Derivatives and Their Corresponding Indefinite Integrals

$$\frac{d}{du}u^{n+1} = (n+1)u^n$$

$$\int u^n \, du = \frac{u^{n+1}}{n+1} + C, \quad n \neq -1$$

$$\frac{d}{du}\ln u = \frac{1}{u}$$

$$\int \frac{1}{u}\, du = \ln|u| + C$$

$$\frac{d}{du}e^u = e^u$$

$$\int e^u \, du = e^u + C$$

$$\frac{d}{du}\sin u = \cos u$$

$$\int \cos u \, du = \sin u + C$$

$$\frac{d}{du}\cos u = -\sin u$$

$$\int \sin u \, du = -\cos u + C$$

$$\frac{d}{du}\tan u = \sec^2 u$$

$$\int \sec^2 u \, du = \tan u + C$$

$$\frac{d}{du}\cot u = -\csc^2 u$$

$$\int \csc^2 u \, du = -\cot u + C$$

$$\frac{d}{du}\sec u = \sec u \tan u$$

$$\int \sec u \tan u \, du = \sec u + C$$

$$\frac{d}{du}\csc u = -\csc u \cot u$$

$$\int \csc u \cot u \, du = -\csc u + C$$

$$\int \tan u \, du = \ln|\sec u| + C$$

$$\int \cot u \, du = \ln|\sin u| + C$$

$$\int \sec u \, du = \ln|\sec u + \tan u| + C$$

$$\int \csc u \, du = \ln|\csc u - \cot u| + C$$

$$\frac{d}{du}\operatorname{Sin}^{-1} u = \frac{1}{\sqrt{1-u^2}}$$

$$\int \frac{1}{\sqrt{1-u^2}}\, du = \operatorname{Sin}^{-1} u + C$$

$$\frac{d}{du}\operatorname{Tan}^{-1} u = \frac{1}{1+u^2}$$

$$\int \frac{1}{1+u^2}\, du = \operatorname{Tan}^{-1} u + C$$

$$\frac{d}{du}\operatorname{Sec}^{-1} u = \frac{1}{u\sqrt{u^2-1}}$$

$$\int \frac{1}{u\sqrt{u^2-1}}\, du = \operatorname{Sec}^{-1} u + C$$

Calculus

Calculus

THIRD EDITION

DENNIS D. BERKEY

Boston University

PAUL BLANCHARD

Boston University

SAUNDERS COLLEGE PUBLISHING

Harcourt Brace Jovanovich College Publishers

Fort Worth   Philadelphia   San Diego   New York
Orlando   Austin   San Antonio   Toronto
Montreal   London   Sydney   Tokyo

Text Typeface: Times Roman
Compositor: York Graphic Services, Inc.
Acquisitions Editor: Robert Stern
Developmental Editor: Alexa Barnes
Managing Editor: Carol Field
Project Editor: Sally Kusch
Copy Editor: Charlotte Nelson
Manager of Art and Design: Carol Bleistine
Text Designer: Rebecca Lemna
Cover Designer: Lawrence R. Didona
Text Artwork: York Graphic Services, Inc.
Director of EDP and Production Manager: Tim Frelick
Marketing Manager: Monica Wilson

About the cover: The cover is a small-scale study of the convergence of Newton's Method applied to the solution of a cubic polynomial in the complex plane. For this particular polynomial $z^3 - 1 = 0$, the initial points that do not lead to a root form the Julia set of this iterative algorithm. The Julia set is the fractal highlighted in this picture. Details and related computer experiments are given in Appendix IV.

Printed in the United States of America

CALCULUS, third edition

0-03-046927-9

Library of Congress Catalog Card Number: 91-053208

345 069 98765432

Preface

This text is intended for use in a traditional three-semester or four-quarter sequence of courses on the calculus, populated principally by mathematics, science and engineering students. It reflects our philosophy that calculus should be taught so as to produce skilled practitioners who understand the mathematical issues underlying the techniques they have acquired. It is written in order to provide students of widely varying interests and abilities with a readable exposition of the principal results, including ample motivation, numerous well-articulated examples, a rich discussion of applications, and a serious description and use of numerical techniques. In particular, we explain how calculators and computers can be used to illustrate the theory, to provide approximate solutions to problems on which more elegant techniques fail, and to relieve the tedium of extensive computations.

Changes for the Third Edition

Content

The Table of Contents of this edition has not changed substantially from that of the Second Edition. The text, however, has been extensively revised and edited. We have expanded our treatment of numerical methods, and we have entirely rewritten the discussion of several key topics. In particular, Chapters 3 and 4 on the derivative and its applications have been significantly revised to place greater emphasis on the derivative as a rate of change and to emphasize its role in extremal problems. Our introduction to the definite integral in Chapter 6 has been entirely rewritten to emphasize the definition as a limit of Riemann sums and to streamline the treatment of the Fundamental Theorem of Calculus. In Chapters 16 and 17 we have reorganized the material on vectors, curves, and surfaces to highlight the geometric issues involved. Finally, we added a brief section on Jacobians at the end of Chapter 19 on multiple integration in response to user requests. For more information about these changes and the order in which various topics can be treated using this text, see our brief guide, *Notes to the Instructor,* immediately following this preface.

Exercises

Over 1000 new exercises designed to challenge the student's understanding have been added to this edition. Many of these are a result of our consideration of the ''lean and lively'' calculus reform movement and a desire on our part for the student to have geometric and numerical as well as algebraic familiarity with the ideas of calculus.

Artwork

Over 20% of the figures in this text are new to this edition. All of the artwork has been redrawn by computer in a four-color format with careful attention to the use of color as a pedagogical tool. (A guide to the pedagogical use of color is given on page xix.) Much of the art has been rendered using the computer-based mathematics system *Mathematica* to provide greater accuracy and more realistic three-dimensional images.

Increased Utilization of Technology

Graphing Calculator Exercises: Over 300 graphing calculator exercises have been added to the third edition. These problems (marked with the logo ▦) can be used with any graphing calculator currently available and have been carefully selected to explore essential concepts of calculus in more depth, not merely to teach students how to push buttons. In fact, many of these problems are actually mini-projects, rather than exercises. Of course, these exercises can also be done with an appropriately equipped personal computer.

For those who need more information about getting started, Appendix V, Calculus and the Graphing Calculator, discusses various graphing calculators and their utility for learning calculus. This appendix includes instructions, explanations, and programs for the Casio, TI-81, HP-28S and HP-48S graphing calculators.

Appendix I on Calculus and the Computer: Appendix I has been significantly revised for this edition. It now contains Pascal versions of the BASIC computer programs previously offered as well as a section that discusses how to use *Mathematica* to illustrate many of the fundamental concepts of calculus.

Appendix IV: We have added a short appendix on Newton's Method in the complex plane. It discusses interesting aspects of the method that are usually ignored in the standard treatment of Newton's Method in calculus courses. More importantly, this appendix gives the interested student a glimpse of an area of current mathematical research.

Features

Even though this revision has been substantial, the distinctive approach, flexibility, and pedagogy of the previous editions have been retained.

Approach

We present the traditional calculus curriculum using informal discussion and geometric arguments whenever possible. Chapter 2 begins with a flexible introduction to the limit, providing both informal and formal discussions of limits. We develop the derivative in Chapter 3 with an emphasis on rates and linear approximation.

Our short Chapter 5 on antidifferentiation precedes the introduction of the definite integral and includes material on differential equations. This chapter is positioned so that antidifferentiation can be treated either as the final application of the derivative or as the first topic in a unit on integration. This ordering reflects our concern that the definite integral is often presented almost simultaneously with the notion of the antiderivative and the Fundamental Theorem of Calculus. As a result, students may leave their first calculus course thinking of a definite integral as a difference between two values of an antiderivative that, coincidentally, might also be identified with the area of a planar region. By introducing antiderivatives first, together with their applications to solving simple differential equations, the distinction between antiderivatives and limits of approximating Riemann sums can be clearly established before the important connection provided by the Fundamental Theorem is introduced. This approach helps students grasp the notion of approximate integration and makes the development of the various applications of the definite integral more accessible.

We define the definite integral as a limit of Riemann sums (Chapter 6), and the natural logarithm as an integral (Chapter 8). We also discuss fully the principal theorems of the calculus, including the theorems of Green and Stokes, as well as the Divergence Theorem (all in Chapter 20).

Flexibility

The organization of this text continues to provide as much flexibility as possible for the individual instructor. We have resisted the current trend towards fewer (and longer) chapters so that instructors can conveniently treat topics in different orders if they so desire. In the Notes to the Instructor, we elaborate on these options.

Examples

Many of the over 800 examples, especially in the early parts of the text, use the two-column Strategy-Solution format. The "Strategy" column gives the student, in abbreviated form, a description of the principal steps involved in the more complete solution. This format helps students identify the more general aspects of the particular solution and develop problem-solving strategies of their own.

Chapter Summaries

Each chapter ends with a brief list of the key concepts of the chapter. Page references are included so that students can easily refer back to the material as they review.

Review Exercises

A comprehensive set of exercises reviewing all of the topics of the chapter follows the chapter summary. Students who work these review exercises can feel comfortable that they have mastered the chapter.

Historical Unit Openers

Each of the seven units of the text begins with an essay describing the key mathematicians whose discoveries underpin the material of that unit. These essays illustrate how mathematics is a human endeavor carried out over centuries and continents.

Accuracy and the Saunders Solution

Because accuracy in mathematics texts is critical, this edition has been subjected to an exhaustive series of accuracy checks. Initially, corrections and suggestions from users of the second edition were incorporated in the reprint of that edition which subsequently became the basis for the third edition. The accuracy of the worked examples and the answers to the odd-numbered exercises were checked during manuscript, galleys and page proof stages by numerous professors as well as by us. All exercises were solved separately by the three professors who wrote the Instructor's Manual and Student Solutions Manual. Finally, a completely independent check of all the exercises was completed by four graduate students at Boston University.

Supplements

The extensive package in support of *Calculus,* Third Edition, reflects Saunders College Publishing's continuing commitment to provide the most helpful, thoughtful, and up-to-date ancillaries.

The **Instructor's Manual,** by Judy Coomes (William Paterson College), Dennis Kletzing (Stetson University), and Michael Motto (Ball State University), is available free to adopters. It contains complete detailed solutions to all the exercises in the text and is arranged in two volumes for convenience. To ensure accuracy, solutions figures are computer-generated, and selected solutions have been checked again using *Mathematica.*

The **Student Solutions Manual,** also by Judy Coomes, Dennis Kletzing, and Michael Motto, is available in two convenient volumes for students who need help with homework. They contain detailed solutions to the odd-numbered exercises; answers to these exercises are also given at the back of the text.

The **Test Bank** by Ken Kramer of CUNY, Queens College, offers over 1300 open-ended test questions corresponding in level and difficulty with the examples and exercises in the text. For each chapter there are four test forms of approximately 15 questions each. The three final exams are divided into two parts (for Chapters 1–11 and 12–21). Answers for each test are also provided and have been checked independently to ensure accuracy.

For adopters with IBM or Macintosh computers, this Test Bank is also available free in computerized form. The *ExaMaster*™ **Computerized Test Bank** software offers two easy, flexible ways to create and edit tests. Each version of the *ExaMaster* software comes with *ExamRecord*™, a gradebook program for each recording, curving, and graphing of grades.

A kit of **transparencies and transparency masters** of the most important figures from the text is available. The 25 four-color overhead transparencies illustrate essential figures that are difficult to draw accurately and effectively. In addition, there are 100 transparency masters of other important and useful figures chosen from throughout the text.

For more material on graphing calculators, we offer two manuals on using the HP-28 and HP-48S graphing calculators to enrich the teaching and learning of calculus. **Calculator Enhancement for Single Variable Calculus** by James H. Nicholson and J.W. Kenelly, consultant, and **Calculator Enhancement for Multivariable Calculus** by J.A. Reneke and D.R. LaTorre, consultant, provide procedures, calculator programs, examples and exercises designed to remove the computational burden normally associated with these courses. Each supplement helps students appreciate the geometrical and graphical aspects of calculus, master its theory and methods, and explore new topics and applications.

Calculus and *Mathematica* by John Emert and Roger Nelson of Ball State University is a computer laboratory manual of 30 projects to supplement the traditional calculus sequence. These experiments use the computer to lead a process of discovery, conjecture, and verification. The experiments cover a wide variety of topics, from the expected Newton's Method and Riemann Sum explorations to adventuresome investigations of symmetric derivatives, chaotic sequences, and global/local behavior.

Calculus and *Derive* by David Olwell and Pat Driscoll of the United States Military Academy contains almost 40 modules (exercises and projects) designed specifically to use the computer to illustrate symbolically and graphically the essential concepts of calculus and to eliminate the drudgery of computation. The average student will take 45 minutes to complete each exercise and four hours to explore each project.

CalcAide by Elizabeth Chang of Hood College provides 17 graphing and numerical utility software modules and is intended for use in the classroom, in a mathematics laboratory, or on students' own computers. CalcAide can draw the graph of most functions and provides numeric and graphic illustrations of many concepts of calculus. The flexible menu system allows you to graph a function on any portion of the coordinate plane, change some features of the graph or function while leaving others unchanged, examine value, slope or other features at a specific point on the graph, and offers many other options. (CalcAide requires an IBM-PC with CGA capabilities.)

Electronic Bulletin Board

The Department of Mathematics at Boston University is committed to excellence in teaching, and as part of this commitment, it has established on its computer system (math.bu.edu) a library of public domain teaching materials available by anonymous login over the Internet. Those who have access to the Internet can use anonymous ftp to retrieve these materials during off-peak hours. Up-to-date materials related to this text will be posted in this library. Those who have access to electronic mail but who do not have access to the Internet can obtain information related to these materials by sending electronic mail to berkey-blanchard@math.bu.edu.

Acknowledgments

Many individuals played instrumental roles in the development of this text. First, we would like to recognize all those who participated in the development of the first two editions. This edition relies heavily on the work of those individuals.

When we began to plan this revision, we already had many helpful suggestions from users of the second edition who kept diaries containing their impressions and comments. These individuals were:

Jake Beard, Tennessee Tech University
Patricia Burgess, Monroe Community College
Peter Collinge, Monroe Community College
M. Hilary Davies, University of Alaska—Anchorage
Edmond D. Dixon, Tennessee Tech University
John Dyer-Bennet, Carleton College
Carol Edwards, St. Louis Community College—Florissant Valley
George Feissner, SUNY–Cortland
Thomas Gruszka, Grand Valley State University
Douglas Kelly, University of North Carolina—Chapel Hill

Calvin Jongsma, Dordt College
Peter M. Knopf, Pace University
Michael Schneider, Belleville Area College
Lowell Stultz, Kalamazoo Valley Community College

These diaries identified topics that needed attention, and they helped us to refine every section in the text.

Ideas for major revisions also came from a focus group that was hosted by Saunders College Publishing in the summer of 1989. At that session, Scott Farrand (California State University–Sacramento), Michael Schneider (Belleville Area College), Bob McFadden (Northern Illinois University), and Ken Kramer (CUNY—Queens College) provided frank evaluations of our plans for this revision as well as a number of excellent suggestions of their own.

We benefited greatly from the considered opinions of the following mathematicians who read parts or all of the initial draft of the revision:

Shih-Chuan Cheng, Creighton University
H. Jay Davis, Rancho Santiago College
Scott Farrand, California State University—Sacramento
Francis G. Florey, University of Wisconsin—Superior
Wayne Dell Gibson, Rancho Santiago College
Fredric T. Howard, Wake Forest University
Elgin Johnston, Iowa State University
Dan Kemp, South Dakota State University
Peter M. Knopf, Pace University
John Kroll, Old Dominion University
Hendrik J. Kuiper, Arizona State University
Jim McKinney, California State Polytechnic University—Pomona
John A. Suvak, Memorial University of Newfoundland
Donald Taranto, California State University—Sacramento
Richard Thompson, University of Arizona
Richard E. Winslow, University of Lowell
James Wiseman, Rochester Institute of Technology

Manuscript, galleys, and page proofs were scrutinized for content and accuracy by:

Alan Candiotti, Drew University
William D. Clark, Steven F. Austin State University
Gloria Langer, University of Colorado
Giles Maloof, Boise State University
Len Miller, Mississippi State University
Allan Mundsack, Los Angeles Mission College
Hugo Sun, California State University—Fresno

All exercises were solved by the three professors who wrote the Instructor's Manual and Student Solutions Manual: Judy Coomes (William Paterson College), Dennis Kletzing (Stetson University), and Michael Motto (Ball State University). Answers to the exercises were checked independently by four graduate students at Boston University: Duff Campbell, Thomas LoFaro, Michèle Nichols, and Mario Casella.

The exercises and answers of the Instructor's Preliminary Edition were checked again by the following instructors. Any errors they found were corrected in the Student Edition.

Paul Allen, University of Alabama
David Crystal, Rochester Institute of Technology
Ed Dixon, Tennessee Technical University
Noal Harbertson, California State University—Fresno
Sven Leukert, University of North Carolina—Chapel Hill
Ed Matzdorf, Chico State University
Dave Obert, University of Minnesota
Ann Ostberg, Central Community College—Platte Campus
Joseph Stephen, Northern Illinois University
Kerry Wyckoff, Brigham Young University

The graphing calculator exercises and appendix were designed and written by James Angelos of Central Michigan University. His knowledge and experience with these remarkable devices is much superior to ours, and we are fortunate to be able to include his materials in this text. David Olwell of the United States Military Academy reviewed every graphing calculator exercise, and his comments helped us to refine several of them. The graphing calculator appendix was reviewed for accuracy by Thomas LoFaro and Duff Campbell of Boston University and Don LaTorre of Clemson University. The Test Bank was checked by Gloria Langer of the University of Colorado.

All of these individuals worked meticulously to ensure the accuracy of the text. Whatever errors might remain are, of course, our sole responsibility. We have worked hard to produce the most accurate text possible, but we welcome your comments on improving or correcting the text. Those who prefer to use electronic mail can reach us at berkey-blanchard@math.bu.edu.

The historical notes, which provide an important human contrast, were written by Duane Deal of Ball State University. They were reviewed by our colleague, Thomas Hawkins of Boston University.

We used *Mathematica* 2.0 to generate many of the three-dimensional figures in this edition, and we thank Wolfram Research, Inc. for providing a beta test copy of version 2.0 so that we could generate these pictures and test the code in the appendix using the current version of *Mathematica*.

The cover of this text is an image obtained in a computer study of Newton's Method, and it was computed at Boston University using *Citool*, a computer program written by Scott Sutherland, Gert Vegter, and Paul Blanchard. Mario Casella helped produce this image, and the final slides were shot by Laura Giannitrapani of the Graphics Lab at Boston University. The images in the appendix on Newton's Method were also produced using *Citool*.

We are greatly appreciative of the support that Saunders College Publishing has shown throughout this lengthy revision. They have enthusiastically provided the resources necessary to complete such a large project in a timely fashion. In particular, we are pleased to recognize the encouragement of Senior Mathematics Editor Robert Stern and Editor-in-Chief Liz Widdicombe. Our schedules made the production of this text especially complicated, and we wish to thank Senior Project Manager Sally Kusch, Manager of Art and Design Carol Bleistine, and Director of Editorial, Design, Production, and Manufacturing Tim Frelick for their extra efforts during the prolonged production of this project. Finally, we must make special note of the work done by Developmental Editor Alexa Barnes, who ably helped us through every step of this revision over a three-year period.

At Boston University, we were assisted by numerous colleagues in addition to those already mentioned. Lisa Doherty, Barbara Leonard, Diurka Rodriguez, and

Angelique Thayer all helped manage the flow of materials between Boston and Philadelphia. Thomas LoFaro and Michèle Taylor helped revise many of the exercise sets, and Reza Behnam produced preliminary versions of many of the three-dimensional figures. Daniel Alexander helped with the production of preliminary *Mathematica* notebooks, which were incorporated into the final version of Appendix I. Thomas LoFaro and Farzan Nadim helped compile the accuracy reports and the answer manuscript.

On a more personal level, the writing of this text was supported by three very special groups of people. First, our students at Boston University, who have helped sharpen our thinking about teaching and calculus for more than a decade. Second, our colleagues in the faculty and administration of Boston University, who believe deeply in the importance of effective teaching. And, most importantly, our families— Cristin, Aaron, Jessica, Eric, and especially Catherine and Dottie, who understood our need to write this book and who shared fully and willingly in the sacrifices that were required. To all we are truly grateful.

DENNIS D. BERKEY
PAUL BLANCHARD
Boston University
January 1992

Notes to the Instructor

These notes provide a brief overview of the organization of this text. They also discuss options regarding possible rearrangement of topics.

Calculator Exercises: Before discussing the individual chapters, it is important to emphasize one issue regarding the calculator and, especially, the graphing calculator exercises. When assigning these exercises, the instructor must remember that these exercises usually take the student more time to complete than standard exercises. In particular, many of the graphing calculator exercises are closer to projects than to exercises. Consequently, we strongly recommend that instructors complete any graphing calculator exercise on their own before assigning it so that they are aware of exactly what is involved. The Instructor's Manual provides annotated solutions to these exercises.

We also note that these exercises are appropriate for personal computers, and in Appendix I we cite the graphing calculator exercises that make effective use of the routines provided there.

Finally, we also note that some of the exercises marked with a graphing calculator icon can also be solved with a programmable calculator.

Overall Organization: The general order of topics in this text is consistent with most popular calculus texts. We begin with a discussion of limits (Chapter 2), develop the derivative and its applications (Chapters 3 and 4), briefly discuss antidifferentiation (Chapter 5), develop the definite integral (Chapter 6), and discuss applications of the integral (Chapter 7). These chapters are covered at Boston University in our standard first-semester calculus course, and we usually must omit a few sections in a 13-week semester.

Next we discuss certain important transcendental functions: logarithms and exponentials (Chapter 8) and the inverse trigonometric functions (Chapter 9). We also develop the standard techniques of integration (Chapters 9 and 10). Then we provide a brief discussion of l'Hôpital's Rule and improper integrals (Chapter 11) in anticipation of their use in the analysis of sequences and series (Chapter 12). Taylor polynomials, Taylor's Theorem, and power series follow immediately (Chapter 13). At Boston University, we finish the second semester with the (optional) topic of conic sections (Chapter 14) and polar coordinates and parametric equations in the plane (Chapter 15). Again, we usually omit a few sections in order to complete these chapters.

Multivariate calculus can either start with a review of the material in Chapter 15 or with the discussion of vectors (Chapter 16). We continue the study of geometry by discussing vector-valued functions, curves, and surfaces (Chapter 17). At that point, we treat the standard topics of multivariate calculus: partial derivatives (Chapter 18), multiple integration (Chapter 19), and vector analysis (Chapter 20). These topics easily fill a complete semester.

Our text ends with a chapter on differential equations (Chapter 21), which is a brief treatment of many of the ideas that are developed in more detail in courses on differential equations.

Here are more specific details to aid with the writing of a syllabus:

Chapter 1:   This chapter is a review of precalculus concepts. The sections that should be covered in detail depend very much on the class. In this edition, we have added a Readiness Test, which the students can use either as a self-study guide to help with their own review or as an indicator that the student is ready to study the calculus after a review of Chapter 1.

Chapter 2:   Our treatment of the limit is designed to accommodate those instructors who desire a rigorous development as well as those instructors who are comfortable with a more intuitive approach. The ϵ-δ definition is discussed in Section 2.3, and the limit theorems are proved rigorously at the end of Sections 2.4, 2.5, and 2.6. Instructors who do not desire an ϵ-δ approach can simply skip Section 2.3 as well as the material at the ends of Sections 2.4 through 2.6.

Chapter 3:   The derivative is developed in the standard manner in Sections 3.1 through 3.5. Then Section 3.6 discusses related rates immediately after the Chain Rule (Section 3.5). We prefer this early exposure to related rate problems for two reasons: it reinforces the fact that the Chain Rule involves rates of change, and it introduces the difficulties associated with setting up and solving word problems as early as possible. We realize that many instructors may prefer to treat this topic after discussing the remainder of Chapter 3. In fact, this topic along with linear approximation in Section 3.7 and Newton's Method in Section 3.10 is optional.

Chapter 4:   We have completely reorganized this chapter so that applied extremum problems are discussed early in the chapter. The theory of critical points and extrema, both absolute and relative, is introduced in Section 4.1, and this theory is related to applied extremal problems in Section 4.2. Then Sections 4.3 through 4.5 discuss the Mean Value Theorem, monotonicity, and concavity. Section 4.6 treats limits at infinity and infinite limits. The material in Sections 4.1 through 4.6 is central to our presentation of calculus. Section 4.7 is a traditional section on curve sketching, and Section 4.8 uses the first derivative version of Taylor's Theorem to analyze the error involved in linear approximation. Either or both of these sections are certainly optional.

Chapter 5:   This is a brief chapter on antidifferentiation and differential equations. The first two sections on antidifferentiation and substitution are crucial to the treatment in Chapter 6, but Section 5.3 (in which differential equations are introduced) can be omitted if desired.

Chapter 6: This chapter introduces the definite integral with a focus on area and Riemann sums and then presents the Fundamental Theorem of Calculus. In order to emphasize that the definite integral is a limit of Riemann sums, we briefly discuss approximation of integrals by the Midpoint Rule in Section 6.2. A more comprehensive discussion of numerical approximation procedures is the topic of the final section of this chapter (Section 6.6). The error analysis in that section is somewhat more geometric than the typical presentation.

Chapter 7: In this chapter, we present eight different applications of the definite integral, and here the instructor has much flexibility about what topics to include or omit.

Chapter 8: The natural logarithm and exponential are defined and developed using the definite integral and the Fundamental Theorem of Calculus. Section 8.1 begins with a brief review of a precalculus definition of the logarithm and moves on to a discussion of inverse functions. The amount of time that should be allotted to this section depends on the background of the class. Section 8.5 presents the application of this theory to exponential growth and decay problems.

Chapter 9: This chapter is a combination of a number of topics that involve the calculus of trigonometric functions. Sections 9.1 and 9.2 describe integration techniques for the trigonometric functions and their powers. Therefore, the instructor has some flexibility about how to treat these two sections based on the manner in which they plan to cover Chapter 10 (techniques of integration). Sections 9.3 and 9.4 present the inverse trigonometric functions and their calculus, so complete coverage of these two sections is important. Section 9.5 discusses the hyperbolic functions, and it can be safely skipped if the instructor so desires.

Chapter 10: We have kept our treatment of techniques of integration to a minimum, but those who are interested in a leaner presentation of the calculus can omit much of this material. However, we believe that integration by parts (Section 10.1) still merits careful treatment.

Chapter 11: This brief chapter discusses l'Hôpital's Rule and improper integrals, and both of these topics are necessary in what follows.

Chapter 12: This chapter on sequences and series is a standard presentation of these two related concepts. Most calculus sequences would contain a complete treatment of this chapter.

Chapter 13: The three concepts of Taylor polynomials, power series, and Taylor series are presented with a complete treatment of Taylor's Theorem. If Section 4.8 was covered earlier, the students benefit from being reminded that the ideas presented here are simply generalizations of the idea of linear approximation. This is another core chapter.

Chapter 14: In this chapter, we review briefly the properties of conics with an emphasis on their analytic description. In what manner and when an instructor covers this material should be determined by the backgrounds of the students. Much of this material can be treated earlier if the instructor so desires.

Chapter 15: There are two basic geometric topics in this chapter: polar coordinates and parametric equations in the plane. Again this is another case where the instructor has a great deal of flexibility as to when and how this material is treated. However, a solid understanding of these topics is crucial for students to succeed in multivariate calculus.

Chapter 16: This chapter has been reorganized to highlight geometric problems involving lines and planes. The entire chapter is crucial to the development of multivariate calculus.

Chapter 17: In this chapter, we introduce vector-valued functions, associated curves in space, and surfaces in space. The section on curvature (Section 17.5) can be safely omitted if the instructor desires. There is also merit to the idea of waiting until triple integrals are discussed in Chapter 19 to introduce cylindrical and spherical coordinates. However, knowledge of these coordinates provides natural examples of the multivariable Chain Rule, which is discussed in Section 18.6.

Chapter 18: This is a standard presentation of partial differentiation. Sections 18.1 through 18.7 are fundamental. Section 18.8 discusses constrained extrema and provides a nice application of the gradient. However, this section can safely be omitted if desired. Section 18.9 discusses the problem of reconstructing a function of two variables from its gradient. This technique is used in the study of vector fields (Section 20.1) and in the study of exact differential equations (Section 21.2). If desired, a discussion of Section 18.9 can be delayed until either of these two topics is presented.

Chapter 19: This is another standard chapter in multivariate calculus. In this edition, we have added an optional section (Section 19.8) on change of variables and Jacobians, and that section has been written so that it has two distinct parts. If an instructor wants to discuss Jacobians for functions of two variables immediately after discussing double integrals in polar coordinates (Section 19.3), the first part of Section 19.8 can be covered at that time. Then the remainder of this section can be completed after triple integrals in cylindrical and spherical coordinates are discussed (Section 19.7).

Chapter 20: This is a standard chapter on vector analysis which should essentially be covered in the order it is written. However, if an instructor has time to cover only Stokes' Theorem or the Divergence Theorem, either topic can be covered independent of the other.

Chapter 21: Our final chapter is a brief discussion of differential equations. Instructors who want to study differential equations earlier can cover many of these sections at appropriate times in the second semester course. For example, first-order linear equations (Section 21.1) can be discussed immediately following Chapter 8. The same is true for Sections 21.3, 21.4, and 21.6. Section 21.5 is an application of the theory of power series in Chapter 13. However, the topic of exact equations (Section 21.2) needs the material presented in Section 18.9 on reconstructing a function from its gradient.

Appendix I: This appendix illustrates ways in which the computer can be used as a tool in the study of the calculus. The first part of the appendix presents short programs written in BASIC and Pascal that are referenced throughout the exercise sets. The *Mathematica* routines in the second part of the appendix illustrate a number of ways the power of a computer-based mathematics system can be used in the teaching of calculus. Relevant graphing calculator exercises from the main part of the text are cited.

Appendix II: Here we complete a number of proofs as promised in the main exposition.

Appendix III: This is a brief summary of the properties of complex numbers.

Appendix IV: As mentioned in the preface, this appendix presents a brief introduction to Newton's Method in the complex plane as well as related computer graphics. Relevant computer projects are also presented.

Appendix V: Many of the graphing calculator exercises are more accessible if certain short programs are available. This appendix by James Angelos of Central Michigan University contains relevant programs and gives a brief introduction to the use of the graphing calculator in calculus. However, it is not meant to replace the more complete manuals on the use of the graphing calculator in calculus (see the descriptions of the Clemson Calculator Enhancement manuals for calculus mentioned in the *Preface*).

Appendixes VI, VII, and VIII: For easy reference, the remaining three appendixes provide tables of transcendental functions, a complete table of integrals, and useful geometry formulas, respectively.

Guide to the Pedagogical Use of Color

The new four-color art program has been computer-generated for increased clarity and accuracy. Color is used pedagogically and consistently to help students grasp key concepts. A guide to the system of colors with references to sample figures is given below.

Color	Purposes	Sample
Blue	primary curve Riemann rectangle above x-axis positive area	Chapter 6, p. 324, Figure 2.5

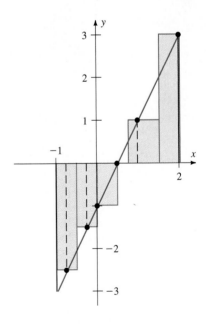

Color	Purposes	Sample
Red	second curve tangent line Riemann rectangle below x-axis negative ''area''	Chapter 2, p. 57, Figure 1.6

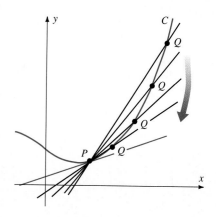

Gray	region in the plane	Chapter 6, p. 324, Figure 2.4

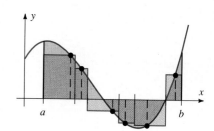

Green, orange, tan, etc.	additional curves and realistic art for application figures	Chapter 13, p. 687, Figure 6.1

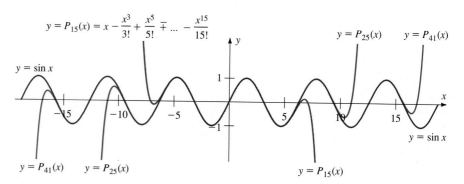

Color   **Purposes**   Sample

Chapter 7, p. 419,
Figure 5.5

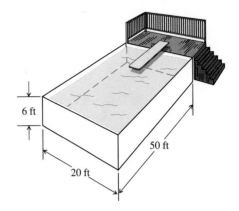

Mathematica   3-dimensional   Chapter 18, p. 885,
                       figures   Figure 1.16

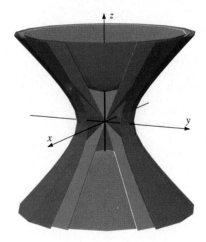

Contents Overview

Contents

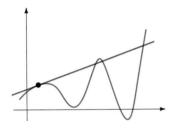

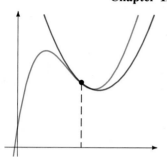
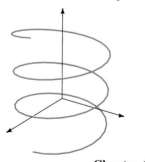

Preliminary Notions

Unit One

Newton and Leibniz—The Unification of the Calculus

More than once in history, a significant mathematical development has been independently discovered at about the same time by two mathematicians widely separated by geography and having no contact with one another. The greatest of these discoveries was that of the calculus. Isaac Newton (1642–1727) in England, and Gottfried Wilhelm Leibniz (1646–1717) in Germany, were completely unaware of each other's work. Newton developed his version of the calculus some ten years before Leibniz, but Newton was generally reluctant to publish his results, and Leibniz published his own version before Newton's publication. These time differences ultimately resulted in an extended controversy between English and Continental mathematicians concerning priority and the possibility of plagiarism. To their credit, however, Newton and Leibniz did not attack one another.

Isaac Newton was born on Christmas day in 1642 in a saddened household, as his father had died about two months earlier. He was a small and frail child, and not expected to live very long. He showed an early interest in mechanical contrivances. He constructed a clock, and a replica of a mill which was powered by a mouse, and he frightened the neighbors with a kite carrying a small fire. His mother wanted him to become a farmer, but he was not interested. His uncle suggested sending him to a university, and at the age of 19 he went to Trinity College at Cambridge, where he was an undistinguished undergraduate student.

Isaac Newton

In the first months of 1665 bubonic plague swept over England, and all public institutions were closed. Newton stayed on at Cambridge to finish his bachelor's degree that spring, and then returned to the family farm. There followed one of the most scientifically productive periods in human history. This 23-year-old youth, in the span of only 18 months, (a) proved that the binomial theorem was valid for fractional and negative exponents as well as for positive integral powers; (b) discovered, using prisms, that white light could be resolved into the colors of the rainbow, and then reconstituted into white light, leading to the theory of optics; (c) discovered the theory of universal gravitation and the inverse square law which was the key to understanding the structure of the solar system; and (d) discovered differential and integral calculus, and more importantly, the inverse relation of the two. The well-known incident of the falling apple which caused him to begin thinking of gravitation occurred during this period on the farm.

It was the infinite series generated by his fractional and negative exponents for binomials which caused young Newton to use these series as the basis for his calculus. He considered a curve as the result of a continuously moving point. The changing quantity he called a *fluent,* and its rate of change he named a *fluxion* (from the Latin *fluere,* to flow). The fluxion was thus what we shall refer to as a derivative, denoted $\dot{y}$ if y were the original fluent, or variable. Similarly, Newton saw that the original fluent y could be thought of as the fluxion of another function, which he designated $\overset{\square}{y}$ or $\grave{y}$. We today call this an integral.

After the plague years, Isaac Newton returned to Cambridge to work on his master's degree, and at the age of only 26 was appointed to the prestigious endowed chair of mathematics, the Lucasian Professorship.

For about twenty years Newton made significant discoveries in many fields, but then the light of creative genius flickered. He suffered a long illness which affected his ability and, while he continued to work, the results were not of the quality of his earlier productivity. He was elected to Parliament, and then, in 1696, was appointed Warden of the Mint. In this capacity, and later as Master of the Mint, his great genius was spent in checking the quality of the metal in British coins. But also in this period, he was elected President of the Royal Society of London successively for 25 years, and presided over the study of the many scientific and mathematical accomplishments of the early eighteenth century.

Gottfried Wilhelm Leibniz was the son of a university professor who died when the boy was only six years old. He was a precocious student, and his teachers sought, unsuccessfully, to permit him to read only material they thought suitable for his age. After his father's death, he was taken from the school and permitted to read in his father's extensive library where he became essentially self-educated. He entered the University of Leipzig at age 15, received his bachelor's degree at 17, and his doctorate (in law) at 20. He became a diplomat by profession, but soon became interested in mathematics as a consuming avocation. By the time Leibniz was 30 in 1676, he had invented his calculus, which is in many ways the calculus as we use it today.

Gottfried Leibniz

About 1673 he came to the realization that areas can be calculated by *summing* "infinitely thin" rectangles, and that tangents to curves involve *differences* in the x and y coordinates. This led him to suspect that the two processes are inverses of one another. In doing these calculations, he soon found himself immersed in infinite series representing functions, as had Newton several years before. Realizing the power of his discoveries, Leibniz set to work developing terminology and notation. On the night of October 29, 1675, he selected the integral sign ∫ as representing the

Latin word *summa,* or sum, and the derivative notation $\frac{dy}{dx}$.

Leibniz's, and the world's, first paper on differential calculus was published in 1684 with the title (translated) *A New Method for Maxima and Minima, and Also for Tangents. …* He gave formulas for differentials of powers, products, and quotients of func-

tions—the same formulas we learn today. Two years later he published his integral calculus, in which the calculation of areas is shown to be the inverse operation of finding tangents.

Leibniz clearly published his calculus (1684–86) before Newton, but Newton made his discoveries (1665–66) before Leibniz (1672–76). The supporters of each man commenced a great controversy concerning the original discoverer of the calculus. Leibniz and Newton corresponded, but it is a tribute to the character of both men that they did not personally enter into the quarrel. The unfortunate result, however, was that the mathematicians on the European continent lost the use of the ideas of Newton, especially concerning his theory of gravitation, and the English mathematicians lost ground due to their refusal to use the more powerful calculus notation of Leibniz.

Leibniz's life was primarily spent in serving a succession of petty German rulers. He became increasingly unhappy, and spent his later years writing about religious and philosophical questions. He never married. His later years were lonely, and he died an embittered man. His secretary was the only mourner at his funeral.

Newton, on the other hand, was respected and revered in his old age. He was knighted by Queen Anne in 1705, and was buried in Westminster Abbey with great honor.

Newton and Leibniz were both creative mathematicians of the first rank. The genius of neither is diminished by that of the other.

(Photographs from the David Eugene Smith Papers, Rare Book and Manuscript Library, Columbia University.)

Chapter 1

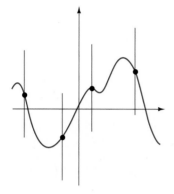

Review of Precalculus Concepts

This chapter summarizes many of the concepts and techniques of precalculus mathematics used throughout the text. You will find little here that you have not seen before. However, there are several reasons you should examine this chapter carefully.

First, you need to become familiar with the notation and style in which this text is written. Studying mathematics is similar to studying a foreign language in that a vocabulary must be mastered, as well as a set of concepts. In mathematics, the vocabulary is quite formal, consisting of symbols and terms that have very precise meanings. However, different authors often do not use exactly the same notation. Let's agree on notation and basic concepts before discussing new ideas. This will make the material that follows less intimidating.

Second, as mathematics has come to be more widely used as a tool in the natural and social sciences, more students than ever before are studying the calculus. This means that the typical calculus class contains students with quite diverse backgrounds in mathematics. This chapter sets the level and scope of precalculus mathematics assumed throughout the text. It therefore provides a common base from which to proceed.

Finally, this chapter provides a gauge by which you can determine whether you are sufficiently well prepared to begin a course on the calculus. If, after carefully reviewing the chapter, you are unable to work most of the exercises successfully, you probably need to consider enrolling in a precalculus course. On the other hand, you may be very familiar with the material in this chapter. If so, we recommend that you start your review by completing the readiness test at the end of this chapter. Those problems will help you identify specific topics you should review.

1.1 The Real Number System

Real Numbers

The **real numbers** may be represented by the set of all points lying on an infinite line. To do so we first select a point 0 on the line that we refer to as the **origin** (Figure 1.1). We then select a **unit** of measurement and locate the point on the line lying one unit to the right of 0. This point is labeled 1 (one). This procedure establishes the **scale** for the **number line** and also uniquely determines the location of each real number. Each positive real number r lies r units to the right of 0, and each negative real number s lies $-s$ units to the left of 0.

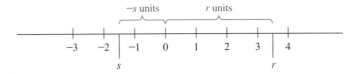

Figure 1.1   The real number line.

There are three important special types of real numbers that can be identified by their decimal forms. The **integers** are the real numbers . . . $-2, -1, 0, 1, 2, \ldots$. Integers may be expressed in decimal form with only zeros to the right of the decimal point. The **rational** numbers are those real numbers that can be expressed as quotients of integers. In other words, if r is a rational number then $r = p/q$ for some integers p and q ($q \neq 0$). The decimal forms of rational numbers repeat eventually. For example, using a bar to identify the digits that repeat, we may write

$$\frac{1}{4} = 0.2500\overline{0} \ldots ,$$

$$\frac{2}{9} = 0.222\overline{2} \ldots ,$$

$$-\frac{25}{11} = -2.272\overline{72} \ldots , \quad \text{and}$$

$$\frac{20}{13} = 1.538461\overline{538461} \ldots .$$

(See Exercise 72 at the end of this section.)

Those real numbers whose decimal forms do not repeat are referred to as **irrational numbers.** The discovery nearly 2500 years ago that not every real number is also a rational number was both profound and traumatic, since the geometric mathematics of the early Greeks assumed that every length could be expressed as the ratio of two integral lengths. We now know that many familiar numbers such as $\sqrt{2}$ and π are irrational.* In fact, there are infinitely many irrational real numbers. (See Exercise 65 at the end of this section to see why $\sqrt{2}$ is irrational.)

Sets of Real Numbers

There are several types of notation that we will use to specify sets of real numbers and operations on these sets. The most familiar notation is probably **set builder notation** in which one specifies a choice of variable and also a rule by which values of that variable are determined. For example, $\{x \mid 2 \leq x \leq 4\}$ is read, "the set of all real numbers equal to or greater than 2 and equal to or less than 4." We will frequently refer to **intervals** of the real number line with the following notation:

$$[a, b] = \{x \mid a \leq x \leq b\},$$
$$[a, b) = \{x \mid a \leq x < b\},$$
$$(a, b] = \{x \mid a < x \leq b\},$$
$$(a, b) = \{x \mid a < x < b\}.$$

*The number $\pi = 3.14159 \ldots$ is the ratio of the circumference of a circle to its diameter. Since this irrational number is often *approximated* by the fraction 22/7, it is sometimes mistakenly assumed that $\pi = 22/7$ and, therefore, that π is rational. The proof that π is irrational is not simple.

We can denote intervals that extend infinitely in one direction by use of the infinity symbol, ∞, as follows:

$$[a, \infty) = \{x \mid a \le x\},$$
$$(-\infty, b) = \{x \mid x < b\}.$$

There are two other cases of intervals that extend infinitely in one direction, namely, (a, ∞) and $(-\infty, b]$; the interval $(-\infty, \infty)$ is the entire number line. It must be clearly understood that ∞ is not a number: it is shorthand for the phrase "continuing infinitely through positive values" (or negative values for $-\infty$). In sketching intervals, we use a solid dot to indicate an endpoint that is included and a hollow dot to indicate an endpoint that is excluded, as illustrated in Figures 1.2 and 1.3.

Figure 1.2 The interval $[a, b) = \{x \mid a \le x < b\}$.

Figure 1.3 The interval $(a, \infty) = \{x \mid a < x < \infty\}$.

An interval of the form $[a, b]$ is called a **closed** interval because it includes both its endpoints. An interval of the form (a, b) is called an **open** interval because it includes neither of its endpoints. Intervals of the form $[a, b)$ or $(a, b]$ are referred to as either **half-open** or **half-closed.** The infinite intervals $[a, \infty)$ and $(-\infty, a]$ are called *closed,* and the infinite intervals (a, ∞) and $(-\infty, a)$ are called *open.*

If x is a number and A is a set of numbers (not necessarily an interval), the statement "$x \in A$," read "x is an element of set A," means that x is one of the numbers in set A. The negation of this statement is "$x \notin A$," read "x is not an element of set A." For example,

$$3 \in \{1, 3, 5\}, \quad 2 \in [0, 7), \quad \text{but} \quad 6 \notin \{x \mid -2 \le x \le 2\}.$$

If A and B are sets of numbers, we say that A is a **subset** of B if every element of A is also an element of B. The notation for this is $A \subseteq B$. When both A and B are intervals, we say that A is a subinterval of B if $A \subseteq B$. For example

$$\{1, 3\} \subseteq \{1, 2, 3, 4\}, \quad \text{and} \quad [-1, 1] \subseteq (-\infty, 3).$$

The symbol for the empty set is $\varnothing$. It is the set with no elements.

There are two **operations** on sets that we shall use frequently. Given two sets, A and B, the **union** of the two sets is the set containing precisely those numbers that are elements of either set A or set B, or both. The symbol for the union of A and B is $A \cup B$. In other words,

$$A \cup B = \{x \mid x \in A \text{ or } x \in B\}.^*$$

For example,

$$\{1, 3\} \cup \{2, 3, 5\} = \{1, 2, 3, 5\}; \quad [-3, 4] \cup (2, 6) = [-3, 6).$$

The **intersection** of two sets A and B is the set containing precisely those numbers that are elements of *both* set A and set B. The symbol for the intersection of A and B is $A \cap B$. Thus,

$$A \cap B = \{x \mid x \in A \text{ and } x \in B\}.$$

---

*In mathematics, the usage of the word "or" is *inclusive*. This means that the statement "U or V" is interpreted as "either U or V, or both." This is in contrast to the *exclusive* usage, in which "U or V" would be interpreted as "either U or V, but *not* both."

For example,

$$\{1, 3\} \cap \{2, 3, 5\} = \{3\}, \quad \text{and} \quad [-3, 4] \cap (2, 6) = (2, 4].$$

Figure 1.4 illustrates these two set operations for the interval examples above.

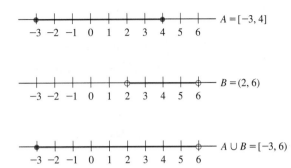

Figure 1.4   Set union and intersection.

Inequalities

There is a natural ordering for the real numbers. If b lies to the right of a on the number line, we say that b is greater than a, and we write $b > a$. Of course, this is the same as saying that a is less than b and writing $a < b$ (Figures 1.5 and 1.6). A purely algebraic way to say that b lies to the right of a is to say that the difference $(b - a)$ is positive. The formal definition of inequality is therefore the following.

Figure 1.5 $a < b$, or $(b - a) > 0$.   **Figure 1.6** $a > b$, or $(a - b) > 0$.

Definition 1

The **inequality** $a < b$ means $b - a$ is positive. The inequality $a \leq b$ means that either $a = b$ or $a < b$.

We will often use the properties given by the following theorem. Their proofs are quite simple, and you are asked to provide them in Exercises 59, 60, and 64.

Theorem 1
Properties of Inequality

Let x, y, z, and c be real numbers. Then

(i) Exactly one of the following holds: $x = y$, $x < y$, or $x > y$.   (trichotomy)
(ii) If $x < y$ and $y < z$, then $x < z$.   (transitivity)
(iii) If $x < y$, then $x + c < y + c$.
(iv) If $x < y$ and $c > 0$, then $cx < cy$.
(v) If $x < y$ and $c < 0$, then $cx > cy$.

Properties (iii)–(v) are frequently used in ''solving'' inequalities, that is, finding the numbers for which the inequalities are true statements. Property (iii) says that a number can be added to both sides of an inequality without changing the sense of the inequality. However, properties (iv) and (v) show that in multiplying both sides of an inequality by c, the sense of the inequality will be retained if $c > 0$ and reversed if $c < 0$. The incorrect use of property (v) is a frequent source of error.

☐ **EXAMPLE 1**

(a) To solve the inequality

$$x + 2 < 5$$

we add the constant -2 to both sides, using property (iii), to obtain the inequality

$$(x + 2) + (-2) < 5 + (-2),$$

or $x < 3$. The **solution set** is, therefore, $\{x \mid x < 3\} = (-\infty, 3)$.

(b) To solve the inequality

$$-\tfrac{3}{2}x \geq 9,$$

we multiply both sides by $-\tfrac{2}{3}$, using property (v). This gives the inequality

$$(-\tfrac{2}{3})(-\tfrac{3}{2}x) \leq (-\tfrac{2}{3})9,$$

or $x \leq -6$. The solution set is $\{x \mid x \leq -6\} = (-\infty, -6]$. ■

The gist of "solving inequalities" is that we attempt to isolate the variable on one side of the inequality, just as we do in solving equations containing a variable. However, you must remember one critical difference—multiplying (or dividing) by a negative number reverses the sense of the inequality.

☐ **EXAMPLE 2**

Solve the inequality $7x - 4 \leq 3x + 8$.

Solution Adding 4 to both sides of the inequality gives

$$7x \leq 3x + 12.$$

We next add $-3x$ to both sides (or, subtract $3x$ from both sides) to obtain

$$4x \leq 12.$$

Finally, we divide both sides by 4 (or, multiplying both sides by $\tfrac{1}{4}$) to conclude that

$$x \leq 3.$$

The solution set is therefore $\{x \mid x \leq 3\} = (-\infty, 3]$. ■

Quadratic inequalities are more difficult to solve than the simple linear inequalities of Examples 1 and 2, since they involve x^2 terms. The following example demonstrates how quadratic inequalities can be solved when we can factor the expression. Note the careful use of the notions of set unions and intersections. Also remember that the product of two positive numbers is positive, that the product of a positive and a negative number is negative, and that the product of two negative numbers is positive.

☐ **EXAMPLE 3**

Solve the inequality $x^2 + x + 1 > 7$.

Strategy · · · · · · · ·
Move all terms to the left side, leaving zero on the right.

Solution
Adding -7 to both sides of the inequality gives

$$x^2 + x - 6 > 0.$$

Factor the left-hand side (if possible) and find its zeros.

Since $x^2 + x - 6 = (x + 3)(x - 2)$, the inequality may be written as

$$(x + 3)(x - 2) > 0.$$

These zeros divide the number line into intervals, on which $x^2 + x - 6$ will have constant sign.

From this factored form we can see that $(x + 3)(x - 2) = 0$ if $x = -3$ or if $x = 2$. We may therefore determine the sign of the left-hand side on the three resulting intervals $(-\infty, -3)$, $(-3, 2)$, and $(2, \infty)$ by examining the signs of the individual factors:

Determine the sign of each factor on each interval. Then, using the signs of the factors, determine the sign of the quadratic expression on each interval.

(i) For $x \in (-\infty, -3)$, $x < -3$, so both

$$x + 3 < 0 \quad \text{and} \quad x - 2 < 0.$$

Thus, $(x + 3)(x - 2) > 0$ if $x \in (-\infty, -3)$.

(ii) For $x \in (-3, 2)$,

    (a) $x > -3$, so $x + 3 > 0$.
    (b) $x < 2$, so $x - 2 < 0$.

Thus, $(x + 3)(x - 2) < 0$ if $x \in (-3, 2)$.

(iii) For $x \in (2, \infty)$, $x > 2$, so both

$$x + 3 > 0 \quad \text{and} \quad x - 2 > 0.$$

Thus, $(x + 3)(x - 2) > 0$ if $x \in (2, \infty)$.

From cases (i)–(iii), we conclude that $(x + 3)(x - 2) > 0$ if

$$x \in (-\infty, -3) \cup (2, \infty). \qquad \blacksquare$$

REMARK Figure 1.7 suggests a simple graphical device for visualizing the procedure in Example 3. Using one number line for each factor and one for the quadratic expression, mark the zeros, and then mark the signs of each factor on each interval. The signs for the quadratic on each interval are then easily recognized.

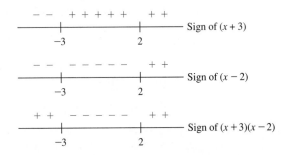

Figure 1.7 Graphical device for determining the sign of a quadratic expression.

Absolute Value and Distance

In order to speak about the distance between two points on the number line, we need a way to refer to the magnitude of a number, regardless of its sign. We use the concept of the **absolute value** of a number x, denoted by $|x|$.

Definition 2

$$|x| = \begin{cases} x & \text{if} & x \geq 0 \\ -x & \text{if} & x < 0. \end{cases}$$

For example, $|3| = 3$, $|-7| = -(-7) = 7$, and $|\pi - 4| = 4 - \pi$. (You should note that the absolute value of a number is *always* nonnegative. Even though the expression $-x$ appears in the second line of the definition of $|x|$, x itself is negative in this case, so $-x$ is positive.) Note that $|x| = \sqrt{x^2}$ for all real numbers x. We shall often simplify expressions involving square roots and squares by rewriting them in terms of their absolute values.

☐ **EXAMPLE 4**

Solve the equation $|2x - 3| = 6$.

Strategy · · · · · · · ·

Write down the two cases from Definition 2.

Solution

From Definition 2, either

$$2x - 3 = 6 \quad \text{or} \quad -(2x - 3) = 6.$$

So, either

$$2x = 9 \quad \text{or} \quad -2x = 3.$$

Solve the two equations that result.

These two equations produce the solutions

$$x_1 = \tfrac{9}{2} \quad \text{and} \quad x_2 = -\tfrac{3}{2}. \qquad \blacksquare$$

The following theorem is useful in solving inequalities involving absolute values. Its proof is left to you as an exercise. (See Exercise 73.)

Theorem 2

Let a be a positive number. The inequality

$$|x| < a$$

is equivalent to the double inequality

$$-a < x < a.$$

The inequality

$$|x| > a$$

holds if either

$$x > a \quad \text{or} \quad x < -a.$$

☐ **EXAMPLE 5**

Solve the inequality $|7 - 2x| \leq 5$.

Strategy · · · · · · · ·

Use Theorem 2 to rewrite the inequality, with the goal of isolating x in the middle.

Solution

By Theorem 2 we can rewrite the inequality as

$$-5 \leq 7 - 2x \leq 5,$$

Subtract 7 from all parts.

so

$$-12 \leq -2x \leq -2,$$

Divide all terms by -2, remembering to reverse sense of all inequalities.

and

$$6 \geq x \geq 1.$$

The solution set is therefore $\{x \mid 1 \leq x \leq 6\} = [1, 6]$. ■

☐ **EXAMPLE 6**

Solve $|4 - 3x| > 1$.

Solution If $|4 - 3x| > 1$, then either

$$4 - 3x > 1 \quad \text{or} \quad 4 - 3x < -1.$$

First, we solve $4 - 3x > 1$ and obtain $x < 1$. Next, we solve $4 - 3x < -1$ and obtain $5/3 < x$. Therefore, the solution set is the union of two intervals

$$\{x \mid x < 1 \text{ or } x > 5/3\} = (-\infty, 1) \cup (5/3, \infty).$$ ■

Using the concept of absolute value, we can define the distance between two numbers (points) on the number line.

Definition 3

The **distance** between the numbers a and b is the number $|b - a|$.

In other words, the distance between the numbers a and b is just the absolute value of their difference (Figure 1.8).

Figure 1.8 The distance between the two real numbers a and b when $a < b$ and when $a > b$.

The following example shows how we may use the notion of distance to interpret certain inequalities.

☐ **EXAMPLE 7**

Find all numbers x that satisfy the inequality

$$|x - 2| < 5.$$

Strategy · · · · · · · ·
Interpret the statement geometrically, using the definition of distance.

Solution
Since $|x - 2|$ is the distance between the numbers x and 2, the inequality is describing precisely those numbers x that lie within 5 units of the number 2. These numbers are described by the inequality

$$-3 < x < 7.$$

Provide a sketch.

The solution set is $\{x \mid -3 < x < 7\} = (-3, 7)$. (See Figure 1.9.) ■

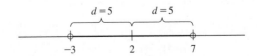

Figure 1.9 Solution of the inequality $|x - 2| < 5$.

We shall make frequent use of the following properties of absolute value.

Theorem 3
Properties of Absolute Value

For any real numbers x and y,

(i) $|x| \geq 0$,

(ii) $|xy| = |x|\,|y|$,

(iii) $|x + y| \leq |x| + |y|$.

Statement (i) simply reminds us that absolute values are never negative. Statement (ii) says that the absolute value of a product is the same as the product of the absolute values. Statement (iii) is referred to as the **triangle inequality** (for reasons we shall see later) and states that the absolute value of the sum of two numbers can never exceed the sum of their absolute values. Notice that statement (iii) is an inequality rather than an equation. An example where equality fails is $|3 + (-5)| = |-2| = 2 < |3| + |-5| = 8$.

Exercise Set 1.1

In each of Exercises 1–9, label the real number as either rational or irrational.

1. $\dfrac{22}{7}$ **2.** $3.141414\ldots$ **3.** $2.1010010001\ldots$

4. $\pi + 3$ **5.** $-6.163\overline{163}\ldots$ **6.** $\dfrac{51}{17}$

7. $\sqrt{256}$ **8.** $\sqrt{2}$ **9.** $1 + \sqrt{2}$

In Exercises 10–15, use the intervals $A = [-2, 5]$, $B = (-1, 6)$, and $C = (-\infty, 0)$ to find the indicated intervals.

10. $A \cap B$ **11.** $A \cup B$ **12.** $A \cup C$

13. $A \cap C$ **14.** $B \cup C$ **15.** $B \cap C$

16. Let $A = [2, 4)$. Find the intervals B so that $A \cup B = [2, 6]$ and $A \cap B = (3, 4)$.

17. True or false? For a circle of positive radius, the radius and the circumference cannot both be rational numbers. What about area and radius? Area and circumference?

In each of Exercises 18–39, solve for x.

18. $2x + 3 < 9$ **19.** $x + 4 \leq 3x$

20. $2x + 7 > 4x - 5$ **21.** $6x - 6 \leq 8x + 8$

22. $-4 \leq 2(x + 2) < 10$ **23.** $(x - 3)(x + 1) < 0$

24. $(x + 7)(2x - 4) > 0$ **25.** $x(x + 6) > -8$

26. $x^2 + x < 0$ **27.** $x^3 + x^2 - 2x > 0$

28. $x^2 + x + 7 > 19$ **29.** $x^4 - 9x^2 < 0$

30. $|x - 7| = 2$ **31.** $|5x - 2| = 0$

32. $x + |x| = 0$ **33.** $2|3x - 1| = 22$

34. $x + |-3| = 7$ **35.** $(|x| + 6)^2 = 49$

36. $x - 6 = 2x - |\pi - 6|$ **37.** $x + 3|x| = 8$

38. $|(x + 2)^2 + 3| = 12$ **39.** $|x - 4| = |x - 7|$

40. True or false? $|a - b| = |b - a|$ for all real numbers a and b.

In Exercises 41–51, solve for x and sketch the solution set.

41. $|x - 3| \leq 2$ **42.** $|x + 5| < 4$

43. $|x + 2| > 1$ **44.** $3|x - 6| \geq 12$

45. $|2x - 7| \leq 3$ **46.** $x > |x|$

47. $|x - 1| = x - 1$ **48.** $|x + 4| - |x - 1| < 4$

49. $|8 - 3x| \geq 5$ **50.** $|x + 2| + |x - 5| \geq 10$

51. $|x + 3| + |x - 2| < 7$

In Exercises 52–55, solve the inequality geometrically by describing in words the meaning of the inequality in terms of distances.

52. $|x - 5| > 2$ **53.** $|x - 1| = |x + 3|$

54. $|x - 1| = 2|x - 3|$ **55.** $2 < |x - 4| < 5$

56. Write down an inequality whose solution is the set of all numbers lying at a distance greater than 4 from the number -5.

57. Write down an equation whose solution is the number whose distance from -2 is twice its distance from 12.

58. The terminology and operations described for sets of real numbers may be applied to more general kinds of sets. **Venn diagrams** may be used to illustrate these concepts, as shown below. The idea is simply that the shaded region represents the particular set.

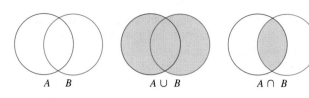

A B   A ∪ B   A ∩ B

Draw Venn diagrams illustrating these sets A, B, and C, so that each pair shares a common area, for the following sets.

a. $A \cup B \cup C$ **b.** $A \cap B \cap C$ **c.** $A \cap (B \cup C)$

59. Prove Theorem 1, part (i), by examining the sign of $(y - x)$.

60. Prove Theorem 1, part (ii) by writing $(z - x) = (z - y) + (y - x)$ and applying Definition 1.

61. True or false? If $a^2 < b^2$ then $a < b$.

62. Prove that if $a^2 < b^2$ and $b > 0$ then $-b < a < b$. (*Hint:* Factor $b^2 - a^2$ as $(b - a)(b + a)$.)

63. Use Exercise 62 to solve the inequality $(x + 6)^2 < 49$.

64. Prove properties (iii)–(v) of Theorem 1.

65. Prove that $\sqrt{2}$ is not a rational number by filling in the details of the following argument:
   a. Argue by contradiction. This means, assume that $\sqrt{2}$ *is* rational and seek a contradiction.
   b. If $\sqrt{2}$ is rational, $\sqrt{2} = p/q$ where p and q have no common factors.
   c. Thus $\sqrt{2}q = p$, so $2q^2 = p^2$.
   d. This means p is divisible by 2 (why?), so $p = 2k$. Thus, $2q^2 = (2k)^2$, so $q^2 = 2k^2$.
   e. This means q is divisible by 2.
   f. Statements d and e contradict the last phrase of statement b. The assumption that $\sqrt{2}$ is rational is therefore false.

66. Prove that $|x| = \sqrt{x^2}$ for all real numbers x.

67. Prove the triangle inequality by noticing that
$$-|x| \le x \le |x| \qquad \text{and} \qquad -|y| \le y \le |y|$$
and adding to obtain $-(|x| + |y|) \le x + y \le |x| + |y|$.

68. True or false? Taking reciprocals reverses the sense of an inequality, i.e., if $a < b$ then $1/a > 1/b$. If false, give a counterexample.

69. State and prove the correct version of the statement in Exercise 68.

70. Prove the inequality $|x| - |y| \le |x - y|$ by writing $x = y + (x - y)$ and applying the triangle inequality.

71. Does the same basic argument as Exercise 65 show that $\sqrt{3}$ is not rational?

72. The following demonstration shows how to recover the fractional form of the repeating decimal (rational number) $2.63\overline{63} \ldots$ Let $x = 2.63\overline{63}$. Then $100x = 263.\overline{63}$. So,
$$99x = (100x - x) = 261.00\overline{00}.$$
Thus, $x = \frac{261}{99} = \frac{29}{11}$.
   a. Use a similar procedure to find a fractional form for the rational number $1.341\overline{341} \ldots$.
   b. Generalize your findings to a statement about how to recover the fractional form for the repeating decimal $x = 0.\overline{a_1 a_2 \ldots a_n}$.

73. Prove Theorem 2 by considering the two cases $x \ge 0$ and $x < 0$ separately.

1.2 The Coordinate Plane, Distance, and Circles

The theory and techniques that we shall develop in the first two thirds of this text are designed for analyzing situations involving two variables. For example, we will consider the relationship between time and distance for particles moving along a line, and the relationship between pressure and temperature for a quantity of an ideal gas.

The Coordinate Plane

In order to illustrate these relationships graphically, we use a two-dimensional coordinate system constructed as follows. If x and y represent the two variables in question, we construct axes for x and for y (known as the **x-axis** and **y-axis,** respectively) so that the axes are perpendicular and so that the two origins lie at the same point. (Also, we follow the convention that the positive half of the horizontal axis extends to the right and the positive half of the vertical axis extends upward.) The plane determined by these two axes is called the **xy-coordinate plane.**

Each point P in the xy-plane is associated with a pair of real numbers, (x_0, y_0), called its coordinates. These coordinates are found by constructing lines through P parallel to both axes. The number x_0 on the x-axis where the vertical line crosses the *x*-axis is called the **x-coordinate** of *P*. Similarly, the number y_0 where the horizontal line crosses the *y*-axis is called the **y-coordinate** of *P*. The coordinate axes divide the plane into four quadrants, numbered as in Figure 2.1. Figure 2.2 shows the coordinates of several points in the plane.

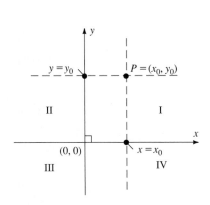

Figure 2.1 The *xy*-coordinate plane.

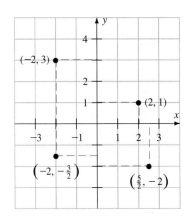

Figure 2.2 Points in the *xy*-plane.

To find the point *Q* in the *xy*-plane corresponding to the given coordinates (x_1, y_1), we reverse the procedure described above, constructing a vertical line through the point $x = x_1$ on the *x*-axis and a horizontal line through $y = y_1$ on the *y*-axis. The point where these lines cross is the desired point *Q*.

Distance

To calculate the distance $d(P, Q)$ between the points $P = (x_1, y_1)$ and $Q = (x_2, y_2)$ in the plane, we make use of the point $R = (x_2, y_1)$. From the Pythagorean theorem, we have

$$[d(P, Q)]^2 = [d(P, R)]^2 + [d(R, Q)]^2$$
$$= |x_2 - x_1|^2 + |y_2 - y_1|^2$$
$$= (x_2 - x_1)^2 + (y_2 - y_1)^2$$

(see Figure 2.3). Therefore, the following definition is consistent with our notion of distance from Euclidean geometry.

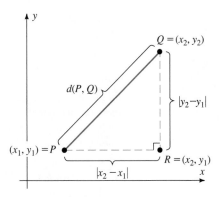

Figure 2.3 Distance between *P* and *Q* is $d(P, Q) = \sqrt{(x_2 - x_1)^2 + (y_2 - y_1)^2}$.

Definition 4

The distance $d(P, Q)$ between the points $P = (x_1, y_1)$ and $Q = (x_2, y_2)$ in the xy-plane is the number

$$d(P, Q) = \sqrt{(x_2 - x_1)^2 + (y_2 - y_1)^2}.$$

(Note that $d(P, Q)$ cannot be negative.)

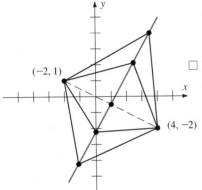

(−2, 1)

(4, −2)

Figure 2.4 Points equidistant from two points lie along a line.

□ **EXAMPLE 1**

The distance between the points $P = (-4, 1)$ and $Q = (1, 3)$ is

$$\begin{aligned}
d(P, Q) &= \sqrt{(1 - (-4))^2 + (3 - 1)^2} \\
&= \sqrt{5^2 + 2^2} \\
&= \sqrt{29}.
\end{aligned}$$

■

□ **EXAMPLE 2**

Find the set of all points lying equidistant from the points $P = (-2, 1)$ and $Q = (4, -2)$.

Strategy · · · · · · · ·

Select an arbitrary point (x, y) satisfying the condition. Try to find a relationship between x and y.

Use Definition 4 to convert the condition of equal distances to an equation in x and y.

Simplify the resulting equation as much as possible.

Solution

We let $T = (x, y)$ be any point lying equidistant from the points $P = (-2, 1)$ and $Q = (4, -2)$. This means that

$$d(P, T) = d(T, Q)$$

or

$$\sqrt{[x - (-2)]^2 + (y - 1)^2} = \sqrt{(x - 4)^2 + [y - (-2)]^2}.$$

Squaring both sides of this equation gives

$$(x + 2)^2 + (y - 1)^2 = (x - 4)^2 + (y + 2)^2,$$

so

$$(x^2 + 4x + 4) + (y^2 - 2y + 1) = (x^2 - 8x + 16) + (y^2 + 4y + 4).$$

This equation simplifies to the equation

$$12x = 6y + 15,$$

or

$$y = 2x - \tfrac{5}{2}.$$

The set of points satisfying this equation is the line sketched in Figure 2.4.

■

Circles

Geometrically, a **circle** may be defined as the set of all points lying at a fixed distance (the **radius**) from a fixed point (the **center**). According to Definition 4, this means that the point (x, y) lies on the circle with radius r and center (h, k) if and only if

$$\sqrt{(x - h)^2 + (y - k)^2} = r, \quad r > 0. \tag{1}$$

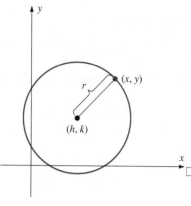

Figure 2.5   The   circle $(x - h)^2 + (y - k)^2 = r^2$.

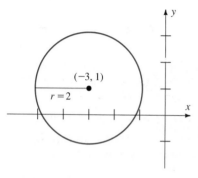

Figure 2.6 Circle $x^2 + y^2 + 6x - 2y = -6$.

Strategy · · · · · · · · ·

Complete the square in x and y terms to try to bring equation into form of equation (1).

Complete the square in x.

Complete the square in y.

Combine the results.

Compare the equation obtained with equation (2).

Squaring both sides of equation (1), we obtain the **standard** form for the equation of the circle with radius r and center (h, k):

$$(x - h)^2 + (y - k)^2 = r^2. \qquad (2)$$

(See Figure 2.5.)

☐ **EXAMPLE 3**

An equation for the circle with center $(1, -2)$ and radius $r = 4$ is, according to equation (2),

$$(x - 1)^2 + [y - (-2)]^2 = 4^2,$$

which simplifies to

$$(x - 1)^2 + (y + 2)^2 = 16,$$

or

$$x^2 - 2x + y^2 + 4y - 11 = 0. \qquad ■$$

Example 3 had to do with finding an equation for a circle, given certain data. Example 4 involves the reverse problem—describing a circle from its equation. The procedure for doing so involves **completing the square.**

☐ **EXAMPLE 4**

Describe and sketch the graph of the equation

$$x^2 + y^2 + 6x - 2y + 6 = 0.$$

Solution

We write the given equation as

$$(x^2 + 6x) + (y^2 - 2y) + 6 = 0. \qquad (3)$$

To complete the square on the first term, we add and subtract the square of half the coefficient of x, that is, we add and subtract $(6/2)^2 = 9$. This gives

$$(x^2 + 6x) = (x^2 + 6x + 9) - 9 = (x + 3)^2 - 9. \qquad (4)$$

Similarly, we complete the square on the second term as

$$(y^2 - 2y) = (y^2 - 2y + 1) - 1 = (y - 1)^2 - 1. \qquad (5)$$

Substituting the expressions in equations (4) and (5) into equation (3) gives the equation

$$[(x + 3)^2 - 9] + [(y - 1)^2 - 1] + 6 = 0$$

or

$$(x + 3)^2 + (y - 1)^2 = 4.$$

The graph is therefore a circle with center $(-3, 1)$ and radius $r = \sqrt{4} = 2$ (see Figure 2.6).   ■

Exercise Set 1.2

1. Find the distance between the following points.
   a. $(2, -1)$ and $(0, 2)$ **d.** $(1, -3)$ and $(6, 6)$
   b. $(3, 1)$ and $(1, 3)$ **e.** $(1, 1)$ and $(-1, -1)$
   c. $(0, 2)$ and $(1, -9)$ **f.** $(-2, -2)$ and $(1, 2)$

2. Find the two points with x-coordinate 5 whose distance from the point $(1, 3)$ is 5.

3. Find two points in the xy-coordinate plane, each of which lies a distance of 4 units from both $(-2, 3)$ and $(2, 3)$.

4. True or false? *Every* point in the plane lying at a distance r from point P must lie on the circle with center P and radius r.

5. Find a so that the triangle with vertices $(-1, 0)$, $(2, 3)$, and $(a, 0)$ is isosceles. (*Hint:* There are four possible answers.)

6. Find b so that the triangle with vertices $(0, 0)$, $(2, 0)$, and $(1, b)$ is equilateral.

7. Find an equation for the set of all points lying equidistant from the points $(-1, -1)$ and $(3, 1)$.

8. Find an equation for the set of all points lying equidistant from the points $(1, 2)$ and $(5, -1)$.

9. Verify that the coordinates of the **midpoint** of the line segment joining the points (x_1, y_1) and (x_2, y_2) are

$$\left(\frac{x_1 + x_2}{2}, \frac{y_1 + y_2}{2} \right).$$

(*Hint:* Use the distance formula.)

10. Find the midpoints of the line segments joining the following pairs of points. (See Exercise 9.)
   a. $(-1, 2)$ and $(6, 4)$
   b. $(1, 1)$ and $(-7, -3)$

In Exercises 11–14, sketch the set of points in the xy-plane satisfying the given property. Then specify an equation or inequality that determines the set.

11. The set of all points lying more than $\sqrt{2}$ units from the point $(-3, 1)$.

12. The set of all points lying at least 3 units to the right of the y-axis and at least 3 units from the x-axis.

13. The set of all points lying no more than 2 units from both axes.

14. The set of all points lying more than 3 but less than 5 units from the point $(1, -2)$.

In Exercises 15–17, write the equation for the circle with the stated properties.

15. The circle with center $(0, 0)$ and radius 3.

16. The circle with center $(-2, 4)$ and radius 3.

17. The circle with center $(-6, -4)$ and radius 5.

18. Find an equation for the circle with center $(2, 3)$ that contains the point $(2, -1)$.

19. Find equations for the circles with radius 5 and containing the points $(-1, -4)$ and $(4, 1)$.

In each of Exercises 20–25, find the center and radius of the circle whose equation is given, and sketch the circle.

20. $x^2 - 2x + y^2 - 8 = 0$

21. $x^2 + y^2 + 4x + 2y - 11 = 0$

22. $x^2 + 14x + y^2 - 10y + 70 = 0$

23. $x^2 + y^2 - 2x - 6y + 3 = 0$

24. $x^2 - 2ax + y^2 + 4ay + 5a^2 - 1 = 0$

25. $x^2 + y^2 - 2by + b^2 - a^2 = 0$

26. Find an equation for the circle with center $(3, 2)$ that is tangent to the y-axis (that is, which intersects the y-axis in a single point).

27. Find the points of intersection of the line $x - y - 5 = 0$ with the circle $x^2 - 8x + y^2 - 4y + 11 = 0$.

28. True or false? Three points determine a circle. Why?

29. Show that the set of all points whose distance from the point $(-2, 1)$ is twice the distance from the point $(4, -2)$ is a circle. Find its radius and center.

30. Find an equation for the circle containing the points $(-2, 1)$, $(1, 4)$, and $(4, 1)$.

31. Find the family of *all* circles containing the two points $(1, 2)$ and $(-1, 2)$.

32. Show that the equation $(x - h)^2 + (y - k)^2 = d$ will have (1) infinitely many solutions (i.e., the points on a circle), (2) exactly one solution, or (3) no solutions, depending on whether d is positive, zero, or negative. (The last two cases are referred to as **degenerate** circles.)

Graphs can be used to suggest solution sets for inequalities. For example, if we want to solve

$$\frac{3x - 2}{x + 3} > 1,$$

we can graph $y = (3x - 2)/(x + 3)$ and $y = 1$.

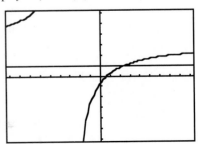

The graphs suggest that

$$\frac{3x - 2}{x + 3} > 1$$

only if $x < -3$ or $x > a$ where a is the solution to the equation

$$\frac{3a - 2}{a + 3} = 1;$$

namely, $a = 5/2$. Thus, the solution set is $(-\infty, -3) \cup (5/2, +\infty)$. In Exercises 33–36, use this technique to find the solution set of the given inequality.

33. $x^2 - x - 6 \leq 0$

34. $\dfrac{2x - 3}{x + 4} > 3$

35. $2x + 3 > -x^2 + 4x + 7$

36. $\left| \dfrac{3 - 2x}{5x + 3} \right| < 1$

In Exercises 37–40, consider the given equation as a quadratic expression in the variable y. Solve for y and graph the equation.

37. $4x^2 - 2xy + 6y^2 = 5$

38. $10x^2 + 8xy - 16y^2 - 16x - 4y = 3$

39. $x^2 + 2xy + y^2 + 2x + 2y = 81$

40. $2x^2 + 6xy + 5y^2 - 4x + 6y = 8$

1.3   Linear Equations

Equations whose graphs are straight lines are very important in the calculus. Indeed, one of the most fundamental problems in the entire subject is that of finding the equation of the line tangent to a given curve at a given point. More generally, lines are important because they represent the simplest relationship between two variables. For example, we shall see that for a freely falling body the relationship between velocity and time is linear. Even when the relationship between two variables is nonlinear, scientists often "linearize" their models. That is, they find the straight line that "best approximates" the true relationship, usually out of a desire to keep their models as simple as possible.

Slope

If $P_1 = (x_1, y_1)$ and $P_2 = (x_2, y_2)$ are two distinct points on a nonvertical line ℓ, then $x_1 \neq x_2$ and the **slope** m of ℓ is defined by the ratio

$$m = \frac{y_2 - y_1}{x_2 - x_1} = \frac{\Delta y}{\Delta x}$$

(1)

where Δx represents the change in x and Δy represents the change in y (Figure 3.1).

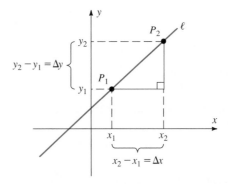

Figure 3.1   The slope $m = \dfrac{y_2 - y_1}{x_2 - x_1} = \dfrac{\Delta y}{\Delta x}$.

For a given line, the slope m is independent of the particular choices for (x_1, y_1) and (x_2, y_2). (See Exercise 44.) Notice that expression (1) for slope is not defined if $x_1 = x_2$. For this reason, we say that a vertical line has no slope. For all other lines slope is a real number, and every real number can occur as the slope of some line. Some illustrations of lines with various slopes appear in Figure 3.2.

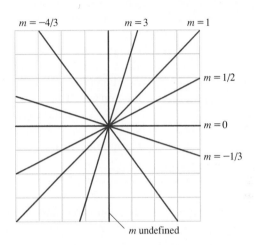

Figure 3.2 Seven lines, all passing through the same point in the plane, along with their associated slopes.

☐ **EXAMPLE 1**

Find the slope of the line through the points

(a) $(-7, 2)$ and $(3, -3)$ and
(b) $(-1, -4)$ and $(7, 16)$.

Solution From the definition of the slope m, we have

(a) $m = \dfrac{-3 - 2}{3 - (-7)} = \dfrac{-5}{10} = -\dfrac{1}{2}$ and

(b) $m = \dfrac{16 - (-4)}{7 - (-1)} = \dfrac{20}{8} = \dfrac{5}{2}.$ ∎

Equations for Lines

Suppose ℓ is the line with slope m that contains the point (x_1, y_1). To find an equation for ℓ we let $P = (x, y)$ be an arbitrary point on ℓ. Then, by the definition of the slope m, we obtain

$$m = \frac{y - y_1}{x - x_1}$$

so

$$\boxed{y - y_1 = m(x - x_1),}$$

(2a)

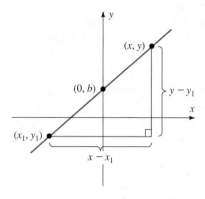

Figure 3.3 $y = mx + b$ has y-intercept b.

or

$$y = mx + b$$

(2b)

where $b = y_1 - mx_1$.

Equation (2a) is called the **point-slope** equation for the line ℓ because it involves the slope and one point of ℓ. Equation (2b) is called the **slope-intercept** equation for ℓ because the constants m and b appearing in equation (2b) are the slope and y-intercept for ℓ, respectively. To see this, notice that if we set $x = 0$ in equation (2b), we obtain the statement $y = b$. This means that the point $(0, b)$ is on the graph of ℓ (Figure 3.3).

Every nonvertical line can be described by an equation of the form (2b). Conversely, if (x_1, y_1) and (x_2, y_2) are any two distinct points whose coordinates satisfy equation (2b), then the slope of the line determined by these two points is

$$\text{Slope} = \frac{y_2 - y_1}{x_2 - x_1} = \frac{(mx_2 + b) - (mx_1 + b)}{x_2 - x_1} = \frac{m(x_2 - x_1)}{x_2 - x_1} = m.$$

It follows that any point (x, y) that satisfies equation (2b) lies on the line with slope m and y-intercept b. The following theorem summarizes our discussion.

Theorem 4

The graph of the equation $y = mx + b$ is a line with slope m and y-intercept b. Conversely, the line with slope m and y-intercept b has this equation.

□ **EXAMPLE 2**

Find an equation in slope-intercept form for the line through $(2, 5)$ with slope $m = 3$.

Solution   Substituting directly into the point-slope form gives

$$y - 5 = 3(x - 2),$$

so

$$y = 3x - 1$$

is the desired equation.   ■

□ **EXAMPLE 3**

Find an equation for the line through $(-3, 4)$ and $(1, -2)$.

Strategy · · · · · · · ·

First, find the slope using

$$m = \frac{\Delta y}{\Delta x}.$$

Then, use slope and one point to write an equation for the line.

Solution

By equation (1), the slope of the line is

$$m = \frac{-2 - 4}{1 - (-3)} = -\frac{6}{4} = -\frac{3}{2}.$$

Using the point $(-3, 4)$ and the slope $m = -\frac{3}{2}$ in the equation (2a) gives

$$y - 4 = (-\tfrac{3}{2})(x + 3),$$

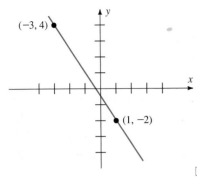

Figure 3.4 Line $y = -\frac{3}{2}x - \frac{1}{2}$.

Strategy · · · · · · · ·

Bring equation to slope-intercept form (equation (2b)).

Read off m and b.
Determine two points on the line.

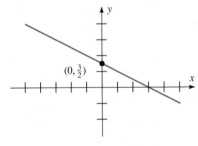

Figure 3.5 Line $y = -\frac{1}{2}x + \frac{3}{2}$.

Vertical and Horizontal Lines

Parallel and Perpendicular Lines

or

$$y = -\tfrac{3}{2}x - \tfrac{1}{2}.$$

(See Figure 3.4.)   ■

In Exercise Set 1.3, you are asked to show that any **general linear equation,** one of the form $ax + by + c = 0$ where a and b are not both zero, describes a line. In the next example, we convert a linear equation to one in slope-intercept form.

☐ **EXAMPLE 4**

Determine the slope of the line $2x + 4y - 6 = 0$ and sketch its graph.

Solution

We put the given equation in slope-intercept form by solving for y. We obtain

$$4y = -2x + 6,$$

so

$$y = -\tfrac{1}{2}x + \tfrac{3}{2}.$$

The line therefore has slope $-1/2$ and y-intercept $3/2$.

To graph the line, we need two points on it, and $b = 3/2$ yields the y-intercept $(0, 3/2)$. For a second point, we use the x-intercept $(3, 0)$, which is obtained by setting $y = 0$ in the original equation. The graph is plotted in Figure 3.5.   ■

As Example 4 illustrates, there are many linear equations that describe a given line. For instance, the linear equations

$$x + 2y - 3 = 0$$
$$2x + 4y - 6 = 0$$
$$y = -\tfrac{1}{2}x + \tfrac{3}{2}$$

are three among the infinitely many different equations that describe the line graphed in Figure 3.5.

A vertical line has the property that all x-coordinates of points on the line are the same, while the values of the y-coordinate(s) are unrestricted. The equation of a vertical line therefore has the form

$$x = a \qquad \text{(vertical line).}$$

A horizontal line has the property that its slope is zero. Setting $m = 0$ in equation (2b) shows that a horizontal line has an equation of the form

$$y = b \qquad \text{(horizontal line).}$$

Figure 3.6 shows the graphs of the vertical line $x = 2$ and the horizontal line $y = 1$.

We will make use of the following theorem from analytic geometry.

Theorem 5

Let ℓ_1 and ℓ_2 be two distinct lines with equations

$$\ell_1: y = m_1 x + b_1$$
$$\ell_2: y = m_2 x + b_2$$

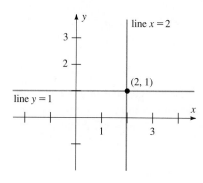

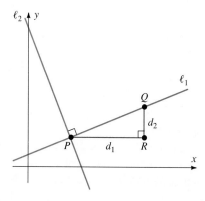

Figure 3.6   Vertical and horizontal lines.

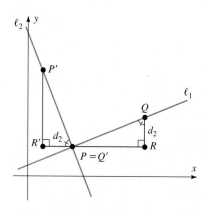

Figure 3.7   Two perpendicular lines.

where neither m_1 nor m_2 is zero. Then

(a) ℓ_1 and ℓ_2 are **parallel** if and only if $m_1 = m_2$.
(b) ℓ_1 and ℓ_2 are **perpendicular** if and only if $m_1 m_2 = -1$.

Of course, horizontal and vertical lines are not considered in Theorem 5 because a vertical line does not have slope. However, if we ignore this exceptional case, we may conclude that parallel lines have the same slope, while the slopes of perpendicular lines are negative reciprocals of each other.

The proof of (a) is treated in Exercise 54 at the end of this section.

Proof of (b):   Since (b) involves the phrase "if and only if," it is actually two logical assertions:

(i) if ℓ_1 and ℓ_2 are perpendicular, then $m_1 m_2 = -1$; and
(ii) if $m_1 m_2 = -1$, then ℓ_1 and ℓ_2 are perpendicular.

Thus, to prove (b), we must prove both (i) and (ii).

To establish (i), suppose that two perpendicular lines ℓ_1 and ℓ_2 intersect at the point $P = (x_0, y_0)$, and assume that the line with positive slope is ℓ_1. Choose any other point $Q = (x_1, y_1)$ on ℓ_1 with $x_1 > x_0$ and form the right triangle $\triangle PQR$ where $R = (x_1, y_0)$ (see Figure 3.7). The slope m_1 of ℓ_1 is the ratio d_2/d_1 of the lengths $d_1 = d(P, R)$ and $d_2 = d(R, Q)$ of the two legs of this triangle.

Let R' be the point $(x_0 - d_2, y_0)$ and P' be the point on ℓ_2 that determines a right triangle $\triangle PR'P'$ with a right angle at R' (see Figure 3.8). In other words, the x-coordinate of P' is the same as the x-coordinate of R'. For notational convenience, we assign P the additional label Q'.

The right triangles $\triangle PRQ$ and $\triangle P'R'Q'$ are congruent. In particular, $\angle PQR$ is equal to $\angle P'Q'R'$ because

$$\angle RPQ + \angle PQR = 90° = \angle RPQ + \angle P'Q'R'.$$

Therefore, $d(Q', R') = d_2$ and $d(P', R') = d_1$. Thus, when we calculate the slope m_2 of ℓ_2 using P' and Q', we get

$$m_2 = \frac{-d_1}{d_2} = -\frac{1}{m_1}.$$

To prove the reverse implication (ii), suppose that ℓ_2 is a line whose slope m_2 satisfies the equation $m_2 = 1/m_1$. Choose a third line ℓ_3 that is perpendicular to ℓ_1. Then, using (i), we know that the slope m_3 of ℓ_3 equals m_2; therefore, ℓ_2 and ℓ_3 are parallel. Since ℓ_3 is perpendicular to ℓ_1, we conclude that ℓ_2 is perpendicular to ℓ_1. ∎

Figure 3.8   The congruent right triangles $\triangle PRQ$ and $\triangle P'R'Q'$.

□ **EXAMPLE 5**

For the line ℓ_1: $3y - 9x = 12$, find

(a) an equation of the line ℓ_2 that is parallel to ℓ_1 and contains the point $(2, 0)$,
(b) an equation of the line ℓ_3 that is perpendicular to ℓ_1 and contains the point $(3, -2)$.

Strategy · · · · · · · ·
(a) Find the slope of ℓ_1 by putting equation in slope-intercept form.

Solution
(a) We first bring the equation for ℓ_1 into slope-intercept form by solving for y:

$$y = 3x + 4.$$

Take $m_2 = m_1$.

The slope of ℓ_1 is therefore $m_1 = 3$, so the slope of ℓ_2 must also be $m_2 = 3$. Since ℓ_2 contains the point $(2, 0)$, we obtain

$$y - 0 = 3(x - 2)$$

Use point-slope form to find equation for ℓ_2.

or

$$\ell_2: y = 3(x - 2).$$

(b) Slope of ℓ_3 is negative reciprocal of m_1.

(b) Line ℓ_3 is perpendicular to ℓ_1, so $m_3 = -1/m_1 = -1/3$. Since ℓ_3 contains the point $(3, -2)$ we obtain

$$y - (-2) = (-\tfrac{1}{3})(x - 3),$$

Use point-slope form to find equation for ℓ_3.

so

$$y + 2 = -\tfrac{1}{3}(x - 3),$$

or

$$\ell_3: y = -\tfrac{1}{3}x - 1.$$

(See Figure 3.9.)

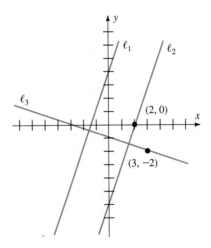

Figure 3.9 Lines in Example 5.

Points of Intersection

Lines in the plane that are not parallel must intersect. If the lines are distinct (that is, if they are not the same line), they intersect at a single point. Since the coordinates of this point of intersection must satisfy the equations for both lines, we can find the point of intersection by solving both equations for y and then equating these two expressions. The result is a first-degree equation in x that is easily solved.

☐ **EXAMPLE 6**

Find the point of intersection for the lines

$$\ell_1: 4y - 2y - 4 = 0,$$
$$\ell_2: x + y - 4 = 0.$$

Strategy · · · · · · ·

Solve both equations for y.

Solution

Solving both equations for y gives

$$\ell_1: y = 2x - 2,$$
$$\ell_2: y = -x + 4.$$

Equate expressions for y. Solve for x.

Substitute value for x into ℓ_1 or ℓ_2 to find y.

Equating expressions for y gives

$$2x - 2 = -x + 4$$

or $3x = 6$. Thus $x = 2$.

From the equation for ℓ_1 we find that $y = 2(2) - 2 = 2$. The point of intersection is therefore $(2, 2)$.   ∎

Of course, there are other ways to solve two linear equations involving two variables. For instance, you could solve both equations for x, or you could add a multiple of one equation to another in order to eliminate one variable. Which method is most efficient depends on the particular equations.

Exercise Set 1.3

1. Find the slope of the line through the following pairs of points
   a. $(3, 1)$ and $(-1, 2)$
   b. $(6, -2)$ and $(-1, -1)$
   c. (a, b) and (b, a), $a \neq b$
   d. $(1, 1)$ and (a, a)

2. True or false? Every line has a slope.

3. True or false? Every slope determines one and only one line.

4. True or false? If ℓ_1 and ℓ_2 are perpendicular lines then ℓ_1 and ℓ_2 have precisely one point of intersection.

5. True or false? If ℓ_1 and ℓ_2 are parallel and distinct, they do not intersect.

6. Find a if the line through $(2, 4)$ and $(-2, a)$ has slope 3.

7. Find b if the line through $(b, 1)$ and $(1, -5)$ has slope 2.

In Exercises 8–22, find an equation for the line determined by the given information.

8. Slope 4 and y-intercept -2.

9. Slope -2 and y-intercept 5.

10. Slope zero and y-intercept -5.

11. Through $(-1, 6)$ and $(4, 12)$.

12. Through $(4, 1)$ and with slope 7.

13. Through $(1, 3)$ and with slope -3.

14. x-intercept -3 and y-intercept 6.

15. Through $(-3, 5)$ and vertical.

16. Through $(-3, 5)$ and horizontal.

17. Through $(-2, 4)$ and $(-6, 8)$.

18. Through $(0, 2)$ and $(-1, -4)$.

19. Through $(1, 4)$ and parallel to the line with equation $2x - 6y + 5 = 0$.

20. Through $(5, -2)$ and parallel to the line with equation $x - y = 2$.

21. Through $(1, 3)$ and perpendicular to the line with equation $3x + y = 7$.

22. Through $(4, -1)$ and perpendicular to the line through the points $(-2, 5)$ and $(-1, 9)$.

23. True or false? If x and y satisfy a linear equation, then y is proportional to x.

24. True or false? If x and y satisfy a linear equation, then y is proportional to $x + a$ for some constant a.

In Exercises 25–32, find the slope, x-intercept, and y-intercept of the line determined by the given equation. Graph the line.

25. $x = 7 - y$

26. $3x - 2y = 6$

27. $x + y + 3 = 0$

28. $2x = 10 - 3y$

29. $y = 5$

30. $y - 2x = 9$

31. $x = 4$

32. $y = x$

In each of Exercises 33–38, find the point(s) of intersection of the two lines, if any.

33. $3x - y - 1 = 0$
    $x + y - 3 = 0$

34. $x - 2y + 3 = 0$
    $x - 2y - 7 = 0$

35. $x - 3y + 3 = 0$
    $2x - 3y + 6 = 0$

36. $x - 2y + 4 = 0$
    $3x + 6y - 12 = 0$

37. $2x - y - 2 = 0$
    $8x - 4y + 2 = 0$

38. $3x - 2y + 2 = 0$
    $3x - y = 0$

39. Explain why any equation of the form $ax + by + c = 0$, where a and b are not both zero, must have a graph that is a line. (*Hint:* Show that if $b \neq 0$, the equation can be put in slope-intercept form. What about the remaining case, $b = 0$?)

40. Because of the result in Exercise 39, the equation $ax + by + c = 0$ is often referred to as the **general linear equation.** If only two points are required to determine a line, why does this equation have three constants?

41. By checking slopes, determine whether the following sets of points lie on a common line.
    a. $(1, 3)$, $(-2, 0)$, and $(4, 6)$
    b. $(2, -7)$, $(-2, -3)$, and $(-1, -4)$
    c. $(5, 15)$, $(0, 3)$, and $(-2, -7)$

42. Determine whether or not the points $(1, 3)$, $(3, 5)$, and $(4, 0)$ form the vertices of a right triangle.

43. The points $(-2, 2)$, $(4, 4)$, and $(0, a)$ are the vertices of a right triangle.
    a. For how many distinct values of a is this condition satisfied?
    b. Find these values of a.

44. Use information about similar triangles to explain why, for a given line, the calculation of slope is independent of the particular points chosen.

45. A market research firm determines that the demand (d) for a certain product, in terms of purchasers per thousand population, is 12 when the product is priced at $P = 20$ dollars, but that the demand is only 6 when the item is priced at \$60. Write down a linear equation that models the relationship between demand (d) and price (P) according to these data. At what price level will demand reach zero?

46. A state has an income tax of 5% on all income over \$5000.
    a. Write a linear equation which determines a person's tax, T, in terms of income, i, for a person earning more than \$5000.
    b. At what income level will a person owe \$800 in state tax?

47. Find a linear equation which determines temperatures in degrees Fahrenheit from temperatures in degrees Celsius.

48. Find the distance from the point $(4, -2)$ to the line ℓ with equation $2x - y + 1 = 0$ as follows:
    a. Find the slope m of the line ℓ.
    b. The slope of the perpendicular form $(4, -2)$ to ℓ is then $-1/m$.
    c. Write the equation for this line.
    d. Find its point of intersection with ℓ?
    e. Compute the desired distance.

49. As a beaker of water is heated the following temperatures are noted:

time (minutes)	0	10	20
temperature (°C)	22	42	62

    a. Find a linear equation that relates time t to temperature T.
    b. According to this equation what will the temperature of the water be after 25 minutes? 35 minutes? (Using an equation obtained from data to predict values of one variable from stated values of the other variable is called **extrapolation**.)

50. Temperatures T on the absolute, or Kelvin, scale are related to temperatures t on the Celsius scale by the equation $T = t + 273$. At the Kelvin temperature 273 (0°C) the volume of a certain quantity of gas is 40 liters.
    a. Find a linear relationship for volume in terms of temperature, according to Charles' Law: $V = kT = k(t + 273)$.
    b. What is the volume of the gas at temperature 323 K?
    c. What is the volume of the gas at temperature 100°C?

51. The length of a rod ℓ is given by the linear equation $\ell = \ell_0(1 + at)$, where a is a constant called the coefficient of thermal expansion, ℓ_0 is the length of the rod at 0°C, and t is the temperature of the rod. Find a if the rod is 100 cm long at temperature 0°C and 100.2 cm long at temperature 50°C.

52. Find an equation for the line through the points of intersection of the circles $x^2 + y^2 - 2y = 0$ and $x^2 - 2x + y^2 = 0$.

53. Find an equation for the line through the centers of the circles with equations $x^2 - 2x + y^2 - 4y + 1 = 0$ and $x^2 + y^2 + 2y = 0$.

54. Let ℓ_1 and ℓ_2 be distinct lines with slopes m_1 and m_2, respectively. Prove that these two lines intersect if and only if $m_1 \neq m_2$.

55. One method of linear approximation is based on the concept of linear **interpolation**. (We estimate a desired quantity using neighboring known quantities.) For example, since $1.4^2 = 1.96$ and $1.5^2 = 2.25$, we know that $\sqrt{1.96} = 1.4$ and $\sqrt{2.25} = 1.5$. Estimate $\sqrt{2}$ as follows:
    a. Determine the slope of the line ℓ that contains the two points $(1.96, 1.4)$ and $(2.25, 1.5)$.
    b. Find the point on ℓ whose x-coordinate is 2.
    c. Using this point, provide an estimate for $\sqrt{2}$.
    d. Graph both $y = \sqrt{x}$ and ℓ on the same plot. Identify the approximation and the actual value of $\sqrt{2}$.

56. The following table is an amortization schedule for a one-year loan of \$4800.00 with an interest rate of 12% and a monthly payment of \$426.47.

Payment no. n	Payment R	Interest $i(n)$	Unpaid balance reduction $r(n)$	Unpaid balance $b(n)$
0				4800.00
1	426.47	48.00	378.47	4421.53
2	426.47	44.22	382.26	4039.27
3	426.47	40.39	386.08	3653.19
4	426.47	36.53	389.94	3263.24
5	426.47	32.63	393.84	2869.40
6	426.47	28.69	397.78	2471.62
7	426.47	24.72	401.76	2069.86
8	426.47	20.70	405.78	1664.09
9	426.47	16.64	409.83	1254.25
10	426.47	12.54	413.93	840.32
11	426.47	8.40	418.07	422.25
12	426.47	4.22	422.25	0
Total	5117.69	317.69	4800.00	

a. Plot the points $(n, i(n))$, $(n, r(n))$, $(n, b(n))$. Do these points appear to fall on straight lines?

b. Find equations of the lines that nearly fit the graphs of part a and draw the points and the lines together on the same graph. *Hint:* For example, use $(i(12) - i(1))/11$ for the slope of the line for the points $(i, i(n))$. Also, assume that this line contains the point $(6.5, \bar{\imath})$ where

$$\bar{\imath} = \frac{\sum\limits_{n=1}^{12} i(n)}{12}.$$

c. Compare the values obtained from the lines computed in part b to the values in the table. Is a straight line the correct function for modelling the amortization table?

1.4 Functions

A linear equation $y = mx + b$ is but one type of rule that assigns a *unique* number y to each number x. More generally, a *function* is *any* rule that assigns a unique *value* to each element of a given set. For example,

(a) The equation $y = x^3$ assigns to each number x its (unique) cube. Thus $y = x^3$ is a function.

(b) The rule "measure his or her height" assigns a unique number (height) to each member of your calculus class. It is also a function.

(c) The equation $x^2 + y^2 = 1$ does *not* determine a function, since it does not determine a unique value of y for a given value of x. This is most readily seen by writing the equation in the form $y = \pm\sqrt{1 - x^2}$. It is then clear, for example, that both $y = 1$ and $y = -1$ are assigned to $x = 0$, violating the property of uniqueness.

Here is a precise definition of the term "function".

Definition 5

A **function** from set A to set B is a rule that assigns to each element $x \in A$ a **unique** element $y \in B$. We write $y = f(x)$ to indicate that the element y is the value assigned by the function f to the element x. The set A is called the **domain** of the function f. The set of all values $\{f(x) \mid x \in A\}$ is called the **range** of the function.

In order to specify a function, we must indicate both the domain of the function and the rule by which values of the function are determined. We will almost always do this by writing an equation of the form $y = f(x)$, where $f(x)$ is an expression involving x's and constants. In this case, the letter f is the name of the function. It simply provides an efficient way to refer to the rule that it represents. Of course, there is nothing special about the letter f, and we often use other names.

When we define a function f using an expression $f(x)$, we usually assume that the domain of f consists of all numbers x for which $f(x)$ makes sense. However, if we wish to specify a special subset of the real numbers as the domain, we do so by writing a statement to this effect following $f(x)$ (see Example 2).

☐ **EXAMPLE 1**

For the function $f(x) = x^2$,

(a) the domain is the whole real line, since nothing has been indicated to restrict the domain;

(b) the range is the interval $[0, \infty)$, since all nonnegative numbers are squares (that is, have square roots), and all squares must be nonnegative.

(See Figure 4.1.)

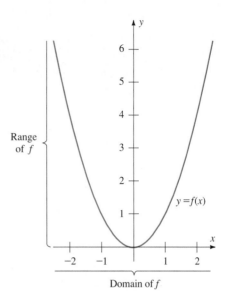

Figure 4.1 For the function $f(x) = x^2$, the domain is $(-\infty, \infty)$ and the range is $[0, \infty)$.

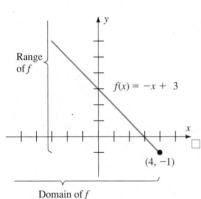

Figure 4.2 For $f(x) = -x + 3$, $x \le 4$, domain of f is $(-\infty, 4]$ and range of f is $[-1, \infty)$.

□ **EXAMPLE 2**

For the function $f(x) = -x + 3$, $x \le 4$,

(a) the domain is specified to be the interval $(-\infty, 4]$ by the inequality $x \le 4$;
(b) the range is the interval $[-1, \infty)$, since $f(x) = -x + 3 \ge -1$ whenever $x \le 4$, and $y = -x + 3$ has a solution in $(-\infty, 4]$ whenever $y \ge -1$.

(See Figure 4.2.)

□ **EXAMPLE 3**

For the function $f(x) = \dfrac{1}{x - 2}$,

(a) the domain is the set $(-\infty, 2) \cup (2, \infty)$. The number $x = 2$ is excluded from the domain since the expression $1/(x - 2)$ is undefined for $x = 2$.
(b) The range is the set $(-\infty, 0) \cup (0, \infty)$ since every value $f(x)$ is a real number other than zero, and the equation $y = 1/(x - 2)$ has a solution for every value y except $y = 0$ (see Figure 4.3.)

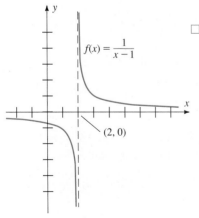

Figure 4.3 For $f(x) = 1/(x - 2)$, domain is $(-\infty, 2) \cup (2, \infty)$ and range is $(-\infty, 0) \cup (0, \infty)$.

Example 3 illustrates one reason certain numbers are not included in the domain of a function even though we do not explicitly exclude them when we define the function. The expression $f(x) = 1/(x - 2)$ has 0 in the denominator when $x = 2$. Hence the number 2 is not in the domain of f even though we did not indicate $x \ne 2$ when we defined $f(x)$. Example 4 illustrates another reason, even real roots only exist for nonnegative real numbers.

□ **EXAMPLE 4**

Find the domain and range of the function $f(x) = \sqrt{x + 3}$.

Solution   Because of the square root, $f(x)$ is defined only when $x + 3 \geq 0$, which gives $x \geq -3$. The domain is therefore $[-3, \infty)$. To determine the range, note that, as x spans the interval $[-3, \infty)$, the quantity $x + 3$ spans the interval $[0, \infty)$. Since $f(x) = \sqrt{x + 3}$, we see that the values $f(x)$ are the (nonnegative) square roots of all of the nonnegative real numbers. Since the set of square roots of all nonnegative real numbers is again the set of all nonnegative real numbers, the range of f is $[0, \infty)$ (see Figure 4.4). ■

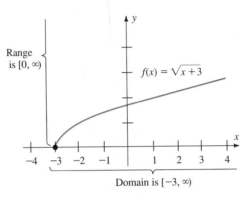

Figure 4.4   For $f(x) = \sqrt{x + 3}$, the domain is $[-3, \infty)$ and the range is $[0, \infty)$.

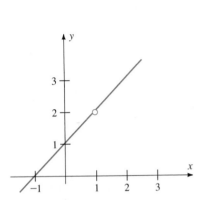

Figure 4.5   Graph of

$$f(x) = \frac{x^2 - 1}{x - 1}.$$

The value $f(1)$ is not defined.

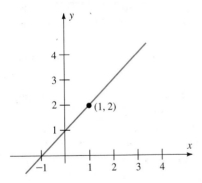

Figure 4.6   Graph of $g(x) = x + 1$. The value $g(1) = 2$.

In the next example, we illustrate the concept of equality for functions. That is, two functions f and g are equal if they have the same domain and if $f(x) = g(x)$ for all x in this domain.

□ **EXAMPLE 5**

The function

$$f(x) = \frac{x^2 - 1}{x - 1}$$

is not defined for $x = 1$. For $x \neq 1$ we can write this function as

$$f(x) = \frac{x^2 - 1}{x - 1} = \frac{(x - 1)(x + 1)}{x - 1} = x + 1, \quad x \neq 1.$$

That is, $f(x) = x + 1$, $x \neq 1$. This function is *not* the same as the function

$$g(x) = x + 1$$

which is defined for *all* x. Thus, even though $f(x) = g(x)$ for $x \neq 1$, the functions are different because their domains are different. (See Figures 4.5 and 4.6.) ■

Graphs of Functions: The Vertical Line Property

The **graph** of the function f is the set of all points (x, y) satisfying the equation $y = f(x)$. That is, the graph is the set

$$\{(x, y) \mid x \text{ is in the domain of } f \text{ and } y = f(x)\}.$$

There is a simple but important property that distinguishes graphs of functions from graphs of equations that are not functions. It is called the **vertical line prop-**

erty, and it means that *any vertical line* (that is, a line parallel to the *y*-axis) *can intersect the graph of a function at most once.* The reason for this is simple: a vertical line has equation $x = a$; but the graph of the *function f* has at most one point, $(a, f(a))$, whose *x*-coordinate is *a*. Figure 4.7 illustrates our previous conclusion, that $x^2 + y^2 = 1$ cannot be the graph of a function—the vertical line $x = a$ can intersect the graph twice. Figure 4.8 illustrates the vertical line property for an arbitrary function.

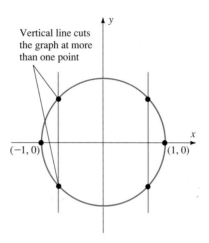

Figure 4.7 Graph of $x^2 + y^2 = 1$ does not have the vertical line property.

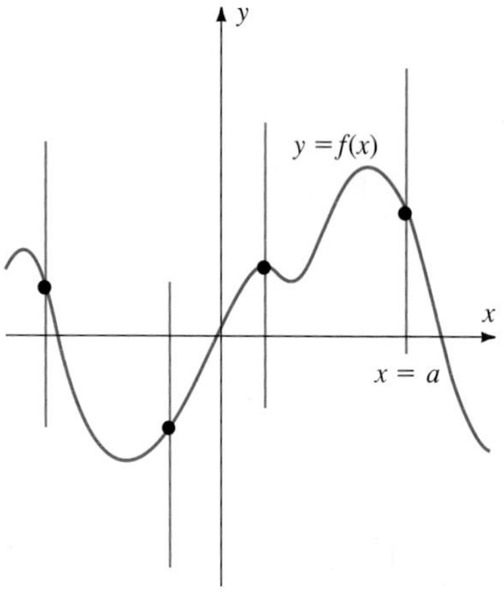

Figure 4.8 Vertical lines intersect the graph of a *function* at most once (Vertical Line Property).

Power and Quadratic Functions

Among the simplest types of functions are the power functions. These have the form

$$f(x) = x^n$$

where *n* is a positive integer. The domain of any power function is the set of all real numbers. If *n* is odd, the range is also the set of all real numbers; if *n* is even, the range is the set of all nonnegative real numbers. Graphs of several power functions appear in Figure 4.9.

Figures 4.10 through 4.12 show the effect of introducing various constants into the equation $y = x^2$.

(a) The graph of $y = ax^2$, $a > 0$, opens upward, but does so more or less quickly, depending on whether the constant *a* is large or small (Figure 4.10).
(b) The graph of $y = ax^2$, $a < 0$, opens downward (Figure 4.10).
(c) The graph of $y = ax^2 + b$ is like the graph of $y = ax^2$, except that it is "shifted" *b* units upward ($b > 0$) or $|b|$ units downward ($b < 0$) (Figure 4.11).
(d) The graph of $y = a(x - c)^2 + b$ is like the graph of $y = ax^2 + b$, except that it is symmetric with respect to the vertical line $x = c$ rather than with respect to the *y*-axis (Figure 4.12).

A **quadratic function** is any function that can be written in the form

$$f(x) = Ax^2 + Bx + C \tag{1}$$

where *A*, *B*, and *C* are constants. By completing the square on the right-hand side of equation (1), we can bring the equation into the form $f(x) = a(x - c)^2 + b$. Thus, the graph of any quadratic function will be of the form illustrated in Figure 4.12. (Such curves are called **parabolas,** and they are discussed in detail in Chapter 14.)

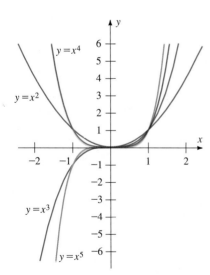

Figure 4.9 The graphs of four basic power functions. Note that the range for those with an even power is $[0, \infty)$ and for those with an odd power is $(-\infty, \infty)$.

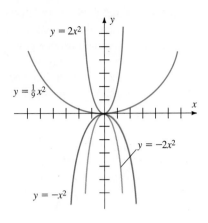

Figure 4.10   Functions of the form $f(x) = ax^2$.

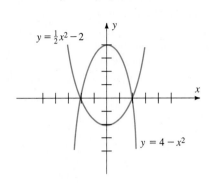

Figure 4.11   Functions of the form $f(x) = ax^2 + b$.

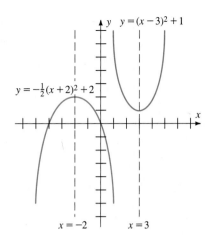

Figure 4.12   Functions of the form $f(x) = a(x - c)^2 + b$.

☐ **EXAMPLE 6**

Sketch the graph of the quadratic function $f(x) = \frac{1}{4}x^2 - x$.

Strategy · · · · · · · ·

Complete the square in x by

(1) factoring by $\frac{1}{4}$ and
(2) adding the square of half the coefficient of x.

Solution

We have

$$f(x) = \tfrac{1}{4}x^2 - x = \tfrac{1}{4}(x^2 - 4x)$$
$$= \tfrac{1}{4}(x^2 - 4x + 4) - \tfrac{1}{4}(4)$$
$$= \tfrac{1}{4}(x - 2)^2 - 1,$$

so

$$f(x) = \tfrac{1}{4}(x - 2)^2 - 1.$$

The graph appears in Figure 4.13.   ■

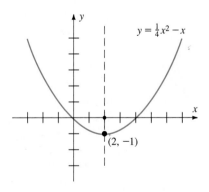

Figure 4.13   Graph of the quadratic function $f(x) = \frac{1}{4}x^2 - x$.

Polynomial Functions

A **polynomial function** is a function that can be written in the form

$$f(x) = a_n x^n + a_{n-1}x^{n-1} + \cdots + a_1 x + a_0$$

where $a_0, a_1, \ldots, a_n$ are constants and n is a positive integer. The integer n is called the **degree** of the polynomial, provided that $a_n \neq 0$. The following are examples of polynomial functions:

$$f(x) = x^3 - 7x + 2 \qquad \text{(degree 3)}$$
$$f(x) = 1 - x^{10} \qquad \text{(degree 10)}$$
$$y = (x - 3)^4 - 2(x + 5)^3 \qquad \text{(degree 4)}.$$

The domain of any polynomial function is the set of all real numbers.

Fractional Power Functions

Fractional powers can be used to define power functions that are more general than the integral power functions mentioned earlier. If m and n are positive integers with no common factors, then the expression $x^{n/m}$ is defined to be the mth root of the nth power of x. Since we cannot take square roots, fourth roots, or in general, roots of even order of negative numbers, the domain of the function

$$f(x) = x^{n/m}$$

depends on whether m is even or odd. For example, the domain of $f(x) = x^{1/3}$ is the set of all real numbers, while the domain of $f(x) = x^{5/6}$ is the set of all nonnegative real numbers. Several fractional power functions are graphed in Figures 4.14 through 4.17. Note how the exponent determines the domain and range of the function.

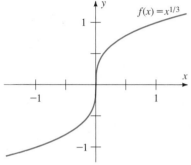

Figure 4.14 The graph of $f(x) = x^{1/3}$. Note that the domain of this power function is the set of all real numbers.

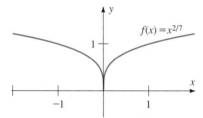

Figure 4.15 The graph of $f(x) = x^{2/7}$. The domain is the set of all real numbers, and since $x^{2/7} = (x^{1/7})^2$, the range is the set of all nonnegative real numbers.

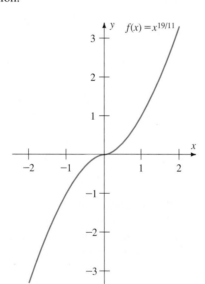

Figure 4.16 The graph of $f(x) = x^{19/11}$. Since the numerator and denominator of the exponent are odd integers, both the domain and range are the set of all real numbers.

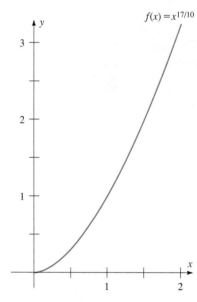

Figure 4.17 The graph of $f(x) = x^{17/10}$. Since the denominator in the exponent is an even integer, the domain is the set of all nonnegative real numbers.

□ **EXAMPLE 7**

Find the domain and range of the function

$$f(x) = (x^2 - x - 2)^{3/2}.$$

(See Figure 4.18.)

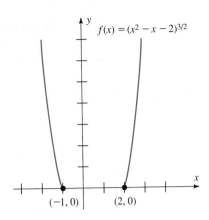

Figure 4.18 For $f(x) = (x^2 - x - 2)^{3/2}$, domain $= (-\infty, -1] \cup [2, \infty)$ and range $= [0, \infty)$.

Strategy · · · · · · · · ·

Note that an even root requires a non-negative base. Write this as an inequality

Factor the left-hand side and find its zeros.

Examine the signs of individual factors between zeros to determine sign of $x^2 - x - 2$ on each of these intervals.

The domain of f is the solution set of inequality (2).

Determine the range by asking, "What values $f(x)$ occur if x is in the domain of f?"

Solution

Since the denominator of the exponent is even, the function is defined only when

$$x^2 - x - 2 \geq 0. \tag{2}$$

To determine the solution set of inequality (2), we write it in factored form as

$$(x - 2)(x + 1) \geq 0.$$

Since $(x - 2)(x + 1) = 0$ for $x = -1$ and $x = 2$, we must examine the signs of the factors on the intervals $(-\infty, -1)$, $(-1, 2)$, and $(2, \infty)$.

(i) If If $x < -1$, then $x - 2 < 0$ and $x + 1 < 0$, so $(x - 2)(x + 1) > 0$.
(ii) If $-1 < x < 2$, then $x - 2 < 0$ but $x + 1 > 0$, so $(x - 2)(x + 1) < 0$.
(iii) If $x > 2$, both $x - 2 > 0$ and $x + 1 > 0$, so $(x - 2)(x + 1) > 0$.

From statements (i)–(iii), it follows that the solution set of inequality (2) is $(-\infty, -1] \cup [2, \infty)$. This is the domain. To determine the range we write $f(x)$ as

$$f(x) = [(x - 2)(x + 1)]^{3/2}.$$

Now consider this expression as x starts at $x = 2$ and increases. The product inside the braces starts at zero and increases through all positive numbers. Since cubes and square roots of positive numbers are positive numbers, $f(x)$ can equal any nonnegative number. Since even roots cannot be negative, the range of f is simply $[0, \infty)$. ∎

Negative powers represent multiplicative inverses. That is, if n/m is a positive rational number, then the function

$$f(x) = x^{-n/m}$$

is defined by

$$f(x) = \frac{1}{x^{n/m}}.$$

Since we cannot divide by zero, we must exclude 0 from the domain of a negative fractional power. Two power functions with negative fractional powers are graphed in Figures 4.19 and 4.20. As with positive fractional powers, the fraction determines the domain and range of the function.

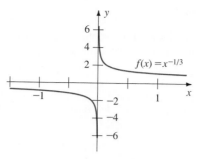

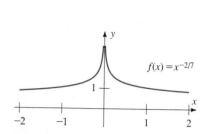

Figure 4.19 The graph of $f(x) = x^{-1/3}$. Note that both its domain and range are the set of all nonzero real numbers.

Figure 4.20 The graph of $f(x) = x^{-2/7}$. Its domain is the set of all nonzero real numbers, and its range is the set of all positive real numbers.

When dealing with combinations of fractional powers, the following **laws for exponents** help simplify the expressions involved. Let x and y be real numbers and m and n be integers. Then

1. $x^n x^m = x^{n+m}$
2. $x^n/x^m = x^{n-m}$, $x \neq 0$
3. $(x^n)^m = x^{nm}$
4. $x^{n/m} = (x^{1/m})^n = (x^n)^{1/m}$
5. $(xy)^{n/m} = (x^{n/m})(y^{n/m})$

Rational Functions

A **rational function** is a quotient of two polynomial functions. A rational function is not defined for numbers x for which its denominator equals zero. For example, the rational function

$$f(x) = \frac{x^2 + 4}{x^3 + x^2 - 12x}$$

is not defined for $x = 0$, 3, or -4 since the denominator $x^3 + x^2 - 12x = x(x-3)(x+4)$ equals zero for these numbers x. Therefore, the domain of f is the union of the three intervals $(-\infty, -4)$, $(-4, 3)$, and $(3, +\infty)$.

Absolute Value Function

The absolute value $|x|$, as defined in Section 1.1, gives rise to the absolute value function $f(x) = |x|$. The graph of $f(x) = |x|$ appears in Figure 4.21.

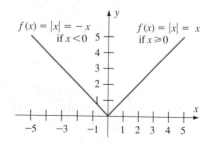

Figure 4.21 Graph of absolute value function.

☐ **EXAMPLE 8**

You can see the effect of the absolute value signs in the function $f(x) = |4 - x^2|$ by comparing Figure 4.22 with the graph of $g(x) = 4 - x^2$ in Figure 4.23. The "legs" of the graph of $4 - x^2$, which otherwise extend below the x-axis, have been turned upward. ■

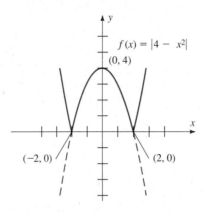

Figure 4.22   Graph of $f(x) = |4 - x^2|$. (Compare with Figure 4.23).

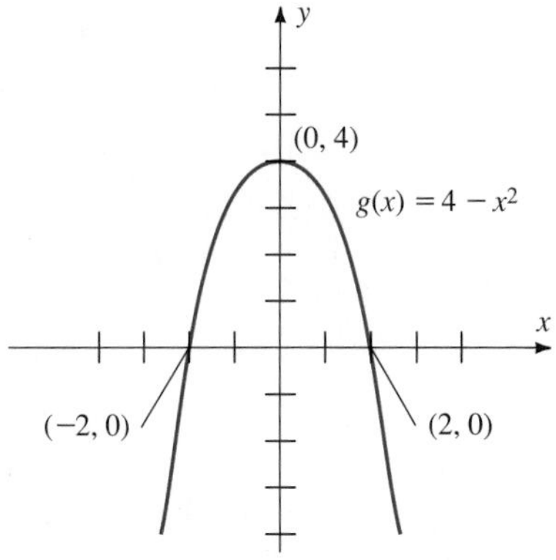

Figure 4.23   Graph of $g(x) = 4 - x^2$.

The Algebra of Functions

It is a simple matter to combine functions algebraically to form new functions that are sums, differences, multiples, products, or quotients of given functions. It is done by calculating the values of the individual terms in the combination separately and then forming the given combination using those values.

Definition 6

Given the functions f and g and the number c, the functions $f + g, f - g, cf, fg,$ and f/g are defined by the following equations:

$$(f + g)(x) = f(x) + g(x)$$
$$(f - g)(x) = f(x) - g(x)$$
$$(cf)(x) = cf(x)$$
$$(fg)(x) = f(x)g(x)$$
$$(f/g)(x) = f(x)/g(x) \qquad \text{(provided } g(x) \neq 0).$$

Of course, these combinations can be formed only for numbers x that are in the domains of both functions. While these definitions may seem obvious, we will need to understand them thoroughly to study limits of functions in Chapter 2.

☐ **EXAMPLE 9**

Let $f(x) = x^3 + 2$ and $g(x) = \sqrt{x + 1}$. Then the sum $f + g$ is defined by the equation

$$(f + g)(x) = f(x) + g(x)$$
$$= [x^3 + 2] + \sqrt{x + 1};$$

for example,

$$(f + g)(3) = f(3) + g(3)$$
$$= [3^3 + 2] + \sqrt{3 + 1}$$
$$= 29 + 2 = 31.$$

But $(f + g)(-2)$ is undefined since $g(-2) = \sqrt{-2 + 1}$ is undefined.

The quotient function $f/(3g)$ is defined by the equation

$$\left(\frac{f}{3g}\right)(x) = \frac{f(x)}{3g(x)}$$
$$= \frac{x^3 + 2}{3\sqrt{x + 1}}$$

so

$$\left(\frac{f}{3g}\right)(0) = \frac{0^3 + 2}{3\sqrt{0 + 1}}$$
$$= \frac{2}{3}.$$

☐ **EXAMPLE 10**

Let $f(x) = \sqrt[3]{x + 2}$ and $g(x) = \sqrt{x + 1}$. Using the laws of exponents, simplify the expression for $h = fg + f^4/g$, where f^4 represents $f \cdot f \cdot f \cdot f$.

Solution   From Definition 6 and the laws of exponents, we have

$$h(x) = \sqrt[3]{x + 2}\sqrt{x + 1} + \frac{(x + 2)^{4/3}}{\sqrt{x + 1}}$$
$$= \frac{\sqrt[3]{x + 2}(x + 1) + (x + 2)^{4/3}}{\sqrt{x + 1}}$$
$$= \frac{\sqrt[3]{x + 2}(x + 1) + \sqrt[3]{x + 2}(x + 2)}{\sqrt{x + 1}}$$
$$= \frac{\sqrt[3]{x + 2}}{\sqrt{x + 1}}[(x + 1) + (x + 2)]$$
$$= \frac{\sqrt[3]{x + 2}}{\sqrt{x + 1}}(2x + 3).$$

Composite Functions

Often, two functions are combined not by an algebraic operation such as addition, but by letting the second function act on values of the first function. The result is called a *composite function,* a composition of the functions.

For example, the function $h(x) = |1 - x^2|$ can be constructed out of the absolute value function and the quadratic function $1 - x^2$. Given a number x we first find $1 - x^2$, and then the absolute value $|1 - x^2|$. Here h is called a composite of the two functions, and we say that the absolute value function acts on values of the quadratic function.

Definition 7

Given two functions f and g, the **composite function** $f \circ g$ is the result of the function f acting on values of the function g. That is,

$$(f \circ g)(x) = f(g(x)).$$

(See Figure 4.24.)

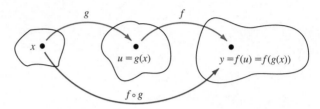

Figure 4.24 The composite function $f \circ g$ is the result of the function f acting on values of the function g.

The distinction between composite functions, or functions formed *by composition,* and algebraic combinations of functions involves the order in which the operations are performed. In an algebraic combination the individual function values are found first and then combined. In the composite function $f \circ g$ we first find the value of the *inside* function $u = g(x)$, and then insert this number u into the *outside* function f to find $f(u) = f(g(x))$.

Thus, the *domain* of the composite function $f \circ g$ is the set of all x in the domain of g for which the number $u = g(x)$ lies in the domain of f.

□ **EXAMPLE 11**

For the functions $f(x) = \sqrt{x}$ and $g(x) = x + 4$,

(a) the composite function $f \circ g$ is

$$(f \circ g)(x) = f(g(x)) = f(x + 4) = \sqrt{x + 4}$$

with domain $[-4, \infty)$, and

(b) the composite function $g \circ f$ is

$$(g \circ f)(x) = g(f(x)) = g(\sqrt{x}) = \sqrt{x} + 4$$

with domain $[0, \infty)$. ∎

□ **EXAMPLE 12**

The function

$$y = \frac{1}{x^2 + 4}$$

may be thought of as a composite function $y = f(g(x))$ where the "inside" function is $u = g(x) = x^2 + 4$ and the "outside" function is $y = f(u) = 1/u$. That is,

$$y = f(u) = \frac{1}{u} = \frac{1}{g(x)} = \frac{1}{x^2 + 4}.$$ ∎

□ **EXAMPLE 13**

The following list shows how several additional functions may be regarded as composite functions of the form $(f \circ g)(x) = f(g(x))$.

Composite function $y = f(g(x))$	Outside function $y = f(u)$	Inside function $u = g(x)$
$y = (x^2 + 7)^{2/3}$	$f(u) = u^{2/3}$	$g(x) = x^2 + 7$
$y = 4 + \sqrt{2 + x^2}$	$f(u) = 4 + \sqrt{u}$	$g(x) = 2 + x^2$
$y = 4 + \sqrt{2 + x^2}$	$f(u) = 4 + u$	$g(x) = \sqrt{2 + x^2}$
$y = \dfrac{8}{|3 + x^4|}$	$f(u) = \dfrac{8}{|u|}$	$g(x) = 3 + x^4$
$y = \dfrac{8}{|3 + x^4|}$	$f(u) = \dfrac{8}{u}$	$g(x) = |3 + x^4|$

The last four entries illustrate that choices for the "inside" and "outside" functions are not necessarily unique. ∎

Finally, we emphasize that the domain of a composite function may not be immediately apparent from its simplified expression.

□ **EXAMPLE 14**

Let $f(x) = x^2 + 1$, $g(x) = \sqrt{x - 4}$, and $h = f \circ g$. Determine the domain of h.

Solution Recall that the domain of $f \circ g$ is the set of all x in the domain of g whose values $g(x)$ lie in the domain of f. Since the domain of f is the set of all real numbers, we see that the domain of h is the same as the domain of g. Thus, the domain of g is

$$\{x \mid x \geq 4\} = [4, \infty).$$ ∎

If we simplify $h(x) = f(g(x))$ in Example 14, we obtain

$$h(x) = (\sqrt{x - 4})^2 + 1 = x - 3.$$

Of course, the domain of the function $H(x) = x - 3$ is the set of all real numbers; consequently, you may be tempted to conclude that the domain of h is the set of all real numbers. To avoid making this mistake, remember that the definition of any particular function must specify both a domain and a rule that determines the values of the function. In this case, both h and H use the same rule, but they are different functions because they are defined on different domains. The function h is a composition of f and g, but H is defined by the rule $H(x) = x - 3$ with the implicit assumption that its domain is the set of all real numbers (see Example 5 earlier in this section).

Exercise Set 1.4

1. Complete the following chart.

$f(x)$	$f(-2)$	$f(0)$	$f(4)$	$f(5)$	$|f(3)|$	$f(f(0))$
$1 - 3x^2$	-11	1	-47	-74	26	-2
$\dfrac{1}{x + 2}$	undefined	$\dfrac{1}{2}$				
$\dfrac{(x - 3)^2}{x^2 + 1}$	5					
$\sqrt{x + 4}$						
$\dfrac{1}{\sqrt{16 - x^2}}$						
$\begin{cases} 1 - x, x < -3 \\ x - 1, x > 1 \end{cases}$						

In Exercises 2–7, determine whether the given equation determines y as a function of x.

2. $2x - y = 7$ **3.** $xy = 5$

4. $xy^2 + 2x = 6$ **5.** $x^2 + 2x + y^2 = 8$

6. $x - y = x + y$ **7.** $y = 5$

8. True or false? The functions

$$f(x) = \frac{x^2 - 4}{x + 2}$$

and $g(x) = x - 2$ are equal. (*Hint:* Do they have the same equations? Domains?)

In Exercises 9–26, state the domain of the given function.

9. $f(x) = x^5 - 5x^3$ **10.** $f(x) = \dfrac{x - 1}{x + 1}, x > 3$

11. $g(x) = \sqrt{x + 4}$ **12.** $h(x) = \sqrt{x(x - 1)}$

13. $f(t) = 1 + t^2, t \ge 0$ **14.** $f(x) = \dfrac{1}{1 - |x|}$

15. $h(s) = \sqrt{16 - s^2}$ **16.** $g(t) = \dfrac{1}{t^2 + 2t - 35}$

17. $f(x) = \begin{cases} 1/x, x > 0 \\ -1/x, x < 0 \end{cases}$ **18.** $f(s) = s(s + 3)\sqrt{1 - s^2}$

19. $h(x) = \dfrac{\sqrt{x^2 + 2x}}{x - 2}$ **20.** $f(x) = \sqrt{6 - |x + 2|}$

21. $f(x) = (x - 2)^{3/2}$ **22.** $g(x) = (x - 2)^{3/5}$

23. $h(s) = (s^2 + 1)^{3/2}$ **24.** $f(t) = (t^2 - 3t + 2)^{3/4}$

25. $g(x) = (x + 1)^{-2/3}$ **26.** $h(t) = (t^2 - 1)^{-7/4}$

27. True or false? Let f be the rational function

$$f(x) = \frac{g(x)}{h(x)}.$$

If $h(x)$ is a polynomial of degree n, then $f(x)$ will be undefined for n distinct values of x.

28. True or false? If a horizontal line intersects the graph of the equation $y = f(x)$ in more than one point, the equation $y = f(x)$ cannot determine y as a function of x. Explain.

For each of the equations in Exercises 29–34, y is determined as a quadratic function of x. Graph each by first completing the square.

29. $3x^2 - 2y - 8 = 0$ **30.** $3x^2 - y - 7 = 0$

31. $y - 3x^2 + x - \frac{1}{2} = 0$ **32.** $y - 2x^2 + 4x - 5 = 0$

33. $4x^2 + y - 24x + 34 = 0$ **34.** $y - \frac{1}{2}x^2 - \frac{1}{2}x - \frac{11}{24} = 0$

An equation of the form $x = ky^2$ determines x as a quadratic function of y. The graph will therefore be a parabola opening in either the positive x direction ($k > 0$) or the negative x direction ($k < 0$). In each of the Exercises 35–40, graph the given parabola.

35. $y^2 + x - 2 = 0$ **36.** $3y^2 - x - 4 = 0$

37. $3x + 2y^2 - 3 = 0$ **38.** $x - 3y^2 + 6y - 4 = 0$

39. $x - 2y^2 + 4y - 2 = 0$ **40.** $4y^2 - x - 24y + 34 = 0$

41. Find the point(s) of intersection of the line $x - y - 2 = 0$ and the parabola $y = 4 - x^2$.

42. Find the points of intersection of the parabolas $y = 6 - x^2$ and $y = x^2 - 2$.

In each of Exercises 43–50, graph the given function.

43. $f(x) = \dfrac{1}{x-1}$

44. $f(x) = \sqrt{2x+1}$

45. $f(x) = \begin{cases} 7 - x, & x \le 2 \\ 2x + 1, & x > 2 \end{cases}$

46. $f(x) = x(4 - 2x)$

47. $f(x) = \dfrac{x^2 - 1}{x - 1}$

48. $f(x) = \begin{cases} \sqrt{x+2}, & -2 \le x \le 2 \\ 4 - x, & x > 2 \end{cases}$

49. $f(x) = (x - 1)^{3/2}$

50. $f(x) = (x + 2)^{2/3}$

In Exercises 51–56, sketch the graph by first noting where the expression inside the absolute value signs changes sign.

51. $f(x) = x + |x|$

52. $y = |1 - x^2|$

53. $y = \dfrac{x+1}{|x-1|}$

54. $f(x) = |x^2 - x + 2|$

55. $f(t) = \dfrac{1}{|t|}$

56. $g(s) = |s^2 - 1|$

57. A function $f(x)$ is called **even** if whenever x is in the domain of f, so is $-x$, and $f(-x) = f(x)$. It is called **odd** if whenever x is in the domain of f, so is $-x$, and $f(-x) = -f(x)$. Determine whether $f(x)$ is even, odd, or neither, and sketch the graph.
a. $f(x) = x^2$
b. $f(x) = 2 - x^2$
c. $f(x) = x^3$
d. $f(x) = 1 - x^3$
e. $f(x) = x^3 + x$
f. $f(x) = 2x^4 + x^2$
g. $f(x) = |x| + 2$
h. $f(x) = x^2 + x$

58. What symmetry properties does the graph of an even function possess? An odd function? (See Exercise 57).

In Exercises 59–66, use the functions
$$f(x) = 3x + 1 \qquad g(x) = x^3 \qquad h(x) = \sqrt{x}$$
to form the indicated composite function.

59. $f \circ g$

60. $f \circ h$

61. $g \circ f$

62. $h \circ g$

63. $h(f(x))$

64. $h(f(g(x)))$

65. $f(g(h(x)))$

66. $g(h(f(x)))$

67. Given the function $f(x) = x^2$ with domain $\{x \mid -1 \le x \le 3\}$ and the function $g(x) = 2x + 6$ find
a. the domain of $f \circ g$
b. the range of $f \circ g$
c. the domain of $g \circ f$
d. the range of $g \circ f$.

68. Given the function $f(x) = x^2$ and the function $g(x) = 2x + 6$ with domain $\{x \mid -5 \le x \le 4\}$ find
a. the domain of $f \circ g$
b. the range of $f \circ g$

c. the domain of $g \circ f$
d. the range of $g \circ f$.

69. Given $f(x) = 1/(3 - x^2)$ and $g(x) = \sqrt{x^2 - 1}$, determine the domains of $f \circ g$ and $g \circ f$.

70. Given $f(x) = x/(x + 1)$ and $g(x) = 1/(x - 1)$, determine the domains of $f \circ g$ and $g \circ f$.

71. Let
$$f(x) = x(x^2 + 1)^{-1/2} - \frac{2\sqrt{x^2 + 1}}{x^3} + \frac{3}{x^3\sqrt{x^2 + 1}}.$$
Determine the domain of f. Does the range of f contain the number 0?

72. Let
$$f(x) = \frac{(x + 4)^{2/3}}{\sqrt[5]{x + 2}} + \frac{(x + 2)^{4/5}}{\sqrt[3]{x + 4}}.$$
Determine the domain of f. Does the range of f contain the number 0?

73. True or false? Every polynomial is a rational function.

74. A rectangle has area 36 cm². Find a function that expresses the width w in terms of the length ℓ of the rectangle. What is the domain of this function?

75. A delivery truck depreciates in value at the rate of 30% per year. Comprehensive theft and damage insurance on the truck costs 8% of its value per year. Find a function C giving the cost of the comprehensive insurance policy on the truck after t years if the original cost of the truck is \$20,000.

76. A water tank has the shape of a right circular cylinder. If the radius of the cylinder is fixed, express its volume as a function of its height.

77. A tool rental agency can rent 500 jackhammers per year at a daily rental of \$30. For each \$1 increase in the daily rental fee 20 fewer jackhammers are rented per year.
a. Express the yearly revenues R from jackhammer rentals as a function of the daily rental price p.
b. At what price will revenues reach zero?

78. Find two nonconstant functions f and g so that the composite function $f \circ g$ is constant.

79. Air quality engineers predict that the level of air pollutants in the center of a particular city will be $c(n) = \sqrt{1 + 0.2n}$ parts per million, where n is the average number of automobiles in the center city area. Traffic engineers predict that the average number of automobiles in the center of the city after t years will be $n(t) = 10,000 + 50t^{2/3}$. Write a composite function $c \circ n$ giving the level of air pollutants after t years.

80. Foxes and rabbits coexist in a certain ecological niche. The number of rabbits r is determined by the number of carrots according to the function $r(c) = 10 + 0.2c + 0.01c^2$. The number of foxes, which prey on the rabbits, is given by the function $f(r) = \sqrt{r + 5}$. Find the composite function f giving the number of foxes as a function of the number of carrots.

81. Find a function f giving the volume of a cube in terms of the length of one of its edges.

82. Find a function giving the length of an edge of a cube in terms of its volume.

83. The frequency of x-rays emitted from a chemical element varies with the atomic number Z according to the equation $\gamma = a(Z - b)^2$, where a and b are positive constants. Discuss the nature of this graph.

84. For a moving automobile, air resistance is proportional to the square of the vehicle's speed. Write this statement in function form and discuss the nature of its graph.

85. Show that the quadratic function $f(x) = Ax^2 + Bx + C$ may be written as $f(x) = A(x - D)^2 + E$ with $D = -B/(2A)$ and $E = C - B^2/(4A)$. (*Hint:* Complete the square.) Using the graph of f, explain the geometric significance of D and E.

86. What is the range of the quadratic function $f(x) = ax^2 + bx + c$ if $a > 0$? What if $a < 0$? (See Exercise 85.)

87. Find the range of the function:
    a. In Exercise 11   e. In Exercise 23
    b. In Exercise 13   f. In Exercise 25
    c. In Exercise 17   g. In Exercise 44
    d. In Exercise 20   h. In Exercise 50

88. Use what you know about the graph of the quadratic function $f(x) = ax^2 + bx + c$ and the method of completing the square to prove the quadratic formula for finding the zeros of the quadratic polynomial $ax^2 + bx + c$.

$$x = \frac{-b \pm \sqrt{b^2 - 4ac}}{2a}$$

In Exercises 89–92, a function f and a graph are given. Find the numbers a, b, c, and d such that the graph of $y = af(b(x + c)) + d$ is the graph specified. (The scale of each graph is provided at the lower left corner.)

89. $f(x) = |x|$

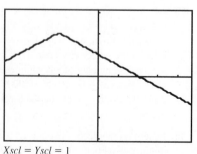

Xscl = Yscl = 1

90. $f(x) = x^2$

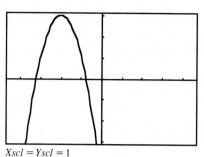

Xscl = Yscl = 1

91. $f(x) = \dfrac{1}{x}$

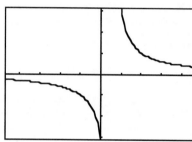

Xscl = Yscl = 1.

92. $f(x) = \sin x$

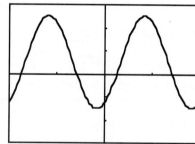

Xscl = $\dfrac{\pi}{2}$, Yscl = 1.

In Exercises 93–96, use the specified functions f and g to graph the functions $f + g$, $f - g$, fg, f/g, g/f, $f \circ g$, and $g \circ f$. Determine the domain and range of each function.

93. $f(x) = 2x + 1$, $g(x) = \sqrt{x}$.

94. $f(x) = x^2 - 4$, $g(x) = |x|$.

95. $f(x) = \sqrt{x + 1}$, $g(x) = x^2 - 2x + 1$.

96. $f(x) = 4x^2 - x$, $g(x) = \dfrac{x + 1}{x}$.

97. Polynomials are sometimes used to produce a function that passes through a given collection of points in the plane. Such a polynomial is said to interpolate the points. For example, $q(x) = -3x^2 + x - 3$ interpolates the points $(-2, -17)$, $(0.5, -\frac{13}{4})$, and $(3, -27)$ because $q(-2) =$

-17, $q(0.5) = -\frac{13}{4}$, and $q(3) = -27$. However, these interpolating polynomials may not faithfully describe the function or the process that generated the points. To see why, consider the polynomial

$$p(x) = -\frac{1}{520}x^6 + \frac{9}{130}x^4 - \frac{59}{104}x^2 + 1.$$

It interpolates the points $(\pm 5, \frac{1}{26})$, $(\pm 3, \frac{1}{10})$, $(\pm 1, \frac{1}{2})$, and $(0, 1)$.

a. Plot the points and then graph the polynomial p.

b. The points given here are generated from the function $f(x) = 1/(1 + x^2)$. Graph p and f. Does p faithfully represent the function f?

98. The following table gives values of the area A and perimeter P of an ellipse with semimajor axis of length a and semiminor axis $b = \sqrt{a}$.

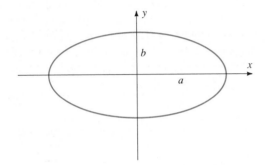

a	A	P
0.5	1.1107	3.8202
1.0	3.1415	6.2832
1.5	5.7715	8.5819
2.0	8.8858	10.8052
2.5	12.4182	12.9843
3.0	16.3242	15.1342
3.5	20.5708	17.2634
4.0	25.1327	19.3769
4.5	29.9895	21.4782
5.0	35.1241	23.5698

a. Plot the points (a, A).

b. Using the plot in part a, determine a function f such that $f(a) = A$.

c. Repeat parts a and b for the points (a, P).

d. Using the equations found in parts b and c, determine a function f such that $f(A) = P$.

99. Rational functions have the ability to produce surprising graphs, and consequently, they are used to model many different types of phenomena. (For example, they are used to design electronic devices that filter out certain types of signals.) To see some of the unusual graphs that occur as graphs of rational functions, graph the following rational functions. You may have to zoom in on some parts of the graph to see all of its details.

a. $f(x) = \dfrac{3x^6 + x^4 + 10x^2 + 1}{-x^6 - 6x^4 - 3x^2 - 1}$.

b. $g(x) = \dfrac{5x^8 + 8x^6 + 5x^4 - 10x^2 + 1}{x^8 + 5}$

100. In the evaluation of the square root function on a computer, a first approximation is sometimes obtained by using a rational function. The function

$$r(x) = \frac{x^2 + 8x + 8}{4x + 8}$$

can be used to find approximations to the values of the function $f(x) = \sqrt{x + 1}$ on the interval $[-\frac{1}{2}, \frac{1}{2}]$, i.e., approximate the square roots of numbers between $\frac{1}{2}$ and $\frac{3}{2}$.

a. Graph both r and f on the interval $[-\frac{1}{2}, \frac{1}{2}]$ and compare the two functions.

b. Determine the maximum of the function $|f(x) - r(x)|$ on the interval $[-\frac{1}{2}, \frac{1}{2}]$. Comment on the accuracy of r as an approximation of f.

c. This rational function can also be used to determine the square root of numbers that are not between $\frac{1}{2}$ and $\frac{3}{2}$. This procedure is outlined below.

Suppose N is a number larger than $\frac{3}{2}$.

(i) Write N as $N = 2^{2n}u$ so that $\frac{1}{2} \le u \le \frac{3}{2}$.

(ii) Set $x = u - 1$. Note that $-\frac{1}{2} \le x \le \frac{1}{2}$.

(iii) Then $\sqrt{N} = 2^n\sqrt{1 + x}$ which is approximated by $2^n r(x)$.

Here's an example of this procedure. Suppose $N = 131$. Then $N = 2^8(\frac{131}{256})$ so $n = 4$ and $u = \frac{131}{256}$. Set $x = u - 1 = -\frac{125}{256}$. Thus

$$\sqrt{131} = 2^4\sqrt{1 + \left(-\frac{125}{256}\right)} \approx 2^4 r\left(-\frac{125}{256}\right)$$
$$= 16(0.7164309795) = 11.46289567.$$

The actual value of $\sqrt{131}$ to 8 decimal places is 11.44552314. Carry out this procedure for several numbers greater than $\frac{3}{2}$. Outline a procedure to estimate the square root of a number that is less than $\frac{1}{2}$.

1.5   Trigonometric Functions

Many problems in mathematics, science, and engineering involve a relationship between angles and lengths. For example, in the study of optics, there is a phenomenon called **refraction,** the bending of a ray of light as it passes from one medium (such as water) to another medium (such as air). Figure 5.1 illustrates the fact that the sizes of the angles associated with the refraction phenomenon are intimately related to the lengths of the sides of certain triangles. The subject of *trigonometry* addresses these kinds of questions, and the relationships that are obtained give rise to the six *trigonometric functions*. In turn, these functions are used to describe numerous periodic and oscillatory phenomena, such as the oscillation of a mass attached to a vibrating spring, the periodic motion of a pendulum, and the flow of electric current in a circuit.

Radian Measure

You are already familiar with the measurement of angles in degrees, and you know that one complete revolution equals 360°. However, the degree system is poorly suited to the needs of calculus. We therefore define a more natural unit of angular measurement, the *radian*.

Consider a unit circle located in the *xy*-coordinate plane, with its center at the origin. Imagine a particle that travels counterclockwise along the circumference of the circle, starting at $(1, 0)$ and carrying with it one end of a line segment whose other end is pivoted at the origin (see Figure 5.2). The angle θ is defined by the positive *x*-axis and the rotating line segment. We define the **radian measure** of the angle θ to be the distance travelled by the particle along the circumference. Since the circumference of the unit circle is 2π we have:

$$2\pi \text{ radians} = 360°.$$

The relation between any angle θ_r measured in radians and the same angle θ_d measured in degrees is therefore

$$\frac{\theta_r}{2\pi} = \frac{\theta_d}{360} \quad \text{or} \quad \theta_r = \frac{2\pi}{360}\theta_d.$$

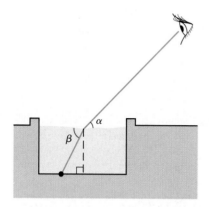

Figure 5.1 The ray of light from the coin in the wishing well to the observer's eye "bends" at the water surface.

□ **EXAMPLE 1**

The following table gives the angular measurement, in both degrees and radians, of some familiar angles.

θ_d	0°	30°	45°	60°	90°	120°	135°	180°	210°	270°	315°	360°
θ_r	0	$\dfrac{\pi}{6}$	$\dfrac{\pi}{4}$	$\dfrac{\pi}{3}$	$\dfrac{\pi}{2}$	$\dfrac{2\pi}{3}$	$\dfrac{3\pi}{4}$	π	$\dfrac{7\pi}{6}$	$\dfrac{3\pi}{2}$	$\dfrac{7\pi}{4}$	2π

■

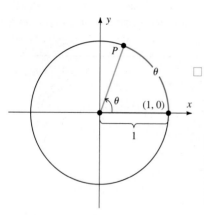

Figure 5.2 Unit circle. Angle measure in radians equals the length of the arc of the unit circle determined by the angle.

We can imagine the point and line segment continuing to rotate for more than one complete revolution, defining angles greater than 360° or 2π. This allows us to identify each real number t with a corresponding point P on the unit circle as follows. If $t > 0$, we allow our travelling point to go a distance t counterclockwise along the unit circle, starting from $(1, 0)$. Let P denote the point on the circle where this motion ends. If $t < 0$, we use the same procedure, except that the point travels in the clockwise direction. This procedure of "wrapping the real number line around the unit circle" allows us to define the *trigonometric functions* for all real numbers.

Sine and Cosine Functions

Let θ be any angle, measured in radians, and let P be the point on the unit circle associated with θ as described above. We define the **sine** of the angle θ, written $\sin \theta$, to be the y-coordinate of P. Similarly, the **cosine** of the angle, $\cos \theta$, is defined to be the x-coordinate of P. In other words,

$$\cos \theta = x$$
$$\sin \theta = y$$

(See Figure 5.3).

Table 5.1 Values of Sine and Cosine for Certain Special Angles (see Example 2 on p. 42).

θ	$\sin \theta$	$\cos \theta$
0	0	1
$\dfrac{\pi}{6}$	$\dfrac{1}{2}$	$\dfrac{\sqrt{3}}{2}$
$\dfrac{\pi}{4}$	$\dfrac{\sqrt{2}}{2}$	$\dfrac{\sqrt{2}}{2}$
$\dfrac{\pi}{3}$	$\dfrac{\sqrt{3}}{2}$	$\dfrac{1}{2}$
$\dfrac{\pi}{2}$	1	0
$\dfrac{2\pi}{3}$	$\dfrac{\sqrt{3}}{2}$	$-\dfrac{1}{2}$
$\dfrac{3\pi}{4}$	$\dfrac{\sqrt{2}}{2}$	$-\dfrac{\sqrt{2}}{2}$
$\dfrac{5\pi}{6}$	$\dfrac{1}{2}$	$-\dfrac{\sqrt{3}}{2}$
π	0	-1
$\dfrac{7\pi}{6}$	$-\dfrac{1}{2}$	$-\dfrac{\sqrt{3}}{2}$
$\dfrac{5\pi}{4}$	$-\dfrac{\sqrt{2}}{2}$	$-\dfrac{\sqrt{2}}{2}$
$\dfrac{4\pi}{3}$	$-\dfrac{\sqrt{3}}{2}$	$-\dfrac{1}{2}$
$\dfrac{3\pi}{2}$	-1	0
$\dfrac{5\pi}{3}$	$-\dfrac{\sqrt{3}}{2}$	$\dfrac{1}{2}$
$\dfrac{7\pi}{4}$	$-\dfrac{\sqrt{2}}{2}$	$\dfrac{\sqrt{2}}{2}$
$\dfrac{11\pi}{6}$	$-\dfrac{1}{2}$	$\dfrac{\sqrt{3}}{2}$

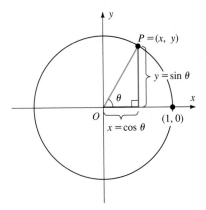

Figure 5.3 Definition of $\sin \theta$ and $\cos \theta$.

In general, values of the sine and cosine are difficult to compute. (We will develop the ability to compute them as part of our work on infinite series later in this text.) However, values of $\sin \theta$ and $\cos \theta$ have been tabulated for many angles (tables of these values appear in Appendix 6). Moreover, most hand calculators are preprogrammed to give values of $\sin \theta$ and $\cos \theta$ with high degrees of precision. There are certain angles, however, for which values of $\sin \theta$ and $\cos \theta$ are easy to compute, and the computation provides precise values, as opposed to approximations. These angles include multiples of $\pi/2$ and the angles involved in 30°-60°-90° and isosceles right triangles. This is because the ratios of sides in such triangles are easily found from elementary geometry (see Figure 5.4).

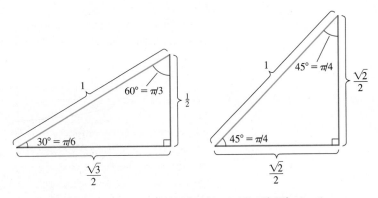

Figure 5.4 Ratios of sides in 30°-60°-90° and 45°-45°-90° triangles.

□ **EXAMPLE 2**

Using the well-known geometric facts about 30°-60°-90° and isosceles right triangles summarized in Figure 5.4, we can determine sine and cosine for 16 "standard" angles between 0 and 2π. For example, Figure 5.5 illustrates sine and cosine for $\pi/6$, $3\pi/4$, $4\pi/3$, and $3\pi/2$.

   In Table 5.1, we list the values of sine and cosine for all angles between 0 and 2π that can be calculated using this technique. You should be able to reproduce this table by drawing triangles such as those illustrated in Figure 5.5 (see Exercise 4).

■

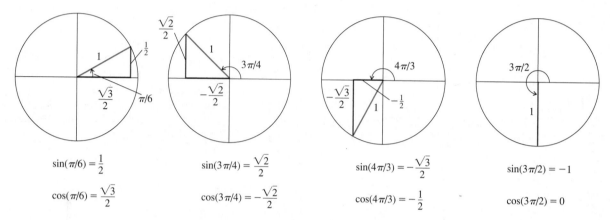

$$\sin(\pi/6) = \frac{1}{2} \qquad \sin(3\pi/4) = \frac{\sqrt{2}}{2} \qquad \sin(4\pi/3) = -\frac{\sqrt{3}}{2} \qquad \sin(3\pi/2) = -1$$

$$\cos(\pi/6) = \frac{\sqrt{3}}{2} \qquad \cos(3\pi/4) = -\frac{\sqrt{2}}{2} \qquad \cos(4\pi/3) = -\frac{1}{2} \qquad \cos(3\pi/2) = 0$$

Figure 5.5 Sin θ and cos θ for various angles.

Note that the definitions of sine and cosine apply to all real numbers (in radian units). For example, $\sin(9\pi/2) = \sin(\pi/2) = 1$, and $\cos(11\pi/4) = \cos(3\pi/4) = -\sqrt{2}/2$. Because all angles with measure $t + 2n\pi$, $n = 0, \pm1, \pm2, \ldots$ are associated with the same point P on the unit circle, we have the identities

$$\sin t = \sin(t + 2n\pi), \qquad n = \pm1, \pm2, \ldots. \tag{1a}$$
$$\cos t = \cos(t + 2n\pi), \qquad n = \pm1, \pm2, \ldots. \tag{1b}$$

Identities (1a) and (1b) say that the values of the sine and cosine are the same at any number located a multiple of 2π radians away from t as they are at the number t. For this reason, we say that these functions are **periodic,** with period $T = 2\pi$.

   Graphs of sin t and cos t appear in Figure 5.6. Notice that, for both functions, the maximum value is 1 and the minimum value is -1. That is, $|\sin t| \le 1$ and $|\cos t| \le 1$.

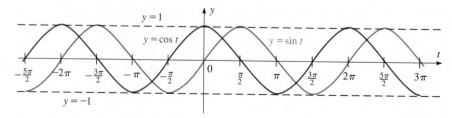

Figure 5.6 Graphs of functions $y = \sin t$ and $y = \cos t$.

The Other Trigonometric Functions

Four additional trigonometric functions are defined as quotients and reciprocals of the sine and cosine functions. They are as follows:

(i) tangent: $\tan t = \dfrac{\sin t}{\cos t}, \qquad t \neq \dfrac{\pi}{2} + n\pi$

(ii) cotangent: $\cot t = \dfrac{\cos t}{\sin t}, \qquad t \neq n\pi$

(iii) secant: $\sec t = \dfrac{1}{\cos t}, \qquad t \neq \dfrac{\pi}{2} + n\pi$

(iv) cosecant: $\csc t = \dfrac{1}{\sin t}, \qquad t \neq n\pi$

where $n = \pm 1, \pm 2, \pm 3, \ldots$.

Graphs of these functions appear in Figure 5.7. (Note that you can sketch these graphs geometrically, using Figure 5.6. For example, the y-coordinates of the points on the graph of $y = \tan t$ are just the quotients of the y-coordinates of the points on the sine and cosine curves.) You can see from the graphs that the tangent and cotangent functions are periodic with period $T = \pi$, whereas the other four trigonometric functions have a period of $T = 2\pi$. (See Exercise 39.)

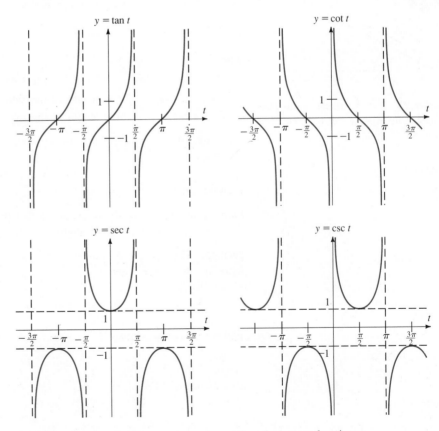

Figure 5.7 Graphs of tangent, cotangent, secant, and cosecant functions.

□ EXAMPLE 3

Find $\tan(\pi/4)$, $\cot(5\pi/3)$, $\sec(-2\pi/3)$, and $\csc(7\pi)$.

Solution

$$\tan(\pi/4) = \frac{\sin(\pi/4)}{\cos(\pi/4)} = \frac{\sqrt{2}/2}{\sqrt{2}/2} = 1$$

$$\cot(5\pi/3) = \frac{\cos(5\pi/3)}{\sin(5\pi/3)} = \frac{1/2}{-\sqrt{3}/2} = -\frac{1}{\sqrt{3}}$$

$$\sec(-2\pi/3) = \sec(4\pi/3) = \frac{1}{\cos(4\pi/3)} = \frac{1}{-1/2} = -2$$

$$\csc(7\pi) = \csc(\pi) = \frac{1}{\sin \pi}, \text{ which is } \textit{undefined} \text{ since } \sin \pi = 0. \quad ■$$

The Trigonometric Functions and Right Triangles

The trigonometric functions can also be viewed as ratios of sides of right triangles. In particular, given a right triangle containing the acute angle θ (as in Figure 5.8), then

$$\sin \theta = \frac{o}{h} \qquad \csc \theta = \frac{h}{o}$$

$$\cos \theta = \frac{a}{h} \qquad \sec \theta = \frac{h}{a}$$

$$\tan \theta = \frac{o}{a} \qquad \cot \theta = \frac{a}{o}$$

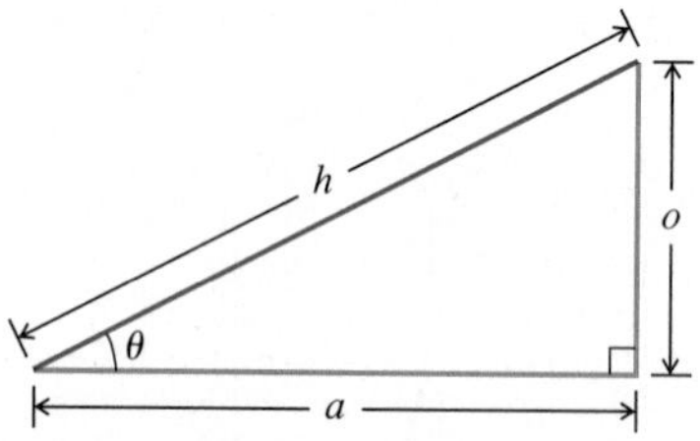

where

a = length of the side adjacent to θ
o = length of the side opposite to θ
h = length of the hypotenuse

Figure 5.8 A right triangle with acute angle θ, its adjacent side a, its opposite side o, and its hypotenuse h.

Therefore, these functions are useful whenever a calculation relates angles and distances.

□ EXAMPLE 4

Find the height of a flagpole given the fact that, at a distance of 10 feet from the pole, one must look up at an angle of $\pi/3$ radians to see its top.

Solution The flagpole is one side of the right triangle sketched in Figure 5.9. Let h represent its height. Then

$$\tan(\pi/3) = \frac{h}{10}. \qquad (2)$$

To find h, we recall that

$$\tan(\pi/3) = \sqrt{3}.$$

So equation (2) becomes

$$\frac{h}{10} = \sqrt{3}.$$

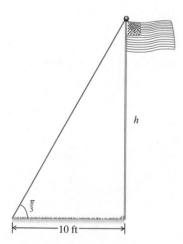

Figure 5.9 The flagpole in Example 4. The flagpole is $10\sqrt{3}$ feet tall. ■

Trigonometric Identities

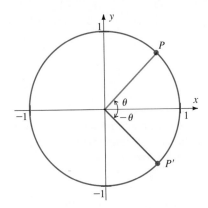

Figure 5.10 The angles θ and $-\theta$ correspond to points on the unit circle that are symmetric with respect to the x-axis.

Because $\sin\theta$ and $\cos\theta$ are the y- and x-coordinates of a point on the unit circle, we have the identity

$$\sin^2\theta + \cos^2\theta = 1. \tag{3}$$

If $\cos\theta \neq 0$, we may divide each term of equation (3) by $\cos^2\theta$ to obtain the identity

$$\tan^2\theta + 1 = \sec^2\theta.$$

Similarly, dividing (3) through by $\sin^2\theta \neq 0$ gives

$$1 + \cot^2\theta = \csc^2\theta.$$

Since the angles θ and $-\theta$ correspond to points $P = (x, y)$ and $P' = (x, -y)$ on the unit circle (Figure 5.10), we see that

$$\sin(-\theta) = -\sin\theta \tag{4a}$$
$$\cos(-\theta) = \cos\theta. \tag{4b}$$

This explains why the graph of $y = \cos\theta$ is symmetric with respect to the y-axis while the graph of $y = \sin\theta$ is symmetric with respect to the origin (see Figure 5.6).

Even though the trigonometric functions are defined geometrically, they satisfy a number of algebraic identities. Many of the identities that we shall need are based on the **addition formulas** (see Exercise 37)

$$\sin(\theta + \phi) = \sin\theta\cos\phi + \cos\theta\sin\phi \tag{5a}$$
$$\cos(\theta + \phi) = \cos\theta\cos\phi - \sin\theta\sin\phi. \tag{5b}$$

Using identities (4a) and (4b), we obtain the corresponding subtraction formulas

$$\sin(\theta - \phi) = \sin\theta\cos\phi - \cos\theta\sin\phi \tag{6a}$$
$$\cos(\theta - \phi) = \cos\theta\cos\phi + \sin\theta\sin\phi. \tag{6b}$$

Then, dividing (5a) by (5b) or (6a) by (6b), we obtain the addition and subtraction formulas for tangent (whenever the denominator on the right-hand side does not equal zero).

$$\tan(\theta \pm \phi) = \frac{\tan\theta \pm \tan\phi}{1 \mp \tan\theta\tan\phi}$$

We can use the subtraction formulas (6a) and (6b) along with the fact that $\cos(\pi/2) = 0$ and $\sin(\pi/2) = 1$ to relate sine and cosine via the **complementary angle formulas**

$$\sin\theta = \cos\left(\frac{\pi}{2} - \theta\right)$$

$$\cos\theta = \sin\left(\frac{\pi}{2} - \theta\right).$$

An important special case of the addition formulas is the **double-angle formulas.** Let $\phi = \theta$ in (5a) and (5b) to obtain

$$\sin 2\theta = 2\sin\theta\cos\theta \tag{7a}$$
$$\cos 2\theta = \cos^2\theta - \sin^2\theta. \tag{7b}$$

Using identity (3) to eliminate either the cosine or the sine term from the right-hand side of (7b) yields the very useful **half-angle formulas**

$$\sin^2 \theta = \frac{1 - \cos 2\theta}{2}$$

$$\cos^2 \theta = \frac{1 + \cos 2\theta}{2}.$$

Finally, we recall two other theorems from trigonometry that relate the angles of a triangle to the lengths of its sides. For any triangle with interior angles α, β, and γ and corresponding opposite sides of length a, b, and c (Figure 5.11), there is a generalization of the Pythagorean theorem called the **Law of Cosines**

$$c^2 = a^2 + b^2 - 2ab \cos \gamma,$$

and a trigonometric version of the theorem of similar triangles called the **Law of Sines**

$$\frac{\sin \alpha}{a} = \frac{\sin \beta}{b} = \frac{\sin \gamma}{c}.$$

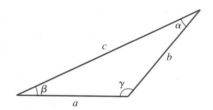

Figure 5.11

Exercise Set 1.5

1. Convert the following angles from degrees to radians.
  a. 30°   **c.** −15°   **e.** $(x + 30)°$
  b. 75°   **d.** 315°   **f.** 2910°

2. Convert the following angles from radians to degrees.
  a. π **c.** $9\pi/2$ **e.** $a + \pi$
  b. 1   **d.** $-21\pi/6$ **f.** $7\pi/16$

3. Complete the following table:

$\sin \theta$	$\cos \theta$	$\tan \theta$	$\sec \theta$
0	1		
$\dfrac{\sqrt{2}}{2}$		1	
$\dfrac{3}{5}$			$\dfrac{5}{4}$
		$\dfrac{5}{12}$	$\dfrac{13}{12}$

4. Complete the following table without referring to Table 5.1:

θ $0 \le \theta < 2\pi$	$\sin \theta$	$\cos \theta$	$\tan \theta$	$\sec \theta$
	0	−1		
$\dfrac{\pi}{3}$				
	$\dfrac{\sqrt{2}}{2}$			−1
	−1	0		
	$\dfrac{1}{2}$	$-\dfrac{\sqrt{3}}{2}$		
			−1	$\sqrt{2}$

5. Find $\tan x$, $\cot x$, $\sec x$, and $\csc x$ for each of the following values of x.

a. $x = \pi/6$ **c.** $x = -13\pi/3$ **e.** $x = -9\pi/4$
b. $x = 5\pi/2$ **d.** $x = 9\pi$ **f.** $x = 7\pi/6$

6. Find all numbers x, with $0 \le x < 2\pi$, in radians, so that
a. $\sin x = 0$
b. $\cos x = \sqrt{2}/2$
c. $\csc x = \sqrt{2}$
d. $\cos(\pi/2 + 2x) = 1$
e. $2\cos^2 x - \cos x - 1 = 0$
f. $\cos 2x + \cos x + 1 = 0$
g. $|\cos x| = \sqrt{3}/2$

7. Find all numbers x, with $-2\pi \le x \le 2\pi$, that satisfy
a. $\cos x = 0$
b. $\sin x = \cos x$
c. $\tan x = -1$
d. $\sin 3x = 0$
e. $\sin 2x = \cos x$
f. $3\cos^2 x - \sin^2 x = 0$
g. $\cos^2 x + \sin x = 0$

In Exercises 8–14, find all numbers t with $0 \le t < 2\pi$, in radians, satisfying all stated properties.

8. $\begin{cases} \sin t = \cos t \\ \tan t > 0 \end{cases}$ **9.** $\begin{cases} \sin t = \sqrt{3}\cos t \\ \cos t < 0 \end{cases}$

10. $\begin{cases} \sec t = 2 \\ \cot t < 0 \end{cases}$ **11.** $\begin{cases} \sin t > 0 \\ \tan t > 0 \end{cases}$

12. $\begin{cases} \sin t \ge \tan t \\ \sin t \ge 0 \end{cases}$ **13.** $\begin{cases} \cos t \ge 0 \\ \sec t \le 0 \end{cases}$

14. $\begin{cases} \cos t < 0 \\ \tan t < 0 \end{cases}$

In Exercises 15–20, use the trigonometric identities to calculate the following values.

15. $\sin(5\pi/12)$ **16.** $\tan(5\pi/12)$

17. $\cos(\pi/12)$ **18.** $\tan(13\pi/12)$

19. $\sec(11\pi/12)$ **20.** $\csc(\pi/12)$

In each of Exercises 21–26, graph the given function by comparing it with the function $y = \sin x$.

21. $y = 2\sin x$ **22.** $f(x) = -2\sin 3x$

23. $y = \pi\sin(\pi x)$ **24.** $f(x) = \sin(2x + \pi/2)$

25. $y = -2\sin(\pi + x)$ **26.** $y = 4\sin(2x + \pi)$

27. State the domains and ranges for each of the six trigonometric functions.

In Exercises 28–33, state whether the given function is even, odd, or neither. (See Exercise 57, Section 1.4.)

28. $y = \sin x$ **29.** $y = \cos x$ **30.** $y = \tan x$

31. $y = \dfrac{\sin x}{x}$ **32.** $y = 3\sec x$ **33.** $y = \sin^2 x$

34. True or false? If the size of an angle is the same in both radians and degrees then that angle must be zero.

35. Let ℓ be a nonvertical line, m be its slope, and θ be its angle of intersection with the x-axis (θ is measured in a counterclockwise direction from the x-axis to ℓ). Show that

$$\tan\theta = m.$$

36. Suppose that C is a circle of radius 2 centered at the origin O and that P and Q are points on C such that the angle POQ measures $7\pi/12$ radians. Calculate the area enclosed by the sector of the circle POQ (see Figure 5.12).

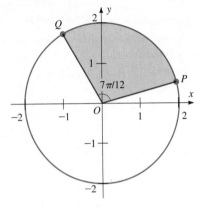

Figure 5.12

37. Prove the addition formula

$$\cos(\theta + \phi) = \cos\theta\cos\phi - \sin\theta\sin\phi$$

in the case where both θ and ϕ are between 0 and $\pi/2$ by filling in the details of the argument sketched below.
a. Let O be the origin and label the following four points on the unit circle.

$$P_1 = (1, 0)$$
$$P_2 = (\cos\phi, \sin\phi)$$
$$P_3 = (\cos(\phi + \theta), \sin(\phi + \theta))$$
$$P_4 = (\cos(-\theta), \sin(-\theta))$$

(See Figure 5.13.)

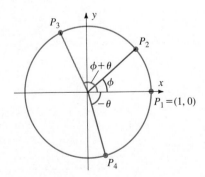

Figure 5.13

b. Show that the triangle OP_1P_3 is congruent to the triangle OP_4P_2. Therefore, the distance of P_3 from P_1 is the same as the distance of P_2 from P_4.

c. Apply the distance formula to obtain the equation

$$[\cos(\theta + \phi) - 1]^2 + \sin^2(\theta + \phi) =$$
$$[\cos(\phi) - \cos(-\theta)]^2 + [\sin(\phi) - \sin(-\theta)]^2.$$

d. Expand and simplify the equation in part c to obtain the equation

$$2 - 2\cos(\theta + \phi) = 2 - 2\cos\theta\cos\phi + 2\sin\theta\sin\phi.$$

e. Derive the addition formula from part d.

38. Explain why the following formula for the area of the triangle in Figure 5.14 is valid.

$$A = \tfrac{1}{2}xh\sin\theta$$

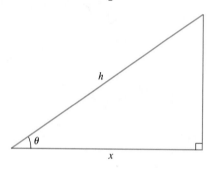

Figure 5.14 $x =$ side adjacent θ; $h =$ hypotenuse.

39. Use the definitions of the tangent and cotangent functions, together with the addition formulas for sine and cosine (equations (5)), to show that the period of the tangent and cotangent is $T = \pi$.

40. True or false? $\sin(x + y) = \sin x + \sin y$.

41. A pendulum consists of a bob hanging at the end of a cord 2 meters long. If the maximum angle of displacement for the pendulum is $\theta = \pi/6$, how far does the bob travel in one complete cycle (measured along the circular arc swept out by the bob)? (Figure 5.15).

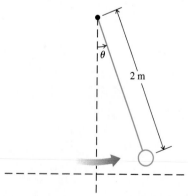

Figure 5.15 Pendulum in Exercise 41.

42. Points A and B are on opposite banks of a straight riverbed, as in Figure 5.16. Point A is directly opposite point B. What is the distance between points B and C if $\theta = \pi/6$?

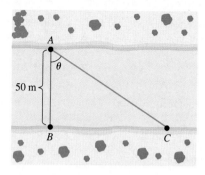

Figure 5.16 Points on opposite sides of a straight riverbed.

43. What is the answer to the question in Exercise 42 if $\theta = \pi/3$?

44. A child stands 300 meters from a point directly below the location of an airplane. If the angle between ground level and the child's line of sight to the airplane is $\pi/6 = 30°$, what is the altitude of the airplane? (Neglect the height of the child.)

45. What is the altitude of the airplane in Exercise 44 if the angle is $\theta = \pi/4$?

46. Use the Law of Sines and Law of Cosines to find the missing sides and angles in the triangles shown in Figures 5.17 and 5.18.

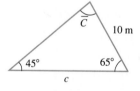

Figure 5.17   **Figure 5.18**

47. How long must a ladder be to reach the top of a 3-meter wall if the angle of inclination between the ladder and the ground is $\pi/3$?

48. Using trigonometry, you can determine the height of a tall building by measuring two different angles of inclination as follows. First, at any particular location, measure the angle α of inclination between the ground and your line of sight to the top of the building. Then walk a distance d towards the building and measure a second angle of inclination β. Then you can establish the height of the building from α, β, and d. Do this calculation in the case where $\alpha = \pi/6$, $\beta = \pi/4$, and $d = 100$ meters.

49. Using the graphs $y = x$ and $y = \cos x$, estimate the solution to the equation $\cos x = x$.

 50. Graph $f_n(x) = f_{n-1}(f(x))$ for $f(x) = \cos x$. For example, $f_2(x) = \cos(\cos x)$, $f_3(x) = \cos(\cos(\cos x))$, etc. Make a conjecture about what happens to $f_n(x)$ as n increases. Compare your conjecture to the solution of the equation in Exercise 49.

51. Using the method of Exercise 49, estimate some of the solutions of the equation $\tan x = x/2$. Provide a simple approximation for these solutions when x is large.

52. Periodic phenomena are often modeled with sums of sine and cosine functions. For $n = 1, 2, 3, 4, 5$ graph the following functions, and conjecture what periodic phenomena the functions are trying to reproduce.

a. $F_n(x) = \dfrac{\pi}{2} + \dfrac{4}{\pi}\left(\cos x + \dfrac{\cos(3x)}{9}\right.$
$\left. + \dfrac{\cos(5x)}{25} + \cdots + \dfrac{\cos[(2n-1)x]}{(2n-1)^2}\right).$

b. $G_n(x) = \dfrac{4}{\pi}\left(\sin x + \dfrac{\sin(3x)}{3}\right.$
$\left. + \dfrac{\sin(5x)}{5} + \cdots + \dfrac{\sin[(2n-1)x]}{2n-1}\right).$

c. $H_n(x) = \dfrac{8}{\pi^2}\left(\sin x - \dfrac{\sin(3x)}{9}\right.$
$\left. + \dfrac{\sin(5x)}{25} - \cdots + \dfrac{(-1)^{n-1}\sin[(2n-1)x]}{(2n-1)^2}\right).$

 53. A pendulum is a bob of mass M that hangs from a massless rod of length l (see Figure 5.19). We know that there are numbers A and ω such that the angle θ varies according to the equation $\theta(t) = A\cos(\omega t)$. In fact, $\omega^2 = g/l$ where $g = 9.8\ m/s^2$ is the force of gravity. The number A is called the amplitude of θ. Supposing that the amplitude $A = \frac{1}{20}$ m,
a. graph $\theta(t)$ for $0 \le t \le 60$ for various values of l ranging from 0.2 m to 900 m, and
b. using these graphs, estimate how long the pendulum must be in order for the period to be 1 s, 2 s, 30 s, and 1 min.

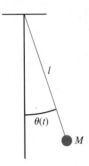

Figure 5.19

Summary Outline of Chapter 1

▌ The **real numbers** correspond to points on a line. The **integers** are the numbers $\ldots -2, -1, 0, 1, 2, \ldots$. **Rational numbers** are quotients of integers. **Decimal expansions** for rational numbers terminate or repeat. Decimal expansions for **irrational** numbers neither repeat nor terminate.   (page 2)

▌ *Interval Notation:*   (page 3)

$$[a, b) = \{x \mid a \le x < b\}; \qquad (a, b) = \{x \mid a < x < b\}$$
$$[a, \infty) = \{x \mid a \le x\}; \qquad [a, b] = \{x \mid a \le x \le b\}$$

▌ *Set Operations:*   (page 4)

$$A \cup B = \{x \mid x \in A \text{ or } x \in B\}$$
$$A \cap B = \{x \mid x \in A \text{ and } x \in B\}$$

▌ The **inequality** $x < y$ means $y - x$ is positive.   (page 5)

▌ *Theorem 1:* (Properties of Inequality)   (page 5)

(i) Either $x = y$, $x < y$, or $x > y$.
(ii) If $x < y$ and $y < z$, then $x < z$.
(iii) If $x < y$, then $x + c < y + c$ for any c.
(iv) If $x < y$ and $c > 0$, then $cx < cy$.
(v) If $x < y$ and $c < 0$, then $cx > cy$.

■ *Absolute Value:* (page 8)

$$|x| = \begin{cases} x \text{ if } x \geq 0 \\ -x \text{ if } x < 0. \end{cases}$$

■ *Theorem 2:* For $a > 0$, $|x| < a$ if and only if $-a < x < a$. (page 8)

■ *Theorem 3:* (Properties of absolute value): For all x (page 10)

  (i) $|x| \geq 0$,

  (ii) $|xy| = |x|\,|y|$,

  (iii) $|x + y| \leq |x| + |y|$.

■ The **distance** d between (x_1, y_1) and (x_2, y_2) in the xy-plane is the nonnegative number (page 13)

$$d = \sqrt{(x_2 - x_1)^2 + (y_2 - y_1)^2}.$$

■ The **circle** with center (h, k) and radius r has equation $(x - h)^2 + (y - k)^2 = r^2$. (page 14)

■ The **slope** of the **line** containing (x_1, y_1) and (x_2, y_2) is (page 16)

$$m = \frac{y_2 - y_1}{x_2 - x_1}, \qquad x_2 \neq x_1.$$

■ An **equation** for the line through (x_1, y_1) with slope m is $(y - y_1) = m(x - x_1)$. (page 17)

■ The **slope-intercept** form for this line is $y = mx + b$, $b = y$-intercept. (page 18)

■ A **vertical** line has no slope and equation $x = a$. A **horizontal** line has slope zero and equation $y = b$. (page 19)

■ Two lines with slope m_1 and m_2 are (a) **parallel** if $m_1 = m_2$, (b) **perpendicular** if $m_1 m_2 = -1$. (page 20)

■ The equation $y = f(x)$ is a **function** if there is a **unique** y corresponding to each x in the **domain** of f. (page 24)

■ The **domain** of f, if not explicitly stated, is the largest set for which $f(x)$ is defined. The **graph** of the function f is the graph of the equation $y = f(x)$. (page 24)

■ The graph of an equation of the form $y = a(x - b)^2 + c$ is a **parabola.** The graph of every **quadratic function** $f(x) = Ax^2 + Bx + C$ is a parabola. A quadratic function can be graphed by **completing the square** in x. (page 27)

■ A **polynomial function** is one that can be written in the form $f(x) = a_n x^n + a_{n-1} x^{n-1} + \cdots + a_1 x + a_0$ where $a_0, a_1, \ldots, a_n$ are constants and n is a positive integer. A **rational function** is a quotient of two polynomials. (page 28)

■ The function $f(x) = x^{n/m}$, where m and n are integers, is the mth root of the nth power of x. Its domain is the set of all real numbers if m is odd and the set of nonnegative real numbers if m is even. (page 29)

■ The **composite function** $f \circ g$ is defined by the equation $(f \circ g)(x) = f(g(x))$. (page 33)

■ **Radian** measure is defined by the equation 2π radians $= 360°$. If $P = (x, y)$ is a point on the unit circle and θ is the angle formed between the radius through P and the positive x-axis, then $\sin \theta = y$ and $\cos \theta = x$. Also, (page 40)

$$\tan \theta = \frac{\sin \theta}{\cos \theta} = \frac{y}{x} \qquad \sec \theta = \frac{1}{\cos \theta} = \frac{1}{x}$$

$$\cot \theta = \frac{\cos \theta}{\sin \theta} = \frac{x}{y} \qquad \csc \theta = \frac{1}{\sin \theta} = \frac{1}{y}$$

when these expressions are defined. The functions $\sin \theta$, $\cos \theta$, $\sec \theta$, and $\csc \theta$ are **periodic** with period $T = 2\pi$. The functions $\tan \theta$ and $\cot \theta$ are periodic with period π.

■ Given a right triangle with acute angle θ, then   (page 44)

$$\sin \theta = o/h \qquad \csc \theta = h/o$$
$$\cos \theta = a/h \qquad \sec \theta = h/a$$
$$\tan \theta = o/a \qquad \cot \theta = a/o$$

■ The trigonometric functions satisfy a number of useful identities: the **addi-**   (page 45)
tion formulas, the **subtraction formulas**, the **complementary angle formu-**
las, the **double-angle formulas**, and the **half-angle formulas**.

■ In any triangle, the sides and angles are related by the **Law of Sines** and the   (page 46)
Law of Cosines.

Review Exercises—Chapter 1

1. For $A = [2, 7)$ and $B = (3, 9)$, find
  a. $A \cup B$ **b.** $A \cap B$

2. List all subsets of the set $\{1, 7, 19, \pi\}$.

3. Solve the inequality $2x - 7 \geq 9$.

4. Find $\{x \mid 6 \leq x^2 - 3 \leq 22\}$.

5. Let $A = \{x \mid x^2 \geq 4\}$, $B = \{x \mid -3 \leq x \leq 3\}$, and $C = \{-4, -3, -2, -1, 0, 1, 2, 3, 4\}$: find
  a. $A \cap B$ **d.** $B \cap C$
  b. $A \cup B$ **e.** $A \cup (B \cap C)$
  c. $A \cap C$ **f.** $A \cap (B \cup C)$

6. Solve the inequality $4 \leq (x + 2)^2 \leq 36$.

7. Solve the inequality $|x - 7| \leq 12$.

8. Find $\{x \mid 2 \leq |x^2 - 2| \leq 7\}$.

9. Let $A = \{x \mid |x| \leq 9\}$ and $B = \{x \mid -2 < x < 2\}$: find $A \cup B$ and $A \cap B$.

In Exercises 10–18, solve the inequality.

10. $|x - 3| \leq 5$ **11.** $-3 \leq |2x + 1| \leq 11$

12. $|x + 2| \geq 6$ **13.** $|2x^2 + 1| \leq 9$

14. $|9 - x^2| \geq 0$

15. $\cos x < \sqrt{3}/2$, $0 \leq x \leq 2\pi$

16. $|\sin x| > \frac{1}{2}$, $0 \leq x \leq 2\pi$

17. $(\sin x - \frac{1}{2})(\cos x + \frac{1}{2}) > 0$, $0 \leq x \leq 2\pi$

18. $\cos 2x + \sin x > 0$, $0 \leq x \leq 2\pi$

19. Find the solution set for the equation $|x - 2| = |x + 2|$.

20. Determine which sets of points lie on a common line.
  a. $\{(0, 0), (-3, 6), (2, 4)\}$
  b. $\{(-4, 2), (-1, 5), (6, 10)\}$
  c. $\{(-3, 2), (-1, 1), (2, -3)\}$
  d. $\{(1, 3), (2, 4), (-2, 0)\}$

21. Find the area of the triangle with vertices $(0, 0)$, $(5, 0)$, and $(2, 3)$.

22. Find a if a right triangle has vertices $(0, 0)$, $(2, 4)$, and $(a, 0)$, and
  a. the right angle is at vertex $(a, 0)$
  b. the right angle is at vertex $(2, 4)$.

23. Find an equation for the circle with center at $(2, -4)$ and diameter $d = 10$.

In Exercises 24–27, find the center and radius of the circle.

24. $x^2 + y^2 = 49$ **25.** $x^2 - 4x + y^2 = 0$

26. $x^2 - 2x + y^2 + 6y = -9$ **27.** $x^2 - 2x + y^2 + 2y = 14$

In Exercises 28–33, find the slope, y-intercept, and x-intercept of the line, if possible.

28. $x + y = 1$ **29.** $x - 3y + 4 = 0$

30. $y = 4$ **31.** $7x - 7y + 21 = 0$

32. $x - y = 5$ **33.** $3x = 6$

34. Find an equation for the line through the points $(-2, 1)$ and $(3, 3)$.

35. Find an equation for the line through the point $(2, -4)$ and parallel to the line with equation $2x - 4y = 14$.

36. The lines $y - ax = 1$ and $3y = 6x + 12$ are parallel. Find a.

37. Find an equation for the line perpendicular to the line with equation $3x - 6y = 8$ and passing through the origin.

38. Find an equation for the line through $(-1, 3)$ parallel to the line through the points $(1, 3)$ and $(6, -4)$.

39. Where does the line $ax + by + c = 0$ cross the x-axis?

40. The line $y = ax + b$ passes through the points $(2, 2)$ and $(4, -5)$. Find a and b.

41. Graph the region bounded by the graph of $y = |4 - x^2|$ and the x-axis.

42. Graph the region R where

$$R = \{(x, y) \mid y \le x + 4\} \cap \{(x, y) \mid y \le -2x + 8\}.$$

43. Sketch the region in the plane whose points satisfy all of the following inequalities.
   a. $x + y \ge 0$
   b. $2y \le 4x + 4$
   c. $x \le 4$

44. Sketch the region R consisting of all points satisfying both of the following inequalities.
   a. $x^2 + y^2 \le 4$
   b. $x + y \ge 2$

45. Find the values of the six trigonometric functions for the given angles (measured in radians):
   a. $\pi/3$ **d.** $9\pi/4$
   b. $-3\pi/4$ **e.** $\pi/12$
   c. $7\pi/3$ **f.** $5\pi/12$

In Exercises 46 and 47, convert the decimals to quotients of integers.

46. $32.61616\overline{1}\ldots$ **47.** $-6.214\overline{214}\ldots$

48. Draw Venn diagrams to illustrate the following statements.
   a. $A \cap (B \cup C) = (A \cap B) \cup (A \cap C)$
   b. $A \cup (B \cap C) = (A \cup B) \cap (A \cup C)$

49. Find the coordinates of the point of intersection of the lines with equations $x + 3y - 6 = 0$ and $x - y = 2$.

50. A beaker of water is being heated. The water is initially at 10°C. After 3 minutes its temperature is 25°C. Using this data, find a linear equation that predicts the temperature of the water at a later time.
   a. What is the predicted temperature after 5 minutes?
   b. What is wrong with extrapolating to predict the temperature after 30 minutes?

51. Find a linear equation that converts temperatures in Kelvin to temperatures in Celsius degrees, given that 100°C = 373 K and 0°C = 273 K.

52. Find an equation that converts temperatures in Kelvin to temperatures in degrees Fahrenheit, given that 212°F = 373 K and 32°F = 273 K.

53. Find the point on the line $x + y = 2$ nearest the point $(4, 1)$.

54. When two chemicals combine to form a solution, the chemical used in greater quantity is called the *solvent*. *Raoult's law* states that in a dilute solution, the vapor pressure P_s of the solvent is linearly proportional to the vapor pressure of P_p of the solvent when pure. Moreover, the constant of proportionality is the ratio of the number n_s of moles of the solvent present to the total number n_T of moles in the solution.
   a. Write an equation expressing Raoult's law.
   b. In a certain solution, the number of moles of solvent is 85% of the total present. Write the equation for Raoult's law in this case, and find the vapor pressure of the solvent

in solution if its vapor pressure when pure is 0.25 atmospheres.

In Exercises 55–58, does the given equation determine y as a function of x?

55. $x^2 + y^2 + 2y = 4$ **56.** $x^2 + y^3 = 3$

57. $y = \pi$ **58.** $x = \sin y$

In Exercises 59–64, state the domain of the given function.

59. $y = \sqrt{x^2 - 1}$ **60.** $y = \dfrac{1}{x + 2}$

61. $f(x) = \sin(1 - x^2)$ **62.** $f(x) = \dfrac{1}{1 + \cos x}$

63. $y = \dfrac{1}{\sqrt{1 - \sin^2 x}}$ **64.** $f(x) = \dfrac{1}{\sqrt{x(x + 2)}}$

In Exercises 65–68, state whether the given function is odd, even, or neither. (See Exercise 57 in Section 1.4.)

65. $f(x) = x^3 \sin x$ **66.** $f(x) = \dfrac{1 - \cos x}{x^2}$

67. $f(x) = (x - 1)(x + 1)$ **68.** $f(x) = \dfrac{x \tan x}{1 + x^3}$

In Exercises 69–72, use the technique of completing the square to graph the function.

69. $y - 2x^2 + 2x - \frac{7}{2} = 0$ **70.** $y^2 - 2x + 4y + 6 = 0$

71. $y^2 + 2x - y + \frac{3}{4} = 0$ **72.** $y - \frac{1}{6}x^2 - x - \frac{5}{2} = 0$

In Exercises 73–76, graph the given function.

73. $f(x) = (x + 1)^{5/2}$ **74.** $f(x) = (x - \pi)^{1/3}$

75. $f(x) = \sin(x - \pi/4)$ **76.** $f(x) = \tan(x + 1)$

77. Find the values of θ such that $0 \le \theta < 2\pi$ and
   a. $\sin \theta = \sqrt{3}/2$ and $\cos \theta < 0$
   b. $\sec \theta = \sqrt{2}$ and $\tan \theta = -1$
   c. $\sin 2\theta = \sin \theta$ and $\sin \theta \ne 0$

78. Sketch a right triangle containing an angle θ for which
   a. $\sec \theta = \sqrt{2}$
   b. $\tan \theta = \sqrt{3}$
   c. $\csc \theta = 2$

In Exercises 79–84, let

$$f(x) = x^3 - x, \qquad g(x) = \frac{1}{1 - x}, \qquad \text{and} \qquad h(x) = \sin x.$$

Form the indicated composite function.

79. $g(h(x))$ **80.** $f(g(x))$ **81.** $h(g(x))$

82. $f(g(h(x)))$ **83.** $g(f(h(x)))$ **84.** $g(h(g(x)))$

85. Prove that the product of two odd functions is an even function. (See Exercise 57 in Section 1.4).

86. Prove that the product of an odd function and an even function is an odd function. (See Exercise 57 in Section 1.4.)

87. Find an equation for the line tangent to the circle $(x - 6)^2 + (y - 4)^2 = 25$ at the point $(3, 8)$.

88. Find numbers a, b, and c so that the parabola $y = ax^2 + bx + c$ will have y-intercept $y = -6$, x-intercept $x = 3$, and vertex at $x = 2$.

89. Given the functions $f(x) = 4 - x^2$, with domain $(-\infty, \infty)$, and $g(x) = \sin x$, with domain $0 \leq x \leq 2\pi$, find the domain and range of the composite function $f \circ g$.

90. Given the functions $f(x) = \sqrt{x - \frac{1}{2}}$ and $g(x) = \sin x$, $0 \leq x \leq 2\pi$, find the domain and range of the composite function $f \circ g$.

91. Find the area of the triangle with vertices $(4, 3)$, $(9, 13)$, and $(20, 15)$.

92. A producer of a certain software product sells the product according to a price structure that encourages volume purchases. For orders of up to five copies the price is $500 per copy. On orders of more than five copies the price per copy is reduced by $10 for each copy in excess of five, except that all orders for more than 25 copies are filled at $300 per copy.
   a. Write a function p giving the price per copy in purchasing x copies of this software.
   b. Graph the function p.

93. Find the function R giving the total revenue received from a single order of x copies of the software described in Exercise 92.

READINESS TEST

The following 25 exercises are designed to test your understanding of the material reviewed in Chapter 1. We suggest that you try all of the problems without consulting the chapter. Then check your answers in the back of the book (the answers to all 25 questions are provided). Along with each answer, we cite the relevant pages and related exercises in the chapter. If you have difficulty with a problem, we recommend that you review the pages cited and do some of the corresponding exercises.

1. Find the center and radius of the circle $x^2 - 2x + y^2 + 3y = 4$.

2. Determine the point of intersection of the two lines
$$x + y = 7$$
$$3x + 2y = 14.$$

3. Find an equation for the line parallel to $3x + 2y = 2$ that contains the point $(-1, 2)$. Express your answer in both point-slope form and slope-intercept form.

4. Sketch a right triangle whose hypotenuse has length 5 and that contains an angle of $\pi/6$ radians. Indicate the lengths of the sides and the sizes (in radians) of the angles.

5. Solve the inequality $|x + 3| > 3$ and express the answer using interval notation.

6. Find an equation for the line that is perpendicular to $x - 2y = 4$ and that has the same y-intercept as $x - 2y = 4$.

7. Graph the quadratic function $f(x) = 2x^2 - 4x + 3$.

8. State the domain of the function $g(x) = \sqrt{x^2 - 2x}$.

9. Express the set $\{x \mid x^2 - 2x - 3 > 0\}$ as a union of intervals using interval notation.

10. Let $f(x) = \sqrt{x}$ and $g(x) = x^2 + 1$. In a–e, specify the domain of the designated function. Then derive an algebraic expression for it.
    a. $2f/g$
    b. g/f
    c. fg
    d. $f \circ g$
    e. $g \circ f$

11. Solve the equation $(|x| - 3)^2 = 9$.

12. Graph the functions $f(x) = 9 - x^2$ and $g(x) = |9 - x^2|$.

13. Find all values of θ such that $\cos \theta = \sqrt{3}/2$ and $\sin \theta < 0$.

14. Find the value of $\cos \theta$ if θ is an angle in the third quadrant and $\cos 2\theta = 1/4$.

15. Sketch a right triangle that contains an angle θ such that $\sin \theta = 2/5$.

16. Solve the inequality $\cos x > 1/2$ for $0 \le x \le 2\pi$. Graph $y = \cos x$ for $0 \le x \le 2\pi$, and illustrate your solution on this graph.

17. State the domain of the function $f(x) = 1/(1 + \sin x)$.

18. State the domain of the function $f(x) = (x + 2)^{3/2}$.

19. Find the set of all points that are equidistant from the points $(-2, 1)$ and $(0, 0)$.

20. Express the set $\{x \mid 0 \le x^2 - 4x < 5\}$ as a union of intervals using interval notation.

21. Give an example of:
    a. a polynomial function of degree 4, and
    b. a rational function that is not a polynomial.

22. The average daily temperature in a certain city rose from 58°F in 1940 to 59.2°F in 1980. If the relationship between temperature and time is assumed to be linear, what temperature do these data project for this city's average daily temperature in 2000?

23. State the domain of the function $f(x) = (x - 2)^{-2/3}$.

24. State the range of the function $f(x) = \sec(x^2)$.

25. Agricultural research shows that a certain type of apple tree will yield 400 pounds of apples per year minus 2 pounds for each tree planted per acre. Write a function Y giving the annual yield in pounds of a one-acre plot containing x such apple trees.

Chapter 2

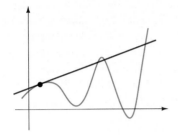

Limits of Functions

Here is where the calculus begins. So that you will have some sense of where we are headed, this chapter opens with a brief discussion of two of the major goals of the calculus. We turn quickly to practical matters, however, by developing in this chapter the concept of the *limit* of a function. This topic is so central to the calculus that it appears frequently throughout every chapter that follows.

2.1   Tangents, Areas, and Limits

Two general problems characterize much of what we shall do in this text:

The Tangent Line Problem: Find the slope of the line ℓ tangent to the graph of the function f at the point P (Figure 1.1).

The Area Problem: Find the area of the region R bounded by the graph of the function f and the x-axis for $a \le x \le b$ (Figure 1.2).

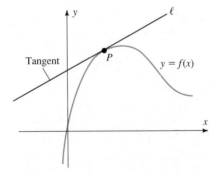

Figure 1.1   The Tangent Line Problem: Find the slope of ℓ.

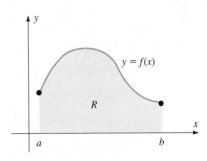

Figure 1.2   The Area Problem: Find the area of R.

The **differential calculus** examines the rate of change of a function, thereby resolving the Tangent Line Problem. The **integral calculus** involves a generalized summation process that solves the Area Problem as well as numerous other applications.

As we begin to develop both the differential and integral calculus, we will use the notation, theory, and techniques of elementary algebra, plane geometry, and analytic geometry, but we shall immediately see that a new concept is required. Whether a study of the calculus is begun with the Tangent Line Problem or the Area Problem, the notion of the **limit** of a function plays a fundamental role. Because this concept underlies every aspect of the calculus, we develop it carefully at the outset. Our plan, therefore, is to discuss the Tangent Line Problem briefly in this section and then to devote the remaining part of this chapter to a careful study of the limit concept that arises here. We will then be prepared for a rigorous study of the differential calculus beginning in Chapter 3.

The Tangent to a Curve

The first step in attacking the Tangent Line Problem is to define clearly what we mean by "the line tangent to the graph of f at point P." From geometry we know that if the graph of f is an arc of a circle, then the tangent at point P may be defined as the (unique) line that intersects the circle only at point P. This is a perfectly adequate definition for arcs of circles, but it fails for more general curves. For example, Figure 1.3 shows several lines each intersecting the graph of f only at P but none we would wish to call a tangent. Moreover, in Figure 1.4, we see that the tangent at P can intersect the graph of f at points other than P.

There is another way, however, of defining the tangent to a circle that does have a satisfactory generalization to more general curves. Figure 1.5 illustrates that a second point Q on the circle determines a *secant* through the points P and Q. As the point Q moves toward P along the circle, the secant rotates into the position of a fixed line through P. This fixed line is the tangent to the circle at P. We use this idea to define the tangent more generally.

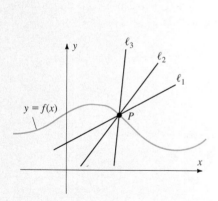

Figure 1.3 Lines intersecting graph of $y = f(x)$ "only at point P," which are not tangents.

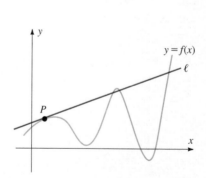

Figure 1.4 A tangent line ℓ that intersects the graph more than once.

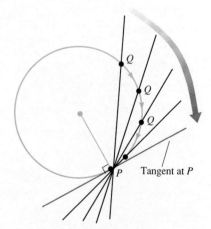

Figure 1.5 As Q "approaches P" along the circle, the secant through P and Q rotates into the tangent at P.

Definition 1

Let P and Q be points on a curve C. The line **tangent** to the curve C at the point P, if it exists, is the limiting position of the secant line through P and Q as Q approaches P along the curve C from either direction. (See Figure 1.6.)

Solving the Tangent Line Problem

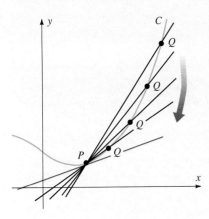

Figure 1.6 Tangent is limiting position of secant as $Q \rightarrow P$.

We can now determine how to define the slope of the line tangent to the graph of a function f at a point P, if it exists, in a way that is consistent with our notion of a tangent to a curve. First, we let x_0 be the x-coordinate of P. Then P has coordinates $(x_0, f(x_0))$ since P is on the graph of f. To find a second point Q on the graph of f we choose any number $h \neq 0$ (positive or negative) and let $Q = (x_0 + h, f(x_0 + h))$. (See Figure 1.7.) Then

$$\begin{bmatrix} \text{Slope of secant} \\ \text{through } P \text{ and } Q \end{bmatrix} = \frac{f(x_0 + h) - f(x_0)}{h} = \frac{\Delta y}{\Delta x} \tag{1}$$

If the graph of f connects the points P and Q in an unbroken arc, the point Q will approach P along the graph of f as the number h approaches zero (see Figure 1.8). Then the tangent to the graph of f at P is the "limiting position" of the secant through P and Q as h approaches zero. Thus, the slope of the tangent should be the "limiting value" of the slope of this secant as h approaches zero:

$$\begin{bmatrix} \text{Slope of the} \\ \text{tangent at } P \end{bmatrix} = \begin{bmatrix} \text{limiting value of the slope of} \\ \text{the secant as } h \text{ approaches zero} \end{bmatrix}$$

$$= \begin{bmatrix} \text{limit as } h \text{ approaches zero of} \\ \dfrac{f(x_0 + h) - f(x_0)}{h} \end{bmatrix}. \tag{2}$$

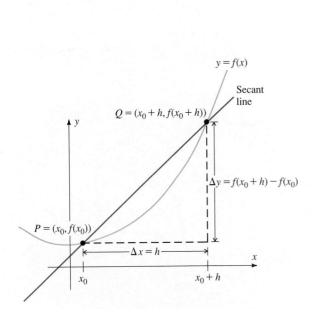

Figure 1.7 The slope $\dfrac{\Delta y}{\Delta x} = \dfrac{f(x_0 + h) - f(x_0)}{h}$ of a secant line.

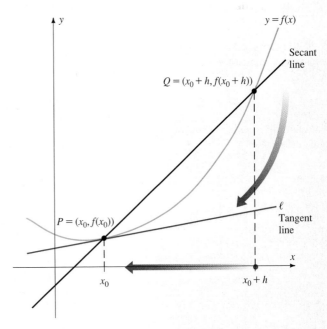

Figure 1.8 The slope of the tangent ℓ is the limit of the slope of the secant as h approaches 0 (and thus $x_0 + h$ approaches x_0).

□ **EXAMPLE 1**

Find the slope of the line tangent to the graph of $f(x) = x^2$ at the point $(2, 4)$.

Before we calculate the slope of the tangent, it is informative to see what happens when we calculate the slopes of selected secants (Table 1.1).

Table 1.1 Selected h values and the slopes of the associated secant lines for Example 1.

h	-1.000	-0.100	-0.010	-0.001	0.001	0.010	0.100	1.000
$\dfrac{f(2+h)-f(2)}{h}$	3.000	3.900	3.990	3.999	4.001	4.010	4.100	5.000

As h approaches 0 for these values, the slopes of the secant lines seem to approach 4. The following calculation indicates how we can justify this assertion.

Strategy · · · · · · · ·

Identify x_0.

Find $f(x_0 + h)$ by substituting $x = x_0 + h$ in equation for $f(x)$.

Solution

Here $x_0 = 2$ and $f(x) = x^2$, so

$$f(x_0 + h) = f(2 + h) = (2 + h)^2 = 4 + 4h + h^2$$

and

$$f(x_0) = f(2) = 2^2 = 4.$$

Thus,

Set up the expression for the slope of the secant; substitute for $f(x_0 + h)$ and $f(x_0)$.

$$\frac{f(x_0 + h) - f(x_0)}{h} = \frac{(4 + 4h + h^2) - 4}{h}$$

$$= \frac{4h + h^2}{h}$$

Simplify; factor h from the numerator. Divide by common factor of h.

$$= \frac{h(4 + h)}{h}$$

$$= 4 + h \qquad (h \neq 0).$$

The slope of the tangent is therefore

Find limit of resulting expression as h approaches zero.

$$m = \lim_{h \to 0} (4 + h) = 4.$$

(See Figure 1.9.)

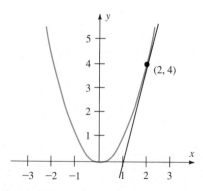

Figure 1.9 Slope of tangent to graph of $f(x) = x^2$ at $(2, 4)$ is $m = 4$.

The "$h \to 0$" notation used in the solution to Example 1 indicates that we are taking the limit of the quotient as the values of h approach 0.

Using equations (1) and (2) we formally define the slope of the line tangent to the graph of a function at a point as follows.

Definition 2	The **slope** of the line tangent to the graph of the function f at the point $(x_0, f(x_0))$, if it exists, is the number

$$m = \lim_{h \to 0} \frac{f(x_0 + h) - f(x_0)}{h}. \tag{3}$$

In the remaining sections of this chapter, we precisely define and discuss the *limit* concept, as represented by the notation

$$\lim_{h \to 0}$$

in equation (3). But for now, we interpret equation (3) as the result of letting the number h in the quotient

$$\frac{f(x_0 + h) - f(x_0)}{h}, \tag{4}$$

approach 0. Of course, if we substitute $h = 0$ in expression (4), we obtain a fraction of the form $\frac{0}{0}$, which is a meaningless algebraic expression. Thus, to determine the limit, we simplify expression (4) in order to eliminate the 0 in the denominator. We obtain an equivalent expression whose limit as $h \to 0$ can be determined ''by inspection.''

□ **EXAMPLE 2**

Find an equation for the line tangent to the graph of $f(x) = x^3 + 3x - 2$ at the point $(1, 2)$.

Solution   We begin by finding the slope of the secant at $(1, 2)$, as in Example 1. With $x_0 = 1$ we have $f(1) = 1^3 + 3 \cdot 1 - 2 = 2$ and

$$\begin{aligned} f(1 + h) &= (1 + h)^3 + 3(1 + h) - 2 \\ &= (1 + 3h + 3h^2 + h^3) + (3 + 3h) - 2 \\ &= h^3 + 3h^2 + 6h + 2. \end{aligned}$$

The slope of the secant through $(1, 2)$ and $(1 + h, f(1 + h))$ is therefore

$$\begin{aligned} \frac{f(1 + h) - f(1)}{h} &= \frac{(h^3 + 3h^2 + 6h + 2) - 2}{h} \\ &= \frac{h^3 + 3h^2 + 6h}{h} \\ &= h^2 + 3h + 6 \end{aligned}$$

when $h \neq 0$. The slope of the tangent at $(1, 2)$ is therefore

$$m = \lim_{h \to 0}(h^2 + 3h + 6) = 6.$$

An equation for the tangent, which has slope $m = 6$ and passes through the point $(1, 2)$, is

$$y - 2 = 6(x - 1). \qquad \blacksquare$$

Exercise Set 2.1

In Exercises 1–17, use equation (3) to find the slope of the line tangent to the graph of the given function at the given point.

1. $f(x) = 3x - 2$, $P = (2, 4)$

2. $f(x) = 7 - 3x$, $P = (1, 4)$

3. $f(x) = 2x^2$, $P = (3, 18)$

4. $f(x) = 9x^2$, $P = (-1, 9)$

5. $f(x) = 2x^2 + 3$, $P = (1, 5)$

6. $f(x) = 5 - x^2$, $P = (-2, 1)$

7. $f(x) = 3x^2 + 4x + 2$, $P = (-2, 6)$

8. $f(x) = x^2 - 6x + 3$, $P = (2, -5)$

9. $f(x) = x^3 + 3$, $P = (2, 11)$

10. $f(x) = x^3 - 2x + 6$, $P = (1, 5)$

11. $f(x) = x^4$, $P = (-2, 16)$

12. $f(x) = ax^2 + bx + c$ where $x = 1$

13. $f(x) = ax^3 + bx^2 + cx + d$ where $x = 1$

14. $f(x) = 1/x$, $P = (1, 1)$

15. $f(x) = 1/(x + 3)$, $P = (-2, 1)$

16. $f(x) = 3/(2x - 1)$, $P = (2, 1)$

17. $f(x) = 4/x^2$, $P = (2, 1)$

18. Find an equation for the line tangent to the graph of the function $f(x) = 2x^2$ at the point $(2, 8)$.

19. Find an equation for the line tangent to the graph of the function $f(x) = 3x^2$ at the point $(-1, 3)$.

20. Find an equation for the line tangent to the graph of the function $f(x) = x^2 - 4x$ at the point $(1/2, -7/4)$.

21. Find an equation for the line tangent to the graph of $f(x) = ax^2$ at the point $(1, a)$.

22. Find an equation for the line tangent to the graph of $f(x) = 5 - x^2$ at the point $(-2, 1)$.

23. Find an equation for the line tangent to the graph of $f(x) = 2x^3 + x$ at the point where $x = 1$.

24. Find an equation for the line tangent to the graph of $f(x) = ax^3$ at the point $(1, a)$.

25. True or false? A tangent to the graph of a function cannot be horizontal. Why or why not?

26. Complete Table 1.2 for the function $f(x) = x^2 - 4x + 3$.

Table 1.2

x_0	$f(x_0)$	$\dfrac{f(x_0 + h) - f(x_0)}{h}$	Slope of tangent at $(x_0, f(x_0))$
-2			
-1			
0			
1			
2			
3			
x_0			

27. By using a calculator we can approximate the slope of the line tangent to the graph of f at $(x_0, f(x_0))$ by computing the slope of the secant

$$\frac{f(x_0 + h) - f(x_0)}{h}$$

for small numbers h.

a. Approximate the slope of the line tangent to the graph of $f(x) = x^3 - 3x$ at the point $(2, 2)$ by completing Table 1.3.

b. Use the method of Examples 1 and 2 to find the slope of the line tangent to the graph of $y = x^3 - 3x$ at the point $(2, 2)$. How does this compare with the results of part a?

Table 1.3

x_0	h	$\dfrac{f(x_0 + h) - f(x_0)}{h}$
	0.1000	
	0.0100	
	0.0010	
	0.0001	
	-0.0001	
	-0.0010	
	-0.0100	
	-0.1000	

28. Let $f(x) = 3x^4$.

a. Approximate the slope of the tangent line to the graph of f at the point $(1, 3)$ using Table 1.3.

b. Use the method of Examples 1 and 2 to verify your prediction in part a.

29. Consider the graph of $f(x) = x^3 + x^2 - 2x$ in Figure 1.10. Note that the tangents at the points with x-coordinates $-3/2$, -1, and $1/2$ are sketched.

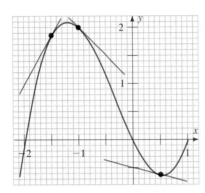

Figure 1.10 The graph of $f(x) = x^3 + x^2 - 2x$ and tangents at $x = -3/2, -1$, and $1/2$.

a. Using the square grid, estimate the slopes of each of these tangents.

b. Use the method of Examples 1 and 2 to calculate these slopes. How close were your estimates?

30. Let $f(x) = 1/x$.

a. Approximate the slope of the tangent line to the graph of f at the point $(1/2, 2)$ using Table 1.3.

b. Use the method of Examples 1 and 2 to verify your prediction in part a.

31.

a. Using Table 1.3, approximate the slope of the tangent line to the graph of $y = \sin x$ at the point $(0, 0)$.

b. Apply the method of Examples 1 and 2 to derive an expression for the limit that must be evaluated to determine the slope of the tangent line. Explain why this expression cannot be evaluated by "inspection." (We will evaluate this limit later in this chapter.)

32. Use Table 1.3 to approximate the slope of the line tangent to the graph $y = \sin x$ at the point $(\pi/4, \sqrt{2}/2)$.

33. Show that the slope of the line tangent to the graph of $f(x) = x^2 + 6x + 1$ at the point where $x = a$ is $m = 2a + 6$. Use this information to find the number x where the slope of the tangent to the graph is 0.

34. Use the method of Exercise 33 to find the point(s) on the graph of $f(x) = x^2 + 2x + 3$ where the tangent line is parallel to the line $y = 6x + 1$. Specify an equation(s) for these tangent line(s).

35. Use the method of Exercise 33 to find the point(s) on the graph of $f(x) = x^2 - 3x + 1$ where the slope of the tangent at the point equals the y-coordinate of the point.

36. Let $f(x) = ax^2 + bx + c$.

a. Use the method of Exercise 33 to show that the slope of the line tangent to the graph of f at $(x, f(x))$ is $m = 2ax + b$.

b. The graph of this function is a parabola. Show that the slope of the tangent at the vertex of the parabola is 0.

37. Find the numbers a, b, and c such that the graph of $f(x) = ax^2 + bx + c$ has y-intercept $(0, 5)$, contains the point $(1, 2)$, and has a tangent with slope 3 when $x = 2$.

38. Prove that the line tangent to the graph of $f(x) = x^3$ at $(x_0, f(x_0))$ is always parallel to the line tangent to this graph at the point $(-x_0, f(-x_0))$. Is this true for any odd function? Why or why not?

39. Let $f(x) = |x|$. Complete Table 1.3 for the point $(0, 0)$. Explain why the data supports the assertion: "The absolute value function does not have a tangent line at the point $(0, 0)$."

40. Knowledge of the slopes of the tangents to the graph of a function often aids in understanding its behavior, even if the exact expression for the function is unknown. Use the indicated values of the slopes in Table 1.4 to sketch tangents at each of the points in Figure 1.11. Then sketch the graph of f as best you can. (Notice that the graph you sketch differs from what you might have sketched without the data of Table 1.4.)

Table 1.4

x	-3	-2	-1	0	1	2	3
Slope	-2	4	0	-1	6	-2	2

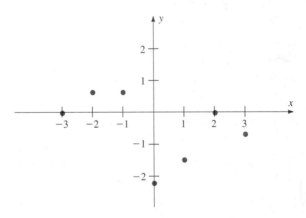

Figure 1.11

In Exercises 41–44, a function f and a number a are given. For $h \neq 0$, let

$$m_h^+ = \frac{f(a + h) - f(a)}{h} \quad \text{and} \quad m_h^- = \frac{f(a) - f(a - h)}{h}.$$

Using $h = 1, 0.1, 0.01, 0.001$, and 0.0001,

a. calculate m_h^+ and m_h^-;

b. graph $y = f(x)$ along with the lines $y = m_h^+(x - a) + f(a)$ and $y = m_h^-(x - a) + f(a)$; and,

c. using these plots, predict the slope of the tangent line to the graph of $y = f(x)$ at $x = a$.

41. $f(x) = x^2 - 2x + 1, \quad a = 3$ **42.** $f(x) = \sin x, \quad a = \pi/2$

43. $f(x) = \begin{cases} x^2 + 2x + 1, & \text{if } x \le 0; \\ x^2 - 2x + 1, & \text{if } x > 0. \end{cases} \quad a = 0$

44. $f(x) = \begin{cases} x^2 + 2x + 1, & \text{if } x \le 0; \\ -x^2 + 2x + 3, & \text{if } x > 0. \end{cases} \quad a = 0$

2.2 Limits of Functions

The slope calculations of Section 2.1 involve special cases of the more general statement

$$L = \lim_{x \to a} f(x) \tag{1}$$

read, "The number L is the limit of the function f as x approaches a." A formal definition of limit, which allows us to actually prove statements in the form of equation (1), is given in the next section. The following intuitive explanation, however, enables us to calculate most limits that we shall encounter.

The statement

$$L = \lim_{x \to a} f(x)$$

means that the values $f(x)$ are as close to L as we desire for all $x \ne a$ sufficiently close to a.

This statement is not a rigorous definition because the phrases "as close to L as we desire" and "sufficiently close" are somewhat imprecise. Nevertheless, once we understand how to interpret them, we shall see how this informal definition characterizes the intuitive calculations we made in Section 2.1.

First, we illustrate this statement in a straightforward example.

☐ **EXAMPLE 1**

Let $f(x) = 2x + 1$. In the next section, we prove that

$$\lim_{x \to 3} f(x) = 7.$$

To see why this function satisfies the intuitive definition given above, we discuss two different choices of "desired closeness" of $f(x)$ to $L = 7$. For instance, suppose that we want $f(x)$ to be within 0.5 of 7. Then, for any x within 0.25 of 3, we have

$$2.75 < x < 3.25$$
$$5.50 < 2x < 6.50$$
$$6.50 < 2x + 1 < 7.50$$
$$6.50 < f(x) < 7.50$$

So, to obtain the "desired" accuracy of ± 0.5 around $L = 7$, we restrict x to the interval $2.75 < x < 3.25$. In other words, if we want to guarantee that the values of

$f(x)$ are within 0.5 of 7 for numbers x that are close to 3, it is sufficient if we restrict x to be within 0.25 of 3.

However, if we want $f(x)$ to be within 0.1 of 7, we must require that x be in a smaller interval around 3. In this case, it suffices for us to require that x be within 0.05 of the number 3. Then

$$2.95 < x < 3.05$$
$$5.90 < 2x < 6.10$$
$$6.90 < 2x + 1 < 7.10$$
$$6.90 < f(x) < 7.10$$

Therefore, if $2.95 < x < 3.05$, then $f(x)$ is within 0.1 of 7.

There is nothing special about the choices of the accuracies 0.5 and 0.1. We chose these two values for illustration purposes only. In fact, given *any* desired accuracy of $f(x)$ to 7, we can find some open interval I centered at 3 such that, if x is contained in I, then the value $f(x)$ differs from 7 by no more than the prescribed accuracy. Thus, we say that

$$\lim_{x \to 3}(2x + 1) = 7.$$

In Section 2.3, we indicate how we determine the appropriate interval I from the desired accuracy. ∎

There are a number of important points regarding limits that you should keep in mind:

REMARK 1 Although the intuitive definition describes what we mean when we say

$$\lim_{x \to a} f(x) = L,$$

it does not indicate how we determine the value of L. In fact, we shall discuss many ways to systematically determine L without using the definition.

REMARK 2 We often use tables such as Tables 2.1 and 2.2 to suggest the limit. The first column contains numbers x that successively approach the number a, and the second column contains the corresponding values $f(x)$. If the limit of $f(x)$ as $x \to a$ is L, the numbers in the $f(x)$ column "approach" L. However, we should emphasize that a single table of this type merely suggests a value for L. In order to *prove* that L is the limit of $f(x)$ as x approaches a, it is necessary to show that the values $f(x)$ approach L for *all* possible choices of x that approach a. In Figure 2.1, we illustrate this condition in terms of the graph of f.

REMARK 3 Sometimes limits do not exist. If we find a table of numbers x approaching a such that the corresponding values $f(x)$ do not approach any particular real number, then the limit does not exist. Also, if we find two different tables of $x \to a$ such that the values $f(x)$ in one table approach the number L_1 and the values $f(x)$ in the other table approach a different number L_2, then the limit does not exist. Examples 6 and 7 at the end of this section are examples of functions that do not have limits as $x \to 0$.

REMARK 4 The function f does not have to be defined at the number a. (That is why Figure 2.1 shows a "hole" in the graph of f at the point (a, L).) **In fact, even if $f(a)$ is defined, this value does not affect the limit of $f(x)$ as $x \to a$.**

Table 2.1 Numerical evidence supporting the assertion that

$$\lim_{x \to 0} \frac{\sin x}{x} = 1.$$

x	$\dfrac{\sin x}{x}$
0.8	0.896695
0.5	0.958851
0.2	0.993347
0.08	0.998934
0.05	0.999583
0.02	0.999933
0.005	0.999996
0.002	0.999999
−0.002	0.999999
−0.005	0.999996
−0.02	0.999933
−0.05	0.999583
−0.08	0.998934
−0.2	0.993347
−0.5	0.958851
−0.8	0.896695

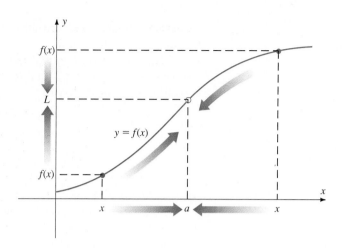

Figure 2.1 The equation $L = \lim_{x \to a} f(x)$ means that $f(x)$ is near L when x is near a.

Table 2.2 Numerical evidence supporting the assertion that
$$\lim_{x \to 0} \frac{(x + 1)^3 - 1}{x} = 3.$$

x	$\dfrac{(x + 1)^3 - 1}{x}$
2.0	13.0000
1.5	9.7500
1.0	7.0000
0.5	4.7500
0.2	3.6400
0.1	3.3100
0.01	3.0301
0.001	3.0030
−0.001	2.9970
−0.01	2.9701
−0.1	2.7100
−0.2	2.4400
−0.5	1.7500
−1.0	1.0000
−1.5	0.7500
−2.0	1.0000

□ **EXAMPLE 2**

The function

$$f(x) = \frac{\sin x}{x}$$

is undefined for $x = 0$. However, the limit

$$\lim_{x \to 0} \frac{\sin x}{x}$$

will be important in our work on trigonometric functions in Chapter 3. Even though $f(x)$ is undefined for $x = 0$, the entries in Table 2.1 (obtained by use of a calculator) suggest that, as x approaches 0, the values $f(x)$ approach the number $L = 1$. Note that this numerical evidence is consistent with the graph of f (see Figure 2.2). Later in this chapter we shall prove that

$$\lim_{x \to 0} \frac{\sin x}{x} = 1. \qquad \blacksquare$$

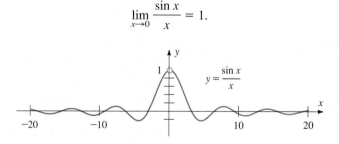

Figure 2.2 The graph $y = (\sin x)/x$, $x \neq 0$. Note that, for x near 0, $(\sin x)/x$ is near 1.

□ **EXAMPLE 3**

Find $\displaystyle\lim_{x \to 0} \frac{(x + 1)^3 - 1}{x}$.

Solution   Here the function

$$f(x) = \frac{(x+1)^3 - 1}{x}$$

is also undefined if $x = 0$. Experimenting with a calculator, however, gives results such as those in Table 2.2 that suggest that for x close to 0 the value $f(x)$ is close to $L = 3$. That is, the data suggest that

$$\lim_{x \to 0} \frac{(x+1)^3 - 1}{x} = 3.$$

We can verify this limit using simple algebra. We write

$$\frac{(x+1)^3 - 1}{x} = \frac{(x^3 + 3x^2 + 3x + 1) - 1}{x}$$

$$= \frac{x^3 + 3x^2 + 3x}{x}$$

$$= x^2 + 3x + 3, \quad x \neq 0.$$

We may conclude therefore that the functions

$$f(x) = \frac{(x+1)^3 - 1}{x} \quad \text{and} \quad g(x) = x^2 + 3x + 3$$

have the same values *except* at $x = 0$, where $f(0)$ is undefined. (See Figures 2.3 and 2.4.) Thus, the limits as x approaches zero of these two functions must be the same. (Remember, we care about only the values $f(x)$ for x **near** zero, not $f(0)$ itself.)

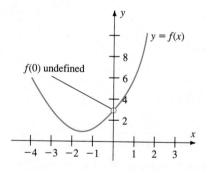

Figure 2.3 $f(x) = \dfrac{(x+1)^3 - 1}{x}$;

$f(0)$ is undefined.

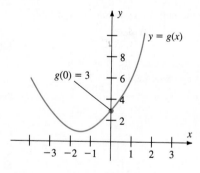

Figure 2.4 $g(x) = x^2 + 3x + 3$;
$g(0) = 3$.

Our limit may be calculated as follows:

$$\lim_{x \to 0} \frac{(x+1)^3 - 1}{x} = \lim_{x \to 0}(x^2 + 3x + 3) = 0 + 0 + 3 = 3. \tag{2}$$

Note that the technique used here to calculate the limit is the same as that of Section 2.1: Factor the term causing the zero from both the numerator and denominator and divide.   ■

REMARK 5: Some limits are quite obvious and may be determined "by inspection." For example, in line (2) the limit

$$\lim_{x \to 0}(x^2 + 3x + 3) = 0 + 0 + 3 = 3$$

is obtained by noting that, if x is close to zero, so are x^2 and $3x$. The limit theorems established in Section 2.4 will justify such calculations. In the following example we use the limit

$$\lim_{x \to 2} \frac{x - 1}{x + 3} = \frac{2 - 1}{2 + 3} = \frac{1}{5},$$

also obtained "by inspection."

☐ **EXAMPLE 4**

Find $\displaystyle\lim_{x \to 2} \frac{x^2 - 3x + 2}{x^2 + x - 6}$.

Strategy · · · · · · · ·

Determine whether limit can be obtained "by inspection" by setting $x = 2$. This yields $\frac{0}{0}$.

Solution

First note that on setting $x = 2$ we obtain the quotient

$$\frac{2^2 - 3 \cdot 2 + 2}{2^2 + 2 - 6} = \frac{4 - 6 + 2}{4 + 2 - 6} = \frac{0}{0}$$

which is undefined. We therefore factor the numerator and denominator, obtaining, for $x \neq 2$,

Factor both numerator and denominator.

$$\frac{x^2 - 3x + 2}{x^2 + x - 6} = \frac{(x - 2)(x - 1)}{(x - 2)(x + 3)}$$

Divide both by common factor $x - 2$.

$$= \frac{x - 1}{x + 3} \quad \left(\text{since } \frac{x - 2}{x - 2} = 1 \text{ if } x \neq 2\right).$$

The limit equals the limit of the resulting expression.

Thus,

$$\lim_{x \to 2} \frac{x^2 - 3x + 2}{x^2 + x - 6} = \lim_{x \to 2} \frac{x - 1}{x + 3} = \frac{2 - 1}{2 + 3} = \frac{1}{5}. \quad ■$$

☐ **EXAMPLE 5**

Find $\displaystyle\lim_{x \to \pi} \frac{\sin^2 x}{1 + \cos x}$.

Solution First note that $\sin^2 \pi = (\sin \pi)^2 = 0$ and $1 + \cos \pi = 1 + (-1) = 0$, so both the numerator and denominator equal zero when x equals π. To find an equivalent expression for

$$\frac{\sin^2 x}{1 + \cos x}$$

we use the identity $\sin^2 x + \cos^2 x = 1$. Then $\sin^2 x = 1 - \cos^2 x$, and

$$\frac{\sin^2 x}{1 + \cos x} = \frac{1 - \cos^2 x}{1 + \cos x} = \frac{(1 - \cos x)(1 + \cos x)}{1 + \cos x} = 1 - \cos x$$

if $\cos x \neq -1$. Thus,

$$\lim_{x \to \pi} \frac{\sin^2 x}{1 + \cos x} = \lim_{x \to \pi}(1 - \cos x) = 1 - (-1) = 2. \quad ■$$

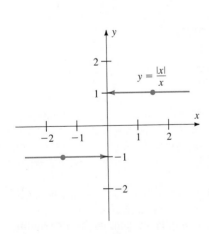

Figure 2.5 $\displaystyle\lim_{x \to 0} \frac{|x|}{x}$ does not exist.

Limits That Fail to Exist

The next two examples show ways in which limits can fail to exist.

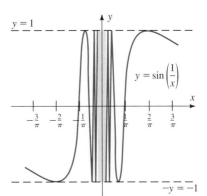

Figure 2.6   The limit

$$\lim_{x \to 0} \sin\left(\frac{1}{x}\right)$$

does not exist.

□ EXAMPLE 6

The limit $\displaystyle\lim_{x \to 0} \frac{|x|}{x}$ does not exist. To see why, we use the definition of absolute value

$$|x| = \begin{cases} x & \text{if } x \geq 0 \\ -x & \text{if } x < 0 \end{cases}$$

to rewrite the function $f(x) = \dfrac{|x|}{x}$ as

$$f(x) = \begin{cases} \dfrac{x}{x} & \text{if } x > 0 \\[2ex] \dfrac{-x}{x} & \text{if } x < 0. \end{cases}$$

(See Figure 2.5.) If x is close to zero *and positive*, $f(x) = 1$. But, if x is close to zero *and negative*, $f(x) = -1$. In order for the limit of $f(x)$ to exist as $x \to 0$, the values $f(x)$ must be close to a single number L when x is close to 0 *on either side*. No matter how close we restrict x to 0, we still include positive x (thus $f(x) = 1$) and negative x (thus $f(x) = -1$). Therefore, the limit does not exist (see Remark 3).   ■

In Example 6, the limit does not exist because the values $f(x)$ approach two different numbers as $x \to 0$ depending upon the sign of x. In Section 2.5, we discuss ''one-sided'' limits, which apply to functions such as the one in Example 6.

In our final example, we discuss another function that does not have a limit as x approaches 0. In this case, the limit does not exist because the values $f(x)$ repeatedly oscillate between 1 and -1 as $x \to 0$.

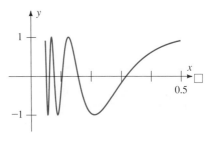

Figure 2.7   The graph $y = \sin(1/x)$ for $0.05 < x < 0.5$.

□ EXAMPLE 7

The limit

$$\lim_{x \to 0} \sin\left(\frac{1}{x}\right)$$

does not exist.

In fact, if we consider the graph of $f(x) = \sin(1/x)$ for x close to 0, as illustrated in Figures 2.6–2.8, we see that $f(x)$ oscillates between 1 and -1 in an increasingly rapid rate as $x \to 0$. These oscillations are caused by the fact that $1/x$ takes the interval $\{x \mid 1 < x < \infty\}$ (on which the sine function oscillates infinitely often) and squeezes it between 0 and 1. Thus, the repeated oscillations of sine cause the graph of f to oscillate infinitely often as $x \to 0$.

Tables 2.3a, 2.3b, and 2.3c illustrate the repeated oscillation of $f(x)$ numerically. Table 2.3a suggests that the limit should be 1, Table 2.3b suggests that the limit should be 0, and Table 2.3c suggests that the limit should be $\sqrt{2}/2$. As we mentioned in Remark 3 earlier, the limit of $\sin(1/x)$ as $x \to 0$ does not exist because the values do not approach a unique number L.   ■

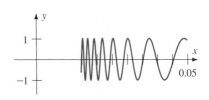

Figure 2.8   The graph $y = \sin(1/x)$ for $0.015 < x < 0.05$.

Table 2.3   Three different tables of numbers x, all approaching 0, along with the corresponding values $\sin(1/x)$. This numerical evidence supports the assertion that $\sin(1/x)$ does not have a limit as x approaches 0.

x	$\sin\left(\dfrac{1}{x}\right)$
$\dfrac{2}{\pi}$	1
$\dfrac{2}{5\pi}$	1
$\dfrac{2}{9\pi}$	1
$\dfrac{2}{13\pi}$	1
$\dfrac{2}{17\pi}$	1

(a)

x	$\sin\left(\dfrac{1}{x}\right)$
$\dfrac{1}{2\pi}$	0
$\dfrac{1}{4\pi}$	0
$\dfrac{1}{6\pi}$	0
$\dfrac{1}{8\pi}$	0
$\dfrac{1}{10\pi}$	0

(b)

x	$\sin\left(\dfrac{1}{x}\right)$
$\dfrac{4}{\pi}$	$\dfrac{\sqrt{2}}{2}$
$\dfrac{4}{9\pi}$	$\dfrac{\sqrt{2}}{2}$
$\dfrac{4}{17\pi}$	$\dfrac{\sqrt{2}}{2}$
$\dfrac{4}{25\pi}$	$\dfrac{\sqrt{2}}{2}$
$\dfrac{4}{33\pi}$	$\dfrac{\sqrt{2}}{2}$

(c)

Exercise Set 2.2

For each of the functions whose graphs are sketched in Exercises 1–6, indicate whether

(i) $\lim_{x \to a} f(x)$ exists and equals $f(a)$,

(ii) $\lim_{x \to a} f(x)$ exists but does not equal $f(a)$, or

(iii) $\lim_{x \to a} f(x)$ does not exist.

1.
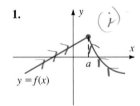

2.
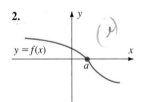

3.
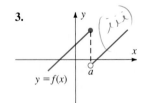

4.
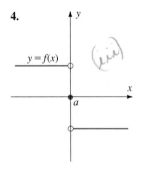

5.
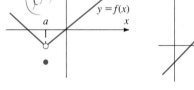

6.
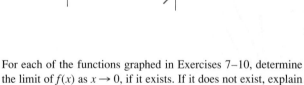

For each of the functions graphed in Exercises 7–10, determine the limit of $f(x)$ as $x \to 0$, if it exists. If it does not exist, explain why.

7.
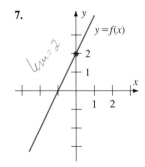

8.
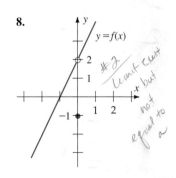

9.

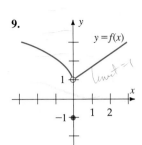

10.

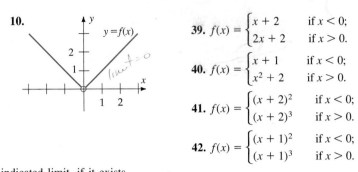

39. $f(x) = \begin{cases} x + 2 & \text{if } x < 0; \\ 2x + 2 & \text{if } x > 0. \end{cases}$

40. $f(x) = \begin{cases} x + 1 & \text{if } x < 0; \\ x^2 + 2 & \text{if } x > 0. \end{cases}$

41. $f(x) = \begin{cases} (x + 2)^2 & \text{if } x < 0; \\ (x + 2)^3 & \text{if } x > 0. \end{cases}$

42. $f(x) = \begin{cases} (x + 1)^2 & \text{if } x < 0; \\ (x + 1)^3 & \text{if } x > 0. \end{cases}$

In Exercises 11–38, find the indicated limit, if it exists.

11. $\lim\limits_{x \to 2}(3 + 7x)$

12. $\lim\limits_{x \to 0}(x^3 - 7x + 5)$

13. $\lim\limits_{x \to 0} \dfrac{(3 + x)^2 - 9}{x}$

14. $\lim\limits_{x \to 3} \dfrac{x^2 - 9}{x - 3}$

15. $\lim\limits_{h \to 0} \dfrac{h^2 - 1}{h - 1}$

16. $\lim\limits_{x \to 0} \dfrac{(x + 3)^3 - 27}{x}$

17. $\lim\limits_{x \to -2} \dfrac{x^2 - x - 6}{x + 2}$

18. $\lim\limits_{h \to 0} \dfrac{(4 + h)^2 - 16}{h}$

19. $\lim\limits_{x \to 0} \dfrac{1 - \cos^2 x}{\sin x \cos x}$

20. $\lim\limits_{x \to 0} \dfrac{\tan x}{\sin x}$

21. $\lim\limits_{h \to 0} \dfrac{\sin 2h}{\sin h}$ (*Hint:* Use identity $\sin 2h = 2 \sin h \cos h$)

22. $\lim\limits_{\theta \to 2} \dfrac{\theta^4 - 2^4}{\theta - 2}$

23. $\lim\limits_{x \to \pi/2} \sin 2x \csc x$

24. $\lim\limits_{x \to \pi/2} \dfrac{\sin 2x}{\cos x}$

25. $\lim\limits_{x \to -1} \dfrac{x^2 - 2x - 3}{x + 1}$

26. $\lim\limits_{x \to 4} \dfrac{x^2 - 2x - 8}{x - 4}$

27. $\lim\limits_{x \to \pi/2} \dfrac{\sec x \cos x}{x}$

28. $\lim\limits_{x \to 1} \dfrac{x^3 + 3x^2 - x - 3}{x^2 - 1}$

29. $\lim\limits_{x \to -1} \dfrac{x^3 + 3x^2 - x - 3}{x^2 - 1}$

30. $\lim\limits_{x \to 1} \dfrac{x^3 - 7x + 6}{x^2 + 2x - 3}$

31. $\lim\limits_{x \to -3} \dfrac{x^3 - 7x + 6}{x^2 + 2x - 3}$

32. $\lim\limits_{x \to 1} \dfrac{x^{5/2} - x^{1/2}}{x^{3/2} - x^{1/2}}$

33. $\lim\limits_{x \to 2} \dfrac{x^{13/4} - 2x^{9/4}}{x^{5/4} - 2x^{1/4}}$

34. $\lim\limits_{x \to -2} \dfrac{x^{7/3} + x^{4/3} - 2x^{1/3}}{x^{4/3} + 2x^{1/3}}$

35. $\lim\limits_{h \to 0} \dfrac{3 - \sqrt{9 + h}}{h}$

36. $\lim\limits_{x \to 4} \dfrac{\sqrt{x} - 2}{x - 4}$

37. $\lim\limits_{x \to 1} \dfrac{(1/x) - 1}{x - 1}$

38. $\lim\limits_{x \to -1} \dfrac{x^{-2} - 1}{x + 1}$

In Exercises 39–46, sketch the graph $y = f(x)$ and determine the limit of $f(x)$ as $x \to 0$, if it exists. If the limit does not exist, explain why.

43. $f(x) = \begin{cases} x^2 - 3x + 3 & \text{if } x < 0; \\ \dfrac{(x - 1)^3 + 1}{x} & \text{if } x > 0. \end{cases}$

44. $f(x) = \begin{cases} 1 & \text{if } x < 0; \\ (\sin x)/x & \text{if } x > 0. \end{cases}$

45. $f(x) = \begin{cases} x^2 + 1 & \text{if } x < 0; \\ (\sin x)/x & \text{if } x > 0. \end{cases}$

46. $f(x) = \begin{cases} |x|/x & \text{if } x < 0; \\ (x - 1)^2 & \text{if } x > 0. \end{cases}$

In Exercises 47–52, complete Table 2.4 using the function specified. Given this numerical evidence, predict the limit of $f(x)$ as $x \to 0$.

Table 2.4

x	$f(x)$
1.000	
0.500	
0.100	
0.050	
0.010	
0.005	
−0.005	
−0.010	
−0.050	
−0.100	
−0.500	
−1.000	

47. $f(x) = \dfrac{x^2}{1 - \cos x}$

48. $f(x) = \dfrac{(\cos 2x) - 1}{x^2}$

49. $f(x) = \dfrac{x - \sin x}{x^3}$ **50.** $f(x) = \dfrac{x^2 - \sin x^2}{x^6}$

51. $f(x) = \dfrac{1 - \cos x^2}{x^4}$ **52.** $f(x) = \dfrac{x^6}{x^2 - \tan x^2}$

53. Let f be the function defined by

$$f(x) = \begin{cases} \dfrac{\sin x}{2x} & \text{if } x < 0; \\ (x + c)^2 & \text{if } x > 0. \end{cases}$$

Find the number(s) c such that the limit of $f(x)$ as $x \to 0$ exists. What is the limit of $f(x)$ as $x \to 0$ in this case?

54. a. Let f be defined by

$$f(x) = \dfrac{(x + 1)^2 - 2}{x} \qquad x \neq 0.$$

Explain why the limit of $f(x)$ as $x \to 0$ does not exist.

b. Let

$$g(x) = \dfrac{(x + 1)^2 - c}{x} \qquad x \neq 0.$$

Find the unique number c such that the limit of $g(x)$ exists as $x \to 0$. What is this limit when it does exist?

55. True or false? If $f(a)$ does not exist then neither does $\lim\limits_{x \to a} f(x)$.

56. Figure 2.9 is a graph of the function $f(x) = x \sin(1/x)$.

a. Does $\lim\limits_{x \to 0} x \sin(1/x)$ exist? Why?

b. What can you say about $\lim\limits_{x \to 0} x^2 \sin(1/x)$?

(Compare these results with that of Example 7.)

 In Exercises 57–60, a function f and a number a are given. Plot the points $(a + h, f(a + h))$ for $h = \pm 1, \pm 0.1, \pm 0.01, \pm 0.001, \pm 0.0001,$ and ± 0.00001. Then predict the limit of $f(x)$ as $x \to a$.

57. $f(x) = \dfrac{2x^2 + 5x - 12}{x + 4}, \quad a = -4$

58. $f(x) = 3x \sin(1/x), \quad a = 0$

59. $f(x) = \dfrac{\sin x}{3x}, \quad a = 0$

60. $f(x) = \begin{cases} 2(x - 1)^2 + 3, & \text{if } x \leq 2; \\ \dfrac{2x + 1}{3x - 5}, & \text{if } x > 2. \end{cases} \quad a = 2$

 In Exercises 61–64, predict the specified limit by repeatedly magnifying the corresponding graph near the appropriate point.

61. $\lim\limits_{x \to 1} \dfrac{2x^2 - x - 1}{x^2 - 1}$ **62.** $\lim\limits_{x \to 1/2} (1 - 2x)\tan \pi x$

63. $\lim\limits_{x \to \sqrt{2}} \dfrac{2x^3 + 3x^2 - 4x - 6}{x - \sqrt{2}}$ **64.** $\lim\limits_{x \to 1/2} \dfrac{\cos \pi x}{2x - 1}$

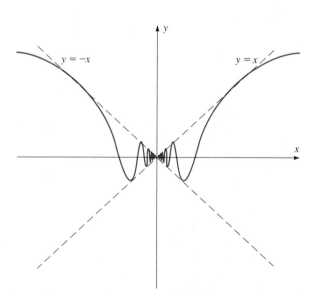

Figure 2.9 $f(x) = x \sin(1/x)$.

2.3   The Formal Definition of Limit

In Section 2.2 we defined the limit of a function informally by saying that

$$L = \lim_{x \to a} f(x)$$

means that the values $f(x)$ are as close to L as we desire for all $x \neq a$ sufficiently close to a.

From the viewpoint of precise mathematics this informal notion of limit is problematic. The difficulty lies in the use of the phrase "close to." A precise mathematical statement can involve constants, variables, equal signs, inequalities, arithmetic operations, and so forth, but not vague references to "closeness."

In order to capture our intuitive notion of limit with precise language, we do the following:

1. In place of the phrase "the values $f(x)$ are as close to L as we desire," we use the inequality

$$|f(x) - L| < \epsilon. \tag{1}$$

2. In place of the phrase "for all $x \neq a$ sufficiently close to a," we use a similar inequality:

$$0 < |x - a| < \delta \tag{2}$$

where δ is a small positive number. (Remember, we do not want to allow $x = a$. That is the reason for the left part of this inequality.)

3. In order to connect these two phrases in the desired way, we say that, no matter what number ϵ is given, we can find a number δ so that if x satisfies inequality (2), then $f(x)$ will satisfy inequality (1). That is, we want to say that

$$|x - a| \text{ small guarantees } |f(x) - L| \text{ small.}$$

These conventions bring us to our formal definition of limit.

Definition 3

Let $f(x)$ be defined for all x in an open interval containing a, except possibly at a. We say that the number L is the **limit** of the function f as x approaches a, written

$$L = \lim_{x \to a} f(x),$$

if and only if, given any number $\epsilon > 0$, there exists a corresponding number $\delta > 0$ so that

$$\text{if} \quad 0 < |x - a| < \delta, \quad \text{then} \quad |f(x) - L| < \epsilon.$$

In other words,

$$L = \lim_{x \to a} f(x)$$

means that the values $f(x)$ are as close to L as we desire (within ϵ units) for all $x \neq a$ sufficiently close (within δ units) to a. (See Figures 3.1 and 3.2.)

Before we discuss examples that illustrate how Definition 3 is used, we repeat an observation made previously (Remark 1 of Section 2.2): Although Definition 3 provides a criterion for proving that L is the limit of $f(x)$ as $x \to a$, it does not provide any information as to how the number L is found. We usually determine L using the techniques that are described in Sections 2.2 and 2.4.

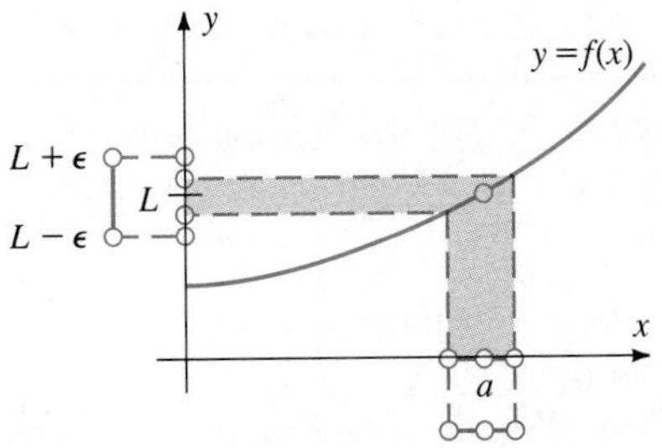

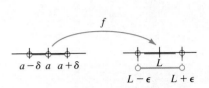

Figure 3.1 If $0 < |x - a| < \delta$, then $|f(x) - L| < \epsilon$.

Figure 3.2 The red intervals on the left consist of those numbers x that satisfy the equation $0 < |x - a| < \delta$. The red interval on the right consists of the corresponding values $f(x)$. Note that these values satisfy $|f(x) - L| < \epsilon$ since they lie within the interval $(L - \epsilon, L + \epsilon)$, which is represented in green.

□ **EXAMPLE 1**

Prove that

$$\lim_{x \to 3}(2x + 1) = 7.$$

This limit was discussed informally in Example 1 of the previous section. Before we establish this limit rigorously using Definition 3, let's examine the meaning of the ϵ and the δ used in the definition. First, note that $f(x) = 2x + 1$, that $a = 3$, and that $L = 7$. Given any positive number ϵ, the inequality

$$|f(x) - L| < \epsilon$$

in Definition 3 is equivalent to the following inequalities:

$$|(2x + 1) - 7| < \epsilon \qquad (3)$$
$$|2x - 6| < \epsilon$$
$$2|x - 3| < \epsilon$$
$$|x - 3| < \epsilon/2. \qquad (4)$$

Since $a = 3$, inequality (4) shows how small the number $|x - a| = |x - 3|$ must be made in order to "force" the equivalent inequality (3) to be true. For example,

(a) if $\epsilon = 0.5$ is given, we choose any x satisfying

$$0 < |x - 3| < \frac{\epsilon}{2} = \frac{0.5}{2} = 0.25.$$

Then inequality (4) holds. Since inequalities (4) and (3) are equivalent, this shows that

$$\text{if} \quad 0 < |x - 3| < 0.25, \quad \text{then} \quad |(2x + 1) - 7| < 0.5.$$

(b) If $\epsilon = 0.1$ is given, we choose x "sufficiently close" to $a = 3$ so that

$$0 < |x - 3| < \frac{\epsilon}{2} = \frac{0.1}{2} = 0.05.$$

By the equivalence of inequalities (3) and (4), we conclude that if $0 < |x - 3| < 0.05$, then $|(2x + 1) - 7| < 0.1$.

The calculations in (a) and (b) are precisely the ones we made in Example 1 of the previous section. In (a), we see that a distance $|x - 3|$ of at most $\delta = 0.25$ from $a = 3$ guarantees an accuracy $|f(x) - 7|$ of $\epsilon = 0.5$ to the limit $L = 7$. In (b), we need a smaller distance of $\delta = 0.05$ to guarantee an accuracy of $\epsilon = 0.1$.

According to Definition 3, we must find an acceptable distance δ for *each* accuracy $\epsilon > 0$ (no matter how small ϵ is) to prove that the limit is 7. The derivation of equation (4) from equation (3) is the key. This computation shows that the inequality

$$|(2x + 1) - 7| < \epsilon \tag{5}$$

is equivalent to the inequality

$$|x - 3| < \epsilon/2. \tag{6}$$

This equivalence shows how to choose δ: with $\delta = \epsilon/2$, we will know that if $0 < |x - 3| < \delta$, then inequality (6) is true. Hence, so is inequality (5). Formally, the proof goes like this:

Let $\epsilon > 0$ be given. Choose $\delta = \epsilon/2$. It follows that if $0 < |x - 3| < \delta$ then

$$\begin{aligned}
|(2x + 1) - 7| &= |2x - 6| \\
&= 2|x - 3| \\
&< 2\delta \quad (\text{since } |x - 3| < \delta) \\
&= 2(\epsilon/2) \quad (\text{since } \delta = \epsilon/2) \\
&= \epsilon.
\end{aligned}$$

That is, if $0 < |x - 3| < \delta$, then $|(2x + 1) - 7| < \epsilon$ as required by Definition 3. (See Figures 3.3 and 3.4.) ∎

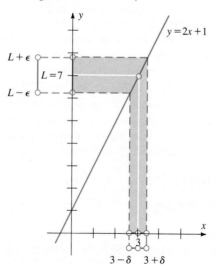

Figure 3.3   We prove

$$\lim_{x \to 3} (2x + 1) = 7.$$

For $\epsilon = 0.8$, $\delta = \epsilon/2 = 0.4$ satisfies Definition 3.

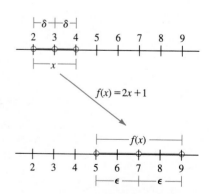

Figure 3.4   For the function $f(x) = 2x + 1$ and the choice of $\epsilon = 2$, then $\delta = \epsilon/2 = 1$.

□ **EXAMPLE 2**

Prove that $\lim\limits_{x \to 2}(4x + 3) = 11$.

Strategy · · · · · · · ·

Identify $f(x)$, L, and a.

Beginning with the inequality

$$|f(x) - L| < \epsilon$$

try to find an equivalent inequality involving $|x - a|$.

The formal proof:

(i) Take $\epsilon > 0$ arbitrary
(ii) Choose $\delta = \epsilon/4$
(iii) Show that

$$0 < |x - a| < \delta$$

guarantees that

$$|f(x) - L| < \epsilon.$$

Solution

Here $f(x) = 4x + 3$, $L = 11$, and $a = 2$. We have the following equivalent inequalities:

$$|f(x) - L| < \epsilon \tag{7}$$
$$|(4x + 3) - 11| < \epsilon$$
$$|4x - 8| < \epsilon$$
$$4|x - 2| < \epsilon$$
$$|x - 2| < \epsilon/4. \tag{8}$$

We will therefore want to take $\delta = \epsilon/4$, since the equivalence of inequalities (7) and (8) shows that if $|x - 2| < \delta = \epsilon/4$, then $|(4x + 3) - 11| < \epsilon$.
Here is the proof:

Given $\epsilon > 0$, choose $\delta = \epsilon/4$. It follows that if

$$0 < |x - 2| < \delta$$

then

$$\begin{aligned}|(4x + 3) - 11| &= |4x - 8| \\ &= 4|x - 2| \\ &< 4\delta \quad \text{(since } |x - 2| < \delta) \\ &= 4(\epsilon/4) \\ &= \epsilon.\end{aligned}$$

Thus, with $\delta = \epsilon/4$, we have that

$$\text{if} \quad 0 < |x - 2| < \delta, \quad \text{then} \quad |(4x + 3) - 11| < \epsilon. \qquad \blacksquare$$

For linear functions, such as those in Examples 1 and 2, the relationship between δ and ϵ in Definition 3 is determined by the slope of the graph of the function. This is because the slope determines the rate at which $f(x)$ changes as x changes.

For nonlinear functions, proofs of particular limits are, in general, more complicated because the relationship between δ and ϵ also depends on the number a at which the limit is being evaluated. The next three examples concern nonlinear functions for which the particular limits in question can be proved in a straightforward manner. Beyond these special types of examples we shall not pursue further the topic of proving individual limits. Rather, we shall determine a set of *properties* of limits, from which individual limits can be deduced.

□ **EXAMPLE 3**

Prove that $\lim\limits_{x \to 2}(x^2 - 4x + 7) = 3$.

Solution In this limit we have $f(x) = x^2 - 4x + 7$, $a = 2$, and $L = 3$. Thus, if ϵ is a given positive number, the following inequalities are equivalent: ·

$$|f(x) - L| < \epsilon$$
$$|(x^2 - 4x + 7) - 3| < \epsilon \tag{9}$$
$$|x^2 - 4x + 4| < \epsilon$$
$$|(x - 2)^2| < \epsilon$$
$$|x - 2| < \sqrt{\epsilon}. \tag{10}$$

The equivalence of inequalities (9) and (10) shows that, if we take $\delta = \sqrt{\epsilon}$, it follows that

$$\text{if} \quad 0 < |x - 2| < \delta, \quad \text{then} \quad |(x^2 - 4x + 7) - 3| < \epsilon.$$

This proves that $\lim_{x \to 2}(x^2 - 4x + 7) = 3$, according to Definition 3. ∎

□ EXAMPLE 4

Prove that $\lim_{x \to 2} x^2 = 4$.

Strategy · · · · · · · ·
Examine $|x^2 - 4|$ to try to find a relationship with $|x - 2|$.

Solution
Here $f(x) = x^2$ and $L = 4$, so we examine

$$|x^2 - 4| = |(x + 2)(x - 2)|$$
$$= |x + 2|\,|x - 2|.$$

Try to bound the factor $|x + 2|$ by some constant.

The factor $|x - 2|$ is the desired quantity $|x - a|$ in Definition 3. But what do we do about the factor $|x + 2|$? Simply this: If x is close to 2, $x + 2$ is close to 4, so we should be able to replace $|x + 2|$ by a constant close to 4.

$|x + 2| < 5$ if $|x - 2| < 1$

More precisely, we observe that if $|x - 2| < 1$, then $|x + 2| < 5$. In this case we will have

$$|x^2 - 4| = |x + 2|\,|x - 2|$$
$$< 5|x - 2|,$$

Take $\delta = \min\left\{\dfrac{\epsilon}{5}, 1\right\}$

Apply Definition 3.

that is, $|x^2 - 4|$ will be less than 5 times $|x - 2|$. We therefore will use $\delta = \epsilon/5$ and recall also that we need to have $|x - 2| < 1$. The formal proof is the following:

Given $\epsilon < 0$, let δ be the smaller of the numbers $\{1, \epsilon/5\}$. Then if $0 < |x - 2| < \delta$, we have both

$$|x - 2| < \epsilon/5 \quad \text{and} \quad |x + 2| < 5.$$

Thus,

$$|x^2 - 4| = |x + 2|\,|x - 2|$$
$$< 5|x - 2|$$
$$< 5(\epsilon/5)$$
$$= \epsilon.$$

This shows that $|x^2 - 4| < \epsilon$ whenever $0 < |x - 2| < \delta$, so

$$\lim_{x \to 2} x^2 = 4.$$

(See Figure 3.5.) ∎

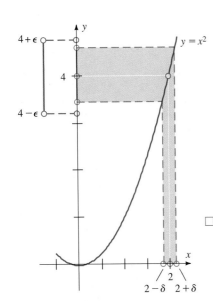

Figure 3.5 We have $|x^2 - 4| < \epsilon$ if $0 < |x - 2| < \delta$, for $\delta = \min\{1, \epsilon/5\}$.

□ EXAMPLE 5

Figure 3.6 illustrates the inequality

$$|\sin x| \le |x|, \qquad -\infty < x < \infty. \tag{11}$$

We may use this inequality to prove that

$$\lim_{x \to 0} \sin x = 0.$$

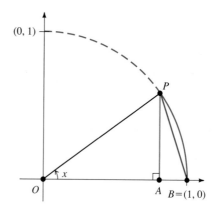

Figure 3.6 If $|x| < \pi/2$, then

$$|\sin x| = \text{length of } AP$$
$$\leq \text{length of } BP$$
$$\leq \text{length of the arc } BP = |x|.$$

In this example we have $f(x) = \sin x$ and $a = L = 0$. Thus, if ϵ is a given positive number, the inequality

$$|f(x) - L| < \epsilon$$

is

$$|\sin x - 0| < \epsilon.$$

From inequality (11) we may conclude that

$$\text{if} \quad |x| < \epsilon, \quad \text{then} \quad |\sin x - 0| = |\sin x| \leq |x| < \epsilon. \tag{12}$$

Since, in this example, $|x - a| = |x - 0| = |x|$, inequality (12) shows that if we take $\delta = \epsilon$, it follows that

$$\text{if} \quad 0 < |x - 0| < \delta, \quad \text{then} \quad |\sin x - 0| < \epsilon.$$

This proves that

$$\lim_{x \to 0} \sin x = 0. \qquad \blacksquare$$

Exercise Set 2.3

Each of Exercises 1–6 presents a limit and asks for certain values of δ related to particular choices of ϵ. These numbers δ are not unique.

1. $\lim_{x \to 2}(2x + 5) = 9$

    a. Show that $|(2x + 5) - 9| < \epsilon$ if and only if $|x - 2| < \epsilon/2$.

    b. Find an appropriate δ for $\epsilon = 2, 0.4, 0.05$.

2. $\lim_{x \to -3}(1 - 4x) = 13$

    a. Show that $|(1 - 4x) - 13| < \epsilon$ if and only if $|x + 3| < \epsilon/4$.

    b. Find an appropriate δ for $\epsilon = 2, 0.4, 0.1$.

3. $\lim_{x \to 0}(x^2 + 3) = 3$

    a. Show that $|(x^2 + 3) - 3| < \epsilon$ if and only if $|x| < \sqrt{\epsilon}$.

    b. Find an appropriate δ for $\epsilon = 2, 1, 0.3$.

4. $\lim_{x \to -2} \dfrac{1}{x - 2} = -\dfrac{1}{4}$

    a. Show that, if $-3 < x < -1$ and $|x + 2| < 12\epsilon$, then

$$\left| \frac{1}{(x - 2)} + \frac{1}{4} \right| < \epsilon.$$

    b. Find an appropriate δ for $\epsilon = 2, 0.4, 0.05$.

5. $\lim\limits_{x\to1}\dfrac{x^2-1}{x-1}=2$

Find an appropriate δ for $\epsilon=2,\ 0.8,\ 0.05$.

6. $\lim\limits_{x\to4}\sqrt{2x+1}=3$

Find an appropriate δ for $\epsilon=2,\ 0.4,\ 0.1$.

In Exercises 7–20 use Definition 3 to prove the stated limit.

7. $\lim\limits_{x\to3}(x+3)=6$

8. $\lim\limits_{x\to1}(2x-3)=-1$

9. $\lim\limits_{x\to4}(7-3x)=-5$

10. $\lim\limits_{x\to2}(3x+5)=11$

11. $\lim\limits_{x\to3}\dfrac{x^2-9}{x-3}=6$

12. $\lim\limits_{x\to2}\dfrac{x^2-4}{x-2}=4$

13. $\lim\limits_{x\to3}x^2=9$

14. $\lim\limits_{x\to2}3x^2=12$

15. $\lim\limits_{x\to1}(x^2-2x+4)=3$

16. $\lim\limits_{x\to3}(x^2-6x+13)=4$

17. $\lim\limits_{x\to4}\sqrt{x}=2$

18. $\lim\limits_{x\to3}|x-3|=0$

19. $\lim\limits_{x\to2}(x^2-2x+2)=2$

20. $\lim\limits_{x\to1}(x^2-4x+1)=-2$

21. Use the technique of Example 5 to prove that $\lim\limits_{x\to0}\sin 2x=0$.

22. Use Definition 3 to prove that $\lim\limits_{x\to0}|x|=0$.

23. Use the trigonometric inequality $|\sin x|\le|x|$ to prove that $\lim\limits_{x\to0}x\sin x=0$. (See Example 5.)

24. Let $f(x)=x\sin(1/x)$ for $x\ne0$ (see Figure 2.9 in Section 2.2).

a. Show that $|f(x)|\le|x|$ for all $x\ne0$.

b. Use this inequality to prove that
$$\lim\limits_{x\to0}x\sin(1/x)=0.$$

25. Let
$$f(x)=\begin{cases}2x+2 & \text{if }x<0;\\3x+2 & \text{if }x>0.\end{cases}$$

Prove that
$$\lim\limits_{x\to0}f(x)=2.$$

26. Let
$$f(x)=\begin{cases}2x-1 & \text{if }x<0;\\x^2-1 & \text{if }x>0.\end{cases}$$

Prove that
$$\lim\limits_{x\to0}f(x)=-1.$$

In Exercises 27–30, given the function f and the number a, predict the limit of $f(x)$ as $x\to a$. Then, by repeatedly magnifying the graph of $y=f(x)$, determine a $\delta>0$ corresponding to $\epsilon=0.5,\ 0.25,\ 0.1$, and 0.05. (See Appendix V for a related program.)

27. $f(x)=2x^2-x-1,\quad a=1/2$

28. $f(x)=-3x^2+4x+1,\quad a=2/3$

29. $f(x)=1/x,\quad a=0.1$

30. $f(x)=\begin{cases}x+1, & \text{if }x<1;\\-x^2+3, & \text{if }x\ge1;\end{cases}\quad a=1.$

31. Let $f(x)=\sin x$ and $\epsilon=0.01$. Is it possible to find a single $\delta>0$ so that $|f(x)-f(a)|<\epsilon$ if $|x-a|<\delta$ for *all* numbers a? Is this possible for the function $f(x)=1/x$, with $a\ne0$? (See Appendix V for a related program.)

2.4 Properties of Limits

The purpose of this section is to establish several properties of limits that will enable us to evaluate more complicated limits from two simple observations:

$$\lim\limits_{x\to a}c=c,\qquad c\text{ constant}\tag{1}$$

and

$$\lim\limits_{x\to a}x=a.\tag{2}$$

While both statements can be proved using the formal definition of limit (see Exercises 46 and 47), you may also regard them as sufficiently obvious to be accepted without proof. Statement (1) says that the limit of the constant function with values $f(x)=c$ for all x is always the number c, regardless of the number a. Statement (2) says that the limit of the linear function $g(x)=x$ as x approaches a is the function value $a=g(a)$.

The following theorem establishes an "algebra" of limits, by which limits of sums, multiples, products, and quotients of functions can be calculated from the limits of the individual terms and factors. Its proof is given at the end of this section.

Theorem 1

Assume that

$$\lim_{x \to a} f(x) = L \qquad \text{and} \qquad \lim_{x \to a} g(x) = M$$

both exist. Let c be any number. Then each of the following limits exists, with the value indicated:

(i) $\lim_{x \to a}[f(x) + g(x)] = \lim_{x \to a} f(x) + \lim_{x \to a} g(x) = L + M$

(ii) $\lim_{x \to a}[cf(x)] = c\left[\lim_{x \to a} f(x)\right] = cL$

(iii) $\lim_{x \to a}[f(x)g(x)] = \left[\lim_{x \to a} f(x)\right]\left[\lim_{x \to a} g(x)\right] = LM$

(iv) $\lim_{x \to a} \dfrac{f(x)}{g(x)} = \dfrac{\lim_{x \to a} f(x)}{\lim_{x \to a} g(x)} = \dfrac{L}{M}$,   provided $M \neq 0$

We can paraphrase Theorem 1 by saying that

(i) the limit of the sum is the sum of the limits;
(ii) the limit of a multiple of a function is the multiple of the limit;
(iii) the limit of a product is the product of the limits;
(iv) the limit of a quotient is the quotient of the limits, when the limit in the denominator is not zero.

☐ **EXAMPLE 1**

Find $\lim_{x \to 2}(3x^2 + 6)$.

Solution

$$\lim_{x \to 2}(3x^2 + 6) = \lim_{x \to 2}(3x^2) + \lim_{x \to 2} 6 \qquad \text{(part i)}$$

$$= 3 \lim_{x \to 2} x^2 + \lim_{x \to 2} 6 \qquad \text{(part ii)}$$

$$= 3\left(\lim_{x \to 2} x\right)\left(\lim_{x \to 2} x\right) + \lim_{x \to 2} 6 \qquad \text{(part iii)}$$

$$= 3 \cdot 2 \cdot 2 + 6 \qquad \text{(equations (1), (2))}$$

$$= 18. \qquad\qquad\qquad \blacksquare$$

☐ **EXAMPLE 2**

Find $\lim_{x \to -1}\left(\dfrac{2x + 3}{1 + x^2}\right)$.

Solution   Since the limit of the denominator is

$$\lim_{x \to -1}(1 + x^2) = \lim_{x \to -1} 1 + \lim_{x \to -1} x^2 \qquad \text{(part i)}$$

$$= \lim_{x \to -1} 1 + \left(\lim_{x \to -1} x\right)\left(\lim_{x \to -1} x\right) \qquad \text{(part iii)}$$

$$= 1 + (-1)(-1) \qquad \text{(equations (1), (2))}$$

$$= 2$$

which is nonzero, we may apply part (iv) to conclude that

$$\lim_{x \to -1}\left(\frac{2x + 3}{1 + x^2}\right) = \frac{\lim_{x \to -1}(2x + 3)}{\lim_{x \to -1}(1 + x^2)} \qquad \text{(part iv)}$$

$$= \frac{2\left(\lim_{x \to -1} x\right) + \lim_{x \to -1} 3}{2} \qquad \text{(parts i, ii)}$$

$$= \frac{2(-1) + 3}{2} = \frac{1}{2}. \qquad \blacksquare$$

Equation (2), part (iii) of Theorem 1, and a straightforward application of mathematical induction give the following extension of Theorem 1 to integral powers of x and "power functions." (See Exercises 38 and 39.)

Theorem 2

Assume that $\lim_{x \to a} f(x) = L$ exists. For any positive integer $n = 1, 2, \ldots$

(i) $\lim_{x \to a} x^n = a^n$

(ii) $\lim_{x \to a}[f(x)]^n = [\lim_{x \to a} f(x)]^n = L^n$.

The following examples show how Theorems 1 and 2 can be used to evaluate limits of polynomials, rational functions, and powers of these kinds of functions. We leave it as an exercise for you to verify which parts of Theorems 1 and 2 are used in each step.

☐ **EXAMPLE 3**

Find $\lim_{x \to 2}(3x^4 + 7x^2 + 4x)$.

Solution

$$\lim_{x \to 2}(3x^4 + 7x^2 + 4x) = 3\left(\lim_{x \to 2} x^4\right) + 7\left(\lim_{x \to 2} x^2\right) + 4\left(\lim_{x \to 2} x\right)$$

$$= 3(2)^4 + 7(2)^2 + 4(2)$$

$$= 84. \qquad \blacksquare$$

□ **EXAMPLE 4**

Find $\lim_{x \to -2}(x^{-3} - 3x^{-2} + 5x^3)$.

Solution

$$\lim_{x \to -2}(x^{-3} - 3x^{-2} + 5x^3) = \lim_{x \to -2}\left(\frac{1}{x^3} + \frac{-3}{x^2} + 5x^3\right)$$

$$= \frac{1}{(-2)^3} + \frac{-3}{(-2)^2} + 5(-2)^3$$

$$= -\frac{327}{8}.$$

□ **EXAMPLE 5**

Find $\lim_{x \to 2}(x^3 - 7x + 1)^3$.

Solution

$$\lim_{x \to 2}(x^3 - 7x + 1)^3 = \left[\lim_{x \to 2}(x^3 - 7x + 1)\right]^3$$

$$= [2^3 - 7(2) + 1]^3 = (-5)^3 = -125.$$

The next theorem enables us to calculate limits of fractional powers of x. Its proof is given in Appendix II.

Theorem 3

Let m and n be positive integers. Then

(i) If m is even, $\lim_{x \to a} x^{n/m} = a^{n/m}$ for $0 < a < \infty$;

(ii) If m is odd, $\lim_{x \to a} x^{n/m} = a^{n/m}$ for $-\infty < a < \infty$.

See Exercise 52 for the case where n/m is negative.

□ **EXAMPLE 6**

Find $\lim_{x \to 4}(3\sqrt{x} + x^{-3/2})$.

Solution Using Theorems 1 and 3 we find that

$$\lim_{x \to 4}(3\sqrt{x} + x^{-3/2}) = \lim_{x \to 4}(3x^{1/2} + x^{-3/2})$$

$$= 3 \lim_{x \to 4} x^{1/2} + \frac{1}{\lim_{x \to 4} x^{3/2}} \qquad \text{(Theorem 1)}$$

$$= 3(4)^{1/2} + \frac{1}{(4)^{3/2}} \qquad \text{(Theorem 3)}$$

$$= 3 \cdot 2 + \frac{1}{8} = \frac{49}{8}.$$

Our final theorem is sometimes called the "Pinching Theorem." This name arises from its geometric interpretation as illustrated in Figure 4.1. Its proof is given in Appendix II.

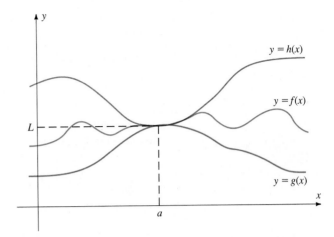

Figure 4.1   The Pinching Theorem.

Theorem 4

Assume that the limit of $g(x)$ and the limit of $h(x)$ both exist as $x \to a$, and that

$$\lim_{x \to a} g(x) = L = \lim_{x \to a} h(x).$$

If the function f satisfies the inequality

$$g(x) \le f(x) \le h(x)$$

for all x in an open interval containing a (except possibly at $x = a$), then

$$\lim_{x \to a} f(x) = L$$

also.

□ **EXAMPLE 7**

In the last section we proved that

$$\lim_{x \to 0} \sin x = 0$$

using the ϵ, δ definition of the limit and the trigonometric inequality

$$|\sin x| \le |x| \tag{3}$$

(see Figure 3.6). If we assume that $|x| \to 0$ as $x \to 0$ (see Exercise 22 in the last section), then we can derive this limit using the Pinching Theorem. From equation (3), we have

$$-|x| \le \sin x \le |x|.$$

Since

$$\lim_{x \to 0} -|x| = 0 = \lim_{x \to 0} |x|,$$

the Pinching Theorem implies that

$$\lim_{x \to 0} \sin x = 0.$$

■

□ **EXAMPLE 8**

We can use the Pinching Theorem to show that

$$\lim_{x \to 0} x \sin \frac{1}{x} = 0. \tag{4}$$

(See Figure 2.9.)

To prove limit (4) we note that since $|\sin u| \le 1$ for all u, we have the inequality

$$0 \le \left| x \sin \frac{1}{x} \right| = |x| \cdot \left| \sin \frac{1}{x} \right| \le |x| \cdot 1 = |x|.$$

Thus,

$$0 \le \left| x \sin \frac{1}{x} \right| \le |x|, \qquad x \ne 0,$$

and therefore,

$$-|x| \le x \sin \frac{1}{x} \le |x|. \tag{5}$$

Since the limit of $|x|$ as $x \to 0$ is 0 (Exercise 22, Section 2.3), limit (4) follows from inequality (5), the Pinching Theorem, and Theorem 1, part (ii).   ■

□ **EXAMPLE 9**

Prove that $\lim_{x \to 0} \cos x = 1$.

Solution   We use the Pinching Theorem and the fact that, if $0 \le a \le 1$, then $0 \le a \le \sqrt{a} \le 1$. On the interval $(-\pi/2, \pi/2)$, we have $\cos x > 0$, so

$$0 < 1 - \sin^2 x \le \sqrt{1 - \sin^2 x} = \cos x \le 1. \tag{6}$$

Since

$$\lim_{x \to 0}(1 - \sin^2 x) = 1 - \left(\lim_{x \to 0} \sin x \right)^2 = 1 - 0 = 1,$$

equation (6) and the Pinching Theorem imply that

$$\lim_{x \to 0} \cos x = 1.$$

■

In our next example we return to the problem of calculating

$$\lim_{x \to 0} \frac{\sin x}{x}.$$

(See Example 2, Section 2.2.) The details of this example require an understanding of Figure 4.2, which concerns an arc of the unit circle subtended by an angle of size x radians, with $0 < x < \pi/2$. The following observations will help you with these details:

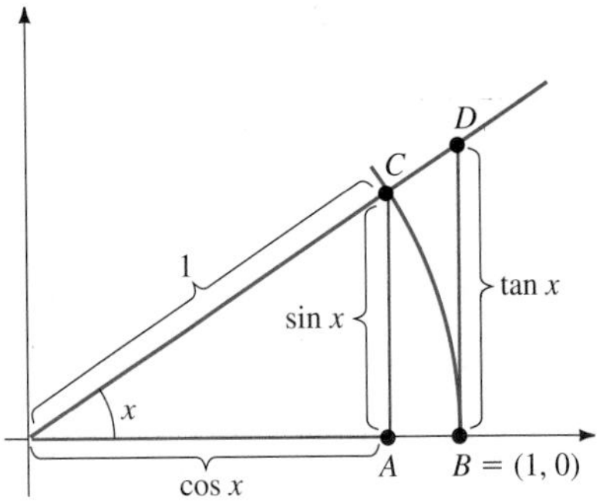

Figure 4.2 An arc of the unit circle subtended by an angle of x radians, with $0 < x < \pi/2$.

(a) The lengths of OC and OB are 1, since each segment is a radius for the unit circle.

(b) $\sin x = \dfrac{\text{length of } AC}{\text{length of } OC} = \text{length of } AC$.

(c) $\cos x = \dfrac{\text{length of } OA}{\text{length of } OC} = \text{length of } OA$.

(d) $\tan x = \dfrac{\text{length of } DB}{\text{length of } OB} = \text{length of } DB$.

□ **EXAMPLE 10**

Show that $\displaystyle\lim_{x \to 0} \frac{\sin x}{x} = 1$.

Solution We assume first that $0 < x < \pi/2$. From Figure 4.2 we observe that

$$\text{Area of triangle } OAC \le \text{Area of sector } OBC \le \text{Area of triangle } OBD. \quad (7)$$

Since

$\text{Area of triangle } OAC = \tfrac{1}{2} \cos x \sin x$

$\text{Area of sector } OBC = \tfrac{1}{2} x \qquad \left(\text{it's } \dfrac{x}{2\pi} \cdot \pi r^2 = \dfrac{x r^2}{2} \text{ with } r = 1 \right)$

$\text{Area of triangle } OBD = \tfrac{1}{2} \tan x$

— area of entire circle

— fractional part of entire circle

inequality (7) becomes

$$\tfrac{1}{2} \cos x \sin x \le \tfrac{1}{2} x \le \tfrac{1}{2} \tan x.$$

Multiplying by 2 and dividing by $\sin x$ (which is nonzero since $0 < x < \pi/2$) gives

$$\cos x \le \frac{x}{\sin x} \le \frac{1}{\cos x}.$$

Inverting all terms then gives

$$\frac{1}{\cos x} \geq \frac{\sin x}{x} \geq \cos x \tag{8}$$

for $0 < x < \pi/2$. Similarly, for $-\pi/2 < x < 0$, inequality (8) also holds (see Exercise 51). Therefore, we can apply the Pinching Theorem with

$$g(x) = \cos x, \quad f(x) = \frac{\sin x}{x}, \quad \text{and} \quad h(x) = \frac{1}{\cos x},$$

since $g(x) \leq f(x) \leq h(x)$. Using Example 9, we have

$$\lim_{x \to 0} \cos x = 1 = \lim_{x \to 0} \frac{1}{\cos x},$$

and thus,

$$\lim_{x \to 0} \frac{\sin x}{x} = 1$$

by the Pinching Theorem. ∎

□ **EXAMPLE 11**

Find $\displaystyle\lim_{x \to 0} \frac{\sin 2x}{x}$.

Strategy · · · · · · · ·

Rewrite

$$\frac{\sin 2x}{x}$$

so that the x in the denominator is paired up with a $\sin x$ term in the numerator.

Use Theorem 1 and facts that

$$\lim_{x \to 0} \frac{\sin x}{x} = 1$$

$$\lim_{x \to 0} \cos x = 1.$$

Solution

Using the identity $\sin 2x = 2 \sin x \cos x$, we obtain

$$\lim_{x \to 0} \frac{\sin 2x}{x} = \lim_{x \to 0} \frac{2 \sin x \cos x}{x}$$

$$= \lim_{x \to 0} \left[2 \left(\frac{\sin x}{x} \right) \cos x \right]$$

$$= 2 \left(\lim_{x \to 0} \frac{\sin x}{x} \right) \left(\lim_{x \to 0} \cos x \right)$$

$$= 2 \cdot 1 \cdot 1 = 2. \qquad ∎$$

REMARK In Section 2.6 (Exercises 40–47), we calculate the more general limit

$$\lim_{x \to 0} \frac{\sin ax}{x}$$

using the notion of a continuous function.

Proof of Theorem 1:

(Part i) To prove that

$$\lim_{x \to a} [f(x) + g(x)] = L + M,$$

using the formal definition of limit, we must show that we can make $|[f(x) + g(x)] - (L + M)|$ small by choosing x with $|x - a|$ sufficiently small. We do this by comparing

$$|[f(x) + g(x)] - (L + M)|$$

with the differences $|f(x) - L|$ and $|g(x) - M|$ as follows:

$$|[f(x) + g(x)] - (L + M)| = |[f(x) - L] + [g(x) - M]|$$
$$\leq |f(x) - L| + |g(x) - M|. \tag{9}$$

Inequality (9) shows that the left side will be smaller than a given number ϵ whenever *each* of the two terms on the right is smaller than $\epsilon/2$. This can be accomplished by choosing $|x - a|$ sufficiently small because

$$\lim_{x \to a} f(x) = L \quad \text{and} \quad \lim_{x \to a} g(x) = M.$$

Here are the details:

Given $\epsilon > 0$, since $\lim_{x \to a} f(x) = L$ there exists a number δ_1 so that

$$\text{if} \quad 0 < |x - a| < \delta_1, \quad \text{then} \quad |f(x) - L| < \epsilon/2. \tag{10}$$

Also, since $\lim_{x \to a} g(x) = M$, there exists a (possibly different) number δ_2 so that

$$\text{if} \quad 0 < |x - a| < \delta_2, \quad \text{then} \quad |g(x) - M| < \epsilon/2. \tag{11}$$

Now let δ be the smaller of the numbers δ_1 and δ_2. Then, if $0 < |x - a| < \delta$, both

$$0 < |x - a| < \delta_1 \quad \text{and} \quad 0 < |x - a| < \delta_2,$$

so by inequality (9) and statements (10) and (11) it follows that

$$|[f(x) + g(x)] - (L + M)| = |[f(x) - L] + [g(x) - M]|$$
$$\leq |f(x) - L| + |g(x) - M|$$
$$< \frac{\epsilon}{2} + \frac{\epsilon}{2} = \epsilon.$$

This proves that $\lim_{x \to a} [f(x) + g(x)] = L + M$. ∎

(Part ii) First note that if $c = 0$, the statement

$$\lim_{x \to a} cf(x) = cL$$

is just $\lim_{x \to a} 0 = 0$, which is obviously true. We may therefore assume $c \neq 0$ in what follows. To prove that the limit of $cf(x)$ as $x \to a$ is cL, we must show that we can make the difference $|cf(x) - cL|$ small by choosing x with $|x - a|$ small. Since

$$|cf(x) - cL| = |c[f(x) - L]| \tag{12}$$
$$= |c| \, |f(x) - L|$$

we can make the left side of this equation smaller than a given number ϵ by requiring that the factor $|f(x) - L|$ on the right side be less than $\epsilon/|c|$. (Remember, $c \neq 0$.)

Here is the proof:

Given $\epsilon > 0$, since $\lim_{x \to a} f(x) = L$ there exists a number $\delta > 0$ so that

$$\text{if} \quad 0 < |x - a| < \delta, \quad \text{then} \quad |f(x) - L| < \epsilon/|c|. \tag{13}$$

Thus, if $0 < |x - a| < \delta$, it follows from equation (12) and line (13) that

$$\begin{aligned}|cf(x) - cL| &= |c[f(x) - L]| \\ &= |c| \, |f(x) - L| \\ &< |c|(\epsilon/|c|) = \epsilon\end{aligned}$$

This shows that $\lim_{x \to a} cf(x) = cL$. ∎

The proofs of parts (iii) and (iv) are similar, but slightly more complicated, and are found in Appendix II.

Exercise Set 2.4

In Exercises 1–20 use Theorems 1–3 to find the limit.

1. $\lim_{x \to 3}(3x - 7)$

2. $\lim_{x \to 2}(x^3 - 4x + 1)$

3. $\lim_{x \to 2} \dfrac{x^2 + 3x}{x - 3}$

4. $\lim_{x \to 5} \dfrac{x^2 - 5x}{x - 5}$

5. $\lim_{x \to 4} \sqrt{x}(1 - x^2)$

6. $\lim_{x \to 2}(4\sqrt{x} - 5x^2)$

7. $\lim_{x \to 1}(3x^9 - \sqrt[3]{x})$

8. $\lim_{x \to 2} \dfrac{x^3 - 6x + 5}{x^2 + 2x + 2}$

9. $\lim_{x \to 3} \dfrac{x^2 - 2x - 3}{x - 3}$

10. $\lim_{x \to 3} \dfrac{x^2 - 10x + 21}{x - 3}$

11. $\lim_{x \to 4} \dfrac{x^{3/2} + 2\sqrt{x}}{x^{5/2} + \sqrt{x}}$

12. $\lim_{x \to 0} \dfrac{\sin x}{\sin 2x}$

13. $\lim_{x \to 0} \dfrac{\tan x}{\sin 2x}$

14. $\lim_{x \to 1} \dfrac{x^3 + 2x^2 + 2x - 5}{x^2 - 1}$

15. $\lim_{x \to -8} \dfrac{x^{2/3} - x}{x^{5/3}}$

16. $\lim_{x \to 4} \dfrac{x^{3/2} - x^{5/2}}{4 + \sqrt{x}}$

17. $\lim_{x \to 4} \dfrac{x - 4}{\sqrt{x} - 2}$

18. $\lim_{x \to -3} \dfrac{x^2 - 6x - 27}{x^2 + 3x}$

19. $\lim_{x \to -1}(x^{7/3} - 2x^{2/3})^2$

20. $\lim_{x \to -1} \dfrac{x^3 + x^2 - x - 1}{x^3 + 2x^2 + 2x + 1}$

In Exercises 21–24, assume that

$$\lim_{x \to a} f(x) = 2, \quad \lim_{x \to a} g(x) = -3, \quad \text{and} \quad \lim_{x \to a} h(x) = 5,$$

and determine the specified limit?

21. $\lim_{x \to a} 3 \cdot f(x) \cdot g(x)$

22. $\lim_{x \to a} \dfrac{f(x) - g(x)}{[h(x)]^2}$

23. $\lim_{x \to a} \dfrac{6f(x) - 4[g(x)]^2}{g(x) - 4f(x)}$

24. $\lim_{x \to a}[f(x) + g(x)]^3$

In Exercises 25–32 use the fact that

$$\lim_{x \to 0} \frac{\sin x}{x} = 1$$

to find the limit.

25. $\lim_{x \to 0} \dfrac{\sin x}{2x}$

26. $\lim_{x \to 0} \dfrac{3 \sin x}{x}$

27. $\lim_{x \to 0} \dfrac{\tan x}{4x}$

28. $\lim_{x \to 0} \dfrac{\sec x \tan x}{x}$

29. $\lim_{x \to 0} x^2 \cot x$

30. $\lim_{x \to 0} \dfrac{\sin^2 x}{x}$

31. $\lim_{x \to 0} \dfrac{\sin x}{\sqrt[5]{x}}$

32. $\lim_{x \to 0} \dfrac{\sin^2 x}{x^{4/3}}$

33. Suppose that $1 - x^2 \le f(x) \le 1 + x^2$ for all x. Find $\lim_{x \to 0} f(x)$.

34. Suppose that $6x - x^2 \le f(x) \le x^2 - 6x + 18$ for all x. Find $\lim_{x \to 3} f(x)$.

35. Suppose that $1 - x^4 \le f(x) \le \sec x$ for $x \in (-\pi/2, \pi/2)$. Find $\lim_{x \to 0} f(x)$.

36. Use Theorem 1, part (i), to show that

$$\lim_{x \to a}[f(x) + g(x) + h(x)] = \lim_{x \to a} f(x) + \lim_{x \to a} g(x) + \lim_{x \to a} h(x)$$

provided each of the limits on the right side exists. (*Hint:* Apply Theorem 1 twice.)

37. Use Theorem 1, part (iii), to show that

$$\lim_{x \to a}[f(x)g(x)h(x)] = \left(\lim_{x \to a} f(x)\right)\left(\lim_{x \to a} g(x)\right)\left(\lim_{x \to a} h(x)\right)$$

provided each of the limits on the right side exists.

38. Use mathematical induction (see Appendix II) and Theorem 1 to prove that $\lim_{x \to a} x^n = a^n$ as follows:

a. Explain why $\lim_{x \to a} x^n = a^n$ if $n = 1$.

b. Assume that

$$\lim_{x \to a} x^m = a^m$$

for a positive integer m and explain why it follows that

$$\lim_{x \to a} x^{m+1} = \lim_{x \to a} x \cdot x^m = \left(\lim_{x \to a} x\right)\left(\lim_{x \to a} x^m\right)$$
$$= a \cdot a^m$$
$$= a^{m+1}.$$

c. Conclude that $\lim_{x \to a} x^n = a^n$ for all positive integers n.

39. Use mathematical induction and Theorem 1 to prove that

$$\text{if} \quad \lim_{x \to a} f(x) = L, \text{ then } \lim_{x \to a}[f(x)]^n = L^n$$

for each positive integer n. (*Hint:* See Exercise 38.)

40. Use Theorem 1, together with mathematical induction, to prove that for the polynomial $Q(x) = a_n x^n + a_{n-1}x^{n-1} + \cdots + a_1 x + a_0$,

$$\lim_{x \to c} Q(x) = a_n c^n + a_{n-1}c^{n-1} + \cdots + a_1 c + a_0.$$

41. Give an example of a function f for which

$$\lim_{x \to a} f(x) = L \quad \text{and} \quad \lim_{x \to b} f(x) = L$$

with $a \neq b$. This shows that a function can have the same limit at more than one number x.

42. Let $f(x) = \dfrac{|x|}{x}$ and $g(x) = -\dfrac{|x|}{x}$, $x \neq 0$.

a. Find the function $h = f + g$.

b. Show that

$$\lim_{x \to 0} h(x) = \lim_{x \to 0}[f(x) + g(x)]$$

exists, but that neither $\lim_{x \to 0^-} f(x)$ nor $\lim_{x \to 0} g(x)$ exists.

43. Find two functions f and g so that $\lim_{x \to 0}[f(x)g(x)]$ exists, but either $\lim_{x \to 0} f(x)$ or $\lim_{x \to 0} g(x)$ fails to exist. (See Exercise 42.)

44. Is it possible to find a function f and a constant c so that $\lim_{x \to 0} cf(x)$ exists but $\lim_{x \to a} f(x)$ does not exist? What if we require $c \neq 0$?

45. Find the number(s) a so that $\lim_{x \to a}(x^2 - 2x - 5) = 10$.

46. Use the formal definition of limit to prove that

$$\lim_{x \to a} c = c$$

for any constant c. (*Hint:* Given $\epsilon > 0$, *any* number $\delta > 0$ will work.)

47. Use the formal definition of limit to prove that

$$\lim_{x \to a} x = a$$

for any number a. (*Hint:* Given $\epsilon > 0$, choose $\delta = \epsilon$.)

48. Show that $\lim_{x \to 0} \dfrac{1 - \cos x}{x} = 0$ as follows:

a. Write

$$\frac{1 - \cos x}{x} = \frac{1 - \cos x}{x}\left(\frac{1 + \cos x}{1 + \cos x}\right)$$
$$= \frac{1 - \cos^2 x}{x(1 + \cos x)}.$$

b. Show that $\dfrac{1 - \cos x}{x} = \dfrac{\sin^2 x}{x(1 + \cos x)}.$

c. Using the fact that, as $x \to 0$, the limit of $\sin x$ is 0, the limit of $\cos x$ is 1, and the limit of $(\sin x)/x$ is 1, obtain the desired limit.

In Exercises 49–50, use the limit obtained in Exercise 48 to determine the specified limit.

49. $\lim_{x \to 0} \dfrac{1 - \cos x}{x^{1/3}}$

50. $\lim_{x \to 0} \dfrac{\cos^2 x - 2 \cos x + 1}{x^{5/3}}$

51. Show that inequality (8) is true for $-\pi/2 < x < 0$ by using the fact that it is true for $0 < x < \pi/2$ and the identities $\sin(-x) = -\sin x$ and $\cos(-x) = \cos x$.

52. Use Theorems 1 and 3 to prove that, for $n/m < 0$,
(i) if m is even, $\lim_{x \to a} x^{n/m} = a^{n/m}$, $0 < a < \infty$

(ii) if m is odd, $\lim_{x \to a} x^{n/m} = a^{n/m}$, $a \neq 0$.

53. Graph the function $f(x) = \sec x$, $g(x) = (\sin x)/x$, and $h(x) = \cos x$ for $-1 < x < 1$, and relate the resulting plot to the solution of Example 10.

2.5   One-Sided Limits

We have defined the statement

$$L = \lim_{x \to a} f(x)$$

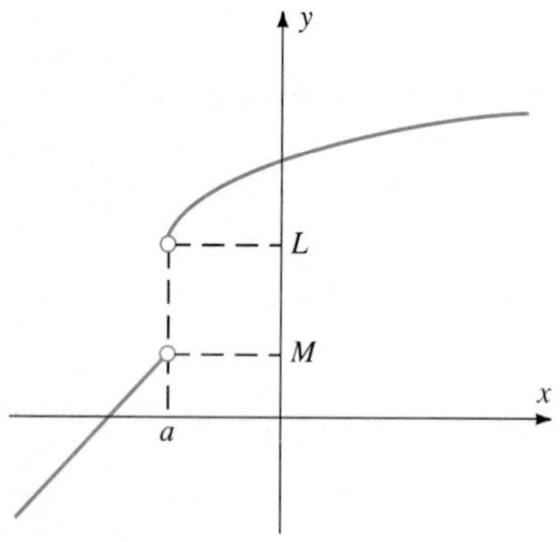

to mean that $|f(x) - L|$ is small whenever $|x - a|$ is small (and nonzero) *regardless of whether x lies to the right of a or to the left*. Because of this "two-sided" property of limits, we say that the limit of $f(x)$ as $x \to a$ does not exist for the function f whose graph is shown in Figure 5.1.

This graph, however, has the property that as x is chosen close to a *on the right*, the corresponding function values $f(x)$ are close to the number L. This is the concept of **right-hand limit,** written

$$\lim_{x \to a^+} f(x) = L.$$

Similarly, we write

$$\lim_{x \to a^-} f(x) = M$$

Figure 5.1 $f(x) \to L$ for $x \to a^+$, $f(x) \to M$ for $x \to a^-$.

to mean that the values $f(x)$ are close to M if x is close to a *on the left*. The superscript $+$ or $-$ on the number a means that we consider only x's lying on the right $(+)$ or left $(-)$ side of a, respectively.

Intuitively, the statement

$$L = \lim_{x \to a^+} f(x)$$

means that the values $f(x)$ are as close to L as we desire for all $x > a$ sufficiently close to a. Similarly, the statement

$$M = \lim_{x \to a^-} f(x)$$

means that the values $f(x)$ are as close to M as we desire for all $x < a$ sufficiently close to a. (See Figure 5.1.)

□ EXAMPLE 1

Because the square root function $f(x) = \sqrt{x}$ is not defined for $x < 0$, the limit of $\sqrt{x}$ as $x \to 0$ does not exist. We may write, however, that

$$\lim_{x \to 0^+} \sqrt{x} = 0.$$

We give a formal proof of this limit at the end of this section.   ■

Since one-sided limits are very much like two-sided limits, it is not surprising that they are computed in the same manner as two-sided limits. In fact, all techniques for computing limits discussed in the preceding section apply to one-sided limits.

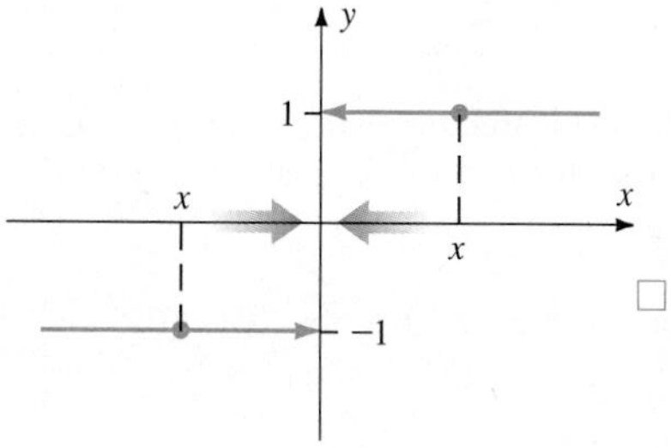

□ EXAMPLE 2

Figure 5.2 shows the graph of the function $f(x) = |x|/x$. In Section 2.2 we concluded that

$$\lim_{x \to 0} \frac{|x|}{x}$$

Figure 5.2 $\displaystyle\lim_{x \to 0^+} \frac{|x|}{x} = 1;$

$\displaystyle\lim_{x \to 0^-} \frac{|x|}{x} = -1.$

does not exist. The one-sided limits of this function at $a = 0$ do exist, however. This is because

(a) for $x > 0$, $|x| = x$, so $\dfrac{|x|}{x} = \dfrac{x}{x} = 1$ and

$$\lim_{x \to 0^+} \frac{|x|}{x} = \lim_{x \to 0^+} 1 = 1, \quad \text{and}$$

(b) for $x < 0$, $|x| = -x$, so $\dfrac{|x|}{x} = \dfrac{-x}{x} = -1$, and

$$\lim_{x \to 0^-} \frac{|x|}{x} = \lim_{x \to 0^-} (-1) = -1. \qquad \blacksquare$$

☐ **EXAMPLE 3**

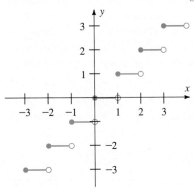

Figure 5.3 shows the graph of the **greatest integer function,** which is defined by

$$[\![x]\!] = \text{the largest integer } n \text{ with } n \le x.$$

Thus, $[\![2.5]\!] = 2$, $[\![-1.2]\!] = -2$, $[\![\pi]\!] = 3$, $[\![7]\!] = 7$, etc.
Although the two-sided limit

$$\lim_{x \to n} [\![x]\!]$$

fails to exist at each integer n, both one-sided limits do exist. For example,

$$\lim_{x \to 2^+} [\![x]\!] = 2; \qquad \lim_{x \to 2^-} [\![x]\!] = 1$$

$$\lim_{x \to -2^-} [\![x]\!] = -3; \qquad \lim_{x \to -2^+} [\![x]\!] = -2. \qquad \blacksquare$$

Figure 5.3 Graph of the greatest integer function $f(x) = [\![x]\!]$.

☐ **EXAMPLE 4**

Figure 5.4 shows the graph of a function f giving the volume, as a function of temperature (in °C), occupied by a fixed mass (weight) of an antifreeze/water solution that freezes at temperature -10°C. If the function f is given on the interval $[-40, 40]$, except at $t = -10$, by

$$f(t) = \begin{cases} 20.2 + .02t, & -10 < t \le 40 \\ 25.1 + .01t, & -40 \le t < -10 \end{cases}$$

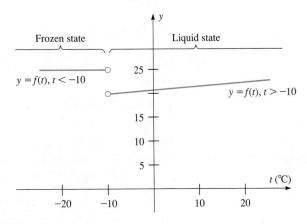

Figure 5.4 Graph of temperature versus volume function in Example 4.

the limit as the temperature approaches the freezing point from the right is

$$\lim_{t \to -10^+} f(x) = \lim_{t \to -10^+} (20.2 + .02t)$$

$$= 20.2 + (.02)(-10) = 20$$

while the limit as t approaches the melting point from the left is

$$\lim_{t \to -10^-} f(x) = \lim_{t \to -10^-} (25.1 + .01t)$$

$$= 25.1 + (.01)(-10) = 25. \qquad \blacksquare$$

Limits vs. One-Sided Limits

In each of Examples 1–4, the two-sided limit of $f(x)$ as $x \to a$ fails to exist although one or both of the one-sided limits exists. No doubt you have noticed that, when the two-sided limit exists, so do both one-sided limits, although the latter convey no information beyond that given by the two-sided limit. The following theorem, whose proof is sketched in Exercise 36, gives the relationship between "two-sided" and "one-sided" limits.

Theorem 5

The two-sided limit of $f(x)$ as $x \to a$ exists if and only if both one-sided limits exist and are equal. That is,

$$\lim_{x \to a} f(x) = L \quad \text{if and only if} \quad \lim_{x \to a^+} f(x) = L = \lim_{x \to a^-} f(x).$$

Theorem 5 provides a way of proving the limit

$$L = \lim_{x \to a} f(x)$$

that is sometimes simpler than applying the formal definition of limit directly: show that

$$\lim_{x \to a^+} f(x) = L = \lim_{x \to a^-} f(x).$$

☐ **EXAMPLE 5**

For the function

$$f(x) = \begin{cases} 4x - 3, & x \geq 1 \\ 2 - x^2, & x < 1 \end{cases}$$

prove that $\lim_{x \to 1} f(x) = 1$. (See Figure 5.5.)

Strategy · · · · · · · ·
Examine the one-sided limits individually.

Solution

To prove that the limit as $x \to 1^+$ of $f(x)$ is 1, we note that if $x > 1$ then $f(x) = 4x - 3$. It follows from the one-sided version of Theorem 1 that

$$\lim_{x \to 1^+} (4x - 3) = 4\left(\lim_{x \to 1^+} x\right) - 3$$

$$= 4(1) - 3 = 1.$$

Figure 5.5 Graph of $f(x) = \begin{cases} 4x - 3, & x \geq 3 \\ 2 - x^2, & x < 1 \end{cases}$.

Use the same strategy to establish the left-hand limit.

Similarly, if $x < 1$, then $f(x) = 2 - x^2$. Using Theorem 1, we obtain

$$\lim_{x \to 1^-} (2 - x^2) = 2 - \left(\lim_{x \to 1^-} x \right)^2$$
$$= 2 - (1)^2 = 1.$$

Use Theorem 5 to conclude that the limit equals the one-sided limits.

Since $\lim_{x \to 1^+} f(x) = 1 = \lim_{x \to 1^-} f(x)$, Theorem 5 gives $\lim_{x \to 1} f(x) = 1$. ∎

Formal Definitions for One-Sided Limits

To formalize our intuitive notion of one-sided limit we modify the "two-sided" condition $0 < |x - a| < \delta$ in the definition of limit (Definition 3) to either the condition

$$a < x < a + \delta \qquad \text{(right-hand limit)}$$

or

$$a - \delta < x < a \qquad \text{(left-hand limit)}.$$

(See Figures 5.6 and 5.7.)

Definition 4	We say that L is the limit of the function f as x approaches *a from the right,* written

$$L = \lim_{x \to a^+} f(x)$$

if and only if: for every $\epsilon > 0$ there exists a corresponding number $\delta > 0$ so that

$$\text{if} \quad a < x < a + \delta, \quad \text{then} \quad |f(x) - L| < \epsilon.$$

Similarly, we say that M is the limit of f as x approaches *a from the left,* written

$$M = \lim_{x \to a^-} f(x)$$

if and only if: for every $\epsilon > 0$ there exists a corresponding $\delta > 0$ so that

$$\text{if} \quad a - \delta < x < a, \quad \text{then} \quad |f(x) - M| < \epsilon.$$

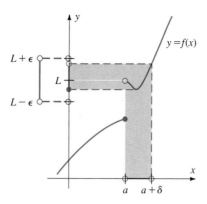

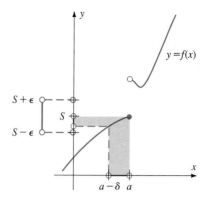

Figure 5.6   The right-hand limit of $f(x)$ as $x \to a^+$ is L. If $a < x < a + \delta$, then $L - \epsilon < f(x) < L + \epsilon$.

Figure 5.7   The left-hand limit of $f(x)$ as $x \to a^-$ is S. If $a - \delta < x < a$, then $S - \epsilon < f(x) < S + \epsilon$.

□ **EXAMPLE 6**

We use Definition 4 to prove that the one-sided limit $\lim_{x \to 0^+} \sqrt{x} = 0$ as follows:

Strategy · · · · · · · ·

Identify f, L, and a.

Examine the difference $|f(x) - L|$ to find a relationship with $x - a = x - 0 = x$.

Use this relationship to determine δ so that $a < x < a + \delta$ guarantees

$$|f(x) - L| < \epsilon.$$

Rewrite findings in the form given by Definition 4 to verify conclusions.

Solution

To prove that $\lim_{x \to 0^+} \sqrt{x} = 0$ we begin by noting that $f(x) = \sqrt{x}$ and $L = 0$, so

$$|f(x) - L| = |\sqrt{x} - 0| = \sqrt{x}.$$

Thus we have $|f(x) - L| < \epsilon$ whenever $\sqrt{x} < \epsilon$. Squaring both sides of this last inequality shows that

$$\sqrt{x} < \epsilon \quad \text{if} \quad 0 < x < \epsilon^2. \tag{1}$$

Since $a = 0$, if we take $\delta = \epsilon^2$, the inequality on the right side of statement (1) is the required condition $a < x < a + \delta$. Thus,

Given $\epsilon > 0$, choose $\delta = \epsilon^2$. Then, if $0 < x < \delta$, we have $\sqrt{x} < \epsilon$, so $|\sqrt{x} - 0| < \epsilon$.

This shows that $\lim_{x \to 0^+} \sqrt{x} = 0$, according to Definition 4. (See Figure 5.8.)   ■

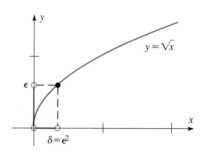

Figure 5.8   In order to prove that

$$\lim_{x \to 0^+} \sqrt{x} = 0,$$

we use $\delta = \epsilon^2$. Then, $\sqrt{x} < \epsilon$ if $0 < x < \delta$.

Exercise Set 2.5

1. For the function f graphed in Figure 5.9, find the following
limits if they exist.
  a. $\lim\limits_{x \to 1^-} f(x)$ 2 **b.** $\lim\limits_{x \to 1} f(x)$ 2

  c. $\lim\limits_{x \to 0} f(x)$ 0

2. For the function f graphed in Figure 5.10, find the following
limits if they exist.
  a. $\lim\limits_{x \to 1^-} f(x)$ -1 **b.** $\lim\limits_{x \to 1^+} f(x)$ 0

  c. $\lim\limits_{x \to 1} f(x)$ DNE

3. For the function f graphed in Figure 5.11, find the following
limits if they exist.
  a. $\lim\limits_{x \to -2} f(x)$ 0 **b.** $\lim\limits_{x \to 2^-} f(x)$ 2

  c. $\lim\limits_{x \to 2^+} f(x)$ -1

4. Let f be the function graphed in Figure 5.9. For what numbers a does:
  a. $\lim\limits_{x \to a^-} f(x) = f(a)$? all R #s   **b.** $\lim\limits_{x \to a^+} f(x) = f(a)$? all R #s

  c. $\lim\limits_{x \to a} f(x)$ exist? 0 **d.** $\lim\limits_{x \to a} f(x)$ exist and equal $f(a)$? 0

5. Let f be the function graphed in Figure 5.10. For what numbers a does:
  a. $\lim\limits_{x \to a^-} f(x) = f(a)$? $a \neq 1$ **b.** $\lim\limits_{x \to a^+} f(x) = f(a)$? all a

  c. $\lim\limits_{x \to a} f(x)$ exist? $a \neq 1$ **d.** $\lim\limits_{x \to a} f(x)$ exist and equal $f(a)$? $a \neq 1$

6. Let f be the function graphed in Figure 5.11. For what numbers a does:
  a. $\lim\limits_{x \to a^-} f(x) = f(a)$?   **b.** $\lim\limits_{x \to a^+} f(x) = f(a)$?

  c. $\lim\limits_{x \to a} f(x)$ exist?   **d.** $\lim\limits_{x \to a} f(x)$ exist and equal $f(a)$?

In Exercises 7–25, find the limit, if it exists.

7. $\lim\limits_{x \to 2^+} \sqrt{x - 2}$

8. $\lim\limits_{x \to 3^-} \dfrac{|x - 3|}{x - 3}$

9. $\lim\limits_{x \to 0^+} (x^{5/2} - 5x^{3/2})$

10. $\lim\limits_{x \to 1^-} \sqrt{1 - x}$

11. $\lim\limits_{x \to 2^-} \dfrac{x^2 - 4x + 4}{x - 2}$

12. $\lim\limits_{x \to 0^+} \dfrac{x - 1}{\sqrt{x} - 1}$

13. $\lim\limits_{x \to 0^+} \dfrac{\sqrt{x} + x^{4/3}}{6 - x^{3/4}}$

14. $\lim\limits_{x \to 3^+} [\![1 + x]\!]$

15. $\lim\limits_{x \to 3^-} [\![1 + x]\!]$

16. $\lim\limits_{x \to 0^+} \dfrac{3\sqrt{x} + 4x^2 - 6}{\sqrt{x} + \sqrt{x + 4}}$

17. $\lim\limits_{x \to 3^-} x[\![x]\!]$

18. $\lim\limits_{x \to 0^-} [\![x^2 - 6x + 5]\!]$

19. $\lim\limits_{x \to 4^+} \dfrac{\sqrt{x} - 4}{x + 2}$

20. $\lim\limits_{x \to 2^-} [\![x^2 - 6x + 5]\!]$

21. $\lim\limits_{x \to 3^+} \dfrac{[\![x - 7]\!]}{[\![x + 4]\!]}$

22. $\lim\limits_{x \to 0^+} \sin\!\left(\dfrac{1}{x}\right)$

23. $\lim\limits_{x \to 0^+} [\![2 - x^2]\!]$

24. $\lim\limits_{x \to 2^-} [\![1 + 4x - x^2]\!]$

25. $\lim\limits_{x \to 0} [\![2 - x^2]\!]$

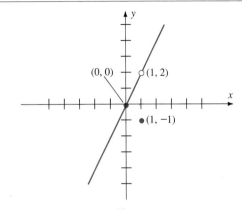

Figure 5.9

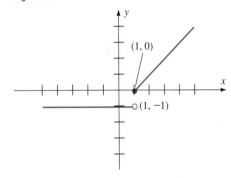

Figure 5.10

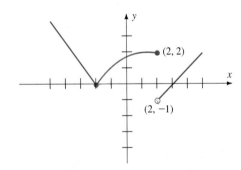

Figure 5.11

26. Let
$$f(x) = \begin{cases} x + 2, & x \le -1 \\ -x, & x > -1. \end{cases}$$

a. Find $\lim\limits_{x \to -1^-} f(x)$.   **b.** Find $\lim\limits_{x \to -1^+} f(x)$.

c. Does $\lim\limits_{x \to -1} f(x)$ exist?

27. Let
$$f(x) = \begin{cases} \cos x, & x \le 0 \\ 1 - x, & x > 0. \end{cases}$$

a. Find $\lim\limits_{x \to 0^-} f(x)$.   **b.** Find $\lim\limits_{x \to 0^+} f(x)$.

c. Does $\lim\limits_{x \to 0} f(x)$ exist?

28. Let
$$f(x) = \begin{cases} (\sin x)/x, & x > 0; \\ [\![x + 2]\!], & x < 0. \end{cases}$$

a. Find $\lim\limits_{x \to 0^-} f(x)$.   **b.** Find $\lim\limits_{x \to 0^+} f(x)$.

c. Does $\lim\limits_{x \to 0} f(x)$ exist?

29. Let
$$f(x) = \begin{cases} [\![x + 3]\!], & x < 0; \\ \dfrac{(x + 1)^3 - 1}{x}, & x > 0. \end{cases}$$

a. Find $\lim\limits_{x \to 0^-} f(x)$.   **b.** Find $\lim\limits_{x \to 0^+} f(x)$.

c. Does $\lim\limits_{x \to 0} f(x)$ exist?

30. Let $f(x) = (\sin x)/|x|$.
a. Find $\lim\limits_{x \to 0^-} f(x)$.   **b.** Find $\lim\limits_{x \to 0^+} f(x)$.

c. Does $\lim\limits_{x \to 0} f(x)$ exist?

31. Let
$$f(x) = \begin{cases} x + 2, & x < 3 \\ 2x - 1, & x > 3. \end{cases}$$

Prove that $\lim\limits_{x \to 3} f(x) = 5$.

32. Let
$$f(x) = \begin{cases} x^2, & x \le 1 \\ x^5, & x > 1. \end{cases}$$

Prove that $\lim\limits_{x \to 1} f(x) = 1$.

33. Let
$$f(x) = \begin{cases} 2, & x \le -1 \\ -x, & -1 < x \le 1 \\ -x^2, & 1 < x. \end{cases}$$

a. Sketch the graph of f.

b. Does $\lim\limits_{x \to -1} f(x)$ exist? Why?

c. Does $\lim\limits_{x \to 1} f(x)$ exist? Why?

d. Prove any limit that exists in parts b and c.

34. Let
$$f(x) = \begin{cases} x + 2, & x \le -1 \\ cx^2, & x > -1. \end{cases}$$

Find c so that $\lim\limits_{x \to -1} f(x)$ exists.

35. Let
$$f(x) = \begin{cases} 3 - x^2, & x \le -2 \\ ax + b, & -2 < x < 2 \\ x^2/2, & x \ge 2. \end{cases}$$

Find a and b so that $\lim\limits_{x \to -2} f(x)$ and $\lim\limits_{x \to 2} f(x)$ exist.

36. Prove the "if" part of Theorem 5 as follows.

a. Assume $\lim\limits_{x \to a^+} f(x) = L = \lim\limits_{x \to a^-} f(x)$.

b. Let ϵ be given. Explain why there exists a number δ_1, so that

$$\text{if} \quad a < x < a + \delta_1 \quad \text{then} \quad |f(x) - L| < \epsilon.$$

c. For this same ϵ, explain why there exists a number δ_2 so that

$$\text{if} \quad a - \delta_2 < x < a \quad \text{then} \quad |f(x) - L| < \epsilon.$$

d. Explain why, for this same ϵ and for $\delta = \min\{\delta_1, \delta_2\}$,

$$\text{if} \quad 0 < |x - a| < \delta, \quad \text{then} \quad |f(x) - L| < \epsilon.$$

37. Graph the functions $f(x) = \sin x$ and $g(x) = (1 - \cos x)/x$ for $0 < x < \pi/2$. What can you conclude about the limit of $g(x)$ as $x \to 0^+$? What happens if you graph f and g for $-\pi/2 < x < 0$? What can you conclude about the limit of $g(x)$ as $x \to 0^-$? What can you conclude about

$$\lim_{x \to 0} \frac{1 - \cos x}{x}?$$

(See Exercise 48 in Section 2.4.)

38. Let
$$f(x) = \begin{cases} \dfrac{\sin(\pi(x + 1))}{x + 1}, & \text{if } x \le -1; \\ \cos(\pi x/2), & \text{if } x > -1. \end{cases}$$

Graph f and predict its limits as $x \to -1^+$ and as $x \to -1^-$.

39. Let
$$f(x) = \begin{cases} x, & \text{if } x \le -2; \\ 2x + 2, & \text{if } -2 < x \le 2; \\ 6/(x + 1), & \text{if } x > 2. \end{cases}$$

Graph f and predict its limits as $x \to -2^-$, as $x \to -2^+$, as $x \to 2^-$, and as $x \to 2^+$.

2.6   Continuity

In defining the limit of $f(x)$ as $x \to a$, we have repeatedly emphasized that this limit need not be equal to $f(a)$. (See Remark 4 of Section 2.2.) Indeed, $f(a)$ may not even be defined. We now turn our attention to the case in which

$$f(a) = \lim_{x \to a} f(x).$$

If this equation holds, we say that the function f is *continuous* at $x = a$.

Definition 5	Suppose that the function f is defined on an open interval containing the number a. Then f is **continuous** at a if $$\lim_{x \to a} f(x) = f(a).$$ Otherwise, f is said to be **discontinuous** at a.

It is important to note that the definition of continuity requires both that *the limit of $f(x)$ as $x \to a$ exists,* and that *the function f is defined at the number a,* in order that the condition

$$\lim_{x \to a} f(x) = f(a)$$

be fulfilled.

We also note that Definition 5 is a definition of continuity at the number a for functions f that are defined on an open interval around a. Continuity at an endpoint of the domain of f is discussed when we discuss continuity on intervals (Definition 6).

Each of the functions whose graphs appear in Figures 6.1–6.3 is discontinuous at a. Geometrically, continuity is a property that assures that the graph of f will not have a hole or otherwise be "broken" at $(a, f(a))$. It rules out each of the possibilities in Figures 6.1–6.3.

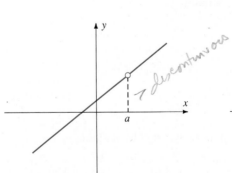

Figure 6.1   The value $f(a)$ is undefined.

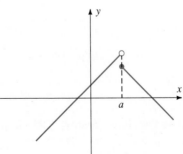

Figure 6.2 $\lim_{x \to a^-} f(x) \neq \lim_{x \to a^+} f(x)$; $\lim_{x \to a} f(x)$ does not exist.

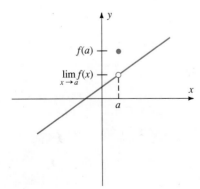

Figure 6.3 $f(a) \neq \lim_{x \to a} f(x)$.

□ EXAMPLE 1

Find the numbers x at which

$$f(x) = \frac{x^2 - 4}{x - 2}$$

is continuous.

Strategy · · · · · · ·
Find the number(s) x for which the denominator equals zero. This is where f is undefined.

Show that

$$\lim_{x \to a} f(x) = f(a)$$

for all other x.

Solution

Since $x - 2 = 0$ when $x = 2$, the value of $f(2)$ is not defined. We say that the function is *discontinuous* at $x = 2$.

For all numbers $a \neq 2$ we have

$$\lim_{x \to a} f(x) = \lim_{x \to a} \frac{x^2 - 4}{x - 2} = \frac{a^2 - 4}{a - 2} = f(a).$$

Thus, f is continuous for all $x \neq 2$. (See Figure 6.4.)   ■

The discontinuity at the number 2 in Example 1 is called a **removable** discontinuity since we can eliminate (i.e., remove) the discontinuity at $x = 2$ by *defining*

$$f(2) = \lim_{x \to 2} f(x) = 4.$$

(See Figure 6.5.) In other words, we add $x = 2$ to the domain of f by defining the new function

$$\hat{f}(x) = \begin{cases} \dfrac{x^2 - 4}{x - 2}, & \text{if } x \neq 2; \\ 4, & \text{if } x = 2. \end{cases}$$

The function $\hat{f}$ is then continuous for all x, and it agrees with f for $x \neq 2$.

Of course, for $x \neq 2$, we have

$$\hat{f}(x) = \frac{x^2 - 4}{x - 2} = \frac{(x + 2)(x - 2)}{x - 2} = x + 2,$$

and therefore, the function $\hat{f}$ is equal to the function g defined by $g(x) = x + 2$. Thus, its graph is a straight line (see Figure 6.5).

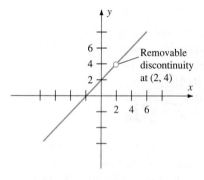

Figure 6.4 The function

$$f(x) = \frac{x^2 - 4}{x - 2}$$

is discontinuous for $x = 2$.

□ EXAMPLE 2

Another function with a removable discontinuity is

$$f(x) = \frac{\sin x}{x}.$$

Although $f(0)$ is undefined, we have shown in Section 2.4 that

$$\lim_{x \to 0} \frac{\sin x}{x} = 1.$$

We may therefore "remove" the discontinuity at $x = 0$ by defining

$$\hat{f}(x) = \begin{cases} \dfrac{\sin x}{x}, & \text{for } x \neq 0; \\ 1, & \text{for } x = 0. \end{cases}$$

The function $\hat{f}$ is then continuous at $x = 0$. (See Figure 2.2 in Section 2.2.)   ■

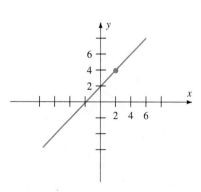

Figure 6.5 The discontinuity in Figure 6.4 is *removed* by defining $f(2) = \lim_{x \to 2} f(x) = 4$.

We leave as exercises the proofs that constant functions and the linear function $f(x) = x$ are continuous at each number x. Because continuity is defined in terms of limits, Theorems 1–3 on limits translate directly into the following theorems on continuity.

Theorem 6

If the functions f and g are continuous at $x = a$ and if c is any real number, then the following functions are also continuous at $x = a$:

(i) the sum $f + g$ of f and g;
(ii) the multiple cf of c times f;
(iii) the product fg of f and g; and
(iv) the quotient f/g, provided that $g(a) \neq 0$.

Proof: To prove that $f + g$ in (i) is continuous at $x = a$, we note first that $(f + g)(a) = f(a) + g(a)$ is defined, since both f and g are continuous at $x = a$. Moreover, $\lim_{x \to a} f(x)$ and $\lim_{x \to a} g(x)$ exist, and

$$\lim_{x \to a} (f + g)(x) = \lim_{x \to a} [f(x) + g(x)] \qquad \text{(Definition of } f + g\text{)}$$

$$= \lim_{x \to a} f(x) + \lim_{x \to a} g(x) \qquad \text{(Theorem 1)}$$

$$= f(a) + g(a) \qquad \text{(Continuity of } f, g \text{ at } a\text{)}$$

$$= (f + g)(a) \qquad \text{(Definition of } f + g\text{)}.$$

Thus $f + g$ satisfies the definition of continuity at $x = a$. The proofs of parts (ii)–(iv), which follow from the corresponding parts of Theorem 1, are left for you as exercises. ∎

☐ **EXAMPLE 3**

Later in this section we shall show that the function $f(x) = \sin x$ is continuous for all x. This fact, together with the continuity of the function $g(x) = x$ and part (iv) of Theorem 6, shows that the function $h(x) = (\sin x)/x$ is continuous for all $x \neq 0$. Thus, the function $\hat{f}$ in Example 2 is continuous for all x. ∎

Theorem 7

For each positive integer $n = 1, 2, \ldots$,

(i) the function $f(x) = x^n$ is continuous at all numbers x.
(ii) if the function g is continuous at $x = a$, the function $f(x) = (g(x))^n$ is continuous at $x = a$.

The proof follows from Theorem 2 and is left as an exercise.

☐ **EXAMPLE 4**

Theorems 6 and 7 combine to show that any polynomial is continuous at all numbers x, and any rational function is continuous wherever its denominator is not zero. In particular

(i) the polynomial $f(x) = x^3 - 2x^2 + 7$ is continuous at all x.

(ii) the rational function

$$g(x) = \frac{x^3 + x + 7}{x - 6}$$

is continuous at all x with $x \neq 6$.

(iii) the fourth power of $g(x)$

$$h(x) = g(x)^4 = \left(\frac{x^3 + x + 7}{x - 6} \right)^4$$

is continuous at all x with $x \neq 6$. ■

Continuity on Intervals

While Definition 5 concerns continuity at a particular number $x = a$, the following definition concerns continuity on an interval.

Definition 6

(i) The function f is **continuous on the open interval** (a, b) if it is continuous at each $x \in (a, b)$.

(ii) The function f is **continuous on the closed interval** $[a, b]$ if it is continuous on (a, b) and, in addition,

$$\lim_{x \to a^+} f(x) = f(a) \quad \text{and} \quad \lim_{x \to b^-} f(x) = f(b).$$

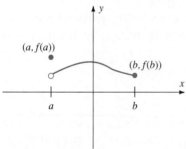

Figure 6.6 The function f is continuous on $(a, b]$ but not on $[a, b]$.

Continuity is defined in a similar way on half-open intervals, such as $(a, b]$, and on infinite intervals, such as $[a, \infty)$, by including the condition that the one-sided limit "from the domain side" equals the function value at any included endpoint. Figure 6.6 illustrates the endpoint condition. *Note that the continuity of f on the closed interval $[a, b]$ does not guarantee the continuity of f at the endpoint $x = a$ or $x = b$ in the sense of Definition 5.*

□ **EXAMPLE 5**

The function $f(x) = \sqrt{x}$ is continuous on the interval $[0, \infty)$. In other words,

$$\lim_{x \to a} \sqrt{x} = \sqrt{a}, \qquad a > 0$$

by Theorem 3 in Section 2.4, and

$$\lim_{x \to 0^+} \sqrt{x} = 0 = f(0)$$

at the included endpoint by Example 1 in Section 2.5. ■

□ **EXAMPLE 6**

The greatest integer function $f(x) = [\![x]\!]$ (defined in Example 3 of Section 2.5) is continuous on each interval of the form $[n, n + 1)$ where n is an integer. That is, f is continuous on the open interval $(n, n + 1)$, and

$$\lim_{x \to n^+} [\![x]\!] = n = f(n).$$

However, this function is discontinuous at each integer $x = n$ because

$$\lim_{x \to n^+} [\![x]\!] \neq \lim_{x \to n^-} [\![x]\!]$$

(see Example 3 in Section 2.5). ∎

The notion of continuity on intervals is required to specify the continuity properties of fractional power functions.

Theorem 8

Let n and m be positive integers. The function $f(x) = x^{n/m}$ is

(i) continuous on $[0, \infty)$ if m is even;
(ii) continuous on $(-\infty, \infty)$ if m is odd.

Proof: To prove part (i) we recall from Theorem 3 that, for m even,

$$\lim_{x \to a} f(x) = \lim_{x \to a} x^{n/m} = a^{n/m} = f(a) \quad \text{if } 0 < a < \infty.$$

Thus f is continuous on $(0, \infty)$. At $a = 0$, the fact that

$$\lim_{x \to 0^+} f(x) = \lim_{x \to 0^+} x^{n/m} = 0 = f(0)$$

follows by the same method of proof used to show

$$\lim_{x \to 0^+} \sqrt{x} = 0. \text{ (See Exercise 71.)}$$

To prove part (ii) we use Theorem 3 again. If m is odd, it guarantees that

$$\lim_{x \to a} f(x) = \lim_{x \to a} x^{n/m} = a^{n/m} = f(a)$$

for all numbers a. ∎

☐ **EXAMPLE 7**

Using Theorems 6–8 we can conclude that the following functions are continuous on the intervals stated:

(a) $f(x) = x^2 - 3\sqrt{x}$ on $[0, \infty)$,

(b) $f(x) = \dfrac{x^2 - x^{2/3}}{1 - x}$ on $(-\infty, 1)$ and $(1, \infty)$,

(c) $f(x) = \dfrac{6x^{2/3} + 5x^{3/2}}{x(x - 2)(x + 3)}$ on $(0, 2)$ and $(2, \infty)$. ∎

Alternative Definition of Continuity

The following theorem gives an alternative characterization of continuity at $x = a$, which is easier to use in certain situations.

Theorem 9

Let f be a function that is defined on an open interval containing the number a. Then f is continuous at a if and only if

$$\lim_{h \to 0} f(a + h) = f(a). \tag{1}$$

Proof: Let $h = x - a$. Then $x \to a$ if and only if $h \to 0$. Also, $a + h = a + (x - a) = x$, so $f(a + h) = f(x)$ and

$$\lim_{h \to 0} f(a + h) = \lim_{x \to a} f(x).$$

Therefore, limit (1) is equivalent to

$$\lim_{x \to a} f(x) = f(a),$$

which is the definition of continuity (Definition 5).   ∎

☐ EXAMPLE 8

Show that the function $f(x) = \sin x$ is continuous at all numbers x.

Strategy · · · · · · · ·

Use Theorem 9 to establish continuity by showing

$$\lim_{h \to 0} \sin(a + h) = \sin a.$$

Rewrite $\sin(a + h)$ using the identity

$$\sin(\alpha + \beta)$$
$$= \sin \alpha \cos \beta + \cos \alpha \sin \beta.$$

Use the facts that

$$\lim_{h \to 0} \cos h = 1$$

$$\lim_{h \to 0} \sin h = 0.$$

Solution

The function $f(x) = \sin x$ is defined for all x. To establish continuity at the number a we use the equivalent definition in part ii of Theorem 9:

$$\lim_{h \to 0} f(a + h) = \lim_{h \to 0} \sin(a + h)$$

$$= \lim_{h \to 0}[\sin a \cos h + \cos a \sin h]$$

$$= (\sin a)\left(\lim_{h \to 0} \cos h\right) + (\cos a)\left(\lim_{h \to 0} \sin h\right)$$

$$= (\sin a)(1) + (\cos a)(0)$$

$$= \sin a$$

$$= f(a). \qquad ■$$

Composite Functions

Recall that the composite function $f \circ g$ is defined by the equation $(f \circ g)(x) = f(g(x))$. For example, the function

$$y = \sin(\pi + x^2)$$

is a composite function: the "inside" function

$$u = g(x) = \pi + x^2$$

is acted on by the "outside" function

$$f(u) = \sin u.$$

The following theorem provides a way of evaluating limits of composite functions if the "outside" function is continuous.

Theorem 10

Let f be a function that is continuous on an open interval containing the number L. If the limit

$$\lim_{x \to a} g(x) = L$$

exists, then

$$\lim_{x \to a} f(g(x)) = f\left(\lim_{x \to a} g(x)\right) = f(L).$$

Theorem 10 says that, if the "outside" function in the composition $f \circ g$ is continuous, we may "pass the limit inside the function f." The proof of Theorem 10 is given at the end of this section.

☐ **EXAMPLE 9**

Theorem 10 may be used as follows:

(a) $\lim_{x \to 0} \sin(\pi + x^2) = \sin[\lim_{x \to 0}(\pi + x^2)] = \sin(\pi + 0) = \sin \pi = 0.$

(b) $\lim_{x \to 0} \sqrt{\dfrac{\sin x}{2x}} = \sqrt{\lim_{x \to 0} \dfrac{\sin x}{2x}} = \sqrt{\dfrac{1}{2} \lim_{x \to 0} \dfrac{\sin x}{x}} = \sqrt{\dfrac{1}{2}} = \dfrac{\sqrt{2}}{2}.$ ■

The next theorem asserts that "the composition of two continuous functions is a continuous function."

Theorem 11	Let g be a function that is continuous on an open interval containing the number a, and let f be a function that is continuous on an open interval containing the number $g(a)$, then the composite function $f \circ g$ is continuous at a.

Proof: By Theorem 10,

$$\lim_{x \to a} (f \circ g)(x) = f\left(\lim_{x \to a} g(x)\right).$$

Since g is continuous at a, we can conclude that

$$\lim_{x \to a} (f \circ g)(x) = f(g(a)).$$ ■

☐ **EXAMPLE 10**

The composite function $y = \sin(\pi + x^2)$ is continuous on $(-\infty, \infty)$ since it has the form $y = f(g(x))$ where the "outside" function $f(u) = \sin u$ is continuous for all u and the "inside" function $g(x) = \pi + x^2$ is continuous for all x. ■

☐ **EXAMPLE 11**

The composite function $y = \sqrt{4 - x^2}$ is continuous on the interval $(-2, 2)$ since

(a) the "inside" function $g(x) = 4 - x^2$ is continuous on $(-2, 2)$ with corresponding values in the interval $(0, 4]$; and

(b) the "outside" function $f(u) = \sqrt{u}$ is continuous on the interval $(0, \infty)$, which contains the interval $(0, 4]$.

By computing the one-sided limits

$$\lim_{x \to -2^+} \sqrt{4 - x^2} = \sqrt{4 - 4} = 0 = y(-2)$$

and

$$\lim_{x \to 2^-} \sqrt{4 - x^2} = \sqrt{4 - 4} = 0 = y(2),$$

we see that $y = \sqrt{4 - x^2}$ is actually continuous on the *closed* interval $[-2, 2]$. ■

The Intermediate Value Theorem

We shall make frequent use of the following property of continuous functions.

Theorem 12

Intermediate Value Theorem

Let f be continuous on the closed interval $[a, b]$ with $f(a) \neq f(b)$. Let d be any number between $f(a)$ and $f(b)$. Then there exists at least one number $c \in (a, b)$ with $f(c) = d$.

The Intermediate Value Theorem states that a continuous function cannot "skip" any numbers in passing from any one of its values to another. Although a completely rigorous proof of this theorem is beyond the scope of this text, we can see that its conclusion is justified by considering the graph of f. If $f(a) \neq f(b)$ and the number d is between $f(a)$ and $f(b)$, then the points $Q_1 = (a, f(a))$ and $Q_2 = (b, f(b))$ lie on opposite sides of the horizontal line $y = d$, as is illustrated in Figure 6.7. Since f is continuous on $[a, b]$, its graph $y = f(x)$, where $a \leq x \leq b$, is an "unbroken" curve from Q_1 to Q_2. Thus, it must intersect the line $y = d$ at some point $P = (c, d)$, where $a < c < b$. Moreover, since P is on the graph of f, we have $f(c) = d$, as claimed.

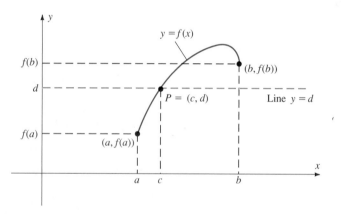

Figure 6.7   The Intermediate Value Theorem in the case where $f(a) < f(b)$.

The Intermediate Value Theorem is one of the several **existence** theorems we shall encounter. It simply guarantees that at least one number c satisfying the stated condition (in this case, $f(c) = d$ with $a < c < b$) exists, but it does not tell us how to find the number. The next example gives one case in which this number can be exactly specified using standard algebraic techniques.

□ **EXAMPLE 12**

The function $f(x) = \sqrt{x^3 + 1}$ is continuous on the interval $[0, 2]$ as a result of Theorems 6, 7, 8, and 11. Since $f(0) = \sqrt{0 + 1} = 1$ and $f(2) = \sqrt{8 + 1} = 3$, the Intermediate Value Theorem guarantees that if d is any "intermediate value" with $1 < d < 3$, there exists a number $c \in (0, 2)$ with $f(c) = d$. In particular, if $d = \sqrt{5}/2$, then $f(c) = \sqrt{5}/2$ gives the equation $\sqrt{c^3 + 1} = \sqrt{5}/2$. We can solve for c as follows:

$$\sqrt{c^3 + 1} = \sqrt{5}/2$$

gives

$$c^3 + 1 = 5/4$$

so

$$c^3 = 1/4$$

and

$$c = 1/\sqrt[3]{4}.$$ ∎

The real power of the Intermediate Value Theorem stems from the fact that it guarantees that certain numbers exist, even if we do not know how to calculate them.

□ EXAMPLE 13

Since the sine function is continuous for all x and since $\sin(0) = 0$ and $\sin(\pi/6) = \sqrt{3}/2$, the Intermediate Value Theorem guarantees that there exists at least one number x between 0 and $\pi/6$ such that

$$\sin x = \sqrt{3}/4.$$ (2)

∎

Without the Intermediate Value Theorem, we have no direct way of knowing that equation (2) has a solution x between 0 and $\pi/6$.

The Intermediate Value Theorem also helps us solve inequalities of the form $f(x) > 0$ or $f(x) < 0$ because it implies that the values of a continuous function do not change sign without passing through 0. More precisely, it implies that the sign of $f(x)$ is either always positive or always negative on any open interval that is bounded by successive zeros of $f(x)$. To see why, suppose that $f(a) = f(b) = 0$, and that $f(x) \neq 0$ for all x between a and b. In addition, suppose that we establish that $f(x_1) > 0$ for one number x_1 between a and b. Then, by the Intermediate Value Theorem, we can conclude that $f(x) > 0$ for all x between a and b. We obtain this conclusion by noting that, if there did exist a number x_2 between a and b such that $f(x_2) < 0$, there would have to exist a third number x_3 between x_1 and x_2 such that $f(x_3) = 0$. But this would contradict our assumption that $f(x) \neq 0$ for x between a and b.

Therefore, we can solve inequalities of the form $f(x) > 0$ or $f(x) < 0$ by first determining where $f(x) = 0$ and then checking the signs of $f(x)$ for selected numbers between these zeros. In other words, we

1. find the zeros of f, and
2. check the sign of $f(x)$ for one number x between each zero.

Step 2 determines the sign of $f(x)$ for all x between the two successive zeros.

□ EXAMPLE 14

Use the Intermediate Value Theorem to solve the inequality $x^3 - 4x^2 > 5x$.

Solution First, we convert the inequality to one in the form of $f(x) > 0$. We obtain $x^3 - 4x^2 - 5x > 0$, and thus we let $f(x) = x^3 - 4x^2 - 5x$. Since f is a polynomial function, it is continuous on $(-\infty, \infty)$, and we can apply the above technique throughout the entire real line. The zeros of f are found by factoring:

$$f(x) = x^3 - 4x^2 - 5x = x(x^2 - 4x - 5) = x(x + 1)(x - 5)$$

so $f(x) = 0$ for $x = -1, 0$, and 5. The following table shows the result of checking the sign of f on each interval.

Interval	Test number x	$f(x)$	Conclusion
$(-\infty, -1)$	$x = -2$	$f(-2) = -14 < 0$	$f(x) < 0$ on $(-\infty, -1)$
$(-1, 0)$	$x = -\frac{1}{2}$	$f(-\frac{1}{2}) = \frac{11}{8} > 0$	$f(x) > 0$ on $(-1, 0)$
$(0, 5)$	$x = 1$	$f(1) = -8 < 0$	$f(x) < 0$ on $(0, 5)$
$(5, \infty)$	$x = 6$	$f(6) = 42 > 0$	$f(x) > 0$ on $(5, \infty)$

The solution of the inequality is therefore $(-1, 0) \cup (5, \infty)$. (See Figures 6.8 and 6.9.) ∎

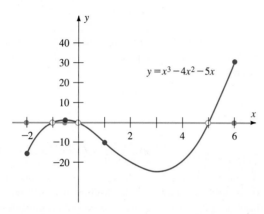

Figure 6.8   The zeros of $f(x) = x^3 - 4x^2 - 5x$ are marked by hollow points on the x-axis. We choose one test number between each successive pair of zeros to determine if the graph of f is above or below the x-axis along the corresponding interval.

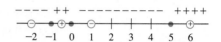

Figure 6.9   Another geometric interpretation of Example 14. In this case, we use a number line and mark the numbers x where $f(x) = 0$. Then we choose test numbers and determine the sign of f at these test numbers. This sign determines the sign of f throughout the interval that contains the test number.

It is important to note that f must be continuous throughout the interval in question in order to apply this technique for solving inequalities.

The ϵ-δ Definition of Continuity

There is a version of Definition 5 whose statement is similar to the formal definition of the limit. If we modify the formal definition of the limit

$$\lim_{x \to a} f(x) = L$$

with $L = f(a)$ to include the possibility that $x = a$, then $f(a)$ must exist and

$$\lim_{x \to a} f(x) = f(a).$$

Theorem 13

Suppose that the function f is defined on an open interval containing the number a. Then f is continuous at a if and only if, given any number $\epsilon > 0$, there exists a corresponding number $\delta > 0$ so that

$$|f(x) - f(a)| < \epsilon$$

if $|x - a| < \delta$.

Using the alternative characterization of continuity given in Theorem 13, we prove Theorem 10.

Proof of Theorem 10:   To prove

$$\lim_{x \to a} f(g(x)) = f(L)$$

we must show that, given a number $\epsilon > 0$, there exists a corresponding number $\delta > 0$ so that

$$\text{if} \quad 0 < |x - a| < \delta, \quad \text{then} \quad |f(g(x)) - f(L)| < \epsilon. \tag{3}$$

Thus we begin by assuming such a number $\epsilon > 0$ is given.

Since f is continuous at L, there exists, by Theorem 13, a number $\delta_1 > 0$ so that

$$\text{if} \quad |u - L| < \delta_1, \quad \text{then} \quad |f(u) - f(L)| < \epsilon. \tag{4}$$

Also, since

$$\lim_{x \to a} g(x) = L$$

and δ_1 is a positive number, we can view δ_1 as another "given number ϵ" and conclude that there exists a number δ_2 so that

$$\text{if} \quad 0 < |x - a| < \delta_2, \quad \text{then} \quad |g(x) - L| < \delta_1. \tag{5}$$

Now let $u = g(x)$. Then, combining statements (4) and (5), we conclude that

$$\text{if} \quad 0 < |x - a| < \delta_2, \quad \text{then} \quad |g(x) - L| < \delta_1,$$
$$\text{so} \quad |f(g(x)) - f(L)| < \epsilon.$$

This is the required statement (3), with $\delta = \delta_2$. ∎

Exercise Set 2.6

1. Let f be the function graphed in Figure 6.10.

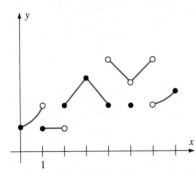

Figure 6.10

a. At what numbers x is f discontinuous? $1, 2, 4, 5, 6$
b. For each pair of successive numbers a and b in part a, is f continuous on the open interval (a, b)? on the half-open interval $[a, b)$? on the half-open interval $(a, b]$? on the closed interval $[a, b]$?

In Exercises 2–24, state the intervals on which the function is continuous.

2. $f(x) = \dfrac{1}{x - 1}$

3. $f(x) = \sec x$

4. $f(x) = \dfrac{1}{4 - x^2}$

5. $f(x) = x \cot x$

6. $f(x) = \dfrac{x}{\cos x}$

7. $f(x) = \dfrac{x + 2}{x^2 - x - 2}$

8. $y = \dfrac{x^2 + x + 1}{x^3 + 2x^2 - 3x}$

9. $y = x^{2/3} - x^{-2/3}$

10. $f(x) = \dfrac{1}{1 + x^{2/3}}$

11. $f(x) = \begin{cases} 1 - x, & x \le 2 \\ x - 1, & x > 2 \end{cases}$

12. $y = \begin{cases} x^2, & x < 0 \\ 3x, & x \ge 0 \end{cases}$

13. $f(x) = \begin{cases} -x, & x \leq -1 \\ 4 - x^2, & -1 < x \leq 2 \\ \frac{1}{2}x - 1, & x > 2 \end{cases}$

14. $y = \begin{cases} ax^2, & x \leq 0 \\ bx^3, & 0 < x \leq 1 \\ bx^4, & x > 1 \end{cases}$

15. $f(x) = \csc x$

16. $f(x) = \sqrt{x + 7}$

17. $f(x) = 1/\sqrt{x + 7}$

18. $f(x) = \sqrt{x^3 - 4x}$

19. $y = x^{2/3} - x^{5/3}$

20. $f(x) = x^{5/2} + x^{-1/3}$

21. $f(x) = \sec x \tan x$

22. $y = \sqrt{\tan x}$

23. $f(x) = \dfrac{\tan x}{x^2 - x - 2}$

24. $f(x) = |9 - x^2|$

In Exercises 25–28 the given function has a removable discontinuity at $x = a$. Determine how to define $f(a)$ so that the function is continuous at a.

25. $f(x) = \dfrac{x^2 - 1}{x - 1}$, $a = 1$

26. $f(x) = \begin{cases} x^2 + 1, & x < 1 \\ \sqrt{3 + x}, & x > 1 \end{cases}$ $a = 1$

27. $f(x) = \dfrac{\cos^2 x - 1}{\sin x}$, $a = 0$

28. $f(x) = \dfrac{x^2 + x - 2}{x^3 - x^2 - 6x}$, $a = -2$

In Exercises 29–31, find the constant k that makes the function continuous at $x = a$.

29. $y = \begin{cases} x^k, & x \leq 2 \\ 10 - x, & x > 2 \end{cases}$ $a = 2$

30. $y = \begin{cases} k, & x \geq 1 \\ 1/\sqrt{kx^2 + k}, & x < 1 \end{cases}$ $a = 1$

31. $h = \begin{cases} (x - k)(x + k), & x \leq 2 \\ kx + 5, & x > 2 \end{cases}$ $a = 2$

In Exercises 32–39, use Theorem 10 to evaluate the limit.

32. $\lim\limits_{x \to 0} \sin\left(\pi - \dfrac{x}{2}\right)$

33. $\lim\limits_{x \to 8}(1 + \sqrt[3]{x})^5$

34. $\lim\limits_{x \to 0} \sqrt{\dfrac{1 - x}{1 + x}}$

35. $\lim\limits_{x \to \pi/2} \cos(\pi + x)$

36. $\lim\limits_{x \to 0}\left(\dfrac{3x + \sin x}{x}\right)^3$

37. $\lim\limits_{x \to 1}(3x^9 - \sqrt[3]{x})^6$

38. $\lim\limits_{x \to 2} \dfrac{\sqrt{4 - x}}{\sqrt{x + 2}}$

39. $\lim\limits_{x \to 0} \cos \pi(x + |x|)$

40. Prove the following equivalent statement of Theorem 10:

"If f is continuous on an open interval containing L, and if $\lim\limits_{x \to a} g(x) = L$, then $\lim\limits_{x \to a} f(g(x)) = \lim\limits_{u \to L} f(u)$."

Use the following steps:

a. The continuity of f at L gives $\lim\limits_{u \to L} f(u) = f(L)$.

b. Theorem 10 gives $\lim\limits_{x \to a} f(g(x)) = f(L)$.

c. Parts a and b give the desired result.

41. Use the result of Exercise 40 to show that

$$\lim_{x \to 0} \dfrac{\sin ax}{x} = a$$

as follows:

a. Define

$$f(u) = \begin{cases} \dfrac{\sin u}{u}, & u \neq 0 \\ 1, & u = 0 \end{cases}$$

and recall that f is continuous for all u (Example 3).

b. Use the fact that

$$\lim_{u \to 0} \dfrac{\sin u}{u} = 1$$

and the result of Exercise 40 to show that

$$\lim_{x \to 0} \dfrac{\sin ax}{x} = \lim_{x \to 0}\left(a\,\dfrac{\sin ax}{ax}\right)$$

$$= a\left(\lim_{x \to 0} \dfrac{\sin ax}{ax}\right)$$

$$= a\left(\lim_{u \to 0} \dfrac{\sin u}{u}\right), \quad u = ax$$

$$= a.$$

In Exercises 42–47, use the fact that

$$\lim_{x \to 0} \dfrac{\sin ax}{x} = a$$

(Exercise 41) to evaluate the limit.

42. $\lim\limits_{x \to 0} \dfrac{\sin 6x}{x}$

43. $\lim\limits_{x \to 0} \dfrac{2x}{\sin x}$

44. $\lim\limits_{x \to 0} \dfrac{\tan 2x}{x}$

45. $\lim\limits_{x \to 0} \dfrac{\sin x}{\sin 3x}$

46. $\lim\limits_{x \to 0} \dfrac{\sin ax}{\sin bx}$

47. $\lim\limits_{x \to 0} x \csc 3x$

In Exercises 48–53, use the Intermediate Value Theorem to find the intervals on which the values of f are strictly positive or strictly negative.

48. $f(x) = x^2 - 4x - 5$

49. $f(x) = 9 - x^2$

50. $f(x) = x^3 - 3x - 2$

51. $f(x) = x^3 + 2x^2 - x - 2$

52. $f(x) = \sin \pi x$

53. $f(x) = \cos(x + \pi)$

In Exercises 54–61, solve the inequality $f(x) > 0$ or $f(x) < 0$ using the Intermediate Value Theorem.

54. $(x - 3)(x + 1) < 0$

55. $x(x + 6) > -8$

56. $x^2 + x < 0$

57. $x^3 + x^2 - 2x > 0$

58. $x^2 + x + 7 > 19$

59. $x^4 - 9x^2 > 0$

60. $x \sec x > 0$, $-2\pi < x < 2\pi$

61. $\sin x \cos x > 0$, $-2\pi < x < 2\pi$

62. Let $f(x) = 1/(x - 2)$. Note that $f(0) = -1/2$ and $f(3) = 1$. Is there a number x between 0 and 3 such that $f(x) = 0$? Does this contradict the Intermediate Value Theorem?

63. Verify the fact that the function $f(x) = x^2$ satisfies the hypothesis of the Intermediate Value Theorem on the interval $[1, 2]$. Use the Intermediate Value Theorem to explain why $\sqrt{2}$ is between 1 and 3/2 and why $\sqrt{3}$ is between 3/2 and 2.

64. Consider the "cash register" function r where $r(x)$ is the closest number to x (rounding up) whose decimal expansion has at most two significant digits to the right of the decimal point. For instance, $r(0.005) = 0.01$, but $r(0.004999) = 0.00$.
   a. At what numbers in the interval $[0, 1]$ is r discontinuous?
   b. Using the greatest integer function $[\![x]\!]$, write down a formula for r.

65. Show that the equation $x + \sin x = 4$ has at least one solution in the interval $[\pi, 2\pi]$.

66. Give an example to show that the composite function $f \circ g$ might be continuous at $x = a$ even though $\lim_{x \to a} g(x) \neq g(a)$.

   Does this contradict Theorem 11?

67. Prove that the constant function with values $f(x) = c$ for all x is continuous at all x.

68. Prove that the linear function $f(x) = x$ is continuous at all x.

69. Prove parts (ii)–(iv) of Theorem 6.

70. Prove Theorem 7.

71. Prove that $\lim_{x \to 0^+} x^{n/m} = 0$ if n and m are positive integers.

72. In this exercise, we explore the continuity of the *roots* of polynomials as functions of the coefficients of the polynomials. Let $f(x) = ax^2 - 4x + 3$.

   a. Graph f with $a = 1 \pm h$ and $h = 0.1, 0.01, 0.001$, and 0.0001. What happens to the roots as h approaches 0?
   b. For what numbers a near 1 will one root be in the interval $(0.95, 1.05)$ and the other root be in the interval $(2.95, 3.05)$?
   c. For $a = 1$, the root is 3/4. Graph f for $h = 0.5, 0.1, 0.01, 0.001$, and 0.0001. What happens to the two roots as h approaches 0?
   d. Repeat parts a–c for the polynomial $g(x) = x^2 + bx + 3$ for $b = -4 \pm h$. What happens to the roots as h approaches 0?
   e. Even though the roots of polynomials are continuous functions of the coefficients, small changes in the coefficients can sometimes cause large changes in the roots. Consider the polynomial

$$p(x) = x^5 - 15x^4 + 85x^3 - 225x^2 + 274x - 120.$$

   It has roots at the numbers 1, 2, 3, 4, and 5. Now replace the coefficient of x^4 with $-15 \pm h$ where $|h| \geq 0.01$, and graph the new polynomial. How many real roots does it have?

The **Bisection Method,** a numerical algorithm for solving equations, is based on the Intermediate Value Theorem. Suppose that f is a continuous function on the interval $[a, b]$ and that $f(a)$ and $f(b)$ have opposite signs (that is, $f(a)f(b) < 0$). Then the Intermediate Value Theorem implies that $f(x) = 0$ for some x between a and b. We can find a root of f (a solution to $f(x) = 0$) by repeatedly halving the interval under consideration. More specifically, we let c_1 be the midpoint of the interval $[a, b]$; that is, $c_1 = (a + b)/2$. We compute $f(c_1)$, and we restrict our attention to the interval $[a, c_1]$ if $f(a)f(c_1) < 0$ or to the interval $[c_1, b]$ if $f(c_1)f(b) < 0$. For example, consider the function f graphed in Figure 6.11. After the first bisection, we restrict our attention to $[a, c_1]$ in this case since $f(a) < 0$ and $f(c_1) > 0$.

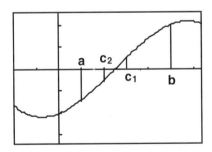

Figure 6.11

Then we apply this halving procedure on the new interval, which we denote $[a_1, b_1]$. We compute $f(c_2)$ where c_2 is the midpoint of the second interval, and using c_2, we obtain a new interval $[a_2, b_2]$, for which $f(a_2)$ and $f(b_2)$ have opposite signs. We repeat this halving procedure until either $|f(c)|$ or the interval under consideration is sufficiently small.

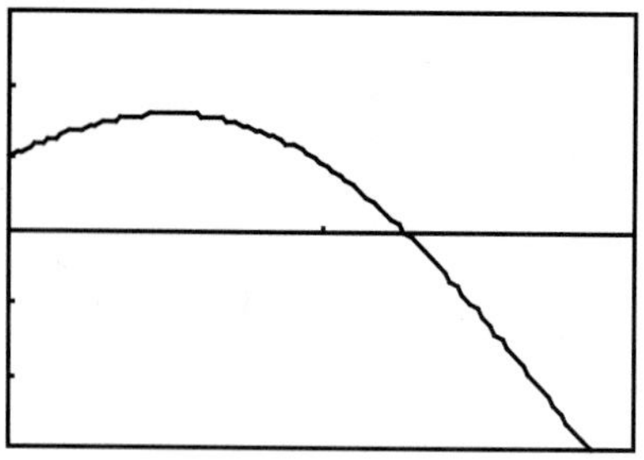

Figure 6.12

n	a_n	b_n	c_n	$f(c_n)$	$(b_n - a_n)/2$
1	1	2	1.5	-1.10888	0.5
2	1	1.5	1.25	0.03597	0.25
3	1.25	1.5	1.375	-0.50896	0.125
4	1.25	1.375	1.3125	-0.22874	0.0625
5	1.25	1.3125	1.28125	-0.09434	0.03125
6	1.25	1.28125	1.26563	-0.02866	0.01563
7	1.25	1.26563	1.25781	0.00379	0.00781
8	1.25781	1.26563	1.26172	-0.01240	0.00391
9	1.25781	1.26172	1.25977	-0.00429	0.00195
10	1.25781	1.25977	1.25879	-0.00025	0.00098
11	1.25781	1.25879	1.25830	0.00177	0.00048
12	1.25830	1.25879	1.25854	0.00076	0.00024
13	1.25854	1.25879	1.25867	0.00025	0.00012
14	1.25867	1.25879	1.25873	0.00000	0.00006

To illustrate the Bisection Method, consider the function $f(x) = 1 - x^2 + \sin 2x$. If we graph f on $[0, 2]$, we obtain Figure 6.12. Consequently, we observe that f has a root between 1 and 2. To approximate this root with an accuracy of 0.0001 using the Bisection Method, we perform the calculations summarized in the following table. The desired root is approximately 1.25873.

In Exercises 73–76, use the Bisection Method to find the roots of the given function f within the specified interval to an accuracy of 0.0001.

73. $f(x) = 2x^3 - 4x^2 - 3$, $[2, 3]$

74. $f(x) = x - \cos x$, $[0, 1]$

75. $f(x) = \sqrt[3]{2x - 7} + \frac{3}{7}x^2 - \frac{3}{2}$, $[2, 3]$

76. $f(x) = (\tan x) - x/2$, (Find the smallest positive root.)

Summary Outline of Chapter 2

■ The line **tangent** to the curve C at point P is the limiting position of the secant through points P and Q as Q approaches P along C. (page 56)

■ The **slope** of the line tangent to the graph of the function f at the point $(x_0, f(x_0))$, if it exists, is the limit (page 59)

$$m = \lim_{h \to 0} \frac{f(x_0 + h) - f(x_0)}{h}.$$

■ Intuitively, $L = \lim_{x \to a} f(x)$ means that the values $f(x)$ are as close to L as we desire for all $x \neq a$ sufficiently close to a. (page 62)

■ Formally, we say that L is the **limit** of the function f as x approaches a, written (page 71)

$$L = \lim_{x \to a} f(x),$$

if and only if given any number $\epsilon > 0$ there exists a corresponding number $\delta > 0$ so that

$$\text{if} \quad 0 < |x - a| < \delta, \quad \text{then} \quad |f(x) - L| < \epsilon.$$

■ **Theorem:** If $\lim_{x \to a} f(x) = L$ and $\lim_{x \to a} g(x) = M$ and c is any constant, (page 78)

(i) $\lim_{x \to a}[f(x) + g(x)] = L + M$

(ii) $\lim_{x \to a}[cf(x)] = cL$

(iii) $\lim_{x \to a}[f(x)g(x)] = LM$

(iv) $\lim_{x \to a} \left(\dfrac{f(x)}{g(x)} \right) = \dfrac{L}{M}$, provided $M \neq 0$

■ **Theorem:**   If $\lim_{x \to a} f(x) = L$ and n is any positive integer,   (page 79)

(i) $\lim_{x \to a} x^n = a^n$

(ii) $\lim_{x \to a} [f(x)]^n = [\lim_{x \to a} f(x)]^n = L^n$.

■ **Theorem:**   Let m and n be positive integers.   (page 80)

(i) If m is even, $\lim_{x \to a} x^{n/m} = a^{n/m}$ for $0 < a < \infty$.

(ii) If m is odd, $\lim_{x \to a} x^{n/m} = a^{n/m}$ for $-\infty < a < \infty$.

■ **Theorem:**   If   (page 81)

$$\lim_{x \to a} g(x) = \lim_{x \to a} h(x) = L$$

exist and $g(x) \leq f(x) \leq h(x)$ for all x in an open interval containing a (except possibly at $x = a$), then $\lim_{x \to a} f(x) = L$ also.

■ **Theorem:**   The limit $L = \lim_{x \to a} f(x)$ exists if and only if $\lim_{x \to a^-} f(x)$ and $\lim_{x \to a^+} f(x)$ (page 90) exist and $\lim_{x \to a^-} f(x) = L = \lim_{x \to a^+} f(x)$.

■ $L = \lim_{x \to a^+} f(x)$ if and only if, given $\epsilon > 0$, there exists a corresponding number   (page 91) $\delta > 0$ so that

$$\text{if}\quad a < x < a + \delta, \quad \text{then}\quad |f(x) - L| < \epsilon.$$

■ $M = \lim_{x \to a^-} f(x)$ if and only if, given $\epsilon > 0$, there exists a number $\delta > 0$ so that   (page 91)

$$\text{if}\quad a - \delta < x < a, \quad \text{then}\quad |f(x) - M| < \epsilon.$$

■ Suppose f is defined on an open interval containing the number a. The func-   (page 95) tion f is **continuous** at a if $\lim_{x \to a} f(x) = f(a)$.

■ **Theorem:**   If f and g are continuous at $x = a$ and if c is any real number, then   (page 97) the following functions are also continuous at $x = a$:

(i) $f + g$
(ii) cf
(iii) fg
(iv) f/g, provided $g(a) \neq 0$.

■ **Theorem:**   For each positive integer $n = 1, 2, \ldots .$ (page 97)

(i) The function $f(x) = x^n$ is continuous at all numbers x.
(ii) If the function g is continuous at $x = a$, so is the function $f(x) = [g(x)]^n$.

■ The function f is **continuous on the open interval** (a, b) if f is continuous at   (page 98) each $x \in (a, b)$. The function f is **continuous on the closed interval** $[a, b]$ if f is continuous at each $x \in (a, b)$ and, in addition,

$$\lim_{x \to a^+} f(x) = f(a) \quad \text{and} \quad \lim_{x \to b^-} f(x) = f(b).$$

■ **Theorem:**   Let m and n be positive integers. The function $f(x) = x^{n/m}$ is   (page 99)

(i) continuous on $[0, \infty)$ if m is even
(ii) continuous on $(-\infty, \infty)$ if m is odd

■ **Theorem:**   Suppose the function f is defined on an open interval containing   (page 99)
the number a. Then f is continuous at a if and only if

$$\lim_{h \to 0} f(a + h) = f(a).$$

■ **Theorem:**   Let $\lim_{x \to a} g(x) = L$ and f be continuous on an open interval con-   (page 100)
taining the number L. Then $\lim_{x \to a} f(g(x)) = f(L)$.

■ **Theorem:**   If g is continuous on an open interval containing the number a (page 101)
and if f is continuous on an open interval containing the number $g(a)$, then the
composite function $f \circ g$ is continuous at $x = a$.

■ **Theorem:**   Let f be continuous on the interval $[a, b]$ with $f(a) \neq f(b)$ and let   (page 102)
d lie between $f(a)$ and $f(b)$. Then there exists at least one number $c \in (a, b)$
with $f(c) = d$.

■ **Theorem:**   Suppose that the function f is defined on an open interval con-   (page 104)
taining the number a. Then f is continuous at a if and only if, given any
number $\epsilon > 0$, there exists a corresponding number $\delta > 0$ so that $|f(x) -
f(a)| < \epsilon$ whenever $|x - a| < \delta$.

Review Exercises—Chapter 2

In Exercises 1–39, find the indicated limit, if it exists.

1. $\lim_{x \to 2}(x^2 - x + 2)$

2. $\lim_{x \to 2} \dfrac{2x - 1}{x + 6}$

3. $\lim_{x \to 3}(x^4 - x - 1)$

4. $\lim_{x \to 2}(3x - 2)^{3/2}$

5. $\lim_{x \to 3}(4 - 4x)^{1/3}$

6. $\lim_{x \to -5} \dfrac{x^2 - 25}{x + 5}$

7. $\lim_{x \to 3/2} \dfrac{4x^2 - 9}{2x - 3}$

8. $\lim_{x \to 3} \dfrac{x - 3}{x^2 - 9}$

9. $\lim_{x \to -2} \dfrac{x^2 + x - 2}{x + 2}$

10. $\lim_{x \to 1} \dfrac{x^2 + 6x - 7}{x - 1}$

11. $\lim_{x \to 1} \dfrac{x^2 + 2x - 3}{x^2 + x - 2}$

12. $\lim_{x \to -2} \dfrac{x^2 + 2x}{x^2 + x - 2}$

13. $\lim_{x \to 1^-} \sqrt{\dfrac{1 - x^2}{1 + x}}$

14. $\lim_{x \to 8} \sqrt{\dfrac{x - 7}{x + 2}}$

15. $\lim_{x \to 7^+} \sqrt{\dfrac{x - 7}{x + 2}}$

16. $\lim_{x \to 0^-} \sqrt{\dfrac{x}{x - 1}}$

17. $\lim_{x \to 0} \dfrac{\tan x}{\sin x}$

18. $\lim_{x \to 0} \dfrac{3x}{\sin 2x}$

19. $\lim_{x \to 0} \dfrac{(2 + x)^2 - 4}{x}$

20. $\lim_{x \to 0} \dfrac{\sqrt{4 + x} - 2}{x}$

21. $\lim_{x \to 0} \dfrac{|x|}{2x}$

22. $\lim_{x \to 2} \dfrac{3x^2 + x + 1}{1 - x^3}$

23. $\lim_{x \to 0} 3x \csc 4x$

24. $\lim_{x \to 0} \dfrac{\sqrt{4x^2 + 1}}{x^2 + 2}$

25. $\lim_{x \to 1} \dfrac{x^3 - 1}{x - 1}$

26. $\lim_{x \to 0} \dfrac{3x + 5x^2}{x}$

27. $\lim_{x \to 4^-} \sqrt{8 - 2x}$

28. $\lim_{x \to 0^+} \dfrac{1 - x}{\sqrt{x} - 1}$

29. $\lim_{x \to 1} \dfrac{|x - 1|}{x - 1}$

30. $\lim_{x \to 2}(1 - [\![2x]\!])$

31. $\lim_{x \to 3}(2[\![x/2]\!] + 3)$

32. $\lim_{x \to 0^+} \sin(1/x)$

33. $\lim_{x \to 1^+} \dfrac{|x - 1|}{x - 1}$

34. $\lim_{x \to 1^-} \dfrac{|x - 1|}{x - 1}$

35. $\lim_{x \to 0^+} \dfrac{2 + \sqrt{x}}{2 - \sqrt{x}}$

36. $\lim_{x \to 1^-} \sqrt{1 - [\![x]\!]}$

37. $\lim_{x \to 1^+} \sqrt{1 - [\![x]\!]}$

38. $\lim_{x \to 0^+} \dfrac{x - [\![x]\!]}{x}$

39. $\lim_{x \to 0} \dfrac{\sqrt[3]{x + 1} - 1}{x}$

In Exercises 40–45, find the slope of the line tangent to the graph
of the given function at the given point, and find an equation for
this tangent.

40. The graph of $y = 4 - 2x$ at $(3, -2)$.

41. The graph of $f(x) = 3x^2 + 4$ at $(1, 7)$.

42. The graph of $y = 9 - x^2$ at $(2, 5)$

43. The graph of $y = 1/x$ at $(2, \frac{1}{2})$.

44. The graph of $f(x) = \dfrac{3}{x + 2}$ at $(1, 1)$

45. The graph of $f(x) = (x + 1)^4$ at $(1, 16)$

In Exercises 46–57, find the intervals on which the given function is continuous.

46. $y = \dfrac{x^2 - 7}{x - 2}$

47. $f(x) = \dfrac{x + 2}{x^2 - x - 6}$

48. $f(x) = \dfrac{|x + 2|}{x + 2}$

49. $y = \sec x, \quad 0 \le x \le 2\pi$

50. $f(x) = \dfrac{\sin 3x}{2x}$

51. $f(x) = \sqrt{x^2 - 3x - 4}$

52. $f(x) = \begin{cases} x - x^2, & x \le 2 \\ -x, & x > 2 \end{cases}$

53. $y = \begin{cases} \sin x, & x \le \pi/4 \\ 1 - \cos x, & x > \pi/4 \end{cases}$

54. $f(x) = \begin{cases} \sqrt{x}, & x \le 4 \\ x - 1, & x > 4 \end{cases}$

55. $f(x) = \begin{cases} 1 - t, & t < -2 \\ -t, & -2 \le t \le 1 \\ 1 - 2t^2, & 1 < t \end{cases}$

56. $f(x) = \begin{cases} 1/(1 - x), & x \ne 1 \\ 0, & x = 1 \end{cases}$

57. $f(x) = \begin{cases} \dfrac{x^2 + 3x - 10}{x + 5}, & x \ne -5 \\ -7, & x = -5 \end{cases}$

58. Use the Intermediate Value Theorem to prove that there exists a number x_0 so that $x_0^3 + x_0 = 3$. Which integer is closest to x_0? Prove your assertion using the Intermediate Value Theorem.

59. Use the function

$$f(x) = \begin{cases} 1, & x \le -1 \\ \sqrt{x + 2}, & -1 < x < 2 \\ \cos \pi x, & 2 \le x \end{cases}$$

to find:

a. $\lim\limits_{x \to -1^-} f(x)$

b. $\lim\limits_{x \to -1^+} f(x)$

c. $\lim\limits_{x \to -1} f(x)$

d. $\lim\limits_{x \to 2^-} f(x)$

e. $\lim\limits_{x \to 2^+} f(x)$

f. $\lim\limits_{x \to 2} f(x)$

g. Is f continuous at $x = -1$?
h. Is f continuous at $x = 2$?

60. Use the function

$$g(x) = \begin{cases} \tan x, & x < 0 \\ \sin x, & 0 < x \le \pi/2 \\ 2x/\pi, & x > \pi/2 \end{cases}$$

to find:

a. $\lim\limits_{x \to 0^-} g(x)$

b. $\lim\limits_{x \to 0^+} g(x)$

c. $\lim\limits_{x \to 0} g(x)$

d. $\lim\limits_{x \to \pi/2^-} g(x)$

e. $\lim\limits_{x \to \pi/2^+} g(x)$

f. $\lim\limits_{x \to \pi/2} g(x)$

g. Is g continuous at $x = 0$?
h. Is g continuous at $x = \pi/2$?

61. The function

$$f(x) = \begin{cases} 4 - x + x^3, & x \le 1 \\ 9 - ax^2, & x > 1 \end{cases}$$

is continuous at $x = 1$. Find a.

62. Suppose that $1 + 4x - x^2 \le f(x) \le x^2 - 4x + 9$ for $x \ne 2$. Find $\lim\limits_{x \to 2} f(x)$.

63. Suppose $1 - |x| \le f(x) \le \sec x$ for $-\frac{1}{2} \le x \le \frac{1}{2}$. Find $\lim\limits_{x \to 0} f(x)$.

In Exercises 64–71, prove the stated limit.

64. $\lim\limits_{x \to 3}(2x + 3) = 9$

65. $\lim\limits_{x \to 1}(4x - 3) = 1$

66. $\lim\limits_{x \to 2}(x^2 + 3) = 7$

67. $\lim\limits_{x \to 4} 1/x = 1/4$

68. $\lim\limits_{x \to 2^+} \sqrt{x - 2} = 0$

69. $\lim\limits_{x \to -3} |x + 3| = 0$

70. $\lim\limits_{x \to 1}(x^2 + 2x) = 3$

71. $\lim\limits_{x \to -2}(x^2 + x + 1) = 3$

72. Given

$$f(x) = \begin{cases} 2ax + b, & x \le 3 \\ ax + 3b, & x > 3 \end{cases}$$

find a and b so that $\lim\limits_{x \to 3} f(x) = 10$.

73. For what real numbers a is

$$\lim\limits_{x \to a}(x^3 - 4x^2 + 3x - 2) = -10?$$

74. For

$$f(x) = \begin{cases} a^2x^2 + bx - 12, & x < 1 \\ ax + b, & x \ge 1 \end{cases}$$

find numbers a and b so that $\lim\limits_{x \to 1} f(x) = 2$.

75. True or false? If $f(x) \le h(x) \le g(x)$ for all x and both f and g are continuous at $x = a$, then h is continuous at $x = a$.

76. Prove that if $\lim\limits_{x \to a} f(x)$ exists, then it is unique. That is, prove that if $\lim\limits_{x \to a} f(x) = L$ and $\lim\limits_{x \to a} f(x) = M$, then $L = M$.

77. Prove that, if $f(x) \geq g(x)$ and if $\lim\limits_{x \to a} g(x) = L$, then $\lim\limits_{x \to a} f(x) \geq L$, if this limit exists.

78. If f is continuous at $x = a$, but $(f + g)$ is discontinuous at $x = a$, what can you say about the continuity of g at $x = a$?

Differentiation

Unit Two

The Origins of Differentiation

Isaac Newton wrote, "If I seem to have seen further than others, it is because I stood on the shoulders of giants." He was referring to the several mathematicians who actually were performing differentiations and integrations in the years before Newton and Leibniz unified calculus into one subject.

One of the principal contributors was Pierre de Fermat (1601–1665). He was a lawyer by profession, and an amateur mathematician in the true sense of the term: a lover of mathematics.

Relatively little is known of his early life, and it is not even known for certain where he received all of his education. He was

financially secure, and as a minor member of the nobility, was allowed to use the "de" in his name. His interests included studying the classics, and he was fluent in a half dozen languages.

Mathematically, he made significant contributions in a number of areas. He was a co-inventor with René Descartes of analytic geometry. He was particularly interested in the theory of numbers; the most famous theorem to which his name is attached is called Fermat's Last Theorem. It states that, for positive integral values of x, y, z, and n, there is no solution to the equation

Pierre de Fermat

$$x^n + y^n = z^n$$

for n larger than 2. After his death, the theorem was found written in the margin of one of his books, to which he added that he had found a marvelous proof, but that the margin was too small to contain it. Thousands of mathematician-hours have been spent on the theorem, but no one has ever been able to prove it.

A friend once wrote Fermat asking if he could determine whether the number 100,895,598,169 is prime. Fermat wrote back immediately to say that the factors are 898,423 and 112,303, and that each of these factors is itself prime. No one has ever been able to figure out how he was able to do it.

The French mathematician/astronomer Laplace called Fermat the discoverer of differential calculus. As early as 1629, before Newton was even born, Fermat was thinking of the differentiation process. Although he was hampered by a lack of the limit concept and of logical procedures, and although his notation was cumbersome, he essentially differentiated. He devised a method of finding maximum and minimum values which is still sometimes called Fermat's Method. He also found the tangent lines to curves of the form $y = f(x)$, noting that the process was similar to his maximum-minimum method.

Fermat determined volumes, centers of gravity, and the lengths of curves, but he did not realize the crucial relation: that integration and differentiation are inverse processes. For this reason, and his reluctance to publish his discoveries, he lost out as an acknowledged creator of the calculus.

Isaac Barrow (1630–1677) was a precocious and rebellious youngster, driving both his parents and his teachers to distraction. His father was heard to pray one time that if God had to take one

of his children (child mortality was common in those times), that Isaac was the one who could most easily be spared. But Isaac Barrow survived, went to Trinity College at Cambridge, and stayed on as a scholar. At the relatively young age of 33 he became the first Lucasian Professor of Mathematics at Cambridge, an endowed chair with relatively few duties. After six years he resigned the chair. One frequently hears that he resigned so that his pupil, Isaac Newton, might fill the professorship, as Barrow recognized Newton's superior genius. Unfortunately, the story appears not to be true. There is no evidence that Barrow was ever Newton's tutor, though Newton may have attended public lectures given by Barrow. Actually, Barrow hoped for an appointment to a different position; he considered himself a theologian rather than a mathematician. Within a year he was appointed chaplain to the king.

Isaac Barrow

Like Pierre de Fermat in France, Barrow performed differentiations much as we do today as he studied the drawing of lines tangent to curves. He also found areas between curved lines, which was close to the integration process. But he was without a theory of limits, and never realized the critically important inverse relationship between differentiation and integration.

Rolle's Theorem is named after Michel Rolle (1652–1719). His education was limited, and he learned most of his mathematics through his own efforts. He married young and had difficulty in supporting his family. But as a result of solving a difficult number theory problem at the age of 30, he was awarded a pension by the French government.

Rolle was at first a strong antagonist to the "new" calculus of Newton and Leibniz. He asserted that it was not based on logical grounds, and that it gave erroneous results. He called the calculus "a collection of ingenious fallacies." Later he became convinced of the usefulness of the calculus. The theorem that bears his name, and which is used as a basis for proving the Mean Value Theorem, appears only incidentally in his writings about finding approximate solutions to equations. Rolle used it only in connection with polynomial functions, although we now know that it is applicable to a much wider variety of functions. His name was not applied to the theorem until 1846.

A nineteenth-century French mathematician, Augustin-Louis Cauchy (1789–1857), is credited with bringing present-day standards of rigor to the calculus. In particular, he defined "limit" in a more precise way than had previously been done, and formally defined "continuous function." Cauchy took the lead in defining the integral as the limit of a sum, rather than as the inverse of differentiation. He also first formally defined the derivative as the limit of the difference quotient, as it is done in today's textbooks.

Cauchy was born during the difficult years of the French

Augustin-Louis Cauchy

Revolution. His early education was provided by his father, although he was fortunate to live not far from Pierre Laplace, who in turn introduced him to Joseph Lagrange; both were famous mathematicians. He became a civil engineer, although he was also interested in the ancient classics. Laplace and Lagrange persuaded him to pursue mathematics, and he became a teacher at the École Polytechnique in Paris. Reports on his teaching range from highest praise to one account from a student that Cauchy spent an entire hour extracting the square root of 17 by a method with which everyone in the class was already familiar. Others said he rambled incessantly.

His mathematical output was prodigious. In twenty years he published 589 papers in just one journal—and submitted many others which they were unable to publish. More theorems and concepts are named after Cauchy than after any other mathematician. His text *Cours d'analyse* (1821), which was based on his own lectures, became the model from which present day calculus text are descended. Death was sudden and abrupt. He was talking with the Archbishop of Paris and commented, "Men pass away, but their deeds abide." With those words, he fell over dead.

(Photographs from the David Eugene Smith Papers, Rare Book and Manuscript Library, Columbia University.)

Chapter 3

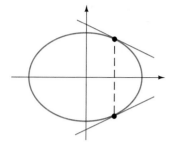

The Derivative

In this chapter we return to the problem of calculating the slope of the line tangent to the graph of a function. Using the properties of *limits,* developed in Chapter 2, we extend this idea to the more general concept of the *derivative* of a function, and we develop the properties of derivatives and the rules by which derivatives can be found. We also interpret the derivative as a rate of change.

3.1   The Derivative as a Function

In Section 2.1 we determined that the slope of the line tangent to the graph of the function f at the point $(x_0, f(x_0))$, if it exists, is the limit

$$m = \lim_{h \to 0} \frac{f(x_0 + h) - f(x_0)}{h}. \tag{1}$$

The quotient

$$\frac{\Delta f}{\Delta x} = \frac{f(x_0 + h) - f(x_0)}{(x_0 + h) - x_0}$$

$$= \frac{f(x_0 + h) - f(x_0)}{h} \tag{2}$$

in equation (1) is the ratio of the *difference* in the values of f and the *difference* in x between the points $(x_0, f(x_0))$ and $(x_0 + h, f(x_0 + h))$ on the graph of f. In other words, it is the change in f divided by the change in x. Since it is a ratio of differences, we call it the **difference quotient.** As we shall see, the limit of (2) as $h \to 0$ has numerous interpretations in addition to its interpretation as the slope of the tangent line. Therefore, we refer this limit more generally as the *derivative* of f at x_0, and we denote it by $f'(x_0)$.

Definition 1

The **derivative of the function f at x_0** is the number

$$f'(x_0) = \lim_{h \to 0} \frac{f(x_0 + h) - f(x_0)}{h} \tag{3}$$

provided this limit exists.

Figure 1.1 shows the interpretation of $f'(x_0)$ as the slope of the line tangent to the graph of f at $(x_0, f(x_0))$. Figure 1.2 reminds us that the limit in equations (1) and (3) is "two-sided." The difference quotient (2) must be defined for $h < 0$ as well as for $h > 0$, and the limits in equations (1) and (3) must exist as h approaches zero from either the right or the left.

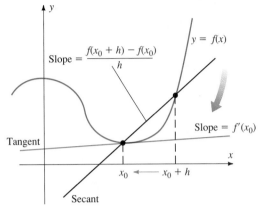

Figure 1.1 Difference quotient (2) is slope of the secant. Derivative $f'(x_0)$ is slope of the tangent at $(x_0, f(x_0))$.

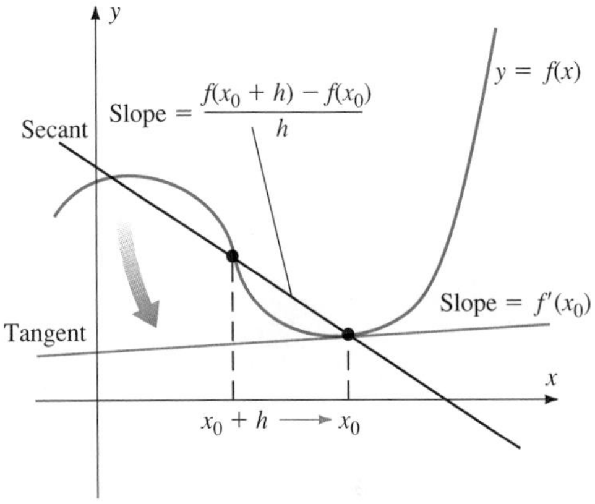

Figure 1.2 Difference quotient (2) with $h < 0$ is also slope of secant.

We say that the function f is *differentiable* at x_0 if the limit $f'(x_0)$ in Definition 1 exists. We refer to the process of calculating the derivative $f'(x_0)$ as *differentiation* of the function f.

When $f'(x_0)$ exists for every x_0 in an interval I, the process of differentiation actually produces a new function f', defined on the interval I, with values $f'(x)$, $x \in I$.

Definition 2

The derivative of the function f on the interval I, denoted by f', is the function with values

$$f'(x) = \lim_{h \to 0} \frac{f(x + h) - f(x)}{h}$$

provided this limit exists for all $x \in I$.

Since the value of the derivative function $f'(x)$ is the slope of the tangent to the graph of f at the point $(x, f(x))$, we sometimes refer to the derivative f' as the *slope function* associated with the function f. Figure 1.3 shows the graph of a function f, and Figure 1.4 shows the graph of its corresponding slope function f'. Note that the values $f'(x)$ are the slopes of the tangents to the graph of f.

In Examples 1–3, we calculate the derivative of three important functions directly from Definition 2.

☐ **EXAMPLE 1**

For the linear function $f(x) = mx + b$, the derivative is

$$f'(x) = \lim_{h \to 0} \frac{f(x + h) - f(x)}{h}$$

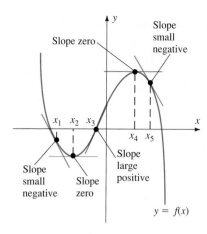

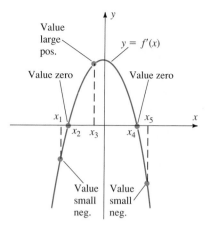

Figure 1.3 Graph of a function f, showing several tangents.

Figure 1.4 Graph of the "slope function" (derivative) f'.

$$= \lim_{h \to 0} \frac{[m(x + h) + b] - [mx + b]}{h}$$

$$= \lim_{h \to 0} \frac{[mx + mh + b] - [mx + b]}{h}$$

$$= \lim_{h \to 0} \frac{mh}{h} = \lim_{h \to 0} m = m.$$

See Figure 1.5. ■

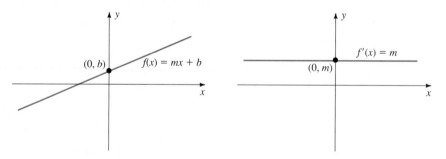

Figure 1.5 For the linear function $f(x) = mx + b$, the slope function (derivative) f' is the constant function $f'(x) = m$.

In some calculations, it is more convenient to use an equivalent form of the difference quotient (2). Since $x_0 + h \to x_0$ as $h \to 0$, we can rewrite the derivative by letting $x = x_0 + h$. Then, $h = x - x_0$, and as $h \to 0$, $x \to x_0$. We obtain

$$f'(x_0) = \lim_{h \to 0} \frac{f(x_0 + h) - f(x_0)}{h} = \lim_{x \to x_0} \frac{f(x) - f(x_0)}{x - x_0}. \tag{4}$$

Thus, we have an alternative way of expressing the limit

$$f'(x_0) = \lim_{\Delta x \to 0} \frac{\Delta f}{\Delta x}.$$

□ **EXAMPLE 2**

The derivative of the function $f(x) = x^2$ is

$$f'(x_0) = \lim_{x \to x_0} \frac{f(x) - f(x_0)}{x - x_0}$$

$$= \lim_{x \to x_0} \frac{x^2 - x_0^2}{x - x_0}$$

$$= \lim_{x \to x_0} \frac{(x + x_0)(x - x_0)}{x - x_0}$$

$$= \lim_{x \to x_0} (x + x_0)$$

$$= x_0 + x_0 = 2x_0.$$

For example, $f'(0) = 0$, $f'(2) = 4$, $f'(-3) = -6$, and $f'(\pi) = 2\pi$. In other words, the derivative of $f(x) = x^2$ is the function $f'(x) = 2x$ (see Figure 1.6). ■

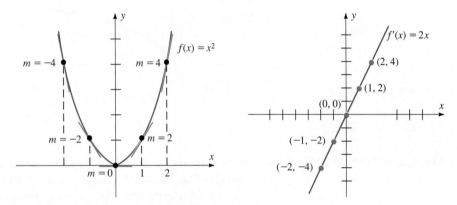

Figure 1.6 For the quadratic function $f(x) = x^2$, the slope function (derivative) f' is $f'(x) = 2x$.

In the following example, we determine the frequently used derivative of the square root function. Note that we must rewrite the difference quotient in a form to which the limit theorems from Chapter 2 apply.

□ **EXAMPLE 3**

The derivative of the function $f(x) = \sqrt{x}$ is

$$f'(x) = \lim_{h \to 0} \frac{f(x + h) - f(x)}{h}$$

$$= \lim_{h \to 0} \frac{\sqrt{x + h} - \sqrt{x}}{h}$$

$$= \lim_{h \to 0} \left(\frac{\sqrt{x + h} - \sqrt{x}}{h} \right)\left(\frac{\sqrt{x + h} + \sqrt{x}}{\sqrt{x + h} + \sqrt{x}} \right)$$

$$= \lim_{h \to 0} \frac{(x + h) - x}{h(\sqrt{x + h} + \sqrt{x})}$$

$$= \lim_{h \to 0} \frac{h}{h(\sqrt{x + h} + \sqrt{x})}$$

$$= \lim_{h \to 0} \frac{1}{(\sqrt{x + h} + \sqrt{x})}$$

$$= \frac{1}{2\sqrt{x}}.$$

That is, for $f(x) = \sqrt{x}$, $f'(x) = 1/(2\sqrt{x})$. Thus, $f'(4) = 1/4$, $f'(9) = 1/6$, $f'(27) = 1/(2\sqrt{27})$, but $f'(0)$ is not defined (see Figure 1.7). ∎

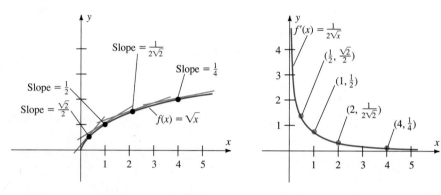

Figure 1.7 For the function $f(x) = \sqrt{x}$, the slope function (derivative) is $f'(x) = 1/(2\sqrt{x})$.

Properties of the Derivative

In this chapter, we determine *properties* of the derivative and *rules* by which derivatives may be calculated. These results will free us from the somewhat cumbersome task of calculating derivatives directly from Definition 2.

Our first result enables us to calculate derivatives of functions that are "built up" from other functions by the processes of addition and multiplication by real numbers.

Theorem 1

If the functions f and g are differentiable at x and c is any real number, then the functions $f + g$ and cf are also differentiable at x, and

(i) $(f + g)'(x) = f'(x) + g'(x)$, and
(ii) $(cf)'(x) = cf'(x)$.

That is, the derivative of a sum of two functions is the sum of their derivatives, and the derivative of the product of a constant c and a function f is the product of the constant and the derivative of f.

Proof: The proof that $f + g$ is differentiable and equation (i) follow from the corresponding property of limits: the limit of the sum is the sum of the limits. (See Theorem 1(i) in Section 2.4.)

$$(f + g)'(x) = \lim_{h \to 0} \frac{(f + g)(x + h) - (f + g)(x)}{h} \qquad \text{(def. of derivative)}$$

$$= \lim_{h \to 0} \frac{[f(x + h) + g(x + h)] - [f(x) + g(x)]}{h} \qquad \text{(def. of } f + g)$$

$$= \lim_{h \to 0} \left[\frac{f(x+h) - f(x)}{h} + \frac{g(x+h) - g(x)}{h} \right] \qquad \text{(rearranging)}$$

$$= \lim_{h \to 0} \frac{f(x+h) - f(x)}{h} + \lim_{h \to 0} \frac{g(x+h) - g(x)}{h} \qquad \begin{array}{l}\text{(property} \\ \text{of limits)}\end{array}$$

$$= f'(x) + g'(x).$$

The proof of part (ii) is left for you as an exercise. ∎

□ **EXAMPLE 4**

For $h(x) = 3x^2 + 4\sqrt{x}$, find $h'(x)$.

Solution   This function has the form

$$h(x) = 3f(x) + 4g(x)$$

with

$$f(x) = x^2 \quad \text{and} \quad g(x) = \sqrt{x}.$$

Using Theorem 1 and the results of Examples 2 and 3, we find that

$$h'(x) = 3f'(x) + 4g'(x) = 3(2x) + 4\left(\frac{1}{2\sqrt{x}}\right) = 6x + \frac{2}{\sqrt{x}}. \qquad ∎$$

Throughout this chapter we shall make frequent use of the relationship between differentiability and continuity, as described in the following theorem.

Theorem 2

Let the function f be defined on an open interval containing x_0. If $f'(x_0)$ exists, then f is continuous at x_0.

Proof: To prove that f is continuous at x_0, we must show that

$$\lim_{x \to x_0} f(x) = f(x_0).$$

Using the assumption that $f'(x_0)$ exists and properties of limits (Theorem 1 in Chapter 2), we have

$$\left(\lim_{x \to x_0} f(x) \right) - f(x_0) = \lim_{x \to x_0} (f(x) - f(x_0))$$

$$= \lim_{x \to x_0} \left(\frac{f(x) - f(x_0)}{x - x_0} (x - x_0) \right)$$

$$= \lim_{x \to x_0} \left(\frac{f(x) - f(x_0)}{x - x_0} \right) \lim_{x \to x_0} (x - x_0)$$

$$= f'(x_0) \cdot 0$$

$$= 0,$$

which proves that

$$\lim_{x \to x_0} f(x) = f(x_0). \qquad ∎$$

It is important to note that the converse of Theorem 2 is not true. That is, a function may be continuous at a number $x = a$ but not differentiable at a. One such example is the absolute value function $f(x) = |x|$ at $x = 0$.

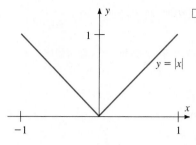

Figure 1.8   The function $f(x) = |x|$ is not differentiable at 0, and its graph does not have a well-defined tangent line at the point $(0, 0)$.

□ **EXAMPLE 5**

We show that $f(x) = |x|$ is not differentiable at 0 by examining its difference quotient at 0. We get

$$\frac{f(0 + h) - f(0)}{h} = \frac{|h|}{h} = \begin{cases} 1, & \text{if } h > 0 \\ -1, & \text{if } h < 0. \end{cases}$$

Thus, the absolute value function would be differentiable at 0 if the limit of $|h|/h$ as $h \to 0$ exists. However, we have already seen that this limit does not exist (Example 6 of Section 2.2). Consequently, the absolute value function is not differentiable at 0.   ■

This lack of differentiability of the absolute value function at 0 corresponds to the fact that its graph has a "corner" at the point $(0, 0)$. Therefore, the graph does not have a well-defined tangent line at $(0, 0)$ (see Figure 1.8).

Exercise Set 3.1

1. Each of the Figures (i)–(v) is the graph of the slope function for one of the functions whose graph is one of Figures (a)–(e). Match the corresponding functions and derivatives.

(a)

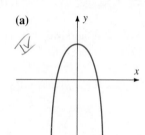

(b)

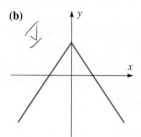

(i)

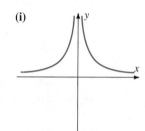

(ii)

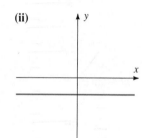

(c)

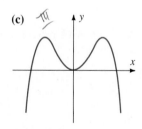

(d)

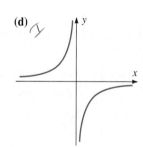

(iii)

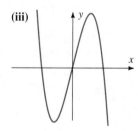

(iv)

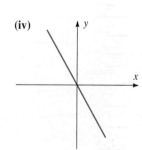

(e)

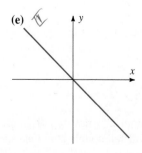

(v)

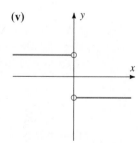

2. For each of the functions graphed in Figures (a)–(d), indicate which of the graphs in Figures (i)–(viii) is the graph of its derivative.

(a)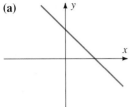

(b)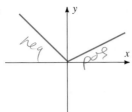

neg   *pos*

(c)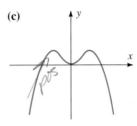

pos

(d)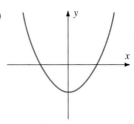

(i)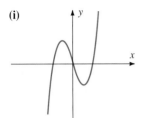

(ii)

b⁻   *pos*

neg

(iii)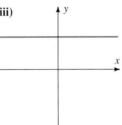

(iv)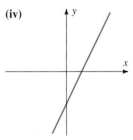

(v)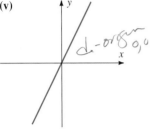

d-orgin 0,0

(vi)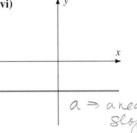

a ⇒ a neg slope

(vii)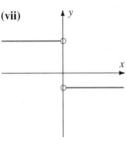

(viii)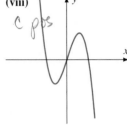

c pos

$$\Rightarrow \lim_{h \to 0} \frac{f(x+h) - f(x)}{h}$$

In Exercises 3–22, use Definition 2 to find f'.

3. $f(x) = 6x + 5$

4. $f(x) = x^2 - 7$

5. $f(x) = 2x^3 + 3$

6. $f(x) = 1 - x^3$

7. $f(x) = ax^3 + bx^2 + cx + d$

8. $f(x) = \dfrac{1}{x - 1}$

9. $f(x) = \dfrac{1}{2x + 3}$

10. $f(x) = \dfrac{1}{x^2 - 9}$

11. $f(x) = \sqrt{x + 1}$

12. $f(x) = \sqrt{2x + 3}$

13. $f(x) = \dfrac{1}{\sqrt{x + 1}}$

14. $f(x) = \dfrac{1}{ax + b}$

15. $f(x) = 1/x^2$

16. $f(x) = x^4$

17. $f(x) = (x + 3)^3$

18. $f(x) = (x - 1)^2$

19. $f(x) = \dfrac{1}{\sqrt{x + 5}}$

20. $f(x) = \dfrac{1}{(x + 2)^2}$

21. $f(x) = \dfrac{3}{(x - 1)^2}$

22. $f(x) = \dfrac{-2}{\sqrt{x + 1}}$

23. Find an equation for the line tangent to the graph of $y = 1/(2x + 3)$ at the point $(0, 1/3)$. *look at #9*

24. Find an equation for the line tangent to the graph of $f(x) = \sqrt{x + 1}$ at the point $(3, 2)$.

25. A **normal** to a curve (see Theorem 5 and Example 5 in Section 1.3) at point P is a line through P perpendicular to the tangent. Find an equation for the normal to the graph of $f(x) = x^2 - 7$ at the point $P = (3, 2)$.

a line through the pt that ⊥ to the tangent line at that pt perpen

26. Find two perpendicular lines that are tangent to the graph of $f(x) = x^2$ and that intersect on the y-axis (see Figure 1.9).

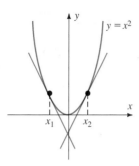

Figure 1.9

27. Find numbers x_1 and x_2 so that the triangle formed by the x-axis and the lines tangent to the graph of $f(x) = 1 - x^2$ at the points $(x_1, f(x_1))$ and $(x_2, f(x_2))$ is equilateral (see Figure 1.10).

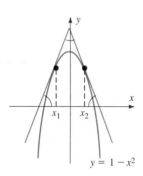

Figure 1.10

28. Find the point where the tangent to the graph of $y = x^2 + 4x + 4$ has slope 4.

29. Find a so that the graph of $y = 2 - ax^2$ has a tangent with slope 6 at $x = -1$.

30. Let $f(x) = ax^2 + bx + 3$. Find a and b so that the tangent to the graph of $y = f(x)$ at $(1, 5)$ has slope 1.

31. Find the constants a and b so that the graph of $y = ax^2 + b$ has tangent $y = 4x$ at the point $P = (1, 4)$.

 32. Sketch the graph of $f(x) = \sin x$ for $0 \le x \le 2\pi$. At each of the values $x = 0, \pi/4, \pi/2, 3\pi/4, \ldots, 2\pi$ estimate the slope of the tangent by evaluating the difference quotient with $h = \pm 0.5, \pm 0.1, \pm 0.05$, and ± 0.01. Using these results sketch the graph of the "slope function" f'.

 33. For $f(x) = \cos x$, estimate the value $f'(x)$ at $x = 0, \pi/4, \pi/2, 3\pi/4, \ldots, 2\pi$ by evaluating the difference quotient with $h = \pm 0.5, \pm 0.1, \pm 0.05$, and ± 0.01. Compare these estimates with the values of $g(x) = \sin x$ at the same angles. What relationship do you observe?

34. Let $f(x) = x^3 + 2x^2 - 4x + 10$. For which numbers x is the slope of the line tangent to the graph of f greater than zero?

35. Let $f(x) = x^2 - 6x + 2$. For which numbers x is the slope of the line tangent to the graph of f less than zero?

36. Let
$$f(x) = \begin{cases} x^2 + 1, & x \ge 3 \\ 6x - 8, & x < 3. \end{cases}$$

a. Is f differentiable at 3? Why?
b. Is f continuous at 3? Why?

37. Let
$$f(x) = \begin{cases} x^2 - x, & x \ge 1 \\ 2x - 2, & x < 1. \end{cases}$$

a. Is f differentiable at 1? Why?
b. Is f continuous at 1? Why?

38. Let
$$f(x) = \begin{cases} 2x^2 - 1, & \text{for } x \ge 0 \\ x - 1, & \text{for } x < 0. \end{cases}$$

a. Is f differentiable at $x = 0$? Why?
b. Is f continuous at $x = 0$? Why?

39. Let $f(x) = \sin x$. Calculate $f'(0)$ directly from Definition 1.

40. Let $f(x) = \sqrt{|x|}$.
a. Sketch the graph of f.
b. For what numbers x is f differentiable?
c. Derive a formula for $f'(x)$.

41. True or false: If the statement is true, explain why. If it is false, give a counterexample.
a. If the limit of $f(x)$ as $x \to a$ exists, then f is continuous at a.
b. If f is continuous at a, then the limit of $f(x)$ as $x \to a$ exists.
c. If the limit of $f(x)$ as $x \to a$ exists, then f is differentiable at a.
d. If f is differentiable at a, then the limit of $f(x)$ as $x \to a$ exists.
e. If f is continuous at a, then f is differentiable at a.
f. If f is differentiable at a, then f is continuous at a.

42. A function f has the following property:
$$f(x_1 + x_2) = f(x_1) + 2x_1x_2 + 3x_2 + x_2^2$$
for all numbers x_1, x_2. Use the definition of the derivative to find $f'(x)$.

In Exercises 43–46, a function f is specified. Given a (fixed) number $h \ne 0$, let m_h be the function
$$m_h(x) = \frac{f(x + h) - f(x)}{h}.$$

Graph $y = m_h(x)$ for $h = 1.0, 0.1, 0.01$, and 0.001. How do these graphs relate to the graph of the derivative of f?

43. $f(x) = 2x + 1$. **44.** $f(x) = x^2$.

45. $f(x) = \sqrt{x}$. **46.** $f(x) = \sin x$.

 In Exercises 47–50, a function f and a number a are given. For $h \neq 0$, let

$$m_h = \frac{f(a + h) - f(a)}{h}.$$

Using $h = \pm 2.0$, ± 1.5, ± 1.0, ± 0.1, ± 0.01, and ± 0.001, graph the secant lines $y = m_h(x - a) + f(a)$. Is f differentiable at a?

47. $f(x) = x^2 - 2x + 1$, $a = 1$

48. $f(x) = \sqrt{x}$, $a = 0$ $(h > 0$ only$)$

49. $f(x) = \begin{cases} \dfrac{\sin x}{x}, & x \neq 0; \\ 1, & x = 0; \end{cases}$ $a = 0$.

50. $f(x) = \begin{cases} 2x - 1, & \text{if } x \leq 1; \\ x^2, & \text{if } x > 1. \end{cases}$ $a = 1$

Formula discovery: In Exercises 51–61, given the function f, graph

$$y = m(x) = \frac{f(x + h) - f(x)}{h}$$

where $h = 0.001$. (The graph of $y = m(x)$ is an approximation to the graph of $y = f'(x)$—see Exercises 43–46.) Compare the graph of $m(x)$ with the graphs of other functions you know, and conjecture a formula for $f'(x)$.

51. $f(x) = x$ **52.** $f(x) = x + 1$

53. $f(x) = 2x + 3$ **54.** $f(x) = x^2$

55. $f(x) = x^2 + 2x + 3$ **56.** $f(x) = x^3$

57. $f(x) = x^3 - 5x^2$ **58.** $f(x) = 1/x$

59. $f(x) = 1/(x + 1)$ **60.** $f(x) = \sin x$

61. $f(x) = \cos 2x$

3.2 Rules for Calculating Derivatives

In this section we begin to develop a list of rules by which derivatives may be calculated. Before doing so, however, we introduce the commonly used alternatives to the notation $f'(x)$.

In addition to the "prime" notation $f'(x)$ for the derivative of f, we shall also find it convenient to use the classical "d by dx" **Leibniz notation**

$$\frac{d}{dx} f(x) \quad \text{and} \quad \frac{df}{dx}$$

(see the historical notes that precede Chapter 1). These two symbols are usually read as "the derivative of f with respect to x," and each is simply another way of writing $f'(x)$. This notation originates with the fact that the derivative is the limit of the ratio of Δf over Δx, and you will often see the expression

$$\frac{df}{dx} = \lim_{\Delta x \to 0} \frac{\Delta f}{\Delta x}$$

as an alternative definition of the derivative.

Moreover, if $y = f(x)$, there are versions of the "prime" notation and "d by dx" notation that involve the variable y. Both

$$y' \quad \text{and} \quad \frac{dy}{dx}$$

represent the derivative function $f'(x)$.

Finally, we should note that the expressions

$$\frac{df}{dx}(x_0) \qquad \frac{df}{dx}\bigg|_{x=x_0} \qquad \frac{dy}{dx}\bigg|_{x=x_0}$$

are frequently used to specify the derivative at a particular number x_0 while using the Leibniz notation. Therefore, each is another way of representing the number $f'(x_0)$.

Now we begin to establish basic rules for the computation of derivatives. Our first result is that the derivative of a constant function is zero.

Theorem 3

Let f be the constant function with values $f(x) = c$ for all x. Then $f'(x) = 0$ for all x. That is,

$$\frac{d}{dx}(c) = 0.$$

Proof: By hypothesis, $f(x + h) = c = f(x)$, so

$$f'(x) = \lim_{h \to 0} \frac{f(x + h) - f(x)}{h} = \lim_{h \to 0} \frac{c - c}{h} = \lim_{h \to 0} \frac{0}{h} = \lim_{h \to 0}(0) = 0. \qquad \blacksquare$$

The Power Rule

In Examples 1 and 2 of Section 3.1 we saw that

$$\text{for} \quad f(x) = x, \quad f'(x) = 1,$$

and

$$\text{for} \quad f(x) = x^2, \quad f'(x) = 2x.$$

Similarly, using the definition of the derivative you can show that

$$\text{for} \quad f(x) = x^3, \quad f'(x) = 3x^2,$$
$$\text{for} \quad f(x) = x^4, \quad f'(x) = 4x^3,$$
$$\text{for} \quad f(x) = x^5, \quad f'(x) = 5x^4,$$

and so forth. Each of these results is a special case of the following rule.

Theorem 4
Power Rule

Let n be any nonzero integer. The function $f(x) = x^n$ is differentiable for all x, and $f'(x) = nx^{n-1}$. That is,

$$\frac{d}{dx}x^n = nx^{n-1}, \qquad n = \pm 1, \pm 2, \ldots.$$

Proof: We shall prove this result for the case $n > 1$ here. The proof for negative integers is given in Example 7. Since

$$x^n - x_0^n = (x - x_0)(x^{n-1} + x^{n-2}x_0 + x^{n-3}x_0^2 + \cdots + xx_0^{n-2} + x_0^{n-1}),$$

we can write the difference quotient as

$$\frac{f(x) - f(x_0)}{x - x_0} = \frac{x^n - x_0^n}{x - x_0}$$

$$= \frac{(x - x_0)(x^{n-1} + x^{n-2}x_0 + x^{n-3}x_0^2 + \cdots + xx_0^{n-2} + x_0^{n-1})}{x - x_0}$$

$$= x^{n-1} + x^{n-2}x_0 + x^{n-3}x_0^2 + \cdots + xx_0^{n-2} + x_0^{n-1}.$$

To compute the derivative, we now take the limit as $x \to x_0$, and we have

$$\lim_{x \to x_0} \frac{f(x) - f(x_0)}{x - x_0} = x_0^{n-1} + x_0^{n-1} + \cdots + x_0^{n-1} \qquad (n \text{ times})$$

$$= nx_0^{n-1}. \qquad \blacksquare$$

A variation of this proof using the Binomial Theorem is given in Appendix II, and a different proof using the Principle of Mathematical Induction is outlined in Exercise 72.

□ **EXAMPLE 1**

(a) For $f(x) = x^{27}$, $f'(x) = 27x^{26}$.

(b) For $y = x^{-4}$, $\dfrac{dy}{dx} = -4x^{-5}$.

□ **EXAMPLE 2**

Using the Power Rule and Theorems 1 and 3, we find that

(a) $\dfrac{d}{dx}(3x^5 + 4) = \dfrac{d}{dx}(3x^5) + \dfrac{d}{dx}(4)$ (Theorem 1, part (i))

$$= 3\frac{d}{dx}(x^5) + \frac{d}{dx}(4) \qquad \text{(Theorem 1, part (ii))}$$

$$= 3 \cdot 5x^4 + 0 \qquad \text{(Power Rule and Theorem 3)}$$

$$= 15x^4.$$

That is, for $f(x) = 3x^5 + 4$, $f'(x) = 15x^4$.

(b) $\dfrac{d}{dx}(7x^4 - 5x^{-3}) = 7\dfrac{d}{dx}(x^4) - 5\dfrac{d}{dx}(x^{-3})$ (Theorem 1, parts (i) and (ii))

$$= 7(4x^3) - 5(-3x^{-4}) \qquad \text{(Power Rule)}$$

$$= 28x^3 + 15x^{-4}.$$

That is, for $f(x) = 7x^4 - 5x^{-3}$, $f'(x) = 28x^3 + 15x^{-4}$.

As Example 2 suggests, the Power Rule may be combined with Theorem 1 to differentiate any polynomial.

□ **EXAMPLE 3**

For $f(x) = 6x^5 - 3x^4 - 2x^3 + 4x^2 - 6x + 5$,

$$f'(x) = 6(5x^4) - 3(4x^3) - 2(3x^2) + 4(2x) - 6(1)$$
$$= 30x^4 - 12x^3 - 6x^2 + 8x - 6.$$

Differentiating Products

We can paraphrase Theorem 1, part (i), by saying that the derivative of the sum is the sum of the derivatives. This observation might lead you to suspect also that the derivative of a product of two functions would be the product of the individual derivatives. However, this is not true. A simple example is found by letting

$$f(x) = x \quad \text{and} \quad g(x) = x^3.$$

Then,

$$f'(x) = 1 \quad \text{and} \quad g'(x) = 3x^2.$$

Now the product is $h(x) = f(x)g(x) = x \cdot x^3 = x^4$, which has derivative $h'(x) = 4x^3$. However, the product of the individual derivatives is $f'(x)g'(x) = (1)(3x^2) = 3x^2$, not $4x^3$.

The correct procedure for differentiating a product is given by the following theorem.

Theorem 5

Product Rule

Let f and g be differentiable at x. Then the product function fg is differentiable at x, and

$$(fg)'(x) = f'(x)g(x) + f(x)g'(x).$$

Before we prove this theorem, we illustrate its use with two examples.

☐ **EXAMPLE 4**

For $f(x) = (2x + 7)(x - 9)$, find $f'(x)$.

Solution   One way to work this problem is to first expand the product,

$$f(x) = (2x + 7)(x - 9) = 2x^2 - 11x - 63,$$

and then differentiate the resulting polynomial:

$$f'(x) = 4x - 11.$$

Another approach is to begin by applying the Product Rule:

$$f'(x) = \left[\frac{d}{dx}(2x + 7) \right](x - 9) + (2x + 7)\left[\frac{d}{dx}(x - 9) \right]$$
$$= 2(x - 9) + (2x + 7)(1)$$
$$= 4x - 11.$$

Of course, both approaches yield the same function $f'(x)$.   ■

☐ **EXAMPLE 5**

For the function $f(x) = (3x^3 - 6x)(9x^4 + 3x^3 + 3)$, it is easier to use the Product Rule to calculate the derivative:

$$f'(x) = \left[\frac{d}{dx}(3x^3 - 6x) \right](9x^4 + 3x^3 + 3) + (3x^3 - 6x)\left[\frac{d}{dx}(9x^4 + 3x^3 + 3) \right]$$
$$= (9x^2 - 6)(9x^4 + 3x^3 + 3) + (3x^3 - 6x)(36x^3 + 9x^2)$$
$$= 189x^6 + 54x^5 - 270x^4 - 72x^3 + 27x^2 - 18.$$ ■

Proof of Product Rule: By the definition of the derivative

$$(fg)'(x) = \lim_{h \to 0} \frac{f(x + h)g(x + h) - f(x)g(x)}{h}.$$

In order to factor the numerator, we subtract and add the quantity $f(x)g(x + h)$ and use the algebra of limits to obtain

$$(fg)'(x) = \lim_{h \to 0} \frac{[f(x + h)g(x + h) - f(x)g(x + h)] + [f(x)g(x + h) - f(x)g(x)]}{h}$$

$$= \lim_{h \to 0}\left[\left(\frac{f(x + h) - f(x)}{h}\right)g(x + h) + f(x)\left(\frac{g(x + h) - g(x)}{h}\right)\right]$$

$$= \left[\lim_{h \to 0} \frac{f(x + h) - f(x)}{h}\right]\left[\lim_{h \to 0} g(x + h)\right]$$

$$+ \left[\lim_{h \to 0} f(x)\right]\left[\lim_{h \to 0} \frac{g(x + h) - g(x)}{h}\right].$$

Now since the function g is differentiable at x it must be continuous at x, by Theorem 2. Thus the limit of $g(x + h)$ as $h \to 0$ equals $g(x)$. Since x is a fixed number in this argument, the limit of $f(x)$ as $h \to 0$ equals $f(x)$. We therefore conclude that

$$(fg)'(x) = f'(x)g(x) + f(x)g'(x)$$

and the proof is complete.   ∎

Our final theorem is the rule for differentiating quotients. It is even more surprising than the rule for products.

Theorem 6

Quotient Rule

Let f and g be differentiable at x with $g(x) \neq 0$. Then the quotient f/g is differentiable at x and

$$\left(\frac{f}{g}\right)'(x) = \frac{f'(x)g(x) - f(x)g'(x)}{[g(x)]^2}.$$

We also delay the proof of this theorem until we complete a few examples.

☐ **EXAMPLE 6**

Find $f'(x)$ for $f(x) = \dfrac{3x^2 + 7x + 1}{9 - x^3}$.

Solution   By the Quotient Rule we have

$$f'(x) = \frac{\left[\dfrac{d}{dx}(3x^2 + 7x + 1)\right](9 - x^3) - (3x^2 + 7x + 1)\left[\dfrac{d}{dx}(9 - x^3)\right]}{(9 - x^3)^2}$$

$$= \frac{(6x + 7)(9 - x^3) - (3x^2 + 7x + 1)(-3x^2)}{(9 - x^3)^2}$$

$$= \frac{3x^4 + 14x^3 + 3x^2 + 54x + 63}{x^6 - 18x^3 + 81}.$$ ■

□ **EXAMPLE 7**

Prove the Power Rule for differentiating $f(x) = x^{-n}$ where n is a positive integer and $x \neq 0$.

Solution   We write $f(x) = 1/x^n$ and apply the Quotient Rule and the Power Rule for the case $n > 0$ to obtain

$$f'(x) = \frac{0 \cdot x^n - 1 \cdot nx^{n-1}}{x^{2n}} = -nx^{n-1-2n} = -nx^{-n-1}.$$ ■

□ **EXAMPLE 8**

Find an equation for the line tangent to the graph of $y = x^2 + x^{-2}$ at the point $(1, 2)$.

Strategy · · · · · · · ·
Find $f'(x)$ for $f(x) = x^2 + x^{-2}$ using the Power Rule in (1).
Slope is $f'(1)$.

Solution

For $f(x) = x^2 + x^{-2}$ we have $f'(x) = 2x - 2x^{-3}$. Then

$$f'(1) = 2 - 2 = 0$$

so the desired line is horizontal. Since this line has slope zero and contains $(1, 2)$, its equation is $y = 2$.   ■

Proof of Quotient Rule: By the definition of the derivative we have

$$\left(\frac{f}{g}\right)'(x) = \lim_{h \to 0} \frac{\dfrac{f(x+h)}{g(x+h)} - \dfrac{f(x)}{g(x)}}{h}$$

$$= \lim_{h \to 0} \frac{f(x+h)g(x) - f(x)g(x+h)}{hg(x)g(x+h)}.$$

We proceed to subtract and add the expression $f(x)g(x)$ in the numerator so as to be able to factor and again use the algebra of limits, to obtain

$$\left(\frac{f}{g}\right)'(x) = \lim_{h \to 0} \frac{[f(x+h)g(x) - f(x)g(x)] + [f(x)g(x) - f(x)g(x+h)]}{hg(x)g(x+h)}$$

$$= \lim_{h \to 0} \frac{\left[\dfrac{f(x+h) - f(x)}{h}\right]g(x) - f(x)\left[\dfrac{g(x+h) - g(x)}{h}\right]}{g(x)g(x+h)}$$

$$= \frac{\left[\lim_{h \to 0} \dfrac{f(x+h) - f(x)}{h}\right]g(x) - f(x)\left[\lim_{h \to 0} \dfrac{g(x+h) - g(x)}{h}\right]}{g(x)\left[\lim_{h \to 0} g(x+h)\right]}$$

$$= \frac{f'(x)g(x) - f(x)g'(x)}{[g(x)]^2}$$

(Notice that we have used the continuity of g in the denominator in the step above, just as we did in the proof of the Product Rule.)   ■

Exercise Set 3.2

In Exercises 1–29, find the derivative of the given function.

1. $f(x) = 8x^3 - x^2$

2. $f(x) = x - x^5$

3. $f(x) = ax^3 + bx$

4. $f(x) = a^5 + 3a^2x^2 + x^3$

5. $f(x) = \dfrac{x^2 + 5}{2}$

6. $f(x) = \frac{2}{3}x^3 + \frac{1}{2}x^2 + x$

7. $f(x) = (x - 1)(x + 2)$

8. $f(x) = (x^2 - 1)(2 - x)$

9. $f(x) = (3x^2 - 8x)(x^2 + 2)$

10. $f(x) = (x^2 + x + 1)(x + 1)$

11. $f(x) = (x^3 - x)^2$

12. $f(x) = \left(x^2 - \dfrac{3}{x^2}\right)^2$

13. $f(x) = 5x^2 + 2x + \dfrac{3}{x} - \dfrac{4}{x^2}$

14. $f(x) = \dfrac{x + 2}{x - 2}$

15. $f(x) = \dfrac{6}{3 - x}$

16. $f(x) = \dfrac{(8x + 2)(x + 1)}{x - 3}$

17. $f(x) = \dfrac{x^4 + 4x + 4}{1 - x^3}$

18. $f(x) = \dfrac{(2x + 1)(3x + 2)}{(x + 1)(x - 1)}$

19. $f(x) = 5x^{-3} - 2x^{-5}$

20. $f(x) = \dfrac{x^2 - 4}{x + 3}$

21. $f(x) = (x - 2)\left(x + \dfrac{1}{x}\right)$

22. $f(x) = \dfrac{1}{(x - 3)^2}$

23. $f(x) = \left(1 + \dfrac{3}{x}\right)^2$

24. $g(t) = \dfrac{t}{t^2 + t + 1}$

25. $f(x) = \left(\dfrac{x + 1}{x - 1}\right)^2$

26. $f(t) = \dfrac{t - 4 + t^2}{t^3 + 3t^2 + 3}$

27. $f(x) = (x^5 + x^{-2})(x^3 - x^{-7})$

28. $f(x) = \dfrac{ax + b}{cx^2 + d}$

29. $f(x) = \dfrac{x^{-3} - x^4}{x^5}$

In each of Exercises 30–38, find f' in two ways—first by the Product Rule, then by first expanding to eliminate the parentheses.

30. $f(x) = x(x + 1)$

31. $f(x) = (x^2 + 2)(x^2 - 2)$

32. $f(x) = \left(\dfrac{1}{x} + 1\right)\left(3 - \dfrac{2}{x^2}\right)$

33. $f(x) = (x^3 + 7)(3x^4 + x + 9)$

34. $f(x) = (1 - x^2)(1 + x^2)$

35. $f(x) = \left(\dfrac{1}{x + 1}\right)\left(\dfrac{2}{x + 2}\right)$

36. $f(s) = \left(s - \dfrac{3}{s^2}\right)\left(s + \dfrac{5}{s^2}\right)$

37. $f(u) = (u^2 + u + 1)(u^2 - u - 1)$

38. $f(x) = (x^{-2} + x^3)(x^2 - x^{-4})$

39. Use the Product Rule to establish the following formula for the derivative of the product of three functions:

$$(fgh)'(x) = f'(x)g(x)h(x) + f(x)g'(x)h(x) + f(x)g(x)h'(x).$$

In Exercises 40–45, use the result of Exercise 39 to find the derivative.

40. $f(x) = x(x + 1)(x + 2)$

41. $f(s) = (2s - 1)(s - 3)(s^2 + 4)$

42. $f(t) = (t^2 - 7)(3t^5 + t)(t^3 - 9)$

43. $f(x) = \left(\dfrac{1}{x}\right)\left(\dfrac{1}{x + 1}\right)\left(\dfrac{1}{x + 2}\right)$

44. $f(u) = (u^2 - 4)^3$

45. $f(x) = (2x^3 - 6x + 9)^3$

46. Use the Product Rule and the result of Exercise 39 to establish the following differentiation rules.
  a. $(f^2)'(x) = 2f(x)f'(x)$ $\quad (f^2(x) = [f(x)]^2)$.
  b. $(f^3)'(x) = 3f^2(x)f'(x)$ $\quad (f^3(x) = [f(x)]^3)$.
  c. Find a similar rule for the derivative of f^4 using the Product Rule and part b. (*Hint:* write f^4 as the product of f and f^3.)
  d. Using parts a–c, formulate a power rule for the derivative of f^n where n is any positive integer. Using mathematical induction, prove your assertion.

47. Find $f'(2)$ for $f(x)$ in Exercise 11.

48. Find $f'(-1)$ for $f(x)$ in Exercise 20.

In Exercises 49–53, use the results of Exercise 46 to find the derivative $f'(x)$.

49. $f(x) = \left(\dfrac{x - 1}{x}\right)^3$

50. $f(s) = \dfrac{1 - s}{(1 + s)^2}$

51. $f(x) = \dfrac{1}{(x - 6)^3}$

52. $f(u) = (u^2 + 4)^3$

53. $f(x) = (x + 2)(x - 1)^2$

In each of Exercises 54–58, find an equation for the line tangent to the graph of the given function at the given point.

54. $f(x) = 3x^3 - 7$ at the point $(1, -4)$

55. $f(x) = \dfrac{x - 1}{x + 1}$ at the point $(1, 0)$

56. $f(x) = (x^2 + x)(1 - x^3)$ at the point $(2, -42)$

57. $f(x) = \left(1 - \dfrac{1}{x}\right)^2$ at the point $(1, 0)$

58. $f(x) = \dfrac{1}{x^3 + x}$ at the point $(2, 1/10)$

In Exercises 59–62, find an equation for the line normal to the graph of the given function at the given point (see Exercise 25 in Section 3.1).

59. $f(x) = 2x^3 + 1$ at the point $(1, 3)$

60. $f(x) = (x - 2)/(x + 1)$ at the point $(2, 0)$

61. $f(x) = 2/x^2$ at the point $(2, 1/2)$

62. $f(x) = \sqrt{x}(x - 1)$ at the point $(3, 2\sqrt{3})$

63. Find the points at which the tangent to the graph of $f(x) = 3x/(2x - 4)$ has slope $m = -3$.

64. Find the constant a if the graph of $y = 1/(ax + 2)$ has tangent $4y + 3x - 2 = 0$ at $(0, 1/2)$. $a = 3$

65. Find the constant b if the graph of $y = b/x^2$ has tangent $4y - bx - 21 = 0$ when $x = -2$.

66. Use the result of Exercise 46 to find a formula for the derivative of $f(x) = \sqrt[3]{x}$. (*Hint:* Use the fact that $f(x)f(x)f(x) = x$.)

67. Use the method of Exercise 66 to find a formula for the derivative of $f(x) = x^{1/n}$, $x \neq 0$, where n is the positive integer.

68. Use the result of Exercises 46 and 67 to derive a power rule for fractional exponents. In other words, if $f(x) = x^{n/m}$, calculate $f'(x)$.

69. Find the numbers x where the tangent to the graph of $f(x) = x^3$ at $(x, f(x))$ is parallel to the secant line (see Section 2.1) through the points $(1, 1)$ and $(2, 8)$. Graph the function f, the secant line, and the tangent lines.

70. Compute

$$\lim_{h \to 0} \frac{(3 + h)^{100} - 3^{100}}{h}.$$

(*Hint:* Use the Power Rule.)

71. Provide a different proof of the Quotient Rule as follows:
   a. Suppose that g is differentiable at x_0 and that $g(x_0) \neq 0$. Show that the difference quotient of $1/g$ at x_0 equals

$$\frac{g(x_0) - g(x_0 + h)}{g(x_0 + h)g(x_0)h}.$$

   b. By taking the limit as $h \to 0$ in part a, prove that

$$\left(\frac{1}{g}\right)'(x) = -\frac{g'(x)}{[g(x)]^2}.$$

   c. Suppose also that f is differentiable at x_0. Derive the Quotient Rule by applying the Product Rule to the product of f with $1/g$ (use part b.).

72. Complete the steps in the following proof of Theorem 4 (the Power Rule):
   a. For $n = 1$, show that the function $f(x) = x$ is differentiable for all x and satisfies the equation $f'(x) = 1 \cdot x^{1-1} = x^0 = 1$. This shows that Theorem 4 holds for $n = 1$.
   b. Assume that Theorem 4 holds for $n = k$ where k is a positive integer greater than 1.
   c. Write $f(x) = x^{k+1} = x(x^k)$ and use assumption (b) and the Product Rule to show that f is differentiable for all x.
   d. Use assumption (b) and the Product Rule to show that $f'(x) = (k + 1)x^k$ for all x.
   e. Apply the Principle of Mathematical Induction to conclude that Theorem 4 holds for all integers $n \geq 1$.

3.3   The Derivative as a Rate of Change

So far we have only interpreted the derivative as the slope of the line tangent to the graph of a function. There is, however, another interpretation that is extremely important—the interpretation of the derivative as a *rate of change*. For example, if $a(t)$ is the surface area of a melting ice cube at time t, then we shall see that $a'(t)$ represents the rate at which its surface area is changing. We begin this section with a discussion of *velocity* as measured by the derivative, and then we discuss rates of change for more general functions of time, functions such as the surface area function $a(t)$.

Velocity

Imagine an object moving along a line, such as a jogger on a footpath or an automobile on a highway. In physics, we define the **velocity** of such an object by the equation

$$\text{velocity} = \frac{\text{change in position}}{\text{change in time}}. \tag{1}$$

Of course, what is meant by equation (1) is really an *average* velocity for the time period in question. For example, if a jogger wearing both a pedometer and a watch finds that she has traveled 12 kilometers in 45 minutes, the velocity association with this time interval is

$$\text{velocity} = \frac{12 \text{ km}}{3/4 \text{ h}} = 16 \text{ km/h}.$$

However, at various times during the jog the jogger is likely to have run both faster and slower than 16 kilometers per hour.

There is a special setting in which we can use the theory of the derivative to define the velocity of an object *at each instant,* rather than having to settle for the average velocity over a finite time interval. First, the motion of the object must be along a line (we call this **rectilinear** motion), rather than along a general curve. Second, we must have available a **position function** s giving the location $s(t)$ of the object along the line at each time t (see Figure 3.1).

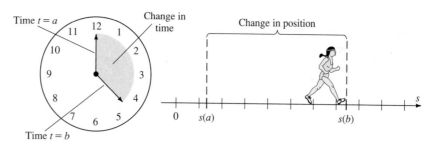

Figure 3.1 A *position function s(t)* gives the location of an object along a (number) line at time t.

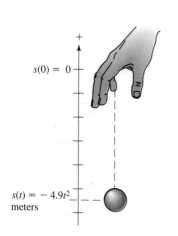

Figure 3.2 A body falling freely from rest falls $4.9t^2$ meters in t seconds.

For example, in physics we learn that, near the surface of the earth, a freely falling body falls $4.9t^2$ meters t seconds after its release. The motion of such an object is therefore along a (vertical) line, and its position function is $s(t) = -4.9t^2$. (The negative sign indicates position *below* the starting point. See Figure 3.2.)

To define the velocity of an object at time t_0 (sometimes referred to as the **instantaneous** velocity at time t_0), we begin by observing that if $h \neq 0$, then $s(t_0 + h) - s(t_0)$ is the change in the position of the object corresponding to the time interval with endpoints t_0 and $t_0 + h$. Thus, the expression

$$\begin{bmatrix} \text{average} \\ \text{velocity} \end{bmatrix} = \frac{s(t_0 + h) - s(t_0)}{h} \tag{2}$$

is precisely the (average) velocity as defined by equation (1).

We now argue that as $h \to 0$, the average velocity corresponding to the (shrinking) time interval with endpoints t_0 and $t_0 + h$ should provide an increasingly accurate measure of the velocity *at the instant* $t = t_0$. For this reason, we define the velocity at time $t = t_0$ to be the limiting value of these average velocities. That is, we *define* the velocity, $v(t_0)$, as

$$v(t_0) = \lim_{h \to 0} \frac{s(t_0 + h) - s(t_0)}{h} = s'(t_0)$$

(3)

whenever this limit exists. Equation (3) simply states that we have *defined* velocity as the derivative of the position function.

Definition 3

If the differentiable function s gives the position at time t of an object moving along a line, then the velocity $v(t)$ at time t is the derivative

$$v(t) = s'(t).$$

That is,

$$v(t) = \frac{d}{dt}s(t).$$

☐ **EXAMPLE 1**

Starting at time $t = 0$, a particle moves along a line so that its position at time t seconds is $s(t) = t^2 - 6t + 8$ meters.

(a) Find its velocity at time t.
(b) When is its velocity zero?

Solution: According to Definition 3, the velocity at time t is

$$v(t) = s'(t) = 2t - 6 \text{ meters per second.}$$

Setting $v(t) = 0$ gives $2t - 6 = 0$, so $t = 3$. The particle has zero velocity at $t = 3$ seconds (see Figures 3.3 and 3.4). ■

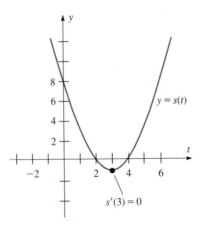

Figure 3.3 Position function $s(t) = t^2 - 6t + 8$.

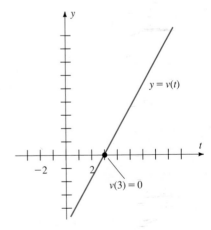

Figure 3.4 Velocity function $v(t) = s'(t) = 2t - 6$.

Notice that the particle in Example 1 lies $s(0) = 8$ meters from the origin at time $t = 0$, but that as t increases from 0 to 3, $s(t)$ decreases from 8 meters to $s(3) = -1$ meter. If we think of the particle as moving along a number line, this says that the particle moves to the left for $0 \leq t < 3$ (Figure 3.5). This corresponds to the observation that $v(t)$ is negative for $t < 3$. For $t > 3$, the position $s(t)$ increases as t increases, so $v(t) > 0$ for $t > 3$. At $t = 3$, the particle reversed direction.

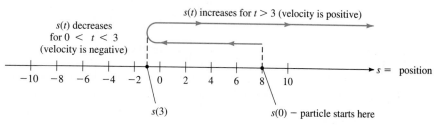

Figure 3.5 For $s(t) = t^2 - 6t + 8$, velocity is negative for $0 < t < 3$ and positive for $t > 3$.

Velocity versus Speed

When we describe motion along a line, the sign ($+$ or $-$) associated with velocity agrees with our choice of which direction along the line corresponds to positive values of position. We shall follow the usual conventions that:

(a) for horizontal motion, positive values of position lie to the *right* of the origin ($s = 0$), so motion to the right corresponds to positive velocity, while negative velocity indicates motion to the left;

(b) for vertical motion, positive values of position lie *above* the origin, so positive velocity indicates motion upward, and negative velocity is motion downward.

Speed, on the other hand, refers to the magnitude of velocity independent of its sign. Thus

$$\text{speed} = |v(t)|.$$

We may summarize the relationship between speed and velocity by saying that velocity indicates both a speed (its absolute value) and a direction (its sign).

□ **EXAMPLE 2**

A pebble is dropped from an open window 100 meters above the ground. Find its velocity and its speed when it strikes the ground.

Solution The motion of the pebble is along a vertical line. Following convention, we take the upward direction as positive. Also, we take the origin $s = 0$ to be at the height of the window (Figure 3.6). As previously stated, at time t seconds the freely falling pebble has position $s(t) = -4.9t^2$ m. It will strike the ground when

$$s(t_0) = -4.9t_0^2 = -100 \text{ m,}$$

or when

$$t_0 = \sqrt{\frac{100}{4.9}} \approx 4.5 \text{ s.}$$

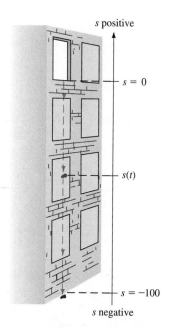

Figure 3.6

Since the corresponding velocity function is

$$v(t) = s'(t) = -2(4.9)t = -9.8t \text{ m/s},$$

the velocity at the time of impact will be

$$v(t_0) \approx v(4.5) = -(9.8)(4.5) = -44.1$$

meters per second (downward). The *speed* at impact will be

$$|v(4.5)| = |-44.1| = 44.1 \text{ m/s}. \qquad \blacksquare$$

□ **EXAMPLE 3**

An object launched vertically upward from ground level with an initial velocity of 72 m/s is located

$$s(t) = 72t - (4.9)t^2$$

meters above ground level at time t seconds. According to this position function,

(a) when does the object stop rising?
(b) what is its maximum height?

Strategy · · · · · · ·

Establish an orientation for the line of motion and locate the origin.

Calculate $v(t) = s'(t)$.

The object stops rising when $v(t_0) = 0$. Solve this equation for t_0.

Maximum height is $s(t_0)$.

Solution

From the statement of the problem we know that the motion is vertical, the origin is at ground level, and the positive direction is upward. The velocity function is

$$v(t) = \frac{d}{dt}(72t - 4.9t^2) = 72 - 9.8t.$$

Setting $v(t_0) = 0$ gives

$$72 - 9.8t_0 = 0$$

or

$$t_0 = \frac{72}{9.8} \approx 7.35 \text{ s}.$$

This is the time at which the object is at maximum height. The maximum height is approximately

$$s(t_0) = 72(7.35) - (4.9)(7.35)^2 = 264.5 \text{ m}. \qquad \blacksquare$$

General Rates of Change

Velocity is one example of the general notion of a *rate of change*. For any function of time, such as the melting ice cube's surface area function $a(t)$ mentioned at the beginning of this section, we can calculate both average and instantaneous rates of change. If $f(t)$ is a function of time t, then the change in time over the interval from time t to time $t + h$ is

$$\Delta t = (t + h) - t = h.$$

Similarly, the change in the function f over the same time interval is

$$\Delta f = f(t + h) - f(t).$$

Then the **average rate of change** in f from time t to time $t + h$ is given by the quotient

$$\frac{\text{change in } f}{\text{change in time}} = \frac{\Delta f}{\Delta t} = \frac{f(t + h) - f(t)}{h}.$$

We define the **instantaneous rate of change** of f to be the limiting value of the average rate of change from time t to time $t + h$ as $h \to 0$, that is,

$$\begin{bmatrix} \text{rate of change of} \\ f \text{ at time } t \end{bmatrix} = \lim_{h \to 0} \frac{f(t + h) - f(t)}{h}.$$

If this limit exists, the right side of the equation is the derivative, that is, we have

$$\begin{bmatrix} \text{rate of change of} \\ f \text{ at time } t \end{bmatrix} = f'(t).$$

In other words, *the derivative $f'(t)$ is the (instantaneous) rate of change of the function f at time t.*

The following example illustrates both average and instantaneous rates of change.

□ **EXAMPLE 4**

Suppose that a spherical balloon is inflated so that, at time t (measured in seconds), its radius is $2t$ centimeters. What is the average rate of change of its surface area from $t = 1$ to $t = 2$? At what (instantaneous) rate is the surface area increasing when $t = 2$?

Solution We know from geometry that the surface area of a sphere of radius r is $4\pi r^2$. Since we are given that $r = 2t$, the surface area $a(t)$ is given by the equation

$$a(t) = 4\pi r^2 = 4\pi(2t)^2 = 16\pi t^2.$$

The average rate of change of the surface area from $t = 1$ to $t = 2$ is

$$\frac{a(2) - a(1)}{2 - 1} = \frac{(64\pi \text{ cm}^2 - 16\pi \text{ cm}^2)}{2 \text{ s} - 1 \text{ s}}$$

$$= \frac{48\pi \text{ cm}^2}{1 \text{ s}}$$

$$= 48\pi \text{ cm}^2/\text{s}.$$

To determine the instantaneous rate of change of the surface area when $t = 2$, we calculate the derivative $a'(t)$. Since

$$a'(t) = 32\pi t,$$

we find that the rate of increase at the desired instant ($t = 2$ s) is

$$a'(2) = 64\pi \text{ cm}^2/\text{s}.$$

Note that the units (cm^2/s) for the instantaneous rate are the same as the units for the average rate of change. The units are the same because the instantaneous rate is the limit of the average rates of change. ■

Exercise Set 3.3

In Exercises 1–4, find the average rate of change of the given function for the given time interval. Then calculate the instantaneous rate of change at the specified time t_0.

1. $s(t) = (1 + t)^2$, $2 \le t \le 3$, $t_0 = 5/2$

2. $s(t) = 1/(1 - t)$, $2 \le t \le 2.5$, $t_0 = 2$

3. $f(t) = \sqrt{t}(1 - t^3)$, $0 \le t \le 1$, $t_0 = 1$

4. $f(t) = (t^{-2} + 2)^3$, $1 \le t \le 2$, $t_0 = 3/2$

In Exercises 5–12, the function s gives the position of a particle moving along a line. Find
  a. the velocity function $v(t)$,
  b. the time(s) $t_0 \geq 0$ at which velocity is zero, and
  c. the intervals in $[0, \infty)$ on which velocity is positive.

5. $s(t) = 3t - 2$

6. $s(t) = t^2 - 6t + 4$

7. $s(t) = \dfrac{1}{1 + t}$

8. $s(t) = t^2 - 3t + 5$

9. $s(t) = t^3 - 9t^2 + 24t + 10$

10. $s(t) = t(t - 1)(t + 2)$

11. $s(t) = t^3 - 6t^2 + 9t + 7$

12. $s(t) = t^4 - 4t + 4$

13. A particle moves along a line so that its position at time t is $s(t) = (t^2 + 2)/(t + 1)$ units. Find its velocity at time $t = 3$.

14. A particle moves along a line so that after t seconds its position is $s(t) = 6 + 5t - t^2$. Find its maximum distance from the origin during the time interval $[0, 6]$.

15. A spherical snowball whose initial radius is 4 inches starts melting, and its radius r decreases according to the formula $r = 4 - t^2$, where t represents time measured in minutes.
  a. What is the average rate of change of its volume over the time intervals $0 \leq t \leq 1$ and $1 \leq t \leq 2$? (Recall that the volume of a sphere of radius r is $\frac{4}{3}\pi r^3$.)
  b. What is the instantaneous rate of change of its volume at $t = 1$?

16. Suppose that water is poured into a cylindrical jar whose radius is 3 inches so that, at time t (measured in minutes), the height of the water is $\sqrt{t}$ inches.
  a. What is the average rate of change of the volume of the water over the time interval $0 \leq t \leq 1$?
  b. What is the instantaneous rate of change of the volume when $t = 1$?

17. A particle moves along a line so that at time t its position is $s(t) = 6t - t^2$.
  a. What is its initial velocity, that is, v_0?
  b. When does it change direction?
  c. How fast is it moving when it reaches the origin the second time?

18. If a particle is projected vertically upward from ground level with an initial velocity v_0, its height after t seconds is $s(t) = v_0 t - 4.9t^2$ meters. Suppose $v_0 = 98$ m/s.
  a. What is the velocity of the particle at time t?
  b. At what time does the particle reach its maximum height? (*Hint:* What is its velocity at maximum height?)
  c. What is the maximum height?
  d. How fast is it moving when it strikes the ground?

19. An object is dropped from rest and strikes the ground with velocity $v = -49$ m/s.
  a. How long did it fall?
  b. From what height was it dropped?
  c. What was its average velocity?

20. When an object is launched vertically from an initial height of s_0 meters with an initial velocity of v_0 m/s, its position after t seconds is

$$s(t) = s_0 + v_0 t - 4.9t^2 \text{ m.}$$

  a. Find its velocity after 4 s.
  b. Find its speed after 4 s.

21. Refer to Exercise 20. A flare is launched vertically from a tower 40 meters above ground level with an initial velocity of $v_0 = 400$ m/s.
  a. Find its position function s.
  b. Find its maximum height.
  c. Find its speed when it strikes the ground (assuming it misses the tower on its way down).

22. A piece of tail section breaks loose from an airliner flying at an altitude of 10,000 meters. Neglecting air resistance,
  a. when does the piece strike the ground?
  b. with what velocity does the piece strike the ground?

23. The height of an object t seconds after it is launched vertically into the air is given by the position function

$$s(t) = 800t - 16t^2.$$

  a. Sketch the graph of the function $v = s(t)$.
  b. Find an equation for its velocity, $v(t)$. *when vel is pos*
  c. For which times t is the object rising? *when vel is neg*
  d. For which times t is the object falling? *when vel is neg*
  e. For which time(s) t is the object at rest before it strikes the ground? *when vel = 0*
  f. What is the maximum height reached? *when vel = 0, when s(25)*
  g. Determine the time required for the object to reach a height of 9600 feet. *s(t) = 9600 2x*
  h. When does the object strike the ground?

24. The following table gives the population of the United States for the years 1900–1980.

Year	Population
1900	75.995
1910	91.972
1920	105.711
1930	123.203
1940	131.669
1950	150.697
1960	179.323
1970	203.212
1980	226.505

Let t represent time (in decades) starting at the year 1900; that is, $t = 0$ corresponds to 1900, $t = 1$ corresponds to 1910, etc. Then the polynomial

$$p(t) = -0.0614t^5 + 1.1502t^4 - 7.0039t^3 + 15.3990t^2 + 6.4932t + 75.995$$

interpolates some of the values in the table, that is, $p(0) = 75.995$, $p(3) = 123.203$, $p(4) = 131.669$, $p(7) = 203.212$, and $p(8) = 226.505$.

a. Graph $p'(t)$.

b. What year between 1900 and 1980 is the population growing most rapidly?

c. At what rate did the population increase between 1950 and 1960?

d. Predict the rate of increase in the population between the years 1980 and 2000. Is this prediction plausible?

e. Repeat parts a–d using the polynomial

$$q(t) = 0.297t^3 - 2.0101t^2 + 17.6901t + 75.995$$

in place of p.

f. Which polynomial, p or q, is the better model for population growth in the United States?

3.4   Derivatives of the Trigonometric Functions

The purpose of this section is to use the differentiation rules of Section 2, together with two limits established in Chapter 2, to find derivatives for all six of the trigonometric functions.

The derivatives for $\sin x$ and $\cos x$ are given by the following theorem. The derivatives of the other four trigonometric functions follow from this theorem and the Quotient Rule.

Theorem 7

The function $f(x) = \sin x$ is differentiable for all x, and $f'(x) = \cos x$. The function $g(x) = \cos x$ is differentiable for all x, and $g'(x) = -\sin x$. That is,

$$\frac{d}{dx} \sin x = \cos x \tag{1}$$

$$\frac{d}{dx} \cos x = -\sin x \tag{2}$$

Figure 4.1 shows the graphs of $f(x) = \sin x$ and $f'(x) = \cos x$ and reminds us of the "*slope function*" interpretation for f': the *slope* of the graph of $f(x) = \sin x$ at $x = x_0$ is given by the *value* $f'(x_0) = \cos x_0$ of the derivative at $x = x_0$. Since the function $f(x) = \sin x$ is *periodic*, with period 2π, this geometric interpretation of equation (1) means that the derivative $f'(x) = \cos x$ must also be periodic with period 2π, a fact that we have already noted in Section 1.5.

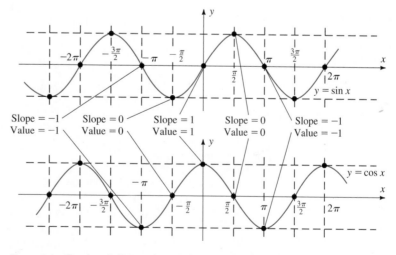

Figure 4.1   Graphs of $f(x) = \sin x$ and its derivative $f'(x) = \cos x$.

Proof of Theorem 7: The proof of this theorem uses the limit

$$\lim_{h \to 0} \frac{\sin h}{h} = 1, \tag{3}$$

established in Section 2.4, and the limit

$$\lim_{h \to 0} \frac{1 - \cos h}{h} = 0, \tag{4}$$

established in Exercise 48, Section 2.4.

To calculate the derivative of $\sin x$, we use the basic definition of the derivative and the addition rule for sines: $\sin(\alpha + \beta) = \sin \alpha \cos \beta + \cos \alpha \sin \beta$:

$$\frac{d}{dx} \sin x = \lim_{h \to 0} \frac{\sin(x + h) - \sin x}{h}$$

$$= \lim_{h \to 0} \frac{(\sin x \cos h + \cos x \sin h) - \sin x}{h}$$

$$= \lim_{h \to 0} \frac{\sin x(\cos h - 1) + \cos x \sin h}{h}$$

$$= \lim_{h \to 0} \left[\sin x \left(\frac{\cos h - 1}{h} \right) + \cos x \left(\frac{\sin h}{h} \right) \right]$$

$$= \sin x \left(\lim_{h \to 0} \frac{\cos h - 1}{h} \right) + \cos x \left(\lim_{h \to 0} \frac{\sin h}{h} \right)$$

$$= (\sin x)(0) + (\cos x)(1) \qquad \text{(limits (3) and (4))}$$

$$= \cos x.$$

To calculate the derivative of $\cos x$, we proceed as above using the addition rule for cosines: $\cos(\alpha + \beta) = \cos \alpha \cos \beta - \sin \alpha \sin \beta$:

$$\frac{d}{dx} \cos x = \lim_{h \to 0} \frac{\cos(x + h) - \cos x}{h}$$

$$= \lim_{h \to 0} \frac{(\cos x \cos h - \sin x \sin h) - \cos x}{h}$$

$$= \lim_{h \to 0} \frac{\cos x(\cos h - 1) - \sin x \sin h}{h}$$

$$= \cos x \left(\lim_{h \to 0} \frac{\cos h - 1}{h} \right) - \sin x \left(\lim_{h \to 0} \frac{\sin h}{h} \right)$$

$$= (\cos x)(0) - (\sin x)(1) \qquad \text{(limits (3) and (4))}$$

$$= -\sin x. \qquad \blacksquare$$

□ **EXAMPLE 1**

For $y = x^3 \sin x$, the derivative is

$$\frac{d}{dx}(x^3 \sin x) = \left(\frac{d}{dx} x^3 \right)(\sin x) + x^3 \left(\frac{d}{dx} \sin x \right) \qquad \text{(Product Rule)}$$

$$= 3x^2 \sin x + x^3 \cos x. \qquad \text{(Equation (1))} \qquad \blacksquare$$

☐ **EXAMPLE 2**

For the function $f(t) = \sin \pi t$, the average rate of change over the interval $[1/4, 1/2]$ is

$$\frac{\Delta f}{\Delta t} = \frac{\sin(\pi/2) - \sin(\pi/4)}{1/2 - 1/4} = \frac{1 - \sqrt{2}/2}{1/4} = 4 - 2\sqrt{2} \approx 1.17,$$

while the instantaneous rate of change at $t = 1/4$ is

$$f'(\pi/4) = \pi \cos(\pi/4) = \frac{\pi\sqrt{2}}{2} \approx 2.22.$$

(See Figure 4.2.)   ■

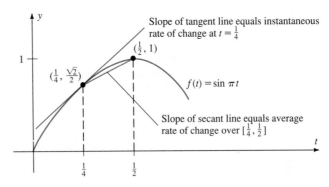

Figure 4.2   Average rate of change versus instantaneous rate of change.

☐ **EXAMPLE 3**

Find $\dfrac{dy}{dx}$ for $y = \dfrac{2 - \cos x}{2 + \cos x}$.

Strategy · · · · · · · ·
Apply Quotient Rule first.

Solution

$$\frac{dy}{dx} = \frac{(2 + \cos x)\left[\dfrac{d}{dx}(2 - \cos x)\right] - \left[\dfrac{d}{dx}(2 + \cos x)\right](2 - \cos x)}{(2 + \cos x)^2}$$

Then apply equation (2).

$$= \frac{(2 + \cos x)[-(-\sin x)] - (-\sin x)(2 - \cos x)}{(2 + \cos x)^2}$$

$$= \frac{4 \sin x}{(2 + \cos x)^2}.$$

■

☐ **EXAMPLE 4**

Find the derivative of $y = \tan x$, where it is defined, and determine for which values of x the function $y = \tan x$ is differentiable.

Strategy · · · · · · · · **Solution**

Write $\tan x = \dfrac{\sin x}{\cos x}$.

$$\frac{d}{dx}\tan x = \frac{d}{dx}\left(\frac{\sin x}{\cos x}\right)$$

Apply Quotient Rule.

$$= \frac{\cos x \cdot \dfrac{d}{dx}(\sin x) - \sin x \cdot \dfrac{d}{dx}(\cos x)}{\cos^2 x}$$

$$= \frac{\cos^2 x + \sin^2 x}{\cos^2 x}$$

Use identity $\cos^2 x + \sin^2 x = 1$.

$\dfrac{1}{\cos x} = \sec x$.

$$= \left(\frac{1}{\cos x}\right)^2$$

$$= \sec^2 x;$$

that is,

$$\frac{d}{dx}\tan x = \sec^2 x.$$

Both $\tan x$ and $\sec x$ are defined for all x for which $\cos x \neq 0$. Since $\cos x$ is the only factor of the denominator in the above calculations, $\tan x$ is differentiable at all values of x with $\cos x \neq 0$ (i.e., except at odd multiples of $\pi/2$). ■

Since each of the remaining trigonometric functions is defined as a ratio involving $\sin x$, $\cos x$, or both, we may proceed as in Example 4 to determine their derivatives (see Exercise 29). The results are the following.

$$\frac{d}{dx}\tan x = \sec^2 x. \tag{5}$$

$$\frac{d}{dx}\cot x = -\csc^2 x. \tag{6}$$

$$\frac{d}{dx}\sec x = \sec x \tan x. \tag{7}$$

$$\frac{d}{dx}\csc x = -\csc x \cot x. \tag{8}$$

□ **EXAMPLE 5**

Find $f'(x)$ for $f(x) = \sec x \tan x$.

Solution   Using the Product Rule and equations (5) and (7) we obtain

$$\frac{d}{dx}(\sec x \tan x) = \left(\frac{d}{dx}\sec x\right)\tan x + \sec x\left(\frac{d}{dx}\tan x\right)$$

$$= (\sec x \tan x)\tan x + \sec x(\sec^2 x)$$

$$= \sec x(\tan^2 x + \sec^2 x). \qquad ■$$

□ **EXAMPLE 6**

Find $\dfrac{dy}{dx}$ for $y = \dfrac{\csc x}{1 + \cot x}$.

Solution   By the Quotient Rule and equations (6) and (8) we obtain

$$\frac{dy}{dx} = \frac{(1 + \cot x)\left(\dfrac{d}{d}\csc x\right) - (\csc x)\left[\dfrac{d}{dx}(1 + \cot x)\right]}{(1 + \cot x)^2}$$

$$= \frac{(1 + \cot x)(-\csc x \cot x) - (\csc x)(-\csc^2 x)}{(1 + \cot x)^2}$$

$$= \frac{\csc^3 x - \csc x \cot x - \csc x \cot^2 x}{(1 + \cot x)^2}$$

$$= \frac{\csc x(\csc^2 x - \cot^2 x - \cot x)}{(1 + \cot x)^2}$$

$$= \frac{\csc x(1 - \cot x)}{(1 + \cot x)^2} \qquad (\csc^2 x - \cot^2 x = 1). \qquad \blacksquare$$

□ **EXAMPLE 7**

Find an equation for the line tangent to the graph of $f(x) = 2x \sec x$ at the point $(0, 0)$.

Strategy · · · · · · · ·

Find $f'(x)$ using the Product Rule.

Solution

The derivative is

$$\frac{d}{dx}(2x \sec x) = \left(\frac{d}{dx}2x\right)(\sec x) + (2x)\left(\frac{d}{dx}\sec x\right)$$

$$= 2 \sec x + 2x \sec x \tan x$$

$$= 2 \sec x(1 + x \tan x)$$

Find the slope: $m = f'(0)$.

So the slope of the line tangent to the graph at $(0, 0)$ is

$$f'(0) = 2(\sec 0)(1 + 0) = 2(1)(1) = 2.$$

Use the slope m and the point $(0, 0)$ to write an equation for the tangent.

An equation for the tangent is therefore

$$y - 0 = 2(x - 0)$$

or

$$y = 2x \qquad \blacksquare$$

□ **EXAMPLE 8**

Figure 4.3 shows a block attached to the end of a spring on a horizontal surface that we shall assume to be frictionless. If the block is oscillating back and forth along the surface so that at time t (in seconds) the pointer is

$$s(t) = 20 + 6 \sin t$$

cm to the right of the fixed end of the spring, find

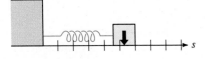

Figure 4.3 Block oscillating on a frictionless surface.

(a) its velocity at time t,

(b) its velocity at time $t = \pi$ s, and

(c) the times at which the block changes direction.

Strategy · · · · · · · · ·   **Solution**

(a) Find $v(t) = s'(t)$.

(a) To find the velocity we find the derivative

$$v(t) = s'(t) = \frac{d}{dt}(20 + 6 \sin t)$$
$$= 6 \cos t.$$

(b) Find $v(\pi)$.

(b) The velocity at time $t = \pi$ is

$$v(\pi) = 6 \cos \pi = 6(-1) = -6 \text{ cm/s}.$$

(c) Set $v(t) = 0$ and solve.

(c) The block changes direction at times t for which $v(t) = 0$. The equation

$$v(t) = 6 \cos t = 0$$

has solutions $t = \pi/2 + n\pi, \quad n = 0, 1, 2, \ldots$. ∎

Exercise Set 3.4

In Exercises 1–20 find $f'(x)$.

1. $f(x) = 4 \cos x$

2. $f(x) = x \sin x$

3. $f(x) = x^3 \tan x$

4. $f(x) = \sin x \cos x$

5. $f(x) = (x^3 - 2) \cot x$

6. $f(x) = \cot x \csc x$

7. $f(x) = \sin x \sec x$

8. $f(x) = \sin^2 x$

9. $f(x) = x \cos x - x \sin x$

10. $f(x) = \cos x(x - \cot x)$

11. $f(x) = \sec x \tan x$

12. $f(x) = \csc^2 x \cot x$

13. $f(x) = \dfrac{x}{2 + \sin x}$

14. $f(x) = \dfrac{\tan x}{1 + x^2}$

15. $f(x) = \dfrac{\sin x - \cos x}{1 + \tan x}$

16. $f(x) = \dfrac{1 - \sin x}{1 + \sin x}$

17. $f(x) = \dfrac{x^2 + 4 \cot x}{x + \tan x}$

18. $f(x) = \dfrac{x^2 + 4}{2 + \sec x}$

19. $f(x) = \dfrac{3 \csc x}{4x^2 - 5 \tan x}$

20. $f(x) = x \csc x - \dfrac{x}{\cot x}$

21. Find an equation for the line tangent to the graph of $y = x \sin x$ at the point $(\pi, 0)$. *first find derivative then slope at π*

22. Find an equation for the line tangent to the graph of $f(x) = \csc x \cot x$ at the point $(\pi/4, \sqrt{2})$.

23. For which number(s) x in the interval $[0, 4\pi]$ is the tangent to the graph of $y = \sec x$ horizontal? *where is slope = 0*

24. Use the identity $\sin 2x = 2 \sin x \cos x$ to find the derivative of the function $f(x) = \sin 2x$.

25. A piston moves up and down in an engine block so that at time t it is $s(t) = 5 - 4 \sin t$ cm from the top of its cylinder. Find the times t at which the piston changes direction.

26. Use the Product Rule to show that $\dfrac{d}{dx} \tan^2 x = \dfrac{d}{dx} \sec^2 x$.

27. Find the number(s) x in the interval $[0, 2\pi]$ for which the tangents to the graphs of $y = \sin x$ and $y = \cos x$ are
a. parallel
b. perpendicular.

28. Find all numbers x in the interval $[0, 2\pi]$ for which the graph of $y = f(x)$ has a horizontal tangent for
a. $f(x) = \cos x$ **b.** $f(x) = \sec x$ **c.** $f(x) = \csc x$

29. Derive equations (6), (7), and (8).

30. Determine where the trigonometric functions $\cot x$, $\sec x$, and $\csc x$ are differentiable on the interval $[0, 2\pi]$.

31. A mass at the end of a vibrating spring is located $s(t) = 30 + 10 \cos t$ cm from the fixed end of the spring t seconds after it is set in motion. If the motion is horizontal, find
a. the velocity at time t.
b. the times at which the mass changes direction.

3.5 The Chain Rule

In this section, we discuss the differentiation of composite functions—functions of the form

$$y = (f \circ g)(x) = f(g(x)).$$

For example, air pressure is a function of altitude. As an airplane takes off, its altitude is a function of time, and since altitude determines air pressure, the air pressure outside of the airplane varies over time. It is reasonable to think that the rate at which the air pressure is decreasing is determined by the rate that the plane is rising and by the rate that air pressure decreases as a function of altitude. The Chain Rule gives the precise relationship among these three rates.

The Power Rule for Functions

Before discussing the Chain Rule in its most general form, we consider the special case of a power g^n of a function g.

If g is a differentiable function of x, the Product Rule may be applied to determine the derivative of its square:

$$(g^2)'(x) = (g \cdot g)'(x) = g'(x)g(x) + g(x)g'(x) = 2g(x)g'(x).$$

Similarly, for integers $n \geq 3$, repeated application of the Product Rule shows that

$$(g^n)'(x) = ng^{n-1}(x)g'(x). \tag{1}$$

Equation (1) is called the **Power Rule for Functions.** Using Leibniz notation where $u = g(x)$ and $y = u^n = g^n(x)$, equation (1) is

$$\frac{dy}{dx} = nu^{n-1}\frac{du}{dx}. \tag{2}$$

In other words, to find the derivative of the function $y = g^n(x) = [g(x)]^n$, let $u = g(x)$ and differentiate the function $y = u^n$ as if y were a function of the independent variable u. Then multiply the result by the derivative $du/dx = g'(x)$.

□ **EXAMPLE 1**

Find $h'(x)$ for $h(x) = (x^4 - 6x^2)^3$.

Solution   This function has the form $h(x) = [g(x)]^3$ with "inside" function $g(x) = x^4 - 6x^2$. Since the derivative of this inside function is

$$g'(x) = 4x^3 - 12x$$

the Power Rule (equation 1) gives

$$h'(x) = 3[g(x)]^2g'(x) = 3(x^4 - 6x^2)^2(4x^3 - 12x).$$

Of course, we could also obtain h' by first writing h as the polynomial

$$h(x) = x^{12} - 18x^{10} + 108x^8 - 216x^6$$

and then differentiating:

$$h'(x) = 12x^{11} - 180x^9 + 864x^7 - 1296x^5.$$

We leave it for you to verify that these two results are the same.   ■

In Example 1 the advantage of the Power Rule solution is that the factors of $h'(x)$ are more easily recognized than in the Polynomial Rule solution, although both methods apply. In Example 2 the Power Rule is the only reasonable method available for finding the derivative.

☐ **EXAMPLE 2**

For $y = \left(\dfrac{x - 3}{x^3 + 7}\right)^9$, find $\dfrac{dy}{dx}$.

Strategy · · · · · · · · ·
Identify the "inside" function u.

Solution

Here we define u to be the function

$$u = \frac{x - 3}{x^3 + 7}.$$

Find $\dfrac{du}{dx}$.

Then, by the Quotient Rule,

$$\frac{du}{dx} = \frac{(x^3 + 7)(1) - (x - 3)(3x^2)}{(x^3 + 7)^2}$$

$$= \frac{-2x^3 + 9x^2 + 7}{(x^3 + 7)^2}.$$

Apply the Power Rule (2) with $n = 9$.

Then, by the Power Rule for Functions,

$$\frac{dy}{dx} = 9u^8 \frac{du}{dx}$$

Substitute back for u in final answer.

$$= 9\left(\frac{x - 3}{x^3 + 7}\right)^8 \left[\frac{-2x^3 + 9x^2 + 7}{(x^3 + 7)^2}\right]$$

$$= \frac{9(-2x^3 + 9x^2 + 7)(x - 3)^8}{(x^3 + 7)^{10}}. \qquad ■$$

☐ **EXAMPLE 3**

Find $\dfrac{dy}{dx}$ for $y = \sin^2 x$.

Solution　We have $y = u^2$ for $u = \sin x$. Applying the Power Rule for Functions, we get

$$\frac{dy}{dx} = 2u\left(\frac{du}{dx}\right)$$

$$= 2 \sin x\left(\frac{d}{dx} \sin x\right)$$

$$= 2 \sin x \cos x. \qquad ■$$

The Chain Rule

Frequently, when we calculate a derivative, we consider a given function h as a *composition* of two more "basic" functions. For example, the function $h(x) = \sqrt{x^3 + 1}$ can be viewed as the composition $h = f \circ g$ where $f(x) = \sqrt{x}$ and $g(x) = x^3 + 1$ (see Figure 5.1). We think of the function h as the application of an "out-

Figure 5.1　The function $h(x) = \sqrt{x^3 + 1}$ can be viewed as a composition $f \circ g$ of the two functions $f(x) = \sqrt{x}$ and $g(x) = x^3 + 1$.

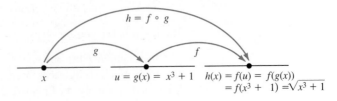

side'' function f acting on values $u = g(x)$ of an ''inside'' function g. Here is a list of functions h together with an indication of how they may be viewed as compositions $f \circ g$. Note that it is possible to view some functions as compositions in more than one way.

$f(g(x))$	$f(u)$	$u = g(x)$
$\sqrt{x^2 + 3}$	$f(u) = \sqrt{u}$	$u = x^2 + 3$
$\dfrac{1}{7 - x}$	$f(u) = \dfrac{1}{u}$	$u = 7 - x$
$\dfrac{3}{(\sqrt{x} + 2)^4}$	$f(u) = \dfrac{3}{u^4}$	$u = \sqrt{x} + 2$
$\left(\dfrac{2x - 9}{2x + 9}\right)^3$	$f(u) = u^3$	$u = \dfrac{2x - 9}{2x + 9}$
$\left(\dfrac{2x - 9}{2x + 9}\right)^3$	$f(u) = \left(\dfrac{u - 9}{u + 9}\right)^3$	$u = 2x$

Since the power function g^n is a special case of a composite function, we would expect the derivative of a composite function $f \circ g$ to be calculated in a manner consistent with the Power Rule for Functions. This is indeed the case, and the result is called the **Chain Rule.**

Theorem 8

Chain Rule

If the function g is differentiable at x and the function f is differentiable at $u = g(x)$, then the composite function $f \circ g$ is differentiable at x, and

$$(f \circ g)'(x) = f'(g(x))g'(x).$$

That is,

$$\frac{d}{dx}f(g(x)) = f'(g(x))g'(x).$$

In Leibniz notation the rule is: If $y = f(u)$ and $u = g(x)$, then

$$\frac{dy}{dx} = \frac{dy}{du}\frac{du}{dx}.$$

In other words, the rate of change $(f \circ g)'(x)$ is the product of two rates, $f'(g(x))$ and $g'(x)$. For instance, if an airplane is climbing at the constant rate of 0.5 km/min and air pressure near the earth's surface is assumed to decrease at the constant rate of 1.6 lb/in.^{2} per kilometer, then the air pressure outside of the plane is dropping at a rate of 0.8 lb/in.^{2} per minute (the product of the two intermediate rates). The Chain Rule indicates that the same is true for any composition as long as one multiplies the appropriate rates. To differentiate the composition $f \circ g$, one first differentiates the ''outside'' function f as a function of $u = g(x)$. Then one multiplies the result by the derivative $du/dx = g'(x)$. We use the Chain Rule in a few examples before we give a more rigorous justification.

☐ **EXAMPLE 4**

For $y = \dfrac{1}{(6x^3 - x)^4}$, find $\dfrac{dy}{dx}$.

Solution Here we let u be the "inside" function $u = 6x^3 - x$. Then

$$y = \frac{1}{(6x^3 - x)^4} = \frac{1}{u^4} = u^{-4},$$

so

$$\frac{dy}{du} = -4u^{-5} \quad \text{and} \quad \frac{du}{dx} = 18x^2 - 1.$$

Then, by the Chain Rule

$$\begin{aligned}
\frac{dy}{dx} &= \frac{dy}{du}\frac{du}{dx} \\
&= -4u^{-5}(18x^2 - 1) \\
&= -4(6x^3 - x)^{-5}(18x^2 - 1) \\
&= \frac{-4(18x^2 - 1)}{(6x^3 - x)^5}.
\end{aligned}$$ ∎

☐ **EXAMPLE 5**

Find $\dfrac{dy}{dx}$ for $y = \sin(6x^2 - x)$.

Solution We can view this function as a composition where the outside function is

$$y = \sin u, \quad \text{with} \quad \frac{dy}{du} = \cos u.$$

The inside function is

$$u = 6x^2 - x, \quad \text{with} \quad \frac{du}{dx} = 12x - 1.$$

The Chain Rule therefore gives

$$\frac{dy}{dx} = \frac{dy}{du}\frac{du}{dx} = (\cos u)(12x - 1) = (12x - 1)\cos(6x^2 - x).$$ ∎

☐ **EXAMPLE 6**

Find $\dfrac{dy}{dx}$ for $y = \dfrac{1}{1 + \sin^2 x}$.

Solution We may regard this as a composition of the form

$$y = \frac{1}{u} = u^{-1} \quad \text{with} \quad u = 1 + \sin^2 x.$$

We know $dy/du = -u^{-2} = -1/u^2$, and we can compute du/dx using the Power Rule for Functions (see Example 3). We obtain

$$\frac{du}{dx} = \frac{d}{dx}(1 + \sin^2 x) = 2 \sin x \cos x.$$

By the Chain rule, we conclude

$$\frac{dy}{dx} = \frac{-1}{u^2}(2 \sin x \cos x)$$

$$= \frac{-2 \sin x \cos x}{(1 + \sin^2 x)^2}.$$ ∎

☐ **EXAMPLE 7**

Find $\dfrac{dy}{dx}$ for $y = \tan^2(4x^3 - 1)$.

Solution   To differentiate this function, we must use the Chain Rule twice. In other words, we let $y = u^2$ where $u = \tan(4x^3 - 1)$. The Power Rule yields

$$\frac{dy}{dx} = 2u\frac{du}{dx}.$$

However, to compute du/dx, we must view $u = \tan(4x^3 - 1)$ as the composition $\tan v$ where $v = 4x^3 - 1$. Applying the Chain Rule a second time yields

$$\frac{du}{dx} = (\sec^2 v)\left(\frac{dv}{dx}\right) = (\sec^2 v)(12x^2).$$

Translating u and v back in terms of the variable x gives

$$\frac{dy}{dx} = 2[\tan(4x^3 - 1)][\sec^2(4x^3 - 1)](12x^2).$$ ∎

Justifying the Chain Rule

The following argument justifies the Chain Rule for many choices of intermediate functions g. We begin with the definition of the derivative of $f \circ g$:

$$(f \circ g)'(x) = \lim_{h \to 0} \frac{f(g(x + h)) - f(g(x))}{h}$$

and use some elementary algebra to obtain the statement

$$(f \circ g)'(x) = \lim_{h \to 0} \left[\frac{f(g(x + h)) - f(g(x))}{g(x + h) - g(x)} \frac{g(x + h) - g(x)}{h} \right]. \tag{3}$$

We next let $u = g(x)$ and $v = g(x + h) - g(x)$. Then $u + v = g(x + h)$, and we can write equation (3) as

$$(f \circ g)'(x) = \lim_{h \to 0} \left[\frac{f(u + v) - f(u)}{v} \frac{g(x + h) - g(x)}{h} \right]. \tag{4}$$

Since g is differentiable at x, it is continuous at x by Theorem 2. Thus, by Theorem 9, Chapter 2

$$\lim_{h \to 0} [g(x + h) - g(x)] = \lim_{h \to 0} v = 0.$$

That is, $v \to 0$ as $h \to 0$. Using this observation and the algebra of limits we have from equation (4) that

$$(f \circ g)'(x) = \left[\lim_{v \to 0} \frac{f(u + v) - f(u)}{v}\right]\left[\lim_{h \to 0} \frac{g(x + h) - g(x)}{h}\right]$$

$$= f'(u)g'(x)$$

$$= f'(g(x))g'(x)$$

as desired. This argument is not completely general, however, since closer attention needs to be paid to the question of whether the denominator v in equation (4) might equal zero for some $h \neq 0$. A complete proof of the Chain Rule is given in Appendix II.

Exercise Set 3.5

In Exercises 1–10 form the composite function $y = f(g(x))$ with $u = g(x)$. Then find

$$\frac{dy}{dx} = \frac{dy}{du}\frac{du}{dx}$$

using the Chain Rule.

1. $f(u) = u^3 + 1$
$g(x) = 1 - x^2$

2. $f(u) = 1/(1 + u)$
$g(x) = 3x^2 - 7$

3. $f(u) = u(1 - u^2)$
$g(x) = 1/x$

4. $f(u) = \dfrac{u + 1}{u - 1}$
$g(x) = \sin x$

5. $f(u) = \tan u$
$g(x) = 1 + x^4$

6. $f(u) = 3u^2 + 5$
$g(x) = \cos x$

7. $f(u) = 1/(1 + u)$
$g(x) = \sin^2 x$

8. $f(u) = u^2 - 1$
$g(x) = 1 + \sec x$

9. $f(u) = \cos u$
$g(x) = \sin x$

10. $f(u) = u^3 - 3u + 1/u$
$g(x) = \tan x$

In Exercises 11–18, determine the function $f(u)$ such that $y = f(u)$. Then find

$$\frac{dy}{dx} = \frac{dy}{du}\frac{du}{dx}$$

using the Chain Rule.

11. $y = \sin(x^2 + x)$
$u = x^2 + x$

12. $y = \tan\left(\dfrac{1 + x}{1 - x}\right)$
$u = \dfrac{1 + x}{1 - x}$

13. $y = (1 + \sin x)/(1 - \sin x)$
$u = \sin x$

14. $y = \sec x/(1 + \sec^2 x)$
$u = \sec x$

15. $y = \left(\dfrac{3x + 7}{3x - 7}\right)^4$
$u = \left(\dfrac{3x + 7}{3x - 7}\right)$

16. $y = (1 + 3\cos x)/(1 - 3\cos x)$
$u = \cos x$

17. $y = \left(\dfrac{3x + 7}{3x - 7}\right)^4$
$u = 3x$

18. $y = (1 + 3\cos x)/(1 - 3\cos x)$
$u = 3\cos x$

In Exercises 19–49, find dy/dx using the Chain Rule.

19. $y = (x^2 + 4)^3$

20. $y = (2 - 3x)^4$

21. $y = (\cos x - x)^6$

22. $y = (x^2 + 8x + 6)^5$

23. $y = \sin(x^3 + 3x)$

24. $y = \cos(\pi - x^2)$

25. $y = \tan(x^2 + x)$

26. $y = \sec(x^2 + 4)$

27. $y = x(x^4 - 5)^3$

28. $y = (x^2 - 7)^3(5 - x)^2$

29. $y = \dfrac{1}{(x^2 - 9)^3}$

30. $y = \dfrac{x + 3}{(x^2 - 6x + 2)^2}$

31. $y = (3\tan x - 2)^4$

32. $y = (x^6 - x^2 + 2)^{-4}$

33. $y = \left(\dfrac{x - 3}{x + 3}\right)^4$

34. $y = \left(\dfrac{1 + \sin x}{1 - \sin x}\right)^{-6}$

35. $y = x\cos(1 - x^2)$

36. $y = (x\cos x - \sin x)^5$

37. $y = \dfrac{\tan^2 x + 1}{1 - x}$

38. $y = \dfrac{x + 2}{1 + \sec(1 + x^2)}$

39. $y = \dfrac{\tan^2(\pi/4) + 1}{\sec(0)}$

40. $y = \dfrac{\sin^2(\pi/6) + 1}{\sin^2(\pi/6) - 1}$

41. $y = \dfrac{x}{(x^2 + x + 1)^6}$

42. $y = \dfrac{2 + \csc(1 + x^2)}{1 - \csc(1 + x^2)}$

43. $y = \left(\dfrac{ax + b}{cx + d}\right)^4$

44. $y = \left(\dfrac{a - bx}{c - dx}\right)^5$

45. $y = \dfrac{1}{1 + \cos^3 x}$

46. $y = \dfrac{x - \sin \pi x}{4 + \cos \pi x}$

47. $y = \dfrac{1}{(1 + x^4)^3}$

48. $y = x \cot\left(\dfrac{1 + x}{1 - x}\right)$

49. $y = \tan(6x) - 6 \tan x$

50. Apply the Chain Rule twice to show that the derivative of the composite function $y = f(g(h(x)))$ is

$$\frac{dy}{dx} = f'(g(h(x))) \cdot g'(h(x)) \cdot h'(x).$$

In Exercises 51–63 find dy/dx using the result of Exercise 50 when it applies.

51. $y = \sec^3 4x$

52. $y = \sin^3(2x - \pi/2)$

53. $y = \tan^2(\pi - x^2)$

54. $y = \dfrac{x - \sin(\pi x)}{\tan^2(\pi x)}$

55. $y = (\sin \pi x - \cos \pi x)^4$

56. $y = \sin^4(x^3 + \pi x)$

57. $y = [1 + (x^2 - 3)^4]^6$

58. $y = \dfrac{1}{a + (bx + c)^4}$

59. $y = \cos^2\left(\dfrac{1 + x}{1 - x}\right)$

60. $y = (\tan \pi x - \cos \pi x)^6$

61. $y = 1 + [1 + (1 + x^2)^2]^2$

62. $y = \cos(\cos(x^2 + 1))$

63. $y = \tan(\sqrt{1 + x^2})$

In each of Exercises 64–69, find an equation for the line tangent to the graph of $y = f(x)$ at point P:

64. $y = (1 - x^2)^3$, $P = (1, 0)$

65. $y = \left(\dfrac{x}{x + 1}\right)^4$, $P = (0, 0)$

66. $y = \sin(\pi + x^3)$, $P = (0, 0)$

67. $y = x \cos(\pi + x^2)$, $P = (0, 0)$

68. $y = \left(\dfrac{3x^2 + 1}{x + 3}\right)^2$, $P = (-1, 4)$

69. $y = \tan\left(\dfrac{(\pi/4) - x}{1 + x}\right)$, $P = (0, 1)$

70. Let $h(x) = f(g(x))$. Find $h'(2)$ if $g(2) = 3$, $g'(2) = 7$ and $f'(3) = -2$.

71. Suppose $f(1) = 1$, $g(1) = 2$, $f'(1) = -2$, $f'(2) = 5$, and $g'(1) = 2$. Find:
a. $(f \circ g)'(1)$
b. $(g \circ f)'(1)$

72. Let $h(t) = [g^2(t) + 1]^3$. Find $h'(3)$ if $g(3) = -2$ and $g'(3) = -1$.

73. Let $h(t) = \sin(g(t))$. Find $h'(2)$ if $g(2) = 3\pi/4$ and $g'(2) = \sqrt{2}$.

74. Let $h(x) = \tan^2(g(x))$. Find $h'(\pi/4)$ if $g(\pi/4) = \pi/6$ and $g'(\pi/4) = \pi$.

75. Find an equation for the line tangent to the graph of the function $f(x) = \sin(x^2)$ at the point $(\sqrt{\pi}/2, \sqrt{2}/2)$.

76. Find an equation for the line tangent to the graph of $f(x) = \csc(\pi x)$ at the point $(\frac{1}{2}, 1)$.

77. A block oscillates at the end of a spring so that at time t (in seconds) it is $s(t) = 20 + 5 \sin(4t)$ cm from the fixed end of the spring. If the motion is along a line, find the velocity $v(t)$ of the block at time t.

78. A particle moves along a line so that at time t (in seconds) it is located at position $s(t) = [t/(t + 1)]^2 + 10 \sin(\pi t)$. If distance is measured in meters, find
a. the velocity of the particle at time t.
b. the velocity $v(0)$ at time 0.

79. Suppose that a spherical balloon is being inflated so that its radius is increasing at a rate of 0.5 in./s. How fast is its volume increasing when its radius is 5 in?

80. Suppose that a coin is tossed into a wishing well and it forms a circular wave whose radius is increasing at the rate of 5 in./s. How fast is the area enclosed by the wave increasing three seconds after the coin hits the water?

81. Let $f(x) = |x|$.
a. Justify the formula $f(x) = \sqrt{x^2}$.
b. Use the Chain Rule to calculate $f'(x)$ for $x \neq 0$.
c. Explain why the Chain Rule does not apply when $x = 0$.

82. In Chapters 1 and 2, we used radian measure when we defined the trigonometric functions, and we derived

$$\lim_{\theta \to 0} \frac{\sin \theta}{\theta} = 1.$$

Moreover, in Section 3.4, this limit was fundamental to our computation of

$$\frac{d}{dx} \sin x = \cos x$$

(see Theorem 7). However, if we decide to measure angles using degrees, we obtain a modified sine function $\text{sindeg}(x)$ which is related to $\sin(x)$ by the formula

$$\text{sindeg}(x) = \sin(\pi x/180).$$

For example, $\text{sindeg}(90) = \sin(\pi/2) = 1$.
a. Calculate the derivative of $\text{sindeg}(x)$.
b. Derive the formula that relates the cosine (in degrees) $\text{cosdeg}(x)$ to $\cos(x)$.
c. Express your answer to part a in terms of $\text{cosdeg}(x)$.
d. Why do we always use radians to measure angles in calculus?

83. Verify equation (1) in the following way, using mathematical induction.

a. For $n = 1$, $(g^n)'(x) = ng^{n-1}(x)g'(x)$ since

$$g'(x) = 1[g(x)]^0 \cdot g'(x).$$

b. Assume that equation (1) holds for the integer $n > 1$.

c. Write $[g^{n+1}]'(x) = (g \cdot g^n)'(x)$ and use the Product Rule to verify that

$$(g^{n+1})'(x) = (n + 1)g^n(x) \cdot g'(x).$$

d. Conclude, from the principle of mathematical induction, that equation (1) holds for all integers $n \geq 1$.

 84. For the pendulum given in Figure 5.2, suppose that the angle $\theta(t)$ varies according to the equation

$$\theta(t) = -\tfrac{1}{10} \sin(\tfrac{5}{2}t) + \tfrac{1}{2}\cos(\tfrac{5}{2}t).$$

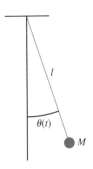

Figure 5.2

a. Using the Bisection Method (Exercises 73–76 in Section 2.6), find the first four positive times when the pendulum is entirely vertical, that is, when $\theta(t) = 0$.

b. Using the graph of $\theta(t)$, determine the first two positive times t when the angle $|\theta(t)|$ is largest.

c. Using the graph of $\theta'(t)$, find the first two times t when $\theta'(t) = 0$. What happens to $\theta(t)$ at these times? Where is the pendulum located?

 85. Suppose that

$$f(x) = \begin{cases} x^2 \sin \dfrac{1}{x}, & \text{if } x \neq 0; \\ 0, & \text{if } x = 0. \end{cases}$$

a. Graph the quotient

$$\frac{f(x)}{x} = x \sin \frac{1}{x}.$$

Repeatedly magnify this graph about the origin. Using these plots, explain why f is differentiable at 0.

b. Calculate $f'(x)$ and graph $y = f'(x)$. Repeatedly magnify the graph about the origin. What do you conclude about $f'(0)$?

c. Evaluate $f'(x)$ at the points $x_n = 1/(n\pi)$ for $n = 1, 2, 3, \dots$. What do the values of $f'(x_n)$ say about the function f' at 0?

86. Suppose that the angle θ at time t (in seconds) of a pendulum of length l is given by the equation

$$\theta(t) = 5 \sin \omega t$$

where $\omega = \sqrt{g/l}$ ($g = 32$ ft/s^2 is the acceleration due to gravity).

a. Graph both $\theta(t)$ and $\theta'(t)$ for $l = 1, 2, 4$, and 8. How do the graphs of θ and θ' change as l increases?

b. If l doubles in length, does the pendulum swing faster or slower? Using the Chain Rule, prove your answer.

3.6 Related Rates

Often we encounter situations in which two or more related quantities are functions of time. For example, as an ice cube melts, its volume, weight, and dimensions change continuously over time, and the *rates* at which each of these quantities change are mutually related. In this section, we determine how to calculate and compare such rates using the Chain Rule.

REMARK 1   If two or more differentiable functions of time are related by a single equation, we usually obtain the relationship between their respective rates of change by differentiating both sides of that equation with respect to time t. In the process we make use of the fact that the equation $f = g$ means that f and g are two forms of the same function. We conclude therefore that the corresponding derivative functions are equal: that is, $f' = g'.$*

---

*For example, if $ax^2 + b = 3x^2 + 4$ we may conclude, by differentiating both sides, that $2ax = 6x$. But we cannot differentiate both sides of the equation $x^2 = 4$ to conclude that $2x = 0$, since the equation $x^2 = 4$ is true only for the two numbers $x = 2$ and $x = -2$. It is not an equation involving *functions*.

□ **EXAMPLE 1**

Suppose two functions f and g are related by the equation

$$f(t) = 3[g(t)]^2 + 10, \quad -\infty < t < \infty. \tag{1}$$

Find the rate at which the function f is changing at $t = 4$ if $g(4) = 2$ and $g'(4) = -5$.

Strategy · · · · · · · ·

First find the rate $f'(t)$ by differentiating both sides of equation (1). Then substitute given data for t, $g(t)$ and $g'(t)$.

Solution

Using the Chain Rule we obtain

$$f'(t) = 3 \cdot 2g(t) \cdot g'(t)$$
$$= 6g(t)g'(t).$$

Substituting $t = 4$, $g(4) = 2$ and $g'(4) = -5$ then gives the rate

$$f'(4) = 6(2)(-5) = -60. \qquad \blacksquare$$

The next example uses this same basic idea. But it involves two geometric variables (volume and radius), each of which are functions of time.

□ **EXAMPLE 2**

A balloon in the shape of a sphere is inflated so that the volume is increasing by 100 cubic centimeters per second. At what rate is the radius increasing when the radius is 9 cm?

Strategy · · · · · · · ·

Name all variables.

Find an equation relating these variables.

Differentiate both sides of the equation, using the Chain Rule where necessary.

Solve for the desired rate, $\dfrac{dr}{dt}$.

Substitute in given values of other rates and variables.

Solution

We let V = volume, r = radius, and t = time. We know that

$$V = \tfrac{4}{3}\pi r^3. \tag{2}$$

We assume that both V and r are differentiable functions of t. Then

$$\frac{dV}{dt} = 4\pi r^2 \frac{dr}{dt},$$

so

$$\frac{dr}{dt} = \frac{1}{4\pi r^2} \frac{dV}{dt}.$$

We are given that $\dfrac{dV}{dt} = 100$ cm³/s. When $r = 9$ cm,

$$\frac{dr}{dt} = \frac{1}{4\pi(9 \text{ cm})^2} \cdot 100 \text{ cm}^3/\text{s}$$

$$= \frac{25}{81\,\pi} \approx .098 \text{ cm/s} \qquad \blacksquare$$

In Example 2, there was no difficulty in determining the basic equation relating the variables. Often, however, related rate problems do not clearly indicate what the basic relationship between the variables might be. We therefore suggest the following *procedure for solving related rate problems:*

1. If the problem is geometric in nature (or can be interpreted geometrically), draw a sketch.
2. Label the important quantities as variables or constants, according to the statement of the problem.
3. From the sketch, together with known relationships among the variables (either given in the problem or known from geometry or trigonometry), write down an equation relating the relevant variables (as functions of time t).
4. Differentiate both sides of the equation with respect to time t, using the Chain Rule where necessary, to obtain an equation relating the rates.
5. Solve for the desired rate. After doing so, substitute given data into this expression to obtain the solution.

REMARK 2   In the solution to Example 2, it was important that we differentiated both sides of equation (2) *before* we used the particular value of r ($r = 9$ cm). If we had reversed these steps, we would not have obtained an equation relating dV/dt and dr/dt. A common error in problems of this kind is to substitute the given values of variables into the equation before the differentiation step. Make sure that you perform step 1 before step 5.

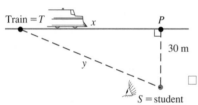

Figure 6.1 Sketch for Doppler effect experiment.

□ **EXAMPLE 3**

A physics student is standing 30 m from a straight section of railroad track in order to perform an experiment on the Doppler effect. A train is approaching, moving along the track at 90 km/h. How fast is the distance between the train and the student decreasing when the train is 50 m from the student?

Strategy · · · · · · · ·

Draw a sketch.

Solution

The situation is sketched in Figure 6.1. We let T denote the location of the train, S the location of the student, and P the point on the tracks nearest the student. Also, we let

$$x = \text{distance from } T \text{ to } P \quad \text{and} \quad y = \text{distance from } T \text{ to } S.$$

The problem is, therefore, to find dy/dt when $y = 50$ m.

Use Pythagorean Theorem to find an equation in x and y.

Since T, P, and S are vertices of a right triangle, we have the equation

$$y^2 = x^2 + 30^2. \tag{3}$$

Differentiate both sides of (3) with respect to time, using the Chain Rule.

Assuming both x and y to be differentiable functions of t, we differentiate both sides of (3) to obtain

$$2y\frac{dy}{dt} = 2x\frac{dx}{dt},$$

so

Solve for $\dfrac{dy}{dt}$.

$$\frac{dy}{dt} = \frac{x}{y}\frac{dx}{dt}. \tag{4}$$

Find x when $y = 50$.

Now when $y = 50$ m, we obtain from equation (3)

$$x = \sqrt{50^2 - 30^2} = \sqrt{40^2} = 40 \text{ m.}$$

Insert given values for all variables and rates.

We are given $dx/dt = -90$ km/h, so when $y = 50$ m, we have from equation (4)

$$\frac{dy}{dt} = \left(\frac{40 \text{ m}}{50 \text{ m}}\right)(-90 \text{ km/h})$$

$$= -72 \text{ km/h.} \qquad \blacksquare$$

□ **EXAMPLE 4**

The top portion of a coffee maker has the shape of a cone 10 cm high. The radius at the top is 4 cm. Coffee is flowing from the top section into the bottom section at a rate of 4 cm³/s. At what rate is the level of coffee in the top section falling when the coffee in the top section is 4 cm deep?

Strategy · · · · · · · · ·
Draw a sketch.

Solution
The coffee maker is sketched in Figure 6.2. An idealized sketch of the top portion appears in Figure 6.3.

Figure 6.2 Coffee maker with the top in the shape of a cone.

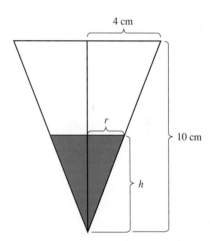

Figure 6.3 Top section of coffee maker.

Label variables.

We let h denote the depth of the coffee in the top section and we let r denote the radius of its surface.

Use equation for volume of a cone.

The formula for the volume of coffee in the upper cone is therefore

$$V = \tfrac{1}{3}\pi r^2 h. \qquad (5)$$

No data given on dr/dt, so r must be eliminated from (5) before differentiation.

Now if we proceed directly to differentiate both sides of equation (5) with respect to t, we will obtain an equation involving dV/dt, dr/dt, and dh/dt, since each of the variables, V, h, and r is a function of time. However, we have no given information on dr/dt. We therefore attempt to find an equation expressing r in terms of V and/or h so that r can be eliminated from equation (5). To do so we observe from Figure 6.3 that, by similar triangles,

Use similar triangles to find equation relating r and h.

$$\frac{r}{h} = \frac{4}{10},$$

so

$$r = \frac{2}{5}h \tag{6}$$

Insert expression for r in (5).

is the desired equation. Substituting this expression for r in (5) gives

$$V = \frac{1}{3}\pi\left(\frac{2}{5}h\right)^2 h = \frac{4\pi}{75}h^3.$$

Differentiating both sides of this equation with respect to t, we obtain

Differentiate both sides.

$$\frac{dV}{dt} = \frac{4\pi}{25}h^2\frac{dh}{dt},$$

so

Solve for $\dfrac{dh}{dt}$.

$$\frac{dh}{dt} = \frac{25}{4\pi h^2}\frac{dV}{dt}.$$

Insert given information. (Note that the sign of dV/dt is negative.)

Substituting the given information $dV/dt = -4\text{ cm}^3/\text{s}$ and $h = 4$ cm then gives

$$\frac{dh}{dt} = \frac{25}{4\pi(4\text{ cm})^2}(-4\text{ cm}^3/\text{s})$$

$$= -\frac{25}{16\pi} \approx -0.497\text{ cm/s.} \qquad \blacksquare$$

REMARK 3 Equation (6) is an example of an **auxiliary** equation. We use it to eliminate one of the variables in the principal equation (5). The strategy in Example (4) was to use the auxiliary equation to eliminate the variable r since no information about r was given. In general, you should look for auxiliary equations in problems of this type when you encounter more variables than can be evaluated with the given data.

□ **EXAMPLE 5**

An airplane is flying a level course due east at a speed of 3 km/min. A second airplane is flying a level course due south at a speed of 2 km/min at an altitude 4 km below that of the first plane. At one instant the second airplane is directly beneath the first. At what rate is the distance between the two airplanes increasing 1 min later?

Solution The situation is sketched in Figure 6.4. In Figure 6.5, points P_1 and P_2 represent the first and second airplanes, respectively, and O_1 and O_2 are the locations of the planes when the first is directly above the second. We name variables as follows:

$$x = \text{distance between } O_1 \text{ and } P_1$$
$$y = \text{distance between } O_2 \text{ and } P_2$$
$$w = \text{distance between } O_2 \text{ and } P_1$$
$$s = \text{distance between } P_1 \text{ and } P_2.$$

The rate to be calculated is ds/dt. Since s is the length of the hypotenuse of the right triangle $P_1O_2P_2$, we have the equation

$$s^2 = w^2 + y^2. \tag{7}$$

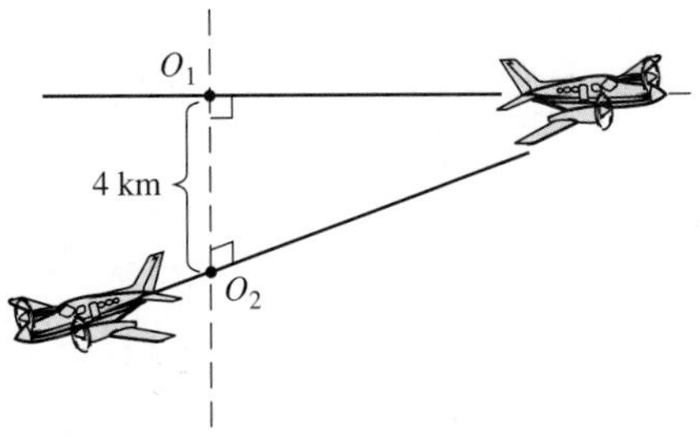

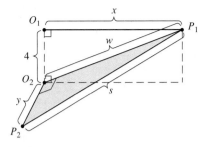

Figure 6.4 Two airplanes.

Figure 6.5 P_1 and P_2 denote the two airplanes.

We are given information about dy/dt, the speed of the second airplane, but we do not have direct information about the variable w. We must therefore find an auxiliary equation relating w to one or more of the other variables. From Figure 6.5 we can see that w is the length of the hypotenuse of triangle $P_1 O_1 O_2$, so we obtain the auxiliary equation

$$w^2 = x^2 + 4^2.$$

Substituting this expression for w^2 into the principal equation (7), we obtain the equation

$$s^2 = x^2 + y^2 + 16.$$

We now differentiate with respect to t. This gives

$$2s\frac{ds}{dt} = 2x\frac{dx}{dt} + 2y\frac{dy}{dt} \qquad \text{so}$$

$$\frac{ds}{dt} = \frac{1}{s}\left(x\frac{dx}{dt} + y\frac{dy}{dt}\right). \tag{8}$$

We are given

$$\frac{dx}{dt} = 3 \text{ km/min} \qquad \text{and} \qquad \frac{dy}{dt} = 2 \text{ km/min}.$$

Therefore, one minute after crossing, we will have $x = 3$ km and $y = 2$ km and $s = \sqrt{3^2 + 2^2 + 16} = \sqrt{29}$ km. Substituting these values into equation (8) gives

$$\frac{ds}{dt} = \left(\frac{1}{\sqrt{29} \text{ km}}\right)[(3 \text{ km})(3 \text{ km/min}) + (2 \text{ km})(2 \text{ km/min})]$$

$$= \frac{13}{\sqrt{29}} \approx 2.414 \text{ km/min.} \qquad \blacksquare$$

Exercise Set 3.6

In Exercises 1–5, find the rate $f'(t)$ for the given values of g, g', and t.

1. $f(t) = 2[g(t)]^3 + 5$, $\quad t = 1$, $\quad g(1) = 3$, $\quad g'(1) = -2$

2. $f(t) = \sqrt{2 + g(t)}$, $\quad t = 0$, $\quad g(0) = 3$, $\quad g'(0) = 4$

3. $f(t) = \dfrac{1}{1 + g(t)}$, $\quad t = 2$, $\quad g(2) = 3$, $\quad g'(2) = -2$

4. $[f(t)]^2 + [g(t)]^3 = 265$, $\quad t = 1$, $\quad g(1) = 6$, $\quad g'(1) = -2$, $\quad f(1) = 7$

5. $\sin(f(t)) = [g(t)]^2$, $\quad t = 0$, $\quad g(0) = 1$, $\quad g'(0) = -2$, $\quad f(0) = \pi/6$

6. A spherical balloon is being inflated so that the radius is increasing at a rate of 3 cm/s. Find the rate at which the volume is increasing when $r = 10$ cm.

7. At what rate is the diagonal of a cube increasing if the edges are increasing at a rate of 2 cm/s?

8. When a pebble is tossed into a still pond, ripples move out from the point where the stone hits in the form of concentric circles. Find the rate at which the area of the disturbed water is increasing when the radius of the outermost circle equals 10 m if this radius is increasing at a rate of 2 m/s.

9. A snowball, in the shape of a sphere, is melting so that the radius is decreasing at a uniform rate of 1 cm/s. How fast is the volume decreasing when the radius equals 6 cm?

10. A point moves along the graph of $y = x^{5/2}$ so that its x-coordinate increases at the constant rate of 2 units per second. Find the rate at which its y-coordinate is increasing as it passes the point $(4, 32)$.

11. A conical water tank with vertex down has a radius of 20 m and a depth of 20 m. Water is being pumped into the tank at the rate of 40 m³/min. How fast is the level of the water rising when the water is 8 m deep?

12. A radio transmitter is located 3 km from a fairly straight section of interstate highway. A truck is travelling away from the transmitter along the highway at a speed of 80 km/h. How fast is the distance between the truck and the transmitter increasing when they are 5 km apart?

13. The area of a rectangle, whose length is twice its width, is increasing at the rate of 8 cm²/s. Find the rate at which the length is increasing when the width is 5 cm.

14. A ladder 5 m long is leaning against a wall. The base of the ladder is sliding away from the wall at a rate of 1 m/s. How fast is the top of the ladder sliding down the wall at the instant when the base is 3 m from the wall?

15. A boat is being pulled to shore by a rope attached to a windlass atop a pier. The height of the windlass above the water is 6 m, and the rope is being wound in at the rate of 5 m/min. How fast is the boat approaching the shore when it is 8 m away?

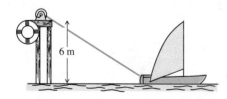

16. The law of cosines, relating the lengths of the three sides of a triangle, is

$$C^2 = A^2 + B^2 - 2AB \cos \theta$$

where θ is the angle opposite side C. Sides $A = 1$ and $B = 2$ are of fixed length, and the angle θ is increasing at the rate of 0.2 radians/minute. Find the rate at which C is increasing at the instant when $C = \sqrt{3}$.

17. A woman 1.6 m tall is walking away from a lamp post 10 m tall. If the woman is walking at a speed of 1.2 m/s, how fast is her shadow increasing when she is 15 m from the lamp post?

18. The lengths of the two equal sides in an isosceles triangle are increasing at the rate of 2 cm/s. The base of the triangle has fixed length 18 cm. Find the rate at which the area of the triangle is increasing when its height equals 12 cm.

19. Following the outbreak of an epidemic, a population of $N(t)$ people can be regarded as being made up of immunes, $I(t)$, and susceptibles, $S(t)$, that is,

$$N(t) = I(t) + S(t).$$

$I(t)$ includes both those who have contracted the disease and those who cannot contract the disease. If the rate of decrease of susceptibles is 20 persons per day, and the rate of increase of immunes is 24 persons per day, how fast is the population growing?

20. Gravel is being poured onto a pile at the rate of 180 m³/min. The pile is in the shape of an inverted cone whose diameter is always three times its height. Find the rate at which the diameter of the base is increasing when the pile is 6 m high.

21. At noon a truck leaves Columbus, Ohio, driving east at a speed of 40 km/hr. An hour later a second truck leaves Columbus driving north at a speed of 60 km/h. At what rate is the distance between the two trucks increasing at 2:00 p.m. that day?

22. A water trough has end pieces in the shape of inverted isosceles triangles with bases 60 cm and heights 40 cm. The trough is 4 m long. Water is being pumped into the trough at a rate of 9000 cm³/s. How fast is the level of the water in the trough rising when the water is 10 cm deep?

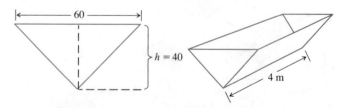

23. Work Exercise 22 if the ends of the trough have the shape of equilateral triangles whose sides have length 60 cm.

24. An observer stands 150 m from a fireworks display rocket which is fired directly upward. When the rocket reaches a height of 200 m, it is travelling at a speed of 12 m/s. At what rate is the angle of elevation formed with the observer increasing at that instant?

25. A point moves along the unit circle $x^2 + y^2 = 1$. Find the points at which the x- and y-coordinates are
   a. changing at the same rates,
   b. changing at opposite rates.

26. For a body moving through air, **drag** is defined as the force opposing the motion of the body. For such a body, the drag D is jointly proportional to the squares of its velocity V and its surface area S, that is,

$$D = kV^2S^2.$$

Find the rate at which drag is increasing if the body is undergoing an acceleration of 8 m/s^2 at the instant when the velocity equals 30 m/s.

27. On a particular autumn day, the sun moves across the sky at the rate of 15°/h. How fast is the shadow cast by a building 30 m high increasing at the moment when the evening sun is 30° above the horizon?

28. For a gas at temperature T confined within a container of volume V at a pressure P, the gas law states that the ratio PV/T is a constant, that is, $PV = kT$. A 1000-cm^3 tank of oxygen is being heated so that the temperature of the oxygen is increasing at a rate of 2°C/min. Find the rate at which the pressure inside the tank is increasing.

29. How fast is the level of the coffee in the bottom of the coffee pot in Example 4 rising at the instant when the coffee in the bottom of the pot is 4 cm deep (assume that the relevant portion of the bottom forms part of a cone whose height is 10 cm and whose radius is 4 cm)?

30. The gravitational force of attraction between two bodies of mass m_1 and m_2 is given by Newton's Law of Gravitation as

$$F = G\frac{m_1m_2}{r^2}$$

where G is the gravitational constant and r is the distance between the two bodies. Find the rate at which the distance between the bodies is changing if the force is changing at the rate α.

31. A lighthouse lies 200 m from a straight shore. The light rotates at a rate of 2 rev/min. A piling marks the spot on the beach nearest the lighthouse. Find the speed at which the light beam is moving along the shore at a point 200 m from the piling.

32. An *idealized heat engine*, designed to provide a model for studying the question of maximum possible efficiency, was proposed by the French engineer Sadi Carnot in 1824. For the Carnot engine the efficiency is defined by the equation

$$E = 1 - \frac{T_c}{T_h}$$

where T_h is the intake temperature and T_c is the exhaust temperature in degrees Kelvin.

a. If T_c is increasing by 3 K/min and the efficiency is not changing, what is the rate of change of T_h?

b. If the efficiency is decreasing by 2%/h and T_c is fixed, what is the rate of change of T_h?

33. If a ray of monochromatic light travelling in a vacuum makes an angle of incidence α with the normal to a surface of a substance a and an angle of refraction β in the substance, then **Snell's Law** states that

$$\frac{\sin \alpha}{\sin \beta} = C_a$$

where C_a is a constant, called the index of refraction for the substance a. For water, the index of refraction is $C_w = 1.33$. If the angle of incidence of a light ray striking water is decreasing at the rate of 0.2 radians per second, find an expression for the rate at which the angle of refraction is decreasing.

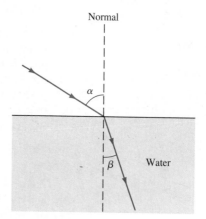

Normal

Water

34. A submarine running due north at a depth of 1 km and a speed of 60 km/h passes directly beneath a ship sailing due west at 30 km/h. At what rate is the distance between the two vessels increasing 20 min later?

35. The velocity of viscous fluid flowing through a circular tube is not the same at all points of a cross section. Provided the velocity is not too great, the flow has maximum velocity at the center and decreases to zero at the walls. For a point at a radial distance r from the center of the tube, the velocity of the flow is given by

$$v = \frac{\alpha}{L}(R^2 - r^2)$$

where R is the radius of the tube, L is its length, and α is a constant. Find the acceleration of the fluid moving at the center of a tube if

a. $L = 25$ cm is fixed and R is increasing at a rate of 0.2 cm/min at the instant when $R = 10$ cm.

b. $R = 10$ cm is fixed and L is increasing at a rate of 0.5 cm/s at the instant when $L = 25$ cm.

3.7 Linear Approximation

Using the derivative to approximate a function is one of the most fundamental ideas in differential calculus. We approximate a differentiable function by a linear function—one whose graph is a line. Although this approach is elementary, it is the starting point for a progression of successively more accurate approximation schemes, as we shall see in Chapter 13. In this section, we introduce linear approximation and the associated error. Analyzing the error in more detail is the topic of Section 4.8.

Before we start our formal discussion, we illustrate the basic idea with a hypothetical example. Suppose that a jet plane flies overhead at a speed of 480 miles per hour (8 miles per minute). How far would you expect the plane to fly in the next 3 minutes? Unless you have some additional information to consider, you probably would predict that the plane would fly at constant speed and, therefore, travel 24 miles. This approximation is linear. You have used the current value of the derivative (the speed of the jet as it passes overhead) to estimate the value of the function (the position of the jet). We now formalize this idea.

Increment Notation

In this section we often use the increment notation Δx to denote a *fixed* nonzero number that is added to a given number x_0 to produce a second number $x = x_0 + \Delta x$. The reason for using increment notation, rather than writing $x = x_0 + h$ as done previously, is that we are *not only* interested here in studying a limit as $h \to 0$. Rather, we also want to approximate the function value $f(x_0 + \Delta x)$ where both x_0 and Δx are fixed numbers. Writing $x = x_0 + \Delta x$ emphasizes the role of the increment Δx in the approximation.

Linear Approximations

Figure 7.1 illustrates how one linearly approximates $f(x_0 + \Delta x)$ using $f(x_0), f'(x_0)$, and Δx. The idea is simply to approximate the value $f(x_0 + \Delta x)$ by the y-coordinate of the point P that is on the *tangent* line to the graph of f at x_0 and that has x-coordinate $x_0 + \Delta x$. Since the slope of this tangent is $f'(x_0)$, the y-coordinate of the point P is $f(x_0) + f'(x_0)\Delta x$. We therefore have the **linear approximation**

$$f(x_0 + \Delta x) \approx f(x_0) + f'(x_0)\Delta x. \tag{1}$$

Of course, approximation (1) makes sense only when $f'(x_0)$ is defined.

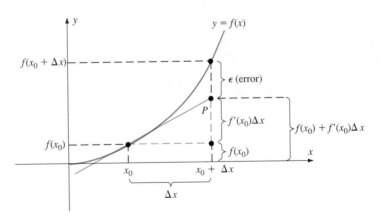

Figure 7.1 $f(x_0) + f'(x_0)\Delta x$ linearly approximates $f(x_0 + \Delta x)$.

□ **EXAMPLE 1**

Use approximation (1) to obtain an approximation for the quantity $\sqrt{36 + \Delta x}$.

Solution Here we take $x_0 = 36$ and $f(x) = \sqrt{x}$. Then $f'(x) = 1/(2\sqrt{x})$, so approximation (1) gives

$$\sqrt{36 + \Delta x} = f(36 + \Delta x) \approx f(36) + f'(36)\Delta x$$

$$= \sqrt{36} + \frac{1}{2\sqrt{36}}\Delta x$$

$$= 6 + \frac{\Delta x}{12},$$

so

$$\sqrt{36 + \Delta x} \approx 6 + \frac{\Delta x}{12}.$$

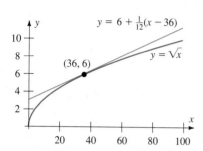

Figure 7.2 The graph of the function $f(x) = \sqrt{x}$ and the graph $y = 6 + \frac{1}{12}(x - 36)$ of its linear approximation near $x_0 = 36$.

For example, using this approach, we can approximate $\sqrt{37}$ by $6 + \frac{1}{12} \approx 6.083$, and this approximation is accurate to three decimal places. Table 7.1 displays values of this approximation for various values of Δx and compares them with the actual values of $\sqrt{36 + \Delta x}$. (See Figure 7.2.) ■

Of course, calculators and computers have essentially eliminated the "by hand" use of linear approximation as a method for estimating the values of standard quantities, such as $\sqrt{37}$. Nevertheless, performing such estimates by hand is a good way to check your understanding of the general procedure.

Absolute and Relative Errors

Table 7.1 also shows the absolute and relative errors in the approximations described in Example 1. In using the approximation

$$y_1 = f(x_0) + f'(x_0)\Delta x$$

to approximate the function value $y = f(x_0 + \Delta x)$,

(a) the **absolute error** is $|y - y_1|$;

Table 7.1 Approximations (to six decimal places) of $\sqrt{36 + \Delta x}$

(a) Δx	(b) Actual value $\sqrt{36 + \Delta x}$	(c) Approximation $6 + \dfrac{\Delta x}{12}$	(d) Absolute error $\lvert (b) - (c) \rvert$	(e) Relative error $\left\lvert \dfrac{(b) - (c)}{(b)} \right\rvert$
4.0	6.324555	6.333333	.008778	.001388
2.0	6.164414	6.166667	.002253	.000365
1.5	6.123724	6.125000	.001276	.000208
1.0	6.082762	6.083333	.000571	.000094
0.5	6.041523	6.041667	.000144	.000024
0.1	6.008328	6.008333	.000006	.000001

(b) the **relative error** is $\left|\dfrac{y - y_1}{y}\right|$;

(c) the **percentage error** is the relative error multiplied by 100%.

The absolute error is the exact amount that an approximation differs from the precise value. The relative error measures the absolute error relative to the size of the number being approximated. For example, the absolute errors in the approximations $1/3 \approx 0.3$ and $3001/3 \approx 1000.3$ are the same while the relative error in the first approximation is $0.1 = 10\%$ and the relative error in the second approximation is less than $0.000034 = 0.0034\%$. In certain applications, the relative error is more important than the absolute error.

□ **EXAMPLE 2**

In the first row of Table 7.1, we see that the linear approximation of $\sqrt{40}$ is $y_1 = 6.333333$, while the actual value (to 6 decimal places) is reported to be $y = 6.324555$. This gives an absolute error of

$$|y - y_1| = |.008778| = .008778,$$

a relative error of

$$\left|\frac{y - y_1}{y}\right| = \frac{.008778}{6.324555} = .001388,$$

and a percentage error of

$$.001388 \times 100\% = 0.1388\%. \qquad \blacksquare$$

□ **EXAMPLE 3**

Find an approximation to $\sin 42°$.

Strategy · · · · · · · · **Solution**

Choose a convenient x_0 near 42°.

We take $x_0 = 45° = \pi/4$ since we are familiar with the fact that

$$\sin(\pi/4) = \sqrt{2}/2.$$

This means that $42° = x_0 + \Delta x$, so

Identify Δx.

$$\Delta x = 42° - 45° = -3°.$$

Convert Δx to radians.

However, since we have defined and differentiated $\sin x$ using radian measure (see Exercise 82 in Section 3.5), we must convert Δx to radians:

$$\Delta x = \frac{-3°}{360°}(2\pi) \approx -.05236 \text{ radians.}$$

Calculate $f'(x)$.

Since $f(x) = \sin x$, $f'(x) = \cos x$. Applying approximation (1) we obtain

Apply (1).

$$\sin(42°) \approx \sin(\pi/4) + \cos(\pi/4)(-.05236)$$

$$= \frac{\sqrt{2}}{2} + \frac{\sqrt{2}}{2}(-.05236) = .6701.$$

The actual value (to four decimal places) is 0.6691. $\qquad \blacksquare$

Table 7.2 presents other approximations to $\sin(\pi/4 + \Delta x)$ for various values of Δx (in radians).

Table 7.2   Approximations to $\sin\left(\dfrac{\pi}{4} + \Delta x\right)$

(a) Δx (in radians)	(b) Actual value $\sin\left(\dfrac{\pi}{4} + \Delta x\right)$	(c) Approximation $\dfrac{\sqrt{2}}{2}(1 + \Delta x)$	(d) Absolute error $\|(b) - (c)\|$	(e) Relative error $\left\|\dfrac{(b) - (c)}{(b)}\right\|$
1.0	0.977062	1.41421	0.437151	0.447414
0.5	0.959549	1.06066	0.101111	0.105374
0.1	0.774165	0.777817	0.003652	0.004718
0.05	0.741562	0.742462	0.000901	0.000121
0.01	0.714140	0.714178	0.000038	0.000053
0.005	0.710631	0.710642	0.000011	0.000016

Geometrically, the error in using approximation (1) to calculate $f(x_0 + \Delta x)$ is represented by the vertical distance labelled ϵ in Figure 7.1. It is

$$\epsilon = f(x_0 + \Delta x) - [f(x_0) + f'(x_0)\Delta x]. \tag{2}$$

As you can see from Figure 7.1, ϵ depends on the increment Δx. As $\Delta x \to 0$, $\epsilon \to 0$. In fact, if we consider ϵ as a function of Δx, then we can derive a limit that justifies the assertion that the approximation

$$f(x) \approx f(x_0) + f'(x_0)\Delta x$$

is the *best* linear approximation to $f(x)$ for x near x_0. We can rewrite equation (2) in the form

$$\frac{f(x_0 + \Delta x) - f(x_0)}{\Delta x} = f'(x_0) + \frac{\epsilon}{\Delta x}, \qquad \Delta x \neq 0. \tag{3}$$

Now if f is differentiable at x_0, then as $\Delta x \to 0$ the limit of the left-hand side of equation (3) is $f'(x_0)$. Thus we may conclude that not only does $\epsilon \to 0$ as $\Delta x \to 0$ but in fact

$$\lim_{\Delta x \to 0} \frac{\epsilon}{\Delta x} = 0. \tag{4}$$

We may interpret equation (4) by saying that as $\Delta x \to 0$ the error approaches zero ''faster'' than Δx. (See Table 7.3).

The preceding discussion is so important in understanding the nature of a differentiable function that we summarize it in the following theorem.

Theorem 9

Let f be differentiable at x_0 and let ϵ be the error function defined by

$$\epsilon = f(x_0 + \Delta x) - [f(x_0) + f'(x_0)\Delta x].$$

Then

$$\lim_{\Delta x \to 0} \frac{\epsilon}{\Delta x} = 0.$$

Table 7.3 The right-hand column illustrates the limit in equation (4) for the function $f(x) = \sqrt{x}$ with $x_0 = 36$.

Δx	$\sqrt{36 + \Delta x}$	$6 + \dfrac{\Delta x}{12}$	ϵ	$\dfrac{\epsilon}{\Delta x}$
4.0	6.32456	6.33333	0.00877801	0.0021945
2.0	6.16441	6.16667	0.00225266	0.00112633
1.5	6.12372	6.125	0.00127564	0.000850429
1.0	6.08276	6.08333	0.000570803	0.000570803
0.5	6.04152	6.04167	0.00014368	0.00028736
0.1	6.00833	6.00833	0.00000577901	0.0000577901

REMARK Although Theorem 9 tells us the limiting behavior of the error ϵ as $\Delta x \to 0$, we have not yet derived any way to estimate ϵ for a fixed nonzero choice of Δx. Without such estimates, we do not have any reason to believe that our approximations are within acceptable bounds.

In order to estimate the error, we need to be able to determine how much the graph of f differs from a straight line. In Chapter 4, we measure this difference. Then we study the accuracy question more carefully.

Approximating Increments in y

We can also use linear approximations to estimate *changes* in the values of a function $y = f(x)$. If we let

$$\Delta y = f(x_0 + \Delta x) - f(x_0) \tag{5}$$

then Δy is the difference in the function values resulting from the increment Δx in the independent variable x from $x = x_0$ to $x = x_0 + \Delta x$. (A shortcoming of the Δy notation, however, is that while this increment depends on both x_0 and Δx, this dependence is not reflected in the notation.)

Using the increment notation of equation (5) we can rewrite approximation (1) as

$$\Delta y \approx f'(x_0)\Delta x. \tag{6}$$

Figure 7.3 illustrates the interpretation of this form of our linear approximation, which says that *the change in the value of the function $y = f(x)$ resulting from an increment Δx in x from $x = x_0$ to $x = x_0 + \Delta x$ is approximated by the product of the derivative $f'(x_0)$ and the increment Δx.*

The form of the approximation in line (6) is useful in approximating the change in a quantity calculated from a measurement if small changes occur in the measurement. The following example is typical.

☐ EXAMPLE 4

A ball bearing in the shape of a perfect sphere of radius 2 cm is to be machined from a metal alloy weighing 9 g/cm^3.

(a) Find the weight of a bearing meeting these specifications.
(b) Approximate the change in this weight resulting from an error in the radius of no more than $\pm.05$ cm.

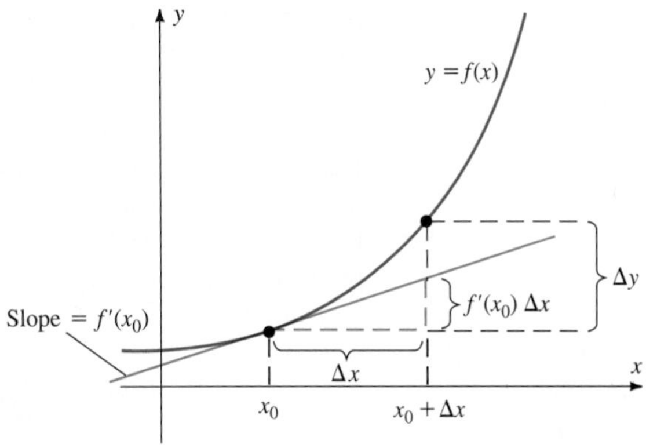

Figure 7.3 Δy is linearly approximated by $f'(x_0)\Delta x$.

Solution Since the volume of a sphere of radius r is $V = \frac{4}{3}\pi r^3$, the weight of a ball bearing of radius r made from this alloy is given by the function

$$W(r) = (\tfrac{4}{3}\pi r^3 \text{ cm}^3)(9 \text{ g/cm}^3) = 12\pi r^3 \text{ g}.$$

(a) The weight of a bearing of radius $r_0 = 2$ cm is therefore

$$W(2) = 12\pi \cdot 2^3 = 96\pi \text{ g}.$$

(b) If the actual measurement of the radius varies from $r_0 = 2$ cm by $\Delta r = \pm 0.05$, approximation (6) gives the approximation to the change ΔW in weight as

$$\Delta W \approx W'(r_0)\Delta r.$$

Since $W'(r) = 12\pi(3r^2) = 36\pi r^2$, this approximation, with $r_0 = 2$ and $\Delta r = \pm 0.05$, is

$$\Delta W \approx 36\pi(2^2)(\pm 0.05) = \pm 7.2\pi \text{ g}.$$

Thus, an error of $\Delta r = \pm 0.05$ cm in the radius will cause a fluctuation of *approximately* $\pm 7.2\pi$ g in the weight.

  To determine the accuracy of this approximation we compare, for example, the *actual* change ΔW in the weight corresponding to $\Delta r = 0.05$ with the *approximate* change, 7.2π. The *actual* change is

$$\begin{aligned} \Delta W &= W(2.05) - W(2) \\ &= 12\pi(2.05)^3 - 12\pi \cdot 2^3 \\ &= 7.3815\pi. \end{aligned}$$

The relative error in this approximation is therefore

$$\frac{(7.3815\pi) - 7.2\pi}{7.3815\pi} = 0.0246,$$

a percentage error of 2.46%. ■

Differential Notation

Frequently the symbol dx is used to denote small changes in x, just as we use the increment notation Δx, and the symbol dy is used to represent *the approximation* to the resulting increment Δy given by the right side of approximation

$$\Delta y \approx f'(x_0)\Delta x. \tag{7}$$

The symbols dx and dy are called **differentials.** If $y = f(x)$, they are related by the equation

$$dy = f'(x)\, dx. \tag{8}$$

Historically they have been used to argue that the derivative can be thought of as the ratio of *infinitesimals* giving the change dy in the function $y = f(x)$ resulting from a tiny change dx in x. The argument is based on the idea that the linear approximation in equation (7) becomes increasingly accurate as Δx approaches zero, so that the derivative

$$f'(x_0) = \lim_{\Delta x \to 0} \frac{f(x_0 + \Delta x) - f(x_0)}{\Delta x} = \lim_{\Delta x \to 0} \frac{\Delta y}{\Delta x}$$

may actually be regarded as the *quotient* of the infinitesimals dy and dx. (This is the origin of the Leibniz notation dy/dx for the derivative $f'(x)$.)

Until we study antidifferentiation in Chapter 5, we shall regard the differentials dx and dy merely as a notational device to help us remember the linear approximation $\Delta y \approx f'(x)\Delta x$. When differentials are used in this way, it is important to remember that the approximation in equation (7) depends on both the value $f'(x_0)$ at the number $x = x_0$ *and* the increment Δx.

□ **EXAMPLE 5**

Differential notation can be used to rewrite derivative formulas from the form $dy/dx = f'(x)$ in the *differential form* $dy = f'(x)\, dx$. While it appears that we do so simply by "multiplying through by dx" we are actually expressing the relationship between y and x in a way that we shall make precise after we discuss the Fundamental Theorem of Calculus in Chapter 6. The following table includes several examples of the differential notation.

Function	Derivative	Differential form
$y = \sin x$	$\dfrac{dy}{dx} = \cos x$	$dy = \cos x\, dx$
$y = x^n$	$\dfrac{dy}{dx} = nx^{n-1}$	$dy = nx^{n-1}\, dx$
$y = \sqrt{x}$	$\dfrac{dy}{dx} = \dfrac{1}{2\sqrt{x}}$	$dy = \dfrac{dx}{2\sqrt{x}}$
$y = \dfrac{1}{x}$	$\dfrac{dy}{dx} = \dfrac{-1}{x^2}$	$dy = -\dfrac{dx}{x^2}$

■

Exercise Set 3.7

In Exercises 1–10, (a) determine the expression

$$f(x_0) + f'(x_0)\Delta x,$$

that you should use to linearly approximate the given function f near the given number x_0. Then (b) approximate the quantity specified.

1. **a.** $f(x) = \sqrt{x}$ near $x_0 = 4$ $\Delta x = .1$
   b. approximate $\sqrt{4.1}$

2. **a.** $f(x) = \sqrt{x}$ near $x_0 = 9$
   b. approximate $\sqrt{9.1}$

3. a. $f(x) = 1/x$ near $x_0 = 2$
 b. approximate $13/30$

4. a. $f(x) = 1/x^2$ near $x_0 = 3$
 b. approximate $4/25$

5. a. $f(x) = x^3$ near $x_0 = 1/2$
 b. approximate $(.48)^3$

6. a. $f(x) = x^{10}$ near $x_0 = 1$
 b. approximate $(.98)^{10}$

7. a. $f(x) = 2 \sin x$ near $x_0 = 0$
 b. approximate $2 \sin(2°)$

8. a. $f(x) = \tan x$ near $x_0 = 0$
 b. approximate $\tan(0.1)$ (note that this angle is being expressed in radians)

9. a. $f(x) = \cos x$ near $x_0 = \pi/6$
 b. approximate $\cos(31°)$

10. a. $f(x) = 2 \sec x$ near $x_0 = \pi/4$
 b. approximate $2 \sec(43°)$

In Exercises 11–18, use linear approximation to approximate the given quantity using nearby "well-known" values.

11. $\sqrt{48}$ **12.** $\sqrt{62}$

13. $1/(4.1)$ **14.** $\sin(46°)$

15. $\cos(59°)$ **16.** $(2.03)^6$

17. $(2.97)^3$ **18.** $\cos(91°)$

19. Use a calculator to find the exact values (to six decimal places) of the quantities in Exercises 11, 13, 15, and 17. In each case, compute the relative and percentage error of the approximation.

20. Use a calculator to find the exact values (to six decimal places) of the quantities in Exercises 12, 14, 16, and 18. In each case, compute the relative and percentage error of the approximation.

21. Find the relative and percentage errors for the approximation in Example 3.

22. Find an approximation to $f(9)$ if $f(10) = 6$ and $f'(10) = -2$.

23. Find an approximation to $f(36)$ if $f(39) = 7$ and $f'(39) = 0.65$.

24. Water flows through a pipe at the rate of 3 L/min/in.² of cross-sectional area.
 a. Find the volume $V(r)$ per minute flowing through the pipe if the cross section is a circle of radius r.
 b. Find the volume per minute, $V(4)$, if the radius of the pipe is $r = 4$ in.
 c. Approximate the change ΔV in the volume per minute resulting from an *increase* in the radius from $r = 4$ to $r = 4.5$ in.

25. A roller bearing has the shape of a cylinder. If the radius of the base is to be $r = 2$ cm and the length is to be $\ell = 5$ cm, approximate the change ΔV in the volume if the length is precise but the radius varies by $\Delta r = 0.02$ cm.

26. A plot of ground is to be laid out in the shape of a square with sides of length $s = 200$ ft. Approximate the change ΔA in the area of the plot, in square feet, if each of the dimensions vary from the intended length by $\Delta s = 2$ ft.

27. The acceleration of a pendulum bob due to gravity is $a(\theta) = -32 \sin \theta$ where θ is the angle between the pendulum cord and the vertical (see Figure 7.4).

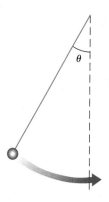

Figure 7.4

 a. Find a linear approximation to the acceleration $a(\theta)$ using $\theta_0 = 0$ degrees (in radians).
 b. Find the approximation to the acceleration when $\theta = \pi/6$ given by your answer to part a.
 c. What is the relative error in the approximation in part b?

28. The motion of a mass connected to the end of a spring is given by the position function
$$s(t) = 10 + 5 \cos t$$
where t is time in seconds.
 a. Find a linear approximation to $s(t)$ using $t_0 = \pi/4$.
 b. Find the approximation to $s(\pi/2)$ given by your answer to part a.
 c. What is the relative error in the approximation in part b?

29. Let $f(x) = \sqrt{x}$.
 a. Using equation (1) with $x_0 = 49$, approximate $\sqrt{57}$.
 b. Using equation (1) with $x_0 = 64$, approximate $\sqrt{57}$.
 c. Which of these two approximations is larger?
 d. Sketch the graph $y = \sqrt{x}$ for $49 \le x \le 64$ along with the two tangent lines that yield the approximations in parts a and b.
 e. Where is $\sqrt{57}$ on the number line relative to the two approximations you computed in parts a and b?

30. Suppose $f(x) = \frac{1}{8}x^3 - x$.
 a. Find an equation $y = t_1(x)$ for the line tangent to the graph of f at $(-2, 1)$.

b. Plot the graphs of f and t_1.

c. Repeatedly magnify this plot about the point $(-2, 1)$. Why do the graphs of t_1 and f seem to coincide?

d. Graph $y = f(x) - t_1(x)$, and find a number $h > 0$ so that, if $-2 - h < x < -2 + h$, then $|f(x) - t_1(x)| < 0.01$.

e. Find an equation $y = t_2(x)$ for the line tangent to the graph of f at $(0, 0)$. Graph $y = f(x) - t_2(x)$ and determine $h > 0$ so that, if $-h < x < h$, then $|f(x) - t_2(x)| < 0.01$.

f. Compare the numbers h found in parts d and e. What features of the graph of f near $x = -2$ and near $x = 0$ account for the difference in the values of h?

 **31.** Repeat Exercise 30 using

$$f(x) = \frac{5}{1 + 10x^2}$$

at $x = 1/2$ and $x = 1$.

3.8 Implicit Differentiation and Rational Power Functions

So far, most of the functions that we have discussed have been defined *explicitly*. For instance, when we write $y = f(x)$, we are providing an explicit rule that determines y from x. Many functions, however, are defined *implicitly* by an equation that relates the two variables. For example, the equation for the ellipse

$$\frac{x^2}{16} + \frac{y^2}{9} = 1 \tag{1}$$

specifies (at most) two y values for each x value between -4 and 4. Therefore if we adopt a convention to determine which of these two y values to pick for each x, we obtain a function. Such a function is said to be implicitly defined by equation (1). We obtain one such function f_1 by specifying that $f_1(x)$ must be nonnegative. In this case, we can write f_1 explicitly by solving for y under the assumption that $y \geq 0$. We obtain

$$y = f_1(x) = 3\sqrt{1 - \frac{x^2}{16}}. \tag{2}$$

If, instead, we specify that $y \leq 0$ along with equation (1), we obtain a second function

$$y = f_2(x) = -3\sqrt{1 - \frac{x^2}{16}} \tag{3}$$

(see Figure 8.1). We say the both f_1 and f_2 are *implicit* in equation (1).

Unfortunately, many equations in x and y cannot be manipulated algebraically into explicit forms such as $y = f(x)$ or $x = g(y)$.* Nevertheless, the graph of such an equation still may have a tangent line at any given point (x_0, y_0), and in order to find equations for these tangents, we need to determine their slopes. As long as a tangent line exists and is not vertical, its slope can often be obtained from the equation using a technique called **implicit differentiation.** As Figure 8.2 indicates, the equation may implicitly define a differentiable function $y = f(x)$ *when it is restricted to a small piece of its graph containing* (x_0, y_0). Thus, the derivative $f'(x_0)$, which equals dy/dx at (x_0, y_0), is the desired slope.

When we differentiate implicitly, we *assume* that y is a function of x, and we derive dy/dx by differentiating both sides of the equation without explicitly solving

---

*In the nineteenth century, Abel, Galois, and Ruffini showed that there are many equations ($y^5 - y - x^2 = -1$, for example) that cannot be explicitly solved for y, *even if one is willing to accept an answer involving extremely complicated combinations of powers and roots!*

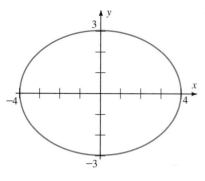

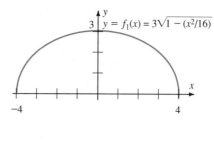

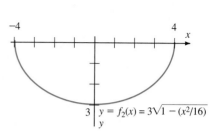

Figure 8.1 (a) The ellipse $\dfrac{x^2}{16} + \dfrac{y^2}{9} = 1$. (b) The graph of the function $f_1(x) = 3\sqrt{1 - (x^2/16)}$. (c) The graph of the function $f(x) = -3\sqrt{1 - (x^2/16)}$.

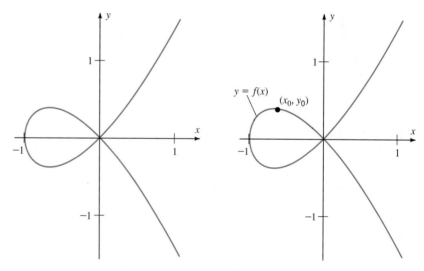

Figure 8.2 (a) The graph of the equation $y^2 = x^3 + x^2$. (b) The red portion of the graph of the equation $y^2 = x^3 + x^2$ is the graph of an implicitly defined function f for x near x_0.

for y. However, when we differentiate a term involving y, we must remember the assumption that y is a function of x. We use the Chain Rule to calculate the appropriate derivative. For example, if one of the terms in the equation were y^2, then its derivative with respect to x would be

$$\frac{d}{dx}(y^2) = \left(\frac{d}{dy}(y^2)\right)\left(\frac{dy}{dx}\right) \qquad \text{(The Chain Rule)}$$

$$= 2y\left(\frac{dy}{dx}\right)$$

☐ **EXAMPLE 1**

Find the slope of the line tangent to the graph of the ellipse

$$\frac{x^2}{16} + \frac{y^2}{9} = 1 \tag{4}$$

at the point $(2, 3\sqrt{3}/2)$ (Figure 8.3).

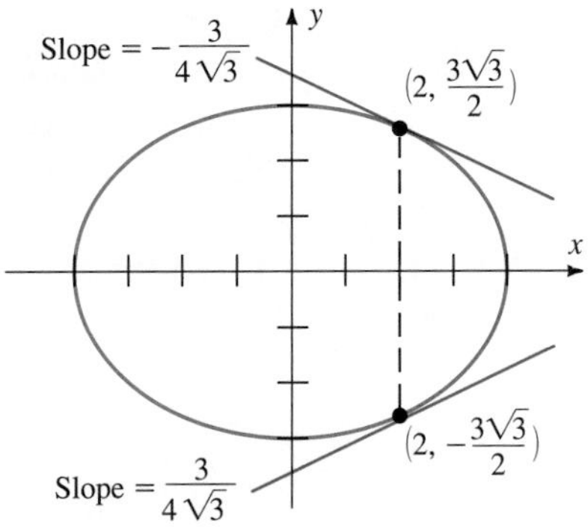

Figure 8.3 Although equation (1) does not define a function, tangents exist at all points on the graph.

Strategy · · · · · · · ·

Assume that y is a function of x.

Differentiate both sides of equation (4), remembering that

$$\frac{d}{dx}\left(\frac{y^2}{9}\right) = \left(\frac{2y}{9}\right)\left(\frac{dy}{dx}\right),$$

according to the Chain Rule.

Solve for dy/dx and substitute values for x and y.

Solution

Differentiating both sides of equation (4) with respect to x, regarding y as an unspecified function of x, we obtain

$$\frac{2x}{16} + \left(\frac{2y}{9}\right)\left(\frac{dy}{dx}\right) = 0,$$

so

$$\frac{dy}{dx} = -\frac{9x}{16y}, \qquad y \neq 0. \tag{5}$$

At the point $(2, 3\sqrt{3}/2)$, we have

$$\frac{dy}{dx} = -\frac{9(2)}{16(3\sqrt{3}/2)} = -\frac{\sqrt{3}}{4},$$

which is the desired slope. (Alternatively, we could have been asked for the slope of the tangent at the point $(2, -3\sqrt{3}/2)$. The result would be $dy/dx = \sqrt{3}/4$ [see Figure 8.3].) ∎

One advantage of this technique is that equation (5) is valid whenever $y \neq 0$. It is not necessary to treat the two cases $y < 0$ and $y > 0$ separately (as we would have had to do if one had used the explicit equations $y = f_1(x)$ and $y = f_2(x)$ [equations (2) and (3)]).

The technique of implicit differentiation, applied to an equation involving the two variables x and y, can be summarized as follows:

(i) Assume that the given equation implicitly defines y as an unspecified differentiable function of x.
(ii) Differentiate both sides of the given equation with respect to x, remembering that the factor dy/dx will occur in the differentiation of any term involving the function y (according to the Chain Rule).
(iii) Solve the resulting equation for dy/dx.

REMARK 1 In this summary, we assumed that y was a function of x. It is equally valid to assume that x is an implicit function of y. Then we could solve for dx/dy, remembering to apply the Chain Rule to terms that include x. In Exercises 60–66, we explore the geometric interpretation of dx/dy.

REMARK 2   Although we can always carry out this procedure symbolically to obtain an expression for dy/dx (assuming that the necessary differentiation formulas are known), we must be careful whenever we use this procedure. We are ignoring the fact that the given equation need not define y as a differentiable function of x. Indeed, in Example 1, we could not obtain a value of dy/dx at the point $(4, 0)$ since the expression $dy/dx = -9x/16y$ is undefined when $y = 0$. Geometrically, this corresponds to the observation that the tangent to the graph of equation (4) at the point $(4, 0)$ is vertical.

A resolution of this difficulty requires a theorem (called the Implicit Function Theorem) that gives precise conditions under which an equation determines y as a differentiable function of x near a particular point (x_0, y_0). Developing this theorem would take us far astray from our main objectives, so we simply refer you to more advanced texts for a discussion of this theorem. (See, for example, *Mathematical Analysis,* 2nd ed., by Tom Apostol, Addison-Wesley, 1974). We will resolve this concern here by cautioning that you should always check to see that both the given equation and the resulting expression for dy/dx are defined at the point (x_0, y_0).

☐ **EXAMPLE 2**

Using implicit differentiation, derive dy/dx for the equation

$$y^2 + x^2y = 3x^2,$$

and find the slope of the line tangent to its graph at the point $(2, 2)$.

Strategy · · · · · · · ·   **Solution**

Differentiate both sides with respect to x.

$$\underbrace{2y \cdot \frac{dy}{dx}}_{\frac{d}{dx}(y^2)} + \underbrace{2xy + x^2 \cdot \frac{dy}{dx}}_{\frac{d}{dx}(x^2y)} = \underbrace{6x}_{\frac{d}{dx}(3x^2)}$$

Collect terms involving $\dfrac{dy}{dx}$.   Thus,

$$(x^2 + 2y)\frac{dy}{dx} = 6x - 2xy,$$

so

Solve for $\dfrac{dy}{dx}$.

$$\frac{dy}{dx} = \frac{6x - 2xy}{x^2 + 2y}.$$

Substitute given values for x and y to   For $x = 2$, $y = 2$ we have
obtain the slope.

$$\frac{dy}{dx} = \frac{6 \cdot 2 - 2 \cdot 2 \cdot 2}{2^2 + 2 \cdot 2} = \frac{1}{2},$$

which is the desired slope.   ∎

☐ **EXAMPLE 3**

Find an equation of the line normal to the graph of

$$y^3 - x^2 = 7$$

at the point $(1, 2)$. (See Figure 8.4 and Exercise 25 in Section 3.1.)

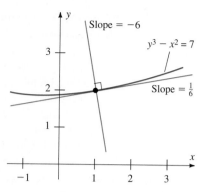

Figure 8.4   Normal to graph of $y^3 - x^2 = 7$ at $(1, 2)$ has slope -6.

Solution   We begin by differentiating implicitly. We obtain

$$3y^2 \cdot \frac{dy}{dx} - 2x = 0$$

so

$$\frac{dy}{dx} = \frac{2x}{3y^2}.$$

At the point $(1, 2)$ the slope of the tangent is therefore

$$m_1 = \frac{2 \cdot 1}{3 \cdot 2^2} = \frac{1}{6},$$

so the slope of the normal, which is perpendicular to the tangent, is

$$m_2 = -\frac{1}{m_1} = -6.$$

(Theorem 5 of Chapter 1.)

An equation for the line is

$$y - 2 = -6(x - 1).$$

(See Figure 8.4.)   ■

□ **EXAMPLE 4**

Find the slope of the line tangent to the graph of the equation

$$\sin(xy) = x^2 \cos y \tag{6}$$

at the point $(2, \pi/2)$.

Solution   We implicitly differentiate both sides of equation (6) with respect to x. We use the Chain Rule and the Product Rule to differentiate the left-hand side of equation (6).

$$\frac{d}{dx}\sin(xy) = \cos(xy)\left(\frac{d}{dx}xy\right)$$

$$= \cos(xy)\left(y + x\frac{dy}{dx}\right)$$

To differentiate the right-hand side of (6), we first use the Product Rule and then the Chain Rule.

$$\frac{d}{dx}(x^2 \cos y) = \left(\frac{d}{dx}x^2\right)(\cos y) + (x^2)\left(\frac{d}{dx}\cos y\right)$$

$$= 2x \cos y - x^2 \sin y \frac{dy}{dx}$$

Equating the two results yields.

$$\cos(xy)\left(y + x\frac{dy}{dx}\right) = 2x \cos y - x^2 \sin y \frac{dy}{dx}$$

After collecting the terms involving dy/dx, we have

$$[x \cos(xy) + x^2 \sin y]\frac{dy}{dx} = 2x \cos y - y \cos(xy),$$

and therefore,

$$\frac{dy}{dx} = \frac{2x \cos y - y \cos(xy)}{x \cos(xy) + x^2 \sin y}$$

whenever $x \cos(xy) + x^2 \sin y \neq 0$.

Now we evaluate the derivative at the point $(2, \pi/2)$ to determine the desired slope.

$$\left.\frac{dy}{dx}\right|_{(x,y)=(2,\pi/2)} = \frac{(2)(2)(0) - (\pi/2)(-1)}{(2)(-1) + (4)(1)} = \frac{\pi}{4} \qquad\blacksquare$$

Differentiating Rational Powers of x.

The following theorem tells us that the Power Rule extends to *all* power functions, $f(x) = x^r$, with r rational.

Theorem 10
Power Rule for the Case of Rational Exponents

Let $f(x) = x^{p/q}$, where p and q are integers with $p \neq 0$ and $q \neq 0$. Then f is differentiable and

$$f'(x) = \frac{p}{q}x^{(p/q)-1} \tag{7}$$

whenever $f(x)$ and the right side of equation (7) are defined. In Leibniz notation,

$$\frac{d}{dx}(x^{p/q}) = \left(\frac{p}{q}\right)x^{(p/q)-1}. \tag{8}$$

Assuming $f(x) = x^{p/q}$ to be a differentiable function of x (a fact that we shall assume but not prove), we can establish equation (7) for $x \neq 0$ using implicit differentiation. Let $x \neq 0$ be in the domain of the function

$$y = x^{p/q}$$

where p and $q \neq 0$ are integers. Then $y \neq 0$, and

$$y^q = (x^{p/q})^q = x^p. \tag{9}$$

If we assume y to be differentiable, then since q is an integer, y^q is differentiable. Applying the Power Rule (for integral powers) to both sides of equation (9) gives

$$qy^{q-1} \cdot \frac{dy}{dx} = px^{p-1}.$$

Recalling that $y = x^{p/q} \neq 0$, we can now solve for $\dfrac{dy}{dx}$ as

$$
\begin{aligned}
\frac{dy}{dx} &= \left(\frac{p}{q}\right) x^{p-1} y^{1-q} \\
&= \left(\frac{p}{q}\right) x^{p-1} (x^{p/q})^{1-q} \\
&= \left(\frac{p}{q}\right) x^{p-1} \cdot x^{(p/q)-p} \\
&= \left(\frac{p}{q}\right) x^{(p-1+(p/q)-p)} \\
&= \left(\frac{p}{q}\right) x^{(p/q)-1},
\end{aligned}
$$

which is the formula in Theorem 10.

REMARK 3:   In Chapter 8, we shall see that the Power Rule applies to all power functions, $f(x) = x^a$, where a is any real number.

□ **EXAMPLE 5**

For $f(x) = \sqrt[4]{x^3 - x^2 + 3}$, find $f'(x)$.

Strategy · · · · · · · ·

Replace the radical sign with a fractional exponent.

Solution

Since $f(x) = (x^3 - x^2 + 3)^{1/4}$,

$$
\begin{aligned}
f'(x) &= \tfrac{1}{4}(x^3 - x^2 + 3)^{-3/4} \cdot \frac{d}{dx}(x^3 - x^2 + 3) \\
&= \tfrac{1}{4}(x^3 - x^2 + 3)^{-3/4}(3x^2 - 2x) \\
&= \frac{3x^2 - 2x}{4\sqrt[4]{(x^3 - x^2 + 3)^3}}.
\end{aligned}
$$

Apply Theorem 10 together with the Chain Rule.

■

□ **EXAMPLE 6**

Find the velocity of an object whose position at time t is

$$s(t) = t^{2/3} \sin(1 + \sqrt{t}).$$

Solution   Using the Product Rule, Power Rule, and Chain Rule, we obtain the velocity function

$$
\begin{aligned}
v(t) = s'(t) &= \left(\tfrac{2}{3}t^{-1/3}\right) \sin(1 + t^{1/2}) + t^{2/3} \cos(1 + t^{1/2}) \cdot \frac{d}{dt}(1 + t^{1/2}) \\
&= \tfrac{2}{3}t^{-1/3} \sin(1 + t^{1/2}) + t^{2/3} \cos(1 + t^{1/2})(\tfrac{1}{2}t^{-1/2}) \\
&= \tfrac{2}{3}t^{-1/3} \sin(1 + t^{1/2}) + \tfrac{1}{2}t^{1/6} \cos(1 + t^{1/2}).
\end{aligned}
$$

(Note that $v(0)$ is undefined.)

■

□ **EXAMPLE 7**

Figure 8.5 shows the graph of $f(x) = x^{1/3}$, with several tangents sketched in. The derivative (slope function) for $f(x) = x^{1/3}$ is $f'(x) = \frac{1}{3}x^{-2/3}$.

Note that $f'(0)$ is undefined and that the slopes of the tangents approach $+\infty$ as x approaches zero from either direction.

However, the graph of $f(x) = x^{1/3}$ *does* have a tangent at $x = 0$ in the sense in which we defined tangents in Section 2.1. As Figure 8.6 illustrates, the slope of the secant line through $(0, 0)$ and $(h, f(h)) = (h, h^{1/3})$ is

$$\frac{f(0 + h) - f(0)}{h} = \frac{h^{1/3} - 0}{h} = \frac{1}{h^{2/3}}.$$

Thus, the secant through $(0, 0)$ and $(h, h^{1/3})$ rotates into the position of the vertical line through $(0, 0)$ as h approaches zero. ■

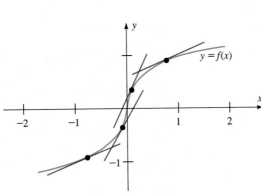

Figure 8.5 Graph of $f(x) = x^{1/3}$ and several tangents.

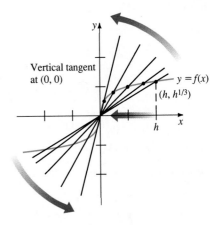

Figure 8.6 Secants through $(0, 0)$ approach the y-axis as $h \to 0$. Tangent to $f(x) = x^{1/3}$ at $(0, 0)$ is vertical.

REMARK 4:   Example 7 illustrates the fact that the graph of a function may have a vertical tangent if its derivative fails to exist.

□ **EXAMPLE 8**

The graph of $f(x) = x^{2/3}$ appears in Figure 8.7. Since $f'(x) = \frac{2}{3}x^{-1/3}$, $f'(0)$ is undefined. Here the secants through $(0, 0)$ have slope

$$\frac{(0 + h)^{2/3} - 0}{h} = \frac{1}{h^{1/3}}.$$

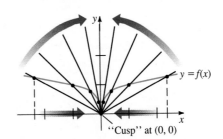

Figure 8.7 Tangent to graph of $f(x) = x^{2/3}$ at $(0, 0)$ is vertical. The graph has a "cusp" at $(0, 0)$.

Again, the secants rotate into the y-axis as $h \to 0$. Thus, the graph of $f(x) = x^{2/3}$ has a vertical tangent at $x = 0$, even though $f'(0)$ is undefined.

Note, however, that the graph of $f(x) = x^{2/3}$ looks quite different from the graph of $g(x) = x^{1/3}$ near $(0, 0)$. This is because $f(x) = x^{2/3} = (x^{1/3})^2$ is an *even* function, while $g(x) = x^{1/3}$ is an *odd* function. (See Exercise 57 in Section 1.4.) Because the tangent to the graph of $f(x) = x^{2/3}$ at $(0, 0)$ is vertical, and because the branch of the graph to the left of $x = 0$ must be the mirror image of the branch to the right of $x = 0$, a "cusp" (sharp point) exists at $(0, 0)$. ∎

Exercise Set 3.8

In Exercises 1–26, find $f'(x)$.

1. $f(x) = (x + 2)^{4/3}$

2. $f(x) = \sqrt[3]{x^2 + 1}$

3. $f(x) = \sqrt{x} + \dfrac{1}{\sqrt{x}}$

4. $f(x) = \dfrac{\sqrt[3]{x + 2}}{1 + \sqrt{x}}$

5. $f(x) = x^{2/3} + x^{-2/3}$

6. $f(x) = (3x^4 + 4x^3)^{-2}$

7. $f(x) = (x^2 - 1)^{-2}(x^2 + 1)^2$

8. $f(x) = (7 - x^2)^{2/3}$

9. $f(x) = \sqrt{1 + \sqrt[3]{x}}$

10. $f(x) = x^{5/2} + 3x^{2/3} + x^{-4/3}$

11. $f(x) = \dfrac{\sqrt[3]{x}}{\sqrt[3]{x} + x}$

12. $f(x) = \sqrt{x^3}(x^{-2} + 2x^{-1} + 1)^4$

13. $f(x) = \sqrt{x} + \sqrt[4]{x} + \sqrt[8]{x}$

14. $f(x) = \dfrac{x^{2/3}}{\sqrt[3]{x^2 + 1}}$

15. $f(x) = \dfrac{(x^2 + 1)^{2/3}}{(1 - x^2)^{4/3}}$

16. $f(x) = \dfrac{(x - 1)^{1/2} + (x + 1)^{1/3}}{(x + 2)^{1/4}}$

17. $f(x) = \sqrt[3]{\sin x^2}$

18. $f(x) = \sqrt{\dfrac{ax^2 + b}{cx + d}}$

19. $f(x) = \dfrac{x^{1/4}(x^{4/3} - x)}{(\sin^3 x - \sqrt{x})^{1/3}}$

20. $f(x) = [x^2 + x \tan(x^{3/2} - x^{1/2})]^{4/3}$

21. $f(x) = \sin^{5/2}(x^{2/3})$

22. $f(x) = (x^{-2/3} - x^{3/2})^{3/5}(6 - x^{5/3})^{-2/5}$

23. $f(x) = \sin \sqrt[3]{x} + \sqrt[3]{\sin x}$

24. $f(x) = \sqrt[3]{\sin^2 x}$

25. $f(x) = \sec \sqrt[4]{3x + 1}$

26. $f(x) = \tan \sqrt[3]{x^2}$

In Exercises 27–50, find $\dfrac{dy}{dx}$ by implicit differentiation.

27. $x^2 + y^2 = 25$

28. $x = \sin y$

29. $x^{1/2} + y^{1/2} = 4$

30. $x^2 + 2xy + y^2 = 8$

31. $x = \tan y$

32. $x^2y + xy^2 = 2$

33. $x \sin y = y \cos x$

34. $x = y(y - 1)$

35. $(xy)^{1/2} = xy - x$

36. $y^2 = \sin^2 x - \cos^2 2x$

37. $x^3 + x^2y + xy^2 + y^3 = 15$

38. $\cos(x + y) = y \sin x$

39. $\sqrt{x + y} = xy - x$

40. $y^2 = \dfrac{x + 1}{x^2 + 1}$

41. $\cot y = 3x^2 + \cot(x + y)$

42. $x^2y^2 = 4$

43. $\sin(xy) = 1/2$

44. $\dfrac{1}{x} + \dfrac{1}{y} + \dfrac{1}{4} = 0$

45. $y^4 = x^5$

46. $\sin(x + y) + \cos(x - y) = 1$

47. $y^2 + 3 = x \sec y$

48. $y^4 + 1 = \tan xy$

49. $x = \sin 2xy$

50. $\cos^2 2y = y - x$

In each of Exercises 51–58, find an equation for the line tangent to the graph determined by the given equation at the given point.

51. $xy = 9$ $(3, 3)$

52. $x^2 + y^2 = 4$ $(\sqrt{2}, \sqrt{2})$

53. $x^3 + y^3 = 16$ $(2, 2)$

54. $x^2y^2 = 16$ $(-1, 4)$

55. $\dfrac{x + y}{x - y} = 4$ $(5, 3)$

56. $(y - x)^2 = x$ $(9, 12)$

57. $y^4 + 3x - x^2 \sin y = 3$ $(1, 0)$

58. $y = x^2 + \sin y$ $(\sqrt{\pi}, \pi)$

59. Assume that an equation determines y as a differentiable function of x. Explain why $dy/dx = 0$ at points where the graph has a horizontal tangent.

60. Assume that an equation determines x as a differentiable function of y. Explain why $dx/dy = 0$ at points where the graph has a vertical tangent.

In Exercises 61–66, use the results of Exercises 59 and 60 to find the equations of all horizontal and vertical tangent lines.

61. $x^2 + y^2 = 2$

62. $x^2 + 4y^2 = 4$

63. $x^2 - y^2 = 1$

64. $(x - 1)^2 + (y + 2)^2 = 9$

65. $x^2 + y^2 + 2x + 4y = -4$

66. $(x + 1)(y - 1) = 1$

67. Find the value of a so that the circles with equations $(x - a)^2 + y^2 = 2$ and $(x + a)^2 + y^2 = 2$ intersect at points where their tangents are perpendicular. perpendicular to Tangent

68. Find an equation for the line normal to the curve $9x^2 + 16y^2 = 144$ at the point $(2, 3\sqrt{3}/2)$.

69. Prove that all normals to the curve $x^2 + y^2 = a^2$ pass through the origin.

70. Explain why the equation $4 + x^2 + y^2 = 2xy$ does not define a function $y = f(x)$ for *any* values of x and y. (*Hint:* Rewrite as $x^2 - 2xy + y^2 = -4$. What is the sign of the left-hand side?)

71. Determine the points on the curve $5x^2 + 6xy + 5y^2 = 8$ where the tangent is parallel to the line $x - y = 1$.

72. Suppose that x and y satisfy the equation for the hyperbola $y^2 - x^2 = 1$.
 a. Derive equations for two implicitly defined, continuous functions $y = f_1(x)$ and $y = f_2(x)$.
 b. Calculate the derivatives of f_1 and f_2.
 c. Calculate dy/dx by differentiating $y^2 - x^2 = 1$ implicitly.
 d. Verify that your answer in part c agrees with those in part b.
 e. Show that $|dy/dx| < 1$ at all points on the hyperbola.

73. Note that the graph of the equation $xy^2 + x^2y = 4x^2$ consists of the vertical line $x = 0$ (y-axis) in addition to the graph of the equation $y^2 + xy = 4x$. Why is this so? What does this say about the existence of a derivative dy/dx "near" $x = 0$?

74. For the equation $x^2 + y^2 = 25$, find the slope of the line tangent to the graph at $(3, 4)$ by
 a. geometry;
 b. solving for y and then finding dy/dx;
 c. finding dy/dx by implicit differentiation.

3.9   Higher Order Derivatives

Early in this chapter we saw that the Tangent Line Problem and the need to calculate instantaneous rates of change lead naturally to the differentiation process. We derive a new function f' from a given function f. If we view the differentiation process as a mathematical operation that produces a new function from a given one, we might wonder if it is useful to repeat this process. In fact, it is extremely useful both for the mathematical theory and for a variety of applications. Repeated differentiation yields the so-called higher order derivatives that often tell us a great deal about the original function. We shall also see that, in many applications, the higher derivatives play a central role.

Given a differentiable function f, we first find its derivative f'. Then, if f' is differentiable, we derive its derivative $(f')'$. This new function is called the **second derivative** of f, and it is denoted by f''. That is,

$$f''(x) = (f')'(x).$$

The Leibniz notation for the second derivative of $y = f(x)$ is

$$\frac{d^2y}{dx^2} = \frac{d}{dx}\left(\frac{dy}{dx}\right).$$

(Note that the 2 that appears twice in the notation indicates that we have differentiated 2 times. It is not an exponent. In particular, we are *not* calculating the derivative with respect to x^2.) Similarly, beginning with f, differentiating three times produces the third derivative, which is denoted by

$$f'''(x), \quad \text{or} \quad \frac{d^3y}{dx^3}.$$

Higher order derivatives are produced analogously. For $n \geq 4$, $f^{(n)}(x)$ denotes the value of the nth derivative of f at x.

□ **EXAMPLE 1**

For the polynomial function

$$f(x) = x^4 - 4x^3 + 3x^2 - 5x + 3$$

we have

$$f'(x) = 4x^3 - 12x^2 + 6x - 5,$$
$$f''(x) = 12x^2 - 24x + 6,$$
$$f'''(x) = 24x - 24,$$
$$f^{(4)}(x) = 24,$$

and $f^{(n)}(x) = 0$ for $n \geq 5$. ■

□ **EXAMPLE 2**

Find the first three derivatives of the function

$$f(x) = x^3 + x \sin x.$$

Solution
$$f'(x) = 3x^2 + \sin x + x \cos x$$
$$f''(x) = 6x + \cos x + \cos x - x \sin x$$
$$= 6x + 2 \cos x - x \sin x$$
$$f'''(x) = 6 - 2 \sin x - \sin x - x \cos x$$
$$= 6 - 3 \sin x - x \cos x$$ ■

□ **EXAMPLE 3**

It is an important property of the functions $f(x) = \sin x$ and $g(x) = \cos x$ that they reappear in the list of their own derivatives:

$f(x) = \sin x$	$g(x) = \cos x$
$f'(x) = \cos x$	$g'(x) = -\sin x$
$f''(x) = -\sin x$	$g''(x) = -\cos x$
$f'''(x) = -\cos x$	$g'''(x) = \sin x$
$f^{(4)}(x) = \sin x$	$g^{(4)}(x) = \cos x$
$f^{(5)}(x) = \cos x$	$g^{(5)}(x) = -\sin x$
etc.	etc.

From these lists you can see that both $\sin x$ and $\cos x$ satisfy the *differential* (or, *derivative*) equations

(i) $f''(x) = -f(x)$
(ii) $f^{(4)}(x) = f(x)$.

Equation (i) is important in the modelling of oscillatory phenomena (such as problems in engineering involving springs), while equation (ii) occurs in problems involving the deflection of beams under heavy loading. We will have much more to say about differential equations later in the text. ■

Given an equation in x and y we will sometimes need to compute d^2y/dx^2. If the equation defines y implicitly as a twice differentiable function of x, we can often compute d^2y/dx^2 by applying the method of implicit differentiation twice.

□ **EXAMPLE 4**

The equation $9x^2 + 4y^2 = 36$ defines y as a twice differentiable function of x near the point $(0, 3)$. Find d^2y/dx^2 for this function.

Strategy · · · · · · · ·
Differentiate both sides of given equation and solve for dy/dx.

Solution

$$18x + 8y \cdot \frac{dy}{dx} = 0$$

$$\frac{dy}{dx} = -\frac{9x}{4y}.$$

Differentiate resulting equation (using the Quotient Rule).

Differentiating both sides of this equation gives

$$\frac{d^2y}{dx^2} = -\frac{(4y)9 - (9x) \cdot 4 \cdot \dfrac{dy}{dx}}{16y^2}$$

$$= -\frac{36y - 36x \cdot \dfrac{dy}{dx}}{16y^2}$$

Substitute for $\dfrac{dy}{dx}$ as found above.

$$= -\frac{36y - 36x\left(-\dfrac{9x}{4y}\right)}{16y^2}$$

$$= -\frac{9}{4y} - \frac{81x^2}{16y^3}$$

$$= -\frac{36y^2 + 81x^2}{16y^3}.$$

Simplify using the original relation, if possible.

Using the original equation $9x^2 + 4y^2 = 36$, we can simplify

$$\frac{d^2y}{dx^2} = -\frac{9(4y^2 + 9x^2)}{16y^3} = -\frac{9(36)}{16y^3} = -\frac{81}{4y^3}.$$ ■

Acceleration

For a particle moving along a line, **acceleration** is defined to be the (instantaneous) rate of change of velocity. Thus, if v is a differentiable velocity function, the acceleration $a(t)$ is defined as

$$a(t) = \lim_{h \to 0} \frac{v(t + h) - v(t)}{h}$$

$$= v'(t).$$

That is, acceleration is the first derivative of velocity. Recall that if $s(t)$ is the position of the particle, then $v(t) = s'(t)$, so acceleration is the second derivative of position:

$$a(t) = v'(t) = s''(t),$$

or

$$a = \frac{dv}{dt} = \frac{d^2s}{dt^2}.$$

REMARK   Since acceleration is the rate of change of velocity with respect to time, its units should be velocity units divided by time units. For example, if we are measuring position in meters and time in seconds, then velocity is measured in units of meters per second. Therefore, acceleration is measured in meters per second per second. These units are usually abbreviated as m/s^2.

☐ **EXAMPLE 5**

A particle moves along a line so that at time t seconds its position is $s(t) = t^3 - 6t^2 + 7t - 2$ m.

(a) Find the acceleration $a(t)$ at time t.
(b) Find the initial acceleration $a(0)$.
(c) Find the velocity $v(t_0)$ at the time t_0 for which the acceleration is zero.

Solution   We have $v(t) = s'(t) = 3t^2 - 12t + 7$, so

$$a(t) = s''(t) = 6t - 12.$$

Thus the initial acceleration is $a(0) = -12$ m/s^2.
   To find the time t_0 for which the acceleration is zero, we solve

$$a(t_0) = 6t_0 - 12 = 0$$

and obtain $t_0 = 2$. At that moment, the velocity is

$$v(2) = 3(2)^2 - 12(2) + 7 = -5 \text{ m/s}. \qquad \blacksquare$$

☐ **EXAMPLE 6**

A projectile is fired vertically upward from ground level with an initial velocity of 100 m/s. In such a case the distance of the particle above ground level is given by the function $s(t) = 100t - 4.9t^2$ m. Find the acceleration of the particle.

Solution   Here $v(t) = s'(t) = 100 - 9.8t$ m/s, so

$$a(t) = v'(t) = -9.8 \text{ m/s}^2.$$

(Note that the acceleration is both constant and downward. It is referred to as the **acceleration due to gravity**.) $\qquad \blacksquare$

   Newton's law of universal gravitation implies that acceleration due to gravity near the surface of the earth is essentially constant. This physical law is one example of why it is useful to consider higher derivatives. For gravitational attraction at modest altitudes (as in many other physical systems), the underlying principles are best expressed in terms of second derivatives. Often it is helpful to formulate the problem in terms of the derivatives involved and then derive the resulting behavior. This reversal of the differentiation operation is the basis of the theory of differential equations, which we shall examine in more detail in later chapters.

Exercise Set 3.9

In each of Exercises 1–16, find $f''(x)$.

1. $f(x) = 2x^3 - 6x$

2. $f(x) = a - bx - cx^2$

3. $f(x) = x^5 - 3x^{-2}$

4. $f(x) = \tan x$

5. $f(x) = \dfrac{1}{1 + x}$

6. $f(x) = \dfrac{ax + b}{cx + d}$

7. $f(x) = \sin^2 x$

8. $f(x) = \sin x \tan x$

9. $f(x) = 1/(1 + \cos x)$

10. $f(x) = x \cos x$

11. $f(x) = \tan^2 x$

12. $f(x) = x^2 \sin^2 x$

13. $f(x) = x - x \sec x$

14. $f(x) = \dfrac{1 - \sin x}{\cos x}$

15. $f(x) = (x^3 + 1)^5$

16. $f(x) = \sec^2 x + \tan^2 x$

In each of Exercises 17–24, find $\dfrac{d^2y}{dx^2}$.

17. $y = x^{-2} - 2x^{-4}$

18. $y = x^2 + 1/x^2$

19. $\dfrac{dy}{dx} = \sec x \tan x$

20. $\dfrac{dy}{dx} = \dfrac{x}{x + \cot x}$

21. $y = x^3 \cos x$

22. $\dfrac{dy}{dx} = x^4(1 - x^4)$

23. $y = 1/(1 - x^2)$

24. $y = x^m + x^n$

In Exercises 25–28, find the indicated derivative.

25. $f''(x)$ for $f(x) = \dfrac{x}{x + 2}$

26. $f''(x)$ for $f(x) = x^3 \sin x$

27. $\dfrac{d^4y}{dx^4}$ for $y = (x^2 - 1)^4$

28. $\dfrac{d^3y}{dx^3}$ for $y = x^3 - 3x^{-2}$

In Exercises 29–32, find $\dfrac{dy}{dx}$ and $\dfrac{d^2y}{dx^2}$.

29. $y^2 = 4x$

30. $x^2 + y^2 = 1$

31. $y^2 - xy = 4$

32. $\sqrt{x} + \sqrt{y} = 1$

33. Explain why a polynomial $p(x) = a_nx^n + \cdots + a_1x + a_0$ has derivatives of all orders (i.e., why $p(x)$ is **infinitely differentiable**).

34. A particle moves along a line so that at time t seconds its position is $s(t) = t^4 - 8t^2 + 2$.
  a. What is the velocity function?
  b. For what time intervals is velocity positive?
  c. What is the acceleration function?
  d. When is the acceleration positive?

35. A particle moves along a line so that at time t its position is $s(t) = t^3 - 6t^2 - 30t$.
  a. What is its velocity function?
  b. When is its velocity negative?
  c. What is its acceleration function?
  d. When is its acceleration positive?

36. A piston moves up and down in an engine block so that, at time t, its position is $s(t) = 4 - 3 \sin t$ cm from the top of its cylinder. Calculate its velocity $v(t)$ and its acceleration $a(t)$. On one graph, plot the position, velocity, and acceleration for $0 \le t \le 4\pi$. What is happening to $s(t)$ when $v(t) > 0$? What is happening to $v(t)$ when $a(t) > 0$?

37. A mass at the end of a vibrating spring is located $s(t) = 10 + 5 \cos t$ cm from the fixed end of the spring t seconds after it is set in motion. Assuming that the motion is horizontal, calculate its velocity $v(t)$ and its acceleration $a(t)$. On one graph, plot the position, velocity, and acceleration for $0 \le t \le 4\pi$. What is happening to $s(t)$ when $v(t) > 0$? What is happening to $v(t)$ when $a(t) > 0$?

38. Calculate $f^{(200)}(x)$ when $f(x) = (x^2 + 1)^{50}$.

39. Calculate $f^{(161)}(x)$ when $f(x) = \sin(-x)$.

40. Calculate $f^{(33)}(x)$ when $f(x) = \cos x$.

41. Calculate $f^{(50)}(x)$ when $f(x) = x \cos(-x)$.

42. Let $f(x) = (x + 1)^{100} = x^{100} + a_{99}x^{99} + \cdots + a_1x + a_0$.
  a. Calculate $f'(0)$.
  b. Calculate $f''(0)$.
  c. Calculate a_1 using part a.
  d. Calculate a_2 using part b.
  e. Calculate a_{50}.

43. Let $f(x) = |x^2 - 1|$. Calculate $f'(x)$ and $f''(x)$. Specify the domains of all three functions.

44. Find a formula for $\dfrac{d^ny}{dx^n}$ if $y = (1 + x)^{-1}$.

45. Find a second-degree polynomial $f(x)$ so that $f(2) = 2$, $f'(2) = 4$, and $f''(2) = 6$.

46. A particle moves with constant acceleration along a line. If $v(2) = 6$ m/s and $v(5) = 15$ m/s, find the acceleration.

47. Examine the first four derivatives of the trigonometric functions $\tan x$, $\cot x$, $\sec x$, and $\csc x$. Do you note any recurring pattern or otherwise see any pattern among these derivatives as happens for $\sin x$ and $\cos x$?

48. The polynomial $q(t) = 0.297t^3 - 2.0101t^2 + 17.6901t + 75.995$ gives the approximate population (in millions) of the

United States from 1900 to 1980, where t represents time (in decades) starting at the year 1900.

a. Graph q' and determine the year when $q'(t)$ is the smallest.

b. Graph q''. When is $q''(t)$ zero? What is the value of $q'(t)$ at this time?

c. Using the graphs of q' and q'', determine how the population is growing when $q''(t) < 0$ and when $q''(t) > 0$.

3.10   Newton's Method

Frequently, in the process of solving a problem or setting up a mathematical model, we encounter an equation of the form $f(x) = 0$, and we need to calculate its solutions—the *zeros* of f. Unfortunately it is often difficult (and sometimes impossible) to calculate these zeros exactly. We are accustomed to the use of algebra and trigonometry to solve equations such as $x^2 - 3x - 2 = 0$ and $\cos x - \sin^2 x = 0$, but there are similar equations for which no such techniques exist.* In these cases, we search for approximate solutions instead.

In this section we discuss an approximation procedure, due to Isaac Newton, for using lines tangent to the graph of f to approximate its zeros. Like the linear approximation procedure of Section 3.7, this method involves "following the tangent." But a new idea also enters, that of **iteration**—repeating the procedure over and over, until a desired degree of accuracy of the approximation is obtained. Although the method described here is now commonly implemented on computers rather than by hand, the mathematical basis for the method and the concept of iteration are important principles to be mastered for the effective use of this idea.

Let us consider a differentiable function f and the problem of approximating a zero c of f (Figure 10.1). Newton's method for approximating c calls for an initial guess x_1 to be made. The line ℓ_1 tangent to the graph of $y = f(x)$ at $(x_1, f(x_1))$ is then constructed. By finding the x-intercept x_2 of ℓ_1, we obtain a second approximation to c. Newton's famous observation is simply that for many functions the second ap-

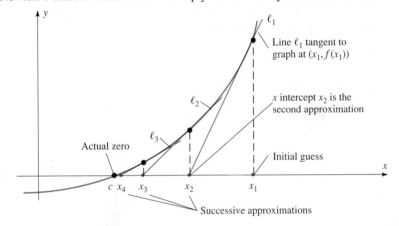

Figure 10.1 Newton's Method for approximating a zero of f.

---

*In advanced algebra courses, it is proved that there are many polynomial equations that are not solvable by traditional algebraic techniques (addition, subtraction, multiplication, division, and taking nth roots). The equation $2x^5 - 10x + 5 = 0$ is one example.

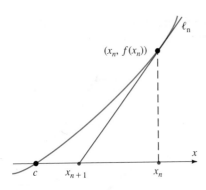

Figure 10.2 Obtaining the approximation x_{n+1} from x_n.

proximation x_2 is better than the first, x_1. If the procedure is then repeated by finding the line ℓ_2 tangent to the graph of f at $(x_2, f(x_2))$, the x-intercept x_3 of ℓ_2 often provides an even closer approximation of c than x_2. The procedure is repeated again and again, until an approximation of sufficiently high accuracy is obtained.

We can obtain a simple equation for the $(n + 1)$st approximation x_{n+1} from the nth approximation x_n by considering Figure 10.2. Since the slope of line ℓ_n tangent to the graph of $y = f(x)$ at $(x_n, f(x_n))$ is given by the derivative $f'(x_n)$, the equation for ℓ_n can be written as

$$y - f(x_n) = f'(x_n)(x - x_n). \tag{1}$$

To find x_{n+1}, the x-intercept of ℓ_n, we set $y = 0$ and solve for x in equation (1). We obtain

$$x = x_n - \frac{f(x_n)}{f'(x_n)}, \quad \text{if} \quad f'(x_n) \neq 0.$$

Since this number is the desired approximation x_{n+1}, we have **the approximation scheme for Newton's Method:**

$$x_{n+1} = x_n - \frac{f(x_n)}{f'(x_n)}, \quad \text{if} \quad f'(x_n) \neq 0. \tag{2}$$

☐ **EXAMPLE 1**

Use Newton's Method to approximate $\sqrt[3]{10}$.

Strategy · · · · · · ·
Specify the function f possessing the desired zero.

Make a rough guess for the first approximation x_1.

Apply (2), using $f'(x) = 3x^2$.

Apply (2) again and so on.

Solution
First, we must express $\sqrt[3]{10}$ as the zero c of some function f. Of course, $\sqrt[3]{10}$ satisfies the equation $x^3 = 10$, so we can consider $\sqrt[3]{10}$ as a zero of the function $f(x) = x^3 - 10$.

Since $f(2) = 2^3 - 10 = -2$ and $f(3) = 3^3 - 10 = 17$, the Intermediate Value Theorem guarantees that $f(x)$ must equal zero for some $x \in (2, 3)$. We use this information to formulate our initial guess, $x_1 = 3$, for the zero c. Then, from equation (2), with $n = 1$, we obtain the second approximation

$$x_2 = 3 - \frac{f(3)}{f'(3)}$$
$$= 3 - \frac{(3^3 - 10)}{3(3)^2}$$
$$= 3 - \frac{17}{27} \approx 2.37.$$

The next approximation, x_3, is obtained from equation (2) using $x_2 = 2.37$ and $n = 2$:

$$x_3 = 2.37 - \frac{f(2.37)}{f'(2.37)}$$
$$= 2.37 - \frac{(2.37)^3 - 10}{3(2.37)^2} \approx 2.17,$$

and so on.

Although the formula for Newton's Method is simple to state, the hand calculations quickly become time-consuming and tedious. However, this is precisely the sort of algorithm that is easy to implement on a hand calculator or on a computer. (Programs for implementing Newton's Method are provided in Appendix I.) Table 10.1 contains the results obtained by using a computer to continue the calculations of this example through $n = 5$ iterations.

Table 10.1

n	x_n	$f(x_n)$	$f'(x_n)$	x_{n+1}
1	3.	17.	27.	2.370370
2	2.370370	3.318295	16.855967	2.173509
3	2.173509	0.267958	14.172419	2.154602
4	2.154602	0.002324	13.926924	2.154435
5	2.154435	0.000004	13.924771	2.154435

The advantage of listing partial calculations in a table such as this is that each approximation x_n (column 2) to the zero can be compared directly to the function value (column 3) at each step. The results obtained in the second and fifth column of Table 10.1 suggest that $x_5 = 2.154435$ is a good approximation to the desired zero. ∎

REMARK   Since Newton's Method is an iterative procedure, it is not necessary to have five columns in tables such as Table 10.1. Note that the second and fifth columns are essentially identical. The only difference is that, after n applications of the method, the second column contains the numbers $x_1, \ldots, x_n$, while the fifth column contains $x_2, \ldots, x_{n+1}$. In subsequent tables, we omit the fifth column.

To use Newton's Method to solve $f(x) = 0$, two questions must be addressed. What is a good initial guess for the method and how many times should the procedure be applied to obtain the desired approximation to the zero?

Unfortunately, we cannot give a general procedure for picking reasonable initial guesses. In fact, this question is a topic of current research among those who study the efficiency of numerical algorithms, and there is still much to be learned. In Appendix IV, we discuss the results of a detailed computer study of the question in the case where $f(x)$ is a cubic polynomial. The results are quite remarkable.

In Example 1, we used the Intermediate Value Theorem to determine that the desired zero was between 2 and 3, and we used 3 as our initial approximation. But the choice $x_1 = 2$ works just as well. Usually, we apply some general technique such as the Intermediate Value Theorem to get a rough estimate for the zero, and then we use Newton's Method to derive an accurate approximation.

In most cases, the second question, regarding the number of iterates, is less problematic. Once the numbers x_n get close to a zero, successive iterates usually tend to the zero quite rapidly. When this happens, we say that the method **converges** to the zero approximated by x_n. If the results are tabulated in a format similar to Table 10.1, then the entries in the third column [the values of $f(x_n)$] are approximately zero. On the other hand, if the entries in the second column do not tend towards a unique number after a few iterations ($n = 10$ or $n = 15$, for example), then you should try a different initial guess x_1.

□ **EXAMPLE 2**

Approximate π by solving the equation $\tan(x/4) = 1$.

Solution   Given our knowledge of the graph of $\tan \theta$ for $0 < \theta < \pi/2$, we know that the graph of the function $f(x) = \tan(x/4) - 1$ crosses the x-axis exactly once for $0 < x < 2\pi$. Thus, π is the only zero of f in $[0, 2\pi)$.

For this function, the Newton iteration scheme is

$$x_{n+1} = x_n - \frac{\tan(x_n/4) - 1}{\frac{1}{4}\sec^2(x_n/4)}$$

$$= x_n - \frac{4\tan(x_n/4) - 4}{\sec^2(x_n/4)}.$$

Using the initial approximation $x_1 = 0$, we obtain the results listed in Table 10.2.

Table 10.2

n	x_n	$f(x_n)$	$f'(x_n)$
1	0.0	-1.0	0.25
2	4.0	0.557408	0.856380
3	3.349112	0.109541	0.557770
4	3.152721	0.005580	0.502798
5	3.141624	0.000016	0.500008
6	3.141593	0.000000	0.500000
7	3.141593		

■

As we mentioned in the beginning of this section, Newton's Method can be used to find approximate solutions to equations on which traditional algebraic techniques either fail or are extremely complicated. The following example falls into the latter category.

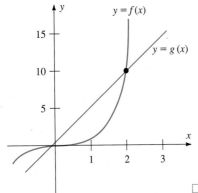

Figure 10.3   The graphs $y = f(x) = 2x^3$ and $y = g(x) = 5x + 1$.

□ **EXAMPLE 3**

Use Newton's Method to find the point in the right half-plane where the graphs of $f(x) = 2x^3$ and $g(x) = 5x + 1$ intersect (see Figure 10.3).

Solution   Finding a number x for which

$$2x^3 = 5x + 1$$

is equivalent to finding a zero of the function $h(x) = 2x^3 - 5x - 1$.

For this function the approximation scheme (2) becomes

$$x_{n+1} = x_n - \frac{2x_n^3 - 5x_n - 1}{6x_n^2 - 5}.$$

Since $h(0) = -1$ and $h(2) = 5$, we know by the Intermediate Value Theorem that a zero must lie in the interval $(0, 2)$. Using a first approximation $x_1 = 2$, we obtain the information contained in Table 10.3 for five iterations of Newton's Method.

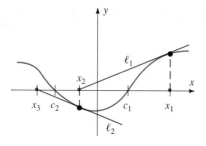

Figure 10.4 Initial approximation x_1 leads to zero c_2 but zero c_1 was desired.

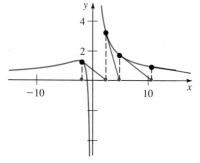

Figure 10.5 Iterates that move away from the zero of the function $f(x) = 4x^{-5/3}(1 + x)$.

Table 10.3

n	x_n	$f(x_n)$	$f'(x_n)$
1	2.0	5.0	19.0
2	1.736842	0.794576	13.099723
3	1.676186	0.037894	11.857560
4	1.672990	0.000103	11.793380
5	1.672982	0.000004	11.793213
6	1.672982		

The desired point of intersection is approximately (1.672982, 9.364914). ■

Depending on the function and the initial approximation, Newton's Method may not converge to the desired zero. We comment on some of the reasons for this here, and pursue some of the details in the Exercise Set. Appendix IV contains a more detailed discussion of the convergence question.

1. The function may have more than one zero. Depending on the initial approximation x_1, the method may converge to a zero other than the one desired (see Figure 10.4).
2. An approximation x_k may be obtained for which $f'(x_k) = 0$. In this case the denominator in the expression for x_{k+1} is zero, so the method fails.
3. The slope of the graph of f may be such that the approximations do not converge, but instead move away from the desired zero (see Figure 10.5 and Exercise 25) or simply oscillate between two or more distinct approximations (see Appendix IV).

Exercise Set 3.10

Most of these exercises are best done with the aid of a calculator or a computer. If a calculator or computer is not available to you, you should calculate only two or three iterations of Newton's Method. On the other hand, if you do have access to either a computer or calculator, your answers should include tables, such as Table 10.2, with at least six decimal places of accuracy.

In Exercises 1–8, use Newton's Method to approximate the zero of the given function lying between the numbers $x = a$ and $x = b$. In each case, first verify that such a zero indeed exists.

1. $f(x) = x^2 - \frac{7}{2}x + \frac{3}{4}$, $a = 0$, $b = 2$

2. $f(x) = 1 - x - x^2$, $a = -2$, $b = -1$

3. $f(x) = 1 - x - x^2$, $a = 0$, $b = 1$

4. $f(x) = x^3 + x - 3$, $a = 1$, $b = 2$

5. $f(x) = \sqrt{x + 3} - x$, $a = 1$, $b = 3$

6. $f(x) = x^3 + x^2 + 3$, $a = -3$, $b = -1$

7. $f(x) = x^4 - 5$, $a = -2$, $b = -1$

8. $f(x) = 5 - x^4$, $a = 1$, $b = 2$

9. Use the results of Exercise 5 to find a point of intersection between the graphs of $f(x) = \sqrt{x + 3}$ and $g(x) = x$.

In Exercises 10–13, use Newton's Method to approximate the number specified.

10. $\sqrt{40}$ **11.** $\sqrt{37}$ **12.** $\sqrt[3]{49}$ **13.** $\sqrt[5]{18}$

14. What are the successive iterates x_n if, when you are applying Newton's method, your first (lucky) guess x_1 is the desired zero?

In Exercises 15–18, use Newton's Method to approximate all solutions of the given equation. You should use either the Intermediate Value Theorem or a graph to help determine initial guesses.

15. $4x^3 - 3x^2 - 11x = -7$ **16.** $3x^3 - 2x^2 - 9x = -4$

17. $x^4 + 4x^3 + 3x^2 - 2x = 1$

18. $6x^4 - 2x^3 - 12x^2 + 6x = -1$

19. Let $f(x) = x \sin 2x$. Approximate the smallest positive number c such that $f'(c) = 0$. What is the geometric significance of c for the graph of f?

20. Consider the three rational numbers $x_1 = 1$, $x_2 = \frac{3}{2}$, and $x_3 = \frac{17}{12}$. Using Newton's Method, explain why these three fractions are successive approximations to $\sqrt{2}$. Express the next iterate x_4 both as a fraction and in decimal form. How accurate an approximation of $\sqrt{2}$ (in terms of the number of decimal places) is x_4? Repeat these calculations for x_5. Note that the number of decimal places of accuracy doubled with one application of Newton's Method. This type of rapid convergence is called **quadratic convergence.**

21. Newton's Method usually exhibits quadratic convergence as described in Exercise 20. However, occasionally the rate of convergence of x_n to a zero is unacceptably slow. To observe this atypical behavior, solve the equation $x^{10} = 0$ using Newton's Method with $x_1 = 0.01$. Note that x_1 agrees with the exact solution to one decimal place. How close (in terms of the number of decimal places of accuracy) is x_{10} to the actual solution? How close would you expect it to be if the convergence had been quadratic?

22. Use a calculator or computer to find an approximation to the point where the graphs of $f(x) = 2x$ and $g(x) = \tan x$ intersect for $0 < x < \pi/2$.

23. Attempt to use Newton's Method to find the zero of $f(x) = (x - 2)^{1/3}$ with initial guess $x_1 = 3$. What happens? Why?

24. The graphs of the functions f and g in Example 3 cross at two other points. Find them.

 25. To determine the temperature of a uniform rod of length L with one end held at a fixed temperature and with the other end radiating heat, we must solve the equation

$$\tan(Lx) = -\frac{x}{h}.$$

where h is a positive constant. Assuming that $L = 2$ m and $h = 3$,

   a. find the first four solutions of this equation using Newton's Method, and

   b. Graph $f(x) = \tan(Lx) + x/h$ using a large scale. Note that the large roots appear to be equally spaced. Explain why.

Summary Outline of Chapter 3

■ The **derivative** of f is defined by   (page 117)

$$f'(x) = \lim_{h \to 0} \frac{f(x + h) - f(x)}{h}.$$

■ The derivative of f at $x = a$ is the slope of the line tangent to the graph of   (page 117)
$y = f(x)$ at $(a, f(a))$. The equation of this line is $y - f(a) = f'(a)(x - a)$.

■ *Theorem:* If f is differentiable at $x = a$, then f is continuous at $x = a$.   (page 121)

■ *Notation:* For $y = f(x)$, $\dfrac{dy}{dx}$ means $f'(x)$; $\dfrac{d^2y}{dx^2}$ means $f''(x)$.   (page 125)

■ If $s(t)$ denotes the location of a particle moving along a line, then the **velocity**   (page 134)
of the particle is $v(t) = s'(t)$, and its **acceleration** is $a(t) = s''(t)$.

■ We have the following differentiation formulas:

If c is a constant, $\dfrac{d}{dx} c = 0$.   (page 126)

If $h(x) = f(x) + g(x)$, $h'(x) = f'(x) + g'(x)$.   (page 120)

If $h(x) = cf(x)$, $h'(x) = cf'(x)$.   (page 120)

If $h(x) = f(x)g(x)$, $h'(x) = f'(x)g(x) + f(x)g'(x)$.   (page 128)

If $h(x) = \dfrac{f(x)}{g(x)}$, $h'(x) = \dfrac{f'(x)g(x) - f(x)g'(x)}{g^2(x)}$.   (page 129)

If $h(x) = x^n$, $h'(x) = nx^{n-1}$.   (page 126)

If $h(x) = \sin x$, $h'(x) = \cos x$.   (page 139)

If $h(x) = \cos x$, $h'(x) = -\sin x$.   (page 139)

If $h(x) = \tan x$, $h'(x) = \sec^2 x$.   (page 142)

If $h(x) = \sec x$, $h'(x) = (\sec x)(\tan x)$.   (page 142)

If $h(x) = \cot x$, $h'(x) = -\csc^2 x$.   (page 142)

If $h(x) = \csc x$, $h'(x) = -\csc x \cot x$.   (page 142)

If $h(x) = f(x)^n$, $h'(x) = nf(x)^{n-1}f'(x)$.   (page 145)

If $h(x) = f(g(x))$, $h'(x) = f'(g(x))g'(x)$.   (page 147)

■ When y is not an explicit function of x we can often calculate dy/dx by the   (page 168)
technique of **implicit differentiation.**

■ By using the tangent to $y = f(x)$ at $(x, f(x))$, we approximate $f(x + \Delta x)$ by the   (page 160)
linear approximation $f(x + \Delta x) \approx f(x) + f'(x)\Delta x$.

■ *Differential Notation:*   If $y = f(x)$, $dy = f'(x)dx$.   (page 166)

■ **Newton's Method** uses the approximation   (page 183)

$$x_{n+1} = x_n - \frac{f(x_n)}{f'(x_n)}$$

to approximate solutions to $f(x) = 0$.

Review Exercises—Chapter 3

In Exercises 1–30, find the derivative of the given function.

1. $f(x) = \dfrac{x}{3x - 7}$

2. $y = \dfrac{\sqrt{x}}{1 + \sqrt{x}}$

3. $y = \dfrac{6x^3 - x^2 + x}{3x^5 + x^3}$

4. $f(x) = \dfrac{x^3 - 1}{x^3 + 1}$

5. $f(t) = \sqrt{t} \sin t$

6. $y = (6x - 4)^{-3}$

7. $y = \dfrac{1}{x + x^2 + x^3}$

8. $f(s) = \left(s + \dfrac{1}{s}\right)^5$

9. $g(t) = (t^{-2} - t^{-3})^{-1}$

10. $y = \left(\dfrac{4x^3 + 3}{x + 2}\right)^4$

11. $y = x^2 \sin x^2$

12. $f(x) = \tan^2 x^2$

13. $y = 1/(x^4 + 4x^2)^3$

14. $f(t) = \tan t \csc t$

15. $f(x) = \dfrac{2x - 7}{\sqrt{x^2 + 1}}$

16. $g(t) = 4t^2\sqrt{t} + t^3\sqrt{t}$

17. $y = [(x^2 + 1)^3 - 7]^5$

18. $y = [\sin(1 + x^2) + x]^3$

19. $f(x) = \sqrt{\dfrac{3x - 9}{x^2 + 3}}$

20. $y = \csc(\cot 3x)$

21. $f(s) = \tan^2 s \cot^5 s$

22. $y = \sqrt[3]{\sin x}$

23. $y = \dfrac{\sqrt{1 + \sin x}}{\cos x}$

24. $f(x) = \dfrac{x + \sqrt{x}}{\sqrt{1 + x^3}}$

25. $f(t) = t^{9/2} - 6t^{5/2}$

26. $y = \dfrac{\sqrt[3]{x}}{x + \sqrt[3]{x}}$

27. $h(t) = \cos^2(t^2 - 1)$

28. $f(s) = \pi^2$

29. $y = \sqrt{x^3}(x^{-3} - 2x^{-1} + 2)^2$

30. $f(x) = x|x|$

In Exercises 31–34, find $f'(x)$ and $f''(x)$.

31. $f(x) = x \sin x$

32. $f(x) = x/(x + 1)$

33. $f(x) = \sec x \tan x$

34. $f(x) = 1/\sqrt{1 + x^2}$

In Exercises 35–38, find the velocity and acceleration of a particle moving along a line if its position along a line after t seconds is $s(t)$.

35. $s(t) = t^2 - 1 + t^{1/2}$

36. $s(t) = \sqrt{t} \sin t$

37. $s(t) = \sqrt{2 + t^2}$

38. $s(t) = \dfrac{\sin t}{1 + \sqrt{t}}$

Use only the definition of the derivative to find dy/dx in Exercises 39–44.

39. $y = 6x - 2$

40. $y = 1/(x + 1)$

41. $y = \sqrt{x + 1}$

42. $y = 1/\sqrt{2x + 1}$

43. $y = x^2 + x + 1$

44. $y = x^3$

In Exercises 45–52, find $\dfrac{dy}{dx}$.

45. $6x^2 - xy - 4y^2 = 0$

46. $6x^3 - 2y^2 = x$

47. $x = y + y^2 + y^3$

48. $\sin \sqrt{x} + \sin \sqrt{y} = 1$

49. $xy = \cot xy$

50. $(2x^2y^3)^{1/3} = 1$

51. $x \sin y = y$

52. $\sqrt{x} + \sqrt{y} = K$

In Exercises 53 and 54, find $\dfrac{d^2y}{dx^2}$.

53. $x^2 + y^2 = 4$

54. $\sin xy = 1/2$

In Exercises 55–59, find an equation for the line tangent to the graph at the indicated point.

55. $\dfrac{1}{x} + \dfrac{1}{y} = 4,$ $(1/2, 1/2)$

56. $y = \tan x - x,$ $(\pi/4, 1 - \pi/4)$

57. $f(x) = \dfrac{\cos x}{\sin x},$ $(\pi/4, 1)$

58. $\sqrt{x} + \sqrt{y} = 1,$ $(1/4, 1/4)$

59. $f(x) = \sec x \tan x,$ $(\pi/4, \sqrt{2})$

In Exercises 60–65, find the differential dy.

60. $y = ax + b$

61. $y = x \sin x$

62. $y = \dfrac{x + 1}{x + 2}$

63. $y = \sqrt{1 + x^2}$

64. $y = \dfrac{\sqrt{x + 1}}{\sqrt{x + 2}}$

65. $y = \cos \sqrt{x}$

In Exercises 66–69, find a linear approximation.

66. $\sqrt{65}$

67. $(8.2)^{2/3}$

68. $\cos 44°$

69. $(0.96)^3$

In Exercises 70–73, determine if the function is differentiable at $x = 2$.

70. $f(x) = |x - 2|$

71. $f(x) = |x^2 - 6x + 8|$

72. $f(x) = \begin{cases} x^3 - 8, & x \le 2 \\ 6x - 12, & x > 2 \end{cases}$

73. $f(x) = \begin{cases} 4 - x^2, & x \le 2 \\ 2x - 4, & x > 2 \end{cases}$

74. A particle moves along a line so that its location at time t is given by a polynomial of degree 2. Prove that it can change direction at most once. Must it change direction?

75. For each of the functions graphed in (a)–(c), determine which of the graphs (i)–(ix) is the graph of its derivative.

(a)
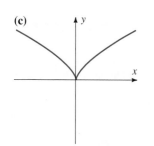

(i)
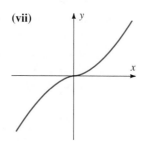

(ii)
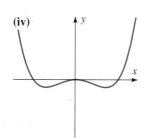

(iii)
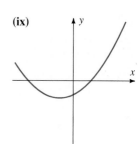

(b)
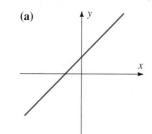

(iv)
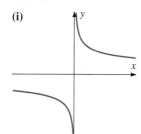

(v)
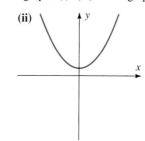

(vi)
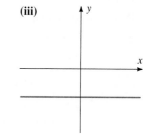

(c)
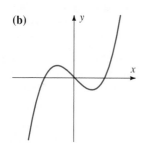

(vii)
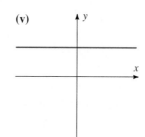

(viii)
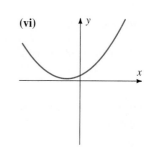

(ix)
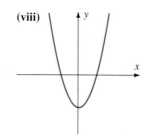

76. A pebble dropped in still water sends out ripples in the shape of concentric circles. If the radius of the outer ripple increases at a rate of 2 m/s, how fast is the area of the disturbance increasing after 5 s?

77. The sides of a regular hexagon are increasing at a rate of 2 cm/s. At what rate is the area increasing when the length of a side is 10 cm?

78. The volume of a sphere is increasing at a rate of 500 cm³/s at the instant when the radius is 25 cm. At what rate is the radius increasing?

79. A ladder 10 m high leans against a wall. The bottom of the ladder is sliding away from the wall at a rate of 1 m/s. At what rate is the area of the triangle formed by the ladder, wall, and ground changing when the base of the ladder is 6 m from the wall?

80. A volume V of oil is spilled at sea. The spill takes the shape of a disc whose radius is increasing at a rate of $a/\sqrt{t}$ m/s where a is constant. Find the rate at which the thickness of the layer of oil is decreasing after 9 s, if the radius at that time is $r(9) = 6a$.

81. A spherical snowball melts at a rate proportional to its surface area. Show that its radius decreases at a constant rate.

82. Calculate the limit

$$\lim_{h \to 0} \frac{(2 + h)^{200} - 2^{200}}{h}.$$

In Exercises 83–86, calculate the equation of the line normal to the graph at the designated point.

83. $y = \sin(x^2)$ at $(\sqrt{\pi/4}, \sqrt{2}/2)$

84. $y = 1/x^2$ at $(1, 1)$

85. $y = x^3 - 2x$ at $(0, 0)$

86. $y = \tan x$ at $(\pi, 0)$

87. Suppose that a spring oscillates according to the position function $s(t) = -4 \cos t$. What is its average velocity during the time interval $0 \le t \le \pi$? What is its instantaneous velocity at $t = \pi/2$?

In Exercises 88–91, specify the expression $f(x_0) + f'(x_0)\Delta x$ that you would use to approximate the given function f near the given number x_0.

88. $f(x) = 1/x$ near $x_0 = 4$

89. $f(x) = \sqrt{x}$ near $x_0 = 8$

90. $f(x) = \sin 2x$ near $x_0 = 0$

91. $f(x) = x^2$ near $x_0 = 1$

92. Prove that all normals to a circle meet at the center. What about ellipses?

93. Find $h'(3)$ if $h(x) = f(g(x))$, $g(3) = 4$, $g'(3) = 2$, $f'(3) = 1$, and $f'(4) = 5$.

94. An isosceles triangle has base 10 cm and base angles $\pi/4$. Approximate the change in the area of the triangle, using linear approximation, if the base angles are increased 0.05 radian.

95. The radius of a sphere is increased from 20 cm to 20.5 cm. Use linear approximation to approximate the change in volume. What is the relative error in this calculation? What is the percentage error?

96. Find the slope of the line tangent to the circle $(x + 2)^2 + (y - 3)^2 = 4$ at the point $(-1, 3 + \sqrt{3})$.

97. At what point(s) is the line tangent to the graph of $y = 4x^3 - 4x + 4$ parallel to the line with equation $y - 8x + 6 = 0$?

98. For $f(x) = \sin 2x$, find the numbers x in the interval $[-\pi, \pi]$ for which the slope of the line tangent to the graph of $y = f'(x)$ equals 2.

99. Give an example of a function f with $f'(x) = 2x^{1/4}$.

100. The graphs of $f(x) = x^2 - 4x + 4$ and its derivative are plotted on the same set of axes. Tangents to the graph of f are drawn at the points of intersection of these two graphs. Find the point of intersection of these two tangents.

101. The position of a particle moving along a horizontal line is given by the function $s(t) = \frac{1}{4}t^4 - 3t^3 + 12t^2 - 20t + 8$.
  a. Find the velocity function $v(t)$.
  b. Find the acceleration function $a(t)$.
  c. For which t is the particle moving to the right?
  d. For which t is the particle moving to the left?
  e. For which t is the particle at rest?
  f. At which t does the particle change direction?
  g. For which t is the acceleration zero?

102. For the function $y = A \sin x + B \tan x$, find A and B if the slope of the tangent to the graph at $x = 0$ is $m_1 = 4$ and the slope of the tangent to the curve at $x = \pi/4$ is $m_2 = 4 + \sqrt{2}$.

103. A particle moves along a line so that after t seconds it is at $s(t) = t^3 + 3t + 3$ on the number line. Find the time at which its velocity and acceleration are of equal magnitude.

104. The adiabatic law for the expansion of air is $PV^{1.4} = C$ where P = pressure, V = volume, and C is a constant. Approximate the percentage change in P caused by a 1% change in V.

105. Prove that the derivative of an odd function is an even function.

106. If the composite function $y = f(g(x))$ is differentiable at $x = a$ must both f and g be differentiable at $g(a)$ and a respectively?

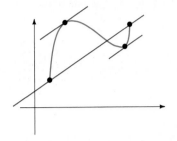

Chapter 4

Applications of the Derivative

In Chapter 3, we discussed two principal interpretations of the derivative of a function f: as the slope of the tangent to the graph $y = f(x)$ and as the instantaneous rate of change of the function f. This chapter is devoted to a more detailed analysis of the relationship between the function f and its derivative f'. In particular, we see how information about f' can be used both to determine certain properties of f and to solve certain types of applied problems.

In Sections 4.1 and 4.2, we see how the derivative can be used to locate the extreme values, both the largest and the smallest, of a function, and we study many applied problems that can be solved by locating the extreme values of a function. Then, in Sections 4.3 through 4.7, we learn how to derive the behavior of f from the values of f' and how to determine the shape of its graph—how it bends on various intervals, its behavior near discontinuities, and its large-scale behavior.

The chapter closes with a discussion of how we use the theory developed here to estimate the error that arises when we approximate using the derivative.

4.1 Extreme Values

There are many situations in which one wishes to find the maximum or minimum value of a function. Here are a few examples.

(a) An aircraft engineer may wish to find the air speed at which an aircraft operates most efficiently.
(b) A chemist may wish to find a temperature at which a certain chemical reaction proceeds most rapidly.
(c) A manufacturing company may wish to determine the production rate at which its average cost per item is a minimum.

In order to solve applied problems of this type we must first determine how to use the derivative to analyze the behavior of a given function. This section presents an important tool, the Extreme Value Theorem, which we shall use in this analysis.

Extreme Values

The extreme values of a function are simply its largest and smallest values. However, to be precise, we must also specify the set of numbers over which we are maximizing or minimizing the function. Usually, the sets that we consider are intervals or unions of intervals (see Section 1.1 for a review of interval notation).

Definition 1	Let the function f be defined for all numbers x in a set S.

Extreme Values

    (i) The **maximum value** of f on the set S is the value M for which

$$f(c) = M \text{ for at least one number } c \in S, \text{ and}$$
$$f(x) \le M \text{ for } all \ x \in S.$$

    (ii) The **minimum value** of f on the set S is the value m for which

$$f(d) = m \text{ for at least one number } d \in S, \text{ and}$$
$$f(x) \ge m \text{ for all } x \in S.$$

    (iii) By the **extreme values** (or **extrema**) of f on S we mean both the maximum and minimum values of f on S.

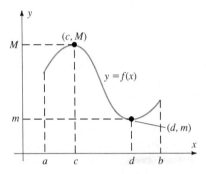

Figure 1.1 Maximum value $M = f(c)$ and minimum value $m = f(d)$ for f on $[a, b]$.

Figure 1.1 illustrates this terminology for a function f where the set S is the closed interval $[a, b]$. Note that the extrema $M = f(c)$ and $m = f(d)$ are *values* of the function f, while c and d are numbers in the interval $[a, b]$ at which these values *occur*. Also, note that maximum or minimum values can occur more than once (Figure 1.2).

It is not necessary that a given function have either a maximum or a minimum value on a given set S. For example, the function $f(x) = x^2$ does not have a maximum value on the set $(-\infty, \infty)$ (Figure 1.3). Similarly, the function $g(x) = x + 1$ does not have a minimum value on the open interval $(-1, \infty)$ because its values form the set of all *positive* real numbers $(0, \infty)$. Since there is no smallest positive real number, g does not have a minimum on $(-1, \infty)$. This example illustrates the significance of the set S in Definition 1. If we consider g on the closed interval $[-1, \infty)$, then it has a minimum value because $g(-1) = 0$ and $0 \le g(x)$ for all $x \in [-1, \infty)$ (Figure 1.4).

An important property of continuous functions, however, is that *a continuous function will always have both a maximum and a minimum value on a closed, finite interval* $[a, b]$. Although the proof of this fact is beyond the scope of this text, it is a theorem on which much of what follows depends because it guarantees the *existence* of the extrema that we shall seek.

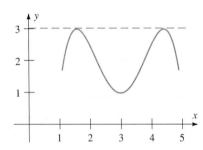

Figure 1.2 The graph of a function defined on the interval $[1, 5]$ for which the maximum value 3 occurs twice.

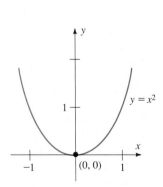

Figure 1.3 The function $f(x) = x^2$ has no maximum value on $S = (-\infty, \infty)$. (The minimum value is $f(0) = 0$.)

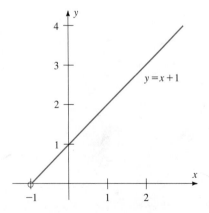

Figure 1.4 The function $g(x) = x + 1$ has no minimum value on the open interval $(-1, \infty)$ because its values form the set of all positive real numbers.

Theorem 1

Existence of Extreme Values

If the function f is continuous on the closed, finite interval $[a, b]$, then f has both a maximum and a minimum value on $[a, b]$.

REMARK 1 The assumptions

(i) f is continuous and

(ii) the set S is a closed and finite interval $[a, b]$

are both necessary as part of any general statement that guarantees the existence of extrema. For example, neither the function $f(x) = x \sin x$ on the unbounded interval $[0, \infty)$ (Figure 1.5) nor the function $g(x) = \tan x$ on the open interval $(-\pi/2, \pi/2)$ (Figure 1.6) possess extrema, and therefore, assumption (ii) regarding the type of set S is appropriate.

To see that assumption (i) requiring continuity is also appropriate, consider the function h defined on the interval $[-1, 1]$ by

$$h(x) = \begin{cases} 1/x, & \text{for } x \neq 0; \\ 0, & \text{for } x = 0 \end{cases}$$

(Figure 1.7). Note that h does not possess any extrema.

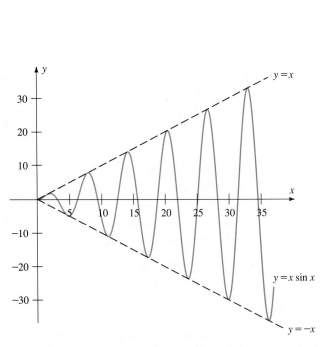

Figure 1.5 The graph of $y = x \sin x$ for $x \geq 0$. This function does not have either a maximum or a minimum value on the closed interval $[0, \infty)$.

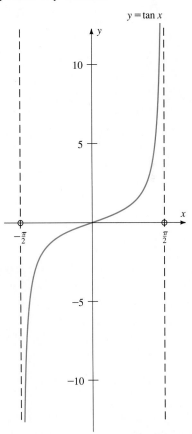

Figure 1.6 The graph of $y = \tan x$ for $-\pi/2 < x < \pi/2$. This function does not have either a maximum or a minimum value on the open interval $(-\pi/2, \pi/2)$.

**Extreme Values and
the Derivative**

Now we pursue the question of how to find the extreme values of a function f over a given interval I. By considering the tangents in Figure 1.8, we find an important relationship between the value of the derivative and the extreme values that occur at *interior points* (numbers that are not endpoints) of the interval I. For instance, if $f(c)$ is the maximum value of f on I, then $(c, f(c))$ is the "high point" on the graph of f. If c is not an endpoint of I, then the tangent to the graph of f at $(c, f(c))$ must be horizontal (assuming the graph has a tangent). Since $f'(c)$ is the slope of this tangent, $f'(c) = 0$. The same reasoning also applies to minimum values.

Theorem 2 and its proof provide a rigorous treatment of this observation.

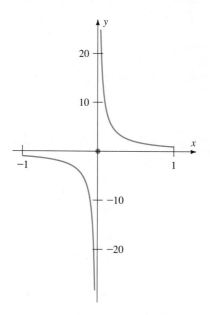

Figure 1.7 The graph of a discontinuous function h that does not have either a maximum or a minimum value on the closed, finite interval $[-1, 1]$.

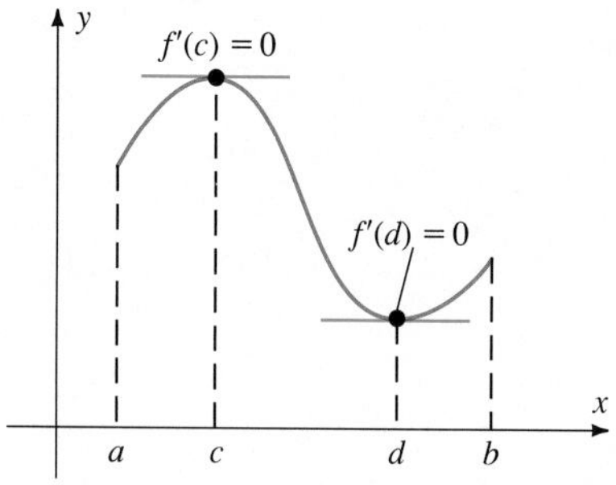

Figure 1.8 If $f(c)$ is the maximum value of f on $[a, b]$, if $c \in (a, b)$, and if $f'(c)$ exists, then $f'(c) = 0$. A similar statement holds for the minimum value $f(d)$.

Theorem 2
Extreme Value Theorem

Suppose $f(c)$ is an extreme value of the function f defined on an interval I. If c is an interior point of I and if f is differentiable at c, then $f'(c) = 0$.

Proof: Consider the case where $f(c)$ is a maximum value of f on I. Then $f(x) \leq f(c)$ for all $x \in I$. Therefore, given h such that $c + h \in I$, we have

$$f(c + h) \leq f(c),$$

so

$$f(c + h) - f(c) \leq 0.$$

Note that the left-hand side of this inequality is the numerator of the difference quotient

$$\frac{f(c + h) - f(c)}{h} \tag{1}$$

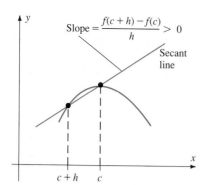

Figure 1.9 If $f(c)$ is a maximum, the slope of a secant line through $(c, f(c))$ and $(c + h, f(c + h))$ is positive for $h < 0$.

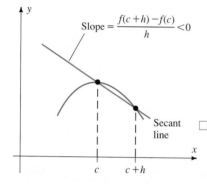

Figure 1.10 If $f(c)$ is a maximum, the slope of a secant line through $(c, f(c))$ and $(c + h, f(c + h))$ is negative for $h > 0$.

involved in the definition of the derivative $f'(c)$. For $h < 0$, the difference quotient (1) is the quotient of a nonpositive number by a negative number. Therefore, for $h < 0$, we have

$$\frac{f(c + h) - f(c)}{h} \geq 0$$

(Figure 1.9). By taking the limit as $h \to 0^-$, we get $f'(c) \geq 0$.

Similarly, if we consider $h > 0$, the difference quotient (1) still has a nonpositive numerator, but now its denominator is positive. Therefore, for $h > 0$, we have

$$\frac{f(c + h) - f(c)}{h} \leq 0$$

(Figure 1.10). By taking the limit as $h \to 0^+$, we get $f'(c) \leq 0$.

Since these two limits must both equal the number $f'(c)$, $f'(c)$ must be zero. The proof in the case where $f(c)$ is a minimum value is similar. ∎

REMARK 2 It is important to note that the Extreme Value Theorem applies only to the case in which the extreme value $f(c)$ occurs *at an interior point c* of I (i.e., *not* at an endpoint) *where the derivative $f'(c)$ exists*.

The following examples illustrate the Extreme Value Theorem.

□ **EXAMPLE 1**

Figure 1.11 shows the graph of the continuous function $f(x) = \sin x$ on the interval $[0, 4\pi]$. Note that

(a) The maximum value $f(\pi/2) = f(5\pi/2) = 1$ occurs where the derivative $f'(x) = \cos x$ equals zero:

$$f'(\pi/2) = \cos(\pi/2) = 0$$
$$f'(5\pi/2) = \cos(5\pi/2) = 0.$$

(b) The minimum value $f(3\pi/2) = f(7\pi/2) = -1$ also occurs where the derivative $f'(x) = \cos x$ equals zero:

$$f'(3\pi/2) = \cos(3\pi/2) = 0$$
$$f'(7\pi/2) = \cos(7\pi/2) = 0.$$ ∎

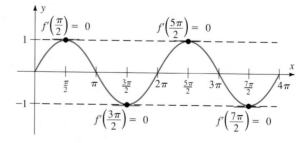

Figure 1.11 Maximum and minimum values of $f(x) = \sin x$ on $[0, 4\pi]$ occur where $f'(x) = \cos x$ equals zero.

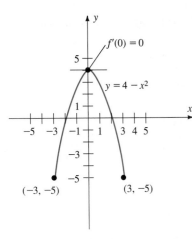

Figure 1.12 The maximum value of $f(x) = 4 - x^2$ on $[-3, 3]$ is $f(0) = 4$, where $f'(0) = 0$. The minimum value is $f(-3) = f(3) = -5$, which occurs at the endpoints of $[-3, 3]$.

☐ **EXAMPLE 2**

Figure 1.12 shows the graph of the function $f(x) = 4 - x^2$ on the interval $[-3, 3]$. Note that the maximum value of f on this interval is

$$f(0) = 4 - 0 = 4$$

and that, since $f'(x) = -2x, f'(0) = 0$, which is consistent with the Extreme Value Theorem.

The minimum value of f on $[-3, 3]$, however, is $f(-3) = f(3) = -5$. Note that we do *not* have $f'(c) = 0$ for $c = -3$ or $c = 3$, since

$$f'(-3) = -2(-3) = 6 \quad \text{and} \quad f'(3) = -2(3) = -6.$$

This is because the minimum value occurs at the *endpoints* of the interval rather than at an interior point. Thus, the Extreme Value Theorem does not apply at these points. ∎

The Extreme Value Theorem is the foundation for a number of similar procedures devised to determine the maximum and minimum values of f. In this section we introduce a version that applies to continuous functions defined on closed, finite intervals $[a, b]$. If $f(c)$ is an extreme value of f, then the Extreme Value Theorem implies that one of three conditions must hold:

(i) c is an endpoint (that is, $c = a$ or $c = b$);
(ii) $f'(c)$ fails to exist (that is, f is not differentiable at c); or
(iii) $f'(c) = 0$ (that is, the graph of f has a horizontal tangent at the point $(c, f(c))$).

It is useful to combine conditions (ii) and (iii) in the following definition.

Definition 2

Let f be a function defined on an interval I and let c be an interior point of I. Then c is a **critical number** for f if either $f'(c) = 0$ or $f'(c)$ fails to exist.

For example, the function graphed in Figure 1.13 possesses five critical numbers, $x_1, x_2, \ldots, x_5$. The derivative of f fails to exist at x_4 (we say that the graph has

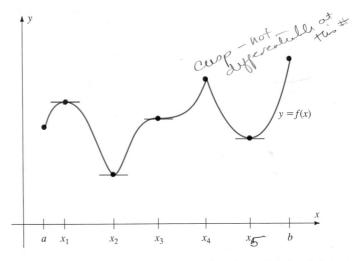

Figure 1.13 The graph $y = f(x)$ of a typical continuous function on the closed, finite interval $[a, b]$. Its minimum value is $f(x_2)$ and its maximum value is $f(b)$.

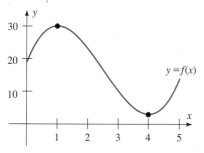

Figure 1.14 The graph $y = f(x)$ for

$$f(x) = 2x^3 - 15x^2 + 24x + 19$$

on the interval $[0, 5]$. The maximum value is 30 at $x = 1$, and the minimum value is 3 at $x = 4$.

a cusp there). At the other four critical numbers, the tangent to the graph is horizontal.

We summarize our first procedure for finding extrema.

Procedure for Finding the Maximum and Minimum Values of a Continuous Function f on a Closed Finite Interval $[a, b]$:

1. Find all numbers $c \in (a, b)$ for which either $f'(c) = 0$ or $f'(c)$ does not exist. (These are the critical numbers.)
2. Compute $f(a), f(b)$, and all values $f(c)$ where c is a critical number for f in (a, b). (That is, check $f(x)$ at the *endpoints* and at all *critical numbers*.)
3. Select the largest and smallest of the values computed in step 2. These are the maximum and minimum values, respectively.

It is important to note that, according to step 2, *you must check the values at the endpoints as well as the values at each of the critical numbers.* Recall that, in Example 2, the maximum value occurred at a critical number but the minimum value occurred at the endpoints.

□ **EXAMPLE 3**

Find the maximum and minimum values of $f(x) = 2x^3 - 15x^2 + 24x + 19$ on the interval $[0, 5]$.

Strategy · · · · · · · · ·
Set $f'(x) = 0$ to find critical numbers.

Solution

We have $f'(x) = 6x^2 - 30x + 24$, so setting $f'(x) = 0$ gives the equation

$$6x^2 - 30x + 24 = 0.$$

We solve for x

$$6(x^2 - 5x + 4) = 0$$
$$6(x - 1)(x - 4) = 0$$

Find remaining critical numbers where $f'(x)$ is undefined. (In this case, there are none.)

and obtain the solutions $x = 1$ and $x = 4$. Note that, since f is a polynomial, it is differentiable everywhere. The critical numbers are, therefore, $x = 1$ and $x = 4$. The values of f at the critical numbers and endpoints are:

Find $f(x)$ for all critical numbers and endpoints.

$$f(0) = 19 \quad \text{(endpoint)}$$
$$f(1) = 30 \quad \text{(critical number)}$$
$$f(4) = 3 \quad \text{(critical number)}$$
$$f(5) = 14 \quad \text{(endpoint)}$$

Find max and min by inspecting these values. Note the corresponding numbers x.

Thus the maximum value of f on $[0, 5]$ is 30, which occurs at $x = 1$. The minimum value is 3, which occurs at $x = 4$. The graph of f appears in Figure 1.14. ■

□ **EXAMPLE 4**

Find the maximum and minimum values of the function $f(x) = 2 + 2x - 3x^{2/3}$ on the interval $[-1, 2]$.

Strategy · · · · · · · ·
Set $f'(x) = 0$ and solve to find critical number(s).

Solution
Since $f'(x) = 2 - 2x^{-1/3}$, setting $f'(x) = 0$ gives the equation

$$2 - 2x^{-1/3} = 0,$$

or

$$x^{1/3} = 1.$$

Determine where

$$f'(x) = 2 - \frac{2}{\sqrt[3]{x}}$$

is undefined to find remaining critical numbers.

Inspect $f(x)$ at critical numbers and endpoints to identify max and min.

The only solution of the equation $f'(x) = 0$ is therefore $x = 1$. Also, $f'(x)$ is undefined at $x = 0$. The critical numbers are therefore $x = 0$ and $x = 1$. Examining $f(x)$ at the critical numbers and endpoints we see that

$$\begin{aligned}
f(-1) &= 2 + 2(-1) - 3(-1)^{2/3} = -3 && \text{(endpoint)} \\
f(0) &= 2 + 2(0) - 3(0)^{2/3} = 2 && \text{(critical number)} \\
f(1) &= 2 + 2(1) - 3(1)^{2/3} = 1 && \text{(critical number)} \\
f(2) &= 2 + 2(2) - 3(2)^{2/3} = 6 - 3\sqrt[3]{4} \approx 1.2378. && \text{(endpoint)}
\end{aligned}$$

We conclude that the maximum value of $f(x)$ on $[-1, 2]$ is $f(0) = 2$ and that the minimum is $f(-1) = -3$. The graph of f appears in Figure 1.15. ∎

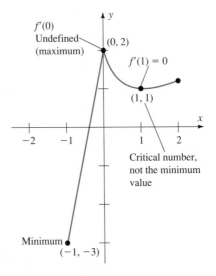

Figure 1.15 For $f(x) = 2 + 2x - 3x^{2/3}$ on $[-1, 2]$, maximum value $= 2$ and minimum value $= -3$.

Relative Extrema

When we use our procedure for finding extrema, we often obtain critical numbers that do not correspond to extreme values (see Example 4). Nevertheless, the behavior of the function near these numbers tells us a great deal about the function. It is therefore useful to introduce additional terminology to help us describe various types of critical numbers. For example, the function graphed in Figure 1.13 has five critical numbers, four of which (x_1, x_3, x_4, and x_5) do not yield extreme values of f on $[a, b]$. However, if we look at the behavior of the function only in the vicinity of these numbers (Figure 1.16), $f(x_1)$ and $f(x_4)$ are maximum values, and $f(x_2)$ and $f(x_5)$ are minima. By contrast, no matter how much we zoom in towards the point $(x_3, f(x_3))$, we find values both larger and smaller than $f(x_3)$. We say that $f(x_1)$ and $f(x_4)$ are *relative maxima* and that $f(x_2)$ and $f(x_5)$ are *relative minima*. The number x_3 is a critical number that does not correspond to any type of extreme point.

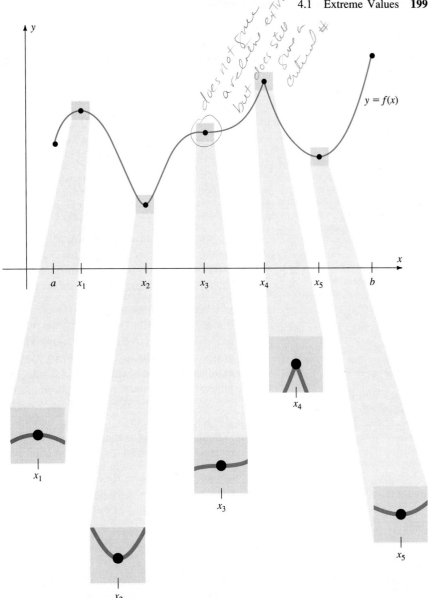

does not have a relative extreme but does still have a critical #

Figure 1.16 The graph $y = f(x)$ from Figure 1.13 along with enlarged views of the graph near the points that correspond to the five critical numbers. The values $f(x_1)$ and $f(x_4)$ are relative maxima, and the values $f(x_2)$ and $f(x_5)$ are relative minima.

Definition 3	Suppose f is a function defined on an interval I and that c is an interior point of I. Then the value $f(c)$ is a **relative maximum** of f if there exists an open interval J containing c such that

$$f(c) \geq f(x) \quad \text{for all} \quad x \in J. \tag{2}$$

Similarly, we say that the value $f(c)$ is a **relative minimum** of f if the inequality in equation (2) is replaced by

$$f(c) \leq f(x) \quad \text{for all} \quad x \in J.$$

A **relative extremum** is a value of f that is either a relative maximum or a relative minimum.

In other words, a relative extremum $f(c)$ as defined in Definition 3 is an extremum (as defined in Definition 1) if we restrict the interval on which the function is considered to a "small" open interval J containing c. Sometimes relative extrema are referred to as local extrema because for all x "near" c, the value $f(c)$ is an extreme value.

REMARK 3   If c is an endpoint of I, then we must adjust our definition to consider only the appropriate half-open interval. For example, if f is a function defined on the closed interval $[a, b]$, then we say that $f(a)$ is a relative maximum of f *on the interval* $[a, b]$ if there exists a number $d > a$ such that

$$f(a) \geq f(x) \quad \text{for all} \quad x \in [a, d).$$

Note that, by indicating the interval $[a, b]$ explicitly, we are clear about the fact that we did not compare the value $f(a)$ to the values $f(x)$ for $x < a$ (assuming that it even makes sense to do so). In Figure 1.16, the value $f(a)$ is a relative minimum for f on $[a, b]$, and the value $f(b)$ is a relative maximum of f on $[a, b]$.

When we want to emphasize the difference between relative extrema and extrema as defined in Definition 1, we refer to a value $f(c)$ that satisfies Definition 1 as an **absolute extremum.** In other words, we call it either an absolute maximum or an absolute minimum. If we think of relative extrema as local extrema, then absolute extrema are global extrema. You should also note that an absolute extremum on an interval I is always a relative extremum on I.

We can use the derivative to locate relative extrema just as we used it to locate absolute extrema. In fact, if c is an interior point of an interval I and $f(c)$ is a relative extremum of f on I, then we can apply the Extreme Value Theorem to f on the interval J. We obtain a relative version of the Extreme Value Theorem.

Theorem 3
Extreme Value Theorem
(Relative Version)

Let f be a function defined on an interval I, c be an interior point of I, and $f(c)$ be a relative extremum of f. If f is differentiable at c, then $f'(c) = 0$.

In other words, if $f(c)$ is a relative extremum at an interior point c, then c is a critical number of f. However, as the critical number x_3 in Figure 1.16 indicates, not all critical numbers correspond to relative extrema.

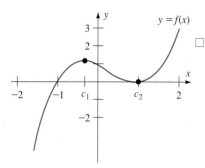

Figure 1.17 The graph $y = f(x) = x^3 - x^2 - x + 1$. Note the relative maximum that occurs at the critical number $c_1 = -1/3$ and the relative minimum that occurs at the critical number $c_2 = 1$.

□ **EXAMPLE 5**

Consider the function

$$f(x) = x^3 - x^2 - x + 1 = (x + 1)(x - 1)^2.$$

Its derivative is

$$f'(x) = 3x^2 - 2x - 1 = (3x + 1)(x - 1).$$

Setting $f'(x) = 0$, we obtain two critical numbers, $c_1 = -1/3$ and $c_2 = 1$. In later sections, we shall learn a systematic way to plot $y = f(x)$, but for now, we simply accept that the graph $y = f(x)$ is as shown in Figure 1.17. The value $f(c_1)$ is a relative maximum, and the value $f(c_2)$ is a relative minimum. Note that, if f is considered as a function of all real numbers, it does not possess any absolute extrema. ∎

☐ EXAMPLE 6

Consider the function $g(x) = x^3$. Since $g'(x) = 3x^2$, we see that g has one critical number, $c = 0$. Moreover, by inspecting the graph $y = x^3$ (Figure 1.18), we see that the value $g(0) = 0$ is not a relative extremum. The function g considered as a function of all real numbers does not possess any absolute extrema. ∎

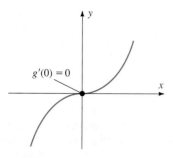

$g'(0) = 0$

Figure 1.18 Critical number $c = 0$ for $g(x) = x^3$ does not yield a relative extremum.

☐ EXAMPLE 7

The function $f(x) = x^{2/3}$ has a relative minimum value of $f(0) = 0$ at $x = 0$. (See Figure 1.19.) This relative minimum is also the minimum value for f throughout its domain since $f(x) = x^{2/3} = (x^{1/3})^2$ is positive for all $x \neq 0$.

Note, however, that *the derivative*

$$f'(x) = \frac{2}{3}x^{-1/3} = \frac{2}{3\sqrt[3]{x}}$$

is undefined for $x = 0$. This fact accounts for the cusp, or point, in the graph of f at $(0, 0)$. ∎

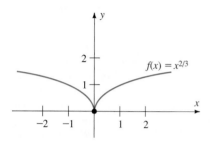

$f(x) = x^{2/3}$

Figure 1.19 Function $f(x) = x^{2/3}$ has a relative minimum value $f(0) = 0$, but $f'(0)$ is undefined.

In this section, we defined three closely related concepts, and throughout the remainder of this chapter, we shall use them repeatedly. Thus, it is important to keep their similarities as well as differences in mind:

(a) Relative and absolute extrema are the maximum or minimum values of a function considered either locally (in the case of a relative extremum) or globally (in the case of an absolute extremum).

(b) In order to locate extrema, we showed that extrema that occur at interior points correspond to critical numbers (the Extreme Value Theorem).

(c) However, you should also remember that some critical numbers do not correspond to extrema.

(d) Finally, remember that extrema can occur at endpoints that are not critical numbers.

Exercise Set 4.1

In Exercises 1–4, a function defined on the interval $[a, b]$ is graphed, and three numbers x_1, x_2, and x_3 between a and b are identified. Determine which of the five numbers a, x_1, x_2, x_3, and b are

a. critical numbers,
b. numbers at which f has a relative minimum on $[a, b]$,
c. numbers at which f has an absolute maximum on $[a, b]$, and
d. numbers at which f has an absolute extremum on $[a, b]$.

1.

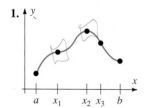

2.

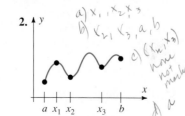

3.

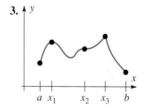

4.

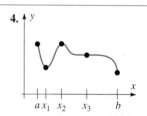

In Exercises 5–10, an extreme value of f on the interval $[a, b]$ occurs at the number c. In each exercise,

a. verify that $f'(c) = 0$,
b. sketch the graph $y = f(x)$ for $x \in [a, b]$, and
c. determine from part b whether $f(c)$ is the (absolute) maximum or minimum value of f on $[a, b]$.

5. $f(x) = x^2 - 4x + 3 \quad [0, 4] \quad c = 2$

6. $f(x) = \sin(x + \pi/4)$ $[0, 2\pi]$ $c = \pi/4$

7. $f(x) = x^2 - 7$ $[-3, 4]$ $c = 0$

8. $f(x) = 6 - (x - 2)^2$ $[0, 5]$ $c = 2$

9. $f(x) = \cos(\pi x)$ $[0, 2]$ $c = 1$

10. $f(x) = 4 + 2x^2$ $[-2, 2]$ $c = 0$

In Exercises 11–28, find all critical numbers and the maximum and minimum values for f on the given interval.

11. $f(x) = x^2(x - 4)$, $x \in [-1, 3]$

12. $f(x) = |x - 2|$, $x \in [0, 5]$

13. $f(x) = \dfrac{1}{x(x - 4)}$, $x \in [1, 3]$

14. $f(x) = x(x^2 - 2)$, $x \in [-1, 2]$

15. $f(x) = x^2(x - 2)$, $x \in [-1, 2]$

16. $f(x) = \sin(x/2)$, $x \in [0, 4\pi]$

17. $f(x) = 1 - \tan^2 x$, $x \in [-\pi/4, \pi/4]$

18. $f(x) = \sin x \cos x$, $x \in [-\pi/2, \pi/2]$

19. $f(x) = \sin x + \cos x$, $x \in [0, \pi]$

20. $f(x) = x^{2/3} + 2$, $x \in [-2, 1]$

21. $f(x) = x + 1/x$, $x \in [1/2, 2]$

22. $f(x) = \left(\dfrac{x + 1}{x - 1}\right)^2$, $x \in [-3, 0]$

23. $f(x) = \sec x$, $x \in [-\pi/4, \pi/4]$

24. $f(x) = \dfrac{x^2}{1 + x}$, $x \in [-1/2, 2]$

25. $f(x) = \dfrac{x^3}{2 + x}$, $x \in [-1, 1]$

26. $f(x) = x \sin x$, $x \in [-\pi/2, \pi/2]$

27. $f(x) = 8x^{1/3} - 2x^{4/3}$, $x \in [-1, 8]$

28. $f(x) = \dfrac{x^{2/3}}{2 + \sqrt[3]{x}}$, $x \in [-1, 8]$

In Exercises 29–32, find the maximum and minimum values for f on the given interval.

29. $f(x) = (\sqrt{x} - x)^2$, $x \in [0, 4]$

30. $f(x) = \sqrt{x}(1 - x^2)$, $x \in [0, 4]$

31. $f(x) = \dfrac{\sqrt{x}}{1 + x}$, $x \in [0, 4]$

32. $f(x) = \dfrac{\sqrt{x}}{1 + \sqrt[3]{x}}$, $x \in [0, 8]$

33. Let c be a nonzero real number. What is the relationship between the critical numbers for the function f on a given interval $[a, b]$ and those of the functions defined by
a. $g(x) = cf(x)$?
b. $h(x) = f(x + c)$?
c. $k(x) = f(cx)$?

34. If f has maximum value M on $[a, b]$ and g has maximum value L on $[a, b]$, is $(L + M)$ necessarily the maximum value of $h = f + g$ on $[a, b]$? Under what conditions will this be true? What relationship will always hold between the maximum value of h on $[a, b]$ and the number $L + M$?

35. Find the numbers b and c if the function $f(x) = x^2 + bx + c$ has minimum value $f(3) = -7$ on the interval $[0, 5]$.

36. Let $f(x) = ax^3 - bx$. Find a and b if $f(2) = 4$ is the maximum value of f on $[0, 4]$.

37. To find the maximum and minimum values of the *periodic* function $f(x) = \sin x$ on the entire number line $(-\infty, \infty)$, we need find these numbers only on $[0, 2\pi]$ (or on *any* interval of length 2π). This is because the graph of $f(x) = \sin x$ on $(-\infty, \infty)$ consists of copies of the graph on the closed interval $[0, 2\pi]$. Generalize this observation to find the maximum and minimum values of the following periodic functions on $(-\infty, \infty)$.
a. $f(x) = \sin x - \cos x$
b. $f(x) = 2 \sin(\pi/2 - x)$
c. $f(x) = 2 \sin x \cos x$.

38. Figure 1.20 shows the graph of $f(x) = |x^2 - 5x + 4|$, which has maximum value $f(0) = 4 = f(5)$ and minimum value $f(1) = f(4) = 0$ for $x \in [0, 5]$. By comparing the graph of f with the graph of $g(x) = x^2 - 5x + 4$ in Figure 1.21, you can see that the effect of the absolute value sign is to reflect the portion of the graph extending below the x-axis through the x-axis.
a. Show that f may be rewritten as

$$f(x) = \begin{cases} x^2 - 5x + 4, & x \le 1 \\ -x^2 + 5x - 4, & 1 < x < 4 \\ x^2 - 5x + 4, & 4 \le x. \end{cases}$$

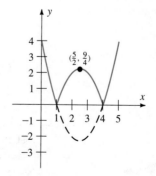

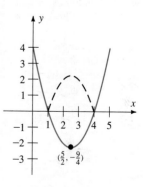

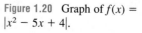

Figure 1.20 Graph of $f(x) = |x^2 - 5x + 4|$.

Figure 1.21 Graph of $g(x) = x^2 - 5x + 4$.

b. Show that f is differentiable for $x \neq 1$ and $x \neq 4$, and that

$$f'(x) = \begin{cases} 2x - 5, & x < 1 \\ 5 - 2x, & 1 < x < 4 \\ 2x - 5, & 4 < x. \end{cases}$$

c. By examining the difference quotient for f, show that $f'(1)$ and $f'(4)$ do not exist.

d. Generalize your findings to a rule for finding the maximum and minimum values for the function $|h(x)|$ in terms of the extrema of the function $h(x)$, $x \in [a, b]$.

In Exercises 39–43, use the results of Exercise 38 to find the maximum and minimum values for the given function on the given interval.

39. $f(x) = |x - 3|, \quad x \in [-2, 6]$

40. $f(x) = |3x - 5|, \quad x \in [0, 4]$

41. $f(x) = |x^2 - x - 6|, \quad x \in [-3, 5]$

42. $f(x) = |\cos x|, \quad x \in [\pi/4, \pi]$

43. $f(x) = |x|/(4 + x), \quad x \in [-2, 4]$.

44. Let $f(x) = 2x^{73} + 7x^{15} + 4x$. Show that the function f does not possess any relative extrema on the interval $(-\infty, \infty)$.

45. Let p_2 be a quadratic polynomial considered as a function on the set of all real numbers. Explain why p_2 has exactly one absolute extremum that occurs at exactly one critical number.

46. Let p_3 be a cubic polynomial considered as a function on the set of all real numbers. What is the maximum number $N_{\max}$ of critical numbers that p_3 can have? What is the minimum number $N_{\min}$? Sketch graphs of cubics to illustrate the fact that, given any integer n such that $N_{\min} \leq n \leq N_{\max}$, there is a cubic with exactly n critical numbers.

47. In this section, we did not discuss the concept of a critical number that is also an endpoint of the interval I under con-

sideration. In this exercise, we study one-sided derivatives so that we can discuss criticality at an endpoint.

Suppose that f is a function defined on the interval $[a, b]$. We define the **right-hand derivative** of f at the endpoint a as

$$f'^{+}(a) = \lim_{h \to 0^+} \frac{f(a + h) - f(a)}{h},$$

if this limit exists.

a. Let $p(x) = x^{3/2}$. Using the definition, calculate the right-hand derivative of p at 0. Explain why this justifies the assertion that p has a critical number at 0.

b. Suppose that g is a function that is defined on a larger interval I that includes the number a as an interior point. Prove that, if $g(x) = f(x)$ for all $x \in [a, b]$ and if g is differentiable at a, then $f'^{+}(a) = g'(a)$.

c. Using part b, show that the function $p(x) = |x|$ has a right-hand derivative at 0 and that $p'^{+}(0) = 1$.

d. Define the **left-hand derivative** of f at the number b.

e. Consider the function $p(x) = x^4 - 2x^2$ defined on the interval $[-1, 1]$. What are the critical numbers of p?

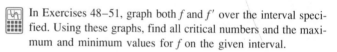 In Exercises 48–51, graph both f and f' over the interval specified. Using these graphs, find all critical numbers and the maximum and minimum values for f on the given interval.

48. $f(x) = 3x^5 - 11x^4 + 2x^3 - 21x^2 + 14x + 123, \quad [-2, 4]$

49. $f(x) = x^3 - 5x + (2x - 1)^{1/3}, \quad [-3, 3]$

50. $f(x) = 2 \sin 2x - \frac{1}{2} \cos(4x - 1), \quad [0, 2\pi]$

51. $f(x) = \begin{cases} x^3 - 3x, & \text{if } x \leq 1 \\ 2(x^2 - 2x), & \text{if } x > 1 \end{cases} \quad [-3, 3]$

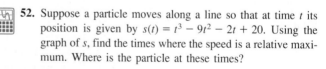 **52.** Suppose a particle moves along a line so that at time t its position is given by $s(t) = t^3 - 9t^2 - 2t + 20$. Using the graph of s, find the times where the speed is a relative maximum. Where is the particle at these times?

4.2 Applied Maximum–Minimum Problems

One of the most important (and difficult) skills needed to become a successful practitioner of the calculus is the ability to formulate a given problem in precise mathematical terms. The examples of this section involve finding the maximum or minimum value of a function on a closed, finite interval. But unlike the examples in the previous section, they are "word problems" stated in prose.

The following diagram illustrates the major conceptual steps involved in applying mathematics to solve problems of this type.

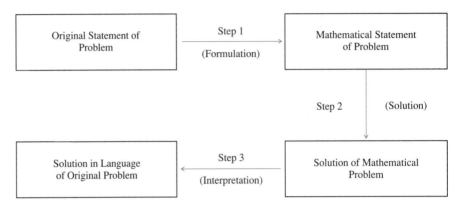

Steps in solving applied problems.

The following procedure refines further these three principal steps.

Procedure for Solving Applied Max–Min Problems

(Formulation)
1. Assign letter names to all variables and, if possible, draw a sketch showing all variables and constants in the problem.
2. Identify the variable for which the extremum is sought. Find an equation for this variable in terms of other variables and constants (principal equation).
3. Find any other equations involving the given variables and constants (auxiliary equations).

(Mathematical Solution)
4. Use auxiliary equation(s) to substitute for variable(s) in the principal equation until it expresses the variable of interest as a *function* of a *single* independent variable.
5. Determine the interval $[a, b]$ on which the function in step 4 is defined.
6. Find the maximum or minimum value for the function on $[a, b]$ by the technique described in Section 4.1.

(Interpretation)
7. Describe the solution found in step 6 in the language of the original problem.

This section consists entirely of examples of problems solved by this procedure. These examples and the exercises at the end of the section provide a ''laboratory'' setting in which you can gain experience with the problem-solving procedure outlined above.

□ **EXAMPLE 1**

Find two nonnegative numbers whose sum is 10 and the sum of whose squares is a maximum.

Strategy · · · · · · ·

Label variables.

Find equation for S by summing squares.

Auxiliary equation states that sum is 10.

Eliminate y in (1), using auxiliary equation.

Find domain of S in the form $[a, b]$ from auxiliary equation and other given information.

Set $S'(x) = 0$ and solve to find critical numbers.

Identify those critical numbers lying in the domain $[0, 10]$.

Inspect $S(x)$ at critical number and endpoints to find maximum.

Solution

Let x and y be the two numbers and let S be the desired sum. We want to maximize

$$S = x^2 + y^2 \qquad \text{(principal equation)} \qquad (1)$$

subject to the condition that

$$x + y = 10 \qquad \text{(auxiliary equation)}. \qquad (2)$$

To eliminate the variable y in equation (1), we solve equation (2) for y to obtain

$$y = 10 - x$$

and substitute into equation (1), which gives

$$S(x) = x^2 + (10 - x)^2.$$

Equation (2) together with the fact that x and y are nonnegative implies that

$$0 \leq x \leq 10.$$

We therefore seek the maximum value of the function S on the interval $[0, 10]$. Setting $S'(x) = 0$ gives the equation

$$S'(x) = 2x - 2(10 - x) = 0,$$

so

$$x = 5.$$

Since $S'(x)$ is defined for all $x \in [0, 10]$, the only critical number in $(0, 10)$ is $x = 5$. The maximum value of S will therefore occur among the following values:

$$\begin{aligned} S(0) &= 0^2 + 10^2 = 100 && \text{(endpoint)} \\ S(5) &= 5^2 + 5^2 = 50 && \text{(critical number)} \\ S(10) &= 10^2 + 0^2 = 100 && \text{(endpoint)}. \end{aligned}$$

The maximum occurs at the endpoints 0 and 10, both of which give the solution as the pair of numbers 0 and 10. ∎

□ **EXAMPLE 2**

An open box is to be made from a square piece of cardboard measuring 12 inches on a side by cutting a square from each corner and folding up the sides as in Figure 2.1. Find the dimensions for which the volume of the resulting box is a maximum.

Strategy · · · · · · ·

Find an equation for the volume V.

Express each dimension in terms of x, the side of the square cut out.

Solution

The volume of the box is

$$V = l \cdot w \cdot h$$

where l = length, w = width, and h = height. If we let x denote the length (and width) of the square cut from each corner, the dimensions of the box can be expressed in terms of x as

$$l = 12 - 2x,$$
$$w = 12 - 2x,$$
$$h = x.$$

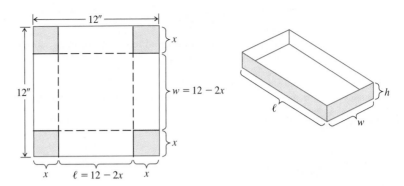

Figure 2.1 Box made from $12'' \times 12''$ square by cutting squares from each corner and folding up flaps.

The volume may now be expressed as a function of x:

Substitute in equation for V to express V as a function of x alone.

$$V(x) = (12 - 2x)(12 - 2x)(x)$$
$$= 4x^3 - 48x^2 + 144x.$$

Find the closed interval $[a, b]$ in which x lies.

Set $V'(x) = 0$ to find its critical numbers in (a, b).

Since the side of the piece cut from a corner can be no longer than $12/2 = 6$ inches, the variable x is restricted to the closed interval $[0, 6]$.

The derivative for V is

$$V'(x) = 12x^2 - 96x + 144$$
$$= 12(x^2 - 8x + 12)$$
$$= 12(x - 2)(x - 6)$$

so the equation $V'(x) = 0$ gives the single critical number $c = 2$ in the interval $(0, 6)$. Since

Compute $V(x)$ at each endpoint and critical number.

$$V(0) = 0, \qquad \text{(endpoint)}$$
$$V(2) = 128, \qquad \text{(critical number)}$$
$$V(6) = 0, \qquad \text{(endpoint)}$$

Select the largest of these values.

the maximum volume of 128 in.^{3} occurs when $x = 2$, which corresponds to the dimensions $l = 12 - 2(2) = 8$ inches, $w = 8$ inches and $h = x = 2$ inches. ∎

□ **EXAMPLE 3**

A water trough is to be constructed from three metal sheets 1 meter wide and 6 meters long plus end panels in the shape of trapezoids. Find the angle at which the long panels should be joined so as to provide a trough of maximum volume (see Figures 2.2 and 2.3).

Strategy · · · · · · ·
Label variables.

Solution
Consider the end view as shown in Figure 2.3. Let θ be the angle between the side panel and the vertical, h the height, and a the length of the side of the triangle opposite the angle θ.

Use

$$h = \frac{h}{1} = \frac{\text{adjacent}}{\text{hypotenuse}}$$
$$= \cos \theta.$$

Then

$$h = \cos \theta, \quad \text{and} \quad a = \sin \theta.$$

The end panel is a trapezoid with small base $b = 1$ and large base $B = 1 + 2a$. The volume of the trough is therefore

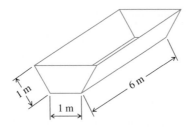

Figure 2.2 Water trough.

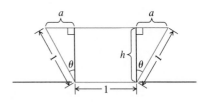

Figure 2.3 End view of the trough.

Find equation for V using formula for area of a trapezoid.

$$V = \tfrac{1}{2}(B + b) \cdot h \cdot 6$$
$$= \tfrac{1}{2}[(1 + 2a) + 1] \cdot h \cdot 6$$
$$= 6(1 + \sin \theta)\cos \theta.$$

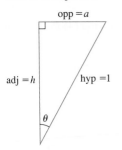

opp $= a$

adj $= h$ hyp $= 1$

θ

Domain of V is $[0, \pi/2]$, from sketch.

Clearly θ must lie within the interval $[0, \pi/2]$. Since V is expressed as a function of θ alone, we may proceed to find the critical numbers for θ:

$$\frac{dV}{d\theta} = -6 \sin \theta + 6 \cos^2 \theta - 6 \sin^2 \theta$$
$$= -6 \sin \theta + 6(1 - \sin^2 \theta) - 6 \sin^2 \theta$$
$$= -12 \sin^2 \theta - 6 \sin \theta + 6$$
$$= -6(2 \sin \theta - 1)(\sin \theta + 1).$$

Set $V'(\theta) = 0$ to find critical numbers.

The equation $dV/d\theta = 0$ therefore yields two critical numbers since

$$2 \sin \theta - 1 = 0 \text{ implies } \sin \theta = 1/2, \quad \text{or} \quad \theta = \pi/6,$$

and

$$\sin \theta + 1 = 0 \text{ implies } \sin \theta = -1, \quad \text{or} \quad \theta = -\pi/2.$$

Remaining critical numbers occur when $V'(\theta) = 0$.

Determine which critical numbers lie in domain $[0, \pi/2]$.

Inspect $V(\theta)$ at critical numbers and endpoints to determine maximum.

Since $dV/d\theta$ is defined for all θ, and the critical number $\theta = -\pi/2$ does not lie within $(0, \pi/2)$, the only critical number for θ in $(0, \pi/2)$ is $\theta = \pi/6$. The maximum value for V must lie among the numbers

$$V(0) = 6(1 + 0)(1) = 6 \qquad \text{(endpoint)}$$
$$V(\pi/6) = 6(1 + 1/2)(\sqrt{3}/2) = 9\sqrt{3}/2 \approx 7.79 \qquad \text{(critical number)}$$
$$V(\pi/2) = 6(1 + 1)(0) = 0 \qquad \text{(endpoint)}$$

The angle corresponding to the maximum volume is therefore $\theta = \pi/6 = 30°$. ∎

☐ **EXAMPLE 4**

Find the right circular cylinder of maximum volume that can be inscribed in a sphere of radius 10 centimeters.

Solution Figure 2.4 shows the cylinder of radius r and height h with edges touching the sphere. The equation for its volume is

$$V = \pi r^2 h. \qquad (3)$$

Since the cross section of the sphere in the xy-plane is a circle of radius 10, it has equation

$$x^2 + y^2 = 10^2. \qquad (4)$$

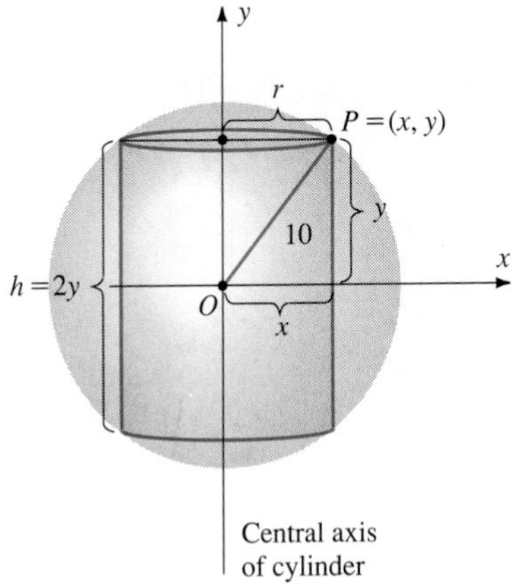

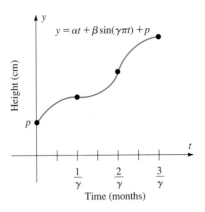

Figure 2.4 Cylinder within a sphere.

Finally, since the edges of the cylinder touch the sphere, we have

$$r = x \qquad (5)$$

and

$$h = 2y. \qquad (6)$$

Substituting in equation (3) for r and h from equations (5) and (6) gives

$$V = \pi x^2 (2y) = 2\pi x^2 y. \qquad (7)$$

To express volume as a function of one variable, we rewrite equation (4) in the form $x^2 = 100 - y^2$ and substitute for x^2 in equation (7):

$$V(y) = 2\pi(100 - y^2)y = 2\pi(100y - y^3), \qquad 0 \le y \le 10.$$

The derivative is

$$V'(y) = 2\pi(100 - 3y^2)$$

so the equation $V'(y) = 0$ gives $3y^2 = 100$, or $y = 10/\sqrt{3}$. This is the only critical number for V. Since $0 \le y \le 10$, checking $V(y)$ at the critical number and endpoints gives

$$V(0) = 2\pi(100 \cdot 0 - 0^3) = 0 \qquad \text{(endpoint)}$$

$$V\left(\frac{10}{\sqrt{3}}\right) = 2\pi\left[100\left(\frac{10}{\sqrt{3}}\right) - \left(\frac{10}{\sqrt{3}}\right)^3\right] = \frac{4000\pi}{3\sqrt{3}} \qquad \text{(critical number)}$$

$$V(10) = 2\pi(100 \cdot 10 - 10^3) = 0. \qquad \text{(endpoint)}$$

The maximum volume of $4000\pi/(3\sqrt{3}) = 4000\pi\sqrt{3}/9$ therefore corresponds to the cylinder of radius $r = x = \sqrt{100 - y^2} = 10\sqrt{6}/3$ and height $h = 2y = 20/\sqrt{3} = 20\sqrt{3}/3$. ∎

□ **EXAMPLE 5**

Researchers who study the rate at which animals (including humans) grow know that growth is not uniform. Periods of rapid growth often occur between periods of very slow growth.

A scientist interested in modelling the growth pattern for a particular species wishes to use an equation of the form

$$y = \alpha t + \beta \sin(\gamma \pi t) + p$$

as a growth model, since this function possesses the growth characteristics described above (see Figure 2.5). Here t represents time in months after birth and y represents height in centimeters. The parameters (constants) α, β, γ, and p are positive numbers calculated by the scientist on the basis of observed data. (This process, called "fitting" the model to the data, is a subject in courses on mathematical statistics.) For a model of this type the growth rate is the derivative dy/dt.

Using this model with $\gamma = 1/4$ and $\alpha = \beta = 1$, we find the maximum and minimum rates of growth and the times at which they occur during the first year of growth.

Figure 2.5 A model for animal growth.

Solution   The growth rate is

$$\frac{dy}{dt} = 1 + \frac{\pi}{4} \cos\left(\frac{\pi t}{4}\right).$$

The mathematical problem is to find the extreme values for the growth rate function on the interval [0, 12].* To obtain the critical numbers for dy/dt, we differentiate, obtaining

$$\frac{d^2y}{dt^2} = -\frac{\pi^2}{16} \sin\left(\frac{\pi t}{4}\right).$$

This derivative equals zero when $\sin\left(\dfrac{\pi t}{4}\right) = 0$, that is, when

$$\frac{\pi t}{4} = 0, \ \pm\pi, \ \pm 2\pi, \ldots$$

or

$$t = 0, \ \pm 4, \ \pm 8, \ \pm 12, \ldots.$$

Now only the critical numbers $t = 4$ and 8 lie within the interval (0, 12). Since d^2y/dt^2 is never undefined, there are no other critical numbers. Table 2.1 (see p. 210) shows the growth rate at each critical number and endpoint.

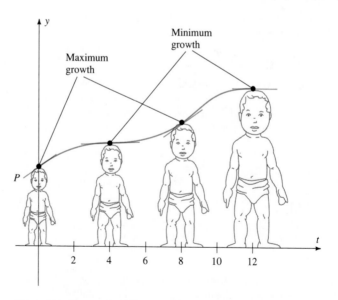

Figure 2.6   Times of minimum and maximum growth rate.

---

*Note here that we are seeking the extreme values of the *derivative dy/dt* rather than the extreme values of y. We therefore examine

$$\frac{d}{dt}\left(\frac{dy}{dt}\right) = \frac{d^2y}{dt^2}$$

for critical numbers.

Table 2.1

t	0	4	8	12
$\dfrac{dy}{dt}$	$1 + \dfrac{\pi}{4}$	$1 - \dfrac{\pi}{4}$	$1 + \dfrac{\pi}{4}$	$1 - \dfrac{\pi}{4}$

The maximum growth rate is observed to be $1 + \pi/4 \approx 1.785$ cm/month, and this rate occurs at birth and at age 8 months. The minimum growth rate $1 - \pi/4 \approx 0.215$ cm/month occurs at ages 4 months and 12 months (see Figure 2.6). ■

Exercise Set 4.2

In Exercises 1–38, solve the given maximum-minimum problem. As part of your answer, specify the function that you are optimizing as well as the closed, finite interval $[a, b]$ that is the relevant domain for the problem.

1. The sum of two nonnegative numbers is 20. Find these numbers if
   a. their product is as small as possible;
   b. the sum of their squares is as small as possible;
   c. the sum of their squares is as large as possible.

2. The sum of two nonnegative numbers is 36. Find these numbers if the first plus the square of the second is
   a. a maximum;
   b. a minimum.

3. A rectangular play yard is to be constructed along the side of a house by erecting a fence on three sides, using the house wall as the fourth wall. Find the dimensions that produce the play yard of maximum area if 20 m of fence is available for the project.

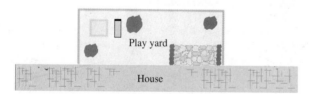

4. A model for the spread of disease assumes that the rate at which a disease spreads is proportional to the product of the number of people infected and the number not infected. Assume the size of the population is a constant N. When is the disease spreading most rapidly?

5. A farmer has 120 m of fencing with which he plans to make a rectangular pig pen. The pen is to have one internal fence that runs parallel to the end fences and divides the pen into two sections. Find the dimensions that produce the pen of maximum area if the length of the larger section is to be twice the length of the smaller section.

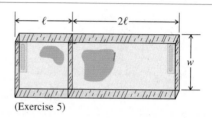

(Exercise 5)

6. Find the minimum and maximum values of the slopes of the lines tangent to the graph
$$y = x^3 - 9x^2 + 7x - 6, \qquad 1 \le x \le 4,$$
and the points where these slopes occur.

7. An open box is to be made from a rectangular sheet of cardboard of dimensions 16 cm by 24 cm by cutting out squares of equal size from each of the four corners and bending up the flaps. Find the dimensions of the box of largest volume that can be made this way.

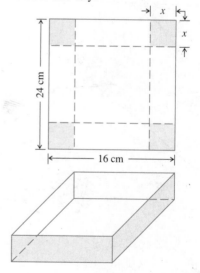

8. A rectangle is inscribed in a right triangle with sides of length 6 cm, 8 cm, and 10 cm, respectively. Find the dimensions of the rectangle of maximum area if two sides of the rectangle lie along two sides of the triangle.

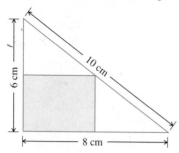

9. A pattern for a rectangular box with a top is to be cut from a sheet of cardboard measuring 10 cm by 16 cm. Find the dimensions of the box for which volume is a maximum.

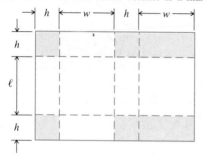

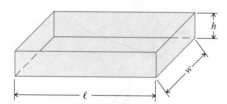

10. A window has the shape of a rectangle surmounted by a semicircle. Find the dimensions that provide maximum area if the perimeter of the window is 10 m.

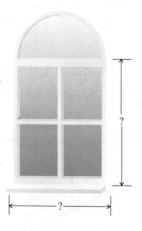

11. A rectangle is inscribed in an isosceles triangle with base 6 cm and height 4 cm. Find the dimensions of the rectangle of maximum area if one side of the rectangle lies along the base of the triangle.

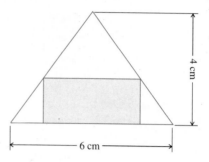

12. A right triangle with hypotenuse 10 cm is rotated about one of its legs to sweep out a right circular cone. Of all such triangles, which generates the cone of maximum volume?

13. A rectangle has two vertices on the x-axis and the other two above the x-axis and on the graph of the equation $y = 4 - x^2$. Find the dimensions for which the area of such a rectangle is a maximum.

14. A triangle has two sides of length a and b. The angle at the vertex where these sides meet is θ. Find the value of θ for which the area of the triangle is a maximum.

15. A sector of a circle of radius r and angle θ, $0 \le \theta \le 2\pi$, is to have fixed perimeter P. Find the dimensions r and θ that maximize the area.

16. Find the dimensions of the right triangle with hypotenuse $h = 2$ and maximum area.

17. An orchard presently has 25 trees per acre. The average yield has been calculated to be 495 apples per tree. It is predicted that for each additional tree planted per acre the yield will be reduced by 15 apples per tree. Should additional trees be planted to increase the yield? If so, how many should be planted to maximize the yield?

18. Find the dimensions of the rectangle of largest area that can be inscribed in a semicircle of radius 8 cm.

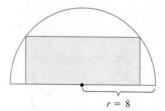

19. Find the dimensions and the area for the rectangle of maximum area that can be inscribed in a circle of radius 4.

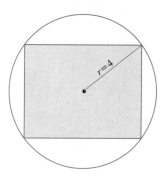

20. A rectangular beam is to be cut from a round log 20 cm in diameter. If the strength of the beam is proportional to the product of its width and the square of its depth, find the dimensions of the cross section for the beam of maximum strength.

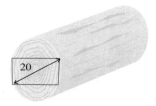

21. A right circular cylinder is inscribed in a right circular cone of radius 3 cm and height 5 cm. Find the dimensions of the cylinder of maximum volume.

22. A right circular cone (shown at the top of the next column) is inscribed in a sphere of radius 10 cm. Find the dimensions for which the volume of the cone is a maximum.

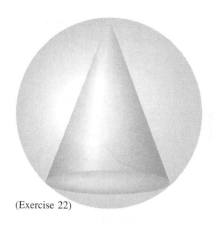

(Exercise 22)

23. For the growth model of Example 5, find the time at which minimum and maximum growth occurs if $\alpha = 3$, $\beta = 2$, $\gamma = 1/3$, *and* $0 \leq t \leq 9$.

24. Prove that $\sin x \leq x$ for $x \geq 0$. (*Hint:* On the interval $[0, 1]$ find the minimum value of the function $f(x) = x - \sin x$. Then argue the case on $[1, \infty)$ from known properties of $\sin x$.)

25. Find the dimensions of the rectangle of maximum area that can be inscribed in the ellipse $x^2 + 4y^2 = 4$.

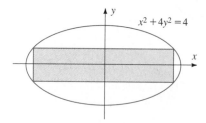

26. Find the points on the ellipse $x^2 + 4y^2 = 4$ nearest the point $(1, 0)$. (*Hint:* Minimize the *square* of the distance.)

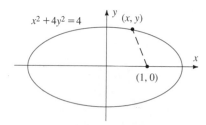

27. A wire 50 cm long is to be cut into two pieces, one of which is to be bent into the shape of a circle and the other of which is to be bent into the shape of a square. Find where the wire should be cut so that the area of the resulting figures is
a. a maximum;
b. a minimum.

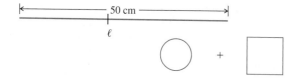

28. United States postal regulations limit the size of a parcel to be mailed parcel post second class according to the rule "length plus girth not to exceed 100 inches." (Girth is the largest perimeter perpendicular to the length.) Find the dimensions of the rectangular box of maximum volume that can meet this restriction. (Neglect the thickness of the material. Then attack the problem in two steps: (1) For fixed girth, find the relationship between width and height that maximizes the area of the cross section. (2) Then find the relationship between girth and length that maximizes volume.)

29. A swimmer is in the ocean 100 meters from a straight shoreline. A person in distress is located on the shoreline 300 meters from the point on the shoreline closest to the swimmer. If the swimmer can swim 3 meters per second and run 5 meters per second, what path should the swimmer follow in order to reach the distressed person as quickly as possible?

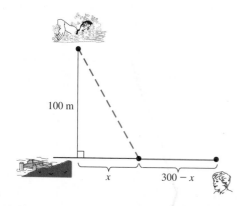

30. Demand for a certain type of electric appliance is related to its selling price p by the equation $x = 2{,}000 - 100p$ where x is the number of appliances that can be sold per month at price p. Find the selling price for which revenues received from sales will be a maximum.

31. For the appliance manufacturer in Exercise 30 the total cost of producing x appliances per month is $C(x) = 500 + 10x$. Find the selling price p at which profits (revenues minus costs) are a maximum.

32. A tool rental company determines that it can achieve 500 daily rentals of jackhammers per year at a daily rental fee of $30. For each $1 increase in rental price 10 fewer jackhammers will be rented. What rental price maximizes revenue?

33. A hotel finds that it can rent 200 rooms per day if it charges $40 per room. For each $1 increase in rental rate 4 fewer rooms will be rented per day. What room rate maximizes revenues?

34. An underground telephone cable is to be laid between two boathouses on opposite banks of a straight river. One boathouse is 600 m downstream from the other. The river is 200 m wide. If the cost of laying the cable is $50 per meter under water and $30 per meter on land, how should the cable be laid to minimize cost?

35. Observations by plant biologists support the thesis that the survival rate for seedlings in the vicinity of a parent tree is proportional to the product of the density of seeds on the ground and their probability of survival against herbivores. (The density of herbivores tends to decrease as distance from the tree increases since the density of the food supply also decreases.) Let x denote the distance in meters from the trunk of the tree. The results of sampling indicate that the density of seeds on the ground for $0 \le x \le 10$ is given by

$$d(x) = \frac{1}{1 + (0.2x)^2}$$

and the probability of survival against herbivores is

$$p(x) = 0.1x.$$

Find the distance, according to the model proposed above, at which the survival rate is a maximum.

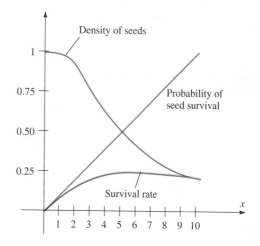

36. In Exercise 35, suppose that the density of seeds for $1 \le x \le 9$ is given by the function

$$d(x) = \frac{1}{x}$$

and that the probability of a seed surviving against herbivores is

$$p(x) = \frac{1}{(x - 10)^2}.$$

Find the distance from the parent tree at which seed survival will be a minimum.

37. Assume that the velocity at which automobiles will travel along a certain section of highway is modeled by the function $v(\rho) = (\rho - 100)^2/100$ where ρ is the density of automobiles (in units of automobiles per kilometer) and $\rho \in [0, 100]$.

a. Find the density for which velocity will be a maximum.

b. If the flow rate is defined to be $q = \rho v(\rho)$, find the density at which flow will be a maximum for the velocity function $v(\rho)$ given above.

38. Fermat's principle states that light travels the path that minimizes the time traveled. Use Fermat's principle to prove the Law of Reflection: a ray of light that strikes a flat reflecting surface at an angle α is reflected away at the same angle. In other words, the angle of incidence equals the angle of reflection.

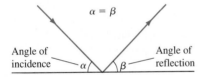

Angle of incidence α $\qquad$ β Angle of reflection

$\alpha = \beta$

39. Starting from a power station on shore, an electrical cable is to be laid along a straight shoreline and then out to an island that is 8 miles down shore and 1.5 miles offshore. The cable will go to a junction box a distance h down shore from the power station and then straight out to the island. Suppose the cost of laying the cable is \$185 per mile on shore and \$275 per mile under water.

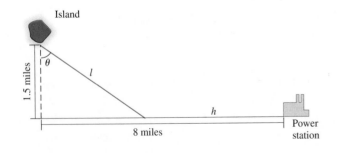

Island

1.5 miles

θ

l

h

8 miles

Power station

a. Find a function $C(\theta)$ that gives the cost of laying the cable from the power station as a function of the angle θ that the cable makes with the direction perpendicular to the shore. (*Hint:* $C(\theta) = 275\ l + 185\ h$.)

b. What is the domain of $C(\theta)$?

c. Graph C and, using the graph, determine if there is an angle θ that minimizes the cost of laying the cable.

d. Graph C' and determine the angle θ where $C'(\theta) = 0$ by magnifying the graph. What is the minimum cost of laying the cable?

40. Suppose you are to design a can that holds 12 fluid ounces (1 fl oz = 1.80469 in.^{3}). Its top is to be made of steel, and the remainder is to be made of aluminum. Moreover, suppose that steel costs 2.5 times as much as aluminum.

a. Determine a function C giving the cost of the can as a function of the radius and the price of aluminum.

b. Draw the graph of C for various prices of aluminum. Do the curves all have a minimum value at the same radius? What radius gives the minimum cost?

c. Find the dimensions of the can that minimize the cost.

41. A horizontal beam 6 m long is imbedded into supports at both ends. Suppose that a load on the beam decreases uniformly from 1500 kg at $x = 0$ to 0 kg at $x = 3$, where x is the distance from one end of the beam. The deflection of the beam in meters is given by

$$y(x) = \alpha \begin{cases} x^2(4x^3 - 60x^2 + 324x - 621), & \text{if } 0 \le x \le 3 \\ -9(4x^3 - 51x^2 + 180x - 108), & \text{if } x > 3 \end{cases}$$

where $\alpha = 3.125 \times 10^{-7}$

(This equation roughly approximates the deflection of a steel beam with this type of load.)

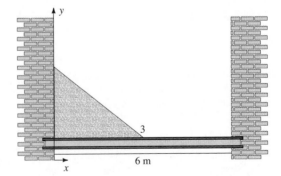

y

3

6 m

x

a. Graph y and, using the graph, determine the maximum deflection of the beam.

b. Show that $y(x)$, $y'(x)$, $y''(x)$, $y^{(3)}(x)$, and $y^{(4)}(x)$ are continuous, but $y^{(5)}(x)$ is not continuous.

4.3   The Mean Value Theorem

Two theorems, the Mean Value Theorem and the Extreme Value Theorem, form the theoretical foundation upon which the techniques in this chapter are based. In Sections 1 and 2 we learned how to use the Extreme Value Theorem to optimize functions. To obtain more information about a function from its derivative, we now present the Mean Value Theorem.

Roughly speaking, the Mean Value Theorem relates the average rate of change of a function to a particular value of the derivative. Consequently, it provides another way for us to use the derivative to predict the behavior of the function. In certain cases, its implications sound so self-evident that one hardly realizes that the Mean Value Theorem is being used. For instance, if you drive 120 miles in 2 hours, then the Mean Value Theorem implies that, at least once, your instantaneous velocity must be exactly 60 miles per hour. Nevertheless, the theorem is so fundamental and so useful that we present a precise mathematical formulation along with a rigorous justification.

Theorem 4

Mean Value Theorem

Let f be continuous on $[a, b]$ and differentiable on (a, b). Then there exists at least one number $c \in (a, b)$ for which

$$f'(c) = \frac{f(b) - f(a)}{b - a}.$$ (1)

(See Figure 3.1.)

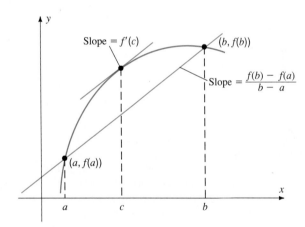

Figure 3.1   Mean Value Theorem: $f'(c) = \dfrac{f(b) - f(a)}{b - a}$.

Before giving a proof of the Mean Value Theorem we comment further on its interpretation and present several examples.

The number

$$\frac{f(b) - f(a)}{b - a}$$

on the right side of equation (1), which is the *slope* of the line through $(a, f(a))$ and $(b, f(b))$, is referred to as the *average change* in f, per unit change in x, over the interval $[a, b]$. The Mean Value Theorem therefore says that *the average* change in f over $[a, b]$ is given by the derivative at one (or more) number c. This ability to represent the change in f *over an interval* by a value of its derivative *at a single number* is the source of the Mean Value Theorem's usefulness.

The following example recalls the concept of average change in f for rectilinear motion (recall Section 3.3).

□ EXAMPLE 1

Consider an object travelling along a line (such as an automobile on a straight highway). If $s(t)$ represents the position of the object at time t, then the *average* velocity of the object between times $t = a$ and $t = b$ is

$$\text{average velocity} = \frac{\text{change in position}}{\text{change in time}} = \frac{s(b) - s(a)}{b - a}.$$

Recall also that we have defined the *instantaneous* velocity of the object at time t to be the derivative

$$v(t) = \lim_{h \to 0} \frac{s(t + h) - s(t)}{h} = s'(t).$$

Thus, for a differentiable position function $y = s(t)$, the Mean Value Theorem says that the *average* velocity from time $t = a$ to time $t = b$ equals the *instantaneous* velocity $v(c) = s'(c)$ for at least one time $t = c$ between a and b.

For example, if an automobile travels 90 miles in 2 hours and $s(t)$ represents the distance travelled after t hours, the *average* velocity during this time period is

$$\text{average velocity} = \frac{s(2) - s(0)}{2 - 0} = \frac{90 - 0}{2} = 45 \text{ mi/h}.$$

We may therefore conclude, by the Mean Value Theorem, that the automobile has velocity $v(c) = 45$ mi/h at at least one instant $t = c$ during this period.   ∎

□ EXAMPLE 2

Figure 3.2 shows the graph of the function $f(x) = \sqrt{x}$ on the interval $[0, 4]$. Since f is continuous on $[0, 4]$ and differentiable on $(0, 4)$, the Mean Value Theorem guarantees the existence of a number c for which

$$f'(c) = \frac{f(4) - f(0)}{4 - 0} = \frac{\sqrt{4} - \sqrt{0}}{4} = \frac{1}{2}. \tag{2}$$

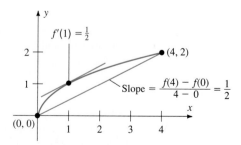

Figure 3.2 For $f(x) = \sqrt{x}$ on $[0, 4]$ the number $c = 1$ satisfies the Mean Value Theorem.

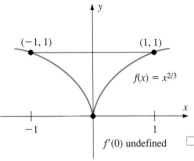

Figure 3.3 The function $f(x) = x^{2/3}$ does not satisfy the conditions of the Mean Value Theorem on $[-1, 1]$ since $f'(0)$ is undefined.

In this example, we can determine all such numbers c by noting that $f'(x) = 1/(2\sqrt{x})$, so equation (2) becomes

$$\frac{1}{2\sqrt{c}} = \frac{1}{2}.$$

Thus, $c = 1$. ∎

□ **EXAMPLE 3**

The function $f(x) = x^{2/3}$ does not satisfy the conditions of the Mean Value Theorem on the interval $[-1, 1]$ because f is not differentiable at 0. In this example we have

$$\frac{f(b) - f(a)}{b - a} = \frac{(1)^{2/3} - (-1)^{2/3}}{2 - (-2)} = 0.$$

Note that, for all $x \neq 0$,

$$f'(x) = \frac{2}{3} x^{-1/3} = \frac{2}{3\sqrt[3]{x}} \neq 0,$$

and therefore, there does not exist any number c between -1 and 1 such that $f'(c) = 0$ (see Figure 3.3). ∎

We prove the Mean Value Theorem by first proving an equivalent, yet apparently more specific theorem—Rolle's Theorem.

Theorem 5

Rolle's Theorem

Let f be continuous on the interval $[a, b]$ and differentiable on (a, b). If $f(a) = f(b)$, then there exists at least one number $c \in (a, b)$ for which $f'(c) = 0$.

Proof: Note first that, if f is a *constant* function on $[a, b]$, $f'(x) = 0$ for all $x \in (a, b)$, so any number $c \in (a, b)$ satisfies the conclusion $f'(c) = 0$ (see Figure 3.4(a)).

If f is not a constant function, let K denote the common value $K = f(a) = f(b)$. Since f is continuous on $[a, b]$, Theorem 1 guarantees that f must have a maximum value M on $[a, b]$ with $M > K$ or a minimum value m with $m < K$. This is because $f(a) = f(b) = K$. (See Figure 3.4(b–d).) In either case we have the existence of an extreme value for f that must occur at a number $c \in (a, b)$. Since $f'(c)$ exists, the Extreme Value Theorem guarantees that $f'(c) = 0$. ∎

We may paraphrase Rolle's Theorem by saying that for a differentiable function f "if the *net* change in the values $f(x)$ between $x = a$ and $x = b$ is zero, then the derivative $f'(x)$ must also equal zero somewhere between a and b." We may also interpret this result geometrically by saying that at some number $c \in (a, b)$ the *tangent to the graph of f must be parallel to the line through $(a, f(a))$ and $(b, f(b))$.*

The Mean Value Theorem generalizes Rolle's Theorem by extending these observations to functions for which $f(a) \neq f(b)$. In order to appreciate how simply this generalization is obtained, think of the graph of f shown in Figure 3.5 as the result of simply "tilting" the graph in Figure 3.4(d) so that $f(b) \neq f(a)$. If this were the case we might expect the tangents at points P and Q still to be parallel to the line through $(a, f(a))$ and $(b, f(b))$. Since the line through $(a, f(a))$ and $(b, f(b))$ has slope

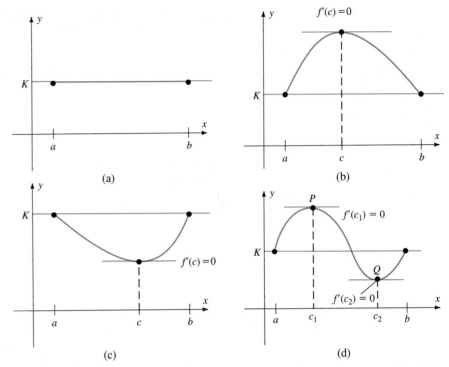

Figure 3.4 Rolle's Theorem. In (a), f is constant and any number c between a and b satisfies the equation $f'(c) = 0$. In (b–d), the extrema of f occur at the number c.

$(f(b) - f(a))/(b - a)$, and since the tangent at a point $(c, f(c))$ has slope $f'(c)$, we would expect to have

$$f'(c) = \frac{f(b) - f(a)}{b - a}$$

when c is the x-coordinate of point P or point Q. We now make this informal argument precise.

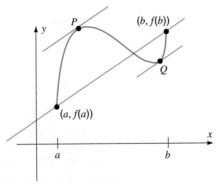

Figure 3.5 If the graph in Figure 3.4(d) is "tilted," the tangents at P and Q are still parallel to the line through $(a, f(a))$ and $(b, f(b))$.

Proof of the Mean Value Theorem

Proof: We seek the number $x = c$ for which the vertical difference $d(x)$ between the graph of f and the line l through the points $(a, f(a))$ and $(b, f(b))$ is an extremum. (See Figure 3.6.)

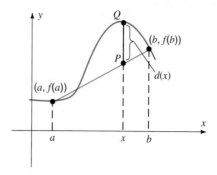

Figure 3.6   Proof of Mean Value Theorem involves finding an extreme value of the difference function d.

Since the line l contains the points $(a, f(a))$ and $(b, f(b))$, the coordinates of a point $P = (x, y)$ on l must satisfy the equation

$$\frac{y - f(a)}{x - a} = \frac{f(b) - f(a)}{b - a} \qquad (= \text{slope of } l)$$

so

$$y = f(a) + \left[\frac{f(b) - f(a)}{b - a}\right](x - a)$$

if $P = (x, y)$ is on l. The vertical difference between the point $P = (x, y)$ on l and the point $Q = (x, f(x))$ on the graph of f is therefore

$$d(x) = f(x) - \left(f(a) + \left[\frac{f(b) - f(a)}{b - a}\right](x - a)\right). \tag{3}$$

We next note the following facts about the function d:

(1) It is continuous on $[a, b]$ and differentiable on (a, b) because f has these properties.

(2) Letting $x = a$ in equation (3) shows that

$$d(a) = f(a) - \left(f(a) + \frac{f(b) - f(a)}{b - a}(0)\right) = 0.$$

(3) Similarly, letting $x = b$ in equation (3) shows that

$$d(b) = f(b) - \left(f(a) + \frac{f(b) - f(a)}{b - a}(b - a)\right) = 0.$$

Thus, the function d satisfies all hypotheses of Rolle's Theorem, so there exists a number $c \in (a, b)$ for which

$$d'(c) = 0. \tag{4}$$

But from equation (3) we see that

$$d'(x) = f'(x) - \frac{f(b) - f(a)}{b - a} \tag{5}$$

so from (4) and (5) we conclude that

$$f'(c) - \frac{f(b) - f(a)}{b - a} = 0,$$

which gives the conclusion of the Mean Value Theorem.   ∎

Exercise Set 4.3

In each of Exercises 1–4, verify that the given function satisfies the hypotheses of Rolle's Theorem on the specified interval $[a, b]$. Find all numbers c, $a < c < b$, that satisfy the equation $f'(c) = 0$.

1. $f(x) = x^2 - 2x + 1$ on $[0, 2]$

2. $f(x) = 1 - x^2$ on $[-1, 1]$

3. $f(x) = \sin 3x$ on $[0, 2\pi]$

4. $f(x) = 1 + \cos 2x$ on $[0, \pi]$

In each of Exercises 5–14, verify that the given function satisfies the hypothesis of the Mean Value Theorem on the specified interval $[a, b]$. Find all numbers c that satisfy both the inequality $a < c < b$ and the equation

$$f'(c) = \frac{f(b) - f(a)}{b - a}.$$

5. $f(x) = 3x + 1$, $x \in [0, 3]$

6. $f(x) = 1 - 2x$, $x \in [-1, 2]$

7. $f(x) = \cos x$, $x \in [\pi/4, 7\pi/4]$

8. $f(x) = 4 - x^2$, $x \in [0, 2]$

9. $f(x) = x^2 + 2x$, $x \in [0, 4]$

10. $f(x) = 3 \sin 2x$, $x \in [0, \pi/2]$

11. $f(x) = x^2 + 2x - 3$, $x \in [-3, 0]$

12. $f(x) = x^3 - 2x + 4$, $x \in [0, 2]$

13. $f(x) = x^3 + x^2 + x$, $x \in [1, 3]$

14. $f(x) = x^3 + x^2$, $x \in [-1, 1]$

15. Sketch the graph of $f(x) = |x|$ for $-1 \le x \le 1$. Does the Mean Value Theorem apply for f on $[-1, 1]$? Why or why not?

16. The function $f(x) = 1/(x - 3)$ is differentiable on the open interval $(0, 3)$, yet the Mean Value Theorem does not apply on $[0, 3]$. Why?

17. Use the Intermediate Value Theorem to show that the equation $x^3 + 3x^2 + 6x + 1 = 0$ has a solution between -1 and 0. Use Rolle's Theorem to show that this solution is the only real number that satisfies this equation.

18. Show that $\tan x > x$ for $0 < x < \pi/2$ as follows.
    a. Let $f(x) = \tan x$ and $g(x) = x$. Note that $f(0) = g(0)$. Show that $f'(x) > g'(x)$ for all x between 0 and $\pi/2$.
    b. Explain why $(f - g)'(x) > 0$ for all x between 0 and $\pi/2$.
    c. Using Rolle's Theorem applied to $f - g$, indicate why 0 is the only number in the interval $[0, \pi/2)$ that satisfies the equation $\tan x = x$.
    d. Using the fact that $\pi < 4$, indicate why $\tan(\pi/4) > \pi/4$.
    e. Use the Intermediate Value Theorem and parts c and d to conclude that $\tan x > x$ for $0 < x < \pi/2$.

19. Show that if f satisfies the conditions of the Mean Value Theorem on $[a, b]$ and $|f'(x)| \le M$ for all $x \in (a, b)$ then

$$|f(b) - f(a)| \le M(b - a).$$

20. Show that if f satisfies the hypotheses of Exercise 19 then

$$f(x) \le f(a) + M(x - a)$$

for all $x \in [a, b]$.

21. Use the results of Exercise 19 to show that

$$|\sin x - \sin y| \le |x - y|$$

for any numbers x and y.

22. Extend the results of Exercise 19 to the case $m \le f'(x) \le M$, $x \in [a, b]$ and obtain the inequality

$$m(b - a) \le |f(b) - f(a)| \le M(b - a)$$

for differentiable functions f.

23. Suppose an automobile travels at speeds between 75 and 90 km/h for 6 h. From Exercise 22 what can you conclude about the distance travelled during this time period?

24. Use the results of Exercise 20 to show that

$$6 < \sqrt{36.2} < 6.02.$$

25. Using the result of Exercise 20, show that

$$\sqrt[3]{1 + \Delta x} < 1 + \frac{\Delta x}{3}$$

for all $\Delta x > 0$.

26. Rolle's Theorem shows that between any two zeros of a differentiable function there must always be a zero of the derivative. Is the following true? If f is differentiable for all x and if $f'(x_1) = f'(x_2) = 0$, f must have a zero in (x_1, x_2).

27. Use the Mean Value Theorem to show that if f and g are continuous on $[a, b]$ and differentiable on (a, b), if $f(a) = g(a)$, and if $f'(x) < g'(x)$ for all $x \in (a, b)$, then $f(b) < g(b)$. (*Hint:* Apply the Mean Value Theorem to the function $h = g - f$.)

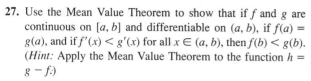 In each of Exercises 28–30, use Newton's Method to approximate the number c satisfying the Mean Value Theorem for the given function on the given interval.

28. $f(x) = x^4 - 3x^2 + 2, \quad x \in [1, 4]$.

29. $f(x) = \tan x, \quad x \in [0, \pi/4]$.

30. $f(x) = \sin \sqrt{x}, \quad x \in [\pi^2/16, \pi^2/4]$.

31. Prove that if f is a quadratic function $f(x) = ax^2 + bx + c$ then the number c in the interval $[x_1, x_2]$ satisfying the Mean Value Theorem is just the midpoint $c = (x_1 + x_2)/2$.

We can interpret the Mean Value Theorem graphically as follows: If f is continuous on $[a, b]$ and differentiable on (a, b), then the graph of $y = f'(x)$ must intersect the line $y = m$ where

$$m = \frac{f(b) - f(a)}{b - a}.$$

In Exercises 32–34, graph f' and the line $y = m$ over the given interval and find their point of intersection c. Then graph the two lines $y = m(x - a) + f(a)$ and $y = f'(c)(x - c) + f(c)$. What do you observe about these two lines?

32. $f(x) = x^2 + 2x - 1, \quad [-2, 2]$

33. $f(x) = x + \cos 3x, \quad [-\pi/2, \pi]$

34. $f(x) = 2x^2 + x^{2/3}, \quad [-1, 2]$
(Note that $f'(0)$ does not exist, but the graph of $y = f'(x)$ still intersects the line $y = m$. Why doesn't this violate the Mean-Value Theorem?)

35. Let $f(x) = x^{2/3}$. Follow the directions for Exercises 32–34 using this function f and the interval $[-2, 1]$. Relate your results to the Mean Value Theorem.

4.4 Increasing and Decreasing Functions

With the Mean Value Theorem, we are now equipped to analyze the behavior of a function on the intervals bordered by its critical numbers. In this section, we establish a relationship between the sign of the derivative of a function and whether its graph is rising or falling. In order to state this relationship precisely, we need the following definition.

Definition 4

Suppose the function f is defined on an interval I. We say that f is **increasing on I** if $f(x)$ increases as x increases. More precisely, f is increasing on I if, for any two numbers x_1 and x_2 in I,

$$x_2 > x_1 \quad \text{implies that} \quad f(x_2) > f(x_1) \qquad \text{(see Figure 4.1).}$$

Similarly, f is **decreasing on I** if, for any two numbers x_1 and x_2 in I,

$$x_2 > x_1 \quad \text{implies that} \quad f(x_2) < f(x_1) \qquad \text{(see Figure 4.2).}$$

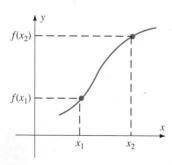

Figure 4.1 f increasing: if $x_2 > x_1$, then $f(x_2) > f(x_1)$.

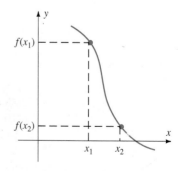

Figure 4.2 f decreasing: if $x_2 > x_1$, then $f(x_2) < f(x_1)$.

Note that we refer to a function as increasing or decreasing only *on intervals,* not at particular numbers. Also, note that we do not specify whether the interval I includes its endpoints.

As Figure 4.3 illustrates, tangents with positive slope suggest a rising graph while tangents with negative slope suggest a falling graph. This relationship is made more precise in the following theorem.

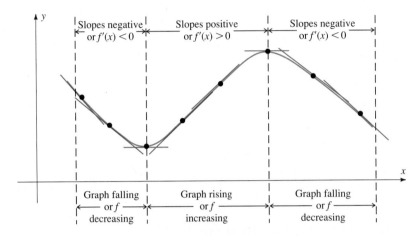

Figure 4.3   Relationship between the graph rising and falling and the sign of $f'(x)$.

Recall that an interior point of an interval I is any number in I that is not an endpoint of I and that the **interior** of I is the set of all interior points of I. In other words, the interior of I is obtained from I by omitting its endpoints.

Theorem 6

Let f be continuous on the interval I and differentiable on the interior of I. Then

(i) If $f'(x) > 0$ for all x in the interior of I, then f is increasing on I.
(ii) If $f'(x) < 0$ for all x in the interior of I, then f is decreasing on I.

Proof: To prove part (i) assume that x_1 and x_2 are any two numbers in I with $x_1 < x_2$. Our hypotheses guarantee that f is continuous on $[x_1, x_2]$ and that $f'(x)$ exists for all $x \in (x_1, x_2)$. Thus, the Mean Value Theorem guarantees the existence of a number $c \in (x_1, x_2)$ for which

$$f'(c) = \frac{f(x_2) - f(x_1)}{x_2 - x_1}. \tag{1}$$

Since we are assuming that $f'(x) > 0$ for all $x \in (x_1, x_2)$, we know that $f'(c) > 0$. Thus, the right side of equation (1) is a positive number. Since $x_2 - x_1 > 0$, it follows that $f(x_2) - f(x_1) > 0$ also. That is, $f(x_1) < f(x_2)$. Since x_1 and x_2 were *any* numbers in I with $x_1 < x_2$, this shows that f is increasing on I, which proves statement (i).

The proof for statement (ii) is the same except, since we are assuming $f'(x) < 0$ for $x \in (x_1, x_2)$, we have $f'(c) < 0$ in equation (1). It follows that $f(x_2) - f(x_1) < 0$, so $f(x_1) > f(x_2)$ and f is decreasing on I.   ∎

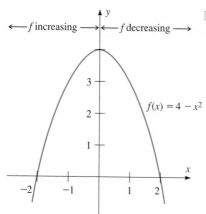

Figure 4.4 $f(x) = 4 - x^2$ is increasing on $(-\infty, 0]$ and decreasing on $[0, \infty)$.

□ **EXAMPLE 1**

Use Theorem 6 to verify that the function $f(x) = 4 - x^2$ is

(a) increasing on the interval $(-\infty, 0]$, and
(b) decreasing on the interval $[0, \infty)$.

Solution   Since $f'(x) = -2x$ exists for all x, f is continuous and differentiable for all x, so Theorem 6 may be applied as follows:

(a) For $x \in (-\infty, 0)$, $f'(x) = -2x$ is *positive*. Thus, f is increasing on $(-\infty, 0]$. (We include the endpoint 0 because f is continuous on the interval $(-\infty, 0]$.)
(b) For $x \in (0, \infty)$, $f'(x) = -2x$ is *negative* so f is decreasing on $[0, \infty)$. (See Figure 4.4.) ■

□ **EXAMPLE 2**

The function $f(x) = \sin x$ satisfies the conditions of Theorem 6 on the interval $[0, 2\pi]$. Theorem 6 shows that f is

(a) increasing on $[0, \pi/2]$, because $f'(x) = \cos x$ is positive for $x \in (0, \pi/2)$.
(b) decreasing on $[\pi/2, 3\pi/2]$ because $f'(x) = \cos x$ is negative on $(\pi/2, 3\pi/2)$.
(c) increasing on $[3\pi/2, 2\pi]$ because $f'(x) = \cos x$ is positive for $x \in (3\pi/2, 2\pi)$. (See Figure 4.5.) ■

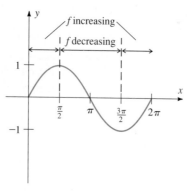

Figure 4.5 $f(x) = \sin x$ is increasing on $[0, \pi/2]$ and $[3\pi/2, 2\pi]$ and decreasing on $[\pi/2, 3\pi/2]$.

□ **EXAMPLE 3**

Figure 4.6 shows the graph of the function f defined as follows:

$$f(x) = \begin{cases} \dfrac{1}{x}, & x \neq 0 \\[2mm] 0, & x = 0. \end{cases}$$

Note that $f(x)$ is defined for all x, but that f is continuous only on the intervals $(-\infty, 0)$ and $(0, \infty)$. Since $f'(x) = -1/x^2$ is negative for all $x \neq 0$, we conclude that f is decreasing both on $(-\infty, 0)$ and on $(0, \infty)$.

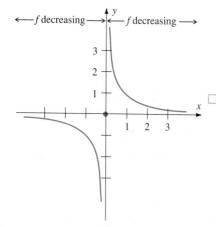

Figure 4.6 $f(x) = 1/x$ is decreasing on $(-\infty, 0)$ and on $(0, \infty)$.

Note that we may *not* conclude that f is decreasing on $(-\infty, 0]$ or on $[0, \infty)$ because f is not continuous on these intervals. In fact, f is *not* decreasing on $(-\infty, 0]$ or on $[0, \infty)$ as you are asked to show in Exercise 85. ∎

REMARK 1:   While Theorem 6 gives conditions that *guarantee* (i.e., are *sufficient*) that f be increasing on an interval I, it is not *necessary* that an increasing function satisfy these conditions. The next example illustrates this point (which is valid also for decreasing functions).

☐ **EXAMPLE 4**

The function $f(x) = x^3$ is continuous and differentiable for all x. Since $x_2^3 > x_1^3$ whenever $x_2 > x_1$, this function is increasing on $(-\infty, \infty)$. We can use Theorem 6 to establish this conclusion in two steps as follows (see Figure 4.7):

(i) For $x < 0$ we have $f'(x) = 3x^2 > 0$. Since f is continuous at $x = 0$, we may conclude that f is increasing on $(-\infty, 0]$.
(ii) Similarly, since $f'(x) = 3x^2 > 0$ for $x > 0$, f is increasing on $[0, \infty)$.

Statements (i) and (ii) combine to tell us that f is increasing on $(-\infty, 0] \cup [0, \infty) = (-\infty, \infty)$. We cannot apply Theorem 6 directly to the interval $(-\infty, \infty)$, however, since $f'(0) = 0$. ∎

REMARK 2   According to Definition 4, if a function f is increasing on an interval I_1 and if I_2 is a "smaller" interval contained in I_1, then f is obviously increasing on I_2. Not much is gained by asserting that a function that is increasing on $[0, 2]$ is also increasing on $[0, 1]$. Therefore, whenever we specify the intervals on which a function is increasing, we mention only those intervals that are the "largest" or "widest" intervals on which f is increasing. This remark also applies to intervals on which a function is decreasing.

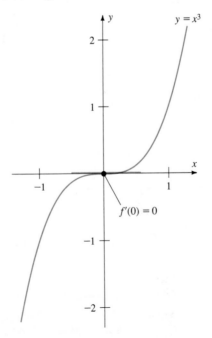

Figure 4.7   The function $f(x) = x^3$ is increasing on $(-\infty, \infty)$ even though $f'(0) = 0$.

The following procedure uses critical numbers to suggest candidates for transition points between intervals on which f is increasing and intervals on which f is decreasing.

> **Procedure for Finding the Intervals on Which f is Either Increasing or Decreasing:**
>
> 1. Locate all critical numbers for f.
> 2. Find the intervals in the domain of f determined by the critical numbers and any numbers x for which $f(x)$ is undefined.
> 3. Apply Theorem 6 in each interval.

Recall that there are critical numbers that do not correspond to transitions between intervals on which f is increasing and intervals on which f is decreasing (Example 4). Thus, you must apply Theorem 6 to each interval as stated in step 3.

☐ **EXAMPLE 5**

Find the intervals on which the function

$$f(x) = \tfrac{1}{3}x^3 - x^2 - 3x + 4$$

is either increasing or decreasing.

Strategy · · · · · · · ·
Find the critical numbers and the numbers x for which $f(x)$ is undefined.

Apply Theorem 6 on each interval.

Solution
Here $f'(x) = x^2 - 2x - 3 = (x - 3)(x + 1)$.

Setting $f'(x) = (x - 3)(x + 1) = 0$ gives the critical numbers $x = -1$ and $x = 3$. There are no numbers x for which $f'(x)$ is undefined or for which $f(x)$ is undefined.

From the factored form of $f'(x)$, we can see (Figure 4.8), that

(i) If $x < -1$, then both $(x - 3) < 0$ and $(x + 1) < 0$. Thus, $f'(x) > 0$ for $x \in (-\infty, -1)$, so f is increasing on $(-\infty, -1]$.

(ii) If $-1 < x < 3$, then $(x - 3) < 0$, but $(x + 1) > 0$. Thus $f'(x) < 0$ for $x \in (-1, 3)$, so f is decreasing on $[-1, 3]$.

(iii) If $x > 3$, then both $(x - 3) > 0$ and $(x + 1) > 0$. Thus, $f'(x) > 0$ for $x \in (3, \infty)$, so f is increasing on $[3, \infty)$.

(See Figure 4.9.) ■

It is often difficult to determine the sign of $f'(x)$ on various intervals, especially if $f'(x)$ is not easily factored. If f' is a continuous function, a procedure based on the Intermediate Value Theorem may be used. The idea is simply that if a and b are successive critical numbers for f, then f' cannot change sign on the interval (a, b). (Otherwise the Intermediate Value Theorem would imply the existence of a zero for f' between a and b.) We can therefore determine the sign of f' on all of (a, b) by

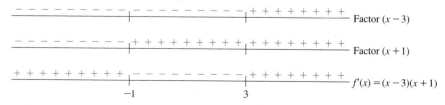

Figure 4.8 Sign analysis for $f'(x) = (x - 3)(x + 1)$ in Example 5.

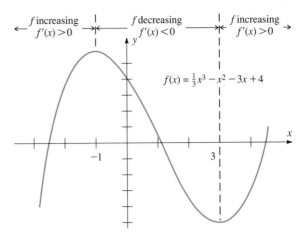

Figure 4.9 Graph of $f(x) = \frac{1}{3}x^3 - x^2 - 3x + 4$.

checking the sign of $f'(t)$ *at any particular* $t \in (a, b)$. We use this idea in the following example.

□ **EXAMPLE 6**

Determine the intervals on which the function

$$f(x) = \frac{x^2}{x - 2}$$

is increasing or decreasing.

Solution   The derivative is found using the Quotient Rule:

$$f'(x) = \frac{(x-2)(2x) - (x^2)(1)}{(x-2)^2} = \frac{x^2 - 4x}{(x-2)^2} = \frac{x(x-4)}{(x-2)^2}.$$

Thus, $f'(x) = 0$ for $x = 0$ and $x = 4$, and $f(2)$ is undefined. The candidates for transition points are therefore $x = 0, 2,$ and 4, and we must inspect the sign of f' on each of the intervals $(-\infty, 0)$, $(0, 2)$, $(2, 4)$, and $(4, \infty)$. Table 4.1 shows the result of checking the sign of $f'(t)$ at an arbitrarily chosen test number t in each of these intervals, and the conclusions from Theorem 6. (See Figure 4.10.) Note that our conclusions include the endpoints 0 and 4, where f is continuous, but not the endpoint 2 because $f(2)$ is undefined.   ■

Table 4.1   Analysis of $f(x) = \frac{x^2}{x - 2}$ for intervals of increase and decrease

Interval I	Test number $t \in I$	Sign of $f'(t)$	Conclusion
$(-\infty, 0]$	$t = -1$	$f'(-1) = \frac{5}{9} > 0$	f increasing on $(-\infty, 0]$
$[0, 2)$	$t = 1$	$f'(1) = -3 < 0$	f decreasing on $[0, 2)$
$(2, 4]$	$t = 3$	$f'(3) = -3 < 0$	f decreasing on $(2, 4]$
$[4, \infty)$	$t = 5$	$f'(5) = \frac{5}{9} > 0$	f increasing on $[4, \infty)$

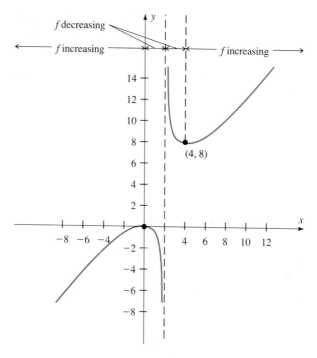

Figure 4.10 Graph of $f(x) = \dfrac{x^2}{x-2}$.

□ **EXAMPLE 7**

Figure 4.11 shows a Ferris wheel of radius $r = 20$ ft mounted 5 ft above ground level. When in motion the Ferris wheel turns at a constant rate of $\pi/4$ radians per second. Figure 4.12 shows that after t seconds the point P that is originally located at

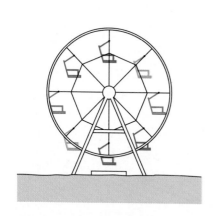

Figure 4.11 Ferris wheel in Example 7.

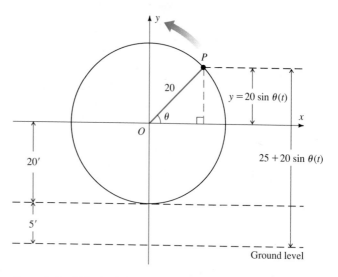

Figure 4.12 Height of point P is $h(t) = 25 + 20 \sin \theta(t)$.

angle $\theta(0) = 0$ will be located at angle $\theta(t) = \pi t/4$. For which times during the first 8 seconds of motion will the seat located at point P be (a) rising and (b) falling?

Strategy · · · · · · · ·

Find the function h giving the altitude of point P at time t.

Solution

With an xy-coordinate system superimposed as in Figure 4.12, the height of point P above ground level at time t is

$$h(t) = 25 + 20 \sin \theta(t).$$

Since $\theta(t) = \dfrac{\pi t}{4}$ radians, we have

$$h(t) = 25 + 20 \sin \frac{\pi t}{4}.$$

Find $h'(t)$.

Then

$$h'(t) = 20\left(\frac{\pi}{4}\right)\cos \frac{\pi t}{4} = 5\pi \cos \frac{\pi t}{4}.$$

Thus, $h'(t) = 0$ if

Set $h'(t) = 0$ and solve to find the critical numbers for h in $[0, 8]$.

$$\cos\left(\frac{\pi t}{4}\right) = 0$$

which occurs when

$$\frac{\pi t}{4} = \frac{\pi}{2} + n\pi, \quad n = 0, \pm1, \pm2, \ldots$$

or

$$t = \frac{4}{\pi}\left(\frac{\pi}{2} + n\pi\right), \quad n = 0, \pm1, \pm2, \ldots$$

$$= 2 + 4n, \quad n = 0, \pm1, \pm2, \ldots.$$

Apply Theorem 6:

P is rising when h is increasing; P is falling when h is decreasing.

The critical numbers for h in the interval $[0, 8]$ are therefore $t = 2$ and $t = 6$. Since h is continuous on $[0, 8]$, we apply Theorem 6 on the intervals $[0, 2]$, $[2, 6]$, and $[6, 8]$. The conclusions are contained in Table 4.2. (See Figure 4.13.) ∎

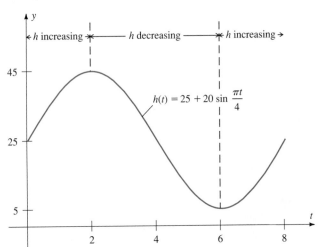

Figure 4.13 Altitude $h(t)$ of point P on Ferris wheel in Example 7.

Table 4.2

Time interval	Test time	$h'(t)$	Conclusion
$[0, 2]$	$t = 1$	$5\pi \cos \dfrac{\pi}{4} = \dfrac{5\pi\sqrt{2}}{2} > 0$	h increasing P rising
$[2, 6]$	$t = 4$	$5\pi \cos \pi = -5\pi < 0$	h decreasing P falling
$[6, 8]$	$t = 7$	$5\pi \cos \dfrac{7\pi}{4} = \dfrac{5\pi\sqrt{2}}{2} > 0$	h increasing P rising

Theorem 7
First Derivative Test

Let f be continuous on $[a, b]$ and suppose c is the only critical number of f in (a, b). Then

   (i) if $f'(x) > 0$ for all $x \in (a, c)$ and if $f'(x) < 0$ for all $x \in (c, b)$, then $f(c)$ is a relative maximum;
  (ii) if $f'(x) < 0$ for all $x \in (a, c)$ and if $f'(x) > 0$ for all $x \in (c, b)$, then $f(c)$ is a relative minimum;
 (iii) if $f'(x) > 0$ for all $x \in (a, c) \cup (c, b)$, then f is increasing on $[a, b]$;
 (iv) if $f'(x) < 0$ for all $x \in (a, c) \cup (c, b)$, then f is decreasing on $[a, b]$.

In cases (iii) and (iv), $f(c)$ is neither a relative maximum nor a relative minimum.

Roughly speaking, we say that $f(c)$ is a relative extremum if f' changes sign at c (see Figures 4.14–4.17).

Proof: To prove statement (i) we first note that, since $f'(x) > 0$ for $x \in (a, c)$ and f is continuous at c, f is increasing on $[a, c]$. Thus, $f(c) \geq f(x)$ for all $x \in [a, c]$. Similarly, since $f'(x) < 0$ for $x \in (c, b)$, f is decreasing on $[c, b]$, so $f(c) \geq f(x)$ for all $x \in [c, b]$. These two observations show that $f(c) \geq f(x)$ for *all* $x \in [a, b]$. Thus, $f(c)$ is a relative maximum.

The proofs of statements (ii)–(iv) follow from the same reasoning and are left as an exercise. ∎

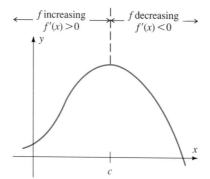

Figure 4.14 First Derivative Test (i): increasing to decreasing implies a relative maximum.

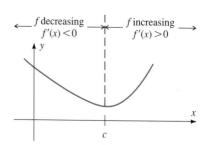

Figure 4.15 First Derivative Test (ii): decreasing to increasing implies a relative minimum.

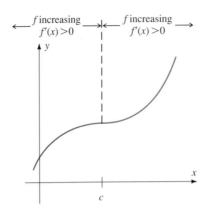

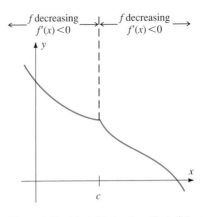

Figure 4.16 First Derivative Test (iii): increasing on the entire interval.

Figure 4.17 First Derivative Test (iv): decreasing on the entire interval.

Taken together, the Extreme Value Theorem (Relative Version—Theorem 3) and the First Derivative Test yield a complete procedure for finding and classifying relative extrema, which is illustrated in the following examples.

☐ **EXAMPLE 8**

For $f(x) = 4x^2(1 - x^2)$ find and classify all relative extrema.

Strategy · · · · · · ·

Find $f'(x)$.

Solution

We first find the critical numbers. The derivative is

$$f'(x) = 8x(1 - x^2) + 4x^2(-2x)$$
$$= 8x - 16x^3$$
$$= 8x(1 - 2x^2).$$

Set $f'(x) = 0$ and solve to find critical numbers.

Identify intervals determined by the critical numbers.

The equation $f'(x) = 0$ yields the critical numbers $x = 0$, $x = -\sqrt{2}/2$, and $x = \sqrt{2}/2$. There are no other critical numbers since $f'(x)$ is defined for all x. We must therefore examine the sign of $f'(x)$ on each of the intervals $(-\infty, -\sqrt{2}/2)$, $(-\sqrt{2}/2, 0)$, $(0, \sqrt{2}/2)$, and $(\sqrt{2}/2, \infty)$. To do so we simply select a convenient number t in each interval and examine the sign of $f'(t)$. The results appear in Table 4.3.

Check the sign of f' on each interval.

Table 4.3

Interval I	Test number $t \in I$	Value $f'(t)$	Sign of $f'(t)$
$\left(-\infty, -\dfrac{\sqrt{2}}{2}\right)$	$t = -1$	$f'(-1) = 8$	$+$
$\left(-\dfrac{\sqrt{2}}{2}, 0\right)$	$t = -\dfrac{1}{2}$	$f'\left(-\dfrac{1}{2}\right) = -2$	$-$
$\left(0, \dfrac{\sqrt{2}}{2}\right)$	$t = \dfrac{1}{2}$	$f'\left(\dfrac{1}{2}\right) = 2$	$+$
$\left(\dfrac{\sqrt{2}}{2}, \infty\right)$	$t = 1$	$f'(1) = -8$	$-$

Apply First Derivative Test to identify relative extrema.

From the results of Table 4.3, as illustrated in Figure 4.18, and from the First Derivative Test, we conclude that $f(-\sqrt{2}/2) = 1$ and $f(\sqrt{2}/2) = 1$ are relative maxima, and that $f(0) = 0$ is a relative minimum. The graph of f appears in Figure 4.19. Note that $f(-\sqrt{2}/2)$ and $f(\sqrt{2}/2)$ are indeed absolute maxima, but no absolute minimum exists. ■

Figure 4.18 Sign of $f'(x)$ in Example 8.

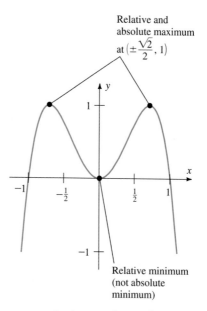

Figure 4.19 Relative extrema for $f(x) = 4x^2(1 - x^2)$.

Finding Absolute Extrema on Unbounded Intervals

For certain types of functions we can apply our techniques for finding relative extrema to the problem of finding the (absolute) maximum or minimum value of the function on its domain. This situation occurs when we know, from physical principles or from the context of a problem, that such a maximum or minimum exists and that it does not occur at an endpoint of the domain. That leaves us with the conclusion that the desired maximum or minimum corresponds to a relative extremum.

The following examples are typical of word problems that can be solved using this idea. Note in both examples that the function for which the extremum is sought must first be expressed as a function of a *single* variable. To do so we eliminate other variables using *auxiliary* equations derived from the given information (recall the procedure described at the beginning of Section 4.2).

□ **EXAMPLE 9**

Find the rectangle with area 64 square inches for which the *perimeter* is a minimum.

Strategy · · · · · · ·
Assign variable names to the two dimensions, length and width. Express perimeter in terms of these variables. Find an *auxiliary equation* involving the two variables.

Solution

If we let x and y denote the length and width of the rectangle, its perimeter is

$$P = 2x + 2y. \tag{2}$$

To express P as a function of a *single* independent variable we use the information that area equals 64 in.^{2} to obtain the *auxiliary* equation

$$xy = 64$$

Solve the auxiliary equation for one variable in terms of the other. Substitute for one of the variables in equation (2), thus *eliminating* the second variable, and obtaining P as a function of *one* variable.

which we can solve for y as

$$y = \frac{64}{x}. \tag{3}$$

Substituting this expression for y in equation (2) gives

$$P = 2x + 2\left(\frac{64}{x}\right)$$

so

$$P(x) = 2x + \frac{128}{x} \tag{4}$$

is the function for which we seek the absolute minimum value.

Find the domain of the function P.

Before proceeding further we note that the *domain* of the function P in equation (4) is $(0, \infty)$. That is because a rectangle with area 64 in.^{2} can have as its length *any* positive number x, as long as its width is $y = 64/x$. (See Figure 4.20.) *We are therefore seeking the minimum value of P on the interval $(0, \infty)$.*

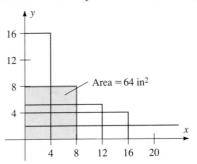

Figure 4.20 Rectangles with area $A = xy = 64$ in^2.

Find the relative extrema for P on its entire domain by setting $P'(x) = 0$ and solving for the critical numbers.

To find the relative extrema for P on $(0, \infty)$, we find the derivative

$$P'(x) = 2 - \frac{128}{x^2}$$

and note that the equation $P'(x) = 0$ gives

$$\frac{128}{x^2} = 2$$

or $x^2 = 128/2 = 64$, which has solutions $x = \pm 8$. Since $P'(x)$ is defined for all $x > 0$, *the only critical number for P in $(0, \infty)$ is $x = 8$.*

Determine whether the critical number yields a relative minimum. If so, determine whether this relative minimum is the absolute minimum by examining the sign of $P'(x)$.

By checking the sign of $P'(x)$ on the intervals $(0, 8)$ and $(8, \infty)$, you can see that

$$P \text{ is } \textit{decreasing} \text{ on } (0, 8], \quad \text{and} \quad P \text{ is } \textit{increasing} \text{ on } [8, \infty).$$

(See Figure 4.21.)

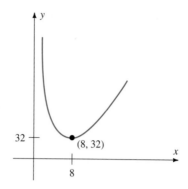

Figure 4.21 Minimum value $P(8) = 32$ of perimeter function $P(x) = 2x + 128/x$ occurs at $x = 8$.

From these observations we may conclude that the minimum value of P on $(0, \infty)$ corresponds to the relative minimum

$$P(8) = 2(8) + 128/8 = 32$$

State the conclusion in terms of the original question.

and that this value of the perimeter occurs when the dimensions of the rectangle are $x = 8$ and $y = 64/8 = 8$. That is, the rectangle of area 64 in.^{2} with the smallest perimeter is a *square* of side $s = 8$ inches. ∎

□ **EXAMPLE 10**

Find the point on the graph of $y = \sqrt{x}$ nearest the point $(4, 0)$.

Solution Following the steps outlined in Example 9, we first express the distance D between a point $P = (x, y)$ on the graph of $y = \sqrt{x}$ and the point $(4, 0)$ as

$$D = \sqrt{(x - 4)^2 + (y - 0)^2}. \tag{5}$$

(See Figure 4.22.) Since the point $P = (x, y)$ lies on the graph of $y = \sqrt{x}$, this is our auxiliary equation and we may substitute $y = \sqrt{x}$ in equation (5) to obtain the distance D as a function of x alone:

$$D(x) = \sqrt{(x - 4)^2 + (\sqrt{x})^2}$$
$$= \sqrt{x^2 - 7x + 16}.$$

Since the square root function $f(x) = \sqrt{x}$ is defined for all $x \geq 0$, we seek the minimum value of the distance function D on the interval $[0, \infty)$.

To find the relative extrema for D we first find

$$D'(x) = \frac{d}{dx}[(x^2 - 7x + 16)^{1/2}]$$
$$= \tfrac{1}{2}(x^2 - 7x + 16)^{-1/2}(2x - 7)$$
$$= \frac{2x - 7}{2\sqrt{x^2 - 7x + 16}}.$$

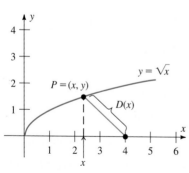

Figure 4.22 $D(x) = \sqrt{x^2 - 7x + 16}$ is the distance from the point $P = (x, y)$ to the point $(4, 0)$.

Setting $D'(x) = 0$ gives $2x - 7 = 0$, or $x = 7/2$. Since $D'(x)$ is defined for all $x \in [0, \infty)$, the number $x = 7/2$ is the only critical number for D in $[0, \infty)$. Checking the sign of $D'(x)$ for x in the intervals $(0, 7/2)$ and $(7/2, \infty)$ gives the conclusions that

D is decreasing on $[0, 7/2]$, and D is increasing on $[7/2, \infty)$.

Thus, $D(7/2) = \sqrt{(7/2)^2 - 7(7/2) + 16} = \sqrt{15}/2$ is a relative minimum for D on $[0, \infty)$ and also the *absolute* minimum for D on this interval.

The point $(7/2, \sqrt{7/2})$ is therefore the point nearest the point $(4, 0)$ on the graph of $y = \sqrt{x}$.   ∎

REMARK 3:   In Exercise 91, we sketch a slightly different solution to Example 10. We minimize the square S of the distance. That is, we minimize the function $S(x) = (D(x))^2$. Using this approach, we avoid the fractional power in the formula for $D(x)$. However, we must justify why we also obtain a minimum for D.

Exercise Set 4.4

In Exercises 1–18, find the intervals on which f is increasing or decreasing. Find all relative extrema for f.

1. $f(x) = x^2 - 4x + 6$

2. $f(x) = x(x - 4)$

3. $f(x) = 2x^3 - 3x^2 - 12x$

4. $f(x) = (x - 3)(x + 5)$

5. $f(x) = 4 + x^{2/3}$

6. $f(x) = 7 - 2x + x^2$

7. $f(x) = x^3 + 4$

8. $f(x) = 9 - x^2$

9. $f(x) = |4 - x^2|$

10. $f(x) = |x^2 - 9|$

11. $f(x) = \sin(x + \pi/4), \quad 0 \le x \le 2\pi$

12. $f(x) = \cos(2x - \pi/2), \quad 0 \le x \le \pi$

13. $f(x) = x^4 - 1$

14. $f(x) = \dfrac{3}{x - 2}$

15. $f(x) = \dfrac{2}{x + 1}$

16. $f(x) = 1/x^2$

17. $f(x) = \begin{cases} 4 - x^2, & -\infty < x \le 1 \\ x + 2, & 1 < x < \infty \end{cases}$

18. $f(x) = |3x - x^3|$

In Exercises 19–54, find the intervals on which f is increasing or decreasing. Classify all relative extrema for f.

19. $f(x) = 4x^3 + 9x^2 - 12x + 7$

20. $f(x) = x^3 + x^2 - 8x + 8$

21. $f(x) = x + \sin x$

22. $f(x) = \tan^2 x$

23. $f(x) = \dfrac{x}{1 + x}$

24. $f(x) = \sqrt{x + 2}$

25. $f(x) = \dfrac{1}{1 + x^2}$

26. $f(x) = \dfrac{x}{1 + x^2}$

27. $f(x) = (x + 3)^{2/3}$

28. $f(x) = 1 - x^{2/3}$

29. $f(x) = \frac{1}{3}x^3 - 3x^2 - 7x + 5$

30. $f(x) = x^3 + 3x^2 + 10$

31. $f(x) = x^3 + 6x^2 + 9x + 1$

32. $f(x) = 2x^3 + 9x^2 - 12x + 2$

33. $f(x) = x^3 + 3x^2 + 6x - 1$

34. $f(x) = x^3 + 6x^2 + 15x + 2$

35. $f(x) = x^4 + 4x^3 + 2x^2 + 1$

36. $f(x) = 3x^4 - 16x^3 + 30x^2 - 24x + 5$

37. $f(x) = 6x^{5/2} - 70x^{3/2} + 15$

38. $f(x) = \sqrt[3]{x}(x - 7)^2$

39. $f(x) = \dfrac{x^2}{1 + x^2}$

40. $f(x) = \dfrac{\sqrt{x}}{4 - x}$

41. $f(x) = \dfrac{x^{1/3}}{x^{2/3} - 4}$

42. $f(x) = x^{2/3}(x + 8)^2$

43. $f(x) = \sec^2 x, \quad 0 \le x \le 2\pi$

44. $f(x) = \sin^2 x + \cos x, \quad 0 \le x \le 2\pi$

45. $f(x) = |\sin x|$

46. $f(x) = |\cos x|$

47. $f(x) = x^{5/3} - 5x^{2/3} + 3$

48. $f(x) = \dfrac{1 - x}{1 + x^2}$

49. $f(x) = \frac{1}{4}x^4 - 2x^3 + \frac{3}{2}x^2 + 10x - 8$

50. $f(x) = \sqrt[3]{8 - x^3}$

51. $f(x) = x + |\sin x|, \quad 0 \le x \le 2\pi$

52. $f(x) = \cos x + |\cos x|, \quad -2\pi \le x \le 2\pi$

53. $f(x) = (x^2 - 4x)^3$

54. $f(x) = x\sqrt{3 - x}$

55. The function $f(x) = 2x^3 - 3ax^2 + 6$ is decreasing only on the interval $[0, 3]$. Find a.

56. Find q so that the function $f(x) = 2x^2 + qx + 5$ is increasing on $[-3, \infty)$ and decreasing on $(-\infty, -3]$.

57. Find q so that the function $f(x) = \frac{1}{3}x^3 - x^2 + qx + 10$ is increasing on $(-\infty, -3]$ and $[5, \infty)$ and decreasing on $[-3, 5]$.

58. A function f has derivative $f'(x) = 2x - 6$. If f is defined for all x, on what interval is f increasing?

59. Explain why a function of the form $y = x^2 + bx + c$ always has precisely one relative minimum and no relative maximum.

60. What is the maximum number of relative extrema that a polynomial of degree n can have? Could such a function have fewer?

61. The function $f(x) = x^2 - ax + b$ has a relative minimum at $x = 2$. Find a.

62. The function $f(x) = x^3 + ax^2 + bx + 7$ has relative extrema at $x = 1$ and $x = -3$.
a. Find a and b.
b. Classify the extrema as relative maxima or minima.

63. The function $f(x) = a(x^2 - bx + 16)$ has a relative extremum at $x = 5$.
a. Find b.
b. Determine the sign of a if $(5, f(5))$ is a relative minimum.

64. The function $f(x) = \sin ax$ has a critical number at $x = \pi/6$. Describe the set of all numbers a for which this is true.

65. A rectangle of variable width and length has a fixed perimeter of 24 inches.
a. Show that the area A of the rectangle is expressed as a function of its width x by the equation $A(x) = x(12 - x)$, $0 \le x \le 12$.
b. What is the largest subinterval of $[0, 12]$ for which $y = A(x)$ is an increasing function of x?
c. What is the maximum value of the function $y = A(x)$ for $x \in [0, 12]$?

66. A salt carton is to be made in the shape of a right circular cylinder and is to contain 500 cm^3 of salt. Find the dimensions for the container that requires the least amount of material. Assume that the container has a top.

67. The position of a particle along a horizontal number line at time t is given by the function $s(t) = -t^2 + 6t - 8$.
a. What is the largest time interval for which s is an increasing function? In which direction is the motion during this time?
b. At what time does the particle change direction?

68. The sum of two positive numbers is 50. If x denotes one of these numbers,
a. for which numbers x is the product of the two numbers an increasing function of x;
b. what is the maximum value of their product?

69. Find the dimensions of the box of capacity V with square top and bottom and rectangular sides that can be constructed from a minimum amount of material.

70. Find the point on the parabola $y = x^2$ nearest the point $(-3, 0)$.

71. A storage bunker is to contain 96 cubic meters and is to have a square base and vertical sides. If the material for the base costs \$10 per square meter and the material for the sides and top costs \$6 per square meter, find the dimensions that minimize cost.

72. Show that of all right circular cylinders of fixed volume V the one with minimum surface area has height equal to the diameter of its base.

73. Prove that the distance from a point $P = (x_0, y_0)$ in the plane to the line with equation $ax + by + c = 0$ is the number

$$d = \frac{|ax_0 + by_0 + c|}{\sqrt{a^2 + b^2}}.$$

74. A format for textbook page layouts is to be chosen so that each printed page has a 4-cm margin at top and bottom and a 2-cm margin on the left and right sides. The rectangular region of printed matter is to have area 800 cm^2. Find the dimensions for the textbook pages that minimize their areas.

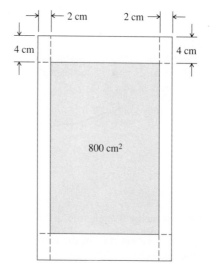

75. Find the length of the longest rod which can be carried horizontally around a (square) corner from a corridor 4 ft wide into another corridor 4 ft wide.

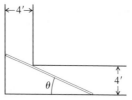

76. Work Problem 75 if the corridors are each 8 ft high and the rod need not be carried horizontally.

77. Use the fact that the lateral surface area of a right circular cone is $S = \pi r\sqrt{r^2 + h^2}$ to find the dimensions of a drinking cup in this shape if the volume of the cup is $36\pi\,\text{cm}^3$ and the surface area is a minimum.

78. Find the point on the graph of $y = \sqrt{x}$ nearest the point $(2, 0)$. (See Remark 3 and Exercise 91.)

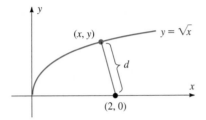

79. Let $f(x) = x + 1/x, \quad x \neq 0$.
   a. Find the intervals on which f is an increasing function.
   b. Use the result of part a to conclude that

$$x + \frac{1}{x} \geq 2$$

   for all $x > 0$.
   c. Can you establish the inequality in part b without using calculus? [*Hint:* Begin with the inequality $(x - 1)^2 \geq 0$.] What about the case $x < 0$?

80. In physics the *total energy* of a particle is defined by the equation

$$E = \frac{mc^2}{\sqrt{1 - \dfrac{v^2}{c^2}}}, \quad 0 < v < c$$

   where v = velocity, m = mass, and c = speed of light. Show that total energy is an increasing function of velocity.

81. In electronics, the ratio $r(\omega)$ of the output voltage to the input voltage in a certain "low-pass" RC circuit is given by the equation

$$r(\omega) = \frac{1/\omega C}{\sqrt{R^2 + \left(\dfrac{1}{\omega C}\right)^2}}$$

   where the constants R and C denote the resistance and capacitance of the circuit, and ω is the frequency of the current. Show that r is a decreasing function of the independent variable ω for $0 < \omega < \infty$.

82. The speed of a satellite moving in a stable orbit about the earth is a function of the altitude h of the satellite above the earth's surface. It is given by the equation

$$S(h) = \sqrt{\frac{GM_e}{R_e + h}} = \sqrt{GM_e}(R_e + h)^{-1/2}$$

where the constants G, M_e, and R_e are defined as follows:

$$G = \text{gravitational constant,}$$
$$M_e = \text{mass of the earth,}$$
$$R_e = \text{radius of the earth.}$$

Show that the speed of the satellite decreases as altitude increases.

83. Use Definition 4 to show that if f is increasing on $[a, b]$ and on $[b, c]$, then f is increasing on $[a, b] \cup [b, c] = [a, c]$.

84. Show that the function f in Example 3 is not decreasing on $(-\infty, 0]$ or on $[0, \infty)$.

85. Prove statements (ii)–(iv) of Theorem 7.

86. Show that $\tan x > x$ for $0 < x < \pi/2$ as follows.
   a. Let $f(x) = \tan x$ and $g(x) = x$. Note that $f(0) = g(0)$. Show that $f'(x) > g'(x)$ for all x between 0 and $\pi/2$. Conclude that $(f - g)'(x) > 0$ on the interval $(0, \pi/2)$.
   b. Use Theorem 6 to conclude that $(f - g)(x) > 0$ on the interval $(0, \pi/2)$.
   c. Derive the inequality $\tan x > x$ for $0 < x < \pi/2$. Compare this argument with the one sketched in Exercise 18 of Section 4.3.

87. Using the technique described in Exercise 86, show that $\sin x < x$ for all $x > 0$.

88. Using Exercise 87 and the technique described in Exercise 86, show that $1 - (x^2/2) < \cos x$ for all $x > 0$.

89. Using Exercises 87 and 88 as well as the technique described in Exercise 86, show that $x - (x^3/6) < \sin x < x$ for all $x > 0$.

90. Sketch the graph of a function f that is increasing on $(-\infty, 0)$ and decreasing on $(0, \infty)$ and for which $f(0)$ is an absolute minimum.

91. Let s be the squaring function $s(x) = x^2$.
   a. Show that s is increasing on $[0, \infty)$.
   b. Given a nonnegative function f, show that $f(c)$ is a minimum if and only if $(f(c))^2$ is a minimum for the square of f. (Note that the square of f is the composition $s \circ f$.)
   c. Rework Example 10 by minimizing the square of D (see Remark 3).
   d. Compare the details of your solution to the solution of Example 10.

In Exercises 92–95, use Theorem 6 to determine the number of real solutions to the polynomial equation specified. Then use Newton's Method to approximate these solutions.

92. $3x^3 + x^2 + x = 3$

93. $4x^5 + 5x^4 = -2$

94. $3x^5 - 10x^3 + 15x = 2$

95. $40x^7 - 140x^6 + 168x^5 - 70x^4 = -1$

96. The path of a projectile fired at an angle θ with an initial velocity v_0 from a height h_0 is given by the equation

$$y = \left(-\frac{16}{v_0^2}\sec^2\theta\right)x^2 + (\tan\theta)x + h_0, \; x \geq 0, \, 0 \leq \theta < \pi/2.$$

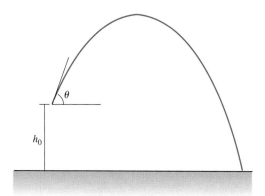

a. Suppose $v_0 = 72$ ft/s and $h_0 = 3.5$ ft. Graph the paths corresponding to $\theta = 25°, 30°, 40°, 45°, 50°, 55°, 60°$, and 65°. Which value of θ yields the maximum horizontal distance?

b. Using magnification, find the maximum height for each of the graphs in part **a**.

c. Find a formula for the maximum horizontal distance $D(\theta)$ travelled by the projectile. Graph D. What value of θ gives the maximum distance?

d. Let $\Delta\theta = 0.001$ and graph

$$\frac{D(\theta + \Delta\theta) - D(\theta)}{\Delta\theta}.$$

What does this graph tell us about the value of θ that produces the greatest horizontal distance travelled by the projectile?

e. Find a formula for the maximum height $M(\theta)$ of the projectile. What value of θ gives the maximum height? (*Hint:* Complete the square.)

f. Let $\Delta\theta = 0.001$ and graph

$$\frac{M(\theta + \Delta\theta) - M(\theta)}{\Delta\theta}.$$

What does this graph tell us about the value of θ that produces the maximum height of the projectile?

4.5 Significance of the Second Derivative: Concavity

We now know several techniques that are useful in determining the properties of a function f. We examine critical numbers to find relative extrema, and we use the first derivative to determine the behavior of the function between these numbers.

However, as illustrated in Figure 5.1, two different functions may have the same relative extrema and may increase and decrease on the same intervals.

Both graphs have a relative maximum at the point B and relative minima at the points A and C. What distinguishes them is the way in which they "bend" between relative extrema and how they "bend" as $|x|$ becomes large. These are the issues of *concavity*, which we discuss here, and *asymptotes*, which we discuss in the next section.

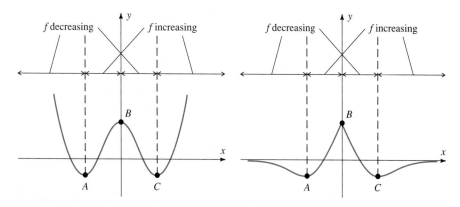

Figure 5.1 Two graphs with the same relative extrema and increasing/decreasing properties may have very different shapes depending on how they bend (their concavity).

Suppose that the function f is differentiable on the interval I. If the derivative f' is an increasing function on I, then the slopes of the lines tangent to the graph of f increase (causing a tangent to rotate counterclockwise) as x increases. This means that the graph will have a "cupped up" shape as illustrated in Figure 5.2.

For this reason we define the *concavity* of the graph of f in terms of whether its derivative f' is an increasing or decreasing function.

Definition 5	Let f be continuous on an interval I and differentiable on the interior of I. We say that the graph of f is

(a) **concave up** on I if f' is an increasing function on the interior of I;
(b) **concave down** on I if f' is a decreasing function on the interior of I.

(See Figures 5.2 and 5.3.)

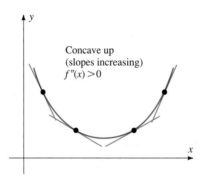

Figure 5.2   Graph of f is concave up if f' is increasing, that is, $f''(x) > 0$.

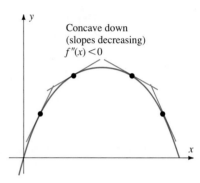

Figure 5.3   Graph of f is concave down if f' is decreasing, that is, $f''(x) < 0$.

REMARK 1:   We are defining concavity in analytic terms (that is, by use of $f'(x)$) rather than geometrically, in order to emphasize the relationship between concavity and the second derivative. A more geometric approach would be to say that the graph of f is concave up on I if, for each $c \in I$, the graph "lies above the tangent at $(c, f(c))$," as is the case in Figure 5.2. In Exercises 57 and 58 you are asked to show that this geometric condition is equivalent to the condition that f' is increasing on I, the condition we have chosen to use. A similar geometric definition can be given for concave down.

The following theorem shows that concavity is determined by the sign of the second derivative.

Theorem 8	Let f be continuous on I and twice differentiable on the interior of I. Then the graph of f is

(a) concave up on I if $f''(x) > 0$ for all x in the interior of I, and
(b) concave down on I if $f''(x) < 0$ for all x in the interior of I.

Proof: To prove part (a) we note that, since $f'' = (f')'$, if $f''(x) > 0$ for all interior points x, it follows from Theorem 6 that f' is an increasing function on I. Thus, the graph of f is concave up on I according to Definition 5. Similar reasoning proves part (b). ∎

If the function f has a continuous derivative f', Theorem 8 provides *a procedure for finding the intervals on which the graph of f is concave up or concave down:*

(i) Find all numbers c for which $f''(c) = 0$ or $f''(c)$ fails to exist.
(ii) Determine the sign of $f''(x)$ for x in each of the resulting intervals and apply Theorem 8.

As in the procedure for finding intervals on which f is increasing or decreasing, we must also be careful to note numbers c at which $f(c)$ or $f'(c)$ fails to exist.

Points on the graph of f that separate arcs of opposite concavity, loosely speaking, are called *inflection points*. Here is a precise definition of this term.

Definition 6

The point $P = (c, f(c))$ is called an **inflection point** for the function f if there exist numbers a and b such that $a < c < b$, f is continuous on (a, b), and the graph of f is either

(a) concave down on (a, c) and concave up on (c, b), or
(b) concave up on (a, c) and concave down on (c, b).

In other words, the point P is an inflection point for f if the direction of the concavity switches at P.

☐ **EXAMPLE 1**

Figure 5.4 shows the graph of $f(x) = \sin x$ on the interval $I = [0, 2\pi]$. For this function we have

$$f'(x) = \cos x, \quad \text{and} \quad f''(x) = -\sin x$$

so $f''(x) = 0$ if $-\sin x = 0$, which occurs on I at $x = 0$, $x = \pi$, and $x = 2\pi$. Since f'' is continuous on I, we check the concavity of the graph of f on the intervals $[0, \pi]$ and $[\pi, 2\pi]$ by checking the sign of $f''(t)$ at a test number t in each of the intervals $(0, \pi)$ and $(\pi, 2\pi)$ and applying Theorem 8. The results are recorded in Table 5.1.

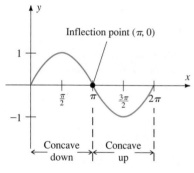

Figure 5.4 Graph of $y = \sin x$ is concave down on $[0, \pi]$ and concave up on $[\pi, 2\pi]$.

Table 5.1 Concavity analysis for $f(x) = \sin x$

Interval	Test number	$f''(t)$	Conclusion
$[0, \pi]$	$t = \dfrac{\pi}{2}$	$-\sin\left(\dfrac{\pi}{2}\right) = -1$	concave down
$[\pi, 2\pi]$	$t = \dfrac{3\pi}{2}$	$-\sin\left(\dfrac{3\pi}{2}\right) = 1$	concave up

Thus, the graph of f is concave down on $[0, \pi]$ and concave up on $[\pi, 2\pi]$. This means that the point $(\pi, f(\pi)) = (\pi, 0)$ is an inflection point, according to Definition 6. ∎

☐ **EXAMPLE 2**

Determine the concavity of the graph of the function $f(x) = x^4 - 3x^2 + 2$.

Strategy
Calculate f''.

Solution
Since $f(x) = x^4 - 3x^2 + 2$ we have

$$f'(x) = 4x^3 - 6x \quad \text{and} \quad f''(x) = 12x^2 - 6.$$

Find all x with $f''(x) = 0$ or $f''(x)$ undefined.

Setting $f''(x) = 0$ gives $12x^2 = 6$, or $x^2 = 1/2$.

The zeros of the second derivative $f''(x)$ are, therefore,

$$x_1 = \sqrt{2}/2 \quad \text{and} \quad x_2 = \sqrt{2}/2.$$

Determine the sign of f'' on each of the resulting intervals by checking $f''(t)$ at a single number. Then apply Theorem 8.

There are no numbers x for which $f''(x)$ is undefined. Since f'' is a continuous function (it's a polynomial), it must have constant sign on each of the intervals determined by its zero. We may therefore determine the sign of f'' on each of these intervals by checking the sign of $f''(t)$ at a "test number" t. The results appear in Table 5.2.

Table 5.2

Interval I	Test number $t \in I$	Value of $f''(t)$	Sign of $f''(t)$	Concavity
$(-\infty, -\sqrt{2}/2)$	$t = -1$	$f''(-1) = 6$	$+$	up
$(-\sqrt{2}/2, \sqrt{2}/2)$	$t = 0$	$f''(0) = -6$	$-$	down
$(\sqrt{2}/2, \infty)$	$t = 1$	$f''(1) = 6$	$+$	up

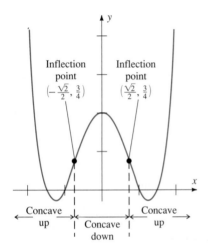

From the results in Table 5.2 and by Theorem 8, we may conclude that the graph of f is

(i) concave up on $(-\infty, -\sqrt{2}/2]$,
(ii) concave down on $[-\sqrt{2}/2, \sqrt{2}/2]$,
(iii) concave up on $[\sqrt{2}/2, \infty)$.

Since the concavity of the graph of $y = f(x)$ changes at $(-\sqrt{2}/2, f(-\sqrt{2}/2)) = (-\sqrt{2}/2, 3/4)$ and at $(\sqrt{2}/2, f(\sqrt{2}/2)) = (\sqrt{2}/2, 3/4)$, both of these points are *inflection points* (see Figure 5.5). ∎

Figure 5.5 Concavity for graph of $f(x) = x^4 - 3x^2 + 2$.

REMARK 2: It is not necessary that $f''(c) = 0$ at an inflection point $(c, f(c))$. Figure 5.6 shows that the graph of $f(x) = |4 - x^2|$ has inflection points $(-2, 0)$ and $(2, 0)$, even though $f''(-2)$ and $f''(2)$ fail to exist.

It is important to note that the point $(x_0, f(x_0))$ need not be an inflection point for the graph of f, even though $f''(x_0) = 0$ or $f''(x_0)$ fails to exist. The following example shows that the concavity can be the same on either side of such a point.

□ **EXAMPLE 3**

Determine the concavity and find the inflection points for the graph of $f(x) = x^{2/3} - \frac{1}{5}x^{5/3} = x^{2/3}(1 - x/5)$.

Solution We have

$$f'(x) = \tfrac{2}{3}x^{-1/3} - \tfrac{1}{3}x^{2/3},$$

so

$$\begin{aligned} f''(x) &= -\tfrac{2}{9}x^{-4/3} - \tfrac{2}{9}x^{-1/3} \\ &= -\tfrac{2}{9}x^{-4/3}(1 + x) \\ &= \frac{-2(1 + x)}{9x^{4/3}}. \end{aligned}$$

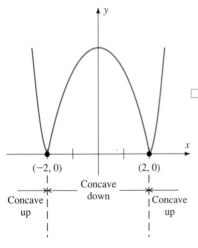

Figure 5.6 Graph of $f(x) = |4 - x^2|$ has inflection points at $(-2, 0)$ and $(2, 0)$ where $f''(x)$ fails to exist.

Thus $f''(x) = 0$ for $x = -1$, and $f''(x)$ is undefined for $x = 0$. Therefore, the sign of f'' must be checked on the intervals $(-\infty, -1)$, $(-1, 0)$, and $(0, \infty)$. This time, rather than substituting particular test numbers and constructing a table, we examine the signs of the numerator and denominator on each interval, as illustrated in Figure 5.7.

From Figure 5.7, we can see that f'' is positive on $(-\infty, -1)$ but negative on both $(-1, 0)$ and $(0, \infty)$. Thus, by Theorem 8, the graph is

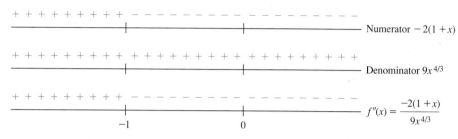

Figure 5.7 Sign analysis for $f''(x) = \dfrac{-2(1 + x)}{9x^{4/3}}$.

(i) concave up on $(-\infty, -1]$,
(ii) concave down on $[-1, 0]$ *and* on $[0, \infty)$.

Hence, the point $(-1, f(-1)) = (-1, 6/5)$ is an inflection point, but the point $(0, f(0)) = (0, 0)$ is *not* an inflection point (see Figure 5.8).

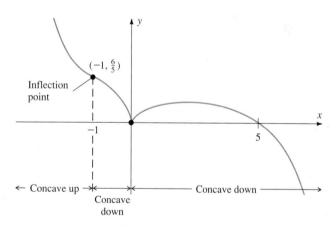

Figure 5.8 Graph of $f(x) = x^{2/3} - \frac{1}{5}x^{5/3} = x^{2/3}(1 - x/5)$. Note that the point $(0, 0)$ is not an inflection point.

The Second Derivative Test for Extrema

There is a direct relationship between concavity and the nature of relative extrema as the following theorem shows.

Theorem 9
Second Derivative Test

Let f be differentiable on an open interval containing the critical number c, with $f'(c) = 0$. Suppose also that $f''(c)$ exists.

(i) If $f''(c) < 0$, then $f(c)$ is a relative maximum.
(ii) If $f''(c) > 0$, then $f(c)$ is a relative minimum.
(iii) If $f''(c) = 0$, there is no conclusion.

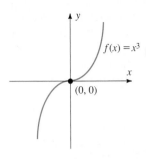

Figure 5.9 $f'(0) = f''(0) = 0$, but $f(0)$ is not a relative extremum.

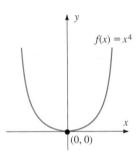

Figure 5.10 $f'(0) = f''(0) = 0$; $f(0)$ is a relative minimum.

Proof: A formal proof of Theorem 9 is outlined in Exercise 59. We prefer to give only an informal sketch of the argument here to avoid clouding the simple idea with details.

In case (i), the condition $f''(c) < 0$ may be interpreted as saying that the graph of f is concave down near $x = c$. This means that f' is a decreasing function near c. Since $f'(c) = 0$, it follows that $f'(x) > 0$ for $x < c$ and $f'(x) < 0$ for $x > c$. Thus, by the First Derivative Test, $f(c)$ is a relative maximum. A similar argument applies for statement (ii). ∎

In applying the Second Derivative Test be sure to note statement (iii)—if $f''(c) = 0$, we obtain no information about whether $f(c)$ is a relative extremum. The following simple examples show why.

☐ **EXAMPLE 4**

For $f(x) = x^3$, $f'(x) = 3x^2$, and $f''(x) = 6x$. Thus $x = 0$ is a critical number, and $f''(0) = 0$. Here $f(0) = 0$ is neither a relative maximum nor a relative minimum (see Figure 5.9). ■

☐ **EXAMPLE 5**

For $f(x) = x^4$, $f'(x) = 4x^3$ and $f''(x) = 12x^2$. Again, $x = 0$ is a critical number, and $f''(0) = 0$. For this function, $f(0) = 0$ is a relative minimum (see Figure 5.10). ■

☐ **EXAMPLE 6**

The function $f(x) = -x^4$ has derivatives $f'(x) = -4x^3$ and $f''(x) = -12x^2$. Here also, $x = 0$ is a critical number, and $f''(0) = 0$. This time $f(0) = 0$ is a relative maximum (see Figure 5.11). ■

REMARK 3: Note that $f''(0) = 0$ in Examples 4–6. In Example 4, $f(0)$ is not a relative extremum. In Example 5, $f(0)$ is a relative minimum, and in Example 6, $f(0)$ is a relative maximum. Taken together, these three examples justify our assertion that, if $f''(c) = 0$ in the Second Derivative Test, we obtain no conclusion.

Figure 5.11 $f'(0) = f''(0) = 0$; $f(0)$ is a relative maximum.

☐ **EXAMPLE 7**

Find all relative extrema for the function $f(x) = x^4 - 3x^2 + 2$ in Example 2.

Strategy · · · · · · · ·

Set $f'(x) = 0$ to find critical numbers.

Classify critical numbers using the Second Derivative Test.

Solution

Setting $f'(x) = 4x^3 - 6x = 0$, we obtain the equation

$$2x(2x^2 - 3) = 0$$

so $x = 0$, $x = -\sqrt{3/2}$, and $x = \sqrt{3/2}$ are critical numbers. We can classify these by the Second Derivative Test.

Since $f''(x) = 12x^2 - 6$ we find that

$f''(-\sqrt{3/2}) = 12 > 0$, so $(-\sqrt{3/2}, -1/4)$ is a relative minimum;

$f''(0) = -6 < 0$, so $(0, 2)$ is a relative maximum;

$f''(\sqrt{3/2}) = 12 > 0$, so $(\sqrt{3/2}, -1/4)$ is a relative minimum (see Figure 5.5). ■

☐ **EXAMPLE 8**

Discuss the relative extrema and concavity for the function $y = x + \sin x$.

Strategy · · · · · · · ·

Find $\dfrac{dy}{dx}$ and $\dfrac{d^2y}{dx^2}$.

Solution

Here

$$\frac{dy}{dx} = 1 + \cos x, \quad \text{so} \quad \frac{d^2y}{dx^2} = -\sin x.$$

Evaluate d^2y/dx^2 at critical numbers to check for extrema by Second Derivative Test. If inconclusive, use First Derivative Test.

The equation $dy/dx = 0$ yields the critical numbers $x = \pm\pi, \pm3\pi, \pm5\pi, \ldots$. At such points $d^2y/dx^2 = 0$, so the Second Derivative Test is inconclusive. However, since

$$\frac{dy}{dx} = 1 + \cos x > 0 \text{ for all } x \neq \pm\pi, \pm3\pi, \pm5\pi, \ldots,$$

the First Derivative Test shows that the function is increasing and that there are no relative extrema.

The equation

$$\frac{d^2y}{dx^2} = -\sin x = 0$$

Examine sign of d^2y/dx^2 between zeros of $d^2y/dx^2 = 0$ to determine concavity by Theorem 8.

has solutions $x = 0, \pm\pi, \pm2\pi, \ldots$. Inspection of the sign of d^2y/dx^2 on the resulting intervals shows that the graph of y is concave up on the intervals $[\pi, 2\pi]$, $[3\pi, 4\pi], \ldots, [(2n-1)\pi, 2n\pi], \ldots$ and concave down on the intervals $[0, \pi]$, $[2\pi, 3\pi], \ldots, [2n\pi, 2(n+1)\pi], \ldots$. Thus all points $(n\pi, n\pi)$, $n = 0, \pm1, \pm2, \ldots$ are inflection points (see Figure 5.12). ■

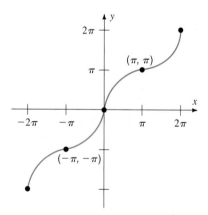

Figure 5.12   Graph of $y = x + \sin x$.

☐ **EXAMPLE 9**

Snell's Law, a law in optics first discovered by the seventeenth-century Dutch mathematician Willebrord Snell, concerns the path of a ray of light that passes from one medium (such as air) to another medium (such as water). We are now able to derive this law using calculus and a principle discovered by Pierre de Fermat, the seventeenth-century French mathematician. Fermat's principle states that a ray of light travels the path of minimum time.

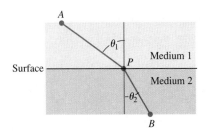

Figure 5.13 Snell's Law of Refraction.

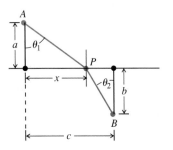

Figure 5.14 Possible path from A to B.

Suppose that the ray travels at a velocity v_1 in the first medium and at a velocity v_2 in the second medium. Then within each medium the path of minimum time is a straight line. If $v_1 \neq v_2$, however, the path followed between a point A in medium 1 and a point B in medium 2 consists of two line segments meeting at a point P on the surface separating the media (see Figure 5.13). This "bending" of the light ray is called refraction, and Snell's Law of Refraction states that

$$\frac{\sin \theta_1}{v_1} = \frac{\sin \theta_2}{v_2}$$

where θ_1 and θ_2 are the angles formed by the ray and the lines normal to the surface at P in each of the media.

To establish this law, we let a and b denote the lengths of the normals from points A and B to the surface, we let c be the distance between these normals, and we let x be the distance along the surface from the normal through A to point P (see Figure 5.14).

Since the time required for the light to travel from point A to point P is

$$T_1 = \frac{\sqrt{a^2 + x^2}}{v_1} \qquad \left(\text{time} = \frac{\text{distance}}{\text{velocity}}\right)$$

and the time required for light to travel from point P to point B is

$$T_2 = \frac{\sqrt{(c - x)^2 + b^2}}{v_2},$$

we seek to minimize the total time

$$T(x) = T_1 + T_2 = \frac{\sqrt{a^2 + x^2}}{v_1} + \frac{\sqrt{(c - x)^2 + b^2}}{v_2}. \tag{1}$$

Since equation (1) for T contains only one independent variable x, we may proceed to calculate $T'(x)$. We obtain

$$T'(x) = \frac{x}{v_1 \sqrt{a^2 + x^2}} - \frac{c - x}{v_2 \sqrt{(c - x)^2 + b^2}}.$$

The function T is differentiable for all x, so the only critical numbers are the solutions to the equation $T'(x) = 0$. This equation gives the condition that

$$\frac{x}{v_1 \sqrt{a^2 + x^2}} = \frac{c - x}{v_2 \sqrt{(c - x)^2 + b^2}}. \tag{2}$$

From Figure 5.14 we see that

$$\frac{x}{\sqrt{a^2 + x^2}} = \sin \theta_1 \quad \text{and} \quad \frac{c - x}{\sqrt{(c - x)^2 + b^2}} = \sin \theta_2.$$

Using these equations, we obtain from equation (2) that

$$\frac{\sin \theta_1}{v_1} = \frac{\sin \theta_2}{v_2}, \tag{3}$$

which is Snell's Law.

We can verify that equation (3) corresponds to a path of minimum time by showing that the graph of the function T is concave up everywhere. A somewhat lengthy calculation (Exercise 60) yields the second derivative

$$T''(x) = \frac{a^2 v_2 [b^2 + (c - x)^2]^{3/2} + b^2 v_1 (a^2 + x^2)^{3/2}}{v_1 v_2 (a^2 + x^2)^{3/2} [(c - x)^2 + b^2]^{3/2}}. \tag{4}$$

By examining equation (4) closely, we see that $T''(x) > 0$ for all real numbers x because the denominator is always positive and both terms in the numerator are positive. Therefore, equation (3) yields the minimum. ∎

Exercise Set 4.5

1. For the function graphed below, determine:
   a. the intervals on which the function is increasing;
   b. the intervals on which the graph is concave up;
   c. the relative minima; $1, 4, 8$
   d. the relative maxima; and $2, 6, 9$ $abs. max\ 6$
   e. the inflection points. 3

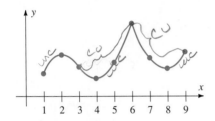

In Exercises 2–12 determine the intervals on which the graph of f is concave up or concave down. Find any inflection points.

2. $f(x) = 3x + 1$

3. $f(x) = x^2 - 4x - 5$

4. $f(x) = 9 - x^3$

5. $f(x) = |1 - x^2|$

6. $f(x) = |x^2 - 4|$

7. $f(x) = \cos x, \quad 0 \le x \le 2\pi$

8. $f(x) = \tan x, \quad -\pi/2 < x < \pi/2$

9. $f(x) = (x + 3)^3$

10. $f(x) = \dfrac{4}{x - 2}$

11. $f(x) = \sqrt{x + 2}$

12. $f(x) = x^{2/3}$

In Exercises 13–28, find the intervals on which the graph of f is concave up or concave down. Find any inflection points.

13. $f(x) = \dfrac{x}{x + 1}$

14. $f(x) = (x + 2)^{1/3}$

15. $f(x) = (2x + 1)^3$

16. $f(x) = (1 - 4x)^3$

17. $f(x) = 2x^3 - 3x^2 + 18x - 12$

18. $f(x) = 2x^3 + 12x^2 + 18x + 12$

19. $f(x) = x^4 + 2x^3 - 36x^2 + 24x - 6$

20. $f(x) = \frac{1}{12}x^4 - \frac{2}{3}x^3 + 2x^2 + 5x - 8$

21. $f(x) = x^4 + 6x^3 + 6x^2 + 12x + 1$

22. $f(x) = 3x^5 + 5x^4 - 20x^3 + 10x + 4$

23. $f(x) = 3x^5 - 25x^4 + 60x^3 - 60x^2 + 2x + 1$

24. $f(x) = 3x^5 + 10x^4 + 20x^3 + 4x + 2$

25. $f(x) = x^{2/3}(119 - x^5)$

26. $f(x) = x/(1 + x^2)$

27. $f(x) = x^{5/3} - 5x^{2/3} + 3$ 28. $f(x) = \dfrac{\sqrt[3]{x}}{1 - x}$

In each of Exercises 29–34, determine whether f has a relative extremum at the given value of x, using either the First or Second Derivative Test.

29. $f(x) = \sin^2 x, \quad x = \pi/2$

30. $f(x) = x^2 + \dfrac{2}{x}, \quad x = 1$

31. $f(x) = x^4 - 4x^3 - 48x^2 + 24x + 20, \quad x = 4$

32. $f(x) = 2x^3 - 3x^2, \quad x = 1$

33. $f(x) = x^4 - 4x^3 + 6x^2 - 4x, \quad x = 1$

34. $f(x) = x^2 + 2 \cos x, \quad x = 0$

In Exercises 35–38, the derivative f' of a function f is specified. Determine if f has a relative extremum at the given x-value using either the First or Second Derivative Test.

35. $f'(x) = \dfrac{x - 1}{x + 1}, \quad x = 1$

36. $f'(x) = \dfrac{1 - \sin 2x}{\cos x}, \quad x = \pi/4$

37. $f'(x) = (x - 1)(x + 2), \quad x = -2$

38. $f'(x) = (x - 2)^3(x + 1), \quad x = 2$

In Exercises 39–42, find all relative extrema, determine the intervals on which the graph is concave up or concave down, and sketch the graph.

39. $f(x) = x^3 - 3x^2 + 6$ **40.** $y = x^4 + 4x^3 - 8x^2 - 48x + 9$

41. $y = |9 - x^2|$ **42.** $f(x) = (x - 8)^{2/3}$

43. The illumination at point Q from a source of light at point P is proportional to the intensity of the light source at P and inversely proportional to the square of the distance from P to Q. Two lamps P_1 and P_2 are 20 meters apart. Find the point Q of minimum illumination on the line joining P_1 and P_2 if one source is twice as strong as the other.

44. A food processing company wants to produce the most economical size can for their pie fillings. Find the size that minimizes the surface area of a can whose volume is 2π cubic inches.

45. Figure 5.15 is the graph of the derivative f' of a function f that is continuous on the interval $[0, 8]$. Determine the intervals on which f is increasing and the intervals on which the graph of f is concave down. Where do the relative extrema occur? Classify all relative extrema.

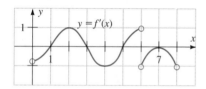

Figure 5.15

46. Does there exist a function that is twice differentiable for all real numbers and that satisfies the three inequalities: (a) $f(x) < 0$ for all x, (b) $f'(x) > 0$ for all x, and (c) $f''(x) > 0$ for all x? If there does exist such a function, sketch its graph. If such a function does not exist, give either an informal explanation or a rigorous proof to justify your assertion.

47. What conditions must hold for the constants a, b, and c in order that the general quadratic function $f(x) = ax^2 + bx + c$ be concave up for all x? Justify your answer using the techniques of this section.

48. Find an example of a function f and a number a so that $f(a)$ is a relative maximum but $f''(a) = 0$.

49. Find an example of a function f and a number b so that $f(b)$ is a relative minimum but $f''(b) = 0$.

50. Find an example of a function f and a number c for which $f'(c) = f''(c) = 0$ but $f(c)$ is not a relative extremum.

51. How many inflection points can a polynomial of degree $n = 2$ possess? $n = 3$? $n = k$?

52. Find an example of a function f that is increasing on $(-\infty, 1]$, decreasing on $[1, \infty)$, concave up on $(-\infty, 0]$ and $[2, \infty)$, and concave down on $[0, 2]$.

53. Find an example of a function that is concave down on $[-\infty, 0]$, concave up on $[0, \infty)$, and decreasing throughout its domain.

54. True or false? If $f(0) = 0$ and f is concave down for all x, then $f(x) \leq 0$ for all x. Justify your answer.

55. Sketch the graph of a function f that is twice differentiable for all real numbers and that satisfies all of the following properties:
(a) $f(x) < 0$ for $x < -2$,
(b) $f(x) > 0$ for $x > -2$,
(c) $f'(x) > 0$ for $x < 0$,
(d) $f'(x) < 0$ for $x > 0$,
(e) $f''(x) < 0$ for $x < -3$ and $-1 < x < 1$,
(f) $f''(x) > 0$ for $-3 < x < -1$ and $x > 1$.

56. Sketch the graph of a function f that is defined and twice differentiable for all real numbers except 0 and that satisfies all of the following properties:
(a) $f(x) < 1$ for $x \neq 0$,
(b) $f(x) < 0$ for $-1 < x < 0$ and for $0 < x < 1$,
(c) $f(x) > 0$ for $x < -1$ and $x > 1$,
(d) $f'(x) > 0$ for $x < -2$ and $x > 0$,
(e) $f'(x) < 0$ for $-2 < x < 0$,
(f) $f''(x) < 0$ for $-3 < x < 0$ and $x > 0$,
(g) $f''(x) > 0$ for $x < -3$.

57. Let f be differentiable on an open interval I. Assume that throughout I "the graph of f is above the tangent." (That is, for each $c \in I$, if $y = g(x)$ is an equation for the tangent at $(c, f(c))$, then $g(x) < f(x)$ for all $x \in I$ with $x \neq c$.) Show that f' is increasing on I.

58. Suppose that f is a differentiable function on an open interval I and that f' is increasing on I. Show that the graph of f is above its tangents (as defined in Exercise 57). (*Hint:* Let the tangent function be g as in Exercise 57. Show that the difference function $d(x) = f(x) - g(x)$ is positive for all $x \neq c$ in I by an application of the Mean Value Theorem.)

59. Prove statement (i) of the Second Derivative Test (Theorem 9) as follows:
a. Assume that $f''(c) = L < 0$.
b. Show that $\lim\limits_{x \to c} \dfrac{f'(x)}{x - c} = L < 0$.

c. Conclude that there exists an interval $I = (a, b)$ so that
$\dfrac{f'(x)}{x - c} < 0$ whenever $x \in I$.

d. Show that $f'(x) > 0$ for $x \in (a, c)$.

e. Show that $f'(x) < 0$ for $x \in (c, b)$.

f. Apply the First Derivative Test to conclude that $(c, f(c))$ is a relative maximum.

60. Verify equation (4) by differentiating $T'(x)$ in Example 9.

4.6   Large-Scale Behavior: Asymptotes

As Figures 6.1 and 6.2 indicate, two functions defined on the interval $(0, \infty)$ can be increasing and can have graphs that are concave down, yet behave differently for large numbers x. The function *(Slope is dec.)*

$$f(x) = \frac{x - 1}{x}$$

is increasing on the interval $(0, \infty)$, so you might think that its values $f(x)$ increase without bound as x increases without bound. However, since $x - 1 < x$ for all x, $f(x) < 1$ for all $x > 0$. The graph of

$$g(x) = \frac{x^2 - 1}{x}$$

is concave down for all $x > 0$, but its values $g(x)$ increase without bound as x increases without bound. Therefore, in order to understand the overall behavior of a function, we need to do more than determine the intervals on which it is increasing or decreasing and the intervals on which its graph is concave up or concave down. We need to understand the limiting properties of $f(x)$ as x tends towards the borders of the domain of f and, if possible, as x increases or decreases without bound.

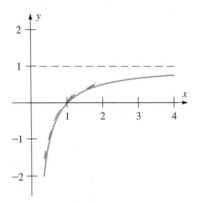

Figure 6.1   The graph of
$$f(x) = \frac{x - 1}{x}$$
for $0 < x < 4$.

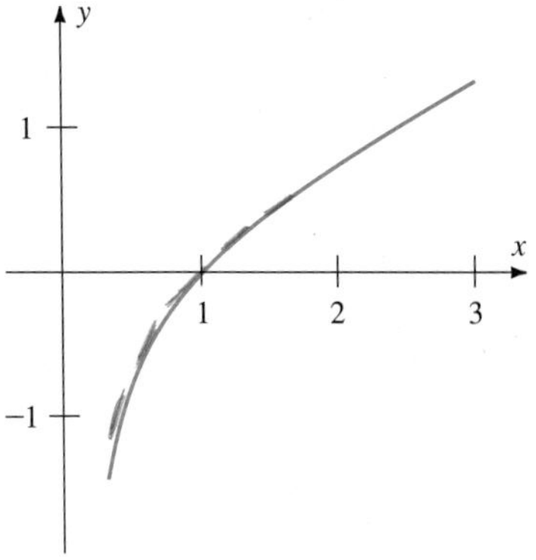

Figure 6.2   The graph of
$$g(x) = \frac{x^2 - 1}{x}$$
for $0 < x < 4$.

Limits at Infinity

The function

$$f(x) = -\frac{2x + 1}{x + 2}$$

is an example of a function whose values approach a constant as x increases without bound. In fact, both the entries of Table 6.1 and the graph of f in Figure 6.3 suggest that the values $f(x)$ approach the number $L = -2$ as x becomes very large.

Table 6.1

x	$f(x) = -\dfrac{2x+1}{x+2}$
0	-0.50000
5	-1.57143
10	-1.75000
20	-1.86364
50	-1.94231
100	-1.97059
250	-1.98810
500	-1.99402
1,000	-1.99701
2,000	-1.99850
5,000	-1.99940
10,000	-1.99997

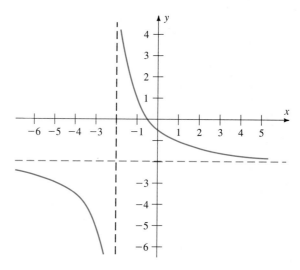

Figure 6.3   Graph of $f(x) = -\dfrac{2x+1}{x+2}$.

In this case we write

$$\lim_{x \to \infty} -\frac{2x+1}{x+2} = -2 \tag{1}$$

and we refer to the line $y = -2$ as a *horizontal asymptote*. More generally, we make the following informal definitions.

Definition 7

The expression

$$\lim_{x \to \infty} f(x) = L$$

means that the values $f(x)$ approach the number L as x increases without bound. The expression

$$\lim_{x \to -\infty} f(x) = M$$

means that the values $f(x)$ approach the number M as x decreases without bound.

Definition 7 is an informal working definition. A formal definition is given in Exercise 64.

An *asymptote* for the graph of f is just any straight line "approached" by the graph of f. Using Definition 7 we may define a horizontal asymptote as follows.

Definition 8

The line $y = L$ is a **horizontal asymptote** for the graph of the function f if either

$$L = \lim_{x \to \infty} f(x) \quad \text{or} \quad L = \lim_{x \to -\infty} f(x).$$

(See Figure 6.4.)

REMARK 1:   The term "asymptote" is commonly used to describe a line that is approached, but not touched, by the curve under consideration. However, Definition 7 includes functions whose graph $y = f(x)$ touches or crosses the line $y = L$. Example 1(e) illustrates this more general use of the term.

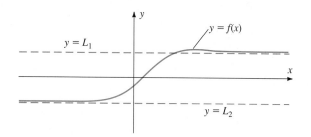

Figure 6.4   Horizontal asymptotes $y = L_1$ and $y = L_2$.

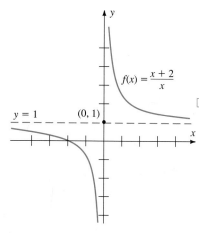

Figure 6.5 $\displaystyle\lim_{x \to \infty}\left(\frac{x+2}{x}\right) = 1$;

$\displaystyle\lim_{x \to -\infty}\left(\frac{x+2}{x}\right) = 1$. Asymptote is $y = 1$.

□ **EXAMPLE 1**

One of the most elementary limits at infinity is

$$\lim_{x \to \infty}\frac{1}{x} = 0$$

(See Exercise 64a.) In this example, we indicate how to use this limit to derive other limits at infinity.

(a) $\displaystyle\lim_{x \to \infty}\frac{x+2}{x} = \lim_{x \to \infty}\left(1 + \frac{2}{x}\right) = (1 - 0) = 1$.

Thus, the line $y = 1$ is a horizontal asymptote for the graph of $f(x) = (x + 2)/x$. (Figure 6.5.)

(b) $\displaystyle\lim_{x \to \infty}\frac{1-x}{1+x} = \lim_{x \to \infty}\left(\frac{1-x}{1+x}\right)\left(\frac{1/x}{1/x}\right) = \lim_{x \to \infty}\left(\frac{\dfrac{1}{x}-1}{\dfrac{1}{x}+1}\right) = \frac{0-1}{0+1} = -1$.

Thus, the line $y = -1$ is a horizontal asymptote for the graph of $f(x) = (1 - x)/(1 + x)$. (Figure 6.6.)

(c) $\displaystyle\lim_{x \to \infty}-\frac{2x+1}{x+2} = \lim_{x \to \infty}\left(\frac{2x+1}{x+2}\right)\left(\frac{1/x}{1/x}\right) = -\lim_{x \to \infty}\left[\frac{2+(1/x)}{1+(2/x)}\right] = -\frac{2+0}{1+0} = -2$.

Thus, the line $y = -2$ is a horizontal asymptote for the graph of $f(x) = -(2x + 1)/(x + 2)$. This technique justifies the limit in equation (1) at the beginning of this section (see Figure 6.3).

(d) The limit of $\sin x$ as $x \to \infty$ does not exist because $\sin x$ oscillates periodically between -1 and 1 as $x \to \infty$. (See Figure 6.7.)

(e) $\displaystyle\lim_{x \to \infty}\frac{\sin x}{x} = 0$ since

$$\left|\frac{\sin x}{x}\right| = \left|\frac{1}{x}\right||\sin x| \le \frac{1}{x} \to 0 \text{ as } x \to \infty.$$

Thus, the line $y = 0$ is a horizontal asymptote for $f(x) = (\sin x)/x$. (See Figure 6.8.)

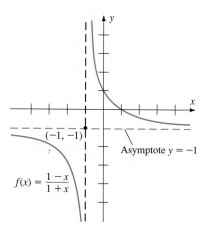

Figure 6.6 $\displaystyle\lim_{x \to \infty}\left(\frac{1-x}{1+x}\right) = 1$;

$\displaystyle\lim_{x \to -\infty}\left(\frac{1-x}{1+x}\right) = -1$. Asymptote is $y = -1$.

Note that, in part (e), the graph crosses the asymptote infinitely many times as $x \to \infty$.   ■

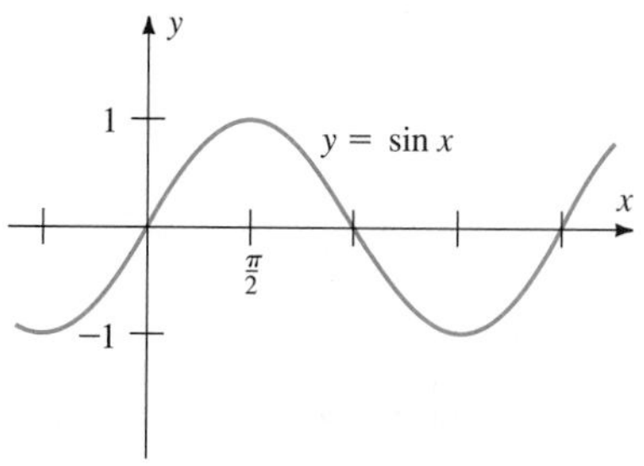

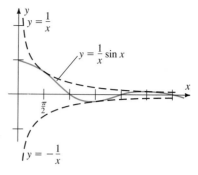

Figure 6.7 $\lim\limits_{x\to\infty} \sin x$ does not exist.

Figure 6.8 $\lim\limits_{x\to\infty} (\sin x)/x = 0$. Asymptote is the x-axis $y = 0$.

Evaluating Limits at Infinity

In Example 1, we used the algebraic properties of limits that we learned in Chapter 2. The following theorem justifies our computations.

Theorem 10

Suppose that $\lim\limits_{x\to\infty} f(x)$ and $\lim\limits_{x\to\infty} g(x)$ both exist. Then

(i) $\lim\limits_{x\to\infty} [f(x) + g(x)] = \lim\limits_{x\to\infty} f(x) + \lim\limits_{x\to\infty} g(x),$

(ii) $\lim\limits_{x\to\infty} [cf(x)] = c \lim\limits_{x\to\infty} f(x),$ c constant,

(iii) $\lim\limits_{x\to\infty} [f(x)g(x)] = [\lim\limits_{x\to\infty} f(x)][\lim\limits_{x\to\infty} g(x)],$

(iv) $\lim\limits_{x\to\infty} \left[\dfrac{f(x)}{g(x)}\right] = \dfrac{\lim\limits_{x\to\infty} f(x)}{\lim\limits_{x\to\infty} g(x)},$ provided $\lim\limits_{x\to\infty} g(x) \neq 0.$

The proof of Theorem 10 is Exercise 52.

To compute limits of rational combinations of power functions, we combine Theorem 10 and algebraic techniques (such as those used in Example 1) with our knowledge of the limiting behavior of elementary power functions. Recall that, in Section 4 of Chapter 1, we graphed functions of the form

$$f(x) = x^{-n/m} = \frac{1}{x^{n/m}}$$

where m and n are positive integers with no common factors. Those graphs suggest that the limit as $x \to \infty$ (and, if it makes sense, the limit as $x \to -\infty$) equals 0. For instance, consider the function $f(x) = x^{-19/11}$ (Figure 6.9) and the function $g(x) = x^{-19/10}$ (Figure 6.10). Since f is defined for all $x \neq 0$, we can calculate both a limit as $x \to \infty$ and a limit as $x \to -\infty$. Both limits can be expressed in one equation as

$$\lim_{x\to\pm\infty} x^{-19/11} = 0 \quad \text{or} \quad \lim_{x\to\pm\infty} \frac{1}{x^{19/11}} = 0.$$

However, $g(x)$ is defined only for $x > 0$, and consequently, we consider only the case where $x \to \infty$. Note that

$$\lim_{x\to\infty} x^{-19/10} = \lim_{x\to\infty} \frac{1}{x^{19/10}} = 0.$$

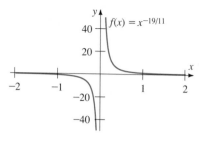

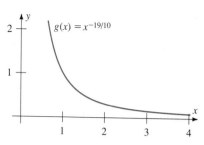

Figure 6.9 The graph of $y = f(x) = x^{-19/11}$.

Figure 6.10 The graph of $y = g(x) = x^{-19/10}$.

The limiting behavior of power functions with negative powers is summarized in the following two equations.

Let $r = p/q$ be any positive rational number. Then

$$\text{(i)} \quad \lim_{x \to \infty} \frac{1}{x^r} = 0 \quad \text{if } q \text{ is even,} \quad \text{and} \tag{2}$$

$$\text{(ii)} \quad \lim_{x \to \pm\infty} \frac{1}{x^r} = 0 \quad \text{if } q \text{ is odd.} \tag{3}$$

Technique for Evaluating Limits at Infinity

To evaluate limits of the form

$$\lim_{x \to \infty} \frac{f(x)}{g(x)} \quad \text{or} \quad \lim_{x \to -\infty} \frac{f(x)}{g(x)}$$

where $f(x)$ and $g(x)$ are polynomials, divide both $f(x)$ and $g(x)$ by the highest power of x present in the denominator. Then use limits (2) and (3).

Although the technique is stated for rational functions only, it works equally well if $f(x)$ and $g(x)$ involve fractional powers of x. Also, note that there is no difficulty in assuring $x \neq 0$ in executing this technique, since we may assume that $|x|$ is large.

□ **EXAMPLE 2**

Using the above technique we obtain the following limits.

Strategy

Solution

Divide numerator and denominator by x^2.

$$\text{(a)} \quad \lim_{x \to \infty} \frac{3x^2 + 7x - 4}{1 - x^2} = \lim_{x \to \infty} \frac{3 + \dfrac{7}{x} - \dfrac{4}{x^2}}{\dfrac{1}{x^2} - 1}$$

Apply Theorem 10 (algebra of limits) and limit (3).

$$= \frac{3 + 7\left(\lim\limits_{x \to \infty} \dfrac{1}{x}\right) - 4\left(\lim\limits_{x \to \infty} \dfrac{1}{x^2}\right)}{\left(\lim\limits_{x \to \infty} \dfrac{1}{x^2}\right) - 1}$$

$$= \frac{3 + 7(0) - 4(0)}{0 - 1} = -3.$$

Divide numerator and denominator by x^3.

(b) $\lim\limits_{x \to -\infty} \dfrac{x - 3}{2x + x^3} = \lim\limits_{x \to -\infty} \dfrac{\dfrac{1}{x^2} - \dfrac{3}{x^3}}{\dfrac{2}{x^2} + 1}$

Apply Theorem 10 and limit (3).

$$= \frac{\left(\lim\limits_{x \to -\infty} \dfrac{1}{x^2}\right) - 3\left(\lim\limits_{x \to -\infty} \dfrac{1}{x^3}\right)}{2\left(\lim\limits_{x \to -\infty} \dfrac{1}{x^2}\right) + 1}$$

$$= \frac{0 - 3(0)}{2(0) + 1} = 0.$$

Divide numerator and denominator by $x^{2/3}$.

(c) $\lim\limits_{x \to \infty} \dfrac{\sqrt{x} + \sqrt[3]{x}}{x^{2/3} + 1} = \lim\limits_{x \to \infty} \dfrac{\dfrac{1}{x^{1/6}} + \dfrac{1}{x^{1/3}}}{1 + \dfrac{1}{x^{2/3}}}$

Apply Theorem 10 and limit (2).

$$= \frac{\left(\lim\limits_{x \to \infty} \dfrac{1}{x^{1/6}}\right) + \left(\lim\limits_{x \to \infty} \dfrac{1}{x^{1/3}}\right)}{1 + \left(\lim\limits_{x \to \infty} \dfrac{1}{x^{2/3}}\right)}$$

$$= \frac{0 + 0}{1 + 0} = 0 \qquad\blacksquare$$

Infinite Limits

Many functions do not have limits as $x \to \pm\infty$. For example, for the parabola $y = x^2$, neither the limit of x^2 as $x \to \infty$ nor the limit of x^2 as $x \to -\infty$ exists in the sense of Definition 7.

We can write, however,

$$\lim_{x \to \infty} x^2 = \infty \tag{4}$$

and

$$\lim_{x \to -\infty} x^2 = \infty, \tag{5}$$

which means that the values $f(x) = x^2$ *increase without bound* as $x \to \infty$ or as $x \to -\infty$ (Figure 6.11).

Since the symbol ∞ represents not a number, but rather a concept, equations (4) and (5) do *not* say that the corresponding limits exist. Rather, they give us information about why the limit fails to exist: $f(x) = x^2$ becomes infinitely large as $x \to \pm\infty$.

Compare this behavior to that of $\sin x$ as $x \to \infty$ (see Example 1(d) and Figure 6.7). The limit of $\sin x$ as $x \to \infty$ fails to exist, but we cannot say that as

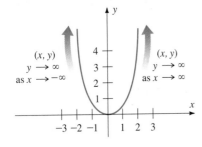

Figure 6.11   For $f(x) = x^2$, both $\lim\limits_{x \to \infty} f(x) = \infty$ and $\lim\limits_{x \to -\infty} f(x) = \infty$.

$x \to \infty$ the limit of $\sin x$ is ∞ since the values of $\sin x$ do not increase without bound as $x \to \infty$. Rather, they simply oscillate periodically between -1 and 1 as $x \to \infty$.

We summarize this discussion of limits at infinity by saying that for the expression

$$\lim_{x \to \infty} f(x)$$

one of three situations occurs:

(i) $\lim_{x \to \infty} f(x)$ exists in the sense of Definition 7. That is, there is a *number L* so that

$$\lim_{x \to \infty} f(x) = L;$$

(ii) $\lim_{x \to \infty} f(x)$ fails to exist but either $\lim_{x \to \infty} f(x) = \infty$ or $\lim_{x \to \infty} f(x) = -\infty$ (such as for

$$f(x) = x^2);$$

(iii) $\lim_{x \to \infty} f(x)$ fails to exist, and neither $\lim_{x \to \infty} f(x) = \infty$ nor $\lim_{x \to \infty} f(x) = -\infty$ (such as

for $f(x) = \sin x$).

Corresponding statements hold for $\lim_{x \to -\infty} f(x)$.

☐ **EXAMPLE 3**

Find

$$\lim_{x \to \infty} \frac{10x^3 - 3x^2 - 4x + 1}{5x^2 - 4x}.$$

Strategy · · · · · · · ·   **Solution**

Divide both numerator and denominator by x^2.

$$\lim_{x \to \infty} \frac{10x^3 - 3x^2 - 4x + 1}{5x^2 - 4x} = \lim_{x \to \infty} \frac{10x - 3 - (4/x) + (1/x^2)}{5 - (4/x)}$$

Examine limits of numerator and denominator separately, using Theorem 10 and limit (3).

In the numerator, we have

$$\lim_{x \to \infty} \left(10x - 3 - \frac{4}{x} + \frac{1}{x^2} \right) = \left(\lim_{x \to \infty} 10x \right) - 3 - 0 + 0$$

$$= \infty;$$

while in the denominator,

$$\lim_{x \to \infty} \left(5 - \frac{4}{x} \right) = 5 - 0.$$

As $x \to \infty$, the denominator approaches the number 5 while the numerator increases without bound. As $x \to \infty$, we have

$$\lim_{x \to \infty} \frac{10x^3 - 3x^2 - 4x + 1}{5x^2 - 4x} = +\infty. \qquad \blacksquare$$

REMARK 2:   The limit in Example 3 can also be determined using division. In other words, we can rewrite the quotient as a sum of a polynomial and a fraction in which the degree of the numerator is less than the degree of the denominator. In Example 3, we obtain

$$\frac{10x^3 - 3x^2 - 4x + 1}{5x^2 - 4x} = 2x + 1 + \frac{1}{5x^2 - 4x}.$$

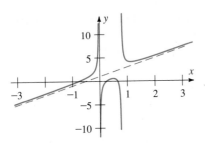

Figure 6.12   The graph of

$$f(x) = \frac{10x^3 - 3x^2 - 4x + 1}{5x^2 - 4x}.$$

The limit of $f(x)$ as $x \to \infty$ is ∞. Note that the graph is asymptotic to the line $y = 2x + 1$ as $x \to \infty$.

Vertical Asymptotes

Since

$$\lim_{x \to \infty} \frac{1}{5x^2 - 4x} = 0,$$

we conclude that the graph of

$$y = \frac{4x^3 + 3x + 3}{2x^2 + x}$$

is asymptotic to the line $y = 2x + 1$ as $x \to \infty$ (see Figure 6.12).

In general, the graph $y = r(x)$ of a rational function r is asymptotic to a line as $x \to \pm\infty$ if the highest power in the numerator is no more than 1 greater than the highest power in the denominator. If so and if the asymptote is not horizontal, we say that r has an **oblique asymptote**. In Exercise 62, we discuss how division can be used to determine curves that are asymptotes.

Infinite limits can also occur for

$$\lim_{x \to a} f(x), \quad \lim_{x \to a^+} f(x), \quad \text{or} \quad \lim_{x \to a^-} f(x),$$

that is, for one- or two-sided limits at the number a. For example, if

$$f(x) = \frac{1}{(x-2)^2},$$

the limit of $f(x)$ as $x \to 2$ is $+\infty$ since $(x-2)^2$ approaches 0 through positive numbers (Figure 6.13 and Table 6.2).

We formalize statements like "the limit of $f(x)$ as $x \to a$ is ∞" with the following working definition.

Definition 9

The statement

$$\lim_{x \to a} f(x) = \infty$$

means that the values $f(x)$ increase without bound as x approaches a from either direction. The statement

$$\lim_{x \to a^+} f(x) = \infty$$

means that the values of $f(x)$ increase without bound as x approaches a from the right. Similar meanings are attached to the statements

$$\lim_{x \to a^-} f(x) = \infty, \quad \lim_{x \to a} f(x) = -\infty,$$

$$\lim_{x \to a^+} f(x) = -\infty, \quad \text{and} \quad \lim_{x \to a^-} f(x) = -\infty.$$

Definition 10

We say that the graph of $y = f(x)$ has a **vertical asymptote** at $x = a$ if any of the infinite limits in Definition 9 exists.

The technique for finding vertical asymptotes is simply to look for those numbers for which the denominator of $f(x)$ becomes zero.

Table 6.2

x	$\dfrac{1}{(x-2)}$	$\dfrac{1}{(x-2)^2}$
1.00	-1	1
1.50	-2	4
1.75	-4	16
1.90	-10	100
2.10	10	100
2.25	4	16
2.50	2	4
3.00	1	1

□ **EXAMPLE 4**

For

$$f(x) = \frac{1}{(x-2)^3}$$

note that if $x - 2$ is positive, so is $(x - 2)^3$. If $(x - 2)$ is negative, $(x - 2)^3$ is also. Thus,

(a) if x approaches 2 from the right, $(x - 2)^3$ approaches zero through positive values, so

$$\lim_{x \to 2^+} \frac{1}{(x-2)^3} = \infty;$$

(b) if, however, x approaches 2 from the left, $(x - 2)^3$ is *negative*, so $(x - 2)^3$ approaches zero through negative values. Thus

$$\lim_{x \to 2^-} \frac{1}{(x-2)^3} = -\infty.$$

The graph of f has a vertical asymptote at $x = 2$. (Figure 6.14.)   ∎

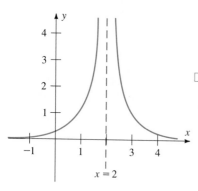

Figure 6.13 $\displaystyle\lim_{x \to 2} \frac{1}{(x-2)^2} = \infty.$

□ **EXAMPLE 5**

Find all vertical asymptotes for the graph of

$$f(x) = \frac{x^2}{4 - x^2}$$

and determine the corresponding limits.

Solution The denominator can be factored as follows:

$$f(x) = \frac{x^2}{4 - x^2} = \frac{x^2}{(2 - x)(2 + x)}.$$

Therefore the denominator equals zero for $x = -2$ and $x = 2$. These numbers are candidates for vertical asymptotes. To find the four corresponding one-sided limits, we must determine the signs of each of the factors of f. This information is easily obtained by use of a marked number line such as Figure 6.15.

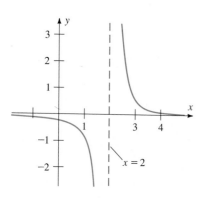

Figure 6.14 $\displaystyle\lim_{x \to 2^+} \frac{1}{(x-2)^3} = \infty;$

$\displaystyle\lim_{x \to 2^-} \frac{1}{(x-2)^3} = -\infty.$ Asymptote is $x = 2$.

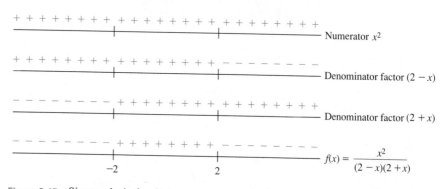

Figure 6.15 Sign analysis for $f(x)$.

Since the numerator of $f(x) = x^2/(4 - x^2)$ approaches 4 and the denominator approaches zero as $x \to \pm 2$, the one-sided limits at $x = \pm 2$ are all infinite. Figure 6.15 allows us to determine the correct signs. We have

$$\lim_{x \to -2^-} \frac{x^2}{4 - x^2} = -\infty, \qquad \lim_{x \to -2^+} \frac{x^2}{4 - x^2} = \infty,$$

$$\lim_{x \to 2^-} \frac{x^2}{4 - x^2} = \infty, \quad \text{and} \quad \lim_{x \to 2^+} \frac{x^2}{4 - x^2} = -\infty.$$

The vertical asymptotes are $x = -2$ and $x = 2$. (See Figure 6.16. Note also that the graph has a *horizontal* asymptote $y = -1$.)   ■

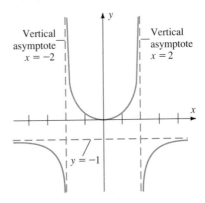

Figure 6.16 The vertical asymptotes of $f(x) = x^2/(4 - x^2)$ are $x = -2$ and $x = +2$. The horizontal asymptote is $y = -1$.

Exercise Set 4.6

In Exercises 1–38, find the indicated limits if they exist.

1. $\lim\limits_{x \to \infty} \dfrac{3x^2 + 2}{10x^2 - 3x}$

2. $\lim\limits_{x \to \infty} \dfrac{6x^4 - 8x}{7 - 3x^4}$

3. $\lim\limits_{x \to \infty} \dfrac{x(4 - x^3)}{3x^4 + 2x^2}$

4. $\lim\limits_{x \to \infty} \left(\dfrac{1}{x}\right)^5$

5. $\lim\limits_{x \to -\infty} x^{-1/3}$

6. $\lim\limits_{x \to -\infty} \dfrac{2x + 6}{x^2 + 1}$

7. $\lim\limits_{x \to \infty} \dfrac{3x^2 + 7x}{1 - x^4}$

8. $\lim\limits_{x \to \infty} \dfrac{x^4 - 4}{x^3 + 7x^2}$

9. $\lim\limits_{x \to -\infty} \dfrac{\sin x}{x}$

10. $\lim\limits_{x \to \infty} \dfrac{x^3 - 2x^2 + 5x - 4}{x - 3}$

11. $\lim\limits_{x \to -\infty} \dfrac{x^4 - 3 \sin x}{3x + 5x^5}$

12. $\lim\limits_{x \to \infty} \dfrac{x^2|2 + x^2|}{4 - x^4}$

13. $\lim\limits_{x \to \infty} \dfrac{|2 - x^2|}{x^2 + 1}$

14. $\lim\limits_{x \to -\infty} \dfrac{(x - 2)^3}{|x^3 - x|}$

15. $\lim\limits_{x \to \infty} \dfrac{\sqrt{x - 1}}{x^2}$

16. $\lim\limits_{x \to \infty} \dfrac{x^{2/3} + x^{4/3}}{x^2}$

17. $\lim\limits_{x \to \infty} \dfrac{x^{2/3} + x}{1 + x^{3/4}}$

18. $\lim\limits_{x \to \infty} \dfrac{\sqrt{x} + 7}{1 - \sqrt[3]{x}}$

19. $\lim\limits_{x \to \infty} x \sin x$

20. $\lim\limits_{x \to -\infty} x + \sin x$

need to find whether + or − ∞

21. $\lim\limits_{x \to 2^+} \dfrac{1}{x - 2}$

22. $\lim\limits_{x \to -2^-} \dfrac{1}{(x + 2)^2}$

23. $\lim\limits_{x \to \pi/2^-} \tan x$

24. $\lim\limits_{x \to \pi/2^+} \tan x$

25. $\lim\limits_{x \to \pi/2} \tan x$

26. $\lim\limits_{x \to -\pi/2^+} x \tan x$

27. $\lim\limits_{x \to 0^-} \dfrac{|x|}{x}$

28. $\lim\limits_{x \to 0^+} \dfrac{|x|}{x}$

29. $\lim\limits_{x \to 5^-} \dfrac{x^2 + 1}{x + 5}$

have to check both − and −

30. $\lim\limits_{x \to 3} \dfrac{x^2 - 4}{x + 3}$

31. $\lim\limits_{x \to 5^-} \dfrac{x^2 - 25}{x - 5}$

32. $\lim\limits_{x \to 3} \dfrac{x^2 - 6x + 9}{x - 3}$

33. $\lim\limits_{x \to 5^-} \dfrac{x - 7}{x - 5}$

34. $\lim\limits_{x \to 3} \dfrac{10 - x^2}{x - 3}$

35. $\lim\limits_{x \to 1^+} \dfrac{1}{(x - 1)^{1/3}}$

36. $\lim\limits_{x \to 1^-} \dfrac{1}{(x - 1)^{1/3}}$

37. $\lim_{x \to 1^+} \dfrac{1}{(x-1)^{2/3}}$

38. $\lim_{x \to 1^-} \dfrac{1}{(x-1)^{2/3}}$

In Exercises 39–48, find the horizontal asymptotes for the given functions, if they exist.

39. $y = \dfrac{1}{1+x}$

40. $y = \dfrac{1+x}{3-x}$

41. $y = \dfrac{2x^2}{x^2+1}$

42. $y = \dfrac{x^2}{1-x}$

43. $y = \dfrac{2x^2}{(x^2+1)^2}$

44. $y = \dfrac{x^2+3}{1-3x^2}$

45. $y = 3 + \dfrac{\sin x}{x}$

46. $y = \dfrac{4x - \sqrt{x}}{x^{2/3}+x}$

47. $y = \dfrac{4-3x}{\sqrt{1+2x^2}}$

48. $y = x \sin(1/x)$

49. The function $y = (ax+7)/(4-x)$ has a horizontal asymptote of $y = 3$. Find a.

In Exercises 50–55, find all vertical asymptotes for the function specified.

50. Exercise 21.

51. Exercise 22.

52. Exercise 35.

53. Exercise 37.

54. Exercise 40.

55. Exercise 42.

56. The function $y = (x^r + 3x)/(7 - x^{4/3})$ has a horizontal asymptote of $y = -1$. Find r.

57. The function $y = (\pi + ax^r)/(1 - 3x^{2/3})$ has a horizontal asymptote of $y = -2$. Find a and r.

58. The function $y = (3x^r + 2x^3)/(rx^3)$ has a horizontal asymptote of $y = 5/3$. Find r.

59. The function $y = (x^2 + ax + b)^{-1}$ has vertical asymptotes of $x = 3$ and $x = 5$. Find a and b.

60. Show that $\lim_{x \to \infty} \dfrac{a_n x^n + a_{n-1}x^{n-1} + \cdots + a_1 x + a_0}{b_m x^m + b_{m-1}x^{m-1} + \cdots + b_1 x + b_0}$ (where $a_n \neq 0$, $b_m \neq 0$) equals
  a. 0 if $n < m$,
  b. ∞ if $n > m$ and a_n and b_m have like signs,
  c. a_n/b_m if $n = m$.

61. Assume that a function f is increasing on $(0, \infty)$, that its graph is convex down on $(0, \infty)$, and that $f(x) < 0$ for all $x > 0$.
  a. Sketch the graph of one such function f.
  b. Find an equation for $f(x)$ that is consistent with these conditions.
  c. Must the limit of $f(x)$ as $x \to \infty$ equal 0? If so, justify your answer. If not, provide a counterexample.

62. Let $f(x) = (2x^3 + 1)/(x - 1)$.
  a. Determine the limit of $f(x)$ as $x \to \infty$.
  b. Let $q(x)$ be the quadratic polynomial $2x^2 + 2x + 2$. Show that the limit of $f(x) - q(x)$ as $x \to \infty$ is 0.
  c. Explain why the limit in part b justifies the assertion that the graph of f approaches the parabola $y = q(x)$ as $x \to \infty$.

63. Suppose g is a function that is twice differentiable for all real numbers x except $x = 1$ and suppose that it is undefined at $x = 1$. Also, assume that it satisfies the following eight conditions:
  (1) $g(x) = 0$ if and only if $x = 0$;
  (2) as $x \to \infty$, the limit of $g(x)$ is 0;
  (3) as $x \to -\infty$, the limit of $g(x)$ is 0;
  (4) as $x \to 1$, the limit of $g(x)$ is ∞;
  (5) $g'(x) < 0$ for $x < -1$ and $x > 1$;
  (6) $g'(x) > 0$ for $-1 < x < 1$;
  (7) $g''(x) < 0$ for $x < -2$; and
  (8) $g''(x) > 0$ for $-2 < x < 1$ and $x > 1$.

Sketch the graph of a function that satisfies these properties. Label the absolute extrema, the relative extrema, and the inflection points.

64. A formal definition of the limit of $f(x)$ is L as $x \to \infty$ is: "Given $\epsilon > 0$ there is an integer N so that if $x > N$ then $|f(x) - L| < \epsilon$." Use this definition to prove that
  a. $\lim_{x \to \infty} \dfrac{1}{x} = 0$,
  b. $\lim_{x \to \infty} \dfrac{3x+1}{x} = 3$,
  c. $\lim_{x \to \infty} \dfrac{\sin x}{x+2} = 0$.

65. State a definition analogous to that of Exercise 64 for $\lim_{x \to -\infty} f(x) = L$ and use it to prove that:
  a. $\lim_{x \to -\infty} \dfrac{1}{x} = 0$,
  b. $\lim_{x \to -\infty} \dfrac{x}{2x+1} = \dfrac{1}{2}$,
  c. $\lim_{x \to -\infty} \dfrac{6x-2}{x+1} = 6$.

66. A formal definition of the limit of $f(x)$ is ∞ as $x \to a$ is: "Given any real number N there exists a number $\delta > 0$ so that if $0 < |x - a| < \delta$, then $f(x) > N$." Use this definition to prove that
  a. $\lim_{x \to 0} \dfrac{1}{|x|} = \infty$,
  b. $\lim_{x \to 0} \dfrac{1}{x^2} = \infty$,
  c. $\lim_{x \to \pi/2} |\tan x| = \infty$.

67. State a formal definition for the statement "the limit of $f(x)$ as $x \to a^-$ is ∞" and use it to prove that $\lim\limits_{x \to \pi/2^-} \tan x = \infty$.

68. State a formal definition for the statement "the limit of $f(x)$ as $x \to a^-$ is $-\infty$" and use it to prove that $\lim\limits_{x \to 0^-} 1/x = -\infty$.

69. Use the definition in Exercise 64 to prove Theorem 10.

 In Exercises 70–73, use the graph of the function f to determine the vertical and horizontal asymptotes as well as the relative extrema.

70. $f(x) = \dfrac{5x^2}{2x^2 + 3x + 1}$

71. $f(x) = \dfrac{x^2 + 4}{x^3 - 9x}$

72. $f(x) = \dfrac{x^2 + 5x + 1}{4x^4 - 5x + 1}$

73. $f(x) = \dfrac{x + \sin 2x}{2x + \cos 5x}$

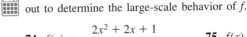 In Exercises 74–77, graph the function f. Then repeatedly zoom out to determine the large-scale behavior of f.

74. $f(x) = -\dfrac{2x^2 + 2x + 1}{x - 1}$

75. $f(x) = \dfrac{x^3 + x - 2}{x^2 - 4}$

76. $f(x) = \dfrac{2x^3 - 3x}{x + 1}$

77. $f(x) = \left(1 + \dfrac{2}{x}\right)^x$

78. A train leaves Kansas City at 1:00 A.M. and travels northeast at 50 mi/h. An hour later a second train leaves Kansas City and travels west at 65 mi/h.

  a. Determine a function $h(t)$ that gives the distance between the trains as a function of time t, in hours. (*Hint:* The Law of Cosines is useful here.)

  b. Graph h. Does the graph approach a straight line as t increases? Why should this be true? Does the graph of h have an oblique asymptote?

  c. Graph h'. Relate the large-scale behavior of this graph to the large-scale behavior of the graph of h.

4.7 Curve Sketching

In Sections 4.1 and 4.4–4.6, we discussed ways in which information about limits and derivatives is used to identify certain properties of a function. In this section, we use these ideas to help us sketch the graph of a function. We see how to determine quickly a function's behavior without having to develop a precise graph.

Procedure for Sketching the Graph of a Function f

1. Determine the domain.
2. If possible, locate the zeros by solving the equation $f(x) = 0$.
3. Determine the large-scale behavior of f: the vertical asymptotes and the limits at $\pm\infty$, if they exist (Section 4.6).
4. Find all critical numbers, determine whether f is increasing or decreasing on the resulting intervals, and classify the extrema (Sections 4.1 and 4.4).
5. Determine the concavity and locate the inflection points (Section 4.5).
6. Calculate the values of the function at a few convenient numbers and locate the corresponding points on the graph. Then sketch in the graph according to this information.

☐ **EXAMPLE 1**

Sketch the graph of the function $f(x) = x^4 - 24x^2 + 20$.

Solution Following the outline above, we have:

(1) Since f is a polynomial, $f(x)$ is defined for all x.

(2) We can determine the zeros of f if we use the quadratic formula to solve for x^2. We obtain $x^2 = 12 \pm 2\sqrt{31}$. The four roots are approximately ± 0.93 and ± 4.81.

(3) Since f is an even degree polynomial whose leading term is positive, $f(x) \to \infty$ as $x \to \pm\infty$. Thus, there are no horizontal asymptotes. There are no vertical asymptotes because $f(x)$ is defined and continuous on $(-\infty, \infty)$.

(4) Since f is a polynomial, it is differentiable everywhere, and its first derivative is $f'(x) = 4x^3 - 48x = 4x(x^2 - 12)$. Thus, to obtain the critical numbers, we solve

$$f'(x) = 4x(x^2 - 12) = 0,$$

which has solutions $x = 0$ and $x = \pm\sqrt{12} = \pm 2\sqrt{3}$. A sign analysis of f' yields the fact that f is decreasing on the intervals $(-\infty, -2\sqrt{3}]$ and $[0, 2\sqrt{3}]$ and increasing on $[-2\sqrt{3}, 0]$ and $[2\sqrt{3}, \infty)$ (see Figure 7.1). The three critical numbers yield the points

$$(-2\sqrt{3}, f(-2\sqrt{3})) = (-2\sqrt{3}, -124)$$
$$(0, f(0)) = (0, 20)$$
$$(2\sqrt{3}, f(2\sqrt{3})) = (2\sqrt{3}, -124).$$

Hence, -124 is the minimum, and $f(0) = 20$ is a relative maximum.

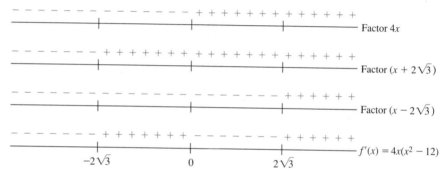

Figure 7.1   Sign analysis of $f'(x) = 4x(x^2 - 12)$.

(5) The second derivative is $f''(x) = 12x^2 - 48 = 12(x^2 - 4)$. Setting $f''(x) = 0$ gives $x = \pm 2$. Checking the sign of f'' on the three intervals determined by ± 2 shows that the graph of f is concave up on both $(-\infty, -2]$ and $[2, \infty)$ and concave down on $[-2, 2]$. Thus, the points $(-2, f(-2)) = (-2, -60)$ and $(2, f(2)) = (2, -60)$ are inflection points.

(6) The graph of f is sketched in Figure 7.2.   ■

☐ **EXAMPLE 2**

Sketch the graph of $f(x) = \dfrac{x}{1 - x^2}$.

Solution   Following the outline stated above, we find that:

(1) The domain of f is all real numbers except $x = 1$ and $x = -1$ since the zeros of the denominator are $x = \pm 1$.

(2) The equation $f(x) = 0$ has the single solution $x = 0$.

(3) The denominator approaches zero as x approaches 1 or -1, but the numerator

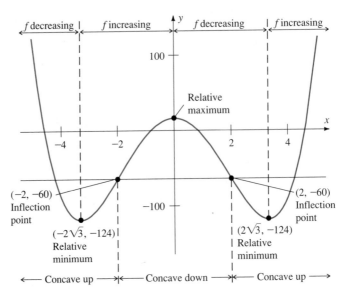

Figure 7.2 The graph of $f(x) = x^4 - 24x^2 + 20$.

does not. Thus, there are vertical asymptotes at $x = \pm 1$. The line $y = 0$ (i.e., the x-axis) is a horizontal asymptote, since

$$\lim_{x \to \pm\infty} \left(\frac{x}{1 - x^2} \right) = \lim_{x \to \pm\infty} \left(\frac{1/x}{(1/x^2) - 1} \right) = \frac{0}{0 - 1} = 0.$$

(4) To find the critical numbers we use the equation

$$f'(x) = \frac{d}{dx} \left(\frac{x}{1 - x^2} \right) = \frac{(1)(1 - x^2) - (x)(-2x)}{(1 - x^2)^2} = \frac{1 + x^2}{(1 - x^2)^2} = 0.$$

This equation has no solutions because $1 + x^2 \neq 0$ for all x. Although $f'(x)$ is undefined for $x = \pm 1$, these numbers are not in the domain of f. There are, therefore, no critical numbers. Since $1 + x^2 \geq 1$ for all x and $(1 - x^2)^2 > 0$ for $x \neq \pm 1$, we can see that $f'(x) > 0$ if $x \neq \pm 1$. Thus, f is increasing on each of the three intervals $(-\infty, -1)$, $(-1, +1)$, and $(+1, +\infty)$.

(5) Since

$$f''(x) = \frac{d}{dx} \left[\frac{1 + x^2}{(1 - x^2)^2} \right] = \frac{2x(1 - x^2)^2 - (1 + x^2)(2)(1 - x^2)(-2x)}{(1 - x^2)^4}$$

$$= \frac{2x^3 + 6x}{(1 - x^2)^3}$$

$$= \frac{2x(x^2 + 3)}{(1 - x^2)^3},$$

the candidate for an inflection point is $x = 0$ ($f''(0) = 0$). We note also that $f''(x)$ is undefined for $x = \pm 1$. The sign of $f''(x)$ on each of the resulting intervals can be obtained from Figure 7.3. From the sign of $f''(x)$ we conclude that the graph of $y = f(x)$ is concave up on $(-\infty, -1)$ and $[0, 1)$ and concave down on $(-1, 0]$ and $(1, \infty)$. The point $(0, f(0)) = (0, 0)$ is an inflection point.

(6) Since $f(-2) = 2/3, f(0) = 0$, and $f(2) = -2/3$, the points $(-2, 2/3)$, $(0, 0)$, and $(2, -2/3)$ are on the graph. The graph is sketched in Figure 7.4. ∎

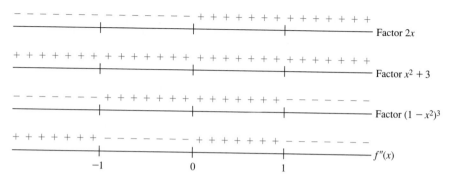

Figure 7.3 Sign analysis for $f''(x) = \dfrac{2x(x^2 + 3)}{(1 - x^2)^3}$.

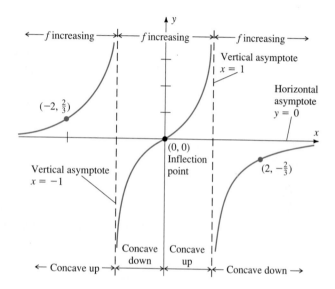

Figure 7.4 Graph of $f(x) = \dfrac{x}{1 - x^2}$.

□ **EXAMPLE 3**

Sketch the graph of $f(x) = \dfrac{\sin x}{2 + \cos x}$ for $-\pi \le x \le \pi$.

Solution Proceeding as in Example 2, we find that

(1) The domain of f is explicitly stated as the interval $[-\pi, \pi]$.
(2) The only solution of the equation

$$f(x) = \frac{\sin x}{2 + \cos x} = 0$$

in $[-\pi, \pi]$ occurs when $\sin x = 0$, that is, at $x = -\pi$, 0, or π.
(3) Since $f(x)$ is defined for all x, there are no vertical asymptotes. Since the domain of f is restricted to a finite interval, we need not consider limits at infinity.

(4) The first derivative is

$$f'(x) = \frac{d}{dx}\left(\frac{\sin x}{2 + \cos x}\right) = \frac{(2 + \cos x)(\cos x) - (\sin x)(-\sin x)}{(2 + \cos x)^2}$$

$$= \frac{2 \cos x + 1}{(2 + \cos x)^2}.$$

Thus $f'(x) = 0$ if $2 \cos x + 1 = 0$, or $\cos x = -1/2$. The solutions of this equation in $[-\pi, \pi]$ are $x = \pm 2\pi/3$. Since $f'(x)$ is defined for all x, these are the critical numbers.

By checking the sign of $f'(x)$ at one "test number" t in each of the resulting intervals, we obtain the information in Table 7.1 regarding the intervals on which f is increasing or decreasing. Consequently $f(-2\pi/3) = -\sqrt{3}/3 \approx -0.58$ is a relative minimum and $f(2\pi/3) = \sqrt{3}/3$ is a relative maximum, by the First Derivative Test.

Table 7.1 Analysis of sign of $f'(x)$.

Interval I	Test number $t \in I$	Sign of $f'(t)$	Conclusion
$\left(-\pi, -\dfrac{2\pi}{3}\right)$	$t = -\dfrac{5\pi}{6}$	$-$	f decreasing on $\left[-\pi, -\dfrac{2\pi}{3}\right]$
$\left(-\dfrac{2\pi}{3}, \dfrac{2\pi}{3}\right)$	$t = 0$	$+$	f increasing on $\left[-\dfrac{2\pi}{3}, \dfrac{2\pi}{3}\right]$
$\left(\dfrac{2\pi}{3}, \pi\right)$	$t = \dfrac{5\pi}{6}$	$-$	f decreasing on $\left[\dfrac{2\pi}{3}, \pi\right]$

(5) $f''(x) = \dfrac{(2 + \cos x)^2(-2 \sin x) - (2 \cos x + 1)(2)(2 + \cos x)(-\sin x)}{(2 + \cos x)^4}$

$$= \frac{2 \sin x(\cos x - 1)}{(2 + \cos x)^3}$$

so $f''(x) = 0$ if $2 \sin x(\cos x - 1) = 0$. The solutions of this equation are $x = 0$ and the endpoints $x = -\pi$ and $x = \pi$. Since $\cos x - 1 \le 0$ and $2 + \cos x \ge 0$ for all x, the factor

$$\frac{2(\cos x - 1)}{(2 + \cos x)^3}$$

is always less than or equal to zero; hence, the sign of $f''(x)$ is the opposite of the sign of the factor $\sin x$. Thus $f''(x) > 0$ on $(-\pi, 0)$ and $f''(x) < 0$ on $(0, \pi)$. The graph is therefore concave up on $[-\pi, 0]$ and concave down on $[0, \pi]$, and the point $(0, 0)$ is an inflection point. The graph appears in Figure 7.5. ∎

REMARK Note that, since $\sin x$ and $\cos x$ are periodic with period 2π, we can use the graph of $f(x)$ for $-\pi \le x \le \pi$ to determine the graph of

$$y = \frac{\sin x}{2 + \cos x} \tag{1}$$

for all x. The graph of equation (1) is simply the graph in Figure 7.5 repeated periodically over every interval of length 2π.

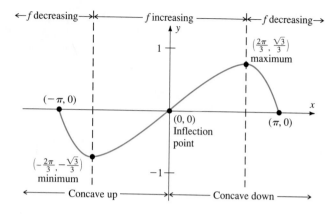

Figure 7.5 Graph of $f(x) = \dfrac{\sin x}{2 + \cos x}$.

□ **EXAMPLE 4**

Sketch the graph of the function $f(x) = \dfrac{x^2 - x - 2}{x - 1}$.

Solution Following the outline above, we begin by writing f in factored form as

$$f(x) = \frac{(x + 1)(x - 2)}{x - 1}.$$

From this equation for f we observe that

(1) The domain of f is all real numbers except $x = 1$.
(2) The zeros of f are $x = -1$ and $x = 2$.
(3) The denominator approaches zero as x approaches 1, but the numerator does not. Thus, the line $x = 1$ is a vertical asymptote. There are no horizontal asymptotes, since

$$\lim_{x \to \pm\infty} \frac{x^2 - x - 2}{x - 1} = \lim_{x \to \pm\infty} \frac{x - 1 - (2/x)}{1 - (1/x)} = \pm\infty.$$

(4) The derivative is

$$f'(x) = \frac{(x - 1)(2x - 1) - (x^2 - x - 2)}{(x - 1)^2}$$

$$= \frac{x^2 - 2x + 3}{(x - 1)^2}$$

$$= \frac{(x - 1)^2 + 2}{(x - 1)^2}$$

$$= 1 + \frac{2}{(x - 1)^2}.$$

Since $f'(x) > 1$ for all x except $x = 1$, where $f(1)$ and $f'(1)$ are undefined, we conclude that

(a) there are no critical numbers for f,
(b) f is increasing on $(-\infty, 1)$ and on $(1, \infty)$, and
(c) there are no relative extrema.

(5) The second derivative is $f''(x) = (-4)/(x - 1)^3$. Since $f''(x)$ is positive for $x < 1$, negative for $x > 1$ and since $f(1)$ is undefined, we conclude that
   (a) there are no inflection points,
   (b) the graph of f is concave up on $(-\infty, 1)$, and
   (c) the graph of f is concave down on $(1, \infty)$.

(6) Noting that $f(-1) = 0$, $f(0) = 2$, and $f(2) = 0$ we obtain the sketch in Figure 7.6. ∎

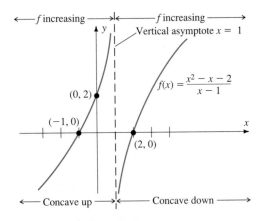

Figure 7.6 Graph of $f(x) = \dfrac{x^2 - x - 2}{x - 1}$.

Exercise Set 4.7

For each of Exercises 1–48, sketch the graph of f following the outline in this section.

1. $f(x) = x^2 - 2x - 8$

2. $f(x) = 9x - x^2$

3. $f(x) = 2x^3 - 3x^2$

4. $f(x) = 3x^4 - 4x^3$

5. $f(x) = x^4 - 8x^3 + 24x^2 - 32x + 19$

6. $f(x) = 3x^5 - 15x^4 + 20x^3$

7. $f(x) = x^3 + x^2 - 8x + 8$

8. $f(x) = x^3 + 2x^2 - x - 2$

9. $f(x) = x^3 - 7x + 6$

10. $f(x) = \dfrac{x - 1}{x + 1}$

11. $f(x) = \dfrac{x + 4}{x - 4}$

12. $f(x) = \sin x + \cos x$

13. $f(x) = 9x - x^{-1}$

14. $f(x) = x + 3x^{2/3}$

15. $f(x) = |4 - x^2|$

16. $f(x) = \sqrt{x + 4}$

17. $f(x) = x^2 + 2/x$

18. $f(x) = (x + 1)^{5/3} - (x + 1)^{2/3}$

19. $f(x) = |x^2 - 3x + 2|$

20. $f(x) = x^{-2} - x^{-1}$

21. $f(x) = \dfrac{1}{x(x - 4)}$

22. $f(x) = x^2 - 9/x^2$

23. $f(x) = \dfrac{x}{(2x + 1)^2}$

24. $f(x) = \sqrt{2x - x^2}$

25. $f(x) = (x - 3)^{2/3} + 1$

26. $f(x) = \sqrt{6x - x^2 - 8}$

27. $f(x) = \frac{1}{3}x^3 - x^2 - 3x + 4$

28. $f(x) = (x^2 + 2)^{-1}$

29. $f(x) = \dfrac{x^2 - 4x + 5}{x - 2}$

30. $f(x) = \frac{3}{2}x^{2/3} - x$

31. $f(x) = x + \cos x$

32. $f(x) = x + \sin(-2x)$

33. $f(x) = \sin x + \cos^2 x$

34. $f(x) = \sin^2 x + \sin x$

35. $f(x) = 4x^2(1 - x^2)$

36. $f(x) = x^4 - 3x^2 + 2$

37. $f(x) = x^{2/3} - \frac{1}{5}x^{5/3}$

38. $f(x) = \dfrac{2x + 1}{x + 2}$

39. $f(x) = \dfrac{x^2}{9 - x^2}$

40. $f(x) = \dfrac{x^2}{x^2 - 16}$

41. $f(x) = 5x^3 - x^5$

42. $f(x) = x^4 - 2x^2 + 1$

43. $f(x) = 16 - 20x^3 + 3x^5$

44. $f(x) = x^4 - 18x^2 + 32$

45. $f(x) = \dfrac{1 - x^2}{x^3}$

46. $f(x) = 3x^{5/3} - 15x^{2/3} + 5$

47. $f(x) = \dfrac{x}{x^2 + 3}$

48. $f(x) = \dfrac{x^2 + 3x}{x^2 + 2x + 1}$

4.8   Linear Approximation Revisited

In Chapter 3, we saw how we can use the derivative to approximate nonlinear functions such as $\sqrt{x}$ and $\sin x$ by *linear* functions. In Theorem 9 of Chapter 3, we characterized the manner in which the error ϵ in such an approximation vanishes as $\Delta x \to 0$. Although Theorem 9 provides an important theoretical property that characterizes the derivative, it does not give us any understanding of the error for a fixed nonzero value of Δx. However, using the second derivative, we can estimate this error.

Recall from Section 3.7 that we approximated the value $f(x_0 + \Delta x)$ by

$$f(x_0 + \Delta x) \approx f(x_0) + f'(x_0)\Delta x, \tag{1}$$

and the error ϵ is the difference between these two numbers

$$\epsilon = f(x_0 + \Delta x) - [f(x_0) + f'(x_0)\Delta x].$$

In practice, we do not know ϵ exactly. If we did, we would consequently know the exact value of $f(x_0 + \Delta x)$, and an approximation would not be necessary. However, using the second derivative, we can determine an upper bound for ϵ.

Theorem 11

Let Δx be a positive real number. Suppose that the function f is continuous on the interval $[x_0, x_0 + \Delta x]$ and twice differentiable on the open interval $(x_0, x_0 + \Delta x)$. If there exists a number M such that

$$|f''(x)| \leq M$$

for all $x \in (x_0, x_0 + \Delta x)$, then the error ϵ is bounded by

$$|\epsilon| \leq \frac{M(\Delta x)^2}{2}. \tag{2}$$

We obtain the same estimate if Δx is negative.

Theorem 11 says that, if we know the maximum of the absolute value of the second derivative on the interval between x_0 and $x_0 + \Delta x$, then we know that our linear approximation differs from the actual value by no more than

$$\frac{M(\Delta x)^2}{2}.$$

It is reasonable to expect that the second derivative would help us predict the error ϵ because the second derivative measures how fast the first derivative is changing. If the first derivative changes slowly, then the graph of the function is almost a straight line, and a linear approximation should be very accurate. On the other hand, if the second derivative is large, then the first derivative is changing rapidly, and the graph is far from linear. In this case, the error involved in any linear approximation is likely to be large.

Before we prove Theorem 11, we illustrate its use with three examples.

☐ **EXAMPLE 1**

Approximate the quantity $(1.3)^2$ using the derivative of $f(x) = x^2$ at $x_0 = 1$. Then use Theorem 11 to bound the associated error.

Strategy · · · · · · · ·

Using $f'(x_0)$, determine the linear equation that approximates the given quantity.

Using the value of Δx, compute the approximation.

Estimate the error by finding an approximate maximum M of the absolute value of the second derivative.

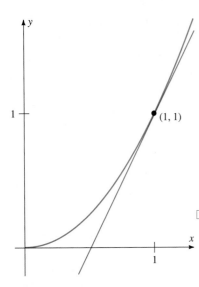

Figure 8.1 The graph of $f(x) = x^2$ and its linear approximation at the point $(1, 1)$.

Solution

To approximate x^2 near $x_0 = 1$, we consider the quantity $(1.3)^2$ as $(1 + \Delta x)^2$ where $\Delta x = 0.3$. The value of the derivative f' at $x_0 = 1$ is 2. Using equation (1), we see that the linear approximation for x^2 near $x_0 = 1$ is

$$(1 + \Delta x)^2 \approx f(1) + f'(1)\Delta x = 1 + 2\Delta x. \tag{3}$$

This expression yields the approximation

$$(1.3)^2 \approx 1 + (2)(0.3) = 1.6.$$

Both $f(x) = x^2$ and the linear approximation (3) about $x_0 = 1$ are graphed in Figure 8.1.

Now, we use Theorem 11 to get an upper bound on the error involved in the estimate $(1.3)^2 \approx 1.6$. Since we already know that $\Delta x = 0.3$, we only need to determine an acceptable value for M. The second derivative of f is $f''(x) = 2$, and therefore, we use $M = 2$. We conclude that

$$|\epsilon| \le \frac{2(0.3)^2}{2} = 0.09.$$

In other words, the linear approximation is within 0.09 of the actual value. ■

REMARK 1   Note that, in Example 1, the bound for the error given by Theorem 11 yields the actual error. This will not usually be the case.

Of course, in this example, it is easy enough to square 1.3 by hand, but if you wanted to compute $(1.22334598734567652346)^2$, the linear approximation method would have some advantages.

□ **EXAMPLE 2**

Approximate the quantity $(1.3)^3$ using a linear approximation of $g(x) = x^3$ around $x_0 = 1$. Bound the error generated by the approximation.

Solution   To approximate x^3 near $x_0 = 1$, we rewrite the quantity $(1.3)^3$ as $(1 + \Delta x)^3$ where $\Delta x = 0.3$. In this case, the linear approximation equation (1) is

$$(1 + \Delta x)^3 \approx 1 + 3\Delta x \tag{4}$$

since $g'(1) = 3$. Using (4), we obtain the approximation

$$(1.3)^3 \approx 1 + 3(0.3) = 1.9.$$

The function $g(x) = x^3$ and the linear approximation (4) are graphed in Figure 8.2.

To analyze the error in this example, we consider the second derivative $g''(x) = 6x$. This function g'' is increasing on $(-\infty, \infty)$, so the largest value of $6x$ for $1 \le x \le 1.3$ occurs when $x = 1.3$. We let $M = 7.8$. A bound on the error is

$$|\epsilon| \le \frac{(7.8)(0.3)^2}{2} = 0.351. \qquad ■$$

If you compare Figures 8.1 and 8.2, you see that it is reasonable to expect the error in Example 2 to be larger. Note that, as Δx increases, the graphs of g and its linear approximation move apart more rapidly than the graph of f and its linear approximation. In Section 4.5, we saw that the second derivative was related to the concavity of the graph. Theorem 11 gives us a way to quantify the relationship between nonlinearity and the second derivative.

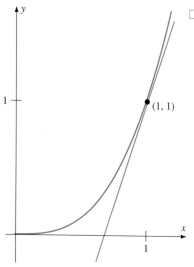

Figure 8.2   The graph of $g(x) = x^3$ and its linear approximation at the point $(1, 1)$.

□ **EXAMPLE 3**

Approximate $\sin(3°)$ using a linear approximation and estimate the accuracy of the resulting approximation.

Solution   Since the differentiation formula for $\sin x$ was derived using radian measure, we start by converting $3°$ to radians. We get

$$3° = \left(\frac{3°}{360°}\right)(2\pi) \text{ radians}$$

$$= \frac{\pi}{60} \text{ radians} \approx 0.05236 \text{ radians}$$

The closest "well-known" angle is $x_0 = 0$. Consequently, we use $x_0 = 0$ to carry out our approximation. Using the fact that the derivative of $\sin x$ is $\cos x$, the linear approximation about $x_0 = 0$ is

$$\sin(\Delta x) \approx \sin 0 + (\cos 0)\Delta x = \Delta x.$$

The resulting approximation is

$$\sin 0.05236 \approx 0.05236. \tag{5}$$

Estimating the accuracy of this approximation is somewhat more complicated in this example. The second derivative of $\sin x$ is $-\sin x$, and so for M, we should find the largest value of $|-\sin x| = \sin x$ for $0 \le x \le 0.05236$. However, since $\sin x$ is increasing on $[0, \pi/2]$, we know that the largest value of $\sin x$ over $[0, 0.05236]$ is $\sin 0.05236$. If we know this value, we would not need to do any approximation.

We resolve this issue by being slightly less efficient in our choice of M. Of course, since $-1 \le \sin x \le 1$ for all x, we could use $M = 1$. However, if we recall the inequality [equation (11) in Section 2.3]

$$|\sin x| \le |x| \quad \text{for all } x,$$

we can use $M = 0.05236$. Therefore, the error in approximation (5) is no larger than

$$\frac{(0.05236)(0.05236)^2}{2}$$

which is 0.000072 (to six decimal places). ■

REMARK 2   When we say that an approximation is accurate to k decimal places, we mean that the error ϵ is no larger than $5 \times 10^{-(k+1)}$. For example, accuracy to two decimal places signifies that $\epsilon \le 0.005$, and accuracy to five decimal places indicates that $\epsilon \le 0.000005$. Therefore, in Example 3, we determined that the approximation $\sin(0.05236) \approx 0.05236$ is accurate to at least three decimal places.

Proof of Theorem 11

We establish the relevance of the second derivative using the first derivative version of Taylor's Theorem.*

---

*Theorem 12 is one instance of a more general topic, the error involved in polynomial approximation, which we shall examine more closely in Chapter 13. Details regarding the development of this topic by the British mathematician Brook Taylor (1685–1731), as well as others, are provided in the historical introduction to Unit 5.

Theorem 12

Taylor's Theorem
(First Derivative Version)

Suppose the function f is continuous on the interval $[a, b]$ and twice differentiable on (a, b). Then there exists a number $c \in (a, b)$ such that

$$f(b) = f(a) + f'(a)(b - a) + \frac{f''(c)}{2}(b - a)^2. \tag{6}$$

Note that the expression $f(a) + f'(a)(b - a)$ on the right-hand side of equation (6) is the linear approximation of $f(b)$ based at a. Therefore, Taylor's Theorem indicates that the error ϵ in a linear approximation is equal to the quantity

$$\frac{f''(c)}{2}(b - a)^2$$

for *some* number c between a and b. Note that this statement is significant because it expresses the error in terms of the second derivative of f at some number between a and b

Proof: This proof is based on two applications of Rolle's Theorem (Theorem 5). We know that there is *some* number k such that

$$f(b) = f(a) + f'(a)(b - a) + k(b - a)^2 \tag{7}$$

because we can find it algebraically. In other words, we solve for k and obtain

$$k = \frac{f(b) - f(a) - f'(a)(b - a)}{(b - a)^2}.$$

Thus, the true significance of the theorem is the fact that this number k is one half of the second derivative $f''(c)$ for some number c between a and b.

Using k as determined by equation (7), we define a second function (a quadratic polynomial)

$$q(x) = f(a) + f'(a)(x - a) + k(x - a)^2, \tag{8}$$

and let d be the difference between f and q,

$$d(x) = f(x) - q(x).$$

The proof is based on the fact that Rolle's Theorem can be applied both to d and to its derivative d'.

First, note that

$$d(a) = f(a) - q(a) = f(a) - f(a) = 0$$

by equation (8) and that

$$d(b) = f(b) - q(b) = f(b) - f(b) = 0$$

by equation (7). Therefore, we can apply Rolle's Theorem to d and obtain a number c_1 between a and b such that $d'(c_1) = 0$.

To calculate the derivative of d at a, we start by calculating the derivative of q at a using equation (8). We differentiate $q(x)$ and get

$$q'(x) = f'(a) + 2k(x - a). \tag{9}$$

Therefore, $q'(a) = f'(a)$, and $d'(a) = f'(a) - q'(a) = f'(a) - f'(a) = 0$. Hence, Rolle's Theorem also applies to d' on the interval $[a, c_1]$, and we obtain a number c between a and c_1 such that $(d')'(c) = d''(c) = 0$.

We use equation (9) to calculate $q''(x) = 2k$. Then

$$d''(x) = f''(x) - q''(x) = f''(x) - 2k.$$

Using the fact that $d''(c) = 0$, we get

$$f''(c) - 2k = 0,$$

and thus, $k = f''(c)/2$, the result claimed in the statement of the theorem. ∎

Recall that Theorem 11 was phrased in terms of a bound M on the absolute value of the second derivative f''. We assumed that

$$|f''(x)| \leq M \tag{10}$$

for all x between x_0 and $x_0 + \Delta x$. Given such a bound, we can estimate the error ϵ by applying Theorem 12 to f on the closed interval between x_0 and $x_0 + \Delta x$:

$$\begin{aligned} |\epsilon| &= |f(x) - (f(x_0) + f'(x_0)\Delta x)| \\ &= \left| \frac{f''(c)}{2}(\Delta x)^2 \right| \\ &= \frac{|f'(c)|}{2}(\Delta x)^2 \end{aligned}$$

for some number c between x_0 and $x_0 + \Delta x$. Using equation (10), we obtain the conclusion of Theorem 11:

$$|\epsilon| \leq \frac{M}{2}(\Delta x)^2.$$

Exercise Set 4.8

In Exercises 1–8, approximate the desired quantity using linear approximation about the specified number x_0. Determine a bound M for Theorem 11, and use Theorem 11 to estimate the error in your approximation.

1. $(2.02)^2$, $x_0 = 2$ **2.** $(7.47)^2$, $x_0 = 7$

3. $(1.97)^2$, $x_0 = 2$ **4.** $(6.5)^2$, $x_0 = 7$

5. $(2.03)^4$, $x_0 = 2$ **6.** $(1.11)^7$, $x_0 = 1$

7. $\sin 46°$, $x_0 = \pi/4$ **8.** $\cos 46°$, $x_0 = \pi/4$

In Exercises 9–12, approximate the desired quantity using linear approximation. Determine a bound M for Theorem 11, and use Theorem 11 to estimate the error in your approximation.

9. $\sin 5°$ **10.** $\cos 89°$

11. $\sqrt{37}$ **12.** $1/(2.1)$

13. Using linear approximation and Theorem 11, show that $\sqrt{50}$ is closer to 7.1 than it is to 7.

14. Approximate $\sqrt{2.5}$ twice, once with $x_0 = 1$ and once with $x_0 = 4$. In light of Theorem 11, which approximation would you expect to be more accurate? Why?

15. Using Theorem 11, specify an interval centered at 0 over which the approximation $\sin x \approx x$ is accurate to two decimal places.

16. Using Theorem 11, determine an interval centered at 1 over which the approximation $\sqrt{1 + \Delta x} \approx 1 + \Delta x/2$ is accurate to one decimal place.

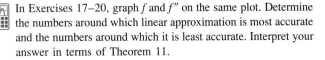 In Exercises 17–20, graph f and f'' on the same plot. Determine the numbers around which linear approximation is most accurate and the numbers around which it is least accurate. Interpret your answer in terms of Theorem 11.

17. $f(x) = x^2 + 2x - 2$, $[-3, 2]$

18. $f(x) = 2x^3 - x^2 - 4x + 2$, $[-2, 2]$

19. $f(x) = x + \cos x$, $[0, \pi]$

20. $f(x) = x^4 - 5x^2 + x$, $[-3, 3]$

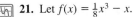 **21.** Let $f(x) = \frac{1}{8}x^3 - x$.
 a. Let T_1 be the linear function that approximates f about the number -2. That is, $T_1(x) = f(-2) + f'(-2)(x + 2)$. By magnifying the graph of f, find a number $h_1 > 0$ such that $|f(x) - T_1(x)| < 0.01$ if $-2 - h_1 < x < -2 + h_1$.

b. Let T_2 be the linear function that approximates f about the number 0. By magnifying the graph of f, find a number $h_2 > 0$ such that $|f(x) - T_2(x)| < 0.01$ if $-h_2 < x < h_2$.

c. Calculate $|f''(-2)|$ and $|f''(0)|$. Relate these two numbers to the results in parts a and b.

d. Graph f and f''. Find the intervals where $|f''(x)| \leq 1$. How well does linear approximation work on these intervals? Interpret your answer in terms of Theorem 11.

 Let h be a small positive number. By applying Theorem 12 to the function f on the interval $[x, x + h]$, we obtain the approximation

$$\frac{2(f(x + h) - f(x) - hf'(x))}{h^2} \approx f''(x)$$

where $b = x + h$ and $a = x$. There also is a version of Theorem 12 that applies to the interval $[x - h, x]$, and it yields the approximation

$$\frac{2(f(x - h) - f(x) + hf'(x))}{h^2} \approx f''(x)$$

where $b = x - h$ and $a = x$. If we average these two approximations we obtain

$$S_h(x) = \frac{f(x + h) - 2f(x) + f(x - h)}{h^2} \approx f''(x).$$

Note that S_h is an approximation of f'' that does not make reference to f'. Thus, we can approximate the graph of f'' by graphing S_h using a (small and fixed) positive number h. We obtain a way to graph f'' without finding f' first. In Exercises 22–25, approximate the graph of f'' by graphing S_h.

22. $f(x) = -x^3 + 4x$, $[-3, 3]$

23. $f(x) = \sin 2x$, $[-\pi, \pi]$

24. $f(x) = x \cos x$, $[-2\pi, 2\pi]$

25. $f(x) = \dfrac{1}{1 + x^2}$, $[-2, 2]$

Summary Outline of Chapter 4

■ To find the maximum and minimum values of a differentiable function f on the interval $[a, b]$ we (page 197)

  (i) find the **critical numbers** by finding all x for which $f'(x) = 0$ or $f'(x)$ is undefined, and

  (ii) inspect $f(x)$ for all critical numbers and both endpoints.

The largest of the values observed in (ii) is the **maximum value of f** on $[a, b]$; the smallest is the **minimum.**

■ The number $f(c)$ is a **relative maximum** if $f(c) \geq f(x)$ for all x in the domain (page 199)
of f near c. The number $f(c)$ is a **relative minimum** if $f(c) \leq f(x)$ for all x in the domain of f near c. Relative maxima and minima are referred to as **relative extrema.**

■ ***Theorem (Mean Value Theorem):*** If f is continuous on $[a, b]$ and differen- (page 215)
tiable on (a, b), there exists $c \in (a, b)$ with

$$f'(c) = \frac{f(b) - f(a)}{b - a}$$

■ The continuous function f is **increasing** on an interval I if $f'(x) > 0$ for all x (page 222)
in the interior of I, and **decreasing** on I if $f'(x) < 0$ for all x in the interior of I.

■ ***Theorem:*** The graph of $y = f(x)$ is (page 238)

  (a) **concave up** on I if $f''(x) > 0$ for all x in the interior of I,
  (b) **concave down** on I if $f''(x) < 0$ for all x in the interior of I.

■ ***Theorem:*** If $f'(c) = 0$ and $f''(c)$ exists, then $f(c)$ is a (page 241)

  (a) relative maximum if $f''(c) < 0$,
  (b) relative minimum if $f''(c) > 0$.

■ **Theorem:** Suppose that the function f is continuous on the interval (page 265) $[x_0, x_0 + \Delta x]$ and twice differentiable on the open interval $(x_0, x_0 + \Delta x)$. If there exists a number M such that $|f''(x)| \le M$ for all $x \in (x_0, x_0 + \Delta x)$, then

$$|\epsilon| \le \frac{M(\Delta x)^2}{2},$$

where ϵ is the error involved in linear approximation of $f(x_0 + \Delta x)$ about x_0.

■ **Theorem:** Suppose the function f is continuous on the interval $[a, b]$ and (page 268) twice differentiable on (a, b). Then there exists a number $c \in (a, b)$ such that

$$f(b) = f(a) + f'(a)(b - a) + \frac{f''(c)}{2}(b - a)^2.$$

Review Exercises—Chapter 4

In Exercises 1–24, find the intervals on which f is increasing and decreasing, find all absolute and relative extrema and points of inflection, determine the concavity, and sketch the graph.

1. $f(x) = 4x - x^2$

2. $f(x) = x(x - 1)(x + 3)$

3. $f(x) = x^2 - 2x + 3$

4. $f(x) = \sin 4x$

5. $f(x) = 2\sin(\pi x)$

6. $f(x) = \dfrac{1 - x}{x}$

7. $f(t) = 2\cos^2(2t)$

8. $f(x) = \sqrt{1 - \sin^2 x}$

9. $f(x) = \dfrac{x - 3}{x + 3}$

10. $f(t) = \dfrac{t^2}{t^2 - 1}$

11. $f(x) = x\sqrt{16 - x^2}$

12. $f(x) = \dfrac{\sqrt{x}}{1 + \sqrt{x}}$

13. $f(t) = \dfrac{t^2}{t^2 + 9}$

14. $f(x) = \dfrac{2}{\sqrt{x}} + \dfrac{\sqrt{x}}{2}$

15. $f(x) = x^4 - 2x^2$

16. $f(x) = (x + 1)(x - 1)^2$

17. $f(x) = x - \cos x$

18. $f(t) = t^3 - 3t^2 + 2$

19. $f(x) = x^2 + (2/x)$

20. $f(x) = \tan x + \cot x$

21. $f(u) = u\sqrt{u^2 + 1}$

22. $f(x) = \dfrac{3x - 1}{\sqrt{2x^2 + 1}}$

23. $f(x) = \cos x + \sin^2 x$

24. $f(x) = x^{2/3} - x^{8/3}$

In Exercises 25–41, find the maximum and minimum values of the function on the given interval. Specify the corresponding numbers x at which these extreme values occur.

25. $f(x) = \dfrac{1}{x^2 - 4}, \quad x \in [-1, 1]$

26. $f(x) = \sin x + \cos x, \quad x \in [-\pi, \pi]$

27. $y = x^3 - 3x^2 + 1, \quad x \in [-1, 1]$

28. $f(x) = x\sqrt{1 - x}, \quad x \in [-3, 0]$

29. $f(x) = x + x^{2/3}, \quad x \in [-1, 1]$

30. $y = x - 2|x|, \quad x \in [-3, 2]$

31. $y = x - \sqrt{1 - x^2}, \quad x \in [-1, 1]$

32. $f(x) = \sqrt[3]{x} - x, \quad x \in [-1, 1]$

33. $f(x) = x - 4x^{-2}, \quad x \in [-3, -1]$

34. $y = \dfrac{x}{1 + x^2}, \quad x \in [-2, 2]$

35. $y = x^4 - 2x^2, \quad x \in [-2, 2]$

36. $f(x) = |x + \sin x|, \quad x \in [-\pi/2, \pi/2]$

37. $f(x) = \dfrac{2 + \cos x}{2 - \cos x}, \quad x \in [-\pi, \pi]$

38. $y = \dfrac{x - 1}{x^2 + 3}, \quad x \in [-2, 4]$

39. $y = \dfrac{x + 2}{\sqrt{x^2 + 1}}, \quad x \in [0, 2]$

40. $f(x) = |x^2 - 1|, \quad x \in [-3, 3]$

41. $f(x) = |x^3 + x^2 - x|, \quad x \in [-3/2, 1]$

In Exercises 42–53, determine the limit specified.

42. $\displaystyle\lim_{x \to \infty} \dfrac{x^4 + 2x + 1}{2x^5 + 3x^2 + 2}$

43. $\displaystyle\lim_{x \to -\infty} \dfrac{3x^2 + x + 1}{x^2 + 1}$

44. $\displaystyle\lim_{x \to 3^+} \dfrac{x^2 + 3x - 4}{x^2 - 2x - 3}$

45. $\displaystyle\lim_{x \to -1^+} \dfrac{x^2 + x + 4}{x^2 - 2x - 3}$

46. $\lim\limits_{x \to 7^-} \dfrac{x - 8}{x^2 - 8x + 7}$

47. $\lim\limits_{x \to 7} \dfrac{x - 8}{x^2 - 8x + 7}$

48. $\lim\limits_{x \to 7^-} \dfrac{x - 8}{x^2 - 14x + 49}$

49. $\lim\limits_{x \to 7} \dfrac{x - 8}{x^2 - 14x + 49}$

50. $\lim\limits_{x \to 0^+} \dfrac{\sin(1/x)}{x}$

51. $\lim\limits_{x \to \infty} \cos(1/x)$

52. $\lim\limits_{x \to 0^+} \cos(1/x)$

53. $\lim\limits_{\theta \to \pi/2^-} \dfrac{\sec \theta}{\tan \theta}$

54. The sum of two positive numbers is 10. For what value(s) will the difference of their squares be a maximum?

55. Sketch a possible graph for a function that is continuous for all x and for which $f(2) = 4, f'(2) = 1$, and $f''(x) < 0$ for all x.

56. Sketch a possible graph for a function that is continuous for all x and for which $f(-2) = -f(2) = -3, f'(-2) = f'(2) = 0, f''(-2) > 0$, and $f''(2) < 0$.

In each of Exercises 57–59, a list of properties for a function f is specified. Produce a graph of $y = f(x)$ that is consistent with these properties. Label the absolute and relative extrema and the inflection points.

57. The function f is twice differentiable for all real numbers x. Also,
  (1) $f(x) > 0$ for all real numbers x;
  (2) $f(x) \to 0$ as $x \to -\infty$;
  (3) $f(x) \to \infty$ as $x \to \infty$;
  (4) $f'(x) > 0$ for $x < 0$ and $x > 2$;
  (5) $f'(x) < 0$ and $0 < x < 2$;
  (6) $f''(x) > 0$ for $x < -1$ and $x > 1$; and
  (7) $f''(x) < 0$ for $-1 < x < +1$.

58. The function f is twice differentiable for all real numbers x. Also,
  (1) $f(x) > 0$ for all real numbers x;
  (2) $f(x) \to 0$ as $x \to \pm\infty$;
  (3) $f'(x) > 0$ for $x < 0$;
  (4) $f'(x) < 0$ for $x > 0$;
  (5) $f''(x) > 0$ for $x < -1$ and $x > 1$; and
  (6) $f''(x) < 0$ for $-1 < x < 1$.

59. The function f is continuous on the interval $[0, 3]$ and it is twice differentiable for all x between 0 and 3 except $x = 1$. Also,
  (1) $f'(1)$ is undefined;
  (2) $f'(x) < 0$ for $0 < x < 1$ and $2 < x < 3$;
  (3) $f'(x) > 0$ for $1 < x < 2$;
  (4) $f''(x) < 0$ for $0 < x < 1$ and $1 < x < 3$; and
  (5) $f(0) = 2, f(1) = 1, f(2) = 3$, and $f(3) = 0$.

In each of Exercises 60–62, a list of properties for a function f is specified. Do the properties contradict one another? If so, explain why. If not, produce a graph $y = f(x)$ that is consistent with these properties. Then label the absolute and relative extrema and the inflection points.

60. The function f is continuous on the closed interval $[0, 3]$ and twice differentiable on the open interval $(0, 3)$. Also,
  (1) $f'(1) = f'(2) = 0$;
  (2) $f'(x) < 0$ for $0 < x < 1$ and $2 < x < 3$;
  (3) $f'(x) > 0$ for $1 < x < 2$;
  (4) $f''(x) > 0$ for $0 < x < 2$; and
  (5) $f''(x) < 0$ for $2 < x < 3$.

61. The function f is continuous on the closed interval $[0, 4]$ and twice differentiable on the open interval $(0, 4)$. Also,
  (1) $f'(1) = f'(3) = 0$;
  (2) $f'(x) > 0$ for $0 < x < 1$ and $1 < x < 3$;
  (3) $f'(x) < 0$ for $3 < x < 4$;
  (4) $f''(x) < 0$ for $0 < x < 1$ and $2 < x < 4$; and
  (5) $f''(x) > 0$ for $1 < x < 2$.

62. The function f is twice differentiable for all real numbers $x \neq 1$ and it is not defined at $x = 1$. Also,
  (1) $f(x) = 0$ if and only if $x = 0$;
  (2) $f'(-1) = 0$;
  (3) $f'(x) > 0$ for $x < -1$;
  (4) $f'(x) < 0$ for $-1 < x < 1$ and $x > 1$;
  (5) $f''(x) > 0$ for $x < -2$ and $x > 1$;
  (6) $f''(x) < 0$ for $-2 < x < 1$;
  (7) $f(x) \to -\infty$ as $x \to 1^-$; and
  (8) $f(x) \to 0$ as $x \to \infty$.

63. A fence 3 m tall is parallel to the wall of a building and 1 m from the building. What is the length of the shortest ladder that can extend from the ground over the fence to the building wall?

64. Use the Intermediate Value Theorem and Rolle's Theorem to prove that if f has a continuous derivative for $x \in [a, b]$, if $f(a)$ and $f(b)$ have opposite signs, and if $f'(x) \neq 0$ for all $x \in [a, b]$, then the equation $f(x) = 0$ has *precisely* one root in $[a, b]$.

65. A publisher currently sells 2000 subscriptions annually for a professional journal at a rate of \$40 per year. Its marketing department predicts that for each \$1 reduction in subscription price an additional 50 subscriptions can be sold. Under this assumption what subscription rate maximizes revenues?

66. Sketch examples of graphs of functions defined on an interval $[a, b]$ for which the conclusion of the Mean Value Theorem is satisfied:
  a. at precisely two points;
  b. at precisely three points.

67. Show that the function $f(t) = t^3 - 6t^2 + 14t + 5$ is increasing for all t. (*Hint:* Complete the square on $f'(t)$.)

68. For $f(x) = x^2 + bx + c$ find a condition involving b and c that will ensure that $f(x) > 0$ for all x. (*Hint:* Ensure that the minimum value is greater than zero.)

69. The function $f(x) = x/(1 + ax^2)$ has minimum values at $x = \pm 3$ for $x \in [-5, 5]$. Find a.

70. Find the number most exceeded by its square root.

71. Prove that the function $f(x) = x^n + ax + b$ cannot have more than 2 zeros if n is even. What if n is odd?

72. A peach orchard has 25 trees per acre, and the average yield is 300 peaches per tree. For each additional tree planted per acre, the average yield per tree will be reduced by approximately 10 peaches. How many trees per acre will give the largest peach crop?

73. A swimming pool is to be constructed in the shape of a sector of a circle with radius r and central angle θ. Find r and θ if the area of the water surface is the constant S and the perimeter is to be a minimum.

74. Find the rectangle of maximum area that can be inscribed within the equilateral triangle of side 10.

75. Find the points on the graph of the hyperbola $x^2 - x + \frac{5}{4} - y^2 = 0$ nearest the origin.

76. Find the dimensions of the rectangle of maximum area that can be inscribed in the ellipse $x^2 + (y^2/4) = 1$.

77. The strength of a wooden beam cut from a log is proportional to the product of its width and the square of its depth. Find the ratio of depth to width of the strongest beam that can be cut from a circular log.

78. A student 1.6 m tall walks directly away from a street light 8 m above the ground at a rate of 1.2 m/s. Find the rate at which the student's shadow is increasing when the student is 20 m from the point directly beneath the light.

79. A propane storage tank is to have the shape of a cylinder capped at both ends by hemispheres. The hemispherical caps are three times as expensive to construct, per square unit of surface area. Find the dimensions that minimize construction costs for a given volume.

80. What is the maximum possible area of an isosceles triangle that can be inscribed in a circle of radius r?

81. Mathematical statisticians often "estimate" the true value x of a parameter in a large population by extracting a sample from the population and calculating the value $\bar{x}$ of the parameter in that sample. For example, we could estimate the average age of all citizens of Chicago by calculating the average age for a sample of 100 citizens, assuming the sample to be representative of the population. Given sample values $x_1, x_2, \ldots, x_n$ of the parameter of interest, statisticians prefer to use the estimate $\bar{x}$ that minimizes the sum of the squares of the errors $(\bar{x} - x_1)^2 + (\bar{x} - x_2)^2 + \cdots + (\bar{x} - x_n)^2$. Show that the number $\bar{x}$ which satisfies this criterion is the simple average

$$\bar{x} = \frac{x_1 + x_2 + \cdots + x_n}{n}$$

82. If a set of data $(x_1, y_1), (x_2, y_2), \ldots, (x_n, y_n)$ appears to be related by an equation of the form $y = mx$, the statistician defines the "best" straight line modeling this relationship as the line $y = mx$ that minimizes the sum of the squares of the errors

$$(mx_1 - y_1)^2 + (mx_2 - y_2)^2 + \cdots + (mx_n - y_n)^2.$$

Show that the value of m producting this "best" straight line is

$$m = \frac{x_1 y_1 + x_2 y_2 + \cdots + x_n y_n}{x_1^2 + x_2^2 + \cdots + x_n^2}.$$

83. A rectangular box is to be constructed so as to hold 1024 cm³. Its base is to have length twice its width. Material for the top costs 12¢ per square centimeter, and material for the sides and bottom costs 6¢ per square centimeter. Find the dimensions that will minimize cost.

84. For $f(x) = x^2 - 2x - 15$:
 a. Find a closed interval on which Rolle's Theorem applies.
 b. Rolle's Theorem guarantees the existence of a critical number c in the interval found in part a. Find c.

85. If the Mean Value Theorem is applied to the function $f(x) = x^3 + qx^2 + 5x - 6$ on the interval $[0, 2]$, the number c determined by the theorem is $c = 2$. What is q?

86. The cost of operating a small aircraft is calculated to be $200 per hour plus the cost of fuel. Fuel costs are $s/100$ dollars per kilometer, where s represents speed in kilometers per hour. Find the speed that minimizes the cost per kilometer.

87. An automobile parts store sells 3000 headlight bulbs per year at a uniform rate. Orders can be placed from the distributor so that shipments arrive just as supplies run out. In ordering bulbs the parts store encounters costs of $20 per order plus 5¢ per bulb. How frequently should the parts store order bulbs if the cost of maintaining one bulb in stock for one year is estimated to be 48¢?

88. A chartered cruise requires a minimum of 100 persons. If 100 people sign up, the cost is $400 per person. For each additional person the cost per person decreases by $1.50. Find the number of passengers that maximizes revenue.

89. A merchant sells shirts uniformly at a rate of about 2000 per year. The cost of carrying a shirt in stock is $3 per year. The cost of ordering shirts is $50 per order plus 20¢ per shirt. Approximately how many times per year should the merchant order shirts to minimize his total yearly costs? (Assume that orders can be placed so that shipment arrives precisely when stock is depleted.)

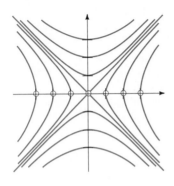

Chapter 5

Antidifferentiation

This brief chapter addresses a single question: Given a function f, what can we say about a function F for which $F' = f$? Does such a function F exist? If so, what are its properties? Is it unique? What can we say about its graph based on what we know about f?

This is the question of finding an *antiderivative* for a given function f. Because the answer involves reversing the process of differentiation, this topic is properly viewed as part of our discussion of the derivative. Yet as we begin our discussion of integration (Chapter 6), the second major topic of the calculus, we shall see that the antiderivative plays a central role.

From an applied point of view, the process of antidifferentiation is also significant, especially to those who model scientific phenomena using mathematics. Since many fundamental laws of nature are expressed by equations involving derivatives, we must be able to pass from the derivative back to the underlying function of interest in order to understand the consequences of these laws.

In this chapter, we initiate our discussion of the general antidifferentiation question. We also include a brief application of antidifferentiation to the solution of certain differential equations. However, the material treated here is simply a beginning. In subsequent chapters, we repeatedly return to the antidifferentiation question. Indeed, we shall see that the concept of an antiderivative is the bridge that links the two major ideas of the calculus—the derivative and the integral.

5.1 Antiderivatives

In Chapter 3 we developed a number of rules for finding derivatives. We now wish to discuss the "inverse" process for differentiation. That is, given a function f, how do we use what we know about differentiation to find a function F *whose derivative is the given function f*? This process is referred to as *antidifferentiation* and the function F is called an antiderivative for f.

Definition 1

Let the function f be defined on an interval I. A function F is called an **antiderivative** for f on the interval I if F is differentiable on I and

$$F'(x) = f(x) \quad \text{for all} \quad x \in I. \tag{1}$$

Obviously, antidifferentiation requires the ability to "think backwards through" (that is, invert) the process of differentiation. Here are several straightforward examples.

☐ **EXAMPLE 1**

The function $F(x) = x^2$ is an antiderivative for the function $f(x) = 2x$ on the interval $(-\infty, \infty)$ since

$$F'(x) = 2x = f(x) \quad \text{for all} \quad x \in (-\infty, \infty).$$ ■

☐ **EXAMPLE 2**

The function $F(x) = \sin x$ is an antiderivative for the function $f(x) = \cos x$ on $(-\infty, \infty)$ since

$$F'(x) = \cos x = f(x) \quad \text{for all} \quad x \in (-\infty, \infty).$$ ■

☐ **EXAMPLE 3**

The absolute value function $F(x) = |x|$ is an antiderivative for the constant function $f(x) = 1$ on the interval $(0, \infty)$. That is because $|x| = x$ if $x > 0$, so

$$F'(x) = \frac{d}{dx}(|x|) = \frac{d}{dx}(x) = 1 = f(x) \quad \text{if} \quad x \in (0, \infty).$$

Note, however, that $F(x) = |x|$ is *not* an antiderivative for f on $(-\infty, \infty)$ since $F'(x) = -1$ for $x < 0$ and $F'(0)$ is not defined.   ■

Example 3 illustrates why we specify the *interval* on which we seek an antiderivative F for f. (If the interval is not explicitly stated, we shall assume that it is the domain for f.)

Antiderivatives Are Not Unique

One important issue is the uniqueness question: How many antiderivatives F of a given function f are there? The following two examples indicate that there can be many.

☐ **EXAMPLE 4**

The function $f(x) = 2x$ in Example 1 also has antiderivatives

$$F_1(x) = x^2 + 1, \quad F_2(x) = x^2 - 10, \quad \text{and} \quad F_3(x) = x^2 + \pi$$

on the interval $(-\infty, \infty)$.   ■

☐ **EXAMPLE 5**

The function $f(x) = \cos x$ in Example 2 also has antiderivatives

$$F_1(x) = \sin x + 5, \quad F_2(x) = \pi + \sin x, \quad \text{and} \quad F_3(x) = \sin x - 6$$

on the interval $(-\infty, \infty)$.   ■

The Geometry of Antiderivatives

Obviously, we need a clearer understanding of what to expect as the antiderivative of a given function. We therefore return to our geometric interpretation of the derivative.

Recall that, starting with the differentiable function F, the process of *differentiation* associates a *slope* $F'(x)$ with each number x at which F is differentiable. (Figure 1.1).

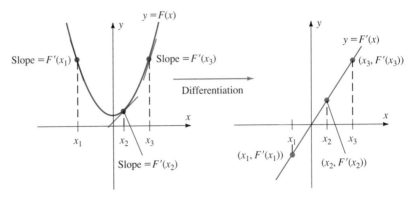

Figure 1.1 Differentiating F produces a slope F'. In other words, the value $F'(x)$ equals the slope of the tangent to the graph of F at x.

Thus, if we begin with the function f and seek F with $F' = f$, we are beginning with a slope function and asking "what function F has slope $f(x)$ at each point $(x, F(x))$ of its graph?" However, there is one difficulty in asking such a question—since we do not know F, we do not know the particular points $(x, F(x))$ on the graph of F. We only know the slope at each point.

Figure 1.2 indicates how we can determine the graph of the antiderivative F from f. The values of f are used to build a **slope portrait** for F. Given a particular number x_0, we interpret the value $f(x_0)$ as the slope of the tangent to the graph of F at x_0. However, since we do not know $F(x_0)$, we sketch infinitely many parallel tangents, all with slope $f(x_0)$ and all centered along the line $x = x_0$ [Figure 1.2(b)].

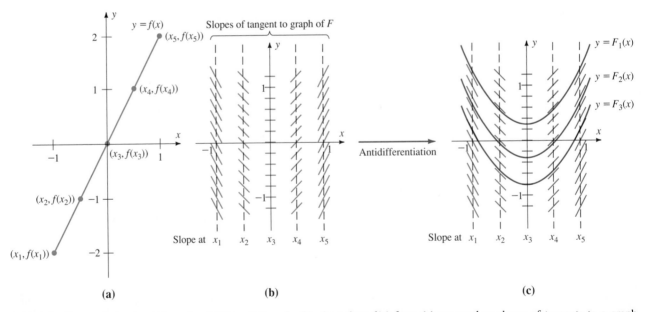

Figure 1.2 Geometric interpretation of antidifferentiation: In (b), the values $f(x)$ from (a) are used as slopes of tangents to a graph $y = F(x)$. In (c), the graphs of three antiderivatives are superimposed on the slope portrait (b).

The equation $F'(x) = f(x)$ implies that the graph of F must be tangent to this slope portrait at all numbers $x \in I$ [Figure 1.2(c)]. Thus, the process of antidifferentiation is equivalent to finding a graph $y = F(x)$ that is tangent to the slope portrait for all $x \in I$.

For example, suppose we ask for an antiderivative for the function $f(x) = \cos x$. That is, we seek F with $F'(x) = \cos x$. Figure 1.3 shows the required slopes $F'(x)$ for various values of x, and Figure 1.4 suggests several functions whose graphs have the required slopes. As Figure 1.4 suggests, we will see that the *graphs of all antiderivatives F have the same shape*. Two such graphs differ only in that one is a "vertical shift" of the other. In other words, if F_1 and F_2 are both antiderivatives for $f(x) = \cos x$, then $F_2(x) = F_1(x) + C$ for some constant C and all x.

Notice this difference between the process of differentiation and that of antidifferentiation: differentiating a particular function produces another (unique) function, while finding an antiderivative results in an entire family of functions. The following theorem confirms that observations.

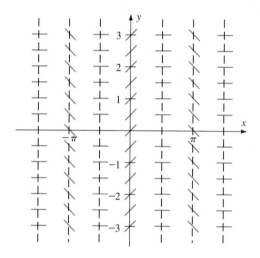

Figure 1.3 Slopes for the graph of F if $F'(x) = \cos x$.

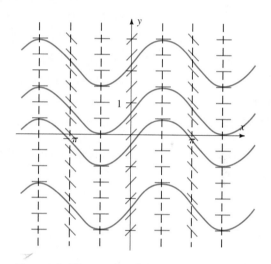

Figure 1.4 The graphs of four antiderivatives for $f(x) = \cos x$.

Theorem 1

Suppose the function F is an antiderivative for the function f on the interval I. That is, assume $F'(x) = f(x)$ for all $x \in I$. Then *any* antiderivative G for f on I satisfies

$$G(x) = F(x) + C, \qquad x \in I$$

for some constant C. Moreover, every function G of this form is an antiderivative for f.

In other words, *any two antiderivatives for the same function on the same interval differ by a constant*. Before proving Theorem 1 we establish a special case using the Mean Value Theorem.

Theorem 2

The only antiderivatives of the zero function are the constant functions. That is, if $F'(x) = 0$ for all x in an interval I, then $F(x) = C$ for some constant C and all $x \in I$.

Proof: If $F'(x) = 0$ for all $x \in I$, then F satisfies the conditions of the Mean Value Theorem on I. Fix a number a in I and let x be any other number in I. The Mean Value Theorem guarantees that

$$\frac{F(x) - F(a)}{x - a} = F'(c) \tag{2}$$

for some number c between a and x. But, since we are assuming that the derivative of F is zero throughout I, we know that $F'(c) = 0$. From equation (2), we obtain $F(x) - F(a) = 0$, or equivalently, $F(x) = F(a)$. If we denote $F(a)$ by C, we conclude that $F(x) = C$ for all $x \in I$.   ∎

We now use Theorem 2 to prove Theorem 1.

Proof of Theorem 1:   First, suppose that F and G differ by a constant C on I. That is, $G(x) = F(x) + C$ for all $x \in I$. Since the derivative of a constant is zero, $G'(x) = F'(x) = f(x)$ for all $x \in I$; therefore G is also an antiderivative of f.

Conversely, suppose that F and G are two antiderivatives for f on I. Let H be the function $G - F$. Then

$$H'(x) = G'(x) - F'(x) = f(x) - f(x) = 0$$

for all $x \in I$. Thus, by Theorem 2 there exists a constant C so that $H(x) = C$ for all $x \in I$. That is, $G(x) - F(x) = C$ or $G(x) = F(x) + C$ for all $x \in I$.   ∎

Notation and Terminology

Once we have found a particular antiderivative F for a function f, Theorem 1 guarantees that *all* antiderivatives for f have the form $F(x) + C$. We shall refer to the family of *all* antiderivatives for f as the *indefinite integral* of f, which we denote by $\int f(x) \, dx$.

Definition 2

The **indefinite integral** of the function f on the interval I is the family of all antiderivatives for f on I and is denoted by $\int f(x) \, dx$. That is,

$$\int f(x) \, dx = F(x) + C \tag{3}$$

where $F'(x) = f(x)$ for all $x \in I$ and C denotes any constant.

The symbol $\int$ in Definition 2 is called an integral sign. It and the symbol dx enclose the function f for which F is an antiderivative. While the symbol dx suggests the differential discussed in Chapter 3, it should be regarded for now as simply part of the notation signifying the indefinite integral for f. The function f in equation (3) is referred to as the *integrand* of the indefinite integral. We refer to the process of finding the indefinite integral for f as *integrating f* or as *integration*.

REMARK 1   In the calculus, you learn two types of integration: indefinite integration (which we also call antidifferentiation) and definite integration. Indefinite integration is simply the differentiation process in reverse. Definite integration is an entirely different mathematical concept—one that is completely distinct from the concept of the derivative. In Chapter 6, we shall explain why we use similar terminology for distinctly different concepts.

☐ **EXAMPLE 6**

Using this notation and Theorem 1 we can give a complete description of the antiderivatives for the functions in Examples 1 and 2 by integration:

$$\int 2x \, dx = x^2 + C$$

and

$$\int \cos x \, dx = \sin x + C.$$ ■

Table 1.1 shows several other indefinite integrals, each resulting from a differentiation fact established in Chapter 3. We refer to equations such as those in the second column as *integration formulas*.

Table 1.1   Some Derivatives and the Corresponding Integration Formulas

$\dfrac{d}{dx} F(x) = f(x)$	$\int f(x) \, dx = F(x) + C$
$\dfrac{d}{dx}(\tfrac{1}{2}x^2) = x$	$\int x \, dx = \tfrac{1}{2}x^2 + C$
$\dfrac{d}{dx}(x^3 + \tfrac{1}{6}x^6) = 3x^2 + x^5$	$\int (3x^2 + x^5) \, dx = x^3 + \tfrac{1}{6}x^6 + C$
$\dfrac{d}{dt}(\cos t) = -\sin t$	$\int (-\sin t) \, dt = \cos t + C$
$\dfrac{d}{d\theta}(\tan \theta) = \sec^2 \theta$	$\int \sec^2 \theta \, d\theta = \tan \theta + C$
$\dfrac{d}{dx}(\sqrt{x}) = \dfrac{1}{2\sqrt{x}}$	$\int \dfrac{1}{2\sqrt{x}} \, dx = \sqrt{x} + C$
$\dfrac{d}{dx}(\tfrac{1}{2}\sin t^2) = t \cos t^2$	$\int t \cos t^2 \, dt = \tfrac{1}{2}\sin t^2 + C$
$\dfrac{d}{dx}(\sqrt{ax^2 + bx + c}) = \dfrac{2ax + b}{2\sqrt{ax^2 + bx + c}}$	$\int \dfrac{2ax + b}{2\sqrt{ax^2 + bx + c}} \, dx = \sqrt{ax^2 + bx + c} + C$

You should not bother remembering Table 1.1 in detail. Its sole purpose is to illustrate how every derivative calculation yields a corresponding integration formula. Since we know how to differentiate infinitely many different functions, we could obviously make a very large table of integration formulas. Instead, we shall learn techniques for deriving the antiderivatives of many functions from a few basic ones using the properties of antiderivatives.

Properties of Antiderivatives

Since antidifferentiation is the inverse process of differentiation, theorems about differentiation often yield corresponding theorems about antidifferentiation. Theorem 3 is one such result.

Theorem 3

Suppose both f and g have antiderivatives on a common interval I. Then, on I,

(a) $\displaystyle\int [f(x) + g(x)] \, dx = \int f(x) \, dx + \int g(x) \, dx$,   and

(b) $\displaystyle\int [cf(x)] \, dx = c \int f(x) \, dx$ for any constant c.

Proof: Let F and G be particular antiderivatives for f and g, respectively. That is, assume that $F'(x) = f(x)$ and $G'(x) = g(x)$ for all $x \in I$. Then on the interval I,

(i) $\displaystyle\frac{d}{dx}[F(x) + G(x)] = F'(x) + G'(x) = f(x) + g(x)$,   and

(ii) $\displaystyle\frac{d}{dx}[cF(x)] = cF'(x) = cf(x)$.

Equation (i) shows that $F + G$ is an antiderivative for $f + g$ and equation (ii) shows that cF is an antiderivative for cf.   ∎

Theorem 3 states that, in finding antiderivatives for sums or constant multiples of functions, we may first find antiderivatives for the individual functions and then form the appropriate sums or multiples, incorporating all arbitrary constants into a single constant.*

Alternatively, we need only find one particular antiderivative for each of the functions involved. Then we form the appropriate sum or multiple and use Theorem 1 to conclude that every other antiderivative differs from the resulting function by a constant.

☐ **EXAMPLE 7**

Find $\displaystyle\int [7x^2 - 2x^3] \, dx$.

Strategy · · · · · · · · ·   **Solution**

Apply Theorem 3 using

$\displaystyle\int x^2 \, dx = \frac{x^3}{3} + C_1$

$\displaystyle\int x^3 \, dx = \frac{x^4}{4} + C_2$

$$\int [7x^2 - 2x^3] \, dx = 7 \int x^2 \, dx - 2 \int x^3 \, dx$$

$$= 7\left[\frac{x^3}{3} + C_1\right] - 2\left[\frac{x^4}{4} + C_2\right]$$

$$= \frac{7}{3}x^3 - \frac{1}{2}x^4 + C.$$

Combine C_1 and C_2.

As indicated before the example, we have combined the various constants into a single constant C.

---

*If C_1 and C_2 denote arbitrary constants, the sum $(C_1 + C_2)$ and the multiple kC_1 are again arbitrary constants. We will therefore always combine arbitrary constants by equations of the form

$$C_1 + C_2 = C \quad \text{and} \quad kC_1 = C.$$

Check result by differentiating.

To verify this result, we check

$$\frac{d}{dx}\left(\frac{7}{3}x^3 - \frac{1}{2}x^4 + C\right) = \frac{7}{3}(3)x^3 - \frac{1}{2}(4)x^3 + 0$$

$$= 7x^2 - 2x^3. \qquad \blacksquare$$

In Example 7, we used the **Power Rule**

$$\int x^n \, dx = \frac{x^{n+1}}{n+1} + C, \quad n \neq -1,$$

which we can derive for all rational exponents $n \neq -1$ using the Power Rule for differentiation (Theorem 10 in Chapter 3):

$$\frac{d}{dx}\left[\frac{x^{n+1}}{n+1} + C\right] = \frac{(n+1)x^n}{n+1} = x^n, \quad n \neq -1.$$

In Table 1.2, we list a number of basic integration formulas that are derived by straightforward differentiation.

Table 1.2 Basic Integral Formulas Derived from Derivatives of Elementary Functions

$\dfrac{d}{dx}F(x) = f(x)$	$\displaystyle\int f(x)\,dx = F(x) + C$
$\dfrac{d}{dx}x^{n+1} = (n+1)x^n$	$\displaystyle\int x^n\,dx = \dfrac{x^{n+1}}{n+1} + C, \quad n \neq -1$
$\dfrac{d}{dx}\sin x = \cos x$	$\displaystyle\int \cos x\,dx = \sin x + C$
$\dfrac{d}{dx}\cos x = -\sin x$	$\displaystyle\int \sin x\,dx = -\cos x + C$
$\dfrac{d}{dx}\tan x = \sec^2 x$	$\displaystyle\int \sec^2 x\,dx = \tan x + C$
$\dfrac{d}{dx}\sec x = \sec x \tan x$	$\displaystyle\int \sec x \tan x\,dx = \sec x + C$
$\dfrac{d}{dx}\cot x = -\csc^2 x$	$\displaystyle\int \csc^2 x\,dx = -\cot x + C$
$\dfrac{d}{dx}\csc x = -\csc x \cot x$	$\displaystyle\int \csc x \cot x\,dx = -\csc x + C$

If a function f has an antiderivative, we have

$$\frac{d}{dx}\left[\int f(x)\,dx\right] = f(x). \tag{4}$$

In other words, *the derivative of an antiderivative for f must be the function f itself.*

In the next example, we illustrate how equation (4) can be used to check the computation of an antiderivative.

☐ **EXAMPLE 8**

Let $f(x) = x^3 + \sin x + \cos x$. Then,

$$\int f(x)\, dx = \int (x^3 + \sin x + \cos x)\, dx$$

$$= \int x^3\, dx + \int \sin x\, dx + \int \cos x\, dx \qquad \text{(Theorem 3)}$$

$$= \frac{x^4}{4} + (-\cos x) + \sin x + C \qquad \text{(Table 1.2)}$$

$$= \frac{x^4}{4} - \cos x + \sin x + C,$$

so

$$\frac{d}{dx}\left[\int f(x)\, dx\right] = \frac{d}{dx}\left[\frac{x^4}{4} - \cos x + \sin x + C\right]$$

$$= \frac{4x^3}{4} - (-\sin x) + \cos x + 0$$

$$= x^3 + \sin x + \cos x$$

$$= f(x)$$

as in equation (4). ■

REMARK 2 In general, computing derivatives is easier than determining antiderivatives. Therefore, we suggest that you develop the habit of checking the validity of any antiderivative calculation by differentiating the resulting antiderivative F, as we did in Example 7. Often a simple derivative calculation exposes an antidifferentiation error.

Position and Velocity as Antiderivatives

Returning to the concepts of position, velocity, and acceleration for motion along a straight line (as discussed in Chapter 3), we use the notation developed here to express position s in terms of velocity v and velocity in terms of acceleration a. All three quantities are expressed as functions of time t.

$$\text{Since} \quad s'(t) = v(t), \quad \int v(t)\, dt = s(t) + C \tag{5}$$

and

$$\text{since} \quad v'(t) = a(t), \quad \int a(t)\, dt = v(t) + C. \tag{6}$$

The following examples illustrate how the constants in equations (5) and (6) are determined in particular situations.

☐ **EXAMPLE 9**

Suppose that, when a hockey player strikes a puck with a certain force, the puck moves along the ice with velocity $v(t) = 64 - \sqrt{t}$ meters per second at time t, where $0 \le t \le 8$ seconds. Let $s(t)$ denote the distance between the puck and the player after t seconds. Find $s(t)$.

Solution Using (5) and the Power Rule we find the position function to be

$$s(t) = \int (64 - t^{1/2})\, dt$$

$$= 64t - \frac{t^{3/2}}{3/2} + C$$

$$= 64t - \tfrac{2}{3}t^{3/2} + C.$$

To determine C we use the fact that $s(0) = 0$, which says that the distance between the player and the puck is zero at the initial time $t = 0$. This gives

$$0 = s(0) = 64(0) - \tfrac{2}{3}(0)^{3/2} + C = C,$$

so

$$C = 0.$$

Thus,

$$s(t) = 64t - \tfrac{2}{3}t^{3/2}. \qquad \blacksquare$$

The condition that $s(0) = 0$ in Example 9 is referred to as an **initial condition** since it provides the value $s(t)$ at a particular number t. (It is not necessary that this value be provided at $t = 0$. Information about $s(t)$ at *any* $t \in [0, 8]$ will allow us to determine C.) In problems involving antidifferentiation, we will need one initial condition to determine the arbitrary constant C for each antidifferentiation performed.

□ **EXAMPLE 10**

When an object is moving freely in the atmosphere it is pulled toward the earth by the force of gravity. Near the surface of the earth the acceleration due to gravity is approximately 9.8 m/s². If a fireworks display rocket is fired vertically from ground level at an initial velocity of 45 m/s and fails to explode, when does it strike the ground?

Strategy · · · · · · ·

Find a from the problem statement.

Solution

Since the only force acting on the rocket is acceleration due to gravity, the acceleration function is the constant function

$$a(t) = -9.8 \text{ m/s}^2.$$

(We take the positive direction to be upward, so the correct sign for $a(t)$ is $-$.) The general form of the velocity function is therefore

Find v by using equation (6).

$$v(t) = \int (-9.8)\, dt$$

$$= -9.8t + C_1.$$

Apply initial condition $v(0) = 45$ to find C_1.

To find the constant C_1, we use the initial condition $v(0) = 45$. Setting $t = 0$ gives

$$v(0) = 45 = (-9.8)(0) + C_1$$

so

$$C_1 = 45.$$

The explicit velocity function is therefore

$$v(t) = -9.8t + 45 \text{ m/s}.$$

Find s from equation (5).

The position function is

$$s(t) = \int (-9.8t + 45) \, dt$$

$$= -4.9t^2 + 45t + C_2.$$

Apply initial condition $s(0) = 0$ to find C_2.

Since the rocket is fired from ground level, the initial condition for s is $s(0) = 0$. Thus

$$s(0) = 0 = -4.9(0)^2 + 45(0) + C_2,$$

so

$$C_2 = 0.$$

The explicit position function is therefore

$$s(t) = -4.9t^2 + 45t \text{ m}.$$

Set $s(t) = 0$ to find desired time.

The rocket strikes the ground when $s(t) = 0$ so we set

$$s(t) = -4.9t^2 + 45t = 0$$

and obtain $t = 0$ (launch), or

$$-4.9t + 45 = 0,$$

which gives

$$t = \frac{45}{4.9} \approx 9.18 \text{ s (after launching).} \qquad \blacksquare$$

Exercise Set 5.1

In Exercises 1–18, find the indefinite integral.

1. $\displaystyle\int (2x^2 + 1) \, dx$

2. $\displaystyle\int (1 - x^3) \, dx$

3. $\displaystyle\int (x^{2/3} + x^{5/2}) \, dx$

4. $\displaystyle\int (\sqrt{x} + 1)^2 \, dx$

5. $\displaystyle\int (2\cos t - \sin t) \, dt$

6. $\displaystyle\int 4 \sec^2 \theta \, d\theta$

7. $\displaystyle\int (\sin x + \sec^2 x) \, dx$

8. $\displaystyle\int \sqrt{t}(t^2 + 1) \, dt$

9. $\displaystyle\int \frac{t^2 + 1}{\sqrt{t}} \, dt$

10. $\displaystyle\int (t - 1)(t + 1) \, dt$

11. $\displaystyle\int (t^2 + 1)(t + 2) \, dt$

12. $\displaystyle\int (\sqrt{t} + 2)(t^3 + 1) \, dt$

13. $\displaystyle\int \frac{1}{\sqrt[3]{x}} \, dx$

14. $\displaystyle\int \frac{1}{1 - \sin^2 x} \, dx$

15. $\displaystyle\int (\tan^2 x + 1) \, dx$

16. $\displaystyle\int (5 \cos x + \sqrt{x}) \, dx$

17. $\displaystyle\int (3x^2 - \sin x + 2 \sec^2 x) \, dx$

18. $\displaystyle\int (6x + 5 - 3 \cot x \csc x) \, dx$

In Exercises 19–26, find the position function s corresponding to the given velocity function and initial condition.

19. $v(t) = \cos t, \quad s(0) = 2$

20. $v(t) = 1 + 2t, \quad s(0) = 0$

21. $v(t) = 2t^2, \quad s(0) = 4$

22. $v(t) = t(t + 2), \quad s(0) = 2$

23. $v(t) = \sqrt{t}(t + 4), \quad s(0) = 0$

24. $v(t) = \dfrac{(\sqrt{t} + 3)^2}{\sqrt{t}}, \quad s(1) = 2/3$

25. $v(t) = (\sqrt{t} + 4)(\sqrt{t} - 4), \quad s(2) = 3$

26. $v(t) = \sin t, \quad s(\pi) = 1$

In Exercises 27–37, find the velocity function v and position function s corresponding to the given acceleration function and initial conditions.

27. $a(t) = 2, \qquad v(0) = 3, \qquad s(0) = 0$

28. $a(t) = -2, \qquad v(0) = 6, \qquad s(0) = 10$

29. $a(t) = 3t, \qquad v(0) = 0, \qquad s(0) = 20$

30. $a(t) = 200, \qquad v(0) = 100, \quad s(0) = 200$

31. $a(t) = 4t + 4, \quad v(0) = 8, \qquad s(0) = 12$

32. $a(t) = \cos t, \qquad v(0) = 1, \qquad s(0) = 9$

33. $a(t) = \sin t, \qquad v(0) = 0, \qquad s(0) = 2$

34. $a(t) = 2t, \qquad v(1) = 1, \qquad s(1) = 1$

35. $a(t) = -2 \sin t, \quad v(\pi) = -3, \quad s(\pi) = -2\pi$

36. $a(t) = -2, \qquad v(2) = -5, \quad s(1) = -2$

37. $a(t) = 0, \qquad v(1) = 1, \qquad s(2) = 3$

38. A particle moves along a line with acceleration

$$a(t) = 2 - t \quad \text{for} \quad 0 \le t \le 4.$$

Must the particle necessarily change direction at some time t_0 between 0 and 4? Support your answer with either a proof that the particle must change direction or with an example of a position function s for which the particle does not change direction.

39. A stone is thrown vertically upward with an initial velocity 15 m/s. How long will it take for the stone to (a) stop rising, (b) strike the ground? (Use $a = -9.8$ m/s^2.)

40. How long will it take for the rocket in Example 10 to reach the highest point of its trajectory?

41. A particle moves along a line with constant acceleration $a = 3$ m/s^2.

    a. How fast is it moving after 6 s if its initial velocity is $v(0) = 10$ m/s?

    b. What is its initial velocity $v(0)$ if its speed after 3 s is 15 m/s?

42. A coin is dropped from the top of a building 150 m high. Using $a = -9.8$ m/s^2,

    a. How long will it take for the coin to strike the ground?

    b. With what velocity will the coin strike the ground?

43. What constant acceleration will enable the driver of an automobile to increase its speed form 20 m/s to 25 m/s in 10 s?

44. What constant negative acceleration (deceleration) is required to bring an automobile travelling at a rate of 72 km/h to a full stop in 100 m? (*Caution:* Be careful to work in common units.)

45. Figures a–d below illustrate the graphs $y = f(x)$ of four particular functions, and Figures i–iv represent the slope portraits for their antiderivatives F. Match each function f with the slope portrait of its antiderivative F. Then, on each of the four slope portraits, illustrate the antidifferentiation process by sketching the graphs $y = F(x)$ of three different antiderivatives.

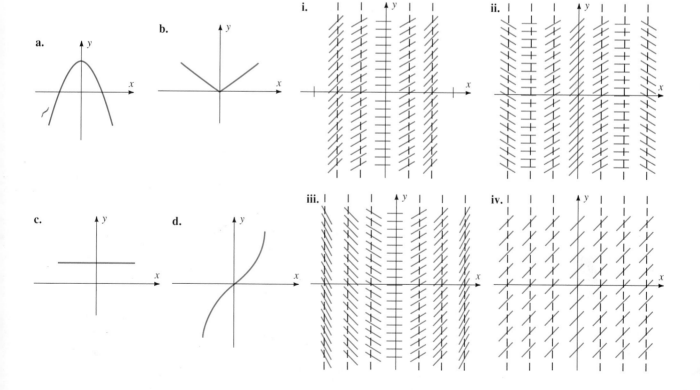

In each of Exercises 46–49, sketch the slope portrait for the antiderivatives F of the given function f. Then, on each portrait, sketch the graphs $y = F(x)$ of three antiderivatives of f. Finally, using the formula for $f(x)$, calculate the indefinite integral of f, and reconcile the graphs $y = F(x)$ with the integral.

46. $f(x) = x$ **47.** $f(x) = 1 - x$

48. $f(x) = 1 - x^2$ **49.** $f(x) = \sin x$

50. Plot the slope portrait that illustrates Theorem 2.

51. Let $f(x) = 1/x$. Note that the antiderivative of f is not determined by the Power Rule. Using f, plot the slope portrait of its antiderivatives F. Sketch the graphs $y = F(x)$ of at least two antiderivatives. From your sketch, what do you conclude about the limit of $F(x)$ as $x \to 0$?

52. (Continuation of Review Exercise 75 in Chapter 3.) For each of the functions f that is graphed in a–c, there is at least one antiderivative F graphed among i–ix. For each f, determine which of the graphs i–ix are graphs of antiderivatives of f.

a.

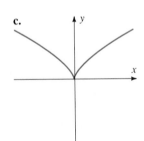

b.

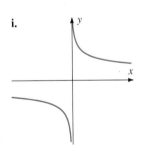

c.

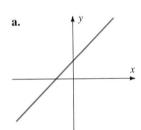

ii.

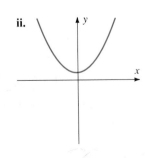

iv.

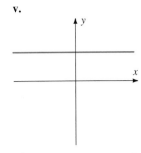

v.

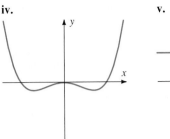

vi.

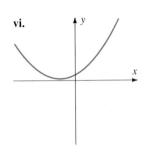

vii.

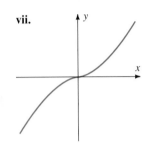

viii.

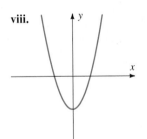

ix.

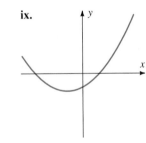

i.

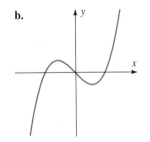

iii.

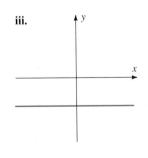

53. Specify a function f that is twice differentiable for all real numbers and that satisfies the three conditions: (1) $f''(x) = 0$ for all x; (2) $f(0) = -2$; and (3) $f(1) = 1$. How many such functions are there? Why?

54. Specify all functions f that satisfy the equation $f^{(3)}(x) = 0$ for all real numbers x.

55. Specify all functions f that satisfy the equation $f''(x) = 2 \sin x$ for all real numbers x.

56. Specify a cubic polynomial $p(x)$ that satisfies all of the following properties:
   (1) p is increasing on the intervals $(-\infty, -1)$ and $(1, \infty)$;
   (2) p is decreasing on the interval $(-1, 1)$;
   (3) the graph of p is concave down on $(-\infty, 0)$;
   (4) the graph of p is concave up on $(0, \infty)$; and
   (5) $p(0) = 2$.

57. Given the graph $y = f(x)$ of a function f below, provide a rough sketch of the graph $y = F(x)$ of its antiderivative F that satisfies the condition $F(0) = 0$. Assume that F is continuous on $[-2, 12]$.

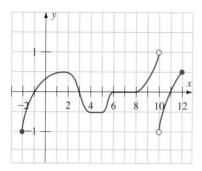

In Exercises 58–63, you are asked to construct approximate graphs of antiderivatives using a **tangent line approximation method**. Relevant programs are provided in Appendix I (for computers) or Appendix V (for graphing calculators).

In Exercises 58–61, graph the function f and its antiderivative over the given interval using the tangent line approximation method, with $n = 10, 20, 30,$ and 50. Use 0 as the arbitrary constant.

58. $f(x) = 2x$, $[-1, 1]$ **59.** $f(x) = \cos x$, $[-2\pi, 2\pi]$

60. $f(x) = 1/x$, $[1/2, 5]$ **61.** $f(x) = 1/(1 + x^2)$, $[-2, 2]$

62. If f is the function that is constantly equal to the number m, that is, $f(x) = m$, then its antiderivative is $F(x) = mx + C$. Using the tangent line approximation method, graph f and its antiderivative F for various numbers m in the interval $[-1, 1]$ using different values of C. Is the approximate graph of $F(x)$ found in this way always the same as the graph of $F(x) = mx + C$? Justify your answer using Theorem 12 in Section 4.8.

63. The following table gives the rate of infection of a contagious disease every two days.

t	$r(t)$
0	.7399
2	1.5387
4	3.1594
6	6.3207
8	12.0080
10	20.7020
12	30.2751
14	34.9569
16	30.7970
18	21.3398
20	12.4823
22	6.6009
24	3.3074
26	1.6127
28	.7759
30	.3709

Suppose that one infected individual initially enters a community of 375 susceptible individuals.
a. Plot the points $(t, r(t))$. From this graph determine the shape of the graph of the number $n(t)$ of individuals that contract the disease every two days.
b. Using the tangent line approximation method, create a table that gives $n(t)$. When will all 375 individuals have the disease?

5.2 Finding Antiderivatives by Substitution

In Chapter 3, we learned how to differentiate typical combinations of "well-known" functions such as power functions, rational functions, and trigonometric functions. Unlike differentiation, however, antidifferentiation is significantly less systematic, and even though we shall present an abstract way to construct an antiderivative for any continuous function, we shall not always be able to express the result in familiar terms. Therefore, we must approach the antidifferentiation problem in stages. First, we identify a small list of functions that are the derivatives of elementary functions such as x^n, $\sin x$, etc. These functions are our basic building blocks (see Table 1.2 in Section 5.1). Then we employ Theorem 3 from Section 5.1 to treat sums and constant multiples. Finally, we try to use the computational theorems from Chapter 3 in reverse. In this section, we discuss the antidifferentiation technique called substitution, which is the Chain Rule in reverse.

The most difficult aspect to learning this method is recognizing when it can be used. We must be able to determine if a given function is the *result* of an application of the Chain Rule, and we must be able to identify the two original functions F and g whose composition $F \circ g$, when differentiated, yields the given function.

Recall the statement of the Chain Rule for differentiating the composite function $F \circ g$:

$$(F \circ g)'(x) = F'(g(x)) \cdot g'(x). \tag{1}$$

Equation (1) tells us that an *antiderivative* for the function $F'(g(x))g'(x)$ on the right side is just the composite function $F \circ g$ that has been differentiated on the left side. We may therefore write

$$\int F'(g(x)) \cdot g'(x)\, dx = F(g(x)) + C$$

or

$$\boxed{\begin{array}{c} \int f(g(x))g'(x)\, dx = F(g(x)) + C \\[2mm] \text{if } F' = f. \end{array}} \tag{2}$$

Equation (2) is useful in finding the indefinite integral for a *product* $(f \circ g)g'$ when the first factor is a composite function $f \circ g$ and the second factor g' is the derivative of the "inside" function in the composition. In this case an antiderivative is the composition $F \circ g$ of an antiderivative F for the "outside" function with the "inside" function g. We emphasize, however, that in order to use equation (2) we must already know an antiderivative F for f.

We also note that $g(x)$ is often written as a new variable, u for instance. Then $f(g(x))$ becomes $f(u)$, $g'(x) = du/dx$, and $F(g(x)) = F(u)$.

☐ **EXAMPLE 1**

Find $\displaystyle\int (x^3 + 2x + 6)^4(3x^2 + 2)dx$.

Strategy · · · · · · · ·

Identify the integrand as the derivative of a composite function:

$$f(g(x))g'(x).$$

Solution

The integrand $(x^3 + 2x + 6)^4(3x^2 + 2)$ has the form $f(g(x))g'(x)$ with

$$f(u) = u^4 \quad \text{and} \quad u = g(x) = x^3 + 2x + 6,$$

since $g'(x) = 3x^2 + 2$ and, thus

$$\underbrace{(x^3 + 2x + 6)^4}_{f(g(x))}\underbrace{(3x^2 + 2)}_{g'(x)} = f(g(x))g'(x).$$

Find an antiderivative F for the outside function f.

An antiderivative for $f(u) = u^4$ is

$$F(u) = \int u^4\, du = \frac{u^5}{5} + C;$$

Apply equation (2).

hence, equation (2) tells us that

$$\int (x^3 + 2x + 6)^4 (3x^2 + 2)\, dx = \frac{(x^3 + 2x + 6)^5}{5} + C,$$

which you can verify by differentiation. ∎

☐ **EXAMPLE 2**

Find $\displaystyle \int \frac{\cos x}{\sqrt{1 + \sin x}}\, dx.$

Solution This integral can be written in the form

$$\int \frac{\cos x}{\sqrt{1 + \sin x}}\, dx = \int (1 + \sin x)^{-1/2} \cos x\, dx,$$

which we recognize as being of the form $\int f(g(x))g'(x)\, dx$ with

$$f(u) = u^{-1/2} \quad \text{and} \quad u = g(x) = 1 + \sin x.$$

An antiderivative for f is $F(u) = \int u^{-1/2}\, du = 2u^{1/2} + C$, so it follows from equation (2) that

$$\int \frac{\cos x}{\sqrt{1 + \sin x}}\, dx = 2\sqrt{1 + \sin x} + C.$$ ∎

The Method of Substitution

There is a way to simplify somewhat the procedure of identifying an integrand in the form $f(g(x))g'(x)$ and finding an antiderivative $F(g(x))$. It is based on the notation for the *differential du* of the function $u = g(x)$. Recall the definition of a differential (Section 3.7):

$$\text{If} \quad u = g(x), \quad \text{then} \quad du = g'(x)\, dx \tag{3}$$

if g is differentiable at x. Using this notation we may write the integrand in equation (2) formally as

$$\underbrace{f(g(x))}_{f(u)}\underbrace{g'(x)\, dx}_{du} = f(u)\, du. \tag{4}$$

With the notation of the line (4), equation (2) may be rewritten as

$$\int f(g(x))g'(x)\, dx = \int f(u)\, du = F(u) + C$$

$$\text{where} \quad u = g(x) \quad \text{and} \quad F' = f. \tag{5}$$

Equation (5) should be regarded as a notational device for remembering equation (2). It is simply equation (2) rewritten using differential notation. The advantage in using equation (5) is that *once we have made the substitution $u = g(x)$ and applied the differential notation $du = g'(x)\, dx$, the antiderivative F may be found by integrating the function f as a function of the independent variable u.* This process is called the **method of substitution.**

In the method of substitution it is important to note that equation (5) results from the *notation du* = $g'(x)$ dx for the differential du and *not* from its interpretation as a linear approximation.

☐ **EXAMPLE 3**

Find $\int \sin^3 x \cos x\, dx$ by the method of substitution.

Strategy · · · · · · · ·
Since $(\sin x)' = \cos x$, we try $u = \sin x$.

Solution
We use the substitution

$$u = \sin x; \quad du = \cos x\, dx.$$

We obtain

Substitute for u and du.

Find antiderivative with respect to u.

Substitute back in terms of x.

$$\int \sin^3 x \cos x\, dx = \int u^3\, du$$

$$= \frac{u^4}{4} + C$$

$$= \frac{\sin^4 x}{4} + C.$$ ■

Often, we encounter indefinite integrals that could be integrated using equation (2) except that they are missing a necessary *constant* factor. Using the fact that

$$\int c \cdot f(x)\, dx = c \int f(x)\, dx,$$

we can always adjust for the "missing" factor using the algebraic manipulation

$$\int f(x)\, dx = \frac{c}{c} \int f(x)\, dx = \frac{1}{c} \int c \cdot f(x)\, dx.$$

The following example illustrates how the method of substitution handles this manipulation automatically.

☐ **EXAMPLE 4**

Find $\int x^2(x^3 + 7)^4\, dx$.

If we set $u = g(x) = x^3 + 7$, we see that $g'(x) = 3x^2$, and in order to use equation (2), we write

$$\int x^2(x^3 + 7)^4\, dx = \tfrac{1}{3} \int (x^3 + 7)^4(3x^2)\, dx.$$

Instead, we prefer to evaluate this integral using the du notation. Note that, when we use this notation, the factor of $\tfrac{1}{3}$ is determined when du is calculated.

Strategy · · · · · · · ·

Take $u = x^3 + 7$ since $du = 3x^2\,dx$ has a factor of x^2. The missing constant factor 3 is not a problem.

Since only $x^2\,dx$ appears in the integrand, solve for this term in terms of du.

Substitute for u and for du.

Apply the Power Rule. Substitute back for u in terms of x.

If desired, check antiderivative by differentiation.

Solution

We make the substitution

$$u = x^3 + 7; \quad du = 3x^2\,dx.$$

Then $x^2\,dx = \frac{1}{3}\,du$, and we obtain

$$\int x^2(x^3 + 7)^4\,dx = \int (x^3 + 7)^4 x^2\,dx$$

$$= \int u^4 (\tfrac{1}{3})\,du$$

$$= \tfrac{1}{3} \int u^4\,du$$

$$= \tfrac{1}{3}\left(\frac{u^5}{5} + C_1\right)$$

$$= \frac{(x^3 + 7)^5}{15} + C.$$

To check this result, we differentiate using the Chain Rule

$$\frac{d}{dx}\left(\frac{(x^3 + 7)^5}{15} + C\right) = \frac{5(x^3 + 7)^4}{15}(3x^2) + 0$$

$$= (x^3 + 7)^4 x^2. \qquad \blacksquare$$

☐ **EXAMPLE 5**

Find $\displaystyle \int \frac{x^2\,dx}{\sqrt{1 + x^3}}$.

Strategy · · · · · · · ·

Given the x^2 factor we try $u = 1 + x^3$.

Solve for the factor $x^2\,dx$ in terms of du.
Make the substitution.

Find the antiderivative with respect to u using Power Rule. Substitute back in terms of x.

Solution

We make the substitution

$$u = 1 + x^3; \quad du = 3x^2\,dx.$$

Then $x^2\,dx = \frac{1}{3}\,du$, so we obtain

$$\int \frac{x^2\,dx}{\sqrt{1 + x^3}} = \int \frac{\frac{1}{3}\,du}{\sqrt{u}}$$

$$= \tfrac{1}{3} \int u^{-1/2}\,du$$

$$= \tfrac{1}{3}\left[(2)u^{1/2} + C_1\right]$$

$$= \tfrac{2}{3}\sqrt{1 + x^3} + C. \qquad \blacksquare$$

REMARK   Remember that the formula

$$\int c \cdot f(x)\,dx = c \int f(x)\,dx$$

only holds for a *constant c*. Therefore, algebraic "adjustments," such as those used in the last two examples, can only be performed with constant factors. For example,

$$\int x\sqrt{x+1}\,dx \neq x\int \sqrt{x+1}\,dx$$

(see Exercise 40).

□ **EXAMPLE 6**

Find $\displaystyle\int \frac{\sin x \cos x\,dx}{\sqrt{1+\sin^2 x}}$.

Solution A first attempt might be to try the substitution

$$u = \sin x; \quad du = \cos x\,dx.$$

Indeed, this works, since we obtain

$$\int \frac{\sin x \cos x\,dx}{\sqrt{1+\sin^2 x}} = \int \frac{u\,du}{\sqrt{1+u^2}}.$$

To handle the antiderivative on the right, we can make a second substitution

$$w = 1 + u^2; \quad dw = 2u\,du,$$

so $u\,du = \frac{1}{2}\,dw$. With this substitution we have

$$\int \frac{\sin x \cos x\,dx}{\sqrt{1+\sin^2 x}} = \int \frac{u\,du}{\sqrt{1+u^2}} = \int \frac{\frac{1}{2}\,dw}{\sqrt{w}} = \int \frac{1}{2}w^{-1/2}\,dw$$
$$= \sqrt{w} + C$$
$$= \sqrt{1+u^2} + C$$
$$= \sqrt{1+\sin^2 x} + C.$$

A shorter solution to this problem is obtained if we simply take

$$u = 1 + \sin^2 x; \quad du = 2\sin x \cos x\,dx.$$

Then $\sin x \cos x\,dx = \frac{1}{2}\,du$, and we obtain

$$\int \frac{\sin x \cos x\,dx}{\sqrt{1+\sin^2 x}} = \int \frac{\frac{1}{2}\,du}{\sqrt{u}} = \int \frac{1}{2}u^{-1/2}\,du = \sqrt{u} + C$$
$$= \sqrt{1+\sin^2 x} + C. \qquad ■$$

As Example 6 illustrates, occasionally one substitution is not enough. More than one substitution may be necessary to determine the desired antiderivative. Example 6 also illustrates the fact that, in some cases, more than one choice of substitution will work.

Identifying a useful substitution $u = g(x)$ is often done by recognizing the differential du of a standard function. Table 2.1 is a list of such differentials.

Table 2.1 Elementary Substitutions and Their Corresponding Differentials

$u = g(x)$	$du = g'(x)\, dx$
$u = x^n$	$du = nx^{n-1}\, dx$
$u = \sin x$	$du = \cos x\, dx$
$u = \cos x$	$du = -\sin x\, dx$
$u = \tan x$	$du = \sec^2 x\, dx$
$u = \sec x$	$du = \sec x \tan x\, dx$
$u = \cot x$	$du = -\csc^2 x\, dx$
$u = \csc x$	$du = -\csc x \cot x\, dx$

In the following example, there are two reasonable choices for u.

☐ **EXAMPLE 7**

Find $\displaystyle\int \tan x \sec^2 x\, dx$.

Solution

Substitution (1): Let $u_1 = \tan x$. Then $du_1 = \sec^2 x\, dx$, and we get

$$\int \tan x \sec^2 x\, dx = \int u_1\, du_1$$

$$= \frac{u_1^2}{2} + C_1$$

$$= \frac{\tan^2 x}{2} + C_1. \tag{6}$$

Substitution (2): Let $u_2 = \sec x$. Then $du_2 = \sec x \tan x\, dx$, and we get

$$\int \tan x \sec^2 x\, dx = \int u_2\, du_2$$

$$= \frac{u_2^2}{2} + C_2$$

$$= \frac{\sec^2 x}{2} + C_2. \tag{7}$$

These two different substitutions seem to result in different answers! However, you should always remember that the indefinite integral is the family of all antiderivatives of a given function. Using the trigonometric identity $\sec^2 x = 1 + \tan^2 x$, we see that the family of functions represented by line (6) is the same as the family in line (7).

Another way to check that both answers are indeed correct is to note that the derivative of (6) is the same as the derivative of (7). ∎

In our final example, we illustrate that occasionally we must solve for x in terms of u as part of the substitution process.

□ **EXAMPLE 8**

Find $\int x^2\sqrt{x+1}\,dx$.

Solution We choose the substitution $u = x + 1$, and consequently $du = dx$. In order to complete the substitution, we must convert the x^2 into terms involving u. Since $u = x + 1$, $x = u - 1$. We obtain

$$\int x^2\sqrt{x+1}\,dx = \int (u-1)^2\sqrt{u}\,du$$

$$= \int (u^{5/2} - 2u^{3/2} + u^{1/2})\,du$$

$$= \frac{2u^{7/2}}{7} - \frac{4u^{5/2}}{5} + \frac{2u^{3/2}}{3} + C$$

$$= \frac{2(x+1)^{7/2}}{7} - \frac{4(x+1)^{5/2}}{5} + \frac{2(x+1)^{3/2}}{3} + C. \quad \blacksquare$$

Exercise Set 5.2

In Exercises 1–6, use the designated substitution to find the specified antiderivatives.

1. $\int x^5 \sin(x^6 + 2)\,dx, \quad u = x^6 + 2$

2. $\int (x + 1)\sqrt{x^2 + 2x}\,dx, \quad u = x^2 + 2x$

3. $\int \frac{\cos\sqrt{x}}{\sqrt{x}}\,dx, \quad u = \sqrt{x}$

4. $\int \frac{\sin x \cos x}{\sqrt{1 + \cos^2 x}}\,dx, \quad u = 1 + \cos^2 x$

5. $\int x\sqrt{x+2}\,dx, \quad u = x + 2$

6. $\int x \sin x^2 \cos x^2\,dx, \quad u = \sin x^2$

In Exercises 7–14, use the method of substitution to find the antiderivative. Check your answer by differentiating.

7. $\int x\sqrt{x^2 + 1}\,dx$

8. $\int x \cos x^2\,dx$

9. $\int \sec 2\theta \tan 2\theta\,d\theta$

10. $\int \csc(\pi x)\cot(\pi x)\,dx$

11. $\int x \csc^2(x^2)\,dx$

12. $\int \frac{\sin\sqrt{x}\cos\sqrt{x}}{\sqrt{x}}\,dx$

13. $\int (1 - t^2)\sqrt{3t^3 - 9t + 9}\,dt$

14. $\int (1 - x)^3\,dx$

In Exercises 15–34, find the antiderivative.

15. $\int \sec^2(x - \pi)\,dx$

16. $\int x \sec^2 x^2\,dx$

17. $\int \frac{(2\sqrt{x} + 3)^2}{\sqrt{x}}\,dx$

18. $\int (t^3 - 6t + 7)^5(2 - t^2)\,dt$

19. $\int (x^2 + 1)^2\,dx$

20. $\int \sqrt{t}(t + 1)\,dt$

21. $\int x\sqrt{x-1}\sqrt{x+1}\,dx$

22. $\int \sqrt{1 - \sin x}\cos x\,dx$

23. $\int x \sec(\pi - x^2)\tan(\pi - x^2)\,dx$

24. $\int \frac{\sec^2\sqrt{2x+1}}{\sqrt{2x+1}}\,dx$

25. $\int (x^5 - 2x^3)(x^6 - 3x^4)^{5/2}\,dx$

26. $\int x^3(1 - x^4)^{10}\,dx$

27. $\int x^2(1 - x^4)^2\,dx$

28. $\int \frac{x^3 + 2x}{\sqrt[3]{x^4 + 4x^2}}\,dx$

29. $\int \frac{t^2 + 2}{\sqrt{t^3 + 6t}}\,dt$

30. $\displaystyle\int \frac{\cos^3 \sqrt{x}\, \sin \sqrt{x}}{\sqrt{x}}\, dx$ **31.** $\displaystyle\int \frac{1}{(1 - 4x)^{2/3}}\, dx$

32. $\displaystyle\int \left(\frac{1 + \tan^2 \sqrt{x}}{\sqrt{x}}\right) dx$

33. $\displaystyle\int x^3 \sqrt{x^2 + 1}\, dx$ **34.** $\displaystyle\int \frac{x}{\sqrt[3]{(1 + x)^2}}\, dx$

In Exercises 35–37, state a substitution appropriate to each part and find the antiderivative.

35. a. $\displaystyle\int (3x + 4)^{10}\, dx$

  b. $\displaystyle\int \sqrt{3x + 4}\, dx$

  c. $\displaystyle\int \sin(3x + 4)\, dx$

  d. $\displaystyle\int \sin^2(3x + 4) \cos(3x + 4)\, dx$

36. a. $\displaystyle\int \sin(3x)\, dx$

  b. $\displaystyle\int \sin(3x) \cos(3x)\, dx$

  c. $\displaystyle\int \sin^4(3x) \cos(3x)\, dx$

37. a. $\displaystyle\int (x^2 + 6)^{21}\, x\, dx$

  b. $\displaystyle\int \frac{x}{\sqrt{x^2 + 6}}\, dx$

  c. $\displaystyle\int x \sec^2(x^2 + 6)\, dx$

38. Some texts include integrals such as $\int \sin nx\, dx$ in their tables of basic integrals. Using substitution, derive an integral formula for $\int \sin nx\, dx$.

39. The integral

$$\int \sin x \cos x\, dx$$

can be calculated using two different substitutions. First, let $u_1 = \sin x$ and calculate the integral. Second, use the trigonometric identity $\sin 2x = 2 \sin x \cos x$ to eliminate the $\cos x$ term. Then use the substitution $u_2 = 2x$. Do you get the same answer both ways?

40. Using the substitution $u = x + 1$, calculate the antiderivatives

$$\int x\sqrt{x + 1}\, dx \quad \text{and} \quad \int \sqrt{x + 1}\, dx.$$

Explain why

$$\int x\sqrt{x + 1}\, dx \neq x \int \sqrt{x + 1}\, dx.$$

41. Four of the following six antiderivatives can be calculated using the methods discussed in Sections 5.1 and 5.2. Which are they? Three of these four can be determined using a substitution. Determine which three and indicate the substitution. How do you calculate the fourth? One of the three amenable to the method of substitution can also be determined without using substitution. Which one? How can this antiderivative be determined if substitution is not used?

  a. $\displaystyle\int \sec^2(2\theta)\, d\theta$ **b.** $\displaystyle\int \sec^2(\theta^2)\, d\theta$

  c. $\displaystyle\int (u^2 + 1)^{10}\, du$ **d.** $\displaystyle\int u(u^2 + 1)^{10}\, du$

  e. $\displaystyle\int \frac{t + 1}{\sqrt{t^2 + 2t}}\, dt$ **f.** $\displaystyle\int \frac{t^2 + 1}{\sqrt{t^3 + t}}\, dt$

5.3 Differential Equations

Calculating functions from their slope functions and finding velocity and position from acceleration are not the only types of problems in which one attempts to determine a function from information about its derivative. For example, a classic experiment in elementary physics involves sprinkling iron filings on a sheet of paper suspended over a horseshoe magnet (Figure 3.1). The iron filings assume the directions of the magnetic force field surrounding the magnet (Figure 3.2). Here we obtain information about the direction (slopes) of the field lines, and we can then ask for the equations of the field lines themselves.

   The general theory for determining a function from certain types of information about its derivative(s) is referred to as the theory of **differential equations**. A differential equation is an equation involving one or more derivatives of an unknown function.

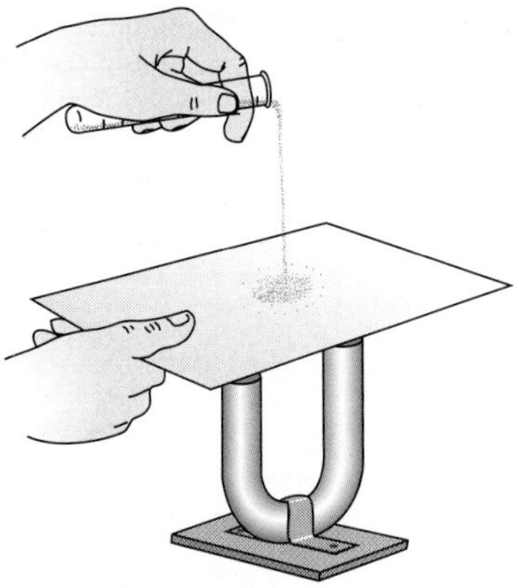

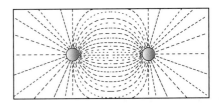

Figure 3.1 Iron filing–magnet experiment.

Figure 3.2 Filings point in direction of magnetic field lines.

For example, since velocity is the derivative of position, we have already seen that the position function s satisfies the differential equation.

$$s'(t) = v(t). \tag{1}$$

Later, we shall see that the function y representing the displacement of the end of an oscillating spring satisfies a differential equation of the form

$$\frac{d^2y}{dt^2} + b\frac{dy}{dt} + cy = 0. \tag{2}$$

Equation (1) is an example of a **first order** differential equation because it involves only a first derivative. Equation (2) is an example of a **second order** differential equation because it involves a second derivative.

A **solution** of a differential equation is a function that is differentiable as many times as the equation requires and that satisfies the equation. For example, we know that the function $y = \sin x$ satisfies the differential equation

$$\frac{dy}{dx} = \cos x.$$

Also, the function $y = x^2$ is a solution to the differential equation

$$x\frac{d^2y}{dx^2} - \frac{dy}{dx} = 0$$

since, if $y = x^2$,

$$x\frac{d^2y}{dx^2} - \frac{dy}{dx} = x(2) - (2x) = 0.$$

☐ **EXAMPLE 1**

The function $y = \sin x$ satisfies the differential equation

$$\frac{d^2y}{dx^2} + y = 0.$$

To show this, we compute the first and second derivatives:

$$\frac{dy}{dx} = \cos x \quad \text{and} \quad \frac{d^2y}{dx^2} = -\sin x.$$

Thus, substitution in the differential equation gives

$$\frac{d^2y}{dx^2} + y = (-\sin x) + \sin x$$

$$= 0$$

as required. ∎

One of the simplest forms for a differential equation is

$$\frac{dy}{dx} = f(x). \tag{3}$$

Any antiderivative $y = F(x)$ for the function f is a solution, since $dy/dx = F'(x) = f(x)$. In fact, once we have identified a particular antiderivative F, Theorem 1 guarantees that all solutions of (3) have the form

$$y = F(x) + C, \qquad C \text{ constant}. \tag{4}$$

We refer to (4) as the **general solution** of differential equation (3) since all particular solutions of (3) may be obtained from (4) by specifying the appropriate constant C.

We will frequently encounter differential equation (3) in the *differential* form

$$dy = f(x)\,dx. \tag{5}$$

In fact, the differential formulation (5) of the differential equation (3) is simply another use of the differential notation introduced in Section 3.7. Recall that, if $y = F(x)$, then we defined $dy = F'(x)\,dx$. When we write $dy = f(x)\,dx$, we are asserting that

$$\frac{dy}{dx} = F'(x) = f(x),$$

which is precisely equation (3).

□ **EXAMPLE 2**

Find the general solution of the differential equation

$$3\,dy - x\,dx = 0.$$

Strategy · · · · · · · · ·

Solve for $\dfrac{dy}{dx}$.

Find y by antidifferentiation.

Solution

Here $\dfrac{dy}{dx} = \dfrac{x}{3}$.

Thus

$$y = \int \frac{x}{3}\,dx = \frac{x^2}{6} + C$$

is the general solution (see Figure 3.3). ∎

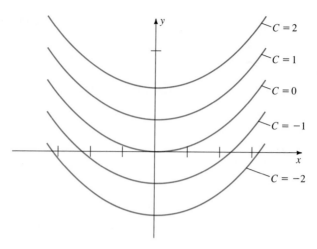

Figure 3.3 Solutions of $3dy - x\,dx = 0$ are $y = \dfrac{x^2}{6} + C.$

As Example 2 illustrates, the general solution of equation (3) is not a function but rather a *family* of functions of the form (4), one such function corresponding to each choice of the constant C. To obtain a *unique* solution to (3) we must also specify an *initial condition*. An **initial condition** is simply a point (x_0, y_0) through which the desired solution curve must pass. For example, if we specify the initial condition $y(1) = 3$ in Example 2, we obtain the equation

$$3 = y(1) = \frac{(1)^2}{6} + C = \frac{1}{6} + C,$$

so $C = 3 - (1/6) = 17/6$. The particular solution satisfying this initial condition is therefore

$$y = \frac{x^2}{6} + \frac{17}{6}.$$

A problem consisting of a differential equation together with an initial condition is called an **initial value problem.**

□ **EXAMPLE 3**

Solve the initial value problem

$$dy = x\sqrt{x^2 + 1}\,dx, \qquad y(0) = 1.$$

Strategy · · · · · · · · · **Solution**

Here $\dfrac{dy}{dx} = x\sqrt{x^2 + 1}$, so

$$y = \int x\sqrt{x^2 + 1}\,dx$$

Find the indefinite integral by the substitution method with

$$u = x^2 + 1, \quad du = 2x\,dx.$$

$$= \int u^{1/2}(\tfrac{1}{2})\,du$$

$$= \tfrac{1}{2}\int u^{1/2}\,du$$

$$= \tfrac{1}{2}(\tfrac{2}{3}u^{3/2} + C_1)$$

$$= \frac{u^{3/2}}{3} + C$$

$$= \frac{(x^2 + 1)^{3/2}}{3} + C.$$

Apply initial condition to find C.

Then $y(0) = 1$ gives the equation

$$1 = y(0) = \frac{(0^2 + 1)^{3/2}}{3} + C$$

$$= \tfrac{1}{3} + C$$

so $C = 2/3$. The desired solution is therefore

$$y = \frac{(x^2 + 1)^{3/2}}{3} + \frac{2}{3}. \qquad \blacksquare$$

Separation of Variables

Another type of differential equation for which we can often find the general solution has the form

$$\frac{dy}{dx} = \frac{f(x)}{g(y)}, \qquad g(y) \neq 0. \tag{6}$$

Differential equations of this type are called **separable** because, if we multiply both sides of equation (6) by $g(y)$, we obtain the equation

$$g(y)\frac{dy}{dx} = f(x). \tag{7}$$

If we can find antiderivatives G for g and F for f, we will have succeeded in **separating the variables** in equation (6): the left side of (7) is the derivative

$$\frac{d}{dx}G(y) = G'(y)\frac{dy}{dx} = g(y)\frac{dy}{dx}$$

of the composite function $G(y)$, and the right side of (7) is the derivative $F'(x) = f(x)$ of the function F. It then follows, from Theorem 1, that

$$G(y) = F(x) + C, \qquad C = \text{constant} \tag{8}$$

on the domain common to $F(x)$ and $G(y)$. Equation (8), in general, defines the solution y of equation (6) implicitly, but equation (8) can often be solved to produce the solution y as an explicit function of x.

REMARK 1 We often use the differential notation

$$dy = \left(\frac{dy}{dx}\right) dx$$

to abbreviate the above discussion by saying that

$$\frac{dy}{dx} = \frac{f(x)}{g(y)} \quad \text{implies that} \quad g(y)\,dy = f(x)\,dx,$$

so

$$\int g(y)\, dy = \int f(x)\, dx,$$

and we obtain the solution y by "integrating both sides." Note, however, that the mathematical justification for the technique described in this remark is the argument, based on the Chain Rule, presented in the preceding paragraph.

☐ **EXAMPLE 4**

Use the technique of separation of variables to find the general solution of the differential equation

$$\frac{dy}{dx} = \frac{x}{y}, \qquad y \neq 0.$$

Strategy · · · · · · ·
Separate variables.

Solution

We rewrite the equation in differential notation and obtain

$$y\, dy = x\, dx$$

Integrate both sides.

so

$$\int y\, dy = \int x\, dx.$$

Thus

$$\frac{y^2}{2} + C_1 = \frac{x^2}{2} + C_2,$$

Solve for y^2, combining all arbitrary constants.

so

$$y^2 + 2C_1 = x^2 + 2C_2,$$

or

$$y^2 = x^2 + C$$

where C is an arbitrary constant.

Graphs of this equation, for various values of C, appear in Figure 3.4. To verify this result we differentiate implicitly in the equation

$$y^2 = x^2 + C$$

to obtain

$$2y\frac{dy}{dx} = 2x$$

so

$$\frac{dy}{dx} = \frac{x}{y}, \qquad y \neq 0. \qquad \blacksquare$$

REMARK 2 As you can see from Figure 3.4, each value of C yields more than one function $y(x)$ that satisfies the differential equation. In order to specify a solution $y(x)$ uniquely, we need an initial condition. For example, if $y(x)$ satisfies the initial

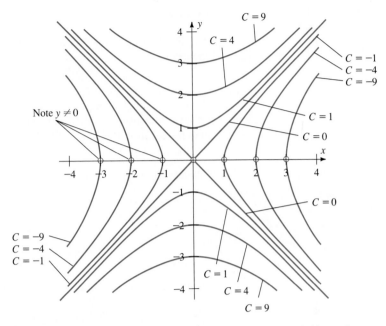

Figure 3.4   Graphs of solutions of the differential equation $dy/dx = x/y$.

condition $y(0) = 1$, then $y(x) = \sqrt{x^2 + 1}$. If, instead, $y(0) = -1$ then $y(x) = -\sqrt{x^2 + 1}$.

☐ EXAMPLE 5

Find the solution of the initial value problem

$$dy = 2xy^2\, dx, \qquad y(0) = 1/2.$$

Strategy · · · · · · · · ·

Separate variables.

Integrate both sides.

Solve for y.

Apply initial condition to solve for C.

Solution

We have

$$y^{-2}\, dy = 2x\, dx,$$

so

$$\int y^{-2}\, dy = \int 2x\, dx$$

and, therefore,

$$-y^{-1} + C_1 = x^2 + C_2.$$

Thus $1/y = -x^2 + C$, so

$$y = \frac{-1}{x^2 + C}.$$

From the initial condition $y(0) = 1/2$ we obtain

$$\frac{1}{2} = y(0) = \frac{-1}{0^2 + C} = \frac{-1}{C}$$

so $C = -2$. The solution is therefore

$$y = \frac{-1}{x^2 - 2}.$$

■

☐ **EXAMPLE 6**

A population of animals in an ecological niche grows according to the differential equation

$$\frac{dP}{dt} = \frac{100}{P}.$$

Here the function $y = P(t)$ represents the number of animals present after t years. Find the function P if initially $P(0) = 50$ animals are present. What is the size of the population after 12 years?

Solution Separating variables in the differential equation

$$\frac{dP}{dt} = \frac{100}{P} \tag{9}$$

gives

$$P \, dP = 100 \, dt.$$

Then

$$\int P \, dP = \int 100 \, dt$$

gives

$$\frac{P^2}{2} + C_1 = 100t + C_2$$

so

$$\frac{P^2}{2} = 100t + C_3$$

or

$$P^2 = 200t + C.$$

Thus

$$P(t) = \sqrt{200t + C} \tag{10}$$

is the general solution of equation (9). (We take the sign of $P(t)$ to be positive because $P(t)$ is the size of a population which cannot be negative.)

To determine the constant C we set $t = 0$ in equation (10) and apply the initial condition $P(0) = 50$. This gives

$$P(0) = \sqrt{200(0) + C} = \sqrt{C} = 50.$$

Thus, $C = 50^2 = 2500$ and

$$P(t) = \sqrt{200t + 2500}$$

is the size of the population after t years.

After $t = 12$ years the population is of size

$$P(12) = \sqrt{2400 + 2500} = \sqrt{4900} = 70. \qquad \blacksquare$$

Differential equations are an important part of applied mathematics, physics, engineering mathematics, mathematical biology, and physical chemistry. In later chapters we shall encounter other types of differential equations, techniques for their solutions, and applications involving such equations.

Exercise Set 5.3

In Exercises 1–14, find the general solution of the differential equation.

1. $\dfrac{dy}{dx} = x - 1$

2. $\dfrac{dy}{dx} = 2 + \sec^2 x$

3. $\dfrac{dy}{dx} = x^2 - \dfrac{1}{x^2}$

4. $\dfrac{dy}{dx} = 4 \cos 2x$

5. $\dfrac{dy}{dx} = \sin x \cos x \sqrt{1 + \sin^2 x}$

6. $\dfrac{dy}{dx} = (x + 1)\sqrt{x^2 + 2x + 2}$

7. $dy = (x - \sqrt{x})\, dx$

8. $dy + 3\, dx = 0$

9. $dy = \dfrac{x}{y}\, dx$

10. $\dfrac{dy}{dx} = \dfrac{x}{y^2}$

11. $dy = -4xy^2\, dx$

12. $dy = \dfrac{\sqrt{x}}{\sqrt{y}}\, dx$

13. $\dfrac{dy}{dx} = -\dfrac{x}{y}$

14. $dy = y^2\, dx$

In Exercises 15–26, find the solution of the initial value problem.

15. $\dfrac{dy}{dx} = \dfrac{x}{y}, \quad y(1) = 4$

16. $\dfrac{dy}{dx} = \dfrac{x}{y}, \quad y(1) = -4$

17. $\dfrac{dy}{dx} = -\dfrac{x}{y}, \quad y(\sqrt{3}) = 1$

18. $2y\, dy = \sqrt{x}\, dx, \quad y(0) = 2$

19. $dy = \dfrac{x^2}{(1 + x^3)^2}\, dx, \quad y(0) = \dfrac{1}{2}$

20. $dy = \dfrac{x}{y\sqrt{1 + x^2}}\, dx, \quad y(0) = 2$

21. $dy = y^2(1 + x^2)\, dx, \quad y(0) = 1$

22. $\dfrac{dy}{dx} = 2x^2 y^2, \quad y(0) = -3$

23. $\dfrac{d^2 y}{dx^2} = x^2, \quad y(0) = 2, \quad y'(0) = 4$

24. $\dfrac{d^2 y}{dx^2} = \sin x, \quad y(0) = \pi, \quad y'(0) = 2$

25. $\dfrac{d^2 y}{dx^2} = -16 \sin 4x, \quad y(\pi) = \pi + 1, \quad y'(\pi) = 1$

26. $\dfrac{d^2 y}{dx^2} = \sqrt{x}, \quad y(1) = 1/3, \quad y'(1) = 1$

27. The slope of the line tangent to the graph of $y = f(x)$ at $(x, f(x))$ is $x - \sqrt{x}, x > 0$. If the point $(1, 2)$ is on the curve, find f.

28. Find a curve in the xy-plane that contains the point $(2, 1)$ and whose normal at (x, y) has slope $y/x, x > 0$.

29. Suppose that y satisfies the differential equation

$$\frac{dy}{dx} = \begin{cases} 2 & \text{for } x \le 1; \\ x + 1 & \text{for } x > 1. \end{cases}$$

Calculate $y(2)$ if $y(0) = 1$.

30. The value of a bottle of a certain rare wine is increasing at a rate of $\sqrt{t + 1}$ dollars per year when the wine is t years old. The wine sold for \$5.00 per bottle when new.
   a. Write an initial value problem that determines the value $V(t)$ of the bottle of wine after t years.
   b. Find the solution of this initial value problem.

31. A population of animals in an ecological niche is growing in time so that its rate of growth dP/dt is related to its current size by the differential equation

$$\frac{dP}{dt} = \frac{900}{P^2}.$$

If time is measured in years and initially there are $P(0) = 10$ animals present, find the population function $P(t)$ giving the size of the population after t years.

32. An automobile manufacturer determines that the acceleration of one of its models under full throttle is related to its velocity $v(t)$ by the differential equation

$$\frac{dv}{dt} = kv^{-1/2}$$

where k is a constant. In this equation v is in units of meters per second. Find the velocity $v(t)$ after t seconds if the velocity of the automobile at time $t = 0$ is $v(0) = 9$ m/s.

33. A population has size $P(t)$ at time t and grows according to the differential equation $dP/dt = 4\sqrt{P}$. Find $P(t), t > 0$, if $P(0) = 10$.

In Exercises 34–39, you are asked to determine approximate graphs of the solutions of differential equations using **Euler's Method.** Relevant programs are provided in Appendix I (for computers) or Appendix V (for graphing calculators).

In Exercises 34–37, use Euler's Method to approximate the graph of the solution of the differential equation. Compare the solutions for $n = 10, 30, 60, 90$, and 120.

34. $\dfrac{dy}{dx} = \dfrac{x}{y + 1}, \quad y(0) = 1, \quad 0 \le x \le 4$

35. $\dfrac{dy}{dx} = (x^2 + 1)y^2, \quad y(-2) = 2, \quad -2 \le x \le 2$

36. $\dfrac{dy}{dx} = x + y, \quad y(0) = -1, \quad 0 \le x \le 6$

37. $\dfrac{dy}{dx} = y \cos 2x, \quad y(0) = 1/2, \quad 0 \le x \le 2\pi$

38. A hemispherical tank of radius R meters is initially full of water. Suppose there is a circular hole at the bottom with radius r meters through which the water drains.

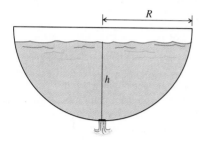

The rate of change dh/dt of the depth $h(t)$ is given by the equation

$$\frac{dh}{dt} = \frac{-r^2\sqrt{2g}}{h(2R - h)}$$

where $g = 9.8$ m/s^2 is the acceleration due to gravity.

a. Given the initial condition $h(0) = R$, solve this differential equation.

b. Suppose $R = 6$ and $r = 0.5$. Graph the solution $h(t)$ using Euler's Method. Using the graph, approximate the amount of time it takes for the tank to empty.

c. Find the time it takes for the tank to empty as a function of R and r.

39. A bicyclist is pedaling down a hill at a velocity of 75 ft/s. Then at the bottom of the hill she coasts. Suppose her rate dv/dt of deceleration due to air resistance is proportional to the square of the velocity; that is,

$$\frac{dv}{dt} = -kv^2$$

where k is a positive constant.

a. Determine v as a function of t. Note that $v(0) = 75$.

b. Suppose at $t = 2$ s her velocity is 45 ft/s. Graph $v(t)$ and determine how long it takes to slow to 5% of her initial velocity.

c. Use Euler's Method to graph $s(t)$ where

$$\frac{ds}{dt} = v, \quad s(0) = 0.$$

How far has she coasted when the velocity is 5% of its initial value?

Summary Outline of Chapter 5

▌ The function F is an **antiderivative** for the function f on the interval I if (page 274) $F'(x) = f(x)$ for all $x \in I$.

▌ *Theorem:* Any two antiderivatives of the same function on the same interval differ by a constant. (page 277)

▌ The **indefinite integral** for f is $\int f(x)\,dx = F(x) + C$ where $F'(x) = f(x)$. (page 278)

▌ *Theorem:* (Properties of antiderivatives) (page 280)
(i) $\int [f(x) + g(x)]\,dx = \int f(x)\,dx + \int g(x)\,dx$.
(ii) $\int cf(x)\,dx = c \int f(x)\,dx$ where C is a constant.

▌ Table 1.2 lists basic **integration formulas.** (page 281)

▌ Use the **method of substitution** to evaluate $\int f(g(x))g'(x)\,dx$ by the substitution $u = g(x)$, $du = g'(x)\,dx$ to obtain (page 289)

$$\int f(g(x))g'(x)\,dx = \int f(u)\,du$$

$$= F(u) + C$$

$$= F(g(x)) + C$$

where $F' = f$.

▌ Solve a **differential equation** of the form (page 299)

$$\frac{dy}{dx} = \frac{f(x)}{g(y)} \quad \text{or} \quad g(y)\,dy = f(x)\,dx$$

by the method of **separation of variables:**

$$\int g(y)\, dy = \int f(x)\, dx$$

so $G(y) = F(x) + C$ where $G' = g$ and $F' = f$.

Review Exercises—Chapter 5

In Exercises 1–18, find the indefinite integral for the given function.

1. $f(x) = 6x^2 - 2x + 1$

2. $f(x) = (x^2 - 6x)^2$

3. $y = \dfrac{\sin \sqrt{t}}{\sqrt{t}}$

4. $f(x) = x\sqrt{9x^4}$

5. $f(x) = 3\sqrt{x} + 3/\sqrt{x}$

6. $y = (t + \sqrt{t})^3$

7. $y = x \sec^2 x^2$

8. $f(s) = s\sqrt{9 - s^2}$

9. $f(x) = \dfrac{x^3 - 7x^2 + 6x}{x}$

10. $y = \dfrac{x^3 + x^2 - x + 2}{x + 2}$

11. $f(t) = t^5\sqrt{t^2 + 1}$

12. $y = x^4(1 + x^3)^2$

13. $f(x) = x \cos(1 + x^2)$

14. $f(x) = \dfrac{\sec x}{1 + \tan^2 x}$

15. $y = (2x - 1)(2x + 1)$

16. $y = \dfrac{x}{4x^4 + 4x^2 + 1}$

17. $f(x) = |x|$

18. $y = |1 - t^2|$

In Exercises 19–24, find the particular function satisfying the stated conditions.

19. $f'(x) = 1 + \cos x, \quad f(0) = 3$

20. $f'(x) = \dfrac{x}{\sqrt{1 + x^2}}, \quad f(0) = 4$

21. $f''(x) = 3, \quad f'(1) = 6, \quad f(0) = 4$

22. $f''(x) = \sin x - \cos x, \quad f'(0) = 3, \quad f(0) = 0$

23. $f'(x) = x \sec x^2 \tan x^2, \quad f(\sqrt{\pi/4}) = 1$

24. $f''(x) = 2 \sec^2 x \tan x, \quad f'(0) = 2, \quad f(0) = 2$

25. Find an equation for the graph whose slope at any point is twice the x-coordinate of that point, and that contains the point (2, 9).

26. Find an equation for the graph containing the point (0, 1) whose slope at any point is the quotient of its x-coordinate by its y-coordinate.

27. From what height must a ball be dropped in order to strike the ground with a velocity of −49 m/s? (The acceleration due to gravity is −9.8 m/s².)

In Exercises 28–31, find the solution of the initial value problem.

28. $\dfrac{dy}{dx} = \sec^2 x, \quad y(0) = 1$

29. $\dfrac{dy}{dx} = \dfrac{x + 1}{y}, \quad y(1) = 2$

30. $dy = 2xy^2\, dx, \quad y(0) = -1/2$

31. $dy = \dfrac{\sqrt{x + 1}}{\sqrt{y}}\, dx, \quad y(0) = 1$

32. A particle moves along the s axis with velocity $v(t) = 2t - (t + 1)^{-2}$.
   a. Find s(t), its position at time t, if s(0) = 0.
   b. Find a(t), its acceleration at time t.

33. A particle moves along the s-axis according to the position function $s(t) = t^3 - 6t^2 + 9t - 4$.
   a. Where does the particle lie at time t = 0?
   b. In which direction is it moving at time t = 0?
   c. When does it change direction?
   d. How many times does it change direction?

34. The value of a certain piece of real estate in San Diego has been increasing at a rate of $V'(t) = \sqrt{36 + t}$ thousand dollars per year since 1975. If its value in 1985 was $200,000, find its value V(t), t years after 1985.

Integration

Unit Three

The Origins of Integration

Historically, the problem of finding the area of a region under a curve, which leads to the integral, was studied earlier and more extensively than the problem of finding the tangent to a curve, which leads to the derivative.

More than two millenia ago, the Greek mathematician Eudoxus (c. 408–355 B.C.) stated what is called the Method of Exhaustion. This principle says that if one successively subtracts from a quantity at least half of it, and from the remainder at least its half, and so on, eventually there will remain something smaller than any preassigned quantity. ("Quantity" may refer to a length, an area, or a volume.) As one example, Eudoxus applied this concept to achieve a difference between the area of a circle and that of an inscribed regular polygon. As the number of sides of the polygon becomes very large, the difference in areas becomes very small. Eudoxus eventually proved, from this result, that the ratio of the areas of two circles equals the ratio of the squares of their diameters. He did not actually determine a formula for the area of a circle, however. He also proved a similar theorem concerning the volumes of spheres. Archimedes wrote that Eudoxus was the first to prove that the volume of a cone is one-third the volume of a cylinder having the same base and altitude. The work of Eudoxus is contained in Book V of Euclid's *Elements*, and many applications are given in Book VI. Little is known of the life of Eudoxus, and none of his original work still exists. He was an authority in many fields, and wrote and lectured on astronomy, geography, music, medicine, and philosophy.

Archimedes of Syracuse (in Sicily)(287–212 B.C.) was the greatest mathematician of antiquity. He used a proof based on the Method of Exhaustion to determine an approximation to π, which he found to be between $3\frac{10}{71}$ and $3\frac{10}{70}$. He though his greatest discovery was that if a right circular cylinder is circumscribed about a sphere, then the area of the sphere is exactly two-thirds the area of the cylinder, and the volume of the sphere is also exactly two-thirds the volume of the cylinder. His desire was that this theorem be inscribed on his tombstone. Three hundred years after his death, when his tomb was long lost, the Roman statesman Cicero hunted for the tomb, and actually found the tombstone with that geometrical figure engraved on it.

Archimedes, using the Method of Exhaustion, came very close to inventing the calculus two thousand years before Newton. It was not until 1906 that the discovery of a manuscript in a Constantinople library showed how he had thought. This

Archimedes

manuscript, copied by a tenth-century scribe, contains several works of Archimedes; among them is a mathematical letter to his friend Eratosthenes, a librarian at Alexandria. The Archimedes material had been washed away in the thirteenth century so that the parchment might be reused for religious texts, but most of the earlier writing can be read. (Such a reused parchment is called a *palimpsest*.) Archimedes explained to his friend that, although the method did not constitute a rigorous proof, it indicated the correct results. He would cut an area or volume to be measured into infinitely many parallel lines or sections, which (in his imagination) he would place at one end of a lever so as to balance an area or volume at the other end whose measure and center of gravity were known. This method approximates the concept of limit and provides the correct results.

In the letter to Eratosthenes mentioned above, Archimedes wrote, "I do believe that men of my time and of the future, and

Bonaventura Cavalieri

through this method, might find still other theorems which have not yet come to my mind." One of these men of the future was Bonaventura Cavalieri (1598–1647). In 1653 he published *Geometria indivisibilibus* in which is introduced his Method of Indivisibles. It was a precursor to integration, and an extension of Archimedes' work. Cavalieri did not define his terms well, but apparently an indivisible of a planar region is a line across that region, and an indivisible of a solid is a section by a plane. The region or solid is made up of infinitely many parallel chords or sections. If these parallel lines (or planes) are slid into other conformations, the total area (or volume) is unchanged. This principle can be used in ingenious ways to find areas and volumes of complex structures. Cavalieri was a brilliant thinker who wrote on astronomy and optics as well as mathematics, and is credited with introducing logarithms in Italy. As a teenager studying geometry, he was so excellent a student that he occasionally substituted for his teacher in mathematics classes at a nearby university.

John Wallis (1616–1703) was a bright child but he showed little aptitude for mathematics. Indeed, mathematics was not even

John Wallis

a part of his grammar school curriculum, although he did learn Latin, Greek, and Hebrew. His acquaintance with mathematics came through reading books his brother had used when studying for a trade. At Cambridge University, mathematics "...was scarce looked upon...[as being]...in fashion." His interest in the subject came through his skill at deciphering captured letters in code written during the British civil war. He was appointed Professor of Geometry (i.e., of Mathematics) at Oxford, and continued in that post for an astonishing 54 years, until his death.

Wallis was one of the first to treat conics as equations of the second degree rather than as sections of a cone, and was the first to explain negative, fractional, and zero exponents in detail. He also created the symbol "∞" still used for infinity. Before Newton, he performed many definite integrations, essentially equivalent to $\int_0^1 x^m dx$, and his work with exponents allowed him to let m take on many values (except −1).

Wallis helped to found the Royal Society of London, and published for the first time since antiquity many editions of the writings of Greek mathematical and musical works. He was also one of the first to attempt to teach deaf mutes to speak. Wallis was very quarrelsome; modesty was not one of his virtues. He attacked violently (in print) some of the finest minds of his day.

During the eighteenth century, a vast amount of work was

Georg Friedrich Bernhard Riemann

Carl Friedrich Gauss

done on the development of the calculus. It was not, however, until the nineteenth century that the foundations of the subject were clearly stated. One of the most important contributors was the German mathematician Georg Friedrich Bernhard Riemann (1826–1866). It was he who finally put the integral onto a firm logical basis, treating it as the limit of upper and lower sums rather than solely as the inverse of differentiation. The modern presentation of the integral is mostly due to Riemann.

Riemann became a lecturer at the University of Göttingen at the age of 28. For his probationary lecture at the university, he had to present three possible titles; the first was almost always chosen. But Carl Friedrich Gauss selected the third title, and Riemann had to work frantically for two months to develop the material, about which he had previously done nothing. The result, however, was a triumph. Riemann's topic was the hypotheses that underlie the foundations of geometry; half a century later the concepts that he presented were used by Albert Einstein as the basis for the theory of general relativity. This lecture was one of the most influential mathematics lectures ever given.

Riemann suffered ill health, and died at only 39 in Italy of tuberculosis. Though he published relatively little, his mathematical influence is still felt in analysis, number theory, and geometry.

(Photographs: Archimedes, Culver Service; Bonaventura Cavalieri and Georg Friedrich Bernhard Riemann, David Eugene Papers, Rare Book and Manuscript Library, Columbia University; John Wallis and Carl Friedrich Gauss, the Library of Congress.)

307

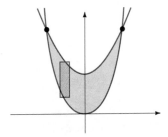

Chapter 6

The Definite Integral

In previous chapters, we used limits to describe the behavior of a function and to calculate its rate of change. Now we employ the limit concept to study an entirely different question: How do we define and calculate the area of a region in the plane? Over the centuries, the study of this question lead to the development of the *definite integral,* a generalized summation procedure with numerous applications in mathematics and the sciences. In this chapter, we present the theory of the definite integral as well as one of the most important theorems in all of mathematics—the Fundamental Theorem of Calculus. With the use of this remarkable result, we shall see how area problems that eluded the efforts of the greatest minds are easily solved with the use of the calculus.

Although we deal mainly with area problems in this chapter, the definite integral has many other applications. We shall discuss several in Chapter 7.

6.1 Area and Summation

The area of a polygonal region is often calculated by first subdividing the region into pieces with known areas, such as triangles or rectangles, and then summing the areas of these pieces. In this section, we generalize this approach for regions with curved boundaries, and we show how area can be computed as a limit of sums.

Consider the region R represented in Figure 1.1. Rather than trying to determine the area of R directly, we will use a version of a procedure used by the ancient Greeks to approximate the area A of R.

If we impose a grid consisting of one-centimeter squares upon the region R (see Figure 1.2) and count the number of squares that (a) lie entirely within R (in this case, none) and (b) contain some portion of R (in this case, eight), we obtain the estimate

$$0 \leq A \leq 8 \text{ cm}^2.$$

We can improve this first estimate by subdividing each square into four smaller squares, each of area 1/4 cm^2 (see Figure 1.3). Doing so, we observe (a) that 7 squares, each of area 1/4 cm^2, lie entirely within R and (b) that 26 squares contain some portion of R. Thus, we arrive at the improved estimate

$$\frac{7}{4} \text{ cm}^2 \leq A \leq \frac{26}{4} \text{ cm}^2.$$

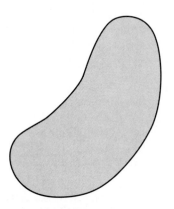

Figure 1.1 Region R.

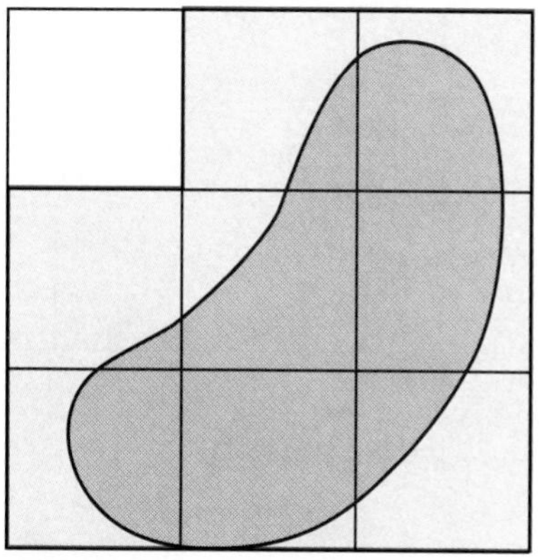

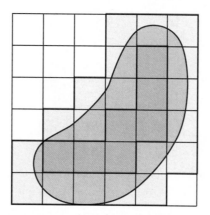

Figure 1.2 No square is contained in R; 8 squares together contain R.

Figure 1.3 7 squares contained in R; 26 squares together contain R.

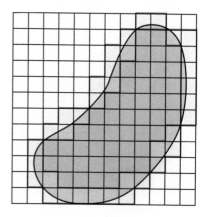

Figure 1.4   48 squares contained in R; 84 squares together contain R.

A further subdivision of the grid into squares of area $1/16$ cm^2 (see Figure 1.4) gives the estimate

$$\frac{48}{16} \text{ cm}^2 \le A \le \frac{84}{16} \text{ cm}^2.$$

Clearly, this procedure can be repeated as many times as we desire, with each halving of the grid dimensions yielding a new estimate. That is, each subdivision yields both a new lower bound and a new upper bound for the desired area A. As we repeat this subdivision procedure again and again, we intuitively see that the lower bounds and upper bounds tend toward the number A ''in the limit'' (see Figure 1.5).

The Area Problem

Using a modification of this procedure, we now define and calculate the area of a region determined by the graph of a nonnegative function. In other words, we address the following area problem.

Area Problem: Let f be a nonnegative continuous function for $x \in [a, b]$. Find the area of the region R bounded by the graph of f, the x-axis, and the lines $x = a$ and $x = b$.

Figure 1.6 represents a typical region R for the case of a nonnegative function f. Figure 1.7 shows our strategy for attacking the Area Problem. In order to define the area A, we approximate the region R with rectangles, and we execute our approxi-

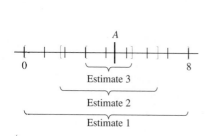

Figure 1.5 Three estimates of the desired area A.

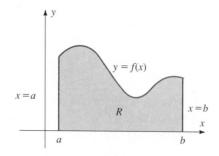

Figure 1.6 Region R bounded by graph of f and x-axis for $a \le x \le b$.

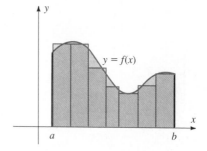

Figure 1.7 Approximating the region R by rectangles.

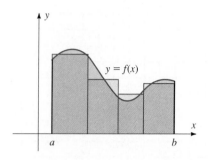
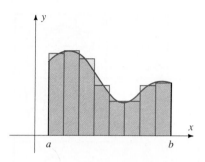
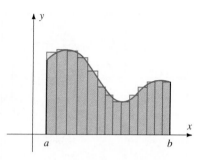

Figure 1.8 As the number n of subintervals increases, the approximation of R by rectangles becomes more precise.

mation scheme in such a way that we will be able to calculate the area A as the limit of this sequence of approximations.

Figure 1.8 illustrates the basic idea behind the approximation of R by rectangles. As the number of rectangles increases and the sizes of the individual rectangles decrease, the union of the set of approximating rectangles more accurately "fits" the region R. Our intention is to define the area A of R to be the limiting value of the areas associated with these approximations. Of course, we must first show that such a limit exists, and that is the purpose of this section.

Lower Approximating Sums

A **lower approximating sum** $\underline{S}_n$ for the region R is an approximation of the area of R by the combined areas of n rectangles of equal width *each of which is entirely contained within* R (see Figures 1.9 and 1.10). Since we use n rectangles of equal width, the width of each is

$$\Delta x = \frac{b - a}{n},$$

and the endpoints of the resulting subintervals are

$$x_0 = a, \ x_1 = a + \Delta x, \ x_2 = a + 2\,\Delta x, \ \dots, \ x_n = a + n\,\Delta x = b. \tag{1}$$

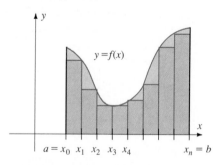

Figure 1.9 A lower approximating sum for R. Union of rectangles lies entirely within R.

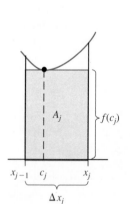

Figure 1.10 The jth approximating rectangle has area $\underline{A}_j = f(c_j)\,\Delta x$.

We want the height of the rectangle constructed over the interval $[x_{j-1}, x_j]$ to be the minimum value of f on $[x_{j-1}, x_j]$. If f is continuous on $[x_{j-1}, x_j]$, there is at least one number $c_j \in [x_{j-1}, x_j]$ with

$$f(c_j) = \min \{f(x) \mid x_{j-1} \le x \le x_j\}$$

(Theorem 1, Chapter 4). Using this notation we can write the area $\underline{A}_j$ of this jth *inscribed* rectangle as

$$\underline{A}_j = f(c_j)\,\Delta x.$$

The lower approximating sum $\underline{S}_n$ is therefore

$$\underline{S}_n = \underline{A}_1 + \underline{A}_2 + \underline{A}_3 + \cdots + \underline{A}_n$$
$$= f(c_1)\,\Delta x + f(c_2)\,\Delta x + f(c_3)\,\Delta x + \cdots + f(c_n)\,\Delta x.$$

☐ EXAMPLE 1

Find the lower approximating sum $\underline{S}_4$ for the area of the region R bounded by the graph of $f(x) = 4 - x^2$ and the x-axis between $x = 0$ and $x = 2$.

Solution Since $n = 4$, the length of each subinterval is

$$\Delta x = \frac{2 - 0}{4} = \frac{1}{2}.$$

The endpoints of the subintervals are therefore

$$x_0 = 0, \quad x_1 = \tfrac{1}{2}, \quad x_2 = 1, \quad x_3 = \tfrac{3}{2}, \quad \text{and} \quad x_4 = 2.$$

Since $f(x) = 4 - x^2$ is decreasing on $[0, 2]$, the minimum value of f on each subinterval occurs at the right endpoint (see Figure 1.11). Thus,

$$c_1 = \tfrac{1}{2}, \quad c_2 = 1, \quad c_3 = \tfrac{3}{2}, \quad \text{and} \quad c_4 = 2.$$

The lower approximating sum is, therefore,

$$\underline{S}_4 = f(\tfrac{1}{2})(\tfrac{1}{2}) + f(1)(\tfrac{1}{2}) + f(\tfrac{3}{2})(\tfrac{1}{2}) + f(2)(\tfrac{1}{2})$$
$$= [4 - (\tfrac{1}{2})^2](\tfrac{1}{2}) + [4 - (1)^2](\tfrac{1}{2}) + [4 - (\tfrac{3}{2})^2](\tfrac{1}{2}) + [4 - (2)^2](\tfrac{1}{2})$$
$$= [\tfrac{15}{4} + 3 + \tfrac{7}{4} + 0](\tfrac{1}{2}) = \tfrac{17}{4}. \qquad \blacksquare$$

A simple but important observation concerning lower sums is that, since each of the associated rectangles lies entirely within the region R, our definition of area should provide that

$$\underline{S}_n \le A \tag{2}$$

for all lower approximating sums $\underline{S}_n$.

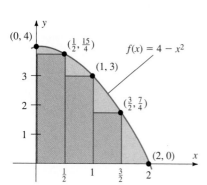

Figure 1.11 Lower approximating sum $\underline{S}_4$ in Example 1 for $f(x) = 4 - x^2$ on the interval $[0, 2]$.

☐ EXAMPLE 2

Find the lower approximating sum $\underline{S}_6$ for the area of the region R bounded by the graph of $\sin x$ and the x-axis between $x = 0$ and $x = \pi$.

Solution Since $n = 6$, the length of each subinterval is $\Delta x = (\pi - 0)/6 = \pi/6$, and the endpoints x_j of the subintervals are

$$x_0 = 0, \ x_1 = \pi/6, \ x_2 = \pi/3, \ x_3 = \pi/2, \ x_4 = 2\pi/3, \ x_5 = 5\pi/6, \ \text{and} \ x_6 = \pi.$$

To determine the c_j, we note that $\sin x$ is increasing on $[0, \pi/2]$ and decreasing on $[\pi/2, \pi]$. Therefore, for $j = 1, 2, 3$, the c_j are the left endpoints of the subintervals, and for $j = 4, 5, 6$, the c_j are the right endpoints of the subintervals (see Figure 1.12). In other words,

$$c_1 = 0, \ c_2 = \pi/6, \ c_3 = \pi/3, \ c_4 = 2\pi/3, \ c_5 = 5\pi/6, \ \text{and} \ c_6 = \pi.$$

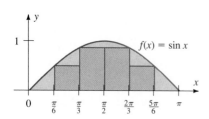

Figure 1.12 Lower approximating sum $\underline{S}_6$ in Example 2 for $f(x) = \sin x$ on the interval $[0, \pi]$.

The lower approximating sum $\underline{S}_6$ is

$$\underline{S}_6 = \sum_{j=1}^{6} \sin(c_j)\,\Delta x$$

$$= (0)(\pi/6) + (1/2)(\pi/6) + (\sqrt{3}/2)(\pi/6) +$$
$$(\sqrt{3}/2)(\pi/6) + (1/2)(\pi/6) + (0)(\pi/6)$$
$$= (1 + \sqrt{3})(\pi/6) \approx 1.4305. \qquad \blacksquare$$

Upper Approximating Sums

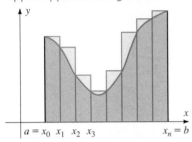

Figure 1.13   An upper approximating sum. The region R lies entirely within the union of the rectangles.

If, instead of using the minimum value of f on each subinterval $[x_{j-1}, x_j]$, we use the maximum value, we obtain what is called an **upper approximating sum** $\overline{S}_n$. That is, we take the height of the rectangle over the interval $[x_{j-1}, x_j]$ to be the value $f(d_j)$, where

$$f(d_j) = \max\,\{f(x)\mid x_{j-1} \le x \le x_j\}.$$

The area of the jth approximating rectangle is, therefore, $\overline{A}_j = f(d_j)\,\Delta x$, and the upper approximating sum is

$$\overline{S}_n = \overline{A}_1 + \overline{A}_2 + \overline{A}_3 + \cdots + \overline{A}_n$$
$$= f(d_1)\,\Delta x + f(d_2)\,\Delta x + f(d_3)\,\Delta x + \cdots + f(d_n)\,\Delta x.$$

(See Figures 1.13 and 1.14.)

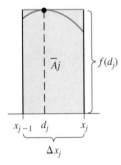

Figure 1.14   The jth approximating rectangle has area $\overline{A}_j = f(d_j)\,\Delta x$.

☐ **EXAMPLE 3**

Find the upper approximating sum $\overline{S}_4$ for the region R in Example 1.

Solution   As in Example 1, we have subintervals of length $\Delta x = 1/2$ and with endpoints

$$x_0 = 0, \quad x_1 = \tfrac{1}{2}, \quad x_2 = 1, \quad x_3 = \tfrac{3}{2}, \quad \text{and} \quad x_4 = 2.$$

However, since $f(x) = 4 - x^2$ is decreasing on $[0, 2]$, the maximum value of f will occur at the *left* endpoint of each subinterval. We will therefore have

$$d_1 = 0, \quad d_2 = \tfrac{1}{2}, \quad d_3 = 1, \quad \text{and} \quad d_4 = \tfrac{3}{2}.$$

The upper approximating sum $\overline{S}_4$ is

$$\overline{S}_4 = f(0)(\tfrac{1}{2}) + f(\tfrac{1}{2})(\tfrac{1}{2}) + f(1)(\tfrac{1}{2}) + f(\tfrac{3}{2})(\tfrac{1}{2})$$
$$= [4 - (0)^2](\tfrac{1}{2}) + [4 - (\tfrac{1}{2})^2](\tfrac{1}{2}) + [4 - (1)^2](\tfrac{1}{2}) + [4 - (\tfrac{3}{2})^2](\tfrac{1}{2})$$
$$= [4 + \tfrac{15}{4} + 3 + \tfrac{7}{4}](\tfrac{1}{2}) = \tfrac{25}{4}.$$

(See Figure 1.15.) $\qquad \blacksquare$

By combining Examples 1 and 3, we conclude that the area A of the region bounded by the graph of $f(x) = 4 - x^2$ and the x-axis between $x = 0$ and $x = 2$ is bounded by $\tfrac{17}{4} < A < \tfrac{25}{4}$.

Summation Notation

In order to compute the desired area precisely, we need to compute the limits of $\overline{S}_n$ and $\underline{S}_n$ as $n \to \infty$. Summation notation is useful in such calculations.

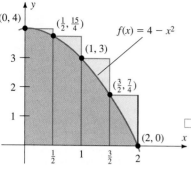

Figure 1.15 Upper approximating sum $\overline{S}_4$ in Example 3 for $f(x) = 4 - x^2$ on the interval $[0, 2]$.

The Greek letter Σ (capital sigma) is used to denote the sum of similar terms when each is a function of the summation index j:

$$\sum_{j=1}^{n} f(j) \quad \text{means} \quad f(1) + f(2) + f(3) + \cdots + f(n).$$

Here are some examples of this notation.

□ **EXAMPLE 4**

$$\sum_{j=1}^{4} (2j + 1) = [2(1) + 1] + [2(2) + 1] + [2(3) + 1] + [2(4) + 1]$$

$$= 3 + 5 + 7 + 9 = 24.$$ ∎

□ **EXAMPLE 5**

$$\sum_{j=1}^{3} (9 - j^2) = (9 - 1^2) + (9 - 2^2) + (9 - 3^2)$$

$$= 8 + 5 + 0 = 13.$$ ∎

□ **EXAMPLE 6**

$$\sum_{j=1}^{4} \sin\left(\frac{j\pi}{4}\right) = \sin\frac{\pi}{4} + \sin\frac{\pi}{2} + \sin\frac{3\pi}{4} + \sin\pi$$

$$= \frac{\sqrt{2}}{2} + 1 + \frac{\sqrt{2}}{2} + 0 = 1 + \sqrt{2}.$$ ∎

Sums indicated by sigma notation need not begin with the index $j = 1$. We may indicate any integer as the starting value of the index, with the understanding that the sum includes terms for each integer between and including the starting value and the terminal value. That is,

$$\sum_{j=k}^{k+p} f(j) \quad \text{means} \quad f(k) + f(k + 1) + f(k + 2) + \cdots + f(k + p).$$

Similarly, variables other than j are frequently used to represent the index.

□ **EXAMPLE 7**

$$\sum_{k=3}^{5} \cos(k\pi) = \cos(3\pi) + \cos(4\pi) + \cos(5\pi)$$

$$= -1 + 1 + (-1) = -1.$$ ∎

We shall need the following formulas, which are established in Appendix II using mathematical induction.

$$\sum_{j=1}^{n} c = c + c + c + \cdots + c = nc \tag{3}$$

$$\sum_{j=1}^{n} j = 1 + 2 + 3 + \cdots + n = \frac{n(n+1)}{2} \tag{4}$$

$$\sum_{j=1}^{n} j^2 = 1 + 4 + 9 + \cdots + n^2 = \frac{n(n+1)(2n+1)}{6} \tag{5}$$

$$\sum_{j=1}^{n} j^3 = 1 + 8 + 27 + \cdots + n^3 = \frac{n^2(n+1)^2}{4} \tag{6}$$

□ **EXAMPLE 8**

Using equation (4) we see that

(a) $1 + 2 + 3 + \cdots + 100 = \sum_{j=1}^{100} j = \frac{100(101)}{2} = 5050 \qquad (n = 100).$

(b) $33 + 34 + 35 + \cdots + 77 = \sum_{j=1}^{77} j - \sum_{j=1}^{32} j = \frac{77(78)}{2} - \frac{32(33)}{2}$

$$= 3003 - 528 = 2475.$$

Using equation (5) we find that

(c) $1 + 2^2 + 3^2 + \cdots + 9^2 = \sum_{j=1}^{9} j^2 = \frac{9(9+1)(2 \cdot 9 + 1)}{6} = 285.$ ■

In Exercise 29, we show how the usual algebraic properties of addition and subtraction as well as the Distributive Law are expressed using summation notation. These properties are used when we calculate sums such as the following example.

□ **EXAMPLE 9**

Determine

$$1 + 4 + 7 + 10 + \cdots + 301 = \sum_{j=1}^{101} (3j - 2).$$

Solution   We split up the sum and evaluate using equations (3) and (4):

$$\sum_{j=1}^{101} (3j - 2) = \sum_{j=1}^{101} 3j + \sum_{j=1}^{101} (-2)$$

$$= 3\sum_{j=1}^{101} j + \sum_{j=1}^{101}(-2)$$

$$= 3\left(\frac{101(102)}{2}\right) + 101(-2)$$

$$= 15{,}453 - 202 = 15{,}251.$$ ∎

The Solution to the Area Problem   Now we are able to complete our discussion of the area problem posed at the beginning of this section. We use limits of upper and lower approximating sums to define the area. Moreover, we use the summation formulas given above to calculate the area for certain functions.

Our next example illustrates the use of the summation formulas to calculate an upper sum with 100 subdivisions.

☐ **EXAMPLE 10**

Find the upper approximating sum $\overline{S}_{100}$ for the area of the region R bounded by the graph of $f(x) = x^2$ and the x-axis between $x = 1$ and $x = 3$.

Strategy · · · · · · · ·   **Solution**

Determine Δx.

To sum the areas of 100 rectangles efficiently, we use summation notation. First, we calculate $\Delta x = (3 - 1)/100 = 1/50$. Thus, when we subdivide the interval $[1, 3]$ into 100 subintervals, the endpoints are

Determine the x_j.

$$x_j = 1 + j(\Delta x) \quad \text{for} \quad j = 0, 1, 2, \dots, 100.$$

Determine the d_j.

Because we are calculating an upper sum and because f is increasing on $[1, 3]$, we use the right endpoints of the subintervals for the d_j. In other words, $d_j = x_j$. We obtain

Form the sum $\overline{S}_{100}$ and evaluate.

$$\overline{S}_{100} = \sum_{j=1}^{100} f(d_j)\,\Delta x$$

$$= \sum_{j=1}^{100} f(1 + j\,\Delta x)\,\Delta x$$

Substitute $f(x) = x^2$.

$$= \sum_{j=1}^{100}(1 + j\,\Delta x)^2\,\Delta x$$

Use summation properties.

$$= \sum_{j=1}^{100} \Delta x + \sum_{j=1}^{100} 2j(\Delta x)^2 + \sum_{j=1}^{100} j^2(\Delta x)^3$$

Use the summation formulas and the fact that $\Delta x = 1/50$.

$$= (100)\,\Delta x + 2\frac{(100)(101)}{2}(\Delta x)^2 + \frac{(100)(101)(201)}{6}(\Delta x)^3$$

$$= (100)\left(\frac{1}{50}\right) + 2\frac{(100)(101)}{2}\left(\frac{1}{50}\right)^2 + \frac{(100)(101)(201)}{6}\left(\frac{1}{50}\right)^3$$

$$= \frac{21{,}867}{2500} = 8.7468.$$ ∎

Since the region R is entirely contained within the union of the approximating rectangles for any upper sum $\overline{S}_n$, the area A of R should satisfy the inequality $A \leq \overline{S}_n$. Similarly, since the union of the approximating rectangles for any lower sum $\underline{S}_n$ is entirely contained within R, we should have $\underline{S}_n \leq A$. In other words, a definition of area should obey

$$\underline{S}_n \leq A \leq \overline{S}_n \tag{7}$$

A

$\underline{S}_n$ $\overline{S}_n$

Figure 1.16   The area A of R is the unique number satisfying $\underline{S}_n \leq A \leq \overline{S}_n$ for all lower and upper approximating sums.

for all lower sums $\underline{S}_n$ and all upper sums $\overline{S}_n$. Now, if both $\underline{S}_n$ and $\overline{S}_n$ approach the same limit* S as $n \to \infty$, we could both satisfy inequality (7) and obtain an unambiguous definition of the area A by taking $A = S$ (see Figure 1.16). Indeed, this is precisely what happens.

Before we summarize this discussion by stating the relevant theorem, we illustrate this limiting behavior in the following example.

□ **EXAMPLE 11**

Let R be the region bounded above by the graph of $f(x) = x^2$, below by the x-axis, on the left by $x = 0$, and on the right by $x = 1$. Show that the number $1/3$ satisfies the inequality

$$\underline{S}_n \leq \tfrac{1}{3} \leq \overline{S}_n$$

for all lower and upper approximating sums, and that

$$\lim_{n \to \infty} \underline{S}_n = \tfrac{1}{3} = \lim_{n \to \infty} \overline{S}_n.$$

Solution   To form the lower approximating sum for $f(x) = x^2$ on $[0, 1]$, we use n subintervals of equal size $\Delta x = 1/n$, and endpoints

$$x_0 = 0, \quad x_1 = \frac{1}{n}, \quad x_2 = \frac{2}{n}, \quad \ldots, \quad x_n = \frac{n}{n} = 1.$$

Since $f(x) = x^2$ is *increasing* on $[0, 1]$, the minimum value of f on each subinterval $[x_{j-1}, x_j]$ will occur at the *left* endpoint x_{j-1} (see Figure 1.17). That is, $c_j = x_{j-1} = (j - 1)/n$ for each $j = 1, 2, \ldots, n$. Thus,

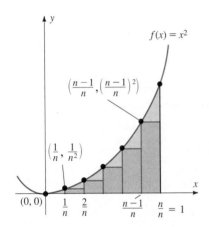

$$f(x) = x^2$$

$$\left(\frac{n-1}{n}, \left(\frac{n-1}{n}\right)^2\right)$$

$$\left(\frac{1}{n}, \frac{1}{n^2}\right)$$

$(0, 0)$ $\dfrac{1}{n}$ $\dfrac{2}{n}$ $\dfrac{n-1}{n}$ $\dfrac{n}{n} = 1$

Figure 1.17   The lower approximating sum $\underline{S}_n$ for $f(x) = x^2$ on $[0, 1]$ is

$$\underline{S}_n = \frac{1}{3} - \left(\frac{3n-1}{6n^2}\right).$$

$$\underline{S}_n = f(0) \cdot \frac{1}{n} + f\left(\frac{1}{n}\right) \cdot \frac{1}{n} + f\left(\frac{2}{n}\right) \cdot \frac{1}{n} + \cdots + f\left(\frac{n-1}{n}\right) \cdot \frac{1}{n}$$

$$= \left[0^2 + \left(\frac{1}{n}\right)^2 + \left(\frac{2}{n}\right)^2 + \cdots + \left(\frac{n-1}{n}\right)^2\right]\left(\frac{1}{n}\right)$$

$$= [1 + 2^2 + 3^2 + \cdots + (n - 1)^2]\left(\frac{1}{n^3}\right).$$

Using formula (5), we can write this sum as

$$\underline{S}_n = \left[\frac{(n - 1)[(n - 1) + 1][2(n - 1) + 1]}{6}\right]\left(\frac{1}{n^3}\right)$$

$$= \frac{(n - 1)(n)(2n - 1)}{6n^3}$$

*The limit as $n \to \infty$ of an approximating sum $\underline{S}_n$ or $\overline{S}_n$ means the same as the limit of $f(x)$ as $x \to \infty$ for a function f, except that the expressions $\underline{S}_n$ and $\overline{S}_n$ are functions defined only for positive integers n. The techniques for evaluating $\lim_{n \to \infty} \underline{S}_n$ and $\lim_{n \to \infty} \overline{S}_n$ are the same as those for evaluating $\lim_{x \to \infty} f(x)$.

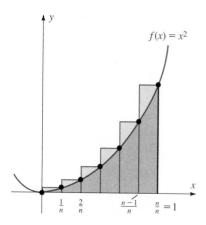

Figure 1.18   The upper approximating sum $\overline{S}_n$ for $f(x) = x^2$ on $[0, 1]$ is $\overline{S}_n = \frac{1}{3} + \left(\frac{3n + 1}{6n^2}\right).$

so

$$\underline{S}_n = \frac{2n^2 - 3n + 1}{6n^2} = \frac{1}{3} - \left(\frac{3n - 1}{6n^2}\right). \tag{8}$$

To obtain the upper approximating sum we use $d_j = x_j$ on the interval $[x_{j-1}, x_j]$, $j = 1, 2, \ldots, n$. That is, d_j is the *right* endpoint of the jth interval (see Figure 1.18). Thus,

$$\overline{S}_n = f\left(\frac{1}{n}\right) \cdot \frac{1}{n} + f\left(\frac{2}{n}\right) \cdot \frac{1}{n} + \cdots + f\left(\frac{n}{n}\right) \cdot \frac{1}{n}$$

$$= \left[\left(\frac{1}{n}\right)^2 + \left(\frac{2}{n}\right)^2 + \cdots + \left(\frac{n}{n}\right)^2\right]\left(\frac{1}{n}\right)$$

$$= [1 + 2^2 + 3^2 + \cdots + n^2]\left(\frac{1}{n^3}\right)$$

$$= \frac{n(n + 1)(2n + 1)}{6n^3} \quad \text{(using formula (5))}$$

$$= \frac{2n^2 + 3n + 1}{6n^2},$$

so

$$\overline{S}_n = \frac{1}{3} + \left(\frac{3n + 1}{6n^2}\right). \tag{9}$$

Now since the numbers $(3n - 1)/6n^2$ and $(3n + 1)/6n^2$ are positive for $n \geq 1$, we have from (8) and (9) that

$$\frac{1}{3} - \left(\frac{3n - 1}{6n^2}\right) = \underline{S}_n < \frac{1}{3} < \overline{S}_n = \frac{1}{3} + \left(\frac{3n + 1}{6n^2}\right) \tag{10}$$

for all $n \geq 1$. Moreover, the number $1/3$ is the *only* number that can occupy the middle position in inequality (10), since

$$\lim_{n \to \infty} \underline{S}_n = \lim_{n \to \infty}\left[\frac{1}{3} - \left(\frac{3n - 1}{6n^2}\right)\right] = \left(\frac{1}{3} - 0\right) = \frac{1}{3},$$

and

$$\lim_{n \to \infty} \overline{S}_n = \lim_{n \to \infty}\left[\frac{1}{3} + \left(\frac{3n + 1}{6n^2}\right)\right] = \left(\frac{1}{3} + 0\right) = \frac{1}{3}.$$

We therefore define the area of the region R to be the number $A = 1/3$.   ∎

The following theorem shows that the result of Example 11 is true for *any* continuous function.

Theorem 1

Let f be continuous and nonnegative on the interval $[a, b]$. Let $\underline{S}_n$ and $\overline{S}_n$ denote the lower and upper approximating sums for f on $[a, b]$. Then $\lim_{n \to \infty} \underline{S}_n$ and $\lim_{n \to \infty} \overline{S}_n$ exist, and

$$\lim_{n \to \infty} \underline{S}_n = \lim_{n \to \infty} \overline{S}_n$$

The proof of Theorem 1 is given in Appendix II. We shall focus here on the significance of this result.

Combining inequality (7) with the statement of Theorem 1, we conclude that the caption on Figure 1.16 is indeed correct: As $n \to \infty$, the lower approximating sums $\underline{S}_n$ and the upper approximating sums $\overline{S}_n$ limit to a single number A. We define this number A to be the area of the region R.

Definition 1

Let R be the region bounded above by the graph of the continuous nonnegative function f, below by the x-axis, on the left by $x = a$, and on the right by $x = b$. The **area** of R is the number A defined by the equations

$$A = \lim_{n \to \infty} \underline{S}_n = \lim_{n \to \infty} \overline{S}_n.$$

Although the idea behind approximating sums is straightforward, the calculations involved in executing a particular problem are both repetitious and time-consuming. This is precisely the kind of situation in which computers can be very helpful. Appendix I contains programs that compute lower and upper sums in various computational environments. Upper and lower approximating sums for the function $f(x) = 3x^2 + 7$ on the interval $[2, 3]$ are shown in Table 1.1.

Table 1.1 Lower and upper approximating sums for $f(x) = 3x^2 + 7$ on $[2, 3]$ obtained on a computer

n	$\underline{S}_n$	$\overline{S}_n$
2	22.3749	29.8750
5	24.5199	27.5200
10	25.2549	26.7550
25	25.7008	26.3008
100	25.9250	26.0750
500	25.9850	26.0150

Exercise Set 6.1

In Exercises 1–8, find the indicated sum.

1. $\displaystyle\sum_{j=1}^{10} j$

2. $\displaystyle\sum_{j=1}^{8} j^2$

3. $\displaystyle\sum_{j=1}^{12} j^3$

4. $\displaystyle\sum_{j=3}^{13} 4$

5. $\displaystyle\sum_{j=1}^{5} \sin\left(\frac{j\pi}{2}\right)$

6. $\displaystyle\sum_{j=1}^{6} \cos\left(\frac{j\pi}{3}\right)$

7. $\displaystyle\sum_{j=2}^{200} (-1)^j$

8. $\displaystyle\sum_{j=3}^{147} j$

In Exercises 9–12, rewrite the given sum in summation notation.

9. $2 + 4 + 6 + 8 + 10$

10. $5 + 8 + 11 + 14 + 17 + 20$

11. $4 + 9 + 16 + 25 + 36$

12. $8 + 18 + 32 + 50 + 72$

In each of Exercises 13–20, find the lower approximating sum $\underline{S}_n$ for the area of the region bounded by the graph of f, the x-axis, the line $x = a$, and the line $x = b$.

13. $f(x) = x$, $\quad a = 0$, $\quad b = 4$, $\quad n = 4$

14. $f(x) = 2x + 5$, $\quad a = 1$, $\quad b = 3$, $\quad n = 4$

15. $f(x) = 2x^2 + 3$, $\quad a = 0$, $\quad b = 4$, $\quad n = 4$

16. $f(x) = 9 - x^2$, $\quad a = -3$, $\quad b = 0$, $\quad n = 3$

17. $f(x) = \sin x$, $\quad a = 0$, $\quad b = \pi$, $\quad n = 4$

18. $f(x) = x^3 + x + 1$, $\quad a = 0$, $\quad b = 6$, $\quad n = 3$

19. $f(x) = 4x + 1$, $\quad a = 1$, $\quad b = 3$, $\quad n = 200$

20. $f(x) = 2 - x$, $\quad a = 0$, $\quad b = 2$, $\quad n = 100$

In each of Exercises 21–28, find the upper approximating sum $\overline{S}_n$ for the area of the region bounded by the graph of f, the x-axis, the line $x = a$, and the line $x = b$.

21. $f(x) = 2x$, $\quad a = 0$, $\quad b = 3$, $\quad n = 3$

22. $f(x) = 6 - x$, $\quad a = 0$, $\quad b = 3$, $\quad n = 6$

23. $f(x) = 3x^2 + 10$, $\quad a = -1$, $\quad b = 3$, $\quad n = 4$

24. $f(x) = 4 - x^2$, $\quad a = 0$, $\quad b = 2$, $\quad n = 4$

25. $f(x) = \cos x$, $\quad a = -\pi/2$, $\quad b = \pi/2$, $\quad n = 4$

26. $f(x) = x^3 + x + 1$, $\quad a = 0$, $\quad b = 6$, $\quad n = 3$

27. $f(x) = 4x - 1$, $\quad a = 1$, $\quad b = 4$, $\quad n = 300$

28. $f(x) = 2x - 1$, $\quad a = 1$, $\quad b = 3$, $\quad n = 200$

29. Explain, using the properties of addition and multiplication for real numbers, why each of the following properties is true:

a. $\displaystyle\sum_{j=1}^{n}(x_j + y_j) = \sum_{j=1}^{n} x_j + \sum_{j=1}^{n} y_j$

b. $\displaystyle\sum_{j=1}^{n} cx_j = c\sum_{j=1}^{n} x_j$

c. $\displaystyle\sum_{j=1}^{n}(x_j - y_j) = \sum_{j=1}^{n} x_j - \sum_{j=1}^{n} y_j$

d. $\displaystyle\sum_{j=1}^{n} c = nc.$

In Exercises 30–35, use properties a–d in Exercise 29 together with equations (3)–(6) to find the sum.

30. $\displaystyle\sum_{j=1}^{10} 2j$

31. $\displaystyle\sum_{j=1}^{7} 3j^2$

32. $\displaystyle\sum_{j=1}^{8} (j^2 + j + 1)$

33. $\displaystyle\sum_{j=1}^{6} (2j^2 + 3j + 5)$

34. $\displaystyle\sum_{j=1}^{n} (j^2 + 2j + 3)$

35. $\displaystyle\sum_{j=1}^{n} (6j^2 - 4j)$

In Exercises 36–40, find the positive integer n that yields the given sum.

36. $\displaystyle\sum_{j=1}^{n} j = 55$

37. $\displaystyle\sum_{j=4}^{n} 2j = 120$

38. $\displaystyle\sum_{j=0}^{n} (3j - 1) = 441$

39. $\displaystyle\sum_{j=n}^{15} (1 - 2j) = -209$

40. $\displaystyle\sum_{j=n}^{12} 6j^2 = 3870$

41. Find the sum of all of the odd integers between 12 and 244.

42. If you want to stack 105 logs in a triangular pile (see Figure 1.19 for a side view), how many logs should you place in the bottom row?

Figure 1.19 A side view of logs stacked in a triangular manner.

In Exercises 43–46, determine the specified limit.

43. $\displaystyle\lim_{n\to\infty} \frac{1}{n^2} \sum_{j=1}^{n} (2j + 1)$

44. $\displaystyle\lim_{n\to\infty} \frac{1}{n^2} \sum_{j=1}^{n} (3j - 2)$

45. $\displaystyle\lim_{n\to\infty} \frac{1}{n^3} \sum_{j=1}^{n} 6j^2$

46. $\displaystyle\lim_{n\to\infty} \frac{1}{n^4} \sum_{j=1}^{n} (3j^3 + 1)$

47. Let $f(x) = 3x + 2$.

a. Find the lower approximating sum $\underline{S}_4$ for f on $[0, 2]$ with $n = 4$ subintervals.

b. Find the lower approximating sum $\underline{S}_{100}$ for f on $[0, 2]$ with $n = 100$ subintervals.

c. Find an expression for the lower approximating sum $\underline{S}_n$ for f on $[0, 2]$ with n subintervals.

d. Using your answer in part c calculate the limit of $\underline{S}_n$ as $n \to \infty$.

e. Find the upper approximating sum $\overline{S}_{100}$ for f on $[0, 2]$ with $n = 100$ subintervals.

f. Find an expression for the upper approximating sum $\overline{S}_n$ for f on $[0, 2]$ with n subintervals.

g. Using your answer in part f, calculate the limit of $\bar{S}_n$ as $n \to \infty$.

h. What do you conclude about the area of the region bounded by the graph of f, the x-axis, the line $x = 0$, and the line $x = 2$?

48. Rework Exercise 47 for the function $f(x) = 8 - 2x$ and the interval $[0, 4]$.

49. Let $f(x) = x^2$.
   a. Let $b > 0$. Find the lower approximating sum $\underline{S}_n$ for f on the interval $[0, b]$.
   b. Using part a, calculate the limit of $\underline{S}_n$ as $n \to \infty$.
   c. Let $0 < a < b$. Find the lower approximating sum $\underline{S}_n$ for f on the interval $[a, b]$.
   d. Using part c, calculate the limit as $n \to \infty$ of the lower approximating sum $\underline{S}_n$ for f on $[a, b]$.
   e. How could you have answered part d using only the answer to part b and intuitive properties of area? (*Hint:* you can use part b to calculate the area of the region under the graph of f from $x = 0$ to $x = a$.)
   f. What is the area of the region bounded by the graph of f, the x-axis, the line $x = 1$, and the line $x = 5$?

50. Rework Exercise 47 the function $f(x) = 3x + 4$ and the interval $[1, 2]$.

51. Use the method of lower approximating sums to calculate the area of the region bounded above by the graph of $y = 3x^2 + 7$, on the left by $x = 2$, on the right by $x = 3$, and below by the x-axis. Compare your results with those of Table 1.1.

52. Rework Exercise 51 using upper approximating sums.

53. Rework Exercise 51 for the function $f(x) = 3x^2 + 6$ and the interval $[0, 1]$.

54. Rework Exercise 51 for the function $f(x) = 2x^2 + x + 3$ and the interval $[0, 2]$.

55. Rework Exercise 51 for the function $f(x) = |1 - x^2|$ and the interval $[0, 2]$.

56. Given a right triangle R with base b and height h, subdivide its base into n equal divisions and, using rectangles, approximate the area A of the region enclosed by R (see Figure 1.20).
   a. What is the width of each rectangle?
   b. What is the height of each rectangle?
   c. Show that the area of the kth rectangle from the left-hand side is $kbh/(n^2)$.
   d. Show that the total area A_n of the n rectangles is

$$A_n = \left(\frac{1}{2} + \frac{1}{2n} \right) bh.$$

   e. Calculate the limit of A_n as $n \to \infty$.

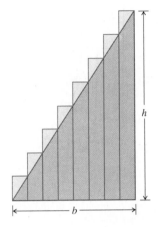

Figure 1.20 A right triangle of base b and height h whose area is approximated by n rectangles.

57. Approximate the area enclosed by the graph of the ellipse

$$\frac{x^2}{9} + \frac{y^2}{4} = 1$$

as follows (Figure 1.21).

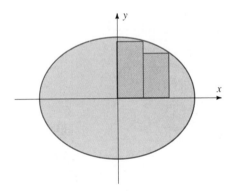

Figure 1.21

   a. Solve the equation for y to obtain

$$y = \pm 2\sqrt{1 - \frac{x^2}{9}}.$$

   Choose the $+$ sign to obtain the equation for the upper half of the figure.
   b. Observe that the total area will be four times the area of the portion of the ellipse lying in the first quadrant.
   c. Use your calculator and the function $f(x) = 2\sqrt{1 - x^2/9}$ to obtain the lower approximating sum $\underline{S}_3$. Multiply by four to estimate the total area.
   d. Obtain an improved estimate by using $n = 6$.

58. Consider the unit circle.
  a. Write down the equation for the function $y = f(x)$ that describes the top half of the graph.
  b. Use your calculator to obtain an estimate for the area of the portion of the circle lying in the first quadrant by calculating the lower approximating sum $\underline{S}_3$.
  c. Calculate $\underline{S}_6$ for the same region.
  d. Compare your estimates with what you know to be the exact value of that area.

59. Here is another procedure for approximating the area of a circle. We construct a regular polygon of n sides circumscribed by the circle of radius r (see Figure 1.22). If we construct radii from the center of the circle to each vertex of the polygon and subdivide each isosceles triangle into two congruent isosceles right triangles, we divide the polygon into $2n$ congruent right triangles.

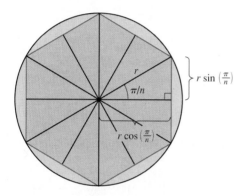

Figure 1.22   Circle of radius r. Approximation of area using inscribed polygon.

  a. Show that the base angle of each triangle is π/n.
  b. Show that the area of each triangle is

$$A_t = \frac{r^2}{4} \sin\left(\frac{2\pi}{n}\right).$$

  (*Hint:* Use the identity $\sin 2\theta = 2 \sin \theta \cos \theta$.)
  c. Show that the resulting approximation to the area of the circle is

$$A_n = \frac{nr^2}{2} \sin\left(\frac{2\pi}{n}\right).$$

  d. Complete Table 1.2 comparing the approximation obtained in part c with the actual value $A = \pi r^2$ for various values of n and r.

60. Here is a third way to approximate the area of a circle. This time we construct the regular polygon on n sides circumscribing the circle of radius r (Figure 1.23). We then construct line segments from the center of the circle to each vertex of the polygon and also construct radii joining the center of the circle with the midpoints of each of the sides of

Table 1.2

r	n	$A_n = \dfrac{nr^2}{2} \sin\left(\dfrac{2\pi}{n}\right)$	$A = \pi r^2$
1	5	2.37764	3.14159
1	20		3.14159
1	50	3.13333	3.14159
4	5		
4	20	49.44271	
4	50	50.13329	
4	200		
10	5	237.76413	
10	20	309.01699	
10	50	313.33308	
10	200		
10	500		
10	1000	314.15720	

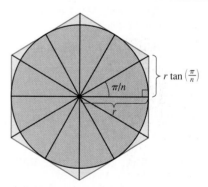

Figure 1.23   Circle of radius r. Approximation of area using circumscribed polygon.

the polygon. This divides the polygon into $2n$ congruent right triangles.
  a. Show that the area of each triangle is

$$A_t = \frac{r^2}{2} \tan\left(\frac{\pi}{n}\right).$$

  b. Show that the resulting approximation to the area of the circle is

$$A_n = nr^2 \tan\left(\frac{\pi}{n}\right).$$

  c. Construct a table like Table 1.2, using the approximation in part b.

61. In Exercises 59 and 60, we derived lower and upper estimates for the area A of a circle of radius r. Explain why

$$\frac{nr^2}{2}\sin\left(\frac{2\pi}{n}\right) < A < nr^2\tan\left(\frac{\pi}{n}\right)$$

for all positive integers n. Prove that $A = \pi r^2$ by taking the limit as $n \to \infty$. (*Hint:* Recall the limits of $(\sin u)/u$ and $(\tan u)/u$ as $u \to 0$.)

62. Verify that $2j - 1 = j^2 - (j-1)^2$. Use this observation to show that

$$\sum_{j=1}^{n}(2j-1) = n^2$$

without using formula (4). (*Hint:* Rewrite the sum in the $n = 4$ case using the formula $2j - 1 = j^2 - (j-1)^2$ on each of the four terms, then simplify. This case is typical of the general case.)

63. Using formula (3) and the result of Exercise 62, derive formula (4).

64. Note that $j^3 - (j-1)^3 = 3j^2 - 3j + 1$. Using this equality as well as formulas (3) and (4), derive formula (5).

65. Establish Lagrange's summation formula for cosine as follows:

a. Using the addition and subtraction identities for sine (Section 1.5), show that

$$\sin(\alpha + \beta) - \sin(\alpha - \beta) = 2\cos\alpha\sin\beta.$$

b. Using induction (Appendix II) and the identity in part a, establish Lagrange's identity

$$\sum_{j=0}^{n-1}\cos j\gamma = \frac{1}{2} + \frac{\sin(n - \frac{1}{2})\gamma}{2\sin\frac{1}{2}\gamma}$$

if $\sin\frac{1}{2}\gamma \ne 0$.

66. Determine a summation formula for

$$\sum_{j=1}^{n}\frac{1}{j(j+1)}.$$

(*Hint:* Note that $1/(j(j+1)) = (1/j) - [1/(j+1)]$.)

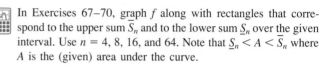 In Exercises 67–70, graph f along with rectangles that correspond to the upper sum $\overline{S}_n$ and to the lower sum $\underline{S}_n$ over the given interval. Use $n = 4, 8, 16$, and 64. Note that $\underline{S}_n < A < \overline{S}_n$ where A is the (given) area under the curve.

67. $f(x) = \frac{1}{2}x + 1$, $\quad -1 \le x \le 2$, $\quad A = 15/4$

68. $f(x) = 9 - x^2$, $\quad 0 \le x \le 3$, $\quad A = 18$

69. $f(x) = \sin^2(\pi x/2)$, $\quad 0 \le x \le 1$, $\quad A = 1/2$

70. $f(x) = \sqrt{x+1}$, $\quad -1 \le x \le 3$, $\quad A = 16/3$

71. Let

$$f(x) = \frac{4}{1+x^2}$$

for $x \in [0, 1]$. Find $\overline{S}_n$ and $\underline{S}_n$ for $n = 1, 2, 4, 8, 16, 32$, and 64. Estimate the rate at which the difference $\overline{S}_n - \underline{S}_n$ decreases as $n \to \infty$. Based on your conjecture, estimate n so that $(\overline{S}_n - \underline{S}_n) < 10^{-5}$. Approximate the area under this curve.

6.2 Riemann Sums: The Definite Integral

To solve the area problem stated in Section 6.1, we used limits of sums of the form

$$\sum_{j=1}^{n}f(t_j)\,\Delta x$$

that were either lower or upper approximating sums depending on the choice of the numbers t_j. In this section, we discuss limits of similar sums in a more general setting. That is, we consider all continuous functions, not just those that are nonnegative, and we are less restrictive in our choice of the numbers t_j and in the regularity of the subdivisions of the interval $[a, b]$. These generalizations are the basis of our definition of the definite integral.

At this point, the extra generality may not seem worth the effort, but in fact, just the opposite is true. The concept of the definite integral, as introduced in this section, is used repeatedly both in the sciences and in engineering. In Chapter 7, we discuss applications of the definite integral to the calculation of quantities as diverse as volume, work, hydrostatic pressure, and centers of gravity.

Partitions, Norms, and Riemann Sums

A **partition** P_n of the closed interval $[a, b]$ is any set of $n + 1$ numbers $\{x_0, x_1, x_2, \ldots, x_n\}$ with

$$a = x_0 < x_1 < x_2 < \cdots < x_n = b.$$

As you can see in Figure 2.1, a partition P_n divides the interval $[a, b]$ into n nonoverlapping subintervals.

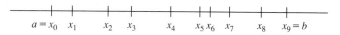

Figure 2.1 A partition P_9 for the interval $[a, b]$.

However, these subintervals are not necessarily of equal length. We define the **norm** of the partition P_n, denoted by $\|P_n\|$, to be the largest of these lengths. That is,

$$\|P_n\| = \max\{(x_1 - x_0), (x_2 - x_1), \ldots, (x_n - x_{n-1})\}.$$

Then

$$x_j - x_{j-1} \le \|P_n\|, \qquad j = 1, 2, 3, \ldots, n. \tag{1}$$

☐ EXAMPLE 1

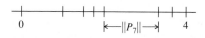

Figure 2.2 The partition P_7 in Example 1.

One partition of $[0, 4]$ is $P_7 = \{0, 1, \frac{3}{2}, \frac{7}{4}, 2, \frac{10}{3}, \frac{11}{3}, 4\}$. The norm of this partition is $\|P_7\| = \frac{10}{3} - 2 = \frac{4}{3}$ (see Figure 2.2). ■

If f is defined on $[a, b]$, we define a *Riemann sum** for f on $[a, b]$ in much the same way as we defined approximating sums in Section 6.1 except that we do not require the subintervals to be of equal length. Nor do we require $f(x) \ge 0$.

Definition 2

A **Riemann sum** R_n for f on $[a, b]$ is any sum of the form

$$R_n = \sum_{j=1}^{n} f(t_j) \, \Delta x_j, \qquad t_j \in [x_{j-1}, x_j]$$

where $\{x_0, x_1, x_2, \ldots, x_n\}$ is a partition of $[a, b]$, $\Delta x_j = x_j - x_{j-1}$, and t_j is an element of $[x_{j-1}, x_j]$, $j = 1, 2, \ldots, n$.

As you can see, the notation R_n for a particular Riemann sum does not indicate all of the choices involved. To be absolutely precise, we should designate the partition P as well as the choices of the numbers t_j within the subintervals determined by P. However, this simplified notation usually suffices.

Figure 2.3 provides a geometric interpretation of a Riemann sum for a partition with six subintervals. Note that, since the values $f(t_j)$ need not be extreme values of f on $[x_{j-1}, x_j]$, we do not know the exact relationship between the combined areas of the rectangles and the area of the region bounded by the x-axis and the graph of f on $[a, b]$. In fact, since we have dropped the condition that $f(x) \ge 0$ for all $x \in [a, b]$, any geometric interpretation of a Riemann sum must allow for the fact that

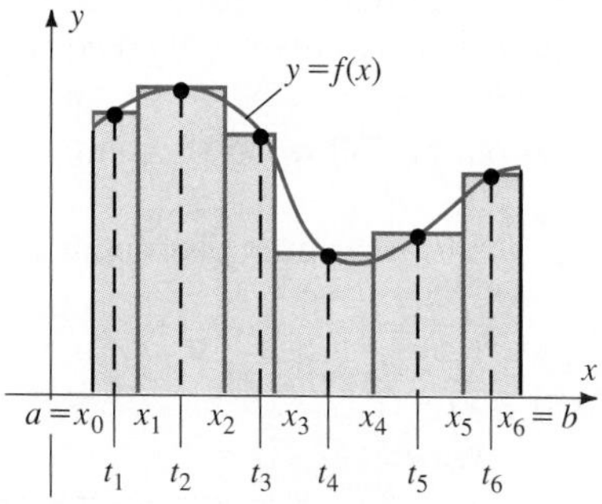

Figure 2.3 A Riemann sum R_6.

---

*Riemann sums and most of what is presented here as the theory of the definite integral are due to the French mathematician Augustin-Louis Cauchy (1826–1866) and the German mathematician Bernhard Riemann, 1826–1866.

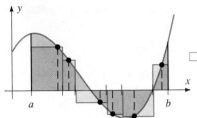

Figure 2.4 A Riemann sum with six terms, three of which are positive and three are negative. The blue rectangles represent the positive terms, and the red rectangles represent the negative terms.

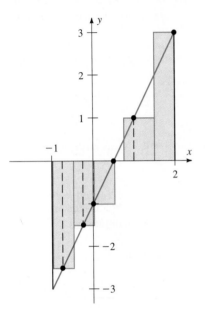

Figure 2.5 A geometric interpretation of the Riemann sum computed in Example 2.

$f(t_j)$ may be negative (see Figure 2.4). Nevertheless, it is important to note that every approximating sum is a Riemann sum, but the latter concept is more general.

□ EXAMPLE 2

Let $f(x) = 2x - 1$ and P_6 be the partition $\{-1, -\frac{1}{2}, 0, \frac{1}{2}, \frac{3}{4}, \frac{3}{2}, 2\}$ of the interval $[-1, 2]$. Determine the Riemann sum corresponding to the choices:

$$t_1 = -\frac{3}{4},\ t_2 = -\frac{1}{4},\ t_3 = 0,\ t_4 = \frac{1}{2},\ t_5 = 1,\ t_6 = 2.$$

Solution First, we calculate the lengths of the subintervals determined by the partition. The first three subintervals and the sixth subinterval each have length $\frac{1}{2}$. The remaining two lengths are $\Delta x_4 = \frac{1}{4}$ and $\Delta x_5 = \frac{3}{4}$. Since the t_j are given, we can evaluate the desired sum.

$$\begin{aligned}
R_6 &= \sum_{j=1}^{6} f(t_j)\,\Delta x_j \\
&= f(t_1)(\tfrac{1}{2}) + f(t_2)(\tfrac{1}{2}) + f(t_3)(\tfrac{1}{2}) + f(t_4)(\tfrac{1}{4}) + f(t_5)(\tfrac{3}{4}) + f(t_6)(\tfrac{1}{2}) \\
&= (-\tfrac{5}{2})(\tfrac{1}{2}) + (-\tfrac{3}{2})(\tfrac{1}{2}) + (-1)(\tfrac{1}{2}) + (0)(\tfrac{1}{4}) + (1)(\tfrac{3}{4}) + (3)(\tfrac{1}{2}) \\
&= -\tfrac{1}{4}
\end{aligned}$$

(see Figure 2.5). ■

As in Section 6.1, we shall use Riemann sums to calculate areas (at least in principle). But in generalizing the notion of approximating sum we have introduced a minor difficulty. We must now be more careful about speaking of the limit of a sequence $\{R_n\}$ of Riemann sums as $n \to \infty$. The reason is illustrated by Figure 2.6.

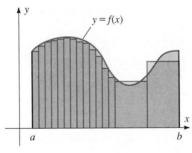

Figure 2.6 As $n \to \infty$, the widths of some rectangles may remain large.

Since the lengths of the subintervals in the partition P_n are not necessarily equal, the lengths of all subintervals need not necessarily become small as n becomes large. If we want this to happen (as we do when approximating regions by rectangles) we must also specify that the P_n are partitions for which $\|P_n\| \to 0$ as $n \to \infty$. By inequality (1) this guarantees that $\Delta x_j \to 0$ for all $j = 1, 2 \ldots, n$.

Our definition of the definite integral is based on the following theorem.

Theorem 2

Suppose f is a continuous function on $[a, b]$. Then there exists a unique number I such that

$$I = \lim_{n \to \infty} R_n = \lim_{n \to \infty} \sum_{j=1}^{n} f(t_j)\,\Delta x_j$$

for all Riemann sums R_n corresponding to partitions P_n for which $\|P_n\| \to 0$ as $n \to \infty$.

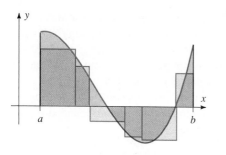
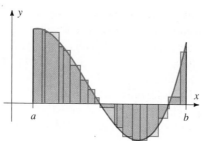
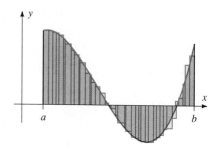

Figure 2.7 As $\|P_n\| \to 0$, the Riemann sums R_n approach a unique number.

In other words, as long as we consider partitions P_n such that $\|P_n\| \to 0$ as $n \to \infty$, the corresponding Riemann sums R_n limit to the same number I *independent* of the way in which we choose the numbers $t_j \in [x_{j-1}, x_j]$ (see Figure 2.7).

Although a precise proof of this result is beyond the scope of this text, the basic ideas in the proof are the same as those given in Appendix II for the proof of Theorem 1.

Definition 3

Let f be continuous on $[a, b]$. The number I designated in Theorem 2 is called the **definite integral** of f from a to b and is denoted

$$\int_a^b f(x)\, dx.$$

The symbol $\int_a^b f(x)\, dx$ is read, "the definite integral from a to b of f with respect to x." As with the symbol $\int f(x)\, dx$ for the collection of all antiderivatives of f, we refer to $\int$ as the **integral sign.** We write the symbol dx following the **integrand** $f(x)$ to indicate that x is the independent variable for f. (We shall later see that the symbol dx has a meaning associated with the differential dx, as suggested by the Riemann sum. But for now simply regard dx as part of the notation identifying the definite integral.) The endpoints a and b are referred to as the **limits** of integration, and their presence at the extremities of the integral sign distinguishes the definite integral $\int_a^b f(x)\, dx$ from the indefinite integral $\int f(x)\, dx$.

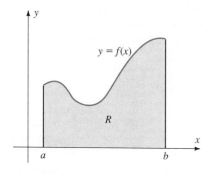

Figure 2.8 Area of $R = \int_a^b f(x)\, dx$ when $f(x) \geq 0$ for all $x \in [a, b]$.

REMARK It is very important to note that the definite integral $\int_a^b f(x)\, dx$ is a *number*, while the indefinite integral $\int f(x)\, dx$ is a family of *functions* (all antiderivatives F of f). That is,

$$\int_a^b f(x)\, dx \in (-\infty, \infty), \quad \text{while} \quad \int f(x)\, dx = F(x) + C$$

where $F'(x) = f(x)$. Although this is a point of possible confusion now, the results of Section 6.3 will reveal a strong relationship between definite integrals and antiderivatives, justifying this similarity of notation.

The Definite Integral and Area

At this point it is important to note that, while approximating sums for areas were defined in Section 6.1 only for nonnegative functions, *our definitions of Riemann sums and the definite integral $\int_a^b f(x)\, dx$ do not require that f be a nonnegative function, nor does Theorem 2*. Thus, the definite integral $\int_a^b f(x)\, dx$ exists whenever f is continuous on $[a, b]$, regardless of the signs of the function values $f(x)$.

The following theorem states, however, that if *f is nonnegative and continuous* on $[a, b]$, the definite integral $\int_a^b f(x)\, dx$ is the area of the region bounded by the graph of f and the x-axis. (See Figure 2.8 on page 325.)

Theorem 3

Let f be continuous on $[a, b]$ with $f(x) \geq 0$ for all $x \in [a, b]$. Then the area A of the region R bounded above by the graph of f, below by the x-axis, on the left by $x = a$, and on the right by $x = b$ is given by the definite integral

$$A = \int_a^b f(x)\, dx.$$

Proof: According to Definition 1, the area A is defined to be the limit

$$A = \lim_{n \to \infty} \sum_{j=1}^n f(t_j)\, \Delta x, \qquad t_j \in [x_{j-1}, x_j] \tag{2}$$

of a sequence of approximating sums for the region R. Since every approximating sum is also a Riemann sum, we may combine Definition 3 with equation (2) to conclude that

$$A = \lim_{n \to \infty} \sum_{j=1}^n f(t_j)\, \Delta x = \int_a^b f(x)\, dx. \qquad \blacksquare$$

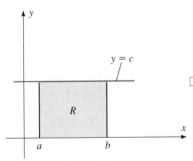

Figure 2.9 Area of R is $A = \int_a^b c\, dx = c(b - a)$.

☐ **EXAMPLE 3**

If the function f is nonnegative we are sometimes able to evaluate $\int_a^b f(x)\, dx$ by identifying the integral with a region whose area we already know.

(a) Figure 2.9 shows that the graph of the constant function $f(x) \equiv c, c > 0$, bounds a rectangle of area $c(b - a)$ over the interval $[a, b]$. Thus,

$$\int_a^b c\, dx = c(b - a), \qquad c > 0.$$

(b) Figure 2.10 shows that the graph of the function $f(x) = x$ bounds a trapezoid over the interval $[a, b]$ if $0 < a < b$. Since this trapezoid has bases of length $B_1 = f(a) = a$ and $B_2 = f(b) = b$, and altitude $h = b - a$, its area is $A = \frac{1}{2}(B_1 + B_2)h = \frac{1}{2}(a + b)(b - a) = \frac{1}{2}(b^2 - a^2)$. Thus,

$$\int_a^b x\, dx = \frac{1}{2}(b^2 - a^2), \qquad 0 < a < b.$$

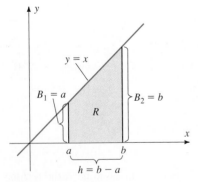

Figure 2.10 Area of R is $A = \int_a^b x\, dx = \frac{1}{2}(b^2 - a^2)$.

(c) Figure 2.11 shows that the graph of the function $f(x) = \sqrt{a^2 - x^2}$ bounds a semicircle of radius $r = a$ over the interval $[-a, a]$ when $a > 0$. Since the area of this semicircle is $A = \frac{1}{2}\pi r^2 = \frac{1}{2}\pi a^2$, we have

$$\int_{-a}^a \sqrt{a^2 - x^2}\, dx = \frac{1}{2}\pi a^2. \qquad \blacksquare$$

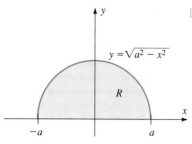

Figure 2.11   Area of R is $A = \int_{-a}^{a} \sqrt{a^2 - x^2}\, dx = \frac{1}{2}\pi a^2$.

☐ **EXAMPLE 4**

In Example 11 of Section 6.1, we determined that the area of the region bounded above by the graph of $f(x) = x^2$, below by the x-axis, on the left by $x = 0$, and on the right by $x = 1$ is $1/3$. Consequently, we have

$$\int_{0}^{1} x^2\, dx = \tfrac{1}{3}.$$ ∎

If $f(x)$ is negative for some $x \in [a, b]$, then a Riemann sum for f may contain terms of the form $f(t_j)\,\Delta x_j$ where $f(t_j) < 0$ (see Example 2 and Figures 2.4 and 2.5). Since Δx_j is positive, the product $f(t_j)\,\Delta x_j$ is negative. Consequently, when we interpret a term in a Riemann sum as the area of a rectangle, we do so with the understanding that a rectangle that is below the x-axis contributes a negative number to the sum. Thus, for a general continuous function f, we can interpret $\int_{a}^{b} f(x)\, dx$ as a "signed" area. In the next two examples, we calculate two definite integrals directly from the definition. In both cases, we can use geometry formulas to confirm that the result is a signed area.

☐ **EXAMPLE 5**

Suppose f is a constant function on the closed interval $[a, b]$, that is, $f(x) = c$ for all $x \in [a, b]$. Show that

$$\int_{a}^{b} f(x)\, dx = c(b - a).$$

Solution   Let P_n be any partition of $[a, b]$. For any Riemann sum R_n formed using P_n, we have

$$R_n = \sum_{j=1}^{n} f(t_j)\,\Delta x_j = \sum_{j=1}^{n} c\,\Delta x_j = c\sum_{j=1}^{n} \Delta x_j, \tag{3}$$

because f is constant. Recall that $\Delta x_j = x_j - x_{j-1}$, so

$$\sum_{j=1}^{n} \Delta x_j = (x_1 - x_0) + (x_2 - x_1) + (x_3 - x_2) + \cdots + (x_n - x_{n-1})$$

$$= x_n - x_0. \tag{4}$$

But $x_0 = a$ and $x_n = b$ for any partition P_n. Thus equations (3) and (4) yield

$$R_n = c(b - a). \tag{5}$$

Since equation (5) holds for *all* Riemann sums, the limit of R_n as $n \to \infty$ must be $c(b - a)$, and therefore, $\int_{a}^{b} f(x)\, dx = c(b - a)$.   ∎

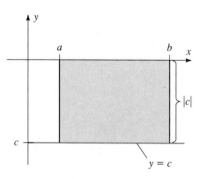

Figure 2.12   For $c < 0$, the integral $\int_{a}^{b} c\, dx$ equals the negative of the area of the rectangle bounded by the x-axis and the lines $y = c$, $x = a$, and $x = b$.

If $c > 0$ in Example 5, we have simply reproduced the result of Example 3(a). If $c < 0$, then $\int_{a}^{b} c\, dx = c(b - a) < 0$, and the area of the rectangle R bounded below by the graph of f, above by the x-axis, on the left by $x = a$, and on the right by $x = b$ is $|c|(b - a) = -\int_{a}^{b} c\, dx$ (see Figure 2.12).

☐ **EXAMPLE 6**

In Exercise 55, we use the methods of Section 6.1 to extend the result of Example 3(b) to the case where $a < 0$. We obtain

$$\int_{a}^{b} x\, dx = \tfrac{1}{2}(b^2 - a^2). \tag{6}$$

In this example, we interpret this formula geometrically in the case where $a < 0 < b$.

The region bounded by the graph of $f(x) = x$, the x-axis, the line $x = a$, and the line $x = b$ consists of two isosceles right triangles T_1 and T_2 (see Figure 2.13). Since $b > 0$, the area of T_2 is $\frac{1}{2}(b)(b) = \frac{1}{2}b^2$. However, since $a < 0$, the area of T_1 is $\frac{1}{2}|a||a| = \frac{1}{2}a^2$. If we compute the difference of the area above the x-axis and the area below the x-axis, we obtain

$$\text{area}(T_2) - \text{area}(T_1) = \frac{1}{2}(b^2 - a^2).$$

Thus, we can interpret equation (6) as a signed area. ∎

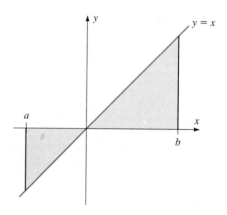

Figure 2.13 Two right triangles determined by the graph of $f(x) = x$ on the interval $[a, b]$ if $a < 0 < b$.

Extending the Definition of $\int_a^b f(x)\, dx$

Since $\int_a^b f(x)\, dx$ has been defined for *intervals* $[a, b]$, it is implicit in Definition 3 that $a < b$. We will also work with definite integrals with $a \geq b$, which we now define.

Definition 4

Let f be continuous on $[a, b]$. Then

(a) $\displaystyle\int_a^a f(x)\, dx = 0$

(b) $\displaystyle\int_b^a f(x)\, dx = -\int_a^b f(x)\, dx.$

These definitions have simple geometric interpretations. Statement (a) says that a region of zero width and finite height has zero area, at least if $f(x) \geq 0$. For (b), note that if $a < b$, the integral on the right is "from left to right." Since the integral on the left is then "from right to left," we are reversing the direction in which we march along the number line adding up terms in the Riemann sums. The effect is to change the sign on each term $\Delta x_j = (x_j - x_{j-1})$, leaving $f(t_j)$ unchanged, and therefore, changing the sign on the Riemann sums associated with $\int_a^b f(x)\, dx$. We may paraphrase (b) by saying that "reversing the direction of integration changes the sign of the integral."

Properties of Definite Integrals

We shall make frequent use of two theorems giving properties of definite integrals. The first states that definite integrals share two properties with derivatives and antiderivatives: The definite integral of a sum of two functions equals the sum of the definite integrals of the individual functions, and the definite integral of a multiple of a function equals that multiple of the definite integral of the function.

Theorem 4

Let f and g be continuous on $[a, b]$ and let c be any constant. Then

(i) $\displaystyle\int_a^b [f(x) + g(x)] \, dx = \int_a^b f(x) \, dx + \int_a^b g(x) \, dx$

(ii) $\displaystyle\int_a^b cf(x) \, dx = c \int_a^b f(x) \, dx.$

We shall not give a rigorous proof of Theorem 4, but here is the basic idea. If $\sum_{j=1}^n f(t_j) \Delta x_j$ and $\sum_{j=1}^n g(t_j) \Delta x_j$ are Riemann sums for f and g, respectively, on the interval $[a, b]$ *with the same partition P*, and the same numbers t_j in each subinterval, then $\sum_{j=1}^n [f(t_j) + g(t_j)] \Delta x_j$, with the same partition P and numbers t_j, is a Riemann sum for the function $f + g$ on $[a, b]$. Since

$$\sum_{j=1}^n [f(t_j) + g(t_j)] \Delta x_j = \sum_{j=1}^n f(t_j) \Delta x_j + \sum_{j=1}^n g(t_j) \Delta x_j$$

the limit as $n \to \infty$ of the sum on the left will exist and equation (i) will result from taking limits as $n \to \infty$ of each of these sums. The argument for equation (ii) is similar.

☐ **EXAMPLE 7**

From Examples 3 and 4 we know that, for $c > 0$,

$$\int_0^1 c \, dx = c, \quad \int_0^1 x \, dx = \frac{1}{2}, \quad \text{and} \quad \int_0^1 x^2 \, dx = \frac{1}{3}.$$

From these facts and Theorem 4 we may conclude that

(a) $\displaystyle\int_0^1 (x + x^2) \, dx = \int_0^1 x \, dx + \int_0^1 x^2 \, dx$ (Property (i))

$$= \frac{1}{2} + \frac{1}{3} = \frac{5}{6}.$$

(b) $\displaystyle\int_0^1 6x^2 \, dx = 6 \int_0^1 x^2 \, dx$ (Property (ii))

$$= 6\left(\frac{1}{3}\right) = 2.$$

(c) $\displaystyle\int_0^1 [3x^2 - 6x + 7] \, dx = \int_0^1 3x^2 \, dx + \int_0^1 (-6x) \, dx + \int_0^1 7 \, dx$

$$= 3 \int_0^1 x^2 \, dx - 6 \int_0^1 x \, dx + \int_0^1 7 \, dx$$

$$= 3\left(\frac{1}{3}\right) - 6\left(\frac{1}{2}\right) + 7 = 5. \qquad \blacksquare$$

The following theorem states that a definite integral may be "split" into the sum of two (or more) definite integrals as follows.

Theorem 5

Let f be continuous on an interval containing the numbers a, b, and c. Then

$$\int_a^b f(x) \, dx = \int_a^c f(x) \, dx + \int_c^b f(x) \, dx$$

regardless of how the numbers a, b, and c are ordered.

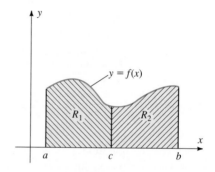

Figure 2.14
$$\int_a^b f(x) \, dx = \text{Area of } R_1 \cup R_2$$

$$= \text{Area of } R_1 + \text{Area of } R_2$$

$$= \int_a^c f(x) \, dx + \int_a^b f(x) \, dx.$$

Figure 2.14 illustrates the statement of Theorem 5 when f is nonnegative on $[a, b]$ and $a < c < b$. In this case the result is obvious from area considerations. Theorem 5 is true, however, whether or not f is nonnegative and whether or not c lies between a and b.

Proof of Theorem 5:   When c lies between a and b the idea behind the proof of Theorem 5 is to require that c be an endpoint in all partitions in a sequence R_n of Riemann sums with limit $\int_a^b f(x) \, dx$. Write $n = m + p$ where m is the number of subintervals in $[a, c]$ and p is the number of subintervals in $[c, b]$. Use the notation $z_k = x_{m+k}$ for the x_j's to the right of c and the notation $u_k = t_{m+k}$ for the t_j's to the right of c. Then

$$\int_a^b f(x) \, dx = \lim_{n \to \infty} \sum_{j=1}^{n} f(t_j) \, \Delta x_j$$

$$= \lim_{n \to \infty} \left[\sum_{j=1}^{m} f(t_j) \, \Delta x_j + \sum_{k=1}^{p} f(u_k) \, \Delta z_k \right] \qquad (n = m + p)$$

$$= \lim_{m \to \infty} \sum_{j=1}^{m} f(t_j) \, \Delta x_j + \lim_{p \to \infty} \sum_{k=1}^{p} f(u_k) \, \Delta z_k$$

$$= \int_{a}^{c} f(x) \, dx + \int_{c}^{b} f(x) \, dx.$$

This is messy, but it is important that you appreciate why we need to allow partitions with unequal subinterval lengths. If all subintervals were of equal length, we could not guarantee that c would correspond to an endpoint, so we could not ''break the sum into two pieces at $x = c$'' as we have done here.

We will next consider the case $c < a < b$, leaving the other four possible orderings for you (Exercise 63).

Since $c < a < b$, we have already explained why

$$\int_{c}^{b} f(x) \, dx = \int_{c}^{a} f(x) \, dx + \int_{a}^{b} f(x) \, dx. \tag{7}$$

But, $\int_{c}^{b} f(x) \, dx = -\int_{a}^{c} f(x) \, dx$ by Definition 4. Solving equation (7) for $\int_{a}^{b} f(x) \, dx$ and using this fact gives

$$\int_{a}^{b} f(x) \, dx = -\int_{c}^{a} f(x) \, dx + \int_{c}^{b} f(x) \, dx = \int_{a}^{c} f(x) \, dx + \int_{c}^{b} f(x) \, dx, \qquad \blacksquare$$

as claimed.

□ **EXAMPLE 8**

Given that $\int_{0}^{2} f(x) \, dx = 4$, $\int_{2}^{3} f(x) \, dx = 2$, and $\int_{2}^{5} f(x) \, dx = 10$ we may conclude from Definition 4 and Theorem 5 that

(a) $\displaystyle\int_{0}^{5} f(x) \, dx = \int_{0}^{2} f(x) \, dx + \int_{2}^{5} f(x) \, dx = 4 + 10 = 14.$

(b) $\displaystyle\int_{5}^{0} f(x) \, dx = -\int_{0}^{5} f(x) \, dx = -14.$

(c) $\displaystyle\int_{3}^{5} f(x) \, dx = \int_{2}^{5} f(x) \, dx - \int_{2}^{3} f(x) \, dx = 10 - 2 = 8.$ $\qquad \blacksquare$

We can use Theorem 5 to justify our geometric interpretation of the definite integral as a signed area (at least for many ''well behaved'' functions). For example, let $a < c < b$ and suppose that f is a continuous function on $[a, b]$ that is positive on $[a, c)$ and negative on $(c, b]$ (see Figure 2.15). Then the graph of f determines two

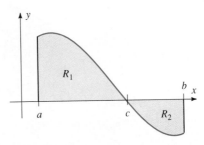

Figure 2.15 The two regions R_1 and R_2 determined by the graph of a function that is continuous on the interval $[a, b]$ and that crosses the x-axis exactly once in $[a, b]$.

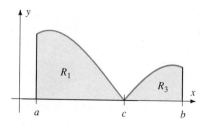

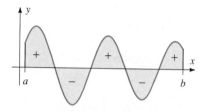

Figure 2.16 The two regions R_1 and R_3 determined by the graph of the nonnegative function that results from taking the absolute value of the function graphed in Figure 2.15.

Figure 2.17 The definite integral $\int_a^b f(x)\,dx$ can be interpreted as the difference between the total area above the x-axis and the total area below the x-axis.

regions R_1 and R_2. The region R_1 is bounded by the graph, the interval $[a, c]$, and the line $x = a$, and R_2 is bounded by the graph, the interval $[c, b]$, and the line $x = b$. Similarly, consider the graph of $|f|$. It also determines two regions—R_1 and R_3 where R_3 is the reflection of R_2 through the x-axis (see Figure 2.16). Using Theorems 3 and 5, we have

$$\int_a^b f(x)\,dx = \int_a^c f(x)\,dx + \int_c^b f(x)\,dx$$

$$= \text{area}(R_1) + (-1)\int_c^b |f(x)|\,dx \qquad (8)$$

because $|f(x)| = -f(x)$ for $x \in [c, b]$. We apply Theorem 3 to $|f|$ on the interval $[c, b]$ to obtain $\int_c^b |f(x)|\,dx = \text{area}(R_3)$. Since R_3 is the reflection of R_2 through the x-axis, we know that $\text{area}(R_3) = \text{area}(R_2)$. Thus, equation (8) becomes

$$\int_a^b f(x)\,dx = \text{area}(R_1) - \text{area}(R_2).$$

Similarly, if f is a continuous function on $[a, b]$ whose graph touches the x-axis a finite number of times, then its graph and the x-axis determine a finite number of regions. We can generalize the previous argument to obtain an interpretation of $\int_a^b f(x)\,dx$ as a signed area: The definite integral $\int_a^b f(x)\,dx$ is equal to the difference between the total area of the regions above the x-axis and the total area of the regions below the x-axis (see Figure 2.17).

In Chapter 4, we observed that a function f that is continuous on a closed finite interval $[a, b]$ has a minimum value m and a maximum value M (Theorem 1 in Section 4.1). We can use these numbers to obtain upper and lower bounds for $\int_a^b f(x)\,dx$.

Theorem 6

Suppose that f is a continuous function on the interval $[a, b]$. Let m be its minimum value and M be its maximum value. Then

$$m(b - a) \le \int_a^b f(x)\,dx \le M(b - a).$$

Figure 2.18 illustrates Theorem 6 in the case where f is nonnegative on $[a, b]$. We leave its proof as an exercise (Exercise 53).

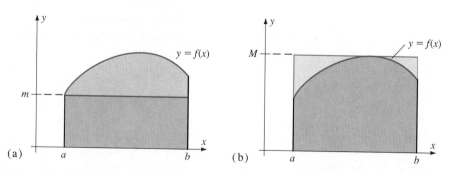

Figure 2.18 (a) If $f(x) > 0$ for $x \in [a, b]$, then Theorem 6 can be interpreted as a statement about areas. The region bounded by the graph of f on $[a, b]$ is larger than the rectangle with base $[a, b]$ and height m. (b) On the other hand, the region is smaller than the rectangle with base $[a, b]$ and height M.

The Midpoint Rule

At this point, the only general means available to us for evaluating the definite integral $\int_a^b f(x)\,dx$ results from Theorem 2: Using any sequence of Riemann sums R_n for f on $[a, b]$, with $\|P_n\| \to 0$ as $n \to \infty$, determine the limit of R_n as $n \to \infty$. The **Midpoint Rule** is a relatively straightforward numerical approximation procedure that uses this approach.

We partition the interval $[a, b]$ into n equal-length subintervals just as we did when we discussed the area problem in Section 6.1. Then $\Delta x_j = \Delta x = (b - a)/n$, and the partition is determined by

$$x_0 = a,\ x_1 = a + \Delta x,\ x_2 = a + 2\Delta x, \ldots, x_n = a + n\,\Delta x = b.$$

To specify a Riemann sum, we must also choose numbers t_j at which to evaluate the function f. The Midpoint Rule uses the midpoint m_j of the subinterval $[x_{j-1}, x_j]$ for t_j. Therefore, we approximate $\int_a^b f(x)\,dx$ by the Riemann sum

$$R_n = \sum_{j=1}^{n} f(m_j)\Delta x,$$

where $m_j = (x_j - x_{j-1})/2$ (see Figure 2.19).

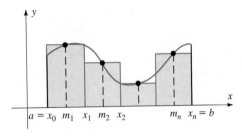

Figure 2.19 Approximation by Midpoint Rule: $m_j = \frac{1}{2}(x_{j-1} + x_j)$.

The Midpoint Rule: Let $\Delta x = \dfrac{b - a}{n}$ and

$$x_0 = a, \quad x_1 = a + \Delta x, \quad x_2 = a + 2\Delta x, \quad \ldots, \quad x_n = a + n\,\Delta x = b.$$

Then

$$\int_a^b f(x)\,dx \approx \Delta x\,(f(m_1) + f(m_2) + \cdots + f(m_n))$$

where m_j is the midpoint of the interval $[x_{j-1}, x_j]$. That is,

$$m_j = \frac{x_{j-1} + x_j}{2}.$$

□ **EXAMPLE 9**

Approximate

$$\int_0^1 \cos(x^2)\,dx$$

using the Midpoint Rule with $n = 5$.

Use $a = 0$, $b = 1$, and $n = 5$ to determine Δx.

Find the midpoints m_j.

Compute the values $f(m_j)$.

The desired approximation is the Riemann sum R_5 given by the Midpoint Rule.

Solution

We subdivide the interval $[0, 1]$ into five equal-length subintervals. Therefore,

$$\Delta x = \frac{b - a}{n} = \frac{1 - 0}{5} = 0.2,$$

and the partition P_5 is $\{0, 0.2, 0.4, 0.6, 0.8, 1\}$. The five midpoints of the five subintervals are

$$m_1 = 0.1, \ m_2 = 0.3, \ m_3 = 0.5, \ m_4 = 0.7, \text{ and } m_5 = 0.9.$$

In order to form the Riemann sum, we compute the values $f(m_j)$ (see Table 2.1). Then we evaluate the sum

$$\int_0^1 \cos(x^2) \, dx \approx (0.2)(0.999950 + 0.995953 + 0.968912 + 0.882333 + 0.689498)$$
$$= (0.2)(4.53665)$$
$$= 0.907329$$

(see Figure 2.20).

Table 2.1   The values $\cos(m_j^2)$ for the midpoint approximation in Example 9.

m_j	$\cos(m_j^2)$
0.1	0.999950
0.3	0.995953
0.5	0.968912
0.7	0.882333
0.9	0.689498

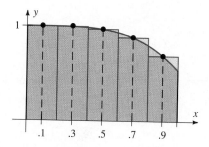

Figure 2.20   The Midpoint Rule applied to the function $f(x) = \cos(x^2)$ on the interval $[0, 1]$ with 5 intervals.

We shall discuss the accuracy of the Midpoint Rule in Section 6.6.

Integrable Functions

From Theorem 2, we know that, if f is continuous on $[a, b]$ and P_n are partitions such that $\|P_n\| \to 0$ as $n \to \infty$, then all corresponding Riemann sums R_n have the limit $\int_a^b f(x) \, dx$. However, continuous functions are not the only class of functions for which all such Riemann sums R_n tend to a unique limit I. In fact, in Exercises 65–69, we show how Riemann sums for functions with "jump" discontinuities approach a unique limit. In general, whenever a function defined on $[a, b]$ has such a limit I, we say that f is **integrable** on $[a, b]$, and we define $\int_a^b f(x) \, dx$ to be the number I. A precise description of the class of all integrable functions is somewhat complicated. However, for our purposes, we usually need only consider the integral of a continuous function or occasionally the integral of a function that is continuous except for a finite number of jump discontinuities.

Exercise Set 6.2

In Exercises 1–4, a function f and a partition $P_n = \{a = x_0, x_1, \ldots, x_n = b\}$ of the interval $[a, b]$ are specified. Let t_j be the left endpoint of the jth subinterval determined by P_n. Calculate the corresponding Riemann sum

$$R_n = \sum_{j=1}^{n} f(t_j)\, \Delta x_j.$$

Sketch the graph of f along with rectangles that provide a geometric interpretation for the number R_n (see Figures 2.4 and 2.5).

1. $f(x) = x + 2, \quad P_3 = \{1, 2, 3, 5\}$

2. $f(x) = 2x + 3, \quad P_4 = \{-1, 0, 2, 3, 4\}$

3. $f(x) = 3 - x, \quad P_4 = \{0, 1, 3/2, 2, 4\}$

4. $f(x) = 4 - 2x, \quad P_3 = \{0, 1/2, 1, 2, 3\}$

In Exercises 5–8, follow the instructions of Exercises 1–4 except let t_j be the right endpoint of the jth subinterval determined by P_n.

5. $f(x) = x^2 - x, \quad P_3 = \{0, 1, 2, 4\}$

6. $f(x) = 4 - x^2, \quad P_3 = \{-1, 0, 2, 3\}$

7. $f(x) = \cos x, \quad P_4 = \{0, \pi/2, \pi, 2\pi, 5\pi/2\}$

8. $f(x) = \sin x, \quad P_3 = \{0, \pi/2, 3\pi/2, 2\pi\}$

In Exercises 9–12, follow the instructions of Exercises 1–4 except let t_j be the midpoint of the jth subinterval determined by P_n. In other words, let $t_j = (x_{j-1} + x_j)/2$.

9. $f(x) = \sqrt{x}, \quad P_3 = \{0, 2, 6, 12\}$

10. $f(x) = \sqrt{1 - x}, \quad P_3 = \{-11, -5, -1, 1\}$

11. $f(x) = \sin x, \quad P_4 = \{0, \pi/2, \pi, 3\pi/2, 2\pi\}$

12. $f(x) = \cos x, \quad P_3 = \{0, \pi/2, 3\pi/2, 2\pi\}$

In Exercises 13–16, partition the interval $[a, b]$ into four subintervals of equal length. Determine both the largest and the smallest Riemann sum associated with this partition for the given function f.

13. $f(x) = 4 - x^2, \quad [a, b] = [-2, 2]$

14. $f(x) = x^2 - 2x, \quad [a, b] = [0, 4]$

15. $f(x) = \sin x, \quad [a, b] = [0, 2\pi]$

16. $f(x) = \sin x, \quad [a, b] = [-\pi, \pi]$

The formulas

$$\int_a^b x\, dx = \tfrac{1}{2}(b^2 - a^2) \quad \text{and} \quad \int_a^b x^2\, dx = \tfrac{1}{3}(b^3 - a^3).$$

are valid for *all* numbers a and b (see either Exercises 55 and 56 or Section 6.3). Use them along with the result of Example 5 to find the value of the integrals in Exercises 17–30.

17. $\int_0^5 6\, dx$

18. $\int_0^3 (2x + 5)\, dx$

19. $\int_{-2}^4 (3x - 2)\, dx$

20. $\int_5^2 (5x + 7)\, dx$

21. $\int_4^0 (9 - 2x)\, dx$

22. $\int_2^5 2x^2\, dx$

23. $\int_3^1 (6 - x^2)\, dx$

24. $\int_{-4}^{-1} (3x^2 - 1)\, dx$

25. $\int_0^1 (x^2 - x + 2)\, dx$

26. $\int_{-2}^2 (2x^2 - 3x + 2)\, dx$

27. $\int_0^4 x(2 + x)\, dx$

28. $\int_3^{-1} (x - 2)(x + 2)\, dx$

29. $\int_{-4}^{-1} 2x(1 - 2x)\, dx$

30. $\int_{-1}^1 (ax^2 + bx + c)\, dx$

31. Given that $\int_0^2 f(x)\, dx = 3, \int_2^5 f(x)\, dx = -2$, and $\int_5^8 f(x)\, dx = 5$, find

a. $\int_2^0 f(x)\, dx$

b. $\int_0^5 f(x)\, dx$

c. $\int_2^8 f(x)\, dx$

d. $\int_0^8 f(x)\, dx$

e. $\int_5^0 f(x)\, dx$

f. $\int_8^2 f(x)\, dx$

32. Calculate

$$\int_0^{2\pi} \sin^2 x\, dx + \int_0^{2\pi} \cos^2 x\, dx.$$

33. Given that $\int_1^3 f(x)\, dx = 5$ and $\int_1^3 g(x)\, dx = -2$, find

a. $\int_1^3 [f(x) + g(x)]\, dx$

b. $\int_1^3 [6g(x) - 2f(x)]\, dx$

c. $\int_3^1 2f(x)\, dx - \int_1^3 g(x)\, dx$

d. $\int_1^3 [2f(x) + 3]\, dx$

e. $\int_3^1 [g(x) - 4f(x) + 5]\, dx$

f. $\int_1^3 [1 - 4g(x) + 3f(x)]\, dx$

In Exercises 34–47, sketch a region in the plane whose area is given by the definite integral. Using geometric formulas, compute the area of that region and, consequently, the integral.

34. $\int_{-2}^3 (x + 2)\, dx$

35. $\int_{-1}^2 (4 - 2x)\, dx$

36. $\int_{1}^{3} (x + 1)\, dx$

37. $\int_{0}^{3} (2x + 1)\, dx$

38. $\int_{-1}^{1} \sqrt{1 - x^2}\, dx$

39. $\int_{0}^{3} \sqrt{9 - x^2}\, dx$

40. $\int_{-2}^{2} |x - 1|\, dx$

41. $\int_{1}^{4} |2x - 7|\, dx$

42. $\int_{0}^{4} (\sqrt{16 - x^2} + 2)\, dx$

43. $\int_{-2}^{0} \sqrt{1 - (x + 1)^2}\, dx$

44. $\int_{-4}^{0} (\sqrt{4 - (x + 2)^2} + 1)\, dx$

45. $\int_{0}^{2} (|x - 1| + 1)\, dx$

46. $\int_{-3}^{3} (4 - \sqrt{9 - x^2})\, dx$

47. $\int_{0}^{2} (x + 2 - \sqrt{4 - x^2})\, dx$

In Exercises 48–51, use the formulas in the instructions for Exercises 17–30 to calculate the integral

$$\int_{0}^{2} f(x)\, dx$$

for the given function f.

48. $f(x) = \begin{cases} x, & \text{for } 0 \le x \le 1; \\ x^2, & \text{for } 1 \le x \le 2. \end{cases}$

49. $f(x) = \begin{cases} 1 - x, & \text{for } 0 \le x \le 1; \\ x^2 - 1, & \text{for } 1 \le x \le 2. \end{cases}$

50. $f(x) = \begin{cases} -x^2, & \text{for } 0 \le x \le 1; \\ x - 2, & \text{for } 1 \le x \le 2. \end{cases}$

51. $f(x) = \begin{cases} x^2 - 1, & \text{for } 0 \le x \le 1; \\ x - 1, & \text{for } 1 \le x \le 2. \end{cases}$

52. a. Prove that if $f(x)$ and $g(x)$ are continuous on $[a, b]$, and if $f(x) \le g(x)$ for all $x \in [a, b]$, then

$$\int_{a}^{b} f(x)\, dx \le \int_{a}^{b} g(x)\, dx.$$

(*Hint:* Write $g(x) = f(x) + h(x)$. What do you know about the sign of $h(x)$? Of $\int_{a}^{b} h(x)\, dx$?)

b. Which is larger, $\int_{0}^{1} \sin(x^2)\, dx$ or $\int_{0}^{1} (x^2 + 1)\, dx$?

53. a. Prove Theorem 6. (*Hint:* Show that *every* Riemann sum R_n is bounded by $m(b - a) \le R_n \le M(b - a)$. Why does this imply the desired result?)

b. Show that

$$\frac{3}{2} \le \int_{1}^{4} \frac{1}{\sqrt{x}}\, dx \le 3.$$

54. a. Prove that, if f is continuous on $[a, b]$, then

$$\left| \int_{a}^{b} f(x)\, dx \right| \le \int_{a}^{b} |f(x)|\, dx.$$

b. Illustrate part a by comparing

$$\left| \int_{0}^{3} (x - 2)\, dx \right| \quad \text{and} \quad \int_{0}^{3} |x - 2|\, dx.$$

55. Derive the formula

$$\int_{a}^{b} x\, dx = \tfrac{1}{2}(b^2 - a^2)$$

by filling in the details in the following steps:

a. Define P_n to be the equal-length partition $\{x_0, \ldots, x_n\}$, where $\Delta x = (b - a)/n$ and $x_j = a + j\, \Delta x$ for $j = 0, 1, 2, \ldots, n$.

b. Choose $t_j = x_j$ for $j = 1, \ldots, n$ and let R_n be the associated Riemann sum.

c. Show that

$$R_n = a(b - a) + \frac{n(n + 1)}{2n^2}(b - a)^2$$

using formula (4) in Section 6.1.

d. Show that

$$\lim_{n \to \infty} R_n = \tfrac{1}{2}(b^2 - a^2).$$

56. Repeat the steps in Exercise 55 to derive

$$\int_{a}^{b} x^2\, dx = \tfrac{1}{3}(b^3 - a^3).$$

In step c, you should obtain

$$R_n = (b - a)a^2 + a(b - a)^2 \frac{n(n + 1)}{n^2}$$
$$+ (b - a)^3 \frac{n(n + 1)(2n + 1)}{6n^3}$$

using formula (5) in Section 6.1.

57. Calculate the limit

$$\lim_{n \to \infty} \sum_{j=1}^{n} \frac{j}{n^2}$$

as follows:

a. Interpret the sum as a Riemann sum. That is, let P_n be the equal-length partition of $[0, 1]$ (namely, $P_n = \{x_j\}$ where $x_j = j/n$). Show that the Riemann sum R_n for $f(x) = x$ using P_n and $t_j = x_j$ is

$$\sum_{j=1}^{n} \left(\frac{j}{n}\right)\left(\frac{1}{n}\right).$$

b. Interpret the limit as $\int_{0}^{1} f(x)\, dx$ and calculate the integral.

In Exercises 58–60, calculate the given limit using the method of Exercise 57.

58. $\displaystyle \lim_{n\to\infty} \sum_{j=1}^{n} \frac{j^2}{n^3}$ **59.** $\displaystyle \lim_{n\to\infty} \sum_{j=1}^{n} \frac{\sqrt{n^2 - j^2}}{n^2}$

60. $\displaystyle \lim_{n\to\infty} \sum_{j=1}^{n} \frac{\sqrt{jn - j^2}}{n^2}$

61. Calculate

$$\int_0^b \cos x \, dx = \sin b$$

as follows:

a. Partition the interval $[0, b]$ into equal-length subintervals using $P_n = \{x_j\}$ where $x_j = j(b/n)$. Using P_n and $t_j = x_{j-1}$, show that the associated Riemann sum R_n equals

$$\left(\frac{b}{n}\right) \sum_{j=0}^{n-1} \cos\left(j\left(\frac{b}{n}\right)\right).$$

b. Using Lagrange's summation formula for cosine (Exercise 65 in Section 6.1), show that

$$R_n = \left(\frac{b}{2n}\right)\left(1 + \frac{\sin\left[\left(n - \frac{1}{2}\right)\left(\frac{b}{n}\right)\right]}{\sin\left(\frac{b}{2n}\right)}\right).$$

c. Calculate the limit of R_n as $n \to \infty$.

62. In this exercise, we verify that

$$\int_0^b \sqrt{x} \, dx = \frac{2}{3} b^{3/2}.$$

a. Sketch the graph $y = \sqrt{x}$ for $0 \le x \le b$ and shade the region R_1 whose area is equal to

$$\int_0^b \sqrt{x} \, dx.$$

(See Figure 2.21.) On the same picture, draw the rectangle R_2 with vertices $(0, 0)$, $(b, 0)$, $(b, \sqrt{b})$, and $(0, \sqrt{b})$.

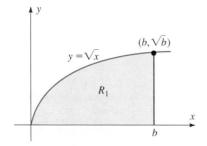

Figure 2.21   Area of R_1 is $A = \int_0^b \sqrt{x} \, dx = \frac{2}{3} b^{3/2}$, $b > 0$.

Let R_3 represent the region that remains when R_1 is removed from R_2.

b. Note that the graph satisfies the equation $y^2 = x$. Explain why the area of R_3 is equal to the integral

$$\int_0^{\sqrt{b}} y^2 \, dy.$$

c. Use the result of Exercise 56 to calculate the area of R_3 and, thus, to deduce the area of R_1.

63. Prove the remaining cases of Theorem 5.

64. Give an argument for statement (ii) of Theorem 4 similar to that given for statement (i).

65. Consider the functions

$$f(x) = x, \qquad 0 \le x \le 1$$

$$g(x) = \begin{cases} x, & 0 \le x < 1 \\ 2, & x = 1 \end{cases}$$

on the interval $[0, 1]$.

a. Note that f is continuous, and therefore integrable, on $[0, 1]$ but that g is *not* continuous on $[0, 1]$, because $g(1) \ne \lim_{x\to 1^-} g(x)$. (See Figure 2.22.)

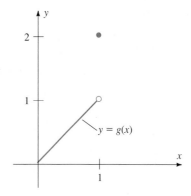

Figure 2.22

b. Let P_n be any partition of $[0, 1]$. Explain why the Riemann sums

$$\sum_{j=1}^{n} f(t_j) \, \Delta x_j \quad \text{and} \quad \sum_{j=1}^{n} g(t_j) \, \Delta x_j$$

over P_n are equal if the numbers t_j in both sums are the same and if we choose t_n with $t_n \ne 1$.

c. Conclude that

$$\lim_{n\to\infty} \sum_{j=1}^{n} g(t_j) \, \Delta x_j = \lim_{n\to\infty} \sum_{j=1}^{n} f(t_j) \, \Delta x_j$$

$$= \int_0^1 x \, dx = \frac{1}{2}.$$

so any limit as $\|P_n\| \to 0$ of Riemann sums for g approaches $1/2$. Thus, g is integrable on $[0, 1]$, and $\int_0^1 g(x)\, dx = 1/2$.

66. Let g be a function defined on $[a, c]$ with a single "jump discontinuity" at the number b. In other words,

$$g(x) = \begin{cases} f_1(x), & \text{for } a \le x \le b; \\ f_2(x), & \text{for } b < x \le c, \end{cases}$$

where f_1 is continuous on $[a, b]$, f_2 is continuous on $(b, c]$, and

$$\lim_{x \to b^+} f_2(x) = l$$

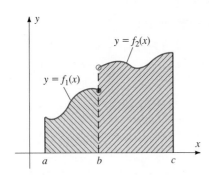

Figure 2.23

exists and does not equal $f_1(b)$ (see Figure 2.23). Generalize the result of Exercise 65 to conclude that f is integrable on $[a, c]$ and

$$\int_a^c g(x)\, dx = \int_a^b f_1(x)\, dx + \int_b^c \hat{f}_2(x)\, dx$$

where $\hat{f}_2$ is the function defined by

$$\hat{f}_2(x) = \begin{cases} l, & \text{if } x = b; \\ f_2(x), & \text{if } b < x \le c. \end{cases}$$

67. Generalize the result of Exercise 66 to conclude that a function g is integrable on $[a, b]$ if it is continuous on $[a, b]$ except at finitely many numbers where it has "jump discontinuities" as in Figure 2.23. (We say that g has a jump discontinuity at $c \in (a, b)$ if the one-sided limits

$$\lim_{x \to c^-} g(x) \quad \text{and} \quad \lim_{x \to c^+} g(x)$$

both exist but are unequal; g has a jump discontinuity at the endpoint a if $g(a) \ne \lim_{x \to a^+} g(x)$ and similarly for the right endpoint b.)

68. Explain how the definite integral for the discontinuous function g in Exercise 67 would be calculated.

69. Let h be the function defined on $[0, 1]$ by

$$h(x) = \begin{cases} 1/x, & \text{for } x \ne 0; \\ 0, & x = 0. \end{cases}$$

Show that h is not integrable on $[0, 1]$. (*Hint:* Consider any Riemann sum R_n associated to the equal-length partition $P_n = \{0, 1/n, 2/n, \ldots, 1\}$ with $t_1 = 1/n^2$. Note that $R_n > n$. Why does this imply that h is not integrable?)

In Exercises 70–75, use the Midpoint Rule with the given value of n to approximate the integral.

70. $\displaystyle\int_1^3 \frac{1}{x}\, dx, \quad n = 2$

71. $\displaystyle\int_1^4 \frac{1}{x}\, dx, \quad n = 3$

72. $\displaystyle\int_0^6 \sqrt{x^3 + 1}\, dx, \quad n = 3$

73. $\displaystyle\int_0^4 \sqrt{x^3 + 1}\, dx, \quad n = 4$

74. $\displaystyle\int_0^1 \frac{1}{1+x^2}\, dx, \quad n = 2$

75. $\displaystyle\int_0^4 \frac{1}{1+x^2}\, dx, \quad n = 4$

In Exercises 76–79, use the Midpoint Rule with $n = 6$ to approximate the integral.

76. $\displaystyle\int_0^3 \frac{1}{\sqrt{1+x^3}}\, dx$

77. $\displaystyle\int_0^1 \sqrt{x^4 + 1}\, dx$

78. $\displaystyle\int_1^2 \frac{1}{x}\, dx$

79. $\displaystyle\int_0^3 \sin\sqrt{x}\, dx$

80. Using $n = 5$ subdivisions with the Midpoint Rule, make a table of estimates of the integral

$$\int_0^b x^3\, dx$$

with $b = 1, 2, 3, 4, 5$. Use these values to sketch a rough graph of the function

$$A(b) = \int_0^b x^3\, dx$$

for b in the interval $[0, 5]$. If you have access to a programmable calculator, you should also estimate the integral for $b = 0.5, 1.5, 2.5, 3.5, 4.5$ in order to obtain a more accurate graph.

81. Using $n = 5$ subdivisions with the Midpoint Rule, make a table of estimates of the integral

$$\int_0^b \sin(x^2)\, dx$$

for $b = 0.5, 1.0, 1.5, 2.0, 2.5, 3.0$. Use these values to sketch a rough graph of the function

$$A(b) = \int_0^b \sin(x^2)\, dx$$

for b in the interval $[0, 3]$. If you have access to a programmable calculator, you should also estimate the integral for additional numbers b between 0 and 3.

82. Let $f(x) = 2x$. Determine the estimate M_n of the integral $\int_0^1 2x\, dx$ obtained by applying the Midpoint Rule with n subdivisions. Why does your answer determine the integral exactly?

In Exercises 83–86, graph the function f as well as the rectangles that correspond to the Riemann sums

$$R_n = \sum_{j=1}^{n} f(t_j)\Delta x,$$

where $\Delta x = (b - a)/n$ and the numbers t_j are the left-hand endpoints of the subintervals. Use $n = 4, 8, 16, 32$, and 64. For the same n, graph f as well as rectangles that correspond to Riemann sums where the numbers t_j are the right-hand endpoints of the subintervals. Compare the sums with the given value I of the integral.

83. $f(x) = -\frac{1}{2}x + 1$, $-4 \le x \le 5$, $I = 27/4$

84. $f(x) = x^3 - 4x$, $-1 \le x \le 2$, $I = -9/4$

85. $f(x) = \begin{cases} 2x, & \text{if } 0 \le x < 1; \\ x^2 - 4x + 5, & \text{if } 1 \le x \le 4. \end{cases}$ $I = 7$

86. $f(x) = 1/x$, $1/8 \le x \le 3$, $I \approx 3.1780538$.

In Exercises 87–90, approximate the given definite integral to 5 decimal places using Riemann sums where the numbers t_j are (a) the left endpoints of the subintervals and (b) the right endpoints of the subintervals.

87. $\displaystyle\int_{0}^{2} \sqrt{4 - x^2}\, dx$

88. $\displaystyle\int_{-1}^{3} (x^2 + 1)^{2/3}\, dx$

89. $\displaystyle\int_{0}^{3} f(x)\, dx$, $f(x) = \begin{cases} \dfrac{\sin x}{x}, & \text{if } x \ne 0; \\ 1, & \text{if } x = 0. \end{cases}$

90. $\displaystyle\int_{0}^{\pi} x \sin 2x\, dx$

6.3   The Fundamental Theorem of Calculus

As its name implies, the Fundamental Theorem of Calculus is the single most important theorem in the calculus. It relates two apparently unrelated concepts—the derivative (a rate of change) and the definite integral (a generalized summation procedure). Theoretically, the Fundamental Theorem is important because it justifies the use of the integral as a mechanism for *defining* functions. Computationally, it is important because it provides a powerful procedure for the evaluation of many definite integrals. In this section, we discuss the theorem and a few of its implications. We elaborate on its use as a technique for the evaluation of definite integrals in the next section.*

The Function

$$A(x) = \int_{a}^{x} f(t)\, dt$$

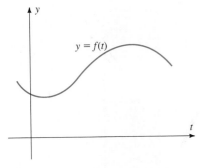

Figure 3.1   The continuous function $y = f(t)$.

We begin with a function f that is continuous on an interval I, and we define a related function A using the definite integral of f *with the upper endpoint of integration as the independent variable for A.* More precisely, suppose that f is continuous on an interval I. Then, by choosing any (fixed) number a in I, we define a function A on I by

$$A(x) = \int_{a}^{x} f(t)\, dt. \tag{1}$$

That is, $A(x)$ equals the definite integral of f from a to x.

If f is nonnegative on I, the value $A(x)$ is simply the area under the graph of $y = f(t)$ for $a \le t \le x$ (see Figures 3.1 and 3.2). In this case, we refer to A as the *area function* determined by f. Consider all $x > a$ in I. As x increases, the value $A(x)$ must increase because we are calculating the area of successively larger regions (see Figure 3.3).

In general, we need not assume that f is nonnegative. The integral $A(x)$ in equation (1) exists for all x in I, regardless of the sign of $f(t)$. Then we interpret $A(x)$ as a "signed" area function.

---

*The history of the development of the Fundamental Theorem of Calculus is intriguing. For more information, see the historical note on Newton and Leibniz at the beginning of Unit 1.

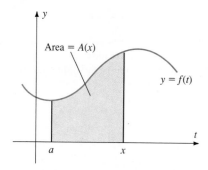

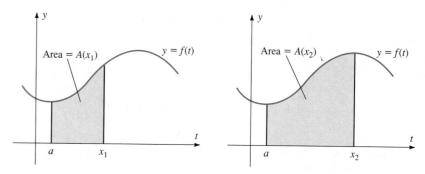

Figure 3.2 If f is nonnegative, then the value $A(x) = \int_a^x f(t)\, dt$ equals the area of the region determined by the graph of f between $t = a$ and $t = x$.

Figure 3.3 If f is nonnegative, the area function A is increasing. That is, $x_2 > x_1$ implies $A(x_2) > A(x_1)$.

Since the calculus involves both integration and differentiation, let's see what happens if we try to differentiate the function A. By definition,

$$A'(x) = \lim_{h \to 0} \frac{A(x + h) - A(x)}{h}. \tag{2}$$

For simplicity, we assume that f is nonnegative and that $x > a$. In order to calculate the limit in equation (2), we consider the difference quotient

$$\frac{A(x + h) - A(x)}{h} \tag{3}$$

for $h > 0$. Since $A(x + h)$ is the area under the graph from $t = a$ to $t = x + h$ and $A(x)$ is the area under the graph from $t = a$ to $t = x$, the difference $A(x + h) - A(x)$ is the area under the graph from $t = x$ to $t = x + h$ (see Figure 3.4). Note that the width of this region is $(x + h) - x = h$. If we can approximate the area by a rectangle of height $f(x)$ and width h (see Figure 3.5), we obtain

$$A(x + h) - A(x) \approx f(x) \cdot h, \tag{4}$$

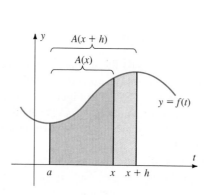

Figure 3.4 Area of pure blue region is $A(x + h) - A(x)$.

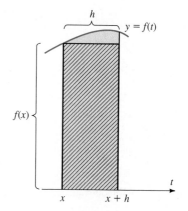

Figure 3.5 Rectangle of area $f(x) \cdot h$ approximates region of area $A(x + h) - A(x)$.

and the difference quotient (3) is approximated by

$$\frac{A(x + h) - A(x)}{h} \approx f(x). \tag{5}$$

As we shall see, the approximation (4) becomes more accurate as $h \to 0^+$, and approximation (5) yields

$$\lim_{h \to 0^+} \frac{A(x + h) - A(x)}{h} = f(x).$$

A similar argument applies for $h < 0$, and we obtain a remarkable fact about A: *It is differentiable, and its derivative is*

$$A'(x) = f(x).$$

In other words, the derivative of the area function A is the original function f! This is the essense of the Fundamental Theorem—the operations of integration (that is, the signed area function A) and differentiation are inverses (as mathematical operations). We shall soon see that the consequences of this fact are extremely significant.

This intuitive argument is not a proof because the approximation in line (4) is not a precise mathematical statement. Nevertheless, from the proof, you will see that our intuition is correct.

A formal presentation of the Fundamental Theorem is now in order.

Theorem 7

Fundamental Theorem of Calculus

(Part i) Let f be continuous on an open interval I containing the number a and let

$$A(x) = \int_a^x f(t)\, dt$$

for each $x \in I$. Then A is differentiable on I, and

$$A'(x) = f(x).$$

That is,

$$\frac{d}{dx}\left(\int_a^x f(t)\, dt\right) = f(x), \qquad x \in I.$$

(Part ii) Let f be continuous on $[a, b]$ and let F be any antiderivative for f on $[a, b]$. Then

$$\int_a^b f(x)\, dx = F(b) - F(a). \tag{6}$$

REMARK 1   In writing $\int_a^b f(x)\, dx$ in equation (6) we have returned to the use of x as the independent variable for f. It does not matter whether we call this "dummy variable" x or t because it is being used simply to fill out the standard notation for the definite integral. As defined in Section 6.2, the definite integral $\int_a^b f(t)\, dt$ is a limit of Riemann sums. This limit depends only on the values of f on the interval $[a, b]$ and not on the notation we use to refer to those values. In other words,

$$\int_a^b f(t)\, dt = \int_a^b f(s)\, ds = \int_a^b f(x)\, dx.$$

However, we are careful to avoid confusing the insignificant dummy variable of integration (t in equation (1)) with the independent variable (x in equation (1)) for $A(x)$.

Proof of the Fundamental Theorem: At first glance, the two parts of the Fundamental Theorem seem unrelated. Consequently, we begin by showing how part (ii) follows from part (i). Then we formalize the argument presented before the statement of the theorem to prove part (i).

The proof of part (ii) assuming part (i): We need to show that the definite integral $\int_a^b f(x)\,dx$ can be evaluated using *any* given antiderivative F of f. By definition, we know that

$$\int_a^b f(x)\,dx = A(b). \tag{7}$$

Since part (i) states that A is one particular antiderivative of f and since we are assuming that F is an arbitrary antiderivative of f, we know that there exists a constant C such that

$$A(x) = F(x) + C \tag{8}$$

(Theorem 1 in Chapter 5). Combining equations (7) and (8), we get

$$\int_a^b f(x)\,dx = A(b) = F(b) + C. \tag{9}$$

To determine the constant C, we evaluate equation (8) at $x = a$. The integral $A(a) = \int_a^a f(x)\,dx$ is 0 by definition, hence (8) yields

$$0 = A(a) = F(a) + C,$$

so

$$C = -F(a).$$

Replacing C by $-F(a)$ in equation (9) establishes part (ii) of the Fundamental Theorem.

The proof of part (i): The basic steps are the same as those presented before the statement of the theorem. We need only replace the approximation in line (4) with a precise estimate. Theorem 6 in Section 6.2 provides the relevant inequalities.

Using the definition of A, we write the difference quotient as

$$\frac{A(x + h) - A(x)}{h} = \frac{1}{h}\left[\int_a^{x+h} f(t)\,dt - \int_a^{x} f(t)\,dt\right]$$

for all h such that $x + h$ is in the interval I. By Theorem 5, we can combine the two integrals and obtain

$$\frac{A(x + h) - A(x)}{h} = \frac{1}{h}\int_x^{x+h} f(t)\,dt. \tag{10}$$

Now we apply Theorem 6 to f on the closed interval J_h between x and $x + h$. Let m_h be the minimum value and M_h be the maximum value of f on J_h. If $h > 0$, we get

$$m_h \cdot h \le \int_x^{x+h} f(t)\,dt \le M_h \cdot h.$$

Thus, the right-hand side of equation (10) satisfies the inequality

$$m_h \leq \frac{1}{h} \int_x^{x+h} f(t)\, dt \leq M_h. \tag{11}$$

As $h \to 0^+$, we know that both m_h and M_h approach $f(x)$ because f is continuous on I and the interval J_h approaches $[x, x] = \{x\}$. Therefore, inequality (11) implies that

$$\lim_{h \to 0^+} \frac{A(x + h) - A(x)}{h} = f(x).$$

Theorem 6 (appropriately modified for the case where $b < a$) also implies inequality (11) when $h < 0$. Therefore, we can conclude that

$$\lim_{h \to 0^-} \frac{A(x + h) - A(x)}{h} = f(x).$$

Hence, A is differentiable and $A'(x) = f(x)$.   ■

It is important to distinguish between the theoretical implications of the Fundamental Theorem and its usefulness for evaluating definite integrals. Part (i) answers an important question posed at the beginning of Chapter 5: does a function f always have an antiderivative? If f is continuous, the answer is yes. The function A is an antiderivative. However, to use part (ii) to evaluate definite integrals, we must already know an antiderivative F that can be evaluated without reference to A. In Chapter 5, we studied methods for determining such antiderivatives, and in later chapters, we shall broaden our repertoire of antidifferentiation techniques.

We now discuss several examples that illustrate the significance of the Fundamental Theorem. We begin with examples that illustrate how part (ii) simplifies the calculation of definite integrals, and we make use of the notation

$$\left[F(x) \right]_a^b = F(b) - F(a).$$

For example,

$$\left[\sin x \right]_{\pi/4}^{\pi/2} = \sin \frac{\pi}{2} - \sin \frac{\pi}{4} = 1 - \frac{\sqrt{2}}{2},$$

and

$$\left[x^2 + 4 \right]_2^3 = (3^2 + 4) - (2^2 + 4) = 5.$$

□ **EXAMPLE 1**

Since $F(x) = 2x^2 + 6x$ is an antiderivative of $f(x) = 4x + 6$,

$$\int_1^2 (4x + 6)\, dx = \left[2x^2 + 6x \right]_1^2 = (8 + 12) - (2 + 6) = 12.$$ ■

Example 1 illustrates the power of the Fundamental Theorem. Try calculating $\int_1^2 (4x + 6)\, dx$ directly from the definition, using the method described in Section 6.1. You will probably find it easier to calculate $F(2) - F(1)$ where $F(x) = 2x^2 + 6x$.

□ **EXAMPLE 2**

Since an antiderivative for $f(x) = x^3 - 3x^2 + 2x - 5$ is

$$F(x) = \frac{x^4}{4} - x^3 + x^2 - 5x,$$

we have

$$\int_{-2}^{2} (x^3 - 3x^2 + 2x - 5)\, dx = \left[\frac{x^4}{4} - x^3 + x^2 - 5x \right]_{-2}^{2}$$
$$= (4 - 8 + 4 - 10) - (4 + 8 + 4 + 10)$$
$$= -36. \qquad\blacksquare$$

□ **EXAMPLE 3**

Since an antiderivative for $f(x) = \cos x$ is $F(x) = \sin x$, we have

$$\int_{0}^{3\pi/2} \cos x\, dx = \left[\sin x \right]_{0}^{3\pi/2} = \sin(3\pi/2) - \sin(0)$$
$$= -1 - 0 = -1. \qquad\blacksquare$$

□ **EXAMPLE 4**

$$\int_{1}^{4} \frac{x+1}{\sqrt{x}}\, dx = \int_{1}^{4} \left(\sqrt{x} + \frac{1}{\sqrt{x}} \right) dx$$
$$= \int_{1}^{4} (x^{1/2} + x^{-1/2})\, dx$$

Since an antiderivative for $f(x) = x^{1/2} + x^{-1/2}$ is $F(x) = \frac{2}{3}x^{3/2} + 2x^{1/2}$, we have

$$\int_{1}^{4} \frac{x+1}{\sqrt{x}}\, dx = \left[\frac{2}{3}x^{3/2} + 2x^{1/2} \right]_{1}^{4}$$
$$= (\tfrac{2}{3} \cdot 8 + 2 \cdot 2) - (\tfrac{2}{3} \cdot 1 + 2) = \tfrac{20}{3}. \qquad\blacksquare$$

REMARK 2   You may be wondering why part (ii) of the Fundamental Theorem works no matter which antiderivative F we use. Since two different antiderivatives F_1 and F_2 of a given function differ only by a constant, $F_1(b) - F_1(a) = F_2(b) - F_2(a)$. In other words, $F_2(x) = F_1(x) + C$, so

$$F_2(b) - F_2(a) = [F_1(b) + C] - [F_1(a) + C]$$
$$= F_1(b) - F_1(a) + (C - C)$$
$$= F_1(b) - F_1(a).$$

REMARK 3   Part (ii) of the Fundamental Theorem explains the use of the integral sign to represent both the antiderivative and the definite integral. If F is any antiderivative for f, we have

$$\int_{a}^{b} f(x)\, dx = \left[F(x) + C \right]_{a}^{b}$$

and the right-hand side can be written

$$\left[\int f(x)\,dx\right]_a^b.$$

However, you should be aware of the conceptual difference between the indefinite integral and definite integral. The indefinite integral is a family of antiderivatives, and the definite integral is a number that is the limit of Riemann sums. It is only by the Fundamental Theorem that we know they are related.

The following example illustrates the use of Fundamental Theorem to compute an integral involving an absolute value.

☐ **EXAMPLE 5**

Find $\displaystyle\int_{-1}^{5} |x - 2|\,dx.$

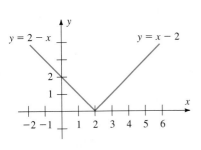

Figure 3.6   Graph of $f(x) = |x - 2|$ (Example 5).

Solution   Applying the definition of absolute value we see that

$$|x - 2| = \begin{cases} x - 2 & \text{if } x - 2 \geq 0 \\ -(x - 2) & \text{if } x - 2 < 0. \end{cases}$$

That is,

$$|x - 2| = \begin{cases} x - 2 & \text{if } x \geq 2 \\ 2 - x & \text{if } x < 2. \end{cases}$$

(See Figure 3.6.)

We therefore evaluate the integral separately on $[-1, 2]$ and on $[2, 5]$, using the corresponding part of the definition of $|x - 2|$ on each interval:

$$\int_{-1}^{5} |x - 2|\,dx = \int_{-1}^{2} |x - 2|\,dx + \int_{2}^{5} |x - 2|\,dx$$

$$= \int_{-1}^{2} (2 - x)\,dx + \int_{2}^{5} (x - 2)\,dx$$

$$= \left[2x - \frac{x^2}{2}\right]_{-1}^{2} + \left[\frac{x^2}{2} - 2x\right]_{2}^{5}$$

$$= \left[\left(2 \cdot 2 - \frac{2^2}{2}\right) - \left(2(-1) - \frac{(-1)^2}{2}\right)\right]$$

$$+ \left[\left(\frac{5^2}{2} - 2 \cdot 5\right) - \left(\frac{2^2}{2} - 2 \cdot 2\right)\right]$$

$$= 9. \qquad\blacksquare$$

REMARK 4   Note that, in Examples 1–5, we needed to know an antiderivative F before we could apply the Fundamental Theorem to evaluate the given definite integral. In later chapters, we shall study several techniques to find antiderivatives, but there are many elementary functions such as $\sin(x^2)$, $\cos(x^2)$, $(\sin x)/x$, and $\sqrt{1 + x^4}$ for which we cannot find suitable antiderivatives. Therefore, although the Fundamental Theorem is the most common way to evaluate definite integrals, there are integrals such as $\int_a^b \cos(x^2)\,dx$ that cannot be evaluated in this way.

Functions Defined by Integrals

As we mentioned earlier in this section, part (i) of the Fundamental Theorem states that any continuous function f has A as one of its antiderivatives. Of course, this fact is significant because it leads directly to part (ii) of the Fundamental Theorem. But we shall also use part (i) to solve the antiderivative problem: Suppose we are given a function f, how do we define a function F whose derivative is f? The next example illustrates the way in which part (i) answers this question.

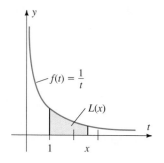

Figure 3.7 The function $L(x)$ is defined using an integral. That is, $L(x) = \int_1^x (1/t)\, dt$.

☐ **EXAMPLE 6**

In Chapter 5, we saw that the Power Rule for antiderivatives,

$$\int x^n\, dx = \frac{x^{n+1}}{n+1} + C,$$

holds for all $n \neq -1$. However, we shall need an antiderivative for $f(x) = 1/x$, and we obtain one by *defining*

$$L(x) = \int_1^x \frac{1}{t}\, dt, \qquad x > 0 \tag{12}$$

(see Figure 3.7). By part (i) of the Fundamental Theorem, we know that $L'(x) = f(x)$. Thus, we now have an antiderivative for $f(x) = 1/x$ for $x > 0$. ∎

Note that we cannot use part (ii) of the Fundamental Theorem to calculate values of L because the only way we know to express an antiderivative of f is by equation (12). Nevertheless, we can produce an accurate approximation to L using techniques to approximate the integral numerically. Figure 3.8 is a graph of L obtained by applying the Midpoint Rule with 20 subdivisions to estimate $L(x)$.

Ironically, even though this function L seems quite abstract, it is one of the most important functions used in applications of the calculus. Essentially we devote an entire chapter (Chapter 8) to a study of L and its applications. Other types of functions involving integrals with the variable x in one or both limits of integration may be differentiated using the Fundamental Theorem, the Chain Rule, and properties of the definite integral. Here are some examples.

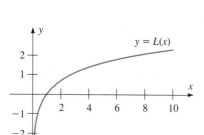

Figure 3.8 The graph of L as obtained by applying the Midpoint Rule using $n = 20$ to estimate $L(x) = \int_1^x (1/t)\, dt$.

☐ **EXAMPLE 7**

For $F(x) = \int_2^x \sqrt{t}\, \sec \pi t\, dt$, find $F'(x)$.

Solution The derivative follows immediately from the Fundamental Theorem, part (i):

$$F'(x) = \sqrt{x}\, \sec \pi x,$$

which is just the integrand evaluated at the upper limit x. ∎

☐ **EXAMPLE 8**

For $F(x) = \int_x^4 \sqrt{1 + t^2}\, dt$, find $F'(x)$.

Strategy · · · · · · · ·   **Solution**

Change order of integration to obtain   Since
x in upper limit, using Definition 4:

$$F(x) = \int_x^4 \sqrt{1 + t^2}\, dt = -\int_4^x \sqrt{1 + t^2}\, dt,$$

$$\int_a^b f(t)\, dt = -\int_b^a f(t)\, dt.$$

we have

Apply Fundamental Theorem, part (i).

$$F'(x) = -\frac{d}{dx}\left[\int_4^x \sqrt{1 + t^2}\, dt\right] = -\sqrt{1 + x^2}. \qquad \blacksquare$$

□ **EXAMPLE 9**

Find $\dfrac{d}{dx}\left[\displaystyle\int_1^{x^3} \dfrac{1}{1 + t^2}\, dt\right]$.

Strategy · · · · · · · ·   **Solution**

Express F as a composite function   The function
$y = G(u)$ with $u = x^3$ as upper limit
of integration.

$$F(x) = \int_1^{x^3} \frac{1}{1 + t^2}\, dt$$

equals $G(u)$ where

$$G(u) = \int_1^u \frac{1}{1 + t^2}\, dt$$

and $u = x^3$. By the Chain Rule we have

Apply Chain Rule and, in the first
factor, the Fundamental Theorem.

$$F'(x) = G'(u)u'(x)$$

$$= \frac{d}{du}\left[\int_1^u \frac{1}{1 + t^2}\, dt\right]\frac{d}{dx}(x^3)$$

$$= \frac{1}{1 + u^2}3x^2$$

$$= \frac{3x^2}{1 + (x^3)^2} = \frac{3x^2}{1 + x^6}. \qquad \blacksquare$$

□ **EXAMPLE 10**

For $F(x) = \displaystyle\int_{\sqrt{x}}^{x^2+3} \cos^3 t\, dt, \quad x > 0, \quad$ find $F'(x)$.

Solution   We first use Theorem 5 to write

$$F(x) = \int_{\sqrt{x}}^a \cos^3 t\, dt + \int_a^{x^2+3} \cos^3 t\, dt$$

for any constant a. Then, proceeding as in Example 8, we reverse limits in the first
integral to obtain

$$F(x) = -\int_a^{\sqrt{x}} \cos^3 t\, dt + \int_a^{x^2+3} \cos^3 t\, dt$$

and apply the Chain Rule and the Fundamental Theorem to obtain

$$F'(x) = (-\cos^3 \sqrt{x})\left(\frac{d}{dx}\sqrt{x}\right) + [\cos^3(x^2 + 3)]\left[\frac{d}{dx}(x^2 + 3)\right]$$

$$= -\frac{\cos^3 \sqrt{x}}{2\sqrt{x}} + 2x\cos^3(x^2 + 3).$$ ∎

The Average Value of a Function

An important step in the proof of the Fundamental Theorem is an analysis of the quantity

$$\frac{A(x + h) - A(x)}{h} = \frac{1}{h}\int_x^{x+h} f(t)\,dt \tag{13}$$

in terms of the function f (see approximation (4) and equation (10) at the beginning of this section). In fact, the expression on the right-hand side of equation (13) is simply the average value of f over the interval between x and $x + h$ in question. We complete this section with a discussion of averages of continuously varying quantities.

For example, consider a temperature function $T(t)$ as a function of time t over a one-week period ($0 \le t \le 7$), and suppose we want to determine the average temperature $\overline{T}$ during the week (see Figure 3.9). There are two reasonable ways to approach the calculation.

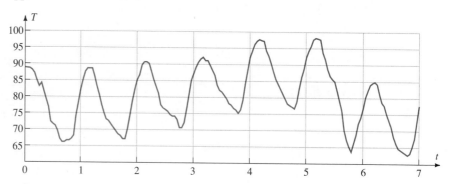

Figure 3.9   A graph of temperature over a one-week period in July in central Illinois.

1. We could approximate $\overline{T}$ by computing an average of the numbers $T(t_j)$, where the t_j are equally spaced times (for example, every hour on the hour). Then, assuming that there are n such numbers,

$$\overline{T} \approx \frac{\sum_{j=1}^{n} T(t_j)}{n}. \tag{14}$$

Then we could repeat this calculation using a smaller time interval Δt. As $\Delta t \to 0$, the approximate averages limit to the average temperature.

2. We could determine the horizontal line ℓ such that the signed area of the region between the graph of T and the line ℓ is zero (see Figure 3.10). In other words, if we write the equation for ℓ as $y = T$, we want the number T such that

$$\int_0^7 (T(t) - \overline{T})\,dt = 0.$$

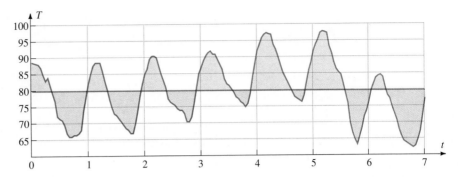

Figure 3.10   The temperature graph from Figure 3.9 along with the line $y = \overline{T}$ corresponding to the average temperature $\overline{T} = 79.94$.

Using this equation, we solve for $\overline{T}$

$$\int_0^7 T(t)\, dt = \int_0^7 \overline{T}\, dt$$
$$= \overline{T}(7 - 0) = 7\overline{T}.$$

In other words,

$$\overline{T} = \frac{1}{7} \int_0^7 T(t)\, dt.$$

The relationship between these two approaches is not immediately apparent. However, using Riemann sums, we can show that they yield identical results. Consider approach (1) in terms of Riemann sums. The time period in question is $0 \le t \le 7$. Therefore, dividing the interval $[a, b] = [0, 7]$ into n equal subintervals yields $\Delta t = (b - a)/n$. The approximate average on the right-hand side of approximation (14) does not look like a Riemann sum as it is written, but by multiplying both the numerator and denominator by $(b - a)$, we obtain the necessary factor of Δt. That is,

$$\overline{T} \approx \frac{\displaystyle\sum_{j=1}^{n} T(t_j)}{n}$$
$$= \frac{(b - a) \displaystyle\sum_{j=1}^{n} T(t_j)}{(b - a)n}$$
$$= \frac{1}{(b - a)} \left[\sum_{j=1}^{n} T(t_j) \frac{(b - a)}{n} \right]$$
$$= \frac{1}{(b - a)} \left[\sum_{j=1}^{n} T(t_j) \Delta t \right].$$

Thus, as the number of sample temperatures $n \to \infty$, we have $\Delta t \to 0$, and we obtain

$$\overline{T} = \frac{1}{(b - a)} \int_a^b T(t)\, dt.$$

This logic is valid for any continuous function f and motivates Definition 5.

Definition 5	Let f be a continuous function on $[a, b]$. The **average value** (or mean value) of f on $[a, b]$ is the number

$$\bar{f} = \frac{1}{b-a} \int_a^b f(x)\,dx.$$

A geometric interpretation of $\bar{f}$ when f is nonnegative on $[a, b]$ is illustrated in Figure 3.11. Since Definition 5 implies

$$\bar{f} \cdot (b-a) = \int_a^b f(x)\,dx,$$

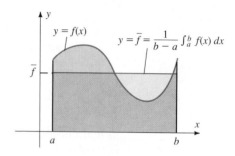

Figure 3.11 The average value $\bar{f}$ satisfies the equation $\bar{f} \cdot (b-a) = \int_a^b f(x)\,dx$.

we see that $\bar{f}$ determines the height of a rectangle with base $[a, b]$ whose area is the same as the area under the graph of f on $[a, b]$.

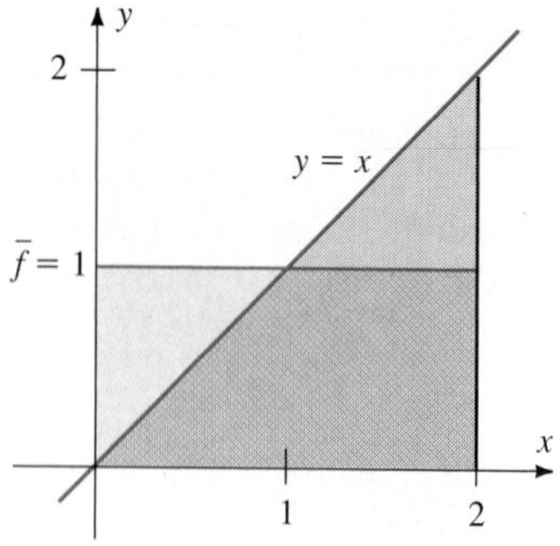

□ **EXAMPLE 11**

Since

$$\int_0^2 x\,dx = \left[\frac{x^2}{2}\right]_0^2 = 2,$$

the mean value for the function $f(x) = x$ on $[0, 2]$ is

$$\bar{f} = \frac{1}{(2-0)} \int_0^2 x\,dx = \frac{1}{2}(2) = 1 \qquad \text{(Figure 3.12)}.$$ ■

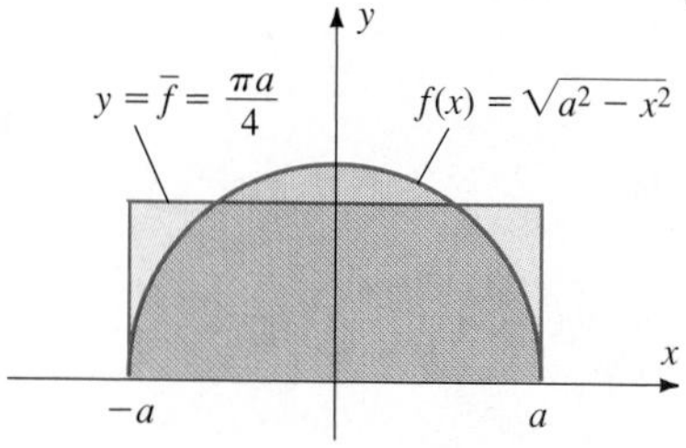

Figure 3.12 Mean value for $f(x) = x$ on $[0, 2]$ is $\bar{f} = 1$.

□ **EXAMPLE 12**

In Example 3, Section 6.2, we showed that

$$\int_{-a}^a \sqrt{a^2 - x^2}\,dx = \frac{\pi a^2}{2}.$$

The mean value for the function $f(x) = \sqrt{a^2 - x^2}$ on $[-a, a]$ is therefore

$$\bar{f} = \frac{1}{a-(-a)} \int_{-a}^a \sqrt{a^2 - x^2}\,dx = \frac{1}{2a}\left(\frac{\pi a^2}{2}\right) = \frac{\pi a}{4}.$$

Figure 3.13 Mean value for $f(x) = \sqrt{a^2 - x^2}$ on $[-a, a]$ is $\bar{f} = \pi a/4$.

(See Figure 3.13.) ■

Suppose f is a continuous function on $[a, b]$ with minimum value m and maximum value M. By Theorem 6, we know that

$$m(b - a) \le \int_a^b f(x)\,dx \le M(b - a).$$

If we divide this equation by $(b - a)$, we obtain

$$m \le \overline{f} \le M.$$

In other words, the average value is between the minimum and maximum values. Using the Intermediate Value Theorem (Section 2.6), we obtain the Mean Value Theorem for Integrals.

Theorem 8

The Mean Value Theorem for Integrals

If f is continuous on $[a, b]$, there exists a number $c \in [a, b]$ such that

$$f(c) = \overline{f} = \frac{1}{b - a} \int_a^b f(x)\,dx.$$

In other words, the graph of f crosses the horizontal line $y = \overline{f}$ at least once over the interval $[a, b]$ (see Figure 3.14).

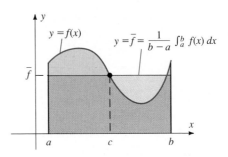

Figure 3.14   The average value $\overline{f}$ is equal to $f(c)$ for some $c \in [a, b]$.

Exercise Set 6.3

In each of Exercises 1–26, evaluate the definite integral using the Fundamental Theorem of Calculus.

1. $\displaystyle\int_{-2}^{2} (x^3 - 1)\,dx$

2. $\displaystyle\int_{0}^{4} (x^2 - 2x + 3)\,dx$

3. $\displaystyle\int_{-1}^{2} \left(\frac{2}{x^3} + 5x\right) dx$

4. $\displaystyle\int_{0}^{3} t(\sqrt[3]{t} - 2)\,dt$

5. $\displaystyle\int_{1}^{3} (t^2 + 2)^2\,dt$

6. $\displaystyle\int_{0}^{1} (x^{3/5} - x^{5/3})\,dx$

7. $\displaystyle\int_{1}^{2} \frac{1 - x}{x^3}\,dx$

8. $\displaystyle\int_{4}^{9} \left(\sqrt{x} - \frac{1}{\sqrt{x}}\right) dx$

9. $\displaystyle\int_{2}^{0} x(\sqrt{x} - 1)\,dx$

10. $\displaystyle\int_{9}^{4} \left(\sqrt{x} + \frac{1}{\sqrt{x}}\right) dx$

11. $\displaystyle\int_{0}^{4} |x - 3|\,dx$

12. $\displaystyle\int_{-2}^{2} |1 - x^2|\,dx$

13. $\displaystyle\int_{0}^{4} |x - \sqrt{x}|\,dx$

14. $\displaystyle\int_{1}^{4} (u^3 - \sqrt{u})\,du$

15. $\displaystyle\int_{4}^{9} x^{1/2}(1 - x^{3/2})\,dx$

16. $\displaystyle\int_{0}^{4} |9 - x^2|\,dx$

17. $\displaystyle\int_{-1}^{4} |x^2 - x - 2|\,dx$

18. $\displaystyle\int_{0}^{\pi/2} \cos t\,dt$

19. $\displaystyle\int_{0}^{\pi} \sin u\,du$

20. $\displaystyle\int_{0}^{\pi/4} \sec x \tan x\,dx$

21. $\displaystyle\int_{-\pi/3}^{\pi/3} \sec^2 \theta\,d\theta$

22. $\displaystyle\int_{0}^{\pi/6} \tan^2 x\,dx$

23. $\displaystyle\int_{-2}^{3} f(x)\,dx,$ where $f(x) = \begin{cases} x^2 - 1 & \text{for } x \le 1 \\ x - 1 & \text{for } x > 1 \end{cases}$

24. $\displaystyle\int_1^5 f(x)\,dx,$ where $f(x) = \begin{cases} \sqrt{x} & \text{for } \;0 \le x \le 4 \\ 2x^2 - 7x - 2 & \text{for } \;x > 4 \end{cases}$

25. $\displaystyle\int_0^\pi \sin 1\,dx$ **26.** $\displaystyle\int_0^{\pi/3} \sec^2 1\,dx$

In Exercises 27–30, find the average value

$$\bar{f} = \frac{1}{b - a}\int_a^b f(x)\,dx$$

of the function f on the given interval.

27. $f(x) = x^2 + x + 1, \quad [0, 1]$

28. $f(x) = \sin x, \quad [0, 2\pi]$

29. $f(x) = \sec^2 x, \quad [0, \pi/4]$

30. $f(x) = \sec x \tan x \quad [0, \pi/4]$

In Exercises 31–34, express F as a function not involving an integral sign using part (ii) of the Fundamental Theorem. Then find $F'(x)$ by differentiation.

31. $F(x) = \displaystyle\int_1^x t\,dt$ **32.** $F(x) = \displaystyle\int_{-2}^x (t^2 + 2t)\,dt$

33. $F(x) = \displaystyle\int_0^x \cos t\,dt$

34. $F(x) = \displaystyle\int_0^x (at^2 + bt + c)\,dt$

In Exercises 35–44, find $F'(x)$ using part (i) of the Fundamental Theorem: $\dfrac{d}{dx}\left[\displaystyle\int_a^x f(t)\,dt\right] = f(x).$

35. $F(x) = \displaystyle\int_2^x t^2 \sin t^2\,dt$ **36.** $F(x) = \displaystyle\int_0^x \sqrt{t^4 + 1}\,dt$

37. $F(x) = \displaystyle\int_x^1 t^3 \cos^2 t\,dt$ **38.** $F(x) = \displaystyle\int_x^5 \sqrt{4 + t^3}\,dt$

39. $F(x) = \displaystyle\int_0^{3x} \sqrt{1 + \sin t}\,dt$ **40.** $F(x) = \displaystyle\int_2^{x^2} \frac{1}{1 + t^3}\,dt$

41. $F(x) = \displaystyle\int_{x^2}^x \cos^3(t + 1)\,dt$ **42.** $F(x) = \displaystyle\int_{-x}^x \sqrt{t^2 + 1}\,dt$

43. $F(x) = x\displaystyle\int_3^{x^2} (t^2 + 1)^{-3}\,dt$

44. $F(x) = \cos x\displaystyle\int_{\sin x}^\pi \sqrt{t^4 + 1}\,dt$

45. Suppose that f and f' are continuous on an open interval I containing the numbers a and b. Simplify the following expressions by eliminating the integral sign.

a. $\dfrac{d}{dx}\left[\displaystyle\int_a^x f(t)\,dt\right]$ **b.** $\dfrac{d}{dx}\left[\displaystyle\int_x^b f(t)\,dt\right]$

c. $\dfrac{d}{dx}\left[\displaystyle\int_a^b f(t)\,dt\right]$ **d.** $\dfrac{d}{dx}\left[\displaystyle\int f(x)\,dx\right]$

e. $\displaystyle\int_a^b f'(x)\,dx$ **f.** $\displaystyle\int f'(x)\,dx$

46. Find all numbers c that satisfy the Mean Value Theorem for Integrals for the given function and interval.
a. $f(x) = 2x + 5, \quad [4, 6]$
b. $f(x) = 3x^2 - 4x + 1, \quad [1, 4]$

47. Find all numbers c that satisfy the Mean Value Theorem for Integrals for the given function and interval.
a. $f(x) = |x - 2|, \quad [1, 4]$
b. $f(x) = 4 - x^2, \quad [-2, 3]$

48. Find the number a so that the average value of the function $f(x) = x^2 + ax + 3$ on the interval $[0, 6]$ is 27.

49. Find the number b so that the average value of the function $f(x) = 2x + 3$ on the interval $[1, b]$ is 11.

50. Find the numbers b such that the average value of the function $f(x) = 4x - x^2$ on the interval $[0, b]$ is 1.

51. Let f be the function graphed in Figure 3.15 and let

$$A(x) = \int_0^x f(t)\,dt, \quad 0 \le x \le 8.$$

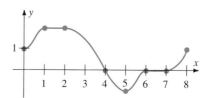

Figure 3.15

a. On which intervals is A increasing?
b. Is A constant on any interval? If so, which?
c. Where do the relative maxima occur?
d. Determine the value $A(0)$.

52. Show that the average value for $f(x) = \sin x$ on $[0, 2\pi]$ is the same as the average value for $f(x) = \cos x$ on $[0, 2\pi]$. Do these average values occur at the same numbers $c \in [0, 2\pi]$?

53. Let f and g be continuous on $[a, b]$. Let $\bar{f}$ be the average value for f on $[a, b]$ and let $\bar{g}$ be the average value for g on $[a, b]$. Show that if $f(x) \le g(x)$ for all $x \in [a, b]$ then $\bar{f} \le \bar{g}$.

54. Show that the converse of the statement in Exercise 53 is false by finding an example of two continuous functions f and g on an interval $[a, b]$ for which $\bar{f} \le \bar{g}$ but for which $f(x) > g(x)$ for some $x \in [a, b]$.

55. Let f be continuous on $[a, b]$ with maximum value M and average value $\bar{f}$. Show that if $\bar{f} = M$, then f is constant on $[a, b]$ and $f(x) = M$ for all $x \in [a, b]$.

56. Determine

$$\lim_{n\to\infty} \left(\frac{\pi}{2n}\right) \sum_{j=1}^{n} \left(\cos\frac{j\pi}{2n}\right)$$

by (i) interpreting the sum as a Riemann sum and then (ii) interpreting the limit as an integral.

57. Determine

$$\lim_{n\to\infty} \left(\frac{x}{n}\right) \sum_{j=1}^{n} \left(\sin\frac{jx}{n}\right)$$

(see Exercise 56).

58. Let $f(x) = \cos(x^2)$ and $A(x) = \int_0^x \cos(t^2)\, dt$.
  a. Graph f over the interval $[0, 3]$. Using this graph, predict the graph of A over $[0, 3]$. Where is A increasing? decreasing? Is A nonnegative on $[0, 3]$?
  b. Using the Midpoint Rule with $n = 10$, approximate the values of $A(x)$ for $x = 0.5, 1.0, 1.5, 2.0,$ and 2.5 (see Example 9 in Section 6.2).

  c. Using the Midpoint Rule with $n = 10$ (for each number x), graph an approximation to the graph of A. Does this graph agree with your predictions in part a?
  d. On the graph of f, identify the numbers x such that $A(x)$ is a relative extremum of A. Then use the First Derivative Test to classify these extrema.

59. Suppose that the rate of change dp/dt of a certain population p of rabbits satisfies the equation

$$\frac{dp}{dt} = \frac{100 - 25t}{t^2 - 8t + 16.1},$$

where t is time measured in years.
  a. Graph dp/dt for $t \geq 0$. From this graph, determine when the rabbit population reaches a maximum.
  b. Assuming that $p(0) = 50$, use the Midpoint Rule to graph

$$p(t) = \int_0^t \frac{dp}{ds}(s)\, ds,$$

  with $n = 20$.
  c. What is the maximum number of rabbits?
  d. Will the population eventually die out? If so, when?

6.4 Substitution in Definite Integrals

Whenever we know an antiderivative F of a function f, we can evaluate the definite integral $\int_a^b f(x)\, dx$ using part (ii) of the Fundamental Theorem of Calculus

$$\int_a^b f(x)\, dx = F(b) - F(a).$$

Thus, any technique for evaluating indefinite integrals is also a technique for evaluating definite integrals.

In Chapter 5, we used the Chain Rule in reverse to produce a useful antidifferentiation technique called substitution. Given the Fundamental Theorem of Calculus, we now interpret substitution as a method for evaluating definite integrals.

Recall that, if we use the differential notation $du = g'(x)\, dx$ when $u = g(x)$, the method of substitution is

$$\int f(g(x))g'(x)\, dx = \int f(u)\, du = F(u) + C = F(g(x)) + C. \tag{1}$$

We can use equation (1) in two different ways to evaluate integrals.

In the first method, we solve the antidifferentiation problem with equation (1) and evaluate it using any one of the resulting antiderivatives.

□ **EXAMPLE 1**

Find $\displaystyle\int_{-1}^{3} \frac{x}{\sqrt{7 + x^2}}\, dx$.

Strategy · · · · · · · ·
Identify that the antiderivative can be determined using the substitution $u = 7 + x^2$.

Solution
To find the antiderivative $\displaystyle\int \frac{x\, dx}{\sqrt{7 + x^2}}$, we use the substitution

$$u = 7 + x^2; \quad du = 2x\, dx.$$

Then $x\,dx = \frac{1}{2}\,du$, and we have

$$\int \frac{x\,dx}{\sqrt{7+x^2}} = \int \frac{\frac{1}{2}\,du}{\sqrt{u}} = \frac{1}{2}\int u^{-1/2}\,du = \frac{1}{2}(2u^{1/2} + C_1)$$
$$= \sqrt{7+x^2} + C.$$

Evaluate $F(3) - F(-1)$ using the antiderivative $F(x) = \sqrt{7+x^2}$.

We therefore use $F(x) = \sqrt{7+x^2}$ and the Fundamental Theorem to find that

$$\int_{-1}^{3} \frac{x\,dx}{\sqrt{7+x^2}} = \left[\sqrt{7+x^2}\right]_{-1}^{3} = \sqrt{7+3^2} - \sqrt{7+(-1)^2}$$
$$= \sqrt{16} - \sqrt{8} = 4 - 2\sqrt{2}. \qquad \blacksquare$$

Note that we expressed the antiderivative in terms of the original variable x as $\sqrt{7+x^2}$ rather than as $\sqrt{u}$. Then we evaluated at $x = -1$ and $x = 3$.

NOTATION   In this example, we wrote the integral as

$$\int_{-1}^{3} \frac{x\,dx}{\sqrt{7+x^2}} \quad \text{rather than as} \quad \int_{-1}^{3} \frac{x}{\sqrt{7+x^2}}\,dx.$$

Writing the dx as part of the fraction is simply a notational convenience.

The second way to use equation (1) is to apply the substitution $u = g(x)$ to the limits of integration as well as to the integrand. This approach is justified by the following theorem.

Theorem 9

Let g' be continuous on $[a, b]$ and let f be continuous on an interval I containing the values $u = g(x)$, $x \in [a, b]$. If f has an antiderivative F on I, then

$$\int_{a}^{b} f(g(x))g'(x)\,dx = \int_{u_a}^{u_b} f(u)\,du = F(u_b) - F(u_a) \qquad (2)$$

where $u_a = g(a)$ and $u_b = g(b)$.

Proof: From equation (1) we see that the composite function $F \circ g$ is an antiderivative for $(f \circ g)g'$. Thus, by the Fundamental Theorem,

$$\int_{a}^{b} f(g(x))g'(x)\,dx = \left[F(g(x))\right]_{a}^{b} = F(g(b)) - F(g(a)) = F(u_b) - F(u_a).$$

But, again using the Fundamental Theorem, we can write this last expression as

$$F(u_b) - F(u_a) = \left[F(u)\right]_{u_a}^{u_b} = \int_{u_a}^{u_b} f(u)\,du. \qquad \blacksquare$$

☐ **EXAMPLE 2**

Use Theorem 9 to evaluate the definite integral $\displaystyle\int_{-1}^{3} \frac{x\,dx}{\sqrt{7+x^2}}$ in Example 1.

Strategy · · · · · · · ·
Identify that the integral can be simplified using (2) and the substitution $u = 7 + x^2$.

Solution
As before, we use the substitution

$$u = 7 + x^2; \quad du = 2x\,dx;$$

so $x \, dx = \frac{1}{2} \, du$. Also, we change limits by noting that

Determine $u_a = u(-1)$ and $u_b = u(3)$.

(i)   when $x = -1$, $u = 7 + (-1)^2 = 8$,   and
(ii) when $x = 3$, $u = 7 + (3)^2 = 16$.

Thus, by Theorem 9,

Substitute in both the limits and the integrand.

$$\int_{-1}^{3} \frac{x \, dx}{\sqrt{7 + x^2}} = \frac{1}{2} \int_{8}^{16} u^{-1/2} \, du$$

$$= \frac{1}{2} \left[2u^{1/2} \right]_{8}^{16}$$

$$= \sqrt{16} - \sqrt{8} = 4 - 2\sqrt{2}. \qquad \blacksquare$$

Compare the solutions in Examples 1 and 2, and note that, in Example 2, we did not convert the antiderivative $\sqrt{u}$ back into a function expressed in terms of the original variable x. Instead, we converted the limits of integration to their corresponding u values, u_a and u_b.

□ **EXAMPLE 3**

Find $\displaystyle\int_{0}^{\pi/4} \tan^2 x \sec^2 x \, dx$.

Solution   To find an antiderivative for the function $\tan^2 x \sec^2 x$, we use the substitution

$$u = \tan x; \quad du = \sec^2 x \, dx.$$

To transform the limits of integration via the substitution, we note that

if $x = 0$,   then $u = \tan(0) = 0$;
if $x = \pi/4$,   then $u = \tan(\pi/4) = 1$.

Thus,

$$\int_{0}^{\pi/4} \tan^2 x \sec^2 x \, dx = \int_{0}^{1} u^2 \, du = \left[\frac{u^3}{3} \right]_{0}^{1} = \frac{1^3}{3} - \frac{0^3}{3} = \frac{1}{3}. \qquad \blacksquare$$

As we saw in Section 5.2, there may be more than one substitution that leads to a successful evaluation of the integral. In Exercise 35, you are asked to evaluate the integral in Example 3 using the substitution $u = \sec x$.

□ **EXAMPLE 4**

Find $\displaystyle\int_{-1}^{2} \frac{x^2 \, dx}{(x^3 + 4)^2}$.

Solution   Here we use the substitution

$$u = x^3 + 4; \quad du = 3x^2 \, dx.$$

Thus, $x^2 \, dx = \frac{1}{3} \, du$. Also,

if $x = -1$, $u = (-1)^3 + 4 = -1 + 4 = 3$;
if $x = 2$, $u = 2^3 + 4 = 8 + 4 = 12$.

Thus,

$$\int_{-1}^{2} \frac{x^2 \, dx}{(x^3 + 4)^2} = \int_{3}^{12} \frac{\frac{1}{3} \, du}{u^2} = \frac{1}{3} \int_{3}^{12} u^{-2} \, du$$

$$= \frac{1}{3} \left[\frac{u^{-1}}{-1} \right]_{3}^{12} = -\frac{1}{3} \left(\frac{1}{12} - \frac{1}{3} \right) = \frac{1}{12}.$$ ∎

Occasionally a substitution transforms an integral with lower limit a and upper limit b, where $a < b$, to an integral with lower limit u_a and upper limit u_b, where $u_a > u_b$. Theorem 9 still applies, and if you evaluate exactly as indicated in equation (2), you obtain the correct result.

□ **EXAMPLE 5**

Evaluate

$$\int_{0}^{\pi/2} \cos^2 x \sin x \, dx.$$

Solution   Let $u = \cos x$. Then $du = -\sin x \, dx$ and $\sin x \, dx = -1 \, du$. Also,

if $x = 0$, $u = \cos 0 = 1$;
if $x = \pi/2$, $u = \cos(\pi/2) = 0$.

Using equation (2), we get

$$\int_{0}^{\pi/2} \cos^2 x \sin x \, dx = \int_{1}^{0} u^2(-1) \, du$$

(note the limits of integration). We bring the constant factor of -1 outside the integral and evaluate

$$-\int_{1}^{0} u^2 \, du.$$

At this point, we can do one of two things. We can evaluate using the Fundamental Theorem

$$-\int_{1}^{0} u^2 \, du = -\left[\frac{u^3}{3} \right]_{1}^{0} = -\left[\frac{0^3}{3} - \frac{1^3}{3} \right] = \frac{1}{3};$$

or we can use Definition 4 to interchange the limits of integration and then apply the Fundamental Theorem

$$-\int_{1}^{0} u^2 \, du = \int_{0}^{1} u^2 \, du = \left[\frac{u^3}{3} \right]_{0}^{1} = \frac{1}{3}.$$ ∎

Exercise Set 6.4

In each of Exercises 1–6, use the Fundamental Theorem of Calculus and the designated substitution to evaluate the integral.

1. $\int_{0}^{\pi/4} \cos 2x \, dx, \quad u = 2x$

2. $\int_{0}^{1} x(x^2 + 1)^{100} \, dx, \quad u = x^2 + 1$

3. $\int_{0}^{\pi/6} \sec(2x) \tan(2x) \, dx, \quad u = 2x$

4. $\int_{0}^{\pi/12} \sec^2 3x \, dx, \quad u = 3x$

5. $\int_{1}^{2} (2 - x^2)(x^3 - 6x + 7)^{50} \, dx, \quad u = x^3 - 6x + 7$

6. $\displaystyle\int_0^3 x\sqrt{9 - x^2}\, dx, \quad u = 9 - x^2$

In Exercises 7–33, evaluate the definite integral.

7. $\displaystyle\int_0^{\pi/4} (1 - \cos 2x)\, dx$

8. $\displaystyle\int_0^{2\pi/3} \sin(2x)\, dx$

9. $\displaystyle\int_0^1 2x(x^2 + 1)^2\, dx$

10. $\displaystyle\int_0^1 x\sqrt{1 - x^2}\, dx$

11. $\displaystyle\int_1^2 x\sqrt{4 - x^2}\, dx$

12. $\displaystyle\int_{-\pi/6}^0 \sec^2(2x)\, dx$

13. $\displaystyle\int_0^{\sqrt{\pi/2}} t \sin(\pi - t^2)\, dt$

14. $\displaystyle\int_0^{\pi/2} \cos^3 t \sin t\, dt$

15. $\displaystyle\int_{-\pi/4}^{\pi/4} \frac{\sin x}{\cos^2 x}\, dx$

16. $\displaystyle\int_0^{\pi/4} \tan \theta \sec^2 \theta\, d\theta$

17. $\displaystyle\int_0^4 \frac{dt}{\sqrt{2t + 1}}$

18. $\displaystyle\int_0^2 \frac{x}{\sqrt{16 + x^2}}\, dx$

19. $\displaystyle\int_{-1}^1 \frac{x}{(1 + x^2)^3}\, dx$

20. $\displaystyle\int_0^3 x\sqrt{9 - x^2}\, dx$

21. $\displaystyle\int_{\pi^2/4}^{\pi^2} \frac{\sin \sqrt{x}}{\sqrt{x}}\, dx$

22. $\displaystyle\int_0^{\pi/4} \sin x\sqrt{\cos x}\, dx$

23. $\displaystyle\int_1^2 \frac{x}{(2x^2 - 1)^3}\, dx$

24. $\displaystyle\int_0^2 t|t^2 - 1|^{10}\, dt$

25. $\displaystyle\int_1^4 s|2 - s|^{12}\, ds$

26. $\displaystyle\int_0^2 x\sqrt{|1 - x^2|}\, dx$

27. $\displaystyle\int_0^1 \frac{1}{(4 - x)^2}\, dx$

28. $\displaystyle\int_0^{\pi/4} \sin^2 x\, dx$ (*Hint:* Use the identity
$\sin^2 x = \frac{1}{2} - \frac{1}{2} \cos 2x$.)

29. $\displaystyle\int_0^{\pi} \cos^2 x\, dx$ (*Hint:* Use the identity
$\cos^2 x = \frac{1}{2} + \frac{1}{2} \cos 2x$.)

30. $\displaystyle\int_0^{\sqrt{\pi/4}} u \sec(u^2) \tan(u^2)\, du$

31. $\displaystyle\int_0^{\pi/4} \sec^2 u\sqrt{\tan u}\, du$

32. $\displaystyle\int_{-\pi/4}^{\pi/4} \tan^2 \theta\, d\theta$

33. $\displaystyle\int_0^{\sqrt{\pi/2}} t \cos^3(t^2) \sin(t^2)\, dt$

34. The integral

$$\int_0^{\pi/2} \cos t\sqrt{1 - \sin^2 t}\, dt$$

can be evaluated in two different ways.
  a. Evaluate it using the identities $\sin^2 t + \cos^2 t = 1$ and
  $\cos^2 t = (1 + \cos 2t)/2$.
  b. Evaluate it using the substitution $u = \sin t$ and the area
  formula for the area of a circle.

35. Evaluate the integral in Example 3 using the substitution
$u = \sec x$. (*Hint:* Remember the identity $\sec^2 x = 1 + \tan^2 x$.)

In Exercises 36–39, find the average value

$$\bar{f} = \frac{1}{b - a} \int_a^b f(x)\, dx$$

of the function f on the given interval.

36. $f(x) = \sqrt{x + 2}, \quad [2, 7]$

37. $f(x) = x\sqrt{x^2 + 1}, \quad [0, 2]$

38. $f(x) = (x + 4)^{2/3}, \quad [-4, 4]$

39. $f(x) = x \cos x^2, \quad [0, \sqrt{\pi/2}]$

In Exercises 40–43, first express F as a function not involving an integral sign using part (ii) of the Fundamental Theorem. Then find $F'(x)$ by differentiation.

40. $\displaystyle F(x) = \int_2^x \sqrt{1 + t}\, dt, \quad x \geq -1$

41. $\displaystyle F(x) = \int_1^x \frac{t}{\sqrt{1 + t^2}}\, dt$

42. $\displaystyle F(x) = \int_0^x t \sin t^2\, dt$

43. $\displaystyle F(x) = \int_x^0 \cos^2 t \sin t\, dt$

6.5   Finding Areas by Integration

In this section, we return to the area problem, and we consider regions more complicated than those discussed in Sections 6.1 and 6.2. In particular, we use the definite integral to calculate the area of a region bounded by the graphs of two continuous functions as well as the area of a region determined by the graph of a continuous function with negative values.

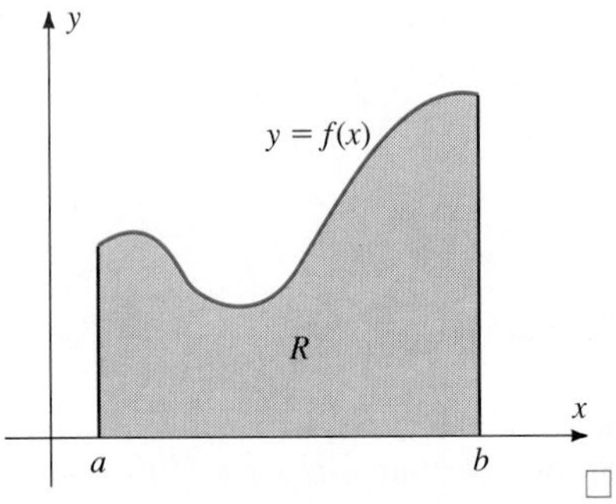

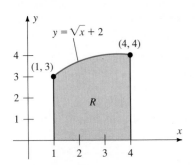

Figure 5.1 If $f(x) \geq 0$ for all x in $[a, b]$, then the area of R equals $\int_a^b f(x)\, dx$.

We begin by recalling Theorem 3 from Section 6.2. The area under the graph of a nonnegative function f over the interval $[a, b]$ equals the definite integral $\int_a^b f(x)\, dx$ (see Figure 5.1).

$$\text{Area of } R = \int_a^b f(x)\, dx \tag{1}$$

$$\text{if } f(x) \geq 0 \text{ for all } x \text{ in } [a, b].$$

☐ **EXAMPLE 1**

Find the area of the region R bounded by the graph of $f(x) = \sqrt{x} + 2$ and the x-axis between $x = 1$ and $x = 4$. (See Figure 5.2.)

Solution   Here $f(x) > 0$ for all x in $[1, 4]$ so we apply equation (1) with $a = 1$ and $b = 4$:

$$\begin{aligned}
\text{Area of } R &= \int_1^4 (\sqrt{x} + 2)\, dx \\
&= \int_1^4 (x^{1/2} + 2)\, dx \\
&= \left[\frac{2}{3} x^{3/2} + 2x \right]_1^4 \\
&= \left(\frac{2}{3} \cdot 4^{3/2} + 2 \cdot 4 \right) - \left(\frac{2}{3} \cdot 1^{3/2} + 2 \cdot 1 \right) \\
&= \frac{2}{3}(8) + 8 - \frac{2}{3} - 2 = \frac{32}{3}.
\end{aligned}$$ ■

Figure 5.2   Region R in Example 1.

In the next example the endpoints a and b defining the region R are not explicitly stated. They must be determined by finding the *zeros* of the function f.

☐ **EXAMPLE 2**

Find the area of the region bounded by the graph of $f(x) = 2 + x - x^2$ and the x-axis.

Strategy · · · · · · · ·

Sketch the graph to locate the region R.

Determine the interval $[a, b]$ defining the region by finding the zeros of f.

Use equation (1) to find the area of R.

Solution

The region bounded by the graph of $f(x) = 2 + x - x^2$ and the x-axis lies above the x-axis, as shown in Figure 5.3. To find the largest and smallest numbers x associated with this region we locate the zeros of f by factoring:

$$f(x) = 2 + x - x^2 = (2 - x)(1 + x).$$

Thus $x = a = 1$ and $x = b = 2$ are the zeros of f. The area of R is thus

$$\begin{aligned}
\text{Area} &= \int_{-1}^2 (2 + x - x^2)\, dx \\
&= \left[2x + \frac{x^2}{2} - \frac{x^3}{3} \right]_{-1}^2 \\
&= \left[2 \cdot 2 + \frac{2^2}{2} - \frac{2^3}{3} \right] - \left[2(-1) + \frac{(-1)^2}{2} - \frac{(-1)^3}{3} \right] \\
&= \frac{9}{2}.
\end{aligned}$$ ■

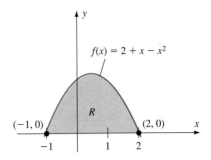

Figure 5.3 Region R in Example 2.

☐ **EXAMPLE 3**

Find the area of the region R bounded above by the graph of $y = |x - 3|$ and below by the x-axis for $1 \leq x \leq 6$.

Strategy · · · · · · · ·
Use definition of absolute value to rewrite $f(x) = |x - 3|$.

Solution

Since we do not have an antiderivative for $f(x) = |x - 3|$, we cannot immediately apply equation (1). However, using the definition of absolute value we may rewrite f as

$$f(x) = |x - 3| = \begin{cases} x - 3 & \text{if } x \geq 3 \\ 3 - x & \text{if } x < 3. \end{cases}$$

Calculate areas of regions over $[1, 3]$ and $[3, 6]$ separately, writing $f(x)$ as a linear function in each interval.

Now we calculate the areas of regions R_1 and R_2 separately as illustrated in Figure 5.4. We obtain

$$\text{Area of } R = \text{Area of } R_1 + \text{Area of } R_2$$

$$= \int_1^3 (3 - x)\, dx + \int_3^6 (x - 3)\, dx$$

$$= \left[\left(3x - \frac{x^2}{2}\right)\right]_1^3 + \left[\left(\frac{x^2}{2} - 3x\right)\right]_3^6$$

$$= \left[\left(3 \cdot 3 - \frac{3^2}{2}\right) - \left(3 \cdot 1 - \frac{1^2}{2}\right)\right]$$

$$+ \left[\left(\frac{6^2}{2} - 3 \cdot 6\right) - \left(\frac{3^2}{2} - 3 \cdot 3\right)\right]$$

$$= \frac{13}{2}.$$

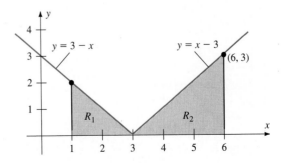

Figure 5.4 Region bounded by the graph of $f(x) = |x - 3|$ and the x-axis for $1 \leq x \leq 6$.

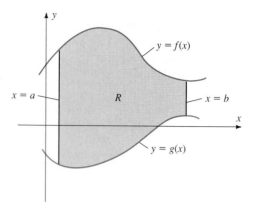

Figure 5.5 Region bounded by two graphs.

The Area of a Region Bounded by Two Curves

Figure 5.5 shows a region R bounded by the graphs of two continuous functions—above by the graph of $y = f(x)$ and below by the graph of $y = g(x)$—for values of x between a and b. We can use the definite integral to compute the area of such regions. That is, if we divide the interval $[a, b]$ into n subintervals of equal length $\Delta x = (b - a)/n$ and select one "test number" t_j in each subinterval, we can approximate the part of the region corresponding to that subinterval by the rectangle with width Δx and height $[f(t_j) - g(t_j)]$. (See Figures 5.6 and 5.7.) Summing the individual approximations then gives the approximation to the area of R:

$$\text{Area of } R \approx \sum_{j=1}^{n} [f(t_j) - g(t_j)] \, \Delta x. \tag{2}$$

Now as n increases without bound the limit of the approximating sum on the right-hand side of equation (2) is the integral

$$\int_{a}^{b} [f(x) - g(x)] \, dx = \lim_{n \to \infty} \sum_{j=1}^{n} [f(t_j) - g(t_j)] \, \Delta x \tag{3}$$

which we define to be the area of R.

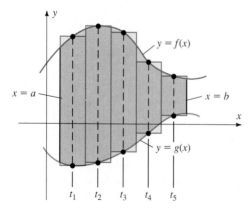

Figure 5.6 Approximating R by rectangles.

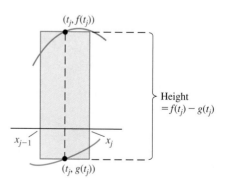

Figure 5.7 Area of jth rectangle is $[f(t_j) - g(t_j)]\Delta x$.

Definition 6	The area of the region R bounded above by the graph of the continuous function f and below by the graph of the continuous function g, between $x = a$ and $x = b$, is

$$\text{Area of } R = \int_a^b [f(x) - g(x)] \, dx. \tag{4}$$

(Note that this requires $f(x) \geq g(x)$ for all x. See Figure 5.5.)

Note that equation (4) generalizes equation (1), since $g(x) = 0$ when the region R is bounded below by the x-axis. Notice also that there are no restrictions on the sign of $f(x)$ or $g(x)$ in equation (4)—either may be positive or negative. *It is essential that $f(x) \geq g(x)$, however,* so that the integrand $[f(x) - g(x)]$ is nonnegative.

□ **EXAMPLE 4**

Find the area of the region R bounded by the graphs of the equations $y = -x^2 + 5x - 4$ and $y = -x - 4$ between $x = 0$ and $x = 6$. (See Figure 5.8.)

Strategy · · · · · · ·
Determine which graph is on top. (To find area we must integrate top curve minus bottom.)

Apply equation (4).

Solution

By graphing the two equations (or by checking individual values) we find that the graph of $f(x) = -x^2 + 5x - 4$ forms the top boundary while the graph of $g(x) = -x - 4$ is the bottom boundary of the region R. Applying equation (4) gives

$$\text{Area of } R = \int_0^6 [(-x^2 + 5x - 4) - (-x - 4)] \, dx$$

$$= \int_0^6 (-x^2 + 6x) \, dx$$

$$= \left[-\frac{x^3}{3} + 3x^2 \right]_0^6$$

$$= -\frac{6^3}{3} + 3 \cdot 6^2 = 36. \qquad ■$$

In Example 4 the region R was completely determined by the graphs of the two functions f and g, even though the numbers $x = 0$ and $x = 6$ were provided in the statement of the problem. If these numbers are not given, it is necessary to solve the equation $f(x) = g(x)$ first in order to determine the horizontal extremities of the region R.

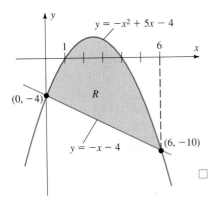

Figure 5.8 Region R in Example 4.

□ **EXAMPLE 5**

Find the area of the region bounded by the graphs of the functions $y = x^4 + 1$ and $y = 2x^2$.

Strategy · · · · · · ·
Sketch region.

Find points of intersection.

Solution

A rough sketch of the region shows that the graphs intersect at two points. To find these points we equate the two functions and obtain

$$x^4 + 1 = 2x^2$$

Solve by factoring (if possible).

or

$$x^4 - 2x^2 + 1 = (x^2 - 1)^2 = 0$$

so $x^2 - 1 = 0$ and $x = \pm 1$. The points of intersection are therefore $(-1, 2)$ and $(1, 2)$.

Determine which curve is on top (see Figure 5.9).

Use Definition 6.

Checking any particular number x in $(-1, 1)$ shows that $f(x) = x^4 + 1$ determines the upper boundary and $g(x) = 2x^2$ determines the lower boundary. Thus by Definition 6

$$\begin{aligned}
A &= \int_{-1}^{1} [(x^4 + 1) - 2x^2] \, dx \\
&= \left[\frac{x^5}{5} + x - 2\frac{x^3}{3} \right]_{x=-1}^{x=1} \\
&= \left(\frac{1}{5} + 1 - \frac{2}{3} \right) - \left(-\frac{1}{5} - 1 + \frac{2}{3} \right) = \frac{16}{15}.
\end{aligned}$$

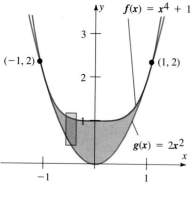

Figure 5.9

The graphs bounding the region may cross several times. In such cases, we apply Definition 6 in each of the resulting subregions, being careful to note the upper and lower boundaries of each region.

□ **EXAMPLE 6**

Find the area of the region between the graphs of $f(x) = \sin x$ and $g(x) = 1/2$ for $0 \le x \le 2\pi$.

Solution   The two graphs cross where $\sin x = 1/2$. The solutions of this equation in $[0, 2\pi]$ are $x = \pi/6$ and $x = 5\pi/6$. By checking particular values of x we can see that, on the resulting intervals,

$$\begin{aligned}
\sin x &\le 1/2 \quad \text{for} \quad x \in [0, \pi/6], \\
\sin x &\ge 1/2 \quad \text{for} \quad x \in [\pi/6, 5\pi/6], \\
\sin x &\le 1/2 \quad \text{for} \quad x \in [5\pi/6, 2\pi].
\end{aligned}$$

(See Figure 5.10.)

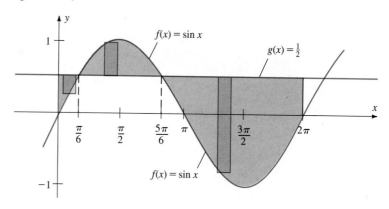

Figure 5.10   Area bounded by $f(x) = \sin x$ and $g(x) = 1/2$, $0 \le x \le 2\pi$.

Thus,

$$A = \int_0^{\pi/6} \left(\frac{1}{2} - \sin x \right) dx + \int_{\pi/6}^{5\pi/6} \left(\sin x - \frac{1}{2} \right) dx + \int_{5\pi/6}^{2\pi} \left(\frac{1}{2} - \sin x \right) dx$$

$$= \left[\frac{x}{2} + \cos x \right]_0^{\pi/6} + \left[-\cos x - \frac{x}{2} \right]_{\pi/6}^{5\pi/6} + \left[\frac{x}{2} + \cos x \right]_{5\pi/6}^{2\pi}$$

$$= \left[\left(\frac{\pi}{12} + \frac{\sqrt{3}}{2} \right) - (0 + 1) \right] + \left[\left(\frac{\sqrt{3}}{2} - \frac{5\pi}{12} \right) - \left(-\frac{\sqrt{3}}{2} - \frac{\pi}{12} \right) \right]$$

$$+ \left[(\pi + 1) - \left(\frac{5\pi}{12} - \frac{\sqrt{3}}{2} \right) \right]$$

$$= \frac{\pi}{3} + 2\sqrt{3} \approx 4.51. \qquad \blacksquare$$

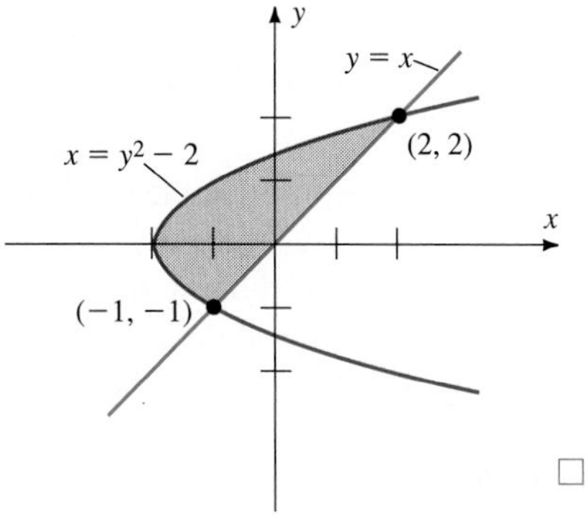

Our next example illustrates that it may sometimes be easier to calculate the desired area by integrating with respect to y rather than with respect to x.

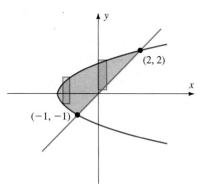

Figure 5.11.

□ **EXAMPLE 7**

Find the area bounded by the graphs of the equations $y = x$ and $x = y^2 - 2$.

Solution   The area in question is illustrated in Figure 5.11. The points of intersection are found by setting

$$y = y^2 - 2$$

or

$$y^2 - y - 2 = (y + 1)(y - 2) = 0,$$

so

$$y = -1 \quad \text{or} \quad y = 2.$$

The points are therefore $(-1, -1)$ and $(2, 2)$.

Figure 5.12 shows that by partitioning the x-axis (i.e., by using vertical rectangles) one encounters a difficulty—some approximating rectangles have both bottoms and tops on the graph of $x = y^2 - 2$, while others have tops on the graphs of $x = y^2 - 2$ and bottoms on the graph of $y = x$.

A simpler approach is to partition the y-axis, thus viewing y as the independent variable. Then all approximating rectangles run from the curve $x = y^2 - 2$ to the line $x = y$ (Figure 5.13). The area is thus calculated as

$$\text{Area} = \int_{y=-1}^{y=2} [y - (y^2 - 2)] \, dy$$

$$= \left[-\frac{y^3}{3} + \frac{y^2}{2} + 2y \right]_{-1}^{2}$$

$$= 9/2. \qquad \blacksquare$$

Figure 5.12

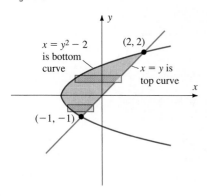

Figure 5.13

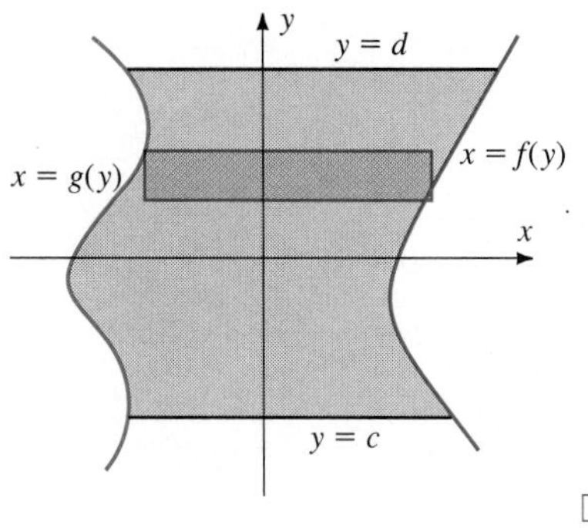

In general, if the region in question is bounded by the graphs of the functions $x = f(y)$ and $x = g(y)$ between $y = c$ and $y = d$ (see Figure 5.14), we have the area formula

$$\text{Area} = \int_c^d [f(y) - g(y)]\, dy$$

$$\text{if}\quad f(y) \geq g(y).$$

Figure 5.14 Region bounded by graphs of functions of y and horizontal lines.

□ **EXAMPLE 8**

Find the area of the region bounded by the graphs of the equations $3y - x = 6$, $x + y = -2$, and $x + y^2 = 4$ as illustrated in Figure 5.15.

Solution   The graphs of the first two equations are lines. The graph of the third is a parabola.

As Figure 5.15 suggests, the simplest solution involves partitioning the y-axis so that the approximating rectangles lie parallel to the x-axis. In this case, all approximating rectangles will have their right edge lying on the parabola. Their left edge will lie on one of the two lines.

To pursue this approach, we express each equation as a function of y, obtaining

$$f(y) = 3y - 6, \quad g(y) = -y - 2, \quad \text{and} \quad h(y) = 4 - y^2.$$

Equating these functions two at a time and solving gives the three points of intersection among these graphs: $(-3, 1)$, $(0, 2)$, and $(0, -2)$.

For $-2 \leq y \leq 1$, the left edges of the approximating rectangles are determined by the graph of $g(y) = -y - 2$, while for $1 \leq y \leq 2$ the left edges intersect the graph of $f(y) = 3y - 6$. The area of the region is therefore

$$A = \int_{-2}^{1} [(4 - y^2) - (-y - 2)]\, dy + \int_{1}^{2} [(4 - y^2) - (3y - 6)]\, dy$$

$$= \int_{-2}^{1} (6 + y - y^2)\, dy + \int_{1}^{2} (10 - 3y - y^2)\, dy$$

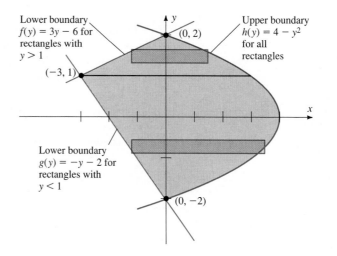

Lower boundary $f(y) = 3y - 6$ for rectangles with $y > 1$

$(-3, 1)$

$(0, 2)$

Upper boundary $h(y) = 4 - y^2$ for all rectangles

Lower boundary $g(y) = -y - 2$ for rectangles with $y < 1$

$(0, -2)$

Figure 5.15   Region in Example 8: Partitioning the interval $[-2, 2]$ on the y-axis.

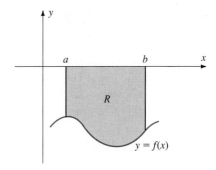

Figure 5.16 Area of $R = -\int_a^b f(x)\,dx$ if $f(x) \le 0$ for $a \le x \le b$.

$$= \left[6y + \frac{y^2}{2} - \frac{y^3}{3}\right]_{-2}^{1} + \left[10y - \frac{3y^2}{2} - \frac{y^3}{3}\right]_{1}^{2}$$

$$= \left[\left(6 + \frac{1}{2} - \frac{1}{3}\right) - \left(-12 + 2 + \frac{8}{3}\right)\right]$$

$$+ \left[\left(20 - 6 - \frac{8}{3}\right) - \left(10 - \frac{3}{2} - \frac{1}{3}\right)\right]$$

$$= \frac{50}{3} \approx 16.67. \qquad\blacksquare$$

Regions Bounded by Negative Functions

We can use Definition 6 to explain the relationship between the integral $\int_a^b g(x)\,dx$ and the area of the region R bounded by the graph of a continuous function g and the x-axis if $g(x) \le 0$ for all $x \in [a, b]$. (See Figure 5.16.) Since the x-axis is the graph of the constant function with values $f(x) = 0$ for all $x \in [a, b]$, the region R is bounded above by the graph of f and below by the graph of g, so

$$\text{Area of } R = \int_a^b [f(x) - g(x)]\,dx \qquad \text{(Definition 6)}$$

$$= \int_a^b [0 - g(x)]\,dx \qquad (f(x) = 0 \text{ for all } x)$$

$$= \int_a^b -g(x)\,dx$$

$$= -\int_a^b g(x)\,dx \qquad \text{(Theorem 4, part ii).}$$

Thus

If g is continuous on $[a, b]$ and $g(x) \le 0$ for all $x \in [a, b]$, then the area of the region R bounded by the graph of g and the x-axis for $a \le x \le b$ is

$$\text{Area of } R = -\int_a^b g(x)\,dx.$$

The following example indicates why we must be careful to ensure nonnegative integrands when calculating areas of regions using the definite integral.

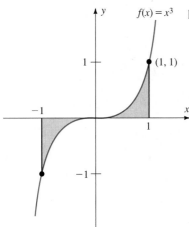

Figure 5.17 Graph of $f(x) = x^3$.

□ **EXAMPLE 9**

Figure 5.17 shows the graph of the function $f(x) = x^3$ for $-1 \le x \le 1$.

(a) To find the area of the region R bounded by the graph of $f(x) = x^3$ and the x-axis for $-1 \le x \le 1$, we must calculate separately the areas of the parts of the regions corresponding to the intervals $[-1, 0]$ and $[0, 1]$. That is because $f(x) \le 0$ for $-1 \le x \le 0$. Thus,

$$\text{Area of } R = -\int_{-1}^{0} x^3 \, dx + \int_{0}^{1} x^3 \, dx$$

$$= -\left[\frac{x^4}{4}\right]_{-1}^{0} + \left[\frac{x^4}{4}\right]_{0}^{1}$$

$$= -\left(0 - \frac{1}{4}\right) + \left(\frac{1}{4} - 0\right) = \frac{1}{2}.$$

(b) The *value of the integral* $\int_{-1}^{1} x^3 \, dx$, however, is simply

$$\int_{-1}^{1} x^3 \, dx = \left[\frac{x^4}{4}\right]_{-1}^{1} = \frac{(1)^4}{4} - \frac{(-1)^4}{4} = 0.$$

Note the difference between parts (a) and (b)! In part (a) the integral was used to calculate the area of a region, and signs were chosen for $\int_{a}^{b} f(x) \, dx$ depending on whether the integrand was positive or negative. Part (b), however, is *not* a calculation of area, but rather simply the evaluation of a definite integral using the Fundamental Theorem. This example emphasizes the difference between evaluating an integral and using the integral to calculate area. ■

REMARK   Using our ability to evaluate definite integrals involving the absolute value function, we can restate the formula for the area A of the region bounded between the graphs of the functions f and g for $a \le x \le b$ as

$$A = \int_{a}^{b} |f(x) - g(x)| \, dx$$

regardless of the sign of $f(x) - g(x)$.

Exercise Set 6.5

In Exercises 1–8, find the area of the region bounded above by the graph of $y = f(x)$ and below by the x-axis for $a \le x \le b$.

1. $f(x) = 2x + 5$, $a = 0$, $b = 2$

2. $f(x) = 9 - x^2$, $a = -3$, $b = 3$

3. $f(x) = \dfrac{1}{\sqrt{x - 2}}$, $a = 3$, $b = 5$

4. $f(x) = \sin 2x$, $a = 0$, $b = \pi/2$

5. $f(x) = \dfrac{x + 1}{(x^2 + 2x)^2}$, $a = 1$, $b = 2$

6. $f(x) = x\sqrt{1 + x^2}$, $a = 0$, $b = 1$

7. $f(x) = \sqrt{5 + x}$, $a = -4$, $b = 4$

8. $f(x) = \dfrac{x}{\sqrt{9 + x^2}}$, $a = 0$, $b = 4$

In Exercises 9–20, sketch the region bounded by the graphs of the given functions between the indicated values of x. Then calculate the area of the region.

9. $f(x) = x + 1$, $g(x) = -2x + 1$, $0 \le x \le 2$

10. $f(x) = 2x + 3$, $g(x) = x^2 - 4$, $-1 \le x \le 1$

11. $f(x) = \sqrt{x}$, $g(x) = -x^2$, $0 \le x \le 4$

12. $f(x) = 1/x^2$, $g(x) = x^{2/3}$, $1 \le x \le 8$

13. $f(x) = \sin x$, $g(x) = \cos x$, $0 \le x \le 2\pi$

14. $f(x) = 1$, $g(x) = \cos x$, $0 \le x \le 2\pi$

15. $f(x) = x - 3$, $g(x) = x^2$, $-2 \le x \le 2$

16. $f(x) = x\sqrt{9 - x^2}$, $g(x) = -x$, $-3 \le x \le 3$

17. $f(x) = |4 - x^2|$, $g(x) = 5$, $-3 \le x \le 3$

18. $f(x) = \sin x$, $g(x) = x$, $0 \le x \le \pi$

19. $f(x) = \dfrac{x^2 - 1}{x^2}$, $g(x) = \dfrac{1 - x^2}{x^2}$, $1 \le x \le 2$

20. $f(x) = x^{2/3}$, $g(x) = x^{1/3}$, $-1 \le x \le 1$

In Exercises 21–29, sketch the region bounded by the graphs of the given equations. Then calculate the area of the region.

21. $y = 4 - x^2$, $y = x - 2$

22. $y = 9 - x^2$, $9y - x^2 + 9 = 0$

23. $y = x^2$, $y = x^3$

24. $y = x^3$, $y = x$

25. $x + y^2 = 4$, $y = x + 2$

26. $x = \sqrt{y}$, $x = \sqrt[3]{y}$

27. $2x = y^2$, $x + 2 = y^2$

28. $y = x^{2/3}$, $y = 2 - x^2$

29. $y = x^{2/3}$, $y = x^2$

In Exercises 30–36, find (a) the definite integral of the given function over the given interval and (b) the area bounded by the graph of the given function and the x-axis over the given interval.

30. $f(x) = 2x + 1$, $[-1, 2]$

31. $f(x) = 3 - x$, $[0, 4]$

32. $f(x) = 3 - x^2$, $[0, 2]$

33. $f(x) = 2x^2 - 1$, $[-1, 1]$

34. $f(x) = 2x^2 + 6x + 1$, $[1, 3]$

35. $f(x) = x^2 + 6$, $[0, 3]$

36. $f(x) = x(x + 1)$, $[-2, 1]$

37. Find the area of the region bounded by the graph of $y = x^{2/3}$ and the line $y = 1$ by partitioning the x-axis.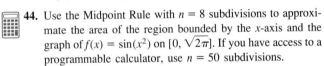

38. Rework Exercise 37, partitioning the y-axis.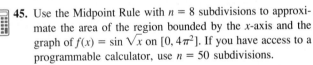

39. Find the area of the region bounded by the graphs of $y = x^3$, $y = -x^3$, $y = 1$ and $y = -1$ by partitioning the y-axis.

40. Rework Exercise 39, partitioning the x-axis.

41. Find a number a so that the line $y = a$ divides the region bounded by the x-axis and the graph of the equation $y = 4 - x^2$ into two regions of equal area.

42. Find the number a so that the line $y = a$ divides the region bounded by the x-axis and the graph of the equation $y = 4x - 4x^2$ into two regions of equal area.

43. Verify that the method of this section produces the same value for the area of the region bounded by the following lines as does the appropriate formula from plane geometry.
    a. $y = x$, $x = 3$, $y = -1$
    b. $y = 2x + 2$, $y = -2x - 1$, $x = 1$, $x = 3$.

44. Use the Midpoint Rule with $n = 8$ subdivisions to approximate the area of the region bounded by the x-axis and the graph of $f(x) = \sin(x^2)$ on $[0, \sqrt{2\pi}]$. If you have access to a programmable calculator, use $n = 50$ subdivisions.

45. Use the Midpoint Rule with $n = 8$ subdivisions to approximate the area of the region bounded by the x-axis and the graph of $f(x) = \sin \sqrt{x}$ on $[0, 4\pi^2]$. If you have access to a programmable calculator, use $n = 50$ subdivisions.

6.6 Numerical Approximation of the Definite Integral

Although our most common method for evaluating the definite integral uses the Fundamental Theorem of Calculus, there are situations in which we need to use other techniques. For example, in many applications of the calculus, we have a table of observed values $f(x)$ for selected numbers x in $[a, b]$ (for example, measurements taken during an experiment), and we need to estimate $\int_a^b f(x)\,dx$. Another situation where the Fundamental Theorem does not help is when we cannot express an antiderivative of the function f in terms of functions we already know. Examples include $f(x) = \sin(x^2)$, $f(x) = \cos(x^2)$, and $f(x) = \sqrt{1 + x^4}$. None of these functions have antiderivatives that can be expressed in terms of the functions we standardly use.

One alternative approach is to approximate the value of the integral just as we did with the Midpoint Rule in Section 6.2. In this section, we introduce two additional procedures for approximating the integral—the Trapezoidal Rule and Simpson's Rule—and we analyze the error involved in each of these three procedures.

First, we recall the Midpoint Rule. Given an integer $n \geq 1$, we partition $[a, b]$ into n equal-length subintervals. If m_j is the midpoint of the jth subinterval, then we approximate the definite integral by

$$\int_a^b f(x)\, dx \approx M_n = \Delta x\, (f(m_1) + f(m_2) + \cdots + f(m_n)) \qquad (1)$$

where $\Delta x = (b - a)/n$.

The following example is a typical application of the Midpoint Rule using a small number of subdivisions ($n = 6$).

□ **EXAMPLE 1**

Approximate

$$\int_1^4 \frac{1}{x}\, dx$$

using the Midpoint Rule with $n = 6$.

Solution With $a = 1$, $b = 4$, and $n = 6$, we have $\Delta x = (4 - 1)/6 = 1/2$, so

$$x_1 = \tfrac{3}{2},\ x_2 = 2,\ x_3 = \tfrac{5}{2},\ x_4 = 3,\ x_5 = \tfrac{7}{2}. \qquad (2)$$

Therefore, the midpoints of the subintervals are

$$m_1 = \tfrac{5}{4},\ m_2 = \tfrac{7}{4},\ m_3 = \tfrac{9}{4},\ m_4 = \tfrac{11}{4},\ m_5 = \tfrac{13}{4},\ m_6 = \tfrac{15}{4},$$

and the Midpoint Rule yields

$$\begin{aligned} M_6 &= \tfrac{1}{2}(\tfrac{4}{5} + \tfrac{4}{7} + \tfrac{4}{9} + \tfrac{4}{11} + \tfrac{4}{13} + \tfrac{4}{15}) \\ &= \frac{62{,}024}{45{,}045} \\ &\approx 1.376934 \end{aligned} \qquad (3)$$

(see Figure 6.1). ■

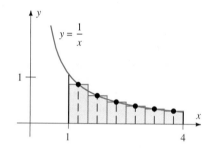

Figure 6.1 The Midpoint Rule applied to $\int_1^4 (1/x)\, dx$ using $n = 6$ subdivisions.

Often when we do these calculations, we use a calculator or a computer, and we skip intermediate steps such as line (3) of Example 1.

REMARK 1 In Figure 3.8 of Section 6.3, we graphed an approximation to the function

$$L(x) = \int_1^x \frac{1}{t}\, dt$$

using the Midpoint Rule with $n = 20$. Example 1 illustrates our calculation of $L(4)$ (with a smaller number n of subintervals).

The Trapezoidal Rule

We can interpret the Midpoint Rule geometrically as a procedure that approximates area using rectangles. There is a similar procedure—the Trapezoidal Rule—that approximates area using trapezoids rather than rectangles. Recall that the area of a trapezoid with base B and parallel heights h_1 and h_2 is

$$B \frac{h_1 + h_2}{2} \tag{4}$$

(see Figure 6.2). We use equation (4) to derive the Trapezoidal Rule.

As with the Midpoint Rule, we subdivide the interval $[a, b]$ into n equal-length subintervals, each with length $\Delta x = (b - a)/n$, and we denote the endpoints of the subintervals by

$$a = x_0 < x_1 < x_2 < \cdots < x_n = b.$$

As illustrated in Figure 6.3, equation (4) yields the approximation

$$\int_{x_{j-1}}^{x_j} f(x)\, dx \approx \Delta x \left[\frac{f(x_{j-1}) + f(x_j)}{2} = \frac{\Delta x}{2} \right] [f(x_{j-1}) + f(x_j)]$$

if $f(x) \geq 0$ on the interval $[a, b]$. Summing these individual approximations yields

$$\int_a^b f(x)\, dx \approx \frac{\Delta x}{2} \Big([f(x_0) + f(x_1)] + [f(x_1) + f(x_2)] + [f(x_2) + f(x_3)]$$
$$+ \cdots + [f(x_{n-2}) + f(x_{n-1})] + [f(x_{n-1}) + f(x_n)] \Big)$$

(see Figure 6.4). Note that each function value, except $f(x_0)$ and $f(x_n)$, occurs twice in the above expression because each such value is involved as a height in two trapezoids, once as a left side and once as a right side. We have

$$\int_a^b f(x)\, dx \approx \frac{\Delta x}{2} [f(x_0) + 2f(x_1) + 2f(x_2) + \cdots + 2f(x_{n-1}) + f(x_n)].$$

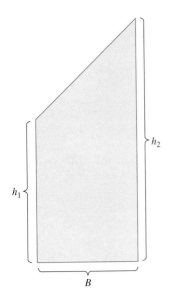

Figure 6.2 The area of a trapezoid is $B(h_1 + h_2)/2$.

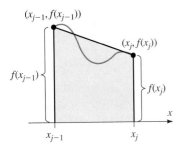

Figure 6.3 Trapezoidal approximation.

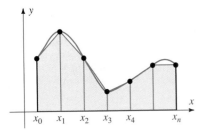

Figure 6.4 The Trapezoidal Rule.

The Trapezoidal Rule: Let $\Delta x = \dfrac{b - a}{n}$ and

$$x_0 = a,\ x_1 = a + \Delta x,\ x_2 = a + 2\Delta x, \ldots, x_n = a + n\,\Delta x = b.$$

Then

$$\int_a^b f(x)\, dx \approx T_n$$

where

$$T_n = \frac{\Delta x}{2} [f(a) + 2f(x_1) + 2f(x_2) + \cdots + 2f(x_{n-1}) + f(b)].$$

In the following example, we approximate the integral from Example 1 using the Trapezoidal Rule with the same number of subdivisions.

□ **EXAMPLE 2**

Approximate

$$\int_1^4 \frac{1}{x}\, dx$$

using the Trapezoidal Rule with $n = 6$.

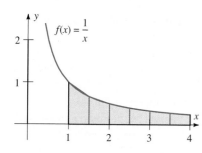

Figure 6.5 Trapezoidal Rule applied to $f(x) = 1/x$ on $[1, 4]$ with $n = 6$.

Solution Since $n = 6$, we have the same partition of $[1, 4]$ as in Example 1

$$x_1 = \tfrac{3}{2}, \; x_2 = 2, \; x_3 = \tfrac{5}{2}, \; x_4 = 3, \; x_5 = \tfrac{7}{2}.$$

The Trapezoidal Rule thus gives the approximation

$$\int_1^4 \frac{1}{x}\,dx \approx \frac{\left(\tfrac{1}{2}\right)}{2}\left(1 + 2 \cdot \tfrac{2}{3} + 2 \cdot \tfrac{1}{2} + 2 \cdot \tfrac{2}{5} + 2 \cdot \tfrac{1}{3} + 2 \cdot \tfrac{2}{7} + \tfrac{1}{4}\right) = \frac{787}{560} \approx 1.405357.$$

(see Figure 6.5). ∎

Later in this section, we shall analyze the errors associated with midpoint and trapezoidal approximation, and we shall see why it is useful to approximate $\int_1^4 (1/x)\,dx$ using both methods.

The following example illustrates how the Trapezoidal Rule is used to approximate the integral of a continuously varying quantity given a table of observations of that quantity.

□ **EXAMPLE 3**

Determine the average temperature over a four-hour time interval using the observed temperatures listed in the following table:

Hourly temperature readings over a four-hour period.

Hour	0	1	2	3	4
Temperature	65	69	72	73	71

Solution Considering temperature T as a function of time t, we want the average value of $T(t)$ for $0 \le t \le 4$. Using Definition 5, we know that the average is

$$\frac{1}{4 - 0} \int_0^4 T(t)\,dt.$$

We use the Trapezoidal Rule to approximate the integral. Since the time interval $[0, 4]$ is divided into four subintervals, we have $\Delta t = (4 - 0)/4 = 1$. Therefore, the trapezoidal approximation is

$$T_4 = \tfrac{1}{2}[65 + 2 \cdot 69 + 2 \cdot 72 + 2 \cdot 73 + 71] = 282.$$

Therefore, the average temperature over the given time interval is approximately $282/4 = 70.5$. ∎

REMARK 2 The formula for T_n assumes that the values $f(x_j)$ correspond to equally spaced numbers x_j throughout $[a, b]$. However, the method of trapezoidal approximation can be modified to handle cases where the observations $f(x_j)$ may not be at regular intervals. See Exercise 29.

Analyzing the Error

Suppose that we approximate an integral $\int_a^b f(x)\,dx$ using both the Midpoint Rule and the Trapezoidal Rule with the same number of subdivisions n. How accurate will our answers be? Which approximation do we expect to be the more accurate?

In order to answer these questions, it is useful to reinterpret the Midpoint Rule using trapezoids. Consider one of the terms

$$f(m_j)\,\Delta x \tag{5}$$

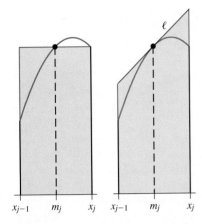

Figure 6.6 Two different geometric interpretations of the product $f(m_j)\,\Delta x$, as the area of a rectangle and as the area of a trapezoid.

used in the Midpoint Rule. We can reinterpret this product (5) as the area of the trapezoid T bounded by ℓ, the x-axis, and the vertical lines $x = x_{j-1}$ and $x = x_j$ where ℓ is the tangent line to the graph of f at the point $(m_j, f(m_j))$ (see Figure 6.6). To see why, let $y = l(x)$ be the linear equation for ℓ with s as its slope. Then

$$l(x_{j-1}) = f(m_j) - s\frac{\Delta x}{2} \quad \text{and} \quad l(x_j) = f(m_j) + s\frac{\Delta x}{2}$$

because $x_{j-1} = m_j - (\Delta x/2)$ and $x_j = m_j + (\Delta x/2)$ (see Figure 6.7). Thus, $l(x_{j-1}) + l(x_j) = 2f(m_j)$. Applying the area formula (4) to T, we obtain

$$\text{area}(T) = \Delta x\left(\frac{l(x_{j-1}) + l(x_j)}{2}\right) = \Delta x(f(m_j)).$$

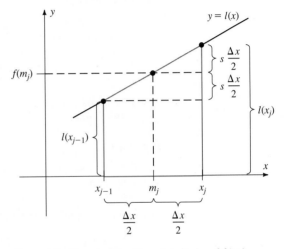

Figure 6.7 The equality $l(x_{j-1}) + l(x_j) = 2f(m_j)$.

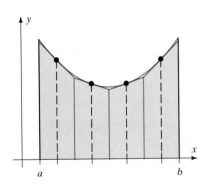

Figure 6.8 The Midpoint Rule underestimates the integral when the function is positive and its graph is concave up.

This interpretation of the Midpoint Rule using trapezoids and tangent lines leads to an interesting observation about the sign of the error

$$E_{M_n} = \int_a^b f(x)\,dx - M_n \tag{6}$$

involved with midpoint approximation. For example, suppose that $f(x) > 0$ for all $x \in [a, b]$ and that the graph of f is concave up throughout $[a, b]$ (see Figure 6.8). Then we conclude that the estimate M_n is smaller than the integral because the tangent lines are below the graph of f.

Moreover, with these assumptions on f, we can likewise analyze the error

$$E_{T_n} = \int_a^b f(x)\,dx - T_n \tag{7}$$

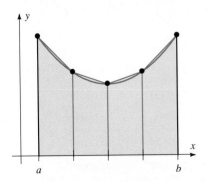

Figure 6.9 The Trapezoidal Rule overestimates the integral when the function is positive and its graph is concave up.

involved with trapezoidal approximation (see Figure 6.9). Note that the trapezoids entirely cover the region determined by the graph of f, and consequently T_n is larger than the integral.

We summarize these observations in the following theorem.

Theorem 10

Suppose f is continuous and nonnegative on the closed interval $[a, b]$.

(i) If the graph of f is concave up on $[a, b]$, then

$$M_n \leq \int_a^b f(x)\, dx \leq T_n.$$

(ii) If the graph of f is concave down on $[a, b]$, then

$$T_n \leq \int_a^b f(x)\, dx \leq M_n.$$

REMARK 3   There is a similar statement for functions f such that $f(x) < 0$ for all $x \in [a, b]$. We recommend that you remember the directions of the inequalities by referring to simple sketches (such as Figures 6.8 and 6.9).

Note that Theorem 10 part (ii) applies to the estimates obtained in Examples 1 and 2. Therefore, we know that

$$1.375 \leq \int_1^4 \frac{1}{x}\, dx \leq 1.406.$$

If we recalculate using $n = 100$, we see that the desired integral is between 1.3862 and 1.3863. In other words, it is approximately 1.386 (to three decimal places).

REMARK 4   It is no coincidence that the estimate from Example 1 is more accurate than the estimate given by Example 2. Typically, the Midpoint Rule is twice as accurate as the Trapezoidal Rule. In fact, under the assumptions of Theorem 10, the Midpoint Rule is *always* more accurate than the Trapezoidal Rule. However, we use both rules in order to determine upper and lower bounds for the integral.

To improve the accuracy of the approximation, we usually increase n so that Δx decreases. Table 6.1 contains the results of both the Midpoint Rule and the Trapezoidal Rule applied to $\int_1^b (1/x)\, dx$ for various numbers b using $n = 10$ subdivisions.

Table 6.1   The Midpoint Rule M_{10} and the Trapezoidal Rule T_{10} applied to the integral $\int_1^b (1/x)\, dx$ using $n = 10$ subdivisions of the interval $[1, b]$

b	M_{10}	T_{10}	$T_{10} - M_{10}$
2	0.69284	0.69377	0.000936
3	1.09714	1.10156	0.004420
4	1.38284	1.39326	0.010426
5	1.60321	1.62204	0.018828
6	1.78204	1.81154	0.029505
7	1.93203	1.97436	0.042332
8	2.06078	2.11796	0.057185
9	2.17322	2.24716	0.073939
10	2.27274	2.36521	0.092474

Table 6.2   The Midpoint Rule M_{100} and the Trapezoidal Rule T_{100} applied to the integral $\int_1^b (1/x)\, dx$ using $n = 100$ subdivisions of the interval $[1, b]$

b	M_{100}	T_{100}	$T_{100} - M_{100}$
2	0.693144	0.693153	0.000009
3	1.0986	1.09864	0.000044
4	1.38626	1.38636	0.000105
5	1.60937	1.60957	0.000192
6	1.79166	1.79196	0.000304
7	1.94576	1.9462	0.000441
8	2.07924	2.07984	0.000603
9	2.19696	2.19775	0.000789
10	2.30225	2.30325	0.001001

From Theorem 10, we see that the difference $T_{10} - M_{10}$ bounds the accuracy of these estimates. Table 6.2 indicates the improved accuracy we obtain when we use 100 subdivisions rather than 10.

Tables 6.1 and 6.2 illustrate two general considerations involved in numerical estimation. For a given interval, increasing the number of subdivisions usually results in a more accurate approximation. On the other hand, larger intervals require more subdivisions in order to maintain a desired accuracy.

Theorem 10 and the resulting error bounds apply whenever the graph of f does not have any inflection points over the interval $[a, b]$. We can also bound the error whenever f is twice differentiable on $[a, b]$. The bounds are

$$|E_{M_n}| \le \frac{M(b - a)^3}{24n^2} \tag{8}$$

and

$$|E_{T_n}| \le \frac{M(b - a)^3}{12n^2} \tag{9}$$

where M is the maximum of $|f''(x)|$ for $a \le x \le b$. (You can find derivations of these bounds in most numerical analysis texts.) Note that, since both the Midpoint Rule and the Trapezoidal Rule are based on approximation by straight line segments, it is not surprising to expect the error to be expressed in terms of the second derivative (see Section 4.8 for a similar use of the second derivative).

☐ **EXAMPLE 4**

Using error bound (8), determine the smallest integer n such that M_n is within 0.0005 (three decimal places of accuracy) of $\int_1^4 (1/x)\, dx$.

Solution   In order to use (8), we need to determine M. We know that $f''(x) = 2/x^3$, which is positive and decreasing for $x > 0$. Therefore, the maximum value of $|f''(x)|$ on $[1, 4]$ occurs at $x = 1$. Thus, $M = 2$. Using the bound given in (8), we want n to satisfy

$$|E_{M_n}| \le \frac{2(4 - 1)^3}{24n^2} \le 0.0005 = \frac{1}{2000}.$$

Thus, we need n large enough so that

$$\frac{54}{24n^2} \le \frac{1}{2000}.$$

Cross multiplying we obtain

$$\frac{54 \cdot 2000}{24} = 4500 \le n^2,$$

or

$$\sqrt{4500} \le n.$$

The smallest integer n that satisfies this inequality is $n = 68$.   ■

Simpson's Rule

Both the Midpoint Rule and the Trapezoidal Rule are based on the use of line segments to approximate the graph of the given function f. If the graph of f is curved, it often makes more sense to approximate the graph (and consequently the integral) using well-known curves. Simpson's Rule uses parabolic arcs. The procedure is based on the fact that, if x_0, x_1, and x_2 are three numbers such that

$$x_1 - x_0 = x_2 - x_1 = \Delta x$$

then the definite integral of the parabola passing through the three points (x_0, y_0), (x_1, y_1), and (x_2, y_2) equals

$$\frac{\Delta x}{3} [y_0 + 4y_1 + y_2]. \tag{10}$$

To see this let $x_0 = -\Delta x$, $x_1 = 0$, and $x_2 = \Delta x$ as in Figure 6.10. Let the parabola through (x_0, y_0), (x_1, y_1), and (x_2, y_2) be denoted by

$$y = ax^2 + bx + c.$$

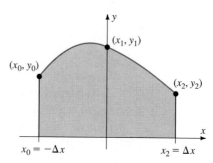

Figure 6.10

Then we have

$$y_0 = a(\Delta x)^2 - b(\Delta x) + c,$$
$$y_1 = c,$$

and

$$y_2 = a(\Delta x)^2 + b(\Delta x) + c.$$

Using these values we find that

$$\int_{-\Delta x}^{\Delta x} (ax^2 + bx + c)\, dx = \left[\frac{1}{3}ax^3 + \frac{1}{2}bx^2 + cx \right]_{-\Delta x}^{\Delta x}$$

$$= \frac{\Delta x}{3} [2a(\Delta x)^2 + 6c]$$

$$= \frac{\Delta x}{3} [y_0 + 4y_1 + y_2].$$

We obtain Simpson's Rule using (10). As before, we divide $[a, b]$ into n subintervals, but now we require that n be an *even* integer. We then approximate the integral

$$\int_{x_{2j-2}}^{x_{2j}} f(x)\, dx$$

over each *pair* of subintervals by expression (10). Thus we are approximating the actual value of the integral of f over $[x_{2j-2}, x_{2j}]$ by the integral of the approximating parabola. We obtain the approximation

$$\int_a^b f(x)\,dx \approx S_n = \frac{\Delta x}{3}\big([f(x_0) + 4f(x_1) + f(x_2)]$$
$$+ [f(x_2) + 4f(x_3) + f(x_4)]$$
$$+ [f(x_4) + 4f(x_5) + f(x_6)] + \cdots$$
$$+ [f(x_{n-4}) + 4f(x_{n-3}) + f(x_{n-2})]$$
$$+ [f(x_{n-2}) + 4f(x_{n-1}) + f(x_n)]\big).$$

Again, notice that each $f(x_{2j})$ is counted twice, except for $f(x_0)$ and $f(x_n)$, for the same reason as in the Trapezoidal Rule. We can state this approximation rule as follows:

Simpson's Rule: Let n be an even integer, $\Delta x = \dfrac{b-a}{n}$, and

$$x_0 = a, x_1 = a + \Delta x, x_2 = a + 2\Delta x, \ldots, x_n = a + n\Delta x = b.$$

Then

$$\int_a^b f(x)\,dx \approx S_n = \frac{b-a}{3n}\Big[f(x_0) + 4f(x_1) + 2f(x_2) + 4f(x_3)$$
$$+ 2f(x_4) + \cdots + 2f(x_{n-2}) + 4f(x_{n-1}) + f(x_n)\Big].$$

(See Figure 6.11.)

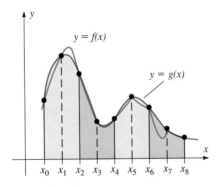

Figure 6.11   With Simpson's Rule, the definite integral $\int_a^b f(x)\,dx$ is approximated by $\int_a^b g(x)\,dx$ where g is a function whose graph consists of parabolic arcs. If $g(x) \geq 0$ on $[a, b]$, the approximation S_n can be interpreted as the area of the region bounded by the graph of g. This region is shaded light and dark blue in this figure.

☐ **EXAMPLE 5**

Approximate

$$\int_0^6 \sqrt{1 + x^4}\,dx$$

using Simpson's Rule with $n = 6$.

Solution   Since we are subdividing the interval $[0, 6]$ into 6 subintervals, we have $\Delta x = 1$. Consequently, $x_j = j$. From Simpson's Rule, we obtain

$$\int_0^6 \sqrt{1 + x^4}\, dx \approx S_6$$

where

$$S_6 = \frac{6 - 0}{3 \cdot 6}\left[\sqrt{1} + 4\sqrt{2} + 2\sqrt{17} + 4\sqrt{82} + 2\sqrt{257} + 4\sqrt{626} + \sqrt{1297}\right]$$

$$\approx 73.0936. \quad \blacksquare$$

There is a bound on the error

$$E_{S_n} = \int_a^b f(x)\, dx - S_n$$

involved in Simpson's Rule similar to the bounds for the Midpoint Rule and the Trapezoidal Rule. It is

$$|E_{S_n}| \leq \frac{M(b - a)^5}{180 n^4} \tag{11}$$

where M is the maximum of $|f^{(4)}(x)|$, the fourth derivative, on the interval $[a, b]$. If you compare this bound to the bounds for E_{M_n} and E_{T_n} (inequalities (8) and (9)), you see why we expect the approximation S_n to be more accurate than either M_n or T_n.

Finally, it is interesting to consider Simpson's Rule in terms of the relative accuracy of M_n and T_n. Recall (Remark 4) that M_n is typically twice as accurate as T_n. Therefore, it is reasonable to compute the average

$$\frac{M_n + M_n + T_n}{3}.$$

In Exercise 30, you are asked to verify that Simpson's Rule S_{2n} is precisely this average. In other words,

$$S_{2n} = \tfrac{2}{3}M_n + \tfrac{1}{3}T_n. \tag{12}$$

Therefore, we can easily add another column to Table 6.2 that contains the estimates S_{200}. See Table 6.3.

Table 6.3   Simpson's Rule S_{200} applied to the integral $\int_1^b (1/x)\, dx$

b	S_{200}
2	0.693147
3	1.09861
4	1.38629
5	1.60944
6	1.79176
7	1.94591
8	2.07944
9	2.19722
10	2.30259

Exercise Set 6.6

In Exercises 1–4, use the Midpoint Rule with the given value of n to approximate the integral.

1. $\int_1^5 \frac{1}{x}\, dx, \quad n = 8$

2. $\int_5^8 \frac{1}{x}\, dx, \quad n = 6$

3. $\int_0^8 \sqrt{1 + x^3}\, dx, \quad n = 4$

4. $\int_\pi^{4\pi} \frac{\sin x}{x}\, dx, \quad n = 3$

In Exercises 5–10, use the Trapezoidal Rule with the given value of n to approximate the integral.

5. $\int_0^4 \sqrt{x^3 + 2}\, dx, \quad n = 4$

6. $\int_1^5 \sqrt{x^3 + 1}\, dx, \quad n = 4$

7. $\int_1^5 \frac{1}{x}\, dx, \quad n = 8$

8. $\int_1^3 \frac{1}{x}\, dx, \quad n = 6$

9. $\int_0^8 \sqrt{1 + x^3}\, dx, \quad n = 4$

10. $\int_0^3 \sqrt{x^4 + 1}\, dx, \quad n = 6$

In Exercises 11–14, use Simpson's Rule with the given value of n to approximate the integral.

11. $\int_1^5 \frac{1}{x}\, dx, \quad n = 8$

12. $\int_1^4 \frac{1}{x}\, dx, \quad n = 6$

13. $\int_1^5 \sqrt{x^3 + 1}\, dx, \quad n = 4$

14. $\int_0^3 \sqrt{x^4 + 1}\, dx, \quad n = 6$

In Exercises 15–18, use the Trapezoidal Rule with $n = 10$ to approximate the integral.

15. $\int_0^2 \sin(x^2)\, dx$

16. $\int_0^1 \frac{1}{\sqrt{x^3 + 1}}\, dx$

17. $\int_0^1 \frac{1}{1 + x^2}\, dx$

18. $\int_0^3 \cos \sqrt{x}\, dx$

In Exercises 19–22, use Simpson's Rule with $n = 10$ to approximate the integral.

19. $\int_0^2 \frac{1}{\sqrt{1 + x^3}}\, dx$

20. $\int_0^1 \sqrt{1 + x^4}\, dx$

21. $\int_1^2 \frac{1}{x}\, dx$

22. $\int_0^4 \sin \sqrt{x}\, dx$

23. In Chapter 9, we shall see that

$$\int_0^1 \frac{4}{1 + x^2}\, dx = \pi.$$

Estimate π using both the Midpoint Rule and the Trapezoidal Rule with $n = 10$. If you have a programmable calculator, repeat the estimates with $n = 100$.

24. Using the results of Exercises 1 and 7, provide upper and lower bounds for the integral

$$\int_1^5 \frac{1}{x}\, dx.$$

25. Using the results of Exercises 3 and 9, provide upper and lower bounds for the integral

$$\int_0^8 \sqrt{x^3 + 1}\, dx.$$

26. During a storm, rainfall rates were recorded once every two hours during an eight-hour period (see Table 6.4). Using the Trapezoidal Rule, estimate the total rainfall during the 8-hour period.

Table 6.4 Rainfall rates at regular intervals during a rain storm

Time (h)	0	2	4	6	8
Rainfall rate (in./h)	0.2	0.6	0.5	1.2	0.3

27. Suppose that you want to cover your irregularly shaped driveway (see Figure 6.12) with crushed stone. Use the

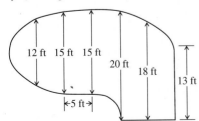

Figure 6.12

Trapezoidal Rule to estimate the area of the driveway so that you can determine how much stone you need.

28. Figure 6.13 is a graph of air temperature for a typical clear day in July for Peoria, Illinois. Using Simpson's Rule with $n = 8$, estimate the average temperature.

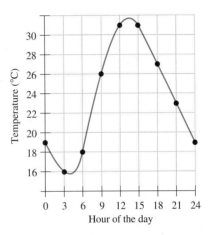

Figure 6.13

29. Suppose that you must estimate the total number of automobiles that travel through a certain traffic intersection during a typical 24-hour day, and the information in Table 6.5 is the only data available.

Table 6.5 Traffic data for Exercise 29

Time of day	5 a.m.	9 a.m.	1 p.m.	3 p.m.	5 p.m.	11 p.m.
Autos per hour	10	100	50	60	110	20

Modify the Trapezoidal Rule and estimate the total number of automobiles that travel through the intersection in one day.

30. Show that Simpson's Rule S_{2n} is equal to the weighted average $\frac{2}{3}M_n + \frac{1}{3}T_n$ of the Midpoint Rule M_n and T_n.

In Exercises 31–34, use the Midpoint Rule, the Trapezoidal Rule, and Simpson's Rule to graph an antiderivative of the given function f.

31. $f(x) = 1/x, \quad 1 \le x \le 5$

32. $f(x) = \cos^2 2x, \quad -\pi \le x \le \pi$

33. $f(x) = 1/(1 + x^2), \quad -4 \le x \le 4$

34. $f(x) = \sqrt{1 + x^2}, \quad 0 \le x \le 8$

35. Suppose that the daily temperature T during a certain year varies according to the function

$$T(t) = 51 \sin\left(\frac{2\pi}{365}(t - 95.2)\right) + 43,$$

where the time t is measured in days.

a. Graph T for $1 \le t \le 365$. During which months is it hottest? coldest? Estimate the average value of the temperature using this graph.

b. Determine the maximum M of T''. Use M to estimate the number n needed to approximate using the Trapezoidal Rule the average value of T with an error of at most 0.1.

c. Using n from part b, estimate the average value of T using the Trapezoidal Rule.

36. The perimeter of an ellipse with a semi-major axis of length a and a semi-minor axis of length b is given by the integral

$$P = 4a \int_0^{\pi/2} \sqrt{1 - e^2 \sin^2 t}\, dt$$

where the positive number e (the eccentricity of the ellipse) satisfies $e^2 = 1 - b^2/a^2$ (see Exercise 28 in Section 14.2). Assume $a = 1$ throughout this exercise.

a. Using Simpson's Rule with $n = 10$, complete the following table.

e	P
0.0	
0.1	
0.2	
0.3	
0.4	
0.5	
0.6	
0.7	
0.8	
0.9	
1.0	

b. Graph the ordered pairs (e, P). (The function that produces this plot is called a complete elliptic function of the second kind.)

c. Using the graph, determine numbers r and s between 0 and 1 so the function of the form

$$f(e) = 2\pi(1 - e)^r + 4e^s,$$

fits the data given in the table. Note that $f(0) = 2\pi$ and $f(1) = 4$.

Summary Outline of Chapter 6

■ The **summation notation** $\sum_{j=1}^{n} f(j)$ means $f(1) + f(2) + \cdots + f(n)$. (page 313)

■ The **lower approximating sum** for the area of the region R bounded by the graph of $f(x) \ge 0$ and the x-axis for $a \le x \le b$ is the sum $\underline{S}_n = \sum_{j=1}^{n} f(c_j)\,\Delta x$, where $\Delta x = (b - a)/n$ and $f(c_j)$ is the minimum value of f on $[x_{j-1}, x_j]$. The **upper approximating sum** is the sum $\overline{S}_n = \sum_{j=1}^{n} f(d_j)\,\Delta x$, where $f(d_j)$ is the maximum value of f on $[x_{j-1}, x_j]$. (page 310)

■ The **area** of R is $A = \lim_{n \to \infty} \underline{S}_n = \lim_{n \to \infty} \overline{S}_n$. (page 318)

■ If f is continuous on $[a, b]$, the **definite integral** of f on $[a, b]$ is defined to be the limit of Riemann sums; that is, as $\|P_n\| \to 0$, (page 325)

$$\int_a^b f(x)\, dx = \lim_{n \to \infty} \sum_{j=1}^{n} f(t_j)\,\Delta x_j, \quad t_j \in [x_{j-1}, x_j].$$

If $f(x) \ge 0$ on $[a, b]$ and f is continuous, then $\int_a^b f(x)\, dx$ is the **area** of the region bounded by the graph of f and the x-axis between $x = a$ and $x = b$.

■ *Properties of the Definite Integral* (page 329)

$$\int_a^b [f(x) + g(x)]\, dx = \int_a^b f(x)\, dx + \int_a^b g(x)\, dx$$

$$\int_a^b [cf(x)]\, dx = c \int_a^b f(x)\, dx$$

$$\int_a^b f(x)\,dx = \int_a^c f(x)\,dx + \int_c^b f(x)\,dx$$

$$\int_a^a f(x)\,dx = 0; \quad \int_b^a f(x)\,dx = -\int_a^b f(x)\,dx$$

▮ The **Fundamental Theorem of Calculus** states that   (page 341)

    a. if f is continuous on an open interval I then

$$\frac{d}{dx}\left(\int_a^x f(t)\,dt\right) = f(x), \quad a \in I$$

    for all $x \in I$, and

    b. if $F'(x) = f(x)$ for all $x \in [a, b]$ then $\int_a^b f(x)\,dx = F(b) - F(a)$.

▮ The **average value** of the continuous function f on $[a, b]$ is   (page 350)

$$\bar{f} = \frac{1}{b - a}\int_a^b f(x)\,dx.$$

▮ ***Theorem:*** $\int_a^b f(x)\,dx = f(c)(b - a)$ for some number c in $[a, b]$ if f is con-   (page 351)
tinuous on $[a, b]$.

▮ The **Method of Substitution** for definite integrals states that   (page 354)

$$\int_a^b f(g(x))g'(x)\,dx = \int_{u_a}^{u_b} f(u)\,du$$

if g' is continuous on $[a, b]$, $u_a = g(a)$, $u_b = g(b)$, and f is continuous on an
interval containing the values $u = g(x)$, $x \in [a, b]$.

▮ The **area** A of the region bounded between the graphs of the functions f and g (page 366)
on the interval $[a, b]$ is

$$A = \int_a^b |f(x) - g(x)|\,dx.$$

▮ The **Midpoint Rule** is   (page 333)

$$\int_a^b f(x)\,dx \approx M_n = \Delta x\,(f(m_1) + f(m_2) + \cdots + f(m_n))$$

where $\Delta x = (b - a)/n$ and m_j is the midpoint of the interval $[x_{j-1}, x_j]$. Its
error E_{M_n} is bounded by the inequality

$$\left|E_{M_n}\right| \le \frac{M(b - a)^3}{24n^2}$$

where M is the maximum of $|f''(x)|$ for $a \le x \le b$.

▮ The **Trapezoidal Rule** is   (page 369)

$$\int_a^b f(x)\,dx \approx T_n = \frac{\Delta x}{2}\left(f(a) + 2f(x_1) + 2f(x_2) + \cdots + 2f(x_{n-1}) + f(b)\right)$$

where $\Delta x = (b - a)/n$. Its error E_{T_n} is bounded by the inequality

$$\left|E_{T_n}\right| \le \frac{M(b - a)^3}{12n^2}$$

where M is the maximum of $|f''(x)|$ for $a \le x \le b$.

■ **Simpson's Rule** is

(page 375)

$$\int_a^b f(x)\,dx \approx S_n = \frac{b-a}{3n}\Big(f(a) + 4f(x_1) + 2f(x_2) + 4f(x_3) + 2f(x_4) + $$

$$\cdots$$

$$+ 4f(x_{n-3}) + 2f(x_{n-2}) + 4f(x_{n-1}) + f(b)\Big)$$

where n is an even integer. Its error E_{S_n} is bounded by the inequality

$$\left| E_{S_n} \right| \le \frac{M(b-a)^5}{180n^4}$$

where M is the maximum of $|f^{(4)}(x)|$ for $a \le x \le b$.

Review Exercises—Chapter 6

In Exercises 1–8, determine **(a)** the lower approximating sum and **(b)** the upper approximating sum for the given function and interval, using n subintervals of equal size.

1. $f(x) = 3x + 1$, $x \in [0, 3]$, $n = 6$

2. $f(x) = 1/x$, $x \in [1, 3]$, $n = 4$

3. $f(x) = 1/(1 + x^2)$, $x \in [-1, 1]$, $n = 6$

4. $f(x) = \sin \pi x$, $x \in [0, 1]$, $n = 6$

5. $f(x) = \cos \pi x$, $x \in [-1/2, 1/2]$, $n = 4$

6. $f(x) = \sec x$, $x \in [-\pi/3, \pi/3]$, $n = 4$

7. $f(x) = 3 - x$, $x \in [0, 2]$, $n = 100$

8. $f(x) = x^2 + 1$, $x \in [-1, 1]$, $n = 200$

In Exercises 9–12, partition the designated interval into n subintervals of equal length and calculate both the largest and smallest Riemann sums of f corresponding to that subdivision.

9. $f(x) = \cos x$, $[-\pi/2, \pi]$, $n = 6$

10. $f(x) = 9 - x^2$, $[0, 3]$, $n = 6$

11. $f(x) = x^2 + 1$, $[-1, 2]$, $n = 6$

12. $f(x) = 2x - x^2$, $[-1, 3]$, $n = 4$

13. Find the area of the region bounded above by the graph of $y = 2x - 2$ and below by the x-axis for $2 \le x \le 4$ by calculating the limit of lower approximating sums.

14. Rework Exercise 13 using upper approximating sums.

In Exercises 15–56, evaluate the definite integral using the Fundamental Theorem of Calculus.

15. $\int_0^9 \sqrt{x}\,dx$

16. $\int_0^3 3\sqrt{x+1}\,dx$

17. $\int_0^1 (x^{2/3} - x^{1/2})\,dx$

18. $\int_1^2 \frac{1-t}{t^3}\,dt$

19. $\int_0^1 x^3(x+1)\,dx$

20. $\int_0^\pi \sin^2 x\,dx$ (*Hint:* $\sin^2\theta = \frac{1}{2} - \frac{1}{2}\cos 2\theta$)

21. $\int_0^1 (3x + 2)\,dx$

22. $\int_1^3 (7 + 3x)\,dx$

23. $\int_{-3}^5 (x^2 + 2)\,dx$

24. $\int_1^5 (3x^2 - 2)\,dx$

25. $\int_1^4 \sqrt{x}\,dx$

26. $\int_0^8 \sqrt{x+1}\,dx$

27. $\int_1^4 \left(\sqrt{x} - \frac{1}{\sqrt{x}}\right)dx$

28. $\int_1^8 (x^{1/3} - 1)\,dx$

29. $\int_0^2 (x + 7)(2x + 2)\,dx$

30. $\int_1^4 (x^2 - 1)(x + 2)\,dx$

31. $\int_0^{\pi/4} \sin x\,dx$

32. $\int_{-\pi/4}^{\pi/4} \cos x\,dx$

33. $\int_0^{\pi/4} \sin(2x)\,dx$

34. $\int_{\pi/4}^{\pi/2} \cos(\pi - 2x)\,dx$

35. $\int_1^2 (x + 4)^{10}\,dx$

36. $\int_0^4 (\sqrt{a} + \sqrt{x})^2\,dx$

37. $\int_2^4 \frac{t^2 - 2t}{5}\,dt$

38. $\int_1^2 \frac{1 + t}{t^3}\,dt$

39. $\int_0^3 \frac{dt}{(t + 1)^2}$

40. $\int_3^8 \frac{1}{\sqrt{x+1}}\,dx$

41. $\int_0^{\pi/4} \frac{1}{\cos^2 x}\,dx$

42. $\int_0^{\pi/3} \frac{\sin x}{\cos^2 x}\,dx$

43. $\int_0^1 (x - \sqrt{x})^2\,dx$

44. $\int_1^4 x^{1/2}(1 + x^{3/2})^5\,dx$

45. $\int_1^8 t(\sqrt[3]{t} - 2t)\,dt$

46. $\int_1^4 \frac{x^2 + 2x + 4}{\sqrt{x}}\,dx$

47. $\int_1^2 \frac{1}{(1 - 2x)^3}\,dx$

48. $\int_{-\pi/4}^{\pi/4} \sec^2 t\,dt$

49. $\int_0^1 \dfrac{x^3 + 8}{x + 2}\, dx$

50. $\int_0^2 x^2 \sqrt{x^3 + 1}\, dx$

51. $\int_0^{\pi/2} \sec(u/2) \tan(u/2)\, du$

52. $\int_{-\pi}^{\pi} \sec^2(v/3)\, dv$

53. $\int_{-1}^3 |x - x^2|\, dx$

54. $\int_0^{2\pi} |\sin 2x|\, dx$

55. $\int_0^1 x\sqrt{x + 1}\, dx$

56. $\int_0^2 x^3 \sqrt{x^2 + 2}\, dx$

57. Determine the positive even integer n such that the sum of the even integers between 21 and $n + 1$ is 1782.

58. Suppose that you exercise daily for 30 days. If you do 10 push-ups the first day and increase the number of push-ups you do by one each day, what is the total number of push-ups you do over the entire 30-day period?

In Exercises 59–65, find the area of the region bounded by the graph of the given function, the x-axis, the line $x = a$, and the line $x = b$.

59. $f(x) = \sqrt{x - 1}, \quad a = 1, \quad b = 5$

60. $f(x) = \dfrac{1}{\sqrt{x - 1}}, \quad a = 2, \quad b = 10$

61. $f(x) = (x^2 + 2)^2, \quad a = 0, \quad b = 1$

62. $f(x) = (x - 1)(x + 2), \quad a = 0, \quad b = 2$

63. $f(x) = 4x - x^2, \quad a = 4, \quad b = 5$

64. $f(x) = \sin \pi x, \quad a = 0, \quad b = 2$

65. $f(x) = 3 + 2x - x^2, \quad a = 1, \quad b = 4$

In Exercises 66–71, find the area of the region bounded above by the graph of $y = f(x)$ and below by the x-axis for the specified intervals.

66. $f(x) = x^3 - x, \quad a = 1, \quad b = 3$

67. $f(x) = x^2 - x - 2, \quad a = 2, \quad b = 4$

68. $f(x) = \sin(2x), \quad a = 0, \quad b = \pi/4$

69. $f(x) = \sqrt{1 - x}, \quad a = -1, \quad b = 1$

70. $f(x) = \dfrac{1}{(x - 2)^2}, \quad a = -2, \quad b = 1$

71. $f(x) = (cx^2 + 3)^2, \quad a = 0, \quad b = 3$

In each of Exercises 72–76, a region is described. Sketch the region, noting where the bounding function f is positive and where it is negative. Then calculate the area of the region using one or more integrals.

72. The region bounded by the graph $f(x) = 9 - x^2$ and the x-axis between $x = -3$ and $x = 3$.

73. The region bounded by the graph of $f(x) = x^2 + 2x - 3$ between $x = -3$ and $x = 1$.

74. The region bounded by the graph of $f(x) = (x - 1)^3$ between $x = 0$ and $x = 2$.

75. The region bounded by the graph of $f(x) = 2x - 4$ and the x-axis for $0 \le x \le 4$.

76. The region bounded by the graph of $f(x) = \sqrt{x} - 2$ and the x-axis for $0 \le x \le 9$.

In Exercises 77–88, find the area of the region bounded by the graphs of the given equations.

77. $y = x^3, \quad y = 0, \quad x = -2, \quad x = 2$

78. $y = 1 - x^2, \quad y = -x - 1, \quad x = -2, \quad x = 1$

79. $y = \sqrt{x}, \quad y = -\sqrt{x}, \quad x = 0, \quad x = 4$

80. $y = x^2, \quad y = 2 - x, \quad x = -2, \quad x = 1$

81. $y = x^2 - 4x + 2, \quad x + y = 6$

82. $y = 2 - 2x - x^2, \quad x = -y$

83. $y = \dfrac{x - 2}{\sqrt{x^2 - 4x + 8}}, \quad x = 0, \quad y = 0$

84. $x - y^2 + 3 = 0, \quad x - 2y = 0$

85. $x + y^2 = 0, \quad x + y + 2 = 0$

86. $x = 2y^2 - 3, \quad x = y^2 + 1$

87. $x = y^{2/3}, \quad x = y^2$

88. $y = 4x^2, \quad 4x + y - 8 = 0$

89. Find the area of the region bounded by the parabola $y = x^2 + x - 2$ and the line through $(-1, -2)$ and $(1, 0)$.

90. Find the area of the region bounded by the parabola $y = 2 - x - x^2$ and the line through $(-1, 2)$ and $(1, 0)$.

91. Find the area of the region bounded by the graphs of the equations $3x + 5y = 23$, $5x - 2y = 28$, and $2x - 7y = -26$.

In Exercises 92–97, find the average value of the given function.

92. $y = x\sqrt{1 - x^2} \quad$ on $\quad [-1, 1]$

93. $y = \sin x \cos x \quad$ on $\quad [0, \pi/2]$

94. $y = x^3 \sin x^2 \quad$ on $\quad [-\pi, \pi] \quad$ (*Hint:* Use the fact that $y = f(x)$ is odd.)

95. $y = \dfrac{x}{\sqrt{1 + x^2}} \quad$ on $\quad [0, 2]$

96. $y = \sqrt{x} + \sqrt[3]{x} \quad$ on $\quad [0, 1]$

97. $y = \sin x \quad$ on $\quad [0, 10\pi]$

98. Show that the average value of $f(x) = \sin x$ on intervals of the form $[2n\pi, 2m\pi]$ is $\bar{f} = 0$, where n and m are integers. What about intervals of the form $[x_0 + 2n\pi, x_0 + 2m\pi]$?

99. Show that the average value of $f(x) = \sin^2 x$ on intervals of the form $[0, n\pi]$ is $\overline{f} = \frac{1}{2}$ where n is a positive integer.

100. Give an argument involving average values to show that
$$\lim_{L \to \infty} \frac{1}{L} \int_0^L \sin^2 x \, dx = \frac{1}{2}. \text{ What about } \lim_{L \to \infty} \frac{1}{L} \int_0^L \cos^2 x \, dx?$$

101. Let $F(x) = \int_0^x t^2 \sqrt{1 + t} \, dt$ for $x > -1$. Find

    a. $F(0)$ **b.** $F'(x)$
    c. $F'(3)$ **d.** $F'(2x)$

102. Let $F(x) = \int_0^{2x} \frac{1}{1 + t^2} \, dt$. Find

    a. $F'(x)$ **b.** $F'(1)$ **c.** $F'(x^2)$

103. True or false? $\dfrac{d}{dx}\left[\int_a^x f(t) \, dt\right] = \int_a^x \dfrac{d}{dt}[f(t)] \, dt.$

104. True or false? If $\int_a^b f(x) \, dx = 0$ then $f(x) \equiv 0$.

105. True or false? If $f(x)$ is continuous on $[a, b]$ and $\int_a^b f(x) \, dx = 0$, then $f(c) = 0$ for at least one $c \in [a, b]$.

106. True or false? If $\int_a^b |f(x)| \, dx = 0$ and $f(x)$ is continuous on $[a, b]$, then $f(x) = 0$ for all $x \in [a, b]$.

107. Find $\dfrac{d}{dx}\left[\int_{2x}^0 \sec t \, dt\right].$

108. Find $\dfrac{d}{dx}\left[\int_x^{x^2} \sqrt{1 + t^2} \, dt\right].$

109. Suppose that f' is continuous on $[a, b]$. Simplify
$$\int_a^b [f(x)]^n f'(x) \, dx$$
by eliminating the integral sign.

In Exercises 110–115, sketch a region in the plane whose area corresponds to the given integral. Then find the value of the integral from area considerations.

110. $\displaystyle\int_1^4 (2x + 1) \, dx$ **111.** $\displaystyle\int_{-1}^1 \sqrt{1 - x^2} \, dx$

112. $\displaystyle\int_0^3 (1 + \sqrt{9 - x^2}) \, dx$ **113.** $\displaystyle\int_{-3}^3 |3 - x| \, dx$

114. $\displaystyle\int_0^4 \sqrt{16 - x^2} \, dx$ **115.** $\displaystyle\int_{-1}^1 (3 + \sqrt{1 - x^2}) \, dx$

In Exercises 116–118, use the Trapezoidal Rule with n subdivisions to approximate the given integral.

116. $\displaystyle\int_{-2}^2 \frac{1}{1 + x^2} \, dx, \quad n = 4$

117. $\displaystyle\int_0^4 \sqrt{1 + x^2} \, dx, \quad n = 4$

118. $\displaystyle\int_{-1}^2 \frac{1}{x + 2} \, dx, \quad n = 6$

119. Using the Trapezoidal Rule and the observations recorded in the following table, estimate the average temperature of a solution during a 10-hour chemical reaction.

Observed temperatures at regular intervals during a chemical reaction

Time (h)	0	2	4	6	8	10
Temperature	72	82	90	94	91	85

120. Using the Midpoint Rule with $n = 4$ subdivisions, estimate
$$\int_0^{2\pi} \sqrt{|\sin x|} \, dx.$$

121. In Chapter 2, we saw that the limit of $f(x) = (\sin x)/x$ is 1 as $x \to 0$. Therefore, if we define $f(0) = 1$, then f is continuous on $(-\infty, \infty)$. Using the Midpoint Rule with $n = 4$ subdivisions, estimate
$$\int_0^{2\pi} f(x) \, dx.$$

Chapter 7

Applications of the Definite Integral

In this chapter, you will encounter a variety of physical and mathematical problems, all of which can be solved using the definite integral. In each instance, we will be led to the solution through a sequence of steps similar to those by which the definite integral arose as the solution to the area problem in Chapter 6.

Because of the strong parallels that exist among the various problems and solutions of this chapter, we begin by summarizing the essential concepts involved in the definition of the integral. You should refer to this introductory summary from time to time, as an aid to understanding the principal idea of this chapter.

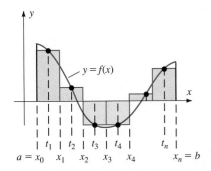

Figure 1 Approximating Riemann sum.

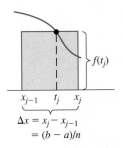

Figure 2 jth approximating rectangle.

Summary Definition of the Integral $\int_a^b f(x)\,dx$:

Let f be continuous on the interval $[a, b]$.

1. For each positive integer n, we partition the interval $[a, b]$ into subintervals of equal* length with endpoints

$$a = x_0 < x_1 < x_2 < \cdots < x_n = b$$

and with $(x_j - x_{j-1}) = \Delta x = \dfrac{b - a}{n}$ for $j = 1, 2, \ldots, n.$

2. A **Riemann sum** for f on $[a, b]$ is an expression of the form

$$\sum_{j=1}^{n} f(t_j)\, \Delta x$$

where t_j is an arbitrary number in the subinterval $[x_{j-1}, x_j]$ for each $j = 1, 2, \ldots, n$ (see Figures 1 and 2).

---

*The most general definition of partition, as given in Chapter 6, allows for subintervals of varying length $\Delta x_j = (x_j - x_{j-1})$. Since all Riemann sums for f on $[a, b]$ have the limit $\int_a^b f(x)\, dx$ and since we will actually be constructing the partitions that are used here, we will work with **regular partitions**—those with subintervals of equal length $\Delta x = (b - a)/n$.

3. The limit of this Riemann sum as $n \to \infty$,

$$\int_a^b f(x)\, dx = \lim_{n \to \infty} \sum_{j=1}^n f(t_j)\, \Delta x,$$

exists independent of how the numbers $t_j \in [x_{j-1}, x_j]$ are chosen.

For each question discussed in this chapter, we will follow the same general problem-solving strategy:

(a) We approximate the desired quantity using Riemann sums of an appropriate function f.

(b) We assume that, as the number of subintervals in the Riemann sum becomes infinite, the approximations approach the desired quantity.

(c) The desired quantity is, therefore, the limiting value of the approximating Riemann sums as $n \to \infty$. That is, it equals the resulting definite integral (point 3 in the preceding summary).

You are urged, in each case, to study carefully the procedure by which the solution is obtained rather than simply accepting the formula that results. Only by striving to understand the ways in which integrals arise from approximation schemes can you gain the insight you will need when encountering new problems whose solutions involve definite integrals.

7.1 Calculating Volumes by Slicing

The volumes of certain types of three-dimensional objects can be calculated as definite integrals. For example, it is common for solid objects to be produced by a process of milling a rotating piece of stock. In using a lathe to produce a wooden table leg, a craftsman presses a chisel against a rapidly rotating block of wood (Figure 1.1). Similarly, a potter works a ball of clay into a vase by using a potter's wheel, which allows the clay to be rotated at a uniform speed about a central axis (Figure 1.2).

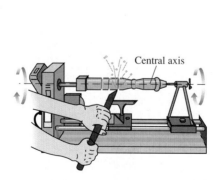

Figure 1.1 Table leg produced on a lathe by pressing chisel against rotating block of wood stock.

Figure 1.2 Pottery produced by shaping clay rotating on a wheel.

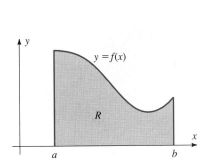

Figure 1.3 The region R bounded by the graph of f.

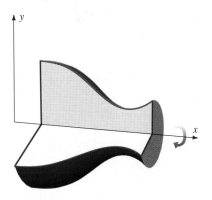

Figure 1.4 As the region R rotates about the x-axis, it sweeps out a solid.

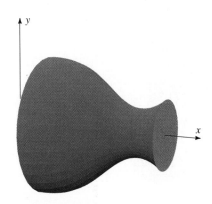

Figure 1.5 The solid of revolution obtained by revolving the region R about the x-axis.

The problem of calculating the volume of such solids is idealized mathematically as follows. Let f be a continuous nonnegative function for $a \leq x \leq b$. Let R denote the region bounded by the graph of f, the x-axis, and the lines $x = a$ and $x = b$ (Figure 1.3). As the region R rotates about the x-axis (Figure 1.4), it sweeps out a **solid of revolution** S (Figure 1.5). Just as for the lathe and pottery wheel illustrations, the cross sections for S taken perpendicular to the x-axis are circles of radius $r = f(x)$. This is because the cross section taken at location x is described by rotating about the x-axis the line segment from $(x, 0)$ to $(x, f(x))$.

To find a formula for the volume V of S, we begin by developing an approximation to the solid S. We do this by partitioning the interval $[a, b]$ into n subintervals of equal length $\Delta x = (b - a)/n$ with endpoints $a = x_0, x_1, \ldots, x_n = b$.

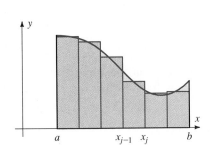

Figure 1.6 Approximating the region R with rectangles.

We arbitrarily select one "test number" t_j in each interval $[x_{j-1}, x_j]$, and we approximate the region R by rectangles with base $[x_{j-1}, x_j]$ and height $f(t_j)$ (Figure 1.6).

To obtain our approximation to the volume V, we rotate these rectangles about the x-axis (Figures 1.7 and 1.8). We get n discs, one for each of the original rectangles (Figures 1.9 and 1.10). The radius of the jth disc is $r_j = f(t_j)$ and its thickness is Δx. Its volume V_j is, therefore,

$$V_j = \pi r_j^2 \, \Delta x = \pi [f(t_j)]^2 \, \Delta x$$

(see Figure 1.8).

Summing the volumes of these individual discs gives the volume of our approximating solid as

$$\sum_{j=1}^{n} V_j = \sum_{j=1}^{n} \pi [f(t_j)]^2 \, \Delta x$$

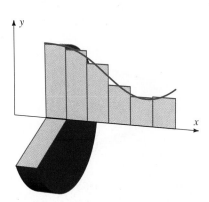

Figure 1.7 As a rectangle rotates about the x-axis, it sweeps out a regular solid.

(see Figure 1.10). Next, we assume that, as $n \rightarrow \infty$ and the thickness Δx of each individual disc becomes small, the volume of our approximating solid approaches the desired volume V (see Figure 1.11). That is,

$$V = \lim_{n \rightarrow \infty} \sum_{j=1}^{n} V_j = \lim_{n \rightarrow \infty} \sum_{j=1}^{n} \pi [f(t_j)]^2 \, \Delta x.$$

Since the sum on the right is a Riemann sum for the function πf^2, we have arrived at the following formula.

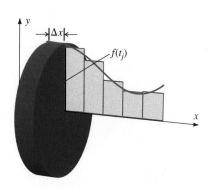

Figure 1.8 After one complete revolution, the rectangle sweeps out a disc. Its thickness is Δx, and the radius is $f(t_j)$. Its volume is $V_j = \pi[f(t_j)]^2 \Delta x$.

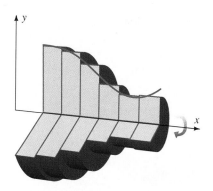

Figure 1.9 As we rotate all of the rectangles, we sweep out a solid consisting of n discs (see Figure 1.4).

Figure 1.10 The n discs yield a solid whose volume approximates the volume of the solid of revolution S (see Figure 1.5).

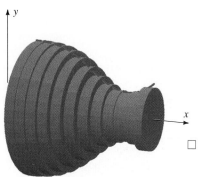

Figure 1.11 As the number of rectangles increases, the volumes of the approximating solids approach the desired volume (compare with Figure 1.10).

Let f be continuous and nonnegative for $a \le x \le b$ and let R denote the region bounded by the graph of $y = f(x)$, the x-axis, and the lines $x = a$ and $x = b$. The volume of the solid obtained by rotating R about the x-axis is

$$V = \int_a^b \pi[f(x)]^2 \, dx. \tag{1}$$

☐ **EXAMPLE 1**

Find the volume of the cone obtained by revolving about the x-axis the region bounded above by the graph of $f(x) = x/3$ and below by the x-axis for $0 \le x \le 3$ (see Figures 1.12–1.14).

Solution Using formula (1) with $f(x) = x/3$ we obtain

$$V = \int_0^3 \pi\left(\frac{x}{3}\right)^2 dx = \int_0^3 \frac{\pi}{9}x^2 \, dx = \left[\frac{\pi}{27}x^3\right]_0^3 = \pi.$$

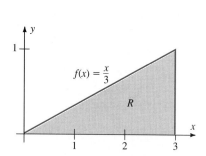

Figure 1.12 The region bounded by the graph of $f(x) = x/3$ for $0 \le x \le 3$.

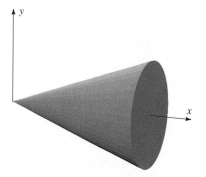

Figure 1.13 Revolving the region in Figure 1.12 about the x-axis produces a cone as a solid of revolution.

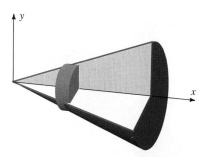

Figure 1.14 A partial rotation of the triangular region R in Example 1 along with one of the discs used to approximate the desired volume.

Since this cone has height $h = 3$ and base of radius $r = 1$, this result agrees with the formula from geometry for the volume of this cone:

$$V = \tfrac{1}{3}\pi r^2 h = \tfrac{1}{3}\pi(1)^2 \cdot 3 = \pi. \qquad \blacksquare$$

☐ **EXAMPLE 2**

Find the volume of the solids obtained by rotating the region bounded by the graphs of $f(x) = \sqrt{x}$ and $g(x) = x^2$ about the x-axis.

Strategy · · · · · · · ·

Find points where the graphs meet.

View solid as difference of two solids of revolution.

Find the volume of each solid using formula (1).

Solution

The two graphs cross at $(0, 0)$ and $(1, 1)$ since the equation $\sqrt{x} = x^2$ implies $x = x^4$ or $x(1 - x^3) = 0$. Since $\sqrt{x} > x^2$ for $0 < x < 1$, the region is bounded above by the graph of $f(x) = \sqrt{x}$ and below by the graph of $g(x) = x^2$ (Figure 1.15). As Figures 1.16–1.18 indicate, we can view the solid S as the "difference" of two other solids. Let S_1 be the solid obtained by rotating the region bounded by the graph of $f(x) = \sqrt{x}$ about the x-axis for $0 \le x \le 1$ and let S_2 be the solid obtained by rotating the region bounded by the graph of $g(x) = x^2$ about the x-axis for $0 \le x \le 1$. Then the original solid S is the solid that results when S_2 is removed from S_1. The desired volume V is, therefore, the difference between the volume of S_1 and the volume of S_2. Using equation (1) twice, we obtain

$$V = \int_0^1 \pi(\sqrt{x})^2 \, dx - \int_0^1 \pi(x^2)^2 \, dx$$

$$= \left[\frac{\pi}{2}x^2\right]_0^1 - \left[\frac{\pi}{5}x^5\right]_0^1 = \frac{3\pi}{10}. \qquad \blacksquare$$

As you may have observed, the volume in Example 2 could have been calculated by the single integral

$$V = \int_0^1 \pi[(\sqrt{x})^2 - (x^2)^2] \, dx.$$

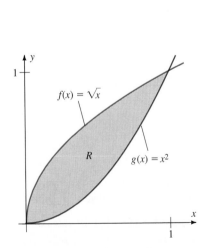

Figure 1.15   The region R bounded by the graphs of $y = x^2$ and $y = \sqrt{x}$.

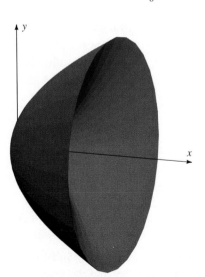

Figure 1.16   The solid of revolution obtained by rotating about the x-axis the region specified in Example 2.

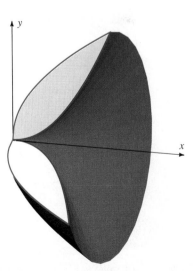

Figure 1.17   A portion of the solid in Example 2 is cut away to expose the entire axis of revolution.

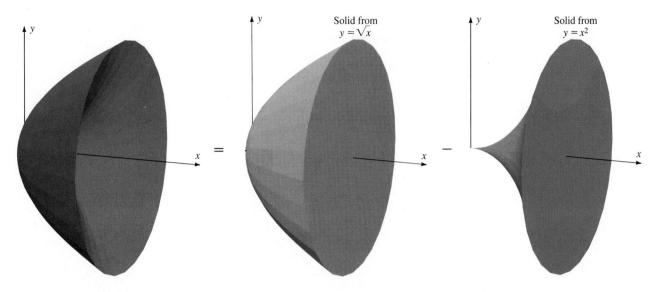

Figure 1.18 The volume of the solid S in Example 2 is the difference of two volumes—one determined by the graph of $y = \sqrt{x}$ and the other determined by the graph of $y = x^2$.

In general, if the region R is bounded above by the graph of $y = f(x)$ and below by the graph of $y = g(x) \geq 0$ and $a \leq x \leq b$, then the formula for volume is

$$V = \int_a^b \pi([f(x)]^2 - [g(x)]^2)\, dx. \qquad (2)$$

However, be careful not to misinterpret the integrand as $[f(x) - g(x)]^2$.

As Figure 1.19 illustrates, equation (2) can also be derived using Riemann sums in much the same way as equation (1) was derived earlier. In this case, the solid is approximated using "washers" with thickness Δx, inside radius $g(t_j)$, and outside radius $f(t_j)$. Then the jth washer has volume $(\pi[f(t_j)]^2 - \pi[g(t_j)]^2)\,\Delta x$, and the desired volume is approximated by the Riemann sum

$$\sum_{j=1}^{n} \pi([f(t_j)]^2 - [g(t_j)]^2)\,\Delta x.$$

As $n \to \infty$, we obtain equation (2).

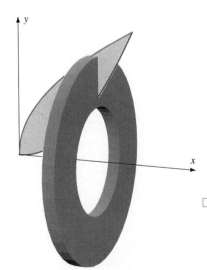

Figure 1.19 The solid in Example 2 can be approximated by a solid made up of "washers" of thickness Δx.

□ **EXAMPLE 3**

The region in the first quadrant bounded by the graph of $y = 4 - x^2$ and the coordinate axes is rotated about the y-axis. Find the volume of the resulting solid.

Solution This problem is similar to that of Example 1 except that the roles of x and y are reversed. Solving the given equation for x as a function of y gives $f(y) = \sqrt{4 - y}$. Since the rotation is about the y-axis, the integration will be with respect to y, and the limits of integration are from $y = 0$ to $y = 4$ (see Figures 1.20 and 1.21).

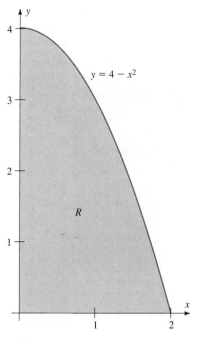

Figure 1.20 The region R in the first quadrant bounded by the graph of $y = 4 - x^2$.

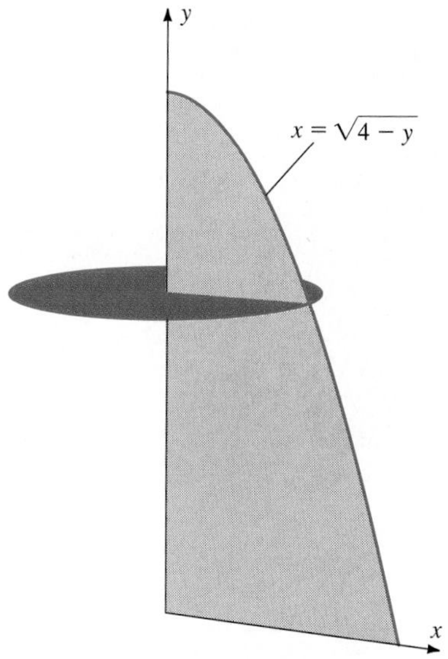

Figure 1.21 The cross sections perpendicular to the y-axis are circles of radius $\sqrt{4 - y}$.

We obtain

$$V = \int_0^4 \pi[\sqrt{4 - y}]^2 \, dy$$

$$= \pi\left[4y - \frac{y^2}{2}\right]_0^4$$

$$= 8\pi.$$

The result of Example 3 generalizes as follows:

> If the region R is bounded by the graph of the continuous function $x = f(y)$ and the y-axis from $y = c$ to $y = d$, then the volume of the solid obtained by rotating R about the y-axis is
>
> $$V = \int_c^d \pi[f(y)]^2 \, dy.$$
>
> (See Figures 1.22 and 1.23.)

Solids of Known Cross-Sectional Area

If we ''slice'' a solid of revolution S in the direction perpendicular to the axis of rotation, we obtain a circular disc. For example, if S is obtained by revolving the region R bounded by the graph of a nonnegative function f about the x-axis for $a \leq x \leq b$, the slice for which $x = x_0$ is a disc with radius $r = f(x_0)$ (see Figure 1.24). Consequently, its area $A(x_0)$ is $\pi[f(x_0)]^2$. In fact, we can interpret formula

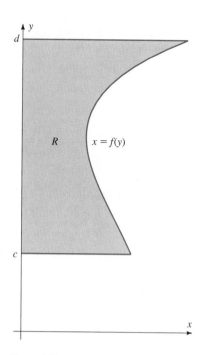

Figure 1.22 A region R bounded by the graph of $x = f(y)$ for $c \leq y \leq d$.

Figure 1.23 A solid of revolution generated by revolving about the y-axis.

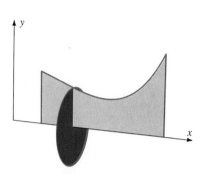

Figure 1.24 A slice perpendicular to the axis of rotation for a solid of revolution is a circular disc.

(1) for the volume of S as the integral of the cross-sectional area function A from $x = a$ to $x = b$. Our next objective is to show that the volume of *any* solid of known cross-sectional area can be calculated in this way.

In particular, suppose that S is the solid for which the area $A(x)$ of each cross section perpendicular to the x-axis is known (S need not be a solid of revolution—see Figure 1.25). Moreover, suppose that S extends from $x = a$ to $x = b$. We partition the interval $[a, b]$ into subintervals of equal length $\Delta x = (b - a)/n$ with endpoints $a = x_0 < x_1 < x_2 < \cdots < x_n = b$. Using the partition, we subdivide S into solids S_j of thickness Δx, where each S_j corresponds to the interval $[x_{j-1}, x_j]$ (see Figure 1.26). By selecting one number t_j in each subinterval $[x_{j-1}, x_j]$, we approximate the volume of S_j by the volume V_j of the cylinder with face area $A(t_j)$ (see Figure 1.27); that is,

$$V_j = A(t_j)\,\Delta x. \tag{3}$$

Summing the V_j for $j = 1, 2, 3, \ldots, n$ yields an approximation to the volume V of S

$$V \approx \sum_{j=1}^{n} V_j = \sum_{j=1}^{n} A(t_j)\,\Delta x$$

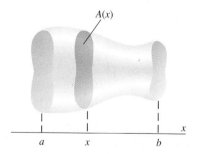

Figure 1.25 A solid S whose cross-sectional areas are given by the area function $A(x)$.

(see Figure 1.28). We assume that, as $n \to \infty$, the Riemann sum on the right-hand side approaches the volume V. Now, if A is continuous for $a \leq x \leq b$, the Riemann sum also approaches the definite integral $\int_a^b A(x)\,dx$. Thus, we obtain the following formula for the volume of solids with known cross-sectional area.

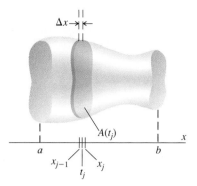

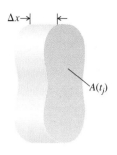

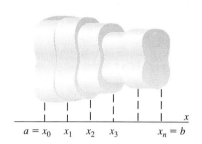

Figure 1.26 The solid S_j of thickness Δx corresponding to the interval $[x_{j-1}, x_j]$.

Figure 1.27 A "volume element" with face area $A(t_j)$ and thickness Δx. Its volume approximates the volume of S_j.

Figure 1.28 The volume of S is approximated by the sum of the volumes of the volume elements.

Let $A(x)$ denote the area of the cross section of S for $a \leq x \leq b$. If the function A is continuous on $[a, b]$, the volume V of S is

$$V = \int_a^b A(x) \, dx. \tag{4}$$

Notice that equation (4) generalizes the familiar formula for the volume of a cylinder with base area A and height h: $V = Ah$. It also contains formula (1) for the volume of a solid of revolution as a special case, since for such solids, $A(x) = \pi[f(x)]^2$.

□ **EXAMPLE 4**

The base of a solid is a circle of radius 4 cm. All cross sections perpendicular to a particular axis are squares (see Figure 1.29). Find the volume of this solid (Figure 1.30).

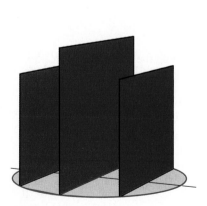

Figure 1.29 Square cross sections sitting on a circular base.

Figure 1.30 The solid in Example 4.

Strategy · · · · · · · ·

Find an expression for the area of a cross section. Begin by finding an equation for the boundary of the base.

Solution

If we impose an xy-coordinate system on the circular base so that the x-axis corresponds to the given axis, then the equation for the boundary of the base is

$$x^2 + y^2 = 16 \qquad \text{(Figure 1.31)}.$$

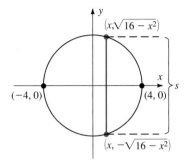

Figure 1.31 Side $s = 2\sqrt{16 - x^2}$.

The equation for the upper semicircle is $y = \sqrt{16 - x^2}$, and the equation for the lower semicircle is $y = -\sqrt{16 - x^2}$. A cross section perpendicular to the x-axis will therefore intersect this circular base in a chord of length $s = 2\sqrt{16 - x^2}$. Since this chord is one side of the square cross section, the area of the cross section is

$$A(x) = s^2 = (2\sqrt{16 - x^2})^2 = 64 - 4x^2.$$

Find the limits of integration.

The smallest and largest values of x are, respectively, -4 and 4. The volume, by equation (4), is therefore

Apply (4).

$$V = \int_{-4}^{4} A(x)\, dx = \int_{-4}^{4} (64 - 4x^2)\, dx$$

$$= \left[64x - \frac{4}{3}x^3 \right]_{-4}^{4}$$

$$= \frac{1024}{3} \text{ cm}^3. \qquad \blacksquare$$

Although the solid in Example 5 is a solid of revolution, the axis of rotation is neither the x- nor the y-axis. We could determine the desired volume using an approach similar to that of Example 2. However, we prefer to use equation (4).

☐ **EXAMPLE 5**

Find the volume of the solid generated by rotating the region bounded by the graph of $f(x) = \sqrt{4 - x}$ and the x-axis for $0 \le x \le 4$ about the line $y = -2$.

Solution As illustrated in Figures 1.32 and 1.33, a cross section taken at location x consists of a circle of radius $R = [\sqrt{4 - x} - (-2)]$ from which a smaller circle of

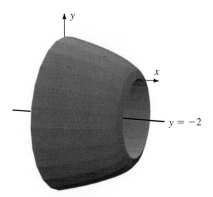

Figure 1.32 The solid generated by rotating about the line $y = -2$ the region in the first quadrant bounded by the graph of $f(x) = \sqrt{4-x}$.

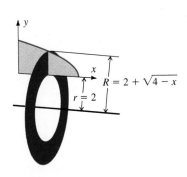

Figure 1.33 A cross section at x consists of a circle of radius $R = 2 + \sqrt{4-x}$ from which a smaller circle of radius $r = 2$ is removed.

radius $r = 2$ has been removed. The cross-sectional area is therefore

$$A(x) = \pi R^2 - \pi r^2$$
$$= \pi(\sqrt{4-x} + 2)^2 - \pi \cdot 2^2$$
$$= \pi(4 - x + 4\sqrt{4-x}).$$

We can now apply equation (4) to find that

$$V = \int_0^4 \pi(4 - x + 4\sqrt{4-x})\,dx$$
$$= \pi\left[4x - \frac{x^2}{2} - \frac{8}{3}(4-x)^{3/2}\right]_0^4 = \frac{88\pi}{3}. \qquad \blacksquare$$

Exercise Set 7.1

In Exercises 1–8,

a. set up an integral to find the volume of the solid obtained by revolving about the x-axis the region bounded by the graph of f and the x-axis, for $a \le x \le b$; and

b. evaluate this integral.

1. $f(x) = 2x + 1, \quad 1 \le x \le 4$

2. $f(x) = \sqrt{4x - 1}, \quad 1 \le x \le 5$

3. $f(x) = \sqrt{\sin x}, \quad 0 \le x \le \pi$

4. $f(x) = \sin x \sqrt{\cos x}, \quad 0 \le x \le \pi/2$

5. $f(x) = x(x^3 - 1)^2, \quad 0 \le x \le 1$

6. $f(x) = 2x^2(x^5 + 1)^5, \quad -1 \le x \le 0$

7. $f(x) = \sqrt{4 - x^2}, \quad 1 \le x \le 2$

8. $f(x) = \dfrac{\sqrt{1 + x}}{x^{3/2}}, \quad 1 \le x \le 2$

In Exercises 9–13, find the volume of the solid obtained by revolving about the x-axis the region bounded by the graph of f and the x-axis, for $a \le x \le b$.

9. $f(x) = \tan x, \quad 0 \le x \le \pi/4$

10. $f(x) = \sin x, \quad 0 \le x \le \pi$ (*Hint:* $\sin^2 x = \frac{1}{2} - \frac{1}{2}\cos 2x$.)

11. $f(x) = \sec x, \quad 0 \le x \le \pi/4$

12. $f(x) = |x - 1|, \quad 0 \le x \le 3$

13. $f(x) = |3x - 6|, \quad 0 \le x \le 5$

In Exercises 14–17,

a. set up an integral to find the volume of the solid obtained by revolving about the x-axis the region bounded by the given curves; and

b. evaluate this integral.

14. $f(x) = \sqrt{\cos x}, \quad g(x) = 1, \quad 0 \le x \le \pi/2$

15. $f(x) = -x^2 + 6$, $g(x) = 2$

16. $f(x) = x^2$, $g(x) = x^3$

17. $f(x) = x^2/4$, $g(x) = x$

In Exercises 18–19, find the volume of the solid obtained by revolving about the *x*-axis the region bounded by the given curves.

18. $f(x) = 1/x$, $g(x) = \sqrt{x}$, $1 \le x \le 4$

19. $f(x) = \sin x$, $g(x) = \cos x$, $0 \le x \le \pi/2$

In Exercises 20–24,

a. set up an integral to find the volume of the solid obtained by revolving about the *y*-axis the region bounded by the given curves; and
b. evaluate this integral.

20. $x + y = 4$, $x = 0$, $0 \le y \le 4$

21. $y = x^2$, $y = 0$, $0 \le x \le 2$ **22.** $y = 4$, $y = x^2$

23. $y = x^3$, $x = 2$, $y = 0$ **24.** $y = x^2$, $y = x^3$

In Exercises 25–27, find the volume of the solid obtained by rotating about the *y*-axis the region bounded by the given curves.

25. $y = x^4$, $y = 0$, $x = 2$ **26.** $y = |x - 2|$, $y = 2$

27. $y = (x - 2)^2$, $y = 4$

28. Find the volume of the right pyramid whose base is a square 10 cm on a side and whose altitude is 8 cm (Figure 1.34).

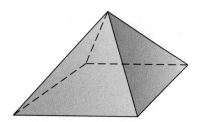

Figure 1.34

29. By integration, derive the formula for the volume of a sphere of radius *r*.

30. The base of a solid is a circle of radius 4. Find the volume of the solid if all cross sections perpendicular to a given axis are equilateral triangles (Figure 1.35).

Figure 1.35 Cross sections perpendicular to a given axis are equilateral triangles (Exercise 30).

31. Find the volume of the solid in Exercise 30 if the cross sections perpendicular to the given axis are isosceles right triangles with bases lying along the base of the solid (Figure 1.36).

Figure 1.36 Cross sections perpendicular to a given axis are isosceles right triangles (Exercise 31).

32. A hemispherical water tank of radius 10 m contains water to a depth of 6 m. How many cubic meters of water does the tank contain?

33. The base of a solid is the region bounded by the graphs of $f(x) = x^2$ and $g(x) = 8 - x^2$. Find the volume of the solid if all cross sections perpendicular to the *x*-axis are squares.

34. Find the volume of the solid in Exercise 33 if the cross sections perpendicular to the *x*-axis are semicircles.

35. Find the formula for the volume of the frustum of a cone with bases of radius *r* and *R*, and height *h* (see Figure 1.37).

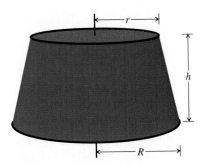

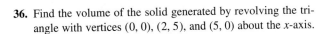

Figure 1.37 The frustum of a cone with bases of radius r and R and with height h (Exercise 35).

36. Find the volume of the solid generated by revolving the triangle with vertices $(0, 0)$, $(2, 5)$, and $(5, 0)$ about the x-axis.

37. Water is running into the tank in Exercise 32 at a rate of 10 m³/min. How fast is the water level rising when its depth is 4 m?

38. A hole of radius 2 cm is drilled through the center of a spherical ball of radius 4 cm. What is the volume of the remaining solid?

39. Find the volume of the solid obtained by rotating the region bounded by the x-axis and the graph of $y = 1 - x^2$ about the line $y = -3$.

40. Find the volume of the solid obtained by rotating the region bounded by the graphs of $y = \sqrt{x}$ and $y = x/2$ about the line $y = 4$.

41. Find the volume of the solid obtained if the region bounded by the graphs of $y = \sqrt{x}$, $y = 0$, and $x = 9$ is rotated about the line $y = -2$.

42. Find the volume of the solid obtained if the graph of the ellipse

$$\frac{x^2}{a^2} + \frac{y^2}{b^2} = 1$$

is rotated about the x-axis.

43. Find the volume of the solid obtained if the region in Exercise 42 is rotated about the y-axis.

44. What formula is obtained from Exercises 42 and 43 if $a = b = r$?

45. When a right circular cylinder of radius r is sliced by two planes, one perpendicular to the central axis of the cylinder and the other at an angle θ with the first, as in Figure 1.38, a wedge results. Find its volume.

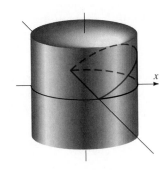

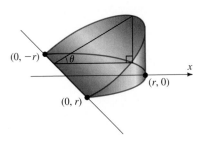

Figure 1.38 A right circular cylinder of radius r cut by two planes—one perpendicular to the central axis and the other at an angle θ with the first. We obtain a wedge (Exercise 45).

46. Use the Trapezoidal Rule with $n = 8$ to approximate the volume of the solid obtained by rotating about the x-axis the region bounded by the graph of

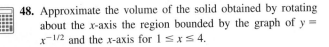

$$f(x) = \frac{1}{\sqrt{1 + x^2}}$$

and the line $y = 0$ for $0 \le x \le 4$.

47. Use Simpson's rule with $n = 12$ to approximate the volume of the solid in Exercise 46.

48. Approximate the volume of the solid obtained by rotating about the x-axis the region bounded by the graph of $y = x^{-1/2}$ and the x-axis for $1 \le x \le 4$.

49. Approximate the volume of the solid obtained by revolving the region in Exercise 42 about the line $y = -2$ if $a = 3$ and $b = 1$. (This exercise can be done either by the use of a numerical approximation procedure or by the use of the formula for the area of a circle.)

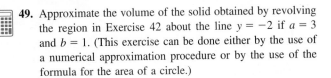

7.2 Calculating Volumes by the Method of Cylindrical Shells

In addition to the slicing methods of Section 7.1, there is another way to calculate the volume of a solid of revolution. While the slicing method is based on the idea of approximating cross sections taken perpendicular to the axis of rotation, the **method of cylindrical shells** uses approximating hollow cylinders centered *about* the axis of rotation.

To describe this method, we let R be the region bounded by the graph of a continuous nonnegative function f and the x-axis for $0 \le a \le x \le b$ (Figure 2.1). If the region R is rotated about the y-axis, a solid S is generated (Figure 2.2).

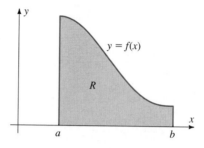

Figure 2.1 The region R bounded by the graph of $y = f(x)$ for $0 \le a \le x \le b$.

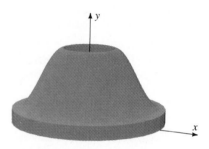

Figure 2.2 The solid generated by rotating the region R in Figure 2.1 about the y-axis.

To calculate the volume V of S, we follow the strategy described in the introduction to this chapter. We partition the interval $[a, b]$ using the equal-length partition

$$a = x_0 < x_1 < x_2 < \cdots < x_n = b$$

where $x_j - x_{j-1} = \Delta x = (b - a)/n$. Then we choose the numbers t_j to be the midpoints of the subintervals. That is, $t_j = (x_{j-1} + x_j)/2$. Using these choices, we approximate the region R by rectangles with base $[x_{j-1}, x_j]$ and height $f(t_j)$ (see Figure 2.3). If we rotate one of these rectangles about the y-axis, we obtain a cylindrical shell (see Figure 2.4). Combining all such shells (Figure 2.5), we obtain a solid whose volume approximates the volume of S.

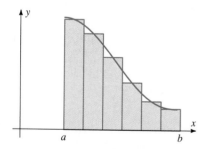

Figure 2.3 Approximating the region R by rectangles using an equal-length partition of the interval $[a, b]$.

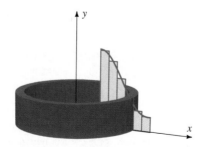

Figure 2.4 A cylindrical shell produced by rotating one of the rectangles about the y-axis.

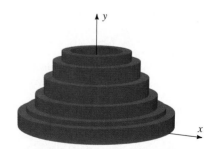

Figure 2.5 Combining all of the shells, we obtain a solid whose volume approximates the desired volume (compare Figure 2.2).

Consider the jth shell (see Figure 2.6). Its height is $f(t_j)$, and the area of its base is $\pi x_j^2 - \pi x_{j-1}^2$. Consequently, its volume V_j is

$$V_j = \pi(x_j^2 - x_{j-1}^2)f(t_j).$$

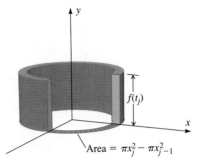

Figure 2.6 A cylindrical shell with a portion cut away to illustrate that its height is $f(t_j)$ and the area of its base is $\pi x_j^2 - \pi x_{j-1}^2$.

Thus, by summing the volumes of all of the shells, we approximate the volume V by

$$V \approx \sum_{j=1}^{n} V_j = \sum_{j=1}^{n} \pi(x_j^2 - x_{j-1}^2)f(t_j). \tag{1}$$

To see that the right-hand side of approximation (1) is a Riemann sum, we factor

$$(x_j^2 - x_{j-1}^2) = (x_j + x_{j-1})(x_j - x_{j-1})$$
$$= (x_j + x_{j-1})\,\Delta x,$$

and rewrite approximation (1) as

$$V \approx \sum_{j=1}^{n} \pi(x_j + x_{j-1})f(t_j)\,\Delta x.$$

Since $2t_j = x_j + x_{j-1}$, we have

$$V \approx \sum_{j=1}^{n} \pi(2t_j)f(t_j)\,\Delta x,$$

and taking the limit as $n \to \infty$, these Riemann sums approach

$$\int_a^b \pi(2x)f(x)\,dx.$$

Assuming that these sums also limit to V, we obtain the following volume formula for S.

If the region R bounded by the graph of the continuous nonnegative function f and the x-axis, for $0 \le a \le x \le b$, is rotated about the y-axis, the volume V of the resulting solid is

$$V = \int_a^b 2\pi x f(x)\,dx. \tag{2}$$

☐ **EXAMPLE 1**

The region bounded by the graph of $y = -2x^2 + 8x - 6$ and the x-axis is rotated about the y-axis (see Figures 2.7 and 2.8). Find the volume V of the resulting solid.

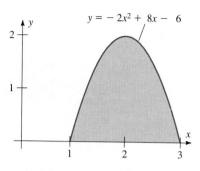

Figure 2.7 The region bounded by the graph of $y = -2x^2 + 8x - 6$ and the x-axis (Example 1).

Figure 2.8 The solid of revolution obtained by revolving the region in Figure 2.7 about the y-axis.

Strategy · · · · · · · ·

Factor equation for y to find limits of integration.

Apply equation (2).

Figure 2.9 A cylindrical shell corresponding to the volume calculation in Example 1.

Solution

Since $y = -2x^2 + 8x - 6 = -2(x - 1)(x - 3)$, the region lies between the lines $x = 1$ and $x = 3$. As Figure 2.9 illustrates,

$$V = \int_1^3 2\pi x[-2x^2 + 8x - 6]\, dx$$

$$= 2\pi \int_1^3 (-2x^3 + 8x^2 - 6x)\, dx$$

$$= 2\pi \left[-\frac{2}{4}x^4 + \frac{8}{3}x^3 - \frac{6}{2}x^2 \right]_1^3 = \frac{32\pi}{3}.$$ ∎

Equation (2) can be generalized to regions bounded below by curves other than the x-axis, as the following example shows.

☐ **EXAMPLE 2**

The region R is bounded by the graphs of $f(x) = -2x^2 + 8x - 6$ and $g(x) = 2x - 6$ (see Figure 2.10). Find the volume of the solid generated by revolving R about the y-axis.

Strategy · · · · · · · ·

Set $f(x) = g(x)$ to find the points where the curves intersect.

Solution

To find the points of intersection, we set $f(x) = g(x)$ and obtain the equation

$$-2x^2 + 8x - 6 = 2x - 6,$$

or

$$-2x^2 + 6x = 0$$

Note which curve bounds the top of the region and which curve bounds the bottom.

so $x = 0$ or $x = 3$. As Figure 2.11 illustrates, when the interval $[0, 3]$ is partitioned, the approximating rectangles are bounded above by $f(x) = -2x^2 + 8x - 6$ and below by $g(x) = 2x - 6$. Since the factor $f(x)$ in equation (2) represents the height of the approximating rectangles, we modify equation (2) to the following:

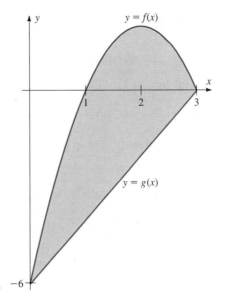

Figure 2.10 The region bounded by the graphs of $f(x) = 2x^2 + 8x - 6$ and $g(x) = 2x - 6$ (Example 2).

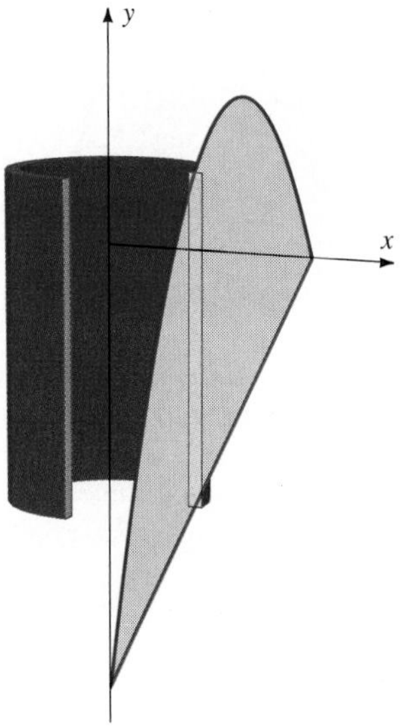

Figure 2.11 A cylindrical shell with a portion cut away to expose one of the approximating rectangles of the region in the xy-plane (Example 2).

Apply equation (2) to both curves.

$$V = \int_0^3 2\pi x[f(x) - g(x)] \, dx$$

$$= \int_0^3 2\pi x[-2x^2 + 8x - 6 - (2x - 6)] \, dx$$

$$= \int_0^3 2\pi x(-2x^2 + 6x) \, dx$$

$$= 2\pi\left[-\frac{2}{4}x^4 + \frac{6}{3}x^3\right]_0^3 = 27\pi. \qquad \blacksquare$$

REMARK 1   We may state the generalization of equation (2) observed in Example 2 as follows: If the region R bounded above by the graph of $y = f(x)$ and below by the graph of $y = g(x)$, for $0 \le a \le x \le b$, is rotated about the y-axis, the volume of the resulting solid is

$$V = \int_a^b 2\pi x[f(x) - g(x)] \, dx. \qquad (3)$$

REMARK 2   You may have noticed that the integrands in equations (2) and (3) have simple geometric interpretations. If the region R to be rotated about the y-axis is sliced vertically at location x and the resulting line segment is rotated about the y-axis, a band of height $[f(x) - g(x)]$ is generated (see Figure 2.12). Since the circumference of this band is $2\pi x$, the surface area of one "side" of the band is precisely the integrand $2\pi x[f(x) - g(x)]$ in equation (3) (Figure 2.13). We may therefore interpret equations (2) and (3) by saying that the volume V is found by integrating the "radial" cross section from $x = a$ to $x = b$.

In the following example, we show how to modify the method of cylindrical shells in cases where equations (2) and (3) do not directly apply.

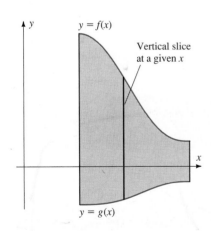

Figure 2.12 The region between the graphs of $f(x)$ and $g(x)$ along with a vertical slice (a line segment) at x.

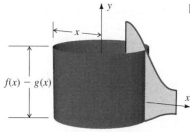

Figure 2.13   The slice in Figure 2.12 generates a band if it is rotated about the y-axis. Its radius is x, and its height is $f(x) - g(x)$. Thus, its circumference is $2\pi x$, and its surface area is $2\pi x(f(x) - g(x))$.

□ **EXAMPLE 3**

Find the volume of the solid obtained by revolving about the line $x = -1$ the region R bounded by the graphs of $f(x) = x$ and $g(x) = (x - 2)^2$.

Solution   Figure 2.14 shows the region R that is bounded above by the graph of $f(x) = x$ and below by the graph of $g(x) = (x - 2)^2 = x^2 - 4x + 4$. To find the points of intersection of these two graphs we set $f(x) = g(x)$, which gives the equation

$$x^2 - 4x + 4 = x$$

or

$$x^2 - 5x + 4 = (x - 1)(x - 4) = 0.$$

The points of intersection are therefore $(1, 1)$ and $(4, 4)$.

Figure 2.15 shows that a vertical line segment through this region generates a circular band of height

$$f(x) - g(x) = x - (x^2 - 4x + 4) = -x^2 + 5x - 4.$$

Since the axis of rotation is the vertical line $x = -1$, the radius of this circular band is $x - (-1) = x + 1$. We must therefore replace the factor x in the integrand in formula (3) by the factor $x + 1$. The volume is therefore

$$V = \int_1^4 2\pi(x + 1)[f(x) - g(x)]\, dx$$

$$= \int_1^4 2\pi(x + 1)(-x^2 + 5x - 4)\, dx$$

$$= \int_1^4 2\pi(-x^3 + 4x^2 + x - 4)\, dx$$

$$= 2\pi\left[-\frac{x^4}{4} + \frac{4}{3}x^3 + \frac{x^2}{2} - 4x\right]_1^4 = \frac{63\pi}{2}. \qquad \blacksquare$$

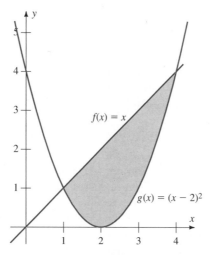

Figure 2.14   The region bounded by the graphs of $f(x) = x$ and $g(x) = (x - 2)^2$.

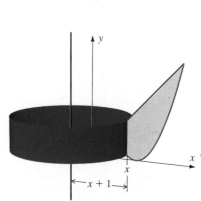

Figure 2.15   Revolving a vertical slice about the line $x = -1$ produces a band whose height is $f(x) - g(x) = -x^2 + 5x - 4$ and whose radius is $x + 1$.

REMARK 3   Many of the problems on volumes of revolution in this section and the preceding section may be solved using either the method of slicing by perpendicular cross sections or the method of cylindrical shells. Usually one method will be simpler to use, depending on the geometry of the region being rotated.

Exercise Set 7.2

In Exercises 1–10,

a. sketch the region bounded by the graphs of the given functions for the specified values of x or y;
b. set up an integral to find the volume of the solid obtained by revolving this region about the y-axis; and
c. evaluate this integral.

1. $x + y = 1$, $x = 0$, $y = 0$

2. $y = \sqrt{x}$, $y = 0$, $1 \le x \le 4$

3. $y = x^3$, $y = 0$, $1 \le x \le 3$

4. $y = \sqrt{1 + x^2}$, $y = 0$, $0 \le x \le 3$

5. $y = \sqrt{x}$, $y = -x$, $x = 4$

6. $y = 1 + x + x^2$, $y = -2$, $1 \le x \le 3$

7. $y = -x^2 - 4x - 3$, $y = 0$

8. $y = 1/x$, $y = 1$, $x = 4$

9. $y = \sin x^2$, $y = 1$, $0 \le x \le \sqrt{\pi/2}$

10. $y = 2$, $y = \sqrt[3]{x}$, $x = 0$

In Exercises 11–15,

a. sketch the region bounded by the graphs of the given functions for the specified values of x or y; and
b. find the volume of the solid obtained by revolving this region about the y-axis.

11. $y = -x \cos x^3$, $0 \le x \le \sqrt[3]{\pi/2}$

12. $y = \dfrac{1}{\sqrt{4 - x^2}}$, $y = 0$, $0 \le x \le 1$

13. $y = \sqrt{9 - x^2}$, $y = 0$, $0 \le x \le 3$

14. $y = \dfrac{1}{\sqrt{x}} + \sqrt{x}$, $y = 0$, $1 \le x \le 4$

15. $y = \dfrac{\sqrt{1 + x^{3/2}}}{\sqrt{x}}$, $y = 0$, $1 \le x \le 4$

In Exercises 16–23, use either the method of cylindrical shells or the method of slicing to find the volume of the solid described.

16. The region bounded by the graphs of $y = \sin x^2$ and $y = -x$, for $0 \le x \le \sqrt{\pi}$, is rotated about the y-axis.

17. The region bounded by the graph of $x = y^2$ and the line $x = 4$ is revolved about the line $x = -1$.

18. The region bounded by the graph of $x = \sqrt{4 + y}$, the line $x = 0$, and the line $y = 0$ is revolved about the line $x = -2$.

19. The region bounded by the graphs of $y = x^2$ and $y = \sqrt{x}$ is revolved about the line $y = -2$.

20. The triangle with vertices $(1, 0)$, $(2, 4)$, and $(4, 0)$ is rotated about the y-axis.

21. The region bounded by the graphs of $y = x$ and $y = x^3$ is revolved about the y-axis.

22. The region bounded by the graphs of $y = 4$ and $y = x^2$ is revolved about the line $y = -2$.

23. The triangle with vertices $(1, 3)$, $(1, 7)$, and $(4, 7)$ is revolved about the line $y = 1$.

24. Find the volume generated by revolving the triangle with vertices $(0, 0)$, $(0, 2)$, and $(2, 0)$ about
    a. the x-axis,
    b. the y-axis,
    c. the line $x = -1$,
    d. the line $y = -1$.

25. Use the method of cylindrical shells to find the volume of the solid obtained by rotating the region bounded by the ellipse

$$\frac{x^2}{a^2} + \frac{y^2}{b^2} = 1$$

about the y-axis.

Figure 2.16 Revolving a disc about an axis to which it is not adjacent produces a torus.

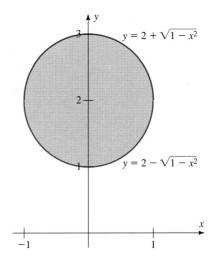

Figure 2.17 The region bounded by the circle of radius 1 centered at $(0, 2)$.

26. Figure 2.16 shows a solid called a *torus*. A **torus** is generated by revolving a disc about an axis to which it is not adjacent. Figure 2.17 shows a region R bounded by the circle of radius $r = 1$ and center $(0, 2)$. Let T be the torus obtained by revolving R about the x-axis.

   a. Show that the volume of T can be expressed as

$$V = \pi \int_{-1}^{1} [2 + \sqrt{1 - x^2}]^2 \, dx - $$
$$\pi \int_{-1}^{1} [2 - \sqrt{1 - x^2}]^2 \, dx.$$

   b. Show that V simplifies to the integral

$$V = 8\pi \int_{-1}^{1} \sqrt{1 - x^2} \, dx.$$

   c. Evaluate the integral in part b by interpreting it as the area of a semicircle.

   d. Find the volume V of T.

 27. Use the method of cylindrical shells and the Trapezoidal Rule to approximate the volume of the solid obtained by revolving about the y-axis the region bounded by the graph of $y = \sin x$ and the x-axis, for $0 \le x \le \pi$.

 28. Use Simpson's Rule to approximate the volume described in Exercise 27.

 29. Approximate the volume of the solid obtained by revolving about the axis $y = -1$ the region bounded by the graph of $y = \sqrt{1 + x^4}$, the line $y = 0$, and the lines $x = 0$ and $x = 2$.

7.3   Arc Length and Surface Area

The problems of computing the length of a curve in the plane and the surface area of a solid of revolution are closely related. We treat both problems in this section and then turn to more physical applications of the definite integral in the remaining sections of the chapter.

Arc Length

The arc length problem is to define and calculate the length of the arc of the graph of the function f from the point $(a, f(a))$ to the point $(b, f(b))$ (Figure 3.1). In the analysis that follows we will discover what specific conditions f needs to satisfy, but at the outset it seems reasonable to assume that f is continuous on the closed interval $[a, b]$.

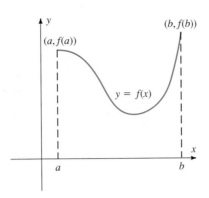

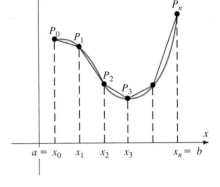

Figure 3.1 Arc whose length is to be calculated

Figure 3.2 Approximating arc by polygonal path.

We approach this problem in a familiar way, by partitioning the interval $[a, b]$ into n subintervals of equal length. As before, we use partitions $a = x_0 < x_1 < x_2 < \cdots < x_n = b$, where $x_j - x_{j-1} = \Delta x = (b - a)/n$ for $j = 1, 2, \ldots, n$. For each subinterval $[x_{j-1}, x_j]$, we approximate the length of the arc by an expression that we can easily calculate. We do this by connecting the endpoints of the arc, $P_{j-1} = (x_{j-1}, f(x_{j-1}))$ and $P_j = (x_j, f(x_j))$, with a line segment ℓ_j (see Figure 3.2).

We approximate the length of the arc over the subinterval $[x_{j-1}, x_j]$ by the length of the line segment ℓ_j. The length of the entire arc is thus approximated by the length of the polygonal path through $P_0, P_1, P_2, \ldots, P_n$. If the limit of this approximation exists as $n \to \infty$, we define the arc length to be this limit.

Using the distance formula, we calculate the length of ℓ_j as

$$\text{Length of } \ell_j = \sqrt{(x_j - x_{j-1})^2 + [f(x_j) - f(x_{j-1})]^2}$$

$$= \left(\sqrt{1 + \left[\frac{f(x_j) - f(x_{j-1})}{x_j - x_{j-1}} \right]^2} \right)(x_j - x_{j-1}).$$

(1)

If f is differentiable on $[x_{j-1}, x_j]$, the Mean Value Theorem guarantees the existence of a number $t_j \in [x_{j-1}, x_j]$ so that

$$f'(t_j) = \frac{f(x_j) - f(x_{j-1})}{x_j - x_{j-1}}. \quad \text{(See Figure 3.3.)}$$

Substituting this expression into equation (1) gives

$$\text{Length of } \ell_j = (\sqrt{1 + [f'(t_j)]^2}) \, \Delta x.$$

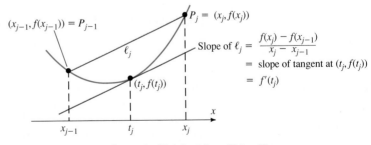

Figure 3.3 Slope of ℓ_j equals $f'(t_j)$ by Mean Value Theorem.

Summing these expressions over all $j = 1, 2, \ldots, n$ gives the approximation

$$\text{Arc length} \approx \sum_{j=1}^{n} \sqrt{1 + [f'(t_j)]^2}\, \Delta x. \tag{2}$$

Now the sum on the right-hand side of approximation (2) is a Riemann sum involving the *derivative* f' of f. Therefore, using the theory of the definite integral from Chapter 6, we know that the limit

$$L = \lim_{n \to \infty} \sum_{j=1}^{n} (\text{length of } \ell_j) = \lim_{n \to \infty} \sum_{j=1}^{n} \sqrt{1 + [f'(t_j)]^2}\, \Delta x$$

equals the integral

$$\int_{a}^{b} \sqrt{1 + [f'(x)]^2}\, dx$$

if f' is a continuous function on $[a, b]$. In this case, we define the arc length to be this limit L, and we can calculate it as follows:

If f has a continuous derivative on $[a, b]$, the length of the graph of $y = f(x)$ from $(a, f(a))$ to $(b, f(b))$ is given by

$$L = \int_{a}^{b} \sqrt{1 + [f'(x)]^2}\, dx. \tag{3}$$

or, in Leibniz notation,

$$L = \int_{a}^{b} \sqrt{1 + \left[\frac{dy}{dx}\right]^2}\, dx.$$

□ **EXAMPLE 1**

Verify that the expression for arc length in equation (3) agrees with the distance formula for the case of a nonvertical line segment joining two points in the plane.

Strategy · · · · · · · · ·

First, determine the length according to the distance formula.

Solution

Let $P = (x_1, y_1)$ and $Q = (x_2, y_2)$ be two points in the plane such that $x_1 \neq x_2$. Then the distance formula gives the length L of the line segment PQ as

$$L = \sqrt{(x_2 - x_1)^2 + (y_1 - y_1)^2}. \tag{4}$$

Find an equation for the line in question. Use form

$$y - y_1 = m(x - x_1)$$

where m = slope.

On the other hand, the equation for the line passing through P and Q is

$$y - y_1 = \frac{y_2 - y_1}{x_2 - x_1}(x - x_1),$$

so

Differentiate to find $\dfrac{dy}{dx}$.

$$\frac{dy}{dx} = \frac{y_2 - y_1}{x_2 - x_1}.$$

Equation (3) therefore gives the length of PQ as

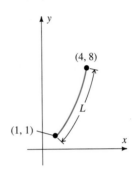

Apply equation (3) to find length by integration.

Note that

$$\sqrt{1 + \left(\frac{y_2 - y_1}{x_2 - x_1}\right)^2}$$

is a constant.

$$L = \int_{x_1}^{x_2} \sqrt{1 + \left(\frac{y_2 - y_1}{x_2 - x_1}\right)^2}\, dx$$

$$= \left(\sqrt{1 + \left(\frac{y_2 - y_1}{x_2 - x_1}\right)^2}\right) \left[x\right]_{x_1}^{x_2}$$

$$= \left(\sqrt{1 + \left(\frac{y_2 - y_1}{x_2 - x_1}\right)^2}\right)(x_2 - x_1)$$

$$= \sqrt{(x_2 - x_1)^2 + (y_2 - y_1)^2},$$

Check that the two results agree.

which agrees with expression (4). ∎

□ **EXAMPLE 2**

Find the arc length of the graph of $y = x^{3/2}$ between $(1, 1)$ and $(4, 8)$. (See Figure 3.4.)

Figure 3.4 The graph of $y = x^{3/2}$ between $(1, 1)$ and $(4, 8)$.

Strategy · · · · · · · ·
Check that dy/dx is continuous.

Apply equation (3).

Use the substitution $u = 1 + \frac{9}{4}x$; $du = \frac{9}{4}\, dx$. Then

$$\int \sqrt{1 + \frac{9}{4}x}\, dx = \int \sqrt{u}\,(\tfrac{4}{9})\, du$$

$$= \tfrac{4}{9}(\tfrac{2}{3})u^{3/2} + C$$
$$= \tfrac{4}{9}(\tfrac{2}{3})(1 + \tfrac{9}{4}x)^{3/2} + C$$

is the antiderivative.

Solution
Here $dy/dx = \frac{3}{2}x^{1/2}$ is continuous on the interval $[1, 4]$, so equation (3) applies. We obtain

$$L = \int_1^4 \sqrt{1 + [\tfrac{3}{2}x^{1/2}]^2}\, dx$$

$$= \int_1^4 \sqrt{1 + \tfrac{9}{4}x}\, dx$$

$$= \left[\tfrac{4}{9}(\tfrac{2}{3})(1 + \tfrac{9}{4}x)^{3/2}\right]_1^4$$

$$= \tfrac{8}{27}[10^{3/2} - (\tfrac{13}{4})^{3/2}] \approx 7.634.$$ ∎

Suppose an arc from the point (a, c) to the point (b, d) with $c < d$ can be expressed in the form $x = f(y)$ where the derivative f' of f is continuous on $[c, d]$. Then we can use the same techniques to derive the arc length formula

$$L = \int_c^d \sqrt{1 + [f'(y)]^2}\, dy. \tag{5}$$

The following example illustrates this approach.

□ **EXAMPLE 3**

Find the length of the arc of the graph of the equation

$$6xy - y^4 - 3 = 0$$

from (19/12, 2) to (14/3, 3).

Strategy · · · · · · · ·

Solve for x rather than y due to presence of y^4 term.

Solution

The form of the equation suggests that we solve for x as a function of y. Doing so we obtain

$$x = f(y) = \frac{y^3}{6} + \frac{1}{2y}.$$

Then

$$f'(y) = \frac{y^2}{2} - \frac{1}{2y^2}.$$

Apply equation (5).

By squaring the binomial term and factoring the result, the term under the radical can be brought into the form of a perfect square.

We therefore obtain

$$
\begin{aligned}
L &= \int_2^3 \sqrt{1 + \left[\frac{y^2}{2} - \frac{1}{2y^2}\right]^2}\, dy \\
&= \int_2^3 \sqrt{1 + \left[\frac{y^4}{4} - \frac{1}{2} + \frac{1}{4y^4}\right]}\, dy \\
&= \int_2^3 \sqrt{\frac{y^4}{4} + \frac{1}{2} + \frac{1}{4y^4}}\, dy \\
&= \int_2^3 \sqrt{\left(\frac{y^2}{2} + \frac{1}{2y^2}\right)^2}\, dy \\
&= \int_2^3 \left(\frac{y^2}{2} + \frac{1}{2y^2}\right) dy \\
&= \left[\frac{y^3}{6} - \frac{1}{2y}\right]_2^3 \\
&= \frac{13}{3} - \frac{13}{12} = \frac{13}{4}.
\end{aligned}
$$

∎

REMARK 1 You may have observed that the integral in equation (3) is, in general, very difficult to evaluate using the Fundamental Theorem. Although we will study additional techniques of integration in Chapter 10, we will not be able to find convenient antiderivatives for most integrands arising from an application of equation (3). Thus, we often need the procedures for approximating integrals in most practical applications of the formulas for arc length.

Surface Area

We saw in Section 7.1 that, if the region bounded by the graph of a continuous function f and the x-axis for $a \le x \le b$ is revolved about the x-axis, a solid of revolution results. Now we determine how to calculate the lateral surface area of such a solid.

As for the arc length problem, we partition $[a, b]$ with numbers $a = x_0 < x_1 < x_2 < \cdots < x_n = b$ where $x_j - x_{j-1} = \Delta x = (b - a)/n$, $j = 1, 2, \ldots, n$. We then connect successive points $P_j = (x_j, f(x_j))$ with line segments to form a polygonal approximation to the graph of $y = f(x)$ from $P_0 = (a, f(a))$ to $P_n = (b, f(b))$ (see

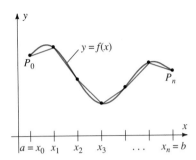

Figure 3.5 A polygon approximation to the graph of $y = f(x)$ from $P_0 = (a, f(a))$ to $P_n =$

Figure 3.6 Revolving the polygonal approximation about the x-axis determines a solid whose surface area approximates the desired surface area. Each subinterval $[x_{j-1}, x_j]$ corresponds to the frustum of a cone.

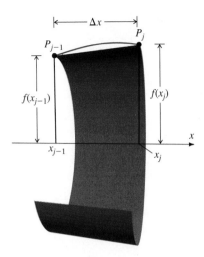

Figure 3.7 One of the frustums with a portion cut away to expose its dimensions.

$(b, f(b))$.

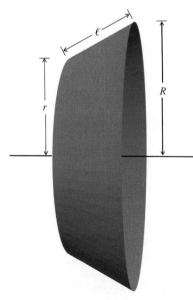

Figure 3.8 The surface area of a frustum of a cone with radii r and R and slant height ℓ is $\pi(r + R)\ell$.

Figure 3.5). As this polygonal path is revolved about the x-axis, it sweeps out a solid whose surface area is an approximation to the surface area S of the solid in question (Figure 3.6).

As the line segment joining P_{j-1} and P_j revolves about the x-axis (Figure 3.7), it sweeps out a frustum of a cone whose radii are $f(x_{j-1})$ and $f(x_j)$ and whose slant height is the length of the line segment from P_{j-1} to P_j; that is,

$$\sqrt{(x_j - x_{j-1})^2 + [f(x_j) - f(x_{j-1})]^2}.$$

Since the formula for the lateral surface area of a frustum of a cone with radii r and R and slant height ℓ is $s = \pi(r + R)\ell$ (Figure 3.8), the surface area of the frustum generated by the line segment ℓ_j from P_{j-1} to P_j is

$$S_j = \pi(f(x_{j-1}) + f(x_j))\sqrt{(x_j - x_{j-1})^2 + [f(x_j) - f(x_{j-1})]^2}.$$

Note that the square root factor in the formula for S_j is the length of ℓ_j. Therefore, we can rewrite it as we did previously in this section (see equation (1)). We obtain

$$S_j = \pi(f(x_{j-1}) + f(x_j)) \sqrt{1 + \left[\frac{f(x_j) - f(x_{j-1})}{x_j - x_{j-1}}\right]^2} \, \Delta x.$$

As in the derivation of the arc length formula, we apply the Mean Value Theorem and get

$$S_j = \pi(f(x_{j-1}) + f(x_j))\sqrt{1 + [f'(t_j)]^2} \, \Delta x$$

for some number t_j in the interval $[x_{j-1}, x_j]$. Summing over all j yields the approximation

$$\text{Surface area} \approx \sum_{j=1}^{n} S_j = \sum_{j=1}^{n} \pi(f(x_{j-1}) + f(x_j))\sqrt{1 + [f'(t_j)]^2} \, \Delta x.$$

It can be shown that, as $n \to \infty$, these approximations approach Riemann sums for the function

$$\pi(2f(x))\sqrt{[1 + [f'(x)]^2]}.$$

Consequently, these estimates motivate the following definition of surface area.

> If the function f has a continuous derivative on the interval $[a, b]$, then the surface area S of the solid of revolution obtained by revolving about the x-axis the region bounded by the graph of $y = f(x) \geq 0$ and the x-axis, for $a \leq x \leq b$, is
>
> $$S = \int_a^b 2\pi f(x)\sqrt{1 + [f'(x)]^2}\, dx \qquad (6)$$
>
> or, in Leibniz notation,
>
> $$S = \int_a^b 2\pi y \sqrt{1 + \left[\frac{dy}{dx}\right]^2}\, dx. \qquad (7)$$

REMARK 2 It is important that $y = f(x) \geq 0$ in equations (6) and (7). If $f(x)$ is negative for some $x \in [a, b]$, we must replace $f(x)$ by $|f(x)|$ in (6) and y by $|y|$ in (7).

□ **EXAMPLE 4**

Find the surface area of the band of the sphere generated by revolving about the x-axis the arc of the circle $x^2 + y^2 = r^2$ lying above the interval $[-a, a]$, $a < r$ (Figure 3.9).

Solution Here $y = \sqrt{r^2 - x^2}$, so

$$\frac{dy}{dx} = \frac{-x}{\sqrt{r^2 - x^2}}.$$

Using equation (7), we obtain

$$\begin{aligned}
S &= \int_{-a}^a 2\pi\sqrt{r^2 - x^2}\sqrt{1 + \left[\frac{-x}{\sqrt{r^2 - x^2}}\right]^2}\, dx \\
&= \int_{-a}^a 2\pi\sqrt{r^2 - x^2}\sqrt{\frac{(r^2 - x^2) + x^2}{r^2 - x^2}}\, dx \\
&= \int_{-a}^a 2\pi\sqrt{r^2}\, dx \\
&= \left[2\pi r x\right]_{-a}^a = 4\pi a r.
\end{aligned}$$ ■

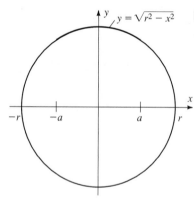

Figure 3.9 Rotating the arc of the circle $x^2 + y^2 = r^2$ lying above $[-a, a]$ about the x-axis produces a band of a sphere.

□ **EXAMPLE 5**

Is it correct in Example 4 to set $a = r$ and conclude that the surface area of the sphere of radius r is $S = 4\pi r^2$?

Solution Although the conclusion is correct, the reasoning is faulty. Applying formula (7) requires that the function f have a continuous derivative for all x in the interval in question. Since

$$\frac{dy}{dx} = \frac{-x}{\sqrt{r^2 - x^2}}$$

is undefined for $x = r$, equation (7) cannot be applied on the interval $[-r, r]$. We will obtain a satisfactory solution to this problem in Chapter 15 when we study curves described by parametric equations. ∎

☐ **EXAMPLE 6**

Find an integral giving the surface area of the "spool" obtained by revolving about the y-axis the region bounded by the graph of $y^2 - x + 1 = 0$ and the y-axis, for $-1 \le y \le 1$ (Figure 3.10).

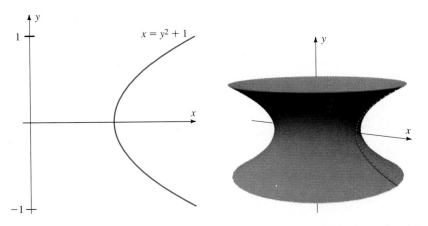

Figure 3.10 The "spool" obtained by revolving about the y-axis the graph of $x = y^2 + 1$ for $-1 \le y \le 1$.

Solution Solving for x as a function of y we obtain

$$x = y^2 + 1$$

so

$$\frac{dx}{dy} = 2y.$$

Since dx/dy is continuous for $-1 \le y \le 1$, we may use the analogue of equation (7) for rotations about the y-axis. We obtain

$$S = \int_{-1}^{1} 2\pi x \sqrt{1 + \left[\frac{dx}{dy}\right]^2} \, dy, \quad \text{or}$$

$$S = \int_{-1}^{1} 2\pi (y^2 + 1)\sqrt{1 + 4y^2} \, dy. \tag{8}$$

As will be the case for many functions to which equation (7) is applied, we cannot find a convenient antiderivative for the integrand in the integral (8) for surface area. Table 3.1, however, shows the results of using the Trapezoidal Rule to approximate the value of this integral. ∎

Table 3.1

n	Trapezoidal approximation
5	27.5483
10	26.5413
25	26.2583
50	26.2179
100	26.2078
200	26.2052

Exercise Set 7.3

In Exercises 1–6,

a. set up an integral to find the length of the arc determined by the given equation; and

b. evaluate this integral.

1. $y = 2x + 3$, from $(-3, -3)$ to $(2, 7)$

2. $y = x^{3/2}$, from $(0, 0)$ to $(4, 8)$

3. $y - 2 = (x + 2)^{3/2}$, from $(-2, 2)$ to $(2, 10)$

4. $3y = (x^2 + 2)^{3/2}$, for $0 \le x \le 2$

5. $3y = 2(x + 1)^{3/2}$, from $(0, 2/3)$ to $(3, 16/3)$

6. $y = \dfrac{x^3}{3} + \dfrac{1}{4x}$, for $1 \le x \le 3$

In Exercises 7–12, find the length of the arc determined by the given equation.

7. $(3y - 6)^2 = (x^2 + 2)^3$, for $0 \le x \le 3$

8. $y = \dfrac{x^4}{4} + \dfrac{1}{8x^2}$, for $1 \le x \le 2$

9. $y = \dfrac{x^3}{6} + \dfrac{1}{2x}$, from $\left(1, \dfrac{2}{3}\right)$ to $\left(4, \dfrac{259}{24}\right)$

10. $y = \dfrac{2}{3}x^{3/2} - \dfrac{1}{2}x^{1/2}$, from $\left(1, \dfrac{1}{6}\right)$ to $\left(3, \dfrac{3\sqrt{3}}{2}\right)$

11. $y = \dfrac{x^4 + 6x + 3}{6x}$, for $2 \le x \le 4$.

12. $y = \dfrac{3x^5}{5} + \dfrac{1}{36x^3}$, for $1 \le x \le 2$.

In Exercises 13–16,

a. set up an integral to find the area of the lateral surface of the solid generated by revolving the stated region about the given axis; and

b. evaluate this integral.

13. The region bounded by the graph of $y = 2x + 3$ and the x-axis for $1 \le x \le 3$ is revolved about the x-axis.

14. The region bounded by the graph of $y = x^3$ and the x-axis for $0 \le x \le 3$ is revolved about the x-axis.

15. The region bounded by the graph of $y = x^{1/3}$ and the y-axis for $0 \le y \le 2$ is revolved about the y-axis.

16. The region bounded by the graph of $x = 2\sqrt{y}$ and the y-axis for $1 \le y \le 4$ is revolved about the y-axis.

In Exercises 17–20, find the area of the lateral surface of the solid generated by revolving the stated region about the given axis.

17. The region bounded by the graph of $y = \sqrt{x}$ and the x-axis for $1 \le x \le 2$ is revolved about the x-axis.

18. The region bounded by the graph of $y = \sqrt{2x - 1}$ and the x-axis for $2 \le x \le 8$ is revolved about the x-axis.

19. The region bounded by the graph of $12xy - 3y^4 = 4$ and the y-axis for $1 \le y \le 3$ is revolved about the y-axis.

20. The region bounded by the graph of

$$y = \frac{x^3}{4} + \frac{1}{3x}$$

and the x-axis for $1 \le x \le 3$ is revolved about the x-axis.

In Exercises 21–22, use the Fundamental Theorem of Calculus to determine the arc length of the graph of f over the specified interval.

21. $f(x) = \displaystyle\int_0^x \sqrt{2t^2 + t^4}\, dt$, $1 \le x \le 2$

22. $f(x) = \displaystyle\int_1^x \sqrt{t^2 - 1}\, dt$, $2 \le x \le 10$

23. Verify using integration that the surface area of a cone with open base is $\pi r l$ where l is the side length and r is the radius of the base.

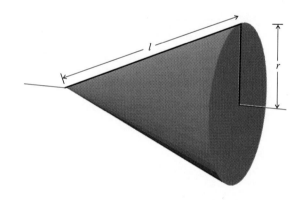

24. The small arc of the circle with equation $x^2 + (y - 1)^2 = 25$ from $(3, 5)$ to $(4, 4)$ is revolved about the x-axis. Use the Trapezoidal Rule to approximate the lateral surface area of the resulting solid.

25. A circle has center $(1, 1)$ and radius 5. The small arc joining the points $(4, 5)$ and $(5, 4)$ on the circle is revolved about the x-axis. Use the Trapezoidal Rule to approximate the lateral surface area of the resulting solid.

26. Find an integral giving the length of the arc of the graph of $f(x) = \sin x$ between $(0, 0)$ and $(\pi/2, 1)$. Use the Trapezoidal Rule with $n = 10$ to approximate the value of this integral.

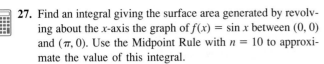

27. Find an integral giving the surface area generated by revolving about the x-axis the graph of $f(x) = \sin x$ between $(0, 0)$ and $(\pi, 0)$. Use the Midpoint Rule with $n = 10$ to approximate the value of this integral.

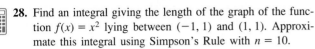

28. Find an integral giving the length of the graph of the function $f(x) = x^2$ lying between $(-1, 1)$ and $(1, 1)$. Approximate this integral using Simpson's Rule with $n = 10$.

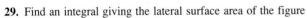

29. Find an integral giving the lateral surface area of the figure

obtained by revolving the upper half of the graph of the ellipse

$$\frac{x^2}{a^2} + \frac{y^2}{b^2} = 1$$

for $-a/2 \le x \le a/2$, about the x-axis. Approximate the value of this integral for $a = 4$ and $b = 3$ using Simpson's Rule with $n = 10$.

7.4 Distance and Velocity

In Chapter 3, we saw that the functions describing the position, velocity, and acceleration of an object moving along a line were related by the process of differentiation. (In particular, $v(t) = s'(t)$ and $a(t) = v'(t)$.) In this section we shall show that the distance D travelled by such an object from time $t = a$ to time $t = b$ can be calculated as a definite integral involving the velocity function v.

Consider an object moving along a line with velocity $v(t)$ at time t. The question we want to answer is: Given that v is a known, continuous function, how can we calculate the distance travelled by the object from time $t = a$ to time $t = b$?

For the romantics among us, here is one centuries-old approach to solving this problem.

> As for distance sailed east or west, early mariners relied solely on dead reckoning, estimating their speed by peering over the side and noting the rate at which foam slipped past the planking. The results they came up with were little more than a compound of hunch, guesswork and intuition. Samuel Eliot Morison calculates that on Columbus' first Atlantic passage, he consistently overestimated the distance of his day's run by about 9 per cent, which scales up to a 90-mile error for every 1,000 miles traveled. However, the remarkable fact is not how far off these navigators were, but how amazingly close they came using such primitive methods.
>
> Progress in the following centuries came slowly. In the late 1500s, some ingenious mariner—name and nationality unknown—invented the common, or chip, log, a simple speed-measuring device. The chip log was a piece of wood shaped like a slice of pie. Attached to it was a light line, knotted at certain equal intervals. To check the ship's speed, a man threw the log overboard, turned a small 30-second sandglass, and began counting the knots that slipped through his fingers as the ship left the log astern. The number of knots he counted in 30 seconds was translated into the nautical miles per hour his ship was traveling. With this device began usage of the term ''knots'' to mean ''nautical miles per hour.''

Life Science Library: The Ships by Edward V. Lewis, Robert O'Brien, and the Editors of LIFE. © 1965 Time-Life Books, Inc.

In our more mathematical setting, we begin by partitioning the time interval $[a, b]$ into n subintervals of equal length, $\Delta t = (b - a)/n$, with endpoints

$$t_0 = a, \quad t_1 = a + \Delta t, \quad t_2 = a + 2\,\Delta t, \quad \ldots, \quad t_n = a + n\,\Delta t = b.$$

Our idea is to assume that $v(t)$ is constant on each subinterval of time, to approximate the distance travelled during each subinterval, and then to sum the individual distances to obtain an approximation to D, the distance travelled by the object.

To do so we let c_j denote any number (time) in the jth interval $[t_{j-1}, t_j]$, and we assume that $v(t) = v(c_j)$ for all $t \in [t_{j-1}, t_j]$. Using the definition

$$\text{distance} = \text{speed} \times \text{time}$$

we approximate the distance D_j travelled by the object during the time subinterval $[t_{j-1}, t_j]$ as

$$D_j = |v(c_j)|(t_j - t_{j-1}) = |v(c_j)| \, \Delta t, \qquad j = 1, 2, \ldots, n.$$

(Recall, if $v(t)$ is velocity, speed $= |v(t)|$. We care only about the rate of the motion, not its direction.) Summing these individual approximations gives an approximation to D as

$$D \approx \sum_{j=1}^{n} D_j = \sum_{j=1}^{n} |v(c_j)| \, \Delta t, \qquad c_j \in [t_{j-1}, t_j]. \tag{1}$$

Now the sum on the right-hand side of approximation (1) is a Riemann sum for the function $|v|$ on the interval $[a, b]$. If v is continuous on $[a, b]$, so is $|v|$. Thus, the Riemann sums on the right-hand side of approximation (1) approach the definite integral of $|v|$ as $n \to \infty$; that is,

$$\int_a^b |v(t)| \, dt = \lim_{n \to \infty} \sum_{j=1}^{n} |v(c_j)| \, \Delta t, \qquad c_j \in [t_{j-1}, t_j].$$

We therefore conclude that

$$\boxed{D = \lim_{n \to \infty} \sum_{j=1}^{n} |v(c_j)| \, \Delta t = \int_a^b |v(t)| \, dt.} \tag{2}$$

☐ **EXAMPLE 1**

Starting at time $t = 0$, a particle moves along a line with velocity $v(t) = 2 + \sqrt{t}$ m/s. How far does it travel during the first 4 seconds?

Solution   Here $v(t) = 2 + \sqrt{t} \geq 0$ for all $t \in [0, 4]$, so $|v(t)| = v(t)$ in (2). Thus, by (2),

$$D = \int_0^4 (2 + \sqrt{t}) \, dt = \left[2t + \frac{2}{3} t^{3/2} \right]_0^4 = \left(2 \cdot 4 + \frac{2}{3} \cdot 8 \right) - (0)$$

$$= \frac{40}{3} \text{ meters.} \qquad \blacksquare$$

☐ **EXAMPLE 2**

A particle moves along a line with velocity $v(t) = 5 - t^2$ m/s. Find the distance travelled from time $t = 3$ s to time $t = 5$ s.

Solution   This time $v(t) = 5 - t^2 < 0$ for $t \in [3, 5]$, so $|v(t)| = -v(t)$ in equation (2). Thus,

$$D = \int_3^5 -(5 - t^2) \, dt = \left[\frac{t^3}{3} - 5t \right]_3^5 = \left(\frac{125}{3} - 25 \right) - \left(\frac{27}{3} - 15 \right)$$

$$= \frac{68}{3} \text{ meters.} \qquad \blacksquare$$

☐ **EXAMPLE 3**

A subway train accelerates at a rate of $a = 0.5$ m/s^2 until reaching its maximum speed of $v = 36$ kilometers per hour. How far does the train travel in the first 2 minutes, assuming maximum acceleration and cruising speed?

Solution   The velocity $v(t)$ of the train while undergoing acceleration $a = 0.5$ m/s^2 is

$$v(t) = \int a(t)\, dt = \int 0.5\, dt = 0.5t + v_0,$$

where the constant v_0 is the initial velocity $v_0 = v(0)$. Since the train begins from rest, we have $v_0 = v(0) = 0$. The velocity function is

$$v(t) = 0.5t, \qquad 0 \le t \le T,$$

where T is the time at which the train reaches its maximum velocity $v_{max} = 36$ km/h. To find T we first convert this maximum velocity to units of m/s:

$$36 \text{ km/h} = 36{,}000 \text{ m/h} = \frac{36{,}000}{(60)^2} \text{ m/s} = 10 \text{ m/s}.$$

The train therefore reaches maximum speed when

$$v(T) = 0.5T = 10 \text{ m/s},$$

that is, at time $T = 10/0.5 = 20$ s.

After time $T = 20$ s, the speed of the train remains constant at 10 m/s. The velocity function for the train during the first 2 minutes ($= 120$ s) is therefore

$$v(t) = \begin{cases} 0.5t, & 0 \le t \le 20; \\ 10, & 20 < t \le 120. \end{cases}$$

The distance travelled during this time is

$$D = \int_0^{120} v(t)\, dt$$

$$= \int_0^{20} (0.5t)\, dt + \int_{20}^{120} 10\, dt$$

$$= \left[0.25t^2 \right]_0^{20} + \left[10t \right]_{20}^{120}$$

$$= 100 + 10(120 - 20)$$

$$= 1100 \text{ m} = 1.1 \text{ km}. \qquad \blacksquare$$

Obviously, a difficulty arises in applying (2) if v has both positive and negative values on $[a, b]$. In such cases, we must first find the zeros of v, and then apply formula (2) separately on each of the subintervals of $[a, b]$ determined by the zeros and the endpoints.

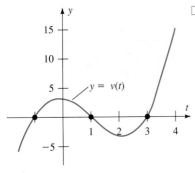

Figure 4.1 $v(t) = t^3 - 3t^2 - t + 3$ in Example 4.

□ **EXAMPLE 4**

If an object moves along a line with velocity $v(t) = t^3 - 3t^2 - t + 3$ m/s, how far does it travel from time $t = 0$ to time $t = 4$ seconds?

Solution In order to determine the sign of $v(t)$ on $[0, 4]$, we must find the solutions of the equation $v(t) = 0$. Now $v(t)$ has factorization

$$v(t) = (t + 1)(t - 1)(t - 3),$$

so $v(t) = 0$ on $[0, 4]$ if $t = 1$ or $t = 3$. By evaluating $v(t)$ at numbers in each of the resulting intervals, you can verify that

$$v(t) \geq 0 \quad \text{for} \quad t \in [0, 1]$$
$$v(t) \leq 0 \quad \text{for} \quad t \in [1, 3]$$
$$v(t) \geq 0 \quad \text{for} \quad t \in [3, 4].$$

(See Figure 4.1.)
Thus,

$$D = \int_0^1 (t^3 - 3t^2 - t + 3) \, dt + \int_1^3 -(t^3 - 3t^2 - t + 3) \, dt$$

$$+ \int_3^4 (t^3 - 3t^2 - t + 3) \, dt$$

$$= \left[\frac{t^4}{4} - t^3 - \frac{t^2}{2} + 3t \right]_0^1 + \left[-\frac{t^4}{4} + t^3 + \frac{t^2}{2} - 3t \right]_1^3$$

$$+ \left[\frac{t^4}{4} - t^3 - \frac{t^2}{2} + 3t \right]_3^4$$

$$= \left(\frac{7}{4} - 0 \right) + \left(\frac{9}{4} - \left(-\frac{7}{4} \right) \right) + \left(4 - \left(-\frac{9}{4} \right) \right)$$

$$= 12 \text{ meters.} \qquad \blacksquare$$

□ **EXAMPLE 5**

The position of the end of a vibrating spring is given by the function $s(t) = 6 \cos \pi t$, where t is in seconds and s is in centimeters. How far does the end of the spring travel between times $t = 0$ and $t = 3$?

Strategy · · · · · · · ·

Differentiate s to find v.

Set $v(t) = 0$ to find zeros in $[0, 3]$.

Determine the sign of $v(t)$ on each of the resulting intervals.

Solution

Since $s(t) = 6 \cos \pi t$ is the *position* of the end of the spring, its velocity is

$$v(t) = \frac{d}{dt}(6 \cos \pi t) = -6\pi \sin \pi t.$$

Then $v(t) = 0$ when $\sin \pi t = 0$, which happens when

$$\pi t = n\pi, \quad \text{or} \quad t = n = 0, \pm 1, \pm 2, \ldots.$$

The zeros for v in $[0, 3]$ are therefore $t = 0, 1, 2,$ and 3. By checking particular numbers in each of the resulting intervals, we find that

$$v(t) \leq 0 \quad \text{for} \quad t \in [0, 1]$$
$$v(t) \geq 0 \quad \text{for} \quad t \in [1, 2]$$
$$v(t) \leq 0 \quad \text{for} \quad t \in [2, 3].$$

(See Figure 4.2.)

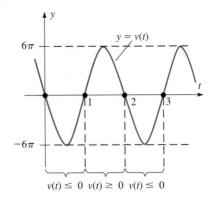

Figure 4.2 Velocity function in Example 5.

Thus, by (2),

Apply (2) on each interval according to the sign of $v(t)$.

$$D = \int_0^1 -v(t)\, dt + \int_1^2 v(t)\, dt + \int_2^3 -v(t)\, dt$$

$$= \int_0^1 6\pi \sin \pi t\, dt + \int_1^2 (-6\pi \sin \pi t)\, dt + \int_2^3 6\pi \sin \pi t\, dt.$$

Since s is an antiderivative of v, we have

$$D = \left[-s(t)\right]_0^1 + \left[s(t)\right]_1^2 + \left[-s(t)\right]_2^3$$

$$= \left[-6 \cos \pi t\right]_0^1 + \left[6 \cos \pi t\right]_1^2 + \left[-6 \cos \pi t\right]_2^3$$

$$= [(-6)(-1) - (-6)(1)] + [(6)(1) - (6)(-1)]$$
$$+ [(-6)(-1) - (-6)(1)]$$

$$= 36 \text{ cm.} \qquad \blacksquare$$

You have undoubtedly noted the similarity between the calculation of area and the calculation of distance. Both are nonnegative quantities, and both are calculated by integrating functions that are nonnegative. In fact, a sketch of a Riemann sum for the integral $D = \int_a^b |v(t)|\, dt$ (Figure 4.3) shows that the sum could just as well be interpreted as an approximation to the area of the region bounded by the graph of $y = |v(t)|$ and the t-axis for $a \le t \le b$.

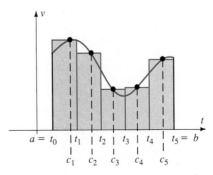

Figure 4.3 Riemann sum R_5 for $D = \int_a^b |v(t)|\, dt$.

For the case of a nonnegative velocity function v, we know from Chapter 5 that the distance travelled from time $t = a$ to time $t = b$ by a particle travelling with velocity $v(t)$ is

$$D = s(b) - s(a) \qquad (3)$$

where s is an antiderivative (a position function) for v. To verify that equations (2) and (3) represent the same quantity, we use the Fundamental Theorem of Calculus. If $s'(t) = v(t)$, we have

$$D = \int_a^b v(t)\,dt = \Big[s(t) \Big]_a^b = s(b) - s(a).$$

The cases $v(t) \le 0$ and $v(t)$ of mixed sign are similar and are left as Exercises 29 and 30.

Exercise Set 7.4

In Exercises 1–11, calculate the distance travelled by a particle moving along a line with the given velocity from time $t = a$ to time $t = b$.

1. $v(t) = 9 - t^2$, $a = 0$, $b = 3$

2. $v(t) = \sqrt{t + 1}$, $a = 0$, $b = 8$

3. $v(t) = \sin t$, $a = 0$, $b = \pi$

4. $v(t) = t - t^3$, $a = 0$, $b = 1$

5. $v(t) = 2 + t^2$, $a = 0$, $b = 5$

6. $v(t) = t^2 + t - 6$, $a = 0$, $b = 2$

7. $v(t) = \cos \pi t$, $a = 1/2$, $b = 3/2$

8. $v(t) = \dfrac{1 - t}{\sqrt{t}}$, $a = 2$, $b = 4$

9. $v(t) = \sin^2 t \cos t$, $a = 0$, $b = \pi/2$

10. $v(t) = \sqrt{t}(t + 2)$, $a = 0$, $b = 1$

11. $v(t) = (2t - 5)(t^2 - 5t + 6)^{1/3}$, $a = 3$, $b = 4$

In Exercises 12–20, the motion of a particle along a line is described by its velocity function. Taking care to note where $v(t)$ changes sign, compute the distance travelled by the particle during the time interval specified.

12. $v(t) = t - 4$, $t = 0$ to $t = 6$

13. $v(t) = 2 - t^2$, $t = 0$ to $t = 4$

14. $v(t) = \sin(2t)$, $t = 0$ to $t = \pi/2$

15. $v(t) = \cos t \sin t$, $t = 0$ to $t = \pi/2$

16. $v(t) = t^{1/3} - t^{2/3}$, $t = 0$ to $t = 8$

17. $v(t) = (t - 1)^5$, $t = 0$ to $t = 2$

18. $v(t) = \dfrac{t^2 - 9}{t + 3}$, $t = 0$ to $t = 4$

19. $v(t) = t(t^2 - 9)^5$, $t = 2$ to $t = 4$

20. $v(t) = t^2(t^3 - 1)^3$, $t = 0$ to $t = 2$

21. The *displacement* of a particle is the net change in position from time $t = a$ to time $t = b$. In other words, for a particle moving along a line its displacement, is the difference $s(b) - s(a)$, where $s(t)$ is the position of the particle at time t. Compare the displacement of the particle to the total distance travelled for the motion specified in
    a. Exercise 12;
    b. Exercise 13;
    c. Exercise 14;
    d. Exercise 15.

22. The motion of a particle along a line is described by the position function $s(t) = t^2 - 7t + 10$.
    a. Find its velocity at time t.
    b. Find its acceleration at time t.
    c. For which times t is the particle at rest?
    d. For which times t is the particle moving to the right?
    e. For which times t is the particle moving to the left?
    f. Find the distance travelled by the particle from $t = 1$ to $t = 8$.
    g. Find the displacement of the particle between times $t = 1$ and $t = 8$.

23. The motion of a particle along a line is described by the velocity function $v(t) = -t^2 + 5t - 6$. The position of the particle at time $t = 0$ is $s(0) = 4$.
    a. Find its position $s(t)$ at time t.
    b. Find its acceleration $a(t)$ at time t.
    c. For which times t is the particle at rest?
    d. For which times t is the particle moving to the right?
    e. For which times t is the particle moving to the left?
    f. Find the distance travelled by the particle from time $t = 0$ to $t = 3$.
    g. Find the displacement of the particle between times $t = 0$ and $t = 3$.

24. The motion of a particle along a line is described by the following equations:

$$a(t) = 2t - 4$$
$$v(0) = 3$$
$$s(0) = 5$$

Here a is acceleration, v velocity, s position.
a. Find the velocity $v(t)$ at time t.
b. Find the position $s(t)$ at time t.
c. For which time(s) t is the particle at rest?
d. For which times t is the particle moving to the right?
e. For which times t is the particle moving to the left?
f. Find the distance travelled by the particle from $t = 0$ and $t = 3$.
g. Find the displacement of the particle between times $t = 0$ and $t = 3$.

25. The acceleration of an object due to gravity is 9.8 m/s². In other words, $a(t) = -9.8$. A ball is thrown straight up from a window 10 m above the ground with initial velocity of 12 m/s.
a. Find $v(t)$.
b. Find $s(t)$.
c. At what time t does the ball reach its highest point?
d. At what time t does the ball hit the ground?

e. What are the displacement and the total distance travelled by the ball?
f. What is the velocity when the ball hits the ground?

26. Find the distance travelled by the subway train in Example 3 if we add the conditions that the train also *decelerates* at the constant rate $a = -0.5$ m/s² and must come to a complete stop at time $t = 2$ minutes.

27. An ultratrain has a maximum cruising speed of 108 km/h and the capability of maximum acceleration of $a_1 = 1.0$ m/s² and maximum deceleration of $a_2 = -1.0$ m/s². How far does the train travel in 10 minutes if it begins from rest, travels under maximum acceleration until reaching its maximum cruising speed, and remains at this speed until time $t = 10$ minutes?

28. What is the maximum distance that the train in Exercise 27 can travel in 10 minutes if we add the condition that it must have come to a complete stop by time $t = 10$ minutes?

29. Verify that $\int_a^b |v(t)|\, dt = -[s(b) - s(a)] = s(a) - s(b)$ if $s'(t) = v(t)$ and $v(t) \le 0$ for all $t \in [a, b]$.

30. What can you say about the relationship between $\int_a^b |v(t)|\, dt$ and the number $s(b) - s(a)$ when $v(t)$ has both positive and negative values on $[a, b]$?

7.5 Hydrostatic Pressure

The word "hydrostatic" refers to properties of still fluids (as opposed to hydrodynamic, which refers to properties of moving fluids). Our interest here is in calculating pressures and forces exerted by water on the walls of containers.

The **pressure** acting on an object is defined to be the force per unit area. For example, if a rain barrel is filled to a depth $h = 4$ feet with water, then above each square foot of the bottom of the barrel extends a rectangular column of water of volume 1 ft × 1 ft × 4 ft = 4 ft³. Since the density of water is 62.4 lb/ft³, the pressure exerted on the bottom of the barrel by the column of water is

$$p = \text{density} \times \text{depth}$$
$$= (62.4 \text{ lb/ft}^3) \times (4 \text{ ft})$$
$$= 249.6 \text{ lb/ft}^2.$$

(Figure 5.1.)*

The total force F exerted on the (horizontal) bottom of the barrel is found by multiplying the area A of the base by the pressure p. If, in this example, the base is a circle of radius 2 ft, then

$$A = \pi r^2 = 4\pi \text{ ft}^2,$$

so

$$F = pA = (249.6 \text{ lb/ft}^2)(4\pi \text{ ft}^2) = 998.4\pi \text{ lb}.$$

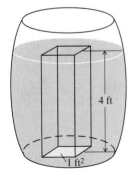

Figure 5.1 Pressure exerted on the bottom of a rain barrel that is filled to a depth of 4 ft.

4 ft

1 ft²

---

*The same principles apply to all fluids. For simplicity, we will concentrate on the most common fluid, water. For a fluid of density ρ lb/ft³, the constant 62.4 is replaced by the constant ρ.

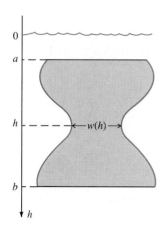

Figure 5.2 Face of submerged object; $w(h)$ is the width at depth h.

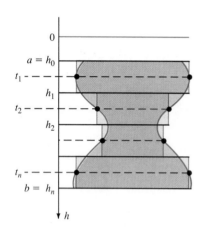

Figure 5.3 Approximating rectangles and test points t_j.

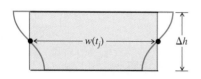

Figure 5.4 jth approximating rectangle.

The problem of calculating the force acting on the side wall of the barrel is more complicated. The preceding discussion shows that pressure depends on depth. However, **Pascal's Principle** states that the pressure exerted by a fluid on a submerged particle acts equally in all directions. Thus, the hydrostatic pressure at any point depends *only* on depth. However, the calculation of the total force acting on a side wall involves more than a simple multiplication, since the wall exists at various depths.

To pursue this question, we consider a submerged object whose face consists of a region bounded above by the line $h = a$, below by the line $h = b$, and whose width at depth h is given by the function $w(h)$. The surface of the water is at depth $h = 0$, the positive h-axis extends vertically downward, and we assume that the face of the object is sitting vertically in the fluid (see Figure 5.2).

Since pressure varies continuously with the depth, we partition the interval $[a, b]$ into n subintervals of equal length with endpoints $a = h_0 < h_1 < h_2 < \cdots < h_n = b$, and we select one arbitrary number t_j in each interval $[h_{j-1}, h_j]$. We then make two types of approximations:

1. We approximate the slice of the face lying between depth h_{j-1} and depth h_j by the *rectangle* with width $w(t_j)$ and height $\Delta h = (b - a)/n$ (see Figures 5.3 and 5.4).
2. We approximate the depth of each point in the jth rectangle by the depth t_j.

That is, we assume the width $w(t)$ to be the constant $w(t_j)$ throughout the jth rectangle, and we assume the pressure throughout the jth rectangle to be constant and equal to $p(t_j) = (62.4)t_j$. Since the jth rectangle has area $A_j = w(t_j)\, \Delta h$, the force F_j acting on the jth rectangle is

$$F_j = (62.4)t_j A_j = (62.4)t_j w(t_j)\, \Delta h, \qquad \Delta h = \frac{b - a}{n}.$$

Finally, we approximate the total force F acting on the face by the sum of these individual forces:

$$F \approx \sum_{j=1}^{n} F_j = \sum_{j=1}^{n} (62.4)t_j w(t_j)\, \Delta h. \tag{1}$$

If the width of the face is a continuous function of depth for $a \le h \le b$, then the function $(62.4)hw(h)$ is a continuous function of h. Consequently, the sum on the right-hand side of approximation (1) is a Riemann sum that must have a limit as $n \to \infty$. We *define* this limit as the total force acting on the face; that is, we define

$$F = \lim_{n \to \infty} \sum_{j=1}^{n} (62.4)t_j w(t_j)\, \Delta h = \int_{a}^{b} (62.4)hw(h)\, dh. \tag{2}$$

Equation (2) states that the total force (in pounds) acting on a submerged vertical face lying between depths $h = a$ and $h = b$ and whose width at depth h is $w(h)$ (all dimensions in feet) is found by integrating the product $(62.4)hw(h)$ over the interval $[a, b]$.

☐ **EXAMPLE 1**

An above-ground swimming pool has the shape of a rectangular box 50 feet long, 20 feet wide, and 6 feet high (Figure 5.5). Find the force exerted on an end wall if the pool is filled to a depth of 5 feet (Figure 5.6).

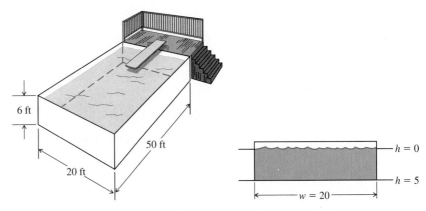

Figure 5.5 Swimming pool.   Figure 5.6 End wall of pool.

Solution   We take the water level to be depth $h = 0$. Then the submerged portion of the wall is a rectangle extending from depth $h = 0$ to depth $h = 5$ feet. The width, $w = 20$ feet, is constant. An application of formula (2) gives

$$F = \int_0^5 (62.4)h(20)\ dh = \left[624h^2 \right]_0^5 = 15{,}600 \text{ pounds.} \qquad \blacksquare$$

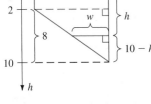

Figure   5.7   Swimming   pool   in Example 2.

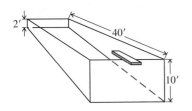

Figure 5.8 $\dfrac{w}{10 - h} = \dfrac{40}{8}$ if $2 \le h < 10$.

☐ **EXAMPLE 2**

Figure 5.7 shows a 40-ft swimming pool 20 ft wide with a floor that slopes from a depth of 2 ft at one end of the pool to a depth of 10 ft at the other end. Find the hydrostatic force on one side wall of this pool when it is full of water.

Solution   The top 2-ft part of this wall is a rectangle of width $w(h) = 40$ ft., $0 \le h \le 2$. The force acting on this part is therefore

$$F_1 = \int_0^2 (62.4)h(40)\ dh = 2496 \left[\frac{h^2}{2} \right]_0^2 = 4992 \text{ lb.}$$

For depths between $h = 2$ ft and $h = 10$ ft the width of a horizontal strip of the wall depends upon its depth. By similar triangles (see Figure 5.8) we have

$$\frac{w}{10 - h} = \frac{40}{8},$$

so

$$w(h) = 5(10 - h), \qquad 2 \le h \le 10.$$

The force acting on this part of the wall is therefore

$$F_2 = \int_2^{10} (62.4)h[5(10 - h)]\, dh$$

$$= 312 \int_2^{10} h(10 - h)\, dh$$

$$= 312 \left[5h^2 - \frac{h^3}{3} \right]_2^{10}$$

$$= 46{,}592 \text{ lb.}$$

The total force on the wall is therefore $F_1 + F_2 = 51{,}584$ lb.   ■

□ **EXAMPLE 3**

Figure 5.9 shows a water trough with semicircular end panels of radius 2 ft. Find the hydrostatic force on one of these end panels when the trough is full of water.

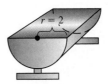

Figure 5.9   Water tank with semicircular end panels.

Solution   Figure 5.10 shows that if we impose an x-h coordinate system with origin at the center of the semicircle, then $x^2 + h^2 = 4$, and the width of a horizontal slice across the end panel at depth h is

$$w(h) = 2x = 2\sqrt{4 - h^2}.$$

According to equation (2) the force on this end panel is

$$F = \int_0^2 (62.4)h \cdot 2\sqrt{4 - h^2}\, dh.$$

To evaluate this integral we use the substitution

$$u = 4 - h^2; \qquad du = -2h\, dh,$$

and we change limits of integration according to the equations:

$$\text{if } h = 0, \quad u = 4 - 0^2 = 4$$
$$\text{if } h = 2, \quad u = 4 - 2^2 = 0.$$

The integral then becomes

$$F = \int_4^0 (62.4)(-1)\sqrt{u}\, du$$

$$= \left[-\tfrac{2}{3}(62.4)u^{3/2} \right]_4^0$$

$$= \tfrac{2}{3}(62.4)(8) = 332.8 \text{ lb.}$$ ■

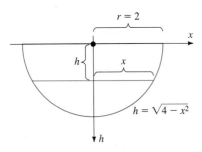

Figure 5.10   The width of a horizontal slice at depth h is $w(h) = 2x = 2\sqrt{4 - h^2}$.

Exercise Set 7.5

1. The vertical ends of a water trough are rectangles 3 ft wide and 18 in. high. Find the hydrostatic force on the end panels when the trough is full of water.

2. A cardboard container for mineral water has a square base and rectangular sides 5 in. wide and 10 in. high. Find the force acting on a side when the carton is full of water.

3. The vertical ends of a water trough are isosceles right triangles with the right angles at the bottom. Find the hydrostatic force acting on an end if the trough is filled to a depth of 2 ft (Figure 5.11).

Figure 5.11 Water trough in Exercise 3.

4. The vertical ends of a water trough are equilateral triangles whose sides are each 2 ft long. Find the hydrostatic force on an end panel when the trough is full of water.

5. The wall of a dam has the shape of the parabola $y = x^2/100$. Find the hydrostatic force acting on the wall if the wall is 80 ft wide across its top and the dam is full of water. (*Hint:* The given equation assumes a coordinate system whose origin lies at the bottom of the wall. You have to find the depth h from the top of the wall.)

6. The end panels of a water trough are rectangles 2 ft wide and 1 ft high. To what depth must the trough be filled so that the force acting on an end panel is exactly half the force acting on the end panel when the trough is full?

7. Find the hydrostatic force on an end panel of a water trough in the shape of the lower half of an ellipse with equation

$$\frac{x^2}{16} + \frac{y^2}{4} = 1$$

(x and y are in units of feet).

8. A submarine has rectangular windows 10 in. wide and 5 in. high. Find the hydrostatic force on such a window when the top of the window lies 30 ft below water level.

9. Find the hydrostatic force on the submarine window in Exercise 8 if it has the shape of an equilateral triangle with side $s = 10$ in., bottom side horizontal, and the top of the window is 30 ft below water level.

10. Suppose an object is submerged in a fluid weighing c pounds per cubic foot. How should equation (2) be modified so as to give the force acting on a face of this object?

11. Use the result of Exercise 10 to find the force acting on one wall of a carton of heavy cream weighing 72 lb/ft^3 if the walls of the carton are rectangles of width 4 in. and height 6 in. and the carton is full of cream.

12. A fuel oil truck has a tank in the shape of a cylinder whose base is described by the ellipse

$$\frac{x^2}{64} + \frac{y^2}{36} = 1.$$

The tank is mounted horizontally on the truck (i.e., the elliptical end panels are vertical). Find the force acting on one end panel when the tank is half filled with fuel oil weighing 58 lb/ft^3 (see Exercise 10).

13. Use Simpson's Rule to approximate the force acting on an end panel of the oil tank in Exercise 12 when the tank is full of oil.

14. If a rectangular plate is suspended in water, the theory of this section enables us to compute the hydrostatic force acting on one face of the plate. Why does the plate not move as a result of this force?

In each of Exercises 15–17, compute the hydrostatic force on a vertical plate of the given shape submerged to the given depth d below water level. In each case, compare your result with the product $F = (62.4)Ah$ where A is the area of the figure and h is the depth of the center of the object below water level.

15. A rectangle of width 15 in. and height 10 in. lying $d = 2$ ft below water level.

16. A square of side 2 ft lying $d = 6$ ft below water level.

17. A rectangle of width 3 ft and height 2 ft lying 7 ft below water level.

18. What do you conjecture, based on the results of Exercises 15–17, about the relationship between the area, depth of center, and hydrostatic force acting on a submerged plate? Do the results of Exercise 13 support this conjecture? (These ideas are pursued in Section 7.7 on Centers of Mass.)

7.6   Work

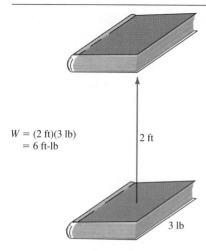

$$W = (2 \text{ ft})(3 \text{ lb}) \\ = 6 \text{ ft-lb}$$

2 ft

3 lb

Figure 6.1 The work done in lifting a 3-pound textbook 2 feet equals 6 foot-pounds.

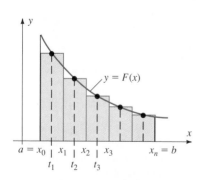

Figure 6.2 Force required to raise anchor depends upon ℓ if we consider the weight of the cable in the calculation.

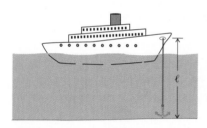

Figure 6.3 Using a Riemann sum to approximate the work done over $[a, b]$ by a variable force $F(x)$.

When a constant force F moves an object through a distance d, the product

$$W = Fd \tag{1}$$

is defined as the **work** done by the force on that object. For example, in lifting a 3-pound textbook 2 feet you do an amount of work equal to $W = (2 \text{ feet}) \times (3 \text{ lb}) = 6$ foot-pound (Figure 6.1).

The choice of the unit foot-pounds reflects the definition of work as the product of force times distance. If force and distance are measured in units other than feet and pounds, the units for work are modified appropriately (inch-pounds or newton-meters, for example).

In many practical situations, the calculation of work cannot be done by a simple multiplication because the force F required to move an object is not constant. For example, the force required to raise a ship's anchor depends on the length of cable, ℓ, between the ship and the anchor (Figure 6.2). This is because the cable usually has substantial weight, and both the cable and the anchor must be raised.

If the motion of the object is along a line, the force required to move the object can sometimes be written as a function F of the object's position x along the line (as we will see in Example 1). In such cases we can generalize equation (1) using the definite integral.

Suppose that an object is to be moved from location $x = a$ to location $x = b$. We can approximate the work done on the object over the entire interval $[a, b]$ by partitioning $[a, b]$ into small subintervals, assuming $F(x)$ to be constant on each subinterval, and then summing the resulting values of work required on each subinterval (Figure 6.3). To do so we choose a partition $a = x_0 < x_1 < x_2 < \cdots < x_n = b$ so that $x_j - x_{j-1} = \Delta x = (b - a)/n$ for each $j = 1, 2, \ldots, n$, and we select one number t_j in each interval $[x_{j-1}, x_j]$. If we approximate the force $F(x)$ throughout the interval $[x_{j-1}, x_j]$ by the constant $F(t_j)$, then by equation (1) the work W_j required to move the object from x_{j-1} to x_j is the product $W_j = F(t_j)(x_j - x_{j-1}) = F(t_j) \Delta x$. By summing these approximations, we obtain the approximation for the work done in moving the object from $x = a$ to $x = b$ as

$$W \approx \sum_{j=1}^{n} W_j = \sum_{j=1}^{n} F(t_j) \Delta x. \tag{2}$$

If the force function F is continuous for $x \in [a, b]$, then the sum on the right-hand side of approximation (2) is a Riemann sum that has a limit as $n \to \infty$. Since our approximation was designed to provide an increasingly accurate measurement of the actual value of the work as $n \to \infty$ (and, hence, as the size of the intervals approaches zero), we *define* work to be the limiting value of this approximating sum, that is,

$$W = \lim_{n \to \infty} \sum_{j=1}^{n} F(t_j) \Delta x = \int_{a}^{b} F(x) \, dx. \tag{3}$$

Equation (3) states that the work done by a variable force in moving an object along a line from location $x = a$ to location $x = b$ is found by integrating the force function from $x = a$ to $x = b$. Figure 6.3 illustrates a typical force function and the geometric interpretation of our approximation scheme. Although the region bounded by the graph of F does not correspond to any physical aspect of the work problem, the area of that region is precisely the work being calculated.

□ **EXAMPLE 1**

A principle from elementary physics known as **Hooke's Law** states that the force required to stretch or compress a spring a distance of x units from its natural length is proportional to that distance; that is,

$$F(x) = kx \qquad \text{(Hooke's Law)}. \qquad (4)$$

Here k is a constant, called the **spring constant,** which depends on the particular spring. Since the force varies with distance, calculating the work done in stretching or compressing a spring requires the use of equation (3) as the following examples illustrate.

(a) A spring has spring constant $k = 20$ lb/ft. The work done in stretching the spring 6 inches beyond its natural length is found using equations (3) and (4):

$$W = \int_0^{1/2} F(x)\, dx$$

$$= \int_0^{1/2} 20x\, dx$$

$$= \left[10x^2 \right]_0^{1/2}$$

$$= 10(\tfrac{1}{4} - 0)$$

$$= 2.5 \text{ ft-lb.}$$

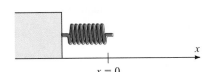

Figure 6.4   Spring in natural state.

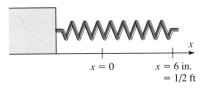

Figure 6.5   Spring stretched 6 in. = 1/2 ft.

(Figures 6.4 and 6.5.)

(b) The work done in stretching the spring in part (a) from 3 inches $= \tfrac{1}{4}$ ft beyond its natural length to 6 in. $= \tfrac{1}{2}$ ft beyond its natural length is

$$W = \int_{1/4}^{1/2} 20x\, dx = \left[10x^2 \right]_{1/4}^{1/2} = \frac{15}{8} \text{ ft-lb.}$$

(c) Suppose that 4 ft-lb of work are required to stretch a spring 1 ft beyond its natural length. We may calculate the spring constant from this information since

$$4 = \int_0^1 kx\, dx = k \left[\frac{x^2}{2} \right]_0^1 = \frac{k}{2}.$$

Thus, $4 = k/2$, so $k = 8$ lb/ft. ■

□ **EXAMPLE 2**

Suppose that a ship's anchor weighs 2 tons (4000 pounds) in water and that the anchor is hanging taut from 100 feet of cable. Find the work required to wind in the anchor if the cable weighs 20 pounds per foot in water.

Solution   This example does not quite fit the general framework within which equation (3) was developed. Although the anchor is moved from depth $h = 100$ to depth $h = 0$, the cable itself is not lifted 100 ft. Rather, each "slice" of cable lying at

$h_0 = 0$
t_1
h_1
t_2
h_2
t_3
h_3

t_j — jth section of cable

$h_n = 100$
$\downarrow h$

Figure 6.6 jth section of cable is assumed to lie uniformly at a depth of t_j feet.

depth h ft is lifted only h ft. To approximate the work done on the entire cable, we partition the interval [0, 100] as usual, with equally spaced numbers $h_0 = 0 < h_1 < h_2 < \cdots < h_n = 100$ (Figure 6.6). Then the force F_j required to lift each section of cable is its weight; that is,

$$F_j = (20 \text{ lb/ft})(\Delta h \text{ ft}).$$

By selecting one number t_j in each interval $[h_{j-1}, h_j]$, we approximate the work required to lift the jth section of cable to the surface by the product of the force F_j and the distance t_j; that is,

$$W_j \approx F_j \cdot t_j = 20t_j \, \Delta h \text{ ft-lb}.$$

The work W_c required to raise the cable is therefore approximated by the sum

$$W_c \approx \sum_{j=1}^{n} W_j = \sum_{j=1}^{n} 20t_j \, \Delta h. \tag{5}$$

Since the expression on the right side of equation (5) is a Riemann sum for the function $f(h) = 20h$, we conclude that

$$W_c = \lim_{n \to \infty} \sum_{j=1}^{n} 20t_j \, \Delta h = \int_0^{100} 20h \, dh. \tag{6}$$

Since the work required to raise a 4000-lb anchor 100 feet is $W_a = (4000 \text{ lb}) \times (100 \text{ ft}) = 400,000$ ft-lb, the total work required is

$$W = W_a + W_c = 400,000 + \int_0^{100} 20h \, dh$$

$$= 400,000 + \left[10h^2 \right]_0^{100}$$

$$= 500,000 \text{ ft-lb.} \qquad \blacksquare$$

REMARK Notice that the integral obtained for the work W_c required to raise the cable in equation (6) cannot be obtained by applying equation (3). The reason is that equation (3) applies to a variable force acting to move an object from location $x = a$ to location $x = b$. However, equation (6) refers to moving a one-dimensional object of uniform weight through a *variable* distance. We could attempt to generalize equation (6), but the wording would be clumsy and, necessarily, vague. Instead, you should note the following cautions:

(i) *Equation (3) may be applied only when a variable force F(x) moves the point at which it is applied from location x = a to location x = b.* (Thus, equation (3) was correctly applied in Example 1, since the point of application corresponded to the end of the spring. However, in Example 2 there was no such "point of application.")

(ii) In all other problems involving the calculation of work, you should
(a) partition the relevant interval,
(b) approximate the work required in each interval,
(c) write the resulting approximation to the total work as a Riemann sum, and
(d) obtain the appropriate integral as the limit of the approximating Riemann sum.

☐ **EXAMPLE 3**

A swimming pool is 40 feet long and 20 feet wide. The floor of the pool has a constant slope from a depth of 2 feet at one end to a depth of 10 feet at the other. Find the work required to pump all the water out through a valve located at the top edge of the pool when the pool is full (Figure 6.7).

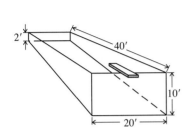

Figure 6.7 Swimming pool with sloping floor.

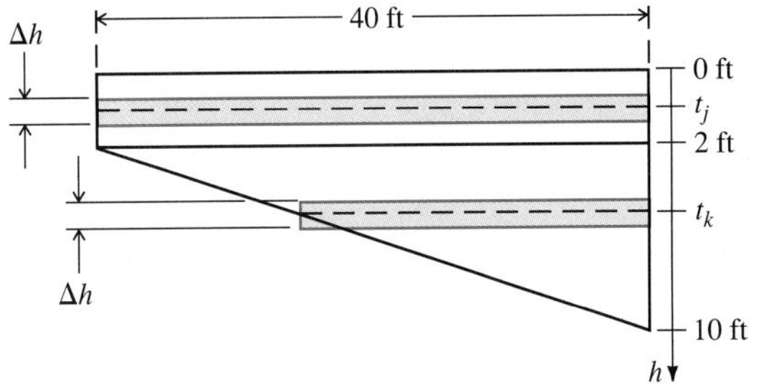

Figure 6.8 Partitioning the depth axis (the h-axis).

Strategy · · · · · · · ·

Label the independent variable.

Treat the problem in two parts according to the geometry.

Partition the interval [0, 2] and approximate the volume of each slab.

$F_j = (62.4) V_j$

$W_j = t \cdot F_j$

Sum the individual approximations. The result is a Riemann sum.

Obtain the integral as the limit of the Riemann sum.

Solution

Let the variable h denote depth. We imagine partitioning the interval [0, 10] and slicing the water volume by horizontal planes into slabs of thickness Δh. As Figure 6.8 illustrates, the resulting slabs are of two types, those with fixed length 40 ft (lying above depth 2 ft) and those with variable length (lying below depth 2 ft). We treat the two types separately.

For depths t_j with $0 \le t_j \le 2$, the slab of thickness Δh has volume

$$V_j = (20 \text{ ft})(40 \text{ ft})(\Delta h \text{ ft}) = 800 \, \Delta h \text{ ft}^3.$$

The force (weight) required to lift this slab is therefore

$$F_j = (62.4 \text{ lb/ft}^3)(800 \, \Delta h \text{ ft}^3) = 49{,}920 \, \Delta h \text{ lb.}$$

If we assume this slab to be concentrated at depth t_j, the work required to lift it to the top of the pool will be

$$W_j = t_j F_j = 49{,}920 t_j \, \Delta h \text{ ft-lb.}$$

Assuming the interval [0, 2] to have been partitioned into n subintervals, the work W_T required to pump out the top 2 ft of water is approximated by

$$W_T \approx \sum_{j=1}^{n} W_j = \sum_{j=1}^{n} 49{,}920 t_j \, \Delta h,$$

so

$$W_T = \int_0^2 49{,}920 h \, dh.$$

For $2 \le h \le 10$, use similar triangles to find length as a function of depth.

For depths t_k with $2 \le t_k \le 10$, the length ℓ_k of the slab depends on its depth. By similar triangles (Figure 6.9), we have

$$\frac{\ell_k}{10 - t_k} = \frac{40}{8},$$

so $\ell_k = 5(10 - t_k)$.

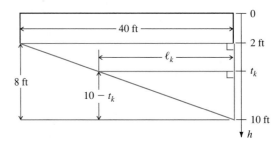

Figure 6.9 Similar triangles give length ℓ_k at depth t_k.

Proceed as in the first part.

A slab of water of thickness Δh assumed to be concentrated at depth t_k therefore has volume

$$V_k = (20 \text{ ft})[5(10 - t_k)\text{ft}](\Delta h \text{ ft}) = 100(10 - t_k) \, \Delta h \text{ ft}^3$$

and requires a lifting force to overcome its weight of

$$F_k = (62.4 \text{ lb/ft}^3)[100(10 - t_k) \, \Delta h \text{ ft}^3]$$
$$= 6240(10 - t_k) \, \Delta h \text{ lb}.$$

The work required to lift this slab to the top edge is therefore

$$W_k = 6240 t_k (10 - t_k) \, \Delta h \text{ ft-lb}.$$

Summing over all slabs lying between depths 2 ft and 10 ft (assuming n such slabs) gives the approximation to W_B, the work required to pump out all water lying below 2 ft:

$$W_B \approx \sum_{k=1}^{n} W_k = \sum_{k=1}^{n} 6240 t_k (10 - t_k) \, \Delta h.$$

Thus

$$W_B = \int_2^{10} 6240 h (10 - h) \, dh.$$

Combine results.

Combining the expressions for W_T and W_B gives the total work W required as

$$W = W_T + W_B$$
$$= \int_0^2 49{,}920 h \, dh + \int_2^{10} 6240 h (10 - h) \, dh$$
$$= \left[24{,}960 h^2 \right]_0^2 + \left[31{,}200 h^2 - 2080 h^3 \right]_2^{10}$$

1 ft-ton = 2000 ft-lb.

$$= 1{,}031{,}680 \text{ ft-lb} = 515.84 \text{ ft-tons.} \qquad \blacksquare$$

Exercise Set 7.6

1. How much work is done in compressing a spring with spring constant $k = 40$ lb/ft a distance of 6 in?

2. A spring has spring constant $k = 5$ newtons per meter. How much work is done in stretching the spring 80 cm beyond its natural length? (Here work will be in units of newtons $\times$ meter = joules.)

3. How much work is done in compressing the spring in Exercise 1 another 6 in?

4. How much work is done in stretching the spring in Exercise 2 another 40 cm?

5. A force of 50 lb stretches a spring 4 in. Find the work required to stretch the spring an additional 4 in.

6. If 40 ft-lb of work are required to stretch a spring 6 in. beyond its natural length, what is its spring constant?

7. True or false? If W_1 is the work done on an object in moving it from point A to point B, and W_2 is the work done on the same object in moving it from point B to point C, then the work W done in moving the object from point A to point C satisfies the equation $W = W_1 + W_2$. What property of definite integrals is involved in this question?

8. The work required to stretch a spring 6 cm beyond its natural length is $W = 2$ ergs. What is the value of the spring constant for this spring? (The units for this spring constant will be dynes per centimeter, where ergs = dynes $\times$ centimeters.)

9. A spring hangs from one end, which is attached to a supporting beam. When a 10-lb weight is attached to the free end of the spring it stretches a distance of 2 ft.
  a. Find the spring constant k.
  b. Find the additional weight that must be added so that the spring will be stretched 3 ft beyond its natural length.
  c. If both weights are removed from the end of the spring, what is the work required to stretch the spring again a distance of 3 ft beyond its natural length?

10. A children's wading pool is 6 ft wide, 10 ft long, and $\frac{1}{2}$ ft deep. How much work is required to pump all the water out through a valve at the top edge of the pool if the pool is full of water?

11. A conical water tank has radius $r = 10$ ft and height $h = 12$ ft and is mounted with its base horizontal and its tip pointing downward. How much work is required to pump the water out through a valve in the top of the tank if the tank is full?

12. A water tank has the shape of a right circular cylinder with radius $r = 10$ ft and depth $h = 20$ ft. It is mounted with its axis vertical. The tank contains 10 ft of water. Find the work done in pumping this water out through a valve in the top of the tank.

13. Find the work done in pumping the water out of the tank in Exercise 12 if the tank is full of water.

14. A water tank has the shape of a right circular cone of height $h = 20$ ft and radius $r = 6$ ft. It is mounted with its axis vertical and tip down. It contains 4 ft of water. Find the work done in pumping this water out through a valve in the top of the tank.

15. Find the work done in pumping the water out of the tank in Exercise 14 if the tank is full of water.

16. Find the work done in Exercise 12 if the water is pumped to a height of 10 ft above the top of the tank.

17. Find the work done in Exercise 14 if the water is pumped to a height of 15 ft above the top of the tank.

18. A hemispherical water tank has radius $r = 5$ ft. The tank is mounted with its circular base on top, lying horizontally. How much work is required to pump the water out through a valve on the top edge of the tank if the tank is full?

19. A 20-ft section of rope weighing 2 lb/ft is hanging from a windlass (crank). How much work is done in winding up the entire length of rope?

20. A chain weighs 40 newtons per meter. How much work is done in winding a 30-m section of chain hanging from a windlass?

21. A crane has a bucket weighing 100 lb that holds 500 lb of concrete. The concrete leaks from the bucket at a rate of 10 lb/s, and the crane lifts the bucket at a rate of 5 ft/s. If we neglect the weight of the cable, how much work is done in lifting the bucket 20 ft? How much work is done if we consider the fact that the cable weighs 3 lb/ft?

22. A woman on the deck of her apartment 30 m above the ground hoists a bucket filled with lemonade from the ground. The bucket weighs 45 newtons if it is empty, but when it leaves the ground, it contains 40 liters of lemonade whose density is 10 newtons/liter. If the lemonade spills at the rate of 0.1 liters/s and the woman lifts at the rate of 1 m/s, how much work will she do getting the lemonade up to her deck? Neglect the weight of the rope.

23. Coulomb's Law governing the force of attraction between two electrically charged bodies states that the force of attraction or repulsion between two point charges is directly proportional to the product of the charges and inversely proportional to the square of the distance between them, that is,

$$F = K\frac{q_1 q_2}{r^2}$$

where q_1 and q_2 are the magnitudes of the charges at points P_1 and P_2, respectively, r is the distance between P_1 and P_2, and K is the constant of proportionality. Suppose that charges of equal magnitude q exist at points P_1 and P_2 that are 4 m apart, and that the force with which the two charges repel each other is 10 newtons.

a. What is the value of the constant K in terms of q?

b. How much work is required to bring one charge from point P_2 to a distance of 2 m from point P_1?

24. Suppose than an ion with charge $+1$ is stationary and located at $(5, 0)$. How much work is required to move a second ion with charge $+1$ from $(0, 0)$ to $(4, 0)$? (See Exercise 23.)

25. What is the answer to Exercise 24 if a third ion with charge $+1$ is located at $(-8, 0)$ and is stationary?

26. The weight of an object varies inversely with the square of its distance from the center of the earth; that is,

$$w(r) = \frac{k}{r^2}$$

where $w(r)$ is the weight of the object if it is located at a distance r from the earth's center. A satellite weighs 5 tons at the earth's surface (4000 miles from the earth's center).

a. Find the value of k for this satellite.

b. Find the amount of work done in propelling this satellite to a height of 100 miles above the earth's surface.

7.7   Moments and Centers of Mass

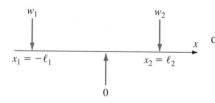

Figure 7.1 A larger weight at a smaller distance balances a smaller weight at a larger distance.

Most children can identify the two factors that determine whether they can "balance" another child seated opposite them on a seesaw: their weight and the distance at which they are seated from the pivot (Figure 7.1). Increasing either of these two quantities increases the tendency of the seesaw to rotate their end downward. We now pursue the mathematical generalizations of these ideas.

We idealize the situation of the seesaw by imagining two objects of weight w_1 and w_2 placed on a flat weightless rod, which in turn is mounted on a fulcrum (pivot). The rod can freely rotate about the fulcrum. Moreover, we let ℓ_1 and ℓ_2 represent the respective distances of the weights w_1 and w_2 from the fulcrum (Figure 7.2).

The tendency of the rod to rotate in the counterclockwise direction is measured by the **torque** $\ell_1 w_1$. Similarly, the torque $\ell_2 w_2$ measures the tendency of the rod to rotate in the clockwise direction. A physical principle called the **law of the lever** states that the rod will balance (equilibrium will occur) if the opposing torques are equal, that is, if

$$\ell_1 w_1 = \ell_2 w_2. \tag{1}$$

We can simplify the model further by introducing a one-dimensional coordinate system along the rod (Figure 7.3). If we denote the point corresponding to the fulcrum as $x = 0$ and take the positive x-axis as the direction in which weights produce clockwise rotation, then for $x_1 = -\ell_1$ and $x_2 = \ell_2$ the equilibrium equation (1) becomes

$$x_1 w_1 + x_2 w_2 = 0. \tag{2}$$

Equation (2) can be generalized to the condition for equilibrium for n weights $w_1, w_2, \ldots, w_n$ located at positions $x_1, x_2, \ldots, x_n$, respectively (Figure 7.4). The condition is

$$x_1 w_1 + x_2 w_2 + \cdots + x_n w_n = 0$$

or

$$\sum_{j=1}^{n} x_j w_j = 0. \tag{3}$$

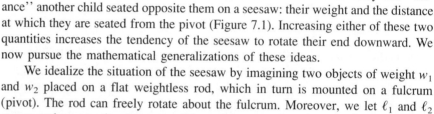

Figure 7.2 Weights balanced on a fulcrum.

Figure 7.3 Assigning coordinates to the lever.

Finally, we observe that, if the fulcrum is located at $x = \bar{x}$ rather than at $x = 0$ (which corresponds to moving the origin of the coordinate system but *does not*

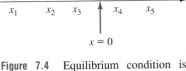

Figure 7.4   Equilibrium condition is $\Sigma x_j w_j = 0$ for fulcrum at $x = 0$.

Figure 7.5   Fulcrum at $\bar{x}$. (Note that all we have done is change the label from 0 to $\bar{x}$.)

change the physical situation), the (signed) distance of the weight w_j to the fulcrum is $(x_j - \bar{x})$ rather than x_j (Figure 7.5). In this case the equilibrium condition (3) is

$$\sum_{j=1}^{n} (x_j - \bar{x}) w_j = 0. \tag{4}$$

Since we can write equation (4) as

$$\sum_{j=1}^{n} x_j w_j - \bar{x} \sum_{j=1}^{n} w_j = 0,$$

we may solve for $\bar{x}$ as

$$\bar{x} = \frac{\displaystyle\sum_{j=1}^{n} x_j w_j}{\displaystyle\sum_{j=1}^{n} w_j}. \tag{5}$$

Equation (5) specifies the location of the fulcrum in our idealized balance problem where equilibrium occurs. Using this version of the equilibrium condition, we can calculate $\bar{x}$ from the weights $w_1, w_2, \ldots, w_n$ and the locations $x_1, x_2, \ldots, x_n$. Not surprisingly, $\bar{x}$ is called the **center of gravity** of the system.

□ **EXAMPLE 1**

A system of weights $w_1 = 10$ lb, $w_2 = 20$ lb, $w_3 = 10$ lb, $w_4 = 20$ lb, and $w_5 = 25$ lb is located along a line at points $x_1 = -6$ ft, $x_2 = -2$ ft, $x_3 = 1$ ft, $x_4 = 3$ ft, and $x_5 = 6$ ft, respectively. Find the center of gravity.

Solution

$$\bar{x} = \frac{[(-6)(10) + (-2)(20) + (1)(10) + (3)(20) + (6)(25)] \text{ ft-lb}}{(10 + 20 + 10 + 20 + 25) \text{ lb}}$$

$$= \frac{24}{17} \text{ ft.} \qquad\blacksquare$$

REMARK 1   The physical definition of the weight w of an object is $w = mg$, where m is the mass of the object and g is the acceleration due to gravity. Since g is a constant, w and m are proportional. Thus, $\bar{x}$ in equation (5) is also the **center of mass** for the n particles whose masses are $m_1 = w_1/g, m_2 = w_2/g, \ldots, m_n = w_n/g$, since all factors of g in equation (5) will cancel. Henceforth we will refer exclusively to centers of mass.

REMARK 2   If we write the equilibrium equation (4) in terms of masses $m_j = w_j/g$, we obtain the equation

$$\sum_{j=1}^{n} (x_j - \bar{x}) m_j = 0. \tag{6}$$

The sum on the left side of equation (6) is called the **first moment** of the mass system about the number $\bar{x}$. The first moment of a system about a number measures the tendency of an idealized rod supported at $\bar{x}$ with masses m_j at locations x_j to rotate, as in the earlier discussion involving torque. Since eqution (5), written for masses $m_j = w_j/g$, becomes

$$\bar{x} = \frac{\sum_{j=1}^{n} x_j m_j}{\sum_{j=1}^{n} m_j}, \tag{7}$$

we may interpret $\bar{x}$ as *the first moment of the mass system about $x = 0$ divided by the total mass*.

The preceding ideas can be used to calculate the center of mass of certain three-dimensional objects.

The Center of Mass of a Rod

Imagine a cylindrical rod of length ℓ and uniform cross-sectional area A. Imagine also that the density ρ of the material is constant across any cross section of the rod, but that the density varies continuously along the rod.

Since the density of the rod is not uniform, the center of mass will not necessarily lie at the midpoint. To approximate the center of mass $\bar{x}$, we position the rod as in Figure 7.6 and we partition the interval $[0, \ell]$ into n subintervals of equal length with endpoints $0 = x_0 < x_1 < x_2 < \cdots < x_n = \ell$. By slicing the rod with planes perpendicular to the axis through each of the points $x_0, x_1, x_2, \ldots, x_n$, we divide the rod into n cylinders, each of length $\Delta x = x_j - x_{j-1} = \ell/n$ and of cross-sectional area A. The volume of each small cylinder is therefore $V_j = A \Delta x$.

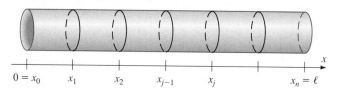

Figure 7.6 Rod of constant cross-sectional area and varying density.

Since the small cylinders are not of uniform density, we select one number t_j in each interval $[x_{j-1}, x_j]$ and *assume* that $\rho(x) = \rho(t_j)$ throughout the interval $[x_{j-1}, x_j]$. This assumption allows us to approximate the mass m_j of the jth cylinder as

$$m_j \approx \rho(t_j)V_j = \rho(t_j)A \Delta x \qquad (\text{mass} = \text{density} \times \text{volume}). \tag{8}$$

Finally, we assume the mass m_j of each cylinder to be concentrated at the point $t_j \in [x_{j-1}, x_j]$. With these assumptions, we have described a system consisting of a finite number of masses, so we may apply the equilibrium condition of equation (6) for the center of mass $\bar{x}$. We obtain, from equation (6) and approximation (8), the condition

$$0 = \sum_{j=1}^{n} (t_j - \bar{x})m_j \approx \sum_{j=1}^{n} (t_j - \bar{x})\rho(t_j)A \Delta x. \tag{9}$$

If ρ is continuous on $[0, \ell]$, the sum in equation (9) is a Riemann sum for the function $f(x) = (x - \bar{x})\rho(x)A$. Thus,

$$0 = \lim_{n \to \infty} \sum_{j=1}^{n} (t_j - \bar{x})\rho(t_j)A \, \Delta x$$

$$= \int_0^\ell (x - \bar{x})\rho(x)A \, dx$$

$$= \int_0^\ell x\rho(x)A \, dx - \bar{x} \int_0^\ell \rho(x)A \, dx.$$

We can now solve for $\bar{x}$. We obtain

$$\bar{x} = \frac{\displaystyle\int_0^\ell x\rho(x)A \, dx}{\displaystyle\int_0^\ell \rho(x)A \, dx}. \tag{10}$$

Equation (10) gives the *center of mass for a rod of constant cross-sectional area A and density $\rho(x)$*. By analogy with the discrete case described by equation (7), the integral in the numerator is called the first moment of the rod about the endpoint $x = 0$, and the integral in the denominator is the total mass. Since A is a constant, we can cancel it from the numerator and denominator, if we so desire.

□ **EXAMPLE 2**

A rod 10 centimeters long has uniform cross-sectional area $A = 4$ cm^2. Find the center of mass of the rod if

(a) the density ρ is uniform, or
(b) the density $\rho(x)$ varies linearly from 3 grams per cubic centimeter at one end to 6 grams per cubic centimeter at the other.

Solution (a) We may apply equation (10) with $\rho(x) = \rho$ (constant) to obtain

$$\bar{x} = \frac{\displaystyle\int_0^{10} x \cdot \rho \cdot 4 \, dx}{\displaystyle\int_0^{10} \rho \cdot 4 \, dx} = \frac{4\rho \left[\dfrac{x^2}{2}\right]_0^{10}}{4\rho \left[x\right]_0^{10}} = \frac{4\rho \cdot 50}{4\rho \cdot 10} = 5,$$

which is the midpoint of the rod, as expected.

(b) For the case of a linearly varying density, we first find an equation representing the density function. The statement that the density $\rho(x)$ varies linearly from $\rho(0) = 3$ to $\rho(10) = 6$ means that $\rho(x) = cx + d$ for some constants c and d.

Since $\rho(0) = c \cdot 0 + d = 3$, we have $d = 3$. Then, from $\rho(10) = c \cdot 10 + 3 = 6$, we obtain $c = \frac{3}{10}$. The density function is therefore $\rho(x) = \frac{3}{10}x + 3$. Thus, from (10),

$$\bar{x} = \frac{\displaystyle\int_0^{10} 4x[\frac{3}{10}x + 3] \, dx}{\displaystyle\int_0^{10} 4[\frac{3}{10}x + 3] \, dx} = \frac{\displaystyle\int_0^{10} (\frac{6}{5}x^2 + 12x) \, dx}{\displaystyle\int_0^{10} (\frac{6}{5}x + 12) \, dx}$$

$$= \frac{\left[\frac{2}{5}x^3 + 6x^2\right]_0^{10}}{\left[\frac{3}{5}x^2 + 12x\right]_0^{10}}$$

$$= \frac{50}{9} \approx 5.56.$$

Because of the variation of the density, the center of mass is closer to the high-density end of the rod. ∎

□ **EXAMPLE 3**

Find an equation for the center of mass $\bar{x}$ of a rod of length ℓ of constant density ρ whose cross-sectional area $A(x)$ is a continuous function for $0 \leq x \leq \ell$.

Solution The differences between this question and the question answered by equation (10) are that (i) the density ρ is now assumed constant and (ii) the cross-sectional area A is now assumed to vary. We use the same analysis as before, except that in approximation (8) we approximate the volume V_j by assuming that the cross-sectional area $A(t_j)$ is uniform throughout the jth cylinder. Thus, the analogue of approximation (8) is

$$m_j = \rho V_j \approx \rho A(t_j)\,\Delta x.$$

The analogue of equation (9) is

$$0 = \sum_{j=1}^{n}(t_j - \bar{x})m_j \approx \sum_{j=1}^{n}(t_j - \bar{x})\rho A(t_j)\,\Delta x,$$

which, as $n \to \infty$, gives

$$\int_0^{\ell}(x - \bar{x})\rho A(x)\,dx = 0.$$

Solving this equation for $\bar{x}$ produces the desired equation

$$\bar{x} = \frac{\displaystyle\int_0^{\ell}\rho x A(x)\,dx}{\displaystyle\int_0^{\ell}\rho A(x)\,dx} \tag{11}$$

for *the center of mass (or gravity) along the x-axis of a rod of uniform density and continuously varying cross-sectional area A(x)* (Figures 7.7 and 7.8). In equation (11), we can cancel the constant factors of ρ, if we so desire. ∎

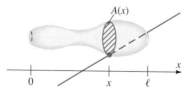

Figure 7.7 Rod of constant density and variable cross-sectional area.

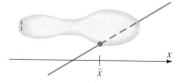

Figure 7.8 Rod balances at center of mass (gravity) $x = \bar{x}$.

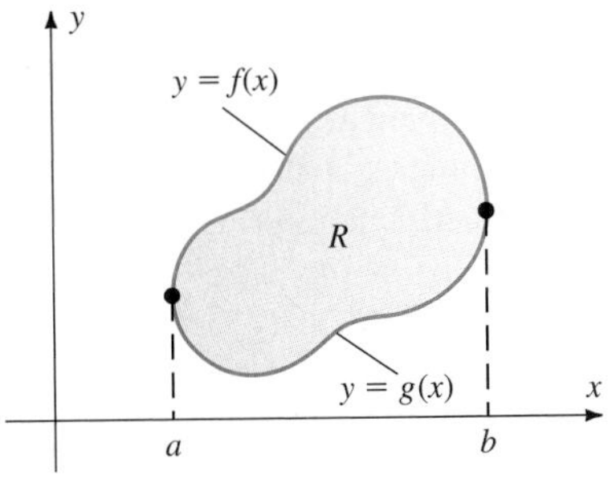

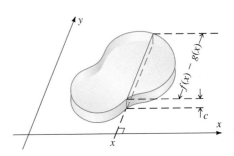

Figure 7.9 Face of thin plate bounded by graphs of $y = f(x)$ and $y = g(x)$.

Figure 7.10 Cross section perpendicular to x-axis has area $A(x) = [f(x) - g(x)]c$.

Center of Mass of a Plate

We may apply the result of Example 3 to a thin plate of uniform density ρ and uniform thickness c. Suppose that the plate lies in the xy-plane and that the face is bounded above by the graph of $y = f(x)$ and below by the graph of $y = g(x)$ for $a \le x \le b$ (Figure 7.9). Then the area of a cross section taken perpendicular to the x-axis is $A(x) = c[f(x) - g(x)]$ (Figure 7.10). Regarding the plate as a rod of constant density and variable cross-sectional area, we apply equation (11) to find the x-coordinate of the center of mass $\bar{x}$:

$$\bar{x} = \frac{\displaystyle\int_a^b c\rho x[f(x) - g(x)]\,dx}{\displaystyle\int_a^b c\rho[f(x) - g(x)]\,dx} = \frac{\displaystyle\int_a^b x[f(x) - g(x)]\,dx}{\displaystyle\int_a^b [f(x) - g(x)]\,dx}. \qquad (12)$$

(See Figure 7.11.)

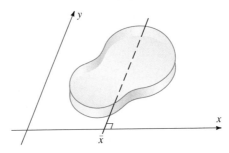

Figure 7.11 Plate balances along perpendicular to x-axis through center of mass $\bar{x}$.

Similarly, a straightforward calculation outlined in Exercise 19 shows that the y-coordinate of the center of mass is

$$\bar{y} = \frac{1}{2} \frac{\displaystyle\int_a^b [f(x)^2 - g(x)^2]\,dx}{\displaystyle\int_a^b [f(x) - g(x)]\,dx}. \qquad (13)$$

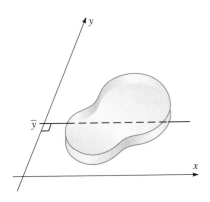

Figure 7.12 Plate "balances" at perpendicular to y-axis through center of mass $\bar{y}$.

(See Figure 7.12.)

The point $(\bar{x}, \bar{y})$ with coordinates given by equations (12) and (13) is called the center of mass, or **centroid,** of the plate. Since the centroid lies on the lines repre-

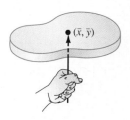

Figure 7.13 Plate balances at centroid $(\bar{x}, \bar{y})$.

senting the centers of gravity with respect to both the x- and y-axes, it is properly referred to as the (two-dimensional) center of gravity of the plate. It is the point where the plate "balances" (see Figure 7.13).

Note two things about equations (12) and (13). First, the constants c and ρ dropped out of both equations completely. This shows that the centroid is a purely geometric property of the face R of the plate, independent of both the uniform thickness and uniform density. Second, the denominators in both equations (12) and (13) are simply the area of the face. We may therefore restate our findings as follows:

> If the region R is bounded above and below by the graphs of the continuous functions f and g for $a \leq x \leq b$, then the coordinates $(\bar{x}, \bar{y})$ of the centroid of R are
>
> $$\bar{x} = \frac{1}{A} \int_a^b x[f(x) - g(x)]\, dx \qquad (14)$$
>
> $$\bar{y} = \frac{1}{2A} \int_a^b [f(x)^2 - g(x)^2]\, dx \qquad (15)$$
>
> where A is the area of R.

□ **EXAMPLE 4**

Find the centroid of the region R bounded by the graphs of $y = \sqrt{x}$ and $y = x^2$.

Strategy · · · · · · · ·

Find the upper and lower bounding functions.

Find $A = \int_a^b [f - g]\, dx$.

Solution

For $0 \leq x \leq 1$, the region is bounded above by $f(x) = \sqrt{x}$ and below by $g(x) = x^2$ (Figure 7.14).

$$A = \int_0^1 (\sqrt{x} - x^2)\, dx = \left[\frac{2}{3}x^{3/2} - \frac{x^3}{3}\right]_0^1 = \frac{1}{3}.$$

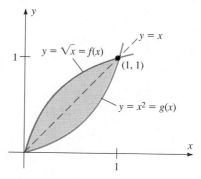

Figure 7.14 The region bounded by $y = \sqrt{x}$ and $y = x^2$ along with the line of symmetry $y = x$.

Find $\bar{x}$ from (14).

We apply equation (14) to obtain

$$\bar{x} = 3 \int_0^1 x[\sqrt{x} - x^2]\, dx$$

$$= 3\left[\frac{2}{5}x^{5/2} - \frac{x^4}{4}\right]_0^1$$

$$= 3\left(\frac{2}{5} - \frac{1}{4}\right) = \frac{9}{20}.$$

Apply (15) to find $\bar{y}$.

By equation (15)

$$\bar{y} = \frac{3}{2}\int_0^1 [(\sqrt{x})^2 - (x^2)^2] \, dx$$

$$= \frac{3}{2}\left[\frac{x^2}{2} - \frac{x^5}{5}\right]_0^1$$

$$= \frac{9}{20}.$$

The centroid of R is therefore

$$(\bar{x}, \bar{y}) = (9/20, 9/20). \qquad \blacksquare$$

REMARK 3   The region R in Example 4 is symmetric with respect to the line $y = x$, meaning that R is its own reflection in the line $y = x$. Notice that the centroid $(9/20, 9/20)$ lies on this line. In fact, *if R is any region that is symmetric with respect to a line ℓ, the centroid of R must lie on ℓ.* If this were not true, reflecting R in the line of symmetry would result in two different ''centers of gravity,'' a physical impossibility. Happily, many of the regions for which one needs to find centroids in practice contain one or more lines of symmetry. In such cases, this observation can simplify the calculation of the centroid greatly.

☐ **EXAMPLE 5**

Find the centroid of the quarter circle R with center $(0, 0)$ and radius $r = 4$ lying in the first quadrant (see Figure 7.15).

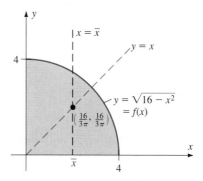

Figure 7.15 Centroid lies on intersection of $x = \bar{x}$ and $y = x$.

Strategy · · · · · · · ·

Find the area A by geometry.

Solution

Since R is a quarter circle, its area A is

$$A = \frac{\pi r^2}{4} = \frac{16\pi}{4} = 4\pi.$$

Find upper and lower bounding functions and find $\bar{x}$ from (14). Centroid lies on line $x = \bar{x}$.

The region R is bounded above by $f(x) = \sqrt{16-x^2}$ and below by $g(x) = 0$. Thus

$$\bar{x} = \frac{1}{4\pi} \int_0^4 x\sqrt{16-x^2}\,dx = \frac{1}{4\pi}\left[\left(-\frac{1}{3}\right)(16-x^2)^{3/2}\right]_0^4 = \frac{16}{3\pi}.$$

Use symmetry to obtain second line containing centroid. Centroid lies on the intersection.

Since R is symmetric about the line $y = x$, the centroid must lie on the line $y = x$. Thus $\bar{x} = \bar{y} = 16/(3\pi)$, so the centroid is $(16/(3\pi), 16/(3\pi))$. ∎

□ **EXAMPLE 6**

Find the centroid of the region R common to both the circle $x^2 + y^2 = 9$ and the circle $(x - 2)^2 + y^2 = 9$ (Figure 7.16).

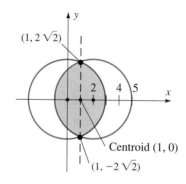

Figure 7.16   Centroid of region common to two circles with axes of symmetry $x = 1$ and $y = 0$.

Strategy · · · · · · · ·

Find points of intersection of the circles.

One line of symmetry is through the centers.

The other is through the points of intersection.

Solution

The circles intersect where $x^2 = (x - 2)^2$. Solving this equation gives $-4x + 4 = 0$ or $x = 1$. The points of intersection are therefore $(1, 2\sqrt{2})$ and $(1, -2\sqrt{2})$. Since both circles have centers on the x-axis, the region R is symmetric with respect to the x-axis. Since the circles have the same radius $r = 3$, the region is also symmetric with respect to the vertical line $x = 1$. The centroid must therefore lie on both lines $x = 1$ and $y = 0$; that is, $(\bar{x}, \bar{y}) = (1, 0)$. ∎

Exercise Set 7.7

1. Find the center of gravity for the system consisting of weights $w_1 = 1$ lb, $w_2 = 31$ lb, and $w_3 = 7$ lb located along a line at positions $x_1 = -4, x_2 = 2$, and $x_3 = 6$, respectively.

2. Find the center of mass for the system consisting of masses $m_1 = 2$ g, $m_2 = 5$ g, $m_3 = 1$ g, and $m_4 = 10$ g located along a line at positions $x_1 = -10, x_2 = -3, x_3 = 1$, and $x_4 = 5$, respectively.

3. An 80-lb child is sitting 3 ft from the pivot on a seesaw. How far from the pivot must a 50-lb child sit so that the seesaw balances?

4. A manufacturing firm ships 100 items per year to city A and 300 items per year to city B. A straight section of interstate highway 200 miles long joins the two cities. The items are shipped by truck along this highway. Where along this high-

way should the manufacturing plant be located if shipping costs are to be equal? What is the relationship between this problem and equation (1)?

5. A system of masses $m_1, m_2, \ldots, m_n$ is located in an xy-plane at points $(x_1, y_1), (x_2, y_2), \ldots, (x_n, y_n)$, respectively. By analogy with the discussion leading up to equations (4) and (6), explain why the center of mass with respect to the x-axis, $\bar{x}$, satisfies the equation

$$\sum_{j=1}^{n}(x_j - \bar{x})m_j = 0.$$

(*Hint:* For the purposes of this calculation, we may regard the points as being distributed along a line parallel to the x-axis.)

6. Let $\bar{y}$ denote the center of mass for the system of Exercise 5. Explain why

$$\sum_{j=1}^{n} (y_j - \bar{y})m_j = 0.$$

7. Conclude from Exercise 5 and Exercise 6 that the center of mass $(\bar{x}, \bar{y})$ of the system of masses $m_1, m_2, \ldots, m_n$ located at points $(x_1, y_1), (x_2, y_2), \ldots, (x_n, y_n)$ is given by the coordinates

$$\bar{x} = \frac{\sum_{j=1}^{n} x_j m_j}{\sum_{j=1}^{n} m_j}, \quad \bar{y} = \frac{\sum_{j=1}^{n} y_j m_j}{\sum_{j=1}^{n} m_j}.$$

8. Use the result of Exercise 7 to find the center of mass of the system consisting of masses $m_1 = 10$ g, $m_2 = 15$ g, $m_3 = 2$ g, and $m_4 = 10$ g located at points $(x_1, y_1) = (0, 0)$, $(x_2, y_2) = (2, -2)$, $(x_3, y_3) = (4, -2)$, and $(x_4, y_4) = (1, 1)$.

9. Find (x_3, y_3) if the system consisting of mass $m_1 = 2$ g at point $(1, 1)$, mass $m_2 = 4$ g at point $(-2, 3)$, and mass $m_3 = 3$ g at point (x_3, y_3) has center of mass $(\bar{x}, \bar{y}) = (0, 1)$.

In Exercises 10–18, find the centroid of the region bounded by the given curves.

10. $y = x^3$ and $y = x^{1/3}$, for $0 \le x \le 1$

11. $y = x^2$ and $y = x^3$

12. $x = 4 - y^2$ and $x = 0$ (*Hint:* Obtain $\bar{y}$ by symmetry considerations. Use the substitution $u = 4 - x$ to find $\bar{x}$.)

13. $y = 4 - x$, $x = 0$, and $y = 0$

14. $y = x$, $y = 4 - x$, and $y = 0$

15. $y = \sqrt{4 - x^2}$ and $y = 0$

16. $x = y^2$ and $x = 4 - y^2$ (*Hint:* This can be done entirely by symmetry considerations.)

17. $y = 2x^2$, $y = x^2 + 1$

18. $y = \frac{1}{3}x^2$, $y = -\frac{2}{3}x^2 + 4$

19. Assuming $f(x)$ and $g(x)$ to be continuous, derive the formula

$$\bar{y} = \frac{\dfrac{1}{2} \displaystyle\int_a^b [f(x)^2 - g(x)^2] \, dx}{\displaystyle\int_a^b [f(x) - g(x)] \, dx}$$

for the y-coordinate of the centroid of the region R bounded above by the graph of $y = f(x)$ and below by the graph of $y = g(x)$, for $a \le x \le b$, as follows:

a. Imagine the region R to be the face of a plate of uniform density ρ and uniform thickness c.

b. Partition the interval $[a, b]$ into n subintervals of equal length $\Delta x = (b - a)/n$ and select one number t_j in each subinterval $[x_{j-1}, x_j]$.

c. Approximate the "slice" of the plate over the subinterval $[x_{j-1}, x_j]$ to be the rectangular "slab" of height $[f(t_j) - g(t_j)]$, with Δx, thickness c, and density ρ. The mass of this slice is therefore approximated by $m_j \approx c\rho[f(t_j) - g(t_j)] \, \Delta x$.

d. Observe that the center of mass of this rectangular slab, with respect to the y-axis, must lie on the line

$$y = y_j = \tfrac{1}{2}[f(t_j) + g(t_j)].$$

e. Write down the analog of equilibrium equation (6) as

$$0 = \sum_{j=1}^{n} (y_j - \bar{y})m_j \approx \sum_{j=1}^{n} (y_j - \bar{y})c\rho[f(t_j) - g(t_j)] \, \Delta x.$$

Substitute for y_j from (d) and apply $\lim_{n \to \infty}$.

In Exercises 20–26, use equations (14) and (15) to find the centroid of the region bounded by the graphs of the given functions.

20. $f(x) = 1 - x^2$, $g(x) = 2$, for $-1 \le x \le 1$

21. $f(x) = x^3$, $g(x) = 0$, for $0 \le x \le 1$

22. $f(x) = \sqrt{x}$, $g(x) = 0$, for $0 \le x \le 4$

23. $f(x) = x$, $g(x) = -x$, for $0 \le x \le 4$

24. $f(x) = \sqrt{9 - x^2}$, $g(x) = 0$, for $-3 \le x \le 3$

25. $f(x) = x^2 + x + 1$, $g(x) = 0$, for $1 \le x \le 3$

26. $f(x) = 4$, $g(x) = 4 - x^2$, for $0 \le x \le 2$

27. Find the center of mass of a 10-cm cylindrical rod if the density of the rod varies uniformly from 2 g/cm³ at one end to 12 g/cm³ at the other.

28. The density of a cylindrical rod of length $\ell = 10$ cm is given by $\rho(x) = 1 + x$ g/cm³, where x represents the distance from the lighter end. Find the center of mass of the rod.

29. Find the center of mass $\bar{x}$ of a 10-cm long cylindrical rod with density $\rho(x) = 1 + x^2$ g/cm³, where x is the distance from the lighter end.

30. A 6-cm long rod has a cross section that is an equilateral triangle with 2-cm sides. Find its center of mass $\bar{x}$ given that its density is $\rho(x) = x + x^2$, where x is the distance from the lighter end.

31. A right circular cone has radius $r = 4$ cm and height $h = 10$ cm. The density of the material from which the cone is made is uniform. Use equation (11) to find the center of mass of the cone.

32. Find the center of mass of the solid obtained by revolving about the x-axis the region bounded by the graph of $y = x^3$ for $0 \le x \le 2$. (Assume the solid to have uniform density.)

33. Find the center of mass of a rod of constant density ρ generated by revolving about the x-axis the graph of $f(x) = x^2$, $0 \le x \le 2$.

34. Find the center of mass of a rod of constant density ρ generated by revolving about the x-axis the graph of $f(x) = 1/x^3$, $1 \le x \le 3$.

35. Give a symmetry argument to show that the centroid of a circle is its center.

36. Use a symmetry argument to find the centroid of the region bounded above by $y = \sin x$ and below by $y = -\sin x$ for $0 \le x \le \pi$.

37. Find the centroid of the region bounded above by the graph of $f(x) = 4 - x^2$ and below by the graph of the function

$$f(x) = \begin{cases} -x - 2 & \text{for} \quad -2 \le x \le 0 \\ x - 2 & \text{for} \quad 0 \le x \le 2 \end{cases}$$

38. From symmetry considerations, find the centroid of the region enclosed by the ellipse $x^2 + y^2 - xy = 6$. (*Hint:* Note that the equation is unchanged under the substitutions $y = x$ and $y = -x$.)

39. Show that the medians of a triangle intersect at the centroid.

7.8 The Theorem of Pappus

You may have noticed a similarity between equation (14) in Section 7.7 and the formula for finding the volume of a solid of revolution by the method of cylindrical shells. To make this observation more precise, suppose that a region R in the xy-plane is bounded above by the graph of $y = f(x)$ and below by the graph of $y = g(x)$, for $a \le x \le b$ (Figure 8.1).

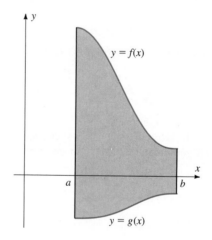

Figure 8.1 The region between the graphs of $y = f(x)$ and $y = g(x)$ for $a \le x \le b$.

If f and g are continuous, the volume V of the solid obtained by revolving R about the y-axis is

$$V = \int_a^b 2\pi x[f(x) - g(x)] \, dx \tag{1}$$

(equation (3) in Section 7.2). Also, the x-coordinate of the centroid $(\bar{x}, \bar{y})$ of R is given by equation (14) of the preceding section as

$$\bar{x} = \frac{1}{A} \int_a^b x[f(x) - g(x)] \, dx. \tag{2}$$

By comparing equations (1) and (2), we see that

$$V = 2\pi A \bar{x}. \tag{3}$$

The quantity $2\pi\bar{x}$ is simply the circumference of the circle swept out by the centroid $(\bar{x}, \bar{y})$ as R is revolved about the y-axis (see Figure 8.2). We have therefore proved a particular case of the following theorem, due to the Greek mathematician Pappus, who lived approximately 300 A.D.

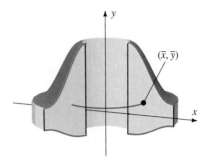

Figure 8.2 The solid obtained by revolving the region in Figure 8.1 about the y-axis with a portion cut away to expose the circumference determined by the centroid as it revolves about the y-axis.

Theorem 1

The Theorem of Pappus

Let R be a region lying in a plane and let ℓ be a line not intersecting R. The volume V of the solid obtained by revolving the region R about the line ℓ is given by the equation $V = cA$ where A is the area of R and c is the circumference of the circle swept out by the centroid of R as R revolves about the line ℓ.

By using this theorem, we can calculate certain volumes for which the previous techniques of this chapter are inadequate.

☐ **EXAMPLE 1**

Find the volume of the torus obtained by revolving the circle $(x - 2)^2 + y^2 = 1$ about the y-axis. (See Figures 8.3 and 8.4.)

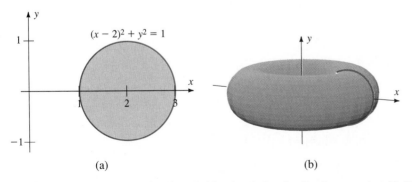

(a)   (b)

Figure 8.3 Revolving the region bounded by the circle of radius 1 centered at $(2, 0)$ about the y-axis produces a torus.

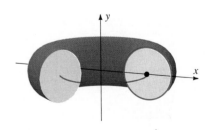

Figure 8.4 The torus from Figure 8.3 with a portion cut away to expose the circumference swept out by the centroid $(2, 0)$ as it revolves about the y-axis.

Solution First we note why the method of shells will not handle this problem, at least with the integration techniques developed thus far. We can write the equation for the upper semicircle as

$$y = \sqrt{1 - (x - 2)^2}$$

and the equation for the bottom semicircle as

$$y = -\sqrt{1 - (x - 2)^2}.$$

Applying the method of cylindrical shells gives the volume as

$$V = \int_1^3 2\pi x [2\sqrt{1 - (x - 2)^2}] \, dx,$$

and we have not yet developed a technique to find an antiderivative for this integrand. (However, we *can* evaluate the definite integral by geometry—see Exercise 11.) But it is easy to see that

(i) the area of the circle is $A = \pi \cdot 1^2 = \pi$,
(ii) the centroid of the circle is $(\overline{x}, \overline{y}) = (2, 0)$, and
(iii) as the circle revolves around the y-axis, the centroid sweeps out a circle of radius $2\pi\overline{x} = 4\pi$.

Therefore, by Theorem 1, $V = (4\pi)\pi = 4\pi^2$. ∎

The Theorem of Pappus can also aid in certain calculations of hydrostatic pressure. Recall from Section 7.5 that if the flat vertical face of a submerged object lying between depths $h = a$ and $h = b$ has width $w(h)$, $a \le h \le b$ (all in feet), then the total hydrostatic force (in pounds) acting on that face is given by the integral

$$F = \int_a^b (62.4) h w(h) \, dh. \tag{4}$$

If the right and left edges of the face are determined by the continuous functions f and g then we can write $w(h) = f(h) - g(h)$, and equation (4) becomes

$$F = \int_a^b (62.4) h [f(h) - g(h)] \, dh. \tag{5}$$

(See Figure 8.5.)
Since the h-coordinate of the centroid of the face is given by

$$\overline{h} = \frac{1}{A} \int_a^b h [f(h) - g(h)] \, dh, \tag{6}$$

we conclude from equations (5) and (6) that

$$F = (62.4) A \overline{h}. \tag{7}$$

Equation (7) states that to calculate the hydrostatic force on a vertical face of a submerged object *we may regard the face as lying horizontally at the depth of its centroid.*

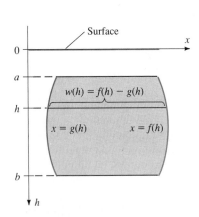

Figure 8.5 Flat vertical face of a submerged object.

☐ **EXAMPLE 2**

A submarine has a circular window of radius 6 inches. Find the hydrostatic force on the window if the submarine is submerged so that the top of the window lies 20 feet below the water level. Assume that the window is mounted vertically (see Figure 8.6).

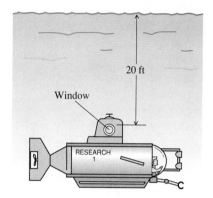

Figure 8.6   Circular window of radius 6 inches in a submerged submarine.

Solution   Since the window is positioned vertically with its top 20 ft below water level, the centroid lies at a depth of $\bar{h} = 20$ ft $+ 6$ in. $= 246$ in., and the area of the window is $\pi(1/2)^2 = \pi/4$ ft$^2 = 36\pi$ in.^{2} Since we are working in units of inches rather than feet, we must convert 62.4 lb/ft^3 to $62.4/1728 = .03611$ lb/in.^{3} From equation (7) we obtain

$$F = (.03611 \text{ lb/in.}^3) \times 36\pi \text{ in.}^2 \times 246 \text{ in.} \approx 1005 \text{ lb.} \qquad ■$$

Exercise Set 7.8

1. Find the volume of the torus obtained by revolving about the x-axis the region enclosed by the circle with the equation $x^2 + (y - 5)^2 = 9$.

2. Find the volume of the torus obtained by revolving about the line $y = 3$ the region enclosed by the circle $(x - 2)^2 + (y + 3)^2 = 4$.

3. Use the Theorem of Pappus to find the volume of the solid obtained by revolving about the y-axis the region bounded by the graphs of $y = x^2$ and $y = \sqrt{x}$ (see Example 4, Section 7.7).

4. Find the volume of the solid obtained by revolving about the line $y = -x$ the region in the first quadrant bounded by the graph of the equation $x^2 + y^2 = 16$ (see Example 5, Section 7.7).

5. The region R common to the circles $x^2 + y^2 = 9$ and $(x - 2)^2 + y^2 = 9$ is revolved about the line $x = -3$. Find the volume of the resulting solid in terms of A, the area of R (see Example 6, Section 7.7).

6. Find the volume of the solid obtained by revolving about the x-axis the region bounded by the graph of $y = 2 + \sqrt{4 - x^2}$ and the x-axis for $-2 \le x \le 2$.

7. Find the force exerted on the end wall of a swimming pool if the wall is a rectangle 10 ft wide and 8 ft deep. Assume the pool to be completely filled.

8. Find the force on a vertical window of a submarine if the window is in the shape of a circle of radius 4 in. and the top of the window is located 30 ft below the surface of the water.

9. Suppose the window in Exercise 8 has the shape of a rectangle 8 in. high and 6 in. long with a semicircle of radius 4 in. attached at either end. Find the force on the window (Figure 8.7).

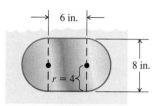

Figure 8.7   Window in submarine (Exercise 9).

10. Find the volume of the solid obtained by revolving the graph of the semicircle $y = \sqrt{4 - x^2}$ about the line $y = 6$ (Figure 8.8).

11. Show that the integral in Example 1 can be evaluated using the u-substitution $u = x - 2$ together with the observation that

$$\int_{-1}^{1} \sqrt{1 - u^2}\, du = \pi/2.$$

(This integral gives the area of the semicircle of radius 1.)

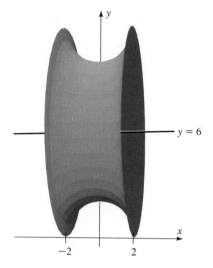

Figure 8.8

Summary Outline of Chapter 7

▌ The **volume** of the solid obtained by revolving about the x-axis the region   (page 386) bounded by the graph of $y = f(x) \geq 0$ and the x-axis for $a \leq x \leq b$ is

$$V = \int_{a}^{b} \pi[f(x)]^2\, dx.$$

▌ The **volume** of the solid that lies between $x = a$ and $x = b$ with cross-   (page 391) sectional area $A(x)$ is

$$V = \int_{a}^{b} A(x)\, dx.$$

▌ The **volume** of the solid obtained by rotating about the y-axis the region   (page 397) bounded by the graph of $y = f(x) \geq 0$ and the x-axis for $a \leq x \leq b$ is

$$V = \int_{a}^{b} 2\pi x f(x)\, dx.$$

▌ The **arc length** of the graph of $y = f(x)$ from $(a, f(a))$ to $(b, f(b))$ is   (page 404)

$$L = \int_{a}^{b} \sqrt{1 + [f'(x)]^2}\, dx.$$

▌ The **lateral surface area** of the solid obtained by rotating about the x-axis the   (page 408) region bounded by the graph of $y = f(x) \geq 0$ and the x-axis, for $a \leq x \leq b$, is

$$S = \int_{a}^{b} 2\pi f(x)\sqrt{1 + [f'(x)]^2}\, dx.$$

■ The **total distance** travelled by an object moving along a line with velocity $v(t)$ at time t, between times $t = a$ and $t = b$, is (page 412)

$$D = \int_a^b |v(t)|\, dt.$$

■ The **hydrostatic force** acting on a submerged face lying between depths $h = a$ and $h = b$ whose width at depth h is $w(h)$ is given by the integral (page 418)

$$F = \int_a^b (62.4)hw(h)\, dh.$$

■ The **work** done by a variable force F in moving an object from point $x = a$ to point $x = b$ is given by the integral (page 422)

$$W = \int_a^b F(x)\, dx.$$

■ The **center of mass** of a rod of constant cross-sectional area A and continuously varying density $\rho(x)$ is (page 431)

$$\bar{x} = \frac{\int_0^\ell x\rho(x)A\, dx}{\int_0^\ell \rho(x)A\, dx}.$$

■ The **center of mass** of a rod of uniform density ρ and continuously varying cross-sectional area $A(x)$ is (page 432)

$$\bar{x} = \frac{\int_0^\ell \rho x A(x)\, dx}{\int_0^\ell \rho A(x)\, dx}.$$

■ The **centroid** of the region R bounded above by the graph of $y = f(x)$ and below by the graph of $y = g(x)$ is $(\bar{x}, \bar{y})$, where (page 434)

$$\bar{x} = \frac{1}{A}\int_a^b x[f(x) - g(x)]\, dx \qquad (A = \text{area})$$

$$\bar{y} = \frac{1}{2A}\int_a^b [f(x)^2 - g(x)^2]\, dx \qquad (A = \text{area})$$

■ ***Theorem of Pappus:*** The volume V of the solid obtained by revolving the region R about the line ℓ (not intersecting R) is $V = cA$, where A is the area of R and c is the circumference of the circle swept out by the centroid of R. (page 439)

Review Exercises—Chapter 7

In Exercises 1–12, use integration to find the volume of the solid S described.

1. The sphere of radius r.

2. The solid obtained by rotating about the x-axis the region bounded by the line $x = 4$ and the parabola $y^2 = x$.

3. The solid obtained by rotating about the x-axis the region bounded by the graph of $y = 1/x$ and the x-axis for $1 \le x \le 4$.

4. The solid obtained by rotating about the x-axis the region bounded by the graph of $y = \sec(\pi x/2)$ and the x-axis for $-1/2 \le x \le 1/2$.

5. The solid obtained by rotating about the x-axis the region bounded by the graph of $y = \sin x$ and the x-axis for $0 \le x \le 3\pi/2$.

6. The solid obtained by rotating about the x-axis the region bounded by the graphs of $f(x) = \sqrt{1 - x^2}$ and $g(x) = 1/2$.

7. The solid obtained by rotating about the y-axis the region bounded by the graph of $f(x) = x^{2/3}$ and the x-axis for $0 \le x \le 2$.

8. The solid obtained by rotating about the y-axis the region bounded by the graph of $y = \sin x^2$ and the x-axis for $0 \le x \le \sqrt{\pi}/2$.

9. The solid obtained by rotating about the y-axis the region bounded by the graphs of $y_1 = \sin x^2$ and $y_2 = \cos x^2$ for $0 \le x \le \sqrt{\pi}/2$.

10. The solid obtained by rotating about the line $y = -2$ the region bounded by the graph of $y = 1/x^2$ and the x-axis for $1 \le x \le 4$.

11. The solid obtained by rotating about the line $y = -3$ the region bounded by the graph of $y = x^{2/3} + 1$ and the x-axis for $0 \le x \le 2$.

12. The solid obtained by rotating about the line $x = 2$ the region bounded by the graph of $y = 3 + x^3$ and the lines $x = 0$ and $y = 0$.

In Exercises 13–18, find the length of the graph of the given equation between the indicated points.

13. $y = 1 + x^{3/2}$ from $(0, 1)$ to $(4, 9)$

14. $x^2 = y^3$ from $(0, 0)$ to $(1, 1)$

15. $y^2 = (x + 1)^3$ from $(0, 1)$ to $(1, \sqrt{8})$

16. $y = \dfrac{1}{8}\left(x^4 + \dfrac{2}{x^2}\right)$ from $(1, 3/8)$ to $(2, 33/16)$

17. $y = x^3 + \dfrac{1}{12x} + 1$ from $(1, 25/12)$ to $(2, 217/24)$

18. $y = x^{5/2} + \dfrac{1}{5\sqrt{x}} + 3$ from $(4, 351/10)$ to $(16, 20{,}541/20)$

19. Find the volume of a triangular pyramid with base area A and height h.

20. The base of the solid is the region bounded by the ellipse

$$x^2 + \frac{y^2}{4} = 1$$

Find the volume given that the cross sections perpendicular to the x-axis are

a. squares,

b. equilateral triangles.

21. A water tank has the shape of a hemisphere of radius r. To what percent capacity is it filled if the water has depth $r/2$?

22. Approximate the length of the graph of $f(x) = \cos^2 x$ from $x = 0$ to $x = \pi$.

23. Approximate the lateral surface area of the solid obtained if one arc ($0 \le x \le \pi$, say) of the graph of $f(x) = \sin x$ is revolved about the x-axis.

24. A spring is stretched 10 in. by a force of 20 lb. How much work is needed to stretch it 15 in?

25. A spring is stretched 20 cm by a force of 10 newtons. How much work is done in stretching the spring 50 cm? How much work is done in stretching the spring from 50 cm to 60 cm?

26. A cylindrical tank 10 ft in diameter and 20 ft high is full of water. How much work will be done in pumping all the water out through a valve in the top?

27. A tank in the shape of an inverted right circular cone has a radius of 8 ft and a depth of 12 ft. The tank is full of water. How much work will be done in pumping all the water to a height 10 ft above the top of the tank?

28. How much work is done in Exercise 27 if only half the water is pumped to a height of 10 ft above the tank?

29. A 400-lb chain 20 ft long hangs from a windlass. How much work is done in winding in the chain?

30. How much work is done in Exercise 29 if a 50-lb hook hangs at the end of the chain?

31. How much work is done in lifting a 1000-lb piano 30 ft vertically using a chain that weighs 6 lb/ft?

32. How much work is done if, after being lifted 15 ft, the piano in Exercise 31 falls and only the chain is hauled the last 15 ft?

33. A water trough has end panels in the shape of trapezoids with lower bases 12 in. and upper bases 18 in. The altitude of the trapezoid is 12 in. Find the hydrostatic pressure on an end panel if the trough is full of water.

34. The vertical wall of a dam has the shape of the region enclosed by the parabola $y = x^2/16$. Find the hydrostatic pressure on the wall when the water level behind it is 16 ft.

35. Show that the hydrostatic force against one face of an object submerged vertically in water is the product of the pressure at the centroid of the face times the area of the face.

36. A semicircular plate of radius 12 in. is submerged vertically in water with its diameter at the top edge and parallel to the surface of the water. Find the hydrostatic force on one face of the plate if the diameter lies 6 in. below water level.

37. A 20-cm rod has uniform cross-sectional area $A = 16\pi$ cm². Find its center of mass if its density varies uniformly from 2 g/cm³ at one end to 8 g/cm³ at the other.

38. A 10-cm rod has uniform density $\rho = 10$ g/cm³. Find the center of mass of the rod if its cross-sectional area at a point x cm from one end is $(2 + .05x^2)$ cm².

39. A 3-m rod has a uniform density of 10 kg/m³. Every cross section is a right triangle with one leg of constant length 20 cm; the length of the other leg varies linearly from 10 cm at one end to 100 cm at the other. Find the center of mass.

40. Three particles of mass 2 g, 5 g, and 7 g are located on the x-axis at points with coordinates -5, 1, and 4, respectively. Find the center of mass of the system.

41. Find the coordinates of the center of mass of the system consisting of four particles with coordinates $(-2, 4)$, $(-2, -2)$, $(0, 6)$, and $(1, -4)$ if the particles are of equal mass.

42. Find the center of mass of a rod of length 100 cm if the cross-sectional area is constant and the density is proportional to $1 + \sqrt{x}$ g/cm³, where x is the distance from one end.

43. Sketch an example of a region R in the plane whose centroid does not lie within R.

44. Find the centroid of the region bounded above by the graph of $y = x$, below by the x-axis, on the left by the line $x = 0$, and on the right by the line $x = 2$.

45. Find the centroid of the region bounded above by the graph of $y = 16 - x^2$ and below by the x-axis.

46. Find the centroid of the region bounded above by the graph of $y = x$ and below by the graph of $y = x^2$.

47. Find the centroid of the region bounded above by the graph of $y = 6x - x^2$ and below by the x-axis.

48. Find the centroid of the ellipse $9x^2 + 4y^2 = 36$.

49. Find the centroid of the region bounded by the graphs of $y = 6x - x^2$ and $y = 3 - |x - 3|$.

50. Find the centroid of the region bounded by the graphs of $y = |2x|$ and $y = 4$.

51. Find the centroid of the region bounded by the graphs of $y = |x - 2|$ and $y = -\frac{1}{5}x + \frac{14}{5}$.

52. Use the Theorem of Pappus to find the volume of the cone obtained by rotating about the y-axis the triangle with vertices $(4, 0)$, $(0, 8)$, and $(0, 0)$.

53. Use the Theorem of Pappus to find the volume of the solid obtained by rotating the region bounded above by the graph of $y = \sqrt{4 - x^2}$ and below by the x-axis about the line
  a. $y = 0$,
  b. $y = -2$,
  c. $y = 6$.

The Transcendental Functions

Unit Four

Discovery of the Transcendental Functions

At the beginning of the eighteenth century, the calculus was still in its infancy. Logically precise foundations of the subject were yet to be developed, and only polynomials could be differentiated or integrated easily. The logarithmic and exponential functions were still to be invented, and the trigonometric functions were not yet really understood as functions.

One of the greatest innovators in mathematics then appeared: **Leonhard Euler** (1707–1783). Born in Basel, Switzerland, Euler spent many years in St. Petersburg , and then taught in Berlin for 25 years. He was hired back to Russia by Catherine the Great and spent the last 17 years of his life there. He was perhaps the most prolific mathematician of all time—he published 520 books and papers and, after his death, the *Proceedings of the St. Petersburg Academy* continued to publish at least one new paper by Euler in every issue for the next 47 years! His output is seen to be all the more incredible by the fact that he was blind for virtually all of his second Russian period, dictating his work to one of his sons.

No one else did more to put calculus and analysis in its present form. Among the symbols Euler initiated are the sigma (Σ) for summation, e to represent the constant 2.71828 . . . , i for the imaginary $\sqrt{-1}$, and even a, b, and c for the sides of a triangle and A, B, and C for the opposite angles. Although William Jones had first used the symbol π for the ratio between the circumference and the diameter of a circle in 1706, Euler made its use standard.

Euler wrote textbooks on differential and integral calculus; the general form of these texts is still in use today. His greatest work was the *Introductio in analysis infinitorum* (1748), in which Euler did for the calculus what Euclid had done for geometry and the theory of numbers in ancient Greece. He systematized differentiation and the method of fluxions (integral calculus), creating a new system which is often called *analysis*—the study of infinite processes. The

Johann Bernoulli

French physicist and astronomer Arago called Euler "analysis incarnate."

Euler transformed the trigonometric ratios into the trigonometric functions as we think of them today, and first used the abbreviations sin, cos, and tan. He treated logarithms and exponents as functions, whereas their creators (John Napier and Henry Briggs) had thought of them merely as tools to aid computation.

Leonhard Euler received both his bachelor's and his master's degrees at the age of 15, studying mathematics primarily under **Johann Bernoulli**. Bernoulli was one of a dozen mathematicians of that name, stretching over six generations, though the most famous are Johann (1667–1748) and his brother Jakob (1654–1705). The two brothers corresponded for many years with Leibniz about the calculus, and made many discoveries with which we are familiar today. Johann first introduced integration by partial fractions. He also discovered a relation between the trigonometric and logarithmic functions, thus paving the way to the realization that there are only two basic types of elementary functions: polynomial, rational, and algebraic functions on the one hand, and the transcendental (trigonometric, logarithmic, exponential, and hyperbolic) functions on the other. Johann and Jakob Bernoulli both studied arc length, the curvature of curves, and points of inflection; however, their ideas seem somewhat naive and inexact today. For example, one postulate states, "Each curved line consists of infinitely many straight lines, these themselves being infinitely small."

Leonhard Euler

Jakob Bernoulli

Jakob Bernoulli was the first to publish on the subject of polar coordinates (although Newton had the general idea some years before). He studied the catenary or hanging chain curve, which is also known as the hyperbolic cosine curve. He suggested the term *calculus integralis*, or integral calculus as we call it today.

Many other mathematicians also contributed to the rapid growth of calculus in the eighteenth century. However, the greatest influence on the modern calculus classroom was that of the functions, symbols, and methods discovered and promoted by Leonhard Euler and the brothers Bernoulli.

Marquis de l'Hôpital

A very useful theorem encountered in this unit is l'Hôpital's Rule, named after a French nobleman, **Guillaume François Antoine de l'Hôpital** (1661–1704). The Marquis de l'Hôpital wrote the first textbook on differential calculus in 1696. It was not what we would think of today as a text for teaching students, but was intended rather for presenting the new subject to mathematicians. The so-called "rule," which enables one to find the limit of a quotient whose numerator and denominator both tend to zero, appears in this book. However, the rule and much of the other material in the book was actually the work of Johann Bernoulli, l'Hôpital's teacher. They had an unusual agreement whereby Bernoulli agreed to turn over to l'Hôpital all of his mathematical discoveries, which l'Hôpital was entitled to claim as his own. L'Hôpital had plans for writing what would also have been the first text in integral calculus, but abandoned this project when Leibniz told him that he was planning such a work.

The first calculus textbook intended for young people was written by an Italian woman, **Maria Gaetana Agnesi** (1718–1799). She was a brilliant child who began learning foreign languages by the age of 4, and by 9 knew Latin, Greek, Hebrew, French, German, and Spanish fluently. When only 10 she published her first book, on a printing press set up for her in her home. The book advocated education for girls and women.

Her father, a professor of mathematics, exploited her brilliance by establishing Sunday afternoon get-togethers at which learned men would sit in a circle around her while she lectured on some topic in areas that included mathematics, physics, logic, chemistry, and philosophy. She usually lectured in Latin, but

Maria Gaetana Agnesi

would respond to questions in the language of the questioner. Her second book was a collection of 190 of these lectures, published when she was 21.

Maria Agnesi was the eldest of 21 children, and hence was expected to supervise the raising of the other 20. To explain mathematics to one of her brothers (and for other teenagers), she wrote a massive two-volume work of 1070 pages. She began with algebra, trigonometry, conic sections, and curve-sketching, but the main part was differential and integral calculus, continuing on into differential equations. It was immediately successful, and was translated from Italian into English. She was only 30 when it was published.

Agnesi was clearly much overworked, and though she had a brilliant mathematical mind, by the age of 40 she had completely lost interest in mathematics. She opened up her home to the hungry and homeless, and eventually she supervised an institution for indigent women. She died at the age of 80 and was buried in a common grave with 15 other elderly women.

(Photograph of Leonhard Euler from the Library of Congress. Photographs of Johann and Jakob Bernoulli from the David Eugene Smith Papers, Rare Book and Manuscript Library, Columbia University. Photo of Maria Gaetana Agnesi from Culver Pictures.)

447

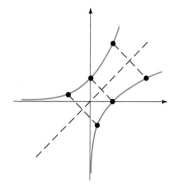

Chapter 8

Logarithmic and Exponential Functions

In previous courses on mathematics you encountered the *logarithm to the base b,* $\log_b x$, which was defined in terms of the *exponential* function $g(y) = b^y$ by the relation

$$y = \log_b x \quad \text{if and only if} \quad x = b^y. \tag{1}$$

Because *logarithm and exponential functions* are used frequently in science to describe the growth and decay of radioactive isotopes, animal populations, etc., we devote this chapter to a careful development of these functions.

The principal difficulty in developing the concept of a logarithm function is that up to this point the exponential b^y in equation (1) has been defined only for exponents y that are rational numbers (that is, numbers that are quotients of integers). In order to apply the calculus to properly defined logarithm and exponential functions they must be continuous and differentiable, and therefore defined for *all* numbers at least on certain intervals.

In order to overcome this difficulty we shall define the *natural logarithm* function as a certain integral, and then show that the exponential function can be obtained as the *inverse* of this integral. Along the way we shall also show that the functions defined in this manner actually are extensions of the functions in line (1). Before beginning on this agenda, however, we review the concepts of $\log_b x$ and inverse function in Section 8.1.

8.1 Review of Logarithms and Inverse Functions

The **logarithm to the base b** of the number x is defined by the statement

$$y = \log_b x \quad \text{if and only if} \quad x = b^y. \tag{1}$$

In other words, $\log_b x$ is the power y to which b must be raised to give x. In equation (1) we assume that b is a *positive* constant not equal to 1 and that $x > 0$. Thus, for example,

$$\log_{10} 100 = 2 \quad \text{because} \quad 100 = 10^2,$$
$$\log_2 8 = 3 \quad \text{because} \quad 8 = 2^3,$$
$$\log_{16} 4 = 1/2 \quad \text{because} \quad 4 = \sqrt{16} = 16^{1/2},$$

and

$$\log_2\left(\frac{1}{4}\right) = -2 \quad \text{because} \quad \frac{1}{4} = \frac{1}{2^2} = 2^{-2}.$$

The logarithm $\log_b x$ defined in statement (1) has the following properties:

$$(L1) \qquad \log_b(uv) = \log_b u + \log_b v,$$

$$(L2) \qquad \log_b\left(\frac{u}{v}\right) = \log_b u - \log_b v,$$

$$(L3) \qquad \log_b(u^r) = r \log_b u, \qquad r \text{ rational.}$$

These properties follow from statement (1) and the laws of exponents. (See Exercises 50–52.)

The Logarithm as an Inverse

It follows from statement (1) that if we calculate the logarithm of the number b^y we obtain simply the number y:

$$\text{If } \quad x = b^y, \quad \text{then} \quad \log_b(b^y) = \log_b x = y.$$

That is,

$$\log_b(b^y) = y. \tag{2}$$

Equation (2) says that the logarithm reverses (or *inverts*) the effect of the exponential function $g(y) = b^y$ acting on the number y. It is also the case that the exponential function "inverts" the application of the logarithm:

$$\text{If } \quad y = \log_b x, \quad \text{then} \quad b^{(\log_b x)} = b^y = x.$$

That is,

$$b^{(\log_b x)} = x. \tag{3}$$

It is important to note that *equations (2) and (3) are valid, however, only if the logarithm and exponential functions have the same base b.*

Inverse Functions

Equations (2) and (3) are examples of identities that hold for pairs of functions f and g if one function is the *inverse* of the other.

Definition 1

Let f be a function with domain D. The **inverse** function for f, written $g = f^{-1}$, is the function g defined by the equation

$$g(f(x)) = x \quad \text{for all } x \in D.$$

That is, g is the inverse of the function f if the composite function $h = g \circ f$ is the identity function $h(x) = (g \circ f)(x) = x$ on the set D. (See Figure 1.1). Thus, g is the inverse of f if g returns the value $y = f(x)$ to the number x for every x in the domain of f.

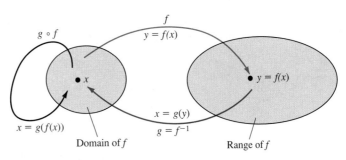

Figure 1.1 g is the inverse of f if $g(f(x)) = x$ for all $x \in D$.

REMARK 1   Be careful to note that we use the notation f^{-1} to denote the inverse function of f, *not* the reciprocal function $1/f$. That is, $f^{-1}(f(x)) = x$ for all x in the domain of f.

REMARK 2   If g is the inverse function of f, then the domain of g is precisely the range of f (nothing more, nothing less) because g is *defined* by the statement

$$g(y) = x \quad \text{if and only if} \quad y = f(x).$$

Even though the *equation* for g may be defined on a larger set, the statement that $g = f^{-1}$ automatically specifies the domain of g to be the range of f.

□ **EXAMPLE 1**

The function $f(x) = 2x$ has domain $(-\infty, \infty)$. Its inverse is the function $g(x) = x/2$ because,

$$\text{if} \quad y = f(x) = 2x, \quad \text{then} \quad g(y) = \frac{y}{2} = \frac{2x}{2} = x$$

for all $x \in (-\infty, \infty)$. That is,

$$g(f(x)) = \frac{f(x)}{2} = \frac{2x}{2} = x, \qquad x \in (-\infty, \infty). \qquad \blacksquare$$

□ **EXAMPLE 2**

The function $f(x) = \sqrt{x}$ has domain $[0, \infty)$. Its inverse is the function $g(x) = x^2$, $x \geq 0$, since

$$\text{if} \quad y = f(x) = \sqrt{x}, \quad \text{then} \quad g(y) = y^2 = (\sqrt{x})^2 = x, \qquad x \in [0, \infty).$$

That is,

$$g(f(x)) = [f(x)]^2 = (\sqrt{x})^2 = x, \qquad x \in [0, \infty).$$

In this case the domain of the inverse function $f^{-1}(x) = g(x) = x^2$ is $[0, \infty)$ because $[0, \infty)$ is the *range* of f. $\qquad \blacksquare$

□ **EXAMPLE 3**

In Chapter 1, we defined $f(x) = 10^x$ for all rational numbers x. In Section 8.4, we shall see how to extend the definition of this exponential function to one that is defined for all real numbers. In any case, equation (2) implies that the inverse function of f is the **common logarithm** function $g(x) = \log_{10} x$. In other words,

$$\text{if} \quad y = f(x) = 10^x \quad \text{then} \quad g(y) = \log_{10} y = \log_{10}(10^x) = x.$$

That is,

$$g(f(x)) = \log_{10}[f(x)] = \log_{10}(10^x) = x.$$

Thus, $f^{-1}(x) = g(x) = \log_{10} x$. $\qquad \blacksquare$

□ **EXAMPLE 4**

The function $f(x) = x^2$ with domain $(-\infty, \infty)$ does *not* have an inverse. That is because each positive number in the range $R = [0, \infty)$ for f corresponds to *two*

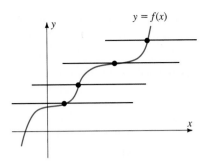

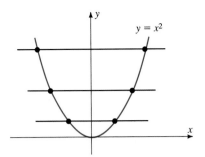

Figure 1.2   A horizontal line intersects the graph of a one-to-one function in at most one point.

Figure 1.3   The function $f(x) = x^2$ is not one-to-one: Some horizontal lines intersect the graph twice.

distinct numbers in the domain of f. For example, for $y = 4$ we have both $f(2) = 4$ and $f(-2) = 4$. It is impossible to define $g(4)$ so that *both* $g(4) = 2$ and $g(4) = -2$. Thus, no function g can satisfy the requirement that $g(f(x)) = x$ for all $x \in (-\infty, \infty)$. ∎

Example 4 shows that *not every function has an inverse.* In general, a function f has an inverse only if each number y in the range of f is the function value for *only one* number x in the domain of f. This allows us to define the inverse function $g = f^{-1}$ by the rule: If $y = f(x)$, then $g(y) = f^{-1}(y) = x$. Functions that have this property are called *one-to-one* functions.

Definition 2

The function f is **one-to-one** if it satisfies the following property for all numbers x_1 and x_2 in its domain:

$$\text{if } x_1 \neq x_2, \quad \text{then} \quad f(x_1) \neq f(x_2).$$

Figure 1.2 illustrates the geometric interpretation of the property of f being one-to-one. Since each number y in the range of f is the value $f(x)$ for only one number x, *any horizontal line* can intersect the graph of a one-to-one function at no more than one point. Figure 1.3 shows that this property fails for the function $f(x) = x^2$ which is not one-to-one (Example 4).

The following theorem summarizes our discussion about the existence of inverse functions.

Theorem 1

If the function f is one-to-one, then

(i) the inverse function f^{-1} exists, and
(ii) the domain of f^{-1} is the range of f.

Proof: Let y be any number in the range of f. Then, since f is one-to-one, there is precisely one number x for which $f(x) = y$. Define $g(y) = x$. Then g is a function defined for all y in the range of f, and $g(f(x)) = x$ for all x in the domain of f. Thus, $g = f^{-1}$ according to Definition 1. ∎

The following theorem gives a useful criterion for f to have an inverse.

Theorem 2

Let f be a function with domain D. Then f^{-1} exists if either

(i) f is increasing on D, or
(ii) f is decreasing on D.

Proof: To prove part (i) we assume that f is increasing on D. We shall show that this guarantees that f is one-to-one. Let x_1 and x_2 be any two numbers in D with $x_1 \neq x_2$. Then, since f is increasing on D,

(a) if $x_1 < x_2$, then $f(x_1) < f(x_2)$;

(b) if $x_1 > x_2$, then $f(x_1) > f(x_2)$.

In either case it follows that $f(x_1) \neq f(x_2)$. Thus, f is one-to-one. It then follows from Theorem 1 that f^{-1} exists.

The proof for part (ii) is similar.   ∎

◻ **EXAMPLE 5**

The function $f(x) = \tan x$ *restricted* to the domain $D = (-\pi/2, \pi/2)$ has an inverse (which we shall study in detail in Chapter 9). To see this we note that the derivative is

$$f'(x) = \frac{d}{dx}(\tan x) = \sec^2 x,$$

which is positive for all $x \in (-\pi/2, \pi/2)$. Thus, f is increasing on $(-\pi/2, \pi/2)$, so f^{-1} exists by Theorem 2.   ∎

Finding Inverse Functions

Given a function f specified by an equation of the form $y = f(x)$, we sometimes find an equation for the inverse function f^{-1} by solving the equation $y = f(x)$ for x in terms of y. This technique succeeds in the next example.

◻ **EXAMPLE 6**

Find the inverse function for $f(x) = 2x - 4$, $x \in (-\infty, \infty)$.

Solution   We solve the equation $y = 2x - 4$ for x:

$$y = 2x - 4$$
$$2x = y + 4$$
$$x = \tfrac{1}{2}y + 2$$

Thus, $f^{-1}(y) = \tfrac{1}{2}y + 2$ since,

$$\text{if} \quad y = 2x - 4, \quad f^{-1}(y) = \tfrac{1}{2}y + 2 = \tfrac{1}{2}(2x - 4) + 2 = x$$

for all $x \in (-\infty, \infty)$. Written as a function of x, the inverse function is

$$f^{-1}(x) = \tfrac{1}{2}x + 2.$$ ∎

Unfortunately, many functions have inverses that cannot be found by this method because the equation $y = f(x)$ cannot be solved algebraically for x. Examples include $y = x + x^3$, $-\infty < x < \infty$, and $y = \tan x$, $-\pi/2 < x < \pi/2$. We may always resort, however, to the defining equation $f^{-1}(f(x)) = x$, $x \in D$, if the inverse function f^{-1} is known to exist.

An Important Identity

Let f be a function with domain D and range R and assume that f^{-1} exists. Then, for each $y \in R$, we know that $x = f^{-1}(y)$ is in D. We may therefore apply f to both sides of this equation and conclude that

$$f(x) = f(f^{-1}(y)), \qquad y \in R. \tag{4}$$

Since $y = f(x)$, equation (4) gives the identity

$$f(f^{-1}(y)) = y, \qquad y \in R. \tag{5}$$

This equation shows that, if f has domain D, range R, and inverse f^{-1}, then *the function f is the inverse for f^{-1}.* (We shall use equation (5) in a critical way in later sections.)

☐ **EXAMPLE 7**

The function $f(x) = x^3$, with domain $(-\infty, \infty)$, has inverse function $f^{-1}(x) = x^{1/3}$ because,

$$\text{if} \quad y = x^3 \quad \text{then} \quad f^{-1}(y) = y^{1/3} = (x^3)^{1/3} = x, \quad x \in (-\infty, \infty).$$

Note also that the *range* of f is $R = (-\infty, \infty)$ and that

$$f(f^{-1}(y)) = f(y^{1/3}) = (y^{1/3})^3 = y, \qquad y \in (-\infty, \infty)$$

as required by equation (5).   ∎

The Graph of the Inverse Function

Figure 1.4 shows the graphs of the functions considered in Examples 2, 6, and 7 with the corresponding inverse function graphed on the same set of axes. Note in each case that the graph of $y = f^{-1}(x)$ is the reflection across the line $y = x$ of the graph of $y = f(x)$. Indeed, this will always be the case, since if $b = f(a)$, then $a = f^{-1}(b)$. This means that the point (b, a) will lie on the graph of $y = f^{-1}(x)$ if (a, b) is on the graph of $y = f(x)$ and conversely.

Properties of Inverse Functions

The fact that the graph of f^{-1} is the reflection across the line $y = x$ of the graph of f suggests that a function and its inverse (if it exists) should have similar continuity and differentiability properties. The following two theorems address this issue. We shall not provide formal proofs, although we shall assume these results in developing rules for differentiating inverses of the logarithm and trigonometric functions.

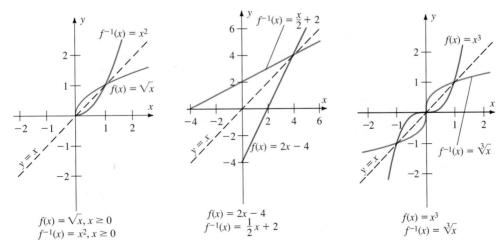

$$f(x) = \sqrt{x}, x \geq 0$$
$$f^{-1}(x) = x^2, x \geq 0$$

$$f(x) = 2x - 4$$
$$f^{-1}(x) = \tfrac{1}{2}x + 2$$

$$f(x) = x^3$$
$$f^{-1}(x) = \sqrt[3]{x}$$

Figure 1.4   The graph of f^{-1} is the reflection across the line $y = x$ of the graph of f. If (a, b) is on the graph of f, (b, a) is on the graph of f^{-1}.

Theorem 3
Continuity of Inverses

If the function f has an inverse f^{-1} on an open interval I and if f is continuous at $a \in I$, then f^{-1} is continuous at $b = f(a)$.

In other words, the inverse of a continuous function is a continuous function where defined.

Theorem 4

Let the function f be continuous on an open interval I and assume that f has an inverse function f^{-1}. Let $a \in I$. The inverse function f^{-1} is differentiable at $b = f(a)$ if

(i) $f'(a)$ exists, and
(ii) $f'(a) \neq 0$.

In this case we have

$$(f^{-1})'(b) = \frac{1}{f'(a)}.$$

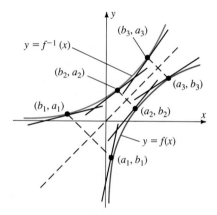

Figure 1.5 Slope of graph of f^{-1} at (b, a) is the reciprocal of slope of graph at (a, b).

Figure 1.5 suggests an intuitive justification for Theorem 4. The existence of the derivative $f'(a)$ means that the graph of f has a non-vertical tangent at (a, b). Reflecting this graph across the line $y = x$ produces a non-vertical tangent to the graph of f^{-1} at (b, a) with reciprocal slope *unless the tangent to the graph of f at (a, b) is horizontal* (in which case the tangent at (b, a) will be vertical with slope, and therefore $(f^{-1})'(b)$, undefined).

Figure 1.6 shows the graph of $f(x) = x^3$ and its inverse function $f^{-1}(x) = x^{1/3}$. Note that f is differentiable for all x, but that $f'(0) = 0$. Thus, the inverse function is *not* differentiable at $f(0) = 0$. This is suggested by the equation for f^{-1} since

$$(f^{-1})'(x) = \frac{d}{dx}(x^{1/3}) = \frac{1}{3x^{2/3}}$$

which is undefined for $x = 0$.

Using implicit differentiation, we can give a simple proof of Theorem 4 in the case where f is differentiable on an open interval I. Let $a \in I$ with $f(a) = b$. Differentiating both sides of the identity

$$f(f^{-1}(x)) = x, \quad x \in I \tag{6}$$

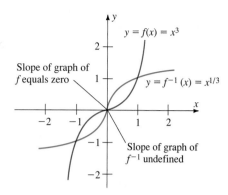

Figure 1.6   For $f(x) = x^3$, inverse function is $f^{-1}(x) = x^{1/3}$, which is not differentiable at $x = 0$ because $f'(0) = 0$.

gives

$$f'(f^{-1}(x)) \cdot (f^{-1})'(x) = 1, \quad x \in I.$$

Let $x = b$. Then $f^{-1}(x) = f^{-1}(b) = a$, and equation (6) becomes

$$f'(a) \cdot (f^{-1})'(b) = 1$$

or, since $f'(a) \neq 0$,

$$(f^{-1})'(b) = \frac{1}{f'(a)}$$

as claimed.

☐ **EXAMPLE 8**

Verify Theorem 4 for the function $f(x) = x^3$ at the point $(2, 8)$.

Solution   Here $f(x) = x^3$, $a = 2$, and $b = 8$. Then $f'(x) = 3x^2$, so $f'(a) = f'(2) = 3 \cdot 2^2 = 12$. According to Theorem 4, we have

$$(f^{-1})'(8) = \frac{1}{f'(2)} = \frac{1}{12}.$$

We can verify this equation directly since we know that $f^{-1}(x) = x^{1/3}$ for $f(x) = x^3$. Thus,

$$(f^{-1})'(x) = \frac{d}{dx}(x^{1/3}) = \frac{1}{3}x^{-2/3}$$

so

$$(f^{-1})'(8) = \frac{1}{3} \cdot 8^{-2/3} = \frac{1}{3 \cdot 8^{2/3}} = \frac{1}{3 \cdot 4} = \frac{1}{12}. \qquad ■$$

Exercise Set 8.1

In Exercises 1–10, find the logarithm.

1. $\log_{10} 1000$

2. $\log_7 343$

3. $\log_8 2$

4. $\log_4 (0.5)$

5. $\log_5\!\left(\dfrac{1}{125}\right)$

6. $\log_2 \sqrt{2}$

7. $\log_4 8$

8. $\log_{343} 7$

9. $\log_{10} 1$ **10.** $\log_b 1$

In Exercises 11–22, solve for x.

11. $\log_x 16 = 4$ **12.** $\log_5 x = 3$

13. $\log_{1/2} x = 2$ **14.** $\log_{125} x = 2/3$

15. $\log_3 x = 1/6$ **16.** $\log_x 243 = 5$

17. $\log_x 27 = 3/2$ **18.** $\log_7 x = 1$

19. $\log_{10} x = 0$ **20.** $\log_7 x = 0$

21. $\log_2 x = -3$ **22.** $\log_5 x = -2$

In Exercises 23–32, find the inverse function f^{-1}.

23. $f(x) = 3x - 2$ **24.** $f(x) = 7 - x$

25. $f(x) = \dfrac{1}{x + 2}$ **26.** $f(x) = \dfrac{2}{4 - x}$

27. $f(x) = \dfrac{x}{x + 3}$ **28.** $f(x) = \dfrac{x}{5 - x}$

29. $f(x) = 4 + \sqrt{x - 1}$ **30.** $f(x) = \sqrt[3]{x + 3}$

31. $f(x) = 3 + x^{5/3}$ **32.** $f(x) = \dfrac{1}{\sqrt{x + 1}}$

In Exercises 33–36, use Theorem 4 to find $(f^{-1})'(b)$ where $b = f(a)$.

33. $f(x) = \dfrac{3}{7 - 2x}$, $\quad a = 1$, $\quad b = 3/5$

34. $f(x) = 2x^2 + x^3$, $\quad a = 2$, $\quad b = 16$

35. $f(x) = \tan x$, $\quad a = \pi/4$, $\quad b = 1$

36. $f(x) = \cos x$, $\quad a = \pi/4$, $\quad b = \sqrt{2}/2$

37. Show that the function $f(x) = 1/x$ is its own inverse.

38. True or false? If f has an inverse, then no horizontal line can intersect the graph of f more than once. Why or why not?

39. Find the inverse of each of the following functions if they are restricted to the indicated domains:
   a. $f(x) = x^2 + 3$, $\quad x \geq 0$
   b. $f(x) = (x + 1)^2$, $\quad x \leq -1$
   c. $f(x) = x^2 + 4x + 4$, $\quad x \geq -2$
   d. $f(x) = x^2 - 2x - 3$, $\quad x \leq 1$

40. Explain why the function $f(x) = ax^2 + bx + c$, $a \neq 0$, has an inverse if it is restricted to the domain $[-b/(2a), \infty)$. (*Hint:* Consider the quadratic formula, or find its relative maximum or minimum.)

41. For the function $f(x) = \sqrt[3]{x + 5}$,
   a. find the derivative $(f^{-1})'(2)$ using Theorem 4;
   b. find the inverse function f^{-1};
   c. find $(f^{-1})'(x)$ from part b;
   d. find $(f^{-1})'(2)$ from part c and verify that it agrees with your answer from part a.

42. Use Theorem 2 to verify that the function $f(x) = \sin x$ has an inverse if it is restricted to the domain $[-\pi/2, \pi/2]$.

43. Let g be the inverse function for $f(x) = \sin x$, $x \in [-\pi/2, \pi/2]$. (See Exercise 42.) Find, using Theorem 4,
   a. $g'(\sqrt{2}/2)$
   b. $g'(0)$
   c. $g'(\sqrt{3}/2)$
   d. $g'(-1/2)$

44. Verify that the function $f(x) = x + x^3$ has an inverse on $(-\infty, \infty)$. (*Hint:* Use Theorem 2.)

45. Let g be the inverse for the function $f(x) = x + x^3$. (See Exercise 44.) Find:
   a. $g(0)$
   b. $g(2)$
   c. $g'(0)$
   d. $g'(2)$

46. Verify that the function $f(x) = -35x - x^5$ has an inverse on $(-\infty, \infty)$.

47. Let g be the inverse for the function $f(x) = -35x - x^5$ (see Exercise 46). Find:
   a. $g(0)$
   b. $g'(0)$
   c. $g(-36)$
   d. $g'(-36)$
   e. $g^{-1}(1)$

48. The exponential function occurs naturally in formulas for the periodic compounding of interest. For example, if an annual rate of interest of r percent is applied to a principal amount P_0 placed in a savings account, the amount $P(1)$ on deposit at the end of one year is

$$P(1) = P_0 + rP_0 = (1 + r)P_0.$$

In the second year the interest rate r is applied to the new principal, $(1 + r)P_0$. Thus, the amount $P(2)$ on deposit at the end of two years is

$$P(2) = (1 + r)P_0 + r[(1 + r)P_0]$$
$$= (1 + r)^2 P_0$$

Let t be a positive integer. Show that the amount $P(t)$ on deposit at the end of t years is given by the exponential function

$$P(t) = (1 + r)^t P_0.$$

49. Show that, if interest is compounded n times per year at an annual percentage rate r, the amount $P(t)$ on deposit in a savings account after t years (t is a positive integer) is

$$P(t) = \left(1 + \frac{r}{n}\right)^{nt} P_0,$$

where P_0 is the amount of the original principle.

50. Prove property (*L1*) of logarithms as follows:
  a. Let $x = \log_b m$ and $y = \log_b n$. Show that $m = b^x$ and $n = b^y$.
  b. By the laws of exponents, show that $mn = b^{x+y}$.
  c. Explain why $\log_b b^z = z$, z rational.
  d. Conclude that
  $$\log_b(mn) = \log_b(b^{x+y}) = x + y = \log_b m + \log_b n.$$

51. Prove property (*L2*)   (see Exercise 50).

52. Prove property (*L3*)   (see Exercise 50).

53. The Beer-Lambert Law relates the absorption of light travelling through a material to the concentration and the thickness of the material. If I_0 and I denote the intensities of light of a particular wavelength before and after passing through the material, respectively, and if x denotes the length of the path followed by the beam of light passing through the material, then
$$\log_{10}\left(\frac{I}{I_0}\right) = kx,$$
where k is a constant depending on the material. Express I as a function of x.

54. True or false? The values of $y = \log_5 x$, as defined in this section, must be rational numbers.

55. True or false? If $0 < x < y$ then $\log_a x < \log_a y$. Why?

Given a function f defined on the interval $[a, b]$, then the graph of its inverse relation is the set
$$\{(f(x), x) \mid a \le x \le b\}$$

in the plane. In Exercises 56–60, graph the function f and its inverse relation over the given interval. Use these graphs to determine if the inverse relation is a function. Justify your answer. (*Hint:* Use a routine that plots parametric equations to plot the inverse relation. See Appendix V.)

56. $f(x) = 2x^2 - x$; $[0, 4]$

57. $f(x) = (x + 1)^{5/2}$; $[-1, 8]$

58. $f(x) = x^3 - 4x$; $[-3, 3]$

59. $f(x) = 2x^2/(x^2 + 10)$; $[-10, 10]$

60. $f(x) = x + \cos x$; $[-2\pi, 2\pi]$

In Exercises 61–63, a function f and an interval $[a, b]$ are specified. Use Theorem 4 to graph $(f^{-1})'$ on the interval between $f(a)$ and $f(b)$. (*Hint:* The points on the graph of $(f^{-1})'$ can be written in the form $(f(x), 1/f'(x))$.)

61. $f(x) = 1/x$; $[1, 3]$ **62.** $f(x) = \sqrt[3]{x - 1}$; $[-1, 3]$

63. $f(x) = \sin x$; $[-\pi/2, \pi/2]$

64. Suppose 1000 is invested in a savings account that pays an annual interest rate of 8.5% compounded n times annually. On one plot, graph both the line $y = 2000$ and the graph of
$$P_n(t) = 1000\left(1 + \frac{0.085}{n}\right)^{nt}$$
for the integers n that correspond to annual, semiannual, quarterly, monthly, daily, and hourly compounding. From each of these plots, approximate the time (in years) it takes to double the initial investment. How do these doubling times differ (in days) for the various compounding times?

8.2   The Natural Logarithm Function

The purpose of this section is to introduce what we shall refer to as the *natural logarithm function*, written $y = \ln x$. We shall begin by answering what seems to be a completely unrelated question, and then show that the resulting function ($y = \ln x$) has the properties *L1–L3* of logarithm functions as discussed in Section 8.1.

An Antiderivative for
$$f(x) = \frac{1}{x}$$

In discussing antiderivatives in Chapters 5 and 6, we were careful to note that the integration formula
$$\int x^n \, dx = \frac{x^{n+1}}{n + 1} + C$$
does *not* apply when $n = -1$. The question of finding an antiderivative of $f(x) = 1/x$ has therefore been addressed only by the Fundamental Theorem of Calculus (see Example 6 and the comments that follow it in Section 6.3). Since $f(x) = 1/x$ is continuous on the interval $(0, \infty)$, the Fundamental Theorem guarantees that the function
$$L(x) = \int_1^x \frac{1}{t} \, dt, \qquad x > 0,$$

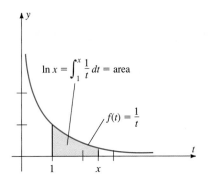

Figure 2.1 $\ln x > 0$ if $x > 1$.

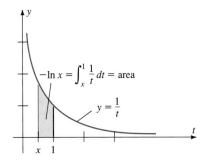

Figure 2.2 $\ln x < 0$ if $0 < x < 1$.

is differentiable on $(0, \infty)$. Moreover, $L'(x) = 1/x$. Consequently, L is an antiderivative of $f(x) = 1/x$. Even though this function is defined in a somewhat abstract fashion (as a function defined by integration), it is one of the most important functions used in applications of the calculus. A large percentage of this chapter is devoted to studying its properties and uses. We shall also see how L is related to the logarithm functions that were discussed in Section 8.1.

Although we define the function L using the definite integral, it is commonly referred to as the *natural logarithm function,* and the natural logarithm of a positive real number x is denoted $\ln x$. That is,

$$\ln x = \int_1^x \frac{1}{t}\, dt, \qquad x > 0, \tag{1}$$

and $y = \ln x$ is a differentiable function for all x in the interval $(0, \infty)$ with

$$\frac{d}{dx}\ln x = \frac{1}{x}, \qquad x > 0. \tag{2}$$

Since a differentiable function is necessarily continuous, this shows that $f(x) = \ln x$ is continuous on $(0, \infty)$.

The geometric interpretation of the expression $y = \ln x$ is simple:

(a) If $x > 1$, $\ln x$ is the area of the region bounded by the graph of $f(t) = 1/t$ and the t-axis between $t = 1$ and $t = x$ (Figure 2.1).
(b) If $x < 1$, $\ln x$ is the *negative* of the area bounded by the graph of $f(t) = 1/t$ and the t-axis between $t = x$ and $t = 1$ (Figure 2.2).
(c) If $x = 1$, $\ln x = \int_1^1 (1/t)\, dt = 0$, so $\ln 1 = 0$.

REMARK   The property that $\ln 1 = 0$ is a property that the logarithm functions, $\log_b x$, discussed in Section 8.1 also satisfy. That is, for any $b > 0$, we know that $\log_b 1 = 0$ since $b^0 = 1$. We shall soon see that $\ln x$ satisfies other properties of $\log_b x$.

The value $\ln x$ can be approximated to any desired accuracy by applying one of the procedures for approximating integrals (the Midpoint Rule, the Trapezoidal Rule, Simpson's Rule, etc.—see Section 6.6) to the integral in equation (1). In fact, in Section 6.6, we obtained approximations to $\ln x$ for $x = 2, 3, 4, \ldots, 10$ that are accurate to five decimal places (see Tables 6.2 and 6.3 in Section 6.6). A table of values for $\ln x$ also appears in Appendix VI, and most scientific calculators and computers are programmed to provide values of $\ln x$. *Note that $\ln x$ is not defined for $x \le 0$.*

If u is a differentiable function of x, we may use the Chain Rule to generalize the differentiation formula of equation (2) to:

$$\frac{d}{dx}\ln u = \frac{1}{u}\frac{du}{dx}, \qquad u > 0. \tag{3}$$

□ **EXAMPLE 1**

Find $f'(x)$ for (a) $f(x) = \ln 3x$, and (b) $f(x) = x\ln(1 + x^2)$.

Solution

(a) $f'(x) = \dfrac{d}{dx}\ln 3x = \dfrac{1}{3x}\cdot 3 = \dfrac{1}{x}$.

(b) By the Product Rule and equation (3) we obtain

$$f'(x) = \frac{d}{dx}[x\ln(1 + x^2)] = (1)\ln(1 + x^2) + x\cdot\frac{d}{dx}[\ln(1 + x^2)]$$

$$= \ln(1 + x^2) + x\left(\frac{1}{1 + x^2}\right)\left[\frac{d}{dx}(1 + x^2)\right]$$

$$= \ln(1 + x^2) + x\left(\frac{1}{1 + x^2}\right)(2x)$$

$$= \ln(1 + x^2) + \frac{2x^2}{1 + x^2}. \qquad\blacksquare$$

But what does the function defined by equation (1) have to do with logarithms? The answer is provided by the following theorem, which shows that *the function $y = \ln x$ satisfies the logarithm properties (L1)–(L3) introduced in Section 8.1.*

Theorem 5	The natural logarithm function satisfies the following properties for all real numbers a, b, and r with $a > 0$, $b > 0$, and r rational:

(L1) $\ln(ab) = \ln a + \ln b$,

(L2) $\ln\left(\dfrac{a}{b}\right) = \ln a - \ln b$,

(L3) $\ln(a^r) = r\ln a$.

We discuss property (L3) in the case where r is irrational in Section 8.4.

The proof of Theorem 5 uses Theorem 1, Chapter 5, which states that *if two functions have the same derivative, they differ at most by a constant.* That is, if $f'(x) = g'(x)$, then $f(x) = g(x) + C$ for some constant C. We prove only property (L1), leaving (L2) and (L3) as exercises.

Proof of (L1): Let f and g be the functions

$$f(x) = \ln ax \quad\text{and}\quad g(x) = \ln x.$$

Then, by equation (3),

$$f'(x) = \frac{1}{ax}\cdot a = \frac{1}{x} = g'(x), \qquad x > 0.$$

Thus, by Theorem 1 of Chapter 5,

$$f(x) = g(x) + C, \qquad x > 0,$$

for some constant C, that is,

$$\ln ax = \ln x + C, \qquad x > 0. \tag{4}$$

To determine the constant C, we use the fact that $\ln 1 = 0$. Setting $x = 1$ in equation (4) gives

$$\ln a = 0 + C,$$

so $C = \ln a$. Equation (4), with $x = b$, now gives the desired result. ∎

The point of Theorem 5 is that the natural logarithm function satisfies the properties of a logarithmic function. Moreover, $\ln x$ is defined for *all* positive, real numbers, both rational and irrational. In Section 8.4 we will show that $\ln x$ can actually be written in the customary form, $\log_e x$, for a particular base number e, and we will justify the terminology *natural* logarithm. The remainder of this section concerns additional properties of $\ln x$.

Graph of $y = \ln x$

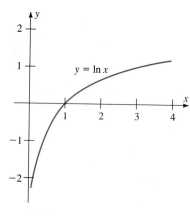

Figure 2.3 Graph of the natural logarithm function.

The graph of the natural logarithm function appears in Figure 2.3. We obtain it either by using numerical approximation techniques (see the comments following Example 6 in Section 6.3) or by using the curve sketching procedure discussed in Chapter 4. Note that the graph has the following properties:

(a) The domain of $y = \ln x$ is $(0, \infty)$.
(b) The function $y = \ln x$ is increasing for all x in its domain. (This follows from the fact that its derivative $1/x$ is positive for $x > 0$.)
(c) The graph of $y = \ln x$ is concave down on $(0, \infty)$. (The second derivative is

$$\frac{d^2}{dx^2} \ln x = \frac{d}{dx}\left(\frac{1}{x}\right) = -\frac{1}{x^2},$$

which is negative for all $x > 0$.)
(d) $\ln(1) = \displaystyle\int_1^1 \frac{1}{t}\, dt = 0.$
(e) $\displaystyle\lim_{x \to \infty} \ln x = +\infty$; $\displaystyle\lim_{x \to 0^+} \ln x = -\infty$. Thus the *range* of $f(x) = \ln x$ is $(-\infty, \infty)$.

(To see the first part of statement (e), we use the fact that $\ln x$ is an increasing function for all x. Thus $\ln 2 > \ln 1 = 0$. Since $\ln 2 > 0$,

$$\lim_{n \to \infty} \ln 2^n = \lim_{n \to \infty} (n \cdot \ln 2) = +\infty$$

for any integer n. Since $\ln x$ is an increasing function, it follows that

$$\lim_{x \to \infty} \ln x = +\infty.$$

The proof that $\displaystyle\lim_{x \to 0^+} \ln x = -\infty$ is similar).

The Number e

Since $y = \ln x$ is a continuous, increasing function whose range is the entire real line, the Intermediate Value Theorem guarantees that there is precisely one number x for which $\ln x = 1$. We denote this number by the letter e, that is, we define the number e by the equation

$$\boxed{\ln e = 1.} \tag{5}$$

The number e has been shown to be an irrational number, and the decimal expansion for e correct to 12 decimal places is known to be

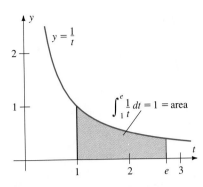

Figure 2.4 The area under $y = 1/t$ for $1 \le t \le e$ is 1.

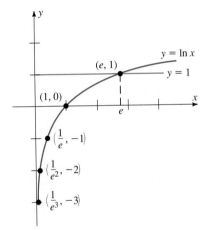

Figure 2.5 Points (e^n, n) on the graph of $y = \ln x$.

Natural Logarithms as Antiderivatives

$$e \approx 2.718281828459.$$

Geometrically speaking, equation (5) says that e is the unique real number greater than 1 such that the area of region bounded by the x-axis, the graph of $y = 1/x$, the line $x = 1$, and the line $x = e$ is 1 (see Figure 2.4).

In the next two sections, we shall see that the number e is the "natural" choice as a base for both the logarithmic and exponential functions. Note also that, by property (L3) and equation (5), we have $\ln e^n = n \cdot \ln e = n \cdot 1 = n$ for all integers n:

$$\ln e^n = n.$$

This equation enables us to locate points on the graph of $y = \ln x$ of the form

$$(e^n, \ln e^n) = (e^n, n), \qquad n = 0, \pm 1, \pm 2, \ldots.$$

(See Table 2.1 and Figure 2.5.)

Table 2.1

n	$x = e^n$	$y = n \cdot \ln e$
-2	$1/e^2 \approx 0.13534$	-2
-1	$1/e \approx 0.36788$	-1
0	$e^0 = 1$	0
1	$e \approx 2.71828$	1
2	$e^2 \approx 7.38906$	2

From the differentiation formula (2), we have the integration formula

$$\int \frac{1}{x} \, dx = \ln |x| + C, \qquad x \ne 0. \tag{6}$$

To see that equation (6) is valid for the case $x < 0$, recall that, if $x < 0$, $|x| = -x$. Thus

$$\frac{d}{dx} \ln |x| = \frac{d}{dx} \ln(-x) = \frac{1}{-x}(-1) = \frac{1}{x},$$

as required.

☐ **EXAMPLE 2**

Find $\displaystyle\int \frac{1}{2x + 3} \, dx$.

Solution We use the substitution

$$u = 2x + 3; \qquad du = \left[\frac{d}{dx}(2x + 3) \right] dx = 2 \, dx.$$

Then $dx = \frac{1}{2} du$, and we may apply equation (6) together with this substitution to obtain

$$\int \frac{1}{2x + 3}\, dx = \int \frac{1}{u}\left(\frac{1}{2}\right) du$$

$$= \frac{1}{2} \int \frac{1}{u}\, du$$

$$= \frac{1}{2} \ln |u| + C$$

$$= \frac{1}{2} \ln |2x + 3| + C. \qquad \blacksquare$$

Using property (L3), we can rewrite the indefinite integral in Example 2 as

$$\int \frac{1}{2x + 3}\, dx = \ln |2x + 3|^{1/2} + C = \ln\sqrt{|2x + 3|} + C,$$

if we so desire.

□ **EXAMPLE 3**

Find $\displaystyle \int \frac{1 + \cos x}{x + \sin x}\, dx, \qquad x + \sin x \neq 0.$

Solution The numerator of the integrand appears to be the derivative of the denominator, so we try the substitution.

$$u = x + \sin x; \qquad du = (1 + \cos x)\, dx.$$

Then

$$\int \frac{1 + \cos x}{x + \sin x}\, dx = \int \frac{1}{u}\, du$$

$$= \ln |u| + C$$

$$= \ln |x + \sin x| + C. \qquad \blacksquare$$

□ **EXAMPLE 4**

Find $\displaystyle \int \frac{x + 1}{x^2 + 2x}\, dx.$

Solution Note that the degree of the numerator is one less than the degree of the denominator. Accordingly, we make the substitution

$$u = x^2 + 2x.$$

Then

$$du = (2x + 2)\, dx = 2(x + 1)\, dx,$$

so

$$(x + 1)\, dx = \frac{1}{2}\, du.$$

With these substitutions we obtain

$$\int \frac{(x+1)\,dx}{x^2+2x} = \int \frac{1}{u}\left(\frac{1}{2}\right)du$$

$$= \frac{1}{2}\int \frac{1}{u}\,du$$

$$= \frac{1}{2}\ln|u| + C$$

$$= \frac{1}{2}\ln|x^2+2x| + C. \qquad \blacksquare$$

☐ **EXAMPLE 5**

Find $\displaystyle\int \frac{x^2+3x}{x+1}\,dx, \qquad x \neq -1.$

Strategy · · · · · · ·
The integrand is an improper fraction. Perform a division to express the integrand as the sum of a polynomial and a constant divided by $x+1$.

Solution
We may write the integrand as

$$\frac{x^2+3x}{x+1} = x+2 - \frac{2}{x+1},$$

so

$$\int \frac{x^2+3x}{x+1}\,dx = \int \left[x+2 - \frac{2}{x+1}\right]dx$$

Apply formula 6 in last term with

$u = x+1, \qquad du = dx.$

$$= \int x\,dx + 2\int dx - 2\int \frac{1}{x+1}\,dx$$

$$= \frac{x^2}{2} + 2x - 2\ln|x+1| + C. \qquad \blacksquare$$

☐ **EXAMPLE 6**

Find the area of the region bounded by the graph of $y = x - \dfrac{1}{x}$ and the x-axis between $x = 1$ and $x = e$ (see Figure 2.6).

Strategy · · · · · · ·
Determine where $f(x) = x - (1/x)$ is nonnegative.

Solution
Since $x \geq 1/x$ for $x \geq 1$, the function $y = x - (1/x)$ is nonnegative for $1 \leq x \leq e$. The area A is therefore

Use formula

$$A = \int_a^b f(x)\,dx.$$

Use facts that $\ln e = 1$ and $\ln 1 = 0$.

$$A = \int_1^e \left(x - \frac{1}{x}\right)dx$$

$$= \left[\frac{x^2}{2} - \ln x\right]_1^e$$

$$= \left(\frac{e^2}{2} - \ln e\right) - \left(\frac{1}{2} - \ln 1\right)$$

$$= \left(\frac{e^2}{2} - 1\right) - \left(\frac{1}{2} - 0\right)$$

$$= \frac{e^2}{2} - \frac{3}{2} \approx 2.19. \qquad \blacksquare$$

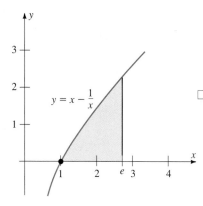

Figure 2.6 Graph of $y = x - 1/x$.

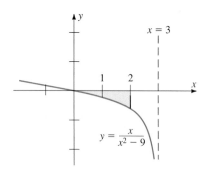

Figure 2.7 Graph of $y = \dfrac{x}{x^2 - 9}$.

In the next example notice how we must take care to observe the absolute value signs in equation (6). The value of the integral is negative, since the function is negative on the interval $(0, 2]$ (see Figure 2.7).

☐ **EXAMPLE 7**

Find $\displaystyle\int_0^2 \frac{x}{x^2 - 9}\, dx$.

Solution We use the substitution

$$u = x^2 - 9; \qquad du = 2x\, dx.$$

Then $x\, dx = \frac{1}{2}\, du$, and the limits of integration change as follows:

$$\text{If } x = 0, \quad u = 0^2 - 9 = -9.$$
$$\text{If } x = 2, \quad u = 2^2 - 9 = -5.$$

Then

$$\int_0^2 \frac{x}{x^2 - 9}\, dx = \int_{-9}^{-5} \frac{1}{u}\frac{1}{2}\, du = \frac{1}{2}\int_{-9}^{-5} \frac{1}{u}\, du$$

$$= \frac{1}{2}\left[\ln|u|\right]_{-9}^{-5}$$

$$= \frac{1}{2}\left[\ln|-5| - \ln|-9|\right]$$

$$= \frac{1}{2}(\ln 5 - \ln 9)$$

$$\approx -0.294. \qquad \blacksquare$$

The next example involves a natural logarithm, although this may not be obvious at first glance.

☐ **EXAMPLE 8**

Find $\displaystyle\int \frac{dx}{\sqrt{x}(1 + \sqrt{x})}$.

Solution We try the substitution

$$u = 1 + \sqrt{x}; \qquad du = \frac{1}{2\sqrt{x}}\, dx.$$

Then $\dfrac{dx}{\sqrt{x}} = 2\, du$, and we obtain

$$\int \frac{dx}{\sqrt{x}(1 + \sqrt{x})} = \int \frac{1}{u}\, 2\, du = 2\int \frac{1}{u}\, du$$

$$= 2\ln|u| + C$$

$$= 2\ln(1 + \sqrt{x}) + C.$$

$(|1 + \sqrt{x}| = 1 + \sqrt{x}$ since $\sqrt{x} \geq 0$.) $\qquad\qquad \blacksquare$

Exercise Set 8.2

1. Use properties of ln x to solve for x. (You will need to use the table in Appendix VI or a calculator.)

   a. ln $2x = 0$
   b. 2 ln $x =$ ln $2x$
   c. ln$(2/x) -$ ln $x = 0$
   d. ln $3^{x^2} = 0$
   e. $\sqrt{\ln \sqrt{x}} = 1$
   f. 3 ln $x + x = 2 +$ ln x^3
   g. $\int_1^x \dfrac{3}{t} dt = 6$
   h. $\int_2^x \dfrac{1}{t} dt = 0$

2. Why may we conclude that the function $y =$ ln x is continuous?

3. True or false? Every number is a natural logarithm.

In Exercises 4–19, find the derivative.

4. $y =$ ln $2x$

5. $y = x$ ln x

6. $y =$ ln$(6 - x^2)$

7. $f(x) =$ ln $\sqrt{x^3 - x}$, $x > 1$

8. $f(t) =$ ln$($ln $t)$, $t > 1$

9. $y =$ sin$($ln $x)$

10. $h(t) = \sqrt{1 +$ ln $t}$

11. $f(x) = \dfrac{x}{1 + \ln x}$

12. $f(x) = x^2$ ln$^2 x$

13. $y = (3$ ln $\sqrt{x})^4$

14. $f(x) = ($sin $x)$ ln$(1 + \sqrt{x})$

15. $f(t) = \dfrac{\ln(a + bt)}{\ln(c + dt)}$

16. $y =$ ln$($sin $t - t$ cos $t)$

17. $f(x) =$ ln cos x

18. $y = x$ ln$(x^2 - x - 3)$, $x > 3$

19. $u(t) = \dfrac{t}{1 + \ln^2 t}$

In Exercises 20–23, find $\dfrac{dy}{dx}$ by implicit differentiation.

20. $y =$ ln(xy)

21. ln$(x + y) +$ ln$(x - y) = 1$

22. x ln $y = 3$

23. x ln $y + y$ ln $x = x$

In Exercises 24–26, find $\dfrac{dy}{dx}$.

24. $y = \int_5^x \dfrac{1}{t} dt$

25. $y = \int_x^1 \dfrac{1}{2t} dt$

26. $y = \int_1^{x^2}$ ln $2t \, dt$

In Exercises 27–30, sketch the graph of the given function, noting all relative extrema.

27. $y = x -$ ln x, $x > 0$

28. ln$(xy) = 1$

29. $y =$ ln$(2 +$ sin $x)$

30. $y =$ ln$(1 + x^2)$

In Exercises 31–53, evaluate the integral.

31. $\displaystyle\int \dfrac{dx}{x + 3}$

32. $\displaystyle\int \dfrac{dx}{2x + 1}$

33. $\displaystyle\int \dfrac{dx}{1 - x}$

34. $\displaystyle\int \dfrac{x \, dx}{x^2 + 1}$

35. $\displaystyle\int \dfrac{x - 1}{x^2 - 2x} dx$

36. $\displaystyle\int \dfrac{x}{1 - 3x^2} dx$

37. $\displaystyle\int \dfrac{x^2 + 3}{x^3 + 9x} dx$

38. $\displaystyle\int$ cot $x \, dx$

39. $\displaystyle\int \dfrac{\sin t \, dt}{4 + 2 \cos t}$

40. $\displaystyle\int \dfrac{dx}{x \ln x}$

41. $\displaystyle\int \dfrac{\ln^2 x}{x} dx$

42. $\displaystyle\int \dfrac{x^2}{x + 1} dx$

43. $\displaystyle\int \dfrac{1}{\sqrt{x}(1 - \sqrt{x})} dx$

44. $\displaystyle\int \dfrac{1 - 2t^2}{1 - t} dt$

45. $\displaystyle\int \dfrac{x^4 + 3x^2 + x + 1}{x + 1} dx$

46. $\displaystyle\int_1^e \dfrac{1}{x} dx$

47. $\displaystyle\int_e^{e^2} \dfrac{1}{x \ln x} dx$

48. $\displaystyle\int_1^e \dfrac{\ln x}{x} dx$

49. $\displaystyle\int_e^{e^2} \dfrac{1}{x \ln(x^2)} dx$

50. $\displaystyle\int_2^3 \dfrac{x}{x^2 + 1} dx$

51. $\displaystyle\int_1^2 \left(\dfrac{1}{1 + x} - \dfrac{1}{2 + x} \right) dx$

52. $\displaystyle\int_1^2 \dfrac{x^2 + 1}{x^3 + 3x} dx$

53. $\displaystyle\int_0^{\pi/3}$ tan $x \, dx$

54. Find the equation of the line tangent to the graph of the equation

$$y = x(\ln x)^2 + \dfrac{x}{\ln x}$$

at the point $(e, 2e)$.

55. Find an equation for the line tangent to the graph $y = x^3$ ln$^2 x$ at $x = 1$.

56. Find the area of the region bounded by the graph of $y = 1/(x - 2)$ and the x-axis for $3 \le x \le 4$.

57. Find the area of the region in the first quadrant bounded by the graphs of $x + y - 6 = 0$ and $xy = 8$.

58. Let ϵ be a small real number.
   a. Using linear approximation, obtain a formula for approximating the quantity ln$(1 + \epsilon)$.
   b. Use the formula obtained in part a to approximate ln(0.8), ln(0.95), ln(1.05), and ln(1.2).

c. Compare results obtained in part (b) with those obtained from a calculator or the table in Appendix VI.

59. True or false? For a given value of $x > 0$, the graphs of $y = \ln x$ and $y = \ln ax$ have the same slope.

60. Find the volume of the solid generated by rotating about the x-axis the region bounded by the graph of

$$y = \sqrt{\frac{2}{x - 1}}$$

and the x-axis, for $3 \le x \le 5$.

61. Find the volume of the solid generated by rotating about the x-axis the region bounded by the graph

$$y = \frac{\ln x}{\sqrt{x}}$$

and the x-axis, for $1 \le x \le 2$.

62. Verify that the function $F(x) = x \ln x - x + C$ is an antiderivative for $f(x) = \ln x$, $x > 0$; that is, show that

$$\int \ln x \, dx = x \ln x - x + C, \qquad x > 0.$$

63. Use the result of Exercise 62 to find the average value of the natural logarithm function on the interval $[1, e]$.

64. A particle moves along a line with acceleration $a(t) = 1/(t + 2) \text{ m/s}^2$. Find the distance travelled by the particle during the time interval $[0, 4]$ if $v(0) = 0$. (*Hint:* Use Exercise 62.)

In Exercises 65–69, solve the differential equation.

65. $\dfrac{dy}{dx} = \dfrac{1}{x}, \quad x > 0$

66. $(x^2 + 3) \, dy - 2x \, dx = 0$

67. $\cos x \, dy = \sin x \, dx$

68. $y^{-1} \, dx = x \, dy$

69. $(2x + 1) \, dy = y^2 \, dx$

70. Economists define the **growth of a function** $y = f(t)$ as the ratio

$$G = \frac{\dfrac{dy}{dt}}{y} = \frac{f'(t)}{f(t)} = \frac{y'}{y}.$$

For example, if $f(2) = 6$ and $f'(2) = 3$, then the growth of the function $y = f(t)$ at time $t = 2$ is $G = 3/6 = 0.5$.
a. Show that the growth of a function may be calculated as the derivative of the natural logarithm of that function, that is, $G = \dfrac{d}{dt} \ln(f(t))$.

b. Show that the growth of the product of two functions is the sum of the individual growths.

c. An oil company determines that the price it obtains for heating oil is increasing at a rate of 15% per year but that the number of gallons of heating oil sold is decreasing by 10% per year. Find the rate at which revenues obtained from the sale of heating oil are increasing.

71. The vapor pressure P, in mm, of a certain fluid is related to its temperature T by the equation

$$\ln P = \frac{-2000}{T} + 5.5.$$

a. Find $\dfrac{dP}{dT}$.

b. Find $\dfrac{dT}{dP}$.

72. Prove statement (*L2*) of Theorem 5.

73. Prove statement (*L3*) of Theorem 5 for the case r rational. (The case r irrational must await the extension of the Power Rule $\dfrac{d}{dx} x^r = r x^{r-1}$ to irrational exponents. This is done in Section 8.4.)

74. Establish the inequality $2 \le e \le 4$ by explaining how each of the following statements leads to the next.
a. For $1 \le x \le 2$, $\quad 1/2 \le 1/x \le 1$
b. $\displaystyle \int_1^2 \frac{1}{2} \, dx \le \int_1^2 \frac{1}{x} \, dx \le \int_1^2 1 \, dx$
c. $1/2 \le \ln 2 \le 1$
d. $1 \le 2 \ln 2 \le 2$
e. $\ln e \le \ln 4 \le \ln e^2$
f. $e \le 4 \le e^2$
(*Hint:* Use the fact that $f(x) = \ln x$ is an increasing function.)
g. $2 \le e \le 4$

75. Approximate $\ln 5$ using the Midpoint Rule with $n = 8$ subdivisions.

76. Approximate $\ln 5$ using the Trapezoidal Rule with $n = 8$ subdivisions.

77. How do the results of Exercises 75 and 76 relate to the precise value of $\ln 5$? Justify your answer without the use of a calculator or a table of values of the natural logarithm function. (*Hint:* Use Theorem 10 in Chapter 6.)

78. Using the Midpoint Rule and the Trapezoidal Rule with $n = 6$ subdivisions, show that $2.7 < e < 2.8$. (*Hint:* Use Theorem 10 in Chapter 6.)

In Exercises 79–81, use your graphing calculator to graph the function f and its derivative f'. Determine the domain and range of f as well as its relative extrema.

79. $f(x) = (\ln x)/x$

80. $f(x) = \ln\left(x + \dfrac{1}{x}\right)$

81. $f(x) = (x^2 - 1) \ln(1 + |x|) - x^2$

82. Using the velocity function v found in Exercise 39 of Section 5.3, determine an exact solution to the differential equation

$$\frac{ds}{dt} = v, \qquad s(0) = 0.$$

Graph the solution s. Compare this to the graph that you obtained using Euler's Method in Chapter 5.

8.3   The Natural Exponential Function

In Section 8.2 we saw that the natural logarithm function is an increasing function for all $x \in (0, \infty)$ with range $(-\infty, \infty)$. These conditions are sufficient to guarantee the existence of the *inverse* of the natural logarithm function (see Theorem 2). We denote this function initially by $\exp(x)$. In other words, we *define* the exponential function as follows:

$$y = \exp(x) \quad \text{if and only if} \quad x = \ln y. \tag{1}$$

(Note that statement (1) implies that $y > 0$.) The definition embodied in statement (1) says that "the value of the exponential function $\exp(x)$ is the number y whose natural logarithm is x."

Now recall from Section 8.1 that if the function g is the inverse of the function f then f is also the inverse of g on the range of f. Thus, $y = \ln x$ is the inverse of $y = \exp(x)$, and we have the identities

$$\exp(\ln x) = x, \quad x > 0 \quad \text{and} \quad \ln(\exp(x)) = x, \quad -\infty < x < \infty.$$

The Graph of $y = \exp(x)$

Since $g(x) = \exp(x)$ is the inverse of $f(x) = \ln x$, the graph of $y = \exp(x)$ is the reflection across the line $y = x$ of the graph of $y = \ln x$. It therefore has the following properties (Figure 3.1):

1. The domain of $y = \exp(x)$ is $(-\infty, \infty)$ because this is the range of $y = \ln x$; that is, the function $y = \exp(x)$ is defined for all numbers.
2. The range of $y = \exp(x)$ is $(0, \infty)$ because the domain of $y = \ln x$ is $(0, \infty)$.
3. Because all points (e^n, n), $n = 0, \pm 1, \pm 2, \ldots$ lie on the graph of $y = \ln x$, the points (n, e^n), $n = 0, \pm 1, \pm 2, \ldots$ lie on the graph of the exponential function.
4. Because $y = \ln x$ is increasing for all $x \in (0, \infty)$, the function $y = \exp(x)$ is increasing for all $x \in (-\infty, \infty)$ (see Exercise 76).

Figures 3.2 and 3.3 show graphs of various functions of the form $y = \exp(kx)$.

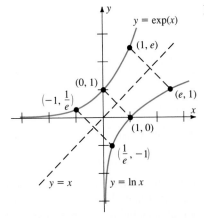

Figure 3.1 Function $y = \exp(x)$ is the inverse of the natural log function $y = \ln x$.

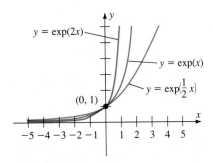

Figure 3.2 Graphs of $y = \exp(kx)$, $k = \frac{1}{2}, 1, 2$.

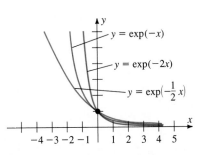

Figure 3.3 Graphs of $y = \exp(kx)$, $k = -2, -1, -\frac{1}{2}$.

Exp(x) is an Exponential Function

At this point we can show that the function $\exp(x)$ is really an exponential function, in the usual sense. In particular, for any rational number x, we have

$$\boxed{\exp(x) = e^x} \tag{2}$$

where e is the number defined by the equation $\ln e = 1$. In fact, applying Theorem 5 to $\ln(e^x)$, we get

$$\ln(e^x) = x \cdot \ln e = x \cdot 1 = x.$$

If we "exponentiate" both sides, we get

$$\exp(\ln(e^x)) = \exp(x),$$

and using the fact that the natural exponential is the inverse of the natural logarithm, we simplify the left-hand side and obtain

$$e^x = \exp(x).$$

From now on we will use the notation e^x rather than $\exp(x)$ to denote the natural exponential function. However, you should keep in mind the fact that statements (1) and (2) together define e^x for *all* real exponents x, both rational and irrational.

The following theorem shows that the function e^x satisfies the characteristic properties of exponential functions for *all* numbers x.

Theorem 6

Let x_1, x_2, and r be any real numbers with r rational. Then

(i) $e^{x_1} \cdot e^{x_2} = e^{x_1 + x_2}$,

(ii) $\dfrac{e^{x_1}}{e^{x_2}} = e^{x_1 - x_2}$,

(iii) $[e^{x_1}]^r = e^{r x_1}$, r rational.

We discuss property (iii) in the case where r is irrational in Section 8.4.

Proof: Let $y_1 = e^{x_1}$ and $y_2 = e^{x_2}$. Then

$$\ln y_1 = x_1 \quad \text{and} \quad \ln y_2 = x_2.$$

By property ($L1$), Section 8.2,

$$x_1 + x_2 = \ln y_1 + \ln y_2 = \ln(y_1 y_2).$$

Thus

$$e^{x_1 + x_2} = e^{\ln(y_1 y_2)} = y_1 y_2 = e^{x_1} \cdot e^{x_2}$$

as required. This proves statement (i). The proofs of statements (ii) and (iii) are similar, and are left as exercises. ∎

The Derivative of $y = e^x$

In asking for the derivative of the exponential function, we uncover one of the most remarkable facts of the calculus. Recall from Section 8.1 that, if the function f has an inverse g, and if f is differentiable in an interval containing x_0, then g is also differentiable at $y_0 = f(x_0)$ if $f'(x_0) \neq 0$. Thus, since $y = \ln x$ is differentiable for all $x > 0$ and $dy/dx = 1/x \neq 0$, we may therefore conclude that its inverse, $y = e^x$, is differentiable for all x in $(-\infty, \infty)$, the range of $y = \ln x$.

To obtain the derivative of $y = e^x$, we apply the Chain Rule to the identity

$$\ln e^x = x, \qquad -\infty < x < \infty. \tag{3}$$

Differentiating both sides of equation (3) to obtain

$$\frac{1}{e^x} \frac{d}{dx} e^x = 1, \qquad -\infty < x < \infty. \tag{4}$$

Multiplying both sides of equation (4) by e^x gives

$$\frac{d}{dx} e^x = e^x. \tag{5}$$

In other words, the exponential function $y = e^x$ is its own derivative! This function, and its multiples, are the only functions in the calculus with this property. (In fact, we could develop the entire theory of exponential and logarithm functions by seeking a function f for which $f'(x) = f(x)$.) Figure 3.4 illustrates the geometric interpretation of formula (5): The slope of the tangent to the graph of $y = e^x$ at (x, e^x) is precisely the y-coordinate e^x.

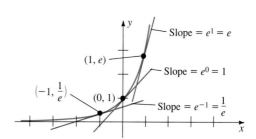

Figure 3.4   Slope of $y = e^x$ equals y-coordinate.

□ **EXAMPLE 1**

Find $\dfrac{dy}{dx}$ for (a) $y = e^{6x}$ and (b) $y = e^{x \sin x}$.

Solution

(a) Equation (5) and the Chain Rule give

$$\frac{d}{dx}(e^{6x}) = e^{6x} \frac{d}{dx}(6x) = 6e^{6x}.$$

(b) Similarly,

$$\frac{d}{dx}(e^{x \sin x}) = e^{x \sin x} \frac{d}{dx}(x \sin x)$$

$$= e^{x \sin x}[\sin x + x \cos x]. \qquad \blacksquare$$

□ **EXAMPLE 2**

Find all relative extrema for the function $f(x) = x^2 e^{-x}$.

Strategy · · · · · · · · · ·

Find f' using the Product Rule.

Solution

The first derivative is

$$f'(x) = \left(\frac{d}{dx} x^2\right) e^{-x} + x^2 \left(\frac{d}{dx} e^{-x}\right)$$
$$= 2xe^{-x} - x^2 e^{-x}$$
$$= x(2 - x)e^{-x}.$$

Set $f'(x) = 0$ and solve to find the critical numbers.

Since $e^{-x} = 1/e^x$ is nonzero for all x, the equation $f'(x) = 0$ gives $x = 0$ or $x = 2$. There are the critical numbers.

The second derivative is

Find f'' using the Product Rule.

$$f''(x) = \frac{d}{dx} (2xe^{-x} - x^2 e^{-x})$$
$$= (2e^{-x} - 2xe^{-x}) - (2xe^{-x} - x^2 e^{-x})$$
$$= (2 - 4x + x^2)e^{-x}.$$

Determine the sign of $f''(c)$ for each critical number c.

Using f'' to check the critical numbers for relative extrema gives

$$f''(0) = 2e^0 = 2 > 0$$

and

$$f''(2) = (2 - 8 + 4)e^{-2} = -\frac{2}{e^2} < 0.$$

Apply the Second Derivative Test for relative extrema.

Thus, the Second Derivative Test shows that $f(0) = 0$ is a relative minimum value for f, and $f(2) = 4e^{-2}$ is a relative maximum.

The graph of f appears in Figure 3.5. ■

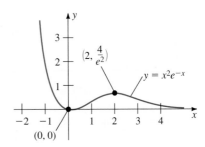

Figure 3.5 Graph of $f(x) = x^2 e^{-x}$ has relative minimum $f(0) = 0$ and relative maximum $f(2) = 4e^{-2}$.

Integrals Involving e^x

The differentiation formula (5) gives the integration formula

$$\int e^x \, dx = e^x + C. \tag{6}$$

☐ **EXAMPLE 3**

Find the following integrals:

(a) $\int e^{-3x}\, dx$

(b) $\int x^2 e^{x^3+1}\, dx.$

Strategy · · · · · · · ·
Make a substitution with

$$u = -3x$$
$$du = -3\, dx$$

Use equation (6).

Solution

(a) For $u = -3x$, $du = -3\, dx$, so $dx = -\frac{1}{3}\, du$. We may therefore write

$$\int e^{-3x}\, dx = \int e^u \left(-\frac{1}{3}\right) du$$

$$= -\frac{1}{3} \int e^u\, du$$

$$= -\frac{1}{3} e^u + C$$

$$= -\frac{1}{3} e^{-3x} + C.$$

Try to view the factor x^2 as a factor of du. This works for the substitution

$$u = x^3 + 1.$$

Use equation (6).

(b) If we use the substitution $u = x^3 + 1$, we have $du = 3x^2\, dx$, so $x^2\, dx = \frac{1}{3}\, du$. Thus

$$\int x^2 e^{x^3+1}\, dx = \int e^u \frac{1}{3}\, du$$

$$= \frac{1}{3} \int e^u\, du$$

$$= \frac{1}{3} e^u + C$$

$$= \frac{1}{3} e^{x^3+1} + C. \qquad \blacksquare$$

☐ **EXAMPLE 4**

Find the area of the region bounded by the graph of $y = xe^{1-x^2}$ and the x-axis for $-2 \le x \le 2$ (see Figure 3.6).

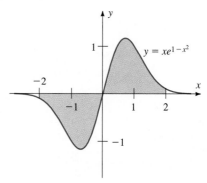

Figure 3.6   Graph of $y = xe^{1-x^2}$ showing region in Example 4.

Strategy · · · · · · · ·

Determine where graph lies above x-axis.

Area on $[-2, 2]$ equals twice the area on $[0, 2]$.

Find a substitution u so that x is a factor of du. The substitution $u = 1 - x^2$ works.

Change limits of integration according to u-substitution.

Apply equation (6).

Apply Fundamental Theorem of Calculus.

Solution

Since $y = xe^{1-x^2}$ is an odd function, its graph is symmetric about the origin. Because $xe^{1-x^2} \geq 0$ for $x > 0$, we may calculate the desired area A as

$$A = 2 \int_0^2 xe^{1-x^2} \, dx.$$

To evaluate this integral, we use the substitution

$$u = 1 - x^2, \qquad du = -2x \, dx.$$

Then

$$x \, dx = -\frac{1}{2} \, du$$

and the limits of integration become

$$u = 1 - 0^2 = 1 \quad \text{if } x = 0$$
$$u = 1 - 2^2 = -3 \quad \text{if } x = 2.$$

With these substitutions, we obtain

$$A = 2 \int_0^2 xe^{1-x^2} \, dx$$

$$= 2 \int_1^{-3} e^u \left(-\frac{1}{2} \right) \, du$$

$$= 2 \left(-\frac{1}{2} \right) \int_1^{-3} e^u \, du$$

$$= -\left[e^u \right]_1^{-3}$$

$$= -[e^{-3} - e]$$

$$= e - \frac{1}{e^3} \approx 2.67.$$

◼

More on the Number e

The number e is difficult to approximate from the defining property

$$\ln e = \int_1^e \frac{1}{x} \, dx = 1,$$

since it appears as a limit of integration. This same number arises in an apparently unrelated manner as the limit

$$e = \lim_{x \to \infty} \left(1 + \frac{1}{x} \right)^x. \tag{7}$$

It is much easier to approximate e as this limit, since we need only calculate values of the function $(1 + 1/x)^x$ for x large. This limit is a special case of the more general statement

$$e^r = \lim_{x \to \infty} \left(1 + \frac{r}{x}\right)^x, \qquad r \text{ constant,} \tag{8}$$

which we shall prove in Chapter 11 (a second proof is outlined in Exercise Set 8.4).

The results of approximating e by the expression $(1 + 1/n)^n$ for various values of n appear in Table 3.1.

Table 3.1 Approximating $e \approx 2.7182818 \ldots$, using $\left(1 + \dfrac{1}{n}\right)^n$

n	$\left(1 + \dfrac{1}{n}\right)^n$	n	$\left(1 + \dfrac{1}{n}\right)^n$
1	2.000000	500	2.715569
5	2.488320	1,000	2.716924
20	2.653298	2,500	2.717738
50	2.691588	5,000	2.718010
100	2.704814	10,000	2.718146
250	2.712865	100,000	2.718268

□ **EXAMPLE 5**

According to equation (8) we have

(a) $\displaystyle \lim_{x \to \infty} \left(1 - \frac{1}{x}\right)^x = e^{-1} = \frac{1}{e}$

(b) $\displaystyle \lim_{x \to \infty} \left(1 + \frac{3}{x}\right)^x = e^3$

(c) $\displaystyle \lim_{x \to \infty} \left(1 + \frac{6}{x}\right)^{2x} = \lim_{x \to \infty} \left[\left(1 + \frac{6}{x}\right)^x\right]^2 = (e^6)^2 = e^{12}$

(d) $\displaystyle \lim_{x \to \infty} \left(1 - \frac{1}{x^2}\right)^x = \lim_{x \to \infty} \left[\left(1 - \frac{1}{x}\right)\left(1 + \frac{1}{x}\right)\right]^x$

$\displaystyle \qquad = \lim_{x \to \infty} \left[\left(1 - \frac{1}{x}\right)^x \left(1 + \frac{1}{x}\right)^x\right]$

$\displaystyle \qquad = e^{-1} e^1 = 1.$ ■

Exercise Set 8.3

1. Simplify the following expressions
   a. $e^{\ln 2}$
   b. $e^{-\ln 4}$
   c. $e^{(\ln x - \ln y)}$
   d. $\ln e^{-x^2}$
   e. $\ln xe^{\sqrt{x}} - \ln x$
   f. $e^{x \ln 2}$
   g. $e^{\ln(1/x)}$
   h. $e^{4 \ln x}$
   i. $\ln xe^{x^2}$
   j. $e^{x - \ln x}$

2. Solve the following equations for x.
   a. $\ln x = 2$
   b. $\ln x^2 = 9$
   c. $e^{x^2} = 5$
   d. $e^{2x} - 2e^x + 1 = 0$

3. Find y if $e^{x-y} = x + 3$.

4. Find y if $e^{(y-1)^2} = x^4 + 1$, $y > 1$.

In Exercises 5–20, find the derivative of the given function.

5. $y = e^{3x}$

6. $f(x) = xe^{-x}$

7. $f(x) = e^{\sqrt{t}}$

8. $y = \dfrac{e^x + 1}{e^x}$

9. $y = e^{x^2 - x}$

10. $f(t) = e^{\sin t}$

11. $f(x) = e^x \sin x$

12. $y = xe^{-3 \ln x}$

13. $f(x) = \ln \dfrac{e^x + 1}{x + 1}$

14. $f(x) = \dfrac{e^x - 1}{e^x + 1}$

15. $f(x) = (2 - e^{x^2})^3$

16. $y = x \ln(e^x + x)$

17. $y = \frac{1}{2}(e^x + e^{-x})$

18. $f(x) = (x^2 + x - 1)e^{x^2 + 3}$

19. $y = e^{\sqrt{x}} \ln \sqrt{x}$

20. $f(t) = te^{(1/t)^2}$

In Exercises 21–24, find $\dfrac{dy}{dx}$ by implicit differentiation.

21. $e^{xy} = x$

22. $e^{x-y} = xy$

23. $\ln(x + 2y) = e^y$

24. $y^2 e^x + y \ln x = 2$

25. True or false? For $y = e^{kx}$, y' is proportional to y.

26. True or false? The equation $y = e^x$ has a solution x for every $y \geq 0$.

27. True or false? You can *always* divide by e^x.

In Exercises 28–41, evaluate the given indefinite integrals.

28. $\displaystyle \int e^{2x} \, dx$

29. $\displaystyle \int e^{-x} \, dx$

30. $\displaystyle \int e^{2x+6} \, dx$

31. $\displaystyle \int xe^{x^2+3} \, dx$

32. $\displaystyle \int x^2 e^{1-x^3} \, dx$

33. $\displaystyle \int e^{2x}(1 + e^{2x})^3 \, dx$

34. $\displaystyle \int \dfrac{e^{\sqrt{x}}}{\sqrt{x}} \, dx$

35. $\displaystyle \int \dfrac{e^{1/x}}{x^2} \, dx$

36. $\displaystyle \int \dfrac{e^x}{\sqrt{e^x + 1}} \, dx$

37. $\displaystyle \int \dfrac{e^x}{1 + e^x} \, dx$

38. $\displaystyle \int \cos x \, e^{\sin x} \, dx$

39. $\displaystyle \int \dfrac{e^x}{(3 + e^x)^2} \, dx$

40. $\displaystyle \int \dfrac{1 + e^{-ax}}{1 - e^{-ax}} \, dx$

41. $\displaystyle \int \dfrac{(1 + e^{\sqrt{x}})e^{\sqrt{x}}}{\sqrt{x}} \, dx$

In Exercises 42–51, evaluate the definite integrals.

42. $\displaystyle \int_0^1 e^{2x} \, dx$

43. $\displaystyle \int_1^{\ln 4} e^{-x} \, dx$

44. $\displaystyle \int_0^1 xe^{1-x^2} \, dx$

45. $\displaystyle \int_0^{\pi/2} \cos x \, e^{\sin x} \, dx$

46. $\displaystyle \int_0^{\ln 2} \dfrac{e^x}{2 + e^x} \, dx$

47. $\displaystyle \int_1^{e^2} \dfrac{\ln x}{x} \, dx$

48. $\displaystyle \int_0^2 \dfrac{e^x + e^{-x}}{2} \, dx$

49. $\displaystyle \int_0^2 e^x \, dx$

50. $\displaystyle \int_0^{\pi/4} \sec^2 x \, e^{\tan x} \, dx$

51. $\displaystyle \int_0^{\pi/3} \sin x \cos x \, e^{\sin^2 x} \, dx$

52. Find the maximum value of the function $y = (3 - x^2)e^x$.

53. Find all relative extrema for the function $y = x^2 e^{1-x^2}$.

54. Find the area of the region bounded by the graph of $y = \ln x$ and the lines $y = 0$, $y = 1$ and the y-axis. (*Hint:* Integrate with respect to y.)

55. For the function $f(x) = xe^{2x}$, find
   a. the interval(s) on which f is increasing,
   b. the interval(s) on which f is decreasing,
   c. all relative extrema,
   d. the interval(s) on which the graph of f is concave up,
   e. the interval(s) on which the graph of f is concave down,
   f. the point(s) of inflection.
   Use this information to sketch the graph of f.

56. Apply the instructions of Exercise 55 to the function $f(x) = xe^{1-x^3}$.

57. Find the length of the graph of $y = \frac{1}{2}(e^x + e^{-x})$ from $(0, 1)$ to $(\ln 2, 5/4)$.

58. Find the lateral surface of the solid obtained by revolving about the x-axis the region bounded by the arc of the graph in Exercise 57 and the x-axis.

59. Solve the differential equation $\dfrac{dy}{dx} = 2xy$.

60. Determine c, $c > 4$, so that $\displaystyle \int_4^c \dfrac{1}{x - 3} \, dx = 1$.

61. The line $y = -1/e$ is tangent to the graph of $y = xe^x$ at the point P. Find P.

62. Find an equation for the line tangent to the graph

$$y = \dfrac{e^x + 1}{e^{3x}}$$

at $x = \ln 2$.

63. Find the area of the region bounded by the graphs of $y = e^x$ and $y = e^{-x}$ for $-1 \leq x \leq 1$.

64. Find the volume of the solid obtained by revolving about the x-axis the region bounded by the graph of $y = e^{2x}$ and the x-axis for $0 \leq x \leq 2$.

65. Show that the function $y = Ae^{kt}$ satisfies the differential equation $y' = ky$ and the initial condition $y(0) = A$. Use this information to solve the following *initial value problems:*
   a. $y' = y$, $y(0) = 1$
   b. $y' = \pi y$, $y(0) = -2$
   c. $y' + 3y = 0$, $y(0) = 2$

66. Find the average value of the function $y = xe^{1-x^2}$ on the interval $[-2, 2]$.

67. Show that $f(x) = (e^x + e^{-x})/2$ is an even function, and that $g(x) = (e^x - e^{-x})/2$ is an odd function.

68. Consider the equation $e^{x-y} = ye^x$.
   a. Using implicit differentiation, show that
   $$\frac{dy}{dx} = \frac{e^{-y} - y}{1 + e^{-y}}.$$
   b. Using part a and the original equation, show that $dy/dx = 0$ for all x. What does this say about y?
   c. Using properties of the exponential function, simplify the original equation and derive the result of part b without differentiating.

69. The Beer-Lambert law states that the intensity of light, after passing through a thickness l of absorbing liquid, is $I = I_0 e^{-kl}$ where I_0 and k are constants. Find k if $dI/dl = 4I$.

70. The function
$$f(x) = \frac{1}{\sigma\sqrt{2\pi}} e^{-(x-\mu)^2/2\sigma^2}$$
is called the *normal* probability density function with mean μ and variance σ^2.
   a. Show that the graph of f is symmetric about the line $x = \mu$.
   b. Show that inflection points exist for $x = \mu \pm \sigma$.
   c. Graph f, obtaining the familiar bell-shaped curve.

71. Human growth in height from age one year to adulthood typically looks like the graph in Figure 3.7. A function that has been used to model this kind of growth is the **double logistic**
$$y(t) = \frac{a}{1 + e^{-b_1(t-c_1)}} + \frac{f - a}{1 + e^{-b_2(t-c_2)}},$$
where the parameters a, b_1, b_2, c_1, c_2, and f are different for each individual. Show that the parameter f is adult height; that is, show that
$$f = \lim_{t \to \infty} y(t).$$

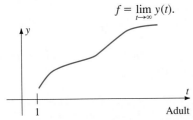

Figure 3.7 Typical double logistic.

72. An ion with n more electrons than protons carries a negative charge of $-np$ where p is the charge on one electron. An ion with m fewer electrons than protons carries a charge of mp. The attraction potential energy between two such ions is
$$E_+ = \frac{-mnp^2}{r}$$
where r is the distance separating the nuclei. However, since the electrons of the two ions repel each other there is a repulsion potential energy of
$$E_- = ae^{-r/b}$$
where a and b are constants. Assume $m = n = p = b = 1$ and $a = 10$, and let $E = E_+ + E_-$ be the total potential energy.
   a. Find the relative maximum and minimum values of the total potential energy on $(0, \infty)$ and the numbers r for which they occur. (*Hint:* You will need to use Newton's Method to solve the equation $dE/dr = 0$. There are two zeros of this equation.)
   b. Graph the function E.

73. Solve the equation $e^{2x} - 8x + 1 = 0$. (*Hint:* Use Newton's Method.)

74. Solve the equation $e^x + x = 5$, $x > 0$.

75. Write a program to calculate entries for the following table, comparing actual values of e^x with those of the approximating polynomials
$$P_1(x) = 1 + x,$$
$$P_2(x) = 1 + x + \frac{x^2}{2!},$$
$$P_3(x) = 1 + x + \frac{x^2}{2!} + \frac{x^3}{3!},$$
$$P_4(x) = 1 + x + \frac{x^2}{2!} + \frac{x^3}{3!} + \frac{x^4}{4!}.$$
(Recall, $n! = n(n - 1)(n - 2) \cdots 2 \cdot 1$.)

x	e^x	$P_1(x)$	$P_2(x)$	$P_3(x)$	$P_4(x)$
1					
0.5					
2					
-2					

76. Prove that, if the function f has an inverse f^{-1} defined for all x and if f is increasing on the interval $[a, b]$, then f^{-1} is increasing on the interval $[f(a), f(b)]$.

77. Prove statement (ii) of Theorem 6.

78. Prove statement (iii) of Theorem 6 for the case r rational.

8.4   Exponentials and Logs to Other Bases

Using the (natural) exponential function e^x, we can extend the domain of the exponential function $y = a^x$, $a > 0$, to *all* real numbers, both rational and irrational. To do so we use the identity

$$a = e^{\ln a}, \qquad a > 0$$

to write

$$a^x = [e^{\ln a}]^x. \tag{1}$$

If x is rational, we can apply Theorem 6 to conclude that

$$[e^{\ln a}]^x = e^{x \ln a}. \tag{2}$$

Combining equations (1) and (2) we obtain

$$\boxed{a^x = e^{x \ln a}} \tag{3}$$

for x rational. However, the right side of equation (3) is defined for all real numbers x and all real numbers $a > 0$. We therefore *define* the function $y = a^x$ by equation (3). Our demonstration shows that this definition agrees with the algebraic definition of a^x if x is rational, so we have now succeeded in extending this definition to all real numbers x. Using equation (3) we can extend the statements of Theorems 5 and 6 as follows:

In Theorem 5, the statement

$$\ln a^r = r \ln a \tag{4}$$

holds for all real numbers $a > 0$ and r, whether or not r is rational.

Proof: Using equation (3) with $x = r$, we have

$$\ln a^r = \ln(e^{r \ln a}) = r \ln a. \qquad \blacksquare$$

In Theorem 6, the statement

$$[e^x]^r = e^{rx} \tag{5}$$

holds for all real numbers x and r, whether or not r is rational.

Proof: Let $y = (e^x)^r$. Then, by equation (4),

$$\ln y = \ln [e^x]^r = r \ln e^x = rx.$$

Thus, $y = e^{rx}$, so $[e^x]^r = e^{rx}$. $\blacksquare$

Property (5) is very useful. For example, we can use it to extend the **Power Rule** for differentiation (Chapter 3–Theorem 10),

$$\frac{d}{dx} x^r = rx^{r-1},$$

to the case where r is an irrational exponent. The details are left as an exercise (Exercise 48).

Another application is the following theorem, which shows that our definition of a^x for irrational x preserves the usual laws of exponents.

Theorem 7

If $a > 0$ is any real number, the equations

$(E1) \quad a^x a^y = a^{x+y}$
$(E2) \quad a^x / a^y = a^{x-y}$
$(E3) \quad (a^x)^y = a^{xy}$

hold for all real numbers x and y.

Proof: We prove only $(E3)$, leaving $(E1)$ and $(E2)$ as exercises:

$$(a^x)^y = (e^{x \ln a})^y \qquad \text{(equation (3))}$$
$$= e^{xy \ln a} \qquad \text{(equation (5))}$$
$$= a^{xy} \qquad \text{(equation (3)).}$$

(See Exercises 32 and 33.)   ∎

Reviewing the Definitions

Before continuing our development of the calculus of logarithms and exponentials, we pause to emphasize how Theorem 7 is the final step in our rigorous development of the function $y = a^x$ defined for *all* real numbers (if $a > 0$). Recall that, in Chapters 1–7, all exponentials were rational. That is, whenever we wrote an expression of the form a^x, the exponent x was a quotient of integers, and a^x was defined algebraically. More precisely, if $x = p/q$ where p and q are integers with no common factors, then a^x is the qth root of the pth power of a. However, in order to study the calculus of exponentials, we needed to extend this definition to all exponents—irrational as well as rational. In this chapter, we succeeded in this extension using our knowledge of the definite integral and the Fundamental Theorem of Calculus as follows:

1. We defined $y = \ln x$ for $x > 0$ *as a definite integral.* Using the Fundamental Theorem of Calculus, we were immediately able to compute its derivative. Moreover, we showed that this function satisfied the logarithm properties $(L1)$–$(L3)$ (Theorem 5). In particular, we established

$$\ln a^r = r \ln a$$

   for all rational numbers r.
2. We defined the number e to be the unique positive real number whose natural logarithm is 1.
3. We defined the natural exponential function $y = \exp(x)$ *as the inverse function of the natural logarithm function,* and we proved that

$$\exp(x) = e^x$$

   for all rational numbers x. Thus, we used $\exp(x)$ to extend the definition of e^x to include irrational x, and we proved that the natural exponential function satisfies the Laws of Exponents (Theorem 6). Consequently, we dropped the notation $\exp(x)$ and replaced it with e^x for *both rational and irrational exponents x.*

4. Finally, we showed that

$$a^x = e^{x \ln a}$$

for x rational. However, since the right-hand side of this equation is defined for all real numbers x, we can use the right-hand side as a *definition of a^x for all x*. Moreover, using this definition, we were able to establish the properties

$$\ln a^r = r \ln a$$

and

$$(e^x)^r = e^{(rx)}$$

for irrational r. Thus, we were able to establish the Laws of Exponents (Theorem 7).

The final result is an exponential function $y = a^x$ that is defined for all real numbers x (assuming only that $a > 0$) and that agrees with our algebraic definition if x is rational.

Differentiating $y = a^x$

Now we continue with our discussion of the calculus of exponentials. Since the exponential function $y = e^x$ and the linear function $y = (\ln a)x$ are both differentiable, equation (3) defines the function $y = a^x$ as the composition of two differentiable functions. As such, $y = a^x$ must be differentiable. To find its derivative we apply the Chain Rule to the right-hand side of equation (3):

$$\frac{d}{dx} a^x = \frac{d}{dx} e^{x \ln a}$$

$$= e^{x \ln a} \frac{d}{dx} (x \ln a)$$

$$= e^{x \ln a} \ln a$$

$$= a^x \ln a.$$

That is,

$$\frac{d}{dx} a^x = a^x \ln a. \tag{6}$$

□ **EXAMPLE 1**

Find $\dfrac{dI}{dx}$ for the Beer-Lambert Law of Exercise 53, Section 8.1.

Solution   Recall that

$$I = I_0 10^{kx}$$

where I_0 and k are constants. Equation (6) gives

$$\frac{dI}{dx} = I_0 10^{kx} (\ln 10) \frac{d}{dx} (kx) = k I_0 10^{kx} \ln 10. \qquad \blacksquare$$

Observe that equation (6) agrees with the differentiation formula for $y = e^x$ when $a = e$, since $\ln e = 1$. This fact explains why the number e is referred to as the *natural* choice for the base of an exponential function: among all bases, e is the one for which the derivative of the exponential function has the simplest form.

The integration formula corresponding to the differentiation formula (6) is

$$\int a^x \, dx = \frac{a^x}{\ln a} + C, \qquad a > 0, \quad a \neq 1. \tag{7}$$

□ **EXAMPLE 2**

Find

$$\int \sin x \cos x \, 10^{\sin^2 x} \, dx.$$

Solution Since the exponent is $\sin^2 x$, we try the substitution.

$$u = \sin^2 x; \qquad du = 2 \sin x \cos x \, dx.$$

Then $\sin x \cos x \, dx = \frac{1}{2} \, du$, and we obtain

$$\int \sin x \cos x \, 10^{\sin^2 x} \, dx = \int 10^u \tfrac{1}{2} \, du$$

$$= \tfrac{1}{2} \int 10^u \, du$$

$$= \frac{10^u}{2 \ln 10} + C$$

$$= \frac{10^{\sin^2 x}}{2 \ln 10} + C. \qquad ∎$$

The Function $y = \log_a x$

For any positive real number $a \neq 1$, equation (3) lets us extend $\log_a x$ to a differentiable function of $x > 0$. To see this we write

$$y = \log_a x \quad \text{if and only if} \quad x = a^y = e^{y \ln a}. \tag{8}$$

Applying the natural logarithm to both sides of the equation $x = e^{y \ln a}$ gives

$$\ln x = \ln(e^{y \ln a}) = y \ln a,$$

so

$$y = \frac{\ln x}{\ln a}, \qquad a \neq 1. \tag{9}$$

Thus the function $\log_a x$ is simply a multiple of the function $\ln x$ and conversely. Statements (8) and (9) together give the equation

$$\log_a x = \frac{\ln x}{\ln a}, \qquad a > 0, \quad a \neq 1. \tag{10}$$

☐ **EXAMPLE 3**

Find formulas for converting common logs to natural logs and vice versa.

Strategy · · · · · · · ·

Find ln 10 from a natural log table or calculator.

Write equation (10) with $a = 10$.

Solve equation (10) for ln x.

Solution

Common logs are logarithms to the base $a = 10$. Since $\ln(10) \approx 2.303$, equation (10) gives

$$\log_{10} x \approx \frac{1}{2.303} \ln x \approx 0.434 \ln x.$$

Solving for ln x we obtain

$$\ln x \approx 2.303 \log_{10} x. \qquad \blacksquare$$

The differentiation formula for $y = \log_a x$ can be found by differentiating both sides of equation (10). We obtain

$$\frac{d}{dx} \log_a x = \frac{1}{x \ln a}, \qquad a > 0, \quad a \neq 1. \tag{11}$$

REMARK   Equation (11) shows why $\ln x = \log_e x$ is called the *natural* logarithm function. Among all logarithm functions, ln x is the one for which the derivative has simplest form, since the factor ln a is just the integer 1 if $a = e$.

☐ **EXAMPLE 4**

Find the minimum value of the function

$$y = \log_{10}(1 + x^2), \qquad -\infty < x < \infty.$$

Solution   To find the critical numbers we set

$$\frac{dy}{dx} = \frac{2x}{(1 + x^2) \ln 10} = 0 \tag{12}$$

and obtain the single critical number $x = 0$. From the form of dy/dx in equation (12) it is easy to see that $dy/dx < 0$ if $x < 0$, and $dy/dx > 0$ if $x > 0$. Thus, $y = \log_{10}(1 + x^2)$ decreases on $(-\infty, 0)$ and increases on $(0, \infty)$. The value $y = 0$ at the point $(0, 0)$ is therefore an absolute minimum on $(-\infty, \infty)$. $\blacksquare$

Logarithmic Differentiation

There are two types of functions for which it is advantageous to calculate the derivative by differentiating the natural logarithm of the function rather than the function itself. The first type is functions of the form

$$y = u(x)^{v(x)}, \qquad u(x) \geq 0. \tag{13}$$

Since the exponent $v(x)$ is not constant, the Power Rule is not applicable. Moreover, the base $u(x)$ is not constant, so rule (6) for differentiating an exponential function does not apply. However, by applying the natural logarithm function to both sides of equation (13) we obtain

$$\ln y = \ln[u(x)^{v(x)}] = v(x) \ln u(x). \tag{14}$$

We may then obtain dy/dx from equation (14) by the technique of implicit differentiation.

□ EXAMPLE 5

Find $\dfrac{dy}{dx}$ for $y = x^{\sin x}$, $\qquad x > 0$.

Strategy · · · · · · · ·

Take natural logs of both sides.

Solution

Since $y = x^{\sin x}$ we use equation (4) to write

$$\ln y = \ln(x^{\sin x}) = \sin x \ln x.$$

Differentiate both sides, using Chain Rule on left, Product Rule on right.

Differentiating implicitly we find that

$$\frac{1}{y}\frac{dy}{dx} = \cos x \ln x + \frac{\sin x}{x},$$

so

Solve for $\dfrac{dy}{dx}$ and substitute for y.

$$\frac{dy}{dx} = y\left[\cos x \ln x + \frac{\sin x}{x}\right]$$

$$= x^{\sin x}\left[\cos x \ln x + \frac{\sin x}{x}\right]. \qquad ■$$

The second type of function for which this technique is useful is one that is complicated by a large number of algebraic operations. Properties $(L1)$–$(L3)$ can often be applied to simplify greatly the form of $\ln y$. The derivative dy/dx is then calculated implicitly as above.

□ EXAMPLE 6

Find $\dfrac{dy}{dx}$ for $y = \dfrac{\sqrt{1 + x^2}(3x + 2)^3}{\sqrt[3]{x^2(x + 1)}}$

Strategy · · · · · · · ·

Take natural logs of both sides.

Solution

We obtain

Use properties $(L1)$–$(L3)$ of natural logs to simplify right-hand side.

$$\ln y = \ln\left[\frac{\sqrt{1 + x^2}(3x + 2)^3}{\sqrt[3]{x^2(x + 1)}}\right]$$

$$= \ln[(1 + x^2)^{1/2}(3x + 2)^3] - \ln[x^2(x + 1)]^{1/3}$$

$$= \tfrac{1}{2}\ln(1 + x^2) + 3\ln(3x + 2) - \tfrac{1}{3}[2\ln x + \ln(x + 1)].$$

Differentiate both sides.

Differentiating both sides then gives

$$\frac{1}{y}\frac{dy}{dx} = \frac{1}{2}\cdot\frac{2x}{1 + x^2} + 3\cdot\frac{3}{3x + 2} - \frac{1}{3}\left[\frac{2}{x} + \frac{1}{x + 1}\right]$$

$$= \frac{x}{1 + x^2} + \frac{9}{3x + 2} - \frac{2}{3x} - \frac{1}{3(x + 1)},$$

so

Solve for $\dfrac{dy}{dx}$.

$$\frac{dy}{dx} = y\left[\frac{x}{1 + x^2} + \frac{9}{3x + 2} - \frac{2}{3x} - \frac{1}{3(x + 1)}\right].$$

If necessary, you could substitute for y in terms of x. ■

Exercise Set 8.4

1. Write each of the following in the form e^u.

   a. $2x$ **b.** π^3 **c.** $7^{\ln x}$

   d. $4^{\sqrt{2}}$ **e.** $3^{\sin x}$ **f.** $2^x 4^{1-x}$

2. Use a calculator with a natural exponential key or a table of values for the exponential function to find decimal expressions for the numbers in Exercise 1, parts b and d.

3. Find a formula for converting powers of 10 to powers of e and vice versa.

4. True or false? $e^{x \ln a} = a^x$ for any a and x. Prove your answer.

In Exercises 5–20, find the derivative of the given function.

5. $y = 3^x$

6. $f(x) = \pi^{x^2 - 1}$

7. $f(x) = \log_{10}(2x - 1)$

8. $y = x 10^{x-1}$

9. $y = \log_{10}(\ln x)$

10. $f(x) = 2^{1-x} \log_2 \sqrt{x}$

11. $g(t) = t^2 \log_2 t$

12. $y = \log_{10} e^{\sqrt{t^2+1}}$

13. $f(x) = \pi^x + x^\pi$

14. $y = \log_{10} 2^{x^2}$

15. $y = x^x$

16. $y = x^{\cos x}$

17. $y = x^{\sqrt{x}}$

18. $y = x^{\ln x}$

19. $y = (\cos x)^{\sin x}, \quad \cos x > 0$

20. $y = [\ln x]^x$

In Exercises 21–30, evaluate the given integral.

21. $\displaystyle \int 5^x \, dx$

22. $\displaystyle \int x 2^{2-x^2} \, dx$

23. $\displaystyle \int \frac{\pi^{\sqrt{x}}}{\sqrt{x}} \, dx$

24. $\displaystyle \int 2^{1-x} 2^{1+x} \, dx$

25. $\displaystyle \int a^{2 \ln x} \, dx$

26. $\displaystyle \int 7^{ax^2+bx}\left(ax + \frac{b}{2}\right) dx$

27. $\displaystyle \int_0^2 x 3^{x^2} \, dx$

28. $\displaystyle \int_0^2 2^{3t-1} \, dt$

29. $\displaystyle \int_0^{\ln 2} 3^{e^{2x}} e^{2x} \, dx$

30. $\displaystyle \int_0^{\pi/4} \sin 2x \, 7^{\cos 2x} \, dx$

31. True or false? The number a^x is nonzero for all x if $a > 0$.

32. Prove statement $(E1)$ in Theorem 7.

33. Prove statement $(E2)$ in Theorem 7.

34. Sketch the graph of the function $y = x 2^{1-x}$ by finding

   a. the relative extrema,

   b. the intervals on which y is increasing,

   c. the inflection points,

   d. the concavity.

35. Find the number $a > 0$ so that the function $y = a^x$ satisfies the differential equation $y' - 2y = 0$.

36. Using linear approximation, obtain a formula for approximating $\log_{10}(10^n + x)$ where n is an integer and x is a small

real number. Use this formula to approximate

   a. $\log_{10}(10.2)$ **b.** $\log_{10}(98)$

   c. $\log_{10}(995)$ **d.** $\log_{10}(.09)$

37. Let R be the region bounded by the graphs of $y = 3^x$, $y = 0$, $x = 0$, and $x = 1$. Find the volume of the solid obtained by revolving this region about the x-axis.

38. Let R be the region bounded by the graph of $f(x) = 2^{x^2}$ and the line $y = 16$. Find the volume of the solid obtained by revolving this region about the y-axis.

39. Find an equation for the line tangent to the graph of $y = \log_{10} x$ at $x = 1$.

40. If P_0 dollars are invested at 10% interest, compounded annually, the amount on deposit after n years is $P(n) = (1.10)^n P_0$ (see Exercise 48, Section 8.1). Find the number r so that

$$P(n) = (1.10)^n P_0 = e^{nr} P_0, \quad n = 1, 2, \ldots .$$

41. The owner of a valuable oil painting estimates that the value of the painting will be approximately $V(t) = (5000) 2^{\sqrt{t}}$ dollars over the next few years. The variable t represents time in years, and the present value of the painting is $V(0) = 5000$ dollars. How fast will the value of the painting be increasing in 4 years?

In Exercises 42–47, find $\dfrac{dy}{dx}$ by logarithmic differentiation.

42. $y = \dfrac{x(x + 1)(x + 2)}{(x + 3)(x + 4)(x + 5)}$

43. $y = \sqrt[3]{\dfrac{x + 2}{x + 3}}$

44. $y = \dfrac{\sqrt[3]{x^2 + 1}}{(x + 1)(x + 2)^2}$

45. $y = \dfrac{(x^2 + 2)^3 (x - 1)^5}{x \sqrt{x + 1} \sqrt{x + 2}}$

46. $y = x^{\sqrt{x+1}}$

47. $y = (x^2 + 1)^x$

48. Prove that the **Power Rule**

$$\frac{d}{dx} x^r = r x^{r-1}$$

is valid for all r, both rational and irrational.

49. The purpose of this exercise is to establish the limit

$$e^r = \lim_{x \to \infty} \left(1 + \frac{r}{x}\right)^x$$

(equation (8) in Section 8.3). Let

$$y = \lim_{x \to \infty} \left(1 + \frac{r}{x}\right)^x.$$

We shall show that ln $y = r$.
a. Let $h = 1/x$. Show that

$$y = \lim_{h \to 0^+} (1 + rh)^{1/h}.$$

b. Let $f(x) = \ln(1 + rx)$. Show that ln $y = f'(0)$.
c. Conclude that ln $y = r$.

 50. Determine the value of e by:
a. Evaluating $f(x) = (1 + 1/x)^x$ for x large. How much accuracy is possible? (The value of e to 20 decimal places is 2.71828182845904523454.)

b. Evaluating $g(x) = (1 + x)^{1/x}$ near zero. How much accuracy is possible?

 51. Given $b \geq 1.5$, let $f(x) = x^b$ and $g(x) = b^x$.
a. For selected values of b, graph f and g for $x > 0$. Using these graphs, determine the number of times that the graph of f and the graph of g intersect.
b. Is there a number b such that the graphs are tangent at the intersection point(s)? If so, find the number b for which this tangency occurs. (Two curves are tangent at a point of intersection if their tangent lines at that point are identical.)

8.5 Exponential Growth and Decay

Imagine the following biological experiment. On a certain day, fruit flies are placed in an enclosed environment such as a large bell jar. If the environment is supportive (e.g., sufficient food supply and proper temperature), the number of fruit flies will increase as times passes. The experiment is to record the number $N(t)$ of fruit flies present after t days and to find a mathematical relationship between time and population size. In experiments of this kind, data such as those presented in Table 5.1 are often obtained (see Figure 5.1).

Table 5.1 Typical data on growth of fruit flies

t days	0	4	8	12	16	20	24
$N(t)$ (fruit flies)	10	18	35	72	107	208	361

On examining these data, one observes that the *rate* of increase of the population increases as the population grows, i.e., the larger the population the more rapid the growth rate. Biologists describe this phenomenon using the **Law of Natural Growth,** which is often stated as:

{Rate of growth of population} $\propto$ {Present population size}. (1)

(The symbol "$\propto$" means "is proportional to." The statement of proportionality, $A \propto B$, is equivalent to the equation $A = kB$, where k is a constant.)

We can develop a mathematical model for populations that grow according to law (1) by assuming that the population size $N(t)$ may be approximated by a differentiable function of time. Since the derivative $N'(t) = dN/dt$ gives the rate of growth (increase or decrease) of the population size, the mathematical formulation of the growth law (1) is

$$\frac{dN}{dt} = kN. \qquad (2)$$

The constant k in equation (2) is called the **growth constant.** Since the mathematical model (2) consists of a *differential equation,* we must find a differentiable function $y = N(t)$ that satisfies the equation.

We have already determined (Section 8.3) that any exponential function of the form $y = Ce^{kt}$ satisfies equation (2), since

$$\frac{dy}{dt} = \frac{d}{dt}(Ce^{kt}) = C\left(\frac{d}{dt} e^{kt}\right) = kCe^{kt} = ky.$$

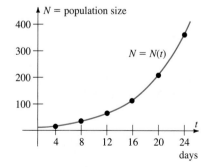

Figure 5.1 Plot of typical data on growth of fruit flies.

In Exercise 35 you will show that, other than the trivial solution $y = 0$, there are no other possible solutions of equation (2). In other words, all solutions of the differential equation (2) must have the form $y = Ce^{kt}$. Finally, we can be more specific about the constant C. Setting $t = 0$ in the equation $y = Ce^{kt}$ gives

$$y(0) = Ce^{k \cdot 0} = C,$$

so $C = y(0)$. We refer to the constant $C = y(0) = y_0$ as the **initial condition** associated with the differential equation (2). We summarize our findings as follows.

Theorem 8

There is a unique solution of the differential equation

$$\frac{dy}{dt} = ky, \tag{3}$$

satisfying the initial condition $y(0) = y_0$. It is the exponential function

$$y = y_0 e^{kt}. \tag{4}$$

Theorem 8 provides the strategy for solving problems involving populations or quantities that satisfy the Law of Natural Growth: If we can safely assume that the size or amount of the population or quantity y can be viewed as a differentiable function of t (which usually represents time), then y has the form given by equation (4). By substituting known data into equation (4) the growth constant k and the initial population size or amount y_0 is obtained. The function y is then completely determined.

Graphs of the growth function (4) are shown in Figure 5.2 for the two general cases $k > 0$ (growth) and $k < 0$ (decay).

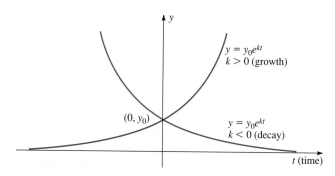

Figure 5.2 Graph of the natural growth function.

☐ **EXAMPLE 1**

Assume that the rate of growth of a population of fruit flies is proportional to the size of the population at each instant of time. If 100 fruit flies are present initially and 300 are present after 10 days, how many will be present after 15 days?

Strategy · · · · · · · ·

Apply Theorem 8 to find the general form of the solution.

Solution

Since the population size N satisfies the differential equation

$$\frac{dN}{dt} = kN,$$

the function N is given by equation (4):

$$N(t) = N_0 e^{kt}, \qquad N(0) = N_0. \tag{5}$$

Find $N_0 = N(0)$ from given data.

We are given that $N = 100$ when $t = 0$, that is, $N_0 = 100$. Also, we are given that $N = 300$ when $t = 10$. Inserting these values in (5) gives

$$300 = 100e^{10k}, \quad \text{so} \quad e^{10k} = 3.$$

Take natural logs of both sides and solve for k.

Thus

$$10k = \ln 3, \quad \text{so} \quad k = \frac{\ln 3}{10}.$$

State explicit formulation of $N(t)$,

The function $N(t)$ is therefore

$$N(t) = 100e^{(\ln 3/10)t},$$

so

Find $N(15)$ using tables or calculator, rounding to nearest integer.

$$N(15) = 100e^{15\ln 3/10} = 100e^{3/2\ln 3} = 100 \cdot 3^{3/2} \approx 520 \text{ fruit flies.} \qquad \blacksquare$$

□ EXAMPLE 2

Certain radioactive elements, such as uranium, decay at a rate proportional to the amount present. If a block of 50 grams of such material decays to 40 grams in 5 days, find the half-life of this material. (The half-life is the amount of time required for the material to decay to half its original amount.)

Strategy · · · · · · · ·

Label the variables.

Solution

Let $A(t)$ denote the amount present at time t. Since A satisfies the differential equation (3), by Theorem 8, A has the form

State form of A from Theorem 8.

$$A(t) = A_0 e^{kt}, \qquad t = \text{time.} \tag{6}$$

Apply initial condition

$$A(0) = 50 = A_0.$$

We are given

$$A_0 = A(0) = 50 \text{ grams.}$$

Use equation $A(5) = 40$ to obtain equation for k.

To find k we use the data $A(5) = 40$ together with equation (6). We obtain

$$40 = A_0 e^{5k} = 50e^{5k}, \quad \text{so} \quad e^{5k} = \tfrac{40}{50} = \tfrac{4}{5}.$$

Solve for k by taking ln's of both sides.

Thus

$$k = \tfrac{1}{5}\ln(\tfrac{4}{5}). \tag{7}$$

Write down condition for T:

$$A(T) = \tfrac{1}{2}A(0).$$

We seek a time T so that

$$A(T) = \tfrac{1}{2}A_0. \tag{8}$$

Combine (6) and (8).

Using (6) we obtain

$$A_0 e^{kT} = \tfrac{1}{2}A_0, \quad \text{so} \quad e^{kT} = \tfrac{1}{2}.$$

Solve for T, taking ln's of both sides and using (7).

Thus

$$T = \frac{\ln(\tfrac{1}{2})}{k}.$$

Using the value for k in (7), we find

$$T = \frac{\ln(\tfrac{1}{2})}{\tfrac{1}{5}\ln(\tfrac{4}{5})} = \frac{5[\ln 1 - \ln 2]}{\ln 4 - \ln 5} \approx 15.53 \text{ days.} \qquad \blacksquare$$

Compound Interest: In Exercise 49, Section 8.1, we showed that, if P_0 dollars are invested at annual interest rate r (in decimal form) compounded n times per year, the amount on deposit after t years is

$$P(t) = \left(1 + \frac{r}{n}\right)^{nt} P_0 \text{ dollars.} \tag{9}$$

However, many banks advertise ''continuous'' compounding of interest. This means that the number of compoundings per year is regarded as having become infinitely large (i.e., interest is added to your account ''each instant''). To obtain an expression for $P(t)$ under continuous compounding of interest we apply $\lim\limits_{n \to \infty}$ to expression (9):

$$P(t) = \lim_{n \to \infty}\left[\left(1 + \frac{r}{n}\right)^{nt} P_0\right] \tag{10}$$

$$= P_0\left[\lim_{n \to \infty}\left(1 + \frac{r}{n}\right)^n\right]^t.$$

We now use equation (8), Section 8.3,

$$e^r = \lim_{n \to \infty}\left(1 + \frac{r}{n}\right)^n$$

in equation (10) to obtain *the equation for continuous compounding of interest*

$$P(t) = P_0 e^{rt}. \tag{11}$$

REMARK   Note that the function P in equation (11) is the solution of the differential equation

$$P'(t) = rP(t) \tag{12}$$

with $P(0) = P_0$. Equation (12) is another way of describing continuous compounding of interest since it says that the rate $P'(t)$ at which the amount $P(t)$ is growing at each instant is the interest rate r applied to the present size of the investment.

□ **EXAMPLE 3**

Find the value of an initial deposit of $1,000 invested at 12% annual interest compounded continuously for 6 years.

Solution   We have $P_0 = 1000$, $r = 0.12$, and $t = 6$. From equation (11),

$$P(6) = 1000e^{6(.12)} \approx 2054.43 \text{ dollars.} \qquad \blacksquare$$

Just as the amount of money available after compounding P_0 dollars continuously for t years can be found using equation (11), the *initial* amount P_0 needed to produce a given amount $P_t = P(t)$ after t years can be calculated. We do this by multiplying both sides of equation (11) by e^{-rt} to obtain

$$P_0 = P_t e^{-rt}. \tag{13}$$

In this case P_0 is called the **present value** of an investment that will yield P_t dollars in t years if compounded continuously at r percent.

□ **EXAMPLE 4**

Find the size of a deposit that, when compounded continuously for 10 years at 12% interest, will yield $20,000.

Solution   From equation (13) we obtain

$$P_0 = e^{-(0.12)10}(20,000) = 6,023.88 \text{ dollars.}$$ ■

□ **EXAMPLE 5**

The owner of a stand of timber estimates that the timber is presently worth $5,000 and that the timber increases in value with time according to the formula

$$V(t) = 5000e^{\sqrt{t}/2} \text{ dollars}$$

where t is time in years. If the prevailing interest rates over the foreseeable future are expected to average 12%, when should the owner sell the timber to maximize profit?

Strategy · · · · · · · · ·

Convert the revenue received at a future date to present value, using (13).

Solution

Since dollars received at different times have different present values, we must discount the changing value of the revenue received for the sale of the timber by converting it to its *present value*. Using equation (13) we find the present value of the revenue received in selling the timber after t years to be

$$R(t) = V(t)e^{-.12t} = 5000e^{(\sqrt{t}/2 - .12t)}$$

Set $R'(t) = 0$.

To maximize R we set $R'(t) = 0$:

$$R'(t) = 5000e^{(\sqrt{t}/2 - .12t)}\left[\frac{1}{4\sqrt{t}} - .12\right] = 0. \tag{14}$$

Solve for t.

Since the exponential function is never zero we must have

$$\frac{1}{4\sqrt{t}} - .12 = 0, \quad \text{or} \quad \sqrt{t} = \left(\frac{1}{.48}\right).$$

Thus,

$$t = \left(\frac{1}{.48}\right)^2 = 4.34 \text{ years,}$$

Verify that the critical point obtained yields a maximum.

or approximately 4 years and 4 months. (We can verify that this value of t indeed produces a profit maximum by applying the First Derivative Test to R' in equation (14).) ■

Exercise Set 8.5

In Exercises 1–6, find an exponential function of the form $y = Ae^{kt}$ that satisfies the given conditions.

1. $y' = 2y$ and $y(0) = 1$

2. $y' = -4y$ and $y(0) = 2$

3. $y' - 5y = 0$ and $y(1) = 1$

4. $y(0) = 2$ and $y(2) = 6$

5. $y(0) = 1$ and $y(1) = e^{-2}$

6. $y(0) = y(1) = 0$

7. True or false? The half-life of an isotope depends upon the amount present. Explain.

8. A radioactive isotope I decays exponentially with a half-life of 20 days. If 50 mg of the isotope remain after 10 days:
   a. How much of the isotope was present initially?
   b. How much will remain after 30 days?
   c. When will 90% of the initial amount have disintegrated?

9. Suppose an object travelling through a fluid is subject only to a force resisting its motion, and this resisting force is proportional to the velocity of the object. Let's write this resisting force as $f_r = cv$ where c is constant and v denotes velocity. By Newton's second law we have $f_r = ma$ where m is the mass of the object and a is its acceleration. Since $a = dv/dt$, we may combine our two equations to obtain

$$m \frac{dv}{dt} = -cv, \quad \text{or} \quad \frac{dv}{dt} = -\left(\frac{c}{m}\right)v.$$

   a. Find a solution $v(t)$ of this differential equation for which $v(0) = v_0$.
   b. Find a solution of the equation for $v_0 = 10$ m/s, $c = 40$, and $m = 10$.
   c. Does $v(t)$ approach a limit as $t \to \infty$? If so, what is this limit?

10. In a **first-order chemical reaction** the rate of disappearance of reactant is proportional to the amount present. Chemists use the notation $[A]$ for the concentration of reactant A in moles per liter. Thus, in first-order reactions, the concentration $[A]$ as a function of time is a solution of the differential equation

$$\frac{d[A]}{dt} = -k[A].$$

   a. Find an expression for the concentration $[A(t)]$ after t seconds if $[A_0] = [A(0)]$.
   b. If the reaction begins with a concentration $[A_0] = 10$ mol/L, and 6 mol/L of reactant A remain after 20 min, find the concentration of reactant A 1 h after the reaction begins.

11. The number of bacteria in a certain culture grows from 50 to 400 in 12 h. Assuming that the rate of increase is proportional to the number of bacteria present:
   a. How long does it take for the number of bacteria present to double?
   b. How many bacteria will be present after 16 h?

12. If we assume temperature to be constant at all altitudes, the rate of decrease in atmospheric pressure p, as a function of altitude, h, is proportional to p, That is,

$$\frac{dp}{dh} = -kp$$

   where the constant k is approximately $k = 0.116$ km^{-1}. Find the pressure in atmospheres 8 km (5 mi) above sea level if the pressure at sea level is 1 atm.

13. Land bought for speculation is expected to be worth $V(t) = 10{,}000(1.2)^{\sqrt{t}}$ dollars after t years. If the cost of money holds constant at 10%, when should the land be sold so as to maximize profits? Does the original price of the land affect this calculation?

14. In a simple model of the growth of a biological population, the rate of increase of N, the number of individuals, is pro-

portional to the difference between the birth rate b and the death rate d. That is, the simple birth-death model of population growth is

$$\frac{dN}{dt} = bN - dN.$$

   a. Find an expression for population size N as a function of the birth rate r, death rate d, and initial population size N_0.
   b. If in a population modelled by this process there are 20 births per year per 100 individuals and 12 deaths per year per 100 individuals, find the anticipated size of the population in 20 years if its present size is $N_0 = 200$ individuals.

15. When a foreign substance is introduced into the body, the body's defense mechanisms move to break down the substance and excrete it. The rate of excretion is usually proportional to the concentration present in the body, and the half-life of the resulting exponential decay is referred to as the **biological half-life** of the substance. If, after 12 h, 30% of a massive dosage of a substance has been excreted by the body, what is the biological half-life of the substance?

16. All living matter contains two isotopes of carbon, ^{14}C and ^{12}C, in its molecules. While the organism is alive the ratio of ^{14}C to ^{12}C is constant, as the ^{14}C is interchanged with fixed levels of $^{14}CO_2$ in the atmosphere. However, when a plant or animal dies this replenishment ceases, and the radioactive ^{14}C present decays exponentially with a half-life of 5760 years. By examining the ^{14}C/^{12}C ratio, archaeologists can determine the percent of the original ^{14}C level remaining and therefore date a fossil.
   a. If 80% of the original amount of ^{14}C remains, how old is the fossil?
   b. A fossil contains 10% of its original amount of ^{14}C. What is its age?

17. The body concentrates iodine in the thyroid gland. This observation leads to the treatment of thyroid cancer by injecting radioactive iodine into the bloodstream. The iodine concentrates in the thyroid and kills the cancerous cells. One isotope of iodine used in this treatment has a half-life of approximately 8 days and decays exponentially in time. If 50 μg of this isotope are injected, what amount remains in the body after three weeks?

18. A chemical dissolves in water at a rate proportional to the amount still undissolved. If 20 g of the chemical are placed in water and 10 g remain undissolved 5 min later, when will 90% of the chemical be dissolved?

19. **Newton's Law of Cooling** states that the *rate* at which an object changes temperature is proportional to the difference between its temperature and the temperature of its environment. In other words, if T_e is the (constant) temperature of the environment, the temperature $T(t)$ of the object at time t satisfies the differential equation

$$\frac{d}{dt}(T(t) - T_e) = -k(T(t) - T_e)$$

for some constant k. Since T_e is assumed to be constant, this equation simplifies to

$$\frac{dT}{dt} = -k(T(t) - T_e). \tag{15}$$

Let T_0 denote the temperature of the object at time $t = 0$, and let $T(t)$ be the function

$$T(t) = T_e + (T_0 - T_e)e^{-kt}.$$

Show that $T(t)$ satisfies the differential equation (15) and has initial value $T(0) = T_0$. What is the limit of $T(t)$ as $t \to \infty$?

20. Let $T(t)$ be the solution to Newton's Law of Cooling, as described in Exercise 19. Show that, if we know the value of $T(t_1)$ at some time $t_1 \neq 0$, we can determine k using t_1, T_0, T_e, and $T(t_1)$.

In Exercises 21–24, use the results of Exercises 19 and 20.

21. How long will it take an object at 100°C in a room of constant temperature 20°C to cool to 30°C if it takes 10 min for it to cool to 50°C?

22. What will be the temperature of the object in Exercise 21 after one hour?

23. Repeat Exercises 21 and 22 for the same object if it is put in a room at 0°C and it takes 5 min for it to cool to 50°C.

24. How long will it take an object at 20°C in a room of constant temperature 100°C to warm to 50°C if it takes 10 min for it to warm to 30°C?

25. The value of a case of a certain French wine t years after being imported to the United States is $V(t) = 100(1.5)^{\sqrt{t}}$. If the cost of money under continuous compounding of interest is expected to remain constant at 12.5% per year, when should the wine be sold so as to maximize profits? (Assume storage costs to be negligible.)

26. In this exercise, we introduce the concept of **effective annual yield**:
  a. Show that, under *continuous* compounding of interest at the rate of $r_c = 10\%$, an initial deposit of $100 will grow in one year to $110.52.
  b. Note that, under a *single annual* compounding of interest, an interest rate of $r_a = 10.52\%$ would be required to produce the same amount of interest in one year. Thus, we say that the *effective annual yield* of $r_c = 10\%$ is $r_a = 10.52\%$. In general, the effective annual yield of an interest rate compounded continuously is the interest rate required, under a *single* annual compounding, to produce the same amount of interest in one year. What is the effective annual yield of the interest rate $r_c = 15\%$?

27. What rate of interest, compounded continuously, will produce an effective annual yield of 8%? (See Exercise 26.)

28. An annuity pays 10% interest, compounded continuously. What amount of money, deposited today, will grow to $2,500 in 8 years?

29. Find the effective annual yield (see Exercise 26) on
  a. an interest rate of 10% compounded semiannually,
  b. an interest rate of 10% compounded quarterly,
  c. an interest rate of 10% compounded daily,
  d. an interest rate of 10% compounded continuously.

30. How long does it take for a deposit of P_0 dollars to double at 10% interest compounded continuously?

31. Show that the effective annual rate of interest i for continuous compounding at rate r is $i = e^r - 1$. (See Exercise 26.)

32. What is the present value of $1000 five years from now if the prevailing rate of interest, when compounded continuously, is 10%? What is the present value if 10% is the prevailing effective annual yield? (See Exercise 26.)

33. True or false? If the population $P(t)$ of a city is increasing at a rate of 5% per year, then the function $P(t)$, if differentiable, satisfies the equation $P'(t) = .05P(t)$. If this is false, what is the correct equation?

34. The population of a certain city is increasing at a rate of 5% per year. If the population $P(t)$ is assumed differentiable, and if $P'(t) = kP(t)$, find k (see Exercise 33).

35. Prove that $y = Ce^{kt}$ is the only nontrivial $(y \neq 0)$ solution of the differential equation $y' = ky$ as follows.
  a. Assume that the differentiable function $y = f(t)$ is also a solution.
  b. Show that the product $g(t) = f(t)e^{-kt}$ is differentiable, and that $g'(t) = 0$.
  c. Conclude that $g(t) = C$ for some constant C, that is, necessarily $f(t) = Ce^{kt}$.

36. To what amount will a deposit of $1 grow in one year at an interest rate of 100% compounded continuously? Interpret your answer in terms of limit (7) in Section 8.3.

37. In this exercise, use the table in Exercise 98 of Section 1.4.
  a. Plot the pairs $(\ln a, \ln A)$. Why does this plot imply that $A = ca^r$ for some numbers c and r?
  b. Determine numbers m and b such that $\ln A = m \ln a + b$. Using m and b, determine the numbers c and r of part a.

38. Suppose that interest is compounded n times per year at an annual percentage rate r (for example, a 9% annual rate corresponds to $r = 0.09$). Show that the doubling time t of an investment satisfies the equation

$$2 = \left(1 + \frac{r}{n}\right)^{nt}.$$

Express t as a function of r and n.
  a. For $n = 1, 2, 4, 12$, and 36, graph this function for $0.001 \leq r \leq 1$. How does the compounding period affect the doubling time for a given interest rate?

b. For $r = 0.05, 0.07, 0.09, 0.11,$ and 0.13, graph this function over the interval $1 \le n \le 365$. How does the interest rate affect the doubling time for a given compounding period.

c. Using your results in parts a and b, which quantity, n or r, most affects the doubling time?

In Exercises 39 and 40, use the table in Exercise 24 of Section 3.3.

39. Assume that the U.S. population $N(t)$ satisfies an exponential growth model; that is, assume

$$N(t) = N_0 e^{kt}.$$

a. Determine N_0 and k if $t_0 = 0$ corresponds to the year 1900 and time t is measured in decades. Graph N and the values in the table. Is the exponential growth model a good one in this case? What does this model predict for the years 1990 and 2000? Do these values seem reasonable?

b. Now let $t = 0$ correspond to the year 1800 when the U.S. population was 5.308 million. Determine new values of N_0 and k. Graph this new function N along with the values in the table. Is this model any better? (Note that, if $t = 0$ is the 1800, then $t = 11$ is the year 1910.)

40. Within most populations there is competition that limits the growth of the population. The **logistic model**

$$\frac{dN}{dt} = kN(1 - bN), \qquad N(0) = N_0$$

is a growth model that considers competition. The solution to the logistic model is given by

$$N(t) = \frac{N_0}{N_0 b + e^{-kt}(1 - N_0 b)}.$$

In this exercise, we measure time t in decades.

a. Use the population of the U.S. in 1900 ($t = 0$), 1920 ($t = 2$) and 1940 ($t = 4$) to determine N_0, k, and b. Graph N and the values in the table. Is this model a good predictor of the U.S. population for this century? What does this model predict for the years 1990 and 2000? Do these values seem plausible? *Hint:* If $N_0 = N(0)$, $N_2 = N(2)$, and $N_4 = N(4)$, show that

$$b = \frac{N_0 - N_2 e^{-2k}}{N_0 N_2 (1 - e^{-2k})} \quad \text{and} \quad e^{-2k} = \frac{N_0(N_4 - N_2)}{N_4(N_2 - N_0)}.$$

b. What will be the limiting value (the so-called carrying capacity) of the U.S. population, i.e., the limit of $N(t)$ as $t \to \infty$? Is this result plausible?

c. Repeat parts a and b using the population of the U.S. in 1900 ($t = 0$), 1960 ($t = 6$), and 1980 ($t = 8$). Then use the model to predict the U.S. population when you are 65 years of age. What percentage of the carrying capacity is this value? Graph N' and N'' as well. When does the population grow the fastest? (*Hint:* Solve the equations needed to find b and k using Newton's Method.)

Summary Outline of Chapter 8

■ **Definition:** $\log_b x = y$ if and only if $b^y = x$.   (page 448)

■ **Properties of Logarithms:**   (page 449)

(L1) $\log_b(mn) = \log_b m + \log_b n$.
(L2) $\log_b(m/n) = \log_b m - \log_b n$.
(L3) $\log_b(n^r) = r \log_b n$.

■ **Definition:** $\ln x = \displaystyle\int_1^x \frac{1}{t}\, dt, \qquad x > 0.$ (page 458)

■ **Theorem:** $\dfrac{d}{dx} \ln x = \dfrac{1}{x}; \qquad \displaystyle\int \frac{1}{x}\, dx = \ln |x| + C.$ (page 458)

■ **Definition:** e is defined by the equation $\ln e = 1$.
$e \approx 2.718281828459.$ (page 460)

■ **Definition:** $y = e^x$ if and only if $x = \ln y$.   (page 467)

■ **Theorem:** $\dfrac{d}{dx} e^x = e^x; \qquad \displaystyle\int e^x\, dx = e^x + C.$ (page 469)

■ **Alternative Definition of e^r:** $e^r = \displaystyle\lim_{k \to \infty} \left(1 + \frac{r}{k}\right)^k, \qquad e = \lim_{k \to \infty} \left(1 + \frac{1}{k}\right)^k.$ (page 472)

▮ **Definition:** $a^x = e^{x \ln a}, \qquad a > 0.$ (page 476)

▮ **Properties of Exponentials:** (page 477)

(E1) $a^x a^y = a^{x+y}$
(E2) $a^x / a^y = a^{x-y}$
(E3) $(a^x)^y = a^{xy}$

▮ **Theorem:** $\dfrac{d}{dx} a^x = a^x \ln a; \qquad \displaystyle\int a^x \, dx = \dfrac{a^x}{\ln a} + C.$ (page 478)

▮ **Definition:** $y = \log_a x$ if and only if $x = a^y = e^{y \ln a}.$ (page 479)

▮ **Conversion Formula:** $\log_a x = \dfrac{\ln x}{\ln a}, \qquad a \neq 1.$ (page 479)

▮ **Model for Natural Growth:** $\dfrac{dy}{dt} = ky.$ (page 484)

Solution: $y = Ce^{kt}, \qquad C = y(0).$

▮ **Formula for Continuous Compounding of Interest:** $P(t) = e^{rt} P_0.$ (page 486)

▮ **Present Value of P_t Dollars in t Years:** $P(0) = e^{-rt} P_t.$ (page 486)

Review Exercises—Chapter 8

1. Simplify the following.
  a. $8^{2/3}$
  b. $36^{-5/2}$
  c. $16^{5/4}$
  d. $4^{-3/4}$

2. Find the following logarithms:
  a. $\log_2 16$
  b. $\log_8 16$
  c. $\ln e^2$
  d. $\log_2 3\sqrt{2}$

3. Find the inverses of the following functions:
  a. $f(x) = x + 2$
  b. $g(x) = x^2, \quad x \geq 0$
  c. $y = 2x + 1$
  d. $y = \sqrt{x}$

4. Solve for x:
  a. $3 \ln x - \ln 3x = 0$
  b. $\ln x^3 = 3$
  c. $\displaystyle\int_1^x \frac{2}{t} \, dt = 4$
  d. $\displaystyle\int_2^x \frac{1}{t} \, dt = 3 + \ln 2$

In Exercises 5–26, find the derivative of each function with respect to its independent variable.

5. $y = x^2 \ln(x - a)$
6. $f(x) = xe^{1-x}$

7. $f(x) = \ln(\ln^2 x)$
8. $y = (2 \ln \sqrt{x})^3$

9. $y = e^{x^2} \tan 2x$
10. $f(t) = \sin^2 t \cos(\ln t)$

11. $f(t) = \dfrac{te^t}{1 + e^t}$
12. $y = x \ln(\sqrt{x} - e^{-x})$

13. $y = \dfrac{\ln(t - a)}{\ln(t^2 + b)}$
14. $f(t) = \ln \cos t$

15. $f(t) = e^{\sqrt{t} - \ln t}$
16. $y = \dfrac{1}{a} \ln\left(\dfrac{a + bx}{x}\right)$

17. $y = \dfrac{1}{3x} + \dfrac{1}{4} \ln\left(\dfrac{1 - 2x}{\sqrt{x}}\right)$

18. $f(t) = \sin(\ln t + te^{\sqrt{t}})$

19. $x \ln y^2 + y \ln x = 1$ $\left(\text{Find } \dfrac{dy}{dx}.\right)$

20. $y = e^{4x}(e^{-x} + \ln x)^2$

21. $x^2 + e^{xy} - y^2 = 2$ $\left(\text{Find } \dfrac{dy}{dx}.\right)$

22. $y = xe^{1 - \sqrt{x} \ln x}$

23. $e^{xy} = \sqrt{xy}$ $\left(\text{Find } \dfrac{dy}{dx}.\right)$

24. $y = \sqrt{e^{-x} + e^x}$

25. $f(x) = \ln \sqrt{e^{2x} + \sin \sqrt{x}}$

26. $y = \displaystyle\int_1^{\sin x} \ln(t + 1) \, dt$

In Exercises 27–44, find the integral.

27. $\displaystyle\int \frac{x \, dx}{1 - x^2}$
28. $\displaystyle\int xe^{1 - x^2} \, dx$

29. $\displaystyle\int \frac{x + 1}{4x + 2x^2} \, dx$
30. $\displaystyle\int \frac{dx}{(x + 1) \ln^3(x + 1)}$

31. $\displaystyle\int \sqrt{e^x} \, dx$
32. $\displaystyle\int (e^x + 1)^2 \, dx$

33. $\displaystyle\int_{e}^{e^2} \frac{1}{x\sqrt{\ln x}}\,dx$

34. $\displaystyle\int \frac{e^x - e^{-x}}{e^x + e^{-x}}\,dx$

35. $\displaystyle\int_{0}^{1} \frac{x^3 - 1}{x + 1}\,dx$

36. $\displaystyle\int_{1}^{4} \frac{e^{\sqrt{x}}}{\sqrt{x}}\,dx$

37. $\displaystyle\int_{0}^{2} x(e^{x^2} + 1)\,dx$

38. $\displaystyle\int_{4}^{9} \frac{1}{\sqrt{x}(1 + \sqrt{x})}\,dx$

39. $\displaystyle\int e^x(1 - e^{2x})^3\,dx$

40. $\displaystyle\int \frac{\cot\sqrt{x}}{\sqrt{x}}\,dx$

41. $\displaystyle\int \frac{2x + 3x^2}{x^3 + x^2 - 7}\,dx$

42. $\displaystyle\int_{1}^{\ln 2} (e^x + 1)(e^x - 1)\,dx$

43. $\displaystyle\int_{-3}^{-2} \frac{x}{x^2 + 1}\,dx$

44. $\displaystyle\int_{0}^{1} \left(\frac{1}{x + 1} - \frac{1}{x + 2}\right)\,dx$

In Exercises 45–48, use logarithmic differentiation to find $\dfrac{dy}{dx}$.

45. $y = x^{\sqrt{x}}$

46. $y = \sqrt{\dfrac{(x - 1)^2(x + 1)}{x(x + 3)^3}}$

47. $y = x^{\sin^2 x}$

48. $y = (\tan x)^{\cos x}$

49. Find all relative extrema for the function $y = x^2 \ln x$.

50. Find the maximum and minimum values of the function $f(x) = e^{x^2/3}$ for x in the interval $[-1, \sqrt{8}]$.

51. On what intervals is the graph of the function $y = \ln(1 + x^2)$ concave up?

52. Find all relative extrema for the function $y = x^2 - \ln x^2$ and sketch the graph.

53. Let $x > 0$ and T be the triangle with vertices $(0, 0)$, $(x, 0)$, and $(0, \ln x)$. If x is increasing at a rate of 4 units/s, how fast is the area of T increasing when $x = 5$ units?

54. The radius of a circle is growing with rate given by $dr/dt = e^r$ in./s. How fast is the area of the circle growing when the radius is 2 in.?

55. Find the area of the region bounded by the graph of $y = x/(x^2 + 1)$ and the x-axis for $-2 \le x \le 2$.

56. Find the area of the region bounded by the graphs of
$$y = \frac{1}{x} \quad \text{and} \quad y = \frac{10x - 21}{4x - 10}.$$
(*Hint:* Divide first.)

57. Find the volume of the solid generated by rotating about the x-axis the portion of the graph of
$$y = \sqrt{\frac{2}{x - 1}}$$
from $x = 3$ to $x = 5$.

58. The function
$$f(x) = \begin{cases} \lambda e^{-\lambda x}, & x > 0 \\ 0, & x \le 0 \end{cases}$$
is called the **exponential probability density function** with parameter λ. For any density function, the probability that x will fall in the interval $[a, b]$ is
$$P\{a \le x \le b\} = \int_{a}^{b} f(x)\,dx.$$
For $\lambda = 1$, find the following probabilities associated with the exponential distribution:
  a. $P\{0 \le x \le 1\}$,
  b. $P\{0 \le x \le \ln 2\}$,
  c. $P\{0 \le x \le 4\}$.

59. Graph the exponential probability density function for $\lambda = 1$ (Exercise 58).

60. (Note Exercise 58.) The **standard normal probability density** function is
$$f(x) = \frac{1}{\sqrt{2\pi}} e^{-x^2/2}.$$
Use a program for numerical integration to approximate
  a. $P\{-1 \le x \le 1\} = \displaystyle\int_{-1}^{1} \frac{1}{\sqrt{2\pi}} e^{-x^2/2}\,dx$
  b. $P\{0 \le x \le 2\}$.

61. Sketch the graph of the standard normal probability density function in Exercise 60.

62. Find the equation of the line tangent to the graph of $y = xe^{2x}$ at the point $(\ln 2, 4 \ln 2)$.

63. Find an equation of the line perpendicular to the graph $y = x^2 \ln x$ at the point (e, e^2).

64. Find an equation of the line tangent to the graph $y = e^{-3x} + 1$ at $x = \ln 3$.

65. Find a formula for approximating the function $y = x \ln x^2$ near $x = e$. Use this formula to approximate $y(2.70) = (2.70)\ln(2.70)^2$.

66. Approximate $e^{\sqrt{4.1}}$ using linear approximation.

67. Let $f(x) = \dfrac{e^x + e^{-x}}{2}$, $g(x) = \dfrac{e^x - e^{-x}}{2}$.
  a. Show that $f'(x) = g(x)$ and that $g'(x) = f(x)$.
  b. Show that both functions satisfy the differential equation $y'' - y = 0$.
  c. Find two solutions of the differential equation $y'' - k^2 y = 0$.

68. A particle, starting from rest, moves along a line with acceleration $a(t) = (t + 2)^{-2}$ m/s^2. Find
  a. the velocity function $v(t)$,
  b. the distance travelled after 10 s.

69. Approximate $\int_0^5 e^{-x^2}\, dx$ using the Midpoint Rule with $n = 10$.

 70. Find r if $2^x = e^{rx}$ for all x.

 71. Use the Midpoint Rule with $n = 10$ to approximate
   a. $\ln 4$ **b.** $\ln 5$ **c.** $\ln(1/2)$

 72. Use Newton's Method to approximate the solution of the equation $2 + \ln x = x$.

In Exercises 73–78, solve the initial value problem.

73. $\dfrac{dy}{dx} = 2y$

   $y(0) = 1$

74. $\dfrac{dy}{dx} + y = 0$

   $y(\ln 2) = 2$

75. $2y' + 4y = 0$

   $y(0) = \pi$

76. $y' + y = 2$

   $y(0) = 1$

77. $4y' - 4y + 4 = 0$

   $y(0) = 1$

78. $(y' + y)^2 = 0$

   $y(0) = 1$

79. Find the average value of the function $y = xe^{x^2+1}$ on the interval $[0, \sqrt{\ln 2}]$.

80. Find the centroid of the region bounded by the curves $y = 1/x$, $y = 0$, $x = e$, and $x = e^2$.

81. Find the centroid of the region bounded by the curves $y = 1/x$, $y = x^2$, $x = 1$, and $x = 3$.

82. Approximate the solution of the equation $e^x + 2x = 0$.

83. The population of the United States in 1980 was 227 million. By 1990 the population had grown to 249 million. Assuming exponential growth, what will the population be in 2000? 2010?

84. In the chemical reaction called the inversion of raw sugar, the inversion rate is proportional to the amount of raw sugar remaining. If 100 kg of raw sugar is reduced to 75 kg in 6 h, how long will it be until
   a. half the raw sugar has been inverted?
   b. 90% of the raw sugar has been inverted?

85. A population of fruit flies grows exponentially. If initially there were 100 flies and if after 10 days there were 500 flies, how many flies were present after 4 days?

86. A snowball melts at a rate proportional to its surface area. If the snowball originally had a radius of 10 cm and, after 20 min, its radius is 8 cm, find
   a. a differential equation for the radius r,
   b. a solution of this differential equation involving two constants,
   c. an explicit solution of this equation, and
   d. the radius of the snowball after one hour.

   (*Hint:* In part a show that

$$\frac{dv}{dt} = s\,\frac{dr}{dt}$$

   where v = volume and s = surface area.)

87. If 100 g of a radioactive substance is reduced to 40 g in 6 h and if the decay is exponential, what is the half-life?

88. Show that the exponential function $f(x) = Ce^{kx}$ satisfies the differential equation $f'(x + a) = kf(x + a)$ for every real number a.

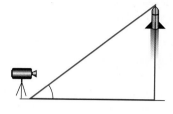

Chapter 9

Trigonometric and Inverse Trigonometric Functions

We have already encountered the six trigonometric functions and their derivatives. The objectives of this chapter are

(i) to obtain antiderivatives for all six trigonometric functions,
(ii) to develop techniques for integrating more complicated expressions involving trigonometric functions,
(iii) to define the inverses of the trigonometric functions and find their derivatives and antiderivatives,
(iv) to define the hyperbolic functions, and
(v) to study some models of physical phenomena involving these functions.

9.1   Integrals of the Trigonometric Functions

We have already determined the following differentiation and integration formulas:

1. $\dfrac{d}{dx}\sin x = \cos x$ 1′. $\displaystyle\int \cos x \, dx = \sin x + C$

2. $\dfrac{d}{dx}\cos x = -\sin x$ 2′. $\displaystyle\int \sin x \, dx = -\cos x + C$

3. $\dfrac{d}{dx}\tan x = \sec^2 x$ 3′. $\displaystyle\int \sec^2 x \, dx = \tan x + C$

4. $\dfrac{d}{dx}\cot x = -\csc^2 x$ 4′. $\displaystyle\int \csc^2 x \, dx = -\cot x + C$

5. $\dfrac{d}{dx}\sec x = \sec x \tan x$ 5′. $\displaystyle\int \sec x \tan x \, dx = \sec x + C$

6. $\dfrac{d}{dx}\csc x = -\csc x \cot x$ 6′. $\displaystyle\int \csc x \cot x \, dx = -\csc x + C$

Identities involving the six trigonometric functions, together with graphs of these six functions, appear in Chapter 1. Remember, all angles are to be measured in *radian* measure when working with trigonometric functions in calculus.

Integrals of tan x and cot x

The problem of finding integrals for sin x and cos x was simple to solve since these functions occur as derivatives of $-\cos x$ and $\sin x$, respectively. However, this is not the case for the remaining four trigonometric functions.

To find the integral for $y = \tan x$ we use the formula

$$\int \frac{1}{u}\, du = \ln |u| + C \qquad (1)$$

from Chapter 8. Since

$$\tan x = \frac{\sin x}{\cos x},$$

we let $u = \cos x$. Then $du = -\sin x\, dx$. Using equation (1) we obtain

$$\int \tan x\, dx = \int \frac{\sin x}{\cos x}\, dx$$

$$= -\int \frac{1}{u}\, du$$

$$= -\ln |u| + C$$

$$= -\ln |\cos x| + C.$$

Since $-\ln |\cos x| = \ln |\cos x|^{-1} = \ln |\sec x|$, we have the formula

$$\int \tan x\, dx = \ln |\sec x| + C, \qquad \cos x \neq 0. \qquad (2)$$

To find the antiderivative for $y = \cot x$, we write

$$\cot x = \frac{\cos x}{\sin x}$$

and let $u = \sin x$. Proceeding as above we obtain

$$\int \cot x\, dx = \ln |\sin x| + C, \qquad \sin x \neq 0. \qquad (3)$$

□ **EXAMPLE 1**

Find $\displaystyle\int x \cot x^2\, dx$.

Strategy · · · · · · · ·
Find a substitution so that x is a factor of du.

Solution
We make the substitution

$$u = x^2; \qquad du = 2x\, dx.$$

Then $x\, dx = \frac{1}{2}\, du$, and we obtain

Make substitution.

$$\int x \cot x^2\, dx = \int \cot u (\tfrac{1}{2})\, du$$

$$= \tfrac{1}{2} \int \cot u\, du$$

Apply equation (3).
Substitute back.

$$= \tfrac{1}{2} \ln |\sin u| + C$$

$$= \tfrac{1}{2} \ln |\sin x^2| + C.$$

□ **EXAMPLE 2**

Find $\int \tan^3 x \, dx$.

Solution We use the identity $\tan^2 x = \sec^2 x - 1$ to write

$$\int \tan^3 x \, dx = \int \tan x [\sec^2 x - 1] \, dx$$

$$= \int \tan x \sec^2 x \, dx - \int \tan x \, dx.$$

In the first integral we use the substitution

$$u = \tan x; \qquad du = \sec^2 x \, dx$$

to obtain

$$\int \tan x \sec^2 x \, dx = \int u \, du$$

$$= \frac{u^2}{2} + C$$

$$= \frac{\tan^2 x}{2} + C.$$

The second integral is given by equation (2). Thus,

$$\int \tan^3 x \, dx = \frac{\tan^2 x}{2} - \ln |\sec x| + C. \qquad ■$$

Integrals of sec x and csc x

The integrals for $\sec x$ and $\csc x$ also involve the natural logarithm function, although the derivations are not so straightforward. To handle $\sec x$ we multiply by

$$\frac{\sec x + \tan x}{\sec x + \tan x}$$

to obtain

$$\int \sec x \, dx = \int \sec x \left(\frac{\sec x + \tan x}{\sec x + \tan x} \right) dx$$

$$= \int \frac{\sec x \tan x + \sec^2 x}{\sec x + \tan x} \, dx$$

$$= \int \frac{1}{\sec x + \tan x} \left[\frac{d}{dx} (\sec x + \tan x) \right] dx$$

$$= \ln |\sec x + \tan x| + C \qquad \text{(by equation (1))}.$$

Thus,

$$\boxed{\int \sec x \, dx = \ln |\sec x + \tan x| + C.}$$

$$(4)$$

Similarly, we can show that

$$\int \csc x \, dx = \ln |\csc x - \cot x| + C. \tag{5}$$

(See Exercise 36.)

□ EXAMPLE 3

Find the length of the graph of the function $y = \ln \sec x$ from $(0, 0)$ to $(\pi/3, \ln 2)$.

Strategy · · · · · · · ·

Find $\dfrac{dy}{dx}$.

Use formula for arc length (Section 7.3)

$$L = \int_a^b \sqrt{1 + \left[\frac{dy}{dx}\right]^2} \, dx.$$

Use identity

$$1 + \tan^2 x = \sec^2 x.$$

Use formula (4).

Solution

Since $\dfrac{dy}{dx} = \tan x$ we have

$$L = \int_0^{\pi/3} \sqrt{1 + \tan^2 x} \, dx$$

$$= \int_0^{\pi/3} \sqrt{\sec^2 x} \, dx$$

$$= \int_0^{\pi/3} \sec x \, dx$$

$$= \left[\ln |\sec x + \tan x| \right]_0^{\pi/3}$$

$$= \ln|2 + \sqrt{3}| - \ln |1 + 0| = \ln(2 + \sqrt{3}) \approx 1.317. \quad \blacksquare$$

□ EXAMPLE 4

Find $\displaystyle\int \frac{\csc \sqrt{x}}{\sqrt{x}} \, dx.$

Strategy · · · · · · · ·

Find a substitution involving $x^{-1/2}$ as factor of du.

Make substitution.

Factor constant.

Apply formula (5).

Substitute back.

Solution

We make the substitution

$$u = \sqrt{x} = x^{1/2}; \qquad du = \frac{dx}{2\sqrt{x}} = \frac{1}{2} x^{-1/2} \, dx.$$

Then $x^{-1/2} \, dx = 2 \, du$, and

$$\int \frac{\csc \sqrt{x}}{\sqrt{x}} \, dx = \int (\csc u) \, 2 \, du$$

$$= 2 \int \csc u \, du$$

$$= 2 \ln |\csc u - \cot u| + C$$

$$= 2 \ln |\csc \sqrt{x} - \cot \sqrt{x}| + C. \quad \blacksquare$$

The exercise set below provides a general review of integration of trigonometric functions as well as exercises concerning equations (3)–(5).

Exercise Set 9.1

In Exercises 1–30, find the integral.

1. $\displaystyle\int \cos 3x \, dx$

2. $\displaystyle\int_0^{\pi/8} \sec^2 2x \, dx$

3. $\displaystyle\int_0^{\pi/8} \sec 2x \tan 2x \, dx$

4. $\displaystyle\int x \csc x^2 \, dx$

5. $\displaystyle\int (\tan^2 x + 1) \, dx$

6. $\displaystyle\int \frac{\tan \sqrt{x}}{\sqrt{x}} \, dx$

7. $\displaystyle\int \frac{x}{\cos x^2} \, dx$

8. $\displaystyle\int \frac{1}{\sqrt{x} \sec \sqrt{x}} \, dx$

9. $\displaystyle\int x \tan(3x^2 - 1) \, dx$

10. $\displaystyle\int_0^{\pi/2} \sin x \sqrt{1 + \cos x} \, dx$

11. $\displaystyle\int_{\pi/4}^{\pi/3} \tan^3 x \sec^2 x \, dx$

12. $\displaystyle\int_{\pi/4}^{\pi/2} \cot x \ln(\sin x) \, dx$

13. $\displaystyle\int_{\pi/6}^{\pi/4} \sec^2 x \, e^{\tan x} \, dx$

14. $\displaystyle\int \frac{\cos^2(2x - 1)}{\sin(2x - 1)} \, dx$

15. $\displaystyle\int \frac{\sec^2 x + 1}{x + \tan x} \, dx$

16. $\displaystyle\int \frac{\csc^2(x + \pi/4)}{\cot^3(x + \pi/4)} \, dx$

17. $\displaystyle\int x \cot^2(1 - x^2) \, dx$

18. $\displaystyle\int \frac{\sec^2 x}{\sqrt{\tan x}} \, dx$

19. $\displaystyle\int (\sec x + \tan x)^2 \, dx$

20. $\displaystyle\int \frac{\sin x + \cos x}{\sin x - \cos x} \, dx$

21. $\displaystyle\int (1 - x)\sec(x^2 - 2x) \, dx$

22. $\displaystyle\int \frac{\sec^5 \sqrt{x} \tan \sqrt{x}}{\sqrt{x}} \, dx$

23. $\displaystyle\int \sqrt{1 - \sin^2 x} \, e^{\sin x} \, dx$

24. $\displaystyle\int \frac{\sec \ln x}{x} \, dx$

25. $\displaystyle\int \frac{dx}{\sqrt{x}(1 + \cos \sqrt{x})}$

26. $\displaystyle\int \csc^4 2x \, dx$

27. $\displaystyle\int \frac{\cos 4x}{\sqrt{1 + \sin 4x}} \, dx$

28. $\displaystyle\int \sin^3 2\theta \, d\theta$

29. $\displaystyle\int \frac{\sec^2 \theta \, d\theta}{\sqrt{\tan \theta + 1}}$

30. $\displaystyle\int \frac{\tan x}{1 + \tan^2 x} \, dx$

31. Find the area bounded by the graph of $y = x \sec x^2$ and the x-axis for $0 \le x \le \sqrt{\pi/3}$.

32. Find the area bounded by the graph of $y = \sec x \tan x$ and the x-axis for $-\pi/3 \le x \le \pi/3$.

33. Find the average value of the function $f(x) = \csc x$ for $\pi/6 \le x \le \pi/3$.

34. Find $\int \sin x \cos x \, dx$ in two ways. (*Hint:* $\sin 2x = 2 \sin x \cos x$.)

35. Find $\int \tan x \sec^2 x \, dx$ in two ways.

36. Derive formula (5).

37. Find the length of the curve $y = \ln \sin x$ from $(\pi/6, -\ln 2)$ to $(\pi/3, \ln(\sqrt{3}/2))$.

38. Find the length of the curve $y = \ln \cos x$ from $(0, 0)$ to $(\pi/4, (\ln 2)/2)$.

39. Find the volume of the solid obtained by revolving about the x-axis the graph of $y = \sec x \tan x$, $-\pi/4 \le x \le \pi/4$.

40. Find the volume of the solid obtained by revolving about the x-axis the region bounded by the graphs of $y = \sec x \tan x$ and $y = \tan x$ for $-\pi/4 \le x \le \pi/4$.

41. Find the area of one of the regions bounded by the graph of $y = \sec x$ and the line $y = -2$.

42. Find the volume of the solid obtained by revolving about the y-axis the graph of $y = \sec x^2$, $0 \le x \le \sqrt{\pi/4}$.

43. Find the volume of the solid obtained by revolving about the y-axis the graph of $y = \tan 2x^2$, $0 \le x \le \sqrt{\pi/8}$.

44. Use Simpson's Rule with $n = 6$ to approximate the length of the graph of $y = \cos x$ for $0 \le x \le 2\pi$.

9.2   Integrals Involving Products of Trigonometric Functions

Integrals that involve products of the trigonometric functions can often be handled using trigonometric identities and appropriate substitutions. For each of the various types of integrals discussed in this section we will present a typical example, followed by a general statement of strategy, followed by more challenging examples. Note that integrating expressions of this kind is rarely straightforward. Indeed, the best recipe for success is knowledge of general strategies coupled with the experience of working many such problems.

**Integrals Involving Odd Powers
of Sine and Cosine**

□ **EXAMPLE 1**

Find $\int \sin^3 x \cos x \, dx$.

Solution   We make the substitution

$$u = \sin x; \qquad du = \cos x \, dx$$

so that $\cos x$ is a factor of du. Then

$$\int \sin^3 x \cos x \, dx = \int u^3 \, du$$

$$= \frac{u^4}{4} + C$$

$$= \frac{\sin^4 x}{4} + C. \qquad ■$$

General Strategy: In an integral of the form

$$\int \sin^n x \cos^m x \, dx, \qquad \text{with } n \text{ or } m \text{ odd:}$$

(i) *If m is odd,* make the substitution $u = \sin x$. Then $du = \cos x \, dx$ involves
one factor of $\cos x$. The remaining even factors of $\cos x$ may be converted
to a function of $\sin x$ by the identity $\cos^2 x = 1 - \sin^2 x$. The integral then
has the form

$$\int \sin^n x (1 - \sin^2 x)^k \cos x \, dx = \int u^n (1 - u^2)^k \, du.$$

(ii) *If n is odd,* make the substitution $u = \cos x$. Then $du = -\sin x \, dx$ and the
remaining even factors of $\sin x$ may be converted to a function of $\cos x$.
Proceed as above.

□ **EXAMPLE 2**

Find $\int \sin^2 x \cos^3 x \, dx$.

Strategy   ·   ·   ·   ·   ·   ·   ·
Use one factor of $\cos x$ for du.

Solution

The exponent on $\cos x$ is odd, so we make the substitution

$$u = \sin x; \qquad du = \cos x \, dx.$$

Then, using the identity $\cos^2 x = 1 - \sin^2 x$, we obtain

Convert remaining factors of $\cos x$ to
functions of $\sin x$.

$$\int \sin^2 x \cos^3 x \, dx = \int \sin^2 x (1 - \sin^2 x) \cos x \, dx$$

$$= \int u^2 (1 - u^2) \, du$$

$$= \frac{u^3}{3} - \frac{u^5}{5} + C$$

$$= \frac{\sin^3 x}{3} - \frac{\sin^5 x}{5} + C. \qquad ■$$

□ **EXAMPLE 3**

Find $\int_0^{\pi/4} \sin^5 2x \, dx$.

Strategy · · · · · · · ·
Use one factor of sin $2x$ for du.

Solution

Even though no factor of $\cos x$ is present, we may make the substitution

$$u = \cos 2x; \qquad du = -2 \sin 2x \, dx.$$

We obtain

$$\int_0^{\pi/4} \sin^5 2x \, dx = \int_0^{\pi/4} \sin^4 2x \sin 2x \, dx$$

Convert remaining $\sin^4 2x$ to $(1 - \cos^2 2x)^2$.

$$= \int_0^{\pi/4} (1 - \cos^2 2x)^2 \sin 2x \, dx$$

Make substitution. If $x = 0$, $u = 1$. If $x = \pi/4$, $u = 0$.

$$= \int_1^0 (1 - u^2)^2 \left(-\tfrac{1}{2}\right) du$$

$$= \tfrac{1}{2} \int_0^1 (1 - 2u^2 + u^4) \, du$$

$$= \left[\frac{u}{2} - \frac{u^3}{3} + \frac{u^5}{10} \right]_0^1$$

$$= \frac{1}{2} - \frac{1}{3} + \frac{1}{10} = \frac{4}{15}.$$

■

Integrals Involving Even Powers of Sine and Cosine

□ **EXAMPLE 4**

Find $\int \sin^2 x \cos^2 x \, dx$.

Strategy · · · · · · · ·
Use half-angle formulas to reduce exponents by half.

Solution

We use the identities

$$\sin^2 x = \frac{1}{2} - \frac{1}{2} \cos 2x$$

$$\cos^2 x = \frac{1}{2} + \frac{1}{2} \cos 2x$$

to obtain

$$\int \sin^2 x \cos^2 x \, dx = \int \left(\frac{1}{2} - \frac{1}{2} \cos 2x \right)\left(\frac{1}{2} + \frac{1}{2} \cos 2x \right) dx$$

$$= \int \left(\frac{1}{4} - \frac{1}{4} \cos^2 2x \right) dx$$

Use identity

$$\cos^2 2x = \frac{1}{2} + \frac{1}{2} \cos 4x$$

(i.e., substitute for $\cos^2$ a second time).

$$= \int \left[\frac{1}{4} - \frac{1}{4}\left(\frac{1}{2} + \frac{1}{2} \cos 4x \right) \right] dx$$

$$= \int \left(\frac{1}{8} - \frac{1}{8} \cos 4x \right) dx$$

$$= \frac{x}{8} - \frac{1}{32} \sin 4x + C.$$

■

General Strategy: In integrals of the form

$$\int \sin^n x \cos^m x \, dx$$

where both n and m are even, use the identities

$$\sin^2 \theta = \frac{1}{2} - \frac{1}{2} \cos 2\theta \qquad (1)$$

$$\cos^2 \theta = \frac{1}{2} + \frac{1}{2} \cos 2\theta \qquad (2)$$

repeatedly until an integrand involving only constants and cosine terms is obtained.

□ **EXAMPLE 5**

Find $\int \sin^4 x \, dx$.

Strategy · · · · · · · ·
Use identity (1) with $\theta = x$.

Solution

$$\int \sin^4 x \, dx = \int (\sin^2 x)^2 \, dx$$

$$= \int \left(\frac{1}{2} - \frac{1}{2} \cos 2x \right)^2 dx$$

Use identity (2) on $\cos^2 2x$ with $\theta = 2x$.

$$= \int \left(\frac{1}{4} - \frac{1}{2} \cos 2x + \frac{1}{4} \cos^2 2x \right) dx$$

$$= \int \left[\frac{1}{4} - \frac{1}{2} \cos 2x + \frac{1}{4} \left(\frac{1}{2} + \frac{1}{2} \cos 4x \right) \right] dx$$

$$= \int \left(\frac{3}{8} - \frac{1}{2} \cos 2x + \frac{1}{8} \cos 4x \right) dx$$

$$= \frac{3x}{8} - \frac{1}{4} \sin 2x + \frac{1}{32} \sin 4x + C. \qquad ■$$

□ **EXAMPLE 6**

Find $\int \cos^2 \left(\frac{x}{2} \right) \sin^4 \left(\frac{x}{2} \right) dx$.

Strategy · · · · · · · ·

Both exponents are even so we use identities (1) and (2) with $\theta = x/2$.

Solution

$$\int \cos^2 \left(\frac{x}{2} \right) \sin^4 \left(\frac{x}{2} \right) dx = \int \cos^2 \left(\frac{x}{2} \right) \left[\sin^2 \left(\frac{x}{2} \right) \right]^2 dx$$

$$= \int \left(\frac{1}{2} + \frac{1}{2} \cos x \right) \left(\frac{1}{2} - \frac{1}{2} \cos x \right)^2 dx$$

$$= \frac{1}{8} \int (1 + \cos x)(1 - 2 \cos x + \cos^2 x) \, dx$$

$$= \frac{1}{8} \int (1 - \cos x - \cos^2 x + \cos^3 x) \, dx$$

Use identity (2) again on $\cos^2 x$.

$$= \frac{1}{8} \int \left(1 - \cos x - \left(\frac{1}{2} + \frac{1}{2} \cos 2x \right) + \cos^3 x \right) dx$$

$$= \frac{1}{8} \int \left(\frac{1}{2} - \cos x - \frac{1}{2} \cos 2x + \cos^3 x \right) dx$$

Handle $\cos^3 x$ term as in Example 2.

$$= \frac{1}{8} \int \left[\frac{1}{2} - \cos x - \frac{1}{2} \cos 2x + (1 - \sin^2 x) \cos x \right] dx$$

$$= \frac{1}{8} \int \left(\frac{1}{2} - \frac{1}{2} \cos 2x - \sin^2 x \cos x \right) dx$$

$$= \frac{x}{16} - \frac{1}{32} \sin 2x - \frac{1}{24} \sin^3 x + C. \qquad \blacksquare$$

Integrals Involving Powers of Secant and Tangent

□ **EXAMPLE 7**

Find $\displaystyle\int \tan^2 x \sec^2 x \, dx$.

Solution   We make the substitution

$$u = \tan x; \qquad du = \sec^2 x \, dx.$$

Then

$$\int \tan^2 x \sec^2 x \, dx = \int u^2 \, du$$

$$= \frac{u^3}{3} + C$$

$$= \frac{\tan^3 x}{3} + C. \qquad \blacksquare$$

□ **EXAMPLE 8**

Find $\displaystyle\int \tan^3 x \sec x \, dx$.

Solution   We use the identity $\tan^2 x = \sec^2 x - 1$ to write

$$\int \tan^3 x \sec x \, dx = \int \tan^2 x \tan x \sec x \, dx$$

$$= \int (\sec^2 x - 1) \tan x \sec x \, dx.$$

We now make the substitution

$$u = \sec x; \qquad du = \tan x \sec x \, dx.$$

With this substitution we have

$$\int \tan^3 x \sec x \, dx = \int (u^2 - 1) \, du$$

$$= \frac{u^3}{3} - u + C$$

$$= \frac{\sec^3 x}{3} - \sec x + C. \qquad \blacksquare$$

General Strategy: **In integrals of the form**

$$\int \sec^n x \tan^m x \, dx:$$

(i) *If n is even,* write $\sec^{n-2} x$ as a function of $\tan x$ using the identity $\sec^2 x = \tan^2 x + 1$. Then make the substitution $u = \tan x$. The remaining factor $\sec^2 x \, dx$ becomes du.

(ii) *If m is odd,* write $\tan^{m-1} x$ as a function of $\sec x$ using the identity above. Make the substitution $u = \sec x$ using one factor of $\sec x$, and using the remaining factor of $\tan x$ as $du = \sec x \tan x \, dx$.

(iii) *If n is odd and m is even,* the technique of integration by parts is needed. (See Section 10.1.)

□ **EXAMPLE 9**

Find $\displaystyle\int \tan^4 x \, dx$.

Strategy · · · · · · · ·   **Solution**

Use identity

$$\tan^2 x = \sec^2 x - 1.$$
Use identity again.

Let $u = \tan x$.

$$du = \sec^2 x \, dx.$$

in first term.

$$\int \tan^4 x \, dx = \int (\sec^2 x - 1) \tan^2 x \, dx$$

$$= \int (\sec^2 x \tan^2 x - \tan^2 x) \, dx$$

$$= \int (\sec^2 x \tan^2 x - \sec^2 x + 1) \, dx$$

$$= \frac{\tan^3 x}{3} - \tan x + x + C.$$ ■

Integrals of the Forms   In these integrals the general strategy is to use the identities

$$\int \sin mx \cos nx \, dx$$

$$\int \sin mx \sin nx \, dx$$

$$\int \cos mx \cos nx \, dx$$

$$\sin mx \cos nx = \tfrac{1}{2}[\sin(m + n) x + \sin(m - n) x] \tag{3}$$

$$\sin mx \sin nx = \tfrac{1}{2}[\cos(m - n) x - \cos(m + n) x] \tag{4}$$

$$\cos mx \cos nx = \tfrac{1}{2}[\cos(m + n) x + \cos(m - n) x]. \tag{5}$$

(See Exercises 45–47.)

□ **EXAMPLE 10**

Find $\displaystyle\int \sin 2x \cos 5x \, dx$.

Solution   We use identity (3) to write

$$\int \sin 2x \cos 5x \, dx = \int \tfrac{1}{2}[\sin(2+5)x + \sin(2-5)x] \, dx$$

$$= \tfrac{1}{2} \int (\sin 7x + \sin(-3x)) \, dx$$

$$= \tfrac{1}{2}[(\tfrac{1}{7})(-\cos 7x) + (-\tfrac{1}{3})(-\cos(-3x))] + C$$

$$= -\tfrac{1}{14} \cos 7x + \tfrac{1}{6} \cos(-3x) + C$$

$$= -\tfrac{1}{14} \cos 7x + \tfrac{1}{6} \cos 3x + C.$$

Exercise Set 9.2

In Exercises 1–36, find the integral.

1. $\int \sin^2 2x \, dx$

2. $\int_0^{\pi/4} \sec^2 x \, dx$

3. $\int_{\pi/4}^{\pi/2} \sin x \cos^2 x \, dx$

4. $\int \sin x \cos^3 x \, dx$

5. $\int \sin^5 x \, dx$

6. $\int \tan x \sec^2 x \, dx$

7. $\int_0^{\pi/2} \sin^2 x \cos^3 x \, dx$

8. $\int_0^{\pi/4} \sin^2 x \cos^2 x \, dx$

9. $\int \sec^3 x \tan^3 x \, dx$

10. $\int \sec^4(2x - 1) \, dx$

11. $\int_0^{\pi} \sin^6 x \, dx$

12. $\int_0^{\pi/4} \tan^3 x \sec^4 x \, dx$

13. $\int \cot^3 x \, dx$

14. $\int \sin^3 x \sqrt{\cos x} \, dx$

15. $\int_0^{\pi/4} \sin x \sin 3x \, dx$

16. $\int_0^{\pi/2} \sin 5x \cos 3x \, dx$

17. $\int_{\pi/2}^{\pi} (\sin x + \cos x)^2 \, dx$

18. $\int_0^{\pi/2} \sin^2 x \cos^5 x \, dx$

19. $\int \tan^4 x \sec^4 x \, dx$

20. $\int \cot^5 x \, dx$

21. $\int \tan^5 x \sec^3 x \, dx$

22. $\int \frac{\sec^2 x}{1 + \tan x} \, dx$

23. $\int \sin x \sin 2x \, dx$

24. $\int \sin 4x \cos 3x \, dx$

25. $\int \frac{\cos^3 x}{\sqrt{\sin x}} \, dx$

26. $\int (\tan x + \cot x)^2 \, dx$

27. $\int \frac{\sec^2 x}{(1 + \tan x)^4} \, dx$

28. $\int \csc x \cot x \, dx$

29. $\int \cos 5x \cos 3x \, dx$

30. $\int \cos 7x \sin 2x \, dx$

31. $\int \frac{\sin^3 \theta}{\cos \theta} \, d\theta$

32. $\int \frac{\tan^3 \theta}{\sec \theta} \, d\theta$

33. $\int \sec^2 x \sqrt{\tan x} \, dx$

34. $\int \sin ax \sin bx \cos cx \, dx$

35. $\int \sin ax \cos bx \cos cx \, dx$

36. $\int \sin(ax + b) \cos(cx + d) \, dx$

In Exercises 37–42, evaluate these definite integrals, which arise often in applied mathematics (n and m are integers).

37. $\dfrac{1}{\pi} \displaystyle\int_{-\pi}^{\pi} \cos n\theta \, d\theta, \quad n \geq 1$

38. $\dfrac{2}{\pi} \displaystyle\int_{0}^{\pi} \sin n\theta \, d\theta, \quad n \geq 1$

39. $\displaystyle\int_{-L}^{L} \cos^2\left(\frac{n\pi x}{L}\right) dx$

40. $\displaystyle\int_{0}^{2\pi} \sin^2 nx \, dx$

41. $\displaystyle\int_{0}^{2\pi} \sin nx \cos mx \, dx$

42. $\displaystyle\int_{0}^{2\pi} \cos nx \cos mx \, dx$

In Exercises 43–44, show how the reduction formulas are obtained.

43. $\displaystyle\int \tan^n x \, dx = \frac{\tan^{n-1} x}{n - 1} - \int \tan^{n-2} x \, dx, \quad n \neq 1$

44. $\displaystyle\int \cot^n x \, dx = -\frac{\cot^{n-1} x}{n - 1} - \int \cot^{n-2} x \, dx, \quad n \neq 1$

45. Prove identity (3) by adding the identities

$$\sin(\theta_1 + \theta_2) = \sin \theta_1 \cos \theta_2 + \cos \theta_1 \sin \theta_2,$$
$$\sin(\theta_1 - \theta_2) = \sin \theta_1 \cos \theta_2 - \cos \theta_1 \sin \theta_2.$$

46. Prove identity (4) by subtracting the identities

$$\cos(\theta_1 - \theta_2) = \cos \theta_1 \cos \theta_2 + \sin \theta_1 \sin \theta_2,$$
$$\cos(\theta_1 + \theta_2) = \cos \theta_1 \cos \theta_2 - \sin \theta_1 \sin \theta_2.$$

47. Prove identity (5).

48. Find the average value of the function $y = \tan x \sec^2 x$ for $-\pi/4 \leq x \leq \pi/4$.

49. Find the average value of the function $f(x) = \sin^2 x$ on $[0, \pi]$.

50. Find the volume of the solid obtained by revolving about the x-axis the region bounded by the graph of $y = \tan x \sec x$ for $0 \leq x \leq \pi/4$.

51. Find the volume of the solid obtained by revolving about the x-axis the region bounded by the graph of $f(x) = \sec x \tan^2 x$ and the x-axis for $0 \leq x \leq \pi/4$.

52. Find the area of the region bounded by the graphs of $f(x) = \tan^2 x$ and $g(x) = -\tan^2 x$ for $0 \leq x \leq \pi/4$.

53. Find the area of the region bounded above by the graph of $y = \sec x$ and below by the x-axis for $0 \leq x \leq \pi/4$.

54. A particle moves along a line with acceleration function $a(t) = \cos^2 t$. Find
  a. its velocity function $v(t)$, if $v(0) = 0$;
  b. its position function $s(t)$, if $s(0) = 4$.

55. Calculate the distance travelled by a particle moving along a line with velocity $v(t) = \sin 3x \cos 2x$ from $t = 0$ to $t = \pi/6$.

56. Calculate the distance travelled by a particle moving along a line with velocity $v(t) = \sin^2 x \cos^3 x$ from $t = 0$ to $t = \pi$.

9.3   The Inverse Trigonometric Functions

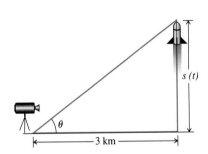

Figure 3.1   Elevation of camera determined by equation (1).

Consider the following problem. A television camera is located 3 kilometers from the launch pad for a rocket that will propel a new satellite into orbit. The problem is to determine the angle $\theta(t)$ at which the camera should be inclined at each instant t so as to track the rocket during the initial phase of its ascent, when its trajectory is nearly vertical.

If the acceleration $a(t)$ of the rocket is known, the height $s(t)$ of the tip of the rocket above the ground may be calculated by integration. Thus, at each instant t we may compute $\tan \theta$ by the equation

$$\tan \theta = \frac{s}{3}. \tag{1}$$

(See Figure 3.1.)

The difficulty with equation (1) is that it describes the *tangent* of the desired angle rather than the angle θ itself. Thus, if we are to solve equation (1) for θ, we must find an *inverse* for the tangent function.

With this simple motivation in mind, we therefore set out to find inverses for each of the six trigonometric functions. We will also obtain some very useful integration formulas.

The Inverse Tangent Function

Consider the graph of $y = \tan x$ (Figure 3.2) and its reflection in the line $y = x$ (Figure 3.3). From these two graphs, we conclude that there is no hope of defining an inverse for the function $y = \tan x$ for all x in the domain of this function. The difficulty is that $y = \tan x$ is not one-to-one. In particular, $\tan(x + n\pi) = \tan x$ for all $n = \pm 1, +2, \ldots$, so each number y in the range of $y = \tan x$ corresponds to *an infinite number* of x's. Thus, the graph of the reflection (Figure 3.3) is not the graph of a *function*. (See Section 8.1 for a general discussion of inverse functions.)

However, not all is lost. Notice in Figure 3.2 that the part of the graph of $y = \tan x$ corresponding to the interval $-\pi/2 < x < \pi/2$ has the property that each value y corresponds to a *unique* number x because the function $y = \tan x$ is *increasing* throughout the interval $(-\pi/2, \pi/2)$. Our strategy is therefore to restrict the domain of the function $y = \tan x$ to the interval $(-\pi/2, \pi/2)$ and then define the *inverse* of the tangent function, denoted by Tan^{-1}, by the equation

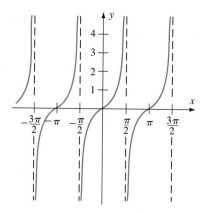

Figure 3.2 Graph of $y = \tan x$.

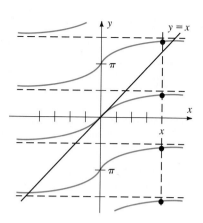

Figure 3.3 Reflection of $y = \tan x$ in line $y = x$. (Many y's correspond to a single number x.)

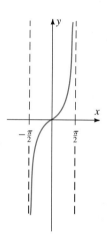

Figure 3.4 Graph of restricted function $y = \operatorname{Tan} x$.

$$\operatorname{Tan}^{-1} y = x \quad \text{if and only if} \quad y = \tan x \quad \text{and} \quad -\pi/2 < x < \pi/2.$$

For example, $\operatorname{Tan}^{-1}(\sqrt{3}) = \pi/3$ since $\pi/3$ is the unique number x between $-\pi/2$ and $\pi/2$ such that $\tan x = \sqrt{3}$. Note that, even though $\tan(4\pi/3) = \sqrt{3}$, $\operatorname{Tan}^{-1}(\sqrt{3}) = \pi/3$.

We denote the restriction of the tangent function to the domain $(-\pi/2, \pi/2)$ by $y = \operatorname{Tan} x$, and we refer to its graph as the **principal branch** of the graph of $y = \tan x$. Figures 3.4 and 3.5 show that the reflection of the graph of $y = \operatorname{Tan} x$ in the line $y = x$ indeed represents the graph of a function. Figure 3.5 is therefore the graph of the inverse of the restricted tangent function $y = \operatorname{Tan} x$.

Since the restricted tangent function $y = \operatorname{Tan} x$ has domain $(-\pi/2, \pi/2)$ and range $(-\infty, \infty)$, the inverse will be defined on $(-\infty, \infty)$ with values in $(-\pi/2, \pi/2)$.

Definition 1

For $x \in (-\infty, \infty)$ the **inverse tangent function** $\operatorname{Tan}^{-1} x$ is defined by

$$y = \operatorname{Tan}^{-1} x \quad \text{if and only if} \quad x = \tan y \text{ with } -\frac{\pi}{2} < y < \frac{\pi}{2}.$$

Be careful to note that $\operatorname{Tan}^{-1} x$ means the inverse of the function $y = \operatorname{Tan} x$, *not* $(\tan x)^{-1}$. The alternatives $y = \operatorname{Arc} \tan x$ and $y = \arctan x$ are also frequently used to represent $y = \operatorname{Tan}^{-1} x$.

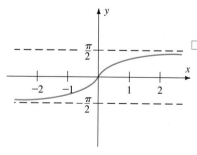

Figure 3.5 Reflection of graph of $y = \operatorname{Tan} x$ is the graph of the inverse $y = \operatorname{Tan}^{-1} x$.

□ **EXAMPLE 1**

Find (a) $\operatorname{Tan}^{-1}(1)$, (b) $\operatorname{Tan}^{-1}(0)$, (c) $\sin(\operatorname{Tan}^{-1}(1/\sqrt{3}))$, (d) $\operatorname{Tan}^{-1}(\tan(\pi/4))$, and (e) $\operatorname{Tan}^{-1}(\tan(5\pi/4))$.

Solution

(a) We seek an angle y so that $-\pi/2 < y < \pi/2$ and $\tan y = 1$. This angle is $y = \pi/4$. Thus, $\operatorname{Tan}^{-1}(1) = \pi/4$ (see Figure 3.6).

(b) $\operatorname{Tan}^{-1}(0) = 0$ since $\tan(0) = 0$.

(c) Since $\tan(\pi/6) = 1/\sqrt{3}$, $\sin(\operatorname{Tan}^{-1}(1/\sqrt{3})) = \sin(\pi/6) = 1/2$ (see Figure 3.7).

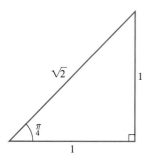

Figure 3.6 $\tan(\pi/4) = 1$;
$\text{Tan}^{-1}(1) = \pi/4$.

(d) Since $\tan(\pi/4) = 1$, solution (a) yields $\text{Tan}^{-1}(\tan(\pi/4)) = \text{Tan}^{-1}(1) = \pi/4$.
(e) Similarly, $\tan(5\pi/4) = 1$. Consequently, $\text{Tan}^{-1}(\tan(5\pi/4)) = \pi/4$. ∎

REMARK   Note that solutions (d) and (e) of Example 1 illustrate the fact that $g(x) = \text{Tan}^{-1} x$ is the inverse to $f(x) = \tan x$ *only* if f is restricted to the domain $(-\pi/2, \pi/2)$. That is,

$$\text{Tan}^{-1}(\tan(x)) = x \quad \text{if} \quad -\pi/2 < x < \pi/2,$$

but

$$\text{Tan}^{-1}(\tan(x)) \neq x \quad \text{if} \quad x < -\pi/2 \text{ or } x > \pi/2.$$

However, Definition 1 implies

$$\tan(\text{Tan}^{-1}(x)) = x$$

for all real numbers x (see Exercise 30).

□ EXAMPLE 2

At what angle should the camera be inclined when the rocket in Figure 3.1 is 7.8 kilometers above the ground?

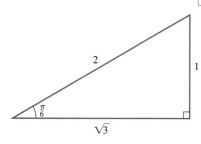

Figure 3.7 $\tan(\pi/6) = 1/\sqrt{3}$;
$\text{Tan}^{-1}(1/\sqrt{3}) = \pi/6$;
$\sin(\text{Tan}^{-1}(1/\sqrt{3})) = 1/2$.

Solution   From equation (1), when $s = 7.8$ km we obtain

$$\tan \theta = \frac{7.8}{3} = 2.6.$$

If you have access to a calculator or computer with the ability to compute values of the inverse tangent, simply compute $\text{Tan}^{-1}(2.6)$. If not, use the table of trigonometric functions in Appendix VI to find an angle whose tangent is approximately 2.6. By either method the answer is

$$\theta = \text{Tan}^{-1}(2.6) \approx 1.20 \text{ radians} \approx 69°. \qquad ∎$$

Other Inverse Trigonometric Functions

To find inverses for the remaining five trigonometric functions, we follow the same method we used to obtain $\text{Tan}^{-1} x$, namely:

(i) Restrict the domain of the trigonometric function $y = f(x)$ to an interval I on which it is one-to-one.
(ii) Define the inverse function by the usual definition

$$y = f^{-1}(x) \quad \text{if and only if} \quad x = f(y) \quad \text{with} \quad y \in I.$$

Definition 2

The remaining five inverse trigonometric functions are defined as follows:

(a) For $-1 \leq x \leq 1$ and $-\pi/2 \leq y \leq \pi/2$, the **inverse sine function** is defined by $y = \text{Sin}^{-1} x$ if and only if $x = \sin y$.
(b) For $-1 \leq x \leq 1$ and $0 \leq y \leq \pi$, the **inverse cosine function** is defined by $y = \text{Cos}^{-1} x$ if and only if $x = \cos y$.
(c) For $|x| \geq 1$ and $y \in [0, \pi/2) \cup [\pi, 3\pi/2)$, the **inverse secant function** is defined by $y = \text{Sec}^{-1} x$ if and only if $x = \sec y$.
(d) For $|x| \geq 1$ and $y \in (0, \pi/2] \cup (\pi, 3\pi/2]$, the **inverse cosecant function** is defined by $y = \text{Csc}^{-1} x$ if and only if $x = \csc y$.
(e) For $-\infty < x < \infty$ and $0 < y < \pi$, the **inverse cotangent function** is defined by $y = \text{Cot}^{-1} x$ if and only if $x = \cot y$.

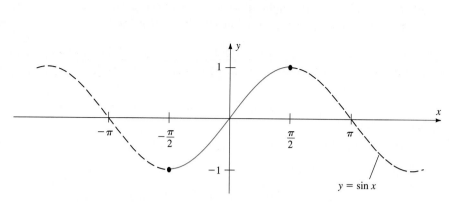

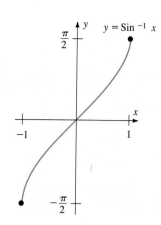

Figure 3.8 Principal branch of $y = \sin x$ is $-\pi/2 \le x \le \pi/2$.

Figure 3.9 Graph of $y = \mathrm{Sin}^{-1} x$ is reflection of graph of principal branch of $y = \sin x$ in line $y = x$.

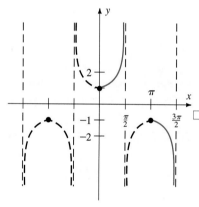

Figure 3.10 Principal branch of $y = \sec x$ is $x \in [0, \pi/2) \cup [\pi, 3\pi/2)$.

(Note that the domains of both $\mathrm{Sec}^{-1} x$ and $\mathrm{Csc}^{-1} x$ are unions of disjoint intervals because the *ranges* of $\sec x$ and $\csc x$ consist of disjoint intervals. It is possible to choose different ranges for $\mathrm{Sec}^{-1} x$ and $\mathrm{Csc}^{-1} x$, such as $[0, \pi/2) \cup (\pi/2, \pi]$ for $\mathrm{Sec}^{-1} x$.)

Figures 3.8 through 3.11 illustrate the principal branches of the sine and secant functions, together with the graphs of the corresponding inverse functions. (The graphs of the other inverse functions are constructed in a similar manner. See Exercise 40.)

☐ **EXAMPLE 3**

Find (a) $\mathrm{Sin}^{-1}(-\sqrt{3}/2)$, (b) $\mathrm{Csc}^{-1}(2)$, and (c) $\tan(\mathrm{Cos}^{-1}(-1/2))$.

Solution

(a) An angle x that satisfies $\sin x = -\sqrt{3}/2$ and $-\pi/2 \le x \le \pi/2$ is $x = -\pi/3$. Thus, $\mathrm{Sin}^{-1}(-\sqrt{3}/2) = -\pi/3$ (Figure 3.12).

(b) An angle x with $\csc x = 2$ and $x \in (0, \pi/2] \cup (\pi, 3\pi/2]$ is $x = \pi/6$. Thus, $\mathrm{Csc}^{-1}(2) = \pi/6$ (Figure 3.13).

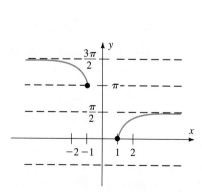

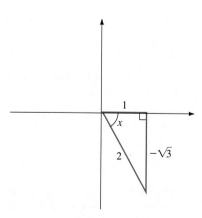

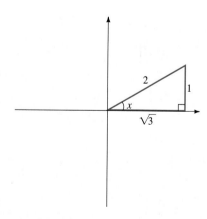

Figure 3.11 Graph of $y = \mathrm{Sec}^{-1} x$ is reflection of principal branch of $y = \sec x$ in line $y = x$.

Figure 3.12 $\sin(-\pi/3) = -\sqrt{3}/2$ so $\mathrm{Sin}^{-1}(-\sqrt{3}/2) = -\pi/3$.

Figure 3.13 $\csc(\pi/6) = 2$ so $\mathrm{Csc}^{-1}(2) = \pi/6$.

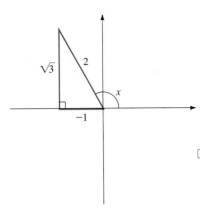

Figure 3.14 $\cos(2\pi/3) = -1/2$ so $\text{Cos}^{-1}(-1/2) = 2\pi/3$, $\tan(\text{Cos}^{-1}(-1/2)) = -\sqrt{3}$.

(c) An angle x with $\cos x = -1/2$ and $0 \le x \le \pi$ is $x = 2\pi/3$. Thus, $\tan(\text{Cos}^{-1}(-1/2)) = \tan(2\pi/3) = -\sqrt{3}$ (Figure 3.14).   ■

The following example demonstrates the interesting fact that a composition of a trigonometric function with an inverse trigonometric function produces an algebraic function.

☐ **EXAMPLE 4**

Let $-1 \le x \le 1$. Express $y = \sin(\text{Cos}^{-1} x)$ as an algebraic function of x.

Solution We use the identity $\sin^2 \theta + \cos^2 \theta = 1$ where $\theta = \text{Cos}^{-1} x$. Since the range of the inverse cosine function is the interval $[0, \pi]$, we know that $\sin \theta \ge 0$. Thus,

$$\sin \theta = \sqrt{1 - \cos^2 \theta}.$$

Then

$$\sin(\text{Cos}^{-1} x) = \sqrt{1 - [\cos(\text{Cos}^{-1} x)]^2}$$
$$= \sqrt{1 - x^2}.$$ ■

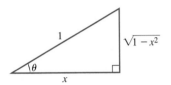

Figure 3.15 $\cos \theta = x$; $\sin \theta = \sqrt{1 - x^2}$.

Figure 3.15 illustrates this simplification.

Exercise Set 9.3

In Exercises 1–19, find the values indicated.

1. $\text{Sin}^{-1}(1/2)$

2. $\text{Cos}^{-1}(-\sqrt{3}/2)$

3. $\text{Tan}^{-1}(0)$

4. $\text{Sec}^{-1}(2)$

5. $\text{Cos}^{-1}(-1)$

6. $\text{Cot}^{-1}(-\sqrt{3})$

7. $\text{Sin}^{-1}(1) + \text{Cos}^{-1}(1)$

8. $\text{Tan}^{-1}(-1/\sqrt{3})$

9. $\tan(\text{Sin}^{-1}(1/2))$

10. $\text{Cos}^{-1}(\sin \pi/4)$

11. $\sec(\text{Tan}^{-1}(\sqrt{3}))$

12. $\text{Cot}^{-1}(\tan(2\pi/3))$

13. $\cos(2\,\text{Tan}^{-1}(\sqrt{3}))$

14. $\sin(2\,\text{Cot}^{-1}(3/4))$

15. $\tan^2(\pi - \text{Sin}^{-1}(\sqrt{2}/2))$

16. $\text{Sec}^{-1}(2 \tan \pi/4)$

17. $\text{Cos}^{-1}(2 - \sqrt{2} \sin(\pi/4))$

18. $\sec(2\,\text{Tan}^{-1} 1)$

19. $\tan(\text{Sec}^{-1} 2)$

In Exercises 20–29, find an algebraic expression for the given function of x.

20. $y = \cos(\text{Sin}^{-1} x)$

21. $y = \tan(\text{Sin}^{-1} x)$

22. $y = \tan(\text{Tan}^{-1} x)$

23. $y = \sin(2 \text{Sin}^{-1}(x))$

24. $y = \sin(\text{Tan}^{-1} x)$

25. $y = \sin(\text{Sec}^{-1} x)$

26. $y = \cos(\text{Tan}^{-1} x)$

27. $y = \tan(\text{Cot}^{-1} x)$

28. $y = \tan(\text{Sec}^{-1} x)$

29. $y = \cos(\text{Csc}^{-1} x)$

30. Explain why the identity $\sin(\text{Sin}^{-1} x) = x$ is valid for $-1 \le x \le 1$. Are similar identities true for the other inverse trigonometric functions? If so, for what values of x do they hold?

31. True or false? $\text{Sec}^{-1}(\cos \pi/4) = 2/\sqrt{2}$.

32. The selection of the principal branch of a trigonometric function is somewhat arbitrary.
    a. Show that the selection of the principal branch of the sine function as $\pi/2 \le x \le 3\pi/2$ leads to a legitimate inverse function for $y = \sin x$.
    b. What would be the disadvantage of choosing the principal branch of $y = \sin x$ to be $0 \le x \le \pi/2$?
    c. What properties should the principal branch of a trigonometric function possess?

33. A lighthouse lies 100 m off a straight shoreline. A boat dock is located on the shoreline at the point nearest the lighthouse. Write an equation describing θ, the angle between the light beam and the line joining the lighthouse and the dock, as a function of the distance x between the dock and the point where the beam of light strikes the shore (see Figure 3.16).

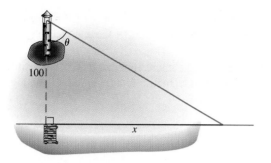

Figure 3.16 Lighthouse problem.

34. Ignoring the effects of gravity, determine the smallest angle at which a football must be kicked to make a 30-yard field goal if the crossbar is 9 ft high? More generally, specify a function that gives the angle θ in terms of the length l of the kick. (The length of a field goal is the distance between the point where the ball is kicked and the base of the goal posts.)

35. A telephone pole is sometimes supported by cables running from near the top of the pole to the ground. Find the angle θ between one of the cables and the ground if the cable is 60 ft long and the telephone pole is 30 ft high.

36. Snell's Law of Refraction states that if a ray of light passes from one medium to another, the angles θ_1 and θ_2 formed between the rays and the normal (perpendicular line) to the boundary are related by the equation

$$\frac{\sin \theta_1}{n_1} = \frac{\sin \theta_2}{n_2}$$

where n_1 and n_1 are constants. Express θ_2 as a function of θ_1 (see Figure 3.17).

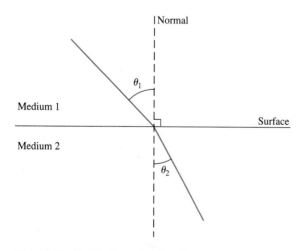

Figure 3.17 Snell's Law of Refraction.

37. A patrol car is parked on an overpass 20 m above a roadway, and the officer is using radar to time the cars moving along the roadway. Ignoring the heights of the cars, express the angle formed between the radar beam and the vertical if a car is x m away from the point on the roadway directly beneath the patrol car.

38. Compute the angle θ in Figure 3.18.

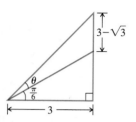

Figure 3.18

39. A person sitting on the ground is flying a kite that is at the end of a $100\sqrt{2}$ ft long kite string. Initially the kite is 100 ft above the ground. Then suddenly it loses one half of its altitude. What is the change in the angle made by the kite string and the ground (assuming that the string is taut at all times)?

40. Graph the inverse functions $y = \text{Cot}^{-1} x$, $y = \text{Cos}^{-1} x$, and $y = \text{Csc}^{-1} x$.

41. Graph the function $f(x) = \text{Cos}^{-1}(1/x)$ for $1 \le x \le \pi$ (use $0 \le y \le 2.1$). Compare this graph to the graph of $g(x) = \sec x$ for $0 \le x < \pi/2$. What do you conclude? (*Hint:* Graph $y = x$ also.) What can you say about $\text{Sin}^{-1}(1/x)$ and $\text{Tan}^{-1}(1/x)$?

42. A courtyard enclosed by a circular wall has only one source of light at night, a floodlight on the wall as shown below.

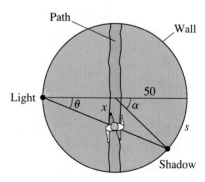

Figure 3.19

A person walking along the path (a 100-ft diameter of the circle) casts a shadow on the opposite wall. Determine the functions that relate the linear displacement x and the angle θ to the circular displacement s. Graph both of these functions and indicate their domains and ranges.

43. The **Tschebyscheff polynomials** are important in the numerical approximation of functions. In this exercise, we see how they arise as compositions of trigonometric and inverse trigonometric functions.

a. Graph the function $T_2(x) = \cos(2 \, \mathrm{Cos}^{-1} x)$. Show that T_2 is a polynomial of degree 2 and determine its domain and range. Do the same for $T_3(x) = \cos(3 \, \mathrm{Cos}^{-1} x)$. (*Hint:* $\cos(3\theta) = \cos(2\theta + \theta)$.) These two functions are Tschebyscheff polynomials of the first kind. In general, a Tschebyscheff polynomial of the first kind is defined by

$T_n(x) = \cos(n \, \mathrm{Cos}^{-1} x)$. Graph $T_n(x)$ for $n = 1, 2, 3, 4$, and 5.

b. Repeat part a for the functions

$$U_n(x) = \frac{\sin((n+1)\,\mathrm{Cos}^{-1} x)}{\sin(\mathrm{Cos}^{-1} x)}.$$

These functions are called the Tschebyscheff polynomials of the second kind.

9.4   Derivatives of the Inverse Trigonometric Functions

Each inverse trigonometric function f^{-1} is differentiable throughout its domain D except at numbers $b = f(a)$ with $f'(a) = 0$, according to Theorem 4, Chapter 8. To obtain these derivatives we shall use the identity

$$f(f^{-1}(x)) = x, \qquad x \in D, \tag{1}$$

and differentiate implicitly to find $\dfrac{dy}{dx} = \dfrac{d}{dx} f^{-1}(x)$.

For the function $y = \mathrm{Tan}^{-1} x$, identity (1) becomes

$$\tan y = x, \qquad -\infty < x < \infty.$$

Differentiating implicitly with respect to x gives

$$\sec^2 y \, \frac{dy}{dx} = 1,$$

so

$$\frac{dy}{dx} = \frac{1}{\sec^2 y}$$

$$= \frac{1}{1 + \tan^2 y} \qquad (\sec^2 y = 1 + \tan^2 y)$$

$$= \frac{1}{1 + x^2} \qquad (\tan y = x).$$

$$\boxed{\frac{d}{dx} \mathrm{Tan}^{-1} x = \frac{1}{1 + x^2}, \qquad -\infty < x < \infty.} \tag{2}$$

□ **EXAMPLE 1**

For the functions

$$\text{(a)} \ f(x) = \mathrm{Tan}^{-1}(3x) \quad \text{and} \quad \text{(b)} \ g(x) = x \, \mathrm{Tan}^{-1} e^x,$$

we have

$$\text{(a)} \ f'(x) = \frac{1}{1 + (3x)^2} \left(\frac{d}{dx} 3x \right) = \frac{3}{1 + 9x^2}, \text{ and}$$

(b) $g'(x) = \left(\dfrac{d}{dx}x\right)\mathrm{Tan}^{-1}e^x + x\left(\dfrac{d}{dx}\mathrm{Tan}^{-1}e^x\right)$

$\qquad = \mathrm{Tan}^{-1}e^x + \dfrac{x}{1+(e^x)^2}\left(\dfrac{d}{dx}e^x\right)$

$\qquad = \mathrm{Tan}^{-1}e^x + \dfrac{xe^x}{1+e^{2x}}.$ ■

To find the derivative of the function $y = \mathrm{Sin}^{-1}x$, we begin with the identity

$$\sin y = x, \quad |x| \le 1, \quad -\pi/2 \le y \le \pi/2.$$

Differentiating implicitly with respect to x gives

$$\cos y \,\frac{dy}{dx} = 1,$$

so

$$\frac{dy}{dx} = \frac{1}{\cos y}$$

$$= \frac{1}{\sqrt{1-(\sin y)^2}} \qquad \begin{array}{l}(\cos y = +\sqrt{1-\sin^2 y}\\ \text{for } -\pi/2 \le y \le \pi/2)\end{array}$$

$$= \frac{1}{\sqrt{1-x^2}} \qquad (\sin y = x)$$

where we must require that $1 - x^2 > 0$.

$$\boxed{\;\frac{d}{dx}\mathrm{Sin}^{-1}x = \frac{1}{\sqrt{1-x^2}}, \qquad |x| < 1.\;}\tag{3}$$

We may establish differentiation formulas for the remaining inverse trigonometric functions by proceeding in the same fashion. The results are:

$$\boxed{\begin{aligned}\frac{d}{dx}\mathrm{Cos}^{-1}x &= \frac{-1}{\sqrt{1-x^2}}, & |x| &< 1 & (4)\\[2mm]\frac{d}{dx}\mathrm{Cot}^{-1}x &= \frac{-1}{1+x^2}, & -\infty &< x < \infty & (5)\\[2mm]\frac{d}{dx}\mathrm{Sec}^{-1}x &= \frac{1}{x\sqrt{x^2-1}}, & |x| &> 1 & (6)\\[2mm]\frac{d}{dx}\mathrm{Csc}^{-1}x &= \frac{1}{x\sqrt{x^2-1}}, & |x| &> 1 & (7)\end{aligned}}$$

Note the restrictions $|x| < 1$ in (4) and $|x| > 1$ in (6) and (7). If these inequalities do not hold, the corresponding formulas are meaningless. Note also that

$$\frac{d}{dx} \text{Cos}^{-1} x = -\frac{d}{dx} \text{Sin}^{-1} x$$

$$\frac{d}{dx} \text{Cot}^{-1} x = -\frac{d}{dx} \text{Tan}^{-1} x$$

$$\frac{d}{dx} \text{Csc}^{-1} x = -\frac{d}{dx} \text{Sec}^{-1} x$$

□ **EXAMPLE 2**

Find $\dfrac{dy}{dx}$ for $y = x^3 \text{Sin}^{-1} x + \text{Cos}^{-1} \sqrt{x}$.

Solution   Using equations (3) and (4) we obtain

$$\frac{dy}{dx} = \left(\frac{d}{dx} x^3\right) \text{Sin}^{-1} x + x^3 \left(\frac{d}{dx} \text{Sin}^{-1} x\right) + \frac{-1}{\sqrt{1 - (\sqrt{x})^2}} \left(\frac{d}{dx} \sqrt{x}\right)$$

$$= 3x^2 \text{Sin}^{-1} x + \frac{x^3}{\sqrt{1 - x^2}} - \frac{1}{2\sqrt{x}\sqrt{1 - x}}.$$

Note that this derivative exists only for $0 < x < 1$.   ■

□ **EXAMPLE 3**

Prove the identity $\text{Tan}^{-1} x = \pi/2 - \text{Tan}^{-1}(1/x)$ by differentiation.

Strategy · · · · · · · ·

Show

$$\frac{d}{dx} \text{Tan}^{-1} x = -\frac{d}{dx} \text{Tan}^{-1} \frac{1}{x}$$

by computing $\dfrac{d}{dx} \text{Tan}^{-1}\left(\dfrac{1}{x}\right)$.

Solution

By equation (2) we have

$$\frac{d}{dx} \text{Tan}^{-1}\left(\frac{1}{x}\right) = \frac{1}{1 + \left(\dfrac{1}{x}\right)^2} \left(-\frac{1}{x^2}\right)$$

$$= \frac{-1}{1 + x^2}$$

$$= -\frac{d}{dx} \text{Tan}^{-1} x.$$

Recall, if $f' = g'$ then $f = g(x) + C$.

Thus

$$\text{Tan}^{-1} x = -\text{Tan}^{-1}\left(\frac{1}{x}\right) + C$$

To find C, substitute particular value of x.

for some constant C. To find C we substitute $x = 1$ to obtain

$$\text{Tan}^{-1} 1 = -\text{Tan}^{-1} 1 + C$$

Use fact that $\text{Tan}^{-1} 1 = \pi/4$.

or $2 \text{Tan}^{-1} 1 = C$. So $C = \pi/2$.   ■

The differentiation formulas (2), (3), and (6) give the following integration formulas:

$$\int \frac{dx}{\sqrt{1 - x^2}} = \mathrm{Sin}^{-1} x + C \qquad (8)$$

$$\int \frac{dx}{1 + x^2} = \mathrm{Tan}^{-1} x + C \qquad (9)$$

$$\int \frac{dx}{x\sqrt{x^2 - 1}} = \mathrm{Sec}^{-1} x + C \qquad (10)$$

(We do not state the integration formulas corresponding to differentiation formulas (4), (5), and (7), since they involve only the negatives of these integrals.)

The next two examples show how we can use substitutions to bring integrals into the form of equations (8–10).

☐ **EXAMPLE 4**

Find $\displaystyle\int \frac{1}{\sqrt{9 - x^2}}\, dx$.

Strategy · · · · · · · ·
Factor out 9 to obtain form of equation (8).

Solution

$$\int \frac{dx}{\sqrt{9 - x^2}} = \int \frac{dx}{\sqrt{9(1 - x^2/9)}} = \int \frac{dx}{3\sqrt{1 - (x/3)^2}}$$

Make substitution.

We make the substitution $u = x/3$; $du = \frac{1}{3}\, dx$. Then $dx = 3\, du$, so the integral becomes

Apply equation (8).

$$\frac{1}{3}\int \frac{3\, du}{\sqrt{1 - u^2}} = \int \frac{du}{\sqrt{1 - u^2}}$$

$$= \mathrm{Sin}^{-1}\left(\frac{x}{3}\right) + C. \qquad ■$$

☐ **EXAMPLE 5**

Find $\displaystyle\int_1^3 \frac{dx}{\sqrt{x}(1 + x)}$.

Strategy · · · · · · · ·
Make a substitution so that $1/\sqrt{x}$ is a factor of du. Here $u = \sqrt{x}$ works.

Solution
We make the substitution

$$u = \sqrt{x}; \qquad du = \frac{1}{2\sqrt{x}}\, dx.$$

Change limits of integration.

Thus $\dfrac{1}{\sqrt{x}}\, dx = 2\, du$. We must also change limits of integration:

$$\text{If } x = 1, \quad u = \sqrt{1} = 1.$$
$$\text{If } x = 3, \quad u = \sqrt{3}.$$

Use formula (9).

With these substitutions we obtain

$$\int_1^3 \frac{dx}{\sqrt{x}(1 + x)} = \int_1^{\sqrt{3}} \frac{2\, du}{1 + u^2}$$

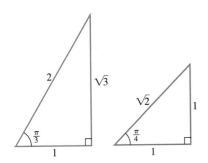

Figure 4.1

$$= 2\left[\operatorname{Tan}^{-1} u\right]_{1}^{\sqrt{3}}$$

$$= 2[\operatorname{Tan}^{-1}\sqrt{3} - \operatorname{Tan}^{-1} 1]$$

$$= 2[\pi/3 - \pi/4] = \pi/6. \qquad \blacksquare$$

The derivatives of inverse trigonometric functions are algebraic functions. Among other things, this remarkable fact yields a simple procedure for approximating the irrational number π.

☐ **EXAMPLE 6**

Use integration formula (9) to approximate π.

Strategy · · · · · · · ·
We select limits to obtain π as the value of the definite integral. We then approximate this integral.

Solution
Since

$$\int \frac{dx}{1 + x^2} = \operatorname{Tan}^{-1} x + C,$$

we have

$$\int_{0}^{1} \frac{dx}{1 + x^2} = \left[\operatorname{Tan}^{-1} x\right]_{0}^{1} = \operatorname{Tan}^{-1}(1) - \operatorname{Tan}^{-1}(0) = \pi/4.$$

Thus

$$\pi = 4\int_{0}^{1} \frac{dx}{1 + x^2}.$$

We may therefore approximate π by approximating the definite integral. The results of using Simpson's Rule with various values of n to approximate this integral appear in Table 4.1. $\qquad \blacksquare$

Table 4.1

n	Approximation to π using Simpson's Rule
4	3.14156863
10	3.14159262
20	3.14159253
50	3.14159251

Exercise Set 9.4

In Exercises 1–14, find the derivative.

1. $y = \operatorname{Sin}^{-1} 3x$

2. $f(x) = x\operatorname{Tan}^{-1}(x - 1)$

3. $f(t) = \operatorname{Sin}^{-1}\sqrt{t}$

4. $y = x^3\operatorname{Csc}^{-1}(1 + x)$

5. $f(x) = \operatorname{Sin}^{-1} e^{-x}$

6. $y = \operatorname{Tan}^{-1}\sqrt{x^2 - 1}$

7. $y = \sqrt{\operatorname{Cos}^{-1} x}$

8. $f(x) = \operatorname{Sin}^{-1} x^2 - x\operatorname{Sec}^{-1}(x + 3)$

9. $f(x) = \ln\operatorname{Tan}^{-1} x$

10. $y = \operatorname{Tan}^{-1}\left(\dfrac{1 - x}{1 + x}\right)$

11. $y = \operatorname{Sec}^{-1}(1/x), \quad 0 < x < 1$

12. $y = e^{\operatorname{Tan}^{-1}\sqrt{x}}$

13. $f(x) = \dfrac{\operatorname{Tan}^{-1} x}{1 + x^2}$

14. $f(x) = x^2\operatorname{Sin}^{-1} 2x$

In Exercises 15–37, find the indicated integral.

15. $\displaystyle\int \frac{1}{4 + x^2}\, dx$

16. $\displaystyle\int_{0}^{\sqrt{3}/2} \frac{dx}{\sqrt{1 - x^2}}$

17. $\displaystyle\int \frac{1}{1 + 9x^2}\, dx$

18. $\displaystyle\int \frac{5}{\sqrt{1 - 4x^2}}\, dx$

19. $\displaystyle\int_{8\sqrt{3}/3}^{8} \frac{1}{2x\sqrt{x^2 - 16}}\, dx$

20. $\displaystyle\int \frac{\cos x}{1 + \sin^2 x}\, dx$

21. $\displaystyle\int \frac{x}{\sqrt{1 - x^2}}\, dx$

22. $\displaystyle\int_{\sqrt{2}}^{2} \frac{1}{x\sqrt{x^2 - 1}}\, dx$

23. $\displaystyle\int \frac{1}{\sqrt{e^{2x} - 1}}\, dx$

24. $\displaystyle\int \frac{x}{\sqrt{1 - x^4}}\, dx$

25. $\displaystyle\int \frac{e^{\sqrt{x}}}{\sqrt{x}(1 + e^{2\sqrt{x}})}\,dx$

26. $\displaystyle\int \frac{x^2}{\sqrt{1 - x^6}}\,dx$

27. $\displaystyle\int \frac{x^2}{1 + x^6}\,dx$

28. $\displaystyle\int \frac{\sin x}{1 + \cos^2 x}\,dx$

29. $\displaystyle\int_0^1 \frac{\mathrm{Tan}^{-1} x}{1 + x^2}\,dx$

30. $\displaystyle\int \frac{\mathrm{Tan}^{-1} 3x}{1 + 9x^2}\,dx$

31. $\displaystyle\int \frac{(\mathrm{Cos}^{-1} 2x)^3}{\sqrt{1 - 4x^2}}\,dx$

32. $\displaystyle\int \frac{1}{(\mathrm{Sin}^{-1} x)\sqrt{1 - x^2}}\,dx$

33. $\displaystyle\int_0^{\pi/2} \frac{\cos x}{\sqrt{4 - \sin^2 x}}\,dx$

34. $\displaystyle\int_1^2 \frac{dx}{x^2 + 2}$

35. $\displaystyle\int_2^3 \frac{dx}{x\sqrt{16x^2 - 25}}$

36. $\displaystyle\int \frac{dx}{2x\sqrt{4x^2 - 1}}$

37. $\displaystyle\int_1^{\sqrt{e}} \frac{1}{x\sqrt{1 - \ln^2 x}}\,dx$

38. Use separation of variables (see Section 5.3) to solve the initial-value problem

$$\frac{dy}{dx} = 4 + y^2, \quad y(0) = 1.$$

39. Solve the initial-value problem

$$\frac{dy}{dx} = \sqrt{1 - y^2}, \quad y(0) = 1.$$

40. In the radar problem of Exercise 37, Section 9.3, how fast must the radar gun rotate to track an automobile 40 m away from the point directly under the radar gun if the automobile is moving at a rate of 120 km/hr?

41. A tapestry 3 m high hangs on a wall so that its lower edge is 1 m above an observer's eye level. How far from the wall should the observer stand so as to maximize the angle subtended in the observer's eye by the tapestry?

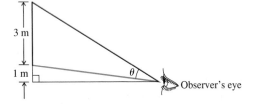

Figure 4.2

42. A ladder 10 m long leans against a wall. If the base of the ladder is slipping away from the wall at a rate of 2 m/s, how fast is the angle between the ladder and the wall increasing when the top of the ladder is 8 m above the ground?

43. A dog is leashed to the top of a 10-ft pole by a flexible leash. If the dog runs away from the pole at a rate of 4 ft/s, how fast is the angle between the leash and the pole changing when the dog is 10 ft from the base of the pole? How fast is the length of the leash changing at that instant?

44. Two boards 10 m long are joined together by a hinge and placed upright on the ground. If the boards fall at the rate of -2 m/s in such a way that the outward movement of each board is identical, how fast is the angle between the boards changing when the hinge is 5 m above the ground?

45. Analysis shows that the first minimum for the diffraction pattern of a circular aperture of diameter d, assuming certain conditions, is given by the equation

$$\sin \theta = 1.22\left(\frac{\lambda}{d}\right), \quad 0 < \theta < \frac{\pi}{2},$$

where λ is the wavelength of light (a constant). Assuming that θ is a differentiable function of d, show that θ is a decreasing function of d.

46. Find an equation for the line tangent to the graph of $y = \mathrm{Sin}^{-1} x^2$ at the point $(\sqrt{2}/2, \pi/6)$.

47. Find the area of the region bounded by the graph of

$$f(x) = \frac{x^2}{1 + x^6}$$

and the x-axis for $0 \le x \le 1$.

48. Find the area bounded by the graphs of $y = 2/(1 + x^2)$ and $y = x$ for $1 \le x \le \sqrt{3}$.

49. Find the volume of the solid generated by revolving about the y-axis the region bounded by the graph of $f(x) = 1/\sqrt{1 - x^4}$ and the x-axis, for $0 \le x \le \sqrt{2}/2$.

50. Find the volume of the solid generated by revolving about the y-axis the region bounded by the graph of

$$f(x) = \frac{1}{x^2\sqrt{x^2 - 1}}$$

and the x-axis, for $\sqrt{2} \le x \le 2$.

51. Find the volume of the solid generated by revolving about the x-axis the region bounded by the graph of $f(x) = 2/\sqrt{1 + 4x^2}$ and the x-axis, for $0 \le x \le \sqrt{3}$.

52. Find the average value of the function

$$f(x) = \frac{xe^{-x^2}}{\sqrt{1 - e^{-2x^2}}}$$

on the interval $[\sqrt{\ln 2}, 1]$.

53. Derive formula (4).

54. Derive formula (5).

55. Derive formula (6).

56. Derive formula (7).

57. What can you conclude about the relationship between the inverse sine function and the inverse cosine function using the equality

$$\frac{d}{dx} \text{Cos}^{-1} x = -\frac{d}{dx} \text{Sin}^{-1} x$$

(see Example 3)? What trigonometric identity from Section 1.5 is relevant to your conclusion?

 58. A radar tracking station is located 250 ft from the base of a 26-ft-high missile launching pad, as illustrated in the following figure.

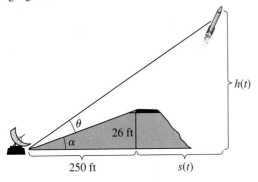

Figure 4.3

Suppose that, at time t, the missile's horizontal distance from the launch pad is given by $s(t) = 8.5t^2$ and its altitude given by $h(t) = 39 + 480t - 24t^2$.

a. Express the angle θ as a function of t.

b. Graph $\theta(t)$ and $\theta'(t)$. Using these graphs, determine the maximum value of $\theta(t)$. What is the altitude h and the distance s of the missile at this time?

c. When will the launch pad block the radar? How high and far away is the missile at this time?

d. How much further will the missile travel after it is no longer visible to the radar?

59. Consider a person walking at the rate of 1.5 ft/s across the courtyard described in Exercise 42 of Section 9.3.

a. Determine the maximum rate of change of the angle θ.

b. Determine the maximum velocity of the shadow.

9.5 The Hyperbolic Functions

Recall from Chapter 8 that the function $y = Ce^x$ is the only nontrivial differentiable function satisfying the differential equation

$$\frac{dy}{dx} = y. \tag{1}$$

We have already seen that equation (1) has various practical interpretations corresponding to the interpretations of the first derivative (e.g., velocity, rate of growth, and slope of tangent). Since the second derivative also has meaningful interpretations (acceleration, concavity, change in rate of growth), it is natural to ask which nonzero functions satisfy the second-order differential equation.

$$\frac{d^2y}{dx^2} = y. \tag{2}$$

A quick check shows that $y_1 = e^x$ and $y_2 = e^{-x}$ are both solutions of equation (2). Moreover, certain combinations of these two functions also satisfy equation (2) and, in addition, satisfy properties quite similar to the properties of the trigonometric functions. We are referring to the following functions.

Definition 3

The **hyperbolic sine and cosine functions,** $\sinh x$ and $\cosh x$, are defined as follows:

$$\sinh x = \frac{e^x - e^{-x}}{2}, \tag{3}$$

$$\cosh x = \frac{e^x + e^{-x}}{2}. \tag{4}$$

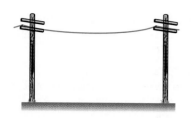

Figure 5.1 Cable suspended from two points hangs in the shape of a *catenary,* which is described by the hyperbolic cosine function.

Roughly speaking, cosh x represents the average of exponential growth and exponential decay, while sinh x represents half the difference between these two phenomena.

The *hyperbolic functions* are important in applied problems in physics and engineering in which the quantities of interest are described not by explicit functions but by *differential equations.* That is why we have come upon these functions by asking for the solutions of certain rather simple differential equations. For example, a cable suspended from two points hangs in the shape of a *catenary* (see Figure 5.1), which is described by the hyperbolic cosine function. (See Exercise 56.) We shall see in Chapter 14 why the term *hyperbolic* is used in conjunction with these functions.

Derivative Formulas

The differentiation formulas for the hyperbolic functions closely resemble those for the trigonometric functions.

For example, from equations (3) and (4) we obtain

$$\frac{d}{dx} \sinh x = \frac{d}{dx}\left(\frac{e^x - e^{-x}}{2}\right)$$
$$= \frac{e^x + e^{-x}}{2}$$
$$= \cosh x,$$

and

$$\frac{d}{dx} \cosh x = \frac{d}{dx}\left(\frac{e^x + e^{-x}}{2}\right)$$
$$= \frac{e^x - e^{-x}}{2}$$
$$= \sinh x.$$

(Note that the derivative of cosh x is sinh x, not $-\sinh x$.)

As a consequence of equations (3) and (4), both sinh x and cosh x are their own second derivatives (just as e^x and e^{-x} are):

$$\frac{d^2}{dx^2}(\sinh x) = \frac{d}{dx}(\cosh x) = \sinh x,$$

and

$$\frac{d^2}{dx^2}(\cosh x) = \frac{d}{dx}(\sinh x) = \cosh x.$$

Therefore, both $y_1 = \sinh x$ and $y_2 = \cosh x$ are solutions of differential equation (2).

REMARK It is easy to show that, if f and g are solutions of differential equation (2), then $c_1 f + c_2 g$ is also a solution for any choice of constants c_1 and c_2. (This is called a **linear combination** of solutions.) The functions sinh x and cosh x are merely special linear combinations of the exponential solutions of (2).

Graphs of sinh x and cosh x

From the definition of sinh x we can see that

$$\sinh(-x) = \frac{e^{-x} - e^{-(-x)}}{2} = \frac{e^{-x} - e^x}{2} = -\frac{e^x - e^{-x}}{2} = -\sinh x.$$

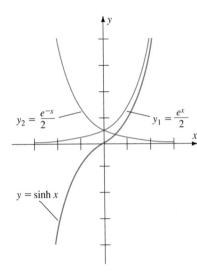

$y_2 = \dfrac{e^{-x}}{2}$ $y_1 = \dfrac{e^x}{2}$

$y = \sinh x$

Figure 5.2 $\sinh x = \dfrac{e^x - e^{-x}}{2}$.

Thus, $\sinh x$ is an *odd* function, and its graph is symmetric with respect to the origin. Moreover, since

$$\frac{d}{dx} \sinh x = \cosh x = \frac{e^x + e^{-x}}{2} > 0$$

for all x, $\sinh x$ is an increasing function on all intervals. Finally, since

$$\frac{d^2}{dx^2} \sinh x = \sinh x \quad \text{is} \quad \begin{cases} \text{positive, if } x > 0 \\ \text{negative, if } x < 0 \end{cases}$$

the graph of $\sinh x$ is concave down on $(-\infty, 0]$ and concave up on $[0, \infty)$ (see Figure 5.2).

A similar analysis shows that

(a) $\cosh x$ is an *even* function: $\cosh(-x) = \cosh x$ for all x. Its graph is therefore symmetric with respect to the y-axis,
(b) $\cosh x > 0$ for all x,
(c) $\cosh x$ is decreasing on $(-\infty, 0]$, increasing on $[0, \infty)$, and
(d) the graph of $\cosh x$ is concave up on $(-\infty, \infty)$.

The graph of $\cosh x$ appears in Figure 5.3.

As with the trigonometric functions, we can define four other hyperbolic functions in terms of $\sinh x$ and $\cosh x$.

Definition 4	The **hyperbolic tangent, cotangent, secant, and cosecant functions** are defined by

$$\tanh x = \frac{\sinh x}{\cosh x} = \frac{e^x - e^{-x}}{e^x + e^{-x}},$$

$$\coth x = \frac{\cosh x}{\sinh x} = \frac{e^x + e^{-x}}{e^x - e^{-x}},$$

$$\operatorname{sech} x = \frac{1}{\cosh x} = \frac{2}{e^x + e^{-x}},$$

$$\operatorname{csch} x = \frac{1}{\sinh x} = \frac{2}{e^x - e^{-x}}.$$

From Definition 4 we obtain the following differentiation formulas:

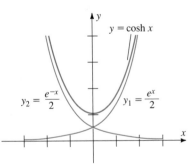

$y = \cosh x$

$y_2 = \dfrac{e^{-x}}{2}$ $y_1 = \dfrac{e^x}{2}$

Figure 5.3 $\cosh x = \dfrac{e^x + e^{-x}}{2}$.

$$\frac{d}{dx} \tanh x = \operatorname{sech}^2 x \tag{5}$$

$$\frac{d}{dx} \coth x = -\operatorname{csch}^2 x \tag{6}$$

$$\frac{d}{dx} \operatorname{sech} x = -\operatorname{sech} x \tanh x \tag{7}$$

$$\frac{d}{dx} \operatorname{csch} x = -\operatorname{csch} x \coth x \tag{8}$$

☐ **EXAMPLE 1**

For $f(x) = x \operatorname{sech}^2 \sqrt{x}$,

$$f'(x) = 1 \cdot \operatorname{sech}^2 \sqrt{x} + x \cdot \frac{d}{dx} (\operatorname{sech}^2 \sqrt{x})$$

$$= \operatorname{sech}^2 \sqrt{x} + x \left[2 \operatorname{sech} \sqrt{x} \, (-\operatorname{sech} \sqrt{x} \tanh \sqrt{x}) \frac{1}{2\sqrt{x}} \right]$$

$$= \operatorname{sech}^2 \sqrt{x} - \sqrt{x} \operatorname{sech}^2 \sqrt{x} \tanh \sqrt{x}.$$ ■

Identities

The hyperbolic functions satisfy identities similar to those for the trigonometric functions. For example,

$$\cosh^2 x = \left[\frac{e^x + e^{-x}}{2} \right]^2$$

$$= \frac{e^{2x} + 2 + e^{-2x}}{4}$$

$$= 1 + \frac{e^{2x} - 2 + e^{-2x}}{4}$$

$$= 1 + \left[\frac{e^x - e^{-x}}{2} \right]^2$$

$$= 1 + \sinh^2 x.$$

Thus,

$$\cosh^2 x - \sinh^2 x = 1. \tag{9}$$

Similarly, we can show that

$$\operatorname{sech}^2 x = 1 - \tanh^2 x \tag{10}$$

and

$$\operatorname{csch}^2 x = \coth^2 x - 1. \tag{11}$$

(Other identities are presented in the exercise set.)

☐ **EXAMPLE 2**

Show that the function $y = \tanh(ax)$ is a solution of the differential equation

$$\frac{dy}{dx} = a(1 - y^2).$$

Solution   By formula (5), the derivative of the given function is

$$\frac{dy}{dx} = \frac{d}{dx} \tanh(ax) = a \operatorname{sech}^2(ax).$$

Thus, with identity (10) we obtain

$$\frac{dy}{dx} = a \operatorname{sech}^2(ax) = a[1 - \tanh^2(ax)] = a(1 - y^2).$$ ■

Inverses

We have already verified that the hyperbolic sine function is increasing for all values of x. This condition guarantees that each value y corresponds to precisely one number x via the equation $y = \sinh x$. We may therefore define the **inverse** of the hyperbolic sine function $\sinh^{-1} x$ to be the function that reverses this correspondence. In other words,

$$y = \sinh^{-1} x \quad \text{if and only if} \quad x = \sinh y. \tag{12}$$

We can obtain an explicit formulation for the inverse hyperbolic sine function by combining equations (3) and (12): If

$$y = \sinh^{-1} x, \quad \text{then} \quad x = \sinh y = \frac{e^y - e^{-y}}{2},$$

so

$$e^y - 2x - e^{-y} = 0. \tag{13}$$

Multiplying both sides of (13) by e^y gives

$$e^{2y} - 2xe^y - 1 = 0. \tag{14}$$

Viewing (14) as a quadratic expression in the variable e^y and applying the quadratic formula gives

$$e^y = \frac{2x + \sqrt{4x^2 + 4}}{2}$$

$$= x \pm \sqrt{x^2 + 1}.$$

Since e^y is never negative and $\sqrt{x^2 + 1} > x$, the ambiguous sign $\pm$ must be $+$. Taking natural logs of both sides of the last equation now gives

$$y = \ln(x + \sqrt{x^2 + 1}),$$

so

$$\boxed{\sinh^{-1} x = \ln(x + \sqrt{x^2 + 1}), \quad -\infty < x < \infty.} \tag{15}$$

By suitably restricting domains where necessary, we may similarly define inverse functions for each of the remaining hyperbolic functions. By rewriting each hyperbolic function in terms of exponential functions, we can then obtain explicit formulas for these inverse functions.

The results are as follows:

$$\cosh^{-1} x = \ln(x + \sqrt{x^2 - 1}), \quad x \geq 1 \tag{16}$$

$$\operatorname{sech}^{-1} x = \ln\left(\frac{1 + \sqrt{1 - x^2}}{x}\right), \quad 0 < x \leq 1 \tag{17}$$

$$\operatorname{csch}^{-1} x = \ln\left(\frac{1}{x} + \frac{\sqrt{1 + x^2}}{|x|}\right), \quad x \neq 0 \tag{18}$$

$$\tanh^{-1} x = \frac{1}{2} \ln\left(\frac{1 + x}{1 - x}\right), \quad |x| < 1 \tag{19}$$

$$\coth^{-1} x = \frac{1}{2} \ln\left(\frac{x + 1}{x - 1}\right), \quad |x| > 1 \tag{20}$$

Derivative and integral formulas follow from equations (15)–(20) by applications of the corresponding rules for differentiation natural logarithms. For example, from (15) we have

$$\frac{d}{dx}\sinh^{-1}x = \frac{d}{dx}\ln(x + \sqrt{x^2 + 1})$$

$$= \frac{1 + \dfrac{1}{2\sqrt{x^2 + 1}}(2x)}{x + \sqrt{x^2 + 1}}$$

$$= \frac{\sqrt{x^2 + 1} + x}{\sqrt{x^2 + 1}(x + \sqrt{x^2 + 1})}$$

$$= \frac{1}{\sqrt{x^2 + 1}}.$$

Thus

$$\frac{d}{dx}\sinh^{-1}x = \frac{1}{\sqrt{x^2 + 1}}. \tag{21}$$

We leave as exercises the formulas

$$\frac{d}{dx}\cosh^{-1}x = \frac{1}{\sqrt{x^2 - 1}}, \qquad x > 1 \tag{22}$$

$$\frac{d}{dx}\tanh^{-1}x = \frac{1}{1 - x^2}, \qquad |x| < 1 \tag{23}$$

$$\frac{d}{dx}\coth^{-1}x = \frac{1}{1 - x^2}, \qquad |x| > 1 \tag{24}$$

$$\frac{d}{dx}\operatorname{sech}^{-1}x = \frac{-1}{x\sqrt{1 - x^2}}, \qquad 0 < x < 1 \tag{25}$$

$$\frac{d}{dx}\operatorname{csch}^{-1}x = \frac{-1}{|x|\sqrt{1 + x^2}}, \qquad x \neq 0 \tag{26}$$

By now you have no doubt realized that the hyperbolic functions and their inverses are really not "new" functions. Rather, they are simply combinations of exponential and logarithmic functions. However, these particular combinations occur often in certain applications. Because the differentiation and integration formulas for the hyperbolic functions are simpler than those for the corresponding exponential and logarithmic forms, it is convenient to use them as shorthand notation.

Exercise Set 9.5

In Exercises 1–10, find the value of the indicated expression.

1. $\sinh 0$

2. $\sinh 1$

3. $\cosh(\ln 2)$

4. $\sinh(\ln 4)$

5. $\tanh(\ln 2)$

6. $\operatorname{sech}(1)$

7. $\coth(\ln 4)$

8. $\operatorname{csch}(\ln \pi^3)$

9. $\sinh^{-1}(1)$

10. $\tanh^{-1}(1/2)$

In Exercises 11–32, find the derivative of the given function.

11. $y = \sinh 2x$

12. $f(x) = x \cosh x$

13. $f(x) = \sinh x \tanh x$

14. $y = \sinh^3(1 - x^2)$

15. $f(x) = \sqrt{\cosh(4x)}$

16. $y = \dfrac{1 - \cosh x}{1 + \cosh x}$

17. $y = 1/\cosh x$

18. $f(x) = \ln(\tanh x^2)$

19. $f(x) = e^x \operatorname{csch} x^2$

20. $f(x) = \cosh^{-1}(1 + x^2)$

21. $y = \sinh^{-1} 2x$

22. $y = \sqrt{\sinh^{-1} x}$

23. $f(s) = \tanh^{-1} s^2$

24. $y = \cosh^{-1} e^{x^2}$

25. $f(x) = \ln \cosh^{-1} \pi x$

26. $f(x) = \ln(1 + \cosh \pi x)$

27. $y = \cosh x \cosh x^2$

28. $y = \sqrt{1 + \cosh^2 3x}$

29. $y = \sinh \ln x^2$

30. $y = x \sinh^{-1}(1 + x^2)$

31. $y = \tanh \sqrt{1 + x^2}$

32. Using implicit differentiation, calculate dy/dx if $x + \cosh xy = y$.

In Exercises 33–48, find the integrals using the differentiation formulas of this section.

33. $\displaystyle \int_1^2 \frac{1}{\sqrt{1 + x^2}}\, dx$

34. $\displaystyle \int_1^2 x \sinh x^2\, dx$

35. $\displaystyle \int \tanh^2(2x)\, dx$

36. $\displaystyle \int e^x \sinh x\, dx$

37. $\displaystyle \int_0^2 \sinh x \cosh x\, dx$

38. $\displaystyle \int_0^1 \frac{1}{4 - x^2}\, dx$

39. $\displaystyle \int_4^6 \frac{dx}{\sqrt{x^2 - 4}}$

40. $\displaystyle \int_2^4 \frac{dx}{2 - x^2}$

41. $\displaystyle \int_0^1 \frac{1}{\sqrt{9x^2 + 25}}\, dx$

42. $\displaystyle \int_1^2 \frac{dx}{x\sqrt{4 + x^2}}$

43. $\displaystyle \int \frac{\sinh x}{\cosh x}\, dx$

44. $\displaystyle \int \frac{1}{\sinh^2 x}\, dx$

45. $\displaystyle \int \frac{\cosh x}{\sqrt{\sinh x}}\, dx$

46. $\displaystyle \int \frac{\sinh x}{1 + \cosh^2 x}\, dx$

47. $\displaystyle \int \tanh^2 x \operatorname{sech}^2 x\, dx$

48. $\displaystyle \int \frac{1 - \tanh^2 x}{1 + \tanh^2 x}\, dx$

49. True or false? The hyperbolic sine function is a periodic function with period 2π.

In Exercises 50–53, verify the identity.

50. $1 - \tanh^2 x = \operatorname{sech}^2 x$

51. $\cosh x \pm \sinh x = e^{\pm x}$

52. $\sinh 2x = 2 \sinh x \cosh x$

53. $\cosh 2x = 2 \sinh^2 x + 1$

In Exercises 54–55, verify the addition formula.

54. $\sinh(x + y) = \sinh x \cosh y + \sinh y \cosh x$.

55. $\cosh(x + y) = \cosh x \cosh y + \sinh x \sinh y$.

56. A flexible cable fastened at both ends hangs in the shape of the **catenary** $y = a \cosh(x/a)$. Find the length of the graph of the catenary $y = \cosh x$ from the point with x-coordinate 0 to the point with x-coordinate b.

57. Consider the situation in which one edge of a rectangular metal plate is held at a constant high temperature, while the opposite edge is held at a constant low temperature; the remaining two edges are insulated. In determining an equation that gives the temperature at each point on the plate, part of the problem involves solving the second-order differential equation

$$\frac{d^2 g(y)}{dy^2} - k^2 g(y) = 0,$$

where k is a nonzero constant. Verify that

$$g(y) = c_1 \sinh(ky) + c_2 \cosh(ky)$$

satisfies the differential equation for any choice of constants c_1 and c_2.

In Exercises 58–59, find as many solutions as you can for the given differential equation.

58. $\dfrac{d^2 y}{dx^2} - 4y = 0$

59. $y'' - k^2\pi^2 y = 0$

60. Verify that the function $y = A \sin x + B \cos x + C \sinh x + D \cosh x$ is a solution of the fourth-order differential equation

$$\frac{d^4 y}{dx^4} = y.$$

61. Find the area of the region bounded by the graphs of $y = \coth x$, $y = \tanh x$, $x = \ln 2$ and $x = \ln 4$.

62. Find the volume of the solid obtained by rotating about the x-axis the region bounded above by the graph of $y = \cosh x$ and below by the x-axis, for $-\ln 2 \le x \le \ln 2$.

In Exercises 63–65, find all relative extrema.

63. $f(x) = \cosh x - \sinh x$

64. $f(x) = 2 \cosh x + 5 \sinh x$

65. $f(x) = 5 \cosh x - 2 \sinh x$

In Exercises 66–68, find a second-order differential equation satisfied by the given function.

66. $f(x) = \cosh 2x$

67. $f(x) = \cosh 3x - 2 \sinh 3x$

68. $f(x) = A \cosh 6x + B \sinh 6x$.

69. Determine which of the hyperbolic trigonometric functions are odd and which are even.

70. Verify differentiation formulas (5)–(8).

71. Verify identity (11).

72. Verify differentiation formulas (22)–(26).

73. According to the definition of $\cosh^{-1} x$ in equation (16), what is the *principal branch* of the graph of $y = \cosh x$?

74. Verify that the function $y = \cosh x$ is increasing on $[0, \infty)$ and therefore invertible on this interval. Sketch the graph of $y = \cosh^{-1} x$ by reflecting the principal branch of $y = \cosh x$ in the line $y = x$.

75. Find the principal branch for each of the functions $\tanh x$, $\coth x$, $\operatorname{sech} x$, and $\operatorname{csch} x$, according to equations (17)–(20), and verify that each function is invertible if it is restricted to its principal branch.

76. Sketch the graphs of the inverse functions defined by equations (17)–(20).

77. A flexible cable hangs in the shape of a catenary $y = a \cosh(x/a)$, where the number a is the height of the lowest point of the cable (the lowest point occurs at $x = 0$). Suppose that this cable is fastened to supports of height 16 ft and 28 ft.

    a. What is the distance between the supports if the lowest point of the cable is 10 ft above the ground? (*Hint:* Graph $y = a \cosh(x/a)$ with $a = 10$ along with the lines $y = 16$ and $y = 28$. Then, either by magnifying the graph or by using Newton's Method, find the distances of the supports from $x = 0$.)

    b. Determine the length of the cable in part a.

    c. Determine the height a for which the supports are farthest apart. Note that a must satisfy $0 < a \le 16$.

78. If an object falls through the air under the force of gravity, its velocity v satisfies the differential equation

$$m\frac{dv}{dt} = -mg + kv^2,$$

where t is in seconds and g is the acceleration due to gravity.

    a. Assume that $v(0) = 0$. By separating variables, show that

$$v(t) = -\sqrt{\frac{mg}{k}} \tanh\left(\sqrt{\frac{kg}{m}}\,t\right).$$

    b. Suppose $m = 80$ kg (about 176 lb), $g = 9.8$, and $k = 0.086$. Graph $v(t)$ and determine the limit of $v(t)$ as $t \to \infty$. This limit is the so-called terminal velocity. Explain the significance of this value for a free-falling skydiver.

    c. Find and graph the exact solution $s(t)$ of

$$\frac{ds}{dt} = v(t) = -\sqrt{\frac{mg}{k}} \tanh\left(\sqrt{\frac{kg}{m}}\,t\right), \quad s(0) = s_0,$$

for the values of m, g, and k given in part b. Determine s_0 so that 95% of terminal velocity is reached at least 1000 ft above the ground.

Summary Outline of Chapter 9

$$\blacksquare \int \sin u \, du = -\cos u + C \qquad \text{(page 494)}$$

$$\int \cos u \, du = \sin u + C$$

$$\int \tan u \, du = \ln|\sec u| + C, \qquad \cos u \neq 0$$

$$\int \cot u \, du = \ln|\sin u| + C, \qquad \sin u \neq 0$$

$$\int \sec u \, du = \ln|\sec u + \tan u| + C$$

$$\int \csc u \, du = \ln|\csc u - \cot u| + C$$

$$\blacksquare \text{ To integrate } \int \sin^n x \cos^m x \, dx: \qquad \text{(page 499)}$$

    (i) If one of m or n is odd (say n), make the substitution $u = \sin x$, so $du = \cos x \, dx$. Change remaining even factors of $\cos x$ to factors of $\sin x$ using the identity $\cos^2 x = 1 - \sin^2 x$.

    (ii) If both n and m are even, use the identities

$$\sin^2 x = \frac{1}{2} - \frac{1}{2}\cos 2x; \qquad \cos^2 x = \frac{1}{2} + \frac{1}{2}\cos 2x.$$

Use a similar strategy in integrals of the form $\int \sec^n x \tan^m x \, dx$.

■ The **inverse trigonometric functions** are defined as follows: (page 506)

$y = \text{Tan}^{-1} x$ if and only if $x = \tan y$, $\quad -\infty < x < \infty, \quad -\pi/2 < y < \pi/2$

$y = \text{Sin}^{-1} x$ if and only if $x = \sin y$, $\quad -1 \le x \le 1, \quad -\pi/2 \le y \le \pi/2$

$y = \text{Cos}^{-1} x$ if and only if $x = \cos y$, $\quad -1 \le x \le 1, \quad 0 \le y \le \pi$

$y = \text{Sec}^{-1} x$ if and only if $x = \sec y$, $\quad |x| \ge 1, \quad y \in [0, \pi/2) \cup [\pi, 3\pi/2)$

$y = \text{Csc}^{-1} x$ if and only if $x = \csc y$, $\quad |x| \ge 1, \quad y \in (0, \pi/2] \cup (\pi, 3\pi/2]$

$y = \text{Cot}^{-1} x$ if and only if $x = \cot y$, $\quad -\infty < x < \infty, \quad 0 \le y \le \pi.$

■ *Differentiation and Integration Formulas for Inverse Trigonometric Func-*   (page 511)
tions:

$$\frac{d}{dx} \text{Tan}^{-1} x = \frac{1}{1 + x^2} \qquad \int \frac{dx}{1 + x^2} = \text{Tan}^{-1} x + C$$

$$\frac{d}{dx} \text{Sin}^{-1} x = \frac{1}{\sqrt{1 - x^2}} \qquad \int \frac{dx}{\sqrt{1 - x^2}} = \text{Sin}^{-1} x + C, \qquad |x| < 1$$

$$\frac{d}{dx} \text{Sec}^{-1} x = \frac{1}{x\sqrt{x^2 - 1}} \qquad \int \frac{dx}{x\sqrt{x^2 - 1}} = \text{Sec}^{-1} x + C, \qquad |x| \ge 1$$

$$\frac{d}{dx} \text{Cot}^{-1} x = \frac{-1}{1 + x^2}$$

$$\frac{d}{dx} \text{Cos}^{-1} x = \frac{-1}{\sqrt{1 - x^2}}$$

$$\frac{d}{dx} \text{Csc}^{-1} x = \frac{-1}{x\sqrt{x^2 - 1}}$$

■ The **hyperbolic functions** are as follows: (page 517)

$$\sinh x = \frac{e^x - e^{-x}}{2} \qquad \coth x = \frac{\cosh x}{\sinh x}$$

$$\cosh x = \frac{e^x + e^{-x}}{2} \qquad \text{sech } x = \frac{1}{\cosh x}$$

$$\tanh x = \frac{\sinh x}{\cosh x} \qquad \text{csch } x = \frac{1}{\sinh x}$$

■ *Differentiation Formulas for Hyperbolic Functions:*   (page 518)

$$\frac{d}{dx} \sinh x = \cosh x; \qquad \frac{d}{dx} \coth x = -\text{csch}^2 x;$$

$$\frac{d}{dx} \cosh x = \sinh x; \qquad \frac{d}{dx} \text{sech } x = -\text{sech } x \cdot \tanh x;$$

$$\frac{d}{dx} \tanh x = \text{sech}^2 x; \qquad \frac{d}{dx} \text{csch } x = -\text{csch } x \cdot \coth x$$

■ The **inverse hyperbolic** functions satisfy the following identities: (page 521)

$$\sinh^{-1} x = \ln(x + \sqrt{x^2 + 1}), \qquad -\infty < x < \infty$$

$$\cosh^{-1} x = \ln(x + \sqrt{x^2 - 1}), \qquad x \ge 1$$

$$\text{sech}^{-1} x = \ln\left(\frac{1 + \sqrt{1 - x^2}}{x}\right), \qquad 0 < x \le 1$$

$$\operatorname{csch}^{-1} x = \ln\left(\frac{1}{x} + \frac{\sqrt{1+x^2}}{|x|}\right), \qquad x \neq 0$$

$$\tanh^{-1} x = \frac{1}{2}\ln\left(\frac{1+x}{1-x}\right), \qquad |x| < 1$$

$$\coth^{-1} x = \frac{1}{2}\ln\left(\frac{x+1}{x-1}\right), \qquad |x| > 1$$

■ Derivatives of the inverse hyperbolic functions: (page 522)

$$\frac{d}{dx}\sinh^{-1} x = \frac{1}{\sqrt{x^2+1}}$$

$$\frac{d}{dx}\cosh^{-1} x = \frac{1}{\sqrt{x^2-1}}, \qquad x > 0$$

$$\frac{d}{dx}\tanh^{-1} x = \frac{1}{1-x^2}, \qquad |x| < 1$$

$$\frac{d}{dx}\coth^{-1} x = \frac{1}{1-x^2}, \qquad |x| > 1$$

$$\frac{d}{dx}\operatorname{sech}^{-1} x = \frac{-1}{x\sqrt{1-x^2}}, \qquad 0 < |x| < 1$$

$$\frac{d}{dx}\operatorname{csch}^{-1} x = \frac{-1}{|x|\sqrt{1+x^2}}, \qquad x \neq 0$$

Review Exercises—Chapter 9

In Exercises 1–9, evaluate the given expression.

1. $\operatorname{Sin}^{-1}(\sqrt{3}/2)$

2. $\operatorname{Cos}^{-1}(-1/2)$

3. $\operatorname{Tan}^{-1}\sqrt{3}$

4. $\operatorname{Sin}^{-1}(\sin \pi/4)$

5. $\operatorname{Cos}^{-1}(\cos(-\pi/4))$

6. $\operatorname{Tan}^{-1}(\sin \pi/2)$

7. $\cot(\operatorname{Sin}^{-1}(\sqrt{2}/2))$

8. $\tan(\operatorname{Sec}^{-1}(-2))$

9. $\sinh(\ln 2)$

In Exercises 10–33, find the derivative of the given function.

10. $y = \operatorname{Tan}^{-1} 3x$

11. $y = \operatorname{Sin}^{-1}\sqrt{x}$

12. $f(x) = \operatorname{Tan}^{-1}\left(\frac{2-x}{2+x}\right)$

13. $f(t) = \operatorname{Cot}^{-1}(1 - t^2)$

14. $y = \cosh(\ln x)$

15. $f(x) = x^2 \sinh(1 - x)$

16. $y = \operatorname{Tan}^{-1}\frac{\sqrt{x}}{2}$

17. $y = \csc(\cot 6x)$

18. $y = \ln^2(\cos^2 x^2)$

19. $f(x) = \operatorname{Sin}^{-1}[\ln(2x + 1)]$

20. $y = \operatorname{Sin}^{-1}(\cos e^{-x})$

21. $f(x) = \frac{\operatorname{Tan}^{-1} x}{1 + x^2}$

22. $y = \frac{\operatorname{Sin}^{-1} e^x}{1 + e^x}$

23. $f(x) = \operatorname{Sec}^{-1}\sqrt{x^2 + 4}$

24. $y = \sqrt{\cosh x^2}$

25. $y = 1/(\pi + \tanh x)$

26. $f(x) = \ln^2(\sinh x)$

27. $y = (\sinh x + \operatorname{Cos}^{-1} 2x)^{1/5}$

28. $y = \ln|\tan x|$

29. $y = \ln(\operatorname{Sin}^{-1} x)$

30. $f(x) = \frac{\tan 2x}{2 + \sec 2x}$

31. $f(x) = x \tanh^{-1}(\ln x)$

32. $y = x^2 \sinh^{-1}(e^x)$

33. $y = \ln\sqrt{\tanh^{-1}(x^2)}$

In Exercises 34–72, find the indicated integral.

34. $\displaystyle\int \frac{dx}{\sqrt{1 + 9x^2}}$

35. $\displaystyle\int \frac{dx}{\sqrt{4x^2 - 1}}$

36. $\displaystyle\int \frac{dx}{x\sqrt{4 + x^2}}$

37. $\displaystyle\int \sin^4 2x \cos 2x \, dx$

38. $\displaystyle\int \sec^5 x \tan x \, dx$

39. $\displaystyle\int_{\pi^2/16}^{\pi^2/4} \frac{\sin^2\sqrt{x}}{\sqrt{x}} \, dx$

40. $\displaystyle\int_0^{\pi/2} \sin^{5/2} x \cos x \, dx$

41. $\displaystyle\int_{-1}^1 \frac{dx}{\sqrt{2 - x^2}}$

42. $\displaystyle\int x \csc^2 x^2 \, dx$

43. $\displaystyle\int_0^{\pi/4} \sqrt{\tan x} \sec^2 x \, dx$

44. $\displaystyle\int \frac{\sec^2\sqrt{x}\tan\sqrt{x}}{\sqrt{x}} \, dx$

45. $\displaystyle\int \frac{1 + \sin^2 2x}{\cos^2 2x} \, dx$

46. $\displaystyle\int \tan(\sec x) \sec x \tan x \, dx$

47. $\displaystyle\int_0^{\pi/8} \tan^2(2x)\sec^2(2x)\,dx$

48. $\displaystyle\int \frac{x}{x^4+1}\,dx$

49. $\displaystyle\int \frac{dx}{\sqrt{4-x^2}}$

50. $\displaystyle\int \frac{e^{-x}}{2+e^{-2x}}\,dx$

51. $\displaystyle\int \frac{\mathrm{Cos}^{-1}x}{\sqrt{1-x^2}}\,dx$

52. $\displaystyle\int \frac{\sqrt{x}}{x^3+4}\,dx$

53. $\displaystyle\int \frac{e^{\sqrt{x}}}{\sqrt{x}(1+e^{2\sqrt{x}})}\,dx$

54. $\displaystyle\int_1^2 \frac{x^2}{x^6+9}\,dx$

55. $\displaystyle\int \frac{\sqrt{\mathrm{Sin}^{-1}x}}{\sqrt{1-x^2}}\,dx$

56. $\displaystyle\int \frac{(\mathrm{Tan}^{-1}2x)^4}{1+4x^2}\,dx$

57. $\displaystyle\int \frac{\mathrm{Sec}^{-1}3x}{x\sqrt{36x^2-4}}\,dx$

58. $\displaystyle\int \frac{\sin x}{\sqrt{2-\cos^2 x}}\,dx$

59. $\displaystyle\int_0^{\pi/9} \sec^3 2x\tan 2x\,dx$

60. $\displaystyle\int_0^{\pi/2} \tan^3(x/2)\,dx$

61. $\displaystyle\int_0^{\pi/3} \sec^4 x\,dx$

62. $\displaystyle\int \sec^3 x\tan^3 x\,dx$

63. $\displaystyle\int_0^{\pi/3} \cos x\cos 5x\,dx$

64. $\displaystyle\int_0^{\pi/4} \sin x\cos 2x\,dx$

65. $\displaystyle\int \sqrt{\cos x}\,\sin^3 x\,dx$

66. $\displaystyle\int \cot^4 2x\,dx$

67. $\displaystyle\int \frac{\sin^3 x}{\cos^2 x}\,dx$

68. $\displaystyle\int_0^{\pi/2} \sin^2 x\cos^4 x\,dx$

69. $\displaystyle\int \frac{\cosh x}{1+\sinh x}\,dx$

70. $\displaystyle\int x\coth x^2\,dx$

71. $\displaystyle\int \sinh^3 x\,dx$

72. $\displaystyle\int \frac{e^{3x}}{\sqrt{9+e^{6x}}}\,dx$

73. Find the area of the region bounded by the graph of $y=\sin^2 x$ and the x-axis between $x=0$ and $x=\pi$.

74. Find the area of the region bounded by the graphs of $y=\sin^2 x$ and $y=\cos^2 x$ for x between $x=0$ and $x=\pi/2$.

75. Find the volume of the solid generated by rotating about the x-axis the region bounded by the graph of $y=\tan x$ and the x-axis for $0\le x\le\pi/4$.

76. Find the volume of the solid generated by rotating about the x-axis the region bounded by the graph of $y=1/\sqrt{1+x^2}$ and the x-axis for $0\le x\le 1$.

77. Find the volume of the solid generated by revolving about the y-axis the region bounded by the graph of $y=1/(1+x^4)$ and the x-axis between $x=0$ and $x=2$.

78. Find the average value of the function $y=\cosh x$ on the interval $[-\ln 2, \ln 2]$.

79. Find dy/dx if $y=\cosh^{-1}x^2y$.

80. Find dy/dx if $y=\mathrm{Cos}^{-1}x^2y$.

81. Find the area bounded by the graph of $y=\cosh x$ and the x-axis for $-1\le x\le 1$.

82. For the function $y=\sinh x\cosh x$, find all relative extrema, determine the concavity, and sketch the graph.

83. A particle moves along a line with velocity $v(t)=\sin^2\pi t$. Find the distance travelled by the particle between times $t=0$ and $t=4$.

84. Using linear approximation about $0.5=1/2$, approximate $\mathrm{Sin}^{-1}(0.48)$.

85. Prove that $\cosh x>\sinh x$ for all x.

86. Find the equation of the line tangent to the graph of $y=\mathrm{Sin}^{-1}x^2$ at the point with x-coordinate $\sqrt{2}/2$.

87. An airplane is flying directly away from a radar station at an altitude of 3 km and a ground speed of 400 km/h. How fast is the angle of elevation of the tracking antenna decreasing when the airplane is directly over a point 4 km from the radar station? (Use inverse trigonometric functions.)

88. A small boat is tied to a rope which is connected to a windlass (crank). The windlass is mounted on the edge of a dock 10 m above the water level. The windlass is pulling the boat ashore by winding in the rope at the rate of 1 m/s. Find the rate at which the angle between the rope and the horizontal is increasing at the instant when 20 m of rope remain out.

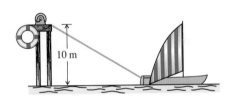

89. Find the points where the line tangent to the graph of $y=\mathrm{Cot}^{-1}x$ is parallel to the line with equation $x+5y-10=0$.

90. Find the x-intercept of the line tangent to the graph $y=\mathrm{Sin}^{-1}x$ at the point $(\sqrt{2}/2, \pi/4)$.

91. A particle moves along a line with acceleration

$$a(t)=-1-\frac{2t}{(1+t^2)^2}\,\text{m/s}^2$$

Find the function $s(t)$ giving the distance travelled after t seconds if the particle starts at the origin with initial velocity 4 m/s.

92. Why is the hyperbolic sine function continuous throughout its domain?

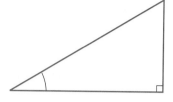

Chapter 10

Techniques of Integration

The goal of this chapter is to develop several additional techniques for finding antiderivatives and evaluating definite integrals. Recall that the function F is an *antiderivative* for the function f on the interval I if $F'(x) = f(x)$ for all $x \in I$. In this case we refer to the most general antiderivative

$$\int f(x)\, dx = F(x) + C \tag{1}$$

as the *indefinite integral* for f. The term *integration* refers to the process of finding the indefinite integral in equation (1).

One reason for developing additional techniques of integration is to expand the class of functions for which we can evaluate *definite integrals* using the Fundamental Theorem of Calculus:

$$\int_a^b f(x)\, dx = F(b) - F(a) \quad \text{if} \quad F'(x) = f(x) \text{ for } x \in [a, b]. \tag{2}$$

This part of the Fundamental Theorem is not always useful for evaluating definite integrals of continuous functions, however, because we must *know* an antiderivative F of f *before* we can apply the theorem. Unfortunately, some continuous functions simply do not have antiderivatives that are expressible in terms of the so-called elementary functions—combinations of the polynomial, rational, and transcendental functions we typically use. For example, the function

$$f(x) = e^{x^2}$$

is continuous on $(-\infty, \infty)$, but there does not exist an elementary function F with $F'(x) = e^{x^2}$.

A second reason for developing additional techniques of integration is to prepare you better to solve the differential equations that arise in advanced courses in science, engineering, and applied mathematics. The techniques required to solve these differential equations involve much of what is discussed in this chapter.

Finally, it is important to note the major role computers now play in both evaluating definite integrals and finding antiderivatives of functions. We have seen in Chapter 6 that numerical methods such as the Midpoint Rule, the Trapezoidal Rule, and Simpson's Rule, when implemented on computers, can be used to evaluate definite integrals to great accuracy without recourse to the Fundamental Theorem. More recently, computer programs have been developed that can now evaluate indefinite integrals symbolically for a wide class of functions (see Appendix I).

However, while the availability of these technological aids reduces the need to calculate integrals "by hand," they cannot eliminate entirely the need for familiarity with the underlying mathematical techniques.

10.1   Integration by Parts

This technique enables us to integrate certain functions that can be interpreted as products. Before discussing the technique generally we illustrate the basic idea with a typical example.

☐ **EXAMPLE 1**

Find $\int x \cos x \, dx$.

Solution   The simple substitution $u = \cos x$ is not helpful because it gives $du = -\sin x \, dx$, introducing a factor $(\sin x)$ that is not present in the integrand and leaving the factor x unaddressed.

The key to finding this integral is to note that the derivative of the *product* $h(x) = x \sin x$ involves the term $x \cos x$ that appears in the integrand:

$$\frac{d}{dx}(x \sin x) = \sin x + x \cos x.$$

Thus,

$$x \cos x = \frac{d}{dx}(x \sin x) - \sin x.$$

Integrating both sides of this equation and using the fact that

$$\int \left[\frac{d}{dx}(x \sin x) \right] dx = x \sin x + C_1$$

gives the desired integral:

$$\int x \cos x \, dx = (x \sin x + C_1) - \int \sin x \, dx$$

$$= (x \sin x + C_1) + (\cos x + C_2)$$

$$= x \sin x + \cos x + C. \qquad \blacksquare$$

The solution to finding the integral in Example 1 was to view the integrand $x \cos x$ as one of the two terms resulting from an application of the Product Rule to a function $h = fg$. To generalize this observation recall that the Product Rule

$$(fg)'(x) = f'(x)g(x) + f(x)g'(x)$$

gives

$$f(x)g'(x) = (fg)'(x) - f'(x)g(x).$$

Integrating both sides of this equation gives

$$\int f(x)g'(x) \, dx = \int (fg)'(x) \, dx - \int f'(x)g(x) \, dx \qquad (1)$$

$$= f(x)g(x) - \int f'(x)g(x) \, dx.$$

Equation (1) uses the fact that $\int (fg)'(x)\,dx = (fg)(x) + C$. We have not written a constant of integration with this term since one will appear when the integral on the right side of the equation is evaluated. Equation (1) is referred to as the **integration by parts formula:**

$$\int f(x)g'(x)\,dx = f(x)g(x) - \int f'(x)g(x)\,dx. \qquad (2)$$

REMARK 1   The general strategy in using equation (2) is to pick the factors f and g' so that

(i) the antiderivative g can be found, and
(ii) the resulting integral $\int f'(x)g(x)\,dx$ is as simple to evaluate as possible.

REMARK 2   There is no need to introduce an arbitrary constant when finding an antiderivative g for g', since (2) holds for *any* antiderivative of g'. The simplest choice is to let the arbitrary constant be zero.

There is a somewhat simpler form by which the integration by parts formula can be remembered. If we make the substitutions

$$u = f(x), \qquad dv = g'(x)\,dx$$

then we can write

$$du = f'(x)\,dx, \qquad v = g(x)$$

and equation (2) becomes

$$\int u\,dv = uv - \int v\,du. \qquad (3)$$

Equation (3) is the Leibniz notation form for the integration by parts formula.

☐ **EXAMPLE 2**

Find $\int x^2 \ln x\,dx$.

Strategy · · · · · · · ·
The derivative of $\ln x$ is a power of x, so take $u = \ln x$, $dv = x^2\,dx$.

Solution
We let

$$u = \ln x, \qquad dv = x^2\,dx,$$

so

Find du, v.

$$du = \frac{1}{x}dx, \qquad v = \frac{x^3}{3}.$$

Apply equation (3).

Integrating by parts then gives

$$\int x^2 \ln x \, dx = \frac{x^3}{3} \ln x - \int \frac{x^3}{3} \frac{1}{x} \, dx$$

$$= \frac{x^3}{3} \ln x - \frac{1}{3} \int x^2 \, dx$$

$$= \frac{x^3}{3} \ln x - \frac{x^3}{9} + C.$$

More than one application of the integration by parts formula may be required, as the following example shows.

□ **EXAMPLE 3**

Find $\int x^2 e^x \, dx$.

Strategy · · · · · · ·
Take $u = x^2$ since differentiation reduces the exponent by one.

Solution
We take

$$u = x^2, \qquad dv = e^x \, dx.$$

Then

$$du = 2x \, dx, \qquad v = e^x,$$

and integration by parts gives

$$\int x^2 e^x \, dx = x^2 e^x - 2 \int x e^x \, dx. \tag{4}$$

Apply equation (3).

In $\int xe^x \, dx$ take $u = x$, since differentiation will yield simply $du = 1 \, dx$.

The integral on the right requires a second application of the parts formula: In

$$\int xe^x \, dx,$$

we take

$$u = x, \qquad dv = e^x \, dx,$$

so

$$du = dx, \qquad v = e^x.$$

Apply equation (3) again.

Then integration by parts gives

$$\int xe^x \, dx = xe^x - \int e^x \, dx \tag{5}$$

$$= xe^x - e^x + C_1.$$

Combine results.

Combining lines (4) and (5) we now have

$$\int x^2 e^x \, dx = x^2 e^x - 2(xe^x - e^x + C_1)$$

$$= e^x(x^2 - 2x + 2) + C.$$

(Since C_1 is an arbitrary constant, C is also an arbitrary constant; we choose to write the answer in the simpler form.)

In the exercise set you will encounter integrals requiring even more than two applications of the parts formula. A different type of problem requiring two applications of the parts formula is the following:

☐ **EXAMPLE 4**

Find $\displaystyle\int e^x \cos x \, dx$.

Solution This time there is no clue as to which function to take as u. We arbitrarily choose

$$u = e^x, \qquad dv = \cos x \, dx.$$

Then

$$du = e^x \, dx, \qquad v = \sin x$$

and the integration by parts formula gives

$$\int e^x \cos x \, dx = e^x \sin x - \int e^x \sin x \, dx. \tag{6}$$

Now the integral on the right side of (6) seems no simpler than the integral on the left. However, let's try another application of the integration by parts formula in the integral $\int e^x \sin x \, dx$. Again taking $u = e^x$ we have

$$u = e^x, \qquad dv = \sin x \, dx$$

and

$$du = e^x \, dx, \qquad v = -\cos x.$$

Thus,

$$\int e^x \sin x \, dx = -e^x \cos x - \int (-\cos x) e^x \, dx \tag{7}$$

$$= -e^x \cos x + \int e^x \cos x \, dx.$$

Note that the desired integral has reappeared on the right side of equation (7)! This happens because the second derivatives of e^x and $\cos x$ are simply multiples of the original functions. Combining equations (6) and (7) gives

$$\int e^x \cos x \, dx = e^x \sin x - \left[-e^x \cos x + \int e^x \cos x \, dx \right]$$

$$= e^x(\sin x + \cos x) - \int e^x \cos x \, dx.$$

We may now add $\int e^x \cos x \, dx$ to both sides to obtain

$$2 \int e^x \cos x \, dx = e^x(\sin x + \cos x) + C_1,$$

so

$$\int e^x \cos x \, dx = \frac{e^x}{2}(\sin x + \cos x) + C. \qquad\blacksquare$$

We can apply the parts formula to any integral of the form $\int f(x)\, dx$ simply by taking $u = f(x)$ and $dv = dx$. The resulting integral may or may not be solvable.

☐ **EXAMPLE 5**

Find $\displaystyle\int \text{Sin}^{-1} x\, dx.$

Strategy · · · · · · ·
Here our only choice is to take

$$u = \text{Sin}^{-1} x, \qquad dv = dx.$$

Apply equation (3). (Handle

$$\int \frac{x}{\sqrt{1 - x^2}}\, dx$$

by a substitution with $u = 1 - x^2$.)

Solution
With $u = \text{Sin}^{-1} x, \qquad dv = dx,$

$$du = \frac{dx}{\sqrt{1 - x^2}}, \quad \text{and} \quad v = x.$$

The parts formula gives

$$\int \text{Sin}^{-1} x\, dx = x\, \text{Sin}^{-1} x - \int \frac{x}{\sqrt{1 - x^2}}\, dx$$

$$= x\, \text{Sin}^{-1} x + \sqrt{1 - x^2} + C. \qquad \blacksquare$$

☐ **EXAMPLE 6**

Show that $\displaystyle\int \ln x\, dx = x \ln x - x + C.$

Solution   We do this by the method of Example 5. Since the only factor of the integrand is $\ln x$, we take

$$u = \ln x, \qquad dv = dx.$$

Then,

$$du = \frac{1}{x}\, dx \quad \text{and} \quad v = x,$$

so an application of the integration by parts formula gives

$$\int \ln x\, dx = x \ln x - \int x\left(\frac{1}{x}\right) dx$$

$$= x \ln x - \int 1\, dx$$

$$= x \ln x - x + C. \qquad \blacksquare$$

The following is one of the trickier integrals involving integration by parts. It should be noted for future reference.

☐ **EXAMPLE 7**

Find $\displaystyle\int \sec^3 x\, dx.$

Strategy · · · · · · ·
Break $\sec^3 x$ into two factors.

Solution
We let

$$u = \sec x, \qquad dv = \sec^2 x\, dx.$$

Then

$$du = \sec x \tan x\, dx, \qquad v = \tan x,$$

so

Apply equation (3).

$$\int \sec^3 x \, dx = \sec x \tan x - \int \tan^2 x \sec x \, dx$$

Use identity $\tan^2 x = \sec^2 x - 1$.

$$= \sec x \tan x - \int (\sec^2 x - 1) \sec x \, dx$$

$$= \sec x \tan x - \int \sec^3 x \, dx + \int \sec x \, dx$$

$$= \sec x \tan x - \int \sec^3 x \, dx + \ln |\sec x + \tan x|.$$

Thus

Add $\int \sec^3 x \, dx$ to both sides and divide by 2.

$$\int \sec^3 x \, dx = \frac{1}{2} \sec x \tan x + \frac{1}{2} \ln |\sec x + \tan x| + C. \qquad \blacksquare$$

Finally, we note that the integration by parts formula can be applied to definite integrals as well as to indefinite integrals. The corresponding formulation of equation (2) is

$$\int_a^b f(x) g'(x) \, dx = \left[f(x) g(x) \right]_a^b - \int_a^b f'(x) g(x) \, dx. \qquad (8)$$

☐ **EXAMPLE 8**

Find $\displaystyle\int_0^{\sqrt{\pi/2}} x^3 \cos x^2 \, dx.$

Strategy ·   ·   ·   ·   ·   ·   ·   ·   ·
We take $u = x^2$ so the remaining factor $dv = x \cos x^2 \, dx$ is a multiple of the derivative of $\sin x^2$.

Solution
Let

$$u = x^2, \qquad dv = x \cos x^2 \, dx.$$

Then

$$du = 2x \, dx, \qquad v = \frac{1}{2} \sin x^2,$$

so the parts formula gives

Apply (8).

$$\int_0^{\sqrt{\pi/2}} x^3 \cos x^2 \, dx = \left[\frac{1}{2} x^2 \sin x^2 \right]_0^{\sqrt{\pi/2}} - \int_0^{\sqrt{\pi/2}} x \sin x^2 \, dx$$

$$= \frac{1}{2} \left(\frac{\pi}{4} \right) \sin \left(\frac{\pi}{4} \right) - \left[-\frac{1}{2} \cos x^2 \right]_0^{\sqrt{\pi/2}}$$

$$= \frac{\sqrt{2}\pi}{16} + \frac{1}{2} \left(\frac{\sqrt{2}}{2} - 1 \right)$$

$$\approx 0.1312. \qquad \blacksquare$$

Exercise Set 10.1

In Exercises 1–36, evaluate the given integral.

1. $\displaystyle\int x\, e^x\, dx$

2. $\displaystyle\int x \ln x\, dx$

3. $\displaystyle\int_{\pi/2}^{\pi} x \cos x\, dx$

4. $\displaystyle\int_{1/2}^{e/2} x^2 \ln 2x\, dx$

5. $\displaystyle\int \mathrm{Tan}^{-1} x\, dx$

6. $\displaystyle\int x \sec x \tan x\, dx$

7. $\displaystyle\int x \sec^2 \pi x\, dx$

8. $\displaystyle\int e^{ax} \sin x\, dx$

9. $\displaystyle\int \sin(\ln x)\, dx$

10. $\displaystyle\int \sin x \sinh x\, dx$

11. $\displaystyle\int_0^1 x(x+2)^8\, dx$

12. $\displaystyle\int_0^4 x\sqrt{x+1}\, dx$

13. $\displaystyle\int_e^{e^3} (\ln x)^2\, dx$

14. $\displaystyle\int_1^e x \ln x^2\, dx$

15. $\displaystyle\int_0^{\ln 2} xe^{2x}\, dx$

16. $\displaystyle\int_1^e x^3 \ln x\, dx$

17. $\displaystyle\int_0^1 x^3 e^{2x}\, dx$

18. $\displaystyle\int \sec^5 x\, dx$

19. $\displaystyle\int_0^1 (3x^2 - 2x + 1)e^{2x}\, dx$

20. $\displaystyle\int_0^{\pi} (4x^2 + 2x)\sin(x/2)\, dx$

21. $\displaystyle\int x \sinh x\, dx$

22. $\displaystyle\int_0^1 \frac{x^3}{\sqrt{x^2+1}}\, dx$

23. $\displaystyle\int_0^{\pi} e^{2x} \cos x\, dx$

24. $\displaystyle\int \mathrm{Cos}^{-1} 2x\, dx$

25. $\displaystyle\int (\mathrm{Cos}^{-1} x)^2\, dx$

26. $\displaystyle\int \sqrt{x} \ln x\, dx$

27. $\displaystyle\int x^3\sqrt{1+x^2}\, dx$

28. $\displaystyle\int x^3\sqrt{9-x^2}\, dx$

29. $\displaystyle\int x^3 e^{ax^2}\, dx$

30. $\displaystyle\int \tanh^{-1} x\, dx$

31. $\displaystyle\int \sec^3 x \tan^2 x\, dx$

32. $\displaystyle\int x \tanh^{-1} x\, dx$

33. $\displaystyle\int x \sinh^{-1} x\, dx$

34. $\displaystyle\int \sqrt{x}\, e^{-\sqrt{x}}\, dx$

35. $\displaystyle\int 2xe^{-\sqrt{x}}\, dx$

36. $\displaystyle\int \sinh^{-1} x\, dx$

37. Find the area of the region bounded by the graph of $y = x \sin \pi x$ and the x-axis between the lines $x = 0$ and $x = 1$.

38. Find the area of the region bounded by the graphs of $y = \ln(1 + x)$, $y = x$, and $x = 1$.

39. Find the volume of the solid obtained by revolving about the x-axis the region bounded by the graph by $y = \ln x$ and the x-axis for $1 \le x \le e$.

40. Find the volume of the solid obtained by revolving the region in Exercise 39 about the y-axis.

41. Find the volume of the solid obtained by revolving about the y-axis the region bounded by the graph of $y = \sin x$ for $0 \le x \le \pi$.

42. Find the volume of the solid obtained by revolving about the y-axis the region bounded by the graph of $y = xe^{2x}$ for $0 \le x \le \ln 2$.

43. Find the average value of the function $f(x) = e^{-x} \sin \pi x$ for $x \in [0, 1]$.

44. Find the average value of the function $y = \mathrm{Sin}^{-1} x$ for $x \in [0, 1]$.

45. A particle moves along a line with velocity $v(t) = t \sin \pi t$ m/s. Find the distance travelled by the particle between times $t = 0$ and $t = 2$ seconds.

46. A particle moves along a line with velocity $v(t) = (t^2 - 1)e^{-t}$ m/s. Find the distance travelled by the particle between times $t = 0$ and $t = 2$ seconds.

47. A water tank has the shape of the solid obtained by revolving about the y-axis the region bounded by the graph of $y = e^x$, the y-axis, and the line $y = 5$. Assuming that the tank is full of water, find the work done in pumping all of the water to the top of the tank.

48. Find the centroid of the region bounded by the graph of $y = e^x$ and the x-axis for $0 \le x \le \ln 2$.

49. Find the centroid of the region bounded by the graphs of $y = x$ and $y = \ln x$ for $1 \le x \le e$.

50. A **reduction formula** is one that reduces the order of the integrand; by repeated applications the order can be reduced to zero or one. Establish the reduction formula

$$\int \sin^n x\, dx = -\frac{1}{n} \sin^{n-1} x \cos x + \frac{n-1}{n} \int \sin^{n-2} x\, dx$$

using integration by parts.

51. Use the formula in Exercise 50 to find $\int \sin^2 x \, dx$.

52. Use the formula in Exercise 50 to find $\int \sin^4 x \, dx$.

53. Establish the reduction formula

$$\int x^n e^x \, dx = x^n e^x - n \int x^{n-1} e^x \, dx$$

using integration by parts.

54. Use the formula in Exercise 53 to find $\int x^4 e^x \, dx$.

55. Establish the reduction formula

$$\int x^n \sin ax \, dx = -\frac{x^n}{a} \cos ax + \frac{n}{a} \int x^{n-1} \cos ax \, dx$$

using integration by parts.

56. Use the formula in Exercise 55 to find $\int x \sin 3x \, dx$.

57. Show that the integral $\int p(x) e^x \, dx$ can be handled by n applications of the integration by parts formula if $p(x)$ is a polynomial of degree n.

10.2 Trigonometric Substitutions

Certain types of integrals may be evaluated by means of a trigonometric substitution of the form $x = a \sin \theta$, $x = a \tan \theta$, or $x = a \sec \theta$. The idea is based on the simple Pythagorean Theorem for right triangles, and it is used to handle integrands involving the factors $\sqrt{x^2 + a^2}$, $\sqrt{a^2 - x^2}$, and $\sqrt{x^2 - a^2}$.

Integrals Involving $\sqrt{x^2 + a^2}$

Construct a right triangle for which the quantity $\sqrt{x^2 + a^2}$ is the length of one of the three sides. The simplest way to do this is to draw a right triangle whose two legs are labelled x and a, respectively (Figure 2.1). The Pythagorean Theorem then states that the length of the hypotenuse is $\sqrt{x^2 + a^2}$.

Next, let θ denote one of the acute angles of this triangle. Since a is constant, θ is a function of x, and vice versa. In fact, you can see that, in Figure 2.1, x and θ are related by the equation

$$\frac{x}{a} = \tan \theta \quad \text{or} \quad x = a \tan \theta, \tag{1}$$

where $-\pi/2 < \theta < \pi/2$. If we make this substitution, the expression $\sqrt{x^2 + a^2}$ becomes

$$\sqrt{x^2 + a^2} = \sqrt{(a \tan \theta)^2 + a^2} = a\sqrt{\tan^2 \theta + 1} = a \sec \theta.$$

(In this case, $\sqrt{\sec^2 \theta} = \sec \theta$ since $\sec \theta > 0$ if $-\pi/2 < \theta < \pi/2$.)*

Consequently, the trigonometric substitution $x = a \tan \theta$ transforms the expression $\sqrt{x^2 + a^2}$, which involves a square root (a radical), to an expression that involves a trigonometric function, but no radical. This is the objective in using trigonometric substitutions. Of course, in using substitution (1) we must also substitute for the differential dx since we will now be integrating with respect to θ. From equation (1) and the definition of the differential we obtain the required equation

$$dx = \frac{d}{d\theta} (a \tan \theta) \, d\theta = a \sec^2 \theta \, d\theta.$$

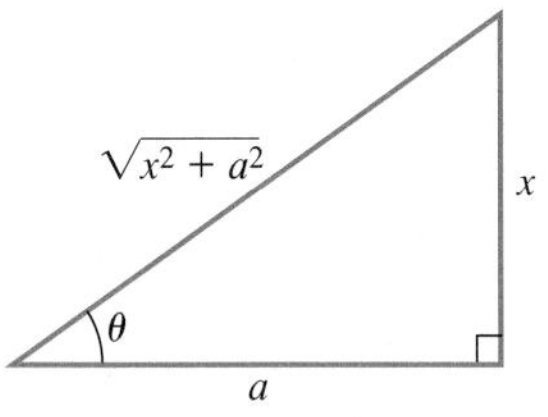

Figure 2.1 Triangle with hypotenuse $\sqrt{x^2 + a^2}$. Then, $\tan \theta = x/a$.

___

*Alternatively, we can establish the equation $\sqrt{x^2 + a^2} = a \sec \theta$ by considering the triangle in Figure 2.1. The side adjacent to the angle θ is a, and the hypotenuse is $\sqrt{x^2 + a^2}$.

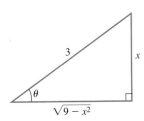

Figure 2.4 Triangle with a side of length $\sqrt{a^2 - x^2}$. Note that $x = a \sin \theta$ and $\sqrt{a^2 - x^2} = a \cos \theta$.

and, therefore, to the substitition

$$x = a \sin \theta, \qquad dx = a \cos \theta \, d\theta.$$

This gives

$$\sqrt{a^2 - x^2} = \sqrt{a^2 - a^2 \sin^2 \theta} = a\sqrt{1 - \sin^2 \theta} = a \cos \theta,$$

since $\cos \theta \geq 0$ for $-\pi/2 \leq \theta \leq \pi/2$.

☐ **EXAMPLE 3**

Find $\displaystyle\int \frac{dx}{x^2\sqrt{9 - x^2}}$.

Solution Here $a = 3$ (Figure 2.5), so we use the substitution

$$x = 3 \sin \theta, \qquad -\pi/2 \leq \theta \leq \pi/2$$
$$dx = 3 \cos \theta \, d\theta.$$

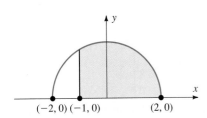

Figure 2.5 Triangle with one side of length $\sqrt{9 - x^2}$. Then, $x = 3 \sin \theta$ and $\sqrt{9 - x^2} = 3 \cos \theta$.

We obtain

$$\int \frac{dx}{x^2\sqrt{9 - x^2}} = \int \frac{3 \cos \theta \, d\theta}{9 \sin^2 \theta \sqrt{9 - 9 \sin^2 \theta}} = \int \frac{3 \cos \theta \, d\theta}{9 \sin^2 \theta \cdot 3 \cdot \sqrt{1 - \sin^2 \theta}}$$

$$= \frac{1}{9} \int \frac{\cos \theta \, d\theta}{\sin^2 \theta \cos \theta}$$

$$= \frac{1}{9} \int \csc^2 \theta \, d\theta$$

$$= -\frac{1}{9} \cot \theta + C$$

$$= -\frac{1}{9} \frac{\sqrt{9 - x^2}}{x} + C. \qquad ■$$

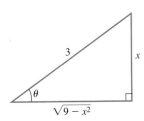

Figure 2.6 Region in Example 4.

☐ **EXAMPLE 4**

Find the area of the region bounded above by the semicircle $y = \sqrt{4 - x^2}$, below by the x-axis, and on the left by the line $x = -1$ (Figure 2.6).

Strategy · · · · · · ·
Write the integral giving the area.

Solution
The area is given by the integral

$$A = \int_{-1}^{2} \sqrt{4 - x^2} \, dx.$$

Use the substitution for form $\sqrt{a^2 - x^2}$.

To evaluate the integral, we use the trigonometric substitution

$$x = 2 \sin \theta, \qquad -\pi/2 \leq \theta \leq \pi/2,$$
$$dx = 2 \cos \theta \, d\theta \qquad \text{(Figure 2.7)}.$$

Change limits of integration, using the function $\text{Sin}^{-1} \theta$.

To find the limits of integration with respect to θ, we note that

$$\sin \theta = -1/2 \quad \text{if} \quad x = -1, \quad \text{and}$$
$$\sin \theta = 1 \quad \text{if} \quad x = 2.$$

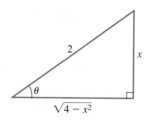

Figure 2.7 A triangle with a side of length $\sqrt{4-x^2}$. Then, $x = 2\sin\theta$ and $\sqrt{4-x^2} = 2\cos\theta$.

To invert these relationships, we recall the principal values of the function $y = \text{Sin}^{-1}x$: $-\pi/2 \le y \le \pi/2$. Thus

$$\theta = \text{Sin}^{-1}(-1/2) = -\pi/6 \quad \text{if} \quad x = -1$$

and

$$\theta = \text{Sin}^{-1}(1) = \pi/2 \quad \text{if} \quad x = 2.$$

From Figure 2.7, we have $\sqrt{4-x^2} = 2\cos\theta$, and therefore our integral becomes

Substitute for x and dx and use new limits of integration.

$$A = \int_{-1}^{2} \sqrt{4-x^2}\, dx = \int_{-\pi/6}^{\pi/2} (2\cos\theta)(2\cos\theta)\, d\theta$$

$$= 4\int_{-\pi/6}^{\pi/2} \cos^2\theta\, d\theta$$

Use identity

$$\cos^2\theta = \frac{1}{2} + \frac{1}{2}\cos 2\theta.$$

$$= 4\int_{-\pi/6}^{\pi/2} \left(\frac{1}{2} + \frac{1}{2}\cos 2\theta\right) d\theta$$

$$= 4\left[\frac{\theta}{2} + \frac{1}{4}\sin 2\theta\right]_{-\pi/6}^{\pi/2}$$

$$= 4\left(\frac{\pi}{3} + \frac{\sqrt{3}}{8}\right) \approx 5.055. \qquad \blacksquare$$

REMARK   Notice that we did not have to substitute back in terms of x in the integral in Example 4. Once we have changed variables and limits of integration in a definite integral, we may evaluate the integral directly.

Integrals Involving $\sqrt{x^2 - a^2}$

The situation for integrals involving $\sqrt{x^2 - a^2}$ is similar to those for the preceding two types except that we must label the hypotenuse x. Thus, one leg of the triangle is a, the other is $\sqrt{x^2 - a^2}$. As suggested by Figure 2.8, we use the substitutions

$$x = a\sec\theta, \qquad 0 \le \theta < \pi/2 \quad \text{or} \quad \pi \le \theta < 3\pi/2.$$
$$dx = a\sec\theta\tan\theta\, d\theta.$$

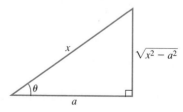

Figure 2.8 A triangle with a side of length $\sqrt{x^2 - a^2}$. Then, $x = a\sec\theta$ and $\sqrt{x^2 - a^2} = a\tan\theta$.

Then the radical $\sqrt{x^2 - a^2}$ simplifies to

$$\sqrt{x^2 - a^2} = \sqrt{a^2 \sec^2 \theta - a^2} = a\sqrt{\tan^2 \theta} = a \tan \theta,$$

since $\tan \theta \geq 0$ when $0 \leq \theta < \pi/2$ or $\pi \leq \theta < 3\pi/2$.

☐ **EXAMPLE 5**

Find $\displaystyle\int \frac{dx}{x^2\sqrt{x^2 - 4}}$.

Strategy   · · · · · · · ·   **Solution**

Noting the radical $\sqrt{x^2 - 4}$, we use the substitution (Figure 2.9)

$$x = 2 \sec \theta, \qquad 0 \leq \theta < \pi/2 \quad \text{or} \quad \pi \leq \theta < 3\pi/2,$$
$$dx = 2 \sec \theta \tan \theta \, d\theta.$$

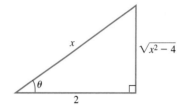

Figure 2.9   A triangle with a side of length $\sqrt{x^2 - 4}$. Then $x = 2 \sec \theta$ and $\sqrt{x^2 - 4} = 2 \tan \theta$.

The integral becomes

Make trigonometric substitution.

$$\int \frac{dx}{x^2\sqrt{x^2 - 4}} = \int \frac{2 \sec \theta \tan \theta \, d\theta}{(2 \sec \theta)^2\sqrt{4 \sec^2 \theta - 4}}$$

Use $\sec^2 \theta - 1 = \tan^2 \theta$.

$$= \int \frac{2 \sec \theta \tan \theta \, d\theta}{4 \sec^2 \theta \cdot 2 \tan \theta}$$

$$= \frac{1}{4} \int \frac{1}{\sec \theta} \, d\theta$$

$\sec \theta = \dfrac{1}{\cos \theta}$.

$$= \frac{1}{4} \int \cos \theta \, d\theta$$

$$= \frac{1}{4} \sin \theta + C$$

From Figure 2.9,

$$\sin \theta = \frac{\sqrt{x^2 - 4}}{x}.$$

$$= \frac{1}{4} \frac{\sqrt{x^2 - 4}}{x} + C.$$ ■

In summary, the following trigonometric substitutions, with the principal branches indicated, are used to handle integrals involving the indicated radicals.

$\sqrt{x^2 + a^2}$	requires	$x = a \tan \theta,$	$-\pi/2 < \theta < \pi/2,$
$\sqrt{a^2 - x^2}$	requires	$x = a \sin \theta,$	$-\pi/2 \leq \theta \leq \pi/2,$
$\sqrt{x^2 - a^2}$	requires	$x = a \sec \theta,$	$0 \leq \theta < \pi/2 \quad$ or
			$\pi \leq \theta < 3\pi/2.$

Other Uses of Trigonometric
Substitution

Trigonometric substitutions are also used to integrate functions that include certain quadratic expressions even if no radical is involved. In Section 10.4, we are more precise about the type of rational function that is amenable to this technique. Here we illustrate the technique with a typical example.

☐ **EXAMPLE 6**

Find $\int \dfrac{1}{(x^2 + 4)^2}\, dx$.

Solution   In this case, the quadratic expression inside the parentheses is $x^2 + 4$. As in Example 2, we use the trigonometric substitution

$$x = 2 \tan \theta \quad \text{with} \quad dx = 2 \sec^2 \theta\, d\theta.$$

We get

$$
\begin{aligned}
x^2 + 4 &= 4 \tan^2 \theta + 4 \\
&= 4(\tan^2 \theta + 1) \\
&= 4 \sec^2 \theta.
\end{aligned}
$$

Consequently,

$$
\begin{aligned}
\int \frac{1}{(x^2 + 4)^2}\, dx &= \int \frac{2 \sec^2 \theta}{16 \sec^4 \theta}\, d\theta \\
&= \frac{1}{8} \int \cos^2 \theta\, d\theta \\
&= \frac{1}{16} \int (1 + \cos 2\theta)\, d\theta \\
&= \frac{1}{16} \left(\theta + \frac{1}{2} \sin 2\theta \right) + C.
\end{aligned}
$$

To complete the integration, we convert θ and $\sin 2\theta$ back to expressions involving the original variable x. First, note that $\theta = \text{Tan}^{-1}(x/2)$. Then, using the techniques of Section 9.3, we convert $\sin 2\theta$ as follows:

$$
\begin{aligned}
\sin 2\theta &= 2 \sin \theta \cos \theta \\
&= 2 \left(\frac{x}{\sqrt{x^2 + 4}} \right) \left(\frac{2}{\sqrt{x^2 + 4}} \right)
\end{aligned}
$$

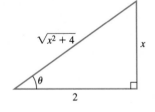

Figure 2.10 A triangle corresponding to the substitution $x = 2 \tan \theta$. Note that

$$\sin \theta = \frac{x}{\sqrt{x^2 + 4}}$$

$$\cos \theta = \frac{2}{\sqrt{x^2 + 4}}.$$

(see Figure 2.10). Thus,

$$\int \frac{1}{(x^2 + 4)^2}\, dx = \frac{1}{16} \text{Tan}^{-1} \left(\frac{x}{2} \right) + \frac{x}{8(x^2 + 4)} + C. \qquad ■$$

Exercise Set 10.2

In Exercises 1–31, evaluate the given integral.

1. $\displaystyle\int \sqrt{4 - x^2}\, dx$

2. $\displaystyle\int \frac{1}{\sqrt{x^2 + 16}}\, dx$

3. $\displaystyle\int_0^{3/2} \frac{dx}{\sqrt{9 - x^2}}$

4. $\displaystyle\int_1^2 \sqrt{x^2 - 1}\, dx$

5. $\displaystyle\int \frac{x^3}{\sqrt{x^2 - 4}}\, dx$

6. $\displaystyle\int \frac{x^2}{\sqrt{9 - x^2}}\, dx$

7. $\displaystyle\int_{1/2}^1 \frac{\sqrt{1 - x^2}}{x}\, dx$

8. $\displaystyle\int_3^5 \frac{1}{x^2 \sqrt{x^2 + 16}}\, dx$

9. $\displaystyle\int \frac{1}{x^3\sqrt{x^2-4}}\,dx$

10. $\displaystyle\int \frac{x}{\sqrt{x^2+4}}\,dx$

11. $\displaystyle\int_0^4 \frac{1}{(16+x^2)^2}\,dx$

12. $\displaystyle\int_0^1 \frac{dx}{(4-x^2)^{3/2}}$

13. $\displaystyle\int \frac{x^2\,dx}{(1-x^2)^{3/2}}$

14. $\displaystyle\int \frac{dx}{(x^2-4)^{3/2}}$

15. $\displaystyle\int_0^1 x^3\sqrt{1-x^2}\,dx$

16. $\displaystyle\int_0^2 \frac{x^2}{\sqrt{4+x^2}}\,dx$

17. $\displaystyle\int \frac{x^2\,dx}{(x^2+3)^{3/2}}$

18. $\displaystyle\int \frac{\sqrt{1-x^2}}{x^4}\,dx$

19. $\displaystyle\int \frac{x^2}{\sqrt{x^2+a^2}}\,dx$

20. $\displaystyle\int \frac{\sqrt{x^2-a^2}}{x}\,dx$

21. $\displaystyle\int \frac{\sqrt{x^2-a^2}}{x^2}\,dx$

22. $\displaystyle\int \frac{x^2}{\sqrt{x^2-a^2}}\,dx$

23. $\displaystyle\int \frac{x\,dx}{\sqrt{(x^2+a^2)^3}}$

24. $\displaystyle\int \frac{\sqrt{x^2+a^2}}{x}\,dx$

25. $\displaystyle\int \frac{dx}{x\sqrt{x^2-a^2}}$

26. $\displaystyle\int \frac{dx}{x\sqrt{x^2+a^2}}$

27. $\displaystyle\int \frac{\sqrt{x^2+a^2}}{x^2}\,dx$

28. $\displaystyle\int \frac{x^2\,dx}{(x^2+a^2)^{3/2}}$

29. $\displaystyle\int \frac{x^2+3}{\sqrt{x^2+9}}\,dx$

30. $\displaystyle\int \frac{x^2-3}{x\sqrt{x^2+4}}\,dx$

31. $\displaystyle\int \frac{x^2-4x+5}{x\sqrt{9-x^2}}\,dx$

32. Use trigonometric substitutions to obtain the formula

$$\int \frac{1}{1+x^2}\,dx = \mathrm{Tan}^{-1}\,x + C.$$

33. Find the area of the region bounded by the graph of $y = 5 - \sqrt{x^2+9}$ and the x-axis.

34. Find the area of the region bounded by the graphs of $y = x^2/5$ and $y = x^2/\sqrt{x^2+16}$.

35. Find the area of the region bounded by the graphs of $y = -(x/60)+(1/4)$ and $y = 1/\sqrt{x^2+16}$ for $0 \le x \le 3$.

36. Find the volume of the solid generated by revolving about the y-axis the region bounded by the graph of $y = x\sqrt{9-x^2}$ and the x-axis.

37. The region bounded above by the graph of $y = x(16-x^2)^{1/4}$ and below by the x-axis is rotated about the x-axis. Find the volume of the resulting solid.

38. Find the volume of the solid generated by revolving about the y-axis the region bounded by the graph of $y = 1/(x^2+9)^{3/2}$ and the x-axis for $0 \le x \le 4$.

39. Find the volume of the solid generated by revolving about the x-axis the region bounded by the graphs of $y = 1/3$, $y = 1/\sqrt{x^2+9}$, and $x = 3$.

40. Find the area of the region enclosed by the ellipse

$$\frac{x^2}{4} + \frac{y^2}{3} = 1.$$

41. A water tank is in the shape of a cylinder with circular end panels of radius 5 m. The tank is mounted with the circular ends vertical. To what percentage capacity is the tank filled if the depth of the water is 6 m?

42. Find the length of the graph of $y = \ln x$ from $(1,0)$ to $(3, \ln 3)$.

43. Find the length of the graph of $y = x^2$ from $(0,0)$ to $(1/2, 1/4)$.

10.3 Integrals Involving Quadratic Expressions

In Section 10.2, we considered a number of integrals that included quadratic expressions with no linear term. In this section, we illustrate how a substitution based on the algebraic operation of completing the square is used to eliminate the linear term (the bx term) of the general quadratic $ax^2 + bx + c$.

In general, integrals of this type involve two or more successive substitutions. As we integrate, it is important to keep a record of the substitutions and the order in which they were performed so that we can correctly state the solution in terms of the original variable.

☐ **EXAMPLE 1**

Find $\displaystyle\int \frac{dx}{\sqrt{x^2 - 2x + 5}}$.

Strategy   · · · · · · ·

First complete the square under the radical.

Solution

By completing the square, we find

$$\sqrt{x^2 - 2x + 5} = \sqrt{(x^2 - 2x + 1) + 4}$$
$$= \sqrt{(x - 1)^2 + 2^2}.$$

Make the substitution

$$u = x - 1.$$

We use the substitution

$$u = x - 1; \qquad du = dx$$

Write the original integral in terms of the substitution.

to write the integral as

$$\int \frac{dx}{\sqrt{x^2 - 2x + 5}} = \int \frac{dx}{\sqrt{(x - 1)^2 + 2^2}}$$
$$= \int \frac{du}{\sqrt{u^2 + 2^2}}.$$

The last integral requires the trigonometric substitution (Figure 3.1)

$$u = 2 \tan \theta, \qquad -\pi/2 < \theta < \pi/2,$$
$$du = 2 \sec^2 \theta \, d\theta.$$

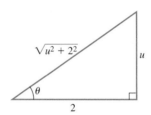

Figure 3.1 $u = 2 \tan \theta$. Thus, $\sqrt{u^2 + 2^2} = 2 \sec \theta$.

With this substitution $\sqrt{u^2 + 2^2} = 2 \sec \theta$, so the integral becomes

$$\int \frac{du}{\sqrt{u^2 + 2^2}} = \int \frac{2 \sec^2 \theta \, d\theta}{2 \sec \theta}$$
$$= \int \sec \theta \, d\theta$$

$\displaystyle\int \sec \theta \, d\theta = \ln |\sec \theta + \tan \theta| + C.$

Substitute back in terms of u.

Substitute back in terms of x.

Use property: $\ln(z/2) = \ln z - \ln 2$.

Incorporate $-\ln 2$ in constant of integration.

$$= \ln |\sec \theta + \tan \theta| + C_1$$
$$= \ln \left| \frac{\sqrt{u^2 + 2^2}}{2} + \frac{u}{2} \right| + C_1$$
$$= \ln \left(\frac{\sqrt{(x - 1)^2 + 2^2}}{2} + \frac{x - 1}{2} \right) + C_1$$
$$= \ln(\sqrt{x^2 - 2x + 5} + x - 1) - \ln 2 + C_1$$
$$= \ln(\sqrt{x^2 - 2x + 5} + x - 1) + C.$$

■

□ **EXAMPLE 2**

Find $\int \dfrac{x + 3}{\sqrt{2x^2 - 8x}}\, dx$.

Strategy · · · · · · ·
Complete the square under the radical.

Solution

We have

$$\sqrt{2x^2 - 8x} = \sqrt{2(x^2 - 4x + 4) - 2(4)}$$
$$= \sqrt{2(x - 2)^2 - 8}.$$

Make a substitution to simplify x term.

We make the substitution

$$u = x - 2; \qquad du = dx.$$

Then

$$x = u + 2, \quad \text{so} \quad x + 3 = u + 5.$$

With these substitutions the integral becomes

$$\int \frac{x + 3}{\sqrt{2x^2 - 8x}}\, dx = \int \frac{x + 3}{\sqrt{2(x - 2)^2 - 8}}\, dx$$

$$= \int \frac{u + 5}{\sqrt{2u^2 - 8}}\, du$$

$$= \frac{1}{\sqrt{2}} \int \frac{u + 5}{\sqrt{u^2 - 4}}\, du$$

Split integrand into two terms.

$$= \frac{1}{\sqrt{2}} \int \frac{u}{\sqrt{u^2 - 4}}\, du + \frac{1}{\sqrt{2}} \int \frac{5}{\sqrt{u^2 - 4}}\, du.$$

First integrand has form

$$\int w^{-1/2}\, dw \quad \text{where} \quad w = u^2 - 4.$$

The first of these two integrals is evaluated using a simple substitution and the Power Rule:

$$\frac{1}{\sqrt{2}} \int \frac{u}{\sqrt{u^2 - 4}}\, du = \frac{1}{2\sqrt{2}} \int \frac{2u}{\sqrt{u^2 - 4}}\, du$$

$$= \frac{1}{\sqrt{2}} \sqrt{u^2 - 4} + C_1.$$

The second integral requires the trigonometric substitution

$$u = 2 \sec \theta, \qquad 0 \le \theta < \pi/2 \quad \text{or} \quad \pi \le \theta < 3\pi/2,$$
$$du = 2 \sec \theta \tan \theta\, d\theta \qquad \text{(Figure 3.2).}$$

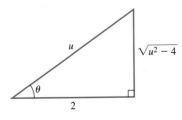

Figure 3.2 $u = 2 \sec \theta$. Thus, $\tan \theta = \dfrac{\sqrt{u^2 - 4}}{2}$.

Apply trigonometric substitution.   It becomes

$$\frac{1}{\sqrt{2}} \int \frac{5}{\sqrt{u^2 - 4}}\, du = \frac{1}{\sqrt{2}} \int \frac{5 \cdot 2 \cdot \sec\theta \tan\theta\, d\theta}{2 \tan\theta}$$

$$= \frac{5}{\sqrt{2}} \int \sec\theta\, d\theta$$

$$= \frac{5}{\sqrt{2}} \ln\left|\sec\theta + \tan\theta\right| + C_2$$

$$= \frac{5}{\sqrt{2}} \ln\left|\frac{u}{2} + \frac{\sqrt{u^2 - 4}}{2}\right| + C_2.$$

Combine results.   Combining these results and recalling that $u = x - 2$ we have

$$\int \frac{x + 3}{\sqrt{2x^2 - 8x}}\, dx = \frac{1}{\sqrt{2}} \sqrt{(x - 2)^2 - 4}$$

$$+ \frac{5}{\sqrt{2}} \ln\left|\frac{x - 2}{2} + \frac{\sqrt{(x - 2)^2 - 4}}{2}\right| + C_3$$

Incorporate factor $-\ln 2$ in C (see Example 1).

$$= \frac{1}{\sqrt{2}} \left[\sqrt{x^2 - 4x} + 5 \ln\left|x - 2 + \sqrt{x^2 - 4x}\right|\right] + C. \quad \blacksquare$$

Exercise Set 10.3

In Exercises 1–18, evaluate the given integral.

1. $\displaystyle\int \frac{dx}{x^2 - 4x + 4}$

2. $\displaystyle\int \frac{dx}{x^2 - 6x + 12}$

3. $\displaystyle\int \frac{dx}{\sqrt{x^2 + 6x + 13}}$

4. $\displaystyle\int \frac{dx}{\sqrt{5 + 4x - x^2}}$

5. $\displaystyle\int \frac{dx}{\sqrt{x^2 - 6x}}$

6. $\displaystyle\int \frac{x\, dx}{x^2 + 6x + 15}$

7. $\displaystyle\int \frac{x\, dx}{\sqrt{6x - x^2}}$

8. $\displaystyle\int \frac{1 + x}{\sqrt{x^2 - 6x + 13}}\, dx$

9. $\displaystyle\int_0^1 \frac{2x + 1}{\sqrt{18 + 6x + x^2}}\, dx$

10. $\displaystyle\int_0^1 \frac{3x - 1}{\sqrt{x^2 + 8x}}\, dx$

11. $\displaystyle\int_{-3}^{-2} \frac{x - 4}{x^2 - 8x - 7}\, dx$

12. $\displaystyle\int_0^{3/2} (2x - 3)\sqrt{16 + 12x - 4x^2}\, dx$

13. $\displaystyle\int_0^1 \frac{x}{(x^2 + 2x + 2)^2}\, dx$

14. $\displaystyle\int \frac{x^2 + 2x + 1}{6x^2 + 12x}\, dx$

15. $\displaystyle\int \frac{dx}{(9 - x)(88 - 18x + x^2)^{3/2}}$

16. $\displaystyle\int \frac{\sqrt{x^2 + 10x + 21}}{x^3 + 15x^2 + 75x + 125}\, dx$

17. $\displaystyle\int \frac{4x^2 + 20x + 25}{\sqrt{4x^2 + 20x + 29}}\, dx$

18. $\displaystyle\int \frac{dx}{(3 - x)\sqrt{18 - 6x + x^2}}$

19. Find the area bounded by the graph of $y = 1/(x^2 - 4x + 8)$ and the x-axis for $2 \le x \le 4$.

20. Find the area bounded by the graph of $y = 5/(2x^2 - 12x + 68)$ and the x-axis for $-2 \le x \le 3$.

21. Find the volume of the solid obtained by revolving about the x-axis the graph of $y = (8x - x^2)^{-1/4}$ for $2 \le x \le 4$.

22. Find the volume of the solid obtained by revolving about the y-axis the graph of $y = 1/\sqrt{8x - x^2}$ for $2 \le x \le 4$.

10.4   The Method of Partial Fractions

In elementary algebra you learned the rule for adding two fractions with different denominators:

$$\frac{a}{b} + \frac{c}{d} = \frac{ad + bc}{bd}.$$

The idea was simply to find a common denominator. For the addition of rational functions of x, the rule is the same. For example,

$$\frac{1}{x + 1} + \frac{3}{x - 2} = \frac{1(x - 2) + 3(x + 1)}{(x + 1)(x - 2)} = \frac{4x + 1}{x^2 - x - 2}. \tag{1}$$

By reversing this procedure, we obtain a technique for integrating rational functions of x such as that on the right side of equation (1). Using equation (1) we see that

$$\int \frac{4x + 1}{x^2 - x - 2} \, dx = \int \left[\frac{1}{x + 1} + \frac{3}{x - 2} \right] dx \tag{2}$$

$$= \ln |x + 1| + 3 \ln |x - 2| + C.$$

Clearly the second integral in equation (2) is easier to handle than the first, since the terms in the integrand involve polynomials of lower degree. The goal of the method of partial fractions is to decompose rational functions into sums of simpler terms, each of which can be integrated by methods we have already developed.

Before describing the method in its full detail, we consider several examples in order to gain some insight into how one goes about finding such a decomposition. We need to keep two points in mind:

POINT 1   In searching for a decomposition of the rational function $p(x)/q(x)$ we must work only with denominators that *are factors of* $q(x)$. Otherwise, the terms obtained cannot possibly add up to $p(x)/q(x)$.

POINT 2   We will work only with *proper fractions*. That is, in all rational expressions $p(x)/q(x)$ we will require that the degree of $p(x)$ be less than the degree of $q(x)$. (If this is not the case we will first simplify using polynomial long division.)

With these two points and our overall goal in mind, we now examine a few examples. Moreover, we assume that such a so-called partial fractions decomposition exists. Later we describe the general procedure for determining the partial fractions decomposition of an arbitrary rational function.

□ **EXAMPLE 1**

Find a partial fractions decomposition for

$$f(x) = \frac{4x - 7}{x^2 - x - 6}.$$

Strategy · · · · · · · ·

Factor the denominator.

Solution

We first factor the denominator as

$$x^2 - x - 6 = (x + 2)(x - 3).$$

Allow one term for each factor of the denominator. (Assume constant numerators for first degree denominators.)

We therefore seek a decomposition of the form

$$\frac{4x - 7}{x^2 - x - 6} = \frac{A}{x + 2} + \frac{B}{x - 3} \tag{3}$$

where A and B are constants. (Neither A nor B can be polynomials of degree higher than zero, or we would be introducing improper fractions, violating Point 2.)

Determine A and B by recombining fractions and equating numerators.

To determine A and B, we recombine the terms on the right of equation (3) according to the rule for adding fractions:

$$\frac{4x - 7}{x^2 - x - 6} = \frac{A}{x + 2} + \frac{B}{x - 3} = \frac{A(x - 3) + B(x + 2)}{(x + 2)(x - 3)} \tag{4}$$

$$= \frac{(A + B)x + (-3A + 2B)}{(x + 2)(x - 3)}.$$

Collect like terms in x in numerator on right-hand side.

Use fact that if two polynomials are equal, coefficients of like powers of x must be the same (see Exercise 42).

Now since the denominators of the two fractions on the left and right sides of equation (4) are equal, the coefficients of their numerators must agree. That is,

$$\begin{cases} A + B = 4 & \text{(coefficients of } x\text{)}, \\ -3A + 2B = -7 & \text{(constants)}. \end{cases}$$

Equate coefficients of like powers of x.

Solve the resulting system either by substitution or by elimination.

The solution of this system of equations may be found by adding three times the top equation to the bottom equation. (Or, by substitution, if you prefer.) Doing so we obtain

$$5B = 5, \quad \text{so} \quad B = 1.$$

From either equation we then obtain $A = 3$. We therefore have

$$\frac{4x - 7}{x^2 - x - 6} = \frac{3}{x + 2} + \frac{1}{x - 3}. \qquad \blacksquare$$

□ **EXAMPLE 2**

Find $\displaystyle\int \frac{4x - 7}{x^2 - x - 6}\, dx$.

Solution   Using the results of Example 1 we find

$$\int \frac{4x - 7}{x^2 - x - 6}\, dx = \int \frac{3}{x + 2}\, dx + \int \frac{1}{x - 3}\, dx$$

$$= 3 \ln |x + 2| + \ln |x - 3| + C. \qquad \blacksquare$$

□ **EXAMPLE 3**

Find a partial fractions decomposition for the function

$$f(x) = \frac{4x^2 - 3x + 1}{x^3 - x^2 + x}.$$

Strategy · · · · · · ·

Factor the denominator.

Solution

The denominator factors as

$$x^3 - x^2 + x = x(x^2 - x + 1).$$

(This is the best we can do since the term $x^2 - x + 1$ does not have real linear factors.*) We therefore seek a partial fractions decomposition of the form

$$\frac{4x^2 - 3x + 1}{x^3 - x^2 + x} = \frac{A}{x} + \frac{Bx + C}{x^2 - x + 1}.$$

Allow one term for each factor of the denominator. (Degree of numerator in each term is one less than the degree of the denominator.)

Here we must allow for a first degree numerator in the second fraction on the right since the denominator has degree two. To find the constants, A, B, and C, we recombine terms to find that

$$\frac{4x^2 - 3x + 1}{x^3 - x^2 + x} = \frac{A}{x} + \frac{Bx + C}{x^2 - x + 1}$$

$$= \frac{A(x^2 - x + 1) + (Bx + C)(x)}{x(x^2 - x + 1)}$$

$$= \frac{(A + B)x^2 + (-A + C)x + A}{x(x^2 - x + 1)}.$$

Recombine terms collecting like terms in x in numerator.

Equate coefficients of like powers of x.

Equating coefficients of the various powers of x, we obtain the equations

$$\begin{cases} A + B & = 4 & \text{(coefficients of } x^2) \\ -A & + C = -3 & \text{(coefficients of } x) \\ A & = 1 & \text{(constants)} \end{cases}$$

Solve resulting system.

The system has solution $A = 1$, $B = 3$, $C = -2$. The partial fractions decomposition is therefore

$$\frac{4x^2 - 3x + 1}{x^3 - x^2 + x} = \frac{1}{x} + \frac{3x - 2}{x^2 - x + 1}.$$ ∎

□ EXAMPLE 4

Find $\displaystyle\int \frac{2x^3 - 8x^2 + 9x + 1}{x^2 - 4x + 4}\, dx.$

Strategy · · · · · · · · ·

Integrand is an improper fraction. Reduce it to a proper fraction, plus a polynomial, by division.

Solution

Before seeking a partial fractions decomposition we note that the integrand is an improper fraction. We therefore perform polynomial long division:

$$\begin{array}{r} 2x \\ x^2 - 4x + 4\overline{)2x^3 - 8x^2 + 9x + 1} \\ \underline{2x^3 - 8x^2 + 8x} \\ x + 1 \end{array}$$

This calculation shows that

$$\frac{2x^3 - 8x + 9x + 1}{x^2 - 4x + 4} = 2x + \frac{x + 1}{x^2 - 4x + 4}.$$

---

*A quadratic polynomial $ax^2 + bx + c$ has real linear factors $ax^2 + bx + c = a(x - A)(x - B)$ if and only if A and B are **roots** of the polynomial. Since the roots are given by the quadratic formula $x = (-b \pm \sqrt{b^2 - 4ac})/2a$, the polynomial has real linear factors only if the **discriminant** $b^2 - 4ac$ is nonnegative. (This statement is called the **Discriminant Test** for the existence of real linear factors.) In this case $a = 1$, $b = -1$, and $c = 1$, so $b^2 - 4ac = (-1)^2 - (4)(1)(1) = 1 - 4 < 0$. Thus there can be no real linear factors.

Since integrating the first term presents no difficulties, we concentrate on finding a partial fractions decomposition for the second term. The denominator factors as

Factor denominator in second term.

$$x^2 - 4x + 4 = (x - 2)^2,$$

and we seek a decomposition of the form

$$\frac{x + 1}{x^2 - 4x + 4} = \frac{A}{x - 2} + \frac{B}{(x - 2)^2}.$$

Allow one term for each factor $(x - 2)$ and $(x - 2)^2$. (Only a constant numerator in the second term is required—see Exercise 41.) Combine terms.

(We need both fractions because neither one alone will suffice. However, as the following computation indicates, the two together are sufficient.)

$$\frac{x + 1}{x^2 - 4x + 4} = \frac{A}{x - 2} + \frac{B}{(x - 2)^2} = \frac{A(x - 2) + B}{(x - 2)^2} = \frac{Ax + (-2A + B)}{(x - 2)^2}.$$

Equate coefficients of like powers of x.

Equating like powers of x we obtain the equations

$$\begin{cases} A \quad\quad = 1 & \text{(coefficients of } x), \\ -2A + B = 1 & \text{(constants).} \end{cases}$$

This system has solution $A = 1$, $B = 3$, so

$$\frac{x + 1}{x^2 - 4x + 4} = \frac{1}{x - 2} + \frac{3}{(x - 2)^2}.$$

Combining our results we now have

$$\int \frac{2x^3 - 8x^2 + 9x + 1}{x^2 - 4x + 4}\, dx = \int 2x\, dx + \int \frac{1}{x - 2}\, dx + \int \frac{3}{(x - 2)^2}\, dx$$

$$= x^2 + \ln|x - 2| - \frac{3}{x - 2} + C. \qquad \blacksquare$$

It is now time to state the formal procedure lurking behind Examples 1–4. The obvious first step in obtaining a partial fractions decomposition for the rational function $p(x)/q(x)$ is to factor the denominator. A theorem about polynomials tells us what to expect as the outcome: *every polynomial $q(x)$ with real coefficients can be factored into the product of only two types of factors:*

(i) powers of linear terms, that is, terms of the form $(x - a)^n$, and/or
(ii) powers of irreducible quadratic terms, that is, terms of the form $(x^2 + bx + c)^m$.

(Thus we are assured, for example, that a cubic polynomial such as $x^3 - 6x^2 + 5x - 9$ can be factored at least into a quadratic term and a linear term, if not three linear terms. Unfortunately, however, this theorem provides no indication of what these factors actually might be.)

We shall not prove this theorem here, but it is central to understanding why the following procedure encompasses all rational integrands.

Procedure for Finding the Partial Fraction Decomposition of the Rational Function

$$f(x) = \frac{p(x)}{q(x)}$$

1. Check that the degree of $p(x)$ is smaller than the degree of $q(x)$. If not, perform polynomial long division.
2. Factor the denominator $q(x)$ into the product of powers of linear and irreducible quadratic terms. That is, obtain the factorization

$$q(x) = d(x - a_1)^{n_1} \cdots (x - a_k)^{n_k}(x^2 + b_1x + c_1)^{m_1} \cdots$$
$$\cdot (x^2 + b_jx + c_j)^{m_j}.$$

3. For each linear factor $(x - a)^n$ include the terms

$$\frac{A_1}{x - a} + \frac{A_2}{(x - a)^2} + \cdots + \frac{A_n}{(x - a)^n}$$

in the partial fractions decomposition. For each factor $(x^2 + bx + c)^m$ include the terms

$$\frac{B_1x + C_1}{x^2 + bx + c} + \frac{B_2x + C_2}{(x^2 + bx + c)^2} + \cdots + \frac{B_mx + C_m}{(x^2 + bx + c)^m}.$$

4. Combine all terms in the partial fractions decomposition and collect coefficients of like powers of x.
5. Equate coefficients of like powers of x between the partial fractions decomposition and the original function $p(x)/q(x)$.
6. Solve the resulting system for the unknown constants.

The procedure is stated in its fullest generality. Do not be intimidated by the notation; usually only a few factors are involved. However, step 3 provides a helpful reference for deciding what types of terms to allow.

□ **EXAMPLE 5**

The following are several examples of rational functions and the forms of their associated partial fractions expansions.

Function	Partial Fractions Expansion
$\dfrac{1}{(x - 1)(x + 2)}$	$\dfrac{A}{x - 1} + \dfrac{B}{x + 2}$
$\dfrac{x + 2}{(x - 1)^2}$	$\dfrac{A}{x - 1} + \dfrac{B}{(x - 1)^2}$
$\dfrac{x^2 - 6x + 1}{x(x^2 + x + 1)}$	$\dfrac{A}{x} + \dfrac{Bx + C}{x^2 + x + 1}$
$\dfrac{x + 7}{x^2(x - 2)^2}$	$\dfrac{A}{x} + \dfrac{B}{x^2} + \dfrac{C}{x - 2} + \dfrac{D}{(x - 2)^2}$
$\dfrac{x^3 + 6x + 1}{x(x^2 + x + 1)^2}$	$\dfrac{A}{x} + \dfrac{Bx + C}{x^2 + x + 1} + \dfrac{Dx + E}{(x^2 + x + 1)^2}$
$\dfrac{x^4 - x^2 + 1}{(x - 1)^3(x^2 + x + 2)^2}$	$\dfrac{A}{x - 1} + \dfrac{B}{(x - 1)^2} + \dfrac{C}{(x - 1)^3} + \dfrac{Dx + E}{x^2 + x + 2} + \dfrac{Fx + G}{(x^2 + x + 2)^2}$

□ **EXAMPLE 6**

Find $\displaystyle\int \frac{3x^4 + 9x^3 + 15x^2 + 10x + 4}{x(x^2 + 2x + 2)^2}\, dx.$

Solution The numerator is fourth degree, and the denominator is fifth degree. Thus, division is not needed. Since $x^2 + 2x + 2$ is an irreducible quadratic, the denominator is in factored form and we proceed to seek the partial fractions decomposition:

$$\frac{3x^4 + 9x^3 + 15x^2 + 10x + 4}{x(x^2 + 2x + 2)^2}$$

$$= \frac{A}{x} + \frac{Bx + C}{x^2 + 2x + 2} + \frac{Dx + E}{(x^2 + 2x + 2)^2}$$

$$= \frac{A(x^2 + 2x + 2)^2 + x(Bx + C)(x^2 + 2x + 2) + x(Dx + E)}{x(x^2 + 2x + 2)^2}$$

$$= \frac{(A + B)x^4 + (4A + 2B + C)x^3 + (8A + 2B + 2C + D)x^2 + (8A + 2C + E)x + 4A}{x(x^2 + 2x + 2)^2}$$

Equating coefficients of like powers of x gives the equations

$$\begin{cases} A + B & = 3 & \text{(coefficients of } x^4) \\ 4A + 2B + C & = 9 & \text{(coefficients of } x^3) \\ 8A + 2B + 2C + D & = 15 & \text{(coefficients of } x^2) \\ 8A \quad + 2C \quad + E = 10 & & \text{(coefficients of } x) \\ 4A & = 4 & \text{(constants)} \end{cases}$$

The solution of this system is $A = 1$, $B = 2$, $C = 1$, $D = 1$, and $E = 0$.
    We therefore conclude that

$$\int \frac{3x^4 + 9x^3 + 15x^2 + 10x + 4}{x(x^2 + 2x + 2)^2}\, dx$$

$$= \int \frac{1}{x}\, dx + \int \frac{2x + 1}{x^2 + 2x + 2}\, dx + \int \frac{x}{(x^2 + 2x + 2)^2}\, dx$$

The first integral produces a logarithm, and the third integral is evaluated by completing the square and performing a trigonometric substitution (see Section 10.3 and Example 6 in Section 10.2; also note Exercise 13 in Section 10.3). To evaluate the second integral, we write

$$\frac{2x + 1}{x^2 + 2x + 2} = \frac{2x + 2}{x^2 + 2x + 2} - \frac{1}{x^2 + 2x + 2}.$$

When we integrate, the first fraction produces a logarithm. That is,

$$\int \frac{2x + 2}{x^2 + 2x + 2}\, dx = \ln(x^2 + 2x + 2) + C_1.$$

Integrating the second fraction involves completing the square and yields an inverse tangent function. In other words,

$$\int \frac{1}{x^2 + 2x + 2}\, dx = \int \frac{1}{(x + 1)^2 + 1}\, dx = \operatorname{Tan}^{-1}(x + 1) + C_2.$$

The result of the original integration is

$$\int \frac{3x^4 + 9x^3 + 15x^2 + 10x + 4}{x(x^2 + 2x + 2)^2} \, dx$$

$$= \ln|x| + \ln(x^2 + 2x + 2) - \frac{3}{2} \mathrm{Tan}^{-1}(x + 1) - \frac{x + 2}{2(x^2 + 2x + 2)} + C. \quad \blacksquare$$

REMARK   At this point, it is useful to compare the various methods used in Examples 2, 4, and 6 to integrate rational functions with quadratics in their denominators. In Example 2, the quadratic factored into two distinct linear factors, and we used logarithms to evaluate the integral. In Example 4, the quadratic was a perfect square, and we used the Power Rule as well as logarithms to evaluate the integral. In Example 6, we integrated two different rational functions with irreducible quadratic factors in their denominators, and in both cases the integration involved multiple steps. We evaluated

$$\int \frac{2x + 1}{x^2 + 2x + 2} \, dx$$

by rewriting the fraction as two fractions where one term can be integrated using logarithms while the other is integrated by completing the square and using the inverse tangent function. Also, we evaluated

$$\int \frac{x}{(x^2 + 2x + 2)^2} \, dx$$

by completing the square and using a trigonometric substitution.

The Heaviside Method[*]   The Heaviside Method is a procedure that often speeds the calculation of the unknown coefficients in a partial fractions expansion. Suppose we wish to find constants A, B, and C so that

$$\frac{p(x)}{(x - a)(x - b)(x - c)} = \frac{A}{x - a} + \frac{B}{x - b} + \frac{C}{x - c}. \tag{5}$$

To find A, we (i) multiply both sides of the equation by $(x - a)$, obtaining

$$\frac{p(x)}{(x - b)(x - c)} = A + B\left(\frac{x - a}{x - b}\right) + C\left(\frac{x - a}{x - c}\right). \tag{6}$$

Then we (ii) set $x = a$, obtaining from equation (6) the statement

$$A = \frac{p(a)}{(a - b)(a - c)}.$$

Similarly, we can find B and C by repeating steps (i) and (ii) with the factors $(x - b)$ and $(x - c)$, respectively. We find that

$$B = \frac{p(b)}{(b - a)(b - c)}; \qquad C = \frac{p(c)}{(c - a)(c - b)}.$$

This procedure, valid only for the case of distinct linear factors, is sometimes described by saying that one "strikes" the factor on the left side of equation (5)

---

[*]This technique is due to Oliver Heaviside (1850–1925).

corresponding to the desired constant and then evaluates the resulting expression at the corresponding value of x.

☐ **EXAMPLE 7**

Find $\displaystyle\int \frac{6x^2 + 6x - 6}{x^3 + 2x^2 - x - 2}\, dx$.

Strategy · · · · · · · ·

Find the factors of $x^3 + 2x^2 - x - 2$.

Solution

First we factor the denominator. We get

$$x^3 + 2x^2 - x - 2 = (x - 1)(x + 1)(x + 2).$$

Write the form of the partial fractions decomposition.

We therefore seek the partial fractions decomposition

$$\frac{6x^2 + 6x - 6}{(x - 1)(x + 1)(x + 2)} = \frac{A}{x - 1} + \frac{B}{x + 1} + \frac{C}{x + 2}. \tag{7}$$

Use the Heaviside method to find A, B, and C.

Using the Heaviside method to find A, we "strike" the factor $(x - 1)$ in the left side of (7) and set $x = 1$ to obtain

$$A = \frac{6(1)^2 + 6(1) - 6}{(1 + 1)(1 + 2)} = \frac{6}{6} = 1.$$

Similarly we find that

$$B = \frac{6(-1)^2 + 6(-1) - 6}{(-1 - 1)(-1 + 2)} = \frac{-6}{-2} = 3$$

and

$$C = \frac{6(-2)^2 + 6(-2) - 6}{(-2 - 1)(-2 + 1)} = \frac{6}{3} = 2.$$

Thus

$$\int \frac{6x^2 + 6x - 6}{x^3 + 2x^2 - x - 2}\, dx = \int \frac{1}{x - 1}\, dx + \int \frac{3}{x + 1}\, dx + \int \frac{2}{x + 2}\, dx$$
$$= \ln |x - 1| + 3 \ln |x + 1| + 2 \ln |x + 2| + C. \qquad ■$$

☐ **EXAMPLE 8**

An important application of the method of partial fractions occurs in the logistic growth model. Recall from Chapter 8 that the Law of Natural Growth

$$\{\text{Rate of growth}\} \propto \{\text{Present size of population}\} \tag{8}$$

leads to the differential equation

$$\frac{dP}{dt} = kP, \tag{9}$$

which has solutions $P(t) = P_0 e^{kt}$. The constant k in equation (9) is referred to as the growth constant. A typical solution curve for equation (9) is depicted in Figure 4.1. One limitation of this growth model is that it assumes an unlimited environment. That is, if $k > 0$, the model assumes that the population can increase at an exponential rate forever.

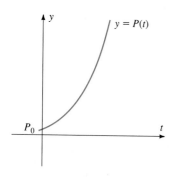

Figure 4.1 Model for natural growth assumes an unlimited environment.

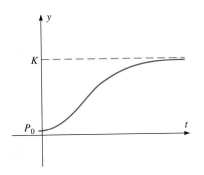

Figure 4.2 Logistic growth model assumes a maximum possible population size K.

The logistic growth model postulates a maximum possible population size K (sometimes called the **carrying capacity** of the system) by assuming that the growth rate is proportional to both present population size and the ''unutilized capacity for growth'' in the system (see Figure 4.2). That is,

$$\text{Rate of Growth} \propto \text{(Percent size) (Unutilized capacity for growth)} \qquad (10)$$

Since the fraction of the carrying capacity K unutilized at population size $P < K$ is $(K - P)/K$, the differential equation corresponding to the logistic growth law in (10) is

$$\frac{dP}{dt} = kP\left(\frac{K - P}{K}\right), \qquad (11)$$

where k is again called the natural growth constant.

To solve (11), we separate variables obtaining

$$\frac{dP}{P(K - P)} = \frac{k}{K}\,dt,$$

so

$$\int \frac{dP}{P(K - P)} = \int \frac{k}{K}\,dt = \left(\frac{k}{K}\right)t + C. \qquad (12)$$

To find the integral on the left side, we use the method of partial fractions: Setting

$$\frac{1}{P(K - P)} = \frac{A}{P} + \frac{B}{K - P} = \frac{A(K - P) + BP}{P(K - P)} = \frac{(-A + B)P + AK}{P(K - P)}$$

gives the equations

$$\begin{cases} -A + B = 0 & \text{(coefficients of } P\text{),} \\ AK = 1 & \text{(constants).} \end{cases}$$

Thus $A = B = \dfrac{1}{K}$. The integral on the left side of (12) is therefore

$$\int \frac{dP}{P(K - P)} = \frac{1}{K}\int \frac{1}{P}\,dP + \frac{1}{K}\int \frac{1}{K - P}\,dP$$

$$= \frac{1}{K}\ln P - \frac{1}{K}\ln(K - P)$$

$$= \frac{1}{K}\ln\left(\frac{P}{K - P}\right).$$

Thus equation (12) becomes

$$\frac{1}{K}\ln\left(\frac{P}{K - P}\right) = \frac{k}{K}t + C$$

or

$$\frac{P}{K - P} = e^{kt + KC}.$$

Solving for P now gives

$$P = \frac{K}{1 + Me^{-kt}} \quad \text{where} \quad M = e^{-KC}. \tag{13}$$

From the form of the solution in line (13) we can indeed see that

$$\lim_{t \to \infty} P(t) = \lim_{t \to \infty} \frac{K}{1 + Me^{-kt}} = K$$

for populations satisfying the logistic growth model. ∎

Exercise Set 10.4

In Exercises 1–34, find the indicated integral.

1. $\displaystyle\int \frac{1}{x(x+1)}\,dx$

2. $\displaystyle\int \frac{3x+2}{x(x+1)}\,dx$

3. $\displaystyle\int_4^6 \frac{2x}{(x-3)(x-1)}\,dx$

4. $\displaystyle\int_7^{11} \frac{-x-8}{(x+1)(x-6)}\,dx$

5. $\displaystyle\int \frac{2x+4}{1-x^2}\,dx$

6. $\displaystyle\int \frac{7x+2}{x^2+x-2}\,dx$

7. $\displaystyle\int \frac{1}{1-x^2}\,dx$

8. $\displaystyle\int \frac{x^2+1}{x^2-1}\,dx$

9. $\displaystyle\int_e^{e^2} \frac{1}{x+x^3}\,dx$

10. $\displaystyle\int_1^e \frac{x^2+3x+3}{x(1+x)}\,dx$

11. $\displaystyle\int \frac{1}{(x+1)(x^2+1)}\,dx$

12. $\displaystyle\int \frac{x^2+x+1}{x+x^3}\,dx$

13. $\displaystyle\int_3^4 \frac{3x^2-2x+2}{(x-2)(x^2+1)}\,dx$

14. $\displaystyle\int_2^5 \frac{-2x^2-x-1}{(3x^2+1)(x-1)}\,dx$

15. $\displaystyle\int \frac{4x^2+x+1}{(x+2)(x^2+1)}\,dx$

16. $\displaystyle\int \frac{1}{x(x+1)^2}\,dx$

17. $\displaystyle\int_e^{e^3} \frac{x^2+5x+1}{x(x+1)^2}\,dx$

18. $\displaystyle\int_1^7 \frac{dx}{(x+1)^2}$

19. $\displaystyle\int \frac{2x^2+3x+2}{x^3+2x^2+x}\,dx$

20. $\displaystyle\int \frac{x^3+x^2+1}{x^4+2x^2+1}\,dx$

21. $\displaystyle\int_1^2 \frac{1}{x^3+3x^2+2x}\,dx$

22. $\displaystyle\int_6^9 \frac{3x^2+6x+2}{x(x+1)(x+2)}\,dx$

23. $\displaystyle\int \frac{5x^3-9x^2+3x-3}{x(x-3)(x^2+1)}\,dx$

24. $\displaystyle\int \frac{6x^3-4x^2+3x-3}{x(x-1)(1+x^2)}\,dx$

25. $\displaystyle\int \frac{4}{x^3+x}\,dx$

26. $\displaystyle\int \frac{dx}{x^3-6x^2+13x-10}$

27. $\displaystyle\int \frac{dx}{(a^2-x^2)^2}$

28. $\displaystyle\int \frac{ax^2+bx+a}{x+x^3}\,dx$

29. $\displaystyle\int \frac{2x^3-4x^2-7x-18}{x^2-2x-8}\,dx$

30. $\displaystyle\int \frac{2x^2-3x-2}{x^3+x^2-2x}\,dx$

31. $\displaystyle\int \frac{5x^2+2x+2}{x^3-1}\,dx$

32. $\displaystyle\int \frac{6x^2-21x-9}{x^3-6x^2+3x+10}\,dx$

33. $\displaystyle\int \frac{9x^2+26x-16}{x^3+2x^2-8x}\,dx$

34. $\displaystyle\int \frac{x^4-x^3+3x^2-10x+8}{x^3-x^2-4}\,dx$

35. Find the area of the region bounded above by the graph of

$$y = \frac{7x+3}{x^2+x}$$

and below by the x-axis for $1 \le x \le 2$.

36. The region bounded above by the graph of $y = x/(x^2+1)$ and below by the x-axis for $0 \le x \le 1$ is revolved about the x-axis. Find the volume of the resulting solid.

37. The region bounded by the graph of

$$y = \frac{1}{(x+1)(x+3)}$$

and the x-axis for $0 \le x \le 2$ is revolved about the y-axis. Find the volume of the resulting solid.

38. Find the average value of the function $f(x) = 4x^2/(1+x^2)^2$ on the interval $[0, 1]$.

39. Find the volume of the solid obtained by revolving about the x-axis the graph of

$$y = \sqrt{\frac{4x^2+1}{x^3+x}}$$

for $1 \le x \le 5$.

40. Find the volume of the solid obtained by revolving about the y-axis the graph of $y = 1/(x - 1)^2$ for $3 \leq x \leq 7$.

41. Show that the expression

$$\frac{A}{x - a} + \frac{B}{(x - a)^2}$$

can always be brought to the form $(Cx + D)/(x - a)^2$ by correctly choosing A and B. (This shows that we need not allow for a first degree numerator in the term $B/(x - a)^2$ when seeking a partial fractions decomposition for $p(x)/(x - a)^2$.)

42. Prove that if

$$p(x) = a_n x^n + a_{n-1} x^{n-1} + \cdots + a_1 x + a_0 \quad \text{and}$$
$$q(x) = b_n x^n + b_{n-1} x^{n-1} + \cdots + b_1 x + b_0$$

are polynomials with $p(x) = q(x)$ for all x, then $a_n = b_n$, $a_{n-1} = b_{n-1}, \ldots, a_0 = b_0$. That is, if two polynomials are

equal then corresponding coefficients of like powers of x must be equal. (*Hint:* Set $x = 0$ to conclude that $a_0 = b_0$. Differentiate.)

43. Find a solution of the logistic differential equation

$$\frac{dP}{dt} = 2P\left(\frac{100 - P}{100}\right).$$

What is the carrying capacity?

44. Verify the evaluation of the integrals in Example 6.

45. Let $P(t)$ be a solution of the differential equation

$$\frac{dP}{dt} = P(1 - P)$$

so that $P(0) > 0$. What is $\lim_{t \to \infty} P(t)$?

10.5 Miscellaneous Substitutions

Except for integration by parts, all integration techniques encountered up to now may be viewed as substitution methods of one sort or another. In this section we present several additional types of substitution techniques.

Radicals of Polynomial Expressions

In integrands involving fractional powers of polynomial expressions, such as $\sqrt[n]{p(x)}$, it is often useful to make a substitution of the form $u^n = p(x)$. This substitution will eliminate the radical sign since $\sqrt[n]{p(x)} = \sqrt[n]{u^n} = u$.

☐ **EXAMPLE 1**

Find $\displaystyle\int \frac{dx}{x\sqrt{x + 1}}, \qquad x > 0$.

Strategy · · · · · · · ·

To obtain a square under the square root we use

$$u^2 = x + 1.$$

Then solve for x and differentiate to find dx.

Substitute.

Form the partial fraction decomposition.

Integrate.

Solve $u^2 = x + 1$ for $u = \sqrt{x + 1}$ and substitute back.

Solution

We make the substitution $u^2 = x + 1$. Then $x = u^2 - 1$, so $dx = 2u\,du$.
    The integral becomes

$$\int \frac{dx}{x\sqrt{x + 1}} = \int \frac{2u\,du}{(u^2 - 1)u}$$

$$= \int \frac{2\,du}{u^2 - 1}$$

$$= \int \left[\frac{1}{u - 1} - \frac{1}{u + 1}\right] du$$

$$= \ln\left|\frac{u - 1}{u + 1}\right| + C$$

$$= \ln\left(\frac{\sqrt{x + 1} - 1}{\sqrt{x + 1} + 1}\right) + C. \qquad \blacksquare$$

□ **EXAMPLE 2**

Find $\displaystyle\int \frac{dx}{\sqrt{x} - \sqrt[3]{x}}$, $\quad x > 0$.

Strategy · · · · · · ·

We need $x = u^n$ where both $n/2$ and $n/3$ are integers. Thus, we use $n = 6$.

Solution

We use the substitution $x = u^6$. Then $dx = 6u^5 \, du$, so

$$\int \frac{dx}{\sqrt{x} - \sqrt[3]{x}} = \int \frac{6u^5 \, du}{\sqrt{u^6} - \sqrt[3]{u^6}}$$

Substitute and simplify.

$$= \int \frac{6u^5 \, du}{u^3 - u^2}$$

$$= 6 \int \frac{u^3 \, du}{u - 1}$$

Reduce integrand by division.

$$= 6 \int \left(u^2 + u + 1 + \frac{1}{u - 1} \right) du$$

Substitute back; $u = x^{1/6}$.

$$= 2u^3 + 3u^2 + 6u + 6 \ln |u - 1| + C$$

$$= 2x^{1/2} + 3x^{1/3} + 6x^{1/6} + 6 \ln |x^{1/6} - 1| + C. \quad \blacksquare$$

□ **EXAMPLE 3**

Find $\displaystyle\int \sqrt{1 + e^x} \, dx$.

Strategy · · · · · · ·

Substitute a square for the quantity $1 + e^x$.

Solution

We use the substitution $u^2 = 1 + e^x$ with $u > 0$. Then

$$e^x = u^2 - 1,$$

so

Note that $u^2 > 1$ for all x, so these expressions are well defined.

$$x = \ln(u^2 - 1) \quad \text{and} \quad dx = \frac{2u \, du}{u^2 - 1}.$$

We obtain

Substitute and simplify.

$$\int \sqrt{1 + e^x} \, dx = \int \sqrt{u^2} \left(\frac{2u}{u^2 - 1} \right) du$$

$$= \int \frac{2u^2}{u^2 - 1} \, du$$

Reduce integrand by division.

$$= \int \left(2 + \frac{2}{u^2 - 1} \right) du$$

Form the partial fraction decomposition.

$$= \int \left(2 + \frac{1}{u - 1} - \frac{1}{u + 1} \right) du$$

Integrate.

$$= 2u + \ln \left| \frac{u - 1}{u + 1} \right| + C$$

Substitute back $u = \sqrt{1 + e^x}$.

$$= 2\sqrt{1 + e^x} + \ln \left(\frac{\sqrt{1 + e^x} - 1}{\sqrt{1 + e^x} + 1} \right) + C. \quad \blacksquare$$

Rational Expressions in Sine and Cosine

There is a special substitution that allows us to handle a number of integrals involving rational functions of $\sin x$ and $\cos x$. It is

$$u = \tan\left(\frac{x}{2}\right). \tag{1}$$

To find the values for $\sin x$ and $\cos x$ corresponding to substitution (1) we observe that

$$\cos\left(\frac{x}{2}\right) = \frac{1}{\sec(x/2)} = \frac{1}{\sqrt{1 + \tan^2(x/2)}} = \frac{1}{\sqrt{1 + u^2}}$$

and

$$\sin\left(\frac{x}{2}\right) = \tan\left(\frac{x}{2}\right)\cos\left(\frac{x}{2}\right) = \frac{u}{\sqrt{1 + u^2}}.$$

Thus

$$\sin x = 2\sin\left(\frac{x}{2}\right)\cos\left(\frac{x}{2}\right) = \frac{2u}{1 + u^2} \tag{2}$$

and

$$\cos x = \cos^2\left(\frac{x}{2}\right) - \sin^2\left(\frac{x}{2}\right) = \frac{1 - u^2}{1 + u^2}. \tag{3}$$

Finally, from (1) we have $x = 2\,\mathrm{Tan}^{-1} u$, so

$$dx = \frac{2}{1 + u^2}\,du. \tag{4}$$

□ **EXAMPLE 4**

Find $\displaystyle\int \frac{dx}{2 + \cos x}$.

Strategy · · · · · · · ·
Use substitution (1) in the form of equations (3) and (4).

Solution
Using equations (3) and (4) we obtain

$$\int \frac{dx}{2 + \cos x} = \int \frac{\dfrac{2}{1 + u^2}}{2 + \left(\dfrac{1 - u^2}{1 + u^2}\right)}\,du$$

Simplify.

$$= \int \frac{\dfrac{2}{1 + u^2}}{\dfrac{2(1 + u^2) + (1 - u^2)}{1 + u^2}}\,du$$

Factor into form

$$\int \frac{dz}{1 + z^2}.$$

$$= \int \frac{2\,du}{3 + u^2}$$

$$= \frac{2}{3}\int \frac{du}{1 + \left(\dfrac{u}{\sqrt{3}}\right)^2}$$

$$= \frac{2}{\sqrt{3}} \int \frac{(1/\sqrt{3})\,du}{1 + (u/\sqrt{3})^2}$$

$$= \frac{2}{\sqrt{3}} \operatorname{Tan}^{-1}\left(\frac{u}{\sqrt{3}}\right) + C$$

Substitute back using (1).

$$= \frac{2}{\sqrt{3}} \operatorname{Tan}^{-1}\left(\frac{1}{\sqrt{3}} \tan \frac{x}{2}\right) + C.$$

■

□ **EXAMPLE 5**

Find $\displaystyle \int \frac{1 + \sin x}{1 + \cos x}\,dx.$

Strategy · · · · · · · · **Solution**
Use substitution (1).   Using equations (2)–(4) we obtain

$$\int \frac{1 + \sin x}{1 + \cos x}\,dx = \int \frac{\left(1 + \dfrac{2u}{1 + u^2}\right)}{\left(1 + \dfrac{1 - u^2}{1 + u^2}\right)} \left(\frac{2}{1 + u^2}\right) du$$

Simplify.

$$= \int \left(\frac{1 + 2u + u^2}{2}\right)\left(\frac{2}{1 + u^2}\right) du$$

Multiply factors in integrand.

$$= \int \frac{1 + 2u + u^2}{1 + u^2}\,du$$

Simplify by division.

$$= \int \left(1 + \frac{2u}{u^2 + 1}\right) du$$

$$= u + \ln(u^2 + 1) + C$$

$$= \tan\left(\frac{x}{2}\right) + \ln\left(\tan^2\left(\frac{x}{2}\right) + 1\right) + C.$$

■

Exercise Set 10.5

Evaluate the following integrals.

1. $\displaystyle \int \frac{\sqrt{x}}{1 + \sqrt{x}}\,dx$

2. $\displaystyle \int \frac{\sqrt{x + 1}}{x}\,dx$

3. $\displaystyle \int \frac{dx}{\sqrt{x} + 2\sqrt[3]{x}}$

4. $\displaystyle \int \frac{\sqrt[3]{x} + 1}{\sqrt[3]{x} - 1}\,dx$

5. $\displaystyle \int_0^7 \frac{x}{\sqrt[3]{x + 1}}\,dx$

6. $\displaystyle \int_3^6 \frac{x + 2}{\sqrt{x - 2}}\,dx$

7. $\displaystyle \int \frac{dx}{2 + \sin x}$

8. $\displaystyle \int \frac{dx}{1 - \sin x}$

9. $\displaystyle \int \frac{dx}{1 - \cos x}$

10. $\displaystyle \int \frac{x^3}{\sqrt{1 + x^2}}\,dx$

11. $\displaystyle \int_0^2 \frac{x^2}{\sqrt[3]{1 + x}}\,dx$

12. $\displaystyle \int_4^{12} \frac{1 + \sqrt{x}}{1 - \sqrt{x}}\,dx$

13. $\displaystyle \int \frac{dx}{\sin x + \cos x}$

14. $\displaystyle \int \frac{1 - \sin x}{1 + \cos x}\,dx$

15. $\displaystyle \int \frac{x^3}{\sqrt{1 - 2x}}\,dx$

16. $\displaystyle \int x^2(x^2 + 1)^{3/2}\,dx$

17. $\displaystyle \int \sqrt{\frac{1 + x}{1 - x}}\,dx$

18. $\displaystyle \int \frac{dx}{1 + e^x}$

19. $\displaystyle \int \frac{dx}{x\sqrt{a + bx}}$

20. $\displaystyle \int_0^3 x\sqrt{1 + x}\,dx$

21. $\displaystyle \int_0^2 \frac{x}{\sqrt{2 + x}}\,dx$

10.6   The Use of Integral Tables

Appendix VII is a *table of integrals*—a list of antiderivatives for various types of functions. While the list in Appendix VII is relatively short, many tables of integrals contain hundreds of entries. Many of the formulas contained in such tables are established using the techniques of this chapter. Such tables are very useful when you are faced with a large number of integrals to evaluate or when you are unwilling to spend the time required to execute the techniques of this chapter.

Using a table of integrals is relatively straightforward. You must match the form of the integrand with those appearing in the table, select an integral matching yours in form, and identify the corresponding values of the constants. For example, the integral

$$\int \frac{dx}{x(3 + 5x^2)}$$

is evaluated using formula (64), which states that

$$\int \frac{du}{u(a + bu^2)} = \frac{1}{2a} \ln \left| \frac{u^2}{a + bu^2} \right| + C.$$

To do so we must take $a = 3$, $b = 5$, and $u = x$. We obtain

$$\int \frac{dx}{x(3 + 5x^2)} = \frac{1}{2 \cdot 3} \ln \left| \frac{x^2}{3 + 5x^2} \right| + C.$$

However, it is often necessary to perform a substitution first to bring an integrand into the form of one of the entries in a table. Be especially careful to carry out the substitution for all variables, including the differential dx. The following examples demonstrate this technique.

□ **EXAMPLE 1**

Use the table of integrals to evaluate

$$\int \frac{dx}{1 + e^{\pi x}}.$$

Solution   We use formula (102):

$$\int \frac{du}{1 + e^u} = \ln \left(\frac{e^u}{1 + e^u} \right) + C.$$

However, first we must use the substitution

$$u = \pi x; \qquad du = \pi \, dx.$$

Then $dx = (1/\pi) \, du$, and we obtain

$$\int \frac{dx}{1 + e^{\pi x}} = \int \frac{(1/\pi) \, du}{1 + e^u} = \frac{1}{\pi} \int \frac{du}{1 + e^u} = \frac{1}{\pi} \ln \left(\frac{e^{\pi x}}{1 + e^{\pi x}} \right) + C. \qquad ■$$

□ **EXAMPLE 2**

Use the table of integrals to evaluate

$$\int x\sqrt{x^2 - 4x + 1}\, dx.$$

Solution This time it is not so clear which formula might correspond to the given integral. Since the table contains no entries involving the general quadratic $ax^2 + bx + c$, we begin by completing the square on the quadratic term:

$$x^2 - 4x + 1 = (x^2 - 4x + 4) - 3 \tag{1}$$
$$= (x - 2)^2 - 3.$$

Our next step is to use the substitution

$$u = x - 2; \qquad du = dx.$$

Then $x = u + 2$, and using equation (1), we obtain

$$\int x\sqrt{x^2 - 4x + 1}\, dx = \int x\sqrt{(x - 2)^2 - 3}\, dx \tag{2}$$

$$= \int (u + 2)\sqrt{u^2 - 3}\, du$$

$$= \int u\sqrt{u^2 - 3}\, du + 2\int \sqrt{u^2 - 3}\, du.$$

To handle the first integral, we use formula (44):

$$\int u\sqrt{u^2 \pm a^2}\, du = \frac{1}{3}(u^2 \pm a^2)^{3/2} + C. \tag{3}$$

The second integral is evaluated using formula (43):

$$\int \sqrt{u^2 \pm a^2}\, du = \frac{1}{2}[u\sqrt{u^2 \pm a^2} \pm a^2 \ln(u + \sqrt{u^2 \pm a^2})] + C. \tag{4}$$

Taking $a = \sqrt{3}$, we combine (2), (3), and (4) to obtain

$$\int x\sqrt{x^2 - 4x + 1}\, dx = \frac{1}{3}[(x - 2)^2 - 3]^{3/2} +$$

$$2 \cdot \frac{1}{2}[(x - 2)\sqrt{(x - 2)^2 - 3} - 3\ln(x - 2 + \sqrt{(x - 2)^2 - 3})] + C. \quad ■$$

Exercise Set 10.6

In Exercises 1–30, use the table of integrals in Appendix VII to evaluate the given integral.

1. $\displaystyle\int \frac{dx}{1 + e^{5x}}$

2. $\displaystyle\int \frac{dx}{x(3 + x)}$

3. $\displaystyle\int_1^2 \frac{dx}{x^2(9 + 2x)^2}$

4. $\displaystyle\int_7^{21/2} \frac{5\, dx}{\sqrt{14x - x^2}}$

5. $\displaystyle\int \frac{x + 3}{2 + 5x^2}\, dx$

6. $\displaystyle\int \frac{3\, dx}{\sqrt{6x + x^2}}$

7. $\displaystyle\int_0^{1/4} \tan^4 \pi x\, dx$

8. $\displaystyle\int_e^{4e} x^2 \ln x\, dx$

9. $\displaystyle\int \sin 6x \sin 3x\, dx$

10. $\displaystyle\int \frac{dx}{8 + 3e^{5x}}$

11. $\displaystyle\int \frac{dx}{x^2(2 - 3x)}$

12. $\displaystyle\int \sqrt{12x - 4x^2}\, dx$

13. $\displaystyle\int \frac{9\, dx}{25 - 4x^2}$

14. $\displaystyle\int \frac{5\, dx}{\sqrt{10x + x^2}}$

15. $\displaystyle\int \tan^3 5x\, dx$

16. $\displaystyle\int e^x \sin 3x\, dx$

23. $\displaystyle\int \frac{3\, dx}{x\sqrt{x^2 + 16}}$

24. $\displaystyle\int \sin \ln \pi x\, dx$

17. $\displaystyle\int \frac{6x^2\, dx}{\sqrt{6 + x}}$

18. $\displaystyle\int \frac{\sqrt{9 - x^2}}{x^2}\, dx$

25. $\displaystyle\int_{-1}^{-2/3} \frac{dx}{9x^2 + 12x + 5}$

26. $\displaystyle\int_{-3}^{0} \frac{dx}{\sqrt{7 - 6x - x^2}}$

19. $\displaystyle\int \frac{6\, dx}{(14 - x^2)^{3/2}}$

20. $\displaystyle\int_{0}^{1/3} x \cos^{-1} 3x\, dx$

27. $\displaystyle\int \frac{7\, dx}{(x + 4)\sqrt{x^2 + 8x + 20}}$

28. $\displaystyle\int \frac{4\, dx}{4x^2 + 20x + 16}$

21. $\displaystyle\int_{0}^{\pi/6} \frac{\pi}{4 + 4 \sin 2x}\, dx$

22. $\displaystyle\int \frac{4\, dx}{x\sqrt{9 - x^2}}$

29. $\displaystyle\int \frac{(2x^2 - 7)\, dx}{\sqrt{16 - x^2}}$

30. $\displaystyle\int \frac{dx}{(2x - x^2)^{3/2}}$

Summary Outline of Chapter 10

■ Integration by parts formula: $\displaystyle\int u\, dv = uv - \int v\, du$. (page 530)

■ Trigonometric substitutions: (page 541)

(a) $\sqrt{x^2 + a^2}$ requires $x = a \tan \theta$, $-\pi/2 < \theta < \pi/2$;
$dx = a \sec^2 \theta\, d\theta$

(b) $\sqrt{a^2 - x^2}$ requires $x = a \sin \theta$, $-\pi/2 \le \theta \le \pi/2$;
$dx = a \cos \theta\, d\theta$

(c) $\sqrt{x^2 - a^2}$ requires $x = a \sec \theta$, $0 \le \theta < \pi/2$ or
$dx = a \sec \theta \tan \theta\, d\theta$ $\pi \le \theta < 3\pi/2$;

■ To handle integrals involving $\sqrt{ax^2 + bx + c}$, complete the square and use (page 543)
the corresponding trigonometric substitution.

■ The method of partial fractions decomposes rational functions into sums of (page 547)
simpler form:

$$\frac{p(x)}{(x - a)(x - b)} = \frac{A}{x - a} + \frac{B}{x - b}$$

$$\frac{p(x)}{(x - a)^n} = \frac{A_1}{x - a} + \cdots + \frac{A_n}{(x - a)^n}$$

$$\frac{p(x)}{(x - a)(x^2 + bx + c)} = \frac{A}{x - a} + \frac{Bx + C}{x^2 + bx + c}$$

$$\frac{p(x)}{(x^2 + bx + c)^n} = \frac{B_1 x + C_1}{x^2 + bx + c} + \cdots + \frac{B_n x + C_n}{(x^2 + bx + c)^n}$$

Review Exercises—Chapter 10

In Exercises 1–81, evaluate the given integral.

1. $\displaystyle\int x\sqrt{x^2 + 9}\, dx$

2. $\displaystyle\int xe^{3x}\, dx$

3. $\displaystyle\int_{0}^{\pi} x^2 \sin 2x\, dx$

4. $\displaystyle\int_{1/2}^{1} x^3 \sin \frac{\pi x}{2}\, dx$

5. $\displaystyle\int \sqrt{x^2 + a^2}\, dx$

6. $\displaystyle\int \frac{dx}{x^2 - 6x + 9}$

7. $\displaystyle\int_{1}^{3} \frac{2x + 3}{x(x + 3)}\, dx$

8. $\displaystyle\int_{3}^{4} \frac{\sqrt{x + 1}}{\sqrt{x}}\, dx$

9. $\displaystyle\int_{-1/2}^{1/2} x \cos \pi x\, dx$

10. $\displaystyle\int_{0}^{1} x \tan^{-1} x\, dx$

11. $\displaystyle\int_{-1/2}^{(\pi - 1)/2} x \cos(2x + 1)\, dx$

12. $\displaystyle\int_{0}^{\ln 2} x^2 e^{3x}\, dx$

13. $\displaystyle\int_{0}^{1} x^2 e^{-5x}\, dx$

14. $\displaystyle\int (x \ln x)^3\, dx$

15. $\displaystyle\int x \ln^2 x\, dx$

16. $\displaystyle\int \frac{dx}{\sqrt{x^2 + a^2}}$

17. $\displaystyle\int \frac{dx}{x^2 - 4x + 9}$

18. $\displaystyle\int \frac{4x + 5}{x^2 - x + 2}\, dx$

19. $\displaystyle\int \frac{\sqrt{x + 4}}{x}\, dx$

20. $\displaystyle\int \frac{1}{\sqrt[3]{x + 1}}\, dx$

21. $\displaystyle\int_0^2 x\sqrt{x + 2}\, dx$

22. $\displaystyle\int_0^1 x(x + 3)^6\, dx$

23. $\displaystyle\int \frac{dx}{\sqrt{a^2 - x^2}}$

24. $\displaystyle\int \frac{dx}{\sqrt{x^2 + 8x + 25}}$

25. $\displaystyle\int \frac{3x^2 + x + 1}{x^2(x + 1)}\, dx$

26. $\displaystyle\int \frac{dx}{2\sqrt{x} + 3\sqrt[3]{x}}$

27. $\displaystyle\int_0^{\pi/4} \tan^2 x\, dx$

28. $\displaystyle\int_1^e x^2 \ln x\, dx$

29. $\displaystyle\int \sqrt{a^2 - x^2}\, dx$

30. $\displaystyle\int \frac{dx}{\sqrt{1 + 4x - x^2}}$

31. $\displaystyle\int_0^1 \frac{4x^2 + x + 2}{(x - 2)(x^2 + 1)}\, dx$

32. $\displaystyle\int_0^1 \frac{2 + \sqrt{x}}{2 - \sqrt{x}}\, dx$

33. $\displaystyle\int \frac{x + 1}{\sqrt[3]{x + 1}}\, dx$

34. $\displaystyle\int x \csc^2 \pi x\, dx$

35. $\displaystyle\int \sqrt{x^2 - a^2}\, dx$

36. $\displaystyle\int \frac{dx}{\sqrt{x^2 - 8x}}$

37. $\displaystyle\int \frac{x^2 + 2x + 3}{(x + 2)(x^2 + x + 1)}\, dx$

38. $\displaystyle\int \frac{x - 3}{\sqrt{x + 2}}\, dx$

39. $\displaystyle\int (x^2 - 4)^{3/2}\, dx$

40. $\displaystyle\int \frac{x\, dx}{x^2 + 4x + 8}$

41. $\displaystyle\int \frac{3x^3 + x + 2}{x(1 + x^3)}\, dx$

42. $\displaystyle\int \frac{dx}{1 + 2\sin x}$

43. $\displaystyle\int \csc^3 x\, dx$

44. $\displaystyle\int \frac{e^{2x}}{\sqrt{1 - e^x}}\, dx$

45. $\displaystyle\int \frac{dx}{x\sqrt{9 + x^2}}$

46. $\displaystyle\int \frac{2 + x}{\sqrt{x^2 - 6x + 13}}\, dx$

47. $\displaystyle\int \frac{3x^2 - x + 10}{(x^2 + 2)(4 - x)}\, dx$

48. $\displaystyle\int \frac{dx}{1 - 2\sin x}$

49. $\displaystyle\int \frac{\cos x}{2 - \cos x}\, dx$

50. $\displaystyle\int \frac{x^2}{\sqrt{1 + x^2}}\, dx$

51. $\displaystyle\int_{\ln 2}^{\ln 3} \frac{dx}{e^x - e^{-x}}$

52. $\displaystyle\int \sec^5 x\, dx$

53. $\displaystyle\int \frac{x}{a^2 + x^2}\, dx$

54. $\displaystyle\int \frac{2x - 1}{\sqrt{17 + 8x + x^2}}\, dx$

55. $\displaystyle\int_2^3 \frac{5x^2 - 2x + 12}{x^3 - x^2 + 4x - 4}\, dx$

56. $\displaystyle\int \frac{dx}{1 + x^3}$

57. $\displaystyle\int \frac{dx}{1 + e^x}$

58. $\displaystyle\int_{-\sqrt{13}}^0 \frac{x}{\sqrt{36 + x^2}}\, dx$

59. $\displaystyle\int \frac{3x + 1}{\sqrt{x^2 + 8x}}\, dx$

60. $\displaystyle\int \frac{6x^2 + 5x - 2}{x^3 + x^2 - 2x}\, dx$

61. $\displaystyle\int \frac{x^5\, dx}{\sqrt{1 + x^2}}$

62. $\displaystyle\int \frac{1}{x^2\sqrt{x^2 - 25}}\, dx$

63. $\displaystyle\int \frac{x^5}{(x^2 + 2)^2}\, dx$

64. $\displaystyle\int \sqrt{1 - \cos x}\, dx$

65. $\displaystyle\int \frac{1}{\sqrt{4x^2 - 25}}\, dx$

66. $\displaystyle\int \sqrt{1 + \sin x}\, dx$

67. $\displaystyle\int_1^e \ln^2 x\, dx$

68. $\displaystyle\int \frac{7\, dx}{13 + x^2}$

69. $\displaystyle\int \frac{3\, dx}{x^2 - 5}, \quad x > \sqrt{5}$

70. $\displaystyle\int \frac{\pi\, dx}{x\sqrt{36 - x^2}}$

71. $\displaystyle\int \frac{dx}{x\sqrt{9 - \pi x}}$

72. $\displaystyle\int \frac{4\, dx}{(7 - 3x)^2}$

73. $\displaystyle\int \frac{5\, dx}{x^2(7 - 2x)}$

74. $\displaystyle\int \frac{dx}{x(\pi + 4x)^2}$

75. $\displaystyle\int \frac{3x\, dx}{\sqrt{9 + 2x}}$

76. $\displaystyle\int \frac{4x^2\, dx}{\sqrt{49 - x^2}}$

77. $\displaystyle\int \frac{dx}{x^2\sqrt{8 - x^2}}$

78. $\displaystyle\int \frac{dx}{\sqrt{5x - x^2}}$

79. $\displaystyle\int \frac{dx}{9 + 4\cos x}$

80. $\displaystyle\int x \operatorname{Sin}^{-1} 2x\, dx$

81. $\displaystyle\int x^3 \ln x\, dx$

82. Find the area bounded by the graph of $y = 1/\sqrt{x^2 - 6x + 5}$ and the x-axis for $-1 \le x \le 0$.

83. Find the area bounded by the graphs of $y = xe^{-x}$ and $y = x/e$.

84. Find the volume of the solid obtained by revolving about the y-axis the graph of $y = x \sin(\pi x/3)$ for $0 \le x \le 3$.

85. Find the volume of the solid obtained by revolving about the x-axis the graph of

$$y = \frac{\sqrt{x+5}}{x\sqrt{x-5}}$$

for $6 \le x \le 7$.

86. Find the arc length of the graph of $y = x^2$ from $(0, 0)$ to $(1, 1)$.

87. Find the arc length of the graph of $y = -\sqrt{1 - x^2}$ from $(0, -1)$ to $(1/2, -\sqrt{3}/2)$.

88. Find the center of mass of a rod 2 m long with uniform cross-sectional area of $\pi\, m^2$ if the density at a point x m from the left end of the rod is given by $\rho(x) = 1/(x^2 + 1)$.

89. Find the center of mass of a rod 4 ft long with uniform cross-sectional area of 3 ft^2 if the density at a point x ft from the left end of the rod is given by $\rho(x) = 4/[(x + 1)(5 - x)]$.

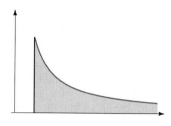

Chapter 11

l'Hôpital's Rule and Improper Integrals

With the exception of asymptotes, which involve limits of the form

$$\lim_{x \to \infty} f(x) = L \quad \text{or} \quad \lim_{x \to a} f(x) = \pm\infty,$$

almost every topic we have considered so far has concerned a continuous function on a bounded interval.

In this chapter we shall develop a method known as l'Hôpital's Rule for evaluating limits that involve the so-called *indeterminate forms* such as

$$\frac{0}{0} \quad \text{or} \quad \frac{\infty}{\infty}.$$

We also extend the notion of the definite integral $\int_a^b f(x)\, dx$ of a continuous function on the closed, finite interval $[a, b]$ to *improper integrals* of the form

$$\int_a^\infty f(x)\, dx$$

where f is continuous on the infinite interval $[a, \infty)$ or improper integrals of the form

$$\int_a^b f(x)\, dx$$

where f is unbounded on the finite interval $[a, b]$. Both l'Hôpital's Rule and improper integrals are used frequently in Chapters 12 and 13, which deal with the theory of infinite sums.

11.1 Indeterminate Forms: l'Hôpital's Rule

On several occasions we have encountered the indeterminate form $\frac{0}{0}$ in attempting to evaluate a limit of the form

$$\lim_{x \to a} \frac{f(x)}{g(x)}. \tag{1}$$

Specifically, this form arises in limit (1) if

$$\lim_{x \to a} f(x) = \lim_{x \to a} g(x) = 0.$$

For example, recall the limit

$$\lim_{x \to 0} \frac{\sin x}{x} \qquad \left(\frac{0}{0} \text{ form} \right),$$

which we determined in Chapter 2 using the Pinching Theorem.* More generally, we have seen that this form will always arise in applying the definition of the derivative:

$$f'(x) = \lim_{h \to 0} \frac{f(x + h) - f(x)}{h} \qquad \left(\frac{0}{0} \text{ form} \right).$$

The theory of the derivative may be applied to establish the following simple procedure for dealing with many limits of type (1) that yield the indeterminate forms 0/0 or ∞/∞. Although the result is credited to the Frenchman G.F.A. l'Hôpital (1661–1704), it was actually discovered by his teacher, Johann Bernoulli (1667–1748).

Theorem 1

l'Hôpital's Rule

Let lim denote one of the symbols

$$\lim_{x \to a}, \ \lim_{x \to a^+}, \ \lim_{x \to a^-}, \ \lim_{x \to \infty}, \ \text{or} \ \lim_{x \to -\infty}$$

and let the functions f and g be differentiable where defined, except possibly at $x = a$. If

$$\lim f(x) = \lim g(x) = 0$$

or

$$\lim f(x) = \lim g(x) = \infty,$$

then

$$\lim \frac{f(x)}{g(x)} = \lim \frac{f'(x)}{g'(x)},$$

provided the limit on the right exists or is infinite.

In other words, l'Hôpital's Rule states that, if we attempt to evaluate the limit

$$\lim_{x \to a^+} \frac{f(x)}{g(x)}$$

and we obtain the indeterminate form 0/0 or ∞/∞, we may separately differentiate both the numerator and denominator and try again. If we then succeed in obtaining a limit, it will also be the limit of the original quotient. (*Caution:* Differentiate f and g *separately*—do *not* apply the Quotient Rule.)

---

*Even though the limit equals 1 *in this case,* you should not assume that the notation $\frac{0}{0}$ represents a valid algebraic expression. If it did, we would have $1 = \frac{0}{0} = \frac{0 + 0}{0} = \frac{0}{0} + \frac{0}{0} = 2$. The notation $\frac{0}{0}$ simply represents the limiting behavior of the two functions f and g involved in limit (1). It does not represent the value of the limit, nor does it even suggest that the limit exists.

REMARK 1   When we say that the limit

$$\lim \frac{f'(x)}{g'(x)}$$

exists or is infinite as part of the statement of l'Hôpital's Rule, we are implicitly assuming that $g'(x) \neq 0$ for all x in an appropriate interval determined by the type of limit we are considering. For example, in the case where we are calculating the two-sided limit

$$\lim_{x \to a} \frac{f(x)}{g(x)},$$

we must have $g'(x) \neq 0$ for all $x \neq a$ in some open interval containing a. If not, we could not calculate $f'(x)/g'(x)$ for certain x near a, and therefore, we could not calculate the limit of this quotient as $x \to a$. We explicitly use this assumption on g' when we prove l'Hôpital's Rule at the end of this section.

REMARK 2   If $f(a) = g(a) = 0$, if f' and g' are continuous on an open interval containing a, and if $g'(a) \neq 0$, then the two-sided version of l'Hôpital's Rule holds because

$$\lim_{x \to a} \frac{f(x)}{g(x)} = \lim_{x \to a} \frac{f(x) - f(a)}{g(x) - g(a)}$$

$$= \frac{\lim_{x \to a} \dfrac{f(x) - f(a)}{x - a}}{\lim_{x \to a} \dfrac{g(x) - g(a)}{x - a}}$$

$$= \frac{f'(a)}{g'(a)} = \lim_{x \to a} \frac{f'(x)}{g'(x)}.$$

We will establish Theorem 1 in the general case after we consider several examples and extensions.

☐ **EXAMPLE 1**

Since both $\lim\limits_{x \to 0} \tan x = 0$ and $\lim\limits_{x \to 0} x = 0$, l'Hôpital's Rule gives

$$\lim_{x \to 0} \frac{\tan x}{x} = \lim_{x \to 0} \frac{\dfrac{d}{dx}(\tan x)}{\dfrac{d}{dx}(x)} = \lim_{x \to 0} \frac{\sec^2 x}{1} = 1. \qquad \blacksquare$$

☐ **EXAMPLE 2**

Find $\lim\limits_{x \to 8} \dfrac{\sqrt[3]{x} - 2}{x - 8}$.

Strategy · · · · · · · ·
Check that both $\sqrt[3]{x} - 2 \to 0$ and $x - 8 \to 0$ as $x \to 8$.

Solution

Since both the numerator and denominator approach zero as $x \to 8$, we apply l'Hôpital's Rule:

$$\lim_{x \to 8} \frac{\sqrt[3]{x} - 2}{x - 8} \qquad \left(\frac{0}{0} \text{ form}\right)$$

Apply l'Hôpital's Rule.

$$= \lim_{x \to 8} \frac{\frac{1}{3} x^{-2/3}}{1} = \frac{\frac{1}{3}(8)^{-2/3}}{1} = \frac{1}{12}. \qquad \blacksquare$$

REMARK 3   Frequently, on applying l'Hôpital's Rule to

$$\lim_{x \to a} \frac{f(x)}{g(x)},$$

we again obtain an indeterminate form. In such cases we simply apply l'Hôpital's Rule again, *provided the hypotheses are fulfilled for the functions f' and g'*.

☐ **EXAMPLE 3**

Find $\displaystyle \lim_{x \to 0} \frac{x - \tan x}{x - \sin x}$.

Strategy · · · · · · · ·
Verify that $\frac{0}{0}$ form results as $x \to 0$.

Solution
Since both $(x - \tan x) \to 0$ and $(x - \sin x) \to 0$ as $x \to 0$, we apply l'Hôpital's Rule to find that

$$\lim_{x \to 0} \frac{x - \tan x}{x - \sin x} \qquad \left(\frac{0}{0} \text{ form}\right)$$

Apply l'Hôpital's Rule.

$$= \lim_{x \to 0} \frac{1 - \sec^2 x}{1 - \cos x} \qquad \left(\frac{0}{0} \text{ form again}\right)$$

Apply l'Hôpital's Rule again!

$$= \lim_{x \to 0} \frac{-2 \sec^2 x \tan x}{\sin x}$$

Simplify using definition of $\sec x$ and $\tan x$.

$$= \lim_{x \to 0} \frac{-2 \left(\dfrac{1}{\cos x}\right)^2 \left(\dfrac{\sin x}{\cos x}\right)}{\sin x}$$

Evaluate limit.

$$= \lim_{x \to 0} \frac{-2}{\cos^3 x} = -2. \qquad ■$$

REMARK 4   Make sure that you check the hypotheses of l'Hôpital's Rule before each application. If the limit (1) is not an indeterminate form, using l'Hôpital's Rule leads to incorrect results. For example, consider

$$\lim_{x \to 0} \frac{\sin x}{x - \sin x} \qquad \left(\frac{0}{0} \text{ form; we can apply l'Hôpital's Rule}\right)$$

$$= \lim_{x \to 0} \frac{\cos x}{1 - \cos x} \qquad (\textit{not} \text{ an indeterminate form})$$

$$= +\infty$$

since $(1 - \cos x)$ approaches 0 *from the positive direction* and $\cos x \to 1$ as $x \to 0$. If we had incorrectly applied l'Hôpital's Rule a second time, we would obtain the limit

$$\lim_{x \to 0} \frac{-\sin x}{\sin x} = 1,$$

which is *not* equal to the original limit.

□ **EXAMPLE 4**

$$\lim_{x \to \pi/2^-} \frac{\tan^2 x}{\sec^2 x} \quad \left(\frac{\infty}{\infty} \text{ form}\right)$$

$$= \lim_{x \to \pi/2^-} \frac{2 \tan x \sec^2 x}{2 \sec^2 x \tan x}$$

$$= \lim_{x \to \pi/2^-} (1) = 1. \qquad \blacksquare$$

□ **EXAMPLE 5**

$$\lim_{x \to \infty} \frac{x^2 + 5}{x + e^x} \quad \left(\frac{\infty}{\infty} \text{ form; apply l'Hôpital's Rule}\right)$$

$$= \lim_{x \to \infty} \frac{2x}{1 + e^x} \quad \left(\text{still } \frac{\infty}{\infty} \text{ form; apply l'Hôpital's Rule again}\right)$$

$$= \lim_{x \to \infty} \frac{2}{e^x} = 0. \qquad \blacksquare$$

□ **EXAMPLE 6**

$$\lim_{x \to 0^+} \frac{\sin x}{x + x^{3/2}} \quad \left(\frac{0}{0} \text{ form; apply l'Hôpital's Rule}\right)$$

$$= \lim_{x \to 0^+} \frac{\cos x}{1 + \frac{3}{2}x^{1/2}}$$

$$= \frac{1}{1 + 0} = 1. \qquad \blacksquare$$

Proof of l'Hôpital's Rule

The justification for l'Hôpital's Rule depends upon a generalization of the Mean Value Theorem due to Augustin L. Cauchy (1789–1857). Recall, the Mean Value Theorem guarantees the existence of a number $c \in (a, b)$ so that

$$f'(c) = \frac{f(b) - f(a)}{b - a} \qquad (2)$$

if f is differentiable on (a, b) and continuous on $[a, b]$. Cauchy's generalization arises by viewing the denominator on the right side of equation (2) as $g(b) - g(a)$, where g is the identity function $g(x) = x$, and then asking what results by allowing g instead to be *any* differentiable function of x. Since $g'(c) = 1$ if g is the identity function, the following result is a direct generalization of the Mean Value Theorem.

Theorem 2
Cauchy's Mean Value Theorem

Let the functions f and g be continuous on the closed interval $[a, b]$ and differentiable on the open interval (a, b). Also, assume $g'(x) \neq 0$ for all x in (a, b). Then there exists a number c in (a, b) so that

$$\frac{f'(c)}{g'(c)} = \frac{f(b) - f(a)}{g(b) - g(a)}.$$

Proof: The Mean Value Theorem was proved by considering the function

$$d(x) = f(x) - \left[f(a) + \frac{f(b) - f(a)}{b - a} (x - a) \right]. \qquad (3)$$

Since Theorem 2 is obtained from the Mean Value Theorem by generalizing the identity function $g(x) = x$ to more general functions g, we consider the function D obtained from (3) by replacing $b - a$ with $g(b) - g(a)$, and $x - a$ with $g(x) - g(a)$:

$$D(x) = f(x) - \left[f(a) + \frac{f(b) - f(a)}{g(b) - g(a)} (g(x) - g(a)) \right]. \tag{4}$$

For this function D to make sense, we must know that $g(b) - g(a) \neq 0$. However since g satisfies the hypotheses of the Mean Value Theorem,

$$\frac{g(b) - g(a)}{b - a} = g'(w)$$

for some number w in (a, b). Since we are assuming $g'(w) \neq 0$ for all w in (a, b) this shows that $g(b) - g(a) \neq 0$. Thus, there is no difficulty in defining $D(x)$ by equation (4).

The function D is continuous on $[a, b]$ and differentiable on (a, b) by our assumptions of f and g. Moreover, $D(a) = D(b) = 0$. We may therefore apply Rolle's Theorem to conclude that for some number $c \in (a, b)$

$$D'(c) = 0. \tag{5}$$

From equation (4) we can see that

$$D'(x) = f'(x) - g'(x) \left(\frac{f(b) - f(a)}{g(b) - g(a)} \right). \tag{6}$$

Letting $x = c$ and combining statements (5) and (6) then gives the desired result.

∎

Proof of l'Hôpital's Rule: We shall prove only the case

$$\lim_{x \to a^+} \frac{f(x)}{g(x)} = \lim_{x \to a^+} \frac{f'(x)}{g'(x)} \tag{7}$$

where the limit of $f(x)$ and the limit of $g(x)$ equals 0 as $x \to a^+$. Since we are assuming the limit on the right to exist or be infinite, the functions f and g are defined and differentiable on an open interval (a, β) with $\beta > a$. We define the functions F and G on $[a, \beta)$ by

$$F(x) = \begin{cases} f(x), & x \neq a \\ 0, & x = a \end{cases} \qquad G(x) = \begin{cases} g(x), & x \neq a \\ 0, & x = a. \end{cases}$$

Then F and G are differentiable on (a, β) and

$$\lim_{x \to a^+} F(x) = \lim_{x \to a^+} f(x) = 0, \tag{8}$$

$$\lim_{x \to a^+} G(x) = \lim_{x \to a^+} g(x) = 0, \tag{9}$$

$$F'(x) = f'(x), \qquad x \in (a, \beta), \tag{10}$$

and

$$G'(x) = g'(x), \qquad x \in (a, \beta). \tag{11}$$

We next let $x > a$ with $a < x < \beta$ and note that the functions F and G satisfy the hypotheses of Cauchy's Mean Value Theorem on the closed interval $[a, x]$. Thus, for each such x there exists a number $c_x \in (a, x)$, which depends on x, with

$$\frac{F(x) - F(a)}{G(x) - G(a)} = \frac{F'(c_x)}{G'(c_x)}. \tag{12}$$

Since $F(a) = G(a) = 0$, equation (12) is just the equation

$$\frac{F(x)}{G(x)} = \frac{F'(c_x)}{G'(c_x)}, \qquad a < c_x < x. \tag{13}$$

Since c_x lies between a and x we must have $c_x \to a^+$ as $x \to a^+$. Thus, equations (8)–(11) and (13) give

$$\lim_{x \to a^+} \frac{f(x)}{g(x)} = \lim_{x \to a^+} \frac{F(x)}{G(x)} = \lim_{x \to a^+} \frac{F'(c_x)}{G'(c_x)}$$

$$= \lim_{c_x \to a^+} \frac{F'(c_x)}{G'(c_x)}$$

$$= \lim_{x \to a^+} \frac{F'(x)}{G'(x)}$$

$$= \lim_{x \to a^+} \frac{f'(x)}{g'(x)}.$$

This proves equation (7).   ∎

Exercise Set 11.1

In Exercises 1–39, find the indicated limit.

1. $\lim_{x \to 1} \dfrac{1 - x}{e^x - e}$

2. $\lim_{x \to 0} \dfrac{\sin 5x}{x}$

3. $\lim_{x \to 0} \dfrac{\sin x^2}{x}$

4. $\lim_{x \to 2} \dfrac{x^3 - x^2 - x - 2}{x - 2}$

5. $\lim_{x \to 0} \dfrac{1 - \cos x}{x^2}$

6. $\lim_{x \to \pi/2} \dfrac{1 - \sin x}{\cos x}$

7. $\lim_{x \to 1^-} \dfrac{\sqrt{1 - x^2}}{x - 1}$

8. $\lim_{x \to 0^+} \dfrac{1 - \cos x}{x^3}$

9. $\lim_{\theta \to 0} \dfrac{\tan \theta - \theta}{\theta - \sin \theta}$

10. $\lim_{x \to \infty} \dfrac{x^3 + 2x + 1}{4x^3 + 1}$

11. $\lim_{x \to 1} \dfrac{\ln x}{x - 1}$

12. $\lim_{x \to 1} \dfrac{\sqrt{x} - \sqrt[4]{x}}{x - 1}$

13. $\lim_{x \to 0^+} \dfrac{1 + \cos \sqrt{x}}{\sin x}$

14. $\lim_{x \to 0} \dfrac{\ln(1 + x)}{1 - e^x}$

15. $\lim_{x \to 0^+} \dfrac{x^2}{x - \sin x}$

16. $\lim_{x \to \infty} \dfrac{\sqrt{1 + x^2}}{x}$

17. $\lim_{x \to \infty} \dfrac{\sqrt[3]{x^3 + 3}}{x^2}$

18. $\lim_{x \to 0} \dfrac{\sin x - x \cos x}{x}$

19. $\lim_{x \to 1} \dfrac{e^{x^2} - e^x}{x^2 - 1}$

20. $\lim_{x \to 0^+} \dfrac{\cos x - x}{\sqrt{x}}$

21. $\lim_{x \to 0} \dfrac{\sin x}{\sqrt[3]{x}}$

22. $\lim_{x \to \pi^+} \dfrac{\sin x}{\sqrt{x - \pi}}$

23. $\lim_{x \to 0} \dfrac{e^{x^2} - 1}{x^2}$

24. $\lim_{x \to 0^+} \dfrac{\cos x - x}{x - \tan x}$

25. $\lim_{x \to \infty} \dfrac{x - \sin x}{x \sin x}$

26. $\lim_{x \to 1^-} \dfrac{x^{5/2} - 1 + \sqrt{1 - x}}{\sqrt{1 - x^2}}$

27. $\lim_{x \to \infty} \dfrac{9 - x^3}{xe^{\pi x}}$

28. $\lim_{x \to \infty} \dfrac{e^{2x}}{1 + x^2}$

29. $\lim_{x \to 0} \dfrac{\sin^{-1} x}{x}$

30. $\lim_{x \to \pi/2} \dfrac{\cos 3x}{\cos x}$

31. $\lim_{x \to \pi/2} \dfrac{\tan(x/2) - 1}{x - \pi/2}$

32. $\lim_{x \to \infty} \dfrac{\tan(1/x)}{1/x}$

33. $\lim_{x \to \infty} \dfrac{\sqrt{x} - 3x^2}{x(6 - x)}$

34. $\lim_{x \to \infty} \dfrac{(ax + b)^3}{(x + c)^3}$

35. $\lim_{x \to \infty} \dfrac{\sqrt{x} - \sqrt{a}}{\sqrt{x} + \sqrt{a}}$

36. $\lim_{x \to \infty} \dfrac{x^3 - 7x^2 + 6x - 5}{(x - 3)(5 - x^2)}$

37. $\lim\limits_{x \to \infty} \dfrac{3x^2 - 4}{4x^2 + 3}$

38. $\lim\limits_{x \to \infty} \dfrac{x^2(x - 1)(x + 3)}{(x^3 - 6)(2x^2 + x + 1)}$

39. $\lim\limits_{x \to \infty} \dfrac{\sin x}{x^2 + \pi}$

In Exercises 40–43, sketch the graph of the given function using l'Hôpital's Rule to find asymptotes and other limiting behavior.

40. $f(x) = \dfrac{x + 4}{x - 2}$

41. $f(x) = \dfrac{x^2 + 3x + 1}{x}$

42. $f(x) = \dfrac{x}{1 - e^x}$

43. $f(x) = \dfrac{\sqrt{x} - 1}{x - 1}$

44. Extend the proof given for l'Hôpital's Rule to the case

$$\lim_{x \to a^-} \frac{f(x)}{g(x)} = \lim_{x \to a^-} \frac{f'(x)}{g'(x)}$$

where $\lim\limits_{x \to a^-} f(x) = \lim\limits_{x \to a^-} g(x) = 0$.

45. Combine the proof given in this section for l'Hôpital's Rule in the case $\lim\limits_{x \to a^+}$ with the proof given in Exercise 44 for the case $\lim\limits_{x \to a^-}$ to prove the case

$$\lim_{x \to a} \frac{f(x)}{g(x)} = \lim_{x \to a} \frac{f'(x)}{g'(x)}$$

if $\lim\limits_{x \to a} f(x) = \lim\limits_{x \to a} g(x) = 0$.

46. Extend the proof given for l'Hôpital's Rule to the case

$$\lim_{x \to \infty} \frac{f(x)}{g(x)} = \lim_{x \to \infty} \frac{f'(x)}{g'(x)}$$

if $\lim\limits_{x \to \infty} f(x) = \lim\limits_{x \to \infty} g(x) = 0$ as follows:

a. Set $t = 1/x$ and conclude that

$$\lim_{x \to \infty} \frac{f(x)}{g(x)} = \lim_{t \to 0^+} \frac{f(1/t)}{g(1/t)}.$$

b. Show that

$$\lim_{t \to 0^+} \frac{f(1/t)}{g(1/t)} = \lim_{t \to 0^+} \frac{(-1/t^2)f'(1/t)}{(-1/t^2)g'(1/t)}$$

$$= \lim_{t \to 0^+} \frac{f'(1/t)}{g'(1/t)} = \lim_{x \to \infty} \frac{f'(x)}{g'(x)}.$$

c. Combine part a and part b to obtain the result.

47. Assume $f(x) = e^x - 1$ and $g(x) = x^3 - 2x$.
a. Graph $f(x)/g(x)$ for $-1 \le x \le 1$. By magnifying the graph determine

$$\lim_{x \to 0} \frac{f(x)}{g(x)}.$$

b. Graph $f'(x)/g'(x)$ for $-1 \le x \le 1$. By magnifying the graph determine

$$\lim_{x \to 0} \frac{f'(x)}{g'(x)}.$$

c. Which method (part a or part b) for determining the limit of $f(x)/g(x)$ as $x \to 0$ is more accurate?

 48. Graph $f(x) = 2x(x - 1)$ and $g(x) = 2x^2(x - 1)^{2/3}$ for $-1/2 \le x \le 3/2$. Using the graphs explain why l'Hôpital's Rule implies that the discontinuity of $f(x)/g(x)$ at $x = 1$ is removable but that the discontinuity at $x = 0$ is not removable (see Section 2.6).

 In Exercises 49–52, are $f(x)/g(x)$ and $g(x)/f(x)$ continuous at $x = a$? If not, use the method of Exercise 48 to determine if the discontinuity is removable.

49. $f(x) = x \sin \pi x$, $g(x) = 2x(x - 1)$; $\quad a = 1$.

50. $f(x) = \tan(\pi x/2)$, $g(x) = \sec(\pi x/2)$; $\quad a = 1$.

51. $f(x) = \ln(2x + 1)$, $g(x) = x \operatorname{Sin}^{-1} x$; $\quad a = 0$.

52. $f(x) = -3 + 6/x$, $g(x) = -x^2/4 + 2x - 3$; $\quad a = 2$.

53. In this exercise, assume that the functions f and g satisfy the hypotheses of l'Hôpital's Rule.
a. Suppose

$$\lim_{x \to +\infty} f(x) = \lim_{x \to +\infty} g(x) = +\infty$$

and

$$\lim_{x \to +\infty} (f(x)/g(x)) = 1.$$

How are the slopes of the tangent lines of $f(x)$ and $g(x)$ related as $x \to +\infty$?
b. Determine two functions $f(x)$ and $g(x)$ that satisfy the conditions of part a and graph them. How do the graphs compare as $x \to +\infty$?
c. Suppose that, in addition to the conditions of part a, we require that f and g satisfy

$$\lim_{x \to +\infty} (f(x) - g(x)) = +\infty.$$

Determine two such functions and graph them. How do their graphs compare as $x \to +\infty$?

11.2 Other Indeterminate Forms

There are five additional indeterminate forms that can often be treated by methods similar to those discussed in Section 11.1. The forms* are

$$\infty - \infty, \quad 0 \cdot \infty, \quad 1^{\infty}, \quad 0^0, \quad \text{and} \quad \infty^0,$$

and here are examples:

$$\lim_{x \to \pi/2^-} (\tan x - \sec x) \qquad (\infty - \infty \text{ form})$$

$$\lim_{x \to \infty} e^{-x} \ln x \qquad (0 \cdot \infty \text{ form})$$

$$\lim_{x \to \infty} \left(1 + \frac{1}{x}\right)^x \qquad (1^{\infty} \text{ form})$$

$$\lim_{x \to 0^+} (\sin x)^x \qquad (0^0 \text{ form})$$

$$\lim_{x \to \pi/2^-} (\tan x)^{(\cos x)} \qquad (\infty^0 \text{ form}).$$

The technique for evaluating limits expressed in any of these five forms is first to rewrite the function in question in a form that yields either $0/0$ or ∞/∞. L'Hôpital's Rule is then applied to yield the desired limit.

In discussing these terms, we shall use the symbol lim to represent one of the limits $\lim\limits_{x \to a}$, $\lim\limits_{x \to a^+}$, $\lim\limits_{x \to a^-}$, $\lim\limits_{x \to \infty}$, or $\lim\limits_{x \to -\infty}$ as in Section 11.1.

The Form $\infty - \infty$

The form $\infty - \infty$ occurs in $\lim(f(x) - g(x))$ when $\lim f(x) = \infty$ and $\lim g(x) = \infty$. In this case the strategy is to rewrite the difference $f - g$ as a quotient to which l'Hôpital's Rule may be applied.

☐ EXAMPLE 1

Find $\lim\limits_{x \to \pi/2^-} (\tan x - \sec x)$.

Strategy · · · · · · · ·
Rewrite difference as quotient by first rewriting $\tan x$ and $\sec x$ in terms of $\sin x$ and $\cos x$. Then combine over common denominator.

Solution

$$\lim_{x \to \pi/2^-} (\tan x - \sec x) \qquad (\infty - \infty \text{ form})$$

$$= \lim_{x \to \pi/2^-} \left(\frac{\sin x}{\cos x} - \frac{1}{\cos x}\right)$$

$$= \lim_{x \to \pi/2^-} \left(\frac{\sin x - 1}{\cos x}\right) \qquad \left(\frac{0}{0} \text{ form}\right)$$

Apply l'Hôpital's Rule to the quotient.

$$= \lim_{x \to \pi/2^-} \left(\frac{\cos x}{-\sin x}\right) = 0$$

since $\cos x \to 0$ and $\sin x \to 1$ as $x \to \pi/2^-$. ∎

---

*Again we emphasize that indeterminate forms are simply notational devices that characterize the limiting behavior of the functions involved in the limit. Do not manipulate them as if they are valid algebraic expressions. For example, do *not* assume that $0 \cdot \infty = 0$ or that $1^{\infty} = 1$. The methods described in this section indicate how you can often calculate limits that can be expressed in this form.

The Form $0 \cdot \infty$

The limit $\lim f(x)g(x)$ yields the indeterminate form $0 \cdot \infty$ if $\lim f(x) = 0$ and $\lim g(x) = \infty$, or vice-versa. Rewriting the product fg as one of the quotients

$$\frac{f}{1/g} \quad \text{or} \quad \frac{g}{1/f}$$

will yield one of the indeterminate forms $0/0$ or ∞/∞.

□ **EXAMPLE 2**

Find $\lim\limits_{x \to \infty} xe^{-x}$.

Strategy · · · · · · · ·
Rewrite factor e^{-x} as $1/e^x$.

Apply l'Hôpital's Rule.

Solution

$$\lim_{x \to \infty} xe^{-x} \qquad (\infty \cdot 0 \text{ form})$$

$$= \lim_{x \to \infty} \frac{x}{e^x} \qquad \left(\frac{\infty}{\infty} \text{ form}\right)$$

$$= \lim_{x \to \infty} \frac{1}{e^x} = 0. \qquad ∎$$

□ **EXAMPLE 3**

Find $\lim\limits_{x \to \infty} x \ln\left(1 + \dfrac{1}{x}\right)$.

Strategy · · · · · · · ·
Rewrite factor x as $1/(1/x)$ to obtain factor $1/x$ in denominator.

Apply l'Hôpital's Rule.

Simplify and evaluate limit.

Solution

$$\lim_{x \to \infty} x \ln\left(1 + \frac{1}{x}\right) \qquad (\infty \cdot 0 \text{ form})$$

$$= \lim_{x \to \infty} \frac{\ln(1 + 1/x)}{1/x} \qquad \left(\frac{0}{0} \text{ form}\right)$$

$$= \lim_{x \to \infty} \frac{\left[\dfrac{1}{(1 + 1/x)}\right](-1/x^2)}{-1/x^2}$$

$$= \lim_{x \to \infty} \left(\frac{1}{1 + 1/x}\right) = 1. \qquad ∎$$

The Form 1^∞

This form occurs in the limit $\lim[f(x)^{g(x)}]$ when $\lim f(x) = 1$ and $\lim g(x) = \infty$. In this case *we* first consider the limit of the *natural logarithm* $\ln y$ of the function $y = f(x)^{g(x)}$. If $L = \lim \ln y$ exists, the continuity of the exponential function guarantees that

$$\lim[f(x)^{g(x)}] = \lim[y] = \lim[e^{\ln y}] = e^{\lim(\ln y)} = e^L.$$

That is, if $L = \lim \ln[f(x)^{g(x)}] = \lim[g(x) \ln f(x)]$, then $\lim[f(x)^{g(x)}] = e^L$.

The following examples show how this procedure enables us to apply l'Hôpital's Rule.

☐ **EXAMPLE 4**

Find $\lim\limits_{x \to \pi/2^-} (\sin x)^{\sec x}$.

Strategy · · · · · · · ·

Verify 1^∞ form.

Use property $\ln y^r = r \ln y$ of logarithms to evaluate limit
$\lim[\ln(\sin x)^{\sec x}]$
$\quad = \lim[\sec x \ln \sin x].$

Apply l'Hôpital's Rule.

Desired limit is e^L.

Solution

Since $\sin x \to 1$ and $\sec x \to +\infty$ as $x \to \pi/2^-$, this limit has the indeterminate form 1^∞. We therefore let $y = (\sin x)^{\sec x}$ and consider

$$
\begin{aligned}
L &= \lim_{x \to \pi/2^-} \ln[(\sin x)^{\sec x}] \\[2mm]
&= \lim_{x \to \pi/2^-} [\sec x \ln \sin x] \qquad (\infty \cdot 0 \text{ form}) \\[2mm]
&= \lim_{x \to \pi/2^-} \frac{\ln \sin x}{\cos x} \qquad \left(\frac{0}{0} \text{ form}\right) \\[2mm]
&= \lim_{x \to \pi/2^-} \frac{\left(\dfrac{\cos x}{\sin x}\right)}{-\sin x} \\[2mm]
&= \lim_{x \to \pi/2^-} \left(-\frac{\cos x}{\sin^2 x}\right) = 0.
\end{aligned}
$$

Thus, the desired limit is

$$
\lim_{x \to \pi/2^-} (\sin x)^{\sec x} = e^L = e^0 = 1. \qquad \blacksquare
$$

☐ **EXAMPLE 5**

Verify that

$$
\lim_{x \to \infty} \left(1 + \frac{r}{x}\right)^x = e^r, \qquad -\infty < r < \infty.
$$

Solution Since this limit has the indeterminate form 1^∞, we let $y = (1 + r/x)^x$ and first investigate the limit $\lim\limits_{x \to \infty} \ln y$. We obtain

$$
\begin{aligned}
\lim_{x \to \infty} \ln y &= \lim_{x \to \infty} \ln[(1 + r/x)^x] \\[2mm]
&= \lim_{x \to \infty} [x \ln(1 + r/x)] \qquad \text{(property of ln's)} \\[2mm]
&= \lim_{x \to \infty} \frac{\ln(1 + r/x)}{1/x} \qquad \left(\frac{0}{0} \text{ form}\right) \\[2mm]
&= \lim_{x \to \infty} \frac{\left(\dfrac{1}{1 + r/x}\right)(-r/x^2)}{(-1/x^2)} \qquad \text{(l'Hôpital's Rule)} \\[2mm]
&= \lim_{x \to \infty} \left(\frac{r}{1 + r/x}\right) = r.
\end{aligned}
$$

The desired limit is therefore

$$
\lim_{x \to \infty} \left(1 + \frac{r}{x}\right)^x = \lim_{x \to \infty} y = e^r.
$$

Setting $r = 1$ we have the important special case

$$e = \lim_{x \to \infty} \left(1 + \frac{1}{x}\right)^x. \qquad \blacksquare$$

The Forms 0^0 and ∞^0

These forms occur in $\lim[f(x)^{g(x)}]$ if $\lim f(x) = \lim g(x) = 0$ or if $\lim f(x) = \infty$ and $\lim g(x) = 0$. We use the same technique as we used with limits of the form 1^∞.

□ **EXAMPLE 6**

Find $\lim_{x \to 0^+} (\sin x)^x$.

Solution Since $\sin x \to 0$ as $x \to 0^+$, this limit yields the indeterminate form 0^0. Let $y = (\sin x)^x$. Then we have

$$\lim_{x \to 0^+} \ln y = \lim_{x \to 0^+} \ln[(\sin x)^x]$$

$$= \lim_{x \to 0^+} [x \ln \sin x].$$

Since $\sin x$ decreases to zero as $x \to 0^+$, the factor $\ln(\sin x)$ decreases toward $-\infty$ as $x \to 0^+$. Thus, the term $-\ln(\sin x)$ increases toward $+\infty$, and to be careful about claiming we have a $0 \cdot \infty$ form, we write

$$\lim_{x \to 0^+} \ln y = (-1) \lim_{x \to 0^+} [x(-\ln \sin x)] \qquad (0 \cdot \infty \text{ form})$$

$$= (-1) \lim_{x \to 0^+} \frac{-\ln \sin x}{1/x} \qquad \left(\frac{\infty}{\infty} \text{ form}\right)$$

$$= (-1) \lim_{x \to 0^+} \frac{-\left(\dfrac{\cos x}{\sin x}\right)}{(-1/x^2)} \qquad (\text{l'Hôpital's Rule})$$

$$= (-1) \lim_{x \to 0^+} \frac{x^2}{\tan x} \qquad \left(\frac{0}{0} \text{ form}\right)$$

$$= (-1) \lim_{x \to 0^+} \frac{2x}{\sec^2 x} \qquad (\text{l'Hôpital's Rule})$$

$$= (-1)(0) = 0.$$

Thus, $\displaystyle\lim_{x \to 0^+} (\sin x)^x = \lim_{x \to 0^+} y = e^0 = 1.$ $\qquad \blacksquare$

□ **EXAMPLE 7**

Find $\lim_{x \to \pi/2^-} (\tan x)^{\cos x}$.

Solution Since $\tan x \to +\infty$ and $\cos x \to 0$ as $x \to \pi/2^-$, this limit yields the indeterminate form ∞^0. Letting $y = (\tan x)^{\cos x}$ we have

$$\lim_{x \to \pi/2^-} \ln y = \lim_{x \to \pi/2^-} \ln(\tan x)^{\cos x}$$

$$= \lim_{x \to \pi/2^-} (\cos x) \ln(\tan x)$$

$$= \lim_{x \to \pi/2^-} \frac{\ln(\tan x)}{\sec x} \qquad \left(\frac{\infty}{\infty} \text{ form}\right)$$

$$= \lim_{x \to \pi/2^-} \frac{\left(\dfrac{\sec^2 x}{\tan x}\right)}{\sec x \tan x} \quad \text{(l'Hôpital's Rule)}$$

$$= \lim_{x \to \pi/2^-} \frac{\sec x}{\tan^2 x} \quad \text{(simplifying)}$$

$$= \lim_{x \to \pi/2^-} \frac{\left(\dfrac{1}{\cos x}\right)}{\left(\dfrac{\sin x}{\cos x}\right)^2}$$

$$= \lim_{x \to \pi/2^-} \frac{\cos x}{\sin^2 x} = \frac{0}{1} = 0.$$

The limit is therefore $\lim_{x \to \pi/2^-} (\tan x)^{\cos x} = e^0 = 1.$ ■

Exercise Set 11.2

In Exercises 1–30, use l'Hôpital's Rule and the techniques of this section to evaluate the given limit.

1. $\lim_{x \to 0^+} \left(\dfrac{1}{x} - \dfrac{1}{x^2} \right)$

2. $\lim_{x \to 0^+} (\csc x - \cot x)$

3. $\lim_{x \to 0} x \cot x$

4. $\lim_{x \to 0} x \sin \dfrac{1}{x}$

5. $\lim_{x \to 0^+} x \ln x$

6. $\lim_{x \to 0} \left(\dfrac{1}{x} - \csc x \right)$

7. $\lim_{x \to 1^+} \left(\dfrac{1}{x - 1} - \dfrac{x}{\sqrt{x - 1}} \right)$

8. $\lim_{x \to \infty} e^{-x} \ln x$

9. $\lim_{x \to 0^+} \left(\dfrac{1}{x} \right) \tan^{-1} x$

10. $\lim_{x \to \infty} \left(\dfrac{1}{x} \right) \ln \dfrac{1}{x}$

11. $\lim_{x \to 0^+} x^x$

12. $\lim_{x \to 0^+} x^{\cos x}$

13. $\lim_{x \to \infty} (1 + x^2) e^{-2x}$

14. $\lim_{x \to 0^+} \sqrt{x^x}$

15. $\lim_{x \to 0^+} \tan x \ln(\tan x)$

16. $\lim_{x \to 0^+} (1 + 2x)^{\cot x}$

17. $\lim_{x \to \infty} \left(1 + \dfrac{2}{x} \right)^{3x}$

18. $\lim_{x \to \infty} \left(1 - \dfrac{2}{x} \right)^{5x}$

19. $\lim_{x \to 0^+} \cot x \, \tan^{-1} x$

20. $\lim_{x \to 0^+} \sqrt{x} \ln x$

21. $\lim_{x \to 0} (2 - e^x)^{1/x}$

22. $\lim_{x \to \infty} [\ln(x^2 + 3) - \ln x]$

23. $\lim_{x \to \infty} [\ln \sqrt{4x + 2} - \ln \sqrt{x + 3}]$

24. $\lim_{x \to 0} (e^x + x)^{1/x}$

25. $\lim_{x \to 0} (e^x + x^2)^{1/x^2}$

26. $\lim_{x \to \infty} \left(\dfrac{x + 1}{x} \right)^{-2x}$

27. $\lim_{x \to \infty} [\ln \sqrt{x^6 + 3x^2} - \ln(2x^3)]$

28. $\lim_{x \to 0^+} x^2 \cot x^2$

29. $\lim_{x \to \infty} \left(\dfrac{x}{x + 1} \right)^{x+1}$

30. $\lim_{x \to 0^+} x^{\sqrt{x}}$

In Exercises 31–34, sketch the graph of the given function using l'Hôpital's Rule to find asymptotes and other limiting behavior.

31. $f(x) = x^2 e^x$

32. $f(x) = x e^{-2x}$

33. $f(x) = \left(\dfrac{1}{x} \right) \ln \dfrac{1}{x}$

34. $f(x) = x \ln^2 x$

35. For which real numbers r is $\lim_{x \to \infty} x^r \ln x = 0$?

36. Explain why $\lim_{x \to \infty} x^n e^{-x} = 0$ for any integer n.

37. This exercise discusses two limits and how well they estimate the natural exponential function.

  a. Using the method of Example 5, show that

$$\lim_{t \to +\infty} \left(1 + \dfrac{1}{t} \right)^{tx} = e^x.$$

  Explain why it follows that

$$\lim_{s \to 0^+} (1 + s)^{x/s} = e^x.$$

  b. From part a, we have $(1 + 1/t)^{tx} e^{-x} \to 1$ as $t \to +\infty$. Using $t = 10$, 100, 1000, and $10,000$, graph $h(x) = 1 - (1 + 1/t)^{tx} e^{-x}$ over the interval $-1 \le x \le 1$. Produce a table of values of $h(x)$ for $x = -1, -1/2, 0, 1/2,$ and 1.

  c. Repeat part b using $g(x) = 1 - (1 + s)^{x/s} e^{-x}$ for $s = 0.1, 0.01, 0.001,$ and 0.0001. Compare your results with those of part b.

38. Recall from Chapter 8, that, if P dollars are initially invested at an annual percentage rate r compounded m times per year, then the amount $A(t)$ at time t (in years) is given by $A(t) = P(1 + r/m)^{mt}$. Using the limit in Example 5, explain why $A(t) = Pe^{rt}$ if the interest is compounded continuously.

a. Suppose $P = \$1000$ and $r = 7.5\% = 0.075$. Graph $A(t)$ both assuming daily compounding and continuous com-

pounding for $0 \le t \le 10$. Do the graphs differ significantly?

b. Compare the amounts $A(10)$ and $A(25)$ in the two cases where the compounding is daily and continuous. How do these values differ? Is there a significant advantage to continuous compounding?

11.3 Improper Integrals

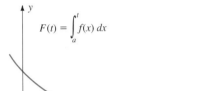

Figure 3.1 $F(t)$ defined by integration.

In Chapter 6 we encountered functions defined by integration of the form

$$F(t) = \int_a^t f(x)\, dx. \qquad (1)$$

If $f(x) \ge 0$ for $x \ge a$, we identified the value $F(t)$ with the area of the region bounded by the graph of $y = f(x)$ and the x-axis between $x = a$ and $x = t$ (Figure 3.1).

Now if f in equation (1) is continuous on $[a, \infty)$, the function F is defined for all $t > a$. We may therefore ask about its limit as $t \to \infty$. In fact, we define the **improper integral**

$$\int_a^\infty f(x)\, dx$$

as this limit, if it exists (Figure 3.2).

Definition 1

Suppose the function f is continuous on the unbounded interval $[a, \infty)$. Then

$$\int_a^\infty f(x)\, dx = \lim_{t \to \infty} \int_a^t f(x)\, dx.$$

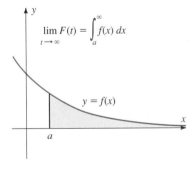

Figure 3.2 $\lim\limits_{t \to \infty} F(t)$ defines an improper integral.

If the limit in Definition 1 exists and is finite, we say that the improper integral **converges**. Otherwise, it is said to **diverge**. Note that we do not assume that f is nonnegative in Definition 1. We need only assume that f is continuous on $[a, \infty)$ to state the definition.

There are several important reasons for studying improper integrals at this point, including the following:

(i) It is not unreasonable to ask whether certain unbounded regions might nevertheless have finite area. Since the definite integral determines area if $f(x) \ge 0$, Definition 1 will give us a way to extend the notion of area to such regions.

(ii) In Chapter 8 we determined that the **present value** P_0 of an asset assumed to be worth P dollars after T years was $P_0 = e^{-rT}P$, assuming continuous compounding at the annual rate r. If the asset itself is income-producing, such as a forest continually harvested for lumber, P is a function of t, and the calculation becomes

$$P_0 = \int_0^T e^{-rt}P(t)\, dt.$$

(See Example 6.) Calculating the present value for an asset with an **infinite** life expectancy then leads to the improper integral

$$P_0 = \int_0^\infty e^{-rt}P(t)\, dt.$$

The calculation of improper integrals of the type given in Definition 1 is simply a matter of evaluating the indicated limit carefully.

□ **EXAMPLE 1**

Find $\displaystyle\int_1^\infty \frac{1}{x^2}\, dx$.

Strategy · · · · · · · ·
Apply Definition 1.

Evaluate the definite integral from $x = 1$ to $x = t$.

Evaluate the limit as $t \to \infty$ of this integral.

Solution

$$\int_1^\infty \frac{1}{x^2}\, dx = \lim_{t\to\infty} \int_1^t \frac{1}{x^2}\, dx$$

$$= \lim_{t\to\infty}\left[-\frac{1}{x}\right]_{x=1}^{x=t}$$

$$= \lim_{t\to\infty}\left[\left(-\frac{1}{t}\right) - (-1)\right]$$

$$= 0 - (-1) = 1.$$

This improper integral converges. ■

□ **EXAMPLE 2**

Find $\displaystyle\int_1^\infty \frac{1}{x}\, dx$.

Strategy · · · · · · ·
Apply Definition 1.

Evaluate the definite integral.

Evaluate the limit using the fact that $\lim_{t\to\infty} \ln t = +\infty$.

Solution

$$\int_1^\infty \frac{1}{x}\, dx = \lim_{t\to\infty} \int_1^t \frac{1}{x}\, dx$$

$$= \lim_{t\to\infty}\left[\ln x\right]_{x=1}^{x=t}$$

$$= \lim_{t\to\infty}[\ln t - \ln 1]$$

$$= +\infty.$$

This improper integral diverges. ■

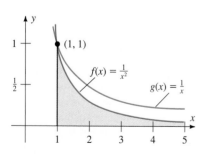

Figure 3.3 Graph of $f(x) = 1/x^2$ approaches x-axis more quickly than graph of $g(x) = 1/x$.

REMARK 1 Note the difference between the improper integrals in Examples 1 and 2. Both are of the form

$$\int_1^\infty \frac{1}{x^p}\, dx,$$

but their behavior is quite different. As Figure 3.3 illustrates, the graph of $y = 1/x^2$ approaches the x-axis more quickly as x increases than does the graph of $y = 1/x$. In an intuitive sense, it is the rate at which the graph of such a function approaches the x-axis that determines whether its improper integral converges. However, this determination cannot be made simply by inspecting a graph. The next example gives a more precise criterion.

□ **EXAMPLE 3**

For which numbers p does the improper integral $\int_1^\infty \frac{1}{x^p}\, dx$ converge?

Strategy · · · · · · · ·

Evaluate the improper integral, carrying along the constant p.

Solution

We have

$$\int_1^\infty \frac{1}{x^p}\, dx = \lim_{t \to \infty} \int_1^t x^{-p}\, dx$$

Assume $p \neq 1$, so the integral is obtained via the Power Rule.

$$= \lim_{t \to \infty} \left[\frac{x^{-p+1}}{-p+1} \right]_{x=1}^{x=t} \quad \text{if} \quad p \neq 1$$

$$= \lim_{t \to \infty} \left[\frac{t^{1-p}}{1-p} - \frac{1}{1-p} \right].$$

The existence of the limit depends upon the sign of the exponent $1 - p$. Examine each case.

Three cases arise:

(a) If $1 - p < 0$, then $t^{1-p} \to 0$ as $t \to \infty$ and the integral converges.

(b) If $1 - p > 0$, then $t^{1-p} \to \infty$ as $t \to \infty$ and the integral diverges.

(c) In the case $1 - p = 0$, excluded above, the integral is $\int_1^\infty (1/x)\, dx$, which diverges (Example 2).

We therefore conclude that

$$\int_1^\infty \frac{1}{x^p}\, dx \text{ converges if and only if } p > 1.$$

■

Using the improper integral, we can extend the notion of area to unbounded regions.

Definition 2

Let f be continuous on $[a, \infty)$ with $f(x) \geq 0$ for all $x \geq a$. The area of the region bounded above by the graph of $y = f(x)$ and below by the x-axis and extending to the right of the line $x = a$ is defined to be

$$A = \int_a^\infty f(x)\, dx$$

if this integral converges.

REMARK 2   Our previous definition of area as

$$A = \int_a^b f(x)\, dx$$

involved A as an infinite limit of approximating sums. Definition 2 actually involves a "double limit," as both the number of approximating rectangles and the interval over which they are constructed become infinite. The notions of volume and surface area for solids of revolution may be extended to unbounded regions in the same way.

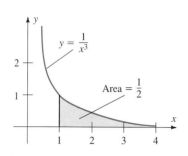

Figure 3.4   An unbounded region with finite area.

□ **EXAMPLE 4**

Find the area of the region bounded above by the graph of $y = 1/x^3$, below by the x-axis, and on the left by the line $x = 1$. (See Figure 3.4.)

Solution   By Definition 2 and Example 3 we have

$$A = \int_1^\infty \frac{1}{x^3}\, dx = \lim_{t\to\infty}\left[\int_1^t x^{-3}\, dx\right] = \lim_{t\to\infty}\left[\frac{t^{-2}}{-2} - \frac{1^{-2}}{-2}\right] = \frac{1}{2}.$$ ■

□ **EXAMPLE 5**

Find the volume of the solid obtained by revolving about the x-axis the region R bounded above by the graph of $y = 1/x$ and below by the x-axis for $1 \le x < \infty$ (see Figures 3.5 and 3.6).

Solution   This volume is

$$V = \int_1^\infty \pi\left[\frac{1}{x}\right]^2 dx = \pi,$$

by Example 1. Thus, although the region R has infinite area, it sweeps out a region of finite volume. (This peculiarity is a result only of the way in which we have extended these definitions to the infinite case. Such an intriguing object cannot be realized.)   ■

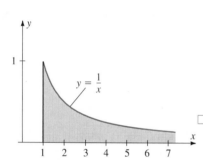

Figure 3.5   The region under the graph of $y = 1/x$ for $1 \le x < \infty$ has infinite area.

□ **EXAMPLE 6**

In Chapter 8 we saw that the **present value** of a sum P of money that will become available after t years is $P_0 = e^{-rt}P$, where r is the rate (in decimal form) of continuous compounding of interest. Closely related to this notion is that of a **revenue stream,** which is simply an anticipated flow of money over time. For example, consider the question of defining the present value of the profits from a small business over the next five years if profits are expected at a rate of $P(t)$ dollars per year, where $P(t)$ is a function of time.

If we partition the time interval $[0, 5]$ into n small intervals of length Δt, then the present value of the profits earned during the time interval $[t_{j-1}, t_j]$ is approximated by

$$P_j = e^{-rs_j}P(s_j)\,\Delta t, \qquad 1 \le j \le n,$$

where s_j is an arbitrary element of the interval $[t_{j-1}, t_j]$. An approximation to the total present value of these profits is then

$$P_0 \approx \sum_{j=1}^n e^{-rs_j}P(s_j)\,\Delta t.$$

As $n \to \infty$ this Riemann sum approaches the integral $\int_0^5 e^{-rt}P(t)\, dt$, so we define

$$P_0 = \int_0^5 e^{-rt}P(t)\, dt.$$

If we assume the business to have an infinite expected lifetime, the present value of all future profits is defined by the improper integral

$$P_0 = \int_0^\infty e^{-rt}P(t)\, dt.$$ ■

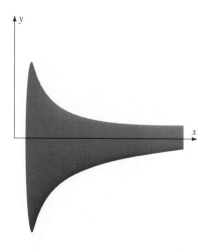

Figure 3.6   Revolving the region in Figure 3.5 about the x-axis results in a solid of finite volume.

Obviously, we can define improper integrals of the form $\int_{-\infty}^{a} f(x)\,dx$ by analogy with Definition 1. We can extend the notion of improper integral to integrals with two infinite limits as follows.

Definition 3

The improper integral $\displaystyle\int_{-\infty}^{\infty} f(x)\,dx$ is defined by

$$\int_{-\infty}^{\infty} f(x)\,dx = \int_{-\infty}^{0} f(x)\,dx + \int_{0}^{\infty} f(x)\,dx.$$

The improper integral in Definition (3) is said to **converge** only if both

$$\int_{-\infty}^{0} f(x)\,dx \quad \text{and} \quad \int_{0}^{\infty} f(x)\,dx$$

converge. Otherwise it is said to **diverge.** Note that there is nothing special about splitting the interval $(-\infty, \infty)$ at 0. We can also determine the convergence of the improper integral by considering

$$\int_{-\infty}^{a} f(x)\,dx + \int_{a}^{\infty} f(x)\,dx$$

for any number a (see Exercise 42).

□ **EXAMPLE 7**

Find $\displaystyle\int_{-\infty}^{\infty} \frac{1}{1 + x^2}\,dx.$

Solution   By Definition 3,

$$\int_{-\infty}^{\infty} \frac{1}{1 + x^2}\,dx = \int_{-\infty}^{0} \frac{1}{1 + x^2}\,dx + \int_{0}^{\infty} \frac{1}{1 + x^2}\,dx$$

$$= \lim_{t \to -\infty} \left[\text{Tan}^{-1} x \right]_{x=t}^{x=0} + \lim_{t \to \infty} \left[\text{Tan}^{-1} x \right]_{x=0}^{x=t}$$

$$= \lim_{t \to -\infty} [-\text{Tan}^{-1}(t)] + \lim_{t \to \infty} [\text{Tan}^{-1}(t)]$$

$$= -\left(-\frac{\pi}{2}\right) + \frac{\pi}{2} = \pi. \qquad \blacksquare$$

A Comparison Test

We will sometimes need to determine whether an improper integral converges or diverges, even though we cannot find an antiderivative in closed form* for its integrand. In such cases we can often resolve the issue of convergence for $\int_{a}^{\infty} f(x)\,dx$ by comparing the integrand f with another function g and using information about $\int_{a}^{\infty} g(x)\,dx$.

---

*By a ''closed form'' antiderivative we mean a function not involving an integral sign. The Fundamental Theorem guarantees that $G(x) = \int_{a}^{x} f(t)\,dt$ is always an antiderivative for f if f is continuous on $[a, x]$, but G is not a ''closed form'' function.

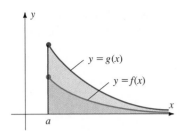

Figure 3.7 Comparison Test: If $0 \leq f(x) \leq g(x)$ and $\int_a^\infty g(x)\,dx$ converges, then $\int_a^\infty f(x)\,dx$ converges.

> **Comparison Test:** Suppose f and g are continuous on $[a, \infty)$ with $0 \leq f(x) \leq g(x)$ for all $x \geq a$. Then
>
> (i) If $\displaystyle\int_a^\infty g(x)\,dx$ converges, so does $\displaystyle\int_a^\infty f(x)\,dx$.
>
> (ii) If $\displaystyle\int_a^\infty f(x)\,dx$ diverges, so does $\displaystyle\int_a^\infty g(x)\,dx$.

Figure 3.7 illustrates statement (i), which we justify intuitively by saying that if the area of the region R_1 bounded by the graph of g and the x-axis, $a \leq x < \infty$, is finite, then the area of the region R_2 bounded by the graph of f and the x-axis, $a \leq x < \infty$, must also be finite, since $R_2 \subseteq R_1$. A similar argument applies to statement (ii): If R_2 is not finite, then R_1 cannot be finite. A more formal proof is given in Chapter 12 for an analogous statement concerning infinite series.

□ **EXAMPLE 8**

Use the Comparison Test to determine whether the improper integral

$$\int_1^\infty e^{-(1+x^2)}\,dx$$

converges.

Solution Since $1 + x^2 > x$ for $x \geq 1$ and e^{-x} is a *decreasing* function, we have the comparison

$$0 < e^{-(1+x^2)} < e^{-x}, \qquad x \geq 1.$$

Since

$$\int_1^\infty e^{-x}\,dx = \lim_{t \to \infty}\left[-e^{-x}\right]_{x=1}^{x=t} = \lim_{t \to \infty}\left(\frac{1}{e} - e^{-t}\right) = \frac{1}{e}$$

converges, so must

$$\int_1^\infty e^{-(1+x^2)}\,dx.$$

(Note, however, that we cannot determine the *value* of this integral by this method—only whether or not it converges.) ■

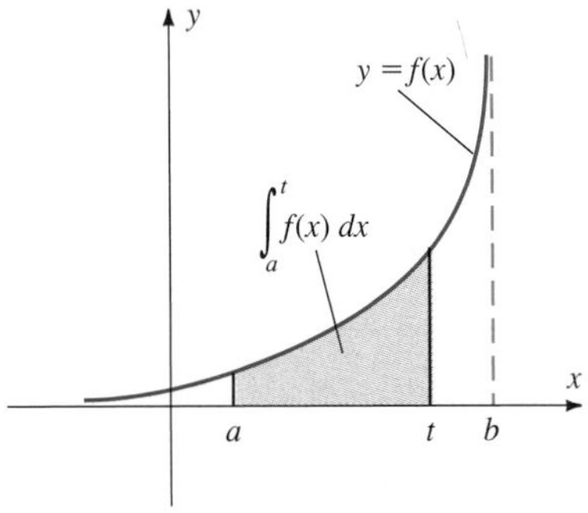

Figure 3.8 $\lim_{x \to b^-} f(x) = +\infty$.

Improper Integrals—Infinite Integrands

A second general type of improper integral occurs if the integrand in a definite integral approaches $\pm\infty$ at one or both limits of integration. For example, suppose that f is defined and continuous on $[a, b)$ but that $f(x) \to \pm\infty$ as $x \to b^-$ (Figure 3.8). In this case the definite integral $\int_a^b f(x)\,dx$ as defined in Chapter 6 does not exist. However, we define this **improper integral** as follows:

Definition 4

If f is continuous on $[a, b)$ but $\lim_{x \to b^-} f(x) = \pm\infty$, we define

$$\int_a^b f(x)\,dx = \lim_{t \to b^-} \int_a^t f(x)\,dx.$$

If the limit on the right exists, the integral is said to **converge.** Otherwise it is said to **diverge** (Figure 3.9).

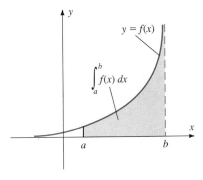

Figure 3.9 $\int_a^b f(x)\,dx = \lim_{t \to b} \int_a^t f(x)\,dx$.

□ **EXAMPLE 9**

Determine whether the improper integral

$$\int_0^1 \frac{x}{\sqrt{1 - x^2}}\,dx$$

converges or diverges.

Strategy · · · · · · · ·

Determine where the integrand fails to exist.

Apply Definition 4.

Use a substitution with

$$u = 1 - x^2, \qquad du = -2x\,dx.$$

Solution

The integrand approaches $+\infty$ as $x \to 1^-$. We have

$$\int_0^1 \frac{x}{\sqrt{1 - x^2}}\,dx = \lim_{t \to 1^-} \left[\int_0^t \frac{x}{\sqrt{1 - x^2}}\,dx \right]$$

$$= \lim_{t \to 1^-} \left[-\sqrt{1 - x^2}\, \right]_{x=0}^{x=t}$$

$$= \lim_{t \to 1^-} \left[-\sqrt{1 - t^2} + 1 \right]$$

$$= 1.$$

The integral converges. ■

If the integrand approaches $\pm\infty$ at the left endpoint, the improper integral is defined by analogy with Definition 4 (see Figure 3.10). That is,

$$\int_a^b f(x)\,dx = \lim_{t \to a^+} \int_t^b f(x)\,dx.$$

If an integrand approaches $\pm\infty$ at a point $c \in (a, b)$, we extend Definition 4 as follows:

$$\int_a^b f(x)\,dx = \int_a^c f(x)\,dx + \int_c^b f(x)\,dx \tag{2}$$

$$= \lim_{t_1 \to c^-} \int_a^{t_1} f(x)\,dx + \lim_{t_2 \to c^+} \int_{t_2}^b f(x)\,dx.$$

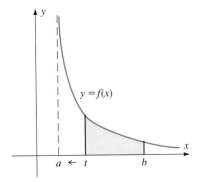

Figure 3.10 $\int_a^b f(x)\,dx = \lim_{t \to a^+} \int_t^a f(x)\,dx$.

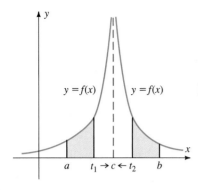

Figure 3.11 $\int_a^b f(x)\,dx = \int_a^c f(x)\,dx + \int_c^b f(x)\,dx$.

(See Figure 3.11.) The improper integral in the left in equation (2) is said to converge only if both improper integrals on the right converge. Otherwise it is said to diverge.

☐ **EXAMPLE 10**

Determine whether the improper integral

$$\int_0^3 (x-2)^{-4/3}\,dx$$

converges or diverges.

Solution   The integrand approaches $+\infty$ as $x \to 2$. Applying (2) we find that

$$\int_0^3 (x-2)^{-4/3}\,dx = \lim_{t_1 \to 2^-}\int_0^{t_1}(x-2)^{-4/3}\,dx + \lim_{t_2 \to 2^+}\int_{t_2}^3 (x-2)^{-4/3}\,dx$$

$$= \lim_{t_1 \to 2^-}\left[-3(x-2)^{-1/3}\right]_0^{t_1} + \lim_{t_2 \to 2^+}\left[-3(x-2)^{-1/3}\right]_{t_1}^3$$

$$= \lim_{t_1 \to 2^-}\left[\frac{-3}{\sqrt[3]{t_1-2}} + \frac{3}{\sqrt[3]{-2}}\right] + \lim_{t_2 \to 2^+}\left[-3 + \frac{3}{\sqrt[3]{t_2-2}}\right]$$

$$= \infty + \infty = \infty.$$

Since both limits fail to exist, the improper integral diverges. (Actually, as soon as we establish that one of these limits fails to exist, we can conclude that the improper integral diverges.)   ■

Exercise Set 11.3

In Exercises 1–28, determine whether the improper integral converges. Evaluate those that converge.

1. $\displaystyle\int_3^\infty \frac{1}{x}\,dx$

2. $\displaystyle\int_2^\infty \frac{1}{(x+1)^2}\,dx$

3. $\displaystyle\int_1^\infty \frac{x}{\sqrt{1+x^2}}\,dx$

4. $\displaystyle\int_1^\infty \frac{dx}{4+x^2}$

5. $\displaystyle\int_0^\infty e^{-x}\,dx$

6. $\displaystyle\int_2^\infty \frac{x}{(4+x^2)^{3/2}}\,dx$

7. $\displaystyle\int_0^\infty e^{-x}\sin x\,dx$

8. $\displaystyle\int_0^\infty e^x\cos x\,dx$

9. $\displaystyle\int_{-\infty}^2 e^{2x}\,dx$

10. $\displaystyle\int_{-\infty}^0 \frac{1}{1+x^2}\,dx$

11. $\displaystyle\int_0^\infty xe^{-x}\,dx$

12. $\displaystyle\int_{-\infty}^\infty \frac{dx}{4+x^2}$

13. $\displaystyle\int_0^1 \frac{1}{x}\,dx$

14. $\displaystyle\int_0^1 \frac{1}{x^2}\,dx$

15. $\displaystyle\int_0^3 \frac{1}{\sqrt{9-x^2}}\,dx$

16. $\displaystyle\int_e^\infty \frac{1}{x\ln x}\,dx$

17. $\displaystyle\int_0^1 x\ln x\,dx$

18. $\displaystyle\int_0^1 \frac{dx}{\sqrt{1-x}}$

19. $\displaystyle\int_0^e x^2\ln x\,dx$

20. $\displaystyle\int_0^1 x^{-2/3}\,dx$

21. $\displaystyle\int_{-1}^1 \frac{1}{x}\,dx$

22. $\displaystyle\int_0^3 \frac{dx}{x^2+2x-3}$

23. $\displaystyle\int_0^\infty e^{-2x}\cos 2x\,dx$

24. $\displaystyle\int_0^\infty e^{-3x}\sin 2x\,dx$

25. $\displaystyle\int_0^\infty \frac{e^{-2\sqrt{x}}}{\sqrt{x}}\,dx$

26. $\displaystyle\int_0^\infty \sqrt{x}\,e^x\,dx$

27. $\displaystyle\int_0^1 x\ln(1+x)\,dx$

28. $\displaystyle\int_0^1 \ln^2 x\,dx$

29. Show that the integral

$$\int_0^\infty \sin x \, dx$$

diverges. Thus, it is not necessary that

$$\lim_{t \to \infty} \int_a^t f(x) \, dx = \pm\infty$$

for an improper integral to diverge.

30. Give a geometric argument that $\int_1^\infty \frac{dx}{x^2} = \int_0^1 \frac{dx}{\sqrt{x}} - 1.$

31. For which numbers p does the integral $\int_0^1 x^p \, dx$ converge?

32. For which numbers p does the integral $\int_0^\infty x^p \, dx$ converge?

33. Give a geometric argument to show that if $0 \le f(x) \le g(x)$ for all $x > a$, then

$$\int_a^\infty g(x) \, dx \text{ diverges} \quad \text{if} \quad \int_a^\infty f(x) \, dx \text{ diverges.}$$

Use the Comparison Test to determine whether the following integrals converge or diverge.

34. $\int_1^\infty \frac{1}{1 + x^3} \, dx$

35. $\int_1^\infty \frac{e^x}{\sqrt{1 + x^2}} \, dx$

36. $\int_1^\infty \frac{1}{\sqrt{1 + x^5}} \, dx$

37. $\int_1^\infty \frac{|\sin x|}{x^2} \, dx$

38. $\int_1^\infty e^{\sqrt{x}} \, dx$

39. $\int_1^\infty \sqrt{e^x + \sin x} \, dx$

40. $\int_1^\infty \frac{dx}{\sqrt{e^{x^2} + x + \cos x}}$

41. Explain why

$$\int_{-\infty}^\infty f(x) \, dx = \lim_{t \to \infty} \int_{-t}^t f(x) \, dx$$

is wrong. (*Hint:* Consider the function $f(x) = x^3$, for example.)

42. Show that if $\int_{-\infty}^\infty f(x) \, dx$ converges then

$$\int_{-\infty}^\infty f(x) \, dx = \int_{-\infty}^a f(x) \, dx + \int_a^\infty f(x) \, dx$$

for all $a \in (-\infty, \infty)$.

43. Consider the region R bounded above by the graph of $y = 1/x$ and below by the x-axis for $1 \le x < \infty$.
   a. Show that the surface area of the solid obtained by revolving R about the x-axis is infinite.

   b. Compare the result of part a with that of Example 5. Comment on the observation that Figure 3.6 represents a horn of infinite length that (i) cannot be painted with a finite amount of paint but (ii) can be filled with a finite amount of paint.

44. Find the volume and the surface area of the solid obtained by revolving about the x-axis the region bounded above by the graph of $f(x) = e^{-x}$ and below by the x-axis for $0 \le x < \infty$.

45. Find the area of the region bounded by the graph of $y = 1/x^4$ and the x-axis for $1 \le x < \infty$.

46. Find the area of the region bounded by the graph of $y = xe^{-4x}$ and the x-axis for $0 \le x < \infty$.

47. Find the area of the region bounded by the graph of $y = \ln x$, the x-axis, and the y-axis.

48. Find the volume of the solid obtained by revolving about the y-axis the region bounded by the graph of $y = x^{-1/4}$ for $0 \le x \le 1$.

49. Find the volume of the solid obtained by revolving about the y-axis the region bounded by the graphs of $y = e^{-x}$ and $y = -1/x^3$ for $1 \le x < \infty$.

50. Find the centroid of the region bounded by the graph of $y = e^{-x}$ and the x-axis for $0 \le x < \infty$.

51. Find the centroid of the region bounded by the graph of $y = 1/x^3$ and the x-axis for $1 \le x < \infty$.

52. The graph of the function

$$f(x) = \frac{x^3 - 1}{x^2}$$

is aysmptotic to the line that is the graph of $g(x) = x$.
   a. Prove this by showing that

$$\lim_{x \to \infty} [f(x) - g(x)] = 0.$$

   b. Find the area of the region bounded by the graphs of $f(x)$ and $g(x)$ for $1 \le x < \infty$.

53. The graph of the function

$$f(x) = \frac{x^2 - 1}{x}$$

is asymptotic to the line that is the graph of $g(x) = x$.
   a. Prove this by showing that

$$\lim_{x \to \infty} [f(x) - g(x)] = 0.$$

   b. Show that the area of the region bounded by the graphs of $f(x)$ and $g(x)$ for $1 \le x < \infty$ is infinite.

54. For what integers n does the integral $\int_0^1 x^n \ln x \, dx$ converge?

55. A business expects to generate one million dollars profit per year indefinitely.

    a. Find the present value of its anticipated profits over the next five-year period, assuming an annual rate of interest of $r = 8\%$.

    b. Find the present value of all anticipated future profits.

56. What is the present value of an investment that will produce revenues at the rate of $R(t) = 1000 + 200t$ dollars per year, assuming an annual rate of continuous compounding of $r = 10\%$.

    a. over the period of the next 5 years?

    b. over an infinite future?

57. Show that

$$\int_0^1 [\ln x]^n \, dx = (-1)^n n!$$

$$(n! = n(n-1)(n-2) \cdot \cdots \cdot 2 \cdot 1).$$

58. Show that $\displaystyle\int_1^\infty \frac{dx}{1 + e^x} = \ln(1 + e) - 1.$

59. Use the notion of an improper integral to explain how you can conclude from the result of Example 4 in Section 7.3 that the surface area of the sphere is $4\pi r^2$.

60. The present value of profits for a business over n years is given by

$$P_n = \int_0^n e^{-rt} P(t) \, dt$$

where $P(t)$ is the expected profit in dollars per year and r is the interest rate compounded continuously. If the interest is compounded m times per year, then the present value of profits is

$$P_{n,m} = \int_0^n \left(1 - \frac{r}{m}\right)^{mt} P(t) \, dt.$$

    a. Assuming that $r = 8.5\%$, that the expected profits are \$215,000 per year, and that interest is compounded yearly, graph P_n and $P_{n,m}$ for $0 \le n \le 20$. What is the difference between P_{20} and $P_{20,1}$? Is the difference significant over a 20-year period? (*Hint:* $\int a^{mt} \, dt = a^{mt}/(m \ln a) + C$.)

    b. What is the difference between the present value of all future anticipated profits (i.e., the limits of P_m and $P_{n,m}$ as $m \to \infty$) if the interest is compounded continuously and if it is compounded yearly? What is this difference if the interest is compounded monthly instead of yearly?

Improper integrals of the form

$$\int_a^{+\infty} f(x) \, dx, \quad a > 0,$$

can be approximated with Simpson's Rule by first applying the change of variable $x = 1/t$. This produces

$$\int_a^{+\infty} f(x) \, dx = \int_{1/a}^0 f(1/t) \frac{-1}{t^2} \, dt = \int_0^{1/a} \frac{f(1/t)}{t^2} \, dt.$$

Note that $t \to 0^+$ as $x \to +\infty$. Then Simpson's Rule can be applied to the last integral, provided the limit of $f(1/t)/t^2$ exists as $t \to 0^+$. For example

$$\int_1^{+\infty} xe^{-2x} \, dx = \int_1^0 \frac{1}{t} e^{-2/t} \frac{-1}{t^2} \, dt = \int_0^1 \frac{e^{-2/t}}{t^3} \, dt.$$

l'Hôpital's Rule implies

$$\lim_{t \to 0} \frac{e^{-2/t}}{t^3} = 0,$$

so we can apply Simpson's Rule to $\int_0^1 f(t) \, dt$ where

$$f(t) = \begin{cases} \dfrac{e^{-2/t}}{t^3}, & \text{if } t > 0; \\ 0, & \text{if } t = 0. \end{cases}$$

With $n = 20$, Simpson's Rule gives 0.1015014675. The exact value is $\frac{3}{4}e^{-2} \approx 0.101501624$. In Exercises 61–64, rewrite the given integral as an integral on a finite interval and use Simpson's Rule with $n = 20$ to approximate the new integral. Compare your answer with the exact value.

61. $\displaystyle\int_2^{+\infty} \frac{dx}{4 + x^2}$, Exact value: $\pi/8$.

62. $\displaystyle\int_1^{+\infty} e^{-\sqrt{x}} \, dx$, Exact value: $4e^{-1}$.

63. $\displaystyle\int_0^{+\infty} \frac{x \, dx}{(1 + x^2)^{3/2}}$, Exact value: 1.

    (*Hint:* Rewrite as two integrals: $\int_0^1 (\ ) \, dx + \int_1^{+\infty} (\ ) \, dx$.)

64. $\displaystyle\int_1^{+\infty} xe^{-2x} \sin \pi x \, dx$, Exact value: $\dfrac{-\pi(\pi^2 + 8)e^{-2}}{(4 + \pi^2)^2}$.

Improper integrals of the form

$$\int_0^a \frac{f(x)}{x^p} \, dx, \quad a > 0, \quad 0 < p < 1,$$

can be approximated with Simpson's Rule by first applying the change of variable $x = t^{1/(1-p)}$. This produces

$$\int_0^a \frac{f(x)}{x^p} \, dx = \int_0^{a^{1-p}} \frac{f(t^{1/(1-p)})}{t^{p/(1-p)}} \frac{t^{p/(1-p)}}{1-p} \, dt$$

$$= \frac{1}{1-p} \int_0^{a^{1-p}} f(t^{1/(1-p)}) \, dt$$

We can apply Simpson's Rule to the last integral provided the limit of $f(t^{1/(1-p)})$ exists as $t \to 0$. For example consider

$$\int_0^\pi \frac{\sin x}{\sqrt{x}} \, dx.$$

Here $p = 1/2$, so the change of variable is $x = t^{1/(1-p)} = t^2$. Therefore

$$\int_0^\pi \frac{\sin x}{\sqrt{x}}\, dx = 2\int_0^{\sqrt{\pi}} \sin t^2\, dt.$$

Applying Simpson's Rule with $n = 20$ to the last integral gives 1.789693687. The true value of this integral to nine decimal places is 1.789662939. In Exercises 65–67, use Simpson's Rule with $n = 20$ to approximate the given integral by applying this change of variable. Compare your answer with the true value (to nine decimal places) given.

65. $\int_0^{1/2} \frac{\cos \pi x}{\sqrt{x}}\, dx$, Value: 1.102935824.

66. $\int_0^4 \frac{e^x}{x^{1/3}}\, dx$, Value: 38.251813991.

67. $\int_0^1 \frac{e^{-x^2/2} \sin 3x}{\sqrt{x}}\, dx$, Value: 0.910274568.

Summary Outline of Chapter 11

▊ **l'Hôpital's Rule** states that if f and g are differentiable (except possibly at (page 567) $x = a$), if lim denotes one of

$$\lim_{x\to a},\ \lim_{x\to a^-},\ \lim_{x\to a^+},\ \lim_{x\to\infty},\ \text{or}\ \lim_{x\to-\infty},$$

and if either $\lim f(x) = 0 = \lim g(x)$ or $\lim f(x) = \infty = \lim g(x)$, then

$$\lim \frac{f(x)}{g(x)} = \lim \frac{f'(x)}{g'(x)}$$

provided that the limit on the right exists or is infinite.

▊ Limits involving the **indeterminate forms**   (page 574)

$$\frac{0}{0},\ \frac{\infty}{\infty},\ \infty - \infty,\ 1^\infty,\ 0^0,\ \text{and}\ \infty^0$$

are evaluated, if possible, by l'Hôpital's Rule.

▊ The **improper integral** $\int_a^\infty f(x)\, dx$ is defined by $\int_a^\infty f(x)\, dx = \lim_{t\to\infty} \int_a^t f(x)\, dx.$ (page 579)

▊ If $\lim_{x\to b^-} f(x) = \pm\infty$, the **improper integral** $\int_a^b f(x)\, dx$ is defined by (page 584)

$$\int_a^b f(x)\, dx = \lim_{t\to b^-} \int_a^t f(x)\, dx.$$

Review Exercises—Chapter 11

In Exercises 1–18, evaluate the limit using l'Hôpital's Rule if necessary.

1. $\lim_{x\to 0} \frac{\sin 3x}{\sin 4x}$

2. $\lim_{x\to 0} \frac{\sin x}{1 - e^x}$

3. $\lim_{x\to 0^+} \frac{\tan x}{x^2}$

4. $\lim_{x\to 2} \frac{x-2}{x^2 + x - 6}$

5. $\lim_{x\to\infty} \frac{\sqrt{x^2 + 2x + 1}}{x}$

6. $\lim_{x\to\infty} \frac{(x^4 + 1)^{3/2}}{x^7}$

7. $\lim_{x\to 0} \frac{x - \sin x}{\tan x}$

8. $\lim_{x\to 0} \frac{\pi - \csc x}{\pi + \cot x}$

9. $\lim_{x\to 0^+} \frac{\sin \sqrt{x}}{\sqrt{x}}$

10. $\lim_{x\to\infty} \frac{\pi/2 - \text{Tan}^{-1} x}{xe^{-x}}$

11. $\lim_{x\to\infty} \frac{\ln \sqrt{x}}{\sqrt{x}}$

12. $\lim_{x\to-\infty} \frac{e^{-x}}{x^2}$

13. $\lim_{x\to\infty} \frac{x^3 + e^x}{xe^{2x}}$

14. $\lim_{x\to\infty} x^2 e^{-x}$

15. $\lim_{x\to\infty} x^{1/x}$

16. $\lim_{x\to 0^+} \tan x \ln x$

17. $\lim_{x\to 0^+} x^{\sin 2x}$

18. $\lim_{x\to\infty} (\ln x)^{e^{-x}}$

In Exercises 19–28, evaluate the improper integral.

19. $\displaystyle\int_1^\infty xe^{-x}\,dx$

20. $\displaystyle\int_{-\infty}^0 \frac{6x}{1+x^2}\,dx$

21. $\displaystyle\int_0^2 \frac{2x+1}{x^2+x-6}\,dx$

22. $\displaystyle\int_0^4 \frac{\ln\sqrt{x}}{\sqrt{x}}\,dx$

23. $\displaystyle\int_{-\infty}^\infty xe^{-x^2}\,dx$

24. $\displaystyle\int_1^\infty \frac{1}{x\sqrt{x^2-1}}\,dx$

25. $\displaystyle\int_0^{\pi/2} \tan x\,dx$

26. $\displaystyle\int_e^\infty \frac{1}{x\ln x}\,dx$

27. $\displaystyle\int_0^4 \frac{1}{\sqrt{16-x^2}}\,dx$

28. $\displaystyle\int_{-\infty}^\infty \mathrm{Tan}^{-1}x\,dx$

In Exercises 29–32, sketch the graph using l'Hôpital's Rule to find asymptotes and other limiting behavior.

29. $y = \dfrac{3x+1}{x-2}$

30. $y = \dfrac{2x^2}{x^2-1}$

31. $y = \dfrac{\ln x}{x}$

32. $y = \dfrac{\ln x}{\sqrt{x}}$

33. Find the area of the region bounded by the graphs $y = (x^5 + 3x + 1)/x^3$ and $y = x^2$ for $1 \le x < \infty$.

34. Find the area of the region bounded by the graph of $y = x^2e^{-x}$ and the x-axis for $0 \le x < \infty$.

35. Find the volume of the solid obtained by revolving about the x-axis the region bounded by the graph of $y = 4/(x^2 + 1)$ for $0 \le x < \infty$.

36. Find the volume of the solid obtained by revolving about the y-axis the region bounded by the graphs of $y = 1/x^3$ and $y = 1/x^4$ for $1 \le x < \infty$.

The Theory of
Infinite Series

The Theory of Infinite Series

It was noted in the introduction to Unit One that Newton and Leibniz both found themselves expressing functions as infinite sums. Later workers in analysis, particularly Leonhard Euler and Johann Bernoulli, also dealt with these infinite series. This early work was not on a firm mathematical footing, however, and operations were sometimes performed that just happened to work because of the circumstances in a particular problem. The development of a logical theory was yet to appear, a process that unfolded over a long period of time.

Brook Taylor

Brook Taylor (1685–1731) and Isaac Newton knew each other through their membership in the Royal Society of London. Newton was its president for many years, and Taylor served for several years as its secretary. Taylor was brought up in a well-to-do home, and music was an important part of his early life. In fact, Taylor and Newton jointly wrote a work entitled *On Musick,* although it was never completed or published. Taylor thought highly of the new Newtonian calculus and attempted to clarify the subject in his writings. His writing was rather murky, however, and his exposition not very successful.

In this unit we write functions in the form of a special type of infinite series called Taylor's series. Taylor developed the series as a result of a chance remark made by a friend in a coffeehouse, and some have questioned whether Taylor should receive full credit. Although he published the concept in 1713, it was so badly written that it had little immediate impact: it took Euler's work 40 years later to make the series concept well known.

Also introduced in this unit is Maclaurin's series, a special case of Taylor's series which was used by Taylor (as was acknowledged by Maclaurin in 1742). Colin Maclaurin (1698–1746), a Scotsman, was the most outstanding British mathematician in the generation following Newton. He was a prodigy, matriculating at the University of Glasgow at the age of eleven. He received his M.A. degree at 15. By age 19 he was a college mathematics teacher, and at 21 he published his first mathemati-

Colin Maclaurin

cal work of importance.

In 1719 Maclaurin met Isaac Newton, and he quickly became a disciple of Newtonian calculus. When Bishop George Berkeley wrote a tract attacking Newton's fluxions, Maclaurin responded in 1742 with his *Treatise on Fluxions,* the first complete and systematic presentation of Newton's calculus. Although it was not a textbook and was still not the final answer to rigor in calculus, it stood as a standard for nearly a century. Maclaurin was not comfortable with the limit concept, and based calculus on geometry instead. An unfortunate consequence of this emphasis was that his writing failed to make clear the useful applications of the subject. Further, the English mathematicians continued to use Newton's inadequate symbolism instead of Leibniz's superior symbolism, and during the 18th century British mathematics lost its preeminence.

Among Maclaurin's mathematical discoveries was what is now known as Cramer's rule for evaluating determinants. There is

Joseph Fourier

some poetic justice to this, for Maclaurin didn't discover Maclaurin's series just as Cramer didn't discover Cramer's rule. (In fairness, it should be noted that Cramer's notation was better than Maclaurin's.)

Joseph Fourier (1768–1830) was the only mathematician ever to serve as Governor of Lower Egypt. He had supported the French Revolution, and was rewarded with an appointment to the École Polytechnique. However, he had always wanted to be an army officer, a career denied him because he was the son of a tailor. When the opportunity came to accompany Napoleon on a military campaign in Egypt, Fourier resigned his teaching position and went along, and was appointed Governor in 1798. When the British took Egypt in 1801, Fourier returned to France.

Fourier studied the flow of heat in metallic plates and rods. The theory that he developed now has applications in industry and in the study of the temperature of the earth's interior. He discovered that many functions could be expressed as infinite sums of sine and cosine terms, now called a trigonometric series, or Fourier series. A paper that he submitted to the Academy of Science in Paris in 1807 was studied by several eminent mathematicians and rejected because he failed to prove his claims. They suggested that he reconsider and refine his paper, and even made heat flow the topic for a prize to be awarded in 1812. Fourier won the prize, but the Academy still declined to publish his paper because of its lack of rigor. (When Fourier became the secretary of the Academy, in 1824, the 1812 paper was published without change.)

As Fourier grew older, he developed at least one peculiar notion. Whether influenced by his stay in the heat of Egypt or by his own studies of the flow of heat in metals, he became obsessed with the idea that extreme heat was the natural condition for the human body. He was always heavily bundled in woolen clothing, and kept his rooms at high temperatures. He died in his sixty-third year, as expressed by Howard Eves in *An Introduction to the History of Mathematics,* "thoroughly cooked."

Most creative mathematicians have shown their genius at an early age. A notable exception was Karl Weierstrass (1816–1897), who did not really become a mathematician until he was 40 years old. He had studied law at the University of Bonn at his father's insistence, but he spent much of his time fencing and drinking. He did not complete his studies, and did not get his degree even after four years. Instead he turned to mathematics, but did not complete his degree in that subject, either. Eventually Weierstrass became a *gymnasium* (high school) teacher in a variety of subjects including not only mathematics

Karl Weierstrass

and physics, but also German, history, botany, geography, gymnastics, and calligraphy. The mathematician in him was struggling to break through, and he did some mathematical research, although he had no contact with other mathematicians. He did manage to publish a few minor papers that attracted some attention. After thirteen years of secondary level teaching, Weierstrass obtained a position at a technical school, and was then called to the University of Berlin. Here his mathematical and pedagogical abilities surfaced, and he is considered the greatest teacher of higher mathematics of the nineteenth century, as judged by the number of his students who became significant researchers. Judged on his own mathematical creativity, he is sometimes called the leading analyst of his time and the father of modern analysis.

Weierstrass worked in several areas, but his principal contribution was the study of functions of complex numbers through power series, which will be mentioned in this unit. This was an extension of the work of such earlier mathematicians as Maclaurin, Taylor, and Euler, but with a new rigor.

With all this creativity, Weierstrass did not publish many papers, and much of what he accomplished is known only through notes taken in class by his students. Indeed, he was uninterested in publicity or fame, and seemingly did not mind that some of his students used material from his lectures as their own. Hence, the record of the mathematical achievements of Karl Weierstrass has never been completely sorted out.

(Photographs from the David Eugene Smith Collection, Rare Book and Manuscript Library, Columbia University.)

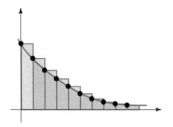

Chapter 12

The Theory of Infinite Series

In the preceding eleven chapters we have seen several instances involving a *sequence* of numbers, one corresponding to each of the positive integers $n = 1, 2, 3, \ldots$. For example,

(1) The definite integral $\int_a^b f(x)\,dx$ has been interpreted as the limit of a sequence S_n of Riemann Sums

$$S_n = \sum_{j=1}^{n} f(t_j)\,\Delta x_j, \qquad n = 1, 2, 3, \ldots.$$

(2) The amount $P(n)$ on deposit in a savings account paying interest annually at the rate r (in decimal form) n years after an initial deposit of P_0 dollars is

$$P(n) = (1 + r)^n P_0, \qquad n = 1, 2, 3, \ldots.$$

(3) Approximations to a zero of a differentiable function f are given by the formula for Newton's Method:

$$x_{n+1} = x_n - \frac{f(x_n)}{f'(x_n)}, \qquad n = 1, 2, 3, \ldots$$

where x_1 is the initial approximation.

One purpose of this chapter is to develop a formal theory for dealing with the limit as $n \to \infty$ of an *infinite sequence*. Given this concept, we are then able to define the sum of an *infinite series*—the sum of an infinite collection of numbers.

In Chapter 13, given the theory of infinite sequences and series developed in this chapter, we are able to discuss two topics of great practical significance: the approximation of differentiable functions by polynomials and the representation of functions by infinite series.

12.1 Infinite Sequences

Up to this point we have informally defined an infinite sequence to be an unending string of numbers of the form

$$a_1, a_2, a_3, \ldots, a_n, \ldots. \tag{1}$$

It is important to note that expression (1) indicates an *order* in which these numbers appear in the string (the subscripts) as well as the numbers themselves (the a_n's). The precise notion of an infinite sequence is given in Definition 1.

Definition 1

An infinite sequence is a function whose domain is the positive integers.

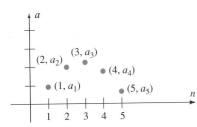

Figure 1.1   Plotting the sequence $\{a_1, a_2, a_3, \ldots\}$ where $a_n = f(n)$.

For example, the infinite sequence $\{1^2, 2^2, 3^2, 4^2, 5^2, \ldots, n^2, \ldots\}$ can be viewed as the set of values of the function $f(n) = n^2$, where $f(1) = 1^2$ is the first term, $f(2) = 2^2$ is the second term, and so on. Using the function concept, we can graph sequences on a coordinate plane. Figure 1.1 shows the graph of an arbitrary sequence $\{a_n\}$.

Usually, a sequence is specified by a rule of the form $a_n = f(n)$ that determines the nth term of the sequence for each integer n, just as functions are usually specified by an equation of the form $y = f(n)$. For example, the rule $a_n = 2^n$ determines the sequence

$$\{2^n\} = \{2^1, 2^2, 2^3, 2^4, 2^5, \ldots\}$$
$$= \{2, 4, 8, 16, 32, \ldots\},$$

while the rule $a_n = (-1)^n$ determines the sequence

$$\{(-1)^n\} = \{-1, 1, -1, 1, -1, 1, -1, \ldots\}.$$

In such cases the term $a_n = f(n)$ is referred to as the **general term** of the sequence. Note that we use braces $\{\ \}$ to denote the entire sequence.

□ **EXAMPLE 1**

Write out the first few terms of the sequences whose general terms are

(a) $a_n = 2n + 1$,

(b) $b_n = 2 + \dfrac{(-1)^n}{n}$.

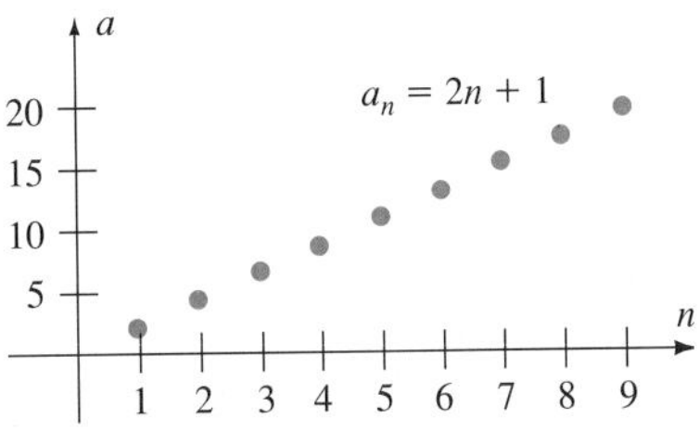

Solution   In part (a) we have

$$a_1 = 2(1) + 1 = 3, \qquad a_2 = 2(2) + 1 = 5, \qquad a_3 = 2(3) + 1 = 7,$$

and so on, so

$$\{2n + 1\} = \{3, 5, 7, 9, 11, 13, 15, \ldots\},$$

while in (b) we have

$$b_1 = 2 + \frac{(-1)}{1} = 1, \qquad b_2 = 2 + \frac{(-1)^2}{2} = \frac{5}{2}, \qquad b_3 = 2 + \frac{(-1)^3}{3} = \frac{5}{3},$$

and so on, so

$$\left\{2 + \frac{(-1)^n}{n}\right\} = \left\{1, \frac{5}{2}, \frac{5}{3}, \frac{9}{4}, \frac{9}{5}, \frac{13}{6}, \frac{13}{7}, \frac{17}{8}, \frac{17}{9}, \ldots\right\}.$$

Graphs of these two sequences appear in Figures 1.2 and 1.3.   ∎

Figure 1.2   Graph of the sequence $\{a_n\} = \{2n + 1\}$.

There is an important difference between the two sequences in Example 1. The terms of the sequence $\{a_n\}$ increase uniformly, not approaching any particular number. In fact, in the language of Chapter 4, we would say that

$$\lim_{n \to \infty} \{2n + 1\} = +\infty$$

since the terms of the sequence increase without bound as $n \to \infty$.

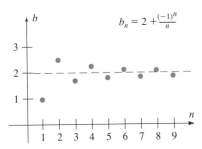

Figure 1.3   Graph of the sequence $\{b_n\} = \left\{2 + \dfrac{(-1)^n}{n}\right\}$.

However, the terms of the sequence $\{b_n\}$ ''approach'' the number $L = 2$ as $n \to \infty$, which we write as

$$2 = \lim_{n \to \infty} \left\{ 2 + \frac{(-1)^n}{n} \right\}.$$

This notion of the *limit* of a sequence is essentially the same as the notion of

$$\lim_{x \to \infty} f(x) = L$$

developed in Chapter 4 for more general functions. That is,

> $L = \lim_{n \to \infty} a_n$ means that the numbers a_n approach
>
> the number L as n increases without bound.

Using this working definition, together with some simple algebra, we can evaluate many types of limits by the same techniques used in Chapter 4 to evaluate $\lim_{x \to \infty} f(x)$.

☐ EXAMPLE 2

Find $\lim_{n \to \infty} \dfrac{6n^3 + 5n^2 + 7}{4n^3 - 2n + 2}$.

Strategy · · · · · · · · **Solution**

Divide all terms by n^3 (the highest power of n in the denominator).

Use fact that, if $k > 0$,

$$\frac{c}{n^k} \to 0$$

as $n \to \infty$.

$$\lim_{n \to \infty} \frac{6n^3 + 5n^2 + 7}{4n^3 - 2n + 2} = \lim_{n \to \infty} \frac{6 + \dfrac{5}{n} + \dfrac{7}{n^3}}{4 - \dfrac{2}{n^2} + \dfrac{2}{n^3}}$$

$$= \frac{6 + 0 + 0}{4 - 0 + 0} = \frac{3}{2}.$$ ∎

☐ EXAMPLE 3

Find $\lim_{n \to \infty} \left[\ln(n + 4) - \dfrac{1}{2} \ln(n) \right]$.

Strategy · · · · · · · · **Solution**

Use properties of $\ln x$ to reduce expression to the logarithm of a single number.

$$\lim_{n \to \infty} \left[\ln(n + 4) - \tfrac{1}{2} \ln(n) \right] = \lim_{n \to \infty} \left[\ln(n + 4) - \ln(n^{1/2}) \right]$$

$$= \lim_{n \to \infty} \ln \left(\frac{n + 4}{\sqrt{n}} \right)$$

Divide both terms in numerator by $\sqrt{n}$.

Use fact that $\ln x \to \infty$ as $x \to \infty$.

$$= \lim_{n \to \infty} \ln \left(\sqrt{n} + \frac{4}{\sqrt{n}} \right)$$

$$= \infty,$$

since $\sqrt{n} \to \infty$ and $\dfrac{4}{\sqrt{n}} \to 0$ as $n \to \infty$.   ∎

□ **EXAMPLE 4**

Find $\displaystyle\lim_{n\to\infty}(-1)^n\left(\frac{n+1}{n}\right)$.

Strategy · · · · · · · · ·
Because of the factor $(-1)^n$, examine even and odd terms separately.

Solution

For even integers n, $(-1)^n = 1$, so we find that

$$\lim_{n\to\infty}(-1)^n\left(\frac{n+1}{n}\right) \qquad \text{(even integers only)}$$

Divide through by n.

$$= \lim_{n\to\infty}(1)\left(1+\frac{1}{n}\right) = 1.$$

However, for odd integers n, $(-1)^n = -1$, so

$$\lim_{n\to\infty}(-1)^n\left(\frac{n+1}{n}\right) \qquad \text{(odd integers only)}$$

Divide through by n.

$$= \lim_{n\to\infty}(-1)\left(1+\frac{1}{n}\right) = -1.$$

These two distinct limits show that the terms do not approach a *single* number as $n \to \infty$. Consequently, the original limit does not exist (see Figure 1.4). ■

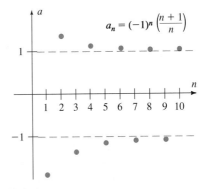

Figure 1.4 Terms of sequence

$$(-1)^n\left(\frac{n+1}{n}\right)$$

approach both 1 and -1 as $n \to \infty$. The limit does not exist.

If the limit as $n \to \infty$ of $\{a_n\}$ exists, we say that the sequence $\{a_n\}$ **converges.** Otherwise the sequence is said to **diverge.** Note from Examples 3 and 4 that a sequence may diverge either because a_n becomes infinite as $n \to \infty$ or because a_n, remaining bounded, fails to approach a *single* number as $n \to \infty$.

We next state a formal definition of the limit of a sequence. Note the similarity to the formal definition for $L = \lim_{x\to\infty} f(x)$ given in Chapter 4.

Definition 2
Formal Definition of Limit

We say that

$$L = \lim_{n \to \infty} a_n$$

if and only if, for each number $\epsilon > 0$, there exists an integer N so that

if $n > N$, then $|a_n - L| < \epsilon$.

Definition 2 says this: If L is the limit of a_n as $n \to \infty$, we will find all terms of the sequence $\{a_n\}$, beyond the Nth term, lying within ϵ units of the number L. Since the integer N, in general, depends upon the number ϵ, we expect to have to look further along the sequence to observe this "closeness" as ϵ decreases in size. Figures 1.5 through 1.7 illustrate choices of N corresponding to three different values of ϵ for a typical sequence $\{a_n\}$.

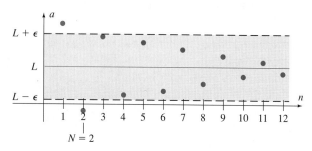

Figure 1.5 Large ϵ.

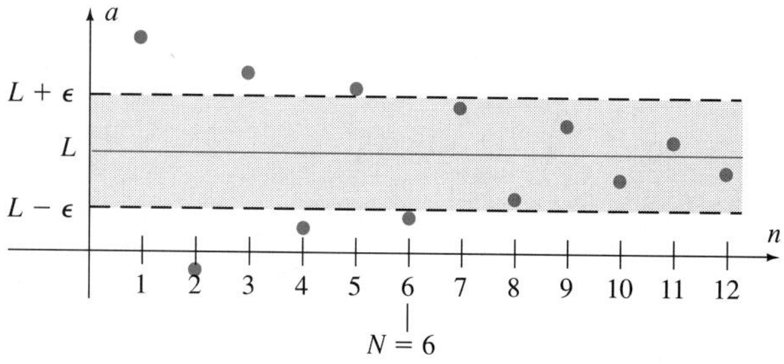

Figure 1.6 Medium ϵ.

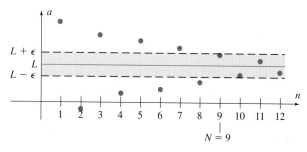

Figure 1.7 Small ϵ.

Definition 2 allows us to prove rigorously statements of the form

$$L = \lim_{n \to \infty} a_n$$

once we have found L. It does not, however, tell us how L is determined from the general term for the sequence $\{a_n\}$. For this, the intuitive notion of limit and familiarity with examples such as Examples 2–4 above are essential.

The following examples illustrate how Definition 2 is used.

□ **EXAMPLE 5**

Prove that $\lim\limits_{n \to \infty} \dfrac{1}{n} = 0$.

Strategy · · · · · · · ·
Set up the inequality

$$|a_n - L| < \epsilon$$

and solve for n to find a relationship between n and ϵ.

Solution
According to Definition 2, we allow $\epsilon > 0$ to be any fixed positive number. Since $a_n = 1/n$ and $L = 0$, we must determine how large to choose n to guarantee that

$$\left|\frac{1}{n} - 0\right| = \frac{1}{n} < \epsilon. \tag{2}$$

Solving inequality (2) for n we see that it is equivalent to the inequality

$$n > \frac{1}{\epsilon}. \tag{3}$$

Choose N large enough that the inequality holds whenever $n > N$.

We therefore take N to be any integer larger* than $1/\epsilon$. Then, inequality (3) holds whenever $n > N$, which guarantees that inequality (2) holds. In other words, if $N > 1/\epsilon$ then

$$\left|\frac{1}{n} - 0\right| < \epsilon \quad \text{whenever} \quad n > N,$$

as required by Definition 2. ■

□ **EXAMPLE 6**

Prove that $\lim\limits_{n \to \infty} \dfrac{2n - 1}{n + 2} = 2$.

Strategy · · · · · · · ·
Set up the inequality

$$|a_n - L| < \epsilon.$$

Solution
We assume $\epsilon > 0$ to be an arbitrary fixed number. Since

$$a_n = \frac{2n - 1}{n + 2}$$

and $L = 2$, we must determine how large to choose n to guarantee that

$$\left|\frac{2n - 1}{n + 2} - 2\right| < \epsilon. \tag{4}$$

Solve this inequality for n.

Inequality (4) is equivalent to the inequality

$$\left|\frac{2n - 1 - 2(n + 2)}{n + 2}\right| < \epsilon,$$

$$\frac{5}{n + 2} < \epsilon$$

or

$$5 < \epsilon(n + 2)$$

$$\frac{5}{\epsilon} < n + 2$$

$$\frac{5}{n + 2} < \epsilon. \tag{5}$$

Solving inequality (5) for n we find that it is equivalent to the inequality

$$n > \frac{5}{\epsilon} - 2.$$

$$n > \frac{5}{\epsilon} - 2. \tag{6}$$

---

*Since ϵ is a positive number, so is $1/\epsilon$. It is a property of the real numbers that, given any positive number a, an integer N can be found with $N > a$.

Choose n sufficiently large that the desired inequality holds.

We therefore take N to be any integer larger than $(5/\epsilon) - 2$. Then, inequality (6) holds whenever $n > N$. Since inequality (6) is equivalent to inequality (4), we may conclude that, for $N > (5/\epsilon) - 2$,

$$\text{if} \quad n > N \quad \text{then} \quad \left| \frac{2n - 1}{n + 2} - 2 \right| < \epsilon,$$

as required by Definition 2.   ■

Definition 2 can be used to prove several theorems that give rules for calculating limits of sequences. Since the proofs of these theorems are similar to the proofs given in Chapters 2 and 4 for the corresponding theorems on limits of functions, we leave them as exercises.

Theorem 1

Properties of Limits of Sequences

If $\lim\limits_{n \to \infty} a_n = L$, $\lim\limits_{n \to \infty} b_n = M$ and c is any real number, then

(i) $\lim\limits_{n \to \infty} (a_n + b_n) = L + M$,

(ii) $\lim\limits_{n \to \infty} (ca_n) = cL$,

(iii) $\lim\limits_{n \to \infty} (a_n b_n) = LM$,

(iv) $\lim\limits_{n \to \infty} \left(\dfrac{a_n}{b_n} \right) = \dfrac{L}{M}$, if $b_n \neq 0$ and $M \neq 0$.

Theorem 1 together with the proof in Example 5 makes legitimate the calculations in Example 2. The next theorem addresses a situation that occurred in Example 3.

Theorem 2

Suppose that

$$\lim_{n \to \infty} a_n = L$$

and each number a_n lies in the domain of the function f. If f is continuous at $x = L$, then

$$\lim_{n \to \infty} f(a_n) = f(L).$$

In other words,

$$\lim_{n \to \infty} f(a_n) = f\left(\lim_{n \to \infty} a_n \right).$$

□ **EXAMPLE 7**

Since $f(x) = \tan x$ is continuous for $-\pi/2 < x < \pi/2$,

$$\lim_{n \to \infty} \tan\left(\frac{\pi n^2 + 1}{3 - 4n^2} \right) = \tan\left[\lim_{n \to \infty} \left(\frac{\pi n^2 + 1}{3 - 4n^2} \right) \right] = \tan\left(-\frac{\pi}{4} \right) = -1.   ■$$

□ **EXAMPLE 8**

Since $f(x) = \sqrt{x}$ is continuous for $x \geq 0$,

$$\lim_{n \to \infty} \sqrt{\frac{4n + 1}{n}} = \sqrt{\lim_{n \to \infty} \frac{4n + 1}{n}} = \sqrt{4} = 2.$$ ■

The following theorem is the analogue for sequences of the Pinching Theorem for functions.

Theorem 3
Pinching Theorem

Let $\{a_n\}$, $\{b_n\}$, and $\{c_n\}$ be sequences. Suppose that

$$a_n \leq b_n \leq c_n$$

for all n greater than some positive integer P. If

$$\lim_{n \to \infty} a_n = L = \lim_{n \to \infty} c_n,$$

then $\lim_{n \to \infty} b_n = L$ also.

□ **EXAMPLE 9**

Show that $\lim_{n \to \infty} \dfrac{1}{n^p} = 0$ if $p \geq 1$.

Solution If $p \geq 1$, $n^p \geq n$ for all $n = 1, 2, 3, \ldots$. Thus

$$0 \leq \frac{1}{n^p} \leq \frac{1}{n}, \qquad n \geq 1.$$

Since $\lim_{n \to \infty} \{0\} = 0 = \lim_{n \to \infty} \left\{ \dfrac{1}{n} \right\}$, the conclusion follows by the Pinching Theorem. ■

□ **EXAMPLE 10**

Find $\lim_{n \to \infty} \dfrac{\sin n}{n}$.

Strategy · · · · · · · ·
Find bounds on $\sin n$.

Solution
We have $|\sin n| \leq 1$ for all n, that is,

$$-1 \leq \sin n \leq 1, \qquad n \geq 1.$$

Thus

Divide by n to find bounds on $\dfrac{\sin n}{n}$.

$$-\frac{1}{n} \leq \frac{\sin n}{n} \leq \frac{1}{n}.$$

Apply Pinching Theorem.

Since $\lim_{n \to \infty} \left(-\dfrac{1}{n} \right) = 0 = \lim_{n \to \infty} \left(\dfrac{1}{n} \right)$,

$$\lim_{n \to \infty} \frac{\sin n}{n} = 0$$

by the Pinching Theorem. ■

We summarize the ideas of this section by noting that finding the limit of sequence $\{a_n\}$ is similar to finding horizontal asymptotes for the function f with $f(n) = a_n$, $n = 1, 2, \ldots$. The principal difference is that $\{a_n\}$ is a function defined only for positive integers. Thus

$$L = \lim_{n \to \infty} a_n \quad \text{if} \quad L = \lim_{x \to \infty} f(x),$$

but the converse need not be true (see Exercise 40).

Exercise Set 12.1

In Exercises 1–33, write out the first four terms of the given sequence and determine whether the sequence converges or diverges. If the sequence converges, find its limit.

1. $\left\{ \dfrac{n}{2n + 1} \right\}$

2. $\left\{ \dfrac{2n - 1}{n + 3} \right\}$

3. $\left\{ \dfrac{n - 4}{n^2 + 2} \right\}$

4. $\left\{ \dfrac{n^2 + 1}{3n(n + 2)} \right\}$

5. $\{e^{1/n}\}$

6. $\left\{ \dfrac{1}{e^n} \right\}$

7. $\{\sqrt{5}\}$

8. $\left\{ \dfrac{(n - 1)(n + 1)}{2n^2 + 2n + 2} \right\}$

9. $\left\{ \sin \dfrac{n\pi}{2} \right\}$

10. $\left\{ \cos \dfrac{n\pi}{4} \right\}$

11. $\left\{ \dfrac{20n}{1 + \sqrt{n}} \right\}$

12. $\left\{ \dfrac{6 - n^{3/2}}{(\sqrt{n} + 1)^2} \right\}$

13. $\left\{ \dfrac{3 + (-1)^n \sqrt{n}}{n + 2} \right\}$

14. $\{(-1)^n \sin(\pi n)\}$

15. $\left\{ \sqrt{1 + \dfrac{1}{n}} \right\}$

16. $\left\{ 1 + \dfrac{(-1)^n}{2^n} \right\}$

17. $\left\{ \cos\left(\dfrac{n - 1}{n^2} \right) \right\}$

18. $\left\{ \dfrac{n + 1}{n} \right\}$

19. $\left\{ \dfrac{n^{3/2} + 2}{2n^{3/2}} \right\}$

20. $\left\{ \dfrac{e^n - e^{-n}}{e^n + e^{-n}} \right\}$

21. $\left\{ \dfrac{1}{n} - \dfrac{1}{n + 1} \right\}$

22. $\left\{ \dfrac{2^n}{5^{n+2}} \right\}$

23. $\{\sqrt{n + 1} - \sqrt{n}\}$

24. $\left\{ \dfrac{\cos^2 \pi n}{n} \right\}$

25. $\left\{ \dfrac{\sqrt{2n^2 + 1}}{n} \right\}$

26. $\left\{ \operatorname{Tan}^{-1}\left(\dfrac{n + 2}{2} \right) \right\}$

27. $\left\{ \operatorname{Tan}^{-1} \dfrac{n + 1}{n} \right\}$

28. $\left\{ n \sin \dfrac{\pi}{2n} \right\}$

29. $\left\{ \dfrac{\cos n\pi}{n} \right\}$

30. $\left\{ \ln \dfrac{n^2 + 1}{(n + 2)(n + 3)} \right\}$

31. $\left\{ \ln \dfrac{2n + 1}{2} \right\}$

32. $\left\{ \left(1 + \dfrac{1}{n} \right)^n \right\}$

33. $\left\{ \left(1 - \dfrac{1}{n} \right)^n \right\}$

In Exercises 34–38, use the definition of limit to prove that $\lim\limits_{n \to \infty} a_n = L$.

34. $a_n = \dfrac{3}{n}$; $L = 0$

35. $a_n = \dfrac{1}{2n + 1}$; $L = 0$ **36.** $a_n = \dfrac{n}{3n + 1}$; $L = \dfrac{1}{3}$

37. $a_n = \dfrac{3n - 1}{n + 1}$; $L = 3$ **38.** $a_n = \dfrac{n^2 + 2n + 3}{1 + n^2}$; $L = 1$

39. Use Theorem 2 to show that $\lim\limits_{n \to \infty} \sqrt[n]{a} = \lim\limits_{n \to \infty} e^{\ln a^{1/n}} = 1$ if $a > 0$.

40. Let $a_n = \sin \pi n$, $n = 1, 2, \ldots$, and let $f(x) = \sin \pi x$. Then $f(n) = a_n$. Show that $\lim\limits_{n \to \infty} a_n = 0$ but that $\lim\limits_{x \to \infty} f(x)$ does not exist.

41. Prove that the limit of a sequence is unique, if it exists. (*Hint:* Assume that both

$$\lim_{n \to \infty} a_n = L \quad \text{and} \quad \lim_{n \to \infty} a_n = M.$$

Then show $L = M$ by examining the inequality

$$|L - M| \le |L - a_n| + |a_n - M|.)$$

42. A sequence is called **bounded** if there is a number M so that $|a_n| \le M$ for all terms a_n of the sequence. Prove that a convergent sequence must be bounded.

43. Give an example showing that a bounded sequence need *not* converge.

44. Prove that the sum of two bounded sequences is again bounded. What about the product of two bounded sequences? The quotient?

45. Prove that $\lim\limits_{n \to \infty} a_n = L$ if and only if $\lim\limits_{n \to \infty} |a_n - L| = 0$.

46. Prove that $\lim\limits_{n\to\infty} \dfrac{1}{a^n} = 0$ if $a > 1$, using the Pinching Theorem.

47. Prove part (i) of Theorem 1.

48. Prove part (ii) of Theorem 1.

49. Prove part (iii) of Theorem 1.

50. Prove part (iv) of Theorem 1.

51. Prove Theorem 2.

52. Prove Theorem 3.

12.2   More on Infinite Sequences

In this section, we derive several limits that we shall use frequently. You should study the derivations carefully because they illustrate a number of techniques that are typically used in the calculation of limits of sequences.

Example 1 illustrates how we can use the techniques for calculating limits at infinity for functions in order to calculate limits of sequences. More precisely, suppose that $\{a_n\}$ is a sequence and $f(x)$ is a function for which $f(n) = a_n$. Then

$$\lim_{n\to\infty} a_n = L \quad \text{if} \quad \lim_{x\to\infty} f(x) = L$$

(see Figure 2.1). Consequently, we can often translate the problem of finding the limit of a sequence into the problem of finding the limit of a function. This reformulation provides the opportunity to use techniques, such as l'Hôpital's Rule, that we have already developed.

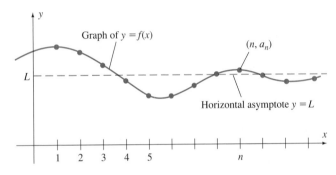

Figure 2.1   If $f(n) = a_n$ and $\lim\limits_{x\to\infty} f(x) = L$, then $\lim\limits_{n\to\infty} a_n = L$ also.

☐ **EXAMPLE 1**

Find $\lim\limits_{n\to\infty} \dfrac{n}{e^n}$.

Solution   The limit is not obvious because both numerator and denominator become infinite as $n \to \infty$. Using the above observation together with l'Hôpital's Rule gives

$$\lim_{n\to\infty} \frac{n}{e^n} = \lim_{x\to\infty} \frac{x}{e^x} \quad \left(\frac{\infty}{\infty} \text{ form; apply l'Hôpital's Rule}\right)$$

$$= \lim_{x\to\infty} \frac{1}{e^x} = 0. \qquad\blacksquare$$

Roughly speaking, the limit in Example 1 is one way of saying that the sequence $\{e^n\}$ goes to infinity "faster" than the sequence $\{n\}$.

Our next sequence is the result of raising a (fixed) number x to positive integral powers n; that is, $\{a_n\} = x^n$.

$$\lim_{n \to \infty} x^n = 0 \quad \text{if} \quad |x| < 1. \tag{1}$$

Proof: Let ϵ be an arbitrary positive number. Then, since

$$\lim_{n \to \infty} \frac{1}{n} = 0,$$

$\epsilon^{1/n} \to \epsilon^0 = 1$ as $n \to \infty$. Thus, given $|x| < 1$, there exists an integer N, by Definition 2, such that $\epsilon^{1/n} > |x|$ whenever $n > N$. Thus

$$|x|^n < (\epsilon^{1/n})^n = \epsilon, \qquad n > N.$$

Equivalently, noting that $|x|^n = |x^n|$, we have

$$|x^n - 0| < \epsilon \quad \text{whenever} \quad n > N.$$

Thus, $\lim\limits_{n \to \infty} x^n = 0$ by Definition 2.   ∎

In Chapter 8, we saw that exponentials such as 10^n rapidly increase without bound as $n \to \infty$. Likewise, $n! \to \infty$ rapidly as $n \to \infty$. The following limit justifies the assertion that, as $n \to \infty$, $n! \to \infty$ more rapidly than any exponential.

$$\lim_{n \to \infty} \frac{x^n}{n!} = 0, \qquad -\infty < x < \infty. \tag{2}$$

Proof: To establish this limit we shall show that

$$\lim_{n \to \infty} \left| \frac{x^n}{n!} \right| = 0, \qquad -\infty < x < \infty. \tag{3}$$

(See Exercise 35.)

Let x be given and let N be an integer such that $N > |x|$. Write

$$J = \left| \left(\frac{x}{1} \right) \left(\frac{x}{2} \right) \left(\frac{x}{3} \right) \cdots \left(\frac{x}{N-1} \right) \right|$$

and note that J is constant since N and x are fixed. Then for $n > N$ we have

$$\left| \frac{x^n}{n!} \right| = \left| \left(\frac{x}{1} \right) \left(\frac{x}{2} \right) \left(\frac{x}{3} \right) \cdots \left(\frac{x}{N-1} \right) \left(\frac{x}{N} \right) \cdots \left(\frac{x}{n} \right) \right|$$

$$= J \left| \left(\frac{x}{N} \right) \left(\frac{x}{N+1} \right) \cdots \left(\frac{x}{n} \right) \right| \qquad (n - N + 1) \text{ factors}$$

$$\leq J \left| \frac{x}{N} \right|^{(n-N+1)}$$

Since $|x| < N$, we have $\left|\dfrac{x}{N}\right| < 1$, so $\displaystyle\lim_{n\to\infty}\left|\dfrac{x}{N}\right|^n = 0$. Thus,

$$\lim_{n\to\infty} J\left|\frac{x}{N}\right|^{(n-N+1)} = J\left|\frac{x}{N}\right|^{(-N+1)} \cdot \lim_{n\to\infty}\left|\frac{x}{N}\right|^n = 0.$$

This establishes (3) by the preceding inequality. ∎

Also in Chapter 8, we learned that $\ln(n) \to \infty$ as $n \to \infty$. Our next limit justifies the assertion that $\ln(n) \to \infty$ "slower" than $n \to \infty$.

$$\lim_{n\to\infty} \frac{\ln(n)}{n} = 0. \tag{4}$$

Proof: $\displaystyle\lim_{n\to\infty} \frac{\ln(n)}{n} = \lim_{x\to\infty} \frac{\ln(x)}{x} = \lim_{x\to\infty} \frac{1/x}{1} = 0$, by l'Hôpital's Rule. ∎

Using logarithms and limit (4), we obtain the following nonzero limit.

$$\lim_{n\to\infty} \sqrt[n]{n} = 1. \tag{5}$$

Proof: $\displaystyle\lim_{n\to\infty} \ln(n^{1/n}) = \lim_{n\to\infty} \frac{\ln(n)}{n} = 0$, by (4). Thus,

$$\lim_{n\to\infty} n^{1/n} = e^0 = 1. \qquad ∎$$

□ **EXAMPLE 2**

$$\lim_{n\to\infty} \left(\frac{5}{n}\right)^{1/n} = \lim_{n\to\infty} \frac{\sqrt[n]{5}}{\sqrt[n]{n}} = \frac{\displaystyle\lim_{n\to\infty} \sqrt[n]{5}}{\displaystyle\lim_{n\to\infty} \sqrt[n]{n}} = \frac{1}{1} = 1$$

because $1 < \sqrt[n]{5} < \sqrt[n]{n}$ if $n > 5$ and, therefore, $\displaystyle\lim_{n\to\infty} \sqrt[n]{5} = 1$ by limit (5) and the Pinching Theorem. ∎

We shall need one more theorem on sequences, which we develop here.

A set S of numbers is said to be *bounded above* if there exists a number M so that $x \le M$ for every number $x \in S$. This number M is called an *upper bound* for S. A fundamental property of the real number system states that among all such upper bounds M there can always be found a smallest bound. This property is referred to as the **completeness axiom,** and it is formally stated as follows.

Completeness Axiom for Real Numbers: If S is a nonempty set of real numbers that is bounded above, then there exists a least upper bound L for S. That is, there exists a number L for which

(i) $x \le L$ for every $x \in S$, and
(ii) if M is any upper bound for S, then $L \le M$.

For example, the set $S = \{x \mid -3 \le x \le 3\}$ is bounded above since $x \le 3$ for any $x \in S$. The number 7 is an upper bound for S since $x < 7$ for all $x \in S$. The numbers π, 37, and 1001 are also upper bounds for S. The number $L = 3$ is the *least* upper bound for S, because any *other* upper bound for S is greater than 3.

The completeness axiom and its role in the definition of the real number system are topics for more advanced courses. We shall use this axiom to establish a theorem about *increasing* sequences. As for functions, we define the sequence $\{a_n\}$ to be **increasing** if $a_n > a_m$ whenever $n > m$. For sequences, this condition is equivalent to the condition $a_{n+1} > a_n$ for all n.

Theorem 4	Every increasing sequence that is bounded above converges.

Proof: We use the definition of the limit of a sequence and the completeness axiom. Let $\epsilon > 0$ be given, and let $\{a_n\}$ denote an increasing sequence that is bounded above. Then the set of numbers $\{a_1, a_2, a_3, \ldots\}$ has a least upper bound L according to the completeness axiom. We prove that $\lim\limits_{n \to \infty} a_n = L$.

Since $\epsilon > 0$, we have $L - \epsilon < L$, so $L - \epsilon$ cannot be an upper bound for the sequence. Thus, there must exist an integer N such that $a_N > L - \epsilon$ (see Figure 2.2).

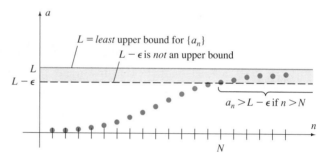

Figure 2.2   An increasing sequence that is bounded above converges to its least upper bound.

But since $\{a_n\}$ is increasing we must have $a_n > a_N$ for all $n > N$. Thus we have the following five numbers in increasing order:

$$L - \epsilon < a_N < a_n \le L < L + \epsilon, \qquad n > N.$$

It now follows from $L - \epsilon < a_n < L + \epsilon$ that

$$|a_n - L| < \epsilon \quad \text{whenever} \quad n > N.$$

Thus, $L = \lim\limits_{n \to \infty} a_n$ according to Definition 2. ∎

There is an obvious extension of Theorem 4 to *decreasing* sequences that are bounded below. If $\{a_n\}$ is such a sequence, then the sequence $\{-a_n\}$ is an *increasing* sequence that is bounded above and which, by Theorem 4, has a limit L. By Theorem 1 we then have

$$\lim_{n \to \infty} a_n = -\lim_{n \to \infty} \{-a_n\} = -L.$$

We summarize these remarks as follows.

Corollary 1   Every decreasing sequence that is bounded below converges.

Recursively Defined Sequences   In many applications, particularly in computer science, sequences are defined *recursively*. One of the most famous recursively defined sequences is the sequence of **Fibonacci numbers**

$$1, 1, 2, 3, 5, 8, 13, 21, 34, \ldots . \tag{6}$$

This sequence was first discovered by the Italian mathematician Leonardo of Pisa (who also went by the name Fibonacci) around the year A.D. 1200 (Fibonacci is regarded by many as the most brilliant of the pre-Renaissance mathematicians.)

The Fibonacci sequence $\{F_n\}$ in (6) is determined by the rules

$$F_1 = 1, \tag{7}$$

$$F_2 = 1, \quad \text{and} \tag{8}$$

$$F_{n+2} = F_n + F_{n+1}, \qquad n = 1, 2, 3, 4, \ldots . \tag{9}$$

That is, every term in the sequence beyond the second is found by adding the two preceding terms. This is what we mean by saying that the sequence $\{a_n\}$ is **recursively defined**: the term a_n is a function of one or more preceding terms, such as a_{n-1}, a_{n-2}, etc. But a_n is *not* written as an explicit function of n.

It is often possible to rewrite a recursively defined sequence as an explicit function of n, and vice versa. For example, the nth term of the Fibonacci sequence can be written

$$F_n = \frac{1}{\sqrt{5}} \left(\frac{1 + \sqrt{5}}{2} \right)^n - \frac{1}{\sqrt{5}} \left(\frac{1 - \sqrt{5}}{2} \right)^n .$$

(See Exercise 28 for a biological interpretation of the Fibonacci sequence.) Another example is the factorial sequence

$$\{a_n\} = \{n!\} = \{1!, 2!, 3!, \ldots\} = \{1, 2, 6, 24, \ldots\}.$$

It can be defined recursively as the sequence $\{f_n\} = \{f_1, f_2, f_3, \ldots\}$ where

$$f_1 = 1$$
$$f_n = nf_{n-1}, \qquad n = 2, 3, 4, \ldots .$$

However, not every sequence $\{a_n\}$ can be defined recursively.

Recursively defined sequences occur frequently in the analysis of computer algorithms. In such situations one is concerned only with how to determine the next term in the sequence given the present term (and, possibly, several preceding terms), not with the correspondence between integers n and the terms a_n. Exercises 28–32 in this section concern recursively defined sequences.

Exercise Set 12.2

In each of Exercises 1–22, find the indicated limit, if it exists.

1. $\lim\limits_{n\to\infty} n \sin\left(\dfrac{2}{n}\right)$

2. $\lim\limits_{n\to\infty} \dfrac{\sin^3 n}{n}$

3. $\lim\limits_{n\to\infty} \sqrt[n]{4n}$

4. $\lim\limits_{n\to\infty} \dfrac{\ln(n)}{\sqrt{n}}$

5. $\lim\limits_{n\to\infty} (n+1)e^{-n}$

6. $\lim\limits_{n\to\infty} \dfrac{n-1}{e^n}$

7. $\lim\limits_{n\to\infty} \dfrac{2^{4n+1}}{9^{n/2}}$

8. $\lim\limits_{n\to\infty} \dfrac{4^{(n+1)/2}}{6^{2n}}$

9. $\lim\limits_{n\to\infty} n^{3/n}$

10. $\lim\limits_{n\to\infty} (n+\pi)^{1/n}$

11. $\lim\limits_{n\to\infty} \left(1-\dfrac{3}{n}\right)^n$

12. $\lim\limits_{n\to\infty} \dfrac{3^n}{(n+3)!}$

13. $\lim\limits_{n\to\infty} \dfrac{7^{2n}}{n!}$

14. $\lim\limits_{n\to\infty} \left(\dfrac{e}{n}\ln\dfrac{e}{n}\right)$

15. $\lim\limits_{n\to\infty} \dfrac{n^2 \ln(n)}{2^n}$

16. $\lim\limits_{n\to\infty} n^{\sin(\pi/n)}$

17. $\lim\limits_{n\to\infty} \dfrac{\cos^2 n\pi}{4^{n+3}}$

18. $\lim\limits_{n\to\infty} \dfrac{\sin^4\left(n+\frac{1}{2}\right)\pi}{3^{1/n}}$

19. $\lim\limits_{n\to\infty} \left(\dfrac{n+3}{n}\right)^n$

20. $\lim\limits_{n\to\infty} \left(1+\dfrac{1}{n^2}\right)^n$

21. $\lim\limits_{n\to\infty} \dfrac{n^n}{n!}$

22. $\lim\limits_{n\to\infty} \dfrac{n-\sin n}{n+\cos n}$

In Exercises 23–26, use Theorem 4 and Corollary 1 to show that the given sequence converges.

23. $\left\{\dfrac{3}{n^2}\right\}$

24. $\left\{\dfrac{2n^2+1}{n^2}\right\}$

25. $\left\{\dfrac{e^{-n}}{3^n}\right\}$

26. $\{\text{Tan}^{-1} n\}$

27. Prove that $\lim\limits_{n\to\infty} x^n$ does not exist if $|x|>1$.

28. The Fibonacci sequence $\{1, 1, 2, 3, 5, 8, 13, 21, 34, \ldots\}$ arises as a mathematical model for the size of a population of rabbits under the following conditions. We assume that we begin with a single pair of rabbits, that this and each other pair of rabbits become fertile one month after birth, that each pair of fertile rabbits gives birth to one new pair of rabbits each month, and that no rabbits die. If F_n represents the number of pairs of rabbits in the population after n months, show that
  a. $F_1 = 1$,
  b. $F_2 = 1$, and

c. $F_{n+2} = 2F_n + (F_{n+1} - F_n)$. (*Hint:* The term $2F_n$ is explained as follows. Every pair of rabbits that was present two months ago is still present along with one pair of offspring. The second term accounts for the fact that those rabbits that were fertile two months ago have produced *two* pairs of offspring since then, one of which is not counted in the first term.)
  d. Conclude from part c that $F_{n+2} = F_n + F_{n+1}$.

29. Let $\{a_n\}$ be a sequence defined recursively by the equations

$$a_0 = 1$$
$$a_n = 2a_{n-1}, \qquad n = 1, 2, \ldots.$$

Show that the general term for this sequence is $a_n = 2^n$.

30. Find the general term for the sequence $\{a_n\}$ defined recursively by the equations

$$a_0 = 4$$
$$a_n = a_{n-1} + 1, \qquad n = 1, 2, \ldots.$$

31. Find the general term for the sequence $\{a_n\}$ defined recursively by the equations

$$a_0 = -5$$
$$a_n = a_{n-1} + 2, \qquad n = 1, 2, \ldots.$$

32. For the Fibonacci sequence (Exercise 28) show that
  a. $F_{n+3} = 2F_{n+1} + F_n$
  b. $F_{n+4} = 3F_{n+1} + 2F_n$
  c. $F_{n+p} = F_p F_{n+1} + F_{p-1}F_n, \qquad p = 3, 4, \ldots.$

33. Give an example of an increasing sequence that does not converge.

34. Must every convergent sequence be bounded and either increasing or decreasing?

35. Prove that if $\lim\limits_{n\to\infty} |a_n| = 0$, then $\lim\limits_{n\to\infty} a_n = 0$.

Let f be a continuous function and x_0 be a number in the domain of f. Define the sequence $\{x_n\}$ by $x_n = f(x_{n-1})$; that is, $x_1 = f(x_0)$, $x_2 = f(x_1)$, $x_3 = f(x_2)$, $\ldots$. Note that, if $x_n \to L$ as $n \to \infty$, then $L = f(L)$. So (L, L) is a point where the graphs of $y = x$ and $y = f(x)$ intersect. For each of the following functions graph $y = f(x)$ and $y = x$ on the given interval and, by magnifying the graph, determine the intersection point(s) to 4 decimal places. Then, using the given number x_0, determine x_1, x_2, $\ldots$, x_{10}. Does $\{x_n\}$ converge to a number L corresponding to an intersection point?

36. $f(x) = 1 + x - \frac{1}{16}x^3$; $[0, 3]$; $x_0 = 1$.

37. $f(x) = (x^2 + 3)/2x$; $[0, 4]$; $x_0 = 3$.

38. $f(x) = 1 - x^2$; $[0, 1]$; $x_0 = 1/2$.

39. $f(x) = 15x(1 - x)/4$; $[0, 1]$; $x_0 = 1/4$.

40. A well-known recursively defined sequence is the arithmetic-geometric sequence defined as follows: For $0 < a_0 < b_0$, let

$$a_n = (a_{n-1} + b_{n-1})/2, \quad b_n = \sqrt{a_{n-1}b_{n-1}}.$$

It is known that the sequence $\{a_n\}$ is increasing and the sequence $\{b_n\}$ is decreasing. Moreover, both sequences are bounded, and therefore both converge.

a. Show that they converge to the same number. This common limit is known as the arithmetic-geometric mean of a_0 and b_0 and is denoted as $AGM(a_0, b_0)$.

b. Complete the following table, and comment on how fast the sequences converge.

a_0, b_0	a_1, b_1	a_2, b_2	a_3, b_3	a_4, b_4	a_5, b_5
1, 2					
1, 50					
7, 8					
7, 150					
200, 500					
150, 2000					

c. In 1818, Gauss proved the following amazing fact about the arithmetic-geometric mean. Suppose

$$K(a, b) = \int_0^{\pi/2} \frac{d\theta}{\sqrt{a^2 \cos^2 \theta + b^2 \sin^2 \theta}}$$

then

$$K(a, b) = \frac{\pi}{2 AGM(a, b)}.$$

This integral is called a **complete elliptic integral of the first kind** and is used to determine the exact period of a pendulum. For each pair a_0, b_0 in the above table, find $K(a_0, b_0)$ using a_5 or b_5 to approximate $AGM(a_0, b_0)$. Compare your answer to the estimate obtained using Simpson's Rule with $n = 20$.

41. Consider the following variation of the arithmetic-geometric sequence. Let $a_0 = \frac{1}{2}$ and $b_0 = \frac{1}{4}$. Define

$$a_n = \sqrt{a_{n-1}b_{n-1}}, \quad b_n = (a_n + b_{n-1})/2.$$

Determine $1/a_{15}$ and $1/b_{15}$. Make a conjecture about the values of the limits

$$\lim_{n \to +\infty} \frac{1}{a_n} \quad \text{and} \quad \lim_{n \to +\infty} \frac{1}{b_n}.$$

(It turns out that $1/a_n$ is the area of an inscribed 2^n-gon, and $1/b_n$ is the area of a circumscribed 2^n-gon about a circle of radius 1.)

12.3 Infinite Series

By an *infinite series* we mean an expression of the form

$$\sum_{k=1}^{\infty} a_k = a_1 + a_2 + a_3 + \cdots. \tag{1}$$

Unlike the situation for finite sums, we cannot associate a "sum" with an infinite series simply by "adding up" the terms $a_1, a_2, \ldots$ because this would require that we perform an infinite number of additions, something not even a supercomputer can accomplish in a finite amount of time.

The method for evaluating improper integrals of the form $\int_a^\infty f(x)\,dx$ provides the idea by which we shall determine whether an infinite series has a sum. We first find the *partial sums*

$$S_n = \sum_{k=1}^{n} a_k$$

of the series in line (1) and then ask whether the limit

$$\lim_{n \to \infty} S_n = \lim_{n \to \infty} \sum_{k=1}^{n} a_k$$

of these partial sums exists. If it does, this limit is what we shall call the sum of the infinite series.

Definition 3

An **infinite series** is an expression of the form

$$\sum_{k=1}^{\infty} a_k = a_1 + a_2 + a_3 + \cdots.$$

The infinite series

$$\sum_{k=1}^{\infty} a_k$$

is said to **converge** to the **sum** S if

$$S = \lim_{n \to \infty} S_n,$$

where S_n denotes the nth **partial sum**

$$S_n = a_1 + a_2 + a_3 + \cdots + a_n = \sum_{k=1}^{n} a_k.$$

If the limit S does not exist, the series $\sum_{k=1}^{\infty} a_k$ is said to **diverge.**

□ **EXAMPLE 1**

The repeating decimal 0.66666 may be interpreted as the infinite series

$$0.66\overline{66} = .6 + (.06) + (.006) + (.0006) + \cdots$$

$$= \frac{6}{10} + \frac{6}{10^2} + \frac{6}{10^3} + \frac{6}{10^4} + \cdots$$

$$= \sum_{k=1}^{\infty} \frac{6}{10^k}.$$

Let us verify that this interpretation is consistent with the usual notion that $0.\overline{666} = 2/3$. The nth partial sum of this series is

$$S_n = \frac{6}{10} + \frac{6}{10^2} + \frac{6}{10^3} + \cdots + \frac{6}{10^n}. \tag{2}$$

Note that each term of the sum is 10 times the following term. Multiplying both sides of equation (2) by $\frac{1}{10}$ gives

$$\frac{1}{10} S_n = \frac{6}{10^2} + \frac{6}{10^3} + \frac{6}{10^4} + \cdots + \frac{6}{10^{n+1}}. \tag{3}$$

Subtracting corresponding sides of equation (3) from those of equation (2) gives

$$S_n - \frac{1}{10} S_n = \frac{6}{10} - \frac{6}{10^{n+1}},$$

so

$$S_n = \frac{10}{9} \left(\frac{6}{10} - \frac{6}{10^{n+1}} \right) = \frac{2}{3} \left(1 - \frac{1}{10^n} \right).$$

According to the definition of an infinite series, the sum of the series is therefore

$$S = \lim_{n \to \infty} S_n = \lim_{n \to \infty} \frac{2}{3}\left(1 - \frac{1}{10^n}\right) = \frac{2}{3}.$$

This shows that

$$\frac{2}{3} = \sum_{k=1}^{\infty} \frac{6}{10^k} = .666\overline{66}.$$ ∎

☐ **EXAMPLE 2**

The infinite series

$$\sum_{k=1}^{\infty} (-1)^k = -1 + 1 - 1 + 1 - 1 + \cdots$$

does not converge. To see why, observe that the partial sums are

$$
\begin{aligned}
S_1 &= -1 \\
S_2 &= -1 + 1 = 0 \\
S_3 &= -1 + 1 - 1 = -1 \\
S_4 &= -1 + 1 - 1 + 1 = 0 \\
&\ \ \vdots \\
S_{2n-1} &= -1 + 1 - 1 + \cdots + 1 - 1 = -1 \qquad \text{(odd partial sums)} \\
S_{2n} &= -1 + 1 - 1 + \cdots + 1 - 1 + 1 = 0 \qquad \text{(even partial sums).}
\end{aligned}
$$

Thus, the terms of the sequence $\{S_n\}$ of partial sums are alternately -1 or 0, so $\lim_{n \to \infty} S_n$ does not exist. ∎

☐ **EXAMPLE 3**

Determine whether the infinite series

$$\sum_{k=1}^{\infty} \frac{1}{k(k+1)}$$

converges. If it does, find its sum.

Solution   By the method of partial fractions we can show that

$$\frac{1}{k(k+1)} = \frac{1}{k} - \frac{1}{k+1}, \qquad k = 1, 2, 3, \ldots.$$

We can therefore write the partial sum S_n for this series as

$$S_n = \sum_{k=1}^{n} \frac{1}{k(k+1)} = \frac{1}{1 \cdot 2} + \frac{1}{2 \cdot 3} + \frac{1}{3 \cdot 4} + \cdots + \frac{1}{(n-1)n} + \frac{1}{n(n+1)}$$

$$= \left[\frac{1}{1} - \frac{1}{2}\right] + \left[\frac{1}{2} - \frac{1}{3}\right] + \left[\frac{1}{3} - \frac{1}{4}\right]$$

$$+ \cdots + \left[\frac{1}{n-1} - \frac{1}{n} \right] + \left[\frac{1}{n} - \frac{1}{n+1} \right]$$

$$= 1 - \frac{1}{n+1}$$

since all intermediate terms in this "telescoping sum" cancel. This formula for S_n shows that

$$\sum_{k=1}^{\infty} \frac{1}{k(k+1)} = \lim_{n \to \infty} S_n = \lim_{n \to \infty} \left(1 - \frac{1}{n+1} \right) = 1.$$

Thus the series converges, and its sum is $S = 1$.   ■

REMARK 1   Up to this point we have written all infinite series so that the first term has index $k = 1$. This is not necessary. For example, we could rewrite

$$\sum_{k=1}^{\infty} a_k \quad \text{as} \quad \sum_{k=2}^{\infty} b_k$$

where $b_k = a_{k-1}$, since in this case we would have

$$\sum_{k=2}^{\infty} b_k = b_2 + b_3 + b_4 + b_5 + \cdots$$

$$= a_{(2-1)} + a_{(3-1)} + a_{(4-1)} + a_{(5-1)} + \cdots$$

$$= a_1 + a_2 + a_3 + a_4 + \cdots.$$

Similarly, we can write

$$\sum_{k=1}^{\infty} a_k \quad \text{as} \quad \sum_{k=0}^{\infty} c_k, \quad \text{where} \quad c_k = a_{k+1}.$$

The remainder of this chapter focuses on the issue of convergence for infinite series, as it is defined in Definition 3. In fact, we usually address the convergence question in two steps:

(1) Given an infinite series, does it converge?
(2) If it converges, what is its sum S?

In Examples 1–3, we were able to answer both questions completely. However, we are often able to answer question (1) but not question (2). In other words, *we frequently determine that a series $\sum_{k=1}^{\infty} a_k$ converges even though we do not establish the sum S to which it converges.*

REMARK 2   One important observation pertaining to question (1) is the fact that *the convergence of an infinite series does not depend on any finite number of terms.* In other words, suppose the series $\sum_{k=1}^{\infty} a_k$ converges and the series $\sum_{k=1}^{\infty} b_k$ differs from $\sum_{k=1}^{\infty} a_k$ in only a finite number of terms (that is, there exists N such that $a_k = b_k$ for all $k > N$). Then $\sum_{k=1}^{\infty} b_k$ converges as well.

To see why, consider the definition of convergence. Let $\{S_n\}$ denote the sequence of partial sums for $\Sigma_{k=1}^{\infty} a_k$ and let $\{T_n\}$ denote the sequence of partial sums for $\Sigma_{k=1}^{\infty} b_k$. If $n > N$, we have

$$S_n = a_1 + a_2 + \cdots + a_n$$
$$= a_1 + \cdots + a_N + a_{N+1} + \cdots + a_n$$
$$= S_N + a_{N+1} + \cdots + a_n,$$

and

$$T_n = b_1 + b_2 + \cdots + b_n$$
$$= b_1 + \cdots + b_N + b_{N+1} + \cdots + b_n$$
$$= T_N + b_{N+1} + \cdots + b_n.$$

Using the fact that $a_k = b_k$ for $k > N$, we have

$$T_n - T_N = S_n - S_N,$$

so

$$T_n = S_n - S_N + T_N.$$

In other words, for $n > N$, the two sequences of partial sums differ by the constant $T_N - S_N$. Consequently, the sequence $\{T_n\}$ converges if and only if the sequence $\{S_n\}$ converges.

This observation is often expressed less formally by saying that the convergence of a series is determined by its "tail"

$$a_N + a_{N+1} + a_{N+2} + \cdots.$$

Consequently, when we are interested solely in answering question (1), we often write an infinite series using the notation Σa_k, which does not explicitly state the initial index k.

Finally, it is useful to note that, under these conditions, the sums to which these series converge differ by $T_N - S_N$.

Geometric Series

In elementary algebra you may have encountered the formula for the sum of a **geometric progression:**

$$1 + x + x^2 + x^3 + \cdots + x^{n-1} = \frac{1 - x^n}{1 - x}, \qquad x \neq 1. \tag{4}$$

(Equation (4) may be verified by multiplying both sides by $1 - x$. On the left side all terms cancel except $1 - x^n$.)

In equation (4) the variable x, referred to as the **ratio term,** can represent any number except $x = 1$. This formula leads directly to the definition of the **geometric series** with ratio term x:

$$\sum_{k=0}^{\infty} x_k = 1 + x + x^2 + x^3 + \cdots + x^k + \cdots.$$

The formula for the partial sum of this series is given by equation (4):

$$S_n = \frac{1 - x^n}{1 - x}, \qquad x \neq 1.$$

To determine the numbers x for which the geometric series converges, we apply Theorem 1 to conclude that

$$\lim_{n\to\infty} S_n = \lim_{n\to\infty} \frac{1 - x^n}{1 - x} = \frac{1 - \lim_{n\to\infty} x^n}{1 - x}. \tag{5}$$

Now by equation (1), Section 12.2,

$$\lim_{n\to\infty} x^n = 0 \quad \text{if} \quad |x| < 1$$

and, by Exercise 27, Section 12.2,

$$\lim_{n\to\infty} x^n \quad \text{does not exist if} \quad |x| > 1.$$

Thus (5) shows that the geometric series $\sum_{k=0}^{\infty} x^k$

(i) converges to $S = \lim_{n\to\infty} S_n = \dfrac{1}{1 - x}$ if $|x| < 1$, and

(ii) diverges if $|x| > 1$.

The two remaining cases are $x = \pm 1$. If $x = 1$, the geometric series becomes the constant series

$$\sum_{k=0}^{\infty} 1^k = 1 + 1 + 1 + \cdots,$$

which diverges. If $x = -1$, the series is

$$\sum_{k=0}^{\infty} (-1)^k = 1 - 1 + 1 - 1 + \cdots,$$

which diverges (see Example 2).

This analysis completely determines the convergence properties of the geometric series, as summarized in the following theorem.

Theorem 5
Convergence of Geometric Series

If $|x| < 1$, the geometric series converges to the sum

$$\sum_{k=0}^{\infty} x^k = \frac{1}{1 - x}.$$

If $|x| \geq 1$, the geometric series $\sum_{k=0}^{\infty} x^k$ diverges.

□ **EXAMPLE 4**

The series

$$1 + \frac{2}{3} + \frac{4}{9} + \frac{8}{27} + \cdots + \frac{2^k}{3^k} + \cdots = \sum_{k=0}^{\infty} \frac{2^k}{3^k}$$

is a geometric series with $x^k = 2^k/3^k = (2/3)^k$ and $x = 2/3$. Since $|2/3| < 1$, the series converges and its sum, by Theorem 5, is

$$S = \sum_{k=0}^{\infty} \left(\frac{2}{3}\right)^k = \frac{1}{1 - (2/3)} = 3. \qquad \blacksquare$$

☐ **EXAMPLE 5**

The series

$$\frac{1}{2} + \frac{1}{4} + \frac{1}{8} + \frac{1}{16} + \cdots + \frac{1}{2^k} + \cdots$$

has the form of the geometric series with $x^k = 1/2^k = (1/2)^k$ except that the term $1 = (1/2)^0$ is missing. In order to use Theorem 5, we write

$$\sum_{k=1}^{\infty} \frac{1}{2^k} = \frac{1}{2} + \frac{1}{4} + \frac{1}{8} + \frac{1}{16} + \cdots + \frac{1}{2^k} + \cdots$$

$$= \left(1 + \frac{1}{2} + \frac{1}{4} + \frac{1}{8} + \frac{1}{16} + \cdots + \frac{1}{2^k} + \cdots\right) - 1$$

$$= \left(\sum_{k=0}^{\infty} \frac{1}{2^k}\right) - 1$$

$$= \left(\frac{1}{1 - (1/2)}\right) - 1 = 1. \qquad ■$$

The Algebra of Convergent Infinite Series

The following theorem shows that a limited amount of algebra may be performed on convergent infinite series.

Theorem 6

Suppose that the series Σa_k and Σb_k both converge, with sums

$$S = \sum a_k \quad \text{and} \quad T = \sum b_k.$$

Then

(i) the series $\Sigma(a_k + b_k)$ converges, with sum

$$\sum (a_k + b_k) = S + T, \quad \text{and}$$

(ii) for any real number c the series $\Sigma c a_k$ converges, with sum

$$\sum c a_k = cS.$$

Proof: To prove part (i) we let S_n denote the nth partial sum for Σa_k and T_n denote the nth partial sum for Σb_k. Then $(S_n + T_n)$ is a partial sum for the series $\Sigma(a_k + b_k)$, since

$$\sum_{k=1}^{n}(a_k + b_k) = (a_1 + b_1) + (a_2 + b_2) + \cdots + (a_n + b_n)$$
$$= (a_1 + a_2 + \cdots + a_n) + (b_1 + b_2 + \cdots + b_n)$$
$$= S_n + T_n.$$

Applying Definition 3 and Theorem 1, part (i), we see that

$$\sum_{k=1}^{\infty}(a_k + b_k) = \lim_{n\to\infty}(S_n + T_n)$$

$$= \lim_{n\to\infty} S_n + \lim_{n\to\infty} T_n$$

$$= S + T.$$

The proof of part (ii) is similar and is left as an exercise (Exercise 42). ∎

REMARK 3   We may paraphrase Theorem 6 by writing

$$\sum(a_k + b_k) = \sum a_k + \sum b_k \tag{6}$$

and

$$\sum ca_k = c\sum a_k. \tag{7}$$

Equations (6) and (7) must be read from right to left—if the sums on the right exist then so do the sums on the left, and equality holds.

☐ **EXAMPLE 6**

To find the sum of the series

$$\sum_{k=0}^{\infty}\frac{3\cdot 2^k + 3^k}{5^k} = 4 + \frac{9}{5} + \frac{21}{25} + \frac{51}{125} + \cdots$$

we use equations (6) and (7) together with Theorem 5:

$$\sum_{k=0}^{\infty}\frac{3\cdot 2^k + 3^k}{5^k} = \sum_{k=0}^{\infty}\left(\frac{3\cdot 2^k}{5^k} + \frac{3^k}{5^k}\right)$$

$$= 3\sum_{k=0}^{\infty}\frac{2^k}{5^k} + \sum_{k=0}^{\infty}\frac{3^k}{5^k}$$

$$= 3\sum_{k=0}^{\infty}\left(\frac{2}{5}\right)^k + \sum_{k=0}^{\infty}\left(\frac{3}{5}\right)^k$$

$$= 3\left(\frac{1}{1 - (2/5)}\right) + \left(\frac{1}{1 - (3/5)}\right)$$

$$= 3\left(\frac{5}{3}\right) + \frac{5}{2} = \frac{15}{2}. \quad ∎$$

Necessary Condition for Convergence

The following theorem gives a condition that every convergent series must satisfy.

Theorem 7
Necessary Condition for Convergence

If the infinite series $\sum a_k$ converges, then $\lim_{k\to\infty} a_k = 0$.

Proof: Let S_k denote the kth partial sum for Σa_k. If Σa_k converges there exists a sum S for the series. That is, $\lim\limits_{k \to \infty} S_k = S$. Then $\lim\limits_{k \to \infty} S_{k-1} = S$ also. But $a_k = S_k - S_{k-1}$, so

$$\lim_{k \to \infty} a_k = \lim_{k \to \infty} (S_k - S_{k-1})$$

$$= \lim_{k \to \infty} S_k - \lim_{k \to \infty} S_{k-1}$$

$$= S - S = 0. \qquad \blacksquare$$

REMARK 4   It is important to note that Theorem 7 is *not* used to demonstrate that the series Σa_k converges. Although the condition

$$\lim_{k \to \infty} a_k = 0$$

is *necessary* for convergence (meaning all convergent series have this property), it is *not sufficient* to guarantee convergence (meaning some divergent series also have this property). However, Theorem 7 *does establish the divergence* of a series Σb_k for which

$$\lim_{k \to \infty} b_k \neq 0.$$

The Venn diagram in Figure 3.1 illustrates the implications of Theorem 7.

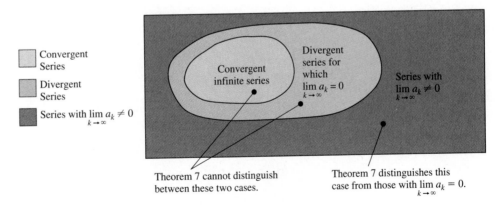

Figure 3.1   Diagram on the applicability of Theorem 7.

The following equivalent formulation of Theorem 7 summarizes these remarks and is the statement you will generally find useful in your work. It is often referred to as the **kth Term Test** (for divergence).

Corollary 2

If $\lim\limits_{k \to \infty} a_k \neq 0$, the series Σa_k diverges.

☐ **EXAMPLE 7**

$$\sum_{k=1}^{\infty} \frac{k}{k+2}$$

diverges, by application of Corollary 2:

$$\lim_{k \to \infty} a_k = \lim_{k \to \infty} \frac{k}{k+2} = 1 \neq 0.$$ ∎

☐ **EXAMPLE 8**

$$\sum_{k=1}^{\infty} \cos(\pi k)$$

diverges, by Corollary 2, since $\lim_{k \to \infty} a_k = \lim_{k \to \infty} \cos(\pi k)$ does not exist. ∎

☐ **EXAMPLE 9**

We show that the series

$$\sum_{k=1}^{\infty} \frac{1}{\sqrt{k}}$$

diverges by analyzing its partial sums $\{S_n\}$. We have

$$S_n = \frac{1}{\sqrt{1}} + \frac{1}{\sqrt{2}} + \frac{1}{\sqrt{3}} + \cdots + \frac{1}{\sqrt{n}}$$
$$> \frac{1}{\sqrt{n}} + \frac{1}{\sqrt{n}} + \frac{1}{\sqrt{n}} + \cdots + \frac{1}{\sqrt{n}}$$
$$= \frac{n}{\sqrt{n}} = \sqrt{n}.$$

Thus, $S_n > \sqrt{n} \to \infty$ as $n \to \infty$. ∎

Note that, in Example 9, the terms $a_k = (1/\sqrt{k}) \to 0$ as $k \to \infty$, but nevertheless, the series diverges. This example illustrates the fact that *the kth Term Test never establishes the convergence of a series*. It can only be used to conclude that a series diverges. In other words, the kth Term Test provides no information about the convergence of a series Σa_k whose terms $a_k \to 0$ as $k \to \infty$.

In this chapter, we encounter relatively few infinite series that yield explicit formulas for their sums (geometric series and those series whose partial sums telescope, as in Example 3, are essentially the only types). Consequently, we concentrate on the convergence question (question (1) earlier). In the remaining sections of this chapter, we discuss a number of techniques, similar to Corollary 2, that resolve the convergence question without the use of an explicit formula for the partial sums.

Exercise Set 12.3

In Exercises 1–6, write out the first four terms of the infinite series

1. $\displaystyle\sum_{k=1}^{\infty} \frac{\cos \pi k}{2^k}$

2. $\displaystyle\sum_{k=1}^{\infty} \frac{\sqrt{k} \sin \pi k}{k+1}$

3. $\displaystyle\sum_{k=1}^{\infty} \frac{2^k + 1}{3^k + 2}$

4. $\displaystyle\sum_{k=0}^{\infty} \tan\left(\frac{\pi k}{3}\right)$

5. $\displaystyle\sum_{k=1}^{\infty} \ln\left(\frac{k}{k+1}\right)$

6. $\displaystyle\sum_{k=1}^{\infty} k^k$

In Exercises 7–28, determine whether the given series converges or diverges. If it converges, find its sum.

7. $\displaystyle\sum_{k=0}^{\infty} \frac{1}{7^k}$

8. $\displaystyle\sum_{k=1}^{\infty} \frac{1}{3^k}$

9. $\displaystyle\sum_{k=0}^{\infty} 4^k$

10. $\displaystyle\sum_{k=1}^{\infty} \frac{7^k + 3^k}{5^k}$

11. $\displaystyle\sum_{k=0}^{\infty} \frac{2^{2k}}{3^{3k}}$

12. $\displaystyle\sum_{k=2}^{\infty} \frac{-1}{3^k}$

13. $\displaystyle\sum_{k=0}^{\infty} \frac{1}{(2+x)^k}, \quad |x| < 1$

14. $\displaystyle\sum_{k=2}^{\infty} \frac{1}{k(k+1)}$

15. $\displaystyle\sum_{k=1}^{\infty} \left[\frac{1}{k+2} - \frac{1}{k+1}\right]$

16. $\displaystyle\sum_{k=2}^{\infty} \frac{2^{k+1} + 2 \cdot 7^k}{9^k}$

17. $\displaystyle\sum_{k=1}^{\infty} \cos \pi k$

18. $\displaystyle\sum_{k=2}^{\infty} \frac{1}{k^2 - 1}$

19. $\displaystyle\sum_{k=1}^{\infty} \frac{1}{k^2 + 5k + 6}$

20. $\displaystyle\sum_{k=4}^{\infty} \frac{1}{k^2 - 9}$

21. $\displaystyle\sum_{k=1}^{\infty} \frac{2}{4k^2 + 8k + 3}$

22. $\displaystyle\sum_{k=1}^{\infty} \ln \frac{k^2}{k^2 + 2k + 1}$

23. $\displaystyle\sum_{k=1}^{\infty} \ln\left(\frac{k}{k+1}\right)$

24. $\displaystyle\sum_{k=1}^{\infty} \frac{2^{k-2} + 3^{k+1}}{5^k}$

25. $\displaystyle\sum_{k=0}^{\infty} \frac{2^{k/2}}{3^k}$

26. $\displaystyle\sum_{k=1}^{\infty} \frac{3^k}{3^{k/2}}$

27. $\displaystyle\sum_{k=1}^{\infty} \frac{e^k}{k^2}$

28. $\displaystyle\sum_{k=0}^{\infty} \frac{2k^4 + 3k^2 - 3}{(k+1)^2}$

In Exercises 29–34, write the given decimal fraction as **(a)** an infinite series, and **(b)** the quotient of two integers.

29. $0.33\overline{3}\ldots$

30. $0.77\overline{7}\ldots$

31. $0.9292\overline{92}\ldots$

32. $0.321515\overline{15}\ldots$

33. $0.412412\overline{412}\ldots$

34. $0.021343\overline{434}\ldots$

In Exercises 35–38, use Theorem 5 on the convergence of geometric series to establish the stated fact.

35. $\displaystyle\sum_{k=0}^{\infty} (-1)^k x^k = \frac{1}{1 + x}$ if $|x| < 1$

36. $\displaystyle\sum_{k=0}^{\infty} x^{2k} = \frac{1}{1 - x^2}$ if $|x| < 1$

37. $\displaystyle\sum_{k=0}^{\infty} \frac{x^k}{y^k} = \frac{y}{y - x}$ if $|x| < |y|$

38. $\displaystyle\sum_{k=1}^{\infty} x^k = \frac{x}{1 - x}$ if $|x| < 1$.

39. When dropped from a height h, a ball rebounds to a height $\frac{2}{3}h$. Write an infinite series expressing the total distance travelled by the ball as it bounces an infinite number of times. What is this distance?

40. Let

$$\sum_{k=0}^{\infty} a_k = \sum_{k=0}^{\infty} \left(1 + \frac{1}{2^k}\right)$$

$$\sum_{k=0}^{\infty} b_k = \sum_{k=0}^{\infty} (-1).$$

Show that the statement $\Sigma(a_k + b_k) = \Sigma a_k + \Sigma b_k$ is false, but that $\Sigma(a_k + b_k)$ converges.

41. If $\Sigma c a_k$ converges for a particular number c, must the series Σa_k converge? Why or why not?

42. Prove Theorem 6, part (ii).

43. Prove that if Σa_k diverges, so must $\Sigma c a_k$, for any $c \neq 0$.

44. Prove that if $\sum_{k=1}^{\infty} a_k$ converges, then $\sum_{k=p}^{\infty} a_k$ converges for any $p \geq 1$. (*Hint:* If S_n is the nth partial sum for $\sum_{k=1}^{\infty} a_k$, then $S_{p+n-1} - S_{p-1}$ is the nth partial sum for $\sum_{k=p}^{\infty} a_k$.)

45. Find a formula that defines the partial sum S_n for the series Σa_k recursively, in terms of the partial sum S_{n-1} and the nth term a_n.

46. Program 6 in Appendix I is a BASIC program that computes partial sums for the geometric series $\sum_{k=p}^{\infty} ax^k$. For example, partial sums for the series $\sum_{k=2}^{\infty} 7(\frac{4}{5})^k$ are obtained by specifying $p = 2$, $a = 7$, and $x = 4/5$. Results appear in Table 3.1.

Table 3.1

n	$S_n = \sum\limits_{k=2}^{n} 7\left(\dfrac{4}{5}\right)^k$
5	13.224959
10	19.393521
25	22.294217
50	22.399598
100	22.399997
200	22.399998
500	22.399998

Show that $\lim\limits_{n\to\infty} S_n = \sum\limits_{k=2}^{\infty} 7\left(\dfrac{4}{5}\right)^k = 22.4$.

In Exercises 47–50, use Program 6 to find the partial sums S_5, S_{10}, S_{20}, S_{50}, and S_{100}. Then find $\lim\limits_{n\to\infty} S_n$.

47. $\sum\limits_{k=0}^{\infty} \left(\dfrac{2}{3}\right)^k$

48. $\sum\limits_{k=3}^{\infty} 4\left(\dfrac{9}{10}\right)^k$

49. $\sum\limits_{k=5}^{\infty} -2\left(\dfrac{5}{7}\right)^k$

50. $\sum\limits_{k=1}^{\infty} 6\left(\dfrac{3}{2}\right)^k$

51. Figure 3.2 illustrates an infinite sequence of inscribed equilateral triangles T_k. The outermost triangle T_1 has sides of length $l_1 = 1$. The second equilateral triangle T_2 is generated from T_1 by joining the midpoints of the sides of T_1; thus the length of each side of T_2 is $l_2 = 1/2$. In general, the kth triangle T_k is generated from T_{k-1} by joining the midpoints of the sides of T_{k-1}. Hence the length of each side of the kth triangle is $l_k = l_{k-1}/2$. Let a_k denote the area of T_k. Determine the sum $\sum_{k=1}^{\infty} a_k$ of the areas of all of the T_k.

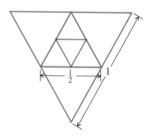

Figure 3.2 Inscribed equilateral triangles in Exercise 51.

52. Starting with the interval $[0, 1]$, we successively remove all middle thirds as follows. First, we remove from $[0, 1]$ the open interval $(1/3, 2/3)$ of length $l_1 = 1/3$. Two intervals $[0, 1/3]$ and $[2/3, 1]$ remain. Next we remove the middle third of each of these two smaller intervals by removing the two intervals $(1/9, 2/9)$ and $(7/9, 8/9)$ (see Figure 3.3). Each middle third removed at this step has length $l_2 = 1/9$. At the kth step, we remove 2^{k-1} intervals, each of length $l_k = 1/3^k$. Show the sum of the lengths of the intervals removed from $[0, 1]$ is 1. Are there any numbers that are not eventually removed?

Figure 3.3 Removing successive middle thirds (Exercise 52).

In Appendix V there are programs that compute the partial sums of an infinite series. For example, by specifying $a_k = 3(2^k)/3^{k-1}$, we obtain the results in the following table.

n	$S_n = \sum\limits_{k=1}^{n} 3\left(\dfrac{2^k}{3^{k-1}}\right)$
5	15.6296293
10	17.68785246
30	17.99990613
50	17.99999997
100	18

Actually, S_{100} is slightly less than 18, but when S_{100} is computed to the accuracy of the calculator, we get $S_{100} \approx 18$. In Exercises 53–56, use these programs to determine S_5, S_{10}, S_{30}, S_{50}, and S_{100}. Find $\lim\limits_{n\to+\infty} S_n$ if it exists.

53. $\sum\limits_{k=0}^{\infty} \dfrac{(-3)^k}{4^k}$

54. $\sum\limits_{k=1}^{\infty} \dfrac{2^k}{k^2}$

55. $\sum\limits_{k=1}^{\infty} \dfrac{k^2}{2^{k-1}}$

56. $\sum\limits_{k=3}^{\infty} \dfrac{3(-4)^k}{6^{k-1}}$

57. The perimeter of an ellipse with semimajor axis of length a and semiminor axis of length b is given by the integral

$$P = 4\int_0^{\pi/2} \sqrt{a^2 \cos^2 t + b^2 \sin^2 t}\, dt.$$

This integral is a so-called **complete elliptic integral of the second kind.** Let $a_0 = a$, $b_0 = b$, $a_n = (a_{n-1} + b_{n-1})/2$, and $b_n = \sqrt{a_{n-1}b_{n-1}}$ as in Exercise 40 in Section 12.2. Then

$$P = 4K(a, b)\left[a_0^2 - \sum_{k=0}^{\infty} 2^{k-1}\, d_k\right]$$

where $d_k = a_k^2 - b_k^2$ and $K(a, b)$ is the value of the complete elliptic integral of the first kind in Exercise 40 in

Section 12.2. Given $a = 5$ and $b = 2$, determine $a_1, \ldots, a_5$, $b_1, \ldots, b_5$, and $d_1, \ldots, d_5$. Then estimate P using a_5 in place of the arithmetic-geometric mean $AGM(5, 2)$ needed to compute $K(a, b)$. Compare your result with the estimate found by approximating the integral using Simpson's Rule. Choose n, the number of points for Simpson's Rule, large enough so that the estimates agree. Which of these two approximation procedures for P is more efficient?

12.4   The Integral Test

In this section and in Sections 12.5 and 12.6 we present five tests for convergence for series *with positive terms.* It is especially important to note that these tests apply only to series Σa_k for which $a_k > 0$ for all k.

The Integral Test

This test exploits the relationship between an infinite series of positive terms and the improper integral of a positive continuous function.

Theorem 8

Integral Test

Suppose $a_k > 0$ for all $k = 1, 2, \ldots$. If f is a function that is continuous and decreasing on $[1, \infty)$ such that $f(k) = a_k$ for each $k = 1, 2, \ldots$, then

$$\sum_{k=1}^{\infty} a_k \text{ converges} \quad \text{if and only if} \quad \int_1^{\infty} f(x)\, dx \text{ converges.}$$

(See Figures 4.1 and 4.2.) That is, the series and the improper integral either both converge or both diverge. Since $f(x) > 0$ for all $x \in [1, \infty)$, we may interpret these two conclusions geometrically by saying that

(i) If the area of the region R_f bounded by the graph of f and the x-axis, for $1 \le x < \infty$, is finite, so is the area R_s of the region enclosed by the rectangles of area $a_1, a_2, a_3, \ldots$ (Figure 4.1).
(ii) If the area of the region R_f is infinite, so is the area of the region R_s (Figure 4.2).

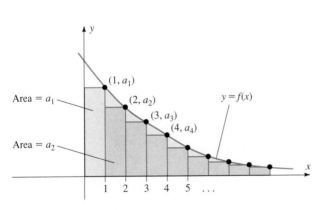

Figure 4.1 Σa_k converges if $\int_1^{\infty} f(x)\, dx < \infty$.

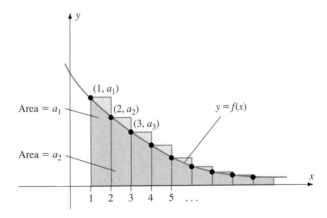

Figure 4.2 Σa_k diverges if $\int_1^{\infty} f(x)\, dx = \infty$.

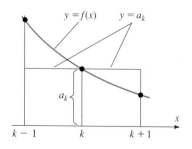

Figure 4.3 $f(x) \geq a_k$, $x \leq k$; $f(x) \leq a_k$, $x \geq k$.

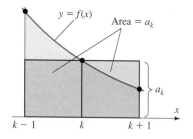

Figure 4.4

$$\int_{k-1}^{k} f(x)\,dx \geq a_k \geq \int_{k}^{k+1} f(x)\,dx.$$

The block of height a_k extending from $k-1$ to k has area a_k; so does the block from k to $k+1$.

Proof of Theorem 8:   Since f is decreasing on $[1, \infty)$, for any integer $k \geq 2$,

$$f(x) \geq f(k) = a_k \quad \text{for all } x \in [k-1, k], \quad \text{and} \tag{1}$$

$$f(x) \leq f(k) = a_k \quad \text{for all } x \in [k, k+1]. \tag{2}$$

(See Figure 4.3.)

Inequalities (1) and (2) imply that

$$\int_{k-1}^{k} f(x)\,dx \geq a_k, \quad \text{and} \tag{3}$$

$$\int_{k}^{k+1} f(x)\,dx \leq a_k, \qquad k = 2, 3, 4, \ldots \quad \text{(Figure 4.4)}. \tag{4}$$

For any integer $n > 2$ we sum inequality (3) from $k = 2$ to n and conclude that

$$\int_{1}^{n} f(x)\,dx = \sum_{k=2}^{n} \int_{k-1}^{k} f(x)\,dx \geq \sum_{k=2}^{n} a_k = S_n - a_1 \tag{5}$$

where S_n denotes the nth partial sum $\sum_{k=1}^{n} a_k$. Thus,

$$S_n \leq a_1 + \int_{1}^{n} f(x)\,dx < a_1 + \int_{1}^{\infty} f(x)\,dx. \tag{6}$$

(The last inequality holds since $f(x) > 0$ on $[1, \infty)$.) Inequality (6) shows that, if the improper integral $\int_{1}^{\infty} f(x)\,dx$ converges, the sequence $\{S_n\}$ of partial sums for the series Σa_k is bounded above. Since $a_k > 0$ for each k, $\{S_n\}$ is also an increasing sequence. Since, by Theorem 4, every bounded increasing sequence converges, the sequence $\{S_n\}$ converges. That is, *if the improper integral $\int_{1}^{\infty} f(x)\,dx$ converges, so does the infinite series Σa_k.*

The converse conclusion is obtained from inequality (4). Summing both sides from $k = 1$ to $k = n$ shows that

$$\int_{1}^{n+1} f(x)\,dx \leq \sum_{k=1}^{n} a_k = S_n. \tag{7}$$

Now $\int_{1}^{n+1} f(x)\,dx$ is an increasing function of n since $f(x) > 0$ for $x \in [1, \infty)$. Thus, if the improper integral $\int_{1}^{\infty} f(x)\,dx$ diverges, we may conclude from inequality (7) that

$$\lim_{n \to \infty} S_n \geq \lim_{n \to \infty} \int_{1}^{n+1} f(x)\,dx = +\infty,$$

so the sequence $\{S_n\}$ diverges. This shows that *if the improper integral diverges, so does the series Σa_k.* This completes the proof. ∎

□ **EXAMPLE 1**

The **harmonic series** is the series

$$\sum_{k=1}^{\infty} \frac{1}{k} = 1 + \frac{1}{2} + \frac{1}{3} + \frac{1}{4} + \cdots + \frac{1}{k} + \cdots.$$

To test this series for convergence, we use the function $f(x) = 1/x$, which is continuous and decreasing on $[1, \infty)$.

Since

$$\int_1^\infty \frac{1}{x}\,dx = \lim_{t \to \infty} \int_1^t \frac{1}{x}\,dx = \lim_{t \to \infty} \ln t = +\infty$$

the harmonic series diverges.   ■

REMARK 1   Note that the kth term $a_k = 1/k$ of the harmonic series satisfies $a_k \to 0$ as $k \to \infty$. Thus, the harmonic series is another example of a divergent series whose terms tend to zero (see Example 9 in Section 12.3).

□ **EXAMPLE 2**

To test the positive term series

$$\sum_{k=1}^\infty k e^{-k^2}$$

for convergence, we use the function $f(x) = xe^{-x^2}$. Since

$$f'(x) = e^{-x^2} - 2x^2 e^{-x^2} = (1 - 2x^2)e^{-x^2} < 0 \quad \text{if} \quad x > \sqrt{2}/2,$$

we are assured that f is decreasing on $[1, \infty)$. Since

$$\int_1^\infty xe^{-x^2}\,dx = \lim_{t \to \infty} \int_1^t xe^{-x^2}\,dx$$

$$= \lim_{t \to \infty} \left[-\frac{1}{2}e^{-x^2} \right]_1^t$$

$$= \lim_{t \to \infty} \left(\frac{1}{2e} - \frac{1}{2e^{t^2}} \right) = \frac{1}{2e},$$

the infinite series $\sum_{k=1}^\infty k e^{-k^2}$ converges, by the Integral Test.   ■

□ **EXAMPLE 3**

A ***p*-series** is a series of the form

$$\sum_{k=1}^\infty \frac{1}{k^p} = 1 + \frac{1}{2^p} + \frac{1}{3^p} + \frac{1}{4^p} + \cdots, \qquad p > 0.$$

Determine the values of p for which the p-series converges.

Strategy · · · · · · · ·   **Solution**

Identify a function f with which to apply the Integral Test.

We use the function $f(x) = \dfrac{1}{x^p} = x^{-p}$.

Since

Verify that f is decreasing on $[1, \infty)$.

$$f'(x) = -px^{-p-1} = \frac{-p}{x^{p+1}}$$

Set up the improper integral and evaluate, carrying along the unknown constant p.

is negative for $x > 0$, the function f is decreasing on $[1, \infty)$.
The improper integral for f is

$$\int_1^\infty x^{-p}\,dx = \lim_{t \to \infty} \int_1^t x^{-p}\,dx$$

$$= \lim_{t \to \infty} \left[\frac{x^{-p+1}}{1-p} \right]_1^t \qquad (p \neq 1)$$

$$= \left[\lim_{t \to \infty} \frac{t^{-p+1}}{1-p} \right] - \frac{1}{1-p}.$$

Determine the values of p for which the improper integral converges.

This limit will exist only if the exponent of t is negative; that is, if

$$-p + 1 < 0, \quad \text{or} \quad p > 1.$$

Handle the remaining case ($p = 1$) directly (see Example 1).

Thus, the improper integral converges if $p > 1$, and diverges if $0 < p < 1$. In the case $p = 1$, (excluded above), the p-series is simply the harmonic series, which diverges.

Apply the Integral Test.

We obtain

> The p-series
>
> $$\sum_{k=1}^\infty \frac{1}{k^p} = 1 + \frac{1}{2^p} + \frac{1}{3^p} + \frac{1}{4^p} + \cdots, \qquad p > 0$$
>
> converges if $p > 1$ and diverges if $0 < p \le 1$.

∎

☐ **EXAMPLE 4**

The series

$$\sum_{k=1}^\infty \frac{1}{k^3} = 1 + \frac{1}{2^3} + \frac{1}{3^3} + \frac{1}{4^3} + \cdots$$

$$= 1 + \frac{1}{8} + \frac{1}{27} + \frac{1}{64} + \cdots$$

converges because it is a p-series with $p = 3 > 1$. ∎

☐ **EXAMPLE 5**

The series

$$\sum_{k=1}^{\infty} \left(\frac{1}{k}\right)^{2/3} = 1 + \left(\frac{1}{2}\right)^{2/3} + \left(\frac{1}{3}\right)^{2/3} + \cdots$$

diverges because it is a *p*-series $\sum_{k=1}^{\infty} \left(\frac{1}{k}\right)^{2/3} = \sum_{k=1}^{\infty} \frac{1}{k^{2/3}}$ with $p = \frac{2}{3} < 1$. ■

REMARK 2   In Remark 2 of Section 12.3, we observed that the question of convergence of a series is determined by the tail of the series. Consequently, we can relax one of the hypotheses of the Integral Test. We need only establish that the function f is decreasing on some interval of the form $[a, \infty)$. That is, f need only be *eventually* decreasing for the Integral Test to apply.

☐ **EXAMPLE 6**

Apply the Integral Test to

$$\sum_{k=1}^{\infty} \frac{k}{2^k}.$$

Solution   Let $f(x) = x/2^x = x2^{-x}$. To check that f is eventually decreasing, we calculate

$$f'(x) = 2^{-x} - x2^{-x} \ln 2 = (1 - x \ln 2)2^{-x}.$$

Since $2^{-x} > 0$ for all x, we see that $f'(x) < 0$ if $(1 - x \ln 2) < 0$. That is, $x > 1/(\ln 2)$. Thus, f is eventually decreasing.

To check the convergence of the improper integral

$$\int_1^{\infty} x2^{-x} \, dx,$$

we integrate by parts with $u = x$ and $dv = 2^{-x} \, dx$. The result is

$$\int_1^{\infty} x2^{-x} \, dx = \lim_{t \to \infty} \left[-\frac{1 + x \ln 2}{2^x (\ln 2)^2} \right]_1^t.$$

After an application of l'Hôpital's Rule (see Example 1 in Section 12.2), we obtain

$$\int_1^{\infty} x2^{-x} \, dx = \frac{1 + \ln 2}{2(\ln 2)^2}.$$

Since the improper integral converges, we conclude that the original series converges. ■

Exercise Set 12.4

In Exercises 1–25, use the Integral Test to determine whether the given series converges or diverges.

1. $\sum \dfrac{1}{2k + 1}$

2. $\sum \dfrac{1}{(3k + 1)^2}$

3. $\sum \dfrac{k + 1}{2k^2 + 4k + 3}$

4. $\sum \dfrac{k}{(k^2 + 2)^3}$

5. $\sum \dfrac{1}{k \ln k}$

6. $\sum \dfrac{1}{k(\ln k)^2}$

7. $\sum \dfrac{1}{1 + k^2}$

8. $\sum k^2 e^{-k^3}$

9. $\sum \dfrac{1}{\sqrt{k + 1}}$

10. $\sum \dfrac{k + 1}{k}$

11. $\sum k^2 2^{-k^3}$

12. $\sum \dfrac{\operatorname{Tan}^{-1} k}{1 + k^2}$

13. $\sum \dfrac{k}{\sqrt{1 + k^2}}$

14. $\sum \dfrac{k^2 - 2}{k^2 + 2}$

15. $\sum \dfrac{1}{\sqrt{k}}$

16. $\sum \dfrac{1}{k^2}$

17. $\sum k(1 + k^2)^{-3}$

18. $\sum k^2 e^{-k}$

19. $\sum \dfrac{\ln k}{k}$

20. $\sum \dfrac{k}{1 + k^2}$

21. $\sum \dfrac{k^2}{k^3 + 1}$

22. $\sum \dfrac{2k^2}{\sqrt{k^3 + 5}}$

23. $\sum \dfrac{\cos^2 \pi k}{k^2 + 1}$

24. $\sum \dfrac{\sin(2k\pi + \pi/2)}{(k + 1)}$

25. $\sum \dfrac{e^{-2\sqrt{k}}}{\sqrt{k}}$

26. Let Σa_k be a convergent infinite series that satisfies the hypotheses of the Integral Test. Show, using inequalities (5) and (7), that

$$\int_{n+1}^{\infty} f(x)\, dx \le \sum_{k=n+1}^{\infty} a_k \le \int_{n}^{\infty} f(x)\, dx \qquad (8)$$

where $f(k) = a_k$ and f is continuous and decreasing on $[1, \infty)$. Inequality (8) gives an estimate for the **truncation error**

$$R_n = \sum_{n+1}^{\infty} a_k$$

which occurs when we use the partial sum

$$S_n = \sum_{k=1}^{n} a_k$$

to approximate the sum of a convergent series

$$S = \sum_{k=1}^{\infty} a_k$$

where $\{a_k\}$ is a decreasing sequence with $\lim\limits_{k \to \infty} a_k = 0$.

27. Use the result of Exercise 26 to estimate the magnitude of the error that occurs in approximating the sum of the p-series

$$\sum_{k=1}^{\infty} \dfrac{1}{k^2}$$

by the partial sum

$$\sum_{k=1}^{10} \dfrac{1}{k^2}.$$

28. Use the result of Exercise 26 to determine n so that the partial sum S_n for the series

$$\sum_{k=2}^{\infty} \dfrac{1}{k \ln^2 k}$$

differs from its sum S by less than 0.1.

29. Use the result of Exercise 26 to determine n so that the partial sum S_n for the series

$$\sum_{k=1}^{\infty} k e^{-k^2}$$

differs from its sum S by less than 0.02.

30. Prove that the harmonic series

$$\sum_{k=1}^{\infty} \dfrac{1}{k}$$

diverges without using the Integral Test:
a. Write the harmonic series as

$$\sum_{k=1}^{\infty} \dfrac{1}{k} = [1] + \left[\dfrac{1}{2}\right] + \underbrace{\left[\dfrac{1}{3} + \dfrac{1}{4}\right]}_{2^1 \text{ terms}} + \underbrace{\left[\dfrac{1}{5} + \dfrac{1}{6} + \dfrac{1}{7} + \dfrac{1}{8}\right]}_{2^2 \text{ terms}}$$

$$+ \underbrace{\left[\dfrac{1}{9} + \cdots + \dfrac{1}{16}\right]}_{(2^3 \text{ terms})}$$

$$+ \left[\frac{1}{17} + \cdots + \frac{1}{32} \right] \quad (2^4 \text{ terms})$$

$$\cdots + \left[\frac{1}{2^n + 1} + \cdots + \frac{1}{2^{n+1}} \right] \quad (2^n \text{ terms})$$

$$+ \cdots.$$

b. Show that the sums of the terms in each of the blocks above are greater than 1/2.

c. Conclude that $S_{2^{n+1}} \geq 1 + \dfrac{(n + 1)}{2}$.

d. From part c, conclude that $\lim\limits_{n \to \infty} S_n = +\infty$.

31. Use the technique of Exercise 30 to prove the "rule of thumb" for the partial sum

$$S_{2^n} = \sum_{k=1}^{2^n} \frac{1}{k}$$

of the harmonic series: $S_{2^n} \geq 1 + \dfrac{n}{2}$.

In Exercises 32–35, use the result of Exercise 26 to determine n so that the partial sum S_n for the given series is within 0.001 of the sum S. Then, using the programs in Appendix V, find S_n. Compare your estimate to the value of S given. Comment on the rate at which S_n converges to S.

32. $\displaystyle\sum_{k=1}^{\infty} \frac{1}{k^4}, \quad S = \dfrac{\pi^4}{90}$

33. $\displaystyle\sum_{k=1}^{\infty} \frac{k}{2^{k^2}}, \quad S = 0.6309205593$

34. $\displaystyle\sum_{k=1}^{\infty} \frac{k}{(k^2 + 4)^3}, \quad S = 0.0141855353$

35. $\displaystyle\sum_{k=1}^{\infty} \frac{1}{(2k + 1)^2}, \quad S = \dfrac{\pi^2}{8} - 1$

12.5   The Comparison Tests

Like the Integral Test, the next test is motivated by area considerations. However, instead of comparing terms of a given series with certain integrals, we compare them with terms of another series.

Theorem 9

Basic Comparison Test

Let Σa_k and Σb_k be infinite series with $0 < a_k \leq b_k$ for each $k = 1, 2, 3, \ldots$. Then

(i) if Σb_k converges, so does Σa_k;
(ii) if Σa_k diverges, so does Σb_k.

Proof: Let

$$S_n = \sum_{k=1}^{n} a_k \quad \text{and} \quad T_n = \sum_{k=1}^{n} b_k$$

be the corresponding partial sums. Since $a_k \leq b_k$ for all k, we have

$$S_n \leq T_n \quad \text{for all} \quad n = 1, 2, \ldots. \tag{1}$$

To prove (i), we note that, since T_n is an increasing sequence and Σb_k converges, we have

$$S_n \leq T_n < \lim_{n \to \infty} T_n < +\infty.$$

Consequently, the sequence $\{S_n\}$ is an increasing sequence that is bounded above; therefore, it converges.

To prove (ii), we note that, since $\{S_n\}$ is an increasing sequence, if Σa_k diverges we must have $\lim\limits_{n\to\infty} S_n = +\infty$. Thus, by inequality (1)

$$\lim_{n\to\infty} T_n \geq \lim_{n\to\infty} S_n = +\infty,$$

so Σb_k diverges.

☐ **EXAMPLE 1**

Use the Basic Comparison Test to show that the series

$$\sum \frac{1}{1 + k^3}$$

converges.

Solution   In order to show that the given series converges, we must compare it to a series whose convergence has already been established. Note that

$$\frac{1}{1 + k^3} < \frac{1}{k^3},$$

and the series

$$\sum \frac{1}{k^3}$$

converges because it is a p-series with $p = 3$.

☐ **EXAMPLE 2**

Apply the Basic Comparison Test to the series

$$\sum \frac{\sqrt{k} - 1}{k^2 + 2}.$$

Solution   Consider the inequalities

$$\frac{\sqrt{k} - 1}{k^2 + 2} < \frac{\sqrt{k}}{k^2 + 2} < \frac{\sqrt{k}}{k^2} = \frac{1}{k^{3/2}}.$$

Since the series

$$\sum \frac{1}{k^{3/2}}$$

is a p-series with $p = 3/2 > 1$, it converges. Consequently, part (i) of the Basic Comparison Test implies that

$$\sum \frac{\sqrt{k} - 1}{k^2 + 2}$$

converges.

REMARK 1   Note that, if the series in Example 2 is changed slightly to

$$\sum \frac{\sqrt{k} + 1}{k^2 + 2},$$

then we cannot use the same comparison as we used in the solution to Example 2. However, a somewhat different comparison leads to the same conclusion. That is, using the fact that

$$\frac{\sqrt{k} + 1}{k^2 + 2} < \frac{\sqrt{k} + \sqrt{k}}{k^2} = \frac{2}{k^{3/2}}$$

we conclude that the series

$$\sum \frac{\sqrt{k} + 1}{k^2 + 2}$$

converges because $\sum(2/k^{3/2}) = 2\sum(1/k^{3/2})$ converges.

☐ **EXAMPLE 3**

Apply the Basic Comparison Test to the series

$$\sum \frac{1 + \ln k}{k}.$$

Solution   In this example, we note that part of the kth term is the kth term of the harmonic series. In other words,

$$\frac{1 + \ln k}{k} > \frac{1}{k}.$$

Since the harmonic series $\sum(1/k)$ diverges (Example 1 in Section 12.4), we conclude that

$$\sum \frac{1 + \ln k}{k}$$

diverges.   ■

REMARK 2   Successful application of the Basic Comparison Test to the series $\sum a_k$ involves finding another series $\sum b_k$ with which you can make a useful comparison. If you suspect that $\sum a_k$ converges, you should look for a *convergent* series $\sum b_k$ with $a_k \leq b_k$. But if you suspect that $\sum a_k$ diverges, look for a divergent series $\sum b_k$ with $a_k \geq b_k$.

REMARK 3   We say that the series $\sum a_k$ is *dominated by* the series $\sum b_k$ (or, that $\sum b_k$ dominates $\sum a_k$) if $a_k \leq b_k$ for all k. Using this terminology we may paraphrase the Basic Comparison Test by saying that a positive series that is dominated by a convergent series must converge, while a series that dominates a positive divergent series must diverge.

It is important to note that two cases remain (dominated by a divergent series and dominating a convergent series) in which no valid conclusions can be drawn. Table 5.1 may help you remember which comparisons yield valid conclusions.

Table 5.1 Conclusions concerning Σa_k when compared with Σb_k.

	Σb_k converges	Σb_k diverges
$0 < a_k \leq b_k$	Σa_k converges	no valid conclusion
$a_k \geq b_k > 0$	no valid conclusion	Σa_k diverges

The Basic Comparison Test is used whenever an appropriate comparison series can be found. As the list of series with which you are familiar expands, your facility with the Basic Comparison Test will grow.

The following variation on the Basic Comparison Test is often easier to apply.

Theorem 10

Limit Comparison Test

Let Σa_k and Σb_k be series with $a_k > 0$, $b_k > 0$ for all $k = 1, 2, \ldots$. If the limit

$$\rho = \lim_{k \to \infty} \frac{a_k}{b_k}$$

exists and $\rho \neq 0$ or $+\infty$, then either both series converge or both series diverge.

The Limit Comparison Test states that if the ratio of the general terms of two positive series tends to a positive limit, then the two series have the same convergence property—either both converge or both diverge.

Before proving this theorem, we consider several examples of its use.

□ **EXAMPLE 4**

The series

$$\sum_{k=1}^{\infty} \frac{1}{2k + 1}$$

diverges, as can be shown by the Limit Comparison Test. Comparing this series with the harmonic series

$$\sum b_k = \sum_{k=1}^{\infty} \frac{1}{k},$$

we find that

$$\rho = \lim_{k \to \infty} \left(\frac{\dfrac{1}{2k + 1}}{\dfrac{1}{k}} \right) = \lim_{k \to \infty} \frac{k}{2k + 1} = \frac{1}{2} > 0.$$

Since the limit $\rho > 0$ exists and is finite and since the harmonic series diverges, the given series diverges as well. ■

□ **EXAMPLE 5**

To apply the Limit Comparison Test to the series

$$\sum_{k=1}^{\infty} \frac{\sqrt{k} + 2}{k^2 + k + 1},$$

we observe that the general term is a quotient containing a highest exponent of 1/2 in its numerator and a highest exponent of 2 in its denominator. This suggests a comparison with the series whose general term is

$$b_k = \frac{k^{1/2}}{k^2} = \frac{1}{k^{3/2}}.$$

We therefore take Σb_k to be the p-series

$$\sum_{k=1}^{\infty} \frac{1}{k^{3/2}}.$$

We obtain

$$\rho = \lim_{k\to\infty} \left(\frac{\dfrac{\sqrt{k}+2}{k^2+k+1}}{\dfrac{1}{k^{3/2}}} \right) = \lim_{k\to\infty} \left(\frac{k^2 + 2k^{3/2}}{k^2 + k + 1} \right) = 1 > 0.$$

We may now conclude that, since the p-series $\displaystyle\sum_{k=1}^{\infty} \frac{1}{k^{3/2}}$ converges, so does the series

$$\sum_{k=1}^{\infty} \frac{\sqrt{k}+2}{k^2 + k + 1}.$$ ∎

☐ **EXAMPLE 6**

Determine whether the series

$$\sum_{k=1}^{\infty} \sin\left(\frac{\pi}{k}\right)$$

converges.

Solution We compare the given series to $\Sigma(\pi/k)$. This comparison leads to the limit

$$\rho = \lim_{k\to\infty} \frac{\sin(\pi/k)}{(\pi/k)}.$$

Since $(\pi/k) \to 0^+$ as $k \to \infty$, we can rewrite ρ in a familiar form. That is,

$$\rho = \lim_{k\to\infty} \frac{\sin(\pi/k)}{(\pi/k)} = \lim_{x\to 0^+} \frac{\sin x}{x} = 1.$$

Thus, by the Limit Comparison Test, the given series has the same convergence property as the series $\Sigma(\pi/k)$, which is a multiple of the (divergent) harmonic series. Therefore,

$$\sum_{k=1}^{\infty} \sin\left(\frac{\pi}{k}\right)$$

diverges. ∎

Proof of Theorem 10:   Since all terms of Σa_k and Σb_k are positive, and $\rho \neq 0$, we must have

$$\rho = \lim_{k \to \infty} \frac{a_k}{b_k} > 0. \tag{2}$$

Now let $\epsilon = \rho/2 > 0$. Then by (2) and the definition of the limit of a sequence there exists an integer N such that

$$\left| \frac{a_k}{b_k} - \rho \right| < \epsilon \quad \text{whenever} \quad k > N. \tag{3}$$

Statement (3) is equivalent to the statement

$$\rho - \epsilon < \frac{a_k}{b_k} < \rho + \epsilon, \qquad k > N. \tag{4}$$

Substituting $\epsilon = \frac{1}{2}\rho$ and multiplying through by $b_k > 0$ in (4) gives

$$\tfrac{1}{2}\rho b_k < a_k < \tfrac{3}{2}\rho b_k, \qquad k > N. \tag{5}$$

Using the left half of inequality (5), we conclude that Σa_k dominates a multiple of Σb_k, since $\Sigma \frac{1}{2}\rho b_k = (\frac{1}{2}\rho)\Sigma b_k$. Thus Σb_k converges if Σa_k converges (Theorem 9). The right half of inequality (5) shows that Σa_k is dominated by $\Sigma \frac{3}{2}\rho b_k = (\frac{3}{2}\rho)\Sigma b_k$, so Σb_k diverges if Σa_k diverges. These two conclusions taken together establish Theorem 10. ∎

Before concluding this section, we repeat an observation that applies to all of the convergence tests discussed in this chapter. In Remark 2 of Section 12.3, we noted that, if the series $\Sigma_{k=N}^{\infty} a_k$ converges, then so does the series $\Sigma_{k=1}^{\infty} a_k$, and vice versa. Consequently, we need not require that the conditions of Theorems 8 through 10 hold on the full series $\Sigma_{k=1}^{\infty} a_k$ but rather only on any "tail" of the form $\Sigma_{k=N}^{\infty} a_k$.

For example, the series

$$\sum_{k=1}^{\infty} \frac{7+k}{1+k^3}$$

may be successfully compared to the series

$$\sum_{k=1}^{\infty} \frac{2}{k^2} = 2 \sum_{k=1}^{\infty} \frac{1}{k^2}$$

for $k > 7$ since

$$\frac{7+k}{1+k^3} < \frac{k+k}{1+k^3} = \frac{2k}{1+k^3} < \frac{2k}{k^3} = \frac{2}{k^2}$$

whenever $k > 7$. The Basic Comparison Test then shows that the "tail"

$$\sum_{k=8}^{\infty} \frac{7+k}{1+k^3}$$

converges, since the p-series $\Sigma_{k=1}^{\infty} (1/k^2)$ converges. Thus the original series

$$\sum_{k=1}^{\infty} \frac{7+k}{1+k^3}$$

also converges.

Exercise Set 12.5

In Exercises 1–10, use the Basic Comparison Test to determine whether the given series converges or diverges.

1. $\sum \dfrac{1}{1 + k^2}$

2. $\sum \dfrac{1}{k^{1/2} + k^{3/2}}$

3. $\sum \dfrac{1}{k - 4}$

4. $\sum \dfrac{3}{k^3 + 3k^2 + 1}$

5. $\sum e^{-3k}$

6. $\sum e^{-2k^2}$

7. $\sum \dfrac{3}{\sqrt{k} + 2}$

8. $\sum (k - 1)e^{-k}$

9. $\sum \dfrac{\sqrt{k}}{1 + k^3}$

10. $\sum \dfrac{\sqrt{k}}{1 + k}$

In Exercises 11–14, use the Limit Comparison Test to determine whether the given series converges or diverges.

11. $\sum \dfrac{k + 3}{2k^2 + 1}$

12. $\sum \dfrac{k^2 - 4}{k^3 + k + 5}$

13. $\sum \dfrac{k + \sqrt{k}}{k + k^3}$

14. $\sum \dfrac{2k + 2}{\sqrt{k^3 + 2}}$

In Exercises 15–31, determine whether the series converges or diverges and state which test you used.

15. $\sum \dfrac{\sin(\pi k)}{k}$

16. $\sum \cos\!\left(\dfrac{\pi k}{4}\right)$

17. $\sum \dfrac{k(k + 1)}{(k + 2)(k^2 + 1)}$

18. $\sum \dfrac{2k^2 + 3k - 1}{k^4 - 6k + 10}$

19. $\sum \dfrac{2^k}{3^k + 1}$

20. $\sum \dfrac{2^k}{3^k - 1}$

21. $\sum \dfrac{1}{\sqrt{4k(k + 1)}}$

22. $\sum \dfrac{2k + 2}{\sqrt{k^3 + 2}}$

23. $\sum \dfrac{1}{\sqrt[3]{k^2 + 2k}}$

24. $\sum \dfrac{k + 1}{2 \ln k}$

25. $\sum \dfrac{\operatorname{Tan}^{-1} k}{k^2}$

26. $\sum \dfrac{\operatorname{Tan}^{-1} \sqrt{k}}{\pi + 6k^2}$

27. $\sum \dfrac{\ln k}{1 + \ln k}$

28. $\sum \dfrac{\ln(k + 1)}{k + 2}$

29. $\sum \dfrac{3^k}{k + 7}$

30. $\sum \dfrac{k + 3}{(k + 2)2^k}$

31. $\sum \dfrac{\sqrt{k}}{\cos(2k - 6) + k^2}$

32. Prove that if Σa_k and Σb_k are positive term series and

$$\lim_{k \to \infty} \frac{a_k}{b_k} = 0,$$

then Σa_k converges if Σb_k converges.

33. Prove that if Σa_k and Σb_k are positive term series and

$$\lim_{k \to \infty} \frac{a_k}{b_k} = +\infty,$$

then Σa_k diverges if Σb_k diverges.

34. Prove that if Σa_k converges, then

$$\lim_{N \to \infty} \sum_{k=N}^{\infty} a_k = 0.$$

That is, if Σa_k converges, its "tails" tend toward zero.

12.6 The Ratio and Root Tests

In this section we discuss two additional tests for series with positive terms. The first involves the limit of the ratio of successive terms of the series.

Theorem 11
Ratio Test

Suppose $a_k > 0$ for all $k = 1, 2, 3, \ldots$ and let

$$\rho = \lim_{k \to \infty} \frac{a_{k+1}}{a_k}. \tag{1}$$

Then, provided this limit exists,

(i) If $\rho < 1$, the series Σa_k converges.
(ii) If $\rho > 1$, the series Σa_k diverges.
(iii) If $\rho = 1$, no conclusion may be drawn.

Before proving Theorem 11, we discuss several examples in which the Ratio Test applies.

□ **EXAMPLE 1**

For the series

$$\sum \frac{k^2}{3^k},$$

the kth term is

$$a_k = \frac{k^2}{3^k}, \quad \text{so} \quad a_{k+1} = \frac{(k+1)^2}{3^{k+1}}.$$

The limit in (1) is therefore

$$\rho = \lim_{k \to \infty} \frac{\dfrac{(k+1)^2}{3^{k+1}}}{\dfrac{k^2}{3^k}} = \lim_{k \to \infty} \frac{3^k(k+1)^2}{3^{k+1}k^2} = \lim_{k \to \infty} \frac{1}{3}\left(\frac{k+1}{k}\right)^2 = \frac{1}{3}.$$

Since $\rho = \dfrac{1}{3} < 1$, the series $\sum \dfrac{k^2}{3^k}$ converges by the Ratio Test. ■

REMARK The Ratio Test is often useful with series involving factorial expressions. Recall that

$$k! = k(k-1)(k-2) \cdots \cdot 3 \cdot 2 \cdot 1.$$

Thus, ratios of factorials can be simplified as follows:

$$\frac{(k+1)!}{k!} = \frac{(k+1)k!}{k!} = k + 1$$

$$\frac{(k+2)!}{k!} = \frac{(k+2)(k+1)k!}{k!} = (k+2)(k+1)$$

$$\frac{k!}{(k+p)!} = \frac{k!}{(k+p)(k+p-1) \cdots \cdot (k+1) \cdot k!}$$

$$= \frac{1}{(k+p)(k+p-1) \cdots \cdot (k+1)}.$$

□ **EXAMPLE 2**

For the series $\sum \dfrac{k^2}{(k+1)!}$ the limit (1) is

$$\rho = \lim_{k \to \infty} \frac{\dfrac{(k+1)^2}{[(k+1)+1]!}}{\dfrac{k^2}{(k+1)!}} = \lim_{k \to \infty} \frac{(k+1)!(k+1)^2}{(k+2)!(k^2)}$$

$$= \lim_{k \to \infty} \left(\frac{1}{k+2}\right)\left(\frac{k+1}{k}\right)^2$$

$$= 0 \cdot 1^2 = 0.$$

Since $\rho < 1$, the series $\sum \dfrac{k^2}{(k+1)!}$ converges, by the Ratio Test. ■

☐ **EXAMPLE 3**

For the series $\sum \dfrac{k!}{k^k}$ the ratio limit is

$$
\rho = \lim_{k \to \infty} \frac{\dfrac{(k+1)!}{(k+1)^{k+1}}}{\dfrac{k!}{k^k}} = \lim_{k \to \infty} \frac{(k+1)!k^k}{k!(k+1)^{k+1}}
$$

$$
= \lim_{k \to \infty} (k+1)\left[\frac{k^k}{(k+1)^{k+1}}\right]
$$

$$
= \lim_{k \to \infty} \left(\frac{k+1}{k+1}\right)\left[\frac{k^k}{(k+1)^k}\right]
$$

$$
= \lim_{k \to \infty} \left(\frac{k}{k+1}\right)^k
$$

$$
= \lim_{k \to \infty} \frac{1}{\left(1 + \dfrac{1}{k}\right)^k} = \frac{1}{e}.
$$

Since $\rho = 1/e < 1$, the series $\sum \dfrac{k!}{k^k}$ converges, by the Ratio Test.   ■

☐ **EXAMPLE 4**

If we attempt to apply the Ratio Test to the series $\sum \dfrac{k}{k^2 + 1}$ we find that

$$
\rho = \lim_{k \to \infty} \frac{\left(\dfrac{k+1}{(k+1)^2 + 1}\right)}{\left(\dfrac{k}{k^2 + 1}\right)}
$$

$$
= \lim_{k \to \infty} \frac{(k+1)(k^2+1)}{k[(k+1)^2 + 1]} = \lim_{k \to \infty} \frac{k^3 + k^2 + k + 1}{k^3 + 2k^2 + 2k} = 1,
$$

so the Ratio Test is inconclusive. However, we can handle this series by a limit comparison with the harmonic series $\Sigma b_k = \Sigma(1/k)$ (Theorem 10). We find

$$
\lim_{k \to \infty} \left(\frac{a_k}{b_k}\right) = \lim_{k \to \infty} \frac{\left(\dfrac{k}{k^2 + 1}\right)}{\left(\dfrac{1}{k}\right)} = \lim_{k \to \infty} \frac{k^2}{k^2 + 1} = 1.
$$

Thus, since the harmonic series diverges, so does the series $\sum \dfrac{k}{k^2 + 1}$.   ■

In Chapter 13 we shall discuss series whose terms contain variables. For such series the Ratio Test is often helpful in determining those values of the variable for which the series converges. The following is one such example.

☐ **EXAMPLE 5**

For which numbers $x > 0$ does the infinite series $\sum \dfrac{x_k}{k}$ converge?

Solution   Applying the Ratio Test we find that

$$\rho = \lim_{k \to \infty} \frac{\left(\dfrac{x^{k+1}}{k+1}\right)}{\left(\dfrac{x^k}{k}\right)} = \lim_{k \to \infty} \frac{x^{k+1}k}{x^k(k+1)}$$

$$= \lim_{k \to \infty} x\left(\frac{k}{k+1}\right)$$

$$= x.$$

Since $\rho = x$, the series will converge for $\rho = x < 1$, by the Ratio Test, and diverge for $x > 1$. The case $\rho = x = 1$ is inconclusive by the Ratio Test, but if $x = 1$ the series is simply the harmonic series, which diverges. Thus, the series $\Sigma x^k/k$, $x > 0$, converges only if $0 < x < 1$.   ∎

In Example 5, we needed the condition $x > 0$ so that all of the terms of the series $\Sigma x_k/k$ are positive. In general, we shall need to consider negative as well as positive values for x, and in the next section, we introduce techniques that eliminate the $x > 0$ assumption.

Proof of Theorem 11:   To prove that $\rho < 1$ implies convergence, we compare the series Σa_k with a geometric series. Since $\rho < 1$, we can find a number γ so that $\rho < \gamma < 1$.
  Since

$$\lim_{k \to \infty} \frac{a_{k+1}}{a_k} = \rho,$$

there exists an integer N so that

$$\frac{a_{k+1}}{a_k} \leq \gamma \quad \text{whenever} \quad k > N. \tag{2}$$

(To see this take $\epsilon = \gamma - \rho > 0$ in Definition 2.)
  Inequality (2) shows that

$$a_{N+1} \leq \gamma a_N$$
$$a_{N+2} \leq \gamma a_{N+1} \leq \gamma^2 a_N$$

and, in general, that

$$a_{N+k} \leq \gamma^k a_N, \quad k = 1, 2, 3, \ldots. \tag{3}$$

Now let $b_k = \gamma^k a_N$ for $k = 1, 2, \ldots$. Then $\Sigma b_k = \Sigma \gamma^k a_N = a_N \Sigma \gamma^k$ is a convergent series since $\gamma < 1$ and the series $\Sigma \gamma^k$ is geometric. Thus, inequality (3) shows that the series $\Sigma_{k=1}^{\infty} a_{N+k}$ is dominated by the convergent series $\Sigma_{k=1}^{\infty} b_k$. The series $\Sigma_{k=1}^{\infty} a_{N+k} = \Sigma_{k=N+1}^{\infty} a_k$ therefore converges, by the Comparison Test. This shows that the series Σa_k converges.
  The proof of statement (ii) is similar (except that $\rho > 1$, so the comparison is with a divergent geometric series) and is left as an exercise.

To prove statement (iii) note that

(a) the harmonic series $\sum \dfrac{1}{k}$ diverges, although

$$\rho = \lim_{k \to \infty} \frac{\left(\dfrac{1}{k+1}\right)}{\left(\dfrac{1}{k}\right)} = \lim_{k \to \infty} \frac{k}{k+1} = 1, \quad \text{and}$$

(b) and p-series $\sum \dfrac{1}{k^2}$ converges, although

$$\rho = \lim_{k \to \infty} \frac{\left(\dfrac{1}{(k+1)^2}\right)}{\left(\dfrac{1}{k^2}\right)} = \lim_{k \to \infty} \frac{k^2}{(k+1)^2} = 1. \qquad \blacksquare$$

The Root Test

The following test for convergence uses information about the kth root of the general term for Σa_k.

Theorem 12
Root Test

Suppose $a_k > 0$ for all $k = 1, 2, 3, \ldots$ and let $\rho = \lim_{k \to \infty} \sqrt[k]{a_k}$. Then

(i)   If $\rho < 1$, the series Σa_k converges.
(ii) If $\rho > 1$, the series Σa_k diverges.
(iii) If $\rho = 1$, no conclusions may be drawn.

The Root Test is primarily applied to series involving kth powers.

$\square$ **EXAMPLE 6**

The series $\sum \dfrac{e^k}{k^k}$ converges. This can be shown using the Root Test since

$$\rho = \lim_{k \to \infty} \left(\frac{e^k}{k^k}\right)^{1/k} = \lim_{k \to \infty} \frac{e}{k} = 0 < 1. \qquad \blacksquare$$

$\square$ **EXAMPLE 7**

For the series $\sum \dfrac{k^2}{2^k}$ we find that

$$\rho = \lim_{k \to \infty} \left(\frac{k^2}{2^k}\right)^{1/k} = \lim_{k \to \infty} \frac{(k^{1/k})^2}{2} = \frac{1}{2} < 1$$

so the series converges. $\qquad \blacksquare$

Proof of Theorem 12:   We proceed as in the proof of the Ratio Test. If $\rho < 1$, there exists a constant γ so that $\rho < \gamma < 1$. Then, since

$$\lim_{k \to \infty} \sqrt[k]{a_k} = \lim_{k \to \infty} a_k^{1/k} = \rho,$$

there exists a constant N so that

$$a_k^{1/k} < \gamma \quad \text{whenever} \quad k > N \tag{4}$$

(take $\epsilon = \gamma - \rho$ in Definition 2). Inequality (4) may be written

$$a_k < \gamma^k, \quad k \geq N,$$

which shows that the series $\sum_{k=N}^{\infty} a_k$ is dominated by the geometric series $\sum_{k=N}^{\infty} \gamma^k$. Since $\gamma < 1$, this geometric series converges. Thus the series $\sum_{k=N}^{\infty} a_k$ converges by comparison with a geometric series. This proves statement (i).

The proof of statement (ii) is similar to that of (i) and is left as an exercise.

The proof of statement (iii) proceeds as in the proof of Theorem 11. We note that

(a) $\rho = \lim\limits_{k \to \infty} \left(\dfrac{1}{k}\right)^{1/k} = \lim\limits_{k \to \infty} \dfrac{1}{\sqrt[k]{k}} = 1$, and the series $\sum \dfrac{1}{k}$ diverges.

(b) $\rho = \lim\limits_{k \to \infty} \left(\dfrac{1}{k^2}\right)^{1/k} = \lim\limits_{k \to \infty} \left(\dfrac{1}{\sqrt[k]{k}}\right)^2 = 1$, and the series $\sum \dfrac{1}{k^2}$ converges. ∎

Exercise Set 12.6

In each of Exercises 1–25, determine whether the given series converges or diverges and state the test you used to verify your conclusion.

1. $\sum \dfrac{1}{(k+1)!}$

2. $\sum \dfrac{1}{(2k)!}$

3. $\sum \dfrac{2^k}{k+2}$

4. $\sum \dfrac{k3^k}{(k+1)!}$

5. $\sum k^{10} e^{-k}$

6. $\sum \dfrac{k!}{2^{k+2}}$

7. $\sum \dfrac{\ln k}{e^k}$

8. $\sum \dfrac{(3k)!}{(k!)^3}$

9. $\sum \dfrac{k+2}{1+k^3}$

10. $\sum \left(\dfrac{k}{2k+1}\right)^k$

11. $\sum \dfrac{1}{(\ln k)^k}$

12. $\sum \left(\dfrac{k}{\ln k}\right)^k$

13. $\sum \left(1 + \dfrac{2}{k}\right)^k$

14. $\sum \left(\dfrac{k}{k+1}\right)^k$

15. $\sum \dfrac{3^k}{k^3 2^{k+2}}$

16. $\sum \dfrac{k^3 2^{k+3}}{2^{2k}}$

17. $\sum \dfrac{(k!)^2}{(3k)!}$

18. $\sum \left(\dfrac{k}{1+k^3}\right)^k$

19. $\sum \dfrac{(k+2)!}{4!k!2^k}$

20. $\sum \dfrac{k!}{e^{3k}}$

21. $\sum \dfrac{k^2 \sin^2 (k\pi/2)}{2k!}$

22. $\sum \dfrac{k \cos^2 k\pi}{k!}$

23. $\sum \left(\dfrac{\sqrt{k}}{3+k^2}\right)^{2k}$

24. $\sum \left(\dfrac{3k^2 + 2}{k^2}\right)^{k/2}$

25. $\sum \dfrac{1}{k\sqrt{\ln k}}$

In Exercises 26–29, find the positive numbers x for which the given series converges.

26. $1 + x^2 + x^4 + \cdots = \sum\limits_{k=0}^{\infty} x^{2k}$

27. $1 + \dfrac{x^2}{2} + \dfrac{x^4}{4} + \cdots = 1 + \sum\limits_{k=1}^{\infty} \dfrac{x^{2k}}{2k}$

28. $1 + \dfrac{x^2}{2} + \dfrac{x^4}{3} + \cdots = \sum\limits_{k=0}^{\infty} \dfrac{x^{2k}}{k+1}$

29. $1 + x + \dfrac{x^2}{2!} + \dfrac{x^3}{3!} + \cdots = \sum\limits_{k=0}^{\infty} \dfrac{x^k}{k!} \quad (0! = 1)$

30. Prove Theorem 11, part (ii). (*Hint:* If $\rho > 1$, let γ be a constant so that $1 < \gamma < \rho$. Then, if $(a_{k+1}/a_k) > \gamma$, it follows that $a_k > \gamma^k$. Proceed as in the proof of part (i).)

31. Prove Theorem 12, part (ii). (*Hint:* If $\rho > 1$, let γ be a constant so that $1 < \gamma < \rho$. Then if $\sqrt[k]{a_k} > \gamma$, it follows that $a_k > \gamma^k$. Proceed as in the proof of part (i).)

32. For the *sequence* $\{a_k\}$ with $a_k > 0$, prove that if $\lim\limits_{k \to \infty} (a_{k+1}/a_k) < 1$, then $\lim\limits_{k \to \infty} a_k = 0$.

33. For the *sequence* $\{a_k\}$ with $a_k > 0$, if $\lim\limits_{k \to \infty} (a_{k+1}/a_k) > 1$, what can you say about $\lim\limits_{k \to \infty} a_k$?

34. In Example 3, we showed that the series

$$\sum_{k=1}^{\infty} \frac{k!}{k^k}$$

converges using the Ratio Test. Using your calculator, investigate the limit we must evaluate if we use the Root Test rather than the Ratio Test. That is, let $a_k = k!/k^k$ and evaluate $a_k^{1/k} = (k!)^{1/k}/k$ for $k = 10, 20, 30, 40, 50,$ and 60. Based on this evidence, do you think that the Root Test applies to this series?

12.7 Absolute and Conditional Convergence

The Integral Test, the Comparison Tests, the Ratio Test, and the Root Test apply only to series with positive terms. In this section we discuss the convergence question for more general series—series containing both positive and negative terms.

Absolute Convergence

One straightforward procedure for testing the series Σa_k for convergence is simply to replace each term by its absolute value and then test the resulting series, $\Sigma |a_k|$, for convergence. This is the idea of *absolute convergence*.

Definition 4

The series Σa_k is said to **converge absolutely** if the series of absolute values, $\Sigma |a_k|$, converges.

□ **EXAMPLE 1**

The series

$$\sum_{k=1}^{\infty} \frac{(-1)^{n+1}}{k^2} = 1 - \frac{1}{4} + \frac{1}{9} - \frac{1}{16} + \frac{1}{25} - \cdots$$

converges absolutely since

$$\sum_{k=1}^{\infty} \left| \frac{(-1)^{n+1}}{k^2} \right| = \sum_{k=1}^{\infty} \frac{1}{k^2}$$

is a convergent *p*-series. ∎

□ **EXAMPLE 2**

The series

$$\sum_{k=1}^{\infty} (-1)^{k+1} = 1 - 1 + 1 - 1 + \cdots$$

does not converge absolutely since $\sum\limits_{k=1}^{\infty} |(-1)^{n+1}| = 1 + 1 + 1 + \cdots$ diverges. ∎

The following theorem shows that the terminology of Definition 4 is well chosen—absolutely convergent series indeed converge.

Theorem 13

If the series $\Sigma |a_k|$ converges then so does the series Σa_k. That is, every absolutely convergent series converges.

Proof: Assume that the series $\Sigma|a_k|$ converges. Since the inequality

$$-|a_k| \le a_k \le |a_k|$$

holds for all terms a_k, we may add $|a_k|$ to all terms in this inequality to get

$$0 \le a_k + |a_k| \le 2|a_k|, \qquad k = 1, 2, 3, \ldots. \tag{1}$$

Since the series $\Sigma|a_k|$ converges, so does the series $2\Sigma|a_k| = \Sigma 2|a_k|$. Thus inequality (1) shows that the series $\Sigma(a_k + |a_k|)$ is a series with positive terms* that is dominated by a convergent series. The Comparison Test therefore guarantees that the series $\Sigma(a_k + |a_k|)$ converges. We may now apply Theorem 6 (Section 12.3) to conclude that the series

$$\Sigma a_k = \Sigma(a_k + |a_k|) - \Sigma|a_k|$$

converges. This completes the proof. ∎

Theorem 13 is simple to apply. If the series Σa_k contains negative terms, we test the positive series $\Sigma|a_k|$ for convergence. If $\Sigma|a_k|$ converges, so does Σa_k. However, if $\Sigma|a_k|$ *diverges, we can draw no conclusions about the original series Σa_k.*

☐ **EXAMPLE 3**

To test the series

$$\sum_{k=0}^{\infty} \frac{(-1)^k k^2}{(k+1)!}$$

for convergence, we apply the Ratio Test to the series

$$\sum_{k=0}^{\infty} \left| \frac{(-1)^k k^2}{(k+1)!} \right| = \sum_{k=0}^{\infty} \frac{k^2}{(k+1)!}.$$

We find that

$$\rho = \lim_{k \to \infty} \frac{a_{k+1}}{a_k} = \lim_{k \to \infty} \frac{\dfrac{(k+1)^2}{(k+2)!}}{\dfrac{k^2}{(k+1)!}} = \lim_{k \to \infty} \frac{(k+1)^2}{k^2(k+2)} = 0.$$

Since $\rho = 0 < 1$, the Ratio Test implies that

$$\sum_{k=0}^{\infty} \frac{k^2}{(k+1)!}$$

converges. Therefore, the original series

$$\sum_{k=0}^{\infty} \frac{(-1)^k k^2}{(k+1)!}$$

converges absolutely, and therefore by Theorem 13, it converges. ∎

---

*If $a_k < 0$, then $a_k + |a_k| = 0$, so this term drops out of the series. This is why $\Sigma(a_k + |a_k|)$ may be regarded as a series with positive terms.

☐ **EXAMPLE 4**

The series

$$\sum_{k=1}^{\infty} \frac{\sin k}{2^k}$$

contains both positive and negative terms. To test for absolute convergence, we consider the series

$$\sum_{k=1}^{\infty} \left| \frac{\sin k}{2^k} \right| = \sum_{k=1}^{\infty} \frac{|\sin k|}{2^k}.$$

This series is positive, so we can apply the Comparison Test. Since $|\sin k| < 1$ for all k, we have

$$\frac{|\sin k|}{2^k} < \frac{1}{2^k}.$$

Using the fact that $\sum(1/2^k)$ is a geometric series that converges (its ratio is $1/2$), we conclude that

$$\sum_{k=1}^{\infty} \frac{|\sin k|}{2^k}$$

converges. The original series

$$\sum_{k=1}^{\infty} \frac{\sin k}{2^k}$$

converges absolutely, and by Theorem 13, it converges. ■

☐ **EXAMPLE 5**

For the series

$$\sum_{k=1}^{\infty} \frac{(-1)^{k+1}}{k} = 1 - \frac{1}{2} + \frac{1}{3} - \frac{1}{4} + \cdots$$

the series of absolute values is the harmonic series $\sum_{k=1}^{\infty} 1/k$, which diverges. Theorem 13 therefore gives no conclusion concerning the convergence of the series

$$\sum_{k=1}^{\infty} \frac{(-1)^{k+1}}{k}.$$

This question will be resolved shortly. ■

□ **EXAMPLE 6**

In Example 5 of Section 12.6, we considered the series

$$\sum_{k=1}^{\infty} \frac{x^k}{k} = x + \frac{x^2}{2} + \frac{x^3}{3} + \cdots$$

with the restriction that $x > 0$. Now, if $x < 0$, this series contains both positive and negative terms. To determine the numbers x for which this series converges we consider the series of absolute values

$$\sum_{k=1}^{\infty} \left| \frac{x^k}{k} \right| = \sum_{k=1}^{\infty} \frac{|x|^k}{k}.$$

Applying the Ratio Test to this series, we find that, for $x \neq 0$,

$$\rho = \lim_{k \to \infty} \frac{|a_{k+1}|}{|a_k|} = \lim_{k \to \infty} \frac{\frac{|x|^{k+1}}{k+1}}{\frac{|x|^k}{k}} = \lim_{k \to \infty} |x| \left(\frac{k}{k+1} \right) = |x|.$$

Thus, $\rho < 1$ if $|x| < 1$. Consequently, by verifying absolute convergence for $|x| < 1$, we know that the given series converges if $|x| < 1$. The $x > 1$ case was also resolved using the Ratio Test (see Example 5 in Section 12.6). However, we obtain no conclusion for $x < -1$ using this approach. ■

We make one final observation regarding absolute convergence. If $\Sigma|a_k|$ converges to a sum S_1, then we know that Σa_k converges to a sum S_2. You should note that $S_2 < S_1$ unless Σa_k is a series consisting entirely of nonnegative terms. In other words, absolute convergence implies convergence, but the two sums involved are different if the original series has negative terms.

Alternating Series

Several types of infinite series containing negative terms can be shown to converge even though they do not converge absolutely. Among these are certain of the *alternating series*.

Definition 5

An **alternating series** is a series of the form

$$\sum_{k=0}^{\infty} (-1)^k a_k = a_0 - a_1 + a_2 - a_3 + a_4 - \cdots$$

where $a_k > 0$ for each $k = 0, 1, 2, 3, \ldots$.

The following theorem gives the conditions that imply convergence for alternating series. We will demonstrate its use before we give a proof.

Theorem 14
Alternating Series Test

Let $a_k > 0$ for each $k = 0, 1, 2, \ldots$. The alternating series

$$\sum_{k=0}^{\infty} (-1)^k a_k$$

converges if both the following hold:

(i) The sequence $\{a_k\}_{k=0}^{\infty}$ is decreasing, and
(ii) $\lim_{k \to \infty} a_k = 0$.

Note that requirement (ii) in the Alternating Series Test is just a restatement of the necessary condition for convergence given by the kth Term Test.

☐ **EXAMPLE 7**

For the series

$$\sum_{k=1}^{\infty} \frac{(-1)^{k+1}}{k} = 1 - \frac{1}{2} + \frac{1}{3} - \frac{1}{4} + \cdots$$

the general term is $(-1)^{k+1} a_k$, where $a_k = 1/k > 0$. Since the sequence $\{1/k\}$ is decreasing and $\lim_{k \to \infty} (1/k) = 0$, the series converges by the Alternating Series Test. ∎

Combining the results of Examples 5 and 7, we see that the **alternating harmonic series**

$$\sum_{k=1}^{\infty} \frac{(-1)^{k+1}}{k} = 1 - \frac{1}{2} + \frac{1}{3} - \frac{1}{4} + \cdots$$

converges, but does not converge absolutely. (This shows that the converse of Theorem 13 is not true.) Such series are said to *converge conditionally*.

Definition 6

The series Σa_k is said to **converge conditionally** if Σa_k converges and $\Sigma |a_k|$ diverges.

☐ **EXAMPLE 8**

Determine whether the series

$$\sum_{k=1}^{\infty} \frac{(-1)^{k+1}k}{1 + k^2}$$

converges absolutely, converges conditionally, or diverges.

Strategy · · · · · · · ·
First, test for absolute convergence.

Solution
The series of absolute values is

$$\sum_{k=1}^{\infty} \frac{k}{1 + k^2}.$$

Use the Integral Test on the series of absolute values.

We may test this series by the Integral Test. Since

$$\int_1^\infty \frac{x}{1+x^2}\, dx = \lim_{t\to\infty} \int_1^t \frac{x}{1+x^2}\, dx$$

$$= \lim_{t\to\infty} \tfrac{1}{2}\ln(1+t^2) - \tfrac{1}{2}\ln 2 = +\infty,$$

the series of absolute values diverges. Thus, the original series does not converge absolutely.

Since the series is alternating but does not converge absolutely, apply the Alternating Series Test.

We next test for conditional convergence using Theorem 14. Since

$$\frac{d}{dx}\left(\frac{x}{1+x^2}\right) = \frac{1+x^2 - x(2x)}{(1+x^2)^2} = \frac{1-x^2}{(1+x^2)^2}$$

is negative for $x > 1$, the sequence $\{k/(1+k^2)\}$ is decreasing. Also,

$$\lim_{k\to\infty} \frac{k}{1+k^2} = 0.$$

Thus, both conditions of Theorem 14 are met, so the series converges. Since it converges but does not converge absolutely, it converges conditionally. ∎

□ **EXAMPLE 9**

For which numbers x does the series

$$\sum_{k=1}^\infty \frac{x^k}{k}$$

converge?

Solution We have already determined that the series converges absolutely for $|x| < 1$ (Example 6). If $|x| > 1$, we may apply l'Hôpital's Rule (differentiating with respect to k) to conclude that

$$\lim_{k\to\infty} \frac{|x|^k}{k} \qquad \left(\frac{\infty}{\infty}\ \text{form}\right)$$

$$= \lim_{k\to\infty} \frac{|x|^k \ln|x|}{1} = +\infty.$$

Thus, the limit of x^k/k as $k \to \infty$ does not exist if $|x| > 1$, so $\sum_{k=1}^\infty x^k/k$ diverges for these values of x, by the kth Term Test. The remaining cases are $x = 1$ and $x = -1$.

If $x = 1$, the series is the harmonic series $\sum_{k=1}^\infty 1/k$, which diverges. If $x = -1$, the series is

$$\sum_{k=1}^\infty \frac{(-1)^k}{k} = -1 + \frac{1}{2} - \frac{1}{3} + \frac{1}{4} - \cdots$$

This is the negative of the alternating harmonic series, which converges conditionally (Example 7).

We have therefore shown that the series $\sum_{k=1}^\infty x^k/k$ converges for $x \in [-1, 1)$ and diverges for all other numbers x. In particular, the series converges absolutely for $x \in (-1, 1)$ and converges conditionally for $x = -1$. ∎

Proof of Theorem 14:   For the alternating series

$$\sum_{k=0}^{\infty}(-1)^{k}a_{k},$$

the even partial sums have the form

$$S_{2n} = (a_0 - a_1) + (a_2 - a_3) + \cdots + (a_{2n-2} - a_{2n-1}) + a_{2n}.$$

Since the sequence $\{a_k\}$ is decreasing, each of the terms in parentheses is positive. This shows that all even partial sums are positive. Also, since $a_{2n+1} > a_{2n+2}$

$$S_{2n+2} = S_{2n} - a_{2n+1} + a_{2n+2} = S_{2n} - (a_{2n+1} - a_{2n+2}) < S_{2n}. \qquad (2)$$

Thus, the sequence of even partial sums $\{S_{2n}\}$ is decreasing. In particular, this shows that the sequence of positive terms $\{S_{2n}\}$ is bounded above by $S_0 = a_0$ and below by zero. Thus $\{S_{2n}\}$ is a bounded decreasing sequence which, by Corollary 1, must converge. We denote the limit of this sequence by $S = \lim\limits_{n\to\infty} S_{2n}$.

Next, since $\lim\limits_{n\to\infty} a_{2n+1} = 0$, we see that

$$\lim_{n\to\infty} S_{2n+1} = \lim_{n\to\infty} (S_{2n} - a_{2n+1}) = S - 0 = S.$$

so both even and odd partial sums converge to S. Thus S is the limit of S_n as $n \to \infty$ and the series $\sum_{k=0}^{\infty}(-1)^{k}a_{k}$ converges. ∎

Estimating Remainders for Alternating Series

For convergent alternating series satisfying the hypotheses of Theorem 14, there is a simple way to obtain both an estimate of the sum and a bound on the error associated with that estimate.

As in line (2) in the proof of the Alternating Series Test, we may show that the odd partial sums of an alternating series $\sum_{k=0}^{\infty}(-1)^{k}a_{k}$ form an *increasing* sequence:

$$S_{2n+1} = S_{2n-1} + (a_{2n} - a_{2n+1}) > S_{2n-1}, \qquad n = 1, 2, \ldots \qquad (3)$$

since $a_{2n+1} < a_{2n}$.

If we write $S = \sum_{k=0}^{\infty}(-1)^{k}a_{k}$, inequalities (2) and (3) together imply that

$$S_{2p-1} < S < S_{2m} \qquad (4)$$

for any positive integers p and m. That is, *any* even partial sum is larger than *any* odd partial sum, and the sum of the series lies between them (see Figure 7.1).

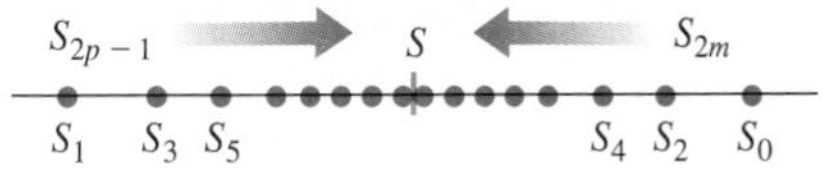

Figure 7.1   Odd partial sums S_{2p-1} increase to S, and even partial sums S_{2m} decrease to S.

Now let

$$S = S_n + R_n \quad \text{where} \quad R_n = \sum_{k=n+1}^{\infty}(-1)^{k}a_{k}. \qquad (5)$$

Then R_n is the **truncation error,** or **remainder,** associated with the approximation of S by the partial sum S_n. From (4) and (5) it follows, by setting $m = n$ and $p = n + 1$, that

$$|R_{2n}| = |S - S_{2n}| < |S_{2n+1} - S_{2n}| = a_{2n+1}. \tag{6}$$

Similarly, by setting $m = p = n$ we find

$$|R_{2n-1}| = |S - S_{2n-1}| < |S_{2n} - S_{2n-1}| = a_{2n}. \tag{7}$$

Thus, regardless of whether we truncate an alternating series after an even or an odd number of terms, *the absolute value of the remainder (or tail) is less than the absolute value of the next term in the series.* This result is sufficiently important that we state it as a theorem.

Theorem 15

Truncation Error

Suppose

$$\sum_{k=0}^{\infty} (-1)^k a_k$$

is an alternating series that satisfies the hypotheses of the Alternating Series Test and converges to the sum S. Then the error $|R_n| = |S - S_n|$ associated with the approximation of S by the nth partial sum $S_n = \sum_{k=0}^{n} (-1)^k a_k$ is less than the absolute value a_{n+1} of the first truncated term. That is,

$$|R_n| < a_{n+1}.$$

☐ **EXAMPLE 10**

Find an approximation to the sum of the alternating harmonic series

$$\sum_{k=0}^{\infty} (-1)^k \left(\frac{1}{k+1} \right) = 1 - \frac{1}{2} + \frac{1}{3} - \frac{1}{4} + \cdots$$

accurate to within 0.05.

Solution If we approximate this sum by the partial sum

$$S_n = \sum_{k=0}^{n} (-1)^k \left(\frac{1}{k+1} \right),$$

the error in this approximation is less than $a_{n+1} = 1/(n+2)$. We therefore choose n sufficiently large to satisfy the inequality

$$\frac{1}{n+2} \leq 0.05 = \frac{1}{20}.$$

This gives

$$n + 2 \geq 20, \quad \text{or} \quad n \geq 18.$$

The partial sum S_{18} therefore has the desired accuracy. It is

$$S_{18} = 1 - \frac{1}{2} + \frac{1}{3} - \frac{1}{4} + \cdots + \frac{1}{19} \approx 0.72. \qquad \blacksquare$$

☐ **EXAMPLE 11**

How many terms must be included in the partial sum for the alternating series

$$\sum_{k=0}^{\infty} \frac{(-1)^k}{1 + k^2} = 1 - \frac{1}{2} + \frac{1}{5} - \frac{1}{10} + \cdots$$

to ensure that the partial sum approximates the sum of the series accurately to within 0.01?

Solution   It is easier to work with S_{n-1}, so that the absolute value of the first truncated term is $a_n = 1/(1 + n^2)$. We therefore need

$$\frac{1}{1 + n^2} \le \frac{1}{100}, \quad \text{or} \quad n^2 \ge 99.$$

The choice $n = 10$ therefore will suffice, and the partial sum S_9 will have the desired accuracy.   ■

Exercise Set 12.7

In each of Exercises 1–30, determine whether the series converges absolutely, converges conditionally, or diverges.

1. $\sum_{k=1}^{\infty} \frac{(-1)^k}{k^3}$

2. $\sum_{k=1}^{\infty} \frac{(-1)^k}{2k + 1}$

3. $\sum_{k=1}^{\infty} \frac{(-1)^{k+1}k^2}{(k + 2)!}$

4. $\sum_{k=2}^{\infty} (-1)^{k+1} \frac{k}{\ln k}$

5. $\sum_{k=1}^{\infty} (-1)^k \frac{k!}{(2k + 1)!}$

6. $\sum_{k=1}^{\infty} \frac{(-k)^3}{3^k}$

7. $\sum_{k=1}^{\infty} \frac{(-1)^k}{1 + \sqrt{k}}$

8. $\sum_{k=1}^{\infty} \frac{\cos \pi k}{k}$

9. $\sum_{k=1}^{\infty} \frac{\sin(\pi k/2)}{k}$

10. $\sum_{k=1}^{\infty} \frac{(-1)^k}{\sqrt{k(k + 2)}}$

11. $\sum_{k=1}^{\infty} \frac{(-1)^k k^3}{2^{k+2}}$

12. $\sum_{k=2}^{\infty} (-1)^k \frac{\sqrt{k}}{\ln k}$

13. $\sum_{k=2}^{\infty} \frac{k^2(-1)^{k+1}}{\ln k}$

14. $\sum_{k=1}^{\infty} \frac{(2k + 1)(-1)^k}{6k + 2}$

15. $\sum_{k=1}^{\infty} \frac{(-1)^k \sqrt{k}}{k + 1}$

16. $\sum_{k=1}^{\infty} \frac{(-1)^{k+1}2^k}{k^3 3^{k+2}}$

17. $\sum_{k=2}^{\infty} \frac{(-1)^k}{k \ln^2 k}$

18. $\sum_{k=1}^{\infty} \frac{(-1)^{k+1}k!}{6^k}$

19. $\sum_{k=1}^{\infty} \frac{(-1)^k}{1 + k^2}$

20. $\sum_{k=1}^{\infty} \sin\left(\frac{k\pi}{4}\right)$

21. $\sum_{k=2}^{\infty} \frac{(-1)^k k}{\ln \sqrt{k}}$

22. $\sum_{k=1}^{\infty} \frac{(-1)^k k^2}{(2k + 1)(k + 3)}$

23. $\sum_{k=1}^{\infty} \frac{(-1)^k \sinh k}{3e^{2k}}$

24. $\sum_{k=1}^{\infty} \frac{(-1)^k \operatorname{Tan}^{-1} k}{\sqrt{k}}$

25. $\sum_{k=1}^{\infty} \frac{\cos \pi k}{k + 2}$

26. $\sum_{k=2}^{\infty} \frac{(-1)^k}{\sqrt[k]{\ln k}}$

27. $\sum_{k=1}^{\infty} \frac{k \sin((k + \frac{1}{2})\pi)}{1 + e^{\sqrt{k}}}$

28. $\sum_{k=2}^{\infty} \frac{(-1)^k(k^{3/2} + 3k)}{7 - k^2 + 2k^{5/2}}$

29. $\sum_{k=1}^{\infty} \frac{\sinh k - \cosh k}{k}$

30. $\sum_{k=2}^{\infty} \frac{(-1)^k \cosh 2k}{ke^k}$

In Exercises 31–34, determine the numbers x for which the given series converges absolutely.

31. $\sum_{k=0}^{\infty} \frac{x^k}{k!} = 1 + x + \frac{x^2}{2!} + \frac{x^3}{3!} + \cdots$

32. $\sum_{k=0}^{\infty} \frac{x^k}{2k + 1} = 1 + \frac{x}{3} + \frac{x^2}{5} + \cdots$

33. $\sum_{k=0}^{\infty} (-1)^k \frac{x^{2k+1}}{(2k + 1)!} = x - \frac{x^3}{3!} + \frac{x^5}{5!} - \frac{x^7}{7!} + \cdots$

34. $\sum_{k=0}^{\infty} (-1)^k \frac{x^{2k}}{(2k)!} = 1 - \frac{x^2}{2!} + \frac{x^4}{4!} - \frac{x^6}{6!} + \cdots$

35. How large must n be taken so that the partial sum

$$S_n = \sum_{k=0}^{n} (-1)^k \left(\frac{1}{1+k} \right)$$

approximates the sum of the alternating harmonic series accurately to within 0.01?

36. With what accuracy does the sum

$$\sum_{k=0}^{10} \frac{(-1)^k}{(k+1)^3}$$

approximate the sum of the series

$$\sum_{k=0}^{\infty} \frac{(-1)^k}{(k+1)^3} ?$$

37. Prove that if Σa_k and Σb_k converge absolutely, then so does $\Sigma(a_k + b_k)$.

38. Prove that if Σa_k converges absolutely and c is any real number, the series Σca_k converges absolutely.

39. Prove for the (not necessarily positive) series Σa_k that, given

$$\rho = \lim_{k \to \infty} \left| \frac{a_{k+1}}{a_k} \right|,$$

(i) Σa_k converges absolutely if $0 \le \rho < 1$;
(ii) Σa_k diverges if $\rho > 1$;
(iii) no conclusions may be drawn if $\rho = 1$.

40. Prove that Σa_k^2 converges if Σa_k converges absolutely. Is the converse true?

41. 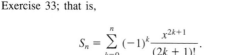 Let $S_n(x)$ represent the nth partial sum of the series given in Exercise 33; that is,

$$S_n = \sum_{k=0}^{n} (-1)^k \frac{x^{2k+1}}{(2k+1)!}.$$

For $n = 1, 2, 3, 4,$ and 5, graph S_n over the interval $[-6, 6]$. What function do you think the infinite series in Exercise 33 represents?

42. Let $C_n(x)$ represent the nth partial sum of the series given in Exercise 34; that is,

$$C_n(x) = \sum_{k=0}^{n} (-1)^k \frac{x^{2k}}{(2k)!}.$$

For $n = 1, 2, 3, 4,$ and 5, graph C_n over the interval $[-6, 6]$. What function do you think the infinite series in Exercise 34 represents?

43. In Example 9, we determined that the series

$$\sum_{k=1}^{\infty} \frac{x^k}{k}$$

converges absolutely on the interal $(-1, 1)$. For $n = 1, 2, 3, 4, 5, 6, 7,$ and 8, graph the nth partial sums

$$L_n(x) = \sum_{k=1}^{n} \frac{x^k}{k}$$

over the interval $(-1, 1)$. What function do you think that the infinite series in Example 9 represents?

In Exercises 44–47, determine n such that the nth partial sum $S_n = \sum_{k=0}^{n} (-1)^k a_k$ of the given alternating series is within 0.001 of the sum $S = \sum_{k=0}^{\infty} (-1)^k a_k$. Using the programs in Appendix V, calculate S_n and compare this estimate to the given value of S.

44. $\displaystyle\sum_{k=0}^{\infty} \frac{(-1)^k}{(k+1)^4}, \ S = \frac{7\pi^4}{720}$ **45.** $\displaystyle\sum_{k=0}^{\infty} (-1)^k \frac{(k+1)}{3^{k+1}}, \ S = \frac{3}{16}$

46. $\displaystyle\sum_{k=0}^{\infty} (-1)^k \frac{2^k}{k!}, \ S = e^{-2}$ **47.** $\displaystyle\sum_{k=1}^{\infty} \frac{(-1)^k}{(4k^2-1)}, \ S = \frac{2-\pi}{4}$

Summary Outline of Chapter 12

■ An **infinite sequence** $\{a_n\}$ is a function whose domain is the set of positive integers.   (page 595)

■ The sequence $\{a_n\}$ is said to **converge** if $L = \lim\limits_{n \to \infty} a_n$ exists.   (page 597)

■ $L = \lim\limits_{n \to \infty} a_n$ if, given $\epsilon > 0$, there exists an integer N so that $|a_n - L| < \epsilon$ (page 598)
whenever $n > N$.

■ The nth **partial sum** S_n of the infinite series $\sum_{k=1}^{\infty} a_k$ is the sum of the first n (page 609)
terms:

$$S_n = \sum_{k=1}^{n} a_k.$$

■ S is called the **sum** of the infinite series if $S = \lim\limits_{n \to \infty} S_n = \lim\limits_{n \to \infty} \sum\limits_{k=1}^{n} a_k$.   (page 610)

■ The series Σa_k is said to **converge** if the sum S exists.   (page 610)

■ The **geometric series** $\Sigma_{k=0}^{\infty} x^k = 1 + x + x^2 + x^3 + \cdots$ converges, if $|x| < 1$,   (page 614) to the sum $S = 1/(1 - x)$, and diverges otherwise.

■ A **necessary condition** for Σa_k to converge is $\lim\limits_{k \to \infty} a_k = 0$. (However, this (page 616) condition does not guarantee convergence.)

■ *kth Term Test*   If $\lim\limits_{k \to \infty} a_k$ does not exist or is not 0, then Σa_k diverges.   (page 617)

■ The **harmonic series** $\sum\limits_{k=1}^{\infty} \dfrac{1}{k} = 1 + \dfrac{1}{2} + \dfrac{1}{3} + \dfrac{1}{4} + \cdots$ diverges.   (page 622)

■ The *p*-series $\sum\limits_{k=1}^{\infty} \dfrac{1}{k^p} = 1 + \dfrac{1}{2^p} + \dfrac{1}{3^p} + \dfrac{1}{4^p} + \cdots$ converges if $p > 1$ and di- (page 623) verges otherwise.

■ *Tests for convergence of Σa_k if $a_k > 0$ for all k:*

(1) **Integral Test**   If $f(k) = a_k$ for all k and f is decreasing on $[1, \infty)$ then (page 621) Σa_k converges if and only if $\int_1^\infty f(x)\, dx$ converges.

(2) **Comparison Test**   Let $0 < a_k \le b_k$ for all k.   (page 627)
    a. If Σb_k converges, Σa_k converges.
    b. If Σa_k diverges, Σb_k diverges.

(3) **Limit Comparison Test**   If $b_k > 0$ for all k and $\rho = \lim\limits_{k \to \infty} \dfrac{a_k}{b_k} > 0$ exists, (page 630) then Σa_k and Σb_k either both converge or both diverge.

(4) **Ratio Test**   If $\rho = \lim\limits_{k \to \infty} \dfrac{a_{k+1}}{a_k}$ exists, then: (page 633)

    a. Σa_k converges if $\rho < 1$.
    b. Σa_k diverges if $\rho > 1$.
    c. No conclusion may be drawn if $\rho = 1$.

(5) **Root Test**   If $\rho = \lim\limits_{k \to \infty} \sqrt[k]{a_k}$ exists, we have the same conclusions as for (page 637) the Ratio Test.

■ The series Σa_k **converges absolutely** if $\Sigma |a_k|$ converges.   (page 639)

■ *Theorem:*   If $\Sigma |a_k|$ converges, then Σa_k converges.   (page 639)

■ An **alternating series** is a series of the form $\sum\limits_{k=0}^{\infty} (-1)^k a_k$ with all $a_k > 0$.   (page 642)

■ *Theorem:*   **Alternating Series Test**   An alternating series converges if (page 643)
    a. $a_k > a_{k+1}$ for all k, and
    b. $\lim\limits_{k \to \infty} a_k = 0$.

■ Suppose $S = \Sigma_{k=0}^{\infty} (-1)^k a_k$ is the sum of an alternating series that satisfies the (page 646) hypotheses of the Alternating Series Test. Then the absolute value of the **error** associated with the approximation of S by the nth partial sum $S_n = \Sigma_{k=0}^{n} (-1)^k a_k$ is less than the absolute value a_{n+1} of the first truncated term.

Review Exercises—Chapter 12

In Exercises 1–10, determine whether the given sequence converges. If it does, find its limit.

1. $\left\{\sin \dfrac{n\pi}{2}\right\}$

2. $\left\{\dfrac{1}{\sqrt{n+1}}\right\}$

3. $\left\{\dfrac{\sqrt{n}}{\ln(\sqrt{n}+1)}\right\}$

4. $\left\{\left(1-\dfrac{2}{n}\right)^{2n}\right\}$

5. $\left\{\dfrac{3^n}{n!}\right\}$

6. $\left\{\dfrac{(-1)^{n+1}}{\sqrt{n^2+1}}\right\}$

7. $\left\{\dfrac{(-1)^{n+1}n^3}{n(1-n+n^2)}\right\}$

8. $\left\{\dfrac{n^2+2n+2}{n^{3/2}}\right\}$

9. $\{e^{2\ln(n)}\}$

10. $\{\ln(n+1)-\ln(n-1)\}$

In Exercises 11–14, find the sum of the given geometric series.

11. $1+\dfrac{1}{6}+\dfrac{1}{36}+\dfrac{1}{216}+\cdots$

12. $10+\dfrac{10}{2}+\dfrac{10}{4}+\dfrac{10}{8}+\cdots$

13. $\dfrac{2}{3}+\dfrac{4}{9}+\dfrac{8}{27}+\cdots$

14. $1-\dfrac{1}{5}+\dfrac{1}{25}-\dfrac{1}{125}+\cdots$

In Exercises 15 and 16, express the repeating decimal as an infinite series and find its rational form.

15. $0.37\overline{37}\ldots$

16. $0.026\overline{3263}\ldots$

In Exercises 17–20, find the sum of the series.

17. $\displaystyle\sum_{k=3}^{\infty}\left(\dfrac{1}{4}\right)^k$

18. $\displaystyle\sum_{k=1}^{\infty}\dfrac{3^k+2^k}{5^k}$

19. $\displaystyle\sum_{k=2}^{\infty}\dfrac{20}{3^{k+1}}$

20. $\displaystyle\sum_{k=1}^{\infty}\left[\dfrac{3}{2^k}+\dfrac{1}{(k+1)(k+2)}\right]$

In Exercises 21–58, determine whether the given series converges or diverges and state which test you used.

21. $\displaystyle\sum_{k=1}^{\infty}\dfrac{1}{k(k+1)}$

22. $\displaystyle\sum_{k=1}^{\infty}\dfrac{3^k}{k^3}$

23. $\displaystyle\sum_{k=3}^{\infty}\dfrac{2^k}{k^5}$

24. $\displaystyle\sum_{k=1}^{\infty}\dfrac{(-1)^k}{2k+1}$

25. $\displaystyle\sum_{k=1}^{\infty}\dfrac{k^3}{k!}$

26. $\displaystyle\sum_{k=0}^{\infty}\dfrac{1}{10k+2}$

27. $\displaystyle\sum_{k=2}^{\infty}\dfrac{\cos \pi k}{\sqrt{k}}$

28. $\displaystyle\sum_{k=2}^{\infty}\dfrac{k^k}{k!}$

29. $\displaystyle\sum_{k=3}^{\infty}\dfrac{\ln k}{k^2}$

30. $\displaystyle\sum_{k=0}^{\infty}\dfrac{2^k+k}{(k+1)!}$

31. $\displaystyle\sum_{k=1}^{\infty}\dfrac{1}{9+k^2}$

32. $\displaystyle\sum_{k=2}^{\infty}\dfrac{(-1)^k k}{\ln k}$

33. $\displaystyle\sum_{k=2}^{\infty}\dfrac{2}{k\ln(k+1)}$

34. $\displaystyle\sum_{k=1}^{\infty}\dfrac{\sqrt{k}}{2^k}$

35. $\displaystyle\sum_{k=0}^{\infty}\dfrac{k^4 2^k}{(k+2)!}$

36. $\displaystyle\sum_{k=1}^{\infty}(-1)^k\dfrac{\ln k}{k}$

37. $\displaystyle\sum_{k=0}^{\infty}\dfrac{1}{\sqrt{4+k^2}}$

38. $\displaystyle\sum_{k=1}^{\infty}\left(\dfrac{k-1}{k+1}\right)^k$

39. $\displaystyle\sum_{k=1}^{\infty}(-1)^{k+1}\dfrac{\sqrt{k}}{2k+1}$

40. $\displaystyle\sum_{k=1}^{\infty}k^3 e^{-k^4}$

41. $\displaystyle\sum_{k=1}^{\infty}\left(\dfrac{k!}{k^k}\right)^k$

42. $\displaystyle\sum_{k=2}^{\infty}(-1)^k\dfrac{k^2+2}{1+k+k^2}$

43. $\displaystyle\sum_{k=1}^{\infty}2ke^{-k}$

44. $\displaystyle\sum_{k=0}^{\infty}\left(\dfrac{k}{1+k}\right)^k$

45. $\displaystyle\sum_{k=0}^{\infty}(-1)^k\sin \pi k$

46. $\displaystyle\sum_{k=2}^{\infty}\dfrac{(2k+1)!}{k^2(k+1)!}$

47. $\displaystyle\sum_{k=1}^{\infty}\dfrac{k!}{e^{2k}}$

48. $\displaystyle\sum_{k=2}^{\infty}\dfrac{1}{(k+6)^{3/2}}$

49. $\displaystyle\sum_{k=1}^{\infty}(-1)^k\dfrac{3}{\sqrt{k}}$

50. $\displaystyle\sum_{k=2}^{\infty}\dfrac{1}{(\ln k)^k}$

51. $\displaystyle\sum_{k=1}^{\infty}\dfrac{2^{2k-1}}{(2k-1)!}$

52. $\displaystyle\sum_{k=2}^{\infty}\dfrac{1}{\sqrt{k}(\ln k)^3}$

53. $\displaystyle\sum_{k=0}^{\infty}\dfrac{(-1)^k}{1+k^2}$

54. $\displaystyle\sum_{k=1}^{\infty}\mathrm{Tan}^{-1}k$

55. $\displaystyle\sum_{k=1}^{\infty}k\left(\dfrac{\pi}{k}\right)^k$

56. $\displaystyle\sum_{k=0}^{\infty}\dfrac{(-1)^k k(k+2)}{(k+1)!}$

57. $\displaystyle\sum_{k=1}^{\infty}\left(\dfrac{k}{2k+1}\right)^k$

58. $\displaystyle\sum_{k=0}^{\infty}\dfrac{(-1)^k \mathrm{Sin}^{-1}\left(\dfrac{1}{k+1}\right)}{k+1}$

In Exercises 59–68, determine whether the given series converges absolutely, converges conditionally, or diverges.

59. $\displaystyle\sum_{k=0}^{\infty}\dfrac{(-1)^{k+1}}{3k+2}$

60. $\displaystyle\sum_{k=2}^{\infty}\dfrac{(-1)^k}{k\ln k}$

61. $\sum_{k=1}^{\infty} (-1)^k \dfrac{k^k}{k!}$

62. $\sum_{k=0}^{\infty} \dfrac{\cos \pi k}{1 + k^2}$

63. $\sum_{k=1}^{\infty} \dfrac{(-1)^{k+1}}{k(k + 1)}$

64. $\sum_{k=0}^{\infty} (-1)^k \left(\dfrac{2}{3}\right)^k$

65. $\sum_{k=1}^{\infty} \dfrac{k!}{(-2)^k}$

66. $\sum_{k=1}^{\infty} \dfrac{\cos(\pi k/3)}{1 + k^2}$

67. $\sum_{k=1}^{\infty} \dfrac{(-1)^k k^2}{(k + 1)!}$

68. $\sum_{k=1}^{\infty} \dfrac{(-1)^{k+1}}{\sqrt{k(k + 1)}}$

In Exercises 69–72, find a bound on the error associated with the approximation of the sum of the given series by the partial sum S_4 consisting of the sum of the first four terms.

69. $\sum_{k=1}^{\infty} \dfrac{2(-1)^k}{1 + k^2} = -1 + \dfrac{2}{5} - \dfrac{2}{10} + \cdots$

70. $\sum_{k=1}^{\infty} \dfrac{(-1)^k}{k^4} = -1 + \dfrac{1}{16} - \dfrac{1}{81} + \cdots$

71. $\sum_{k=1}^{\infty} \dfrac{(-1)^{k+1}}{2k + 1} = \dfrac{1}{3} - \dfrac{1}{5} + \dfrac{1}{7} - \dfrac{1}{9} + \cdots$

72. $\sum_{k=1}^{\infty} \dfrac{(-1)^{k+1} \sin^2\left(\dfrac{\pi}{2k}\right)}{k + 1} = \dfrac{1}{2} - \dfrac{1}{6} + \dfrac{1}{16} - \cdots$

73. Define the sequence $\{a_k\}$ by $a_1 = c$, $a_{k+1} = a_1 \cdot a_k$, $k \geq 1$. For what values of c does $\{a_k\}$ converge?

74. Prove that if the sequence $\{a_n\}$ is not bounded it cannot converge.

75. If Σa_k converges but $\Sigma(a_k + b_k)$ diverges, what can you conclude about Σb_k?

76. Use an infinite series to determine the time between 1 p.m. and 2 p.m. at which the minute hand on a clock is directly over the hour hand. (*Hint:* The minute hand moves 12 times as fast as the hour hand.)

77. A ball is dropped from a height of 10 m. Each time it strikes the ground, it rebounds to 5/6 of the height from which it fell. Find the total distance travelled by the ball
   a. when it has struck the ground 10 times;
   b. when it has come to rest.

78. Prove that, if Σa_k diverges, then $\Sigma c a_k$ diverges for any number $c \neq 0$.

79. Let Σa_k and Σb_k be series with positive terms. If $\Sigma(a_k + b_k)^2$ converges, what can you conclude about
   a. Σa_k^2?
   b. Σb_k^2?
   c. $\Sigma a_k b_k$?

80. Let $p(k)$ and $q(k)$ be polynomials in k with positive coefficients. Let n be the degree of $p(k)$ and let m be the degree of $q(k)$. Show that
   a. $\sum \dfrac{p(k)}{q(k)}$ converges if $m > n + 1$
   b. $\sum \dfrac{p(k)}{q(k)}$ diverges if $m \leq n + 1$.

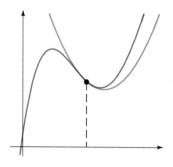

Chapter 13

Taylor Polynomials and Power Series

We begin this chapter with an *approximation problem:* Given a differentiable function f, how can we approximate it using only polynomials? We answer this question by developing the notion of a *Taylor polynomial,* and we discuss various applications of *Taylor's Theorem,* a result that determines a bound for the error involved in these approximations.

Then we return to the topic of infinite series and discuss *power series,* infinite series whose terms include powers of an independent variable. Applying the theory and techniques of Chapter 12, we show that a convergent power series determines a differentiable function. Moreover, we show that Taylor polynomials may be regarded as partial sums for power series, and we see that many familiar functions, such as trigonometric, exponential, and logarithmic functions may be represented as power series.

13.1 The Approximation Problem and Taylor Polynomials

In the first half of this chapter we consider the following problem:

The Approximation Problem: Given a function f and a number a, how do we use polynomials to approximate the function's values $f(x)$ for x near a? How accurate are the resulting approximations?

Of course, if f is a polynomial, the answer is easy. We simply use f to approximate itself, with complete accuracy. In fact, we are really interested in approximating more complicated functions such as transcendental functions ($\sin x$ or e^x, for example) or algebraic functions (like $\sqrt[5]{x}$). Even though we have developed a rich theory of differentiation and integration for these functions, we do not yet have a general means for calculating their values, except in special cases. For instance, how do we calculate $\sin 53°$, $e^{0.05}$, or $\sqrt[5]{216}$? Even though calculators can provide accurate approximations to these values, an appreciation of the underlying mathematics of approximation is essential to understanding many of the applications of calculus.

We have already discussed linear approximation in Sections 3.7 and 4.8. That is, we used the degree one polynomial

$$P_1(x) = f(a) + f'(a)(x - a), \tag{1}$$

whose graph is a line, to approximate the function f near a. Now we consider higher degree polynomials P_n that provide more accurate approximations.

Taylor Polynomials

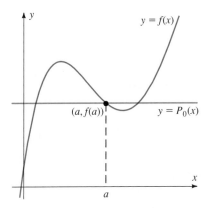

Figure 1.1 The constant polynomial $P_0(x) = f(a)$ approximating $f(x)$ near a.

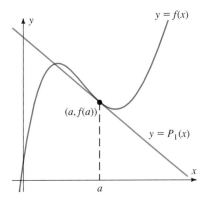

Figure 1.2 The first degree polynomial $P_1(x) = f(a) + f'(a)(x - a)$ approximating $f(x)$ near a.

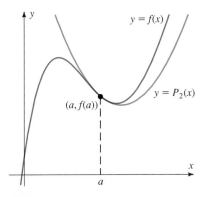

Figure 1.3 The second degree polynomial P_2 approximating f near a.

Given the function f, we first ask for the polynomial P_0 of lowest degree (degree zero) that best approximates $f(x)$ for x near the number a. Since polynomials of degree zero are constant functions, the obvious answer to this question is the constant polynomial

$$P_0(x) \equiv f(a) \qquad \text{(see Figure 1.1).} \qquad (2)$$

Next, let us look for the polynomial P_1 of degree one (that is, a linear function) that best approximates f near a. Using the same basic approach as we used when we discussed linear approximation (Section 3.7), we write this polynomial in the form

$$P_1(x) = a_0 + a_1(x - a)$$

and derive the coefficients a_0 and a_1. Again we want $P_1(a) = f(a)$, so $a_0 = f(a)$. Now, if we suppose that f is differentiable at a, the line through $(a, f(a))$ that best approximates the graph of f near a has slope $f'(a)$. Thus, we also specify that $f'(a) = P_1'(a) = a_1$. The polynomial P_1 is therefore

$$P_1(x) = f(a) + f'(a)(x - a) \qquad \text{(see Figure 1.2).} \qquad (3)$$

Thus, we obtain P_1 as defined in equation (1).

Continuing in this way, we write the second degree polynomial P_2 approximating $f(x)$ as

$$P_2(x) = a_0 + a_1(x - a) + a_2(x - a)^2.$$

As before, if we assume that $P_2(a) = f(a)$, we obtain $a_0 = f(a)$, and if we assume that $P_2'(a) = f'(a)$, we obtain $a_1 = f'(a)$. To determine the remaining constant a_2, we require that P_2 and f have the same measure of concavity at a; that is, we set $P_2''(a) = f''(a)$. This equation is equivalent to

$$P_2''(a) = 2a_2 = f''(a).$$

Thus, $a_2 = f''(a)/2$, and P_2 is therefore

$$P_2(x) = f(a) + f'(a)(x - a) + \frac{f''(a)}{2}(x - a)^2 \qquad \text{(see Figure 1.3).} \qquad (4)$$

To find the general form of these polynomials, we must carry this analysis one step further. We derive a polynomial P_3 of degree three such that $P_3(a) = f(a)$, $P_3'(a) = f'(a)$, $P_3''(a) = f''(a)$, and $P_3'''(a) = f'''(a)$. To do so, we write

$$P_3(x) = a_0 + a_1(x - a) + a_2(x - a)^2 + a_3(x - a)^3,$$

and we obtain

$$\begin{aligned}
P_3(a) &= a_0 = f(a); & a_0 &= f(a) \\
P_3'(a) &= a_1 = f'(a); & a_1 &= f'(a) \\
P_3''(a) &= 2a_2 = f''(a); & a_2 &= f''(a)/2 \\
P_3'''(a) &= 3 \cdot 2 \cdot a_3 = f'''(a); & a_3 &= f'''(a)/3!.
\end{aligned}$$

(Recall that $n! = n(n - 1)(n - 2) \cdots 3 \cdot 2 \cdot 1$. We also define $0! = 1$.) The polynomial P_3 is therefore

$$P_3(x) = f(a) + f'(a)(x - a) + \frac{f''(a)}{2}(x - a)^2 + \frac{f'''(a)}{3!}(x - a)^3 \qquad (5)$$

(see Figure 1.4).

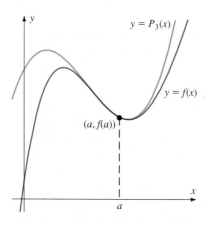

Figure 1.4 The third degree polynomial P_3 approximating f near a.

We generalize equations (2)–(5) by stating that the approximating polynomial of degree n for the function f near the number a is

$$P_n(x) = f(a) + f'(a)(x - a) + \frac{f''(a)}{2}(x - a)^2$$
$$+ \frac{f'''(a)}{3!}(x - a)^3 + \cdots + \frac{f^{(n)}(a)}{n!}(x - a)^n. \tag{6}$$

The polynomial P_n in (6) is called the nth **Taylor polynomial*** for f, expanded about $x = a$. The ideas we discuss here were developed by the English mathematician Brook Taylor (1686–1731) early in the eighteenth century (see the historical notes that begin Unit 5).

Figures 1.1 through 1.4 suggest that the polynomials P_n approximate the function f near $x = a$ more closely as the degree of P_n increases. In this section, we verify this observation for a number of examples. In Sections 13.2 and 13.3, we present a general method for determining a bound on the error involved in the approximation of $f(x)$ by $P_n(x)$.

Before discussing the examples, we note that the kth coefficient of P_n (the coefficient of the $(x - a)^k$ term) is

$$\frac{f^{(k)}(a)}{k!}$$

if $k \le n$. Consequently, the first $n - 1$ terms of the nth Taylor polynomial agree with the $(n - 1)$st Taylor polynomial; that is,

$$P_n(x) = P_{n-1}(x) + \frac{f^{(n)}(a)}{n!}(x - a)^n.$$

□ **EXAMPLE 1**

Find P_0, P_1, P_2, and P_3 for the function $f(x) = e^x$ at $a = 0$ and use these polynomials to approximate the number e.

Solution The values of $f(x) = e^x$ and the first three derivatives at $a = 0$ are

$$f(x) = e^x; \qquad f(0) = e^0 = 1$$
$$f'(x) = e^x; \qquad f'(0) = 1$$
$$f''(x) = e^x; \qquad f''(0) = 1$$
$$f'''(x) = e^x; \qquad f'''(0) = 1.$$

From this information and equation (6) we obtain

$$P_0(x) = 1$$

$$P_1(x) = 1 + x$$

---

*There is a slight abuse of terminology here. Since it may be the case that $f^{(n)}(a) = 0$, an nth Taylor polynomial as defined above may actually be a polynomial of lower degree (see Example 2).

$$P_2(x) = 1 + x + \frac{x^2}{2}$$

$$P_3(x) = 1 + x + \frac{x^2}{2} + \frac{x^3}{3!}.$$

Using these polynomials we may approximate the value $f(1) = e^1 = e$ by the numbers

$$P_0(1) = 1$$

$$P_1(1) = 1 + 1 = 2$$

$$P_2(1) = 1 + 1 + \frac{1^2}{2} = \frac{5}{2} = 2.5$$

$$P_3(1) = 1 + 1 + \frac{1^2}{2} + \frac{1^2}{6} = \frac{8}{3} = 2.666\overline{6}\ldots.$$

Figure 1.5 shows graphs of these Taylor polynomials and the approximations $P_n(1)$ to the value $f(1) = e^1 = e$. Table 1.1 compares, to four decimal place accuracy, values of these Taylor polynomials and the function $f(x) = e^x$ for several different numbers x.

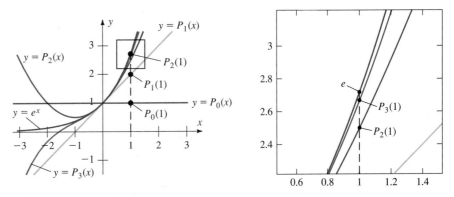

Figure 1.5   The Taylor polynomials P_0, P_1, P_2, and P_3 expanded about $a = 0$. The numbers $P_0(1)$, $P_1(1)$, $P_2(1)$, and $P_3(1)$ are approximations to $f(1) = e$.

Table 1.1   Values of Taylor approximations to e^x near $a = 0$

x	$P_0(x)$	$P_1(x)$	$P_2(x)$	$P_3(x)$	e^x
-1.5	1.0	-0.5	0.6250	0.0625	0.2231
-1.0	1.0	0	0.5	0.3333	0.3679
0	1.0	1.0	1.0	1.0	1.0
0.5	1.0	1.5	1.625	1.6458	1.6487
1.0	1.0	2.0	2.5	2.6667	2.7183
1.5	1.0	2.5	3.6250	4.1875	4.4817

☐ **EXAMPLE 2**

Find the Taylor polynomials P_0, P_1, P_2, ..., P_5 for $f(x) = \sin x$ expanded about $a = 0$.

Solution   Here

$$
\begin{aligned}
f(x) &= \sin x; & f(0) &= 0 \\
f'(x) &= \cos x; & f'(0) &= 1 \\
f''(x) &= -\sin x; & f''(0) &= 0 \\
f'''(x) &= -\cos x; & f'''(0) &= -1 \\
f^{(4)}(x) &= \sin x; & f^{(4)}(0) &= 0 \\
f^{(5)}(x) &= \cos x; & f^{(5)}(0) &= 1.
\end{aligned}
$$

Thus, by (6),

$$P_0(x) = 0$$

$$P_1(x) = 0 + 1x = x$$

$$P_2(x) = 0 + 1x + \frac{0}{2}x^2 = x$$

$$P_3(x) = 0 + 1x + \frac{0}{2}x^2 + \frac{-1}{3!}x^3 = x - \frac{x^3}{3!}$$

$$P_4(x) = 0 + 1x + \frac{0}{2}x^2 + \frac{-1}{3!}x^3 + \frac{0}{4!}x^4 = x - \frac{x^3}{3!}$$

$$P_5(x) = 0 + 1x + \frac{0}{2}x^2 + \frac{-1}{3!}x^3 + \frac{0}{4!}x^4 + \frac{1}{5!}x^5 = x - \frac{x^3}{3!} + \frac{x^5}{5!}.$$

(See Figure 1.6)   ■

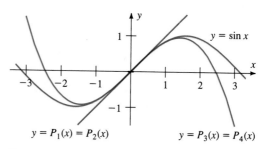

$$y = P_1(x) = P_2(x) \qquad\qquad y = P_3(x) = P_4(x)$$

Figure 1.6   The graphs of four Taylor polynomials for $f(x) = \sin x$ expanded about $a = 0$. Note that $P_1(x)$ and $P_2(x)$ coincide, as do $P_3(x)$ and $P_4(x)$.

☐ **EXAMPLE 3**

In Example 2 the even powers of x in the Taylor polynomials for $f(x) = \sin x$ at $a = 0$ dropped out because even derivatives of $\sin x$ are zero at $a = 0$. For comparison, we next determine the fourth Taylor polynomial for $f(x) = \sin x$ expanded about $a = \pi/6$.

$$
\begin{aligned}
f(x) &= \sin x; & f(\pi/6) &= 1/2 \\
f'(x) &= \cos x; & f'(\pi/6) &= \sqrt{3}/2
\end{aligned}
$$

$$f''(x) = -\sin x; \qquad f''(\pi/6) = -1/2$$
$$f'''(x) = -\cos x; \qquad f'''(\pi/6) = -\sqrt{3}/2$$
$$f^{(4)}(x) = \sin x; \qquad f^{(4)}(\pi/6) = 1/2.$$

By (6),

$$P_4(x) = \frac{1}{2} + \frac{\sqrt{3}}{2}\left(x - \frac{\pi}{6}\right) + \left(\frac{1}{2}\right)\left(-\frac{1}{2}\right)\left(x - \frac{\pi}{6}\right)^2$$

$$+ \left(\frac{1}{3!}\right)\left(-\frac{\sqrt{3}}{2}\right)\left(x - \frac{\pi}{6}\right)^3 + \left(\frac{1}{4!}\right)\left(\frac{1}{2}\right)\left(x - \frac{\pi}{6}\right)^4$$

$$= \frac{1}{2} + \frac{\sqrt{3}}{2}\left(x - \frac{\pi}{6}\right) - \frac{1}{4}\left(x - \frac{\pi}{6}\right)^2 - \frac{\sqrt{3}}{12}\left(x - \frac{\pi}{6}\right)^3$$

$$+ \frac{1}{48}\left(x - \frac{\pi}{6}\right)^4. \qquad \blacksquare$$

Comparing P_4 in Example 2 with P_4 in Example 3, we see that the form of the nth Taylor polynomial depends a great deal on the number a about which the polynomial is expanded.

□ **EXAMPLE 4**

Find the Taylor polynomials $P_0, P_1, \ldots, P_4$ for the function $f(x) = \cos x$ expanded about $a = 0$.

Solution   The required derivatives are as follows:

$$f(x) = \cos x; \qquad f(0) = 1$$
$$f'(x) = -\sin x; \qquad f'(0) = 0$$
$$f''(x) = -\cos x; \qquad f''(0) = -1$$
$$f'''(x) = \sin x; \qquad f'''(0) = 0$$
$$f^{(4)} = \cos x; \qquad f^{(4)}(0) = 1.$$

The Taylor polynomials are

$$P_0(x) = 1$$
$$P_1(x) = 1 + 0x = 1$$
$$P_2(x) = 1 + 0x + \frac{-1}{2}x^2 = 1 - \frac{x^2}{2}$$
$$P_3(x) = 1 + 0x + \frac{-1}{2}x^2 + \frac{0}{3!}x^3 = 1 - \frac{x^2}{2}$$
$$P_4(x) = 1 + 0x + \frac{-1}{2}x^2 + \frac{0}{3!}x^3 + \frac{1}{4!}x^4 = 1 - \frac{x^2}{2} + \frac{x^4}{4!}. \qquad \blacksquare$$

□ **EXAMPLE 5**

Find the Taylor polynomials $P_0, P_1, \ldots, P_5$ expanded about $a = 1$ for the function $f(x) = \ln x$ and use these polynomials to approximate $\ln 2$.

Solution   We have

$$f(x) = \ln x; \qquad f(1) = \ln 1 = 0$$
$$f'(x) = 1/x = x^{-1}; \qquad f'(1) = 1^{-1} = 1$$

$$f''(x) = -x^{-2}; \qquad f''(1) = -1^{-2} = -1$$
$$f'''(x) = 2x^{-3}; \qquad f'''(1) = 2 \cdot 1^{-3} = 2$$
$$f^{(4)}(x) = -6x^{-4}; \qquad f^{(4)}(1) = -6 \cdot 1^{-4} = -6$$
$$f^{(5)}(x) = 24x^{-5}; \qquad f^{(5)}(1) = 24 \cdot 1^{-5} = 24.$$

Thus,

$$P_0(x) = 0$$
$$P_1(x) = 0 + 1(x - 1) = x - 1$$
$$P_2(x) = 0 + 1(x - 1) + \frac{-1}{2}(x - 1)^2 = (x - 1) - \frac{1}{2}(x - 1)^2$$
$$P_3(x) = 0 + 1(x - 1) + \frac{-1}{2}(x - 1)^2 + \frac{2}{3!}(x - 1)^3$$
$$= (x - 1) - \frac{1}{2}(x - 1)^2 + \frac{1}{3}(x - 1)^3$$
$$P_4(x) = 0 + 1(x - 1) + \frac{-1}{2}(x - 1)^2 + \frac{2}{3!}(x - 1)^3 + \frac{-6}{4!}(x - 1)^4$$
$$= (x - 1) - \frac{1}{2}(x - 1)^2 + \frac{1}{3}(x - 1)^3 - \frac{1}{4}(x - 1)^4$$

and

$$P_5(x) = 0 + 1(x - 1) + \frac{-1}{2}(x - 1)^2 + \frac{2}{3!}(x - 1)^3 + \frac{-6}{4!}(x - 1)^4$$
$$+ \frac{24}{5!}(x - 1)^5$$
$$= (x - 1) - \frac{1}{2}(x - 1)^2 + \frac{1}{3}(x - 1)^3 - \frac{1}{4}(x - 1)^4 + \frac{1}{5}(x - 1)^5.$$

Since $\ln 2 = f(2)$, we obtain approximations to $\ln 2$ from these polynomials by setting $x = 2$:

$$P_0(2) = 0$$
$$P_1(2) = (2 - 1) = 1$$
$$P_2(2) = (2 - 1) - \tfrac{1}{2}(2 - 1)^2 = 1 - \tfrac{1}{2} = \tfrac{1}{2} = 0.5$$
$$P_3(2) = (2 - 1) - \tfrac{1}{2}(2 - 1)^2 + \tfrac{1}{3}(2 - 1)^3 = 1 - \tfrac{1}{2} + \tfrac{1}{3} = \tfrac{5}{6} \approx 0.8333$$
$$P_4(2) = (2 - 1) - \tfrac{1}{2}(2 - 1)^2 + \tfrac{1}{3}(2 - 1)^3 - \tfrac{1}{4}(2 - 1)^4$$
$$= 1 - \tfrac{1}{2} + \tfrac{1}{3} - \tfrac{1}{4} = \tfrac{7}{12} \approx 0.5833$$
$$P_5(2) = (2 - 1) - \tfrac{1}{2}(2 - 1)^2 + \tfrac{1}{3}(2 - 1)^3 - \tfrac{1}{4}(2 - 1)^4 + \tfrac{1}{5}(2 - 1)^5$$
$$= 1 - \tfrac{1}{2} + \tfrac{1}{3} - \tfrac{1}{4} + \tfrac{1}{5} = \tfrac{47}{60} \approx 0.7833.$$

By comparison, the exact value of $\ln 2$, to four decimal places, is 0.6931. ∎

Exercise Set 13.1

In Exercises 1–22, find the nth Taylor polynomial for the function f expanded about $x = a$.

1. $f(x) = e^{-x}, \quad a = 0, \quad n = 4$

2. $f(x) = e^x, \quad a = 1, \quad n = 4$

3. $f(x) = \cos x, \quad a = \pi/4, \quad n = 6$

4. $f(x) = \sin x, \quad a = \pi/3, \quad n = 5$

5. $f(x) = \ln(1 + x), \quad a = 0, \quad n = 4$

6. $f(x) = \dfrac{1}{1+x}$, $a = 0$, $n = 4$

7. $f(x) = \tan x$, $a = 0$, $n = 3$

8. $f(x) = \cos 2x$, $a = 0$, $n = 6$

9. $f(x) = \sin(-\pi x)$, $a = 0$, $n = 5$

10. $f(x) = \sec x$, $a = 0$, $n = 3$

11. $f(x) = e^{x^2}$, $a = 0$, $n = 3$

12. $f(x) = \sqrt{x}$, $a = 2$, $n = 2$

13. $f(x) = \sqrt{1-x}$, $a = 0$, $n = 3$

14. $f(x) = x \sin x$, $a = 0$, $n = 4$

15. $f(x) = \sec x$, $a = \pi/4$, $n = 3$

16. $f(x) = \dfrac{1}{1+x^2}$, $a = 0$, $n = 2$

17. $f(x) = \text{Tan}^{-1} x$, $a = 0$, $n = 3$

18. $f(x) = \dfrac{1}{1-x^3}$, $a = 1/2$, $n = 1$

19. $f(x) = \dfrac{2x}{2+x^2}$, $a = 1$, $n = 2$

20. $f(x) = \dfrac{1}{1+e^x}$, $a = 0$, $n = 3$

21. $f(x) = 3x^3 + 2x + 1$, $a = -1$, $n = 3$

22. $f(x) = x^4$, $a = 2$, $n = 4$

23. Compare the results of Exercises 5 and 6 and those of Exercises 16 and 17. Formulate a conjecture on the relationship between the Taylor polynomial for f and that for f' and test your conjecture on a few examples.

24. What is the relationship between the Taylor polynomial for f at $a = 0$ and that for $h(x) = xf(x)$? Can you prove this?

25. True or false? If f is an even function, the Taylor polynomial for f expanded about $a = 0$ will contain only even powers of x. Explain.

26. The Taylor polynomial P_n is defined for f if each of the first n derivatives of f exists at $x = a$. Does this imply that $f(x)$ is defined for all x for which $P_n(x)$ is defined? (*Hint:* Consider this question for the function in Example 5.)

In Exercises 27–30 graph the given function f along with its Taylor polynomials P_1, P_2, P_3, P_4, and P_5 about the given number a over the specified interval. How well do these Taylor polynomials approximate f on this interval?

27. $f(x) = e^{2x}$, $a = 0$, $[-1/2, 1/2]$

28. $f(x) = \ln x$, $a = 1$, $[1/2, 2]$

29. $f(x) = \sqrt{x+1}$, $a = 1$, $[0, 2]$

30. $f(x) = \cos 2x$, $a = \pi/4$, $[0, \pi/2]$

31. The Taylor polynomial P_{2n} for $f(x) = 1/(1+x^2)$ is

$$P_{2n}(x) = \sum_{k=0}^{n} (-1)^k x^{2k}.$$

  a. For $n = 0, 1, 2, 3, 4$, and 5, graph f and P_{2n} over the interval $[-2, 2]$. How well do these polynomials approximate f on the intervals $[-2, -1]$ and $[1, 2]$? Justify your answer. (*Hint:* $\sum_{k=0}^{\infty} (-1)^k x^{2k}$ is a geometric series. See Exercise 36 in Section 12.3.)

  b. Graph $F(x) = \text{Tan}^{-1} x$ and $Q_{2n+1}(x) = \int_0^x P_{2n}(t)\, dt$ for $n = 0, 1, 2, 3, 4$, and 5. Formulate a conjecture for the relationship between the Taylor polynomials of f and the Taylor polynomials of F.

32. Graph the difference $f(x) - P_n(x)$ for $-\pi/2 \le x \le \pi/2$ where $f(x) = \cos x$ and $P_n(x)$ is the Taylor polynomial of $f(x)$ about $a = 0$. Determine the numbers x in this interval at which $|f(x) - P_n(x)|$ is a maximum. How does the maximum of $|f(x) - P_n(x)|$ decrease as n increases?

33. The Taylor polynomial about $x = a$ is a good approximation to $f(x)$ as long as x is near a. However, there are different, more accurate polynomial approximations over an entire interval. Compare the graphs of $f(x) = e^x$ and its Taylor polynomials of degrees 1, 2, and 3 about $a = 0$ to the graphs of the polynomials

$$q_1(x) = 1.26428 + 1.1752x$$
$$q_2(x) = 0.98904 + 1.12018x + 0.55404x^2$$
$$q_3(x) = 0.99458 + 0.99567x + 0.54297x^2 + 0.17953x^3$$

over the interval $-1 \le x \le 1$. These polynomials are called the "best" **uniform approximations** to e^x on the interval $[-1, 1]$.

13.2 Taylor's Theorem

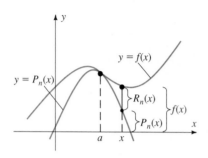

Figure 2.1 The remainder $R_n(x) = f(x) - P_n(x)$.

In the previous section, we derived a formula for the Taylor polynomial P_n of degree n about the number a by matching derivatives at a. That is, we defined P_n for the function f by requiring

$$P_n^{(k)}(a) = f^{(k)}(a)$$

for $k = 0, 1, 2, 3, \ldots, n$. In this section, we state and prove a theorem that provides bounds on the accuracy of the approximation of $f(x)$ by $P_n(x)$. In the next section, we use this theorem to derive such bounds.

Let

$$R_n(x) = f(x) - P_n(x) \tag{1}$$

denote the *error* in the approximation of $f(x)$ by $P_n(x)$ (see Figure 2.1). We call $R_n(x)$ the **remainder** in a Taylor approximation.* Taylor's Theorem provides a way to estimate the size of $R_n(x)$ when we approximate using Taylor polynomials.

Theorem 1

Taylor's Theorem

Let f be a function that is $(n + 1)$-times differentiable on the open interval I and let a be any number in I. Then, for each $x \in I$, there exists a number c between a and x so that

$$f(x) = f(a) + f'(a)(x - a) + \frac{f''(a)}{2}(x - a)^2 + \cdots + \frac{f^{(n)}(a)}{n!}(x - a)^n$$
$$+ \frac{f^{(n+1)}(c)}{(n + 1)!}(x - a)^{n+1}.$$

The formula for $f(x)$ given in Theorem 1 is called **Taylor's formula with remainder** or just **Taylor's formula.** If we write

$$f(x) = P_n(x) + R_n(x),$$

then Taylor's formula states that the remainder $R_n(x)$ is given by

$$R_n(x) = \frac{f^{(n+1)}(c)}{(n + 1)!}(x - a)^{n+1} \tag{2}$$

for *some* number c between a and x. Note that Theorem 12 in Section 4.8 is a special case of this result.

Roughly speaking, Taylor's Theorem says that the error involved in approximating a value of an $(n + 1)$-times differentiable function by the value of its Taylor polynomial $P_n(x)$ is just the term we would add to $P_n(x)$ to obtain $P_{n+1}(x)$, *except* that the coefficient

$$\frac{f^{(n+1)}(c)}{(n + 1)!}$$

is evaluated at some number c between a and b rather than at a. In the next section, we see how to determine a bound for $R_n(x)$.

---

*The error ϵ discussed in Section 4.8 is essentially the same as $R_1(x)$. In the notation of Section 4.8 with $x_0 = a$, $\epsilon = R(a + \Delta x)$.

☐ **EXAMPLE 1**

Write Taylor's formula using the third Taylor polynomial for $f(x) = e^x$ expanded about $a = 0$.

Solution   From Example 1 of Section 13.1 we have

$$P_3(x) = 1 + x + \frac{x^2}{2!} + \frac{x^3}{3!}.$$

For $f(x) = e^x$, $f^{(4)}(x) = e^x$, so $f^{(4)}(c) = e^c$. Thus, by Taylor's Theorem,

$$R_3(x) = \frac{e^c}{4!}(x - 0)^4$$

where c lies between 0 and x. Taylor's formula is therefore

$$e^x = 1 + x + \frac{x^2}{2} + \frac{x^3}{3!} + \frac{e^c}{4!}x^4. \qquad\blacksquare$$

Since Taylor's Theorem is both important and useful, it is unfortunate that no easily motivated proof has been found. We present one of the standard proofs in the remainder of this section.

Proof of Theorem 1:   In addition to the number a, choose a number $b \in I$. Then we can define a number C (which depends on a, b, and n) by the equation

$$f(b) = f(a) + f'(a)(b - a) + \cdots + \frac{f^{(n)}(a)}{n!}(b - a)^n + C. \qquad (3)$$

That is, C is the difference $C = f(b) - P_n(b)$, where $P_n(b)$ is the Taylor polynomial of degree n expanded about a.

First note that if $b = a$, then (3) collapses to $f(a) = f(a)$, and $C = 0$. Thus, we consider $b \neq a$ (it makes no difference whether $a < b$ or $b < a$) and prove that C has the form of remainder $R_n(x)$ given in the theorem.

To do this, we construct a rather odd-looking function of x, in order to apply Rolle's Theorem on the interval between a and b:

$$F(x) = f(b) - \left[f(x) + f'(x)(b - x) + \cdots + \frac{f^{(n)}(x)}{n!}(b - x)^n \right.$$

$$\left. + C\frac{(b - x)^{n+1}}{(b - a)^{n+1}} \right]. \qquad (4)$$

Now we show the following facts about F:

(i) If $x = a$, then

$$F(a) = f(b) - \left[f(a) + f'(a)(b - a) + \cdots \right.$$

$$\left. + \frac{f^{(n)}(a)}{n!}(b - a)^n + C\frac{(b - a)^{n+1}}{(b - a)^{n+1}} \right]$$

$$= 0 \qquad \text{(by (3))}.$$

(ii) If $x = b$, then

$$F(b) = f(b) - \left[f(b) + f'(b)(b - b) + \cdots \right.$$
$$\left. + \frac{f^{(n)}(b)}{n!}(b - b)^n + C\frac{(b - b)^{n+1}}{(b - a)^{n+1}} \right]$$
$$= f(b) - f(b) = 0.$$

(iii) Because $f^{(n+1)}(x)$ exists for each $x \in I$, F is differentiable on the entire interval, and

$$F'(x) = -\frac{f^{(n+1)}(x)}{n!}(b - x)^n + (n + 1)C\frac{(b - x)^n}{(b - a)^{n+1}}.$$

To see that (iii) is true, we differentiate F in equation (4), using the Product Rule, and note that many terms cancel:

$$F'(x) = 0 - \left[f'(x) + f''(x)(b - x) - f'(x) \right.$$
$$+ \frac{f'''(x)}{2}(b - x)^2 - f''(x)(b - x)$$
$$+ \frac{f^{(4)}(x)}{3!}(b - x)^3 - \frac{f'''(x)}{2}(b - x)^2$$
$$+ \cdots + \frac{f^{(n)}(x)}{(n - 1)!}(b - x)^{n-1} - \frac{f^{(n-1)}(x)}{(n - 2)!}(b - x)^{n-2}$$
$$+ \frac{f^{(n+1)}(x)}{n!}(b - x)^n - \frac{f^{(n)}(x)}{(n - 1)!}(b - x)^{n-1} - (n + 1)C\frac{(b - x)^n}{(b - a)^{n+1}} \right]$$
$$= -\frac{f^{(n+1)}(x)}{n!}(b - x)^n + (n + 1)C\frac{(b - x)^n}{(b - a)^{n+1}}.$$

From statements (i) and (ii) and Rolle's Theorem, it follows that there is a number c between a and b such that $F'(c) = 0$. Using statement (iii), we may write this condition as

$$\frac{f^{(n+1)}(c)}{n!}(b - c)^n = (n + 1)C\frac{(b - c)^n}{(b - a)^{n+1}}.$$

Solving for C now gives

$$C = \frac{f^{(n+1)}(c)}{(n + 1)!}(b - a)^{n+1}. \tag{5}$$

Since we have placed no restriction on b other than $b \neq a$ (and we already know that $C = 0$ if $b = a$), we can substitute x for b in (5), which yields exactly $C = R_n(x)$. This completes the proof. ∎

Exercise Set 13.2

In Exercises 1–14, write Taylor's formula with remainder (Theorem 1) for f using the nth Taylor polynomial expanded about $x = a$.

1. $f(x) = e^{-x}$, $n = 3$, $a = 0$

2. $f(x) = \sin x$, $n = 3$, $a = 0$

3. $f(x) = \cos x$, $n = 4$, $a = 0$

4. $f(x) = \sin x$, $n = 3$, $a = \pi/4$

5. $f(x) = \mathrm{Tan}^{-1} x$, $n = 3$, $a = 0$

6. $f(x) = \sqrt{x}$, $n = 3$, $a = 4$

7. $f(x) = 1/(1 + x^2)$, $n = 2$, $a = 1$

8. $f(x) = 3x^4 + 2x + 2$, $n = 3$, $a = 2$

9. $f(x) = \sec x$, $n = 2$, $a = \pi/4$

10. $f(x) = \ln(1 + x^2)$, $n = 3$, $a = 0$

11. $f(x) = \sinh x$, $n = 4$, $a = 0$

12. $f(x) = x \sinh x$, $n = 3$, $a = 0$

13. $f(x) = \cosh x$, $n = 3$, $a = \ln 2$

14. $f(x) = (1 + x)^{3/2}$, $n = 3$, $a = 0$

15. Let f be a polynomial of degree n. Use Theorem 1 to prove that $f = P_n$.

16. Refer to the precise hypotheses of Rolle's Theorem in Chapter 4. Then show that the result of Theorem 1 still holds if $f^{(n)}$ is continuous on the closed interval $[a, x]$ and if f is $(n + 1)$-times differentiable on the open interval (a, x).

17. Use Taylor's Theorem to justify Newton's Method for approximating the root of a function.

18. Let P_n be the nth Taylor polynomial for $f(x) = \sin x$ expanded about $a = 0$. Show that $\lim_{x \to \infty} R_n(x) = 0$.

19. Let $f(x) = 1/(1 - x)$.
   a. Show that the nth Taylor polynomial of f about $a = 0$ is

$$P_n(x) = 1 + x + x^2 + \cdots + x^n.$$

   b. Show that

$$\frac{1}{1 - x} = 1 + x + x^2 + \cdots + x^n + \frac{x^{n+1}}{1 - x}, \qquad x \neq 1,$$

   by multiplying both sides of the equation by $1 - x$.
   c. Conclude that

$$R_n(x) = \frac{x^{n+1}}{1 - x}.$$

   Does this contradict Taylor's Theorem?

20. By substituting $-x$ for x in the equation in Exercise 19 conclude that

$$\frac{1}{1 + x} = 1 - x + x^2 - x^3 + \cdots$$
$$+ (-1)^n x^n + \frac{(-1)^{n+1} x^{n+1}}{1 + x}, \qquad x \neq -1.$$

Using Taylor's Theorem with $a = 0$, verify that the nth Taylor polynomial for $f(x) = 1/(1 + x)$ is

$$P_n(x) = 1 - x + x^2 - x^3 + \cdots + (-1)^n x^n.$$

21. Replace x by x^2 in Exercise 20 to conclude that

$$\frac{1}{1 + x^2} = 1 - x^2 + x^4 - \cdots$$
$$+ (-1)^n x^{2n} + \frac{(-1)^{n+1} x^{2n+2}}{1 + x^2}.$$

What is the nth Taylor polynomial P_n, expanded about $a = 0$, for $f(x) = 1/(1 + x^2)$?

22. Replace x by t in Exercise 20 and integrate between 0 and x to conclude that

$$\int_0^x \frac{1}{1 + t} \, dt = \int_0^x \left(1 - t + t^2 - t^3 + \cdots \right.$$
$$\left. + (-1)^{n-1} t^{n-1} + \frac{(-1)^n t^n}{1 + t} \right) dt$$
$$= x - \frac{x^2}{2} + \frac{x^3}{3} - \frac{x^4}{4} + \cdots + \frac{(-1)^{n-1} x^n}{n}$$
$$+ (-1)^n \int_0^x \frac{t^n}{1 + t} \, dt.$$

What is the nth Taylor polynomial, expanded about $a = 0$, for $f(x) = \ln(1 + x)$?

23. Use Taylor's Theorem to prove the Binomial Theorem:

$$(x + a)^n = a^n + na^{n-1}x + \frac{n(n - 1)}{2} a^{n-2} x^2 + \cdots$$
$$+ \binom{n}{r} a^{n-r} x^r + \cdots + nax^{n-1} + x^n$$

where n is a positive integer and

$$\binom{n}{r} = \frac{n!}{r!(n - r)!}$$

is the **binomial coefficient.**

24. (Refer to Exercise 33 in Section 13.1.) The maximum error between $f(x) = e^x$ and $q_3(x) = 0.99458 + 0.99567x + 0.54297x^2 + 0.17953x^3$ for $-1 \le x \le 1$ is 0.00553. Determine n so that $|e^x - P_n(x)| \le 0.005$ on $[-1, 1]$ where $P_n(x)$ is the nth Taylor polynomial about $a = 0$. Graph $e^x - q_3(x)$ and $e^x - P_n(x)$. Which polynomial is the more accurate approximation throughout the entire interval?

25. Let r be the rational function

$$r(x) = \frac{x^3 + 9x^2 + 36x + 60}{3x^2 - 24x + 60}.$$

   a. Graph $f(x) = e^x$ and $r(x)$ over the interval $-1 \le x \le 1$. Is there any difference between $f(x)$ and $r(x)$ over this interval? The function r is called a **Padé approximation** and is a generalization of the Taylor polynomial.
   b. Find the maximum of $|e^x - r(x)|$ on $-1 \le x \le 1$. What Taylor polynomial of e^x about $a = 0$ approximates e^x with the same accuracy as $r(x)$ over this interval?

13.3 Applications of Taylor's Theorem

According to Taylor's Theorem, if we approximate $f(x)$ by the value of the nth Taylor polynomial $P_n(x)$ expanded about $x = a$, the error (remainder) in the calculation is

$$R_n(x) = \frac{f^{(n+1)}(c)}{(n+1)!}(x-a)^{n+1} \tag{1}$$

where c lies between a and x. Since x is given, the size of the error depends on the numbers a and n. Obviously, we will want to choose a close to x so that the factors $(x - a)$ in $R_n(x)$ are small. Also, we will want to choose a so that the function value $f(a)$ and the various derivatives $f^{(k)}(a)$ are easy to compute.

Once a convenient value for a is chosen, the magnitude of the error depends only on the integer n. Since we seldom know the number c, we usually cannot determine $R_n(x)$ precisely. However, in most applications we are concerned only with knowing the approximate size of $R_n(x)$. For example, to approximate $\ln(1.2)$ to within 0.01, we need only demonstrate that $|R_n(1.2)| < 0.01$.

In general, to achieve a Taylor approximation for $f(x)$ to within ϵ, we find n sufficiently large to guarantee that the inequality

$$\left| \frac{f^{(n+1)}(c)}{(n+1)!}(x-a)^{n+1} \right| < \epsilon$$

holds. By (1) this estimate assures the desired accuracy.

□ EXAMPLE 1

Use the Taylor polynomial of degree 3 for $f(x) = \ln x$ expanded about $a = 1$ to approximate $\ln(1.5)$ and estimate the accuracy of this approximation.

Strategy · · · · · · · · ·
Find P_3 as in Section 13.1.

Solution

We have

$$f(x) = \ln x; \qquad f(1) = 0$$

$$f'(x) = \frac{1}{x}; \qquad f'(1) = 1$$

$$f''(x) = -\frac{1}{x^2}; \qquad f''(1) = -1$$

$$f'''(x) = \frac{2}{x^3}; \qquad f'''(1) = 2.$$

Thus

$$P_3(x) = 1(x-1) - \frac{1}{2}(x-1)^2 + \frac{2}{3!}(x-1)^3.$$

Evaluate P_3 at $x = 1.5$ to obtain approximation $P_3(1.5)$.

The approximation is therefore

$$\ln(1.5) \approx P_3(1.5) = (.5) - \frac{1}{2}(.5)^2 + \frac{1}{3}(.5)^3$$

$$= .416\overline{6}$$

Since $f^{(4)}(x) = -6x^{-4}$ for $f(x) = \ln x$, by (1) the error in the approximation is

Write expression for $|R_3(1.5)|$ using (1).

$$|R_3(1.5)| = \left| \frac{-6c^{-4}}{4!}(1.5 - 1)^4 \right| = \left| \frac{-6c^{-4}}{24}(.5)^4 \right|$$

$$= |c^{-4}| \frac{(.5)^4}{4}.$$

Since c is unknown, we must determine the largest possible value for $|f^{(4)}(c)|$ to obtain an upper bound on the error. We know that c is between $a = 1$ and $x = 1.5$.

Since $|c^{-4}| < 1$ for $1 < c < 1.5$, we have the inequality

$$|R_3(1.5)| < \frac{(.5)^4}{4} = .015625.$$

We may conclude only that the approximation is accurate to one decimal place. ∎

REMARK When we say that a number is accurate to k decimal places, we mean that the error is less than $5 \times 10^{-(k+1)}$. That is, accuracy to one decimal place means an error less than $5 \times 10^{-2} = .05$, accuracy to two decimal places means an error less than .005, and so forth. Thus we could claim that the approximation in Example 1 was accurate to one decimal place, and the approximation in Example 2 (below) is accurate to three decimal places.

☐ **EXAMPLE 2**

Suppose we use the approximation

$$\sin x \approx x$$

for numbers x satisfying the inequality $|x| < \pi/45$. Use Taylor's Theorem to find a bound for the maximal error.

Strategy · · · · · · · ·
Determine the Taylor polynomial $P_n(x)$ associated with the approximation.

Once $f(x)$, n, and a are known, use equation (1).

Estimate the maximum possible size of $|f''(c)|$. Use to obtain bound on $|R_2(x)|$.

Solution
For $f(x) = \sin x$, both $P_1(x) = x$ and $P_2(x) = x$ when $a = 0$. We may therefore take $n = 2$ in equation (1). Since $f'''(c) = -\cos(c)$, the error is

$$|R_2(x)| = \left| \frac{-\cos(c)}{3!}(x - 0)^3 \right| = \frac{|\cos(c)|}{3!}|x|^3.$$

Since $|\cos(c)| \le 1$ for all numbers c and since $|x| < \pi/45$, we have

$$|R_2(x)| \le \frac{1}{3!}|x|^3 < \frac{1}{3!}\left(\frac{\pi}{45}\right)^3 < .00006.$$

The error in the approximation is less than $.00006 = 6 \times 10^{-5}$. Note that if we had used $n = 1$ in these calculations, we would have had $|f''(c)| = |-\sin(c)| \le 1$ and we could have claimed only that

$$|R_1(x)| \le \frac{1}{2!}|x|^2 < \frac{1}{2}\left(\frac{\pi}{45}\right)^2 < .0025. \quad ∎$$

☐ **EXAMPLE 3**

How large must n be so that $\cos 48°$ is approximated with four decimal place accuracy using the Taylor polynomial P_n for $f(x) = \cos x$ expanded about $a = 45° = \pi/4$?

Strategy · · · · · · · · ·

Set up the expression for $|R_n(x)|$ using $x = 48° = 4\pi/15$.

Find an upper bound for $|R_n(x)|$.

Require that the upper bound be less than the desired degree of accuracy.

Solution

For $f(x) = \cos x, f^{(n+1)}(c)$ is either $\pm \sin c$ or $\pm \cos c$. In either case, $|f^{(n+1)}(c)| \leq 1$. Using this inequality we obtain

$$\left| R_n\left(\frac{4\pi}{15}\right) \right| = \left| \frac{f^{(n+1)}(c)}{(n+1)!}\left(\frac{4\pi}{15} - \frac{\pi}{4}\right)^{n+1} \right| < \frac{1}{(n+1)!}\left(\frac{\pi}{60}\right)^{n+1}. \tag{2}$$

To ensure that $\left| R_n\left(\frac{4\pi}{15}\right) \right| < 5 \times 10^{-5}$, we need to find n large enough so that

$$\frac{1}{(n+1)!}\left(\frac{\pi}{60}\right)^{n+1} < 5 \times 10^{-5}. \tag{3}$$

We now proceed to find n sufficiently large that (3) holds. Values of the left side of inequality (3) for various integers n are as follows

n	$\dfrac{1}{(n+1)!}\left(\dfrac{\pi}{60}\right)^{n+1}$
1	.0013708
2	.0000240
3	.0000003

Thus $n = 2$ is sufficient to give the desired accuracy.   ■

☐ **EXAMPLE 4**

Find a bound on the magnitude of $|x|$ so that the approximation

$$e^x \approx P_3(x) = 1 + x + \frac{x^2}{2!} + \frac{x^3}{3!}$$

is accurate to within .001.

Strategy · · · · · · · · ·

Set up the expression for $|R_3(x)|$ using (1).

Find an expression for the maximum size of the factor e^c.

Since e^x is what is being approximated, the factor $e^{|x|}$ must be replaced by a "safe" upper bound.

Solution

First note that the form of the given approximation is the Taylor polynomial P_3 expanded about $a = 0$.

Since $f^{(n)}(c) = e^c$ for all n, we have

$$|R_3(x)| = \left| \frac{e^c}{4!}x^4 \right| = \frac{e^c}{24}x^4.$$

Now since e^c is an increasing function and c lies between 0 and x, the maximum value of e^c for $-|x| < c < |x|$ is $e^{|x|}$. Thus

$$|R_3(x)| \leq \frac{e^{|x|}}{24}x^4. \tag{4}$$

For $|R_3(x)|$ in (4) to be less than .001, we will clearly need to take $|x| < 1$. Thus we may safely bound the term $e^{|x|}$ by, say, 4 since $e^1 \approx 2.718 < 4$.* We therefore need to find the maximum value of x for which

---

*See Exercise 74, Section 8.2, where the inequality $e < 4$ was established. Of course, we must also obtain a valid estimate using the fact that $e < 3$, but we have not *proved* that $e < 3$.

Solve the resulting inequality for x^n.

$$\frac{4}{24}x^4 < .001 \quad \text{or} \quad x^4 < .006.$$

By trial and error (or by extracting the fourth root), find a value of x satisfying the inequality.

Since $(.25)^4 < .0039 < .006$, the bound $|x| < .25$ will assure the desired accuracy.

■

We conclude this section with remarks on two generalizations suggested by Example 3.

Since $(\pi/60) < 1$, we may write inequality (2) of Example 3 as

$$\left| R_n\left(\frac{4\pi}{15}\right) \right| < \frac{1}{(n+1)!}\left(\frac{\pi}{60}\right)^{n+1} < \frac{1}{(n+1)!}.$$

This inequality shows that $R_n(4\pi/15) \to 0$ as $n \to \infty$. In other words, $\cos 48°$ can be approximated to any desired degree of accuracy by simply taking n sufficiently large. Since increasing n corresponds to taking polynomials of higher degree, this means that the *sequence* of estimates

$$P_0\left(\frac{4\pi}{15}\right) = \frac{\sqrt{2}}{2}$$

$$P_1\left(\frac{4\pi}{15}\right) = \frac{\sqrt{2}}{2} - \frac{\sqrt{2}}{2}\left(\frac{\pi}{60}\right)$$

$$P_2\left(\frac{4\pi}{15}\right) = \frac{\sqrt{2}}{2} - \frac{\sqrt{2}}{2}\left(\frac{\pi}{60}\right) - \frac{\sqrt{2}}{4}\left(\frac{\pi}{60}\right)^2$$

$$P_3\left(\frac{4\pi}{15}\right) = \frac{\sqrt{2}}{2} - \frac{\sqrt{2}}{2}\left(\frac{\pi}{60}\right) - \frac{\sqrt{2}}{4}\left(\frac{\pi}{60}\right)^2 + \frac{\sqrt{2}}{12}\left(\frac{\pi}{60}\right)^3$$

approaches the number $\cos(4\pi/15)$. In the language of Chapter 12 we will say that

(i) the *sequence* of approximations

$$\left\{ P_0\left(\frac{4\pi}{15}\right), P_1\left(\frac{4\pi}{15}\right), P_2\left(\frac{4\pi}{15}\right), \dots \right\}$$

approaches $\cos(4\pi/15)$ *in the limit,* that is,

$$\lim_{n\to\infty} P_n\left(\frac{4\pi}{15}\right) = \cos\left(\frac{4\pi}{15}\right);$$

(ii) the *series* of terms

$$\sum_{j=0}^{n} \frac{f^{(j)}(\pi/4)}{j!}\left(\frac{\pi}{60}\right)^j$$

$$= \frac{\sqrt{2}}{2} - \frac{\sqrt{2}}{2}\left(\frac{\pi}{60}\right) - \frac{\sqrt{2}}{4}\left(\frac{\pi}{60}\right)^2 + \cdots + \frac{f^{(n)}(\pi/4)}{n!}\left(\frac{\pi}{60}\right)^n$$

approaches $\cos(4\pi/15)$ *in the limit* as $n \to \infty$, that is,

$$\sum_{j=0}^{\infty} \frac{f^{(j)}(4\pi/15)}{j!}\left(\frac{\pi}{60}\right)^j = \cos(4\pi/15).$$

These observations give a glimpse of what is ahead: the limit

$$\lim_{x \to \infty} P_n(x) = \sum_{k=0}^{\infty} \frac{f^{(k)}(a)}{k!}(x-a)^k$$

will be viewed as an infinite series containing powers of the variable $(x-a)$ that when it converges, "represents" the function f. Before developing this notion of *Taylor series,* we discuss the more general notion of *power series* in Section 13.4.

Exercise Set 13.3

In Exercises 1–10, use Taylor's Theorem to make the indicated approximation and estimate the accuracy using equation (1).

1. $\ln(1.5)$, $f(x) = \ln(x+1)$, $a = 0$, $n = 3$

2. $\cos 36°$, $f(x) = \cos x$, $a = \pi/4$, $n = 2$

3. $\sin 80°$, $f(x) = \sin x$, $a = \pi/2$, $n = 3$

4. $\cos 1$, $f(x) = \cos x$, $a = \pi/3$, $n = 2$

5. $e^{0.2}$, $f(x) = e^x$, $a = 0$, $n = 3$

6. $\sqrt{3.91}$, $f(x) = \sqrt{x}$, $a = 4$, $n = 2$

7. $\sqrt{9.2}$, $f(x) = \sqrt{x}$, $a = 9$, $n = 2$

8. $\text{Sin}^{-1}(0.2)$, $f(x) = \text{Sin}^{-1} x$, $a = 0$, $n = 1$

9. $\sqrt[3]{10}$, $f(x) = \sqrt[3]{x}$, $a = 8$, $n = 2$

10. $\text{Tan}^{-1}(1/2)$, $f(x) = \text{Tan}^{-1} x$, $a = 0$, $n = 2$

In Exercises 11–17, determine a bound on the accuracy of the given approximation for the indicated range of x.

11. $\sin x \approx x$, $|x| < .05$

12. $\sin x \approx x - \dfrac{x^3}{3!}$, $|x| < .15$

13. $\cos x \approx \dfrac{1}{2} - \dfrac{\sqrt{3}}{2}\left(x - \dfrac{\pi}{3}\right)$, $\left|x - \dfrac{\pi}{3}\right| < .05$

14. $\tan x \approx 1 + 2\left(x - \dfrac{\pi}{4}\right)$, $\left|x - \dfrac{\pi}{4}\right| < \dfrac{\pi}{36}$

15. $\sqrt[3]{1+x} \approx 1 + \dfrac{x}{3}$, $|x| < .025$

16. $\ln x \approx (x-1) - \dfrac{1}{2}(x-1)^2 + \dfrac{1}{3}(x-1)^3$, $|x-1| < 0.1$

17. $\sqrt{1+x} \approx 1 + \dfrac{x}{2}$, $0 < x < .02$

In Exercises 18–23, determine how large n must be taken to ensure accuracy to four decimal places in approximating the given quantity.

18. $\sqrt{38}$ using $f(x) = \sqrt{x}$, $a = 36$

19. $\ln 1.3$ using $f(x) = \ln(x+1)$, $a = 0$

20. $\sin 9°$ using $f(x) = \sin x$, $a = 0$

21. $\cos 42°$ using $f(x) = \cos x$, $a = \pi/4$

22. $e^{0.3}$ using $f(x) = e^x$, $a = 0$.

23. $\sqrt[3]{9.2}$ using $f(x) = \sqrt[3]{x}$, $a = 8$

24. Approximate $\ln 1.25$ to 4 decimal places.

25. Approximate e to four decimal places.

26. (Another way to approximate $\ln x$.)
   a. Find the third Taylor polynomial with $a = 0$ for

$$f(x) = \ln\left(\frac{1+x}{1-x}\right)$$

   including the remainder term.

   b. Find the number x for which $\dfrac{1+x}{1-x} = 1.5$.

   c. Find the accuracy in using the polynomial in part a to approximate $\ln(1.5)$.

   d. Compare this accuracy with that obtained in Example 1 using the third Taylor polynomial for $f(x) = \ln(1+x)$.

27. A scientist needing to calculate $\cos x$ for small angles, say $|x| < 6°$, wonders how much accuracy is lost in simply using the approximation $\cos x \approx 1$. What is the answer?

28. Suppose that you need to make many hand calculations of the function $f(x) = x^5 + 3x^3 + 2x + 6$ for numbers x between 0.8 and 1.0. Explain how you might obtain the approximation

$$f(x) \approx 12 + 16(x-1).$$

What accuracy can you expect from such approximations? What if, instead, you use the approximation

$$f(x) \approx 12 + 16(x-1) + 19(x-1)^2?$$

29. Use Taylor's Theorem to prove the Second Derivative Test for relative extrema.

30. Show that

$$\sin x = x - \frac{x^3}{3!} + R_4(x)$$

where $|R_4(x)| \leq |x|^5/5!$. Use this to show that

$$\int_0^1 \sin x \, dx = \int_0^1 \left(x - \frac{x^3}{3!} + R_4(x) \right) dx$$

$$= \frac{11}{24} + \int_0^1 R_4(x) \, dx.$$

Conclude that the number 11/24 provides an approximation to the integral

$$\int_0^1 \sin x \, dx$$

with an error E no greater than

$$E = \left| \int_0^1 \sin x \, dx - \frac{11}{24} \right| = \left| \int_0^1 R_4(x) \, dx \right|$$

$$\leq \int_0^1 |R_4(x)| \, dx$$

$$\leq \int_0^1 \frac{x^5}{5!} \, dx$$

$$\leq \frac{1}{6!} \approx 0.0014.$$

31. Second degree Taylor polynomials can be used to find the maximum of a function f. For a first guess a_0 at the desired maximum of f, find a number a_0 for which $f''(a_0) < 0$. (Why is this inequality relevant?) Then calculate the second degree Taylor polynomial

$$P_2(x) = f(a_0) + f'(a_0)(x - a_0) + \frac{f''(a_0)}{2}(x - a_0)^2$$

about a_0 and determine the x-coordinate a_1 of the vertex of the graph of P_2. Iterate this procedure where a_{n+1} is determined using the vertex of the second degree Taylor polynomial of f about a_n. For example, we illustrate this algorithm for $f(x) = e^{-2x} + \sin x$ starting with $a_0 = 1$.

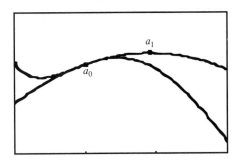

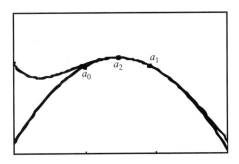

For this example, we obtain the numbers $a_1 = 1.8983836087$, $a_2 = 1.47058242099$, and $a_3 = 1.46348576842$. The actual number at which the maximum occurs is 1.46346875793.

For each of the following functions, graph f and choose an initial guess a_0 for which $f''(a_0) < 0$. Then, using this algorithm, find a_1, a_2, and a_3. Graph f and P_2 at each step of the iteration process.

a. $f(x) = e^{-x} + \cos 2x$

b. $f(x) = xe^{-x^2}$

c. $f(x) = \ln \dfrac{2x + 1}{x^2 + 1}$

32. Show that the algorithm discussed in Exercise 31 is actually Newton's Method applied to f'.

13.4   Power Series

Because the geometric series $\sum_{k=0}^{\infty} x^k$ converges for $|x| < 1$ (Theorem 5 in Section 12.3), it actually represents a *function* with domain $(-1, 1)$. From the formula for its sum, we know that this function has equation $f(x) = 1/(1 - x)$. That is,

$$f(x) = \frac{1}{1 - x} = \sum_{k=0}^{\infty} x^k, \qquad -1 < x < 1.$$

In the remaining sections of this chapter we study certain generalizations of the geometric series, called *power series,* to determine when they converge and the properties of the functions they represent. Among these, Taylor series will have the special form corresponding to Taylor polynomials, as discussed in Sections 13.1–13.3.

Definition 1

A **power series** in powers of $(x - a)$ is an expression of the form

$$\sum_{k=0}^{\infty} a_k(x - a)^k = a_0 + a_1(x - a) + a_2(x - a)^2 + \cdots \tag{1}$$

$$+ a_k(x - a)^k + \cdots$$

where the coefficients $a_0, a_1, a_2, \ldots$ are constants and x is regarded as an independent variable.*

Often power series of the form $\sum_{k=0}^{\infty} a_k(x - a)^k$ are said to be expanded about the number a (the same terminology that was used for Taylor polynomials).

Given a power series expanded about a, we can always use the change of variable $u = (x - a)$ to transform the power series in equation (1) to one of the form

$$\sum_{k=0}^{\infty} a_k u^k = a_0 + a_1 u + a_2 u^2 + \cdots + a_k u^k + \cdots, \tag{2}$$

which is a power series about 0. Consequently, we work almost exclusively with power series about 0.

Here are several examples of power series:

$$\sum_{k=0}^{\infty} x^k = 1 + x + x^2 + x^3 + \cdots + x^k \cdots.$$

$$\sum_{k=0}^{\infty} \frac{x^k}{k!} = 1 + x + \frac{x^2}{2!} + \frac{x^3}{3!} + \cdots + \frac{x^k}{k!} + \cdots \qquad (0! = 1). \tag{3}$$

$$\sum_{k=0}^{\infty} \frac{(-1)^k}{1 + k} x^k = 1 - \frac{x}{2} + \frac{x^2}{3} + \cdots + \frac{(-1)^k}{1 + k} x^k + \cdots. \tag{4}$$

We can use the techniques of Chapter 12 to determine the numbers x for which these series converge. For example, if we use the Ratio Test to test the series in equation (3) for absolute convergence, we find that

$$\rho = \lim_{k \to \infty} \frac{\left| \dfrac{x^{k+1}}{(k+1)!} \right|}{\left| \dfrac{x^k}{k!} \right|} = \lim_{k \to \infty} \frac{|x|}{k+1} = 0, \qquad x \neq 0,$$

for all $x \neq 0$, so the series in equation (3) converges for all x.

---

*Also it should be noted that any infinite series $\sum a_k(bx - c)^k$ involving powers of $(bx - c)$ is a power series, since the general term may be written

$$a_k(bx - c)^k = a_k \left[b \left(x - \frac{c}{b} \right) \right]^k = b^k a_k \left(x - \frac{c}{b} \right)^k = A_k(x - C)^k$$

with $A_k = b^k a_k$ and $C = c/b$.

To determine the numbers x for which the series in equation (4) converges, we again apply the Ratio Test to test for absolute convergence: Since

$$\rho = \lim_{k \to \infty} \frac{\left| \dfrac{(-1)^{k+1} x^{k+1}}{k+2} \right|}{\left| \dfrac{(-1)^k x^k}{k+1} \right|} = \lim_{k \to \infty} \left(\frac{k+1}{k+2} \right) |x| = |x|, \qquad x \neq 0, \qquad (5)$$

this power series converges absolutely if $|x| < 1$. If $x = 1$ we obtain the alternating series

$$\sum_{k=0}^{\infty} \frac{(-1)^k}{1+k},$$

which converges, and if $x = -1$ we obtain the harmonic series

$$\sum_{k=0}^{\infty} \frac{1}{1+k},$$

which diverges. However, we still need to determine convergence for $|x| > 1$. Rather than pursue specific examples in this way, we now establish two theorems that tell us the convergence properties of power series in general.

Theorem 2

(i) If the power series $\Sigma a_k x^k$ converges for $x = c \neq 0$, then it converges absolutely for all x with $|x| < |c|$.
(ii) If the power series $\Sigma a_k x^k$ diverges for $x = d$, then it diverges for all x with $|x| > |d|$.

Proof: To prove part (i) we assume that $\Sigma a_k c^k$ converges. It is a necessary condition (Theorem 7, Chapter 12) that

$$\lim_{k \to \infty} a_k c^k = 0.$$

Thus, there exists an integer N so that $|a_k c^k| < 1$ whenever $k > N$. Now let x be any number such that $|x| < |c|$, and let $\gamma = |x|/|c| < 1$. Then whenever $k > N$ we have

$$|a_k x^k| = \frac{|a_k c^k x^k|}{|c^k|} = |a_k c^k| \left(\frac{|x|}{|c|} \right)^k < \gamma^k.$$

This shows that the series $\Sigma_{k=N}^{\infty} |a_k x^k|$ is dominated by the convergent geometric series $\Sigma_{k=N}^{\infty} \gamma^k$. Thus, by the Comparison Test, the series $\Sigma_{k=N}^{\infty} |a_k x^k|$ converges. The series $\Sigma a_k x^k$ therefore converges absolutely if $|x| < |c|$.

To prove part (ii), we assume that the series $\Sigma a_k d^k$ diverges and we let x be any number so that $|x| > |d|$. Then $\Sigma a_k x^k$ cannot converge, since the convergence of $\Sigma a_k x^k$ would imply convergence of $\Sigma a_k d^k$, by part (i). Thus $\Sigma a_k x^k$ diverges whenever $|x| > |d|$. This completes the proof. ∎

If we want to consider a power series $\Sigma a_k x^k$ as a function, we need to understand the set S of all numbers x for which it converges. We can use Theorem 2 to determine the geometry of S as a subset of the real line.

Now suppose that there exists a number $d \notin S$. Then $\Sigma a_k d^k$ diverges, so by Theorem 2, $|x| \leq |d|$ for every $x \in S$. This shows that the set S is bounded if it is not the entire real line, $(-\infty, \infty)$. By the Completeness Axiom there exists a least upper bound r for S. Let us look at two cases:

(i) If $|x| < r$, then $|x|$ is not an upper bound for S, so there exists an element $c \in S$ with $|x| < c$. Since $c \in S$, $\Sigma a_k c^k$ converges. Thus $\Sigma a_k x^k$ converges absolutely, by Theorem 2.
(ii) If $|x| > r$, then $x \notin S$ so $\Sigma a_k x^k$ diverges.

Case (i) shows that S contains the interval $(-r, r)$, and that $\Sigma a_k x^k$ converges absolutely for every $x \in (-r, r)$. Case (ii) shows that the only other possible elements of S are the endpoints r and $-r$. For this reason the number r is called the **radius of convergence** of the power series. The interval $(-r, r)$, $[-r, r)$, $(-r, r]$, or $[-r, r]$ on which $\Sigma a_k x^k$ converges is called the **interval of convergence**. We summarize these findings in the following theorem.

Theorem 3

Given the power series $\Sigma a_k x^k$, precisely one of the following holds:

(a) The power series converges only for $x = 0$.
(b) There exists a positive number r so that the power series converges absolutely for $|x| < r$ and diverges for $|x| > r$. (The series may or may not converge for $x = \pm r$.)
(c) The power series converges for all x.

(See Figure 4.1).

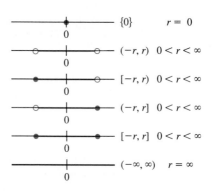

Figure 4.1   The six possible types of intervals of convergence for power series $\Sigma a_k x^k$ expanded about 0. The radius of convergence is r.

☐ **EXAMPLE 1**

(a) The radius of convergence for the geometric power series Σx^k is $r = 1$, and the interval of convergence is $(-1, 1)$.
(b) The radius of convergence for the power series $\Sigma x^k/k!$ in equation (3) is $r = \infty$, and the interval of convergence is $(-\infty, \infty)$.
(c) To determine the radius of convergence of the power series

$$\sum \frac{(-1)^k}{1 + k} x^k$$

in equation (4) we test for absolute convergence using the Ratio Test. We found in equation (5) that $\rho = |x|$. Thus, $\rho < 1$ if and only if $|x| < 1$. So $r = 1$ is the radius of convergence, and the power series converges absolutely for $|x| < 1$. We also checked the two values $x = r = 1$ and $x = -r = -1$ and found that the power series converges for $x = 1$ and diverges for $x = -1$. The interval of convergence is therefore $(-1, 1]$. ∎

□ **EXAMPLE 2**

Find the radius and interval of convergence for the power series

$$\sum_{k=1}^{\infty} \frac{1}{k2^k} x^k.$$

Strategy · · · · · · · ·

First, find the radius of convergence, using the Ratio Test. (The Root Test also works well in this example.)

Solution

We test for absolute convergence using the Ratio Test:

$$\rho = \lim_{k \to \infty} \frac{\left| \dfrac{1}{(k+1)2^{k+1}} x^{k+1} \right|}{\left| \dfrac{1}{k2^k} x^k \right|}$$

$$= \lim_{k \to \infty} \left(\frac{k}{k+1} \right)\left(\frac{1}{2} \right)|x| = \frac{1}{2}|x|.$$

Thus $\rho < 1$ if $|x| < 2$. The radius of convergence is therefore $r = 2$.

To determine the interval of convergence check the endpoints $x = r = 2$ and $x = -r = -2$.

If $x = 2$ we obtain the harmonic series

$$\sum_{k=1}^{\infty} \frac{1}{k2^k}(2^k) = \sum_{k=1}^{\infty} \frac{1}{k} = 1 + \frac{1}{2} + \frac{1}{3} + \cdots,$$

which diverges. If $x = -2$ we obtain the *alternating* harmonic series

$$\sum_{k=1}^{\infty} \frac{1}{k2^k}(-2)^k = \sum_{k=1}^{\infty} \frac{(-1)^k}{k} = -1 + \frac{1}{2} - \frac{1}{3} + \cdots,$$

which converges. The interval of convergence is therefore $[-2, 2)$. ∎

□ **EXAMPLE 3**

Find the interval of convergence for the power series

$$\sum_{k=0}^{\infty} \frac{x^k}{3^k}.$$

Solution   Testing for absolute convergence using the Root Test we find that

$$\rho = \lim_{k \to \infty} \left| \frac{x^k}{3^k} \right|^{1/k} = \lim_{k \to \infty} \frac{|x|}{3} = \frac{|x|}{3}.$$

Thus, $\rho < 1$ if $|x| < 3$. The radius of convergence is $r = 3$. (The Ratio Test also leads to this conclusion.)

If $x = 3$, the series becomes

$$\sum_{k=0}^{\infty} \frac{3^k}{3^k} = \sum_{k=0}^{\infty} 1 = 1 + 1 + 1 + \cdots,$$

which diverges. When $x = -3$, we obtain the divergent series

$$\sum_{k=0}^{\infty} \frac{(-3)^k}{3^k} = \sum_{k=0}^{\infty} (-1)^k = 1 - 1 + 1 - 1 + \cdots.$$

The interval of convergence is therefore $(-3, 3)$. ■

☐ **EXAMPLE 4**

The radius of convergence for the power series

$$\sum_{k=0}^{\infty} k! x^k$$

is $r = 0$ since, by the Ratio Test,

$$\rho = \lim_{k \to \infty} \frac{|(k + 1)! x^{k+1}|}{|k! x^k|} = \lim_{k \to \infty} (k + 1)|x| = \begin{cases} 0, & x = 0 \\ \infty, & x \neq 0 \end{cases}.$$

The "interval" of convergence is simply $\{0\}$. ■

The next example involves a power series of the form $\Sigma a_k(x - a)^k$. As we stated at the beginning of this section, we can use the change of variable $u = x - a$ to bring the power series into the form $\Sigma a_k u^k$. Note, however, that the interval of convergence will be centered at $x = a$ rather than at $x = 0$.

☐ **EXAMPLE 5**

Find the interval of convergence for the power series

$$\sum_{k=1}^{\infty} \frac{3^k}{k} (2x - 1)^k.$$

Strategy · · · · · · · · · ·
Use the substitution $u = 2x - 1$ to bring power series to the form of equation (2). (The interval of convergence will therefore be centered about $x = 1/2$ because $u = 0$ corresponds to $x = 1/2$.)

Apply Ratio Test to find radius of convergence in the variable u.

Solution

Letting $u = 2x - 1$ we obtain the series

$$\sum_{k=1}^{\infty} \frac{3^k}{k} u^k.$$

Testing this series for absolute convergence using the Ratio Test we find that

$$\rho = \lim_{k \to \infty} \frac{\left| \dfrac{3^{k+1} u^{k+1}}{(k + 1)} \right|}{\left| \dfrac{3^k u^k}{k} \right|}$$

$$= \lim_{k \to \infty} 3 \left(\frac{k}{k + 1} \right) |u| = 3|u|.$$

Rewrite the inequality for $|u|$ in terms of the original variable x.

The series converges absolutely for $|u| < 1/3$ or for $|2x - 1| < 1/3$. This last inequality is equivalent to

$$-\tfrac{1}{3} < 2x - 1 < \tfrac{1}{3},$$

or

$$\tfrac{1}{3} < x < \tfrac{2}{3}.$$

Test the endpoints individually.

The radius of convergence (about $x = 1/2$) is therefore $r = 1/6$. At the endpoint $x = 2/3$ we obtain the series

$$\sum_{k=1}^{\infty} \frac{3^k}{k} \left(\frac{1}{3}\right)^k = \sum_{k=1}^{\infty} \frac{1}{k},$$

which diverges. At the endpoint $x = 1/3$ we obtain

$$\sum_{k=1}^{\infty} \frac{3^k}{k} \left(-\frac{1}{3}\right)^k = \sum_{k=1}^{\infty} \frac{(-1)^k}{k},$$

which converges. The interval of convergence is therefore $[1/3, 2/3)$. ∎

Exercise Set 13.4

In Exercises 1–39, find the interval of convergence of the given power series.

1. $\displaystyle\sum_{k=0}^{\infty} \frac{x^k}{k + 2}$

2. $\displaystyle\sum_{k=1}^{\infty} \frac{x^k}{2k}$

3. $\displaystyle\sum_{k=0}^{\infty} \frac{(-1)^{k+1}}{k!} x^k$

4. $\displaystyle\sum_{k=0}^{\infty} \frac{2^k x^k}{(k + 1)!}$

5. $\displaystyle\sum_{k=1}^{\infty} \frac{(k^2 + 1)}{k!} x^k$

6. $\displaystyle\sum_{k=0}^{\infty} \frac{kx^k}{2^k}$

7. $\displaystyle\sum_{k=2}^{\infty} \frac{x^k}{\ln k}$

8. $\displaystyle\sum_{k=1}^{\infty} \frac{(-1)^k e^k}{k^2} x^k$

9. $\displaystyle\sum_{k=1}^{\infty} \frac{\cos \pi k}{1 + k} x^k$

10. $\displaystyle\sum_{k=0}^{\infty} \frac{\sin((k + \tfrac{1}{2})\pi)}{k^2 + 1} x^k$

9. $\displaystyle\sum_{k=1}^{\infty} \frac{\cos \pi k}{1 + k} x^k$

12. $\displaystyle\sum_{k=0}^{\infty} \frac{k^2}{k + 1} x^{3k}$

11. $\displaystyle\sum_{k=1}^{\infty} \frac{(2k + 1)!}{(2k)!} x^{2k}$

14. $\displaystyle\sum_{k=1}^{\infty} k^2 c^k x^k$

15. $\displaystyle\sum_{k=1}^{\infty} \frac{(-1)^k}{k!} (x - 3)^k$

16. $\displaystyle\sum_{k=1}^{\infty} \frac{k}{3^k} (x - \pi)^k$

17. $\displaystyle\sum_{k=0}^{\infty} k!(x - 1)^k$

18. $\displaystyle\sum_{k=1}^{\infty} \frac{3^k}{k^2} (2x - 1)^k$

19. $\displaystyle\sum_{k=2}^{\infty} \frac{(-1)^k}{\ln k} (3x - 2)^k$

20. $\displaystyle\sum_{k=2}^{\infty} \frac{(x - 1)^k}{(\ln k)^k}$

21. $\displaystyle\sum_{k=1}^{\infty} \frac{x^{2k+1}}{k + 2}$

22. $\displaystyle\sum_{k=0}^{\infty} \frac{x^{2k+1}}{\pi^k}$

23. $\displaystyle\sum_{k=1}^{\infty} \frac{(x - 3)^k}{k(k + 1)}$

24. $\displaystyle\sum_{k=1}^{\infty} \frac{(x + 2)^k}{(k + 1)3^k}$

25. $\displaystyle\sum_{k=0}^{\infty} \frac{(2x)^k}{\sqrt{k^3 + 2}}$

26. $\displaystyle\sum_{k=0}^{\infty} \frac{(2x + 5)^k}{\sqrt{2k + 8}}$

27. $\displaystyle\sum_{k=1}^{\infty} \frac{k(x - 2)^k}{e^k}$

28. $\displaystyle\sum_{k=0}^{\infty} \frac{(3x + 1)^k}{\sqrt{k^4 + 1}}$

29. $\displaystyle\sum_{k=2}^{\infty} \frac{k(7x + 1)^k}{2^k}$

30. $\displaystyle\sum_{k=1}^{\infty} \frac{(x - 2)^k}{3^k k^2}$

31. $\displaystyle\sum_{k=3}^{\infty} kx^k$

32. $\displaystyle\sum_{k=1}^{\infty} \frac{x^k}{\ln(k + 1)}$

33. $\displaystyle\sum_{k=1}^{\infty} \frac{(x - 1)^k}{3^k \sqrt{k + 1}}$

34. $\displaystyle\sum_{k=0}^{\infty} \frac{(2x - 1)^k}{5^k}$

35. $\displaystyle\sum_{k=1}^{\infty} \frac{(k + 4)x^k}{(k + 1)(k + 2)e^k}$

36. $\displaystyle\sum_{k=0}^{\infty} \frac{k^3 (x - 2)^k}{3^k}$

37. $\displaystyle\sum_{k=0}^{\infty} \frac{(-1)^k x^{2k+1}}{2(k+1)!}$

38. $\displaystyle\sum_{k=0}^{\infty} \frac{(-1)^k x^{2k}}{(2k)!}$

39. $\displaystyle\sum_{k=1}^{\infty} \frac{x^k}{k(k+1)}$

40. Show that the radius of convergence of the power series

$$\sum_{k=1}^{\infty} \frac{1}{k}(2x-3)^k \text{ is } r = 1/2.$$

41. Prove that if the interval of convergence of the power series $\Sigma a_k x^k$ is $[-r, r)$, then the power series is conditionally convergent, but not absolutely convergent, for $x = -r$.

42. Prove that if the power series $\Sigma a_k x^k$ has radius of convergence r, then the power series $\Sigma a_k x^{ck}$ has radius of convergence $r^{1/c}$, $c > 0$.

43. Prove that if the power series $\Sigma a_k x^k$ has radius of convergence r_a and if the power series $\Sigma b_k x^k$ has radius of convergence r_b, then the series $\Sigma(a_k + b_k)x^k$ converges absolutely for $|x| < c$, where c is the smaller of r_a and r_b.

44. Show that the power series $\Sigma a_k(x-a)^k$ always converges at $x = a$.

 In Exercises 45–47, a function f and a Taylor series for it are specified. The radius of convergence R of this series is also given. Graph f and the nth degree Taylor polynomial $P_n(x) = \sum_{k=0}^{n} a_k(x-a)^k$ for $n = 3, 6,$ and 9. From these graphs, form a conjecture about the convergence of the Taylor series at the endpoints of the interval of convergence. Then prove your conjecture.

45. $f(x) = \dfrac{\ln(3-2x)}{2(1-x)},$ $\displaystyle\sum_{k=0}^{\infty} \frac{2^k(x-1)^k}{(k+1)},$ $R = 1/2$

46. $f(x) = \dfrac{x^2-5x+6}{(4-x)^3},$ $\displaystyle\sum_{k=0}^{\infty} k^2(x-3)^k,$ $R = 1$

47. $f(x) = \ln(x+\sqrt{x^2+1}),$ $x + \displaystyle\sum_{k=1}^{\infty}(-1)^k \frac{(2k)! x^{2k+1}}{(2k+1)(2^k k!)^2},$
$R = 1.$ (*Hint:* $(2k)! = (1 \cdot 3 \cdot 5 \cdot \ldots \cdot (2k-1))2^k k!$ and $2^k k! = 2 \cdot 4 \cdot 6 \cdot \ldots \cdot (2k).$)

13.5   Differentiation and Integration of Power Series

Within its interval of convergence, a power series represents a perfectly legitimate function. Thus, if the power series $\Sigma a_k x^k$ converges on the interval I_a, we may define a function f by

$$f(x) = \sum a_k x^k, \qquad x \in I_a.$$

If a function f is defined in this way, we say that $\Sigma a_k x^k$ is a **power series representation** of f or that f is represented by the power series $\Sigma a_k x^k$.

If $\Sigma b_k x^k$ is a second power series with interval of convergence I_b, then $g(x) = \Sigma b_k x^k$ is again a function, but g is defined on a (possibly) different interval than f is. However, if the intervals I_a and I_b overlap (that is, if $I_a \cap I_b$ is not empty), we may form sums and differences of f and g as follows:

$$(f+g)(x) = f(x) + g(x)$$
$$= \sum a_k x^k + \sum b_k x^k = \sum (a_k + b_k)x^k, \qquad x \in I_a \cap I_b$$
$$(f-g)(x) = f(x) - g(x)$$
$$= \sum a_k x^k - \sum b_k x^k = \sum (a_k - b_k)x^k, \qquad x \in I_a \cap I_b.$$

(The concepts of multiplication and division for power series are more complicated and will not be pursued here.)

□ **EXAMPLE 1**

The formula for the sum of a geometric series shows that the function

$$f(x) = \frac{1}{1-x}$$

may be represented as a power series for $|x| < 1$:

$$\frac{1}{1 - x} = \sum_{k=0}^{\infty} x^k = 1 + x + x^2 + x^3 + \cdots. \tag{1}$$

Multiplying both sides of equation (1) by x shows that

$$\frac{x}{1 - x} = \sum_{k=0}^{\infty} x^{k+1} = x + x^2 + x^3 + x^4 + \cdots, \qquad |x| < 1. \tag{2}$$

Replacing x by $-x$ in (1) gives

$$\frac{1}{1 + x} = \sum_{k=0}^{\infty} (-x)^k = 1 - x + x^2 - x^3 + \cdots, \qquad |x| < 1. \tag{3}$$

Adding equations (1) and (3) (which is allowed because their intervals of convergence are identical) shows that

$$\frac{2}{(1 - x)(1 + x)} = \frac{1}{1 - x} + \frac{1}{1 + x}$$

$$= \sum_{k=0}^{\infty} x^k + \sum_{k=0}^{\infty} (-x)^k = 2 + 2x^2 + 2x^4 + 2x^6 + \cdots$$

$$= 2 \sum_{k=0}^{\infty} x^{2k}, \qquad |x| < 1. \qquad \blacksquare$$

If you are beginning to suspect that within its interval of convergence the power series

$$f(x) = a_0 + a_1 x + a_2 x^2 + \cdots \tag{4}$$

behaves just like an "infinitely long polynomial," that is precisely our point. In fact, the following theorem shows that we may even differentiate expression (4) term by term, just as for polynomials, obtaining

$$f'(x) = a_1 + 2a_2 x + 3a_3 x^2 + \cdots,$$

which is the power series representation for the derivative of the function in equation (4). Its proof is given in more advanced courses.

Theorem 4
Differentiation of Power Series

Suppose that the power series $\Sigma a_k x^k$ has a radius of convergence $r \neq 0$ and that the function f is defined to be its sum:

$$f(x) = \sum_{k=0}^{\infty} a_k x^k = a_0 + a_1 x + a_2 x^2 + a_3 x^3 + \cdots, \qquad |x| < r.$$

Then

(i) the function f is differentiable for $x \in (-r, r)$,

(ii) the power series $\displaystyle\sum_{k=0}^{\infty} k a_k x^{k-1}$ converges absolutely for each $x \in (-r, r)$, and

(iii) $f'(x) = \displaystyle\sum_{k=0}^{\infty} k a_k x^{k-1} = a_1 + 2a_2 x + 3a_3 x^2 + 4a_4 x^3 + \cdots, \qquad |x| < r.$

□ **EXAMPLE 2**

Applying Theorem 4 to the geometric series

$$\frac{1}{1-x} = \sum_{k=0}^{\infty} x^k = 1 + x + x^2 + x^3 + \cdots, \qquad |x| < 1,$$

we conclude that

$$\frac{d}{dx}\left[\frac{1}{1-x}\right] = \frac{1}{(1-x)^2} = \sum_{k=0}^{\infty} kx^{k-1}$$

$$= 1 + 2x + 3x^2 + 4x^3 + \cdots, \qquad |x| < 1. \qquad ■$$

□ **EXAMPLE 3**

Find a power series representation for the function

$$f(x) = \frac{x}{(1-x^2)^2}.$$

Solution   Since f does not resemble any of the functions whose power series representations we already know, we attack the problem in another way. It is easy to find an antiderivative of f; that is, $f(x) = F'(x)$ where

$$F(x) = \frac{1}{2}\left(\frac{1}{1-x^2}\right).$$

Now we note that the expression in the parentheses is the sum of a geometric series with x^2 in place of x in equation (1). In other words,

$$\frac{1}{1-x^2} = \sum_{k=0}^{\infty} (x^2)^k = 1 + x^2 + x^4 + x^6 + \cdots,$$

which converges for $|x^2| < 1$. Thus, we have a power series representation of $F(x)$ for $|x| < 1$.

We conclude from Theorem 4 that

$$\frac{x}{(1-x^2)^2} = \frac{1}{2}\frac{d}{dx}\left[\sum_{k=0}^{\infty} x^{2k}\right] = \frac{1}{2}\sum_{k=0}^{\infty} 2kx^{2k-1} = x + 2x^3 + 3x^5 + \cdots.$$

This series converges absolutely for $|x| < 1$. ■

□ **EXAMPLE 4**

The power series

$$\sum_{k=0}^{\infty} \frac{x^k}{k!} = 1 + x + \frac{x^2}{2!} + \frac{x^3}{3!} + \cdots$$

converges for all x, as noted in Example 1, Section 13.4. On differentiating this series we find that

$$\frac{d}{dx}\left[\sum_{k=0}^{\infty}\frac{x^k}{k!}\right] = \frac{d}{dx}\left[1 + x + \frac{x^2}{2!} + \frac{x^3}{3!} + \frac{x^4}{4!} + \cdots\right]$$

$$= \left[1 + \frac{2x}{2!} + \frac{3x^2}{3!} + \frac{4x^3}{4!} + \cdots\right]$$

$$= \left[1 + x + \frac{x^2}{2!} + \frac{x^3}{3!} + \cdots\right]$$

$$= \sum_{k=0}^{\infty}\frac{x^k}{k!}.$$

That is, if $f(x) = \sum_{k=0}^{\infty} x^k/k!$, then $f'(x) = f(x)$ for all x. In Chapter 8 we proved that the only nonzero functions that satisfy this differential equation are functions of the form $f(x) = Ce^x$. Thus

$$\sum_{k=0}^{\infty}\frac{x^k}{k!} = 1 + x + \frac{x^2}{2!} + \frac{x^3}{3!} + \cdots = Ce^x$$

for some constant C. Setting $x = 0$ shows that $C = 1$, so we conclude that the power series representation for the function e^x is

$$e^x = \sum_{k=0}^{\infty}\frac{x^k}{k!} = 1 + x + \frac{x^2}{2!} + \frac{x^3}{3!} + \cdots$$

and that this representation is valid for all x. ∎

The statement of Theorem 4 raises an obvious question about integrals of functions represented by power series. The answer is just what you might suspect.

Theorem 5

Integration of Power Series

Suppose that the power series $\Sigma a_k x^k$ has radius of convergence $r \neq 0$ and that the function f is defined as its sum:

$$f(x) = \sum_{k=0}^{\infty} a_k x^k = a_0 + a_1 x + a_2 x^2 + a_3 x^3 + \cdots, \qquad |x| < r.$$

Then

(i) the power series $\displaystyle\sum_{k=0}^{\infty}\left(\frac{a_k}{k+1}\right)x^{k+1}$ converges absolutely for each $x \in (-r, r)$,

and

(ii) $\displaystyle\int f(x)\,dx = \sum_{k=0}^{\infty}\left(\frac{a_k}{k+1}\right)x^{k+1} + C$

$$= \left[a_0 x + \frac{a_1}{2}x^2 + \frac{a_2}{3}x^3 + \frac{a_3}{4}x^4 + \cdots\right] + C, \qquad |x| < r.$$

In other words, a power series may be integrated term by term within its radius of convergence. The proof of Theorem 5 is left for more advanced courses. Note that, rather than writing a constant of integration for each term, we collect all constants into a single number C.

□ **EXAMPLE 5**

Since $\ln(1 + x) = \int \dfrac{1}{1 + x}\,dx$, we integrate equation (3) according to Theorem 5.

$$\ln(1 + x) = \sum_{k=0}^{\infty} \left[\int (-x)^k\,dx \right]$$

$$= \sum_{k=0}^{\infty} -\frac{(-x)^{k+1}}{k + 1} + C$$

$$= \sum_{k=0}^{\infty} \frac{(-1)^k x^{k+1}}{k + 1} + C$$

$$= \left[x - \frac{x^2}{2} + \frac{x^3}{3} - \frac{x^4}{4} + \cdots \right] + C, \qquad |x| < 1.$$

Setting $x = 0$ gives $\ln 1 = 0 = C$, so

$$\ln(1 + x) = \sum_{k=0}^{\infty} -\frac{(-x)^{k+1}}{k + 1}$$

$$= x - \frac{x^2}{2} + \frac{x^3}{3} - \frac{x^4}{4} + \cdots, \qquad |x| < 1. \tag{5}$$

■

REMARK   Equation (5) provides a practical means for calculating values of $\ln a$ for $0 < a < 2$. For example, to approximate $\ln(1.2)$ we set $x = 0.2$ and apply (5) to obtain

$$\ln(1.2) \approx .2 - \frac{(.2)^2}{2} + \frac{(.2)^3}{3} - \frac{(.2)^4}{4} = .182266$$

with an error of less than $\dfrac{.2^5}{5} = .000064$ (Theorem 15, Section 12.7).

□ **EXAMPLE 6**

Find a power series representation for $\text{Tan}^{-1} x$.

Solution   Replacing x by x^2 in equation (3) shows that

$$\frac{1}{1 + x^2} = \sum_{k=0}^{\infty} (-x^2)^k = 1 - x^2 + x^4 - x^6 + \cdots, \qquad |x| < 1.$$

Since $\int \dfrac{1}{1 + x^2}\,dx = \text{Tan}^{-1} x + C$, Theorem 5 gives

$$\mathrm{Tan}^{-1} x = \sum_{k=0}^{\infty} \left[\int (-x^2)^k \, dx \right] + C$$

$$= \sum_{k=0}^{\infty} \frac{(-1)^k}{2k+1} x^{2k+1} + C$$

$$= \left[x - \frac{x^3}{3} + \frac{x^5}{5} - \cdots \right] + C, \qquad |x| < 1.$$

Setting $x = 0$ gives $\mathrm{Tan}^{-1}(0) = 0 = C$. Thus

$$\mathrm{Tan}^{-1} x = \sum_{k=0}^{\infty} \frac{(-1)^k}{2k+1} x^{2k+1} = x - \frac{x^3}{3} + \frac{x^5}{5} - \cdots, \qquad |x| < 1. \qquad \blacksquare$$

Exercise Set 13.5

In Exercises 1–14, find a power series representation for the given function using equation (1). State the radius of convergence for the power series obtained.

1. $\dfrac{1}{1 - 2x}$

2. $\dfrac{1}{1 + 3x}$

3. $\dfrac{x^2}{1 - x}$

4. $\dfrac{x^3}{1 + x}$

5. $\dfrac{1}{1 + 4x^2}$

6. $\dfrac{1}{1 - 9x^2}$

7. $\dfrac{x}{1 + x^2}$

8. $\dfrac{x}{1 - x^2}$

9. $\dfrac{1}{1 - x^4}$

10. $\dfrac{3}{1 + x^3}$

11. $\dfrac{1}{9 - x^2}$

12. $\dfrac{x}{4 + x^2}$

13. $\dfrac{x - 1}{x + 1}$

14. $\dfrac{x - 1}{1 + x^2}$

In Exercises 15–23, find a power series representation for the given function using Theorem 4. State the radius of convergence.

15. $f(x) = \dfrac{2}{(1 + x)^2}$ $\left(\textit{Hint: } f(x) = -2 \dfrac{d}{dx} \left(\dfrac{1}{1 + x} \right). \right)$

16. $f(x) = \dfrac{2}{(1 - x)^3}$ $\left(\textit{Hint: } f(x) = \dfrac{d^2}{dx^2} \left(\dfrac{1}{1 - x} \right). \right)$

17. $f(x) = \dfrac{x}{(1 + x^2)^2}$

18. $f(x) = \dfrac{1 - x^2}{(1 + x^2)^2}$ $\left(\textit{Hint: } f(x) = \dfrac{d}{dx} \left(\dfrac{x}{1 + x^2} \right). \right)$

19. $f(x) = \dfrac{1 + x^2}{(1 - x^2)^2}$ $\left(\textit{Hint: } f(x) = \dfrac{d}{dx} \left(\dfrac{x}{1 - x^2} \right). \right)$

20. $f(x) = \dfrac{1}{(1 + 4x)^2}$

21. $f(x) = \dfrac{8x}{(1 + 4x^2)^2}$ (*Hint:* Use Exercise 5.)

22. $f(x) = \dfrac{1 + 2x - x^2}{(1 + x^2)^2}$ (*Hint:* Use Exercise 14.)

23. $f(x) = \dfrac{2}{(x + 1)^2}$ (*Hint:* Use Exercise 13.)

In Exercises 24–31, find a power series representation for the given function using Theorem 5. State the radius of convergence.

24. $f(x) = \ln(1 + x)$

25. $f(x) = \ln(1 - x)$

26. $f(x) = x \ln(1 + x)$

27. $f(x) = \mathrm{Tan}^{-1}(2x)$

28. $f(x) = x \, \mathrm{Tan}^{-1} x$

29. $f(x) = \ln(4 + x)$

30. $\displaystyle \int \dfrac{dx}{1 + x^4}$

31. $\ln(1 + x^2)$

32. Use the results of Exercises 24 and 25 to show that

$$\ln \left(\frac{1 + x}{1 - x} \right) = 2 \left(x + \frac{x^3}{3} + \frac{x^5}{5} + \cdots \right), \qquad |x| < 1.$$

33. For the power series

$$\sum_{k=0}^{\infty} (-1)^k \frac{x^{2k}}{(2k)!} = 1 - \frac{x^2}{2!} + \frac{x^4}{4!} - \frac{x^6}{6!} + \cdots :$$

a. Use the Ratio Test to show that the series converges absolutely for all values of x.

b. Using Theorem 4, show that the function

$$f(x) = \sum_{k=0}^{\infty} (-1)^k \frac{x^{2k}}{(2k)!}$$

satisfies the differential equation $f''(x) = -f(x)$.

c. Show that $f(0) = 1$.

d. What function have you seen previously that satisfies both properties b and c?

34. For the power series

$$\sum_{k=0}^{\infty}(-1)^k\frac{x^{2k+1}}{(2k+1)!} = x - \frac{x^3}{3!} + \frac{x^5}{5!} - \frac{x^7}{7!} + \cdots:$$

a. Use the Ratio Test to show that the series converges absolutely for all values of x.

b. Using Theorem 4, show that the function

$$g(x) = \sum_{k=0}^{\infty}(-1)^k\frac{x^{2k+1}}{(2k+1)!}$$

satisfies the differential equation $g''(x) = -g(x)$.

c. Show that $g(0) = 0$.

d. Show that $g'(x) = f(x)$ and $f'(x) = -g(x)$ where f is the function in Exercise 33. For what common function is $g(x)$ a power series representation?

35. Use the Chain Rule to show that if the power series

$$f(x) = \sum_{k=0}^{\infty} a_k(bx + c)^k$$

converges for $|bx + c| < r$, then

$$f'(x) = \sum_{k=0}^{\infty} kba_k(bx + c)^{k-1}, \qquad |bx + c| < r.$$

13.6 Taylor and Maclaurin Series

We are about to uncover a remarkable relationship between Taylor polynomials and power series, one that will provide a precise solution to the approximation problem posed in Section 13.1.

Recall that if the function f has $n + 1$ derivatives in an interval containing $x = a$, then the nth Taylor polynomial for f is

$$P_n(x) = f(a) + f'(a)(x - a) + \frac{f''(a)}{2!}(x - a)^2 + \cdots + \frac{f^{(n)}(a)}{n!}(x - a)^n \quad (1)$$

$$= \sum_{k=0}^{n}\frac{f^{(k)}(a)}{k!}(x - a)^k.$$

We refer to the coefficient

$$\frac{f^{(k)}(a)}{k!}$$

of $(x - a)^k$ as the **kth Taylor coefficient**. Also, if we write

$$f(x) = P_n(x) + R_n(x), \quad (2)$$

then, by Taylor's Theorem, the remainder term $R_n(x)$ is bounded by

$$|R_n(x)| \le \left|\frac{f^{(n+1)}(c)}{(n + 1)!}(x - a)^{n+1}\right| \quad (3)$$

where c lies between a and x.

Now note that the Taylor polynomial $P_n(x)$ is precisely the nth partial sum for the power series.

$$\sum_{k=0}^{\infty}\frac{f^{(k)}(a)}{k!}(x - a)^k = f(a) + f'(a)(x - a) + \frac{f''(a)}{2!}(x - a)^2 + \frac{f'''(a)}{3!}(x - a)^3 + \cdots. \quad (4)$$

Series (4) is referred to as the **Taylor series** for f, expanded about $x = a$. It is simply the result of allowing the Taylor polynomial $P_n(x)$ to become "infinitely long." In other words,

$$\sum_{k=0}^{\infty} \frac{f^{(k)}(a)}{k!}(x-a)^k = \lim_{n \to \infty} \sum_{k=0}^{n} \frac{f^{(k)}(a)}{k!}(x-a)^k = \lim_{n \to \infty} P_n(x). \qquad (5)$$

Of course, the expression in equation (4) makes sense only if the function f is infinitely differentiable. That is, the derivative $f^{(n)}(a)$ must exist for all integers $n = 1, 2, \ldots$ in order that all Taylor coefficients be defined.

To determine the numbers x for which a Taylor series converges to $f(x)$, we use the remainder term $R_n(x)$. Taking the limit as $n \to \infty$ on both sides of equation (2) and using (5) we see that

$$f(x) = \lim_{n \to \infty} P_n(x) + \lim_{n \to \infty} R_n(x)$$

$$= \sum_{k=0}^{\infty} \frac{f^{(k)}(a)}{k!}(x-a)^k + \lim_{n \to \infty} R_n(x).$$

This shows that *the Taylor series (4) converges for f(x) if and only if* $\lim_{n \to \infty} R_n(x) = 0$.

□ **EXAMPLE 1**

Find the Taylor series for the function $f(x) = \sin x$ expanded about $a = \pi/4$ and determine the numbers x for which the series converges to $f(x)$.

Strategy · · · · · · · · ·
Find the Taylor coefficients

$$\frac{f^{(k)}(a)}{k!}$$

with $a = \pi/4$.

Solution
We have

$$f(x) = \sin x; \qquad\qquad f(\pi/4) = \sqrt{2}/2$$
$$f'(x) = \cos x; \qquad\qquad f'(\pi/4) = \sqrt{2}/2$$
$$f''(x) = -\sin x; \qquad\qquad f''(\pi/4) = -\sqrt{2}/2$$
$$f'''(x) = -\cos x; \qquad\qquad f'''(\pi/4) = -\sqrt{2}/2$$
$$f^{(4)}(x) = \sin x; \qquad\qquad f^{(4)}(\pi/4) = \sqrt{2}/2$$
$$\vdots$$
$$f^{(2k)}(x) = (-1)^k \sin x; \qquad f^{(2k)}(\pi/4) = (-1)^k\sqrt{2}/2$$
$$f^{(2k+1)}(x) = (-1)^k \cos x; \qquad f^{(2k+1)}(\pi/4) = (-1)^k\sqrt{2}/2$$
$$\vdots$$

Insert the Taylor coefficients and $a = \pi/4$ in (4) to obtain the Taylor series.

The Taylor series is therefore

$$\sin x = \frac{\sqrt{2}}{2} + \frac{\sqrt{2}}{2}\left(x - \frac{\pi}{4}\right) - \frac{\sqrt{2}}{2 \cdot 2!}\left(x - \frac{\pi}{4}\right)^2$$
$$- \frac{\sqrt{2}}{2 \cdot 3!}\left(x - \frac{\pi}{4}\right)^3 + \frac{\sqrt{2}}{2 \cdot 4!}\left(x - \frac{\pi}{4}\right)^4 + \cdots.$$

Determine where

$$\lim_{n \to \infty} |R_n(x)| = 0$$

to find the values of x for which the series converges to $f(x)$.

Use fact that

$$\lim_{n \to \infty} \frac{x^n}{n!} = 0$$

(limit (2), Section 12.2).

Since $f^{(n+1)}(c)$ is either $\pm \sin c$ or $\pm \cos c$, we know that $|f^{(n+1)}(c)| \leq 1$ for all n. Thus

$$\lim_{n \to \infty} |R_n(x)| = \lim_{n \to \infty} \left| \frac{f^{(n+1)}(c)}{(n+1)!} \left(x - \frac{\pi}{4} \right)^{n+1} \right|$$

$$\leq \lim_{n \to \infty} \frac{\left| x - \dfrac{\pi}{4} \right|^{n+1}}{(n+1)!}$$

$$= 0$$

for all x. Thus the series converges to $\sin x$ for all x. ∎

The special case of a Taylor series expanded about $a = 0$ is referred to as a **Maclaurin series:**

$$\sum_{k=0}^{\infty} \frac{f^{(k)}(0)}{k!} x^k = f(0) + f'(0)x + \frac{f''(0)}{2!} x^2 + \cdots + \frac{f^{(k)}(0)}{k!} x^k + \cdots. \qquad (6)$$

□ **EXAMPLE 2**

Find a Maclaurin series for $f(x) = e^x$ and determine the numbers x for which it converges to $f(x)$.

Solution Every derivative of $f(x) = e^x$ is the same:

$$f(x) = e^x; \qquad f(0) = 1$$
$$f'(x) = e^x; \qquad f'(0) = 1$$
$$\vdots$$
$$f^{(n)}(x) = e^x; \qquad f^{(n)}(0) = 1$$
$$\vdots$$

Thus

$$e^x = 1 + x + \frac{x^2}{2!} + \frac{x^3}{3!} + \cdots + \frac{x^k}{k!} + \cdots = \sum_{k=0}^{\infty} \frac{x^k}{k!}.$$

Here, for all n,

$$|f^{(n+1)}(c)| = |e^c| = e^c \leq e^{|c|} \leq e^{|x|}$$

since $|c| \leq |x|$, so

$$\lim_{n \to \infty} |R_n(x)| = \lim_{n \to \infty} \left| \frac{e^c}{(n+1)!} x^{n+1} \right| \leq \lim_{n \to \infty} e^{|x|} \frac{|x|^{n+1}}{(n+1)!} = 0$$

for all x. The Maclaurin series for e^x therefore converges to e^x for all x. (Note that this result was proved in Example 4, Section 13.5, by a different method.) ∎

☐ **EXAMPLE 3**

Find the Maclaurin series for $f(x) = \ln(1 + x)$.

Solution   Here

$$f(x) = \ln(1 + x); \qquad\qquad f(0) = 0$$
$$f'(x) = (1 + x)^{-1}; \qquad\qquad f'(0) = 1$$
$$f''(x) = -(1 + x)^{-2}; \qquad\qquad f''(0) = -1$$
$$f'''(x) = 2(1 + x)^{-3}; \qquad\qquad f'''(0) = 2$$
$$\vdots$$
$$f^{(k)}(x) = (-1)^{k+1}(k - 1)!(1 + x)^{-k}; \qquad f^{(k)}(0) = (-1)^{k+1}(k - 1)!$$
$$\vdots$$

Using (6) we obtain

$$\ln(1 + x) = 0 + 1x - \frac{1}{2!}x^2 + \frac{2}{3!}x^3 - \cdots + \frac{(-1)^{k+1}(k - 1)!}{k!}x^k + \cdots$$

$$= x - \frac{x^2}{2} + \frac{x^3}{3} - \frac{x^4}{4} + \cdots + \frac{(-1)^{k+1}}{k}x^k + \cdots$$

$$= \sum_{k=1}^{\infty} \frac{(-1)^{k+1}}{k}x^k.$$

In Example 5 of Section 13.5, we showed that this power series converges to $\ln(1 + x)$ for $|x| < 1$. ■

Obviously, every Taylor or Maclaurin series is a power series. We now show that, if the function f is represented as a power series expanded about $x = a$ with radius of convergence $r > 0$, then it must be the Taylor series for f expanded about $x = a$. First, suppose $a = 0$ and

$$f(x) = a_0 + a_1 x + a_2 x^2 + \cdots + a_k x^k + \cdots, \qquad |x| < r \neq 0. \tag{7}$$

Using Theorem 4, we repeatedly differentiate both sides of (7), obtaining

$$f'(x) = a_1 + 2a_2 x + 3a_3 x^2 + \cdots + ka_k x^{k-1} + \cdots$$
$$f''(x) = 2a_2 + 2 \cdot 3a_3 x + \cdots + (k - 1)ka_k x^{k-2} + \cdots$$
$$\vdots$$
$$f^{(k)}(x) = k!a_k + (k + 1)!a_{k+1}x + \frac{(k + 2)!}{2!}a_{k+2}x^2 + \cdots. \tag{8}$$

Setting $x = 0$ in each equation causes all terms containing powers of x to vanish, so

$$f(0) = a_0, \qquad a_0 = \frac{f(0)}{0!}$$

$$f'(0) = a_1, \qquad a_1 = \frac{f'(0)}{1!}$$

$$f''(0) = 2a_2, \qquad a_2 = \frac{f''(0)}{2!}$$

$$\vdots \qquad\qquad \vdots$$

$$f^{(k)}(0) = k!a_k, \qquad a_k = \frac{f^{(k)}(0)}{k!}.$$

In other words, if f is represented by a convergent power series, as in equation (7), then the coefficients a_k must be precisely the Maclaurin coefficients $a_k = f^{(k)}(0)/k!$.

If we replace x by $(x - a)$ in equation (7) and evaluate the derivatives at $x = a$, we prove the following theorem for Taylor series.

Theorem 6

If the function f can be represented by a power series of the form

$$f(x) = a_0 + a_1(x - a) + a_2(x - a)^2 + \cdots + a_k(x - a)^k + \cdots$$

$$= \sum_{k=0}^{\infty} a_k(x - a)^k, \qquad |x - a| < r$$

with a radius of convergence $r > 0$, then the coefficients a_k in the power series expansion must be the Taylor coefficients

$$a_k = \frac{f^{(k)}(a)}{k!}, \qquad k = 0, 1, 2, \ldots.$$

REMARK 1   The precise formulation of Theorem 6 is important. If we *know* that $f(x) = \Sigma a_k(x - a)^k$, then this series is the Taylor series for f. There are examples, however, for which the Taylor series $\Sigma a_k(x - a)^k$ for f converges, but $f(x) \neq \Sigma a_k(x - a)^k$. That is, *a convergent Taylor series for f need not converge to the value $f(x)$*. (See Exercise 45.)

☐ **EXAMPLE 4**

In Exercise 3 you are asked to obtain the Maclaurin series expansion

$$\sin x = x - \frac{x^3}{3!} + \frac{x^5}{5!} - \frac{x^7}{7!} + \cdots + \frac{(-1)^k x^{2k+1}}{(2k + 1)!} + \cdots. \tag{9}$$

To show that this series converges to $\sin x$ for all x, we must show that $\lim_{n \to \infty} |R_n(x)| = 0$ for all x.

We do so, using Taylor's Theorem. Note that the derivative $f^{(n+1)}$ for $f(x) = \sin x$ is one of $\pm \sin x$ or $\pm \cos x$. Thus, for any x, we know that

$$|f^{(n+1)}(c)| \leq 1$$

for all c between 0 and x. By Taylor's Theorem,

$$|R_n(x)| \leq \frac{1}{(n + 1)!} |x|^{n+1}, \qquad -\infty < x < \infty,$$

so

$$\lim_{n \to \infty} |R_n(x)| \leq \lim_{n \to \infty} \frac{|x|^{n+1}}{(n + 1)!} = 0, \qquad -\infty < x < \infty$$

according to equation (2), Section 12.2. (See Figure 6.1.)

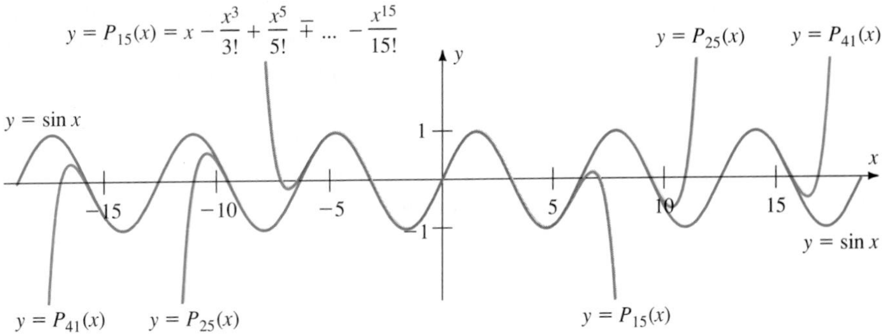

Figure 6.1 The graphs of $f(x) = \sin x$ along with the graphs of three of its Taylor polynomials P_{15}, P_{25}, and P_{41}, about $a = 0$. These graphs are consistent with the limit.

$$\lim_{n \to \infty} R_n(x) = 0$$

for all x.

Since the Maclaurin series (9) is a power series that converges for all x, we may differentiate this series term by term, to obtain a power series for $\cos x$:

$$\cos x = 1 - \frac{3x^2}{3!} + \frac{5x^4}{5!} - \frac{7x^6}{7!} + \cdots + \frac{(-1)^k(2k+1)x^{2k}}{(2k+1)!} + \cdots$$

$$= 1 - \frac{x^2}{2!} + \frac{x^4}{4!} - \frac{x^6}{6!} + \cdots + \frac{(-1)^k x^{2k}}{(2k)!} + \cdots,$$

which converges for all x. Theorem 6 then assures that this series is, in fact, the Maclaurin series for $f(x) = \cos x$. ∎

□ **EXAMPLE 5**

Find a Maclaurin series for $x^3 e^{x^2}$.

Strategy · · · · · · · · · ·
Do *not* calculate Taylor coefficients for $f(x) = x^3 e^{x^2}$. Work directly with the Maclaurin series for e^x and substitute x^2 for x.

Solution
Replacing x by x^2 in the series for e^x in Example 2 shows that

$$e^{x^2} = 1 + x^2 + \frac{x^4}{2!} + \frac{x^6}{3!} + \cdots = \sum_{k=0}^{\infty} \frac{x^{2k}}{k!},$$

which converges for all x. Thus

Multiply by x^3.

$$x^3 e^{x^2} = x^3 \left[1 + x^2 + \frac{x^4}{2!} + \frac{x^6}{3!} + \cdots \right]$$

$$= x^3 + x^5 + \frac{x^7}{2!} + \frac{x^9}{3!} + \cdots$$

$$= \sum_{k=0}^{\infty} \frac{x^{2k+3}}{k!}$$

converges for all x. By Theorem 6 this must be the Maclaurin series for $x^3 e^{x^2}$. ∎

REMARK 2   The strategy to the solution of Example 5 suggests a convenient method for calculating Taylor polynomials for certain functions. Since the nth Taylor polynomial P_n is the nth partial sum of the Taylor series, we can use Taylor series representations of functions to find P_n. For example, consider the function $f(x) = x^3 e^{x^2}$ in Example 5. Calculating $P_9(x)$ using the method of Section 13.1 involves the calculation of the derivatives of $f^{(k)}$ for $k = 0, 1, 2, \ldots, 9$. Alternatively, we can use the Maclaurin series for f to provide P_9. That is,

$$P_9(x) = x^3 + x^5 + \frac{x^7}{2!} + \frac{x^9}{3!}.$$

The following examples show how Taylor and Maclaurin series may be used to calculate certain constants and integrals.

☐ **EXAMPLE 6**

Since the Maclaurin series

$$e^x = \sum_{k=0}^{\infty} \frac{x^k}{k!} = 1 + x + \frac{x^2}{2!} + \frac{x^3}{3!} + \cdots$$

converges to e^x for all x, we may use it to approximate e^x for any number x. Setting $x = 1$ gives

$$e = 1 + 1 + \frac{1}{2!} + \frac{1}{3!} + \cdots + \frac{1}{k!} + \cdots = \sum_{k=0}^{\infty} \frac{1}{k!},$$

and setting $x = -1$, we obtain

$$\frac{1}{e} = 1 - 1 + \frac{1}{2!} - \frac{1}{3!} + \frac{1}{4!} - \cdots + \frac{(-1)^k}{k!} + \cdots = \sum_{k=0}^{\infty} \frac{(-1)^k}{k!}.$$

Since this last series is an alternating series, we may approximate $1/e$ by, say,

$$\frac{1}{e} \approx \frac{1}{2!} - \frac{1}{3!} + \frac{1}{4!} - \frac{1}{5!} = .3667.$$

with accuracy no worse than $1/6! = .0014$. ∎

☐ **EXAMPLE 7**

We may approximate the integral

$$\int_0^1 e^{-x^2}\, dx$$

using Maclaurin series as follows. Since

$$e^x = 1 + x + \frac{x^2}{2!} + \frac{x^3}{3!} + \cdots + \frac{x^k}{k!} + \cdots$$

converges for all x, the series

$$e^{-x^2} = 1 - x^2 + \frac{x^4}{2!} - \frac{x^6}{3!} + \cdots + \frac{(-x^2)^k}{k!} + \cdots$$

also converges for all x. By Theorem 5 an antiderivative for e^{-x^2} is represented as

$$\int e^{-x^2}\, dx = \int 1\, dx - \int x^2\, dx + \int \frac{x^4}{2!}\, dx - \int \frac{x^6}{3!}\, dx + \cdots + C$$

$$= x - \frac{x^3}{3} + \frac{x^5}{5 \cdot 2!} - \frac{x^7}{7 \cdot 3!} + \cdots + \frac{(-1)^k x^{2k+1}}{(2k+1)k!} + \cdots + C.$$

Thus,

$$\int_0^1 e^{-x^2}\, dx = 1 - \frac{1}{3} + \frac{1}{5 \cdot 2!} - \frac{1}{7 \cdot 3!} + \cdots + \frac{(-1)^k}{(2k+1)k!} + \cdots.$$

Since this is an alternating series that satisfies the Alternating Series Test, we may approximate this series to any desired degree of accuracy by taking sufficiently many terms. For example, to obtain accuracy of .01, we find by trial and error that for $k = 4$,

$$\frac{1}{(2 \cdot 4 + 1)4!} = \frac{1}{9 \cdot 24} = \frac{1}{216} < \frac{1}{100}.$$

Thus, we use terms up through $k = 3$ to obtain

$$\int_0^1 e^{-x^2}\, dx \approx 1 - \frac{1}{3} + \frac{1}{5 \cdot 2!} - \frac{1}{7 \cdot 3!} \approx .74,$$

accurate to within .01. ∎

The Binomial Series

According to the Binomial Theorem, we know that

$$(1 + x)^n = 1 + nx + \frac{n(n-1)}{2} x^2 + \cdots$$

$$+ \binom{n}{k} x^k + \cdots + nx^{n-1} + x^n \tag{10}$$

if n is a positive integer. In writing (10) we have used the notation

$$\binom{n}{k}$$

for the **binomial coefficient**

$$\binom{n}{k} = \frac{n!}{k!(n-k)!} = \frac{n(n-1)(n-2)\cdots(n-k+1)}{k!}.$$

We can extend equation (10) to the case where n is not a positive integer. However, the result involves an infinite series, rather than a polynomial in x. The result is called the **Binomial Series:**

$$(1 + x)^r = 1 + rx + \frac{r(r-1)}{2} x^2 + \frac{r(r-1)(r-2)}{3!} x^3 + \cdots \tag{11}$$

$$= 1 + \sum_{k=1}^{\infty} \frac{r(r-1)(r-2)\cdots(r-k+1)}{k!} x^k, \qquad |x| < 1.$$

(Note that the infinite series in (11) reduces to the polynomial in (10) if $r = n$ is a positive integer, since the term $(r - k + 1)$ becomes zero when k reaches $r + 1$.)

Proving the validity of (11) involves showing that the right-hand side is precisely the Maclaurin series for $f(x) = (1 + x)^r$, and that the radius of convergence for this series is one. To do so we note that

$$f(x) = (1 + x)^r \quad \text{gives} \quad f(0) = 1$$
$$f'(x) = r(1 + x)^{r-1} \quad \text{gives} \quad f'(0) = r$$
$$f''(x) = r(r - 1)(1 + x)^{r-2} \quad \text{gives} \quad f''(0) = r(r - 1)$$
$$\vdots$$
$$f^{(k)}(x) = r(r - 1) \cdots (r - k + 1)(1 + x)^{r-k}$$
$$\text{gives} \quad f^{(k)}(0) = r(r - 1) \cdots (r - k + 1).$$

Thus, the kth Maclaurin coefficient for $f(x) = (1 + x)^r$ is indeed

$$\frac{f^{(k)}(0)}{k!} = \frac{r(r - 1)(r - 2) \cdots (r - k + 1)}{k!}$$

as in the Binomial Series (11).

To verify that the series converges for $|x| < 1$, we let a_k denote the kth term

$$a_k = \frac{r(r - 1) \cdots (r - k + 1)}{k!} x^k.$$

Then

$$\rho = \lim_{k \to \infty} \left| \frac{a_{k+1}}{a_k} \right| = \lim_{k \to \infty} \left| \frac{r(r - 1) \cdots (r - k)}{r(r - 1) \cdots (r - k + 1)} \frac{k!}{(k + 1)!} \frac{x^{k+1}}{x^k} \right|$$
$$= \lim_{k \to \infty} \left| \frac{(r - k)}{1} \frac{1}{(k + 1)} x \right|$$
$$= \lim_{k \to \infty} \left| \frac{r - k}{k + 1} \right| |x|$$
$$= |x|.$$

Thus, if $|x| < 1$ we have $\rho < 1$, and the series converges absolutely by the Ratio Test. We assert, but do not prove, that this series converges to $(1 + x)^r$ for $|x| < 1$.

□ **EXAMPLE 8**

The Binomial Series, with $r = \frac{1}{2}$, gives

$$\sqrt{1 + x} = (1 + x)^{1/2} = 1 + \frac{1}{2}x + \frac{\frac{1}{2}(\frac{1}{2} - 1)}{2}x^2 + \frac{\frac{1}{2}(\frac{1}{2} - 1)(\frac{1}{2} - 2)}{3!}x^3$$
$$+ \frac{\frac{1}{2}(\frac{1}{2} - 1)(\frac{1}{2} - 2)(\frac{1}{2} - 3)}{4!}x^4 + \cdots$$
$$= 1 + \frac{1}{2}x - \frac{1}{8}x^2 + \frac{1}{16}x^3 - \frac{5}{128}x^4 + \cdots, \qquad |x| < 1. \quad \blacksquare$$

□ **EXAMPLE 9**

Replacing x by x^3 in Example 8 gives

$$\sqrt{1 + x^3} = (1 + x^3)^{1/2}$$
$$= 1 + \frac{1}{2}x^3 - \frac{1}{8}x^6 + \frac{1}{16}x^9 - \frac{5}{128}x^{12} + \cdots, \qquad |x| < 1. \quad \blacksquare$$

□ **EXAMPLE 10**

With $r = -\frac{1}{3}$ and $-x$ in place of x, (11) gives

$$\frac{1}{\sqrt[3]{1-x}} = [1 + (-x)]^{-1/3}$$

$$= 1 + (-\tfrac{1}{3})(-x) + \frac{(-\frac{1}{3})(-\frac{1}{3} - 1)}{2}(-x)^2$$

$$+ \frac{(-\frac{1}{3})(-\frac{1}{3} - 1)(-\frac{1}{3} - 2)}{3!}(-x)^3$$

$$+ \frac{(-\frac{1}{3})(-\frac{1}{3} - 1)(-\frac{1}{3} - 2)(-\frac{1}{3} - 3)}{4!}(-x)^4 + \cdots$$

$$= 1 + \tfrac{1}{3}x + \tfrac{2}{9}x^2 + \tfrac{14}{81}x^3 + \tfrac{35}{243}x^4 + \cdots, \qquad |x| < 1. \qquad ■$$

The Summary Outline includes a list of frequently used Maclaurin series.

Exercise Set 13.6

In Exercises 1–23, find the Taylor or Maclaurin series for the given function expanded about the given point and determine the numbers x for which the series converges.

1. $f(x) = e^{2x}, \quad a = 0$ **2.** $f(x) = e^{-x}, \quad a = 0$

3. $f(x) = \sin x, \quad a = 0$ **4.** $f(x) = \sin \pi x, \quad a = 0$

5. $f(x) = \cos x, \quad a = \pi/4$ **6.** $f(x) = \sin x, \quad a = \pi/6$

7. $f(x) = 1 + x^2, \quad a = 2$ **8.** $f(x) = 3x^2 + 2x - 4, \quad a = 3$

9. $f(x) = 1/(1 + x), \quad a = 0$

10. $f(x) = \ln(3 + x), \quad a = 0$

11. $f(x) = x \sin 2x, \quad a = 0$
   (*Hint:* First find the series for $\sin 2x$.)

12. $f(x) = 3^{-x}, \quad a = 0$ **13.** $f(x) = 2^x, \quad a = 0$

14. $f(x) = (1 + x)^{2/5}$ **15.** $f(x) = (1 + x)^{3/2}, \quad a = 0$

16. $f(x) = (1 + x)^{5/2}, \quad a = 0$ **17.** $f(x) = \sqrt{x}, \quad a = 4$

18. $f(x) = \sqrt{x + 1}, \quad a = 0$ **19.** $f(x) = 1/x, \quad a = 2$

20. $f(x) = (\sin x)/x, \quad a = 0$ **21.** $f(x) = x^2 e^{-x}, \quad a = 0$

22. $f(x) = x \sin x, \quad a = 0$ **23.** $f(x) = x^2 \ln(1 + x), \quad a = 0$

24. Find a Maclaurin series for $f(x) = \cos^2 x$ using the identity $\cos^2 x = \frac{1}{2}[1 + \cos 2x]$.

25. Find the Maclaurin series for $f(x) = \sin^2 x$ (see Exercise 24).

26. Find the first four terms of the Maclaurin series for $f(x) = \tan x$.

27. Find the first three terms of the Maclaurin series for $f(x) = \sec^2 x$ using the result of Exercise 26.

28. Find the Maclaurin series for $x \sin x^2$ using the Maclaurin series for $\sin x$.

29. Find a Maclaurin series for $\cos x^2$ using the technique of Exercise 28.

30. Find the Maclaurin series for $f(x) = \sinh x$ from the Maclaurin series for e^x and e^{-x}.

31. Find the Maclaurin series for $x \cosh x$.

In Exercises 32–40, use Taylor series to approximate the given quantity accurately to three decimal places.

32. $\sin 2°$ **33.** e^2

34. $\ln(1.1)$ **35.** $\displaystyle\int_0^{\pi/4} \sin x^2 \, dx$

36. $\displaystyle\int_0^1 \cos x^2 \, dx$ **37.** $\displaystyle\int_0^1 \frac{\sin x}{x} \, dx$

38. $\displaystyle\int_0^{1/2} \frac{1}{1 + x^3} \, dx$ **39.** $\displaystyle\int_0^1 e^{-x^2} \, dx$

40. $\displaystyle\int_0^1 e^{-x^4} \, dx$

In Exercises 41–44, use the Binomial Series to find a power series for the given function and the radius of convergence.

41. $f(x) = \sqrt{1 + 2x}$ **42.** $f(x) = \sqrt[3]{27 + x}$

43. $f(x) = (9 + 3x)^{3/2}$ **44.** $f(x) = 1/\sqrt[3]{8 - x^2}$

45. Define the function $f(x)$ by

$$f(x) = \begin{cases} e^{-(1/x^2)}, & x \neq 0 \\ 0, & x = 0. \end{cases}$$

a. Show that $f^{(n)}(0)$ exists and equals zero for every integer $n \geq 1$.

b. Show that the Maclaurin series for f is identically equal to zero for all values of x. Thus the function f does not equal its Maclaurin series in any interval containing zero. (*Remark:* This example shows that the existence of all derivatives is not a sufficient condition for a function to equal its Taylor series. The condition that $\lim_{n\to\infty} R_n(x) = 0$ is essential.)

46. Prove that power series representations are unique. That is, prove that if

$$f(x) = \sum_{k=0}^{\infty} a_k x^k \quad \text{and} \quad f(x) = \sum_{k=0}^{\infty} b_k x^k,$$

then $a_0 = b_0, a_1 = b_1, \cdots, a_k = b_k, \cdots$. (*Hint:* Set $x = 0$ to obtain $a_0 = b_0$. Then differentiate and set $x = 0$ again, and so on.)

47. Find a decimal approximation to π as follows:
a. Determine the Maclaurin series for $(1 - x^2)^{-1/2}$ using the Binomial Series
b. Using part a, find the Maclaurin series for $\text{Sin}^{-1} x$. Graph the first 5 partial sums of this series along with the graph of $\text{Sin}^{-1} x$. Do these partial sums converge rapidly on the interval $-1/2 \leq x \leq 1/2$? (*Hint:* The identity $(2k)! = 2^k k![1 \cdot 3 \cdot 5 \cdot \ldots \cdot (2k - 1)]$ is helpful.)
c. Let $x = 1/2$ in the series of part b. Use the resulting series to determine the first 8 digits of π.

Summary Outline of Chapter 13

▮ The **nth Taylor polynomial** for f, expanded about $x = a$, is (page 654)

$$P_n(x) = f(a) + f'(a)(x - a) + \frac{f''(a)}{2!}(x - a)^2 + \cdots + \frac{f^{(n)}(a)}{n!}(x - a)^n$$

$$= \sum_{j=0}^{n} \frac{f^{(j)}(a)}{j!}(x - a)^j.$$

▮ *Theorem:* If f is $(n + 1)$-times differentiable and $f(x) = P_n(x) + R_n(x)$, then (page 660)

$$R_n(x) = \frac{f^{(n+1)}(c)}{(n + 1)!}(x - a)^{n+1},$$

where c lies between a and x.

▮ A **power series** has the form $\sum_{k=0}^{\infty} a_k(x - a)^k$. (page 670)

▮ *Theorem:* If the power series $\sum a_k x^k$ converges for $x = c$, it does so for all x (page 671) with $|x| < |c|$.

▮ *Theorem:* The set of all x for which a power series converges is either (i) (page 672) $\{a\}$, (ii) an interval with midpoint $x = a$ and radius $r \neq 0$, or (iii) $(-\infty, \infty)$. (The radius r is called the **radius of convergence.**)

▮ *Theorem:* If $f(x) = \sum a_k x^k$ with radius of convergence r, then $f'(x)$ exists, (page 677) and $f'(x) = \sum k a_k x^{k-1}$ with radius of convergence r.

▮ *Theorem:* If $f(x) = \sum a_k x^k$ with radius of convergence r, then (page 679)

$$\int f(x)\, dx = \sum \left(\frac{a_k}{k + 1}\right) x^{k+1} + C$$

with radius of convergence r.

▌A **Taylor series** is a power series of the form   (page 682)

$$\sum_{k=0}^{\infty} \frac{f^{(k)}(a)}{k!}(x-a)^k = f(a) + f'(a)(x-a)$$
$$+ \frac{f''(a)}{2!}(x-a)^2 + \cdots + \frac{f^{(k)}(a)}{k!}(x-a)^k + \cdots.$$

If $a = 0$, this series is called a **Maclaurin series.**

▌Frequently used Maclaurin series are: (page 677)

$$\frac{1}{1-x} = 1 + x + x^2 + \cdots + x^k + \cdots, \qquad |x| < 1$$

$$= \sum_{k=0}^{\infty} x^k, \qquad |x| < 1$$

$$\frac{1}{1+x} = 1 - x + x^2 \mp \cdots + (-1)^k x^k + \cdots, \qquad |x| < 1 \qquad \text{(page 677)}$$

$$= \sum_{k=0}^{\infty} (-1)^k x^k, \qquad |x| < 1$$

$$\frac{1}{1+x^2} = 1 - x^2 + x^4 \mp \cdots + (-1)^{k+1} x^{2k} + \cdots, \qquad |x| < 1 \qquad \text{(page 680)}$$

$$= \sum_{k=0}^{\infty} (-1)^k x^{2k}, \qquad |x| < 1$$

$$\ln(1+x) = x - \frac{x^2}{2} + \frac{x^3}{3} \mp \cdots + (-1)^{k+1} \frac{x^k}{k} + \cdots, \qquad |x| < 1 \qquad \text{(page 680)}$$

$$= \sum_{k=1}^{\infty} (-1)^{k+1} \frac{x^k}{k}, \qquad |x| < 1$$

$$\text{Tan}^{-1} x = x - \frac{x^3}{3} + \frac{x^5}{5} \mp \cdots + (-1)^k \frac{x^{2k+1}}{2k+1} + \cdots, \qquad |x| < 1 \qquad \text{(page 680)}$$

$$= \sum_{k=1}^{\infty} (-1)^{k+1} \frac{x^k}{k}, \qquad |x| < 1$$

$$e^x = 1 + x + \frac{x^2}{2!} + \cdots + \frac{x^k}{k!} + \cdots \qquad \text{(page 684)}$$

$$= \sum_{k=0}^{\infty} \frac{x^k}{k!}$$

$$\sin x = x - \frac{x^3}{3!} + \frac{x^5}{5!} \mp \cdots + (-1)^k \frac{x^{2k+1}}{(2k+1)!} + \cdots \qquad \text{(page 683)}$$

$$= \sum_{k=0}^{\infty} (-1)^k \frac{x^{2k+1}}{(2k+1)!}$$

$$\cos x = 1 - \frac{x^2}{2!} + \frac{x^4}{4!} \mp \cdots + (-1)^k \frac{x^{2k}}{(2k)!} + \cdots \qquad \text{(page 687)}$$

$$= \sum_{k=0}^{\infty} (-1)^k \frac{x^{2k}}{(2k)!}$$

■ The **Binomial Series** is   (page 689)

$$(1 + x)^r = 1 + \sum_{k=0}^{\infty} \frac{r(r-1)(r-2)\cdots(r-k+1)}{k!} x^k, \quad |x| < 1.$$

Review Exercises—Chapter 13

In Exercises 1–10, find the Taylor polynomial of degree n for $f(x)$ expanded about $x = a$.

1. $f(x) = \sin 2x$, $a = 0$, $n = 5$

2. $f(x) = \ln(1 + x^2)$, $a = 0$, $n = 2$

3. $f(x) = x \cos x$, $a = \pi/4$, $n = 3$

4. $f(x) = \sqrt{1 + x^2}$, $a = 0$, $n = 3$

5. $f(x) = e^{x^2}$, $a = 0$, $n = 3$

6. $f(x) = \sqrt{2x + 3}$, $a = 11$, $n = 3$

7. $f(x) = \mathrm{Tan}^{-1} x$, $a = 1$, $n = 3$

8. $f(x) = x \ln(1 + x)$, $a = 0$, $n = 4$

9. $f(x) = 2^x$, $a = 0$, $n = 3$

10. $f(x) = 1/(1 + x^3)$, $a = 0$, $n = 2$

In Exercises 11–15, find the accuracy of the approximation of the given quantity by the nth Taylor polynomial for the function $f(x)$ expanded about $x = a$.

11. $\sin 34°$, $f(x) = \sin x$, $a = \pi/6$, $n = 3$

12. $\tan(\pi/12)$, $f(x) = \tan x$, $a = 0$, $n = 2$

13. $\sqrt[5]{35}$, $f(x) = \sqrt[5]{x}$, $a = 32$, $n = 2$

14. $\sec(3\pi/16)$, $f(x) = \sec x$, $a = \pi/4$, $n = 2$

15. $e^{.25}$, $f(x) = e^x$, $a = 0$, $n = 3$

16. Show that the fourth Taylor polynomial for $f(x) = \cosh x^2$ expanded about $a = 0$ is $P_4(x) = 1 + x^4/2$.

17. By integrating the $(n - 1)$st Taylor polynomial for $f(x) = 1/(1 - x)$, show that

$$\ln\frac{1}{1 - x} = x + \frac{x^2}{2} + \frac{x^3}{3} + \cdots + \frac{x^n}{n} + \int_0^x \frac{t^n \, dt}{1 - t}.$$

18. Show that, if $P_n(x)$ is the nth Taylor polynomial for $f(x) = \sinh x$ expanded about $a = 0$, and if $Q_{n+1}(x)$ is the Taylor polynomial of degree $n + 1$ for $\cosh x$ expanded about $a = 0$, then

$$\frac{d}{dx} Q_{n+1}(x) = P_n(x).$$

What if $a \neq 0$?

19. Let $P_n(x)$ be the nth Taylor polynomial, expanded about $a = 0$, for $f(x) = x^{7/2}$. Show that $P_0(0) = P_1(0) = P_2(0) = P_3(0) = 0$, but that $P_n(0)$ is undefined for $n \geq 4$.

20. Let $P_n(x)$ be the nth Taylor polynomial for f expanded about $x = a$. If $P_n(x) = 0$ for all n for which $P_n(x)$ is defined, must f be the zero function? (*Hint:* See Exercise 19.)

In Exercises 21–26, find the interval of convergence for the given series.

21. $\sum k(x - 3)^k$

22. $\sum (-1)^k k^2 (x - 1)^k$

23. $\sum \frac{\sqrt{k}(x - 2)^k}{k + 3}$

24. $\sum \frac{(x + 2)^k}{2^k}$

25. $\sum \frac{(-1)^k x^k}{k \ln k}$

26. $\sum \frac{(\ln k)(x - 1)^k}{e^k}$

In Exercises 27–32, find the Maclaurin series for f and determine its radius of convergence.

27. $f(x) = \dfrac{x}{1 - x^2}$

28. $f(x) = \dfrac{1}{8 + x^3}$

29. $f(x) = \ln(4 + x)$

30. $f(x) = \dfrac{\sin x^3}{x}$

31. $f(x) = \sin x^2 + 2x^2 \cos x^2$

$$\left(Hint: f(x) = \frac{d}{dx}(x \sin x^2)\right)$$

32. $f(x) = 2x \cos x^3 - 3x^4 \sin x^3$

$$\left(Hint: f(x) = \frac{d}{dx}(x^2 \cos x^3)\right)$$

33. Show that $\displaystyle\sum_{k=0}^{\infty} a_k x^{k+1} = \sum_{k=1}^{\infty} a_{k-1} x^k$.

34. Show that $\displaystyle\sum_{k=1}^{\infty} k(k + 1)x^k = \sum_{k=2}^{\infty} (k - 1)kx^{k-1}$.

35. Find numbers $a_1, a_2, \ldots$ so that

$$\sum k a_k x^{k-1} + c \sum a_k x^k = 0, \quad c \neq 0.$$

What is the function f whose power series is $\Sigma a_k x^k$?

36. Find the numbers $a_2, a_3, a_4, \ldots$ so that

$$\sum_{n=2}^{\infty} n(n-1)a_n x^{n-2} + 4\sum_{n=0}^{\infty} a_n x^n = 0.$$

37. Find a power series representation for the function $f(x) = 1/(1 + x^4)$.

38. Find a Maclaurin series for the function $f(x) = \sin x \cos x$.

39. Find a Maclaurin series for $f(x) = \sqrt[3]{1 + x^2}$.

40. Find a power series for the function

$$f(x) = \frac{1}{(1-x)(2-x)}$$

using the geometric series. What is the radius of convergence?

41. Find the first 3 nonzero terms of the Taylor series for $f(x) = \sin \sqrt{x}$ expanded about $x = \pi^2/4$.

42. What is the radius of convergence of a power series for $e^{\sqrt{x}}$ expanded about $a = 0$?

43. Find a Maclaurin series for $f(x) = x^2 e^{x^2}$ using the series for e^x.

44. Find the Taylor series for $f(x) = \sqrt{x + 4}$ expanded about $a = 2$. For what values of x does it converge?

45. Use a Taylor series to approximate $\sqrt{e}$ accurate to three decimal places.

Geometry in the Plane and in Space

Unit Six

Geometry in the Plane and in Space

René Descartes

René Descartes (1596–1650) and Pierre de Fermat (1601–1665) had at least two things in common: both were French, and neither was a professional mathematician. They shared something else of far greater importance—each independently discovered analytic geometry.

Descartes left school at the age of 16 and went to Paris, where he occasionally studied mathematics. At 21 he became a soldier-for-pay, hiring out to the armies of various nations and minor political subdivisions. He alternated military service with study and travel throughout Europe and met a number of mathematicians. His soldiering was apparently successful; he was offered a commission as lieutenant general, but declined in since he preferred having less responsibility and more time to think. After nine years of this life, his growing interest in matters of philosophy led him to the relative peace of the Netherlands, where he studied and wrote for 20 years.

on the night of November 10, 1619, Descartes had three dreams that profoundly affected his life and the development of mathematics. He claimed that in one of these dreams he was given the key to understanding nature. Although he never revealed precisely what this was, it is widely believed that he was referring to the relation between algebra and geometry. This may therefore be said to be the founding date of analytical geometry, although publication was not to occur for 18 years.

In 1637 Descartes published *Discours de la méthode…*, or *Discourse on the Method of Rightly Conducting the Reason and Seeking Truth in the Sciences*. This was a program for conducting philosophical research—an attempt to reach valid conclusions in many fields by systematic reasoning. The *Discourse* included three appendices. Each was brilliant illustration of the application of the method to a different field. The first, *La dioptrique*, included the first published statement of the law of refraction (although the Dutch mathematician Willebrord Snell (1591–1626) had discovered it earlier). The second appendix, on meteorology,

included an explanation of the colors of the rainbow. The third, and most famous, was *La géométrie*. Starting with an ancient problem from the Greek mathematician Pappus (third century A.D.), Descartes applied algebra to geometry and vice versa. This was not quite today's analytic geometry, however. Coordinate axes were implied but not used explicitly, and were thought of as oblique rather than the perpendicular axes that we now label Cartesian. The notions of distance, angle between lines, and slope were not included. In no case was a curve plotted from an equation. Descartes did not consider negative numbers of negative coordinates. Only one equation was considered in detail, the general second degree equation, and that was not discussed until Chapter 15. Descartes gave the conditions on the coefficients of this equation for the curve to be an ellipse, parabola, or hyperbola. The details were omitted, and indeed the entire work was difficult to read. Readers often found it difficult to see how the appendices were related to the philosophy set forth in the main text. In spite of these inadequacies, the concepts of analytic geometry were introduced in this little work, the only book about mathematics that Descartes ever produced. It was a remarkable and striking achievement.

Descartes was lured away from the Netherlands to a position with Queen Christina of Sweden, who was interested in philosophy and wanted him to establish an academy of sciences. Unfortunately, she wanted her tutoring in philosophy at 5:00 A.M. which seriously conflicted with Descartes' lifelong practice of late rising. He was unable to maintain his health in the cold, damp Swedish climate, and died of pneumonia at 54 years of age.

Pierre de Fermat

Pierre de Fermat has already been mentioned in the historical note preceding Unit Two, in which his early work on differentiation (before Newton and Leibniz) is discussed. As a lawyer, politician, and judge, he apparently had sufficient time to devote to the study of his real love, mathematics. His work on analytic geometry was actually done several years before that of Descartes, but he refused to publish his mathematical discoveries. What few items were printed during his lifetime were published by friends but without Fermat's name attached. Most of his work is known either through his notes, often written in the margins of books, or through the many letters he wrote to other mathematicians. The letters frequently contained original mathematics of such merit that copies were made and then recopied for circulation to other mathematicians. But Fermat was so casual about his work that he usually failed even to keep copies of his letters and essays for himself.

Fermat's work in analytic geometry was actually written several years before Descartes' discovery. It was not published, though, until 1679, 14 years after Fermat's death and a half century after its completion. Descartes wrote in more modern notation than did Fermat, but the latter's ideas much more closely resemble modern mathematics. The 1679 volume, titled (in translation) *Introduction to Plane and Solid Loci,* present in archaic terminology an astonishing amount of material which is familiar to us today. His linear equation written *"D in A acequitur B in E"* we would write as $Dx = By$, and Fermat sketched it as a line—or rather, as a half-line, since he did not accepted negative coordinates. He dealt with circles, parabolas, ellipses, and hyperbolas. He even extended his analytic geometry to three-dimensional space, a generalization that had not occurred to Descartes.

Fermat considered the family of equations of the form $y = x^n$ where n is a positive or negative integer. In an attempt to find the maxima and minima of the associated curves, he found the derivatives of these special functions. He also found the tangents to these curves by using what was essentially the differentiation process, though without the idea of the limit. Later Fermat found the area between such a curve and a line, which corresponds to the integral of x^n. It is curious that he did not seem to have recognized the inverse nature of these processes. Although Fermat is sometimes called a discoverer of the calculus, he worked with only a few polynomial functions, he lacked the concept of the limit, and he failed to note the Fundamental Theorem of Calculus. Thus, he cannot be said to have reached the intellectual summit that might have allowed him to be ranked with Newton and Leibniz.

Fermat's chief mathematical interest was neither the calculus nor analytic geometry. It involved such topics as prime numbers, perfect numbers, and magic squares, which fall under the heading of the theory of numbers. He further was one of the earliest to develop the theory of probability. He was a classical scholar and could read and write fluently in several languages: French, Spanish, Italian, Greek, and Latin. He even composed poetry in Latin. But his refusal to publish his work resulted in his brilliance not being recognized until long after his death.

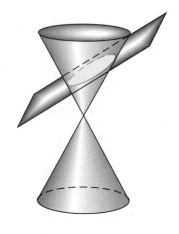

Chapter 14

The Conic Sections

The purpose of this chapter is to develop a broader familiarity with the types of planar curves that arise as graphs of equations of the form

$$Ax^2 + Bxy + Cy^2 + Dx + Ey + F = 0. \qquad (1)$$

If at least one of A, B, and C is nonzero, equation (1) is referred to as the general second degree equation in the two variables x and y. We have already examined one special case of this equation rather carefully.

If $A = C \neq 0$ and $B = 0$ the equation can be written

$$x^2 + y^2 + dx + ey + f = 0,$$

where $d = D/A$, $e = E/A$, $f = F/A$. By completing the squares in x and in y this equation can be brought to the standard form for the equation of a circle:

$$(x - h)^2 + (y - k)^2 = r^2.$$

(See Section 1.2.)

In the following three sections we shall see that if $B = 0$, the remaining nondegenerate possibilities for the graph of equation (1) are limited to the parabola, the ellipse, or the hyperbola. The case $B \neq 0$ is handled in Section 14.4, where we

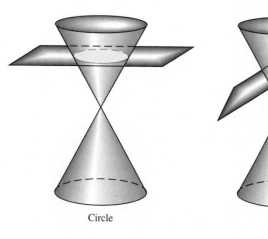

Circle

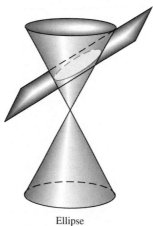

Ellipse

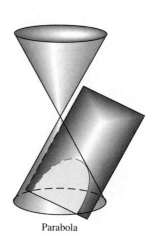

Parabola

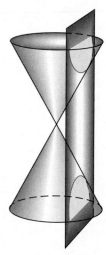

Hyperbola

show that rotating the x- and y-axes produces new coordinates for the plane in which the xy term in equation (1) is not present. Thus, we will show that the graph of any second degree equation of the form (1) must correspond to one of the four shown on page 698.

The circle, parabola, ellipse, and hyperbola are referred to as **conic sections** because each arises as the curve of intersection of an infinite cone and a plane, as illustrated on page 698.

We shall not pursue this three-dimensional interpretation of the conic sections. Rather, we shall use the techniques of analytic geometry in the plane to analyze equation (1) in the cases corresponding to each of these figures.

14.1 Parabolas

We begin with the purely geometric definition of the parabola.

Definition 1

A **parabola** is the set of all points P in the plane lying equidistant from a fixed line ℓ and a fixed point F (not on the line).

The fixed point F is called the **focus** of the parabola, and the fixed line ℓ is called the **directrix.** The line through the focus perpendicular to the directrix is called the **axis** of the parabola.

It is easy to see from Definition 1 that the point V on the axis midway between the focus and directrix lies on the parabola. This point is called the **vertex** of the parabola. A typical parabola, illustrating the requirement of Definition 1, is sketched in Figure 1.1.

To determine the equation satisfied by the coordinates of a point $P = (x, y)$ on the graph of a parabola, we position the parabola in the xy-plane so that the vertex V lies at the origin and so that the focus F lies at the point $(0, c)$ (see Figure 1.2). The directrix then has equation $\ell: y = -c$. The number c, representing the (signed) distance from the vertex to the focus, is referred to as the **focal length** of the parabola.

Now let $P = (x, y)$ be any point on the parabola, and let Q be the point on ℓ nearest P. Then $Q = (x, -c)$, and the distance from P to ℓ is the same as the distance from P to Q. Using the distance formula we can state the requirement $d_1 = d_2$ of Definition 1 as

$$\sqrt{(x - 0)^2 + (y - c)^2} = \sqrt{(x - x)^2 + (y - (-c))^2},$$

or

$$\sqrt{x^2 + (y - c)^2} = |y + c|.$$

Squaring both sides gives

$$x^2 + (y - c)^2 = (y + c)^2, \quad \text{or}$$
$$x^2 + (y^2 - 2cy + c^2) = y^2 + 2cy + c^2,$$

$$\boxed{x^2 = 4cy, \quad \text{or} \quad y = \frac{x^2}{4c}.} \tag{1}$$

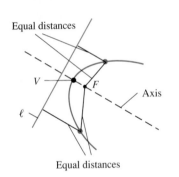

Figure 1.1 Parabola (ℓ = directrix, F = focus, V = vertex).

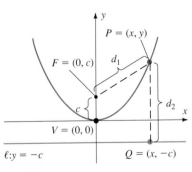

Figure 1.2 Parabola in standard position.

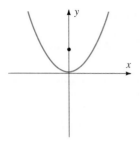

Figure 1.3 $x^2 = 4cy, c > 0$.

Figure 1.4 $x^2 = 4cy, c < 0$.

Equation (1) is the **standard form** for the parabola with vertex at the origin and focus on the y-axis. In Figure 1.2 we have assumed $c > 0$. However, the calculations above did not depend on this assumption, so equation (1) is also valid if $c < 0$. In the case $c < 0$ the parabola opens in the negative y-direction (see Figures 1.3 and 1.4).

REMARK   The above demonstration shows that any point (x, y) satisfying the *geometric* condition of Definition 1 satisfies the *algebraic* condition of equation (1). In Exercise 39 you are asked to show the converse—that any point (x, y) satisfying the algebraic condition (1) also satisfies the geometric condition of Definition 1.

For a parabola with vertex at $(0, 0)$, the axis may lie along the x-axis rather than the y-axis. In this case the focus is labelled $(c, 0)$, the directrix is $x = -c$, and the requirement of Definition 1 is that

$$\sqrt{(x - c)^2 + (y - 0)^2} = \sqrt{(x - (-c))^2 + (y - y)^2},$$

which simplifies to the equation

$$y^2 = 4cx, \quad \text{or} \quad x = \frac{y^2}{4c}. \tag{2}$$

(See Figure 1.5.) Equation (2) is the standard form for the parabola with vertex at $(0, 0)$ and focus on the x-axis. If $c > 0$, the parabola opens in the positive x-direction. If $c < 0$ it opens in the negative x-direction (Figure 1.6).

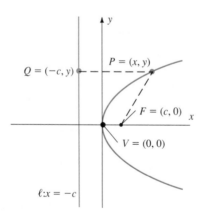

Figure 1.5 $y^2 = 4cx, c > 0$.

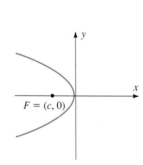

Figure 1.6 $y^2 = 4cx, c < 0$.

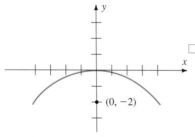

Figure 1.7 $x^2 = -8y$.

□ **EXAMPLE 1**

The equation $x^2 + 8y = 0$ may be written in the form

$$x^2 = -8y, \quad \text{or} \quad y = -x^2/8$$

corresponding to equation (1). Its graph is therefore a parabola with vertex $(0, 0)$, with focal length $c = -2$, and opening in the negative y-direction. Its focus is at $(0, -2)$ (see Figure 1.7).   ∎

□ **EXAMPLE 2**

The graph of the equation $\frac{1}{2}y^2 - x = 0$ can be written in the form of equation (2) as

$$y^2 = 2x, \quad \text{or} \quad x = \frac{y^2}{2}.$$

Its graph is a parabola opening in the positive x-direction with vertex $(0, 0)$, focal length $c = 1/2$, and focus $(1/2, 0)$ (Figure 1.8). ■

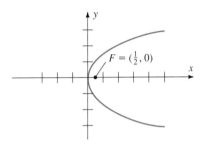

Figure 1.8 $y^2 = 2x$.

Translating the Parabola

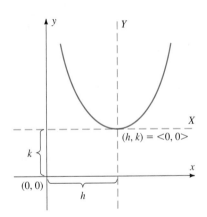

Figure 1.9 A translated parabola: $(\,,\,) = xy$-coordinates; $\langle\,,\,\rangle = XY$-coordinates.

If the vertex of a parabola lies at a point (h, k) rather than at $(0, 0)$, we may apply the technique of **translation of axes** to analyze its graph. To do so we introduce new XY-coordinates, via the substitutions

$$X = x - h \quad \text{and} \quad Y = y - k. \tag{3}$$

These substitutions determine new X- and Y-axes, parallel to the x- and y-axes, but with the origin in XY-coordinates located at the vertex of the parabola (see Figure 1.9).

If, after translating axes, the parabola opens along the Y-axis with focal length c, the equation for the parabola in XY-coordinates is

$$X^2 = 4cY, \tag{4}$$

according to equation (1). Combining equation (4) with substitutions (3) we conclude that

$$\boxed{(x - h)^2 = 4c(y - k)} \tag{5}$$

is the *standard form for the equation for the parabola with vertex at (h, k), focal length c, and axis parallel to the y-axis.* Similarly, from equation (2) and substitutions (3) it follows that

$$\boxed{(y - k)^2 = 4c(x - h)} \tag{6}$$

is the *standard form for the equation of the parabola with vertex at (h, k), focal length c, and axis parallel to the x-axis.*

As was the case for equations of circles in Section 1.2, the usual technique for analyzing the equation of a parabola is to complete the square first, if necessary, and then to bring the resulting equation into the form of equation (5) or equation (6).

□ **EXAMPLE 3**

Describe the graph of the equation.

$$(y - 1)^2 = 6(x + 2).$$

Strategy · · · · · · · ·

Identify the form of the equation as that of equation (6).

Since the parabola opens in the x direction we have

> vertex = (h, k),
> focal length = c,
> axis: $y = k$,
> focus: $(h + c, k)$.

Solution

The equation has the form of equation (6) with

$$h = -2 \quad \text{and} \quad k = 1,$$

so the graph is a parabola with axis parallel to the x-axis. Since $6 = 4c$, we have $c = 3/2$. The vertex is $(-2, 1)$, the focal length is $3/2$, the axis is $y = 1$, and the focus is $(-1/2, 1)$ (see Figure 1.10). ■

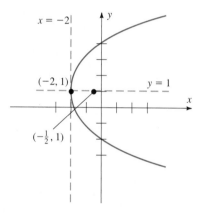

Figure 1.10 $(y - 1)^2 = 6(x + 2)$.

□ **EXAMPLE 4**

Find the vertex, focus, axis, and focal length for the parabola with equation $x^2 - 6x + 6y - 3 = 0$.

Strategy · · · · · · · ·

Complete the square in x by adding 9 to both sides.

Move all terms except the squared term to the right side.

Identify the form of the resulting equation.

Solution

We first observe that

$$x^2 - 6x + 6y - 3 = 0$$

gives

$$[x^2 - 6x + 9] + 6y - 3 = 9,$$
$$(x - 3)^2 + 6y = 12,$$
$$(x - 3)^2 = -6y + 12$$
$$= -6(y - 2)$$
$$= 4(-3/2)(y - 2).$$

This equation has the the form of equation (5) with

$$h = 3, \quad k = 2, \quad c = -3/2.$$

Thus the vertex is $(3, 2)$, the focal length is $c = -3/2$, the axis is $x = h = 3$, and the focus is $(h, k + c) = (3, 1/2)$. The graph appears in Figure 1.11. ■

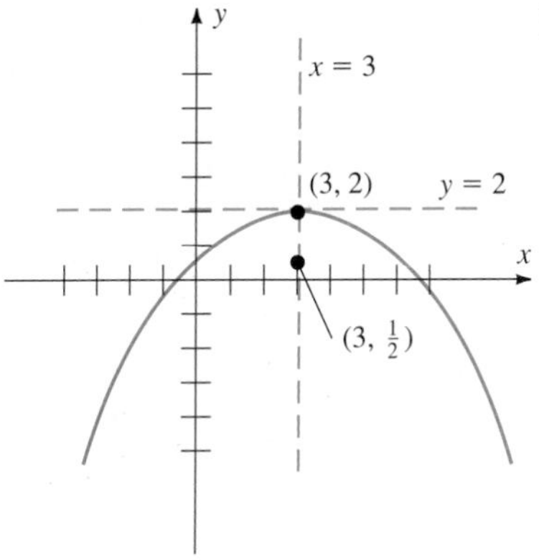

y

x = 3

(3, 2)   y = 2

x

(3, ½)

Figure 1.11 $x^2 - 6x + 6y - 3 = 0$.

Reflecting Properties of Parabolas

☐ **EXAMPLE 5**

Find the equation of the parabola with directrix $y = -2$ and focus $F = (2, 5)$.

Solution   Since the axis lies perpendicular to the directrix, the axis is parallel to the y-axis. The standard equation for the parabola is therefore that of equation (5). Since the vertex lies midway between the directrix and the focus, the vertex has coordinates $(2, -2 + \frac{1}{2}(5 - (-2))) = (2, 3/2)$. Thus, $h = 2$ and $k = 3/2$. Since the distance from the focus to the vertex is $5 - 3/2 = 7/2$, the focal length is $c = 7/2$. The desired equation is therefore

$$(x - 2)^2 = 4(\tfrac{7}{2})(y - 3/2),$$

or

$$(x - 2)^2 = 14(y - 3/2).$$ ■

An important application of parabolic arcs follows from the physical law that when a ray of light strikes a reflecting surface the angle of incidence equals the angle of reflection. The relationship between this law and parabolas lies in the fact that the line tangent to the graph of a parabola at a point P makes equal angles with the ray from P to the focus and the line through P parallel to the axis of the parabola (Figure 1.12 and Exercise 31).

Thus, rays of light parallel to the axis of a parabolic reflector will all be reflected precisely to the focus of the parabola (Figure 1.13). This principle is used in reflecting telescopes, where the parallel rays of light from the stars are reflected off a parabolic mirror to the eyepiece located at the focus. The same principle applies to radar signals, which are gathered by means of parabolic reflecting "dishes."

The same principle in reverse explains the design of automobile headlamps. By placing the bulb at the focus of a parabolic mirror, one ensures that the beams of light will leave the headlamp parallel to the axis of the parabolic mirror.

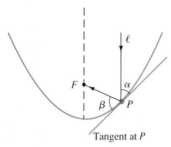

ℓ

F   α
β   P

Tangent at P

Figure 1.12 Reflecting property: $\alpha = \beta$ if ℓ is parallel to axis.

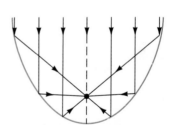

Figure 1.13 Rays parallel to the axis are all reflected through the focus of a parabola, and conversely.

Exercise Set 14.1

In Exercises 1–20, sketch the parabola whose equation is given, noting the vertex, focal length, axis, and focus.

1. $y = 2x^2$

2. $y = 4x^2 + 8$

3. $y = -2x^2 + 2$

4. $x = -y^2$

5. $x = 3y^2 + 1$

6. $x + 2y^2 = 4$

7. $8y + x^2 = 2$

8. $y - x^2 = 4$

9. $(x - 2)^2 = 8(y - 1)$

10. $x^2 = 12(y + 2)$

11. $(y - 2)^2 = 6(x + 3)$

12. $(y - 1)^2 = -7x$

13. $y^2 - 2y - 4x - 7 = 0$

14. $y^2 + 4y + x + 7 = 0$

15. $x^2 + 2x - 6y - 11 = 0$

16. $x^2 - 4x - 4y = 0$

17. $x^2 + 6x - 10y + 19 = 0$

18. $y^2 - 4y - 8x + 20 = 0$

19. $y^2 + 4y - 2x + 6 = 0$

20. $y^2 + 6y - 3x + 6 = 0$.

In Exercises 21–27, find an equation for the parabola with the stated properties.

21. Vertex $(0, 0)$ and focus $(0, 3)$.

22. Vertex $(-1, 2)$ and focus $(-1, -2)$.

23. Vertex $(1, 3)$ and directrix $y = -5$.

24. Focus $(-1, -1)$ and directrix $x = 3$.

25. Focus $(3, -6)$ and vertex $(-1, -6)$.

26. Vertex $(0, 4)$ and directrix $x = 2$.

27. Axis $x = 0$, directrix $y = -3$, and containing the point $(0, 1)$.

28. State a procedure for sketching the graph of a parabola according to Definition 1, assuming that you are given the equation for the directrix and the location of the focus and assuming that you can use only a ruler and compass.

29. The chord through the focus of a parabola perpendicular to its axis is called the **latus rectum** of the parabola (see Figure 1.14). Prove that the length of the latus rectum of a parabola is $|4c|$ where $|c|$ is the focal length.

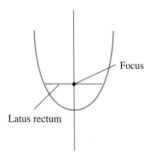

Figure 1.14 Latus rectum.

30. Prove that a circle that has a diameter corresponding to the latus rectum of a parabola must be tangent to the directrix of that parabola.

31. In Figure 1.12, α is the angle formed between the tangent at P and the line ℓ through P parallel to the axis of the parabola; β is the angle formed between the tangent at P and the line segment from P to the focus F. Prove that $\alpha = \beta$. (*Hint:* Construct a coordinate system with origin at the vertex of the parabola and extend ℓ through P toward the directrix. Also draw a line through F, perpendicular to the tangent at P, until it intersects ℓ. What can you show about the triangle formed in this way?)

32. Prove that the graph of the second degree equation

$$Ax^2 + Bxy + Cy^2 + Dx + Ey + F = 0$$

is a parabola with axis parallel to either the x- or y-axis whenever $B = 0$ and either $A = 0, CD \neq 0$ or $C = 0$, $AE \neq 0$.

33. Find the area of the region bounded by the graph of $y = p(x)$ and the x-axis if the graph of $y = p(x)$ is a parabola with focus at the origin and directrix $y = -2$.

34. Find the volume of the solid obtained by revolving about the y-axis the area of the region bounded by the y-axis and the parabola with vertex $(-1, 0)$ and directrix $x = -3/2$.

35. Find the area of the region bounded by the x-axis and the parabola with vertex $(4, 2)$ and directrix $y = 4$.

36. Find the volume of the solid obtained by revolving about the y-axis the region bounded by the y-axis and the parabola with focus $(0, 4)$ and directrix $x = 4$.

37. Find the average value, on the interval $[0, 5]$, of the function whose graph is a parabola with focus $(2, 3)$ and vertex $(2, -1)$.

38. Find the area of the region bounded by the graph of the function in Exercise 34 and the line $x = 5$.

39. Show that if a point (x, y) satisfies the equation $x^2 = 4cy$ then it has the property that its distance from the directrix $y = -c$ equals its distance from the focus $(0, c)$. Thus, any point satisfying the algebraic condition $x^2 = 4cy$ for a parabola also satisfies the geometric condition of Definition 1.

14.2 The Ellipse

The geometric definition of the ellipse is the following.

Definition 2

An **ellipse** is the set of all points in the plane the sum of whose distances from two fixed points is a constant.

The two fixed points in Definition 2 are referred to as the **foci** of the ellipse. We may interpret Definition 2 by imagining a length of string whose ends are tied to two tacks driven into a flat surface. The tacks represent the foci. If a pencil is held

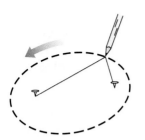

Figure 2.1 Tracing out an ellipse.

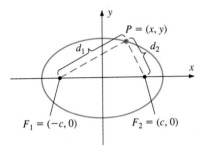

Figure 2.2 Ellipse in standard position.

against the surface as in Figure 2.1, so that the string is kept taut, the figure traced out by the possible locations for the pencil tip is an ellipse. This is because the sum of the distances of the pencil tip from the foci equals the length of the string, which is constant.

To determine the equation for the ellipse, we construct a coordinate system so that the foci for the ellipse are located at points $F_1 = (-c, 0)$ and $F_2 = (c, 0)$ (Figure 2.2). We let $P = (x, y)$ be an arbitrary point on the ellipse, and we let d_1 and d_2 be the distances from P to F_1 and from P to F_2, respectively. The condition that P be on the ellipse is that $d_1 + d_2$ equal a constant, which we take to be $2a$. (Note that we must have $a > c$, since the distance between the foci is $2c$.)

Using the distance formula, we can now write the requirement of Definition 2 as

$$d_1 + d_2 = 2a,$$

or

$$\sqrt{(x + c)^2 + y^2} + \sqrt{(x - c)^2 + y^2} = 2a.$$

To simplify this equation we move the second radical to the right-hand side and square both sides. We obtain

$$(x + c)^2 + y^2 = 4a^2 - 4a\sqrt{(x - c)^2 + y^2} + (x - c)^2 + y^2.$$

After the squared terms are expanded, this simplifies to

$$2cx = 4a^2 - 4a\sqrt{(x - c)^2 + y^2} - 2cx,$$

so

$$a\sqrt{(x - c)^2 + y^2} = a^2 - cx.$$

Squaring both sides again gives

$$a^2[(x - c)^2 + y^2] = c^2x^2 - 2a^2cx + a^4,$$

or

$$(a^2 - c^2)x^2 + a^2y^2 = a^2(a^2 - c^2). \tag{1}$$

To simplify (1), we define the constant b by

$$b = \sqrt{a^2 - c^2}. \tag{2}$$

Substituting (2) into (1) then gives

$$b^2x^2 + a^2y^2 = a^2b^2,$$

or

$$\boxed{\frac{x^2}{a^2} + \frac{y^2}{b^2} = 1, \qquad a > b.} \tag{3}$$

Equation (3) is the *standard form for the equation of the ellipse with center at* $(0, 0)$ *and foci on the x-axis*. (The **center** of an ellipse is the midpoint of the line segment joining the foci.) In Exercise 39 you are asked to show the converse—that any point satisfying equation (3) also satisfies the geometric condition of Definition 2.

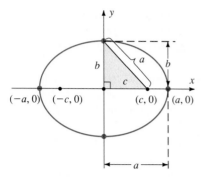

Figure 2.3   The ellipse

$$\frac{x^2}{a^2} + \frac{y^2}{b^2} = 1$$

with $a > b$.

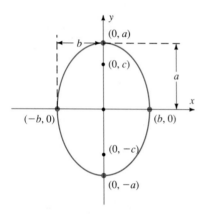

Figure 2.4   The ellipse

$$\frac{x^2}{b^2} + \frac{y^2}{a^2} = 1$$

with $a > b$.

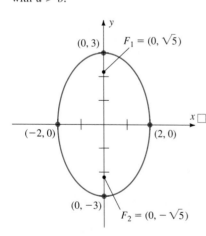

Figure 2.5 $\dfrac{x^2}{4} + \dfrac{y^2}{9} = 1.$

The geometry associated with equation (3) is worth noting. Setting $y = 0$ in (3) shows that the x-intercepts of the ellipse are $x = \pm a$ (see Figure 2.3). The line segment joining the points $(-a, 0)$ and $(a, 0)$ is referred to as the *major* axis of the ellipse. Its length is $2a$. In general, the **major axis** of an ellipse is the chord passing through both foci.

Setting $x = 0$ in (3) shows that the y-intercepts are $y = \pm b$. The line segment joining $(0, b)$ and $(0, -b)$ is called the *minor* axis of the ellipse. Its length is $2b$. More generally, the **minor axis** of an ellipse is the chord through the center and perpendicular to the major axis.

Figure 2.4 illustrates the ellipse that results if the foci are located along the y-axis at $(0, c)$ and at $(0, -c)$. The equation describing this ellipse may be developed just as for the previous case. However, it is much simpler to note that in passing from Figure 2.3 to Figure 2.4 we have merely interchanged the roles of x and y. Thus,

$$\frac{x^2}{b^2} + \frac{y^2}{a^2} = 1, \qquad a > b \tag{4}$$

is *the standard form for the equation of the ellipse with center at $(0,0)$ and foci along the y-axis.*

Note in equation (4) that the constant a still represents half the length of the major axis and that b is again half the length of the minor axis. For either equation (3) or equation (4), the distance between the origin and a focus is found by solving equation (2) for c:

$$c = \sqrt{a^2 - b^2}. \tag{5}$$

□ **EXAMPLE 1**

Describe the graph of the equation $\dfrac{x^2}{4} + \dfrac{y^2}{9} = 1$.

Solution   The equation has the form of equation (4) with $b^2 = 4$, $a^2 = 9$. Thus, $b = 2$ and $a = 3$. The figure is an ellipse with center at $(0, 0)$, with major axis vertical and of length $2a = 6$, and with minor axis of length $2b = 4$. From equation (5) we find that $c = \sqrt{9 - 4} = \sqrt{5}$, so the foci are located at $(0, \sqrt{5})$ and $(0, -\sqrt{5})$ (see Figure 2.5). ∎

□ **EXAMPLE 2**

Describe the graph of the equation $4x^2 + 16y^2 - 64 = 0$.

Solution   To bring the equation into standard form, we move the constant term to the right and divide both sides by this constant. We obtain $4x^2 + 16y^2 = 64$, so

$$\frac{x^2}{4^2} + \frac{y^2}{2^2} = 1.$$

By setting $y = 0$, we obtain the x-intercepts, $x = \pm 4$. The endpoints of the major axis are therefore $(-4, 0)$ and $(4, 0)$. Setting $x = 0$ gives the y-intercepts, $y = \pm 2$.

The endpoints of the minor axis are $(0, 2)$ and $(0, -2)$. Since the equation is of the form of equation (3), with $a = 4$ and $b = 2$, we have from equation (5) that $c = \sqrt{4^2 - 2^2} = \sqrt{12}$. The foci are therefore $(\pm\sqrt{12}, 0)$ (Figure 2.6). ■

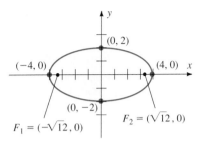

Figure 2.6 $4x^2 + 16y^2 - 64 = 0$.

Translating the Ellipse

When the center of an ellipse lies at the point (h, k) rather than at the origin, we can analyze its graph by first translating the coordinate axes by the substitutions

$$X = x - h \quad \text{and} \quad Y = y - k \quad \text{(Figure 2.7)}. \tag{6}$$

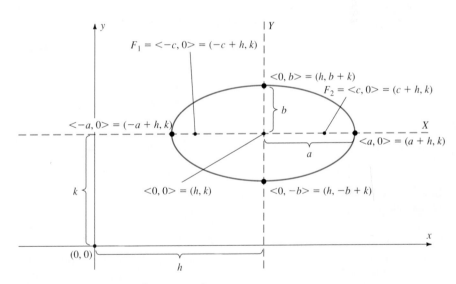

Figure 2.7 Ellipse $\dfrac{(x - h)^2}{a^2} + \dfrac{(y - k)^2}{b^2} = 1$, $a > b$. $\langle , \rangle = XY$-coordinates; $(,) = xy$-coordinates.

In XY-coordinates, the center of the ellipse lies at the origin. If the major axis of the ellipse is parallel to the x-axis, the equation of the ellipse in XY-coordinates is, by equation (3),

$$\frac{X^2}{a^2} + \frac{Y^2}{b^2} = 1. \tag{7}$$

Using equation (6), we conclude that

$$\frac{(x-h)^2}{a^2} + \frac{(y-k)^2}{b^2} = 1, \qquad a > b \tag{8}$$

is *the standard form of the equation for the ellipse with center at (h, k) and major axis parallel to the x-axis.* In (8), the constant *a* represents half the length of the major axis while *b* is half the length of the minor axis. Similarly,

$$\frac{(x-h)^2}{b^2} + \frac{(y-k)^2}{a^2} = 1, \qquad a > b \tag{9}$$

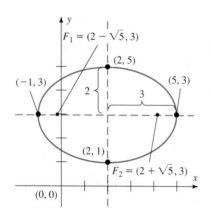

Figure 2.8 Graph of equation $4x^2 + 9y^2 - 16x - 54y + 61 = 0$.

is *the standard form of the equation for the ellipse with center at (h, k), with major axis parallel to the y-axis and of length 2a, and with minor axis of length 2b.*

For either equation (8) or equation (9), the distance between the center of the ellipse and either focus is given by *c* in equation (5).

□ **EXAMPLE 3**

Describe the graph of the equation

$$4x^2 + 9y^2 - 16x - 54y + 61 = 0.$$

Strategy · · · · · · · ·
Complete the square in both variables.

Solution

$$4x^2 + 9y^2 - 16x - 54y + 61$$
$$= 4(x^2 - 4x + 4) + 9(y^2 - 6y + 9) + 61 - 16 - 81$$
$$= 4(x - 2)^2 + 9(y - 3)^2 - 36.$$

Write the original equation in squared form

The equation is therefore equivalent to

$$4(x - 2)^2 + 9(y - 3)^2 = 36,$$

Bring equation to form (8) or (9). Identify all constants.

or

$$\frac{(x - 2)^2}{3^2} + \frac{(y - 3)^2}{2^2} = 1.$$

Center is (h, k).
Major axis has length $2a$.
Minor axis has length $2b$.
Foci are at $(h \pm c, k)$ where $c = \sqrt{a^2 - b^2}$.

This equation has the form of equation (8) with $h = 2$, $k = 3$, $a = 3$, $b = 2$, and $c = \sqrt{3^2 - 2^2} = \sqrt{5}$. The graph is an ellipse with center (2, 3), major axis parallel to the *x*-axis and of length $2a = 6$, and minor axis of length $2b = 4$. Foci are at $(2 - \sqrt{5}, 3)$ and $(2 + \sqrt{5}, 3)$. The graph appears in Figure 2.8. ∎

□ **EXAMPLE 4**

Find the equation for the ellipse with center $(-4, 3)$, with minor axis of length 6, and with foci at $(-4, 3 \pm 4)$.

Solution The distance from the center of the ellipse to one of the foci is $c = 4$. The length of the minor axis is $2b = 6$, so $b = 3$. Thus, by equation (5),

$$a^2 = b^2 + c^2 = 3^2 + 4^2 = 5^2,$$

so $a = 5$. Also, $h = -4$ and $k = 3$ are the coordinates of the center. Since the major axis is vertical, we use equation (9) to write

$$\frac{(x + 4)^2}{3^2} + \frac{(y - 3)^2}{5^2} = 1.$$

(See Figure 2.9.) ∎

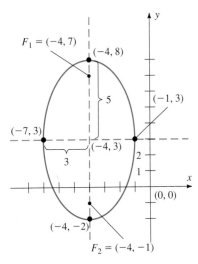

Figure 2.9 The ellipse with center $(-4, 3)$, minor axis of length 6, and foci at $(-4, 3 \pm 4)$.

Reflecting Property of Ellipses

In Exercise 32 you are asked to show that the rays $\overline{PF_1}$ and $\overline{PF_2}$, from a point P on an ellipse to the foci, make equal angles with the line tangent to the ellipse at P (see Figure 2.10). Applying the physical law that the angle of incidence equals the angle of reflection, we conclude that if a ray of light or a sound wave emanates from the focus F_1 of an elliptical reflecting surface, it will be reflected through the second focus F_2. Visitors to the United States Capitol Building hear tales of just such a phenomenon. In the days when congressmen were seated on the floor beneath the beautiful elliptical dome, the unfortunate politician seated at a focus risked losing whispered secrets to the statesman seated at the other focus (see Figure 2.11).

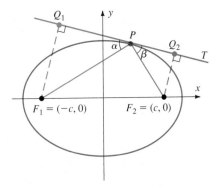

Figure 2.10 $\alpha = \beta$ if T is the tangent to the ellipse at P.

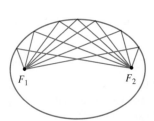

Figure 2.11 Reflecting property of ellipses.

Exercise Set 14.2

In Exercises 1–16, find the center and foci for the given ellipse. State the length of the major and minor axes and sketch the graph.

1. $\dfrac{x^2}{9} + \dfrac{y^2}{1} = 1$

2. $\dfrac{x^2}{16} + \dfrac{y^2}{49} = 1$

3. $4x^2 + 16y^2 = 64$

4. $9x^2 + 4y^2 = 36$

5. $50x^2 + 18y^2 - 450 = 0$

6. $8x^2 + 18y^2 - 72 = 0$

7. $\dfrac{(x-2)^2}{3} + \dfrac{(y-1)^2}{4} = 1$

8. $\dfrac{(x-2)^2}{3} + \dfrac{(y+2)^2}{4} = 25$

9. $(x+3)^2 + 4(y-1)^2 = 16$

10. $5(x+2)^2 + 4(y-3)^2 = 25$

11. $9x^2 + 4y^2 - 18x + 8y = 23$

12. $x^2 + 4y^2 + 16y + 12 = 0$

13. $9x^2 + 25y^2 + 54x - 50y - 119 = 0$

14. $9x^2 + 4y^2 - 36x + 48y + 144 = 0$

15. $2x^2 + 3y^2 + 6y - 4x - 1 = 0$

16. $4x^2 + 2y^2 + 24x + 12y + 46 = 0$

In Exercises 17–23, find the equation for the ellipse with the stated properties.

17. Center at $(0, 0)$, major axis of length 8, minor axis of length 6, foci on x-axis.

18. Foci at $(\pm 4, 0)$, major axis of length 10.

19. Foci at $(\pm 4, 0)$, minor axis of length 6.

20. Foci at $(0, \pm\sqrt{3})$, minor axis of length 2.

21. Foci at $(4, 5)$ and $(4, -1)$, minor axis of length 8.

22. Center at $(2, -6)$, foci at $(2, -3)$, and $(2, -9)$, major axis of length 10.

23. Foci at $(-2, 3)$ and $(-2, 9)$, minor axis of length 8.

24. Find the area of the region enclosed by the ellipse

$$\frac{x^2}{4} + \frac{y^2}{9} = 1.$$

25. Find the volume of the solid obtained by revolving about the x-axis the region in the first quadrant bounded by the graph of the ellipse $25x^2 + 9y^2 = 225$.

26. Find the volume of the solid obtained by revolving about the y-axis the region described in Exercise 25.

27. A chord through a focus of an ellipse and parallel to the minor axis is called a **latus rectum.** Find the length of a latus rectum for the ellipse

$$\frac{x^2}{a^2} + \frac{y^2}{b^2} = 1, \qquad a^2 > b^2.$$

28. The **eccentricity** of the ellipse

$$\frac{x^2}{a^2} + \frac{y^2}{b^2} = 1$$

is defined to be the ratio

$$e = \frac{c}{a} = \frac{\sqrt{a^2 - b^2}}{a}.$$

   a. Show that $0 \le e < 1$ for every ellipse.
   b. Discuss the change in the nature of the ellipse as e approaches zero.
   c. Discuss the change in the nature of the ellipse as e approaches one.

29. True or false? A circle is a special case of an ellipse. Explain.

30. Suppose that the numbers A and C are nonzero and have the same sign. Show that the graph of the second degree equation

$$Ax^2 + Cy^2 + Dx + Ey + F = 0$$

is either empty, a point, or an ellipse. What conditions on the coefficients distinguish the three cases?

31. Consider the equation for an ellipse:

$$\frac{x^2}{a^2} + \frac{y^2}{b^2} = 1.$$

   a. Differentiate implicitly with respect to x to show that

$$\frac{dy}{dx} = \frac{-b^2 x}{a^2 y}, \qquad y \neq 0.$$

   b. Use part a to show that the equation for the line tangent to the graph of this ellipse at (x_0, y_0) can be written

$$(b^2 x_0)x + (a^2 y_0)y - a^2 b^2 = 0.$$

32. Use the result of Exercise 31 to prove that triangle $F_1 Q_1 P$ is similar to triangle $F_2 Q_2 P$ in Figure 2.10. Conclude that $\alpha = \beta$.

33. Find the equation for the set of all points the sum of whose distances from $(2, 3)$ and $(-4, 3)$ is 12.

34. Find the equation for the set of all points whose distance from the point $(4, 0)$ is half the distance from the line $y = 3$.

35. Find the volume of the solid obtained by revolving about the y-axis the region bounded by the ellipse $16x^2 + 9y^2 = 144$.

36. A 10-ft long water tank has the shape of a cylinder with elliptical end panels. The end panels have major axes horizontal, of length 32 ft, and minor axes vertical, of length 18 ft. The tank is full. Find the work done in pumping the water out through a valve in the top of the tank.

37. Find the answer to Exercise 36 if the tank is only half full.

38. State the domain and range of the function

$$f(x) = \sqrt{b^2\left(1 - \frac{x^2}{a^2}\right)}.$$

What is the relationship between the graph of this function and the ellipse with equation $b^2x^2 + a^2y^2 = a^2b^2$?

39. Show that if (x, y) is a point whose coordinates satisfy the equation

$$\frac{x^2}{a^2} + \frac{y^2}{b^2} = 1,$$

then (x, y) satisfies the geometric condition of Definition 2, that the sum of its distances from the foci $(-c, 0)$ and $(c, 0)$ is $2a$, where $c^2 = a^2 - b^2$.

14.3   The Hyperbola

As for the parabola and the ellipse, we begin with the geometric definition of the hyperbola and then develop the corresponding equations.

Definition 3

A **hyperbola** is the set of all points in the plane the difference of whose distances from two fixed points is a constant.

To interpret Definition 3, we let F_1 and F_2 denote the two fixed points (called **foci**) in the plane and we let k denote the constant difference in the distances. Then a point P is on the hyperbola if either $d(P, F_1) - d(P, F_2) = k$ or $d(P, F_2) - d(P, F_1) = k$ (see Figure 3.1).

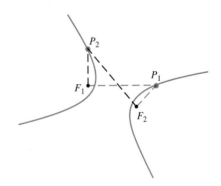

Figure 3.1   A point P is on the hyperbola if either

$$d(P, F_1) - d(P, F_2) = k$$

or

$$d(P, F_2) - d(P, F_1) = k.$$

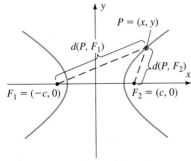

Figure 3.2   Hyperbola with center at $(0, 0)$.

To determine the equations for the hyperbola, we construct a pair of xy-coordinate axes so that the foci are located at $F_1 = (-c, 0)$ and $F_2 = (c, 0)$ (Figure 3.2). (We refer to the midpoint of the line segment $\overline{F_1F_2}$ as the **center** of the hyperbola, so the center in this case is the origin.) Let $P = (x, y)$ be a point on the hyperbola, and let $2a = k$ be the constant difference referred to in Definition 3. Then P is on the hyperbola if and only if

$$\left| d(P, F_2) - d(P, F_1) \right| = 2a.$$

Using the distance formula and the coordinates for P, F_1, and F_2, we can rewrite this equation as

$$\left| \sqrt{(x-c)^2 + y^2} - \sqrt{(x+c)^2 + y^2} \right| = 2a, \qquad c > a. \tag{1}$$

By squaring both sides of this equation and simplifying, just as we did in Section 14.2 for the ellipse, we can obtain the *standard form of the equation for the hyperbola with center at $(0,0)$ and foci on the x-axis:*

$$\boxed{\frac{x^2}{a^2} - \frac{y^2}{b^2} = 1, \qquad b = \sqrt{c^2 - a^2}.} \tag{2}$$

By setting $y = 0$ in equation (2), we obtain the x-intercepts of the hyperbola $x = \pm a$. The points $V_1 = (-a, 0)$ and $V_2 = (a, 0)$ are called the **vertices** of the hyperbola. (By setting $x = 0$ you will observe that the graph of equation (2) has no y-intercepts. This observation will help you to determine whether the foci of a given hyperbola lie along the x-axis or along the y-axis.)

In graphing equation (2) it is helpful to note that *the lines $y = \pm (b/a)x$ are asymptotes for the graph of equation (2)*. This means that the points on the graph of the hyperbola approach the lines $y = \pm(b/a)x$ as $|x| \to \infty$. To see this we solve (2) for y:

$$\frac{y^2}{b^2} = \frac{x^2}{a^2} - 1,$$

so

$$y = \pm b \sqrt{\frac{x^2}{a^2} - 1} = \pm \frac{b}{a} x \sqrt{1 - \frac{a^2}{x^2}}. \tag{3}$$

Since $\displaystyle \lim_{|x| \to \infty} \sqrt{1 - \frac{a^2}{x^2}} = \sqrt{1 - 0} = 1$, equation (3) shows that

$$\boxed{y = \pm \frac{b}{a} x} \tag{4}$$

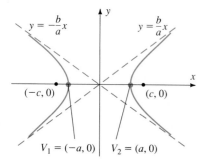

Figure 3.3　The hyperbola

$$\frac{x^2}{a^2} - \frac{y^2}{b^2} = 1$$

with asymptotes $y = \pm \left(\dfrac{b}{a} \right) x.$

are the equations for asymptotes to the graph of equation (2) (see Figure 3.3).

It is worth noting the geometric significance of b in equation (2). If a rectangle is constructed with vertical sides passing through the vertices $(-a, 0)$ and $(a, 0)$ and with horizontal sides passing through the points $(0, b)$ and $(0, -b)$, the diagonals of this rectangle pass through $(0, 0)$ and have slope $\pm b/a$ (Figure 3.4).

Thus, the asymptotes (4) may be sketched by extending the diagonals of this rectangle. Also, the second equation in line (2) may be solved for c to give $c^2 = a^2 + b^2$. Thus, c may be interpreted as the length of the hypotenuse of one of the eight triangles determined by the diagonals of this rectangle.

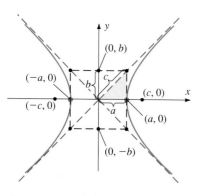

Figure 3.4　Relationships among the constants a, b, and c.

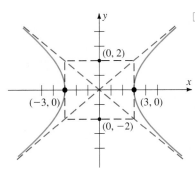

Figure 3.5 $\dfrac{x^2}{9} - \dfrac{y^2}{4} = 1$.

□ **EXAMPLE 1**

Sketch the hyperbola $\dfrac{x^2}{9} - \dfrac{y^2}{4} = 1$.

Solution   This equation has the form of equation (2) with $a = 3$ and $b = 2$. The vertices are therefore $V_1 = (-3, 0)$ and $V_2 = (3, 0)$, and the asymptotes are $y = \pm(2/3)x$. The graph appears in Figure 3.5.   ■

If the foci of a hyperbola lie along the y-axis at $F_1 = (0, c)$ and $F_2 = (0, -c)$, the condition

$$|d(P, F_2) - d(P, F_1)| = 2a$$

becomes

$$\left|\sqrt{x^2 + (y + c)^2} - \sqrt{x^2 + (y - c)^2}\right| = 2a, \tag{5}$$

which simplifies to the *standard form of the equation for the hyperbola with center at $(0, 0)$ and foci along the y-axis*

$$\frac{y^2}{a^2} - \frac{x^2}{b^2} = 1, \qquad b = \sqrt{c^2 - a^2}, \qquad c > a. \tag{6}$$

As in the previous case, we may solve for y to determine the equations for the asymptotes:

$$y^2 = a^2\left(\frac{x^2}{b^2} + 1\right),$$

so

$$y = \pm a\sqrt{\frac{x^2}{b^2} + 1} = \pm\frac{a}{b}x\sqrt{1 + \frac{b^2}{x^2}}. \tag{7}$$

Since

$$\lim_{x \to \pm\infty} \sqrt{1 + \frac{b^2}{x^2}} = 1,$$

equation (7) shows that *the asymptotes for the graph of equation (6) are given by*

$$y = \pm\frac{a}{b}x. \tag{8}$$

(See Figure 3.6.)

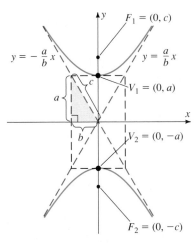

Figure 3.6   Hyperbola $\dfrac{y^2}{a^2} - \dfrac{x^2}{b^2} = 1$.

REMARK   A simple way to avoid confusion in working with equations (4) and (8) for the asymptotes associated with the standard form for a hyperbola is to note that we obtain these equations by replacing the 1 on the right-hand side of the standard forms with 0.

□ **EXAMPLE 2**

Graph the hyperbola $25x^2 - 9y^2 + 225 = 0$.

Solution   Subtracting the constant term from both sides and dividing by -225 gives

$$\frac{y^2}{25} - \frac{x^2}{9} = 1, \quad \text{or} \quad \frac{y^2}{5^2} - \frac{x^2}{3^2} = 1.$$

This equation has the form of equation (6) with $a = 5$ and $b = 3$. By (8) the asymptotes are $y = \pm(5/3)x$. The vertices are $(0, a) = (0, 5)$ and $(0, -a) = (0, -5)$. The graph appears as Figure 3.7.   ■

Translating the Hyperbola

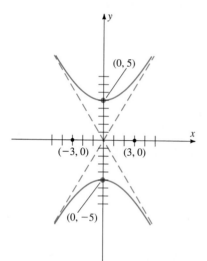

Figure 3.7   Graph of the equation
$$25x^2 - 9y^2 + 225 = 0.$$

If the center of a hyperbola is located at (h, k) rather than at $(0, 0)$, we translate axes using the equations

$$X = x - h \quad \text{and} \quad Y = y - k. \tag{9}$$

If the foci of the hyperbola lie on the X-axis, the equation in XY-coordinates, from (2), is

$$\frac{X^2}{a^2} - \frac{Y^2}{b^2} = 1, \tag{10}$$

and the equations for the asymptotes are

$$Y = \pm\frac{b}{a}X. \tag{11}$$

Using (9), we may write these equations in xy-coordinates. The equation

$$\boxed{\frac{(x - h)^2}{a^2} - \frac{(y - k)^2}{b^2} = 1, \qquad b = \sqrt{c^2 - a^2} \qquad a < c} \tag{12}$$

is *the standard form of the equation for the hyperbola with center at (h, k) and foci on a line parallel to the x-axis.* The asymptotes for this hyperbola are

$$\boxed{y - k = \pm\frac{b}{a}(x - h)} \qquad \text{(see Figure 3.8).} \tag{13}$$

Similarly, *the standard form of the equation for the hyperbola with center at (h, k) and foci on a line parallel to the y-axis* is

$$\boxed{\frac{(y - k)^2}{a^2} - \frac{(x - h)^2}{b^2} = 1, \qquad b = \sqrt{c^2 - a^2}, \qquad a < c,} \tag{14}$$

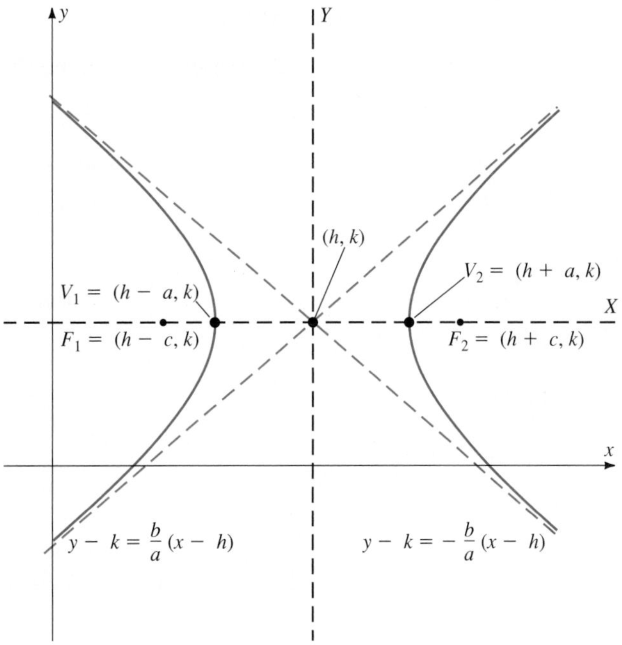

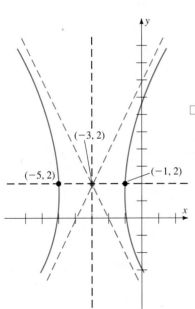

Figure 3.8 Hyperbola with center (h, k).

and the asymptotes for this hyperbola are

$$y - k = \pm \frac{a}{b}(x - h). \tag{15}$$

□ **EXAMPLE 3**

Describe the graph of the hyperbola

$$16(x + 3)^2 - 4(y - 2)^2 = 64.$$

Solution   Dividing both sides by 64 shows that the standard form for the hyperbola is

$$\frac{(x + 3)^2}{2^2} - \frac{(y - 2)^2}{4^2} = 1.$$

This equation has the form of equation (12) with $h = -3$, $k = 2$, $a = 2$, $b = 4$, and $c = \sqrt{a^2 + b^2} = \sqrt{20} = 2\sqrt{5}$. The hyperbola therefore has center $(-3, 2)$, vertices $(-5, 2)$ and $(-1, 2)$, foci $(-3 - 2\sqrt{5}, 2)$ and $(-3 + 2\sqrt{5}, 2)$, and asymptotes $y - 2 = \pm 2(x + 3)$. The graph appears as Figure 3.9. ∎

Figure 3.9   Graph of the equation

$$16(x + 3)^2 - 4(y - 2)^2 = 64.$$

□ **EXAMPLE 4**

Describe the graph of the equation

$$25y^2 - 9x^2 - 50y - 54x - 281 = 0.$$

Solution   We complete the square in both variables on the left-hand side to find that

$$25(y^2 - 2y + 1) - 9(x^2 + 6x + 9) - 281 = 25 - 81,$$

so

$$25(y - 1)^2 - 9(x + 3)^2 = 225,$$

or

$$\frac{(y - 1)^2}{3^2} - \frac{(x + 3)^2}{5^2} = 1.$$

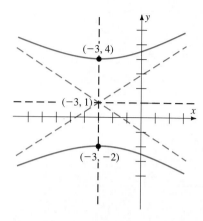

Figure 3.10   Graph of equation $25y^2 - 9x^2 - 50y - 54x - 281 = 0$.

This equation has the form of equation (14) with $h = -3$, $k = 1$, $a = 3$, $b = 5$, and $c = \sqrt{a^2 + b^2} = \sqrt{34}$. Its graph is a hyperbola with center $(-3, 1)$, vertices $(-3, -2)$ and $(-3, 4)$, foci $(-3, 1 - \sqrt{34})$ and $(-3, 1 + \sqrt{34})$, and asymptotes $y - 1 = \pm(3/5)(x + 3)$ (see Figure 3.10).   ■

Exercise Set 14.3

In Exercises 1–20, graph the hyperbola corresponding to the given equation. State the center and the equations for the asymptotes.

1. $\dfrac{x^2}{2^2} - \dfrac{y^2}{3^2} = 1$ **2.** $\dfrac{y^2}{4^2} - \dfrac{x^2}{2^2} = 1$

3. $\dfrac{x^2}{5} - \dfrac{y^2}{9} = 1$ **4.** $\dfrac{x^2}{9} - \dfrac{y^2}{2} = 1$

5. $16x^2 - 4y^2 - 64 = 0$ **6.** $x^2 - 9y^2 = 9$

7. $4y^2 - 7x^2 - 28 = 0$ **8.** $5x^2 - 9y^2 - 10x = 25$

9. $9x^2 - 4y^2 + 54x + 32y + 119 = 0$

10. $16y^2 - 4x^2 - 32y - 8x - 52 = 0$

11. $9x^2 - 4y^2 - 36x - 24y - 36 = 0$

12. $4y^2 - x^2 - 2x = 2$

13. $x^2 - 4y^2 = 4$

14. $4x^2 - y^2 + 24x + 4y + 28 = 0$

15. $2x^2 - y^2 - 4\sqrt{2}x + 2\sqrt{2}y = 6$

16. $3x^2 - 5y^2 + 6x - 20y = -8.$

17. $16x^2 - 25y^2 + 64x + 150y = 561$

18. $9x^2 - y^2 + 36x + 4y + 23 = 0$

19. $9y^2 - 16x^2 - 126y - 160x = 184$

20. $4y^2 - 16y - x^2 - 4x + 11 = 0$

In Exercises 21–27, find the equation for the hyperbola with the stated properties.

21. Foci at $(-2, 0)$ and $(2, 0)$, vertices at $(-1, 0)$ and $(1, 0)$.

22. Foci at $(2, 5)$ and $(2, -3)$, vertices at $(2, 3)$ and $(2, -1)$.

23. Vertices at $(0, 4)$ and $(0, 0)$, asymptotes $y = 2 \pm 2x$.

24. Foci at $(-3, 0)$ and $(3, 0)$, asymptotes $y = \pm\frac{3}{2}x$.

25. Foci at $(-5, 2)$ and $(3, 2)$, asymptotes $y - 2 = \pm 2(x + 1)$.

26. Center at the origin, one vertex $(0, 4)$, and containing the point $(2, 8)$.

27. Center at the origin, one vertex $(0, 4)$, and one focus $(0, -5)$.

28. Find an equation for the line tangent to the graph of the hyperbola $4x^2 - 16y^2 = 64$ at the point $(4\sqrt{2}, 2)$.

29. What properties do the graph of the hyperbola

$$\frac{x^2}{a^2} - \frac{y^2}{b^2} = 1$$

and its **conjugate** $\dfrac{y^2}{b^2} - \dfrac{x^2}{a^2} = 1$ have in common?

30. Show that the line tangent to the graph of the hyperbola $b^2x^2 - a^2y^2 = a^2b^2$ at the point (x_0, y_0) has slope

$$m = \frac{b^2x_0}{a^2y_0}.$$

31. Using the result of Exercise 30, find the equation for the line tangent to the graph of $b^2x^2 - a^2y^2 = a^2b^2$ at the point (x_0, y_0).

32. Suppose that the numbers A and C are nonzero and have opposite signs. Show that the graph of the second degree equation

$$Ax^2 + Cy^2 + Dx + Ey + F = 0$$

is either two intersecting lines or a hyperbola. What conditions on the coefficients distinguish the two cases?

33. A chord of a hyperbola through a focus perpendicular to the principal axis is called a **latus rectum.** Show that the length of a latus rectum is $2b^2/a$.

34. Find an equation for the hyperbola with center at the origin, latus rectum of length 20, and one vertex at $(0, 4)$.

35. Show that any point (x, y) satisfying equation (2) satisfies the geometric condition of Definition 3.

14.4 Rotation of Axes

Up to this point, we have discussed all possible second degree equations of the form

$$Ax^2 + Bxy + Cy^2 + Dx + Ey + F = 0 \tag{1}$$

except those involving an xy term. In this section we complete that discussion by considering the $B \neq 0$ case. On encountering such an expression, we find a suitable change of variables from (x, y)-coordinates to (x', y')-coordinates so that in the new coordinates equation (1) contains no $x'y'$ term.

The simplest change of variables that accomplishes this simplification is a rotation of the xy-axes through an angle θ. After we construct new x'- and y'-axes by rotating the x- and y-axes counterclockwise through an angle θ, a point P with xy-coordinates (x, y) has $x'y'$-coordinates given by the equations

$$x' = d \cos \phi, \tag{2a}$$

$$y' = d \sin \phi, \tag{2b}$$

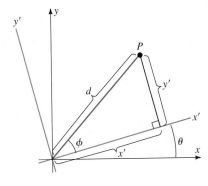

where d is the distance from the origin O to P and ϕ is the angle between $\overline{OP}$ and the x'-axis (Figure 4.1).

Similarly, we can express the xy-coordinates of P as

$$x = d \cos(\theta + \phi), \tag{3a}$$

$$y = d \sin(\theta + \phi). \tag{3b}$$

Using the addition laws for sine and cosine we can combine equations (2) and (3) to obtain the equations

$$x = d(\cos \theta \cos \phi - \sin \theta \sin \phi) = x' \cos \theta - y' \sin \theta, \tag{4a}$$

$$y = d(\sin \theta \cos \phi + \cos \theta \sin \phi) = x' \sin \theta + y' \cos \theta. \tag{4b}$$

Solving this system for x' and y', we obtain

$$x' = x \cos \theta + y \sin \theta, \tag{5a}$$

$$y' = -x \sin \theta + y \cos \theta. \tag{5b}$$

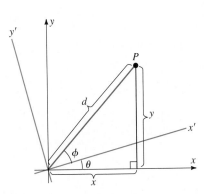

Figure 4.1 Rotating the axes by an angle θ produces $x'y'$ coordinates that are related to the xy coordinates by equations (2) and (3).

By using equations (4) and (5), we can pass back and forth from xy- to $x'y'$-coordinates. In particular, if we substitute the right-hand sides of equations (4) for x and y into equation (1), we find that the coefficient of the $x'y'$ term is

$$2(C - A) \sin \theta \cos \theta + B(\cos^2 \theta - \sin^2 \theta),$$

which equals

$$(C - A) \sin 2\theta + B \cos 2\theta.$$

Thus, the coefficient of the $x'y'$ term is zero if

$$(C - A) \sin 2\theta + B \cos 2\theta = 0.$$

Since we are assuming that $B \neq 0$, we can rewrite this equation as

$$\frac{\cos 2\theta}{\sin 2\theta} = \frac{A - C}{B}$$

or as

$$\cot 2\theta = \frac{A - C}{B}, \qquad 0 < \theta < \pi/2. \qquad (6)$$

If we choose θ so that equation (6) is satisfied, then equation (1), when it is written in $x'y'$-coordinates, does not have an $x'y'$ term. Thus, this equation in x' and y' can be analyzed using the techniques of Sections 14.1–14.3.

□ **EXAMPLE 1**

For the equation $5x^2 + 6xy + 5y^2 - 8 = 0$,

(a) find a rotation that eliminates the xy term, and
(b) sketch the graph.

Solution   We have $A = 5$, $B = 6$, and $C = 5$, so equation (6) becomes $\cot 2\theta = (5 - 5)/6 = 0$. The equation $\cot 2\theta = 0$ has solution $2\theta = \pi/2$, or $\theta = \pi/4$. With this value of θ, equations (4a) and (4b) become

$$x = \frac{\sqrt{2}}{2}x' - \frac{\sqrt{2}}{2}y',$$

$$y = \frac{\sqrt{2}}{2}x' + \frac{\sqrt{2}}{2}y'.$$

Inserting these expressions for x and y in the original equation gives

$$5\left(\frac{\sqrt{2}}{2}x' - \frac{\sqrt{2}}{2}y'\right)^2 + 6\left(\frac{\sqrt{2}}{2}x' - \frac{\sqrt{2}}{2}y'\right)\left(\frac{\sqrt{2}}{2}x' + \frac{\sqrt{2}}{2}y'\right)$$

$$+ 5\left(\frac{\sqrt{2}}{2}x' + \frac{\sqrt{2}}{2}y'\right)^2 - 8 = 0$$

or, on simplifying,

$$x'^2 + \frac{y'^2}{4} = 1.$$

The graph of this ellipse is shown in Figure 4.2.

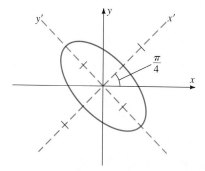

Figure 4.2 Graph of the equation $5x^2 + 6xy + 5y^2 - 8 = 0$ is a rotated ellipse.

□ **EXAMPLE 2**

Graph the equation $xy = 1$.

Strategy · · · · · · · ·

Identify form of the equation.

Rotate axes to eliminate xy term.

Find the rotation angle from (6).

Solution

This is an equation of the form of the general second degree equation (1) with $B = 1$ and $A = C = D = E = 0$. Although we are already familiar with the general nature of its graph, we may use the method of rotation of axes to verify that the graph is, in fact, a hyperbola.

From equation (6) we find that

$$\cot 2\theta = \frac{A - C}{B} = 0,$$

so $\theta = \pi/4$, as in Example 1. From equations (4) we then obtain

Express x and y as functions of x' and y'. Insert these expressions into the given equation.

$$x = \frac{\sqrt{2}}{2}x' - \frac{\sqrt{2}}{2}y',$$

$$y = \frac{\sqrt{2}}{2}x' + \frac{\sqrt{2}}{2}y'.$$

Inserting these expressions in $xy = 1$ gives

$$\left(\frac{\sqrt{2}}{2}x' - \frac{\sqrt{2}}{2}y'\right)\left(\frac{\sqrt{2}}{2}x' + \frac{\sqrt{2}}{2}y'\right) = 1,$$

or

Bring resulting equation to standard form.

$$\frac{x'^2}{2} - \frac{y'^2}{2} = 1.$$

The graph is therefore a hyperbola in $x'y'$-coordinates, with center at $(0, 0)$ and asymptotes $y' = \pm x'$ (see Figure 4.3). ∎

We have now completed our analysis of the general second degree equation

$$Ax^2 + Bxy + Cy^2 + Dx + Ey + F = 0, \tag{7}$$

where at least one of A and C is nonzero.

If $B \neq 0$, we can employ a rotation of axes, as described above, to obtain an equation in $x'y'$-coordinates of the form

$$A'x'^2 + C'y'^2 + D'x' + E'y' + F' = 0. \tag{8}$$

Equation (8) can describe a degenerate conic. For example, the graph of equation

$$x'^2 + y'^2 + 1 = 0$$

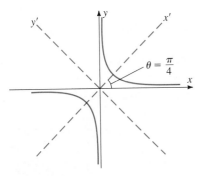

Figure 4.3 Graph of $xy = 1$ is a rotated hyperbola.

($A' = C' = F' = 1$ and $D' = E' = 0$) is empty. Similarly, the graph of the equation

$$x'^2 - 1 = 0$$

($A' = 1$, $C' = D' = E' = 0$, and $F' = -1$) consists of two parallel lines. However, if the graph of Equation (8) is nondegenerate, then it must fall under one of the following cases:

(i) If $A' = C' \neq 0$, the graph of (8) is a circle (Section 1.2).
(ii) If $A' = 0$ or $C' = 0$ (but not both), the graph of (8) is a parabola (Exercise 32, Section 14.1).

(iii) If $A'C' \neq 0$ and $A' \neq C'$, the graph of (8) is
(a) an ellipse, if A' and C' have the same sign (Exercise 30, Section 14.2),
(b) a hyperbola, if A' and C' have opposite signs (Exercise 32, Section 14.3).

Exercise Set 14.4

In each of Exercises 1–6, find a rotation θ that eliminates the xy term.

1. $3x^2 + 2xy + y^2 + 6x + 9 = 0$

2. $6x^2 + 2xy + 2y^2 + 8y + 8 = 0$

3. $x^2 - 3xy + y^2 - x + y + 10 = 0$

4. $4x^2 + \sqrt{3}xy + 3y^2 + 2x + 7y - 8 = 0$

5. $5x^2 + 4xy + 2y^2 = 1$

6. $9x^2 - 24xy + 16y^2 - 40x - 30y + 100 = 0$

In each of Exercises 7–16, find a rotation θ that eliminates the xy term. Then use equations (4) to obtain an equation in $x'y'$-coordinates. Finally, graph the equation.

7. $5x^2 + 6xy + 5y^2 - 32 = 0$

8. $3x^2 + 10xy + 3y^2 - 32 = 0$

9. $3x^2 + 2\sqrt{3}xy + y^2 + 2x - 2\sqrt{3}y + 16 = 0$

10. $5x^2 + 4xy + 5y^2 = 21$

11. $9x^2 - 24xy + 16y^2 - 40x - 30y + 100 = 0$

12. $x^2 - 6xy + y^2 + 16 = 0$

13. $7x^2 - 2\sqrt{3}xy + 5y^2 - 16 = 0$

14. $23x^2 - 26\sqrt{3}xy - 3y^2 + 144 = 0$

15. $x^2 + 2xy + y^2 + \sqrt{2}x - \sqrt{2}y = 0$

16. $31x^2 + 10\sqrt{3}xy + 21y^2 = 144$

17. Find the endpoints of the major axis of the ellipse in Exercise 16.

18. Find the coordinates of the focus of the parabola in Exercise 15.

19. Find the vertices of the hyperbola in Exercise 14.

20. Find the endpoints of the minor axis of the ellipse in Exercise 13.

Summary Outline of Chapter 14

◼ The standard forms of the equations for the **parabola** with vertex at (h, k) and focal length c are   (page 701)

$$(x - h)^2 = 4c(y - k) \quad \text{(axis vertical)},$$
$$(y - k)^2 = 4c(x - h) \quad \text{(axis horizontal)}.$$

◼ The standard form of the equation for the **ellipse** with center at (h, k) is   (page 708)

$$\frac{(x - h)^2}{a^2} + \frac{(y - k)^2}{b^2} = 1.$$

◼ The standard forms of the equations for the **hyperbola** with center at (h, k) are   (page 714)

$$\frac{(x - h)^2}{a^2} - \frac{(y - k)^2}{b^2} = 1 \quad \text{(foci on line parallel to } x\text{-axis)},$$

$$\frac{(y - k)^2}{a^2} - \frac{(x - h)^2}{b^2} = 1 \quad \text{(foci on line parallel to } y\text{-axis)}.$$

◼ A **rotation of axes** to eliminate the xy term in $Ax^2 + Bxy + Cy^2 + Dx +$ (page 717) $Ey + F = 0$ is accomplished by the substitutions

$$x = x' \cos \theta - y' \sin \theta, \quad y = x' \sin \theta + y' \cos \theta$$

where θ is determined by the equation $\cot 2\theta = \dfrac{A - C}{B}$, $0 < \theta < \pi/2$.

Review Exercises—Chapter 14

1. Write an equation for the circle with center $(3, -4)$ and radius 6.

2. Write an equation for the circle with center $(-2, -3)$ and radius 5.

3. Write an equation for the circle with center $(2, 3)$ that contains the point $(2, -1)$.

4. Write equations for the circles with radius 5 containing the points $(-1, -4)$ and $(4, 1)$. How many such circles exist?

5. Write an equation for the parabola with vertex $(0, 4)$ and directrix $x = -2$.

6. Write equations for parabolas with axis horizontal, vertex on the y-axis, and containing the points $(2, 1)$ and $(18, 3)$.

7. Write an equation for the parabola with axis vertical and containing the points $(0, 5)$, $(2, 5)$, and $(3, 14)$.

8. Find an equation for the circle that passes through the focus and vertex of the parabola $y = 8x^2$ and that is tangent to the parabola at $(0, 0)$.

9. Find an equation of the ellipse with center $(-1, 1)$, focus $(1, 1)$, and vertex $(2, 1)$.

10. Find an equation for the ellipse with focus $(0, 1)$ and vertices $(0, 0)$ and $(0, 8)$.

11. Does the origin lie inside or outside the ellipse with equation $(x - 4)^2 + 4(y - 3)^2 = 4$?

Find the points of intersection of each of the following pairs of graphs.

12. The line $x + y - 4 = 0$ and the circle $x^2 + y^2 = 16$.

13. The line $x - y - 2 = 0$ and the parabola $y = 4 - x^2$.

14. The line $4y - 3x + 12 = 0$ and the ellipse $9x^2 + 16y^2 = 144$.

15. The parabola $y = 6 - x^2$ and the parabola $y = x^2 - 2$.

In Exercises 16–44, write the equation in standard form, identify the graph, and sketch.

16. $3y^2 - x - 4 = 0$

17. $x^2 - 2x + y^2 + 6y + 2 = 0$

18. $16x^2 + 7y^2 - 112 = 0$

19. $16x^2 - 7y^2 + 112 = 0$

20. $x^2 - 3y^2 + 3 = 0$

21. $16x^2 + 12y^2 - 192 = 0$

22. $3x^2 + 3y^2 - 3 = 0$

23. $2x^2 - y^2 + 4 = 0$

24. $9x^2 + 4y^2 - 36 = 0$

25. $4x^2 - 9y^2 - 36 = 0$

26. $4x^2 + 9y^2 - 36 = 0$

27. $y^2 + x - 2 = 0$

28. $7y^2 - 2x + 6 = 0$

29. $x^2 + 14x + y^2 - 10y + 70 = 0$

30. $9x^2 + 8y^2 - 72 = 0$

31. $9x^2 - 8y^2 - 72 = 0$

32. $4x^2 + y + 8 = 0$

33. $4x^2 + 25y^2 - 100 = 0$

34. $x^2 + 3y^2 - 3 = 0$

35. $3x^2 - 3y^2 + 12 = 0$

36. $x^2 - 6x + y^2 - 4y + 8 = 0$

37. $x^2 - 2ax + y^2 + 4ay + 5a^2 - 1 = 0$

38. $x^2 - 2x + y^2 - 8 = 0$

39. $y^2 - x^2 - 2 = 0$

40. $3x + 2y^2 - 3 = 0$

41. $x^2 + y^2 - 2x - 6y + 3 = 0$

42. $4x^2 - 25y^2 + 100 = 0$

43. $9y^2 - 25x^2 - 225 = 0$

44. $2x - 16y^2 - 12 = 0.$

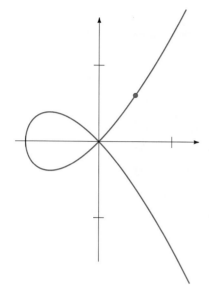

Chapter 15

Polar Coordinates and Parametric Equations

Up to this point we have studied curves in the plane by regarding them as graphs of functions plotted in Cartesian (rectangular) coordinates. This chapter presents two additional ways to describe such curves: as graphs of functions of the form $r = f(\theta)$ plotted in *polar coordinates* and as graphs determined by *parametric equations* for the x- and y-coordinates of points in Cartesian coordinates.

15.1 The Polar Coordinate System

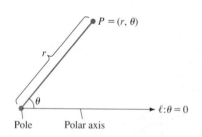

Figure 1.1 The polar coordinates (r, θ) for point P.

Figure 1.2 For a point charge (shown here as positive), equipotential lines are circles $r = c$ (=constant). Lines of force are rays $\theta = d$ (=constant).

Cartesian coordinates provide a convenient scheme to use when graphing functions whose variables represent naturally perpendicular quantities (length versus height, for example) or when attempting to approximate areas by rectangles. However, many plane curves can be described more easily in a coordinate system in which the coordinates of a point P (r, θ) measure its (radial) distance r from the origin along a ray that makes an angle θ with some fixed reference ray (usually the positive x-axis; see Figure 1.1).

These coordinates are especially convenient for describing curves that are symmetric with respect to the origin. An example of this situation occurs in the theory of electric potential fields. For a point charge, the lines representing equal electric potentials (called equipotentials) in any plane containing the point charge are circles with the point as center (Figure 1.2). These equipotential circles can most easily be described simply by stating their radius. Similarly, the lines of force emanating from such a point charge are rays, which are described completely by the angle they form with the positive x-axis.

Let O be any given point in the plane, called the **pole,** and let ℓ be any ray emanating from O. The ray ℓ is called the **polar axis** for the plane. The choice of a pole and a polar axis determines a **polar coordinate system** for the plane in which any point P may be assigned **polar coordinates** (r, θ), where

(i) r, called the **radial variable,** is the distance from O to P, and
(ii) θ, called the **angular variable,** is the angle formed between the polar axis and the ray OP, measured in the counterclockwise direction (Figure 1.1).

Figure 1.3 illustrates various points plotted in polar coordinates. As in most applications of trigonometry in the calculus, we shall measure angles using radian measure. Although statements (i) and (ii) above imply that both r and θ are positive, we shall soon extend their definitions to include negative values.

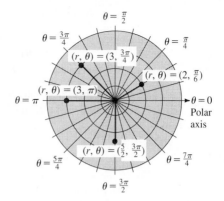

Figure 1.3   Polar coordinates (r, θ) for four points in the plane.

In the xy-coordinate plane, the graphs of equations of the form $x =$ constant or $y =$ constant are lines. In the polar coordinate plane, graphs of equations of the form $r =$ constant are circles, while graphs of equations of the form $\theta =$ constant are rays (see Figure 1.4).

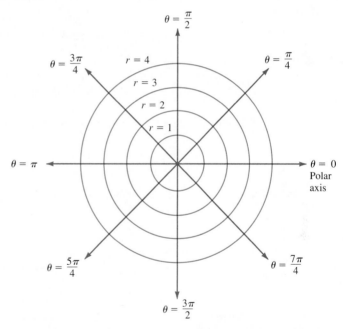

Figure 1.4   In polar coordinates, graphs of equations of the form $r =$ constant are circles, while graphs of equations of the form $\theta =$ constant are rays.

You may have already noticed that polar coordinates are not unique. For example, the point with polar coordinates $(r, \theta) = (2, \pi/4)$ is also represented by the polar coordinates $(2, 9\pi/4)$. In fact, since all angles of the form $\theta + 2n\pi$ for $n = 0, \pm 1, \pm 2, \ldots$ determine the same ray, we have

$$(2, \pi/4) = (2, 9\pi/4) = (2, 17\pi/4) = (2, -7\pi/4) = \ldots,$$

where we use the equal sign to indicate that these polar coordinates represent the same point in the plane.

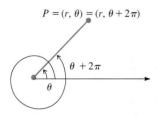

Figure 1.5 The point P with polar coordinates $(r, \theta + 2\pi)$ is the same as the point with polar coordinates (r, θ).

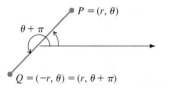

Figure 1.6 The point Q with polar coordinates $(-r, \theta)$ is the same as the point with polar coordinates $(r, \theta + \pi)$.

In general,

$$(r, \theta) = (r, \theta + 2n\pi), \qquad n = \pm 1, \pm 2, \pm 3, \dots . \tag{1}$$

(See Figure 1.5). Moreover, it is customary to extend the range of values for the radial variable r to include negative numbers via the identity

$$(-r, \theta) = (r, \theta + \pi). \tag{2}$$

(See Figure 1.6). In other words, the point with polar coordinates $(-r, \theta)$ lies r units along the ray pointing in the direction opposite that specified by the angular variable θ.

☐ EXAMPLE 1

By identities (1) and (2) we have

(a) $(3, \pi/6) = (3, 13\pi/6) = (3, -11\pi/6)$,
(b) $(-4, \pi/4) = (4, 5\pi/4) = (4, -3\pi/4)$,
(c) $(7, -\pi/2) = (-7, \pi/2) = (-7, 5\pi/2) = (-7, -3\pi/2)$. ■

The *pole O* in the polar coordinate plane has coordinates $(0, \theta)$ for every angle θ. It is the only point in the plane not associated with a unique angle θ in $[0, \pi)$.

Relationships Between Polar and Rectangular Coordinates

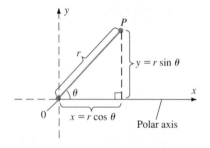

Figure 1.7 The relationship between rectangular and polar coordinates. The point P has rectangular coordinates (x, y) and polar coordinates (r, θ).

The relationships between polar and xy-coordinates are easily determined when the origin in xy-coordinates is the pole and the positive x-axis is the polar axis (see Figure 1.7). If the point P has rectangular coordinates (x, y) and polar coordinates (r, θ), and if $r > 0$, the definitions of $\sin \theta$ and $\cos \theta$ give

$$x = r \cos \theta \quad \text{and} \quad y = r \sin \theta. \tag{3}$$

In Exercise 62, you are asked to show that equations (3) hold for $r < 0$ as well. It follows from the equations in (3) (or from Figure 1.7) that

$$r^2 = x^2 + y^2, \qquad \tan \theta = \frac{y}{x}, \qquad x \neq 0. \tag{4}$$

The equations in line (3) are used to change from polar to rectangular coordinates, and the equations in line (4) allow us to change from rectangular to polar coordinates.

☐ EXAMPLE 2

To find the rectangular coordinates corresponding to the polar coordinates $(2, 2\pi/3)$, we use equations (3) with $r = 2$, $\theta = 2\pi/3$:

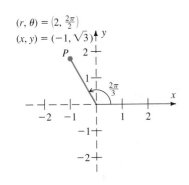

$(r, \theta) = \left(2, \frac{2\pi}{2}\right)$
$(x, y) = (-1, \sqrt{3})$

Figure 1.8 The point P with polar coordinates $(r, \theta) = (2, 2\pi/3)$ has rectangular coordinates $(-1, \sqrt{3})$.

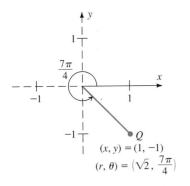

$(x, y) = (1, -1)$
$(r, \theta) = \left(\sqrt{2}, \frac{7\pi}{4}\right)$

Figure 1.9 The point Q with rectangular coordinates $(x, y) = (1, -1)$ has polar coordinates $(r, \theta) = (\sqrt{2}, 7\pi/4)$. Of course, there are infinitely many other pairs (r, θ) of polar coordinates for Q.

$$x = 2 \cos \frac{2\pi}{3} = 2\left(-\frac{1}{2}\right) = -1,$$

$$y = 2 \sin \frac{2\pi}{3} = 2\left(\frac{\sqrt{3}}{2}\right) = \sqrt{3}.$$

(See Figure 1.8.)   ∎

□ **EXAMPLE 3**

To find the polar coordinates for the point Q with rectangular coordinates $(1, -1)$, we use the equations in (4):

$$r^2 = x^2 + y^2 = 1^2 + 1^2 = 2 \tag{5}$$

$$\tan \theta = \frac{y}{x} = \frac{-1}{1} = -1. \tag{6}$$

From equation (5), we know that $r = \pm\sqrt{2}$. Moreover, equation (6) has two solutions in the interval $[0, 2\pi)$: $\theta = 3\pi/4$ and $\theta = 7\pi/4$. The proper value of θ depends upon our choice for r. If we use $r = +\sqrt{2}$, we know that $\theta = 7\pi/4$ since Q lies in the fourth quadrant (see Figure 1.9).   ∎

REMARK 1   It is important to note, as in Example 3, that equations (4) do not immediately determine r and θ for given values of x and y. The value of r can be positive or negative, and each value of $\tan \theta$ occurs twice for $\theta \in [0, 2\pi)$. This difficulty is overcome most simply by noting the quadrant in which the point (x, y) lies. An alternative approach is to use the positive value of r from equation (4) and then determine θ as the (unique) simultaneous solution of equations (3): $x = r \cos \theta$; $y = r \sin \theta$.

□ **EXAMPLE 4**

Graph the polar equation $r = 1 + \cos \theta$ and find the rectangular form for the equation.

Solution   One approach to graphing polar equations is to select several (convenient) values of θ (beginning with $\theta = 0$), calculate the corresponding values of r, and plot the points (r, θ) obtained. The graph is then completed by sketching a curve connecting these points. For the equation $r = 1 + \cos \theta$, we obtain the following points.

θ	0	$\frac{\pi}{6}$	$\frac{\pi}{4}$	$\frac{\pi}{3}$	$\frac{\pi}{2}$	$\frac{2\pi}{3}$	$\frac{3\pi}{4}$	$\frac{5\pi}{6}$	π
r	2	$\frac{2+\sqrt{3}}{2}$	$\frac{2+\sqrt{2}}{2}$	$\frac{3}{2}$	1	$\frac{1}{2}$	$\frac{2-\sqrt{2}}{2}$	$\frac{2-\sqrt{3}}{2}$	0

θ	$\frac{7\pi}{6}$	$\frac{5\pi}{4}$	$\frac{4\pi}{3}$	$\frac{3\pi}{2}$	$\frac{5\pi}{3}$	$\frac{7\pi}{4}$	$\frac{11\pi}{6}$	2π
r	$\frac{2-\sqrt{3}}{2}$	$\frac{2-\sqrt{2}}{2}$	$\frac{1}{2}$	1	$\frac{3}{2}$	$\frac{2+\sqrt{2}}{2}$	$\frac{2+\sqrt{3}}{2}$	2

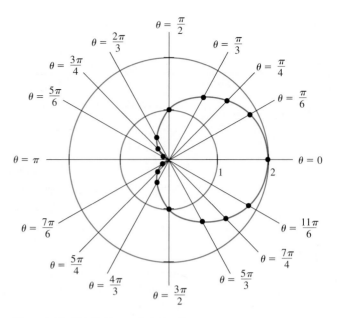

Figure 1.10   Cardioid $r = 1 + \cos \theta$.

The graph is the **cardioid** in Figure 1.10. In Section 15.2 we shall discuss a technique for graphing polar equations that addresses the problem of ensuring that points obtained as above are connected in the correct manner.

To find the rectangular form of the equation $r = 1 + \cos \theta$, we multiply both sides by r to obtain

$$r^2 = r + r \cos \theta.$$

Since $r = 1 + \cos \theta \geq 0$ for all θ, we know that $r = \sqrt{x^2 + y^2}$. Using the equations in (3) and (4), we obtain

$$x^2 + y^2 = \sqrt{x^2 + y^2} + x.$$

Clearly the original equation is simpler to graph than its rectangular counterpart. ∎

REMARK 2   In Example 4, we considered angles θ between 0 and 2π only. For this example, restricting θ to the interval $[0, 2\pi]$ is appropriate because the function $r(\theta) = 1 + \cos \theta$ is a periodic function with a period of 2π. That is,

$$r(\theta + 2\pi) = r(\theta). \tag{7}$$

If r is any function that satisfies equation (7) and if $\theta_2 = \theta_1 + 2n\pi$ for some integer n, then $r(\theta_2) = r(\theta_1)$. Therefore, the point with polar coordinates $(r(\theta_2), \theta_2)$ is the same as the point with polar coordinates $(r(\theta_1), \theta_1)$ since it has the same radius and the angles θ_1 and θ_2 determine the same ray from the pole.

In general, if a function r satisfies equation (7), we need only plot points corresponding to θ in some interval of length 2π. The intervals $[0, 2\pi]$ and $[-\pi, \pi]$ are commonly used.

Symmetry in Polar Coordinates

Perhaps you noticed in Example 4 that the graph of the equation $r = 1 + \cos \theta$ is symmetric about the x-axis. This symmetry is due to the identity $\cos(-\theta) = \cos \theta$, or $\cos(2\pi - \theta) = \cos \theta, 0 \leq \theta \leq \pi$. More generally, graphs of polar equations will be

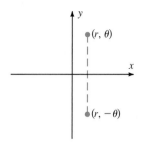

Figure 1.11 Symmetry about the *x*-axis.

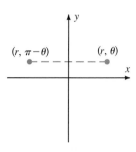

Figure 1.12 Symmetry about the *y*-axis.

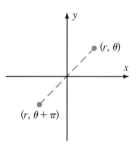

Figure 1.13 Symmetry about the origin.

(i) *symmetric about the x-axis* if $(r, -\theta)$ lies on the graph whenever (r, θ) does (Figure 1.11);

(ii) *symmetric about the y-axis* if $(r, \pi - \theta)$ lies on the graph whenever (r, θ) does (Figure 1.12);

(iii) *symmetric about the origin* if $(r, \theta + \pi)$ lies on the graph whenever (r, θ) does (Figure 1.13).

Additional tests for symmetry are discussed in Exercise 16.

□ **EXAMPLE 5**

Graph the polar equation $r = \sin 2\theta$.

θ	0	$\dfrac{\pi}{6}$	$\dfrac{\pi}{4}$	$\dfrac{\pi}{3}$	$\dfrac{\pi}{2}$	$\dfrac{2\pi}{3}$	$\dfrac{3\pi}{4}$	$\dfrac{5\pi}{6}$	π
r	0	$\dfrac{\sqrt{3}}{2}$	1	$\dfrac{\sqrt{3}}{2}$	0	$\dfrac{-\sqrt{3}}{2}$	-1	$\dfrac{-\sqrt{3}}{2}$	0

Plotting these points gives the two leaves shown in Figure 1.14 (note that values of r are negative for $\pi/2 < \theta < \pi$).

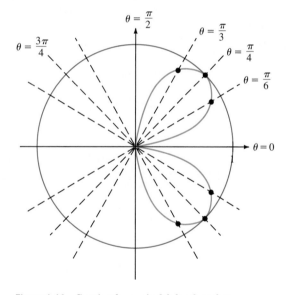

Figure 1.14   Graph of $r = \sin 2\theta$ for $0 \le \theta \le \pi$.

We can use symmetry considerations to obtain the remaining portion of the curve. Using the double angle identity, we write

$$r(\theta) = \sin 2\theta = 2 \sin \theta \cos \theta.$$

Since $\sin(\theta + \pi) = -\sin \theta$ and $\cos(\theta + \pi) = -\cos \theta$, we obtain

$$r(\theta + \pi) = 2 \sin(\theta + \pi) \cos(\theta + \pi) = 2 \sin \theta \cos \theta = r(\theta).$$

This equality shows that $(r, \theta + \pi)$ is on the graph whenever (r, θ) is. That is, we may sketch a leaf in the third quadrant symmetric with the leaf in the first quadrant, and a leaf in the second quadrant symmetric with the leaf in the fourth quadrant

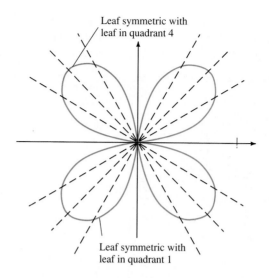

Figure 1.15   Graph of $r = \sin 2\theta$ for $0 \le \theta \le 2\pi$.

(Figure 1.15). (Actually we could obtain the entire graph by applying symmetry arguments just to the leaf obtained for $0 \le \theta \le \pi/2$. Can you see how this could be done?) The resulting figure is called a four-leaved rose. ■

□ **EXAMPLE 6**

For the polar equation $r = 2 \cos \theta$,

(a) sketch the graph;
(b) note any symmetries of the graph;
(c) find the rectangular form of the equation.

Solution

(a) For $0 \le \theta \le \pi$, we obtain the following points:

θ	0	$\dfrac{\pi}{6}$	$\dfrac{\pi}{4}$	$\dfrac{\pi}{3}$	$\dfrac{\pi}{2}$	$\dfrac{2\pi}{3}$	$\dfrac{3\pi}{4}$	$\dfrac{5\pi}{6}$	π
r	2	$\sqrt{3}$	$\sqrt{2}$	1	0	-1	$-\sqrt{2}$	$-\sqrt{3}$	-2

These points produce the curve shown in Figure 1.16.

For $\pi < \theta < 2\pi$, we note that $r(\theta_0 + \pi) = 2 \cos(\theta_0 + \pi) = -2 \cos(\theta_0) = -r(\theta_0)$ where $0 < \theta_0 < \pi$. Thus, these values simply trace over the same points a second time.

(b) Figure 1.16 suggests symmetry about the x-axis. This is indeed the case, since, for $r(\theta) = 2 \cos \theta$,

$$r(-\theta) = 2 \cos(-\theta) = 2 \cos \theta = r(\theta).$$

Thus, $(r, -\theta)$ is on the graph whenever (r, θ) is.

(c) To find the rectangular form for $r = 2 \cos \theta$, we multiply by r to obtain

$$r^2 = 2r \cos \theta.$$

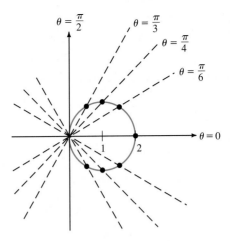

Figure 1.16   Graph of $r = 2 \cos \theta$.

Using equations (3) and (4) we obtain

$$x^2 + y^2 = 2x \quad \text{or} \quad x^2 - 2x + y^2 = 0.$$

Completing the square in x gives

$$(x^2 - 2x + 1) + y^2 = 1$$

or

$$(x - 1)^2 + y^2 = 1.$$

The graph is therefore a circle of radius $r = 1$ with center $(1, 0)$.   ■

REMARK 3   In Example 6, we only considered values of θ in the interval $[0, \pi]$. For this example, not only do the observations of Remark 2 apply, but the function $r = r(\theta) = 2 \cos \theta$ also has the property

$$r(\theta + \pi) = -r(\theta). \tag{8}$$

Given convention (2), which states that $(-r, \theta)$ is the same point as $(r, \theta + \pi)$, equation (8) implies that

$$
\begin{aligned}
(r(\theta + \pi), \theta + \pi) &= (-r(\theta), \theta + \pi) &&\text{(equation (8))} \\
&= (r(\theta), \theta + 2\pi) &&\text{(convention (2))} \\
&= (r(\theta), \theta)
\end{aligned}
$$

Thus, we only need to use values of θ in an interval of length π. In fact, if we plot all points corresponding to angles θ in the interval $[0, 2\pi]$, we plot each point on the circle twice.

Exercise Set 15.1

In Exercises 1–8, find rectangular coordinates for the point given in polar coordinates.

1. $(1, \pi/2)$ **2.** $(3, \pi/6)$ **5.** $(\sqrt{2}, -\pi/4)$ **6.** $(-1, -3\pi/2)$

3. $(0, \pi)$ **4.** $(-2, \pi/4)$ **7.** $(-3, \pi)$ **8.** $(\pi, -\pi)$

In Exercises 9–15, find polar coordinates, subject to the stated restrictions on θ, for the point given in rectangular coordinates.

9. $(1, 1)$, $0 < \theta < \pi$ **10.** $(1, 1)$, $\pi < \theta < 2\pi$

11. $(-3, 0)$, $-\pi/2 < \theta < \pi/2$ **12.** $(1, -\sqrt{3})$, $\pi < \theta < 2\pi$

13. $(1, -\sqrt{3})$, $0 < \theta < \pi$ **14.** $(-2, 2)$, $0 < \theta < \pi$

15. $(-2, 2)$, $3\pi < \theta < 4\pi$

16. In this exercise, we derive additional tests of symmetry for the graph of a polar equation.

  a. Using Figure 1.17 show that, if the point with polar coordinates $(-r, -\theta)$ lies on the graph whenever the point (r, θ) is on the graph, then the graph is symmetric with respect to the y-axis.

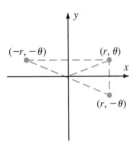

Figure 1.17

  b. Apply this test for symmetry to the graph of $r = \theta$. Graph this curve for $-4\pi \le \theta \le 4\pi$.
  c. In general, if the point with polar coordinates $(-r, \pi - \theta)$ lies on a graph whenever the point (r, θ) does, what type of symmetry does the graph have?
  d. In general, if the point with polar coordinates $(-r, \theta)$ lies on a graph whenever the point (r, θ) does, what type of symmetry does the graph have?

In Exercises 17–24, identify all symmetries (about the x-axis, the y-axis, or the origin) possessed by the graph of the given equation. (See Exercise 16.)

17. $r = 4 \cos \theta$ **18.** $r = 2 \sin \theta$

19. $r = 1 + \sin \theta$ **20.** $r^2 = \cos \theta$

21. $r^2 = \cos 2\theta$ **22.** $r = \sin 3\theta$

23. $r = 2$ **24.** $r = 4 \sin 2\theta$

In Exercises 25–38, find an equation in polar coordinates for the given equation.

25. $x^2 + y^2 = 4$ **26.** $y^2 = 16x$

27. $4x^2 + y^2 = 4$ **28.** $xy = 2$

29. $x^2 + 2y^2 = 1$ **30.** $(x^2/a^2) + (y^2/b^2) = 1$

31. $x^2 - y^2 = 1$ **32.** $y = 3x^2$

33. $x = 6$ **34.** $y = 4$

35. $x^2 + y^2 + 2y = 0$ **36.** $y = x$

37. $y = x/2$ **38.** $y = |x|$

In Exercises 39–48, find an equation in rectangular coordinates for the given polar equation.

39. $r = 4 \sin \theta$ **40.** $r^2 = 3r \cos \theta$

41. $r = \tan \theta$ **42.** $r = 6$

43. $r = 4 \sec \theta$ **44.** $r = \csc \theta$

45. $r = 1/(1 - \cos \theta)$ **46.** $r = 1 + \sin \theta$

47. $r^2 = 4 \sec \theta$ **48.** $r = a \tan \theta \sec \theta$

In Exercises 49–58, sketch the graph of the given polar equation.

49. $r = \theta/2$ (spiral of Archimedes)

50. $r = 2 \sin \theta$ (circle)

51. $r = 1 + \sin \theta$ (cardioid)

52. $r = \cos 2\theta$ (4-leaved rose)

53. $r = 1 + 2 \cos \theta$ (limaçon)

54. $r^2 = \cos 2\theta$ (lemniscate)

55. $r = e^{\theta}$, $\theta \ge 0$ (spiral)

56. $r = e^{-\theta}$, $\theta \ge 0$

57. $r = 1 - 2 \cos \theta$ (limaçon)

58. $r = \sin 4\theta$ (8-leaved rose)

59. Prove that the graph of $r = 2 \sin \theta - 2 \cos \theta$ is a circle. Find its center and radius.

60. Prove that the graph of $r = a \sin \theta$ is a circle. Sketch the graph.

61. Prove that the graph of $r = a \cos \theta$ is a circle. Sketch the graph.

62. Show that equations (3) are valid for the case $r < 0$.

63. Show that the equation $y = ax$ of a nonvertical line through the origin is, in polar coordinates, $\tan \theta = a$.

64. The line $r = 4 \sec \theta$ is tangent to the graph of $r = a(1 + \cos \theta)$. Find a.

65. Find the area of the region enclosed by the graph of $r = 3 \cos \theta$. (*Hint:* See Exercise 61.)

66. Find the area of the region enclosed by the graph of $r^2 = 4 \cos^2 \theta$. (*Hint:* See Exercise 65.)

67. Find the vertices of the triangle determined by the three lines

$$\theta = 0, \quad \tan \theta = 1, \quad \text{and} \quad r = 4 \sec \theta.$$

68. Find the area of the triangle determined by the three lines

$$\tan \theta = 2, \quad \tan \theta = -2, \quad \text{and} \quad r = 4 \csc \theta.$$

15.2 Graphing Techniques for Polar Equations

In Section 15.1 we graphed polar equations by plotting a few points (r_j, θ_j) and sketching in a curve. A slightly different approach is often simpler to use. We put the polar equation in the form $r = f(\theta)$, if possible, and then "think dynamically," noting the behavior of r as θ sweeps out one or more revolutions about the pole.

For example, to graph the polar equation $r = a(1 + \cos \theta)$ where $a > 0$, we first note that the maximum value of $r = a(1 + 1) = 2a$ occurs when $\cos \theta = 1$, that is, when $\theta = 0$. The curve is therefore contained within the circle $r = 2a$ (Figure 2.1). To plot the curve we begin at the point $(2a, 0)$ and note that as θ in-

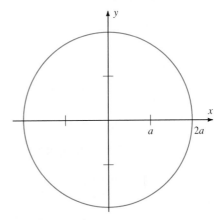

Figure 2.1 Circle $r = 2a$.

creases from 0 to $\pi/2$, r *decreases* from length $r(0) = 2a$ to length $r(\pi/2) = a(1 + \cos(\pi/2)) = a$. This observation enables us to sketch the arc labelled 1 in Figure 2.2. We next note the behavior of r as θ increases from $\pi/2$ to π: r decreases from $r(\pi/2) = a$ to $r(\pi) = a(1 + (-1)) = 0$. This produces the arc labelled 2 in Figure 2.3. Similarly, arcs 3 and 4 are "swept out" as θ increases from π to 2π. Allowing θ to increase beyond 2π simply retraces the arcs already obtained, so the complete curve corresponding to $r = a(1 + \cos \theta)$ is obtained from $0 \le \theta \le 2\pi$ (Figure 2.3).

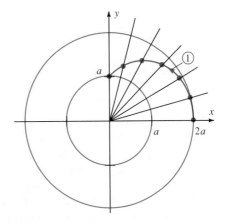

Figure 2.2 As θ increases toward $\pi/2$, r decreases from $2a$ to a.

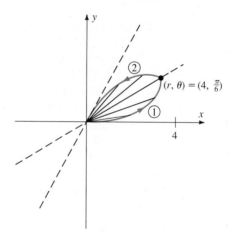

Figure 2.3 Graph of $r = a(1 + \cos \theta)$. The points are labelled with their polar coordinates.

☐ **EXAMPLE 1**

Sketch the graph of the polar equation $r = 4 \sin 3\theta$.

Solution   Since $|\sin 3\theta| \leq 1$ for all θ, the curve must lie within the circle $r = 4$. Beginning at the point with $\theta = 0$, we note the following:

(a) As θ turns from 0 to $\pi/6$, 3θ turns from 0 to $\pi/2$, so r *increases* from $r(0) = 4 \sin 0 = 0$ to $r(\pi/6) = 4 \sin 3(\pi/6) = 4$. This produces arc 1 in Figure 2.4.

Figure 2.4 $r = 4 \sin 3\theta$, $0 \leq \theta \leq \pi/3$.

(b) As θ turns from $\pi/6$ to $\pi/3$, 3θ turns from $\pi/2$ to π, so r *decreases* from $r(\pi/6) = 4$ to $r(\pi/3) = 0$ (arc 2 in Figure 2.4).
(c) As θ turns from $\pi/3$ to $\pi/2$, 3θ turns from π to $3\pi/2$, so r *decreases* from $r(\pi/3) = 0$ to $r(\pi/2) = 4 \sin 3(\pi/2) = -4$, producing arc 3 in Figure 2.5.
(d) As θ increases through the next quarter turn, from $\pi/2$ to π, 3θ increases from $3\pi/2$ to 3π. Thus, r increases from -4 to 4 for $\pi/2 \leq \theta \leq 5\pi/6$, producing arcs 4 and 5, and decreases from 4 to 0 for $5\pi/6 \leq \theta \leq \pi$, producing arc 6 (Figure 2.6).

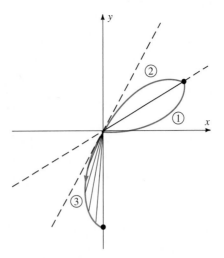

Figure 2.5 $r = 4 \sin 3\theta$, $0 \le \theta \le \pi/2$.

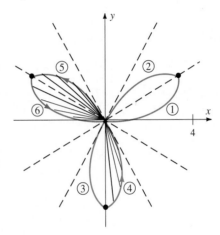

Figure 2.6   Three-leaved rose $r = 4 \sin 3\theta$, $0 \le \theta \le \pi$.

By noting the behavior of r for $\theta > \pi$, we can see that the entire curve (a three-leaved rose) has been obtained with only $0 \le \theta \le \pi$. (Figure 2.7 recalls the graph of the function $f(\theta) = 4 \sin 3\theta$. Note how the arcs numbered 1–6 correspond to the arcs in Figures 2.4–2.6.)   ■

☐ **EXAMPLE 2**

Graph the lemniscate $r^2 = a \cos 2\theta$, $a > 0$.

Solution   Taking square roots of both sides shows that the given equation corresponds to the *pair* of equations

$$r = r_1(\theta) = \sqrt{a \cos 2\theta} \tag{1}$$

and

$$r = r_2(\theta) = -\sqrt{a \cos 2\theta}. \tag{2}$$

Thus, some values of θ will correspond to two values r, while others will not correspond to any real value r.

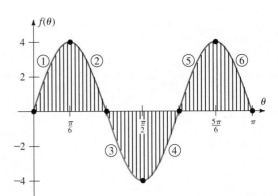

Figure 2.7 The graph of $f(\theta) = 4\sin(3\theta)$. Note that each choice of θ yields a vertical line segment on this graph, which in turn corresponds to a radial line segment when we sketch the polar equation $r = 4\sin(3\theta)$. Match the vertical segments in this figure with the radial segments in Figures 2.4–2.6.

Beginning with $\theta = 0$, we observe that:

(i) As θ increases from 0 to $\pi/4$, 2θ increases from 0 to $\pi/2$. Thus,

(a) r_1 decreases from $r_1(0) = \sqrt{a}$ to $r_1(\pi/4) = 0$ (arc 1 in Figure 2.8),
(b) r_2 increases from $r_2(0) = -\sqrt{a}$ to $r_2(\pi/4) = 0$ (arc 2).

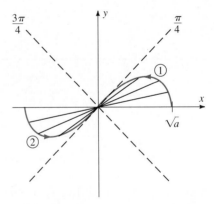

Figure 2.8 $r^2 = a\cos 2\theta$, $0 \le \theta \le \pi/4$.

(ii) As θ increases from $\pi/4$ to $3\pi/4$, 2θ increases from $\pi/2$ to $3\pi/2$. Since $\cos 2\theta < 0$ for $\pi/2 < 2\theta < 3\pi/2$, equations (1) and (2) have no solutions in this interval.

(iii) As θ increases from $3\pi/4$ to π, 2θ increases from $3\pi/2$ to 2π. Thus,

(a) r_1 increases from $r_1(3\pi/2) = 0$ to $r_1(2\pi) = \sqrt{a}$ (arc 3 in Figure 2.9),
(b) r_2 decreases from $r_2(3\pi/2) = 0$ to $r_2(2\pi) = -\sqrt{a}$ (arc 4).

Checking equations (1) and (2) for values of $\theta > \pi$ shows that all points on the graph are obtained by using only $0 \le \theta \le \pi/4$ and $3\pi/4 \le \theta \le \pi$. ∎

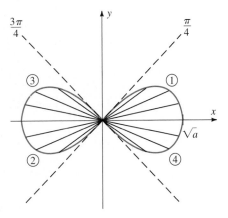

Figure 2.9 $r^2 = a \cos 2\theta$.

Intersections of Graphs in Polar Coordinates

In Section 15.3, we will need to determine the points of intersection for two curves specified in terms of polar coordinates as

$$r = r_1(\theta) \quad \text{and} \quad r = r_2(\theta). \tag{3}$$

As for equations in rectangular coordinates, we can find points of intersection by equating the right-hand sides of equations (3) and solving the resulting equation for θ. However, unlike the situation for rectangular equations, *this method will not necessarily produce all points of intersection,* as the following examples show.

☐ **EXAMPLE 3**

Find all points of intersection for the graphs of the equations $r = 1 + \cos \theta$ and $r = 3 \cos \theta$.

Solution   In this example, we have $r_1(\theta) = 1 + \cos \theta$ and $r_2(\theta) = 3 \cos \theta$. Setting $r_1 = r_2$ gives the equation

$$1 + \cos \theta = 3 \cos \theta,$$

so

$$\cos \theta = 1/2.$$

The solutions of this equation for $-\pi < \theta < \pi$ are $\theta_1 = \pi/3$ and $\theta_2 = -\pi/3$. The corresponding points of intersection are $(3/2, \pi/3)$ and $(3/2, -\pi/3)$. However, as Figure 2.10 illustrates, the *pole* $(0, \theta)$ is also common to both graphs. Setting $r_1(\theta) = r_2(\theta)$ misses this point of intersection, since the pole corresponds to $\theta = \pi$ in the graph of r_1, but it corresponds to $\theta = \pi/2$ in the graph of r_2.   ■

☐ **EXAMPLE 4**

Find all points of intersection of the four-leaved rose $r = 4 \cos 2\theta$ and the circle $r = 2$.

Solution   Let $r_1(\theta) = 4 \cos 2\theta$ and $r_2(\theta) = 2$. Setting $r_1 = r_2$ gives $4 \cos 2\theta = 2$, or $\cos 2\theta = 1/2$. Thus $2\theta = \pm \pi/3$ or $2\theta = \pm 7\pi/3$. The four distinct solutions of this equation in the interval $(0, 2\pi)$ are $\theta = \pi/6, 5\pi/6, 7\pi/6$, and $11\pi/6$. These values of θ give the four points of intersection $(2, \pi/6)$, $(2, 5\pi/6)$, $(2, 7\pi/6)$, and $(2, 11\pi/6)$.

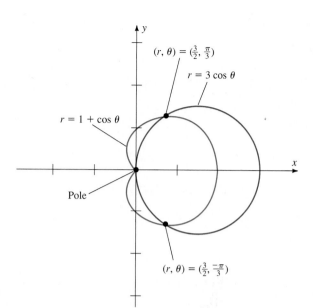

Figure 2.10 The graphs of $r = r_1(\theta)$ and $r = r_2(\theta)$ intersect in 3 points.

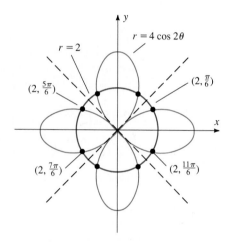

Figure 2.11 Graphs of $r = r_1(\theta)$ and $r = r_2(\theta)$ intersect in 8 points.

However, the method of setting $r_1 = r_2$ has failed to detect *four* additional points of intersection $(2, \pi/3)$, $(2, 2\pi/3)$, $(2, 4\pi/3)$, and $(2, 5\pi/3)$ (see Figure 2.11).

In fact, we determine these points by observing that the circle $r = 2$ is also the graph of $r = r_3(\theta) = -2$. Setting $r_1 = r_3$ gives $\cos 2\theta = -1/2$. This equation also has four distinct solutions in the interval $(0, 2\pi)$. They are $\theta = \pi/3$, $\theta = 2\pi/3$, $\theta = 4\pi/3$, and $\theta = 5\pi/3$. Thus, we obtain the four points of intersection $(-2, \pi/3) = (2, 4\pi/3)$, $(-2, 2\pi/3) = (2, 5\pi/3)$, $(-2, 4\pi/3) = (2, \pi/3)$, and $(-2, 5\pi/3) = (2, 2\pi/3)$. ∎

Examples 3 and 4 show that simply solving $r_1(\theta) = r_2(\theta)$ for θ will not necessarily yield all points common to the graphs of equations (3). The reason for this difficulty is the nonuniqueness of polar coordinates. Rather than trying to find an algorithm allowing for all possible ways in which points of intersection of two polar equations can arise, *you should simply develop the habit of always sketching the two curves to ensure that all such points are found.*

Exercise Set 15.2

In Exercises 1–14, sketch the graph of the given polar equation using the method of Examples 1 and 2.

1. $r = 4 \cos \theta$

2. $r = a \cos \theta$

3. $r = 3 \sin \theta$

4. $r = a \sin \theta$

5. $r = 5 \sin 2\theta$

6. $r = a \sin 2\theta$

7. $r = a(1 + \cos \theta)$

8. $r = a(1 - \sin \theta)$

9. $r = 1 + 2 \sin \theta$

10. $r = 2\theta$

11. $r^2 = 4 \cos 2\theta$

12. $r^2 = \sin^2 \theta$

13. $r = \sin \theta + \cos \theta$

14. $r = a(1 - \cos \theta)$

In Exercises 15–26, find all points of intersection of the graphs of the two given equations.

15. $r = 2 \cos \theta$
$r = 2 \sin \theta$

16. $r = 2 \cos \theta$
$r = 1$

17. $r = 1 + \sin \theta$
$r = 3 \sin \theta$

18. $r = a \sin 3\theta$
$r = a$

19. $r = a(1 + \cos \theta)$
$r = 3a \cos \theta$

20. $r = 2a \cos 2\theta$
$r = a$

21. $r = 1 + \sin \theta$
$r = 1 - \sin \theta$

22. $r = a(1 + \cos \theta)$
$r = a(1 - \cos \theta)$

23. $r = \cos 2\theta$
$r = \sqrt{2}/2$

24. $r = 1/\theta$
$\theta = \pi/4$

25. $r = \sin \theta$
$r = \sin^2 \theta$

26. $r = a \sin 4\theta$
$r = a$

27. Show that the graphs of the polar equations

$$r = \frac{a}{1 \pm \cos \theta}; \quad r = \frac{a}{1 \pm \sin \theta}$$

are parabolas. (*Hint:* Convert to rectangular equations.)

28. Show that the graphs of the polar equations

$$r = \frac{ab}{1 \pm b \cos \theta}; \quad r = \frac{ab}{1 \pm b \sin \theta}$$

are ellipses if $0 < b < 1$. (*Hint:* Convert to rectangular equations.)

In Exercises 29–32, use the results of Exercises 27 and 28 to convert the given equations to polar coordinates.

29. $y = \frac{1}{8} - 2x^2$

30. $x = 7y^2 - \frac{1}{28}$

31. $\dfrac{9(x - 2/3)^2}{16} + \dfrac{3y^2}{4} = 1$

32. $\dfrac{2x^2}{9} + \dfrac{16(y + 3/4)^2}{81} = 1$

In Exercises 33–38, find a rectangular form for the given polar equation, identify the graph as either a parabola or an ellipse, and sketch the graph. (See Exercises 27 and 28.)

33. $r = \dfrac{1}{1 + \cos \theta}$

34. $r = \dfrac{4}{1 - \sin \theta}$

35. $r = \dfrac{2}{1 + \frac{1}{2}\cos \theta}$

36. $2r = \dfrac{4}{2 - \sin \theta}$

37. $r - r \cos \theta = 2$

38. $3r + r \cos \theta - 6 = 0$

 In Exercises 39–42, graph the given polar equation using your calculator.

39. $r = 1 + \tan \theta$

40. $r = 2(1 - \sin 2\theta)$

41. $r = \sin 2\theta + 2 \cos \theta$

42. $r = \sec 2\theta$

 43. The general equation for a limaçon is $r = a \pm b \cos \theta$ or $r = a \pm b \sin \theta$. Using your calculator, investigate the various shapes of limaçons. Make sure that you try examples in which $|a| = |b|$, $|a| < |b|$, and $|a| > |b|$. There are limaçons with inner loops or dimples. Others are cardioids, and some, bound convex regions in the plane.

 44. Consider the polar equation $r = 1 + a \sin m\theta$. For $a = \frac{1}{8}, \frac{1}{4}, \frac{1}{2}$, and $\frac{9}{10}$, graph this equation for $m = \frac{1}{3}, \frac{1}{2}, 1, 2$, and 5.

 If $r = f(\theta)$ is a polar equation in which f is an increasing function of θ, its graph is a spiral. Similarly, if f is decreasing, it also generates a spiral. In Exercises 45–47, graph the given spirals.

45. $r = \theta/2$

46. $r = e^{\theta/20}$

47. $r = 5/\theta$

 48. Graph the polar equations $r = \frac{1}{10}\theta \cos m\theta$, where $m = 1, 2, 3$. Compare the results to the graphs obtained in Exercises 45–47. Explain the similarities as well the differences.

 With the aid of your calculator, find the points of intersection of each pair of curves in Exercises 49–51.

49. $r = 1 + 2 \cos \theta$
$r = 1 - 2 \sin \theta$

50. $r = 1$
$r^2 = 2 \cos 3\theta$

51. $r = 1 + \sin \theta$
$r = \sin 2\theta$

 52. On the same plot, graph the two curves

$$r = \cos \theta + \sin^3 \left(\frac{5}{2}\theta\right)$$
$$r = \frac{1}{2}(\cos 3\theta)^{3/2}.$$

Note that this plot resembles a butterfly. Produce other shapes by changing the constants in these equations.

15.3 Calculating Area in Polar Coordinates

If a region R in the plane is determined by the graph of an equation written in the *polar* form $r = f(\theta)$, the area of R cannot be calculated by a direct application of the equations developed in Chapter 6. The difficulty lies in the fact that these formulas were obtained for functions $y = g(x)$ written in *rectangular* coordinates. To calculate areas of regions determined by the graphs of polar equations, we might try rewriting these equations in rectangular coordinates, using the substitutions $x = r \cos \theta$ and $y = r \sin \theta$, as in Section 15.1. However, this strategy often leads to undesirable consequences. Either the resulting rectangular equations are much more complicated than the original polar equations, or, even worse, the resulting equations might not describe either variable as a *function* of the other.

A more satisfactory approach is to return to the basic theory of the definite integral and determine the formulas appropriate for calculating area in polar coordinates. The small amount of time spent in this effort will allow us to avoid the step of converting to rectangular coordinates for each area problem in polar coordinates.

Suppose that a region R is bounded by the graph of the function $r = f(\theta)$ for θ between the numbers $\theta = a$ and $\theta = b$ (Figure 3.1). Since θ is the independent variable, we partition the interval $[a, b]$ into n equal subintervals using the increment $\Delta\theta = (b - a)/n$ and the endpoints

$$\theta_0 = a, \quad \theta_1 = a + \Delta\theta, \quad \theta_2 = a + 2\,\Delta\theta, \ldots, \quad \theta_n = a + n\,\Delta\theta = b.$$

The rays $\theta = \theta_j, j = 0, 1, 2, \ldots, n$, "slice" the region R into n wedge-shaped subregions (Figure 3.2).

We now approximate the area of each of the subregions. To do so we choose one number t_j in each interval $[\theta_{j-1}, \theta_j], j = 1, 2, \ldots, n$ (Figure 3.3). We then approximate the area of the jth subregion by the area of the circular **sector** of constant radius $r_j = f(t_j)$ and angle $\Delta\theta$ (see Figure 3.4).

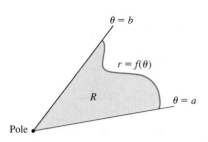

Figure 3.1 Region determined by $r = f(\theta)$, $a \le \theta \le b$.

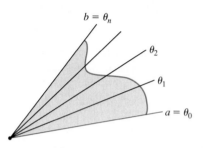

Figure 3.2 Partition $\theta_0 < \theta_1 < \cdots < \theta_n$ slices R into wedge-shaped regions.

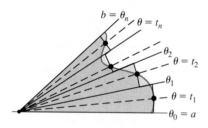

Figure 3.3 Area of R approximated by areas of sectors of circles.

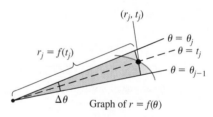

Figure 3.4 $\Delta A_j = \frac{1}{2} r_j^2 \Delta\theta$.

Since the area of the circle of radius r_j is πr_j^2, the area of the circular sector comprising the fractional part $\Delta\theta/(2\pi)$ of the entire circle is

$$\Delta A_j = \pi r_j^2 \left(\frac{\Delta\theta}{2\pi} \right) = \frac{1}{2} r_j^2 \Delta\theta = \frac{1}{2} [f(t_j)]^2 \Delta\theta.$$

Summing these approximations, one for each subregion, gives an approximation to the area A of the whole region R of the form

$$A \approx \sum_{j=1}^{n} \Delta A_j = \sum_{j=1}^{n} \frac{1}{2} [f(t_j)]^2 \Delta\theta. \tag{1}$$

If the function $r = f(\theta)$ is continuous for $a \le \theta \le b$, the sum on the right-hand side of approximation (1) is a *Riemann sum* for the function $\frac{1}{2}[f(\theta)]^2$, which "converges" to the definite integral

$$\int_a^b \frac{1}{2} [f(\theta)]^2 \, d\theta = \lim_{n \to \infty} \sum_{j=1}^{n} \frac{1}{2} [f(t_j)]^2 \Delta\theta \tag{2}$$

as $n \to \infty$. We therefore argue that as $n \to \infty$, the circular sectors provide an increasingly accurate approximation to the region R and that the area of R is correctly defined as follows.

> The area A of the region bounded by the graph of the continuous function $r = f(\theta)$ between the rays $\theta = a$ and $\theta = b$, $a < b$, is given by the definite integral
>
> $$A = \int_a^b \frac{1}{2} [f(\theta)]^2 \, d\theta. \tag{3}$$

The same cautions should be observed in using equation (3) as in calculating areas in rectangular coordinates. First, sketch the region carefully to determine the proper limits of integration.

☐ **EXAMPLE 1**

Find the area of the region bounded by the rays $\theta = -\pi/4$ and $\theta = \pi/4$, and the graph of the equation $r = 1 + \sin \theta$.

Solution   The region is the portion of the cardioid swept out by the radius of length $r = 1 + \sin \theta$ as θ increases from $\theta = -\pi/4$ to $\theta = \pi/4$ (Figure 3.5). By equation (3) the area is

$$
\begin{aligned}
A &= \int_{-\pi/4}^{\pi/4} \frac{1}{2} [1 + \sin \theta]^2 \, d\theta \\
&= \int_{-\pi/4}^{\pi/4} \frac{1}{2} [1 + 2 \sin \theta + \sin^2 \theta] \, d\theta \\
&= \int_{-\pi/4}^{\pi/4} \frac{1}{2} \left[1 + 2 \sin \theta + \left(\frac{1}{2} - \frac{1}{2} \cos 2\theta \right) \right] d\theta \qquad \left(\text{using the identity} \right. \\
&\qquad\qquad\qquad\qquad\qquad\qquad\qquad\qquad \left. \sin^2 \theta = \frac{1}{2} - \frac{1}{2} \cos 2\theta \right) \\
&= \int_{-\pi/4}^{\pi/4} \frac{1}{2} \left[\frac{3}{2} + 2 \sin \theta - \frac{1}{2} \cos 2\theta \right] d\theta
\end{aligned}
$$

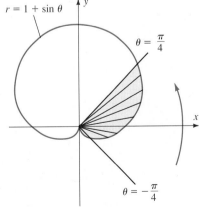

Figure 3.5 $-\pi/4 \leq \theta \leq \pi/4$.

$$= \frac{1}{2}\left[\frac{3\theta}{2} - 2\cos\theta - \frac{1}{4}\sin 2\theta\right]_{-\pi/4}^{\pi/4}$$

$$= \frac{1}{2}\left[\left(\frac{3\pi}{8} - \sqrt{2} - \frac{1}{4}\right) - \left(-\frac{3\pi}{8} - \sqrt{2} + \frac{1}{4}\right)\right]$$

$$= \frac{3\pi}{8} - \frac{1}{4} \approx 0.928.$$

■

□ EXAMPLE 2

Find the area of the region enclosed by the graph of the cardioid $r = 1 + \cos\theta$.

Strategy · · · · · · · ·

Sketch the region.

Find a pair of smallest and largest values of θ required to sweep out the region. These are the limits of integration. (Note that $0 \le \theta \le 2\pi$ works just as well as $-\pi \le \theta \le \pi$.)

Set up the integral given by equation (3).

Square $f(\theta)$ in the integrand and simplify, using the identity

$$\cos^2\theta = \frac{1}{2} + \frac{1}{2}\cos 2\theta.$$

Integrate.

Solution

The cardioid is sketched in Figure 3.6. Since the cardioid is the graph of the function $f(\theta) = 1 + \cos\theta$ as θ increases from $\theta = -\pi$ to $\theta = \pi$, the limits of integration are $a = -\pi$ and $b = \pi$. By equation (3) the area is

$$A = \int_{-\pi}^{\pi} \frac{1}{2}[1 + \cos\theta]^2 \, d\theta$$

$$= \frac{1}{2}\int_{-\pi}^{\pi} [1 + 2\cos\theta + \cos^2\theta] \, d\theta$$

$$= \frac{1}{2}\int_{-\pi}^{\pi} \left[1 + 2\cos\theta + \left(\frac{1}{2} + \frac{1}{2}\cos 2\theta\right)\right] d\theta$$

$$= \frac{1}{2}\int_{-\pi}^{\pi} \left[\frac{3}{2} + 2\cos\theta + \frac{1}{2}\cos 2\theta\right] d\theta$$

$$= \frac{1}{2}\left[\frac{3\theta}{2} + 2\sin\theta + \frac{1}{4}\sin 2\theta\right]_{-\pi}^{\pi}$$

$$= \frac{1}{2}\left[\left(\frac{3\pi}{2} + 2\cdot 0 + \frac{1}{4}\cdot 0\right) - \left(\frac{3}{2}(-\pi) + 2\cdot 0 + \frac{1}{4}\cdot 0\right)\right]$$

$$= \frac{3\pi}{2}.$$

■

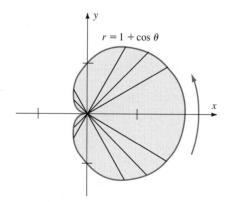

$$r = 1 + \cos\theta$$

Figure 3.6 $-\pi \le \theta \le \pi$.

Often, we can use symmetry considerations to simplify the calculation of area, as the following example illustrates.

☐ **EXAMPLE 3**

Find the area of the region enclosed by the graph of the polar equation $r = a \sin 3\theta$.

Solution   The region is a three-leaved rose, as illustrated in Figure 3.7. Since the 3 leaves are congruent and since the leaf determined by $0 \le \theta \le \pi/3$ is symmetric about the ray $\theta = \pi/6$, we may obtain the area of the entire figure by calculating the area of the half leaf determined by $0 \le \theta \le \pi/6$ and multiplying the result by 6. We obtain

$$A = 6 \int_0^{\pi/6} \frac{1}{2} [a \sin 3\theta]^2 \, d\theta$$

$$= 3a^2 \int_0^{\pi/6} \sin^2 3\theta \, d\theta$$

$$= 3a^2 \int_0^{\pi/6} \left[\frac{1}{2} - \frac{1}{2} \cos 6\theta \right] d\theta$$

$$= 3a^2 \left[\frac{\theta}{2} - \frac{1}{12} \sin 6\theta \right]_0^{\pi/6}$$

$$= \frac{a^2 \pi}{4}. \qquad \blacksquare$$

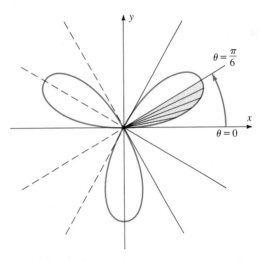

Figure 3.7   Area of three-leaved rose is 6 times area of region determined by $0 \le \theta \le \pi/6$.

As happens with equations in rectangular coordinates, an equation in polar coordinates may not express r as a function of θ. The following example shows how symmetry can be used to overcome this difficulty.

☐ **EXAMPLE 4**

Find the area of the region enclosed by the lemniscate $r^2 = \cos 2\theta$.

Strategy   · · · · · · · ·

Sketch the figure. (Recall Example 2, Section 15.2.)

Take square roots to solve for r. Obtain a function $r = f(\theta)$.

Solution

The lemniscate is sketched in Figure 3.8. By extracting square roots on both sides of the equation $r^2 = \cos 2\theta$ we obtain the equation $r = \pm \sqrt{\cos 2\theta}$. Choosing the positive sign gives the function

$$f(\theta) = \sqrt{\cos 2\theta}.$$

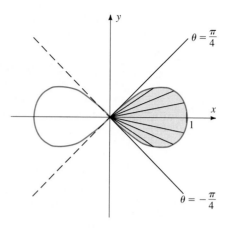

Figure 3.8 $-\pi/4 \le \theta \le \pi/4$ determines half the area of the lemniscate $r^2 = \cos 2\theta$.

Compare region determined by the function with that of the original equation.

This function is defined only for

$$-\frac{\pi}{4} \le \theta \le \frac{\pi}{4} \quad \text{and} \quad \frac{3\pi}{4} \le \theta \le \frac{5\pi}{4}.$$

Use symmetry to simplify calculation.

However, as θ ranges through these two intervals the two lobes of the lemniscate are traced out. We may therefore integrate over $-\pi/4 \le \theta \le \pi/4$ to obtain the area of one lobe and double the result. We obtain

Apply equation (3).

$$A = 2\int_{-\pi/4}^{\pi/4} \frac{1}{2}[\sqrt{\cos 2\theta}]^2 \, d\theta$$

$$= \int_{-\pi/4}^{\pi/4} \cos 2\theta \, d\theta$$

$$= \left[\frac{1}{2}\sin 2\theta\right]_{-\pi/4}^{\pi/4} = 1. \qquad \blacksquare$$

Area Between Two Curves

If a region R lies between the graphs of two polar equations, as in Figure 3.9, we may calculate the area of R by subtracting the area enclosed by the inner curve from the area enclosed by the outer curve. That is, if R lies inside the graph of $r = f(\theta)$ and outside the graph of $r = g(\theta)$, for $a \le \theta \le b$, the area A of R is given by the integral.

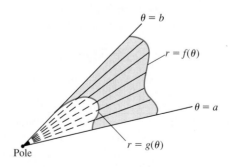

Figure 3.9   Region between graphs of $r = f(\theta)$ and $r = g(\theta)$ for $a \le \theta \le b$.

$$A = \int_a^b \frac{1}{2}[f(\theta)]^2 \, d\theta - \int_a^b \frac{1}{2}[g(\theta)]^2 \, d\theta. \tag{4}$$

☐ **EXAMPLE 5**

Find the area of the region lying inside the circle $r = 3 \cos \theta$ and outside the cardioid $r = 1 + \cos \theta$.

Strategy · · · · · · · · ·

Sketch the region.

Find the points of intersection to determine the limits of integration.

Solution

The graphs of the two equations are sketched in Figure 3.10. By solving the two equations simultaneously, we find the points of intersection to be $(3/2, \pm \pi/3)$ (see Example 3, Section 15.2). The limits of integration are therefore $-\pi/3 \leq \theta \leq \pi/3$.

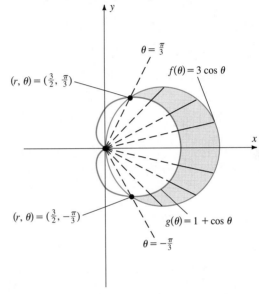

Figure 3.10 The region inside the circle $r = f(\theta) = 3 \cos \theta$ and outside the cardioid $r = g(\theta) = 1 + \cos \theta$.

Determine which is the outer curve.

In this interval, the outer (greater) function is $f(\theta) = 3 \cos \theta$. The inner function is $g(\theta) = 1 + \cos \theta$. By equation (4) the area is

Apply equation (4).

$$A = \int_{-\pi/3}^{\pi/3} \frac{1}{2}[3 \cos \theta]^2 \, d\theta - \int_{-\pi/3}^{\pi/3} \frac{1}{2}[1 + \cos \theta]^2 \, d\theta$$

Simplify integrand.

$$= \int_{-\pi/3}^{\pi/3} \frac{1}{2}[9 \cos^2 \theta - (1 + 2 \cos \theta + \cos^2 \theta)] \, d\theta$$

$$= \int_{-\pi/3}^{\pi/3} \frac{1}{2}[8 \cos^2 \theta - 1 - 2 \cos \theta] \, d\theta$$

$$\cos^2 \theta = \frac{1}{2} + \frac{1}{2} \cos 2\theta.$$

$$= \int_{-\pi/3}^{\pi/3} \frac{1}{2}\left[8\left(\frac{1}{2} + \frac{1}{2} \cos 2\theta\right) - 1 - 2 \cos \theta\right] d\theta$$

$$= \int_{-\pi/3}^{\pi/3} \left(\frac{3}{2} + 2 \cos 2\theta - \cos \theta\right) d\theta$$

$$= \left[\frac{3\theta}{2} + \sin 2\theta - \sin \theta\right]_{-\pi/3}^{\pi/3} = \pi. \qquad ■$$

Exercise Set 15.3

In Exercises 1–8, find the area of the region determined by the given equations and inequalities.

1. $r = 1 + \sin \theta, \quad -\pi/2 < \theta < \pi/2$

2. $r = a \sin 2\theta, \quad 0 \le \theta \le \pi/2$

3. $r = 2 \sin \theta, \quad 0 \le \theta \le \pi$

4. $r = \theta, \quad 0 \le \theta \le \pi$

5. $r = 2 + \sin \theta, \quad 0 \le \theta \le \pi$

6. $r = a \cos 3\theta, \quad \pi/2 \le \theta \le 2\pi/3$

7. $r = 4 + \sin \theta, \quad \pi/4 \le \theta \le 3\pi/4$

8. $r = 3 \cos 2\theta, \quad -\pi/4 \le \theta \le \pi/4$

In Exercises 9–19, find the area of the region enclosed by the graph of the given equation.

9. $r = 2 \cos \theta$ **10.** $r = 3 \sin \theta$

11. $r = 1 + \sin \theta$ **12.** $r = -1 + \cos \theta$

13. $r = 4 \sin 2\theta$ **14.** $r = a \sin 2\theta$

15. $r = \cos 2\theta$ **16.** $r = a \cos 3\theta$

17. $r = a \sin 4\theta$ **18.** $r = \sin \theta + \cos \theta$

19. $r^2 = \cos \theta$

In Exercises 20–27, find the area of the region described.

20. The region inside the cardioid $r = 1 + \sin \theta$ and outside the circle $r = 2 \sin \theta$.

21. The regions inside the cardioid $r = 1 - \cos \theta$ and outside the circle $r = -2 \cos \theta$.

22. The region inside the cardioid $r = 1 + \cos \theta$ and outside the circle $r = 3 \cos \theta$. (*Hint:* You will need to treat the intervals $[\pi/3, \pi/2]$ and $[\pi/2, \pi]$ separately.)

23. The regions common to both cardioids $r = a(1 + \cos \theta)$ and $r = a(1 - \cos \theta)$.

24. The region outside the three-leaved rose $r = 4 \sin 3\theta$ and inside the circle $r = 4$.

25. The region common to the circles $r = 2 \cos \theta$ and $r = 2 \sin \theta$.

26. The region inside the circle $r = 2$ and outside the circle $r = 2 \cos \theta$.

27. The region common to the circle $r = \sqrt{2}/2$ and the lemniscate $r^2 = \cos 2\theta$.

28. Find the area of the region bounded by the graphs of the spirals $r_1 = \theta$ and $r_2 = e^\theta$ for $0 \le \theta \le \pi$.

29. Find the area of the region inside the graph of $r = 3 + \sin \theta$ and outside the graph of $r = 4 \sin \theta$.

30. Find the area of the region outside the graph of $r = 1 + \cos \theta$ and inside the graph of $r = 2 - \cos \theta$.

31. Find the area of the region bounded by the graph of the equation $r - r \cos \theta = 2$ and the line $\theta = \pi/2$.

32. Show that the result of calculating the area of the region enclosed by the graph of $r = 2 \cos \theta$ in polar coordinates agrees with the result of calculating the area of the same region in rectangular coordinates.

33. Let $f(\theta) \ge g(\theta)$ for all θ. Sketch the region whose area is given by the integral

$$\int_a^b \frac{1}{2} [f(\theta) - g(\theta)]^2 \, d\theta.$$

Compare this region with the region whose area is given by the integral in equation (4). Show that, in general, the two integrals are not equal.

34. Program 7 in Appendix I is a BASIC program that approximates

$$\int_a^b \frac{1}{2} [f(\theta)]^2 \, d\theta$$

for the function $f(\theta) = 1 + \sqrt{\sin \theta}$.
a. Use Program 7 to approximate the integral

$$\int_0^{\pi/2} \frac{1}{2} [1 + \sqrt{\sin \theta}]^2 \, d\theta$$

with $n = 10$, 100, and 200.
b. Use Program 7 to approximate the area of the region enclosed by the graph of $f(\theta) = 1 + \sqrt{\sin \theta}$ for $0 \le \theta \le \pi$.
c. Modify Program 7 to approximate the area of the region enclosed by the graph of $f(\theta) = 1 - \sqrt{\cos \theta}$, $0 \le \theta \le \pi/2$.

35. The polar curves

$$r^4 + b^4 - 2r^2 b^2 \cos 2\theta = k^4, \quad b < k,$$

called the **Ovals of Cassini,** are used to model the shape of red blood cells. Graph the oval that corresponds to $b = 2$ and $k = 17/8$. Using Simpson's Rule, estimate the area bounded by this curve to 4 decimal places of accuracy. (*Hint:* The equation is quadratic in r^2.)

15.4 Parametric Equations in the Plane

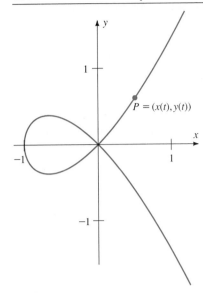

Figure 4.1 A curve in the plane described by parametric equations. The equations for this curve are given in Exercise 53.

Parametric equations provide a means of using functions to describe a curve in the plane, even though the curve may not represent the graph of a function of the form $y = f(x)$. Rather than describing either coordinate as a function of the other, we allow both coordinates to be functions of a third (independent) variable, called a **parameter,*** which is usually denoted by t. That is, we describe the curve as the collection of points $P(t) = (x(t), y(t))$ where the parameter t ranges through some specified domain (see Figure 4.1).

For example, the graph of the equation

$$x^2 + y^2 = 1 \tag{1}$$

is the familiar unit circle in the plane. However, equation (1) does not determine y as a function of x, since two distinct values of y correspond to each x with $-1 < x < 1$. Nevertheless, we may use our knowledge of trigonometry to obtain parametric equations, or a **parameterization,** for the unit circle. Let t denote the angle formed between the positive x-axis and the radius from $(0, 0)$ to (x, y), measured in the counterclockwise direction. Then

$$x = \cos t \quad \text{and} \quad y = \sin t$$

are the coordinate functions of the point (x, y) (see Figure 4.2). By inspection we can see that as t increases from $t = 0$ to $t = 2\pi$, the point $(x, y) = (\cos t, \sin t)$ traverses the unit circle once in the counterclockwise direction. We therefore say that a parameterization of the unit circle is given by the equations

$$\begin{cases} x(t) = \cos t \\ y(t) = \sin t \end{cases} \quad 0 \le t < 2\pi. \tag{2a} \tag{2b}$$

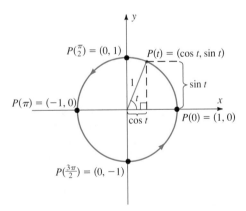

Figure 4.2 Parameterization of unit circle: $(x(t), y(t)) = (\cos t, \sin t)$, $0 \le t < 2\pi$.

---

*The term *parameter* denotes a variable that often has no geometric or physical interpretation. Here the parameter t has no geometric interpretation with respect to the curve it helps describe. In some applications the parameter t represents time, while the variables x and y represent coordinates of points in the plane.

Note that parameterizations for curves are not unique. For example, each of the following pairs of equations also represents a parameterization of the unit circle

$$\begin{cases} x(t) = \cos 2t \\ y(t) = \sin 2t \end{cases} \quad 0 \le t < \pi,$$

$$\begin{cases} x(t) = \cos t \\ y(t) = \sin(-t) \end{cases} \quad 0 \le t < 2\pi.$$

Later in this section, we shall see that parametric equations can describe motion. When used in this manner, the parameter t usually represents time, and the three different parameterizations of the unit circle given above correspond to three different types of circular motion. In all three cases, the curve swept out by the motion is the unit circle. The three motions differ either in speed or direction.

☐ **EXAMPLE 1**

Find parametric equations for the line ℓ containing the point $P_0 = (2, 4)$ and with slope 3.

Solution   We first find the equation for ℓ in xy-coordinates, using the point-slope form. We obtain

$$y - 4 = 3(x - 2), \tag{3}$$

so

$$y = 3x - 2. \tag{4}$$

If we set $x = t$ in equation (4), we find that $y = 3t - 2$, and we obtain the parametric equations

$$\begin{cases} x(t) = t \\ y(t) = 3t - 2 \end{cases} \quad -\infty < t < \infty.$$

To illustrate the various possible ways to parameterize a given curve, we also parameterize ℓ using the parameter $t = \sqrt[3]{x - 2}$. Then, $x - 2 = t^3$, and equation (3) yields $y - 4 = 3t^3$. The corresponding parameterization is

$$\begin{cases} x = t^3 + 2 \\ y = 3t^3 + 4. \end{cases} \qquad \blacksquare$$

Up to this point, we have considered only the problem of finding parameterizations for curves described by equations in xy-coordinates. Just as important is the problem of describing a curve that is presented in parametric form. One technique that is always available is to sketch the curve by selecting various values for t, calculating the corresponding coordinates of the point $P(t) = (x(t), y(t))$, and sketching a curve passing through the various points obtained (in order!). However, through a familiarity with the xy-coordinate forms of various curves, you can often convert the parametric equations directly into their xy counterparts.

☐ **EXAMPLE 2**

A point moves along a path with coordinates

$$(x(t), y(t)) = (a \cos t, b \sin t)$$

at time t. Describe the path.

Solution By squaring both coordinate functions we observe that

$$[x(t)]^2 = a^2 \cos^2 t; \qquad [y(t)]^2 = b^2 \sin^2 t.$$

We obtain

$$\frac{[x(t)]^2}{a^2} + \frac{[y(t)]^2}{b^2} = \cos^2 t + \sin^2 t = 1.$$

The curve is, therefore, the ellipse $\dfrac{x^2}{a^2} + \dfrac{y^2}{b^2} = 1$ seen in Figure 4.3. ■

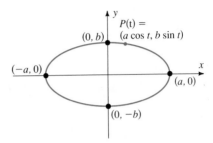

Figure 4.3 Parameterization for the ellipse

$$\frac{x^2}{a^2} + \frac{y^2}{b^2} = 1.$$

☐ **EXAMPLE 3**

To convert the parametric equations

$$\begin{cases} x(t) = t^2 - 2 \\ y(t) = t^2 + 1 \end{cases}$$

to an equation in x and y alone, we begin by noting that

$$y = t^2 + 1 = (t^2 - 2) + 3 = x + 3.$$

But the resulting curve is not simply the line $y = x + 3$. To see why, consider the equation $x = t^2 - 2$. Since $t^2 \geq 0$ for all t, $x \geq -2$. Therefore, we describe the resulting curve as

$$y = x + 3, \qquad x \geq -2.$$

(See Figure 4.4.) ■

☐ **EXAMPLE 4**

Parametric equations are useful in describing the motion of objects in space. For example, we may use them to determine the motion of a bullet fired horizontally from an altitude of 2 meters with an initial velocity of v_0 m/s.

Let $(x(t), y(t))$ denote the position of the bullet t seconds after it is fired, as in Figure 4.5. We will show in Chapter 17 that (neglecting the effect of air resistance) the coordinate functions x and y are

$$\begin{cases} x(t) = v_0 t, \\ y(t) = 2 - 4.9t^2. \end{cases}$$

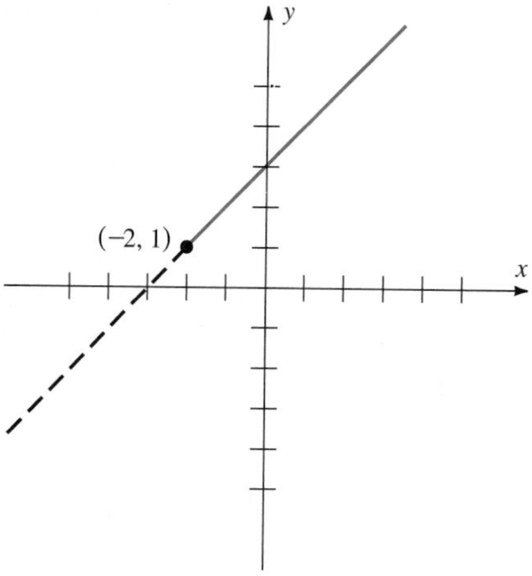

Figure 4.4 Graph of parametric equations $x(t) = t^2 - 2$, $y(t) = t^2 + 1$.

To see that these equations, for $t \geq 0$, describe an arc of a parabola, we note that

$$y(t) = 2 - 4.9t^2$$
$$= 2 - \frac{4.9}{v_0^2}(v_0 t)^2$$
$$= 2 - \left(\frac{4.9}{v_0^2}\right)[x(t)]^2.$$

That is, $y = kx^2 + 2$ with $k = -4.9/v_0^2$. ∎

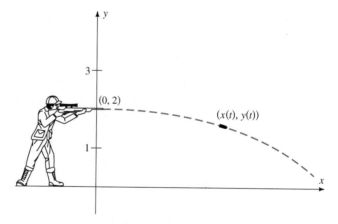

Figure 4.5 Bullet fired by a marksman follows a parabolic path.

So far, all of our examples have been familiar curves rewritten in parametric form. The next example illustrates how parametric equations can be used to describe curves and motions that are difficult to describe without the use of a parameter.

☐ **EXAMPLE 5**

Consider the motion of a pebble wedged in the tread of an automobile tire as the automobile moves along a straight road at constant speed. For simplicity, assume that the tire has a radius of 1 ft and that the automobile is moving at the constant rate of k ft/s. Describe this motion using parametric equations.

Solution   Let the x-axis represent the road surface and assume that the pebble P is initially located at the origin $O = (0, 0)$. Then t seconds later, the point Q of contact of the tire with the road has coordinates $(kt, 0)$. Likewise, since the radius is 1, the center C of the tire is located at the point $(kt, 1)$, and the angle between the radial arcs $\overline{CQ}$ and $\overline{CP}$ is kt (see Figure 4.6). Let α represent the angle of $\overline{CP}$ from the horizontal, as shown in Figure 4.6. Then we have

$$\begin{cases} x(t) = kt + \cos\alpha & \text{(5a)} \\ y(t) = 1 + \sin\alpha. & \text{(5b)} \end{cases}$$

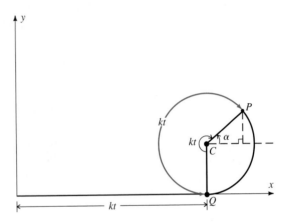

Figure 4.6   The point P on the circumference of the tire at time t.

To eliminate α, we observe that

$$\alpha = \frac{3\pi}{2} - kt.$$

Using trigonometric identities for sine and cosine (Section 1.5), we obtain

$$\cos\alpha = \cos\left(\frac{3\pi}{2} - kt\right) = -\sin kt$$

and

$$\sin\alpha = \sin\left(\frac{3\pi}{2} - kt\right) = -\cos kt.$$

Thus, equations (5a) and (5b) become

$$\begin{cases} x(t) = kt - \sin kt \\ y(t) = 1 - \cos kt. \end{cases}$$

The resulting motion is shown in Figure 4.7.   ∎

The curve specified in Example 5 is called a **cycloid.** Note that, using the equation $x(t) = kt - \sin kt$, we have

$$x'(t) = k - k\cos kt = k(1 - \cos kt) > 0$$

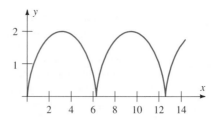

Figure 4.7   The path of the pebble in Example 5 is a cycloid.

except when kt is an integer multiple of 2π. Therefore, the function $x(t)$ is increasing for all real numbers t. In other words, the pebble is always moving forwards *even during those times when it is located in the back half of the tire*.

Now that we have studied parametric equations somewhat, we pause to sharpen our use of the terminology **curve.** In general, a curve C in the plane is the set of all points of the form

$$C = \{(x(t), y(t)) \mid t \in I\} \tag{6}$$

where I is some interval and the coordinate functions x and y are **continuous** functions of $t \in I$.

The requirement that x and y be continuous ensures that the curve will itself be continuous (unbroken) in the sense of the graph of a continuous function $y = f(x)$. The curve C in (6) is called **smooth** if the derivatives x' and y' are continuous functions of $t \in I$ and $[x'(t)]^2 + [y'(t)]^2 \neq 0$ for all $t \in I$. The reason for this last condition is explained below.

Equations of Tangents

Suppose that the curve C is given in the parametric form of equation (6) and that we wish to find the slope of the line tangent to the curve at the point $P(t_0) = (x(t_0), y(t_0))$. (For example, the curve C might be the trajectory of a rocket, and we might wish to describe its angle of elevation as a function of time.) To do so, we begin by letting h be a small real number. Then $P(t_0 + h) = (x(t_0 + h), y(t_0 + h))$ is, in general, a second point on C (see Figure 4.8). If $x(t_0) \neq x(t_0 + h)$, we can write the slope of the secant through $P(t_0)$ and $P(t_0 + h)$ as

$$\text{Slope of secant} = \frac{y(t_0 + h) - y(t_0)}{x(t_0 + h) - x(t_0)}. \tag{7}$$

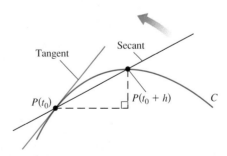

Figure 4.8   Slope of tangent is limit of slopes of secants.

the limit of this slope as h tends to zero is the number m, the desired slope of the tangent at $P(t_0)$. By dividing both numerator and denominator in line (7) by $h \neq 0$, we find that

$$m = \lim_{h \to 0} \frac{\left(\dfrac{y(t_0 + h) - y(t_0)}{h} \right)}{\left(\dfrac{x(t_0 + h) - x(t_0)}{h} \right)} = \frac{y'(t_0)}{x'(t_0)}. \tag{8}$$

Obviously, m in equation (8) makes sense only if both $y'(t_0)$ and $x'(t_0)$ exist and $x'(t_0) \neq 0$. Also, recall that we needed to assume that $x(t_0 + h) \neq x(t_0)$. However, this assumption will be valid for all h near zero whenever $x'(t_0)$ exists and is not zero (see Exercise 49). Thus, we have shown that *if $x'(t_0)$ and $y'(t_0)$ exist, and if $x'(t_0) \neq 0$, the slope of the line tangent to the graph of the curve C at $P(t_0) = (x(t_0), y(t_0))$ is*

$$\boxed{m = \frac{y'(t_0)}{x'(t_0)}, \qquad x'(t_0) \neq 0.} \tag{9}$$

The equation for this tangent line is therefore

$$y - y(t_0) = \frac{y'(t_0)}{x'(t_0)}(x - x(t_0))$$

or

$$\boxed{x'(t_0)[y - y(t_0)] - y'(t_0)[x - x(t_0)] = 0.} \tag{10}$$

If $x'(t_0) = 0$ and $y'(t_0) \neq 0$, equation (10) reduces to the equation $x = x(t_0)$ so the tangent is a vertical line through $(x(t_0), y(t_0))$. If both $x'(t_0) = 0$ and $y'(t_0) = 0$, no conclusion can be drawn about the tangent at $(x(t_0), y(t_0))$. This lack of a well-defined tangent explains why we require $[x'(t)]^2 + [y'(t)]^2 \neq 0$ in the definition of a smooth curve.

Of course, the condition $[x'(t)]^2 + [y'(t)]^2 \neq 0$ is simply another way to say that at least one of the derivatives $x'(t)$ and $y'(t)$ is nonzero. It is interesting to return to the parametric equations for the cycloid derived in Example 5 to see how this condition relates to the shape of the curve. In this example,

$$\begin{cases} x(t) = kt - \sin kt \\ y(t) = 1 - \cos kt, \end{cases}$$

so we have

$$\begin{cases} x'(t) = k(1 - \cos kt) \\ y'(t) = k \sin kt. \end{cases}$$

Thus, $[x'(t)]^2 + [y'(t)]^2 = 0$ only if kt is an integer multiple of 2π. Those values of t correspond to the times when the pebble touches the road. Figure 4.7 illustrates the lack of a well-defined tangent at those times.

□ **EXAMPLE 6**

Find the equation of the line tangent to the curve given by the parametric equations

$$\begin{cases} x(t) = 2 \cos t \\ y(t) = 4 \sin t \end{cases} \quad 0 \le t < 2\pi$$

at the point $(\sqrt{2}, 2\sqrt{2})$.

Strategy · · · · · · · · ·

Find t_0.

Solution

To find the number t_0 for which $P(t_0) = (\sqrt{2}, 2\sqrt{2})$, we set

$$x(t_0) = 2 \cos t_0 = \sqrt{2},$$
$$y(t_0) = 4 \sin t_0 = 2\sqrt{2}.$$

This gives $\cos t_0 = \sqrt{2}/2$ and $\sin t_0 = \sqrt{2}/2$, so $t_0 = \pi/4$.

Find $x'(t_0), y'(t_0)$.

Then

$$x'(t_0) = -2 \sin(\pi/4) = -\sqrt{2}$$

and

$$y'(t_0) = 4 \cos(\pi/4) = 2\sqrt{2}.$$

Use equation (9) to determine the slope.

Thus, the slope m is

$$\frac{2\sqrt{2}}{-\sqrt{2}} = -2.$$

An equation for the tangent line is

$$y - 2\sqrt{2} = -2(x - \sqrt{2}),$$

or

$$y = -2x + 4\sqrt{2}.$$

(See Figure 4.9.)

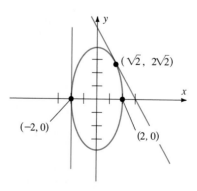

Figure 4.9 Tangents to curve in Examples 6 and 7.

□ **EXAMPLE 7**

Determine the points for which the curve in Example 6 has vertical tangents.

Strategy · · · · · · · ·
Find t_0 so that $x'(t_0) = 0$.

Solution
Since $x'(t) = -2 \sin t$, the equation

$$x'(t) = -2 \sin t = 0$$

has solutions $t_0 = 0$, π for $0 \leq t < 2\pi$.

Determine if $y'(t_0) \neq 0$ for each such t_0.

Since

$$y'(0) = 4 \cos 0 = 4 \neq 0 \quad \text{and}$$
$$y'(\pi) = 4 \cos \pi = -4 \neq 0,$$

If so, the points

$$P(t_0) = (x(t_0), y(t_0))$$

yield vertical tangents.

the points

$$P(0) = (2 \cos 0, 4 \sin 0) = (2, 0)$$

and

$$P(\pi) = (2 \cos \pi, 4 \sin \pi) = (-2, 0)$$

yield vertical tangents (see Figure 4.9). ■

REMARK   The graph $y = f(x)$ of a continuous function f for $x \in [a, b]$ can always be parameterized in a straightforward way: simply let $x = t$. Then the parametric equations

$$\begin{cases} x(t) = t \\ y(t) = f(t) \end{cases} \qquad a \leq t \leq b$$

describe the graph. Moreover, if f' exists and is continuous on (a, b), then the curve is smooth.

This simple observation tells us that any statement or property regarding parametric equations generalizes some fact about graphs of the form $y = f(x)$. For instance, if $x = t$ and $y = f(t)$, equation (9) implies $m = f'(t)/1$, which is the standard interpretation of the derivative as the slope of the tangent.

Tangents to Polar Curves

If a curve in the plane is the graph of a polar equation of the form $r = f(\theta)$, we can use the equations

$$x = r \cos \theta \tag{11a}$$

$$y = r \sin \theta \tag{11b}$$

to obtain parametric equations for the curve. Multiplying the equation $r = f(\theta)$ on both sides by $\cos \theta$ gives the equation $r \cos \theta = f(\theta) \cos \theta$. Using (11a), we obtain

$$x(\theta) = f(\theta) \cos \theta. \tag{12a}$$

Similarly, multiplying both sides of $r = f(\theta)$ by $\sin \theta$ and using (11b), we obtain

$$y(\theta) = f(\theta) \sin \theta. \tag{12b}$$

If the derivative $f'(\theta) = dr/d\theta$ exists, we may differentiate both sides of equation (12a) and use equation (11b) to conclude that

$$x'(\theta) = f'(\theta)\cos\theta - f(\theta)\sin\theta \tag{13a}$$

$$= \frac{dr}{d\theta}\cos\theta - r\sin\theta.$$

Similarly, differentiating both sides of (12b) and using (11a), we obtain

$$y'(\theta) = f'(\theta)\sin\theta + f(\theta)\cos\theta \tag{13b}$$

$$= \frac{dr}{d\theta}\sin\theta + r\cos\theta.$$

Finally, we may apply equation (9) (with parameter θ rather than t) and equations (13a) and (13b) to conclude that *the slope of the line tangent to the graph of the polar equation $r = f(\theta)$ at the point (r, θ) is*

$$m = \frac{\dfrac{dr}{d\theta}\sin\theta + r\cos\theta}{\dfrac{dr}{d\theta}\cos\theta - r\sin\theta}. \tag{14}$$

As noted for equation (9), m in equation (14) is defined only if

$$x'(\theta) = \frac{dr}{d\theta}\cos\theta - r\sin\theta \neq 0.$$

If $x'(\theta) = 0$ and $y'(\theta) \neq 0$, the graph has a vertical tangent at (r, θ). If both $x'(\theta) = 0$ and $y'(\theta) = 0$, no conclusions may be drawn.

□ **EXAMPLE 8**

Find the points where the four-leaved rose with equation

$$r = \sin 2\theta, \qquad 0 \le \theta < 2\pi$$

has vertical tangents.

Strategy · · · · · · · ·
Find a parameterization for the curve.

Solution

By equations (12a) and (12b), a parameterization for $f(\theta) = \sin 2\theta = 2\sin\theta\cos\theta$ is

$$\begin{cases} x(\theta) = 2\sin\theta\cos^2\theta \\ y(\theta) = 2\sin^2\theta\cos\theta. \end{cases}$$

Find the angles θ for which $x'(\theta) = 0$.

Thus,

$$x'(\theta) = 2\cos^3\theta - 4\sin^2\theta\cos\theta = 0$$

implies

$$\cos\theta(\cos^2\theta - 2\sin^2\theta) = 0.$$

Thus, $x'(\theta) = 0$ whenever

$$\cos\theta = 0 \quad \text{or} \quad \cos^2\theta = 2\sin^2\theta.$$

The equation $\cos \theta = 0$ has solutions $\theta = \pi/2$ and $\theta = 3\pi/2$ in the interval $[0, 2\pi)$. The equation $\cos^2 \theta = 2 \sin^2 \theta$ gives $\tan^2 \theta = \frac{1}{2}$ or $\tan \theta = \pm\sqrt{2}/2$, which has four solutions in the interval $[0, 2\pi)$, $\theta = \pm\text{Tan}^{-1}(\pm\sqrt{2}/2)$. The six solutions of $x'(\theta) = 0$ are therefore $\theta = \pi/2$, $\theta = 3\pi/2$, and $\theta = \pm\text{Tan}^{-1}(\pm\sqrt{2}/2)$.

Find $y'(\theta)$.

To determine which of these yield vertical tangents we must examine

$$y'(\theta) = 4 \sin \theta \cos^2 \theta - 2 \sin^3 \theta$$
$$= 2 \sin \theta[2 \cos^2 \theta - \sin^2 \theta].$$

Determine if $y'(\theta) \neq 0$ for each θ with $x'(\theta) = 0$.

If $\theta = \pi/2$ or $3\pi/2$, then

$$y'(\theta) = -2 \sin^3 \theta \neq 0.$$

Similarly, if $\theta = \pm\text{Tan}^{-1}\left(\pm\dfrac{\sqrt{2}}{2}\right)$, then

$$y'(\theta) = 2\left(\pm\frac{\sqrt{3}}{3}\right)\left[2 \cdot \frac{2}{3} - \frac{1}{3}\right] \neq 0.$$

If so, (r, θ) is a point that yields a vertical tangent. (See Figure 4.10.)

Thus, all six values of θ yield vertical tangents. (Note in Figure 4.10 that both $\theta = \pi/2$ and $\theta = 3\pi/2$ correspond to the origin.) ∎

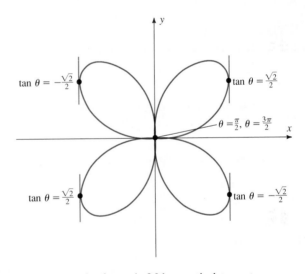

Figure 4.10 Points where graph of $r = \sin 2\theta$ has vertical tangents.

Exercise Set 15.4

In Exercises 1–12, sketch the curve described by the given parametric equations and find an equation in xy-coordinates whose graph includes the given curve.

1. $\begin{cases} x(t) = t \\ y(t) = t + 6 \end{cases}$

2. $\begin{cases} x(t) = t + 2 \\ y(t) = 3 - t \end{cases}$

3. $\begin{cases} x(t) = 1 + 3t \\ y(t) = 2t + 2 \end{cases}$

4. $\begin{cases} x(t) = t \\ y(t) = t^2 + 2 \end{cases}$

5. $\begin{cases} x(t) = t^2 + 1 \\ y(t) = t^2 - 1 \end{cases}$

6. $\begin{cases} x(t) = \sin t \\ y(t) = \cos^2 t \end{cases}$

7. $\begin{cases} x(t) = e^t \\ y(t) = e^{3t} \end{cases}$

8. $\begin{cases} x(t) = \sqrt{t} \\ y(t) = 3t - 2 \end{cases}$

9. $\begin{cases} x(t) = \sec t \\ y(t) = \tan t \end{cases}$

10. $\begin{cases} x(t) = 3 \sin t \\ y(t) = 5 \cos t \end{cases}$

11. $\begin{cases} x(t) = \sin t \\ y(t) = \sin 2t \end{cases}$

12. $\begin{cases} x(t) = \cos t \\ y(t) = \sec t \end{cases}$

In Exercises 13–24, find both the slope (if defined) of the line tangent to the curve at the indicated point and an equation for the tangent line.

13. $\begin{cases} x(t) = t \\ y(t) = t^2 + 1 \end{cases}$
$t = 2$

14. $\begin{cases} x(t) = t^2 \\ y(t) = 1 - t \end{cases}$
$t = 2$

15. $\begin{cases} x(t) = \sin t \\ y(t) = \cos t \end{cases}$
$t = \pi/3$

16. $\begin{cases} x(t) = \cos t \\ y(t) = \sin t \end{cases}$
$t = \pi/4$

17. $\begin{cases} x(t) = 1/t \\ y(t) = 3t^2 - 7 \end{cases}$
$t = 2$

18. $\begin{cases} x(t) = 3t^3 \\ y(t) = \sin \pi t \end{cases}$
$t = 1$

19. $\begin{cases} x(t) = 1 + \sqrt{t} \\ y(t) = 1 - \sqrt{t} \end{cases}$
$t = 4$

20. $\begin{cases} x(t) = \sec t \\ y(t) = \tan t \end{cases}$
$t = \pi/4$

21. $\begin{cases} x(t) = t^3 \\ y(t) = t^2 \end{cases}$
$(x, y) = (8, 4)$

22. $\begin{cases} x(t) = t^2 + 2t - 3 \\ y(t) = 1 - t \end{cases}$
$(x, y) = (0, 4)$

23. $\begin{cases} x(t) = e^t \\ y(t) = \ln t^2 \end{cases}$
$(x, y) = (e, 0)$

24. $\begin{cases} x(t) = (1 - 2\cos t)\cos t \\ y(t) = (1 - 2\cos t)\sin t \end{cases}$
$(x, y) = (-3, 0)$

In Exercises 25–30, find the slope, if defined, of the line tangent to the graph of the polar equation at the point corresponding to the given value of θ.

25. $r = \cos \theta$
$\theta = \pi/4$

26. $r = 1 + \cos \theta$
$\theta = \pi/6$

27. $r = a \sin 2\theta$
$\theta = \pi/6$

28. $r = 2 \sin \theta$
$\theta = \pi/4$

29. $r = \theta$
$\theta = \pi/2$

30. $r = a \sin 3\theta$
$\theta = \pi/3$

In Exercises 31–36, find all points at which the curve described by the parametric equations has **(a)** a vertical tangent, **(b)** a horizontal tangent.

31. $\begin{cases} x(t) = \cos t \\ y(t) = \sin t \end{cases}$

32. $\begin{cases} x(t) = \sin 2t \\ y(t) = \sin t \end{cases}$

33. $\begin{cases} x(t) = t^2 + 4 \\ y(t) = 3t^2 - 6t + 2 \end{cases}$

34. $\begin{cases} x(t) = 5 + 2\sin t \\ y(t) = 3 - \cos t \end{cases}$

35. $\begin{cases} x(t) = 3t^2 + 6 \\ y(t) = t - t^2 \end{cases}$

36. $\begin{cases} x(t) = t^{3/2} \\ y(t) = t + \cos t \end{cases}$

37. Find the point on the curve

$$C_1: \begin{cases} x(t) = t \\ y(t) = 2t^2 + 3 \end{cases}$$

where the tangent is parallel to the line

$$C_2: \begin{cases} x(t) = t + 3 \\ y(t) = 4t - 10. \end{cases}$$

38. Find parametric equations for the cardioid $r = 1 + \cos \theta$.

39. Find parametric equations for the three-leaved rose $r = a \sin 3\theta$.

40. Find parametric equations for the spiral $r = \theta$.

41. Find the points at which the cardioid $r = 1 + \sin \theta$ has vertical tangents.

42. Find the points at which the cardioid $r = 1 + \cos \theta$ has horizontal tangents.

43. Find the points at which the cardioid $r = 1 + \cos \theta$ has vertical tangents.

44. Show that the curve determined by the parametric equations

$$\begin{cases} x(t) = a \cos t + h \\ y(t) = b \sin t + k \end{cases}$$

is an ellipse with center at (h, k).

In Exercises 45–48, a particle moves about the plane so that at time t its position is $(x_1(t), y_1(t))$. A second particle also moves in the plane so that at time t its position is $(x_2(t), y_2(t))$.
a. Find an equation in x and y alone for each of the curves.
b. Do the paths of the particles cross? If so, where?
c. Do the particles collide? If so, where and at what time?

45. $x_1(t) = t + 4$, $\quad y_1(t) = 8 - t^2$
$x_2(t) = t + 4$, $\quad y_2(t) = t + 6$

46. $x_1(t) = t$, $\quad y_1(t) = t^2$
$x_2(t) = 2t$, $\quad y_2(t) = 2t + 2$

47. $x_1(t) = \cos t$, $\quad y_1(t) = \sin t$
$x_2(t) = 0$, $\quad y_2(t) = \sin 2t$

48. $x_1(t) = \cos 2t \cos t$, $\quad y_1(t) = \cos 2t \sin t$
$x_2(t) = \cos t$, $\quad y_2(t) = \sin t$
(*Hint:* Consider $(x_1(t), y_1(t))$ in terms of polar coordinates and recall Exercise 52 in Section 15.1.)

49. Prove that if $x(t)$ is differentiable at t_0 and if $x'(t_0) \neq 0$, then $x(t_0 + h) \neq x(t_0)$ for all sufficiently small h.

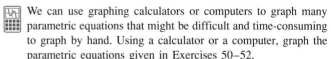 We can use graphing calculators or computers to graph many parametric equations that might be difficult and time-consuming to graph by hand. Using a calculator or a computer, graph the parametric equations given in Exercises 50–52.

50. $\begin{cases} x(t) = 2 \cos t + \cos 2t \\ y(t) = 2 \sin t - \sin 2t \end{cases}$ $\quad 0 \leq t \leq 2\pi$

51. $\begin{cases} x(t) = 3 \cos t + \cos 3t \\ y(t) = 3 \sin t - \sin 3t \end{cases}$ $\quad 0 \leq t \leq 2\pi$

52. $\begin{cases} x(t) = \cos t \sin 8t \\ y(t) = \sin t \sin 8t \end{cases}$ $0 \le t \le 2\pi$

53. Consider the curve $y^2 = x^3 + x^2$ in the plane.
 a. Show that the equations
$$\begin{cases} x(t) = t^2 - 1 \\ y(t) = t(t^2 - 1) \end{cases}$$
 parameterize this curve.
 b. As in Exercises 50–52, use a calculator or a computer to graph this curve. The result is Figure 4.1.

In Exercises 54–61, graph the parameterized curve. Then identify it.

54. $\begin{cases} x(t) = 2 \sin t \\ y(t) = 2 \cos t \end{cases}$ $0 \le t \le 2\pi$

55. $\begin{cases} x(t) = 1 + \sqrt{t} \\ y(t) = 2 - \sqrt{t} \end{cases}$ $0 \le t \le 36$

56. $\begin{cases} x(t) = \sec t \\ y(t) = \cos t \end{cases}$ $0 \le t \le 2\pi$

57. $\begin{cases} x(t) = \sin t \\ y(t) = \csc t \end{cases}$ $0 \le t \le 2\pi$

58. $\begin{cases} x(t) = 3 \sin t \\ y(t) = 8 \cos t \end{cases}$ $0 \le t \le 2\pi$

59. $\begin{cases} x(t) = 2t/(1 + t^2) \\ y(t) = (1 - t^2)/(1 + t^2) \end{cases}$ $-10 \le t \le 10$

60. $\begin{cases} x(t) = \sin t \\ y(t) = 2 \cos 2t \end{cases}$ $-\pi \le t \le \pi$

61. $\begin{cases} x(t) = 5 \cot t \\ y(t) = 5 \sin^2 t \end{cases}$ $0 \le t \le \pi$

In Exercises 62–65, graph the given parametric curve.

62. $\begin{cases} x(t) = 3t - 5 \sin t \\ y(t) = 3 - 5 \cos t \end{cases}$ $-10 \le t \le 10$,   (Prolate Cycloid).

63. $\begin{cases} x(t) = 3.5 \cos t + 0.5 \cos 7t \\ y(t) = 3.5 \sin t - 0.5 \sin 7t \end{cases}$ $0 \le t \le 2\pi$,   (Hypocycloid).

64. $\begin{cases} x(t) = 3t/(1 + t^3) \\ y(t) = 3t^2/(1 + t^3) \end{cases}$ $-10 \le t \le 10$,   (Folium of Descartes).

65. $\begin{cases} x(t) = -3t(4 - t^6) \\ y(t) = -3t^2(4 - t^6) \end{cases}$ $-2 \le t \le 2$,   (Bifolium).

In computer-aided design special parametric curves called **Bezier curves** are used extensively. A cubic Bezier curve is determined by an ordered set of four points in the plane, the control points of the curve. That is, given four points $P_1 = (x_1, y_1)$, $P_2 = (x_2, y_2)$, $P_3 = (x_3, y_3)$, and $P_4 = (x_4, y_4)$, the cubic Bezier curve is defined by the parametric equations
$$\begin{cases} x = x_1(1 - t)^3 + x_2(3t(1 - t)^2) + x_3(3t^2(1 - t)) + x_4t^3 \\ y = y_1(1 - t)^3 + y_2(3t(1 - t)^2) + y_3(3t^2(1 - t)) + y_4t^3. \end{cases}$$

By carefully choosing the control points, we can draw many different shapes. For example, using the set of control points
$$\{P_1, P_2, P_3, P_4\} = \{(30, 10), (90, 60), (10, 60), (60, 10)\},$$
we obtain the Bezier curve shown in Figure 4.11.

Figure 4.11

Similarly, using the three sets of control points
$$\{(30, 50), (60, 50), (60, 30), (30, 30)\}$$
$$\{(30, 30), (60, 30), (60, 10), (30, 10)\}$$
$$\{(30, 50), (30, 40), (30, 20), (30, 10)\}$$
we can draw curves that approximate a letter (see Figure 4.12).

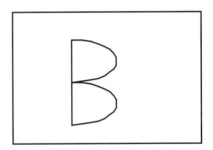

Figure 4.12

In Exercises 66–71, graph the Bezier curves that correspond to the given set(s) of control points.

66. $\{(20, 10), (20, 60), (80, 60), (80, 10)\}$

67. $\{(20, 10), (80, 60), (20, 60), (80, 10)\}$

68. $\{(50, 60), (20, 30), (60, 60), (50, 40)\}$

69. $\{(50, 60), (80, 20), (20, 20), (50, 60)\}$

70. $\{(50, 35), (20, 10), (20, 60), (50, 35)\}$
 $\{(50, 35), (80, 60), (80, 20), (50, 35)\}$

71. $\{(20, 20), (25, 30), (23, 40), (20, 45)\}$
 $\{(22, 40), (30, 50), (40, 40), (40, 20)\}$
 $\{(35, 40), (45, 50), (50, 40), (55, 20)\}$

15.5 Arc Length and Surface Area Revisited

In Section 7.3 we developed formulas for arc length and surface area calculations associated with the graph of a differentiable function f. Since polar equations and parametric equations often describe curves that are not graphs of functions, we should ask whether the concepts of arc length and surface area can be extended to general curves. In both cases the answer is yes, and the development here parallels that of Section 7.3 very closely.

Arc Length

Suppose that C is a curve in the plane that is determined by the parametric equations

$$C: \begin{cases} x = x(t) \\ y = y(t) \end{cases} \quad t \in I,$$

where I is some interval. Suppose further that a and b are numbers in I, with $a < b$, so that the arc of the curve from $P = (x(a), y(a))$ to $Q = (x(b), y(b))$ does not intersect itself, except possibly if $P = Q$. The problem is to define and calculate the length L of the arc of C connecting P and Q (see Figure 5.1).

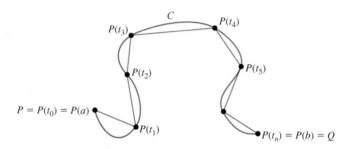

Figure 5.1 The polygonal path joining points $P(t_j) = (x(t_j), y(t_j))$ approximates C.

As in Section 7.3, we begin by partitioning the interval $[a, b]$ for the independent variable (parameter) t into n subintervals of equal length $\Delta t = (b - a)/n$ with endpoints

$$a = t_0 < t_1 < t_2 < \cdots < t_n = b.$$

For each integer $j = 0, 1, 2, \ldots, n$, we let $P(t_j) = (x(t_j), y(t_j))$. Then each $P(t_j)$ is a point on the arc of C joining P and Q. In each interval $[t_{j-1}, t_j]$, we use the length of the line segment from $P(t_{j-1})$ to $P(t_j)$ to approximate the length of the arc C_j connecting these two points. Using the distance formula, we can write this distance as

$$\Delta L_j = \sqrt{\Delta x_j^2 + \Delta y_j^2}$$
$$= \sqrt{[x(t_j) - x(t_{j-1})]^2 + [y(t_j) - y(t_{j-1})]^2}, \quad j = 1, 2, \ldots, n. \quad (1)$$

(See Figure 5.2.)

If x is a differentiable function of t in each interval $[t_{j-1}, t_j]$, we may apply the Mean Value Theorem to conclude that there exists a number $c_j \in [t_{j-1}, t_j]$ so that

$$x'(c_j) = \frac{x(t_j) - x(t_{j-1})}{t_j - t_{j-1}} = \frac{x(t_j) - x(t_{j-1})}{\Delta t}.$$

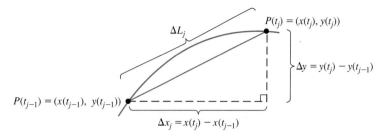

Figure 5.2 ΔL_j approximates length of arc from $P(t_{j-1})$ to $P(t_j)$.

Thus, we can write

$$x(t_j) - x(t_{j-1}) = x'(c_j)\,\Delta t, \qquad j = 1, 2, \ldots, n. \tag{2}$$

Similarly, if y is a differentiable function of t, there exist numbers $d_j \in [t_{j-1}, t_j]$ so that

$$y(t_j) - y(t_{j-1}) = y'(d_j)\,\Delta t, \qquad j = 1, 2, \ldots, n. \tag{3}$$

Combining equations (1)–(3), we obtain the length of the approximating line segment as

$$\Delta L_j = \sqrt{[x'(c_j)\,\Delta t]^2 + [y'(d_j)\,\Delta t]^2} \tag{4}$$

$$= \sqrt{[x'(c_j)]^2 + [y'(d_j)]^2}\,\Delta t, \qquad j = 1, 2, \ldots, n.$$

Finally, we approximate the length L of the arc of C from P to Q by the length of the polygonal path connecting the points $P(t_0), P(t_1), \ldots, P(t_n)$. The latter is simply the sum of the lengths ΔL_j in line (4). We obtain the approximation

$$L \approx \sum_{j=1}^{n} \Delta L_j = \sum_{j=1}^{n} \sqrt{[x'(c_j)]^2 + [y'(d_j)]^2}\,\Delta t. \tag{5}$$

The expression on the right-hand side of equation (5) is "almost" a Riemann sum. The difficulty is that the functions x' and y' are being evaluated at two (possibly) different points in each subinterval. However, in more advanced courses, it is shown that, if both x' and y' are continuous on $[a, b]$, this difficulty can be overcome and that as $n \to \infty$ the approximating sum converges to the integral

$$L = \int_a^b \sqrt{[x'(t)]^2 + [y'(t)]^2}\,dt. \tag{6}$$

As in Section 7.3, we argue that as $n \to \infty$ the polygonal paths more closely approximate the curve C, so that the limiting value of the length of the polygonal paths provides a reasonable definition of the length of the arc. It is important to keep in mind that *both x' and y' must be continuous on $[a, b]$ for formula (6) to be valid.*

☐ **EXAMPLE 1**

Find the length of the arc of the curve

$$C: \begin{cases} x(t) = \cos t + t \sin t \\ y(t) = \sin t - t \cos t \end{cases}$$

connecting the points $P = (x(0), y(0)) = (1, 0)$ and $Q = (x(\pi/2), y(\pi/2)) = (\pi/2, 1)$.

Solution Here

$$x'(t) = -\sin t + \sin t + t \cos t = t \cos t,$$
$$y'(t) = \cos t - \cos t + t \sin t = t \sin t.$$

By formula (6)

$$L = \int_0^{\pi/2} \sqrt{(t \cos t)^2 + (t \sin t)^2} \, dt$$

$$= \int_0^{\pi/2} \sqrt{t^2(\cos^2 t + \sin^2 t)} \, dt$$

$$= \int_0^{\pi/2} t \, dt = \left[\frac{t^2}{2}\right]_0^{\pi/2} = \frac{\pi^2}{8}. \qquad \blacksquare$$

Arc Length in Polar Coordinates

In Section 15.4 we showed that the graph of the polar equation $r = f(\theta)$ can be parameterized, using θ as the parameter, as

$$\begin{cases} x(\theta) = f(\theta) \cos \theta, \\ y(\theta) = f(\theta) \sin \theta. \end{cases}$$

If $f'(\theta)$ is continuous, so are the functions

$$\begin{cases} x'(\theta) = f'(\theta) \cos \theta - f(\theta) \sin \theta, \\ y'(\theta) = f'(\theta) \sin \theta + f(\theta) \cos \theta, \end{cases}$$

so we may apply equation (6) to determine a formula for arc length in polar coordinates. Before doing so we note that

$$[x'(\theta)]^2 = [f'(\theta)]^2 \cos^2 \theta - 2f'(\theta)f(\theta) \sin \theta \cos \theta + [f(\theta)]^2 \sin^2 \theta,$$
$$[y'(\theta)]^2 = [f'(\theta)]^2 \sin^2 \theta + 2f'(\theta)f(\theta) \sin \theta \cos \theta + [f(\theta)]^2 \cos^2 \theta,$$

so

$$[x'(\theta)]^2 + [y'(\theta)]^2 = [f'(\theta)]^2 + [f(\theta)]^2. \tag{7}$$

From equations (6) and (7) it follows that

$$L = \int_a^b \sqrt{[f'(\theta)]^2 + [f(\theta)]^2} \, d\theta \tag{8}$$

gives the length of the arc of the graph of $r = f(\theta)$ from $\theta = a$ to $\theta = b$.

☐ **EXAMPLE 2**

Find the length of the cardioid $r = 1 + \cos \theta$.

Strategy · · · · · · ·

Determine the limits of integration.

Verify that equation (8) applies.

Apply (8).

Solution

Here $f(\theta) = r = 1 + \cos \theta$, and the cardioid is swept out as θ makes one complete revolution. Appropriate limits of integration are therefore $\theta = 0$ to $\theta = 2\pi$.

Since $f'(\theta) = -\sin \theta$ is continuous on $[0, 2\pi]$, we may apply formula (8):

$$L = \int_0^{2\pi} \sqrt{(1 + \cos \theta)^2 + (-\sin \theta)^2} \, d\theta$$

$$= \int_0^{2\pi} \sqrt{(1 + 2 \cos \theta + \cos^2 \theta) + \sin^2 \theta} \, d\theta$$

$$= \int_0^{2\pi} \sqrt{2 + 2 \cos \theta} \, d\theta$$

Use identity

$$\cos^2 \phi = \frac{1}{2} + \frac{1}{2} \cos 2\phi$$

in reverse to simplify integrand.

Use the symmetry of the cosine function to handle the absolute value signs. (Alternatively, we could have used the symmetry of the cardioid and integrated over $[0, \pi]$.)

$$= \int_0^{2\pi} \sqrt{4\left(\frac{1}{2} + \frac{1}{2} \cos \theta\right)} \, d\theta$$

$$= \int_0^{2\pi} \sqrt{4 \cos^2\left(\frac{\theta}{2}\right)} \, d\theta$$

$$= \int_0^{2\pi} \left| 2 \cos\left(\frac{\theta}{2}\right) \right| \, d\theta$$

$$= 2 \int_0^{\pi} 2 \cos\left(\frac{\theta}{2}\right) \, d\theta$$

$$= \left[8 \sin\left(\frac{\theta}{2}\right) \right]_0^{\pi} = 8. \qquad ■$$

Surface Area

Given a differentiable function y such that $y(t) > 0$ for $a \le t \le b$, suppose that the region bounded by the smooth curve C with equations

$$C: \begin{cases} x = x(t) \\ y = y(t) \end{cases} \qquad a \le t \le b$$

and the x-axis is to be revolved about the x-axis. In order to calculate the surface area of the resulting solid we must require that $x'(t) \ne 0$ for all $t \in [a, b]$. (This last condition ensures that the curve will have at most one point with any given x-coordinate (see Exercise 39).)

We begin with the usual step of partitioning the interval $[a, b]$ into n equal subintervals of length $\Delta t = (b - a)/n$ with endpoints

$$a = t_0 < t_1 < t_2 < \cdots < t_n = b.$$

Let $P(t_j) = (x(t_j), y(t_j))$. We approximate C by the polygonal path joining these consecutive points on C (see Figure 5.3). As each of these line segments is revolved about the x-axis, it generates the frustum of a cone, as in Figure 5.4. The radii of the jth frustum are $y(t_{j-1})$ and $y(t_j)$, so the lateral surface area of the jth frustum is

$$\Delta S_j = \pi[y(t_{j-1}) + y(t_j)] \, \Delta L_j, \qquad j = 1, 2, \ldots, n$$

where

$$\Delta L_j = \sqrt{(\Delta x_j)^2 + (\Delta y_j)^2}$$

$$= \sqrt{[x(t_j) - x(t_{j-1})]^2 + [y(t_j) - y(t_{j-1})]^2}. \qquad \text{(See Figures 5.2 and 5.4.)}$$

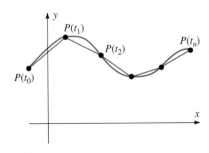

Figure 5.3 Curve to be revolved about the x-axis.

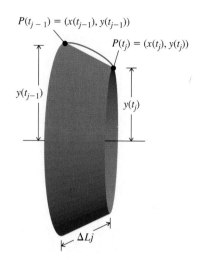

As in our development of the formula for arc length, we obtain the following approximation to the lateral surface area S of the volume of revolution:

$$S \approx \sum_{j=1}^{n} \Delta S_j = \sum_{j=1}^{n} \pi[y(t_{j-1}) + y(t_j)]\sqrt{[x'(c_j)]^2 + [y'(d_j)]^2}\, \Delta t$$

which converges to the definite integral

$$S = \int_{a}^{b} 2\pi y(t)\sqrt{[x'(t)]^2 + [y'(t)]^2}\, dt. \tag{9}$$

Figure 5.4 The frustum generated by revolving the line segment $\overline{P(t_{j-1})P(t_j)}$ about the x-axis. (See Section 7.3 for the area formula for a frustum of a cone.)

□ **EXAMPLE 3**

In Example 4, Section 7.3, we showed that the surface area of the band of the sphere obtained by revolving the arc of the circle $x^2 + y^2 = r^2$ lying above the interval $[-a, a]$, $a < r$, about the x-axis is $S = 4\pi ar$ (Figure 5.5). Moreover, we noted in Example 5 of that section that we could not conclude from this calculation that the surface area of a sphere is $S = 4\pi r^2$.

We may now use equation (9) to demonstrate that the surface area of the sphere of radius r is indeed $S = 4\pi r^2$ (Figure 5.6). We parameterize the upper semicircle by

$$\begin{cases} x(t) = r\cos t \\ y(t) = r\sin t \end{cases} \quad 0 \le t \le \pi.$$

Then

$$x'(t) = -r\sin t; \qquad y'(t) = r\cos t.$$

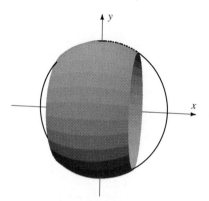

Figure 5.5 The surface area of a band of a sphere is $4\pi ar$.

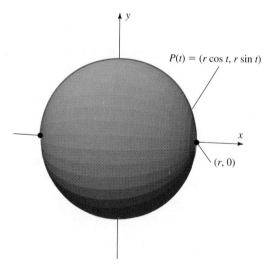

Figure 5.6 A sphere of radius r obtained by revolving about the x-axis the upper semicircle of $x^2 + y^2 = r^2$. Its surface area is $4\pi r^2$.

Both of these derivatives are continuous for all t, in contrast to dy/dx (which is undefined at $x = r$ and $x = -r$). Thus,

$$S = \int_0^{\pi} 2\pi(r \sin t)\sqrt{(-r \sin t)^2 + (r \cos t)^2}\, dt$$

$$= \int_0^{\pi} 2\pi r^2 \sin t\, dt$$

$$= \left[-2\pi r^2 \cos t\right]_0^{\pi} = -2\pi r^2(-1 - 1) = 4\pi r^2. \qquad \blacksquare$$

Exercise Set 15.5

In Exercises 1–15, a curve is described by a pair of parametric equations. Find the length of the given curve.

1. $\begin{cases} x(t) = t \\ y(t) = t^{3/2} \end{cases} \quad 0 \le t \le 4/9$

2. $\begin{cases} x(t) = t + 1 \\ y(t) = \frac{2}{3}t^{3/2} \end{cases} \quad 0 \le t \le 3$

3. $\begin{cases} x(t) = 3t^2 \\ y(t) = 2t^3 \end{cases} \quad 0 \le t \le \sqrt{2}$

4. $\begin{cases} x(t) = \sin t \\ y(t) = \cos t \end{cases} \quad 0 \le t \le 2\pi$

5. $\begin{cases} x(t) = 2(2t + 3)^{3/2} \\ y(t) = 3(t + 1)^2 \end{cases} \quad 0 \le t \le 2$

6. $\begin{cases} x(t) = 2(1 - \cos t) \\ y(t) = 2 \sin t \end{cases} \quad 0 \le t \le \pi$

7. $\begin{cases} x(t) = \cos^3 t \\ y(t) = \sin^3 t \end{cases} \quad 0 \le t \le \pi$

8. $\begin{cases} x(t) = 3t^3 \\ y(t) = 3t^2 \end{cases} \quad 0 \le t \le \sqrt{5}$

9. $\begin{cases} x(t) = e^t \cos t \\ y(t) = e^t \sin t \end{cases} \quad 0 \le t \le \pi$

10. $\begin{cases} x(t) = \ln \cos t \\ y(t) = t \end{cases} \quad 0 \le t \le \pi/3$

11. $\begin{cases} x(t) = t^3 - 3t^2 \\ y(t) = 3t^2 \end{cases} \quad 0 \le t \le 1$

12. $\begin{cases} x(t) = t^2/2 \\ y(t) = \frac{1}{3}(2t + 1)^{3/2} \end{cases}$ from $(0, 1/3)$ to $(8, 9)$

13. $\begin{cases} x(t) = \frac{1}{3}(2t + 1)^{3/2} \\ y(t) = t \end{cases}$ from $(8/3, 3/2)$ to $(64/3, 15/2)$

14. $\begin{cases} x(t) = t^3 \\ y(t) = 2t^2 \end{cases}$ from $(0, 0)$ to $(8, 8)$

15. $\begin{cases} x(t) = \sqrt{2}t^2 \\ y(t) = \frac{2}{3}t^3 - t \end{cases}$ from $(0, 0)$ to $(2\sqrt{2}, \sqrt{2}/3)$

In Exercises 16–22, a curve is described by a pair of parametric equations. Find the surface area of the solid generated by revolving this curve about the x-axis.

16. $\begin{cases} x(t) = t \\ y(t) = t^2 \end{cases} \quad 0 \le t \le 1$

17. $\begin{cases} x(t) = t^2 \\ y(t) = 4t \end{cases} \quad 0 \le t \le 2$

18. $\begin{cases} x(t) = t \\ y(t) = \dfrac{t^4}{4} + \dfrac{1}{8t^2} \end{cases} \quad 1 \le t \le 2$

19. $\begin{cases} x(t) = t \\ y(t) = \sqrt{t} \end{cases} \quad 1 \le t \le 2$

20. $\begin{cases} x(t) = t \\ y(t) = (t^2 - 1)/2 \end{cases} \quad 0 \le t \le 1$

21. $\begin{cases} x(t) = 3t^2 \\ y(t) = 2t^3 \end{cases} \quad 1 \le t \le 2$

22. $\begin{cases} x(t) = \cos^3 t \\ y(t) = \sin^3 t \end{cases} \quad 0 \le t \le \pi$

23. Find the length of the cardioid $r = \cos^2(\theta/2)$.

24. Find the length of the cardioid $r = 1 + \cos \theta$.

25. Find the length of the spiral $r = e^{2\theta}, \quad 0 \le \theta \le \pi$.

26. Find the area of the surface generated if one arch of the cycloid $x(t) = t - \sin t$, $y(t) = 1 - \cos t$ is revolved about the x-axis.

27. Find the length of the circle $x(t) = 2 \cos t$, $y(t) = 2 \sin t$, $0 \le t \le 2\pi$.

28. Find the length of the spiral $r = 2\theta, \quad 0 \le \theta \le \pi$.

29. Using equation (8), show that the circumference of the circle $r = 3 \cos \theta$ is 3π.

30. Find the length of the cardioid $r = a(1 + \sin \theta)$.

31. Find the length of the graph of $r = e^{\theta}$ for $0 \le \theta \le 2\pi$.

32. Find the length of the graph of the equation $r = 4 \sec \theta$ for $0 \le \theta \le \pi/4$.

33. Find the surface area of the solid obtained by revolving about the x-axis the region bounded by the graph of $r = e^{\theta}$, $0 \le \theta \le \pi$, and the x-axis.

34. Find the area of the surface of the solid obtained by revolving about the line $\theta = 0$ the region bounded by the graph of $r = 4 \sin \theta$.

35. Suppose that the parametric equations $x = x(t)$, $y = y(t)$ describe a smooth curve C for $a \le t \le b$. Moreover, assume that $x(t) > 0$ and that $y'(t) > 0$ for $a \le t \le b$. Determine the formula for the surface area of the surface generated if C is revolved about the y-axis.

36. Use the result of Exercise 35 to find the surface area of the solid generated if the curve given in Exercise 17 is revolved about the y-axis.

37. Use the result of Exercise 35 to find the surface area of the solid obtained by revolving about the y-axis the curve in Exercise 19.

38. Use Simpson's Rule to approximate the length of the ellipse

$$\begin{cases} x(t) = 4 \cos t, \\ y(t) = 6 \sin t. \end{cases}$$

39. Show that the condition $x'(t) \ne 0$ is sufficient to guarantee that a differentiable curve C will have at most one point with any particular x-coordinate.

Summary Outline of Chapter 15

■ The **polar coordinates** (r, θ) for the point whose rectangular coordinates are (x, y) are determined by the equations $r = \sqrt{x^2 + y^2}$, $\tan \theta = y/x$, with $x = r \cos \theta$, $y = r \sin \theta$.   (page 722)

■ The following identities hold for polar coordinates: $(-r, \theta) = (r, \theta + \pi)$, $(r, \theta + 2n\pi) = (r, \theta)$, $n = \pm 1, \pm 2, \ldots$.   (page 724)

■ The graph of the polar equation $r = r(\theta)$ is   (page 727)

1. symmetric with respect to the x-axis if $r(-\theta) = r(\theta)$ or if $r(\pi - \theta) = -r(\theta)$;

2. symmetric with respect to the y-axis if $r(\pi - \theta) = r(\theta)$ or if $r(-\theta) = -r(\theta)$;

3. symmetric with respect to the origin if $r(\theta + \pi) = r(\theta)$.

■ The area A of the region bounded by the graph of the function $r = f(\theta)$ and the rays $\theta = a$ and $\theta = b$ is   (page 739)

$$A = \int_a^b \frac{1}{2} [f(\theta)]^2 \, d\theta.$$

■ The slope of the line tangent to the curve determined by the **parametric equations** $x = x(t)$, $y = y(t)$, at the point $(x(t_0), y(t_0))$ is   (page 751)

$$m = \frac{y'(t_0)}{x'(t_0)}$$

provided $x'(t_0)$ and $y'(t_0)$ exist and $x'(t_0) \ne 0$.

■ The slope of the line tangent to the graph of the polar equation $r = f(\theta)$ at the point (r, θ) is   (page 754)

$$m = \frac{\dfrac{dr}{d\theta} \sin \theta + r \cos \theta}{\dfrac{dr}{d\theta} \cos \theta - r \sin \theta}.$$

■ For the arc of the smooth curve C determined by the parametric equations (page 759)
$x = x(t)$, $y = y(t)$, between the points $(x(a), y(a))$ and $(x(b), y(b))$:

1. The length of the arc is

$$L = \int_a^b \sqrt{[x'(t)]^2 + [y'(t)]^2} \, dt.$$

2. The surface area of the volume obtained by revolving the arc about the
   x-axis is

$$S = \int_a^b 2\pi y(t) \sqrt{[x'(t)]^2 + [y'(t)]^2} \, dt.$$

■ The length of the arc of the graph of the polar equation $r = f(\theta)$, from $\theta = a$ (page 760)
to $\theta = b$, is

$$L = \int_a^b \sqrt{[f'(\theta)]^2 + [f(\theta)]^2} \, d\theta.$$

Review Exercises—Chapter 15

In Exercises 1–6, sketch the curve described by the given parametric equations.

1. $\begin{cases} x = 5\cos\theta \\ y = 5\sin\theta \end{cases}$ $0 \le \theta \le 2\pi$

2. $\begin{cases} x = 3 + 2t \\ y = 8 - 6t \end{cases}$ $-\infty < t < \infty$

3. $\begin{cases} x = t\cos\pi t \\ y = t\sin\pi t \end{cases}$ $0 \le t \le 6$

4. $\begin{cases} x = t^2 + 1 \\ y = t^4 - 4 \end{cases}$ $-\infty < t < \infty$

5. $\begin{cases} x = \cos t \\ y = \sin 2t \end{cases}$ $0 \le t \le \pi$

6. $\begin{cases} x = 3\sqrt{t} + 1 \\ y = 1 - \sqrt{t} \end{cases}$ $0 \le t$

7. Eliminate the parameter in Exercise 3 to find an equation in
   x and y that represents the given curve.

8. Repeat Exercise 7 for the curve given in Exercise 4.

9. Repeat Exercise 7 for the curve given in Exercise 5.

In each of Exercises 10–13, find parametric equations for the
given curves.

10. $x^2 + y^2 = 9$ 11. $4x^2 + 9y^2 = 36$

12. $x + y = 7$ 13. $x = 2y^2 + 4y + 5$

14. Show that the parametric equations $x(t) = t$, $y(t) = t^2$ de-
    scribe the graph of the equation $y = x^2$ as well as the graph
    of the equation $y^3 = x^6$, but *not* the graph of $y^2 = x^4$.

15. Sketch the curve given parametrically in polar coordinates
    by the equations $r = 2t$, $\theta = \pi t^2$.

16. Sketch the curve given parametrically by the equations $\theta = t$, $r = e^t$.

17. Show that the points on the curve determined by the para-
    metric equations $x(t) = t^2 - 1$, $y(t) = 3t$ lie on a parabola.
    Find the equation for this parabola in xy-coordinates.

18. Find an equation in polar coordinates for the circle with cen-
    ter at the origin and radius 5.

In Exercises 19–28, sketch the graph of the given polar equation.

19. $r = \cos 2\theta$ 20. $r = 3\theta$

21. $r = 5\cos 3\theta$ 22. $r = |\sin 2\theta|$

23. $r = 6\sin\theta$ 24. $r = a\tan\theta$

25. $r = 2 + \sin 2\theta$ 26. $r = \dfrac{1}{2 - \cos\theta}$

27. $\theta = \pi/2$ 28. $r = 1 - 2\sin\theta$

In Exercises 29–32, write the equations of the given circles in
polar coordinates.

29. The circle centered at $(a, 0)$ of radius a.

30. The circle centered at $(-a, 0)$ of radius a.

31. The circle centered at $(0, a)$ of radius a.

32. The circle centered at $(0, -a)$ of radius a.

In Exercises 33–38, find the area of the region enclosed by the
given curve.

33. $r = 2(1 + \sin 2\theta)$ 34. $r = 2 - \cos\theta$

35. $r = a(1 + \sin\theta)$ 36. $r = \sqrt{1 - \sin\theta}$

37. $r = 6\cos 3\theta$ 38. $r = 2 + 4\sin\theta\cos\theta$

In Exercises 39–44, find the length of the curve given by the parametric equations.

39. $x = 1 + t$, $y = (1 + t)^{3/2}$, $0 \le t \le 1$.

40. $x = 2t + 1$, $y = t^2$, $0 \le t \le 1$

41. $x = 2t^2$, $y = 4t + 1$, $0 \le t \le 1$

42. $x = \sin t - t$, $y = \cos t$, $0 \le t \le \pi$

43. $x = 1 - \cos t$, $y = \sin t$, $0 \le t \le \pi$

44. $x = 9t^2$, $y = 3t^3 - 9t^2$, $0 \le t \le 1$

In Exercises 45–48, find the length of the curve given by the polar equation.

45. $r = \cos \theta$, $-\pi/4 \le \theta \le \pi/4$

46. $r = \theta^2$, $1 \le \theta \le 2$

47. $r = 1 + \sin \theta$, $0 \le \theta \le \pi/2$

48. $r = 1 - \cos \theta$, $0 \le \theta \le \pi$

49. Find the area of the region common to the circles $r = \sin \theta$ and $r = \cos \theta$.

50. Find the area of the region enclosed by the lemniscate $r^2 = 4 \cos 2\theta$.

51. Find the area of the region inside the circle $r = 6 \cos \theta$ and outside the cardioid $r = 2(1 + \cos \theta)$.

52. Find the area of the region inside the circle $r = 4 \cos \theta$ and outside the circle $r = 2$.

53. Find the area of the region common to the circle $r = 4 \sin \theta$ and the circle $r = 2$.

54. Find the slope of the line tangent to the curve given by the parametric equations $x(t) = \sin t$, $y(t) = \sin 2t$ at the point where $t = \pi/4$.

55. Find the equation of the line tangent to the curve given by

$$\begin{cases} x(t) = t^2 + 2t + 1 \\ y(t) = 4t - 3 \end{cases}$$

at the point $(1, -11)$.

56. The motions of two particles p and q moving in the plane are described parametrically by

$$\begin{cases} x_p(t) = 4e^{-t} \\ y_p(t) = 4e^{-2t} \end{cases} \text{and} \begin{cases} x_q(t) = -3e^{-t} \\ y_q(t) = -3e^{-2t}. \end{cases}$$

Do the paths of these particles ever cross? If so, where? Do the particles ever collide? If so, where and when? What happens to p and q as $t \to \infty$?

57. Repeat Exercise 56 with

$$\begin{cases} x_p(t) = -t \\ y_p(t) = -2t \end{cases} \text{and} \begin{cases} x_q(t) = -t^2 \\ y_q(t) = -t \end{cases}$$

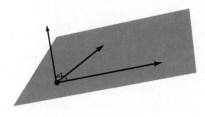

Chapter 16

Vectors, Lines, and Planes

Up to this point, we have dealt exclusively with the calculus of functions of one variable, functions f that specify one real number $f(x)$ for each real number x in their domains. In other words, our calculus has been essentially one-dimensional. Although we have explored multidimensional concepts such as area and volume, those discussions covered only special cases that could be treated using a one-variable function. Now we broaden our point of view and consider multivariable functions as well.

The fundamental notions of calculus—the limit, the derivative, and the integral—are equally valid for functions of several variables. Indeed, they are often most useful in this more general context. In this chapter, we begin our study of multivariate functions by discussing vectors, the multidimensional generalization of real numbers. In later chapters, we use the vector algebra introduced here to develop the calculus of multivariate functions.

16.1   Vectors in the Plane and in Space

Certain quantities, such as area, mass, or time, are measured by a single real number. Others, such as velocity or force, involve a direction as well as a magnitude. For example, the path of a hurricane is described by specifying both its speed and direction. Similarly, at any given moment, the gravitational force on the earth due to the sun involves both an intensity (determined by the distance between the earth and the sun) and a direction (towards the sun). Quantities that involve both a magnitude and a direction are described mathematically using vectors.

In this section, we introduce vectors and their fundamental operations. We begin with a geometric interpretation that applies both to vectors in the plane and to vectors in space. Then we present the formal algebraic definition in two stages. First, we define two-dimensional vectors. Then we discuss three-dimensional rectangular coordinates in order to define vectors in space. Finally, we conclude the section with a discussion of the properties that are common to vectors either in the plane or in space.

Figure 1.1   The displacement $\overrightarrow{PQ}$.

Displacements as Vectors

Geometrically, we interpret vectors as directed line segments. For example, consider two points P and Q. Then the **displacement** vector $\overrightarrow{PQ}$ is represented by an arrow consisting of the line segment $\overline{PQ}$ with an arrowhead at Q. This arrowhead distinguishes the *terminal point* Q from the *initial point* P (see Figure 1.1). The displace-

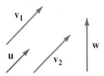

Figure 1.2   The vectors v_1 and v_2 are equal. They are the only two that are equal in this figure.

ment $\overrightarrow{PQ}$ has both a *length* $|\overrightarrow{PQ}|$, the length of the line segment $\overline{PQ}$, and a direction, the direction of Q from P.

The two essential attributes of any displacement vector are its length and its direction. This is also true for quantities such as velocities or forces. Therefore, a vector is often considered as independent of any particular initial point. When the initial point is not significant, we prefer to represent a vector by boldface notation (for example, **v**) rather than as a displacement $\overrightarrow{PQ}$, and we say that two vectors **v** and **w** are **equal** if they have the same length and same direction. In other words, if $\mathbf{v} = \overrightarrow{PQ}$, then the displacement $\overrightarrow{PQ}$ is only one of the possible representations for **v**. Two directed line segments with the same length and the same direction represent the same vector (see Figure 1.2).

Using this notion of equality, we can now describe the basic vector operations geometrically.

Operations on Vectors

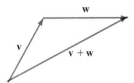

Figure 1.3   Sum of vectors **v** and **w**.

Let **v** and **w** be vectors positioned so that the terminal point of **v** lies at the initial point of **w** (Figure 1.3). The vector **v** + **w** is defined to be the vector that originates at the initial point of **v** and terminates at the terminal point of **w**. This definition agrees with our concept of vectors as displacements—the result of the displacement **v** followed by the displacement **w** is precisely the displacement **v** + **w**.

You should note that this definition of addition applies only if the vectors **v** and **w** are positioned with the initial point of **w** at the terminal point of **v**. Otherwise, we must first translate **w** to the terminal point of **v** (Figure 1.4). By positioning **v** and **w** to originate at the same point, we can see that **v** + **w** may also be interpreted as a diagonal of the parallelogram determined by **v** and **w** (Figure 1.5).

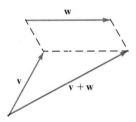

Figure 1.4   Translating **w** into position to form **v** + **w**.

Figure 1.5 illustrates the fact that **v** + **w** = **w** + **v**, since the same diagonal is obtained regardless of the order in which the two adjacent sides of the parallelogram are listed.

For any real number c, we define the multiple $c\mathbf{v}$ to be the vector whose length is $|c|$ times the length of **v** and whose direction is

(i) the same as that of **v** if $c > 0$, or
(ii) opposite that of **v** if $c < 0$ (Figure 1.6).

Two special cases are important: If $c = -1$, then $c\mathbf{v} = -\mathbf{v}$ is called the negative of **v** (Figure 1.7); it has the same length as **v** but the opposite direction. If $c = 0$, the result is a vector of zero length, called the **zero vector** and denoted by **0**. The zero vector has many properties similar to those of the real number zero, and it is the only vector for which a direction is not defined.

Finally, we define the operation of **subtraction for vectors** by the equation

$$\mathbf{v} - \mathbf{w} = \mathbf{v} + (-\mathbf{w}). \tag{1}$$

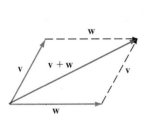

Figure 1.5   **v** + **w** is the diagonal of the parallelogram whose sides are **v** and **w**.

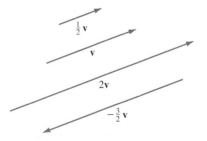

Figure 1.6   Multiples of the vector **v**.

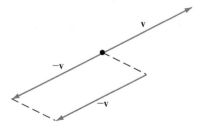

Figure 1.7   The negative of **v**.

That is, the difference **v** − **w** is defined to be the *sum* of **v** and the negative of **w.** Figure 1.8 shows how **v** − **w** may be obtained using equation (1) and the head-to-tail rule for vector addition. Figure 1.9 shows that the vector **v** − **w** is the diagonal opposite **v** + **w** in the parallelogram whose adjacent sides are **v** and **w** (compare Figure 1.5). Figure 1.10 illustrates the simplest method for constructing the vector **v** − **w**: if **v** and **w** originate at the same point, **v** − **w** is the vector originating at the terminal point of **w** and terminating at the terminal point of **v**. This follows from the equation **w** + (**v** − **w**) = **v**.

Note that **v** − **v** = **0**; that is, the result is the **zero vector,** not the real number zero.

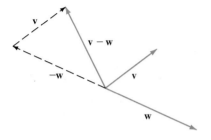

Figure 1.8   **v** − **w** = **v** + (−**w**).

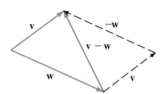

Figure 1.9   The vector **v** − **w** is the diagonal opposite **v** + **w** in the parallelogram whose adjacent sides are **v** and **w**.

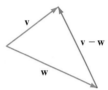

Figure 1.10   Using the equation **w** + (**v** − **w**) = **v** to construct **v** − **w**.

Applications of Vectors

These vector operations are useful in a variety of applications. The following example illustrates how vector addition is used to calculate velocities.

□ **EXAMPLE 1**

Suppose that a wind is blowing due west at 40 mph. In what direction should a pilot steer an airplane flying with an air speed of 200 mph in order to maintain a course that is due north?

Solution   The velocity $\mathbf{v}_g$ of the plane relative to the ground is the vector sum of its velocity relative to the air $\mathbf{v}_a$ and the velocity of the wind **w**. That is, $\mathbf{v}_g = \mathbf{v}_a + \mathbf{w}.$ Moreover, the vector **w** points west, and we want the vector $\mathbf{v}_g$ to point north (see Figure 1.11). Consequently, the triangle formed by the three vectors $\mathbf{v}_a$, **w**, and $\mathbf{v}_g$ is a right triangle.

Suppose θ is the angle between the vectors $\mathbf{v}_a$ and $\mathbf{v}_g$. We have

$$\sin \theta = \frac{|\mathbf{w}|}{|\mathbf{v}_a|} = \frac{40}{200} = \frac{1}{5}.$$

Consequently, the pilot should steer the plane at an angle

$$\theta = \text{Sin}^{-1}(1/5) \approx 11.54°$$

to the east of north.   ■

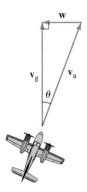

Figure 1.11   The velocity $\mathbf{v}_g$ relative to the ground is $\mathbf{v}_a + \mathbf{w}$.

In problems in which an object moves along an inclined plane, the force $\mathbf{w}_g$ due to gravity must be ''resolved'' into the component forces acting parallel and perpendicular to the plane (see Figure 1.12). That is, we must find two vectors, one parallel to the plane and one perpendicular to the plane, whose sum is $\mathbf{w_g}$. We can do this using trigonometry.

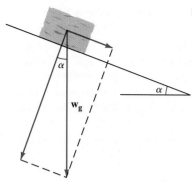

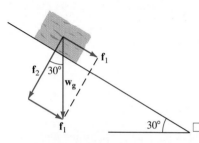

Figure 1.12 Force due to gravity is resolved into forces parallel and perpendicular to the inclined surface.

Figure 1.13 The force due to gravity of a block on a plane inclined 30° from the horizontal.

□ **EXAMPLE 2**

A block weighing 10 lb is placed on a plane inclined 30° from the horizontal. Find

(a) the magnitude of the force $\mathbf{f}_1$, acting parallel to the plane, and
(b) the magnitude of the force $\mathbf{f}_2$, acting perpendicular to the plane.

Solution   Since the desired vectors $\mathbf{f}_1$ and $\mathbf{f}_2$ are perpendicular, they form two sides of a right triangle with the force $\mathbf{w_g}$ due to gravity as the hypotenuse (see Figure 1.13). Since the weight of the block is precisely this force, $|\mathbf{w_g}| = 10$ lb. Using the right triangle, we have

$$|\mathbf{f}_1| = |\mathbf{w_g}| \sin 30° = 10(1/2) = 5 \text{ lb}$$
$$|\mathbf{f}_2| = |\mathbf{w_g}| \cos 30° = 10(\sqrt{3}/2) = 5\sqrt{3} \text{ lb.}$$

The physical significance of $\mathbf{f}_1$ and $\mathbf{f}_2$ is: The vector $\mathbf{f}_2$, perpendicular to the surface of the plane, is exactly counteracted by a "reaction" force exerted by the plane. Thus, $\mathbf{f}_2$ causes no acceleration of the block. The vector $\mathbf{f}_1$, if it is not opposed by some external force, will cause acceleration down the plane, according to Newton's second law. If an external force equal to $-\mathbf{f}_1$ is applied to the block, the system will be in equilibrium.   ∎

The following example shows how vector methods can be used in plane geometry.

□ **EXAMPLE 3**

Use vectors to prove that the diagonals of a parallelogram bisect each other.

Solution   Let the vertices of the parallelogram be A, B, C, and D, as in Figure 1.14. Since $\overrightarrow{AC} = \overrightarrow{AB} + \overrightarrow{AD}$, the vector from A to the midpoint of diagonal $\overrightarrow{AC}$ is

$$\mathbf{v} = \tfrac{1}{2}(\overrightarrow{AB} + \overrightarrow{AD}).$$

Since $\overrightarrow{DB} = \overrightarrow{AB} - \overrightarrow{AD}$, the vector from A to the midpoint of diagonal $\overrightarrow{DB}$ is

$$\mathbf{w} = \overrightarrow{AD} + \tfrac{1}{2}(\overrightarrow{AB} - \overrightarrow{AD})$$
$$= \overrightarrow{AD} + \tfrac{1}{2}\overrightarrow{AB} - \tfrac{1}{2}\overrightarrow{AD}$$
$$= \tfrac{1}{2}(\overrightarrow{AB} + \overrightarrow{AD})$$
$$= \mathbf{v.}$$

Since $\mathbf{v} = \mathbf{w}$, these midpoints are the same.   ∎

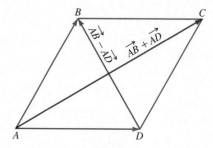

Figure 1.14 The diagonals of a parallelogram bisect each other.

The Analytic Definition of Vectors

Our geometric interpretation of the vector operations apply both to vectors in the plane and vectors in space. Now, we develop an analytic description of vectors using coordinates, and to do so, we treat the two- and three-dimensional cases separately.

Given a vector $\mathbf{v}$ in the plane, there exists a unique point P in the plane such that the vector $\overrightarrow{OP}$ (whose initial point is the origin O) equals $\mathbf{v}$ (see Figure 1.15). Consequently, we can use the usual rectangular coordinates (a, b) for P to associate a unique ordered pair $\langle a, b \rangle$ to $\mathbf{v}$. Moreover, we use the notation $\langle a, b \rangle$ to represent any vector that is equal to $\overrightarrow{OP}$. This observation is the basis of our analytic definition.

Definition 1

The set of all ordered pairs $\langle a, b \rangle$ of real numbers, together with the rules

(i) $\langle a_1, b_1 \rangle + \langle a_2, b_2 \rangle = \langle a_1 + a_2, b_1 + b_2 \rangle$
(ii) $c\langle a, b \rangle = \langle ca, cb \rangle$

for addition and for multiplication by real numbers c, is called the set of **vectors in the plane.** The numbers a and b are referred to as the **components** of the vector $\langle a, b \rangle$.

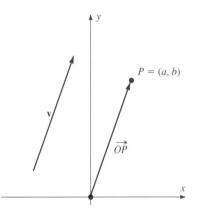

Figure 1.15   The vector $\mathbf{v} = \overrightarrow{OP} = \langle a, b \rangle$.

It is important to note that the vector $\langle a, b \rangle$ is an *ordered pair of numbers,* not a point (a, b) in the plane. We shall sometimes wish to associate the vector $\langle a, b \rangle$ with the point (a, b), but the concept of a vector is more general.

The multiplication of the vector $\langle a, b \rangle$ by the real number c in equation (ii) is called **scalar multiplication,** and the number c is referred to as a **scalar.**

When we motivated Definition 1 using our geometric description of vectors, we considered the vector $\overrightarrow{OP}$ whose initial point is located at the origin (see Figure 1.15). Such vectors are called **position vectors** because they provide a unique correspondence between points (positions) and vectors. In subsequent sections and chapters, we shall frequently use position vectors to describe geometric objects and to study motion in the plane and in space.

EXAMPLE 4

For $\mathbf{v} = \langle 2, 5 \rangle$ and $\mathbf{w} = \langle -3, 2 \rangle$, we have

(a) $\mathbf{v} + \mathbf{w} = \langle 2 + (-3), 5 + 2 \rangle = \langle -1, 7 \rangle$,
(b) $3\mathbf{v} + \mathbf{w} = \langle 3 \cdot 2, 3 \cdot 5 \rangle + \langle -3, 2 \rangle = \langle 3, 17 \rangle$,
(c) $4\mathbf{v} + 2\mathbf{w} = \langle 4 \cdot 2, 4 \cdot 5 \rangle + \langle 2(-3), 2 \cdot 2 \rangle = \langle 2, 24 \rangle$.   ∎

We define the vector $-\mathbf{v}$ as $(-1)\mathbf{v}$. That is, if $\mathbf{v} = \langle a, b \rangle$, then

$$-\mathbf{v} = \langle -a, -b \rangle.$$

Also, we define $\mathbf{v} - \mathbf{w}$ as $\mathbf{v} + (-\mathbf{w})$. That is, if $\mathbf{v} = \langle a_1, b_1 \rangle$ and $\mathbf{w} = \langle a_2, b_2 \rangle$, then

$$\mathbf{v} - \mathbf{w} = \mathbf{v} + (-\mathbf{w}) = \langle a_1 - a_2, b_1 - b_2 \rangle.$$

The **zero vector** is defined to be the vector $\mathbf{0} = \langle 0, 0 \rangle$. Finally, we define two vectors to be **equal** if and only if they have the same components. That is,

$$\langle a_1, b_1 \rangle = \langle a_2, b_2 \rangle \quad \text{if and only if} \quad a_1 = a_2 \text{ and } b_1 = b_2.$$

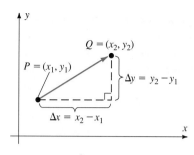

Figure 1.16 $\overrightarrow{PQ} = \langle \Delta x, \Delta y \rangle$
$= \langle x_2 - x_1, y_2 - y_1 \rangle.$

□ **EXAMPLE 5**

For $\mathbf{v} = \langle 1, -3 \rangle$ and $\mathbf{w} = \langle 2, 5 \rangle$,

(a) $\mathbf{v} - \mathbf{w} = \langle 1 - 2, -3 - 5 \rangle = \langle -1, -8 \rangle.$
(b) $2\mathbf{v} - 3\mathbf{w} = \langle 2, -6 \rangle - \langle 6, 15 \rangle = \langle -4, -21 \rangle.$
(c) $\mathbf{v} - \mathbf{v} = \langle 1 - 1, -3 + 3 \rangle = \langle 0, 0 \rangle = \mathbf{0}.$ ■

To make the connection between our geometric description and our analytic definition of a vector, we must determine the components of a displacement $\overrightarrow{PQ}$. If the point $P = (x_1, y_1)$ and the point $Q = (x_2, y_2)$, then

$$\overrightarrow{PQ} = \langle \Delta x, \Delta y \rangle, \qquad (2)$$

where $\Delta x = x_2 - x_1$ and $\Delta y = y_2 - y_1$ (see Figure 1.16).

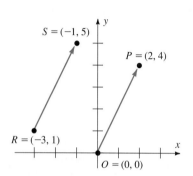

Figure 1.17 $\overrightarrow{RS} = \overrightarrow{OP} = \langle 2, 4 \rangle.$

□ **EXAMPLE 6**

Let $O = (0, 0)$, $P = (2, 4)$, $R = (-3, 1)$, and $S = (-1, 5)$. Then

$$\overrightarrow{OP} = \langle 2 - 0, 4 - 0 \rangle = \langle 2, 4 \rangle$$

and

$$\overrightarrow{RS} = \langle -1 - (-3), 5 - 1 \rangle = \langle 2, 4 \rangle,$$

so $\overrightarrow{OP} = \overrightarrow{RS}$ (see Figure 1.17). Thus, $\overrightarrow{OP}$ is the unique position vector that is equal to $\overrightarrow{RS}$. ■

Using equation (2), we see that the geometric interpretation of vector addition agrees with our analytic definition. If the vector $\mathbf{v} = \langle a_1, b_1 \rangle$ is represented as an arrow originating at point $P = (x_1, y_1)$, this arrow has terminal point $Q = (x_2, y_2)$ with

$$x_2 = x_1 + a_1,$$
$$y_2 = y_1 + b_1.$$

If the vector $\mathbf{w} = \langle a_2, b_2 \rangle$ is represented as an arrow originating at Q, this arrow has terminal point $R = (x_3, y_3)$ with

$$x_3 = x_2 + a_2 = x_1 + (a_1 + a_2),$$
$$y_3 = y_2 + b_2 = y_1 + (b_1 + b_2).$$

Since Definition 1 requires that

$$\mathbf{v} + \mathbf{w} = \langle a_1, b_1 \rangle + \langle a_2, b_2 \rangle = \langle a_1 + a_2, b_1 + b_2 \rangle,$$

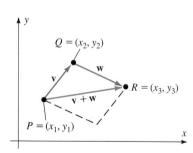

Figure 1.18 Head-to-tail addition for vectors as arrows in the plane.

this shows that $\mathbf{v} + \mathbf{w}$ is represented by the arrow originating at P and terminating at R. (See Figure 1.18.) Thus, the head-to-tail rule for adding arrows (displacements) corresponds to our formal definition of vector addition.

Length of Vectors

If we represent the vector $\mathbf{v} = \langle a, b \rangle$ by a position vector, it terminates at the point (a, b). The obvious way to define the length of $\mathbf{v}$ is to use the distance formula in the plane (see Figure 1.19).

Definition 2

The **length of the vector** $\mathbf{v} = \langle a, b \rangle$ is $|\mathbf{v}| = \sqrt{a^2 + b^2}$.

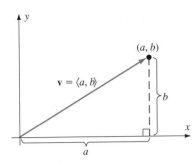

Figure 1.19 $|\mathbf{v}| = \sqrt{a^2 + b^2}$.

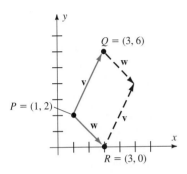

Figure 1.20 The vectors $\mathbf{v}$ and $\mathbf{w}$ in Example 8.

Note that Definition 2 applies to all vectors in the plane, not just to position vectors. For example, given the displacement vector $\overrightarrow{PQ} = \langle \Delta x, \Delta y \rangle$ from $P = (x_1, y_1)$ to $Q = (x_2, y_2)$ where $\Delta x = x_2 - x_1$ and $\Delta y = y_2 - y_1$, Definition 2 yields

$$|\overrightarrow{PQ}| = \sqrt{(\Delta x)^2 + (\Delta y)^2},$$

which equals the length of the segment $\overline{PQ}$ as given by the distance formula in the plane.

☐ **EXAMPLE 7**

For $\mathbf{v} = \langle 2, 1 \rangle$ and $\mathbf{w} = \langle -1, 4 \rangle$,

(a) $|\mathbf{v}| = \sqrt{2^2 + 1^2} = \sqrt{5}$,

(b) $|\mathbf{w}| = \sqrt{(-1)^2 + 4^2} = \sqrt{17}$,

(c) $|2\mathbf{v} - \mathbf{w}| = \sqrt{(2 \cdot 2 - (-1))^2 + (2 \cdot 1 - 4)^2} = \sqrt{5^2 + 2^2} = \sqrt{29}$. ■

☐ **EXAMPLE 8**

Let $P = (1, 2)$, $Q = (3, 6)$, and $R = (3, 0)$. Let $\mathbf{v} = \overrightarrow{PQ}$ and $\mathbf{w} = \overrightarrow{PR}$ (see Figure 1.20). Using the distance formula, we can see that

$$|\mathbf{v}| = \text{length of } \mathbf{v} = \text{length of } \overline{PQ} = \sqrt{(3 - 1)^2 + (6 - 2)^2} = 2\sqrt{5},$$

$$|\mathbf{w}| = \text{length of } \mathbf{w} = \text{length of } \overline{PR} = \sqrt{(3 - 1)^2 + (0 - 2)^2} = 2\sqrt{2}.$$

Also, $\mathbf{v} = \langle 2, 4 \rangle$ and $\mathbf{w} = \langle 2, -2 \rangle$. From this information it follows that

(a) $\mathbf{v}$, when originating at $(3, 0)$, terminates at $(5, 4)$,

(b) $\mathbf{w}$, when originating at $(3, 6)$, terminates at $(5, 4)$,

(c) $\mathbf{v} + \mathbf{w}$, when originating at $(1, 2)$, terminates at $(5, 4)$,

(d) $\mathbf{v} - \mathbf{w}$, when originating at $(3, 0)$, terminates at $(3, 6)$,

(e) $-\mathbf{w}$, when originating at $(1, 2)$, terminates at $(-1, 4)$,

(f) $\frac{1}{2}\mathbf{v} - \mathbf{w}$, when originating at $(1, 2)$, terminates at $(0, 6)$.

(See Figures 1.21 and 1.22.) ■

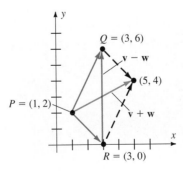

Figure 1.21 The vectors $\mathbf{v} + \mathbf{w}$ and $\mathbf{v} - \mathbf{w}$ in Example 8.

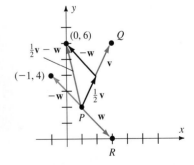

Figure 1.22 The vectors $-\mathbf{w}$ and $\frac{1}{2}\mathbf{v} - \mathbf{w}$ in Example 8.

The upshot of the preceding discussion is that a vector in the plane may be regarded as an ordered pair $\langle a, b \rangle$ of numbers, subject to the laws of addition and scalar multiplication given by Definition 1. Moreover, the vector $\langle a, b \rangle$ may be *represented* geometrically by an arrow originating at any point (x, y) and terminating at the point $(x + a, y + b)$ (Figure 1.23).

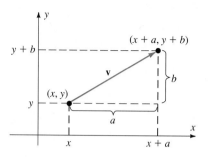

Figure 1.23 The vector $\mathbf{v} = \langle a, b \rangle$.

Rectangular Coordinates in Space

In order to work with vectors in space analytically, we need a coordinate system for points in space, and in this case, it is best to use rectangular coordinates. We usually represent the three coordinates by the variables x, y, and z, and we construct three mutually perpendicular axes, one for x, one for y, and one for z. All three have their origins at the common point of intersection. When sketching figures, we typically follow the convention that the positive x-axis points towards the reader, the positive y-axis points to the right, and the positive z-axis points upwards (see Figure 1.24). Occasionally, it is also convenient to surround the sketch with a box on which we label the ranges of x, y, and z.

We may identify each point P in space with a unique triple of numbers $P = (x_0, y_0, z_0)$ as follows: the plane through P perpendicular to the x-axis intersects the x-axis at x_0; the plane through P perpendicular to the y-axis intersects the y-axis at y_0; the plane through P perpendicular to the z-axis intersects the z-axis at z_0 (see Figure 1.25). Examples of several points in space are shown in Figure 1.27.

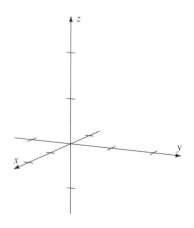

Figure 1.24 The coordinate axes for rectangular coordinates in space.

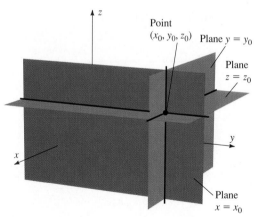

Figure 1.25 The point common to these three planes, has coordinates (x_0, y_0, z_0).

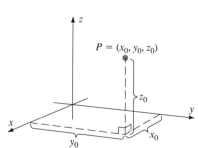

Figure 1.26 The coordinates of P are (x_0, y_0, z_0).

Figure 1.28 illustrates that the set of all points obtained by specifying two of the three coordinates (and allowing the third to range unrestricted) is a line parallel to the axis of the unrestricted variable. If we specify only one variable, allowing all

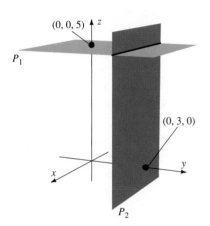

Figure 1.27   Three points in space along with their coordinates. The boxes are shown simply to highlight the relative positions of the points.

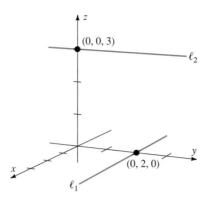

Figure 1.28 ℓ_1: the line $y = 2$, $z = 0$.
ℓ_2: the line $x = 0$, $z = 3$.

values of the other two, we obtain a plane perpendicular to the axis of the restricted variable (Figure 1.29). Equations for lines and planes not parallel or perpendicular to coordinate axes are more complicated and are discussed in detail in Sections 16.2 through 16.4.

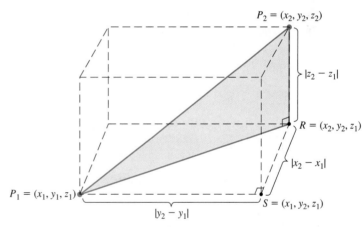

Figure 1.29 P_1: the plane $z = 5$.
P_2: the plane $y = 3$.

Figure 1.30 $d(P_1, P_2) = \sqrt{(x_2 - x_1)^2 + (y_2 - y_1)^2 + (z_2 - z_1)^2}$.

The Distance Formula

Let $P_1 = (x_1, y_1, z_1)$ and $P_2 = (x_2, y_2, z_2)$ be two points in space. To find the distance $d(P_1, P_2)$ between these points, we use the Pythagorean theorem twice.

Let $S = (x_1, y_2, z_1)$ and $R = (x_2, y_2, z_1)$, as in Figure 1.30. Then, since the segment $\overline{P_1 R}$ is the hypotenuse of the right triangle $\Delta P_1 R S$, the Pythagorean theorem gives

$$[d(P_1, R)]^2 = [d(S, R)]^2 + [d(P_1, S)]^2$$
$$= (x_2 - x_1)^2 + (y_2 - y_1)^2.$$

Similarly, the segment $\overline{P_1 P_2}$ is the hypotenuse of the right triangle $\Delta P_1 R P_2$. Thus, a second application of the Pythagorean theorem, together with the above equation, gives

$$[d(P_1, P_2)]^2 = [d(P_1, R)]^2 + [d(R, P_2)]^2$$
$$= (x_2 - x_1)^2 + (y_2 - y_1)^2 + (z_2 - z_1)^2.$$

Taking square roots of both sides of this equation gives the desired formula:

> The distance between $P_1 = (x_1, y_1, z_1)$ and $P_2 = (x_2, y_2, z_2)$ is
> $$d(P_1, P_2) = \sqrt{(x_2 - x_1)^2 + (y_2 - y_1)^2 + (z_2 - z_1)^2}.$$

☐ **EXAMPLE 9**

The distance between $P = (-3, 0, 4)$ and $Q = (2, 5, -2)$ is
$$d = \sqrt{(2 - (-3))^2 + (5 - 0)^2 + (-2 - 4)^2}$$
$$= \sqrt{5^2 + 5^2 + (-6)^2} = \sqrt{86}. \qquad ∎$$

Vectors in Space

To define vectors in space we must work with **ordered triples,** since points in space are determined by three coordinates.

Definition 3

The set of all ordered triples $\langle x, y, z \rangle$ together with the operations

(i) $\langle x_1, y_1, z_1 \rangle + \langle x_2, y_2, z_2 \rangle = \langle x_1 + x_2, y_1 + y_2, z_1 + z_2 \rangle$
(ii) $c\langle x, y, z \rangle = \langle cx, cy, cz \rangle, \qquad c \in (-\infty, \infty)$

for addition and multiplication by real numbers is the **set of vectors in space.**

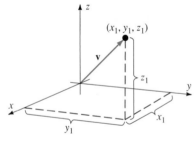

Figure 1.31 The vector $\mathbf{v} = \langle x_1, x_2, x_3 \rangle$ represented as a position vector.

For the vector $\mathbf{v} = \langle x_1, y_1, z_1 \rangle$, the numbers $x_1, y_1,$ and z_1 are again referred to as the **components** of $\mathbf{v}$. You have probably noticed that Definition 3 states that vector addition and scalar multiplication are to be carried out for vectors in space just as for vectors in the plane, except that now there is a third component. Similarly, we define

(a) the **negative** of the vector $\mathbf{v} = \langle x, y, z \rangle$ as $-\mathbf{v} = (-1)\mathbf{v} = \langle -x, -y, -z \rangle,$
(b) the **difference** of two vectors $\mathbf{v} = \langle x_1, y_1, z_1 \rangle$ and $\mathbf{w} = \langle x_2, y_2, z_2 \rangle$ as
$$\mathbf{v} - \mathbf{w} = \mathbf{v} + (-\mathbf{w}) = \langle x_1 - x_2, y_1 - y_2, z_1 - z_2 \rangle,$$
(c) the **zero vector** as $\mathbf{0} = \langle 0, 0, 0 \rangle,$
(d) **equality** of two vectors $\mathbf{v} = \langle x_1, y_1, z_1 \rangle$ and $\mathbf{w} = \langle x_2, y_2, z_2 \rangle$ by

$$\mathbf{v} = \mathbf{w} \quad \text{if and only if} \quad x_1 = x_2, y_1 = y_2 \text{ and } z_1 = z_2.$$

The vector $\mathbf{v} = \langle x_1, y_1, z_1 \rangle$ can be represented by a position vector terminating at the point (x_1, y_1, z_1). More generally, if (a, b, c) is any point in space, the vector $\mathbf{v}$ can be represented by a vector originating at (a, b, c) and terminating at the point $(a + x_1, b + y_1, c + z_1)$ (see Figures 1.31 and 1.32).

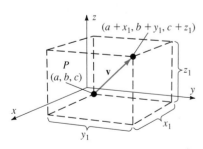

Figure 1.32 The vector $\mathbf{v} = \langle x_1, y_1, z_1 \rangle$ represented as a displacement with initial point $P = (a, b, c)$ and terminal point $(a + x_1, b + y_1, c + z_1)$.

As with vectors in the plane, we shall not concern ourselves with the distinction between vectors defined as ordered triples and their representations as arrows in space. For example, the vector originating at the point $P = (x_1, y_1, z_1)$ and terminating at the point $Q = (x_2, y_2, z_2)$ is written in component form as
$$\overrightarrow{PQ} = \langle x_2 - x_1, y_2 - y_1, z_2 - z_1 \rangle.$$

We define the length of a vector $\mathbf{v}$ in space by applying the distance formula (see Figures 1.31 and 1.32).

Definition 4	The **length of the vector** $\mathbf{v} = \langle x, y, z \rangle$ is

$$|\mathbf{v}| = \sqrt{x^2 + y^2 + z^2}.$$

☐ **EXAMPLE 10**

For the vectors $\mathbf{v} = \langle 2, -1, 4 \rangle$ and $\mathbf{w} = \langle 0, 3, -5 \rangle$,

(a) $\mathbf{v} + \mathbf{w} = \langle 2 + 0, -1 + 3, 4 + (-5) \rangle = \langle 2, 2, -1 \rangle$

(b) $3\mathbf{v} = \langle 3 \cdot 2, 3(-1), 3 \cdot 4 \rangle = \langle 6, -3, 12 \rangle$

(c) $\mathbf{v} - \mathbf{w} = \langle 2 - 0, -1 - 3, 4 - (-5) \rangle = \langle 2, -4, 9 \rangle$

(d) $|2\mathbf{v} + \mathbf{w}| = |\langle 2 \cdot 2 + 0, 2(-1) + 3, 2 \cdot 4 + (-5) \rangle|$
$= |\langle 4, 1, 3 \rangle| = \sqrt{4^2 + 1^2 + 3^2} = \sqrt{26}.$　■

Properties of Vectors

Vectors, both in the plane and in space, satisfy many of the same properties as do real numbers. For example, vector addition is **commutative.** This means that $\mathbf{v} + \mathbf{w} = \mathbf{w} + \mathbf{v}$ for any two vectors $\mathbf{v}$ and $\mathbf{w}$. However, not all properties of real numbers are shared by vectors. For example, we shall not encounter a useful way to define the *product* of two vectors in the plane as another vector in the plane.

The following theorem specifies the legitimate properties of vector addition and scalar multiplication.

Theorem 1	Let $\mathbf{u}$, $\mathbf{v}$, and $\mathbf{w}$ be vectors and let a and b be real numbers. Then

    (i) $\mathbf{v} + \mathbf{w} = \mathbf{w} + \mathbf{v}$　　　　　　　　(commutativity),

    (ii) $(\mathbf{u} + \mathbf{v}) + \mathbf{w} = \mathbf{u} + (\mathbf{v} + \mathbf{w})$　(associativity),

    (iii) $\mathbf{0} + \mathbf{v} = \mathbf{v}$　　　　　　　　　　(identity for vector addition),

    (iv) $\mathbf{v} + (-1)\mathbf{v} = \mathbf{0}$　　　　　　　(inverse for vector addition)

    (v) $1\mathbf{v} = \mathbf{v}$　　　　　　　　　　　(identity for scalar multiplication),

    (vi) $a(\mathbf{v} + \mathbf{w}) = a\mathbf{v} + a\mathbf{w}$　　　(scalar multiplication distributes over vector addition).

  (vii) $(a + b)\mathbf{v} = a\mathbf{v} + b\mathbf{v}.$

  (viii) $a(b\mathbf{v}) = (ab)\mathbf{v}$

To prove part (i) where $\mathbf{v}$ and $\mathbf{w}$ are vectors in the plane, we write the vectors $\mathbf{v}$ and $\mathbf{w}$ in component form

$$\mathbf{v} = \langle x_1, y_1 \rangle, \qquad \mathbf{w} = \langle x_2, y_2 \rangle$$

and apply Definition 1:

$$\mathbf{v} + \mathbf{w} = \langle x_1, y_1 \rangle + \langle x_2, y_2 \rangle = \langle x_1 + x_2, y_1 + y_2 \rangle, \tag{3}$$

$$\mathbf{w} + \mathbf{v} = \langle x_2, y_2 \rangle + \langle x_1, y_1 \rangle = \langle x_2 + x_1, y_2 + y_1 \rangle. \tag{4}$$

Since addition for real numbers is commutative, $x_1 + x_2 = x_2 + x_1$, and $y_1 + y_2 = y_2 + y_1$. Thus, the right-hand sides of equations (3) and (4) represent the same components, so $\mathbf{v} + \mathbf{w} = \mathbf{w} + \mathbf{v}.$

You are asked to prove statements (ii) through (viii) in the exercise set. Each statement is proved by writing the vector(s) in component form and using the corresponding properties of real numbers.

The following theorem summarizes the properties of length for vectors.

Theorem 2

Let $\mathbf{v}$ and $\mathbf{w}$ be vectors and let c be a real number. Then

(i) $|\mathbf{v}| \geq 0;$ $|\mathbf{v}| = 0$ if and only if $\mathbf{v} = \mathbf{0}$,

(ii) $|c\mathbf{v}| = |c|\,|\mathbf{v}|$,

(iii) $|\mathbf{v} + \mathbf{w}| \leq |\mathbf{v}| + |\mathbf{w}|$.

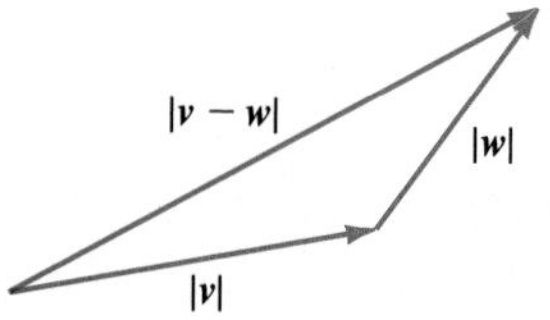

Interpreted geometrically, properties (i) and (ii) are obvious. To prove them for vectors in the plane according to Definition 1, we write $\mathbf{v}$ in the component form $\mathbf{v} = \langle a, b \rangle$. Then

$$|\mathbf{v}| = \sqrt{a^2 + b^2} \geq 0 \quad \text{for all } a \text{ and } b,$$

which proves the first part of (i). The second part of (i) is proved by noting that

$$|\mathbf{v}| = \sqrt{a^2 + b^2} = 0 \quad \text{if and only if} \quad a = b = 0,$$

in which case $\mathbf{v} = \langle a, b \rangle = \langle 0, 0 \rangle = \mathbf{0}$. Statement (ii) is just as easy:

$$|c\mathbf{v}| = |\langle ca, cb \rangle| = \sqrt{(ca)^2 + (cb)^2} = \sqrt{c^2}\sqrt{a^2 + b^2} = |c|\,|\mathbf{v}|.$$

Statement (iii) is referred to as the **triangle inequality.** It may be interpreted geometrically as saying that the length of one side of a triangle cannot exceed the sum of the lengths of the other two sides (Figure 1.33). A proof of statement (iii) is given in Section 16.3.

Figure 1.33 $|\mathbf{v} + \mathbf{w}| \leq |\mathbf{v}| + |\mathbf{w}|$.

Unit Vectors

A vector $\mathbf{u}$ with $|\mathbf{u}| = 1$ is called a **unit vector.** In what follows we shall frequently need to find a unit vector $\mathbf{u}$ in the same direction as a given vector $\mathbf{v}$. We can do so using property (ii) of Theorem 2. Since

$$\left|\left(\frac{1}{|\mathbf{v}|}\right)\mathbf{v}\right| = \frac{1}{|\mathbf{v}|}|\mathbf{v}| = 1,$$

the vector

$$\mathbf{u} = \frac{1}{|\mathbf{v}|}\mathbf{v} \tag{5}$$

is a unit vector. It points in the same direction as $\mathbf{v}$ because it is a *positive* scalar multiple of $\mathbf{v}$. Thus, the unit vector $\mathbf{u}$ given in equation (5) is frequently referred to as the **direction** of $\mathbf{v}$. Of course, equation (5) makes sense only if $|\mathbf{v}| \neq 0$.

□ **EXAMPLE 11**

Find a unit vector in the same direction as the vector $\mathbf{v} = \langle 6, -4 \rangle$.

Solution

$$|\mathbf{v}| = \sqrt{6^2 + (-4)^2} = \sqrt{52} = 2\sqrt{13}.$$

Thus, by equation (5), the desired unit vector is

$$\mathbf{u} = \frac{1}{2\sqrt{13}}\langle 6, -4 \rangle = \left\langle \frac{3}{\sqrt{13}}, \frac{-2}{\sqrt{13}} \right\rangle.$$

(Use Definition 2 to verify that $|\mathbf{u}| = 1$.)   ■

A similar problem concerning unit vectors in the plane is that of finding a unit vector making an angle α with a given line or vector. For example, Figure 1.34 illustrates that *the unit vector forming an angle α measured counterclockwise from the positive x-axis is*

$$\mathbf{u} = \langle \cos \alpha, \sin \alpha \rangle. \tag{6}$$

Equation (6) follows from the definition of the trigonometric functions $\sin \theta$ and $\cos \theta$.

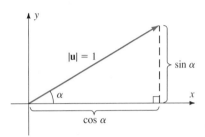

Figure 1.34 $\mathbf{u} = \langle \cos \alpha, \sin \alpha \rangle$ is a unit vector.

□ **EXAMPLE 12**

Find

(a) a unit vector $\mathbf{u}$ in the plane making an angle of $60°$ with the positive x-axis.
(b) a vector in the plane of length 5 making an angle of $45°$ with the positive x-axis.

Strategy · · · · · · · ·
(a) Use equation (6) and the facts

$$\cos 60° = \cos(\pi/3) = 1/2,$$
$$\sin 60° = \sin(\pi/3) = \sqrt{3}/2.$$

(b) First, use equation (6) to find a unit vector $\mathbf{u}$ in the given direction.

The solution is then

$$\mathbf{v} = 5\mathbf{u}.$$

Solution
(a) Using equation (6) we obtain

$$\mathbf{u} = \langle \cos 60°, \sin 60° \rangle = \langle 1/2, \sqrt{3}/2 \rangle.$$

(b) By equation (6) a unit vector making an angle of $45°$ with the positive x-axis is

$$\mathbf{u} = \langle \cos 45°, \sin 45° \rangle = \langle \sqrt{2}/2, \sqrt{2}/2 \rangle.$$

The desired vector is therefore

$$\mathbf{v} = 5\mathbf{u} = \langle 5\sqrt{2}/2, 5\sqrt{2}/2 \rangle. \qquad ■$$

Unit Coordinate Vectors in the Plane

Two special vectors enable us to develop yet another notation for representing vectors in the plane. They are the **unit coordinate vectors**

$$\mathbf{i} = \langle 1, 0 \rangle \quad \text{and} \quad \mathbf{j} = \langle 0, 1 \rangle.$$

(See Figure 1.35.) Using these two vectors, we may represent the vector $\langle a, b \rangle$ as

$$\langle a, b \rangle = a\langle 1, 0 \rangle + b\langle 0, 1 \rangle = a\mathbf{i} + b\mathbf{j}.$$

Geometrically, $a\mathbf{i}$ and $b\mathbf{j}$ represent the adjacent sides of the rectangle whose diagonal is $\langle a, b \rangle$ (Figure 1.36). We refer to the numbers a and b as the $\mathbf{i}$ and $\mathbf{j}$ components of the vector $a\mathbf{i} + b\mathbf{j}$, respectively.

The principal advantage of this vector notation is that it helps avoid confusion between points and vectors. We shall make frequent use of this notation.

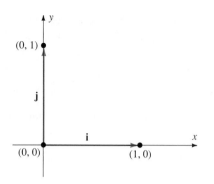

Figure 1.35 Unit coordinate vectors.

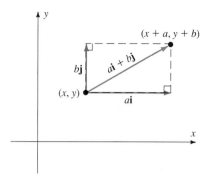

Figure 1.36 $\langle a, b \rangle = a\mathbf{i} + b\mathbf{j}$.

□ **EXAMPLE 13**

For $\mathbf{v} = \langle -2, 1 \rangle$ and $\mathbf{w} = \langle 3, 4 \rangle$,

(a) $\mathbf{v} = -2\mathbf{i} + \mathbf{j}; \qquad \mathbf{w} = 3\mathbf{i} + 4\mathbf{j}$,
(b) $\mathbf{v} + \mathbf{w} = (-2\mathbf{i} + \mathbf{j}) + (3\mathbf{i} + 4\mathbf{j}) = (-2 + 3)\mathbf{i} + (1 + 4)\mathbf{j} = \mathbf{i} + 5\mathbf{j}$,
(c) $3\mathbf{v} - 2\mathbf{w} = (-6\mathbf{i} + 3\mathbf{j}) - (6\mathbf{i} + 8\mathbf{j}) = (-6 - 6)\mathbf{i} + (3 - 8)\mathbf{j} = -12\mathbf{i} - 5\mathbf{j}$,
(d) $|\mathbf{v}| = |-2\mathbf{i} + \mathbf{j}| = \sqrt{(-2)^2 + 1^2} = \sqrt{5}$. ∎

Unit Coordinate Vectors in Space   Similarly we define the three unit coordinate vectors in space by

$$\mathbf{i} = \langle 1, 0, 0 \rangle, \qquad \mathbf{j} = \langle 0, 1, 0 \rangle, \qquad \mathbf{k} = \langle 0, 0, 1 \rangle.$$

(See Figure 1.36.) We may write the vector $\mathbf{v} = \langle x, y, z \rangle$ in terms of $\mathbf{i}$, $\mathbf{j}$, and $\mathbf{k}$ as

$$\mathbf{v} = \langle x, y, z \rangle = x\langle 1, 0, 0 \rangle + y\langle 0, 1, 0 \rangle + z\langle 0, 0, 1 \rangle$$
$$= x\mathbf{i} + y\mathbf{j} + z\mathbf{k} \qquad \text{(Figure 1.37)}.$$

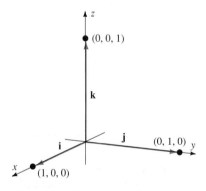

Figure 1.37 The unit coordinate vectors in space.

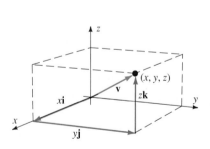

Figure 1.38 The vector $\mathbf{v}$ is equal to $x\mathbf{i} + y\mathbf{j} + z\mathbf{k}$.

Exercise Set 16.1

In Exercises 1–10, sketch the given vector originating at the given point P.

1. $\langle 1, 3 \rangle$ at point $P = (3, 2)$

2. $\langle -6, 2 \rangle$ at point $P = (-1, 1)$

3. $\langle -4, -2 \rangle$ at point $P = (2, 6)$

4. $\langle -2, 4 \rangle$ at point $P = (1, 5)$

5. $\langle 3, -5 \rangle$ at point $P = (1, 1)$

6. $\mathbf{i} + \mathbf{j}$ at point $P = (3, 5)$

7. $-\mathbf{i} + 2\mathbf{j}$ at point $P = (6, 0)$

8. $5\mathbf{i} + 2\mathbf{j}$ at point $P = (-3, 4)$

9. $-3\mathbf{i} + 2\mathbf{j}$ at point $P = (3, -2)$

10. $4\mathbf{j}$ at point $P = (0, 2)$

In Exercises 11–16, let $P = (2, 5)$, $Q = (-1, 2)$, and $R = (2, -6)$. Find

11. $\mathbf{v} = \overrightarrow{PQ}$

12. $\mathbf{v} = \overrightarrow{PQ} + \overrightarrow{QR}$

13. $\mathbf{v} = 2\overrightarrow{PQ} - \overrightarrow{RQ}$

14. $\mathbf{v} = \overrightarrow{PR} + 4\overrightarrow{RQ}$

15. $\mathbf{v} = 2\overrightarrow{RQ} - 2\overrightarrow{RP}$

16. $\mathbf{v} = \overrightarrow{PQ} + 2\overrightarrow{QR} - 3\overrightarrow{RP}$

In Exercises 17–22, let $\mathbf{u} = \langle 3, 1 \rangle$, $\mathbf{v} = \langle -2, 4 \rangle$, and $\mathbf{w} = \langle -4, -2 \rangle$. Find

17. $\mathbf{u} + 2\mathbf{v}$

18. $\mathbf{v} - 2\mathbf{u}$

19. $2\mathbf{u} + 2\mathbf{v} - 2\mathbf{w}$

20. $-3\mathbf{u} + 2\mathbf{v}$

21. $\mathbf{v} - \mathbf{u} - \mathbf{w}$

22. $7\mathbf{u} + 3\mathbf{w} - 6\mathbf{v}$

In Exercises 23–30, let $\mathbf{u} = 3\mathbf{i} - \mathbf{j}$, $\mathbf{v} = 2\mathbf{i} + 6\mathbf{j}$, and $\mathbf{w} = -\mathbf{i} + \mathbf{j}$. Find the indicated vector.

23. $\mathbf{u} + 2\mathbf{v}$

24. $\mathbf{u} - 4\mathbf{w}$

25. $\mathbf{u} + \mathbf{v} + \mathbf{w}$

26. $\mathbf{u} - \mathbf{v} - \mathbf{w}$

27. $3\mathbf{u} + 4\mathbf{v} - 2\mathbf{w}$

28. $6\mathbf{u} - 5\mathbf{w} + \mathbf{v}$

29. $3\mathbf{u} + 3\mathbf{v} + 3\mathbf{w}$

30. $-\mathbf{u} + \mathbf{v} + 2\mathbf{w}$

31. Plot the following points in space: $P_1 = (-2, 1, 5)$, $P_2 = (3, 0, 2)$, $P_3 = (1, 1, 5)$, $P_4 = (2, -3, 6)$, $P_5 = (-3, -4, -5)$.

In Exercises 32–37, let $\mathbf{u} = \langle 2, -1, 5 \rangle$, $\mathbf{v} = \langle -3, 5, 0 \rangle$, and $\mathbf{w} = \langle 3, 3, 1 \rangle$. Find the indicated vector or number.

32. $\mathbf{v} + \mathbf{w}$

33. $\mathbf{u} + 6\mathbf{v}$

34. $\mathbf{v} - \mathbf{w}$

35. $3\mathbf{u} - 2\mathbf{v}$

36. $\dfrac{\mathbf{v}}{|\mathbf{v}|}$

37. $|\mathbf{u} + 4\mathbf{v}|$

In Exercises 38–45, find the length of the given vector.

38. $\langle 6, -1 \rangle$

39. $\langle -3, 4 \rangle$

40. $\langle a, a^2 \rangle$

41. $\mathbf{i}$

42. $\mathbf{i} + \mathbf{j}$

43. $3\mathbf{i} + 4\mathbf{j}$

44. $6\mathbf{i} - 3\mathbf{j}$

45. $4(\mathbf{i} - 3\mathbf{j})$

46. Find the distance between the given pair of points.
a. $P = (1, 0, 1)$ $Q = (3, 2, 1)$
b. $P = (1, 2, -3)$, $Q = (-1, 4, 5)$
c. $P = (-3, 6, 2)$, $Q = (0, 3, 0)$
d. $P = (a, b, c)$, $Q = (2a, 2b, 2c)$

47. Find the point D so that $\overrightarrow{AB} = \overrightarrow{CD}$ if $A = (3, 1, 6)$, $B = (-2, 1, 5)$, and $C = (6, -2, 2)$.

48. Find the vector $\mathbf{v} = \overrightarrow{PQ}$ for
a. $P = (1, 0, 1)$ $Q = (3, 1, 3)$
b. $P = (-3, 2, 6)$, $Q = (1, 1, 1)$
c. $P = (1, -5, 3)$, $Q = (7, -2, 2)$

49. Let $P = (1, 2, -2)$, $Q = (3, 1, 1)$, and $R = (2, 2, 3)$. Determine the lengths of the following vectors.
a. $\overrightarrow{PQ}$ **b.** $\overrightarrow{QP}$
c. $\overrightarrow{PR}$ **d.** $\overrightarrow{QR}$
e. $\overrightarrow{PQ} + \overrightarrow{QR}$ **f.** $-2\overrightarrow{PR}$
g. $\overrightarrow{PQ} + 2\overrightarrow{QR}$

50. Find the point D so that $\overrightarrow{AB} = 2\overrightarrow{CD}$ if $A = (2, 1, 1)$, $B = (4, 7, 6)$, and $C = (-1, 1, 2)$.

51. Find a unit vector pointing in the direction opposite the positive x-axis.

52. Find a unit vector in the direction of $\mathbf{v} = 3\mathbf{i} + 4\mathbf{j}$.

53. Find a unit vector in the plane making an angle of $120°$ with the positive x-axis.

54. Find a vector of length 3 in the direction opposite of $\mathbf{v} = 2\mathbf{i} - 3\mathbf{j}$.

55. Find a unit vector in the direction of $\mathbf{w} = \mathbf{i} + 4\mathbf{j} + 3\mathbf{k}$.

In Exercises 56–59, let $\mathbf{u} = \mathbf{i} + 2\mathbf{j} - \mathbf{k}$, $\mathbf{v} = 3\mathbf{i} - 2\mathbf{j} + 2\mathbf{k}$, and $\mathbf{w} = 5\mathbf{i} - \mathbf{j} + 3\mathbf{k}$. Find the indicated vector or number.

56. $\mathbf{u} + 2\mathbf{v} + \mathbf{w}$

57. $\mathbf{v} - 3\mathbf{w}$

58. $|3\mathbf{v} + \mathbf{w}|$

59. $\dfrac{\mathbf{v}}{|\mathbf{v}|} + \dfrac{\mathbf{w}}{|\mathbf{w}|}$

60. Find a unit vector in space that makes an angle of $30°$ with the positive x-axis and whose $\mathbf{j}$ component is zero.

61. Find two unit vectors in the plane that originate at $(1, 1)$ and that are tangent to the graph of $y = x^3$.

62. Show that the length of the vector originating at $P = (x, y)$ and terminating at $Q = (x + a, y + b)$ is $\sqrt{a^2 + b^2}$.

63. The points $A = (1, 1)$, $B = (5, 1)$, and $C = (6, 3)$ form 3 vertices of a parallelogram. Find the fourth vertex D if
a. A and C lie on a diagonal,
b. A and C lie on a common side.

64. Let $A = (0, 2)$, $B = (b, 5)$, $C = (5, 2)$, $D = (7, 5)$. Find b if $\overrightarrow{AB} + \overrightarrow{AC} = \overrightarrow{AD}$.

65. In the triangle with vertices A, B, and C, let D be the midpoint of side $\overline{AB}$ and let E be the midpoint of side $\overline{BC}$. Prove that $\overrightarrow{DE}$ is parallel to $\overrightarrow{AC}$ and half as long.

66. Prove that the midpoints of the sides of any quadrilateral form the vertices of a parallelogram.

67. Prove statements (ii) through (viii) of Theorem 1.

68. Find the midpoint of the line segment joining the points $P = (-4, 6, 1)$ and $Q = (-1, -3, -11)$.

69. Find the point one third of the distance from $P = (-4, 6, 1)$ to $Q = (-1, -3, -11)$.

70. Let $v_1 = i + j$ and $v_2 = -i + j$. Show that for any vector w in the plane one can find constants c_1 and c_2 so that

$$w = c_1 v_1 + c_2 v_2.$$

(*Hint:* Express w in component form and obtain two linear equations for the unknowns c_1 and c_2.)

71. Generalize Exercise 70 by showing that if v_1 and v_2 are any nonzero and nonparallel vectors in the plane, then any vector w in the plane can be expressed as

$$w = c_1 v_1 + c_2 v_2$$

for appropriate c_1 and c_2.

72. In this exercise, we use the concept of a position vector in the plane to derive a vector equation for a line ℓ in the plane. Recall that a position vector is a vector whose initial point is the origin. Consequently, there is a unique correspondence between points P and position vectors $\overrightarrow{OP}$. Given two points P_1 and P_2 on ℓ, let b be the displacement vector $\overrightarrow{P_1 P_2}$. Moreover, let a be the position vector $\overrightarrow{OP_1}$.
  a. For any scalar t, let $P(t)$ be the terminal point of the position vector $a + tb$. In other words,

$$\overrightarrow{OP(t)} = a + tb.$$

  Show that $P(t)$ is a point on ℓ.
  b. For any point P on ℓ, show that there exists a scalar t such that $P = P(t)$.
  c. Note that parts a and b imply that the line ℓ is the set of terminal points of the position vectors

$$r(t) = \overrightarrow{OP(t)} = a + tb, \qquad -\infty < t < \infty.$$

  Let $r(t) = x(t)i + y(t)j$, $a = a_1 i + a_2 j$, and $b = b_1 i + b_2 j$. Using this notation, derive parametric equations for ℓ.

73. We say that two vectors v and w in space are **parallel** if there is a scalar c so that $v = cw$. The vectors point in the same direction if $c > 0$, and they point in opposite directions if $c < 0$.

a. Find two vectors of length 2 parallel to the vector $v = 2i - 4j + 4k$.
b. Find the constant a so that the vectors $v = i - 3j + 4k$ and $w = ai + 9j - 12k$ are parallel.
c. Find a vector of length 5 in the direction opposite that of $v = i - 2j + 3k$.
d. Find a and b so that the vectors $3i - j + 4k$ and $ai + bj - 2k$ are parallel.

74. Under what geometric conditions does equality hold in the triangle inequality (statement (iii) of Theorem 2), that is, when does $|v + w| = |v| + |w|$?

75. Let $v_1 = i + j$, $v_2 = j + k$, $v_3 = i + 2j + 3k$. Show that given any vector w there exist scalars c_1, c_2, and c_3 so that $w = c_1 v_1 + c_2 v_2 + c_3 v_3$. (*Hint:* Express w in component form and obtain three linear equations for the scalars c_1, c_2, and c_3.)

76. A woman rows a boat at a rate of 30 meters per minute across a stream 150 meters wide. Assuming that she always rows in a direction perpendicular to the banks of the stream, how far downstream does she land if the stream is flowing at a rate of 10 meters per minute?

77. How long does it take the woman in Exercise 76 to reach the opposite bank?

78. At what angle should the woman in Exercise 76 row if she wishes to reach the opposite bank at the point directly opposite her starting point?

79. An airplane flies at a heading of due north at an air speed of 200 km/h in a wind blowing due east at 30 km/h. (The heading is the direction in which the airplane is pointed, and its air speed is its speed with respect to the air, rather than the ground.)
  a. What is the location of the airplane after 2 h?
  b. How far does the airplane travel in 2 h?
  c. At which heading should the airplane fly so as to be moving due north?

80. What force must be overcome in pushing a 200-lb motorbike up a 30° incline, assuming that the only force to be overcome is that due to gravity? (See Example 2.)

81. A force of 50 lb is required to hold a block on a 45° frictionless incline. How much does the block weigh?

16.2 Lines in Space

As you know, a line in the plane is determined by its slope and one point on the line. However, a point and a slope do not provide sufficient information to determine a line in space. Indeed, even the term "slope" becomes ambiguous in space. For example, think about the problem of writing an equation for the line determined by the beam from a stationary searchlight. The beam is determined not only by a *point* (the location of the searchlight) but also by the *direction* in which the searchlight points (Figure 2.1).

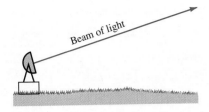

Figure 2.1 Beam from a searchlight.

One of the simplest ways to think about how to find the general form for a line in space is to remember that two points determine a line. If P and Q are two points in space, then the vector $\mathbf{b} = \overrightarrow{PQ}$ is parallel to the line ℓ containing P and Q. Moreover, any multiple $t\mathbf{b}$ is also parallel to ℓ (Figure 2.2).

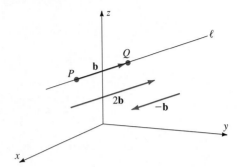

Figure 2.2 Multiples of $\mathbf{b} = \overrightarrow{PQ}$ are parallel to ℓ.

Using the point P, the vector $\mathbf{b}$, and basic vector operations, we describe the line ℓ as a set of **position vectors,** vectors that originate at the origin and terminate on ℓ. We start with the position vector $\mathbf{a} = \overrightarrow{OP}$, which represents the point P. Then, since $t\mathbf{b}$ is parallel to ℓ, the position vector $\mathbf{r}(t) = \mathbf{a} + t\mathbf{b}$ also terminates on the line ℓ (Figure 2.3). As the values of t increase from 0 to ∞, the position vector $\mathbf{a} + t\mathbf{b}$ terminates successively at all points on ℓ lying on one side of P, while as t decreases from 0 to $-\infty$ the position vector $\mathbf{a} + t\mathbf{b}$ determines all points on ℓ lying on the opposite side of P (Figure 2.4). (See Exercise 72 of Section 16.1 for a demonstration of this concept for a line in the plane.)

Thus, a line ℓ in space is determined by a *position vector* $\mathbf{a}$, originating at the origin, and a *direction vector* $\mathbf{b}$:

$$\ell: \mathbf{r}(t) = \mathbf{a} + t\mathbf{b}, \qquad -\infty < t < \infty. \tag{1}$$

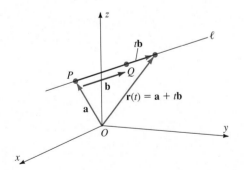

Figure 2.3 If $\mathbf{a} = \overrightarrow{OP}$ and $\mathbf{b} = \overrightarrow{PQ}$, the position vector $\mathbf{r}(t) = \mathbf{a} + t\mathbf{b}$ terminates on ℓ.

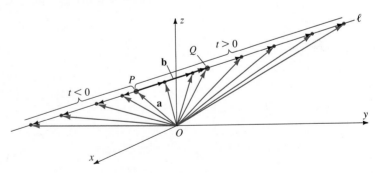

Figure 2.4 The points on the line ℓ are the terminal points of the position vector $\mathbf{r}(t) = \mathbf{a} + t\mathbf{b}$ for $-\infty < t < \infty$.

Equation (1) is an equation for ℓ written in **vector form.** If P and Q are distinct points on ℓ we can always take

$$\mathbf{a} = \overrightarrow{OP} \tag{2}$$

and

$$\mathbf{b} = \overrightarrow{PQ} \tag{3}$$

$\mathbf{b} = \mathbf{i} - \mathbf{j} + 2\mathbf{k}$

$P = (2, -1, 4)$

$\mathbf{a} = 2\mathbf{i} - \mathbf{j} + 4\mathbf{k}$

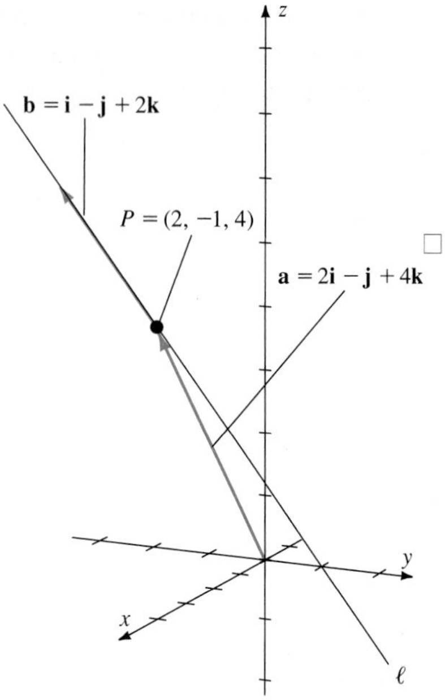

Figure 2.5 The line ℓ in Example 1.

as in the preceding discussion. In this case equation (1) becomes

$$\ell: \mathbf{r}(t) = \overrightarrow{OP} + t\overrightarrow{PQ}, \qquad -\infty < t < \infty.$$

More generally, equation (1) allows us to interpret the line ℓ as an infinite collection of position vectors, each terminating at a point on ℓ.

□ **EXAMPLE 1**

Find an equation for the line containing the point $P = (2, -1, 4)$ and parallel to the vector $\mathbf{v} = \mathbf{i} - \mathbf{j} + 2\mathbf{k}$.

Solution   A position vector is

$$\mathbf{a} = \overrightarrow{OP} = 2\mathbf{i} - \mathbf{j} + 4\mathbf{k}$$

and a direction vector is given as

$$\mathbf{b} = \mathbf{v} = \mathbf{i} - \mathbf{j} + 2\mathbf{k}.$$

The line is, by equation (1), the set of all position vectors of the form

$$\mathbf{r}(t) = (2\mathbf{i} - \mathbf{j} + 4\mathbf{k}) + t(\mathbf{i} - \mathbf{j} + 2\mathbf{k}) \qquad \text{(Figure 2.5).} \qquad \blacksquare$$

□ **EXAMPLE 2**

Find an equation for the line containing the points $P = (4, -1, 2)$ and $Q = (-3, 5, 1)$.

Solution   By equation (2) a position vector is

$$\mathbf{a} = \overrightarrow{OP} = 4\mathbf{i} - \mathbf{j} + 2\mathbf{k}$$

and, by (3), a direction vector is

$$\mathbf{b} = \overrightarrow{PQ} = (-3 - 4)\mathbf{i} + (5 - (-1))\mathbf{j} + (1 - 2)\mathbf{k} = -7\mathbf{i} + 6\mathbf{j} - \mathbf{k}.$$

The line has vector equation

$$\mathbf{r}(t) = (4\mathbf{i} - \mathbf{j} + 2\mathbf{k}) + t(-7\mathbf{i} + 6\mathbf{j} - \mathbf{k}) \qquad \text{(Figure 2.6).} \qquad \blacksquare$$

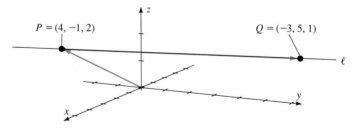

Figure 2.6   The line ℓ in Example 2.

REMARK 1   Equation (1) does not produce a unique representation of a line ℓ. For example, both

$$\mathbf{r}_1(t_1) = t_1\mathbf{i} \quad \text{and} \quad \mathbf{r}_2(t_2) = (2t_2 + 1)\mathbf{i}$$

are vector equations that describe the x-axis. In other words, any point of the form $(x, 0, 0)$, if written as a position vector $\langle x, 0, 0 \rangle$, can be expressed either as $\langle t_1, 0, 0 \rangle$ where $t_1 = x$ or as $\langle 2t_2 + 1, 0, 0 \rangle$ where $t_2 = (x - 1)/2$.

Even though the direction vector **b** plays an important role in the description of the line, it is not unique. In fact, since any two points P and Q on the line determine a direction vector $\mathbf{b} = \overrightarrow{PQ}$, every line actually has infinitely many different direction vectors. However, any two direction vectors $\mathbf{b}_1$ and $\mathbf{b}_2$ for the same line must be nonzero scalar multiples of each other ($\mathbf{b}_1 = c\mathbf{b}_2$ for some scalar c). In Exercise 73 of Section 16.1, we noted that two vectors **v** and **w** are **parallel** if $\mathbf{v} = c\mathbf{w}$ for some scalar c. Therefore, any two direction vectors for a line are parallel. Similarly, parallel lines have parallel direction vectors.

□ EXAMPLE 3

Find an equation for the line ℓ that is parallel to the line in Example 2 and that contains the point $(1, 2, 1)$.

Solution   Since ℓ is parallel to a line with direction vector

$$\mathbf{b} = -7\mathbf{i} + 6\mathbf{j} - \mathbf{k},$$

b is also a direction vector for ℓ. Writing the point $(1, 2, 1)$ as a position vector

$$\mathbf{a} = \mathbf{i} + 2\mathbf{j} + \mathbf{k},$$

we obtain

$$\begin{aligned}\mathbf{r}(t) &= \mathbf{a} + t\mathbf{b} \\ &= (\mathbf{i} + 2\mathbf{j} + \mathbf{k}) + t(-7\mathbf{i} + 6\mathbf{j} - \mathbf{k}).\end{aligned}$$ ■

□ EXAMPLE 4

Find the point of intersection, if it exists, for the lines

$$\ell_1 \colon \mathbf{r}_1(t) = (3\mathbf{i} + 2\mathbf{j} - \mathbf{k}) + t(-6\mathbf{i} + 4\mathbf{j} + 3\mathbf{k})$$

$$\ell_2 \colon \mathbf{r}_2(t) = (5\mathbf{i} + 4\mathbf{j} + 7\mathbf{k}) + t(14\mathbf{i} - 6\mathbf{j} + 2\mathbf{k}).$$

Solution   If ℓ_1 and ℓ_2 intersect at point P, it is not necessary that P is given by the same number t for both $\mathbf{r}_1$ and $\mathbf{r}_2$. The condition of intersection is therefore that

$$\mathbf{r}_1(t_1) = \mathbf{r}_2(t_2) \tag{4}$$

for some numbers t_1 and t_2. The vector equation (4) is equivalent to the three scalar equations obtained by equating the **i**, **j**, and **k** components of $\mathbf{r}_1$ and $\mathbf{r}_2$. Using the given equations for ℓ_1 and ℓ_2, we have

$$3 - 6t_1 = 5 + 14t_2 \tag{5}$$

$$2 + 4t_1 = 4 - 6t_2 \tag{6}$$

$$-1 + 3t_1 = 7 + 2t_2. \tag{7}$$

If we solve (5) and (6) simultaneously, we obtain $t_1 = 2$ and $t_2 = -1$. Then we check that equation (4) holds by verifying equation (7) with these two numbers t_1 and t_2.

To determine the coordinates of the point of intersection, we can use either $\mathbf{r}_1$ or $\mathbf{r}_2$. Using $\mathbf{r}_1$ and $t_1 = 2$, we obtain

$$\mathbf{r}_1(2) = (3\mathbf{i} + 2\mathbf{j} - \mathbf{k}) + 2(-6\mathbf{i} + 4\mathbf{j} + 3\mathbf{k}).$$

Thus the lines intersect at the point $(-9, 10, 5)$.   ■

Parametric Equations for Lines

Suppose that the vector $\mathbf{a} = a_1\mathbf{i} + a_2\mathbf{j} + a_3\mathbf{k}$ and the vector $\mathbf{b} = b_1\mathbf{i} + b_2\mathbf{j} + b_3\mathbf{k}$. Then, the vector equation $\mathbf{r}(t) = \mathbf{a} + t\mathbf{b}$ can be written in component form as

$$x(t)\mathbf{i} + y(t)\mathbf{j} + z(t)\mathbf{k} = (a_1\mathbf{i} + a_2\mathbf{j} + a_3\mathbf{k}) + t(b_1\mathbf{i} + b_2\mathbf{j} + b_3\mathbf{k})$$
$$= (a_1 + tb_1)\mathbf{i} + (a_2 + tb_2)\mathbf{j} + (a_3 + tb_3)\mathbf{k}.$$

Equating components in this equation gives an equation in **parametric form** for the line ℓ with position vector $\mathbf{a} = a_1\mathbf{i} + a_2\mathbf{j} + a_3\mathbf{k}$ and direction vector $\mathbf{b} = b_1\mathbf{i} + b_2\mathbf{j} + b_3\mathbf{k}$:

$$\begin{cases} x(t) = a_1 + tb_1 \\ y(t) = a_2 + tb_2 \\ z(t) = a_3 + tb_3. \end{cases} \qquad (8)$$

The point (x, y, z) is on ℓ if and only if the coordinates satisfy the equations in (8) for some single value of t.

☐ EXAMPLE 5

Find parametric equations for the line in Example 1.

Solution If we collect the coefficients of $\mathbf{i}$, $\mathbf{j}$, and $\mathbf{k}$ of $\mathbf{r}(t)$ in Example 1, we get

$$\mathbf{r}(t) = (2 + t)\mathbf{i} + (-1 - t)\mathbf{j} + (4 + 2t)\mathbf{k}.$$

Thus, by (8)

$$\begin{cases} x(t) = 2 + t \\ y(t) = -1 - t \\ z(t) = 4 + 2t \end{cases}$$

are parametric equations for ℓ. ■

☐ EXAMPLE 6

Find a vector parallel to the line with parametric equations

$$x = -4 + t, \qquad y = 3 - 6t, \quad \text{and} \quad z = 2 + 2t.$$

Solution A vector equation for the line is

$$\mathbf{r}(t) = (-4 + t)\mathbf{i} + (3 - 6t)\mathbf{j} + (2 + 2t)\mathbf{k}$$
$$= (-4\mathbf{i} + 3\mathbf{j} + 2\mathbf{k}) + t(\mathbf{i} - 6\mathbf{j} + 2\mathbf{k})$$

Using equation (1), we see that the vector $\mathbf{i} - 6\mathbf{j} + 2\mathbf{k}$ is a direction vector for the line. ■

☐ EXAMPLE 7

Show that the two lines

$$\ell_1: \begin{cases} x = 3t_1 \\ y = t_1 \\ z = t_1 \end{cases} \qquad \ell_2: \begin{cases} x = 3t_2 \\ y = 2t_2 \\ z = t_2 + 1 \end{cases}$$

do not intersect and are not parallel.

Solution First, we proceed as in the solution to Example 4 to determine the point of intersection. Equating the equations for x, y, and z, we obtain

$$3t_1 = 3t_2 \tag{9}$$

$$t_1 = 2t_2 \tag{10}$$

$$t_1 = t_2 + 1. \tag{11}$$

From (9) and (10), we see that $t_1 = t_2 = 0$. However, these solutions are inconsistent with equation (11). Thus, the lines do not intersect.

To determine if the lines are parallel, we obtain direction vectors from the parametric equations. The coefficients of t_1 in the equations for ℓ_1 yield the direction vector

$$\mathbf{b}_1 = 3\mathbf{i} + \mathbf{j} + \mathbf{k}$$

for ℓ_1. Similarly, we determine the direction vector

$$\mathbf{b}_2 = 3\mathbf{i} + 2\mathbf{j} + \mathbf{k}$$

for ℓ_2. Since the vectors $\mathbf{b}_1$ and $\mathbf{b}_2$ are not parallel, the two lines are not parallel. ∎

REMARK 2 Lines in the plane either intersect or are parallel. Example 7 shows that, for lines in space, there is a third possibility. Lines can be **skew,** that is, both nonparallel and nonintersecting.

Symmetric Equations for Lines

Solving each of the equations in (8) for t gives the equations

$$t = \frac{x(t) - a_1}{b_1}; \qquad t = \frac{y(t) - a_2}{b_2}; \qquad t = \frac{z(t) - a_3}{b_3}.$$

Equating the right-hand sides of each of these equations gives an equation in **symmetric** (or **rectangular**) form for the line ℓ with position vector $\mathbf{a} = a_1\mathbf{i} + a_2\mathbf{j} + a_3\mathbf{k}$ and direction vector $\mathbf{b} = b_1\mathbf{i} + b_2\mathbf{j} + b_3\mathbf{k}$:

$$\boxed{\frac{x - a_1}{b_1} = \frac{y - a_2}{b_2} = \frac{z - a_3}{b_3}.} \tag{12}$$

Since the position vector $\mathbf{a}$ terminates at the point (a_1, a_2, a_3) on ℓ, we may also refer to equation (12) as an equation in symmetric form for the line containing the point (a_1, a_2, a_3) and having direction vector $\mathbf{b} = b_1\mathbf{i} + b_2\mathbf{j} + b_3\mathbf{k}.$

□ **EXAMPLE 8**

Find symmetric equations for the line determined by the parametric equations in Example 5.

Solution Solving for t in the parametric equations in Example 5, we obtain three equations for t, and thus,

$$x - 2 = \frac{y + 1}{-1} = \frac{z - 4}{2}. \qquad \blacksquare$$

☐ **EXAMPLE 9**

Find symmetric equations for the line containing the points $P = (7, 3, -1)$ and $Q = (4, -5, -2)$.

Strategy · · · · · · · ·

Find the direction vector $\mathbf{b} = \overrightarrow{PQ}$.

Solution

The direction vector for the line is

$$\mathbf{b} = \overrightarrow{PQ} = (4 - 7)\mathbf{i} + (-5 - 3)\mathbf{j} + (-2 - (-1))\mathbf{k}$$
$$= -3\mathbf{i} - 8\mathbf{j} - \mathbf{k}.$$

Use $P = (a_1, a_2, a_3)$ and substitute into equations in (12).

Using the point $P = (7, 3, -1)$ on the line and the direction vector $\mathbf{b}$ we obtain from the equations in (12) that

$$\frac{x - 7}{-3} = \frac{y - 3}{-8} = \frac{z + 1}{-1}.$$ ∎

Occasionally, a line described parametrically as in (8) has one or two of the components b_i equal to zero. If this happens, the symmetric representation of the line cannot be written exactly as displayed in equation (12).

☐ **EXAMPLE 10**

Find an equation in symmetric form for the line containing the two points $P = (3, 4, 1)$ and $Q = (2, 5, 1)$.

Solution   First we calculate a direction vector

$$\mathbf{b} = \overrightarrow{PQ} = -\mathbf{i} + \mathbf{j}.$$

Since the $\mathbf{k}$-component of $\mathbf{b}$ is zero, we see that the third coordinate (the z-coordinate) of any point on the line must be 1. Thus, we use the first half of equation (12) as well as the equation $z = 1$ to describe the line. Using $P = (3, 4, 1)$ and $\mathbf{b}$, we get

$$\frac{x - 3}{-1} = \frac{y - 4}{1} \quad \text{and} \quad z = 1.$$

In other words,

$$y = 7 - x \quad \text{and} \quad z = 1.$$ ∎

Exercise Set 16.2

In Exercises 1–7, find a parametric form of the equations for the line described.

1. The line with direction vector $\mathbf{b} = \mathbf{i} + \mathbf{j} - \mathbf{k}$ and containing the point $P = (1, 2, 3)$.

2. The line with direction vector $\mathbf{b} = 2\mathbf{i} - \mathbf{j} + 3\mathbf{k}$ and containing the point $P = (3, -6, 2)$.

3. The line with position and direction vectors $\mathbf{a} = \mathbf{b} = 3\mathbf{i} - \mathbf{j} + 5\mathbf{k}$.

4. The line containing the points $P = (1, 3, -1)$ and $Q = (7, -2, 5)$.

5. The line containing the points $P = (-4, 2, 1)$ and $Q = (-3, 5, 3)$.

6. The line with position vector $\mathbf{a} = \mathbf{i} + 2\mathbf{j} - 6\mathbf{k}$ and direction vector $\mathbf{b} = 2\mathbf{i} - \mathbf{j} + 4\mathbf{k}$.

7. The line that contains the origin and the point $(\pi, 2, \sqrt{2})$.

In Exercises 8–11, find equations in symmetric form for the line determined by the given parametric equations.

8. $x(t) = 2t$
   $y(t) = 3 - t$
   $z(t) = t$

9. $x(t) = t + 7$
   $y(t) = 4t - 6$
   $z(t) = 3 - 2t$

10. $x(t) = 2t + 5$
$y(t) = t - 6$
$z(t) = 5t + 2$

11. $x(t) = t$
$y(t) = 8t - 6$
$z(t) = 4t + 4$

In Exercises 12–14, find symmetric equations for

12. The line in Exercise 2.

13. The line in Exercise 3.

14. The line in Exercise 5.

In Exercises 15–18, find parametric equations for the line determined by the given symmetric equations.

15. $\dfrac{x}{3} = \dfrac{y}{2} = \dfrac{z}{5}$

16. $\dfrac{x-2}{3} = \dfrac{y+1}{-4} = \dfrac{z}{3}$

17. $\dfrac{x+4}{4} = \dfrac{y-2}{-2} = \dfrac{z+3}{3}$

18. $x = \dfrac{y-3}{2} = z + 1$

19. Find a vector parallel to the line in Exercise 9.

20. Find a vector parallel to the line in Exercise 16.

21. Find a direction vector for the line in Exercise 17.

22. Find a direction vector for the line in Exercise 18.

In Exercises 23–28, find the point, if it exists, at which each of the following pairs of lines intersect.

23. $\mathbf{r}_1(t) = (2\mathbf{i} + \mathbf{j} + 2\mathbf{k}) + t(5\mathbf{i} + \mathbf{j} + 3\mathbf{k})$
$\mathbf{r}_2(t) = (-4\mathbf{i} + 7\mathbf{j} + 10\mathbf{k}) + t(3\mathbf{i} - 3\mathbf{j} - 4\mathbf{k})$

24. $\mathbf{r}_1(t) = (3\mathbf{i} + \mathbf{j} + 3\mathbf{k}) + t(\mathbf{i} + \mathbf{j} + \mathbf{k})$
$\mathbf{r}_2(t) = (5\mathbf{i} - 6\mathbf{j} - \mathbf{k}) + t(-\mathbf{i} + 2\mathbf{j} + \mathbf{k})$

25. ℓ_1: $x(t) = 1 + t$, $y(t) = 2 - 2t$, $z(t) = t + 5$
ℓ_2: $x(t) = 2 + 2t$, $y(t) = 5 - 9t$, $z(t) = 2 + 6t$

26. ℓ_1: $x(t) = 3 - t$, $y(t) = 2t$, $z(t) = t$
ℓ_2: $x(t) = 3 + t$, $y(t) = 6 + t$, $z(t) = 6 + 2t$

27. $3 - x = y + 2 = z + 3$
$x + 1 = y = \dfrac{z+1}{2}$

28. $x - 4 = y - 5 = z - 4$
$x - 3 = -y = \dfrac{z+3}{-2}$

29. Find parametric equations for the line that contains the point $(1, 6, -5)$ and is parallel to the line in Exercise 11.

30. Find parametric equations for the line that contains the point $(5, 6, -2)$ and is parallel to the line in Exercise 10.

31. Find symmetric equations for the line that contains the point $(1, 2, 1)$ and is parallel to the line in Exercise 15.

32. Find symmetric equations for the line that contains the point $(-2, \pi, 1)$ and is parallel to the line in Exercise 17.

33. Find a vector equation for the line that contains the point $(1, 2, 4)$ and is parallel to the line in Exercise 9.

34. Find a vector equation for the line that contains the point $(1, 0, 0)$ and is parallel to the line in Exercise 17.

35. Which of the following five lines are parallel? Are any equal?

ℓ_1: $\dfrac{x-3}{4} = \dfrac{y-1}{-2} = \dfrac{z-2}{2}$

ℓ_2: $x(t) = t + 3$, $y(t) = t + 4$, $z(t) = 2 - t$

ℓ_3: $x - 1 = y - 2 = z + 1$

ℓ_4: $\mathbf{r}(t) = (\mathbf{i} + 2\mathbf{j} + \mathbf{k}) + t(2\mathbf{i} - \mathbf{j} + \mathbf{k})$

ℓ_5: $x(t) = 2t + 3$, $y(t) = 2t + 4$, $z(t) = 2t + 2$

36. Show that the following two lines are skew:

$$\dfrac{x+1}{3} = \dfrac{y-2}{5} = \dfrac{z+1}{2}$$

$$x + 2 = \dfrac{y-3}{2} = z + 2.$$

16.3 The Dot Product

Just as the geometric concepts of length and direction are expressed analytically as vectors, the concept of angle is expressed in terms of a **scalar product** of two vectors. To determine the angle between two directions in the plane or in space, we represent the directions by vectors $\mathbf{v}$ and $\mathbf{w}$, and we calculate their **dot product** $\mathbf{v} \cdot \mathbf{w}$. The resulting scalar (real number) is not the desired angle, but we can easily determine the angle from this number.

Definition 5

The **dot product** of the two vectors $\mathbf{v}$ and $\mathbf{w}$ is the number

$$|\mathbf{v}||\mathbf{w}| \cos \theta$$

where θ is the angle between $\mathbf{v}$ and $\mathbf{w}$.

The dot product is also referred to as the **scalar product** or the **inner product** of $\mathbf{v}$ and $\mathbf{w}$.

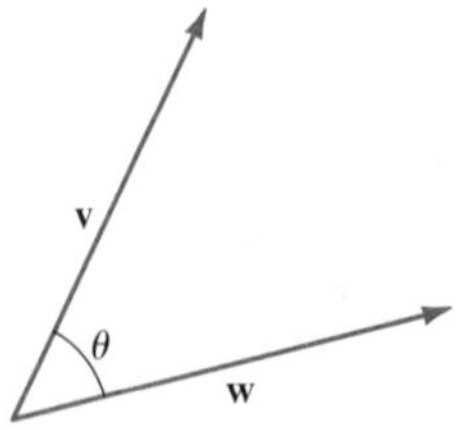

REMARK 1   If two vectors have the same direction, then the angle θ between them is zero. If they point in opposite directions, then $\theta = \pi$ (in radians). Otherwise, we think of the two vectors $\mathbf{v}$ and $\mathbf{w}$ as having the same initial point. Then θ is the angle between 0 and π determined by $\mathbf{v}$ and $\mathbf{w}$ (see Figure 3.1). As usual, we say that $\mathbf{v}$ and $\mathbf{w}$ are **perpendicular** if the angle between them is $\pi/2$. Perpendicular vectors are also said to be **orthogonal.**

Figure 3.1 Angle between two vectors.

□ **EXAMPLE 1**

The dot product

$$\mathbf{i} \cdot (\mathbf{i} + \mathbf{j})$$

equals

$$|\mathbf{i}||\mathbf{i} + \mathbf{j}| \cos(\pi/4) = (1)(\sqrt{2})(\sqrt{2}/2) = 1$$

since we know that the angle between $\mathbf{i}$ and $\mathbf{i} + \mathbf{j}$ is $\pi/4$. ∎

Usually, we use the dot product to *determine* the angle θ between two given vectors, and to do so, we must have another way to compute it.

Theorem 3

If $\mathbf{v} = \langle x_1, y_1 \rangle$ and $\mathbf{w} = \langle x_2, y_2 \rangle$, then

$$\mathbf{v} \cdot \mathbf{w} = x_1 x_2 + y_1 y_2.$$

If $\mathbf{v} = \langle x_1, y_1, z_1 \rangle$ and $\mathbf{w} = \langle x_2, y_2, z_2 \rangle$, then

$$\mathbf{v} \cdot \mathbf{w} = x_1 x_2 + y_1 y_2 + z_1 z_2.$$

Proof: According to our geometric concept of vectors, the vectors $\mathbf{v}$ and $\mathbf{w}$ determine a triangle in which the side opposite the angle θ has length $|\mathbf{v} - \mathbf{w}|$ (Figure 3.2). Thus, according to the Law of Cosines (see Section 1.5),

$$|\mathbf{v} - \mathbf{w}|^2 = |\mathbf{v}|^2 + |\mathbf{w}|^2 - 2|\mathbf{v}||\mathbf{w}| \cos \theta. \tag{1}$$

If $\mathbf{v} = \langle x_1, y_1 \rangle$ and $\mathbf{w} = \langle x_2, y_2 \rangle$, then $\mathbf{v} - \mathbf{w} = \langle x_1 - x_2, y_1 - y_2 \rangle$, so equation (1) may be written in the form

$$(x_1 - x_2)^2 + (y_1 - y_2)^2 = (x_1^2 + y_1^2) + (x_2^2 + y_2^2) - 2|\mathbf{v}||\mathbf{w}| \cos \theta$$

which simplifies to the equation

$$-2x_1 x_2 - 2y_1 y_2 = -2|\mathbf{v}||\mathbf{w}| \cos \theta,$$

or

$$x_1 x_2 + y_1 y_2 = |\mathbf{v}||\mathbf{w}| \cos \theta = \mathbf{v} \cdot \mathbf{w}.$$

If $\mathbf{v}$ and $\mathbf{w}$ are vectors in space, the final step of the proof is essentially the same. ∎

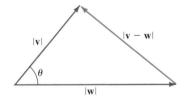

Figure 3.2 Law of Cosines:
$|\mathbf{v} - \mathbf{w}|^2 = |\mathbf{v}|^2 + |\mathbf{w}|^2 - 2|\mathbf{v}||\mathbf{w}| \cos \theta.$

□ **EXAMPLE 2**

For $\mathbf{v} = \langle 2, 3 \rangle$ and $\mathbf{w} = \langle -4, 5 \rangle$,

$$\mathbf{v} \cdot \mathbf{w} = (2)(-4) + (3)(5) = 7. \quad ∎$$

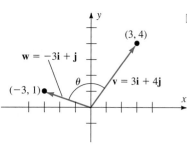

Figure 3.3 The vectors **v** and **w** in Example 4.

□ **EXAMPLE 3**

For **v** = 3**i** + 2**j** − **k** and **w** = **i** + 4**j** + **k**,

$$\mathbf{v} \cdot \mathbf{w} = (3)(1) + (2)(4) + (-1)(1) = 10.$$ ■

If both **v** ≠ **0** and **w** ≠ **0**, we may write Definition (5) in the equivalent form

$$\cos \theta = \frac{\mathbf{v} \cdot \mathbf{w}}{|\mathbf{v}||\mathbf{w}|}.$$ (2)

Equation (2) is useful in calculating angles between vectors, as the following example shows.

□ **EXAMPLE 4**

Find the angle between the vectors **v** = 3**i** + 4**j** and **w** = −3**i** + **j** (Figure 3.3).

Solution We use equation (2). Since

$$\mathbf{v} \cdot \mathbf{w} = 3(-3) + 4 \cdot 1 = -5,$$
$$|\mathbf{v}| = \sqrt{3^2 + 4^2} = \sqrt{25} = 5,$$
$$|\mathbf{w}| = \sqrt{3^2 + 1^2} = \sqrt{10},$$

we have

$$\cos \theta = \frac{-5}{5\sqrt{10}} \approx -0.3162.$$

Since we want an angle θ such that $0 \le \theta \le \pi$, we have

$$\theta = \text{Cos}^{-1}\left(\frac{-5}{5\sqrt{10}}\right) \approx 0.602\pi \ (\approx 108.4°).$$ ■

The dot product provides a useful criterion for determining whether two nonzero vectors are orthogonal. If $|\mathbf{v}| \ne 0$ and $|\mathbf{w}| \ne 0$, equation (2) shows that $\mathbf{v} \cdot \mathbf{w} = 0$ if and only if $\cos \theta = 0$. Since θ is the angle between **v** and **w**, we know that $0 \le \theta \le \pi$. Thus, $\cos \theta = 0$ if and only if $\theta = \pi/2$. Combining these observations gives the following useful observation.

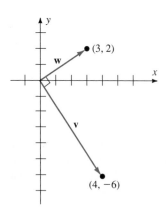

Figure 3.4 The vectors **v** and **w** in Example 5.

The nonzero vectors **v** and **w** are orthogonal if and only if $\mathbf{v} \cdot \mathbf{w} = 0$.

□ **EXAMPLE 5**

Show that the vectors **v** = 4**i** − 6**j** and **w** = 3**i** + 2**j** are orthogonal (Figure 3.4).

Solution Since

$$\mathbf{v} \cdot \mathbf{w} = 4 \cdot 3 + (-6) \cdot 2 = 0,$$

the vectors are orthogonal. ■

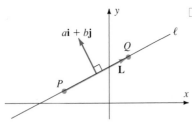

Figure 3.5   The vector $a\mathbf{i} + b\mathbf{j}$ in the plane is perpendicular to the line ℓ with equation $ax + by + c = 0$.

☐ **EXAMPLE 6**

Show that the vector $a\mathbf{i} + b\mathbf{j}$ is orthogonal to the line ℓ with equation

$$ax + by + c = 0.$$

(See Figure 3.5.)

Solution   Our strategy is first to find a vector $\mathbf{L}$ parallel to the line ℓ and then to show that $(a\mathbf{i} + b\mathbf{j}) \cdot \mathbf{L} = 0$. Let $P = (x_1, y_1)$ and $Q = (x_2, y_2)$ be distinct points on ℓ. Then $\overrightarrow{PQ}$ is a segment of ℓ, so $\mathbf{L} = \overrightarrow{PQ}$ represents a vector parallel to ℓ, which has the component form

$$\mathbf{L} = (x_2 - x_1)\mathbf{i} + (y_2 - y_1)\mathbf{j}.$$

Now, since P and Q lie on ℓ, both

$$ax_1 + by_1 + c = 0 \quad \text{and} \quad ax_2 + by_2 + c = 0.$$

Subtracting corresponding sides of these equations gives the equation

$$a(x_2 - x_1) + b(y_2 - y_1) = 0,$$

which we can rewrite as $(a\mathbf{i} + b\mathbf{j}) \cdot \mathbf{L} = 0$. This shows that $a\mathbf{i} + b\mathbf{j}$ is orthogonal to $\mathbf{L}$, and, hence, to ℓ. ■

We shall make frequent use of the following properties of the dot product.

Theorem 4
Properties of the Dot Product

Let $\mathbf{u}$, $\mathbf{v}$, and $\mathbf{w}$ be vectors and let c be a real number. Then

(i) $\mathbf{v} \cdot \mathbf{v} = |\mathbf{v}|^2$,

(ii) $\mathbf{v} \cdot \mathbf{w} = \mathbf{w} \cdot \mathbf{v}$ ($\cdot$ is commutative),

(iii) $\mathbf{u} \cdot (\mathbf{v} + \mathbf{w}) = \mathbf{u} \cdot \mathbf{v} + \mathbf{u} \cdot \mathbf{w}$ ($\cdot$ distributes over vector addition),

(iv) $(c\mathbf{v}) \cdot \mathbf{w} = c(\mathbf{v} \cdot \mathbf{w})$ (scalars may be factored),

(v) $\mathbf{0} \cdot \mathbf{v} = 0$,

(vi) $|\mathbf{v} \cdot \mathbf{w}| \leq |\mathbf{v}||\mathbf{w}|$ (Schwarz inequality).

Proof: The proofs of statements (i) through (v) follow either from the definition of the dot product or from Theorem 3. To prove statement (iii) for vectors $\mathbf{u}$, $\mathbf{v}$, and $\mathbf{w}$ in the plane, we write them in component form as

$$\mathbf{u} = \langle x_1, y_1 \rangle, \quad \mathbf{v} = \langle x_2, y_2 \rangle, \quad \text{and} \quad \mathbf{w} = \langle x_3, y_3 \rangle.$$

Then

$$\begin{aligned}
\mathbf{u} \cdot (\mathbf{v} + \mathbf{w}) &= x_1(x_2 + x_3) + y_1(y_2 + y_3) \\
&= x_1 x_2 + x_1 x_3 + y_1 y_2 + y_1 y_3 \\
&= (x_1 x_2 + y_1 y_2) + (x_1 x_3 + y_1 y_3) \\
&= \mathbf{u} \cdot \mathbf{v} + \mathbf{u} \cdot \mathbf{w}.
\end{aligned}$$

To prove the Schwarz inequality, we use the definition of the dot product. If either $\mathbf{v} = \mathbf{0}$ or $\mathbf{w} = \mathbf{0}$, the inequality holds, since both sides are zero. Otherwise, an angle θ is determined between $\mathbf{v}$ and $\mathbf{w}$. Since $|\cos \theta| \leq 1$ for all θ, we conclude that

$$|\mathbf{v} \cdot \mathbf{w}| = |\mathbf{v}||\mathbf{w}||\cos \theta| \leq |\mathbf{v}||\mathbf{w}|.$$

The proofs of the remaining properties are similar and are left as exercises. ■

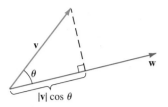

Figure 3.6 Component of **v** in the direction of **w** is

$$|\mathbf{v}| \cos \theta = \frac{\mathbf{v} \cdot \mathbf{w}}{|\mathbf{w}|}.$$

We can use the Schwarz inequality to prove the triangle inequality (Theorem 2, part (iii), Section 16.1):

$$|\mathbf{v} + \mathbf{w}| \le |\mathbf{v}| + |\mathbf{w}|. \tag{3}$$

The strategy will be to prove the inequality

$$|\mathbf{v} + \mathbf{w}|^2 \le (|\mathbf{v}| + |\mathbf{w}|)^2 \tag{4}$$

from which inequality (3) is obtained by taking square roots.

Applying Theorem 4 parts (i) through (iii), we find that

$$\begin{aligned} |\mathbf{v} + \mathbf{w}|^2 &= (\mathbf{v} + \mathbf{w}) \cdot (\mathbf{v} + \mathbf{w}) \\ &= \mathbf{v} \cdot \mathbf{v} + 2\mathbf{v} \cdot \mathbf{w} + \mathbf{w} \cdot \mathbf{w} \\ &= |\mathbf{v}|^2 + 2\mathbf{v} \cdot \mathbf{w} + |\mathbf{w}|^2. \end{aligned} \tag{5}$$

Now the Schwarz inequality shows that

$$2\mathbf{v} \cdot \mathbf{w} \le 2|\mathbf{v} \cdot \mathbf{w}| \le 2|\mathbf{v}|\,|\mathbf{w}|. \tag{6}$$

Combining (5) and (6), we obtain the inequality

$$\begin{aligned} |\mathbf{v} + \mathbf{w}|^2 &\le |\mathbf{v}|^2 + 2|\mathbf{v}|\,|\mathbf{w}| + |\mathbf{w}|^2 \\ &= (|\mathbf{v}| + |\mathbf{w}|)^2, \end{aligned}$$

which proves inequality (4).

Components

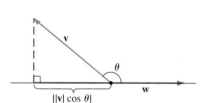

Figure 3.7 The component $|\mathbf{v}| \cos \theta$ is negative if $\pi/2 < \theta < \pi$. Its absolute value is the length of the projection of **v** onto the line determined by **w**.

Suppose **v** is an arbitrary vector, **w** is a nonzero vector, and let θ be the angle formed between **v** and **w**. The number $|\mathbf{v}| \cos \theta$ is called **the component of v with respect to w,** written $\text{comp}_{\mathbf{w}}\,\mathbf{v} = |\mathbf{v}| \cos \theta$ (Figure 3.6). The number $\text{comp}_{\mathbf{w}}\,\mathbf{v}$ can be interpreted as the change, in the direction of **w**, represented by the vector **v**. Using the dot product we write

$$|\mathbf{v}| \cos \theta = \frac{|\mathbf{v}|\,|\mathbf{w}| \cos \theta}{|\mathbf{w}|} = \frac{\mathbf{v} \cdot \mathbf{w}}{|\mathbf{w}|}$$

so

$$\boxed{\ \text{comp}_{\mathbf{w}}\,\mathbf{v} = \frac{\mathbf{v} \cdot \mathbf{w}}{|\mathbf{w}|}.\ }$$

Notice that

$$\text{comp}_{\mathbf{i}}\,\mathbf{v} = \frac{\mathbf{v} \cdot \mathbf{i}}{|\mathbf{i}|} = \mathbf{v} \cdot \mathbf{i}.$$

Since $\mathbf{v} \cdot \mathbf{i}$ is the first component (the **i**-component) of **v**, we see that $\text{comp}_{\mathbf{w}}\,\mathbf{v}$ is a generalization of the concept of the **i**-component. We also note that, if $\theta > \pi/2$, then $\text{comp}_{\mathbf{w}}\,\mathbf{v}$ is negative because $\cos \theta < 0$ (see Figure 3.7).

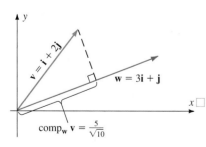

Figure 3.8 The component $\text{comp}_{\mathbf{w}}\,\mathbf{v}$ in Example 7.

☐ **EXAMPLE 7**

For $\mathbf{v} = \mathbf{i} + 2\mathbf{j}$ and $\mathbf{w} = 3\mathbf{i} + \mathbf{j}$,

$$\text{comp}_{\mathbf{w}}\,\mathbf{v} = \frac{1 \cdot 3 + 2 \cdot 1}{\sqrt{3^2 + 1^2}} = \frac{5}{\sqrt{10}} \qquad \text{(Figure 3.8).} \qquad ■$$

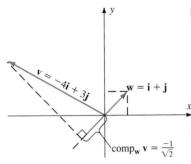

Figure 3.9 The component $\text{comp}_{\mathbf{w}}\mathbf{v}$ in Example 8.

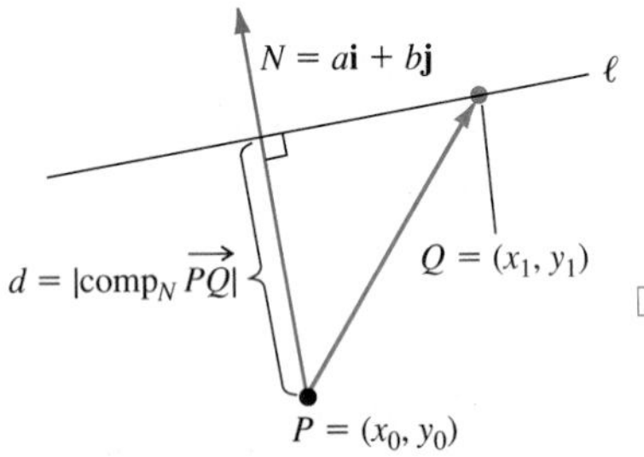

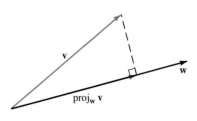

Figure 3.10 Distance from (x_0, y_0) to $\ell: ax + by + c = 0$ is $|\text{comp}_{\mathbf{N}}\overrightarrow{PQ}|$.

□ **EXAMPLE 8**

For $\mathbf{v} = -4\mathbf{i} + 3\mathbf{j}$ and $\mathbf{w} = \mathbf{i} + \mathbf{j}$

$$\text{comp}_{\mathbf{w}}\mathbf{v} = \frac{(-4)(1) + 3 \cdot 1}{\sqrt{1^2 + 1^2}} = \frac{-1}{\sqrt{2}} \qquad \text{(Figure 3.9)}. \qquad ∎$$

□ **EXAMPLE 9**

Let $\mathbf{v} = 3\mathbf{i} + 2\mathbf{j} - \mathbf{k}$ and $\mathbf{w} = \mathbf{i} + 4\mathbf{j} + \mathbf{k}$. Then

$$\text{comp}_{\mathbf{w}}\mathbf{v} = \frac{\mathbf{v} \cdot \mathbf{w}}{|\mathbf{w}|} = \frac{10}{\sqrt{18}} = \frac{10}{3\sqrt{2}}. \qquad ∎$$

The notion of component can be used to find the distance from the point $P = (x_0, y_0)$ to the line ℓ in the plane with equation $\ell: ax + by + c = 0$. The idea is summarized by Figure 3.10: if $\mathbf{N}$ is a vector orthogonal to ℓ, and if $Q = (x_1, y_1)$ is any point on ℓ, then $|\text{comp}_{\mathbf{N}}\overrightarrow{PQ}|$ is the required distance.

□ **EXAMPLE 10**

Find the distance from the point $P = (4, -2)$ to the line in the plane with equation $3x - y + 4 = 0$.

Solution From Example 6, we note that the vector $\mathbf{N} = 3\mathbf{i} - \mathbf{j}$ is perpendicular to the line. In order to use the observation above (see Figure 3.10), we need one point Q on the line. Setting $x = 0$, we obtain $y = 4$, so we use $Q = (0, 4)$. The distance is

$$\begin{aligned}
|\text{comp}_{\mathbf{N}}\overrightarrow{PQ}| &= \frac{|\overrightarrow{PQ} \cdot \mathbf{N}|}{|\mathbf{N}|} \\
&= \frac{|(-4\mathbf{i} + 6\mathbf{j}) \cdot (3\mathbf{i} - \mathbf{j})|}{|3\mathbf{i} - \mathbf{j}|} \\
&= \frac{|-18|}{\sqrt{10}} = \frac{18}{\sqrt{10}}. \qquad ∎
\end{aligned}$$

Projections

The component $\text{comp}_{\mathbf{w}}\mathbf{v}$ is a *number* that indicates the length and direction of the projection of $\mathbf{v}$ onto $\mathbf{w}$. In order to obtain the *vector* that is the **orthogonal projection** of $\mathbf{v}$ onto $\mathbf{w}$ (Figure 3.11), we multiply the unit vector $\mathbf{w}/|\mathbf{w}|$ by the scalar $\text{comp}_{\mathbf{w}}\mathbf{v}$. We obtain the vector

$$\text{proj}_{\mathbf{w}}\mathbf{v} = \text{comp}_{\mathbf{w}}\mathbf{v}\left(\frac{\mathbf{w}}{|\mathbf{w}|}\right) = \left(\frac{\mathbf{v} \cdot \mathbf{w}}{|\mathbf{w}|}\right)\left(\frac{\mathbf{w}}{|\mathbf{w}|}\right) = \left(\frac{\mathbf{v} \cdot \mathbf{w}}{|\mathbf{w}|^2}\right)\mathbf{w}. \qquad (7)$$

Since $\mathbf{w}/|\mathbf{w}|$ is a unit vector, equation (7) indicates that

$$|\text{proj}_{\mathbf{w}}\mathbf{v}| = \left|\frac{\mathbf{w} \cdot \mathbf{v}}{|\mathbf{w}|}\right| = |\text{comp}_{\mathbf{w}}\mathbf{v}|.$$

That is, the length of the vector $\text{proj}_{\mathbf{w}}\mathbf{v}$ equals the absolute value of the number $\text{comp}_{\mathbf{w}}\mathbf{v}$.

Figure 3.11 Projection of $\mathbf{v}$ onto $\mathbf{w}$.

The vector

$$\text{proj}_{\perp \mathbf{w}}\, \mathbf{v} = \mathbf{v} - \text{proj}_{\mathbf{w}}\, \mathbf{v} \qquad (8)$$

is called the **orthogonal projection of v perpendicular to w.**

It is obvious from equation (8) that

$$\mathbf{v} = \text{proj}_{\mathbf{w}}\, \mathbf{v} + \text{proj}_{\perp \mathbf{w}}\, \mathbf{v} \qquad \text{(Figure 3.12)}.$$

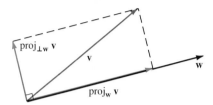

Figure 3.12 $\mathbf{v} = \text{proj}_{\mathbf{w}}\, \mathbf{v} + \text{proj}_{\perp \mathbf{w}}\, \mathbf{v}$.

In other words, $\text{proj}_{\mathbf{w}}\, \mathbf{v}$ and $\text{proj}_{\perp \mathbf{w}}\, \mathbf{v}$ may be interpreted as adjacent sides of a parallelogram for which **v** is the diagonal. In fact, the parallelogram is actually a rectangle. That is,

$$(\text{proj}_{\mathbf{w}}\, \mathbf{v}) \cdot (\text{proj}_{\perp \mathbf{w}}\, \mathbf{v}) = 0. \qquad (9)$$

The proof of (9) is simply a calculation:

$$(\text{proj}_{\mathbf{w}}\, \mathbf{v}) \cdot (\text{proj}_{\perp \mathbf{w}}\, \mathbf{v}) = \left[\left(\frac{\mathbf{v} \cdot \mathbf{w}}{|\mathbf{w}|^2} \right) \mathbf{w} \right] \left[\mathbf{v} - \left(\frac{\mathbf{v} \cdot \mathbf{w}}{|\mathbf{w}|^2} \right) \mathbf{w} \right]$$

$$= \left(\frac{\mathbf{v} \cdot \mathbf{w}}{|\mathbf{w}|^2} \right) \mathbf{w} \cdot \mathbf{v} - \left(\frac{\mathbf{v} \cdot \mathbf{w}}{|\mathbf{w}|^2} \right)^2 \mathbf{w} \cdot \mathbf{w}$$

$$= \left(\frac{\mathbf{v} \cdot \mathbf{w}}{|\mathbf{w}|} \right)^2 - \left(\frac{\mathbf{v} \cdot \mathbf{w}}{|\mathbf{w}|^2} \right)^2 |\mathbf{w}|^2$$

$$= 0.$$

Equation (9) explains the use of the terminology "orthogonal" projection in defining $\text{proj}_{\perp \mathbf{w}}\, \mathbf{v}$. The projections $\text{proj}_{\mathbf{w}}\, \mathbf{v}$ and $\text{proj}_{\perp \mathbf{w}}\, \mathbf{v}$ allow us to express a given vector **v** as the sum of a vector parallel to a given vector or line and a vector orthogonal to the given vector or line. This concept is particularly useful in physics, where one often needs to "resolve" a force or velocity vector into **component vectors** parallel and perpendicular to a given force or velocity vector (recall Example 2 in Section 16.1).

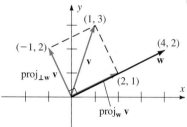

Figure 3.13   Projections of **v** onto and orthogonal to **w**.

□ **EXAMPLE 11**

For $\mathbf{v} = \mathbf{i} + 3\mathbf{j}$ and $\mathbf{w} = 4\mathbf{i} + 2\mathbf{j}$,

(a) $\text{proj}_{\mathbf{w}}\, \mathbf{v} = \left[\dfrac{(\mathbf{i} + 3\mathbf{j}) \cdot (4\mathbf{i} + 2\mathbf{j})}{|4\mathbf{i} + 2\mathbf{j}|^2} \right] (4\mathbf{i} + 2\mathbf{j})$

$\qquad\qquad = \left[\dfrac{1 \cdot 4 + 3 \cdot 2}{4^2 + 2^2} \right] (4\mathbf{i} + 2\mathbf{j})$

$\qquad\qquad = 2\mathbf{i} + \mathbf{j},$

and

$$\text{proj}_{\perp \mathbf{w}}\, \mathbf{v} = (\mathbf{i} + 3\mathbf{j}) - (2\mathbf{i} + \mathbf{j}) = -\mathbf{i} + 2\mathbf{j} \qquad \text{(Figure 3.13)}.$$

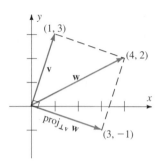

Figure 3.14 $\mathbf{v} = \text{proj}_{\mathbf{v}}\,\mathbf{w}$.

(b) $\text{proj}_{\mathbf{v}}\,\mathbf{w} = \left[\dfrac{(4\mathbf{i} + 2\mathbf{j}) \cdot (\mathbf{i} + 3\mathbf{j})}{|\mathbf{i} + 3\mathbf{j}|^2}\right](\mathbf{i} + 3\mathbf{j})$

$\qquad\qquad = \left[\dfrac{4 \cdot 1 + 2 \cdot 3}{1^2 + 3^2}\right](\mathbf{i} + 3\mathbf{j})$

$\qquad\qquad = \mathbf{i} + 3\mathbf{j} \quad (=\mathbf{v}),$

and

$\qquad \text{proj}_{\perp\mathbf{v}}\,\mathbf{w} = (4\mathbf{i} + 2\mathbf{j}) - (\mathbf{i} + 3\mathbf{j}) = 3\mathbf{i} - \mathbf{j} \qquad \text{(Figure 3.14).}$ ■

□ **EXAMPLE 12**

Find the force required to hold a 5000-kilogram automobile motionless on a 30° incline, assuming that the only force that must be overcome is that due to gravity.

Solution   We may represent the force due to gravity as a vector $\mathbf{F}$ pointing vertically downward. According to Newton's second law, the magnitude of this force vector is $|\mathbf{F}| = mg = 5000g$, where $g = 9.8$ m/s^2 is the acceleration due to gravity. Thus

$$\mathbf{F} = (-5000g)\mathbf{j}.$$

By equation (6), Section 16.1, a unit vector $\mathbf{u}$ in the direction of the 30° incline is represented in Figure 3.15 and by the equation

$$\mathbf{u} = \cos 30°\mathbf{i} + \sin 30°\mathbf{j} = \frac{\sqrt{3}}{2}\mathbf{i} + \frac{1}{2}\mathbf{j}.$$

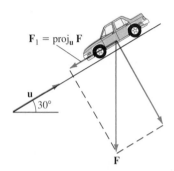

Figure 3.15 $|\mathbf{F}_1| = |\text{proj}_{\mathbf{u}}\,\mathbf{F}|$ is the force required to prevent auto from rolling down the incline.

The force due to $\mathbf{F}$ acting in the direction of the incline is therefore represented by the vector

$$\mathbf{F}_1 = \text{proj}_{\mathbf{u}}\,\mathbf{F} = \frac{\mathbf{F} \cdot \mathbf{u}}{|\mathbf{u}|^2}\mathbf{u},$$

which equals $(\mathbf{F} \cdot \mathbf{u})\mathbf{u}$ because $\mathbf{u}$ is a unit vector. Consequently,

$$\mathbf{F}_1 = -2500g\left(\frac{\sqrt{3}}{2}\mathbf{i} + \frac{1}{2}\mathbf{j}\right).$$

The magnitude of this force is

$$|\mathbf{F}_1| = |\text{comp}_{\mathbf{u}}\,\mathbf{F}| = 2500g.$$

This is the force due to gravity that must be overcome to hold the automobile in place. ■

□ **EXAMPLE 13**

When a constant force of magnitude F moves the point at which it is applied a distance d *in the direction in which the force is applied,* the work done by the force is defined to be the product.

$$W = Fd. \qquad (10)$$

However, if the point of application is confined to move along a line in a direction different from that of the applied force, equation (10) does not apply. If the

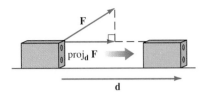

Figure 3.16 Work done by force **F** over the displacement **d** is $|\text{proj}_{\mathbf{d}} \mathbf{F}||\mathbf{d}| = \mathbf{F} \cdot \mathbf{d}$.

vector **F** represents the applied force and the vector **d** represents the resulting displacement, the work done is defined to be the dot product

$$W = \mathbf{F} \cdot \mathbf{d}.$$

Figure 3.16 illustrates this definition. The vector component of **F** acting in the direction of **d** is $\text{proj}_{\mathbf{d}} \mathbf{F}$, whose magnitude is $\text{comp}_{\mathbf{d}} \mathbf{F}$. Since this is the force acting in the direction of motion, equation (10) gives

$$W = (\text{comp}_{\mathbf{d}} \mathbf{F})(|\mathbf{d}|) = \frac{\mathbf{F} \cdot \mathbf{d}}{|\mathbf{d}|}(|\mathbf{d}|) = \mathbf{F} \cdot \mathbf{d}. \qquad ■$$

Equations for Planes

We often think of a plane in space as being determined by two intersecting lines or by three noncollinear points. However, to find the equation of a plane, we use a less familiar geometric fact: through any point P on a line ℓ, there is exactly one plane σ perpendicular to ℓ (see Figure 3.17). In other words, a point $Q \neq P$ lies on the plane σ if and only if the vector $\overrightarrow{PQ}$ is perpendicular to ℓ. This observation is the basis for the following analytic description of a plane.

If **n** is a direction vector for the line ℓ, the vector $\overrightarrow{PQ}$ is perpendicular to the line ℓ if and only if the vectors $\overrightarrow{PQ}$ and **n** are orthogonal. This last condition is equivalent to the condition that $\overrightarrow{PQ} \cdot \mathbf{n} = 0$. We may therefore write *the plane σ determined by the point P and the vector* **n** *as*

$$\sigma = \{Q \mid \overrightarrow{PQ} \cdot \mathbf{n} = 0\}; \qquad (11)$$

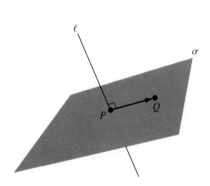

Figure 3.17 The plane σ is the unique plane that is perpendicular to the line ℓ at the point P.

that is, the plane σ is the set of points Q whose displacement vector $\overrightarrow{PQ}$ from the point P is perpendicular to the vector **n**. The vector **n** in equation (11) is referred to as a **normal vector** for the plane σ (Figure 3.18).

To develop an equation in *xyz*-coordinates for the plane σ, we express **n** in component form as $\mathbf{n} = a\mathbf{i} + b\mathbf{j} + c\mathbf{k}$, and we assume the point P to have coordinates $P = (x_0, y_0, z_0)$. Let the point Q have coordinates $Q = (x, y, z)$. Then the vector $\overrightarrow{PQ}$ is

$$\overrightarrow{PQ} = (x - x_0)\mathbf{i} + (y - y_0)\mathbf{j} + (z - z_0)\mathbf{k}, \qquad (12)$$

and the condition in line (11) is that

$$[(x - x_0)\mathbf{i} + (y - y_0)\mathbf{j} + (z - z_0)\mathbf{k}] \cdot (a\mathbf{i} + b\mathbf{j} + c\mathbf{k}) = 0. \qquad (13)$$

Equation (13) simplifies to the equation

$$a(x - x_0) + b(y - y_0) + c(z - z_0) = 0,$$

or

$$\boxed{ax + by + cz = d} \qquad (14)$$

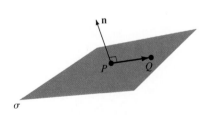

Figure 3.18 The plane σ is determined by the point P and the normal vector **n**.

where d is the constant $ax_0 + by_0 + cz_0$.

Equation (14) is an *equation for the plane with normal vector* $\mathbf{n} = a\mathbf{i} + b\mathbf{j} + c\mathbf{k}$ *containing the point* (x_0, y_0, z_0). Notice that the coefficients of x, y, and z are just the respective components of the normal vector **n**, while the right-hand side is a *constant* determined from **n** and P.

☐ **EXAMPLE 14**

Find an equation for the plane that is perpendicular to the vector $\mathbf{n} = \mathbf{i} + 4\mathbf{j} + 2\mathbf{k}$ and that contains the point $P = (6, -3, 4)$.

Solution   Let $Q = (x, y, z)$ be an arbitrary point in the desired plane. Then

$$\mathbf{n} \cdot \overrightarrow{PQ} = 0.$$

Thus

$$(\mathbf{i} + 4\mathbf{j} + 2\mathbf{k}) \cdot [(x - 6)\mathbf{i} + (y + 3)\mathbf{j} + (z - 4)\mathbf{k}] = 0$$

and

$$(x - 6) + 4(y + 3) + 2(z - 4) = 0.$$

The desired equation is therefore

$$x + 4y + 2z = 2. \qquad \blacksquare$$

REMARK 2   Using equation (14), we know that any plane can be represented by an equation of the form

$$ax + by + cz = d, \tag{15}$$

given a vector $\mathbf{n} = a\mathbf{i} + b\mathbf{j} + c\mathbf{k}$ that is normal (perpendicular) to the plane. In Exercise 70, we shall see that *any* equation in the form of equation (15) is the equation of a plane (assuming that one of the numbers a, b, or c is nonzero) and that the vector $\mathbf{n} = a\mathbf{i} + b\mathbf{j} + c\mathbf{k}$ is perpendicular to that plane.

☐ **EXAMPLE 15**

Find a vector normal to the plane with equation $3x + 5y - z = 9$.

Solution   Using the remark directly above, the components of a normal vector $\mathbf{n}$ are simply the coefficients of x, y, and z. Thus, the vector

$$\mathbf{n} = 3\mathbf{i} + 5\mathbf{j} - \mathbf{k}$$

is a normal vector for the plane. $\qquad \blacksquare$

We can use normal vectors and the dot product to determine the distance of a point to a plane.

☐ **EXAMPLE 16**

Find the distance between the point $P = (3, -1, 2)$ and the plane with equation $2x - y + z = 4$.

Strategy · · · · · · · ·   **Solution**

Note that, given any point Q in the plane, $|\text{comp}_{\mathbf{n}} \overrightarrow{PQ}|$ is the desired distance (see Figure 3.19).

Find a normal $\mathbf{n}$ for the plane.   A normal vector for the plane is

$$\mathbf{n} = 2\mathbf{i} - \mathbf{j} + \mathbf{k}.$$

Set $x = y = 0$ and solve for z to find   A point Q in the plane is $Q = (0, 0, 4)$. A vector $\mathbf{v}$ parallel to $\mathbf{n}$ originating at P and
any convenient point Q in the plane.   terminating in the plane is

$$\mathbf{v} = \text{proj}_{\mathbf{n}} \overrightarrow{PQ}$$

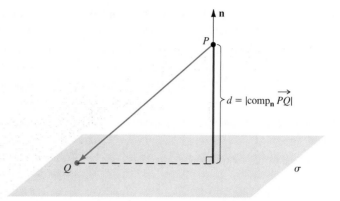

Figure 3.19 Distance from a point P to a plane σ.

Project $\overrightarrow{PQ}$ onto **n**. Recall that

$$|\text{proj}_\mathbf{n}\, \overrightarrow{PQ}| = |\text{comp}_\mathbf{n}\, \overrightarrow{PQ}|.$$

The desired distance is the length of **v**

$$|\mathbf{v}| = |\text{comp}_\mathbf{n}\, \overrightarrow{PQ}| = \frac{|\overrightarrow{PQ} \cdot \mathbf{n}|}{|\mathbf{n}|}$$

$$= \frac{|(-3\mathbf{i} + \mathbf{j} + 2\mathbf{k}) \cdot (2\mathbf{i} - \mathbf{j} + \mathbf{k})|}{|2\mathbf{i} - \mathbf{j} + \mathbf{k}|}$$

$$= \frac{5}{\sqrt{6}}$$

You may wish to try another convenient point Q, say $(2, 0, 0)$, and verify that the result is the same. ∎

Exercise Set 16.3

In Exercises 1–8, find the dot product $\mathbf{v} \cdot \mathbf{w}$.

1. $\mathbf{v} = \mathbf{i} + 2\mathbf{j}, \quad \mathbf{w} = 3\mathbf{i} - \mathbf{j}$

2. $\mathbf{v} = \langle -3, 5 \rangle, \quad \mathbf{w} = \langle 1, 4 \rangle$

3. $\mathbf{v} = \langle -3, 5 \rangle, \quad \mathbf{w} = \langle 6, -2 \rangle$

4. $\mathbf{v} = a\mathbf{i} + b\mathbf{j}, \quad \mathbf{w} = c\mathbf{i} + d\mathbf{j}$

5. $\mathbf{v} = 2\mathbf{i} + \mathbf{j} - \mathbf{k}, \quad \mathbf{w} = \mathbf{i} - 2\mathbf{j} + 3\mathbf{k}$

6. $\mathbf{v} = \langle 2, -3, 1 \rangle, \quad \mathbf{w} = \langle 2, 4, 1 \rangle$

7. $\mathbf{v} = \mathbf{i} + 4\mathbf{j} - \mathbf{k}, \quad \mathbf{w} = \mathbf{j} + 2\mathbf{k}$

8. $\mathbf{v} = \mathbf{i} + \mathbf{j}, \quad \mathbf{w} = -2\mathbf{j} + 2\mathbf{k}$

9. Find the cosine of the angle between the vectors $\mathbf{v} = \mathbf{i} - 3\mathbf{j}$ and $\mathbf{w} = -4\mathbf{i} + \mathbf{j}$.

10. Find the cosine of the angle between the vectors $\mathbf{v} = \langle 3, 2 \rangle$ and $\mathbf{w} = \langle -2, 2 \rangle$.

11. Find the cosine of the angle between the vectors $\mathbf{v} = 2\mathbf{i} - \mathbf{j} + 2\mathbf{k}$ and $\mathbf{w} = \mathbf{i} - \mathbf{j} + \mathbf{k}$.

12. Find the cosine of the angle between the vectors $\mathbf{v} = \langle 1, 1, 1 \rangle$ and $\mathbf{w} = \langle 2, 1, 2 \rangle$.

13. For the vectors $\mathbf{u} = -2\mathbf{i} + 4\mathbf{j}$, $\mathbf{v} = 3\mathbf{i} + 5\mathbf{j}$, and $\mathbf{w} = 6\mathbf{i} - 4\mathbf{j}$ find

    a. $\mathbf{u} \cdot \mathbf{v}$ b. $\mathbf{u} \cdot \mathbf{w}$

    c. $\mathbf{u} \cdot (\mathbf{v} + \mathbf{w})$ d. $\mathbf{u} \cdot (\mathbf{v} - \mathbf{w})$

    e. $(\mathbf{u} + \mathbf{v}) \cdot (\mathbf{v} - \mathbf{w})$ f. $\mathbf{u} \cdot (2\mathbf{v} + 3\mathbf{w})$

    g. $\text{comp}_\mathbf{v}\, \mathbf{u}$ h. $\text{comp}_\mathbf{w}\, \mathbf{v}$

    i. $\text{proj}_\mathbf{u}\, \mathbf{v}$ j. $\text{proj}_{\perp \mathbf{w}}\, \mathbf{u}$

14. For $\mathbf{v} = 2\mathbf{i} - 3\mathbf{j}$ and $\mathbf{w} = 5\mathbf{i} + \mathbf{j}$, find

    a. $\text{comp}_\mathbf{w}\, \mathbf{v}$ b. $\text{comp}_\mathbf{i}\, \mathbf{v}$

    c. $\text{comp}_\mathbf{v}\, \mathbf{w}$ d. $\text{proj}_\mathbf{w}\, \mathbf{v}$

    e. $\text{proj}_{\perp \mathbf{w}}\, \mathbf{v}$ f. $\text{proj}_{\perp \mathbf{v}}\, \mathbf{w}$

15. Let $\mathbf{v} = a\mathbf{i} + \mathbf{j}$ and $\mathbf{w} = 3\mathbf{i} + 4\mathbf{j}$. Find a so that

    a. $\mathbf{v} \cdot \mathbf{w} = 0$ b. $\mathbf{v} \cdot \mathbf{w} = 10$

    c. $\text{proj}_\mathbf{w}\, \mathbf{v} = \mathbf{v}$ d. $\text{proj}_\mathbf{w}\, \mathbf{v} = \mathbf{0}$

In Exercises 16–23, let $\mathbf{u} = \langle 2, -1, 5 \rangle$, $\mathbf{v} = \langle -3, 5, 0 \rangle$, and $\mathbf{w} = \langle 3, 3, 1 \rangle$. Find the indicated vector or number.

16. $\mathbf{v} \cdot \mathbf{w}$ 17. $\mathbf{v} \cdot (\mathbf{u} + 2\mathbf{w})$

18. $\text{proj}_\mathbf{v}\, \mathbf{w}$ 19. $\text{proj}_\mathbf{w}\, \mathbf{v}$

20. $\text{comp}_\mathbf{v}\, \mathbf{w}$ 21. $\text{comp}_\mathbf{w}\, \mathbf{v}$

22. $\text{comp}_\mathbf{u}\, \mathbf{u}$ 23. $\text{proj}_\mathbf{u}\, \mathbf{u}$

In Exercises 24–29, let $\mathbf{u} = \mathbf{i} + 2\mathbf{j} - \mathbf{k}$, $\mathbf{v} = 3\mathbf{i} - 2\mathbf{j} + 2\mathbf{k}$, and $\mathbf{w} = 5\mathbf{i} - \mathbf{j} + 3\mathbf{k}$. Find the indicated vector or number.

24. $\mathbf{u} \cdot (2\mathbf{v} + 3\mathbf{w})$ **25.** $|\mathbf{v} \cdot \mathbf{w} + \mathbf{w} \cdot \mathbf{u}|$

26. $\mathbf{v} \cdot \mathbf{w} - |\mathbf{v}||\mathbf{w}|$ **27.** $\text{proj}_{\mathbf{w}}\, \mathbf{v}$

28. $\text{comp}_{\mathbf{u}}\, \mathbf{w}$ **29.** $\text{comp}_{\mathbf{u}}\, \mathbf{v}$

30. Find the interior angles of the triangle with vertices $(0, 0)$, $(3, 0)$, and $(3, 4)$.

31. Determine which of the following pairs of vectors are orthogonal
  a. $\mathbf{v} = 2\mathbf{i} + 4\mathbf{j}$ **b.** $\mathbf{v} = 3\mathbf{i} - 2\mathbf{j}$
      $\mathbf{w} = 5\mathbf{i} - \mathbf{j}$ $\mathbf{w} = 10\mathbf{i} + 15\mathbf{j}$

  c. $\mathbf{v} = \mathbf{i} - 3\mathbf{j}$ **d.** $\mathbf{v} = \mathbf{i} + \mathbf{j}$
      $\mathbf{w} = 6\mathbf{i} + 2\mathbf{j}$ $\mathbf{w} = -\mathbf{i} - \mathbf{j}$

32. Find the number a so that the vectors $\mathbf{v} = 3\mathbf{i} - 5\mathbf{j}$ and $\mathbf{w} = a\mathbf{i} + 3\mathbf{j}$ are the orthogonal.

33. Find two unit vectors in the plane that are orthogonal to $\mathbf{v} = 2\mathbf{i} + \mathbf{j}$.

34. Find an example of three nonzero vectors $\mathbf{u}$, $\mathbf{v}$, and $\mathbf{w}$ for which $\mathbf{u} \cdot \mathbf{v} = \mathbf{u} \cdot \mathbf{w}$, but $\mathbf{v} \neq \mathbf{w}$. This shows that a "cancellation law" for dot products cannot hold.

35. Which of the following sets of points are vertices of right triangles?
  a. $P = (1, 3, 2)$, $Q = (4, 1, 4)$, $R = (6, 5, 5)$
  b. $P = (0, 2, 5)$, $Q = (1, 3, 1)$, $R = (1, 4, 5)$
  c. $P = (-3, 1, 2)$, $Q = (1, -3, 2)$, $R = (2, -2, 2)$

36. Let $\mathbf{u} = \mathbf{i} - 2\mathbf{j} + 3\mathbf{k}$, $\mathbf{v} = \mathbf{i} + 2\mathbf{j} + \mathbf{k}$,   and $\mathbf{w} = \mathbf{i} - 4\mathbf{j} + \mathbf{k}$. Find t so that the vector $\mathbf{u} + t\mathbf{v}$ is orthogonal to $\mathbf{w}$.

In each of Exercises 37–41, find an equation for the plane with the given normal containing the given point.

37. $\mathbf{n} = \mathbf{i} + 2\mathbf{j} - \mathbf{k}$, $P = (1, 2, -3)$

38. $\mathbf{n} = 4\mathbf{i} - \mathbf{j} + \mathbf{k}$, $P = (-1, 3, 5)$

39. $\mathbf{n} = 2\mathbf{i} - 2\mathbf{j} + 3\mathbf{k}$, $P = (0, -2, 5)$

40. $\mathbf{n} = \mathbf{i} + \mathbf{j}$ $P = (4, 2, -3)$

41. $\mathbf{n} = \mathbf{i} + 2\mathbf{j} + \mathbf{k}$ $P = (0, -1, 3)$

42. Find a vector normal to the plane with equation $2x - 3y + z = 5$.

43. Find all unit vectors perpendicular to the plane with equation $x + 2y + z = 4$.

44. Express the vector $\mathbf{v} = 3\mathbf{i} + 4\mathbf{j}$ as the sum of a vector parallel to $\mathbf{w} = 3\mathbf{i} + \mathbf{j}$ and a vector orthogonal to $\mathbf{w}$.

45. Express the unit coordinate vector $\mathbf{i}$ as the sum of a vector parallel to the vector $\mathbf{v} = 2\mathbf{i} - \mathbf{j}$ and a vector perpendicular to $\mathbf{v}$.

46. Find the distance from the point $(-4, 3)$ to the line in the plane with equation $2x - y - 3 = 0$.

47. Find the distance from the point $(3, 6)$ to the line in the plane with equation $x + y - 1 = 0$.

48. Find the distance from the origin to the line in the plane with vector equation $\mathbf{r}(t) = \mathbf{i} + 2\mathbf{j} + t(\mathbf{i} - \mathbf{j})$.

49. Find an equation for the plane perpendicular to the line with symmetric equation

$$\frac{x - 2}{3} = \frac{1 - y}{6} = \frac{z + 2}{2}$$

and containing the point $(3, -2, 5)$.

50. Find an equation for the plane containing the point $(3, -1, -1)$ and perpendicular to the line with parametric equations $x = 3 + t$, $y = -1 + 4t$, $z = 5 + 3t$.

51. The angle between two planes is equal to the angle between vectors normal to the two planes. Find the angle between the planes with equations $x + y + z = 6$ and $3x - y + 2z = 5$.

52. Find the angle between the planes with equations $x - y + z = 2$ and $x + 3y + 3z = -4$ (see Exercise 51).

53. Find the distance from the point $P = (3, -6, 5)$ to the plane with equation $6x - y + z = 2$.

54. Find symmetric equations for the line that contains the point $(2, 1, -2)$ and is perpendicular to the plane with equation $3x - y + z = -3$.

55. Find parametric equations for the line containing the point $(2, 4, -3)$ and perpendicular to the plane with equation $2x + 3y - 7z = 9$.

56. Prove statements (i), (ii), (iv), and (v) of Theorem 4.

57. A child pulls a sled by exerting a 20-lb force on a rope that makes an angle of $45°$ with the horizontal. Find the work done by the child in pulling the sled a distance of 100 ft.

58. Let $\mathbf{v} = x\mathbf{i} + y\mathbf{j} + z\mathbf{k}$ be a nonzero vector, and let α, β, and γ denote the angles formed between $\mathbf{v}$ and the unit coordinate vectors $\mathbf{i}$, $\mathbf{j}$, and $\mathbf{k}$, respectively (see Figure 3.20). The angles α, β, and γ are called the **direction angles** for $\mathbf{v}$. The numbers $\cos \alpha$, $\cos \beta$, and $\cos \gamma$ are called the **direction cosines** for the vector $\mathbf{v}$.

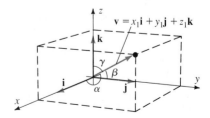

Figure 3.20 Direction angles α, β, γ for the vector $\mathbf{v}$.

a. Using the dot product, derive the equation $x = |\mathbf{v}| \cos \alpha$. Derive similar equations for y and z in terms of the other two direction cosines.

b. Derive the representation $\mathbf{v} = |\mathbf{v}|[(\cos \alpha)\mathbf{i} + (\cos \beta)\mathbf{j} + (\cos \gamma)\mathbf{k}]$.

c. Using part b, show that the vector $\mathbf{u} = (\cos \alpha)\mathbf{i} + (\cos \beta)\mathbf{j} + (\cos \gamma)\mathbf{k}$ is a unit vector with the same direction as $\mathbf{v}$.

59. Find the direction cosines for the given vector.
  a. $\mathbf{v} = 3\mathbf{i} - \mathbf{j} + 2\mathbf{k}$
  b. $\mathbf{v} = 6\mathbf{i} - 2\mathbf{j} + \mathbf{k}$
  c. $\overrightarrow{PQ}$, where $P = (2, 1, 5)$ and $Q = (1, 3, 1)$

60. Show that a nonzero vector is completely determined by its direction cosines and its length.

61. Using Exercise 60, find all unit vectors in space that make an angle of $45°$ with $\mathbf{i}$ and an angle of $60°$ with $\mathbf{j}$.

62. Which of the following triples can be direction angles of a single vector?
  a. $45°$, $45°$, $60°$
  b. $30°$, $45°$, $60°$
  c. $45°$, $60°$, $60°$

63. Show that $\mathbf{v} + \mathbf{w}$ and $\mathbf{v} - \mathbf{w}$ are perpendicular if $|\mathbf{v}| = |\mathbf{w}| \neq 0$. Conclude that the diagonals of a rhombus are perpendicular. (A rhombus is a parallelogram with all sides of equal length.)

64. Prove the converse of Exercise 63: In a parallelogram, if the diagonals are perpendicular, then the parallelogram is a rhombus.

65. Assume $\mathbf{v} \neq \mathbf{0}$ and $\mathbf{w} \neq \mathbf{0}$. Under what conditions is $\mathbf{v} \cdot \mathbf{w} = |\mathbf{v}||\mathbf{w}|$?

66. Use the dot product to prove the following statements.
  a. $|\mathbf{v} + \mathbf{w}| = |\mathbf{v} - \mathbf{w}|$ if and only if $\mathbf{v}$ and $\mathbf{w}$ are orthogonal.
  b. $|\mathbf{v} + \mathbf{w}|^2 = |\mathbf{v}|^2 + |\mathbf{w}|^2$ if and only if $\mathbf{v}$ and $\mathbf{w}$ are orthogonal.

67. Let $\mathbf{v}_1$, $\mathbf{v}_2$, and $\mathbf{v}_3$ be mutually orthogonal vectors in space. Use the dot product to show that if c_1, c_2, and c_3 are scalars so that $c_1\mathbf{v}_1 + c_2\mathbf{v}_2 + c_3\mathbf{v}_3 = \mathbf{0}$, then $c_1 = c_2 = c_3 = 0$.

68. Use the dot product to show that if $P = (a_1, b_1)$ and $Q = (a_2, b_2)$ are endpoints of the diameter of a circle and if $X = (x, y)$ is a point on the circle, then $\overrightarrow{XP}$ and $\overrightarrow{XQ}$ are orthogonal.

69. Use the result of Exercise 68 and the dot product to find an equation for the circle having (a_1, b_1) and (a_2, b_2) as endpoints of a diameter.

70. Let a, b, c, and d be (fixed) numbers and suppose that at least one of a, b, or c is nonzero. Let σ represent the set of all points (x, y, z) in space that satisfy the equation $ax + by + cz = d$. We show that σ is a plane as follows:
  a. Show that the set σ contains at least one point $P = (x_0, y_0, z_0)$.
  b. Let $\mathbf{n}$ be the vector $a\mathbf{i} + b\mathbf{j} + c\mathbf{k}$. Show that, for any point Q in σ, $\overrightarrow{PQ} \cdot \mathbf{n} = 0$.
  c. Using part b, show that σ coincides with the plane through P that is normal to $\mathbf{n}$.

71. Which of the following five planes are parallel? Are any equal?

$$\sigma_1: x + y - 2z = 4$$
$$\sigma_2: x - 2y + z = 4$$
$$\sigma_3: -x + 4y + z = -8$$
$$\sigma_4: 2x - 4y + 2z = 8$$
$$\sigma_5: 3x + 3y - 6z = 5$$

72. The distance between two parallel planes is the same as the distance from one of the planes to any point in the other plane. Determine equations for the two planes that are 4 units away from the plane with equation $3x + y + z = 5$.

16.4 The Cross Product

When we throw a Frisbee or a stick for a dog to fetch, the resulting motion involves rotation as well as displacement. Likewise, when we tighten a bolt with a wrench, we exert torque on the bolt. In these and similar physical situations, the rotational aspect of the motion is measured using a vector product—the **cross product**—of vectors in space.

The cross product also has geometric properties that we use throughout Chapters 17 through 20. Therefore, in this section we introduce the cross product and discuss these properties. We give a geometric definition, and then we derive an equivalent algebraic formulation. We also discuss applications of the cross product to problems involving lines and planes in space.

To define the cross product, we need a convention, the **right-hand rule,** that consistently associates a perpendicular direction to any pair of nonparallel vectors $\mathbf{v}$

and **w.** To determine this direction, translate **v** and **w** so that their initial points coincide and consider the unique plane σ in which these vectors lie. The right-hand rule consistently chooses one of the two directions that are perpendicular to σ.

Right-hand Rule: Given two nonparallel vectors **v** and **w,** then the vector **N(v, w)** is the unique unit vector that:

 (i) is perpendicular to the plane containing **v** and **w** and
 (ii) points in the same direction as the thumb of the right hand if the index finger points in the direction of **v** and the middle finger points in the direction of **w.**

In other words, the two fingers determine the angle $\theta, 0 < \theta < \pi$, between **v** and **w** (see Figures 4.1 and 4.2).

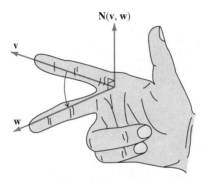

Figure 4.1 Right-hand rule.

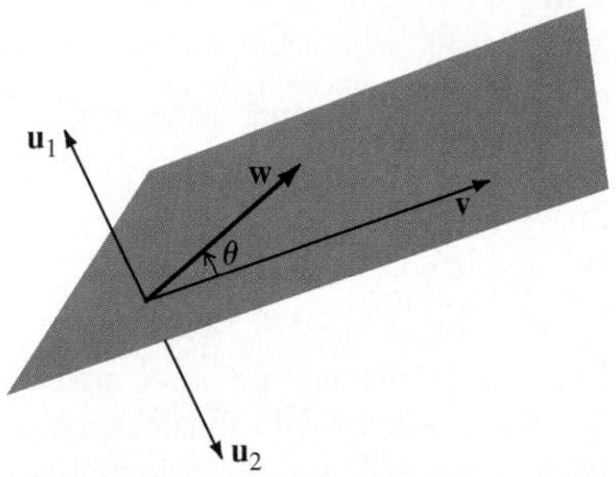

Figure 4.2 Both units vectors $\mathbf{u}_1$ and $\mathbf{u}_2$ are orthogonal to the plane determined by **v** and **w**. Which one corresponds to **N(v, w)**?

REMARK 1   The result of the rule *depends on the order in which we choose* **v** *and* **w!** If we interchange that order, the wrist of the right hand must rotate 180°. Therefore,

$$\mathbf{N(w, v)} = -\mathbf{N(v, w)}.$$

□ **EXAMPLE 1**

As Figures 4.3 and 4.4 illustrate,

$$\mathbf{N(-i - j, i - j)} = \mathbf{k} \quad \text{and} \quad \mathbf{N(i - j, -i - j)} = -\mathbf{k}. \qquad ■$$

Given the right-hand rule, we can now define the cross product.

Definition 6

Let **v** and **w** be two nonzero vectors. Then

$$\mathbf{v} \times \mathbf{w} = (|\mathbf{v}||\mathbf{w}|\sin \theta)\mathbf{N(v, w)}$$

where $0 < \theta < \pi$ is the angle between **v** and **w.**

REMARK 2   Be careful to use the angle θ *between 0 and* π determined by **v** and **w** in Definition 6. Since $\sin(2\pi - \theta) = -\sin(\theta)$, the use of the larger angle $2\pi - \theta$ leads to an incorrect result, one that differs by a minus sign from the correct result.

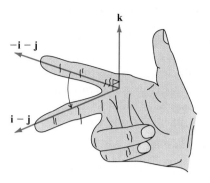

Figure 4.3 $\mathbf{N}(-\mathbf{i}-\mathbf{j}, \mathbf{i}-\mathbf{j}) = \mathbf{k}.$

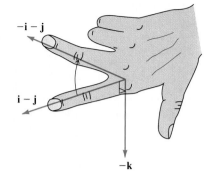

Figure 4.4 $\mathbf{N}(\mathbf{i}-\mathbf{j}, -\mathbf{i}-\mathbf{j}) = -\mathbf{k}.$

REMARK 3   Since $\mathbf{N}(\mathbf{w}, \mathbf{v}) = -\mathbf{N}(\mathbf{v}, \mathbf{w})$, we see that

$$\mathbf{w} \times \mathbf{v} = -\mathbf{v} \times \mathbf{w}. \tag{1}$$

In other words, the cross product is a vector multiplication that results in another vector, but it is not commutative. However, even though it is not commutative, equation (1) tells us that the result changes by only a minus sign if we interchange the order of multiplication.

REMARK 4   If $\mathbf{v}$ and $\mathbf{w}$ are parallel, then $\sin \theta = 0$ and $\mathbf{v} \times \mathbf{w} = \mathbf{0}.$

☐ **EXAMPLE 2**

Since the unit coordinate vectors are orthogonal, we can calculate cross products among them simply by keeping track of the right-hand rule. We obtain

$$\mathbf{i} \times \mathbf{j} = \mathbf{k} \qquad \text{(Figure 4.5)}. \tag{2}$$

Similarly, you can verify that

$$\mathbf{j} \times \mathbf{k} = \mathbf{i} \qquad \text{(Figure 4.6)} \tag{3}$$

and

$$\mathbf{k} \times \mathbf{i} = \mathbf{j} \qquad \text{(Figure 4.7)}. \tag{4}$$

The curved arrows in Figures 4.5 through 4.7 indicate the order of the vectors in the cross products.

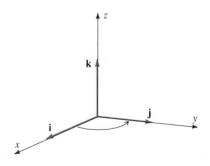

Figure 4.5 $\mathbf{i} \times \mathbf{j} = \mathbf{k}.$

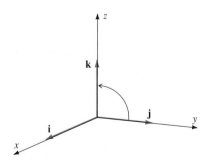

Figure 4.6 $\mathbf{j} \times \mathbf{k} = \mathbf{i}.$

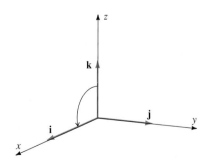

Figure 4.7 $\mathbf{k} \times \mathbf{i} = \mathbf{j}.$

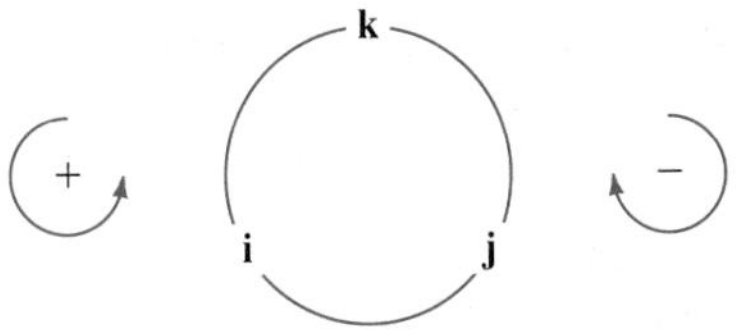

Figure 4.8   Device for remembering cross products of unit coordinate vectors.

Using either Definition 6 or equation (1), you can also verify that

$$\mathbf{j} \times \mathbf{i} = -\mathbf{k}, \quad \mathbf{k} \times \mathbf{j} = -\mathbf{i}, \quad \text{and} \quad \mathbf{i} \times \mathbf{k} = -\mathbf{j}. \tag{5}$$

Figure 4.8 presents a device for remembering these results. By beginning at the first factor and traversing the circle in the direction leading directly to the second factor, we determine the cross product: the product is $+$ or $-$ the remaining unit coordinate vector and the sign is $+$ if the direction is counterclockwise (equations (2) $-$ (4)) and $-$ if the direction is clockwise.   ■

Just as with the dot product, this vector operation is useful because it has an analytic formulation in addition to its geometric definition.

Theorem 5

If $\mathbf{v} = x_1\mathbf{i} + y_1\mathbf{j} + z_1\mathbf{k}$ and $\mathbf{w} = x_2\mathbf{i} + y_2\mathbf{j} + z_2\mathbf{k}$, then

$$\mathbf{v} \times \mathbf{w} = (y_1z_2 - z_1y_2)\mathbf{i} + (z_1x_2 - x_1z_2)\mathbf{j} + (x_1y_2 - y_1x_2)\mathbf{k}. \tag{6}$$

Although equation (6) seems somewhat complicated, we shall soon see that it can be nicely reformulated in terms of determinants.

Proof: Let $\mathbf{c}$ represent the vector defined by the right-hand side of equation (6). That is,

$$\mathbf{c} = (y_1z_2 - z_1y_2)\mathbf{i} + (z_1x_2 - x_1z_2)\mathbf{j} + (x_1y_2 - y_1x_2)\mathbf{k}.$$

First, we show that $\mathbf{c}$ is orthogonal to both $\mathbf{v}$ and $\mathbf{w}$ using the dot product. To determine orthogonality with $\mathbf{v}$, we compute

$$\mathbf{c} \cdot \mathbf{v} = (y_1z_2 - z_1y_2)x_1 + (z_1x_2 - x_1z_2)y_1 + (x_1y_2 - y_1x_2)z_1$$
$$= 0.$$

Orthogonality with $\mathbf{w}$ is computed similarly.

Next, we express the length of $\mathbf{c}$ in terms of the lengths of $\mathbf{v}$ and $\mathbf{w}$. We get

$$\begin{aligned}
|\mathbf{c}|^2 &= (y_1z_2 - z_1y_2)^2 + (z_1x_2 - x_1z_2)^2 + (x_1y_2 - y_1x_2)^2 \\
&= (x_1^2 + y_1^2 + z_1^2)(x_2^2 + y_2^2 + z_2^2) - (x_1x_2 + y_1y_2 + z_1z_2)^2 \\
&= |\mathbf{v}|^2|\mathbf{w}|^2 - (\mathbf{v} \cdot \mathbf{w})^2 \\
&= |\mathbf{v}|^2|\mathbf{w}|^2 - |\mathbf{v}|^2|\mathbf{w}|^2 \cos^2 \theta \\
&= |\mathbf{v}|^2|\mathbf{w}|^2(1 - \cos^2 \theta) \\
&= |\mathbf{v}|^2|\mathbf{w}|^2 \sin^2 \theta \\
&= |\mathbf{v} \times \mathbf{w}|^2.
\end{aligned}$$

Thus, the vector $\mathbf{c}$ has the desired length.

To complete the proof, we would need to establish that the direction of $\mathbf{c}$ agrees with the direction determined by the right-hand rule. Certainly, this fact can be checked explicitly for special cases such as $\mathbf{v} = \mathbf{i}$ and $\mathbf{w} = \mathbf{j}$. Using techniques from linear algebra not discussed in this book, we can show that verifying the general case reduces to checking this special case.   ■

□ **EXAMPLE 3**

For $\mathbf{v} = 2\mathbf{i} + \mathbf{j} + 3\mathbf{k}$ and $\mathbf{w} = 4\mathbf{i} - \mathbf{j} + 2\mathbf{k}$,

$$\begin{aligned}
\mathbf{v} \times \mathbf{w} &= [1 \cdot 2 - 3(-1)]\mathbf{i} + [3 \cdot 4 - 2 \cdot 2]\mathbf{j} + [2(-1) - 1 \cdot 4]\mathbf{k} \\
&= 5\mathbf{i} + 8\mathbf{j} - 6\mathbf{k}.
\end{aligned}$$

■

Properties of the Cross Product

The cross product satisfies many standard properties, which are summarized in the following theorem. We have already discussed properties (i), (iv), and (v). Properties (ii) and (iii) can be verified using Theorem 5.

Theorem 6
Properties of the Cross Product

Let **u, v,** and **w** be vectors and let c be a scalar. Then

(i) $\mathbf{v} \times \mathbf{w} = -\mathbf{w} \times \mathbf{v}$ ($\times$ is anticommutative)
(ii) $\mathbf{u} \times (\mathbf{v} + \mathbf{w}) = \mathbf{u} \times \mathbf{v} + \mathbf{u} \times \mathbf{w}$ ($\times$ distributes over addition)
(iii) $c(\mathbf{v} \times \mathbf{w}) = (c\mathbf{v}) \times \mathbf{w} = \mathbf{v} \times (c\mathbf{w})$
(iv) $(\mathbf{v} \times \mathbf{w}) \perp \mathbf{v}; \ (\mathbf{v} \times \mathbf{w}) \perp \mathbf{w}$
(v) $\mathbf{v} \times \mathbf{v} = \mathbf{0}$ ($\times$ is self-annihilating)

The Determinant Notation

A useful formula for remembering equation (6) involves the concept of **determinants.** The determinant of the 2×2 matrix

$$\begin{bmatrix} a_1 & b_1 \\ a_2 & b_2 \end{bmatrix}$$

is the number

$$\det \begin{bmatrix} a_1 & b_1 \\ a_2 & b_2 \end{bmatrix} = a_1 b_2 - b_1 a_2.$$

For example,

$$\det \begin{bmatrix} 3 & 2 \\ 4 & 1 \end{bmatrix} = 3 \cdot 1 - 2 \cdot 4 = 3 - 8 = -5.$$

The determinant of the 3×3 matrix

$$\begin{bmatrix} a_1 & b_1 & c_1 \\ a_2 & b_2 & c_2 \\ a_3 & b_3 & c_3 \end{bmatrix}$$

is the number

$$\det \begin{bmatrix} a_1 & b_1 & c_1 \\ a_2 & b_2 & c_2 \\ a_3 & b_3 & c_3 \end{bmatrix} = a_1 \cdot \det \begin{bmatrix} b_2 & c_2 \\ b_3 & c_3 \end{bmatrix} - b_1 \cdot \det \begin{bmatrix} a_2 & c_2 \\ a_3 & c_3 \end{bmatrix} + c_1 \cdot \det \begin{bmatrix} a_2 & b_2 \\ a_3 & b_3 \end{bmatrix}$$

$$= a_1(b_2 c_3 - c_2 b_3) - b_1(a_2 c_3 - c_2 a_3) + c_1(a_2 b_3 - b_2 a_3). \quad (7)$$

For example,

$$\det \begin{bmatrix} 1 & 2 & 3 \\ 4 & 5 & 6 \\ 7 & 8 & 9 \end{bmatrix} = 1 \cdot \det \begin{bmatrix} 5 & 6 \\ 8 & 9 \end{bmatrix} - 2 \cdot \det \begin{bmatrix} 4 & 6 \\ 7 & 9 \end{bmatrix} + 3 \cdot \det \begin{bmatrix} 4 & 5 \\ 7 & 8 \end{bmatrix}$$

$$= (5 \cdot 9 - 6 \cdot 8) - 2(4 \cdot 9 - 6 \cdot 7) + 3(4 \cdot 8 - 5 \cdot 7)$$

$$= 0.$$

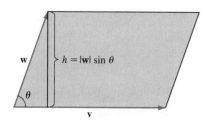

Figure 4.9 Area of parallelogram with base $|\mathbf{v}|$ and adjacent side $\mathbf{w}$ is $|\mathbf{v} \times \mathbf{w}|$.

Our reason for mentioning determinants is that by comparing equations (6) and (7) you can see the memory device

$$\mathbf{v} \times \mathbf{w} = \det \begin{bmatrix} \mathbf{i} & \mathbf{j} & \mathbf{k} \\ x_1 & y_1 & z_1 \\ x_2 & y_2 & z_2 \end{bmatrix}$$

$$= (y_1 z_2 - z_1 y_2)\mathbf{i} + (z_1 x_2 - x_1 z_2)\mathbf{j} + (x_1 y_2 - y_1 x_2)\mathbf{k} \qquad (8)$$

provides a convenient way to remember equation (6). (Of course, equation (8) is not really a proper use of the determinant defined in equation (7), since the top row of the matrix consists of vectors rather than numbers.) Determinants play an important role in the theory of systems of linear equations. Here we use determinants simply as a way to remember the result of Theorem 5.

□ **EXAMPLE 4**

For $\mathbf{v} = 3\mathbf{i} - \mathbf{j} - 4\mathbf{k}$ and $\mathbf{w} = -2\mathbf{i} + 2\mathbf{j} + \mathbf{k}$,

$$\mathbf{v} \times \mathbf{w} = \det \begin{bmatrix} \mathbf{i} & \mathbf{j} & \mathbf{k} \\ 3 & -1 & -4 \\ -2 & 2 & 1 \end{bmatrix}$$

$$= ((-1)(1) - (-4)(2))\mathbf{i}$$
$$+ ((-4)(-2) - (3 \cdot 1))\mathbf{j} + (3 \cdot 2 - (-1)(-2))\mathbf{k}$$

$$= 7\mathbf{i} + 5\mathbf{j} + 4\mathbf{k}. \qquad ■$$

Figure 4.9 illustrates a geometric consequence of Definition 6. If θ is the angle between $\mathbf{v}$ and $\mathbf{w}$, the altitude of the parallelogram determined by the vectors $\mathbf{v}$ and $\mathbf{w}$ is $h = |\mathbf{w}| \sin \theta$. Since the base has length $b = |\mathbf{v}|$, the area A must be

$$A = bh = |\mathbf{v}||\mathbf{w}| \sin \theta = |\mathbf{v} \times \mathbf{w}|. \qquad (9)$$

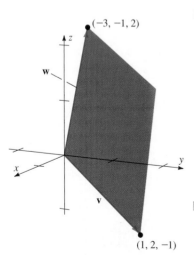

Figure 4.10 The parallelogram determined by the vectors $\mathbf{v}$ and $\mathbf{w}$ in Example 5.

□ **EXAMPLE 5**

For $\mathbf{v} = \mathbf{i} + 2\mathbf{j} - \mathbf{k}$ and $\mathbf{w} = -3\mathbf{i} - \mathbf{j} + 2\mathbf{k}$,

$$\mathbf{v} \times \mathbf{w} = \det \begin{bmatrix} \mathbf{i} & \mathbf{j} & \mathbf{k} \\ 1 & 2 & -1 \\ -3 & -1 & 2 \end{bmatrix} = (4 - 1)\mathbf{i} + (3 - 2)\mathbf{j} + (-1 + 6)\mathbf{k}$$

$$= 3\mathbf{i} + \mathbf{j} + 5\mathbf{k}.$$

The area of the parallelogram determined by $\mathbf{v}$ and $\mathbf{w}$ is, by equation (9),

$$A = |3\mathbf{i} + \mathbf{j} + 5\mathbf{k}| = \sqrt{9 + 1 + 25} = \sqrt{35} \qquad \text{(Figure 4.10)}. \qquad ■$$

□ **EXAMPLE 6**

Find the sine of the angle between the vectors $\mathbf{v}$ and $\mathbf{w}$ in Example 5.

Solution Solving equation (9) for $\sin \theta$ gives

$$\sin \theta = \frac{|\mathbf{v} \times \mathbf{w}|}{|\mathbf{v}||\mathbf{w}|}.$$

Since $|\mathbf{v} \times \mathbf{w}| = \sqrt{35}$, $|\mathbf{v}| = \sqrt{1^2 + 2^2 + 1^2} = \sqrt{6}$, and $|\mathbf{w}| = \sqrt{3^2 + 1^2 + 2^2} = \sqrt{14}$, we have

$$\sin \theta = \frac{\sqrt{35}}{\sqrt{6}\sqrt{14}} = \frac{1}{2}\sqrt{\frac{5}{3}}.$$

We can also use equation (9) to learn something about 2×2 determinants. Let $\mathbf{v} = x_1\mathbf{i} + y_2\mathbf{j}$ and $\mathbf{w} = x_2\mathbf{i} + y_2\mathbf{j}$. Then $\mathbf{v}$ and $\mathbf{w}$ are vectors in the plane $z = 0$, and the area of the parallelogram they determine is

$$A = |\mathbf{v} \times \mathbf{w}| = \det \begin{bmatrix} \mathbf{i} & \mathbf{j} & \mathbf{k} \\ x_1 & y_1 & 0 \\ x_2 & y_2 & 0 \end{bmatrix}$$
$$= |(x_1y_2 - y_1x_2)\mathbf{k}|$$
$$= |x_1y_2 - y_1x_2|$$
$$= \left|\det \begin{bmatrix} x_1 & y_1 \\ x_2 & y_2 \end{bmatrix}\right|.$$

That is, the absolute value of the determinant of the matrix

$$\begin{bmatrix} x_1 & y_1 \\ x_2 & y_2 \end{bmatrix}$$

is the area of the parallelogram determined by the vectors $x_1\mathbf{i} + y_1\mathbf{j}$ and $x_2\mathbf{i} + y_2\mathbf{j}$. Since a diagonal of a parallelogram bisects the parallelogram into two congruent triangles, another interpretation of the 2×2 determinant is that the absolute value of the determinant

$$\det \begin{bmatrix} x_1 & y_1 \\ x_2 & y_2 \end{bmatrix}$$

is twice the area of the triangle determined by the vectors $x_1\mathbf{i} + y_1\mathbf{j}$ and $x_2\mathbf{i} + y_2\mathbf{j}$ (Figure 4.11).

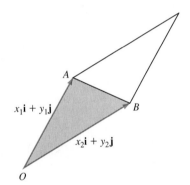

Figure 4.11 Area of $\triangle OAB$ is $\frac{1}{2}\left|\det \begin{bmatrix} x_1 & y_1 \\ x_2 & y_2 \end{bmatrix}\right|$

□ **EXAMPLE 7**

Find the area of the triangle with vertices $(2, 3)$, $(4, 7)$, and $(5, 1)$.

Solution Let $P = (2, 3)$, $Q = (4, 7)$, and $R = (5, 1)$. Two of the sides may be interpreted as the vectors

$$\mathbf{v} = \overrightarrow{PQ} = 2\mathbf{i} + 4\mathbf{j}$$

and

$$\mathbf{w} = \overrightarrow{PR} = 3\mathbf{i} - 2\mathbf{j} \quad \text{(Figure 4.12).}$$

By the preceding observation, the area of the triangle is

$$A = \frac{1}{2}\left|\det \begin{bmatrix} 2 & 4 \\ 3 & -2 \end{bmatrix}\right| = \frac{1}{2}|2(-2) - 4(3)| = \frac{16}{2} = 8.$$

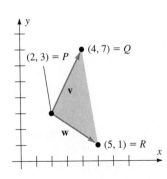

Figure 4.12 The triangle in Example 7.

Another geometric application of the cross product is to find a vector that is mutually perpendicular to two given vectors. This is particularly important if we want to determine the equation of a plane given two nonparallel lines or three noncollinear points in the plane. In particular, if we can find two vectors, $\mathbf{v}_1$ and $\mathbf{v}_2$, lying in the plane, then the cross product $\mathbf{n} = \mathbf{v}_1 \times \mathbf{v}_2$ is a normal to the plane, since it is orthogonal both to $\mathbf{v}_1$ and to $\mathbf{v}_2$ (Figure 4.13).

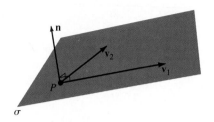

Figure 4.13 The cross product $\mathbf{n} = \mathbf{v}_1 \times \mathbf{v}_2$ is orthogonal to the plane σ that contains $\mathbf{v}_1$ and $\mathbf{v}_2$.

□ **EXAMPLE 8**

Find an equation for the plane determined by the three points $P = (-4, 0, 2)$, $Q = (1, -3, 1)$, and $R = (2, -2, 6)$.

Strategy · · · · · · · ·
Find two vectors $\mathbf{v}_1 = \overrightarrow{PQ}$, $\mathbf{v}_2 = \overrightarrow{PR}$ in the plane.

Solution

Two vectors in the plane are

$$\mathbf{v}_1 = \overrightarrow{PQ} = 5\mathbf{i} - 3\mathbf{j} - \mathbf{k}$$

and

$$\mathbf{v}_2 = \overrightarrow{PR} = 6\mathbf{i} - 2\mathbf{j} + 4\mathbf{k}.$$

Find a normal $\mathbf{n} = \mathbf{v}_1 \times \mathbf{v}_2$.

A normal vector is therefore

$$\mathbf{n} = \mathbf{v}_1 \times \mathbf{v}_2 = \det \begin{bmatrix} \mathbf{i} & \mathbf{j} & \mathbf{k} \\ 5 & -3 & -1 \\ 6 & -2 & 4 \end{bmatrix}$$
$$= -14\mathbf{i} - 26\mathbf{j} + 8\mathbf{k}.$$

Find the equation of the plane, using $\mathbf{n}$ and P.

Using $\mathbf{n}$ and the point P, the equation for the plane is

$$\mathbf{n} \cdot \overrightarrow{PQ} = 0$$

where $Q = (x, y, z)$ is an arbitrary point on the plane. We obtain

$$(-14\mathbf{i} - 26\mathbf{j} + 8\mathbf{k}) \cdot [(x + 4)\mathbf{i} + y\mathbf{j} + (z - 2)\mathbf{k}] = 0.$$

The desired equation is

$$-14(x + 4) - 26y + 8(z - 2) = 0,$$

or

$$-14x - 26y + 8z = 72. \qquad \blacksquare$$

The Distance from a Point to a Line in Space

The problem of finding the distance d from a point $P = (x_0, y_0, z_0)$ to a line in space with equation $\mathbf{r}(t) = \mathbf{a} + t\mathbf{b}$ can also be solved using the cross product.

Figure 4.14 illustrates that the desired distance d is given by the expression

$$d = |\overrightarrow{QP}||\sin \theta| \qquad (10)$$

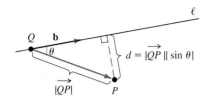

Figure 4.14

$$d = |\overrightarrow{QP}||\sin \theta| = \frac{|\overrightarrow{QP} \times \mathbf{b}|}{|\mathbf{b}|}.$$

where Q is any point on ℓ and θ is the angle formed between $\overrightarrow{QP}$ and the direction vector **b.**

The right side of equation (10) is suggestive of the cross product. In fact, we may rewrite equation (10) as

$$d = |\overrightarrow{QP}||\sin \theta| = \frac{|\overrightarrow{QP}||\mathbf{b}||\sin \theta|}{|\mathbf{b}|} = \frac{|\overrightarrow{QP} \times \mathbf{b}|}{|\mathbf{b}|}.$$

□ **EXAMPLE 9**

Find the distance from the point $P = (1, -1, 2)$ to the line with vector equation

$$\ell: \mathbf{r}(t) = 3\mathbf{i} - 2\mathbf{j} + 4\mathbf{k} + t(\mathbf{i} - 2\mathbf{j} + 2\mathbf{k}).$$

Strategy · · · · · · · ·

Find a direction vector **b.**

Solution

A direction vector for ℓ is

$$\mathbf{b} = \mathbf{i} - 2\mathbf{j} + 2\mathbf{k}.$$

Find a point Q on the line ℓ.

Setting $t = 0$ shows that the point

$$Q = (3, -2, 4)$$

lies on ℓ. Thus,

Form the vector $\overrightarrow{QP}$.

$$\overrightarrow{QP} = (1 - 3)\mathbf{i} + (-1 - (-2))\mathbf{j} + (2 - 4)\mathbf{k}$$
$$= -2\mathbf{i} + \mathbf{j} - 2\mathbf{k}.$$

Find the distance using

$$d = \frac{|\overrightarrow{QP} \times \mathbf{b}|}{|\mathbf{b}|}.$$

Since $\overrightarrow{QP} \times \mathbf{b} = -2\mathbf{i} + 2\mathbf{j} - 3\mathbf{k}$, the distance is

$$d = \frac{|\overrightarrow{QP} \times \mathbf{b}|}{|\mathbf{b}|} = \frac{\sqrt{17}}{3}. \qquad \blacksquare$$

Exercise Set 16.4

In Exercises 1–6, let $\mathbf{u} = \mathbf{i} + \mathbf{j} + \mathbf{k}$, $\mathbf{v} = 2\mathbf{i} - \mathbf{j} + 2\mathbf{k}$, and $\mathbf{w} = 3\mathbf{i} + 4\mathbf{j} - \mathbf{k}$. Find the indicated vector or number.

1. $\mathbf{v} \times \mathbf{w}$ **2.** $\mathbf{u} \times \mathbf{w}$

3. $\mathbf{w} \times \mathbf{u}$ **4.** $\mathbf{u} \times \mathbf{v}$

5. $\mathbf{u} \cdot (\mathbf{v} \times \mathbf{w})$ **6.** $\mathbf{v} \cdot (\mathbf{w} \times \mathbf{u})$

In Exercises 7–14, let $\mathbf{u} = 2\mathbf{i} - \mathbf{j} + 4\mathbf{k}$, $\mathbf{v} = \mathbf{i} - 3\mathbf{k}$, and $\mathbf{w} = 2\mathbf{i} + 3\mathbf{j} + 4\mathbf{k}$. Find the indicated vector or number.

7. $\mathbf{u} \times \mathbf{v}$ **8.** $\mathbf{v} \times \mathbf{w}$

9. $\mathbf{u} \times \mathbf{w}$ **10.** $3\mathbf{u} \times 2\mathbf{w}$

11. $\mathbf{u} \times (\mathbf{v} \times \mathbf{w})$ **12.** $(\mathbf{u} \times \mathbf{v}) \times \mathbf{w}$

13. $\mathbf{u} \cdot (\mathbf{v} \times \mathbf{w})$ **14.** $\mathbf{w} \cdot (\mathbf{v} \times \mathbf{u})$

In Exercises 15–18, find the area of the parallelogram determined by the given vectors.

15. $\mathbf{v} = 2\mathbf{i} + \mathbf{j} + \mathbf{k}$, $\mathbf{w} = 3\mathbf{i} - \mathbf{j} - 2\mathbf{k}$

16. $\mathbf{v} = -3\mathbf{i} + \mathbf{k}$, $\mathbf{w} = 2\mathbf{i} + 2\mathbf{j} + \mathbf{k}$

17. $\mathbf{v} = 2\mathbf{i} - \mathbf{j}$, $\mathbf{w} = 4\mathbf{i} + 2\mathbf{j}$

18. $\mathbf{v} = -6\mathbf{i} + 4\mathbf{j} - 3\mathbf{k}$, $\mathbf{w} = \mathbf{i} - \mathbf{j} - \mathbf{k}$

In Exercises 19–21, find the area of the triangle in space with the given vertices.

19. $A = (3, 0, 1)$, $B = (2, -1, 2)$, $C = (1, 3, -2)$

20. $A = (5, 2, -2)$, $B = (-1, 4, 2)$, $C = (-4, 5, 3)$

21. $A = (0, 2, 1)$, $B = (-4, 1, -2)$, $C = (1, 1, -2)$

22. Find the area of the triangle with vertices $(2, 4)$, $(-3, 8)$, and $(-5, -2)$.

23. Find the area of the triangle with vertices $(1, 5)$, $(-3, -4)$, and $(4, 6)$.

24. Find a unit vector orthogonal to both $\mathbf{v} = 2\mathbf{i} + \mathbf{j} - \mathbf{k}$ and $\mathbf{w} = \mathbf{i} + \mathbf{j} + 4\mathbf{k}$.

25. Find a unit vector orthogonal to both $\mathbf{v} = \mathbf{i} + 3\mathbf{j}$ and $\mathbf{w} = 4\mathbf{i} - \mathbf{j}$.

26. Show that the cross product is not associative. That is, find vectors **u**, **v**, and **w** so that $\mathbf{u} \times (\mathbf{v} \times \mathbf{w}) \neq (\mathbf{u} \times \mathbf{v}) \times \mathbf{w}$.

27. Find two unit vectors orthogonal to $\mathbf{v} = \mathbf{i} + \mathbf{j}$ and $\mathbf{w} = 2\mathbf{i} - \mathbf{j} + 3\mathbf{k}$.

28. Find parametric equations for the line that contains the point $(1, 2, 0)$ and is orthogonal both to $\mathbf{i} + 3\mathbf{j} + \mathbf{k}$ and to $\mathbf{j} - 2\mathbf{k}$.

29. Find symmetric equations for the line that contains the point $(1, 2, 2)$ and is orthogonal both to $\mathbf{i} + \mathbf{j}$ and to $\mathbf{i} + \mathbf{j} + 2\mathbf{k}$.

30. Find the distance from the origin to the line with symmetric equations

$$x - 1 = \frac{2 - y}{2} = \frac{z}{-2}.$$

31. Find the distance from the point $P = (2, -1, 4)$ to the line with equation $\mathbf{r}(t) = (3\mathbf{i} - \mathbf{j} - \mathbf{k}) + t(\mathbf{i} + 2\mathbf{j} + \mathbf{k})$.

32. Find the distance from the point $P = (1, 3, -2)$ to the line with equations $x(t) = 1 + t$, $y(t) = 3 - 2t$, $z(t) = 2t - 2$.

33. Find the distance from the point $P = (0, -2, 1)$ to the line with equations

$$\frac{x - 1}{4} = \frac{y + 3}{-2} = \frac{z + 1}{5}.$$

34. Find the distance d between the lines

$$\ell_1 : \frac{x - 2}{3} = \frac{y + 1}{2} = \frac{z - 3}{5}$$

and

$$\ell_2 : \frac{x + 4}{1} = \frac{y - 3}{2} = \frac{z + 1}{2}$$

as follows:
  a. Find direction vectors $\mathbf{b}_1$ and $\mathbf{b}_2$ for lines ℓ_1 and ℓ_2, respectively.
  b. Observe that a vector perpendicular to both lines ℓ_1 and ℓ_2 is the cross product $\mathbf{n} = \mathbf{b}_1 \times \mathbf{b}_2$.
  c. Find a point P on ℓ_1 and a point Q on ℓ_2.
  d. Obtain the distance d as

$$d = |\text{comp}_\mathbf{n} \overrightarrow{PQ}|.$$

35. Use Exercise 34 to find the distance between the lines ℓ_1 and ℓ_2:

$$\ell_1 : x(t) = 1 + t, \quad y(t) = 5t, \quad z(t) = 1 - t,$$
$$\ell_2 : x(t) = 2 + t, \quad y(t) = 2 - 3t, \quad z(t) = 1 + 5t.$$

In Exercises 36–40, find an equation for the plane containing the given points.

36. $P = (0, 1, -2), \quad Q = (1, 1, 1), \quad R = (3, 5, 1)$

37. $P = (-2, 3, -4), \quad Q = (1, 5, 1), \quad R = (-2, -2, -2)$

38. $P = (7, 0, -2), \quad Q = (1, -3, 4), \quad R = (5, 2, -3)$

39. $P = (-1, 1, 1), \quad Q = (0, 2, -1), \quad R = (3, 5, -2)$

40. $P = (-2, 1, 5), \quad Q = (2, 0, -2), \quad R = (-1, -1, 3)$

41. Find a vector normal to the plane containing $(-2, 1, 3)$, $(5, 1, 5)$, and $(0, 0, 2)$.

42. Find a vector that is perpendicular to the plane containing the two lines

$$\frac{x - 1}{2} = \frac{y - 1}{3} = z - 1$$

and

$$x - 1 = \frac{y - 1}{-2} = \frac{z - 1}{2}.$$

43. Find parametric equations for the line of intersection of the planes with equations $x + y - 2z = 6$ and $2x - 4y + z = 3$.

44. Show that if $\mathbf{u} = x_1\mathbf{i} + y_1\mathbf{j} + z_1\mathbf{k}$, $\mathbf{v} = x_2\mathbf{i} + y_2\mathbf{j} + z_2\mathbf{k}$, and $\mathbf{w} = x_3\mathbf{i} + y_3\mathbf{j} + z_3\mathbf{k}$, then

$$\mathbf{u} \cdot (\mathbf{v} \times \mathbf{w}) = \det \begin{bmatrix} x_1 & y_1 & z_1 \\ x_2 & y_2 & z_2 \\ x_3 & y_3 & z_3 \end{bmatrix}.$$

45. Determine which of the following sets of points are coplanar.
  a. $(2, 1, 0), (3, 0, 5), (1, 1, 1), (2, 3, -12)$
  b. $(1, -6, 2), (3, -5, 11), (4, 0, 4), (1, 5, -2)$
  c. $(1, 7, 2), (3, 4, -2), (1, 5, 1), (1, 9, 3)$

46. Find an equation for the plane containing the point $(3, -1, 3)$ and the line with symmetric equations

$$\frac{x - 2}{3} = \frac{y + 1}{5} = \frac{z}{4}.$$

47. Find an equation for the plane that contains the two lines

$$\frac{x}{5} = y = z \quad \text{and} \quad x = y = z.$$

48. Find an equation for the plane that contains the two lines

$$x = y = z \quad \text{and} \quad x = y + 1 = z + 2.$$

49. Show that $|\mathbf{v} \times \mathbf{w}| = |\mathbf{v}||\mathbf{w}|$ if $\mathbf{v}$ and $\mathbf{w}$ are orthogonal.

50. Show that the volume of the parallelepiped formed by $\mathbf{u}$, $\mathbf{v}$, and $\mathbf{w}$ is $|\mathbf{u} \cdot (\mathbf{v} \times \mathbf{w})|$. (*Hint:* $|\mathbf{v} \times \mathbf{w}|$ is the area of the base formed by $\mathbf{v}$ and $\mathbf{w}$. Thus,

$$|\mathbf{u} \cdot (\mathbf{v} \times \mathbf{w})| = |\mathbf{v} \times \mathbf{w}|(|\mathbf{u}||\cos \theta|)$$

is the area of the base times the altitude. Why?)

51. Use Exercises 49 and 50 to find the volume of the parallelepiped determined by the vectors $\mathbf{u} = 2\mathbf{i} + \mathbf{j} - 4\mathbf{k}$, $\mathbf{v} = -\mathbf{i} + 3\mathbf{j} - 2\mathbf{k}$, and $\mathbf{w} = 4\mathbf{i} - \mathbf{j} + 3\mathbf{k}$.

52. Find the volume of the parallelepiped determined by the vectors $\mathbf{u} = -3\mathbf{i} + \mathbf{j} - \mathbf{k}$, $\mathbf{v} = 2\mathbf{i} + 3\mathbf{j} + 2\mathbf{k}$, and $\mathbf{w} = \mathbf{i} + 4\mathbf{j} + 2\mathbf{k}$.

53. Prove that $\mathbf{u} \times (\mathbf{v} \times \mathbf{w}) = (\mathbf{u} \cdot \mathbf{w})\mathbf{v} - (\mathbf{u} \cdot \mathbf{v})\mathbf{w}$.

54. Prove that $\mathbf{u} \cdot (\mathbf{v} \times \mathbf{w}) = \mathbf{v} \cdot (\mathbf{w} \times \mathbf{u})$.

55. Prove that $\mathbf{u} \cdot (\mathbf{v} \times \mathbf{w}) = \mathbf{w} \cdot (\mathbf{u} \times \mathbf{v})$.

56. Prove that $(\mathbf{u} \times \mathbf{v}) \times \mathbf{w} = \mathbf{u} \times (\mathbf{v} \times \mathbf{w})$ only if $(\mathbf{u} \times \mathbf{w}) \times \mathbf{v} = \mathbf{0}$.

57. Suppose that $\mathbf{u} + \mathbf{v} + \mathbf{w} = \mathbf{0}$. Show that $\mathbf{u} \times \mathbf{v} = \mathbf{v} \times \mathbf{w} = \mathbf{w} \times \mathbf{u}$. What is the geometric interpretation of this result?

Summary Outline of Chapter 16

■ A **vector in the plane** is an ordered pair of numbers $\mathbf{v} = \langle a, b \rangle$ subject to the laws   (page 768)

$$\langle a_1, b_1 \rangle + \langle a_2, b_2 \rangle = \langle a_1 + a_2, b_1 + b_2 \rangle,$$
$$c \langle a, b \rangle = \langle ca, cb \rangle, \quad c \in (-\infty, \infty).$$

■ A **vector in space** is an ordered triple $\mathbf{v} = \langle a, b, c \rangle$ subject to the laws   (page 776)

$$\langle a_1, b_1, c_1 \rangle + \langle a_2, b_2, c_2 \rangle = \langle a_1 + a_2, b_1 + b_2, c_1 + c_2 \rangle,$$
$$d\langle a, b, c \rangle = \langle da, db, dc \rangle, \quad d \in (-\infty, \infty).$$

■ The **unit coordinate vectors** are $\mathbf{i} = \langle 1, 0, 0 \rangle$, $\mathbf{j} = \langle 0, 1, 0 \rangle$, $\mathbf{k} = \langle 0, 0, 1 \rangle$.   (page 779)

■ *Notation:* $\mathbf{v} = \langle a, b, c \rangle = a\mathbf{i} + b\mathbf{j} + c\mathbf{k}$.   (page 780)

■ The distance d between two points $P = (x_1, y_1, z_1)$ and $Q = (x_2, y_2, z_2)$ in   (page 776)
space is

$$d = \sqrt{(x_2 - x_1)^2 + (y_2 - y_1)^2 + (z_2 - z_1)^2}.$$

■ The **length** of the vector $\mathbf{v} = a\mathbf{i} + b\mathbf{j} + c\mathbf{k}$ is $|\mathbf{v}| = \sqrt{a^2 + b^2 + c^2}$.   (page 777)

■ Vector length has the following properties: (page 778)

(i) $|\mathbf{v}| \geq 0$; $|\mathbf{v}| = 0$ if and only if $\mathbf{v} = \mathbf{0}$,
(ii) $|c\mathbf{v}| = |c||\mathbf{v}|$,
(iii) $|\mathbf{v} + \mathbf{w}| \leq |\mathbf{v}| + |\mathbf{w}|$.

■ A **unit vector** in the direction of $\mathbf{v}$ is $\mathbf{u} = \dfrac{\mathbf{v}}{|\mathbf{v}|}$.   (page 778)

■ The **line** with position vector $\mathbf{a} = a_1\mathbf{i} + a_2\mathbf{j} + a_3\mathbf{k}$ and direction vector $\mathbf{b} =$ (page 784)
$b_1\mathbf{i} + b_2\mathbf{j} + b_3\mathbf{k}$ has

(i) vector equation $\mathbf{r}(t) = \mathbf{a} + t\mathbf{b}$, $-\infty < t < \infty$.
(ii) parametric equations $x(t) = a_1 + tb_1$, $y(t) = a_2 + tb_2$, $z(t) = a_3 + tb_3$.

(iii) symmetric equations $\dfrac{x - a_1}{b_1} = \dfrac{y - a_2}{b_2} = \dfrac{z - a_3}{b_3}$.

■ The **dot product** of the vectors $\mathbf{v}$ and $\mathbf{w}$ is the number $|\mathbf{v}||\mathbf{w}| \cos \theta$, where θ (page 789)
is the angle between $\mathbf{v}$ and $\mathbf{w}$.

■ The vectors $\mathbf{v}$ and $\mathbf{w}$ are **orthogonal** (perpendicular) if and only if $\mathbf{v} \cdot \mathbf{w} = 0$.   (page 791)

■ The **component** of the vector $\mathbf{v}$ in the direction of the vector $\mathbf{w}$ is the number   (page 793)

$$\text{comp}_\mathbf{w}\, \mathbf{v} = \frac{\mathbf{v} \cdot \mathbf{w}}{|\mathbf{w}|}.$$

■ The **projection** of the vector $\mathbf{v}$ along the vector $\mathbf{w}$ is the **vector**   (page 794)

$$\text{proj}_\mathbf{w}\, \mathbf{v} = \left(\frac{\mathbf{v} \cdot \mathbf{w}}{|\mathbf{w}|^2} \right) \mathbf{w}.$$

■ **Theorem:**   If $\mathbf{v} = x_1\mathbf{i} + y_1\mathbf{j} + z_1\mathbf{k}$ and $\mathbf{w} = x_2\mathbf{i} + y_2\mathbf{j} + z_2\mathbf{k}$, then   (page 790)

$$\mathbf{v} \cdot \mathbf{w} = x_1x_2 + y_1y_2 + z_1z_2.$$

■ The **plane** with normal vector $\mathbf{n}$ containing the point P is the set of all points   (page 797) Q for which $\overrightarrow{PQ} \cdot \mathbf{n} = 0$.

■ The **cross product** of the vectors $\mathbf{v}$ and $\mathbf{w}$ is the **vector** $(|\mathbf{v}||\mathbf{w}| \sin \theta)\mathbf{N}$, where   (page 802) $\mathbf{N}$ is a unit vector orthogonal to both $\mathbf{v}$ and $\mathbf{w}$ in the direction determined by the right-hand rule.

■ $|\mathbf{v} \times \mathbf{w}|$ is the area of the parallelogram determined by $\mathbf{v}$ and $\mathbf{w}$.   (page 806)

■ **Theorem:**   (page 806)

$$\mathbf{v} \times \mathbf{w} = \det \begin{bmatrix} \mathbf{i} & \mathbf{j} & \mathbf{k} \\ x_1 & y_1 & z_1 \\ x_2 & y_2 & z_2 \end{bmatrix} = (y_1z_2 - z_1y_2)\mathbf{i} + (z_1x_2 - x_1z_2)\mathbf{j} + (x_1y_2 - y_1x_2)\mathbf{k}.$$

■ The **distance** d from the point P to the line with direction vector $\mathbf{b}$ containing   (page 809) the point Q is

$$d = |\overrightarrow{QP}| \sin \theta = \frac{|\overrightarrow{QP} \times \mathbf{b}|}{|\mathbf{b}|}$$

where θ is the angle between $\overrightarrow{QP}$ and $\mathbf{b}$.

Review Exercises—Chapter 16

1. Find the distance between the points $P = (1, 2, -4)$ and $Q = (2, -5, 2)$.

2. Find an equation for the set of all points $P = (x, y, z)$ equidistant from the fixed points $P_1 = (1, -2, 1)$ and $Q = (3, 4, -3)$.

3. Let $\mathbf{u} = 2\mathbf{i} + \mathbf{j} - \mathbf{k}$, $\mathbf{v} = \mathbf{i} + \mathbf{j} + \mathbf{k}$, and $\mathbf{w} = \mathbf{j} - \mathbf{k}$. Find
   a. $\mathbf{u} \cdot \mathbf{v}$
   b. $\mathbf{u} \times \mathbf{v}$
   c. $\mathbf{v} \times \mathbf{u}$
   d. $\text{comp}_\mathbf{w}\,\mathbf{u}$
   e. $\text{proj}_\mathbf{w}\,\mathbf{u}$
   f. $\mathbf{u} \cdot (\mathbf{v} \times \mathbf{w})$

4. Let $\mathbf{u} = \mathbf{i} + 2\mathbf{j} - 3\mathbf{k}$, $\mathbf{v} = 2\mathbf{i} + 2\mathbf{j} + 6\mathbf{k}$, $\mathbf{w} = \mathbf{i} - 4\mathbf{j} + 3\mathbf{k}$. Find
   a. $\mathbf{u} + 2\mathbf{v}$
   b. $\mathbf{u} - \mathbf{v}$
   c. $|3\mathbf{u} + \mathbf{v}|$
   d. $\mathbf{u} \cdot \mathbf{v}$
   e. $\mathbf{u} \cdot (\mathbf{v} \times \mathbf{w})$
   f. $|\mathbf{u} + \mathbf{v} - \mathbf{w}|$

5. Find the cosine of the angle between the vectors $\mathbf{v} = \mathbf{i} - 3\mathbf{j} + 2\mathbf{k}$ and $\mathbf{w} = 3\mathbf{i} + 3\mathbf{j} + 2\mathbf{k}$.

6. Determine whether the following pairs of lines intersect and, if so, at what point.
   a. $x = t$, $y = t + 2$, $z = 2t - 4$;
      $x = 1 - t$, $y = 3 + t$, $z = 4t$
   b. $x = 1 + 2t$, $y = t - 2$, $z = 1 + t$;
      $x = 2t + 1$, $y = 4 - 2t$, $z = 5 - t$

7. Find parametric equations for the line containing the point $(2, 1, -3)$ that is perpendicular to both of the lines:

   $$\ell_1: x = 2t, \quad y = 3 + t, \quad z = 5t$$
   $$\ell_2: x = t + 5, \quad y = 6 - 4t, \quad z = 3t + 4$$

8. Find the direction cosines for the vector $\mathbf{v} = 3\mathbf{i} + 4\mathbf{j} + 5\mathbf{k}$ (See Exercise 58 in Section 16.3).

9. Find an equation for the plane containing the points $(-1, 2, 1)$, $(2, 5, 3)$, and $(-4, 0, 2)$.

10. Find an equation for the plane containing the line

    $$\ell_1: \mathbf{r}(t) = \mathbf{i} - 3\mathbf{j} + \mathbf{k} + t(4\mathbf{i} + 2\mathbf{j} - \mathbf{k})$$

    and the point $(1, 2, 1)$.

11. Find a vector normal to the plane with equation $8x + y - 2z = 5$.

12. Find the distance from the point $P = (1, 2, -4)$ to the plane with equation $3x + 2y - 5z = 5$.

13. Find a vector equation for the line of intersection of the planes with equations $2x + 3y - z = 4$ and $x - 3y + 5z = 2$.

14. Show that if $\mathbf{v} \times \mathbf{w} = 0$ and $\mathbf{w} \neq 0$ then $\mathbf{v} = c\mathbf{w}$ for some constant c.

15. Show that $(\mathbf{u} + \mathbf{v}) \times (\mathbf{u} - \mathbf{v}) = 2\mathbf{v} \times \mathbf{u}$.

16. Find parametric equations for the line containing the points $(3, 5, -6)$ and $(-2, 3, -1)$.

17. Find symmetric equations for the line containing the points $(1, 2, -1)$ and $(2, 3, 1)$.

18. Find a vector equation for the line that is perpendicular to the plane with equation $3x + y + z = 0$ at the origin.

19. Find an equation for the plane containing the three points $(1, 1, 1)$, $(2, 1, 4)$, and $(3, 2, 3)$.

20. Find an equation for the plane containing the point $(2, -5, 3)$ and perpendicular to the line with parametric equations $x = 4 + 2t$, $y = 3 - t$, $z = 5 + 6t$.

21. Find an equation for the plane that contains the lines

$$\frac{x}{2} = y = 1 - z \quad \text{and} \quad x = y = 2z - 2.$$

22. For $\mathbf{v} = \mathbf{i} + 2\mathbf{j} + 4\mathbf{k}$ and $\mathbf{w} = 3\mathbf{j} + 4\mathbf{k}$, find
  a. $\text{comp}_{\mathbf{w}} \mathbf{v}$ **b.** $\text{proj}_{\mathbf{w}} \mathbf{v}$

23. Find the distance from the point $(-4, 2, 5)$ to the line with equation $x = 3 + t$, $y = 4 - 2t$, $z = 5 + 5t$.

24. Find the area of the triangle with vertices $(1, 3, -2)$, $(1, 1, 1)$, and $(4, 0, 3)$.

25. Find a unit vector in the same direction as $\mathbf{v} = \mathbf{i} - 3\mathbf{j} + 4\mathbf{k}$.

26. Find an equation for the plane containing the points $(3, -2, 6)$ and $(4, -2, 2)$ that is perpendicular to the xz-plane.

27. Find symmetric equations for the line with parametric equations $x = 1 + 3t$, $y = 2 + 7t$, $z = 3 - t$.

28. Find the vector of length 3 with direction cosines $\sqrt{2}/2$, 0, and $\sqrt{2}/2$ (see Exercise 58 in Section 16.3).

29. For $\mathbf{v} = 2\mathbf{i} + 3\mathbf{j} - \mathbf{k}$ and $\mathbf{w} = \mathbf{i} + 3\mathbf{j} + \mathbf{k}$ find
  a. $\text{comp}_{\mathbf{w}} \mathbf{v}$ **b.** $\text{proj}_{\mathbf{v}} \mathbf{w}$

30. Find the distance from the point $(1, 2, 1)$ to the plane containing the points $(1, 1, 1)$, $(2, -1, 5)$, and $(3, 1, -2)$.

31. Find the distance of the point $(1, 2, 2)$ to the plane with equation $x + y + z = 0$.

32. Find the cosine of the angle of intersection of the planes with equations $x + 2y + z = 1$ and $x + y + z = 3$ (see Exercise 51 in Section 16.3).

33. Find symmetric equations for the line of intersection of the two planes with equations

$$x + y + z = 4 \quad \text{and} \quad 2x + y - z = 1$$

34. Find an equation for the plane that intersects the x-axis at $x = 3$, the y-axis at $y = 4$, and the z-axis at $z = 5$.

35. Find an equation for the plane perpendicular to the line with symmetric equations

$$\frac{x - 2}{3} = \frac{y + 1}{2} = \frac{z - 3}{-2}$$

and containing the point $(3, -2, 3)$.

36. Find parametric equations for the line parallel to the line of intersection of the two planes with equations $x + y + 2z = 6$ and $x - y - 2z = 4$ and containing the point $(5, 2, -3)$.

37. Find all vectors of length 2 orthogonal to both $\mathbf{v} = 3\mathbf{i} + 2\mathbf{j} + \mathbf{k}$ and $\mathbf{w} = \mathbf{i} + \mathbf{j} - 2\mathbf{k}$.

38. Find the area of the parallelogram determined by the vectors $\mathbf{v} = 2\mathbf{i} + 3\mathbf{j} + \mathbf{k}$ and $\mathbf{w} = \mathbf{i} - \mathbf{j} + 2\mathbf{k}$.

39. Prove that the planes with equations $a_1 x + b_1 y + c_1 z = d_1$ and $a_2 x + b_2 y + c_2 z = d_2$ are perpendicular if and only if $a_1 a_2 + b_1 b_2 + c_1 c_2 = 0$.

40. Show that the points $(1, 1, -1)$, $(0, 1, -1/2)$, $(-1, 1, 0)$, and $(0, 0, 1/4)$ all lie in the same plane.

Chapter 17

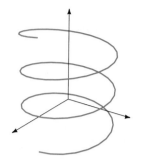

Curves and Surfaces

In previous chapters, we studied curves in the plane as well as a special type of surface, the surface of revolution. Now, in order to develop the calculus of multivariate functions, we need a more general theory of curves, one that applies both to curves in the plane and curves in space. Similarly, we need to be able to work with surfaces that are more general than those obtained by revolving a planar curve around an axis in space. In this chapter, we discuss both topics.

We study curves by introducing a new type of function—the *vector-valued* function. Like the functions we have previously considered, the domains of vector-valued functions are sets of real numbers; however, their values are vectors (either in the plane or in space) rather than real numbers. Studying curves as vector-valued functions is useful because the concept of a vector-valued function is more general than the concept of a curve. For example, the study of the motion of a planet is best described in terms of a continuously varying family of displacement vectors, and this approach easily yields important quantities such as velocity and acceleration. Much of this chapter is devoted to the theory and applications of vector-valued functions.

We also begin to study the geometry of surfaces. We discuss methods for understanding the structure of a surface from an equation that represents it, and we introduce two alternative spacial coordinate systems in order to simplify the representation of some commonly used surfaces.

17.1   Vector-Valued Functions; Curves in Space

A **vector-valued function** is a function

$$\mathbf{f}(t) = x(t)\mathbf{i} + y(t)\mathbf{j} + z(t)\mathbf{k} \tag{1}$$

whose domain is a set of real numbers and whose values $\mathbf{f}(t)$ are vectors. Just as with parametric equations (Chapter 15), each of the functions $x(t)$, $y(t)$, and $z(t)$ is real-valued. If they are combined as in equation (1), they form a vector.

We refer to the functions x, y, and z as the "component functions," to the values $x(t)$, $y(t)$, and $z(t)$ as the "scalar components," and to the vectors $x(t)\mathbf{i}$, $y(t)\mathbf{j}$, and $z(t)\mathbf{k}$ as the "vector components." When we use the term "components" alone, it will always be clear from the context which type we mean.

The **curve** generated by a vector-valued function $\mathbf{f}$ is the set of all points in space that are terminal points of the arrows representing the vectors $\mathbf{f}(t)$ where $\mathbf{f}(t)$ *originates at the origin.* In other words, we consider $\mathbf{f}(t) = x(t)\mathbf{i} + y(t)\mathbf{j} + z(t)\mathbf{k}$ as

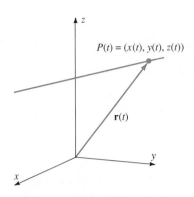

Figure 1.1   A line in space specified as the curve generated by the vector-valued function $\mathbf{r}(t)$.

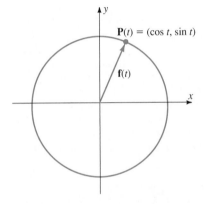

Figure 1.2   The unit circle is the curve generated by the vector-valued function

$$\mathbf{f}(t) = \cos t\,\mathbf{i} + \sin t\,\mathbf{j}$$

(see Section 15.4).

a position vector, and we associate the point $(x(t), y(t), z(t))$ in space with the vector $\mathbf{f}(t)$.

We shall also discuss vector-valued functions whose values are vectors in the plane. In this case, we do not need the function $z(t)$. Thus, equation (1) becomes

$$\mathbf{f}(t) = x(t)\mathbf{i} + y(t)\mathbf{j}.$$

We have already seen two examples of vector-valued functions:

(i) The **line** in space with vector equation

$$\begin{aligned}
\mathbf{r}(t) &= (a_1\mathbf{i} + a_2\mathbf{j} + a_3\mathbf{k}) + t(b_1\mathbf{i} + b_2\mathbf{j} + b_3\mathbf{k}) \\
&= (a_1 + tb_1)\mathbf{i} + (a_2 + tb_2)\mathbf{j} + (a_3 + tb_3)\mathbf{k}
\end{aligned}$$

is the curve generated by the vector-valued function with $x(t) = (a_1 + tb_1)$, $y(t) = (a_2 + tb_2)$, and $z(t) = (a_3 + tb_3)$. The vector $\mathbf{r}(t)$ is interpreted as extending from the origin to the point $P(t)$ whose coordinates are the components of the vector $\mathbf{r}(t)$. If t denotes time, the vector function represents the motion of

a particle moving in space (see Figure 1.1).

(ii) The **curve** C in the plane determined by the parametric equations

$$C: \quad \begin{cases} x = x(t) \\ y = y(t) \end{cases} \qquad a \le t \le b$$

may be interpreted as the curve generated by the vector function

$$\mathbf{f}(t) = x(t)\mathbf{i} + y(t)\mathbf{j}, \qquad a \le t \le b$$

where the vector $\mathbf{f}(t)$ extends from the origin to the point $(x(t), y(t))$ on C (see Figure 1.2).

The concept of a vector-valued function allows us to generalize both of these examples in a unified way. By interpreting $\mathbf{f}(t)$ as a vector from the origin to a particle in space, we may study the motion of that particle, addressing the usual issues of position, speed, velocity, acceleration, distance, and elapsed time.

In studying vector-valued functions we shall make frequent use of the fact that the vector function $\mathbf{f}(t)$ is the sum of three vector components, each of which involves a function of a single independent variable (Figure 1.3). The laws of vector algebra allow us to perform algebraic operations with the components, while the theory of one-variable calculus enables us to address many of the issues associated with derivatives and integrals of vector functions.

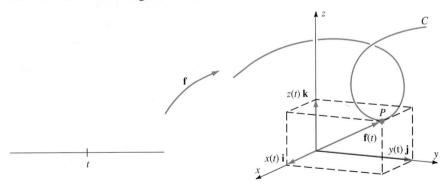

Figure 1.3   The curve C generated by the vector-valued function $\mathbf{f}$. We have $\mathbf{f}(t) = x(t)\mathbf{i} + y(t)\mathbf{j} + z(t)\mathbf{k}$, and $P(t) = (x(t), y(t), z(t))$.

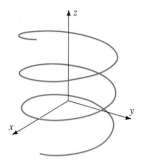

Figure 1.4 The circular helix generated by the vector-valued function

$$\mathbf{f}(t) = a \cos t\mathbf{i} + a \sin t\mathbf{j} + bt\mathbf{k}.$$

□ **EXAMPLE 1**

Sketch the curve generated by the vector function

$$\mathbf{f}(t) = a \cos t\mathbf{i} + a \sin t\mathbf{j} + bt\mathbf{k}$$

where a and b are positive constants.

Solution We note that in any plane parallel to the xy-plane the distance from the point $(a \cos t, a \sin t, bt)$ to the point $(0, 0, bt)$ on the z-axis is

$$d = \sqrt{a^2 \cos^2 t + a^2 \sin^2 t} = |a|.$$

As t increases, the z-coordinate of the point determined by $\mathbf{f}(t)$ increases uniformly. For every point $(x(t), y(t), z(t))$ on the curve, infinitely many points $(x(t), y(t), z(t) + 2\pi nb)$, $n = \pm 1, \pm 2, \ldots$ also lie on the curve. The curve is the **circular helix** in Figure 1.4. ∎

Algebra, Limits, and Continuity

Given two vector-valued functions

$$\mathbf{f}(t) = x_1(t)\mathbf{i} + y_1(t)\mathbf{j} + z_1(t)\mathbf{k} \quad \text{and} \quad \mathbf{g}(t) = x_2(t)\mathbf{i} + y_2(t)\mathbf{j} + z_2(t)\mathbf{k},$$

we may form the following functions, as indicated.

(i) The **sum** of $\mathbf{f}$ and $\mathbf{g}$ is the vector-valued function $\mathbf{f} + \mathbf{g}$ where

$$\begin{aligned}(\mathbf{f} + \mathbf{g})(t) &= \mathbf{f}(t) + \mathbf{g}(t) \\ &= [x_1(t) + x_2(t)]\mathbf{i} + [y_1(t) + y_2(t)]\mathbf{j} + [z_1(t) + z_2(t)]\mathbf{k}.\end{aligned}$$

In other words, we add vector functions component by component.

(ii) The **scalar multiple** of $\mathbf{f}$ by the real number c is the vector-valued function $c\mathbf{f}$ where

$$(c\mathbf{f})(t) = c\mathbf{f}(t) = [cx_1(t)]\mathbf{i} + [cy_1(t)]\mathbf{j} + [cz_1(t)]\mathbf{k}.$$

That is, multiplication of vector functions by scalars is done component by component.

(iii) The **dot product** of $\mathbf{f}$ and $\mathbf{g}$ is the **real-valued function** $\mathbf{f} \cdot \mathbf{g}$ where

$$(\mathbf{f} \cdot \mathbf{g})(t) = \mathbf{f}(t) \cdot \mathbf{g}(t) = x_1(t)x_2(t) + y_1(t)y_2(t) + z_1(t)z_2(t).$$

The values of the dot product function are *numbers* rather than vectors.

(iv) The **cross product** of $\mathbf{f}$ and $\mathbf{g}$ is the vector-valued function $\mathbf{f} \times \mathbf{g}$ where

$$\begin{aligned}(\mathbf{f} \times \mathbf{g})(t) &= \mathbf{f}(t) \times \mathbf{g}(t) \\ &= [y_1(t)z_2(t) - z_1(t)y_2(t)]\mathbf{i} + [z_1(t)x_2(t) - x_1(t)z_2(t)]\mathbf{j} \\ &\quad + [x_1(t)y_2(t) - y_1(t)x_2(t)]\mathbf{k}.\end{aligned}$$

The values of the cross product function are *vectors*.

□ **EXAMPLE 2**

From the vector-valued functions

$$\mathbf{f}(t) = t\mathbf{i} + t^2\mathbf{j} + \sqrt{t}\mathbf{k}, \qquad t \geq 0$$

and

$$\mathbf{g}(t) = \cos t\mathbf{i} + \sin t\mathbf{j} + t\mathbf{k},$$

the following functions may be formed:

(a) $(\mathbf{f} + \mathbf{g})(t) = (t + \cos t)\mathbf{i} + (t^2 + \sin t)\mathbf{j} + (\sqrt{t} + t)\mathbf{k}, \qquad t \geq 0.$

(b) $3\mathbf{f}(t) = 3t\mathbf{i} + 3t^2\mathbf{j} + 3\sqrt{t}\mathbf{k}, \qquad t \geq 0.$

(c) $(\mathbf{f} - 2\mathbf{g})(t) = (t - 2\cos t)\mathbf{i} + (t^2 - 2\sin t)\mathbf{j} + (\sqrt{t} - 2t)\mathbf{k}, \qquad t \geq 0.$

(d) $(\mathbf{f} \times \mathbf{g})(t) = \mathbf{f}(t) \times \mathbf{g}(t)$

$$= \det \begin{bmatrix} \mathbf{i} & \mathbf{j} & \mathbf{k} \\ t & t^2 & \sqrt{t} \\ \cos t & \sin t & t \end{bmatrix}$$

$$= (t^3 - \sqrt{t}\sin t)\mathbf{i} + (\sqrt{t}\cos t - t^2)\mathbf{j} + (t\sin t - t^2\cos t)\mathbf{k}. \qquad ■$$

□ **EXAMPLE 3**

The function

$$\mathbf{f}(t) = [3\cos(t) + \cos(4t)]\mathbf{i} + [3\sin(t) + \sin(4t)]\mathbf{j}$$

is a simplified model of a moon orbiting a planet that is itself orbiting a sun. To see why, we note that the vector $\mathbf{f}(t)$ is the sum of two vectors $\mathbf{v}_1(t)$ and $\mathbf{v}_2(t)$ where

$$\mathbf{v}_1(t) = 3\cos(t)\mathbf{i} + 3\sin(t)\mathbf{j}$$
$$\mathbf{v}_2(t) = \cos(4t)\mathbf{i} + \sin(4t)\mathbf{j}.$$

In Section 15.4, we saw that the corresponding parametric equations determine circles (see Figure 1.2 in this section). Thus, the position determined by $\mathbf{f}(t)$ can be viewed as the result of adding a displacement vector $\mathbf{v}_2(t)$, which corresponds to the circular motion of the moon around the planet, to the position vector $\mathbf{v}_1(t)$, which represents the position of the planet relative to the sun. In this case, the factor of 4 in the equation for $\mathbf{v}_2$ indicates that the moon revolves exactly four times around the planet during one revolution of the planet around the sun. The resulting motion of the moon relative to the sun is illustrated in Figure 1.5. ■

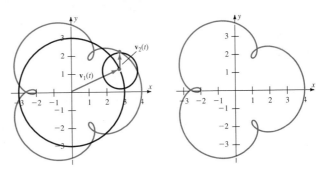

Figure 1.5 The curve generated by the vector-valued function

$$\mathbf{f}(t) = \mathbf{v}_1(t) + \mathbf{v}_2(t)$$

in Example 3 viewed both as a vector sum and as a subset of the plane.

Limits of Vector-Valued Functions

The limit of a vector-valued function is defined just as the limit of a real-valued function:

$$\lim_{x \to a} \mathbf{f}(t) = \mathbf{L}$$

informally means that, as the number t approaches the number a, the vector $\mathbf{f}(t)$

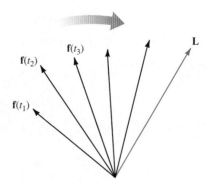

Figure 1.6 The vectors $\mathbf{f}(t_i)$ approach the limiting vector $\mathbf{L}$ as $t_i \to a$.

approaches the vector $\mathbf{L}$ (see Figure 1.6). In other words, $|\mathbf{f}(t) - \mathbf{L}|$ is small whenever $|t - a|$ is small. Note, however, that since $\mathbf{f}(t)$ and $\mathbf{L}$ are vectors, the expression $|\mathbf{f}(t) - \mathbf{L}|$ is the *length* of the vector $\mathbf{f}(t) - \mathbf{L}.$

Here is the formal definition.

Definition 1

Let $\mathbf{f}$ be a vector-valued function that is defined on an open interval containing the number a, except possibly at $t = a$. We say that the vector $\mathbf{L}$ is the *limit* of the function $\mathbf{f}$ as t approaches a, written

$$\mathbf{L} = \lim_{t \to a} \mathbf{f}(t),$$

if, corresponding to each number $\epsilon > 0$ there is a number $\delta > 0$ so that

$$\text{if} \quad 0 < |t - a| < \delta, \quad \text{then} \quad |\mathbf{f}(t) - \mathbf{L}| < \epsilon.$$

(See Figure 1.7.)

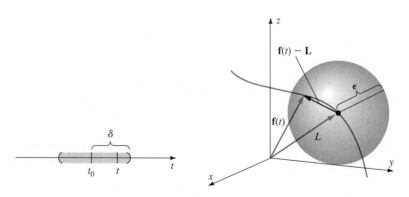

Figure 1.7 If $0 < |t - t_0| < \delta$, then $|\mathbf{f}(t) - \mathbf{L}| < \epsilon$.

The following theorem shows that Definition 1 is equivalent to the existence of limits for each of the component functions of $\mathbf{f}$. This equivalence allows us to evaluate limits of vector-valued functions by using the techniques of Chapter 2 for limits of real-valued functions. The proof of Theorem 1 is given at the end of this section.

Theorem 1

Let $\mathbf{f}$ be the vector-valued function $\mathbf{f}(t) = x(t)\mathbf{i} + y(t)\mathbf{j} + z(t)\mathbf{k}$ and let $\mathbf{L} = L_1\mathbf{i} + L_2\mathbf{j} + L_3\mathbf{k}$ be a vector. Let $\mathbf{f}$ be defined on an open interval containing the number a except possibly at $t = a$. Then

$$\mathbf{L} = \lim_{t \to a} \mathbf{f}(t)$$

if and only if

$$L_1 = \lim_{t \to a} x(t), \qquad L_2 = \lim_{t \to a} y(t), \quad \text{and} \quad L_3 = \lim_{t \to a} z(t).$$

Theorem 1 states that the limit of a vector-valued function is the vector whose components are the limits of the individual components. If one or more of these component limits fails to exist we say that the limit of the vector-valued function fails to exist.

□ **EXAMPLE 4**

Find $\lim_{t \to \pi/3} (t^2\mathbf{i} + \sin t\mathbf{j} + \cos t\mathbf{k})$.

Solution The limits of the components are

$$L_1 = \lim_{t \to \pi/3} x(t) = \lim_{t \to \pi/3} t^2 = \frac{\pi^2}{9},$$

$$L_2 = \lim_{t \to \pi/3} y(t) = \lim_{t \to \pi/3} \sin t = \frac{\sqrt{3}}{2} \qquad \text{(Figure 1.8)},$$

$$L_3 = \lim_{t \to \pi/3} z(t) = \lim_{t \to \pi/3} \cos t = \frac{1}{2} \qquad \text{(Figure 1.8)}.$$

Thus, by Theorem 1,

$$\lim_{t \to \pi/3} (t^2\mathbf{i} + \sin t\mathbf{j} + \cos t\mathbf{k}) = \frac{\pi^2}{9}\mathbf{i} + \frac{\sqrt{3}}{2}\mathbf{j} + \frac{1}{2}\mathbf{k}. \qquad ■$$

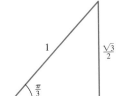

Figure 1.8

Continuity

Continuity for vector-valued functions is defined in terms of limits, just as for real-valued functions. Let $\mathbf{f}$ be a vector-valued function, let a be a number in the domain of $\mathbf{f}$, and suppose that $\mathbf{f}$ is defined for all numbers t near a. Then $\mathbf{f}$ is continuous at a if

$$\lim_{t \to a} \mathbf{f}(t) = \mathbf{f}(a).$$

The following definition is an equivalent statement expressed in language similar to that of the formal definition of the limit.

Definition 2

Let $\mathbf{f}$ be a vector-valued function that is defined on an open interval I containing the number $t = a$. We say that $\mathbf{f}$ is *continuous* at a if, corresponding to each number $\epsilon > 0$ there is a number $\delta > 0$ so that

$$\text{if} \quad |t - a| < \delta, \quad \text{then} \quad |\mathbf{f}(t) - \mathbf{f}(a)| < \epsilon.$$

We say that $\mathbf{f}$ is continuous on I if $\mathbf{f}$ is continuous at each number $a \in \mathbf{I}$.

Like the definition of limit, the definition of continuity for a vector-valued function is equivalent to the requirement that each of the component functions be continuous. We state this result as a corollary of Theorem 1 and we leave its proof, which is similar to the proof of Theorem 1, as an exercise.

Corollary 1

The vector-valued function $\mathbf{f}(t) = x(t)\mathbf{i} + y(t)\mathbf{j} + z(t)\mathbf{k}$ is **continuous** at $t = t_0$ if and only if each of the component functions $x(t)$, $y(t)$, and $z(t)$ is continuous at t_0.

Note that all components of $\mathbf{f}$ must be continuous at t_0 if $\mathbf{f}$ is continuous at t_0. Thus, a vector-valued function is **discontinuous** at t_0 if one or more of its component functions is discontinuous at t_0.

☐ **EXAMPLE 5**

Find the intervals on which the vector-valued function

$$\mathbf{f}(t) = \frac{1}{t}\mathbf{i} + t^2\mathbf{j} + \frac{2}{t^2 - 4}\mathbf{k}$$

is continuous.

Strategy · · · · · · ·
Determine the numbers t where one or more of the component functions is undefined or discontinuous.

Solution
The $\mathbf{i}$-component $x(t) = 1/t$ is undefined at $t = 0$. It is continuous for all $t \neq 0$.

The $\mathbf{j}$-component $y(t) = t^2$ is continuous for all t.

The $\mathbf{k}$-component

$$z(t) = \frac{2}{t^2 - 4} = \frac{2}{(t - 2)(t + 2)}$$

is undefined for $t = -2, 2$. It is continuous for all $t \neq -2, 2$.

The vector-valued function is continuous for all other values of t.

The vector-valued function $\mathbf{f}$ is therefore undefined at $t = -2, 0$, and 2. It is defined and continuous at all other real numbers, so $\mathbf{f}$ is continuous on the intervals

$$(-\infty, -2), (-2, 0), (0, 2), \text{ and } (2, \infty). \qquad \blacksquare$$

Proof of Theorem 1: First, let us assume that

$$L_1 = \lim_{t \to t_0} x(t), \tag{2}$$

$$L_2 = \lim_{t \to t_0} y(t), \quad \text{and} \tag{3}$$

$$L_3 = \lim_{t \to t_0} z(t). \tag{4}$$

Let $\epsilon > 0$ be given. According to the definition of the limit of a function of a single variable and equation (2), there exists a number δ_1 so that

$$\text{if} \quad 0 < |t - t_0| < \delta_1, \quad \text{then} \quad |x(t) - L_1| < \epsilon/3. \tag{5}$$

(You will see in a few lines why we use $\epsilon/3$ rather than just ϵ.) Similarly, it follows from equations (3) and (4) that there exist positive numbers δ_2 and δ_3 so that

$$\text{if} \quad 0 < |t - t_0| < \delta_2, \quad \text{then} \quad |y(t) - L_2| < \epsilon/3 \tag{6}$$

and

$$\text{if} \quad 0 < |t - t_0| < \delta_3, \quad \text{then} \quad |z(t) - L_3| < \epsilon/3. \tag{7}$$

If we take δ to be the smallest of the numbers δ_1, δ_2, and δ_3, then each of the inequalities on the left-hand sides of (5) through (7) is fulfilled when $0 < |t - t_0| < \delta$. In this case, using inequalities (5) through (7) and the triangle inequality, we find that

$$|\mathbf{f}(t) - \mathbf{L}| = |(x(t)\mathbf{i} + y(t)\mathbf{j} + z(t)\mathbf{k}) - (L_1\mathbf{i} + L_2\mathbf{j} + L_3\mathbf{k})|$$
$$= |(x(t) - L_1)\mathbf{i} + (y(t) - L_2)\mathbf{j} + (z(t) - L_3)\mathbf{k}|$$
$$\leq |x(t) - L_1| + |y(t) - L_2| + |z(t) - L_3|$$
$$< \frac{\epsilon}{3} + \frac{\epsilon}{3} + \frac{\epsilon}{3} = \epsilon.$$

That is,

$$\text{if} \quad 0 < |t - t_0| < \delta, \quad \text{then} \quad |\mathbf{f}(t) - \mathbf{L}| < \epsilon. \tag{8}$$

To prove the converse, we assume that whenever $\epsilon > 0$ is given we can find a number $\delta > 0$ so that statement (8) holds, and we show that this guarantees that statements (2)–(4) hold. If t is any number for which $0 < |t - t_0| < \delta$ (with δ as in equation (8)) we have

$$|x(t) - L_1| = \sqrt{|x(t) - L_1|^2}$$
$$\leq \sqrt{|x(t) - L_1|^2 + |y(t) - L_2|^2 + |z(t) - L_3|^2}$$
$$= |\mathbf{f}(t) - \mathbf{L}|$$
$$< \epsilon.$$

That is, if $0 < |t - t_0| < \delta$, then $|x(t) - L_1| < \epsilon$. In other words, the limit as $t \to t_0$ of $x(t)$ is L_1.

Similar comparisons show that, as $t \to t_0$, the limit of $y(t)$ is L_2 and the limit of $z(t)$ is L_3. Consequently, if $\mathbf{L}$ is the limit of $\mathbf{f}(t)$ as $t \to t_0$, then statements (2)–(4) hold, and the proof is complete.   ∎

Exercise Set 17.1

In Exercises 1–8, sketch the curve generated by the given vector-valued function.

1. $\mathbf{f}(t) = \mathbf{i} + \mathbf{j} + t\mathbf{k}$

2. $\mathbf{f}(t) = \cos t\mathbf{i} + \sin t\mathbf{j} + t\mathbf{k}$

3. $\mathbf{f}(t) = t\mathbf{i} + \cos t\mathbf{j} + \sin t\mathbf{k}$

4. $\mathbf{f}(t) = t\mathbf{i} + t^2\mathbf{j}$

5. $\mathbf{f}(t) = t\mathbf{i} + t^2\mathbf{j} + t^3\mathbf{k}$

6. $\mathbf{f}(t) = \sin t\mathbf{i} + \mathbf{j} + \mathbf{k}$

7. $\mathbf{f}(t) = e^t\mathbf{i} + t\mathbf{j} + 2t\mathbf{k}$

8. $\mathbf{f}(t) = t\mathbf{i} + 3t^2\mathbf{j} + 3\mathbf{k}$

The **implicit domain** of a vector-valued function $\mathbf{f}$ is the largest set of numbers t for which each of the component functions is defined. In Exercises 9–16, state the implicit domain for the given vector-valued function.

9. $\mathbf{f}(t) = t\mathbf{i} + \sqrt{t}\mathbf{j} + \sin t\mathbf{k}$

10. $\mathbf{f}(t) = t^2\mathbf{i} - t\mathbf{j} + \tan t\mathbf{k}$

11. $\mathbf{f}(t) = \dfrac{1}{\sqrt{1 - t}}\mathbf{i} + 2t\mathbf{j} + \sin^2 t\mathbf{k}$

12. $\mathbf{f}(t) = \dfrac{1}{9 - t^2}\mathbf{i} + \dfrac{2}{1 + t}\mathbf{j} + \sqrt{t}\mathbf{k}$

13. $\mathbf{f}(t) = (t^2 - 1)\mathbf{i} + \sec t\mathbf{j} + t^3\mathbf{k}$

14. $\mathbf{f}(t) = t^2\mathbf{i} + t^{3/2}\mathbf{j} + t^{2/3}\mathbf{k}$

15. $\mathbf{f}(t) = \ln t\mathbf{i} + \cos t\mathbf{j} + \sqrt{9 - t^2}\mathbf{k}$

16. $\mathbf{f}(t) = \text{Sin}^{-1} t\mathbf{i} + \sqrt{t}\mathbf{j} + t^2\mathbf{k}$

In Exercises 17–22, let $\mathbf{f}(t) = t\mathbf{i} + 3t\mathbf{j} + t^2\mathbf{k}$ and $\mathbf{g}(t) = \sin t\mathbf{i} + \cos t\mathbf{j} + \mathbf{k}$. Find the indicated functions.

17. $\mathbf{f} + \mathbf{g}$ **18.** $\mathbf{f} - \mathbf{g}$

19. $3\mathbf{f} + 2\mathbf{g}$ **20.** $4\mathbf{f} - 3\mathbf{g}$

21. $\mathbf{f} \cdot \mathbf{g}$ **22.** $\mathbf{f} \times \mathbf{g}$

In Exercises 23–28, let $\mathbf{f}(t) = (1 + t)\mathbf{i} + (1 - t^2)\mathbf{j} + t\mathbf{k}$, $\mathbf{g}(t) = \sqrt{t}\,\mathbf{i} + (1 - t)\mathbf{j} + e^t\mathbf{k}$, and $h(t) = \sin t$. Find the indicated function.

23. $\mathbf{f} + 3\mathbf{g}$ **24.** $h\mathbf{f}$

25. $h[\mathbf{f} - \mathbf{g}]$ **26.** $2\mathbf{f} - 3h\mathbf{g}$

27. $\mathbf{f} \cdot \mathbf{g}$ **28.** $3\mathbf{f} \times 2\mathbf{g}$

In Exercises 29–38, find the indicated limit, if it exists.

29. $\lim\limits_{t \to \pi/4} (\sin t\mathbf{i} + \tan t\mathbf{j} + \cos t\mathbf{k})$

30. $\lim\limits_{t \to 2}(t^2\mathbf{i} + (3t - 2)\mathbf{j} + 6t\mathbf{k})$

31. $\lim\limits_{t \to 0}\left[\left(\dfrac{\sin t}{t}\right)\mathbf{i} + \left(\dfrac{1 - \cos t}{t}\right)\mathbf{j} + e^{2t}\mathbf{k}\right]$

32. $\lim\limits_{t \to 0}\left[\dfrac{\sin 3t}{t}\mathbf{i} + \dfrac{\tan 2t}{3t}\mathbf{j} + \ln(1 + t)\mathbf{k}\right]$

33. $\lim\limits_{t \to 0}\left[\dfrac{|t|}{t}\mathbf{i} + \sin 2t\mathbf{j} - |t|\mathbf{k}\right]$

34. $\lim\limits_{t \to -2}\left[\ln t^2\mathbf{i} + \dfrac{t^2 - t - 6}{t + 2}\mathbf{j} + \dfrac{t^2 - 6t + 9}{t + 2}\mathbf{k}\right]$

35. $\lim\limits_{t \to 3}\left[\left(\dfrac{9 - t^2}{3 - t}\right)\mathbf{i} + \left(\dfrac{t^2 + t - 12}{t - 3}\right)\mathbf{j} + \left(\dfrac{t^3 - 13t + 12}{t - 3}\right)\mathbf{k}\right]$

36. $\lim\limits_{t \to 2}\left[(t + 3)\mathbf{i} + \left(\dfrac{t^2 - 4}{t - 2}\right)\mathbf{j} + \left(\dfrac{t^3 + t^2 - 4t - 4}{t - 2}\right)\mathbf{k}\right]$

37. $\lim\limits_{t \to \infty}\left[\dfrac{3t^2 + 1}{t - 4}\mathbf{i} + \dfrac{4t - 3}{t - 4}\mathbf{j} + \dfrac{2t + 1}{(t - 4)^3}\mathbf{k}\right]$

38. $\lim\limits_{t \to \infty}\left[e^{-2t}\mathbf{i} + \dfrac{6t - 7}{t - 2}\mathbf{j} + \dfrac{\sin t}{t}\mathbf{k}\right]$

In Exercises 39–47, determine the intervals on which the given vector function is continuous.

39. $\mathbf{f}(t) = 2t\mathbf{i} + \cos t\mathbf{j} + \tan t\mathbf{k}$

40. $\mathbf{f}(t) = \sqrt{9 - t^2}\,\mathbf{i} + \dfrac{1}{t}\mathbf{j} + \sin t\mathbf{k}$

41. $\mathbf{f}(t) = \dfrac{1}{1 + \sqrt{t}}\mathbf{i} + \dfrac{1}{1 - \sqrt{t}}\mathbf{j} + t^{3/2}\mathbf{k}$

42. $\mathbf{f}(t) = \left(\dfrac{1}{t^2 - t + 12}\right)\mathbf{i} + \ln(2t)\mathbf{j} + e^{-t}\mathbf{k}$

43. $\mathbf{f}(t) = \ln(1 + t^2)\mathbf{i} + \ln(1 - t^2)\mathbf{j} + \sqrt{t}\,\mathbf{k}$

44. $\mathbf{f}(t) = \begin{cases} t\mathbf{i} + (3t - 2)\mathbf{j} + (3 - t)\mathbf{k}, & -\infty < t < 2 \\ 2\mathbf{i} + (2 + t)\mathbf{j} + t\mathbf{k}, & 2 \le t < \infty \end{cases}$

45. $\mathbf{f}(t) = \begin{cases} t^2\mathbf{i} + (3 - t)\mathbf{j} + 4t\mathbf{k}, & -\infty < t < 1 \\ t\mathbf{i} + (4 + t)\mathbf{j} + (1 - t^2)\mathbf{k}, & 1 \le t < \infty \end{cases}$

46. $\mathbf{f}(t) = \dfrac{|t|}{t}\mathbf{i} + \dfrac{|t + 1|}{t + 1}\mathbf{j} + \dfrac{|t + 2|}{t + 2}\mathbf{k}$

47. $\mathbf{f}(t) = t\mathbf{i} + t^2\mathbf{j} + \sqrt{\sin t}\,\mathbf{k}$

48. An automobile assembly robot turns a machine screw at a constant rate of 10π radians per second. The **pitch** of the screw (the number of threads per millimeter of length) is such that the screw advances 0.5 mm for each complete revolution.

   a. Write a vector function for the motion of a paint spot on one thread of the screw.

   b. At what rate, in mm/s, must the robot arm advance?

49. The **polarization** of a light wave is determined by the motion of the tip of the associated "electric vector" $\mathbf{E}(t)$. If the motion follows a circular helix, the light is said to be circularly polarized. If a particular light wave is circularly polarized according to $\mathbf{E}(t) = \cos(10^3 t)\mathbf{i} + \sin(10^3 t)\mathbf{j} + (3 \times 10^8)t\mathbf{k}$, how far does the wave advance along the z-axis during one complete revolution of the electric vector?

50. Describe the curve generated by the vector function $\mathbf{f}(t) = 3 \cos t\mathbf{i} + 4 \sin t\mathbf{j} + t\mathbf{k}$.

51. Prove that if
$$\lim\limits_{t \to t_0} \mathbf{f}(t) = \mathbf{L} \quad \text{and} \quad \lim\limits_{t \to t_0} \mathbf{g}(t) = \mathbf{M},$$
then $\lim\limits_{t \to t_0} (\mathbf{f} + \mathbf{g})(t) = \mathbf{L} + \mathbf{M}$ as follows.
   a. Write $\mathbf{f}(t)$ and $\mathbf{g}(t)$ in component form.
   b. Obtain $(\mathbf{f} + \mathbf{g})(t)$ in component form.

52. Prove that if $\mathbf{f}$ and $\mathbf{g}$ are continuous at $t = t_0$ then so are
   a. $\mathbf{f} + \mathbf{g}$,
   b. $c\mathbf{f}$,
   c. $\mathbf{f} \times \mathbf{g}$.

53. Prove that if $\mathbf{f}$ and $\mathbf{g}$ are continuous vector-valued functions then so are the real-valued functions
   a. $\mathbf{f} \cdot \mathbf{g}$,
   b. $|\mathbf{f}|$.

54. For each set of vector-valued functions, find the intervals on which $\mathbf{f}$, $\mathbf{g}$, $\mathbf{h}$, and d are continuous.
   a. $\mathbf{f}(t) = \sqrt{t}\,\mathbf{i} - \dfrac{1}{t}\mathbf{j} + t\mathbf{k}$, $\mathbf{g}(t) = -2\sqrt{t}\,\mathbf{i} + \dfrac{2}{t}\mathbf{j} + 3t\mathbf{k}$,
       $\mathbf{h}(t) = (\mathbf{f} + \mathbf{g})(t)$

   b. $\mathbf{f}(t) = \dfrac{1}{t}\mathbf{i} + \ln(1 + t)\mathbf{j} + \sqrt{t}\,\mathbf{k}$, $\mathbf{g}(t) = t\mathbf{i} + \sqrt{t}\,\mathbf{k}$,
       $d(t) = (\mathbf{f} \cdot \mathbf{g})(t)$

55. If $\mathbf{f} + \mathbf{g}$ is continuous, what can you say about the continuity of $\mathbf{f}$ and $\mathbf{g}$? What conclusions can you draw from the continuity of the dot product $\mathbf{f} \cdot \mathbf{g}$?

56. Prove Corollary 1.

57. The vector-valued function
$$\mathbf{f}(t) = (2 \cos ht + \cos kt)\mathbf{i} + 2 \sin ht\mathbf{j} + \sin kt\mathbf{k}$$
describes the motion of a particle on a torus, and certain properties of this motion depend upon the choices of h and k. For example, for some choices the motion is periodic, and for others the motion is not periodic. Use a computer program capable of plotting three-dimensional graphics to plot the curve generated by $\mathbf{f}$ for various values of h and k.

17.2   Derivatives and Integrals of Vector-Valued Functions

We can differentiate and integrate vector-valued functions in much the same way as we do for real-valued functions. Moreover, many of the techniques we use in the real-valued case also apply to the vector-valued case. In this section, we develop the theories of differentiation and integration for vector-valued functions.

Let **f** be a vector-valued function and let t be a number in the domain of **f**. Also, assume that the numbers $t + h$ lie in the domain of **f** for h sufficiently small. The vector

$$\left(\frac{1}{h}\right)[\mathbf{f}(t + h) - \mathbf{f}(t)], \qquad h \neq 0 \tag{1}$$

is referred to as a **difference quotient** for the vector function $\mathbf{f}(t)$. The vector in equation (1) may be interpreted as the average change in the vector $\mathbf{f}(t)$ over the interval between t and $t + h$. As for functions of a single variable, we want to know about the limiting value of this vector as $h \to 0$, which we refer to as the *derivative* of the vector-valued function **f.** The next three sections focus on the geometric and physical interpretations of this derivative. Our main interest here is the definition and calculation of this particular limit.

Definition 3

The vector-valued function **f** is said to be **differentiable** at the number t if the limit

$$\mathbf{f}'(t) = \lim_{h \to 0} \frac{1}{h}[\mathbf{f}(t + h) - \mathbf{f}(t)] \tag{2}$$

exists. The vector $\mathbf{f}'(t)$ is called the **derivative** of the vector-valued function **f** at t.

As for functions of a single variable, the vector-valued function **f** is said to be differentiable on an open interval I if $\mathbf{f}'(t)$ in (2) exists for every $t \in I$. Thus, the derivative of a vector-valued function is again a vector-valued function, and the domain of $\mathbf{f}'$ is the subset of the domain of **f** for which the limit in line (2) exists.

Tangents for Curves

Even though most of this section concerns arbitrary vector-valued functions, we begin with a brief interpretation of Definition 3 in the case where a vector-valued function **r** is used to parameterize a curve C in space. That is, we interpret the vector $\mathbf{r}(t)$ as a position vector, and C is the curve generated by the function **r** for $t \in I$. Consider the limit

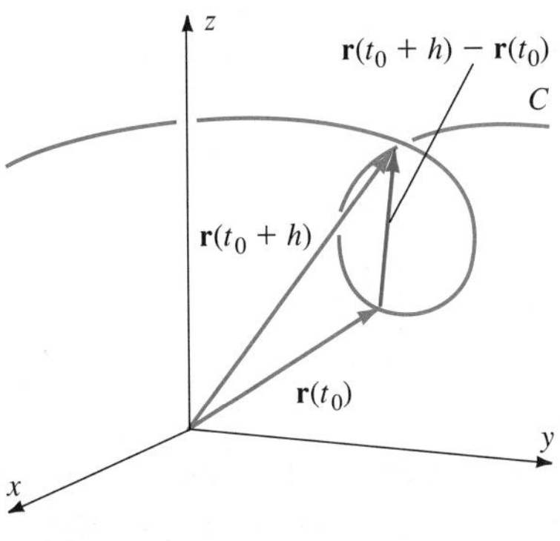

$$\lim_{h \to 0} \frac{1}{h}[\mathbf{r}(t_0 + h) - \mathbf{r}(t_0)],$$

and observe that the vector

$$\mathbf{r}(t_0 + h) - \mathbf{r}(t_0) \tag{3}$$

originates at the tip of $\mathbf{r}(t_0)$ and terminates at the tip of $\mathbf{r}(t_0 + h)$ (see Figure 2.1). This vector therefore determines a secant vector, joining the points $P(t_0)$ and $P(t_0 + h)$ on C. As $h \to 0$ the secant through these same points approaches a *tangent* to C at $\mathbf{r}(t_0)$. However, we cannot obtain a direction vector for the tangent line simply by letting $h \to 0$ in equation (1) since the length of this vector will approach zero. Therefore we multiply the vector in (3) by the scalar $1/h$ (changing its length,

Figure 2.1 The secant vector $\mathbf{r}(t_0 + h) - \mathbf{r}(t_0)$.

but not its direction) before applying the limit. Thus, a direction vector for the line tangent to C at $\mathbf{r}(t_0)$ is

$$\mathbf{r}'(t_0) = \lim_{h \to 0}\left[\frac{1}{h}[\mathbf{r}(t_0 + h) - \mathbf{r}(t_0)]\right]. \tag{4}$$

(See Figure 2.2.) If the limit $\mathbf{r}'(t_0)$ exists and is not $\mathbf{0}$, we obtain a vector that is tangent to the curve. In Section 17.3, we shall see that its length $|\mathbf{r}(t_0)|$ also has an important geometric interpretation.

Computing Derivatives

Now we return to the general case of an arbitrary vector-valued function $\mathbf{f}$ and develop techniques for the calculation of its derivative $\mathbf{f}'$. The following theorem shows that the derivative $\mathbf{f}'$ may be calculated directly from the components of $\mathbf{f}$.

Theorem 2

Let $\mathbf{f}(t) = x(t)\mathbf{i} + y(t)\mathbf{j} + z(t)\mathbf{k}$. The vector-valued function $\mathbf{f}$ is differentiable if and only if each of the component functions x, y, and z is differentiable. In this case

$$\mathbf{f}'(t) = x'(t)\mathbf{i} + y'(t)\mathbf{j} + z'(t)\mathbf{k}. \tag{5}$$

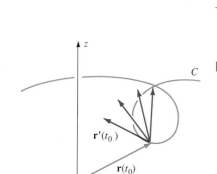

Figure 2.2 As $h \to 0$, the limit of

$$\left(\frac{1}{h}\right)[\mathbf{r}(t_0 + h) - \mathbf{r}(t_0)]$$

tends to a vector $\mathbf{r}'(t_0)$ that is tangent to C (assuming that the limit exists and is nonzero).

Theorem 2 is proved by examining the difference quotient in Definition 3.

Proof: According to Theorem 1,

$$\mathbf{f}'(t) = \lim_{h \to 0}\frac{1}{h}[\mathbf{f}(t + h) - \mathbf{f}(t)]$$

$$= \lim_{h \to 0}\frac{1}{h}([x(t + h)\mathbf{i} + y(t + h)\mathbf{j} + z(t + h)\mathbf{k}] - [x(t)\mathbf{i} + y(t)\mathbf{j} + z(t)\mathbf{k}])$$

$$= \lim_{h \to 0}\left[\left(\frac{x(t + h) - x(t)}{h}\right)\mathbf{i} + \left(\frac{y(t + h) - y(t)}{h}\right)\mathbf{j} + \left(\frac{z(t + h) - z(t)}{h}\right)\mathbf{k}\right]$$

$$= \left[\lim_{h \to 0}\left(\frac{x(t + h) - x(t)}{h}\right)\right]\mathbf{i} + \left[\lim_{h \to 0}\left(\frac{y(t + h) - y(t)}{h}\right)\right]\mathbf{j}$$

$$+ \left[\lim_{h \to 0}\left(\frac{z(t + h) - z(t)}{h}\right)\right]\mathbf{k}.$$

This calculation shows that $\mathbf{f}$ is differentiable if and only if each of the component functions is differentiable. Applying each of the indicated limits and using Definition 3 establishes equation (5). ∎

☐ **EXAMPLE 1**

Let $\mathbf{f}(t) = t^2\mathbf{i} + \cos t\mathbf{j} + e^{3t}\mathbf{k}$. Find $\mathbf{f}'(t)$.

Solution $\mathbf{f}'(t) = \left[\dfrac{d}{dt}(t^2)\right]\mathbf{i} + \left[\dfrac{d}{dt}(\cos t)\right]\mathbf{j} + \left[\dfrac{d}{dt}(e^{3t})\right]\mathbf{k}$

$$= 2t\mathbf{i} - \sin t\mathbf{j} + 3e^{3t}\mathbf{k}. \qquad ■$$

If the derivative $\mathbf{f}'$ is itself differentiable, we define its derivative to be the *second* derivative of $\mathbf{f}$. That is, $\mathbf{f}'' = (\mathbf{f}')'$. The second derivative of $\mathbf{f}$ is again a vector-valued function. Just as for real-valued functions, some or all of the higher-order derivatives may exist as well.

☐ **EXAMPLE 2**

Let $\mathbf{f}(t) = \sin 2t\mathbf{i} + \cos 2t\mathbf{j} + \sqrt{t}\mathbf{k}$. Find both $\mathbf{f}'$ and $\mathbf{f}''$.

Solution   For $t > 0$ we have

$$\mathbf{f}'(t) = 2\cos 2t\mathbf{i} - 2\sin 2t\mathbf{j} + \tfrac{1}{2}t^{-1/2}\mathbf{k}$$

and

$$\mathbf{f}''(t) = -4\sin 2t\mathbf{i} - 4\cos 2t\mathbf{j} - \tfrac{1}{4}t^{-3/2}\mathbf{k}.$$ ■

☐ **EXAMPLE 3**

Show that the function $\mathbf{f}(t) = A\sin \omega t\mathbf{i} + B\cos \omega t\mathbf{j}$ satisfies the **vector differential equation**

$$\mathbf{f}''(t) + \omega^2\mathbf{f}(t) = \mathbf{0}. \tag{6}$$

Strategy · · · · · · · ·
Calculate $\mathbf{f}'(t)$ and $\mathbf{f}''(t)$.

Solution
Here

$$\mathbf{f}'(t) = \omega A\cos \omega t\mathbf{i} - \omega B\sin \omega t\mathbf{j}$$

and

$$\mathbf{f}''(t) = -\omega^2 A\sin \omega t\mathbf{i} - \omega^2 B\cos \omega t\mathbf{j}.$$

Substitute $\mathbf{f}$ and $\mathbf{f}''$ into the left-hand side of (6) and verify that $\mathbf{0}$ is obtained.

Substituting into equation (4) shows that

$$\mathbf{f}''(t) + \omega^2\mathbf{f}(t) = [-\omega^2 A\sin \omega t\mathbf{i} - \omega^2 B\cos \omega t\mathbf{j}] + \omega^2[A\sin \omega t\mathbf{i} + B\cos \omega t\mathbf{j}]$$
$$= \mathbf{0}.$$ ■

The following theorem shows how derivatives of various other vector-valued functions may be calculated.

Theorem 3

Let the vector-valued functions $\mathbf{f}$ and $\mathbf{g}$ and the scalar-valued function h be differentiable on appropriate intervals and let c be a number. Then the vector-valued functions $\mathbf{f} + \mathbf{g}$, $c\mathbf{f}$, $h\mathbf{f}$, $\mathbf{f} \times \mathbf{g}$, and $\mathbf{f} \circ h$, and the scalar-valued function $\mathbf{f} \cdot \mathbf{g}$ are differentiable, and

  (i) $(\mathbf{f} + \mathbf{g})'(t) = \mathbf{f}'(t) + \mathbf{g}'(t)$,
 (ii) $(c\mathbf{f})'(t) = c\mathbf{f}'(t)$,
(iii) $(h\mathbf{f})'(t) = h(t)\mathbf{f}'(t) + h'(t)\mathbf{f}(t)$,
 (iv) $(\mathbf{f} \times \mathbf{g})'(t) = [\mathbf{f}(t) \times \mathbf{g}'(t)] + [\mathbf{f}'(t) \times \mathbf{g}(t)]$,
  (v) $(\mathbf{f} \circ h)'(t) = h'(t)\mathbf{f}'(h(t))$ (Chain Rule),
 (vi) $(\mathbf{f} \cdot \mathbf{g})'(t) = \mathbf{f}(t) \cdot \mathbf{g}'(t) + \mathbf{f}'(t) \cdot \mathbf{g}(t)$.

Before commenting on the proof of Theorem 3, we consider two examples of its use.

☐ **EXAMPLE 4**

Let $\mathbf{f}(t) = t^2\mathbf{i} + e^t\mathbf{k}$ and $\mathbf{g}(t) = \sin t\mathbf{i} + \cos t\mathbf{j} + \mathbf{k}$. Find the derivative of the function $\mathbf{r} = \mathbf{f} \times \mathbf{g}$.

Solution   The derivatives of the given functions are

$$\mathbf{f}'(t) = 2t\mathbf{i} + e^t\mathbf{k} \quad \text{and} \quad \mathbf{g}'(t) = \cos t\mathbf{i} - \sin t\mathbf{j}.$$

According to part (iv) of Theorem 3 we have

$$\mathbf{r}'(t) = [\mathbf{f}(t) \times \mathbf{g}'(t)] + [\mathbf{f}'(t) \times \mathbf{g}(t)]$$

$$= \det \begin{bmatrix} \mathbf{i} & \mathbf{j} & \mathbf{k} \\ t^2 & 0 & e^t \\ \cos t & -\sin t & 0 \end{bmatrix} + \det \begin{bmatrix} \mathbf{i} & \mathbf{j} & \mathbf{k} \\ 2t & 0 & e^t \\ \sin t & \cos t & 1 \end{bmatrix}$$

$$= [e^t \sin t\mathbf{i} + e^t \cos t\mathbf{j} - t^2 \sin t\mathbf{k}]$$
$$\quad + [-e^t \cos t\mathbf{i} + (e^t \sin t - 2t)\mathbf{j} + 2t \cos t\mathbf{k}]$$

$$= e^t(\sin t - \cos t)\mathbf{i} + [e^t(\sin t + \cos t) - 2t]\mathbf{j}$$
$$\quad + (2t \cos t - t^2 \sin t)\mathbf{k}.$$

This same result may be obtained directly by noting that

$$\mathbf{r}(t) = \mathbf{f}(t) \times \mathbf{g}(t)$$

$$= \det \begin{bmatrix} \mathbf{i} & \mathbf{j} & \mathbf{k} \\ t^2 & 0 & e^t \\ \sin t & \cos t & 1 \end{bmatrix}$$

$$= -e^t \cos t\mathbf{i} + (e^t \sin t - t^2)\mathbf{j} + t^2 \cos t\mathbf{k}$$

and differentiating component by component, according to Theorem 2. ■

□ **EXAMPLE 5**

Let **f** and **g** be as in Example 4. Find the derivative of the scalar-valued function $s = \mathbf{f} \cdot \mathbf{g}$.

Solution By Theorem 3, part (vi),

$$s'(t) = \mathbf{f}(t) \cdot \mathbf{g}'(t) + \mathbf{f}'(t) \cdot \mathbf{g}(t)$$
$$= [(t^2\mathbf{i} + e^t\mathbf{k}) \cdot (\cos t\mathbf{i} - \sin t\mathbf{j})] + [(2t\mathbf{i} + e^t\mathbf{k}) \cdot (\sin t\mathbf{i} + \cos t\mathbf{j} + \mathbf{k})]$$
$$= t^2 \cos t + 2t \sin t + e^t.$$

To calculate this derivative directly, we first compute

$$s(t) = \mathbf{f}(t) \cdot \mathbf{g}(t) = t^2 \sin t + (0)(\cos t) + e^t$$
$$= t^2 \sin t + e^t.$$

Differentiating then gives the above result. ■

Examples 4 and 5 suggest that Theorem 3 may be proved by writing the vector-valued functions **f** and **g** in component form and applying Theorem 2. This is indeed the case. We prove part (v) here and leave the other statements as exercises. If

$$\mathbf{f}(t) = x(t)\mathbf{i} + y(t)\mathbf{j} + z(t)\mathbf{k},$$

then

$$\mathbf{f}(h(t)) = x(h(t))\mathbf{i} + y(h(t))\mathbf{j} + z(h(t))\mathbf{k}.$$

According to Theorem 2 and the Chain Rule of Chapter 3,

$$(\mathbf{f} \circ h)'(t) = [(x \circ h)'(t)]\mathbf{i} + [(y \circ h)'(t)]\mathbf{j} + [(z \circ h)'(t)]\mathbf{k}$$
$$= [x'(h(t))h'(t)]\mathbf{i} + [y'(h(t))h'(t)]\mathbf{j} + [z'(h(t))h'(t)]\mathbf{k}$$
$$= h'(t)[x'(h(t))\mathbf{i} + y'(h(t))\mathbf{j} + z'(h(t))\mathbf{k}]$$
$$= h'(t)\mathbf{f}'(h(t)).$$

The result of the following example will be useful in later calculations.

☐ EXAMPLE 6

Let **f** be a vector-valued function and let c be a constant. Prove that if **f** is differentiable on some interval I and if $|\mathbf{f}(t)| = c$ for all $t \in I$, then $\mathbf{f}(t)$ and $\mathbf{f}'(t)$ are orthogonal for all $t \in I$.

Strategy · · · · · · · ·
Express $|\mathbf{f}(t)|^2$ as a dot product using $|\mathbf{v}|^2 = \mathbf{v} \cdot \mathbf{v}$.

Solution

By Theorem 4, Chapter 16,

$$|\mathbf{f}(t)|^2 = \mathbf{f}(t) \cdot \mathbf{f}(t).$$

The resulting expression equals c^2.

Thus, since $|\mathbf{f}(t)| = c$, we have

$$\mathbf{f}(t) \cdot \mathbf{f}(t) = c^2, \qquad t \in I.$$

Differentiate both sides of this equation using part (vi) of Theorem 3.

Differentiating both sides of this equation with respect to t gives

$$\mathbf{f}(t) \cdot \mathbf{f}'(t) + \mathbf{f}'(t) \cdot \mathbf{f}(t) = 0$$

or

$$2\mathbf{f}(t) \cdot \mathbf{f}'(t) = 0.$$

Use the fact that **v** is orthogonal to **w** if and only if $\mathbf{v} \cdot \mathbf{w} = 0$.

Thus $\mathbf{f}'(t) \cdot \mathbf{f}(t) = 0$, $t \in I$, so $\mathbf{f}(t)$ and $\mathbf{f}'(t)$ are orthogonal for all $t \in I$. ∎

Integrals of Vector-Valued Functions

Just as for real-valued functions, we say that the vector-valued function **F** is an *antiderivative* for the vector-valued function **f** on the interval I if $\mathbf{F}'(t) = \mathbf{f}(t)$ for all t in I.

If the vector-valued function

$$\mathbf{F}(t) = X(t)\mathbf{i} + Y(t)\mathbf{j} + Z(t)\mathbf{k}$$

is an antiderivative for

$$\mathbf{f}(t) = x(t)\mathbf{i} + y(t)\mathbf{j} + z(t)\mathbf{k}$$

on the interval I, it follows that any other antiderivative **G** for **f** on I must have the form

$$
\begin{aligned}
\mathbf{G}(t) &= [X(t) + c_1]\mathbf{i} + [Y(t) + c_2]\mathbf{j} + [Z(t) + c_3]\mathbf{k} \\
&= X(t)\mathbf{i} + Y(t)\mathbf{j} + Z(t)\mathbf{k} + (c_1\mathbf{i} + c_2\mathbf{j} + c_3\mathbf{k}) \\
&= \mathbf{F}(t) + \mathbf{C}, \qquad \mathbf{C} = (c_1\mathbf{i} + c_2\mathbf{j} + c_3\mathbf{k}).
\end{aligned}
$$

We refer to the family of *all* antiderivatives for **f** on I as the *indefinite integral* for **f** on I, denoted by $\int \mathbf{f}(t) \, dt$, the component form of which is given by the following theorem.

Theorem 4

The **indefinite integral** of the vector-valued function $\mathbf{f}(t) = x(t)\mathbf{i} + y(t)\mathbf{j} + z(t)\mathbf{k}$ on the interval I is

$$\int \mathbf{f}(t) \, dt = \left[\int x(t) \, dt \right]\mathbf{i} + \left[\int y(t) \, dt \right]\mathbf{j} + \left[\int z(t) \, dt \right]\mathbf{k} \tag{7}$$

provided that the integrals of the component functions exist on I.

REMARK It is important to note in applying equation (7) that *three* constants of integration will arise—one for each of the antiderivatives of the respective components. That is, if $X'(t) = x(t)$, $Y'(t) = y(t)$, and $Z'(t) = z(t)$, we can write equation (7) as either

$$\int \mathbf{f}(t)\, dt = [X(t) + c_1]\mathbf{i} + [Y(t) + c_2]\mathbf{j} + [Z(t) + c_3]\mathbf{k} \tag{8}$$

or

$$\int \mathbf{f}(t)\, dt = X(t)\mathbf{i} + Y(t)\mathbf{j} + Z(t)\mathbf{k} + \mathbf{C} \tag{9}$$

where

$$\mathbf{C} = c_1\mathbf{i} + c_2\mathbf{j} + c_3\mathbf{k}.$$

A common mistake is to write the constant $\mathbf{C}$ in equation (9) as a number rather than as a vector.

☐ **EXAMPLE 7**

Find the indefinite integral $\int \mathbf{f}(t)\, dt$ for

$$\mathbf{f}(t) = 2t\mathbf{i} + e^{-3t}\mathbf{j} + \sec^2 t\mathbf{k}.$$

Solution By Theorem (4)

$$\int \mathbf{f}(t)\, dt = \left[\int 2t\, dt\right]\mathbf{i} + \left[\int e^{-3t}\, dt\right]\mathbf{j} + \left[\int \sec^2 t\, dt\right]\mathbf{k}$$

$$= (t^2 + c_1)\mathbf{i} + (-\tfrac{1}{3}e^{-3t} + c_2)\mathbf{j} + (\tan t + c_3)\mathbf{k}$$

$$= t^2\mathbf{i} - \tfrac{1}{3}e^{-3t}\mathbf{j} + \tan t\mathbf{k} + \mathbf{C}, \qquad \mathbf{C} = c_1\mathbf{i} + c_2\mathbf{j} + c_3\mathbf{k}. \qquad ■$$

☐ **EXAMPLE 8**

Find the vector-valued function $\mathbf{f}$ for which

$$\mathbf{f}'(t) = \mathbf{i} + t^3\mathbf{j} + \left(\frac{1}{1 + t^2}\right)\mathbf{k},$$

and $\mathbf{f}(0) = 2\mathbf{i} - \mathbf{j} + 3\mathbf{k}$.

Solution By integration we obtain

$$\mathbf{f}(t) = \int \mathbf{f}'(t)\, dt$$

$$= \int \left[\mathbf{i} + t^3\mathbf{j} + \left(\frac{1}{1 + t^2}\right)\mathbf{k}\right] dt$$

$$= (t + c_1)\mathbf{i} + \left(\frac{t^4}{4} + c_2\right)\mathbf{j} + (\text{Tan}^{-1} t + c_3)\mathbf{k}.$$

Setting $t = 0$ and using the given *initial condition* $\mathbf{f}(0) = 2\mathbf{i} - \mathbf{j} + 3\mathbf{k}$ gives the equation

$$\mathbf{f}(0) = c_1\mathbf{i} + c_2\mathbf{j} + c_3\mathbf{k} = 2\mathbf{i} - \mathbf{j} + 3\mathbf{k}.$$

Thus, $c_1 = 2$, $c_2 = -1$, and $c_3 = 3$, so

$$\mathbf{f}(t) = (t + 2)\mathbf{i} + \left(\frac{t^4}{4} - 1\right)\mathbf{j} + (\text{Tan}^{-1}\, t + 3)\mathbf{k}$$

is the desired solution.   ∎

Finally, the definite integral of the vector-valued function $\mathbf{f}$ is defined as the vector whose components are the respective definite integrals of the components of $\mathbf{f}$.

Definition 4

Let $\mathbf{f}(t) = x(t)\mathbf{i} + y(t)\mathbf{j} + z(t)\mathbf{k}$. If the definite integral of each of the components exists on the interval $[a, b]$, we define

$$\int_a^b \mathbf{f}(t)\, dt = \left[\int_a^b x(t)\, dt\right]\mathbf{i} + \left[\int_a^b y(t)\, dt\right]\mathbf{j} + \left[\int_a^b z(t)\, dt\right]\mathbf{k}.$$

☐ **EXAMPLE 9**

For $\mathbf{f}(t) = \sin 3t\,\mathbf{i} + \cos t\,\mathbf{j} + \mathbf{k}$, find

$$\int_0^{\pi/2} \mathbf{f}(t)\, dt.$$

Solution   By Definition 4,

$$\int_0^{\pi/2} \mathbf{f}(t)\, dt = \left[\int_0^{\pi/2} \sin 3t\, dt\right]\mathbf{i} + \left[\int_0^{\pi/2} \cos t\, dt\right]\mathbf{j} + \left[\int_0^{\pi/2} 1\, dt\right]\mathbf{k}$$

$$= \left[-\frac{\cos 3t}{3}\right]_0^{\pi/2}\mathbf{i} + \left[\sin t\right]_0^{\pi/2}\mathbf{j} + \left[t\right]_0^{\pi/2}\mathbf{k}$$

$$= [0 - (-1/3)]\mathbf{i} + [1 - 0]\mathbf{j} + [\pi/2 - 0]\mathbf{k}$$

$$= (1/3)\mathbf{i} + \mathbf{j} + (\pi/2)\mathbf{k}.$$ ∎

We may summarize the results of this section by saying that for the purposes of calculating limits, derivatives, or integrals, the vector-valued function $\mathbf{f}(t) = x(t)\mathbf{i} + y(t)\mathbf{j} + z(t)\mathbf{k}$ is regarded as a sum of three component functions of a single variable. In the remaining sections of this chapter, we shall see how these results may be combined with a broader interpretation of $\mathbf{f}(t)$ as a vector in space to develop the concepts of tangent, normal, velocity, acceleration, and curvature associated with curves in space.

Exercise Set 17.2

In Exercises 1–4, state the intervals on which the given vector-valued function is differentiable.

1. $\mathbf{f}(t) = t\mathbf{i} + \cos t\mathbf{j} + \sqrt{t}\,\mathbf{k}$

2. $\mathbf{f}(t) = t^{-2}\mathbf{i} + |t - 3|\mathbf{j} + t^3\mathbf{k}$

3. $\mathbf{f}(t) = \ln(3 + t)\mathbf{i} + \left(\dfrac{1}{t + 2}\right)\mathbf{j} + e^t\mathbf{k}$

4. $\mathbf{f}(t) = \sqrt{1 - t^2}\,\mathbf{i} + \ln t\mathbf{j} + \dfrac{1}{1 - t}\mathbf{k}$

In Exercises 5–12, find the derivative of the given vector-valued function.

5. $\mathbf{f}(t) = t\mathbf{i} + \sqrt{t}\,\mathbf{j}$

6. $\mathbf{f}(t) = \mathbf{i} + \sin t\mathbf{j} + \cos 2t\mathbf{k}$

7. $\mathbf{f}(t) = \sqrt{t}\,\mathbf{i} + t^{-3/2}\mathbf{j} + \ln(2t - 1)\mathbf{k}$

8. $\mathbf{f}(t) = \dfrac{3t^3 - 5t^2 + 4t - 1}{t^2}\mathbf{i} + (t - 2)(3t + 7)\mathbf{j} + \mathbf{k}$

9. $\mathbf{f}(t) = e^t(t^2 - 1)\mathbf{i} + t^2 \sin t\mathbf{j} + \sqrt{t^3 - 2t - 1}\,\mathbf{k}$

10. $\mathbf{f}(t) = \cos^2 t\mathbf{i} + \sec t\mathbf{j} + \text{Tan}^{-1} t\mathbf{k}$

11. $\mathbf{f}(t) = \text{Sin}^{-1} t\mathbf{i} + \sqrt{1 + t^2}\mathbf{j} + e^{-t^3}\mathbf{k}$

12. $\mathbf{f}(t) = e^{\sqrt{t}}\mathbf{i} + 3\mathbf{j} - \text{Cos}^{-1} 2t\mathbf{k}$

13. Find $\mathbf{f}''$ for the function $\mathbf{f}$ in Exercise 7.

14. Find $\mathbf{f}''$ for the function $\mathbf{f}$ in Exercise 12.

15. Show that the function $\mathbf{f}(t) = Ae^{\omega t}\mathbf{i} + Be^{-\omega t}\mathbf{j}$ satisfies the vector differential equation

$$\mathbf{f}''(t) - \omega^2 \mathbf{f}(t) = \mathbf{0}.$$

16. Show that the function

$$\mathbf{f}(t) = C \sinh \omega t\mathbf{i} + D \cosh \omega t\mathbf{j}$$

satisfies the vector differential equation in Exercise 15.

In Exercises 17–22, let

$$\mathbf{f}(t) = \sin t\mathbf{i} + \mathbf{j} + t^2\mathbf{k},$$
$$\mathbf{g}(t) = t\mathbf{i} + \cos t\mathbf{k},$$
$$h(t) = e^{3t}.$$

Find the derivative of the given function.

17. $\mathbf{r}(t) = (3\mathbf{f} - 2\mathbf{g})(t)$
18. $\mathbf{r}(t) = h(t)\mathbf{f}(t)$

19. $\mathbf{r}(t) = (\mathbf{f} \times \mathbf{g})(t)$
20. $\mathbf{r}(t) = (\mathbf{f} \cdot \mathbf{g})(t)$

21. $\mathbf{r}(t) = \mathbf{f}(h(t))$
22. $\mathbf{r}(t) = |\mathbf{g}(t)|^2$

In Exercises 23–26, let

$$\mathbf{f}(2) = 2\mathbf{i} + 8\mathbf{j} + 4\mathbf{k} \qquad \mathbf{f}'(2) = 4\mathbf{i} + 12\mathbf{j}$$
$$\mathbf{g}(2) = -\mathbf{i} + 9\mathbf{j} + 4\mathbf{k} \qquad \mathbf{g}'(2) = \mathbf{i} + 2\mathbf{j} + 4\mathbf{k}$$
$$h(2) = 4 \qquad h'(2) = 4.$$

Use Theorem 3 to evaluate the derivative of the given function at $t = 2$.

23. $\mathbf{r}(t) = (\mathbf{f} + \mathbf{g})(t)$
24. $\mathbf{r}(t) = (h\mathbf{f})(t)$

25. $\mathbf{r}(t) = (\mathbf{f} \cdot \mathbf{g})(t)$
26. $\mathbf{r}(t) = (\mathbf{f} \times \mathbf{g})(t)$

In Exercises 27–28, let

$$\mathbf{f}(t) = t\mathbf{i} + t^2\mathbf{j} + e^t\mathbf{k}$$
$$\mathbf{g}(t) = t\mathbf{i} + t\mathbf{j} + \mathbf{k}.$$

27. Find $(\mathbf{f} \cdot \mathbf{g})'(t_0)$ if $\mathbf{f}(t_0) = \mathbf{k}$.

28. Find $(\mathbf{g} \times \mathbf{f})'(t_0)$ if $\mathbf{f}(t_0) = -2\mathbf{i} + 4\mathbf{j} + e^{-2}\mathbf{k}$.

29. Verify that $\mathbf{f}(t)$ and $\mathbf{f}'(t)$ are orthogonal if $\mathbf{f}$ is the vector function $\mathbf{f}(t) = \cos 2t\mathbf{i} + \sin 2t\mathbf{j} + 3\mathbf{k}$.

In Exercises 30–34, find the indefinite integral for the given vector-valued function.

30. $\mathbf{f}(t) = t\mathbf{i} + \cos t\mathbf{j} + \sin t\mathbf{k}$

31. $\mathbf{f}(t) = \sqrt{t}\mathbf{i} + e^{2t}\mathbf{j} + \dfrac{1}{t}\mathbf{k}$

32. $\mathbf{f}(t) = t \sin t\mathbf{i} + \cos^2 t\mathbf{j} + (1 - t)\mathbf{k}$

33. $\mathbf{f}(t) = \ln t\mathbf{i} + \dfrac{1}{t \ln t}\mathbf{j}$

34. $\mathbf{f}(t) = t \sec^2(1 - t^2)\mathbf{i} + \sqrt{1 - t}\mathbf{j} + \sqrt{t}\mathbf{k}$

35. Find $\mathbf{f}$ if $\mathbf{f}'(t) = \mathbf{i} + t^2\mathbf{j}$ and $\mathbf{f}(0) = 3\mathbf{i} + 5\mathbf{j}$

36. Find $\mathbf{f}$ if

$$\mathbf{f}'(t) = \cos 2t\mathbf{i} + \left(\frac{1}{1 + t^2}\right)\mathbf{j} + \frac{t}{\sqrt{1 + t^2}}\mathbf{k}$$

and $\mathbf{f}(0) = -4\mathbf{i} + \mathbf{k}$.

37. Find $\mathbf{f}$ if $\mathbf{f}'(t) = \cos^2 t\mathbf{i} + \sin^2 t\mathbf{j} + e^{-t}\mathbf{k}$ and $\mathbf{f}(0) = 3\mathbf{i} - 6\mathbf{j} + 2\mathbf{k}$.

38. Find a solution of the vector differential equation

$$\mathbf{f}'(t) = \mathbf{f}(t)$$

for which $\mathbf{f}(0) = \mathbf{i} + 2\mathbf{j} - \mathbf{k}$.

39. Find a solution of the vector differential equation

$$\mathbf{f}'(t) + 4\mathbf{f}(t) = \mathbf{0}$$

satisfying the initial condition $\mathbf{f}(0) = \mathbf{i} + 4\mathbf{k}$.

40. Find a family of solutions to the vector differential equation

$$\mathbf{f}''(t) - 2\mathbf{f}(t) = \mathbf{0}.$$

41. Find $\displaystyle\int_a^b (t\mathbf{i} + t^2\mathbf{j} - t^3\mathbf{k})\,dt$.

42. Find $\displaystyle\int_0^1 \mathbf{f}(t)\,dt$ if $\mathbf{f}(t) = t^2\mathbf{i} - 2t\mathbf{j} + \sqrt{t}\mathbf{k}$.

43. Find $\displaystyle\int_0^{\pi/4} \mathbf{f}(t)\,dt$ if $\mathbf{f}(t) = \cos t\mathbf{i} + \sin 2t\mathbf{j} + \cos^2 t\mathbf{k}$.

44. Find $\displaystyle\int_1^4 \mathbf{f}(t)\,dt$ if $\mathbf{f}(t) = e^t\mathbf{i} + t^2\mathbf{j} + \ln 2t\mathbf{k}$.

45. Show that if $\mathbf{f}'(t) = \mathbf{0}$ for all $t \in I$, then $\mathbf{f}$ is constant on the interval I.

46. Show that if $\mathbf{f}'(t) = \mathbf{g}'(t)$ for all $t \in I$ then $\mathbf{f}(t) = \mathbf{g}(t) + \mathbf{C}$ for some vector $\mathbf{C}$ and all $t \in I$.

47. True or false? If $\mathbf{f}$ is differentiable on $[a, b]$, there exists a constant $c \in (a, b)$ so that

$$\mathbf{f}'(c) = \frac{1}{b - a}[\mathbf{f}(b) - \mathbf{f}(a)].$$

48. Prove statements (i) through (iv) and (vi) of Theorem 3 by first writing the vector functions in component form.

49. Complete the following alternative proof of differentiation formula (iii) of Theorem 3:

$$(h\mathbf{f})'(t) = h'(t)\mathbf{f}(t) + h(t)\mathbf{f}'(t).$$

a. Show that we can write

$$(h\mathbf{f})'(t) = \lim_{s \to 0} \frac{1}{s}[h(t + s)\mathbf{f}(t + s) - h(t)\mathbf{f}(t)]$$

$$= \lim_{s \to 0} \frac{1}{s}[h(t + s) - h(t)]\mathbf{f}(t + s)$$

$$+ \lim_{s \to 0} \frac{1}{s}[\mathbf{f}(t + s) - \mathbf{f}(t)]h(t).$$

b. Show that the limit in part a is

$$(h\mathbf{f})'(t) = h'(t)\mathbf{f}(t) + h(t)\mathbf{f}'(t).$$

50. Use a method similar to that of Exercise 49 to prove part (vi) of Theorem 3:

$$(\mathbf{f} \cdot \mathbf{g})'(t) = \mathbf{f}'(t) \cdot \mathbf{g}(t) + \mathbf{f}(t) \cdot \mathbf{g}'(t).$$

17.3   Curves: Tangents and Arc Length

In Sections 17.1 and 17.2, we saw how a curve C (in the plane or in space) can be parameterized using a vector-valued function $\mathbf{r}(t)$, and we interpreted the derivative $\mathbf{r}'(t_0)$ as a vector tangent to this curve. In this section, we focus on the use of vector-valued functions to study curves.

Suppose we are given a curve C and a vector-valued function

$$\mathbf{r}(t) = x(t)\mathbf{i} + y(t)\mathbf{j} + z(t)\mathbf{k}, \qquad t \in I \tag{1}$$

that parameterizes C. We say that the curve C is **differentiable** if the function $\mathbf{r}$ is differentiable on I, which in turn requires that each of the component functions x, y, and z is differentiable (Theorem 2). The curve is **smooth** if $\mathbf{r}'$ is continuous and nonzero on I.

As we discussed in Section 17.2, if the derivative $\mathbf{r}'(t)$ exists and is nonzero, we refer to it as a *tangent vector*.

Definition 5

For the differentiable curve C in equation (1), if $\mathbf{r}'(t_0) \neq \mathbf{0}$, the vector

$$\mathbf{r}'(t_0) = x'(t_0)\mathbf{i} + y'(t_0)\mathbf{j} + z'(t_0)\mathbf{k}, \qquad t_0 \in I$$

is called a **tangent vector** for C at the tip of the vector $\mathbf{r}(t_0)$ (Figure 3.1).

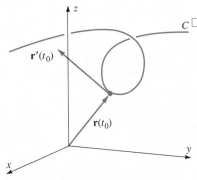

Figure 3.1 The vector $\mathbf{r}(t_0)$ determines a point on the curve C, and the derivative $\mathbf{r}'(t_0)$ is tangent to C at that point.

□ EXAMPLE 1

The vector-valued function

$$\mathbf{r}(t) = a \cos t\mathbf{i} + a \sin t\mathbf{j}, \qquad a > 0$$

parameterizes the circle in the xy-plane with center $(0, 0)$ and radius a. For this function the tangent vector at $\mathbf{r}(t)$ is

$$\mathbf{r}'(t) = -a \sin t\mathbf{i} + a \cos t\mathbf{j}.$$

Note that

$$\mathbf{r}(t) \cdot \mathbf{r}'(t) = -a^2(\cos t)(\sin t) + a^2(\sin t)(\cos t) = 0,$$

which confirms the familiar fact that the tangent to a circle at point P is perpendicular to the radius at that point. (See Figure 3.2.) ∎

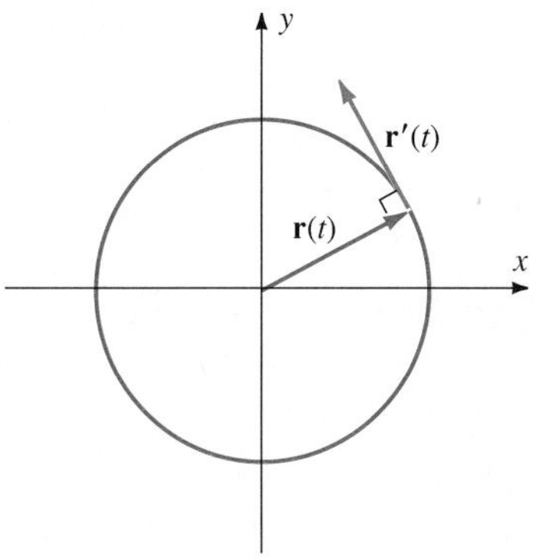

Figure 3.2 For the circle with parameterization

$$\mathbf{r}(t) = a \cos t\mathbf{i} + a \sin t\mathbf{j},$$

the tangent $\mathbf{r}'(t)$ is always orthogonal to the radius vector $\mathbf{r}(t)$.

If $\mathbf{r}'(t_0) \neq \mathbf{0}$, a unit vector $\mathbf{T}$ tangent to the curve C at $\mathbf{r}(t_0)$ is given by

$$\mathbf{T} = \frac{\mathbf{r}'(t_0)}{|\mathbf{r}'(t_0)|} \qquad \textbf{(unit tangent).} \tag{2}$$

☐ **EXAMPLE 2**

Find a unit vector tangent to the circular helix $\mathbf{r}(t) = a \cos t\mathbf{i} + a \sin t\mathbf{j} + bt\mathbf{k}$ at the point $(0, a, b\pi/2)$.

Strategy · · · · · · · ·
Find t_0 for which $\mathbf{r}(t_0)$ terminates at the point $(0, a, b\pi/2)$.

Solution
Setting $\mathbf{r}(t_0) = 0\mathbf{i} + a\mathbf{j} + \dfrac{b\pi}{2}\mathbf{k}$ gives the equations

$$a \cos t_0 = 0, \qquad a \sin t_0 = a, \qquad bt_0 = b\pi/2,$$

so $t_0 = \pi/2$.

Find $\mathbf{r}'(t_0)$.

According to Definition 5 a tangent vector at $\mathbf{r}(t_0) = \mathbf{r}(\pi/2)$ is

$$\mathbf{r}'(\pi/2) = -a \sin(\pi/2)\mathbf{i} + a \cos(\pi/2)\mathbf{j} + b\mathbf{k}$$
$$= -a\mathbf{i} + b\mathbf{k}.$$

Find $|\mathbf{r}'(t_0)|$.

This vector has length

$$|\mathbf{r}'(\pi/2)| = \sqrt{a^2 + b^2}.$$

Apply (2).

The desired unit tangent is

$$\mathbf{T} = \frac{\mathbf{r}'(\pi/2)}{|\mathbf{r}'(\pi/2)|} = \frac{-a\mathbf{i} + b\mathbf{k}}{\sqrt{a^2 + b^2}}.$$

(See Figure 3.3.) ∎

To find an equation for the *line* ℓ tangent to C at $\mathbf{r}(t_0)$, we use the fact that $\mathbf{r}(t_0)$ is a position vector for ℓ and $\mathbf{r}'(t_0)$ is a direction vector for ℓ (see Figure 3.4). Since t is the parameter for C, we should use a different variable to parametrize ℓ, say ω.

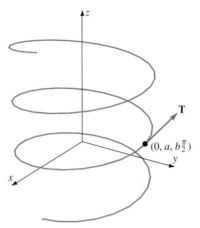

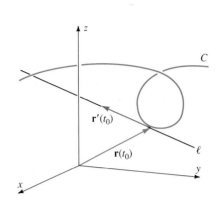

Figure 3.3 The unit tangent vector **T** to the circular helix at $(0, a, b\pi/2)$.

Figure 3.4 The vector $\mathbf{r}'(t_0)$ is a direction vector for the tangent line ℓ to C at $\mathbf{r}(t_0)$.

We may then use equation (1), Section 16.2, to write a parameterization for ℓ as

$$\ell: \quad \mathbf{R}(\omega) = \mathbf{r}(t_0) + \omega \mathbf{r}'(t_0). \tag{3}$$

(Be careful when using equation (3): $\mathbf{R}(\omega)$ is the position vector of a point on ℓ, while $\mathbf{r}(t_0)$ and $\mathbf{r}'(t_0)$ are fixed vectors determined by the curve C.)

☐ **EXAMPLE 3**

Find vector, parametric, and symmetric equations for the line ℓ tangent to the curve

$$C: \quad \mathbf{r}(t) = t^2 \mathbf{i} + (3 - t)\mathbf{j} + t^3 \mathbf{k}$$

at the point $(4, 1, 8)$.

Strategy · · · · · · · ·

Find t_0 for which $\mathbf{r}(t_0)$ terminates at the point $(4, 1, 8)$. $\mathbf{r}(t_0)$ is a position vector for ℓ.

Find $\mathbf{r}'(t_0)$. This is a direction vector for ℓ.

Use equation (3) to write the vector form of ℓ.

The parametric equations for x, y, and z are the components of $\mathbf{R}(\omega)$.

Solution

Setting $\mathbf{r}(t_0) = 4\mathbf{i} + \mathbf{j} + 8\mathbf{k}$ gives the equations

$$t_0^2 = 4, \quad 3 - t_0 = 1, \quad \text{and} \quad t_0^3 = 8$$

for which the solution is $t_0 = 2$.

Since $\mathbf{r}'(t) = 2t\mathbf{i} - \mathbf{j} + 3t^2 \mathbf{k}$,

$$\mathbf{r}'(t_0) = \mathbf{r}'(2) = 4\mathbf{i} - \mathbf{j} + 12\mathbf{k}.$$

The vector form of ℓ is therefore

$$\ell: \quad \mathbf{R}(\omega) = (4\mathbf{i} + \mathbf{j} + 8\mathbf{k}) + \omega(4\mathbf{i} - \mathbf{j} + 12\mathbf{k}).$$

To find parametric equations for ℓ, we let $P(\omega) = (x(\omega), y(\omega), z(\omega))$ be a point on ℓ. Then

$$x(\omega) = 4 + 4\omega \quad \text{(\textbf{i}-component)}$$
$$y(\omega) = 1 - \omega \quad \text{(\textbf{j}-component)}$$
$$z(\omega) = 8 + 12\omega \quad \text{(\textbf{k}-component)}$$

The symmetric equations are found from the parametric equations as in Section 16.2.

are the parametric equations for the coordinates of P. Solving each of these equations for ω and equating the results gives the symmetric equations

$$\frac{x-4}{4} = 1 - y = \frac{z-8}{12}$$

for ℓ.   ∎

Using the vector operations discussed in Chapter 16 and the tangent vector $\mathbf{r}'(t_0)$, we can answer many geometric questions about curves in the plane or in space. In the following example, we use the dot product to determine the angle at which two curves intersect.

□ **EXAMPLE 4**

Consider the curves

$$C_1: \quad \mathbf{r}_1(t) = e^t \mathbf{i} + 2 \sin t \mathbf{j} + (t^2 - 2)\mathbf{k}$$
$$C_2: \quad \mathbf{r}_2(t) = t\mathbf{i} + (t^2 - 1)\mathbf{j} + (t - 3)\mathbf{k}.$$

Determine their points of intersection and the angles at which they intersect.

Solution   Since a point of intersection may correspond to one value of the parameter on one curve and a different value of the parameter on the other curve, we must use two different variables as parameters for the two curves. We keep the variable t as the parameter for $\mathbf{r}_1$, and we use u as the parameter for $\mathbf{r}_2$. In other words, we rewrite the formula for $\mathbf{r}_2$ as

$$\mathbf{r}_2(u) = u\mathbf{i} + (u^2 - 1)\mathbf{j} + (u - 3)\mathbf{k}.$$

A point of intersection corresponds to values of t and u such that the $\mathbf{i}$-, $\mathbf{j}$-, and $\mathbf{k}$-components of $\mathbf{r}_1(t)$ and $\mathbf{r}_2(u)$ are equal. We obtain the three scalar equations

$$u = e^t \tag{4}$$

$$u^2 - 1 = 2 \sin t \tag{5}$$

$$u - 3 = t^2 - 2. \tag{6}$$

Equation (6) is equivalent to $u = t^2 + 1$. Combining this equation with equation (4) yields $e^t = t^2 + 1$, which has only one solution, $t = 0$. (The proof that $t = 0$ is the only solution is left as an exercise.) If $t = 0$, then $u = 1$. In order to be sure that we have obtained a point of intersection, we must also verify that equation (5) holds for $t = 0$ and $u = 1$, and it does. Using either $\mathbf{r}_1(t)$ or $\mathbf{r}_2(u)$, we find that the point of intersection is $(1, 0, -2)$.

To determine the angle at which these two curves intersect, we calculate the tangent vectors for both curves at $(1, 0, -2)$ and use the dot product. Using $\mathbf{r}_1(t)$, we have

$$\mathbf{r}_1'(t) = e^t \mathbf{i} + 2 \cos t \mathbf{j} + 2t\mathbf{k},$$

so $\mathbf{r}_1'(0) = \mathbf{i} + 2\mathbf{j}$. Using $\mathbf{r}_2(u)$, we get

$$\mathbf{r}_2'(u) = \mathbf{i} + 2u\mathbf{j} + \mathbf{k}$$

and $\mathbf{r}_2'(1) = \mathbf{i} + 2\mathbf{j} + \mathbf{k}$. Finally, the angle of intersection θ is computed using the dot product

$$\cos\theta = \frac{\mathbf{r}_1'(0) \cdot \mathbf{r}_2'(1)}{|\mathbf{r}_1'(0)||\mathbf{r}_2'(1)|} = \frac{5}{\sqrt{5}\sqrt{6}} = \sqrt{5/6}.$$

Then $\theta \approx 24.09°$. ∎

Arc Length

If C is a smooth curve in space parameterized by the vector function

$$\mathbf{r}(t) = x(t)\mathbf{i} + y(t)\mathbf{j} + z(t)\mathbf{k}$$

an expression for the length L of the arc of C from $\mathbf{r}(a)$ to $\mathbf{r}(b)$ is obtained in a manner similar to that for plane curves (see Sections 7.3 and 15.5). First, the interval $[a, b]$ is partitioned into n subintervals of equal length $\Delta t = (b - a)/n$ with endpoints $a = t_0 < t_1 < t_2 < \cdots < t_n = b$. This partition determines a polygonal path connecting the tips of the vectors $\mathbf{r}(t_0), \mathbf{r}(t_1), \ldots, \mathbf{r}(t_n)$ of total length

$$\sum_{j=1}^{n} L_j = \sum_{j=1}^{n} \sqrt{\Delta x_j^2 + \Delta y_j^2 + \Delta z_j^2} \tag{7}$$

where $\Delta x_j = x(t_j) - x(t_{j-1})$, $\Delta y_j = y(t_j) - y(t_{j-1})$, $\Delta z_j = z(t_j) - z(t_{j-1})$ and L_j is the length of the polygonal path from the tip of $\mathbf{r}(t_{j-1})$ to $\mathbf{r}(t_j)$ (see Figure 3.5). As in the two previous discussions of arc length (Sections 7.3 and 15.5), applications of the Mean Value Theorem (one application for each component on each subinterval for each component) lead to an approximation of the form

$$L \approx \sum_{j=1}^{n} \sqrt{[x'(u_j)]^2 + [y'(v_j)]^2 + [z'(w_j)]^2}\, \Delta t. \tag{8}$$

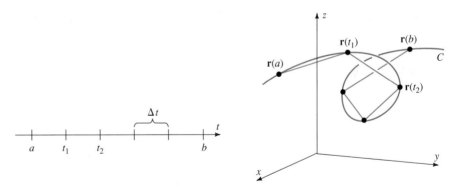

Figure 3.5 The interval $[a, b]$ partitioned into an equal length partition along with the corresponding points $\mathbf{r}(t_i)$ on the curve C. These position vectors determine a polygonal path whose length approximates the arc length of C from $\mathbf{r}(a)$ to $\mathbf{r}(b)$.

As $n \rightarrow \infty$, $\Delta t \rightarrow 0$ and we obtain

$$L = \int_a^b \sqrt{[x'(t)]^2 + [y'(t)]^2 + [z'(t)]^2}\, dt \tag{9}$$

which we take as the definition of the arc length L. (Since this is familiar ground, we do not pursue the details of this development here.) Since the expression under the radical in equation (9) is equal to $\mathbf{r}'(t) \cdot \mathbf{r}'(t) = |\mathbf{r}'(t)|^2$, the arc length may be expressed more compactly as

$$L = \int_a^b |\mathbf{r}'(t)|\, dt. \tag{10}$$

□ **EXAMPLE 5**

Find the length of the curve

$$C: \quad \mathbf{r}(t) = \cos \pi t \mathbf{i} + \sin \pi t \mathbf{j} + t^{3/2} \mathbf{k}$$

from $\mathbf{r}(0) = \mathbf{i}$ to $\mathbf{r}(4) = \mathbf{i} + 8\mathbf{k}$.

Solution　Here

$$\mathbf{r}'(t) = -\pi \sin \pi t \mathbf{i} + \pi \cos \pi t \mathbf{j} + \tfrac{3}{2}\sqrt{t}\,\mathbf{k}$$

so

$$\begin{aligned}
|\mathbf{r}'(t)| &= \sqrt{\pi^2 \sin^2 \pi t + \pi^2 \cos^2 \pi t + \tfrac{9}{4}t} \\
&= \sqrt{\pi^2 + \tfrac{9}{4}t}.
\end{aligned}$$

By formula (9),

$$\begin{aligned}
L &= \int_0^4 \sqrt{\pi^2 + \tfrac{9}{4}t}\, dt \\
&= \tfrac{8}{27}\left[(\pi^2 + \tfrac{9}{4}t)^{3/2}\right]_0^4 \\
&= \tfrac{8}{27}[(\pi^2 + 9)^{3/2} - \pi^3] \approx 15.1
\end{aligned}$$

(Figure 3.6). ■

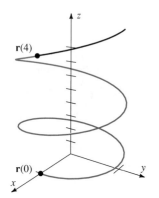

Figure 3.6　The curve C whose length is calculated in Example 5.

At first, formula (10) seems like an entirely new formula for arc length, but it is simply a more general version of the formulas we obtained in Sections 7.3 and 15.5. In Section 7.3, we compute the length L_1 of the graph $y = f(x)$ from $(a, f(a))$ to $(b, f(b))$ by

$$L_1 = \int_a^b \sqrt{1 + [f'(x)]^2}\, dx. \tag{11}$$

To see that formula (11) is a special case of formula (10), we parameterize the graph by

$$\mathbf{r}(t) = t\mathbf{i} + f(t)\mathbf{j}$$

where $a \le t \le b$. Then $\mathbf{r}'(t) = \mathbf{i} + f'(t)\mathbf{j}$, and formula (10) becomes

$$\int_a^b \sqrt{1 + [f'(t)]^2}\, dt,$$

which is the same integral as in formula (11).

In Section 15.5, we computed the arc length L_2 of a parameterized curve

$$C: \quad \begin{cases} x = x(t) \\ y = y(t) \end{cases}$$

for $a \le t \le b$ by

$$L_2 = \int_a^b \sqrt{[x'(t)]^2 + [y'(t)]^2}\, dt. \tag{12}$$

To see that this formula is another special case of formula (10), note that the parametric description of C is equivalent to the vector-valued formulation

$$\mathbf{r}(t) = x(t)\mathbf{i} + y(t)\mathbf{j}.$$

Then $|\mathbf{r}'(t)| = \sqrt{[x'(t)]^2 + [y'(t)]^2}$ and, therefore, formula (12) is precisely formula (10).

Parameterization by Arc Length

Given two vectors $\mathbf{a}$ and $\mathbf{b}$ in space, consider the vector-valued functions $\mathbf{f}(t) = \mathbf{a} + t\mathbf{b}$ and $\mathbf{g}(t) = \mathbf{a} + 2t\mathbf{b}$. In Section 16.2, we saw that both $\mathbf{f}$ and $\mathbf{g}$ generate the same line in space. However, a particle located at the tip of $\mathbf{g}(t)$ moves along the line at twice the speed as a particle located at the tip of $\mathbf{f}(t)$. In this case, we say that $\mathbf{g}$ generates the line twice as fast as $\mathbf{f}$ does.

Up to this point, we have not considered the *rate* at which a vector-valued function $\mathbf{f}$ generates its corresponding curve. However, in order to proceed with our analysis of the geometry of curves, we shall need a standard way to parameterize a given curve. We do so by *(re)parameterizing the curve by arc length*. Intuitively, this notion says that, as the parameter t varies through an interval of length L on the t-axis, the vector-valued function $\mathbf{f}$ generates a curve whose arc length is L (see Figure 3.7). We now give a precise definition.

Definition 6

Let the curve C be parameterized by the differentiable function $\mathbf{r}(t) = x(t)\mathbf{i} + y(t)\mathbf{j} + z(t)\mathbf{k}$ defined on an interval I. The curve C (and the function $\mathbf{r}$) is said to be **parameterized by arc length** if

$$\int_{t_1}^{t_2} |\mathbf{r}'(\omega)|\, d\omega = t_2 - t_1 \tag{13}$$

for all t_1 and t_2 in the interval I.

Since the integral in equation (13) is the length of the arc of C from $\mathbf{r}(t_1)$ to $\mathbf{r}(t_2)$, Definition 6 agrees with the intuitive description of parameterization by arc length given above. If we think of t_1 as a fixed number in I and replace t_2 by the variable t, equation (13) becomes

$$\int_{t_1}^{t} |\mathbf{r}'(\omega)|\, d\omega = t - t_1. \tag{14}$$

Using the Fundamental Theorem of Calculus, we differentiate both sides of equation (14) and obtain

$$\boxed{|\mathbf{r}'(t)| = 1 \quad \text{for all} \quad t \in I,} \tag{15}$$

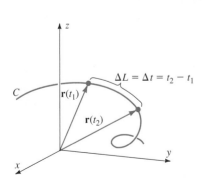

Figure 3.7 $\mathbf{r}$ is parameterized by arc length if ΔL always equals Δt.

which is equivalent to C being parameterized by arc length. Thus, a curve is *parameterized by arc length if and only if the length of its derivative is constantly equal to one.*

☐ **EXAMPLE 6**

The unit circle

$$\mathbf{r}(t) = \cos t\mathbf{i} + \sin t\mathbf{j}, \qquad t \in [0, 2\pi)$$

is parameterized by arc length, since

$$|\mathbf{r}'(t)| = |-\sin t\mathbf{i} + \cos t\mathbf{j}| = \sqrt{\sin^2 t + \cos^2 t} = 1$$

for all $t \in [0, 2\pi)$. (This should not be surprising. Both the circumference of the unit circle and the length of the interval $[0, 2\pi)$ are 2π. In fact, we defined radian measure for angles using the arc length of the unit circle (see Section 1.5).) ■

☐ **EXAMPLE 7**

The circle of radius ρ centered at the origin is traced out by the vector function

$$\mathbf{r}(t) = \rho \cos \alpha t\mathbf{i} + \rho \sin \alpha t\mathbf{j}, \qquad t \in [0, 2\pi/\alpha), \qquad \alpha > 0.$$

Find the value of α for which this circle is parameterized by arc length.

Solution   We first find $|\mathbf{r}'(t)|$:

$$\mathbf{r}'(t) = -\alpha\rho \sin \alpha t\mathbf{i} + \alpha\rho \cos \alpha t\mathbf{j}.$$

Thus

$$|\mathbf{r}'(t)| = \sqrt{(-\alpha\rho \sin \alpha t)^2 + (\alpha\rho \cos \alpha t)^2} = |\alpha\rho|.$$

In order for equation (13) to hold, we must have

$$|\mathbf{r}'(t)| = |\alpha\rho| = 1.$$

We therefore take $\alpha = 1/\rho$. The circle of radius ρ is parameterized by arc length by the function

$$\mathbf{r}(t) = \rho \cos (t/\rho)\mathbf{i} + \rho \sin (t/\rho)\mathbf{j}. \qquad ■$$

NOTATION   If a curve C is parameterized by arc length, it is common to use the variable s as the parameter. This is a convention we shall follow. Consequently, if we parameterize a curve C by $\mathbf{r}(s)$, the arc length of C from $\mathbf{r}(s_1)$ to $\mathbf{r}(s_2)$ is $s_2 - s_1$.

REMARK   In theory, any smooth curve C parameterized by a vector-valued function $\mathbf{r}_1(t)$ can also be given a parameterization $\mathbf{r}_2(s)$ by arc length. However, in many cases, the method that determines $\mathbf{r}_2$ from $\mathbf{r}_1$ involves definite integrals that cannot be evaluated using the Fundamental Theorem of Calculus. Thus, it may be impossible to determine an elementary formula that relates s to t. A simple example is the ellipse

$$\mathbf{r}_1(t) = 2 \cos t\mathbf{i} + \sin t\mathbf{j}.$$

Calculation of its arc length leads to an *elliptic integral*

$$\int \sqrt{1 - (\tfrac{3}{4})\sin^2 t} \, dt,$$

which cannot be evaluated using the Fundamental Theorem. Nevertheless, it is often theoretically useful to know that we can describe a smooth curve using arc length as a parameter even if we cannot write an explicit formula for $\mathbf{r}(s)$.

We conclude by highlighting the comment following equation (15). If the curve C determined by the vector function $\mathbf{r}$ is parameterized by arc length, the *unit tangent* $\mathbf{T}(s)$ at $\mathbf{r}(s)$ is

$$\mathbf{T}(s) = \mathbf{r}'(s). \tag{16}$$

Exercise Set 17.3

In Exercises 1–6, find a vector tangent to the given curve at the given point.

1. $\mathbf{r}(t) = t^3\mathbf{i} + \sqrt{t}\,\mathbf{j} + 2\mathbf{k}, \quad t = 4$

2. $\mathbf{r}(t) = \sin t\mathbf{i} + \cos 2t\mathbf{j} + e^{3t}\mathbf{k}, \quad t = 0$

3. $\mathbf{r}(t) = \sec t\mathbf{i} + \tan t\mathbf{j}, \quad t = \pi/4$

4. $\mathbf{r}(t) = \sqrt{1+t^2}\mathbf{i} + \mathrm{Tan}^{-1} t\mathbf{j} + \ln(1+t)\mathbf{k}, \quad t = 0$

5. $\mathbf{r}(t) = (1+t^2)\mathbf{i} + (t-4)\mathbf{j} + 6t\mathbf{k}$, at the point $(2, -5, -6)$

6. $\mathbf{r}(t) = a\mathbf{i} + bt\mathbf{j} + ct^2\mathbf{k}$, at the point (a, b, c).

In Exercises 7–10, find a unit vector tangent to the given curve at the given point.

7. $\mathbf{r}(t) = (t^2 - 6)\mathbf{i} + (t^3 - 9t + 4)\mathbf{j} + (t^2 - 4t + 1)\mathbf{k}$, at the point $(-2, -6, -3)$.

8. $\mathbf{r}(t) = (9t - 4)\mathbf{i} + (t^2 - t + 3)\mathbf{j} - (90/\pi)\cos(\pi t/10)\mathbf{k}$, at the point $(41, 23, 0)$

9. $\mathbf{r}(t) = \sin t\mathbf{i} + \cos t\mathbf{j} + (t+4)^{3/2}\mathbf{k}$, at the point $(0, 1, 8)$

10. $\mathbf{r}(t) = a\cos \omega t\mathbf{i} + b\sin \omega t\mathbf{j}, \quad \omega t = \pi/4$

In Exercises 11–13, let $\mathbf{r}(t) = (t^2 + 2)\mathbf{i} + 3t\mathbf{j}$.

11. Find the number(s) t for which $\mathbf{r}(t)$ and $\mathbf{r}'(t)$ are orthogonal.

12. Find the number(s) t for which $\mathbf{r}(t)$ and $\mathbf{r}'(t)$ have the same direction.

13. Find the number(s) t for which $\mathbf{r}(t)$ and $\mathbf{r}'(t)$ have opposite direction.

14. Find an equation in vector form for the line tangent to the curve in Exercise 1 at the given point.

15. Find an equation in parametric form for the line tangent to the curve in Exercise 2 at the given point.

16. Find an equation in symmetric form for the line tangent to the curve in Exercise 5 at the given point.

17. Show that every tangent to the unit circle is orthogonal to the radius vector through the point of tangency.

18. Let C be the plane curve parameterized by the function $\mathbf{r}(t) = t\mathbf{i} + (2t^2 + 3)\mathbf{j}$ and let ℓ be the line parameterized by $\mathbf{L}(t) = (t + 3)\mathbf{i} + (t/4 - 10)\mathbf{j}$. Find the point on C where the tangent line is perpendicular to ℓ.

19. Let C be the plane curve parameterized by the function $\mathbf{r}(t) = (t^2 + 1)\mathbf{i} + 8t\mathbf{j}$ and let ℓ be the line parameterized by $\mathbf{L}(t) = (-2t + 3)\mathbf{i} + (8t + 4)\mathbf{j}$. Find the point on C where the tangent line is parallel to ℓ.

In Exercises 20–23, find the equation of the plane perpendicular to the given curve at the specified point.

20. $\mathbf{r}(t) = \ln t\mathbf{i} + t\mathbf{j} + t^2\mathbf{k}, \quad (1, e, e^2)$

21. $\mathbf{r}(t) = \ln(2t + 1)\mathbf{i} + \ln(t^2 - 2)\mathbf{j} + \ln(t^3)\mathbf{k},$ $(\ln 5, \ln 2, \ln 8)$

22. $\mathbf{r}(t) = \sin 2t\mathbf{i} + \cos 2t\mathbf{j} + e^t\mathbf{k}, \quad (1, 0, e^{\pi/4})$

23. $\mathbf{r}(t) = -\sin 2\pi t\mathbf{i} + \cos 4\pi t\mathbf{j} + t\mathbf{k}, \quad (0, 1, 1)$

The angle of intersection θ of a curve and a plane in space is defined to be $(\pi/2) - \alpha$, where α is the angle between the curve and the normal line to the plane at the point of intersection. In Exercises 24 and 25, calculate the cosine of the angle of intersection θ between the plane $x + y + z = 3$ and the curve generated by the given vector-valued function $\mathbf{r}$.

24. $\mathbf{r}(t) = t\mathbf{i} + t^2\mathbf{j} + t^3\mathbf{k}$

25. $\mathbf{r}(t) = t^3\mathbf{i} + (1 - 2t^3/3)\mathbf{j} - (1 + 2t^3/9)\mathbf{k}$

In Exercises 26–29, determine all points of intersection of the two curves specified. Calculate the cosine of the angle of intersection at each point of intersection.

26. $\mathbf{r}_1(t) = t\mathbf{i} + t\mathbf{j} + t\mathbf{k}, \quad \mathbf{r}_2(t) = t^3\mathbf{i} + \mathbf{j} + \mathbf{k}$

27. $\mathbf{r}_1(t) = t^2\mathbf{i} + t\mathbf{j} - \mathbf{k},$ $\mathbf{r}_2(t) = (1 - t)\mathbf{i} + (1 - t)\mathbf{j} + (t - 2)\mathbf{k}$

28. $\mathbf{r}_1(t) = \sin 2t\mathbf{i} + \cos 2t\mathbf{j} + 2t\mathbf{k}, \quad \mathbf{r}_2(t) = t\mathbf{i} + \mathbf{j} + 8\pi\mathbf{k}$

29. $\mathbf{r}_1(t) = \sin t\mathbf{i} + t\mathbf{j} + \sin t\mathbf{k},$ $\mathbf{r}_2(t) = \mathbf{i} + t\mathbf{j} + [(t - \pi/2)^2 + 1]\mathbf{k}$

In Exercises 30–36, find the length of the indicated arc.

30. $\mathbf{r}(t) = \cos t\mathbf{i} + \sin t\mathbf{j}, \quad 0 \le t \le \pi/4$

31. $\mathbf{r}(t) = \cos t\mathbf{i} + (\sqrt{2}/2) \sin t\mathbf{j} + (\sqrt{2}/2) \sin t\mathbf{k},$
$0 \le t \le \pi/2$

32. $\mathbf{r}(t) = e^t \cos t\mathbf{i} + e^t \sin t\mathbf{j}, \quad 0 \le t \le \pi/2$

33. $\mathbf{r}(t) = t\mathbf{i} + (t^2 + 1)\mathbf{j} + t\mathbf{k}, \quad 0 \le t \le 1$

34. $\mathbf{r}(t) = t^3\mathbf{i} + 3t^2\mathbf{j} + 6t\mathbf{k}, \quad 0 \le t \le 3$

35. $\mathbf{r}(t) = 2t^2\mathbf{i} + (t^3/3)\mathbf{k}, \quad$ from $(0, 0, 0)$ to $(18, 0, 9)$.

36. $\mathbf{r}(t) = (2t - 1)\mathbf{i} + 2t^{3/2}\mathbf{j} + \sqrt{5}t\mathbf{k},$
from $(-1, 0, 0)$ to $(1, 2, \sqrt{5})$.

37. Find a unit tangent $\mathbf{T}(t)$ to the graph of $y = x^2 + 3$ at the point $(2, 7)$. (*Hint:* Let $x = t$. Then $y = t^2 + 3$, and a vector parameterization for the curve is $\mathbf{r}(t) = t\mathbf{i} + (t^2 + 3)\mathbf{j}$.)

38. Use the method of Exercise 37 to find a unit tangent $\mathbf{T}(t)$ to the graph of $y^3 = 1 - x^2$ at the point $(3, -2)$.

39. Let C be the curve parameterized by $\mathbf{r}(t) = (1 + t^2)\mathbf{i} + (t - 4)\mathbf{j} + 6t\mathbf{k}$. Find an equation in symmetric form of the line tangent to C at the point $(2, -5, -6)$.

40. Let C be the curve parameterized by $\mathbf{r}(t) = \sin 2t\mathbf{i} + \cos 2t\mathbf{j} + t\mathbf{k}$. Find an equation in vector form for the line tangent to C at the point $(1, 0, \pi/4)$.

41. Determine the constant α so that the curve

$\mathbf{r}(t) = \cos \alpha t\mathbf{i} + \sin \alpha t\mathbf{j} + \alpha t\mathbf{k}, \quad 0 \le t \le 2\pi/\alpha, \quad \alpha > 0,$

is parameterized by arc length.

42. Find a parameterization by arc length for the curve $y = 3x - 2$ so that $x = 0$ and $y = -2$ if $s = 0$.

43. Show that if $\mathbf{r}(t) = x(t)\mathbf{i} + y(t)\mathbf{j}$ and $x'(t_0) \ne 0$, Definition 5 for the *tangent* at $\mathbf{r}(t_0)$ agrees with our earlier definition of tangent for the curve C determined by the parametric equations

$$C: \begin{cases} x = x(t) \\ y = y(t) \end{cases}$$

(see Section 15.4).

44. Use Example 6, Section 17.2, to show that if

$$\mathbf{T}(t) = \frac{\mathbf{r}'(t)}{|\mathbf{r}'(t)|}$$

is differentiable on I, then $\mathbf{T}'(t)$ is orthogonal to $\mathbf{T}(t)$ for $t \in I$. If $\mathbf{T}'(t) \ne 0$, we define

$$\mathbf{N}(t) = \frac{\mathbf{T}'(t)}{|\mathbf{T}'(t)|}$$

to be the **principal unit normal** to the curve $C = \{\mathbf{r}(t) \mid t \in I\}$.

45. Find the principal unit normals to the curve with parameterization

$$\mathbf{r}(t) = \sqrt{2} \cos t\mathbf{i} + \sqrt{2} \sin t\mathbf{j}, \quad 0 \le t \le 2\pi.$$

(See Exercise 44.)

46. Show that the line with vector equation

$$\mathbf{R}(\omega) = \mathbf{r}(t_0) + \omega \mathbf{r}''(t_0)$$

is orthogonal to the tangent vector $\mathbf{T}(t_0)$ if $\mathbf{r}(t)$ is a parameterization by arc length, $\mathbf{r}(t)$ is twice differentiable at $\mathbf{r}(t_0)$ and $\mathbf{r}''(t_0) \ne \mathbf{0}$. (We refer to this line as the line **normal** to the curve at $\mathbf{r}(t_0)$.)

47. Find an equation for the line normal to the curve generated by

$$\mathbf{r}(t) = 2 \cos t\mathbf{i} + 2 \sin t\mathbf{j} + 2t\mathbf{k}$$

at the point $(\sqrt{2}, \sqrt{2}, \pi/2)$. (*Hint:* Use Exercise 46. Also remember that in Exercise 46 we assumed that the curve is parameterized by arc length.)

48. Give an example to show that $\mathbf{r}(t)$ is not always orthogonal to $\mathbf{r}'(t)$.

49. Use Simpson's Rule to estimate the arc length of

$$\mathbf{r}(t) = (2 \cos t + \cos 2t)\mathbf{i} + (2 \sin t - \sin 2t)\mathbf{j}$$

for $0 \le t \le 2\pi$.

17.4 Velocity and Acceleration

If a particle moves in space so that its location at time t is the point $P(t) = (x(t), y(t), z(t))$, its motion can be described by the **position vector** function

$$\mathbf{r}(t) = x(t)\mathbf{i} + y(t)\mathbf{j} + z(t)\mathbf{k}. \tag{1}$$

That is, we think of the position of the particle as the tip of the vector $\mathbf{r}(t)$ originating at the origin and terminating at the point $P(t)$.

In such cases the **average velocity** from time t to time $t + h$ is

$$\text{average velocity} = \frac{\text{change in position}}{\text{change in time}}$$

$$= \frac{1}{h} \Delta \mathbf{r}$$

$$= \frac{1}{h} [\mathbf{r}(t + h) - \mathbf{r}(t)].$$

(See Figure 4.1.)

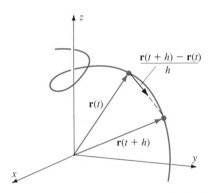

Figure 4.1   Average velocity is the vector

$$\frac{1}{h} [\mathbf{r}(t + h) - \mathbf{r}(t)].$$

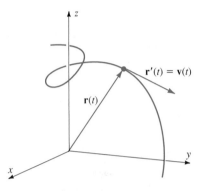

Figure 4.2   Velocity is the vector

$$\mathbf{v}(t) = \mathbf{r}'(t) = \lim_{h \to 0} \frac{1}{h} [\mathbf{r}(t + h) - \mathbf{r}(t)],$$

and $|\mathbf{v}(t_0)| = $ speed.

In order to extend our earlier concept of (instantaneous) velocity as the limit of average velocity as $h \to 0$ to motion in space, we should define

$$\text{velocity at time } t = \lim_{h \to 0} \frac{1}{h} [\mathbf{r}(t + h) - \mathbf{r}(t)]. \tag{2}$$

(See Figure 4.2.) We have already seen that the limit on the right-hand side of equation (2), if it exists, is the derivative $\mathbf{r}'(t)$ of the vector-valued function $\mathbf{r}$. According to Theorem 2, this limit exists precisely when each of the component functions x, y, and z is differentiable. The *velocity* of a particle moving in space is therefore defined as follows:

Definition 7

Let the vector function

$$\mathbf{r}(t) = x(t)\mathbf{i} + y(t)\mathbf{j} + z(t)\mathbf{k}$$

be the position function for a particle moving in space and let the parameter t denote time. If $\mathbf{r}$ is differentiable at time t_0, the **velocity** of the particle at t_0 is the vector

$$\mathbf{v}(t_0) = \mathbf{r}'(t_0) = x'(t_0)\mathbf{i} + y'(t_0)\mathbf{j} + z'(t_0)\mathbf{k}. \tag{3}$$

If $\mathbf{r}$ is differentiable on an interval I, equation (3) determines the **velocity function** for $\mathbf{r}$ on the interval I.

Since velocity is just the derivative of the position function, we have already determined that $\mathbf{v}(t_0)$ is *tangent* to the graph of $\mathbf{r}$ at $\mathbf{r}(t_0)$ and points in the direction of increasing t, provided $\mathbf{v}(t_0) \neq \mathbf{0}$ (Section 17.2).

As with real-valued functions, the length of the velocity vector $|\mathbf{v}(t_0)| = |\mathbf{r}'(t_0)|$ is defined to be the **speed** (rate of change of distance along the path) of the particle at time t_0. This interpretation of $|\mathbf{r}'(t)|$ is consistent with the arc length formula

$$s(t) = \int_{t_0}^{t} |\mathbf{r}'(\omega)| \, d\omega,$$

derived in Section 17.3. By the Fundamental Theorem of Calculus, the derivative $s'(t)$ of arc length is $|\mathbf{r}'(t)|$. If $\mathbf{r}$ represents motion, then arc length corresponds to distance traveled, and the derivative of distance traveled is speed.

□ EXAMPLE 1

Show that a particle moving about the unit circle in the xy-plane with position function

$$\mathbf{r}(t) = \cos \alpha t \mathbf{i} + \sin \alpha t \mathbf{j}$$

moves at a constant speed.

Solution   By Definition 7, the velocity function is

$$\mathbf{v}(t) = \mathbf{r}'(t) = -\alpha \sin \alpha t \mathbf{i} + \alpha \cos \alpha t \mathbf{j}.$$

The speed at time t is therefore

$$|\mathbf{v}(t)| = \sqrt{(\alpha \sin \alpha t)^2 + (\alpha \cos \alpha t)^2} = |\alpha|,$$

which is constant. For this reason, we call this **uniform circular motion**. ∎

As in the case of motion along a line, we define the *acceleration* of a particle moving in space to be the rate of change of velocity with respect to time.

Definition 8

Let the vector function

$$\mathbf{r}(t) = x(t)\mathbf{i} + y(t)\mathbf{j} + z(t)\mathbf{k}$$

be the position function of a particle moving in space and let the parameter t denote time. If $\mathbf{r}$ is twice differentiable at time t_0, the **acceleration** of the particle at t_0 is the vector

$$\mathbf{a}(t_0) = \mathbf{r}''(t_0) = x''(t_0)\mathbf{i} + y''(t_0)\mathbf{j} + z''(t_0)\mathbf{k}.$$

Thus,

$$\mathbf{a}(t_0) = \mathbf{v}'(t_0). \tag{4}$$

□ EXAMPLE 2

Show that, for the particle moving about the unit circle in Example 1, the acceleration vector always points toward the center of the circle and is of constant magnitude.

Solution   From the solution to Example 1 and equation (4) we have

$$\begin{aligned}
\mathbf{a}(t) = \mathbf{v}'(t) &= -\alpha^2 \cos \alpha t \mathbf{i} - \alpha^2 \sin \alpha t \mathbf{j} \\
&= -\alpha^2 (\cos \alpha t \mathbf{i} + \sin \alpha t \mathbf{j}) \\
&= -\alpha^2 \mathbf{r}(t).
\end{aligned}$$

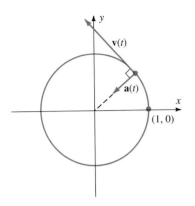

Figure 4.3   Velocity and acceleration vectors associated with uniform circular motion.

Thus, $\mathbf{a}(t)$ points in the direction opposite $\mathbf{r}(t)$ and has length

$$|\mathbf{a}(t)| = \sqrt{(\alpha^2 \cos \alpha t)^2 + (\alpha^2 \sin \alpha t)^2} = \alpha^2.$$

(See Figure 4.3.) Note that $\mathbf{v}(t)$ and $\mathbf{a}(t)$ are orthogonal, an observation that follows from Example 6, Section 17.2, since $|\mathbf{v}(t)| = |\alpha| = $ constant.   ∎

□ **EXAMPLE 3**

Find the velocity and acceleration functions associated with the position function

$$\mathbf{r}(t) = \cos t\mathbf{i} + \sin t\mathbf{j} + t^2\mathbf{k}$$

and show that $|\mathbf{a}(t)|$ is constant.

Solution

$$\mathbf{v}(t) = \mathbf{r}'(t) = -\sin t\mathbf{i} + \cos t\mathbf{j} + 2t\mathbf{k}$$
$$\mathbf{a}(t) = \mathbf{v}'(t) = -\cos t\mathbf{i} - \sin t\mathbf{j} + 2\mathbf{k}.$$

Since $|\mathbf{a}(t)| = \sqrt{\cos^2 t + \sin^2 t + 4} = \sqrt{5}$, $|\mathbf{a}(t)|$ is constant. The vectors $\mathbf{r}(t)$, $\mathbf{v}(t)$, and $\mathbf{a}(t)$ are sketched in Figure 4.4.   ∎

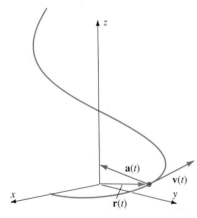

Figure 4.4   Velocity and acceleration vectors for

$$\mathbf{r}(t) = \cos t\mathbf{i} + \sin t\mathbf{j} + t^2\mathbf{k}.$$

Just as velocity and acceleration can be determined by differentiating the position function $\mathbf{r}$, the functions for position and velocity can be obtained by integrating the acceleration function, provided antiderivatives of each component function can be found. More precisely, equation (4) gives

$$\mathbf{v}(t) = \int \mathbf{a}(t)\, dt, \tag{5}$$

and equation (3) shows that

$$\mathbf{r}(t) = \int \mathbf{v}(t)\, dt. \tag{6}$$

In using equations (5) and (6), you must remember that one constant of integration will appear in each component in each integration. These constants may be determined if *initial conditions* are specified for the functions $\mathbf{v}$ and $\mathbf{r}$.

□ **EXAMPLE 4**

A particle starts from rest at the point $P_0 = (2, 0, -3)$ and moves with an acceleration of $\mathbf{a}(t) = 2\mathbf{i} + 6t\mathbf{j}$ m/s^2. Find

(a) the velocity function for the particle,
(b) the position function for the particle,
(c) the location of the particle after 3 seconds.

Strategy · · · · · · · ·
Identify the initial conditions.

Solution
(a) Since the particle starts at $P_0 = (2, 0, -3)$, we have

$$\mathbf{r}_0 = \mathbf{r}(0) = 2\mathbf{i} - 3\mathbf{k}.$$

Since the particle starts from rest, we have

$$\mathbf{v}_0 = \mathbf{v}(0) = \mathbf{0}.$$

Obtain **v** by integrating **a**. Don't forget that

$$\left[\int 0 \, dt \right] \mathbf{k} = (0 + c_3) \mathbf{k}.$$

Using equation (5), we have

$$\mathbf{v}(t) = \int (2\mathbf{i} + 6t\mathbf{j}) \, dt$$

$$= (2t - c_1)\mathbf{i} + (3t^2 + c_2)\mathbf{j} + c_3\mathbf{k}.$$

Apply the initial condition $\mathbf{v}_0 = \mathbf{0}$ to determine c_1, c_2, and c_3.

Setting $t = 0$ and applying the initial condition $\mathbf{v}(0) = \mathbf{0}$ gives

$$\mathbf{0} = c_1\mathbf{i} + c_2\mathbf{j} + c_3\mathbf{k},$$

so $c_1 = c_2 = c_3 = 0$. The velocity function is therefore

$$\mathbf{v}(t) = 2t\mathbf{i} + 3t^2\mathbf{j}.$$

Obtain **r** by integrating **v**.

(b) Using equation (6), we find that

$$\mathbf{r}(t) = \int (2t\mathbf{i} + 3t^2\mathbf{j}) \, dt$$

$$= (t^2 + d_1)\mathbf{i} + (t^3 + d_2)\mathbf{j} + d_3\mathbf{k}.$$

Determine the constants d_1, d_2, and d_3 by applying the initial condition

$$\mathbf{r}(0) = 2\mathbf{i} - 3\mathbf{k}.$$

Setting $t = 0$ and applying the initial condition $\mathbf{r}(0) = 2\mathbf{i} - 3\mathbf{k}$ gives

$$2\mathbf{i} - 3\mathbf{k} = d_1\mathbf{i} + d_2\mathbf{j} + d_3\mathbf{k}$$

so $d_1 = 2$, $d_2 = 0$, and $d_3 = -3$. Thus,

$$\mathbf{r}(t) = (t^2 + 2)\mathbf{i} + t^3\mathbf{j} - 3\mathbf{k}.$$

The location of the particle after 3 seconds is $\mathbf{r}(3)$.

(c) The location of the particle after 3 seconds is

$$\mathbf{r}(3) = (3^2 + 2)\mathbf{i} + 3^3\mathbf{j} - 3\mathbf{k} = 11\mathbf{i} + 27\mathbf{j} - 3\mathbf{k},$$

or $(11, 27, -3)$. Note that, because there is no component of **a** in the z direction, the entire motion takes place in the plane $z = -3$.   ■

Projectile Motion

A frequent application of equations (5) and (6) concerns the trajectory followed by a projectile launched with a prescribed initial velocity $\mathbf{v}_0$. The term "projectile" refers to any object, such as a missile, a flare, or a baseball, which is not self-propelling. Thus, the trajectory followed by a projectile is completely determined by the speed and direction (i.e., the velocity) at which it is launched. In the discussion that follows, we shall assume that air resistance is negligible, so that the only force acting on the projectile after launch is the force due to gravity. Also, we shall take the y-axis as vertical and the x-axis as horizontal, with the positive x-axis pointing in the direction of the horizontal component of the initial velocity. The motion is then restricted entirely to the xy-plane.

The motion of projectiles (and, for that matter, of all moving objects) is governed by the vector form of Newton's second law of motion:

$$\mathbf{F} = m\mathbf{a}. \tag{7}$$

Here **F** is the vector sum of all forces acting on the projectile, m is its mass, and **a** is its acceleration.

According to our use of the term projectile, the only force acting on the object in flight is the force due to gravity,

$$\mathbf{F} = -mg\mathbf{j} \tag{8}$$

where $g = 9.81$ m/s^2 is the gravitational constant. (We use a negative sign since this force acts downward.)

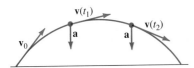

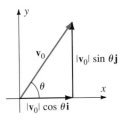

Figure 4.5 Motion of a projectile with initial velocity $\mathbf{v}_0$.

Figure 4.6 Components of initial velocity are

$$\mathbf{v}_0 = |\mathbf{v}_0| = \cos \theta \mathbf{i} + |\mathbf{v}_0| \sin \theta \mathbf{j}.$$

Combining equations (7) and (8) we find that a projectile in flight experiences the acceleration

$$\mathbf{a} = -g\mathbf{j}. \tag{9}$$

Figure 4.5 illustrates a typical trajectory for a projectile fired with initial velocity $\mathbf{v}_0$.

Now consider the problem of determining the trajectory of a projectile fired from the origin at an angle θ with the horizontal and with an initial speed s_0. This means that the initial velocity $\mathbf{v}_0$ has length $|\mathbf{v}_0| = s_0$. We can therefore write $\mathbf{v}_0$ in component form as

$$\mathbf{v}_0 = s_0 \cos \theta \mathbf{i} + s_0 \sin \theta \mathbf{j}. \tag{10}$$

(See Figure 4.6.)

Applying equation (5) with $\mathbf{a}(t)$ as in equation (9), we find

$$\mathbf{v}(t) = \int \mathbf{a}\, dt = \int -g\mathbf{j}\, dt = c_1 \mathbf{i} + (c_2 - gt)\mathbf{j}.$$

Setting $t = 0$ and applying the initial condition given by equation (10), we obtain

$$\mathbf{v}(0) = c_1 \mathbf{i} + c_2 \mathbf{j} = s_0 \cos \theta \mathbf{i} + s_0 \sin \theta \mathbf{j}.$$

Thus,

$$c_1 = s_0 \cos \theta; \qquad c_2 = s_0 \sin \theta,$$

and therefore

$$\mathbf{v}(t) = s_0 \cos \theta \mathbf{i} + (s_0 \sin \theta - gt)\mathbf{j}.$$

According to equation (6), the position function is

$$\mathbf{r}(t) = \int \mathbf{v}(t)\, dt = \int [s_0(\cos \theta)\mathbf{i} + (s_0 \sin \theta - gt)\mathbf{j}]\, dt$$

$$= [s_0(\cos \theta)t = d_1]\mathbf{i} + [s_0(\sin \theta)t - \tfrac{1}{2}gt^2 + d_2]\mathbf{j}.$$

Since the projectile is fired from the origin, we have the initial condition $\mathbf{r}(0) = \mathbf{0}$. Setting $t = 0$ in the above expression for $\mathbf{r}(t)$ and applying this initial condition gives

$$\mathbf{r}(0) = d_1 \mathbf{i} + d_2 \mathbf{j} = \mathbf{0},$$

so $d_1 = d_2 = 0$. The position function for the projectile motion is therefore

$$\mathbf{r}(t) = [s_0(\cos \theta)t]\mathbf{i} + [s_0(\sin \theta)t - \tfrac{1}{2}gt^2]\mathbf{j}.$$

We summarize our findings as follows:

> The trajectory of a projectile fired from the origin with initial speed s_0 and angle of elevation θ is parameterized by the vector-valued function
>
> $$\mathbf{r}(t) = [s_0(\cos \theta)t]\mathbf{i} + [s_0(\sin \theta)t - \tfrac{1}{2}gt^2]\mathbf{j}. \tag{11}$$

□ **EXAMPLE 5**

A flare is fired from ground level at an elevation of 60° with an initial speed of 50 m/s. Find

(a) the maximum height of the trajectory,
(b) the length of time during which the flare is airborne, and
(c) the distance from the point of launch to the point of impact.

Strategy · · · · · · · ·

Write the expression for $\mathbf{r}(t) = x(t)\mathbf{i} + y(t)\mathbf{j}$ using (11).

Solution

(a) With $\theta = 60° = \pi/3$ and $s_0 = 50$ m/s, $\mathbf{r}(t)$ is

$$\mathbf{r}(t) = 50(\cos \pi/3)t\mathbf{i} + (50(\sin \pi/3)t - \tfrac{1}{2}gt^2)\mathbf{j}$$
$$= 25t\mathbf{i} + (25\sqrt{3}t - \tfrac{1}{2}gt^2)\mathbf{j}.$$

Find the time t_{end} of maximum height by maximizing the $\mathbf{j}$-component of $\mathbf{r}(t)$.

The height of the flare is given by the $\mathbf{j}$-component

$$y(t) = 25\sqrt{3}t - \tfrac{1}{2}gt^2.$$

The maximum height occurs when

$$y'(t) = 25\sqrt{3} - gt = 0,$$

or

$$t_{\text{max}} = \frac{25\sqrt{3}}{g} = \frac{25\sqrt{3}}{9.81} \approx 4.4 \text{ seconds.}$$

$y(t_{\text{max}})$ is the maximum height.

The maximum height is

$$y(t_{\text{max}}) = 25\sqrt{3}\left(\frac{25\sqrt{3}}{9.81}\right) - \tfrac{1}{2}(9.81)\left(\frac{25\sqrt{3}}{9.81}\right)^2$$
$$= \frac{(\tfrac{1}{2})25^2 \cdot 3}{9.81} \approx 95.6 \text{ meters.}$$

Find the time t_{end} at which the projectile lands by setting $y(t) = 0$.

(b) The projectile lands when the $\mathbf{j}$-component reaches zero, that is, when

$$y(t) = 25\sqrt{3}t - \tfrac{1}{2}gt^2 = 0.$$

This equation has solutions $t = 0$ and $t = 50\sqrt{3}/g$.

The time $t = 0$ corresponds to launch. The time at which the projectile returns to ground level is therefore

$$t_{\text{end}} = \frac{50\sqrt{3}}{g} = \frac{50\sqrt{3}}{9.81} \approx 8.8 \text{ seconds.}$$

Distance from launch point to impact point is $x(t_{\text{end}})$.

(c) The distance from the point of launch to the point of impact is the $\mathbf{i}$-component of $\mathbf{r}(t_{\text{end}})$.

$$x(t_{\text{end}}) = 25(8.8) = 220 \text{ meters.} \qquad ■$$

□ **EXAMPLE 6**

The following experiment is performed in an elementary physics class. A target is hung on a wall of a gymnasium at a height of h meters above the floor. A small cannon, which fires a lead slug and whose muzzle velocity is adjustable, is located on the floor d meters from the wall. The cannon is aimed at the bull's-eye on the target. By means of a trip mechanism, the target is released from its hanger at the precise instant at which the cannon is fired and falls toward the floor. The result of

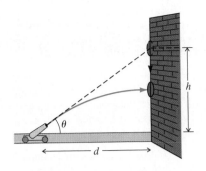

repeated trials of the experiment is that no matter what muzzle velocity is chosen, the slug always strikes the bull's-eye (assuming that d, θ, and the muzzle velocity are such that the slug does not strike the floor before reaching the wall). (See Figure 4.7.) How can this be explained?

Solution   Let θ be the angle of elevation of the barrel of the cannon when aimed at the bull's-eye, and let s_0 be the muzzle velocity of the cannon. We assume the cannon to be located at the origin of an xy-coordinate system. The trajectory of the slug is then given by $\mathbf{r}(t)$ in equation (11).

The slug reaches the wall when the **i**-component of $\mathbf{r}(t)$ satisfies the equation

$$x(t) = s_0(\cos \theta)\, t = d$$

which occurs at time $t_f = \dfrac{d}{s_0 \cos \theta}$. At this time the **j**-component of the slug is

$$y(t_f) = s_0 \sin \theta \left(\frac{d}{s_0 \cos \theta} \right) - \frac{1}{2} g \left(\frac{d}{s_0 \cos \theta} \right)^2$$

$$= d \tan \theta - \tfrac{1}{2} g t_f^2.$$

Since $\dfrac{h}{d} = \tan \theta$, $h = d \tan \theta$. Thus

$$y(t_f) = h - \tfrac{1}{2} g t_f^2$$

is the **j**-component of $\mathbf{r}(t)$ when the slug reaches the wall.

Since the target is a freely-falling body, after t_f seconds it has fallen a distance $\tfrac{1}{2} g t_f^2$, so the **j**-component of the bull's-eye's position vector after t_0 seconds is also $h - \tfrac{1}{2} g t_f^2$. Since the slug and the bull's-eye have the same position vector, $\mathbf{r}(t_f) = d\mathbf{i} + (h - \tfrac{1}{2} g t_f^2)\mathbf{j}$, at time $t = t_f$, the slug must always score a direct hit on the target.   ■

Exercise Set 17.4

In Exercises 1–7, find the velocity and acceleration functions corresponding to the given position functions.

1. $\mathbf{r}(t) = \mathbf{i} + 2t\mathbf{j}$

2. $\mathbf{r}(t) = 3t\mathbf{i} + t^3\mathbf{j} + (2t + 3)\mathbf{k}$

3. $\mathbf{r}(t) = \cos 3t\mathbf{i} + \sin 3t\mathbf{j}$

4. $\mathbf{r}(t) = e^t\mathbf{i} + e^{-t}\mathbf{j} + \sqrt{t}\mathbf{k}, \quad t \geq 1$

5. $\mathbf{r}(t) = \ln t^2\mathbf{i} + \cos 2t\mathbf{j} + (1/t)\mathbf{k}, \quad t \geq 1$

6. $\mathbf{r}(t) = t(t - 1)\mathbf{i} + \mathrm{Tan}^{-1} t\mathbf{j} + te^t\mathbf{k}$

7. $\mathbf{r}(t) = e^t \cos t\mathbf{i} + e^t \sin t\mathbf{j} + \cos t\mathbf{k}$

8. A particle moves with position function $\mathbf{r}(t) = \cos \pi t\mathbf{i} + \sin \pi t\mathbf{j}$. Find its speed at time $t = 1/4$.

9. A particle moves with position vector $\mathbf{r}(t) = t^2\mathbf{i} + 2t\mathbf{j} + t^3\mathbf{k}$. Find its speed at time $t = 2$.

In Exercises 10–13, find the position function $\mathbf{r}$ from the given velocity function and the initial condition $\mathbf{r}(0) = \mathbf{r}_0$.

10. $\mathbf{v}(t) = 2t\mathbf{i} + t^2\mathbf{j}, \quad \mathbf{r}_0 = \mathbf{i} + 4\mathbf{j}$

11. $\mathbf{v}(t) = \cos t\mathbf{i} + \sin t\mathbf{j}, \quad \mathbf{r}_0 = 3\mathbf{i} + 2\mathbf{j}$

12. $\mathbf{v}(t) = e^t\mathbf{i} + \sqrt{t}\mathbf{j} + 2t\mathbf{k}, \quad \mathbf{r}_0 = 6\mathbf{i} + 4\mathbf{k}$

13. $\mathbf{v}(t) = (1/(1 + t^2))\mathbf{i} + te^{t^2}\mathbf{j}, \quad \mathbf{r}_0 = -2\mathbf{i} + \mathbf{j} + 4\mathbf{k}$

In Exercises 14–18, find the velocity function $\mathbf{v}$ and the position function $\mathbf{r}$ given the acceleration function $\mathbf{a}$ and the initial conditions $\mathbf{v}_0 = \mathbf{v}(0)$ and $\mathbf{r}_0 = \mathbf{r}(0)$.

14. $\mathbf{a} = \mathbf{i} + \mathbf{j}, \quad \mathbf{v}_0 = \mathbf{i} + \mathbf{j}, \quad \mathbf{r}_0 = 0$

15. $\mathbf{a} = 2\mathbf{i} + \mathbf{k}, \quad \mathbf{v}_0 = 0, \quad \mathbf{r}_0 = 3\mathbf{i} - \mathbf{j} + 4\mathbf{k}$

16. $\mathbf{a} = 6t\mathbf{i} + e^t\mathbf{j}, \quad \mathbf{v}_0 = 0, \quad \mathbf{r}_0 = 4\mathbf{i} + \mathbf{j} + 2\mathbf{k}$

17. $\mathbf{a} = \cos t\mathbf{i} + \sin t\mathbf{j}, \quad \mathbf{v}_0 = \mathbf{k}, \quad \mathbf{r}_0 = 3\mathbf{i} - \mathbf{j} + 2\mathbf{k}$

18. $\mathbf{a} = e^t\mathbf{i} + t\mathbf{j} + e^{-t}\mathbf{k}, \quad \mathbf{v}_0 = \mathbf{i} + \mathbf{k}, \quad \mathbf{r}_0 = \mathbf{i} + 4\mathbf{j} + \mathbf{k}$

19. Verify that the vectors $\mathbf{v}(t)$ and $\mathbf{a}(t)$ of Example 2 are orthogonal for each t.

20. Prove that if a particle moves with constant speed its velocity and acceleration vectors are orthogonal.

21. A particle moves about a circle in the xy-plane with position function $\mathbf{r}(t) = \rho \cos \alpha t \mathbf{i} + \rho \sin \alpha t \mathbf{j}$. Find a relationship between the radius of the circle and the speed of the particle.

22. A projectile is fired from ground level at an elevation angle of $\theta = 45°$ with an initial speed of 60 m/s. Find
a. the position function $\mathbf{r}$ of the projectile,
b. the maximum altitude of the projectile,
c. the flight time,
d. the distance from the launch point to the impact point,
e. the speed on impact.

23. Find each of the quantities in Exercise 22 for a projectile fired at an elevation angle of $\theta = 30°$ with an initial speed of 100 m/s.

24. A bale of hay is dropped from an airplane flying at a ground speed of 200 km/h and an elevation of 1000 m. How far from the drop point does the bale strike the ground?

25. A handgun used in a physics experiment simultaneously fires one slug through the barrel and drops a second slug mounted on the side of the barrel. Independent of the muzzle velocity, the two slugs are observed to strike the ground at the same instant when the gun is fired horizontally. Why?

26. How far along the path $\mathbf{r}(t)$ does the particle of Example 4 travel during the first 3 seconds of its motion? What is its distance from the starting point P_0 at $t = 3$ seconds?

27. a. Write the integral that expresses the arc length of the motion of the flare in Example 5.
b. Use Simpson's Rule with $n = 8$ to approximate the value of the integral in part a.

17.5   Curvature

Curvature is a measure of the rate at which a curve in space turns or twists. To make this idea more precise, we consider a curve C determined parametrically by the vector function

$$C: \quad \mathbf{r}(s) = x(s)\mathbf{i} + y(s)\mathbf{j} + z(s)\mathbf{k}, \qquad a \le s \le b. \tag{1}$$

We assume that C is parameterized by arc length, so that the tip of $\mathbf{r}(s)$ moves along C uniformly. That is, we assume that $|\mathbf{r}'(s)| = 1$, $a \le s \le b$.

Since $\mathbf{r}'(s)$ is a unit tangent at each point $\mathbf{r}(s)$, the second derivative $\mathbf{r}''$ measures the rate at which the unit tangent changes (turns or twists) as s increases. This is analogous to the one-variable case in which the second derivative f'' measures the rate at which the slope function f' is changing. However, the derivatives here are vectors rather than numbers. We therefore make the following definition.

Definition 9

Let the curve C in (1) be parameterized by arc length, and let $\mathbf{r}$ in (1) be twice differentiable for $a \le s \le b$. The vector $\mathbf{r}''(s)$ is called the **curvature vector** at $\mathbf{r}(s)$. Its length, $|\mathbf{r}''(s)|$, is called the **curvature** of C at $\mathbf{r}(s)$.

REMARK: If C were not parameterized by arc length, then $|\mathbf{r}''|$ would depend upon both the direction and the magnitude of $\mathbf{r}'$. By insisting on parameterization by arc length, we have removed the dependence of the curvature on the ''speed of the curve $\mathbf{r}'$.''

The Greek letter κ (kappa) is usually used to denote curvature. Thus, Definition 9 states that

$$\kappa(s) = |\mathbf{r}''(s)|, \qquad a \le s \le b. \tag{2}$$

Figures 5.1 and 5.2 illustrate the essence of curvature: $\kappa(s)$ is large if C turns sharply (rapid change in $\mathbf{r}'(s)$), and $\kappa(s)$ is small if C turns slowly (small change in $\mathbf{r}'(s)$).

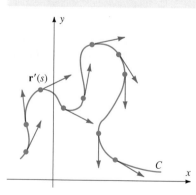

Figure 5.1 $\kappa(s)$ is large when C turns sharply.

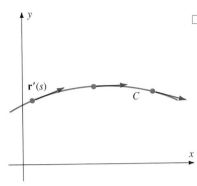

☐ **EXAMPLE 1**

In Example 7, Section 17.3, we determined that the circle of radius ρ centered at the origin is parameterized by arc length by the vector function

$$\mathbf{r}(s) = \rho \cos\left(\frac{s}{\rho}\right)\mathbf{i} + \rho \sin\left(\frac{s}{\rho}\right)\mathbf{j}, \qquad 0 \le s < 2\rho\pi.$$

The curvature vector is therefore

$$\mathbf{r}''(s) = -\frac{1}{\rho}\cos\left(\frac{s}{\rho}\right)\mathbf{i} - \frac{1}{\rho}\sin\left(\frac{s}{\rho}\right)\mathbf{j} = -\frac{1}{\rho^2}\mathbf{r}(s)$$

and the curvature at $\mathbf{r}(s)$ is

$$\kappa(s) = |\mathbf{r}''(s)| = \frac{1}{\rho}.$$ ∎

If the circle is allowed to have an arbitrary center, a slight generalization of this argument shows that *the curvature of a circle of radius ρ is $\kappa = 1/\rho$ at each point on the circle* (see Exercise 33). Figure 5.3 shows the curvature vectors for circles of various radii.

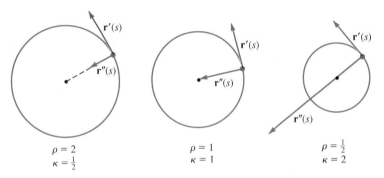

Figure 5.3   Tangent and curvature vectors for various circles.

For more general plane curves C parameterized by arc length, the expression $\rho(s) = 1/\kappa(s)$ is called the **radius of curvature** at $\mathbf{r}(s)$. Since $\rho(s)$ is the radius of the circle whose curvature is also $\kappa(s)$, the number $\rho(s)$ determines a circle that

(i) is tangent to C at $\mathbf{r}(s)$,
(ii) has radius $\rho(s) = 1/\kappa(s)$,
(iii) has the same curvature vector as C at $\mathbf{r}(s)$.

This circle is called the **osculating circle,** or **circle of curvature,** at $\mathbf{r}(s)$. The center of this circle is called the **center of curvature.** Figures 5.4 and 5.5 show the osculating circles at various points on a curve C.

For plane curves that are not parameterized by arc length, the following theorem indicates how we compute curvature.

Theorem 5

Let the curve C be determined parametrically by the twice differentiable function

$$\mathbf{r}(t) = x(t)\mathbf{i} + y(t)\mathbf{j}, \qquad a \le t \le b. \tag{3}$$

The curvature $\kappa(t)$ of C at the point $(x(t), y(t))$ is

$$\kappa(t) = \frac{|x'(t)y''(t) - y'(t)x''(t)|}{[(x'(t))^2 + (y'(t))^2]^{3/2}}. \tag{4}$$

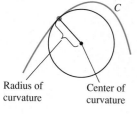

Radius of
curvature

Center of
curvature

Figure 5.4   Circle of curvature for a plane curve C.

Before we discuss the proof of Theorem 5, we consider three examples.

☐ **EXAMPLE 2**

For the curve C given parametrically by the function

$$\mathbf{r}(t) = (2t^2 + 1)\mathbf{i} + (t^3 - 3)\mathbf{j}, \qquad 1 \le t \le 3$$

find

(a) the curvature function κ, and
(b) the radius of curvature at the point $(3, -2)$.

Solution   Here $x(t) = 2t^2 + 1$ and $y(t) = t^3 - 3$. Thus

$$\begin{aligned} x'(t) &= 4t & y'(t) &= 3t^2, \\ x''(t) &= 4 & y''(t) &= 6t. \end{aligned}$$

According to Theorem 5, the curvature is

$$\kappa(t) = \frac{|(4t)(6t) - (3t^2)(4)|}{[(4t)^2 + (3t^2)^2]^{3/2}} = \frac{12t^2}{(16t^2 + 9t^4)^{3/2}}.$$

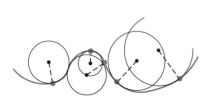

Figure 5.5   Osculating circles at various points on C.

The point $(3, -2)$ corresponds to $t = 1$. The curvature at this point is therefore

$$\kappa(1) = \frac{12}{(16 + 9)^{3/2}} = \frac{12}{125}.$$

The radius of curvature at this point is

$$\rho(1) = \frac{1}{\kappa(1)} = \frac{125}{12}. \qquad \blacksquare$$

☐ **EXAMPLE 3**

Find the curvature function κ for the graph of the function $f(t) = \sin t$.

Solution   We can write the function $f(t) = \sin t$ in parametric (vector) form as

$$\mathbf{r}(t) = t\mathbf{i} + \sin t\mathbf{j}.$$

Thus, $x(t) = t$ and $y(t) = \sin t$, so

$$\begin{aligned} x'(t) &= 1 & y'(t) &= \cos t, \\ x''(t) &= 0 & y''(t) &= -\sin t. \end{aligned}$$

By Theorem 5,

$$\kappa(t) = \frac{|(1)(-\sin t) - (0)(\cos t)|}{[1^2 + (\cos t)^2]^{3/2}} = \frac{|\sin t|}{(1 + \cos^2 t)^{3/2}}.$$

Notice that $\kappa(t)$ is a minimum (zero) at $t = 0$ $\pm \pi$, $\pm 2\pi$, ... and that $\kappa(t)$ is a maximum (one) at $t = \pi/2 \pm n\pi$, $n = 1, 2, \ldots$ ∎

Example 3 shows how Theorem 5 may be used to calculate the curvature for the graph of a twice differentiable function $y = f(x)$. Such functions can always be written in vector form as

$$\mathbf{r}(x) = x\mathbf{i} + f(x)\mathbf{j}.$$

Then $x' = 1$, $x'' = 0$, $y' = f'(x)$, and $y'' = f''(x)$, so the curvature $\kappa(x)$, according to Theorem 5, is

$$\kappa(x) = \frac{|(0)(f'(x)) - (f''(x))(1)|}{[1 + (f'(x))^2]^{3/2}}.$$

We state this result as a corollary of Theorem 5.

Corollary 2

Let f be a twice differentiable function of x. The curvature $\kappa(x)$ at the point $(x, f(x))$ on the graph of $y = f(x)$ is

$$\kappa(x) = \frac{|f''(x)|}{[1 + (f'(x))^2]^{3/2}}.$$

☐ **EXAMPLE 4**

Show that the curvature of a nonvertical line in the plane is zero.

Solution   The equation for a nonvertical line in the plane can be written in the form $f(x) = mx + b$ where m and b are constants. Thus $f'(x) = m$ and $f''(x) = 0$. So, $\kappa(x) = 0$, according to Corollary 2. ∎

Proof of Theorem 5:   We will demonstrate the proof up to a point at which you can complete the argument with a calculation (Exercise 34).

The difficulty to be overcome is that the parameterization given by $\mathbf{r}(t)$ in equation (3) is not necessarily a parameterization by arc length. We address this problem by letting l denote the length of the arc of C from $\mathbf{r}(a)$ to $\mathbf{r}(b)$. Then the *arc length function*

$$s(t) = \int_a^t |\mathbf{r}'(\omega)| d\omega \tag{5}$$

(see Section 17.3) yields the length of the arc of C from $\mathbf{r}(a)$ to $\mathbf{r}(t)$. Consequently, it has the properties that $s(a) = 0$, $s(b) = l$, and

$$s'(t) = |\mathbf{r}'(t)|, \qquad a \le t \le b. \tag{6}$$

Now, suppose that the vector function

$$\mathbf{R}(s) = X(s)\mathbf{i} + Y(s)\mathbf{j}, \qquad 0 \le s \le l$$

is an arc length parameterization for C. Then the composition of $\mathbf{R}$ with s in (5) is

$$\mathbf{R}(s(t)) = X(s(t))\mathbf{i} + Y(s(t))\mathbf{j}, \qquad a \le t \le b.$$

It is important to note that $\mathbf{R}(s(t)) = \mathbf{r}(t)$ for each $t \in [a, b]$, since both vectors terminate at a point $s(t)$ units along C from $\mathbf{r}(a)$ (see Figure 5.6).

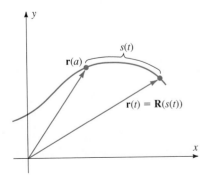

Figure 5.6 $\mathbf{R}(s(t)) = \mathbf{r}(t)$, $a \le t \le b$.

Since $\mathbf{R}(s)$ is a parameterization by arc length,

$$\mathbf{T}(t) = \mathbf{R}'(s) = \frac{d}{ds}\mathbf{R}(s(t)) \tag{7}$$

is a unit tangent for C at $\mathbf{r}(t) = \mathbf{R}(s(t))$ (equation (16), Section 17.3). Differentiating both sides of equation (7) with respect to t, gives

$$\begin{aligned}
\mathbf{T}'(t) &= \frac{d}{dt}\left(\frac{d}{ds}\mathbf{R}\right) \\
&= \frac{d}{ds}\left(\frac{d}{ds}\mathbf{R}\right)\frac{ds}{dt} \qquad \text{(by Chain Rule)} \\
&= \mathbf{R}''(s)\frac{ds}{dt} \\
&= \mathbf{R}''(s)\,|\mathbf{r}'(t)| \qquad \text{(by equation (6)),}
\end{aligned}$$

so

$$|\mathbf{R}''(s)| = \frac{|\mathbf{T}'(t)|}{|\mathbf{r}'(t)|}. \tag{8}$$

The expression in equation (8) is the curvature of C at $\mathbf{r}(t) = \mathbf{R}(s(t))$. It is left for you to show (Exercise 34) that

$$\frac{|\mathbf{T}'(t)|}{|\mathbf{r}'(t)|} = \frac{|x''(t)y'(t) - x'(t)y''(t)|}{[(x'(t))^2 + (y'(t))^2]^{3/2}}, \tag{9}$$

which will complete the proof. ∎

Curvature for Space Curves

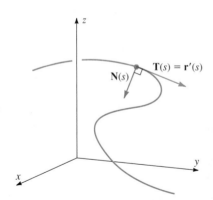

If C is a space curve parameterized by arc length by the vector function

$$C: \quad \mathbf{r}(s) = x(s)\mathbf{i} + y(s)\mathbf{j} + z(s)\mathbf{k}, \qquad a \le s \le b, \tag{10}$$

the vector $\mathbf{T}(s) = \mathbf{r}'(s)$ is a unit tangent to the curve C at $\mathbf{r}(s)$. Since $|\mathbf{r}'(s)| = 1$, the curvature vector $\mathbf{r}''(s)$ is orthogonal to the tangent $\mathbf{T}(s)$ (Example 6, Section 17.2). Thus, the curvature vector $\mathbf{r}''(s)$ is **normal** to C at $\mathbf{r}(s)$, and the **principal unit normal** to C at $\mathbf{r}(s)$ is the vector

$$\mathbf{N}(s) = \frac{1}{|\mathbf{r}''(s)|}\mathbf{r}''(s). \tag{11}$$

Of course, the principal unit normal in (11) is defined only if $\mathbf{r}$ is twice differentiable and $|\mathbf{r}''(s)| \ne 0$ (see Figure 5.7). If C is a plane curve, the vector $\mathbf{N}(s)$ points toward the center of curvature. Multiplying (11) through by $|\mathbf{r}''(s)| = \kappa(s)$ and writing $\mathbf{r}'(s) = \mathbf{T}(s)$ gives the equation

Figure 5.7 Principal unit normal is

$$\mathbf{N}(s) = \frac{\mathbf{r}''(s)}{|\mathbf{r}''(s)|}.$$

$$\mathbf{T}'(s) = \kappa(s)\mathbf{N}(s), \tag{12}$$

and equation by which we could alternatively have defined curvature for space curves.

Normal and Tangential Components of Acceleration

We may apply these ideas to obtain a useful expression for the velocity of a particle moving in space with position function

$$\mathbf{r}(t) = x(t)\mathbf{i} + y(t)\mathbf{j} + z(t)\mathbf{k}. \tag{13}$$

We assume that **r** is twice differentiable, but we do not require that **r** be an arc length parameterization for the path. If s denotes arc length, then we know that the vector $\mathbf{T} = d\mathbf{r}/ds$ is the unit tangent vector to the path. Applying the Chain Rule, we obtain the velocity function **v** as

$$\mathbf{v}(t) = \frac{d\mathbf{r}}{dt} = \frac{d\mathbf{r}}{ds}\frac{ds}{dt} = \frac{ds}{dt}\mathbf{T}.$$

Similarly, the acceleration function is

$$\mathbf{a}(t) = \frac{d\mathbf{v}}{dt} = \frac{d}{dt}\left(\frac{d\mathbf{r}}{ds}\frac{ds}{dt}\right) \tag{14}$$

$$= \frac{d^2s}{dt^2}\frac{d\mathbf{r}}{ds} + \frac{ds}{dt}\left(\frac{d^2\mathbf{r}}{ds^2}\frac{ds}{dt}\right)$$

$$= \frac{d^2s}{dt^2}\frac{d\mathbf{r}}{ds} + \left(\frac{ds}{dt}\right)^2\left[\frac{d}{ds}\left(\frac{d\mathbf{r}}{ds}\right)\right]$$

$$= \frac{d^2s}{dt^2}\mathbf{T}(s) + \left(\frac{ds}{dt}\right)^2\kappa(s)\mathbf{N}(s) \qquad \text{(equation (12)).}$$

That is, we can express acceleration as the sum of a vector parallel to the unit tangent $\mathbf{T}(s)$ and a vector parallel to the unit normal $\mathbf{N}(s)$. It is customary to refer to the **tangential component of acceleration** as the coefficient

$$a_T = \frac{d^2s}{dt^2}$$

and to the **normal component of acceleration** as the coefficient

$$a_N = \left(\frac{ds}{dt}\right)^2\kappa(s).$$

We may then write

$$\mathbf{a} = a_T\mathbf{T} + a_N\mathbf{N}. \tag{15}$$

The components a_T and a_N have straightforward interpretations. The tangential component is

$$a_T = \frac{d^2s}{dt^2} = \frac{d}{dt}\left(\frac{ds}{dt}\right) = \frac{d}{dt}(|\mathbf{v}(t)|),$$

so that a_T is just the rate at which speed is changing. For example, if the speed of the particle is constant, then $a_T = 0$. The normal component of acceleration can be written

$$a_N = \left(\frac{ds}{dt}\right)^2\kappa(s) = |\mathbf{v}|^2\kappa.$$

Thus, a_N is influenced both by speed and by curvature. Since **N** is orthogonal to **T,** this explains why the force tending to pull an object (such as an automobile) from its path is influenced both by the speed of the object and by the curvature of the path.

Since $a_T = d|\mathbf{v}|/dt$, computing a_T from $\mathbf{r}$ is usually simple. However, calculating a_N can be considerably more difficult. One method uses equation (15) and the fact that $\mathbf{T}$ and $\mathbf{N}$ are orthogonal unit vectors:

$$|\mathbf{a}|^2 = \mathbf{a} \cdot \mathbf{a} = (a_T\mathbf{T} + a_N\mathbf{N}) \cdot (a_T\mathbf{T} + a_N\mathbf{N})$$
$$= (a_T)^2 |\mathbf{T}|^2 + a_N|\mathbf{N}|^2$$
$$= a_T^2 + a_N^2.$$

Thus

$$a_N = \sqrt{|\mathbf{a}|^2 - a_T^2}. \tag{16}$$

□ **EXAMPLE 5**

Find the tangential and normal components of acceleration for a particle moving along the helical path

$$\mathbf{r}(t) = \cos t^2\mathbf{i} + \sin t^2\mathbf{j} + t\mathbf{k}.$$

Strategy · · · · · · · · · ·

Find $\mathbf{v}(t) = \mathbf{r}'(t)$.

Find $|\mathbf{v}(t)|$.

$a_T = \dfrac{d}{dt}(|\mathbf{v}(t)|).$

Find $\mathbf{a}$.

Find $|\mathbf{a}(t)|$.

Apply equation (16).

Solution

Here

$$\mathbf{v}(t) = -2t \sin t^2\mathbf{i} + 2t \cos t^2\mathbf{j} + \mathbf{k},$$

so

$$|\mathbf{v}(t)| = \sqrt{4t^2 \sin^2 t^2 + 4t^2 \cos^2 t^2 + 1}$$
$$= \sqrt{4t^2 + 1}.$$

Thus

$$a_T = \frac{d}{dt}(\sqrt{4t^2 + 1}) = \frac{4t}{\sqrt{4t^2 + 1}}.$$

To find $\mathbf{a}$ we differentiate $\mathbf{v}$:

$$\mathbf{a}(t) = (-2 \sin t^2 - 4t^2 \cos t^2)\mathbf{i} + (2 \cos t^2 - 4t^2 \sin t^2)\mathbf{j},$$

so

$$|\mathbf{a}(t)| = (4 \sin^2 t^2 + 16t^2 \sin t^2 \cos t^2 + 16t^4 \cos^2 t^2 + 4 \cos^2 t^2$$
$$- 16t^2 \sin t^2 \cos t^2 + 16t^4 \sin^2 t^2)^{1/2}$$
$$= \sqrt{4 + 16t^4}.$$

Thus

$$a_N = \sqrt{|\mathbf{a}|^2 - a_T^2} = \sqrt{(4 + 16t^4) - \frac{16t^2}{4t^2 + 1}}$$
$$= \sqrt{\frac{64t^6 + 16t^4 + 4}{4t^2 + 1}}.$$

Notice that $\lim_{t \to \infty} a_T = 2$, while $\lim_{t \to \infty} a_N = +\infty$. Can you explain why this is so?

∎

Vector Form of Curvature

In general, the curvature $\kappa(s)$ of a curve C is difficult to compute, using Definition 9, when C determined by $\mathbf{r}$ is not parameterized by arc length. We shall now develop a formula for κ that is easier to apply. Imagine the parameterization $\mathbf{r}$ for C as the position function of a particle moving in space. Then

$$\mathbf{v}(t) = \frac{d\mathbf{r}}{dt} = \frac{d\mathbf{r}}{ds}\frac{ds}{dt} = \left(\frac{ds}{dt}\right)\mathbf{T}.$$

Using equation (14) we find that

$$\mathbf{v} \times \mathbf{a} = \left(\frac{ds}{dt}\right)\mathbf{T} \times \left[\left(\frac{d^2s}{dt^2}\right)\mathbf{T} + \kappa\left(\frac{ds}{dt}\right)^2\mathbf{N}\right]$$

$$= \kappa\left(\frac{ds}{dt}\right)^3 (\mathbf{T} \times \mathbf{N})$$

since $\mathbf{T} \times \mathbf{T} = \mathbf{0}$. Also, since $\mathbf{T}$ and $\mathbf{N}$ are orthogonal unit vectors,

$$|\mathbf{T} \times \mathbf{N}| = |\mathbf{T}||\mathbf{N}||\sin\theta| = 1.$$

Thus,

$$|\mathbf{v} \times \mathbf{a}| = \kappa\left(\frac{ds}{dt}\right)^3 = \kappa|\mathbf{v}|^3,$$

so

$$\boxed{\kappa = \frac{|\mathbf{v} \times \mathbf{a}|}{|\mathbf{v}|^3}.} \tag{17}$$

☐ **EXAMPLE 6**

Find the curvature of the helix

$$C: \quad \mathbf{r}(t) = \cos t\mathbf{i} + \sin t\mathbf{j} + t\mathbf{k}.$$

Solution   Here

$$\mathbf{v}(t) = -\sin t\mathbf{i} + \cos t\mathbf{j} + \mathbf{k}$$

and

$$\mathbf{a}(t) = -\cos t\mathbf{i} - \sin t\mathbf{j}.$$

Thus

$$\mathbf{v} \times \mathbf{a} = \det\begin{bmatrix} \mathbf{i} & \mathbf{j} & \mathbf{k} \\ -\sin t & \cos t & 1 \\ -\cos t & -\sin t & 0 \end{bmatrix}$$

$$= \sin t\mathbf{i} - \cos t\mathbf{j} + (\sin^2 t + \cos^2 t)\mathbf{k},$$

so

$$|\mathbf{v} \times \mathbf{a}| = \sqrt{\sin^2 t + \cos^2 t + 1} = \sqrt{2}.$$

Also,

$$|\mathbf{v}| = \sqrt{\sin^2 t + \cos^2 t + 1} = \sqrt{2}.$$

Thus, by (17),

$$\kappa = \frac{\sqrt{2}}{(\sqrt{2})^3} = \frac{1}{2}.$$

Exercise Set 17.5

In Exercises 1–4, verify that the given curve is parameterized by arc length. Then find (a) the unit tangent $\mathbf{T}(s)$, (b) the curvature vector $\mathbf{r}''(s)$, and (c) the curvature $\kappa(s)$.

1. $\mathbf{r}(s) = \dfrac{1}{2}s\mathbf{i} + \dfrac{\sqrt{3}}{2}s\mathbf{j}$

2. $\mathbf{r}(s) = (4 + \cos s)\mathbf{i} + (2 + \sin s)\mathbf{j}$

3. $\mathbf{r}(s) = \dfrac{\sqrt{2}}{2}\cos s\,\mathbf{i} + \dfrac{\sqrt{2}}{2}\sin s\,\mathbf{j} + \dfrac{\sqrt{2}}{2}s\mathbf{k}$

4. $\mathbf{r}(s) = \sin\left(\dfrac{s}{2}\right)\mathbf{i} + \dfrac{\sqrt{3}}{2}s\mathbf{j} + \cos\left(\dfrac{s}{2}\right)\mathbf{k}$

In Exercises 5–10, find the curvature function $\kappa(t)$ for the given plane curves.

5. $\mathbf{r}(t) = 3\mathbf{i} + t^2\mathbf{j}$ **6.** $\mathbf{r}(t) = t\mathbf{i} + (t^2 + 3)\mathbf{j}$

7. $\mathbf{r}(t) = t\mathbf{i} + e^t\mathbf{j}$ **8.** $\mathbf{r}(t) = e^t\mathbf{i} + e^{-t}\mathbf{j}$

9. $\mathbf{r}(t) = (3 + 2\sin t)\mathbf{i} + (5 + 2\cos t)\mathbf{j}$

10. $\mathbf{r}(t) = (t - \sin t)\mathbf{i} + (1 - \cos t)\mathbf{j}$

In Exercises 11–16, find the curvature of the graph of the given function.

11. $f(x) = 3 + x^2$ **12.** $f(x) = \sqrt{x + 4}$

13. $f(x) = x^3$ **14.** $f(x) = \ln x$

15. $f(x) = \cos x$ **16.** $f(x) = e^{x^2}$

In Exercises 17–20, find the curvature $\kappa(t)$ for the space curve determined by $\mathbf{r}(t)$.

17. $\mathbf{r}(t) = \mathbf{i} + t\mathbf{j} + t^2\mathbf{k}$

18. $\mathbf{r}(t) = \cos t\,\mathbf{i} + \sin t\,\mathbf{j} + \mathbf{k}$

19. $\mathbf{r}(t) = e^t\cos t\,\mathbf{i} + e^t\sin t\,\mathbf{j} + t\mathbf{k}$

20. $\mathbf{r}(t) = \ln t\,\mathbf{i} + t\mathbf{j} + t\mathbf{k}, \quad t > 0$

21. Find the center of curvature for the graph of $y = \sin x$ at the point $(\pi/2, 1)$.

22. Find the point on the graph of $y = \ln x$ where curvature is a maximum.

23. For the graph of $y = \sqrt{x}$, find
 a. the curvature at $(1, 1)$,
 b. the center of curvature at $(1, 1)$,
 c. $\lim\limits_{x \to 0^+} \kappa(x)$.

24. Find the equation for the osculating circle (circle of curvature) for the graph of $y = e^{-x}$ at the point where $x = 1$.

25. Show that the curvature of a line in space is zero.

26. Show that if the curve C is the graph of the polar equation $r = f(\theta)$ and if $f''(\theta)$ exists, then the curvature $\kappa(\theta)$ is given by

$$\kappa(\theta) = \frac{|f(\theta)f''(\theta) - 2[f'(\theta)]^2 - [f(\theta)]^2|}{[(f'(\theta))^2 + (f(\theta))^2]^{3/2}}.$$

(*Hint:* C can be written as $\mathbf{r}(\theta) = x(\theta)\mathbf{i} + y(\theta)\mathbf{j}$, with $x(\theta) = f(\theta)\cos\theta$, $y(\theta) = f(\theta)\sin\theta$.)

In Exercises 27–30, use the result of Exercise 26 to find the curvature function $\kappa(\theta)$.

27. $r = 1 + \sin\theta$ **28.** $r = \theta$

29. $r = 1 - \cos\theta$ **30.** $r = e^\theta$

31. For a curve C in the plane, explain why the curvature vector $\mathbf{r}''(s)$ always points in the direction of the concave side of the curve.

32. Find the curvature of the ellipse $x^2 + 2y^2 = 4$ at the point $(2, 0)$.

33. Prove that the curvature of a circle of radius ρ is $\kappa = \dfrac{1}{\rho}$.

34. Complete the proof of Theorem 5.

35. Find the tangential and normal components of acceleration for a particle moving along the helical path $2t\mathbf{i} + \sin t^2\mathbf{j} + \cos t^2\mathbf{k}$.

17.6   Surfaces

The best mathematical description for certain objects (for example, a piece of twisted wire with a small diameter) is as a parameterized curve C. We specify its shape using a vector-valued function $\mathbf{r}$ that parameterizes C. Then we can discuss its length and the manner in which it curves through space, but we ignore its diameter.

Similarily, we use *surfaces* to model objects that have area but whose thickness we neglect. For instance, soap bubbles, membranes, thin sheets of metal, and pieces of plastic wrap are described as surfaces (see Figure 6.1). In addition, we shall soon see that surfaces arise naturally in the study of multivariate calculus.

In this section we begin our discussion of surfaces by considering a number of examples. Then we discuss graphing techniques that help us understand the geometry of a surface from an algebraic representation.

☐ **EXAMPLE 1**

The simplest of all types of surfaces is the plane. In Section 16.3, we saw that planes are described by equations of the form

$$ax + by + cz = d, \tag{1}$$

where at least one of the numbers a, b, or c is nonzero. Consequently, any equation in the form of equation (1) yields a surface (see Figure 6.2). ∎

Figure 6.1 A soap bubble. (Yoav Levy/ Phototake)

Spheres

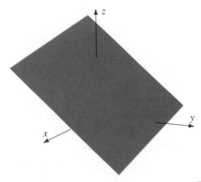

Another basic type of surface is the **sphere.** Given a point $C = (a, b, c)$ in space and a positive real number $r > 0$, the sphere centered at C of radius r is the set of points P whose distance from C to P is exactly r. Using the distance formula, we may write this condition as

$$\sqrt{(x - a)^2 + (y - b)^2 + (z - c)^2} = r,$$

or, in standard form, as

$$\boxed{(x - a)^2 + (y - b)^2 + (z - c)^2 = r^2.} \tag{2}$$

Figure 6.2 A plane in space is a surface.

☐ **EXAMPLE 2**

The sphere with center $C = (2, -3, 1)$ and radius $r = 3$ has equation

$$(x - 2)^2 + (y + 3)^2 + (z - 1)^2 = 9 \qquad \text{(Figures 6.3 and 6.4).} \quad ∎$$

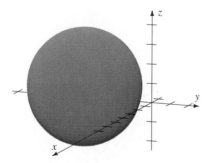

Figure 6.3 The sphere in Example 2.

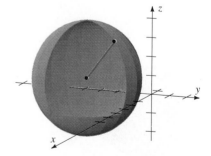

Figure 6.4 The sphere in Example 2 with a portion cut away to indicate the center $(2, -3, 1)$ and a radial segment of length 3.

□ **EXAMPLE 3**

To determine whether the equation

$$x^2 + y^2 + z^2 - 4x + 6z - 3 = 0$$

is that of a sphere, we complete the square in each variable:

$$
\begin{aligned}
0 &= x^2 + y^2 + z^2 - 4x + 6z - 3 \\
&= (x^2 - 4x) + y^2 + (z^2 + 6z) - 3 \\
&= (x^2 - 4x + 4) + y^2 + (z^2 + 6z + 9) - 3 - 4 - 9.
\end{aligned}
$$

Thus,

$$(x - 2)^2 + y^2 + (z + 3)^2 = 16.$$

The graph is a sphere with center at $(2, 0, -3)$ and radius $r = 4$. ■

Note that the sphere described by equation (2) is the surface of the solid ball B consisting of all points $P = (x, y, z)$ whose distance from C is at most r. That is,

$$B = \{(x, y, z) \mid (x - a)^2 + (y - b)^2 + (z - c)^2 \le r^2\}.$$

Many of the surfaces that we study arise as the surface or the boundary of a solid region in space.

Surfaces of Revolution

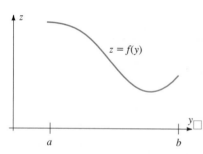

A third familiar type of surface is the **surface of revolution.** In Sections 7.1 through 7.3 and in Section 15.5, we discussed solids of revolution, which are obtained by revolving a planar region about an axis. The same construction yields a surface of revolution. For example, consider the graph of a function $z = f(y)$ for $a \le y \le b$ (see Figure 6.5). For simplicity, we assume that $f(y) > 0$ for all y in $[a, b]$. If we revolve this graph about the y-axis, we obtain a surface (see Figure 6.6). As the following example indicates, this idea can be extended to curves that are not graphs of functions.

Figure 6.5 The graph of $z = f(y)$ for $a \le y \le b$.

□ **EXAMPLE 4**

Consider the circle C with radius 1 and center $(0, 3, 0)$ in the yz-plane. We can revolve C about the z-axis to obtain a surface of revolution T called a **torus** (see Figures 6.7 and 6.8). This surface is the mathematical idealization of the surface of a donut. In the exercise set, we obtain a precise description of T. ■

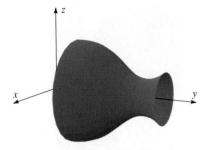

Figure 6.6 The surface of revolution obtained by revolving the graph in Figure 6.5 about the y-axis.

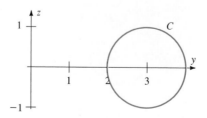

Figure 6.7 The circle C with center $(0, 3, 0)$ and radius 1.

Figure 6.8 The torus that we obtain if the circle C in Figure 6.7 is revolved about the z-axis.

Before we start a systematic study of certain specific surfaces, we introduce one more example—the Möbius strip, which is also discussed when we study vector analysis in Chapter 20.

☐ **EXAMPLE 5**

In the review exercises at the end of this chapter, we develop a precise description of the Möbius strip. For now, we simply give instructions on how you can make a model. Take a long, thin strip of paper (1 in. by 11 in., for example), and tape the two small ends together *after* twisting the strip by a single half-turn (a twist of 180°, see Figure 6.9). The resulting surface is a Möbius strip (see Figure 6.10), and it is interesting because it has only one side! To prove this to yourself, place the strip on a desk and hold it down using the point of a pen (see Figure 6.11). Without lifting the pen from the desk, move it along the center of the strip until the pen comes back to where it started. Inspect the strip and note that you cannot find an unmarked side. Now take a pair of scissors and cut the strip along the central curve that you drew. What happens? ■

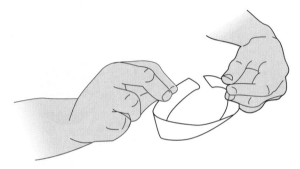

Figure 6.9 Making a Möbius band by twisting a strip of paper one-half turn and joining the ends.

Figure 6.10 A Möbius strip.

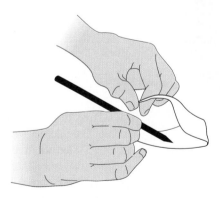

Figure 6.11 The tip of a pen placed on the Möbius strip.

Slicing Surfaces

So far, our discussion of surfaces has been primarily descriptive. However, in order to work with them, we must be able to translate back and forth between their geometric and algebraic descriptions. The remainder of this section is devoted to the technique of determining a geometric description of a surface from an equation by building it out of planar slices.

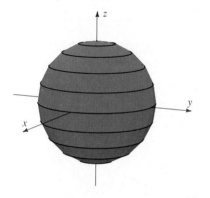

Figure 6.12 The three coordinate planes.

Given a surface we can often determine its shape by finding its *traces* in the coordinate planes and in planes parallel to the coordinate planes. To understand what this means, we first recall the equations for the three coordinate planes:

$$z = 0 \text{ is the equation for the } xy\text{-plane,} \tag{3a}$$

$$y = 0 \text{ is the equation for the } xz\text{-plane,} \tag{3b}$$

$$x = 0 \text{ is the equation for the } yz\text{-plane.} \tag{3c}$$

(See Figure 6.12).

The **trace** of a surface in any plane is simply the intersection of the surface and the plane. The equation for this trace is obtained by substituting the constant value determining the plane for the corresponding variable in the equation for the surface. For example, we know that the equation for the sphere with center $(0, 0, 0)$ and radius r is

$$x^2 + y^2 + z^2 = r^2. \tag{4}$$

Combining equations (3a) and (4), we find that the equation for the trace of the sphere in the xy-plane is $x^2 + y^2 = r^2$. The trace is therefore the circle with center $(0, 0)$ and radius r in the xy-plane (Figure 6.13).

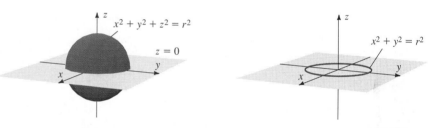

Figure 6.13 The trace of a sphere $x^2 + y^2 + z^2 = r^2$ in the plane $z = 0$ is the circle $x^2 + y^2 = r^2$.

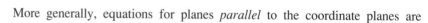

More generally, equations for planes *parallel* to the coordinate planes are

$$z = d, \quad \text{a plane parallel to the } xy\text{-plane,} \tag{5a}$$

$$y = d, \quad \text{a plane parallel to the } xz\text{-plane,} \tag{5b}$$

$$x = d, \quad \text{a plane parallel to the } yz\text{-plane.} \tag{5c}$$

Thus, to find the equation of the trace of the sphere in, say, the plane $z = d$, we combine equations (4) and (5a) to get $x^2 + y^2 + d^2 = r^2$, or $x^2 + y^2 = r^2 - d^2$. If $d^2 < r^2$, then the trace is again a circle (although smaller than the one in the xy-plane); if $d^2 > r^2$, there are no points of intersection (the plane is above or below the sphere; see Figure 6.14).

Figure 6.14 If $d^2 < r^2$, the trace of the sphere $x^2 + y^2 + z^2 = r^2$ in the plane $z = d$ is a circle. A number of these traces are shown in this figure.

Generalized Cylinders

Now we apply this slicing technique to the surface

$$\{(x, y, z) \mid x^2 + y^2 = 1\}.$$

All slices by planes of the form $z = d$, where d is any (fixed) number, produce the equation $x^2 + y^2 = 1$. Therefore, this surface is made up of horizontal circles, all of whose centers lie on the z-axis. Thus the surface is a circular cylinder (see Figure 6.15).

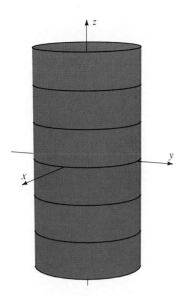

Figure 6.15   The circular cylinder $x^2 + y^2 = 1$. All traces by planes of the form $z = d$ are circles.

Our analysis of these horizontal traces indicates that the circular cylinder can be viewed as the parallel translation of the familiar unit circle in the two directions that are perpendicular to the xy-plane. Indeed, parallel translation of a plane curve is another method for producing a surface using a plane curve. Instead of revolving the plane curve about an axis in that plane, we translate the curve in the two directions perpendicular to that plane. Such surfaces are called **generalized cylinders.**

The absence of the z variable in the equation $x^2 + y^2 = 1$ for the cylinder suggests that it is a generalized cylinder. Whenever an equation involving two variables is used to describe a subset of xyz-space, the traces obtained by holding the third variable constant are all identical. The resulting set can therefore be thought of as an "infinite stack" of identical curves. Other examples of generalized cylinders are

(a) elliptic cylinders of the form

$$\frac{x^2}{a^2} + \frac{y^2}{b^2} = 1$$

(see Figure 6.16);

(b) a plane whose equation contains only two variables, such as $x + y = 2$ (Figure 6.17);

(c) the parabolic cylinder $y = x^2$ (Figure 6.18);

(d) the hyperbolic cylinder $\dfrac{y^2}{a^2} - \dfrac{x^2}{b^2} = 1$ (Figure 6.19).

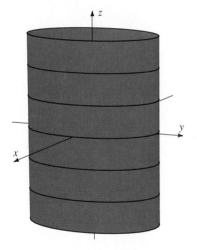

Figure 6.16   An elliptic cylinder of the form

$$\frac{x^2}{a^2} + \frac{y^2}{b^2} = 1$$

In this figure, $b^2 > a^2$.

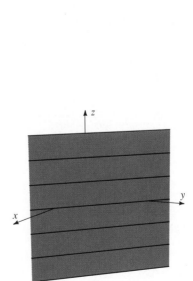

Figure 6.17   The plane $x + y = 2$ is a generalized cylinder.

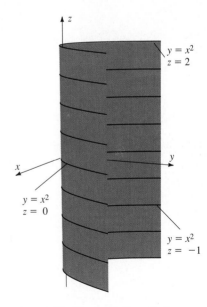

Figure 6.18   The parabolic cylinder $y = x^2$ is a generalized cylinder.

□ **EXAMPLE 6**

Sketch the cylinder $9x^2 + 4z^2 - 18x - 16z = 11$.

Solution   First we complete the square in x and z:

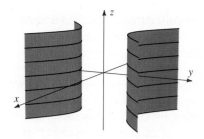

Figure 6.19 The hyperbolic cylinder
$$\frac{y^2}{a^2} - \frac{x^2}{b^2} = 1.$$

Quadric Surfaces

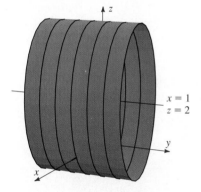

Figure 6.20 The elliptic cylinder

$$9x^2 - 18x + 4z^2 - 16z = 11.$$

Its central axis is the line $x = 1$ and $z = 2$.

$$9x^2 - 18x + 4z^2 - 16z = 11$$
$$9(x^2 - 2x) + 4(z^2 - 4z) = 11$$
$$9(x^2 - 2x + 1) + 4(z^2 - 4z + 4) = 11 + 9 \cdot 1 + 4 \cdot 4$$
$$9(x - 1)^2 + 4(z - 2)^2 = 36$$
$$\frac{(x - 1)^2}{2^2} + \frac{(z - 2)^2}{3^2} = 1.$$

This is the equation for an ellipse with center $(1, 2)$ in the xz-plane. Since the variable y is missing, the figure is an elliptic cylinder with central axis parallel to the y-axis (Figure 6.20). ∎

Just as nondegenerate second-degree equations in the plane describe lines, parabolas, circles, ellipses, and hyperbolas, nondegenerate second-degree equations in three variables

$$Ax^2 + By^2 + Cz^2 + Dxy + Exz + Fyz + Gx + Hy + Iz + J = 0 \qquad (6)$$

describe surfaces in space. These surfaces are called **quadric surfaces,** and there are six basic types:

1. the ellipsoid,
2. the elliptic paraboloid,
3. the elliptic cone,
4. the hyperboloid of one sheet,
5. the hyperboloid of two sheets, and
6. the hyperbolic paraboloid.

We can use traces to understand the general shape of each type, but we shall not attempt a complete analysis of all possibilities. Instead, we illustrate how the slicing technique can be used to help understand the geometry of a given quadric surface.

The Ellipsoid: $\quad \dfrac{x^2}{a^2} + \dfrac{y^2}{b^2} + \dfrac{z^2}{c^2} = 1$

Setting $z = 0$ shows that the trace in the xy-plane is the ellipse

$$\frac{x^2}{a^2} + \frac{y^2}{b^2} = 1.$$

Similarly, setting $y = 0$ and then $x = 0$ shows that the traces in the xz- and yz-plane are also ellipses. These three traces are shown in Figure 6.21.

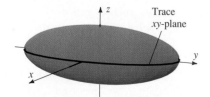
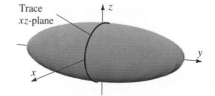
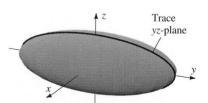

Figure 6.21 The traces of the ellipsoid $\dfrac{x^2}{a^2} + \dfrac{y^2}{b^2} + \dfrac{z^2}{c^2} = 1$ in the three coordinate planes.

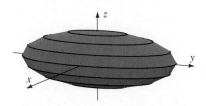

Figure 6.22 The traces of the ellipsoid in the planes $z = d$.

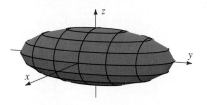

Figure 6.23 Various traces of the ellipsoid in the planes $y = d$ and $z = d$.

Using Equations (5a) through (5c), we find that the traces in planes that are parallel to the coordinate planes and that intersect the ellipsoid are again ellipses. For example, setting $z = d$ gives the trace

$$\frac{x^2}{a^2} + \frac{y^2}{b^2} = 1 - \frac{d^2}{c^2}. \tag{7}$$

The graph of equation (7) is an ellipse if $0 \le |d| < |c|$, it is a pair of points $(0, 0, \pm d)$ if $|d| = |c|$, and the equation has no solutions if $|d| > |c|$ (Figure 6.22). Figure 6.23 shows traces of the ellipsoid for various planes determined by equations (5a) through (5c).

The ellipsoid is symmetric with respect to each of the coordinate planes, since replacing x by $-x$, y by $-y$, or z by $-z$ does not change the equation. If $a = b = c$, the ellipsoid is, of course, a sphere.

An Elliptic Paraboloid: $z = \dfrac{x^2}{a^2} + \dfrac{y^2}{b^2}$

Setting $x = 0$ shows that the trace of this figure in the yz-plane is the parabola $z = y^2/b^2$. Similarly, the trace in the xz-plane is the parabola $z = x^2/a^2$. The trace in the xy-plane ($z = 0$) is simply the origin $(0, 0)$ (Figure 6.24).

The traces in the planes $x = d_1$ and $y = d_2$ are again parabolas:

$$z = \frac{y^2}{b^2} + \frac{d_1^2}{a^2}; \qquad z = \frac{x^2}{a^2} + \frac{d_2^2}{b^2} \qquad \text{(Figure 6.25)}.$$

However, traces in the plane $z = d$ are ellipses with equations

$$\frac{x^2}{a^2 d} + \frac{y^2}{b^2 d} = 1$$

if $d > 0$. If $d < 0$ there are no traces. (See Figure 6.26.)

The elliptic paraboloid is symmetric with respect to the xz- and yz-planes, as you can see by replacing x by $-x$, or y by $-y$ in its equation.

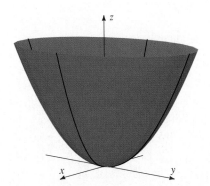

Figure 6.24 For the elliptic paraboloid

$$z = \frac{x^2}{a^2} + \frac{y^2}{b^2},$$

the traces in the xz- and the yz-planes are parabolas.

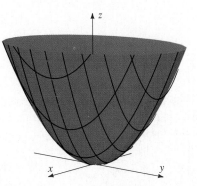

Figure 6.25 The elliptic paraboloid with a number of traces in the planes $x = d$ and $y = d$.

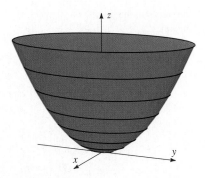

Figure 6.26 The traces of the elliptic paraboloid

$$z = \frac{x^2}{a^2} + \frac{y^2}{b^2}$$

in the planes $z = d$ $(d > 0)$ are ellipses.

An Elliptic Cone: $z^2 = \dfrac{x^2}{a^2} + \dfrac{y^2}{b^2}$.

The difference between the equation of the elliptic cone and that of the elliptic paraboloid is the presence of z^2 rather than z. As a result, the traces in the xz-plane ($y = 0$) are the lines $z = \pm x/a$, and the traces in the yz-plane ($x = 0$) are the lines $z = \pm y/b$. The trace in the xy-plane ($z = 0$) is simply the origin $(0, 0, 0)$. (See Figure 6.27.)

In the planes $x = d$, the traces are the hyperbolas

$$\frac{a^2 z^2}{d^2} - \frac{a^2 y^2}{b^2 d^2} = 1.$$

Similarly, the traces in the planes $y = d$ are hyperbolas (Figure 6.28). However, in the planes $z = d$, the traces are the ellipses with equations

$$\frac{x^2}{a^2 d^2} + \frac{y^2}{b^2 d^2} = 1 \qquad \text{(Figure 6.29)}.$$

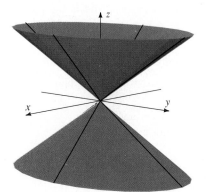

Figure 6.27 For the elliptic cone

$$z^2 = \frac{x^2}{a^2} + \frac{y^2}{b^2},$$

the traces in the planes $x = 0$ and $y = 0$ are lines.

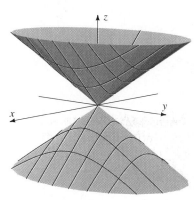

Figure 6.28 The traces of an elliptic cone

$$z^2 = \frac{x^2}{a^2} + \frac{y^2}{b^2}$$

in planes of the form $x = d$ and $y = d$ are hyperbolas.

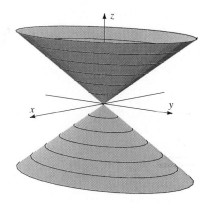

Figure 6.29 The z-traces of an elliptic cone

$$z^2 = \frac{x^2}{a^2} + \frac{y^2}{b^2}$$

are ellipses.

The elliptic cone is symmetric with respect to each of the three coordinate planes. It differs from the elliptic paraboloid in that it is unbounded in both the positive and negative z directions. If $a = b$, the cone is called a *right* or *circular* cone.

A Hyperboloid of One Sheet: $\dfrac{x^2}{a^2} + \dfrac{y^2}{b^2} - \dfrac{z^2}{c^2} = 1$

The trace in the xy-plane ($z = 0$) is the ellipse

$$\frac{x^2}{a^2} + \frac{y^2}{b^2} = 1,$$

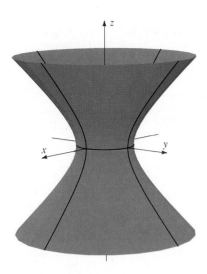

while traces in the other two coordinate planes are the hyperbolas

$$\frac{x^2}{a^2} - \frac{z^2}{c^2} = 1 \quad \text{and} \quad \frac{y^2}{b^2} - \frac{z^2}{c^2} = 1$$

(Figure 6.30). Similarly, setting $z = d$ shows that traces parallel to the xy-plane are ellipses, while the traces in planes parallel to the xz- or yz-planes are hyperbolas. The hyperboloid of one sheet is symmetric with respect to each of the coordinate axes. (See Figure 6.31.)

A Hyperboloid of Two Sheets: $\quad \dfrac{x^2}{a^2} + \dfrac{y^2}{b^2} - \dfrac{z^2}{c^2} = -1$

Setting $z = 0$ shows that this surface has no trace in the xy-plane. The traces in the xz- and yz-planes are the hyperbolas

$$\frac{z^2}{c^2} - \frac{x^2}{a^2} = 1 \quad \text{and} \quad \frac{z^2}{c^2} - \frac{y^2}{b^2} = 1,$$

Figure 6.30 For this hyperboloid of one sheet, the traces in the planes $x = 0$ and $y = 0$ are hyperbolas, and the trace in the plane $z = 0$ is an ellipse.

respectively. The traces in planes parallel to the xz- and yz-planes are also hyperbolas. However, traces in planes parallel to the xy-plane ($z = d$) are ellipses

$$\frac{x^2}{a^2} + \frac{y^2}{b^2} = \frac{d^2}{c^2} - 1$$

provided $|d| > |c|$ (Figure 6.32). The hyperboloid of two sheets is symmetric with respect to each of the coordinate planes (Figure 6.33).

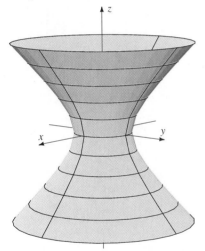

Figure 6.31 The shape of this hyperboloid of one sheet can be determined using the traces in the coordinate planes and traces in planes of the form $z = d$.

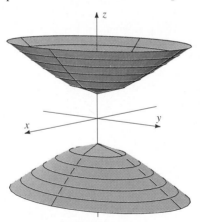

Figure 6.32 For this hyperboloid of two sheets, the traces in the xz- and yz-planes are hyperbolas. Traces in planes of the form $z = d$ are either empty, a single point, or an ellipse.

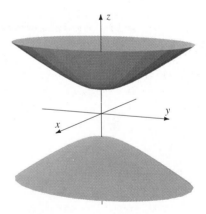

Figure 6.33 The hyperboloid of two sheets that consists of the traces shown in Figure 6.32.

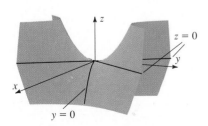

Figure 6.34 For this hyperbolic paraboloid, the $z = 0$ trace consists of two lines. The $y = 0$ trace is a parabola.

A Hyperbolic Paraboloid: $\quad z = \dfrac{y^2}{b^2} - \dfrac{x^2}{a^2}$

Setting $z = 0$ shows that the trace in the xy-plane is the pair of intersecting straight lines $y = \pm |b/a| x$. In the xz-plane ($y = 0$) the trace is the parabola $z = -x^2/b^2$ (Figure 6.34). The traces in planes parallel to the xz- and yz-planes are also parabolas. However, the traces in planes parallel to the xy-plane ($z = d$) are hyperbolas. The hyperbolic paraboloid is symmetric with respect to the xz- and yz-planes. (See Figure 6.35.)

In the next example, we illustrate how the algebraic technique of completing the square indicates that a **translation of the axes** can be used to simplify the characterization of a quadric surface. In other words, we introduce new variables X, Y, and Z defined by

$$X = x - h, \tag{8a}$$

$$Y = y - k, \tag{8b}$$

$$Z = z - l. \tag{8c}$$

As in the two-variable case, these substitutions correspond to a relocation of the coordinate axes so that the origin lies at (h, k, l) and so that each of the coordinate axes lies parallel to its original position (see Figure 6.36).

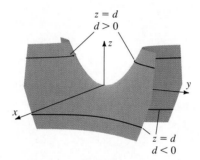

Figure 6.35 For this hyperbolic paraboloid, the traces in planes of the form $z = d$ ($d \neq 0$) are hyperbolas.

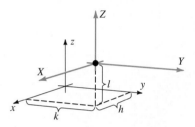

Figure 6.36 Translation of axes:
$$X = x - h,$$
$$Y = y - k,$$
$$Z = z - l.$$

☐ **EXAMPLE 7**

Describe the graph of the equation

$$x^2 + y^2 + 3z^2 - 2x + 4y - 4 = 0.$$

Strategy · · · · · · · ·
Complete the square in x and in y.

Solution

$$x^2 - 2x + y^2 + 4y + 3z^2 - 4 = 0$$
$$(x^2 - 2x + 1) + (y^2 + 4y + 4) + 3z^2 - 4 - 1 - 4 = 0$$
$$(x - 1)^2 + (y + 2)^2 + 3z^2 - 9 = 0$$

Use translated variables (8a) through (8c) to simplify equation.

The given equation therefore has the form

$$X^2 + Y^2 + 3Z^2 = 9,$$

or

Identify the form of the equation obtained.

$$\frac{X^2}{3^2} + \frac{Y^2}{3^2} + \frac{Z^2}{(\sqrt{3})^2} = 1$$

with $X = x - 1$, $Y = y + 2$, and $Z = z$. The graph of this equation is an ellipsoid with center at $(1, -2, 0)$. (See Figure 6.37.) ■

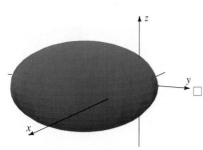

Figure 6.37 The ellipsoid

$$x^2 + y^2 + 3z^2 - 2x + 4y - 4 = 0.$$

Its center is the point $(1, -2, 0)$.

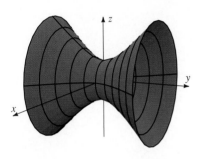

Figure 6.38 The hyperboloid of one sheet $x^2 - y^2 + z^2 = 1$ discussed in Example 8.

Our final example illustrates how the slicing technique works on a quadric surface whose equation is somewhat different from those discussed above.

☐ **EXAMPLE 8**

Identify and sketch the surface

$$x^2 - y^2 + z^2 = 1.$$

Solution First we pick a particular direction in which we determine many slices. Moving the y^2 term to the right-hand side of the equation, we get

$$x^2 + z^2 = 1 + y^2;$$

this form suggests that we slice by planes parallel to the xz-plane. That is, we slice by planes of the form $y = d$ for various choices of the constant d. We obtain

$$x^2 + z^2 = 1 + d^2,$$

which is the equation of a circle whose radius is at least 1. If we also slice by the other coordinate planes, we get

$$x^2 - y^2 = 1 \qquad (\text{if } z = 0)$$

and

$$z^2 - y^2 = 1 \qquad (\text{if } x = 0).$$

These two slices are hyperbolas. Thus we have the traces of a hyperboloid of one sheet. When we sketch the surface, we note that the circles are all parallel to the xz-plane. The result is shown in Figure 6.38. ∎

Exercise Set 17.6

In Exercises 1–5, find the center and radius for the sphere with the given equation.

1. $x^2 + y^2 + z^2 - 4z = 5$

2. $x^2 + y^2 + z^2 - 2y - 4z = 4$

3. $x^2 + y^2 + z^2 - 4x + 2y - 6z = 2$

4. $x^2 + y^2 + z^2 - 4x + 4z = -4$

5. $x^2 + y^2 + z^2 - 6x + 2y + 4z = 11$

6. Write an equation for the sphere with
   a. center $(0, 0, 0)$ and radius $r = 2$,
   b. center $(1, -1, 0)$ and radius $r = 1$,
   c. center $(-2, 3, -5)$ and radius $r = 3$,
   d. center $(4, 6, -2)$ and radius $r = 10$.

In Exercises 7–26, describe the quadric surface that is the graph of the given equation.

7. $x^2 + y^2 + z^2 = 10$

8. $\dfrac{x^2}{4} + \dfrac{y^2}{9} + \dfrac{z^2}{16} = 1$

9. $\dfrac{x^2}{4} + \dfrac{y^2}{9} = z^2$

10. $6x^2 + 12y^2 - 8z^2 - 24 = 0$

11. $4x^2 + 9y^2 - 36z^2 + 36 = 0$

12. $9y^2 - 6x^2 = 54$

13. $2x^2 - 3y^2 + 8x + 6y - 6z + 5 = 0$

14. $8x^2 + 4y^2 - 2z^2 + 24y - 4z + 44 = 0$

15. $x^2 + 2y^2 - 2z^2 + 6x - 4y + 7 = 0$

16. $x^2 + 4y^2 - 4z^2 + 6x - 8y - 8z + 9 = 0$

17. $6x^2 + 9y^2 + 36y - 54z + 36 = 0$

18. $4x^2 + 3y^2 + 3z^2 - 12y + 18z + 27 = 0$

19. $9z^2 + 4y^2 - 36x = 0$

20. $x^2 - 4y^2 + 4z^2 = 0$

21. $z^2 + y^2 - x^2 - 1 = 0$

22. $6x^2 - 12y^2 + 8z^2 + 24 = 0$

23. $x^2 + y^2 + z^2 - 6x + 2y + 6z + 18 = 0$

24. $9x^2 + 4y^2 + 9z^2 - 36x + 8y - 18z + 13 = 0$

25. $x^2 + y^2 + 4x - 2y - 36z + 175 = 0$

26. $y^2 - x^2 - 6x - 2y - 4z - 16 = 0$

In Exercises 27–42, describe the given generalized cylinder in *xyz*-space.

27. $x = 3$ **28.** $y = 6$

29. $x + y = 6$ **30.** $x^2 + y^2 = 5$

31. $\dfrac{x^2}{4} - \dfrac{y^2}{9} = 1$ **32.** $\dfrac{x^2}{2} + \dfrac{y^2}{4} = 1$

33. $xy = 1$ **34.** $y = \sin x$

35. $y = 4z^2$ **36.** $x = \sqrt{9 - z^2}$

37. $x^2 - z^2 = 1$ **38.** $z^2 - y^2 = 1$

39. $z = 1 - y^2$ **40.** $(x - 1)^2 + (z + 3)^2 = 9$

41. $3x^2 + 12x - z + 16 = 0$

42. $9x^2 + 4z^2 + 18x - 16z - 11 = 0$

43. Write the equation for the surface obtained by revolving about the *y*-axis the graph of $y = 2x$.

44. Describe the set of all points P for which the distance from P to the *y*-axis is twice the distance from P to the *xz*-plane.

45. Find an equation for the surface obtained by revolving about the *y*-axis the graph of $y = z^2$.

46. Each of the quadric surfaces (i)–(viii) is given by one of the equations (a)–(h). Match the corresponding surfaces and equations.

a. $y = x^2 + z^2$ **e.** $x^2 - y^2 + z^2 = -1$
b. $x^2 + y^2 - z^2 = 1$ **f.** $4x^2 + y^2 + z^2 = 1$
c. $z = x^2 + 2y^2$ **g.** $4x^2 + y^2 + 4z^2 = 1$
d. $z = 2x^2 + y^2$ **h.** $z = 2y^2 - x^2$

(ii)

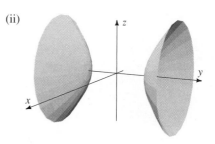

(iii)

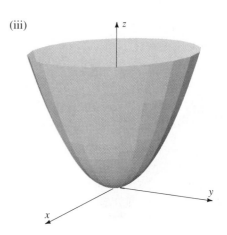

(iv)

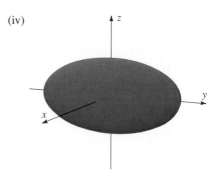

(i)

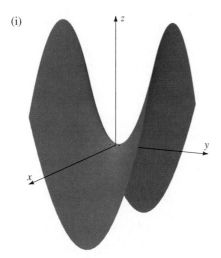

(v)

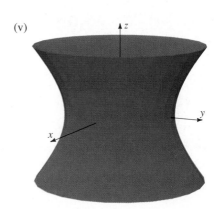

(vi)

(vii)

(viii)

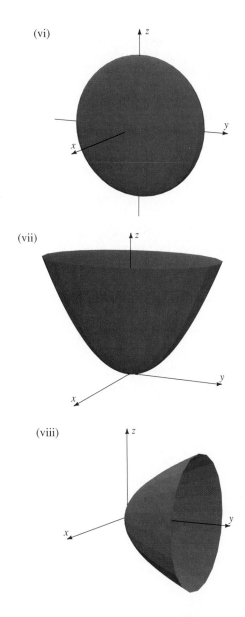

b. The two surfaces are shown below. Indicate which surface is Q_1 and which is Q_2. Draw the traces S_1, S_2, and S_3 on the appropriate surface.

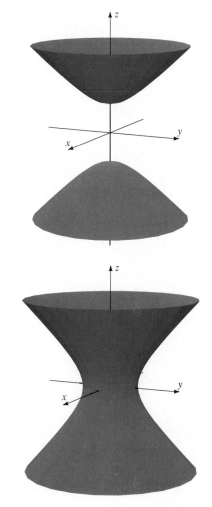

47. Let Q_1 be the quadric surface $x^2 + y^2 - z^2 = -1$ and Q_2 be the quadric surface $x^2 + y^2 - z^2 = 1$. Let S_1 be the equation $y^2 - z^2 = 1$, S_2 be the equation $z^2 - y^2 = 0$, and S_3 be the equation $x^2 + y^2 = \frac{1}{2}$. Each of these three equations results from slicing one of the two surfaces by one or more planes.

a. For each trace S_1, S_2, and S_3, determine the surface, Q_1 or Q_2, that contains it. Specify an equation for the plane or planes that yield these traces.

48. Consider the circle C with radius 1 and center $(0, 3, 0)$ in the yz-plane as described in Example 4.

a. Specify an equation in y and z that describes C.

b. Let $r = \sqrt{x^2 + y^2}$. Show that r is the distance of a point (x, y, z) to the z-axis.

c. Specify an equation in r and z that describes the torus T in Example 4.

d. Using part c, determine an equation in x, y, and z that describes T.

17.7   Cylindrical and Spherical Coordinates

In Chapter 15, we saw that certain curves in the plane are represented most simply using polar coordinates. Similarly, certain surfaces in space are more easily described using *cylindrical* or *spherical* coordinates in place of rectangular coordinates. In fact, some surfaces that are easily specified in one of these coordinate systems are almost impossible to describe in rectangular coordinates. In this section, we introduce cylindrical and spherical coordinates so that we may use them in our development of multivariable calculus.

Cylindrical Coordinates

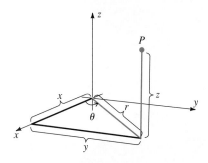

The cylindrical coordinate system in space uses polar coordinates in the xy-plane with a third coordinate that is the usual rectangular z-coordinate. That is, if the point P has cylindrical coordinates $P = (r, \theta, z)$ and rectangular coordinates $P = (x, y, z)$, the equations

$$x = r \cos \theta, \qquad r \geq 0 \tag{1a}$$

$$y = r \sin \theta, \qquad r \geq 0 \tag{1b}$$

$$z = z \tag{1c}$$

Figure 7.1 The rectangular coordinates (x, y, z) and the cylindrical coordinates (r, θ, z) of the point P.

give the rectangular coordinates in terms of the cylindrical coordinates. The equations

$$r = \sqrt{x^2 + y^2} \tag{2a}$$

$$\tan \theta = \frac{y}{x} \tag{2b}$$

$$z = z \tag{2c}$$

give the cylindrical coordinates in terms of the rectangular coordinates (Figure 7.1). Note that we now require $r \geq 0$, unlike the case for polar coordinates in the plane and that equation (2a) implies

$$r^2 = x^2 + y^2 \tag{3}$$

Figure 7.2 shows why cylindrical coordinates are named as they are. The graph of the equation $r = r_0$, for r_0 constant, is a cylinder in space. The graph of $\theta = \theta_0$, for θ_0 constant, is a half plane, as in Figure 7.3. The graph of $z = z_0$, for z_0 constant,

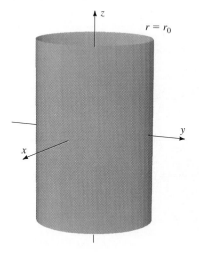

Figure 7.2 The graph of $r = r_0$ is a circular cylinder whose central axis is the z-axis.

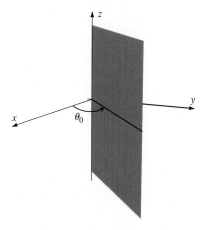

Figure 7.3 The graph of $\theta = \theta_0$ is a half-plane that borders the z-axis.

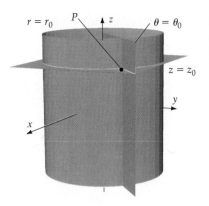

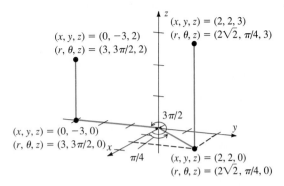

Figure 7.4 The cylindrical coordinates $(r, \theta, z) = (r_0, \theta_0, z_0)$ correspond to the point P that lies on the intersection of the cylinder $r = r_0$, the half-plane $\theta = \theta_0$, and the plane $z = z_0$.

is a horizontal plane, as in rectangular coordinates. Figure 7.4 indicates how these surfaces determine the cylindrical coordinates of a point. Several points, expressed in terms of both rectangular and cylindrical coordinates, appear in Figure 7.5.

Figure 7.5 Four points expressed both in terms of rectangular and cylindrical coordinates.

☐ EXAMPLE 1

Find cylindrical coordinates for the point with rectangular coordinates $(1, \sqrt{3}, 4)$.

Solution   Here $x = 1$, $y = \sqrt{3}$, $z = 4$. By equations (2a) through (2c),

$$r = \sqrt{1^2 + (\sqrt{3})^2} = 2,$$

$$\tan \theta = \sqrt{3} \quad \text{so} \quad \theta = \pi/3 \qquad \text{(since } x \text{ and } y \text{ place the point in the first quadrant),}$$

$$z = 4.$$

The cylindrical coordinates are $(2, \pi/3, 4)$. ■

Equations (1) do not specify θ uniquely. In Example 1 we had infinitely many choices for θ, and we chose the unique angle θ between 0 and 2π corresponding to this point.

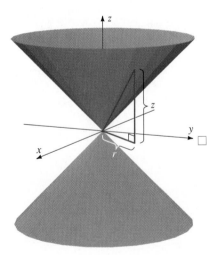

Figure 7.6 The graph of $z = \pm r$ is a cone.

☐ EXAMPLE 2

Express the surface $z^2 = x^2 + y^2$ in cylindrical coordinates.

Solution   Using (3), we get

$$z^2 = x^2 + y^2$$
$$= r^2.$$

The surface is therefore expressed in cylindrical coordinates as

$$z = \pm r,$$

Thus the surface is the cone in Figure 7.6. ■

☐ EXAMPLE 3

Graph the equation $r^2 = a^2 \sin \theta$ in cylindrical coordinates.

Solution   Since the variable z is missing, the graph is a generalized cylinder in the *z*-direction whose trace in the *xy*-plane is the graph of the lemniscate $r^2 = a^2 \sin \theta$ (see Figure 7.7). ∎

Spherical Coordinates

In spherical coordinates, a point is determined by the ordered triple (ρ, θ, ϕ) where $\rho = |\overrightarrow{OP}|$ is the distance of the point P from the origin, θ is the polar angle used in cylindrical coordinates, and ϕ is the (tilt) angle between the vector $\overrightarrow{OP}$ and the positive *z*-axis (Figure 7.8). By convention, we require $\rho \geq 0$, $0 \leq \theta < 2\pi$, and $0 \leq \phi \leq \pi$.

Figure 7.9 illustrates the reason for the terminology "spherical coordinates." The graph of the equation $\rho = \rho_0$, for ρ_0 constant, is a sphere of radius ρ_0 centered at the origin. The graph of $\theta = \theta_0$, for θ_0 constant, is a half plane, as in cylindrical coordinates. The graph of $\phi = \phi_0$, for ϕ_0 constant, is a cone (Figure 7.10). Figure 7.11 illustrates how these three surfaces determine the spherical coordinates of a point.

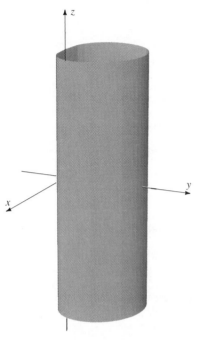

Figure 7.7 The graph of $r^2 = a^2 \sin \theta$ is a generalized cylinder.

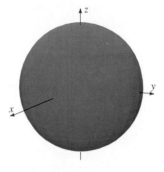

Figure 7.9 The graph of $\rho = \rho_0$ is a sphere centered at the origin.

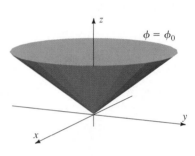

Figure 7.10 The graph of $\phi = \phi_0$ is one half of a circular cone. If $0 < \phi_0 < \pi/2$, then the half-cone lies above the *xy*-plane. If $\pi/2 < \phi_0 < \pi$, then the half-cone lies below the *xy*-plane.

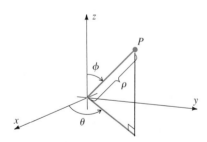

Figure 7.8 The spherical coordinates (ρ, θ, ϕ) of a point P.

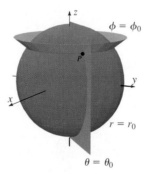

Figure 7.11 The spherical coordinates $(\rho, \theta, \phi) = (\rho_0, \theta_0, \phi_0)$ correspond to the point P that lies on the intersection of the sphere $\rho = \rho_0$, the half-plane $\theta = \theta_0$, and the half-cone $\phi = \phi_0$.

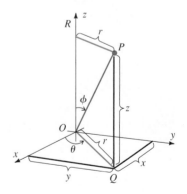

Figure 7.12 The relationships between rectangular, cylindrical, and spherical coordinates.

Figure 7.12 illustrates the relationships between the spherical coordinates (ρ, θ, ϕ), the rectangular coordinates (x, y, z), and the cylindrical coordinates

(r, θ, z) for a point P. Since the vector $\overrightarrow{OP}$ is a diagonal of the rectangle $OQPR$,

$$r = \rho \sin \phi \quad \text{and} \quad z = \rho \cos \phi. \tag{4}$$

Also,

$$x = r \cos \theta \quad \text{and} \quad y = r \sin \theta. \tag{5}$$

Combining equations (4) and (5) gives

$$x = r \cos \theta = \rho \sin \phi \cos \theta, \qquad \rho \geq 0 \tag{6a}$$

$$y = r \sin \theta = \rho \sin \phi \sin \theta, \qquad \rho \geq 0 \tag{6b}$$

$$z = \rho \cos \phi, \qquad \rho \geq 0. \tag{6c}$$

Also, from the distance formula we find

$$\rho = \sqrt{x^2 + y^2 + z^2}. \tag{7}$$

Equations for θ and ϕ may be easily derived from equations (4) and (5).

□ **EXAMPLE 4**

The point P has spherical coordinates $(\rho, \theta, \phi) = (3, \pi/3, \pi/4)$. Find rectangular coordinates for P.

Solution   By equations (6a) through (6c),

$$x = 3 \sin\left(\frac{\pi}{4}\right) \cos\left(\frac{\pi}{3}\right) = 3\left(\frac{\sqrt{2}}{2}\right)\left(\frac{1}{2}\right) = \frac{3\sqrt{2}}{4},$$

$$y = 3 \sin\left(\frac{\pi}{4}\right) \sin\left(\frac{\pi}{3}\right) = 3\left(\frac{\sqrt{2}}{2}\right)\left(\frac{\sqrt{3}}{2}\right) = \frac{3\sqrt{6}}{4},$$

$$z = 3 \cos\left(\frac{\pi}{4}\right) = \frac{3\sqrt{2}}{2}.$$

The rectangular coordinates are $P = \left(\dfrac{3\sqrt{2}}{4}, \dfrac{3\sqrt{6}}{4}, \dfrac{3\sqrt{2}}{2}\right)$. ■

□ **EXAMPLE 5**

Express the surface $x^2 - y^2 + z^2 = 4$ in spherical coordinates.

Solution   Using equations (6a) through (6c) the equation becomes

$$(\rho \sin \phi \cos \theta)^2 - (\rho \sin \phi \sin \theta)^2 + (\rho \cos \phi)^2 = 4,$$
$$\rho^2 \sin^2 \phi [\cos^2 \theta - \sin^2 \theta] + \rho^2 \cos^2 \phi = 4,$$
$$\rho^2 \sin^2 \phi [1 - 2 \sin^2 \theta] + \rho^2 \cos^2 \phi = 4,$$
$$\rho^2 - 2\rho^2 \sin^2 \phi \sin^2 \theta = 4.$$

■

Exercise Set 17.7

1. The following points are given in rectangular coordinates. Find their cylindrical coordinates.
   a. $(1, 1, 0)$ c. $(-1, 1, -2)$
   b. $(\sqrt{3}, 1, 3)$ d. $(-1, \sqrt{3}, 4)$

2. The following points are given in rectangular coordinates. Find their cylindrical coordinates.
   a. $(0, 3, -5)$ c. $(4\sqrt{3}, 12, -2)$
   b. $(-\sqrt{2}, \sqrt{2}, \sqrt{2})$ d. $(-2, -2, 1)$

3. The following points are given in cylindrical coordinates. Find their rectangular coordinates.
   a. $(2, \pi/4, -3)$ c. $(4, \pi/3, -5)$
   b. $(1, \pi/6, 4)$ d. $(2, 4\pi/3, 2)$

4. The following points are given in cylindrical coordinates. Find their rectangular coordinates.
   a. $(1, \pi, 1)$ c. $(2, 3\pi/2, -3)$
   b. $(5, 5\pi/3, 5)$ d. $(3, 7\pi/4, 0)$

5. The following points are given in rectangular coordinates. Find their spherical coordinates.
   a. $(1, 0, 0)$ c. $(1, -1, \sqrt{2})$
   b. $(1, 1, \sqrt{2})$ d. $(1, \sqrt{3}, 2)$

6. The following points are given in rectangular coordinates. Find their spherical coordinates.
   a. $(-\sqrt{3}, 1, -2)$ c. $(-1, -1, \sqrt{2})$
   b. $(-2, 2, 2\sqrt{2})$ d. $(0, 2, 2\sqrt{3})$

7. The following points are given in spherical coordinates. Find their rectangular coordinates.
   a. $(2, \pi/4, \pi/3)$ c. $(2, 3\pi/4, \pi/4)$
   b. $(1, \pi/2, \pi)$ d. $(3, 3\pi/2, 2\pi/3)$

8. The following points are given in spherical coordinates. Find their rectangular coordinates.
   a. $(5, \pi/6, 5\pi/6)$ c. $(4, \pi/2, \pi/6)$
   b. $(2, 5\pi/3, 3\pi/4)$ d. $(\sqrt{17}, \pi/4, \mathrm{Tan}^{-1}(3/2))$

9. The following points are given in cylindrical coordinates. Find their spherical coordinates.
   a. $(1, 0, 0)$ c. $(2, \pi/3, 2)$
   b. $(\sqrt{2}, -\pi/4, \sqrt{2})$ d. $(2, \pi/4, 2)$

10. The following points are given in cylindrical coordinates. Find their spherical coordinates.
    a. $(2, 5\pi/3, 0)$ c. $(4\sqrt{2}, 5\pi/6, 8)$
    b. $(2, \pi/6, 2)$ d. $(\sqrt{3}, 7\pi/6, -1)$

11. The following points are given in spherical coordinates. Find their cylindrical coordinates.
    a. $(2, \pi/4, \pi/2)$ c. $(3, 2\pi/3, \pi/2)$
    b. $(2, \pi/2, \pi/4)$ d. $(1, \pi/2, 2\pi/3)$

12. The following points are given in spherical coordinates. Find their cylindrical coordinates.
    a. $(4, \pi/4, 0)$ c. $(2, 3\pi/4, \pi/3)$
    b. $(2, \pi/3, \pi/2)$ d. $(4, \pi/4, \pi/4)$

In Exercises 13–20, an equation in cylindrical coordinates is given. Write the equation in rectangular coordinates and sketch the graph.

13. $z = 3$ 14. $r = 2$

15. $z = 2r$ 16. $z = r \sin \theta$

17. $r^2 + z^2 = 4$ 18. $z^2 = (r^2 \cos^2 \theta) - 1$

19. $\cos^2 \theta - \sin^2 \theta = a^2/r^2$ 20. $z = r^2 \cos^2 \theta$

In Exercises 21–28, an equation in rectangular coordinates is given. Write the equation in cylindrical coordinates.

21. $x^2 + y^2 = 9$ 22. $x^2 + y^2 + z^2 = 9$

23. $x^2 + y^2 = 9z$ 24. $y^2 + z^2 = 1$

25. $x^2 + z^2 = 4$ 26. $x + y + z = 4$

27. $xy - ax = 4$ 28. $x^2 - y^2 = 4$

In Exercises 29–36, rewrite the equation in the given exercise in spherical coordinates.

29. Exercise 21   30. Exercise 22

31. Exercise 23   32. Exercise 24

33. Exercise 25   34. Exercise 26

35. Exercise 27   36. Exercise 28

Summary Outline of Chapter 17

■ A **vector-valued function f** has the form   (page 814)

$$\mathbf{f}(t) = x(t)\mathbf{i} + y(t)\mathbf{j}, \quad \text{or} \quad \mathbf{f}(t) = x(t)\mathbf{i} + y(t)\mathbf{j} + z(t)\mathbf{k}$$

where $x(t)$, $y(t)$, and $z(t)$ are real-valued functions.

■ For $\mathbf{f}(t) = x(t)\mathbf{i} + y(t)\mathbf{j} + z(t)\mathbf{k}$,   (page 819)

(i) $\displaystyle\lim_{t \to t_0} \mathbf{f}(t) = \left[\lim_{t \to t_0} x(t)\right]\mathbf{i} + \left[\lim_{t \to t_0} y(t)\right]\mathbf{j} + \left[\lim_{t \to t_0} z(t)\right]\mathbf{k}$

(ii) $\mathbf{f}'(t) = x'(t)\mathbf{i} + y'(t)\mathbf{j} + z'(t)\mathbf{k}$

(iii) $\displaystyle\int \mathbf{f}(t)\,dt = \left[\int x(t)\,dt\right]\mathbf{i} + \left[\int y(t)\,dt\right]\mathbf{j} + \left[\int z(t)\,dt\right]\mathbf{k}$

(iv) $\displaystyle\int_a^b \mathbf{f}(t)\,dt = \left[\int_a^b x(t)\,dt\right]\mathbf{i} + \left[\int_a^b y(t)\,dt\right]\mathbf{j} + \left[\int_a^b z(t)\,dt\right]\mathbf{k}.$

(v) $\mathbf{f}$ is continuous at t_0 if and only if $x(t)$, $y(t)$, and $z(t)$ are all continuous at t_0.

■ If $\mathbf{r}'(t_0) \neq \mathbf{0}$, the **tangent** to the curve generated by $\mathbf{r}$ at $\mathbf{r}(t_0)$ is $\mathbf{r}'(t_0) =$ (page 823) $x'(t_0)\mathbf{i} + y'(t_0)\mathbf{j} + z'(t_0)\mathbf{k}$.

■ Derivatives of vector-valued functions satisfy: (page 825)

(i) $(\mathbf{f} + \mathbf{g})'(t) = \mathbf{f}'(t) + \mathbf{g}'(t)$
(ii) $(c\mathbf{f})'(t) = c\mathbf{f}'(t)$
(iii) $(h\mathbf{f})'(t) = h(t)\mathbf{f}'(t) + h'(t)\mathbf{f}(t)$
(iv) $(\mathbf{f} \times \mathbf{g})'(t) = [\mathbf{f}(t) \times \mathbf{g}'(t)] + [\mathbf{f}'(t) \times \mathbf{g}(t)]$
(v) $(\mathbf{f} \circ h)'(t) = h'(t)\mathbf{f}'(h(t))$
(vi) $(\mathbf{f} \cdot \mathbf{g})'(t) = \mathbf{f}(t) \cdot \mathbf{g}'(t) + \mathbf{f}'(t) \cdot \mathbf{g}(t)$.

■ The **unit tangent** to the curve generated by $\mathbf{r}$ at $\mathbf{r}(t_0)$ is $\mathbf{T} = \dfrac{1}{|\mathbf{r}'(t_0)|}\mathbf{r}'(t_0)$,   (page 832) $\mathbf{r}'(t_0) \neq \mathbf{0}$.

■ The curve C is **parameterized by arc length** on the interval I if   (page 837)

$$\int_0^t |\mathbf{r}'(\omega)|\,d\omega = t$$

for all t in I. Equivalently, $|\mathbf{r}'(t)| = 1, \qquad t \in I.$

■ For a particle moving in space with the twice differentiable **position**   (page 841) **function r,**

(i) the **velocity** function is $\mathbf{v} = \mathbf{r}'$
(ii) the **speed** is $|\mathbf{v}(t)| = |\mathbf{r}'(t)|$
(iii) the **acceleration** function is $\mathbf{a} = \mathbf{v}' = \mathbf{r}''$.

■ The **trajectory** of a projectile fired from the origin with initial speed s_0 and   (page 845) angle of elevation θ is

$$\mathbf{r}(t) = s_0(\cos\theta)t\mathbf{i} + (s_0\sin\theta t - \tfrac{1}{2}gt^2)\mathbf{j}.$$

■ If a curve is parameterized by arc length by the vector-valued function $\mathbf{r}$, the   (page 848) **curvature** at $\mathbf{r}(s)$ is $\kappa(s) = |\mathbf{r}''(s)|$.

■ The **radius of curvature** at $\mathbf{r}(s)$ is   (page 849)

$$\rho(s) = \frac{1}{\kappa(s)}, \quad \kappa(s) \neq 0.$$

■ If the plane curve C is parameterized by the equations $x = x(t)$, $y = y(t)$, the   (page 850) **curvature** at $(x(t), y(t))$ is

$$\kappa(t) = \frac{|x'(t)y''(t) - y'(t)x''(t)|}{[(x'(t))^2 + (y'(t))^2]^{3/2}}.$$

■ If the curve C is the graph of $y = f(x)$,   (page 851)

$$\kappa(x) = \frac{|f''(x)|}{[1 + (f'(x))^2]^{3/2}}.$$

■ The **principal unit normal** to $\mathbf{r}(s)$ is $\mathbf{N}(s) = \dfrac{1}{|\mathbf{r}''(s)|}\mathbf{r}''(s)$ if $\mathbf{r}(s)$ is   (page 852) parameterized by arc length.

■ If **r** is a twice differentiable position function, its acceleration function can be   (page 853) written $\mathbf{a}(t) = a_T\mathbf{T} + a_N\mathbf{N}$, where the **tangential component of acceleration** is

$$a_T = \frac{d^2s}{dt^2}$$

and the **normal component of acceleration** is

$$a_N = \left(\frac{ds}{dt}\right)^2 \kappa(s) = |\mathbf{v}|^2\,\kappa.$$

■ If the curve C is determined by the twice differentiable **position function r,**   (page 855) then

$$\kappa = \frac{|\mathbf{v} \times \mathbf{a}|}{|\mathbf{v}|^3}$$

at each point on C.

■ There are six types of **quadric surfaces:**   (page 862)

   (i) ellipsoids,
   (ii) elliptic paraboloids,
   (iii) elliptic cones,
   (iv) hyperboloids of one sheet,
   (v) hyperboloids of two sheets,
   (vi) hyperbolic paraboloids.

■ The **cylindrical coordinates** (r, θ, z) and the rectangular coordinates (x, y, z) (page 870) for a point P are related via the equations

$$\begin{cases} x = r\cos\theta, & r \geq 0, \\ y = r\sin\theta, & r \geq 0, \\ z = z \end{cases} \quad \text{and} \quad \begin{cases} r = \sqrt{x^2 + y^2} \\ \tan\theta = y/x, \\ z = z. \end{cases}$$

■ The **spherical coordinates** (ρ, θ, ϕ) and the rectangular coordinates (x, y, z) (page 872) for a point P are related via the equations

$$\begin{cases} x = \rho\sin\phi\cos\theta, & \rho \geq 0 \\ y = \rho\sin\phi\sin\theta, & \rho \geq 0 \\ z = \rho\cos\phi, & \rho \geq 0 \end{cases} \quad \text{and} \quad \begin{cases} \rho = \sqrt{x^2 + y^2 + z^2} \\ \tan\theta = y/x, & x \neq 0 \\ \tan\phi = \dfrac{\sqrt{x^2 + y^2}}{z}, & z \neq 0. \end{cases}$$

Review Exercises—Chapter 17

1. Find an equation in rectangular coordinates for the curve generated by the vector function $\mathbf{r}(t) = a\cos t\mathbf{i} + b\sin t\mathbf{j}$.

2. Sketch the curve generated by the vector-valued function $\mathbf{r}(t) = \mathbf{i} + \cos t\mathbf{j} + \mathbf{k}$.

3. Find the implicit domain of the vector function

$$\mathbf{r}(t) = t\mathbf{i} + \sqrt{1 - t^2}\mathbf{j} + \frac{1}{t}\mathbf{k}.$$

4. Show that the curve generated by the vector-valued function $\mathbf{r}(t) = \tan t\mathbf{i} + \sec t\mathbf{j}$, $-\pi/2 < t < \pi/2$, is one branch of the hyperbola with rectangular equation $y^2 - x^2 = 1$. What are the asymptotes for this curve?

5. Sketch the **cycloid** given by the vector-valued function $\mathbf{r}(t) = a(t - \sin t)\mathbf{i} + a(1 - \cos t)\mathbf{j}$, $t \geq 0$. Find the numbers t for which the tangent vector is zero. (The cycloid gives the path of a point on the rim of a wheel of radius a as the wheel rolls along a horizontal surface.)

In Exercises 6–8, find the indicated limit.

6. $\displaystyle\lim_{t \to 0^+}\left[\cos 2t\mathbf{i} + t\cos t\mathbf{j} + \frac{t}{\sqrt{t} + 2}\mathbf{k}\right]$

7. $\lim\limits_{t\to 0}\left[\left(\dfrac{\sin 3t}{t}\right)\mathbf{i} + \sqrt{t+2}\,\mathbf{j} + \dfrac{\tan 2t}{t}\mathbf{k}\right]$

8. $\lim\limits_{t\to 2^+}\left[\sqrt{2+t}\,\mathbf{i} + \ln\sqrt{t}\,\mathbf{j} + \dfrac{t^3-8}{t-2}\mathbf{k}\right]$

In Exercises 9–11, determine the numbers t for which the function is discontinuous.

9. $\mathbf{r}(t) = \sin t\mathbf{i} + \sec t\mathbf{j}, \quad 0 \le t \le 2\pi$

10. $\mathbf{r}(t) = t^{-1/2}\mathbf{i} + \tan t\mathbf{j} + \ln(1+t)\mathbf{k}, \qquad 0 \le t \le \pi$

11. $\mathbf{r}(t) = \dfrac{1}{t^2-9}\mathbf{i} + \dfrac{t+2}{t^2-4}\mathbf{j} + t\mathbf{k}, \qquad -4 \le t \le 4$

In Exercises 12–15, find $\mathbf{r}'$ and $\mathbf{r}''$

12. $\mathbf{r}(t) = e^{-2t}\mathbf{i} + \cos(\pi t/4)\mathbf{j} + \mathbf{k}$

13. $\mathbf{r}(t) = te^t\mathbf{i} + te^{-t}\mathbf{j}$

14. $\mathbf{r}(t) = \text{Tan}^{-1} 2t\mathbf{i} + \sqrt{t}\,\mathbf{j} + t\mathbf{k}$

15. $\mathbf{r}(t) = \ln 3t\mathbf{i} + e^{\sqrt{t}}\mathbf{j}$

16. Find $\displaystyle\int_0^1 \mathbf{r}(t)\,dt$ for $\mathbf{r}$ in Exercise 12.

17. Find $\displaystyle\int_0^1 \mathbf{r}(t)\,dt$ for $\mathbf{r}$ in Exercise 13.

18. Find a vector function $\mathbf{r}$ for which $\mathbf{r}'(t) = 4\mathbf{r}(t)$ and $\mathbf{r}(0) = 4\mathbf{i} - \mathbf{j} + \pi\mathbf{k}$.

19. Find a vector function $\mathbf{r}$ for which $\mathbf{r}''(t) = 9\mathbf{r}(t)$ and $\mathbf{r}(0) = 4\mathbf{i} + 3\mathbf{j}$.

20. Find the numbers α for which the curve generated by the vector-valued function

$$\mathbf{r}(t) = \alpha t\mathbf{i} + \cos(2\alpha t)\mathbf{j} + \sin(2\alpha t)\mathbf{k}, \qquad 0 \le t \le \pi$$

is parameterized by arc length.

21. Find an arc length parameterization for the circle $x^2 + y^2 = 25$.

22. Find the unit tangent $\mathbf{T}(t)$ for the ellipse $\mathbf{r}(t) = a\cos t\mathbf{i} + b\sin t\mathbf{j}, 0 \le t < 2\pi$.

23. Describe the graph of the equation $x^2 + y^2 + z^2 - 6x + 4y - 2z + 10 = 0$.

24. Sketch the graph of the equation $x^2 + z^2 = 1 + y$.

25. Sketch the graph of the equation $y^2 = 9 + z^2$.

26. Find an equation for the sphere with center on the z-axis and containing the points $(5, 0, 0)$ and $(0, 0, 4)$.

27. Find rectangular coordinates for the point with spherical coordinates $(2, \pi/4, \pi/3)$.

28. Find cylindrical coordinates for the point with rectangular coordinates $(2, 2, 5)$.

In Exercises 29–32, find a rectangular equation for the given equation and sketch the graph.

29. $\phi = \pi/4$

30. $r = \cos\theta$

31. $\rho\sin\phi = 2\cos\theta$

32. $\rho = 2\sec\phi$

33. For the curve generated by the vector-valued function $\mathbf{r}(t) = t\mathbf{i} + \cos t\mathbf{j} + \sin t\mathbf{k}$, find
   a. the curvature $\kappa(t)$,
   b. the unit tangent $\mathbf{T}(t)$.

34. A particle moves in space with position function $\mathbf{r}(t) = a\cos t\mathbf{i} + tb\mathbf{j} + a\sin t\mathbf{k}$. Show that
   a. the velocity vector $\mathbf{v}(t)$ and the acceleration vector $\mathbf{a}(t)$ have constant lengths,
   b. $\mathbf{v}(t)$ and $\mathbf{a}(t)$ are orthogonal for each t.

35. Find the length of the helix $\mathbf{r}(t) = 3t\mathbf{i} + \cos 2t\mathbf{j} + \sin 2t\mathbf{k}$ between the points $\mathbf{r}(0)$ and $\mathbf{r}(3)$.

36. Find the length of the curve determined by the position function $\mathbf{r}(t) = \ln t\mathbf{i} + t\mathbf{j}$ between $\mathbf{r}(1)$ and $\mathbf{r}(2)$.

37. Find the curvature of the cubic curve $y = x^3$ at the point $(2, 8)$.

38. Find the curvature of the curve determined by the vector-valued function $\mathbf{r}(t) = 3\mathbf{i} + \sqrt{t}\,\mathbf{j} + t^2\mathbf{k}$ at the point $\mathbf{r}(4)$.

39. Let C be a curve in space determined by the differentiable vector function $\mathbf{r}, a \le t \le b$. Show that if $t_0 \neq a, t_0 \neq b$, and $\mathbf{r}(t_0)$ is a point on C either nearest or farthest from the origin, then $\mathbf{r}(t_0)$ and $\mathbf{r}'(t_0)$ are orthogonal. (*Hint:* Consider $|\mathbf{r}(t)|^2 = \mathbf{r}(t) \cdot \mathbf{r}(t)$.)

In Exercises 40–46, describe the graph.

40. $9x - 4y^2 + 36z^2 = -36$

41. $6x^2 + y^2 - 2z^2 = 6$

42. $36(x-1)^2 + 18(y+3)^2 + 8(z+1)^2 = 72$

43. $-x^2 + y^2 = 9z$

44. $y^2 = 4x^2 + 2z$

45. $9x^2 + 9y^2 - 4z^2 + 18x - 16z - 43 = 0$

46. $(x+1)(y-3) = 1$

47. Use Newton's law $\mathbf{F} = m\mathbf{a}$ to show that if a particle moves in space subject to zero net force, then
   a. the motion is along a line, and
   b. both the tangential and normal components of acceleration are zero.

48. Prove that for a particle moving in space, the speed of the particle is constant if and only if its velocity and acceleration vectors are orthogonal at all points along its path.

49. Find the radius of curvature and an equation for the circle of curvature at the point $(0, 1)$ on the graph of $y = \cos x$.

50. Find the point on the graph of $y = e^x$ at which the radius of curvature is a minimum.

51. A particle moves along a curve with position function $\mathbf{r}(t) = t \cos t \mathbf{i} + t \sin t \mathbf{j}$. Find its speed as a function of t.

52. Let the curve C be parameterized by $\mathbf{r}(t) = t^3 \mathbf{i} + 3t^2 \mathbf{j} + 6t \mathbf{k}$. Find the arc length of C from $(0, 0, 0)$ to $(8, 12, 12)$.

53. A projectile is fired with an angle of elevation $\alpha = 30°$ and an initial speed $|\mathbf{v}(0)| = 20$ m/s at a vertical wall 10 m away. At what height does it strike the wall?

54. The curve parameterized by $\mathbf{r}(t) = t\mathbf{i} + t^2\mathbf{j} + t^3\mathbf{k}$ intersects the plane $z = 1$ exactly once. What is the cosine of the angle θ between the normal to the plane and the tangent to the curve at the point of intersection?

55. In this exercise, we develop a description of the Möbius strip (Example 5 in Section 17.6) using cylindrical coordinates (r, θ, z).

  a. For a fixed angle θ between 0 and 2π, consider the circle C, with center at $(r, \theta, z) = (3, \theta, 0)$ and radius 1, in the half plane determined by θ. Let P_θ be the point $(r, \theta, z) =$

$(3 + \cos(\theta/2), \theta, \sin(\theta/2))$ on C, and let Q_θ be the point on C that is diametrically opposite to P_θ. Determine the cylindrical coordinates of Q_θ.

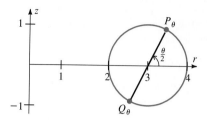

  b. Show that the diameter determined by P_θ and Q_θ has cylindrical coordinates $(r(t), \theta, z(t))$, where $r(t) = 3 + t \cos(\theta/2)$ and $z(t) = t \sin(\theta/2)$ for $-1 \le t \le 1$.

  c. Now let θ vary from 0 to 2π. Note that the radial segment returns to its original position, but with a half twist. Determine equations for x, y, and z in terms of t and θ that describe the Möbius strip in rectangular coordinates.

Calculus in Higher Dimensions and Differential Equations

Calculus in Higher Dimensions and Differential Equations

A host of mathematicians contributed to the several topics included in this unit: partial differentiation, multiple integration, and vector analysis. Several of these people have been discussed in other units.

Newton differentiated functions of two variables by means of formulas that we now obtain by partial differentiation. This work is recorded in his personal papers, but was not published. Leibniz also differentiated functions of two variables, but made little use of them. Jakob Bernoulli and his nephew Nicolaus used what were essentially partial derivatives around 1720, although they seemingly did not recognize that there is a difference between differentiation of a function of one variable and that for a function of two or more variables. The same symbol was used by many early writers for regular and partial derivatives, which of course led to much confusion. The "rounded d" symbol, ∂, was first used by Euler in 1776, but not in the way that we use it today. The first to use $\partial y/\partial x$ was the Frenchman Adrien-Marie Legendre in 1786, but it was more than a century before that notation came into general use.

Multiple integration was first used by Newton, but his arguments were geometrical and somewhat unclear. In the first half of the eighteenth century Euler used repeated integrations in order to integrate over a bounded domain. Joseph Louis Lagrange used a triple integral in a work on gravitation involving ellipsoids around 1775. By the nineteenth century the use of multiple integrals had become fairly common.

The concept of a vector was known in antiquity. Aristotle represented forces by vectors, and knew that two forces acting in different directions could be summed by what we call the parallelogram law. Representation of complex numbers in the plane was done by Wessel in Denmark, Argand in France, and Gauss in Germany between 1798 and 1806. It was fairly well known by 1830 that vectors could be nicely expressed as complex numbers. However, many physical situations involve more than two factors which are not all in the same plane. For many years mathematicians searched for a three-dimensional vector system which preserved all of the properties of real-number algebra, but their efforts were unsuccessful.

The problem was resolved by Ireland's greatest mathematician, William Rowan Hamilton (1805–1865). Hamilton was a largely self-taught youthful prodigy whose ability first manifested itself, as with several other mathematicians, in language. He read

Greek, Latin, and Hebrew by the age of 5, and by the age of 14 he knew 14 languages including Sanskrit and Arabic. He became interested in mathematics through meeting an American calculating prodigy, and had soon read all four volumes of Newton's *Principia* and Laplace's *Celestial Mechanics*, in which he found a mathematical error. At the age of 22, still an undergraduate, he was appointed Professor of Astronomy at an Irish university and Royal Astronomer of Ireland.

While walking to Dublin, Hamilton suddenly realized the impossibility of a three-dimensional system with the desired properties, but that a *four*-dimensional system is possible. He stopped to carve the basic relations on the handrail of a bridge, and that carving can still be seen. He postulated a system with four mutually independent unit vectors, called *quaternions*. Quaternions do not obey the commutative law (that is, $A \cdot B \neq B \cdot A$); this was the first algebra in which such behavior was studied. The use of quaternions was advocated throughout the nineteenth century by Hamilton and others, and courses in the subject were taught in British and American graduate schools well into the twentieth century. Eventually, however, it was superseded by the vector analysis pioneered by J. W. Gibbs. Hamilton did not lead a happy life; his marriage was unpleasant and his quaternions were not

Josiah Willard Gibbs

well-received on the Continent. He was knighted early, in 1835, but became an alcoholic in the last twenty years of his life.

Josiah Willard Gibbs (1839–1903) was an American mathematician and scientist in a day when American science was held in little esteem. Born in New Haven, Connecticut, he attended Yale College and obtained a Ph.D. in physics in 1863, one of the first doctorates to be granted in the United States. He then went to Europe for further study. Returning in 1871, he became Professor of Mathematical Physics at Yale, though he received no salary for the first nine years. He remained at Yale for the rest of his life, living in the house in which he had grown up, less than a block from the campus. In 1881 he had printed, at his own expense, a little pamphlet called *Elements of Vector Analysis* for the use of his students. In it, he created the subject much as it is known today. The pamphlet became well-known, and E. B. Wilson published a book called *Vector Analysis*, based on Gibbs' lectures, in 1901.

Oliver Heaviside (1850–1925) was another contributor to the development of vector analysis, and in fact he essentially created the subject independently of Gibbs. He had only an elementary school education in his native England, but embarked on a regimen of reading which resulted in his becoming a pragmatically successful mathematical physicist. He began his career as a telegrapher, but abandoned work as a result of the deafness which bothered him all of his life. His creative work was disdained by many professional mathematicians particularly because of his lack of formal education and unorthodox methods. His work in electromagnetic theory was particularly brilliant when he generalized Maxwell's theory. Heaviside used ideas from Hamilton's theory of quaternions, and developed both scalar and vector multiplication. He never married, but lived with his parents until they died. He spent his last years in poverty and died in a nursing home.

Green's Theorem and Stokes' Theorem are two applications of vectors which are found in this unit. George Green (1793–1841), like Heaviside, had only an elementary school English education, and was largely self-taught. He was one of the first to treat static electricity and magnetism in mathematical terms. At the age of 36 Green decided to get a university education. He studied for four years, was admitted to Cambridge at the age of 40, and graduated four years later. His marks were not high, largely because he spent more time writing original mathematics than in studying for classes. In 1828 he had printed at his own expense his most important work, in which his mathematical treatment of electricity and magnetism appears. Here too is found what we now call Green's Theorem. It remained unknown until four years after his death when the British physicist Lord Kelvin became aware of the paper and had it reprinted in a German journal. By this time the Russian mathematician Michel Ostrogradski had also discovered the theorem; it is known by his name in the Soviet Union.

George Gabriel Stokes

George Gabriel Stokes (1819–1903) was an Irishman who was Lucasian Professor of Mathematics at Cambridge, the same post held by Newton a century and a half earlier. Though prestigious, the position did not pay well, and Stokes had to teach at a School of Mines in order to make ends meet. Also like Newton, he served the Royal Society of London as an officer for many years—he was either president or secretary for 45 years. He studied various physical phenomena from a mathematical point of view, and through a brilliant paper established the foundations of hydrodynamics. The theorem named after him is a three-dimensional vector version of Green's Theorem.

Chapter 18

Differentiation for Functions of Several Variables

Multivariable functions abound in nature. For example, the ideal gas law in chemistry relates the pressure, the volume, and the temperature of an ideal gas by the equation

$$PV = nRT$$

where n is the number of moles of the gas present and R is a constant. Consequently, we can consider the pressure P as a function f of the three independent quantities n, T, and V. In other words,

$$P = (nR)\left(\frac{T}{V}\right) = f(n, T, V).$$

Similarly, in economics, the simple equation

$$R = Px$$

gives the revenue obtained from selling x items at a price of P dollars per item. That is, the revenue R is a function $f(x, P)$ of the two independent variables x and P; we must know both the selling price and the number of sales in order to calculate revenue.

In this chapter, we begin our discussion of the calculus of multivariable functions by extending the theory of differentiation to real-valued functions with more than one independent variable.

18.1 Functions of Several Variables

A **function of two variables** is a rule that assigns a *unique* number to each *ordered pair* (x, y) of numbers for which the rule is defined. Generally, we use the notation

$$z = f(x, y)$$

to indicate that the value z is determined by the pair (x, y). If the function f is defined for all ordered pairs (x, y), we may represent this situation symbolically as

$$f : \mathbb{R}^2 \to \mathbb{R} \quad \text{or} \quad (x, y) \xrightarrow{f} z$$

where $\mathbb{R}^2$ represents the set of all pairs (x, y) with both $x \in \mathbb{R}$ and $y \in \mathbb{R}$. (You may think of $\mathbb{R}^2$ as simply the coordinate plane.) If the function f is defined only for (x, y) in a subset D of $\mathbb{R}^2$ (called the **domain** of the function), we write

$$f : D \to \mathbb{R}, \qquad D \subset \mathbb{R}^2.$$

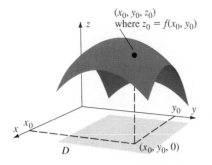

Figure 1.1 The graph of a function $z = f(x, y)$ of two variables with domain D.

As for functions of a single variable, the set of values $f(x, y)$, for $(x, y) \in D$, is called the **range** of the function f.

Figure 1.1 shows how we graph functions of two variables. We interpret the ordered pairs (x, y) as points in the xy-plane. We indicate the value of the function, $z = f(x, y)$, by plotting the point (x, y, z) in space. Then, the height of the point (x, y, z) above or below the point $(x, y, 0)$ represents the number (or **value**) z assigned by the function to the ordered pair (x, y). The set of all such points is called the **graph of the function** $z = f(x, y)$. Notice that while graphs of functions of a single variable are *curves in the plane* (in general), graphs of functions of two variables are *surfaces in space*.

An example of a function of two variables is given by the equation

$$z = x^2 + y^2.$$

The graph of this function is the paraboloid in Figure 1.2. However, the equation

$$z^2 = x^2 + y^2 \tag{1}$$

does *not* describe a function. The reason is that each pair of independent variables $(x, y) \neq (0, 0)$ corresponds to *two* values of z. This can be seen by writing equation (1) in the form

$$z = \pm\sqrt{x^2 + y^2}.$$

The graph of equation (1) is the cone in Figure 1.3.

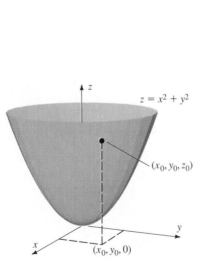

Figure 1.2 The circular paraboloid $z = x^2 + y^2$ is the graph of the function $f(x, y) = x^2 + y^2$.

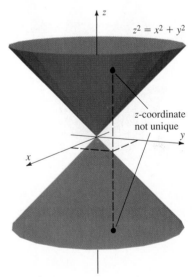

Figure 1.3 The cone $z^2 = x^2 + y^2$ is not the graph of a function $f(x, y)$ because two z-values correspond to each ordered pair (x_0, y_0) except $(0, 0)$.

Functions of three or more independent variables are defined in the analogous way. A function w of the three independent variables x, y, and z takes the form

$$w = f(x, y, z),$$

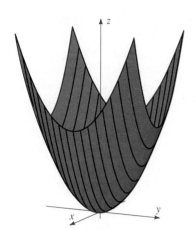

Figure 1.4   The traces of $z = x^2 + y^2$ in the planes $y = c$. Note that all of these traces are parabolas.

Graphing Techniques: Traces and Level Curves

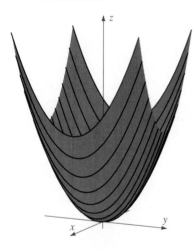

Figure 1.5   The traces of $z = x^2 + y^2$ in the planes $x = c$. All of these traces are parabolas.

and we write

$$f:\mathbb{R}^3 \to \mathbb{R} \quad \text{or} \quad (x, y, z) \xrightarrow{\,f\,} w$$

An example of a function of three variables is the temperature function for the room in which you are sitting. If you use a rectangular coordinate system to describe the space within the room, the number $w_0 = T(x_0, y_0, z_0)$ is the temperature at the point in the room with coordinates (x_0, y_0, z_0). Unfortunately, we cannot sketch a graph for a function of three variables, since all three axes are required just to plot the points in its domain.

More generally, a function of the n independent variables $x_1, x_2, \ldots, x_n$ has the form

$$w = f(x_1, x_2, \ldots, x_n).$$

Symbolically, we describe such functions by writing

$$f:\mathbb{R}^n \to \mathbb{R} \quad \text{or} \quad (x_1, x_2, \ldots, x_n) \xrightarrow{\,f\,} w.$$

Graphing a function of two variables is difficult, at best. Two techniques frequently give a general idea of the appearance of the graph of the function $z = f(x, y)$. The first is simply setting one of the two independent variables equal to a constant, so that we obtain a function of a single variable, whose graph can be sketched in the appropriate plane in space. Setting $x = c$ in $z = f(x, y)$ gives the equation $z = f(c, y) = g(y)$, whose graph is the intersection of the desired graph with the plane $x = c$; setting $y = c$ in $z = f(x, y)$ gives the equation $z = f(x, c) = h(x)$, whose graph is the intersection of the desired graph with the plane $y = c$.

We refer to the intersections of the graph of f with the planes $x = c$ or $y = c$ as the **traces** of f in the respective planes. By sketching traces of f in several planes, we can sometimes gain a fairly accurate picture of the graph of f. (Note that this technique is the same as the technique discussed in Section 17.6.)

☐ **EXAMPLE 1**

Sketch several traces for the graph of $z = x^2 + y^2$.

Solution   Tables 1.1 and 1.2 show the results of setting one of the independent variables equal to one of several constants.

Traces in each of these planes are indicated in Figures 1.4 and 1.5. They are combined in Figure 1.6. Figure 1.2 is a sketch of the corresponding paraboloid. ■

Table 1.1

Plane $y = c$	Function $z = f(x, c)$
0	$z = x^2$
1	$z = x^2 + 1$
2	$z = x^2 + 4$
−1	$z = x^2 + 1$
−2	$z = x^2 + 4$

Table 1.2

Plane $x = c$	Function $z = f(c, y)$
0	$z = y^2$
1	$z = y^2 + 1$
2	$z = y^2 + 4$
−1	$z = y^2 + 1$
−2	$z = y^2 + 4$

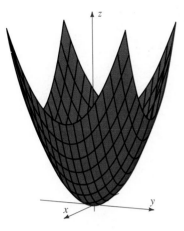

Figure 1.6 The graph of $z = x^2 + y^2$ with traces in planes that are parallel to the xz-plane and with traces in planes that are parallel to the yz-plane.

☐ **EXAMPLE 2**

For the function $f(x, y) = y^2 - x^2$, the equations of various traces are given in Tables 1.3 and 1.4. They are all parabolas.

Table 1.3

Plane $y = c$	Equation $z = f(x, c)$
0	$z = -x^2$
1	$z = 1 - x^2$
2	$z = 4 - x^2$
-1	$z = 1 - x^2$
-2	$z = 4 - x^2$

Table 1.4

Plane $x = c$	Equation $z = f(c, y)$
0	$z = y^2$
1	$z = y^2 - 1$
2	$z = y^2 - 4$
-1	$z = y^2 - 1$
-2	$z = y^2 - 4$

Figure 1.7 shows both types of traces. Figure 1.8 shows the graph of $f(x, y) = y^2 - x^2$. ∎

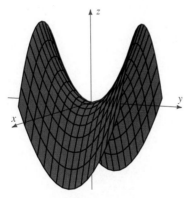

Figure 1.7 The graph of $f(x, y) = y^2 - x^2$ along with traces in planes of the form $x = c$ and $y = c$. Traces in planes of the form $x = c$ are parabolas that open upward, and traces in planes of the form $y = c$ are parabolas that open downward.

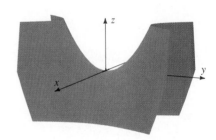

Figure 1.8 The graph of $f(x, y) = y^2 - x^2$. No traces are shown.

Level Curves and Level Surfaces

The surface $z = f(x, y)$ is one way to represent a function of two variables. Another approach involves the drawing of a purely two-dimensional sketch that conveys certain information about the graph. In this case, we plot the pairs (x, y) that correspond to certain prescribed values of the function; that is, we plot the points that satisfy the equation $c = f(x, y)$ for various choices of c. Each choice of c yields the **level curve** of level c.

The use of level curves to represent functions of two variables is a common technique. For example, topographical maps use curves of constant altitude to indicate the "lay of the land" (see Figures 1.9 and 1.10). Similarly, weather maps often use bands of color to represent temperature ranges. In this case, the curves that separate the different bands are level curves of the temperature function (see Figure 1.11).

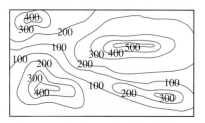

Figure 1.9 A picture of a mountainous region.

Figure 1.10 A topographical map of the region.

Average sea level temperature °F (July)

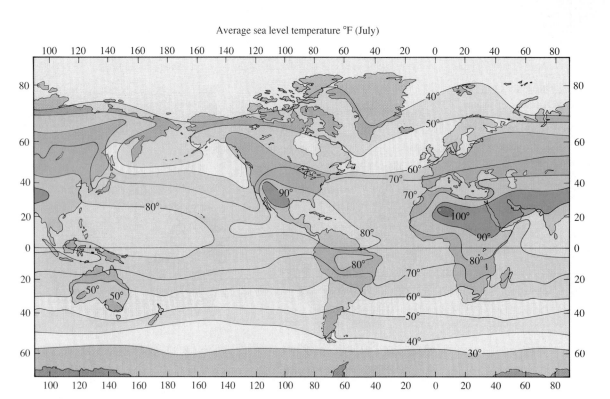

Figure 1.11 Average sea level temperatures in July (°F). Longitude

To use level curves to represent $z = f(x, y)$, we set $z = c$ and graph the resulting equation $f(x, y) = c$ in the xy-plane. Doing so for several values of c conveys the same type of information about the graph of f as does a topographical map. Note that although setting $z = c$ produces a trace of the surface $z = f(x, y)$, we make no attempt to produce a three-dimensional sketch by this method. Also, you should be sure to label the value of z corresponding to each level curve sketched.

☐ **EXAMPLE 3**

Consider the function $z = x^2 + y^2$ in Example 1. Setting $z = c$ gives the equation $x^2 + y^2 = c$, valid for $c \geq 0$. The level curves are concentric circles, as sketched in Figure 1.12.

■

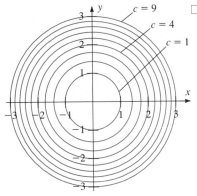

Figure 1.12 The level curves of level $c > 0$ of the function $z = x^2 + y^2$ are circles centered at $(0, 0)$. In this figure, we have sketched the level curves $x^2 + y^2 = c$ for $c = 1, 2, 3, \ldots, 9$.

□ **EXAMPLE 4**

For the function $f(x, y) = y^2 - x^2$ in Example 2, the level curves have equation $y^2 - x^2 = c$ or $y = \pm\sqrt{x^2 + c}$. Graphs of several such curves appear in Figure 1.13. ■

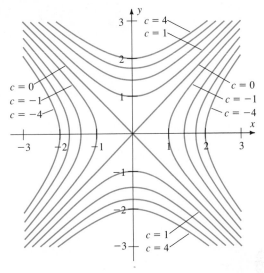

Figure 1.13 The level curves $y^2 - x^2 = c$ for $c = -4, -3, -2, -1, 0, 1, 2, 3, 4$.

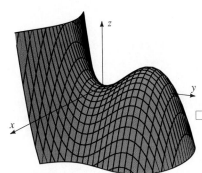

Figure 1.14 The graph of

$$z = \frac{y^2}{2} - \frac{y^3}{12} - \frac{x^2}{4}.$$

Traces in planes of the form $y = c$ are parabolas that open downward. Traces in planes of the form $z = c$ are graphs of cubics with one relative minimum and one relative maximum.

□ **EXAMPLE 5**

Figure 1.14 illustrates the x and y traces of the function

$$z = \frac{y^2}{2} - \frac{y^3}{12} - \frac{x^2}{4}.$$

If we set $x = c$, we obtain $z = y^2/2 - y^3/12 - c^2/4$ whose graph (in the $x = c$ plane) is a cubic with one relative maximum and one relative minimum. On the other hand, if we set $y = c$, we get $z = c^2/2 - c^3/12 - x^2/4$ whose graph (in the $y = c$ plane) is a parabola that opens downward. Figure 1.15 illustrates the correspondence between the level curves for this function and the z traces of its graph. ■

Although we cannot sketch graphs of functions of three variables, we can sketch their **level surfaces**. If $w = f(x, y, z)$, we set $w = c$ and graph the resulting surface. For example, the level surfaces for the function $w = x^2 + y^2 + z^2$ are the spheres $x^2 + y^2 + z^2 = c$, $c \geq 0$. In the following example, we determine the level surfaces of a function using the methods we discussed in Section 17.6.

□ **EXAMPLE 6**

Consider the function

$$w = x^2 + y^2 - z^2.$$

The three level surfaces $w = 1$, $w = 0$, and $w = -1$ are three different quadric surfaces. Using the methods of Section 17.6, we have

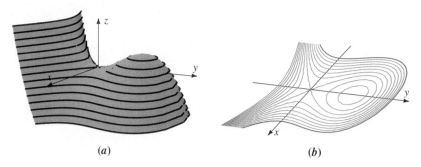

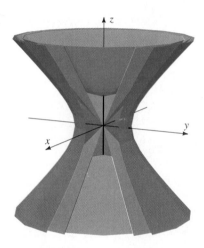

Figure 1.15 The relationship between the graph of

$$z = f(x, y) = \frac{y^2}{2} - \frac{y^3}{12} - \frac{x^2}{4}$$

and the level curves of f.

$$
\begin{aligned}
w = 1 &= x^2 + y^2 - z^2 && \text{(hyperboloid of one sheet)}\\
w = 0 &= x^2 + y^2 - z^2 && \text{(cone)}\\
w = -1 &= x^2 + y^2 - z^2 && \text{(hyperboloid of two sheets).}
\end{aligned}
$$

(see Figure 1.16). ∎

REMARK   Later in this chapter we shall exploit a general observation regarding graphs of functions of two variables. Any graph of the form $z = f(x, y)$ also can be regarded as the level surface (of level 0) for the function $g(x, y, z)$ defined by

$$g(x, y, z) = z - f(x, y).$$

Figure 1.16 Three level surfaces of the function $w = x^2 + y^2 - z^2$. The hyperboloid of one sheet is level 1, the cone is level 0, and the hyperboloid of two sheets is level -1. The outer two surfaces are cut away to show the third.

Consequently, any technique that we develop for level sets applies to graphs of functions.

Finally, we note that one simple physical example of a level surface (of level T_0) is the set of all points in your classroom at which the temperature $T(x, y, z)$ equals T_0. Such surfaces of constant temperature are called **isothermal surfaces.** Another example occurs in the theory of electricity and magnetism, where surfaces on which an electric potential is constant are called **equipotential surfaces.**

Limits

Just as limits play a fundamental role in the differentiation of functions of one variable, the limit concept is again fundamental in the multivariable case. We conclude this section with a discussion of limits and continuity for functions of more than one variable, and to do so, we introduce the concept of a neighborhood of a point.

A **neighborhood** of a point $P_0 = (x_0, y_0)$ in the plane is an open **disc** N with center (x_0, y_0). That is,

$$N = \{(x, y) \mid \sqrt{(x - x_0)^2 + (y - y_0)^2} < r\} \tag{2}$$

where r is the radius of the disc N. (By analogy with open intervals on the real number line, the term "open" means that the boundary points are not included in the set.) Similarly, a neighborhood of a point $Q = (x_0, y_0, z_0)$ in space is an open **ball** N with center (x_0, y_0, z_0).

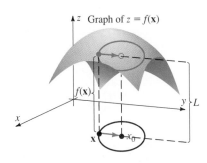

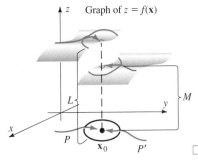

Figure 1.17 If $f(x) \to L$ as $\mathbf{x} \to \mathbf{x}_0$ along every possible path that approaches x_0, then

$$\lim_{x \to x_0} f(x) = L.$$

Figure 1.18 $f(\mathbf{x}) \to L$ as $\mathbf{x} \to \mathbf{x}_0$ along path P, but $f(\mathbf{x}) \to M$ as $\mathbf{x} \to \mathbf{x}_0$ along path P', so $\lim_{\mathbf{x} \to \mathbf{x}_0} f(\mathbf{x})$ does not exist.

$$N = \{(x, y, z) \mid \sqrt{(x - x_0)^2 + (y - y_0)^2 + (z - z_0)^2} < r\} \tag{3}$$

where r is the radius of the ball. Using vector notation, we may generalize both (2) and (3) by saying that a neighborhood of the vector $\mathbf{x}_0$ is a set of vectors

$$N = \{\mathbf{x} \mid |\mathbf{x} - \mathbf{x}_0| < r\}. \tag{4}$$

In equation (4), r is called the **radius** of the neighborhood.

Finally, we define the term **deleted neighborhood** to mean all points in a neighborhood of $\mathbf{x}_0$ except the vector $\mathbf{x}_0$ itself. We shall use this terminology in defining the limit of $f(x)$ as $x \to x_0$ where we shall require $f(\mathbf{x})$ to be defined for all $\mathbf{x}$ in a neighborhood of $\mathbf{x}_0$ except possibly at $\mathbf{x}_0$ itself.

Just as for functions of a single variable, the statement

$$\lim_{\mathbf{x} \to \mathbf{x}_0} f(\mathbf{x}) = L \tag{5}$$

means that the values $f(\mathbf{x})$ of the function f "approach" the number L as the vector (point) $\mathbf{x}$ approaches the fixed vector $\mathbf{x}_0$. However, in formulating this definition we must realize that $\mathbf{x}$ may approach $\mathbf{x}_0$ along many different paths. This observation points out a significant difference between functions of one variable and functions of several variables. In the one-variable case, a variable can approach a number only from one of two directions along the number line. In writing statement (5) we shall mean that $f(\mathbf{x}) \to L$ as $\mathbf{x} \to \mathbf{x}_0$ *regardless of the path* along which $\mathbf{x}$ approaches $\mathbf{x}_0$. Figure 1.17 illustrates this concept. Figure 1.18 suggests a situation in which the limit in (5) fails to exist.

Before formulating a precise definition for the limit of a function of several variables, we discuss several examples.

□ **EXAMPLE 7**

Consider the function $f(x, y) = 4 - x - y$. We claim that

$$\lim_{(x, y) \to (1, 1)} f(x, y) = 2. \tag{6}$$

Notice first that we have not used the vector notation of statement (5) in statement (6), since the number and names of the independent variables are known explicitly. From the form of $f(x, y)$ it is natural to conclude that $f(x, y) \to (4 - 1 - 1) = 2$ as $x \to 1$ and $y \to 1$. Furthermore, we can see that this result is independent of the path along which (x, y) approaches $(1, 1)$ by setting $z = f(x, y)$ and recalling that the graph of the equation $z = 4 - x - y$ is a plane (see Figure 1.19). Regardless of how (x, y) approaches $(1, 1)$, the corresponding point (x, y, z) on the plane approaches $(1, 1, 2)$. ■

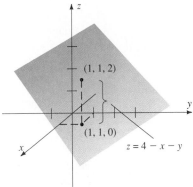

Figure 1.19 $\lim_{(x, y) \to (1, 1)} (4 - x - y) = 2.$

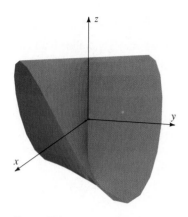

Figure 1.20 The graph of

$$f(x, y) = \frac{-xy}{x^2 + y^2}.$$

Note that the limit of $f(x, y)$ as $(x, y) \to (0, 0)$ does not exist. If $(x, y) \to (0, 0)$ along either the x- or y-axes, then $f(x, y) \to 0$. If $(x, y) \to (0, 0)$ along the line $y = x$, then $f(x, y) \to -1/2$.

☐ **EXAMPLE 8**

Consider the function $f(x, y) = \dfrac{-xy}{x^2 + y^2}$. We claim that

$$\lim_{(x, y) \to (0, 0)} \frac{-xy}{x^2 + y^2} \tag{7}$$

does not exist. On first glance this conclusion may not be obvious. For example, if we set $y = 0$ and allow (x, y) to approach $(0, 0)$ along the x-axis, we obtain

$$\lim_{(x, 0) \to (0, 0)} \frac{-xy}{x^2 + y^2} = \lim_{x \to 0} \frac{-x \cdot 0}{x^2 + 0^2} = 0 \qquad \text{(see Figure 1.20).} \tag{8}$$

Similarly, allowing (x, y) to approach $(0, 0)$ along the y-axis $(x = 0)$ gives

$$\lim_{(0, y) \to (0, 0)} \frac{-xy}{x^2 + y^2} = \lim_{y \to 0} \frac{-0 \cdot y}{0^2 + y^2} = 0. \tag{9}$$

However, if we allow (x, y) to approach $(0, 0)$ along the line $y = x$, we find that

$$\lim_{(x, x) \to (0, 0)} \frac{-xy}{x^2 + y^2} = \lim_{x \to 0} \frac{-x^2}{x^2 + x^2} = -\frac{1}{2}. \tag{10}$$

Since the result in (10) does not agree with that in (8) or in (9), we conclude that the limit in (7) cannot exist. (See Exercise 66 for a different look at this function.) ∎

☐ **EXAMPLE 9**

Find $\displaystyle \lim_{(x, y, z) \to (2, \pi/3, 3)} x \cos yz$.

Solution Since $f(x, y, z) = x \cos yz$ is a function of three variables, we cannot rely on a graph to determine this limit. We must, instead, rely on our intuition about both the cosine function and the operation of multiplication. As $y \to \pi/3$ and $z \to 3$, $yz \to (\pi/3)(3) = \pi$, so $\cos yz \to \cos \pi = -1$. Thus, as $x \to 2$, $x \cos yz \to 2(-1) = -2$. That is,

$$\lim_{(x, y, z) \to (2, \pi/3, 3)} x \cos yz = -2. \qquad \blacksquare$$

We now give a formal definition for the limit in line (5).

Definition 1

Let $\mathbf{x}$ be the position vector for the point $(x_1, x_2, \ldots, x_n)$ in $\mathbb{R}^n$. Let f be a function of n variables defined for all $\mathbf{x}$ in a deleted neighborhood of $\mathbf{x}_0$. Let L be a real number. We say that L is the **limit of the function f** as $\mathbf{x}$ approaches $\mathbf{x}_0$, written

$$\lim_{\mathbf{x} \to \mathbf{x}_0} f(\mathbf{x}) = L,$$

if and only if, for every $\epsilon > 0$, there exists a number $\delta > 0$ so that

$$\text{if } 0 < |\mathbf{x} - \mathbf{x}_0| < \delta, \quad \text{then} \quad |f(\mathbf{x}) - L| < \epsilon. \tag{11}$$

Note in statement (11) that the second inequality involves the *absolute value* of the *number $f(\mathbf{x}) - L$*, while the first concerns the *length* of the *vector $\mathbf{x} - \mathbf{x}_0$*. When applying Definition 1 in a particular situation, you should rewrite the first inequality in (11) in the appropriate component form.

☐ **EXAMPLE 10**

Use Definition 1 to prove that

$$\lim_{(x, y)\to(0, 0)} \sqrt{9 - x^2 - y^2} = 3.$$

Solution   Here $L = 3$, and $\mathbf{x}_0 = \mathbf{0}$ is the position vector associated with the origin. We begin by rewriting the first inequality in (11) as

$$0 < \sqrt{(x - 0)^2 + (y - 0)^2} = \sqrt{x^2 + y^2} < \delta. \tag{12}$$

Also, the second inequality in (11) becomes

$$\left|\sqrt{9 - x^2 - y^2} - 3\right| < \epsilon, \tag{13}$$

or

$$3 - \sqrt{9 - x^2 - y^2} < \epsilon$$

since $\sqrt{9 - x^2 - y^2} \le 3$. Solving this inequality for the radical term, we obtain

$$\sqrt{9 - x^2 - y^2} > 3 - \epsilon,$$

which holds if and only if the following chain of equivalent inequalities holds:

$$9 - (x^2 + y^2) > (3 - \epsilon)^2 \quad \text{(assuming } \epsilon < 3\text{)},$$
$$x^2 + y^2 < 6\epsilon - \epsilon^2,$$
$$\sqrt{x^2 + y^2} < \sqrt{6\epsilon - \epsilon^2}. \tag{14}$$

Now let's assume that $\epsilon > 0$ is a small ($\epsilon < 3$) positive given number. If we define the number δ to be $\delta = \sqrt{6\epsilon - \epsilon^2}$, then (12) holds whenever inequality (14) holds. Since inequality (14) is equivalent to inequality (13), this shows that

$$\text{if } 0 < \sqrt{x^2 + y^2} < \delta, \quad \text{then} \quad \left|\sqrt{9 - x^2 - y^2} - 3\right| < \epsilon.$$

This proves the stated limit (see Figure 1.21).   ■

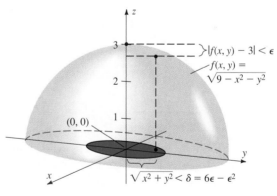

Figure 1.21 $\lim_{(x, y)\to(0, 0)} \sqrt{9 - x^2 - y^2} = 3.$

The algebra of limits for functions of several variables is analogous to that for functions of a single variable. The following theorem may be proved using the same ideas used to prove Theorem 1, Chapter 2 (see Exercises 68 and 69).

Theorem 1

Let f and g be functions of two or three variables defined in a deleted neighborhood of $\mathbf{x}_0$. Suppose that both $\lim_{\mathbf{x}\to\mathbf{x}_0} f(\mathbf{x})$ and $\lim_{\mathbf{x}\to\mathbf{x}_0} g(\mathbf{x})$ exist. Let c be any constant. Then

(i) $\lim\limits_{\mathbf{x} \to \mathbf{x}_0} [f(\mathbf{x}) + g(\mathbf{x})] = \lim\limits_{\mathbf{x} \to \mathbf{x}_0} f(\mathbf{x}) + \lim\limits_{\mathbf{x} \to \mathbf{x}_0} g(\mathbf{x})$,

(ii) $\lim\limits_{\mathbf{x} \to \mathbf{x}_0} cf(\mathbf{x}) = c \lim\limits_{\mathbf{x} \to \mathbf{x}_0} f(\mathbf{x})$,

(iii) $\lim\limits_{\mathbf{x} \to \mathbf{x}_0} f(\mathbf{x}) g(\mathbf{x}) = \left(\lim\limits_{\mathbf{x} \to \mathbf{x}_0} f(\mathbf{x}) \right) \left(\lim\limits_{\mathbf{x} \to \mathbf{x}_0} g(\mathbf{x}) \right)$,

(iv) $\lim\limits_{\mathbf{x} \to \mathbf{x}_0} \dfrac{f(\mathbf{x})}{g(\mathbf{x})} = \dfrac{\lim\limits_{\mathbf{x} \to \mathbf{x}_0} f(\mathbf{x})}{\lim\limits_{\mathbf{x} \to \mathbf{x}_0} g(\mathbf{x})}$, if $\lim\limits_{\mathbf{x} \to \mathbf{x}_0} g(\mathbf{x}) \neq 0$.

Continuity

Continuity for functions of several variables is defined just as for functions of a single variable.

Definition 2

Let f be a function that is defined on a neighborhood of $x_0 \in \mathbb{R}^n$. Then f is **continuous** at $\mathbf{x}_0$ if $\lim\limits_{\mathbf{x} \to \mathbf{x}_0} f(\mathbf{x})$ exists and

$$\lim\limits_{\mathbf{x} \to \mathbf{x}_0} f(\mathbf{x}) = f(\mathbf{x}_0).$$

In other words, f is continuous at $\mathbf{x}_0$ if $f(\mathbf{x}) \to f(\mathbf{x}_0)$ as $\mathbf{x} \to \mathbf{x}_0$, regardless of the manner in which $\mathbf{x}$ approaches $\mathbf{x}_0$. Here are some examples.

1. The function $f(x, y) = 4 - x - y$ in Example 7 is continuous at $(1, 1)$ since

$$\lim\limits_{(x, y) \to (1, 1)} (4 - x - y) = 2 = f(1, 1).$$

2. The function in Figure 1.18 is discontinuous at $\mathbf{x}_0$, since $\lim\limits_{\mathbf{x} \to \mathbf{x}_0} f(\mathbf{x})$ does not exist.

3. The function $f(x, y, z) = x \cos yz$ in Example 9 is continuous at $(2, \pi/3, 3)$ since

$$\lim\limits_{(x, y, z) \to (2, \pi/3, 3)} x \cos yz = -2 = f(2, \pi/3, 3).$$

4. The function $f(x, y) = \sqrt{9 - x^2 - y^2}$ in Example 10 is continuous at $(0, 0)$ since

$$\lim\limits_{(x, y) \to (0, 0)} \sqrt{9 - x^2 - y^2} = 3 = f(0, 0).$$

If a function is defined at a point $\mathbf{x}_0$ but is not defined on a neighborhood of $\mathbf{x}_0$, we must modify the definition of continuity so that we consider only those $\mathbf{x} \to \mathbf{x}_0$ *within the domain* of f. For example, consider the function

$$f(x, y) = \sqrt{9 - x^2 - y^2}$$

defined on the disc

$$D = \{(x, y) \mid x^2 + y^2 \leq 9\}.$$

If a point (x_0, y_0) lies on the circle $x^2 + y^2 = 9$ that bounds D, we need only consider the limiting values of $f(x, y)$ for points $(x, y) \to (x_0, y_0)$ that lie in D. In this case, we have $f(x, y) \to 0 = f(x_0, y_0)$, so f is continuous at (x_0, y_0). Since f is continuous at every point of D, we say that f is continuous on D.

As is the case with limits of functions of several variables, continuity is a property that is often difficult to prove. We usually establish continuity by knowing how continuous functions combine to form other continuous functions. For example, since powers, products, and sums of continuous functions of a single variable are again continuous, we would expect a polynomial function such as

$$p(x, y) = x^3y^2 + 6xy + xy^4$$

to be a continuous function of two variables. Indeed, using Definition 2 and the ideas of Chapter 2, we can prove that sums, multiples, products, powers, and quotients (where denominators are not zero) of continuous functions of several variables are continuous (see Exercises 70 and 71). Moreover, the composition of a continuous function of a single variable with a continuous function of n variables is a continuous function of n variables (see Exercise 72).

Points of discontinuity for functions of several variables occur for the usual reasons: a denominator becomes zero, an expression underneath a radical sign of even order becomes negative, and so forth.

Exercise Set 18.1

In Exercises 1–13, state the implicit domain for the function f.

1. $f(x, y) = \dfrac{1}{x^2 + y^2}$

2. $f(x, y) = \dfrac{x - y}{x + y}$

3. $f(x, y) = \sqrt{y - x}$

4. $f(x, y) = \sqrt{1 - xy^2}$

5. $f(x, y) = \sin xy^2$

6. $f(x, y) = \dfrac{1 + e^x}{(1 - e^y) \ln x^2}$

7. $f(x, y) = \dfrac{\cos xy}{\text{Sin}^{-1}(x^2 + y^2)}$

8. $f(x, y, z) = \sqrt{4 - x^2 - y^2 - z^2}$

9. $f(x, y, z) = xyz$

10. $f(x, y) = \dfrac{1}{x - 3} + \dfrac{2}{x - y}$

11. $f(x, y, z) = \dfrac{\sin y}{xyz}$

12. $f(x, y, z) = \dfrac{z^3 - 10}{(4 - y^2) \tan x}$

13. $f(x, y, z) = \dfrac{1}{\ln(x^2yz^2)}$

14. Let $f(x, y) = x^2 + y^2$, and $g(z) = 2z + 3$. Write the composite function $h(x, y) = g(f(x, y))$ as an explicit function of x and y.

15. Let $f(x, y) = x + y^2$, and $g(z) = \sqrt{z}$.
a. Write the composite function $h(x, y) = g(f(x, y))$ as an explicit function of x and y.
b. Find the domain of the function h.

16. Let V be the volume of a right circular cone. Write $V = V(r, h)$ as a function of the radius r and height h.

17. Let S be the total exterior surface area of a right circular cylinder with top and bottom. Express S as a function $S(r, h)$ of the radius and the height of the cylinder.

18. An object is dropped from rest h meters above the ground. Express the time $T(g, h)$ required for it to fall to the ground as a function of h and g, where g is the acceleration due to gravity (ignore air resistance).

19. Let V be the volume of a right circular cone capped by a hemisphere of radius r (an "ice cream" cone). Write V as a function of the radius r and the height h of the cone.

20. Let S be the total surface area of a rectangular box (with six sides). Write S as a function of the length l, the width w, and the volume V of the box.

In Exercises 21–30, evaluate the given limit.

21. $\displaystyle\lim_{(x, y) \to (1, 3)} (x^2 - 2y)$

22. $\displaystyle\lim_{(x, y) \to (1, -1)} \sqrt{x - y}$

23. $\displaystyle\lim_{(x, y) \to (3, -1)} \dfrac{1}{\sqrt{x + y}}$

24. $\displaystyle\lim_{(x, y) \to (1, -2)} \dfrac{x + y^3}{x^2 + 2xy + y^2}$

25. $\displaystyle\lim_{(x, y) \to (\pi/2, 1)} x \cos xy$

26. $\displaystyle\lim_{(x, y) \to (1, 0)} xe^{y-x}$

27. $\displaystyle\lim_{(x, y) \to (\pi/2, 1)} \ln \sin xy$

28. $\displaystyle\lim_{(x, y) \to (2, 2)} \dfrac{\text{Tan}^{-1}(y/x)}{1 + xy}$

29. $\displaystyle\lim_{(x, y) \to (0, 0)} \dfrac{\sin(xy)}{xy}$

30. $\displaystyle\lim_{(x, y) \to (0, 0)} \dfrac{e^{2xy} - e^x}{y}$

Figures (i)–(vi) are the graphs of six different functions of two variables, and Figures (a)–(f) are graphs of level curves for the same six functions. In Exercises 31–36, determine which of the graphs (i)–(vi) is the graph of $z = f(x, y)$ and which of the graphs (a)–(f) is the graph of the level curves of f. Assume that the level curves correspond to equally spaced levels.

31. $f(x, y) = y^2 - y^3$

32. $f(x, y) = 3x^2 + y^2$

33. $f(x, y) = x^2 + 3y^2$

34. $f(x, y) = 2ye^{-x^2 - y^2}$

35. $f(x, y) = 1/(1 + x^2 + 4y^2)$

36. $f(x, y) = -x^2 \sin \pi y$

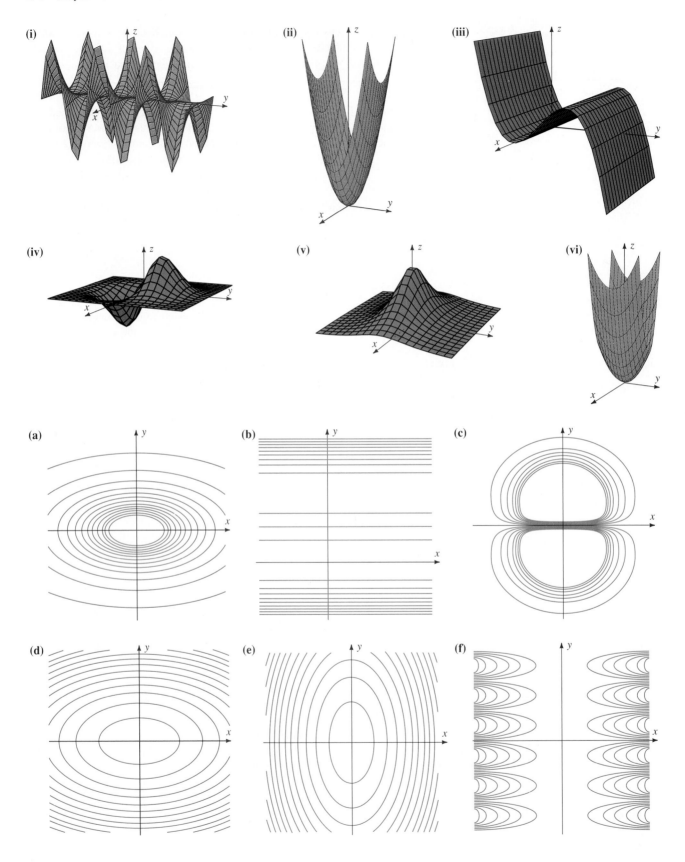

In Exercises 37–44, sketch several level curves for the given functions.

37. $f(x, y) = y - x^2$

38. $f(x, y) = x^2 + 4y^2$

39. $f(x, y) = 2xy$

40. $f(x, y) = x^2 - y^2$

41. $f(x, y) = xe^y$

42. $f(x, y) = x \cos y$

43. $f(x, y) = \sqrt{x + y}$

44. $f(x, y) = \sqrt{y^2 - x^2}$

In Exercises 45–52, sketch the graph of $z = f(x, y)$ by first sketching several traces of $z = f(x, y)$ in the planes $x = c$ and $y = c$.

45. $f(x, y) = x$

46. $f(x, y) = y^2$

47. $f(x, y) = x + y$

48. $f(x, y) = x \cos y$

49. $f(x, y) = x^2 + 4y^2$

50. $f(x, y) = y - x$

51. $f(x, y) = y \sin x$

52. $f(x, y) = y/x$

In Exercises 53–56, sketch the graph of $z = f(x, y)$ by first sketching several level curves and several traces in the planes $x = c$ and $y = c$.

53. $f(x, y) = x^3$

54. $f(x, y) = 2x^2 + 3y^2$

55. $f(x, y) = \sin xy$

56. $f(x, y) = \dfrac{x + y}{x - y}$

57. Find $\displaystyle\lim_{(x, y) \to (1, 0)} \frac{xy - y}{x^2 + y^2 - 2x + 1}$, if it exists.

58. Let

$$f(x, y) = \begin{cases} \dfrac{4xy}{x^2 + y^2}, & (x, y) \neq (0, 0); \\ 0, & (x, y) = (0, 0). \end{cases}$$

Is f continuous at $(0, 0)$? Let

$$g(x, y) = \begin{cases} \dfrac{4xy}{\sqrt{x^2 + y^2}}, & (x, y) \neq (0, 0); \\ 0, & (x, y) = (0, 0). \end{cases}$$

Is g continuous at $(0, 0)$?

59. Is the function

$$f(x, y) = \begin{cases} \dfrac{x^2 y}{x^3 + y^3}, & (x, y) \neq (0, 0) \\ 0, & (x, y) = (0, 0) \end{cases}$$

continuous at $(0, 0)$? (*Hint:* Consider the path $y = x$.)

60. Show that the function

$$f(x, y) = \begin{cases} \dfrac{x^3 y}{x^3 + y^5}, & (x, y) \neq (0, 0) \\ 0, & (x, y) = (0, 0) \end{cases}$$

is not continuous at $(0, 0)$.

61. Is the function

$$f(x, y) = \begin{cases} \dfrac{x^3}{x^2 + y^2}, & (x, y) \neq (0, 0) \\ 0, & (x, y) = (0, 0) \end{cases}$$

continuous at $(0, 0)$?

62. Let

$$f(x, y) = \begin{cases} \dfrac{\sin x \cos y}{x}, & (x, y) \neq (0, 0) \\ 0, & (x, y) = (0, 0) \end{cases}$$

Find $\displaystyle\lim_{(x, y) \to (0, 0)} f(x, y)$. (*Hint:* Write

$$f(x, y) = g(x, y)h(x, y), \qquad (x, y) \neq (0, 0),$$

with $g(x, y) = \cos y$, $h(x, y) = (\sin x)/x$, and apply Theorem 1.)

63. Let

$$f(x, y) = \begin{cases} x \csc x \tan y, & (x, y) \neq (0, 0) \\ 0, & (x, y) = (0, 0). \end{cases}$$

Find $\displaystyle\lim_{(x, y) \to (0, 0)} f(x, y)$ (see Exercise 62).

64. True or false? If $\displaystyle\lim_{(x, y) \to (a, b)} f(x, y) = L$, then $\displaystyle\lim_{x \to a} f(x, b) = L$. Explain.

65. True or false? If $\displaystyle\lim_{x \to a} f(x, b) = L$ and $\displaystyle\lim_{y \to b} f(a, y) = L$, then

$$\lim_{(x, y) \to (a, b)} f(x, y) = L.$$

66. Consider the function

$$f(x, y) = \frac{-xy}{x^2 + y^2}$$

whose graph appears in Figure 1.20.

a. Letting $x = r \cos \theta$ and $y = r \sin \theta$, show that f may be expressed in polar coordinates as

$$f(r, \theta) = -\tfrac{1}{2} \sin 2\theta.$$

b. Conclude from part a that the values $f(r, \theta)$ are independent of r, depending only on θ. Explain why this shows that the limit of $f(x, y)$ as $(x, y) \to (0, 0)$ does not exist.

67. Prove that $\displaystyle\lim_{(x, y) \to (0, 0)} (x^2 + y^2) = 0$ using Definition 1.

68. Prove part (i) of Theorem 1.

69. Prove part (ii) of Theorem 1.

70. Prove that sums and constant multiples of continuous functions of several variables are continuous.

71. Prove that the product of two continuous functions of several variables is continuous.

72. Prove that if g is a continuous function of a single variable and if $y = f(\mathbf{x})$ is a continuous function of several variables, then the composite function $R(\mathbf{x}) = g(f(\mathbf{x}))$ is a continuous function of several variables.

18.2   Partial Differentiation

The *derivative* of the function f of one variable is the limit

$$f'(x) = \lim_{h \to 0} \frac{f(x + h) - f(x)}{h}, \tag{1}$$

which measures the *rate of change* of $f(x)$ with respect to change in x. In the case of a function of two or more independent variables we have already seen that the question of calculating a rate of change for $z = f(x, y)$ at (x_0, y_0) using limits is a complicated issue, since the ''nearby'' point (x, y) may approach (x_0, y_0) along an infinite number of distinct paths.

We begin by examining rates at which $f(x, y)$ changes along paths parallel to the coordinate axes. This is the concept of **partial differentiation.**

Definition 3

Let $f(x, y)$ be defined in a neighborhood of (x_0, y_0). The **partial derivative of f with respect to x** at (x_0, y_0) is the number

$$\frac{\partial f}{\partial x}(x_0, y_0) = \lim_{h \to 0} \frac{f(x_0 + h, y_0) - f(x_0, y_0)}{h}, \tag{2}$$

if this limit exists. Similarly, the **partial derivative of f with respect to y** at (x_0, y_0) is the number

$$\frac{\partial f}{\partial y}(x_0, y_0) = \lim_{h \to 0} \frac{f(x_0, y_0 + h) - f(x_0, y_0)}{h}, \tag{3}$$

provided this limit exists.

Comparing equations (1) and (2), we see that the partial derivative $\partial f / \partial x$ at (x_0, y_0) is simply the result of holding the variable y constant and differentiating the function $z = f(x, y_0)$ as a function of x alone. Similarly, the partial derivative $\partial f / \partial y$ at (x_0, y_0) results from treating the variable x as the constant $x = x_0$ and differentiating $z = f(x_0, y)$ as a function of y alone. Thus, partial derivatives may be calculated by the rules developed earlier for differentiating functions of a single variable.

☐ **EXAMPLE 1**

Calculate the partial derivatives with respect to x and y for the function $f(x, y) = x^2 y^3 + e^x + \ln y$ and evaluate each at $(1, 4)$.

Solution   When we compute the partial derivative with respect to x, we consider y (and any expression containing y alone) to be a constant. Thus,

$$\frac{\partial f}{\partial x}(x, y) = \left(\frac{d}{dx} x^2\right) y^3 + \frac{d}{dx} e^x + \frac{d}{dx} \ln y$$
$$= 2xy^3 + e^x + 0.$$

Similarly, when we compute the partial derivative with respect to y, we consider x to be a constant. Thus,

$$\frac{\partial f}{\partial y}(x, y) = x^2 \left(\frac{d}{dy} y^3\right) + \frac{d}{dy} e^x + \frac{d}{dy} \ln y$$
$$= 3x^2 y^2 + 0 + \frac{1}{y}, \qquad y \neq 0.$$

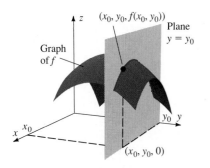

Figure 2.1 The intersection of the plane $y = y_0$ with the graph of f.

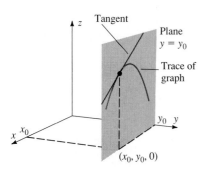

Figure 2.2 The partial derivative $\partial f/\partial x$ at (x_0, y_0) is the slope of the line tangent to the trace of the graph of f in the plane $y = y_0$.

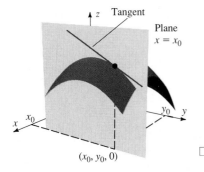

Figure 2.3 The partial derivative $\partial f/\partial y$ at (x_0, y_0) is the slope of the line tangent to the trace of the graph of f in the plane $x = x_0$.

Consequently,

$$\frac{\partial f}{\partial x}(1, 4) = 2 \cdot 1 \cdot 4^3 + e^1 = 128 + e$$

and

$$\frac{\partial f}{\partial y}(1, 4) = 3 \cdot 1^2 \cdot 4^2 + \frac{1}{4} = 48 + \frac{1}{4} = \frac{193}{4}. \quad \blacksquare$$

□ **EXAMPLE 2**

For $f(x, y) = \sin xy$, we apply the rule

$$\frac{d}{dt} \sin at = a \cos at,$$

with one variable playing the role of a and the other playing the rule of t:

$$\frac{\partial f}{\partial x}(x, y) = y \cos xy \quad \text{and} \quad \frac{\partial f}{\partial y}(x, y) = x \cos xy. \quad \blacksquare$$

Figures 2.1 through 2.3 illustrate the geometric interpretations of the partial derivatives in Definition 3. To interpret $\partial f/\partial x$ at (x_0, y_0), we note that the set of points (x, y, z) in $\mathbb{R}^3$ with $y = y_0$ is a plane. The intersection of this plane with the graph of f is the *trace* of f in the plane $y = y_0$ (Figure 2.1). This curve may be viewed as the graph of the function of one variable

$$h(x) = f(x, y_0), \qquad y_0 \text{ fixed.}$$

Since

$$h'(x_0) = \frac{\partial f}{\partial x}(x_0, y_0),$$

the partial derivative $\partial f/\partial x$ at (x_0, y_0) gives the slope of the line tangent to this trace at the point $(x_0, y_0, f(x_0, y_0))$ (Figure 2.2). A similar interpretation is valid for the partial derivative $\partial f/\partial y$ at (x_0, y_0) (Figure 2.3).

□ **EXAMPLE 3**

Calculate $\partial f/\partial x$ and $\partial f/\partial y$ at $(1, 2)$ for

$$f(x, y) = \sqrt{9 - x^2 - y^2},$$

and interpret these numbers geometrically.

Solution Differentiating with respect to x, holding y constant, gives

$$\frac{\partial f}{\partial x}(x, y) = \tfrac{1}{2}(9 - x^2 - y^2)^{-1/2}(-2x)$$

$$= \frac{-x}{\sqrt{9 - x^2 - y^2}}.$$

Thus, $\partial f/\partial x$ at $(1, 2)$ is $-1/2$.

Similarly, the partial derivative with respect to y is

$$\frac{\partial f}{\partial y} = \frac{-y}{\sqrt{9 - x^2 - y^2}}.$$

Thus, $\partial f / \partial x$ at $(1, 2)$ is -1.

Figure 2.4 illustrates the geometric interpretation of these two numbers. The graph of $z = f(x, y)$ is a hemisphere, and if $x = 1$ and $y = 2$, then $z = 2$. The partial derivative $\partial f / \partial y$ at $(1, 2)$ is the slope of the line that lies in the plane $y = 2$ and is tangent to this hemisphere at $(1, 2, 2)$. Similarly, the partial derivative $\partial f / \partial y$ at $(1, 2)$ is the slope of the line that lies in the plane $x = 1$ and is tangent to this hemisphere at $(1, 2, 2)$. ■

Notation for Partial Derivatives

For a function of two variables $z = f(x, y)$, the symbol $\partial / \partial x$ denotes the partial derivative with respect to x just as the Leibniz notation d / dx denotes the derivative of the function of a single variable. We also use subscripts to denote partial derivatives, such as $f_x(x, y)$ or z_x for $\partial f / \partial x$ at (x, y).

We may summarize the various types of notation for partial derivatives as follows: If $z = f(x, y)$, then

$$\frac{\partial f}{\partial x}(x, y) = \frac{\partial}{\partial x} f(x, y) = z_x(x, y) = z_x,$$

and

$$\frac{\partial f}{\partial y}(x, y) = \frac{\partial}{\partial y} f(x, y) = z_y(x, y) = z_y.$$

Functions of More than Two Variables

Partial derivatives are defined for functions of more than two variables by the same idea used in Definition 3: Hold all variables constant except one, and differentiate the resulting function of a single variable as before. For example, if $w = f(x, y, z)$ is a function of the three independent variables x, y, and z, the three partial derivatives are defined as follows:

$$\frac{\partial f}{\partial x}(x, y, z) = \lim_{h \to 0} \frac{f(x + h, y, z) - f(x, y, z)}{h}$$

$$\frac{\partial f}{\partial y}(x, y, z) = \lim_{h \to 0} \frac{f(x, y + h, z) - f(x, y, z)}{h}$$

$$\frac{\partial f}{\partial z}(x, y, z) = \lim_{h \to 0} \frac{f(x, y, z + h) - f(x, y, z)}{h}$$

Partial derivatives for functions of more than three independent variables are defined and calculated in an analogous way.

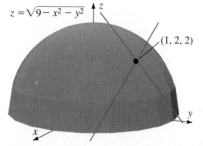

Figure 2.4 The hemisphere $z = f(x, y)$ where $f(x, y) = \sqrt{9 - x^2 - y^2}$. The partial derivatives $\partial f / \partial x$ and $\partial f / \partial y$ at $(1, 2)$ are the slopes of two lines tangent to the hemisphere at $(1, 2, 2)$.

☐ **EXAMPLE 4**

Let $f(x, y, z) = \sqrt{x} e^{y/z}$, $z \ne 0$, $x \ge 0$. Then,

$$\frac{\partial f}{\partial x}(x, y, z) = \left[\frac{d}{dx} \sqrt{x} \right] e^{y/z} = \frac{1}{2\sqrt{x}} e^{y/z},$$

$$\frac{\partial f}{\partial y}(x, y, z) = \sqrt{x}\left[\frac{\partial}{\partial y}(e^{y/z})\right]\sqrt{x}e^{y/z}\frac{1}{z} = \frac{\sqrt{x}}{z}e^{y/z},$$

$$\frac{\partial f}{\partial z}(x, y, z) = \sqrt{x}\left[\frac{\partial}{\partial z}e^{y/z}\right] = \sqrt{x}e^{y/z}\left(\frac{-y}{z^2}\right) = \frac{-y\sqrt{x}}{z^2}e^{y/z}. \qquad \blacksquare$$

□ **EXAMPLE 5**

The radial rate of heat flow in a substance between two concentric spheres is given by the function

$$H(r, R, t, T) = \frac{(t - T)4\pi kRr}{R - r}$$

where k is a constant, and where the inner sphere has radius r and temperature t and the outer sphere has radius R and temperature T. Thus (using the Quotient Rule to differentiate with respect to r and R),

$$H_r = \frac{\partial}{\partial r}H(r, R, t, T) = \frac{(R - r)[(t - T)4\pi kR] - (-1)[(t - T)4\pi kRr]}{(R - r)^2}$$

$$= \frac{(t - T)4\pi kR^2}{(R - r)^2},$$

$$H_R = \frac{\partial}{\partial R}H(r, R, t, T) = \frac{(R - r)[(t - T)4\pi kr] - (1)[(t - T)4\pi kRr]}{(R - r)^2}$$

$$= \frac{-(t - T)4\pi kr^2}{(R - r)^2},$$

$$H_t = \frac{\partial}{\partial t}H(r, R, t, T) = \frac{\partial}{\partial t}\left[t\frac{4\pi kRr}{R - r} - T\frac{4\pi kRr}{R - r}\right] = \frac{4\pi kRr}{R - r},$$

$$H_T = \frac{\partial}{\partial T}H(r, R, t, T) = \frac{\partial}{\partial T}\left[t\frac{4\pi kRr}{R - r} - T\frac{4\pi kRr}{R - r}\right] = -\frac{4\pi kRr}{R - r} \qquad \blacksquare$$

□ **EXAMPLE 6**

For the vectors $\mathbf{x} = x_1\mathbf{i} + x_2\mathbf{j} + x_3\mathbf{k}$ and $\mathbf{y} = y_1\mathbf{i} + y_2\mathbf{j} + y_3\mathbf{k}$, the dot product

$$\mathbf{x} \cdot \mathbf{y} = x_1y_1 + x_2y_2 + x_3y_3$$

may be viewed as a function of the six independent variables (components) $x_1, x_2, x_3, y_1, y_2, y_3$. Thus,

$$\frac{\partial}{\partial x_1}(\mathbf{x} \cdot \mathbf{y}) = y_1; \qquad \frac{\partial}{\partial y_1}(\mathbf{x} \cdot \mathbf{y}) = x_1,$$

$$\frac{\partial}{\partial x_2}(\mathbf{x} \cdot \mathbf{y}) = y_2; \qquad \frac{\partial}{\partial y_2}(\mathbf{x} \cdot \mathbf{y}) = x_2,$$

$$\frac{\partial}{\partial x_3}(\mathbf{x} \cdot \mathbf{y}) = y_3; \qquad \frac{\partial}{\partial y_3}(\mathbf{x} \cdot \mathbf{y}) = x_3.$$

Thus, the rate at which $\mathbf{x} \cdot \mathbf{y}$ changes with respect to change in the component x_1 is just the component y_1, and so forth. $\qquad \blacksquare$

Limitations of Partial Derivatives

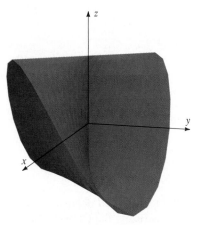

Figure 2.5 From the graph of the function

$$f(x, y) = \frac{-xy}{x^2 + y^2},$$

$$(x, y) \neq (0, 0), \quad \text{and} \quad f(0, 0) = 0$$

we see that both $\partial f / \partial x$ and $\partial f / \partial y$ at $(0, 0)$ are 0. However, the function changes in a discontinuous fashion as (x, y) leaves $(0, 0)$ along the line $y = x$.

Before proceeding further, we should clarify a frequent misconception. Although knowledge of the derivative $f'(x_0)$ completely determines the rate at which the function f changes at $x = x_0$, *knowledge of the partial derivatives $f_x(x_0, y_0)$ and $f_y(x_0, y_0)$ is not always sufficient to determine the rate at which the function of two variables $z = f(x, y)$ is changing at (x_0, y_0).* The analogous statement for functions of more than two variables is also true.

To see why, consider the function

$$f(x, y) = \begin{cases} \dfrac{-xy}{x^2 + y^2}, & (x, y) \neq (0, 0); \\ 0, & (x, y) = (0, 0); \end{cases}$$

(see Example 8 in Section 18.1). Using Definition 3, we find that

$$\frac{\partial f}{\partial x}(0, 0) = \lim_{h \to 0} \frac{f(0 + h, 0) - f(0, 0)}{h} = \lim_{h \to 0} \left(\frac{0}{h} \right) = 0$$

and

$$\frac{\partial f}{\partial y}(0, 0) = \lim_{h \to 0} \frac{f(0, 0 + h) - f(0, 0)}{h} = \lim_{h \to 0} \left(\frac{0}{h} \right) = 0.$$

Thus, both partial derivatives at $(0, 0)$ are zero. But it is false to conclude that the rate of change of $f(x, y)$ at $(0, 0)$ in *every direction* is zero, as can be seen from Figure 2.5. For example, a small change from the point $(0, 0)$ to the point (h, h) causes an abrupt change in the value $f(x, y)$ from $f(0, 0) = 0$ to $f(h, h) = -1/2$, no matter how small h may be.

It is important to understand that partial derivatives give information about functions *only* in the directions of the coordinate axes. In Sections 18.5 and 18.7 we develop a richer theory of differentiation in order to obtain general conclusions about the rate of change of a function of several variables near a particular point.

Higher Order Partial Derivatives

Repeated application of partial differentiation leads to **higher order partial derivatives.** Just as the higher order derivatives of a function of one variable indicate how that function's graph curves, the higher order partials for a function of two variables indicate how its graph bends in space. In fact, in Section 18.4, we develop a general method for determining the shape of the graph $z = f(x, y)$ using the higher order partials of f.

Since we now have more than one independent variable, we encounter **mixed partial derivatives,** in which one differentiation is performed with respect to a particular variable, followed by another differentiation with respect to a different variable. We use the following notation:

$$\frac{\partial^2 f}{\partial x^2}(x, y) = \frac{\partial^2}{\partial x^2} f(x, y) \quad \text{means} \quad \frac{\partial}{\partial x} \left(\frac{\partial f}{\partial x}(x, y) \right).$$

$$\frac{\partial^2 f}{\partial y \partial x}(x, y) = \frac{\partial^2}{\partial y \partial x} f(x, y) \quad \text{means} \quad \frac{\partial}{\partial y} \left(\frac{\partial f}{\partial x}(x, y) \right).$$

$$\frac{\partial^2 f}{\partial x \partial y}(x, y) = \frac{\partial^2}{\partial x \partial y} f(x, y) \quad \text{means} \quad \frac{\partial}{\partial x} \left(\frac{\partial f}{\partial y}(x, y) \right).$$

$$\frac{\partial^2 f}{\partial y^2}(x, y) = \frac{\partial^2}{\partial y^2} f(x, y) \quad \text{means} \quad \frac{\partial}{\partial y} \left(\frac{\partial f}{\partial y}(x, y) \right).$$

Note that the order in which the differentiations are performed is indicated by reading the "denominator" of the derivative notation from *right* to *left*. Similar definitions hold for third and higher order partial derivatives.

☐ **EXAMPLE 7**

For the function $f(x, y) = x^2y^3 + \cos x \sin y$,

$$\frac{\partial^2 f}{\partial x^2}(x, y) = \frac{\partial}{\partial x}\left(\frac{\partial f}{\partial x}(x, y)\right) = \frac{\partial}{\partial x}(2xy^3 - \sin x \sin y) = 2y^3 - \cos x \sin y,$$

$$\frac{\partial^2 f}{\partial y \partial x}(x, y) = \frac{\partial}{\partial y}\left(\frac{\partial f}{\partial x}(x, y)\right) = \frac{\partial}{\partial y}(2xy^3 - \sin x \sin y) = 6xy^2 - \sin x \cos y,$$

$$\frac{\partial^2 f}{\partial x \partial y}(x, y) = \frac{\partial}{\partial x}\left(\frac{\partial f}{\partial y}(x, y)\right) = \frac{\partial}{\partial x}(3x^2y^2 + \cos x \cos y) = 6xy^2 - \sin x \cos y,$$

$$\frac{\partial^2 f}{\partial y^2}(x, y) = \frac{\partial}{\partial y}\left(\frac{\partial f}{\partial y}(x, y)\right) = \frac{\partial}{\partial y}(3x^2y^2 + \cos x \cos y) = 6x^2y - \cos x \sin y,$$

$$\frac{\partial^3 f}{\partial y^3}(x, y) = \frac{\partial}{\partial y}\left(\frac{\partial^2 f}{\partial y^2}(x, y)\right) = \frac{\partial}{\partial y}\left(\frac{\partial^2 f}{\partial y^2}\right) = \frac{\partial}{\partial y}(6x^2y - \cos x \sin y)$$

$$= 6x^2 - \cos x \cos y. \qquad ∎$$

We also use subscript notation to indicate higher order partial derivatives. If $z = f(x, y)$,

$$f_{xx}(x, y) \text{ or } z_{xx} \quad \text{means} \quad \frac{\partial^2 f}{\partial x^2}(x, y),$$

$$f_{xy}(x, y) \text{ or } z_{xy} \quad \text{means} \quad \frac{\partial^2 f}{\partial y \partial x}(x, y),$$

$$f_{yx}(x, y) \text{ or } z_{yx} \quad \text{means} \quad \frac{\partial^2 f}{\partial x \partial y}(x, y),$$

$$f_{yy}(x, y) \text{ or } z_{yy} \quad \text{means} \quad \frac{\partial^2 f}{\partial y^2}(x, y),$$

with similar statements holding for functions of more than two variables and for higher order derivatives.

It is important to note that, when subscripts are used, the differentiations are performed *in the order indicated by the subscripts, read from left to right*. This is just the opposite of the order in which the differentiations are indicated in Leibniz notation.

☐ **EXAMPLE 8**

For the function $f(x, y, z) = x^2y^3z^4$,

$$f_x = 2xy^3z^4; \qquad f_y = 3x^2y^2z^4; \qquad f_z = 4x^2y^3z^3$$

$$f_{xy} = \frac{\partial}{\partial y}(2xy^3z^4) = 6xy^2z^4; \qquad f_{yx} = \frac{\partial}{\partial x}(3x^2y^2z^4) = 6xy^2z^4$$

$$f_{yz} = \frac{\partial}{\partial z}(3x^2y^2z^4) = 12x^2y^2z^3; \qquad f_{zy} = \frac{\partial}{\partial y}(4x^2y^3z^3) = 12x^2y^2z^3$$

$$f_{xz} = \frac{\partial}{\partial z}(2xy^3z^4) = 8xy^3z^3; \qquad f_{zx} = \frac{\partial}{\partial x}(4x^2y^3z^3) = 8xy^3z^3$$

$$f_{xx} = 2y^3z^4; \qquad f_{yy} = 6x^2yz^4; \qquad f_{zz} = 12x^2y^3z^2$$

$$f_{xyz} = \frac{\partial}{\partial z}(6xy^2z^4) = 24xy^2z^3; \qquad f_{yzx} = \frac{\partial}{\partial x}(12x^2y^2z^3) = 24xy^2z^3$$

$$f_{xxx} = 0; \qquad f_{yyy} = 6x^2z^4; \qquad f_{zzz} = 24x^2y^3z. \qquad \blacksquare$$

Equality of Mixed Partials

You have no doubt observed, in Example 7, that $f_{yx} = f_{xy}$ and, in Example 8, that $f_{xy} = f_{yx}, f_{yz} = f_{zy}$, and $f_{xz} = f_{zx}$. This is *not* true for all functions. However, if the function f and various of its partial derivatives are continuous, these mixed partials will be equal. The following theorem makes this precise. It is typically proved in courses on advanced calculus.

Theorem 2
Equality of Mixed Partials

If the function $z = f(x, y)$ and the partial derivatives

$$\frac{\partial f}{\partial x}, \quad \frac{\partial f}{\partial y}, \quad \frac{\partial^2 f}{\partial x \partial y}, \quad \text{and} \quad \frac{\partial^2 f}{\partial y \partial x}$$

are all continuous in a neighborhood of the point (x_0, y_0), then

$$\frac{\partial^2 f}{\partial x \partial y}(x_0, y_0) = \frac{\partial^2 f}{\partial y \partial x}(x_0, y_0). \qquad (4)$$

Equation (4) plays an important role in the theory of differentiation that follows.

Exercise Set 18.2

In Exercises 1–20, find all first order partial derivatives.

1. $f(x, y) = xy$

2. $z = \sqrt{x + y^2}$

3. $z = x \tan y^2$

4. $f(x, y) = xy^3 + \sqrt{y}$

5. $f(x, y) = e^{x^2 + y^2}$

6. $z = \mathrm{Tan}^{-1}(y/x)$

7. $f(r, \theta) = r \cos \theta$

8. $f(s, t) = \dfrac{s - t}{s + t}$

9. $z = \ln(xy^2 + x - y)$

10. $h(u, v) = e^{u-v} + e^{v-u}$

11. $f(x, y) = x^y$

12. $z = 2^x y^2$

13. $f(r, \theta) = r^2 \cos \theta$

14. $f(x, y, z) = x^3 e^y \ln z$

15. $f(x, y, z) = xy^3 - yz^2$

16. $w = \ln(x^2 + y^2 + z^2)$

17. $f(x, y, z) = \left(\dfrac{x - y}{x + y}\right)^z$

18. $f(u, v, w) = \dfrac{\sin u}{v^3 \, \mathrm{Tan}^{-1} w}$

19. $f(r, s, t) = \dfrac{\sqrt{r} \, s \ln t}{\sqrt{s^2 - 2r + t}}$

20. $f(u, v, w) = \dfrac{u e^{vw}}{\sin^2 u + \tan^2 w}$

21. Find $\dfrac{\partial f}{\partial x}(2, 5)$ for $f(x, y) = xy^3 - y$.

22. Find $\dfrac{\partial f}{\partial y}(1, 0)$ for $f(x, y) = e^{x-y^2}$.

23. Find $z_x(2, 1)$ for $z = \sqrt{x + y^2}$.

24. Find f_{xx}, f_{xy}, f_{yx}, and f_{yy} for $f(x, y)$ in Exercise 4.

25. Find $\dfrac{\partial^2 f}{\partial r^2}, \dfrac{\partial^2 f}{\partial r \partial \theta}$, and $\dfrac{\partial^2 f}{\partial \theta^2}$ for $f(r, \theta)$ in Exercise 13.

26. Find $f_{xx}, f_{xy}, f_{xz}, f_{yz}, f_{yy}$, and f_{zz} for $f(x, y, z)$ in Exercise 15.

27. Find $w_{xx} + w_{yy} + w_{zz}$ for $w(x, y, z)$ in Exercise 16.

28. For a particle traveling in a circular orbit of radius r, the relation between angular speed ω and velocity v is $v = \omega r$. Find $\partial v/\partial r$ and $\partial v/\partial \omega$.

29. For the constant volume flow of an incompressible fluid through a tube of varying cross-sectional area, the equation $A_1v_1 = A_2v_2$ expresses the relationship between the respective cross-sectional areas and velocities at two points in the tube.

a. Express v_2 as a function of A_2, A_1, and v_1.

b. Find $\partial v_2/\partial A_2$, the rate at which v_2 changes with respect to change in A_2 alone.

c. Suppose that $A_1 = 5 \text{ cm}^2$, $A_2 = 3 \text{ cm}^2$, and $v_1 = 20 \text{ cm/s}$. Find the rate of change of v_2 with respect to A_2 if A_1 and v_1 are held constant.

d. With A_1, A_2, and v_1 as in part c, find the rate of change of A_2 with respect to change in v_1 if A_1 and v_2 are held constant.

In Exercises 30–36, use the equations

$$x = \rho \sin \phi \cos \theta$$
$$y = \rho \sin \phi \sin \theta$$
$$z = \rho \cos \phi$$

for changing from rectangular to spherical coordinates to find the indicated partial derivative.

30. $\dfrac{\partial x}{\partial \phi}$ **31.** $\dfrac{\partial y}{\partial \theta}$ **32.** $\dfrac{\partial z}{\partial \rho}$

33. $\dfrac{\partial x}{\partial \rho}$ **34.** $\dfrac{\partial y}{\partial \phi}$ **35.** $\dfrac{\partial z}{\partial \phi}$

36. $\dfrac{\partial \phi}{\partial z}$

37. For $f(x, y) = \displaystyle\int_x^{x+y} \cos t^2 \, dt$ find

a. $\dfrac{\partial f}{\partial x}$ **b.** $\dfrac{\partial f}{\partial y}$

38. For $f(x, y, z) = \displaystyle\int_z^x \sqrt{t^3 + 1} \, dt - \int_y^z \sqrt{t^3 + 1} \, dt$ find

a. $\dfrac{\partial f}{\partial x}$ **b.** $\dfrac{\partial f}{\partial y}$ **c.** $\dfrac{\partial f}{\partial z}$

39. Show that the function

$$f(x, y) = \text{Sin}^{-1}\left(\frac{x - y}{x + y}\right)$$

is a solution of the differential equation

$$x\frac{\partial z}{\partial x} + y\frac{\partial z}{\partial y} = 0.$$

40. Show that the function $w = \ln(e^x + e^y + e^z)$ satisfies the partial differential equation

$$\frac{\partial w}{\partial x} + \frac{\partial w}{\partial y} + \frac{\partial w}{\partial z} = 1.$$

41. The **electrical potential** at an axial point for a charged disc is give by

$$V = \frac{\sigma}{2\epsilon_0}(\sqrt{a^2 + r^2} - r)$$

where σ and ϵ_0 are constants, a is the radius of the disc, and r is the distance from the point to the disc.

a. Find the rate of change of V with respect to a if r is held constant.

b. Find the rate of change of V with respect to r if a is held constant.

42. The national unemployment rate u may be viewed as the dot product $u = \mathbf{v} \cdot \mathbf{w}$ where the components of the vector $\mathbf{v} = \langle v_1 \quad \ldots, v_n \rangle$ are the percentages of employable citizens in each of the n job categories and the components of the vector $\mathbf{w} = \langle w_1, w_2, \ldots, w_n \rangle$ are the unemployment rates within each category.

a. Find $\partial u/\partial v_j$ and interpret this rate.

b. Find $\partial u/\partial w_j$ and interpret this rate.

43. Let $f(x, y, z) = e^{x+y+z}$. Show that all partial derivatives of all orders equal $f(x, y, z)$.

44. Show that a function of the form $f(x, y) = e^{kx}g(y)$ satisfies the differential equation

$$\frac{\partial f}{\partial x}(x, y) = kf(x, y).$$

45. Show that the function $f(x, y) = \sin xy$ satisfies the differential equation

$$x\frac{\partial f}{\partial x}(x, y) - y\frac{\partial f}{\partial y}(x, y) = 0.$$

46. Show that the function $f(x, y) = \sin xy$ satisfies the differential equation

$$x^2\frac{\partial^2 f}{\partial x^2}(x, y) - y^2\frac{\partial^2 f}{\partial y^2}(x, y) = 0.$$

47. Find a solution of the differential equation

$$x^n\frac{\partial^n f}{\partial x^n} = y^n\frac{\partial^n f}{\partial y^n}, \qquad n = 1, 2, 3, \ldots.$$

(See Exercises 45 and 46.)

48. Show that the function $y(x, t) = g(t - cx)$ satisfies the **one-dimensional wave equation**

$$\frac{\partial^2 y}{\partial t^2} = \frac{1}{c^2}\frac{\partial^2 y}{\partial x^2}.$$

49. Laplace's equation for the function $f(x, y)$ is

$$\frac{\partial^2 f}{\partial x^2} + \frac{\partial^2 f}{\partial y^2} = 0.$$

Show that the following functions satisfy Laplace's equation.

a. $f(x, y) = e^x \sin y$

b. $f(x, y) = e^{-x} \cos y$

c. $f(x, y) = \ln \sqrt{x^2 + y^2}$

50. Show that a polynomial $P(x, y)$ in x and y (like $P(x, y) = 3x^2y^3 + xy^6 - 5x^3y$) has the property that

$$\frac{\partial^2 P}{\partial x \partial y} = \frac{\partial^2 P}{\partial y \partial x}.$$

51. Find the slope of the line that lies in the plane $y = 2$ and is tangent to the surface $z = x^3 + y^2$ at the point $(1, 2, 5)$. Find the point at which this tangent line intersects the xy-plane.

52. Suppose that the altitude z of a mountainous terrain satisfies the equation $z = xy^2 + x^3 - 2xy$. In which direction, the positive x direction or the positive y direction, is the mountain rising most rapidly at the point $(2, 1, 6)$?

18.3 Tangent Planes

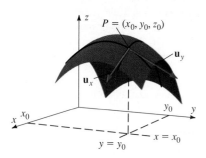

Figure 3.1 The partial derivatives $\partial f/\partial x$ and $\partial f/\partial y$ at (x_0, y_0) determine the vectors $\mathbf{u}_x$ and $\mathbf{u}_y$ that are tangent to the graph of $z = f(x, y)$ at the point P.

For a function f of one variable, knowledge of $f(a)$ and the derivative $f'(a)$ enables us to write an equation for the line tangent to the graph of f at the point $(a, f(a))$. In this brief section, we see how we can obtain an equation for a plane tangent to the graph of a function of two variables $z = f(x, y)$ from knowledge of its partial derivatives. We assume that such a tangent plane exists, and we focus on how we can describe it using the partial derivatives of f. The question of the existence of a tangent plane is addressed in Section 18.5.

Suppose that f is a function of two variables that is defined in a neighborhood of the point (x_0, y_0). Assume also that the partial derivatives $\partial f/\partial x$ and $\partial f/\partial y$ both exist at (x_0, y_0). Using the geometric interpretation of the partials discussed in Section 18.2, we may conclude the following (see Figure 3.1):

(i) The partial derivative $\partial f/\partial x$ at (x_0, y_0) gives the slope of the line tangent to the trace of f in the plane $y = y_0$. Since the plane $y = y_0$ is parallel to the xz-coordinate plane, a vector parallel to this tangent line is

$$\mathbf{u}_x = \mathbf{i} + \frac{\partial f}{\partial x}(x_0, y_0)\mathbf{k}.$$

(To check this statement, note that the slope of the vector $\mathbf{u}_x$, thought of as a line segment in the plane $y = y_0$, equals the value of $\partial f/\partial x$ at (x_0, y_0), as asserted.)

(ii) The partial derivative $\partial f/\partial y$ at (x_0, y_0) gives the slope of the line tangent to the trace of f in the plane $x = x_0$. Since the plane $x = x_0$ is parallel to the yz-coordinate plane, a vector parallel to this tangent line is

$$\mathbf{u}_y = \mathbf{j} + \frac{\partial f}{\partial y}(x_0, y_0)\mathbf{k}.$$

To find an equation for the tangent plane, we use the cross product along with the techniques discussed in Sections 16.3 and 16.4 (see Example 8 in Section 16.4). Since both $\mathbf{u}_x$ and $\mathbf{u}_y$ must lie in the tangent plane, they together determine a normal to the plane as

$$\mathbf{N} = \mathbf{u}_y \times \mathbf{u}_x = \det \begin{bmatrix} \mathbf{i} & \mathbf{j} & \mathbf{k} \\ 0 & 1 & \dfrac{\partial f}{\partial y}(x_0, y_0) \\ 1 & 0 & \dfrac{\partial f}{\partial x}(x_0, y_0) \end{bmatrix},$$

which yields

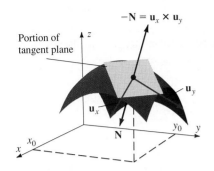

Figure 3.2 The vectors $\mathbf{u}_x$ and $\mathbf{u}_y$ determine the normal $\mathbf{N} = \mathbf{u}_y \times \mathbf{u}_x$ for the plane tangent to the graph of f at the point P.

$$\mathbf{N} = \frac{\partial f}{\partial x}(x_0, y_0)\mathbf{i} + \frac{\partial f}{\partial y}(x_0, y_0)\mathbf{j} - \mathbf{k}. \tag{1}$$

(See Figure 3.2). Since the plane with normal vector $\mathbf{N}$ that contains the point $P = (x_0, y_0, z_0) = (x_0, y_0, f(x_0, y_0))$ and the arbitrary point $Q = (x, y, z)$ satisfies the equation

$$\mathbf{N} \cdot \overrightarrow{PQ} = 0,$$

we obtain the following equations for the tangent plane to the graph of f at P:

$$\frac{\partial f}{\partial x}(x_0, y_0)(x - x_0) + \frac{\partial f}{\partial y}(x_0, y_0)(y - y_0) - (z - z_0) = 0 \tag{2}$$

or

$$z = \frac{\partial f}{\partial x}(x_0, y_0)(x - x_0) + \frac{\partial f}{\partial y}(x_0, y_0)(y - y_0) + z_0. \tag{3}$$

Either equation (2) or equation (3) may be used to write the equation of the plane tangent to the graph of f at the point (x_0, y_0, z_0), where $z_0 = f(x_0, y_0)$. More generally, you may simply remember the idea used to develop these equations: the partial derivatives $\partial f/\partial x$ and $\partial f/\partial y$ at (x_0, y_0) determine **direction vectors, $\mathbf{u}_x$ and $\mathbf{u}_y$,** whose cross product $\mathbf{N} = \mathbf{u}_y \times \mathbf{u}_x$ is a normal to the desired plane.

□ **EXAMPLE 1**

Find an equation for the plane tangent to the graph of $f(x, y) = x^2 + 4y^2$ at the point $(2, 1, 8)$.

Solution   The required partial derivatives are

$$\frac{\partial f}{\partial x}(2, 1) = 2x \Big|_{\substack{x=2 \\ y=1}} = 4,$$

$$\frac{\partial f}{\partial y}(2, 1) = 8y \Big|_{\substack{x=2 \\ y=1}} = 8,$$

and $x_0 = 2$, $y_0 = 1$, $z_0 = 8$. Thus, by equation (3), the equation is

$$z = 4(x - 2) + 8(y - 1) + 8$$

or

$$z = 4x + 8y - 8.$$

□ EXAMPLE 2

Find an equation for the line of intersection of the plane tangent to the graph of $z = 9 - x^2 - y^2$ at $(1, 2, 4)$ and the xy-coordinate plane.

Solution   Since

$$z_x(1, 2) = -2x \Big|_{\substack{x=1 \\ y=2}} = -2$$

and

$$z_y(1, 2) = -2y \Big|_{\substack{x=1 \\ y=2}} = -4,$$

the equation of the tangent plane, according to (3), is

$$z = -2(x - 1) - 4(y - 2) + 4$$

or

$$z = -2x - 4y + 14.$$

Setting $z = 0$ gives the line of intersection of this plane with the xy-plane as $2x + 4y - 14 = 0$ (see Figure 3.3). ∎

Normal Lines

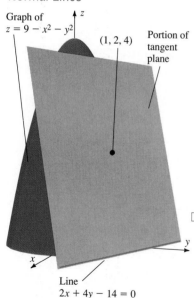

Figure 3.3 Plane tangent to the graph of $z = 9 - x^2 - y^2$ at $(1, 2, 4)$ intersects the xy-plane along the line $2x + 4y - 14 = 0$.

If a surface has a tangent plane at point P, we say that a line through P is **normal** to the surface if it is normal to the tangent plane at P. Using $\mathbf{N}$ in (1) as a direction vector normal to the tangent plane for f at (x_0, y_0, z_0), we may write the normal line in vector form as

$$\ell : \mathbf{r}(t) = \mathbf{r}_0 + t\mathbf{N} \tag{4}$$

where $\mathbf{r}_0$ is the position vector $\mathbf{r}_0 = x_0\mathbf{i} + y_0\mathbf{j} + z_0\mathbf{k}$. Writing $\mathbf{r}(t)$ in component form as $\mathbf{r}(t) = x(t)\mathbf{i} + y(t)\mathbf{j} + z(t)\mathbf{k}$ and using (1), we may write (4) in the component form

$$x(t)\mathbf{i} + y(t)\mathbf{j} + z(t)\mathbf{k}$$
$$= (x_0\mathbf{i} + y_0\mathbf{j} + z_0\mathbf{k}) + t\left(\frac{\partial f}{\partial x}(x_0, y_0)\mathbf{i} + \frac{\partial f}{\partial y}(x_0, y_0)\mathbf{j} - \mathbf{k}\right). \tag{5}$$

□ EXAMPLE 3

Find equations for the line normal to the graph of $z = e^{y - x^2}$ at the point $(1, \ln 2, 2/e)$.

Solution   For the function $f(x, y) = e^{y - x^2}$,

$$\frac{\partial f}{\partial x}(1, \ln 2) = -2xe^{y - x^2} \Big|_{\substack{x=1 \\ y=\ln 2}} = -2e^{\ln 2 - 1} = \frac{-4}{e}$$

and

$$\frac{\partial f}{\partial y}(1, \ln 2) = e^{y - x^2} \Big|_{\substack{x=1 \\ y=\ln 2}} = e^{\ln 2 - 1} = \frac{2}{e}.$$

Also, $x_0 = 1$, $y_0 = \ln 2$, and $z_0 = 2/e$. The vector form for the normal line is therefore

$$\ell : \mathbf{r}(t) = \mathbf{i} + \ln 2\mathbf{j} + \frac{2}{e}\mathbf{k} + t\left(-\frac{4}{e}\mathbf{i} + \frac{2}{e}\mathbf{j} - \mathbf{k}\right). \quad ∎$$

Exercise Set 18.3

In Exercises 1–16, find an equation for the plane tangent to the graph of the given function at point P, assuming it exists.

1. $f(x, y) = x^2 + y^2$; $P = (1, 3, 10)$

2. $f(x, y) = x^2 + 2xy + y^2$; $P = (3, -1, 4)$

3. $z = x^2 + y^2 - xy - 4x - 2y$; $P = (1, -1, 1)$

4. $f(x, y) = 2x^2 - y^2$; $P = (2, \sqrt{5}, 3)$

5. $f(x, y) = \dfrac{x - 2}{y + 2}$; $P = (4, -1, 2)$

6. $f(x, y) = \sqrt{9 - x^2 - y^2}$; $P = (1, -2, 2)$

7. $f(x, y) = \ln xy$; $P = (1, 1, 0)$

8. $f(x, y) = e^{-x} \sin y$; $P = (0, \pi/6, 1/2)$

9. $z = \dfrac{x}{x^2 + y^2}$; $P = (1, 1, 1/2)$

10. $f(x, y) = \mathrm{Tan}^{-1}(y/x)$; $P = (2, 2, \pi/4)$

11. $z = \ln y^x$; $P = (1, 1, 0)$

12. $z = \dfrac{y - 6}{3x^2 - 1}$; $P = (1, 4, -1)$

13. $f(x, y) = \cos x \sin y$; $P = (0, \pi/4, \sqrt{2}/2)$

14. $f(x, y) = \sec xy$; $P = (0, 0, 1)$

15. $f(x, y) = \ln\left(\dfrac{y - x}{y + x}\right)$, $P = (0, e, 0)$

16. $f(s, t) = \dfrac{1 - \sqrt{s}}{t(s - \sqrt{t})}$, $P = (4, 1, -1/3)$

In Exercises 17–22, find a vector equation for the line normal to the graph of the given function at the point P, as described in the stated exercise.

17. Exercise 2.

18. Exercise 3.

19. Exercise 6.

20. Exercise 8.

21. Exercise 11.

22. Exercise 16.

23. Find the point on the graph of $z = -x^2 + xy + 2y^2$ where the tangent plane is parallel to the plane with equation $x - 14y + z = 4$.

24. Find the point(s) on the graph of $z = y \cos x$ where the tangent plane is parallel to the plane $x - \sqrt{3}y + 2z = -2$.

25. Find a vector equation for the line of intersection of the plane tangent to the graph of $z = x^2 + 2y^2 - 4y + 2$ at $(2, 1, 4)$ and the xy-coordinate plane.

26. Find a vector equation for the line of intersection of the plane tangent to $z = e^x \sin y$ at $(0, \pi/2, 1)$ and the plane tangent to $z = x^2 + 2y^2$ at $(1, 2, 9)$.

27. Show that all lines normal to the graph of $f(x, y) = x \sin y$ at points $(x, \pi/2)$ are parallel.

28. Show that at all points $(x, \pi/2)$, the planes tangent to the graph of $f(x, y) = x \sin y$ are the same. Find an equation for this plane.

29. Find an equation for the plane tangent to the paraboloid $z = 9x^2 + 4y^2$ at the point $(1, 2, 25)$.

30. Show that the volume of the tetrahedron formed by the planes $x = 0$, $y = 0$, and $z = 0$, and any tangent to the graph of the function $f(x, y) = c/(xy)$, $c > 0$, is $V = 9c/2$.

31. Find an equation for the plane tangent to the sphere $x^2 + y^2 + z^2 = r^2$ at the point (x_0, y_0, z_0), $z_0 \neq 0$.

32. Find an equation for the plane tangent to the graph of the function $y = f(x, z)$ at the point $(x_0, z_0, z_0) = (x_0, f(x_0, z_0), z_0)$.

33. Use the result of Exercise 32 to find the equation of the plane tangent to the graph of $y = \mathrm{Tan}^{-1}(z/x)$ at the point where $x = z = 1$.

34. Use the result of Exercise 32 to find a vector equation for the line of intersection of the plane tangent to the graph of $y = x^2 + 2xz - z^2$ at the point $(2, -8, -2)$ and the xz-coordinate plane.

In Exercises 35 and 36, let the angle of intersection θ between a surface and a curve at an intersection point P be $(\pi/2) - \alpha$, where α is the angle between the tangent vector to the curve at P and the normal vector to the surface at P.

35. Find the angle of intersection between the surface $z = 2x + 4y - x^2 - y^2$ and the curve $\mathbf{r}(t) = -\cos(\pi t)\mathbf{i} + 2\sin(\pi t/2)\mathbf{j} + 5t\mathbf{k}$ at the point $(1, 2, 5)$.

36. Find the angle of intersection between the surface $z = x^2 - y^2 - 1$ and the helix $\mathbf{r}(t) = (\cos t)\mathbf{i} + (\sin t)\mathbf{j} + t\mathbf{k}$ at the point $(1, 0, 0)$.

In Exercises 37 and 38, let the angle of intersection between two surfaces at an intersection point P be the angle between their two normal vectors at P.

37. Find the angle of intersection between the surface $z = 2x + 4y - x^2 - y^2 - 1$ and the surface $z = \sqrt{(x - 1)^2 + (2y + 1)^2}$ at the point $(1, 1, 3)$.

38. Find the angle of intersection between the surfaces $z = y^2$ and $z = x^2 + y^2/2$ at the point $(1, \sqrt{2}, 2)$.

39. Prove that every normal to a sphere passes through the center.

18.4   Relative and Absolute Extrema

One of the principal applications of the derivative for functions of a single variable is in finding relative and absolute extrema. In this section, we show how partial derivatives may be used to find relative extrema for functions of several variables. Although the ideas discussed here are not confined to functions of just two independent variables, we will deal mainly with this case because of the opportunity to interpret the results geometrically.

We begin by defining what we mean by relative extrema.

Definition 4

Let f be a function defined on a set S in $\mathbb{R}^2$. The number $z_0 = f(x_0, y_0)$ is a **relative maximum** for f if there exists a neighborhood N of (x_0, y_0) such that

$$f(x_0, y_0) \geq f(x, y)$$

for all $(x, y) \in N \cap S$.

The number $z_0 = f(x_0, y_0)$ is a **relative minimum** for f if there exists a neighborhood N of (x_0, y_0) such that

$$f(x_0, y_0) \leq f(x, y)$$

for all $(x, y) \in N \cap S$.

The number $z_0 = f(x_0, y_0)$ is a **relative extremum** for f if it is either a relative maximum or a relative minimum.

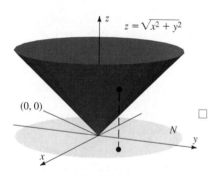

Figure 4.1 The value $f(0, 0) \leq f(x, y)$ for all (x, y) near $(0, 0)$; $f(0, 0)$ is a relative minimum.

Thus, a relative maximum for f is the precise analogue of a relative maximum for a function of a single variable: $z_0 = f(x_0, y_0)$ is the largest value $z = f(x, y)$ for all (x, y) "near" (x_0, y_0). A similar interpretation holds for relative minima.

☐ **EXAMPLE 1**

The function $f(x, y) = \sqrt{x^2 + y^2}$ has a relative minimum of $z_0 = 0$ at the point $(x_0, y_0) = (0, 0)$. This is easy to see, since

$$f(x, y) = \sqrt{x^2 + y^2} \geq 0 = f(0, 0)$$

for all points (x, y) (see Figure 4.1).   ■

One way to verify that the number $z_0 = f(x_0, y_0)$ is a relative extremum is simply to compare the number $f(x_0, y_0)$ with values of the function f for points (x, y) near (x_0, y_0). While such comparisons are sometimes difficult, if not impossible, to make, this method works for a variety of polynomial functions in two variables. The idea is to write $x = x_0 + h$ and $y = y_0 + k$ and then to examine the sign of the difference:

$$f(x_0, y_0) - f(x_0 + h, y_0 + k). \tag{1}$$

If this difference is nonnegative for all small values of h and k, we conclude that $f(x_0, y_0)$ is a relative maximum. Similarly, if the difference in line (1) is nonpositive for all small values of h and k, we conclude that $f(x_0, y_0)$ is a relative minimum.

□ **EXAMPLE 2**

Verify that $f(1, 2) = 4$ is a relative maximum for the function $f(x, y) = 2x + 4y - x^2 - y^2 - 1$.

Solution Here $x_0 = 1$ and $y_0 = 2$, so the nearby point (x, y) is written $(x, y) = (x_0 + h, y_0 + k) = (1 + h, 2 + k)$. The difference in (1) is

$$f(1, 2) - f(1 + h, 2 + k) \qquad (2)$$
$$= 4 - [2(1 + h) + 4(2 + k) - (1 + h)^2 - (2 + k)^2 - 1]$$
$$= 4 - 2 - 2h - 8 - 4k + 1 + 2h + h^2 + 4 + 4k + k^2 + 1$$
$$= h^2 + k^2.$$

Since $h^2 + k^2 \geq 0$ for all h and k, this calculation shows that the difference in (2) is nonnegative for all small h and k. Thus, $f(1, 2) = 4$ is a relative maximum for the function f (see Figure 4.2). ■

Finding Relative Extrema

The preceding discussion addressed the issue of verifying that $f(x_0, y_0)$ is a relative extremum. But how do we find (x_0, y_0) to begin with? Recall the one-variable case again. If the function f has a relative extremum at $x = x_0$, then either $f'(x_0) = 0$ or $f'(x_0)$ fails to exist. In the former case, the line tangent to the graph of f at $(x_0, f(x_0))$ is horizontal (see Figure 4.3). In the latter case, no tangent exists at $(x_0, f(x_0))$.

From a strictly geometric viewpoint, it seems that the same relationship should hold between the points (x_0, y_0), at which the function f has a relative extremum, and the *plane* tangent to the graph of f at $(x_0, y_0, f(x_0, y_0))$: At such points the tangent plane should either be horizontal or fail to exist (see Figure 4.4).

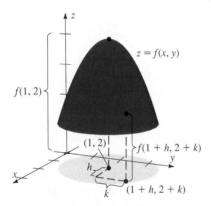

Figure 4.2 For the function f in Example 2, $f(1, 2) \geq f(1 + h, 2 + k)$ for all small values of h and k; $f(1, 2)$ is a relative maximum.

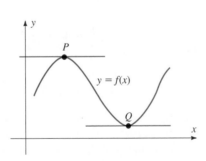

Figure 4.3 At relative extremum P or Q, $f'(x_0) = 0$ and tangent line is horizontal.

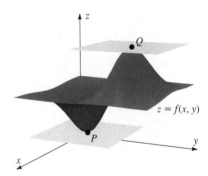

Figure 4.4 At the relative extrema P and Q,

$$\frac{\partial f}{\partial x}(x_0, y_0) = \frac{\partial f}{\partial y}(x_0, y_0) = 0,$$

and the tangent plane is horizontal.

Since a tangent plane is determined at the point $(x_0, y_0, f(x_0, y_0))$ by the partial derivatives $\partial f / \partial x$ and $\partial f / \partial y$ at (x_0, y_0), these geometric observations lead to the following theorem.

Theorem 3

If the number $z_0 = f(x_0, y_0)$ is a relative extremum for the function f at the point (x_0, y_0), one of the following two conditions must hold:

(i) $\dfrac{\partial f}{\partial x}(x_0, y_0) = \dfrac{\partial f}{\partial y}(x_0, y_0) = 0$, or

(ii) either one or both of $\dfrac{\partial f}{\partial x}(x_0, y_0)$ and $\dfrac{\partial f}{\partial y}(x_0, y_0)$ fail to exist.

Proof: First consider the case where $f(x_0, y_0)$ is a relative maximum. We assume that both $\partial f/\partial x$ and $\partial f/\partial y$ exist at (x_0, y_0), and we show that

$$\frac{\partial f}{\partial x}(x_0, y_0) = \frac{\partial f}{\partial y}(x_0, y_0) = 0.$$

Holding $y = y_0$ fixed, we define the function (of a single variable) g by

$$g(x) = f(x, y_0).$$

Since $f(x_0, y_0)$ is a relative maximum, $f(x_0, y_0) \geq f(x, y)$ for all (x, y) near (x_0, y_0). Thus,

$$g(x_0) = f(x_0, y_0) \geq f(x, y_0) = g(x)$$

for all x near x_0. This shows that $g(x_0)$ is a relative maximum for the function g. Since the derivative of g is

$$\begin{aligned} g'(x) &= \lim_{h \to 0} \frac{g(x + h) - g(x)}{h} \\ &= \lim_{h \to 0} \frac{f(x + h, y_0) - f(x, y_0)}{h} \\ &= \frac{\partial f}{\partial x}(x, y_0), \end{aligned}$$

we know that $g'(x_0) = \partial f/\partial x$ at (x_0, y_0) exists (Figure 4.5). But, if $g(x_0)$ is a relative maximum and $g'(x_0)$ exists, then $g'(x_0) = 0$. This shows that

$$\frac{\partial f}{\partial x}(x_0, y_0) = g'(x_0) = 0.$$

By repeating this argument with $x = x_0$ fixed, we can show that $\partial f/\partial y$ at (x_0, y_0) is 0 as well. Finally, we note that we have made no special use of the assumption that the extremum $f(x_0, y_0)$ was a relative maximum—the argument applies for relative minima as well. ∎

Theorem 3 determines a procedure for finding relative extrema for f: Find all points (x_0, y_0) where

$$\frac{\partial f}{\partial x}(x_0, y_0) = \frac{\partial f}{\partial y}(x_0, y_0) = 0$$

or where

either $\dfrac{\partial f}{\partial x}(x_0, y_0)$ or $\dfrac{\partial f}{\partial y}(x_0, y_0)$ fails to exist.

(We call these points **critical points**.) Then test each critical point to determine whether it yields a relative extremum.

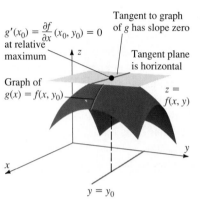

$g'(x_0) = \dfrac{\partial f}{\partial x}(x_0, y_0) = 0$ at relative maximum

Tangent to graph of g has slope zero

Tangent plane is horizontal

Graph of $g(x) = f(x, y_0)$

$z = f(x, y)$

$y = y_0$

Figure 4.5   A relative maximum: If $\partial f/\partial x$ exists at (x_0, y_0), then $\partial f/\partial x = 0$ at (x_0, y_0).

☐ **EXAMPLE 3**

For the function $f(x, y) = 2x + 4y - x^2 - y^2 - 1$, setting both partial derivatives equal to zero gives the equations

$$\frac{\partial f}{\partial x}(x, y) = 2 - 2x = 0, \qquad \frac{\partial f}{\partial y}(x, y) = 4 - 2y = 0.$$

The (simultaneous) solution of this pair of equations is $x = 1$, $y = 2$. Thus, condition (i) in Theorem 3 yields the single critical point $(1, 2)$. We have verified that $f(1, 2)$ is a relative maximum in Example 2. Since the partial derivatives are defined for all (x, y), there are no points satisfying condition (ii) of Theorem 3. Thus, the only relative extremum for this function is the relative maximum at $(1, 2)$ (see Figure 4.2). ■

☐ **EXAMPLE 4**

Find all relative extrema for the function $f(x, y) = \sqrt{x^2 + y^2}$.

Solution   The partial derivatives are

$$\frac{\partial f}{\partial x}(x, y) = \frac{x}{\sqrt{x^2 + y^2}}; \qquad \frac{\partial f}{\partial y}(x, y) = \frac{y}{\sqrt{x^2 + y^2}}.$$

Both partial derivatives are undefined at $(x, y) = (0, 0)$. For all other points at least one of the partial derivatives is nonzero. Since it is easy to see that $f(0, 0) = 0 < f(x, y)$ for all $(x, y) \neq (0, 0)$, $f(0, 0) = 0$ is a relative minimum. (See Figure 4.1.) ■

It is important to understand that the conditions of Theorem 3 do not *guarantee* that $f(x_0, y_0)$ is a relative extremum. Theorem 3 merely provides *necessary* conditions for an extremum. Without satisfying condition (i) or (ii) of Theorem 3, $f(x_0, y_0)$ cannot be a relative extremum. However, some critical points do not yield extrema. The following is a typical example.

☐ **EXAMPLE 5**

For the function $f(x, y) = y^2 - x^2$, the partial derivatives are

$$\frac{\partial f}{\partial x} = -2x; \qquad \frac{\partial f}{\partial y} = 2y.$$

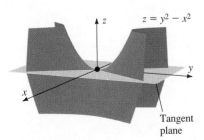

Figure 4.6   Graph of $z = y^2 - x^2$ has a saddle point at $(0, 0)$. Note that the tangent plane is horizontal.

Thus, both $\partial f/\partial x$ and $\partial f/\partial y$ are 0 at $(0, 0)$. However, the number $f(0, 0) = 0$ is neither a relative maximum nor a relative minimum. To see this, we compare $f(0, 0)$ with $f(h, k)$ where (h, k) is a point near $(0, 0)$. Since

$$f(0, 0) - f(h, k) = 0 - (k^2 - h^2) = h^2 - k^2,$$

we see that the sign of this difference depends only on the relative sizes of $|h|$ and $|k|$. Since this difference is not of constant sign, $f(0, 0)$ is neither a maximum nor a minimum.

Figure 4.6 illustrates why $f(x, y) = y^2 - x^2$ does not have an extremum at $(0, 0)$. Although both partial derivatives are zero, the function $g(x) = f(x, 0) = -x^2$ reaches a relative *maximum* at $(0, 0)$ while the function $h(y) = f(0, y) = y^2$ reaches a relative *minimum* at $(0, 0)$. ■

In Example 5, the point $(0, 0)$ is called a *saddle point* for rather obvious reasons. The surface bows upward along one axis and downward along another. More generally, a point (x_0, y_0) in the domain of a function of two variables is called a **saddle point** if (x_0, y_0) is a critical point and if $f(x_0, y_0)$ is neither a relative maximum nor a relative minimum. Figure 4.7 shows the graph of another function that has a saddle point at $(0, 0)$ (see Exercise 26).

Second Derivative Test

There is a theorem that helps sort out extrema from saddle points. It may be regarded as the analogue of the Second Derivative Test for functions of a single variable. A proof of this theorem may be found in texts on advanced calculus.

Theorem 4
Second Derivative Test

Let f be a function of two variables. Suppose that all second order partial derivatives of f are continuous in a neighborhood of (x_0, y_0) and that

$$\frac{\partial f}{\partial x}(x_0, y_0) = \frac{\partial f}{\partial y}(x_0, y_0) = 0.$$

Let

$$A = \frac{\partial^2 f}{\partial x^2}(x_0, y_0), \qquad B = \frac{\partial^2 f}{\partial y \partial x}(x_0, y_0), \qquad C = \frac{\partial^2 f}{\partial y^2}(x_0, y_0)$$

and

$$D = B^2 - AC.$$

Then

(i) If $D < 0$ and $A < 0$, $f(x_0, y_0)$ is a relative maximum.
(ii) If $D < 0$ and $A > 0$, $f(x_0, y_0)$ is a relative minimum.
(iii) If $D > 0$, (x_0, y_0) is a saddle point.
(iv) If $D = 0$, no conclusions may be drawn.

Theorem 4 verifies that the critical point $(0, 0)$ in Example 5 is a saddle point, since

$$A = \frac{\partial^2 f}{\partial x^2}(0, 0) = -2, \qquad B = \frac{\partial^2 f}{\partial y \partial x}(0, 0) = 0, \qquad C = \frac{\partial^2 f}{\partial y^2}(0, 0) = 2$$

and

$$D = B^2 - AC = 4 > 0.$$

Similarly, for the function f in Example 3 and the critical point $(1, 2)$, we have

$$A = \frac{\partial^2 f}{\partial x^2}(1, 2) = -2, \qquad B = \frac{\partial^2 f}{\partial y \partial x}(1, 2) = 0, \qquad C = \frac{\partial^2 f}{\partial y^2}(1, 2) = -2$$

and

$$D = B^2 - AC = -4.$$

Thus, since $D < 0$ and $A < 0$, the Second Derivative Test agrees with the conclusion of Example 3, that $f(1, 2)$ is a relative maximum.

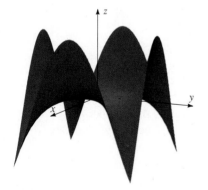

Figure 4.7 The graph of $f(x, y) = x^4 - 3x^2y^2 + y^4$ has a saddle point at $(0, 0)$.

☐ **EXAMPLE 6**

Find and classify all relative extrema for the function $f(x, y) = x^4 + y^4 - 4xy$.

Solution The partial derivatives are

$$\frac{\partial f}{\partial x} = 4x^3 - 4y; \qquad \frac{\partial f}{\partial y} = 4y^3 - 4x.$$

Since both partial derivatives are defined for all (x, y), the extrema can occur only at points where

$$\frac{\partial f}{\partial x} = 4x^3 - 4y = 0, \quad \text{or} \quad y = x^3 \tag{3}$$

and

$$\frac{\partial f}{\partial y} = 4y^3 - 4x = 0, \quad \text{or} \quad x = y^3 \tag{4}$$

Substituting for x in (3) using (4) gives the equation $y = y^9$. Thus, either $y = 0$, or $y^8 = 1$, which gives $y = \pm 1$. Using (4) we find that $x = 0$ if $y = 0$, $x = 1$ if $y = 1$, and $x = -1$ if $y = -1$. The three critical points are therefore $(0, 0)$, $(1, 1)$, and $(-1, -1)$.

Next, we calculate the second order partials:

$$\frac{\partial^2 f}{\partial x^2} = 12x^2; \qquad \frac{\partial^2 f}{\partial y \partial x} = -4; \qquad \frac{\partial^2 f}{\partial x^2} = 12y^2.$$

At the critical point $(1, 1)$ we have

$$A = \frac{\partial^2 f}{\partial x^2}(1, 1) = 12, \qquad B = \frac{\partial^2 f}{\partial y \partial x}(1, 1) = -4, \qquad C = \frac{\partial^2 f}{\partial y^2}(1, 1) = 12,$$

and

$$D = B^2 - AC = 16 - 12 \cdot 12 = -128 < 0.$$

Thus, since $D < 0$ and $A > 0$, $f(1, 1) = -2$ is a relative minimum, according to Theorem 4.

At the critical point $(-1, -1)$ the values of A, B, C, and D are the same as for $(1, 1)$ so $f(-1, -1) = -2$ is also a relative minimum.

At the critical point $(0, 0)$, $A = C = 0$ and $B = -4$, so $D = B^2 - AC = 16 > 0$. Thus, $(0, 0)$ is a saddle point (see Figure 4.8). ■

In each of Examples 7, 8, and 9, we obtain $D = 0$ at the critical point. These three examples show that any of the three possible outcomes (relative maximum, relative minimum, saddle point) may result if $D = 0$.

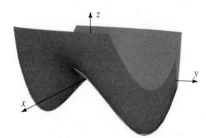

Figure 4.8 The graph of $f(x, y) = x^4 + y^4 - 4xy$. It has relative minima at $(1, 1)$ and $(-1, -1)$ and a saddle point at $(0, 0)$.

☐ **EXAMPLE 7**

For the function $f(x, y) = e^{-(x^4 + y^4)}$, we have

$$\frac{\partial f}{\partial x} = -4x^3 e^{-(x^4 + y^4)} \qquad\qquad \frac{\partial f}{\partial y} = -4y^3 e^{-(x^4 + y^4)}$$

$$\frac{\partial^2 f}{\partial x^2} = (16x^6 - 12x^2)e^{-(x^4 + y^4)} \qquad \frac{\partial^2 f}{\partial y^2} = (16y^6 - 12y^2)e^{-(x^4 + y^4)}$$

$$\frac{\partial^2 f}{\partial y \partial x} = 16y^3 x^3 e^{-(x^4 + y^4)}.$$

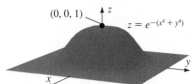

Figure 4.9 The function $f(x, y) = e^{-(x^4+y^4)}$ has a relative maximum at $(0, 0)$.

The point $(0, 0)$ is a critical point since both $\partial f/\partial x$ and $\partial f/\partial y$ are 0 at $(0, 0)$. At this critical point

$$A = \frac{\partial^2 f}{\partial x^2}(0, 0) = 0; \qquad B = \frac{\partial^2 f}{\partial y \partial x}(0, 0) = 0; \qquad C = \frac{\partial^2 f}{\partial y^2}(0, 0) = 0.$$

Thus, the Second Derivative Test yields no conclusion about the critical point $(0, 0)$. However, it is easy to see that the expression $x^4 + y^4$ has a minimum at $(0, 0)$, so

$$f(x, y) = e^{-(x^4+y^4)} = \frac{1}{e^{x^4+y^4}}$$

has a relative maximum at $(0, 0)$ (see Figure 4.9). ■

□ **EXAMPLE 8**

The function $f(x, y) = x^4 + y^4$ obviously has a relative minimum of $f(0, 0) = 0$, since $f(x, y) > 0$ at all other points. However, we find

$$A = \frac{\partial^2 f}{\partial x^2}(0, 0) = 12x^2 \Big|_{x=0} = 0, \qquad B = \frac{\partial^2 f}{\partial y \partial x}(0, 0) = 0,$$

$$C = \frac{\partial^2 f}{\partial y^2}(0, 0) = 12y^2 \Big|_{y=0} = 0, \qquad D = B^2 - AC = 0.$$

Thus, the Second Derivative Test fails to classify this critical point. ■

□ **EXAMPLE 9**

For the function $f(x, y) = x^3 - y^3$, the only simultaneous solution of the two equations

$$\frac{\partial f}{\partial x}(x, y) = 3x^2 = 0 \quad \text{and} \quad \frac{\partial f}{\partial y}(x, y) = -3y^2 = 0$$

is $x = y = 0$, so $(0, 0)$ is the only critical point. Since

$$\frac{\partial^2 f}{\partial x^2} = 6x, \qquad \frac{\partial^2 f}{\partial y \partial x} = 0, \qquad \frac{\partial^2 f}{\partial y^2} = -6y,$$

we have $A = B = C = D = 0$ at the critical point $(0, 0)$.

Thus, the test gives no information as to the nature of this critical point. However, since $f(x, y) = x^3 - y^3$ takes on both positive and negative values in every neighborhood of $(0, 0)$, the critical point $(0, 0)$ is a saddle point. ■

Functions of More Than Two Variables

This section has focused on finding relative extrema for functions of two variables. For functions of more than two variables, the statement of Theorem 3 generalizes directly:

(a) If the function f of three variables has a relative extremum at (x_0, y_0, z_0), then either

$$\frac{\partial f}{\partial x} = \frac{\partial f}{\partial y} = \frac{\partial f}{\partial z} = 0$$

at this point, or else one or more of these partial derivatives fails to exist at (x_0, y_0, z_0).

(b) More generally, the function f of n variables can have a relative extremum at the point $(a_1, a_2, \ldots, a_n)$ only if one or more of the partial derivatives fails to exist, or if

$$\frac{\partial f}{\partial x_1} = \frac{\partial f}{\partial x_2} = \cdots = \frac{\partial f}{\partial x_n} = 0$$

at this point.

However, the Second Derivative Test does not generalize quite so easily. For functions of more than two variables, the classification of critical points is considerably more difficult and will not be pursued here.

Extrema on Closed Sets

Finally, we note that we have not addressed the question of *absolute* extrema, as we did for functions of a single variable in Chapter 4. In a manner analogous to the behavior of a function of one variable on a closed interval, a continuous function f of two variables will assume both an absolute maximum and an absolute minimum on any **closed bounded set** S in the plane. By a closed set in the plane we mean (roughly speaking) a set that includes its boundary, such as the disc $D = \{(x, y) \mid x^2 + y^2 \leq 1\}$ or the square $S = \{(x, y) \mid 0 \leq x \leq 1, 0 \leq y \leq 1\}$. As you might suspect, such absolute extrema will occur either at critical points or at points on the boundary of S. Checking for extrema among boundary points can be a quite complicated task, so we have chosen to defer this topic to more advanced courses where appropriate theory and techniques are developed. (However, this issue is addressed in simple settings in Exercises 35–38.)

Exercise Set 18.4

In Exercises 1–22, find all critical points of the function f. Classify each as a relative maximum, relative minimum, or saddle point.

1. $f(x, y) = x^2 + y^2 + 4y + 4$

2. $f(x, y) = x^2 + y^2 + 4x - 2y + 11$

3. $f(x, y) = x^2 - y^2 + 6x + 4y + 5$

4. $f(x, y) = 7 - 2x + 2y - x^2 - y^2$

5. $f(x, y) = xy + 9$

6. $f(x, y) = x^2 + y^4 - 2x - 4y^2 + 5$

7. $f(x, y) = 5x^2 + y^2 - 10x - 6y + 15$

8. $f(x, y) = x^2 + y^3 - 3y$

9. $f(x, y) = x^3 - y^3$

10. $f(x, y) = e^{x^2 - 2x + y^2 + 4}$

11. $f(x, y) = x^2 - xy$

12. $f(x, y) = e^x \cos y$

13. $f(x, y) = x^3 + y^3 + 4xy$

14. $f(x, y) = x^4 + y^4 - 4xy$

15. $f(x, y) = x \cos y$

16. $f(x, y) = \dfrac{y}{x} - \dfrac{x}{y}$

17. $f(x, y) = \sin(x - y)$

18. $f(x, y) = \sin x - \cos y$

19. $f(x, y) = \ln(x^2 + y^2 + 1)$

20. $f(x, y) = \ln(x^2 + 2x + y^2 - 4y + 6)$

21. $f(x, y) = e^{\left(\frac{1}{x^2 + y^2 + 1}\right)}$

22. $f(x, y) = e^{x^2 + y^2}$

23. Show that the function $z = 4 - \sqrt{x^2 + y^2}$ has a relative maximum at $(0, 0)$. (Note that none of the partial derivatives exist at this point.)

24. Show that the function $z = x^2 + y^2 - 4x - 2y + 9$ has a relative minimum at $(2, 1)$ using the method of Example 2.

25. Find the highest point on the graph of $z = 2y^3 - 3x^2 - 3xy + 9x$ if the positive z-axis is upward.

26. Show that the function $f(x, y) = x^4 - 3x^2y^2 + y^4$ has a saddle point at $(0, 0)$ (see Figure 4.7).

27. Show that the function $f(x, y) = x^2 - y^2 + 2x + 4y - 3$ has a saddle point at $(-1, 2)$ by the method of Example 2.

28. Consider the function $f(x, y) = \dfrac{x^2 + y^2}{(x + y)^2}$.

a. Show that

$$\frac{\partial f}{\partial x}(x, y) = \frac{\partial f}{\partial y}(x, y) = 0$$

if and only if $y = x$ and $x \neq 0$.

b. Show that f has a relative minimum at each point (x, x) with $x \neq 0$.

c. Show that the function

$$g(\lambda) = f(x, \lambda x) = \frac{1 + \lambda^2}{(1 + \lambda)^2},$$

$\lambda \neq -1$, has a relative minimum at $\lambda = 1$.

d. Explain the geometric and algebraic relationships between the functions g and f.

29. A rectangular box, with a top, is to hold 16 cubic meters. Find the dimensions that produce the least expensive box if the material for the side walls is half as expensive as the material for the top and the bottom.

30. Find the point on the plane $2x + 3y + z - 14 = 0$ nearest the origin.

31. Find the point on the plane with equation $ax + by + cz = d$ nearest the origin.

32. Find the dimensions of the closed rectangular box of volume $V = 8000 \text{ cm}^3$ and of minimum surface area.

33. Find the dimensions of the rectangular package of largest volume that can be mailed under the restrictions that length plus girth cannot exceed 84 inches. (Girth is the perimeter of the cross section taken perpendicular to the length.)

34. A manufacturer of tape recorders intends to market x recorders through retail outlets and y recorders through wholesale outlets. The manufacturer estimates that the revenue per recorder sold retail will be

$$r(x, y) = 250 - \frac{x}{20} - \frac{y}{5},$$

while the revenue per recorder on the wholesale market is estimated to be

$$w(x, y) = 200 - \frac{x}{50} - \frac{y}{20}.$$

The cost of producing the recorders is known to be $100 each. Using the model Profit = Revenue − Cost, find the number of recorders that should be marketed retail and the number that should be marketed wholesale so as to maximize profit.

35. Find the *absolute* maximum and minimum values of the function $f(x, y) = 2x + 2y + 4$ on the *closed* disc $D = \{(x, y) \mid x^2 + y^2 \leq 1\}$. (*Hint:* To check for absolute extrema along the boundary $C = \{(x, y) \mid x^2 + y^2 = 1\}$ use the

parameterization $x = \cos t$, $y = \sin t$, and optimize $f(x(t), y(t))$ as a function of t.)

36. Find the absolute maximum and minimum values of the function $f(x, y) = 3x - 2y^3 + 2$ on the (closed) square $S = \{(x, y) \mid 0 \leq x \leq 1, 0 \leq y \leq 1\}$ (see Exercise 35).

37. Find the absolute maximum and minimum values of the function $z = x + 2y + 6$ on the closed set consisting of the ellipse $4x^2 + 2y^2 = 4$ and its interior (see Exercise 35).

38. Find the absolute maximum and minimum values for the function $f(x, y) = x^2 + 2y^2 - 2xy - 6x + 4y$ on the rectangle $0 \leq x \leq 8$, $-2 \leq y \leq 2$.

39. A company manufactures two models of a portable stereo ("boom box")—the standard model and the deluxe model. Manufacturing costs are $65 for the standard model and $80 for the deluxe model. Market research indicates that, if the standard model is priced at x dollars and the deluxe model is priced at y dollars, then the company will sell $1200 - 40x + 20y$ standard models and $2650 + 10x - 30y$ deluxe models. What prices x and y maximize profits?

40. Find three positive numbers x, y, and z such that $x + y + z = 16$ and x^2yz^2 is maximized.

41. Given a set of points $(x_1, y_1), (x_2, y_2), \ldots, (x_n, y_n)$, one way to define a line $y = mx + b$ that "best fits" these points is the **Method of Least Squares.** As illustrated by Figure 4.10, the distance from the data point (x_j, y_j) to the point on the line $y = mx + b$ with x-coordinate x_j is

$$|y_j - (mx_j + b)| = |y_j - mx_j - b|.$$

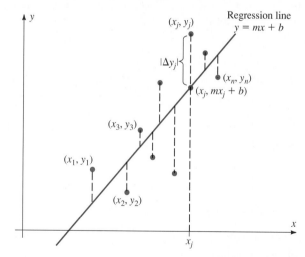

Figure 4.10 The Method of Least Squares determines the line that "best fits" the data points $(x_1, y_1), (x_2, y_2), \ldots, (x_n, y_n)$.

The Method of Least Squares defines the best fitting line $y = mx + b$ (called the **regression line**) to be the line that minimizes the sum of the squares of these individual distances:

$$S(m, b) = \sum_{j=1}^{n} (y_j - mx_j - b)^2.$$

Viewing $x_1, x_2, \ldots, x_n$ and $y_1, y_2, \ldots, y_n$ as fixed constants and m and b as independent variables, show that the values of m and b for which $S(m, b)$ is a minimum are given by the formulas

$$m = \frac{n \sum\limits_{j=1}^{n} x_j y_j - \left(\sum\limits_{j=1}^{n} x_j\right)\left(\sum\limits_{j=1}^{n} y_j\right)}{n \sum\limits_{j=1}^{n} x_j^2 - \left(\sum\limits_{j=1}^{n} x_j\right)^2},$$

$$b = \frac{\left(\sum\limits_{j=1}^{n} x_j^2\right)\left(\sum\limits_{j=1}^{n} y_j\right) - \left(\sum\limits_{j=1}^{n} x_j\right)\left(\sum\limits_{j=1}^{n} x_j y_j\right)}{n \sum\limits_{j=1}^{n} x_j^2 - \left(\sum\limits_{j=1}^{n} x_j\right)^2}.$$

(*Hint*: Set

$$\frac{\partial S}{\partial m}(m, b) = 0 \quad \text{and} \quad \frac{\partial S}{\partial b}(m, b) = 0$$

and solve the resulting system of equations.)

42. For a group of $n = 6$ calculus students, achievement test scores x_j and final course averages in calculus y_j were as follows:

j	1	2	3	4	5	6
x_j	52	46	69	54	61	48
y_j	74	66	94	91	84	80

The regression line for these data appears in Figure 4.11. Use the result of Exercise 41 to show that this regression line has equation $y = 0.95x + 29.31$.

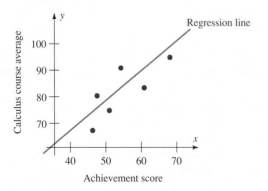

Figure 4.11 Achievement scores versus calculus course averages for six students.

43. A regression line corresponding to a set of data points $(x_1, y_1), (x_2, y_2), \ldots, (x_n, y_n)$ provides a model for *predicting* a value of y corresponding to any particular value of x. Use the regression line obtained in Exercise 42 to predict the course average for a student whose achievement score is $x = 60$.

44. Given the five points whose coordinates are:

x	0	1	1	2	3
y	2	4	3	6	6

  a. Find the regression line for the data using the Method of Least Squares.
  b. Plot the data points and the regression line.
  c. Find the predicted value for $x = 4$.

45. Repeat Exercise 44 for the data below.

x	6	8	9	10	12
y	2	5	5	7	9

46. For data consisting of ordered triples $(x_1, y_1, z_1), \ldots, (x_n, y_n, z_n)$, the analog of the regression line is the (regression) **plane** $z = ax + by + c$. This plane is defined as the plane that minimizes the sum of squares

$$S = \sum_{j=1}^{n} [z_j - (ax_j + by_j + c)]^2.$$

Show that the coefficients a, b, and c in this equation satisfy the equations

(i) $a \sum\limits_{j=1}^{n} x_j + b \sum\limits_{j=1}^{n} y_j + nc = \sum\limits_{j=1}^{n} z_j,$

(ii) $a \sum\limits_{j=1}^{n} x_j^2 + b \sum\limits_{j=1}^{n} x_j y_j + c \sum\limits_{j=1}^{n} x_j = \sum\limits_{j=1}^{n} x_j z_j,$

(iii) $a \sum\limits_{j=1}^{n} x_j y_j + b \sum\limits_{j=1}^{n} y_j^2 + c \sum\limits_{j=1}^{n} y_j = \sum\limits_{j=1}^{n} y_j z_j.$

47. Recall, from Section 7.7, that the *centroid* of the n data points $(x_1, y_1), (x_2, y_2), \ldots, (x_n, y_n)$ is the point $(\bar{x}, \bar{y})$ where

$$\bar{x} = \frac{1}{n} \sum_{j=1}^{n} x_j, \qquad \bar{y} = \frac{1}{n} \sum_{j=1}^{n} y_j.$$

Show that the centroid lies on the regression line determined by the Method of Least Squares.

48. Suppose that you suspected that the n data points $(x_1, y_1), (x_2, y_2), \ldots, (x_n, y_n)$ could be "fitted" well by a curve of the form $y = m \ln x + b$ (a common situation in biology and chemistry). Explain how the Method of Least Squares could be used to find the constants m and b.

49. Prove Theorem 4 for functions of the form $f(x, y) = ax^2 + bxy + cy^2$.

18.5   Approximation and Differentiability

For the function of a single variable $y = f(x)$, the existence of the derivative $f'(x)$ leads to the linear approximation formula

$$f(x + \Delta x) \approx f(x) + f'(x)\, \Delta x \tag{1}$$

and to the definition of the differential

$$df = f'(x)\, dx. \tag{2}$$

In this section, we take up the approximation question for functions of several variables. The key to the entire discussion (and to the discussions of the sections that follow) is the Approximation Theorem (the multivariable generalization of Theorem 9 in Section 3.7).

Theorem 5

Approximation Theorem

Let f be a function of two variables. Suppose f and its first partial derivatives

$$\frac{\partial f}{\partial x}(x, y) \quad \text{and} \quad \frac{\partial f}{\partial y}(x, y)$$

are continuous in an open rectangle

$$R = \{(x, y) \mid a_1 < x < a_2,\, b_1 < y < b_2\}$$

in the xy-plane. Let the point (x_0, y_0) and the point $(x_0 + \Delta x, y_0 + \Delta y)$ both lie in R. Then

$$f(x_0 + \Delta x, y_0 + \Delta y) = f(x_0, y_0) + \frac{\partial f}{\partial x}(x_0, y_0)\, \Delta x + \frac{\partial f}{\partial x}(x_0, y_0)\, \Delta y \tag{3}$$

$$+ \,\epsilon_1\, \Delta x + \epsilon_2\, \Delta y$$

where $\lim\limits_{\substack{\Delta x \to 0 \\ \Delta y \to 0}} \epsilon_1 = 0$ and $\lim\limits_{\substack{\Delta x \to 0 \\ \Delta y \to 0}} \epsilon_2 = 0$.

Theorem 5 characterizes the relationship between the value of the function f at (x_0, y_0) and the value of this function at the "nearby" point $(x_0 + \Delta x, y_0 + \Delta y)$. Ignoring the error term $(\epsilon_1\, \Delta x + \epsilon_2\, \Delta y)$ in equation (3) leads to the approximation

$$f(x_0 + \Delta x, y_0 + \Delta y) \approx f(x_0, y_0) + \frac{\partial f}{\partial x}(x_0, y_0)\, \Delta x + \frac{\partial f}{\partial y}(x_0, y_0)\, \Delta y \tag{4}$$

for functions of two variables.

Figure 5.1 provides a geometric interpretation of approximation (4). Since both $\partial f / \partial x$ and $\partial f / \partial y$ exist at (x_0, y_0), they determine a "tangent" plane with equation

$$z = \frac{\partial f}{\partial x}(x_0, y_0)(x - x_0) + \frac{\partial f}{\partial y}(x_0, y_0)(y - y_0) + f(x_0, y_0) \tag{5}$$

(equation (3), Section 18.3). To find the point on this plane above $(x, y) = (x_0 + \Delta x, y_0 + \Delta y)$, we set $x = x_0 + \Delta x$ and $y = y_0 + \Delta y$. Then $x - x_0 = \Delta x$ and $y - y_0 = \Delta y$, so (5) gives the z-coordinate as

$$z = \frac{\partial f}{\partial x}(x_0, y_0)\, \Delta x + \frac{\partial f}{\partial y}(x_0, y_0)\, \Delta y + f(x_0, y_0),$$

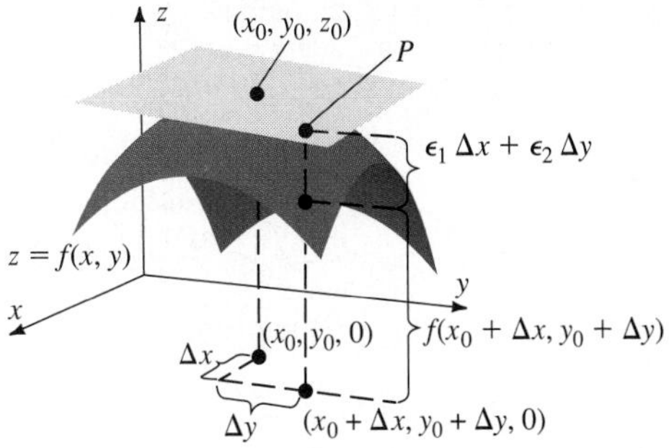

Figure 5.1   The point P on the tangent plane approximates the point on the graph of $z = f(x, y)$ corresponding to $(x_0 + \Delta x, y_0 + \Delta y)$. The coordinates of P are $(x_0 + \Delta x, y_0 + \Delta y, f_x \Delta x + f_y \Delta y)$.

which is precisely the right-hand side of approximation (4). Thus, the left-hand side of approximation (4) is the z-coordinate of the point above $(x_0 + \Delta x, y_0 + \Delta y)$ *on the graph of f,* the right-hand side of (4) is the z-coordinate of the point above $(x_0 + \Delta x, y_0 + \Delta y)$ *on the tangent plane,* and the difference between these two numbers is the "error term," $\epsilon_1 \Delta x + \epsilon_2 \Delta y$.

The proof of Theorem 5 is given at the end of this section.

☐ **EXAMPLE 1**

To see how this works in a particular example, consider the function $f(x, y) = 2x^2 + 4y^2$ and the problem of calculating $f(1 + \Delta x, 2 + \Delta y)$. Here

$$(x_0, y_0) = (1, 2)$$
$$f(x_0, y_0) = f(1, 2) = 2 \cdot 1^2 + 4 \cdot 2^2 = 18$$
$$\frac{\partial f}{\partial x}(x_0, y_0) = 4x \bigg|_{x=1} = 4$$
$$\frac{\partial f}{\partial y}(x_0, y_0) = 8y \bigg|_{y=2} = 16$$

and

$$
\begin{aligned}
f(1 + \Delta x, 2 + \Delta y) &= 2(1 + \Delta x)^2 + 4(2 + \Delta y)^2 \qquad\qquad (6) \\
&= 2(1 + 2\,\Delta x + \Delta x^2) + 4(4 + 4\,\Delta y + \Delta y^2) \\
&= 18 + 4\,\Delta x + 16\,\Delta y + 2\,\Delta x^2 + 4\,\Delta y^2 \\
&= 18 + 4\,\Delta x + 16\,\Delta y + (2\,\Delta x)\Delta x + (4\,\Delta y)\Delta y.
\end{aligned}
$$

$$\underbrace{}_{f(1,2)} \quad \underbrace{\phantom{\frac{\partial f}{\partial x}(1,2)}}_{\frac{\partial f}{\partial x}(1,2)} \quad \underbrace{\phantom{\frac{\partial f}{\partial y}(1,2)}}_{\frac{\partial f}{\partial y}(1,2)} \quad \underbrace{}_{\epsilon_1} \ \underbrace{}_{\epsilon_2}$$

Thus, the approximation (4) is

$$f(1 + \Delta x, 2 + \Delta y) \approx 18 + 4\,\Delta x + 16\,\Delta y. \qquad\qquad (7)$$

By comparing (6) and (7), we can see that the error in this approximation is

$$\epsilon_1 \Delta x + \epsilon_2 \Delta y = 2(\Delta x)^2 + 4(\Delta y)^2.$$

Thus, $\epsilon_1 = 2\,\Delta x$ and $\epsilon_2 = 4\,\Delta y$ in this example.   ■

□ **EXAMPLE 2**

Use approximation (4) to estimate the value of the expression $\sqrt{(3.04)^2 + (3.95)^2}$.

Solution If we let $f(x, y) = \sqrt{x^2 + y^2}$, we must approximate $f(3.04, 3.95)$. For $x_0 = 3$ and $y_0 = 4$ it is easy to compute

$$f(x_0, y_0) = \sqrt{3^2 + 4^2} = \sqrt{25} = 5.$$

We therefore write

$$3.04 = x_0 + \Delta x \quad \text{and} \quad 3.95 = y_0 + \Delta y,$$

where

$$x_0 = 3, \quad \Delta x = 0.04, \quad y_0 = 4, \quad \text{and} \quad \Delta y = -0.05.$$

Also

$$\frac{\partial f}{\partial x}(x_0, y_0) = \frac{\partial}{\partial x}(\sqrt{x^2 + y^2})\Big|_{(3, 4)} = \frac{3}{\sqrt{3^2 + 4^2}} = 0.6$$

and

$$\frac{\partial f}{\partial y}(x_0, y_0) = \frac{\partial}{\partial y}(\sqrt{x^2 + y^2})\Big|_{(3, 4)} = \frac{4}{\sqrt{3^2 + 4^2}} = 0.8.$$

Thus, by (4),

$$\sqrt{(3.04)^2 + (3.95)^2} \approx 5 + (0.6)(0.04) + (0.8)(-0.05) = 4.984.$$

The actual value of this expression, to four decimal places, is 4.9844. The *relative error* (see Section 3.7) in the approximation is therefore

$$\frac{4.9844 - 4.984}{4.9844} \approx .00008,$$

a *percentage error* of less than 0.01%. ∎

Approximation (4) is written in a form that is useful in approximating a particular value $f(x_0 + \Delta x, y_0 + \Delta y)$. By using the notation

$$\Delta f = f(x_0 + \Delta x, y_0 + \Delta y) - f(x_0, y_0),$$

we may rewrite approximation (4) as

$$\boxed{\Delta f \approx \frac{\partial f}{\partial x}(x_0, y_0)\,\Delta x + \frac{\partial f}{\partial y}(x_0, y_0)\,\Delta y.} \qquad (8)$$

Approximation (8) is useful in approximating the *change* in the value of the function f due to changes Δx and Δy in the independent variables (see Figure 5.2).

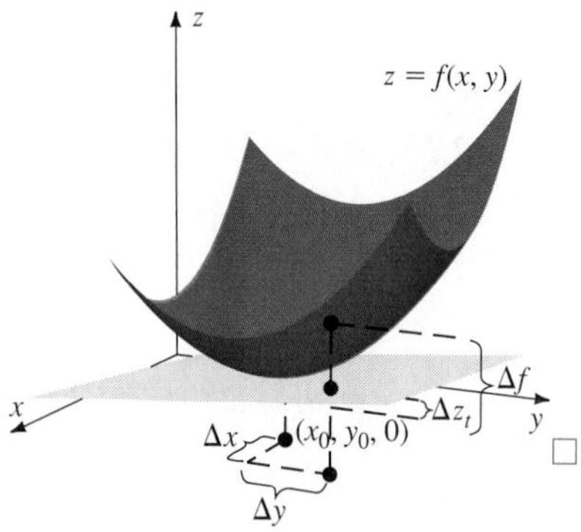

Figure 5.2 The change Δf in the function f from (x_0, y_0) to $(x_0 + \Delta x, y_0 + \Delta y)$ is approximated by the change Δz_t in z along the tangent plane.

□ **EXAMPLE 3**

The ideal gas law is $PV = nRT$, where P is pressure, V is volume, n is the number of moles of gas present, R is constant, and T is temperature. Approximate, according to the ideal gas law, the change in the volume of 1000 cc of gas at temperature 300 K and pressure 780 mm of mercury if the gas is heated by 10 K and the pressure is increased by 5 mm of mercury.

Solution   We first simplify the gas law by solving for the constant nR and using the given data. We find that

$$nR = \frac{PV}{T} = \frac{780 \cdot 1000}{300} = 2600.$$

Using this constant, we may solve the gas law for V as a function of T and P:

$$V(T, P) = \frac{nRT}{P} = \frac{2600T}{P}.$$

According to approximation (8),

$$\Delta V \approx \frac{\partial}{\partial T}\left(\frac{2600T}{P}\right)\Delta T + \frac{\partial}{\partial P}\left(\frac{2600T}{P}\right)\Delta P$$

$$= \frac{2600\,\Delta T}{P} - \frac{2600T\,\Delta P}{P^2}.$$

Using the data $T = 300$, $P = 780$, $\Delta T = 10$, and $\Delta P = 5$, we obtain

$$\Delta V \approx \frac{2600(10)}{780} - \frac{2600(300)(5)}{780^2} \tag{9}$$

$$= 26.92 \text{ cc.}$$

The actual value of V corresponding to the temperature $T = 310$ and pressure $P = 785$ mm of mercury is

$$V = \frac{2600(310)}{785} = 1026.75,$$

so the actual value of ΔV is 26.75. The approximation in (9) is therefore in error by $|26.75 - 26.92| = 0.17$ cc, a percentage error of only

$$\left(\frac{0.17}{26.75}\right) \times 100\% = 0.64\%. \qquad \blacksquare$$

Differentials

The preceding ideas may be written more compactly using the terminology of **differentials.** Recall that in the one-variable case the differential df for the differentiable function f is defined to be

$$df = f'(x)\,dx. \tag{10}$$

The differential in (10) provides an approximation to the change Δf in $f(x)$, corresponding to small changes in x (see Section 3.7).

For precisely the same reason, we define the differential of the function f of two variables to be

$$df = \frac{\partial f}{\partial x}(x, y)\,dx + \frac{\partial f}{\partial y}(x, y)\,dy. \tag{11}$$

The differential df provides an approximation to the change Δf corresponding to small changes in x and y. The differential in equation (11) is sometimes referred to as the **total differential** for the function $z = f(x, y)$.

□ **EXAMPLE 4**

In economic theory, the Cobb-Douglas production function relating output production y to the input of labor L and capital K has the form

$$y = \lambda L^\alpha K^{1-\alpha}$$

where λ and α are positive constants. The change Δy in production resulting from a change dL in labor input and a change dK in capital input is therefore approximated by the total differential

$$dy = \frac{\partial}{\partial L}(\lambda L^\alpha K^{1-\alpha})\, dL + \frac{\partial}{\partial K}(\lambda L^\alpha K^{1-\alpha})\, dK$$

$$= \alpha \lambda L^{\alpha-1} K^{1-\alpha}\, dL + (1-\alpha)\lambda L^\alpha K^{-\alpha}\, dK. \qquad ■$$

Differentiability

In Section 18.3, we derived an equation for the tangent plane to the graph of $z = f(x, y)$, assuming that such a plane existed. Using the ideas discussed in this section, we can now be more precise about the existence of a tangent plane. In fact, we can also generalize the notion of differentiability to functions of more than one variable. In Section 18.2, we discussed the function

$$f(x, y) = \begin{cases} \dfrac{-xy}{x^2 + y^2}, & (x, y) \neq (0, 0); \\ 0, & (x, y) = (0, 0); \end{cases}$$

whose partial derivatives $\partial f/\partial x$ and $\partial f/\partial y$ exist at (x_0, y_0) even though the function f itself is not continuous at (x_0, y_0). The concept of differentiability precludes this exceptional behavior.

Recall Theorem 9 from Section 3.7: If f is a function of one variable that is differentiable at x_0, then the error function

$$\epsilon = f(x_0 + \Delta x) - [f(x_0) + f'(x_0)(x - x_0)]$$

satisfies

$$\lim_{\Delta x \to 0} \frac{\epsilon}{\Delta x} = 0.$$

It is precisely this limiting property of ϵ that generalizes the notion of differentiability. Consider a function f of two variables. If the partial derivatives $\partial f/\partial x$ and $\partial f/\partial y$ exist at (x_0, y_0), then we consider the error function

$$\epsilon = f(x_0 + \Delta x, y_0 + \Delta y) -$$
$$\left[f(x_0, y_0) + \frac{\partial f}{\partial x}(x_0, y_0)(x - x_0) + \frac{\partial f}{\partial y}(x_0, y_0)(y - y_0) \right].$$

Since the distance between (x_0, y_0) and $(x_0 + \Delta x, y_0 + \Delta y)$ is $\sqrt{\Delta x^2 + \Delta y^2}$, we consider the limit

$$\lim_{\substack{\Delta x \to 0 \\ \Delta y \to 0}} \frac{\epsilon}{\sqrt{\Delta x^2 + \Delta y^2}}. \qquad (12)$$

If this limit is 0, we say that f is **differentiable** at (x_0, y_0). It is a fact, which we shall not prove, that a function is continuous at (x_0, y_0) if it is differentiable there.

Since the graph of

$$z = f(x_0, y_0) + \frac{\partial f}{\partial x}(x_0, y_0)(x - x_0) + \frac{\partial f}{\partial y}(x_0, y_0)(y - y_0) \tag{13}$$

is a plane, we can interpret the error ϵ as the vertical difference between this plane and the graph of f. If the limit (12) is zero, we say that the plane given by equation (13) is tangent to the graph of f. In advanced calculus texts, it is shown that the tangent plane to a graph of a function is unique, if it exists.

Now we relate the concept of differentiability to the Approximation Theorem (Theorem 5). If the hypothesis of Theorem 5 is satisfied, we know that the error function ϵ can be written as

$$\epsilon = \epsilon_1 \, \Delta x + \epsilon_2 \, \Delta y$$

where $\epsilon_1 \to 0$ and $\epsilon_2 \to 0$ as Δx and $\Delta y \to 0$. It follows from Exercise 32 of this section that

$$\lim_{\substack{\Delta x \to 0 \\ \Delta y \to 0}} \frac{\epsilon_1 \, \Delta x + \epsilon_2 \, \Delta y}{\sqrt{\Delta x^2 + \Delta y^2}} = 0.$$

Consequently, if a function f and its partial derivatives $\partial f/\partial x$ and $\partial f/\partial y$ are continuous on an open rectangle R, then f is differentiable at all points of R. Moreover, the graph of f has a tangent plane at each point $(x_0, y_0, f(x_0, y_0))$ corresponding to a point (x_0, y_0) in R. Since it is usually easier to verify the hypothesis of Theorem 5 than to verify that limit (12) is 0, we mostly consider continuous functions whose partial derivatives are also continuous. Differentiability then follows from Theorem 5.

Functions of Three Variables

The ideas of this section generalize directly to functions of three variables. The corresponding version of the Approximation Theorem is the following.

Theorem 5′

Let f and its partial derivatives be continuous in the open rectangular box

$$B = \{(x, y, z) \mid a_1 < x < a_2, b_1 < y < b_2, c_1 < z < c_2\}$$

in $\mathbb{R}^3$. If (x_0, y_0, z_0) and $(x_0 + \Delta x, y_0 + \Delta y, z_0 + \Delta z)$ both lie in B, then

$$f(x_0 + \Delta x, y_0 + \Delta y, z_0 + \Delta z) = f(x_0, y_0, z_0) + \frac{\partial f}{\partial x}(x_0, y_0, z_0)\Delta x$$

$$+ \frac{\partial f}{\partial y}(x_0, y_0, z_0)\Delta y + \frac{\partial f}{\partial z}(x_0, y_0, z_0)\Delta z$$

$$+ \epsilon_1 \, \Delta x + \epsilon_2 \, \Delta y + \epsilon_3 \, \Delta z$$

where $\epsilon_1 \to 0$, $\epsilon_2 \to 0$, and $\epsilon_3 \to 0$ as $\Delta x \to 0$, $\Delta y \to 0$, and $\Delta z \to 0$.

The proof of Theorem 5′ is similar to that of Theorem 5 (see Exercise 31). Finally, the definition of the differential, suggested by Theorem 5′, is

$$\boxed{df = \frac{\partial f}{\partial x}(x, y, z) \, dx + \frac{\partial f}{\partial y}(x, y, z) \, dy + \frac{\partial f}{\partial z}(x, y, z) \, dz}$$

or

$$dw = f_x \, dx + f_y \, dy + f_z \, dz$$

where $w = f(x, y, z)$.

□ **EXAMPLE 5**

According to Newton's Law of Universal Gravitation, the force of attraction between two bodies of mass m_1 and m_2 is

$$F(m_1, m_2, r) = \frac{Gm_1m_2}{r^2},$$

where r is the distance between the bodies and G is the universal gravitation constant. The change ΔF in this force caused by changes dm_1 and dm_2 in the masses and dr in the distance is approximated by the differential

$$dF = \frac{\partial}{\partial m_1}\left(\frac{Gm_1m_2}{r^2}\right) dm_1 + \frac{\partial}{\partial m_2}\left(\frac{Gm_1m_2}{r^2}\right) dm_2 + \frac{\partial}{\partial r}\left(\frac{Gm_1m_2}{r^2}\right) dr$$

$$= \frac{Gm_2}{r^2} \, dm_1 + \frac{Gm_1}{r^2} \, dm_2 - \frac{2Gm_1m_2}{r^3} \, dr. \qquad ■$$

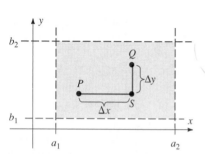

Figure 5.3 Diagram for proof of Theorem 5.

Proof of Theorem 5: Let $P = (x_0, y_0)$ and $Q = (x_0 + \Delta x, y_0 + \Delta y)$. Since R is a rectangle, the point $S = (x_0 + \Delta x, y_0)$ also lies in R, as do the line segments PS and SQ (see Figure 5.3).

Along the line segment PS the variable $y = y_0$ is fixed, so $g(x) = f(x, y_0)$ is a function of x alone, and $g'(x) = \partial f/\partial x$ exists at (x, y_0) for all $x \in (x_0, x_0 + \Delta x)$. Thus, by the Mean Value Theorem, there exists a number $c \in (x_0, x_0 + \Delta x)$ so that

$$\frac{f(x_0 + \Delta x, y_0) - f(x_0, y_0)}{\Delta x} = \frac{\partial f}{\partial x}(c, y_0). \qquad (14)$$

Similarly, along the line segment SQ, the function $h(y) = f(x_0 + \Delta x, y)$ is a function of y alone, and $h'(y) = \partial f/\partial y$ at $(x_0 + \Delta x, y)$ for each $y \in (y_0, y_0 + \Delta y)$. Again by the Mean Value Theorem, there exists a number $d \in (y_0, y_0 + \Delta y)$ so that

$$\frac{f(x_0 + \Delta x, y_0 + \Delta y) - f(x_0 + \Delta x, y_0)}{\Delta y} = \frac{\partial f}{\partial y}(x_0 + \Delta x, d). \qquad (15)$$

Next we multiply both sides of equation (14) by Δx and both sides of equation (15) by Δy. Adding the resulting equations gives

$$f(x_0 + \Delta x, y_0 + \Delta y) - f(x_0, y_0) = \frac{\partial f}{\partial x}(c, y_0) \, \Delta x + \frac{\partial f}{\partial y}(x_0 + \Delta x, d) \, \Delta y. \qquad (16)$$

We now invoke the continuity of the partial derivatives. Since c lies between x_0 and $x_0 + \Delta x$, we have $c \to x_0$ as $\Delta x \to 0$. Thus, since $\partial f/\partial x$ is continuous,

$$\lim_{\Delta x \to 0} \frac{\partial f}{\partial x}(c, y_0) = \frac{\partial f}{\partial x}(x_0, y_0). \qquad (17)$$

Another way to write equation (17) is simply

$$\frac{\partial f}{\partial x}(c, y_0) = \frac{\partial f}{\partial x}(x_0, y_0) + \epsilon_1 \qquad (18)$$

where $\epsilon_1 \to 0$ as $\Delta x \to 0$. Similarly, since d lies between y_0 and $y_0 + \Delta y$, $d \to y_0$ as $\Delta y \to 0$. Thus, since $\partial f/\partial y$ is continuous, we conclude that

$$\lim_{\substack{\Delta x \to 0 \\ \Delta y \to 0}} \frac{\partial f}{\partial y}(x_0 + \Delta x, d) = \frac{\partial f}{\partial y}(x_0, y_0),$$

or that

$$\frac{\partial f}{\partial y}(x_0 + \Delta x, d) = \frac{\partial f}{\partial y}(x_0, y_0) + \epsilon_2 \tag{19}$$

where $\epsilon_2 \to 0$ as $\Delta x \to 0$ and $\Delta y \to 0$.

Combining statements (16), (17), and (19) now gives

$$f(x_0 + \Delta x, y_0 + \Delta y) = f(x_0, y_0) + \frac{\partial f}{\partial x}(x_0, y_0)\,\Delta x + \frac{\partial f}{\partial y}(x_0, y_0)\,\Delta y$$

$$+ \epsilon_1\,\Delta x + \epsilon_2\,\Delta y$$

where $\epsilon_1 \to 0$ and $\epsilon_2 \to 0$ as $\Delta x \to 0$ and $\Delta y \to 0$ as desired.   ∎

Exercise Set 18.5

In Exercises 1–12, find the total differential df.

1. $f(x, y) = x^2 y^4$

2. $f(x, y) = x^{2/3} y^{5/2}$

3. $f(x, y) = \sqrt{x^2 + y^4}$

4. $f(x, y) = x \sin y^2$

5. $f(x, y) = e^{\sqrt{x}} \cos y$

6. $f(x, y) = \text{Tan}^{-1}(y/x)$

7. $f(x, y, z) = x^2 y z^3$

8. $f(x, y, z) = \ln(x^2 + 3y + z^3)$

9. $f(x, y, z) = \dfrac{x}{y + 3^z}$

10. $f(x, y, z) = \sqrt{\dfrac{x}{y^2 + z^2}}$

11. $f(x, y, z) = \dfrac{x - y}{x^2 + y^2 + z^2}$

12. $f(x, y, z) = z e^{x^2 - y^3}$

13. Let $f(x, y) = 2xy^2 + x^2 y$, $x_0 = 2$, $y_0 = 3$, $\Delta x = 0.1$, and $\Delta y = 0.2$.
  a. Calculate $f(x_0, y_0)$ and $f(x_0 + \Delta x, y_0 + \Delta y)$.
  b. Calculate $\Delta f = f(x_0 + \Delta x, y_0 + \Delta y) - f(x_0, y_0)$.
  c. Approximate Δf using the linear approximation given in equation (8).
  d. Determine the relative and percentage error involved in this approximation.

14. Let $f(x, y) = \sin x \cos y$, $x_0 = \pi/2$, $y_0 = \pi/4$, $\Delta x = \pi/16$, and $\Delta y = -\pi/16$. Answer the questions in Exercise 13.

In Exercises 15–22, use the linear approximation (8) to approximate the given number.

15. $\sqrt{(3.02)^2 + (4.08)^2}$

16. $\sqrt{9.4} + \sqrt{15.6}$

17. $\sin 88° \cos 42°$

18. $(2.02)^3 + (3.96)^3$

19. $(5.03)^2(1.02)^3$

20. $(5.94)\ln(1.15)$

21. $(6.04)(3.1)(2.96)$

22. $\sqrt{(12.1)^2 + (4.05)^2 + (2.96)^2}$

23. A cylindrical bearing of radius 2 cm and length 6 cm is dipped in a molten brass solution to produce a brass coating. If the thickness of the coating increases from 0.02 cm to 0.06 cm, approximate the volume of the coated bearing.

24. The money supply in the economy is defined by the equation $M = C + D$ where C is the currency outstanding in banks and D is the amount of money in demand and term deposits. Use differentials to approximate the percentage change in the money supply caused by a 10% increase in the outstanding currency and a 5% decrease in total deposits.

25. For the Cobb-Douglas production function $P = \lambda L^3 K^{3/2}$, use differentials to approximate the percentage change in P caused by a 4% increase in labor L and a 10% decrease in capital K.

26. For the Cobb-Douglas production function $P = \lambda L^\alpha K^\beta$, what can you say about α and β if a percentage decrease in labor of γ percent, combined with a percentage increase in capital of γ percent, produces no change in P?

27. For small oscillations, the period of a simple pendulum is given by the equation

$$T = 2\pi\sqrt{s/g}$$

where s is the length of the pendulum, g is the acceleration due to gravity, and T is time. (If the units for s and g involve meters and seconds, the units for T are seconds.) What is the approximate error in the calculation of T resulting from its calculation using $s = 10$ m and $g = 9.8$ m/s^2 if the actual values of these quantities are $s = 10.2$ and $g = 9.75$?

28. Approximate $f(x_0, y_0, z_0)$ for

$$f(x, y, z) = \frac{\sqrt{x^2 - y^2}}{\tan z}$$

and $(x_0, y_0, z_0) = (5.04, 3.97, 0.78)$.
(*Hint:* $\pi/4 \approx 0.785398$.)

29. When three resistors with resistances R_1, R_2, and R_3 are connected in parallel, the resistance of the resulting circuit is

$$\frac{1}{R} = \frac{1}{R_1} + \frac{1}{R_2} + \frac{1}{R_3}.$$

What is the maximum percentage change in R resulting from a change of no more than 10% in each of R_1, R_2, and R_3?

30. The work involved in an adiabatic (no heat lost to the surroundings) expansion of an ideal gas is given by the equation

$$w = nC_v(T_2 - T_1)$$

where C_v is the average heat capacity of the gas, and T_1 and T_2 are the initial and final temperatures. Show that if both T_1 and T_2 are increased by λ percent, the amount of work involved is also increased by λ percent.

31. Prove Theorem 5', using the Mean Value Theorem three times.

32. Show that $\displaystyle\lim_{\substack{\Delta x \to 0 \\ \Delta y \to 0}} \frac{\epsilon_1 \Delta x + \epsilon_2 \Delta y}{\sqrt{\Delta x^2 + \Delta y^2}} = 0$ if

$$\lim_{\substack{\Delta x \to 0 \\ \Delta y \to 0}} \epsilon_1 = 0 \quad \text{and} \quad \lim_{\substack{\Delta x \to 0 \\ \Delta y \to 0}} \epsilon_2 = 0.$$

(*Hint:* Write this as two separate fractions and use the fact that $\left| \dfrac{\Delta x}{\sqrt{\Delta x^2 + \Delta y^2}} \right| \le 1$.)

18.6   Chain Rules

In this section, we use the Approximation Theorem to establish a generalization of the Chain Rule for functions of a single variable:

$$(f \circ g)'(x) = f'(g(x))g'(x).$$

Our generalization concerns the *composite* function

$$w(t) = f(x(t), y(t), z(t)), \qquad t \in (a, b).$$

It is important to note that this composite function is just a function of a single variable. Our interest is in knowing when the derivative $w'(t)$ exists and, if it does, how it may be calculated.

Before stating the theorem that answers this question, we describe a simple setting in which a composite function of this type occurs. Imagine a region of airspace, such as that over the state of California, through which an airplane is flying. The region of airspace can be described by xyz-coordinates (which represent latitude, longitude, and altitude, if you prefer), and the path of the airplane can be thought of as a curve parameterized by the vector function $\mathbf{r}(t) = x(t)\mathbf{i} + y(t)\mathbf{j} + z(t)\mathbf{k}$, $a \le t \le b$, giving the location of the airplane at time t. (See Figure 6.1.)

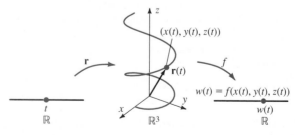

Figure 6.1   If $r{:}\mathbb{R} \to \mathbb{R}^3$ and $f{:}\mathbb{R}^3 \to \mathbb{R}$, the composite function $w(t) = f(r(t)) = f(x(t), y(t), z(t))$ maps $\mathbb{R}$ to $\mathbb{R}$.

Assuming a steady state temperature distribution throughout the region, we let $T(x, y, z)$ be the temperature at location (x, y, z) in the airspace. *The composite function*

$$w(t) = T(x(t), y(t), z(t))$$

gives the temperature outside the airplane at time t.

What factors determine the rate dw/dt at which the temperature outside the airplane changes? Obviously dx/dt, dy/dt, and dz/dt are involved, since they are the components of the velocity of the airplane. But dw/dt will also depend upon the partial derivatives $\partial T/\partial x$, $\partial T/\partial y$, and $\partial T/\partial z$, since these determine the rate of change of temperature with respect to change in position. The precise relationship between these rates is given by the following theorem.

Theorem 6

Chain Rule

Let f be a continuous function with continuous partial derivatives for all (x, y, z) in an open "box"

$$Q = \{(x, y, z) \mid a_1 < x < b_1, a_2 < y < b_2, a_3 < z < b_3\}$$

in space. Assume that x, y, and z are functions of t so that $x'(t)$, $y'(t)$, and $z'(t)$ exist for each $t \in (a, b)$, and so that the point $(x(t), y(t), z(t))$ lies in Q for each $t \in (a, b)$. Then the composite function

$$w(t) = f(x(t), y(t), z(t))$$

is a differentiable function of $t \in (a, b)$, and

$$\frac{dw}{dt} = \frac{\partial f}{\partial x}\frac{dx}{dt} + \frac{\partial f}{\partial y}\frac{dy}{dt} + \frac{\partial f}{\partial z}\frac{dz}{dt}. \tag{1}$$

The proof of Theorem 6 is given at the end of this section.

□ **EXAMPLE 1**

Let $f(x, y, z) = \sqrt{x}\, y^2 e^{2z}$, $x(t) = 3t^2 + 2$, $y(t) = 6t$, $z(t) = 1 - t^3$, and $w(t) = f(x(t), y(t), z(t))$. Find $w'(t)$.

Solution We have

$$\frac{\partial f}{\partial x} = \frac{\partial}{\partial x}(\sqrt{x}\, y^2 e^{2z}) = \frac{y^2 e^{2z}}{2\sqrt{x}}$$

$$\frac{\partial f}{\partial y} = \frac{\partial}{\partial y}(\sqrt{x}\, y^2 e^{2z}) = 2\sqrt{x}\, y e^{2z}$$

and

$$\frac{\partial f}{\partial z} = \frac{\partial}{\partial z}(\sqrt{x}\, y^2 e^{2z}) = 2\sqrt{x}\, y^2 e^{2z}.$$

Also,

$$x'(t) = 6t, \quad y'(t) = 6, \quad \text{and} \quad z'(t) = -3t^2.$$

According to Theorem 6 we must have

$$w'(t) = \frac{y^2 e^{2z}}{2\sqrt{x}}(6t) + 2\sqrt{x}\, y e^{2z}(6) + 2\sqrt{x}\, y^2 e^{2z}(-3t^2)$$

$$= \frac{108t^3 e^{2(1-t^3)}}{\sqrt{3t^2 + 2}} + 72t\sqrt{3t^2 + 2}\, e^{2(1-t^3)} - 216t^4\sqrt{3t^2 + 2}\, e^{2(1-t^3)}$$

$$= \sqrt{3t^2 + 2}\, e^{2(1-t^3)}\left[\frac{108t^3}{3t^2 + 2} + 72t - 216t^4\right]. \qquad \blacksquare$$

Of course, we can also calculate $w'(t)$ in Example 1 by determining $w(t)$ as an explicit function of t and then differentiating using our techniques for functions of one variable. Nevertheless, it is important to be competent with the Chain Rule, as stated in Theorem 6 and as used in the solution to Example 1. It plays an important role in the calculus of multivariable functions.

☐ **EXAMPLE 2**

A function giving the temperature $T(x, y, z)$ at a point (x, y, z) in space is

$$T(x, y, z) = \lambda\sqrt{x^2 + y^2 + z^2}$$

where λ is constant. Find the rate of change of temperature with respect to t along the elliptical helix

$$\mathbf{r}(t) = a\cos t\mathbf{i} + b\sin t\mathbf{j} + ct\mathbf{k}.$$

Solution The temperature function along the helix is the composite function $T(x(t), y(t), z(t))$ where

$$x(t) = a\cos t, \qquad y(t) = b\sin t, \qquad z(t) = ct.$$

Using equation (1), we obtain the desired rate as

$$\frac{dT}{dt} = \frac{\partial}{\partial x}(\lambda\sqrt{x^2 + y^2 + z^2})\frac{d}{dt}(a\cos t)$$

$$+ \frac{\partial}{\partial y}(\lambda\sqrt{x^2 + y^2 + z^2})\frac{d}{dt}(b\sin t) + \frac{\partial}{\partial z}(\lambda\sqrt{x^2 + y^2 + z^2})\frac{d}{dt}(ct)$$

$$= \frac{\lambda x(-a\sin t)}{\sqrt{x^2 + y^2 + z^2}} + \frac{\lambda y b\cos t}{\sqrt{x^2 + y^2 + z^2}} + \frac{\lambda cz}{\sqrt{x^2 + y^2 + z^2}}$$

$$= \frac{\lambda(-a^2 + b^2)\sin t\cos t + \lambda c^2 t}{\sqrt{a^2\cos^2 t + b^2\sin^2 t + c^2 t^2}}. \qquad \blacksquare$$

☐ **EXAMPLE 3**

The **electric potential** at an axial point for a charge disc is given by the equation

$$V = \frac{\sigma}{2\epsilon_0}(\sqrt{a^2 + r^2} - r)$$

where a is the radius of the disc, r is the distance from the point to the disc, and σ and ϵ_0 are constants. If the radius of the disc is increasing at a rate of 2 cm/s and the point is moving away from the disc at a rate of 5 cm/s, find the rate at which the electric potential is changing when $a = 3$ cm and $r = 4$ cm.

Solution Since V is a function of only two variables, a and r, we simply consider the third variable in the Chain Rule (1) to be $z = 0$, so $dz/dt = 0$. Applying the Chain Rule we find

$$\frac{dV}{dt} = \frac{\partial}{\partial a}\left[\frac{\sigma}{2\epsilon_0}(\sqrt{a^2 + r^2} - r)\right]\frac{da}{dt} \tag{2}$$

$$+ \frac{\partial}{\partial r}\left[\frac{\sigma}{2\epsilon_0}(\sqrt{a^2 + r^2} - r)\right]\frac{dr}{dt}$$

$$= \frac{\sigma}{2\epsilon_0}\left(\frac{a}{\sqrt{a^2 + r^2}}\right)\frac{da}{dt} + \frac{\sigma}{2\epsilon_0}\left(\frac{r}{\sqrt{a^2 + r^2}} - 1\right)\frac{dr}{dt}.$$

We are given

$$\frac{da}{dt} = 2 \text{ cm/s}, \quad \frac{dr}{dt} = 5 \text{ cm/s}, \quad a = 3 \text{ cm}, \quad \text{and} \quad r = 4 \text{ cm}.$$

Substituting these values into (2) gives

$$\frac{dV}{dt} = \frac{\sigma}{2\epsilon_0}\left(\frac{3}{\sqrt{3^2 + 4^2}}\right)(2) + \frac{\sigma}{2\epsilon_0}\left(\frac{4}{\sqrt{3^2 + 4^2}} - 1\right)(5)$$

$$= \frac{\sigma}{10\epsilon_0}. \qquad\blacksquare$$

Differentiating Other Types of Composite Functions

Various other sorts of composite functions can be formed using functions of several variables, each of which may be differentiated using the Chain Rule. For example, if

$$x = x(s, t), \quad y = y(s, t), \quad \text{and} \quad z = z(s, t)$$

are functions of two variables, and if

$$w = f(x, y, z)$$

is a function of three variables, the composite function

$$w(s, t) = f(x(s, t), y(s, t), z(s, t))$$

is again a function of two variables. To obtain $\partial w/\partial s$, we hold t constant and differentiate the resulting function of a single variable $g(s) = w(s, t)$ according to Theorem 6. Similarly, the partial derivative $\partial w/\partial t$ is obtained by holding s constant and differentiating with respect to t.

☐ **EXAMPLE 4**

Let $f(x, y) = x^2 y^3$ where the variables x and y are functions of the polar variables r and θ:

$$x(r, \theta) = r \cos \theta \quad \text{and} \quad y(r, \theta) = r \sin \theta.$$

To find the partial derivative $\partial f/\partial r$, we treat θ as a constant, leaving x and y as functions of the single independent variable r. We then apply Theorem 6:

$$\frac{\partial f}{\partial r} = \frac{\partial f}{\partial x}\frac{\partial x}{\partial r} + \frac{\partial f}{\partial y}\frac{\partial y}{\partial r}$$

$$= (2xy^3)(\cos \theta) + (3x^2 y^2)(\sin \theta)$$

$$= (2r^4 \cos \theta \sin^3 \theta)(\cos \theta) + (3r^4 \cos^2 \theta \sin^2 \theta)(\sin \theta)$$

$$= 5r^4 \cos^2 \theta \sin^3 \theta. \qquad\blacksquare$$

Higher Order Derivatives

Using Theorem 6, we can determine formulas for higher order ordinary and partial derivatives of composite functions. For instance, in the next example, we consider the standard rectangular coordinates x and y on $\mathbb{R}^2$ as functions of the polar coordinates r and θ, and we determine an expression for $\partial^2 f/\partial r^2$ in terms of the second order partials of f with respect to x and y.

□ **EXAMPLE 5**

Let f be an arbitrary function of two variables with continuous second order partial derivatives. Express $\partial^2 f/\partial r^2$ in terms of the second order partial derivatives of f with respect to x and y.

Solution Our first application of the Chain Rule yields

$$\frac{\partial f}{\partial r} = \frac{\partial f}{\partial x}\frac{\partial x}{\partial r} + \frac{\partial f}{\partial y}\frac{\partial y}{\partial r}$$

$$= \left[\frac{\partial f}{\partial x}\right](\cos\theta) + \left[\frac{\partial f}{\partial y}\right](\sin\theta). \tag{3}$$

Now we must apply the Chain Rule both to $\partial f/\partial x$ and $\partial f/\partial y$. Differentiating $\partial f/\partial x$ with respect to r, we obtain

$$\frac{\partial}{\partial r}\left(\frac{\partial f}{\partial x}\right) = \frac{\partial^2 f}{\partial x^2}\frac{\partial x}{\partial r} + \frac{\partial^2 f}{\partial y\partial x}\frac{\partial y}{\partial r}$$

$$= \frac{\partial^2 f}{\partial x^2}\cos\theta + \frac{\partial^2 f}{\partial y\partial x}\sin\theta. \tag{4}$$

Similarly, we obtain

$$\frac{\partial}{\partial r}\left(\frac{\partial f}{\partial y}\right) = \frac{\partial^2 f}{\partial x\partial y}\cos\theta + \frac{\partial^2 f}{\partial y^2}\sin\theta. \tag{5}$$

Since θ is held constant in these calculations and since the mixed partials of f are equal, we combine equations (3)–(5) to obtain

$$\frac{\partial^2 f}{\partial r^2} = \frac{\partial^2 f}{\partial x^2}\cos^2\theta + 2\frac{\partial^2 f}{\partial y\partial x}\sin\theta\cos\theta + \frac{\partial^2 f}{\partial y^2}\sin^2\theta. \qquad ■$$

Proof of Theorem 6: Let $t_0 \in (a, b)$ and let $\Delta t \neq 0$ be sufficiently small so that $(t_0 + \Delta t)$ also lies in (a, b). Let $x_0 = x(t_0)$, $y_0 = y(t_0)$, $z_0 = z(t_0)$, and

$$\Delta x = x(t_0 + \Delta t) - x(t_0)$$
$$\Delta y = y(t_0 + \Delta t) - y(t_0)$$
$$\Delta z = z(t_0 + \Delta t) - z(t_0).$$

Then, according to the Approximation Theorem (Theorem 5′),

$$w(t_0 + \Delta t) = w(t_0) + \frac{\partial f}{\partial x}(x_0, y_0, z_0)\,\Delta x + \frac{\partial f}{\partial y}(x_0, y_0, z_0)\,\Delta y \tag{6}$$

$$+ \frac{\partial f}{\partial z}(x_0, y_0, z_0)\,\Delta z + \epsilon_1\Delta x + \epsilon_2\Delta y + \epsilon_3\Delta z$$

where $\epsilon_1 \to 0$, $\epsilon_2 \to 0$, and $\epsilon_3 \to 0$ as $\Delta x \to 0$, $\Delta y \to 0$, and $\Delta z \to 0$.

Subtracting $w(t_0)$ from both sides of equation (6) and dividing by Δt gives the equation

$$\frac{w(t_0 + \Delta t) - w(t_0)}{\Delta t} = \frac{\partial f}{\partial x}(x_0, y_0, z_0)\frac{\Delta x}{\Delta t} + \frac{\partial f}{\partial y}(x_0, y_0, z_0)\frac{\Delta y}{\Delta t} \tag{7}$$

$$+ \frac{\partial f}{\partial z}(x_0, y_0, z_0)\frac{\Delta z}{\Delta t} + \epsilon_1\frac{\Delta x}{\Delta t} + \epsilon_2\frac{\Delta y}{\Delta t} + \epsilon_3\frac{\Delta z}{\Delta t}.$$

Now according to our definitions, we have

$$\lim_{\Delta t \to 0} \frac{\Delta x}{\Delta t} = \lim_{\Delta t \to 0} \frac{x(t_0 + \Delta t) - x(t_0)}{\Delta t} = x'(t_0).$$

Thus,

$$\lim_{\Delta t \to 0}\left[\frac{\partial f}{\partial x}(x_0, y_0, z_0)\frac{\Delta x}{\Delta t}\right] = \frac{\partial f}{\partial x}(x_0, y_0, z_0)\lim_{\Delta t \to 0}\frac{\Delta x}{\Delta t} = \frac{\partial f}{\partial x}(x_0, y_0, z_0)x'(t_0) \tag{8}$$

and

$$\lim_{\Delta t \to 0}\left(\epsilon_1\frac{\Delta x}{\Delta t}\right) = \left(\lim_{\Delta t \to 0}\epsilon_1\right)\left(\lim_{\Delta t \to 0}\frac{\Delta x}{\Delta t}\right) = 0 \cdot x'(t_0) = 0. \tag{9}$$

Equations analogous to (8) and (9) hold for the variables y and z.

Finally, from equation (7) and the equations of the form (8) and (9) for each of the variables x, y, and z, we conclude that

$$w'(t_0) = \lim_{\Delta t \to 0}\frac{w(t_0 + \Delta t) - w(t_0)}{\Delta t}$$

$$= \frac{\partial f}{\partial x}(x_0, y_0, z_0)x'(t_0) + \frac{\partial f}{\partial y}(x_0, y_0, z_0)y'(t_0)$$

$$+ \frac{\partial f}{\partial z}(x_0, y_0, z_0)z'(t_0),$$

as required. ∎

Exercise Set 18.6

In Exercises 1–10, use the Chain Rule (Theorem 6) to find the rate of change df/dt of f along the given curves.

1. $f(x, y) = x^2 + y^2$, $x(t) = 2t$, $y(t) = 6 - t^2$

2. $f(x, y) = x - y^2$, $x(t) = 3t^2 + 2$, $y(t) = 4 + t$

3. $f(x, y) = xy^2$, $\mathbf{r}(t) = \cos t\mathbf{i} + \sin t\mathbf{j}$

4. $f(x, y) = x^2 - y^2$, $\mathbf{r}(t) = e^t\mathbf{i} + e^{-t}\mathbf{j}$

5. $f(x, y) = \sqrt{x + y}$, $x(t) = \sqrt{t}$, $y(t) = e^{t^2}$

6. $f(x, y) = y^2 - x^2$, $x(t) = a\cos t$, $y(t) = a\sin t$

7. $f(x, y, z) = xy - x + z^2$, $\mathbf{r}(t) = t^2\mathbf{i} - 2t\mathbf{j} + \sin t\mathbf{k}$

8. $f(x, y, z) = x^2 + y^2 + z^2$,
$\mathbf{r}(t) = a\cos \pi t\mathbf{i} + b\sin \pi t\mathbf{j} - t^2\mathbf{k}$

9. $f(x, y, z) = z(y^2 - x^2)$, $x(t) = a\cosh t$,
$y(t) = b\sinh t$, $z(t) = e^{-2t}$

10. $f(x, y, z) = e^{xyz}$, $x(t) = \ln \sqrt{t}$, $y(t) = t\sin t$, $z(t) = 2^t$.

In Exercises 11–22, use the Chain Rule (Theorem 6) to find the indicated derivative(s).

11. $f(x, y) = x^2y^3$, $x(t) = \cos t$, $y(t) = t\sin t$. Find df/dt.

12. $f(x, y) = x^2y^3$, $x(s, t) = st$, $y(s, t) = s^2 - t^2$. Find

a. $\dfrac{\partial f}{\partial s}$ **b.** $\dfrac{\partial f}{\partial t}$

13. $f(x, y, z) = x^2 + y^2 - z^2$, $x(s, t) = e^{st}$,
$y(s, t) = st$, $z(s, t) = s - t$. Find

a. $\dfrac{\partial f}{\partial s}$ **b.** $\dfrac{\partial f}{\partial t}$

14. $f(x, y) = xy - y^2$, $x(r, s, t) = rst$, $y(r, s, t) = e^{rst}$. Find

a. $\dfrac{\partial f}{\partial r}$ **b.** $\dfrac{\partial f}{\partial s}$ **c.** $\dfrac{\partial f}{\partial t}$

15. $f(x, y) = \sin(xy^2) - x^2y$, $x(s, t) = s^2 - st$,
$y(s, t) = t^2s^2$. Find

a. $\dfrac{\partial f}{\partial s}$ **b.** $\dfrac{\partial f}{\partial t}$

16. $f(x, y) = x^2 e^{y-x}$, $x(s, t) = s - t$, $y(s, t) = \sqrt{s + t}$. Find

a. $\dfrac{\partial f}{\partial s}$ **b.** $\dfrac{\partial f}{\partial t}$

17. $f(x, y, z) = xy^3z^2$, $x(s, t) = s \sin t$,
$y(s, t) = t \cos(s)$, $z(s, t) = t^2 - s^2$. Find

a. $\dfrac{\partial f}{\partial s}$ **b.** $\dfrac{\partial f}{\partial t}$

18. $f(x, y) = e^{xy}$, $x(r, s) = \sqrt{s^2 - r^2}$,
$y(r, s) = \mathrm{Tan}^{-1}(r/s)$. Find

a. $\dfrac{\partial f}{\partial r}$ **b.** $\dfrac{\partial f}{\partial s}$

19. $f(u, v) = e^{u^2} - e^{2v}$, $u(x, y) = xy^2$, $v(x, y) = x^2y$. Find

a. $\dfrac{\partial f}{\partial x}$ **b.** $\dfrac{\partial f}{\partial y}$

20. $f(x, y, z) = x^2 - xy + yz$, $x(r, \theta) = r \cos \theta$,
$y(r, \theta) = 2r$, $z(r, \theta) = r \sin \theta$. Find

a. $\dfrac{\partial f}{\partial r}$ **b.** $\dfrac{\partial f}{\partial \theta}$

21. $f(r, \theta) = r^2(1 - \cos \theta)$, $r(t) = 1 + t^3$,
$\theta(t) = \sqrt{1 + t^2}$. Find df/dt.

22. $f(x, y, z) = \ln(x^2 + y^2 + z^2)$, $x(u, v, w) = u \cos v$,
$y(u, v, w) = v \sin u$, $z(u, v, w) = uvw$. Find

a. $\dfrac{\partial f}{\partial u}$ **b.** $\dfrac{\partial f}{\partial v}$ **c.** $\dfrac{\partial f}{\partial w}$

23. The radius of the base of a cone is 6 cm, and it is increasing at a rate of 2 cm/s. The height of the cone is 10 cm, and it is increasing at a rate of 3 cm/s. At what rate is the volume increasing?

24. The radius of a cylinder is 8 cm, and it is increasing at a rate of 2 cm/s. The height of the cylinder is 20 cm, and it is increasing at a rate of 4 cm/s.
a. Find the rate at which the volume is increasing.
b. Find the rate at which the lateral surface area is increasing.

25. A rectangle is 3 in. wide and 5 in. long. Suppose its width is increasing at the rate of 1 in./h and its length is increasing by 0.5 in./h. How fast is its area increasing 2 h later?

26. Consider a right elliptical cylinder whose height is increasing at a rate of 1 cm/s and whose base area is increasing at a rate of 2 cm²/s. How fast is the volume of the cylinder increasing when the height is 5 cm and the area of the base is 10 cm²?

27. The radius of a right circular cone is increasing at a rate of 1 cm/s and the height is increasing at a rate of 3 cm/s. How fast is its surface area S increasing when the height is 10 cm and the radius is 3 cm? (*Hint:* $S = \pi r \sqrt{r^2 + h^2}$.)

28. The radius of a right circular cylinder is decreasing at a rate of 1 ft/min and the height is increasing at a rate of 3 ft/min. Is the volume increasing or decreasing when the height is 10 ft and the radius is 2 ft?

29. A particle moves along a helix with position function $\mathbf{r}(t) = \cos t\mathbf{i} + \sin t\mathbf{j} + t\mathbf{k}$.
a. Find the function $D(x, y, z)$ giving the distance from the particle to the origin.
b. Find the rate dD/dt at which the distance from the particle to the origin changes as a function of time.
c. Show that $\lim\limits_{t \to \infty} D'(t) = 1$; $\lim\limits_{t \to -\infty} D'(t) = -1$.

30. In economic theory, the **rate of growth** (as opposed to rate of change) of a function f is defined to be the ratio

$$\frac{f'(t)}{f(t)} = \frac{d}{dt}[\ln(f(t))] = \frac{\text{marginal function}}{\text{total function}}.$$

Show, according to this definition, that if the rate of growth of consumption C is a, and if the rate of growth of population P is b, then the rate of growth of per capita consumption, $R = C/P$, is $a - b$.

31. The money supply in the economy is defined by the equation $M = C + D$ where C is the cash on deposit in banks and D is the total of all time and demand deposits. If m, c, and d are the respective rates of growth of M, C, and D, show that

$$m = \frac{cC}{C + D} + \frac{dD}{C + D}.$$

(See Exercise 30 for the definition of rate of growth.)

32. Let $f(x, y) = xe^{y^2}$, $x(r, \theta) = r \cos \theta$,
$y(r, \theta) = r \sin \theta$. Find

a. $\dfrac{\partial^2 f}{\partial r^2}$ **b.** $\dfrac{\partial^2 f}{\partial \theta^2}$

33. Let $f(x, y) = x^2 + y^4$, $x(s, t) = s^2t$,
$y(s, t) = t^2 - s^2$. Find

a. $\dfrac{\partial^2 f}{\partial s^2}$ **b.** $\dfrac{\partial^2 f}{\partial t^2}$

34. Find $\dfrac{d^2 f}{dt^2}$ for the function $f(x, y)$ in Exercise 11.

35. Find $\dfrac{\partial^2 f}{\partial r^2}$ for the function $f(x, y)$ in Exercise 14.

36. Find an expression for the mixed second order partial derivative $\partial^2 f / \partial s \partial t$ for the composite function

$$f(x, y) = f(x(s, t), y(s, t)).$$

37. Use the result of Exercise 36 to find $\partial^2 f / \partial r \partial \theta$ for the function f in Exercise 32.

38. Use the result of Exercise 36 to find $\partial^2 f / \partial s \partial t$ for the function f in Exercise 33.

39. By use of the Chain Rule, show that Laplace's equation

$$\frac{\partial^2 v}{\partial x^2} + \frac{\partial^2 v}{\partial y^2} + \frac{\partial^2 v}{\partial z^2} = 0$$

in cylindrical coordinates is

$$\frac{\partial^2 v}{\partial r^2} + \frac{1}{r}\frac{\partial v}{\partial r} + \frac{1}{r^2}\frac{\partial^2 v}{\partial \theta^2} + \frac{\partial^2 v}{\partial z^2} = 0.$$

40. Use the Chain Rule to show that Laplace's equation

$$\frac{\partial^2 v}{\partial x^2} + \frac{\partial^2 v}{\partial y^2} + \frac{\partial^2 v}{\partial z^2} = 0$$

in spherical coordinates is

$$\frac{\partial^2 v}{\partial \rho^2} + \frac{2}{\rho}\frac{\partial v}{\partial \rho} + \frac{1}{\rho^2}\frac{\partial^2 v}{\partial \phi^2} + \frac{\cot \phi}{\rho^2}\frac{\partial v}{\partial \phi}$$
$$+ \frac{\csc^2 \phi}{\rho^2}\frac{\partial^2 v}{\partial \theta^2} = 0.$$

18.7   Directional Derivatives; The Gradient

If f is a function of two variables defined in a neighborhood of a point (x_0, y_0), the partial derivatives $\partial f / \partial x$ and $\partial f / \partial y$ at (x_0, y_0) measure the rate of change of f in the directions of the x- and y-coordinate axes, respectively. But how do we calculate the rate of change of f in an arbitrary direction? The answer to this question is provided by the *directional derivative*.

Suppose that we want to calculate the rate of change of f at the point $P_0 = (x_0, y_0)$ in the direction determined by the *unit* vector $\mathbf{u} = u_1 \mathbf{i} + u_2 \mathbf{j}$ (see Figure 7.1). Using $\mathbf{u}$, we parameterize the line in the xy-plane containing P_0 with direction vector $\mathbf{u}$ by

$$\begin{cases} x(t) = x_0 + tu_1 \\ y(t) = y_0 + tu_2. \end{cases}$$

Note that $(x(0), y(0)) = (x_0, y_0) = P_0$ and that $(x(1), y(1)) = (x_0 + u_1, y_0 + u_2)$, which is the point P_1 that results from adding the displacement $\mathbf{u}$ to the point P_0. In addition, note that, as t decreases from 1 to 0, $(x(t), y(t))$ generates the line segment from P_1 to P_0.

For each number t, we calculate the change

$$\Delta f = f(x(t), y(t)) - f(x_0, y_0)$$

in f in the direction of $\mathbf{u}$ over the interval of length

$$\sqrt{(x(t) - x_0)^2 + (y(t) - y_0)^2} = \sqrt{(tu_1)^2 + (tu_2)^2}$$
$$= |t|\sqrt{u_1^2 + u_2^2}$$
$$= |t|,$$

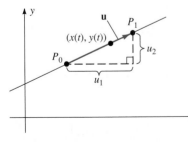

Figure 7.1   The direction vector $\mathbf{u}$ determines a line through the point $P_0 = (x_0, y_0)$.

since $|\mathbf{u}| = \sqrt{u_1^2 + u_2^2} = 1$. For $t > 0$, the rate of change of f over the line segment from (x_0, y_0) to $(x(t), y(t))$ is

$$\frac{\Delta f}{\Delta t} = \frac{f(x(t), y(t)) - f(x_0, y_0)}{t - 0}$$

$$= \frac{f(x_0 + tu_1, y_0 + tu_2) - f(x_0, y_0)}{t}$$

(see Figure 7.2). The directional derivative is obtained by taking the limit of this quotient as $t \to 0^+$.

Definition 5	Let f be a function of two variables defined in a neighborhood of (x_0, y_0). Let $\mathbf{u} = u_1\mathbf{i} + u_2\mathbf{j}$ be a unit vector. The **directional derivative** $D_\mathbf{u}f(x_0, y_0)$ in the direction of the unit vector $\mathbf{u}$ is the limit

$$D_\mathbf{u}f(x_0, y_0) = \lim_{t \to 0^+} \frac{f(x_0 + tu_1, y_0 + tu_2) - f(x_0, y_0)}{t}. \qquad (1)$$

Figure 7.3 illustrates the fact that $D_\mathbf{u}f(x_0, y_0)$ is the slope of a line tangent to the graph of f. That is, the direction vector $\mathbf{u}$, along with the standard coordinate vector $\mathbf{k}$, determines a vertical plane through the point $(x_0, y_0, 0)$. The directional derivative $D_\mathbf{u}f(x_0, y_0)$ is the slope of the graph of f in this plane at the point $(x_0, y_0, f(x_0, y_0))$.

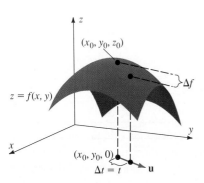

Figure 7.2 The rate of change of f over the line segment from (x_0, y_0) to $(x_0 + tu_1, y_0 + tu_2)$ is $\Delta f/\Delta t = \Delta f/t$.

REMARK 1 Definition 5 has an obvious generalization to functions of more than two variables. If $\mathbf{u} = \langle u_1, u_2, \ldots, u_n \rangle$ is a unit vector in $\mathbb{R}^n$ and f is a function of n variables, the directional derivative of f at $(x_1, x_2, \ldots x_n)$ in the direction of $\mathbf{u}$ is the limit

$$D_\mathbf{u}f(x_1, x_2, \ldots, x_n) \qquad (2)$$
$$= \lim_{t \to 0^+} \frac{f(x_1 + tu_1, x_2 + tu_2, \ldots, x_n + tu_n) - f(x_1, x_2, \ldots, x_n)}{t}.$$

Unfortunately, equations (1) and (2) fail to provide a simple procedure for calculating the directional derivative. If f satisfies the hypotheses of the Approximation Theorem (Theorem 5), we can use this result to establish such a procedure. First, we write the numerator in (1) as

$$f(x_0 + tu_1, y_0 + tu_2) - f(x_0, y_0) = \frac{\partial f}{\partial x}(x_0, y_0)(tu_1) + \frac{\partial f}{\partial y}(x_0, y_0)(tu_2)$$
$$+ \epsilon_1(tu_1) + \epsilon_2(tu_2)$$

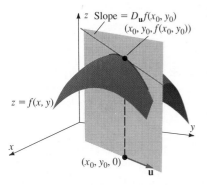

Figure 7.3 The directional derivative $D_\mathbf{u}f(x_0, y_0)$ is the rate of change of f in the direction of the vector $\mathbf{u}$. It equals the slope of the tangent line to the graph of $f(x, y)$ determined by the vector $\mathbf{u}$.

where $\epsilon_1 \to 0$ and $\epsilon_2 \to 0$ as $t \to 0$. Dividing both sides by $t \neq 0$ and taking the limit as $t \to 0^+$ yields

$$D_\mathbf{u}f(x_0, y_0) = \lim_{t \to 0^+} \frac{\frac{\partial f}{\partial x}(x_0, y_0)(tu_1) + \frac{\partial f}{\partial y}(x_0, y_0)(tu_2) + \epsilon_1(tu_1) + \epsilon_2(tu_2)}{t}$$

$$= \lim_{t \to 0^+} \left[\frac{\partial f}{\partial x}(x_0, y_0)u_1 + \frac{\partial f}{\partial y}(x_0, y_0)u_2 + \epsilon_1 u_1 + \epsilon_2 u_2 \right]$$

$$= \frac{\partial f}{\partial x}(x_0, y_0)u_1 + \frac{\partial f}{\partial y}(x_0, y_0)u_2 + \lim_{t \to 0^+} \epsilon_1 u_1 + \lim_{t \to 0^+} \epsilon_2 u_2.$$

Since $\epsilon_1 \to 0$ and $\epsilon_2 \to 0$ as $t \to 0$, the last two terms on the right side of the above equation are zero. We have therefore established the following theorem.

Theorem 7

If the function f and its first partial derivatives are continuous in a neighborhood of (x_0, y_0), the directional derivative $D_{\mathbf{u}} f(x_0, y_0)$ in the direction of the unit vector $\mathbf{u} = u_1 \mathbf{i} + u_2 \mathbf{j}$ is given by

$$D_{\mathbf{u}} f(x_0, y_0) = \frac{\partial f}{\partial x}(x_0, y_0) u_1 + \frac{\partial f}{\partial y}(x_0, y_0) u_2. \tag{3}$$

REMARK 2   The corresponding result for functions of n variables, $n \geq 3$, is that the directional derivative in (2) may be expressed as

$$D_{\mathbf{u}} f(x_1, x_2, \ldots, x_n) = \frac{\partial f}{\partial x_1}(x_1, \ldots, x_n) u_1 \tag{4}$$

$$+ \frac{\partial f}{\partial x_2}(x_1, x_2, \ldots, x_n) u_2 + \cdots$$

$$+ \frac{\partial f}{\partial x_n}(x_1, x_2, \ldots, x_n) u_n$$

if f and its first order partial derivatives are continous in a neighborhood of $(x_1, x_2, \ldots, x_n)$ in $\mathbb{R}^n$.

REMARK 3   It is important to note that the vector $\mathbf{u}$ in expressions (1) through (4) is a *unit* vector. If you wish to calculate the directional derivative of f in the direction of a vector $\mathbf{w}$ with $|\mathbf{w}| \neq 1$, you must first obtain the unit vector $\mathbf{u} = \mathbf{w}/|\mathbf{w}|$ in the direction of $\mathbf{w}$.

REMARK 4   Note that, if $\mathbf{u} = \mathbf{i}$ (so that $u_1 = 1$ and $u_2 = 0$) and if the partial derivative $\partial f/\partial x$ exists at (x, y) then the directional derivative $D_{\mathbf{i}} f(x, y)$ is just $\partial f/\partial x$ at (x, y). Similarly, if $\mathbf{u} = \mathbf{j}$ and $\partial f/\partial y$ exists at (x, y), then $D_{\mathbf{j}} f(x, y)$ is equal to $\partial f/\partial y$ at (x, y). This is most easily seen from the form for $D_{\mathbf{u}} f(x, y)$ in Theorem 7. However, $D_{\mathbf{u}} f(x, y)$ may exist, as defined in Definition 5, for all $\mathbf{u}$ even if one or both of the partial derivatives do not exist (see Exercise 61).

☐ **EXAMPLE 1**

Find the directional derivative $D_{\mathbf{u}} f(2, 1)$ for

$$f(x, y) = x^2 e^{3y} \quad \text{and} \quad \mathbf{u} = \frac{1}{\sqrt{5}} \mathbf{i} + \frac{2}{\sqrt{5}} \mathbf{j}.$$

Solution   The partial derivatives are

$$\frac{\partial f}{\partial x}(2, 1) = 2x e^{3y} \bigg|_{(2, 1)} = 4e^3$$

and

$$\frac{\partial f}{\partial y}(2, 1) = 3x^2 e^{3y} \bigg|_{(2, 1)} = 12e^3.$$

Since $|\mathbf{u}| = \sqrt{\frac{1}{5} + \frac{4}{5}} = 1$, we have from (3) that

$$D_{\mathbf{u}} f(2, 1) = (4e^3)\left(\frac{1}{\sqrt{5}}\right) + 12e^3\left(\frac{2}{\sqrt{5}}\right) = \frac{28e^3}{\sqrt{5}} \approx 251.5. \qquad \blacksquare$$

□ **EXAMPLE 2**

Find the directional derivative of the function $f(x, y, z) = e^x \cos y + xz$ at the point $(1, \pi, -1)$ in the direction of the vector $\mathbf{w} = \mathbf{i} - 3\mathbf{j} + 4\mathbf{k}$.

Solution The partial derivatives of f are

$$\frac{\partial f}{\partial x}(1, \pi, -1) = e^x \cos y + z \bigg|_{(1, \pi, -1)} = -e - 1,$$

$$\frac{\partial f}{\partial y}(1, \pi, -1) = -e^x \sin y \bigg|_{(1, \pi, -1)} = 0,$$

and

$$\frac{\partial f}{\partial z}(1, \pi, -1) = x \bigg|_{(1, \pi, -1)} = 1.$$

Since $|\mathbf{w}| = \sqrt{1^2 + 3^2 + 4^2} = \sqrt{26}$, a *unit* vector in the direction of $\mathbf{w}$ is

$$\mathbf{u} = \frac{1}{\sqrt{26}}(\mathbf{i} - 3\mathbf{j} + 4\mathbf{k}),$$

so $u_1 = 1/\sqrt{26}$, $u_2 = -3/\sqrt{26}$, and $u_3 = 4/\sqrt{26}$. According to (4),

$$D_{\mathbf{u}}f(1, \pi, -1) = (-e - 1)\left(\frac{1}{\sqrt{26}}\right) + (0)\left(\frac{-3}{\sqrt{26}}\right) + (1)\left(\frac{4}{\sqrt{26}}\right)$$

$$= \frac{-e + 3}{\sqrt{26}} \approx 0.055. \quad\blacksquare$$

If $\mathbf{u} = u_1\mathbf{i} + u_2\mathbf{j}$ is a unit vector and θ is the angle formed between $\mathbf{u}$ and the positive x-axis, then

$$u_1 = \frac{u_1}{|\mathbf{u}|} = \cos\theta, \quad\text{and}\quad u_2 = \frac{u_2}{|\mathbf{u}|} = \sin\theta.$$

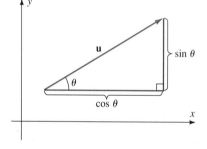

Figure 7.4 If $\mathbf{u} = u_1\mathbf{i} + u_2\mathbf{j}$ with $|\mathbf{u}| = 1$, then $u_1 = \cos\theta$, $u_2 = \sin\theta$.

(See Figure 7.4.) We may use these equations to write the directional derivative $D_{\mathbf{u}}f(x_0, y_0)$ in (3) in the form

$$D_{\mathbf{u}}f(x_0, y_0) = \frac{\partial f}{\partial x}(x_0, y_0)\cos\theta + \frac{\partial f}{\partial y}(x_0, y_0)\sin\theta. \tag{5}$$

Equation (5) implies that, if f and its first partial derivatives are continuous, the directional derivative depends only on the partial derivatives and the direction of the unit vector $\mathbf{u}$.

□ **EXAMPLE 3**

Let $f(x, y) = xy - y^3$. Find the unit vector $\mathbf{u}$ for which the directional derivative $D_{\mathbf{u}}f(2, 1)$ is a maximum.

Solution We begin by calculating the partial derivatives for f:

$$\frac{\partial f}{\partial x}(2, 1) = y \bigg|_{(2, 1)} = 1,$$

$$\frac{\partial f}{\partial y}(2, 1) = x - 3y^2 \bigg|_{(2, 1)} = -1.$$

According to (5), the directional derivative $D_{\mathbf{u}}f(2, 1)$ is

$$D_{\mathbf{u}}f(2, 1) = (1) \cos \theta + (-1) \sin \theta = \cos \theta - \sin \theta.$$

We must therefore find the value of θ for which the function

$$g(\theta) = \cos \theta - \sin \theta$$

is a maximum. To do so we set

$$g'(\theta) = -\sin \theta - \cos \theta = 0$$

and obtain the equation

$$\sin \theta = -\cos \theta, \quad \text{or} \quad \tan \theta = -1,$$

which has solutions $\theta = 3\pi/4$ and $\theta = 7\pi/4$ for $0 \le \theta \le 2\pi$. Since

$$g''(3\pi/4) = -\cos(3\pi/4) + \sin(3\pi/4) = \sqrt{2} > 0$$

and

$$g''(7\pi/4) = -\cos(7\pi/4) + \sin(7\pi/4) = -\sqrt{2} < 0,$$

the angle $\theta = 7\pi/4$ corresponds to the maximum. For this angle,

$$\mathbf{u} = (\cos \theta)\mathbf{i} + (\sin \theta)\mathbf{j} = \frac{\sqrt{2}}{2}\mathbf{i} - \frac{\sqrt{2}}{2}\mathbf{j}$$

is the required unit vector.   ∎

The Gradient

The form of the directional derivative given by Theorem 7 can be written as a dot product:

$$D_{\mathbf{u}}f(x_0, y_0) = \frac{\partial f}{\partial x}(x_0, y_0)u_1 + \frac{\partial f}{\partial y}(x_0, y_0)u_2 \tag{6}$$

$$= \left[\frac{\partial f}{\partial x}(x_0, y_0)\mathbf{i} + \frac{\partial f}{\partial y}(x_0, y_0)\mathbf{j} \right] \cdot [u_1\mathbf{i} + u_2\mathbf{j}].$$

The second factor in this dot product is just the unit vector $\mathbf{u} = u_1\mathbf{i} + u_2\mathbf{j}$. The first factor is called the *gradient* of f at (x_0, y_0). It is usually written as

$$\nabla f(x_0, y_0) = \frac{\partial f}{\partial x}(x_0, y_0)\mathbf{i} + \frac{\partial f}{\partial y}(x_0, y_0)\mathbf{j} \tag{7}$$

or just

$$\nabla f = \frac{\partial f}{\partial x}\mathbf{i} + \frac{\partial f}{\partial y}\mathbf{j}.$$

For functions of three variables, the gradient is

$$\nabla f = \frac{\partial f}{\partial x}\mathbf{i} + \frac{\partial f}{\partial y}\mathbf{j} + \frac{\partial f}{\partial z}\mathbf{k}.$$

Thus, the gradient ∇f is a *vector* whose components are the partial derivatives of f.

□ **EXAMPLE 4**

For $f(x, y) = x \cos y$, the gradient is

$$\nabla f(x, y) = \left[\frac{\partial}{\partial x}(x \cos y)\right]\mathbf{i} + \left[\frac{\partial}{\partial y}(x \cos y)\right]\mathbf{j}$$
$$= \cos y\mathbf{i} - x \sin y\mathbf{j}$$

and the gradient at $(2, \pi/4)$ is

$$\nabla f(2, \pi/4) = \frac{\sqrt{2}}{2}\mathbf{i} - \sqrt{2}\mathbf{j}. \qquad \blacksquare$$

□ **EXAMPLE 5**

For $f(x, y, z) = \sqrt{x}\, e^y \, \text{Tan}^{-1} z$, the gradient is

$$\nabla f(x, y, z) = \frac{e^y \, \text{Tan}^{-1} z}{2\sqrt{x}}\mathbf{i} + \sqrt{x}\, e^y \, \text{Tan}^{-1} z\mathbf{j} + \frac{\sqrt{x}\, e^y}{1 + z^2}\mathbf{k},$$

and the gradient at $(4, 0, 1)$ is

$$\nabla f(4, 0, 1) = \frac{\pi}{16}\mathbf{i} + \frac{\pi}{2}\mathbf{j} + \mathbf{k}. \qquad \blacksquare$$

Vector Notation and Gradients

Using vector notation, we may express the directional derivative as simply

$$\boxed{D_{\mathbf{u}} f(\mathbf{x}) = \nabla f(\mathbf{x}) \cdot \mathbf{u}.} \qquad (8)$$

Equation (8) provides insight into the geometry associated with $\nabla f(\mathbf{x})$. From the definition of the dot product, we know that

$$\nabla f(\mathbf{x}) \cdot \mathbf{u} = |\nabla f(\mathbf{x})|\, |\mathbf{u}| \cos \theta \qquad (9)$$
$$= |\nabla f(\mathbf{x})| \cos \theta \qquad (\text{since } |\mathbf{u}| = 1)$$

where θ is the angle between the vectors $\nabla f(\mathbf{x})$ and $\mathbf{u}$. Combining equations (8) and (9) gives

$$D_{\mathbf{u}} f(\mathbf{x}) = |\nabla f(\mathbf{x})| \cos \theta. \qquad (10)$$

Thus, since $-1 \le \cos \theta < 1$ for all θ, equation (10) shows that

(i) $-|\nabla f(\mathbf{x})| \le D_{\mathbf{u}} f(\mathbf{x}) \le |\nabla f(\mathbf{x})|$, and
(ii) $D_{\mathbf{u}} f(\mathbf{x})$ assumes its maximum value if $\cos \theta = 1$, that is, if $\theta = 0$.

Since the case $\theta = 0$ occurs precisely when $\nabla f(\mathbf{x})$ and the direction vector $\mathbf{u}$ point in the same direction, conclusion (ii) says that $\nabla f(\mathbf{x})$ points in the direction of the maximum value of $D_{\mathbf{u}} f(\mathbf{x})$. That is,

$$\boxed{\nabla f(\mathbf{x}) \text{ points in the direction of most rapid increase for } f \text{ at } \mathbf{x}} \qquad (11)$$

(assuming, of course, that the hypotheses of Theorem 7 hold).

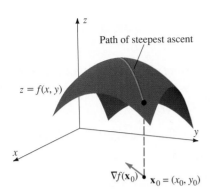

Figure 7.5 The gradient $\nabla f(\mathbf{x})$ points in the direction of most rapid increase of the function f at $\mathbf{x}$.

This observation has several important applications. If $\mathbf{x} = (x, y)$ is a point in the domain of the function f, then $\nabla f(x, y)$ points in the direction in which a path through $(x, y, f(x, y))$ on the graph will rise most rapidly (see Figure 7.5). If a path on a surface has the property that the tangent at each point of the path is parallel to the gradient of the function defining the surface, it is called a **path of steepest ascent.** This concept has many important uses. For example, it is used to develop procedures for approximating extrema for complicated functions of several variables. (It should be geometrically obvious that proceeding in the opposite direction, $-\nabla f(x, y)$, leads to the **path of steepest descent.** In Exercise 53, you are asked to prove this analytically.)

A second application of observation (11) concerns particles moving so as to maximize an attribute of the medium through which they are moving. Examples include insects flying toward a light source (maximizing the intensity of light), heat-seeking missiles (maximizing temperature), and sharks in water (maximizing blood concentration). In such situations, if $A(\mathbf{x})$ is the value of the attribute at location $\mathbf{x}$, the particle will move so that its tangent vector points in the direction of $\nabla A(\mathbf{x})$ (see Example 7).

☐ **EXAMPLE 6**

For the function $f(x, y) = 9 - \dfrac{x^2 + y^2}{4}$, the gradient at $\mathbf{x} = (x, y)$ is

$$\nabla f(\mathbf{x}) = -\frac{x}{2}\mathbf{i} - \frac{y}{2}\mathbf{j} = -\frac{1}{2}(x\mathbf{i} + y\mathbf{j}) = -\frac{1}{2}\mathbf{x}.$$

Since $\mathbf{x} = x\mathbf{i} + y\mathbf{j}$ is the position vector of the point (x, y), the vector $\nabla f(\mathbf{x}) = -\frac{1}{2}\mathbf{x}$ points toward the origin for all $(x, y) \neq (0, 0)$. In particular,

$$\begin{aligned}
\text{if } \mathbf{x} &= (2, 2), & \nabla f &= -\mathbf{i} - \mathbf{j}, \\
\text{if } \mathbf{x} &= (1, 4), & \nabla f &= -\tfrac{1}{2}\mathbf{i} - 2\mathbf{j}, \\
\text{if } \mathbf{x} &= (2, -1), & \nabla f &= -\mathbf{i} + \tfrac{1}{2}\mathbf{j}.
\end{aligned}$$

This should not be surprising, since the graph of f is a circular paraboloid. At any point on this surface, the z-coordinate is increased most rapidly by moving directly toward the z-axis (see Figure 7.6). ■

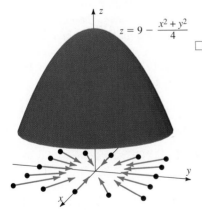

Figure 7.6 For $f(x, y) = 9 - (\tfrac{1}{4})(x^2 + y^2)$, $\nabla f(\mathbf{x}) = -\tfrac{1}{2}\mathbf{x}$. The gradient always points towards the origin for this function f.

☐ **EXAMPLE 7**

The temperature distribution across the surface of a rectangular plate is given by the function $T(x, y) = 100 - x^2 - 2y^2$. Find the path followed by a heat-seeking particle placed on the plate at the point $(4, 2)$.

Solution Let the path followed by the particle have parameterization

$$\mathbf{r}(t) = x(t)\mathbf{i} + y(t)\mathbf{j}.$$

Assuming the components x and y to be differentiable functions of t, we recall from Chapter 17 that the tangent vector $\mathbf{r}'(t)$ at each point along the path is

$$\mathbf{r}'(t) = x'(t)\mathbf{i} + y'(t)\mathbf{j}.$$

Since the particle is heat-seeking, this tangent vector should point in the same direction as the gradient

$$\nabla T(x, y) = -2x\mathbf{i} - 4y\mathbf{j}.$$

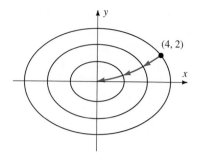

Figure 7.7 Heat-seeking particle approaches origin along parabolic path.

This will be the case if

$$x'(t) = -2x(t) \quad \text{and} \quad y'(t) = -4y(t).$$

These are the familiar differential equations for exponential decay (see Chapter 8). Their solutions are

$$x(t) = x(0)e^{-2t} \quad \text{and} \quad y(t) = y(0)e^{-4t}.$$

Since the particle begins at $(4, 2)$, we have $x(0) = 4$ and $y(0) = 2$. Thus

$$x(t) = 4e^{-2t}, \qquad y(t) = 2e^{-4t}$$

are the components of a parameterization of the path. We may eliminate the parameter t by noting that

$$y(t) = 2e^{-4t} = \tfrac{1}{8}(4e^{-2t})^2 = \tfrac{1}{8}[x(t)]^2.$$

The particle therefore approaches the origin along the parabola $y = \tfrac{1}{8}x^2$ (Figure 7.7). ∎

Level Curves and the Gradient

Let $\mathbf{r}(t) = x(t)\mathbf{i} + y(t)\mathbf{j} + z(t)\mathbf{k}$ be a parameterization for a curve for $\mathbb{R}^3$, and let $w = f(x, y, z)$ be a function of three variables. Using the gradient and assuming that all necessary derivatives exist, we may write the Chain Rule (Theorem 6)

$$\frac{d}{dt}f(x, y, z) = \frac{\partial f}{\partial x}\frac{dx}{dt} + \frac{\partial f}{\partial y}\frac{dy}{dt} + \frac{\partial f}{\partial z}\frac{dz}{dt}$$

as

$$\frac{d}{dt}f(\mathbf{r}(t)) = \nabla f(\mathbf{r}(t)) \cdot \mathbf{r}'(t). \tag{12}$$

Equation (12) says that the derivative of the composite function $f(\mathbf{r}(t))$ is the dot product of the gradient $\nabla f(\mathbf{r}(t))$ with the tangent vector $\mathbf{r}'(t) = x'(t)\mathbf{i} + y'(t)\mathbf{j} + z'(t)\mathbf{k}$ for each t. Equation (12) also holds for curves $\mathbf{r}(t) = x(t)\mathbf{i} + y(t)\mathbf{j}$ in the plane composed with functions $z = f(x, y)$ of two variables, as does Theorem 6. (The beauty of the vector notation is that we do not need to distinguish between these two cases in writing equation (12).)

Equation (12) reveals an important relationship between gradients and level curves. Suppose that f is a functon of two variables for which the equation

$$f(x, y) = k, \qquad k \text{ constant} \tag{13}$$

determines a curve C in the xy-plane with parameterization

$$\mathbf{r}(t) = x(t)\mathbf{i} + y(t)\mathbf{j}.$$

Then along this level curve we have

$$f(\mathbf{r}(t)) = f(x(t), y(t)) = k, \tag{14}$$

according to equation (13).

Now let $\mathbf{x}_0 = \mathbf{r}(t_0) = (x(t_0), y(t_0))$ be a point on C and differentiate both sides of equation (14). Since the right-hand side is a constant k, the derivative is zero. If the functions on the left-hand side satisfy the hypotheses of the Chain Rule, we can

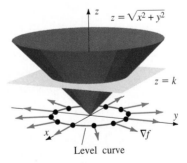

Figure 7.8 The $\nabla f(\mathbf{x}_0)$ is orthogonal to the level curve through $\mathbf{x}_0$.

use equation (12) to write the derivative of the left-hand side as a dot product. We obtain

$$\frac{d}{dt} f(\mathbf{r}(t)) \bigg|_{t=t_0} = \nabla f(\mathbf{r}(t_0)) \cdot \mathbf{r}'(t_0) = 0. \tag{15}$$

Since $\mathbf{r}'(t_0)$ is a vector tangent to the level curve C, equation (15) shows that *the gradient $\nabla f(\mathbf{r}(t_0))$ is orthogonal to the level curve C at $\mathbf{x}_0$* (see Figure 7.8). That is:

> Let f be a function of two variables. If f and its first partial derivatives are continuous, then at each point in the domain of f the gradient vector, if nonzero, is orthogonal to the level curve through that point. $\tag{16}$

□ **EXAMPLE 8**

The graph of the function $z = \sqrt{x^2 + y^2}$ is a cone. The level curves associated with this graph are the circles

$$\sqrt{x^2 + y^2} = k, \quad \text{or} \quad x^2 + y^2 = k^2.$$

The gradient for this function is

$$\nabla f(x, y) = \frac{x}{\sqrt{x^2 + y^2}} \mathbf{i} + \frac{y}{\sqrt{x^2 + y^2}} \mathbf{j} = \frac{x\mathbf{i} + y\mathbf{j}}{\sqrt{x^2 + y^2}}.$$

In vector notation, the gradient may be written as

$$\nabla f(\mathbf{x}) = \frac{\mathbf{x}}{|\mathbf{x}|}.$$

This expression shows that the gradient vectors are parallel to radius vectors for the circular level curves and, therefore, are orthogonal to the level curves (Figure 7.9). ■

Figure 7.9 For $f(x, y) = \sqrt{x^2 + y^2}$, the level curves are circles centered at the origin; $\nabla f(\mathbf{x})$ is parallel to the corresponding radii of the circle.

Statement (16) has many useful geometric applications. For example, we can use the gradient to solve a tangent line problem that we solved using implicit differentiation in Chapter 3 (Example 1 in Section 3.8).

☐ **EXAMPLE 9**

Find a vector equation for the line tangent to the ellipse

$$\frac{x^2}{16} + \frac{y^2}{9} = 1$$

at the point $P = (2, 3\sqrt{3}/2)$.

Solution Consider the ellipse in question as the level curve of level 1 for the function

$$f(x, y) = \frac{x^2}{16} + \frac{y^2}{9}.$$

Then we can use statement (16) to find a normal direction to the ellipse at P. We have

$$\nabla f(x, y) = \frac{x}{8}\mathbf{i} + \frac{2y}{9}\mathbf{j}.$$

Hence, the vector $\mathbf{n} = \nabla f(2, 3\sqrt{3}/2) = (1/4)\mathbf{i} + (1/\sqrt{3})\mathbf{j}$ is normal to the ellipse at P.

To find an equation for the tangent line, we need a direction vector $\mathbf{d}$ for the line. Given $\mathbf{n}$, we obtain an orthogonal vector $\mathbf{d}$ by interchanging the $\mathbf{i}$ and $\mathbf{j}$ coefficients of n and then multiplying one of these coefficients by -1 (see Section 16.3). Thus, the vector

$$\mathbf{d} = \frac{1}{\sqrt{3}}\mathbf{i} + \frac{-1}{4}\mathbf{j}$$

is a direction vector for the line.

Using P and $\mathbf{d}$, we obtain the vector equation

$$\mathbf{r}(t) = \left[2\mathbf{i} + \frac{3\sqrt{3}}{2}\mathbf{j}\right] + t\left[\frac{1}{\sqrt{3}}\mathbf{i} + \frac{-1}{4}\mathbf{j}\right]$$

$$= \left[2 + \frac{t}{\sqrt{3}}\right]\mathbf{i} + \left[\frac{3\sqrt{3}}{2} - \frac{t}{4}\right]\mathbf{j}$$

for the tangent line. ■

In Section 3.8, we found the slope of this tangent line using implicit differentiation. We recommend that you compare that solution with the one given in Example 9. The two techniques are related by the Chain Rule.

Level Surfaces and the Gradient

A similar relationship exists between gradients for functions of three variables and level surfaces. Suppose that

$$f(x, y, z) = k$$

is the equation of a level surface S for the function f, and suppose that the point $\mathbf{x}_0$ lies on S. Choose a curve C on the surface S with parameterization

$$\mathbf{r}(t) = x(t)\mathbf{i} + y(t)\mathbf{j} + z(t)\mathbf{k}$$

for which $\mathbf{r}(t_0) = \mathbf{x}_0$ (see Figure 7.10). Then, along this curve

$$f(\mathbf{r}(t)) = f(x(t), y(t), z(t)) = k, \tag{17}$$

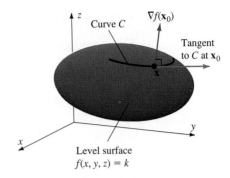

Figure 7.10 The gradient $\nabla f(\mathbf{x}_0)$ is orthogonal to the curve C at the point $\mathbf{x}_0$.

that is, the composite function $f \circ \mathbf{r}$ is constant. As before, if the function f and the component functions x, y, and z satisfy the hypotheses of Theorem 6 in a neighborhood of $\mathbf{x}_0 = \mathbf{r}(t_0)$, we may differentiate both sides of equation (17), using the Chain Rule, to conclude that

$$\frac{d}{dt}f(\mathbf{r}(t))\bigg|_{t=t_0} = \nabla f(\mathbf{r}(t_0)) \cdot \mathbf{r}'(t_0) = 0. \tag{18}$$

Equation (18) shows that the gradient $\nabla f(\mathbf{x}_0)$ is orthogonal to the tangent $\mathbf{r}'(t_0)$ to the curve C through $\mathbf{x}_0$ on the surface S. Since C was an *arbitrary* curve on S, it follows that $\nabla f(\mathbf{x}_0)$ is orthogonal to *every* tangent to S at $\mathbf{x}_0$. In particular, $\nabla f(\mathbf{x}_0)$ is orthogonal to the tangent plane determined by $\partial f/\partial x$ and $\partial f/\partial y$ at $\mathbf{x}_0$ and therefore to the level surface S itself.

Let f be a function of three variables. If f and its first partial derivatives are continuous, then at each point in the domain of f the gradient vector, if nonzero, is orthogonal to the level surface containing that point. (19)

Thus, if the level surface has a tangent plane at the point $\mathbf{x}_0$, $\nabla f(\mathbf{x}_0)$ is a normal vector for this plane (see Figure 7.11). Let $\mathbf{x} = (x, y, z)$ be an arbitrary point on the tangent plane. An equation for the tangent plane is

$$\nabla f(\mathbf{x}_0) \cdot (\mathbf{x} - \mathbf{x}_0) = 0,$$

which becomes

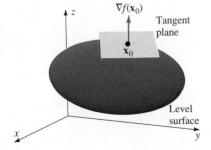

Figure 7.11 The gradient $\nabla f(\mathbf{x}_0)$ is normal to the tangent plane of the level surface at $\mathbf{x}_0$.

$$\frac{\partial f}{\partial x}(x_0, y_0, z_0)(x - x_0) + \frac{\partial f}{\partial y}(x_0, y_0, z_0)(y - y_0)$$
$$+ \frac{\partial f}{\partial z}(x_0, y_0, z_0)(z - z_0) = 0. \tag{20}$$

☐ **EXAMPLE 10**

Find equations for a vector normal to the graph of the ellipsoid $2x^2 + 4y^2 + z^2 = 21$ at the point $(2, 1, 3)$, and find an equation for the plane tangent to the graph at that point.

Solution   We regard the graph of the equation $2x^2 + 4y^2 + z^2 = 21$ as the level surface $f(x, y, z) = 21$ for the function

$$f(x, y, z) = 2x^2 + 4y^2 + z^2.$$

Using the gradient vector $\nabla f(2, 1, 3) = 8\mathbf{i} + 8\mathbf{j} + 6\mathbf{k}$, we derive the equation for the tangent plane as

$$8(x - 2) + 8(y - 1) + 6(z - 3) = 0$$

$$8x + 8y + 6z = 42. \qquad ■$$

REMARK 5   We could have worked Example 10 by solving the equation $2x^2 + 4y^2 + z^2 = 21$ for z as $z = \pm\sqrt{21 - 2x^2 - 4y^2}$. Since the point in question is on the upper half of the ellipsoid, we could then apply the method of Section 18.3 to the function $g(x, y) = \sqrt{21 - 2x^2 - 4y^2}$. However, using the gradient of f is a more direct method.

Exercise Set 18.7

In Exercises 1–9, find the gradient of the given function at the point P.

1. $f(x, y) = x^2y$, $P = (3, 1)$

2. $f(x, y) = xy^2 - ye^x$, $P = (0, 2)$

3. $f(x, y) = x\cos(y - x)$, $P = (\pi/2, \pi/4)$

4. $f(x, y) = \dfrac{2x}{y - x}$, $P = (2, 1)$

5. $f(x, y, z) = x^2y + xz^2$, $P = (1, 1, 2)$

6. $f(x, y, z) = xe^{yz}$, $P = (2, 0, 1)$

7. $f(x, y, z) = x^2y + xz^3 - y^2z$, $P = (1, 2, 1)$

8. $f(x, y, z) = e^x\cos y - e^y\sin z$, $P = (0, \pi/4, \pi/3)$

9. $f(x, y, z) = x\sqrt{y}\cosh z$, $P = (2, 4, 1)$

In Exercises 10–19, find the directional derivative of the given function at the given point in the direction of the given vector.

10. $f(x, y) = 3x^2 - y^2$, $P = (1, 2)$, $\mathbf{w} = \mathbf{i} + \mathbf{j}$

11. $f(x, y) = x^2y^2$, $P = (-2, 3)$, $\mathbf{w} = \mathbf{i} - \mathbf{j}$

12. $f(x, y) = \sin(y - x)$, $P = (\pi/2, \pi/4)$, $\mathbf{w} = \mathbf{i} + 2\mathbf{j}$

13. $f(x, y) = \dfrac{x}{x + y}$, $P = (1, 2)$, $\mathbf{w} = \sqrt{3}\mathbf{i} + \mathbf{j}$

14. $f(x, y) = xe^{y^2} - y$, $P = (1, 3)$, $\mathbf{w} = \sqrt{2}\mathbf{i} + \sqrt{2}\mathbf{j}$

15. $f(x, y, z) = xy + xz + yz$, $P = (1, 2, 1)$, $\mathbf{w} = \mathbf{i} + \mathbf{j} - \mathbf{k}$

16. $f(x, y, z) = \cosh x - \sinh y + \mathrm{Tan}^{-1}z$, $P = (-1, 1, 3)$, $\mathbf{w} = \mathbf{i} - \mathbf{j} - 2\mathbf{k}$

17. $f(x, y, z) = xe^{yz}$, $P = (2, 0, 1)$, $\mathbf{w} = \sqrt{3}\mathbf{i} + \sqrt{3}\mathbf{j} - \sqrt{5}\mathbf{k}$

18. $f(x, y, z) = xy^2 - x^2z + (yz)^3$, $P = (1, 0, 1)$, $\mathbf{w} = \mathbf{i} + 2\mathbf{j} + \mathbf{k}$

19. $f(x, y, z) = x\cos y - y\sinh z$, $P = (6, \pi/4, -1)$, $\mathbf{w} = \mathbf{i} - 2\mathbf{j} - 2\mathbf{k}$

In Exercises 20–28, write vector equations for the normal and tangent lines to the given curve at the given point.

20. $2x^2 + 3y^2 = 11$, $P = (2, 1)$

21. $4x^2 - y^2 = 7$, $P = (2, 3)$

22. $3x^2 - y^4 + x = 13$, $P = (2, 1)$

23. $\sqrt{x} + \sqrt{y} = 4$, $P = (4, 4)$

24. $6x^2 - 4y^2 = 18$, $P = (3, -3)$

25. $\sin x - \cos y = 1$, $P = (\pi/2, \pi/2)$

26. $ye^x = 1$, $P = (0, 1)$

27. $\ln x + 2\ln y = 1$, $P = (1, \sqrt{e})$

28. $\ln x + \sin y = 5$, $P = (e^5, 0)$

29. For $f(x, y) = xy^2 + ye^x$, find the directional derivative at $(0, 1)$ in the direction of most rapid increase of f.

30. For $f(x, y) = x^2 + 2y^3$, find the directional derivative at $(2, 3)$ in the direction toward the origin.

31. For the function f given in Exercise 29, find $D_{\mathbf{u}} f(0, 1)$ in the direction of the origin.

32. For $f(x, y) = x/(x + y)$, find $D_{\mathbf{u}} f(1, 1)$ in the direction of the point $(2, 3)$.

33. For $f(x, y) = e^x \sin y$, find $D_{\mathbf{u}} f(0, \pi/4)$ in the direction of the point $(1, \pi/2)$.

34. For $f(x, y) = \ln(x + y)$, find $D_{\mathbf{u}} f(1/2, 1/2)$ in the direction of the origin.

35. For the function f given in Exercise 32, find $D_{\mathbf{u}} f(1, 1)$ in the direction of most rapid increase of f.

36. Use the gradient to find the point(s) on the graph of the hyperbola $3x^2 - 2y^2 - 6x + 8y = 3$ where the tangent is horizontal.

37. Use the gradient to find the point(s) on the graph of the ellipse $x^2 - 6x + 2y^2 - 4y = -7$ where the tangent is
   a. horizontal,   **b.** vertical.

38. Use the gradient to find the point(s) on the graph of the parabola $x^2 - 2xy + y^2 - x - y = -1$ where the tangent is
   a. vertical,   **b.** horizontal.

39. Use the gradient to find the point(s) on the graph of $x \ln y = 1$ where the tangent is parallel to the line $y = -x$.

In Exercises 40–46, find a normal vector and an equation for the tangent plane for the given surface at the given point.

40. $x^2 + 4y^2 + 2z^2 = 9$, $P = (-1, 0, 2)$

41. $xyz = 6$, $P = (2, 1, 3)$

42. $x^2 + y^2 - 3yz = 8$, $P = (1, 1, -2)$

43. $z = x^2 - y^3 + xy$, $P = (2, 1, 5)$

44. $xy - zx + xz^2 = 1$, $P = (1, -1, 2)$

45. $y = \sin x$, $P = (\pi/2, 1, 5)$

46. $\sqrt{x} + \sqrt{y} + \sqrt{z} = 10$, $P = (1, 16, 25)$

47. Find a unit normal to the graph of the equation $xy^2 + x^2y - xz = 4$ at the point $(1, 2, 3)$.

48. Find a vector equation for the normal to the graph of the equation $xyz = 6$ at the point $(1, 3, 2)$.

49. Show that the cylinder $x^2 + y^2 = 4$ and the sphere $x^2 + y^2 + z^2 - 8y - 6z + 21 = 0$ are tangent at the point $(0, 2, 3)$.

50. Show that the ellipsoid $9x^2 + 4y^2 + 9z^2 = 36$ and the sphere $x^2 + y^2 + z^2 - 10z + 16 = 0$ are tangent at $(0, 0, 2)$.

51. Show that the sphere $x^2 + y^2 + z^2 - 2\sqrt{2}\, z + 1 = 0$ and the cone $z = \sqrt{x^2 + y^2}$ share common tangent planes along a circle. Find an equation for that circle.

52. Use equation (8) to show that the directional derivative of f in the direction of the unit vector $\mathbf{u}$ is the component of ∇f in the direction of $\mathbf{u}$.

53. Use equation (8) to show that the negative gradient, $-\nabla f(\mathbf{x})$, points in the direction of most rapid decrease of the function f at $\mathbf{x}$.

54. The temperature distribution in a room is given by the function $T(x, y, z) = 30 - (x^2 + 2y^2 + 3z^2)$. An insect flies so as to experience the most rapid decrease in temperature. In what direction does it move when located at point $(2, 1, 1)$?

55. Show that $|\nabla f(\mathbf{x}_0)|$ is the maximum value of $D_{\mathbf{u}} f(\mathbf{x}_0)$ if f and its first partials are continuous at $\mathbf{x}_0$.

56. Show that if $\cos \alpha$, $\cos \beta$, and $\cos \gamma$ are the direction cosines for the unit vector $\mathbf{u}$, then

$$D_{\mathbf{u}} f = \frac{\partial f}{\partial x} \cos \alpha + \frac{\partial f}{\partial y} \cos \beta + \frac{\partial f}{\partial z} \cos \gamma.$$

57. Find the direction of steepest ascent at the point above $(4, 2)$ on the graph of $z = e^{x^2 + y^2}$.

58. Find the direction of steepest ascent at the point above $(1, 1)$ on the graph of $z = xe^y + ye^x$.

59. Find the direction of steepest *descent* at the point above $(1, 2)$ on the graph of $z = 4x^2 - 2y^2$.

60. The temperature distribution on a metal plate is $T(x, y) = 200 - (x^2 + 4y^2)$. Find the path followed by a heat-seeking particle placed at the point $(0, 2)$.

61. Show that if the partial derivative $\partial f/\partial x$ exists at (x_0, y_0), then $D_{\mathbf{i}} f(x_0, y_0)$ exists and

$$D_{\mathbf{i}} f(x_0, y_0) = \frac{\partial f}{\partial x}(x_0, y_0).$$

Give an example to show that $D_{\mathbf{u}} f(x_0, y_0)$ can exist for all $\mathbf{u}$ while $\partial f/\partial x$ at (x_0, y_0) (and, therefore, $\nabla f(x_0, y_0)$) fails to exist.

62. Use statement (19) in this section to derive the equation for the tangent plane to the graph of $z = f(x, y)$ [equation (3) in Section 18.3]. (*Hint:* Consider the function $g(x, y, z) = f(x, y) - z$.)

63. Using the definition of the directional derivative, show that the function

$$f(x, y) = \begin{cases} \dfrac{xy^2}{x^2 + y^4}, & (x, y) \neq (0, 0); \\ 0, & (x, y) = (0, 0) \end{cases}$$

has directional derivatives in all directions at $(0, 0)$. Is f differentiable at $(0, 0)$? (*Hint:* Consider $(x, y) \rightarrow (0, 0)$ along the parabola $y^2 = x$.)

18.8   Constrained Extrema: The Method of Lagrange Multipliers

One application of the theory of the gradient is a method due to the French mathematician Joseph L. Lagrange (1736–1813) for finding extrema of functions of several variables subject to constraints. Examples of this kind of problem are:

□ **EXAMPLE 1**

Find the maximum and minimum values of the function $f(x, y) = 2x^2 + 4y^2$, given that $x^2 + y^2 = 1$.

□ **EXAMPLE 2**

Find the point(s) on the hyperbolic paraboloid $z = (y + 1)^2 - (x - 2)^2 + 1$ nearest the point $(2, -1, 2)$.

□ **EXAMPLE 3**

A cylindrical tin can, with a top and bottom, is to be manufactured using 100 cm² of tin, ignoring waste. What dimensions produce the can of maximum volume?

Each of these examples involves finding the extreme values of a function

$$w = f(\mathbf{x}), \qquad \mathbf{x} \in \mathbb{R}^2 \text{ or } \mathbb{R}^3 \qquad \text{(function to be optimized)} \qquad (1)$$

subject to a **constraint** of the form

$$g(\mathbf{x}) = 0 \qquad \text{(constraint equation)}. \qquad (2)$$

In Example 1 the function to be optimized is $f(x, y) = 2x^2 + 4y^2$, and the constraint is $x^2 + y^2 = 1$. The latter equation can be put in the form of equation (2) by writing it as

$$g(x, y) = x^2 + y^2 - 1 = 0.$$

In Example 2, we are to minimize the distance function $D(x, y, z) = \sqrt{(x - 2)^2 + (y + 1)^2 + (z - 2)^2}$ subject to the constraint that (x, y, z) lie on the hyperboloid. We can write this constraint in the form of equation (2) as

$$g(x, y, z) = z - (y + 1)^2 + (x - 2)^2 - 1 = 0.$$

In Example 3, the dimensions of the tin can are r (radius) and h (height), so we need to maximize the volume $V = \pi r^2 h$ subject to the constraint that the total surface area equals 100 cm². This constraint may by expressed as the equation

$$g(r, h) = 2\pi r^2 + 2\pi rh - 100 = 0.$$

Each of these examples illustrates the important difference between finding *relative* extrema for functions f of several variables and finding *extrema in the presence of constraints,* or *constrained extrema.* The constraint equation $g(\mathbf{x}) = 0$ restricts the points $\mathbf{x}$ that may be considered to just those satisfying this constraint. Constrained extrema usually do not correspond to relative extrema.

Figure 8.1 illustrates the constrained extrema problem of Example 1. Notice how the constraint $x^2 + y^2 = 1$ restricts the graph of f to the curve C above the unit circle. Note also that neither the constrained maxima nor the constrained minima correspond to relative extrema for the function f.

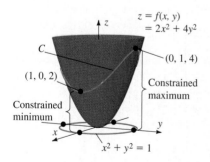

Figure 8.1   The constraint $g(x, y) = x^2 + y^2 - 1 = 0$ restricts the graph of f to the curve C. There are two constrained extrema on the back of the elliptical paraboloid. They correspond to the points $(-1, 0, 2)$ and $(0, -1, 4)$.

It is sometimes possible to solve a problem involving a constraint by solving the constraint equation (2) for one of the independent variables and then substituting for this variable in equation (1). The method due to Lagrange is more general, in that it does not depend on our ability to solve the constraint equation for any particular variable. It is based on the following theorem.

Theorem 8

Let $\mathbf{x}$ denote a point (position vector) in either $\mathbb{R}^2$ or $\mathbb{R}^3$, and let f and g be functions of either two or three variables. Assume that both f and g have continuous partial derivatives in a neighborhood of the point $\mathbf{x}_0$. If $\mathbf{x}_0$ maximizes or minimizes the function

$$w = f(\mathbf{x}) \tag{3a}$$

subject to the constraint

$$g(\mathbf{x}) = 0, \tag{3b}$$

and if $\nabla g(\mathbf{x}_0) \neq \mathbf{0}$, then

$$\nabla f(\mathbf{x}_0) = \lambda \nabla g(\mathbf{x}_0) \tag{4}$$

for some real number λ. In other words, $\nabla f(\mathbf{x}_0)$ and $\nabla g(\mathbf{x}_0)$ are parallel.

We sketch the proof of Theorem 8 at the end of this section.

Theorem 8 states that the gradients of the function f and the constraint g are parallel at points that correspond to constrained extrema. Therefore, the **method of Lagrange multipliers** is simply to determine all points $\mathbf{x}_0$

1. that satisfy the constraint and
2. for which $\nabla f(\mathbf{x}_0)$ and $\nabla g(\mathbf{x}_0)$ are parallel.

Then the constrained extrema are the extrema of the corresponding values $f(\mathbf{x}_0)$ among all $\mathbf{x}_0$ that satisfy conditions (1) and (2).

The multiplier λ in equation (4) is a scalar (real number) whose only role is to indicate that the two vectors ∇f and ∇g are parallel. The value of λ is not significant for the method.

The following is a more detailed description of the method of Lagrange.

---

1. Find all simultaneous solutions of the equations

$$\nabla f(\mathbf{x}_0) = \lambda \nabla g(\mathbf{x}_0) \tag{5}$$

and

$$g(\mathbf{x}_0) = 0. \tag{6}$$

   a. If $\mathbf{x} = (x, y) \in \mathbb{R}^2$, equations (5) and (6) are equivalent to the three equations

$$\begin{cases} \dfrac{\partial f}{\partial x}(x_0, y_0) = \lambda \dfrac{\partial g}{\partial x}(x_0, y_0), \\[2mm] \dfrac{\partial f}{\partial y}(x_0, y_0) = \lambda \dfrac{\partial g}{\partial y}(x_0, y_0), \\[2mm] g(x_0, y_0) = 0. \end{cases} \tag{7}$$

b. If $\mathbf{x} = (x, y, z) \in \mathbb{R}^3$, equations (5) and (6) are equivalent to the four equations

$$\begin{cases} \dfrac{\partial f}{\partial x}(x_0, y_0, z_0) = \lambda \dfrac{\partial g}{\partial x}(x_0, y_0, z_0), \\[2mm] \dfrac{\partial f}{\partial y}(x_0, y_0, z_0) = \lambda \dfrac{\partial g}{\partial y}(x_0, y_0, z_0), \\[2mm] \dfrac{\partial f}{\partial z}(x_0, y_0, z_0) = \lambda \dfrac{\partial g}{\partial z}(x_0, y_0, z_0), \\[2mm] g(x_0, y_0, z_0) = 0. \end{cases} \tag{8}$$

2. Calculate $f(\mathbf{x}_0)$ for all $\mathbf{x}_0$ obtained in step 1.
3. Using geometric, analytic, or physical reasoning, determine which of the numbers obtained in step 2 correspond to constrained extrema.

Solution to Example 1   We wish to find the extreme values of

$$f(x, y) = 2x^2 + 4y^2$$

subject to

$$g(x, y) = x^2 + y^2 - 1 = 0.$$

The three equations corresponding to equations (7) are:

$$4x = 2\lambda x \qquad (f_x = \lambda g_x), \tag{9}$$

$$8y = 2\lambda y \qquad (f_y = \lambda g_y), \tag{10}$$

and

$$x^2 + y^2 - 1 = 0 \qquad (g = 0). \tag{11}$$

To solve this system of three equations, we begin with equation (9). If $x = 0$ equation (9) is satisfied, equation (11) then becomes simply $y^2 - 1 = 0$, so $y = \pm 1$. We therefore obtain the two points $(0, 1)$ and $(0, -1)$ that must be checked for extrema.

On the other hand, if $x \neq 0$ in (9), we may divide both sides of (9) by x to obtain $4 = 2\lambda$, so $\lambda = 2$. Substituting this value of λ into (10) then gives $8y = 4y$, so y must equal zero. With $y = 0$, equation (11) becomes $x^2 - 1 = 0$, so $x = \pm 1$. We have therefore obtained two additional points, $(1, 0)$ and $(-1, 0)$.

Finally, we calculate the value of $f(x, y)$ for each of the four points $(0, 1)$, $(0, -1)$, $(1, 0)$, and $(-1, 0)$. We find

$$f(0, 1) = 4 \qquad \text{(maximum)},$$

$$f(0, -1) = 4 \qquad \text{(maximum)},$$

$$f(1, 0) = 2 \qquad \text{(minimum)},$$

$$f(-1, 0) = 2 \qquad \text{(minimum)}.$$

As Figure 8.1 illustrates, the constrained maximum is 4, occurring at $(0, 1)$ and $(0, -1)$, and the constrained minimum is $2 = f(1, 0) = f(-1, 0)$. ∎

REMARK 1   In these examples, note that the method we use to solve the systems of equations (7) and (8) varies from problem to problem. The great versatility of the method of Lagrange multipliers is that it applies to a wide variety of functions and constraints. However, the resulting systems of equations will therefore be of various types, and often most or all of the resulting equations will be nonlinear. Your success in using this method will depend on your ability to solve these various sorts of systems of equations.

REMARK 2   Note that although we obtained the value $\lambda = 2$ at one point in the solution of Example 1, the value of λ did not appear in the solution. This will always be the case, since the Lagrange multiplier λ is an "artificial" variable that is introduced merely to indicate that ∇f and ∇g are parallel at the desired points.

The solution of Example 2 is similar to that of Example 1, and we leave it to you as Exercise 19.

Solution to Example 3   Since the volume of the cylinder is $V(r, h) = \pi r^2 h$, the areas of the top and borrom are each πr^2, and the lateral surface area is $2\pi rh$, we must maximize the function

$$V(r, h) = \pi r^2 h$$

subject to the side condition

$$g(r, h) = 2\pi r^2 + 2\pi rh - 100 = 0.$$

The equations corresponding to equations (7) are therefore

$$2\pi rh = \lambda(4\pi r + 2\pi h) \qquad (V_r = \lambda g_r), \tag{12}$$

$$\pi r^2 = \lambda(2\pi r) \qquad (V_h = \lambda g_h), \tag{13}$$

and

$$2\pi r^2 + 2\pi rh - 100 = 0 \qquad (g = 0). \tag{14}$$

Solving equation (12) for λ gives

$$\lambda = \frac{rh}{2r + h}, \tag{15}$$

while solving equation (13) for λ gives

$$\lambda = \frac{r}{2}. \tag{16}$$

Equating the right sides of (15) and (16) we find

$$\frac{rh}{2r + h} = \frac{r}{2} \quad \text{so} \quad 2r + h = 2h,$$

or $h = 2r$. Substituting $h = 2r$ for h in (14) then gives

$$2\pi r^2 + 2\pi r(2r) - 100 = 0,$$

which has solution $r = 10/\sqrt{6\pi}$. Since it is clear from the geometry of the situation that at least one pair (r, h) must produce a maximum volume, we conclude that the maximum volume of

$$V = \pi\left(\frac{10}{\sqrt{6\pi}}\right)^2\left(\frac{20}{\sqrt{6\pi}}\right) = \frac{1000}{3\sqrt{6\pi}} \text{ cm}^3$$

corresponds to the dimensions $r = 10/\sqrt{6\pi}$ cm and $h = 2r = 20/\sqrt{6\pi}$ cm.   ∎

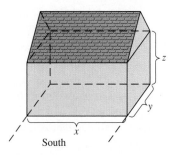

Figure 8.2 The house in Example 4. The rectangular portion determined by the dimensions x, y, and z is the heated space.

□ **EXAMPLE 4**

A builder wishes to design a rectangular house containing V cubic meters of heated space, so as to minimize heating costs. One wall of the building is to face south. The annual heating costs are estimated to be \$4 per square meter of floor space, \$3 per square meter for all exterior walls not facing south, and \$2 per square meter for exterior wall space facing south. What dimensions will produce the most energy-efficient building?

Solution As in Figure 8.2, let x denote the length of the wall facing south. We refer to this dimension as the width of the house. Let y be the depth, and let z be the height of the heated portion of the house. Since the area of the floor is xy, the area of each side wall is yz, and the area of each front and rear wall is xz, the annual heating cost is

$$C(x, y, z) = 4xy + 3(2yz) + 3xz + 2xz$$

Cost for south wall

Cost for rear (north) wall

Costs for side walls (both)

Costs for roof (floor space)

$$= 4xy + 6yz + 5xz.$$

The constraint is that the volume of the building is to equal the constant V. That is,

$$g(x, y, z) = xyz - V = 0.$$

Applying equations (8), we find

$$4y + 5z = \lambda yz \qquad (C_x = \lambda g_x), \tag{17}$$

$$4x + 6z = \lambda xz \qquad (C_y = \lambda g_y), \tag{18}$$

$$5x + 6y = \lambda xy \qquad (C_z = \lambda g_z), \tag{19}$$

and

$$xyz - V = 0 \qquad (g = 0). \tag{20}$$

To solve these equations, it is helpful to multiply both sides of equation (17) by x, both sides of (18) by y, and both sides of (19) by z. The result is the equations

$$4xy + 5xz = \lambda xyz, \tag{21}$$

$$4xy + 6yz = \lambda xyz, \tag{22}$$

and

$$5xz + 6yz = \lambda xyz. \tag{23}$$

Equating the left sides of (21) and (22) then gives $4xy + 5xz = 4xy + 6yz$, so $y = 5x/6$. Similarly, equating the left sides of (22) and (23) gives $4xy + 6yz = 5xz + 6yz$, or $z = 4y/5 = 2x/3$. With these substitutions (20) becomes

$$x(\tfrac{5}{6}x)(\tfrac{2}{3}x) = \tfrac{5}{9}x^3 = V.$$

so

$$x = \sqrt[3]{\frac{9V}{5}}, \quad y = \frac{5}{6}\sqrt[3]{\frac{9V}{5}}, \quad \text{and} \quad z = \frac{2}{3}\sqrt[3]{\frac{9V}{5}}. \qquad ■$$

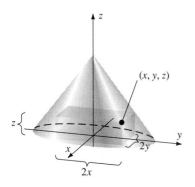

Figure 8.3 Box inscribed within a cone.

□ **EXAMPLE 5**

A rectangular box is to be inscribed in the cone $z = 9 - \sqrt{x^2 + y^2}$, $z \geq 0$ (see Figure 8.3). Find the dimensions for the box that maximize its volume.

Solution It is clear that to achieve maximum volume we should position the box with one face lying in the xy-plane. If (x, y, z) denotes one corner of the box lying in the first octant (that is, $x > 0$, $y > 0$, and $z > 0$), the dimensions of the box are

$$\text{length} = 2x, \quad \text{width} = 2y, \quad \text{and} \quad \text{height} = z.$$

The problem is therefore to maximize the function

$$V(x, y, z) = 4xyz$$

subject to the constraint

$$g(x, y, z) = z + \sqrt{x^2 + y^2} - 9 = 0.$$

Applying the Lagrange criterion (8), we find

$$4yz = \frac{\lambda x}{\sqrt{x^2 + y^2}} \qquad (f_x = \lambda g_x), \tag{24}$$

$$4xz = \frac{\lambda y}{\sqrt{x^2 + y^2}} \qquad (f_y = \lambda g_y), \tag{25}$$

$$4xy = \lambda, \tag{26}$$

and

$$z + \sqrt{x^2 + y^2} - 9 = 0. \tag{27}$$

Multiplying (24) through by x, (25) by y, and (26) by z gives

$$4xyz = \frac{\lambda x^2}{\sqrt{x^2 + y^2}}; \qquad 4xyz = \frac{\lambda y^2}{\sqrt{x^2 + y^2}}; \qquad 4xyz = \lambda z.$$

Equating the right-hand sides of these equations gives

$$\frac{\lambda x^2}{\sqrt{x^2 + y^2}} = \frac{\lambda y^2}{\sqrt{x^2 + y^2}} \tag{28}$$

and

$$\lambda z = \frac{\lambda x^2}{\sqrt{x^2 + y^2}}. \tag{29}$$

From (28) it follows that $y = x$, since neither x nor y can be negative. With $y = x$, (29) gives $z = x^2/\sqrt{2x^2} = x/\sqrt{2}$. Substituting these expressions for y and z in (27) then gives

$$\frac{x}{\sqrt{2}} + \sqrt{x^2 + x^2} - 9 = \left(\frac{1}{\sqrt{2}} + \sqrt{2}\right)x - 9 = 0,$$

so

$$x = \frac{9}{\dfrac{1}{\sqrt{2}} + \sqrt{2}} = \frac{9\sqrt{2}}{1 + 2} = 3\sqrt{2}.$$

Thus, $y = 3\sqrt{2}$ and $z = 3$. The maximum volume is therefore

$$V = 4(3\sqrt{2})(3\sqrt{2})(3) = 216.$$ ∎

Proof of Theorem 8 (Sketch): First consider the planar case $\mathbf{x} = (x, y) \in \mathbb{R}^2$. Then the constraint equation $g(\mathbf{x}) = g(x, y) = 0$ determines a curve C in the xy-plane on which the point $\mathbf{x}_0$ must lie. This curve C is a *level curve* for the function g. From statement (16), Section 18.7, if $\nabla g(\mathbf{x}_0) \neq \mathbf{0}$ then $\nabla g(\mathbf{x}_0)$ is orthogonal to C at $\mathbf{x}_0$. We complete the proof for the planar case by showing that $\nabla f(\mathbf{x}_0)$ is also orthogonal to C at $\mathbf{x}_0$.

Let $\mathbf{r}(t) = x(t)\mathbf{i} + y(t)\mathbf{j}$ be a parameterization for the curve C so that $\mathbf{r}(t_0) = \mathbf{x}_0$ and so that $\mathbf{r}'(t_0) \neq \mathbf{0}$.* Since $\mathbf{x}_0 = \mathbf{r}(t_0)$ maximizes $f(\mathbf{x})$ on C, the scalar function $h = f \circ \mathbf{r}$ has a relative extremum (as discussed in Chapter 4) at $t = t_0$. Thus, from the Extreme Value Theorem in Chapter 4, we know that $h'(t_0) = 0$. However, we can also compute $h'(t_0)$ using the multivariable Chain Rule (see equation (12) in Section 18.7). Therefore,

$$h'(t_0) = \nabla f(\mathbf{r}(t_0)) \cdot \mathbf{r}'(t_0) = 0. \tag{30}$$

This equation shows that $\nabla f(\mathbf{r}(t_0)) = \nabla f(\mathbf{x}_0)$ is orthogonal to $\mathbf{r}'(t_0)$. Since $\mathbf{r}'(t_0)$ is *tangent* to C at $\mathbf{x}_0$, it follows that $\nabla f(\mathbf{x}_0)$ is orthogonal to C at $\mathbf{x}_0$.

Since $\nabla f(\mathbf{x}_0)$, $\nabla g(\mathbf{x}_0)$, and C all lie in the xy-plane, we may now conclude, since $\nabla f(\mathbf{x}_0)$ and $\nabla g(\mathbf{x}_0)$ are both orthogonal to C at $\mathbf{x}_0$, that $\nabla f(\mathbf{x}_0)$ and $\nabla g(\mathbf{x}_0)$ are parallel. That is, $\nabla f(\mathbf{x}_0) = \lambda \nabla g(\mathbf{x}_0)$ for some scalar λ.

The case of $\mathbf{x} \in \mathbb{R}^3$ is similar to the planar case. Now, the constraint equation $g(\mathbf{x}) = 0$ determines a surface S in $\mathbb{R}^3$ that is a level surface for the function $u = g(\mathbf{x})$. By statement (19), Section 18.7, if $\nabla g(\mathbf{x}_0) \neq \mathbf{0}$ then $\nabla g(\mathbf{x}_0)$ is orthogonal to S at $\mathbf{x}_0$.

As in the planar case, we show that $\nabla f(\mathbf{x}_0)$ is also orthogonal to S at $\mathbf{x}_0$. We begin by letting C be an arbitrary curve on S with parameterization $\mathbf{r}(t) = x(t)\mathbf{i} + y(t)\mathbf{j} + z(t)\mathbf{k}$ for which $\mathbf{r}(t_0) = \mathbf{x}_0$ and $\mathbf{r}'(t_0) \neq \mathbf{0}$. Again, the scalar function $h = f \circ \mathbf{r}$ has a relative extremum at t_0, so equation (30) holds. Since $\mathbf{r}'(t_0)$ is tangent to C at $\mathbf{x}_0$, it is also tangent to S at $\mathbf{x}_0$. This shows that $\nabla f(\mathbf{x}_0)$ is orthogonal to the tangent $r'(t_0)$ at $\mathbf{x}_0$. But since the curve C was arbitrary, so is its tangent $\mathbf{r}'(t_0)$. Thus, $\nabla f(\mathbf{x}_0)$ is orthogonal to *all* tangents to S at $\mathbf{x}_0$, and $\nabla f(\mathbf{x}_0)$ is orthogonal to S. Since two vectors orthogonal to a (smooth) surface at a common point must be parallel, it follows that $\nabla f(\mathbf{x}_0) = \lambda \nabla g(\mathbf{x}_0)$ for some scalar λ. ∎

Exercise Set 18.8

In Exercises 1–18, find the maximum and minimum values of the given function subject to the given constraint.

1. $f(x, y) = 2x^2 + 4y^2$ subject to $x^2 + y^2 = 1$

2. $f(x, y) = x^2 + y$ subject to $x^2 + y^2 = 9$

3. $f(x, y) = x^3 - y^3$ subject to $x - y = 2$

4. $f(x, y) = xy$ subject to $x^2 + y^2 - 4y = 5$

5. $f(x, y) = xy$ subject to $x^2 + y^2 = 1$

6. $f(x, y) = y - x$ subject to $x^2 + y^2 = 2$

7. $f(x, y) = x^2 + 4x + 4y^2$ subject to $x^2 + 2y^2 = 4$

8. $f(x, y, z) = x + y + z$ subject to $x^2 + y^2 + z^2 = 4$

9. $f(x, y) = x^2 - 4x + 4y^2$ subject to $x^2 + y^2 = 1$

---

*$\mathbf{r}(t)$ can always be chosen so that $\mathbf{r}'(t_0) \neq \mathbf{0}$ if $\nabla g(\mathbf{x}_0) \neq \mathbf{0}$. This follows from the Implicit Function Theorem, a result usually discussed in courses on advanced calculus. Actually, the Implicit Function Theorem underlies several of the seemingly obvious statements being made here, which is why this argument is only a sketch of a proof.

10. $f(x, y, z) = xyz$ subject to $2x^2 + y^2 + 4z^2 = 9$

11. $f(x, y, z) = x + 2y - z$ subject to $x^2 + y^2 + z^2 = 1$

12. $f(x, y, z) = \sqrt{xyz}$ subject to $x + y + z = 4$

13. $f(x, y, z) = x + y + z$ subject to $x^2 + y^2 + z^2 = 12$

14. $f(x, y, z) = x^2 + 2y^2 + 4z^2$ subject to $x^2 + y^2 + z^2 = 1$

15. $f(x, y) = x \sin y + y \sin x$ subject to $x - y = 0$

16. $f(x, y) = ye^x$ subject to $x^2 + 2y = 4$

17. $f(x, y, z) = x + y^2 + z$ subject to $z + \ln x = 0$

18. $f(x, y, z) = x - 2y + z$ subject to $x^2 + y^2 + (z - 3)^2 = 1$

19. Find the solution of Example 2.

20. Find the point(s) on the ellipsoid $9x^2 + 36y^2 + 4z^2 = 36$ nearest the origin.

21. Find the point on the circle $(x - 3)^2 + (y + 2)^2 = 9$ nearest the origin.

22. Find the point on the sphere $x^2 + y^2 + z^2 = 1$ furthest from the point $(3, 2, 1)$.

23. Find the point on the ellipsoid

$$(x - 1)^2 + \frac{(y - 2)^2}{4} + \frac{(z - 1)^2}{9} = 1$$

nearest the point $(1, -1, 1)$.

24. Find the point on the cone $z = \sqrt{x^2 + y^2}$ nearest the point $(3, 1, 0)$.

25. Find the point(s) on the hyperbolic paraboloid $z = (y + 1)^2 - (x - 2)^2 + 1$ nearest the point $(2, -1, 1)$.

26. The plane $2y - 3z = 8$ intersects the cone $z^2 = 4x^2 + 4y^2$ in an ellipse. Find the highest and lowest points of intersection.

27. Find the dimensions of the rectangular box of maximum volume that can be inscribed in a sphere of radius r.

28. Find the dimensions of the rectangular box of maximum volume that can be inscribed in the ellipsoid $x^2 + 4y^2 + 2z^2 = 8$.

29. Find the dimensions for the cylindrical barrel, with volume 2 ft^3, that has minimum exterior surface area. (Assume that the barrel has a top and that the thickness of its side is negligible.)

30–34. Rework Exercises 29–33 of Section 18.4 using the method of Lagrange multipliers.

18.9 Reconstructing a Function from Its Gradient

In several ways, the gradient is the analogue of the derivative for functions of a single variable. It is therefore reasonable to ask whether the concept of antidifferentiation makes sense for gradients and, if so, whether this concept is useful. The goal of this section is to define what we mean by the "antiderivative" of a gradient (we don't really use this particular terminology), to show how such functions can be found, and to prove a theorem indicating if such functions exist. We will restrict our discussion to functions of two variables, although analogous results may be developed in more general settings. The results of this section will be used extensively in Chapters 20 and 21.

Potential Functions

We begin by recalling that a gradient is a vector-valued function

$$\nabla f(x, y) = \frac{\partial f}{\partial x}(x, y)\mathbf{i} + \frac{\partial f}{\partial y}(x, y)\mathbf{j}.$$

The antidifferentiation question for gradients is therefore: Given a vector-valued function of the form

$$\mathbf{F}(x, y) = M(x, y)\mathbf{i} + N(x, y)\mathbf{j},$$

is there a function f of two variables for which

$$\mathbf{F}(x, y) = \nabla f(x, y)?$$

We prefer not to refer to f as the antiderivative for the vector-valued function **F**. Instead, we call it a *potential function,* or simply a *potential.* (The reason for this particular terminology has to do with the roles played by such functions in mechanics and in the theory of electricity and magnetism.)

Given the vector-valued function **F**, the task of finding a function f for which $\mathbf{F}(x, y) = \nabla f(x, y)$ is therefore referred to as finding a potential for **F**. This is also what we mean by "reconstructing the function f from its gradient." As Example 1 shows, potentials are not unique.

Definition 6

The function f of two variables is called a **potential** for the vector-valued function

$$\mathbf{F}(x, y) = M(x, y)\mathbf{i} + N(x, y)\mathbf{j}$$

in a rectangle $Q \subseteq \mathbb{R}^2$ if $\dfrac{\partial f}{\partial x}(x, y)$ and $\dfrac{\partial f}{\partial y}(x, y)$ exist and if

$$\mathbf{F}(x, y) = \nabla f(x, y)$$

for all (x, y) in Q.

☐ **EXAMPLE 1**

If $f(x, y) = xe^{2y}$, then

$$\nabla f(x, y) = e^{2y}\mathbf{i} + 2xe^{2y}\mathbf{j}.$$

Thus, the function

$$\mathbf{F}(x, y) = e^{2y}\mathbf{i} + 2xe^{2y}\mathbf{j}$$

has the function $f(x, y) = xe^{2y}$ as a potential. However, the function $g(x, y) = xe^{2y} + C$ is also a potential for **F,** since

$$\nabla g(x, y) = e^{2y}\mathbf{i} + 2xe^{2y}\mathbf{j} = \mathbf{F}(x, y).$$

Thus, if f is a potential for **F**, so is $f + C$ for any constant C. ∎

Finding Potentials

If a potential f for **F** exists, how is it found? According to Definition 6 the following must be true.

If $\mathbf{F}(x, y) = M(x, y)\mathbf{i} + N(x, y)\mathbf{j} = \nabla f(x, y)$, then

$$M(x, y) = \frac{\partial f}{\partial x}(x, y) \tag{1}$$

and

$$N(x, y) = \frac{\partial f}{\partial y}(x, y). \tag{2}$$

Equations (1) and (2) are the keys to finding f. Beginning with equation (1) and "integrating partially with respect to x" we find

$$f(x, y) = \int \left(\frac{\partial f}{\partial x}(x, y) \right) dx = \int M(x, y) \, dx = G(x, y) + h(y). \tag{3}$$

By "integrating partially with respect to x" we mean treating y as a constant and integrating $M(x, y)$ as a function of x alone—just the reverse of partial differentiation. We must remember, however, that functions of y alone vanish entirely when differentiated partially with respect to x, so we must allow for the appearance of a function $h(y)$ of y alone when integrating partially with respect to x.

To determine $h(y)$, we use equation (2). Differentiating equation (3) with respect to y yields

$$\frac{\partial f}{\partial y} = \frac{\partial G}{\partial y} + h'(y) = N(x, y). \tag{4}$$

If **F** has a potential function, then equation (4) leads to an expression for $h'(y)$ that does not involve x. If so, we determine $f(x, y)$ by finding an antiderivative for h'.

Of course, we can also begin by "partially integrating equation (2) with respect to y," if we prefer. Then we obtain a function $h(x)$, and we differentiate with respect to x to find f.

☐ **EXAMPLE 2**

Let **F** be the vector-valued function

$$\mathbf{F}(x, y) = (3x^2 + 2y^2)\mathbf{i} + 4xy\mathbf{j}.$$

Find a function f for which $\mathbf{F}(x, y) = \nabla f(x, y)$ for all $f(x, y)$.

Strategy · · · · · · · ·
Identify M and N.

Solution
Here $\mathbf{F}(x, y) = M(x, y)\mathbf{i} + N(x, y)\mathbf{j}$ with

$$M(x, y) = 3x^2 + 2y^2; \qquad N(x, y) = 4xy.$$

Integrate M partially with respect to x to find an expression for $f(x, y)$.

Partially integrating M with respect to x according to equation (3) gives

$$f(x, y) = \int M(x, y)\, dx$$

$$= \int (3x^2 + 2y^2)\, dx$$

$$= x^3 + 2xy^2 + h(y) \tag{5}$$

Find $h(y)$ by differentiating (5) with respect to y.

where h is a function of y alone. To determine h, we differentiate with respect to y and obtain

$$\frac{\partial f}{\partial y} = 4xy + h'(y)$$

Set $\partial f/\partial y = N$.

Equating $\partial f/\partial y$ with N yields

$$4xy + h'(y) = 4xy.$$

Determine $h'(y)$ and then $h(y)$.

Consequently, $h'(y) = 0$, and $h(y)$ is a constant C.
Since $h(y) = C$ in equation (5), we have

$$f(x, y) = x^3 + 2xy^2 + C.$$

Check by calculating $\nabla f(x, y)$.

Using partial differentiation, we can check that

$$\nabla f(x, y) = \mathbf{F}(x, y). \qquad \blacksquare$$

☐ **EXAMPLE 3**

Find a potential for the function

$$\mathbf{F}(x, y) = (ye^{xy} - 2x \sin x^2)\mathbf{i} + \left(\frac{1}{\sqrt{y}} + xe^{xy}\right)\mathbf{j}.$$

Solution   Here $\mathbf{F}(x, y) = M(x, y)\mathbf{i} + N(x, y)\mathbf{j}$ with

$$M(x, y) = ye^{xy} - 2x \sin x^2; \qquad N(x, y) = \frac{1}{\sqrt{y}} + xe^{xy}.$$

Integrating M partially with respect to x gives

$$f(x, y) = \int M(x, y)\, dx = \int (ye^{xy} - 2x \sin x^2)\, dx$$

$$= e^{xy} + \cos x^2 + h(y) \qquad (6)$$

where h is a function of y alone. Differentiating with respect to y, we get

$$\frac{\partial f}{\partial y} = xe^{xy} + h'(y).$$

Setting $\partial f/\partial y = N$ yields

$$xe^{xy} + h'(y) = \frac{1}{\sqrt{y}} + xe^{xy}.$$

Thus, $h'(y) = 1/\sqrt{y} = y^{-1/2}$, and $h(y) = 2\sqrt{y} + C$. Combining this expression for $h(y)$ with equation (6) yields the desired potential function

$$f(x, y) = e^{xy} + \cos x^2 + 2\sqrt{y} + C. \qquad ■$$

☐ **EXAMPLE 4**

Show that the function

$$\mathbf{F}(x, y) = y\mathbf{i} - x\mathbf{j}$$

has no potential. That is, $\mathbf{F}$ cannot be the gradient of a function f of two variables.

Solution   On the contrary, let us assume that there does exist a function f with $\nabla f = \mathbf{F}$. Then

$$\frac{\partial f}{\partial x} = y \quad \text{and} \quad \frac{\partial f}{\partial y} = -x.$$

Differentiating once again, we have

$$\frac{\partial^2 f}{\partial y \partial x} = 1 \quad \text{and} \quad \frac{\partial^2 f}{\partial x \partial y} = -1,$$

which contradicts the theorem on equality of mixed partial derivatives.

Since the assumption that $\mathbf{F}(x, y) = y\mathbf{i} - x\mathbf{j}$ has a potential leads to a contradiction, we conclude that $\mathbf{F}$ has no potential.   ■

The method of solution to Example 4 provides a condition that guarantees the existence of a potential function.

Theorem 9	Let M and N have continuous first partial derivatives in an open rectangle R in the xy-plane. Then the vector-valued function

$$\mathbf{F}(x, y) = M(x, y)\mathbf{i} + N(x, y)\mathbf{j}$$

has a potential in R if and only if

$$\frac{\partial M}{\partial y}(x, y) = \frac{\partial N}{\partial x}(x, y) \tag{7}$$

for all (x, y) in R.

Before giving the proof, note that condition (7) is satisfied for $\mathbf{F}$ in Example 2, since

$$\frac{\partial M}{\partial y}(x, y) = \frac{\partial}{\partial y}(3x^2 + 2y^2) = 4y = \frac{\partial N}{\partial x}(x, y).$$

Also, equation (7) holds for $\mathbf{F}$ in Example 3:

$$\frac{\partial M}{\partial y}(x, y) = \frac{\partial}{\partial y}(ye^{xy} - 2x \sin x^2)$$

$$= e^{xy} + xye^{xy}$$

$$= \frac{\partial}{\partial x}\left(\frac{1}{\sqrt{y}} + xe^{xy}\right)$$

$$= \frac{\partial N}{\partial x}(x, y).$$

However, condition (7) fails for $\mathbf{F}$ in Example 4 since

$$\frac{\partial M}{\partial y} = \frac{\partial}{\partial y}(y) = 1 \neq -1 = \frac{\partial}{\partial x}(-x) = \frac{\partial N}{\partial x}.$$

Proof of Theorem 9:   First we assume that $\mathbf{F}$ has a potential f and show that equation (7) holds. This means that

$$\mathbf{F}(x, y) = \nabla f(x, y)$$

where $M = \partial f/\partial x$ and $N = \partial f/\partial y$. Since both M and N are assumed to have continuous first partial derivatives in R, it follows that

$$\frac{\partial^2 f}{\partial y \partial x} = \frac{\partial}{\partial y}\left(\frac{\partial f}{\partial x}\right) = \frac{\partial M}{\partial y}(x, y) \tag{8}$$

and

$$\frac{\partial^2 f}{\partial x \partial y} = \frac{\partial}{\partial x}\left(\frac{\partial f}{\partial y}\right) = \frac{\partial N}{\partial x}(x, y). \tag{9}$$

Moreover, by Theorem 2, the mixed partial derivatives on the left-hand sides of equations (8) and (9) are equal. Thus, the right-hand sides of these equations are equal, and condition (7) is obtained.

The remaining part of the proof involves assuming that condition (7) holds and actually finding a potential for **F**. We begin by fixing a point (x_0, y_0) in R and defining the function f as

$$f(x, y) = \int_{x_0}^{x} M(t, y_0)\, dt + \int_{y_0}^{y} N(x, s)\, ds. \tag{10}$$

Figure 9.1 illustrates the line segments that correspond to these integrals. Since we want to construct f in such a way that $\partial f/\partial x = M$ and $\partial f/\partial y = N$, the idea is to integrate M along a path involving only change in x and to integrate N along a path involving only change in y. The reason we require R to be a rectangle is simply to ensure that such a path lies entirely within R.

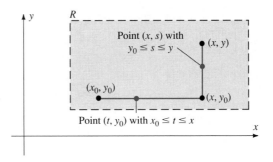

Figure 9.1 Rectangle R in Theorem 9.

Since M and N are continuous in R, and since R is a rectangle, $f(x, y)$ is defined for all $(x, y) \in R$. Moreover, by the Fundamental Theorem of Calculus, it follows that

$$\frac{\partial f}{\partial y}(x, y) = \frac{\partial}{\partial y}\left[\int_{x_0}^{x} M(t, y_0)\, dt + \int_{y_0}^{y} N(x, s)\, ds\right]$$

$$= \frac{\partial}{\partial y}\int_{y_0}^{y} N(x, s)\, ds = N(x, y) \tag{11}$$

since the first integral inside the braces does not involve y. We also want to calculate $\partial f/\partial x$, but this is a bit harder. We find that

$$\frac{\partial f}{\partial x}(x, y) = \frac{\partial}{\partial x}\left[\int_{x_0}^{x} M(t, y_0)\, dt + \int_{y_0}^{y} N(x, s)\, ds\right]$$

$$= M(x, y_0) + \frac{\partial}{\partial x}\int_{y_0}^{y} N(x, s)\, ds. \tag{12}$$

In courses on advanced calculus it is proved that

$$\frac{\partial}{\partial x}\int_{y_0}^{y} N(x, s)\, ds = \int_{y_0}^{y} N(x, s)\, ds. \tag{13}$$

That is, we may pass the partial differentiation with respect to x under the y-integral sign in the integral on the right-hand side of equation (12). Using (13) we return to (12) and use condition (7) to find

$$\frac{\partial f}{\partial x}(x, y) = M(x, y_0) + \int_{y_0}^{y} \frac{\partial}{\partial x} N(x, s) \, ds$$

$$= M(x, y_0) + \int_{y_0}^{y} \frac{\partial}{\partial s} M(x, s) \, ds \qquad \left(\frac{\partial N}{\partial x}(x, s) = \frac{\partial M}{\partial s}(x, s) \right)$$

$$= M(x, y_0) + \left[M(x, s) \right]_{s=y_0}^{s=y}$$

$$= M(x, y_0) + [M(x, y) - M(x, y_0)] = M(x, y). \tag{14}$$

Reviewing equations (11) and (14), we find that the function f defined in line (10) has the properties

$$\frac{\partial f}{\partial x}(x, y) = M(x, y); \qquad \frac{\partial f}{\partial y}(x, y) = N(x, y).$$

That is, $\nabla f = \mathbf{F}$, so $\mathbf{F}$ indeed has a potential. ∎

Exercise Set 18.9

In Exercises 1–12, determine whether the given vector function has a potential. If so, find a potential for $\mathbf{F}$.

1. $\mathbf{F}(x, y) = \mathbf{i} - \mathbf{j}$

2. $\mathbf{F}(x, y) = 6y\mathbf{i} + 6x\mathbf{j}$

3. $\mathbf{F}(x, y) = \pi y\mathbf{i} - \pi x\mathbf{j}$

4. $\mathbf{F}(x, y) = 3x\mathbf{i} + 4y\mathbf{j}$

5. $\mathbf{F}(x, y) = (x^2 - y)\mathbf{i} + (x + y^2)\mathbf{j}$

6. $\mathbf{F}(x, y) = \sin y\mathbf{i} + x \cos y\mathbf{j}$

7. $\mathbf{F}(x, y) = (3x^2 \cos y - 1)\mathbf{i} + (2y - x^3 \sin y)\mathbf{j}$

8. $\mathbf{F}(x, y) = (2xy^3 - 2x)\mathbf{i} + (3x^2y^2 + 1)\mathbf{j}$

9. $\mathbf{F}(x, y) = (2xe^{xy} + x^2ye^{xy})\mathbf{i} + \left(x^3e^{xy} - \frac{1}{2\sqrt{y}} \right)\mathbf{j}$

10. $\mathbf{F}(x, y) = (x^3 - y \sin x)\mathbf{i} + (y^3 - \cos x)\mathbf{j}$

11. $\mathbf{F}(x, y) = (2xy^2 - y \sin x)\mathbf{i} + (2x^2y + \cos x)\mathbf{j}$

12. $\mathbf{F}(x, y) = 2x \operatorname{Tan}^{-1} y\mathbf{i} + \frac{x^2}{1 + y^2}\mathbf{j}$

Summary Outline of Chapter 18

■ A **level curve** for the function $z = f(x, y)$ is the graph of the equation (page 885) $f(x, y) = c$ in the xy-plane. A **level surface** for the function $w = f(x, y, z)$ is the graph of an equation $f(x, y, z) = c$ in space.

■ If $\mathbf{x} \in \mathbb{R}^n$ and $f: \mathbb{R}^n \to \mathbb{R}$, $\lim_{\mathbf{x} \to \mathbf{x}_0} f(\mathbf{x}) = L$ means $|f(\mathbf{x}) - L| \to 0$ as $\mathbf{x} \to \mathbf{x}_0$. (page 890)

■ The function $f: \mathbb{R}^n \to \mathbb{R}$ is **continuous** at $\mathbf{x}_0$ if $\lim_{\mathbf{x} \to \mathbf{x}_0} f(\mathbf{x}) = f(\mathbf{x}_0)$. (page 892)

■ The **partial derivatives** of f are the limits (page 896)

$$\frac{\partial f}{\partial x}(x, y, z) = \lim_{h \to 0} \frac{f(x + h, y, z) - f(x, y, z)}{h},$$

$$\frac{\partial f}{\partial y}(x, y, z) = \lim_{h \to 0} \frac{f(x, y + h, z) - f(x, y, z)}{h},$$

and

$$\frac{\partial f}{\partial z}(x, y, z) = \lim_{h \to 0} \frac{f(x, y, z + h) - f(x, y, z)}{h}.$$

■ **Theorem:** $\dfrac{\partial^2 f}{\partial y \partial x}(x, y) = \dfrac{\partial^2 f}{\partial x \partial y}(x, y)$ if f, $\dfrac{\partial f}{\partial x}$, $\dfrac{\partial f}{\partial y}$, $\dfrac{\partial^2 f}{\partial x \partial y}$, and $\dfrac{\partial^2 f}{\partial y \partial x}$ are   (page 902)
continuous.

■ The plane tangent to the graph of $z = f(x, y)$ are (x_0, y_0, z_0) has equation   (page 905)

$$\frac{\partial f}{\partial x}(x_0, y_0)(x - x_0) + \frac{\partial f}{\partial y}(x_0, y_0)(y - y_0) - (z - z_0) = 0.$$

■ **Theorem:** If f has a relative extremum at (x_0, y_0) then either   (page 909)

(i) $\dfrac{\partial f}{\partial x}(x_0, y_0) = \dfrac{\partial f}{\partial y}(x_0, y_0) = 0$,   or

(ii) one or both of $\dfrac{\partial f}{\partial x}(x_0, y_0)$ and $\dfrac{\partial f}{\partial y}(x_0, y_0)$ fail to exist.

■ **Theorem:** If $\dfrac{\partial f}{\partial x}(x_0, y_0) = \dfrac{\partial f}{\partial y}(x_0, y_0) = 0$ and if $A = \dfrac{\partial^2 f}{\partial x^2}(x_0, y_0)$, $B =$ (page 912)
$\dfrac{\partial^2 f}{\partial y \partial x}(x_0, y_0)$, $C = \dfrac{\partial^2 f}{\partial y^2}(x_0, y_0)$, and $D = B^2 - AC$, then

(i) If $D < 0$, and $A < 0$, $f(x_0, y_0)$ is a relative maximum.
(ii) If $D < 0$, and $A > 0$, $f(x_0, y_0)$ is a relative minimum.
(iii) If $D > 0$, (x_0, y_0) is a saddle point.
(iv) If $D = 0$ there is no conclusion.

■ **Theorem:** (**Linear Approximation**) If f and its first partials are continuous,   (page 918)
then

$$f(x_0 + \Delta x, y_0 + \Delta y) = f(x_0, y_0) + \frac{\partial f}{\partial x}(x_0, y_0)\, \Delta x + \frac{\partial f}{\partial y}(x_0, y_0)\, \Delta y$$
$$+ \epsilon_1 \Delta x + \epsilon_2 \Delta y$$

where $\epsilon_1 \to 0$ and $\epsilon_2 \to 0$ as $\Delta x \to 0$ and $\Delta y \to 0$.

■ **Chain Rule:** If $w = f(x(t), y(t))$ then   (page 927)

$$\frac{dw}{dt} = \frac{\partial f}{\partial x}\frac{dx}{dt} + \frac{\partial f}{\partial y}\frac{dy}{dt}.$$

■ The **directional derivative** $D_{\mathbf{u}} f(x_0, y_0)$ in the direction of the unit vector   (page 934)
$\mathbf{u} = u_1 \mathbf{i} + u_2 \mathbf{j}$ is

$$D_{\mathbf{u}} f(x_0, y_0) = \lim_{t \to 0^+} \frac{f(x_0 + tu_1, y_0 + tu_2) - f(x_0, y_0)}{t}.$$

■ The **gradient** $\nabla f(x_0, y_0)$ of the function f at (x_0, y_0) is the *vector*   (page 937)

$$\nabla f(x_0, y_0) = \frac{\partial f}{\partial x}(x_0, y_0)\mathbf{i} + \frac{\partial f}{\partial y}(x_0, y_0)\mathbf{j}$$

or

$$\nabla f(x_0, y_0, z_0) = \frac{\partial f}{\partial x}(x_0, y_0, z_0)\mathbf{i} + \frac{\partial f}{\partial y}(x_0, y_0, z_0)\mathbf{j} + \frac{\partial f}{\partial z}(x_0, y_0, z_0)\mathbf{k}.$$

■ If $\dfrac{\partial f}{\partial x}(x, y)$ and $\dfrac{\partial f}{\partial y}(x, y)$ are continuous, then   (page 938)

$$D_{\mathbf{u}} f(x_0, y_0) = \frac{\partial f}{\partial x}(x_0, y_0)u_1 + \frac{\partial f}{\partial x}(x_0, y_0)u_2$$
$$= \nabla f(x_0, y_0) \cdot u.$$

■ $\nabla f(\mathbf{x}_0)$ points in the direction of most rapid increase of the function f at $\mathbf{x}_0$.   (page 938)

■ $\nabla f(\mathbf{x}_0)$ is orthogonal to the level curve for $z = f(x, y)$ at $\mathbf{x}_0 = (x_0, y_0)$.   (page 941)

■ $\nabla f(\mathbf{x}_0)$ is orthogonal to the level surface for $z = f(x, y, z)$ at $\mathbf{x}_0 = (x_0, y_0, z_0)$.   (page 943)

■ The extreme values of the function $w = f(\mathbf{x})$ subject to the constraint $g(\mathbf{x}) =$ (page 947)
   0 occur at points $\mathbf{x}_0$ where $\nabla f(\mathbf{x}_0) = \lambda \nabla g(\mathbf{x}_0)$, where λ is a scalar.

■ The vector-valued function $\mathbf{F}(x, y) = M(x, y)\mathbf{i} + N(x, y)\mathbf{j}$ is said to have the   (page 954)
   potential f if $\mathbf{F}(x, y) = \nabla f(x, y)$.

■ ***Theorem:*** If M, N, and their first partial derivatives are continuous in a   (page 957)
   rectangle R, then

$$\mathbf{F}(x, y) = M(x, y)\mathbf{i} + N(x, y)\mathbf{j}$$

has a potential in R if and only if $\dfrac{\partial M}{\partial y} = \dfrac{\partial N}{\partial x}$ for all $(x, y) \in R$.

Review Exercises—Chapter 18

1. Does $\displaystyle\lim_{(x, y) \to (0, 0)} \frac{x^2 - y^2}{x^2 + y^2}$ exist? If so, find it.

2. Find $\displaystyle\lim_{(x, y) \to (0, 0)} \frac{2x^2 y^3}{(x^2 + y^2)^2}$.

3. Sketch level curves for the function

$$f(x, y) = \frac{x}{x^2 + y^2}$$

corresponding to levels $z = -2, -1, 1, 2, 4$.

4. Find df/dt if $f(x, y) = x \operatorname{Tan}^{-1}(y/x)$, $x = 1 + t^2$, and $y = 1 - t$.

5. Find a vector normal to the curve $x^3 - 2xy^2 + y + 5 = 0$ at the point $(1, 2)$.

6. Find an equation for the plane tangent to the graph of the equation $x^3 + y^3 - 6xy + z = 0$ at the point $(2, 2, 8)$.

7. Find a vector normal to the surface $e^x \cos y - z = 4$ at the point $(0, \pi, -5)$.

8. What is the z-coordinate of the point $P = (1, 3, z)$ if P lies on the plane tangent to the ellipsoid $4x^2 + y^2 + 9z^2 = 17$ at the point $(1, 2, 1)$?

9. Find both first order partial derivatives for the function $f(x, y) = x \sin \sqrt{x^2 + y^2}$.

10. Find an equation for the plane tangent to the graph of $x^3 + y^3 + xz^2 + z^3 - 9 = 0$ at the point $(2, 1, -2)$.

11. Let $f(x, y, z) = xy^3 + x^2\sqrt{y^2 + z^2}$. Find the directional derivative of f in the direction of the vector $2\mathbf{i} + \mathbf{j} + 2\mathbf{k}$ at the point $(2, 3, 4)$.

12. Show that $f(x, y, z) = (ax + by + cz)^3$ satisfies the partial differential equation

$$x \frac{\partial f}{\partial x} + y \frac{\partial f}{\partial y} + z \frac{\partial f}{\partial z} = 3f.$$

13. Find $\partial f/\partial r$ and $\partial f/\partial s$ if $f(x, y) = 3x^3 + 2xy^2 - y^2$, $x = 2r + 5s$, and $y = r - 2s^2$.

14. Suppose that $w = f(x, y)$ has partial derivatives with respect to both variables. Show that the function $z = f(x - y, y - x)$ satisfies the partial differential equation

$$\frac{\partial z}{\partial x} + \frac{\partial z}{\partial y} = 0.$$

15. Find an equation for the line tangent to the graph of $x \cos \pi y + x^2 e^y = 6$ at the point $(2, 0)$.

16. Find the directional derivative $D_{\mathbf{u}} f$ of the function $f(x, y) = y^2 \operatorname{Sin}^{-1} x + ye^x$ at the point $(0, 1)$ in the direction of the vector $\mathbf{w} = \mathbf{i} + \sqrt{e}\mathbf{j}$.

17. Let $f(x, y)$ and $g(x, y)$ have partial derivatives with respect to both variables. Show that $\nabla(\alpha f + \beta g) = \alpha \nabla f + \beta \nabla g$ where α and β are constants.

18. Find the maximum value of the directional derivative for the function $w = xy^3 + z \cos y - \ln(x^2 + y)$ at the point $(1, 0, 4)$.

19. Find a vector pointing in the direction of most rapid decrease for the function $f(x, y, z) = xyz - z \operatorname{Tan}^{-1}(y/x)$ at the point $(1, 2, 3)$.

20. Find an equation for the plane tangent to the surface $z^2 x - 2zy + e^{xy} = 9$ at the point $(2, 0, 2)$.

21. Show that the function $f(x, t) = e^{x+ct}$ is a solution of the wave equation

$$\frac{\partial^2 f}{\partial t^2} = c^2 \frac{\partial^2 f}{\partial x^2}.$$

22. Find the maximum and minimum values of the function $f(x, y) = 4x^2 + xy + 2y^2$ on the square $S = \{(x, y) \,|\, -1 \le x \le 1, -1 \le y \le 1\}$.

23. Use the method of Lagrange multipliers to find the rectangular box of largest volume that can be inscribed in the ellipsoid

$$x^2 + \frac{y^2}{9} + \frac{z^2}{4} = 1.$$

In Exercises 24–30, find and classify all relative extrema.

24. $f(x, y) = 4y^2 - 2x^2$

25. $f(x, y) = 3x^2 + xy - 6y^2$

26. $f(x, y) = x^2y + xy^2 + 4x + 4y$

27. $f(x, y) = e^{x^2 - 4xy}$

28. $f(x, y) = \ln(1 + x^2 + y^2)$

29. $f(x, y) = e^{1 + x^2 - y^2}$

30. $f(x, y) = 6x^2 - 2x - 3xy + y^2 + 5y + 5$

31. The directional derivative of $w = f(x, y)$ at P in the direction of the vector $\mathbf{u}_1 = \mathbf{j}$ is $D_{\mathbf{u}_1} f(P) = 3$, and the directional derivative of f at P in the direction of $\mathbf{u}_2 = 3\mathbf{i} + 4\mathbf{j}$ is $D_{\mathbf{u}_2} f(P) = 3$ also. Find

a. $\dfrac{\partial f}{\partial x}(P)$,   **b.** $\dfrac{\partial f}{\partial y}(P)$,   and   **c.** $\nabla f(P)$.

32. The directional derivative of f at (x_0, y_0) in the direction of the unit vector $\mathbf{u}$ is 6. What is
a. $D_{2\mathbf{u}} f(x_0, y_0)$?
b. $D_{-\mathbf{u}} f(x_0, y_0)$?

33. Find a function f for which $\nabla f(x, y) = 2xe^y \mathbf{i} + (x^2 e^y - \sin y)\mathbf{j}$.

34. Find a potential for the vector function $F(x, y) = (ye^x + e^y)\mathbf{i} + (1 + e^x + xe^y)\mathbf{j}$.

35. If three resistors r_1, r_2, and r_3 are connected in parallel, the net resistance R is determined by the equation

$$\frac{1}{R} = \frac{1}{r_1} + \frac{1}{r_2} + \frac{1}{r_3}.$$

Suppose $r_1 = 10$ ohms, $r_2 = 20$ ohms, and $r_3 = 25$ ohms. Approximate the change in the value of R that results from each of r_1, r_2, and r_3 being increased by 10%.

36. Is the function

$$f(x, y) = \begin{cases} \dfrac{6xy}{x^2 + y^2}, & (x, y) \ne (0, 0) \\ 0, & (x, y) = (0, 0) \end{cases}$$

differentiable at $(0, 0)$? Why or why not?

37. Show that the function

$$w = e^{x-y} + \cos(y - z) + \sqrt{z - x}$$

satisfies the partial differential equation

$$\frac{\partial w}{\partial x} + \frac{\partial w}{\partial y} + \frac{\partial w}{\partial z} = 0.$$

38. An airplane flying due north at a speed of 200 km/h and an altitude of 2 km passes directly over an automobile traveling due east along a straight highway at a speed of 100 km/h. At what rate are the plane and the automobile moving apart after 6 minutes?

39. Find the maximum and minimum values of the function

$$f(x, y) = 3x - 2y + 5$$

on or inside the ellipse $\dfrac{x^2}{4} + \dfrac{y^2}{9} = 1$.

40. A closed rectangular box is to contain 1000 cm^3. If the material for the top and bottom costs 2¢/cm^2 and the material for the sides costs 3¢/cm^2, find the dimensions that minimize cost.

41. Let $\mathbf{u}$ and $\mathbf{v}$ be distinct unit vectors. Is $D_{\mathbf{u}+\mathbf{v}} f(x_0, y_0) = D_{\mathbf{u}} f(x_0, y_0) + D_{\mathbf{v}} f(x_0, y_0)$? Why or why not?

42. Find an equation for the plane through the point $(1, 2, 1)$ with positive x-, y-, and z-intercepts that bounds the solid of least volume in the first octant.

43. Find the point on the surface.

$$x^2 + y^2 + z^2 = 16$$

where the function $f(x, y, z) = x + 2y - 3z + 1$ is a maximum.

44. Suppose that the level surfaces $f(x, y, z) = K$ and $g(x, y, z) = L$ intersect in a curve C. Suppose $h(x, y, z)$ is maximized at x_0 on C, and assume that f, g, and h have continuous partial derivatives. Show that $\nabla f(\mathbf{x}_0)$, $\nabla g(\mathbf{x}_0)$, and $\nabla h(\mathbf{x}_0)$ lie in a common plane if they are based at $\mathbf{x}_0$.

45. Prove that if f is differentiable for all (x, y) in a closed and bounded disc D, then f is bounded on D. (This means that there exists a number M with $|f(x, y)| \le M$ for all $(x, y) \in D$.)

46. The function $f(x, y, z)$ is called homogeneous of degree n if $f(tx, ty, tz) = t^n f(x, y, z)$ for all $x, y, z,$ and t. Show that if n is an integer and f has continuous partial derivatives, then

$$x\frac{\partial f}{\partial x} + y\frac{\partial f}{\partial y} + z\frac{\partial f}{\partial z} = nf$$

when f is homogeneous of degree n.

47. For the function $f(x, y) = \sqrt{x^2 + y^2}$, show that all directional derivatives exist at $(0, 0)$, but that neither partial derivative exists at $(0, 0)$.

48. Show that the spheres with equations $x^2 + (y - 1)^2 + z^2 = 2$ and $x^2 + (y + 1)^2 + z^2 = 2$ intersect in a circle. Then show that at each point on the circle of intersection the tangent planes to the two circles are orthogonal.

49. Use the method of Lagrange multipliers to find the point on the plane $x - 3y + 2z = 6$ nearest the origin.

50. Find a vector in the direction of most rapid increase of the function $f(x, y, z) = e^{x-z} \operatorname{Tan}^{-1}(y + z)$ at the point $(3, 1, 2)$.

51. Find an equation for the plane tangent to the surface $x^2 + 4y^2 + z^2 = 12$ at the point $(2, 1, 2)$.

52. Find $\dfrac{d^2 f}{dt^2}$ if $f(x, y) = e^{y^2 - x^2}$, $x(t) = 4 - t^2$, $y(t) = t \cos \pi t$.

53. Find the minimum value of the function $f(x, y) = 3y^2 - xy + x^2$ subject to the constraint that $x^2 + 3y^2 = 3$.

54. Find the maximum volume for a rectangular solid inscribed in the ellipsoid

$$\frac{x^2}{4} + \frac{y^2}{9} + \frac{z^2}{4} = 1.$$

55. Show that $u = f(xy)$ satisfies the partial differential equation

$$x\frac{\partial u}{\partial x} = y\frac{\partial u}{\partial y}.$$

56. Show that $u = f(x + 2y^2)$ satisfies the partial differential equation

$$\frac{\partial u}{\partial y} = 4y\frac{\partial u}{\partial x}.$$

57. Show that $u = f(t - x)$ is a solution of the partial differential equation

$$\frac{\partial^2 u}{\partial x^2} + 2\frac{\partial^2 u}{\partial x \partial t} + \frac{\partial^2 u}{\partial t^2} = 0.$$

58. Show that $u = f(x^2 + y^2)$ satisfies the partial differential equation

$$\frac{\partial u}{\partial y} = \frac{y}{x}\frac{\partial u}{\partial x}.$$

59. Show that if f is any differentiable function, the composite function $u(x, t) = f(bx - t)$ is a solution of the partial differential equation

$$\frac{\partial u}{\partial x} = -b\frac{\partial u}{\partial t}.$$

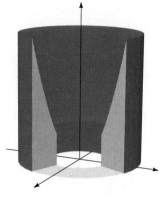

Chapter 19

Double and Triple Integrals

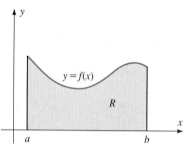

Figure 1.1 Region bounded by the graph of a continuous nonnegative function f.

In principle, the definite integral of a function of two or three variables is a straightforward generalization of the definite integral for a function of a single variable. It is the appropriate limit of Riemann sums. However, since the regions over which we integrate are now subsets of the plane or space, we shall rely extensively on our study of curves and surfaces in order to evaluate the integrals discussed here.

We begin by generalizing the definition of the definite integral to a function of two variables. Next we discuss methods for evaluating the integral, all based on the concept of iterated integration. We reduce the problem of evaluation to the operation of repeated integration of integrals involving only one variable. We also illustrate a few of the applications of this theory by discussing the calculation of surface area and the calculation of centers of mass. Finally we generalize the integral to functions of three variables.

19.1 The Double Integral over a Rectangle

We began our study of the definite integral $\int_a^b f(x)\,dx$ of a continuous function f of one variable by discussing an area problem. That is, for a nonnegative function f, we computed the area of the region R bounded by the graph of $y = f(x)$ and the x-axis for $a \le x \le b$ (Figure 1.1). To do so, we partitioned the interval $[a, b]$ into subintervals of length $\Delta x_j = x_j - x_{j-1}$. After choosing one number t_j arbitrarily in each interval, we formed the approximating Riemann sum

$$S_n = \sum_{j=1}^{n} f(t_j)\,\Delta x_j$$

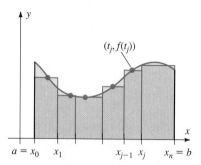

Figure 1.2 Riemann sum approximates region R by rectangles.

representing the sum of the areas of the rectangles illustrated in Figure 1.2. We then proved that the limit of this Riemann sum, as $n \to \infty$ and as the norm of the partition $\|P_n\| \to 0$, is the desired area. This led to the definition of the definite integral

$$\int_a^b f(x)\,dx = \lim_{n \to \infty} \sum_{j=1}^{n} f(t_j)\,\Delta x_j$$

as a limit of Riemann sums (see Section 6.2). Not only did the definite integral solve this area problem, but more generally it provided a summation procedure by which we can compute any quantity that can be interpreted as a sum of a continuously

964

varying function. In Chapter 7, we calculated quantities such as arc length, distance, hydrostatic pressure, and work using definite integrals.

We now follow the same approach to define the definite integral of a function of two variables. We start with a rectangular region R in the domain of a nonnegative function f of two variables. The rectangle R and the graph $z = f(x, y)$ over R determine a solid in space (see Figure 1.3), and we approximate the volume of this solid by rectangular prisms (Figure 1.4). This approximation is a Riemann (double) sum.

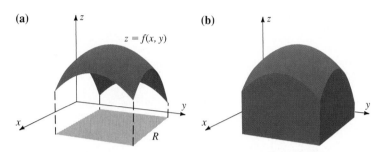

Figure 1.3   The graph of $z = f(x, y)$ over the rectangle R determines a solid in space.

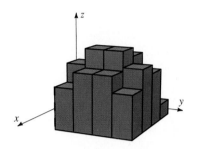

Figure 1.4   The volume of the solid bounded by the graph of f over R is approximated using rectangular prisms.

Developing the Double Integral over a Rectangle

More precisely, we calculate the volume V of a solid bounded above by the graph of the continuous nonnegative function $z = f(x, y)$, below by the rectangle

$$R = \{(x, y) \mid a \le x \le b, c \le y \le d\}$$

in the xy-plane, and on four sides by the vertical planes $x = a$, $x = b$, $y = c$, and $y = d$.

Using the same terminology as in the one-variable case, we let $P_1 = \{a = x_0, x_1, x_2, \ldots, x_n = b\}$ be a partition of the interval $[a, b]$, and we let $P_2 = \{c = y_0, y_1, y_2, \ldots, y_m = d\}$ be a partition of the interval $[c, d]$. Also, we let

$$\Delta x_j = x_j - x_{j-1}, \qquad j = 1, 2, \ldots, n$$

and

$$\Delta y_k = y_k - y_{k-1}, \qquad k = 1, 2, \ldots, m.$$

As Figure 1.5 illustrates, these partitions determine a grid dividing the region R into rectangles R_{jk} of area $\Delta A_{jk} = \Delta x_j \, \Delta y_k$ for $j = 1, 2, \ldots, n$ and $k = 1, 2, \ldots, m$.

We refer to this grid as the *partition P of R* determined by the partitions P_1 and P_2. We define the norm $\|P\|$ of this partition to be the larger of the norms $\|P_1\|$ and $\|P_2\|$ of the partitions P_1 and P_2. That is,

$$\|P\| = \max\{\|P_1\|, \|P_2\|\}$$
$$= \max\{\Delta x_1, \Delta x_2, \ldots, \Delta x_n, \Delta y_1, \Delta y_2, \ldots, \Delta y_m\}.$$

We approximate the volume of the region above the rectangle R_{jk} and below the graph of f with the volume of the rectangular prism with base of area $\Delta A_{jk} = \Delta x_j \, \Delta y_k$. For the height of this prism we use the function value $f(s_j, t_k)$, where the

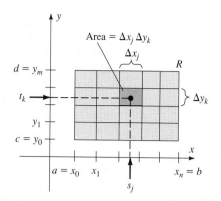

Figure 1.5 The partitions of $[a, b]$ and of $[c, d]$ determine a grid dividing R into rectangles R_{jk}. The area of R_{jk} is $\Delta A_{jk} = \Delta x_j \, \Delta y_k$.

"test point" (s_j, t_k) is chosen arbitrarily in the rectangle R_{jk} (see Figure 1.6). This leads to the double sum

$$S_{n, m} = \sum_{j=1}^{n} \sum_{k=1}^{m} f(s_j, t_k) \, \Delta A_{jk}, \qquad \Delta A_{jk} = \Delta x_j \, \Delta y_k, \tag{1}$$

which is called a *Riemann sum* for the function f on the rectangle R. Just as in the one-variable case, we obtain the definite integral as the limit of these Riemann sums as the Δx_j and Δy_k approach 0 (see Figure 1.7).

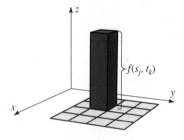

Figure 1.6 The volume of this rectangular prism is $f(s_j, t_k) \, \Delta x_j \, \Delta y_k$, where (s_j, t_k) is a point in the rectangle R_{jk}.

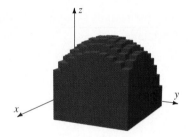

Figure 1.7 As Δx_j and $\Delta y_k \to 0$, the corresponding Riemann sums approach the desired volume.

Theorem 1

If f is continuous on the rectangle R, then there exists a unique number I such that

$$I = \lim_{n, m \to \infty} S_{n, m} = \lim_{n, m \to \infty} \sum_{j=1}^{n} \sum_{k=1}^{m} f(s_j, t_k) \, \Delta A_{jk}, \qquad \Delta A_{jk} = \Delta x_j \, \Delta y_k,$$

for all Riemann sums $S_{n, m}$ corresponding to partitions $P_{n, m}$ for which $\|P_{n, m}\| \to 0$ as both m and $n \to \infty$.

Consequently, the arbitrary choice of test points (s_j, t_k) for any particular Riemann sum does not matter. In the limit, all Riemann sums corresponding to partitions such that $\|P_{n, m}\| \to 0$ approach the number I.

Since the basic ideas in the proof of Theorem 1 are similar to those discussed in Appendix II for the one-variable case, we omit its proof. Nevertheless, we emphasize that Theorem 1 provides the foundation for the theory of the definite integral as we see in Definition 1.

Definition 1

Let f be a continuous function of two variables on the rectangle R. Then the number I designated in Theorem 1 is called the **definite integral of f over the rectangle R.** It is usually written as

$$\iint\limits_{R} f(x, y)\, dA.$$

We use two integral signs to indicate that this integral represents the result of a double limit process—the x-interval $[a, b]$ and the y-interval $[c, d]$ have both been partitioned into increasingly small subintervals. The subscript R denotes the rectangle over which the integral is evaluated. For now, the symbol dA (which may also be written $dx\, dy$) indicates that the Riemann sum has been obtained by partitioning R into rectangles of area $\Delta A_{jk} = \Delta x_j\, \Delta y_k$. As before, the function f is referred to as the **integrand.**

According to the development that led to the Riemann sum, we conclude that the volume V of the solid bounded above by the graph of the continuous nonnegative function f and below by the rectangle R in the xy-plane is

$$V = \iint\limits_{R} f(x, y)\, dA. \tag{2}$$

In the one-variable case, if $f(x) < 0$ for all $x \in [a, b]$, then the area bounded by the graph $y = f(x)$ and the x-axis equals $-\int_a^b f(x)\, dx$. An analogous statement also holds for negative functions $f(x, y)$ of two variables. If $f(x, y) < 0$ for all $(x, y) \in R$, then the volume of the solid determined by R and the graph $z = f(x, y)$ equals $-\iint_R f(x, y)\, dA$.

Indeed, Theorem 1 and Definition 1 apply to *all* continuous functions f of two variables. For such f, the value of the resulting integral can be interpreted geometrically as the difference between the volume of the solid determined by that portion of the graph of $z = f(x, y)$ that is above the xy-plane and the volume of the solid determined by that portion of the graph that is below the xy-plane.

Iterated Integrals

Although Theorem 1 provides both the theoretical foundation for the definite integral and a means by which the integral can be approximated, we would like to have some method similar to the Fundamental Theorem of Calculus for computing the value of the integral. By considering the case where f is nonnegative over R, we can derive one such method.

Recall from Section 7.1 that the volume V in equation (2) is given by the definite integral

$$V = \int_a^b A(x)\, dx, \qquad a < b, \tag{3}$$

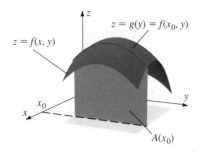

Figure 1.8 The area $A(x_0)$ of the cross section at x_0 is

$$A(x_0) = \int_c^d f(x_0, y) \, dy.$$

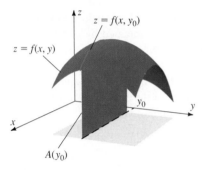

Figure 1.9 The area $A(y_0)$ of the cross section at y_0 is

$$A(y_0) = \int_a^b f(x, y_0) \, dx.$$

where $A(x)$ is the area of the cross section taken perpendicular to the x-axis. Now if $x_0 \in [a, b]$ is fixed, the area of this cross section is just

$$A(x_0) = \int_c^d f(x_0, y) \, dy, \qquad c < d, \tag{4}$$

since the cross section is bounded above by the continuous function $g(y) = f(x_0, y)$ (see Figure 1.8).

Combining equations (3) and (4) we conclude that

$$V = \int_a^b \left[\int_c^d f(x, y) \, dy \right] dx, \qquad a < b, \quad c < d. \tag{5}$$

Equation (5) indicates that the volume V is calculated by first integrating f with respect to y (treating x as a constant) from c to d, and then integrating the resulting function of x from a to b. As Figure 1.9 illustrates, we can also begin by fixing y_0 and obtaining the area of the cross section perpendicular to the y-axis at $y = y_0$ as

$$A(y_0) = \int_a^b f(x, y_0) \, dx, \qquad a < b.$$

The resulting calculation for volume is

$$V = \int_c^d \left[\int_a^b f(x, y) \, dx \right] dy, \qquad a < b, \quad c < d. \tag{6}$$

The integrals in (5) and (6) are called **iterated integrals** because they involve the composition of two successive integrations, each with respect to a single variable. We usually omit the brackets and simply write

$$\int_a^b \int_c^d f(x, y) \, dy \, dx = \int_a^b \left[\int_c^d f(x, y) \, dy \right] dx \tag{7}$$

and

$$\int_c^d \int_a^b f(x, y) \, dx \, dy = \int_c^d \left[\int_a^b f(x, y) \, dx \right] dy. \tag{8}$$

It is important to note that iterated integrals are interpreted ''from inside out.'' First the inside integral is evaluated. Then the outside integral is determined.

Finally, we note that our assumption that the function f was nonnegative is not essential. Iterated integration applies to any function f that is continuous on the rectangle R.

Theorem 2

Let $R = \{(x, y) \mid a \leq x \leq b, c \leq y \leq d\}$ be a rectangle in the plane and f be a continuous function on R. Then

$$\iint_R f(x, y) \, dA = \int_a^b \int_c^d f(x, y) \, dy \, dx$$

$$= \int_c^d \int_a^b f(x, y) \, dx \, dy.$$

Either of these two iterated integrals can be used to determine the value of the double integral, and changing from one to the other is called **interchanging the order of integration**.

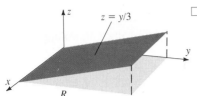

Figure 1.10 The graph of $z = y/3$ over the rectangle R determines a solid wedge.

□ **EXAMPLE 1**

Let $R = \{(x, y) \mid 0 \le x \le 2, 0 \le y \le 3\}$. Evaluate the double integral

$$\iint_R \frac{y}{3} \, dA$$

and interpret the result geometrically.

Solution   The graph of $z = y/3$ is a plane, and therefore this integral corresponds to the volume of a solid wedge determined by R and the plane $z = y/3$ (see Figure 1.10).

Using equation (7), we have

$$\iint_R \frac{y}{3} \, dA = \int_0^2 \int_0^3 \frac{y}{3} \, dy \, dx$$

$$= \int_0^2 \left[\frac{y^2}{6} \right]_{y=0}^{y=3} dx$$

$$= \int_0^2 \frac{3}{2} \, dx \qquad (9)$$

$$= \frac{3}{2} \left[x \right]_{x=0}^{x=2} = 3.$$

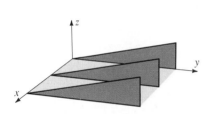

Figure 1.11 All x-slices of the wedge in Example 1 are triangles of area 3/2.

Geometrically, this calculation corresponds to slicing the wedge into slices that are parallel to the yz-plane (x-slices). Note that, since the graph of $z = y/3$ does not involve the variable x, all of the slices are equally sized triangles (see Figure 1.11). Each is a right triangle with a base of 3 and a height of 1. Thus its area is 3/2. The fact that the area of each of the slices is 3/2 is reflected in the above computation in equation (9). The integrand of the outermost integral is the constant function $c(x) = 3/2$.

It is instructive to evaluate this integral a second time using the opposite order of integration. From equation (8), we have

$$\iint_R \frac{y}{3} \, dA = \int_0^3 \int_0^2 \frac{y}{3} \, dx \, dy$$

$$= \int_0^3 \left[\frac{xy}{3} \right]_{x=0}^{x=2} dy$$

$$= \int_0^3 \left[\frac{2y}{3} \right] dy \qquad (10)$$

$$= \left[\frac{y^2}{3} \right]_{y=0}^{y=3} = 3.$$

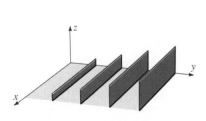

Figure 1.12 All y-slices of the wedge in Example 1 are rectangles. The area of these rectangles depends on the choice of y.

In this case, we are slicing the solid by y-slices (parallel to the xz-plane). Note that these slices are rectangles whose area depends on y (see Figure 1.12). Thus the integrand in equation (10) depends on y.   ■

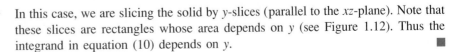

□ **EXAMPLE 2**

Let R be the square $\{(x, y) \mid -1 \le x \le 1, -1 \le y \le 1\}$. Calculate the volume of the solid region determined by the graph of $f(x, y) = 8 - x^2 - y^2$ over R.

Solution The desired volume may be calculated as

$$\iint_R (8 - x^2 - y^2) \, dA = \int_{-1}^{1} \left[\int_{-1}^{1} (8 - x^2 - y^2) \, dx \right] dy$$

$$= \int_{-1}^{1} \left[8x - \frac{x^3}{3} - xy^2 \right]_{x=-1}^{x=1} dy$$

$$= \int_{-1}^{1} \left(\frac{46}{3} - 2y^2 \right) dy$$

$$= \left[\frac{46y}{3} - \frac{2y^3}{3} \right]_{y=-1}^{y=1} = \frac{88}{3}.$$

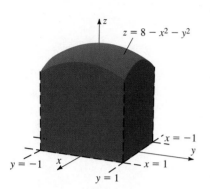

Figure 1.13 The solid whose volume is calculated in Example 2.

(See Figure 1.13.) ∎

Our final example illustrates how, in some cases, one order of integration is preferable to the other.

□ **EXAMPLE 3**

Calculate

$$\iint_R x \cos(xy) \, dA$$

where R is the rectangle $\{(x, y) \mid 0 \le x \le \pi/4, 0 \le y \le 2\}$.

Solution Theorem 2 provides two choices:

$$\int_0^{\pi/4} \int_0^2 x \cos(xy) \, dy \, dx \quad \text{or} \quad \int_0^2 \int_0^{\pi/4} x \cos(xy) \, dx \, dy.$$

Both can be evaluated using the Fundamental Theorem of Calculus, but the first is more convenient because it is easier to find an antiderivative for its inside integral. Since $\partial \sin(xy)/\partial y = x \cos(xy)$,

$$\int_0^{\pi/4} \int_0^2 x \cos(xy) \, dy \, dx = \int_0^{\pi/4} \left[\sin(xy) \right]_{y=0}^{y=2} dx$$

$$= \int_0^{\pi/4} \sin(2x) \, dx$$

$$= -\left[\frac{\cos(2x)}{2} \right]_0^{\pi/4}$$

$$= -\frac{0 - 1}{2} = \frac{1}{2}. \quad \blacksquare$$

Note that, for the other order of integration, the inside integral is

$$\int_0^{\pi/4} x \cos(xy)\, dx,$$

which is evaluated using integration by parts. Consequently, our solution is more immediate.

REMARK Even though we know that Theorem 2 holds for all continuous functions, the technique of iterated integration is successful only if we can evaluate the corresponding iterated integrals. Alternatively, if we cannot find the necessary antiderivatives, then we can approximate the desired integral using techniques that approximate the integral via Riemann sums. Exercises 40 through 43 illustrate this approach.

Exercise Set 19.1

In Exercises 1–16, evaluate the iterated integral.

1. $\displaystyle\int_0^1 \int_0^2 xy\, dx\, dy$

2. $\displaystyle\int_{-1}^1 \int_1^3 (y - x)\, dy\, dx$

3. $\displaystyle\int_1^3 \int_1^2 (4 + x - y)\, dx\, dy$

4. $\displaystyle\int_0^2 \int_{-1}^1 (x + y)^2\, dx\, dy$

5. $\displaystyle\int_0^1 \int_0^{\pi/2} x \sin y\, dy\, dx$

6. $\displaystyle\int_0^1 \int_0^1 ye^{x-y^2}\, dy\, dx$

7. $\displaystyle\int_0^2 \int_1^e y^2 \ln x\, dx\, dy$

8. $\displaystyle\int_0^9 \int_1^4 \sqrt{y/x}\, dx\, dy$

9. $\displaystyle\int_0^{\pi/2} \int_0^3 y \sin(xy)\, dx\, dy$

10. $\displaystyle\int_0^{\pi/4} \int_0^1 x \sec^2(xy)\, dy\, dx$

11. $\displaystyle\int_0^2 \int_0^2 xye^{xy^2}\, dy\, dx$

12. $\displaystyle\int_0^1 \int_0^1 x \cosh y\, dx\, dy$

13. $\displaystyle\int_0^1 \int_0^{\pi/2} xy \sin x\, dx\, dy$

14. $\displaystyle\int_0^\pi \int_0^1 \sinh x \cosh(\pi - y)\, dx\, dy$

15. $\displaystyle\int_0^1 \int_0^4 \frac{\sqrt{y}}{1 + x^2}\, dy\, dx$

16. $\displaystyle\int_0^{\pi/4} \int_0^{\pi/4} \tan x \sec^2 y\, dy\, dx$

In Exercises 17–25, evaluate the double integral over the rectangle R.

17. $\displaystyle\iint_R (x + y^2)\, dA, \quad R = \{(x, y)\,|\,0 \le x \le 1,\ 0 \le y \le 1\}$

18. $\displaystyle\iint_R (x^2 + y^2)\, dA, \quad R = \{(x, y)\,|\,0 \le x \le a,\ 0 \le y \le b\}$

19. $\displaystyle\iint_R x \cos y\, dA, \quad R = \{(x, y)\,|\,0 \le x \le 4,\ 0 \le y \le \pi/2\}$

20. $\displaystyle\iint_R \frac{xy}{\sqrt{x^2 + y^2}}\, dA, \quad R = \{(x, y)\,|\,1 \le x \le 2,\ 1 \le y \le 2\}$

21. $\displaystyle\iint_R xy \sec^2(xy^2)\, dA,\quad R = \{(x, y)\,|\,0 \le x \le \pi/4,\ 0 \le y \le 1\}$

22. $\displaystyle\iint_R xy \cos(xy^2)\, dA, \quad R = \{(x, y)\,|\,0 \le x \le \pi/4,\ 0 \le y \le \sqrt{2}\}$

23. $\displaystyle\iint_R ye^{xy}\, dA, \quad R = \{(x, y)\,|\,0 \le x \le 2,\ 0 \le y \le 1\}$

24. $\displaystyle\iint_R y \cos(x + y)\, dA, \quad R = \{(x, y)\,|\,0 \le x \le \pi/4,\ 0 \le y \le \pi/4\}$

25. $\displaystyle\iint_R \frac{1}{\sqrt{x + y}}\, dA, \quad R = \{(x, y)\,|\,4 \le x \le 8,\ 0 \le y \le 4\}$

In Exercises 26–31, use a double integral to calculate the volume of the solid bounded above by the graph of f and below by the rectangle R in the xy-plane.

26. $f(x, y) = 16 - 4x - 2y, \quad R = \{(x, y)\,|\,0 \le x \le 2,\ 0 \le y \le 1\}$

27. $f(x, y) = x, \quad R = \{(x, y)\,|\,0 \le x \le 2,\ 0 \le y \le 3\}$

28. $f(x, y) = 9 - x^2 - y^2$, $R = \{(x, y) \mid -1 \le x \le 1,$
$-1 \le y \le 1\}$

29. $f(x, y) = x \sin y$, $R = \{(x, y) \mid 0 \le x \le 1,$
$0 \le y \le \pi/2\}$

30. $f(x, y) = \dfrac{\sqrt{x}}{1 + y^2}$, $R = \{(x, y) \mid 0 \le x \le 4,$
$0 \le y \le 1\}$

31. $f(x, y) = \dfrac{e^{\sqrt{x}}}{\sqrt{xy}}$, $R = \{(x, y) \mid 1 \le x \le 4, 1 \le y \le 9\}$

In Exercises 32–35, find the volume of the solid bounded by the four surfaces given.

32. $z = 1 - x^2$
$z = 0$
$y = 1$
$y = 3$

33. $z = 2 - y^2$
$z = 1$
$x = -1$
$x = 1$

34. $z = 1 - y^2$
$z = -1$
$x = 0$
$x = 3$

35. $z = x^2 - 1$
$z = 1$
$y = 0$
$y = 2$

36. Show that if f is continuous on the rectangle R, then

$$\iint_R cf(x, y) \, dA = c \iint_R f(x, y) \, dA$$

for any constant c.

37. Show that if f and g are continuous on the rectangle R then

$$\iint_R [f(x, y) + g(x, y)] \, dA =$$

$$\iint_R f(x, y) \, dA + \iint_R g(x, y) \, dA.$$

38. Show that if $f(x, y) \le 0$ for all $(x, y) \in R = \{(x, y) \mid a \le x \le b, c \le y \le d\}$ and if V is the volume of the solid bounded by R and the graph of $z = f(x, y)$, then

$$V = -\iint_R f(x, y) \, dA.$$

39. Show that, if f is continuous on the rectangle $R = \{(x, y) \mid a \le x \le b, c \le y \le d\}$, then

$$\int_a^b \int_c^d f(x, y) \, dy \, dx = -\int_b^a \int_c^d f(x, y) \, dy \, dx$$

$$= -\int_a^b \int_d^c f(x, y) \, dy \, dx$$

$$= \int_b^a \int_d^c f(x, y) \, dy \, dx.$$

40. Using the programs in Appendix I, approximate the volume of the solid bounded above by the graph of $z = \sin \sqrt{xy}$ and below by the rectangle

$$R = \{(x, y) \mid 0 \le x \le 1, 0 \le y \le 1\}.$$

41. Using the programs in Appendix I, approximate the double integral

$$\iint_R e^{x^2 + y^2} \, dA$$

where R is the rectangle $R = \{(x, y) \mid 0 \le x \le 1, 0 \le y \le 2\}$.

42. Using the programs in Appendix I, approximate the iterated integral

$$\int_0^1 \int_1^2 \sin(x^2 + y^2) \, dx \, dy.$$

43. Using the programs in Appendix I, approximate the iterated integral

$$\int_1^3 \int_2^4 \sqrt{x^3 + y^3} \, dx \, dy.$$

44. The purpose of this exercise is to enable you to sketch a proof of the fact that

$$\iint_R f(x, y) \, dA = \int_a^b \int_c^d f(x, y) \, dy \, dx$$

where f is continuous and $R = \{(x, y) \mid a \le x \le b, c \le y \le d\}$.

a. Begin with the partitions $a = x_0 < x_1 < \cdots < x_n = b$ and $c = y_0 < y_1 < \cdots < y_m = d$ and the Riemann sum

$$S_{n, m} = \sum_{j=1}^{n} \sum_{k=1}^{m} f(s_j, t_k) \, \Delta y_k \, \Delta x_j$$

$$= \sum_{j=1}^{n} \left[\sum_{k=1}^{m} f(s_j, t_k) \, \Delta y_k \right] \Delta x_j.$$

b. With j fixed, pick t_k in each interval $[y_{k-1}, y_k]$ to be the number for which

$$f(s_j, t_k) \, \Delta y_k = \int_{y_{k-1}}^{y_k} f(s_j, t) \, dt.$$

Why can this be done?

c. Conclude that

$$\sum_{k=1}^{m} f(s_j, t_k) \, \Delta y_k = \sum_{k=1}^{m} \int_{y_{k-1}}^{y_k} f(s_j, y) \, dy = \int_c^d f(s_j, y) \, dy.$$

Why is this true?

d. From a and c conclude that

$$S_{n, m} = \sum_{j=1}^{n} \left[\int_c^d f(s_j, y) \, dy \right] \Delta x_j.$$

e. Note that $S_{n, m}$ in d has the form $S_{n, m} = \displaystyle\sum_{j=1}^{n} G(s_j) \, \Delta x_j$

where

$$G(s_j) = \int_c^d f(s_j, t) \, dt$$

Conclude that

$$\lim_{\|P\| \to 0} S_{n, m} = \lim_{n \to \infty} \sum_{j=1}^{n} G(s_j) \, \Delta x_j = \int_a^b G(x) \, dx$$

$$= \int_a^b \left[\int_c^d f(x, y) \, dy \right] dx.$$

45. Show that if f is continuous on the interval $[a, b]$, and if g is continuous on the interval $[c, d]$, then

$$\iint\limits_R f(x) g(y) \, dA = \left[\int_a^b f(x) \, dx \right]\left[\int_c^d g(y) \, dy \right]$$

where $R = \{(x, y) \mid a \le x \le b, \, c \le y \le d\}$.

19.2   Double Integrals over More General Regions

In many applications, we must integrate functions over regions that are not rectangular. For example, computing the volume of a hemisphere of radius 1 corresponds to computing the double integral of $f(x, y) = \sqrt{1 - (x^2 + y^2)}$ over the circular disk $D = \{(x, y) \mid x^2 + y^2 \le 1\}$ of radius 1. In this section, we extend our definition of the double integral to more general regions, and we develop the related computational techniques.

More precisely, we extend the definition of the definite integral to functions f that are continuous on bounded regions Q with piecewise smooth boundaries. That is, a region Q in the plane is *bounded* if it is contained in a finite rectangle. Its boundary is *piecewise smooth* if the boundary consists of a finite number of smooth curves that do not cross either themselves or one another. A finite rectangle is one such region, as is a pentagon, a triangle, and an ellipse. In general, we consider regions such as the one in Figure 2.1.

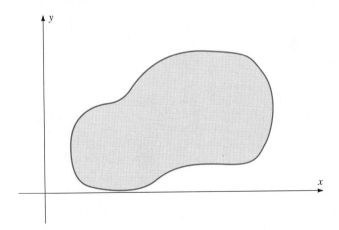

Figure 2.1   A bounded region Q with a smooth boundary.

Suppose that f is a continuous function of two variables defined on Q and that Q is contained in the rectangle $R = \{(x, y) \mid a \le x \le b, \, c \le y \le d\}$. We construct a grid over R (as in Section 19.1), which partitions R into smaller rectangles (see Figure 2.2), and we let $\|P\|$ denote the norm of this partition P.

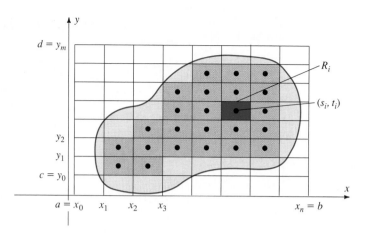

Figure 2.2 Partitioning the region Q.

Let $R_1, R_2, \ldots, R_m$ be a list of all of these smaller rectangles that lie entirely within Q. From each such rectangle R_i, we choose an arbitrary point (s_i, t_i). Finally, we form the Riemann sum

$$S = \sum_{i=1}^{m} f(s_i, t_i) \, \Delta A_i, \qquad \Delta A_i = \text{area of } R_i. \tag{1}$$

By considering successively finer partitions ($\|P_n\| \to 0$ as $n \to \infty$), we obtain a sequence of Riemann sums S_n. Under the conditions stated on f and Q, these sums converge to a number I that is independent of the choices of the partitions P_n and the points (s_i, t_i). This number I is the **double integral of f over Q.** That is,

$$\iint\limits_{Q} f(x, y) \, dA = \lim_{\|P\| \to 0} \sum_{i=1}^{m} f(s_i, t_i) \, \Delta A_i. \tag{2}$$

Before looking at ways to evaluate such integrals, we wish to emphasize four points concerning this definition.

(i) Although only one summation sign appears in equation (2), this definition agrees with Definition 1 if Q is a rectangle. In Definition 1, we used a two-dimensional counting procedure (rows by columns). In equation (2), we have simply listed all included rectangles $R_1, R_2, \ldots, R_m$ using a single index.

(ii) If $f(x, y) \geq 0$ for all $(x, y) \in Q$, the double integral (2) gives the volume V of the solid bounded by the graph of $z = f(x, y)$ and the region Q just as in the case of the double integral over a rectangle.

(iii) If f assumes both positive and negative values over the region Q, then the double integral (2) geometrically corresponds to the difference between the volume determined by that portion of the graph of $z = f(x, y)$ that is above the xy-plane and the volume determined by the portion of the graph that is below the xy-plane.

(iv) Finally, it is important to note that the double integral is a generalized summation procedure. We use it whenever we must sum a continuously varying quantity over a planar region. For example, in Section 9.4, we use the double integral to compute the mass and the center of mass of planar objects (for instance, a thin sheet of metal).

Regular Regions

There are two special types of regions for which the double integral in (2) can be evaluated as an iterated integral: x-simple and y-simple regions.

Definition 2

A region Q in the xy-plane is called **y-simple** if there exist continuous functions g_1 and g_2 so that

$$Q = \{(x, y) \mid a \leq x \leq b, g_1(x) \leq y \leq g_2(x)\}.$$

The region Q is called **x-simple** if there exist continuous functions h_1 and h_2 so that

$$Q = \{(x, y) \mid c \leq y \leq d, h_1(y) \leq x \leq h_2(y)\}.$$

The region Q is called **regular** if it is both x-simple and y-simple.

Figure 2.3 gives two illustrations of y-simple regions. The condition that $g_1(x) \leq y \leq g_2(x)$ for all $x \in [a, b]$ simply means that *the vertical line segment connecting $(x, g_1(x))$ and $(x, g_2(x))$ lies entirely within the region Q.* Thus, we see that Q is made up of infinitely many x-slices, each of which is a vertical line segment from the point $(x, g_1(x))$ to the point $(x, g_2(x))$. Moreover, lines of the form $x = x_0$ for $a < x_0 < b$ intersect the boundary of Q at most twice.

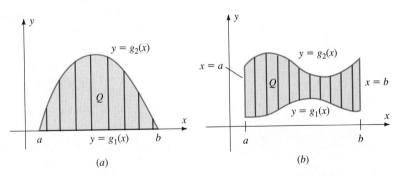

(a) (b)

Figure 2.3 Two y-simple regions: vertical lines intersect the boundary of Q at most twice.

Figure 2.4 shows two x-simple regions. Lines of the form $y = y_0$ for $c < y_0 < d$ intersect the boundary of Q at most twice. Note also that the regions in Figures 2.3(a) and 2.4(a) are regular (both x-simple and y-simple). However, the region in Figure 2.3(b) is not x-simple, and the region in Figure 2.4(b) is not y-simple.

The following theorem shows how double integrals over x-simple or y-simple regions may be evaluated as iterated integrals.

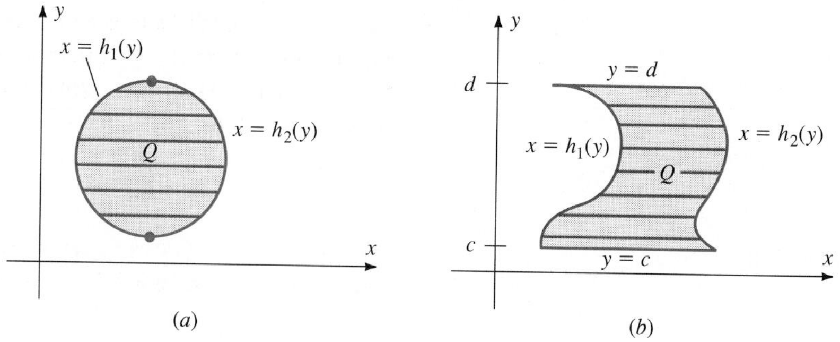

Figure 2.4 Two x-simple regions: horizontal lines intersect the boundary of Q at most twice.

Theorem 3	Let f be a continuous function of two variables on the region Q.

(i) If $Q = \{(x, y) \,|\, a \leq x \leq b, g_1(x) \leq y \leq g_2(x)\}$ is y-simple, then

$$\iint\limits_{Q} f(x, y) \, dA = \int_a^b \int_{g_1(x)}^{g_2(x)} f(x, y) \, dy \, dx = \int_a^b \left[\int_{g_1(x)}^{g_2(x)} f(x, y) \, dy \right] dx. \qquad (3)$$

(ii) If $Q = \{(x, y) \,|\, c \leq y \leq d, h_1(y) \leq x \leq h_2(y)\}$ is x-simple, then

$$\iint\limits_{Q} f(x, y) \, dA = \int_c^d \int_{h_1(y)}^{h_2(y)} f(x, y) \, dx \, dy = \int_c^d \left[\int_{h_1(y)}^{h_2(y)} f(x, y) \, dx \right] dy. \qquad (4)$$

In other words, if Q is y-simple, we can evaluate the integral by first holding x constant and integrating $f(x, y)$ as a function of y from $g_1(x)$ to $g_2(x)$. The result is a function solely of x, which we then integrate from a to b.

If Q is x-simple, we first hold y constant and integrate $f(x, y)$ as a function of x with limits $h_1(y)$ and $h_2(y)$. This result is a function of y, which is then integrated from c to d.

We shall not prove Theorem 3. However, Figure 2.5 illustrates why equation (4) holds if f is nonnegative on an x-simple region Q. If we fix $y = y_0$, then the integral

$$A(y_0) = \int_{h_1(y_0)}^{h_2(y_0)} f(x, y_0) \, dx \qquad (5)$$

equals the area of the $y = y_0$ slice of the solid determined by Q and the graph $z = f(x, y)$. As we saw in Section 7.1, the volume V of this solid is the integral of $A(y)$ over the interval $[c, d]$. More precisely,

$$\iint\limits_{Q} f(x, y) \, dA = V = \int_c^d A(y) \, dy$$

$$= \int_c^d \int_{h_1(y)}^{h_2(y)} f(x, y) \, dx \, dy \qquad \text{(by equation (5))}.$$

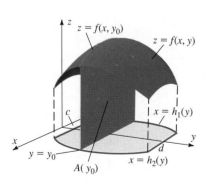

Figure 2.5 If the region Q is x-simple, the area of a cross section perpendicular to the y-axis at y_0 is

$$A(y_0) = \int_{h_1(y_0)}^{h_2(y_0)} f(x, y_0) \, dx.$$

This yields equation (4).

In using Theorem 3 to evaluate double integrals, it is important first to sketch the region Q to determine whether it is x-simple or y-simple. This determination will

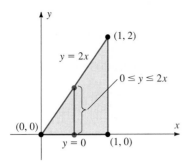

Figure 2.6 Treating the triangle Q as y-simple.

often dictate the order of integration in the iterated integral. (Of course, if Q is a regular region, you may proceed in either order. Just be careful that the limits of integration correspond to the chosen order of integration.)

□ **EXAMPLE 1**

Evaluate the double integral

$$\iint\limits_{Q} (2xy + y^2)\, dA$$

where Q is the triangle with vertices $(0, 0)$, $(1, 0)$, and $(1, 2)$.

Solution The triangular region Q is sketched in Figure 2.6. Since Q is y-simple, we find the equation of the line segment joining $(0, 0)$ and $(1, 2)$. It is simply $y = 2x$. An x-slice through Q extends from $y = g_1(x) = 0$ to $y = g_2(x) = 2x$. That is, the limits of integration are determined by the inequalities

$$0 \le x \le 1 \quad \text{and} \quad 0 \le y \le 2x. \tag{6}$$

According to equation (3), the double integral is evaluated as

$$\iint\limits_{Q} (2xy + y^2)\, dA = \int_0^1 \int_0^{2x} (2xy + y^2)\, dy\, dx$$

$$= \int_0^1 \left[xy^2 + \frac{y^3}{3} \right]_{y=0}^{y=2x} dx$$

$$= \int_0^1 \frac{20x^3}{3}\, dx$$

$$= \left[\frac{5}{3} x^4 \right]_0^1 = \frac{5}{3}.$$

The region Q is also x-simple, and thus we can also compute this integral using y-slices (equation (4)). To use equation (4) we must first write the equation for the line segment joining $(0, 0)$ and $(1, 2)$ as a function of y. Solving $y = 2x$ for x gives $x = h_1(y) = y/2$ as the left boundary. As Figure 2.7 illustrates, $h_2(y) = 1$ is the right boundary of Q. The limits of integration are therefore

$$0 \le y \le 2 \quad \text{and} \quad y/2 \le x \le 1. \tag{7}$$

Equation (4) then gives

$$\iint\limits_{Q} (2xy + y^2)\, dA = \int_0^2 \int_{y/2}^1 (2xy + y^2)\, dx\, dy$$

$$= \int_0^2 \left[x^2 y + xy^2 \right]_{x=y/2}^{x=1} dy$$

$$= \int_0^2 \left(-\frac{3}{4} y^3 + y^2 + y \right) dy$$

$$= \left[-\frac{3}{16} y^4 + \frac{y^3}{3} + \frac{y^2}{2} \right]_0^2 = \frac{5}{3}. \quad ■$$

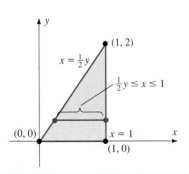

Figure 2.7 Treating the triangle Q as x-simple.

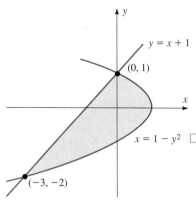

Figure 2.8 The region Q bounded by graphs of $y = x + 1$ and $x = 1 - y^2$.

REMARK 1 Notice that the order of integration is entirely determined by whether we describe Q as y-simple, using inequalities (6), or as x-simple, using inequalities (7). Whichever integration is performed last must involve only constant limits of integration, since the definite integral is a particular real number determined only by the function f and the region Q.

□ **EXAMPLE 2**

Evaluate the double integral

$$\iint_Q 4xy \, dA$$

where Q is the region bounded by the graphs of the equations $y = x + 1$ and $x = 1 - y^2$.

Solution The region Q is sketched in Figure 2.8. This region, like that in Example 1, is both x-simple and y-simple. However, we choose to work with Q as an x-simple region, since every horizontal line through Q originates on the line $y = x + 1$ and terminates on the parabola $x = 1 - y^2$ (see Figure 2.9). (Viewing Q as y-simple would be much messier—some vertical lines terminate on the line and others terminate on the parabola; see Figure 2.10.)

Solving the equation $y = x + 1$ for x gives $x = h_1(y) = y - 1$ as the left boundary of Q. The right boundary is $h_2(y) = 1 - y^2$. The limits of integration are therefore

$$-2 \leq y \leq 1 \quad \text{and} \quad y - 1 \leq x \leq 1 - y^2.$$

The double integral is evaluated as

$$\iint_Q 4xy \, dA = \int_{-2}^{1} \int_{y-1}^{1-y} 4xy \, dx \, dy$$

$$= \int_{-2}^{1} \left[2x^2 y \right]_{x=y-1}^{x=1-y^2} dy$$

$$= \int_{-2}^{1} (2y^5 - 6y^3 + 4y^2) \, dy$$

$$= \left[\tfrac{1}{3} y^6 - \tfrac{3}{2} y^4 + \tfrac{4}{3} y^3 \right]_{-2}^{1} = \tfrac{27}{2}. \qquad ■$$

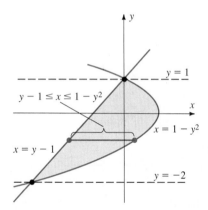

Figure 2.9 Considering the region Q as x-simple.

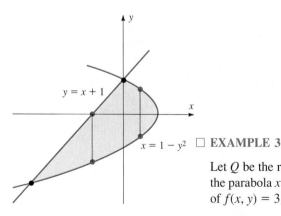

Figure 2.10 Considering the region Q as y-simple.

□ **EXAMPLE 3**

Let Q be the region in the xy-plane bounded by the lines $x = 0$, $y = -1$, $y = 1$, and the parabola $x = 1 + y^2$. Find the volume V of the solid bounded above by the graph of $f(x, y) = 3/2 - x/7 - y/2$ and below by the region Q.

Solution The region Q is sketched in Figure 2.11. Note that Q is x-simple but not y-simple.

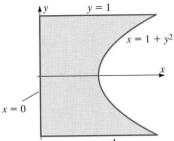

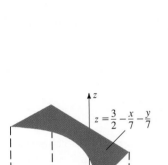

Figure 2.11   The region Q in Example 3. It is x-simple but not y-simple.

Figure 2.12   The solid determined by the region Q and the function $f(x, y) = 3/2 - x/7 - y/2$ in Example 3.

Since $f(x, y) \geq 0$ for all $(x, y) \in Q$, the volume V is given by the definite integral of f over Q. Since Q is x-simple, we slice it by y-slices and obtain the limits of integration

$$-1 \leq y \leq 1 \quad \text{and} \quad 0 \leq x \leq 1 + y^2.$$

The volume is

$$V = \iint\limits_{Q} \left(\frac{3}{2} - \frac{x}{7} - \frac{y}{2} \right) dA$$

$$= \int_{-1}^{1} \int_{0}^{1+y^2} \left(\frac{3}{2} - \frac{x}{7} - \frac{y}{2} \right) dx\, dy$$

$$= \int_{-1}^{1} \left[\frac{3x}{2} - \frac{x^2}{14} - \frac{xy}{2} \right]_{x=0}^{x=1+y^2} dy$$

$$= \int_{-1}^{1} \left(\frac{3(1 + y^2)}{2} - \frac{(1 + y^2)^2}{14} - \frac{(1 + y^2)y}{2} \right) dy$$

$$= \int_{-1}^{1} \left(\frac{10}{7} - \frac{y}{2} + \frac{19y^2}{14} - \frac{y^3}{2} - \frac{y^4}{14} \right) dy$$

$$= \left[\frac{10y}{7} - \frac{y^2}{4} + \frac{19y^3}{42} - \frac{y^4}{8} - \frac{y^5}{70} \right]_{-1}^{1} = \frac{56}{15}.$$

The solid is shown in Figure 2.12.   ■

□ **EXAMPLE 4**

Find the volume V of the solid bounded by the graph of $f(x, y) = e^{x+y^2}$ and the region

$$Q = \{(x, y) \mid \ln y \leq x \leq \ln 2y,\ 1 \leq y \leq 2\}$$

in the xy-plane.

Solution   The region Q is sketched in Figure 2.13. Although Q is both x-simple and y-simple, this time the order of integration is determined by the integrand $f(x, y) = e^{x+y^2}$. We must attempt to integrate first with respect to x, since there is no hope of finding an antiderivative with respect to y. Treating Q as an x-simple region, we use the limits

$$\ln y \leq x \leq \ln 2y \quad \text{and} \quad 1 \leq y \leq 2$$

and obtain

$$V = \iint\limits_{Q} e^{x+y^2}\, dA = \int_{1}^{2} \int_{\ln y}^{\ln 2y} e^{x+y^2}\, dx\, dy$$

$$= \int_{1}^{2} \int_{\ln y}^{\ln 2y} e^x e^{y^2}\, dx\, dy$$

$$= \int_{1}^{2} e^{y^2} \int_{\ln y}^{\ln 2y} e^x\, dx\, dy$$

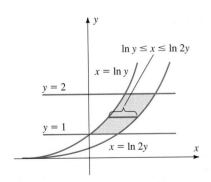

Figure 2.13   Region Q in Example 4.

$$= \int_1^2 e^{y^2} \left[e^x \right]_{x=\ln y}^{x=\ln 2y} dy$$

$$= \int_1^2 e^{y^2} [2y - y] \, dy$$

$$= \int_1^2 y e^{y^2} \, dy$$

$$= \left[\frac{e^{y^2}}{2} \right]_1^2 = \frac{e^4 - e}{2} \approx 25.94. \qquad \blacksquare$$

Finding Areas by Double Integration

The value of the definite integral of the function f that is identically 1 over the region Q is equal to the area of Q. To see why, we consider the definition of the double integral of f over Q (equation (2) in this section). Because $f(x, y) = 1$ for all $(x, y) \in Q$, equation (2) yields

$$\iint\limits_Q 1 \, dA = \lim_{\|P\| \to 0} \sum_i \Delta A_i,$$

and this limit converges to the area of Q as $\|P\| \to 0$.

Figure 2.14 illustrates this assertion geometrically. The area of Q is the same as the volume of the generalized cylinder with base Q and uniform height $h = 1$. Thus,

$$\boxed{\text{Area of } Q = \iint\limits_Q 1 \, dA.} \qquad (8)$$

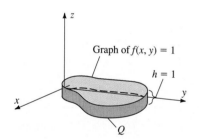

Figure 2.14 Area of $Q = \iint\limits_Q 1 \, dA$.

□ **EXAMPLE 5**

Use a double integral to find the area of the region Q lying inside the circle $x^2 + y^2 = 4$ and above the line $y = 1$.

Solution The points on the circle $x^2 + y^2 = 4$ with y-coordinate equal to 1 have x-coordinates $x = \pm\sqrt{4 - 1} = \pm\sqrt{3}$. The region may therefore be described by the inequalities

$$-\sqrt{3} \leq x \leq \sqrt{3}, \qquad 1 \leq y \leq \sqrt{4 - x^2}.$$

(See Figure 2.15.) According to (8),

$$\text{Area of } Q = \iint\limits_Q 1 \, dA$$

$$= \int_{-\sqrt{3}}^{\sqrt{3}} \int_1^{\sqrt{4-x^2}} 1 \, dy \, dx$$

$$= \int_{-\sqrt{3}}^{\sqrt{3}} \left[y \right]_{y=1}^{y=\sqrt{4-x^2}} dx$$

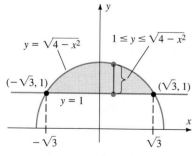

Figure 2.15 Region inside circle $x^2 + y^2 = 4$ and above $y = 1$.

$$= \int_{-\sqrt{3}}^{\sqrt{3}} (\sqrt{4 - x^2} - 1) \, dx$$

$$= 2 \int_{0}^{\sqrt{3}} \sqrt{4 - x^2} \, dx - 2\sqrt{3}.$$

Using a trigonometric substitution, we can determine that

$$\int \sqrt{4 - x^2} \, dx = 2 \, \text{Sin}^{-1} \left(\frac{x}{2} \right) + \frac{x\sqrt{4 - x^2}}{2} + C.$$

The desired area is therefore

$$2 \left[2 \, \text{Sin}^{-1} \left(\frac{\sqrt{3}}{2} \right) + \frac{\sqrt{3}}{2} \right] - 2\sqrt{3} \approx 2.457. \qquad \blacksquare$$

Interchanging the Order of Integration

There are iterated integrals for which we need to interchange the order of integration because an antiderivative cannot be found with respect to the "inside" variable. For example, in the iterated integral

$$\int_{0}^{1} \int_{y^2}^{1} y e^{x^2} \, dx \, dy = \int_{0}^{1} y \int_{y^2}^{1} e^{x^2} \, dx \, dy, \qquad (9)$$

we cannot express the antiderivative $\int e^{x^2} \, dx$ in closed form.

However, the region over which we are integrating is also x-simple. Thus this integral is equal to a second iterated integral, one that is derived using y-slices of the region. The following three-step procedure indicates how we may interchange the order of integration.

To reverse the order of integration in the iterated integral

$$\int_{c}^{d} \int_{h_1(y)}^{h_2(y)} f(x, y) \, dx \, dy:$$

(i) Identify (and sketch, if possible) the region Q for which the iterated integral can be written as the double integral

$$\int_{c}^{d} \int_{h_1(y)}^{h_2(y)} f(x, y) \, dx \, dy = \iint_{Q} f(x, y) \, dA.$$

(ii) Find constants a and b, and continuous functions g_1 and g_2, so that the region Q can be expressed as

$$Q = \{(x, y) \mid a \leq x \leq b, \, g_1(x) \leq y \leq g_2(x)\}.$$

(iii) Rewrite the iterated integral as

$$\int_{c}^{d} \int_{h_1(y)}^{h_2(y)} f(x, y) \, dx \, dy = \iint_{Q} f(x, y) \, dA$$

$$= \int_{a}^{b} \int_{g_1(x)}^{g_2(x)} f(x, y) \, dy \, dx.$$

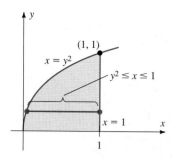

Figure 2.16 Considering the region Q in Example 6 as x-simple.

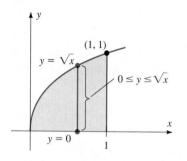

Figure 2.17 Considering the region Q in Example 6 as y-simple.

Obviously, this procedure can be applied only if Q is both x-simple and y-simple. Also, though the procedure is stated for reversing the order of integration from $dx\,dy$ to $dy\,dx$, the procedure for changing from order $dy\,dx$ to order $dx\,dy$ is analogous. Finally, there is no guarantee that the resulting iterated integral is any easier to evaluate than the original integral.

□ EXAMPLE 6

Use the procedure for reversing order of integration to evaluate the iterated integral

$$\int_0^1 \int_{y^2}^1 y e^{x^2}\, dx\, dy.$$

Solution   From the given limits of integration, the region Q is described by the inequalities

$$y^2 \le x \le 1, \qquad 0 \le y \le 1.$$

That is, Q is the region bounded between the graphs of $x = y^2$ and $x = 1$ for $0 \le y \le 1$. From Figure 2.16, we can see that Q is regular and can also be described by the inequalities

$$0 \le x \le 1, \qquad 0 \le y \le \sqrt{x} \qquad \text{(Figure 2.17)}.$$

Beginning with the given integral, we therefore reverse the order of integration as follows:

$$\int_0^1 \int_{y^2}^1 y e^{x^2}\, dx\, dy = \iint_Q y e^{x^2}\, dA = \int_0^1 \int_0^{\sqrt{x}} y e^{x^2}\, dy\, dx$$

$$= \int_0^1 \left[\frac{y^2}{2} e^{x^2} \right]_{y=0}^{y=\sqrt{x}} dy\, dx$$

$$= \int_0^1 \frac{x}{2} e^{x^2}\, dx$$

$$= \left[\frac{e^{x^2}}{4} \right]_0^1$$

$$= \frac{e-1}{4} \approx 0.43. \qquad \blacksquare$$

REMARK 2   Note that we cannot simply interchange limits of integration when we interchange the order of integration. There is no alternative to working out the new limits of integration from knowledge of the boundary of Q.

□ EXAMPLE 7

Evaluate the iterated integral

$$\int_0^1 \int_0^{\sqrt{1-x}} x y^2\, dy\, dx$$

by first reversing the order of integration.

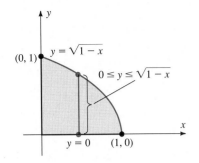

Figure 2.18 Considering the region Q in Example 7 as y-simple.

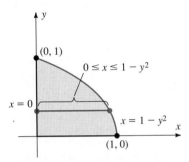

Figure 2.19 Considering the region Q in Example 7 as x-simple.

Properties of Double Integrals

Solution   The given limits of integration are

$$0 \le x \le 1, \qquad 0 \le y \le \sqrt{1-x},$$

which describe the region Q bounded above by the graph of $y = \sqrt{1-x}$ and below by the x-axis for $0 \le x \le 1$ (see Figure 2.18). By solving the equation $y = \sqrt{1-x}$ for x, we find that this region may also be described by the inequalities

$$0 \le x \le 1 - y^2, \qquad 0 \le y \le 1.$$

(See Figure 2.19.) We may therefore evaluate the iterated integral as

$$\int_0^1 \int_0^{\sqrt{1-x}} xy^2 \, dy \, dx = \iint_Q xy^2 \, dA = \int_0^1 \int_0^{1-y^2} xy^2 \, dx \, dy$$

$$= \int_0^1 \left[\frac{x^2}{2} y^2 \right]_{x=0}^{x=1-y^2} dy$$

$$= \int_0^1 \frac{(1-y^2)^2}{2} y^2 \, dy$$

$$= \frac{1}{2} \int_0^1 (1-y^2)^2 y^2 \, dy$$

$$= \frac{1}{2} \int_0^1 y^2 - 2y^4 + y^6 \, dy$$

$$= \frac{1}{2} \left[\frac{y^3}{3} - \frac{2y^5}{5} + \frac{y^7}{7} \right]_0^1 = \frac{4}{105}.$$

As defined in this section, the double integral satisfies the following properties:

(i) $\displaystyle \iint_Q [f(x, y) + g(x, y)] \, dA = \iint_Q f(x, y) \, dA + \iint_Q g(x, y) \, dA,$

(ii) $\displaystyle \iint_Q cf(x, y) \, dA = c \iint_Q f(x, y) \, dA, \qquad c = \text{constant},$

(iii) $\displaystyle \iint_{Q_1 \cup Q_2} f(x, y) \, dA = \iint_{Q_1} f(x, y) \, dA + \iint_{Q_2} f(x, y) \, dA.$

In (iii) we mean that $Q = Q_1 \cup Q_2$ is the union of two nonoverlapping regions Q_1 and Q_2, each satisfying the properties required of Q in this section. Property (iii) extends to unions of finitely many nonoverlapping regions $Q_1, Q_2, \ldots, Q_n$ of this type.

The proofs of these properties are analogous to those in the one-variable case, and we omit them. Nevertheless, as in the one-variable case, we use these properties frequently. In particular, property (iii) is especially useful when we must calculate a double integral over a region Q that is neither x-simple nor y-simple. If we can divide Q into a finite number of disjoint regions that are either x-simple or y-simple, then we can calculate the integral using property (iii) and the methods of this section.

Exercise Set 19.2

In Exercises 1–14, sketch the region Q determined by the limits of integration and evaluate the iterated integral.

1. $\displaystyle\int_0^1\int_0^x (y^2 - x^2)\, dy\, dx$

2. $\displaystyle\int_0^1\int_0^{1-x} 2xy\, dy\, dx$

3. $\displaystyle\int_{-1}^0\int_{-1}^{y+1} (xy - x)\, dx\, dy$

4. $\displaystyle\int_0^1\int_0^y xye^{x^2}\, dx\, dy$

5. $\displaystyle\int_{-1}^1\int_0^{1-x^2} xy\, dy\, dx$

6. $\displaystyle\int_0^1\int_{-x}^{\sqrt{x}} \frac{y}{1+x}\, dy\, dx$

7. $\displaystyle\int_0^1\int_{x^3}^{x^2} x\, dy\, dx$

8. $\displaystyle\int_0^{\sqrt{\pi/2}}\int_0^{\sqrt{y}} x\sin y^2\, dx\, dy$

9. $\displaystyle\int_0^{\pi/2}\int_0^{\sin x} 2y\cos x\, dy\, dx$

10. $\displaystyle\int_1^e\int_0^{\ln y} e^{x+y}\, dx\, dy$

11. $\displaystyle\int_0^1\int_0^{y^2} e^{x/y}\, dx\, dy$

12. $\displaystyle\int_0^{\pi/2}\int_0^{\cos y} x\sin y\, dx\, dy$

13. $\displaystyle\int_0^{\pi/4}\int_0^{\tan x} 3y\sec^2 x\, dy\, dx$

14. $\displaystyle\int_0^{\pi/6}\int_1^{1+\cos y} \cos y\, dx\, dy$

In Exercises 15–24, evaluate the double integral.

15. $\displaystyle\iint_Q y\sqrt{x}\, dA, \quad Q = \{(x,y)\,|\,0 \le x \le y^2, 0 \le y \le 1\}$

16. $\displaystyle\iint_Q x\cos\pi y\, dA, \quad Q = \{(x,y)\,|\,0 \le x \le 1, 0 \le y \le x\}$

17. $\displaystyle\iint_Q (x+2)\sqrt{1+e^y}\, dA,$

$Q = \{(x,y)\,|\,0 \le x \le e^{y/2}, 0 \le y \le 1\}$

18. $\displaystyle\iint_Q \frac{x^2}{1+y}\, dA, \quad Q = \{(x,y)\,|\,0 \le x \le 1, 0 \le y \le e^x - 1\}$

19. $\displaystyle\iint_Q xy\, dA, \quad Q = \{(x,y)\,|\,y \le x \le \sqrt{y}, 0 \le y \le 1\}$

20. $\displaystyle\iint_Q (x^2 + y^2)\, dA,$

$Q = \{(x,y)\,|\,-2 \le x \le 2, -3 \le y \le 1 - x^2\}$

21. $\displaystyle\iint_Q y\, dA, \quad Q = \{(x,y)\,|\,-1 \le x \le 1, e^x \le y \le e\}$

22. $\displaystyle\iint_Q ye^x\, dA, \quad Q = \{(x,y)\,|\,-\ln y \le x \le \ln y, 1 \le y \le 2\}$

23. $\displaystyle\iint_Q x\, dA, \quad Q = \{(x,y)\,|\,0 \le x \le \sqrt{1 - y^2}, 0 \le y \le 1\}$

24. $\displaystyle\iint_Q xy\, dA, \quad Q = \{(x,y)\,|\,0 \le x^2 + y^2 \le 1\}$

In Exercises 25–28, evaluate the given double integral.

25. $\displaystyle\iint_Q 3\sec x^3\, dA$

where Q is the region bounded by the x-axis, the curve $y = x^2$, and the line $x = \sqrt[3]{\pi/3}$.

26. $\displaystyle\iint_Q xy^2\, dA$

where Q is the region in the first quadrant that is inside the unit circle centered at the origin.

27. $\displaystyle\iint_Q e^x e^{e^x}\, dA$

where Q is the region bounded by the y-axis, the curve $x = \ln y$, and the line $y = 4$.

28. $\displaystyle\iint_Q (1 - x^3)y^2\, dA$

where Q is the region bounded by the curves $y = x^2$ and $x = y^2$.

In Exercises 29–34, sketch the region Q determined by the limits of integration, interchange the order of integration, and evaluate the given integral, where possible.

29. $\displaystyle\int_{-1}^1\int_0^{x+1} (x+y)\, dy\, dx$

30. $\displaystyle\int_0^1\int_{x^2}^1 xe^{y^2}\, dy\, dx$

31. $\displaystyle\int_0^1\int_0^y xy^2\, dx\, dy$

32. $\displaystyle\int_{-2}^0\int_{x^2}^4 xe^{y^2}\, dy\, dx$

33. $\displaystyle\int_1^e\int_0^{\ln x} f(x,y)\, dy\, dx$

34. $\displaystyle\int_0^1\int_0^{\sin^{-1}x} f(x,y)\, dy\, dx$

In Exercises 35–39, use a double integral to find the area of Q.

35. Q is the region bounded by the graphs of $y = 4 - x^2$ and the line $y = x + 2$.

36. Q is the region bounded by the graphs of $y = x^2$ and $y = x^3$.

37. Q is the region bounded by the graphs of $y = \sin x$ and $y = \cos x$ for $0 \le x \le \pi/4$.

38. Q is the region bounded by the graphs of $y = x^3$ and $y = 4x$.

39. Q is the region bounded by the graphs of $y = \sqrt{x}$ and $y = x^2$.

40. Find the volume of the solid bounded by the planes $x = z$, $x = 4$, $y = 7$, the xy-plane, and the xz-plane.

41. Let Q be the region in the xy-plane that is bounded by the curve $y = 1 + x^2$ and the line $y = 1 + x$. Find the volume of the solid determined by the graph of $f(x, y) = 1/y^2$ over Q.

42. Use a double integral to find the volume of the tetrahedron with vertices $(1, 0, 0)$, $(0, 1, 0)$, $(0, 0, 0)$, and $(0, 0, 1)$.

43. Find the volume of the solid bounded by the coordinate planes and the plane $6x + 3y + 2z = 6$.

44. Find the area of the ellipse

$$\frac{x^2}{4} + \frac{y^2}{9} = 1$$

using a double integral.

45. Sketch the region in the first octant common to the two cylinders $z^2 = 1 - x^2$ and $z^2 = 1 - y^2$. Find its volume.

19.3 Double Integrals in Polar Coordinates

Given a double integral

$$\int\int_Q f(x, y)\, dA$$

in which both the function f and the region Q have simple descriptions in terms of polar coordinates, we usually find it convenient to *convert the integral to polar coordinates*. The resulting iterated integral is often easier to compute. In this section, we describe the conversion process beginning with functions f that are already expressed in terms of polar coordinates. Later we describe how a function f defined in terms of the Cartesian (rectangular) coordinates can be rewritten in terms of polar coordinates.

Consider a region Q (see Figure 3.1) that is bounded by the rays $\theta = a$ and $\theta = b$ as well as by the graph of the nonnegative function $r = g(\theta)$. Given a continuous function $f(r, \theta)$ on the region Q expressed in terms of polar coordinates, how do we calculate the definite integral of f over Q?

Of course, we could use the techniques of Section 19.2 if Q is either x-simple or y-simple, but then we would not be taking advantage of the fact that Q has a simple description in terms of polar coordinates. We now derive a different representation for the integral using a *polar partition* of Q.

As Figure 3.2 illustrates, the circular arcs $r = r_j$ and the rays $\theta = \theta_k$ partition the region Q into wedge-shaped subregions R_{jk}. To determine the area of R_{jk} we use two facts (see Figure 3.3):

(i) The subregion R_{jk} lies between the concentric circles with radii $r = r_{j-1}$ and $r = r_j$. The area of the entire annular region lying between these two circles is

$$\text{Area of annulus} = \pi r_j^2 - \pi r_{j-1}^2.$$

(ii) The angle $\Delta\theta_k$, measured in radians, determines the proportion of the entire annulus lying in the region R_{jk}. That is,

$$\text{Proportion of annulus in } R_{jk} = \frac{\Delta\theta_k}{2\pi}.$$

From statements (i) and (ii), you can see that the area of the region R_{jk} is

$$\Delta A_{jk} = (\pi r_j^2 - \pi r_{j-1}^2)\left(\frac{\Delta\theta_k}{2\pi}\right) = \frac{r_j^2 - r_{j-1}^2}{2}\, \Delta\theta_k. \tag{1}$$

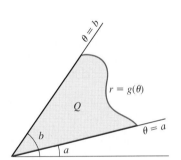

Figure 3.1 The region Q, bounded by the graph of $r = g(\theta)$ and the rays $\theta = a$ and $\theta = b$.

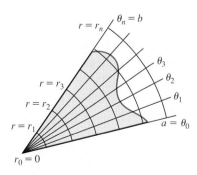

Figure 3.2 The region Q is partitioned into subregions by arcs $r = r_j$ and rays $\theta = \theta_k$.

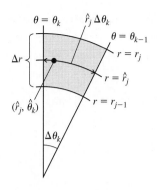

Figure 3.3 Area of region R_{jk} is $\Delta A_{jk} = \hat{r}_j \, \Delta r_j \, \Delta \theta_k$.

Since $r_j^2 - r_{j-1}^2 = (r_j + r_{j-1})(r_j - r_{j-1}) = (r_j + r_{j-1})\Delta r_j$ where $\Delta r_j = r_j - r_{j-1}$, we may rewrite equation (1) as

$$\Delta A_{jk} = \frac{r_j + r_{j-1}}{2} \, \Delta r_j \, \Delta \theta_k. \tag{2}$$

Let $\hat{r}_j$ denote the average of the two radii r_{j-1} and r_j; that is,

$$\hat{r}_j = \frac{r_j + r_{j-1}}{2}.$$

Then, from (2), the area of the region R_{jk} is

$$\Delta A_{jk} = \hat{r}_j \, \Delta r_j \, \Delta \theta_k. \tag{3}$$

Now let f be a continuous function defined on the region Q, and let $\hat{\theta}_k$ be any number with $\theta_{k-1} \le \hat{\theta}_k \le \theta_k$. Then the number $f(\hat{r}_j, \hat{\theta}_k)$ is the value of the function f at the point $(\hat{r}_j, \hat{\theta}_k)$ in the region R_{jk}. If $f(r, \theta) \ge 0$ for all $(r, \theta) \in Q$, then the product $f(\hat{r}_j, \hat{\theta}_k) \, \Delta A_{jk}$ approximates the volume of the solid bounded by the graph of f over R_{jk}, and the sum

$$\sum_{j=1}^{n} \sum_{k=1}^{m} f(\hat{r}_j, \hat{\theta}_k) \, \Delta A_{jk} = \sum_{j=1}^{n} \sum_{k=1}^{m} f(\hat{r}_j, \hat{\theta}_k) \, \hat{r}_j \, \Delta r_j \, \Delta \theta_k \tag{4}$$

approximates the volume of the solid bounded by the graph of f over the entire region Q. (In line (4) the sums are taken only over those regions R_{jk} that lie entirely within Q.) We therefore obtain the double integral of $f(r, \theta)$ over Q as

$$\iint\limits_{Q} f(r, \theta) \, dA = \lim_{\|P\| \to 0} \sum_{j=1}^{n} \sum_{k=1}^{m} f(\hat{r}_j, \hat{\theta}_k) \, \hat{r}_j \, \Delta r_j \, \Delta \theta_k \tag{5}$$

where $\|P\|$ is an appropriately defined norm for the partition P that we have described above.

Of course, equation (5) is of little value unless we know how to evaluate the limit on its right-hand side. The following theorem, analogous to Theorem 3, shows how this may be done. It is proved by interpreting the right-hand side of equation (5) as a limit of Riemann sums for the function $f(r, \theta) \, r$ over a different region Q' (see Section 19.8 for more details).

Theorem 4

Let Q be the region bounded by the graph of the continuous function $r = g(\theta)$ and the rays $\theta = a$ and $\theta = b$, as in Figure 3.1. Let f be a continuous function defined on the region Q. Then

$$\iint\limits_{Q} f(r, \theta) \, dA = \int_{a}^{b} \int_{0}^{g(\theta)} f(r, \theta) \, r \, dr \, d\theta. \tag{6}$$

Theorem 3 says that the double integral in (5) may be evaluated as an iterated integral, where we integrate first with respect to r alone, for $0 \le r \le g(\theta)$, and then integrate the resulting function of θ over the limits $a \le \theta \le b$. However, *it is impor-*

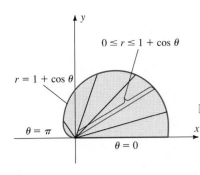

Figure 3.4 Region bounded by graph of $r = 1 + \cos \theta$ for $0 \leq \theta \leq \pi$.

tant to note that the integrand in the iterated integral is the product $f(r, \theta)$ r, not just $f(r, \theta)$. The reason for this factor of r can be seen from equation (3): The area of the region R_{jk} is $\Delta A_{jk} = \hat{r}_j \, \Delta r_j \, \Delta \theta_k$. The factor r in the iterated polar double integral appears because of this formula for the area of the approximating region R_{jk}.

□ **EXAMPLE 1**

Evaluate the double integral

$$\iint\limits_{Q} r \sin \theta \, dA$$

where Q is the region inside the upper half of the cardioid $r = 1 + \cos \theta$ (see Figure 3.4).

Solution Here $f(r, \theta) = r \sin \theta$, and the region Q may be described by the inequalities

$$0 \leq r \leq 1 + \cos \theta, \qquad 0 \leq \theta \leq \pi.$$

Thus, $g(\theta) = 1 + \cos \theta$ in equation (6), and

$$\iint\limits_{Q} r \sin \theta \, dA = \int_0^\pi \int_0^{1+\cos \theta} r \sin \theta \cdot r \, dr \, d\theta \quad \text{← note extra factor } r$$

$$= \int_0^\pi \int_0^{1+\cos \theta} r^2 \sin \theta \, dr \, d\theta$$

$$= \int_0^\pi \left[\frac{r^3}{3} \sin \theta \right]_{r=0}^{r=1+\cos \theta} d\theta$$

$$= \int_0^\pi \frac{(1 + \cos \theta)^3}{3} \sin \theta \, d\theta$$

$$= \left[-\frac{(1 + \cos \theta)^4}{12} \right]_{\theta=0}^{\theta=\pi}$$

$$= -\frac{0 - 2^4}{12} = \frac{4}{3}. \qquad \blacksquare$$

□ **EXAMPLE 2**

Find the area of the region enclosed by the three-leaved rose $r = \sin 3\theta$ (see Figure 3.5).

Solution As illustrated in Figure 3.5, one sixth of the entire region Q lies between the rays $\theta = 0$ and $\theta = \pi/6$. Using the symmetry of the region and equation (6), we find that

$$\text{Area of } Q = \iint\limits_{Q} 1 \, dA$$

$$= 6 \int_0^{\pi/6} \int_0^{\sin 3\theta} 1 \cdot r \, dr \, d\theta \quad \text{← note extra factor } r$$

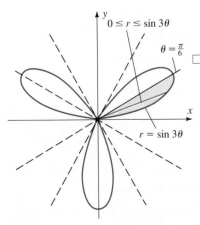

Figure 3.5 Region enclosed by three-leaved rose $r = \sin 3\theta$.

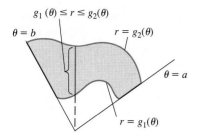

Figure 3.6 An r-simple region Q bounded by two polar curves, $r = g_1(\theta)$ and $r = g_2(\theta)$.

$$= 6 \int_0^{\pi/6} \left[\frac{r^2}{2} \right]_{r=0}^{r=\sin 3\theta} d\theta$$

$$= 3 \int_0^{\pi/6} \sin^2 3\theta \, d\theta$$

$$= \tfrac{3}{2} \int_0^{\pi/6} (1 - \cos 6\theta) \, d\theta \qquad (\sin^2 3\theta = \tfrac{1}{2}(1 - \cos 6\theta))$$

$$= \tfrac{3}{2} \left[\theta - \tfrac{1}{6} \sin 6\theta \right]_{\theta=0}^{\theta=\pi/6} = \pi/4. \qquad \blacksquare$$

Figure 3.6 shows a region Q that is **r-simple**: it lies *between* the graphs of the polar equations $r = g_1(\theta)$ and $r = g_2(\theta)$ for $a \leq \theta \leq b$. That is,

$$Q = \{(r, \theta) \mid g_1(\theta) \leq r \leq g_2(\theta), a \leq \theta \leq b\}.$$

If we let

$$Q_1 = \{(r, \theta) \mid 0 \leq r \leq g_1(\theta), a \leq \theta \leq b\},$$

and

$$Q_2 = \{(r, \theta) \mid 0 \leq r \leq g_2(\theta), a \leq \theta \leq b\},$$

then $Q = Q_2 - Q_1$. Accordingly, we have

$$\iint_Q f(r, \theta) \, dA = \iint_{Q_2} f(r, \theta) \, dA - \iint_{Q_1} f(r, \theta) \, dA$$

$$= \int_a^b \int_0^{g_2(\theta)} f(r, \theta) \, r \, dr \, d\theta - \int_a^b \int_0^{g_1(\theta)} f(r, \theta) \, r \, dr \, d\theta$$

$$= \int_a^b \left[\int_0^{g_2(\theta)} f(r, \theta) \, r \, dr - \int_0^{g_1(\theta)} f(r, \theta) \, r \, dr \right] d\theta$$

$$= \int_a^b \int_{g_1(\theta)}^{g_2(\theta)} f(r, \theta) \, r \, dr \, d\theta.$$

Thus, if Q is described by the polar inequalities

$$g_1(\theta) \leq r \leq g_2(\theta), \qquad a \leq \theta \leq b,$$

then

$$\iint_Q f(r, \theta) \, dA = \int_a^b \int_{g_1(\theta)}^{g_2(\theta)} f(r, \theta) \, r \, dr \, d\theta. \tag{7}$$

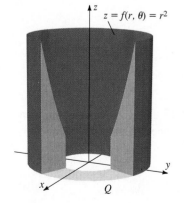

Figure 3.7 The bearing sleeve of Example 3 with a portion cut away to illustrate the region Q.

□ **EXAMPLE 3**

A bearing sleeve has the shape of the solid bounded above by the graph of $f(r, \theta) = r^2$ and below by the xy-plane for $1 \leq r \leq 2$. Find the volume of the sleeve.

Solution The solid is illustrated in Figure 3.7. The region Q over which the solid is defined is described using polar coordinates by the inequalities

$$1 \leq r \leq 2, \qquad 0 \leq \theta \leq 2\pi.$$

The volume is therefore

$$\iint\limits_{Q} f(r, \theta)\, dA = \int_0^{2\pi} \int_1^2 r^2 \cdot r \, dr \, d\theta$$

$$= \int_0^{2\pi} \int_1^2 r^3 \, dr \, d\theta$$

$$= \int_0^{2\pi} \left[\frac{r^4}{4} \right]_{r=1}^{r=2} d\theta$$

$$= \int_0^{2\pi} \frac{15}{4} \, d\theta = \frac{15\pi}{2}.$$

Changing from Cartesian to Polar Coordinates

Often, a double integral in Cartesian coordinates is more easily evaluated by first changing to polar coordinates. For example, to find the volume of the sphere $x^2 + y^2 + z^2 = 4$ lying above the plane $z = 1$, we must calculate the double integral

$$V = \int_{-\sqrt{3}}^{\sqrt{3}} \int_{-\sqrt{3-x^2}}^{\sqrt{3-x^2}} (\sqrt{4 - (x^2 + y^2)} - 1) \, dy \, dx. \tag{8}$$

(See Figure 3.8.)

We can calculate this volume much more simply using polar coordinates, together with the equations

$$x = r \cos \theta \tag{9a}$$

$$y = r \sin \theta \tag{9b}$$

for changing from Cartesian coordinates to polar coordinates.

The region $Q = \{(x, y) \mid -\sqrt{3} \le x \le \sqrt{3}, -\sqrt{3 - x^2} \le y \le \sqrt{3 - x^2}\}$ can be described in polar coordinates as

$$Q = \{(r, \theta) \mid 0 \le r \le \sqrt{3}, 0 \le \theta \le 2\pi\}.$$

Using equations (9a) and (9b), the function $f(x, y) = \sqrt{4 - (x^2 + y^2)}$ in polar coordinates is simply

$$f(r \cos \theta, r \sin \theta) = \sqrt{4 - (r^2 \cos^2 \theta + r^2 \sin^2 \theta)} = \sqrt{4 - r^2}.$$

The desired volume is then obtained using equation (6).

$$V = \iint\limits_{Q} f(r \cos \theta, r \sin \theta) \, dA = \int_0^{2\pi} \int_0^{\sqrt{3}} (\sqrt{4 - r^2} - 1) \, r \, dr \, d\theta$$

$$\overset{\text{note extra } r}{\underset{}{}}$$

$$= \int_0^{2\pi} \left[-\frac{1}{3}(4 - r^2)^{3/2} - \frac{r^2}{2} \right]_{r=0}^{r=\sqrt{3}} d\theta$$

$$= \int_0^{2\pi} \left[-\frac{1}{3}(1 - 4^{3/2}) - \frac{3}{2} \right] d\theta$$

$$= \int_0^{2\pi} \frac{5}{6} \, d\theta = \frac{5\pi}{3}.$$

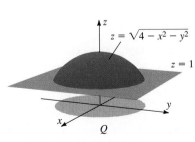

Figure 3.8   Calculating the volume of the part of the solid sphere $x^2 + y^2 + z^2 \le 4$ that lies above the plane $z = 1$ involves an integral over the circular region $Q = \{(x, y) \mid x^2 + y^2 \le 3\}$.

This example is typical of a more general problem: Given an iterated integral in Cartesian coordinates, how can we evaluate the integral using an iterated integral in polar coordinates (without, of course, changing the value of the integral)? The answer is the following:

To express the iterated integral

$$\int_c^d \int_{h_1(y)}^{h_2(y)} f(x, y)\, dx\, dy \qquad \text{(or equivalent)}$$

in polar coordinates,

(i) express the region $Q = \{(x, y) \mid h_1(y) \le x \le h_2(y),\, c \le y \le d\}$ in polar coordinates as

$$Q = \{(r, \theta) \mid g_1(\theta) \le r \le g_2(\theta),\, a \le \theta \le b\}, \text{ and}$$

(ii) using the substitutions $x = r \cos\theta$ and $y = r \sin\theta$, replace the integrand

$$f(x, y) \quad \text{by} \quad f(r \cos\theta, r \sin\theta) \cdot r.$$

The result of steps (i) and (ii) is the iterated integral

$$\int_a^b \int_{g_1(\theta)}^{g_2(\theta)} f(r \cos\theta, r \sin\theta)\, r\, dr\, d\theta.$$

Another way to write this is simply

$$\boxed{\iint_Q f(x, y)\, dx\, dy = \iint_Q f(r \cos\theta, r \sin\theta)\, r\, dr\, d\theta.} \qquad (10)$$

Equation (10) is referred to as a **change of variables** formula.

We shall not prove statement (10). However, it may be justified by comparing Theorem 3, which expresses

$$\iint_Q f\, dA$$

as an iterated integral in Cartesian coordinates, with Theorem 4, which gives

$$\iint_Q f\, dA$$

as an iterated integral in polar coordinates. We may paraphrase equation (10) by saying that in changing from Cartesian to polar coordinates we replace the *element of area*

$$dA = dx\, dy$$

in rectangular coordinates with the element of area

$$dA = r\, dr\, d\theta \qquad (11)$$

in polar coordinates. It is very important to include the factor r in (11), and to remember that, in the integral on the right side of equation (10), Q must be described using polar coordinates.

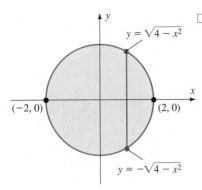

Figure 3.9 If $-2 \le x \le 2$ and $-\sqrt{4 - x^2} \le y \le \sqrt{4 - x^2}$, then the region Q is circular.

□ **EXAMPLE 4**

Evaluate the integral

$$\int_{-2}^{2} \int_{-\sqrt{4-x^2}}^{\sqrt{4-x^2}} e^{x^2+y^2} \, dy \, dx$$

by first changing to polar coordinates.

Solution   From the limits of integration, we see that the integral is being evaluated over the disc-shaped region

$$Q = \{(x, y) \mid -2 \le x \le 2, \, -\sqrt{4 - x^2} \le y \le \sqrt{4 - x^2}\}.$$

(See Figure 3.9.) This region can be described in polar coordinates by the inequalities

$$0 \le r \le 2, \qquad 0 \le \theta \le 2\pi \qquad \text{(Figure 3.10)}.$$

With $x = r \cos \theta$ and $y = r \sin \theta$, the function $f(x, y) = e^{x^2+y^2}$ becomes

$$f(r \cos \theta, r \sin \theta) = e^{r^2 \cos^2 \theta + r^2 \sin^2 \theta} = e^{r^2}.$$

Using (10), we obtain

$$\int_{-2}^{2} \int_{-\sqrt{4-x^2}}^{\sqrt{4-x^2}} e^{x^2+y^2} \, dy \, dx = \int_{0}^{2\pi} \int_{0}^{2} e^{r^2} \underset{\uparrow}{r} \, dr \, d\theta \quad \text{note the extra factor } r$$

$$= \int_{0}^{2\pi} \left[\tfrac{1}{2} e^{r^2} \right]_{r=0}^{r=2} d\theta$$

$$= \int_{0}^{2\pi} \tfrac{1}{2}(e^4 - 1) \, d\theta$$

$$= \pi(e^4 - 1). \qquad ■$$

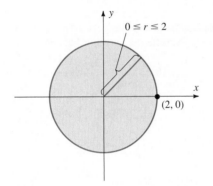

Figure 3.10 The circular region Q in Example 4 can be expressed in polar coordinates as $0 \le r \le 2$ and $0 \le \theta \le 2\pi$.

REMARK   Note that the iterated integral in Example 4 could not be evaluated in rectangular coordinates, since we would not be able to find an antiderivative for $e^{x^2+y^2}$ with respect to either x or y. Thus, it is sometimes essential that we change to polar coordinates, if possible. The next example involves a volume that could be calculated using Cartesian coordinates, but the calculation in polar coordinates is much easier. It is typical of the kind of problem we will encounter in Chapter 20.

□ **EXAMPLE 5**

Interpret the iterated integral

$$\int_{0}^{2} \int_{0}^{\sqrt{4-x^2}} \sqrt{5 - x^2 - y^2} \, dy \, dx$$

geometrically, and evaluate the integral by first changing to polar coordinates.

Solution   Since the integrand is nonnegative, the integral may be interpreted as the volume of the solid bounded above by the graph of $f(x, y) = \sqrt{5 - x^2 - y^2}$ and below by the quarter disc $Q = \{(x, y) \mid 0 \le x \le 2, \, 0 \le y \le \sqrt{4 - x^2}\}$ of radius 2 (Figure 3.11). In polar coordinates, Q is determined by the inequalities

$$0 \le r \le 2, \qquad 0 \le \theta \le \pi/2.$$

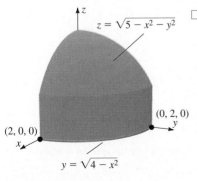

Figure 3.11 The solid whose volume is computed in Example 5.

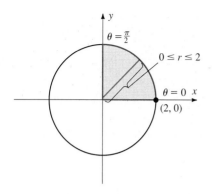

Figure 3.12 The region Q of integration in Example 5 is a quarter disc.

(See Figure 3.12.) Thus

$$\int_0^2 \int_0^{\sqrt{4-x^2}} \sqrt{5 - x^2 - y^2}\, dy\, dx$$

$$= \int_0^{\pi/2} \int_0^2 \sqrt{5 - (r\cos\theta)^2 - (r\sin\theta)^2} \cdot r\, dr\, d\theta$$

note the extra factor r

$$= \int_0^{\pi/2} \int_0^2 \sqrt{5 - r^2}\, r\, dr\, d\theta$$

$$= \int_0^{\pi/2} \left[-\frac{1}{3}(5 - r^2)^{3/2} \right]_{r=0}^{r=2} d\theta$$

$$= \int_0^{\pi/2} -\frac{1}{3}(1 - 5^{3/2})\, d\theta = \frac{\pi}{6}(5^{3/2} - 1).$$
∎

Exercise Set 19.3

In Exercises 1–7, sketch the region Q and evaluate the double integral of the given function f over the given region Q.

1. $f(r, \theta) = r, \quad Q = \{(r, \theta) \mid 0 \leq r \leq 1, 0 \leq \theta \leq 2\pi\}$

2. $f(r, \theta) = 1 - r, \quad Q = \{(r, \theta) \mid 0 \leq r \leq 2, 0 \leq \theta \leq \pi\}$

3. $f(x, y) = e^{x^2+y^2}, \quad Q = \{(r, \theta) \mid r \leq a\}$

4. $f(x, y) = e^{x^2+y^2}, \quad Q = \{(r, \theta) \mid 1 \leq r \leq 2, \ 0 \leq \theta \leq \pi/4\}$

5. $f(r, \theta) = 3r^2, \quad Q = \{(r, \theta) \mid 1 \leq r \leq 2, 0 \leq \theta \leq \pi/4\}$

6. $f(r, \theta) = 2 - r, \quad Q = \{(r, \theta) \mid 0 \leq r \leq (1 + \cos\theta), \ 0 \leq \theta \leq \pi/2\}$

7. $f(r, \theta) = r, \quad Q = \{(r, \theta) \mid 0 \leq r \leq \sin 3\theta, \ 0 \leq \theta \leq \pi/3\}$

In Exercises 8–16, find the area of the given region Q using a double integral.

8. Q is the region enclosed by the circle $r = 2\cos\theta$.

9. Q is the region enclosed by the three-leaved rose $r = \sin 3\theta$.

10. Q is the region enclosed by the cardioid $r = a(1 + \cos\theta)$.

11. Q is the region inside the cardioid $r = 1 + \cos\theta$ and outside the circle $r = 1$.

12. Q is the region inside the four-leaved rose $r = 4\sin 2\theta$ and outside the circle $r = 2$.

13. Q is the region inside the circle $r = 5$ and outside the circle $r = 10\sin\theta$.

14. Q is the region bounded by the graph of $r^2 = a^2 \sin 2\theta$.

15. Q is the region enclosed by the graph of $r = 2 - \sin\theta$.

16. Q is the region enclosed by the graph of $r^2 = 4\cos^2\theta$.

In Exercises 17–26, change the iterated integral from Cartesian to polar coordinates. Then evaluate the resulting integral.

17. $\int_0^1 \int_0^{\sqrt{1-x^2}} 2\, dy\, dx$

18. $\int_{-2}^{\sqrt{2}} \int_x^{\sqrt{4-x^2}} 1\, dy\, dx$

19. $\int_0^1 \int_0^{\sqrt{1-y^2}} e^{x^2+y^2}\, dx\, dy$

20. $\int_0^2 \int_{\sqrt{2y-y^2}}^{\sqrt{2y-y^2}} \sqrt{x^2 + y^2}\, dx\, dy$

21. $\int_1^2 \int_{-\sqrt{4-x^2}}^{\sqrt{4-x^2}} \frac{x^2}{x^2 + y^2}\, dy\, dx$

22. $\int_0^1 \int_x^{\sqrt{2-x^2}} (x^2 + y^2)\, dy\, dx$

23. $\int_0^2 \int_0^{\sqrt{2x-x^2}} \frac{1}{\sqrt{x^2 + y^2}}\, dy\, dx$

24. $\int_0^1 \int_0^{\sqrt{4-x^2}} (x^2 + y^2)^{3/2}\, dy\, dx$

25. $\int_{-2}^2 \int_0^{\sqrt{4-x^2}} \frac{x}{\sqrt{x^2 + y^2}}\, dy\, dx$

26. $\int_0^1 \int_0^{\sqrt{1-y^2}} (x^2 + y^2)^{3/2}\, dx\, dy$

27. Find the volume of the portion of the cylinder $x^2 + (y - 1)^2 = 4$ bounded above by the plane $z = x + 4$ and below by the xy-plane.

28. Find the area of the region outside the spiral $r = \theta$ and inside the spiral $r = 2\theta$ for $0 \leq \theta \leq 2\pi$.

29. Find the volume of the region lying inside the sphere $x^2 + y^2 + z^2 = 4$ and outside the cylinder $x^2 + y^2 = 1$.

30. Find the volume of the region lying inside both the cone $z^2 = x^2 + y^2$ and the sphere $x^2 + y^2 + z^2 = 2$.

31. Find the volume of the solid bounded above by the plane $z = y + 2$ and below by the region inside the cardioid $r = 1 + \cos \theta$.

32. Use a double integral in polar coordinates to obtain the formula for the volume of a right circular cylinder of radius r and height h.

33. Use a double integral to establish the formula for the volume of a right circular cone of radius r and height h.

34. Find the volume of the solid bounded by the paraboloid $z = \frac{1}{4}(x^2 + y^2)$ and the paraboloid $z = 5 - x^2 - y^2$.

35. Find the volume of the solid bounded above by the graph of $z = 9 - x^2 - y^2$ and below by the graph of $z = 1 + x^2 + y^2$.

36. A hole 2 cm in diameter is drilled through the center of a spherical bearing of radius 3 cm. Find the volume of the remaining solid.

37. Change the order of integration in

$$\int_0^{\pi/2} \int_0^{\sin \theta} \sin \theta \, dr \, d\theta$$

and evaluate the integral.

38. Find the volume of the solid inside both the ellipsoid $z^2 + 4r^2 = 4$ and the cylinder $r = \sin \theta$.

39. Find the volume of the solid bounded by the cone $z = \sqrt{x^2 + y^2}$, the cylinder $x^2 + y^2 = 4$, and the xy-plane.

19.4 Calculating Mass and Centers of Mass

In this section we use the double integral to calculate mass and centroids for lamina (thin flat objects) lying in the plane. The difference between the discussion of Chapter 7 and what we do here is that we previously had assumed the density ρ of the material to be constant. Here we will treat the more general case of a variable density ρ. We assume the density function ρ to be continuous throughout the planar region Q that describes the lamina.

More specifically, let Q be a region in the plane (which we think of as the base of the lamina) and let $\rho(x, y)$ be the **mass per unit area** of the lamina at point (x, y). This is what we mean by density. (Thus, $\rho(x, y)$ is affected both by the thickness of the material and by its mass per unit volume. We will not be concerned about these two factors individually.) If R_j is a rectangle in Q of area ΔA, and if (s_j, t_j) is a point in R_j, then the product

$$\Delta M_j = \rho(s_j, t_j) \, \Delta A \qquad \text{(mass} = \text{mass per unit area} \times \text{area)}$$

is an approximation to the mass of the lamina over the rectangle R_j. If, as in Section 19.2, the region Q is partitioned by a rectangular grid* and $R_1, R_2, \ldots, R_n$ is a list of all rectangles lying within Q, then

$$M \approx \sum_{j=1}^{n} \Delta M_j = \sum_{j=1}^{n} \rho(s_j, t_j) \, \Delta A \tag{1}$$

is an approximation to the mass of the lamina over Q. Since, as $n \to \infty$, the union of the rectangles R_j provides an increasingly accurate approximation to the region Q, we obtain M, the mass of Q, as the double integral

$$M = \iint_Q \rho(x, y) \, dA. \tag{2}$$

---

*As we did in Chapter 7, we shall henceforth use *regular* grids to partition the region Q into rectangles of equal dimensions Δx and Δy, and areas $\Delta A = \Delta x \, \Delta y$.

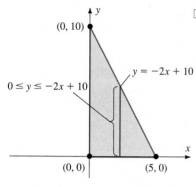

Figure 4.1 The triangular region in Example 1.

□ **EXAMPLE 1**

A thin plate has the shape of a right triangle with legs of length 5 cm and 10 cm. The density, in terms of mass per unit area, at each point on the plate is proportional to the square of the distance from the vertex corresponding to the right angle. What is the mass of the plate?

Solution With the triangle positioned as in Figure 4.1, the region it occupies is described by the inequalities

$$0 \le x \le 5, \qquad 0 \le y \le -2x + 10.$$

Since the right angle is at the origin, the density function is

$$\rho(x, y) = \lambda(x^2 + y^2)$$

where λ is constant. By (2), the mass is

$$M = \iint_Q \lambda(x^2 + y^2)\, dA = \int_0^5 \int_0^{-2x+10} \lambda(x^2 + y^2)\, dy\, dx$$

$$= \int_0^5 \lambda \left[x^2 y + \frac{y^3}{3} \right]_{y=0}^{y=-2x+10} dx$$

$$= \int_0^5 \lambda \left(-\frac{14}{3}x^3 + 50x^2 - 200x + \frac{1000}{3} \right) dx$$

$$= \lambda \left[-\frac{7}{6}x^4 + \frac{50}{3}x^3 - 100x^2 + \frac{1000}{3}x \right]_0^5$$

$$= \frac{3125\lambda}{6}.$$

□ **EXAMPLE 2**

A machine part has the shape of a half annulus—the region between two concentric semicircles of radius $r = 1$ and $r = 2$ (see Figure 4.2). Find the mass of the part if the density at each point is proportional to the distance from that point to the common center of the two semicircles.

Solution The density function can be written as

$$\rho(x, y) = \lambda\sqrt{x^2 + y^2}$$

where λ is constant. However, it is most convenient to describe the region Q in polar coordinates using the inequalities

$$0 \le \theta \le \pi, \qquad 1 \le r \le 2,$$

in which case the density function is $\rho(r, \theta) = \lambda r$.
    The mass is therefore

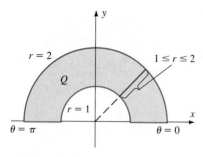

Figure 4.2 Annular region of Example 2.

$$M = \iint_Q \lambda r\, dA = \int_0^\pi \int_1^2 \lambda r \cdot r\, dr\, d\theta$$

$$= \int_0^\pi \int_1^2 \lambda r^2\, dr\, d\theta = \int_0^\pi \left[\lambda \frac{r^3}{3} \right]_{r=1}^{r=2} d\theta$$

$$= \int_0^\pi \frac{7\lambda}{3}\, d\theta = \frac{7\lambda\pi}{3}. \qquad\blacksquare$$

Centers of Mass

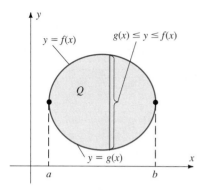

$y = f(x)$

$g(x) \le y \le f(x)$

Q

$y = g(x)$

a b

Figure 4.3   A y-simple region Q.

In Chapter 7 we determined that the x-coordinate of the *centroid* of the lamina with shape Q and constant density ρ is

$$\bar{x} = \frac{\int_a^b \rho x [f(x) - g(x)]\, dx}{\int_a^b \rho [f(x) - g(x)]\, dx}. \tag{3}$$

In equation (3), we assume that the region Q can be described by the inequalities

$$a \le x \le b, \qquad g(x) \le y \le f(x) \qquad \text{(Figure 4.3)}.$$

In other words, Q is y-simple.

To generalize to the case of a variable density function ρ, we begin with the integral in the numerator of $\bar{x}$ in (3), which we write as

$$\int_a^b \rho x [f(x) - g(x)]\, dx = \int_a^b \left[\rho xy \right]_{y=g(x)}^{y=f(x)} dx \tag{4}$$

$$= \int_a^b \int_{g(x)}^{f(x)} \rho x\, dy\, dx.$$

The right-hand side of (4) allows us to define the first moment of mass for Q about the y-axis if ρ is nonconstant. We simple replace ρ by $\rho(x, y)$ and define

$$M_y = \int_a^b \int_{g(x)}^{f(x)} x\rho(x, y)\, dy\, dx = \iint_Q x\rho(x, y)\, dA.$$

Similarly, the first moment of mass for Q about the x-axis is defined to be

$$M_x = \iint_Q y\rho(x, y)\, dA.$$

Finally, the denominator of $\bar{x}$ can be written

$$\int_a^b \rho [f(x) - g(x)]\, dx = \int_a^b \left[\rho y \right]_{y=g(x)}^{y=f(x)} dx$$

$$= \int_a^b \int_{g(x)}^{f(x)} \rho\, dy\, dx$$

$$= \iint_Q \rho\, dy\, dx,$$

which is just the mass (area $\times$ density) of Q. We are therefore ready to generalize the concept of centroid to that of *center of mass*.

Definition 3

Let Q be a region in the xy-plane that is either x-simple or y-simple, and let ρ be a continuous density function defined on Q. Then

(i) The **mass** of Q is the number

$$M = \iint\limits_{Q} \rho(x, y) \, dA.$$

(ii) The **first moment of mass for Q about the y-axis** is the number

$$M_y = \iint\limits_{Q} x\rho(x, y) \, dA.$$

(iii) The **first moment of mass for Q about the x-axis** is the number

$$M_x = \iint\limits_{Q} y\rho(x, y) \, dA.$$

(iv) The **center of mass** of Q is the point $(\bar{x}, \bar{y})$, where

$$\bar{x} = \frac{M_y}{M}, \qquad \bar{y} = \frac{M_x}{M}.$$

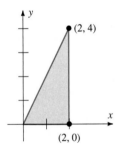

Figure 4.4 The triangle corresponding to the lamina in Example 3.

Because of the way the concept of centroid is generalized by Definition 3, the physical interpretation of the center of mass is the same as before—the lamina associated with Q will "balance" at the point $(\bar{x}, \bar{y})$.

□ **EXAMPLE 3**

A thin lamina has the shape of a triangle with vertices at $(0, 0)$, $(2, 0)$, and $(2, 4)$. The density function associated with the lamina has equation $\rho(x, y) = 4x + 2y + 2$. Find the mass and center of mass of the lamina (see Figure 4.4).

Solution The region Q associated with the lamina can be described by the inequalities

$$0 \le x \le 2, \qquad 0 \le y \le 2x.$$

Thus, by Definition 3,

$$M = \iint\limits_{Q} (4x + 2y + 2) \, dA = \int_0^2 \int_0^{2x} (4x + 2y + 2) \, dy \, dx$$

$$= \int_0^2 \left[4xy + y^2 + 2y \right]_{y=0}^{y=2x} dx$$

$$= \int_0^2 (12x^2 + 4x) \, dx$$

$$= \left[4x^3 + 2x^2 \right]_0^2$$

$$= 40,$$

$$M_y = \iint\limits_Q x(4x + 2y + 2)\, dA = \int_0^2 \int_0^{2x} (4x^2 + 2xy + 2x)\, dy\, dx$$

$$= \int_0^2 \left[4x^2 y + xy^2 + 2xy \right]_{y=0}^{y=2x} dx$$

$$= \int_0^2 (12x^3 + 4x^2)\, dx = \left[3x^4 + \frac{4x^3}{3} \right]_0^2$$

$$= \frac{176}{3},$$

and

$$M_x = \iint\limits_Q y(4x + 2y + 2)\, dA = \int_0^2 \int_0^{2x} (4xy + 2y^2 + 2y)\, dy\, dx$$

$$= \int_0^2 \left[2xy^2 + \frac{2y^3}{3} + y^2 \right]_{y=0}^{y=2x} dx$$

$$= \int_0^2 \left(\frac{40}{3} x^3 + 4x^2 \right) dx = \left[\frac{10}{3} x^4 + \frac{4x^3}{3} \right]_0^2$$

$$= \frac{192}{3}.$$

Thus,

$$\bar{x} = \frac{M_y}{M} = \frac{\left(\frac{176}{3}\right)}{40} = \frac{22}{15}$$

and

$$\bar{y} = \frac{M_x}{M} = \frac{\left(\frac{192}{3}\right)}{40} = \frac{8}{5}.$$

The center of mass is therefore $(\bar{x}, \bar{y}) = \left(\frac{22}{15}, \frac{8}{5}\right)$. ∎

□ **EXAMPLE 4**

Find the center of mass for the lamina described in Example 2.

Solution The density function in Example 2 is

$$\rho(x, y) = \lambda \sqrt{x^2 + y^2}$$

and the region Q is described in polar coordinates as

$$0 \le \theta \le \pi, \qquad 1 \le r \le 2.$$

We evaluate the integral for M_y in polar coordinates as

$$M_y = \iint\limits_Q x\rho(x, y)\, dA = \int_0^\pi \int_1^2 (r\cos\theta)\lambda r \cdot \underbrace{r\, dr\, d\theta}_{dA}$$

$$\rho = \lambda\sqrt{x^2 + y^2} = \lambda r$$
$$x = r\cos\theta$$

$$= \lambda \int_0^\pi \int_1^2 r^3 \cos\theta\, dr\, d\theta$$

$$= \lambda \int_0^\pi \left[\frac{r^4}{4}\cos\theta\right]_{r=1}^{r=2} d\theta$$

$$= \lambda \int_0^\pi \frac{15}{4}\cos\theta\, d\theta$$

$$= \lambda \left[\frac{15}{4}\sin\theta\right]_0^\pi = 0.$$

(This should not surprise you since both Q and ρ are symmetric with respect to the y-axis.) The integral for M_x is

$$M_x = \iint\limits_Q y\rho(x, y)\, dA = \int_0^\pi \int_1^2 (r\sin\theta)\,\lambda r \cdot \underbrace{r\, dr\, d\theta}_{dA}$$

$$\rho = \lambda\sqrt{x^2 + y^2} = \lambda r$$
$$y = r\sin\theta$$

$$= \lambda \int_0^\pi \int_1^2 r^3 \sin\theta\, dr\, d\theta$$

$$= \lambda \int_0^\pi \left[\frac{r^4}{4}\sin\theta\right]_{r=1}^{r=2} d\theta$$

$$= \lambda \int_0^\pi \frac{15}{4}\sin\theta\, d\theta$$

$$= \lambda \left[-\frac{15}{4}\cos\theta\right]_0^\pi = \frac{15\lambda}{2}.$$

Since we found in Example 2 that $M = 7\lambda\pi/3$, we have

$$\bar{x} = \frac{M_y}{M} = 0$$

and

$$\bar{y} = \frac{M_x}{M} = \frac{(15\lambda/2)}{(7\lambda\pi/3)} = \frac{45}{14\pi} \approx 1.02.$$

The center of mass is therefore $(0, 45/(14\pi))$. ■

Exercise Set 19.4

In Exercises 1–10, find the mass of a lamina with shape given by the region Q and density function ρ.

1. $\rho(x, y) = x + y$,
$Q = \{(x, y) \mid 0 \le x \le 2, 0 \le y \le 1\}$

2. $\rho(x, y) = x^2 + y$,
$Q = \{(x, y) \mid 0 \le x \le 2, 0 \le y \le x\}$

3. $\rho(x, y) = 6 + x$,
$Q = \{(x, y) \mid -1 \le x \le 1, 0 \le y \le 1 - x^2\}$

4. $\rho(x, y) = xy$,
$Q = \{(x, y) \mid 0 \le x \le 1 - y^2, 0 \le y \le 1\}$

5. $\rho(x, y) = \sin(x + y)$,
$Q = \{(x, y) \mid 0 \le x \le \pi/4, 0 \le y \le \pi/4\}$

6. $\rho(x, y) = xy$,
$Q = \{(x, y) \mid 0 \le x \le 1, 0 \le y \le 1\}$

7. $\rho(x, y) = x^2 + y^2$,
$Q = \{(x, y) \mid -1 \le x \le 1, 0 \le y \le \sqrt{1 - x^2}\}$

8. $\rho(x, y) = \sqrt{x^2 + y^2}$,
$Q = \{(r, \theta) \mid 0 \le r \le 2, 0 \le \theta \le \pi/2\}$

9. $\rho(x, y) = \dfrac{1}{\sqrt{x^2 + y^2}}$, Q is the annulus $1 \le r \le 2$

10. $\rho(x, y) = xy$,
$Q = \{(r, \theta) \mid 0 \le r \le 1, 0 \le \theta \le \pi/2\}$

11. Find the center of mass of the lamina in Exercise 1.

12. Find the center of mass of the lamina in Exercise 3.

13. Find the center of mass of the lamina in Exercise 8.

14. A lamina has the shape of a triangle with vertices $(0, 0)$, $(0, 4)$, and $(1, 0)$. The density at each point (x, y) is $\rho(x, y) = y - x + 8$. Find the mass of the lamina.

15. Find the center of mass of the lamina in Exercise 14.

16. Find the centroid of the region bounded by the graph of $r = \sin 2\theta$ for $0 \le \theta \le \pi/2$. (Assume $\rho(x, y) = 1$ throughout the region.)

17. Find the centroid of the planar region bounded by the parabola $y = 4 - x^2$ and the x-axis.

18. Find the centroid of the region bounded by the graph of the function $f(x) = 1/x$ and the line $2x + 2y = 5$.

19. Find the centroid of the region obtained by connecting the points $(0, 0)$, $(4, 0)$, $(4, 4)$, $(2, 4)$, $(2, 1)$, $(0, 1)$, and $(0, 0)$ in order by line segments.

20. Find the centroid of the region bounded by the graph of $y = x^2$ and $y = x^3$.

21. Show that the centroid of the region
$$R = \{(x, y) \mid 0 \le x \le a, 0 \le y \le \sqrt{a^2 - x^2}\}$$
is $(\bar{x}, \bar{y}) = \left(\dfrac{4a}{3\pi}, \dfrac{4a}{3\pi} \right)$.

22. Find the centroid of the region bounded by the graph of $f(x) = \sinh x$ and the x-axis for $0 \le x \le 1$.

23. Find the centroid of the region bounded by the graph of $x = y(4 - y)$ and the y-axis.

24. A lamina has the shape of a circle of radius R. The density at any point P is proportional to the distance from the center. What is the mass?

25. A lamina has the shape of a right triangle with legs of length 2 and 6. The density at any point is proportional to the distance from the longer leg. What is the mass?

26. Find the distance from the vertex at the right angle to the center of mass for the triangle in Exercise 25.

27. Assume that the region in Example 1 is oriented with the 10-cm leg of the triangle along the x-axis, again with the right angle at the origin. Show that the mass calculated in this way is the same as that found in Example 1.

28. A lamina has the shape of a four-leaved rose (see Section 15.2) with petals of length 2. What is its mass if its density at any point is inversely proportional to the point's distance from the origin?

29. Find the center of mass of the region outside the cardioid $r = 1 + \cos \theta$ and inside the circle $r = 3 \cos \theta$ if its density function is $\rho(r, \theta) = 1 + r$.

30. Find the center of mass of a lamina with density function $\rho(r, \theta) = \lambda(1 + \cos^2 \theta)$ over the half annulus $Q = \{(r, \theta) \mid 1 \le r \le 2, 0 \le \theta \le \pi\}$. Locate the center of mass on a sketch of the region Q. State (in words) how the mass of the lamina is distributed, and what relation this has to the location of the center of mass.

19.5   Surface Area

As our final application of double integrals, we consider the problem of calculating the area of a surface in space. We shall restrict our discussion to the case of surfaces that are graphs of functions of two variables, although the ideas discussed here can be extended to more general types of surfaces.

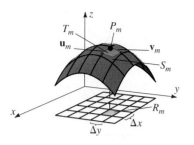

Figure 5.1 The tangent parallelogram T_m that lies directly over the patch S_m and the rectangle R_m.

Let Q be a region in the plane that is either x-simple or y-simple, and let the surface S be the graph of the continuous function $z = f(x, y)$ on Q. We partition the region Q with a rectangular grid, and we denote by $R_1, R_2, \ldots, R_n$ the rectangles lying entirely within Q. This grid is constructed so that each of the rectangles R_m has dimensions Δx and Δy (see Figure 5.1). The grid on Q partitions the surface S into patches $S_1, S_2, \ldots, S_n$. Specifically, by the patch S_m we mean the portion of the surface S lying above the rectangle R_m.

As Figure 5.1 suggests, the patch S_m is nearly a parallelogram, but not quite, since S_m is part of the (generally) curved surface S. We therefore construct a parallelogram T_m that approximates the patch S_m in the following way. Let P_m be the point on the corner of S_m nearest the z-axis and let (x_m, y_m) be the vertex of the rectangle R_m directly beneath the point P_m. Then the vector

$$\mathbf{u}_m = \Delta x\, \mathbf{i} + \frac{\partial f}{\partial x}(x_m, y_m)\, \Delta x\, \mathbf{k}, \tag{1}$$

when originating at point P_m, terminates at some point above the vertex $(x_m + \Delta x, y_m)$ of R_m. Moreover, $\mathbf{u}_m$ is tangent to S at P_m, by definition of the partial derivative $\partial f/\partial x$ at (x_m, y_m). Similarly, the vector

$$\mathbf{v}_m = \Delta y\, \mathbf{j} + \frac{\partial f}{\partial y}(x_m, y_m)\, \Delta y\, \mathbf{k}, \tag{2}$$

when originating at P_m, terminates above the vertex $(x_m, y_m + \Delta y)$ of R_m and is tangent to S at P_m (see Figure 5.2).

From the properties noted for the vectors $\mathbf{u}_m$ and $\mathbf{v}_m$, we conclude that $\mathbf{u}_m$ and $\mathbf{v}_m$ determine a parallelogram T_m that

(i) lies directly above the rectangle R_m (and, hence, the patch S_m), and
(ii) lies tangent to the surface S at point P_m.

The idea is therefore to use the area of the parallelogram T_m as an approximation to the area of the patch S_m. From Section 16.4, we recall that the area of the parallelogram T_m is equal to the length of the cross product of the vectors $\mathbf{u}_m$ and $\mathbf{v}_m$, or

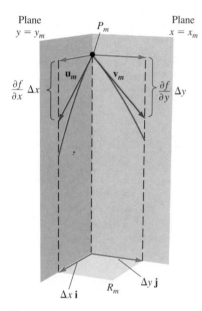

Figure 5.2 An enlarged view of the vectors $\mathbf{u}_m$ and $\mathbf{v}_m$ with the patch S_m removed. The blue curves are the traces of the graph of $z = f(x, y)$ in the planes $x = x_m$ and $y = y_m$.

$$\text{Area of } T_m = |\mathbf{u}_m \times \mathbf{v}_m|$$

$$= \left| \left(\Delta x\, \mathbf{i} + \frac{\partial f}{\partial x}(x_m, y_m)\, \Delta x\, \mathbf{k} \right) \times \left(\Delta y\, \mathbf{j} + \frac{\partial f}{\partial y}(x_m, y_m)\, \Delta y\, \mathbf{k} \right) \right|$$

$$= (\Delta x\, \Delta y) \left| \det \begin{bmatrix} \mathbf{i} & \mathbf{j} & \mathbf{k} \\ 1 & 0 & \dfrac{\partial f}{\partial x}(x_m, y_m) \\ 0 & 1 & \dfrac{\partial f}{\partial y}(x_m, y_m) \end{bmatrix} \right|$$

$$= (\Delta x\, \Delta y) \left| -\frac{\partial f}{\partial x}(x_m, y_m)\mathbf{i} - \frac{\partial f}{\partial y}(x_m, y_m)\mathbf{j} + \mathbf{k} \right|$$

$$= \sqrt{\left[\frac{\partial f}{\partial x}(x_m, y_m) \right]^2 + \left[\frac{\partial f}{\partial y}(x_m, y_m) \right]^2 + 1}\, \Delta x\, \Delta y.$$

Summing these approximations over all patches $S_1, S_2, \ldots, S_n$ gives the approximation to the area of S as

$$\text{Area of } S \approx \sum_{m=1}^{n} \sqrt{\left[\frac{\partial f}{\partial x}(x_m, y_m)\right]^2 + \left[\frac{\partial f}{\partial y}(x_m, y_m)\right]^2 + 1} \, \Delta x \, \Delta y.$$

If the partial derivatives $\partial f/\partial x$ and $\partial f/\partial y$ are continuous on Q, this Riemann sum converges to an integral as $n \to \infty$ and as $\Delta x \to 0$ and $\Delta y \to 0$. This integral provides our definition of surface area.

Definition 4

Let Q be a region in the plane that is either x-simple or y-simple, and let S be the graph of the function $z = f(x, y)$ for $(x, y) \in Q$. If $\partial f/\partial x$ and $\partial f/\partial y$ are continuous on Q, the **area of the surface S** is defined to be

$$A_S = \iint_Q \sqrt{\left[\frac{\partial f}{\partial x}(x, y)\right]^2 + \left[\frac{\partial f}{\partial y}(x, y)\right]^2 + 1} \, dA. \tag{3}$$

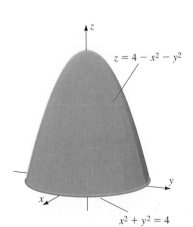

$z = f(x, y) = x^2$

Figure 5.3 The cylinder $z = x^2$ over the square $Q = \{(x, y) \mid -1 \le x \le 1, -1 \le y \le 1\}$.

□ **EXAMPLE 1**

Find the area of the surface that is the graph of the equation $f(x, y) = x^2$ lying above the rectangle $Q = \{(x, y) \mid -1 \le x \le 1, -1 \le y \le 1\}$.

Solution The surface is the generalized cylinder sketched in Figure 5.3. Using the method of integration by trigonometric substitution, we may evaluate the surface area integral (3) as

$$A_S = \iint_Q \sqrt{(2x)^2 + 1} \, dA$$

$$= \int_{-1}^{1} \int_{-1}^{1} \sqrt{4x^2 + 1} \, dx \, dy$$

$$= \int_{-1}^{1} \left[\tfrac{1}{2}x \sqrt{4x^2 + 1} + \tfrac{1}{4}\ln|2x + \sqrt{4x^2 + 1}|\right]_{x=-1}^{x=1} dy$$

$$= \int_{-1}^{1} (\sqrt{5} + \tfrac{1}{4}[\ln(2 + \sqrt{5}) - \ln(\sqrt{5} - 2)]) \, dy$$

$$= 2\sqrt{5} + \tfrac{1}{2}[\ln(2 + \sqrt{5}) - \ln(\sqrt{5} - 2)] \approx 5.92. \quad ∎$$

$z = 4 - x^2 - y^2$

$x^2 + y^2 = 4$

Figure 5.4 The portion of the paraboloid $z = 4 - x^2 - y^2$ that lies above the xy-plane.

□ **EXAMPLE 2**

Find the surface area of the portion of the paraboloid

$$z = 4 - x^2 - y^2$$

lying above the xy-plane.

Solution The paraboloid intersects the xy-plane in the circle $4 - x^2 - y^2 = 0$, or $x^2 + y^2 = 4$. (See Figure 5.4.) In xy-coordinates, this region may be described by the inequalities

$$-2 \le x \le 2, \qquad -\sqrt{4 - x^2} \le y \le \sqrt{4 - x^2}.$$

With $f(x, y) = 4 - x^2 - y^2$, we have

$$\frac{\partial f}{\partial x}(x, y) = -2x, \quad \text{and} \quad \frac{\partial f}{\partial y}(x, y) = -2y.$$

By (3) the integral giving the surface area is

$$A_S = \iint_Q \sqrt{4x^2 + 4y^2 + 1} \, dA = \int_{-2}^{2} \int_{-\sqrt{4-x^2}}^{\sqrt{4-x^2}} \sqrt{4x^2 + 4y^2 + 1} \, dy \, dx.$$

This integral is most easily evaluated in polar coordinates. With $x = r \cos \theta$ and $y = r \sin \theta$, and the region Q described as

$$0 \le \theta \le 2\pi, \qquad 0 \le r \le 2,$$

we have

$$A_S = \int_{0}^{2\pi} \int_{0}^{2} \sqrt{4(r \cos \theta)^2 + 4(r \sin \theta)^2 + 1} \, r \, dr \, d\theta$$

$$= \int_{0}^{2\pi} \int_{0}^{2} \sqrt{4r^2 + 1} \, r \, dr \, d\theta$$

$$= \int_{0}^{2\pi} \left[\tfrac{1}{12}(4r^2 + 1)^{3/2} \right]_{r=0}^{r=2} d\theta$$

$$= \frac{\pi}{6}(17^{3/2} - 1) \approx 36.13.$$ ∎

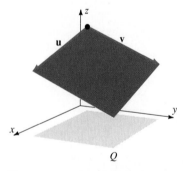

Figure 5.5 The portion of the plane that lies over the rectangle Q is a parallelogram with the vectors **u** and **v** as adjacent sides.

□ **EXAMPLE 3**

Show that Definition 4 of surface area agrees with the usual definition of area if S is a flat surface over a rectangle in the plane.

Solution If S is a flat surface in space, then S is a portion of a plane. Thus, we let

$$z = f(x, y) = Ax + By + C$$

be the equation for the plane that contains S, and we let

$$Q = \{(x, y) \mid a \le x \le b, c \le y \le d\}$$

be a rectangle in the xy-plane.

As in Figure 5.5, the portion of the plane $Ax + By + C$ lying over the rectangle Q is a parallelogram. From (1) and (2) we see that the vectors

$$\mathbf{u} = \Delta x \, \mathbf{i} + \frac{\partial f}{\partial x} \Delta x \, \mathbf{k} = (b - a)\mathbf{i} + A(b - a)\mathbf{k}$$

and

$$\mathbf{v} = \Delta y \, \mathbf{j} + \frac{\partial f}{\partial y} \Delta y \, \mathbf{k} = (d - c)\mathbf{j} + B(d - c)\mathbf{k},$$

when positioned at the vertex $(a, c, f(a, c))$, form two adjacent sides of the parallelogram. Using the vector methods in Section 16.4, we know that the area of the parallelogram is

$$A_S = |\mathbf{u} \times \mathbf{v}|$$

$$= \left| \det \begin{bmatrix} \mathbf{i} & \mathbf{j} & \mathbf{k} \\ b - a & 0 & A(b-a) \\ 0 & d - c & B(d-c) \end{bmatrix} \right|$$

$$= |-A(b-a)(d-c)\mathbf{i} - B(b-a)(d-c)\mathbf{j} + (b-a)(d-c)\mathbf{k}|$$

$$= (b-a)(d-c)\sqrt{A^2 + B^2 + 1}.$$

Using Definition 4 with $\partial f / \partial x = A$ and $\partial f / \partial y = B$ we find

$$A_S = \int_c^d \int_a^b \sqrt{A^2 + B^2 + 1} \, dx \, dy = (b-a)(d-c)\sqrt{A^2 + B^2 + 1}.$$

Thus, Definition 4 agrees with our usual concept of area for flat surfaces.   ■

REMARK   In Chapter 20, we shall find it useful to employ the differential version of the surface area formula. If dS represents the differential of surface area, then equation (3) can be expressed as

$$dS = \sqrt{\left[\frac{\partial f}{\partial x}\right]^2 + \left[\frac{\partial f}{\partial y}\right]^2 + 1} \, dA.$$

It is also useful to recall from Section 18.3 that a normal vector $\mathbf{N}$ to the graph of f is

$$\mathbf{N} = \frac{\partial f}{\partial x}\mathbf{i} + \frac{\partial f}{\partial y}\mathbf{j} - \mathbf{k}.$$

Note that the square root factor in the surface area formula equals $\|\mathbf{N}\|$. Consequently, the differential form of the surface area formula can be written as

$$dS = \|\mathbf{N}\| \, dA.$$

The general notion of calculating surface area by double integration will be put to important use in the next chapter.

Exercise Set 19.5

In Exercises 1–12, find the surface area of the graph of $z = f(x, y)$ above the region Q in the plane.

1. $f(x, y) = x + y + 6$, $\quad Q = \{(x, y) \mid 0 \le x \le 1, 0 \le y \le 1\}$

2. $f(x, y) = 9 - x + 2y$, $\quad Q = \{(x, y) \mid 0 \le x^2 + y^2 \le 1\}$

3. $f(x, y) = 9 - x^2 - y^2$, $\quad Q = \{(x, y) \mid 0 \le x^2 + y^2 \le 3\}$

4. $f(x, y) = 4 + y^2$, $\quad Q = \{(x, y) \mid 0 \le x \le 1, 0 \le y \le 2\}$

5. $f(x, y) = 2 - x - y$,
$\quad Q = \{(x, y) \mid 0 \le x \le 2, 0 \le y \le 2 - x\}$

6. $f(x, y) = 3 + y^2$, $\quad Q = \{(x, y) \mid 0 \le x \le 2, 0 \le y \le 2\}$

7. $f(x, y) = \sqrt{x^2 + y^2}$, $\quad Q = \{(x, y) \mid 1 \le x^2 + y^2 \le 4\}$

8. $f(x, y) = x + y^2$, $\quad Q = \{(x, y) \mid 0 \le x \le 1, 0 \le y \le 2\}$

9. $f(x, y) = \sqrt{3}\, y - x^2$, $\quad Q = \{(x, y) \mid 0 \le x \le 1, 0 \le y \le 1\}$

10. $f(x, y) = x^2 + y$, $\quad Q = \{(x, y) \mid 0 \le x \le 1, 0 \le y \le x\}$

11. $f(x, y) = 6 + \ln(\sec x)$,
$\quad Q = \{(x, y) \mid 0 \le x \le \pi/4, 0 \le y \le \sec^2 x\}$

12. $f(x, y) = \sqrt{x^2 - y^2}$,
$\quad Q = \{(x, y) \mid 1 \le x \le y, \sqrt{2}/2 \le y \le 1\}$

13. Find the surface area of the portion of the graph of $z = y + 2x^2$ over the triangular region with vertices $(0, 0)$, $(0, 1)$, and $(1, 1)$.

14. Find the area of the part of the plane $x + y + z = 4$ bounded by the cylinder $x^2 + y^2 = 4$.

15. Find the surface area of the portion of the paraboloid $z = 16 - x^2 - y^2$ lying between the planes $z = 4$ and $z = 9$.

16. Find the surface area of the part of the sphere $x^2 + y^2 + z^2 = 4$ lying above the plane $z = 1$.

17. Find the surface area of the part of the hemisphere $z = \sqrt{4 - x^2 - y^2}$ lying inside the cylinder $x^2 + y^2 = 1$.

18. Find the surface area of the part of the paraboloid $z = 4 - x^2 - y^2$ lying inside the cylinder $x^2 + y^2 = 1$.

19. Find the surface area of the portion of the hyperbolic paraboloid (saddle) $z = y^2 - x^2$ that lies inside the cylinder $x^2 + y^2 = 25$ and above the xy-plane.

20. Find the surface area of the portion of the cone $z^2 = x^2 + y^2$ that lies inside the sphere $x^2 + y^2 + z^2 = 8$.

21. Find the surface area of the portion of the cylinder $y^2 + z^2 = 1$ that lies inside the sphere $x^2 + y^2 + z^2 = 50$.

22. Find the area of the portion of the plane $z = x + 2$ that lies inside the surface $z^2 = 2(x^2 + y^2)$.

23. Use Simpson's Rule to approximate the surface area of the portion of the paraboloid $z = 3 - x^2 - y^2 + 2y$ lying above the plane $z = 2y + 2$.

24. Use Simpson's Rule to approximate the surface area of the portion of the sphere $x^2 + y^2 + z^2 - 4z = 0$ that lies above the cardioid $r = 1 + \cos\theta$ in the xy-plane.

25. Develop the formula for the surface area of a sphere using a double integral.

26. What relationship exists between the formula for the surface area of a solid of revolution and Definition 4?

27. Find the surface area of the portion of the surface $x = 1 - y^2$ that is bounded by the planes $y = 0$, $z = 0$, and $y + z = 1$.

28. Find the area of the portion of the plane $x = 4 - y - z$ that lies in the first octant.

29. Find the area of the portion of the surface $y = x^2 + z^2$ that lies in the first octant and is bounded by the planes $x = 0$ and $z = 0$ and by the cylinder $x^2 + z^2 = 4$. First set up the integral in rectangular coordinates and then convert it to polar coordinates.

30. Intersecting the sphere $x^2 + y^2 + z^2 = 9$ with the plane $x = 2$ splits the sphere into two parts. Find the surface area of the smaller part. First set up the integral in rectangular coordinates and then convert it to polar coordinates.

31. (A coordinate-free formula for surface area) Let $\mathbf{u}_m$ and $\mathbf{v}_m$ be the vectors in (1) and (2). Let ΔT_m be the area of the parallelogram tangent to S at P_m, over the rectangle R_m of area ΔA, as before.
a. Show that $\mathbf{N}_m = \mathbf{u}_m \times \mathbf{v}_m$ is normal to S at P_m.
b. Show that $\mathbf{N} \cdot \mathbf{k} = \Delta x \, \Delta y = \Delta A$.
c. Show that, also, $\mathbf{N} \cdot \mathbf{k} = |\mathbf{u}_m \times \mathbf{v}_m| \cos\theta$, where θ is the angle between $\mathbf{N}$ and $\mathbf{k}$.
d. Conclude from b and c that

$$\Delta T_m = |\mathbf{u}_m \times \mathbf{v}_m| = \frac{\Delta A}{\cos\theta} = \sec\theta \, \Delta A.$$

e. Conclude from d that $A_S = \iint\limits_{Q} \sec\theta \, dA.$

19.6 Triple Integrals

In this section, we define the triple integral for a continuous function of three independent variables. As we did for double integrals, we shall first carry out this development for special types of regions Q (namely, boxes) and then indicate how the concept extends to more general regions.

The Triple Integral Over a Box

Let Q be the box-shaped region in $\mathbb{R}^3$ defined by the inequalities

$$a \le x \le b, \qquad c \le y \le d, \qquad p \le z \le q.$$

(See Figure 6.1.) Let f be a continuous function defined on Q. By constructing planes perpendicular to the x-axis at $x_0, x_1, \ldots, x_n$, planes perpendicular to the y-axis at $y_0, y_1, \ldots, y_m$, and planes perpendicular to the z-axis at $z_0, z_1, \ldots, z_\ell$, we partition the box Q into smaller rectangular boxes Q_{ijk}, each of which has volume $\Delta V_{ijk} = \Delta x_i \, \Delta y_j \, \Delta z_k$ (Figure 6.2).

Next, we select one point (s_i, t_j, u_k) in each box Q_{ijk}, and we form the approximating sum

$$S_n = \sum_{i=1}^{n} \sum_{j=1}^{m} \sum_{k=1}^{\ell} f(s_i, t_j, u_k) \, \Delta V_{ijk}. \tag{1}$$

By analogy with the one- and two-variable cases, this approximating sum is called a Riemann sum for f on Q and the set of rectangular boxes constitutes a *partition P*

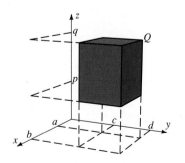

Figure 6.1 The rectangle box $Q = \{(x, y, z) \mid a \le x \le b, \; c \le y \le d, \; p \le z \le q\}$.

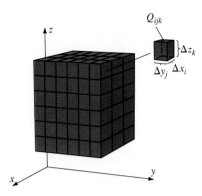

Figure 6.2 Partitioning the box Q into smaller boxes Q_{ijk} of volume $\Delta V_{ijk} = \Delta x_i \, \Delta y_j \, \Delta z_k$.

of Q. As in the one- and two-variable cases, if f is continuous on Q, this sum approaches a limit as $\|P\| \to 0$. This limit is defined to be the **triple integral** of f on the box Q:

$$\iiint_Q f(x, y, z) \, dV = \lim_{\|P\| \to 0} \sum_{i=1}^{n} \sum_{j=1}^{m} \sum_{k=1}^{\ell} f(s_i, t_j, u_k) \, \Delta V_{ijk}. \tag{2}$$

In the special case $f(x, y, z) \equiv 1$, we can give a geometric interpretation of the triple integral in (1). The terms in the approximating Riemann sum (1) are just

$$f(s_i, t_j, u_k) \, \Delta V_{ijk} = 1 \cdot \Delta V_{ijk} = \text{volume of } Q_{ijk}.$$

Since the union of the boxes Q_{ijk} is the box Q, it follows that

$$\iiint_Q 1 \, dV = \text{Volume of } Q. \tag{3}$$

That is, the triple integral of the function $f(x, y, z) \equiv 1$ over the box Q is just the volume of Q.

As for double integrals, triple integrals over boxes can be evaluated as iterated integrals. In particular, with Q as above, we have

$$\iiint_Q f(x, y, z) \, dV = \int_p^q \int_c^d \int_a^b f(x, y, z) \, dx \, dy \, dz. \tag{4}$$

In evaluating the iterated integral in (4), there is no reason why the first integration must be performed with respect to x. Since each of the variables ranges between constant limits, the order of integration can be any of the six possible orders xyz, xzy, yxz, yzx, zxy, or zyx.

We shall not prove equation (4), although it is easy to explain in the case $f(x, y, z) \equiv 1$. If v is any number in the z-interval $[p, q]$, the plane $z = v$ determines a rectangular cross section of Q of area

$$A(z) = \int_c^d \int_a^b 1 \, dx \, dy.$$

Thus,

$$\text{Volume of } Q = \int_p^q A(z) \, dz \tag{5}$$

$$= \int_p^q \left[\int_c^d \int_a^b 1 \, dx \, dy \right] dz$$

$$= \int_p^q \int_c^d \int_a^b 1 \, dx \, dy \, dz.$$

Combining equations (3) and (5) results in equation (4) in this special case.

☐ **EXAMPLE 1**

Evaluate the triple integral

$$\iiint\limits_{Q} xe^y \cos z \, dV$$

where Q is the box $\{(x, y, z) \mid 0 \leq x \leq 2, 0 \leq y \leq \ln 2, 0 \leq z \leq \pi/2\}$.

Solution Using equation (4), we find

$$\iiint\limits_{Q} xe^y \cos z \, dV = \int_0^{\pi/2} \int_0^{\ln 2} \int_0^2 xe^y \cos z \, dx \, dy \, dz$$

$$= \int_0^{\pi/2} \int_0^{\ln 2} \left[\frac{x^2}{2} e^y \cos z \right]_{x=0}^{x=2} dy \, dz$$

$$= \int_0^{\pi/2} \int_0^{\ln 2} 2e^y \cos z \, dy \, dz$$

$$= \int_0^{\pi/2} \left[2e^y \cos z \right]_{y=0}^{y=\ln 2} dz$$

$$= \int_0^{\pi/2} 2 \cos z \, dz$$

$$= \left[2 \sin z \right]_{z=0}^{z=\pi/2} = 2. \qquad \blacksquare$$

Triple Integrals Over More General Regions

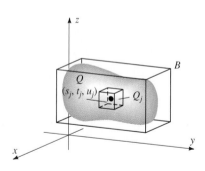

Figure 6.3 More general region Q enclosed by the box B.

If Q is a bounded region in space (not necessarily a box), we define the triple integral of f over Q as follows. First, we find a box B containing the region Q. (See Figure 6.3.) Next, we partition the box B into smaller rectangular boxes, just as we did when Q itself was a box, and we let $Q_1, Q_2, \ldots, Q_n$ be a list of all such rectangular boxes *lying entirely within the region Q*. For each such box we select a point $(s_j, t_j, u_j) \in Q_j$ and we denote by ΔV_j the volume of Q_j. The limit of this sequence of approximating sums, if it exists, is called the **triple integral** of f over Q:

$$\iiint\limits_{Q} f(x, y, z) \, dV = \lim_{\|P\| \to 0} \sum_{j=1}^{n} f(s_j, t_j, u_j) \, \Delta V_j. \qquad (6)$$

The triple integral in (6) exists if f is continuous on Q and Q is a "sufficiently nice" region in space. Rather than worry too much about the precise meaning of this last phrase, we shall state a theorem showing how triple integrals may be evaluated for regions of the type encountered in this text and in most applications. Before doing so, however, we need to make one observation concerning the approximating sum in equation (6). In the special case $f(x, y, z) \equiv 1$, the terms in the approximating sum are, as before, the volumes of the approximating boxes. The sum therefore approximates the volume of Q, and in the limit we obtain

$$\iiint\limits_{Q} 1 \, dV = \text{Volume of } Q \tag{7}$$

just as in the case when Q itself is a box.

Evaluating Triple Integrals

Intuitively, we say that a region Q in $\mathbb{R}^3$ is **z-simple** if every vertical line (that is, a line parallel to the z-axis) intersects the boundary of Q at most twice. More precisely, let Q' be the projection of the region Q into the xy-plane. Then the region Q is z-simple if there exist two continuous functions $g_1(x, y)$ and $g_2(x, y)$ defined on Q' such that

$$Q = \{(x, y, z) \mid (x, y) \in Q' \quad \text{and} \quad g_1(x, y) \le z \le g_2(x, y)\}.$$

(see Figure 6.4). We also define x-simple and y-simple regions Q in $\mathbb{R}^3$ by considering Q from the point of view of its projections in the yz- and xz-planes, respectively. The solid torus in Figure 6.5 is a region that is z-simple, but neither x-simple nor y-simple.

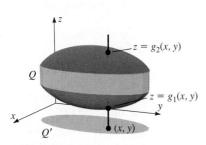

Figure 6.4   The solid region Q in $\mathbb{R}^3$ is z-simple.

Given a z-simple region Q, we can evaluate

$$\iiint\limits_{Q} f(x, y, z) \, dV$$

by considering the integral

$$\int_{g_1(x, y)}^{g_2(x, y)} f(x, y, z) \, dz$$

as a continuous function defined on the region Q' in the xy-plane. As such, we can integrate it over Q'. The resulting value equals the desired triple integral. That is,

$$\iiint\limits_{Q} f(x, y, z) \, dV = \iint\limits_{Q'} \left[\int_{g_1(x, y)}^{g_2(x, y)} f(x, y, z) \, dz \right] dA \tag{8}$$

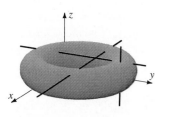

Figure 6.5   This solid torus is z-simple. However, it is neither x-simple nor y-simple.

where dA corresponds to area in the xy-plane. The right-hand side of equation (8) is evaluated by integrating the innermost integral. Then the resulting double integral is calculated using the methods discussed in Section 19.2. The following theorem is a more explicit version of equation (8) in the case where the region Q' is y-simple (considered as a region in the xy-plane).

Theorem 5

Suppose Q is a region in $\mathbb{R}^3$ described by the inequalities

$$a \le x \le b, \qquad h_1(x) \le y \le h_2(x), \qquad g_1(x, y) \le z \le g_2(x, y)$$

where h_1, h_2, g_1, and g_2 are continuous functions. If f is continuous on Q, then

$$\iiint\limits_{Q} f(x, y, z) \, dV = \int_{a}^{b} \int_{h_1(x)}^{h_2(x)} \int_{g_1(x, y)}^{g_2(x, y)} f(x, y, z) \, dz \, dy \, dx.$$

A proof of Theorem 5 requires a deeper treatment of multiple integrals than we have given here and is left to more advanced courses. Before proceeding to apply this result, we need to emphasize two points.

REMARK 1 Theorem 5 remains true if the roles of the independent variables are interchanged. For example, if the region Q is described by the inequalities

$$c \leq y \leq d, \qquad h_1(y) \leq z \leq h_2(y), \qquad g_1(y, z) \leq x \leq g_2(y, z),$$

then the iterated integral formula is

$$\iiint\limits_Q f(x, y, z)\, dV = \int_c^d \int_{h_1(y)}^{h_2(y)} \int_{g_1(y, z)}^{g_2(y, z)} f(x, y, z)\, dx\, dz\, dy.$$

REMARK 2 Theorem 5 may be paraphrased this way. To evaluate the triple integral

$$\iiint\limits_Q f(x, y, z)\, dV:$$

(i) Find the constants and/or functions that bound the region Q in each of the three directions corresponding to the coordinate axes. (Write these down!)

(ii) Evaluate the integral as an iterated integral, integrating first with respect to a variable whose bounds depend on the other one or two variables, integrating second with respect to a variable whose limits involve the remaining variable, and integrating last with respect to a variable whose limits involve only constants.

It is important to note that the outermost limits of integration are always constants, that the intermediate limits of integration can only depend on the outermost variable of integration, and that the innermost limits of integration depend only on the two outer variables of integration. For example, iterated integrals of the form

$$\int_a^b \int_{g_1(x, y)}^{g_2(x, y)} \int_{h_1(y)}^{h_2(y)} f(x, y, z)\, dz\, dy\, dx$$

and

$$\int_{h_1(x)}^{h_2(x)} \int_{g_1(x, y)}^{g_2(x, y)} \int_a^b f(x, y, z)\, dx\, dy\, dz$$

are nonsense.

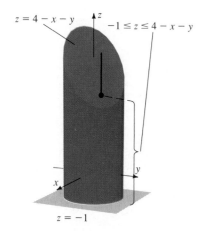

$z = 4 - x - y$
$-1 \leq z \leq 4 - x - y$
$z = -1$

Figure 6.6 The z-simple region Q in Example 2.

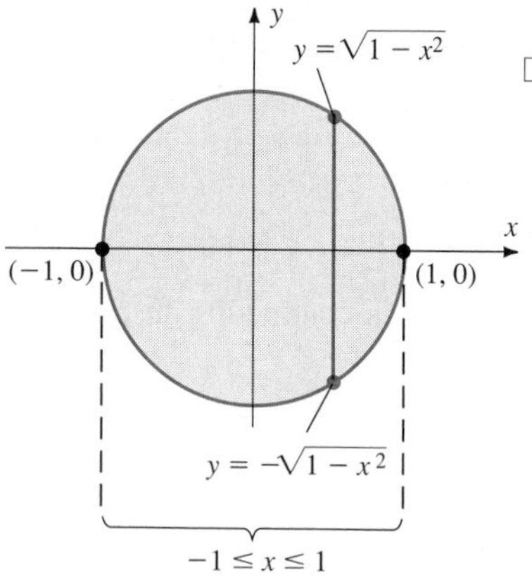

Figure 6.7 Projection of Q into the xy-plane.

$y = \sqrt{1 - x^2}$
$(-1, 0)$
$(1, 0)$
$y = -\sqrt{1 - x^2}$
$-1 \leq x \leq 1$

☐ **EXAMPLE 2**

Evaluate the triple integral

$$\iiint\limits_Q 2xy\, dV,$$

where Q is the region inside the cylinder $x^2 + y^2 = 1$ bounded above by the plane $x + y + z = 4$ and below by the plane $z = -1$.

Solution The region is sketched in Figure 6.6. Figure 6.7 shows the projection of Q into the xy-plane and illustrates how the inequalities involving x and y are obtained. The equation $x + y + z = 4$ gives $z = 4 - x - y$, so $-1 \leq z \leq 4 - x - y$. The region is therefore completely described by the inequalities

$$-1 \leq x \leq 1, \qquad -\sqrt{1 - x^2} \leq y \leq \sqrt{1 - x^2}, \qquad -1 \leq z \leq 4 - x - y.$$

Thus, according to Theorem 5,

$$\iiint\limits_{Q} 2xy \, dV = \int_{-1}^{1} \int_{-\sqrt{1-x^2}}^{\sqrt{1-x^2}} \int_{-1}^{4-x-y} 2xy \, dz \, dy \, dx$$

$$= \int_{-1}^{1} \int_{-\sqrt{4-x^2}}^{\sqrt{4-x^2}} \left[2xyz \right]_{z=-1}^{z=4-x-y} dy \, dx$$

$$= \int_{-1}^{1} \int_{-\sqrt{4-x^2}}^{\sqrt{4-x^2}} (10xy - 2x^2y - 2xy^2) \, dy \, dx$$

$$= \int_{-1}^{1} \left[5xy^2 - x^2y^2 - \tfrac{2}{3}xy^3 \right]_{y=-\sqrt{4-x^2}}^{y=\sqrt{4-x^2}} dx$$

$$= \int_{-1}^{1} -\tfrac{4}{3}x(4 - x^2)^{3/2} \, dx$$

$$= \left[\tfrac{4}{15}(4 - x^2)^{5/2} \right]_{x=-1}^{x=1} = 0.$$

(This result is explained by the fact that both the integrand and the region Q are symmetric with respect to the plane $y = x$.) ∎

REMARK 3 The integral in Example 2 could also have been evaluated as

$$\iiint\limits_{Q} 2xy \, dV = \int_{-1}^{1} \int_{-\sqrt{1-y^2}}^{\sqrt{1-y^2}} \int_{-1}^{4-x-y} 2xy \, dz \, dx \, dy.$$

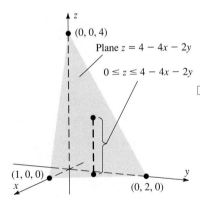

Plane $z = 4 - 4x - 2y$

$0 \le z \le 4 - 4x - 2y$

(0, 0, 4)

(1, 0, 0)

(0, 2, 0)

Figure 6.8 Tetrahedron of Example 3.

□ **EXAMPLE 3**

Evaluate the triple integral

$$\iiint\limits_{Q} (x - y + z) \, dV$$

where Q is the tetrahedron with vertices $(0, 0, 0)$, $(1, 0, 0)$, $(0, 2, 0)$, and $(0, 0, 4)$.

Solution The tetrahedron is sketched in Figure 6.8. Substituting the given points into the equation $z = Ax + By + C$ shows that the plane that bounds Q from above has equation $z = 4 - 4x - 2y$. As Figure 6.9 illustrates, the base of Q is a triangle in the xy-plane bounded by the lines $x = 0$, $y = 0$, and $y = -2x + 2$. The region Q may therefore be described by the inequalities

$$0 \le x \le 1, \qquad 0 \le y \le -2x + 2, \qquad 0 \le z \le 4 - 4x - 2y.$$

The integral is therefore

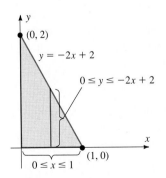

(0, 2)

$y = -2x + 2$

$0 \le y \le -2x + 2$

(1, 0)

$0 \le x \le 1$

Figure 6.9 Projection of tetrahedron into the xy-plane.

$$\iiint\limits_{Q} (x - y + z) \, dV = \int_{0}^{1} \int_{0}^{2-2x} \int_{0}^{4-4x-2y} (x - y + z) \, dz \, dy \, dx$$

$$= \int_{0}^{1} \int_{0}^{2-2x} \left[(x - y)z + \frac{z^2}{2} \right]_{z=0}^{z=4-4x-2y} dy \, dx$$

$$= \int_0^1 \int_0^{2-2x} (4x^2 - 12x + 10xy - 12y + 4y^2 + 8) \, dy \, dx$$

$$= \int_0^1 \left[(4x^2 - 12x + 8)y + 5xy^2 - 6y^2 + \frac{4y^3}{3} \right]_{y=0}^{y=2-2x} dx$$

$$= \int_0^1 (\tfrac{4}{3}x^3 - 4x + \tfrac{8}{3}) \, dx$$

$$= \left[\tfrac{1}{3}x^4 - 2x^2 + \tfrac{8}{3}x \right]_0^1 = 1. \qquad ∎$$

REMARK 4 The integral in Example 3 could also have been evaluated as

$$\int_0^2 \int_0^{1-y/2} \int_0^{4-4x-2y} (x - y + z) \, dz \, dx \, dy,$$

or as

$$\int_0^1 \int_0^{4-4x} \int_0^{2-2x-z/2} (x - y + z) \, dy \, dz \, dx,$$

or as one of three other iterated integrals (see Exercise 35).

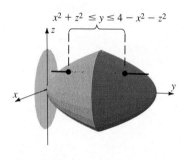

$x^2 + z^2 \le y \le 4 - x^2 - z^2$

Figure 6.10 The solid Q bounded by the paraboloids $y = x^2 + z^2$ and $y = 4 - x^2 - z^2$ is y-simple.

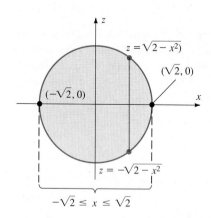

Figure 6.11 Projection Q' of Q into xz-plane.

□ **EXAMPLE 4**

Find the volume of the solid bounded by the paraboloids $y = 4 - x^2 - z^2$ and $y = x^2 + z^2$.

Solution The region is sketched in Figure 6.10. From the description of the region we can see that

$$x^2 + z^2 \le y \le 4 - x^2 - z^2. \qquad (9)$$

We therefore seek inequalities for the variables x and z. Equating the two expressions for y gives

$$4 - x^2 - z^2 = x^2 + z^2$$

or

$$x^2 + z^2 = 2.$$

This is the equation of the circle (in the plane $y = 2$) where the paraboloids intersect. The projection of this circle into the xz-plane is shown in Figure 6.11. From this sketch we see that the inequalities on x and on z are

$$-\sqrt{2} \le x \le \sqrt{2}, \qquad -\sqrt{2 - x^2} \le z \le \sqrt{2 - x^2}. \qquad (10)$$

Using statement (7) and inequalities (9) and (10), we obtain the desired volume as

$$V = \iiint_Q 1 \, dV = \int_{-\sqrt{2}}^{\sqrt{2}} \int_{-\sqrt{2-x^2}}^{\sqrt{2-x^2}} \int_{x^2+z^2}^{4-x^2-z^2} 1 \, dy \, dz \, dx$$

$$= \int_{-\sqrt{2}}^{\sqrt{2}} \int_{-\sqrt{2-x^2}}^{\sqrt{2-x^2}} \left[y \right]_{y=x^2+z^2}^{y=4-x^2-z^2} dz \, dx$$

$$= \int_{-\sqrt{2}}^{\sqrt{2}} \int_{-\sqrt{2-x^2}}^{\sqrt{2-x^2}} (4 - 2x^2 - 2z^2) \, dz \, dx$$

$$= \int_{-\sqrt{2}}^{\sqrt{2}} \left[(4 - 2x^2)z - \frac{2z^3}{3} \right]_{z=-\sqrt{2-x^2}}^{z=\sqrt{2-x^2}} dx$$

$$= \int_{-\sqrt{2}}^{\sqrt{2}} \tfrac{8}{3}(2 - x^2)^{3/2} \, dx = 4\pi. \tag{11}$$

(The integral in line (11) is evaluated by means of a trigonometric substitution.) ∎

REMARK 5   The triple integral in Example 4 could also have been evaluated as

$$\int_{-\sqrt{2}}^{\sqrt{2}} \int_{-\sqrt{2-z^2}}^{\sqrt{2-z^2}} \int_{x^2+z^2}^{4-x^2-z^2} 1 \, dy \, dx \, dz.$$

Density

The triple integral may also be used to calculate the mass of a solid object if we know the *density* of the material in units of mass per unit volume (such as grams/cm³) as a continuous *density function* ρ. If Q is an object with a density function ρ of this type, we may approximate the mass of Q by partitioning Q into approximating boxes $Q_1, Q_2, \dots, Q_n$, as before. If (s_j, t_j, u_j) is a point in Q_j, and if the volume ΔV_j of Q_j is small, then the quantity

$$M_j = \rho(s_j, t_j, u_j) \, \Delta V_j \qquad \text{(mass = density × volume)}$$

provides an approximation to the mass of the jth box Q_j. Summing these approximations over all boxes contained within Q gives the approximating sum

$$M \approx \sum_{j=1}^{n} M_j = \sum_{j=1}^{n} \rho(s_j, t_j, u_j) \, \Delta V_j.$$

Thus, the **mass** of Q is defined by the triple integral

$$M = \iiint_Q \rho(x, y, z) \, dV. \tag{12}$$

That is, mass is the integral of the density function over the region Q, just as in the one- and two-variable cases.

□ **EXAMPLE 5**

A small wedge has the shape of the region

$$Q = \{(x, y, z) \mid -1 \le x \le 1, 0 \le y \le 2, 0 \le z \le y\}.$$

Find its mass if the density at any point (x, y, z) is given by the density function $\rho(x, y, z) = 1 + x^2 + y^2$ grams/cm³ and the dimensions for Q are in centimeters (Figure 6.12).

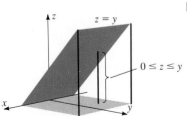

Figure 6.12 The wedge in Example 5.

Solution   According to equation (12) and Theorem 5, the mass is

$$M = \iiint\limits_{Q} (1 + x^2 + y^2) \, dV = \int_{-1}^{1} \int_{0}^{2} \int_{0}^{y} (1 + x^2 + y^2) \, dz \, dy \, dx$$

$$= \int_{-1}^{1} \int_{0}^{2} \left[(1 + x^2 + y^2)z \right]_{z=0}^{z=y} \, dy \, dx$$

$$= \int_{-1}^{1} \int_{0}^{2} (y + x^2 y + y^3) \, dy \, dx$$

$$= \int_{-1}^{1} \left[\frac{y^2}{2} + \frac{x^2 y^2}{2} + \frac{y^4}{4} \right]_{y=0}^{y=2} \, dx$$

$$= \int_{-1}^{1} (6 + 2x^2) \, dx$$

$$= \left[6x + \frac{2x^3}{3} \right]_{-1}^{1} = \frac{40}{3} \text{ grams.} \qquad \blacksquare$$

Moments and Center of Mass

Let Q denote a solid in $\mathbb{R}^3$, and let ρ be a continuous function giving the density $\rho(x, y, z)$ of Q at each point (x, y, z). As before, let Q be partitioned into rectangular boxes $Q_1, Q_2, \ldots, Q_n$ and let (s_j, t_j, u_j) be a point in $Q_j, j = 1, 2, \ldots, n$. Since s_j is the distance from the point (s_j, t_j, u_j) to the yz-plane, the product $s_j \rho(s_j, t_j, u_j) \, \Delta V_j$ may be interpreted as an approximation to the product of the mass of Q_j and the length of its "lever arm" extending from the yz-plane. By analogy with our previous discussions on moments (see Section 19.4), we define *the first moment, M_{yz}, of the solid Q with respect to the yz-plane* to be

$$M_{yz} = \iiint\limits_{Q} x\rho(x, y, z) \, dV = \lim_{n \to \infty} \sum_{j=1}^{n} s_j \rho(s_j, t_j, u_j) \, \Delta V_j.$$

Similarly, *the moments of Q with respect to the xz- and xy-planes* are

$$M_{xz} = \iiint\limits_{Q} y\rho(x, y, z) \, dV$$

and

$$M_{xy} = \iiint\limits_{Q} z\rho(x, y, z) \, dV.$$

Finally, if M denotes the mass of Q, the **center of mass** of Q is the point $(\bar{x}, \bar{y}, \bar{z})$ where

$$\bar{x} = \frac{M_{yz}}{M}; \qquad \bar{y} = \frac{M_{xz}}{M}; \qquad \bar{z} = \frac{M_{xy}}{M}.$$

□ **EXAMPLE 6**

Find the three first moments and the center of mass for the solid in Example 5.

Solution   With $\rho(x, y, z) = 1 + x^2 + y^2$ and Q as given in Example 5, we have

$$M_{yz} = \iiint_Q x(1 + x^2 + y^2) \, dV = \int_{-1}^{1} \int_{0}^{2} \int_{0}^{y} (x + x^3 + xy^2) \, dz \, dy \, dx$$

$$= \int_{-1}^{1} \int_{0}^{2} \Big[(x + x^3 + xy^2)z \Big]_{z=0}^{z=y} \, dy \, dx$$

$$= \int_{-1}^{1} \int_{0}^{2} [(x + x^3)y + xy^3] \, dy \, dx$$

$$= \int_{-1}^{1} \Big[(x + x^3)\frac{y^2}{2} + \frac{xy^4}{4} \Big]_{y=0}^{y=2} \, dx$$

$$= \int_{-1}^{1} (6x + 2x^3) \, dx$$

$$= \Big[3x^2 + \frac{x^4}{2} \Big]_{-1}^{1} = 0.$$

Similar calculations show that

$$M_{xz} = \iiint_Q y(1 + x^2 + y^2) \, dV$$

$$= \int_{-1}^{1} \int_{0}^{2} \int_{0}^{y} (y + x^2y + y^3) \, dz \, dy \, dx = \frac{896}{45}$$

and that

$$M_{xy} = \iiint_Q z(1 + x^2 + y^2) \, dV$$

$$= \int_{-1}^{1} \int_{0}^{2} \int_{0}^{y} (z + x^2z + y^2z) \, dz \, dy \, dx = \frac{448}{45}.$$

Since we have previously calculated the mass to be $M = 40/3$, the coordinates of the center of mass $(\bar{x}, \bar{y}, \bar{z})$ are

$$\bar{x} = \frac{M_{yz}}{M} = \frac{0}{(40/3)} = 0,$$

$$\bar{y} = \frac{M_{xz}}{M} = \frac{(896/45)}{(40/3)} = \frac{112}{75} \approx 1.49,$$

and

$$\bar{z} = \frac{M_{xy}}{M} = \frac{(448/45)}{(40/3)} = \frac{56}{75} \approx 0.75.$$ ■

Exercise Set 19.6

In Exercises 1–8, evaluate the iterated integral.

1. $\displaystyle\int_0^1 \int_0^1 \int_0^1 xyz \, dx \, dy \, dz$

2. $\displaystyle\int_0^2 \int_{-\pi/2}^{\pi/2} \int_1^2 xe^z \cos y \, dx \, dy \, dz$

3. $\displaystyle\int_0^1 \int_0^y \int_0^x 3 \, dz \, dx \, dy$

4. $\displaystyle\int_1^3 \int_0^{\pi/4} \int_0^x \cos(x+y) \, dy \, dx \, dz$

5. $\displaystyle\int_0^2 \int_0^x \int_0^{x+y} z \, dz \, dy \, dx$

6. $\displaystyle\int_{-1}^1 \int_{-\sqrt{1-x^2}}^{\sqrt{1-x^2}} \int_0^{\sqrt{1-x^2-y^2}} 1 \, dz \, dy \, dx$ (*Hint:* Use geometry.)

7. $\displaystyle\int_{-1}^1 \int_0^y \int_0^x ye^{x^2+y^2} \, dz \, dx \, dy$

8. $\displaystyle\int_0^2 \int_0^x \int_{x+y}^{x^2+y^2} 1 \, dz \, dy \, dx$

9. Interchange the order of integration in

$$\int_0^1 \int_0^{2x} \int_0^{x+y} f(x, y, z) \, dz \, dy \, dx$$

from order $dz \, dy \, dx$ to order $dz \, dx \, dy$.

10. Interchange the order of integration in

$$\int_{-2}^2 \int_{-\sqrt{4-x^2}}^{\sqrt{4-x^2}} \int_{-\sqrt{16-x^2-y^2}}^{\sqrt{16-x^2-y^2}} f(x, y, z) \, dz \, dy \, dx$$

from order $dz \, dy \, dx$ to order $dy \, dz \, dx$.

11. Interchange the order of integration in

$$\int_0^4 \int_{-\sqrt{z}}^{\sqrt{z}} \int_{-\sqrt{z-y^2}}^{\sqrt{z-y^2}} f(x, y, z) \, dx \, dy \, dz$$

from order $dx \, dy \, dz$ to order $dz \, dy \, dx$.

12. Use a triple integral to find the volume of the tetrahedron with vertices $(0, 0, 0)$, $(1, 0, 0)$, $(1, 1, 0)$, and $(1, 1, 1)$.

13. Sketch the solid whose volume is given by the iterated integral

$$\int_0^2 \int_0^{2x} \int_0^{x+y} dz \, dy \, dx.$$

14. Sketch the solid whose volume is given by the iterated integral

$$\int_{-1}^1 \int_{-\sqrt{1-x^2}}^{\sqrt{1-x^2}} \int_{\sqrt{x^2+y^2}}^{2-\sqrt{x^2+y^2}} dz \, dy \, dx.$$

15. Find the volume of the region in Exercise 14.

16. Find the volume of the tetrahedron bounded by the plane $x + y + z = 1$ and the coordinate planes $x = 0$, $y = 0$, and $z = 0$.

17. Find the volume of the region lying above the xy-plane, inside the cylinder $x^2 + y^2 = 9$, and below the plane $z = y + 3$.

18. Sketch the solid whose volume is given by the integral

$$V = \int_0^2 \int_0^{\sqrt{2x-x^2}} \int_0^{2-x} dz \, dy \, dx$$

and find the volume.

19. Evaluate the integral

$$\iiint_Q (3x + xz) \, dV$$

where Q is the region bounded by the cylinder $x^2 + z^2 = 9$, the plane $y + z = 3$, and the plane $y = 0$.

20. Sketch the solid bounded by the cone $x = y^2 + z^2$ and the plane $x = 3$. Find its volume using a triple integral.

21. Sketch the solid bounded by the plane $x = 1$ and the graph of $x = 9 - (y^2 + z^2)/4$. Find its volume using a triple integral.

22. Interchange the order of integration in the integral

$$\int_{-2}^2 \int_0^{\sqrt{4-x^2}} \int_{-\sqrt{4-x^2-y^2}}^{\sqrt{4-x^2-y^2}} f(x, y, z) \, dz \, dy \, dx$$

from $dz \, dy \, dx$ to $dy \, dz \, dx$.

23. Find the volume of the region bounded by the paraboloids $x = y^2 + z^2$ and $x = 2 - y^2 - z^2$.

24. Find the volume of the solid bounded above by the paraboloid $z = 2 - x^2 - y^2$ and below by the plane $z = 2 - 2x$.

25. Find the volume of the solid bounded by the graphs of $z = x^2$, $z = y$, and $y = 4$.

26. Find the volume of the solid bounded by the graphs of $x = z^3$, $y = 2x$, and $y = 4z^2$.

27. Let Q be the solid bounded by the cylinder $x^2 + y^2 = 9$ and the planes $z = 0$ and $x + z = 3$. Find the mass of Q if the density at each point (x, y, z) is given by the function $\rho(x, y, z) = z$.

28. Find the volume of the region common to the cylinders $x^2 + z^2 = 1$ and $y^2 + z^2 = 1$.

29. The density at each point of the box $Q = \{(x, y, z) \mid 0 \le x \le 1, 0 \le y \le 2, 0 \le z \le 2\}$ is proportional to the square of the distance from the origin. Find the mass.

30. Find the center of mass of the region in Exercise 16 if the density is constant.

31. Find the center of mass of the solid in Exercise 17 if the density is constant.

32. Find the center of mass of the solid in Exercise 27.

33. Find the center of mass of the part of the region in the first octant enclosed by the sphere $x^2 + y^2 + z^2 = 4$ if the density is constant.

34. Find the center of mass of the cube $Q = \{(x, y, z) \mid 0 \le x \le 1, 0 \le y \le 1, 0 \le z \le 1\}$ if the density function is $\rho(x, y, z) = xyz$.

35. Find five other iterated integrals by which the triple integral of Example 3 may be evaluated.

36. Find the volume, mass, and center of gravity for the solid in Example 5 if the density is uniform $\rho(x, y, z) = 1$ gram/cm^3. Compare your answers with those in Examples 5 and 6, and account for the differences in physical terms.

19.7 Triple Integrals in Cylindrical and Spherical Coordinates

The purpose of this section is to calculate the triple integral for functions written in cylindrical or spherical coordinates. In part, we want to know how to calculate volumes and masses for solids described in these coordinate systems. Another reason for studying these topics is that certain triple integrals, originally expressed in Cartesian coordinates, are more easily evaluated by changing to either cylindrical or spherical coordinates.

Triple Integrals in Cylindrical Coordinates

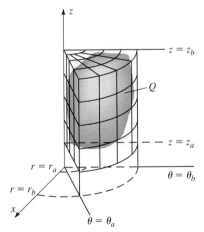

Figure 7.1 Partitioning the region Q using a grid specified in cylindrical coordinates.

Recall the relationship between cylindrical and Cartesian coordinates: If a point P has cylindrical coordinates $P = (r, \theta, z)$ and Cartesian coordinates $P = (x, y, z)$, then (r, θ) are the polar coordinates for the point (x, y) in the xy-plane (Figure 7.1). Now suppose that $\mathbb{R}^3$ is coordinatized by cylindrical coordinates, that Q is a region in $\mathbb{R}^3$, and that f is a continuous function defined on Q. We shall calculate the triple integral of f over Q in the usual way—by partitioning the region Q, forming an approximating sum, and obtaining the limit of the approximating sum.

Suppose that the region Q lies within the "cylindrical box" determined by the inequalities

$$r_a \le r \le r_b, \qquad \theta_a \le \theta \le \theta_b, \qquad z_a \le z \le z_b.$$

(See Figure 7.1.) We divide the region containing Q into small cylindrical blocks

$$Q_{ijk} = \{(r, \theta, z) \mid r_{i-1} \le r \le r_i, \theta_{j-1} \le \theta \le \theta_j, z_{k-1} \le z \le z_k\}$$

as illustrated in Figure 7.2. According to equation (3), Section 19.3, the area of the base of block Q_{ijk} is

$$\Delta A_{ijk} = \hat{r}_i \Delta r_i \Delta \theta_j$$

where $\hat{r}_i = (r_{i-1} + r_i)/2$. Since the block Q_{ijk} has height Δz_k, the volume of the block Q_{ijk} is

$$\Delta V_{ijk} = \hat{r}_i \Delta r_i \Delta \theta_j \Delta z_k. \qquad (1)$$

Now let θ_j^* be any number in the interval $[\theta_{j-1}, \theta_j]$ and let z_k^* be any number in the interval $[z_{k-1}, z_k]$. Then, since $\hat{r}_i \in [r_{i-1}, r_i]$, the point $(\hat{r}_i, \theta_j^*, z_k^*)$ lies in the block Q_{ijk}. The sum

$$\sum_i \sum_j \sum_k f(\hat{r}_i, \theta_j^*, z_k^*)\, \Delta V_{ijk} = \sum_i \sum_j \sum_k f(\hat{r}_i, \theta_j^*, z_k^*) r_i \Delta r_i \Delta \theta_j \Delta z$$

is the approximating Riemann sum for f over Q, where the su~
boxes Q_{ijk} lying entirely within Q. If the sizes of all of the

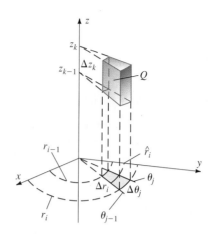

Figure 7.2 Volume of circular wedge is $\Delta V_{ijk} = \hat{r}_i \, \Delta r_i \, \Delta \theta_j \, \Delta z_k$ where $\hat{r}_i = (r_{i-1} + r_i)/2$.

(written $\|P\| \to 0$, as before), if f is continuous, and if Q is as described in Theorem 6, then the limit of this Riemann sum is the triple integral of f over Q:

$$\iiint\limits_Q f(r, \theta, z) \, dV = \lim_{\|P\| \to 0} \sum_i \sum_j \sum_k f(\hat{r}_i, \theta_j^*, z_k^*) r_i \Delta r_i \Delta \theta_j \Delta z_k. \qquad (3)$$

The following theorem shows how this triple integral may be evaluated for the types of regions encountered in most applications.

Theorem 6

Let Q be a region in $\mathbb{R}^3$ of the form

$$Q = \{(r, \theta, z) \mid a \le \theta \le b, h_1(\theta) \le r \le h_2(\theta), g_1(r, \theta) \le z \le g_2(r, \theta)\}$$

where g_1, g_2, h_1, and h_2 are continuous functions. Let f be continuous on Q. Then

$$\iiint\limits_Q f(r, \theta, z) \, dV = \int_a^b \int_{h_1(\theta)}^{h_2(\theta)} \int_{g_1(r, \theta)}^{g_2(r, \theta)} f(r, \theta, z) \, r \, dz \, dr \, d\theta. \qquad (4)$$

REMARK 1 Note the factor r appearing in the iterated integral in (4). The reason for its appearance is the same as in Section 19.3: it can be considered a result of converting dV to cylindrical coordinates. It arises from the factor r_i in equation (1).

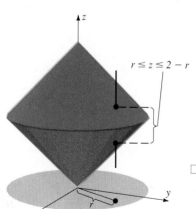

$r \le z \le 2 - r$

re 7.3 The solid "top" Q in Exam-

☐ **EXAMPLE 1**

A solid "top" has the shape of the region

$$Q = \{(r, \theta, z) \mid 0 \le r \le 1, 0 \le \theta \le 2\pi, r \le z \le 2 - r\}$$

as illustrated in Figure 7.3. Find the mass of the object if the density at any point is proportional to the distance from the z-axis.

Solution   The density function described here is $\rho(r, \theta, z) = \lambda r$, where λ is the constant of proportionality. By equation (12), Section 19.6, and Theorem 6, we have

$$\text{Mass} = \iiint\limits_{Q} \rho(r, \theta, z) \, dV = \iiint\limits_{Q} \lambda r \, dV$$

$$= \int_0^{2\pi} \int_0^1 \int_r^{2-r} \lambda r^2 \, dz \, dr \, d\theta \qquad \text{— note extra factor } r$$

$$= \int_0^{2\pi} \int_0^1 \left[\lambda r^2 z \right]_{z=r}^{z=2-r} dr \, d\theta$$

$$= \int_0^{2\pi} \int_0^1 (2\lambda r^2 - 2\lambda r^3) \, dr \, d\theta$$

$$= \int_0^{2\pi} \left[\frac{2\lambda r^3}{3} - \frac{\lambda r^4}{2} \right]_{r=0}^{r=1} d\theta$$

$$= \int_0^{2\pi} \left(\frac{\lambda}{6} \right) d\theta = \frac{\pi \lambda}{3}. \qquad \blacksquare$$

Changing to Cylindrical Coordinates

Sometimes a triple integral written in Cartesian coordinates is more easily evaluated by first changing to cylindrical coordinates. Doing so is analogous to changing a double integral from Cartesian to polar coordinates. Specifically, to write the iterated integral

$$\int_a^b \int_{h_1(x)}^{h_2(x)} \int_{g_1(x, y)}^{g_2(x, y)} f(x, y, z) \, dz \, dy \, dx$$

in cylindrical coordinates, we do the following:

(i) Express the region

$$Q = \{(x, y, z) \,|\, a \le x \le b, h_1(x) \le y \le h_2(x), g_1(x, y) \le z \le g_2(x, y)\}$$

in cylindrical coordinates as

$$Q = \{(r, \theta, z) \,|\, c \le \theta \le d, h_3(\theta) \le r \le h_4(\theta), g_3(r, \theta) \le z \le g_4(r, \theta)\}.$$

(ii) Using the substitutions $x = r \cos \theta$ and $y = r \sin \theta$, replace the integrand $f(x, y, z)$ by $f(r \cos \theta, r \sin \theta, z) \, r$. (Do not forget the factor r.)

(iii) Obtain the equation

$$\int_a^b \int_{h_1(x)}^{h_2(x)} \int_{g_1(x, y)}^{g_2(x, y)} f(x, y, z) \, dz \, dy \, dx$$

$$= \int_c^d \int_{h_3(\theta)}^{h_4(\theta)} \int_{g_3(r, \theta)}^{g_4(r, \theta)} f(r \cos \theta, r \sin \theta, z) \, r \, dz \, dr \, d\theta. \qquad (5)$$

Equation (5) is verified by comparing Theorems 5 and 6, each of which expresses the triple integral

$$\iiint\limits_{Q} f \, dV$$

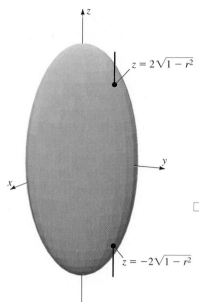

Figure 7.4

as one of the two iterated integrals in (5). One way to paraphrase equation (5) is to say that *in changing from Cartesian coordinates to cylindrical coordinates, the* **volume element**

$$dV = dz\, dy\, dx$$

in Cartesian coordinates is replaced by the volume element

$$dV = r\, dz\, dr\, d\theta \tag{6}$$

in cylindrical coordinates. It is very important to note the extra factor r that appears in the integrand on the right-hand side of (5) and in equation (6).

□ **EXAMPLE 2**

Calculate the volume of the ellipsoid

$$4x^2 + 4y^2 + z^2 = 4.$$

Solution   The ellipsoid is sketched in Figure 7.4. Since the ellipsoid is symmetric with respect to the xy-plane, we may calculate the volume as twice the volume of the region Q lying above the xy-plane. Since this region is described by the inequalities

$$0 \le z \le 2\sqrt{1 - x^2 - y^2}, \qquad -\sqrt{1 - x^2} \le y \le \sqrt{1 - x^2}, \qquad -1 \le x \le 1,$$

the volume is given by the iterated integral

$$V = 2 \int_{-1}^{1} \int_{-\sqrt{1-x^2}}^{\sqrt{1-x^2}} \int_{0}^{2\sqrt{1-x^2-y^2}} 1\, dz\, dy\, dx. \tag{7}$$

Clearly, this integral will be difficult to evaluate, so we try switching to cylindrical coordinates. In cylindrical coordinates the region Q is described by the inequalities

$$0 \le r \le 1, \qquad 0 \le \theta \le 2\pi, \qquad 0 \le z \le 2\sqrt{1 - r^2}$$

since $2\sqrt{1 - x^2 - y^2} = 2\sqrt{1 - (x^2 + y^2)} = 2\sqrt{1 - r^2}$. Using equation (5) we may rewrite the integral (7) as

$$V = 2 \int_{0}^{2\pi} \int_{0}^{1} \int_{0}^{2\sqrt{1-r^2}} r\, dz\, dr\, d\theta \underset{\text{\small note the extra factor } r}{\underbrace{}}$$

$$= 2 \int_{0}^{2\pi} \int_{0}^{1} \Big[zr \Big]_{z=0}^{z=2\sqrt{1-r^2}} dr\, d\theta$$

$$= 2 \int_{0}^{2\pi} \int_{0}^{1} 2r\sqrt{1 - r^2}\, dr\, d\theta$$

$$= 2 \int_{0}^{2\pi} \Big[-\tfrac{2}{3}(1 - r^2)^{3/2} \Big]_{r=0}^{r=1} d\theta = \frac{8\pi}{3}. \qquad ■$$

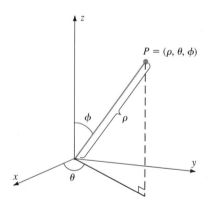

Figure 7.5   Spherical coordinates (ρ, θ, ϕ) for P.

Spherical Coordinates

Recall that the point P has spherical coordinates $P = (\rho, \theta, \phi)$ if $\rho \ge 0$ is its distance from the origin, θ is its rotation angle (as in cylindrical coordinates), and ϕ is its angle of inclination from the vertical (see Figure 7.5). Figure 7.6 illustrates the "spherical block" determined by the inequalities

$$\rho_{i-1} \le \rho \le \rho_i, \qquad \theta_{j-1} \le \theta \le \theta_j, \qquad \phi_{k-1} \le \phi \le \phi_k \tag{8}$$

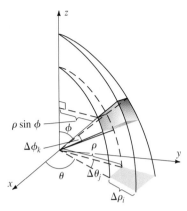

Figure 7.6 The spherical block.

where $\Delta\rho_i = \rho_i - \rho_{i-1}$, $\Delta\theta_j = \theta_j - \theta_{j-1}$, and $\Delta\phi_k = \phi_k - \phi_{k-1}$. A fact, which we shall not prove, is that the volume of this spherical block is given by

$$\Delta V_{ijk} = \rho_i^{*2} \sin \phi_k^* \, \Delta\rho_i \Delta\theta_j \Delta\phi_k \qquad (9)$$

for some $\rho_i^* \in [\rho_{i-1}, \rho_i]$ and $\phi_k^* \in [\phi_{k-1}, \phi_k]$ (see Exercise 39).

In order to calculate the integral of a continuous function f defined on Q, we select one point $(\rho_i^*, \theta_j^*, \phi_k^*)$ in each block Q_{ijk}, form the approximating sum, and evaluate the limit of this sum as the sizes of the blocks approach zero. We obtain the triple integral of f over Q:

$$\iiint\limits_Q f(\rho, \theta, \phi) \, dV$$

$$= \lim_{n\to\infty} \sum_i \sum_j \sum_k f(\rho_i^*, \theta_j^*, \phi_k^*)\rho_i^{*2} \sin \phi_k^* \, \Delta\rho_i \Delta\theta_j \Delta\phi_k.$$

This representation of the triple integral leads to the following theorem regarding its evaluation as an iterated integral.

Theorem 7

Let Q be a region in $\mathbb{R}^3$ of the form

$$Q = \{(\rho, \theta, \phi) \mid a \le \theta \le b, h_1(\theta) \le \phi \le h_2(\theta), g_1(\theta, \phi) \le \rho \le g_2(\theta, \phi)\}$$

where h_1, h_2, g_1, and g_2 are continuous functions. Let f be continuous on Q. Then,

$$\iiint\limits_Q f(\rho, \theta, \phi) \, dV = \int_a^b \int_{h_1(\theta)}^{h_2(\theta)} \int_{g_1(\theta, \phi)}^{g_2(\theta, \phi)} f(\rho, \theta, \phi) \, \rho^2 \sin \phi \, d\rho \, d\phi \, d\theta.$$

REMARK 2 Note the factor of $\rho^2 \sin \phi$ in the integrand of the iterated integral. In other words, when we convert from Cartesian to spherical coordinates, the volume element $dV = dz \, dy \, dx$ is replaced by $dV = \rho^2 \sin \phi \, d\rho \, d\theta \, d\phi$. Equation (9) is the key ingredient in the proof of this fact. See Section 19.8 for more details.

☐ **EXAMPLE 3**

Find the mass of the solid occupying the region

$$Q = \{(\rho, \theta, \phi) \mid 0 \le \rho \le 2, 0 \le \theta \le 2\pi, 0 \le \phi \le \pi/6\}$$

if the density at any point is proportional to its distance from the origin.

Solution The region Q is the "ice cream cone" region sketched in Figure 7.7. As before, we obtain the mass of the solid by integrating the density function $D(\rho, \theta, \phi) = \lambda\rho$. (Note that we are now using ρ for radial distance, not density.) Thus, by Theorem 7

$$\text{Mass} = \iiint\limits_Q \lambda\rho \, dV = \int_0^{2\pi} \int_0^{\pi/6} \int_0^2 (\lambda\rho)(\rho^2 \sin \phi) \, d\rho \, d\phi \, d\theta$$

note extra factor
$\rho^2 \sin \phi$

$$= \int_0^{2\pi} \int_0^{\pi/6} \int_0^2 \lambda\rho^3 \sin \phi \, d\rho \, d\phi \, d\theta$$

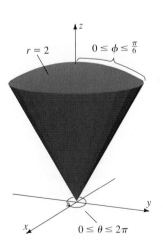

Figure 7.7 The "ice cream cone" region of Example 3.

$r = 2$ $0 \le \phi \le \dfrac{\pi}{6}$

$0 \le \theta \le 2\pi$

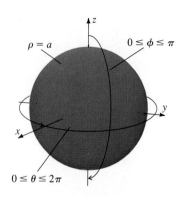

Figure 7.8 The solid sphere of radius a is determined by the inequalities

$$0 \le \rho \le a,$$
$$0 \le \phi \le \pi,$$
$$0 \le \theta \le 2\pi.$$

$$= \int_0^{2\pi} \int_0^{\pi/6} \left[\lambda \frac{\rho^4}{4} \sin \phi \right]_{\rho=0}^{\rho=2} d\phi \, d\theta$$

$$= \int_0^{2\pi} \int_0^{\pi/6} 4\lambda \sin \phi \, d\phi \, d\theta$$

$$= \int_0^{2\pi} \left[-4\lambda \cos \phi \right]_{\phi=0}^{\phi=\pi/6} d\theta$$

$$= \int_0^{2\pi} 4\lambda(1 - \sqrt{3}/2) \, d\theta$$

$$= 8\lambda\pi(1 - \sqrt{3}/2). \qquad \blacksquare$$

□ **EXAMPLE 4**

Use spherical coordinates to obtain the formula for the volume of a sphere of radius a.

Solution   The sphere of radius a is described by the spherical inequalities

$$0 \le \rho \le a, \qquad 0 \le \phi \le \pi, \qquad 0 \le \theta \le 2\pi \qquad \text{(Figure 7.8)}.$$

Thus

$$\text{Volume} = \iiint_Q 1 \, dV = \int_0^{2\pi} \int_0^{\pi} \int_0^{a} (\rho^2 \sin \phi) \, d\rho \, d\phi \, d\theta$$

note extra factor $\rho^2 \sin \phi$

$$= \int_0^{2\pi} \int_0^{\pi} \left[\frac{\rho^3}{3} \sin \phi \right]_{\rho=0}^{\rho=a} d\phi \, d\theta$$

$$= \int_0^{2\pi} \int_0^{\pi} \frac{a^3}{3} \sin \phi \, d\phi \, d\theta$$

$$= \int_0^{2\pi} \left[-\frac{a^3}{3} \cos \phi \right]_{\phi=0}^{\phi=\pi} d\theta$$

$$= \int_0^{2\pi} \frac{2a^3}{3} \, d\theta = \frac{4}{3} \pi a^3. \qquad \blacksquare$$

Changing From Cartesian to Spherical Coordinates

Given a function f written in Cartesian coordinates for which we wish to calculate a triple integral over a spherical region, we rewrite the function in spherical coordinates using the substitutions

$$x = \rho \sin \phi \cos \theta, \qquad y = \rho \sin \phi \sin \theta, \qquad z = \rho \cos \phi.$$

(See Section 17.7 for a derivation of these equations.) We shall not write out a general formula for converting triple integrals from Cartesian coordinates to spherical coordinates since the situations in which such changes are possible are limited almost exclusively to functions involving the expression

$$x^2 + y^2 + z^2 = \rho^2 \sin^2 \phi \cos^2 \theta + \rho^2 \sin^2 \phi \sin^2 \theta + \rho^2 \cos^2 \phi$$
$$= \rho^2 \sin^2 \phi(\cos^2 \theta + \sin^2 \theta) + \rho^2 \cos^2 \phi$$
$$= \rho^2.$$

☐ **EXAMPLE 5**

A hemispherical solid is described in Cartesian coordinates as

$$Q = \{(x, y, z) \mid 0 \le x^2 + y^2 \le 1, 0 \le z \le \sqrt{1 - x^2 - y^2}\}.$$

Find the mass of the object if the density at the point (x, y, z) is given by the function

$$D(x, y, z) = e^{-(x^2 + y^2 + z^2)^{3/2}}.$$

Solution   The hemisphere may be described by the inequalities

$$0 \le \rho \le 1, \qquad 0 \le \phi \le \pi/2, \qquad 0 \le \theta \le 2\pi,$$

and the density function, in spherical coordinates, is

$$D(\rho, \theta, \phi) = e^{-\rho^3}.$$

Thus,

$$\text{Mass} = \int_0^{2\pi} \int_0^{\pi/2} \int_0^1 e^{-\rho^3} \underbrace{\rho^2 \sin \phi}_{} \, d\rho \, d\phi \, d\theta \qquad \text{note extra factor } \rho^2 \sin \phi$$

$$= \int_0^{2\pi} \int_0^{\pi/2} \left[-\frac{e^{-\rho^3}}{3} \sin \phi \right]_{\rho=0}^{\rho=1} d\phi \, d\theta$$

$$= \int_0^{2\pi} \int_0^{\pi/2} \frac{1}{3} \left(1 - \frac{1}{e} \right) \sin \phi \, d\phi \, d\theta$$

$$= \int_0^{2\pi} \left[\left(-\frac{e-1}{3e} \right) \cos \phi \right]_{\phi=0}^{\phi=\pi/2} d\theta$$

$$= \int_0^{2\pi} \left(\frac{e-1}{3e} \right) d\theta$$

$$= \frac{2\pi(e-1)}{3e}.$$

Exercise Set 19.7

In Exercises 1–10, use cylindrical coordinates.

1. Find the volume of the solid bounded by the graphs of $z = x^2 + y^2$ and $z = 9$.

2. Find the volume of the solid bounded by the graphs of the equations $z = 9 - x^2 - y^2$, $x^2 + y^2 = 3$, and $z = 9$.

3. Find the center of mass of the solid in Exercise 1 if the density is constant.

4. Find the volume of the region that lies inside both the cylinder $x^2 + y^2 = 1$ and the sphere $x^2 + y^2 + z^2 = 4$.

5. A solid is bounded by the cylinder $y^2 + z^2 = 4$ and the planes $x = 0$ and $x = 4$. Find its mass if the density at each point is proportional to the distance of that point from the central axis of the cylinder.

6. Find the volume of the region that lies inside the cylinder $x^2 + y^2 = 1$, above the xy-plane, and below the hyperboloid of two sheets $x^2 + y^2 - z^2 = -1$.

7. Find the volume of the solid remaining if the region inside the cylinder $r = 3 \sin \theta$ is removed from the solid region bounded by the sphere $x^2 + y^2 + z^2 = 9$.

8. Find the volume of the region lying inside both the cylinder $x^2 - 2x + y^2 = 0$ and the sphere $x^2 + y^2 + z^2 = 4$.

9. Find the volume of the solid bounded above and below by the cone $z^2 = 2x^2 + 2y^2$ and on the sides by the cylinder $x^2 + y^2 - 4y = 0$.

10. Find the volume of the solid bounded above by the plane $z = x$ and below by the paraboloid $z = x^2 + y^2$.

In Exercises 11–15, evaluate the iterated integral by first changing to cylindrical coordinates.

11. $\int_{-1}^{1} \int_{-\sqrt{1-x^2}}^{\sqrt{1-x^2}} \int_{\sqrt{x^2+y^2}}^{1} x^2 \, dz \, dy \, dx$

12. $\int_{-2}^{2} \int_{-\sqrt{4-x^2}}^{\sqrt{4-x^2}} \int_{0}^{y+2} xy \, dz \, dy \, dx$

13. $\int_{-1}^{1} \int_{0}^{\sqrt{1-x^2}} \int_{x^2+y^2}^{1} z \, dz \, dy \, dx$

14. $\int_{0}^{2} \int_{-\sqrt{2x-x^2}}^{\sqrt{2x-x^2}} \int_{0}^{\sqrt{x^2+y^2}} 1 \, dz \, dy \, dx$

15. $\int_{0}^{3} \int_{0}^{\sqrt{3y-y^2}} \int_{\sqrt{x^2+y^2}}^{4} 2 \, dz \, dx \, dy$

16. Sketch the solid whose mass is given by

$$8 \int_{0}^{1} \int_{0}^{\sqrt{1-x^2}} \int_{0}^{\sqrt{x^2+y^2}} D(x, y, z) \, dz \, dy \, dx.$$

Find the mass if the density D is given by $D(x, y, z) = \sqrt{x^2 + y^2}$.

In Exercises 17–24, use spherical coordinates.

17. Find the volume of the solid that lies between the spheres $x^2 + y^2 + z^2 = r^2$ and $x^2 + y^2 + z^2 = R^2$ (where $r < R$).

18. Find the mass of a sphere of radius $r = 2$ if the density at each point is proportional to the square of the distance from the center.

19. Find the volume of the solid bounded above by the sphere $x^2 + y^2 + z^2 - 2z = 0$ and below by the cone $z = \sqrt{x^2 + y^2}$.

20. Find the mass and the center of mass of a hemisphere of radius 1 whose density at each point is proportional to its distance from the base.

21. Find the mass of the solid lying between the spheres $x^2 + y^2 + z^2 = 1$ and $x^2 + y^2 + z^2 = 4$ if the density at each point is proportional to the reciprocal of the distance from the center of the spheres.

22. Find the volume of the solid remaining if the region lying inside the cone $z^2 = x^2 + y^2$ is removed from the region bounded by the sphere $x^2 + y^2 + z^2 = 4$.

23. Find the center of mass of the quarter sphere

$$Q = \{(\rho, \theta, \phi) \mid 0 \le \rho \le 1, 0 \le \phi \le \pi/2, 0 \le \theta \le \pi\}$$

if the density is constant.

24. Use spherical coordinates to obtain the formula for the volume of a right circular cone of radius r and height h.

In Exercises 25–28, evaluate the iterated integral by first changing to spherical coordinates.

25. $\int_{-1}^{1} \int_{-\sqrt{1-x^2}}^{\sqrt{1-x^2}} \int_{0}^{\sqrt{1-x^2-y^2}} 3 \, dz \, dy \, dx$

26. $\int_{-2}^{2} \int_{-\sqrt{4-x^2}}^{\sqrt{4-x^2}} \int_{-\sqrt{4-x^2-y^2}}^{\sqrt{4-x^2-y^2}} (x^2 + y^2) \, dz \, dy \, dx$

27. $\int_{0}^{1} \int_{0}^{\sqrt{1-x^2}} \int_{\sqrt{x^2+y^2}}^{\sqrt{2-x^2-y^2}} \sqrt{x^2 + y^2 + z^2} \, dz \, dy \, dx$

28. $\int_{0}^{3} \int_{-\sqrt{9-x^2}}^{0} \int_{\sqrt{x^2+y^2}}^{\sqrt{18-x^2-y^2}} e^{(x^2+y^2+z^2)^{3/2}} \, dz \, dy \, dx$

29. Two circular cylinders of radius R meet at right angles. Use triple integration to find the volume of the solid common to both cylinders.

30. Find the volume of the region lying inside the sphere $x^2 + y^2 + z^2 = 4$ and outside the cylinder $x^2 + z^2 = 1$.

31. Evaluate the triple integral

$$\iiint\limits_{Q} \frac{z}{(x^2 + y^2)^{3/2}} \, dV$$

where Q is the region

$$Q = \{(x, y, z) \mid 1 \le x^2 + y^2 \le 3, 0 \le z \le 3\}.$$

32. Evaluate the triple integral

$$\iiint\limits_{Q} \cos \pi y \sqrt{x^2 + z^2} \, dV$$

where Q is the region bounded by the cylinders $x^2 + z^2 = 1$ and $x^2 + z^2 = 4$, and the planes $y = -1$ and $y = 2$.

33. Evaluate the triple integral

$$\iiint\limits_{Q} \sqrt{\frac{x}{y^2 + z^2}} \, dV$$

where Q is the region bounded by the cone $y^2 + z^2 = x^2$, the cylinder $y^2 + z^2 = 4$, and the planes $x = 0$ and $x = 2$.

34. Find the volume of the solid that remains if the cone $3y^2 = x^2 + z^2$ is removed from the sphere $x^2 + y^2 + z^2 = 4$.

35. Find the volume of the smaller of the two parts of the sphere $\rho = 4$ determined by the plane $y = 2$.

36. Evaluate the integral

$$\iiint\limits_{Q} (x^2 + y^2) \, dV$$

where Q is the sphere $x^2 + y^2 + z^2 \le 4$.

37. Find the volume of the solid lying between the spheres $x^2 + y^2 + z^2 = 1$ and $x^2 + y^2 + z^2 = 9$ and inside the cone $y^2 = x^2 + z^2$.

38. Find the volume of the solid bounded below by the graph of $x^2 + y^2 + z^2 + 2z = 0$ and above by the graph of $z^2 = x^4 + 2x^2y^2 + y^4$.

39. Obtain the approximation $\Delta V_{ijk} \approx \rho_i^2 \sin \phi_k \, \Delta\rho_i \, \Delta\theta_j \, \Delta\phi_k$ as follows: Assume that the spherical block in Figure 7.9 is

nearly a parallelepiped. We therefore approximate its volume by the product

$$(\text{length of } PQ)(\text{length of } PS)(\text{length of } PR).$$

a. Show that the length of the arc PS is $\rho_i \, \Delta\phi_k$. (*Hint:* It lies on a circle of radius ρ_i.)
b. Show that the length of the arc PR is $\rho_i \sin \phi_k \, \Delta\theta_j$. (*Hint:* It lies on a horizontal circle of radius $\rho_i \sin \phi_k$. Why?)
c. Derive the desired approximation.

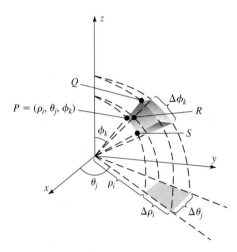

Figure 7.9 Spherical block of volume ΔV_{ijk}.

19.8 Change of Variables and Jacobians

When we evaluate integrals involving functions of one variable, we frequently substitute variables; that is, we use the equality

$$\int_a^b f(g(x)) \, g'(x) \, dx = \int_{u_a}^{u_b} f(u) \, du \tag{1}$$

where $u = g(x)$ (Theorem 9 in Section 6.4). In this chapter, we have also simplified the evaluation of multiple integrals by changing variables. For example, in Section 19.3, we evaluated certain double integrals by changing from rectangular to polar coordinates using the change of variables formula

$$\iint_Q f(x, y) \, dA = \int_a^b \int_{g_1(\theta)}^{g_2(\theta)} f(r \cos \theta, r \sin \theta) \, r \, dr \, d\theta \tag{2}$$

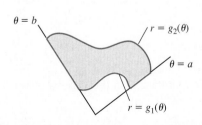

Figure 8.1 The region of integration Q of equation (2).

(equation (10) in Section 19.3) where Q is the region that is expressed in polar coordinates by $Q = \{(r, \theta) \mid a \le \theta \le b, g_1(\theta) \le r \le g_2(\theta)\}$ (see Figure 8.1). In this section, we indicate how each of these change of variables formulas is a special case of the general change of variables formula in the theory of integration. We begin by focusing on changing variables in double integrals, and at the end of the section, we indicate how the concepts generalize to triple integrals.

Transformations

The general change of variables formula for double integrals is based on the concept of a **transformation** T from the plane $\mathbb{R}^2$ with one pair of coordinates (often written as u and v) to the plane $\mathbb{R}^2$ with another pair of coordinates (usually the standard x- and y-coordinates). More precisely, a transformation T is a function defined on a planar region Q' that assigns to each point (u, v) in Q' a unique point (x, y) in $\mathbb{R}^2$. Therefore, both x and y are functions of u and v, and we often express the transformation $T(u, v) = (x, y)$ as

$$x = x(u, v)$$

$$y = y(u, v)$$

or as

$$T(u, v) = (x(u, v), y(u, v)). \tag{3}$$

The point (x, y) is called the **image** of the point (u, v).

One way to visualize a transformation T is to plot the image curves under T of a standard rectangular grid in the uv-plane. That is, for a fixed number u_0, the transformation T generates the curve $T(u_0, v)$ in the xy-plane that is the image of the vertical line $u = u_0$. Of course, we can also plot the image of a horizontal line $v = v_0$. If we plot these curves for a number of equally spaced horizontal and vertical lines, we produce a "grid" of curves in the xy-plane that help us understand the transformation (see Figure 8.2).

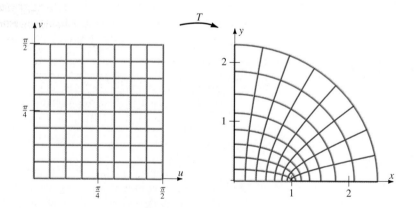

Figure 8.2 The transformation T "transforms" the square in the uv-plane into a curved region in the xy-plane.

An important example to keep in mind throughout this section is the transformation from (r, θ) to (x, y) determined by polar coordinates. In this case, we have

$$x = x(r, \theta) = r \cos \theta$$

$$y = y(r, \theta) = r \sin \theta.$$

This transformation by polar coordinates is illustrated in Figure 8.3.

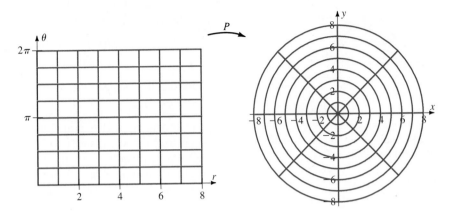

Figure 8.3 The polar coordinate transformation $P(r, \theta) = (x(r, \theta), y(r, \theta))$ where $x(r, \theta) = r \cos \theta$ and $y(r, \theta) = r \sin \theta$.

□ **EXAMPLE 1**

Determine the image of the rectangle

$$R = \{(u, v) \mid 0 \le u \le 1, 0 \le v \le 2\}$$

under the transformation T given by

$$x = 3u + v$$
$$y = u + v.$$

Solution　First we determine the image of the u-axis under the transformation. Since the u-axis is the line $v = 0$, we set $v = 0$ in the transformation and obtain

$$x(u, 0) = 3u$$
$$y(u, 0) = u,$$

which is an equation in parametric form for a line with slope $1/3$. Since $T(0, 0) = (0, 0)$ and $T(1, 0) = (3, 1)$, we see that the line segment determined by $0 \le u \le 1$ on the u-axis is transformed to the line segment from $(0, 0)$ to $(3, 1)$ in the xy-plane. Therefore, the image of one side of the rectangle R is the line segment from $(0, 0)$ to $(3, 1)$ in the xy-plane.

Similarly, to determine the image of the side of the rectangle on which $u = 1$, we set $u = 1$ and obtain

$$x(1, v) = 3 + v$$
$$y(1, v) = 1 + v,$$

which is an equation in parametric form for a line with slope 1. Since $T(1, 0) = (3, 1)$ and $T(1, 2) = (5, 3)$, the image of a second side of the rectangle R is the line segment from $(3, 1)$ to $(5, 3)$ in the xy-plane.

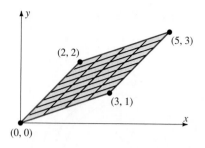

In general, all horizontal lines in the uv-plane (lines determined by holding v constant) are transformed to lines of slope $1/3$ in the xy-plane, and all vertical lines in the uv-plane (lines determined by holding u constant) are transformed to lines of slope 1 in the xy-plane. Since

$$T(0, 0) = (0, 0) \quad T(1, 0) = (3, 1)$$
$$T(0, 2) = (2, 2) \quad T(1, 2) = (5, 3),$$

we conclude that the image of the rectangle R is the parallelogram P in the xy-plane with vertices $(0, 0)$, $(3, 1)$, $(5, 3)$, and $(2, 2)$ (see Figure 8.4). ∎

A transformation $(x, y) = T(u, v)$ is a **one-to-one** transformation of the region Q' to its image Q if every point (x, y) in Q is the image of no more than one point (u, v) in Q'. In addition, we say that T is a C^1 **transformation** if the component functions $x(u, v)$ and $y(u, v)$ have continuous first-order partial derivatives. We now present the change of variables formula for one-to-one C^1 transformations.

Area and Changing Variables

Suppose $f(x, y)$ is a continuous function defined on a region Q in the xy-plane and that $(x, y) = T(u, v)$ is a one-to-one C^1 transformation from a region Q' to Q. We want a generalization of equation (2) where the polar coordinate transformation is replaced by the arbitrary transformation T. In other words, suppose we are given the function f, the region Q, and the transformation T from Q' to Q. What function do we integrate over Q' to obtain the double integral of f over Q? Certainly, this function should include the function f written in terms of u and v [that is, the composition $f(x(u, v), y(u, v))$]. We need only identify the additional factor $g(u, v)$ (such as the factor r in equation (2)) that is necessary to obtain the equality

$$\iint\limits_{Q} f(x, y)\, dx\, dy = \iint\limits_{Q'} f(x(u, v), y(u, v))\, g(u, v)\, du\, dv.$$

The correct form of g is determined by considering the constant function $f(x, y) = 1$ for all (x, y) in Q. Since

$$\iint\limits_{Q} f(x, y)\, dx\, dy = \iint\limits_{Q} 1\, dA = \text{Area of } Q,$$

the function $g(u, v)$ must satisfy

$$\text{Area of } Q = \iint\limits_{Q'} g(u, v)\, du\, dv. \tag{4}$$

Since we can interpret the right-hand side of equation (4) as the product

$$(\text{the average of } g \text{ over } Q')(\text{Area of } Q'),$$

we interpret $g(u, v)$ as the factor that measures the amount that T transforms area from Q' to Q.

To find g, we approximate this "area conversion" factor as follows. Consider a small rectangle R_i in the uv-plane with opposite corners (u_0, v_0) and $(u_0 + \Delta u, v_0 + \Delta v)$. In general, the image of R_i is a region Q_i in the xy-plane whose boundary consists of four curves. We approximate Q_i using tangents to these curves at the point $(x_0, y_0) = T(u_0, v_0)$ (see Figure 8.5). Holding $v = v_0$, the function $T(u, v_0) = (x(u, v_0), y(u, v_0))$ determines a curve in the xy-plane with tan-

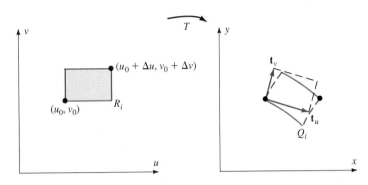

Figure 8.5 The transformation T takes the rectangle R_i to the region Q_i whose area is approximately $|\mathbf{t}_u \times \mathbf{t}_v|$.

gent vector $(\partial x/\partial u)\mathbf{i} + (\partial y/\partial u)\mathbf{j}$. Therefore, we approximate the side of Q that corresponds to holding $v = v_0$ by the vector

$$\mathbf{t}_u = \Delta u\left(\frac{\partial x}{\partial u}\mathbf{i} + \frac{\partial y}{\partial u}\mathbf{j}\right).$$

Similarly, by holding $u = u_0$, we approximate an adjacent side of Q_i by

$$\mathbf{t}_v = \Delta v\left(\frac{\partial x}{\partial v}\mathbf{i} + \frac{\partial y}{\partial v}\mathbf{j}\right).$$

Finally, we approximate the area of Q_i by the area of the parallelogram formed by $\mathbf{t}_u$ and $\mathbf{t}_v$. Using the cross product, we obtain

$$\text{Area of } Q_i \approx |\mathbf{t}_u \times \mathbf{t}_v| = \left|\frac{\partial x}{\partial u}\frac{\partial y}{\partial v} - \frac{\partial x}{\partial v}\frac{\partial y}{\partial u}\right|.$$

Since $\Delta u\,\Delta v$ is the area of the rectangle R_i, we see that the areas of R_i and Q_i are approximately related by the factor

$$\left|\frac{\partial x}{\partial u}\frac{\partial y}{\partial v} - \frac{\partial x}{\partial v}\frac{\partial y}{\partial u}\right|.$$

This informal argument motivates the following definition.

Definition 5

Suppose the transformation $T(u, v) = (x(u, v), y(u, v))$ is a C^1 transformation. The determinant

$$\det\begin{bmatrix} \dfrac{\partial x}{\partial u} & \dfrac{\partial x}{\partial v} \\[2mm] \dfrac{\partial y}{\partial u} & \dfrac{\partial y}{\partial v} \end{bmatrix} = \frac{\partial x}{\partial u}\frac{\partial y}{\partial v} - \frac{\partial x}{\partial v}\frac{\partial y}{\partial u}$$

is called the **Jacobian** of the transformation T, which is written as

$$\frac{\partial(x, y)}{\partial(u, v)}.$$

The Jacobian is named after the German mathematician C. G. J. Jacobi (1804–1851).

Changing Variables in Double Integrals

Using the Jacobian we can give a precise statement of the change of variables formula for double integrals. We assume that all regions involved are bounded and have piecewise smooth boundaries.

Theorem 8

Change of Variables in Double Integrals

Let

$$T(u, v) = (x(u, v), y(u, v))$$

be a C^1 transformation that takes the region Q' to the region Q and is one-to-one on the interior of Q. If the Jacobian of T is nonzero on Q' and if f is continuous on Q, then

$$\iint_Q f(x, y)\, dx\, dy = \iint_{Q'} f(x(u, v), y(u, v)) \left| \frac{\partial(x, y)}{\partial(u, v)} \right| du\, dv. \tag{5}$$

The same statement holds if the order of integration is interchanged on either the left-hand or the right-hand side of equation (5).

Note that, if $f(x, y) = 1$ for all (x, y) in Q, we obtain

$$\boxed{\text{Area of } Q = \iint_{Q'} \left| \frac{\partial(x, y)}{\partial(u, v)} \right| du\, dv,}$$

which is consistent with the approximation we discussed above.

☐ **EXAMPLE 2**

Show that the change of variables formula (equation (2)) for polar coordinates follows immediately from Theorem 8.

Solution   Converting an integral to polar coordinates corresponds to changing variables by the transformation

$$x(r, \theta) = r \cos \theta$$
$$y(r, \theta) = r \sin \theta.$$

Therefore, to apply Theorem 8, we calculate the Jacobian

$$\frac{\partial(x, y)}{\partial(r, \theta)} = \det \begin{bmatrix} \dfrac{\partial x}{\partial r} & \dfrac{\partial x}{\partial \theta} \\[2mm] \dfrac{\partial y}{\partial r} & \dfrac{\partial y}{\partial \theta} \end{bmatrix}$$

$$= \det \begin{bmatrix} \cos \theta & -r \sin \theta \\ \sin \theta & r \cos \theta \end{bmatrix}$$

$$= r \cos^2 \theta + r \sin^2 \theta = r.$$

Thus, equation (2) is a special case of Theorem 8.   ■

The following example illustrates how Theorem 8 can simplify an integration.

☐ **EXAMPLE 3**

Let P be the parallelogram with vertices $(0, 0)$, $(3, 1)$, $(5, 3)$, and $(2, 2)$ discussed in Example 1. Calculate

$$\iint_P xy \, dx \, dy$$

using the change of variables given in Example 1.

Solution   The transformation given in Example 1 is

$$x = 3u + v$$
$$y = u + v.$$

Thus, its Jacobian is

$$\frac{\partial(x, y)}{\partial(u, v)} = \det \begin{bmatrix} 3 & 1 \\ 1 & 1 \end{bmatrix} = 2.$$

From Example 1, we know that the parallelogram P is the image of the rectangle $R = \{(u, v) \mid 0 \le u \le 1, 0 \le v \le 2\}$ (see Figure 8.4). Therefore, using this change of variables, we have

$$\iint_P xy \, dx \, dy = \iint_R (3u + v)(u + v)2 \, du \, dv$$

$$= 2 \int_0^2 \int_0^1 (3u^2 + 4uv + v^2) \, du \, dv$$

$$= 2 \int_0^2 \left[u^3 + 2u^2v + v^2u \right]_{u=0}^{u=1} dv$$

$$= 2 \int_0^2 (1 + 2v + v^2) \, dv$$

$$= 2 \left[v + v^2 + \frac{v^3}{3} \right]_0^2 = \frac{52}{3}. \qquad \blacksquare$$

Change of Variables for Triple Integrals

There is a similar change of variables formula for triple integrals. Let

$$T(u, v, w) = (x(u, v, w), y(u, v, w), z(u, v, w))$$

be a one-to-one C^1 transformation that takes a region Q' in uvw-space to a region Q in xyz-space. Then the Jacobian of T is

$$\frac{\partial(x, y, z)}{\partial(u, v, w)} = \det \begin{bmatrix} \dfrac{\partial x}{\partial u} & \dfrac{\partial x}{\partial v} & \dfrac{\partial x}{\partial w} \\[2mm] \dfrac{\partial y}{\partial u} & \dfrac{\partial y}{\partial v} & \dfrac{\partial y}{\partial w} \\[2mm] \dfrac{\partial z}{\partial u} & \dfrac{\partial z}{\partial v} & \dfrac{\partial z}{\partial w} \end{bmatrix}$$

With this definition of the Jacobian, Theorem 8 generalizes immediately to triple integrals.

Theorem 9

Change of Variables in Triple
Integrals

Let

$$T(u, v, w) = (x(u, v, w), y(u, v, w), z(u, v, w))$$

be a C^1 transformation that takes the region Q' to the region Q and is one-to-one on the interior of Q. If the Jacobian of T is nonzero on Q' and f is continuous on Q, then

$$\iiint\limits_{Q} f(x, y, z)\, dx\, dy\, dz =$$

$$\iiint\limits_{Q'} f(x(u, v, w), y(u, v, w), z(u, v, w)) \left| \frac{\partial(x, y, z)}{\partial(u, v, w)} \right| du\, dv\, dw.$$

☐ **EXAMPLE 4**

Using the transformation

$$x = au$$
$$y = bv$$
$$z = cw$$

and Theorem 9, calculate the volume of the solid ellipsoid Q

$$\frac{x^2}{a^2} + \frac{y^2}{b^2} + \frac{z^2}{c^2} \leq 1.$$

Solution   The ellipsoid Q' is the image of the solid sphere S

$$u^2 + v^2 + w^2 \leq 1$$

of radius 1 under the given transformation. To apply Theorem 9, we calculate the Jacobian and obtain

$$\frac{\partial(x, y, z)}{\partial(u, v, w)} = \det \begin{bmatrix} a & 0 & 0 \\ 0 & b & 0 \\ 0 & 0 & c \end{bmatrix} = abc.$$

Using Theorem 9, the desired volume is

$$\iiint\limits_{Q} 1\, dx\, dy\, dz = \iiint\limits_{S} abc\, du\, dv\, dw$$

$$= abc \iiint\limits_{S} du\, dv\, dw$$

$$= abc(\text{Volume of } S).$$

In Example 4 of Section 19.7, we calculated the volume of S to be $\frac{4}{3}\pi$. Therefore, the ellipsoid has volume $\frac{4}{3}\pi\, abc$.   ■

Exercise Set 19.8

In Exercises 1–6, sketch the image of the square

$$S = \{(u, v) \mid 0 \le u \le 1, 0 \le v \le 1\}$$

under the given transformation.

1. $x = 2u + v$
$y = u + v$

2. $x = 3u - v$
$y = u + v$

3. $x = 2u$
$y = 3v$

4. $x = -v$
$y = u$

5. $x = u^2 - v^2$
$y = 2uv$

6. $x = u \cos v$
$y = u \sin v$

In Exercises 7–10, evaluate the double integral using the given change of variables.

7. $\displaystyle\iint_P xy \, dx \, dy$

where P is the parallelogram with vertices $(0, 0)$, $(3, 1)$, $(3, 2)$, $(0, 1)$ and $x = 3u$, $y = u + v$.

8. $\displaystyle\iint_S y \, dx \, dy$

where S is the square with vertices $(0, 0)$, $(1, 1)$, $(0, 2)$, $(-1, 1)$ and $x = u - v$, $y = u + v$.

9. $\displaystyle\iint_E \left(\frac{x^2}{9} + \frac{y^2}{4} \right) dx \, dy$

where E is the elliptical region

$$\frac{x^2}{36} + \frac{y^2}{16} \le 1$$

and $x = 3u$ and $y = 2v$.

10. $\displaystyle\iint_Q y^2 \, dx \, dy$

where Q is the region in the first quadrant bounded by the curves $xy = 1$, $xy = 2$, $y = 2$, and $y = 3$ and $x = u/v$, $y = v$.

11. Using a triple integral, find the volume of the elliptical cylinder

$$C = \left\{ (x, y, z) \,\middle|\, \frac{x^2}{4} + \frac{y^2}{9} \le 1, 0 \le z \le 2 \right\}.$$

12. Find the volume of the solid

$$Q = \left\{ (x, y, z) \,\middle|\, x^2 + \frac{y^2}{4} \le 1, 0 \le z \le x^2 + y^2 \right\}.$$

13. Find the volume of the solid

$$Q = \left\{ (x, y, z) \,\middle|\, x \ge 0, y \ge 0, z \ge 0, 0 \le z \le 1 - \frac{x^2}{4} - \frac{y^2}{9} \right\}.$$

14. Sketch a region Q' in the $r\theta$-plane whose image under the polar coordinate change of variables is the region Q in Figure 8.1.

15. Derive equation (3) in Section 15.3 from the change of variables formula for polar coordinates. (See Exercise 14.)

16. Derive equation (5) in Section 19.7 as a special case of Theorem 9.

17. Using Theorem 9, derive a change of variables formula for spherical coordinates assuming the hypotheses of Theorem 7 in Section 19.7.

18. Interpret equation (1) in terms of the change of variables formulas stated in this section under the assumption that the function $g'(x) \ne 0$ for $a \le x \le b$. (*Hint:* Interchange the roles of x and u.)

Summary Outline of Chapter 19

■ If f is continuous on the planar region   (page 976)

$$Q = \{(x, y) \mid a \le x \le b, g_1(x) \le y \le g_2(x)\}$$

then

(i) $\displaystyle\iint_Q f(x, y) \, dA = \int_a^b \int_{g_1(x)}^{g_2(x)} f(x, y) \, dy \, dx.$

(ii) Area of $Q = \displaystyle\iint_Q 1 \, dA.$

■ If f is continuous on the region Q described in polar coordinates as   (page 986)

$$Q = \{(r, \theta) \mid g_1(\theta) \le r \le g_2(\theta), a \le \theta \le b\},$$

then

(i) $\displaystyle\iint_Q f(r, \theta)\, dA = \int_a^b \int_{g_1(\theta)}^{g_2(\theta)} f(r, \theta)\, r\, dr\, d\theta.$

(ii) Area of $Q = \displaystyle\int_a^b \int_{g_1(\theta)}^{g_2(\theta)} r\, dr\, d\theta$ if $g_1(\theta) \ge 0$ for $a \le \theta \le b.$

■ In changing an iterated double integral from Cartesian to polar coordinates,   (page 990) the differential for area $dA = dx\, dy$ is replaced by the polar differential for area $dA = r\, dr\, d\theta.$

■ If a thin lamina with density ρ has the shape of the planar region Q, then   (page 995)

(i) Mass $= M = \displaystyle\iint_Q \rho(x, y)\, dA.$

(ii) Center of mass $= (\bar{x}, \bar{y})$ where $\bar{x} = \dfrac{M_y}{M}$, $\bar{y} = \dfrac{M_x}{M}$,

$$M_y = \iint_Q x\rho(x, y)\, dA, \quad \text{and} \quad M_x = \iint_Q y\rho(x, y)\, dA.$$

■ The **surface area** of the graph of $z = f(x, y)$ over the region Q is   (page 1001)

$$A_S = \iint_Q \sqrt{\left[\frac{\partial f}{\partial x}(x, y)\right]^2 + \left[\frac{\partial f}{\partial y}(x, y)\right]^2 + 1}\, dA.$$

■ If the region Q in $\mathbb{R}^3$ is described by the inequalities $a \le x \le b$, $h_1(x) \le y$ (page 1007) $\le h_2(x)$, $g_1(x, y) \le z \le g_2(x, y)$, then

(i) $\displaystyle\iiint_Q f(x, y, z)\, dV = \int_a^b \int_{h_1(x)}^{h_2(x)} \int_{g_1(x, y)}^{g_2(x, y)} f(x, y, z)\, dz\, dy\, dx.$

(ii) Volume of $Q = \displaystyle\iiint_Q 1\, dV.$

■ If a solid has the shape of the region Q in $\mathbb{R}^3$ and density function ρ,   (page 1011) then

(i) Mass $= M = \displaystyle\iiint_Q \rho(x, y, z)\, dV.$

(ii) Center of mass $= (\bar{x}, \bar{y}, \bar{z})$ where $\bar{x} = \dfrac{M_{yz}}{M}$, $\bar{y} = \dfrac{M_{xz}}{M}$, $\bar{z} = \dfrac{M_{xy}}{M}$,

$$M_{yz} = \iiint_Q x\rho(x, y, z)\, dV, \qquad M_{xz} = \iiint_Q y\rho(x, y, z)\, dV,$$

$$M_{xy} = \iiint_Q z\rho(x, y, z)\, dV.$$

■ If the region Q is described in cylindrical coordinates by the inequali-   (page 1016) ties $a \le \theta \le b$, $h_1(\theta) \le r \le h_2(\theta)$, $g_1(r, \theta) \le z \le g_2(r, \theta)$, then

$$\iiint_Q f(r, \theta, z)\, dV = \int_a^b \int_{h_1(\theta)}^{h_2(\theta)} \int_{g_1(r, \theta)}^{g_2(r, \theta)} f(r, \theta, z)\, r\, dz\, dr\, d\theta.$$

- In changing an iterated triple integral from Cartesian coordinates to cylin- (page 1017) drical coordinates, the Cartesian differential for volume $dV = dx\,dy\,dz$ is replaced by the cylindrical differential for volume $dV = r\,dz\,dr\,d\theta$.

- If the region Q in $\mathbb{R}^3$ is described in spherical coordinates by the inequali- (page 1019) ties $a \le \theta \le b$, $h_1(\theta) \le \phi \le h_2(\theta)$, $g_1(\theta, \phi) \le \rho \le g_2(\theta, \phi)$, then

$$\iiint\limits_{Q} f(\rho, \theta, \phi)\,dV = \int_a^b \int_{h_1(\theta)}^{h_2(\theta)} \int_{g_1(\theta, \phi)}^{g_2(\theta, \phi)} f(\rho, \theta, \phi)\rho^2 \sin\phi\,d\rho\,d\phi\,d\theta.$$

- In changing an iterated triple integral from Cartesian to spherical coordi- (page 1019) nates, the Cartesian differential for volume $dV = dx\,dy\,dz$ is replaced by the spherical differential for volume $dV = \rho^2 \sin\phi\,d\rho\,d\phi\,d\theta$.

- The **Jacobian** of a C^1 transformation $T(u, v) = (x(u, v), y(u, v))$ is (page 1027)

$$\frac{\partial(x, y)}{\partial(u, v)} = \det \begin{bmatrix} \dfrac{\partial x}{\partial u} & \dfrac{\partial x}{\partial v} \\[2mm] \dfrac{\partial y}{\partial u} & \dfrac{\partial y}{\partial v} \end{bmatrix} = \frac{\partial x}{\partial u}\frac{\partial y}{\partial v} - \frac{\partial x}{\partial v}\frac{\partial y}{\partial u}.$$

- The change of variables formula for double integrals is (page 1028)

$$\iint\limits_{Q} f(x, y)\,dx\,dy = \iint\limits_{Q'} f(x(u, v), y(u, v)) \left| \frac{\partial(x, y)}{\partial(u, v)} \right| du\,dv.$$

- The **Jacobian** of a C^1 transformation $T(u, v, w) = (x(u, v, w), y(u, v, w), z(u, v, w))$ is (page 1029)

$$\frac{\partial(x, y, z)}{\partial(u, v, w)} = \det \begin{bmatrix} \dfrac{\partial x}{\partial u} & \dfrac{\partial x}{\partial v} & \dfrac{\partial x}{\partial w} \\[2mm] \dfrac{\partial y}{\partial u} & \dfrac{\partial y}{\partial v} & \dfrac{\partial y}{\partial w} \\[2mm] \dfrac{\partial z}{\partial u} & \dfrac{\partial z}{\partial v} & \dfrac{\partial z}{\partial w} \end{bmatrix}$$

- The change of variables formula for triple integrals is (page 1030)

$$\iiint\limits_{Q} f(x, y, z)\,dx\,dy\,dz =$$

$$\iiint\limits_{Q'} f(x(u, v, w), y(u, v, w), z(u, v, w)) \left| \frac{\partial(x, y, z)}{\partial(u, v, w)} \right| du\,dv\,dw.$$

Review Exercises—Chapter 19

1. Find the volume of the solid bounded above by the graph of $z = xy^2$ and below by the triangle with vertices $(0, 0)$, $(2, 0)$, and $(0, 1)$.

2. Evaluate the double integral

$$\iint\limits_{Q} \frac{x}{xy + 2}\,dA$$

where Q is the rectangle

$$Q = \{(x, y)\,|\,0 \le x \le 1, 0 \le y \le 1\}.$$

3. Evaluate

$$\iint\limits_{Q} e^{-x^2/2}\,dy\,dx$$

where Q is the region

$$Q = \{(x, y)\,|\,0 \le x \le 1, 0 \le y \le 2x\}.$$

4. Evaluate

$$\iint_Q (x - 4y)\, dA$$

where Q is the rectangle

$$Q = \{(x, y) \mid -1 \le x \le 1, 0 \le y \le 2\}.$$

5. Find the volume of the solid in the first octant bounded by the graphs of $z = y^2$, $y = x$, $z = 0$, and $y = 4$.

6. Change the order of integration:

$$\int_{-1}^{2} \int_{x^2-2}^{x} f(x, y)\, dy\, dx.$$

7. Find the surface area of the part of the sphere $x^2 + y^2 + z^2 = 4$ that lies inside the cylinder $x^2 + y^2 - 2x = 0$.

8. Find the center of mass of a solid that is bounded above by the paraboloid $4z = 4 - (x^2 + y^2)$ and below by the xy-plane if the density of the material is uniform.

9. Find the volume of the solid bounded by the cylinder $r = 2 \cos \theta$, the paraboloid $z = 2r^2$ and the plane $z = 0$ (use cylindrical coordinates).

10. Rewrite the integral

$$\int_{-\pi/2}^{\pi/2} \int_{0}^{2\cos\theta} \int_{-\sqrt{4-r^2}}^{\sqrt{4-r^2}} r\, dz\, dr\, d\theta$$

in rectangular coordinates.

11. Find the volume of the solid bounded by the graph of

$$\frac{x}{a} + \frac{y}{b} + \frac{z}{c} = 1 \qquad (a > 0, b > 0, c > 0)$$

and the three coordinate planes.

12. Sketch the region Q corresponding to the iterated integral, reverse the order of integration, and evaluate the resulting integral:

$$\int_{-2}^{0} \int_{-\sqrt{x+2}}^{\sqrt{x+2}} y^2\, dy\, dx.$$

13. Calculate the volume of the solid lying inside the cylinder $x^2 + y^2 = 4$ and between the planes $y + z = 9$ and $z = 0$.

14. Use a double integral to calculate the area of the region lying between the graph of $\sqrt{x} + \sqrt{y} = 1$ and the graph of $x + y = 1$.

15. Find the center of mass of a solid bounded by the cylinder $r = 2$, the cone $z = r$, and the plane $z = 0$ if the density of the material is uniform.

16. Calculate the volume of the ellipse

$$\frac{x^2}{4} + \frac{y^2}{9} + \frac{z^2}{4} = 1.$$

17. Find the volume of the region common to the sphere $r^2 + z^2 = a^2$ and the cylinder $r = a \cos \theta$.

18. Find the surface area of the paraboloid $z = x^2 + y^2$ lying between the planes $z = 1$ and $z = 9$.

19. Find the area of the region cut from the plane $z = 4y$ by the cylinder $x^2 + y^2 = 4$.

20. Find the volumes of the two regions cut from the sphere $\rho = 4$ by the plane $x = 2$.

21. Find the volume of the region common to the sphere $x^2 + y^2 + z^2 = 16$ and the cylinder $x^2 + z^2 = 4$.

22. Use a double integral to find the area enclosed by the lemniscate $r^2 = 2 \cos 2\theta$.

23. A solid is bounded below by the region bounded by the graphs of $y = x$ and $y = x^2 - 2$. It is bounded above by the plane $z - x + 2y = 10$. Find its volume.

24. Find

$$\iint_Q \frac{\sin x}{x}\, dA$$

where Q is the triangle with vertices $(0, 0)$, $(2, 0)$ and $(2, 2)$.

25. Find the centroid of the half disc $r = 4$, $0 \le \theta \le \pi$.

26. Find the mass of a right circular cone of radius r and height h if the density at each point is proportional to the distance of that point from the vertex.

27. Find

$$\iiint_Q e^{(x^2+y^2+z^2)^{3/2}}\, dV$$

where Q is the half sphere

$$Q = \{(r, \theta, \phi) \mid 0 \le r \le 1, 0 \le \phi \le \pi/2, 0 \le \theta \le 2\pi\}.$$

28. Show that if a thin lamina of constant density has the shape of the region Q described in polar coordinates, then the coordinates $(\bar{x}, \bar{y})$ of the center of mass may be calculated by the formulas

$$\bar{x} = \frac{1}{\text{area}} \iint_Q r^2 \cos \theta\, dr\, d\theta;$$

$$\bar{y} = \frac{1}{\text{area}} \iint_Q r^2 \sin \theta\, dr\, d\theta.$$

29. Use the result of Exercise 28 to calculate the center of mass for a thin lamina of constant density whose shape is the cardioid $r = 1 + \cos \theta$.

30. Find the volume of the solid bounded above by the plane $z - x = 2$ and below by the paraboloid $z = x^2 + y^2$.

31. Find the volume of the region bounded by the paraboloids

$$z = 4 - x^2 + 2x - y^2 - 4y$$

and

$$z = x^2 - 2x + y^2 + 4y + 5.$$

32. Find the volume of the solid bounded above by the cylinder $z = 4 - x^2$ and below by the paraboloid $3x^2 + y^2 = z$.

33. A solid corresponds to the region bounded by the cylinder $x^2 + y^2 = 4$ and the planes $z = 0$ and $z = 4$. Calculate the mass of the solid if the density at each point is proportional to the distance from the xy-plane.

34. Evaluate the integral

$$\iint_Q \cos \sqrt{x^2 + y^2} \, dA$$

where Q is the disc $Q = \{(x, y) \mid 0 \le x^2 + y^2 \le 4\}$.

35. Evaluate the integral

$$\iint_Q \frac{1}{\sqrt{1 + x^2 + y^2}} \, dA$$

where Q is the quarter circle

$$Q = \{(x, y) \mid 0 \le x \le 1, 0 \le y \le \sqrt{1 - x^2}\}.$$

36. Find the volume of the solid bounded above by the graph of $z = 4 - r$ and below by the region in the $r\theta$-plane bounded by the graph of $r = 3 \sin \theta$.

37. Find the volume of the solid bounded above by the cone $z^2 = x^2 + y^2$, on the sides by the cylinder $x^2 + y^2 - 4y = 0$, and below by the xy-plane.

38. Find the volume of the solid inside the cylinder $x^2 - 4x + y^2 = 0$ lying above the xy-plane and below the plane $z - x = 4$.

39. Use the programs in Appendix I to approximate

$$\int_0^2 \int_1^3 \sin^2 \sqrt{x - y^2} \, dy \, dx.$$

40. Find the area of the region bounded by the graphs of $y = \sqrt{x + 2}$ and $x - 3y + 2 = 0$ by double integration.

41. Find the volume of the solid bounded by the cone $z^2 = x^2 + y^2$ and the cylinder $x^2 + y^2 = 9$.

42. Find the centroid of the region bounded by the graph of $r = \cos 2\theta$.

43. Find the surface area of the part of the sphere $x^2 + y^2 + z^2 = 4$ lying outside the cylinder $x^2 + y^2 = 1$.

Chapter 20

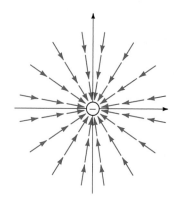

Vector Analysis

Vector fields are vector-valued functions defined in the plane or in space, and vector analysis is the calculus of these functions. In this chapter, we introduce vector fields and briefly discuss a few of their many applications (for example, the study of gravitation as well as other forces, and the study of fluid flows). Then we present the integrals used to analyze these fields. In particular, we emphasize the theorems of vector analysis that generalize the Fundamental Theorem of Calculus to these integrals.

20.1 Vector Fields

In Chapter 17, we studied vector-valued functions defined on subsets of the real numbers. In this chapter, we study vector-valued functions defined on subsets of the plane or space.

Definition 1

Let Q be a subset of $\mathbb{R}^3$. A **vector field** on Q is a function

$$\mathbf{F}(x, y, z) = M(x, y, z)\mathbf{i} + N(x, y, z)\mathbf{j} + P(x, y, z)\mathbf{k} \tag{1}$$

that assigns a vector $\mathbf{w} = \mathbf{F}(x, y, z)$ to each point (x, y, z) in Q.

If Q is a subset of $\mathbb{R}^2$, a vector field on Q is a function of the form

$$\mathbf{F}(x, y) = M(x, y)\mathbf{i} + N(x, y)\mathbf{j}. \tag{2}$$

The real-valued functions M, N, and P in equation (1) are referred to as the **component** functions of the vector field $\mathbf{F}$. A vector field is **continuous** if each of its component functions is continuous. Using the position vector

$$\mathbf{r} = x\mathbf{i} + y\mathbf{j} + z\mathbf{k}, \qquad (x, y, z) \in Q,$$

we may write the vector field in equation (1) simply as $\mathbf{w} = \mathbf{F}(\mathbf{r})$. Functions of the form $w = f(x, y, z)$, which assign real numbers to vectors, will now be referred to as **scalar functions** or scalar fields.

The following examples indicate the types of vector fields and scalar functions that we shall study in this chapter. Note that, when we sketch a vector field $\mathbf{F}$ as in equation (1) or (2), we place the initial point of the vector $\mathbf{F}(\mathbf{r}_0)$ at the point $\mathbf{r}_0$. The examples indicate why vector fields are best interpreted in this fashion.

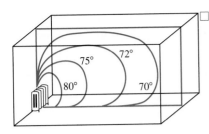

Figure 1.1 Isothermal curves in a cross section of a room with a single heat source.

EXAMPLE 1

Figure 1.1 represents a room with a single radiator. The function T that gives the temperature $T(x, y, z)$ at the point in the room with coordinates (x, y, z) is an example of a *scalar* function. The curves drawn in a cross section of the room are locations of points of equal temperature and are called **isothermal curves.**

Since warmer air rises and cooler air falls, the presence of a single heat source in a room causes air to flow about the room in **convection currents.** If the vector $\mathbf{F}(x, y, z)$ represents the velocity vector of the convection current passing through the point (x, y, z), the function $\mathbf{w} = \mathbf{F}(x, y, z)$ is an example of a vector field defined throughout the room (Figure 1.2). ∎

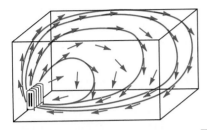

Figure 1.2 Velocity vectors $\mathbf{F}(x, y, z)$ for convection currents determine a vector field in the room.

EXAMPLE 2

An electric current of magnitude I flowing through a thin wire induces a **magnetic field** around the wire. (The statement of Ampère's Law, giving the relationship between the electric current I and its magnetic effect, involves the idea of a *line integral,* introduced in Section 20.2.) The set of vectors $\mathbf{F}(x, y, z)$, giving the direction and intensity of the magnetic field at location (x, y, z), determines a vector field in the space surrounding the wire (Figure 1.3). ∎

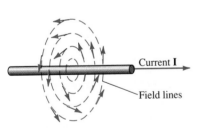

Figure 1.3 Vectors that give the direction and intensity of the magnetic field form a (magnetic) vector field in space.

EXAMPLE 3

An example of a force field is a **gravitational field.** According to Newton's Law of Gravitation, the magnitude of the force of attraction exerted on a particle P_2 of mass m_2 by a particle P_1 of mass m_1 is given by the equation

$$F = \frac{Gm_1m_2}{r^2} \qquad (3)$$

where r is the distance between the particles and G is a constant. If we assume the particle P_1 is located at the origin in xyz-space, then the force in (3) is a function of the coordinates (x, y, z) of the location of the particle P_2. That is,

$$F(x, y, z) = \frac{Gm_1m_2}{x^2 + y^2 + z^2}. \qquad (4)$$

Since the force whose magnitude is given by (4) acts toward the origin, it acts in the direction opposite that of the position vector $\mathbf{r} = x\mathbf{i} + y\mathbf{j} + z\mathbf{k}$ of P_2. We may therefore write the force vector $\mathbf{F}(x, y, z)$ as

$$\mathbf{F}(x, y, z) = \left(\frac{Gm_1m_2}{(x^2 + y^2 + z^2)}\right)\left(-\frac{x\mathbf{i} + y\mathbf{j} + z\mathbf{k}}{\sqrt{x^2 + y^2 + z^2}}\right) \qquad (5)$$

$$= \left(\frac{-Gm_1m_2}{(x^2 + y^2 + z^2)^{3/2}}\right)(x\mathbf{i} + y\mathbf{j} + z\mathbf{k}).$$

Using the position vector $\mathbf{r} = x\mathbf{i} + y\mathbf{j} + z\mathbf{k}$, we may write the force field in (5) as

$$\mathbf{F}(\mathbf{r}) = -\frac{Gm_1m_2}{|\mathbf{r}|^3}\mathbf{r}, \qquad \mathbf{r} \neq \mathbf{0} \qquad \text{(Figure 1.4).} \qquad (6)$$

Since the vector $\mathbf{F}$ in (5) and (6) is defined for all position vectors $\mathbf{r} \neq \mathbf{0}$, either equation defines a vector field at all points of $\mathbb{R}^3$ except the origin (Figure 1.5). ∎

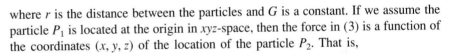

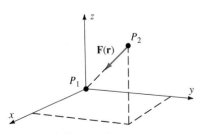

Figure 1.4 Gravitational force vector $\mathbf{F}(\mathbf{r})$ exerted on particle P_2 by particle P_1.

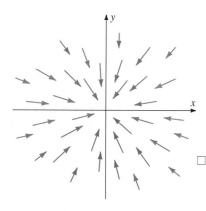

Figure 1.5 Gravitational force field (drawing represents xy-plane only).

The gravitational field of Example 3 is an example of a **central force field** since the force vector at each point points toward the origin. More generally, a central force field in space has the form

$$\mathbf{F}(x, y, z) = f(x, y, z)(x\mathbf{i} + y\mathbf{j} + z\mathbf{k})$$

where f is a scalar function of three variables.

□ **EXAMPLE 4**

Another example of a central force field is obtained by letting $\mathbf{F}(x, y, z)$ be the vector representing the force exerted on a particle P_2 with electric charge q_2 at location (x, y, z) by a particle P_1 with electric charge q_1 located at the origin. According to **Coulomb's Law,** this force vector is

$$\mathbf{F}(\mathbf{r}) = \frac{kq_1q_2}{|\mathbf{r}|^3}\mathbf{r}, \qquad \mathbf{r} \neq \mathbf{0} \tag{7}$$

where $\mathbf{r}$ is the position vector $\mathbf{r} = x\mathbf{i} + y\mathbf{j} + z\mathbf{k}$ of P_2, and k is a constant that depends on the choice of units for $\mathbf{r}$, q_1, and q_2. Figure 1.6 represents the **electric force field F** (for positive charges P_2) due to a positive charge at the origin. Figure 1.7 represents the force field **F** (for positive charges P_2) due to a negative charge at the origin. (Although the force field is defined in space, we show only the force vectors lying in the xy-plane in these figures.) ■

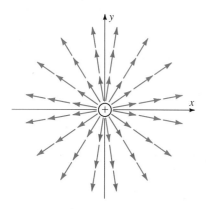

Figure 1.6 Electric force field due to a positive charge at the origin.

If charge is considered to be analogous to mass, it is clear from a comparison of equations (6) and (7) that the gravitational and electric force fields have the same form; both forces obey an *inverse square law*. The major difference is that the direction of $\mathbf{F}(\mathbf{r})$ may be either toward or away from the origin for the electric field, but it can only be toward the origin for the gravitational field (since all masses are positive).

Electric force fields satisfy the **principle of superposition,** which states that the force acting on a charge at a point P due to two separate charges is the vector sum of the individual forces acting on the charge at P. Figure 1.8 represents the electric field, acting on a positive charge, that results from a positive charge at $(-1, 0)$ and a negative charge of equal magnitude at $(1, 0)$. Note that while the individual force fields in Figure 1.6 and 1.7 are central, their sum is not (see Figure 1.8).

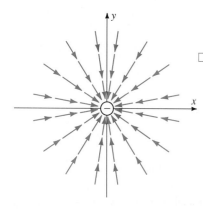

Figure 1.7 Electric field due to a negative charge at the origin.

□ **EXAMPLE 5**

If f is a differentiable scalar function of three variables, the gradient function

$$\mathbf{F}(x, y, z) = \nabla f(x, y, z) = f_x(x, y, z)\mathbf{i} + f_y(x, y, z)\mathbf{j} + f_z(x, y, z)\mathbf{k} \tag{8}$$

defines a vector field on $\mathbb{R}^3$. If f is a differentiable scalar function of two variables, the corresponding gradient field on $\mathbb{R}^2$ is

$$\mathbf{F}(x, y) = \nabla f(x, y) = f_x(x, y)\mathbf{i} + f_y(x, y)\mathbf{j}. \tag{9}$$

For example, the scalar function $f(x, y, z) = xe^{2y} \sin z$ yields the gradient vector field

$$\mathbf{F}(x, y, z) = e^{2y} \sin z\,\mathbf{i} + 2xe^{2y} \sin z\,\mathbf{j} + xe^{2y} \cos z\,\mathbf{k}. \qquad ■$$

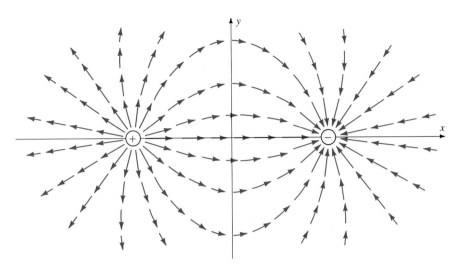

Figure 1.8 Electric force field due to two opposite charges is the vector sum of the individual central force fields.

Table 1.1 Selected vectors from the vector field $\mathbf{F}(x, y) = -y\mathbf{i} + x\mathbf{j}$*

(x, y)	$\mathbf{F}(x, y) = -y\mathbf{i} + x\mathbf{j}$
$(1, 0)$	$\mathbf{j}$
$(0, 1)$	$-\mathbf{i}$
$(-1, 0)$	$-\mathbf{j}$
$(0, -1)$	$\mathbf{i}$
$(\sqrt{2}, \sqrt{2})$	$-\sqrt{2}\mathbf{i} + \sqrt{2}\mathbf{j}$
$(-\sqrt{2}, \sqrt{2})$	$-\sqrt{2}\mathbf{i} - \sqrt{2}\mathbf{j}$

*These vectors are among those plotted in Figure 1.9.

As in Section 18.9, we will refer to the function f in equation (8) as a **scalar potential** for the vector field $\mathbf{F}$. Although every differentiable scalar function yields a gradient vector field, a given vector field $\mathbf{F}$ need not be the gradient of a scalar potential. For reasons that are discussed in Section 20.3, those vector fields that can be written as gradients of scalar potentials are called **conservative vector fields.**

There are many other familiar examples of vector fields. The flow of air currents around a moving automobile or airplane determines a **velocity field.** The study of such vector fields is called *aerodynamics.* Similarly, the velocity field determined by the flow of water through a container, such as a pipe or a dam, is the subject of *hydrodynamics.*

Our final example is a vector field that corresponds to circular motion about the origin.

□ **EXAMPLE 6**

Sketch the vector field

$$\mathbf{F}(x, y) = -y\mathbf{i} + x\mathbf{j}$$

in the plane and indicate why it corresponds to circular motion about the origin.

Solution   First we calculate $\mathbf{F}(x, y)$ for a sufficient number of vectors so that we get a feel for the nature of the vector field. For each point (x, y) that we consider, we plot the vector $\mathbf{F}(x, y)$ with its initial point at (x, y). From Table 1.1, we obtain the sketch in Figure 1.9.

Note that these vectors are tangent to circles centered at the origin. We establish this fact by calculating the dot product of the position vector $x\mathbf{i} + y\mathbf{j}$ with $\mathbf{F}(x, y)$. We obtain

$$(x\mathbf{i} + y\mathbf{j}) \cdot (-y\mathbf{i} + x\mathbf{j}) = -xy + xy = 0.$$

Note also that the length $|\mathbf{F}(x, y)|$ is $\sqrt{y^2 + x^2}$, which equals the distance of the point (x, y) to the origin. Therefore, we can interpret this vector field as the velocity vectors for points on a disc that is rotating about its center at constant angular velocity.   ■

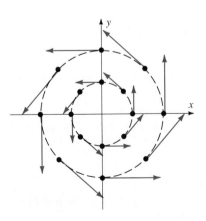

Figure 1.9 Selected vectors from the vector field $\mathbf{F}(x, y) = -y\mathbf{i} + x\mathbf{j}$.

Exercise Set 20.1

In Exercises 1–10, sketch enough vectors in the given vector field to get a sense of the nature of the vector field.

1. $\mathbf{F}(x, y) = x\mathbf{i} - y\mathbf{j}$ **2.** $\mathbf{F}(x, y) = (x - y)\mathbf{i} + (y - x)\mathbf{j}$

3. $\mathbf{F}(x, y) = 3\mathbf{i}$ **4.** $\mathbf{F}(x, y) = -x\mathbf{i} - y\mathbf{j}$

5. $\mathbf{F}(x, y) = \dfrac{x}{\sqrt{x^2 + y^2}}\mathbf{i} + \dfrac{y}{\sqrt{x^2 + y^2}}\mathbf{j}$

6. $\mathbf{F}(x, y) = \dfrac{x}{\sqrt{x^2 + y^2}}\mathbf{i} - \dfrac{y}{\sqrt{x^2 + y^2}}\mathbf{j}$

7. $\mathbf{F}(x, y, z) = \mathbf{j} + \mathbf{k}$ **8.** $\mathbf{F}(x, y, z) = x\mathbf{i} + y\mathbf{j}$

9. $\mathbf{F}(x, y, z) = 2x\mathbf{i} + 2y\mathbf{j} + 2z\mathbf{k}$

10. $\mathbf{F}(x, y, z) = \mathbf{i} - \mathbf{j} + \mathbf{k}$

11. Which of the vector fields in Exercises 1–10 are central?

Find the gradient vector field $\mathbf{F} = \nabla f$ for each of the functions in Exercises 12–17.

12. $f(x, y) = xy$ **13.** $f(x, y) = x^2 - y^2$

14. $f(x, y) = x \tan xy$

15. $f(x, y, z) = \sqrt{x^2 + y^2 + z^2}$

16. $f(x, y, z) = x \ln(y^2 + z^2)$ **17.** $f(x, y, z) = ze^{x-y}$

In Exercises 18–27, determine whether the given vector field is a gradient vector field. If so, find a potential ϕ with $\mathbf{F} = \nabla\phi$ (see Section 18.9).

18. $\mathbf{F}(x, y) = \sin y\mathbf{i} + \cos y\mathbf{j}$

19. $\mathbf{F}(x, y) = (\tan xy + xy \sec^2 xy)\mathbf{i} + x^2 \sec^2 xy\mathbf{j}$

20. $\mathbf{F}(x, y) = \left[\ln(y - x) - \dfrac{x}{y - x}\right]\mathbf{i} + \dfrac{x}{y - x}\mathbf{j}$

21. $\mathbf{F}(x, y) = \left[\ln(y - x) + \dfrac{x}{y - x}\right]\mathbf{i} + \left(\dfrac{x}{y - x}\right)\mathbf{j}$

22. $\mathbf{F}(x, y, z) = x^2\mathbf{i} + x^2 \sin z\mathbf{j} - \cos yz\mathbf{k}$

23. $\mathbf{F}(x, y, z) = yze^{xyz}\mathbf{i} + xze^{xyz}\mathbf{j} + xye^{xyz}\mathbf{k}$

24. $\mathbf{F}(x, y, z) = 2xe^y\mathbf{i} + (x^2e^y - \sin y)\mathbf{j} + \dfrac{1}{\sqrt{z^2 + 36}}\mathbf{k}$

25. $\mathbf{F}(x, y, z) = \sqrt{y}\cosh z\mathbf{i} + \dfrac{x}{2\sqrt{y}}\cosh z\mathbf{j} + x\sqrt{y}\sinh z\mathbf{k}$

26. $\mathbf{F}(x, y, z) = -\dfrac{z}{y^2}\ln\left(\dfrac{y + \sqrt{x^2 + y^2}}{x}\right)\mathbf{i} - $

$\dfrac{z}{x^2}\ln\left(\dfrac{x + \sqrt{x^2 + y^2}}{y}\right)\mathbf{j} + \dfrac{1}{xy\sqrt{x^2 + y^2}}\mathbf{k}$

27. $\mathbf{F}(x, y, z) = 2xye^{x^2y}\ln z\mathbf{i} + x^2e^{x^2y}\ln z\mathbf{j} + $

$\left(\dfrac{e^{x^2y}}{z} - \dfrac{1}{z^2 + 2z + 2}\right)\mathbf{k}$

28. The **flow lines** for a vector field $\mathbf{F}$ are curves $\boldsymbol{\alpha}(t)$ tangent to the vector field. That is, $\boldsymbol{\alpha}'(t) = \mathbf{F}(\boldsymbol{\alpha}(t))$ for every t and

every flow line. (An example of flow lines associated with a vector field in the plane is shown in Figure 1.10.) Sketch several flow lines for each of the vector fields in Exercises 1–10.

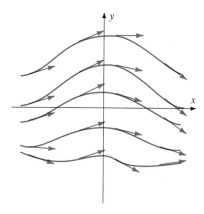

Figure 1.10   Flow lines for a vector field.

29. Imagine water flowing through a pipe as in Figure 1.11. At each point (x, y, z) within the pipe let $\mathbf{F}(x, y, z)$ be the velocity vector for the water at that point. This determines a vector field within the pipe. Why are the vectors in the narrow part of the pipe larger in magnitude?

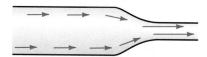

Figure 1.11   Velocity field in a water pipe.

30. Show that the central force field

$$\mathbf{F}(\mathbf{r}) = -\left(\dfrac{k}{|\mathbf{r}|^3}\right)\mathbf{r}$$

is the gradient of the scalar field $f(\mathbf{r}) = k/|\mathbf{r}|$. ($\mathbf{r}$ denotes the position vector $\mathbf{r} = x\mathbf{i} + y\mathbf{j} + z\mathbf{k}$ associated with the point (x, y, z).)

31. Let $\mathbf{F}(x, y, z)$ be the electric field describing the force on a particle with charge $q = +1$ at location (x, y, z) due to the combined effect of a charge $q_1 = +3$ at $(-1, 0, 0)$ and a charge $q_2 = -1$ at $(1, 0, 0)$.

a. Find the force vector $\mathbf{F}(0, 0, 0)$.

b. Find the force vector $\mathbf{F}(-2, 0, 0)$.

c. Sketch enough vectors $\mathbf{F}(x, y, 0)$ to get an idea of the force field in the xy-plane.

d. Sketch several of the flow lines for the vector field (see Exercise 28).

20.2 Work and Line Integrals

When a force of magnitude F moves an object d units along a line, the **work** done on the object by the force is the product $W = Fd$. That is,

$$\text{Work} = (\text{force}) \times (\text{distance}) \qquad (1)$$

for constant forces applied along a line.

In Chapter 7 we applied the theory of the definite integral to calculate the work done by a continuously varying force f in moving an object from location $x = a$ to location $x = b$ along a line:

$$W = \lim_{n \to \infty} \sum_{j=1}^{n} f(t_j)\,\Delta x = \int_a^b f(x)\,dx,$$

which agrees with (1) if f is a constant force (Figures 2.1 and 2.2).

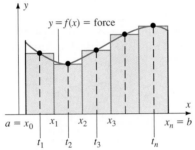

Figure 2.1 $W \approx \sum_{j=1}^{n} f(t_j)\,\Delta x.$

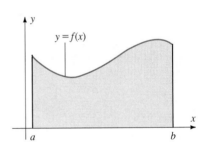

Figure 2.2 $W = \int_a^b f(x)\,dx.$

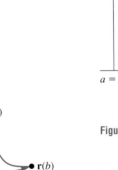

Figure 2.3 The curve C: $\mathbf{r}(t) = x(t)\mathbf{i} + y(t)\mathbf{j}$.

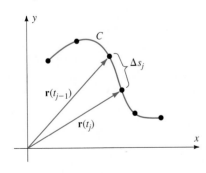

Figure 2.4 Arc length
$$\Delta s_j = \int_{t_{j-1}}^{t_j} |\mathbf{r}'(t)|\,dt.$$

In this section we wish to generalize these concepts to calculate the work done on a particle by a force that moves the particle along a curve C lying in the plane or in space. Examples of this situation include small charged particles moving in electric fields, objects moving in gravitational fields, and others suggested by the discussion of Section 20.1.

To formulate the problem more precisely, assume that $\mathbf{F}$ is a vector field defined on an open box D of $\mathbb{R}^3$ that determines a force vector $\mathbf{F}(\mathbf{r})$ for each position vector $\mathbf{r} = x\mathbf{i} + y\mathbf{j} + z\mathbf{k}$ in D. Also, let C be a curve lying within D that is parameterized by the vector function.

$$C: \quad \mathbf{r}(t) = x(t)\mathbf{i} + y(t)\mathbf{j} + z(t)\mathbf{k}, \qquad a \le t \le b \qquad (\text{see Figure 2.3}). \qquad (2)$$

If we partition the parameter interval $[a, b]$ into n subintervals of equal length $\Delta t = (b - a)/n$ with endpoints $a = t_0 < t_1 < t_2 < \cdots < t_n = b$, the position vectors $\mathbf{r}(t_0), \mathbf{r}(t_1), \dots, \mathbf{r}(t_n)$ divide the curve C into n arcs. If the curve C is smooth (meaning that $\mathbf{r}'(t)$ is continuous and that $|\mathbf{r}'(t)| \neq 0$ for $t \in [a, b]$), equation (10) of Section 17.3 gives the length of the jth arc as

$$\Delta s_j = \int_{t_{j-1}}^{t_j} |\mathbf{r}'(t)|\,dt \qquad (\text{see Figure 2.4}).$$

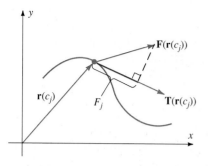

Figure 2.5 F_j is the component of the force field in the direction of the unit tangent $\mathbf{T}$ at $\mathbf{r}(c_j)$.

Under this assumption, the Mean Value Theorem for Integrals guarantees the existence of a number c_j in the interval $[t_{j-1}, t_j]$ for which

$$\Delta s_j = \int_{t_{j-1}}^{t_j} |\mathbf{r}'(t)|\, dt = |\mathbf{r}'(c_j)|(t_j - t_{j-1}) \tag{3}$$

$$= |\mathbf{r}'(c_j)|\, \Delta t.$$

To approximate the work ΔW_j done by the force field in moving a particle along the jth arc, we assume that the magnitude F_j of the force acting throughout the jth arc is the tangential component of $\mathbf{F}(\mathbf{r})$ at the point with position vector $\mathbf{r}(c_j)$. That is, we use

$$F_j = \operatorname{comp}_{\mathbf{T}(\mathbf{r}(c_j))} \mathbf{F}(\mathbf{r}(c_j)) = \mathbf{F}(\mathbf{r}(c_j)) \cdot \mathbf{T}(\mathbf{r}(c_j)) \tag{4}$$

where $\mathbf{T}(\mathbf{r}(c_j))$ is the unit tangent to C at $\mathbf{r}(c_j)$ (see Figure 2.5).

Multiplying (3) and (4) and summing over all n arcs, we obtain the approximation

$$W \approx \sum_{j=1}^{n} F_j\, \Delta s_j = \sum_{j=1}^{n} \mathbf{F}(\mathbf{r}(c_j)) \cdot \mathbf{T}(\mathbf{r}(c_j)) |\mathbf{r}'(c_j)|\, \Delta t$$

for the work done by the force field $\mathbf{F}$ in moving the particle over the curve C. We therefore define the work W as the limit as $n \to \infty$ of this approximating sum:

$$W = \lim_{n \to \infty} \sum_{j=1}^{n} \mathbf{F}(\mathbf{r}(c_j)) \cdot \mathbf{T}(\mathbf{r}(c_j)) |\mathbf{r}'(c_j)|\, \Delta t \tag{5}$$

$$= \int_{a}^{b} \mathbf{F}(\mathbf{r}(t)) \cdot \mathbf{T}(\mathbf{r}(t)) |\mathbf{r}'(t)|\, dt.$$

Equation (5) may be simplified somewhat by recalling that the unit tangent $\mathbf{T}(\mathbf{r}(t))$ to C at $\mathbf{r}(t)$ is given by

$$\mathbf{T}(\mathbf{r}(t)) = \frac{\mathbf{r}'(t)}{|\mathbf{r}'(t)|}. \tag{6}$$

Combining equations (5) and (6) gives the desired definitions of work done by a (continuous) force field in moving a particle along a smooth curve C.

Definition 2

Let $\mathbf{F}$ be a continuous force field defined in some open box containing the smooth curve

$$C: \quad \mathbf{r}(t) = x(t)\mathbf{i} + y(t)\mathbf{j} + z(t)\mathbf{k}, \qquad a \le t \le b, \qquad |\mathbf{r}'(t)| \ne 0.$$

The **work done by the force field $\mathbf{F}$** in moving a particle along C from $\mathbf{r}(a)$ to $\mathbf{r}(b)$ is given by the definite integral

$$W = \int_{a}^{b} \mathbf{F}(\mathbf{r}(t)) \cdot \mathbf{r}'(t)\, dt. \tag{7}$$

REMARK 1   For the curve C as in Definition 2,

$$\mathbf{r}'(t) = \frac{d}{dt}(\mathbf{r}(t)) = \frac{dx}{dt}\mathbf{i} + \frac{dy}{dt}\mathbf{j} + \frac{dz}{dt}\mathbf{k},$$

which suggests the notation

$$d\mathbf{r} = \mathbf{r}'(t)\,dt. \tag{8}$$

Using (8), we sometimes abbreviate the integral in (7) as

$$W = \int_C \mathbf{F} \cdot d\mathbf{r}. \tag{9}$$

In using equation (9), it is important to remember that one must first find a smooth parameterization $\mathbf{r}(t)$ for C and then proceed as in equation (7). It is a fact, which we shall not prove, that the value of the integral in (9) does not depend on the particular parameterization chosen for C.

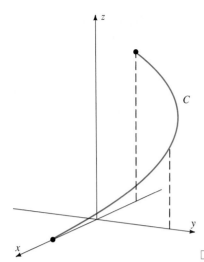

Figure 2.6   The helix $C: \mathbf{r}(t) = \cos t\mathbf{i} + \sin t\mathbf{j} + t\mathbf{k}$ for $0 \le t \le \pi$.

□ **EXAMPLE 1**

Find the work done by the force field

$$\mathbf{F}(x, y, z) = 2x\mathbf{i} + 3y\mathbf{j} + z\mathbf{k} \qquad \text{(see Figures 2.6 and 2.7)}$$

in moving a particle along the circular helix

$$C: \quad \mathbf{r}(t) = \cos t\mathbf{i} + \sin t\mathbf{j} + t\mathbf{k}$$

from point $\mathbf{r}(0) = \mathbf{i}$ to point $\mathbf{r}(\pi) = -\mathbf{i} + \pi\mathbf{k}$.

Solution   The parameterization given for C is

$$x(t) = \cos t; \qquad y(t) = \sin t; \qquad z(t) = t,$$

so

$$\mathbf{F}(\mathbf{r}(t)) = 2\cos t\mathbf{i} + 3\sin t\mathbf{j} + t\mathbf{k}.$$

Also,

$$\mathbf{r}'(t) = -\sin t\mathbf{i} + \cos t\mathbf{j} + \mathbf{k}.$$

Thus, by Definition 2,

$$
\begin{aligned}
W &= \int_0^\pi [2\cos t\mathbf{i} + 3\sin t\mathbf{j} + t\mathbf{k}] \cdot [-\sin t\mathbf{i} + \cos t\mathbf{j} + \mathbf{k}]\,dt \\
&= \int_0^\pi (-2\cos t\sin t + 3\sin t\cos t + t)\,dt \\
&= \int_0^\pi (\sin t\cos t + t)\,dt \\
&= \left[\frac{\sin^2 t}{2} + \frac{t^2}{2}\right]_0^\pi = \frac{\pi^2}{2}.
\end{aligned}
$$

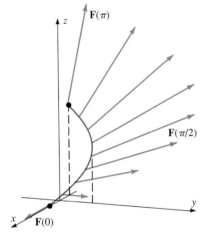

Figure 2.7   Selected vectors in the force field $\mathbf{F}(x, y, z) = 2x\mathbf{i} + 3y\mathbf{j} + z\mathbf{k}$ that correspond to points on C.

□ **EXAMPLE 2**

Find the work done by the force field

$$\mathbf{F}(x, y) = 3y\mathbf{i} - x^2\mathbf{j}$$

in moving a particle along the plane curve $y = \sqrt{x}$ from the point $(1, 1)$ to the point $(4, 2)$.

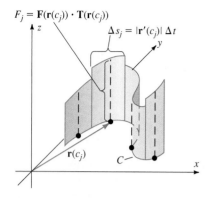

$F_j = \mathbf{F}(\mathbf{r}(c_j)) \cdot \mathbf{T}(\mathbf{r}(c_j))$

$\Delta s_j = |\mathbf{r}'(c_j)| \Delta t$

Figure 2.8 The terms $\mathbf{F}(\mathbf{r}(c_j)) \cdot \mathbf{T}(\mathbf{r}(c_j))|\mathbf{r}'(c_j)| \Delta t$ in (5) can be interpreted as the surface area of a curved "rectangle" that sits on the curve C.

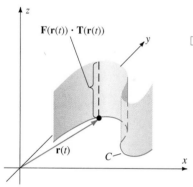

$\mathbf{F}(\mathbf{r}(t)) \cdot \mathbf{T}(\mathbf{r}(t))$

Figure 2.9 If the curve C lies in the xy-plane, the limit of sums in (5) can be interpreted as the area of a surface.

Solution Here we must first find a parameterization for the arc C of the graph of $y = \sqrt{x}$ from $(1, 1)$ to $(4, 2)$. We do so by setting $x(t) = t$, $y(t) = \sqrt{t}$. Then

$$\mathbf{r}(t) = t\mathbf{i} + \sqrt{t}\mathbf{j}, \qquad 1 \le t \le 4,$$

$$\mathbf{r}'(t) = \mathbf{i} + \frac{1}{2\sqrt{t}}\mathbf{j},$$

and

$$\mathbf{F}(\mathbf{r}(t)) = 3\sqrt{t}\mathbf{i} - t^2\mathbf{j}.$$

Thus, by (7)

$$W = \int_1^4 [3\sqrt{t}\mathbf{i} - t^2\mathbf{j}] \cdot \left[\mathbf{i} + \frac{1}{2\sqrt{t}}\mathbf{j}\right] dt$$

$$= \int_1^4 (3t^{1/2} - \tfrac{1}{2}t^{3/2})\,dt$$

$$= \left[2t^{3/2} - \tfrac{1}{5}t^{5/2}\right]_1^4 = \tfrac{39}{5}. \qquad\blacksquare$$

□ **EXAMPLE 3**

Show that the result of Example 2 remains unchanged if the parameterization $x(t) = t^2$, $y(t) = t$, $1 \le t \le 2$, is used for C.

Solution Here $\mathbf{r}(t) = t^2\mathbf{i} + t\mathbf{j}$, $\mathbf{r}'(t) = 2t\mathbf{i} + \mathbf{j}$, and $\mathbf{F}(\mathbf{r}(t)) = 3t\mathbf{i} - t^4\mathbf{j}$. Thus,

$$W = \int_1^2 [3t\mathbf{i} - t^4\mathbf{j}] \cdot [2t\mathbf{i} + \mathbf{j}]\,dt$$

$$= \int_1^2 (6t^2 - t^4)\,dt = \left[2t^3 - \frac{t^5}{5}\right]_1^2 = \frac{39}{5}.$$

This result illustrates our earlier remark that the value of the integral in (7) depends only on $\mathbf{F}$ and on C, not on the particular parameterization chosen for C. $\blacksquare$

Figure 2.8 gives a geometric interpretation of the approximating sum for the integral in equation (5). As Figure 2.9 illustrates, the integral in equation (7) may be interpreted as surface areas of curved vertical "slabs" of variable height $\mathbf{F}(\mathbf{r}(t)) \cdot \mathbf{T}(\mathbf{r}(t))$.

Line Integrals

The existence of the integral in Definition 2 depends only on the mathematical properties of the force field $\mathbf{F}$ and the curve C. We may therefore generalize Definition 2 to apply to any vector field $\mathbf{F}$, not necessarily a force field. The result is called a *line integral*.

Definition 3

Let $\mathbf{F}$ be a continuous vector field in some open box D and let C be a smooth curve lying within D with parameterization $C = \{\mathbf{r}(t) \,|\, a \le t \le b\}$. The **line integral** of $\mathbf{F}$ over C is defined as

$$\int_C \mathbf{F} \cdot d\mathbf{r} = \int_a^b \mathbf{F}(\mathbf{r}(t)) \cdot \mathbf{r}'(t)\,dt. \tag{10}$$

Definitions 2 and 3 are the same except that Definition 2 refers to the more specific situation of calculating work when **F** is a force field.

The line integral in (10) may be written in various other ways. If the vector field **F** has the form

$$\mathbf{F}(x, y, z) = M(x, y, z)\mathbf{i} + N(x, y, z)\mathbf{j} + P(x, y, z)\mathbf{k},$$

we use the notation

$$d\mathbf{r} = dx\,\mathbf{i} + dy\,\mathbf{j} + dz\,\mathbf{k}$$

to write **F** · $d\mathbf{r}$ as

$$\mathbf{F} \cdot d\mathbf{r} = M(x, y, z)\,dx + N(x, y, z)\,dy + P(x, y, z)\,dz.$$

If the parameterization for C has the form

$$\mathbf{r}(t) = x(t)\mathbf{i} + y(t)\mathbf{j} + z(t)\mathbf{k},$$

then

$$\mathbf{r}'(t) = x'(t)\mathbf{i} + y'(t)\mathbf{j} + z'(t)\mathbf{k}.$$

With this notation, the line integral in (10) can be written

$$\int_C \mathbf{F} \cdot d\mathbf{r} = \int_a^b [M(x(t), y(t), z(t))\,x'(t) \\ + N(x(t), y(t), z(t))\,y'(t) \\ + P(x(t), y(t), z(t))\,z'(t)]\,dt. \tag{11}$$

Equation (11) may be abbreviated simply as

$$\int_C \mathbf{F} \cdot d\mathbf{r} = \int_C M(x, y, z)\,dx + N(x, y, z)\,dy + P(x, y, z)\,dz. \tag{12}$$

If the vector field **F** and the curve C are defined in the plane by

$$\mathbf{F}(x, y) = M(x, y)\mathbf{i} + N(x, y)\mathbf{j}$$

and

$$\mathbf{r}(t) = x(t)\mathbf{i} + y(t)\mathbf{j}, \qquad a \le t \le b,$$

the line integral in Definition 3 takes the form

$$\int_C \mathbf{F} \cdot d\mathbf{r} = \int_a^b [M(x(t), y(t))\,x'(t) + N(x(t), y(t))\,y'(t)]\,dt \tag{13}$$

or

$$\int_C \mathbf{F} \cdot d\mathbf{r} = \int_C M(x, y)\,dx + N(x, y)\,dy. \tag{14}$$

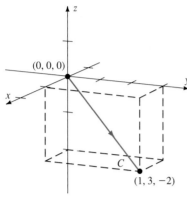

Figure 2.10   Path C from $(0, 0, 0)$ to $(1, 3, -2)$.

□ **EXAMPLE 4**

Evaluate the line integral $\int_C \mathbf{F} \cdot d\mathbf{r}$ where $\mathbf{F}$ is the vector field

$$\mathbf{F}(x, y, z) = x^2\mathbf{i} + xy\mathbf{j} + xz\mathbf{k}$$

and C is the line segment from $(0, 0, 0)$ to $(1, 3, -3)$ (see Figure 2.10).

Solution   Here $\mathbf{F} = M\mathbf{i} + N\mathbf{j} + P\mathbf{k}$ with

$$M(x, y, z) = x^2, \qquad N(x, y, z) = xy, \qquad P(x, y, z) = xz. \tag{15}$$

The line segment C from $(0, 0, 0)$ to $(1, 3, -2)$ may be parameterized with

$$x(t) = t, \qquad y(t) = 3t, \qquad z(t) = -2t \tag{16}$$

for $0 \leq t \leq 1$ (Figure 2.10). Thus,

$$x'(t) = 1, \qquad y'(t) = 3, \qquad z'(t) = -2. \tag{17}$$

Combining equations (15) and (16) shows that

$$M(x(t), y(t), z(t)) = t^2; \qquad N(x(t), y(t), z(t)) = 3t^2; \tag{18}$$
$$P(x(t), y(t), z(t)) = -2t^2.$$

Using equations (11), (17), and (18), we find

$$\int_C \mathbf{F} \cdot d\mathbf{r} = \int_0^1 [(t^2)(1) + (3t^2)(3) + (-2t^2)(-2)]\, dt$$
$$= \int_0^1 14t^2\, dt = \left[\frac{14t^3}{3}\right]_0^1 = \frac{14}{3}. \qquad \blacksquare$$

□ **EXAMPLE 5**

Evaluate the line integral

$$\int_C xy\, dx - 2y^2\, dy$$

where C is the arc of the unit circle from $(1, 0)$ to $(0, 1)$ traversed counterclockwise (Figure 2.11).

Solution   The line integral has the form $\mathbf{F} = M\mathbf{i} + N\mathbf{j}$ where

$$M(x, y) = xy, \qquad N(x, y) = -2y^2.$$

A parameterization for the curve C is given by the equations

$$x(t) = \cos t, \qquad y(t) = \sin t, \qquad 0 \leq t \leq \pi/2 \qquad \text{(Figure 2.11)}.$$

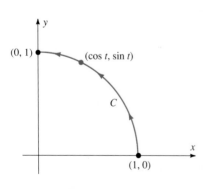

Figure 2.11   Path C from $(1, 0)$ to $(0, 1)$.

Thus

$$x'(t) = -\sin t, \qquad y'(t) = \cos t.$$

Equating the right sides of equations (13) and (14) and using the above functions we obtain

$$\int_C xy \, dx - 2y^2 \, dy = \int_0^{\pi/2} [(\cos t)(\sin t)(-\sin t) - 2(\sin^2 t)(\cos t)] \, dt.$$

$$= \int_0^{\pi/2} -3 \sin^2 t \cos t \, dt$$

$$= \left[-\sin^3 t \right]_0^{\pi/2} = -1. \qquad \blacksquare$$

REMARK 2   Note that the value of the line integral in Definition 3 depends on the *direction* in which the curve C is traced out by the parameterization $\mathbf{r}(t)$. *Reversing the direction along C changes the sign of the line integral.* The reason for this can be most easily seen in equation (5): Reversing the direction in which C is traversed changes the sign of the unit tangent vector $\mathbf{T}(\mathbf{r}(c_j))$.

To illustrate this remark, we evaluate the line integral in Example 5 over the same curve C, but traversed in the opposite direction, from $(0, 1)$ to $(1, 0)$. (We will refer to the curve C with its new orientation as $-C$). A parameterization for $-C$ is

$$x(t) = \cos\left(\frac{\pi}{2} - t\right), \qquad y(t) = \sin\left(\frac{\pi}{2} - t\right), \qquad 0 \le t \le \frac{\pi}{2},$$

so

$$x'(t) = \sin\left(\frac{\pi}{2} - t\right), \qquad y'(t) = -\cos\left(\frac{\pi}{2} - t\right),$$

and

$$\int_{-C} xy \, dx - 2y^2 \, dy = \int_0^{\pi/2} \left[\cos\left(\frac{\pi}{2} - t\right) \sin^2\left(\frac{\pi}{2} - t\right) \right.$$

$$\left. + 2 \sin^2\left(\frac{\pi}{2} - t\right) \cos\left(\frac{\pi}{2} - t\right) \right] dt$$

$$= \int_0^{\pi/2} 3 \sin^2\left(\frac{\pi}{2} - t\right) \cos\left(\frac{\pi}{2} - t\right) dt$$

$$= \left[-\sin^3\left(\frac{\pi}{2} - t\right) \right]_0^{\pi/2} = 1.$$

This is the negative of the result obtained in Example 5. We summarize this remark by writing

$$\boxed{\int_{-C} \mathbf{F} \cdot d\mathbf{r} = -\int_C \mathbf{F} \cdot d\mathbf{r}.}$$

If the curve C is a line segment parallel to the x-axis, the line integral in (11) and (12) reduces to

$$\int_C \mathbf{F} \cdot d\mathbf{r} = \int_C M(x, y, z) \, dx = \int_a^b M(x(t), y(t), z(t)) \, x'(t) \, dt.$$

Such integrals are called **line integrals with respect to x.** Line integrals with respect to y and with respect to z are defined similarly. Using these definitions we may write the line integral in (12) as

$$\int_C \mathbf{F} \cdot d\mathbf{r} = \int_C M(x, y, z)\, dx + \int_C N(x, y, z)\, dy + \int_C P(x, y, z)\, dz.$$

Piecewise Smooth Curves

Up to this point we have defined the line integral only for smooth curves C and continuous vector fields $\mathbf{F}$. (Recall that the curve $C = \{\mathbf{r}(t) \mid a \leq t \leq b\}$ is *smooth* if $\mathbf{r}'(t)$ is continuous and $|\mathbf{r}'(t)| \neq 0$ for all t in the interval $[a, b]$.) The reason why we have assumed that C is smooth is that the integrand in the definite integral in line (10) involves the derivative $\mathbf{r}'(t)$. Thus, when $\mathbf{F}$ and $\mathbf{r}'$ are continuous, the theory developed in Chapter 6 guarantees that the definite integral in equation (10) exists.

However, this integral exists under slightly less restrictive conditions. In particular, it was stated in Chapter 6 that the definite integral exists if the integrand is merely piecewise continuous. Thus, we need only require that $\mathbf{r}'$ be piecewise continuous for the line integral in Definition 3 to exist. Curves of this type are called *piecewise smooth*. We shall refer to such curves as *paths* (see Figures 2.12 and 2.13).

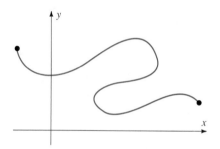

Figure 2.12 A smooth curve ($\mathbf{r}'$ continuous and nonzero).

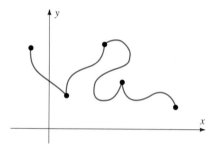

Figure 2.13 A piecewise smooth curve, or *path* ($\mathbf{r}'$ piecewise continuous).

Definition 4

A curve $C = \{\mathbf{r}(t) \mid a \leq t \leq b\}$ is **piecewise smooth** if there exist numbers $a = t_0 < t_1 < t_2 < \cdots < t_n = b$ so that $\mathbf{r}$ is continuous on $[a, b]$ and so that $\mathbf{r}'(t)$ is continuous and $|\mathbf{r}'(t)| \neq 0$ on each of the subintervals (t_0, t_1), (t_1, t_2), . . . , (t_{n-1}, t_n). A piecewise smooth curve is called a **path.**

REMARK 3 For piecewise smooth arcs, we evaluate the line integral by first integrating over each of the subarcs on which $\mathbf{r}$ is smooth and then summing the results:

$$\int_C \mathbf{F} \cdot d\mathbf{r} = \int_a^b \mathbf{F}(\mathbf{r}(t)) \cdot \mathbf{r}'(t)\, dt \tag{19}$$

$$= \int_{t_0}^{t_1} \mathbf{F}(\mathbf{r}(t)) \cdot \mathbf{r}'(t)\, dt + \cdots + \int_{t_{n-1}}^{t_n} \mathbf{F}(\mathbf{r}(t)) \cdot \mathbf{r}'(t)\, dt$$

$$= \int_{C_1} \mathbf{F} \cdot d\mathbf{r} + \int_{C_2} \mathbf{F} \cdot d\mathbf{r} + \cdots + \int_{C_n} \mathbf{F} \cdot d\mathbf{r}$$

where C_j denotes the arc $C_j = \{\mathbf{r}(t) \mid t_{j-1} \leq t \leq t_j\}$.

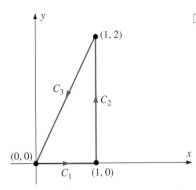

Figure 2.14 The path consisting of three line segments in Example 6.

□ **EXAMPLE 6**

Evaluate the line integral

$$\int_C (x + y)\, dx + xy\, dy$$

over the path C consisting of the line segments

$$C_1: \quad \text{from } (0, 0) \text{ to } (1, 0),$$
$$C_2: \quad \text{from } (1, 0) \text{ to } (1, 2),$$
$$C_3: \quad \text{from } (1, 2) \text{ to } (0, 0),$$

(See Figure 2.14).

Solution   Here the vector field is $\mathbf{F}(x, y) = M(x, y)\mathbf{i} + N(x, y)\mathbf{j}$ with

$$M(x, y) = x + y, \qquad N(x, y) = xy.$$

A parameterization for C_1 is

$$C_1: \begin{cases} x(t) = t \\ y(t) = 0 \end{cases} \quad 0 \le t \le 1.$$

On this arc, $x'(t) = 1$ and $y'(t) = 0$. The line integral over C_1 is therefore

$$\int_{C_1} (x + y)\, dx + xy\, dy = \int_0^1 [(t + 0)(1) + (t \cdot 0)(0)]\, dt$$
$$= \int_0^1 t\, dt = \tfrac{1}{2}.$$

A parameterization for C_2 is

$$C_2: \begin{cases} x(t) = 1 \\ y(t) = t \end{cases} \quad 0 \le t \le 2.$$

On this arc, $x'(t) = 0$ and $y'(t) = 1$. Thus,

$$\int_{C_2} (x + y)\, dx + xy\, dy = \int_0^2 [(1 + t)(0) + (1 \cdot t)(1)]\, dt$$
$$= \int_0^2 t\, dt = 2.$$

Finally, a parameterization for C_3 is

$$C_3: \begin{cases} x(t) = 1 - t \\ y(t) = 2 - 2t \end{cases} \quad 0 \le t \le 1.$$

On this arc, $x'(t) = -1$ and $y'(t) = -2$. Thus, by (13),

$$\int_{C_3} (x + y)\, dx + xy\, dy = \int_0^1 ([(1 - t) + (2 - 2t)](-1)$$
$$+ (1 - t)(2 - 2t)(-2))\, dt$$
$$= \int_0^1 (-4t^2 + 11t - 7)\, dt = -\tfrac{17}{6}.$$

Applying equation (19), we conclude

$$\int_C (x + y)\, dx + xy\, dy = \int_{C_1} (x + y)\, dx + xy\, dy$$

$$+ \int_{C_2} (x + y)\, dx + xy\, dy$$

$$+ \int_{C_3} (x + y)\, dx + xy\, dy$$

$$= \tfrac{1}{2} + 2 - \tfrac{17}{6} = -\tfrac{1}{3}. \qquad ■$$

Line Integrals with Respect to Arc Length

Thus far we have defined line integrals only for vector fields. Another type of line integral, called a **line integral with respect to arc length,** concerns a scalar-valued function, say f on $\mathbb{R}^2$. Let C be a path with parameterization

$$C = \{\mathbf{r}(t) = x(t)\mathbf{i} + y(t)\mathbf{j} \mid a \le t \le b\}.$$

The usual partitioning of the interval $[a, b]$ produces arcs $C_1, C_2, \ldots, C_n$ whose lengths Δs_j, as before, are

$$\Delta s_j = \sqrt{[x'(c_j)]^2 + [y'(c_j)]^2}\, \Delta t, \qquad j = 1, 2, \ldots, n,$$

where c_j is a number in the jth subinterval of $[a, b]$. If the function f is continuous on an open rectangle containing the path C, the Riemann sum

$$\sum_{j=1}^{n} f(x(c_j), y(c_j))\, \Delta s_j \qquad (20)$$

$$= \sum_{j=1}^{n} f(x(c_j), y(c_j))\sqrt{[x'(c_j)]^2 + [y'(c_j)]^2}\, \Delta t$$

converges to the definite integral

$$\int_C f(x, y)\, ds = \int_a^b f(x(t), y(t))\sqrt{[x'(t)]^2 + [y'(t)]^2}\, dt. \qquad (21)$$

The integral in equation (21) is called the *line integral of f over C with respect to arc length.* The difference between the integral in (21) and the line integral in Definition 3 is that (21) is an integral of a scalar function f with respect to change in arc length, while the integral in (10) is an integral of the dot product of a vector field and the tangent with respect to change in the parameter t. By using the differential for arc length

$$ds = \sqrt{[x'(t)]^2 + [y'(t)]^2}\, dt = |\mathbf{r}'(t)|\, dt,$$

we may interpret the line integral in Definition 3 as a special case of the integral in (21) with $f(x(t), y(t)) = \mathbf{F}(\mathbf{r}(t)) \cdot \mathbf{T}(\mathbf{r}(t))$. That is,

$$\int_C \mathbf{F} \cdot d\mathbf{r} = \int_C \mathbf{F} \cdot \mathbf{T}\, ds. \qquad (22)$$

This can be seen by comparing the approximating sums in lines (5) and (20).

Although we shall work almost exclusively with line integrals as defined in Definition 3, equation (22) will be used in Section 20.4 to interpret certain statements about line integrals.

☐ **EXAMPLE 7**

Find $\int_C xy^3 \, ds$ where C is the quarter circle $C = \{\cos t\mathbf{i} + \sin t\mathbf{j} \mid 0 \le t \le \pi/2\}$.

Solution   Here $x(t) = \cos t$, $y(t) = \sin t$, $x'(t) = -\sin t$, and $y'(t) = \cos t$. Thus, by (21)

$$
\int_C xy^3 \, ds = \int_0^{\pi/2} (\cos t)(\sin^3 t)\sqrt{[-\sin t]^2 + [\cos t]^2} \, dt
$$
$$
= \int_0^{\pi/2} \sin^3 t \cos t \, dt
$$
$$
= \left[\frac{\sin^4 t}{4} \right]_0^{\pi/2} = \frac{1}{4}. \qquad \blacksquare
$$

It is important to note that, unlike line integrals of the form $\int_C \mathbf{F} \cdot d\mathbf{r}$, line integrals with respect to arc length are independent of the direction along which the curve C is traversed (see Exercise 32).

An application of line integrals with respect to arc length concerns finding the mass M of a thin wire of variable density whose shape is given by the curve $C = \{\mathbf{r}(t) = x(t)\mathbf{i} + y(t)\mathbf{j} \mid a \le t \le b\}$. If the mass density (in units of mass per unit length) is given by the continuous function f, the expression

$$
\Delta m_j = f(x(c_j), y(c_j)) \, \Delta s_j
$$

approximates the mass of a section of the wire of length Δs_j, one point of which has coordinates $(x(c_j), y(c_j))$. The sum in equation (20) therefore approximates the total mass of the wire, which is given precisely by the line integral with respect to arc length in equation (21).

For curves in space, line integrals with respect to arc length are defined in an entirely analogous manner. That is, if f is continuous in an open box containing the piecewise smooth curve

$$
C = \{\mathbf{r}(t) = x(t)\mathbf{i} + y(t)\mathbf{j} + z(t)\mathbf{k} \mid a \le t \le b\},
$$

then

$$
\int_C f(x, y, z) \, ds
$$
$$
= \int_a^b f(x(t), y(t), z(t))\sqrt{[x'(t)]^2 + [y'(t)]^2 + [z'(t)]^2} \, dt.
$$

Exercise Set 20.2

In Exercises 1–10, calculate the work done by the force field **F** in moving a particle along the specified path C.

1. $\mathbf{F}(x, y) = x\mathbf{i} + y\mathbf{j}$,
 C: $\mathbf{r}(t) = t^2\mathbf{i} + (3 + t)\mathbf{j}$, $0 \le t \le 2$

2. $\mathbf{F}(x, y) = xy\mathbf{i} + x^2\mathbf{j}$,
 C: $\mathbf{r}(t) = \cos t\mathbf{i} + \sin t\mathbf{j}$, $0 \le t \le \pi$

3. $\mathbf{F}(x, y) = x^2\mathbf{i} + y^2\mathbf{j}$,
 C: $\mathbf{r}(t) = \sqrt{t}\mathbf{i} + 3t\mathbf{j}$, $1 \le t \le 4$

4. $\mathbf{F}(x, y) = -y\mathbf{i} + x\mathbf{j}$,
 C: $\mathbf{r}(t) = (t + 1)^2\mathbf{i} + (t - 1)^2\mathbf{j}$, $0 \le t \le 2$

5. $\mathbf{F}(x, y) = (x^2 + y^2)\mathbf{i} + xy\mathbf{j}$,
 C: $\mathbf{r}(t) = t\mathbf{i} + t^2\mathbf{j}$, $0 \le t \le 2$

6. $\mathbf{F}(x, y) = x^2 y\mathbf{i} + xy^2\mathbf{j}$,
 C: $\mathbf{r}(t) = t\mathbf{i} + 2t^2\mathbf{j}$, $0 \le t \le 1$

7. $\mathbf{F}(x, y, z) = xy\mathbf{i} + xz\mathbf{j} + yz\mathbf{k}$,
 C: $\mathbf{r}(t) = \cos t\mathbf{i} + \sin t\mathbf{j} + t\mathbf{k}$, $0 \le t \le \pi$

8. $\mathbf{F}(x, y, z) = x^2\mathbf{i} + y^2\mathbf{j} + z^2\mathbf{k}$,
 C: $\mathbf{r}(t) = t\mathbf{i} + 3t^2\mathbf{j} - 2t\mathbf{k}$, $0 \le t \le 2$

9. $\mathbf{F}(x, y, z) = (2xy + z^2)\mathbf{i} + (x^2 + 6yz^2)\mathbf{j} + (2xz + 6y^2 z)\mathbf{k}$,
 C: $\mathbf{r}(t) = 3t^2\mathbf{i} - t\mathbf{j} + 6\mathbf{k}$, $-1 \le t \le 1$

10. $\mathbf{F}(x, y, z) = \cos x\mathbf{i} - 3\sin z\mathbf{j} + 6\sec^2(z - y)\mathbf{k}$
 C: $\mathbf{r}(t) = 2\sqrt{\pi}t\mathbf{i} + t^2\mathbf{j} + 9t^2\mathbf{k}$, $0 \le t \le \sqrt{\pi}$

In Exercises 11–19, evaluate the line integral of the vector field **F** over the indicated path C.

11. $\mathbf{F}(x, y) = y\mathbf{i} + 2x\mathbf{j}$, C is the line segment from $(0, 0)$ to $(4, 2)$.

12. $\mathbf{F}(x, y) = -x^2\mathbf{i} + y^2\mathbf{j}$, C is the upper unit semicircle from $(1, 0)$ to $(-1, 0)$.

13. $\mathbf{F}(x, y) = (y - x)\mathbf{i} + xy\mathbf{j}$, C is the unit circle traversed counterclockwise.

14. $\mathbf{F}(x, y) = xy^2\mathbf{i} + (x + y)\mathbf{j}$, C is the triangular path from $(0, 0)$ to $(4, 0)$ to $(4, 2)$ to $(0, 0)$.

15. $\mathbf{F}(x, y) = (x + y)\mathbf{i} + (y^2 - x^2)\mathbf{j}$, C is the triangle with vertices $(-1, 0)$, $(1, -4)$, $(0, 2)$ traversed counterclockwise.

16. $\mathbf{F}(x, y, z) = xy\mathbf{i} + y\mathbf{j} + xz\mathbf{k}$, C is the line segment from $(0, 0, 0)$ to $(2, 4, -6)$.

17. $\mathbf{F}(x, y, z) = y\mathbf{i} - x\mathbf{j} + 2z\mathbf{k}$, C is the arc of the circular helix $\mathbf{r}(t) = \cos t\mathbf{i} + \sin t\mathbf{j} + t\mathbf{k}$ from $(1, 0, 0)$ to $(0, 1, \pi/2)$.

18. **F** is the vector field in Exercise 9, C is the circular path from $(0, 0, 1)$ to $(0, 0, 1)$ traversed counterclockwise in the plane $z = 1$.

19. **F** is the vector field in Exercise 8, C is the line segment from $(0, 0, 0)$ to $(2, 12, -4)$.

20. Evaluate

$$\int_C xy\, dx$$

where C is the arc of the unit circle from $(0, 1)$ to $(-1, 0)$ traversed counterclockwise.

21. Evaluate

$$\int_C (x^2 + y^2)\, dy$$

where C is the path in Exercise 20.

22. Evaluate

$$\int_C \mathbf{F} \cdot d\mathbf{r}$$

where $\mathbf{F}(x, y) = (x^2 + y^2)\mathbf{i} + 2xy\mathbf{j}$ and C is the arc of the circle $x^2 + y^2 = 1$ from $(1, 0)$ to $(0, 1)$, followed by the line segment from $(0, 1)$ to $(-1, 1)$ (see Figure 2.15).

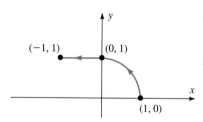

Figure 2.15   The path in Exercise 22.

23. Evaluate

$$\int_C (x^2 + y^2 + z^2)\, dy$$

along the line segment from $(0, 1, 0)$ to $(-1, 2, 1)$.

24. Find the work done on a particle by the force field $\mathbf{F}(x, y) = (x - y)\mathbf{i} + xy\mathbf{j}$ in moving a particle counterclockwise around the ellipse $9x^2 + 4y^2 = 36$ from $(2, 0)$ to $(-2, 0)$.

25. Evaluate

$$\int_C xy\, dx + xy^2\, dy$$

where C is the curve in Exercise 22.

26. Find the work done by the force field $\mathbf{F}(x, y) = (x\mathbf{i} - y\mathbf{j})$ in moving a particle along the path in Figure 2.16.

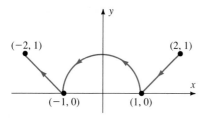

Figure 2.16 The path in Exercise 26.

27. Evaluate

$$\int_C (x + y) \, dx + (x^2 - y^2) \, dy$$

where C is the arc of the parabola $y = 2x^2$ from $(0, 0)$ to $(2, 8)$.

28. Evaluate

$$\int_C (x - y) \, dx + xy \, dy$$

where C is the path consisting of the line segments from $(0, 1)$ to $(2, -1)$, from $(2, -1)$ to $(2, 5)$, and from $(2, 5)$ to $(4, 1)$.

29. Evaluate

$$\int_C x^2 y \, dx + xy^2 \, dy$$

where C is the path from $(0, 2)$ to $(4, 1)$ consisting of the three line segments from $(0, 2)$ to $(0, 0)$ to $(4, 0)$ and then to $(4, 1)$.

30. Let $\mathbf{F}$ be the central force field

$$\mathbf{F}(\mathbf{r}) = \frac{2}{|\mathbf{r}|^3} \mathbf{r}, \mathbf{r} \neq \mathbf{0}.$$

Find the work done by $\mathbf{F}$ in moving a particle along the upper half of the unit circle from $(1, 0)$ to $(-1, 0)$.

31. Show that Definition 2 agrees with the definition of work when a force $f(x)$ is applied to a particle moving along the x-axis.

32. Show that the line integral with respect to arc length in line (21) is independent of the direction along which C is traversed.

In Exercises 33–37, evaluate the line integral with respect to arc length.

33. $\displaystyle\int_C (x + 2y^2) \, ds$ where $C = \{\mathbf{r}(t) = 2t\mathbf{i} - 4t\mathbf{j} \mid 0 \le t \le 1\}$

34. $\displaystyle\int_C xe^y \, ds$ where $C = \{\mathbf{r}(t) = \cos t\mathbf{i} + \sin t\mathbf{j} \mid 0 \le t \le \pi\}$

35. $\displaystyle\int_C (8x + y - 3) \, ds$

where $C = \{\mathbf{r}(t) = t\mathbf{i} + 3\mathbf{j} + t^2\mathbf{k} \mid 0 \le t \le 1\}$

36. $\displaystyle\int_C (3x^2 - 6y + z^3) \, ds$

where $C = \{\mathbf{r}(t) = 2\cos t\mathbf{i} - 2\sin t\mathbf{j} - 2t\mathbf{k} \mid 0 \le t \le \pi/3\}$

37. $\displaystyle\int_C (x^2 + y^2 - 9) \, ds$

where $C = \{\mathbf{r}(t) = 3\sin t\mathbf{i} + 3\cos t\mathbf{j} - 3t\mathbf{k} \mid 0 \le t \le 100\}$

38. Give an argument to support the conclusion that if the curve C describes the shape of a thin wire of mass density $\rho(x, y, z)$ then the total mass of the wire is

$$M = \int_C \rho \, ds$$
$$= \int_a^b \rho(x(t), y(t), z(t)) \sqrt{[x'(t)]^2 + [y'(t)]^2 + [z'(t)]^2} \, dt.$$

39. Refer to Exercise 38. Find the total mass of a wire with shape C: $\mathbf{r}(t) = \cos t\mathbf{i} - \sin t\mathbf{j} + 2\mathbf{k}$, $0 \le t \le \pi$ if the density is $\rho(x, y, z) = x^2 + y^2 + z^2$.

40. Refer to Exercise 38. Find the mass of a wire of constant density $\rho(x, y, z) = k$ shaped like the helix $\mathbf{r}(t) = 4\cos t\mathbf{i} + 4\sin t\mathbf{j} + 3t\mathbf{k}$ for $0 \le t \le 2\pi$.

20.3 Line Integrals: Independence of Path

The Fundamental Theorem of Calculus allows us to evaluate definite integrals of functions of a single variable by means of an antiderivative. In this section, we develop a similar theorem for line integrals. To do so requires the concept of a *scalar potential* ϕ for $\mathbf{F}$. Recall that the differentiable scalar function ϕ is a scalar potential for the vector function $\mathbf{F}$ if $\nabla\phi = \mathbf{F}$ (see Sections 18.9 and 20.1).

Our Fundamental Theorem for line integrals is

Let $C = \{r(t) \mid a \leq t \leq b\}$ be a piecewise smooth curve in an open rectangle D. Let $\mathbf{F}$ be a vector field on D and let ϕ be a differentiable scalar function for which $\mathbf{F}(\mathbf{r}) = \nabla\phi(\mathbf{r})$ for all $\mathbf{r}$ in D. Then

$$\int_C \mathbf{F} \cdot d\mathbf{r} = \phi(\mathbf{r}(b)) - \phi(\mathbf{r}(a)).$$

In other words, if the vector field $\mathbf{F}$ is conservative, meaning that it is the gradient of some scalar function ϕ, then the line integral of $\mathbf{F}$ over C from $A = \mathbf{r}(a)$ to $B = \mathbf{r}(b)$ depends only on the values of the scalar potential ϕ at A and at B. Thus, if the vector field $\mathbf{F}$ is *conservative,* we can determine the value of the line integral by finding a scalar potential ϕ for $\mathbf{F}$. Before proving Theorem 1, we present two examples.

□ **EXAMPLE 1**

Find

$$\int_C \mathbf{F} \cdot d\mathbf{r}$$

where $\mathbf{F}$ is the vector field $\mathbf{F}(x, y) = 2xy\mathbf{i} + x^2\mathbf{j}$ and C is the line segment from $(-1, 2)$ to $(4, -3)$.

Solution   Here $\phi(x, y) = x^2y$ is a potential for $\mathbf{F}$ since

$$\nabla\phi(x, y) = 2xy\mathbf{i} + x^2\mathbf{j} = \mathbf{F}(x, y).$$

Thus, by Theorem 1,

$$\int_C \mathbf{F} \cdot d\mathbf{r} = \phi(4, -3) - \phi(-1, 2) = 4^2(-3) - (-1)^2 2 = -50. \quad \blacksquare$$

□ **EXAMPLE 2**

Evaluate the line integral

$$\int_C 2xye^{x^2}\, dx + e^{x^2}\, dy$$

over a path C from $(0, 1)$ to $(1, 3)$.

Solution   This integral has the form $\int_C \mathbf{F} \cdot d\mathbf{r}$ where

$$\mathbf{F}(x, y) = 2xye^{x^2}\mathbf{i} + e^{x^2}\mathbf{j}$$

and $d\mathbf{r} = dx\,\mathbf{i} + dy\,\mathbf{j}$. A potential for $\mathbf{F}$ is $\phi(x, y) = ye^{x^2}$ since

$$\nabla\phi(x, y) = 2xye^{x^2}\mathbf{i} + e^{x^2}\mathbf{j} = \mathbf{F}(x, y).$$

Thus, by Theorem 1

$$\int_C 2xye^{x^2}\, dx + e^{x^2}\, dy = \phi(1, 3) - \phi(0, 1) = 3e - e^0 = 3e - 1. \quad \blacksquare$$

Proof of Theorem 1:   We will use the Chain Rule: If x and y are differentiable component functions, then

$$\frac{d}{dt}\phi(x(t), y(t)) = \phi_x(x(t), y(t))\, x'(t) + \phi_y(x(t), y(t))\, y'(t)$$

$$= \nabla\phi(x(t), y(t)) \cdot [x'(t)\mathbf{i} + y'(t)\mathbf{j}].$$

Using the position vector notation, this can be written

$$\frac{d}{dt}\phi(\mathbf{r}(t)) = \nabla\phi(\mathbf{r}(t)) \cdot \mathbf{r}'(t). \tag{1}$$

Assume first that C is a smooth curve. Then, under the hypotheses of Theorem 1, equation (1) holds. Thus,

$$\int_C \mathbf{F} \cdot d\mathbf{r} = \int_a^b \mathbf{F}(\mathbf{r}(t)) \cdot \mathbf{r}'(t)\, dt$$

$$= \int_a^b \nabla\phi(\mathbf{r}(t)) \cdot \mathbf{r}'(t)\, dt$$

$$= \int_a^b \frac{d}{dt}[\phi(\mathbf{r}(t))]\, dt \qquad \text{(equation (1))}$$

$$= \left[\phi(\mathbf{r}(t))\right]_a^b = \phi(\mathbf{r}(b)) - \phi(\mathbf{r}(a)).$$

The extension to the case C piecewise smooth is easy and is left as an exercise. ∎

In trying to determine whether the vector field $\mathbf{F}$ is conservative, it is helpful to recall Theorem 9, Section 18.9: If the functions M and N are continuously differentiable in an open rectangle D, *the vector function*

$$\mathbf{F}(x, y) = M(x, y)\mathbf{i} + N(x, y)\mathbf{j}$$

is conservative if and only if

$$\boxed{M_y(x, y) = N_x(x, y)} \tag{2}$$

for all (x, y) in D. Equation (2) provides a quick check of whether Theorem 1 applies for a vector field $\mathbf{F}$ in the plane. If equation (2) holds, the method of Section 18.9 can often be applied to find the potential ϕ.

□ **EXAMPLE 3**

Find the work done by the vector field

$$\mathbf{F}(x, y) = (e^y + 3x^2y^2)\mathbf{i} + (xe^y + 2x^3y + 2)\mathbf{j}$$

in moving a particle from the point $(0, 0)$ to the point $(3, 2)$ in the plane.

Solution   Here $\mathbf{F}(x, y) = M(x, y)\mathbf{i} + N(x, y)\mathbf{j}$ where

$$M(x, y) = e^y + 3x^2y^2, \qquad N(x, y) = xe^y + 2x^3y + 2.$$

Then

$$M_y(x, y) = e^y + 6x^2y = N_x(x, y),$$

so the vector field **F** is conservative. To find a potential ϕ with

$$\nabla\phi(x, y) = \phi_x(x, y)\mathbf{i} + \phi_y(x, y)\mathbf{j} \tag{3}$$
$$= M(x, y)\mathbf{i} + N(x, y)\mathbf{j} = \mathbf{F}(x, y),$$

we first equate **i**-components and integrate partially with respect to x to find that

$$\phi(x, y) = \int (e^y + 3x^2y^2) \, dx = xe^y + x^3y^2 + f(y) \tag{4}$$

where f is a function of y alone. To determine f, we differentiate both sides of equation (4) with respect to y and obtain

$$\phi_y(x, y) = xe^y + 2x^3y + f'(y).$$

Since we require that $\phi_y = N$, we have $f'(y) = 2$. Thus,

$$f(y) = 2y + C. \tag{5}$$

From equations (4) and (5), we know that

$$\phi(x, y) = xe^y + x^3y^2 + 2y$$

is a scalar potential for **F,** and we apply Theorem 1. The desired work is

$$W = \int_C \mathbf{F} \cdot d\mathbf{r} = \phi(3, 2) - \phi(0, 0)$$
$$= (3e^2 + 3^3 \cdot 2^2 + 2 \cdot 2) - 0 = 3e^2 + 112. \qquad \blacksquare$$

Whether or not a potential ϕ can explicitly be determined, Theorem 1 guarantees that the value of a line integral for a conservative vector field depends only on the endpoints of the path C and not on the path itself. (Physicists paraphrase this statement by saying that "work is a function of position, not of path.") The terminology we choose to use is that the line integral of a conservative force field is **independent of path.** The following theorem characterizes this notion.

Theorem 2	Let C be a piecewise smooth curve in an open rectangle D. The line integral $$\int_C \mathbf{F} \cdot d\mathbf{r}$$ is independent of path if and only if the vector field **F** is conservative in D.

The "if" part of this theorem follows directly from Theorem 1. The "only if" part is an important observation in more advanced courses in mathematics but will not be used here in a direct way. Consequently we omit its proof.

The following corollary is a useful formulation of Theorem 2 and the statement concerning equation (2) for vector fields in the plane.

Corollary 1

Let D be an open rectangle in the plane and let C be any path from point A to point B lying within D. Let M and N be continuously differentiable in D. The line integral

$$\int_C M(x, y)\, dx + N(x, y)\, dy \tag{6}$$

is independent of path if and only if

$$M_y(x, y) = N_x(x, y) \tag{7}$$

for all (x, y) in D.

☐ **EXAMPLE 4**

Evaluate the line integral

$$\int_C (\cos xy - xy \sin xy)\, dx - x^2 \sin xy\, dy$$

where C is the upper unit semicircle from $(1, 0)$ to $(-1, 0)$ (see Figure 3.1).

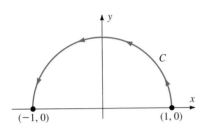

Figure 3.1 Circular path C from $(1, 0)$ to $(-1, 0)$.

Solution   This integral has the form (6) with

$$M(x, y) = \cos xy - xy \sin xy, \qquad N(x, y) = -x^2 \sin xy,$$

so

$$M_y(x, y) = -2x \sin xy - x^2 y \cos xy = N_x(x, y).$$

Thus, equation (7) holds, so the integral is independent of path. However, instead of applying the method of Example 3 to actually find the potential ϕ, we demonstrate how to apply Corollary 1 in evaluating this line integral.

Since the integral is independent of path, we may choose a path over which the line integral is easy to compute. In this case, we choose the path C_1 consisting of the (straight) line segment from $(1, 0)$ to $(-1, 0)$ (see Figure 3.2). A parameterization for C_1 is

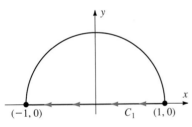

Figure 3.2 Straight path C_1 from $(1, 0)$ to $(-1, 0)$.

$$C_1: \begin{cases} x(t) = 1 - t \\ y(t) = 0 \end{cases} \quad 0 \le t \le 2.$$

Thus $dx = x'(t)\, dt = (-1)\, dt$ and $dy = y'(t)\, dt = (0)\, dt$. Calculating the line integral according to the definition (as expressed in equation (13) in Section 20.2) yields

$$\int_C (\cos xy - xy \sin xy)\, dx - x^2 \sin xy\, dy$$

$$= \int_C (\cos xy - xy \sin xy)\, dx - \int_C x^2 \sin xy\, dy$$

$$= \int_0^2 [\cos(0) - 0 \cdot \sin(0)](-1)\, dt - \int_0^2 [(1 - t)^2 \sin(0)](0)\, dt$$

$$= \int_0^2 (-1)\, dt = -2.$$

Now we note that a scalar potential for this vector field is $\phi(x, y) = x \cos xy$. With this information the line integral could have been evaluated using Theorem 1 as

$$\phi(-1, 0) - \phi(1, 0) = (-1)\cos(0) - (1)\cos(0) = -2. \qquad ■$$

REMARK   Since a line integral over a path from point A to point B for a conservative force field $\mathbf{F}$ depends only on the points A and B, we sometimes use the notation

$$\int_C \mathbf{F} \cdot d\mathbf{r} = \int_A^B \mathbf{F} \cdot d\mathbf{r}.$$

If $\mathbf{F} = \nabla \phi$, Theorem 1 is stated in this notation as

$$\int_A^B \mathbf{F} \cdot d\mathbf{r} = \phi(B) - \phi(A).$$

Note that this notation is meaningful only if the line integral is independent of path.

☐ **EXAMPLE 5**

In the above notation, the result of Example 4 is written as

$$\int_{(1, 0)}^{(-1, 0)} (\cos xy - xy \sin xy) \, dx - x^2 \sin xy \, dy = -2.$$ ■

☐ **EXAMPLE 6**

Although all of the preceding examples have dealt with paths in the plane, Theorems 1 and 2 apply also to paths and vector fields in space. The line integral

$$\int_C yz \, dx + xz \, dy + xy \, dz$$

over the helical path

$$C: \quad \mathbf{r}(t) = \cos t \mathbf{i} + \sin t \mathbf{j} + t \mathbf{k}, \qquad 0 \le t \le \pi/4$$

may be handled by means of Theorem 1 by noting that the vector field

$$\mathbf{F}(x, y, z) = yz\mathbf{i} + xz\mathbf{j} + xy\mathbf{k}$$

is the gradient of the scalar potential

$$\phi(x, y, z) = xyz.$$

Since the endpoints of C are $A = \mathbf{r}(0) = (1, 0, 0)$ and $B = \mathbf{r}(\pi/4) = (\sqrt{2}/2, \sqrt{2}/2, \pi/4)$, we have

$$\int_C yz \, dx + xz \, dy + xy \, dz = \int_{(1, 0, 0)}^{(\sqrt{2}/2, \sqrt{2}/2, \pi/4)} yz \, dx + xz \, dy + xy \, dz$$

$$= \phi(\sqrt{2}/2, \sqrt{2}/2, \pi/4) - \phi(1, 0, 0)$$

$$= \pi/8.$$ ■

Line Integrals Over Closed Paths

The path (piecewise smooth curve) $C = \{\mathbf{r}(t) \mid a \le t \le b\}$ is called **closed** if $\mathbf{r}(a) = \mathbf{r}(b)$. Thus, circles, ellipses, triangles, squares, and rectangles are all examples of closed paths in the plane. There is a very simple situation governing line integrals over closed paths. *If the vector field $\mathbf{F}$ is conservative, the line integral*

$$\int_C \mathbf{F} \cdot d\mathbf{r} = 0$$

if the path C is closed. This statement follows directly from Theorem 1: Since a closed path $C = \{\mathbf{r}(t) \mid a \le t \le b\}$ satisfies the equation $\mathbf{r}(a) = \mathbf{r}(b)$,

$$\int_C \mathbf{F} \cdot d\mathbf{r} = \phi(\mathbf{r}(b)) - \phi(\mathbf{r}(a)) = 0.$$

This fact is well known in physics. In terms of work, it says that the work done by a conservative force field (meaning that no energy is lost in the process) in moving a particle around a closed path is zero.

Conservation of Energy

Let's take the physics one step further. Let $\mathbf{F}$ be a conservative force field. If a particle of mass m moves in the vector field $\mathbf{F}$ with a velocity $\mathbf{v}(t) = \mathbf{r}'(t)$, we define the **kinetic energy** of the particle at time t as

$$K(t) = \tfrac{1}{2}m|\mathbf{v}(t)|^2. \tag{8}$$

Also, the change in **potential energy** U for the particle as it is moved from point A to point B is defined as the negative of the work done by the force field $\mathbf{F}$ in moving it from A to B:

$$U(B) - U(A) = -\int_A^B \mathbf{F} \cdot d\mathbf{r}. \tag{9}$$

Now if $\mathbf{a}(t)$ denotes the acceleration of the particle at time t, Newton's second law of motion states that

$$\mathbf{F}(\mathbf{r}(t)) = m\mathbf{a}(t) = m\mathbf{r}''(t),$$

so we can rewrite the integral in Definition 2 as

$$\begin{aligned}
\mathbf{F}(\mathbf{r}(t)) \cdot \mathbf{r}'(t) &= m\mathbf{r}''(t) \cdot \mathbf{r}'(t) \tag{10}\\
&= \tfrac{1}{2}m[2\mathbf{r}''(t) \cdot \mathbf{r}'(t)]\\
&= \tfrac{1}{2}m\,\frac{d}{dt}[\mathbf{r}'(t) \cdot \mathbf{r}'(t)]\\
&= \frac{d}{dt}[\tfrac{1}{2}m|\mathbf{v}(t)|^2].
\end{aligned}$$

Using (8) and (10) and the fact that $\mathbf{F}$ is conservative, we find that

$$\begin{aligned}
\text{Work} = \int_A^B \mathbf{F} \cdot d\mathbf{r} &= \int_a^b \mathbf{F}(\mathbf{r}(t)) \cdot \mathbf{r}'(t)\,dt \tag{11}\\
&= \int_a^b \frac{d}{dt}[\tfrac{1}{2}m|\mathbf{v}(t)|^2]\,dt\\
&= \tfrac{1}{2}m|\mathbf{v}(b)|^2 - \tfrac{1}{2}m|\mathbf{v}(a)|^2\\
&= K(B) - K(A)
\end{aligned}$$

where $A = \mathbf{r}(a)$, $B = \mathbf{r}(b)$. That is,

$$\boxed{\text{Work} = K(B) - K(A).} \tag{12}$$

Equation (12) may be interpreted as the principle that, *in a conservative system, the work done on an object (by a force) equals the change in its kinetic energy.*

Finally, combining equations (9) and (11) gives the equation

$$U(A) - U(B) = K(B) - K(A)$$

or

$$K(A) + U(A) = K(B) + U(B). \tag{13}$$

Equation (13) is the famous **principle of conservation of energy:** In a conservative system, the sum of kinetic and potential energy of a moving particle remains constant from point to point.

Exercise Set 20.3

In Exercises 1–14, evaluate the line integral by verifying that the vector field is conservative and applying Theorem 1.

1. $\displaystyle\int_C \mathbf{F} \cdot d\mathbf{r}, \quad \mathbf{F}(x, y) = y\mathbf{i} + x\mathbf{j},$

C is a path from $(0, 0)$ to $(3, 1)$.

2. $\displaystyle\int_C \mathbf{F} \cdot d\mathbf{r}, \quad \mathbf{F}(x, y) = 3x^2y^2\mathbf{i} + 2x^3y\mathbf{j},$

C is a path from $(-3, 1)$ to $(2, 2)$.

3. $\displaystyle\int_C \mathbf{F} \cdot d\mathbf{r}, \quad \mathbf{F}(x, y) = ye^{xy}\mathbf{i} + xe^{xy}\mathbf{j},$

C is a path from $(0, 0)$ to $(1, 2)$.

4. $\displaystyle\int_C \mathbf{F} \cdot d\mathbf{r}, \quad \mathbf{F}(x, y, z) = yze^{xyz}\mathbf{i} + xze^{xyz}\mathbf{j} + xye^{xyz}\mathbf{k},$

C is a path from $(0, 0, 0)$ to $(1, 0, 1)$.

5. $\displaystyle\int_C \mathbf{F} \cdot d\mathbf{r}, \quad \mathbf{F}(x, y) = 2x\mathbf{i} + 2y\mathbf{j},$

C is the upper unit semicircle traversed counterclockwise.

6. $\displaystyle\int_{(1, 1)}^{(2, 3)} (1 + 2xy^2) \, dx + 2x^2y \, dy$

7. $\displaystyle\int_{(0, 0)}^{(2, 1)} e^{y^2} \, dx + 2xye^{y^2} \, dy$

8. $\displaystyle\int_{(0, 0)}^{(\pi/2, 1)} y \sin xy \, dx + x \sin xy \, dy$

9. $\displaystyle\int_{(0, 0)}^{(1, \pi/4)} e^x \sin y \, dx + e^x \cos y \, dy$

10. $\displaystyle\int_{(0, 0, 0)}^{(0, \pi/4, \pi/4)} e^x \sin y \cos z \, dx$

$\qquad + e^x \cos y \cos z \, dy - e^x \sin y \sin z \, dz$

11. $\displaystyle\int_{(1, 3, -4)}^{(7, 14, 0)} (2xy + z^2) \, dx + (x^2 - 2) \, dy + (2xz + 1) \, dz$

12. $\displaystyle\int_{(0, \pi/2, 6)}^{(1, 0, 2)} (e^x + 2xy) \, dx + (x^2 - \sin y) \, dy + 4z^3 \, dz$

13. $\displaystyle\int_B^A \frac{dx + dy + dz}{(x + y + z)\sqrt{x^2 + y^2 + z^2 + 2xy + 2xz + 2yz - 1}}$

where $A = (1, 1, 0)$ and $B = (0, 0, 2\sqrt{3}/3)$. (*Hint:* Let $u = x + y + z$.)

14. $\displaystyle\int_{(1, 1, 1)}^{(e, e^2, e^3)} \left(yze^{xyz} + \frac{1}{x}\right) dx + \left(xze^{xyz} + \frac{1}{y}\right) dy$

$\qquad + \left(xye^{xyz} + \frac{1}{z}\right) dz$

15. Let $\mathbf{F}(x, y)$ define a conservative vector field throughout the plane. Let C_1 be the path along the upper semicircle from $(1, 0)$ to $(-1, 0)$. Let C_2 be the path along the x-axis from $(-1, 0)$ to $(1, 0)$. If

$$\int_{C_1} \mathbf{F} \cdot d\mathbf{r} = a,$$

what is

$$\int_{C_2} \mathbf{F} \cdot d\mathbf{r}?$$

Why?

16. Evaluate

$$\int_C \sin y \, dx + x \cos y \, dy$$

where C is the curve with parameterization C: $\mathbf{r}(t) = \cos t\mathbf{i} + \sin t\mathbf{j}, \quad 0 \le t \le \pi.$

17. Use Theorem 2 to prove that the vector field $\mathbf{F}$ is conservative in the open set D if and only if

$$\int_C \mathbf{F} \cdot d\mathbf{r} = 0$$

for every closed path C in D.

18. Use Theorem 2 and Exercise 17 to prove that the line integral

$$\int_C \mathbf{F} \cdot d\mathbf{r}$$

is independent of path in the open set D if and only if

$$\int_C \mathbf{F} \cdot d\mathbf{r} = 0$$

for every closed path C' in D.

19. Let $\mathbf{F}(x, y) = 2xye^{x^2y}\mathbf{i} + x^2 e^{x^2y}\mathbf{j}$. Let C_1 be the path from $A = (-1, 0)$ to $B = (1, 0)$ clockwise along the unit circle, and let C_2 be the path consisting of the line segment from A to B. Use the method of Section 20.2 to evaluate each of the line integrals

$$\int_{C_1} \mathbf{F} \cdot d\mathbf{r} \quad \text{and} \quad \int_{C_2} \mathbf{F} \cdot d\mathbf{r}.$$

Are they equal? Why?

20. Repeat Exercise 19 for the vector field $\mathbf{F}(x, y) = x^3 y^2 \mathbf{i} + x^3 y \mathbf{j}$.

21. Show that, in a conservative force field, the force is equal to the negative gradient of the potential.

22. Find a force field $\mathbf{F}$ so that

$$\int_{(0, 0, 0)}^{(x, y, z)} \mathbf{F} \cdot d\mathbf{r} = xyz$$

for all $(x, y, z) \in \mathbb{R}^3$, independent of path.

23. Are line integrals of $\mathbf{F}(x, y) = (x - y)\mathbf{i} + (x + y)\mathbf{j}$ independent of path in $\mathbb{R}^2$? Why or why not?

24. a. Show that

$$\int_C \frac{1}{x^2 + y^2}(x \, dy - y \, dx) = 2\pi$$

where C is the unit circle oriented counterclockwise.

b. Show that the vector field

$$\mathbf{F}(x, y) = \frac{-y}{x^2 + y^2}\mathbf{i} + \frac{x}{x^2 + y^2}\mathbf{j}.$$

is not conservative.

c. Show that, for

$$M(x, y) = \frac{-y}{x^2 + y^2} \quad \text{and} \quad N(x, y) = \frac{x}{x^2 + y^2},$$

$M_y(x, y) = N_x(x, y)$ if $(x, y) \neq (0, 0)$.

d. Explain why parts a through c do not contradict the statement involving equation (2).

25. Prove Theorem 1 in the case where C is piecewise smooth.

20.4   Green's Theorem

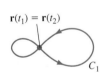

$\mathbf{r}(t_1) = \mathbf{r}(t_2)$

C_1

Figure 4.1 The curve C_1 is not a simple closed curve.

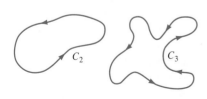

C_2 C_3

Figure 4.2 The curves C_2 and C_3 are simple closed curves.

The Fundamental Theorem of Calculus states that, if $F'(x) = f(x)$ for all x in $[a, b]$, the definite integral $\int_a^b f(x) \, dx$ may be evaluated by means of the formula

$$\int_a^b f(x) \, dx = F(b) - F(a). \tag{1}$$

In other words, the value of the integral in (1) is completely determined by the associated antiderivative at the endpoints of the interval $[a, b]$. In this section, we establish a similar result for certain functions of two variables: The value of a double integral of such a function over a region R in the plane is the same as an associated line integral taken around the boundary of R. This remarkable result is due to George Green, an English mathematician and physicist (1793–1841).

The statement of Green's Theorem involves the concept of **simple closed curves**. A closed curve $C = \{\mathbf{r}(t) \mid a \le t \le b\}$ is called simple if $\mathbf{r}(t_1) \neq \mathbf{r}(t_2)$ for all numbers t_1 and t_2 in (a, b) with $t_1 \neq t_2$. (Remember, we must have $\mathbf{r}(a) = \mathbf{r}(b)$ if C is closed.) In other words, *a simple closed curve does not cross itself* (see Figures 4.1 and 4.2). Be careful not to confuse the notion of simple closed *curves* with our earlier concepts of vertically simple or horizontally simple *regions*.

Theorem 3

Green's Theorem

Let C be a piecewise smooth simple closed curve in the plane, oriented counter-clockwise, and let Q be the region enclosed by C. If the functions $M(x, y)$ and $N(x, y)$ have continuous first partial derivatives in an open region containing Q, then

$$\int_C M(x, y)\, dx + N(x, y)\, dy = \iint_Q \left(\frac{\partial N}{\partial x} - \frac{\partial M}{\partial y} \right) dA. \tag{2}$$

Before discussing the proof of Green's Theorem, we illustrate its significance with a few examples and applications. We begin by expressing the line integral in (2) in vector form. If we write

$$\mathbf{F}(x, y) = M(x, y)\mathbf{i} + N(x, y)\mathbf{j}$$

and

$$C = \{\mathbf{r}(t)\,|\,a \le t \le b\},$$

then the conclusion (2) may be written as

$$\int_C \mathbf{F}(\mathbf{r}) \cdot d\mathbf{r} = \iint_Q \left(\frac{\partial N}{\partial x} - \frac{\partial M}{\partial y} \right) dA \tag{3}$$

Now, if $\mathbf{F}$ is a conservative vector field, equation (3) is not surprising since both integrals are zero. The line integral is zero since C is closed, and the double integral is zero since $\partial N/\partial x - \partial M/\partial y = 0$ if $\mathbf{F}$ is conservative. Thus, Green's Theorem is most significant for evaluating line integrals for *nonconservative* vector fields.

In Section 20.2, we presented a geometric interpretation of the line integral as the area of a surface with base C and height $\mathbf{F}(\mathbf{r}(t)) \cdot \mathbf{T}(\mathbf{r}(t))$ (assuming $\mathbf{F}(\mathbf{r}(t)) \cdot \mathbf{T}(\mathbf{r}(t)) > 0$). Interpreted similarly, Green's Theorem states that this surface area equals the volume of the solid that is based on the region Q enclosed by C and whose height is $\partial N/\partial x - \partial M/\partial y$ (see Figure 4.3).

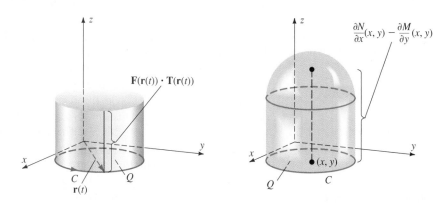

Figure 4.3 The statement of Green's Theorem interpreted geometrically.

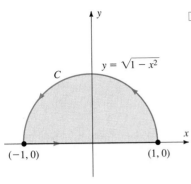

Figure 4.4 Region Q in Example 1.

□ **EXAMPLE 1**

Use Green's Theorem to evaluate the line integral

$$\int_C xy^2 \, dx + 2x^3y \, dy$$

where C is the path consisting of the upper unit semicircle traversed counterclockwise and the directed line segment from $(-1, 0)$ to $(1, 0)$ (see Figure 4.4).

Solution Here $M(x, y) = xy^2$, $N(x, y) = 2x^3y$. The region Q enclosed by the path C can be described by the inequalities

$$Q: \quad -1 \le x \le 1; \quad 0 \le y \le \sqrt{1 - x^2}.$$

With $dA = dx \, dy$, Green's Theorem says that

$$\int_C xy^2 \, dx + 2x^3y \, dy = \int_{-1}^{1} \int_0^{\sqrt{1-x^2}} \left[\frac{\partial}{\partial x}(2x^3y) - \frac{\partial}{\partial y}(xy^2) \right] dy \, dx$$

$$= \int_{-1}^{1} \int_0^{\sqrt{1-x^2}} (6x^2y - 2xy) \, dy \, dx$$

$$= \int_{-1}^{1} \left[(3x^2 - x)y^2 \right]_0^{\sqrt{1-x^2}} dx$$

$$= \int_{-1}^{1} (3x^2 - x)(1 - x^2) \, dx$$

$$= \int_{-1}^{1} (-3x^4 + x^3 + 3x^2 - x) \, dx = \tfrac{4}{5}. \qquad \blacksquare$$

□ **EXAMPLE 2**

Use Green's Theorem to evaluate the line integral

$$\int_C xy \, dx + e^y \, dy$$

where C is the path from $(0, 0)$ to $(1, 1)$ along the graph of $y = x^2$ and from $(1, 1)$ to $(0, 0)$ along the graph of $y = \sqrt{x}$.

Solution Note in Figure 4.5 that C is oriented counterclockwise, as required. The region Q enclosed by C can be described by the inequalities

$$Q: \quad 0 \le x \le 1; \quad x^2 \le y \le \sqrt{x}.$$

Since $M(x, y) = xy$ and $N(x, y) = e^y$, an application of Green's Theorem gives

$$\int_C xy \, dx + e^y \, dy = \int_0^1 \int_{x^2}^{\sqrt{x}} \left[\frac{\partial}{\partial x}(e^y) - \frac{\partial}{\partial y}(xy) \right] dy \, dx$$

$$= \int_0^1 \int_{x^2}^{\sqrt{x}} -x \, dy \, dx$$

$$= \int_0^1 \left[-xy \right]_{x^2}^{\sqrt{x}} dx$$

$$= \int_0^1 (-x^{3/2} + x^3) \, dx = -\tfrac{3}{20}. \qquad \blacksquare$$

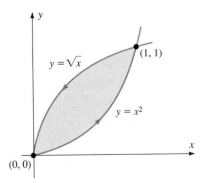

Figure 4.5 Region Q in Example 2.

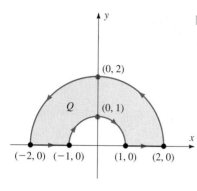

Figure 4.6 The path C in Example 3 that encloses the region Q.

□ **EXAMPLE 3**

Use Green's Theorem to evaluate the line integral

$$\int_C xy\,dx + (x + y)\,dy$$

where C is the path indicated in Figure 4.6.

Solution   The region Q enclosed by C can be described in polar coordinates by the inequalities

$$Q: \quad 1 \le r \le 2; \quad 0 \le \theta \le \pi.$$

Since $M(x, y) = xy$ and $N(x, y) = x + y$, we have

$$\frac{\partial N}{\partial x} - \frac{\partial M}{\partial y} = \frac{\partial}{\partial x}(x + y) - \frac{\partial}{\partial y}(xy) = 1 - x. \tag{4}$$

Since we are using polar coordinates, we set

$$x = r\cos\theta, \quad y = r\sin\theta, \quad \text{and} \quad dA = r\,dr\,d\theta. \tag{5}$$

Combining (4) and (5) with the statement of Green's Theorem, we obtain

$$\int_C xy\,dx + (x + y)\,dy = \int_0^\pi \int_1^2 [1 - r\cos\theta]r\,dr\,d\theta$$

$$= \int_0^\pi \left[\frac{r^2}{2} - \frac{r^3}{3}\cos\theta\right]_{r=1}^{r=2} d\theta$$

$$= \int_0^\pi \left(\frac{3}{2} - \frac{7}{3}\cos\theta\right) d\theta$$

$$= \left[\frac{3\theta}{2} - \frac{7}{3}\sin\theta\right]_0^\pi = \frac{3\pi}{2}. \quad ■$$

There should be no doubt in your mind that the Green's Theorem solution to this problem is much easier than a direct evaluation of the line integral, since the latter would require four separate parameterizations, one for each arc of the curve C.

Calculating Area by Line Integrals

So far we have seen examples in which the double integral over Q has been easier to evaluate than the line integral over C in Green's Theorem. Sometimes it is the other way around. Since areas of certain regions in the plane may be calculated by means of a double integral, Green's Theorem enables us to find expressions for the area of Q (in Theorem 3) in terms of line integrals over C. For example, if we use the functions $M(x, y) = 0$ and $N(x, y) = x$, we obtain

$$\int_C x\,dy = \iint_Q \left[\frac{\partial}{\partial x}(x) - \frac{\partial}{\partial y}(0)\right] dA = \iint_Q 1\,dA = \text{Area of } Q. \tag{6}$$

Similarly, by choosing $M(x, y) = y$ and $N(x, y) = 0$, we obtain

$$\int_C y\,dx = \iint_Q \left[\frac{\partial}{\partial x}(0) - \frac{\partial}{\partial y}(y)\right] dA = \iint_Q (-1)\,dA = -(\text{Area of } Q). \tag{7}$$

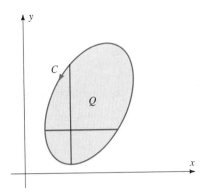

Figure 4.7 The region Q is both x- simple and y-simple.

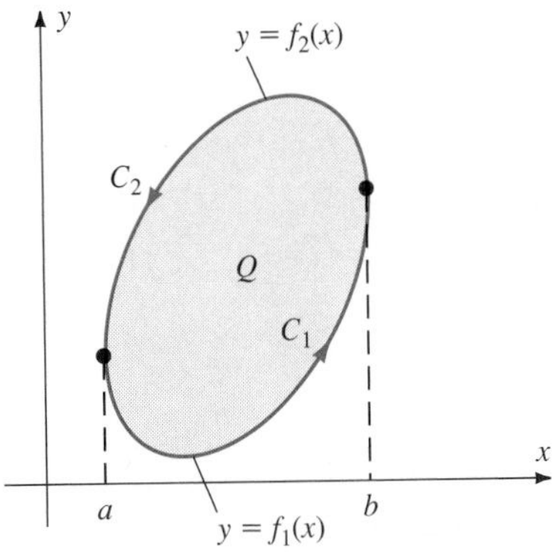

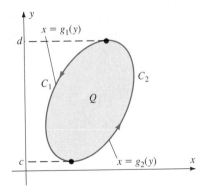

Figure 4.8 The curve C_j is graph of $y = f_j(x)$, for $j = 1, 2$.

Subtracting the corresponding sides of equations (6) and (7) and dividing by 2 gives

$$\text{Area of } Q = \tfrac{1}{2} \int_C x \, dy - y \, dx. \tag{8}$$

Any of equations (6), (7), or (8) may be used in calculating the area of Q. In the following example, equation (8) provides an easy solution to a problem that otherwise requires the technique of integration by trigonometric substitution.

□ **EXAMPLE 4**

Calculate the area A of the region enclosed by the ellipse

$$\frac{x^2}{a^2} + \frac{y^2}{b^2} = 1.$$

Solution A parameterization for the ellipse with counterclockwise orientation is

$$x(t) = a \cos t, \qquad y(t) = b \sin t, \qquad 0 \le t \le 2\pi.$$

Thus

$$x'(t) = -a \sin t, \quad \text{and} \quad y'(t) = b \cos t.$$

By equation (8),

$$A = \tfrac{1}{2} \int_C x \, dy - y \, dx$$

$$= \tfrac{1}{2} \int_0^{2\pi} [(a \cos t)(b \cos t) - (b \sin t)(-a \sin t)] \, dt$$

$$= \tfrac{1}{2} \int_0^{2\pi} ab(\cos^2 t + \sin^2 t) \, dt = \pi ab. \qquad ■$$

Proof of Theorem 3: We will prove Theorem 3 only for regions that are both x-simple and y-simple and then comment on how the proof can be extended to more general regions.

Specifically, let us assume that Q is a region such as that in Figure 4.7 that can be described both as

$$Q = \{(x, y) \mid a \le x \le b, f_1(x) \le y \le f_2(x)\} \qquad \text{(Figure 4.8)} \tag{9}$$

and as

$$Q = \{(x, y) \mid g_1(y) \le x \le g_2(y), c \le y \le d\} \qquad \text{(Figure 4.9)} \tag{10}$$

where each of the functions f_1, f_2, g_1, and g_2 is continuous and piecewise differentiable.

We will prove Theorem 3 by showing both that

$$\int_C M(x, y) \, dx = -\iint_Q \frac{\partial M}{\partial y} \, dA \tag{11}$$

Figure 4.9 The curve C_j is graph of $x = g_j(y)$, for $j = 1, 2$.

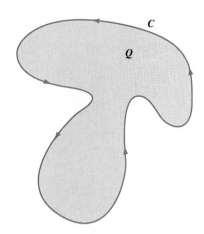

Figure 4.10 Q is neither x-simple nor y-simple.

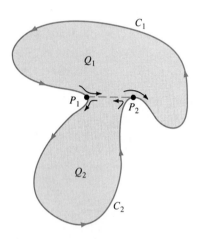

Figure 4.11 Q_1 and Q_2 are both x-simple and y-simple.

Extending Green's Theorem to More General Regions

and that

$$\int_C N(x, y)\, dy = \iint_Q \frac{\partial N}{\partial x}\, dA. \tag{12}$$

Adding the corresponding sides of equations (11) and (12) then establishes equation (2), which is what we are trying to prove.

To prove (11), we first use Figure 4.8 and equation (9) to find that

$$\int_C M(x, y)\, dx = \int_{C_1} M(x, y)\, dx + \int_{C_2} M(x, y)\, dx$$

$$= \int_a^b M(x, f_1(x))\, dx + \int_b^a M(x, f_2(x))\, dx$$

$$= \int_a^b M(x, f_1(x))\, dx - \int_a^b M(x, f_2(x))\, dx.$$

Thus,

$$\int_C M(x, y)\, dx = \int_a^b [M(x, f_1(x)) - M(x, f_2(x))]\, dx. \tag{13}$$

On the other hand, we also have

$$\iint_Q \frac{\partial M}{\partial y}\, dA = \int_a^b \int_{f_1(x)}^{f_2(x)} \frac{\partial}{\partial y} M(x, y)\, dy\, dx \tag{14}$$

$$= \int_a^b \left[M(x, y) \right]_{y=f_1(x)}^{y=f_2(x)}\, dx$$

$$= \int_a^b [M(x, f_2(x)) - M(x, f_1(x))]\, dx.$$

Equations (13) and (14) yield equation (11). Similarly, using Figure 4.9 and equation (10) we can show that equation (12) holds. (This step is left to you as Exercise 22.) This completes the proof for regions Q that are both x-simple and y-simple. ∎

Figure 4.10 shows a region that is neither x-simple nor y-simple, although it does satisfy the hypotheses of Theorem 3. To extend the preceding proof to this region, we make a "cut" (that is, we draw a line) from point P_1 to point P_2, chosen in such a way that the resulting subregions Q_1 and Q_2 are both x-simple and y-simple (See Figure 4.11.) Thus, if C_1 denotes the boundary of Q_1, and C_2 the boundary of Q_2, Green's Theorem on the individual subregions shows that

$$\int_{C_1} M(x, y)\, dx + N(x, y)\, dy = \iint_{Q_1} \left[\frac{\partial N}{\partial x} - \frac{\partial M}{\partial y} \right] dA \tag{15}$$

and

$$\int_{C_2} M(x, y)\, dx + N(x, y)\, dy = \iint_{Q_2} \left[\frac{\partial N}{\partial x} - \frac{\partial M}{\partial y} \right] dA. \tag{16}$$

Adding the corresponding sides of these two equations then gives equation (2). More specifically, the contribution to the line integral in (15) resulting from the path

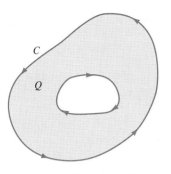

Figure 4.12 *C* consists of two arcs.

from P_1 to P_2 is precisely the negative of the contribution of this same path to the line integral in (16), since the path is traversed in opposite directions by the two line integrals. Thus, the two line integrals taken over the path P_1P_2 "cancel."

Green's Theorem may be extended to regions even more general than those in our statement of Theorem 3. In particular, Green's Theorem holds for regions *Q* containing "holes," such as that in Figure 4.12. For such regions, the curve *C* is the entire boundary—both the "inner" and "outer" curves. The meaning of "counter-clockwise" for such regions is that each arc of *C* is traversed so as to keep the region *Q* on the left-hand side, as indicated in Figure 4.12. Green's Theorem is extended to such regions by making "cuts" (see Figure 4.13) that divide *Q* into subregions without holes, to which Theorem 3 applies. As before, the contributions to the line integral along each cut cancel, so equation (2) is obtained adding the corresponding sides of the equations holding in each subregion.

Other Formulations of Green's Theorem

Green's Theorem can be stated in two special forms, each of which is a two-dimensional version of more general results yet to come. We will interpret each of these statements in terms of the flow of a thin layer of fluid, such as that very near the surface of the water in a swimming pool or in a cross section of a water pipe.

Let $\mathbf{F}(x, y) = M(x, y)\mathbf{i} + N(x, y)\mathbf{j}$ be the vector field giving the velocity (speed and direction) of the fluid flow at each point (x, y) in the thin layer. Using equation (22), Section 20.2, we may write the vector form (3) of Green's Theorem as

$$\int_C \mathbf{F} \cdot \mathbf{T} \, ds = \iint_Q \left(\frac{\partial N}{\partial x} - \frac{\partial M}{\partial y} \right) dA \tag{17}$$

where *C* is any smooth closed curve in the fluid layer. The integrand $\mathbf{F} \cdot \mathbf{T}$ in the line integral in (17) is just the component of the vector field (i.e., the flow) in the direction of the unit tangent $\mathbf{T}$ to the curve *C* at each point (see Figure 4.14). If the fluid has a tendency to circulate around the curve *C* (such as what you see when you pull the plug in your bathtub), we would intuitively expect the line integral in (17) to be large, since the velocity vector $\mathbf{F}$ and the tangent vector $\mathbf{T}$ should point consistently either in approximately the same directions (if the fluid rotates counterclockwise) or in opposite directions (if the fluid rotates clockwise), as illustrated in Figure 4.15. If the fluid has little tendency to rotate, such as occurs in a constant flow field, we expect the line integral in (17) to be small since the component of $\mathbf{F}$

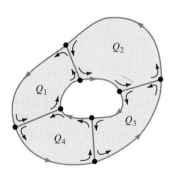

Figure 4.13 Cuts divide *Q* into regions without holes.

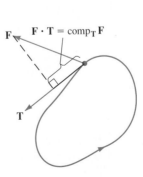

Figure 4.14 The dot product $\mathbf{F} \cdot \mathbf{T}$ is the component of $\mathbf{F}$ in the direction of $\mathbf{T}$.

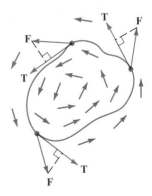

Figure 4.15 In a rotating fluid, $\left| \int_C \mathbf{F} \cdot \mathbf{T} \, ds \right|$ is large.

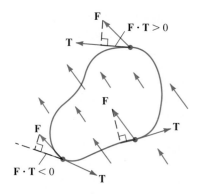

Figure 4.16 In a fluid with little rotation, $\left|\int_C \mathbf{F} \cdot \mathbf{T}\, ds\right|$ is small.

on $\mathbf{T}$ will be of opposite sign on opposite sides of C (Figure 4.16). For these reasons we say that the line integral in (17) is a measure of the tendency of the fluid layer to circulate or to rotate. A vector field in the plane for which this line integral is always zero is called **irrotational.**

Since the double integral in equation (17) has the same value as the line integral, it too measures the tendency of the fluid to rotate. However, it does so by integrating the scalar function

$$\frac{\partial N}{\partial x} - \frac{\partial M}{\partial y}$$

over the region Q enclosed by C. For this reason we refer to this function as the **scalar curl** of the vector field $\mathbf{F}$. That is,

$$\text{curl } \mathbf{F} = \frac{\partial N}{\partial x} - \frac{\partial M}{\partial y}. \tag{18}$$

Using this definition, we may rewrite the formulation of Green's Theorem given in line (17) as

$$\boxed{\int_C \mathbf{F} \cdot \mathbf{T}\, ds = \iint\limits_Q \text{curl } \mathbf{F}\, dA} \qquad \text{(Circulation of } \mathbf{F} \text{ around } C\text{).} \tag{19}$$

Equation (19) is referred to as **Stokes' Theorem in the plane.** In Section 20.6 we shall study the generalizations of curl and of Stokes' Theorem to three dimensions.

A second vector formulation of Green's Theorem concerns the outward unit normal to the closed smooth simple curve C. To obtain this vector we let $\mathbf{T}$ be the unit tangent at a point on the curve C (oriented counterclockwise) and we let θ be the angle formed between $\mathbf{T}$ and the unit vector $\mathbf{i}$ (Figure 4.17). Then

$$\mathbf{T} = \cos\theta\,\mathbf{i} + \sin\theta\,\mathbf{j}. \tag{20}$$

The unit vector $\mathbf{n}$ obtained by rotating $\mathbf{T}$ $90°$ in the counterclockwise direction is the **unit normal**

$$\mathbf{n} = \cos(\theta + \pi/2)\mathbf{i} + \sin(\theta + \pi/2)\mathbf{j} = -\sin\theta\,\mathbf{i} + \cos\theta\,\mathbf{j}.$$

However, this normal points inward for a closed curve C oriented counterclockwise, so the desired **outward unit normal** $\mathbf{N}$ is the vector

$$\mathbf{N} = -\mathbf{n} = \sin\theta\,\mathbf{i} - \cos\theta\,\mathbf{j}. \tag{21}$$

Now recall from Section 17.3 that the unit tangent vector $\mathbf{T}$ may be written

$$\mathbf{T} = \frac{dx}{ds}\mathbf{i} + \frac{dy}{ds}\mathbf{j} \tag{22}$$

where s is the arc length parameter and $\mathbf{r}(s) = x(s)\mathbf{i} + y(s)\mathbf{j}$ is a parameterization for C by arc length. From equations (20), (21), and (22) it follows that the outward unit normal $\mathbf{N}$ may be written

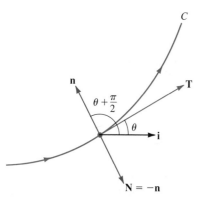

Figure 4.17 $\mathbf{N}$ is the outward unit normal.

$$\mathbf{N} = \frac{dy}{ds}\mathbf{i} - \frac{dx}{ds}\mathbf{j}. \tag{23}$$

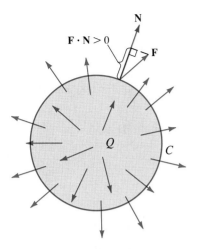

$\mathbf{F} \cdot \mathbf{N} > 0$

Figure 4.18 $\int_C \mathbf{F} \cdot \mathbf{N} \, ds > 0$ if net flow across C is outward.

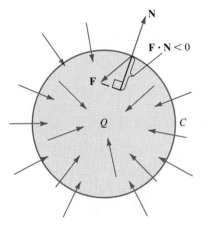

$\mathbf{F} \cdot \mathbf{N} < 0$

Figure 4.19 $\int_C \mathbf{F} \cdot \mathbf{N} \, ds < 0$ if net flow across C is inward.

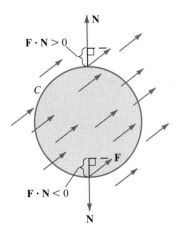

$\mathbf{F} \cdot \mathbf{N} > 0$

$\mathbf{F} \cdot \mathbf{N} < 0$

Figure 4.20 $\int_C \mathbf{F} \cdot \mathbf{N} \, ds = 0$ if net flow across C is zero.

For the vector field $\mathbf{F}(x, y) = M(x, y)\mathbf{i} + N(x, y)\mathbf{j}$, the expression (23) for $\mathbf{N}$ suggests that

$$\int_C \mathbf{F} \cdot \mathbf{N} \, ds = \int_C M(x, y) \, dy - N(x, y) \, dx, \tag{24}$$

and Green's Theorem, applied to the right-hand side of equation (24) gives

$$\int_C M(x, y) \, dy - N(x, y) \, dx = \iint_Q \left(\frac{\partial M}{\partial x} + \frac{\partial N}{\partial y} \right) dA. \tag{25}$$

Finally, combining (24) and (25), we obtain

$$\int_C \mathbf{F} \cdot \mathbf{N} \, ds = \iint_Q \left(\frac{\partial M}{\partial x} + \frac{\partial N}{\partial y} \right) dA. \tag{26}$$

Equation (26) is referred to as the **Divergence Theorem in the plane** for the following reasons. Since the scalar quantity $\mathbf{F} \cdot \mathbf{N}$ in the line integral (with respect to arc length) is the component of the vector field (fluid flow) in the direction of the outward normal, the line integral in (26) measures the net flow of fluid outward across the boundary C of the region Q. The term **flux** is used to denote this net rate at which fluid flows across a boundary, and the line integral in (26) is called a **flux integral.** If the line integral is *positive,* more fluid is leaving Q than is entering. (Think of an open faucet somewhere inside Q.) If the line integral in (26) is *negative,* more fluid is flowing into Q than is leaving. (Think of an open drain somewhere inside Q.) If the line integral is small or zero, there is little or no net gain or loss of fluid in Q. (Think of fluid moving uniformly, or of motionless fluid. See Figures 4.18 through 4.20.)

Since the double integral in (26) equals the flux integral, the integrand $\partial M/\partial x + \partial N/\partial y$ is referred to as the **divergence** of the vector field $\mathbf{F}$, written div $\mathbf{F}$. That is

$$\text{div } \mathbf{F} = \frac{\partial M}{\partial x} + \frac{\partial N}{\partial y}. \tag{27}$$

Using (27), we may rewrite the Divergence Theorem in the plane as

$$\int_C \mathbf{F} \cdot \mathbf{N} \, ds = \iint_Q \text{div } \mathbf{F} \, dA. \qquad \text{(Flux of } \mathbf{F} \text{ across } C\text{).} \tag{28}$$

In Section 20.7 we generalize both the concept of divergence of a vector field and the Divergence Theorem to three dimensions and apply these generalizations to spacial fluid flows.

Exercise Set 20.4

(In all exercises the closed curves C are oriented counterclockwise.) In Exercises 1–10, use Green's Theorem to evaluate the given line integral.

1. $\int_C xy\,dx + (x + y)\,dy$, C is the square with vertices $(0, 0)$, $(0, 1)$, $(1, 1)$, and $(1, 0)$.

2. $\int_C xy\,dx + (y - x)\,dy$, C is the unit circle.

3. $\int_C x^2 y^3\,dx + (x^2 + 1)\,dy$, C is the square in Exercise 1.

4. $\int_C \sqrt{y}\,dx + \sqrt{x}\,dy$, C is the rectangle with vertices $(1, 1)$, $(1, 2)$, $(4, 1)$, and $(4, 2)$.

5. $\int_C xy^2\,dy - x^2 y\,dx$, C is the quarter circular path from $(0, 0)$ to $(1, 0)$ to $(0, 1)$ to $(0, 0)$.

6. $\int_C e^x \tan y\,dx + e^x \sec^2 y\,dy$, C is the triangle with vertices $(-1, 2)$, $(3, 0)$, and $(1, 6)$.

7. $\int_C (x^3 + y)\,dx + (y - x^2)\,dy$, C is the boundary of the region bounded by the graphs of $y = x^2$ and $y = x^3$.

8. $\int_C (\mathrm{Tan}^{-1} x + y^2)\,dx + (\ln^2 y - x^2)\,dy$, C is the circle $x^2 + y^2 = 4$.

9. $\int_C (\cos^5 x + \sqrt{x})\,dx + \mathrm{Tan}^{-1} y\,dy$, C is the ellipse $4x^2 + y^2 = 1$.

10. $\int_C \sqrt{x}\,dx + \ln(x^2 + y^2)\,dy$, C is the curve in Figure 4.6.

In Exercises 11–14, use Green's Theorem to find $\int_C \mathbf{F} \cdot d\mathbf{r}$.

11. $\mathbf{F}(x, y) = 2y\mathbf{i} - 3x\mathbf{j}$, C is the unit circle.

12. $\mathbf{F}(x, y) = xy\mathbf{i} + x^2\mathbf{j}$, C is the square with vertices $(0, 0)$, $(0, 2)$, $(2, 2)$, and $(2, 0)$.

13. $\mathbf{F}(x, y) = x^2(y^2 - x^2)\mathbf{i} + \frac{2}{3}x^3 y\mathbf{j}$, C is the triangle with vertices $(1, 1)$, $(4, 1)$, and $(3, 5)$.

14. $\mathbf{F}(x, y) = e^x \sin y\mathbf{i} + e^x \cos y\mathbf{j}$, C is the ellipse $x^2 + 9y^2 = 1$.

15. Find the work done by the force field $\mathbf{F}(x, y) = (e^{x^2} + 2y)\mathbf{i} + (ye^y - x)\mathbf{j}$ in moving a particle once around the path indicated in Figure 4.6.

16. Use equation (8) to find the area of the region bounded by the graphs of $y = x$ and $y = x^2$.

17. Use equations (8) to find the area of the region bounded by the graphs of $y = x$ and $y = \sqrt{x}$.

18. Find the area of the region enclosed by the graph of the parametric equations $x(t) = \sin t \cos t$, $y(t) = \sin t$ for $0 \le t \le \pi$.

19. Find the area of the region bounded by the graph of the parametric equations $x(t) = a \cos^3 t$, $y(t) = a \sin^3 t$, $0 \le t \le 2\pi$.

20. Let Q be a region in the plane whose boundary is a simple closed piecewise smooth curve C. Let A be the area of Q. Show that the coordinates of the centroid of Q are

$$\bar{x} = \frac{1}{2A} \int_C x^2\,dy, \qquad \bar{y} = -\frac{1}{2A} \int_C y^2\,dx.$$

21. Use Green's Theorem to prove that if $\mathbf{F}$ is a conservative vector field in the plane, then any line integral

$$\int_C \mathbf{F} \cdot d\mathbf{r}$$

over a piecewise smooth curve C is independent of path.

22. Complete the proof of Theorem 3 by showing that equation (12) holds.

23. Let $\mathbf{F}(x, y) = M(x, y)\mathbf{i} + N(x, y)\mathbf{j}$ where M and N have continuous first partial derivatives in some open rectangle D. Show that $\mathbf{F}$ is conservative in D if and only if curl $\mathbf{F} = 0$ for all $(x, y) \in D$.

20.5 Surface Integrals

The primary motivation for the remainder of this chapter will be to extend the concept of flux, introduced in Section 20.4, to the flow of fluids in and out of regions in space. Since regions in space are bounded by surfaces rather than curves, we must begin by developing the concept of integrals over surfaces, or *surface integrals*. Although we are primarily interested in defining the integral of a *vector field* **F** over a surface S (Figure 5.1), we begin by defining the integral of a scalar function f over a surface S (Figure 5.2).

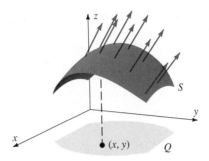

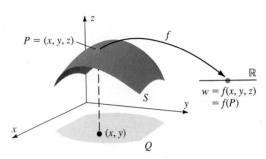

Figure 5.1   A vector field $\mathbf{F}(x, y, z)$ assigns a vector to each point on the surface S.

Figure 5.2   A scalar function $f(x, y, z)$ assigns a number to each point P on the surface S.

In our discussion, we focus on the case where the surface S is the graph $z = g(x, y)$ of a function of two variables over some regular region Q in the xy-plane (see Figure 5.3). (Recall that a region is regular if it is both x- and y-simple.) Even though many surfaces cannot be described in this way, the techniques we develop can be modified to treat most surfaces encountered in practice.

Suppose that the continuous scalar function f is defined for all points $(x, y, g(x, y))$ on the surface S. Recall from Section 19.5 that a partition of the region Q into n rectangles of area $\Delta A_j = \Delta x\, \Delta y$ partitions "most" of the surface S into curvilinear "patches" S_j of area ΔS_j (Figure 5.4). We form the product of the surface area ΔS_j of the path S_j and the value $f(x_j, y_j, z_j)$ of the scalar function f at an arbitrary point (x_j, y_j, z_j) on the patch. Summing these products over all n patches gives the approximating sum

$$\sum_{j=1}^{n} f(x_j, y_j, z_j)\, \Delta S_j, \qquad (x_j, y_j, z_j) \in S_j.$$

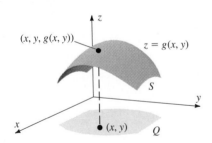

Figure 5.3   The surface S is the graph of a continuous function $z = g(x, y)$ over the region Q.

As the norm of the partition P approaches zero ($\|P\| \to 0$), the limit of this approximating sum, if it exists, is defined as the surface integral of f over S:

$$\iint_S f(x, y, z)\, dS = \lim_{\|P\| \to 0} \sum_{j=1}^{n} f(x_j, y_j, z_j)\, \Delta S_j. \tag{1}$$

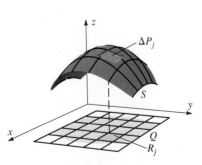

Figure 5.4   A parallelogram of area ΔP_j over the rectangle R_j approximates the area ΔS_j of the "patch" S_j.

We shall determine several ways to calculate this surface integral. The first of these involves recalling that the area ΔS_j of the patch S_j may be approximated by the area ΔP_j of the parallelogram over R_j tangent to the surface S at one corner (Figure 5.4). That is, we have the approximation

$$\Delta S_j \approx \Delta P_j = \sqrt{\left(\frac{\partial g}{\partial x}\right)^2 + \left(\frac{\partial g}{\partial y}\right)^2 + 1}\ \Delta x\, \Delta y \tag{2}$$

where the partial derivatives $\partial g/\partial x$ and $\partial g/\partial y$ are evaluated at one corner (x_j, y_j) of the rectangle R_j. Using approximation (2) in equation (1) leads to the following definition.

Definition 5

Let Q be a regular region in the plane and let g be a continuous function defined on some open rectangle containing Q on which both $\partial g/\partial x$ and $\partial g/\partial y$ are continuous. Let S be the graph of g over Q, and let f be a continuous scalar function defined on S. The **surface integral** of f over S is

$$\iint_S f(x, y, z)\, dS = \iint_Q f(x, y, g(x, y)) \sqrt{\left(\frac{\partial g}{\partial x}\right)^2 + \left(\frac{\partial g}{\partial y}\right)^2 + 1}\, dx\, dy. \qquad (3)$$

☐ **EXAMPLE 1**

Evaluate the surface integral

$$\iint_S (x^2 + y + z)\, dS$$

where S is the graph of the function $g(x, y) = \sqrt{3}\, y - x^2$ over the unit square

$$Q = \{(x, y) \mid 0 \le x \le 1, 0 \le y \le 1\}.$$

Solution   The integrand is the function

$$f(x, y, z) = x^2 + y + z.$$

Also,

$$\frac{\partial g}{\partial x} = -2x, \quad \text{and} \quad \frac{\partial g}{\partial y} = \sqrt{3}.$$

Substituting into equation (3) gives

$$\iint_S (x^2 + y + z)\, dS = \iint_Q [x^2 + y + (\sqrt{3}\, y - x^2)]\sqrt{(-2x)^2 + (\sqrt{3})^2 + 1}\, dx\, dy$$

$$= \int_0^1 \int_0^1 (1 + \sqrt{3})y\sqrt{4x^2 + 4}\, dy\, dx$$

$$= (1 + \sqrt{3})\int_0^1 \left[y^2\sqrt{x^2 + 1}\right]_{y=0}^{y=1} dx$$

$$= (1 + \sqrt{3})\int_0^1 \sqrt{x^2 + 1}\, dx \qquad \text{(trigonometric substitution)}$$

$$= (1 + \sqrt{3})\left(\frac{\sqrt{2}}{2} + \frac{1}{2}\ln(1 + \sqrt{2})\right) \approx 3.14. \qquad \blacksquare$$

☐ **EXAMPLE 2**

Evaluate the surface integral

$$\iint_S (x^2 + y^2 + z)\, dS$$

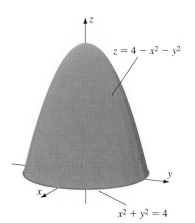

Figure 5.5 The surface S is the graph of $z = 4 - x^2 - y^2$.

where S is the portion of the graph of the function $z = 4 - x^2 - y^2$ bounded below by the xy-plane.

Solution   The integrand is $f(x, y, z) = x^2 + y^2 + z$, and the surface is the graph of the function $g(x, y) = 4 - x^2 - y^2$. Thus

$$\frac{\partial g}{\partial x} = -2x, \quad \text{and} \quad \frac{\partial g}{\partial y} = -2y.$$

The region Q in the xy-plane over which S is defined is

$$Q = \{(x, y) \mid x^2 + y^2 \le 4\} \qquad \text{(Figure 5.5)}.$$

According to Definition 5, the surface integral is

$$\iint_S (x^2 + y^2 + z) \, dS = \iint_Q [x^2 + y^2 + (4 - x^2 - y^2)]\sqrt{(-2x)^2 + (-2y)^2 + 1} \, dx \, dy$$

$$= \iint_Q 4\sqrt{4x^2 + 4y^2 + 1} \, dx \, dy.$$

Switching to polar coordinates allows us to evaluate this integral as

$$\iint_S (x^2 + y^2 + z) \, dS = \int_0^{2\pi} \int_0^2 4\sqrt{4r^2 + 1} \, r \, dr \, d\theta$$

$$= \int_0^{2\pi} \left[\tfrac{1}{3}(4r^2 + 1)^{3/2} \right]_{r=0}^{r=2} d\theta$$

$$= \frac{2\pi}{3}(17^{3/2} - 1) \approx 144.71. \qquad \blacksquare$$

REMARK   For surfaces that are graphs of functions of the form $y = g(x, z)$ or $x = g(y, z)$, we simply interchange roles among the variables x, y, and z as required. The following example presents one such situation. (Notice that the given surface cannot be described as the graph of a function of the form $z = g(x, y)$.)

□ **EXAMPLE 3**

Evaluate the surface integral

$$\iint_S yz \, dS$$

where S is the portion of the graph of $x = 1 - z^2$ bounded by the yz-plane and the planes $y = -2$ and $y = 2$ (Figure 5.6).

Solution   In order to describe S as the graph of a function, we must take x as the dependent variable. Interchanging roles of x and z in Definition 5, we find that

$$f(x, y, z) = yz \quad \text{and} \quad g(y, z) = 1 - z^2.$$

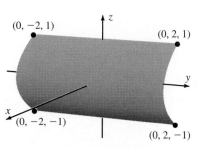

Figure 5.6   The surface S determined by the graph of $x = 1 - z^2$ bounded by the yz-plane and the planes $y = -2$ and $y = 2$.

Thus,

$$\frac{\partial g}{\partial z} = -2z, \quad \text{and} \quad \frac{\partial g}{\partial y} = 0.$$

The region Q in the yz-plane over which S is defined is the rectangle

$$Q = \{(y, z) \mid -2 \le y \le 2, -1 \le z \le 1\}.$$

Definition 5 now gives

$$\iint_S yz \, dS = \iint_Q yz \sqrt{(-2z)^2 + 1} \, dz \, dy$$

$$= \int_{-2}^{2} \int_{-1}^{1} yz \sqrt{4z^2 + 1} \, dz \, dy$$

$$= \int_{-2}^{2} \left[\frac{y}{12} (4z^2 + 1)^{3/2} \right]_{z=-1}^{z=1} dy$$

$$= \int_{-2}^{2} \frac{y}{12} (5^{3/2} - 5^{3/2}) \, dy = 0.$$

The result should not be surprising since the surface is symmetric with respect to the xy-plane and $f(x, y, -z) = -f(x, y, z)$. ■

Calculating Mass

A straightforward application of surface integrals of the form (1) occurs in the calculation of the total mass of a hollow object (such as a basketball or a vase) whose mass density $\rho(x, y, z)$ per unit surface area varies continuously over the surface of the object. That is, we assume that $\rho(x, y, z)$ gives the mass density (in grams/cm^2, say) of the object at location (x, y, z). Then, if S_j is a small patch of area ΔS_j on the surface of the object, the total mass ΔM_j of S_j may be approximated by

$$\Delta M_j \approx \rho(x_j, y_j, z_j) \, \Delta S_j \qquad \left(\text{mass} = \frac{\text{mass}}{\text{unit area}} \times \text{area} \right) \tag{4}$$

where (x_j, y_j, z_j) is a point on S_j. Summing (4) over all patches in a partition covering S, we obtain the approximation to total mass:

$$M \approx \sum_{j=1}^{n} \rho(x_j, y_j, z_j) \, \Delta S_j. \tag{5}$$

Comparing approximation (5) with equation (1), we conclude that, in the limit as $\Delta S_j \to 0$ assuming appropriate conditions on the shape of the object and on the density function ρ,

$$\boxed{M = \iint_S \rho(x, y, z) \, dS.} \tag{6}$$

In other words, **the total mass of the object** is simply the integral, over the surface of the object, of the function giving the mass per unit area.

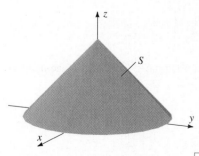

Figure 5.7 The cone described by $z = 1 - r$ in cylindrical coordinates.

☐ **EXAMPLE 4**

Let S be the idealized cone described in cylindrical coordinates by the equation $z = 1 - r, 0 \le r \le 1$. Find the mass of S if the density at any point is proportional to the square of the distance between the point and the origin (see Figure 5.7).

Solution   Here $\rho(x, y, z) = \lambda(x^2 + y^2 + z^2)$ is the mass density, where λ is a constant. We will work with ρ in cylindrical coordinates:

$$\rho(r, \theta, z) = \lambda(r^2 + z^2).$$

The surface S is described by the function

$$z = g(r, \theta) = 1 - r.$$

Using the result of Exercise 25 of this section (for evaluating a surface integral in cylindrical coordinates) together with equation (6), we find that

$$
\begin{aligned}
M &= \iint\limits_{S} \rho(r, \theta, z) \, dS \\
&= \iint\limits_{Q} \rho(r, \theta, g(r, \theta)) \sqrt{r^2 + r^2\left(\frac{\partial g}{\partial r}\right)^2 + \left(\frac{\partial g}{\partial \theta}\right)^2} \, dr \, d\theta \\
&= \int_0^{2\pi} \int_0^1 \lambda(r^2 + (1 - r)^2)\sqrt{r^2 + r^2(-1)^2 + 0^2} \, dr \, d\theta \\
&= \int_0^{2\pi} \int_0^1 \sqrt{2}\lambda(2r^3 - 2r^2 + r) \, dr \, d\theta \\
&= \frac{2\sqrt{2}\pi\lambda}{3}.
\end{aligned}
$$

■

The concept of surface integral can be extended to various types of surfaces that do not fulfill the hypotheses of Definition 5. If a surface S is a union of a finite number of surfaces $S = S_1 \cup S_2 \cup \cdots \cup S_n$, each of which satisfies the hypotheses of Definition 5, we define the surface integral of $f(x, y, z)$ over S to be the sum of the individual surface integrals

$$\iint\limits_{S} f(x, y, z) \, dS = \sum_{j=1}^{n} \iint\limits_{S_j} f(x, y, z) \, dS_j, \tag{7}$$

each of which is evaluated by use of Definition 5.

□ **EXAMPLE 5**

Evaluate

$$\iint\limits_{S} (x^2 + y^2 + z) \, dS$$

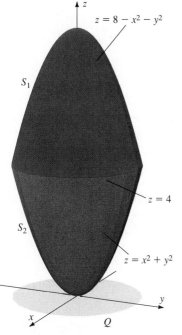

Figure 5.8 The surface S in Example 5 is the union of the two surfaces S_1 and S_2.

where the surface S is the boundary of the region bounded by the graphs of $g_1(x, y) = 8 - (x^2 + y^2)$ and $g_2(x, y) = x^2 + y^2$ (Figure 5.8).

Solution   The two graphs intersect in the circle $x^2 + y^2 = 4$ in the plane $z = 4$. Therefore, we can describe the surface S as $S = S_1 \cup S_2$ where S_1 corresponds to the equation $z = 8 - (x^2 + y^2)$ with $z \geq 4$ and S_2 corresponds to the equation $z = x^2 + y^2$ with $z \leq 4$. Both S_1 and S_2 project onto the region $Q = \{(x, y) \mid x^2 + y^2 \leq 4\}$ in the xy-plane. Thus,

$$
\begin{aligned}
S_1 &= \{(x, y, 8 - (x^2 + y^2)) \mid (x, y) \in Q\}, \\
S_2 &= \{(x, y, x^2 + y^2) \mid (x, y) \in Q\}.
\end{aligned}
$$

For the surface S_1, with $f(x, y, z) = x^2 + y^2 + z$ and $g_1(x, y) = 8 - (x^2 + y^2)$, Definition 5 gives

$$\iint\limits_{S_1} (x^2 + y^2 + z) \, dS$$

$$= \iint\limits_{Q} [x^2 + y^2 + (8 - (x^2 + y^2))]\sqrt{(-2x)^2 + (-2y)^2 + 1} \, dx \, dy$$

$$= \int_0^{2\pi} \int_0^2 8\sqrt{4r^2 + 1} \, r \, dr \, d\theta \qquad \text{(switching to polar coordinates)}$$

$$= \frac{4\pi}{3}(17^{3/2} - 1).$$

For the surface S_2, with $f(x, y, z) = x^2 + y^2 + z$ and $g_2(x, y) = x^2 + y^2$, we obtain

$$\iint\limits_{S_2} (x^2 + y^2 + z) \, dS = \iint\limits_{Q} [x^2 + y^2 + (x^2 + y^2)]\sqrt{(2x)^2 + (2y)^2 + 1} \, dx \, dy$$

$$= \int_0^{2\pi} \int_0^2 2r^2\sqrt{4r^2 + 1} \, r \, dr \, d\theta \qquad \text{(let } u = 4r^2 + 1\text{)}$$

$$= \frac{\pi}{20}(17^{5/2} - 1) - \frac{\pi}{12}(17^{3/2} - 1).$$

Combining these results according to equation (7) gives

$$\iint\limits_{S} (x^2 + y^2 + z) \, dS = \frac{\pi}{20}(17^{5/2} - 1) + \frac{5\pi}{4}(17^{3/2} - 1) \approx 458.3. \qquad \blacksquare$$

Definition 5 could not be applied directly to the entire surface S in Example 5 for two reasons. First, S is not the graph of a *function* $z = g(x, y)$ since there are two distinct points corresponding to some (x, y)-coordinates. Second, the surface S is not differentiable (meaning that $\partial g/\partial x$ and $\partial g/\partial y$ are not continuous) along the "seam" where S_1 meets S_2.

Another difficulty that can arise in attempting to apply Definition 5 is that the integrand in the double integral has a singularity (i.e., becomes infinite) along the boundary of the region Q. In such cases we define

$$\iint\limits_{S} f(x, y, z) \, dS$$

to be the corresponding improper integral if this integral exists. When combined with the notion of equation (7), this idea allows us to evaluate surface integrals over spheres, an important aspect of many applications of surface integrals.

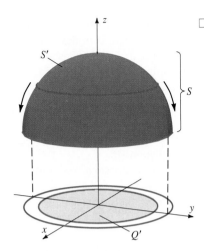

Figure 5.9 The surface integral over the hemisphere $z = 1 + \sqrt{1 - x^2 - y^2}$ is evaluated as an improper integral.

□ **EXAMPLE 6**

Evaluate the surface integral

$$\iint_S z \, dS$$

where S is the hemisphere $z = 1 + \sqrt{1 - x^2 - y^2}$ (Figure 5.9).

Solution: Here $g(x, y) = 1 + \sqrt{1 - x^2 - y^2}$ and $f(x, y, z) = z$. Equation (3) becomes

$$\iint_S z \, dS = \iint_Q [1 + \sqrt{1 - x^2 - y^2}]$$

$$\cdot \sqrt{\left(\frac{-x}{\sqrt{1 - x^2 - y^2}}\right)^2 + \left(\frac{-y}{\sqrt{1 - x^2 - y^2}}\right)^2 + 1} \, dx \, dy$$

$$= \iint_Q \left[\frac{1}{\sqrt{1 - x^2 - y^2}} + 1 \right] dx \, dy$$

where Q is the unit circle $x^2 + y^2 = 1$. However, this double integral is improper, since the denominator of the first term in the integrand vanishes as (x, y) approaches the boundary of Q. We evaluate this improper integral as suggested in Figure 5.9— by evaluating it over a disc Q^- of radius $\hat{r} < 1$ and then calculating the limit as $\hat{r}$ approaches 1. To do so we switch to polar coordinates and find that

$$\iint_S z \, dS = \iint_Q \left[\frac{1}{\sqrt{1 - r^2}} + 1 \right] r \, dr \, d\theta$$

$$= \lim_{\hat{r} \to 1^-} \int_0^{\hat{r}} \int_0^{2\pi} \left[\frac{r}{\sqrt{1 - r^2}} + r \right] d\theta \, dr$$

$$= \lim_{\hat{r} \to 1^-} \int_0^{\hat{r}} 2\pi \left[\frac{r}{\sqrt{1 - r^2}} + r \right] dr$$

$$= \lim_{\hat{r} \to 1^-} 2\pi \left[-\sqrt{1 - r^2} + \frac{r^2}{2} \right]_{r=0}^{r=\hat{r}}$$

$$= \lim_{\hat{r} \to 1^-} 2\pi \left[-\sqrt{1 - \hat{r}^2} + \frac{\hat{r}^2}{2} - (-\sqrt{1}) \right] = 3\pi. \quad ■$$

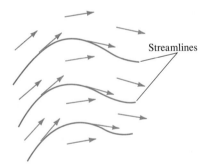

Figure 5.10 A velocity field and its streamlines.

Flux Integrals

Returning now to our primary motivation concerning fluid flow, we consider the motion of a fluid, such as water or a gas, through a region R in space. At each point (x, y, z) in the region, let $\mathbf{v}(x, y, z)$ be the velocity vector for the motion of the fluid. That is, $\mathbf{v}(x, y, z)$ is a vector pointing in the direction of motion whose length is the speed of the fluid at (x, y, z). We shall only consider the case of **steady state fluid motion.** This means that the velocity vector $\mathbf{v}(x, y, z)$ does not change as time changes. Thus, $\mathbf{v}(x, y, z)$ is a *vector field* defined on the region R. (The curves that are tangent to the velocity field at each point are called the **streamlines** of the velocity field. See Figure 5.10.)

Now suppose that S is a permeable surface suspended in the fluid. (In other words, the fluid can flow through the surface S, as coffee flows through a paper

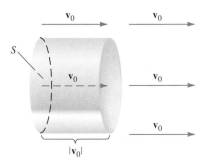

Figure 5.11 Volume of fluid flowing across S in unit time is $\Delta V = |\mathbf{v}_0|\,\Delta S$ if $\mathbf{v}_0$ is perpendicular to S.

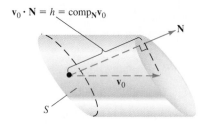

Figure 5.12 Volume of fluid crossing S in a unit of time is $\Delta V = (\mathbf{v}_0 \cdot \mathbf{N})\,\Delta S$.

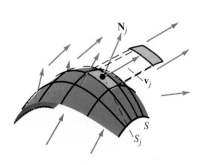

Figure 5.13 Volume of fluid crossing patch S_j is approximated by $\Delta V \approx (\mathbf{v}_j \cdot \mathbf{N}_j)\,\Delta S_j$.

filter.) For a steady state velocity field $\mathbf{v}$, we define the **flux** of $\mathbf{v}$ across S as the rate at which mass is flowing across the surface S. Our interest here is in developing a formula by which flux can be calculated for a given surface and a given vector field.

We can simplify this problem somewhat by assuming that the density δ (mass per unit volume) is constant throughout the fluid. Since

$$\text{Mass} = (\text{density}) \times (\text{volume}) \tag{8}$$

we need only find the rate at which volume (i.e., fluid) is flowing across S. Multiplying by δ then gives the desired flux, according to equation (8).

The simplest case of this problem occurs if $\mathbf{v}$ is a constant vector field, $\mathbf{v}(x, y, z) \equiv \mathbf{v}_0$, and S is a flat surface perpendicular to $\mathbf{v}_0$ (Figure 5.11). Since the speed at which fluid crosses S is $|\mathbf{v}_0|$, in one unit of time the quantity

$$\Delta V = |\mathbf{v}_0|\,\Delta S \tag{9}$$

crosses S, where ΔS is the area of S. If $\mathbf{v}(x, y, z) = \mathbf{v}_0$ is constant and S is flat, but not orthogonal to $\mathbf{v}_0$, the situation is only slightly more complicated. Let $\mathbf{N}$ be a unit vector normal to S. (There are two possible choices for $\mathbf{N}$. Take the one for which $0 \le \theta \le 90°$, where θ is the angle between $\mathbf{N}$ and $\mathbf{v}_0$.) The volume of fluid that crosses S in a unit of time can be thought of as a prism with base of area ΔS and height $h = \text{comp}_{\mathbf{N}}\,\mathbf{v}_0 = \mathbf{v}_0 \cdot \mathbf{N}$, as illustrated in Figure 5.12. The desired volume is thus

$$\Delta V = (\mathbf{v}_0 \cdot \mathbf{N})\,\Delta S. \tag{10}$$

The more general case of a curved surface S and a nonconstant velocity field $\mathbf{v}$ is handled in a familiar way. First, we assume that S has a normal vector $\mathbf{N}(x, y, z)$ at each point and that $\mathbf{N}(x, y, z)$ is continuous on S. Also, we assume that S can be recognized as having two sides and that $\mathbf{N}(x, y, z)$ always remains on the same side of S. All these assumptions are met if S is the graph of a differentiable function $z = g(x, y)$ over a regular region Q in the plane.

By partitioning Q, as in Figure 5.4, we divide S into nonoverlapping patches S_1, $S_2, \ldots, S_n$. For each patch we choose one point $(x_j, y_j, z_j) \in S_j$, and we let ΔS_j denote the surface area of S_j. Also, we approximate $\mathbf{v}(x, y, z)$ for every $(x, y, z) \in S_j$ by the vector $\mathbf{v}_j = \mathbf{v}(x_j, y_j, z_j)$ and we let $\mathbf{N}_j = \mathbf{N}(x_j, y_j, z_j)$. Using equation (10), we approximate the volume ΔV_j of fluid crossing the patch S_j in a unit of time as

$$\Delta V_j \approx (\mathbf{v}_j \cdot \mathbf{N}_j)\,\Delta S_j \qquad \text{(Figure 5.13)}.$$

Summing these approximations over all patches gives the approximation

$$\Delta V \approx \sum_{j=1}^{n} (\mathbf{v}_j \cdot \mathbf{N}_j)\,\Delta S_j. \tag{11}$$

Finally, we argue that as the number of patches becomes large (and their individual sizes become small) the approximation in (11) should converge to the desired rate. Since the right-hand side of approximation (11) has the form of the approximating sum in equation (1), we conclude from equations (8) and (11) that the rate at which the mass of the fluid is flowing across S (that is, the flux) is given by the surface integral

$$\frac{dM}{dt} = \iint_S \delta\mathbf{v}(x, y, z) \cdot \mathbf{N}(x, y, z)\,dS. \tag{12}$$

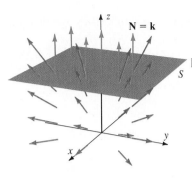

Figure 5.14 Flow across the square S for a central velocity field.

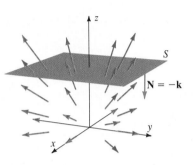

Figure 5.15 This velocity field flows in a direction opposite to $N = -k$.

The integral in (12) is sometimes called the **rate of mass transport** across S in the direction of **N.**

☐ **EXAMPLE 7**

A fluid with mass density δ is emanating from the origin and flowing according to the central velocity field

$$\mathbf{v}(x, y, z) = \lambda(x\mathbf{i} + y\mathbf{j} + z\mathbf{k})$$

where λ is constant. Find the rate dM/dt of mass transport across the square region

$$S = \{(x, y, z) \,|\, z = 2, -1 \le x \le 1, -1 \le y \le 1\}.$$

Solution   The square S is horizontal, so we use the unit normal $\mathbf{N} = \mathbf{k}$ at each point. Then $\mathbf{v}(x, y, z) \cdot \mathbf{N} = \lambda z$, so, by (12),

$$\frac{dM}{dt} = \iint_S \delta\lambda z \, dS = \int_{-1}^{1} \int_{-1}^{1} 2\delta\lambda \, dx \, dy = 8\delta\lambda.$$

Note that the choice of $\mathbf{N} = \mathbf{k}$ was arbitrary. If we had instead used $\mathbf{N} = -\mathbf{k},$ the resulting integral would have yielded $dM/dt = -8\delta\lambda.$ This simply says that the flow is in the direction *opposite* the vector $\mathbf{N} = -\mathbf{k}$ (see Figures 5.14 and 5.15).   ■

We can generalize equation (12) by letting $\mathbf{F}$ denote any vector field and $\mathbf{N} = \mathbf{N}(x, y, z)$ be a unit normal defined at each point on the surface S as described above. The flux of the vector field $\mathbf{F}$ across S in the direction of $\mathbf{N}$ is defined to be the surface integral

$$\text{Flux of } \mathbf{F} \text{ across } S = \iint_S \mathbf{F} \cdot \mathbf{N} \, dS. \tag{13}$$

Thus (12) is a special case of (13) with $\mathbf{F} = \delta\mathbf{v}.$ However, the vector field in (13) need not be a velocity field for a fluid. It may, for example, be an electric field induced by one or more point charges, or a gradient field for a temperature function.

Recall that, for a surface of the form $z = g(x, y)$, the vector

$$\mathbf{V} = -\frac{\partial g}{\partial x}\mathbf{i} - \frac{\partial g}{\partial y}\mathbf{j} + \mathbf{k}$$

is normal to the surface (equation (1) of Section 18.3). Consequently, in equation (13), we use either

$$\mathbf{N}_1 = \frac{\mathbf{V}}{|\mathbf{V}|} = \frac{-\dfrac{\partial g}{\partial x}\mathbf{i} - \dfrac{\partial g}{\partial y}\mathbf{j} + \mathbf{k}}{\sqrt{\left(\dfrac{\partial g}{\partial x}\right)^2 + \left(\dfrac{\partial g}{\partial y}\right)^2 + 1}} \tag{14}$$

or

$$\mathbf{N}_2 = \frac{\mathbf{V}}{|\mathbf{V}|} = \frac{\dfrac{\partial g}{\partial x}\mathbf{i} + \dfrac{\partial g}{\partial y}\mathbf{j} - \mathbf{k}}{\sqrt{\left(\dfrac{\partial g}{\partial x}\right)^2 + \left(\dfrac{\partial g}{\partial y}\right)^2 + 1}} \tag{15}$$

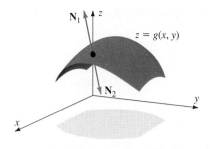

Figure 5.16 There are two unit normal vectors to the graph of $z = g(x, y)$. This choice affects the sign of the number given by equation (13).

(See Figure 5.16.) The choice between $\mathbf{N}_1$ and $\mathbf{N}_2$ is determined by whether you wish $\mathbf{N}$ to point upward ($\mathbf{N}_1$) or downward ($\mathbf{N}_2$).

Before proceeding to our final example, we use equations (14) and (15) to express the flux integral (13) in a more manageable form. If the surface S is the graph of $z = g(x, y)$ and the vector field $\mathbf{F}$ has the component form

$$\mathbf{F}(x, y, z) = M(x, y, z)\mathbf{i} + N(x, y, z)\mathbf{j} + P(x, y, z)\mathbf{k},$$

then the flux integral (13) corresponding to the upward unit normal $\mathbf{N}_1$ in (14) is, according to Definition 5,

$$\iint_S \mathbf{F} \cdot \mathbf{N}\, dS = \iint_Q (M\mathbf{i} + N\mathbf{j} + P\mathbf{k})$$

$$\cdot \left(\frac{-\dfrac{\partial g}{\partial x}\mathbf{i} - \dfrac{\partial g}{\partial y}\mathbf{j} + \mathbf{k}}{\sqrt{\left(\dfrac{\partial g}{\partial x}\right)^2 + \left(\dfrac{\partial g}{\partial y}\right)^2 + 1}} \right) \left(\sqrt{\left(\dfrac{\partial g}{\partial x}\right)^2 + \left(\dfrac{\partial g}{\partial y}\right)^2 + 1} \right) dx\, dy$$

where Q is the region in the xy-plane over which S is defined. This simplifies to

$$\iint_S \mathbf{F} \cdot \mathbf{N}\, dS = \iint_Q \left[-M(x, y, g(x, y))\frac{\partial g}{\partial x} - N(x, y, g(x, y))\frac{\partial g}{\partial y} \right.$$
$$\left. + P(x, y, g(x, y)) \right] dx\, dy. \tag{16}$$

Similarly, the flux integral (13) corresponding to the downward unit normal $\mathbf{N}_2$ in (15) is

$$\iint_S \mathbf{F} \cdot \mathbf{N}\, dS = \iint_Q \left[M(x, y, g(x, y))\frac{\partial g}{\partial x} + N(x, y, g(x, y))\frac{\partial g}{\partial y} \right.$$
$$\left. - P(x, y, g(x, y)) \right] dx\, dy. \tag{17}$$

□ **EXAMPLE 8**

Calculate the flux over the sphere S with center $(0, 0, 0)$ and radius R associated with the (central) inverse square field

$$\mathbf{F}(\mathbf{r}) = \frac{\lambda \mathbf{r}}{|\mathbf{r}|^3}, \qquad \mathbf{r} \neq \mathbf{0}, \qquad \lambda \text{ constant.}$$

That is,

$$\mathbf{F}(x, y, z) = \lambda \frac{x\mathbf{i} + y\mathbf{j} + z\mathbf{k}}{(x^2 + y^2 + z^2)^{3/2}}, \qquad x^2 + y^2 + z^2 \neq 0.$$

Solution Since the force field $\mathbf{F}$ has the same symmetry with respect to the origin as the sphere S does, we will calculate the flux over the upper hemisphere only and

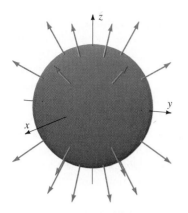

Figure 5.17 The flux of a *central* force field over a sphere is twice the flux over the upper hemisphere $z = \sqrt{R^2 - x^2 - y^2}$.

then double the result (see Figure 5.17). Note that we are dealing with an improper integral as in Example 6. Using equation (16) with

$$M(x, y, z) = \frac{\lambda x}{(x^2 + y^2 + z^2)^{3/2}} = \frac{\lambda x}{R^3},$$

$$N(x, y, z) = \frac{\lambda y}{(x^2 + y^2 + z^2)^{3/2}} = \frac{\lambda y}{R^3},$$

$$P(x, y, z) = \frac{\lambda z}{(x^2 + y^2 + z^2)^{3/2}} = \frac{\lambda z}{R^3},$$

and

$$g(x, y) = \sqrt{R^2 - x^2 - y^2} = z$$

gives

$$\iint_S \mathbf{F} \cdot \mathbf{N} \, dS = 2 \iint_Q \left[-\frac{\lambda x}{R^3}\left(\frac{-x}{\sqrt{R^2 - x^2 - y^2}} \right) \right.$$

$$\left. - \frac{\lambda y}{R^3}\left(\frac{-y}{\sqrt{R^2 - x^2 - y^2}} \right) + \frac{\lambda \sqrt{R^2 - x^2 - y^2}}{R^3} \right] dx \, dy$$

$$= 2 \iint_Q \frac{\lambda}{R\sqrt{R^2 - x^2 - y^2}} \, dx \, dy$$

$$= 2 \int_0^{2\pi} \int_0^R \frac{\lambda}{R\sqrt{R^2 - r^2}} \, r \, dr \, d\theta \qquad \text{(switch to polar coordinates)}$$

$$= 2 \int_0^{2\pi} \left. -\left[\frac{\lambda}{R}\sqrt{R^2 - r^2} \right] \right|_{r=0}^{r=R} d\theta$$

$$= 4\pi\lambda. \qquad \blacksquare$$

The surprising result of Example 8 is that *the flux of the inverse square field* **F** *over the sphere S is independent of the radius of S!* This result is true for any inverse square field, such as a gravitational field or an electric field. In the latter case the result of Example 8, together with another property of electric fields (Coulomb's Law), gives **Gauss's Law:**

> The flux of the electric field **E** through any closed surface S is
>
> $$\iint_S \mathbf{E} \cdot \mathbf{N} \, dS = 4\pi \sum_{j=1}^{n} q_j$$
>
> where $q_1, q_2, \ldots, q_n$ are the point charges contained within S.

The Orientation of a Surface

If the surface S is the graph of a single function $z = g(x, y)$, there is no ambiguity about what is meant by the *upward* unit normal (as opposed to the downward unit normal, so a flux integral is easily determined to be of the form (16) or of the form (17). Also, in the cases where S is the graph of the function $y = g(x, z)$ or $x = g(y, z)$, the corresponding meaning of upward normal and the modifications of equa-

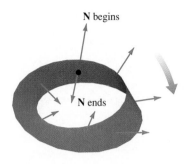

N begins

N ends

Figure 5.18 The Möbius band is not orientable.

tions (16) and (17) should be clear. However, the situation becomes less clear for surfaces of the form $S = S_1 \cup S_2 \cup \cdots \cup S_n$ where each S_j is as above. We need to discuss the meaning of **N** in the flux integral (13) for these more general cases.

According to our original motivation, flux is a measure of the *net* (positive minus negative) rate of flow across S. If more fluid is flowing upward across $S: z = g(x, y)$ than is flowing downward, flux is positive, although fluid may actually be flowing upward across some portions of S and downward across others. It is therefore crucial that **N** always point on the same "side" of S, since **N** provides the direction used to measure the rate of flow at each point. Thus, in order for the flux integral in (13) to make sense, the surface S must be *orientable,* that is, we must be able to identify two distinct sides of S and to assign a *unique* unit normal $\mathbf{N}(x, y, z)$ that varies continuously as (x, y, z) varies across S.

We have already seen that this is simple to do for surfaces of the form

$$S = \{(x, y, z) \mid z = g(x, y), (x, y) \in Q.$$

We just choose the unit normal with positive z-component (upward) or negative z-component (downward). For closed surfaces (such as a sphere or an ellipsoid) the two sides are the inside and the outside, and we must choose between the *inner* unit normal or the *outer* unit normal. (Figure 5.17 shows outer unit normals for a sphere.)

These two general situations (upper versus lower, or inner versus outer) encompass most of the surfaces with which you will need to deal, and we refer to any such surface as **orientable.** We shall not attempt to give a precise definition for this term (as is done in more advanced courses). However, you should know that not all surfaces are orientable. A classic example is the Möbius band, which is obtained by twisting one end of a long rectangular strip $180°$ and "gluing" it to the other end. As Figure 5.18 shows, a unit normal **N** beginning at point P and moving continuously once "around" the Möbius band arrives back at P pointing in the opposite direction.

As a final observation, note that the surface integral in Definition 5 does not involve the notion of orientation for S. However, the flux integral in (13) does depend on an orientation for S. Reversing the orientation for S (and, hence, replacing **N** by $-\mathbf{N}$) changes the sign of the integral in (13).

Exercise Set 20.5

1. Evaluate

$$\iint_S (x + 3y + z) \, dS$$

over the portion of the plane $x - y + 2z = 4$ lying above the rectangle $Q = \{(x, y) \mid 0 \le x \le 1, 0 \le y \le 2\}$.

2. Evaluate the surface integral in Exercise 1 where S is the portion of the plane $x - y + 2z = 4$ lying above the unit circle.

3. Evaluate

$$\iint_S x^2 z \, dS$$

where S is the portion of the plane $2x + 3y + z = 6$ lying in the first octant ($x \ge 0, y \ge 0, z \ge 0$).

4. Evaluate

$$\iint_S y^2 z \, dS$$

where S is the portion of the cone $z^2 = x^2 + y^2$ lying between the planes $z = 1$ and $z = 4$.

5. Evaluate the surface integral

$$\iint_S (x^2 + 5y - z) \, dS$$

where S is the graph of the cylinder $z = x^2$ over the square $-1 \le x \le 1, -1 \le y \le 1$.

6. Evaluate

$$\iint\limits_S (x^2 - y^2 + 1)\, dS$$

where S is the part of the plane $z = x + 2$ inside the cylinder $x^2 + y^2 = 4$.

7. Find the mass of the part of a sphere $x^2 + y^2 + z^2 = 9$ lying inside the cylinder $x^2 + y^2 = 4$ if the density per unit area δ is constant.

8. Find

$$\iint\limits_S \mathbf{F} \cdot \mathbf{N}\, dS$$

where $\mathbf{F} = 2x\mathbf{i} - y\mathbf{j} + 3z\mathbf{k}$, S is the part of the plane $x - y + 2z = 4$ bounded by the planes $y = 0$, $y = -4$, $x = 0$, and $x = 4$, and $\mathbf{N}$ is the upward unit normal.

9. Find

$$\iint\limits_S (x + y + z^2)\, dS$$

where S is the hemisphere $z = \sqrt{4 - x^2 - y^2}$.

10. Evaluate

$$\iint\limits_S (y - x)\, dS$$

where S is the part of the plane $6x + 4y + 2z = 8$ lying inside the cylinder $x^2 + y^2 = 4$.

11. Find the flux

$$\iint\limits_S \mathbf{F} \cdot \mathbf{N}\, dS$$

where $\mathbf{F} = x\mathbf{i} + y\mathbf{j} + z\mathbf{k}$, S is the upper half sphere $z = \sqrt{1 - x^2 - y^2}$, and $\mathbf{N}$ is the upward unit normal.

12. Use the result of Exercise 11 to find

$$\iint\limits_S \mathbf{F} \cdot \mathbf{N}\, dS$$

where $\mathbf{F} = x\mathbf{i} + y\mathbf{j} + z\mathbf{k}$, S is the entire unit sphere, and $\mathbf{N}$ is the outward unit normal.

13. Calculate the flux integral

$$\iint\limits_S \mathbf{F} \cdot \mathbf{N}\, dS$$

where $\mathbf{F} = 2x\mathbf{i} + y\mathbf{j} + z\mathbf{k}$, S is the surface of the paraboloid $z = x^2 + y^2$ bounded by the planes $z = 0$ and $z = 4$, and $\mathbf{N}$ is the outward unit normal.

14. Let $\mathbf{F} = 2\mathbf{i} - x\mathbf{j} + y\mathbf{k}$. Find the flux integral

$$\iint\limits_S \mathbf{F} \cdot \mathbf{N}\, dS$$

where S is the portion of the cylinder $x^2 + y^2 = 1$ lying between the planes $z = 0$ and $z = 1$ and $\mathbf{N}$ is the outward unit normal.

15. Find the flux outward through the sphere $x^2 + y^2 + z^2 = a^2$ for the vector field $\mathbf{F}(x, y, z) = z\mathbf{k}$. What is the flux if $\mathbf{F}(x, y, z) = y\mathbf{j}$? Can you give a geometric argument for these results?

16. Find

$$\iint\limits_S \mathbf{F} \cdot \mathbf{N}\, dS$$

where S is the ellipsoid $4x^2 + y^2 + z^2 = 4$, $\mathbf{F} = x^2\mathbf{i} + y^2\mathbf{j} + z^2\mathbf{k}$, and $\mathbf{N}$ is the outward unit normal. (This will require a geometric argument.)

17. Find the flux

$$\iint\limits_S \mathbf{F} \cdot \mathbf{N}\, dS$$

where $\mathbf{F} = x\mathbf{i} + y\mathbf{j} + z\mathbf{k}$, $\mathbf{N}$ is the outward unit normal, and S is the surface of the unit cube $\{(x, y, z) \mid 0 \le x \le 1, 0 \le y \le 1, 0 \le z \le 1\}$.

18. Suppose that $\mathbf{F}(x, y, z) = a x\mathbf{i} + b y\mathbf{j} + c z\mathbf{k}$ and that S is part of the plane $ax + by + cz = d$. Show that the flux integral

$$\iint\limits_S \mathbf{F} \cdot \mathbf{N}\, dS$$

is $(a^2 + b^2 + c^2)A$ where A is the area of S, if $\mathbf{N}$ is the unit normal with $\mathbf{N} \cdot c\mathbf{k} \ge 0$.

19. Is the result of Example 8 the same if $\mathbf{F}$ is an inverse *cube* field

$$\mathbf{F}(\mathbf{r}) = \frac{\mathbf{r}}{|\mathbf{r}|^4}?$$

What is the result in this case?

20. Show that the *flux* of the vector field $\mathbf{F} = M\mathbf{i} + N\mathbf{j} + P\mathbf{k}$ across a surface S can be written

$$\iint\limits_S \mathbf{F} \cdot \mathbf{N}\, dS = \iint\limits_S (P \cos \alpha + Q \cos \beta + R \cos \gamma)\, dS$$

where α, β, and γ are the direction numbers for the unit normal $\mathbf{N}$.

For a surface S, the coordinates $(\bar{x}, \bar{y}, \bar{z})$ of the *centroid* are defined to be

$$\bar{x} = \frac{1}{A} \iint_S x \, dS, \quad \bar{y} = \frac{1}{A} \iint_S y \, dS, \quad \bar{z} = \frac{1}{A} \iint_S z \, dS$$

where A is the surface area of S. Use these formulas in Exercises 21–23.

21. Find the centroid of the hemisphere $z = \sqrt{a^2 - x^2 - y^2}$, $z \geq 0$. (*Hint:* Use symmetry as much as possible.)

22. Find the centroid of the cylinder $z = 1 - x^2$, $-2 \leq y \leq 2$, $-1 \leq x \leq 1$.

23. Find the centroid of the part of the spherical surface $x^2 + y^2 + z^2 = 1$ lying in the first octant.

24. Show that the surface integral, as defined by equation (1), can be written in the form

$$\iint_S f(x, y, z) \, dS = \iint_Q f(x, y, g(x, y)) \sec \gamma \, dx \, dy$$

where $\gamma = \gamma(x, y)$ is the angle between the upward normal to the graph of $z = g(x, y)$ at (x, y, z) and the vector $\mathbf{k}$. (*Hint:* Refer to Section 19.5, Exercise 31.)

25. Show that if the function $f(r, \theta, z)$ is written using cylindrical coordinates for $\mathbb{R}^3$ and if the surface S is the graph of the function $z = g(r, \theta)$ in polar coordinates, then the formula in Definition 5 becomes

$$\iint_S f(r, \theta, z) \, dS =$$

$$\iint_Q f(r, \theta, g(r, \theta)) \cdot \sqrt{r^2 + r^2 \left(\frac{\partial g}{\partial r}\right)^2 + \left(\frac{\partial g}{\partial \theta}\right)^2} \, dr \, d\theta.$$

20.6 Stokes' Theorem

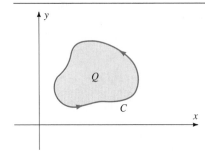

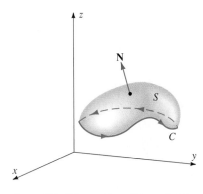

Figure 6.1 Curve C and region Q in Green's Theorem.

Figure 6.2 The surface S bounded by the space curve C in Stokes' Theorem.

In this section we encounter a generalization of Green's Theorem, which we have stated earlier as

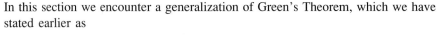

$$\int_C \mathbf{F} \cdot d\mathbf{r} = \iint_Q \left(\frac{\partial N}{\partial x} - \frac{\partial M}{\partial y}\right) dA = \iint_Q (\text{curl } \mathbf{F}) \, dA. \tag{1}$$

Recall the assumptions of equation (1): $\mathbf{F}(x, y) = M(x, y)\mathbf{i} + N(x, y)\mathbf{j}$ is a vector field in the plane (with continuously differentiable components), C is a piecewise smooth simple closed curve enclosing the region Q, and curl $\mathbf{F}$ is the scalar function

$$\text{curl } \mathbf{F} = \frac{\partial N}{\partial x} - \frac{\partial M}{\partial y}. \tag{2}$$

All this takes place in the plane $\mathbb{R}^2$.

The generalization we discuss in this section is Stokes' Theorem, which differs from Green's Theorem in three ways. First, the setting for Stokes' Theorem is $\mathbb{R}^3$. The curve C will now be a closed space curve, and the vector field $\mathbf{F}$ will have the form

$$\mathbf{F}(x, y, z) = M(x, y, z)\mathbf{i} + N(x, y, z)\mathbf{j} + P(x, y, z)\mathbf{k}. \tag{3}$$

Second, whereas the integral on the right-hand side of equation (1) is a double integral over the plane region Q, the right-hand side of Stokes' Theorem will involve a surface integral evaluated over a smooth surface S bounded by the closed space curve C. Figures 6.1 and 6.2 illustrate the geometry associated with these two theorems.

Stokes' Theorem (which we shall state in more precise terms later) is

$$\int_C \mathbf{F} \cdot d\mathbf{r} = \iint\limits_S (\text{curl } \mathbf{F}) \cdot \mathbf{N} \, dS \qquad (4)$$

where $\mathbf{N}$ is a unit normal for the surface S.

The strong similarity between equations (1) and (4) is obvious. However, the integrand in the surface integral in (4) has not yet been defined. This brings us to the third major difference between these two theorems.

Recall that the line integral in (1) was interpreted in Section 20.4 as the *circulation* around C of a fluid flowing with velocity field $\mathbf{F}$. This interpretation followed from the equation

$$\int_C \mathbf{F} \cdot d\mathbf{r} = \int_C \mathbf{F} \cdot \mathbf{T} \, ds \qquad (5)$$

discussed earlier. Now equation (5) holds for vector fields of the form (3) defined on space curves as well as in the planar case. Using fluid flow as our primary motivation, we would like to define curl $\mathbf{F}$ for $\mathbf{F}$ in (3) in such a way that equation (4) can be interpreted as a statement about fluid flow (and, of course, so that equation (4) is true!) The following definition of curl $\mathbf{F}$ does this. We shall discuss its physical interpretation at the end of this section.

Definition 6	Let $\mathbf{F}$ be a vector field of the form $$\mathbf{F}(x, y, z) = M(x, y, z)\mathbf{i} + N(x, y, z)\mathbf{j} + P(x, y, z)\mathbf{k}.$$ If all first order partial derivatives for each component function M, N, and P exist at (x, y, z), **curl $\mathbf{F}(x, y, z)$** is defined to be the vector $$\text{curl } \mathbf{F}(x, y, z) = \left(\frac{\partial P}{\partial y} - \frac{\partial N}{\partial z}\right)\mathbf{i} + \left(\frac{\partial M}{\partial z} - \frac{\partial P}{\partial x}\right)\mathbf{j} + \left(\frac{\partial N}{\partial x} - \frac{\partial M}{\partial y}\right)\mathbf{k} \qquad (6)$$ where each partial derivative is evaluated at (x, y, z).

Note that curl $\mathbf{F}(x, y, z)$ is a *vector* in $\mathbb{R}^3$ while curl $\mathbf{F}(x, y)$ in (2) is a *number*. Now curl $\mathbf{F}(x, y, z) \cdot \mathbf{N}$ (the integrand in the surface integral in Stokes' Theorem) is a generalization of curl $\mathbf{F}(x, y)$ (the integrand in the double integral in Green's Theorem) in the following sense. If we restrict $\mathbf{F}$ in (3) to the xy-plane by setting $z \equiv 0$ and $P(x, y, z) \equiv 0$, equation (6) gives

$$\text{curl } \mathbf{F}(x, y, 0) = \left(\frac{\partial N}{\partial x} - \frac{\partial M}{\partial y}\right)\mathbf{k}. \qquad (7)$$

Since a unit normal to the xy-plane is $\mathbf{N} = \mathbf{k}$, we obtain

$$[\text{curl } \mathbf{F}(x, y, 0)] \cdot \mathbf{N} = \left[\left(\frac{\partial N}{\partial x} - \frac{\partial M}{\partial y}\right)\mathbf{k}\right] \cdot \mathbf{k}$$

$$= \frac{\partial N}{\partial x} - \frac{\partial M}{\partial y} = \text{curl } \mathbf{F}(x, y),$$

as claimed.

You will often see the notation

$$\text{curl } \mathbf{F} = \nabla \times \mathbf{F}. \tag{8}$$

This results from thinking of the gradient ∇ as an "operator"

$$\nabla = \frac{\partial}{\partial x}\mathbf{i} + \frac{\partial}{\partial y}\mathbf{j} + \frac{\partial}{\partial z}\mathbf{k}$$

and forming the cross product*

$$\nabla \times \mathbf{F} = \det \begin{bmatrix} \mathbf{i} & \mathbf{j} & \mathbf{k} \\ \dfrac{\partial}{\partial x} & \dfrac{\partial}{\partial y} & \dfrac{\partial}{\partial z} \\ M & N & P \end{bmatrix} \tag{9}$$

$$= \left(\frac{\partial P}{\partial y} - \frac{\partial N}{\partial z}\right)\mathbf{i} + \left(\frac{\partial M}{\partial z} - \frac{\partial P}{\partial x}\right)\mathbf{j} + \left(\frac{\partial N}{\partial x} - \frac{\partial M}{\partial y}\right)\mathbf{k}.$$

□ **EXAMPLE 1**

For the vector field $\mathbf{F}(x, y, z) = y\mathbf{i} - z^2\mathbf{j} + 2x\mathbf{k}$, $M = y$, $N = -z^2$, and $P = 2x$. From equations (8) and (9) we find that

$$\text{curl } \mathbf{F} = \det \begin{bmatrix} \mathbf{i} & \mathbf{j} & \mathbf{k} \\ \dfrac{\partial}{\partial x} & \dfrac{\partial}{\partial y} & \dfrac{\partial}{\partial z} \\ y & -z^2 & 2x \end{bmatrix} = 2z\mathbf{i} - 2\mathbf{j} - \mathbf{k}.$$
∎

One final detail needs to be addressed before we state Stokes' Theorem. We must require that the surface S be **simply connected.** Basically, this means that S contains no holes. Alternatively, given any point P_0 on S, the boundary curve C can be "continuously contracted" across S to an arbitrarily small closed curve containing P_0 (see Figure 6.3). (If S had a hole you could not do this since the contraction of C would get "hung up" at the hole.)

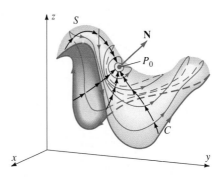

Figure 6.3 An oriented simply connected surface S. Contracting C continuously across S reveals the orientation of C induced by the orientation of S.

---

*As in the mnemonic for remembering how to compute cross products, the determinant here is only formal notation, since only the third row of the matrix actually contains numbers.

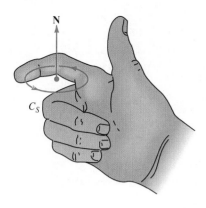

Figure 6.4 Right-hand rule determines orientation of C_S from direction of **N**.

This notion of contracting C across S toward a point also allows us to clarify the issue of orientation. We shall require that the surface S be oriented. As defined in Section 20.5, this means that S has a unit normal **N** that varies continuously across S, always remaining on the same side of S. In the statement of Stokes' Theorem we must require that the orientation of the boundary curve C be the orientation *induced* on C by the orientation of S. Here's what we mean. Let **N** be the unit normal to S at the point P_0. If S is smooth, then very near P_0 the surface of S may be approximated by a plane tangent to S at P_0. Think of a very small circle C_S on this tangent plane with center P_0. This small circle approximates a small closed curve on S, and we can continuously contract C across S so that it takes the shape of this small closed curve. The orientation on C is the orientation induced by the orientation of S if the orientation of all three curves (C, the small curve on S, and the small circle C_S on the tangent plane) are the same. The normal **N** determines an orientation for C_S by a variation of the right-hand rule: Think of wrapping the index finger of your right hand around C_S, keeping your thumb parallel to **N**. Your index finger points out the orientation of C_S (see Figure 6.4). This is the orientation induced by **N** on C.

We are now ready to state Stokes' Theorem.

Theorem 4
Stokes' Theorem

Let S be a smooth, simply connected, oriented surface bounded by a piecewise smooth simple closed curve C with orientation induced by the orientation of S. Let

$$\mathbf{F}(x, y, z) = M(x, y, z)\mathbf{i} + N(x, y, z)\mathbf{j} + P(x, y, z)\mathbf{k}$$

be a vector field defined on an open box D containing S and C so that each of the component functions M, N, and P is continuous and has continuous partial derivatives on D. Then

$$\int_C \mathbf{F} \cdot d\mathbf{r} = \int\int_S (\text{curl } \mathbf{F}) \cdot \mathbf{N} \, dS \tag{10}$$

where **N** is the unit normal to the oriented surface S.

Using equation (3), we may write equation (10) in the alternative form

$$\int_C M \, dx + N \, dy + P \, dz = \int\int_S (\text{curl } \mathbf{F}) \cdot \mathbf{N} \, dS.$$

We may paraphrase Stokes' Theorem by saying that ''the line integral of the vector field around the boundary of S equals the integral of the normal component of curl **F** over the surface S.''

Note that the particular shape of S is unimportant—the surface integral over S is determined entirely by the line integral along its boundary. For example, the surface integrals of (curl **F**) · **N** over each of the surfaces in Figure 6.5 are the same, since each has the same boundary C.

A proof of Stokes' Theorem is beyond the scope of this text. (If you've noticed that some of the concepts of this chapter appear to be less precise than in earlier chapters, you are correct. We are now discussing ideas that require considerably more advanced mathematics to treat rigorously than we can present in a first calculus course.) However, we can discuss why Stokes' Theorem should hold.

First consider the case of the tetrahedron sketched in Figure 6.6. We shall take the surface S to be the union of the three faces ABE, BDE, and DAE. Then S is

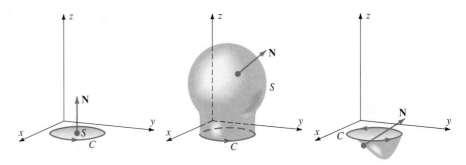

Figure 6.5 The integrals over all of these surfaces of (curl **F**) · **N** are the same:

$$\iint_S (\text{curl } \mathbf{F}) \cdot \mathbf{N} \, dS = \int_C \mathbf{F} \cdot d\mathbf{r}.$$

bounded by the triangle ABD, which is the curve C. We orient S by taking the outward unit normal **N** at each point, which induces the orientation ABD on C.

Since each face of the surface is a plane region, we apply Green's Theorem to each face and obtain the three equations

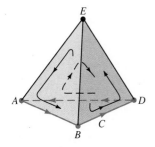

Figure 6.6 Stokes' Theorem holds for a tetrahedral surface.

$$\int_{ABE} \mathbf{F} \cdot d\mathbf{r} = \iint_{ABE} (\text{curl } \mathbf{F}) \cdot \mathbf{N} \, dS, \tag{11}$$

$$\int_{BDE} \mathbf{F} \cdot d\mathbf{r} = \iint_{BDE} (\text{curl } \mathbf{F}) \cdot \mathbf{N} \, dS, \tag{12}$$

$$\int_{DAE} \mathbf{F} \cdot d\mathbf{r} = \iint_{DAE} (\text{curl } \mathbf{F}) \cdot \mathbf{N} \, dS. \tag{13}$$

(In each of equations (11) through (13), we have used the remark concerning equation (7) to write the scalar curl in vector form.)

Next, we wish to add the corresponding sides in equations (11) through (13). When we do so, the contributions to the resulting line integral along the edges AE, BE, and DE will be zero, since each of these edges is traversed twice, one in each direction. Also, the three surface integrals should sum to the surface integral over S. The result, then, is

$$\int_{ABD} \mathbf{F} \cdot d\mathbf{r} = \iint_S (\text{curl } \mathbf{F}) \cdot \mathbf{N} \, dS$$

which is equation (10).

Now consider a more complex polyhedron S, such as that in Figure 6.7. The idea is the same. If we apply Green's Theorem to each face and sum the corresponding sides of the resulting equations, the only nonzero contributions to the line integral occur along the bottom edges. Along all other edges, the line integral will have been evaluated twice, once in each direction. Referring to the bottom edge as C, we again obtain equation (10).

For a smooth surface S we approximate S by a sequence of polyhedra, each of whose vertices lies on the surface S. As the number of vertices in the polyhedra becomes large, the polyhedra become increasingly accurate approximations to the

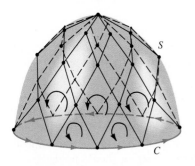

Figure 6.7 A polyhedral approximation to the surface S.

surface. Since equation (10) holds for each polyhedron, it holds for the smooth surface "in the limit."

☐ **EXAMPLE 2**

Find

$$\int_C \mathbf{F} \cdot d\mathbf{r}$$

where $\mathbf{F}$ is the vector field $\mathbf{F}(x, y, z) = y\mathbf{i} - z^2\mathbf{j} + 2x\mathbf{k}$ and C is the unit circle $C = \{(x, y, 0) \mid x^2 + y^2 = 1\}$ oriented in the counterclockwise direction.

Solution We could evaluate the line integral directly (and painfully) by parameterizing C by $x(t) = \cos t$, $y(t) = \sin t$, $0 \le t \le 2\pi$. Instead, we apply Stokes' Theorem using any smooth surface S bounded by C. The simplest choice, of course, is the unit disc

$$S = \{(x, y, 0) \mid 0 \le x^2 + y^2 \le 1\}.$$

From Example 1 we know that

$$\text{curl } \mathbf{F} = 2z\mathbf{i} - 2\mathbf{j} - \mathbf{k}$$

and a unit normal giving S the required orientation is $\mathbf{N} = \mathbf{k}$. Thus

$$(\text{curl } \mathbf{F}) \cdot \mathbf{N} = (2z\mathbf{i} - 2\mathbf{j} - \mathbf{k}) \cdot \mathbf{k} = -1.$$

Thus, by Stokes' Theorem,

$$\int_C \mathbf{F} \cdot d\mathbf{r} = \iint_S (-1) \, dS = -\pi,$$

since π is the area of the unit disc. ∎

☐ **EXAMPLE 3**

Verify Stokes' Theorem if $\mathbf{F}$ is the vector field

$$\mathbf{F}(x, y, z) = y\mathbf{i} - x^2\mathbf{j} + 2z^2\mathbf{k}$$

and S is the portion of the paraboloid $z = 4 - x^2 - y^2$ lying above the xy-plane. Take $\mathbf{N}$ to be the upward unit normal to S.

Solution The curve C bounding S is the circle

$$C = \{(x, y, 0) \mid x^2 + y^2 = 4\}$$

which has parameterization $x(t) = 2 \cos t$, $y(t) = 2 \sin t$, $z(t) = 0$, $0 \le t \le 2\pi$. The line integral on the left side of equation (10) is therefore

$$\int_C \mathbf{F} \cdot d\mathbf{r} = \int_C y \, dx - x^2 \, dy + 2z^2 \, dz$$

$$= \int_0^{2\pi} (-4 \sin^2 t - 8 \cos^3 t) \, dt$$

$$= -4\pi.$$

To calculate the surface integral in (10), we first use (9) to find

$$\text{curl }\mathbf{F} = \nabla \times \mathbf{F} = \det \begin{bmatrix} \mathbf{i} & \mathbf{j} & \mathbf{k} \\ \dfrac{\partial}{\partial x} & \dfrac{\partial}{\partial y} & \dfrac{\partial}{\partial z} \\ y & -x^2 & 2z^2 \end{bmatrix} = -(2x + 1)\mathbf{k}.$$

Then, using equation (16), Section 20.5, we evaluate the surface integral as

$$\iint\limits_{S} (\text{curl }\mathbf{F}) \cdot \mathbf{N}\, dS = \iint\limits_{S} [-(2x + 1)\mathbf{k}] \cdot \mathbf{N}\, dS$$

$$= \iint\limits_{Q} -(2x + 1)\, dx\, dy$$

$$= \int_{0}^{2\pi} \int_{0}^{2} -(2r \cos\theta + 1)\, r\, dr\, d\theta \qquad \text{(switch to polar coordinates)}$$

$$= -4\pi,$$

which agrees with the value of the line integral. ■

REMARK   Notice that the calculation of the surface integral involved the same double integral as we would have obtained had the problem simply been to calculate

$$\iint\limits_{S'} \text{curl }\mathbf{F} \cdot \mathbf{N}\, dS$$

over the planar disc $S' = \{(x, y, 0) \mid 0 \leq x^2 + y^2 \leq 4\}$ (Figure 6.8). This is a direct illustration of the statement of Stokes' Theorem—the value of the surface integral depends only on the line integral around the boundary.

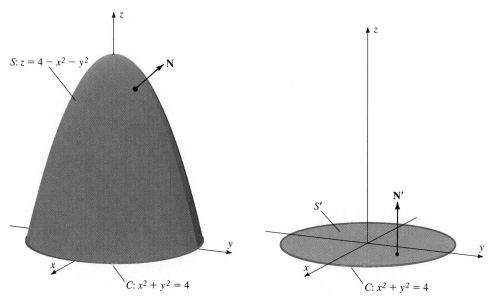

According to Stokes' Theorem,

$$\iint\limits_{S} (\text{curl }\mathbf{F}) \cdot \mathbf{N}\, dS = \iint\limits_{S'} (\text{curl }\mathbf{F}) \cdot \mathbf{N}'\, dS.$$

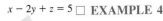

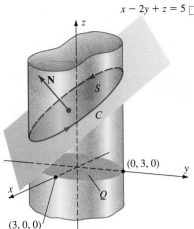

Figure 6.9 Line integral $\int_C \mathbf{F} \cdot d\mathbf{r}$ is evaluated as a surface integral

$$\iint\limits_{S} (\text{curl } \mathbf{F}) \cdot \mathbf{N} \, dS$$

over S.

□ **EXAMPLE 4**

Find

$$\int_C \mathbf{F} \cdot d\mathbf{r}$$

where the space curve C is the intersection of the plane $x - 2y + z = 5$ with the cylinder $x^2 + y^2 = 9$, oriented as in Figure 6.9, and $\mathbf{F}$ is the vector field $\mathbf{F}(x, y, z) = (x^2 - 3y^2)\mathbf{i} + (z^2 + y)\mathbf{j} + (x + 2z^2)\mathbf{k}$.

Solution Attempting to evaluate the line integral directly would be difficult at best, so we resort to Stokes' Theorem, taking S to be the portion of the plane enclosed by C. From the equation of the plane and the orientation of C, we see that a unit normal to S is

$$\mathbf{N} = \frac{1}{\sqrt{6}}(\mathbf{i} - 2\mathbf{j} + \mathbf{k})$$

which is upward. Also,

$$\text{curl } \mathbf{F} = \det \begin{bmatrix} \mathbf{i} & \mathbf{j} & \mathbf{k} \\ \dfrac{\partial}{\partial x} & \dfrac{\partial}{\partial y} & \dfrac{\partial}{\partial z} \\ (x^2 - 3y^2) & (z^2 + y) & (x + 2z^2) \end{bmatrix} = -2z\mathbf{i} - \mathbf{j} + 6y\mathbf{k}.$$

The equation for S becomes $z = g(x, y) = 5 - x + 2y$. By Stokes' Theorem and equation (16), Section 20.5, we have

$$\int_C \mathbf{F} \cdot d\mathbf{r} = \iint\limits_{S} (\text{curl } \mathbf{F}) \cdot \mathbf{N} \, dS$$

$$= \iint\limits_{Q} (-[-2(5 - x + 2y)](-1) - (-1)(2) + 6y) \, dx \, dy$$

$$= \iint\limits_{Q} (2x + 2y - 8) \, dx \, dy$$

$$= \int_0^{2\pi} \int_0^3 (2r \cos\theta + 2r \sin\theta - 8)r \, dr \, d\theta$$

$$= -72\pi.$$

(Note that we did not use the components of the unit normal $\mathbf{N}$—only the fact that it is oriented upward.) ∎

□ **EXAMPLE 5**

Calculate

$$\iint\limits_{S} (\text{curl } \mathbf{F}) \cdot \mathbf{N} \, dS$$

where S is the enclosed surface shown in Figure 6.10 and $\mathbf{F}$ is a vector field satisfying the hypotheses of Stokes' Theorem.

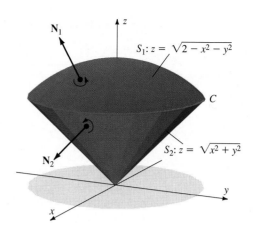

Figure 6.10   For the surface $S = S_1 \cup S_2$ in Example 5,

$$\iint\limits_{S} (\text{curl } \mathbf{F}) \cdot \mathbf{N} \, dS = 0.$$

Solution   The closed surface S may be written $S = S_1 \cup S_2$ where S_1 is the hemisphere $z = \sqrt{2 - x^2 - y^2}$ and S_2 is the graph of the cone $z = \sqrt{x^2 + y^2}$ over the unit disc. These two surfaces share the common boundary curve

$$C = \{(x, y, 1) \mid x^2 + y^2 = 1\}.$$

However, the two surfaces induce opposite orientations on C. This is because the unit normal to S_1 is upward, inducing a counterclockwise orientation on C, while the unit normal to S_2 is downward, inducing a clockwise orientation on C. Thus, by Stokes' Theorem,

$$\iint\limits_{S} (\text{curl } \mathbf{F}) \cdot \mathbf{N} \, dS = \iint\limits_{S_1} (\text{curl } \mathbf{F}) \cdot \mathbf{N}_1 \, dS + \iint\limits_{S_2} (\text{curl } \mathbf{F}) \cdot \mathbf{N}_2 \, dS$$

$$= \int_{C} \mathbf{F} \cdot d\mathbf{r} + \int_{-C} \mathbf{F} \cdot d\mathbf{r}$$

$$= \int_{C} \mathbf{F} \cdot d\mathbf{r} - \int_{C} \mathbf{F} \cdot d\mathbf{r} = 0. \qquad \blacksquare$$

Curl F and Conservative Force Fields

For conservative planar vector fields $\mathbf{F}(x, y) = M\mathbf{i} + N\mathbf{j}$, we have seen that the condition that $\mathbf{F}$ be *conservative* in an open rectangle D is equivalent to the condition that curl $\mathbf{F}(x, y) = 0$ for all $(x, y) \in D$ (see Exercise 23, Section 20.4). We now sketch an argument for the same result in three dimensions.

First, suppose that curl $\mathbf{F}(x, y, z) = \mathbf{0}$ for all (x, y, z) in some simply connected region D in $\mathbb{R}^3$. Then if C is any closed path in D,

$$\int_{C} \mathbf{F} \cdot d\mathbf{r} = \iint\limits_{S} \text{curl } \mathbf{F} \cdot \mathbf{N} \, dS = \iint\limits_{S} \mathbf{0} \cdot \mathbf{N} \, dS = 0$$

where S is any smooth surface in D bounded by C. Thus,

$$\int_{C} \mathbf{F} \cdot d\mathbf{r} = 0$$

for any closed path in D, so $\mathbf{F}$ is conservative (Exercise 17, Section 20.3).

Conversely, suppose that **F** is conservative in D. Then (again by Exercise 17, Section 20.3)

$$\int_C \mathbf{F} \cdot d\mathbf{r} = 0$$

for any closed curve C in D. Now let (x_0, y_0, z_0) be any point in D, let C_r be a small circle of radius and center (x_0, y_0, z_0), and let **N** be any unit normal to the plane containing C_r. Let S_r be the disc enclosed by C_r. Then, by Stokes' Theorem

$$\int_{C_r} \mathbf{F} \cdot d\mathbf{r} = \iint_{S_r} (\text{curl } \mathbf{F}) \cdot \mathbf{N} \, dS. \tag{14}$$

Now the line integral in (14) is zero, since **F** is conservative. Thus,

$$\iint_{S_r} (\text{curl } \mathbf{F}) \cdot \mathbf{N} \, dS = 0,$$

no matter how small r is chosen. From this statement we conclude that

$$(\text{curl } \mathbf{F}(x_0, y_0, z_0)) \cdot \mathbf{N} = 0. \tag{15}$$

(If the expression in (15) were either positive or negative, we could invoke the continuity of **F** to contract C_r to a small circle on which $(\text{curl } \mathbf{F}) \cdot \mathbf{N}$ is either always positive or always negative. The resulting surface integral in (14) would then be nonzero, a contradiction of equation (14).

Since **N** was arbitrary, it follows from (15) that curl $\mathbf{F}(x_0, y_0, z_0) = \mathbf{0}$. Finally, since (x_0, y_0, z_0) was chosen arbitrarily in D, we conclude that curl $\mathbf{F}(x, y, z) = \mathbf{0}$ for all $(x, y, z) \in D$ if F is conservative.

The preceding argument is not a rigorous proof. However, it does convey the general flavor of the arguments linking the ideas of conservative vector fields, independence of path for line integrals, and the condition curl $\mathbf{F} = \mathbf{0}$. These relationships are summarized by the following theorem.

Theorem 5

Let $\mathbf{F}(x, y, z) = M(x, y, z)\mathbf{i} + N(x, y, z)\mathbf{j} + P(x, y, z)\mathbf{k}$ be a vector field for which all of the component functions, together with all their first partial derivatives, are continuous in some open box D in $\mathbb{R}^3$. The following conditions are all equivalent:

(i) **F** is conservative in D.

(ii) $\displaystyle\int_C \mathbf{F} \cdot d\mathbf{r}$ is independent of path C in D.

(iii) $\displaystyle\int_C \mathbf{F} \cdot d\mathbf{r} = 0$ for every closed path C in D.

(iv) curl $\mathbf{F}(x, y, z) = \mathbf{0}$ for all $(x, y, z) \in D$.

Theorem 5 gives the following condition for determining if a vector field $\mathbf{F} = M\mathbf{i} + N\mathbf{j} + P\mathbf{k}$ is conservative.

Corollary 2

Let $\mathbf{F}$ and D be as in Theorem 5. Then $\mathbf{F}$ is conservative in D if and only if each of the following equations holds for all (x, y, z) in D:

(i) $\dfrac{\partial P}{\partial y}(x, y, z) = \dfrac{\partial N}{\partial z}(x, y, z),$

(ii) $\dfrac{\partial M}{\partial z}(x, y, z) = \dfrac{\partial P}{\partial x}(x, y, z),$

(iii) $\dfrac{\partial N}{\partial x}(x, y, z) = \dfrac{\partial M}{\partial y}(x, y, z).$

The proof of Corollary 2 uses Theorem 5, parts (i) and (iv) and the definition of curl $\mathbf{F}$ (Definition 6).

A Physical Interpretation of Curl

As a last look at Stokes' Theorem, we use equation (5) to write it in the form

$$\int_C \mathbf{F} \cdot \mathbf{T}\, ds = \iint_S (\operatorname{curl} \mathbf{F}) \cdot \mathbf{N}\, dS. \qquad (16)$$

Returning to our primary motivation of fluid flow, we recall from Section 20.4 that if $\mathbf{F}$ is the velocity field of a moving fluid in which the curve C is submerged, the line integral in (16), called the **circulation** of $\mathbf{F}$ around C, is a measure of the tendency of the fluid to rotate, or circulate, around the curve C. Now let $P_0(x_0, y_0, z_0)$ be a point in the fluid, and let C_r be a circle with radius r and center P_0 (see Figure 6.11). Applying the Mean Value Theorem for double integrals (an obvious generalization of the Mean Value Theorem for functions of a single variable which we have not previously stated) to equation (16), we find that for some point $P_r = (x_r, y_r, z_r)$ in S_r (the disc enclosed by C_r)

$$\int_{C_r} \mathbf{F} \cdot \mathbf{T}\, ds = \iint_{S_r} (\operatorname{curl} \mathbf{F}) \cdot \mathbf{N}\, dS = \pi r^2 [\operatorname{curl} \mathbf{F}(x_r, y_r, z_r)] \cdot \mathbf{N}.$$

Letting $r \to 0$, we conclude that

$$[\operatorname{curl} \mathbf{F}(x_0, y_0, z_0)] \cdot \mathbf{N} = \lim_{r \to 0} \frac{1}{\pi r^2} \int_{C_r} \mathbf{F} \cdot \mathbf{T}\, dS. \qquad (17)$$

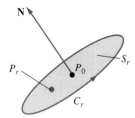

Figure 6.11 The circle C_r of radius r centered at P_0.

We interpret equation (17) as follows. The right-hand side, as described earlier, is a measure of the tendency of the fluid to circulate, or rotate, about the small circle C_r. Since the normal $\mathbf{N}$ to the disc enclosed by this circle appears on the left-hand side of (17), this tendency to rotate will be the largest if C_r is positioned so that $\mathbf{N}$ is parallel to curl $\mathbf{F}(x_0, y_0, z_0)$. Thus, curl $\mathbf{F}(x_0, y_0, z_0)$ is a vector whose *magnitude* is a measure of the tendency of the fluid at (x_0, y_0, z_0) to rotate (as about the drain hole in a bathtub), and whose *direction* is along the axis about which the fluid has the maximal tendency to rotate.

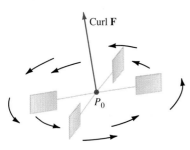

Figure 6.12 Paddle wheel interpretation of curl $\mathbf{F}$. The wheel rotates most rapidly if the axis is parallel to curl $\mathbf{F}$.

Figure 6.12 shows an interpretation of equation (17) in terms of a paddle wheel. The motion of the fluid will cause the paddle wheel, based at P_0, to rotate most quickly if the axis points parallel to curl $\mathbf{F}(x_0, y_0, z_0)$. For obvious reasons, fluid fields for which curl $\mathbf{F} = \mathbf{0}$ (such as conservative fields) are called *irrotational*.

Note that these last few observations interpret curl $\mathbf{F}(x, y, z)$ as a *local* property of the vector field—the tendency of the fluid *at location* (x, y, z) to rotate. In general, this tendency will vary from point to point within the fluid. Stokes' Theorem says that the collective measure of this tendency taken over the entire surface S (that is, the value of the surface integral) is completely determined by, and equals, the tendency of the fluid to circulate around the boundary C (the value of the line integral).

Exercise Set 20.6

In Exercises 1–4, calculate curl **F**.

1. $\mathbf{F}(x, y, z) = x\mathbf{i} - y\mathbf{j} + z^2\mathbf{k}$

2. $\mathbf{F}(x, y, z) = xy^2\mathbf{i} + xz^2\mathbf{j} + yz^2\mathbf{k}$

3. $\mathbf{F}(x, y, z) = xyz\mathbf{i} - \cos(xy)\mathbf{j} + \sin(yz)\mathbf{k}$

4. $\mathbf{F}(x, y, z) = e^{xy}\mathbf{i} - x^2z^2\mathbf{j} + \sqrt{xy}\,\mathbf{k}$

In Exercises 5–8, determine whether the given vector field is conservative.

5. $\mathbf{F}(x, y, z) = yz\mathbf{i} + xz\mathbf{j} + xy\mathbf{k}$

6. $\mathbf{F}(x, y, z) = 2xyz^2\mathbf{i} + x^2z^2\mathbf{j} + 2x^2yz\mathbf{k}$

7. $\mathbf{F}(x, y, z) = \sin y\mathbf{i} + \cos x\mathbf{j} + \sqrt{xy}\,\mathbf{k}$

8. $\mathbf{F}(x, y, z) = yze^{xyz}\mathbf{i} + xze^{xyz}\mathbf{j} + xye^{xyz}\mathbf{k}$

9. Verify Stokes' Theorem for the vector field $\mathbf{F}(x, y, z) = z\mathbf{i} + x\mathbf{j} + y\mathbf{k}$ and the surface of the hemisphere $z = \sqrt{1 - x^2 - y^2}$.

10. Use Stokes' Theorem to calculate the line integral

$$\int_C \mathbf{F} \cdot d\mathbf{r}$$

where $\mathbf{F}(x, y, z) = x^2y^2\mathbf{i} + x^2z^2\mathbf{j} + y^2z^2\mathbf{k}$ and C is the perimeter of the rectangle with vertices $(1, 1, 0)$, $(1, 5, 0)$, $(3, 5, 0)$, and $(3, 1, 0)$ traversed in this order.

11. Verify Stokes' Theorem for $\mathbf{F}(x, y, z) = x^2y\mathbf{i} + y^2z\mathbf{j} + xz\mathbf{k}$ where C is the boundary of the rectangle $Q = \{(x, y, 0) \,|\, 0 \le x \le 1, 0 \le y \le 2\}$ oriented counterclockwise.

12. Use Stokes' Theorem to calculate the line integral

$$\int_C \mathbf{F} \cdot d\mathbf{r}$$

where $\mathbf{F}(x, y, z) = x^2y\mathbf{i} + y^2z\mathbf{j} + xz\mathbf{k}$ and C is the intersection of the plane $x + 3y + z = 4$ with the cylinder $x^2 + y^2 = 1$ with orientation induced by the upward unit normal to the plane.

13. Verify Stokes' Theorem for the portion of the plane $x + 2y + z = 2$ in the first octant, the vector field $\mathbf{F}(x, y, z) = z\mathbf{i} + x\mathbf{j} + y\mathbf{k}$, and the upward unit normal.

14. Calculate

$$\int_C \mathbf{F} \cdot d\mathbf{r}$$

where $\mathbf{F}(x, y, z) = xz\mathbf{i} + 2z\mathbf{j} - xy\mathbf{k}$ and C is the intersection of the plane $y = z + 2$ and the cylinder $x^2 + y^2 = 4$. The orientation on C is that induced by the upward unit normal to the plane.

15. Evaluate

$$\int_C \mathbf{F} \cdot d\mathbf{r}$$

around the unit circle in the xy-plane, counterclockwise, where

$$\mathbf{F}(x, y, z) = (\sqrt{x} + y)\mathbf{i} + (e^y - x)\mathbf{j} + (\sin z + y)\mathbf{k}.$$

16. Calculate

$$\int_C \mathbf{F} \cdot d\mathbf{r}$$

where $\mathbf{F}(x, y, z) = xyz\mathbf{i} + xz\mathbf{k}$ and C is the intersection of the paraboloid $z = x^2 + y^2$ and the plane $z = 4$. The orientation on C is that induced by the upward unit normal on the paraboloid.

17. Verify Stokes' Theorem for the surface $z = 9 - x^2 - y^2$, $z \ge 0$ and the vector field $\mathbf{F}(x, y, z) = x^2\mathbf{i} + y^2\mathbf{j} + z^2\mathbf{k}$.

18. Show that curl $\mathbf{F} = \mathbf{0}$ for $\mathbf{F}(x, y, z) = x^2\mathbf{i} + y^2\mathbf{j} + z^2\mathbf{k}$. Can you give a geometric interpretation for this result?

19. Let $\mathbf{F}(x, y) = M(x, y)\mathbf{i} + N(x, y)\mathbf{j}$ be a conservative vector field. Let $\mathbf{G}(x, y, z) = \mathbf{F}(x, y) + \lambda z\mathbf{k}$ where λ is constant. Show that $\mathbf{G}(x, y, z)$ is conservative. Is the result true if $z\mathbf{k}$ is replaced by $g(z)\mathbf{k}$?

20. Let $\mathbf{F}(x, y, z)$ be a vector field satisfying the hypotheses of Stokes' Theorem. Show that

$$\iint_Q (\text{curl } \mathbf{F}) \cdot \mathbf{N} \, dS = 0$$

where S is the unit sphere.

21. Use Stokes' Theorem to show that a line integral is independent of parameterization if orientation is preserved.

20.7   The Divergence Theorem

In the last section, we generalize equation (28) of Section 20.4,

$$\int_C \mathbf{F} \cdot \mathbf{N}\, ds = \iint_Q \operatorname{div} \mathbf{F}(x, y)\, dA \qquad \text{(flux of } \mathbf{F} \text{ across } C), \tag{1}$$

to an appropriate equation for vector fields in $\mathbb{R}^3$. However, this time we begin with the physical interpretation and then see where the mathematics leads us.

Recall the interpretation for equation (1) in terms of fluid flow: $\mathbf{F}$ is a vector field in the plane giving the velocity $\mathbf{v}(x, y) = \mathbf{F}(x, y)$ for the steady state motion of a thin layer of fluid; C is a closed curve in the fluid layer; and $\mathbf{N}$ is the outward unit normal to C. In this setting the line integral $\int_C \mathbf{F} \cdot \mathbf{N}\, ds$ in (1) is interpreted as the **flux** of the fluid outward across C, that is, the net rate at which fluid is crossing the boundary C of Q (Figure 7.1). In Section 20.5 we encountered the notion of the flux of a vector field $\mathbf{F}$ across a closed surface S in $\mathbb{R}^3$

$$\text{Flux} = \iint_S \mathbf{F} \cdot \mathbf{N}\, dS, \tag{2}$$

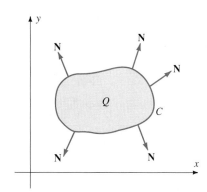

Figure 7.1 $\int_C \mathbf{F} \cdot \mathbf{N}\, ds$ is the flux of $\mathbf{F}(x, y)$ across C in $\mathbb{R}^2$.

where $\mathbf{N}$ is the outward unit normal to S at (x, y, z). That is, the natural generalization of the line integral in (1), at least in terms of fluid flow, is the surface integral in (2) (see Figure 7.2).

Now the right-hand side of equation (1) is a double integral of the scalar function div $\mathbf{F}$ evaluated over Q, the region bounded by C. We can paraphrase equation (1) by saying that "div $\mathbf{F}$ is a measure of the local behavior of $\mathbf{F}$ that, when integrated over the enclosed region Q, determines the flux of $\mathbf{F}$ across the boundary C."

Thus, to generalize equation (1) properly for a closed surface S in $\mathbb{R}^3$, we need to find a scalar function div $\mathbf{F}$ that is a measure of the local behavior of $\mathbf{F}$ and that, when integrated over the region R enclosed by S, gives the flux of $\mathbf{F}$ across the surface S. That is, div $\mathbf{F}$ must satisfy the equation

$$\iint_S \mathbf{F} \cdot \mathbf{N}\, dS = \iiint_R \operatorname{div} \mathbf{F}\, dV. \tag{3}$$

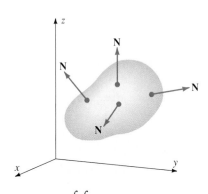

Figure 7.2 $\iint_S \mathbf{F} \cdot \mathbf{N}\, dS$ is the flux of $\mathbf{F}(x, y, z)$ across S in $\mathbb{R}^3$.

Although you can probably guess how we're going to define div $\mathbf{F}$, let's proceed "experimentally" by investigating the surface integral in (3). We begin by expressing $\mathbf{F}$ in component form as

$$\mathbf{F}(x, y, z) = M(x, y, z)\mathbf{i} + N(x, y, z)\mathbf{j} + P(x, y, z)\mathbf{k}. \tag{4}$$

Also, we assume that S is a closed surface that is the union $S = S_1 \cup S_2$ of two smooth surfaces, which are graphs of the functions

$$S_1: \quad z = g_1(x, y), \qquad (x, y) \in Q$$
$$S_2: \quad z = g_2(x, y), \qquad (x, y) \in Q$$

with $g_1(x, y) \le g_2(x, y)$ for all $(x, y) \in Q$ (see Figure 7.3).

Using (4), we write the left side of equation (3) as

$$\iint_S \mathbf{F} \cdot \mathbf{N}\, dS = \iint_S (M\mathbf{i}) \cdot \mathbf{N}\, dS + \iint_S (N\mathbf{j}) \cdot \mathbf{N}\, dS + \iint_S (P\mathbf{k}) \cdot \mathbf{N}\, dS. \tag{5}$$

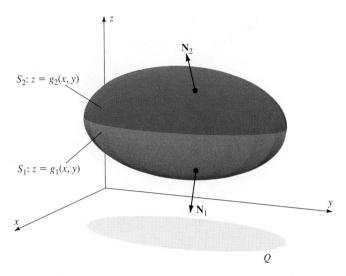

Figure 7.3   A closed surface S that is the union of two graphs of the form $z = g(x, y)$.

Let's now focus on the last integral on the right side of (5). Since the unit normal to S_2 is upward and the unit normal to S_1 is downward, we use both equation (16) and equation (17) of Section 20.5 to find that

$$\iint\limits_S [P(x, y, z)\mathbf{k}] \cdot \mathbf{N} \, dS \tag{6}$$

$$= \iint\limits_{S_2} [P(x, y, z)\mathbf{k}] \cdot \mathbf{N}_2 \, dS + \iint\limits_{S_1} [P(x, y, z)\mathbf{k}] \cdot \mathbf{N}_1 \, dS$$

$$= \iint\limits_Q P(x, y, g_2(x, y)) \, dx \, dy + \iint\limits_Q -P(x, y, g_1(x, y)) \, dx \, dy$$

$$= \iint\limits_Q \left[P(x, y, z) \right]_{z=g_1(x, y)}^{z=g_2(x, y)} dx \, dy$$

$$= \iint\limits_Q \left[\int_{g_1(x, y)}^{g_2(x, y)} \frac{\partial}{\partial z} P(x, y, z) \, dz \right] dx \, dy$$

$$= \iiint\limits_R \frac{\partial}{\partial z} P(x, y, z) \, dx \, dy \, dz.$$

(Note that we have changed the order of integration from $dz \, dx \, dy$ to $dx \, dy \, dz$.) That is,

$$\iint\limits_S (P\mathbf{k}) \cdot \mathbf{N} \, dS = \iiint\limits_R \frac{\partial P}{\partial z} \, dx \, dy \, dz. \tag{7}$$

Similarly, if S can be written as in Figure 7.4, we can show that

$$\iint\limits_S (N\mathbf{j}) \cdot \mathbf{N} \, dS = \iiint\limits_R \frac{\partial N}{\partial y} \, dx \, dy \, dz, \tag{8}$$

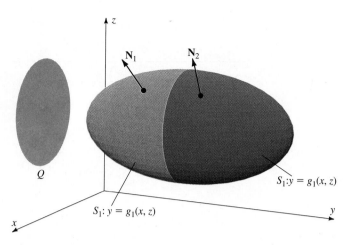

Figure 7.4 The closed surface S viewed as the union of two graphs of the form $y = g(x, z)$.

and if S can be written as in Figure 7.5, we obtain

$$\iint\limits_{S} (M\mathbf{i}) \cdot \mathbf{N}\, dS = \iiint\limits_{R} \frac{\partial M}{\partial x}\, dx\, dy\, dz. \tag{9}$$

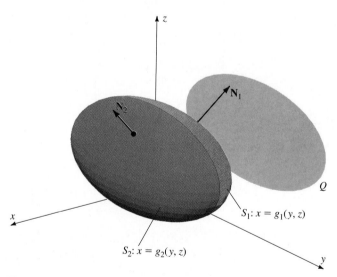

Figure 7.5 The closed surface S viewed as the union of two graphs of the form $x = g(y, z)$.

Adding the corresponding sides of equations (7), (8), and (9), using equation (5), we obtain the equation

$$\iint\limits_{S} \mathbf{F} \cdot \mathbf{N}\, dS = \iiint\limits_{R} \left(\frac{\partial M}{\partial x} + \frac{\partial N}{\partial y} + \frac{\partial P}{\partial z} \right) dx\, dy\, dz. \tag{10}$$

Comparing equations (3) and (10) leads directly to the following definition.

Definition 7	The **divergence of the vector field**

$$\mathbf{F}(x, y, z) = M(x, y, z)\mathbf{i} + N(x, y, z)\mathbf{j} + P(x, y, z)\mathbf{k}, \tag{11}$$

written div **F,** is the scalar function

$$\text{div } \mathbf{F}(x, y, z) = \frac{\partial}{\partial x}M(x, y, z) + \frac{\partial}{\partial y}N(x, y, z) + \frac{\partial}{\partial z}P(x, y, z). \tag{12}$$ |

With this definition of div **F,** the preceding discussion shows that equation (3) holds for smooth closed surfaces that are simultaneously x-simple, y-simple, and z-simple. This is our desired generalization of the planar equation (1). A more general and precisely stated version of equation (3) is our final result.

Theorem 6 Divergence Theorem	Let S be a closed piecewise smooth surface enclosing the region R. Let $\mathbf{F}(x, y, z) = M(x, y, z)\mathbf{i} + N(x, y, z)\mathbf{j} + P(x, y, z)\mathbf{k}$ be a vector field for which all of the components, together with all their first partial derivatives, are continuous throughout R. Then

$$\iint\limits_{S} \mathbf{F} \cdot \mathbf{N} \, dS = \iiint\limits_{R} \text{div } \mathbf{F}(x, y, z) \, dx \, dy \, dz \tag{13}$$

where $\mathbf{N}$ is the outward unit normal to S. |

REMARK 1   The Divergence Theorem was first discovered by the German mathematician Carl Friedrich Gauss and is often referred to as Gauss' Theorem.

REMARK 2   Using the operator notation

$$\nabla = \frac{\partial}{\partial x}\mathbf{i} + \frac{\partial}{\partial y}\mathbf{j} + \frac{\partial}{\partial z}\mathbf{k}$$

we can write

$$\text{div } \mathbf{F} = \nabla \cdot \mathbf{F} = \frac{\partial M}{\partial x} + \frac{\partial N}{\partial y} + \frac{\partial P}{\partial z}.$$

With this notation, the Divergence Theorem may be written either as

$$\iint\limits_{S} \mathbf{F} \cdot \mathbf{N} \, dS = \iiint\limits_{R} \nabla \cdot \mathbf{F} \, dV, \qquad dV = dx \, dy \, dz \tag{14}$$

or as

$$\iint\limits_{S} \mathbf{F} \cdot \mathbf{N} \, dS = \iiint\limits_{R} \left(\frac{\partial M}{\partial x} + \frac{\partial N}{\partial y} + \frac{\partial P}{\partial z} \right) dx \, dy \, dz. \tag{15}$$

The appeal of equation (14) is that it expresses Theorem 6 in what is referred to as "coordinate-free" form. On the other hand, equation (15) is the most explicit form to use when working in Cartesian (rectangular) coordinates.

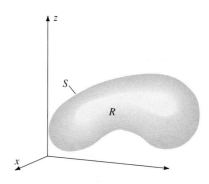

Figure 7.6   A region R that is not y-simple.

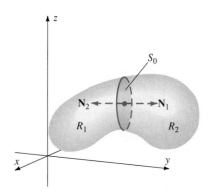

Figure 7.7 R_1 and R_2 are both y-simple. N_j is the outward unit normal for $R_j, j = 1, 2$ along the surface S_0.

REMARK 3   Our proof of the Divergence Theorem addressed only the case of a surface enclosing a region that is simultaneously x-simple, y-simple, and z-simple. Figures 7.6 and 7.7 illustrate how this proof may be extended to include more general regions. By constructing the surface S_0, we partition the region R into two regions, R_1 and R_2, both of which are simultaneously x-simple, y-simple, and z-simple. If S_1 denotes the boundary of R_1 and S_2 the boundary of R_2, the previous proof shows that

$$\iint_{S_1} \mathbf{F} \cdot \mathbf{N} \, dS = \iiint_{R_1} \operatorname{div} \mathbf{F} \, dx \, dy \, dz \tag{16}$$

and

$$\iint_{S_2} \mathbf{F} \cdot \mathbf{N} \, dS = \iiint_{R_2} \operatorname{div} \mathbf{F} \, dx \, dy \, dz. \tag{17}$$

On the face S_0, the outward unit normal to S_1 is the *opposite* of the outward unit normal to S_2. Thus, the contributions over the face S_0 to the surface integrals in (16) and (17) cancel. Adding the corresponding sides of (16) and (17) then gives

$$\iint_{S} \mathbf{F} \cdot \mathbf{N} \, dS = \iint_{S_1} \mathbf{F} \cdot \mathbf{N} \, dS + \iint_{S_2} \mathbf{F} \cdot \mathbf{N} \, dS$$

$$= \iiint_{R_1} \operatorname{div} \mathbf{F} \, dx \, dy \, dz + \iiint_{R_2} \operatorname{div} \mathbf{F} \, dx \, dy \, dz$$

$$= \iiint_{R} \operatorname{div} \mathbf{F} \, dx \, dy \, dz.$$

The same argument can be applied to more complicated surfaces.

Before we discuss a few examples, let's clarify what the Divergence Theorem says about fluid flow. Of course, the left-hand side of equation (13) is the **net flux** (rate out minus rate in) of the fluid outward across the surface S. The volume integral on the right-hand side of (13) can be thought of as a sum taken throughout R of the product $\operatorname{div} \mathbf{F} \, dV$, where dV is an infinitesimal element of volume containing the point (x, y, z). It is this local property of $\operatorname{div} \mathbf{F}$ that we wish to understand better.

To do so, let $\Delta V_\epsilon = \frac{4}{3} \pi \epsilon^3$ be the volume of a small sphere S_ϵ of radius ϵ and center (x_0, y_0, z_0) contained within S. According to Theorem 6

$$\text{Flux of } \mathbf{F} \text{ out of } S_\epsilon = \iiint_{R_\epsilon} \operatorname{div} \mathbf{F} \, dx \, dy \, dz \tag{18}$$

where R_ϵ is the region enclosed by the sphere S_ϵ. We now present an argument analogous to that of Section 20.6 for the interpretation of curl $\mathbf{F}$. The mean value property for triple integrals guarantees the existence of a point $(x_\epsilon, y_\epsilon, z_\epsilon)$ in R_ϵ for which

$$\iiint_{R_\epsilon} \operatorname{div} \mathbf{F} \, dx \, dy \, dz = \operatorname{div} \mathbf{F}(x_\epsilon, y_\epsilon, z_\epsilon) \, \Delta V_\epsilon. \tag{19}$$

Combining (18) and (19) we conclude that

$$\text{div } \mathbf{F}(x_\epsilon, y_\epsilon, z_\epsilon) = \frac{\text{Flux of } \mathbf{F} \text{ out of } S_\epsilon}{\Delta V_\epsilon}. \tag{20}$$

Since $(x_\epsilon, y_\epsilon, z_\epsilon) \to (x_0, y_0, z_0)$ as $\epsilon \to 0$, we conclude that

$$\text{div } \mathbf{F}(x_0, y_0, z_0) = \lim_{\epsilon \to 0} \frac{\text{Flux of } \mathbf{F} \text{ out of } S_\epsilon}{\Delta V_\epsilon}. \tag{21}$$

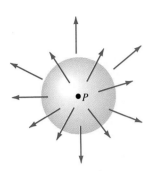

Figure 7.8 div $\mathbf{F} > 0$; P is a source.

In other words, div $\mathbf{F}(x_0, y_0, z_0)$ is the **flux per unit volume** at the point (x_0, y_0, z_0). Thus,

(i) If div $\mathbf{F}(x_0, y_0, z_0) > 0$, more fluid is flowing out across a small sphere centered at (x_0, y_0, z_0) than is flowing in. In this case we say that the fluid is **expanding** at (x_0, y_0, z_0), or that (x_0, y_0, z_0) is a **source** for the vector field $\mathbf{F}$ (see Figure 7.8).

(ii) If div $\mathbf{F}(x_0, y_0, z_0) < 0$, more fluid is flowing into a small sphere centered at (x_0, y_0, z_0) than is flowing out. The fluid is said, therefore, to be **contracting** at (x_0, y_0, z_0), and (x_0, y_0, z_0) is called a **sink** for $\mathbf{F}$ (see Figure 7.9).

(iii) If div $\mathbf{F}(x_0, y_0, z_0) = 0$, the amount of fluid flowing out of the small sphere centered at (x_0, y_0, z_0) equals the amount flowing in. In this case the fluid is said to be **incompressible** at (x_0, y_0, z_0) (see Figure 7.10).

The Divergence Theorem states that knowledge of the local property div $\mathbf{F}(x, y, z)$, summed throughout the interior of S, determines the flux of $\mathbf{F}$ across the surface S, and conversely.

Figure 7.9 div $\mathbf{F} < 0$; P is a sink.

☐ **EXAMPLE 1**

Calculate the flux of the vector field

$$\mathbf{F}(x, y, z) = (x - y^2)\mathbf{i} + 3y\mathbf{j} + (x^3 - z)\mathbf{k}$$

outward over the sphere $S = \{(x, y, z) \mid x^2 + y^2 + z^2 = 9\}$.

Solution Here

$$\text{div } \mathbf{F} = \frac{\partial}{\partial x}(x - y^2) + \frac{\partial}{\partial y}(3y) + \frac{\partial}{\partial z}(x^3 - z)$$

$$= 1 + 3 - 1 = 3.$$

By the Divergence Theorem,

$$\text{Flux across } S = \iint_S \mathbf{F} \cdot \mathbf{N} \, dS = \iiint_R 3 \, dx \, dy \, dz = 3 \cdot \tfrac{4}{3}\pi \cdot 3^3 = 108\pi,$$

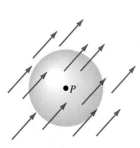

Figure 7.10 div $\mathbf{F} = 0$; $\mathbf{F}$ is *incompressible* at P.

since the volume of the sphere of radius r is $\tfrac{4}{3}\pi r^3$. ∎

☐ **EXAMPLE 2**

Evaluate the surface integral

$$\iint\limits_{S} \mathbf{F} \cdot \mathbf{N} \, dS$$

where $\mathbf{F}(x, y, z) = xy\mathbf{i} + xz\mathbf{j} + yz\mathbf{k}$ and $\mathbf{N}$ is the outward unit normal to the ellipsoid $S = \{(x, y, z) \mid x^2 + 4y^2 + z^2 = 1\}$.

Solution We use the Divergence Theorem rather than trying to evaluate the surface integral directly. We have

$$\text{div } \mathbf{F} = \frac{\partial}{\partial x}(xy) + \frac{\partial}{\partial y}(xz) + \frac{\partial}{\partial z}(yz) = 2y.$$

Letting R denote the region enclosed by the ellipsoid, we obtain

$$\iint\limits_{S} \mathbf{F} \cdot \mathbf{N} \, dS = \iiint\limits_{R} 2y \, dx \, dy \, dz$$

$$= \int_{-1}^{1} \int_{-(1/2)\sqrt{1-x^2}}^{(1/2)\sqrt{1-x^2}} \int_{-\sqrt{1-x^2-4y^2}}^{\sqrt{1-x^2-4y^2}} 2y \, dz \, dy \, dx.$$

Now this iterated integral can be evaluated directly. However, since the integrand satisfies the equation

$$\text{div } \mathbf{F}(x, -y, z) = -2y = -\text{div } \mathbf{F}(x, y, z),$$

and since the ellipsoid S is symmetric about the plane $y = 0$, the value of this integral is zero. ∎

☐ **EXAMPLE 3**

Calculate the value of the surface integral

$$\iint\limits_{S} \mathbf{F} \cdot \mathbf{N} \, dS$$

where $\mathbf{F}(x, y, z) = x^2y\mathbf{i} + xy\mathbf{j} + y^2z^3\mathbf{k}$, S is the cube formed by the planes $x = \pm 1$, $y = \pm 1$, and $z = \pm 1$, and $\mathbf{N}$ is the outward unit normal on S.

Solution Rather than calculate the surface integral directly for each of the six faces, we use the Divergence Theorem. Since

$$\text{div } \mathbf{F} = \frac{\partial}{\partial x}(x^2y) + \frac{\partial}{\partial y}(xy) + \frac{\partial}{\partial z}(y^2z^3) = 2xy + x + 3y^2z^2$$

we have, from Theorem 6,

$$\iint\limits_{S} \mathbf{F} \cdot \mathbf{N} \, dS = \int_{-1}^{1} \int_{-1}^{1} \int_{-1}^{1} (2xy + x + 3y^2z^2) \, dx \, dy \, dz$$

$$= \int_{-1}^{1} \int_{-1}^{1} \left[x^2y + \frac{x^2}{2} + 3y^2z^2x \right]_{x=-1}^{x=1} dy \, dz$$

$$= \int_{-1}^{1} \int_{-1}^{1} 6y^2 z^2 \, dy \, dz$$

$$= \int_{-1}^{1} \left[2y^3 z^2 \right]_{y=-1}^{y=1} dz$$

$$= \int_{-1}^{1} 4z^2 \, dz = \tfrac{8}{3}.$$

◻ **EXAMPLE 4**

Find the flux of the vector field

$$\mathbf{F}(x, y, z) = y^2 z^3 \mathbf{i} + 4x^2 yz^2 \mathbf{j} + x^2 z^3 \mathbf{k}$$

outward across the unit sphere $S = \{(x, y, z) \mid x^2 + y^2 + z^2 = 1\}$.

Solution   Let R be the unit ball $R = \{(x, y, z) \mid x^2 + y^2 + z^2 \le 1\}$. We have

$$\text{div } \mathbf{F} = \frac{\partial}{\partial x}(y^2 z^3) + \frac{\partial}{\partial y}(4x^2 yz^2) + \frac{\partial}{\partial z}(x^2 z^3) = 7x^2 z^2.$$

According to the Divergence Theorem,

$$\text{Flux across } S = \iint_S \mathbf{F} \cdot \mathbf{N} \, dS = \iiint_R 7x^2 z^2 \, dx \, dy \, dz.$$

To evaluate the triple integral, we use spherical coordinates:

$$x = \rho \cos \theta \sin \phi \qquad 0 \le \rho \le 1,$$
$$y = \rho \sin \theta \sin \phi \qquad 0 \le \theta \le 2\pi,$$
$$z = \rho \cos \phi \qquad\quad 0 \le \phi \le \pi,$$

We obtain

$$\iiint_R 7x^2 z^2 \, dx \, dy \, dz = 7 \int_0^1 \int_0^{2\pi} \int_0^\pi [\rho \cos \theta \sin \phi]^2 [\rho \cos \phi]^2 \rho^2 \sin \phi \, d\phi \, d\theta \, d\rho$$

$$= 7 \int_0^1 \int_0^{2\pi} \int_0^\pi \rho^6 \cos^2 \theta \sin^3 \phi \cos^2 \phi \, d\phi \, d\theta \, d\rho$$

$$= 7 \int_0^1 \int_0^{2\pi} \int_0^\pi \rho^6 \cos^2 \theta [1 - \cos^2 \phi] \cos^2 \phi \sin \phi \, d\phi \, d\theta \, d\rho$$

$$= 7 \int_0^1 \int_0^{2\pi} \rho^6 \cos^2 \theta \left[-\frac{\cos^3 \phi}{3} + \frac{\cos^5 \phi}{5} \right]_0^\pi d\theta \, d\rho$$

$$= 7 \cdot \tfrac{4}{15} \int_0^1 \int_0^{2\pi} \rho^6 \cos^2 \theta \, d\theta \, d\rho$$

$$= 7 \cdot \tfrac{4}{15} \cdot \pi \int_0^1 \rho^6 \, d\rho = \frac{4\pi}{15} \qquad\blacksquare$$

The Divergence Theorem has numerous applications in physics, engineering, and applied mathematics. We have restricted our discussion to the example of fluid flow to keep the discussion straightforward and to focus primarily on the mathematics. The subject area of electricity and magnetism makes considerable use of each of the major theorems presented in this chapter (and involves another theorem due to Gauss).

Exercise Set 20.7

In Exercises 1–6, find div **F**.

1. $\mathbf{F}(x, y, z) = x^2\mathbf{i} + y^2\mathbf{j} + z^2\mathbf{k}$

2. $\mathbf{F}(x, y, z) = x^2 z\mathbf{i} + y^2 x\mathbf{j} + xz^2\mathbf{k}$

3. $\mathbf{F}(x, y, z) = x\mathbf{i} + xy\mathbf{j} + xyz\mathbf{k}$

4. $\mathbf{F}(x, y, z) = (y - x)\mathbf{i} + (y - z)\mathbf{j} + (x - y)\mathbf{k}$

5. $\mathbf{F}(x, y, z) = \cos xy\mathbf{i} + e^{xyz}\mathbf{j} + y \sin(xz)\mathbf{k}$

6. $\mathbf{F}(x, y, z) = x \operatorname{Tan}^{-1} y\mathbf{i} - \sqrt{yz}\mathbf{j} - z \sec y\mathbf{k}$

7. Find the flux of the vector field $\mathbf{F}(x, y, z) = z\mathbf{i} + x\mathbf{j} + y\mathbf{k}$ out of the ellipsoid $x^2 + 9y^2 + 4z^2 = 1$.

8. Find the value of the surface integral
$$\iint\limits_S \mathbf{F} \cdot \mathbf{N} \, dS$$
where **F** is the vector field $\mathbf{F}(x, y, z) = -y^2\mathbf{i} + x\mathbf{j} + z\mathbf{k}$, S is the sphere $x^2 + y^2 + z^2 = 5$ and **N** is the outward unit normal.

9. Find the flux of the vector field $\mathbf{F}(x, y, z) = 3x\mathbf{i} - z^2 y\mathbf{j} + 2z\mathbf{k}$ outward across the rectangular box with vertices $(1, 0, 0)$, $(1, 3, 0)$, $(-2, 0, 0)$, $(-2, 3, 0)$, $(1, 0, 5)$, $(1, 3, 5)$, $(-2, 0, 5)$, and $(-2, 3, 5)$.

10. Find the value of the surface integral
$$\iint\limits_S \mathbf{F} \cdot \mathbf{N} \, dS$$
where **F** is the vector field $\mathbf{F}(x, y, z) = 3x\mathbf{i} - 4y\mathbf{j} + 5z\mathbf{k}$ and V is the volume of the solid enclosed by a smooth surface S.

11. Find the flux of the vector field $\mathbf{F}(x, y, z) = 2xy\mathbf{i} + z^2 y\mathbf{j} + xz\mathbf{k}$ over the cube formed by the coordinate planes and the planes $x = 1$, $y = 1$, and $z = 1$.

12. Find the value of the surface integral
$$\iint\limits_S \mathbf{F} \cdot \mathbf{N} \, dS$$
where $\mathbf{F}(x, y, z) = (x + e^y)\mathbf{i} + (e^{xz} - y)\mathbf{j} + (xy + z)\mathbf{k}$, S is the cylinder $\{(x, y, z) \mid x^2 + y^2 = 4, 0 \le z \le 2\}$ and **N** is the outward unit normal.

13. Find the flux of the vector field $\mathbf{F}(x, y, z) = x^2 y^2\mathbf{i} + xy^3\mathbf{j} + xy\mathbf{k}$ outward across the cylinder $\{(x, y, z) \mid x^2 + y^2 = 4, 0 \le z \le 2\}$. (*Hint:* Use cylindrical coordinates.)

14. Find the value of the surface integral
$$\iint\limits_S \mathbf{F} \cdot \mathbf{N} \, dS$$
where **F** is the vector field $\mathbf{F}(x, y, z) = x^2 y\mathbf{i} + xy^2\mathbf{j} + xyz\mathbf{k}$ and S is the surface of the quarter cylinder $C = \{(r, \theta, z) \mid 0 \le r \le 1, 0 \le \theta \le \pi/2, 0 \le z \le 1\}$.

15. Find the flux of the vector field $\mathbf{F}(x, y, z) = x^3\mathbf{i} + xz^2\mathbf{j} + x^2 z\mathbf{k}$ outward across the sphere $S = \{(x, y, z) \mid x^2 + y^2 + z^2 = 4\}$. (Use spherical coordinates.)

16. Find the value of the surface integral
$$\iint\limits_S \mathbf{F} \cdot \mathbf{N} \, dS$$
where **F** is the vector field $\mathbf{F}(x, y, z) = x^2 y\mathbf{i} + xy^2\mathbf{j} + xyz\mathbf{k}$, S is the unit sphere $x^2 + y^2 + z^2 = 1$, and N is the outward unit normal. (Use spherical coordinates.)

17. Find the flux of $\mathbf{F}(x, y, z) = 6x\mathbf{i} - y\mathbf{j} + 4\mathbf{k}$ across the ellipsoid $x^2 + 4y^2 + z^2 = 4$. (Calculate the volume of the ellipsoid as a volume of revolution.)

18. Verify the Divergence Theorem for the vector field $\mathbf{F}(x, y, z) = x\mathbf{i} + y\mathbf{j} + z\mathbf{k}$ and the closed surface S of the cylinder $\{(x, y, z) \mid x^2 + y^2 \le 1, 0 \le z \le 2\}$.

In Exercises 19–21, verify the stated identities for differentiable vector fields $\mathbf{F}(x, y, z)$ and $\mathbf{G}(x, y, z)$.

19. $\nabla \times (\mathbf{F} + \mathbf{G}) = \nabla \times \mathbf{F} + \nabla \times \mathbf{G}$

20. $\nabla \cdot (\mathbf{F} + \mathbf{G}) = \nabla \cdot \mathbf{F} + \nabla \cdot \mathbf{G}$

21. $\nabla \cdot (\mathbf{F} \times \mathbf{G}) = (\nabla \times \mathbf{F}) \cdot \mathbf{G} - (\nabla \times \mathbf{G}) \cdot \mathbf{F}$

The **Laplacian of the scalar field** $\phi = \phi(x, y, z)$ is defined by the equation
$$\nabla^2 \phi = \nabla \cdot (\nabla \phi) = \frac{\partial^2 \phi}{\partial x^2} + \frac{\partial^2 \phi}{\partial y^2} + \frac{\partial^2 \phi}{\partial z^2}.$$

The equation $\nabla^2 \phi = 0$ is called **Laplace's equation.** Functions which satisfy Laplace's equation are called **harmonic functions.**

22. Show that if $\mathbf{F} = \nabla \phi$, then ϕ is harmonic if and only if div $\mathbf{F} = 0$.

23. Determine which of the following vector fields are gradients of harmonic scalar functions
   a. $\mathbf{F}(x, y, z) = yz\mathbf{i} + xz\mathbf{j} + xy\mathbf{k}$
   b. $\mathbf{F}(x, y, z) = 2xye^z\mathbf{i} + x^2 e^z\mathbf{j} + x^2 ye^z\mathbf{k}$
   c. $\mathbf{F}(x, y, z) = x \sin yz\mathbf{i} + z \cos yz\mathbf{j} + y \cos yz\mathbf{k}$

24. For the inverse square field

$$\mathbf{F}(\mathbf{r}) = \frac{\mathbf{r}}{|\mathbf{r}|^3}, \quad |\mathbf{r}| \neq 0,$$

show that div $\mathbf{F} = 0$.

25. For the inverse square field of Exercise 24 show that

$$\iint_S \mathbf{F} \cdot \mathbf{N} \, dS = 0$$

if S is a surface containing a region R, as in the statement of the Divergence Theorem, so that $(0, 0, 0) \notin S \cup R$.

26. Let S be a sphere with center $(0, 0, 0)$. Show that

$$\iint_S \mathbf{F} \cdot \mathbf{N} \, dS = 4\pi$$

where $\mathbf{F}$ is the inverse square field of Exercise 24, and S is the outward unit normal. Why does this result not contradict Theorem 6?

27. Show that

$$\iint_S \mathbf{F} \cdot \mathbf{N} \, dS = 0$$

if the vector field $\mathbf{F}$ is incompressible for all $(x, y, z) \in S$. $(S,$ $\mathbf{F},$ and $\mathbf{N}$ are as in Theorem 6.)

28. Show that, if

$$\iint_S \mathbf{F} \cdot \mathbf{N} \, dS > 0,$$

then S must contain at least one source in its interior. $(S, \mathbf{F},$ and $\mathbf{N}$ are as in Theorem 6.) What if

$$\iint_S \mathbf{F} \cdot \mathbf{N} \, dS < 0?$$

29. True or false? If

$$\iint_S \mathbf{F} \cdot \mathbf{N} \, dS = 0,$$

S may contain neither sources nor sinks. $(S, \mathbf{F},$ and $\mathbf{N}$ are as in Theorem 6.) Explain.

30. Let $\mathbf{F}, S, \mathbf{N},$ and R satisfy the hypotheses of Theorem 6 in a region Ω. If

$$\iint_S \mathbf{F} \cdot \mathbf{N} \, dS = 0$$

for all closed surfaces S within Ω must $\mathbf{F} \equiv \mathbf{0}$ for all $(x, y, z) \in \Omega$? Why or why not?

Summary Outline of Chapter 20

■ A **vector field** on $\mathbb{R}^3$ is a function   (page 1036)

$$\mathbf{F}(x, y, z) = M(x, y, z)\mathbf{i} + N(x, y, z)\mathbf{j} + P(x, y, z)\mathbf{k}.$$

■ The **work** done by the force field $\mathbf{F}$ in moving an object along a curve   (page 1042)
$C = \{\mathbf{r}(t) \mid a \leq t \leq b\}$ is

$$W = \int_C \mathbf{F} \cdot d\mathbf{r} = \int_a^b \mathbf{F}(\mathbf{r}(t)) \cdot \mathbf{r}'(t) \, dt.$$

■ The **line integral** of $\mathbf{F}$ over $C = \{\mathbf{r}(t) = x(t)\mathbf{i} + y(t)\mathbf{j} + z(t)\mathbf{k} \mid a \leq t \leq b\}$ is (page 1044)

$$\int_C \mathbf{F} \cdot d\mathbf{r} = \int_a^b \mathbf{F}(\mathbf{r}(t)) \cdot \mathbf{r}'(t) \, dt$$

$$= \int_a^b [M(x(t), y(t), z(t))x'(t) + N(x(t), y(t), z(t))y'(t) +$$

$$P(x(t), y(t), z(t))z'(t)] \, dt$$

$$= \int_C M \, dx + N \, dy + P \, dz$$

■ The **line integral** of the scalar function f **with respect to arc length** over   (page 1050)
the path C (above) is

$$\int_C f(x, y, z) \, ds = \int_a^b f(x(t), y(t), z(t))\sqrt{[x'(t)]^2 + [y'(t)]^2 + [z'(t)]^2} \, dt.$$

▮ **Theorem 1:** Under appropriate hypotheses, if $\mathbf{F} = \nabla\phi$ and $C = \{\mathbf{r}(t) \mid$ (page 1054)
$a \le t \le b\}$, then

$$\int_C \mathbf{F} \cdot d\mathbf{r} = \phi(\mathbf{r}(b)) - \phi(\mathbf{r}(a)).$$

▮ **Theorem 2:** $\int_C \mathbf{F} \cdot d\mathbf{r}$ is independent of path if and only if $\mathbf{F}$ is **conser-**   (page 1056)
vative (i.e., $\mathbf{F} = \nabla\phi$ for some scalar potential ϕ).

▮ **Greens' Theorem:** Let C be a simple closed path, oriented counter-   (page 1062)
clockwise, that encloses Q. Then

$$\int_C M\,dx + N\,dy = \iint_Q \left(\frac{\partial N}{\partial x} - \frac{\partial M}{\partial y}\right) dA.$$

▮ The **surface integral** of the scalar function f over the surface   (page 1072)
$S = \{(x, y, g(x, y)) \mid (x, y) \in Q\}$ is

$$\iint_S f(x, y, z)\,dS = \iint_Q f(x, y, g(x, y)) \sqrt{\left(\frac{\partial g}{\partial x}\right)^2 + \left(\frac{\partial g}{\partial y}\right)^2 + 1}\,dx\,dy.$$

▮ The **flux** of the vector field $\mathbf{F}$ across the surface S in the direction indicated   (page 1079)
by the unit normal $\mathbf{N}$ is

$$\text{Flux} = \iint_S \mathbf{F} \cdot \mathbf{N}\,dS.$$

▮ The **curl** of the vector field $\mathbf{F} = M\mathbf{i} + N\mathbf{j} + P\mathbf{k}$ is the **vector**   (page 1086)

$$\text{curl } \mathbf{F} = \nabla \times \mathbf{F} = \left(\frac{\partial P}{\partial y} - \frac{\partial N}{\partial z}\right)\mathbf{i} + \left(\frac{\partial M}{\partial z} - \frac{\partial P}{\partial x}\right)\mathbf{j} + \left(\frac{\partial N}{\partial x} - \frac{\partial M}{\partial y}\right)\mathbf{k}.$$

▮ **Stokes' Theorem:** If the surface S is bounded by the curve C, then   (page 1087)

$$\int_C \mathbf{F} \cdot d\mathbf{r} = \iint_S (\text{curl } \mathbf{F}) \cdot \mathbf{N}\,dS.$$

▮ The **divergence** of the vector field $\mathbf{F} = M\mathbf{i} + N\mathbf{j} + P\mathbf{k}$ is the **scalar**   (page 1099)

$$\text{div } \mathbf{F} = \frac{\partial M}{\partial x} + \frac{\partial N}{\partial y} + \frac{\partial P}{\partial z}.$$

▮ **Divergence Theorem (Gauss):** If the surface S encloses the region R,   (page 1099)

$$\iint_S \mathbf{F} \cdot \mathbf{N}\,dS = \iiint_R \text{div } \mathbf{F}\,dV.$$

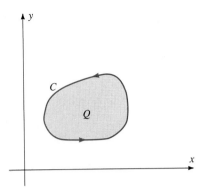

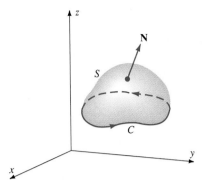

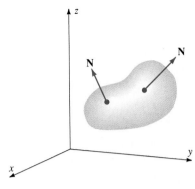

Green's Theorem

$$\int_C M\,dx + N\,dy = \iint_Q \left(\frac{\partial N}{\partial x} - \frac{\partial M}{\partial y} \right) dA$$

Stokes' Theorem

$$\int_C \mathbf{F} \cdot d\mathbf{r} = \iint_S (\text{curl } \mathbf{F}) \cdot \mathbf{N}\,dS$$

Divergence Theorem

$$\iint_S \mathbf{F} \cdot \mathbf{N}\,dS = \iiint_R \text{div } \mathbf{F}\,dV$$

Review Exercises—Chapter 20

1. Compute the gradient vector field of the given function.
   a. $f(x, y, z) = y(\cos x)\ln z$
   b. $f(x, y, z) = x^3 yz^2 + \sin xy$

2. Calculate the work done by the force field $\mathbf{F}(x, y) = x^2\mathbf{i} + 2xy\mathbf{j}$ when a particle is moved once around the triangle with vertices $(-1, 2)$, $(1, 0)$, and $(3, 2)$.

3. Show that the force field

   $$\mathbf{F}(x, y) = \frac{y}{x^2 + y^2}\mathbf{i} - \frac{x}{x^2 + y^2}\mathbf{j}$$

   is conservative.

4. Show that curl $\mathbf{F} = \mathbf{0}$ where $\mathbf{F}$ is the force field in Exercise 3.

5. Let C be the unit circle. Show that

   $$\int_C \mathbf{F} \cdot d\mathbf{r} \neq 0$$

   where $\mathbf{F}$ is the vector field in Exercise 3. Does this result together with the result of Exercise 4 contradict Stokes' Theorem? Why or why not?

6. For $\mathbf{F}(x, y, z) = xy\mathbf{i} + (y^2 - x^2)\mathbf{j} + x^2 z^2\mathbf{k}$ find
   a. curl $\mathbf{F}$ b. div $\mathbf{F}$

7. Is the vector field $\mathbf{F}(x, y, z) = y^2 z\mathbf{i} + 2xyz\mathbf{j} + xy^2\mathbf{k}$ conservative? Why or why not?

8. Show that

   $$\iint_S \mathbf{F} \cdot \mathbf{N}\,dS = 0$$

   if S is the unit sphere and $\mathbf{F}$ is the vector field $\mathbf{F}(\mathbf{r}) = \mathbf{r}_0$ where $\mathbf{r}_0$ is constant.

9. Find the flux of the vector field $\mathbf{F}(x, y, z) = zx\mathbf{i} + xy\mathbf{j} - x^2 z\mathbf{k}$ outward over the tetrahedron with vertices $(0, 0, 0)$, $(0, 1, 0)$, $(1, 0, 0)$, and $(0, 0, 1)$.

10. Show that the vector field $\mathbf{F}(x, y, z) = x\mathbf{i} + y^2\mathbf{j} + z^3\mathbf{k}$ is irrotational.

11. Let $f(x, y, z)$ be a differentiable scalar function. Show that curl $(\nabla f) = \mathbf{0}$.

12. Let $\mathbf{F}(x, y, z)$ be a differentiable vector field. Show that div (curl $\mathbf{F}) = 0$.

13. True or false? If curl $\mathbf{F} = \mathbf{0}$ then $\mathbf{F} = \nabla\phi$ for some differentiable scalar function $\phi(x, y, z)$. Explain.

14. Find the value of the surface integral

    $$\iint_S \mathbf{F} \cdot \mathbf{N}\,dS$$

    where $\mathbf{F}(x, y, z) = xy\mathbf{i} + x^2 z\mathbf{j} + xz\mathbf{k}$, S is the unit sphere and $\mathbf{N}$ is the outward unit normal.

15. Let $\mathbf{F}(x, y, z) = ax\mathbf{i} + by\mathbf{j} + cz\mathbf{k}$. Find the flux of $\mathbf{F}$ over the closed surface S in terms of the volume V of the region enclosed by S.

16. Find $\int_C \mathbf{F} \cdot d\mathbf{r}$ where $\mathbf{F}(x, y) = x\mathbf{i} - 3xy\mathbf{j}$ and C is the square with vertices $(0, 0)$, $(2, 0)$, $(2, 2)$, $(0, 2)$ oriented counterclockwise.

17. Find $\int_C \mathbf{F} \cdot d\mathbf{r}$ where $\mathbf{F}(x, y) = x^2\mathbf{i} - xy\mathbf{j}$ and C is the upper unit semicircle from $(1, 0)$ to $(-1, 0)$.

18. Find $\displaystyle\int_{(1, 2)}^{(3, 2)} \mathbf{F} \cdot d\mathbf{r}$ where $\mathbf{F}(x, y) = 4x^3 y^2\mathbf{i} + 2x^4 y\mathbf{j}$.

In Exercises 19–23, evaluate the given line integral.

19. $\int_C xz\,dx + yz\,dy + z\,dz$ where C is parameterized by $\mathbf{r}(t) = t\mathbf{i} + \mathbf{j} + \sin t\mathbf{k},\ 0 \le t \le \pi$.

20. $\int_C y\,dx + x\,dy$, C is the triangle with vertices $(0, 0)$, $(4, 0)$, and $(2, 4)$, taken in that order.

21. $\int_C \mathbf{F} \cdot d\mathbf{r}$, $\mathbf{F}(x, y, z) = 2y\mathbf{i} - x\mathbf{j} + 3z^2\mathbf{k}$, $\mathbf{r}(t) = t\mathbf{i} - 3t\mathbf{j} + \mathbf{k},\ 0 \le t \le 1$.

22. $\int_C \mathbf{F} \cdot d\mathbf{r}$, $\mathbf{F}(x, y) = e^{2y}\mathbf{i} + \cos \pi x\mathbf{j}$, C is the triangle with vertices $(0, 0)$, $(1, 0)$, and $(0, 1)$ taken in that order.

23. $\int_C \mathbf{F} \cdot d\mathbf{r}$, $\mathbf{F}(x, y) = xy^2\mathbf{i} + x^3y\mathbf{j}$, C is the upper unit semicircle taken counterclockwise.

In Exercises 24–28, evaluate the given surface integral.

24. $\iint_S (x^2 + 2)\,dS$,

S is the portion of the plane $x + y + z = 6$ lying inside the cylinder $x^2 + y^2 = 9$.

25. $\iint_S (x + y)\,dS$,

S is the portion of the plane $x - 2y + z = 8$ lying above the square with vertices $(0, 0, 0)$, $(1, 0, 0)$, $(1, 1, 0)$, $(0, 1, 0)$.

26. $\iint_S (x^2 + y^2 + 1)\,dS$,

S is the part of the plane $z = x + 4$ inside the cylinder $x^2 + y^2 = 4$.

27. $\iint_S x^2z\,dS$,

S is the portion of the cone $z^2 = x^2 + y^2$ lying between the planes $z = 1$ and $z = 4$.

28. $\iint_S \sqrt{x^2 + y^2}\,dS$,

S is the part of the graph of $z = 2xy$ lying inside the cylinder $x^2 + y^2 = 1$.

In Exercises 29–33, find the flux

$$\iint_S \mathbf{F} \cdot \mathbf{N}\,dS$$

of the given vector field outward over the given surface.

29. $\mathbf{F}(x, y, z) = xy\mathbf{i} + y\mathbf{j} + 2z\mathbf{k}$, S is the graph of the paraboloid $z = x^2 + y^2$ for $1 \le z \le 4$, $\mathbf{N}$ is the downward unit normal.

30. $\mathbf{F}(x, y, z) = y\mathbf{i} + x\mathbf{j}$, S is the unit sphere $x^2 + y^2 + z^2 = 1$.

31. $\mathbf{F}(x, y, z) = y\mathbf{i} + x\mathbf{j}$, S is the cylinder $x^2 + y^2 = 1$, $0 \le z \le 1$, including the top and bottom discs.

32. $\mathbf{F}(x, y, z) = (x + y^3)\mathbf{i} + (z^2 - y)\mathbf{j} + (x^3 - y^2)\mathbf{k}$, S is the unit sphere $x^2 + y^2 + z^2 = 1$.

33. $\mathbf{F}(x, y, z) = ax\mathbf{i} + by\mathbf{j} + cz\mathbf{k}$, S is the graph of $z = 4 - x^2 - y^2$, $z \ge 0$, $\mathbf{N}$ is the upward unit normal.

In Exercises 34–37, evaluate the given line integral.

34. $\int_C \mathbf{F} \cdot d\mathbf{r}$, $\mathbf{F}(x, y, z) = (x + y)\mathbf{i} + (z - y)\mathbf{j} + z\mathbf{k}$, C is the circle $x^2 + y^2 = 4$ oriented counterclockwise in the plane $z = 1$.

35. $\int_C \mathbf{F} \cdot d\mathbf{r}$, $\mathbf{F}(x, y, z) = 2y^2\mathbf{i} - 3z\mathbf{j} + 2x\mathbf{k}$, C is the triangle with vertices $(0, 0, 0)$, $(2, 4, 2)$, and $(0, 4, 0)$, taken in that order.

36. $\int_C \mathbf{F} \cdot d\mathbf{r}$, $\mathbf{F}(x, y, z) = 3y\mathbf{i} - z\mathbf{j} + 4x\mathbf{k}$, C is the boundary of the part of the unit sphere $x^2 + y^2 + z^2 = 1$ lying in the first quadrant with orientation induced by the outward unit normal to the sphere.

37. $\int_C \mathbf{F} \cdot d\mathbf{r}$, $\mathbf{F}(x, y, z) = (x^3 - y)\mathbf{i} + (\cos y - x)\mathbf{j} + \sin z\mathbf{k}$, C is the triangle with vertices $(0, 0, 0)$, $(4, 2, 0)$, and $(2, 4, 2)$, taken in that order.

38. Let $\mathbf{F}(x, y, z) = x^3\mathbf{i} + z\mathbf{j} + y\mathbf{k}$. Find the work done by $\mathbf{F}$ on an object that moves from $(1, 0, 0)$ to $(-1, 0, \pi)$ along a straight line.

39. Find the work done by $\mathbf{F}$ in Exercise 38 if the motion is along the helix $\mathbf{r}(t) = \cos t\mathbf{i} + \sin t\mathbf{j} + t\mathbf{k}$.

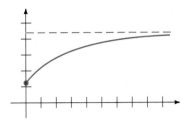

Chapter 21

Differential Equations

In this chapter we briefly indicate how to use the theory and techniques of the calculus to solve certain types of ordinary differential equations. (See Section 5.3, where the basic definitions associated with differential equations, and the method of separation of variables, were first introduced.)

For a comprehensive treatment of differential equations, you are referred to courses and texts with this title. The discussion here, however, should help you appreciate the use made of differential equations in elementary courses in science and engineering.

21.1   First Order Linear Differential Equations

Recall that a differential equation is an equation involving one or more derivatives of an unknown function. Two examples of differential equations are

$$\frac{dy}{dt} = ky \qquad (1)$$

and

$$f''(t) + 4f'(t) + f(t) = \cos t. \qquad (2)$$

A *solution* of a differential equation on an interval I is a function that is differentiable on I as many times as the equation requires and that satisfies the equation. For example, in Chapter 8 we saw that the function $y = e^{kt}$ is a solution of equation (1) on $I = (-\infty, \infty)$ because $y = e^{kt}$ is differentiable for all $t \in (-\infty, \infty)$ and

$$\frac{dy}{dt} = \frac{d}{dt}(e^{kt}) = k(e^{kt}) = ky, \qquad -\infty < t < \infty.$$

The *general solution* of a differential equation is a function involving one or more constants that is a solution of the differential equation and from which all particular solutions can be obtained by specifying the appropriate values for these constants. For example, the general solution of equation (1) is $y = Ce^{kt}$, since we showed in Chapter 8 that all solutions of equation (1) are of this form.

The *order* of a differential equation is the order of the highest derivative in the equation. Thus, equation (1) is a first order equation and equation (2) is a second order equation.

In Section 5.3 we showed that a first order differential equation of the form

$$\frac{dy}{dt} = \frac{f(t)}{g(y)}$$

may be solved by the method of *separation of variables*. This method involves rewriting the equation as

$$g(y)\frac{dy}{dt} = f(t)$$

and "integrating both sides:"

$$\int g(y)\left(\frac{dy}{dt}\right)dt = \int f(t)\,dt$$

or

$$\int g(y)\,dy = \int f(t)\,dt.$$

In the remaining part of this section we show how this technique and the method of integrating factors may be used to solve first order linear equations, equations of the general form

$$\frac{dy}{dt} + p(t)y = g(t)$$

where the functions p and g (which may be constants, including zero) are continuous on the interval of interest.

Separation of Variables

First, suppose we arbitrarily add a constant b to the right side of equation (1). We obtain the differential equation

$$\frac{dy}{dt} = ky + b.$$

By letting $a = -b/k$ we can write this equation as

$$\frac{dy}{dt} = k(y - a). \tag{3}$$

The interpretation of equation (3) is that the rate of increase or decrease of y is proportional to the *difference* between y and the constant a. Before turning to practical examples, we observe that equation (3) may be solved by the method of **separation of variables.** The following example illustrates the use of this method.

□ **EXAMPLE 1**

Find a solution of the differential equation

$$\frac{dy}{dt} = 2 - ay, \qquad a \neq 0. \tag{4}$$

Strategy · · · · · · ·
Separate variables.

Solution

Recall that *separating variables* means isolating y terms on one side of the equation and t terms on the other. To do so we must assume $2 - ay \neq 0$. We then divide both sides by $(2 - ay)$ and multiply both sides by dt to obtain

$$\frac{dy}{2 - ay} = dt.$$

Integrate both sides.

Integrating both sides then gives

$$\int \frac{dy}{2 - ay} = \int dt,$$

so

(Only one constant of integration need be displayed.)

$$-\frac{1}{a} \ln |2 - ay| = t + C \qquad (C \text{ arbitrary})$$

or

$$\ln |2 - ay| = -at - aC.$$

Thus

Solve for y by applying the exponential function to both sides.

$$|2 - ay| = e^{(-at - aC)}$$
$$2 - ay = \pm e^{-at - aC}$$
$$-ay = -2 \pm e^{-at - aC}$$
$$y = \frac{2}{a} + Ae^{-at}, \qquad A = \pm \frac{1}{a} e^{-aC} \qquad (5)$$

A is determined by initial conditions.

where the constant A is determined from an initial condition. For example, if $y(0) = 1$ we obtain

$$1 = \frac{2}{a} + Ae^0, \quad \text{so} \quad A = \left(1 - \frac{2}{a} \right).$$

Verify that assumption on $2 - ay$ holds for solutions obtained.

To check that the assumption $2 - ay \neq 0$ holds, notice that from (5)

$$2 - ay = 2 - a\left[\frac{2}{a} + Ae^{-at} \right]$$
$$= -aAe^{-at},$$

which is not zero unless $A = 0$. However, if $A = 0$ equation (5) still gives the valid solution $y = 2/a$. Thus, in all cases equation (5) represents a solution of (4). ∎

Mixing Problems

Examples 2 and 3 illustrate how differential equations of the above type arise as models in certain "mixing" problems.

□ **EXAMPLE 2**

A tank in which chocolate milk is being mixed initially contains a mixture of 460 liters of milk and 40 liters of chocolate syrup. Syrup and milk are then added to the tank at the rates of 2 liters per minute of syrup and 8 liters per minute of milk. Simultaneously, the mixture is withdrawn at the rate of 10 liters per minute. Assuming perfect mixing of the milk and syrup, find the function giving the amount of syrup in the tank at time t.

Strategy · · · · · · ·
Label variables.

Solution
If y represents the amount of syrup in the tank, then dy/dt is the rate at which the amount of syrup is changing. This rate is the rate at which syrup enters the tank (2 liters per minute) minus the rate at which syrup leaves the tank. We calculate the rate that it leaves the tank by multiplying the fraction of syrup in the tank ($y/500$) by the rate that the mixture leaves the tank (10 liters per minute). Thus, syrup leaves the tank at a rate of $.02y$ liters per minute. The function y therefore satisfies the differential equation

$$\frac{dy}{dt} = 2 - .02y.$$

Find y by solving this differential equation.

This is a particular form of equation (4) treated in Example 1, with $a = .02$. From (5) the solution is

$$y = Ae^{-.02t} + 100.$$

Determine the constant A by using the initial conditions.

To determine A we use the initial condition that $y = 40$ when $t = 0$. We obtain the equation

$$40 = A + 100,$$

so

$$A = -60.$$

State the solution.

The desired solution is therefore

$$y = 100 - 60e^{-.02t} \text{ liters.} \tag{6}$$

Verify that restrictions on y are met.

Note that the requirement of Example 1 that

$$y \neq \frac{2}{a} = \frac{2}{.02} = 100$$

is met since $60e^{-0.2t} > 0$ for all t. The graph of the solution appears in Figure 1.1 and values for $y(t)$ are listed in Table 1.1. ■

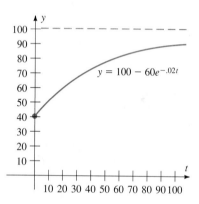

Figure 1.1 Solution of mixing problem in Example 2.

Table 1.1 Calculated values of $y(t)$ in Figure 1.1

t	0	1	2	3	5	10	20	100	200
$y = 100 - 60e^{-.02t}$	40	41.2	42.4	43.5	45.7	50.9	59.8	91.9	98.9

REMARK Figure 1.1 illustrates an important property of solutions of differential equations of the form of equation (3), namely, that the solution approaches a horizontal asymptote (constant limit) as $t \to \infty$. For the solution y of Example 2 it is easy to see that

$$\lim_{t \to \infty} y = \lim_{t \to \infty} (100 - 60e^{-.02t}) = 100,$$

as Figure 1.1 suggests. This asymptote $y = 100$ is called the **steady state** part of the solution, while the term $60e^{-.02t}$ is referred to as the **transient** part of the solution. The result of Example 2 is precisely what intuition suggests—regardless of the initial concentration of syrup, the steady state concentration will be $100/500 = .2$, since 2 liters of syrup enter for every 8 liters of milk.

☐ **EXAMPLE 3**

Newton's law of cooling states that the rate at which an object gains or loses heat is proportional to the difference in temperature between the object and its surroundings. A bottle of soda at temperature 6°C is removed from a refrigerator and placed in a room at temperature 22°C. If, after 10 minutes, the temperature of the soda is 14°C, find its temperature after 20 minutes according to Newton's law.

Strategy · · · · · · ·

Label variables.

Solution

Let $y(t)$ denote the temperature of the soda t minutes after it is removed from the refrigerator. Newton's law of cooling states that

Interpret Newton's law as a differential equation for y.

$$\frac{dy}{dt} = k(22 - y).$$

Thus

Separate variables.

$$\frac{dy}{22 - y} = k \, dt,$$

so

$$\int \frac{dy}{22 - y} = \int k \, dt.$$

Integrate.

Integrating both sides we obtain

Apply exponential function to both sides.

$$-\ln |22 - y| = kt + C$$
$$|22 - y| = e^{-kt - C}$$
$$22 - y = \pm e^{-kt - C}.$$

Thus

Solve for y.

$$y = 22 - Ae^{-kt}, \qquad A = \pm e^{-C}. \tag{7}$$

Use initial data to find A.

To find A we use the initial data* that $y_0 = 6°$ when $t = 0$. Inserting this information in (7) gives

$$6 = 22 - Ae^0,$$

so

$$A = 16. \tag{8}$$

Use additional data to find k.

To find k we use (7), (8), and the data $y(10) = 14$. We obtain the equation

$$14 = 22 - 16e^{-10k}.$$

Thus

$$e^{-10k} = \frac{-8}{-16} = \frac{1}{2},$$

so

$$\ln(e^{-10k}) = -10k = \ln(\tfrac{1}{2})$$
$$= -\ln 2,$$

so

$$k = \frac{\ln 2}{10}.$$

$$k = \frac{-\ln 2}{-10} = \frac{\ln 2}{10}.$$

The function y is now completely determined as

$$y = 22 - 16e^{-\left(\frac{\ln 2}{10}\right)t}. \tag{9}$$

The temperature after 20 minutes is therefore

Find $y(20)$

$e^{\ln 2^{-2}} = 2^{-2} = \frac{1}{4}.$

$$y(20) = 22 - 16e^{-\left(\frac{\ln 2}{10}\right)20}$$
$$= 22 - 16e^{-2\ln 2}$$
$$= 22 - 16e^{\ln 2^{-2}}$$
$$= 22 - 16 \cdot \tfrac{1}{4} = 18°.$$

(See Table 1.2 and Figure 1.2 on page 1114.)

■

*Initial data refer to information about a function at *any* value of the independent variable, not necessarily $t = 0$.

Integrating Factors

Table 1.2 Values of the solution $y(t)$ in Example 3.

t	$y(t)$
0	6.00
4	9.87
8	12.81
12	15.04
16	16.72
20	18.00
24	18.97
28	19.70
32	20.26
36	20.68

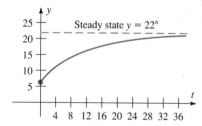

Figure 1.2 Graph of
$$y = 22 - 16e^{-\left(\frac{\ln 2}{10}\right)t}.$$

Equation (1) can also be generalized to linear first order differential equations of the form

$$\frac{dy}{dt} + p(t)y = g(t) \tag{10}$$

where p and g are continuous functions of t. The differences between equation (10) and equation (1) involve these two functions. In equation (1) we have $p(t) \equiv -k$, and $g(t) \equiv 0$.

To find solutions to equations of the form (10), we let $u(t) = \int p(t)\, dt$ and note that *if we multiply both sides of* (10) *by the integrating factor*

$$e^{u(t)} = e^{\int p(t)\, dt} \tag{11}$$

it is easy to find the integral of the left side of equation (10). (In (11) the expression $\int p(t)\, dt$ denotes any *particular* antiderivative for p.) Note that

$$\left[\frac{dy}{dt} + p(t)y\right] e^{u(t)} = \frac{dy}{dt} e^{\int p(t)\, dt} + yp(t)e^{\int p(t)\, dt} \tag{12}$$

$$= \frac{d}{dt}[ye^{\int p(t)\, dt}]$$

$$= \frac{d}{dt}[ye^{u(t)}].$$

To find a solution of the differential equation (10), we begin by multiplying both sides by an integrating factor e^u in (11) to obtain

$$\left[\frac{dy}{dt} + p(t)y\right] e^{u(t)} = g(t)e^{u(t)}.$$

Using (12) we can write this equation as

$$\frac{d}{dt}[ye^{u(t)}] = g(t)e^{u(t)},$$

so

$$ye^{u(t)} = \int g(t)e^{u(t)}\, dt + C$$

or

$$\boxed{y = e^{-u(t)}\left[\int g(t)e^{u(t)}\, dt + C\right].} \tag{13}$$

The function y in (13) is the **general solution** of equation (10) because, as is shown in more advanced courses, every solution of (10) has the form (13) for some choice of the constant C.

Rather than memorizing formula (13), you should simply remember that solving a differential equation of the form (10) requires the integrating factor $e^{u(t)} = e^{\int p(t)\, dt}$.

☐ **EXAMPLE 4**

Find the general solution of the differential equation

$$\frac{dy}{dt} + 2ty = t. \qquad (14)$$

Solution   This equation has the form (10) with $p(t) = 2t$, $g(t) = t$. We therefore use the integrating factor

$$e^{u(t)} = e^{\int 2t\, dt} = e^{t^2}.$$

Multiplying both sides of (14) by e^{t^2} gives

$$e^{t^2}\frac{dy}{dt} + e^{t^2}(2ty) = te^{t^2}.$$

We next rewrite the left side as a derivative

$$\frac{d}{dt}[e^{t^2} \cdot y] = e^{t^2}\frac{dy}{dt} + (e^{t^2} \cdot 2t)y$$

and integrate both sides:

$$e^{t^2}y = \int te^{t^2}\, dt$$

$$= \tfrac{1}{2}e^{t^2} + C.$$

Finally, we multiply both sides by e^{-t^2} to obtain

$$y = e^{-t^2}[\tfrac{1}{2}e^{t^2} + C]$$

$$= Ce^{-t^2} + \tfrac{1}{2}. \qquad ■$$

☐ **EXAMPLE 5**

Find the solution of

$$\frac{dy}{dt} - \frac{2}{t}y = t^3, \qquad t > 0,$$

that satisfies the initial condition $y(1) = 2$.

Strategy · · · · · · ·

Find an integrating factor.

Solution

We use the integrating factor

$$e^{u(t)} = e^{\int -2/t\, dt} = e^{-2 \ln t} = e^{\ln t^{-2}} = t^{-2} = 1/t^2.$$

Multiply both sides of the equation by the integrating factor.

We then obtain

$$\frac{1}{t^2}\frac{dy}{dt} - \frac{2}{t^3}y = t^3\left(\frac{1}{t^2}\right).$$

We can write this equation as

Write the left side as a derivative.

$$\frac{d}{dt}\left[\frac{1}{t^2}y\right] = t,$$

so

Integrate both sides.

$$\frac{1}{t^2} y = \int t \, dt = \frac{t^2}{2} + C.$$

Solve for y.   Thus

$$y = t^2 \left[\frac{t^2}{2} + C \right]$$

$$= \frac{t^4}{2} + Ct^2.$$

Apply initial conditions to solve for C.

The initial condition $y(1) = 2$ gives

$$2 = \tfrac{1}{2} + C,$$

so

$$C = 2 - \tfrac{1}{2} = \tfrac{3}{2}.$$

The desired solution is

$$y(t) = \frac{t^4}{2} + \frac{3t^2}{2}. \qquad \blacksquare$$

Exercise Set 21.1

In Exercises 1–4, find a particular solution of the initial value problem. Identify the steady state and transient parts of the solution.

1. $\dfrac{dy}{dt} = y + 1, \quad y(0) = 1$

2. $\dfrac{dy}{dt} = 2y + 2, \quad y(0) = -1$

3. $\dfrac{dy}{dt} = -y + \pi, \quad y(0) = \pi/2$

4. $\dfrac{dy}{dx} + y = 2, \quad y(0) = 2$

In Exercises 5–14 find the general solution of the differential equation.

5. $y' + 2y = 4$ **6.** $y' + ay = b$

7. $y' - \dfrac{2}{t} y = t^4$ **8.** $y' = \dfrac{1}{t^2} y$

9. $y' - 2y = e^{2t} \sin t$ **10.** $y' = 2ty + 3t$

11. $y' + y = e^t$ **12.** $ty' - 2y = -2t$

13. $ty' + y = t$ **14.** $y' = y - 2e^{-t}$

15. Use the method of separation of variables to find a solution of the differential equation $dy/dx = axy$ where a is constant and $y > 0$.

16. A curve in the xy-plane has the property that at each point (x, y) the slope of the tangent equals xy, the product of the coordinates. Find an equation for such a curve containing the point $(0, 1)$ and sketch its graph. (*Hint:* Use the result of Exercise 15.)

17. A tank contains 200 L of brine (salt and water). The initial concentration of salt is 0.5 kg/L of brine. Fresh water is then added at a rate of 3 L/min and the brine is drawn off from the bottom of the tank at the same rate. Assume perfect mixing.
   a. Find a differential equation for the number $p(t)$ of kg of salt in the solution at time t.
   b. How much salt is present after 20 min?
   c. What is the steady state concentration of salt?

18. A bottle of water whose temperature is 26°C is placed in a room at temperature 12°C. If the temperature of the water after 30 min is 20°C, find its temperature after 2 h.

19. A person initially places $100 in a savings account that pays interest at a rate of 10%/yr compounded continuously. If the person makes additional deposits of $100/yr continuously throughout each year, find:
   a. a differential equation for $P(t)$, the amount on deposit after t years, and
   b. the amount on deposit after 7 years.

20. An electrical circuit contains a battery supplying E volts, a resistance R, and an inductance L. When the battery is connected, the resulting current I satisfies the differential equation

$$L\frac{dI}{dt} + RI = E, \qquad I(0) = 0.$$

a. Find the solution I as a function of t.
b. Obtain Ohm's Law, $I = E/R$, as the steady state part of the solution.

21. If, in the circuit described in Exercise 20, the inductance is replaced by a capacitance C, the equation becomes

$$R\frac{dI}{dt} + \frac{I}{C} = \frac{dE}{dt}.$$

Find the solution $I(t)$, if the battery is switched off (i.e., $dE/dt = 0$ at the time $t = 0$.)

22. In the chemical reaction called inversion of raw sugar, the inversion rate is proportional to the amount of raw sugar remaining. If 100 kg of raw sugar is reduced to 75 kg in 6 h, how long will it be until
a. half the raw sugar has been inverted?
b. 90% of the raw sugar has been inverted?

23. A pan of boiling water is removed from a burner and cools to 80°C in 3 min. Find its temperature after 10 min, according to Newton's law of cooling, if the surrounding temperature is 30°C.

24. A room containing 1000 ft^3 of air is initially free of pollutants. Air containing 100 parts of pollutants per cubic foot is pumped into the room at a rate of 50 ft^3/min. The air in the room is kept perfectly mixed, and air is removed from the room at the same rate of 50 ft^3/min. Find the function $P(t)$ giving the number of parts of pollutants per cubic foot in the room air after t min.

25. A tank contains a mixture of 500 gal of salt brine (salt and water) initially containing 2 lb of salt/gal. A second mixture, containing 6 lb of salt/gal, is then pumped in at a rate of 10 gal/min. The resulting mixture is drawn off at the same rate. Assuming perfect mixing, find the function that describes the concentration of salt (lb/gal) in the mixture in the tank t min after mixing begins.

26. Influenza spreads through a university community at a rate proportional to the product of the proportion of those already infected and the proportion of those not yet infected. Assuming that those infected remain infected, find the proportion $P(t)$ infected after t days if initially 10% were infected and after 3 days 30% were infected.

27. An annuity is initially set up containing $10,000 to which interest is compounded continuously at an annual rate of 10%. The owner of the annuity withdraws $2000 per year in a continuous manner. When does the value of the annuity reach zero?

28. Certain biological populations grow in size according to the Gompertz growth equation

$$\frac{dy}{dt} = -ky \ln y$$

where k is a positive constant. Find the general solution of this differential equation.

29. The Weber-Fechner law in psychology is a model for the rate of change of the reaction y to a stimulus of strength s. It is given by the differential equation

$$\frac{dy}{ds} = k\frac{y}{s}$$

where k is a positive constant. Find the general solution of this differential equation.

21.2   Exact Equations

Certain types of first order differential equations may be solved by a technique discussed in Chapter 18. These are *exact* differential equations.

Definition 1

Let the functions M and N be defined in an open rectangle R in $\mathbb{R}^2$. The differential equation

$$M(x, y) + N(x, y)\frac{dy}{dx} = 0, \qquad (x, y) \in R \qquad (1)$$

is called **exact** if there exists a function f of two variables for which

$$M(x, y) = \frac{\partial f}{\partial x}(x, y) \text{ and } N(x, y) = \frac{\partial f}{\partial y}(x, y)$$

for all $(x, y) \in R$.

If a differential equation is exact, then we can use methods from Chapter 18 to determine its solutions. To see why, consider the equation

$$f(x, y) = C, \qquad (C \text{ constant}), \tag{2}$$

which determines a level curve for f. If (x_0, y_0) is a point on this curve at which the tangent is not parallel to the y-axis, then equation (2) determines a function $y = y(x)$ of x for all x near x_0. Then we rewrite equation (2) as

$$f(x, y(x)) = C. \tag{3}$$

Using the Chain Rule for functions of two variables, we differentiate both sides of equation (3) with respect to x and obtain

$$\frac{\partial f}{\partial x}(x, y(x)) + \frac{\partial f}{\partial y}(x, y(x))\frac{dy}{dx} = 0. \tag{4}$$

Note that equation (4) is an exact differential equation. From this calculation we conclude that, if a differential equation is exact, then the graphs of the solutions lie on level curves of the function f. Thus, we solve an exact differential equation by determining the function f such that $\partial f/\partial x = M$ and $\partial f/\partial y = N$.

☐ **EXAMPLE 1**

The differential equation

$$e^{2y} + 2xe^{2y}\frac{dy}{dx} = 0$$

is exact since, for the function $f(x, y) = xe^{2y}$,

$$M(x, y) = e^{2y} = \frac{\partial}{\partial x}(xe^{2y}) = \frac{\partial f}{\partial x}$$

and

$$N(x, y) = 2xe^{2y} = \frac{\partial}{\partial y}(xe^{2y}) = \frac{\partial f}{\partial y}.$$

The general solution is therefore determined by the level curves for the function $f(x, y) = xe^{2y}$,

$$xe^{2y} = C. \qquad\qquad ■$$

We also use the language of differentials to write equation (1) in the form

$$M(x, y)\, dx + N(x, y)\, dy = 0$$

(see Section 5.3).

☐ **EXAMPLE 2**

The differential equation

$$(3x^2 + 2y^2)\, dx + 4xy\, dy = 0$$

is exact, since the function $f(x, y) = x^3 + 2xy^2$ has the property that

$$M(x, y) = 3x^2 + 2y^2 = \frac{\partial}{\partial x}(x^3 + 2xy^2) = \frac{\partial f}{\partial x}$$

and

$$N(x, y) = 4xy = \frac{\partial}{\partial y}(x^3 + 2xy^2) = \frac{\partial f}{\partial y}.$$

The general solution is therefore

$$x^3 + 2xy^2 = C,$$

or

$$y = \pm\left(\frac{C - x^3}{2x}\right)^{1/2}.$$ ∎

Since Definition 1 may be paraphrased by saying that equation (1) is exact if there exists a function f such that $\nabla f = M(x, y)\mathbf{i} + N(x, y)\mathbf{j}$, Theorem 9 of Chapter 18 provides the following criterion for equation (1) to be exact.

Theorem 1

Let the functions M and N have continuous partial derivatives in an open rectangle R in $\mathbb{R}^2$. The differential equation

$$M(x, y) + N(x, y)\frac{dy}{dx} = 0$$

is exact in R if and only if

$$\frac{\partial M}{\partial y}(x, y) = \frac{\partial N}{\partial x}(x, y)$$

for all (x, y) in R.

☐ **EXAMPLE 3**

The differential equation

$$xy^3 \, dx + x^3y \, dy = 0$$

is not exact, since

$$\frac{\partial M}{\partial y}(x, y) = \frac{\partial}{\partial y}(xy^3) = 3xy^2 \neq 3x^2y = \frac{\partial}{\partial x}(x^3y) = \frac{\partial N}{\partial x}(x, y).$$ ∎

☐ **EXAMPLE 4**

Test the differential equation

$$(3x^2y + \cos x) \, dx + x^3 \, dy = 0$$

for exactness. If it is exact, find a solution passing through the point $(1, \pi)$.

Solution Here

$$M(x, y) = 3x^2y + \cos x; \qquad N(x, y) = x^3.$$

Since both M and N have continuous partial derivatives throughout $\mathbb{R}^2$, we may apply Theorem 1 to test for exactness. Since

$$\frac{\partial M}{\partial y}(x, y) = 3x^2 = \frac{\partial N}{\partial x}(x, y),$$

the equation is exact.

To solve the differential equation, we must find a function f such that $\partial f/\partial x = M$ and $\partial f/\partial y = N$. To do so we integrate $M(x, y)$ partially with respect to x to obtain

$$f(x, y) = \int M(x, y)\, dx = \int (3x^2 y + \cos x)\, dx$$

$$= x^3 y + \sin x + h(y).$$

To determine $h(y)$, we differentiate this expression for f with respect to y and obtain

$$\frac{\partial f}{\partial y} = x^3 + h'(y).$$

Since we require $\partial f/\partial y = N$, we have $h'(y) = 0$. Thus, h is constant, and

$$f(x, y) = x^3 y + \sin x + K$$

is the desired function for any constant K. We therefore choose $K = 0$ and obtain the family of level curves

$$x^3 y + \sin x = C \tag{5}$$

as the general solution for the differential equation. To find the particular curve passing through the point $(x_0, y_0) = (1, \pi)$, we set $x = 1$ and $y = \pi$ in (5) to find

$$1^3 (\pi) + \sin(\pi) = \pi = C,$$

so $C = \pi$. A solution passing through $(1, \pi)$ is therefore determined by the equation

$$x^3 y + \sin x = \pi$$

as

$$y = \frac{\pi - \sin x}{x^3}, \qquad x^3 \neq 0. \qquad \blacksquare$$

Example 4 demonstrates the general technique for solving exact equations. Once the equation is shown to be exact by use of Theorem 1, the solution $f(x, y) = C$ is found by reconstructing the function f from its gradient (see Section 18.9).

☐ **EXAMPLE 5**

Solve the differential equation

$$\cos y\, dx + (3y^2 - x \sin y)\, dy = 0.$$

Solution Writing the equation in the form

$$\cos y + (3y^2 - x \sin y)\frac{dy}{dx} = 0$$

and letting

$$M(x, y) = \cos y; \qquad N(x, y) = 3y^2 - x \sin y,$$

we verify that the equation is exact by noting that

$$\frac{\partial M}{\partial y} = -\sin y = \frac{\partial N}{\partial x}.$$

To find the solution $f(x, y) = C$, we first integrate M partially with respect to x to find that

$$f(x, y) = \int M(x, y)\, dx = \int \cos y\, dx = x \cos y + g(y).$$

To determine $g(y)$, we differentiate this expression for f with respect to y and obtain

$$\frac{\partial f}{\partial y} = -x \sin y + g'(y).$$

Since we require $\partial f/\partial y = N$, $g'(y) = 3y^2$, so $g(y) = y^3 + K$ where K is an arbitrary constant. (We choose $K = 0$ for simplicity.) Thus, $f(x, y) = y^3 + x \cos y$ and the general solution is

$$y^3 + x \cos y = C. \qquad\blacksquare$$

Exercise Set 21.2

In Exercises 1–10, determine whether the given equation is exact. If it is, find a family of solution (level) curves of the form $f(x, y) = C$.

1. $2xy\, dx + x^2\, dy = 0$

2. $y^2 + xy\dfrac{dy}{dx} = 0$

3. $\dfrac{x\, dx}{\sqrt{x^2 + y^2}} + \dfrac{y\, dy}{\sqrt{x^2 + y^2}} = 0$

4. $\cos y^2 - 2xy \sin y^2 \dfrac{dy}{dx} = 0$

5. $\dfrac{x}{y^2}dx + \dfrac{y}{x^2}dy = 0$

6. $\dfrac{y}{1 + x^2y^2} + \dfrac{x}{1 + x^2y^2}\dfrac{dy}{dx} = 0$

7. $(x - 2xy^2)\, dx + (y + x^2y)\, dy = 0$

8. $(y^2 - 2xy)\, dx + (2xy - x^2)\, dy = 0$

9. $\dfrac{x\, dx}{\sqrt{x + y^2}} + \dfrac{y\, dy}{\sqrt{x + y^2}} = 0$

10. $\dfrac{1}{\sqrt{x - y^2}} - \dfrac{2y}{\sqrt{x - y^2}}\dfrac{dy}{dx} = 0$

11. Find the solution of the differential equation in Exercise 1 that passes through the point $(1, 3)$.

12. Find the solution of the differential equation in Exercise 4 that passes through the point $(2, 0)$.

13. Find the solution of the differential equation in Exercise 8 that passes through the point $(1, 2)$.

14. Show that a separable differential equation $f(x)\, dx + g(y)\, dy = 0$ is always exact. What is the "solution" function $f(x, y)$?

15. Show that the differential equation $(1 + x^2y^2 + y)\, dx + x\, dy = 0$ is not exact. Then show that, if this equation is multiplied through by the integrating factor

$$p(x, y) = \frac{1}{1 + x^2y^2},$$

an exact differential equation results. Is a solution of this exact equation also a solution of the original equation?

21.3   Second Order Linear Equations

By adding a term involving the second derivative to the first order linear differential equation $y' = ky$ for natural growth, we obtain one type of *second order linear differential equation*,*

$$\frac{d^2y}{dt^2} + a\frac{dy}{dt} + by = 0 \qquad (1)$$

---

*More precisely, equation (1) is referred to as a second order, constant-coefficient, linear, homogeneous equation. The term homogeneous refers to the fact that the right-hand side is zero. This is but one type of second order equation. There are many others, but we shall not pursue them here.

where a and b are constants. From our experience with first order equations we might suspect that equation (1) has a solution of the form $y = e^{rt}$ where r is a constant. To check, let $y = e^{rt}$. Then,

$$\frac{dy}{dt} = re^{rt} \quad \text{and} \quad \frac{d^2y}{dt^2} = r^2 e^{rt}.$$

Substituting into equation (1) would therefore give

$$r^2 e^{rt} + a(re^{rt}) + b(e^{rt}) = 0,$$

or

$$(r^2 + ar + b)e^{rt} = 0. \tag{2}$$

Since values of the exponential function $y = e^{rt}$ are always nonzero, the only way equation (2) holds is if the *characteristic polynomial* $r^2 + ar + b$ equals zero. That is, r must be a (real) *root* of the quadratic polynomial $r^2 + ar + b$.

Let r be a real number. The function $y = e^{rt}$ is a solution of the differential equation

$$\frac{d^2y}{dt^2} + a\frac{dy}{dt} + by = 0$$

if and only if the constant r is a real root of the characteristic polynomial

$$r^2 + ar + b = 0.$$

Thus, we expect to find two, one, or zero solutions of the differential equation (1) of the form $y = e^{rt}$, depending on the number of real roots r of the associated characteristic polynomial.

□ **EXAMPLE 1**

For the differential equation

$$\frac{d^2y}{dt^2} - \frac{dy}{dt} - 2y = 0$$

the constant coefficients are $a = -1$ and $b = -2$, so the characteristic polynomial is

$$r^2 - r - 2 = (r - 2)(r + 1).$$

This characteristic polynomial has real roots $r_1 = 2$ and $r_2 = -1$. Two solutions of the differential equation are therefore

$$y_1 = e^{r_1 t} = e^{2t} \quad \text{and} \quad y_2 = e^{r_2 t} = e^{-t},$$

as you can verify. (We shall say more about the *general* solution of this equation later.) ■

☐ **EXAMPLE 2**

The differential equation

$$\frac{d^2y}{dt^2} + 4\frac{dy}{dt} + 4y = 0$$

has characteristic polynomial $r^2 + 4r + 4 = (r + 2)^2$, which has only the single real root $r = -2$. Thus, the function $y_1 = e^{-2t}$ is a solution.   ∎

REMARK 1   If the characteristic polynomial for equation (1) has only the single real root r, equation (1) will have both the functions $y_1 = e^{rt}$ and $y_2 = te^{rt}$ as solutions. Thus, the function $y_2 = te^{-2t}$ is also a solution of the equation in Example 2, as you can verify.

REMARK 2   If the characteristic polynomial $r^2 + ar + b$ has the *complex** root $r_1 = \alpha + i\beta$, $\beta \neq 0$, then the complex number $r_2 = \alpha - i\beta$ is also a root, and the differential equation (1) has the two solutions

$$y_1 = e^{\alpha t} \sin \beta t \quad \text{and} \quad y_2 = e^{\alpha t} \cos \beta t.$$

☐ **EXAMPLE 3**

For the differential equation

$$\frac{d^2y}{dt^2} - 2\frac{dy}{dt} + 5y = 0$$

the characteristic polynomial is

$$r^2 - 2r + 5.$$

The roots of this polynomial, according to the quadratic formula, are

$$r = \frac{2 \pm \sqrt{(-2)^2 - 4(1)(5)}}{2}$$

$$= 1 \pm \tfrac{1}{2}\sqrt{-16}$$

$$= 1 \pm 2i, \quad i^2 = -1.$$

Thus, the roots are $r_1 = \alpha + i\beta = 1 + 2i$ and $r_2 = \alpha - i\beta = 1 - 2i$ and $\alpha = 1$ and $\beta = 2$, so two solutions are

$$y_1 = e^t \sin 2t \quad \text{and} \quad y_2 = e^t \cos 2t,$$

which we leave for you to verify.   ∎

The General Solution of Equation (1)

If y_1 and y_2 are any two solutions of equation (1), then so is any *linear combination* of these solutions of the form

$$y(t) = Ay_1(t) + By_2(t) \tag{3}$$

where A and B are constants. (See Exercise 36.) If the functions y_1 and y_2 are not multiples of each other, the function y is called the *general solution* of equation (1) since it can be shown that any solution can be written in this form for an appropriate choice of the constants A and B.

---

*An introduction to complex numbers is presented in Appendix III. The ideas contained in that appendix are assumed in this discussion.

□ **EXAMPLE 4**

(i) The general solution of the differential equation

$$\frac{d^2y}{dt^2} - \frac{dy}{dt} - 2y = 0$$

in Example 1 is

$$y = Ae^{2t} + Be^{-t}.$$

(ii) The general solution of the differential equation

$$\frac{d^2y}{dt^2} + 4\frac{dy}{dt} + 4y = 0$$

in Example 2 is

$$y = Ae^{-2t} + Bte^{-2t}.$$

(iii) The general solution of the differential equation

$$\frac{d^2y}{dt^2} - 2\frac{dy}{dt} + 5y = 0$$

in Example 3 is

$$y = Ae^t \sin 2t + Be^t \cos 2t = e^t(A \sin 2t + B \cos 2t). \qquad ■$$

Initial Conditions

In general, two initial conditions *specified at the same number* t_0 are required to determine a unique solution of a second order differential equation, one for the solution y and one for its derivative dy/dt. Since the general solution of equation (1) has the form given by equation (3), two initial conditions will produce two linear equations for the constants A and B, from which these numbers may be determined.

□ **EXAMPLE 5**

Find a solution of the initial value problem consisting of the differential equation

$$y'' - 2y' - 3y = 0 \qquad (4)$$

and the initial conditions.

$$y(0) = 0$$
$$y'(0) = 4.$$

Strategy · · · · · · · ·
First, find the general solution by

1. finding the characteristic polynomial
2. identifying its roots r_1 and r_2
3. combining the solutions $e^{r_1 t}$ and $e^{r_2 t}$ as in equation (3).

Solution

The characteristic polynomial for equation (4) is

$$r^2 - 2r - 3 = (r - 3)(r + 1),$$

which has roots $r_1 = 3$ and $r_2 = -1$. The general solution of equation (4) is therefore

$$y(t) = Ae^{3t} + Be^{-t}, \qquad (5)$$

so the derivative y' has the form

$$y'(t) = 3Ae^{3t} - Be^{-t}. \qquad (6)$$

Next, substitute initial conditions into general solution y and its derivative y', to obtain two equations in A and B.

Substituting the initial condition $y(0) = 0$ into equation (5) gives the equation

$$A + B = 0. \tag{7}$$

Substituting the initial condition $y'(0) = 4$ into equation (6) gives the equation

$$3A - B = 4. \tag{8}$$

Solve the two equations for A and B by substitution or by elimination.

Substituting $B = -A$ from equation (7) into equation (8) gives

$$3A - (-A) = 4$$

or

$$4A = 4.$$

Find particular solution by substituting for A and B in general solution.

Thus, $A = 1$ and $B = -A = -1$. Equation (5) now becomes

$$y(t) = e^{3t} - e^{-t},$$

which is the desired solution.   ∎

Modelling Oscillatory Motion

A version of equation (1) that occurs frequently in applications is

$$\frac{d^2y}{dt^2} + by = 0, \qquad b > 0. \tag{9}$$

The characteristic polynomial for equation (9) is $r^2 + b$, which has the complex roots $r_1 = \sqrt{b}\,i$ and $r_2 = -\sqrt{b}\,i$. Since two solutions of this equation are $e^{\alpha t} \sin \beta t$ and $e^{\alpha t} \cos \beta t$ with $\alpha = 0$ and $\beta = \sqrt{b}$, we have:

> The differential equation
>
> $$\frac{d^2y}{dt^2} + by = 0, \qquad b > 0, \tag{10}$$
>
> has the general solution
>
> $$y = A \sin \sqrt{b}\,t + B \cos \sqrt{b}\,t. \tag{11}$$

□ **EXAMPLE 6**

Find the general solution of the differential equation

$$y'' + 9y = 0.$$

Solution   Since the differential equation has the form of equation (10), with $b = 9$ the general solution is

$$y = A \sin 3t + B \cos 3t.$$ ∎

Because solutions of equation (9) are made up of sine and cosine functions, versions of this equation occur in models for the motion of vibrating physical systems involving springs, pendulums, etc. A typical example is the harmonic oscillator.

The Harmonic Oscillator

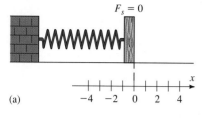

(a)

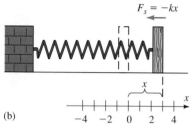

(b)

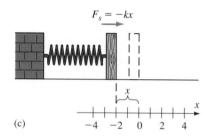

(c)

Figure 3.1 Mass-spring system (a) at equilibrium, (b) stretched, and (c) compressed. Displacement x is measured with respect to a particular point on the block.

Imagine a mass m lying on a frictionless surface and attached to a spring (see Figure 3.1). When the mass is moved x units from its natural rest position, the spring will exert a restoring force F_s proportional to the displacement x (Hooke's Law). Since the restoring force acts in the direction opposite to the motion of the mass, we write

$$F_s = -kx \qquad (12)$$

where the constant k is called the **spring constant** and depends on the particular spring.

Newton's second law of motion states that the sum of all forces acting on the mass must equal the product of the mass m and acceleration a. Since the net force* acting on the mass after it is displaced and released is F_s, we can write Newton's law as

$$F_s = ma, \qquad a = \text{acceleration.}$$

Since the variable x represents the displacement of the mass from its rest position, the time derivative dx/dt is the velocity of the mass, and the second derivative d^2x/dt^2 is its acceleration. Assuming the displacement x to be a twice differentiable function of t, we can rewrite Newton's law as

$$F_s = m\frac{d^2x}{dt^2}. \qquad (13)$$

Combining equations (12) and (13) now gives the differential equation

$$m\frac{d^2x}{dt^2} = -kx,$$

or

$$\boxed{\frac{d^2x}{dt^2} + \frac{k}{m}x = 0.} \qquad (14)$$

Equation (14) is called the **harmonic oscillator equation,** and it provides the desired mathematical model for the motion of the mass for the situation described above. The solution of equation (14) is given by equation (11):

$$x(t) = A \cos \sqrt{\frac{k}{m}}\, t + B \sin \sqrt{\frac{k}{m}}\, t. \qquad (15)$$

☐ EXAMPLE 7

An 8-kg mass is attached to a spring as in Figure 3.1. The spring constant is $k = 2$ N/m. Find the resulting motion if the mass is displaced 10 cm from its equilibrium position and released from rest (zero initial velocity).

Strategy · · · · · · · ·
Write differential equation (14), using given values of k and m.

Solution

Here $k/m = 2/8 = 1/4$, so the differential equation (14) is

$$\frac{d^2x}{dt^2} + \frac{1}{4}x = 0.$$

---

*The force of gravity is exactly counteracted by the supporting force exerted by the surface, so the net *vertical* force is zero.

Write general solution from (15).

From (15), with $\sqrt{k/m} = 1/2$, the general solution is

$$x(t) = A \cos \frac{t}{2} + B \sin \frac{t}{2}. \tag{16}$$

Set $x(0) = 10$ cm $= \frac{1}{10}$ m to find one equation in A and B.

To find A and B, we apply the given *initial conditions*

(i) $x(0) = \frac{1}{10}$ (initial displacement 10 cm $= \frac{1}{10}$ m), and
(ii) $x'(0) = 0$ (initial velocity zero).

From equation (16) and condition (i) we have

$$\tfrac{1}{10} = A \cos 0 + B \sin 0 = A,$$

so $A = \frac{1}{10}$. Since, from (16),

Differentiate $x(t)$ to obtain equation for $x'(t)$.

$$\frac{dx}{dt} = -\frac{A}{2} \sin \frac{t}{2} + \frac{B}{2} \cos \frac{t}{2}, \tag{17}$$

condition (ii) gives

Set $x'(0) = 0$ to obtain second equation in A and B.

$$0 = -\frac{A}{2} \sin 0 + \frac{B}{2} \cos 0 = \frac{B}{2}.$$

Thus, $B = 0$. The solution is therefore

$$x(t) = \frac{1}{10} \cos \frac{t}{2} \text{ (meters)}. \qquad \blacksquare$$

□ **EXAMPLE 8**

Repeat Example 7, except that the mass now is released with an initial velocity of 3 m/s.

Strategy · · · · · · · ·
Same solution as for Example 7, except that the condition $x'(0) = 0$ is replaced by the condition $x'(0) = 3$.

Solution
As in Example 7 we have

$$x(t) = A \cos \frac{t}{2} + B \sin \frac{t}{2}.$$

Also, we still have $x(0) = \frac{1}{10}$, so $A = \frac{1}{10}$.
    However, the condition that the initial velocity is 3 m/s means that $x'(0) = 3$. Equation (17) therefore gives

Set $x'(0) = 3$ and solve for B.

$$3 = -\frac{A}{2} \sin 0 + \frac{B}{2} \cos 0 = \frac{B}{2},$$

so $B = 6$. In this case the solution is therefore

$$x(t) = \frac{1}{10} \cos \frac{t}{2} + 6 \sin \frac{t}{2}. \qquad \blacksquare$$

Exercise Set 21.3

In Exercises 1–6 find the general solution of the differential equation in the form $y = Ae^{r_1t} + Be^{r_2t}$ by finding two distinct real roots of the characteristic polynomial.

1. $\dfrac{d^2y}{dt^2} + 5\dfrac{dy}{dt} + 6y = 0$ **2.** $y'' + 3y' - 4y = 0$

3. $y'' - y = 0$ **4.** $\dfrac{d^2y}{dt^2} - \dfrac{dy}{dt} - 6y = 0$

5. $\dfrac{d^2y}{dt^2} + 3\dfrac{dy}{dt} - 10y = 0$ **6.** $y'' - 4y = 0$

In Exercises 7–10 find the general solution of the differential equation in the form $y = Ae^{rt} + Bte^{rt}$ by finding the single root r of the characteristic polynomial.

7. $y'' + 2y' + 1y = 0$ **8.** $y'' - 4y' + 4y = 0$

9. $\dfrac{d^2y}{dt^2} + 6\dfrac{dy}{dt} + 9y = 0$ **10.** $y'' - 2y' + 1y = 0$

In Exercises 11–14 find the general solution of the differential equation in the form $y = A \sin \sqrt{b}\, t + B \cos \sqrt{b}\, t$.

11. $\dfrac{d^2y}{dt^2} + 4y = 0$ **12.** $y'' + 16y = 0$

13. $y'' = -5y$ **14.** $\dfrac{d^2y}{dt^2} = -25y$

In Exercises 15–24 find the general solution of the differential equation.

15. $y'' - 3y' - 10y = 0$ **16.** $y'' + 8y' + 16 = 0$

17. $y'' - 4y' + y = 0$ **18.** $y'' + 6y' + 5y = 0$

19. $y'' - 2y' + 6y = 0$ **20.** $y'' - 6y' + 9y = 0$

21. $y'' - y' - 12y = 0$ **22.** $y'' - 2y' + 3y = 0$

23. $y'' + 2y' + 4y = 0$ **24.** $y'' + 4y' + 4y = 0$

In Exercises 25–30 find the solution of the initial value problem.

25. $y'' - y' - 6y = 0$ **26.** $y'' - 5y' + 6y = 0$
$\quad y(0) = 0$ $\quad y(0) = 2$
$\quad y'(0) = 5$ $\quad y'(0) = 0$

27. $y'' + 2y' + y = 0$ **28.** $y'' - 4y' + 4y = 0$
$\quad y(0) = 3$ $\quad y(0) = 2$
$\quad y'(0) = 1$ $\quad y'(0) = 5$

29. $y'' + 9y = 0$ **30.** $y'' + y = 0$
$\quad y(0) = 3$ $\quad y(0) = 0$
$\quad y'(0) = -3$ $\quad y'(0) = 2$

In Exercises 31–32 find the general solution by first converting the system of two first order equations into a second order equation.

31. $\dfrac{dy}{dt} = 4x$ **32.** $\dfrac{dy}{dt} = 2(x - 3)$

$\quad \dfrac{dx}{dt} = -y$ $\quad \dfrac{dx}{dt} = 8(y + 5)$

33. A 0.5 kg mass is attached to a spring with spring constant $k = 2$ N/m, as in Figure 3.1. The mass is pulled 10 cm from the equilibrium position and released with zero velocity. Find an equation describing the resulting motion.

34. A 0.5 kg mass is attached to a spring with spring constant $k = 100$ N/m, as in Figure 3.1. The mass is tapped while in its equilibrium position so as to give it an initial velocity of 10 cm/s.
 a. Find a function describing the resulting motion.
 b. When will the mass return to its equilibrium point for the first time?

35. Sketch the graph of the solution of the initial value problem

$$\dfrac{d^2x}{dt^2} + kx = 0$$
$$x(0) = 1$$
$$x'(0) = 0$$

for $k = 0,\ \pm 1,\ \pm 4,\ \pm 9$. What can you conclude about this solution as
 a. $k \to \infty$?   **b.** $k \to -\infty$?
 c. $k \to 0^+$?   **d.** $k \to 0^-$?

36. Verify that if y_1 and y_2 are solutions of the differential equation

$$y'' + ay' + by = 0,$$

then so is any function of the form

$$y(t) = Ay_1(t) + By_2(t).$$

21.4 Nonhomogeneous Linear Equations

In Section 21.3 we noted that if y_1 and y_2 are solutions of the linear differential equation

$$\dfrac{d^2y}{dt^2} + a\dfrac{dy}{dt} + by = 0, \tag{1}$$

then so is any function y of the form

$$y(t) = C_1 y_1(t) + C_2 y_2(t) \tag{2}$$

where C_1 and C_2 are constants. We say that the solutions y_1 and y_2 are *linearly independent* if they are not multiples of each other. It can be shown that, if y_1 and y_2 are linearly independent solutions of equation (1), the general solution of equation (1) is given by the function y in line (2). Equation (1) is called *homogeneous* because the term on the right-hand side is simply zero.

We next wish to determine the general solution of the *nonhomogeneous* differential equation of the form

$$\frac{d^2y}{dt^2} + a\frac{dy}{dt} + by = f(t) \tag{3}$$

for certain types of nonzero functions f. Note that the nonhomogeneous equation (3) does *not* have the property that the sum of two solutions of (3) is again a solution. (See Exercise 21.) The following theorem shows how the general solution of equation (3) is related to the general solution of the associated homogeneous equation (1). Its proof is sketched in Exercise 19.

Theorem 2

Let y_p be any particular solution of the nonhomogeneous linear differential equation (3) and let y_g be the general solution of the associated homogeneous linear equation (1). Then the general solution of the nonhomogeneous equation (3) is

$$y(t) = y_g(t) + y_p(t). \tag{4}$$

Theorem 2 says that to find the general solution of the nonhomogeneous equation (3) we must do two things. First, find *any particular* solution y_p of equation (3). Then, find the *general* solution y_g of the "homogeneous part" given by equation (1). The general solution of (3) is the sum of these two solutions.

□ **EXAMPLE 1**

The nonhomogeneous equation

$$y'' + y' - 2y = -4t \tag{5}$$

has the particular solution $y_p = 2t + 1$. To verify this, note that

$$y_p' = 2 \quad \text{and} \quad y_p'' = 0,$$

so

$$y_p'' + y_p' - 2y_p = 0 + 2 - 2(2t + 1) = -4t$$

as required.

To find the *general* solution we next note that the associated homogeneous equation

$$y'' + y' - 2y = 0 \tag{6}$$

has characteristic polynomial $r^2 + r - 2 = (r + 2)(r - 1)$, with roots $r_1 = -2$ and $r_2 = 1$, so the general solution of the homogeneous equation (6) is

$$y_g = C_1 e^{-2t} + C_2 e^t.$$

The general solution of equation (5) is therefore

$$y = C_1 e^{-2t} + C_2 e^t + 2t + 1$$

according to Theorem 2.

■

The Method of Undetermined Coefficients

Although Theorem 2 specifies the form of the general solution of equation (3), it does not tell us how to find the particular solution y_p required to form the general solution.

For certain types of functions f, a simple observation enables us to find y_p. The idea is to consider which types of functions, when differentiated and combined according to the left side of equation (3), produce the function f on the right side of equation (3). We then substitute the "general forms" of such functions into the equation and see what happens.

For example, in the nonhomogeneous equation

$$y'' + 3y' - 4y = e^{2t} \tag{7}$$

we suggest that functions of the form $y = Ae^{2t}$ are candidates for a solution since derivatives of Ae^{2t} are again of the form Ae^{2t}. With $y = Ae^{2t}$ we have

$$y' = 2Ae^{2t} \quad \text{and} \quad y''(t) = 4Ae^{2t},$$

so substituting into the left side of equation (7) gives

$$y'' + 3y' - 4y = 4Ae^{2t} + 3(2Ae^{2t}) - 4Ae^t = 6Ae^{2t}. \tag{8}$$

Comparing the right sides of equations (7) and (8), we conclude that $y = Ae^{2t}$ is a solution of equation (7) is $e^{2t} = 6Ae^{2t}$. Thus, $6A = 1$, so $A = \frac{1}{6}$. A particular solution of equation (7) is therefore $y_p(t) = \frac{1}{6}e^{2t}$, as you can verify.

By the "general form" for a function f, we mean the most general form of a function y which, when combined with its derivatives dy/dt and d^2y/dt^2 according to the left side of a linear second order differential equation

$$\frac{d^2y}{dt^2} + a\frac{dy}{dt} + by = f(t), \tag{9}$$

can produce the function f on the right side. The method we are describing works well for the following list of functions f. Their general forms are also listed.

Function f	General Form
$f(t) = t$	$y = At + B$
$f(t) = t^2$	$y = At^2 + Bt + C$
$f(t) = e^{kt}$	$y = Ae^{kt}$
$f(t) = \sin kt$	$y = A \sin kt + B \cos kt$
$f(t) = \cos kt$	$y = A \sin kt + B \cos kt$

The *method of undetermined coefficients* for finding a particular solution of equation (9) is to insert the general form for the function f into equation (9) and to solve for the "undetermined coefficients" A, B, etc.

☐ **EXAMPLE 2**

In the differential equation

$$y'' + 3y' + 2y = t^2 \tag{10}$$

the general form for the function $f(t) = t^2$ is

$$y = At^2 + Bt + C. \tag{11}$$

Then $y' = 2At + B$ and $y'' = 2A$. Inserting these expressions in equation (10) gives

$$2A + 3(2At + B) + 2(At^2 + Bt + C) = t^2$$

or

$$2At^2 + (6A + 2B)t + (2A + 3B + 2C) = t^2.$$

We therefore obtain the system of equations

$$
\begin{aligned}
2A &= 1 \quad &\text{(coefficients of } t^2\text{)} \\
6A + 2B &= 0 \quad &\text{(coefficients of } t\text{)} \\
2A + 3B + 2C &= 0 \quad &\text{(constants)}
\end{aligned}
$$

which has solution $A = \frac{1}{2}$, $B = -\frac{3}{2}$, $C = \frac{7}{4}$. A particular solution of equation (10), given by equation (11), is therefore

$$y_p(t) = \tfrac{1}{2}t^2 - \tfrac{3}{2}t + \tfrac{7}{4}.$$

Since the homogeneous equation

$$y'' + 3y' + 2y = 0 \tag{12}$$

has characteristic polynomial $r^2 + 3r + 2 = (r + 2)(r + 1)$ with roots $r_1 = -2$ and $r_2 = -1$, the general solution of equation (12) is

$$y_g(t) = C_1 e^{-2t} + C_2 e^{-t},$$

and the general solution of equation (10) is

$$y(t) = C_1 e^{-2t} + C_2 e^{-t} + \tfrac{1}{2}t^2 - \tfrac{3}{2}t + \tfrac{7}{4}. \qquad \blacksquare$$

☐ **EXAMPLE 3**

In the differential equation

$$y'' - 2y' + y = \cos t \tag{13}$$

the general form for the function $f(t) = \cos t$ is

$$y = A \sin t + B \cos t.$$

Then

$$y' = A \cos t - B \sin t$$

and

$$y'' = -A \sin t - B \cos t.$$

Inserting these expressions into equation (13) gives

$$(-A \sin t - B \cos t) - 2(A \cos t - B \sin t) + (A \sin t + B \cos t) = \cos t$$

or

$$2B \sin t - 2A \cos t = \cos t.$$

Thus, we obtain the equations

$$
\begin{aligned}
2B &= 0 \qquad \text{(coefficients of } \sin t) \\
-2A &= 1 \qquad \text{(coefficients of } \cos t)
\end{aligned}
$$

with solution $A = -\frac{1}{2}$ and $B = 0$. A particular solution of equation (13) is therefore

$$y_p(t) = -\tfrac{1}{2} \sin t.$$

Since the corresponding homogeneous equation

$$y'' - 2y' + y = 0 \tag{14}$$

has characteristic polynomial $r^2 - 2r + 1 = (r - 1)^2$ with the repeated root $r = 1$, the general solution of equation (14) is

$$y_g(t) = C_1 e^t + C_2 t e^t,$$

and the general solution of equation (13) is

$$y = C_1 e^t + C_2 t e^t - \tfrac{1}{2} \sin t. \qquad \blacksquare$$

Superposition of Solutions

If the function f on the right side of the nonhomogeneous equation

$$\frac{d^2 y}{dt^2} + a \frac{dy}{dt} + by = f(t)$$

is a *linear combination* of two or more functions of the form t^n, e^{kt}, $\sin kt$, or $\cos kt$, the following theorem enables us to obtain a particular solution as a linear combination of solutions of equations involving only one of these functions on their right-hand sides.

Theorem 3

Let y_1 be a solution of the equation

$$\frac{d^2 y}{dt^2} + a \frac{dy}{dt} + by = f_1(t), \tag{15}$$

and let y_2 be a solution of the equation

$$\frac{d^2 y}{dt^2} + a \frac{dy}{dt} + by = f_2(t). \tag{16}$$

Then the function

$$y(t) = \alpha y_1(t) + \beta y_2(t) \tag{17}$$

is a solution of the equation

$$\frac{d^2 y}{dt^2} + a \frac{dy}{dt} + by = \alpha f_1(t) + \beta f_2(t). \tag{18}$$

Theorem 3 says that we may find a solution of equation (18) by first finding solutions y_1 and y_2 of equations (15) and (16), and then forming the linear combination $y(t) = \alpha y_1(t) + \beta y_2(t)$ of these solutions. The proof of this theorem is given in Exercise 20.

☐ **EXAMPLE 4**

We leave it as an exercise for you to verify that the equation

$$\frac{d^2y}{dt^2} + 4y = e^t$$

has the particular solution $y_1(t) = \frac{1}{5}e^t$ and that the equation

$$\frac{d^2y}{dt^2} + 4y = t^2$$

has the particular solution $y_2(t) = \frac{1}{4}t^2 - \frac{1}{8}$.

Thus, according to Theorem 3,

(i) the equation

$$\frac{d^2y}{dt^2} + 4y = 6e^t$$

has solution

$$y(t) = 6y_1(t) = \frac{6}{5}e^t;$$

(ii) the equation

$$\frac{d^2y}{dt^2} + 4y = -4t^2$$

has solution

$$y(t) = -4y_2(t) = -4(\tfrac{1}{4}t^2 - \tfrac{1}{8}) = -t^2 + \tfrac{1}{2};$$

(iii) the equation

$$\frac{d^2y}{dt^2} + 4y = 10e^t + 8t^2 \tag{19}$$

has solution

$$\begin{aligned}
y &= 10y_1(t) + 8y_2(t) \\
&= 10(\tfrac{1}{5}e^t) + 8(\tfrac{1}{4}t^2 - \tfrac{1}{8}) \\
&= 2e^t + 2t^2 - 1.
\end{aligned}$$

We could also have found a particular solution of equation (19) directly by beginning with the general form

$$y = Ae^t + Bt^2 + Ct + D$$

and applying the method of undetermined coefficients. ■

A Complicating Issue

There is a situation in which the method of undetermined coefficients, as it has been described thus far, fails to produce a particular solution of the nonhomogeneous equation

$$\frac{d^2y}{dt^2} + a\frac{dy}{dt} + by = f(t). \tag{20}$$

This occurs if the function $y = f(t)$ is a solution of the *homogeneous* equation

$$\frac{d^2y}{dt^2} + a\frac{dy}{dt} + by = 0. \tag{21}$$

In this case, inserting the general form for the function $y = f(t)$ in the left side of equation (20) will simply yield the zero function, because of equation (21). In this case we modify the method of undetermined coefficients as follows:

If the function $y = f(t)$ in equation (20) is a solution of the homogeneous equation (21), we seek a particular solution of equation (20) in the form $y = tg(t)$ where $g(t)$ is the general form for the function f.

We shall not prove this statement, because we have described the method of undetermined coefficients only as a way to *seek* particular solutions of equation (20). The following examples give you some indication, however, as to why this technique works.

□ **EXAMPLE 5**

For the nonhomogeneous equation

$$y'' + 2y' - 3y = e^t \tag{22}$$

the corresponding homogeneous equation is

$$y'' + 2y' - 3y = 0. \tag{23}$$

The characteristic polynomial for equation (23) is

$$r^2 + 2r - 3 = (r + 3)(r - 1),$$

which has roots $r_1 = -3$ and $r_2 = 1$, so the general solution of equation (23) is

$$y_g(t) = C_1 e^{-3t} + C_2 e^t. \tag{24}$$

Thus, the function $f(t) = e^t$ on the right side of equation (22) is a solution of equation (23) (with $C_1 = 0$ and $C_2 = 1$ in equation (24)).

If we were to seek a particular solution of equation (22) using the general form

$$y = Ae^t,$$

we would have $y' = Ae^t$ and $y'' = Ae^t$. Inserting these functions in equation (22) would give the ''equation''

$$Ae^t + 2Ae^t - 3Ae^t = e^t. \tag{25}$$

Since ''equation'' (25) simplifies to the false statement $0 = e^t$, there is no number A for which $y = Ae^t$ is a solution of equation (22).

If, however, we begin with the general form

$$y = Ate^t \tag{26}$$

we obtain

$$y' = A(1 + t)e^t \quad \text{and} \quad y'' = A(2 + t)e^t.$$

Substituting these functions in equation (22) then gives

$$A(2 + t)e^t + 2A(1 + t)e^t - 3Ate^t = e^t,$$

so

$$A(2 + 2)e^t + A(1 + 2 - 3)te^t = e^t$$

or

$$4Ae^t = e^t.$$

Thus $A = \frac{1}{4}$, and a particular solution of equation (22) is

$$y_p(t) = \frac{1}{4}te^t$$

in equation (26). From equations (24) and (26) we may conclude that the *general solution* of equation (22) is

$$y(t) = C_1e^{-3t} + C_2e^t + \frac{1}{4}te^t. \qquad \blacksquare$$

□ **EXAMPLE 6**

For the nonhomogeneous equation

$$y'' + y = \sin t \qquad (27)$$

the associated homogeneous equation $y'' + y = 0$ has general solution

$$y_g(t) = C_1 \sin t + C_2 \cos t.$$

To find a particular solution of the nonhomogeneous equation (27) we must therefore use the general form

$$y = t(A \sin t + B \cos t) = At \sin t + Bt \cos t.$$

We leave it for you to verify that substituting this general form into equation (27) leads to the conclusion $A = 0$ and $B = -\frac{1}{2}$, so a particular solution of (27) is

$$y_p(t) = -\frac{1}{2}t \cos t,$$

and the general solution of (27) is

$$y(t) = C_1 \sin t + C_2 \cos t - \frac{1}{2}t \cos t. \qquad \blacksquare$$

□ **EXAMPLE 7**

In the nonhomogeneous equation

$$y'' - 4y' + 4y = e^{2t} \qquad (28)$$

the associated homogeneous equation

$$y'' - 4y' + 4y = 0 \qquad (29)$$

has characteristic polynomial $r^2 - 4r + 4 = (r - 2)^2$. Since $r = 2$ is a *repeated* root, the general solution of (29) is

$$y_g(t) = C_1e^{2t} + C_2te^{2t}.$$

This means that not only is the function $f(t) = e^{2t}$ a solution of (29), *but so is the function $tf(t) = te^t$*. To find a particular solution of (28) we must therefore begin with the general form obtained by multiplying by yet another factor of t:

$$y = t(Ate^{2t}) = At^2e^{2t} \qquad (30)$$

gives

$$y' = 2Ate^{2t} + 2At^2e^{2t}$$

and

$$y'' = 2Ae^{2t} + 8Ate^{2t} + 4At^2e^{2t},$$

so

$$y'' - 4y' + 4y = (4A - 8A + 4A)t^2e^{2t} + (8A - 8A)te^{2t} + 2Ae^{2t}$$

and we obtain the equation $2Ae^{2t} = e^{2t}$. Thus, $A = \frac{1}{2}$ and the particular solution of (28) of the form in line (30) is

$$y_p(t) = \frac{1}{2}t^2e^{2t},$$

and the general solution of (28) is

$$y(t) = C_1e^{2t} + C_2te^{2t} + \frac{1}{2}t^2e^{2t}.$$ ∎

Exercise Set 21.4

In Exercises 1–12, find the general solution y of the given second order differential equation.

1. $y'' + 2y' - y = \sin 2t$

2. $y'' - 3y' + 4y = t$

3. $y'' + y = e^{-t}$

4. $y'' - 9y = t^2$

5. $y'' - 4y = 4t$

6. $y'' + 7y' + 12y = e^{2t}$

7. $y'' - 3y' - 10y = 5e^{2t}$

8. $y'' + y = t^2$

9. $y'' + 9y = \cos 3t$

10. $y'' + 2y' + y = 4e^{-t}$

11. $y'' - 3y' - 4y = 2e^{-t}$

12. $y'' + 5y' + 6y = 3e^{-3t}$

In Exercises 13–17, use Theorem 3 and the method of undetermined coefficients to find the general solution of the given differential equation.

13. $y'' + 2y' + y = 3e^t + t^2$

14. $y'' - 2y' - 3y = 2e^{2t} + 3\sin t$

15. $y'' - 2y' + y = 2\cos 2t + \sin 2t + 2e^{-t}$

16. $y'' + y' = t^3 + 2t^2$

17. $y'' + y' + y = t^3 + 2t^2$

18. Solve the initial value problem

$$y'' - 2y' - 3y = 2e^t - 10\sin t$$
$$y(0) = 2$$
$$y'(0) = 4.$$

19. Prove Theorem 2 by proving the following statements.
  a. If y_p and z_p are any two particular solutions of the nonhomogeneous equation

$$y'' + ay' + by = f(t),$$

then the function $y = y_p - z_p$ is a solution of the *homogeneous* equation

$$y'' + ay' + by = 0.$$

(*Hint:* Simply substitute the function $y = y_p - z_p$ into the left side of the homogeneous equation.)

  b. Let $y_g(t) = C_1y_1(t) + C_2y_2(t)$ be the general solution for the homogeneous equation. Conclude from part a that for some choice of the constants C_1 and C_2.

$$z_p(t) = C_1y(t) + C_2y_2(t) + y_p(t).$$

  c. Conclude from parts a and b that *any* particular solution z_p of the nonhomogeneous equation must have the form

$$z_p = y_g + y_p$$

where y_g is the general solution of the homogeneous equation.

20. Prove Theorem 3 by substituting $y(t) = \alpha y_1(t) + \beta y_2(t)$ into equation (18) and verifying the equality.

21. Demonstrate that the sum $y = y_1 + y_2$ of two distinct solutions of the nonhomogeneous equation

$$y'' + ay' + by = f(t), \qquad f(t) \neq 0$$

is not again a solution of this equation. For what differential equation is y a solution?

22. The method of undetermined coefficients may also be applied to certain *first* order nonhomogeneous equations. Verify the following facts required in its application.
  a. If a is a constant, the general solution of the first order homogeneous equation

$$y' + ay = 0$$

is $y_g(t) = Ce^{-at}$.

  b. If y_p is any particular solution of the nonhomogeneous equation

$$y' + ay = f(t),$$

the general solution of the nonhomogeneous equation is

$$y = Ce^{-at} + y_p.$$

(*Hint:* As in Exercise 19, verify that the difference of any two particular solutions is a solution of the homogeneous equation in part a.)

c. Conclude that, if $f(t)$ is a function of the form t^n, e^{kt}, $\sin kt$, or $\cos kt$, and the method of undetermined coefficients can be used to find a particular solution of the equation

$$y' + ay = f(t),$$

then the general solution of this equation is as in part b.

Use the method of undetermined coefficients and the results of Exercise 22 to find the general solution of the differential equations in Exercises 23–30.

23. $y' - 3y = 6t - 14$ **24.** $y' + 2y = 3t$

25. $y' - 4y = 1 - 4t^2$ **26.** $y' + 5y = 10$

27. $y' - 4y = 2e^{3t}$ **28.** $y' + 2y = 2 \sin t + 11 \cos t$

29. $y' - 3y = 4e^t + 6t - 14$

30. $y' - 2y = -4t + 8 + 5 \sin t$

In Exercises 31–34, find the solution of the given initial value problem.

31. $y' + 4y = 0$
     $y(0) = 3$

32. $y' - 2y = 12 - 4t$
     $y(0) = 8$

33. $y' + 3y = 5e^{2t}$
     $y(0) = 7$

34. $y' - 2y = 2 \sin 4t + 16 \cos 4t$
     $y(0) = 8$

21.5   Power Series Solutions of Differential Equations

The theory of power series provides a method for solving certain types of differential equations. For differential equations of the form

$$f(x, y, y') = 0 \tag{1}$$

the idea is to express the (unknown) solution y as a power series in the independent variable x:

$$y = \sum_{k=0}^{\infty} a_k x^k. \tag{2}$$

Then, assuming (2) has a nonzero radius of convergence, we apply the theorem on differentiating power series to conclude that

$$y' = \sum_{k=0}^{\infty} k a_k x^{k-1}. \tag{3}$$

We next insert representations (2) and (3) into the differential equation (1). For each integer k, coefficients of x^k will appear in various places in equation (1). By equating coefficients of x^k on either side of equation (1) we can often find a set of equations that determine the coefficients $a_0, a_1, a_2, a_3, \ldots$ completely. In such cases we obtain a power series representation for the solution.

□ **EXAMPLE 1**

Use the power series method to solve the differential equation

$$y' = y. \tag{4}$$

Solution   We assume that $y = \sum_{k=0}^{\infty} a_k x^k$, so that $y' = \sum_{k=0}^{\infty} k a_k x^{k-1}$, and both of these series have the same radius of convergence. Inserting these expansions in the differential equation (4) gives

$$\sum_{k=0}^{\infty} k a_k x^{k-1} = \sum_{k=0}^{\infty} a_k x^k$$

or

$$a_1 + 2a_2x + 3a_3x^2 + \cdots + ka_kx^{k-1} + \cdots$$
$$= a_0 + a_1x + a_2x^2 + \cdots + a_{k-1}x^{k-1} + \cdots. \quad (5)$$

Since the two series in equation (5) are equal, the coefficients of like powers of x must be the same. Thus,

$$
\begin{aligned}
a_1 &= a_0 && \text{(constant terms)} \\
2a_2 &= a_1 && (x \text{ terms)} \\
3a_3 &= a_2 && (x^2 \text{ terms)} \\
&\vdots \\
ka_k &= a_{k-1} && (x^{k-1} \text{ terms)} \\
&\vdots
\end{aligned}
$$

Using these equations, we may solve for each coefficient $a_2, a_3, a_4, \ldots$ in terms of the one preceding and, therefore, in terms of a_0:

$$
\begin{aligned}
a_1 &= a_0 \\
a_2 &= \frac{1}{2}a_1 = \frac{1}{2}a_0 = \frac{1}{2!}a_0 \\
a_3 &= \frac{1}{3}a_2 = \frac{1}{3 \cdot 2}a_0 = \frac{1}{3!}a_0 \\
&\vdots \\
a_k &= \frac{1}{k}a_{k-1} = \frac{1}{k(k-1)\cdot \, \cdots \, \cdot 3 \cdot 2}a_0 = \frac{1}{k!}a_0. \quad (6)
\end{aligned}
$$

Returning to the expansion for y, we can now write

$$y = a_0 + a_0x + \frac{1}{2!}a_0x^2 + \frac{1}{3!}a_0x^3 + \cdots + \frac{1}{k!}a_0x^k + \cdots \quad (7)$$
$$= a_0\left[1 + x + \frac{x^2}{2!} + \frac{x^3}{3!} + \cdots + \frac{x^k}{k!} + \cdots\right],$$

which is the power series representation for the function $y = a_0e^x$. Since this power series converges for all values of x, the solution $y = a_0e^x$ is valid for all x. The constant a_0 is determined by an initial condition for the equation (4). ∎

REMARK   The result of Example 1 should not have been unexpected, since we determined in Chapter 8 that the solution of the differential equation $y' = y$ is $y = Ce^x$. However, here we see that the use of power series provides an entirely different approach to solving differential equations.
   The equation

$$a_k = \frac{1}{k}a_{k-1}$$

on the left side of equation (6) warrants special attention. It is referred to as the **recurrence relation** for the differential equation because it gives the general formula by which each coefficient in the series expansion for the solution (except the first) may be determined from its predecessor. If you can succeed in finding a recurrence relation, you will be able to generate all coefficients in the expansion for the solution of a differential equation. However, two issues then remain to be addressed:

(1) For which values of x does the power series obtained from the recurrence relation converge? A power series solution is a legitimate solution of the differential equation *only* if it has a nonzero radius of convergence.
(2) Can a closed form expression* for the power series solution be recognized, as in Example 1 where we recognized the solution as the power series for $a_0 e^x$? The answer to this question is not critical if you are willing to accept solutions in the form of power series. In fact, many important functions in mathematical physics arise as series solutions of differential equations and do not have closed form expressions (see Example 3).

□ **EXAMPLE 2**

Use the power series method to solve the initial value problem

$$y' = xy$$
$$y(0) = 1.$$

Strategy · · · · · · · ·

Assume a power series form for the solution y. Differentiate to find series form for y'.

Solution

We assume that

$$y = \sum_{k=0}^{\infty} a_k x^k, \quad \text{so} \quad y' = \sum_{k=0}^{\infty} k a_k x^{k-1}.$$

Insert the expansions for y and y' into the differential equation.

This gives

$$\sum_{k=0}^{\infty} k a_k x^{k-1} = x \sum_{k=0}^{\infty} a_k x^k$$

$$= \sum_{k=0}^{\infty} a_k x^{k+1}$$

Write out the first few terms, including the general terms on both sides *for the same power of x.* (Here we arbitrarily picked x^{k-1}.)

or

$$a_1 + 2a_2 x + 3a_3 x^2 + \cdots + k a_k x^{k-1} + \cdots$$
$$= a_0 x + a_1 x^2 + a_2 x^3 + \cdots + a_{k-2} x^{k-1} + \cdots,$$

Equate coefficients of like powers of x.

so

$$a_1 = 0 \qquad \text{(constant terms)}$$

$$a_2 = \frac{1}{2} a_0 \qquad \text{(}x\text{ terms)}$$

$$a_3 = \frac{1}{3} a_1 = 0 \qquad \text{(}x^2\text{ terms)}$$

$$\vdots$$

The recurrence relation is obtained by equating coefficients of x^{k-1}.

$$a_k = \frac{1}{k} a_{k-2} \qquad \text{(}x^{k-1}\text{ terms)}$$

$$\vdots$$

---

*A closed form expression is one that does not involve an infinite process, such as summation. $f(x) = 1/(1 - x)$ is expressed in closed form, while $f(x) = \sum_{k=0}^{\infty} x^k$ is not.

Determine the form of all coefficients using the recurrence relation.

From the recurrence relation

$$a_k = \frac{1}{k} a_{k-2}$$

and the fact that $a_1 = 0$, we see that all odd coefficients are zero: $0 = a_1 = a_3 = a_5 = \cdots$. The even coefficients are

$$a_2 = \frac{1}{2} a_0$$

$$a_4 = \frac{1}{4} a_2 = \frac{1}{4 \cdot 2} a_0 = \frac{1}{2^2} \left(\frac{1}{2 \cdot 1} \right) a_0 = \frac{1}{2^2} \cdot \frac{1}{2!} a_0$$

$$a_6 = \frac{1}{6} a_4 = \frac{1}{6 \cdot 4 \cdot 2} a_0 = \frac{1}{2^3} \left(\frac{1}{3 \cdot 2 \cdot 1} \right) a_0 = \frac{1}{2^3} \cdot \frac{1}{3!} a_0$$

$$\vdots$$

$$a_{2k} = \frac{1}{2k} a_{2k-2} = \frac{1}{2k(2k-2) \cdot \cdots \cdot 2} a_0 = \frac{1}{2^k} \cdot \frac{1}{k!} a_0.$$

Write the series for y using coefficients found above.

The power series for the solution y is therefore

$$y = a_0 + a_1 x + a_2 x^2 + a_3 x^3 + a_4 x^4 + \cdots$$

Try to bring the series for y into the form of a known power series. (Here we use the fact that

$$e^u = \sum_{k=0}^{\infty} \frac{u^k}{k!}$$

for all u.)

$$= a_0 + \frac{1}{2} a_0 x^2 + \frac{1}{2^2} \cdot \frac{1}{2!} a_0 x^4 + \frac{1}{2^3} \cdot \frac{1}{3!} a_0 x^6 + \cdots + \frac{1}{2^k} \cdot \frac{1}{k!} a_0 x^{2k} + \cdots$$

$$= a_0 \left[1 + \frac{x^2}{2} + \frac{(x^2/2)^2}{2!} + \frac{(x^2/2)^3}{3!} + \cdots + \frac{(x^2/2)^k}{k!} + \cdots \right]$$

$$= a_0 \sum_{k=0}^{\infty} \frac{(x^2/2)^k}{k!}.$$

This is the power series expansion for $y = a_0 e^{x^2/2}$, which converges for all values of x. The general solution for the differential equation $y' - xy = 0$ is therefore $y = a_0 e^{x^2/2}$. The constant a_0 is determined from the initial condition:

Apply initial condition to determine a_0.

$$1 = y(0) = a_0 e^0 = a_0.$$

The solution of the initial value problem is $y = e^{x^2/2}$.   ■

Power series methods may be used in higher order differential equations as well. Note in the following example that the power series for y must be differentiated twice since the differential equation is of order 2.

☐ **EXAMPLE 3**

Find a power series solution for the differential equation

$$xy'' - y = 0.$$

Solution   We shall work with the equation in the form

$$xy'' = y. \tag{8}$$

We assume the solution y to have the form

$$y = \sum_{k=0}^{\infty} a_k x^k.$$

Then

$$y' = \sum_{k=0}^{\infty} k a_k x^{k-1} \quad \text{and} \quad y'' = \sum_{k=0}^{\infty} k(k-1) a_k x^{k-2}.$$

Equation (8) becomes

$$x \sum_{k=0}^{\infty} k(k-1) a_k x^{k-2} = \sum_{k=0}^{\infty} a_k x^k,$$

or

$$2a_2 x + 3 \cdot 2a_3 x^2 + 4 \cdot 3 \cdot a_4 x^3 + \cdots + k(k-1) a_k x^{k-1} + \cdots$$
$$= a_0 + a_1 x + a_2 x^2 + a_3 x^3 + \cdots + a_{k-1} x^{k-1} + \cdots.$$

Thus

$$a_0 = 0$$

$$2a_2 = a_1 \text{ gives } a_2 = \frac{1}{2} a_1$$

$$3 \cdot 2 \cdot a_3 = a_2 \text{ gives } a_3 = \frac{1}{3 \cdot 2} a_2 = \frac{1}{3 \cdot 2^2} a_1 = \frac{1}{3(2!)^2} a_1$$

$$4 \cdot 3 \cdot a_4 = a_3 \text{ gives } a_4 = \frac{1}{4 \cdot 3} a_3 = \frac{1}{4 \cdot 3^2 \cdot 2^2} a_1 = \frac{1}{4(3!)^2} \cdot a_1$$

$$\vdots$$

$$k(k-1) a_k = a_{k-1} \text{ gives } a_k = \frac{1}{k(k-1)} a_{k-1} = \cdots = \frac{1}{k[(k-1)!]^2} \cdot a_1$$

The solution y is therefore

$$y = a_1 \left[x + \frac{1}{2} x^2 + \frac{1}{3 \cdot 2^2} x^3 + \frac{1}{4 \cdot (3!)^2} x^4 + \cdots + \frac{1}{k[(k-1)!]^2} x^k + \cdots \right]. \quad (9)$$

To determine the radius of convergence for (9) we apply the Ratio Test.

$$\rho = \lim_{k \to \infty} \left| \frac{\dfrac{a_1}{(k+1)(k!)^2} x^{k+1}}{\dfrac{a_1}{k[(k-1)!]^2} x^k} \right| = \lim_{k \to \infty} \left(\frac{k}{k+1} \right) \left(\frac{1}{k^2} \right) |x| = 0$$

for all values of x, so the series in (9) converges for all x. We have therefore obtained a legitimate solution for the differential equation (8), although the solution is expressed in power series form and is not immediately recognizable as the power series for a known function in closed form. ∎

As you might suspect at this point, the theory associated with the use of power series to solve differential equations is not simple. In fact, the examples presented here were carefully chosen to convey the basic idea while avoiding the difficulties surrounding this method. These difficulties include the following:

(i) Often we can obtain only the first few terms of the series solution rather than the general term. If so, we cannot address the convergence question, although in many cases one can obtain at least an approximation to the solution.

(ii) The assumption that the solution y has a power series representation contains the implicit assumption that y has derivatives of all orders. Since the solution to a differential equation of degree n need have only n derivatives, the power series approach necessarily fails to identify solutions to certain equations because it assumes too much.

Further work on power series methods in differential equations is left to more specialized courses. If you take such a course you will find that the material in Chapter 13 constitutes an important foundation on which much of the work of that course depends.

Exercise Set 21.5

In Exercises 1–8, use the method of this section to find a power series form of the solution of the given differential equation or initial value problem. Check your work by solving the equation by the method of separation of variables.

1. $y' + y = 0$

2. $y' + 2y = 0$

3. $y' - 6y = 0$

4. $y' + xy = 0$

5. $y' - 2xy = 0$

6. $y' + ax = 0$

7. $y' + 4y = 0$

$y(0) = 1$

8. $y' + xy = 0$

$y(0) = 2$

9. Find a power series solution for the second order differential equation $xy'' + y = 0$.

10. Find a power series solution for the initial value problem

$$y'' + y = 0$$
$$y(0) = 0$$
$$y'(0) = 1.$$

(*Hint:* Let $y = \sum_{k=0}^{\infty} a_k x^k$. Since $y(0) = 0$, $a_0 = 0$. This observation simplifies the recurrence relation.)

11. Find a power series solution for the initial value problem

$$y'' + y = 0$$
$$y(0) = 1$$
$$y'(0) = 0.$$

12. Conclude from Exercises 10 and 11 that the function $y = A \sin x + B \cos x$ is a solution of the differential equation $y'' + y = 0$.

21.6   Approximating Solutions of Differential Equations

There are many differential equations for which simple solutions cannot be found. In such cases, however, we can resort to certain approximation procedures to obtain information about the solution, just as we can use the Trapezoidal Rule or Simpson's Rule to approximate a definite integral if the corresponding antiderivative cannot be found.

To give you an idea of how such approximation procedures work, we discuss Euler's method. It is named for the Swiss mathematician Leonhard Euler (1707–1783).

Euler's Method

Euler's method can be used to approximate the solution to an initial value problem of the form

$$\frac{dy}{dt} = f(t, y) \tag{1}$$

$$y(a) = y_0$$

on some interval $[a, b]$. (We shall assume that f and its partial derivatives are continuous for all t and y, although less restrictive conditions can be given.)

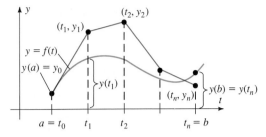

Figure 6.1   Approximation by Euler's method to the solution y of the initial value problem

$$y' = f(t, y), \qquad y(a) = y_0.$$

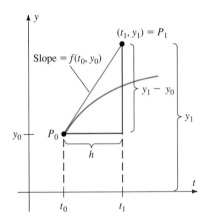

Figure 6.2 $y_1 = y_0 + f(t_0, y_0)h.$

Figure 6.1 illustrates the basic idea. Assume that $y = f(t)$ is a particular solution of the given differential equation. We divide the interval $[a, b]$ into n subintervals of equal length $h = (b - a)/n$ and obtain the $(n + 1)$ endpoints

$$a = t_0 < t_1 < t_2 < \cdots < t_n = b.$$

Beginning at the left endpoint $t_0 = a$ we use the given information about the solution y (namely, the initial value $y(a) = y_0$) and its derivative to approximate the value $y(t_1)$. We call the approximation y_1. (Note, as suggested by Figure 6.1, that, in general, the approximation y_1 will not equal the (unknown) function value $y(t_1)$). Next, we use the assumed value y_1 and the number $t_1 = a + h$ to approximate $y(t_2)$. We call this approximation y_2. Continuing in this way we finally approximate $y(t_n) = y(b)$ from the approximate value y_{n-1} and the number $t_{n-1} = a + (n - 1)h$. We therefore obtain *approximations* $y_1, y_2, \ldots, y_n$ to the (unknown) values of the solution $y(t_1), y(t_2), \ldots, y(t_n)$ at finitely many numbers t_j in $[a, b]$. The choice of the endpoint b and the number n of intervals depend upon which values of the solution you wish to approximate and with what accuracy.

Figure 6.2 illustrates how we start Euler's method. To find the approximation y_1 to the value $y(t_1)$, we follow the line tangent to the graph of y at the point $P_0 = (t_0, y_0)$ until we reach the point $P_1 = (t_1, y_1)$ with t-coordinate t_1. Since the slope of this line is $y'(t_0) = f(t_0, y_0)$, it follows that

$$\frac{y_1 - y_0}{t_1 - t_0} = f(t_0, y_0),$$

and since $h = t_1 - t_0$,

$$y_1 = y_0 + f(t_0, y_0)h. \tag{2}$$

Once we obtain the approximation y_1 from equation (2), we *pretend* that y_1 is the actual value of the solution y at $t = t_1$. Since we don't know the actual value $y(t_1)$, the approximation y_1 is the ''next best thing.'' This ''pretending'' is necessary in order that we may carry out the same procedure to obtain the approximation y_2 for $y(t_2)$. Starting with the point (t_1, y_1) and the slope $f(t_1, y_1)$,* we obtain the y-coordinate of the point $P_2 = (t_2, y_2)$ from the equation

$$\frac{y_2 - y_1}{t_2 - t_1} = f(t_1, y_1).$$

---

*The pretending occurs here in two places. Since we do not know $y(t_j)$, we do not know the actual slope $f(t_j, y(t_j))$, which we approximate by $f(t_j, y_j)$. Neither do we know the point of tangency, $(t_j, y(t_j))$, so we use (t_j, y_j) instead.

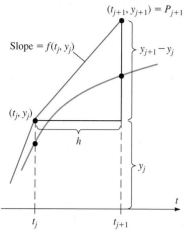

Figure 6.3 $y_{j+1} = y_j + f(t_j, y_j)h.$

That is, since $h = t_2 - t_1$,

$$y_2 = y_1 + f(t_1, y_1)h.$$

Continuing in this way (see Figure 6.3), we obtain each succeeding approximation y_{j+1} from the preceding y_j and t_j by the equation

$$y_{j+1} = y_j + f(t_j, y_j)h \qquad j = 0, 1, \ldots, n - 1 \tag{3}$$

which is the summary statement of Euler's method.

The following examples show how a first order differential equation may be put in the form of equation (1), and how the solution of an associated initial value problem may be approximated using Euler's method.

□ **EXAMPLE 1**

Use Euler's method to approximate the solution of the initial value problem

$$y' = y(4 - y) \tag{4}$$
$$y(0) = 1$$

on the interval $[0, 1]$ using 4 subintervals of equal length.

Solution   With $[0, 1]$ divided into 4 subintervals of length

$$h = \frac{1 - 0}{4} = \frac{1}{4},$$

we have endpoints

$$t_0 = 0, \qquad t_1 = \tfrac{1}{4}, \qquad t_2 = \tfrac{1}{2}, \qquad t_3 = \tfrac{3}{4}, \quad \text{and} \quad t_4 = 1.$$

With $y_0 = y(0) = 1$ and $f(t, y) = y(4 - y)$, equation (3) gives

$$\begin{aligned}
y_1 &= y_0 + y_0(4 - y_0)h \\
&= 1 + 1(4 - 1)(\tfrac{1}{4}) \\
&= 1 + \tfrac{3}{4} = 1.75
\end{aligned}$$

$$\begin{aligned}
y_2 &= y_1 + y_1(4 - y_1)h \\
&= 1.75 + 1.75(4 - 1.75)(0.25) \approx 2.7344,
\end{aligned}$$

etc. Table 6.1 shows the approximations y_1, y_2, y_3, y_4 together with the actual values $y(t_1), y(t_2), y(t_3), y(t_4)$ of the solution

$$y = \frac{4}{1 + 3e^{-4t}}$$

of the initial value problem. (This solution is obtained by the method of Section 21.1.) ■

Table 6.1 Approximate and actual values of the solution to the initial value problem in Example 1 ($h = 0.25$).

t_j	0	.25	.50	.75	1.0
y_j	1	1.75	2.7344	3.5995	3.9599
$y(t_j)$	1	1.9015	2.8449	3.4802	3.7917

☐ **EXAMPLE 2**

Approximate the solution to the initial value problem in Example 1 on the interval $[0, 1]$ using $n = 10$ subintervals of equal size.

Solution This time we have

$$h = \frac{1 - 0}{10} = \frac{1}{10},$$

so Euler's equation becomes

$$y_{j+1} = y_j + y_j(4 - y_j)(0.10), \qquad j = 0, 1, 2, \ldots, 9.$$

Table 6.2 shows the resulting approximations, together with the actual values of the solution, to four decimal place accuracy. (Note the improvement in accuracy compared with Example 1.) ∎

Table 6.2 Values of the approximations and actual solution to the initial value problem in Example 2 ($h = 0.10$).

t_j	0	.1	.2	.3	.4	.5	.6	.7	.8	.9	1.0
y_j	1	1.3	1.651	2.0388	2.4387	2.8194	3.1523	3.4195	3.6180	3.7562	3.8478
$y(t_j)$	1	1.3285	1.7036	2.1013	2.4911	2.8449	3.1443	3.3829	3.5642	3.6970	3.7917

☐ **EXAMPLE 3**

Use Euler's method to approximate the solution of the initial value problem

$$y' + 2y = e^t \tag{5}$$
$$y(0) = 1$$

on the interval $[0, 1]$.

Solution We begin by moving all terms in equation (5) except the derivative y' to the right-hand side, obtaining

$$\frac{dy}{dt} = e^t - 2y.$$

Thus, $f(t, y) = e^t - 2y$ in equation (3) for Euler's method. We calculate two approximations corresponding to two different choices of n.

(a) Using $n = 4$ subdivisions, we have

$$h = \frac{1 - 0}{4} = 0.25,$$

and equation (3) becomes

$$y_{j+1} = y_j + (e^{t_j} - 2y_j)(0.25), \qquad j = 0, 1, 2, 3.$$

Thus since $y_0 = y(0) = 1$ is given,

$$
\begin{aligned}
y_1 &= y_0 + (e^{t_0} - 2 \cdot y_0)(0.25) \\
&= 1 + (e^0 - 2 \cdot 1)(0.25) \\
&= 1 - 0.25 = 0.75,
\end{aligned}
$$

$$
\begin{aligned}
y_2 &= y_1 + (e^{t_1} - 2y_1)(0.25) \\
&= 0.75 + [e^{.25} - 2(0.75)](0.25) = 0.6960.
\end{aligned}
$$

etc. Table 6.3 shows the same approximations y_1, y_2, y_3, and y_4 to four decimal place accuracy, as well as the actual values $y(t_1)$, $y(t_2)$, $y(t_3)$, and $y(t_4)$ for the actual solution $y = \frac{1}{3}(2e^{-2t} + e^t)$.

Table 6.3 Approximate and actual values for the initial value problem in Example 3 ($h = 0.25$).

t_j	0	.25	.5	.75	1.0
y_j	1	.75	.6960	.7602	.9093
$y(t_j)$	1	.8324	.7948	.8544	.9963

(b) The same solution, approximated using 10 subintervals of $[0, 1]$ of length

$$h = \frac{1 - 0}{10} = 0.1,$$

is obtained using the Euler equation

$$y_{j+1} = y_j + (e^{t_j} - 2y_j)(0.1), \qquad j = 0, 1, \dots, 9.$$

Table 6.4 shows the results of this approximation. ■

Table 6.4 Approximate and actual values of the solution of the initial value problem in Example 3 ($h = 0.1$).

t_j	0	.1	.2	.3	.4	.5	.6	.7	.8	.9	1.0
y_j	1	.9	.8305	.7866	.7642	.7606	.7733	.8009	.8421	.8962	.9629
$y(t_j)$	1	.9142	.8540	.8158	.7968	.7948	.8082	.8356	.8764	.9301	.9963

Euler's method is one procedure for approximating the solution of an initial value problem involving a first order differential equation. While we have used examples in which an actual solution can be obtained for purposes of comparison, Euler's method is usually applied to differential equations whose solutions are not known.

Obviously there is a fair amount of tedium in applying Euler's method. Calculators and especially computers can reduce this tedium to a minimum. Appendix I contains computer programs that were used to perform the calculations in Tables 6.1–6.4. You may find them useful in working the exercises.

Several sources of error enter into the calculations that are involved in Euler's method. Since we obtain only approximations y_j instead of the actual value $y(t_j)$ at each step, the slope calculation $dy/dt = f(t_j, y_j)$ contains an error due to the difference $|y_j - y(t_j)|$, and the point (t_j, y_j) of assumed tangency contains an error due to the same factor. Finally, roundoff error in such calculations is inescapable. In general, however, we can say that accuracy decreases as we approximate values of the solution further away from the number t_0 where the initial condition is prescribed and that accuracy increases as the stepsize h decreases. These important issues of accuracy are pursued in more advanced courses.

Exercise Set 21.6

In Exercises 1–10, use Euler's method to approximate the value of the solution to the given initial value problem at the endpoints of 4 subintervals of $[0, 1]$ of size $h = 0.25$.

1. $y' = 2y$
   $y(0) = 1$

2. $y' = 9y(1 - y)$
   $y(0) = \frac{1}{2}$

3. $y' + 2y = 4$
   $y(0) = 1$

4. $\dfrac{dy}{dx} = \dfrac{y + 1}{x + 1}$
   $y(0) = 0$

5. $\dfrac{dy}{dx} = x(y + 1)$
   $y(0) = 1$

6. $\dfrac{dy}{dx} = 2 - 4y$
   $y(0) = 1$

7. $\dfrac{dy}{dt} + 2ty = t$
   $y(0) = \frac{3}{2}$

8. $e^y \dfrac{dy}{dt} - t^2 = 0$
   $y(0) = 0$

9. $\dfrac{dy}{dt} = 2y - 2t$
   $y(0) = \frac{3}{2}$

10. $\dfrac{dy}{dt} - 2y = e^{2t} \sin t$
   $y(0) = 0$

11. Find the actual solution of the initial value problem in Exercise 3. Compute the actual value of the solution at each endpoint and compare it with the value from Euler's method.

12. An investor opens a savings account paying 10% interest compounded continuously with an initial deposit of $1000. The investor makes deposits of $1000 per year in a continuous manner.
   a. Show that the function P, giving the amount on deposit after t years, satisfies the differential equation
   $$P'(t) = 0.10P(t) + 1000.$$
   b. Use Euler's method to approximate the values $P(1), P(2), P(3),$ and $P(4)$.

Summary Outline of Chapter 21

■ First order differential equations of the form   (page 1110)
$$\frac{dy}{dt} = ky + b$$
can be solved by the method of *separation of variables*.

■ First order differential equations of the form   (page 1114)
$$\frac{dy}{dt} + p(t)y = g(t)$$
can be solved using the *integrating factor* $e^{\int p(t)\, dt}$

■ The differential equation   (page 1117)
$$M(x, y) + N(x, y)\frac{dy}{dx} = 0, \qquad (x, y) \in R \qquad (1)$$
is *exact* if there exists a function f for which
$$M(x, y) = \frac{\partial f}{\partial x}(x, y), \qquad N(x, y) = \frac{\partial f}{\partial y}(x, y).$$

■ *Theorem:* Equation (1) is exact in an open rectangle R if and only if   (page 1119)
$$\frac{\partial M}{\partial y}(x, y) = \frac{\partial N}{\partial x}(x, y), \qquad (x, y) \in R.$$

■ The second order homogeneous linear differential equation $y'' + ay' + by =$ (page 1122)
0 has characteristic polynomial $r^2 + ar + b$ and general solution

   (i) $y = C_1 e^{r_1 t} + C_2 e^{r_2 t}$ if $r^2 + ar + b$ has two distinct real roots r_1 and r_2;
   (ii) $y = C_1 e^{rt} + C_2 t e^{rt}$ if $r^2 + ar + b$ has a repeated real root r;
   (iii) $y = e^{\alpha t}[C_1 \sin \beta t + C_2 \cos \beta t]$ if $r^2 + ar + b$ has the complex roots
        $r = \alpha \pm \beta i$; $\beta \neq 0$.

■ The second order linear homogeneous equation $y'' + by = 0$, $b > 0$, has gen- (page 1125)
eral solution $y = C_1 \sin \sqrt{b}t + C_2 \cos \sqrt{b}t$.

■ The nonhomogeneous second order linear equation $y'' + ay' + by = f(t)$ can (page 1130)
be solved by the *method of undetermined coefficients* if $f(t)$ is of the form t^n,
e^{kt}, $\sin kt$, or $\cos kt$.

■ A differential equation can sometimes be solved by assuming a *power series* (page 1137)
solution of the form $y = \Sigma a_k x^k$, substituting into the differential equation,
and solving the resulting equations for the coefficients a_k.

■ *Euler's Method* for approximating solutions of the differential equation $y' =$ (page 1144)
$f(t, y)$ with $y(a) = y_0$ is based on the formula $y_{j+1} = y_j + f(t_j, y_j)h$, which
gives the approximation y_{j+1} to the value $y(t_{j+1})$.

Review Exercises—Chapter 21

In Exercises 1–30, find the general solution of the given differential equation.

1. $\dfrac{dy}{dx} = x - 3$

2. $\dfrac{dy}{dt} = 4 - \sqrt{t}$

3. $\dfrac{dy}{dt} = \dfrac{\sec \sqrt{t} \tan \sqrt{t}}{\sqrt{t}}$

4. $\dfrac{d^2 y}{dx^2} = \sqrt{1 + x}$

5. $\dfrac{dy}{dt} = 3ty^2$

6. $\dfrac{dy}{dt} = \dfrac{\sqrt{t+1}}{y}$

7. $\dfrac{dy}{dx} = (1 + x)(2 + y)$

8. $ty' = \dfrac{t^2 + 3}{y}$

9. $y' + 2y = 4$

10. $y' = y(3 - y)$

11. $y' = y(4 + y)$

12. $ty' = y \ln t$

13. $y' + y \cos t = 0$

14. $y' = 4y \ln y$

15. $y' = 4y + 8$

16. $y' = 4y + 4t + 8$

17. $y'' + 4y' + 4y = 0$

18. $y'' - 3y' - 10y = 0$

19. $y'' - 2y' - 15y = \cos 2t$

20. $y'' + 9y = 5 + e^{2t}$

21. $y'' - 9y = 5 + e^{2t}$

22. $y' - 3y = -9 \sin 2t$

23. $t^2 y' + y = 0$

24. $\dfrac{dy}{dx} = y(2 - y)$

25. $\dfrac{d^2 y}{dt^2} - 5\dfrac{dy}{dt} - 14y = 2 + t$

26. $y'' + y' - 6y = 6 - e^{2t}$

27. $2xy^3 + 3x^2 y^2 \dfrac{dy}{dx} = 0$

28. $\cos y - x \sin y \dfrac{dy}{dx} = 0$

29. $2y - y^2 - 2x\dfrac{dy}{dx} = 0$

30. $(e^{xy} + xye^{xy})\dfrac{dy}{dx} + y^2 e^{xy} = 0$

In Exercises 31–38, find the solution of the given initial value problem.

31. $\dfrac{dy}{dx} = \dfrac{x}{\sqrt{1 + x^2}}$
  $y(0) = 3$

32. $y' = 6y$
  $y(0) = 2$

33. $\dfrac{dy}{dt} = ty$
  $y(0) = 3$

34. $y' = 4(1 - y)$
  $y(0) = 1$

35. $\dfrac{d^2 y}{dt^2} = -9y$
  $y(0) = 3$
  $y'(0) = 9$

36. $y'' - 16y = 0$
  $y(0) = 2$
  $y'(0) = 8$

37. $y'' - 6y' + 9y = 0$
  $y(0) = 0$
  $y'(0) = 6$

38. $y'' - 4y = -2e^{3x}$
  $y(0) = 3$
  $y'(0) = 10$

39. Find a differential equation satisfied by the function
  a. $y = Ce^{2t}$
  b. $y = C_1 e^t + C_2 e^{-t}$.

40. A tank contains 500 L of brine with an initial concentration of salt of 1 kg of salt per liter. A second mixture of brine containing 2 kg of salt per liter is then added at a rate of 50 L/min. The tank is kept well mixed, and brine is drawn off the bottom at the same rate, 50 L/min. Find the concentration of salt t min after the second mixture begins entering the first.

41. An automobile radiator contains five gallons of pure antifreeze. The owner begins adding fresh water at the rate of 1 gal/min, with the engine running to ensure complete mixing, and draining the radiator at the same rate of 1 gal/min. Find the concentration $y(t)$ of antifreeze/gal t minutes after the owner begins this process.

42. An investor places $5000 in a savings account paying 10% interest compounded continuously and pledges to make additional deposits of $2000 per year in a continual manner.
   a. Find a differential equation for $P(t)$, the amount on deposit in this account t years after it is opened.
   b. Use Euler's method to approximate $P(1)$, $P(2)$, $P(3)$, and $P(4)$.

43. A cold drink is removed from a refrigerator at temperature 40°F and is placed on a sunporch where the surrounding temperature is 90°F. After 5 min the temperature of the drink is 50°F. Find its temperature after 10 min.

44. Use a power series to find a solution to the initial value problem

$$y' - xy = 0$$
$$y(0) = 2.$$

45. Find a power series solution for the second order differential equation $xy'' - y = 0$.

46. Find a power series solution for the initial value problem

$$xy'' + y = 0$$
$$y(0) = 0$$
$$y'(0) = 1.$$

Why is the condition $y(0) = 0$ *necessary?*

47. Find the general form of solutions to the differential equation

$$y'' - 4y' + 6y = 0.$$

48. Find the solution for the initial value problem

$$y'' - 4y' + 8y = 0$$
$$y(0) = 0$$
$$y'(0) = 2.$$

Appendix I

Calculus and the Computer

In this appendix, we illustrate the use of the computer as an aid to the study of calculus in two different ways. First, we provide nine computer programs that are referenced in various examples and exercises throughout the text. Implemented in both BASIC and Pascal, they are presented as ''bare-bones'' prototypes, which those with access to computing facilities (personal computers, programmable calculators, or large computers) can use to design programs that actually operate on particular machines. Second, we illustrate the use of one of the sophisticated mathematics-based computer systems, *Mathematica,* as an aid in the study of calculus.

I.1   BASIC and Pascal Programs

BASIC Program 1: Newton's Method for $f(x) = x^3 - 7$

```
 10 DEF FNF(T) = T^3 - 7
 20 DEF FND(T) = 3*(T^2)
 30 PRINT "how many iterations?"
 40 INPUT N
 50 PRINT "what is your first guess?"
 60 INPUT X
 70 FOR I = 1 TO N
 80   LET Z1 = FNF(X)
 90   LET Z2 = FND(X)
100   LET W = X - (Z1/Z2)
110   PRINT I,W
120   LET X = W
130 NEXT I
140 END
```

Comment: This program implements Newton's Method to locate a zero of the function $f(x) = x^3 - 7$. The value of the function is computed in line 10, the derivative $f'(x) = 3x^2$ is computed in line 20, and the formula for Newton's Method is implemented in line 100. The user supplies the number of iterations and a first guess at the root.

A.1

Pascal Program 1: Newton's Method for $f(x) = x^3 - 7$

```pascal
program newton;

var
   maxits, numits, i: integer;
   x: real;

function f(x: real): real;
begin
   f:= x*x*x - 7;   (* the function *)
end;

function df(x: real): real;
begin
   df:= 3*x*x;    (* the derivative of the function *)
end;

begin
   write('How many iterations? ');
   readln(numits);
   write('Your initial guess for a root? ');
   readln(x);
   for i:= 1 to numits do
   begin
      x:= x - f(x)/df(x);    (* Newton iteration *)
      writeln(i:3, '    ', x:8:5);
   end;
   write('Press <enter> to finish ...');
   readln;
end.
```

Comment: This program implements Newton's Method to locate a zero of the function $f(x) = x^3 - 7$. The function and its derivative are specified in the routines **f(x)** and **df(x)**.

BASIC Program 2: Lower Riemann Sums for $f(x) = 3x^2 + 7$

```basic
10 DEF FNF(T) = 3*(T^2) + 7
20 PRINT "enter interval endpoints a,b"
30 INPUT A,B
40 PRINT "how many subintervals?"
50 INPUT N
60 LET D = (B - A)/N
70 LET S = 0
80 FOR I = 1 TO N
90   LET X = A + (I - 1)*D
100  LET S = S + FNF(X)*D
110 NEXT I
120 PRINT S
130 END
```

Comment: This program computes a lower Riemann sum for the function $f(x) = 3x^2 + 7$ on the interval $[a, b]$, if $a > 0$, using n equal subintervals. The numbers a, b, and n are supplied by the user. The program computes the function value at left endpoints of the resulting subintervals, since f is an increasing function on the interval $[0, \infty)$.

Pascal Program 2: Lower Riemann Sums for $f(x) = 3x^2 + 7$

```
program lower_Riemann_sum;

var
   numintervals, i: integer;
   sum, x, deltax, a, b: real;

function f(x: real): real;
begin
   f:= 3*x*x + 7;   (* the function *)
end;

begin
   write('Enter the left endpoint of the interval: ');
   readln(a);
   write('Enter the right endpoint of the interval: ');
   readln(b);
   write('How many subintervals? ');
   readln(numintervals);
   deltax:= (b - a)/numintervals;
   sum:= 0;
   x:= a;
   for i:= 1 to numintervals do
   begin
      sum:= sum + f(x)*deltax;
      x:= x + deltax;
   end;
   writeln('The left-hand endpoint Riemann sum is ',
           sum:8:4);
   write('Press <enter> to finish ...');
   readln;
end.
```

Comment: This program computes a lower Riemann sum for the function $f(x) = 3x^2 + 7$ on the interval $[a, b]$, if $a > 0$, using n equal subintervals. It computes the function value at the left endpoints of the resulting subintervals, since f is an increasing function on the interval $[0, \infty)$. The function is specified in the routine **f(x)**.

BASIC Program 3: Upper Riemann Sums for $f(x) = 3x^2 + 7$

```
10 DEF FNF(T) = 3*(T^2) + 7
20 PRINT "enter interval endpoints a,b"
30 INPUT A,B
40 PRINT "how many subintervals?"
```

```
50 INPUT N
60 LET D = (B - A)/N
70 LET S = 0
80 FOR I = 1 TO N
90   LET X = A + I*D
100   LET S = S + FNF(X)*D
110 NEXT I
120 PRINT S
130 END
```

Comment: This program computes an upper Riemann sum for the function $f(x) = 3x^2 + 7$ on the interval $[a, b]$, if $a > 0$, using n equal subintervals. The numbers a, b, and n are supplied by the user. The program computes the function values at right endpoints of the resulting subintervals, since f is an increasing function on the interval $[0, \infty)$.

Pascal Program 3: Upper Riemann Sums for $f(x) = 3x^2 + 7$

```
program upper_Riemann_sum;

var
   numintervals, i: integer;
   sum, x, deltax, a, b: real;

function f(x: real): real;
begin
  f:= 3*x*x + 7;   (* the function *)
end;

begin
  write('Enter the left endpoint of the interval: ');
  readln(a);
  write('Enter the right endpoint of the interval: ');
  readln(b);
  write('How many subintervals? ');
  readln(numintervals);
  deltax:= (b - a)/numintervals;
  sum:= 0;
  x:= a + deltax;
  for i:= 1 to numintervals do
  begin
    sum:= sum + f(x)*deltax;
    x:= x + deltax;
  end;
  writeln('The right-hand endpoint Riemann sum is ',
          sum:8:4);
  write('Press <enter> to finish ...');
  readln;
end.
```

Comment: This program computes an upper Riemann sum for the function $f(x) = 3x^2 + 7$ on the interval $[a, b]$, *if $a > 0$*, using n equal subintervals. It computes the function value at the right endpoints of the resulting subintervals, since f is an increasing function on the interval $[0, \infty)$. The function is specified in the routine **f(x)**.

BASIC Program 4: Trapezoidal Rule Applied to $f(x) = \dfrac{1}{x}$

```
 10 DEF FNF(T) = 1/T
 20 PRINT "enter interval endpoints a,b"
 30 INPUT A,B
 40 PRINT "how many subintervals?"
 50 INPUT N
 60 LET D = (B - A)/N
 70 LET S = FNF(A)
 80 FOR I = 1 TO (N - 1)
 90   LET X = A + I*D
100   LET S = S + 2*FNF(X)
110 NEXT I
120 LET S = S + FNF(B)
130 LET S = S*(D/2)
140 PRINT S
150 END
```

Comment: This program implements the Trapezoidal Rule for approximate integration of the function $f(x) = 1/x$ on the interval $[a, b]$ with $a > 0$. The numbers a, b, and n (the number of subintervals) are supplied by the user.

Pascal Program 4: Trapezoidal Rule Applied to $f(x) = \dfrac{1}{x}$

```
program Trapezoidal_Rule;
var
   numintervals, i: integer;
   trapapprox, x, deltax, a, b: real;
function f(x: real): real;
begin
   f:= 1/x;   (* the function *)
end;

begin
   write('Enter the left endpoint of the interval: ');
   readln(a);
   write('Enter the right endpoint of the interval: ');
   readln(b);
   write('How many subintervals? ');
   readln(numintervals);
```

```
            deltax:= (b - a)/numintervals;
            trapapprox:= f(a) + f(b);
            x:= a;
            for i:= 1 to (numintervals - 1) do
            begin
              x:= x + deltax;
              trapapprox:= trapapprox + 2*f(x);
            end;
            trapapprox:= trapapprox * (deltax/2);
            writeln('The Trapezoidal Approximation is ',
                    trapapprox:8:4);
            write('Press <enter> to finish ...');
            readln;
          end.
```

Comment: This program implements the Trapezoidal Rule for approximate integration of the function $f(x) = 1/x$ on the interval $[a, b]$ with $a > 0$. The numbers a, b, and n (the number of subintervals) are supplied by the user. The function is specified in the routine **f(x)**.

BASIC Program 5: Simpson's Rule Applied to the Function $f(x) = \dfrac{1}{x}$

```
 10 DEF FNF(T) = 1/T
 20 PRINT "enter interval endpoints a,b"
 30 INPUT A,B
 40 PRINT "how many subintervals (an even number)?"
 50 INPUT N
 60 LET C = (B - A)/N
 70 LET D = C/3
 80 LET S = FNF(A)
 90 FOR I = 1 TO (N - 1) STEP 2
100   LET S = S + 4*FNF(A + I*C)
110 NEXT I
120 FOR I = 2 TO (N - 2) STEP 2
130   LET S = S + 2*FNF(A + I*C)
140 NEXT I
150 LET S = S + FNF(B)
160 LET S = S*D
170 PRINT S
180 END
```

Comment: This program implements Simpson's Rule for approximate integration of the function $f(x) = 1/x$ on the interval $[a, b]$ with $a > 0$. The numbers a, b, and n (the number of subintervals, *an even number*) are supplied by the user.

Pascal Program 5: Simpson's Rule Applied to $f(x) = \dfrac{1}{x}$

```pascal
program Simpsons_Rule;

var
  numintervals, i: integer;
  simpsonapprox, x, deltax, a, b: real;

function f(x: real): real;
begin
  f:= 1/x;  (* the function *)
end;

begin
  write('Enter the left endpoint of the interval: ');
  readln(a);
  write('Enter the right endpoint of the interval: ');
  readln(b);
  write('How many subintervals (an even number)? ');
  readln(numintervals);
  deltax:= (b - a)/numintervals;
  simpsonapprox:= f(a) + f(b);
  x:= a;
  for i:= 1 to ((numintervals div 2) - 1) do
  begin
    x:= x + 2*deltax;
    simpsonapprox:= simpsonapprox + 2*f(x);
  end;
  x:= a + deltax;
  for i:=1 to (numintervals div 2) do
  begin
    simpsonapprox:= simpsonapprox + 4*f(x);
    x:= x + 2*deltax;
  end;
  simpsonapprox:= simpsonapprox * (delta/3);
  writeln('Simpson''s Approximation is ',
          simpsonapprox:8:4);
  write('Press <enter> to finish ...');
  readln;
end.
```

Comment: This program implements Simpson's Rule for approximate integration of the function $f(x) = 1/x$ on the interval $[a, b]$ with $a > 0$. The numbers a, b, and n (the number of subintervals, *an even number*) are supplied by the user. The function is specified in the routine **f(x)**.

BASIC Program 6: Partial Sums of the Geometric Series

```
10 PRINT "enter p,a,x,n"
20 INPUT P,A,X,N
30 LET S = 0
40 FOR K = P TO N
50   LET T = A*X^K
60   LET S = S + T
70 NEXT K
80 PRINT S
90 END
```

Comment: This program computes the partial sum

$$\sum_{k=p}^{n} ax^k$$

where the constants p, a, x, and n are supplied by the user.

Pascal Program 6: Partial Sums of a Geometric Series

```
program partial_sums;

var
   lowindex, highindex, i: integer;
   sum, term, ratio, a: real;

begin
   write('Enter the beginning index: ');
   readln(lowindex);
   write('Enter the ending index: ');
   readln(highindex);
   write('Enter the constant factor a: ');
   readln(a);
   write('Enter the ratio x: ');
   readln(ratio);
   term:= a;
   if lowindex > 0 then
      for i:=1 to lowindex do term:= term * ratio;
   sum:= term;
   for i:= (lowindex + 1) to highindex do
     begin
     term:= term * ratio;
     sum:= sum + term;
     end;

   writeln('The partial sum is ', sum:8:4);
   write('Press <enter> to finish ...');
   readln;
end.
```

Comment: This program computes the partial sum

$$\sum_{k=p}^{n} ax^k$$

where the constants p, a, x, and n are supplied by the user.

BASIC Program 7: A Riemann Sum in Polar Coordinates for $f(\theta) = 1 + \sqrt{\sin \theta}$

```
10 DEF FNF(T) = 1 + SQR(SIN(T))
20 DEF FNG(T) = 0.5 * (T^2)
30 PRINT "enter interval endpoints a,b"
40 INPUT A,B
50 PRINT "how many subintervals?"
60 INPUT N
70 LET D = (B - A)/N
80 LET S = 0
90 FOR I = 1 TO N
100   LET Y = FNF(A + I*D)
110   LET S = S + FNG(Y)*D
120 NEXT I
130 PRINT S
140 END
```

Comment: This program computes a Riemann sum, using n subintervals of equal length, for the integral

$$\int_{a}^{b} \tfrac{1}{2}(1 + \sqrt{\sin \theta})^2 \, d\theta$$

which approximates the area bounded by the curve $r = f(\theta) = 1 + \sqrt{\sin \theta}$. The parameters a, b, and n are supplied by the user.

Pascal Program 7: A Riemann Sum in Polar Coordinates for $f(\theta) = 1 + \sqrt{\sin \theta}$

```
program polar_area;

var
   numintervals, i: integer;
   sum, theta, deltatheta, a, b: real;

function f(theta: real): real;
begin
   f:= 1 + sqrt(sin(theta));   (* the function
                                   f(theta) *)
end;

begin
   write('Enter left endpoint of theta interval: ');
   readln(a);
```

```
                         write('Enter right endpoint of theta interval: ');
                         readln(b);
                         write('How many subintervals? ');
                         readln(numintervals);
                         deltatheta:= (b - a)/numintervals;
                         sum:= 0;
                         theta:= a + deltatheta;
                         for i:= 1 to numintervals do
                         begin
                            sum:= sum + 0.5*f(theta)*f(theta)*deltatheta;
                            theta:= theta + deltatheta;
                         end;
                         writeln('The Riemann sum is ', sum:8:4);
                         write('Press <enter> to finish ...');
                         readln;
                      end.
```

Comment: This program computes a Riemann sum, using n subintervals of equal length, for the integral

$$\int_a^b \tfrac{1}{2}(1 + \sqrt{\sin\theta})^2 \, d\theta$$

which approximates the area bounded by the curve $r = f(\theta) = 1 + \sqrt{\sin\theta}$. The parameters a, b, and n are supplied by the user. The function $f(\theta)$ is specified in the routine **f(theta)**.

BASIC Program 8: A Riemann Sum for a Double Integral

```
 10 PRINT "enter a,b,c,d"
 20 INPUT A,B,C,D
 30 PRINT "how many subintervals?"
 40 INPUT N
 50 LET D1 = (B - A)/N
 60 LET D2 = (D - C)/N
 70 Let D3 = D1*D2
 80 LET S = 0
 90 FOR J = 1 TO N
100   FOR K = 1 TO N
110     LET S = S + (8 - 2*(A + J*D1) - 4*(C + K*D2))
120   NEXT K
130 NEXT J
140 LET S = S*D3
150 PRINT S
160 END
```

Comment: This program computes a Riemann sum for the double integral

$$\int_c^d \int_a^b (8 - 2x - 4y) \; dx \; dy$$

using a grid consisting of n^2 rectangles, each with area $(b - a)(d - c)/n^2$. The parameters a, b, c, d, and n are supplied by the user.

Pascal Program 8: A Riemann Sum for a Double Integral

```
program double_integral_approx;

var
  numintervals, i, j: integer;
  sum, a, b, c, d: real;
  x, y, deltax, deltay, deltaxy: real;

function f(x,y: real): real;
begin
  f:= 8 - 2*x - 4*y;   (* the function f(x,y) *)
end;

begin
  write('Enter a (the minimum x): ');
  readln(a);
  write('Enter b (the maximum x): ');
  readln(b);
  write('Enter c (the minimum y): ');
  readln(c);
  write('Enter d (the maximum y): ');
  readln(d);
  write('How many subintervals? ');
  readln(numintervals);
  deltax:= (b - a)/numintervals;
  deltay:= (d - c)/numintervals;
  deltaxy:= deltax * deltay;
  sum:= 0;
  x:= a;
  for i:= 1 to numintervals do
    begin
    x:= x + deltax;
    y:= c;
    for j:= 1 to numintervals do
      begin
      y:= y + deltay;
      sum:= sum + f(x,y)*deltaxy;
      end;
    end;
  writeln('The Riemann sum is ', sum:8:4);
  write('Press <enter> to finish ...');
  readln;
end.
```

Comment: This program computes a Riemann sum for the double integral

$$\int_{c}^{d}\int_{a}^{b}(8 - 2x - 4y)\,dx\,dy$$

using a grid consisting of n^2 rectangles, each with area $(b - a)(d - c)/n^2$. The parameters a, b, c, d, and n are supplied by the user. The function $f(x, y) = 8 - 2x - 4y$ is specified in the routine **f(x,y)**.

BASIC Program 9: Euler's Method Applied to the Initial Value Problem

$$\frac{dy}{dt} = e^t - 2y$$
$$y(0) = 1$$

```
10 LET Y = 1
20 LET T = 0
30 FOR I = 1 TO 4
40 T = T + 0.25
50 Y = Y + (EXP(T) - 2*Y)*(0.25)
60 PRINT T,Y
70 NEXT I
80 END
```

Comment: This program calculates an approximation to the given initial value problem on the interval [0, 1] using $n = 4$ steps. The equation for Euler's method is implemented in line 50. Note that the right side of this equation involves current values of both T and Y.

Pascal Program 9: Euler's Method Applied to the Initial Value Problem

$$\frac{dy}{dt} = e^t - 2y$$
$$y(0) = 1$$

```
program Eulers_Method;

var
   numsteps, i: integer;
   y, t, a, b, deltat: real;

function f(y,t: real): real;
begin
   f:= exp(t) - 2*y;   (* the function f(y,t) *)
end;

begin
   write('Enter the initial time a: ');
   readln(a);
   write('Enter the final time b: ');
   readln(b);
   write('Enter the initial value y0: ');
   readln(y);
```

```
write('Enter the number of steps: ');
readln(numsteps);
deltat:= (b - a)/numsteps;
t:= a;
for i:= 1 to numsteps do
  begin
  y:= f(y,t) * deltat;
  t:= t + deltat;
  writeln('t= ', t:8:4, '  y = ', y:8:4);
  end;
write('Press <enter> to finish ...');
readln;
end.
```

Comment: This program calculates an approximation to the given initial value problem on the interval [*a*, *b*]. The numbers *a*, *b*, y_0, and *n* (the number of steps) are supplied by the user. The function *f* on the right-hand side of the differential equation is given in the routine **f(y,t)**.

I.2   Calculus and *Mathematica*

The second half of this appendix is devoted to the use of *Mathematica* as an aid to the study of calculus. *Mathematica* is a computer system designed to perform high-level mathematical computations, both numerically and symbolically. It also has extensive graphics capabilities.

We provide a number of commands and programs written in *Mathematica* that can be used to enrich traditional exercise sets. Our goal is to show how a mathematics-based computer language can be an effective tool in the study of calculus. We focus on only a few of the many ways that *Mathematica* can be used in the classroom. However, at the end of this appendix, we provide a bibliography for those who are interested in reading more about the use of mathematics-based computer languages as an aid to the study of calculus. Although the commands and routines given here are written in *Mathematica,* most of the examples we discuss can be treated with other mathematics-based computer systems (for example, *Derive, Maple,* and *Macsyma*).

The examples given here were tested with version 2.0 of *Mathematica.* However, with minor modifications, all commands and routines will work with earlier versions of *Mathematica.*

Getting Started

If you are just getting started with *Mathematica,* we recommend that you begin by trying out some of the commands that are described in the documentation that comes with the program. For example, the *Mathematica* distribution includes files grouped under the category of **Kernel Help.** These files contain many elementary examples. We recommend that you begin with the file called ''Getting Started'' and then try examples from other **Kernel Help** files.

Another source of introductory material is Stephen Wolfram's reference book [W]. It begins with a tour of *Mathematica.* You should enter some of those examples to get a feeling for interacting with *Mathematica.* Finally, you can use *Mathematica*'s **Plot** command to solve many of the graphing calculator problems in Chapter 1 of this text. Here are some examples of how you can use **Plot** to graph functions (in this appendix, we often suppress the output of a command in order to save space):

1. Graph $\sin x$ over the interval $[0, 2\pi]$:

   ```
   Plot[Sin[x], {x, 0, 2 Pi}]
   ```

2. Graph of $\sin x$ with same scale on both axes:

   ```
   Plot[Sin[x], {x, 0, 2 Pi}, AspectRatio -> Automatic]
   ```

3. Graph of both $\sin x$ and $\cos x$ on one plot:

   ```
   Plot[{Sin[x], Cos[x]}, {x, 0, 2 Pi}]
   ```

When you work with the program, it is important to keep the following *Mathematica* conventions in mind:

1. Arithmetic operations are specified using the traditional symbols:
   + is used for addition;
   − is used for subtraction;
   * is used for multiplication;
   / is used for division; and
   ^ is used for exponentiation.
   Moreover, the product of **x** and **y** can also be written **x y** (with a blank space in place of *****). Since variable names can be more than one character long, **xy** is considered to be a new variable rather than the product **x*y**. However, variable names must start with a letter, and therefore, **2x** is interpreted as the product **2*x** even though **x2** is the name of a variable.
2. The names of built-in constants and functions always begin with capital letters, and function arguments are always enclosed in square braces. For example, π is **Pi,** and e is **E.** The sine function $\sin x$ is written **Sin[x].**
3. We follow the convention of writing *Mathematica* input in boldface type.

Tangent Line Problem

In Section 2.1, we studied the tangent line problem by considering the tangent line as a limit of secant lines. This limiting process can be visualized using *Mathematica*'s ability to display animations. Using the following four commands, we can illustrate this limiting process for the tangent line to $y = \sin x$ at $(\pi/2, 1)$. We discuss some of the notation after we finish with the animation.

1. Define the function f:

   ```
   f[x_] := Sin[x]
   ```

2. Set the number a that corresponds to the desired tangent:

   ```
   a = Pi/2;
   ```

3. Define the slope of the secant line that corresponds to a given h:

   ```
   m[h_,a_] := (f[a + h] - f[a])/h
   ```

4. Display 19 different plots ($h = 1, 0.95, 0.90, 0.85, \ldots, 0.05$) of the graph of f along with the graph of the secant line that corresponds to the given number h:

   ```
   Do[Plot[{f[x], m[h, a] (x - a) + f[a]},
       {x, 0, Pi},
       (* use the next line with a color monitor *)
   ```

```
    PlotStyle -> {RGBColor[0,0,1],RGBColor[0,1,0]},
    PlotRange -> {0, 1.1}
    ],
{h, 1, 0.05, -0.05}];
```

This Do command takes a few minutes to complete. After it is done, you can animate these plots. On the Macintosh, this is done by scrolling so that the first graphics cell is completely visible. Then double-click on first picture to see the animation. Controls for the animation appear in the bottom left-hand corner of the notebook.

When we defined the function *f* in command 1, we used two special *Mathematica* symbols that warrant explanation. First, we used **x_** on the left-hand side of the equals sign and **x** on the right-hand side. The underbar indicates that **x** is a variable in the definition of the function. We use it only on the left-hand side of the definition. Also, we used **: =** as the equals sign when we defined the functions, and we used **=** when we defined the number *a*. In general, we use **: =** whenever we are defining a function, and we use **=** whenever we are specifying a constant in a calculation. The difference between the standard equals sign **=** and the delayed assignment sign **: =** is somewhat complicated. It is discussed in Part 2 of the *Mathematica* book.

Warning: When using *Mathematica,* remember that, once you have defined a function or a variable, that definition holds throughout that session unless you redefine it or explicitly remove it using the **Remove** command.

Exercises

1. Animate the tangent line problem for:

  a. $f(x) = \cos x$ at $a = 0$
  b. $f(x) = 1/x$ at $a = 1$
  c. $f(x) = \sin x$ at $a = \pi/4$

Limits

If you prefer to examine the limit of the slopes of secant lines using tables of numbers as discussed in Sections 2.1 and 2.2, you can do so using *Mathematica*'s **N** function.

The following command makes a table with two columns, one for **h** and one for the slope **m[h,a]** of the corresponding secant line. We are assuming that the function **f**, the constant **a**, and the slope function **m[h,a]** have already been defined.

```
TableForm[Table[{h,N[m[h,a],10]},{h,1,0.05,-0.05}]]
```

In general, the **TableForm** and **Table** commands can be used to investigate any limit.

  1. Specify the function under consideration:

```
f[x_]:= (1 - 2x) Tan[Pi x]
```

  2. Set the number at which the limit is being considered:

```
a = 1/2;
```

3. Make the table:

```
TableForm[Table[{x,N[f[x],10]},
                {x,a+0.1,a+0.005,-0.005}]]
```

The table that we obtain in this case corresponds to a choice of numbers x that approach the number a from above. Consult the documentation of the **Table** command to see how this command can be modified so that the numbers x approach a from below.

Limits can also be studied using the **Plot** command (don't worry about the error messages):

```
Plot[Sin[x]/x, {x, -15, 15},
                PlotRange -> {-1.5,1.5}]
```

Relevant Exercises

Section 2.2: Exercises 47–52 and 57–64.

Section 2.4: Exercise 53.

Section 2.5: Exercises 37–39.

The Derivative Function

We can visualize the graph of the derivative function using animation in much the same way as we visualized the tangent line as a limit of secant lines.

1. Define the function f:

```
f[x_]:= x^2
```

2. Define the slope of the secant line that corresponds to a given h at a given number x:

```
m[h_,x_]:= (f[x + h] - f[x])/h
```

3. Plot m using a small h (for example, $h = 0.05$):

```
Plot[m[0.05,x], {x, 0, 2}]
```

Can you guess the derivative of f from this graph? (In Section 3.1, we calculated that the derivative of f, so we already know the answer in this case.)

4. Display 10 different plots ($h = 0.5, 0.45, \ldots, 0.05$) of the graph of m along with the graph of our guess for the derivative:

```
Do[Plot[{m[h,x],2x},
        {x, 0, 2},
        (* use the next line with a color monitor *)
        PlotStyle -> {RGBColor[1,0,0],RGBColor[0,0,1]},
        PlotRange -> {0,4}
        ],
    {h,0.5,0.05,-0.05}];
```

Relevant Exercises

Section 3.1: Exercises 43–46 and 51–61.

Newton's Method

We present two ways to implement Newton's Method. One simply computes the iterations. The other illustrates the method graphically. Note that both functions take advantage of *Mathematica*'s ability to calculate exact derivatives.

1. In the following routine, we provide the function whose zeros we desire, the variable, the initial value of that variable, and the maximum number of iterates desired:

```
NewtonIterates[exp_, {x_, x0_}, maxits_]:=
  Module[{f,df,iterates, values},
    f = Function[x, exp];
    df = f';
    iterates =
      FixedPointList[(# - f[#]/df[#])&,
                       N[x0], maxits];
    values = Map[Chop[f[#],10^(-6)]&,iterates];
    PaddedForm[TableForm[
        Transpose[{iterates,values}]], {8,5}]
  ]
```

For example, we estimate $\sqrt[3]{10}$ using an initial guess of 4. The first column contains the iterates, and the second column contains the values of the function corresponding to those iterates.

```
NewtonIterates[x^3 - 10, {x, 4}, 5]
```

4.00000	54.00000
2.87500	13.76367
2.31994	2.48625
2.16596	0.16137
2.15450	0.00085
2.15443	0.00000

2. The following routine uses *Mathematica*'s ability to display graphics to illustrate the convergence of Newton's method:

```
NewtonGraphics[exp_, {x_, a_, b_}, x0_, maxits_]:=
  Module[{graph, iterates, t, f, df, pts1, pts2,
          lines},
          f = Function[x, exp];
          df = f';
          graph = Plot[f[t], {t, a, b},
                        DisplayFunction -> Identity,
                        PlotStyle -> RGBColor[0,0,1]];
          iterates = FixedPointList[
                  (# - f[#]/df[#])&, N[x0], maxits];
          pts1 = Map[{#,f[#]}&, Drop[iterates,-1]];
          pts2 = Map[{# ,0}&, Drop[iterates, 1]];
          lines = Map[Line,Transpose[{pts1,pts2}]];
          lines = Prepend[lines, RGBColor[1,0,0]];
```

```
points = Map[Point,Join[pts1,pts2]];
points = Prepend[points, PointSize[0.015]];
points = Prepend[points, RGBColor[0,0,0]];
Show[graph, Graphics[Join[lines, points]],
     DisplayFunction -> $DisplayFunction];
]
```

Again we calculate $\sqrt[3]{10}$ via Newton's Method, but now we see the corresponding tangents:

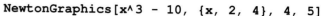

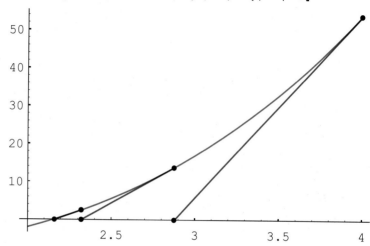

Relevant Exercises

Exercise Set 3.10.

Section 4.3: Exercises 28–30.

Section 4.4: Exercises 92–95.

Slope Portraits and Antiderivatives

Here we show how to construct a slope portrait with graphs of the corresponding antiderivatives superimposed upon it. The steps we use can easily be combined to form a general routine for plotting slope portraits. The following code was used to generate part of Figure 1.2 in Chapter 5. That figure illustrates the slope portrait for the antiderivatives of $f(x) = 2x$.

1. We define the function f:

   ```
   f[x_]:=2 x
   ```

2. We form a grid of points in the plane at which we will display the tangent line segments:

   ```
   grid = Flatten[Table[{x,y}, {x,-1,1,0.5},
                              {y,-1,1.6,0.2}],
                  1];
   ```

3. We specify a small number **xwidth** that determines the sizes of the tangent segments, and we write a simple function that produces a tangent at the point {x,y}:

```
xwidth = 0.1;
Tangent[{x_,y_}]:= Graphics[{RGBColor[1,0,0],
            Line[{{x - xwidth, y - f[x] xwidth},
                  {x + xwidth, y + f[x] xwidth}}]
       }]
```

4. We produce the slope portrait by applying the function **Tangent** to the points of the grid. We display it in Step 6 below.

```
slopeportrait = Map[Tangent, grid];
```

5. We produce the graphs of three antiderivatives of f, but at this point, they are not displayed. We will display these graphs in Step 6.

```
graph1 = Plot[x^2 + 0.3, {x, -1.2,1.2},
            PlotStyle -> RGBColor[0,0,1],
            DisplayFunction -> Identity];

graph2 = Plot[x^2 - 0.3, {x, -1.2,1.2},
            PlotStyle -> RGBColor[0,0,1],
            DisplayFunction -> Identity];

graph3 = Plot[x^2 - 0.8, {x, -1.2,1.2},
            PlotStyle -> RGBColor[0,0,1],
            DisplayFunction -> Identity];
```

6. We combine all of the graphics we have produced and we display the result:

```
Show[slopeportrait,graph1,graph2,graph3,
            AspectRatio -> Automatic,
            Ticks -> {{-1,1},{-1,1}},
            DisplayFunction -> $DisplayFunction];
```

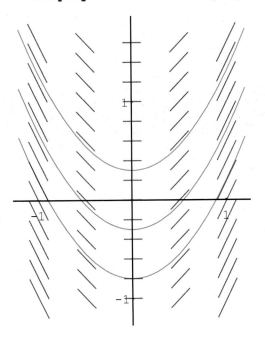

Tangent Line Approximation of Antiderivatives

We can approximate the graph of an antiderivative of a given function f using line segments that are tangent to the graph. Once again, the steps we follow can be combined to form a general routine for graphing antiderivatives.

1. Specify the function, the interval under consideration, and the step size (determined by the number of steps):

```
f[x_]:= 2x

a = -1; b = 1; n = 50; deltax = (b-a)/n;
```

2. Generate a list of points on the graph of the approximation to the antiderivative of f. Start the list at the point (a, C) where C is the arbitrary constant of integration. In this example, we pick $C = 0$.

```
points = {{a,0}};

For[x = a + deltax, x <= b, x += deltax,
    points = Append[points,
        {x,Last[points][[2]] + f[x] deltax}]
    ];
```

3. Plot the graph of the approximation:

```
ListPlot[points, PlotJoined -> True,
                PlotStyle -> RGBColor[0,0,1]];
```

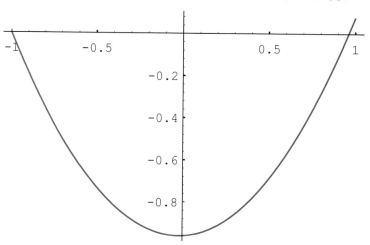

Relevant Exercises

Section 5.1: Exercises 58–63.

Euler's Method

Euler's Method for approximating the solutions of a differential equation can be implemented in much the same way as the method for plotting graphs of antiderivatives.

1. Specify the differential equation, the interval under consideration, and step size (determined by the number of steps):

```
f[x_,y_]:= x(y+1)

a = 0; b = 1; n = 10; deltax = (b-a)/n;
```

2. Generate a list of points on the graph of the approximation to the solution. Start the list at the point (a, C) where C is the initial value. In this example, we are given $C = 1$.

```
points = {{a,1}};

For[x = a + deltax, x <= b, x += deltax,
   points = Append[points,
                     {x,Last[points][[2]] +
                       f[x,Last[points][[2]]] deltax}]
   ];
```

3. Plot the graph of the approximate solution:

```
ListPlot[points, PlotJoined -> True,
                 PlotStyle -> RGBColor[0,0,1]];
```

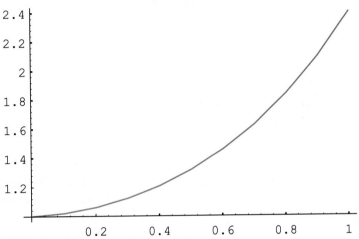

Relevant Exercises

Section 5.3: Exercises 34–39.

Exercise Set 21.6.

Riemann Sums

The following procedure calculates the Riemann Sum of a function f given a partition of the interval and a selection of numbers, one from each subinterval. It also illustrates this sum using rectangles.

```
RiemannSumGraphics[exp_,{x_,partition_},tnumbers_]:=
   Module[{f,rsum,i,j,rectangles,lines,tdots,value,graph},
           f = Function[x, exp];

           (* compute Riemann sum *)
           rsum=0;
           For[i=2,i<=Length[partition],i++,
            value[i]= f[tnumbers[[i-1]]];
            rsum=rsum +
                  value[i] (partition[[i]]-partition[[i-1]]);
           ];
```

```
                                    Print["Riemann Sum = ",Simplify[rsum]];
                                    Print["which is approximately ",N[rsum]];
                                    (* produce graphics *)
                                    rectangles=Table[
                                            Rectangle[{partition[[j-1]],0},
                                                    {partition[[j]],value[j]}],
                                                    {j,2,Length[partition]}
                                                    ];
                                    lines=Table[
                                                Line[{{partition[[j-1]],0},
                                                        {partition[[j-1]],value[j]},
                                                        {partition[[j]],value[j]},
                                                        {partition[[j]],0}}],
                                                    {j,2,Length[partition]}
                                                    ];
                                    tdots=Map[Point,
                                            Join[
                                                Table[{tnumbers[[j]],0},
                                                        {j,1,Length[tnumbers]}],
                                                Table[{tnumbers[[j]],value[j+1]},
                                                        {j,1,Length[tnumbers]}]
                                                ]
                                            ];
                                    tdots=Prepend[tdots,PointSize[0.01]];
                                    rectangles=Prepend[rectangles,
                                                    RGBColor[0,0.7,1.0]];
                                    lines=Prepend[lines,GrayLevel[0]];
                                    graph=Plot[f[x],
                                            {x,First[partition],Last[partition]},
                                            DisplayFunction->Identity];
                                    Show[
                                        Graphics[Join[rectangles,lines,tdots]],
                                        graph,
                                        DisplayFunction->$DisplayFunction
                                        ];
                                    ]
```

To implement the Midpoint Rule along with illustrative graphics, we use the function **RiemannSumGraphics** along with two additional functions—one that produces an equal length partition of the interval and one that produces the midpoints of the subintervals.

```
EqualLengthPartition[a_,_,b_,n_]:=
        Table[a + j (b-a)/n ,{j, 0, n}]

Midpoints[a_,b_,n_]:=
        Table[a + (j - 1/2) (b-a)/n,{j,1,n}]
```

We combine **RiemannSumGraphics, EqualLengthPartition,** and **Midpoints** to produce the procedure **MidpointApproxGraphics**

```
MidpointApproxGraphics[f_, {x,_,a_,b_}, n_]:=
        RiemannSumGraphics[f,
                    {x,EqualLengthPartition[a,b,n]},
                    Midpoints[a,b,n]]
```

Here is an example of **MidpointApproxGraphics** applied to $f(x) = 1/x$ on the interval $[1, 4]$ with 6 subdivisions.

```
MidpointApproxGraphics[1/x, {x, 1, 4}, 6]
```

Riemann Sum = $\dfrac{62024}{45045}$

which is approximately 1.37693

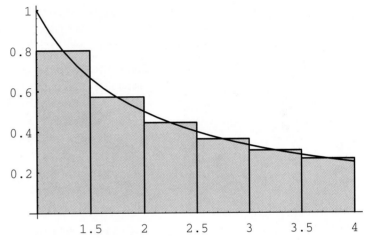

Of course, if we only want an approximate value of the Riemann sum, then we can replace **RiemannSumGraphics** by a much simpler routine:

```
RiemannSum[exp_,{x_,partition_},tnumbers_]:=
Module[{f,rsum,i,j,rectangles,lines,tdots,value,graph},
        f = Function[x, exp];

        rsum=0;
        For[i=2,i<=Length[partition],i++,
            value[i]= N[f[tnumbers[[i-1]]]];
            rsum=rsum +
                value[i] (partition[[i]]-partition[[i-1]]);
        ];
        rsum
        ]
```

Then if we want to calculate the Riemann sum that corresponds to a particular partition or to a particular type of partition, we simply use the approach illustrated in **MidpointApproxGraphics** above. For example, to define a routine that calculates the Riemann sum corresponding to an equal length partition using the left-hand endpoints, we define:

```
LeftRiemannSum[f_, {x_,a_,b_}, n_]:=
        RiemannSum[f,
                    {x,EqualLengthPartition[a,b,n]},
                    Drop[EqualLengthPartition[a,b,n],-1]]
```

Relevant Exercises

Section 6.1: Exercises 67–70.

Section 6.2: Exercises 87–90.

Numerical Approximation of the Definite Integral

In Chapter 6, we discussed three numerical methods that approximate the integral—the Midpoint Rule, the Trapezoidal Rule, and Simpson's Rule. Here we implement the Midpoint Rule and the Trapezoidal Rule.

```
MidpointRule[exp_, {x_, a_, b_}, n_] :=
   Module[{f, midpts, values, deltax, xi},
        f = Function[x, exp];
        deltax = (b - a)/n;
        midpts = Table[N[xi],
                    {xi,a + deltax/2,b,deltax}];
        values = Map[f,midpts];
        Apply[Plus, values] deltax
   ]

TrapezoidalRule[exp_, {x_, a_, b_}, n_] :=
   Module[{f, partition, values, deltax, xi},
        f = Function[x, exp];
        deltax = (b - a)/n;
        partition = Table[N[xi], {xi, a, b, deltax}];
        values = Map[f,partition];
        (Apply[Plus, values] -
            (values[[1]] + values[[n+1]])/2) deltax
   ]
```

These routines can be used to reproduce Table 6.1 in Section 6.6.

```
TableForm[Table[{b,
                MidpointRule[1/x,{x,1,b},10],
                TrapezoidalRule[1/x,{x,1,b},10]},
                {b, 2, 10}
            ]]
```

2	0.692835	0.693771
3	1.09714	1.10156
4	1.38284	1.39326
5	1.60321	1.62204
6	1.78204	1.81154

7	1.93203	1.97436
8	2.06078	2.11796
9	2.17322	2.24716
10	2.27274	2.36521

Relevant Exercises

Section 6.2: Exercises 70–81.

Section 6.5: Exercises 44–45.

Exercise Set 6.6.

Fundamental Theorem of Calculus

The Fundamental Theorem of Calculus is based on the concept of a function that is defined by varying the upper limit of integration. That is,

$$A(x) = \int_a^x f(t)\, dt.$$

We can either use one of the numerical approximation rules mentioned above or *Mathematica*'s built-in numerical approximation routine to graph A given f. Here we illustrate the idea using the built-in routine. We graph A over the interval [0, 3] for $f(x) = \cos x^2$.

```
f[x_]:= Cos[x^2]; a = 0; b = 3;

(* this command will take a few minutes to execute *)
(* we could use the routine MidpointRule in place
    of NIntegrate *)
Plot[NIntegrate[f[t], {t, a, x}],
    {x, a, b},
    PlotPoints -> 10,
    PlotStyle -> RGBColor[0,0,1]];
```

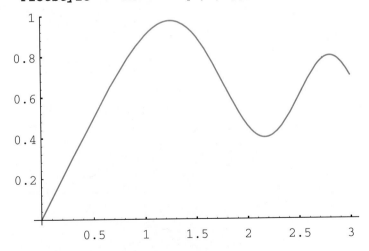

It is interesting to compare this graph with the graph of f:

```
Plot[f[x], {x, a, b},
    PlotStyle -> RGBColor[0,0,1]];
```

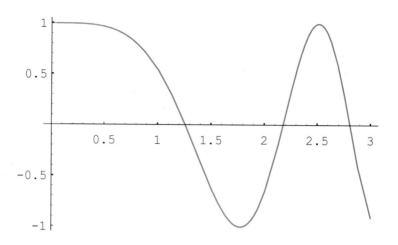

Relevant Exercises

Section 6.2: Exercises 80–81.

Section 6.3: Exercise 58.

The Graph of the Natural Logarithm

In Section 8.2, the natural logarithm function is defined by

$$\ln x = \int_1^x \frac{1}{t}\, dt.$$

Using the methods described above, we can sketch an approximation of its graph.

1. We define an approximation to $\ln x$ with

```
ln[x_]:= MidpointRule[1/t, {t, 1, x}, 50] /; x > 1
ln[x_]:= -MidpointRule[1/t, {t, x, 1}, 50] /;
                                    0 < x && x < 1
ln[1] = 0;
```

2. We graph this approximation to the natural logarithm function:

```
Plot[ln[x], {x, 0.1, 10},
    PlotStyle -> RGBColor[0,0,1]];
```

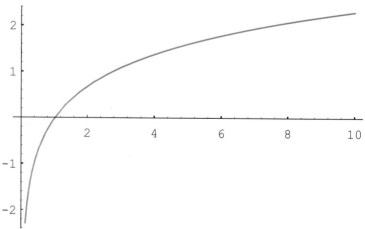

Methods of Integration

In *Mathematica,* you can define rules for mathematical operations in very general ways. Therefore, you can define your own integration function and add to it throughout your study of methods of integration. The basic idea is discussed in Section 2.3.13 of the *Mathematica* reference book [W] by Wolfram. For example, start by calling your integration function **integrate** (as opposed to *Mathematica*'s built-in integration function **Integrate**).

1. The basic integral formula

$$\int x^n \, dx = \frac{x^{n+1}}{n+1} + C, \quad n \neq -1,$$

can be defined in *Mathematica* as

```
integrate[x_^n_., x_] := x^(n+1)/(n+1) /;
                         FreeQ[n,x] && n != -1
```

2. The fact that the integral of a sum is the sum of the integrals is written as

```
integrate[y_ + z_, x_] :=
                    integrate[y,x] + integrate[z,x]
```

3. Now our integration function knows how to calculate the integral of a simple sum:

```
integrate[t^2 + t^5, t]
```
$$\frac{t^3}{3} + \frac{t^6}{6}$$

See Wolfram's book [W] for more details. Allan Hayes [H] has written a very nice *Mathematica* notebook that treats methods of integration from this point of view.

Taylor Polynomials

It is easy to study Taylor approximation using *Mathematica.*

1. We define a function that gives us the *n*th degree Taylor polynomial of $\sin x$ about the number *a*:

```
a = 0;
```

```
TaylorSin[x_,n_]:=Normal[Series[Sin[x],{x,a,n}]];
```

2. Now we use the **Plot** command to compare the graph of the sine function with the graph of one of its Taylor polynomials:

```
Plot[Evaluate[{Sin[x],TaylorSin[x,5]}],
     {x, -2 Pi, 2 Pi},
       PlotStyle ->
          {RGBColor[0,0,1], (* Sin[x] is blue *)
           RGBColor[1,0,0]} (* Taylor poly is red *)
   ];
```

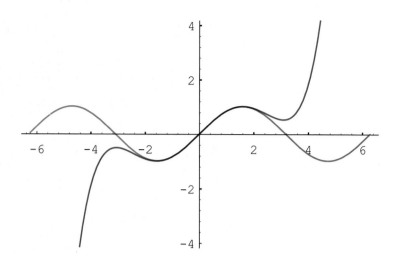

Exercises

1. Find the lowest degree n such that the nth degree Taylor polynomial for $\sin x$ about 0 is a good approximation over the interval $[-2\pi, 2\pi]$. What happens if the graph of this polynomial is compared to the graph of the sine function over the interval $[-3\pi, 3\pi]$? Is it possible to find an n such that the graph of the Taylor polynomial of degree n is a good approximation to the graph of the sine function over the entire real line?

2. Repeat Exercise 1 using the inverse tangent function $\text{Tan}^{-1} x$ in place of the sine function. (Note that *Mathematica*'s notation for the inverse tangent function is **ArcTan[ ]**.)

References

[C] Child, J. Douglas: *A Guide to Calculus T/L.* Brooks/Cole Publishing Company, 1990. *Calculus T/L* is an interesting offshoot of *Maple* intended for use in calculus courses.

[CR] Crooke, Philip, and Ratcliffe, John: *A Guidebook to Calculus with Mathematica.* Wadsworth Publishing Company, 1991. A *Mathematica*-based laboratory manual for calculus.

[EL] Ellis, Wade, and Lodi, Ed: *A Tutorial Introduction to Mathematica.* Brooks/Cole Publishing Company, 1990. An elementary tutorial intended for students who want to learn the basics of *Mathematica.*

[EN] Emert, John, and Nelson, Roger: *Calculus and Mathematica.* Saunders College Publishing, 1992. A *Mathematica*-based laboratory manual for calculus. It is an ancillary to this text.

[GG] Gray, Theodore, and Glynn, Jerry: *Exploring Mathematics with Mathematica.* Addison-Wesley, 1991. Explorations of various mathematical topics using *Mathematica.*

[H] Hayes, Allan: *Integration Techniques. Mathematica* notebook available by ftp—see Preface of this text.

[M] Maeder, Roman: *Programming in Mathematica,* 2e. Addison-Wesley, 1991. An advanced source that concentrates on *Mathematica* programming.

[OD] Olwell, D., and Driscoll, P: *Calculus and Derive.* Saunders College Publishing, 1992. A *Derive*-based laboratory manual for calculus. It is an ancillary to this text.

[S] Solow, Anita, ed.: *Learning by Discovery.* Available from Wayne Roberts, Department of Mathematics, Macalester College, St. Paul, Minnesota. A laboratory manual for calculus developed by the ACM-GLCA Calculus Reform Project.

[SH] Small, Donald B., and Hosack, John M.: *Explorations in Calculus with a Computer Algebra System.* McGraw-Hill, Inc., 1991. A laboratory manual for calculus that uses various computer algebra systems.

[T] *Elementary Tutorial Notes.* Available from Wolfram Research, Inc. Tutorial notes from the 1991 *Mathematica* conference.

[Wa] Wagon, Stan: *Mathematica in Action.* W. H. Freeman and Company, 1991. A discussion of various mathematical topics illustrated with the use of *Mathematica.*

[W] Wolfram, Stephen: *Mathematica: A System for Doing Mathematics by Computer,* 2e. Addison-Wesley, 1991. The definitive reference on *Mathematica.*

Appendix II

Some Additional Proofs

Mathematical Induction

The *principle of mathematical induction* is a property of the real number system (we will think of it as an axiom) that allows us to prove statements about positive integers. Such statements, represented generally as $P(n)$, are usually formulas involving some or all of the integers $1, 2, 3, \ldots, n$. Four particular formulas that we wish to prove are equations (3)–(6) of Section 6.1:

$$\sum_{j=1}^{n} c = c + c + c + \cdots + c = nc \tag{1a}$$

$$\sum_{j=1}^{n} j = 1 + 2 + 3 + \cdots + n = \frac{n(n+1)}{2} \tag{1b}$$

$$\sum_{j=1}^{n} j^2 = 1 + 4 + 9 + \cdots + n^2 = \frac{n(n+1)(2n+1)}{6} \tag{1c}$$

$$\sum_{j=1}^{n} j^3 = 1 + 8 + 27 + \cdots + n^3 = \frac{n^2(n+1)^2}{4} \tag{1d}$$

We use the following principle to prove these formulas.

Principle of Mathematical Induction

If $P(n)$ represents a mathematical statement involving the positive integer n, then $P(n)$ is true for all positive integers n if both of the following conditions hold:

(i) $P(1)$ is true.
(ii) If $P(n)$ is true, $P(n+1)$ is also true.

The principle of mathematical induction (PMI) is not surprising. If a statement $P(1)$ is true, then setting $n = 1$ shows that $P(2) = P(1 + 1)$ is true by condition (ii). Since $P(2)$ is then known to be true, we next set $n = 2$ and conclude, again by (ii), that $P(3) = P(2 + 1)$ is true. Continuing in this way we verify that $P(n)$ is true for each of the positive integers $1, 2, 3, \ldots$.

Proof of Formula (1a):   We prove formula (1a) by applying the PMI to the formula

$$P(n)\colon \sum_{j=1}^{n} c = c + c + c + \cdots + c = nc.$$

Setting $n = 1$ yields the statement

$$P(1): c = 1 \cdot c,$$

which is certainly valid. Thus, condition (i) of the PMI holds. To verify condition (ii) of the PMI, we assume that $P(n)$ holds, and we verify $P(n + 1)$. That is,

$$\sum_{j=1}^{n+1} c = c + \sum_{j=1}^{n} c$$

$$= c + nc = (n + 1)c$$

since $P(n)$ holds. We have verified condition (ii) of the PMI, and therefore, equation (1a) holds for all positive integers. ∎

Proof of Formula (1b): We prove formula (1b) by applying the PMI to the formula

$$P(n): 1 + 2 + 3 + \cdots + n = \frac{n(n + 1)}{2}. \tag{2}$$

First, setting $n = 1$ gives the statement

$$P(1): 1 = \frac{1(1 + 1)}{2},$$

which is true, since $\frac{1(1 + 1)}{2} = \frac{1 \cdot 2}{2} = 1$. Thus, condition (i) of the PMI holds.

To show that condition (ii) of the PMI holds, we assume that $P(n)$ in (2) is true, and we show that

$$P(n + 1): 1 + 2 + 3 + \cdots + (n + 1) = \frac{(n + 1)[(n + 1) + 1]}{2} \tag{3}$$

is true. Using statement (2), we write

$$1 + 2 + 3 + \cdots + n + (n + 1) = [1 + 2 + 3 + \cdots + n] + n + 1$$

$$= \frac{n(n + 1)}{2} + n + 1$$

$$= \frac{n(n + 1) + 2(n + 1)}{2}$$

$$= \frac{(n + 1)(n + 2)}{2}$$

$$= \frac{(n + 1)[(n + 1) + 1]}{2},$$

which establishes (3). Thus, by the PMI, statement (2) holds for all positive integers. This proves formula (1b). ∎

Proof of Formula (1c): As in the preceding proof, we use the PMI. The statement to be proved is

$$P(n): 1^2 + 2^2 + 3^2 + \cdots + n^2 = \frac{n(n + 1)(2n + 1)}{6}. \tag{4}$$

The statement corresponding to $n = 1$ is

$$P(1): 1^2 = \frac{1(1 + 1)(2 \cdot 1 + 1)}{6},$$

which is true, so condition (i) of the PMI holds.

Next, we assume that $P(n)$ in (4) holds, and we prove the statement $P(n + 1)$:

$$1^2 + 2^2 + 3^2 + \cdots + (n + 1)^2 = \frac{(n + 1)[(n + 1) + 1][2(n + 1) + 1]}{6}. \tag{5}$$

We use statement (4):

$$\begin{aligned}
1^2 + 2^2 + 3^2 + \cdots + (n + 1)^2 &= [1^2 + 2^2 + 3^2 + \cdots + n^2] + (n + 1)^2 \\
&= \frac{n(n + 1)(2n + 1)}{6} + (n + 1)^2 \\
&= \frac{n(n + 1)(2n + 1) + 6(n + 1)^2}{6} \\
&= \frac{(n + 1)[n(2n + 1) + 6(n + 1)]}{6} \\
&= \frac{(n + 1)[2n^2 + 7n + 6]}{6} \\
&= \frac{(n + 1)[(n + 2)(2n + 3)]}{6} \\
&= \frac{(n + 1)[(n + 1) + 1][2(n + 1) + 1]}{6}.
\end{aligned}$$

Thus, $P(n + 1)$ in (5) is true if $P(n)$ is true. Thus, condition (ii) of the PMI holds, so formula (1c) is proved by the PMI. ∎

The proof of formula (1d) is similar to the proofs of formulas (1b) and (1c). We leave the details as an exercise.

Chapter 2, Theorem 3
parts (i) and (ii)

Let m and n be positive integers.

(i) If n is even, $\lim\limits_{x \to a} x^{m/n} = a^{m/n}$ whenever $0 < a < \infty$.

(ii) If n is odd, $\lim\limits_{x \to a} x^{m/n} = a^{m/n}$ for all $-\infty < a < \infty$.

Proof: To prove part (i), we need only prove that

$$\lim_{x \to a} \sqrt[n]{x} = \sqrt[n]{a}$$

from which part (i) follows by Theorem 2 of Chapter 2. According to Definition 3, Chapter 2, we must show that if $\epsilon > 0$ is given, there can be found a corresponding number δ so that

$$|\sqrt[n]{x} - \sqrt[n]{a}| < \epsilon \quad \text{if} \quad 0 < |x - a| < \delta. \tag{6}$$

The first inequality holds the key to determining how to choose δ, so we begin by rewriting

$$|\sqrt[n]{x} - \sqrt[n]{a}| < \epsilon \tag{7}$$

as

$$-\epsilon < \sqrt[n]{x} - \sqrt[n]{a} < \epsilon.$$

The strategy now is to isolate the expression $x - a$ in the middle position, which we do as follows: From the above we have

$$\sqrt[n]{a} - \epsilon < \sqrt[n]{x} < \sqrt[n]{a} + \epsilon, \tag{8}$$

so

$$(\sqrt[n]{a} - \epsilon)^n < x < (\sqrt[n]{a} + \epsilon)^n, \tag{9}$$

so

$$(\sqrt[n]{a} - \epsilon)^n - a < x - a < (\sqrt[n]{a} + \epsilon)^n - a.$$

(There is a potential difficulty here. Going from step (8) to step (9) is valid only if $\sqrt[n]{a} - \epsilon$ is nonnegative. We can ensure that this holds by requiring that $\epsilon < \sqrt[n]{a}$ initially. There is no problem in making this assumption since, if statement (6) holds for a smaller value of ϵ than that originally given, it also must hold for the given value.)

Finally, we rewrite (9) as

$$-[a - (\sqrt[n]{a} - \epsilon)^n] < x - a < [(\sqrt[n]{a} + \epsilon)^n - a] \tag{10}$$

in which both expressions inside brackets are positive. Then, subject to the assumption that $\epsilon < \sqrt[n]{a}$, inequality (10) is equivalent to inequality (7). Now suppose we take $\delta > 0$ to be small enough so that both

$$\delta < [a - (\sqrt[n]{a} - \epsilon)^n] \quad \text{and} \quad \delta < [(\sqrt[n]{a} + \epsilon)^n - a]. \tag{11}$$

Then the inequality

$$0 < |x - a| < \delta \tag{12}$$

can be rewritten as

$$-\delta < x - a < \delta. \tag{13}$$

From (11) it follows that if (13) holds, so does (10). Since (13) is equivalent to (12), and (10) is equivalent to (7), this shows that

$$|\sqrt[n]{x} - \sqrt[n]{a}| < \epsilon \quad \text{if} \quad 0 < |x - a| < \delta,$$

which completes the proof of part (i).

To prove (ii), we assume that n is odd. As in part (i) we show only that

$$\lim_{x \to a} \sqrt[n]{x} = \sqrt[n]{a}.$$

If $a > 0$ the proof for part (i) applies. Thus, we must only concern ourselves with the case $a < 0$. But in this case $-a > 0$, so we may apply the proof of part (i) to conclude that

$$\lim_{x \to a} \sqrt[n]{-x} = \sqrt[n]{-a}. \tag{14}$$

Of course, we now wish simply to "factor out" the -1's in (14), but we must do it rigorously. By Definition 3, Chapter 2, equation (14) means that, given $\epsilon > 0$, there exists a $\delta > 0$ so that

$$|\sqrt[n]{-x} - \sqrt[n]{-a}| < \epsilon \quad \text{if} \quad 0 < |-x - (-a)| < \delta. \tag{15}$$

Since, for odd n,

$$|\sqrt[n]{-x} - \sqrt[n]{-a}| = |\sqrt[n]{-1}\sqrt[n]{x} - \sqrt[n]{-1}\sqrt[n]{a}| = |(-1)(\sqrt[n]{x} - \sqrt[n]{a})|$$
$$= |\sqrt[n]{x} - \sqrt[n]{a}|$$

and

$$|-x - (-a)| = |(-1)(x - a)| = |x - a|,$$

inequality (15) is equivalent to the statement

$$|\sqrt[n]{x} - \sqrt[n]{a}| < \epsilon \quad \text{if} \quad 0 < |x - a| < \delta,$$

which shows that $\lim_{x \to a} \sqrt[n]{x} = \sqrt[n]{a}$ in the present case. This completes the proof. ∎

Chapter 2, Theorem 1
parts (iii) and (iv)

If $\lim_{x \to a} f(x) = L$ and $\lim_{x \to a} g(x) = M$, then

(iii) $\lim_{x \to a} f(x)g(x) = LM$, and

(iv) $\lim_{x \to a} \dfrac{f(x)}{g(x)} = \dfrac{L}{M}, \qquad M \neq 0.$

Proof: To prove (iii), we must show that, given $\epsilon > 0$, there exists a $\delta > 0$ so that

$$|f(x)g(x) - LM| < \epsilon \quad \text{if} \quad 0 < |x - a| < \delta.$$

We do this by using the triangle inequality to write

$$|f(x)g(x) - LM| = |f(x)g(x) - f(x)M + f(x)M - LM| \tag{16}$$
$$\leq |f(x)g(x) - f(x)M| + |f(x)M - LM|$$
$$= |f(x)||g(x) - M| + |M||f(x) - L|.$$

We need to make three observations about the terms on the right side of inequality (16).

(a) Since $\lim_{x \to a} f(x) = L$, there exists a number $\delta_1 > 0$ so that

$$|f(x) - L| < 1 \quad \text{if} \quad 0 < |x - a| < \delta_1.$$

Thus,

$$|f(x)| < |L| + 1 \quad \text{if} \quad 0 < |x - a| < \delta_1. \tag{17}$$

(b) Also, since $\lim_{x \to a} f(x) = L$, there exists a number $\delta_2 > 0$ so that

$$|f(x) - L| < \frac{\epsilon}{2(|M| + 1)} \quad \text{if} \quad 0 < |x - a| < \delta_2. \tag{18}$$

(c) Finally, since $\lim_{x \to a} g(x) = M$, there exists a number δ_3 so that

$$|g(x) - M| < \frac{\epsilon}{2(|L| + 1)} \quad \text{if} \quad 0 < |x - a| < \delta_3. \tag{19}$$

Now, we take δ to be the smallest of the numbers δ_1, δ_2, and δ_3. Then each of the conditions in statements (17), (18), and (19) is fulfilled if $0 < |x - a| < \delta$. Combining inequality (16) with (17) through (19) then shows that

$$|f(x)g(x) - LM| \le |f(x)|\,|g(x) - M| + |M|\,|f(x) - L|$$

$$< (|L| + 1)\left[\frac{\epsilon}{2(|L| + 1)}\right] + |M|\left[\frac{\epsilon}{2(|M| + 1)}\right]$$

$$< \frac{\epsilon}{2} + \frac{\epsilon}{2}$$

$$= \epsilon$$

if $0 < |x - a| < \delta$. This proves part (iii).

To prove (iv), we first show that

$$\lim_{x \to a} \frac{1}{g(x)} = \frac{1}{M}, \qquad M \ne 0. \tag{20}$$

Statement (iv) then follows from part (iii) and statement (20). To prove (20), we must show that, given $\epsilon > 0$, there exists a $\delta > 0$ so that

$$\left|\frac{1}{g(x)} - \frac{1}{M}\right| < \epsilon \quad \text{if} \quad 0 < |x - a| < \delta.$$

Since

$$\left|\frac{1}{g(x)} - \frac{1}{M}\right| = \left|\frac{M - g(x)}{Mg(x)}\right| = \left(\frac{1}{|M|\,|g(x)|}\right)|M - g(x)| \tag{21}$$

and $\lim_{x \to a} g(x) = M$, there exists a number δ_1 so that

$$|g(x) - M| < \frac{|M|}{2} \quad \text{if} \quad 0 < |x - a| < \delta_1. \tag{22}$$

From (22) it follows that $|g(x)| > \dfrac{|M|}{2}$, or

$$\frac{1}{|g(x)|} < \frac{2}{|M|} \quad \text{if} \quad 0 < |x - a| < \delta_1. \tag{23}$$

Next, again since $\lim_{x \to a} g(x) = M$, there exists a number δ_2 so that

$$|g(x) - M| < \frac{|M|^2 \epsilon}{2} \quad \text{if} \quad 0 < |x - a| < \delta_2. \tag{24}$$

Finally, we take δ to be the smaller of δ_1 and δ_2. Then from statements (21), (23), and (24) it follows that

$$\left|\frac{1}{g(x)} - \frac{1}{M}\right| = \left(\frac{1}{|M|\,|g(x)|}\right)|M - g(x)|$$

$$< \left(\frac{1}{|M|}\right)\left(\frac{2}{|M|}\right)\left(\frac{|M|^2 \epsilon}{2}\right)$$

$$= \epsilon$$

if $0 < |x - a| < \delta$. This proves statement (iv). ∎

Chapter 2, Theorem 4

Assume that $\lim_{x \to a} g(x)$ and $\lim_{x \to a} h(x)$ exist and that

$$\lim_{x \to a} g(x) = L = \lim_{x \to a} h(x).$$

If the function f satisfies the inequality

$$g(x) \leq f(x) \leq h(x)$$

for all x in an open interval containing a (except possibly at $x = a$), then $\lim_{x \to a} f(x) = L$.

Proof of Theorem 4:   Since

$$\lim_{x \to a} g(x) = L,$$

given $\epsilon > 0$, there is a number δ_1 so that if $0 < |x - a| < \delta_1$,

$$-\epsilon < g(x) - L < \epsilon$$

or, adding L to each term, so that

$$L - \epsilon < g(x) < L + \epsilon. \tag{25}$$

Similarly, since

$$\lim_{x \to a} h(x) = L,$$

there is a (possibly different) number δ_2 so that when $0 < |x - a| < \delta_2$,

$$L - \epsilon < h(x) < L + \epsilon. \tag{26}$$

Now since $g(x) \leq f(x) \leq h(x)$, we may use the left side of inequality (25) and the right side of inequality (26) to conclude that

$$L - \epsilon < g(x) \leq f(x) \leq h(x) < L + \epsilon$$

if both $0 < |x - a| < \delta_1$ and $0 < |x - a| < \delta_2$. This shows that if we take $\delta = \min \{\delta_1, \delta_2\}$, we will have

$$L - \epsilon < f(x) < L + \epsilon$$

or

$$-\epsilon < f(x) - L < \epsilon,$$

if $0 < |x - a| < \delta$, as required. ∎

We now use the Binomial Theorem to give a proof of Theorem 4 in Chapter 3.

Chapter 3, Theorem 4
Power Rule

Let n be any nonzero integer. The function $f(x) = x^n$ is differentiable for all x, and $f'(x) = nx^{n-1}$. That is,

$$\frac{d}{dx} x^n = nx^{n-1}, \qquad n = \pm 1, \pm 2, \ldots .$$

Proof: We shall prove this result for the case $n > 1$ here. Using the Binomial Theorem we write

$$(x + h)^n = x^n + nx^{n-1}h + \frac{n(n-1)}{2}x^{n-2}h^2 + \cdots$$

$$+ \frac{n(n-1)}{2}x^2h^{n-2} + nxh^{n-1} + h^n.$$

We can therefore write the required difference quotient as follows:

$$\frac{f(x + h) - f(x)}{h} = \frac{(x + h)^n - x^n}{h}$$

$$= nx^{n-1} + \text{(terms with factors of } h^P, P \geq 1).$$

We now observe that in the limit as $h \to 0$, all terms involving factors of h^P approach zero. Thus, we conclude that

$$f'(x) = \lim_{h \to 0} \frac{(x + h)^n - x^n}{h} = nx^{n-1}.$$ ∎

Now we give a complete proof of the Chain Rule.

Chapter 3, Theorem 8

Chain Rule

If the function g is differentiable at x and the function f is differentiable at $u = g(x)$, then the composite function $f \circ g$ is differentiable at x, and

$$(f \circ g)'(x) = f'(g(x))g'(x).$$

Proof: Let x_0 be in the domain of g. To prove that

$$(f \circ g)'(x_0) = f'(g(x_0))g'(x_0),$$

we begin by defining a new function F for t in the domain of f:

$$F(t) = \begin{cases} \dfrac{f(t) - f(g(x_0))}{t - g(x_0)} & \text{if} \quad t \neq g(x_0) \\ f'(g(x_0)) & \text{if} \quad t = g(x_0) \end{cases}$$

The function F is continuous at $t_0 = g(x_0)$, since

$$\lim_{t \to g(x_0)} F(t) = \lim_{t \to g(x_0)} \frac{f(t) - f(g(x_0))}{t - g(x_0)} = f'(g(x_0)) = F(g(x_0)).$$

The function F is also continuous at numbers $t = g(x) \neq g(x_0)$ since we have, for $s \neq g(x_0)$,

$$\lim_{s \to t} F(s) = \lim_{s \to t} \frac{f(s) - f(g(x_0))}{s - g(x_0)} = \frac{f(t) - f(g(x_0))}{t - g(x_0)} = F(t).$$

We now work with the composite function $F \circ g$. Since F is continuous and g is differentiable (hence, continuous), the composite function $F \circ g$ is continuous on the domain of g. Thus, according to the definition of F,

$$\lim_{x \to x_0} F(g(x)) = F(g(x_0)) = f'(g(x_0)). \tag{27}$$

Next, we note that, if $x \neq x_0$, we can write

$$\frac{f(g(x)) - f(g(x_0))}{x - x_0} = F(g(x)) \left[\frac{g(x) - g(x_0)}{x - x_0} \right]. \tag{28}$$

To see this, note that if $t = g(x) \neq g(x_0)$, then

$$F(g(x)) = \frac{f(g(x)) - f(g(x_0))}{g(x) - g(x_0)}$$

$$F(g(x))[g(x) - g(x_0)] = f(g(x)) - f(g(x_0))$$

$$F(g(x)) \left[\frac{g(x) - g(x_0)}{x - x_0} \right] = \frac{f(g(x)) - f(g(x_0))}{x - x_0}.$$

If $g(x) = g(x_0)$ for $x \neq x_0$, both sides of (28) are zero. Using equations (27) and (28) together with the continuity of F and g, we may now conclude that

$$(f \circ g)'(x_0) = \lim_{x \to x_0} \frac{f(g(x)) - f(g(x_0))}{x - x_0}$$

$$= \lim_{x \to x_0} F(g(x)) \left[\frac{g(x) - g(x_0)}{x - x_0} \right]$$

$$= f'(g(x_0))g'(x_0). \qquad \blacksquare$$

We conclude this appendix with a rough sketch of the proof of Theorem 1 in Chapter 6.

Chapter 6, Theorem 1

Let f be continuous and nonnegative on the interval $[a, b]$. Let $\underline{S}_n$ and $\overline{S}_n$ denote the lower and upper approximating sums for f on $[a, b]$. Then $\lim_{n \to \infty} \underline{S}_n$ and $\lim_{n \to \infty} \overline{S}_n$ exist, and

$$\lim_{n \to \infty} \underline{S}_n = \lim_{n \to \infty} \overline{S}_n.$$

Proof: This argument is only a sketch of a proof, since a complete proof involves the concept of *uniform continuity,* a topic for more advanced courses in analysis. However, the basic notion is a straightforward calculation.

To show that the lower sums $\underline{S}_n$ approach a limit as $n \to \infty$, we fix an integer m and note that $\underline{S}_{km} > \underline{S}_m$ for all positive integers k since each subinterval corresponding to $\underline{S}_{km}$ lies entirely within a subinterval corresponding to $\underline{S}_m$. (Thus, the minimum values on these smaller subintervals increase.) Therefore, the sequence $\underline{S}_{k^n m}$ is an increasing sequence that is bounded above (by an upper sum), and such sequences always have limits, as is proved in Chapter 12. Using uniform continuity, one can show that the original sequence $\underline{S}_n$ has the same limit. Therefore, we assume that this limit exists. Similarly, one can show that the upper sums $\overline{S}_n$ approach a limit as $n \to \infty$.

To prove $\lim_{n \to \infty} \underline{S}_n = \lim_{n \to \infty} \overline{S}_n$, it is sufficient to prove that

$$\lim_{n \to \infty} (\overline{S}_n - \underline{S}_n) = 0. \tag{29}$$

According to the definitions of $\underline{S}_n$ and $\overline{S}_n$ we have

$$\lim_{n \to \infty} (\overline{S}_n - \underline{S}_n) = \lim_{n \to \infty} \left[\sum_{j=1}^{n} f(d_j) \, \Delta x - \sum_{j=1}^{n} f(c_j) \, \Delta x \right] \tag{30}$$

$$= \lim_{n \to \infty} \sum_{j=1}^{n} [f(d_j) - f(c_j)] \, \Delta x.$$

Now recall that $\Delta x = (x_j - x_{j-1}) = (b - a)/n$, so as $n \to \infty$ the width of each of the subintervals approaches zero. Since f is a continuous function, the difference between the maximum value $f(d_j)$ and the minimum value $f(c_j)$ must approach zero as the width of the interval $[x_{j-1}, x_j]$ approaches zero.* Thus, if we let

$$M_n = \max \{f(d_j) - f(c_j) \mid 1 \le j \le n\},$$

we have both

$$f(d_j) - f(c_j) \le M_n \qquad \text{for all } j = 1, 2, \ldots, n, \tag{31}$$

(that is, M_n is the largest difference) and

$$\lim_{n \to \infty} M_n = 0 \tag{32}$$

(that is, the differences approach zero as the interval width approaches zero).

Returning to equation (30) and using inequality (31) we obtain the statement

$$\lim_{n \to \infty} (\overline{S}_n - \underline{S}_n) = \lim_{n \to \infty} \sum_{j=1}^{n} [f(d_j) - f(c_j)] \, \Delta x \tag{33}$$

$$\le \lim_{n \to \infty} \sum_{j=1}^{n} M_n \, \Delta x$$

$$= \lim_{n \to \infty} M_n \sum_{j=1}^{n} \Delta x.$$

Since

$$\sum_{j=1}^{n} \Delta x = \sum_{j=1}^{n} \left(\frac{b - a}{n}\right) = \frac{n(b - a)}{n} = b - a,$$

equation (32) and inequality (33) give

$$\lim_{n \to \infty} (\overline{S}_n - \underline{S}_n) = (b - a) \lim_{n \to \infty} M_n = 0,$$

which establishes the desired equation (29). ∎

---

*A rigorous proof of this statement uses uniform continuity.

Appendix III

Complex Numbers

In the set of real numbers, the quadratic equation

$$x^2 = -1 \qquad (1)$$

has no solution, since the square of a real number is never negative. The *complex numbers* constitute a number system containing the real number system (just as the real numbers contain the integers. In this system equations such as equation (1) have solutions.

To form the complex number system, we first define the complex number i by the equation

$$i^2 = -1. \qquad (2)$$

Another way to write (2) is to say that $i = \sqrt{-1}$. Thus, the complex number i is a solution to equation (1), as is the number $-i = -\sqrt{-1}$. However, i and $-i$ are not the only complex numbers. The following definition determines the complex number system fully.

Definition 1

The **complex numbers** are all numbers of the form

$$z = a + bi,$$

where a and b are real numbers, together with the following operations:

(i) Scalar multiplication: $\lambda(a + bi) = \lambda a + (\lambda b)i, \qquad \lambda \in \mathbb{R}$
(ii) Addition: $(a_1 + b_1 i) + (a_2 + b_2 i) = (a_1 + a_2) + (b_1 + b_2)i$
(iii) Multiplication: $(a_1 + b_1 i) \cdot (a_2 + b_2 i) = (a_1 a_2 - b_1 b_2) + (a_1 b_2 + a_2 b_1)i$

Notice the following about Definition 1:

(a) If $b = 0$, the complex number $z = a + 0i = a$ is a real number. Moreover, the definitions of addition and multiplication agree with the definitions of addition and multiplication of real numbers if $b_1 = b_2 = 0$. Thus, the set of complex numbers of the form $\{a + bi \mid a \in \mathbb{R}, b = 0\}$ is just the set $\mathbb{R}$ of real numbers.
(b) The definition of multiplication is obtained by assuming that the usual distribu-

tive and commutative laws hold for all expressions involving a_1, a_2, b_1, b_2, and i:

$$(a_1 + b_1 i) \cdot (a_2 + b_2 i) = a_1(a_2 + b_2 i) + (b_1 i)(a_2 + b_2 i)$$
$$= a_1 a_2 + a_1 b_2 i + a_2 b_1 i + b_1 b_2 i^2$$
$$= a_1 a_2 - b_1 b_2 + (a_1 b_2 + a_2 b_1)i.$$

For $z = a + bi$, we say that z is **pure imaginary** if $a = 0$.

☐ **EXAMPLE 1**

If $z_1 = 3 + 2i$ and $z_2 = 4 - i$, then

(i) $4z_1 = 4(3 + 2i) = 4 \cdot 3 + (4 \cdot 2)i = 12 + 8i.$

(ii) $z_1 + z_2 = (3 + 2i) + (4 - i) = (3 + 4) + (2 + (-1))i = 7 + i.$

(iii) $z_1 - z_2 = z_1 + (-z_2) = (3 + 2i) + (-1)(4 - i)$
$$= (3 + 2i) + (-4 + i)$$
$$= [3 + (-4)] + (2 + 1)i$$
$$= -1 + 3i.$$

(iv) $z_1 z_2 = (3 + 2i)(4 - i) = 3 \cdot 4 - (2)(-1) + [3(-1) + 2 \cdot 4]i$
$$= 14 + 5i.$$ ■

Geometric and Vector Interpretations

Since the complex number $z = a + bi$ is determined by a pair of real numbers, we refer to a as the **real part** (or real component) of z, and we refer to b as the **imaginary part** (imaginary component) of z. (Remember that b is a real number.) Plotting the real part of z on the x-axis (called the real axis) and the imaginary part of z on the y-axis (called the imaginary axis), we can plot the complex number $z = a + bi$ as the ordered pair (a, b) in the xy-plane (see Figure III.1). Note that the real numbers (considered as a subset of the complex numbers) correspond to points on the real axis, while pure imaginary numbers correspond to points on the imaginary axis.

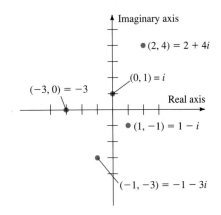

Figure III.1 Complex numbers $a + bi$ can be represented as points (a, b) in the complex plane.

Since points (a, b) in the plane can be regarded as position vectors $\langle a, b \rangle$ originating at the origin $(0, 0)$, we can interpret the complex number $a + bi$ as the position vector $\langle a, b \rangle$. As Figure III.2 illustrates, the definition of addition for com-

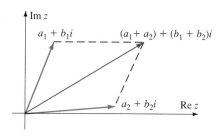

Figure III.2   Addition of complex numbers corresponds to vector addition.

plex numbers agrees with the definition of vector addition. You can verify that the corresponding definitions of scalar multiplication agree as well. The vector interpretation leads directly to the concept of the *modulus* (also called the length or absolute value) of a complex number.

Definition 2

The **modulus of the complex number** $z = a + bi$ is the real number $|z| = \sqrt{a^2 + b^2}$.

Note that the modulus of a complex number is just its absolute value if the number is real.

A concept related to the modulus is the *conjugate* of a complex number.

Definition 3

The **conjugate of the complex number** $z = a + bi$ is the complex number $\bar{z} = a - bi$.

The relationhip between the modulus and the conjugate is that

$$z\bar{z} = (a + bi)(a - bi) = a^2 + b^2 = |z|^2.$$

Figure III.3 shows that the conjugate $\bar{z}$ is just the reflection in the real axis of the complex number z.

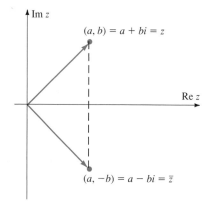

Figure III.3   The conjugate $\bar{z}$ of z is the reflection of z in the real axis.

Division for Complex Numbers

The concepts of modulus and conjugate allow us to define the operation of division. Since

$$z\bar{z} = |z|^2$$

it follows that

$$\left(\frac{1}{|z|^2}\right)(z\bar{z}) = 1,$$

so the reciprocal $\frac{1}{z}$, $z \neq 0$, is defined as

$$\frac{1}{z} = \frac{\bar{z}}{|z|^2}, \qquad z \neq 0.$$

Division is then defined as multiplication by a reciprocal.

Definition 4

Let z_1 and z_2 be complex numbers with $z_2 \neq 0$. The quotient $\dfrac{z_1}{z_2}$ is defined to be

$$\frac{z_1}{z_2} = z_1\left(\frac{1}{z_2}\right) = \frac{z_1\bar{z}_2}{|z_2|^2}.$$

Again, this definition agrees with the definition of division in $\mathbb{R}$ when z_1 and z_2 are real.

☐ **EXAMPLE 2**

Let $z_1 = 1 + 3i$ and $z_2 = 2 - 4i$. Then

(i) $\bar{z}_1 = 1 - 3i$.

(ii) $z_1\bar{z}_1 = (1 + 3i)(1 - 3i) = (1 + 9) + (3 - 3)i = 10 = |z_1|^2$.

(iii) $\dfrac{1}{z_1} = \dfrac{\bar{z}_1}{|z_1|^2} = \frac{1}{10}(1 - 3i) = \frac{1}{10} - \frac{3}{10}i$.

(iv) $\dfrac{z_1}{z_2} = z_1\left(\dfrac{1}{z_2}\right) = \dfrac{z_1\bar{z}_2}{|z_2|^2} = \frac{1}{20}(1 + 3i)(2 + 4i)$

$\qquad = \frac{1}{20}[(2 - 12) + (6 + 4)i]$

$\qquad = -\frac{1}{2} + \frac{1}{2}i$. ∎

Solutions of Quadratic Equations

We conclude by recalling that the solutions of the quadratic equation

$$ax^2 + bx + c = 0$$

are given by the quadratic formula as

$$x = -\frac{b}{2a} \pm \frac{\sqrt{b^2 - 4ac}}{2a}$$

if $b^2 - 4ac \geq 0$. Using complex numbers we may now write the roots as

$$x = -\frac{b}{2a} \pm \left(\frac{\sqrt{4ac - b^2}}{2a}\right)i$$

if $b^2 - 4ac < 0$.

□ **EXAMPLE 3**

For the quadratic equation

$$x^2 - 2x + 3 = 0$$

the solutions are

$$x = -\frac{(-2)}{2} \pm \frac{\sqrt{(-2)^2 - 4(1)(3)}}{2}$$

$$= 1 \pm \frac{\sqrt{-8}}{2}$$

$$= 1 \pm \sqrt{2}\, i.$$

■

Appendix IV

Newton's Method in the Complex Plane

In this appendix, we return to questions that arose when we discussed Newton's Method in Section 3.10. Recall that, when we use Newton's Method to find the roots of a function f, we begin by making an initial guess x_1. Then we apply the Newton iteration

$$x_{k+1} = x_k - \frac{f(x_k)}{f'(x_k)} \tag{1}$$

which produces a sequence of numbers $x_2, x_3, x_4, \ldots$ that often converges to a root of f. In this appendix, we use computer graphics to examine a number of questions, all related to the issue of how to choose the initial guess. As we shall see, the answers to these questions are surprisingly involved even if we confine our examination to the case where f is a polynomial.

Of course, an arbitrary function may not have any roots. However, if f is a polynomial, we take advantage of our knowledge of complex numbers (see Appendix III) and the Fundamental Theorem of Algebra. That is, we know that every polynomial

$$f(x) = a_n x^n + a_{n-1} x^{n-1} + \cdots + a_1 x + a_0$$

of degree n ($a_n \neq 0$) has at least one root and at most n roots in the complex plane. We therefore search for roots in the complex plane.

We provide this appendix for two reasons. First, we want to discuss the mathematical significance of this book's cover. We are pleased that the cover is based on one of the topics discussed in the text, and as this section indicates, much can be learned about Newton's Method by a careful examination of this picture. Moreover, using Newton's Method to produce beautiful graphics has a nice by-product. If you perform the experiments at the end of this appendix, you will learn more about Newton's Method, *and* you will be able to produce your favorite alternatives to our cover.

Second, this appendix illustrates the fact that the frontiers of mathematical research are not as remote as is frequently imagined. Often, important mathematical research originates by the consideration of fundamental questions in new and important ways. Indeed, much of what is described here was discovered in the first half of the 1980s.

Brief History

Newton's Method does indeed go back to Newton (1669), although equation (1) is due to Raphson (1690). However, the question of how best to choose initial guesses was first studied extensively by Schröder and Cayley in the 1870s. More generally, comprehensive studies of the global convergence of analytic iterative procedures (of which equation (1) is one example) were done in the decade surrounding the first World War by Fatou and Julia in Paris. With the aid of computer graphics, the subject underwent a renaissance in the 1980s. Detailed studies of Newton's Method were conducted by Curry, Garnett, and Sullivan (1983) and by Douady and Hubbard (1985). Almost everything that is discussed in this appendix was brought to our attention in conversations with these five mathematicians.

Newton's Method as an Iterated Function

As mentioned earlier, we now consider equation (1) as an iterative procedure where the number x is a complex number. Since we typically represent complex numbers with variables such as z and w, we rewrite equation (1) as

$$z_{k+1} = N(z_k) = z_k - \frac{f(z_k)}{f'(z_k)} \tag{2}$$

where the function N is often called the *Newton function* corresponding to the function f. For the complex polynomial f given by

$$f(z) = a_n z^n + a_{n-1} z^{n-1} + \cdots + a_1 z + a_0,$$

the derivative f' is given by

$$f'(z) = na_n z^{n-1} + (n-1)a_{n-1} z^{n-2} + \cdots + a_1.$$

Therefore, given an initial quess z_1 for a root of f, we calculate a sequence of approximations as

$$z_2 = N(z_1)$$
$$z_3 = N(z_2) = N(N(z_1))$$
$$z_4 = N(z_3) = N(N(z_2)) = N(N(N(z_1)))$$
$$\cdot$$
$$\cdot$$
$$\cdot$$

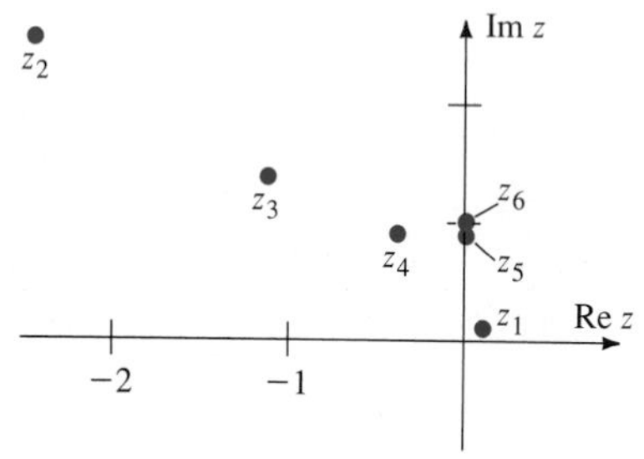

Figure IV.1 Six iterates of $z_1 = 0.1 + 0.1i$ for Newton's Method applied to $z^2 + 1 = 0$. Note that this sequence of iterates converges to i.

The numbers z_k are called the *iterates* of z_1 under the iterative procedure $z_{k+1} = N(z_k)$. For example, if $f(z) = z^2 + 1$, then

$$N(z) = z - \frac{z^2 + 1}{2z} = \frac{z^2 - 1}{2z}. \tag{3}$$

Table IV.1 contains the iterates z_k and their corresponding function values $f(z_k)$ for the Newton function in equation (3) applied to the starting value $z_1 = 0.1 + 0.1i$. The iterates z_k are also plotted in Figure IV.1.

Basin Portraits

Given a polynomial whose roots are known, we can use computer graphics to help us understand the question of choosing "good" starting points—numbers z_1 whose iterates z_k tend to a root relatively quickly. We do this in the following way:

1. We pick a rectangular region R in the complex plane. (The starting points that we sample will be equally spaced within R.) Then we subdivide this region into a grid of equally sized subrectangles R_{ij}. (See Section 19.1 where we show how to carry out such a subdivision.)
2. We select an accuracy ϵ about each root, and we use it to determine that the iterates have converged to a root. That is, once an iterate z_k is within ϵ of a root, we stop applying Newton's Method.

Table IV.1 The iterates of $z_1 = 0.1 + 0.1i$ for Newton's Method applied to the equation $z^2 + 1 = 0$.

k	z_k	$f(z_k)$
1	$0.100000 + 0.100000i$	$1.000000 + 0.020000i$
2	$-2.450000 + 2.550000i$	$0.500000 - 12.495000$
3	$-1.127040 + 1.376960i$	$0.374201 - 3.103770i$
4	$-0.385542 + 0.905924i$	$0.327944 - 0.698543i$
5	$0.006097 + 0.920251i$	$0.153176 + 0.011222i$
6	$-0.000551 + 1.003430i$	$-0.006875 - 0.001106i$
7	$0.000000 + 1.000000i$	$0.000000 + 0.000000i$

3. We pick a maximum for the number of iterates we compute for any particular starting number.
4. For each subrectangle R_{ij} we let z_1 be the complex number that corresponds to the midpoint of R_{ij}. We calculate the iterates z_k of z_1 until either z_k is within ϵ of one of the roots or we reach the maximum number of iterates.
5. If we reach the maximum number of iterates and the iterates have not converged to a root, then we color the rectangle R_{ij} black. Otherwise, we color the rectangle according to some coloring scheme. One common scheme is to use a spectrum of colors to indicate the relative number of iterates it takes the initial point to get within ϵ of the root. When we use this scheme, the color red represents fast convergence (a small number of iterates) and purple represents slow convergence (a large number of iterates). In this appendix, we use a spectrum that goes from red through yellow through green through blue through purple to black (no convergence).

In our figures, we subdivide R using a large number of rectangles. Consequently, we cannot distinguish between a subrectangle R_{ij} and its midpoint.

□ **EXAMPLE 1**

If we apply equation (2) to the polynomial equation

$$f(z) = z^3 - 1 = 0$$

we obtain the Newton function

$$N(z) = z - \frac{z^3 - 1}{3z^2} = \frac{2z^3 + 1}{3z^2}.$$

Figure IV.2 illustrates the convergence of the iterates of N for starting points in the rectangle

$$R = \{z \mid -2 \le \text{Re } z \le 2, -2 \le \text{Im } z \le 2\}.$$

In this case, we are approximating the three solutions to the equation $z^3 - 1 = 0$, namely, the three cube roots of 1. These cube roots are

$$1, -\frac{1}{2} + \frac{\sqrt{3}}{2}i, \quad \text{and} \quad -\frac{1}{2} - \frac{\sqrt{3}}{2}i.$$

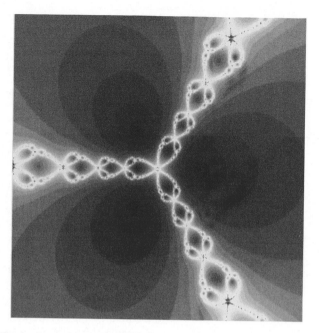

Figure IV.2 The basin portrait of Newton's Method applied to the solution of $z^3 - 1 = 0$.

Note that initial points that are close to these numbers are colored red while points that are farther away get different colors from the spectrum. The colors are completely determined by the number of applications of Newton's Method to find one of the roots using the given initial point.* ■

For polynomials such as $f(z) = z^3 - 1$ in Example 1, we can ask a related question which leads to a different coloring scheme. Given an initial point whose iterates converge to a root, to which of the three roots do they converge? To answer this question, we repeat the same computer experiment using a coloring scheme that indicates the root that is found. In Figure IV.3, we use shades of the three primary colors to indicate the root. For example, all initial guesses whose iterates converge to the root 1 are colored with a shade of blue. The color red corresponds to the root $(-1 + \sqrt{3})/2$, and the color green corresponds to the root $(-1 - \sqrt{3})/2$. Note that in certain parts of Figure IV.3, the iterates of two initial guesses that are quite close together converge to two different roots. Indeed, the red, green, and blue sets in Figure IV.3 intermingle in an extremely complicated fashion.

Figure IV.3 illustrates why we call these pictures *basin portraits*. We think of the roots of f as "attracting" the iterates of nearby starting guesses. Therefore, for a given root r, we consider the set of all starting points whose iterates are attracted to r. This set is called the *basin* of r. More precisely, we define the basin of the root r to be the set

$$\text{basin}(r) = \{z_1 \,|\, \text{the iterates } z_k \text{ of } z_1 \text{ converge to } r\}.$$

Thus, in Figure IV.3, the basin of the root 1 is colored blue.

---

*In addition to the book's cover, all figures in this appendix were made with the use of *Citool*, a window-based computer graphics program designed to produce images of complex-analytic iterative procedures. *Citool* was written at Boston University by Scott Sutherland, Gert Vegter, and Paul Blanchard. Mario Casella helped with the production of many of these images, including the cover of this book.

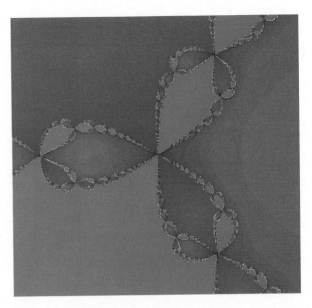

Figure IV.3 This is the same basin portrait as Figure IV.2 except that the coloring scheme indicates which of the three roots to $z^3 - 1 = 0$ was found.

In Figures IV.4 and IV.5, we use two other coloring schemes that highlight the boundaries of the basins in Figure IV.3. We magnify selected subrectangles of Figure IV.3, and as in Figure IV.2, we use a coloring scheme that highlights the common boundary of these three basins. The colors simply indicate the number of

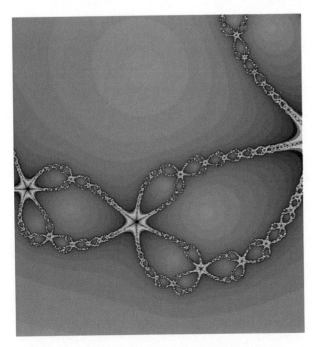

Figure IV.4 An enlargement of the basin portrait for $z^3 - 1$ with a coloring scheme that highlights the set that divides the three basins.

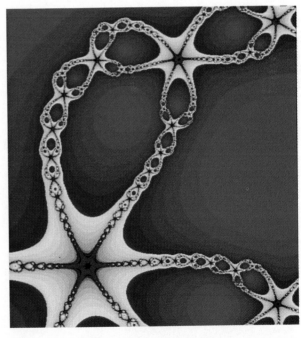

Figure IV.5 An enlargement of another part of the basin portrait for $z^3 - 1 = 0$. This image is very close to the image used to produce the cover of this book.

iterates that are necessary to come within the prescribed accuracy of one of the roots.

In Figures IV.6 through IV.8, we return to the original coloring scheme of Figure IV.2 and illustrate some of the various basin portraits that arise for polynomials of higher degree.

Figure IV.6   The basin portrait of $z^4 - 1 = 0$.

Figure IV.7   The basin portrait for $z^8 - 1 = 0$.

Figure IV.8   The basin portrait for $z^8 + 3z^4 - 4 = 0$.

In Section 3.10, we indicated that it was also possible for the iterates of an initial guess to oscillate rather than to converge to one of the roots. Table IV.2 illustrates this phenomenon for the polynomial $f(z) = \frac{1}{2}z^3 - z + 1$. The iterates corresponding to the initial guess $z_1 = 0.1$ approach a sequence that oscillates between 0 and 1. Note that neither 0 nor 1 are roots of the polynomial f.

Table IV.2 The iterates of $z_1 = 0.1$ for Newton's Method applied to the polynomial $f(z) = \frac{1}{2}z^3 - z + 1$.

k	z_k	$f(z_k)$
1	0.100000	0.900500
2	1.014210	0.507411
3	0.079656	0.920597
4	1.009100	0.504674
5	0.052227	0.947845
6	1.003970	0.502006
7	0.023329	0.976677
8	1.000800	0.500403
9	0.004807	0.995193
10	1.000030	0.500017
11	0.000207	0.999793
12	1.000000	0.500000
13	0.000000	1.000000
14	1.000000	0.500000
15	0.000000	1.000000

Figures IV.9 and IV.10 show the basin portrait of another cubic polynomial that displays this same lack of convergence for certain starting guesses. The rectangle R in Figure IV.9 contains all three roots. Note that a small portion of R is black. These black points correspond to starting guesses whose iterates converge to an oscillating sequence consisting of four complex numbers. In Figure IV.10, we enlarge a subrectangle of R that contains one of the black sets. We note that the set of black points (that is, the set of initial points that do not lead to one of the three roots) is rather distinctive. These points form a set whose structure remains complicated even under repeated magnification.* It is remarkable that a classical numerical method such as Newton's Method leads to the study of complicated and fascinating sets such as this one.

Problematic Polynomials

In Figures IV.2 and IV.9, we see two different types of basin portraits. For the polynomial $f(z) = z^3 - 1$ studied in Figure IV.2, we see that a random choice of initial guess results in a sequence of iterates that converges to one of the three roots.

---

*Sets with this property—the property of remaining complicated under repeated magnification—come under the category of *fractal* sets. The study of fractal sets is a topic of widespread current research in both mathematics and the natural sciences.

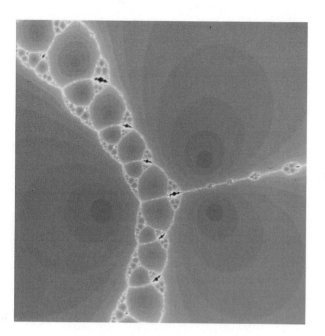

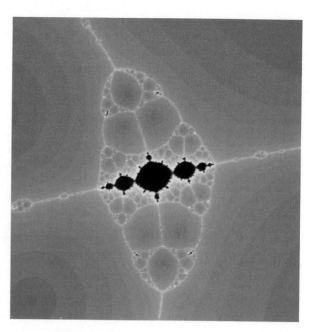

Figure IV.9 The basin portrait for a cubic polynomial for which a number of starting points yield sequences that oscillate instead of converging to one of the roots.

Figure IV.10 An enlargement of a portion of the black set in Figure IV.9. Its structure remains complicated under repeated magnification.

However, for the polynomial studied in Figure IV.9, we see that there is a small percentage of starting guesses whose iterates do not converge to any of the three roots. In general, it is interesting to try to determine polynomials that possess this lack of general convergence under Newton's Method. We conclude this appendix with a discussion of how we can use computer graphics to help locate such polynomials.

The idea is simple but clever. Rather than considering a grid of initial guesses for a given polynomial, we use one fixed initial guess ($z_1 = 0$, for example) for a *grid of polynomials*. We apply Newton's Method with $z_1 = 0$ to the *family* of cubic polynomials

$$f_a(z) = z^3 - (3a^2 + 3a + 1)z + (2a^3 + 3a^2 + a). \tag{4}$$

That is, we use a grid in the a-plane of complex numbers, and we color the point a according to the number of iterates that are necessary for the starting point $z_1 = 0$ to converge to a root of the polynomial. If the starting guess does not converge within the maximum number of iterates specified, we color it black. The result of this experiment is shown in Figure IV.11.

Note that there is very little black in Figure IV.11. In other words, the iterates of the initial guess $z_1 = 0$ converge to one of the roots for most polynomials in this family. Nevertheless, there are small black regions in Figure IV.11. In Figure IV.12, we enlarge one of these regions. At the top of Figure IV.12 there is an unusual-looking black set. The points in this set correspond to polynomials for which $z_1 = 0$ is an ineffective starting guess for Newton's Method. We examine this black set more closely in Figure IV.13. Note that it looks remarkably similar to a set that has received widespread attention over the last decade—the Mandelbrot set. In 1985, A. Douady and J. H. Hubbard published an important paper in which they develop a theory that explains the existence of sets like the Mandelbrot set in this context.

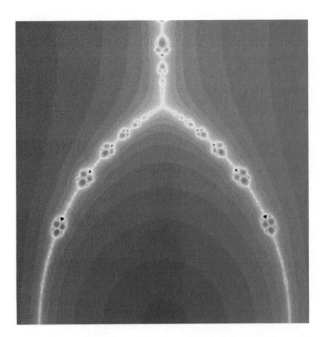

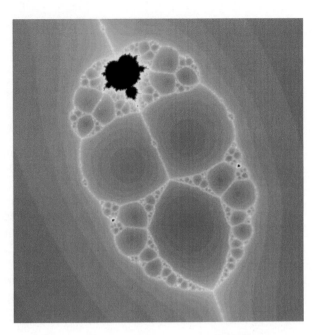

Figure IV.11 Following the sequence of iterates for $z_1 = 0$ corresponding to a *grid* of cubic polynomials. The black set corresponds to cubics that have a basin portrait similar to that of Figure IV.9.

Figure IV.12 An enlargement of a small region in Figure IV.11. Note the unusual black set in the top of the figure.

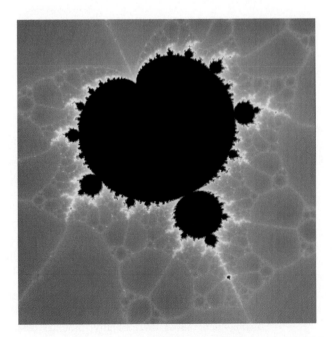

Figure IV.13 A further enlargement of the black set in Figure IV.12. The cubic polynomial whose basin portrait is shown in Figure IV.9 corresponds to a complex number a in this set.

Unfortunately, the details are beyond the scope of this text. Nonetheless, the significance of the large black set in Figure IV.13 is that the corresponding cubic polynomials possess many starting values that do not lead to a root under Newton's Method. In fact, we obtained the polynomial whose basin portrait is Figure IV.9 by selecting one of the numbers a in the black region in Figure IV.13.

Related Computer Experiments

1. Write a computer program that plots basin portraits using the five-part algorithm described in this appendix. Use a coloring scheme that is similar to the one used in Figure IV.2 and in Figures IV.6 through IV.10. In other words, given:
   a. a polynomial f whose roots are known,
   b. a desired accuracy ϵ for the location of the root,
   c. a maximum number of iterates,
   d. a rectangular region in the complex plane,
   e. the number of subdivisions of the rectangle (usually n^2, where n is number of subdivisions of each side of the region),
   plot the basin portrait for Newton's Method applied to the solution of $f(z) = 0$ in the complex plane. Check your program by reproducing the basin portraits shown in this appendix.

2. Modify the program produced in Experiment 1 for the particular polynomial $f(z) = z^3 - 1$ to use the coloring scheme employed in Figure IV.3. In other words, the color should represent the root to which the iterates of the starting guess converge. Then modify this program so that it produces the basin portrait for Newton's Method applied to the equation $z^3 - z = 0$. Finally, modify these programs so that they can compute the basin portraits for $f_a(z) = 0$ where f_a is the cubic polynomial given in equation (4). Note that the roots of f_a are the three complex numbers a, $1 + a$, $-1 - 2a$.

3. Using the program developed in Experiment 1, examine the solutions to the equation $z^n - 1 = 0$ for various integers $n \geq 2$. How does the basin portrait $n = 2$ differ from all of the other portraits?

4. Using the program developed in Experiment 1, compare the basin portrait for $z^2 - 1 = 0$ to the basin portrait for $z^2 + 1 = 0$. What happens to the iterates of a starting guess z_1 for $z^2 + 1 = 0$ if that guess is a real number (that is, if Im $z = 0$)?

5. Write a computer program that calculates the iterates of a given starting point under Newton's Method applied to $f(z) = 0$. (See Appendix I: The BASIC and Pascal programs must be modified so that they use the arithmetic of complex numbers. The *Mathematica* program works as given.) Consider Newton's Method applied to the equation $x^3 - 5x = 0$. What are the iterates if we use the starting guess $x_1 = 1$? What are the iterates if we use the starting guess $x_1 = 0.99$? Using the computer programs developed in Experiments 1 and 2, plot the basin portrait for $z^3 - 5z = 0$. Analyze the various results of these computations. How well does Newton's Method work for solving the equation $z^5 - 5z = 0$?

6. Repeat the computations described in Experiment 5 for the equation $\frac{1}{2}z^3 - z + 1 = 0$. What happens if the starting value $z_1 = 0$ is used? Reproduce the iterates displayed in Table IV.2. Finally, plot the basin portrait for Newton's Method applied to this equation. How do the results of this experiment differ from those of Experiment 5?

Calculus and the Graphing Calculator

The purpose of this appendix is to introduce the use of graphing calculators in solving problems in calculus. You are probably familiar with scientific calculators and their operation, but graphing calculators add a large screen and have the ability to produce the graph of virtually any function that comes to mind. Some of these calculators have a limited amount of symbol manipulation ability, and all are programmable. All of these devices are relatively inexpensive and provide a tool to visualize most of the concepts and applications of calculus, removing the drudgery of computations and paper-and-pencil graphing. This appendix will not serve as a substitute for the user manuals that accompany your calculator. You should know the user manuals well enough to consult them for specific information when needed.

Four companies currently produce graphing calculators: Casio, Hewlett-Packard, Sharp, and Texas Instruments. Our focus here will be on the graphing calculators made by Casio, Hewlett-Packard, and Texas Instruments.

V.1 Basic Operations

Casio

All Casio graphing calculators operate in the algebraic mode, i.e., a calculation of the form "$3+4 \div 6 \times (-3)$" is entered exactly as it is written here. You will notice the absence of an "$=$" key. In its place is the $\boxed{\text{EXE}}$ key, which in effect operates in the same way; since programmed commands can be used, it is more appropriate to have this **EXE**cute key.

There are 26 named storage locations, which are accessed via the $\boxed{\text{ALPHA}}$ key. For example, $\boxed{\text{ALPHA}}$ $\boxed{+}$ $\boxed{\text{EXE}}$ recalls the value stored in location **X**. The **ALPHA** shifted option of the plus key is the storage location **X**. From now on we will denote this by $\boxed{\text{ALPHA}}$ $\boxed{\text{X}}$, using the letter instead of the primary key name. One advantage of these storage locations is that they can be used as if they were variables. For example, $\boxed{(-)}$ $\boxed{2}$ $\boxed{\rightarrow}$ $\boxed{\text{ALPHA}}$ $\boxed{\text{X}}$ $\boxed{\text{EXE}}$ stores -2 in location **X** and $\boxed{4}$ $\boxed{\text{ALPHA}}$ $\boxed{\text{X}}$ $\boxed{\text{X}^2}$ $\boxed{+}$ $\boxed{2}$ $\boxed{\text{ALPHA}}$ $\boxed{\text{X}}$ $\boxed{-}$ $\boxed{3}$ $\boxed{\text{EXE}}$ evaluates the expression $4x^2 + 2x - 3$ at $x = -2$, the value of which is 9.

The truly novel feature of the Casio is that it can produce the graphs of functions given by expressions that can be written in the form "$y = f(x)$". A brief synopsis of how to produce and analyze a graph follows.

Range

Press $\boxed{\text{Range}}$ and you will see a screen that sets the coordinate system of the graphics screen. In order these values are **Xmin**, the left edge of the screen, **Xmax**, the right edge of the screen, **Xscl**, the distance between tick marks on the x-axis, **Ymin**, the bottom edge of the screen, **Ymax**, the top edge of the screen, and **Yscl**, the distance between tick marks on the y-axis. Simply move the cursor to the particular line with the arrow keys, $\boxed{\Leftarrow}$, $\boxed{\Rightarrow}$, $\boxed{\Uparrow}$, $\boxed{\Downarrow}$ or with $\boxed{\text{EXE}}$ and enter the value. The Casio has a default setting obtained by pressing $\boxed{\text{SHIFT}}$ $\boxed{\text{DEL}}$ or $\boxed{\text{Reset}}$ on the fx-7700. (On the fx-7700 there is a menu bar at the bottom of the screen. The items on the bar identify various procedures; each is activated by pressing the key below that item. This will be denoted by $\boxed{\quad}$.) This setting produces a coordinate system equally scaled on both x- and y-axes. To leave the range screen press the $\boxed{\text{Range}}$ key again.

Graphs

Drawing a graph is accomplished by pressing $\boxed{\text{Graph}}$, entering the expression using the variable **X**, and pressing $\boxed{\text{EXE}}$. For example, entering $\boxed{\text{Graph}}$ $\boxed{\text{ALPHA}}$ $\boxed{\text{X}}$ $\boxed{\text{X}^2}$ $\boxed{-}$ $\boxed{1}$ $\boxed{\text{EXE}}$ will draw the graph of $f(x) = x^2 - 1$. If you want more than one graph on the screen, simply put a $\boxed{:}$ or $\boxed{\text{SHIFT}}$ $\boxed{\text{EXE}}$ between the graph commands. Either of these will concatenate command lines, and the latter moves the cursor to a new line. To clear the graphics screen press $\boxed{\text{SHIFT}}$ $\boxed{\text{Cls}}$ $\boxed{\text{EXE}}$. Clear the graphics screen, set the range to the default setting, enter **Graph Y = X² − 1:Graph Y = cos X** and press $\boxed{\text{EXE}}$. [Be certain that the calculator is in the radian mode ($\boxed{\text{MODE}}$ $\boxed{5}$ $\boxed{\text{EXE}}$). Check this by pressing and holding $\boxed{\text{M Disp}}$ and looking at the angle mode.] You will see both of the graphs $y = x^2 - 1$ and $y = \cos x$. If you press $\boxed{\Leftarrow}$ or $\boxed{\Rightarrow}$ you will recover the last command executed; this is very helpful when you want to change the last function graphed or correct an error. When an error is made, pressing these same keys will move the cursor to the point where the error occurred and using $\boxed{\text{DEL}}$ will delete the command under the cursor. The sequence $\boxed{\text{SHIFT}}$ $\boxed{\text{INS}}$ is used to insert commands immediately to the left of the cursor position.

Tracing and Zooming

Using the default range setting, graph $f(x) = x^2 - 1$ and $g(x) = \cos x$. After the graphs are drawn, press $\boxed{\text{SHIFT}}$ $\boxed{\text{Trace}}$. You will see a flashing dot at the left end of the screen that moves when you press $\boxed{\Leftarrow}$ or $\boxed{\Rightarrow}$. The x-coordinate is given at the bottom, and the y-coordinate is viewed by pressing $\boxed{\text{SHIFT}}$ $\boxed{\text{X} \leftrightarrow \text{Y}}$. (The fx-7700 displays both simultaneously). Move the dot to the first point where the graphs intersect and press $\boxed{\text{SHIFT}}$ $\boxed{\times}$. You have now zoomed in on the intersection point by a factor of 2. (On the fx-7700 there is a zoom menu that will do the same thing, but the zoom factors can be set as well.) Using the trace again, move to the intersection point and press $\boxed{\text{SHIFT}}$ $\boxed{\div}$ to zoom out by a factor of 2. The zoom factors can be changed by using the **Factor** command. The following will draw the same two graphs and halt to allow you to move the dot to the intersection point; then, upon pressing $\boxed{\text{EXE}}$ twice, you zoom in on the point by a factor of 4 and the graph is redrawn.

> **Graph Y = X² − 1:Graph Y = cos X◢**
> **Factor 4**

The ◢ command ($\boxed{\text{SHIFT}}$ $\boxed{◢}$) halts the execution so that the trace can be used. This command is also used to display values that have been computed in a program. The x and y zoom factors can be set independently as well; for example, $\boxed{\text{SHIFT}}$ $\boxed{\text{Factor}}$

4 SHIFT , 1 6 will zoom in on a point by a factor of 4 in the x direction and a factor of 16 in the y direction. This setting is very useful for zooming in on the maximum or minimum points of a graph. If the zoom factors are set to be values that are less than 1, then you zoom out. Be aware that the trace function will work only on functions graphed with the **Graph** command and is active only on the last graph drawn. If you have an fx-7700 then you can trace more than one graph by switching graphs with the up arrow key.

Programming

It is not our intention to give details of all the syntax of the programming language of the Casio, but only how to enter, edit, and execute a program. To program the Casio, first press MODE 2. Across the bottom of the screen will be **Prog 0123456789**. These numbers mark the 10 program locations available for storing programs. You will also see a number that tells how many bytes are free. Consult your owner's manual to determine how much is available for programs. Moving the cursor under one of the numbered locations and pressing EXE gives a text screen where you enter the program. Programming lines can be strung together by using the : key or pressing EXE to move the cursor to a new line. To execute a program, first return to the **RUN** mode by entering MODE 1, and then enter Prog (location number) EXE.

As an example we offer the following progrAm that should be placed in program location 9, where it can be kept indefinitely. This program sets the range so that it is centered on any given point and so that the x-axis and y-axis are equally scaled. The number **S** sets the distance between pixels or dots on the screen. For example, if **S** is equal to 0.1 then each pixel is 0.1 of a unit.

```
"X"?→X
"Y"?→Y
"S"?→S
Range X−47S,X+47S,10S,Y−31S,Y+31S,10S
```

Execute this program and enter 0 for both **X** and **Y** and 1 for **S**. Then key in the sequence SHIFT Plot EXE. (If you have an fx-7500 then key in Plot EXE.) This places a dot at the screen center, with the x-coordinate displayed at the bottom of the screen. Move this dot around the screen with the arrow keys and note how the x- and y-coordinates change.

Texas Instruments

As with the other types of TI calculators the TI-81 is an algebraic entry type calculator. This means that a calculation like "$3+4\div2+6\times(-3)$" is entered exactly as it is written here. You will notice that there is no "$=$" key like on many other calculators. The TI-81 calculator has an ENTER key which in effect operates the same way, but since it is possible to use programmed commands it is more appropriate to have this key. There are 27 storage locations **A–Z** and $\boldsymbol{\theta}$. These storage locations are accessed with the ALPHA key. For example, X|T ENTER or ALPHA STO▷ ENTER recalls the value stored in the location **X**. The **ALPHA** shifted value of the STO▷ key is the memory location **X**. This key sequence will be denoted from now on as ALPHA X, i.e., giving ALPHA followed by the letter instead of the primary key name. The following example shows how to use these storage locations in an arithmetic computation: (−) 2 STO▷ X|T ENTER 4 X|T X² + 2 X|T − 3 ENTER evaluates the expression $4x^2 + 2x - 3$ at $x = -2$, the value of which is 9.

Probably the most useful feature of the TI-81 is its capability to produce the graph of any function that can be expressed in the form "$y = f(x)$". A brief synopsis of how to produce and analyze the graph of a function follows.

Range

Press RANGE and you will see a screen that is used to set the coordinate system of the graphics screen. In order, these values are **Xmin**, the left edge of the screen, **Xmax**, the right edge of the screen, **Xscl**, the distance between tick marks on the x-axis, **Ymin**, the bottom edge of the screen, **Ymax**, the top edge of the screen, and **Yscl**, the distance between tick marks on the y-axis. Simply move the cursor to the particular line with the arrow keys ◁, ▷, ▽, △ or with ENTER and enter the value. For example, set the range to be **Xmin = −4.8**, **Xmax = 4.7**, **Xscl = 1**, **Ymin = −3.2**, **Ymax = 3.1**, **Yscl = 1**, and **Xres = 1**. The last value is the resolution of the graph of a function, i.e., the number of pixels, horizontally, that will be darkened. If **Xres** is 1, then every pixel will be darkened when a graph is drawn; if **Xres** is 2, then every other pixel will be darkened. This range setting is quite useful since it scales both the x- and y-axes equally. It also sets the distance between pixels to be 0.1. In the **ZOOM** menu there are some default range settings: **Square**, **Standard**, **Trig**, and **Integer**; each is quite useful, but we will not discuss the details of these. Consult your owner's manual for details. To get out of the range screen press 2nd QUIT. This key sequence is used to exit from any menu screen.

Graphs

To draw the graph of a function, press Y=. A list appears with four lines. Here you can enter up to four expressions labeled **Y₁**, **Y₂**, **Y₃**, and **Y₄**, all of which must use the variable **X** in order to draw their graphs. Enter on the first line the function **X² − 1**. Press GRAPH and the graph of the function $y = x^2 − 1$ will be drawn. To draw more than one function, simply add another expression to the **Y** variables list. Enter on the **Y₂ =** line **cos X** and press GRAPH again. Now both curves are drawn on the screen. (Be certain that the calculator is in the radian mode. Check this by pressing MODE and selecting and entering **Rad**.) The graphics screen is cleared by selecting and entering **ClrDraw** from the **DRAW** menu. If you make an error while entering an expression in the **Y =** menu, simply move the cursor over the error and press DEL to delete it. The INS key can be used to enter commands immediately to the left of the cursor position.

Tracing and Zooming

Using the range setting given above, graph $f(x) = x^2 − 1$ and $g(x) = \cos x$. After the graphs are drawn, press TRACE. You will see a flashing × on the screen, centered on the graph of $y = x^2 − 1$, and which moves when you press ◁ or ▷. The x- and y-coordinates of the point are given at the bottom of the screen. This trace can be switched to the other graph by pressing △ or ▽. Move the cursor to the first point where the graphs intersect and then press ZOOM. Select and enter the fourth item, **Set Factors**. A screen now appears where the zoom factors can be set. Enter **2** for both of these and press ZOOM again. Select the second item **Zoom In** and press ENTER. The graph is redrawn, and you have zoomed in on the intersection point by a factor of 2. Trace again to the intersection point, select **Zoom Out** from the **ZOOM** menu, and press ENTER. The graphs are redrawn again, but now zoomed out by a factor of 2. You have probably noticed the first command in the **ZOOM** menu, **Box**. This is used for drawing a box around a point. Simply select **Box** from the menu and press ENTER. A flashing cross is placed on the screen center. Move this cross to one corner of the box and press [ENTER]. The cursor now changes to a flashing box. As this flashing box is moved a box is generated. Move this so it encloses a point of interest on the graph. After the box is drawn and [ENTER] is pressed, the graph is redrawn with the box scaled up to the full screen, thus zooming in on the points in the box. It is especially useful for zooming in on a maximum or minimum point of a graph. This can also be done by setting the *x* and *y* zoom factors to be different values. A good choice is **XFact = 4** and **YFact = 16**.

Programming

The intention here is not to give complete details of the syntax of the programming language, but simply how to enter, edit, and execute a program. To program the TI-81, first press [PRGM]. You will see the titles **EXEC**, **EDIT**, and **ERASE** across the top of the screen and a list of the program numbers **1–0**, **A–Z**, and **θ**. To enter a program, use the right arrow key to select **EDIT** and then the number or letter of the program you want to create or edit. This gives an edit screen with a flashing **A** (**ALPHA** mode) on the first line. Here you enter the program's name, which can contain up to eight characters. After entering the name, press [ENTER] and the cursor is now on the first executable line of a program. Errors are corrected in the same way as in the **Y =** list, using the [DEL] and [INS] keys. To execute a program, you first must return to the text screen by pressing [2nd] [QUIT]. Then press [PRGM], enter the program number or letter, press [ENTER] and the program is executed.

As an example we offer the following program that should be placed in program location **R**, where it can be kept indefinitely and easily remembered. This program sets the screen range so that it is centered on any given point and so that the *x*-axis and *y*-axis are equally scaled. The number **S** sets the distance between the pixels. For example, if **S** is equal to 0.1 then each pixel is 0.1 of a unit. The range variables are accessed from the [VARS] menu under the **RNG** title, and the → command is simply the [STO▷] key. After each line is keyed in, press [ENTER] to move to a new line. All special commands must be entered from a menu and cannot be entered by simply typing in the letters one at a time. The **Disp** and **Input** commands are found by pressing [PRGM] and selecting the **I/O** title.

```
PrgmR:RANGE
Disp "X"
Input X
Disp "Y"
Input Y
Disp "S"
Input S
X − 48S→Xmin
X + 47S→Xmax
10S→Xscl
Y − 32S→Ymin
Y + 31S→Ymax
10S→Yscl
```

Execute this program by pressing [PRGM] [ALPHA] [R] [ENTER] and enter 0 for both **X** and **Y** and 1 for **S**. Then go to the graph ([GRAPH]) and press any one of the arrow keys. A flashing crosshair is placed at the screen center. Move the crosshair around the screen with the arrow keys and note how the *x*- and *y*-coordinates change.

Hewlett-Packard

Hewlett-Packard makes two graphing calculator models, the HP-28S and the HP-48S or HP-48SX. What distinguishes these calculators from others is that they do not use algebraic entry but postfix or reverse Polish entry, typical of most Hewlett-Packard calculators. The other distinguishing feature of these calculators is they are much more powerful than the others because they can do a certain amount of symbol manipulation. For example, these calculators can symbolically compute a derivative of a single-formula function.

The postfix notation that these calculators use takes time to learn at first, but if you own one of these calculators it is to your advantage to learn to use it well. Consider for example the calculation ''$2 \times 5 = 10$''. This is the usual way that we think about multiplication, but in the postfix notation it would be written as ''2 5 $\times = 10$''. This is precisely how the HP calculators operate. To perform this calculation you would enter the sequence of keystrokes: [2] [ENTER] [5] [×]. The value of 10 would then be displayed. This calculation can be entered in an algebraic mode as well by enclosing the item in single quotes and pressing [EVAL], i.e., ['] [2] [×] [5] [ENTER] [EVAL].

Functions or numbers can be entered and stored as names on the menu bar, and evaluated by pressing this menu item and then evaluating. Consider the following example, where you will enter the function $y = x^2 - 5x + 4$ and evaluate it at $x = -3$. First enter ['] [X] [ENTER] [■] [PURGE] (HP-28S) or ['] [α] [X] [ENTER] [←] [PURGE] (HP-48SX). This procedure simply purges the value of **X** from the user menu if it exists. For the HP-48SX we will use [←] for the (orange) left shift and [→] for the (blue) right shift. Expressions like $x^2 - 5x + 4$ can be entered by either postfix or algebraic entry. The following describes each of these.

Postfix Entry for HP-28S

[X] [■] [X²] [5] [ENTER] [X] [×] [−] [4] [+]

['] [F] [STO]

Postfix Entry for HP-48SX

[α] [X] [←] [X²] [5] [ENTER] [α] [X] [×] [−] [4] [+]

['] [α] [F] [STO]

Algebraic Entry for HP-28S

['] [X] [■] [×] [2] [−] [5] [×] [X] [+] [4] [ENTER]

['] [F] [STO]

Algebraic Entry for HP-48SX

Make sure the calculator is in the **INS** mode. You can check this by pressing ['] and watching the blinking cursor. If the cursor is an arrow, you are in the **INS** mode. If the cursor is a square you must change it to an arrow. This is done by entering ['] and then pressing [←] [EDIT]. The edit menu will then appear; press [||| INS |||] . The cursor will change from a square to an arrow.

['] [α] [X] [yˣ] [2] [−] [5] [×] [α] [X] [+] [4] [ENTER]

['] [α] [F] [STO]

Now that we have the function $y = x^2 - 5x + 4$ stored in the variable **F**, it is a simple matter to evaluate the function. Enter -3 3 CHS (HP-28S) or 3 +/− (HP-48SX). Then enter ' X STO (HP-28S) or ' α X STO (HP-48SX) to store the value -3 in the variable **X**. Press the key under |||F||| and the expression defining the function is displayed. If the menu key that shows **F** is not visible, press USER and NEXT (HP-28S), or press VAR and NEXT (HP-48SX) a few times until the variable **F** is visible. Once the expression for **F** is displayed press EVAL and the function is evaluated. The value 28 is displayed. From now on ||| ||| will denote that a white key under a menu item is to be pressed.

We now describe how you can draw a graph with these calculators and analyze functions using some of the special built-in features.

You have probably noticed that there are numbers along the left of the screen. These number the levels of what is called the *stack*, and it is this structure that these calculators utilize in all operations. It's worth your time to get used to "stack" ideas in order to operate the calculator efficiently.

HP-28S

Range

The HP-28S has a built-in range setting that sets the screen coordinate system so that the x- and y-axes are equally scaled. The default screen range of the HP-28S is $-6.8 \le x \le 6.8$ and $-1.5 \le y \le 1.6$; this makes each pixel a width of 0.1. Enter the **PLOT** menu by pressing ■ PLOT and go to the next page of this menu with NEXT. Now press ' |||PPAR||| ■ PURGE. This sets the independent variable to be **X** and will also set the default range.

Graphs

Let's go through the steps used to draw the graph of a function. The first thing to do is to purge **X** from the user menu and set the default screen range as described above. Now enter '**X^2 − 1**' onto the stack and activate the **PLOT** menu by entering ■ PLOT. There you will see the menu item |||STEQ||| which stands for store equation. With the function on the stack, press the key under this item and the function $y = x^2 - 1$ is now stored in the plotting utility. Now press |||DRAW|||. Soon you will see the graph drawn, but because the screen of the HP-28S is so long and narrow you don't see much of it. Now press ON (**ATTN**) and then NEXT, followed by |||PPAR|||. This puts on the stack a list of the plot parameters. In order, these values are the lower left point, the upper right point, the independent variable, the resolution, and the screen center. All of these can be changed by using the following commands in the **PLOT** menu: |||PMIN||| sets the lower left point, |||PMAX||| sets the upper right point, |||INDEP||| sets the independent variable, |||RES||| sets the resolution, and |||CENTR||| sets the screen center. For example, **(−1,−1)** |||PMIN||| will set the lower left corner of the screen to be $(-1, -1)$.

Redraw the graph with the default screen range. After the graph is drawn, press one of the arrow keys below the screen; you will see a crosshair moving around the screen. To see the coordinates of the crosshair position press and hold ◁⊕▷; at the bottom of the screen you will see an ordered pair. Unfortunately, there is no facility for tracing along the graph of a function on the HP-28S, but an ordered pair can be "digitized," i.e., placed on level 1 of the stack, by pressing INS. Once the point is digitized press ON (**ATTN**) to leave the graph and go back to the stack. There you will see the ordered pair on level 1 of the stack. To get only the x-coordinate of the point, go to the **COMPLEX** menu (■ COMPLEX) and press |||C→R|||, which separates the coordinates into two values on levels 1 and 2 of the stack. The value of level 2 is the *x*-coordinate. To get rid of the *y*-coordinate, press ‖DROP‖. Store the *x*-coordinate in the variable **X** as described earlier, and then you can evaluate the function by pressing ‖EQ‖ ‖EVAL‖. (‖EQ‖ is found in the **USER** menu.) Whenever a function is drawn, the plotting utility stores the expression in **EQ** in the user menu.

To graph two functions on the screen, simply put both expressions on the stack, enter ‖=‖, and store the equation in the plot utility. Pressing ‖DRAW‖ will draw both of the functions. For example, enter '**X^2−1**' and '**COS(X)**' and then press ‖=‖ ‖ENTER‖. You now have '**X^2−1=COS(X)**' on the stack. Activate the plot menu and press ‖STEQ‖ followed by ‖DRAW‖. The calculator responds by drawing both functions simultaneously.

Zooming

To zoom in on a point we use the commands ‖*W‖, ‖*H‖, and ‖CENTER‖, which multiply the width and height of the screen by a given positive number and set the coordinates of the center of the screen. Draw the graphs of $y = x^2 - 1$ and $y = \cos x$ on the default screen. When the graph is finished, move the crosshair to the first point on the left where the functions intersect and digitize it (‖INS‖). Now get out of the graphics screen (‖ON‖) and search the plot menu for **CENTR** and press it. The screen is now centered on this point. Enter **.5** twice, press ‖*H‖ followed by ‖*W‖, and redraw the graph with ‖DRAW‖. You have now zoomed in by a factor of 2 on the point of intersection. If you had entered **2** twice and pressed ‖*H‖ and ‖*W‖ you would have zoomed out on the point by a factor of 2.

Programming

It can take a considerable amount of time to learn all of the programming commands and how to use them. Our intention here is not to delve deeply into all the syntax of the programming language but simply to give the basic ideas on how to enter, edit, and execute a program. All programs begin with the delimiter ≪ and end with ≫. While you are writing a program you can move the cursor to any place using the arrow keys while the menu bar is hidden. To hide the menu bar press ‖◁⊕▷‖. A space is entered with ‖SPACE‖ and a new line with ▮ ‖NEWLINE‖. Any command can be typed in letter for letter or entered directly from its menu. The key ‖⇐‖ can be used to backspace over a character, causing it to be deleted, and ‖INS‖ is used to insert characters or commands immediately to the left of the cursor. The ‖DEL‖ key deletes the character beneath the cursor.

Once a program is finished being keyed in, press ‖ENTER‖; if there are no errors, the program will be placed on level 1 of the stack. If there are errors, the calculator will refuse to put the program on the stack and will return the message "**Syntax Error.**" With a correct program on the stack, enter the name in single quotes and press ‖STO‖ to save the program in user memory. To execute a program, simply press the white key under its name. To edit the contents of an item in **USER** memory, including a program, enter '**(name)**' ‖ENTER‖ ▮ ‖VISIT‖.

We now give a collection of programs that set the range so that the *x*- and *y*-axes are scaled equally, graph any number of functions on the same screen, and recall a graph to the screen without redrawing it. Lowercase letters are obtained by pressing ‖LC‖ before entering the letter. Pressing ‖LC‖ a second time returns the uppercase letters. The name at the top of each program is not to be put in the program, but is the name that the program should be stored under.

'RANGE'	'MGRAPH'	'RCGRAPH'
≪	≪	≪
{STO X Y S}	DEPTH →LIST	SCRN →LCD
MENU HALT X Y	DUP 'EQS' STO	DGTIZ
R→C CENTR	LIST→ → n	≫
X 68 S * −	≪ CLLCD 1 n	
Y 15 S * − R→C	START STEQ DRAW	
PMIN X 68 S * +	NEXT LCD→	
Y 16 S * + R→C	'SCRN' STO	
PMAX { X Y S }	DGTIZ	
PURGE	≫	
≫	≫	

The program **RANGE** puts a custom menu on the screen calling for input of the screen center coordinates **X** and **Y** and the value of the scaling **S** (which must be positive), and then halts. Enter the desired numbers and press the appropriate key under the item to store the values. Once this is done, press ▮ CONT to continue the program. The value of **S** sets the distance between pixels. For example, if **X** = 0, **Y** = 0, and **S** = 0.1 then the range will be the default, which has a distance of 0.1 between pixels.

For **MGRAPH**, put the functions to be plotted on the stack, one on each level, and execute the program. Make sure that nothing else is on stack. The functions are stored in the list named **EQS**. After the graphs are drawn the screen is put into a string variable and stored under the name **SCRN**, where it is used by **RCGRAPH**. The command **DGTIZ** puts the crosshair on the screen so points can be digitized. **RCGRAPH** recalls to the screen the last graphs drawn by **MGRAPH** by putting back to the screen the string stored in the variable **SCRN**. **RCGRAPH** also puts the crosshair on the screen so points may be digitized. The functions can all be redrawn by putting **EQS** on the stack, issuing the commands **LIST** → **DROP**, and executing **MGRAPH** again. This is useful if you want to change the range and redraw the functions. (The command **LIST**→ is found in the **LIST** menu.)

HP-48SX

Range
The default range of the HP-48SX is $-6.5 \le x \le 6.5$, $-3.1 \le y \le 3.2$. The current range is listed on the screen when entering the **PLOT** menu and accessing the ▮▮▮ **PLOTR** ▮▮▮ submenu. The values of the range can be put on the stack by pressing → ▮▮▮ **XRNG** ▮▮▮, giving the values of the x range, pressing → ▮▮▮ **YRNG** ▮▮▮, and giving the values of the y range. These keys can also be used to set the range: simply enter the two values onto the stack and press ▮▮▮ **XRNG** ▮▮▮ or ▮▮▮ **YRNG** ▮▮▮.

Graphs
The programs needed on the HP-28S to draw several graphs on the screen at once are unnecessary on the HP-48SX, since this capability is already built in. Enter the **PLOT** menu and select the plot type **FUNCT**. Return to the main plot menu, enter a function, and press ▮▮▮ **NEW** ▮▮▮. You will be prompted to give the function a name and press ENTER. In this way you can enter as many functions as you wish. To graph all of these you press ▮▮▮ **CAT** ▮▮▮, and a list appears of all the functions stored in the **VAR** menu. Select one of the functions with the up or down arrow keys and press ▮▮▮ **EQ+** ▮▮▮. This takes the name of a function and adds it to a list that is displayed at

the top of screen. Press ||| **PLOTR** ||| and the list becomes the active equation. The graphs can now be drawn by first pressing ||| **ERASE** ||| (to erase any previous graph) and then ||| **DRAW** |||. Once the graphs are drawn, there are many features that can be used to analyze the functions. We list only a few. Note that these can be used only if the function is of the '$y = f(x)$' kind, i.e., a single formula.

||| **ROOT** ||| Finds a root of the current equation.

||| **ISECT** ||| Finds a point of intersection of two graphs.

||| **COORD** ||| Displays the coordinate of the crosshair.

||| **FNC** ||| Produces another menu in which the current function can be evaluated at a coordinate. Also provides a facility for moving from equation to equation, and one that will graph the derivative of the current equation.

The last graph can always be recalled to the screen by pressing **GRAPH** (the ◁ key).

Zooming

Immediately after the graph is drawn you will see ||| **ZOOM** ||| and ||| **CENT** ||| on the menu bar. These are the keys used to zoom in or out on a point. Move the crosshair to the point on the graph where you want to zoom in and press ||| **CENT** |||. The graph will be redrawn with this point as the new center of the graph. When the graph is completed press ||| **ZOOM** ||| and you will have a menu of features to choose. These are **XY** to scale both x and y, **X** to scale x only, **Y** to scale y only, and **XAUTO**. This last feature will automatically scale the y-axis. When any of these features is selected, you are prompted to put in the zoom factor. It is also possible to box in a point and zoom in on it. This is done with ||| **Z-BOX** |||. First move the crosshair to a corner of the box and press ||| **Z-BOX** |||. Then move the crosshair to the opposite corner of the box and press ||| **Z-BOX** ||| again. The graph will be redrawn with the region enlarged to fit the screen.

Programming

It can take a considerable amount of time to learn all of the programming commands and how to use them. Our intention here is not to delve deeply into all the syntax of the programming language but simply to give the basic ideas on how to enter, edit, and execute a program. All programs begin with the delimiter ≪ and end with ≫. Once a program is correctly keyed onto the command line, press **ENTER**; if there are no errors, the program will be put on level 1 of the stack. If an error does occur, the point where the error occurred will be flagged. With a correct program on the stack, enter the program name in single quotes and press **STO**. The program will now be stored in the user memory, and the menu that holds the name is accessed by pressing **VAR**. A program is executed by pressing the white key under the menu name. If a program needs to be edited, enter its name in single quotes onto level 1 and press '**(name)**' **ENTER** → **VISIT**.

As an example we offer the following program, which sets the range so that the x- and y-axes are equally scaled and the graph is centered on any given coordinate. The name **RANGE** is not to be keyed into the program, but is the name under which it is stored.

```
'RANGE'
≪
"Enter X,Y,S"
{ ":X:
:Y:
:S:" { 0 1 }}
INPUT OBJ→
'S' STO 'Y' STO
'X' STO X Y R→C
CENTR X 65 S * −
X 65 S * + XRNG
Y 31 S * − Y 32
S * + YRNG
{ X Y S }
PURGE
≫
```

When the program is executed you will see a prompt to enter **X**, **Y**, **X**. Following the colon, enter the desired values, moving from line to line with the up or down arrow keys. The value of **S** (which must be positive) sets the distance between pixels. For example, if **S** is set equal to 0.1 you will get the default range, for which pixels are 0.1 unit apart.

V.2   Piecewise Defined Functions

Casio

The only way to draw piecewise defined functions on the fx-7000, fx-7500, fx-8000, and fx-8500 models is with the following point plotting program. Keep this program at all times, since it is used over and over. Program location **0** will always be used to evaluate a function and store the value in location **Y**. This point plotting program should be placed in program location **4**. The names of the programs are not to be entered into the program. The following programs will produce the graph of the function

$$f(x) = \begin{cases} -2x - 4, & \text{if } x \le -2 \\ 4 - x^2, & \text{if } -2 < x \le 2 \\ -\frac{1}{2}x + 1, & \text{if } 2 < x \end{cases}$$

PTPLOT

```
"A"?→A:"B"?→B
"N"?→N
(B − A) ÷ N→H
0→I:Lbl 1
A + IH→X
Prog 0
Plot X,Y
Isz I
I≤N⇒Goto 1
```

Function Evaluator

```
X≤−2⇒Goto 1
X≤2⇒Goto 2
−X÷2+1→Y:Goto 3
Lbl 1
−2X−4→Y:Goto 3
Lbl 2
4−X²→Y:Goto 3
Lbl 3
```

Before you execute this program, the range must first be set so that the values of **A** (left endpoint) and **B** (right endpoint) are the range values **Xmin** and **Xmax**, respectively. The program **PTPLOT** prompts for the input of the endpoints; for this example **A = −4.7** and **B = 4.7** could be used if the default range is used. There are 95 pixels across the screen, so a good value for **N** is 95. This will be the number of points the program plots. The function evaluator should be placed in **Prog 0** since this is the program called by **PTPLOT** to evaluate the function. Unfortunately, it is not possible to use the trace with this type of graph. However, if you enter $\boxed{\text{SHIFT}}$ $\boxed{\text{Plot}}$ $\boxed{\text{EXE}}$, a flashing dot can be put on the screen and moved to the part of the graph of interest, where its approximate x- and y-coordinates can be read. (If you have an fx-7500 then $\boxed{\text{Plot}}$ $\boxed{\text{EXE}}$ will put a flashing dot on the screen.)

The Casio fx-7700 can graph piecewise functions. This is accomplished by a command like the following.

> **Graph Y = −2x−4,[−4.7,−2]**
> **:Graph Y = 4−X²,[−2,2]**
> **:Graph Y = −X÷2+1,[2,4.7]**

On the fx-7700 it is possible to trace each piece of the function by pressing the up arrow key to switch to the different pieces.

TI-81

Piecewise defined functions can be drawn from the **Y** variables list or with the **DrawF** command. Enter the definition of each piece in parentheses followed by its interval in parentheses; all of these are combined with plus signs:

> **:Y₁ = (−2X−4)(X≤−2)+(4−x²)(−2<X)(X ≤ 2)+(−X/2+1)(X>2)**

The expression **(−2X−4)(X≤−2)** is a product of the two expressions **(−2X−4)** and **(X≤−2)**. The expression **(X≤−2)** is evaluated by the calculator as 1 if $x \leq -2$ and as 0 if $x > -2$. So this has the effect of turning on the function $-2x - 4$ if $x \leq -2$ and making it zero if $x > -2$. It is also possible to divide by **(X≤−2)**. This causes the calculator not to draw anything for $x > -2$. Just as with ordinary functions, it is possible to trace along this function. However, not all piecewise functions can be drawn in this way. For example, suppose we replace the center portion of this example with $\sqrt{4 - x^2}$. Since this is not defined outside the interval $[-2, 2]$ the calculator ignores all parts of the function outside this interval. This can be fixed by putting the first piece of the function in **Y₁**, the second piece in **Y₂**, and the third piece in **Y₃**. Now, instead of multiplying by the expressions with the inequalities, divide by them:

> **:Y₁ = (−2X−4)/(X≤−2)**
> **:Y₂ = (√(4−x²))/((−2<X)(X≤2))**
> **:Y₃ = (−X/2+1)/(X>2)**

Hewlett-Packard

A simple program written with the **IF . . . THEN** program structure can be used to evaluate a piecewise function. If the program is written to take a value of x, evaluate the function, and leave the y value on level 1 of the stack, then the plotting utility can take this program as an "equation" to draw. This is a unique feature of the HP calculators. Let's illustrate this with an example. The **IF . . . ELSE** commands are in the **BRANCH** menu of the HP-28S and in the **BRCH** submenu under the **PRG** menu on the HP-48SX.

≪ IF X − 2 ≤ THEN '−2*X−4' EVAL END
IF X − 2 > X 2 ≤ AND THEN '4−SQ(X)' EVAL END
IF X 2 > THEN '−X/2+1' EVAL END ≫

This program can be used just as any other function, except that the facility in the plotting utility that graphs the derivative cannot handle a function defined this way. It is possible to create a piecewise function that can be differentiated by the calculator and used with the other built-in features of the HP-48SX. This technique uses the **IFTE** (**IF..THEN..ELSE**) command found in the **BRANCH** (HP-28S) or **BRCH** (HP-48SX) menu. The syntax for this command is '**IFTE**(*test,true clause, false clause*)'. The following shows what is to be entered into the calculator.

'IFTE(X≤ − 2, − 2*X − 4,IFTE(X ≤ 2,4 − X^2, − X/2 + 1))'

This definition seems a bit odd at first, but let's see why it produces the correct function. If $x \leq -2$ is true then $y = -2x - 4$ is evaluated and the next **IFTE** is ignored. If $x > -2$ then the second **IFTE** is executed. At this point we know that $x > -2$, so a check is made to see if $x \leq 2$. If $x \leq 2$ then $y = 4 - x^2$ is evaluated. If it turns out that $x > 2$ then $y = -x/2 + 1$ is evaluated. Thus this structure evaluates the correct pieces for the various values of x.

V.3   Epsilon/Delta Computations

With the following programs it is possible to determine a value of δ for a given ϵ. In the definition of the limit $\lim_{x \to a} f(x) = L$, a value of $\epsilon > 0$ is given, and a value of $\delta > 0$ is to be found so that if $0 < |x - a| < \delta$ then $|f(x) - L| < \epsilon$. Graphically this implies that if the screen range is set to be **Ymin** $= L - \epsilon$, **Ymax** $= L + \epsilon$ then we must find δ so that if **Xmin** $= a - \delta$ and **Xmax** $= a + \delta$ then the graph of $f(x)$ must enter the left edge of the screen and leave the right edge of the screen. If this happens then a suitable value of δ has been found. The name **EPSID** is not to be keyed into the program for the Casio. For the HP-28S or HP-48SX, this is a suggested name to store the program under. On the HP-48SX, lower case letters are entered by pressing $\boxed{\alpha}$ $\boxed{\leftarrow}$ followed by the desired letter.

Casio	TI-81	HP-28S	HP-48SX
EPSID	PrgmE:EPSID	'EPSID'	'EPSID'
"A"?→A	All-Off	≪	≪
"L"?→L	Y₁−On	→fale	→fale
"E"?→E	Disp "A"	≪a 1−	≪a 1 − a 1 +
1→D	Input A	le − R→C	XRNG l e − l e +
Lbl 1	Disp "L"	PMIN	YRNG 1 'D' STO
Range A−D,A+D,	Input L	a 1 + l e +	f STEQ 1 'W' STO
D,L−E,L+E,E	Disp "E"	R→C PMAX	WHILE W 0 ≠
Prog 1◢	Input E	1 'D' STO	REPEAT
"CONTINUE?"	L−E→Ymin	1 'W' STO	ERASE DRAX DRAW
"1−YES,0−NO"	L+E→Ymax	f STEQ	GRAPH D 2 / 'D'
?→W	E→Yscl	WHILE W 0 ≠	.5 *W STO
W=0⇒Goto 2	1→D	REPEAT	"Continue:1−Yes,0−No"
D÷2→D	Lbl 1	CLLCD DRAX DRAW	{": :"} INPUT

Casio	TI-81	HP-28S	HP-48SX
Goto 1	A − D→Xmin	D 2 / 'D' STO	OBJ→ 'W' STO
Lbl 2	A + D→Xmax	.5 *W	END D 2 * 'D' STO
"D =":D◢	D→Xscl	"Continue?"	D 'D' →TAG
	DispGraph	"1 − Yes,0 − No"	'W' PURGE
	Pause	HALT 'W' STO	≫
	Disp "CONTINUE?"	DROP2 END	≫
	Disp "1 − YES,0 − NO"	D 2 * 'D' STO	
	Input W	"D =" D →STR	
	If W = 0	'W' PURGE	
	Goto 2	≫	
	D/2→D	≫	
	Goto 1		
	Lbl 2		
	Disp "D ="		
	Disp D		

Explanation

The explanations that follow will use the example $\lim_{x \to 1}(x^2 − x − 2) = −2$ with $\epsilon = 0.1$.

Casio: Program location **1** contains the commands that graph the function. For this example, program location **1** would contain **Graph Y = X² − X − 2**. When the program is executed you are prompted to enter **A**, which is 1 for the example, the limit **L**, −2 for the example, and **E** (ϵ), 0.1 for the example. After the graph is drawn, press EXE and you will be asked to continue. If the graph did not enter the left edge and leave the right edge, then enter 1 to continue. For this example, continuing four times (pressing EXE each time after the graph is drawn and entering 1) will produce the correct graph; then, when 0 is entered, the program responds by outputting **D = 0.0625**. This is the estimate for δ.

TI-81:   The expression defining the function should be entered in Y_1 before you execute the program. For this example, Y_1 should contain **X² − X − 2**. When the program is executed you are prompted to enter **A**, 1 for the example, the limit **L**, −2 for the example, and **E** (ϵ), 0.1 for the example. After the graph is drawn, press ENTER and you will be asked to continue. If the graph did not enter the left edge and leave the right edge, then enter 1 to continue. For this example, continuing four times (pressing ENTER each time after the graph is drawn and entering 1) will produce the correct graph; then, when 0 is entered, the program responds by outputting **D = 0.0625**. This is the estimate for δ.

HP-28S: Before the program is executed, place the following items on the stack: the expression defining the function on level 4, the value of a on level 3, the value of L on level 2, and the value of ϵ on level 1. For the example considered here, the stack would look like the following before the program is executed, and there should be nothing else on the stack.

4:	'X^2 − X − 2'
3:	1
2:	−2
1:	.1

Once these values are entered, execute the program. After the graph is drawn, press ON (**ATTN**) and you will be asked to continue. If the graph did not enter the left edge and leave the right edge, then enter 1 and press ▉ CONT to continue. For this example, continuing four times (pressing ON each time after the graph is drawn and entering 1 and pressing ▉ CONT) will produce the correct graph; when 0 is entered, the program responds by outputting **D = 0.0625**. This is the estimate for δ. The value of **D** is also stored in user memory.

HP-48SX: Before the program is executed, the stack should be the same as shown above for the HP-28S. After the graph is drawn, press ON (**ATTN**) and you will be asked to continue. If the graph did not enter the left edge and leave the right edge, then enter 1 and press ENTER to continue. For this example, continuing four times (pressing ON each time after the graph is drawn, entering 1 and pressing ENTER) will produce the correct graph; when 0 is entered, the program responds by outputting **D:0.0625**. This is the estimate for δ. The value of **D** is also stored in user memory.

V.4 Bisection

The following programs can be used to implement the Bisection Method for approximating the zeros of a function. These programs use the fact that, if $f(x)$ is continuous on the interval $[a, b]$ and $f(a)$ and $f(b)$ have opposite signs, then the Intermediate Value Theorem guarantees that $f(z) = 0$ for some point z, where $a \le z \le b$. The method halves the interval and determines in which half the zero of $f(x)$ lies. By repeating this procedure the method determines smaller and smaller intervals in which the zero lies. The titles above the programs serve the same purpose as they did for the programs in Section V.3.

Casio	TI-81	HP-28S	HP-48SX
BISCT	PrgmB:BISCT	'BISCT'	'BISCT'
"A"?→A	All-Off	≪ → f a b err	≪ → f a b err
"B"?→B	Disp "A"	≪a 'X' STO f	≪a 'X' STO f
"ERROR"?→E	Input A	EVAL →NUM b 'X' STO	EVAL→NUM b 'X' STO
A→X:Prog 0:Y→L	Disp "B"	f EVAL →NUM	f EVAL →NUM
B→X:Prog 0:Y→R	Input B	→ l r	→ l r
LR≥0⇒Goto 4	Disp "ERROR"	≪IF l r * 0	≪IF l r * 0
Lbl 1	Input E	≥ THEN "NO ROOT"	≥ THEN "NO ROOT"
(B−A)÷2→D	A→X	KILL END	KILL END
(A+B)÷2→C	Y₁→L	DO b a − 2 / 'D' STO	DO b a − 2 / 'D' STO
C→X	B→X	a b + 2 / DUP 'C'	a b + 2 / DUP 'C'
Prog 0:Y→M	Y₁→R	STO 'X' STO	STO 'X' STO
"C":C◢	If LR≥0	f EVAL →NUM	f EVAL →NUM
"F(C)":M◢	Goto 4	'M' STO	'M' STO
D<E⇒Goto 3	Lbl 1	"C=" C →STR +	C 'C' →TAG
Abs M<1ᴇ−11⇒Goto 3	(B−A)/2→D	"F(C)=" M →STR +	M 'FC' →TAG
LM < 0⇒Goto 2	(A+B)/2→C	HALT CLEAR	HALT CLEAR
C→A:M→L	C→X	IF l M * 0 <	IF l M * 0 <
Goto 1	Y₁→M	THEN C 'b' STO	THEN C 'b' STO
Lbl 2	Disp "C"	M 'r' STO ELSE	M 'r' STO ELSE
C→B:M→T	Disp C	C 'a' STO M 'l'	C 'a' STO M 'l'

Casio	TI-81	HP-28S	HP-48SX
Goto 1	Disp "F(C)"	STO END	STO END
Lbl 3	Disp M	UNTIL	UNTIL
"DONE C =":C⬕	Pause	D err < M ABS	D err < M ABS
"F(C) = ":M⬕	If D<E	1E − 11 < OR END	1E − 11 < OR END
Goto 5	Goto 3	"DONE"	"DONE"
Lbl 4	If abs M<1ᴇ−11	"C =" C →STR +	C 'C' →TAG
"NO ROOT"	Goto 3	"F(C) =" M →STR +	M 'FC' →TAG
Lbl 5	If LM<0	{ D X M }	{ D X M }
	Goto 2	PURGE ≫	PURGE ≫
	C→A	≫	≫
	M→L	≫	≫
	Goto 1		
	Lbl 2		
	C→B		
	M→R		
	Goto 1		
	Lbl 3		
	Disp "DONE C ="		
	Disp C		
	Disp "F(C) ="		
	Disp M		
	Goto 5		
	Lbl 4		
	Disp "NO ROOT"		
	Lbl 5		
	End		

Explanation

For the explanations that follow, we will use the example $f(x) = \cos x - x$ on the interval $[0, 1]$. The maximum error in the value of x will be 0.0001. Note that $f(0) = 1$ and $f(1) < 0$.

Casio: Program location **0** contains the commands that evaluate the function. For the example, program location **0** should contain **cos X − X→Y**. Upon execution you are prompted to enter the left endpoint **A** (0 for the example), the right endpoint **B** (1 for the example), and the error **E** (0.0001 for example). The midpoints and the function values at these midpoints are displayed at every step. After the values are displayed, press $\boxed{\text{EXE}}$ to continue. The program ends when the interval containing the zero is sufficiently narrow or the value of the function at a midpoint is nearly zero. For the example, 14 intermediate values will be displayed before the program says it is done. The final values are $c = 0.739074707$, $f(c) = 1.74493462 \times 10^{-5}$. You can remove the pieces of the program that display the intermediate values, in which case the program will display only the approximate zero and the function value when done.

TI-81:   The expression defining the function should be entered in $\mathbf{Y_1}$ before execution. For the example, $\mathbf{Y_1}$ should contain **cos X − X**. Upon execution you are prompted to enter the left endpoint **A** (0 for the example), the right endpoint **B** (1 for the example), and the error **E** (0.0001 for the example). The midpoints and the function values at these midpoints are displayed at every step. After the values are displayed, press ENTER to continue. The program ends when the interval containing the zero is sufficiently narrow or the value of the function at a midpoint is nearly zero. For the example, 14 intermediate values will be displayed before the program says it is done. The final values are $c = 0.739074707$, $f(c) = 1.74493462 \times 10^{-5}$. You can remove the pieces of the program that display the intermediate values, in which case the program will display only the approximate zero and the function value when done.

HP-28S: Before the program is executed, place on the stack the expression defining the function on level 4, the value of a on level 3, the value of b on level 2, and the value of the error on level 1. For the example considered here, the stack would look like the following before the program is executed. There should be nothing else on the stack.

4:	'COS(X) − X'
3:	0
2:	1
1:	.0001

Once these values are entered, execute the program. The midpoints and the function values at these midpoints are displayed at every step. After the values are displayed, press ■ CONT to continue. The program ends when the interval containing the zero is sufficiently narrow or the value of the function at a midpoint is nearly zero. For the example, 14 intermediate values will be displayed before the program says it is done. The final values are $c = 0.73907470703$, $f(c) = .000017449349$. You can remove the pieces of the program that display the intermediate values. Simply delete the commands: "**C =**" **C** →**STR** + "**F(C)=**" **M** →**STR** + **HALT CLEAR**. The program will display only the approximate zero and the function value when done.

HP-48SX: Before the program is executed, the stack should be the same as shown above for the HP-28S. The midpoints and the function values at these midpoints are displayed at every step. After the values are displayed, press ← CONT to continue. The number of intermediate values displayed and the final values are the same as those given for the HP-28S. You can remove the pieces of the program that display the intermediate values. Simply delete the commands: **C 'C'** →**TAG M 'FC'** →**TAG HALT CLEAR**. The program will display only the approximate zero and the function value when done.

V.5 Slopes and Derivatives

The evaluation of the slope of a secant line and its limit is most easily done on the Casio and TI-81 with short programs. The first set evaluates the quotients $m_+(h) = (f(a + h) − f(a))/h$ and $m_-(h) = (f(a) − f(a − h))/h$ for smaller and smaller values of h.

Casio	TI-81
SLOPE	**PrgmS:SLOPE**
"A"?→A	**Disp "A"**
Lbl 1	**Input A**
"H"?→H	**Lbl 1**
H=0⇒Goto 2	**Disp "H"**
A+H→X:Prog 0	**Input H**
Y→M:A→X	**If H=0**
Prog 0:Y→F	**Goto 2**
(M−F)÷H→M	**A+H→X**
A−H→X:Prog 0	**Y₁→M**
(F−Y)÷H→N	**A→X**
"M+=":M◢	**Y₁→F**
"M−=":N◢	**(M−F)/H→M**
Goto 1	**A−H→X**
Lbl 2	**(F−Y₁)/H→N**
	Disp "M+="
	Disp M
	Disp "M−="
	Disp N
	Pause
	Goto 1
	Lbl 2

Explanation

For the explanations that follow we will use the example

$$f(x) = \begin{cases} x^2/2 - 1, & \text{if } x \le 1 \\ x - 3/2, & \text{if } 1 < x. \end{cases}$$

Casio: Program location **0** contains the expressions that evaluate the function. For this example, program location **0** should contain the following.

```
X≤1⇒Goto 1
X−3÷2→Y
Goto 2
Lbl 1
X²÷2−1→Y
Lbl 2
```

Upon execution of the program you are prompted to input **A** (1 for the example) and **H** (0.01 for the example). The program computes the two quotients $m_+(h)$ and $m_-(h)$ at $a = 1$ for $h = 0.01$. The values displayed for this example are **M+ = 1.** and **M− = 0.995.** This program will continue prompting for a value of **H** until the value **H = 0** is entered. The values of $m_+(h)$ and $m_-(h)$ are stored in **M** and **N**, respectively.

TI-81: The expressions needed to evaluate the function are assumed to be in **Y₁**. For the example, **Y₁** should contain **(X²/2 − 1)(X≤1) + (X − 3/2)(X>1)**. Upon execution of the program you are prompted to input **A** (1 for the example) and **H** (0.01 for the example). The program computes the two quotients $m_+(h)$ and $m_-(h)$ at $a = 1$ for $h = 0.01$. The values displayed for this example are **M+ = 1.** and **M− = 0.995.**

This program will continue prompting for a value of **H** until the value **H=0** is entered. The values of $m_+(h)$ and $m_-(h)$ are stored in **M** and **N**, respectively.

Hewlett-Packard: With the HP calculators this computation can be accomplished easily when the function is defined by a single expression. Enter the function '**X^2 − 2*X**' and store it as **F**. Now store '**A+H**' in '**X**' and put **F** on the stack and evaluate it. The expression on the stack now is $f(a + h)$. Store '**A**' in '**X**' and put **F** on the stack and evaluate it. Subtract this expression from the previous one and then divide it by '**H**'. You now have an expression that will evaluate the quotient $(f(a + h) - f(a))/h$. A nice way to evaluate this is to activate the ⌷SOLV⌷ utility and store this equation there by pressing ⌷⌷⌷STEQ⌷⌷⌷. Once this is done, press ⌷⌷⌷SOLVR⌷⌷⌷ and you will see a menu with the listings ⌷⌷⌷A⌷⌷⌷, ⌷⌷⌷H⌷⌷⌷, and ⌷⌷⌷EXPR=⌷⌷⌷. Enter a value for **A** and press ⌷⌷⌷A⌷⌷⌷; do the same thing for **H** and then press ⌷⌷⌷EXPR=⌷⌷⌷. The value on the stack is the quotient $(f(a + h) - f(a))/h$. A similar procedure can be used to evaluate the quotient $(f(a) - f(a - h))/h$. Unfortunately, this procedure will not work with a piecewise function defined by a program or with the '**IFTE**' command. For a function like this, a short program is necessary.

> HP-28S, HP-48SX
>
> ```
> 'DIFQUO'
> ≪ → a h
> ≪a 'X' STO F
> EVAL →NUM DUP
> a h + 'X' STO F
> EVAL →NUM SWAP
>  − h / SWAP a h −
> 'X' STO F EVAL
> →NUM − h /
> ≫'X' PURGE
> ≫
> ```

Explanation

Let us use the piecewise defined function given above. In the user variable **F**, store the expression '**IFTE(X≤1,X^2/2 − 1,X − 3/2)**'. The program assumes that the function is stored under the variable name '**F**'. Before the program is executed, enter the value of the point a on level 2 and the value of h on level 1. For the example, you would put 1 on level 1 and .01 on level 2, but nothing else should be on the stack. The program takes the two numbers from the stack, and evaluates the two quotients: $(f(a + h) - f(a))/h$, whose value is left on level 2, and $(f(a) - f(a - h))/h$, whose value is left on level 1.

Graphing the Derivative

Casio: Producing the graph of the derivative on the Casio requires the program **PTPLOT** given in Section V.2. The commands required to approximate the derivative should be placed in program location **2**. In this case, the quotient $(f(x + h) - f(x - h))/(2h)$ will be used because it is more accurate than $(f(x + h) - f(x))/h$. The second derivative can be drawn by using a similar program and evaluating the quotient $(f(a + h) - 2f(a) + f(a - h))/h^2$, and the commands for this should be placed in program location **3**. In each case, the value of h is chosen to be 0.001. As always, program location **0** evaluates the function and program location **1** draws the graph of $f(x)$.

Derivative Evaluator   Second Derivative Evaluator

X→T:T − .001→X   X→T:T − .001→X
Prog 0:Y→M   **Prog 0:Y→M**
T + .001→X:Prog 0   T + .001→X:Prog 0
(Y − M) ÷ .002→M   Y + M→M:T→X
T→X:M→Y   **Prog 0:(M − 2Y) ÷ .000001→Y**

The explanation that follows will use the example $f(x) = x^3 - 4x$. Place in program location **0** the expressions needed to evaluate the function. For the example, program location **0** will be **Xx^{y}3 − 4X→Y**. To draw the graph of $f(x)$ as well as the graphs of its first and second derivatives, program location **1** should contain **Graph Y = Xx^{y}3 − 4X**. The range must be set first; also, **A** must be given the value of **Xmin** and **B** the value of **Xmax**. You will notice that the value of x is placed in location **T**. This is necessary to evaluate $f(x)$ at $x - h$ and $x + h$. Also, when the **Plot X,Y** command is used, the closest screen coordinate to **X** is put in location **X**, thus changing its value. The necessary changes to **PTPLOT** follow.

PTPLOT

"A"?→A:"B"?→B
"N"?→N
(B − A) ÷ N→H
Prog 1
0→I:Lbl 1
A + IH→X
Prog 2
Plot X,Y
T→X:Prog 3
Plot X,Y
Isz I
I ≤ N⇒Goto 1

It is certainly possible to graph $f'(x)$ and $f''(x)$ by computing these functions symbolically and then using the **Graph** command. It should be pointed out that if a function is not defined outside the interval $[a, b]$, such as for $f(x) = \sqrt{1 - x^2}$, $-1 \le x \le 1$, then the value for **A** should be $-1 + 0.001$ and the value for **B** should be $1 - 0.001$. This will avoid an error when the program evaluates $f(a - 0.001)$ and $f(b + 0.001)$.

TI-81: In the **MATH** menu, the eighth item under the **MATH** title is the **NDeriv(** command. This command has the syntax **NDeriv('function', 'positive number')** and is used to evaluate the quotient $(f(a + h) - f(a - h))/2h$. So the graph of the derivative of $f(x) = x^3 - 4x$ can be obtained by entering **X^3 − 4X** for **Y$_1$** and **NDeriv(Y$_1$,.001)** for **Y$_2$**. The second derivative can be graphed as well by entering **NDeriv(Y$_2$,.001)** for **Y$_3$** where **Y$_2$** contains **NDeriv(Y$_1$,.001)**. The graphs of these functions can be traced as well.

Hewlett-Packard: Each of the Hewlett-Packard calculators can symbolically compute derivatives of functions that are defined by single expressions. This is done by entering the function on level 2 of the stack and the variable name on level 1, and pressing ■ d/dx on the HP-28S or pressing → ∂ on the HP-48SX. With this facility it is a simple matter to graph $f(x)$, $f'(x)$, and $f''(x)$ on the same screen. For piecewise defined functions the command **IFTE** should be used. For a description of how this is done, see Section V.2.

V.6 Newton's Method

The following programs can be used to implement Newton's Method. It is known that Newton's Method will not always converge to a zero of the function. If this occurs in any of the programs, stop the program and restart it with a different initial guess. If initial guesses are estimated from the graph of the function, the possibility of the method not converging is reduced.

Casio	TI-81	HP-28S and HP-48SX
NEWTON	PrgmN:NEWTON	'NEWTON'
"X0"?→X	Disp "X0"	≪ → x
Lbl 1	Input X	≪x DUP 'X' STO F
Prog 0	Lbl 1	EVAL →NUM DF EVAL
X−Y÷D→X▲	X−Y$_1$/Y$_2$→X	→NUM / − DUP 'X'
Abs Y < 1ᴇ−11⇒Goto 2	Disp "X"	STO ≫ ≫
Goto 1	Disp X	
Lbl 2	Pause	
"DONE X=":X▲	If abs Y$_1$ <1ᴇ−11	
Prog 0	Goto 2	
"Y=":Y▲	Goto 1	
	Lbl 2	
	Disp "DONE X="	
	Disp X	
	Disp "Y="	
	Disp Y$_1$	
	Y$_2$→D	

Explanation
In what follows, the example $f(x) = x^3 - x + 1$ will be used.

Casio: Program location **0** should contain the commands that evaluate both $f(x)$ and $f'(x)$. For the example, program location **0** will be as follows.

$$Xx^y3 - X + 1 → Y$$
$$3X^2 - 1 → D$$

Upon execution of the program you are prompted to enter the initial guess for the zero. For the example, we see from the graph that the zero is near -1, so enter -1 for the initial guess. The intermediate values of x are displayed at each step and so the EXE key must be pressed after each value is displayed in order for the program to continue. The program stops when $|f(x)| < 10^{-11}$. For the example, six values will be displayed and the program ends displaying $x = -1.324717957$ and $y = 0$. If at any time during the execution $f'(x)$ is zero, the program will stop and the calculator will display **Ma error**. For example, if the initial guess $-1/\sqrt{3}$ is entered, this error will occur. If it happens that $f'(x)$ is nearly zero, the estimate displayed will be very large in absolute value. When either of these occurs, stop the program and restart it with another guess. The final value of the root is stored in the location **X**.

TI-81:   The expressions needed to evaluate $f(x)$ and $f'(x)$ are assumed to be in **Y₁** and **Y₂**, respectively. For the example, you should enter **X³ − X + 1** for **Y₁**, and enter **3X² − 1** for **Y₂**. Upon execution of the program you are prompted to enter the initial guess for the zero. For the example, from the graph we see that the zero is near -1, so enter -1 for the initial guess. The intermediate values of x are displayed at each step and so the $\boxed{\text{ENTER}}$ key must be pressed after each value is displayed in order for the program to continue. The program stops when $|f(x)| < 10^{-11}$. For the example, five values will be displayed and the program ends displaying $x = -1.324717957$ and $y = -10^{-12}$. If at any time during the execution $f'(x)$ is zero, the program will stop and the calculator will display **ERROR 02 MATH** because of an attempt to divide by zero. If it happens that $f'(x)$ is nearly zero, the estimate displayed will be very large in absolute value. For example, if the initial guess $-1/\sqrt{3}$ is entered, the next estimate will be $-3.846944943 \times 10^{10}$. When either of these occurs, stop the program and restart it with another guess. The final value of the root is stored in the location **X**.

Hewlett-Packard: The expressions needed to evaluate the function and its derivative are assumed to be stored under the names **F** and **DF**, respectively. These should be algebraic expressions. For the example, **F** should be '**X^3 − X + 1**' and **DF** should be '**3*X^2 − 1**'. To execute the program, first enter the initial guess on level 1 of the stack, and then press the white key under the menu label that holds the program (this should be **NEWTON**). The program simply takes the value of x on the stack, stores it under the name '**X**', and evaluates the expression $x - f(x)/f'(x)$ once and leaves its value on the stack. To continue, reexecute the program until the values left on the stack are the same. For the example, enter -1 as the initial guess and execute **NEWTON**. The subsequent values on the stack are -1.5, -1.34782608696, -1.32520039895, -1.324718174, -1.32471795725. All of the values after this are the same. To see that this is the zero of $f(x)$ execute $\boxed{|||}\boxed{\text{F}}\boxed{|||}$ $\boxed{\text{EVAL}}$. The value displayed will be -0.00000000002. If at any time it happens that $f'(x)$ is zero, then the error **/ Error: Infinite Result** is displayed. If it happens that $f'(x)$ is nearly zero, the estimate displayed will be very large in absolute value. For example, if the initial guess $-1/\sqrt{3}$ is entered, the next estimate will be $1.38490017946 \times 10^{12}$. When either of these occurs, drop the value from the stack and enter a new initial guess.

V.7   Antiderivatives and Euler's Method

The following programs can be used to graph the antiderivative of a function. The programs use the tangent line approximation to approximate the graph of the antiderivative.

Casio	TI-81	HP-28S	HP-48SX
ANTID	**PrgmA:ANTID**	**'ANTID'**	**'ANTID'**
Cls	**ClrDraw**	≪ → f y a b n	≪ → f y a b n
"A"?→A:"B"?→B	**All-Off**	≪b a − n / → h	≪b a − n / → h
"N"?→N:"F(A)"?→F	**Y₁ − On**	≪ f STEQ CLLCD DRAX	≪ f STEQ ERASE DRAX
(B − A) ÷ N→H	**DispGraph**	DRAW a y R→C PIXEL	DRAW a y R→C PIXON
Prog 1:Plot A,F	Disp "A"	1 n FOR i a i h * +	1 n FOR i a i h * +
1→I:Lbl 1	Input A	DUP 'X' STO f EVAL	DUP 'X' STO f EVAL
A + IH→X:Prog 0	Disp "B"	→NUM h * y + 'y'	→NUM h * y + 'y'
F + HY→F	Input B	STO y R→C PIXEL	STO y R→C PIXON
Plot X,F:Isz I	Disp "N"	NEXT 'X' PURGE	NEXT GRAPH 'X' PURGE
I≤N⇒Goto 1	Input N	≫ ≫	≫ ≫
	Disp "F(A)"	≫	≫
	Input F		
	(B − A)/N→H		
	PT − On(A,F)		
	1→I		
	Lbl 1		
	A + IH→X		
	F + HY₁→F		
	PT − On(X,F)		
	IS > (I,N)		
	Goto 1		

Explanation

In the following explanations we will consider the example of graphing the antiderivative of $f(x) = \sqrt{x + 1}$ on the interval $[-1, 1]$.

Casio: In program location **0** enter the expressions that evaluate $f(x)$. Program location **1** is used to draw the graph of $y = f(x)$. For the example, program location **0** will contain $\sqrt{(X+1)}→Y$ and program location **1** will contain **Graph** $Y = \sqrt{(X+1)}$. The range must be set before the program is executed, so for this example set **Xmin = −1, Xmax = 1, Ymin = 0** and **Ymax = 3**. Upon execution of the program you are prompted to enter the endpoints **A** and **B** (−1 and 1 for the example), **N**, and an initial value of the antiderivative, **F(A)**. A good value for **F(A)** most of the time is zero. (This value can also be thought of as the arbitrary constant.) Changing this value will simply raise or lower the graph of the antiderivative. Since there are 95 pixels across the screen, this is a good value for **N**, initially. A larger value may be needed to obtain a more accurate graph.

TI-81:   The expressions that evaluate $f(x)$ are assumed to be placed in **Y₁**. For the example, **Y₁** will contain $\sqrt{(X+1)}$. The range must be set before the program is executed, so for this example set **Xmin = −1, Xmax = 1, Ymin = 0** and **Ymax = 3**. Upon execution of the program you are prompted to enter the endpoints **A** and **B** (−1 and 1 for the example), **N**, and an initial value of the antiderivative, **F(A)**. A good value for **F(A)** most of the time is zero. (This value can also be thought of as the arbitrary constant.) Changing this value will simply raise or lower the graph of the antiderivative. Since there are 96 pixels across the screen, this is a good value for **N**, initially. A larger value may be needed to obtain a more accurate graph.

Hewlett-Packard: Before the program is executed, the screen range must be set. For the example this is done as follows:

HP-28S: **(−1,0)** |||| PMIN |||| **(1,3)** |||| PMAX ||||

HP-48SX: **−1 1** |||| XRNG |||| **0 3** |||| YRNG ||||

The program assumes that prior to execution level 5 has the function in algebraic form, level 4 has the initial value of the antiderivative, level 3 has the left endpoint, level 2 has the right endpoint, and level 1 has the number of points to plot. For the example this takes the following form.

5:	'$\sqrt{(X+1)}$'
4:	0
3:	−1
2:	1
1:	60

There should be nothing else on the stack. Upon execution of the program, the screen is cleared and a point graph of the antiderivative is drawn. The graph is held on the screen until ON (ATTN) is pressed.

The following programs can be used to graph the solution of $y' = f(x, y)$, $y(a) = y_0$. Each program uses Euler's Method, which is based on a tangent line approximation.

Casio	TI-81	HP-28S	HP-48SX
EULER	PrgmF:EULER	'EULER'	'EULER'
Cls	ClrDraw	≪ → f y a b n	≪ → f y a b n
"A"?→A:"B"?→B	All-Off	≪b a − n / → h	≪b a − n / → h
"N"?→N:"Y(A)"→T	Disp "A"	≪CLLCD DRAX	≪ERASE DRAX
(B−A)÷N→H	Input A	a y R→C PIXEL	a y R→C PIXON
Plot A,T:1→I	Disp "B"	1 n FOR i a i h * +	1 n FOR i a i h * +
Lbl 1	Input B	DUP 'X' STO y 'Y'	DUP 'X' STO y 'Y' STO
A+IH→X:T→Y	Disp "N"	STO f EVAL →NUM	f EVAL →NUM h *
Prog 0:T+HF→T	Input N	h * y + DUP 'y' STO	y + DUP 'y' STO
Plot X,T:Isz I	Disp "Y(A)"	R→C PIXEL NEXT	R→C PIXON NEXT
I≤N⇒Goto 1	Input T	{ X Y } PURGE	{ X Y } PURGE
	(B−A)/N→H	≫ ≫	GRAPH
	PT−On(A,T)	≫	≫ ≫
	1→I		≫
	Lbl 1		
	A+IH→X		
	T→Y		
	T+HY₁→T		
	PT−On(X,T)		
	IS>(I,N)		
	Goto 1		

Explanation

For the explanations that follow we will use the example $y' = 4x/y$, $y(-2) = 4$, on the interval $[-2, 2]$.

Casio: The expression needed to evaluate $f(x, y)$ must be placed in program location **0**. It is assumed that the value of $f(x, y)$ is in **F**. For the example, program location **0** would have **4X ÷ Y→F**. Before execution the range must be set. If you have no idea about how to set the vertical range, a first try might be $y(a) - 5$ and $y(a) + 5$ where $y(a)$ is the initial value. For the example, the range should be set at **Xmin = −2, Xmax = 2, Ymin = 0, Ymax = 4**. Upon execution of the program, you will be prompted to enter the endpoints **A** and **B**, the number of points to be plotted **N**, and the initial value **Y(A)**. A point graph of the approximate solution is then drawn.

TI-81:   The expression needed to evaluate $f(x, y)$ must be placed in **Y₁**. For the example, **Y₁** would have **4X/Y**. Before execution the range must be set. If you have no idea about how to set the vertical range, a first try might be $y(a) - 5$ and $y(a) + 5$ where $y(a)$ is the initial value. For the example, the range should be set at **Xmin = −2, Xmax = 2, Ymin = 0, Ymax = 4**. Upon execution of the program you will be prompted to enter the endpoints **A** and **B**, the number of points to be plotted **N**, and the initial value **Y(A)**. A point graph of the approximate solution is then drawn.

Hewlett-Packard: Just as in the program **'ANTID'**, the screen range must be set before execution. If you have no idea how to set the vertical range, a first try might be $y(a) - 5$ for the bottom edge and $y(a) + 5$ for the top, where $y(a)$ is the initial value. For the example, the range should be set as follows.

HP-28S: **(−2,0}** ▯▯ **PMIN** ▯▯ **(2,4)** ▯▯ **PMAX** ▯▯

HP-48SX: **−2 2** ▯▯ **XRNG** ▯▯ **0 4** ▯▯ **YRNG** ▯▯

The program assumes that prior to execution level 5 has the function $f(x, y)$ in algebraic form, level 4 has the initial value, level 3 has the left endpoint, level 2 has the right endpoint, and level 1 has the number of points to plot. For the example this takes the following form.

5:	**'4*X/Y'**
4:	**4**
3:	**−2**
2:	**2**
1:	**60**

There should be nothing else on the stack. On the HP-28S, when the program is executed, the screen is cleared and a point graph of the approximate solution is drawn. On the HP-48S, when the program is executed, after several seconds the graph appears in complete form. The graph is held on the screen until $\boxed{\text{ON}}$ **(ATTN)** is pressed.

V.8 Integration

The first set of programs draw the graph of a function and a Riemann sum approximation of its definite integral. The graph is the set of rectangles that make up the Riemann sum $\sum_{j}^{n} f(t_j)\Delta x$ where $\Delta x = (b-a)/n$. The points t_j in the interval $[x_{j-1}, x_j]$ can be chosen to be the left endpoints (type 1), the right endpoints (type 2), the midpoints (type 3), or random points (type 4).

Casio	TI-81	HP-28S	HP-48SX
RSUMS	PrgmS:RSUMS	'RSUMS'	'RSUMS'
Cls	ClrDraw	≪ → f a b n typ	≪ → f a b n typ
"A"?→A:"B"?→B	All-Off	≪b a − n / → h	≪b a − n / → h
"N"?→N	Y₁-On	≪0 'R' STO f	≪0 'R' STO f STEQ
"POINT TYPE"?→W	Disp "A"	STEQ CLLCD DRAX	a b XRNG ERASE AUTO
(B−A)÷N→H	Input A	DRAW 1 n FOR j	DRAX DRAW 1 n FOR j
Prog 1	Disp "B"	IF typ 1 = = THEN	IF typ 1 = = THEN
1→J:0→R	Input B	a j 1 − h * + DUP	a j 1 − h * + DUP 'X'
Lbl 0	Disp "N"	'X' STO DUP h + 'T2' STO	STO DUP h + 'T2' STO 'T1'
W=1⇒Goto 1	Input N	'T1' STO END	STO END
W=2⇒Goto 2	Disp "POINT TYPE"	IF typ 2 = = THEN	IF typ 2 = = THEN
W=3⇒Goto 3	Input W	a j h * + DUP 'X'	a j h * + DUP 'X' STO
W=4⇒Goto 4	(B−A)/N→H	STO DUP h − 'T1' STO	DUP h − 'T1' STO 'T2'
Lbl 1	0→R	'T2' STO END	STO END
A+(J−1)H→X	1→J	IF typ 3 = = THEN	IF typ 3 = = THEN
X→P:X+H→Q	Lbl 0	a j h * + h 2 /−	a j h * + h 2 / − DUP
Goto 5	If W=1	DUP 'X' STO DUP h	'X' STO DUP h 2 / DUP
Lbl 2	Goto 1	2 / DUP ROT +'T2'	ROT + 'T2' STO − 'T1'
A+JH→X:X→Q	If W=2	STO − 'T1' STO END	STO END
X−H→P	Goto 2	IF typ 4 = = THEN	IF typ 4 = = THEN
Goto 5	If W=3	a j 1 − h * + DUP	a j 1 − h * + DUP 'T1'
Lbl 3	Goto 3	'T1' STO DUP h +	STO DUP h +'T2' STO
A+JH−H÷2→X	If W=4	'T2' STO h RAND *	h RAND * + 'X' STO
X−H÷2→P	Goto 4	+ 'X' STO END	END
X+H÷2→Q	Lbl 1	f EVAL DUP 'R' STO+	f EVAL DUP 'R' STO+
Goto 5	A+(J−1)H→X	'Y' STO T1 0 R→C	'Y' STO T1 0 R→C
Lbl 4	X→P	T1 Y R→C DUP ROT	T1 Y R→C DUP ROT
A+(J−1)H→X	X+H→Q	LINE† T2 Y	LINE T2 Y
X→P:X+H→Q	Goto 5	R→C DUP ROT LINE	R→C DUP ROT LINE
HRan# +X→X	Lbl 2	T2 0 R→C LINE	T2 0 R→C LINE

---

†**LINE** is a separate program that the HP-28S program uses to draw a line between two points on the screen. The code for this follows:

```
≪
→ p1 p2
≪ 0 1 FOR t
'p1*(1−t)+p2*t'
EVAL PIXEL
.0625 STEP ≫
≫
```

Casio	TI-81	HP-28S	HP-48SX
Goto 5	**A + JH→X**	**NEXT h 'R' STO***	**NEXT h 'R' STO***
Lbl 5:Prog 0	**X − H→P**	**{ T1 T2 Y X }**	**{ T1 T2 Y X }**
Y→F:F + R→R	**X→Q**	**PURGE ≫**	**PURGE GRAPH ≫**
Plot P,0	**Goto 5**	**≫ ≫**	**≫ ≫**
Plot P,F:Line	**Lbl 3**		
Plot Q,F:Line	**A + JH − H/2→X**		
Plot Q,0:Line	**X − H/2→P**		
Isz J	**X + H/2→Q**		
J≤N⇒Goto 0	**Goto 5**		
Line◢	**A + (J − 1)H→X**		
"RS = ":HR→R◢	**X→P**		
	X + H→Q		
	HRand + X→X		
	Goto 5		
	Lbl 5		
	Y₁→F		
	F + R→R		
	Line(P,0,P,F)		
	Line(P,F,Q,F)		
	Line(Q,F,Q,0)		
	IS > (J,N)		
	Goto 0		
	Pause		
	HR→R		
	Disp "RS = "		
	Disp R		

Explanation

The following explanations will use the function $f(x) = \sin x$ on the interval $[0, \pi/2]$. The goal is to graph Riemann sum approximations to the area bounded by the x-axis and the curve $y = \sin x$ for $0 \le x \le \pi/2$.

Casio: Before executing the program, the screen range must be set. For this example set **Xmin = 0**, **Xmax = 1.5707963** ($\pi/2$), **Ymin = 0**, and **Ymax = 1**. Program location **0** should contain the expression that evaluates the function; for this example, this would be **sin X→Y**. Program location **1** should contain the commands to graph the function; for this example, this would be **Graph Y = sin X**. Upon execution you will be prompted to enter the endpoints **A** and **B** (0 and $\pi/2$ for the example), and the number of points **N**. This is then followed by a prompt to enter the point type: 1 for left endpoints, 2 for right endpoints, 3 for midpoints, and 4 for random points. For the example, if you enter point type 1 and **N** = 10, the Riemann sum will be that for left endpoints. The graph is held on the screen until ⟦EXE⟧ is pressed and the value of the Riemann sum is displayed. The value for this example will be 0.91940317. Notice that since the function is increasing on $[0, \pi/2]$, the Riemann sum using left endpoints is a lower sum and the sum using right endpoints is an upper sum.

TI-81:   Before executing the program, the screen range must be set. For this example, set **Xmin = 0**, **Xmax = 1.5707963** ($\pi/2$), **Ymin = 0**, and **Ymax = 1**. The expression that evaluates the function should be placed in **Y₁**; for this example, this would be **sin X**. Upon execution you will be prompted to enter the endpoints **A** and **B** (0 and $\pi/2$ for the example), and the number of points **N**. This is then followed by a prompt to enter the point type: 1 for left endpoints, 2 for right endpoints, 3 for midpoints, and 4 for random points. For the example, if you enter point type 1 and **N** = 10, the Riemann sum will be that for left endpoints. The graph is held on the screen until [ENTER] is pressed and the value of the Riemann sum is displayed. The value for this example will be 0.91940447. Notice that since the function is increasing on [0, $\pi/2$], the Riemann sum using left endpoints is a lower sum and the sum using right endpoints is an upper sum.

Hewlett-Packard: Before executing the program on the HP-28S, the screen range must be set. For this example, that would be (0, 0) for **PMIN** and (1.5707963268, 1) for **PMAX**. (The exact procedure for this is given after the program '**ANTID**'.) Setting the range is not required for the HP-48SX since the endpoints a and b are put into **XRNG** by the program and the autoscaling utility in the **PLOTR** submenu is used. Before executing the program the function must be on level 5, the left endpoint a on level 4, the right endpoint b on level 3, the number of points on level 2, and the point type on level 1: 1 for left endpoints, 2 for right endpoints, 3 for midpoints, and 4 for random points. For the example considered here, the stack should look like the following prior to execution.

5:	'SIN(X)'
4:	0
3:	1.5707963268
2:	10
1:	1

When the program is executed the screen is cleared, the function is graphed, and the rectangles that approximate the area are drawn. After the graph is complete, press [ON] (**ATTN**) and then press |||| R |||| for the value of the sum to be displayed. For the values entered for this example ($n = 10$, point type: 1) the Riemann sum is 0.919403170021. Because of the lack of a built-in utility that draws a line between points, the HP-28S can take some time to produce the graph.

The next set of programs approximate the definite integral by Riemann sums that use left endpoints and right endpoints as in the previous program, the Midpoint Rule, the Trapezoidal Rule, and Simpson's Rule with $2n$ points. All of these approximations are found every time the programs are executed.

Casio	TI-81	HP-28S and HP-48SX
INTAP	Prgml:INTAP	'INTAP'
"A"?→A:"B"?→B	Disp "A"	≪ → f a b n
Lbl 0:"N"?→N	Input A	≪b a − n / DUP 2 /
N=0⇒Goto 2	Disp "B"	→ h k
(B−A)÷N→H	Input B	≪0 DUP DUP 'L' STO
0→L:0→R:0→M	Lbl 0	'R' STO 'M' STO
1→I	Disp "N"	1 n FOR i
Lbl 1	Input N	a i 1 − h * + 'X' STO
A+(I−1)H→X	If N=0	f EVAL →NUM 'L' STO+
Prog 0:Y+L→L	Goto 2	X h + 'X' STO f EVAL

Casio	TI-81	HP-28S and HP-48SX
X + H→X:Prog 0	(B − A)/N→H	→NUM 'R' STO +
Y + R→R	0→L	X k − 'X' STO f EVAL
X − H ÷ 2→X:Prog 0	0→R	→NUM 'M' STO + NEXT
Y + M→M	0→M	h 'L' STO* h 'R' STO*
Isz I	1→I	h 'M' STO* L R + 2 /
I≤N⇒Goto 1	Lbl 1	'T' STO 2 M * T + 3 /
"L = ":HL→L◢	A + (I − 1)H→X	'S' STO 'X' PURGE
"R = ":HR→R◢	Y₁ + L→L	≫ ≫
"M = ":HM→M◢	X + H→X	≫
"T = ":(L + R) ÷ 2→T◢	Y₁ + R→R	
"S = ":(2M + T) ÷ 3→S◢	X − H/2→X	
Goto 0	Y₁ + M→M	
Lbl 2	IS>(I,N)	
	Goto 1	
	HL→L	
	HR→R	
	HM→M	
	Disp "L = "	
	Disp L	
	Disp "R = "	
	Disp R	
	Disp "M = "	
	Disp M	
	Pause	
	(L + R)/2→T	
	(2M + T)/3→S	
	Disp "T = "	
	Disp T	
	Disp "S = "	
	Disp S	
	Pause	
	Goto 0	
	Lbl 2	
	End	

Explanation

For the explanations that follow, the example $f(x) = x^2\sqrt{x + 1}$, $-1 \le x \le 2$ will be used.

Casio: Program location **0** should contain the expressions needed to evaluate the function. For the example this will be $X^2\sqrt{(X + 1)} \rightarrow Y$. When the program is executed you will be prompted to enter the endpoints **A** and **B** (for the example these would be -1 and 2), and the number of points **N**. If **N** = 10 for the example, the values displayed will be: left endpoint Riemann sum **L** = 3.344727175, right endpoint Riemann sum **R** = 5.423188144, Midpoint Rule **M** = 4.332849077, Trapezoidal Rule **T** = 4.383957659, and Simpson's Rule for 20 points **S** = 4.349885271. The EXE key must be pressed after each value is displayed in order to continue. After **S** is displayed and EXE is pressed, you are prompted to enter **N**. This way you can enter larger values of **N** without executing the program again. If at this point you enter **N** = 0 the program will stop.

TI-81: The expressions needed to evaluate the function should be placed in **Y₁**; for the example this will be **X²√(X + 1)**. When the program is executed you will be prompted to enter the endpoints **A** and **B** (for the example these would be −1 and 2), and the number of points **N**. If **N** = 10 for the example, the values displayed will be: left endpoint Riemann sum **L** = 3.344727175, right endpoint Riemann sum **R** = 5.423188144, Midpoint Rule **M** = 4.332849077, Trapezoidal Rule **T** = 4.383957659, and Simpson's Rule for 20 points **S** = 4.349885271. The ENTER key must be pressed after the values of **L**, **R**, and **M** are displayed in order to continue. After **S** is displayed and ENTER is pressed, you are prompted to enter **N**. This way you can enter larger values of **N** without executing the program again. If at this point you enter **N** = 0 the program will stop.

Hewlett-Packard: Before the program is executed, the function must be on level 4, the left endpoint on level 3, the right endpoint on level 2, and the number of points on level 1. Nothing else should be on the stack. For the example, the stack should look like the following.

4:	'X^2*√(X + 1)'
3:	−1
2:	2
1:	10

After the program has finished, the values of **L**, **R**, **M**, **T**, and **S** are stored in user memory. The values can be recalled to the stack by pressing the white key under the appropriate menu item in the **USER** menu (HP-28S) or the **VAR** menu (HP-48SX).

The Hewlett-Packard calculators have a built-in numerical method for approximating definite integrals. The description of how to use these follows.

HP-28S: To use the integration utility, put the function or program on level 3, a list of the form {**'variable' a b** } on level 2, and the prescribed error on level 1. Once this is entered, press ▉ ∫ and the integrator is started. When the integration is complete, the approximate value of the integral is on level 2 and the computed error is on level 1. The following example will help make this clear.

3:	'SIN(SQ(X))'
2:	{ X 0 1 }
1:	0.00001

Invoking the integration utility as described above results in **.310268297696** on level 2 and **3.101601635E-6** on level 1.

HP-48SX: The HP-48SX has the utility EquationWriter that makes it very easy to enter equations or integrals. Enter this utility and, using the right arrow key, you can enter the formula $\int_0^1 \mathbf{SIN(X^2)}\, \mathbf{dX}$ just as it is written here. The **'dX'** is obtained by pressing the right arrow key twice after the left parenthesis is keyed in. To invoke the numerical integration utility press EVAL . If it is possible for the calculator to find an exact value (not possible in this case), it will do so; otherwise it will return to the stack the expression **'∫(0,1,SIN(X^2),X)'**. A numerical value is computed upon pressing $\boxed{\rightarrow}$ $\boxed{\rightarrow \text{NUM}}$. When the integration is complete the numerical approximation will be on level 1 and the value for the error will be **IERR** in the **VAR** menu. For this example, if the calculator is in **STD** model, the approximate value will be **.310268301723**. To get a value with greater error put the calculator in the **FIX** mode with the number of digits of accuracy desired.

Now that we have the methods that approximate a definite integral, a few changes can be made in these programs so that an antiderivative can be drawn. Suppose we are given a continuous function $f(x)$ defined on the interval $[a, b]$. Our task is to draw the graph of $F(x) = \int_a^x f(t) \, dt$. The technique is to partition $[a, b]$ into k subintervals, say $[a, x_1], [x_1, x_2], \ldots, [x_{k-1}, b]$, and apply the Midpoint Rule, Trapezoidal Rule, or Simpson's Rule to each of the subintervals. That is, since

$$F(x_1) = \int_a^{x_1} f(t) \, dt, \ F(x_2) = F(x_1) + \int_{x_1}^{x_2} f(t) \, dt, \ldots,$$

$$F(x_i) = F(x_{i-1}) + \int_{x_{i-1}}^{x_i} f(t) \, dt$$

we apply one of these rules with n points to each of the integrals $\int_{x_{j-1}}^{x_j} f(t) \, dt, j = 1, 2, \ldots, i$, and sum the results to approximate $F(x_i)$. The following programs do just that. The explanation of how to use the programs is similar to that for **'ANTID'**. The only difference is that **K** is the number of subintervals (the number of points plotted) and **N** is the number of points used for the Midpoint Rule, the Trapezoidal Rule, or Simpson's Rule on each of the subintervals. For the Hewlett-Packard calculators the programs require the function to be on level 6, the initial value on level 5, the left endpoint on level 4, the right endpoint on level 3, the number of subintervals k on level 2, and the number of points n on level 1. The range must be set before execution on all of the programs except those for the HP-48SX.

The following programs draw a function $f(x)$ and its antiderivative using the Midpoint Rule.

Casio	TI-81	HP-28S	HP-48SX
ADMPT	PrgmC:ADMPT	'ADMPT'	'ADMPT'
Cls: Prog 1	**ClrDraw**	**≪ → f m a b k n**	**≪ → f m a b k n**
"A"?→A:"B"?→B	**All-Off**	**≪b a − k / DUP 2 n**	**≪b a − k / DUP 2 n**
"K"?→K:"N"?→N	**Y₁−On**	*** / → h r**	*** / → h r**
"FA"?→M	**Disp "A"**	**≪f STEQ CLLCD**	**≪a b XRNG f STEQ**
(B−A)÷K→H	**Input A**	**DRAX DRAW**	**ERASE AUTO DRAX DRAW**
H÷N→R	**Disp "B"**	**1 k FOR i 0 'S'**	**1 k FOR i 0 'S'**
1→I:Lbl 1:0→S	**Input B**	**STO a i h * + 'U' STO**	**STO a i h * + 'U' STO**
A+IH→U	**Disp "FA"**	**1 n FOR j U h − 2**	**1 n FOR j U h − 2 j * 1**
1→J:Lbl 2	**Input M**	**j * 1 − r * + 'X'**	**− r * + 'X' STO f EVAL**
U−H+(2J−1)R÷2→X	**Disp "N"**	**STO f EVAL →NUM**	**→NUM 'S' STO+ NEXT**
Prog 0:S+Y→S	**Input N**	**'S' STO+ NEXT 2 r**	**2 r S * * m + 'm' STO**
Isz J	**Disp "K"**	**S * * m + 'm' STO**	**U m R→C PIXON NEXT**
J ≤ N⇒Goto 2	**Input K**	**U m R→C PIXEL**	**{ X U S EQ }**
RS + M→M	**(B−A)/K→H**	**NEXT { X U S EQ }**	**PURGE GRAPH ≫**
Plot U,M	**H/N→R**	**PURGE DGTIZ ≫**	**≫ ≫**
Isz I	**1→I**	**≫ ≫**	
I ≤ K⇒Goto 1	**Lbl 1**		
	0→S		

TI-81

```
A + IH→U
1→J
Lbl 2
U − H + (2J − 1)R/2→X
S + Y₁→S
IS > (J,N)
Goto 2
RS + M→M
PT − On(U,M)
IS > (I,K)
Goto 1
```

The following programs draw a function $f(x)$ and its antiderivative using the Trapezoidal Rule.

Casio	TI-81	HP-28S	HP-48SX
ADTRP	PrgmT:ADTRP	'ADTRP'	'ADTRP'
Cls:Prog 1	ClrDraw	≪ → f t a b k n	≪ → f t a b k n
"A"?→A:"B"?→B	All-Off	≪b a − k / DUP n	≪b a − k / DUP n
"K"?→K:"N"?→N	Y₁ − On	/ → h r	/ → h r
"FA"?→T	Disp "A"	≪f STEQ CLLCD	≪f STEQ a b XRNG
(B − A) ÷ K→H	Input A	DRAX DRAW 1 k	ERASE AUTO DRAX DRAW
H ÷ N→R	Disp "B"	FOR i 0 'S' STO	1 k FOR i
1→I:Lbl 1:0→S	Input B	a i h * + 'U' STO	0 'S' STO a i h * +
A + IH→U	Disp "FA"	1 n 1 − FOR j U h − j r *	'U' STO 1 n 1 − FOR j U h −
1→J:Lbl 2	Input T	+ 'X' STO f EVAL	j r * + 'X' STO f
U − H + JR→X	Disp "N"	→NUM 'S' STO +	EVAL →NUM 'S' STO +
Prog 0:S + Y→S	Input N	NEXT U h − 'X' STO	NEXT U h − 'X' STO
Isz J	Disp "K"	f EVAL →NUM 2 'S'	f EVAL →NUM 2 'S'
J≤N − 1⇒Goto 2	Input K	STO* 'S' STO + U 'X'	STO* 'S' STO + U 'X'
U − H→X:Prog 0	(B − A)/K→H	STO f EVAL →NUM S	STO f EVAL →NUM S
Y + 2S→S:U→X	H/N→R	+ r * 2 / t + 't' STO	+ r * 2 / t + 't' STO
Prog 0	1→I	U t R→C PIXEL	U t R→C PIXON NEXT
R(Y + S) ÷ 2 + T→T	Lbl 1	NEXT { X S U EQ}	{ X S U EQ }
Plot U,T	0→S	PURGE DGTIZ ≫	PURGE GRAPH ≫
Isz I	A + IH→U	≫ ≫	≫ ≫
I ≤ K⇒Goto 1	1→J		
	Lbl 2		
	U − H + JR→X		
	S + Y₁→S		
	IS>(J,N − 1)		
	Goto 2		
	U − H→X		
	Y₁ + 2S→S		
	U→X		
	R(Y₁ + S)/2 + T→T		
	PT − On(U,T)		
	IS>(I,K)		
	Goto 1		

The following programs draw a function $f(x)$ and its antiderivative using Simpson's Rule.

Casio	TI-81	HP-28S	HP-48SX
ADSMP	PrgmU:ADSMP	'ADSMP'	'ADSMP'
Cls:Prog 1	ClrDraw	≪ → f s a b k n	≪ → f s a b k n
"A"?→A:"B"?→B	All-Off	≪ b a − k / DUP n	≪ b a − k / DUP n
"K"?→K:"N"?→N	Y₁ − On	/ → h r	/ → h r
"FA"?→S	Disp "A"	≪ f STEQ CLLCD	≪ f STEQ a b XRNG
(B − A) ÷ K→H	Input A	DRAX DRAW 1 k	ERASE AUTO DRAX DRAW
H ÷ N→R	Disp "B"	FOR i 0 'T' STO 0	1 k FOR i
1→I:Lbl 1:0→T	Input B	'M' STO a i h * +	0 'T' STO 0 'M' STO
0→M:A + IH→U	Disp "FA"	'U' STO 1 n 1 −	a i h * + 'U' STO
1→J:Lbl 2	Input S	FOR j U h − j r *	1 n 1 − FOR j U h −
U − H + JR→X	Disp "N"	+ 'X' STO f EVAL	j r * + 'X' STO
Prog 0:T + Y→T	Input N	→NUM 'T' STO+ X	f EVAL →NUM 'T'
X − R ÷ 2→X	Disp "K"	r 2 / − 'X' STO	STO+ X r 2 / − 'X'
Prog 0:Y + M→M	Input K	f EVAL →NUM	STO f EVAL →NUM
Isz J	(B − A)/K→H	'M' STO+ NEXT U h	'M' STO+ NEXT U h −
J≤N − 1⇒Goto 2	H/N→R	− 'X' STO f EVAL	'X' STO f EVAL →NUM
U − H→X:Prog 0	1→I	→NUM 2 'T' STO*	2 'T' STO* 'T' STO+
Y + 2T→T:U→X	Lbl 1	'T' STO+ U 'X' STO	U 'X' STO f EVAL
Prog 0	0→T	f EVAL →NUM T	→NUM T + r * 2 /
R(Y + T) ÷ 2→T	0→M	+ r * 2 / 'T' STO	'T' STO X r 2 / −
U − R ÷ 2→X:Prog 0	A + IH→U	X r 2 / − 'X' STO	'X' STO f EVAL →NUM
R(Y + M)→M	1→J	f EVAL →NUM	'M' STO+ r 'M' STO*
(T + 2M) ÷ 3 + S→S	Lbl 2	'M' STO+ r 'M' STO*	T M 2 * + 3 / s + 's'
Plot U,S	U − H + JR→X	T M 2 * + 3 / s + 's'	STO U s R→C
Isz I	T + Y₁→T	STO U s R→C	PIXON NEXT
I≤K⇒Goto 1	X − R/2→X	PIXEL NEXT	{ X U T M EQ }
	Y₁ + M→M	{ X U T M EQ }	PURGE GRAPH ≫
	IS>(J,N − 1)	PURGE DGTIZ ≫	≫ ≫
	Goto 2	≫ ≫	
	U − H→X		
	Y₁ + 2T→T		
	U→X		
	R(Y₁ + S)/2→T		
	U − R/2→X		
	R(Y₁ + M)→M		
	(T + 2M)/3 + S→S		
	PT − On(U,S)		
	IS>(I,K)		
	Goto 1		

V.9 Series and Taylor Polynomials

The following programs compute the partial sums of an infinite series $\sum_{k=j}^{m} a_k$. Note that they sum the series in reverse order; this helps control round-off error.

Casio	TI-81	HP-28S and HP-48SX
PSUM	PrgmP:PSUM	'PSUM'
"J"?→J:"M"?→M	Disp "J"	≪
M→K:0→S	Input J	→ ak j m
Lbl 1	Disp "M"	≪
Prog 0:Y+S→S	Input M	0 'S' STO m j FOR k
K−1→K	0→S	k 'K' STO
K≥J⇒Goto 1	M→K	ak EVAL →NUM
"S=":S◢	Lbl 1	'S' STO+ −1 STEP
	Y₁+S→S	S { K S }
	DS<(K,J)	PURGE ≫
	Goto 1	≫
	Disp "S="	
	Disp S	

Explanation

Casio: Program location **0** should contain the expression needed to evaluate a_k. For example, to evaluate the partial sums of $\sum_{k=1}^{\infty} (1/k^2)$, program **0** should be **1÷K²→Y**. For $\sum_{k=1}^{\infty} ((-1)^k/k!)$, program **0** should be **(−1)x^yK÷K!→Y**. Upon execution you are prompted to enter the starting index **J** and the ending index **M**. When the program is complete, the value of the sum is displayed. For the latter series the value would be -0.6321205588 for **J** = 1 and **M** = 20.

TI-81: In the variables list, **Y₁** should contain the expression needed to evaluate a_k. For example, to evaluate the partial sums of $\sum_{k=1}^{\infty} (1/k^2)$, **Y₁** should be **1/K²**. For $\sum_{k=1}^{\infty} ((-1)^k/k!)$, **Y₁** should be **(−1)^K/K!**. Upon execution you are prompted to enter the starting index **J** and the ending index **M**. When the program is complete, the value of the sum is displayed. For the latter series the value would be -0.6321205588 for **J** = 1 and **M** = 20.

Hewlett-Packard: Before executing the program you must create a function that evaluates a_k and name it whatever you like (e.g., **AK**) or just put it on the stack. For example, to evaluate the partial sums of $\sum_{k=1}^{\infty} (1/k^2)$, **AK** should be **'INV(SQ(K))'**. For $\sum_{k=1}^{\infty} ((-1)^k/k!)$, **AK** should be **'(−1)^K/FACT(K)'**. The program assumes that **AK** is on level 3 of the stack, the initial index **j** on level 2, and the value of **m** on level 1. So before execution the stack should look like

'(−1)^K/FACT(K)'
2:                              1
1:                             20

This would evaluate the 20th partial sum of the series $\sum_{k=1}^{\infty} ((-1)^k/k!)$. When the program is finished, the value of the sum is on level 1 of the stack. For this example the value would be -0.632120558829.

There are no special commands that will compute Taylor polynomials on the Casio or the TI-81. Thus to graph Taylor polynomials and the functions they approximate, it is necessary first to find the polynomial by hand.

The Hewlett-Packard calculators have a utility that will compute the Taylor polynomial at 0 of any function. This is done by placing on the stack the function, the name of the independent variable, and the degree and pressing ||| **TAYLR** ||| found in the **ALGEBRA** menu. For example, entering on the stack.

3:           'SIN(X)'
2:               'X'
1:                5

and pressing ||| **TAYLR** ||| will produce **'X − .166666666667*X^3 + 8.33333333333E − 3*X^5'** (HP-28S) or **'X − 1/3!*X^3 + 1/5!*X^5** (HP-48SX). In order to produce a Taylor polynomial centered at some other point, you should find the following program useful.

'TYLRA'
≪ → f x a n
≪ Y a + x STO f
EVAL 'Y' n TAYLR
x a − 'Y'
STO EVAL 'Y' PURGE
≫
≫

To execute this program the stack must have the function on level 4, the variable name on level 3, the point where the polynomial is centered on level 2, and the degree on level 1. For example, entering on the stack

4:           'SIN(X)'
3:               'X'
2:              'π/4'
1:                3

and executing the program **TYLRA** will produce the following.

'SIN(π/4) + COS(π/4)*
(X − π/4) − SIN(π/4)/2*
(X − π/4)^2 − COS(π/4)/6*
(X − π/4)^3'

V.10   Polar and Parametric Graphs

Producing the graphs of polar and parametric equations must be done with a program on the Casio and the HP-28S. However, the Casio fx-7700 has built-in facilities for producing these graphs, as do the TI-81 and the HP-48SX. We will describe these one at a time.

Casio: On the Casio all that is needed is a slight variation of the program **PTPLOT** given in Section V.2. That program is given below, along with the commands in program location **0** that will evaluate a polar equation and a parametric equation.

PTPLOT	Polar Equation	Parametric Equation
"A"?→A:"B"?→B	cos 2Tcos T→X	2cos T→X
"N"?→N	cos 2Tsin T→Y	4sin T→Y
(B − A) ÷ N→H		
A→T:Prog 0		
Plot X,Y		
1→I:Lbl 1		
A + IH→T		
Prog 0		
Plot X,Y:Line		
Isz I		
I≤N⇒Goto 1		

The values **A** and **B** should be the range of the independent variable, θ for polar equations and t for parametric equations. The range must be set prior to executing the program. This is where the program **RANGE** in Section V.1 is most important. Setting the range with this program will scale the screen so that the polar graphs will look as they should; i.e., circles will look like circles. The two examples here will graph the polar equation $r = \cos 2\theta$ or the parametric equations $x(t) = 2 \cos t$, $y(t) = 4 \sin t$. Note that the program that evaluates the polar equation is written in parametric form.

TI-81:   The TI-81 has two different graphing modes, **Function** for graphing single-variable functions and **Param** for graphing parametrically defined functions. These modes are set by accessing the **MODE** screen and selecting and entering the desired mode. When in this mode, the **Y** variables list contains X_{1t}, Y_{1t}, ... The expressions entered here must contain the variable **T**. To draw the polar graph $r = \cos 2\theta$, enter for X_{1t} the expression **cos 2Tcos T** and for Y_{1t} the expression **cos 2Tsin T**. To draw the graph of the parametric equations $x(t) = 2 \cos t$, $y(t) = 4 \sin t$, enter for X_{1t} the expression **2cos T** and for Y_{1t} the expression **4sin T**. Note that in either case the equations are treated as parametric equations. In this **Param** mode the range setting screen has a different look. The first three lines are now **Tmin =**, **Tmax =**, and **Tstep =**, which are used to set the range of the independent variable **T**. For example, if you are graphing the polar curve $r = \cos 2\theta$ you would set **Tmin** to be 0, **Tmax** to be 6.283185307 (or 2π), and **Tstep** to be 0.1256637061 (or $2\pi/50$). The resolution of the graph is controlled by **Tstep**. (In the degree mode you could use **Tmin = 0, Tmax = 360**, and **Tstep = 5**.) These variables can also be accessed from the **VARS** menu under the **RNG** title. It is important when graphing polar or parametric type equations that you use the **RANGE** program of Section V.1. Setting the range with this program will scale the screen so that polar graphs will look as they should; i.e., circles will look like circles.

HP-28S: The HP-28S needs a point plotting program in order to draw parametric and polar graphs. These two programs are given below.

'POLPLT'	'PARPLT'
≪ → a b tstep	≪ → a b tstep
≪ DEPTH →LIST	≪ DEPTH →LIST
DUP SIZE → req n	DUP SIZE → peq n
≪ CLLCD DRAX	≪ CLLCD DRAX
1 n FOR i req i GET	1 n FOR i peq i GET
'R' STO	'X' STO peq i 1 + GET
a b FOR t t DUP 'T'	'Y' STO a b FOR t
STO R EVAL →NUM	t 'T' STO X EVAL
SWAP R→C P→R	→NUM Y EVAL
PIXEL tstep	→NUM R→C
STEP NEXT	PIXEL tstep STEP
{ T R } PURGE	2 STEP { T X Y }
DGTIZ	PURGE DGTIZ
≫ ≫	≫ ≫
≫	≫

The polar plotter program, **POLPLT**, uses the command **P→R** to convert the polar coordinate to a rectangular coordinate. Before execution the left endpoint of the range of θ must be on level 3, the right endpoint of the range of θ on level 2, and **tstep** (which defines the resolution of the graph) on level 1. For example, if **tstep** is $2\pi/50 = .125663706144$, then 50 points will be plotted if, say $\mathbf{a} = 0$ and $\mathbf{b} = 2\pi$. On levels 4 or higher should be the polar equations with variable **'T'** for θ.

Before executing parametric plotter, **PARPLT**, the left endpoint of the parameter range must be on level 3, the right endpoint of the parameter range on level 2, and **tstep** on level 1. On levels 4 and higher the x and y coordinate functions defining the curves must be placed: the x-coordinate function first, followed by the y-coordinate function for each curve to be drawn. It is assumed that these are functions of the variable **'T'**. It is important when graphing these type of equations that you use the **RANGE** program of Section V.1. Setting the range with this program will scale the screen so that polar graphs will look as they should; i.e., circles will look like circles. Below we give examples of the what the stack should look like before these programs are executed. For the examples, **POLPLT** will draw the graph of the four-petaled rose $r = \cos 2\theta$ and the cardioid $r = 1 + \cos \theta$, and **PARPLT** will draw the graphs of the ellipse $x(t) = 4 \cos t$, $y(t) = 1.5 \sin t$ and the circle $x(t) = 1.5 \cos t$, $y(t) = 1.5 \sin t$.

	'POLPLT'		'PARPLT'
5:	'COS(2*T)'	7:	'4*COS(T)'
4:	'1 + COS(T)'	6:	'1.5*SIN(T)'
3:	0	5:	'1.5*COS(T)'
2:	6.28318530718	4:	'1.5*SIN(T)'
1:	.125663706144	3:	0
		2:	6.28318530718
		1:	.125663706144

HP-48SX: For polar equations the correct plot type must be selected, which in this case is **POLAR**. Polar equations can be entered as they are written; the variable θ is obtained by $\boxed{\alpha}$ $\boxed{\rightarrow}$ $\boxed{\text{F}}$. For example, to graph $r = \cos 2\theta$ enter '**COS(2*θ)**' and set the independent variable to be θ.

The Hewlett-Packard calculators treat the ordered pair (a, b) as the complex number $a + bi$, and it is this way that a parametric equation must be entered. First, though, you must be in the correct plot type, which in this case is **PARA**. Once the plot type is entered, a parametric equation is drawn by entering the equations in the format '**x(t)+i*y(t)**'. For example, to graph the parametric equations $x(t) = 2 \cos t$, $y = 4 \sin t$, you would enter '**2*COS(T)+i*4*SIN(T)**' and the independent variable must be set to be **T**. It is important when graphing these types of equations that you use the **RANGE** program of Section V.1. Setting the range with this program will scale the screen so that polar graphs will look as they should; i.e., circles will look like circles. The range of the independent variable for these types of plots can be changed by entering the two values on the stack and pressing $\boxed{||||\,\text{INDEP}\,||||}$ in the **PLOTR** submenu.

JAMES D. ANGELOS
Central Michigan University

Appendix VI

Tables of Transcendental Functions

Trigonometric functions (x in radians)

x	$\sin x$	$\cos x$	$\tan x$
0.0	.00000	1.00000	0.00000
0.1	.09983	.99500	.10033
0.2	.19867	.98007	.20271
0.3	.29552	.95534	.30934
0.4	.38942	.92106	.42279
0.5	.47943	.87758	.54630
0.6	.56464	.82534	.68414
0.7	.64422	.76484	.84229
0.8	.71736	.69671	1.02964
0.9	.78333	.62161	1.26016
1.0	.84147	.54030	1.55741
1.1	.89121	.45360	1.96476
1.2	.93204	.36236	2.57215
1.3	.96356	.26750	3.60210
1.4	.98545	.16997	5.79788
1.5	.99749	.07074	14.10142
$\pi/2$	1.00000	.00000	undefined
1.6	.99957	−.02920	−34.23254
1.7	.99166	−.12884	−7.69660
1.8	.97385	−.22720	−4.28626
1.9	.94630	−.32329	−2.92710
2.0	.90930	−.41615	−2.18504
2.1	.86321	−.50485	−1.70985
2.2	.80850	−.58850	−1.37382
2.3	.74571	−.66628	−1.11921
2.4	.67546	−.73739	−.91601
2.5	.59847	−.80114	−.74702
2.6	.51550	−.85689	−.60160
2.7	.42738	−.90407	−.47273
2.8	.33499	−.94222	−.35553
2.9	.23925	−.97096	−.24641
3.0	.14112	−.98999	−.14255
3.1	.04158	−.99914	−.04162
π	.00000	−1.00000	.00000
3.2	−.05837	−.99829	.05847
3.3	−.15775	−.98748	.15975

x	$\sin x$	$\cos x$	$\tan x$
3.4	−.25554	−.96680	.26432
3.5	−.35078	−.93646	.37459
3.6	−.44252	−.89676	.49347
3.7	−.52984	−.84810	.62473
3.8	−.61186	−.79097	.77356
3.9	−.68777	−.72593	.94742
4.0	−.75680	−.65364	1.15782
4.1	−.81828	−.57482	1.42353
4.2	−.87158	−.49026	1.77778
4.3	−.91617	−.40080	2.28585
4.4	−.95160	−.30733	3.09632
4.5	−.97753	−.21080	4.63733
4.6	−.99369	−.11215	8.86017
4.7	−.99992	−.01239	80.71271
$3\pi/2$	−1.00000	−.00000	undefined
4.8	−.99616	.08750	−11.38487
4.9	−.98245	.18651	−5.26749
5.0	−.95892	.28366	−3.38051
5.1	−.92581	.37798	−2.44939
5.2	−.88345	.46852	−1.88564
5.3	−.83227	.55437	−1.50127
5.4	−.77276	.63469	−1.21754
5.5	−.70554	.70867	−.99558
5.6	−.63127	.77557	−.81394
5.7	−.55069	.83471	−.65973
5.8	−.46460	.88552	−.52467
5.9	−.37388	.92748	−.40311
6.0	−.27942	.96017	−.29101
6.1	−.18216	.98327	−.18526
6.2	−.08309	.99654	−.08338
2π	.00000	1.00000	.00000
6.3	.01681	.99986	.01682
6.4	.11655	.99318	.11735
6.5	.21512	.97659	.22028

Exponential functions

x	e^x	e^{-x}
0.00	1.00000	1.00000
.05	1.05127	.95123
.10	1.10517	.90484
.15	1.16183	.86071
.20	1.22140	.81873
.25	1.28403	.77880
.30	1.34986	.74082
.35	1.41907	.70469
.40	1.49182	.67032
.45	1.56831	.63763
.50	1.64872	.60653
.55	1.73325	.57695
.60	1.82212	.54881
.65	1.91554	.52205
.70	2.01375	.49659

x	e^x	e^{-x}
.75	2.11700	.47237
.80	2.22554	.44933
.85	2.33965	.42741
.90	2.45960	.40657
.95	2.58571	.38674
1.00	2.71828	.36788
2.00	7.38906	.13534
3.00	20.08554	.04979
4.00	54.59815	.01832
5.00	148.41316	.00674
6.00	403.42879	.00248
7.00	1,096.63316	.00091
8.00	2,980.95799	.00034
9.00	8,103.08393	.00012
10.00	22,026.46579	.00005

Note: $e^{a+x} = e^a e^x$

Natural logarithms

x	$\ln x$	x	$\ln x$	x	$\ln x$	x	$\ln x$
.1	−2.30258	2.6	.95551	5.1	1.62924	7.6	2.02815
.2	−1.60943	2.7	.99325	5.2	1.64866	7.7	2.04122
.3	−1.20396	2.8	1.02962	5.3	1.66771	7.8	2.05412
.4	−.91628	2.9	1.06471	5.4	1.68640	7.9	2.06686
.5	−.69314	3.0	1.09861	5.5	1.70475	8.0	2.07944
.6	−.51082	3.1	1.13140	5.6	1.72277	8.1	2.09186
.7	−.35666	3.2	1.16315	5.7	1.74047	8.2	2.10413
.8	−.22313	3.3	1.19392	5.8	1.75786	8.3	2.11626
.9	−.10535	3.4	1.22378	5.9	1.77495	8.4	2.12823
1.0	0.00000	3.5	1.25276	6.0	1.79176	8.5	2.14007
1.1	.09531	3.6	1.28093	6.1	1.80829	8.6	2.15176
1.2	.18232	3.7	1.30833	6.2	1.82455	8.7	2.16332
1.3	.26236	3.8	1.33500	6.3	1.84055	8.8	2.17475
1.4	.33647	3.9	1.36098	6.4	1.85630	8.9	2.18605
1.5	.40547	4.0	1.38629	6.5	1.87180	9.0	2.19722
1.6	.47000	4.1	1.41099	6.6	1.88707	9.1	2.20827
1.7	.53063	4.2	1.43508	6.7	1.90211	9.2	2.21920
1.8	.58779	4.3	1.45862	6.8	1.91692	9.3	2.23001
1.9	.64185	4.4	1.48160	6.9	1.93152	9.4	2.24071
2.0	.69315	4.5	1.50408	7.0	1.94591	9.5	2.25129
2.1	.74194	4.6	1.52606	7.1	1.96009	9.6	2.26176
2.2	.78846	4.7	1.54756	7.2	1.97408	9.7	2.27213
2.3	.83291	4.8	1.56862	7.3	1.98787	9.8	2.28238
2.4	.87547	4.9	1.58924	7.4	2.00148	9.9	2.29253
2.5	.91629	5.0	1.60944	7.5	2.01490	10.0	2.30259

Note: $\ln 10x = \ln x + \ln 10$

Appendix VII

Table of Integrals

1. $\displaystyle\int af(u)\,du = a\int f(u)\,du$

2. $\displaystyle\int [af(u) + bg(u)]\,du = a\int f(u)\,du + b\int g(u)\,du$

3. $\displaystyle\int g'(f(u))\,f'(u)\,du = g(f(u)) + C$

4. $\displaystyle\int u\,dv = uv - \int v\,du$

5. $\displaystyle\int a\,du = au + C$

6. $\displaystyle\int (au + b)\,du = \frac{a}{2}u^2 + bu + C$

7. $\displaystyle\int u^n\,du = \frac{u^{n+1}}{n+1} + C, \quad n \neq -1$

8. $\displaystyle\int \frac{1}{u}\,du = \ln|u| + C$

9. $\displaystyle\int \frac{f'(u)}{f(u)}\,du = \ln|f(u)| + C$

10. $\displaystyle\int \sin u\,du = -\cos u + C$

11. $\displaystyle\int \cos u\,du = \sin u + C$

12. $\displaystyle\int \tan u\,du = \ln|\sec u| + C$

13. $\displaystyle\int \cot u\,du = \ln|\sin u| + C$

14. $\displaystyle\int \sec u\,du = \ln|\sec u + \tan u| + C$

15. $\displaystyle\int \csc u\,du = \ln|\csc u - \cot u| + C$

16. $\displaystyle\int e^u\,du = e^u + C$

17. $\displaystyle\int a^u\,du = \frac{a^u}{\ln a} + C, \quad a \neq 1$

18. $\displaystyle\int \ln u\,du = u\ln u - u + C$

19. $\displaystyle\int \frac{du}{1 + u^2} = \mathrm{Tan}^{-1}\,u + C$

20. $\displaystyle\int \frac{du}{a^2 + u^2} = \frac{1}{a}\mathrm{Tan}^{-1}\left(\frac{u}{a}\right) + C$

21. $\displaystyle\int \frac{du}{\sqrt{1 - u^2}} = \mathrm{Sin}^{-1}\,u + C = -\mathrm{Cos}^{-1}\,u + C$

22. $\displaystyle\int \frac{du}{\sqrt{a^2 - u^2}} = \mathrm{Sin}^{-1}\left(\frac{u}{a}\right) + C = -\mathrm{Cos}^{-1}\left(\frac{u}{a}\right) + C$

23. $\displaystyle\int \frac{du}{u\sqrt{u^2 - 1}} = \mathrm{Sec}^{-1}\,u + C$

24. $\displaystyle\int \frac{du}{u\sqrt{u^2 - a^2}} = \frac{1}{a}\mathrm{Sec}^{-1}\left(\frac{u}{a}\right) + C, \quad a > 0$

Forms Involving $a + bu$

25. $\displaystyle\int (a + bu)^n \, du = \frac{(a + bu)^{n+1}}{(n + 1)b} + C, \quad n \neq -1$

26. $\displaystyle\int \frac{du}{a + bu} = \frac{1}{b} \ln|a + bu| + C$

27. $\displaystyle\int \frac{du}{(a + bu)^2} = -\frac{1}{b(a + bu)} + C$

28. $\displaystyle\int \frac{u \, du}{(a + bu)^2} = \frac{1}{b^2}\left[\ln(a + bu) + \frac{a}{a + bu}\right] + C$

29. $\displaystyle\int \frac{du}{u(a + bu)} = -\frac{1}{a} \ln\left|\frac{a + bu}{u}\right| + C$

30. $\displaystyle\int \frac{du}{u(a + bu)^2} = \frac{1}{a(a + bu)} - \frac{1}{a^2} \ln\left|\frac{a + bu}{u}\right| + C$

31. $\displaystyle\int \frac{du}{u^2(a + bu)} = -\frac{1}{au} + \frac{b}{a^2} \ln\left|\frac{a + bu}{u}\right| + C$

32. $\displaystyle\int \frac{du}{u^2(a + bu)^2} = -\frac{a + 2bu}{a^2 u(a + bu)} + \frac{2b}{a^3} \ln\left|\frac{a + bu}{u}\right| + C$

33. $\displaystyle\int u\sqrt{a + bu} \, du = -\frac{2(2a - 3bu)\sqrt{(a + bu)^3}}{15b^2} + C$

34. $\displaystyle\int \frac{du}{u\sqrt{a + bu}} = \begin{cases} \dfrac{1}{\sqrt{a}} \ln\left|\dfrac{\sqrt{a + bu} - \sqrt{a}}{\sqrt{a + bu} + \sqrt{a}}\right| + C, & a > 0 \\[3mm] \dfrac{2}{\sqrt{-a}} \mathrm{Tan}^{-1} \sqrt{\dfrac{a + bu}{-a}} + C, & a < 0 \end{cases}$

35. $\displaystyle\int u^2\sqrt{a + bu} \, du =$
$$\frac{2(8a^2 - 12abu + 15b^2 u^2)\sqrt{(a + bu)^3}}{105 b^3} + C$$

36. $\displaystyle\int \frac{u \, du}{\sqrt{a + bu}} = -\frac{2(2a - bu)}{3b^2}\sqrt{a + bu} + C$

37. $\displaystyle\int \frac{u^2 \, du}{\sqrt{a + bu}} = \frac{2(8a^2 - 4abu + 3b^2 u^2)}{15 b^3}\sqrt{a + bu} + C$

Forms Containing $u^2 \pm a^2$

38. $\displaystyle\int \frac{du}{a^2 - u^2} = \frac{1}{2a} \ln\left|\frac{a + u}{a - u}\right| + C$

39. $\displaystyle\int \frac{du}{u^2 - a^2} = \frac{1}{2a} \ln\left|\frac{u - a}{u + a}\right| + C$

40. $\displaystyle\int \frac{du}{\sqrt{u^2 \pm a^2}} = \ln|u + \sqrt{u^2 \pm a^2}| + C$

41. $\displaystyle\int \frac{du}{u\sqrt{a^2 \pm u^2}} = -\frac{1}{a} \ln\left|\frac{a + \sqrt{a^2 \pm u^2}}{u}\right| + C$

42. $\displaystyle\int \frac{du}{au^2 + c} = \frac{1}{\sqrt{ac}} \mathrm{Tan}^{-1}\left(u\sqrt{\frac{a}{c}}\right) + C, \quad ac > 0$

43. $\displaystyle\int \sqrt{u^2 \pm a^2} \, du =$
$$\frac{1}{2}\left[u\sqrt{u^2 \pm a^2} \pm a^2 \ln(u + \sqrt{u^2 \pm a^2})\right] + C$$

44. $\displaystyle\int u\sqrt{u^2 \pm a^2} \, du = \frac{1}{3}(u^2 \pm a^2)^{3/2} + C$

45. $\displaystyle\int u^2\sqrt{a^2 + u^2} \, du =$
$$\frac{u}{8}(a^2 + 2u^2)\sqrt{a^2 + u^2} - \frac{a^4}{8} \ln|u + \sqrt{a^2 + u^2}| + C$$

46. $\displaystyle\int \frac{du}{\sqrt{u^2 \pm a^2}} = \ln(u + \sqrt{u^2 \pm a^2}) + C$

47. $\displaystyle\int \frac{du}{u\sqrt{u^2 - a^2}} = \frac{1}{a} \mathrm{Sec}^{-1}\left(\frac{u}{a}\right) + C$

48. $\displaystyle\int \frac{du}{u\sqrt{u^2 + a^2}} = -\frac{1}{a} \ln\left(\frac{a + \sqrt{u^2 + a^2}}{u}\right) + C$

49. $\displaystyle\int \frac{\sqrt{u^2 + a^2}}{u} \, du = \sqrt{u^2 + a^2} - a \ln\left(\frac{a + \sqrt{u^2 + a^2}}{u}\right) + C$

50. $\displaystyle\int \frac{\sqrt{u^2 - a^2}}{u} \, du = \sqrt{u^2 - a^2} - a\, \mathrm{Sec}^{-1}\left(\frac{u}{a}\right) + C, \quad a > 0$

51. $\displaystyle\int \frac{du}{\sqrt{(u^2 \pm a^2)^3}} = \frac{\pm u}{a^2 \sqrt{u^2 \pm a^2}} + C$

52. $\displaystyle\int \frac{\sqrt{u^2 \pm a^2}}{u^2} \, du = -\frac{\sqrt{u^2 \pm a^2}}{u} + \ln(u + \sqrt{u^2 \pm a^2}) + C$

53. $\displaystyle\int \frac{\sqrt{u^2 - a^2}}{u^3} \, du =$
$$-\frac{\sqrt{u^2 - a^2}}{2u^2} + \frac{1}{2a} \mathrm{Sec}^{-1}\left(\frac{u}{a}\right) + C, \quad a > 0$$

54. $\displaystyle\int \sqrt{a^2 - u^2} \, du =$
$$\frac{1}{2}\left[u\sqrt{a^2 - u^2} + a^2 \mathrm{Sin}^{-1}\left(\frac{u}{a}\right)\right] + C, \quad a > 0$$

55. $\displaystyle\int \frac{du}{\sqrt{a^2 - u^2}} = \mathrm{Sin}^{-1}\left(\frac{u}{a}\right) + C$

56. $\displaystyle \int \frac{du}{u\sqrt{a^2 - u^2}} = -\frac{1}{a} \ln\left(\frac{a + \sqrt{a^2 - u^2}}{u}\right) + C$

57. $\displaystyle \int \frac{du}{\sqrt{(a^2 - u^2)^3}} = \frac{u}{a^2\sqrt{a^2 - u^2}} + C$

58. $\displaystyle \int u^2\sqrt{a^2 - u^2}\, du = -\frac{u}{4}\sqrt{(a^2 - u^2)^3}$
$$+ \frac{a^2}{8}\left(u\sqrt{a^2 - u^2} + a^2\,\text{Sin}^{-1}\frac{u}{a}\right) + C, \quad a > 0$$

59. $\displaystyle \int \frac{u^2\, du}{\sqrt{a^2 - u^2}} = -\frac{u}{2}\sqrt{a^2 - u^2} + \frac{a^2}{2}\,\text{Sin}^{-1}\left(\frac{u}{a}\right) + C, \quad a > 0$

60. $\displaystyle \int \frac{du}{u^2\sqrt{a^2 - u^2}} = -\frac{\sqrt{a^2 - u^2}}{a^2 u} + C$

61. $\displaystyle \int \frac{\sqrt{a^2 - u^2}}{u^2}\, du = -\frac{\sqrt{a^2 - u^2}}{u} - \text{Sin}^{-1}\left(\frac{u}{a}\right) + C, \quad a > 0$

Forms Containing $a + bu^2$

62. $\displaystyle \int \frac{du}{a + bu^2} = \frac{1}{\sqrt{ab}}\,\text{Tan}^{-1}\frac{u\sqrt{ab}}{a} + C$

63. $\displaystyle \int \frac{du}{(u^2 + a^2)^2} = \frac{1}{2a^3}\,\text{Tan}^{-1}\frac{u}{a} + \frac{u}{2a^2(u^2 + a^2)} + C$

64. $\displaystyle \int \frac{du}{u(a + bu^2)} = \frac{1}{2a}\ln\left|\frac{u^2}{a + bu^2}\right| + C$

Forms Involving $2au - u^2$

65. $\displaystyle \int \sqrt{2au - u^2}\, du =$
$$\frac{1}{2}\left[(u - a)\sqrt{2au - u^2} + a^2\,\text{Sin}^{-1}\left(\frac{u - a}{a}\right)\right] + C$$

66. $\displaystyle \int \frac{du}{\sqrt{2au - u^2}} = 2\,\text{Sin}^{-1}\sqrt{\frac{u}{2a}} + C, \quad a > 0$

67. $\displaystyle \int \frac{du}{(2au - u^2)^{3/2}} = \frac{u - a}{a^2\sqrt{2au - u^2}} + C$

68. $\displaystyle \int \frac{du}{\sqrt{2au + u^2}} = \ln|u + a + \sqrt{2au + u^2}| + C$

Forms Involving Trigonometric Functions

69. $\displaystyle \int \sin^2 u\, du = \frac{u}{2} - \frac{1}{4}\sin 2u + C$

70. $\displaystyle \int \sin^3 u\, du = -\frac{1}{3}\cos u(\sin^2 u + 2) + C$

71. $\displaystyle \int \cos^2 u\, du = \frac{1}{2}u + \frac{1}{4}\sin 2u + C$

72. $\displaystyle \int \cos^3 u\, du = \frac{1}{3}\sin u(\cos^2 u + 2) + C$

73. $\displaystyle \int \frac{du}{1 \pm \sin u} = \mp\tan\left(\frac{\pi}{4} \mp \frac{u}{2}\right) + C$

74. $\displaystyle \int \frac{du}{1 + \cos u} = \tan\left(\frac{u}{2}\right) + C$

75. $\displaystyle\int \frac{du}{a + b \sin u} = \begin{cases} \dfrac{2}{\sqrt{a^2 - b^2}} \text{Tan}^{-1} \dfrac{a \tan\left(\dfrac{u}{2}\right) + b}{\sqrt{a^2 - b^2}} + C, \quad a^2 > b^2 \\[3em] \dfrac{1}{\sqrt{b^2 - a^2}} \ln \dfrac{a \tan\left(\dfrac{u}{2}\right) + b - \sqrt{b^2 - a^2}}{a \tan\left(\dfrac{u}{2}\right) + b + \sqrt{b^2 - a^2}} + C, \quad a^2 < b^2 \end{cases}$

76. $\displaystyle\int \frac{du}{a + b \cos u} = \begin{cases} \dfrac{2}{\sqrt{a^2 - b^2}} \text{Tan}^{-1} \dfrac{\sqrt{a^2 - b^2} \tan\left(\dfrac{u}{2}\right)}{a + b} + C, \quad a^2 > b^2 \\[3em] \dfrac{1}{\sqrt{b^2 - a}} \ln\left(\dfrac{\sqrt{b^2 - a^2} \tan\left(\dfrac{u}{2}\right) + a + b}{\sqrt{b^2 - a^2} \tan\left(\dfrac{u}{2}\right) - a - b} \right) + C, \quad a^2 < b^2 \end{cases}$

77. $\displaystyle\int \sin(mu)\sin(nu)\, du = \frac{\sin(m - n)u}{2(m - n)} - \frac{\sin(m + n)u}{2(m + n)} + C, \quad m^2 \neq n^2$

78. $\displaystyle\int \cos(mu)\cos(nu)\, du = \frac{\sin(m - n)u}{2(m - n)} + \frac{\sin(m + n)}{2(m + n)} + C, \quad m^2 \neq n^2$

79. $\displaystyle\int \tan^3 u\, du = \frac{1}{2} \tan^2 u + \ln|\cos u| + C$

80. $\displaystyle\int \tan^4 u\, du = \frac{1}{3} \tan^3 u - \tan u + u + C$

81. $\displaystyle\int \sin(mu)\cos(nu)\, du = -\frac{\cos(m - n)u}{2(m - n)} - \frac{\cos(m + n)u}{2(m + n)} + C, \quad m^2 \neq n^2$

82. $\displaystyle\int \sin^2 u \cos^2 u\, du = -\frac{1}{32} \sin 4u + \frac{u}{8} + C$

Forms Involving Inverse Trigonometric Functions

83. $\displaystyle\int \text{Sin}^{-1} u\, du = u \,\text{Sin}^{-1} u + \sqrt{1 - u^2} + C$

84. $\displaystyle\int \text{Cos}^{-1} u\, du = u \,\text{Cos}^{-1} u - \sqrt{1 - u^2} + C$

85. $\displaystyle\int \text{Tan}^{-1} u\, du = u \,\text{Tan}^{-1} u - \frac{1}{2} \ln(1 + u^2) + C$

86. $\displaystyle\int \text{Cot}^{-1} u\, du = u \,\text{Cot}^{-1} u + \frac{1}{2} \ln(1 + u^2) + C$

87. $\displaystyle\int \text{Sec}^{-1} u\, du = u \,\text{Sec}^{-1} u - \ln|u + \sqrt{u^2 - 1}| + C$

88. $\displaystyle\int \text{Csc}^{-1} u\, du = u \,\text{Csc}^{-1} u + \ln|u + \sqrt{u^2 - 1}| + C$

89. $\displaystyle\int (\text{Sin}^{-1} u)^2\, du =$
$$u(\text{Sin}^{-1} u)^2 - 2u + 2\sqrt{1 - u^2}(\text{Sin}^{-1} u) + C$$

90. $\displaystyle\int (\text{Cos}^{-1} u)^2\, du =$
$$u(\text{Cos}^{-1} u)^2 - 2u - 2\sqrt{1 - u^2}(\text{Cos}^{-1} u) + C$$

91. $\displaystyle\int u\,\text{Sin}^{-1}(au)\, du =$
$$\frac{1}{4a^2}[(2a^2u^2 - 1)\text{Sin}^{-1}(au) + au\sqrt{1 - a^2u^2}] + C$$

92. $\displaystyle\int u\,\text{Cos}^{-1}(au)\, du =$
$$\frac{1}{4a^2}[(2a^2u^2 - 1)\text{Cos}^{-1}(au) - au\sqrt{1 - a^2u^2}] + C$$

Forms Involving Logarithms

93. $\displaystyle\int \ln u \, du = u \ln u - u + C$

94. $\displaystyle\int u \ln u \, du = \frac{u^2}{2} \ln u - \frac{u^2}{4} + C$

95. $\displaystyle\int u^2 \ln u \, du = \frac{u^3}{3} \ln u - \frac{u^3}{9} + C$

96. $\displaystyle\int (\ln u)^2 \, du = u(\ln u)^2 - 2u \ln u + 2u + C$

97. $\displaystyle\int \frac{du}{u \ln u} = \ln|\ln u| + C$

98. $\displaystyle\int u^m \ln u \, du = u^{m+1} \left[\frac{\ln u}{m+1} - \frac{1}{(m+1)^2} \right] + C$

99. $\displaystyle\int \sin \ln u \, du = \frac{1}{2} u \sin \ln u - \frac{1}{2} u \cos \ln u + C$

100. $\displaystyle\int \cos \ln u \, du = \frac{1}{2} u \sin \ln u + \frac{1}{2} u \cos \ln u + C$

Forms Involving Exponential Functions

101. $\displaystyle\int u e^{au} \, du = \frac{e^{au}}{a^2}(au - 1)$

102. $\displaystyle\int \frac{du}{1 + e^u} = \ln \left(\frac{e^u}{1 + e^u} \right) + C$

103. $\displaystyle\int \frac{du}{a + be^{cu}} = \frac{u}{a} - \frac{1}{ac} \ln|a + be^{cu}| + C$

104. $\displaystyle\int e^{au} \sin bu \, du = \frac{e^{au}[a \sin(bu) - b \cos(bu)]}{a^2 + b^2} + C$

105. $\displaystyle\int e^{au} \cos bu \, du = \frac{e^{au}[a \cos(bu) + b \sin(bu)]}{a^2 + b^2} + C$

Appendix VIII

Geometry Formulas

(A = area, C = circumference, V = volume, S = surface area, r = radius, b = base, h = height)

Triangle

$A = \frac{1}{2}bh$

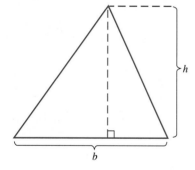

Parallelogram

$A = bh$

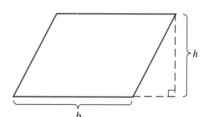

Rectangle

$A = bh$

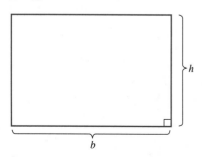

Trapezoid

$A = \frac{1}{2}(b_1 + b_2)h$

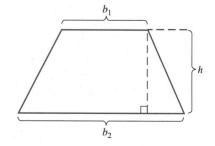

Circle

$A = \pi r^2$
$C = \pi d = 2\pi r$

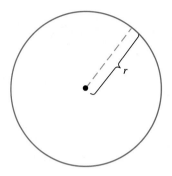

Cylinder with Parallel Bases

$V = Bh$

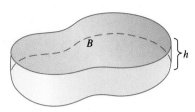

Right Circular Cylinder

$V = \pi r^2 h$
$S = 2\pi rh + 2\pi r^2$

Sphere

$V = \frac{4}{3}\pi r^3$
$S = 4\pi r^2$

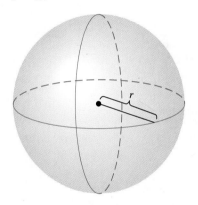

Right Circular Cone

$V = \frac{1}{3}\pi r^2 h$

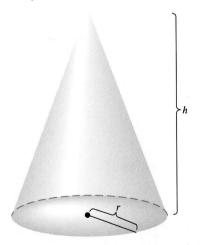

Answers to Selected Exercises

Chapter 1

1. Rational   **3.** Irrational

5. Rational   **7.** Rational

9. Irrational   **11.** $[-2, 6)$

13. $[-2, 0)$ **15.** $(-1, 0)$

17. True, true, true   **19.** $x \geq 2$

21. $x \geq -7$ **23.** $-1 < x < 3$

25. $x < -4$ or $x > -2$ **27.** $-2 < x < 0$ or $x > 1$

29. $-3 < x < 3$ and $x \neq 0$
Equivalently $-3 < x < 0$ or $0 < x < 3$

31. $x = 2/5$

33. $x = 4$ or $x = -10/3$ **35.** $x = 1$ or $x = -1$

37. $x = 2$ or $x = -4$ **39.** $x = 11/2$

41. $1 \leq x \leq 5$ **43.** $x > -1$ or $x < -3$

   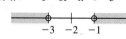

45. $2 \leq x \leq 5$ **47.** $x \geq 1$

49. $x \leq 1$ or $x \geq \frac{13}{3}$

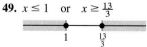

51. $-4 < x < 3$ **53.** $x = -1$

55. x is more than 2 units, but less than 5 units, from 4.

57. $|x + 2| = 2|x - 12|$

61. False   **63.** $-13 < x < 1$ **71.** Yes

1. a. $\sqrt{13}$ **b.** $2\sqrt{2}$ **c.** $\sqrt{122}$
   d. $\sqrt{106}$ **e.** $2\sqrt{2}$ **f.** 5

3. $(0, 3 + 2\sqrt{3})$ and $(0, 3 - 2\sqrt{3})$

5. $a \in \{-1 + 3\sqrt{2}, -1 - 3\sqrt{2}, 5, 2\}$

7. $2x + y - 2 = 0$

11. $\{(x, y) \mid (x + 3)^2 + (y - 1)^2 > 2\}$

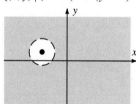

13. $\{(x, y) \mid |x| \leq 2 \text{ and } |y| \leq 2\}$

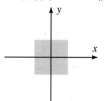

15. $x^2 + y^2 = 9$

17. $(x + 6)^2 + (y + 4)^2 = 25$

19. $(x - 4)^2 + (y + 4)^2 = 25$
     $(x + 1)^2 + (y - 1)^2 = 25$

21. Center $(-2, -1)$, radius 4

23. Center $(1, 3)$, radius $\sqrt{7}$

25. Center $(0, b)$, radius $|a|$

27. $(7, 2)$ and $(4, -1)$

29. Center $(6, -3)$, radius $2\sqrt{5}$

31. $\{x^2 + (y - a)^2 = 1 + (a - 2)^2 \mid a \in \mathbb{R}\}$

Exercise Set 1.3

1. a. $-1/4$ **b.** $-1/7$ **c.** -1 **d.** 1

3. False

5. True

7. $b = 4$

9. $y = -2x + 5$

11. $6x - 5y = -36$

13. $y = -3x + 6$

15. $x = -3$

17. $y = -x + 2$

19. $x - 3y + 11 = 0$

21. $x - 3y = -8$

23. False

25. $m = -1$, x-intercept $= 7$, y-intercept $= 7$

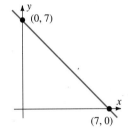

27. $m = -1$, x-intercept $= -3$, y-intercept $= -3$

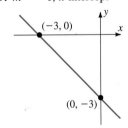

29. $m = 0$, no x-intercept, y-intercept $= 5$

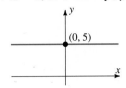

31. No slope, x-intercept $= 4$, no y-intercept

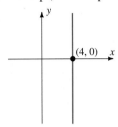

33. $(1, 2)$ **35.** $(-3, 0)$ **37.** No intersection

39. $y = -\dfrac{a}{b}x - \dfrac{c}{b}$; if $b = 0$, $x = -\dfrac{c}{a}$

41. a. Yes   **b.** Yes   **c.** No

43. a. 4 **b.** $-4, 6, 16, 0$

45. $P = 100 - \frac{20}{3}d$, $P = \$100$

47. $F = \frac{9}{5}C + 32$

49. a. $T = 2t + 22$ **b.** $72°C$, $92°C$

51. $a = 1/25{,}000$

53. $y = 3x - 1$

55. a. $m = \frac{10}{29}$ **b.** $\frac{3}{2} - \frac{25}{290}$ **c.** $1.41379 \ldots$

Exercise Set 1.4

1.

$f(x)$	$f(-2)$	$f(0)$	$f(4)$	$f(5)$	$\lvert f(3) \rvert$	$f(f(0))$
$1 - 3x^2$	-11	1	-47	-74	26	-2
$\dfrac{1}{x + 2}$	undefined	$\dfrac{1}{2}$	$\dfrac{1}{6}$	$\dfrac{1}{7}$	$\dfrac{1}{5}$	$\dfrac{2}{5}$
$\dfrac{(x - 3)^2}{x^2 + 1}$	5	9	$\dfrac{1}{17}$	$\dfrac{2}{13}$	0	$\dfrac{18}{41}$
$\sqrt{x + 4}$	$\sqrt{2}$	2	$2\sqrt{2}$	3	$\sqrt{7}$	$\sqrt{6}$
$\dfrac{1}{\sqrt{16 - x^2}}$	$\dfrac{1}{2\sqrt{3}}$	$\dfrac{1}{4}$	undefined	undefined	$\dfrac{1}{\sqrt{7}}$	$\dfrac{4\sqrt{255}}{255}$
$\begin{cases} 1 - x, & x < -3 \\ x - 1, & x > 1 \end{cases}$	undefined	undefined	3	4	2	undefined

3. Yes

5. No

7. Yes

9. $x \in (-\infty, \infty)$

11. $x \geq -4$

13. $t \in [0, \infty)$

15. $-4 \leq s \leq 4$

17. $x \neq 0$

19. $x \leq -2$ or $0 \leq x < 2$ or $2 < x$

21. $x \geq 2$

23. $x \in (-\infty, \infty)$

25. $x \neq -1$

27. False

29. $y = \frac{3}{2}x^2 - 4$

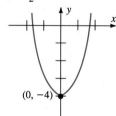

31. $y = 3(x - \frac{1}{6})^2 + \frac{5}{12}$

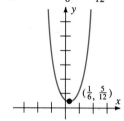

33. $y = -4(x - 3)^2 + 2$

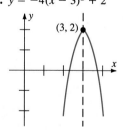

35. $x = -y^2 + 2$

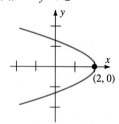

37. $x = -\frac{2}{3}y^2 + 1$

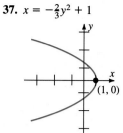

39. $x = 2(y - 1)^2$

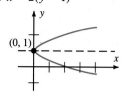

41. $(-3, -5), (2, 0)$

43.

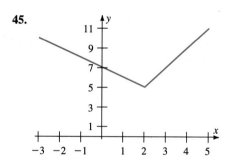

45.

47.

49.

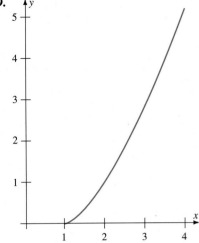

51.

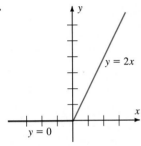

53.

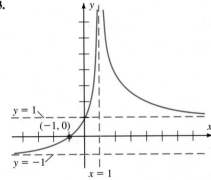

55.

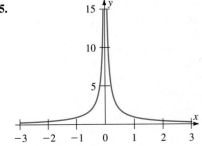

57. a. Even

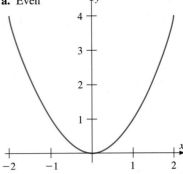

b. Even

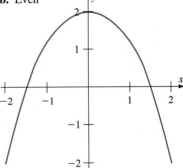

c. Odd

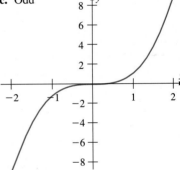

d. Neither

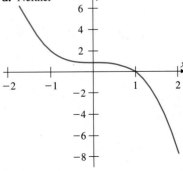

e. Odd

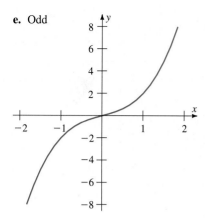

f. Even

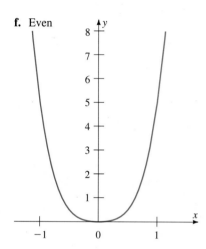

g. Even

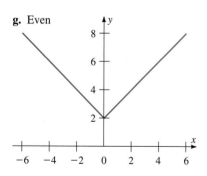

h. Neither

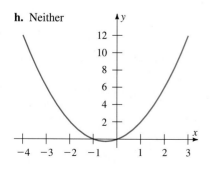

59. $3x^3 + 1$

61. $(3x + 1)^3$

63. $\sqrt{3x + 1}$

65. $3x^{3/2} + 1$

67. a. $[-7/2, -3/2]$ **b.** $[0, 9]$
   c. $[-1, 3]$ **d.** $[6, 24]$

69. Domain $f \circ g$: $(-\infty, -2) \cup (-2, -1] \cup [1, 2) \cup (2, \infty)$
   Domain $g \circ f$: $[-2, -\sqrt{3}) \cup (-\sqrt{3}, -\sqrt{2}] \cup [\sqrt{2}, \sqrt{3}) \cup (\sqrt{3}, 2]$

71. Domain f: $(-\infty, 0) \cup (0, \infty)$; yes

73. True

75. $C(t) = 0.08\,[20,000(0.7)^t]$

77. a. $R(p) = 1100p - 20p^2$ **b.** $p = \$55$

79. $C(n(t)) = \sqrt{2001 + 10t^{2/3}}$

81. $f(s) = s^3$

83. A parabola with vertex $(b, 0)$

87. a. $R = [0, \infty)$ **b.** $R = [1, \infty)$
   c. $R = (0, \infty)$ **d.** $R = [0, \sqrt{6}]$
   e. $R = [0, \infty)$

Exercise Set 1.5

1. a. $\pi/6$ **b.** $5\pi/12$ **c.** $-\pi/12$
   d. $7\pi/4$ **e.** $\dfrac{2\pi}{360}(x + 30)$ **f.** $97\pi/6$

3.

$\sin\theta$	$\cos\theta$	$\tan\theta$	$\sec\theta$
0	1	0	1
$\sqrt{2}/2$	$\sqrt{2}/2$	1	$\sqrt{2}$
3/5	4/5	3/4	5/4
5/13	12/13	5/12	13/12

5.

	$\tan x$	$\cot x$	$\sec x$	$\csc x$
a.	$\sqrt{3}/3$	$\sqrt{3}$	$2\sqrt{3}/3$	2
b.	undefined	0	undefined	1
c.	$-\sqrt{3}$	$-\sqrt{3}/3$	2	$-2\sqrt{3}/3$
d.	0	undefined	-1	undefined
e.	-1	-1	$\sqrt{2}$	$-\sqrt{2}$
f.	$\sqrt{3}/3$	$\sqrt{3}$	$-2\sqrt{3}/3$	-2

7. a. $-3\pi/2,\ -\pi/2,\ \pi/2,\ 3\pi/2$
   b. $-7\pi/4,\ -3\pi/4,\ \pi/4,\ 5\pi/4$
   c. $-5\pi/4,\ -\pi/4,\ 3\pi/4,\ 7\pi/4$
   d. $\pm 2\pi,\ \pm 5\pi/3,\ \pm 4\pi/3,\ \pm\pi,\ \pm 2\pi/3,\ \pm\pi/3,\ 0$
   e. $-11\pi/6,\ -3\pi/2,\ -7\pi/6,\ -\pi/2,\ \pi/6,\ \pi/2,\ 5\pi/6,\ 3\pi/2$
   f. $\pm 5\pi/3,\ \pm 4\pi/3,\ \pm 2\pi/3,\ \pm\pi/3$
   g. $-5\pi/6,\ 7\pi/6,\ 11\pi/6$

9. $t = 4\pi/3$

11. $0 < t < \pi/2$

13. No t

15. $\dfrac{\sqrt{2+\sqrt{3}}}{2}$

17. $\dfrac{\sqrt{2+\sqrt{3}}}{2}$

19. $\dfrac{-2\sqrt{2+\sqrt{3}}}{2+\sqrt{3}}$

21.

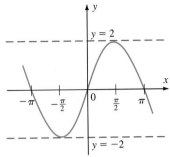

23.

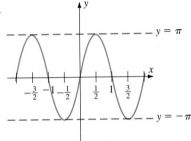

25.

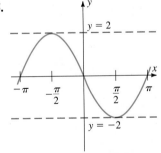

27.

	Domain	Range
$\sin x$	all x	$[-1, 1]$
$\cos x$	all x	$[-1, 1]$
$\tan x$	$x \neq \pi/2 + n\pi$	$(-\infty, \infty)$
$\sec x$	$x \neq \pi/2 + n\pi$	$(-\infty, -1] \cup [1, \infty)$
$\csc x$	$x \neq n\pi$	$(-\infty, -1] \cup [1, \infty)$
$\cot x$	$x \neq n\pi$	$(-\infty, \infty)$

29. Even   **31.** Even   **33.** Even

41. $4\pi/3$ meters   **43.** $50\sqrt{3}$ meters

45. 300 meters   **47.** $2\sqrt{3}$ meters

Review Exercises—Chapter 1

1. $[2, 9)$ **3.** $x \geq 8$

5. a. $2 \leq |x| \leq 3$ **b.** $-\infty < x < \infty$
   c. $\{-4, -3, -2, 2, 3, 4\}$ **d.** $\{-3, -2, -1, 0, 1, 2, 3\}$
   e. $\{x \mid 2 \leq |x|\} \cup \{-1, 0, 1\}$ **f.** $\{x \mid 2 \leq |x| \leq 3 \text{ or } |x| = 4\}$

7. $-5 \leq x \leq 19$ **9.** $A \cup B = A;\ A \cap B = B$

11. $-6 \leq x \leq 5$ **13.** $-2 \leq x \leq 2$

15. $\pi/6 < x < 11\pi/6$

17. $\pi/6 < x < 2\pi/3$ or $5\pi/6 < x < 4\pi/3$

19. $x = 0$ **21.** $15/2$

23. $(x-2)^2 + (y+4)^2 = 25$

25. Center $(2, 0)$, radius 2

27. Center $(1, -1)$, radius 4

29. $1/3,\ 4/3,\ -4$ **31.** $1,\ 3,\ -3$

33. Undefined, none, 2

35. $x - 2y = 10$ **37.** $y = -2x$

39. If $a \neq 0$ then at $x = -c/a$, otherwise never

41.

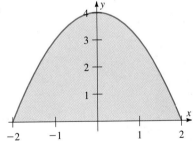

43.

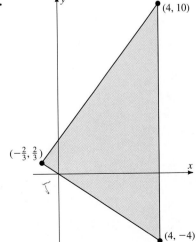

45.

	sin x	cos x	tan x	cot x	sec x	csc x
a.	$\sqrt{3}/2$	$1/2$	$\sqrt{3}$	$\sqrt{3}/3$	2	$2\sqrt{3}/3$
b.	$-\sqrt{2}/2$	$-\sqrt{2}/2$	1	1	$-\sqrt{2}$	$-\sqrt{2}$
c.	$\sqrt{3}/2$	$1/2$	$\sqrt{3}$	$\sqrt{3}/3$	2	$2\sqrt{3}/3$
d.	$\sqrt{2}/2$	$\sqrt{2}/2$	1	1	$\sqrt{2}$	$\sqrt{2}$
e.	$\frac{1}{2}\sqrt{2-\sqrt{3}}$	$\frac{1}{2}\sqrt{2+\sqrt{3}}$	$2-\sqrt{3}$	$2+\sqrt{3}$	$2(2-\sqrt{3})\sqrt{2+\sqrt{3}}$	$2(2+\sqrt{3})\sqrt{2-\sqrt{3}}$
f.	$\frac{1}{2}\sqrt{2+\sqrt{3}}$	$\frac{1}{2}\sqrt{2-\sqrt{3}}$	$2+\sqrt{3}$	$2-\sqrt{3}$	$2(2+\sqrt{3})\sqrt{2-\sqrt{3}}$	$2(2-\sqrt{3})\sqrt{2+\sqrt{3}}$

47. $-6208/999$

49. $(3, 1)$

51. $C = K - 273$

53. $(\frac{5}{2}, -\frac{1}{2})$

55. No

57. Yes

59. $x \le -1$ or $x \ge 1$

61. All x

63. $x \ne \pi/2 + n\pi$

65. Even

67. Even

69. $y = 2(x - \frac{1}{2})^2 + 3$

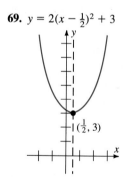

71. $x = -\frac{1}{2}(y - \frac{1}{2})^2 - \frac{1}{4}$

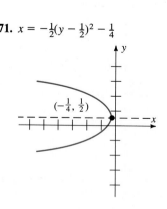

73.

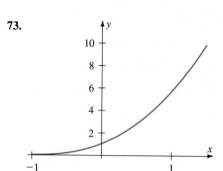

75.

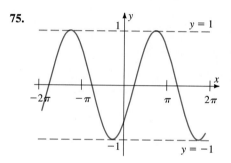

77. a. $2\pi/3$ **b.** $7\pi/4$ **c.** $\pi/3$ or $5\pi/3$

79. $\dfrac{1}{1 - \sin x}$

81. $\sin\left(\dfrac{1}{1 - x}\right)$

83. $\dfrac{1}{1 - \sin^3 x + \sin x}$

87. $3x - 4y + 23 = 0$

89. Domain $= [0, 2\pi]$, range $= [3, 4]$

91. 50

93. $P(x) = \begin{cases} 500 & 0 \le x \le 5 \\ 500 - 10(x - 5) & 5 < x \le 25 \\ 300 & x > 25 \end{cases}$

Readiness Test

1. Center: $(1, -3/2)$; radius: $\sqrt{29}/2$. See pages 13–14 and Exercises 15–26 in Section 2.

2. $(x, y) = (0, 7)$. See pages 21–22 and Exercises 33–38 in Section 3.

3. Point-slope: $(y - 2) = (-\frac{3}{2})(x + 1)$. Slope-intercept: $y = (-\frac{3}{2})x + \frac{1}{2}$. See pages 17–21 and Exercises 8–32 in Section 3.

4.

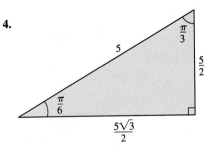

See page 41 and Exercises 42–45 in Section 5.

5. $x \in (-\infty, -6) \cup (0, \infty)$. See pages 6–9 and Exercises 30–55 in Section 1.

6. One equation for the line is $y = -2x - 2$. See pages 19–21 and Exercises 8–32 in Section 3.

7. $y = f(x) = 2(x - 1)^2 + 1$

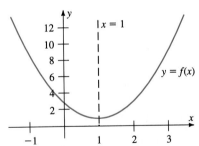

See pages 27–28 and Exercises 29–34 in Section 4.

8. The domain of g is $(-\infty, 0] \cup [2, \infty)$. See pages 24–27 and Exercises 9–27 in Section 4.

9. $x \in (-\infty, -1) \cup (3, \infty)$. See pages 3–7 and Exercises 10–15 and 23–29 in Section 1.

10. a. $2\sqrt{x}/(x^2 + 1)$; domain: $\{x \mid x \geq 0\}$
b. $(x^2 + 1)/\sqrt{x}$; domain: $\{x \mid x > 0\}$
c. $x^{5/2} + x^{1/2}$; domain: $\{x \mid x \geq 0\}$
d. $\sqrt{x^2 + 1}$; domain consists of all real numbers x.
e. $x + 1$; domain: $\{x \mid x \geq 0\}$
See pages 32–35 and Exercises 59–70 in Section 4.

11. $x = -6, 0, 6$. See page 8 and Exercises 30–40 in Section 1.

12.

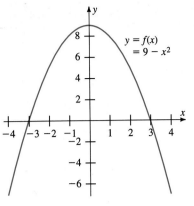

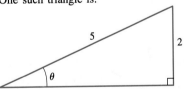

See pages 31–32 and Exercises 51–56 in Section 4.

13. $\theta = -\pi/6 + 2k\pi$ where k is any integer. See pages 41–42 and Exercises 5–7 in Section 5.

14. $\cos \theta = -\sqrt{5/8}$. See pages 45–46 and Exercises 7 and 15–20 in Section 5.

15. One such triangle is:

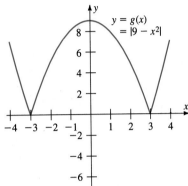

See page 44 and Exercises 42–45 and 47 in Section 5.

16. $x \in [0, \pi/3) \cup (5\pi/3, 2\pi]$.

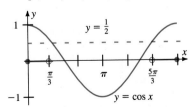

See pages 41–42 and Exercises 8–14 in Section 5.

17. $x \neq 3\pi/2 + 2k\pi$ where k is any integer. See pages 41–42 and Exercises 4–7 in Section 5.

18. $x \geq -2$. See pages 29–31 and Exercises 11, 12, 15, and 18–20 in Section 4.

19. All (x, y) such that $y = 2x + 5/2$. See page 13 and Exercises 7–9 in Section 2.

20. $(-1, 0] \cup [4, 5)$. See page 7 and Exercises 18–39 in Section 1.

21. There are many such examples. Here are two arbitrary ones:
   a. $2x^4 - x^3 - \pi x + \sqrt{2}$
   b. $\dfrac{x^2 + 1}{x + 2}$
   See pages 27–29 and page 31.

22. $59.8°$. See pages 16–18 and Exercises 45–47, 49, and 50 in Section 3.

23. $x \neq 2$. See pages 29–31 and Exercises 11, 12, 15, and 18–20 in Section 4.

24. $y \in (-\infty, -1]$ and $[1, \infty)$. See page 43 and Exercises 5 and 32 in Section 5.

25. $Y(x) = 400x - 2x^2$. See Exercises 74–77 and 79–84 in Section 4.

Chapter 2

Exercise Set 2.1

1. 3   **3.** 12   **5.** 4

7. -8 **9.** 12   **11.** -32

13. $3a + 2b + c$ **15.** -1 **17.** -1

19. $y = -6x - 3$ **21.** $y = 2ax - a$ **23.** $y = 7x - 4$

25. False. The tangent to $y = x^2$ at $(0, 0)$ is horizontal.

27. a.

x_0	h	$\dfrac{f(x_0 + h) - f(x_0)}{h}$
2	0.1000	9.61
2	0.0100	9.0601
2	0.0010	9.006001
2	0.0001	9.00060001
2	−0.0001	8.99940001
2	−0.0010	8.994001
2	−0.0100	8.9401
2	−0.1000	8.41

b. Slope is 9.

29. a. $f'(-3/2) \approx 7/4$, $f'(-1) \approx -1$, $f'(1/2) \approx -1/4$
   b. $f'(-3/2) = 7/4$, $f'(-1) = -1$, $f'(1/2) = -1/4$

31. a.

x_0	h	$\dfrac{f(x_0 + h) - f(x_0)}{h}$
0	0.1000	0.9983341665
0	0.0100	0.9999833334
0	0.0010	0.9999998333
0	0.0001	0.9999999983
0	−0.0001	0.9999999983
0	−0.0010	0.9999998333
0	−0.0100	0.9999833334
0	−0.1000	0.9983341665

b. $\lim\limits_{h \to 0} \dfrac{\sin h}{h}$, "Inspection" gives $\dfrac{0}{0}$, which is undefined.

33. $a = -3$

35. $(4, 5)$ and $(1, -1)$

37. $a = 2$, $b = -5$, $c = 5$

39.

x_0	h	$\dfrac{f(x_0 + h) - f(x_0)}{h}$
0	0.1000	1
0	0.0100	1
0	0.0010	1
0	0.0001	1
0	−0.0001	−1
0	−0.0010	−1
0	−0.0100	−1
0	−0.1000	−1

Exercise Set 2.2

1. (i)   **3.** (iii)   **5.** (ii)   **7.** 2   **9.** 1

11. 17   **13.** 6   **15.** 1   **17.** -5 **19.** 0

21. 2   **23.** 0   **25.** -4 **27.** $2/\pi$ **29.** 2

31. -5 **33.** 4   **35.** $-1/6$ **37.** -1 **39.** 2

41. Does not exist   **43.** 3   **45.** 1

47.

x	$f(x)$
1.000	2.17534
0.500	2.04219
0.100	2.00167
0.050	2.00042
0.010	2.000017
0.005	2.000004
−0.005	2.000004
−0.010	2.000017
−0.050	2.00042
−0.100	2.00167
−0.500	2.04219
−1.000	2.17534

49.

x	$f(x)$
1.000	0.158529
0.500	0.164596
0.100	0.166583
0.050	0.166646
0.010	0.166666
0.005	0.166666
−0.005	0.166666
−0.010	0.166666
−0.050	0.166646
−0.100	0.166583
−0.500	0.164596
−1.000	0.158529

51.

x	$f(x)$
1.000	0.459698
0.500	0.497401
0.100	0.499996
0.050	0.4999997
0.010	0.5
0.005	0.5
−0.005	0.5
−0.010	0.5
−0.050	0.4999997
−0.100	0.499996
−0.500	0.497401
−1.000	0.459698

53. $c = \pm\sqrt{2}/2$; $1/2$

55. False

Exercise Set 2.3

1. b. 1, 0.2, 0.025

3. b. $\sqrt{2}$, 1, 0.547

5. b. 2, 0.8, 0.05

Exercise Set 2.4

1. 2   **3.** −10   **5.** −30   **7.** 2   **9.** 4

11. $\frac{6}{17}$ **13.** $\frac{1}{2}$ **15.** $-\frac{3}{8}$ **17.** 4   **19.** 9

21. −18   **23.** $\frac{24}{11}$ **25.** $\frac{1}{2}$ **27.** $\frac{1}{4}$ **29.** 0

31. 0   **33.** 1   **35.** 1

41. $\lim\limits_{x\to 0} (x^2 - x) = \lim\limits_{x\to 1} (x^2 - x) = 0$

43. $f(x) = g(x) = \dfrac{|x|}{x}$

45. $a = 5, -3$

49. 0

Exercise Set 2.5

1. a. 2   **b.** 2   **c.** 0   **3. a.** 0   **b.** 2   **c.** −1

5. a. $a \neq 1$ **b.** all a **c.** $a \neq 1$ **d.** $a \neq 1$

7. 0   **9.** 0   **11.** 0   **13.** 0   **15.** 3

17. 6   **19.** 0   **21.** −4/7   **23.** 1   **25.** 1

27. a. 1   **b.** 1   **c.** Yes, 1

29. a. 2   **b.** 3   **c.** No

33. a.

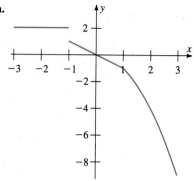

b. No
c. Yes

35. $a = 3/4$, $b = 1/2$

Exercise Set 2.6

1. a. 1, 2, 4, 5, 6
  b. Yes for all (a, b)
      Yes for $[1, 2)$
      No for $[4, 5)$, $[5, 6)$
      No for all $(a, b]$
      No for all $[a, b]$

3. $(k\pi - \pi/2, k\pi + \pi/2)$
   $k = \ldots, -1, 0, 1, \ldots$

5. $(k\pi, (k + 1)\pi)$
   $k = \ldots, -1, 0, 1, \ldots$

7. $(-\infty, -1)$ and $(-1, 2)$ and $(2, \infty)$

9. $(-\infty, 0)$ and $(0, \infty)$

11. $(-\infty, 2]$ and $(2, \infty)$

13. $(-\infty, -1]$ and $(-1, \infty)$ **15.** $(k\pi, (k + 1)\pi)$

17. $(-7, \infty)$ **19.** $(-\infty, \infty)$

21. $[(k - \frac{1}{2})\pi, (k + \frac{1}{2})\pi]$

23. $\cdots, (-3\pi/2, -\pi/2), (-\pi/2, -1), (-1, \pi/2),$
   $(\pi/2, 2), (2, 3\pi/2), (3\pi/2, 5\pi/2), \cdots$

25. $f(1) = 2$ **27.** $f(0) = 0$ **29.** $k = 3$ **31.** $k = -1$

33. 243   **35.** 0   **37.** 64   **39.** 1

43. 2   **45.** 1/3   **47.** 1/3

49. Positive: $(-3, 3)$; negative: $(-\infty, -3)$, $(3, \infty)$

51. Positive: $(-2, -1)$, $(1, \infty)$; negative: $(-\infty, -2)$, $(-1, 1)$

53. Positive: $((2k + \frac{1}{2})\pi, (2k + \frac{3}{2})\pi)$;
   negative: $((2k - \frac{1}{2})\pi, (2k + \frac{1}{2})\pi)$

55. $x < -4$ or $x > -2$

57. $-2 < x < 0$ or $x > 1$

59. $x < -3$ or $x > 3$

61. $-2\pi < x < -3\pi/2$ or $-\pi < x < -\pi/2$ or
   $0 < x < \pi/2$ or $\pi < x < 3\pi/2$

Review Exercises—Chapter 2

1. 4   **3.** 77   **5.** −2

7. 6   **9.** −3   **11.** 4/3

13. 0   **15.** 0   **17.** 1

19. 4   **21.** Does not exist   **23.** 3/4

25. 3   **27.** 0   **29.** Does not exist

31. 5   **33.** 1   **35.** 1

37. 0   **39.** 1/3   **41.** 6, $y = 6x + 1$

43. −1/4, $x + 4y = 4$ **45.** 32, $y = 32x - 16$

47. $(-\infty, -2)$, $(-2, 3)$, $(3, \infty)$

49. $[0, \pi/2)$, $(\pi/2, 3\pi/2)$, $(3\pi/2, 2\pi]$

51. $(-\infty, -1]$, $[4, \infty)$ **53.** $(-\infty, \pi/4]$, $(\pi/4, \infty)$

55. $(-\infty, -2)$, $[-2, \infty)$ **57.** $(-\infty, \infty)$

59. a. 1   **b.** 1   **c.** 1   **d.** 2   **e.** 1
   f. Does not exist   **g.** Yes   **h.** No

61. 5   **63.** 1   **73.** $a = -1$

75. False

Chapter 3

Exercise Set 3.1

1. a. (iv)   **b.** (v)   **c.** (iii)   **d.** (i)   **e.** (ii)

3. 6   **5.** $6x^2$

7. $3ax^2 + 2bx + c$ **9.** $-2/(2x + 3)^2$

11. $\dfrac{1}{2\sqrt{x + 1}}$ **13.** $-\frac{1}{2}(x + 1)^{-3/2}$

15. $-2x^{-3}$ **17.** $3(x + 3)^2$

19. $-\frac{1}{2}(x + 5)^{-3/2}$ **21.** $-6(x - 1)^{-3}$

23. $2x + 9y - 3 = 0$ **25.** $x + 6y = 15$

27. $x_1 = -\sqrt{3}/2$, $x_2 = \sqrt{3}/2$ **29.** $a = 3$

31. $a = 2$, $b = 2$

33.

h \ x	0	$\pi/4$	$\pi/2$	$3\pi/4$	π	$5\pi/4$	$3\pi/2$	$7\pi/4$	2π
-0.5	0.244835	-0.504886	-0.958851	-0.851135	-0.244835	0.504886	0.958851	0.851135	0.244835
-0.1	0.0499583	-0.670603	-0.998334	-0.741255	-0.0499583	0.670603	0.998334	0.741255	0.0499583
-0.05	0.0249948	-0.689138	-0.999583	-0.724486	-0.0249948	0.689138	0.999583	0.724486	0.0249948
-0.01	0.00499996	-0.703559	-0.999983	-0.710631	-0.00499996	0.703559	0.999983	0.710631	0.00499996
Est.	0	-0.707	-1	-0.707	0	0.707	1	0.707	0
0.01	-0.00499996	-0.710631	-0.999983	-0.703559	0.00499996	0.710631	0.999983	0.703559	-0.00499996
0.05	-0.0249948	-0.724486	-0.999583	-0.689138	0.0249948	0.724486	0.999583	0.689138	-0.0249948
0.1	-0.0499583	-0.741255	-0.998334	-0.670603	0.0499583	0.741255	0.998334	0.670603	-0.049958
0.5	-0.244835	-0.851135	-0.958851	-0.504886	0.244835	0.851135	0.958851	0.504886	-0.244835

35. $x < 3$

37. a. No   **b.** Yes

39. 1

41. a. False   **b.** True   **c.** False
d. True   **e.** False   **f.** True

Exercise Set 3.2

1. $24x^2 - 2x$

3. $3ax^2 + b$

5. x

7. $2x + 1$

9. $12x^3 - 24x^2 + 12x - 16$ **11.** $6x^5 - 8x^3 + 2x$

13. $10x + 2 - 3x^{-2} + 8x^{-3}$ **15.** $6(3 - x)^{-2}$

17. $(1 - x^3)^{-2}(-x^6 + 12x^3 + 12x^2 + 4)$

19. $-15x^{-4} + 10x^{-6}$ **21.** $2x - 2 + 2x^{-2}$

23. $-6x^{-2} - 18x^{-3}$

25. $(x - 1)^{-3}(-4x - 4)$

27. $8x^7 + 1 + 2x^{-3} + 9x^{-10}$

29. $-8x^{-9} + x^{-2}$

31. $4x^3$ **33.** $21x^6 + 88x^3 + 27x^2 + 7$

35. $-2(x^2 + 3x + 2)^{-2}(2x + 3)$

37. $4u^3 - 2u - 2$ **41.** $8s^3 - 21s^2 + 22s - 28$

43. $(x^3 + 3x^2 + 2x)^{-2}(-3x^2 - 6x - 2)$

45. $(18x^2 - 18)(2x^3 - 6x + 9)^2$

47. 132   **49.** $3x^{-4}(x - 1)^2$

51. $-3(x - 6)^{-4}$ **53.** $3x^2 - 3$

55. $x - 2y - 1 = 0$ **57.** $y = 0$

59. $y = -\frac{1}{6}x + \frac{19}{6}$ **61.** $y = 2x - 7/2$

63. $(3, 9/2)$, $(1, -3/2)$ **65.** $b = 7$

67. $\dfrac{1}{n}x^{(1/n)-1}$

69. $\pm\sqrt{7/3}$

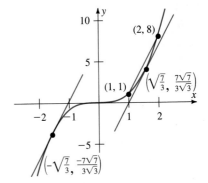

Exercise Set 3.3

1. 7, 7

3. 0, -3

5. a. 3   **b.** Never   **c.** $[0, \infty)$

7. a. $-(1 + t)^{-2}$ **b.** Never   **c.** Never

9. a. $3(t - 2)(t - 4)$ **b.** 2, 4   **c.** $[0, 2) \cup (4, \infty)$

11. a. $3(t - 3)(t - 1)$ **b.** 1, 3   **c.** $[0, 1) \cup (3, \infty)$

13. 13/16

15. a. $-148\pi/3$, -36π **b.** -72π

17. a. 6   **b.** 3   **c.** -6

19. a. 5 s   **b.** 122.5 m   **c.** -24.5 m/s

21. a. $40 + 400t - 4.9t^2$ **b.** 8200 m   **c.** 401 m/s

23. a.

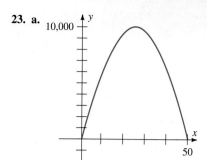

b. $32(25 - t)$ **c.** $0 < t < 25$ **d.** $25 < t < 50$
e. $t = 25$ **f.** $10,000$ **g.** $t = 20, 30$
h. $t = 50$

Exercise Set 3.4

1. $-4 \sin x$

3. $3x^2 \tan x + x^3 \sec^2 x$

5. $3x^2 \cot x - (x^3 - 2) \csc^2 x$

7. $\sec^2 x$

9. $(1 - x) \cos x - (1 + x) \sin x$

11. $\sec x \tan^2 x + \sec^3 x$

13. $\dfrac{2 - x \cos x + \sin x}{(2 + \sin x)^2}$

15. $\dfrac{(\cos x + \sin x)(1 + \tan x) - (\sin x - \cos x) \sec^2 x}{(1 + \tan x)^2}$

17. $\dfrac{(2x - 4 \csc^2 x)(x + \tan x) - (x^2 + 4 \cot x)(1 + \sec^2 x)}{(x + \tan x)^2}$

19. $\dfrac{(-3 \csc x \cot x)(4x^2 - 5 \tan x) - 3 \csc x(8x - 5 \sec^2 x)}{(4x^2 - 5 \tan x)^2}$

21. $y = -\pi x + \pi^2$

23. $x = 0, \ \pi, \ 2\pi, \ 3\pi, \ 4\pi$

25. $t = \pi/2 + n\pi, \ n = 0, \ \pm 1, \ \pm 2, \ \ldots$

27. a. $3\pi/4, \ 7\pi/4$
   b. Never

31. a. $-10 \sin t$
   b. $k\pi, \ k = 0, \ \pm 1, \ \pm 2, \ \ldots$

Exercise Set 3.5

1. $-6x(1 - x^2)^2$

3. $-x^{-2} + 3x^{-4}$

5. $4x^3 \sec^2(1 + x^4)$

7. $-(2 \sin x \cos x)(1 + \sin^2 x)^{-2}$

9. $-\sin(\sin x) \cos x$

11. $(2x + 1) \cos (x^2 + x)$

13. $2 \cos x(1 - \sin x)^{-2}$

15. $-168(3x + 7)^3(3x - 7)^{-5}$

17. $-168(3x + 7)^3(3x - 7)^{-5}$

19. $6x(x^2 + 4)^2$

21. $-6(\sin x + 1)(\cos x - x)^5$

23. $3(x^2 + 1) \cos (x^3 + 3x)$

25. $(2x + 1) \sec^2 (x^2 + x)$

27. $(x^4 - 5)^2(13x^4 - 5)$

29. $-6x(x^2 - 9)^{-4}$

31. $12 \sec^2 x(3 \tan x - 2)^3$

33. $24(x - 3)^3(x + 3)^{-5}$

35. $\cos(1 - x^2) + 2x^2 \sin(1 - x^2)$

37. $[2 \tan x \sec^2 x(1 - x) + (\tan^2 x + 1)](1 - x)^{-2}$

39. 0

41. $(1 - 5x - 11x^2)(x^2 + x + 1)^{-7}$

43. $4(ad - bc)(ax + b)^3(cx + d)^{-5}$

45. $3 \cos^2 x \sin x(1 + \cos^3 x)^{-2}$

47. $-12x^3(1 + x^4)^{-4}$

49. $6 \sec^2(6x) - 6 \sec^2 x$

51. $12 \sec^3 4x \tan 4x$

53. $-4x \tan(\pi - x^2) \sec^2 (\pi - x^2)$

55. $4\pi(\sin \pi x - \cos \pi x)^3(\cos \pi x + \sin \pi x)$

57. $48x(x^2 - 3)^3[1 + (x^2 - 3)^4]^5$

59. $\dfrac{-4}{(1 - x)^2} \cos\!\left(\dfrac{1 + x}{1 - x}\right) \sin\!\left(\dfrac{1 + x}{1 - x}\right)$

61. $8x(1 + x^2)[1 + (1 + x^2)^2]$

63. $\dfrac{x}{\sqrt{1 + x^2}} \sec^2 \sqrt{1 + x^2}$

65. Tangent: $y = 0$

67. Tangent: $y = -x$

69. Tangent: $y = 1 - (2 + \pi/2)x$

71. a. 10 **b.** -4

73. -1

75. Tangent: $y = \sqrt{\pi/2}(x - \sqrt{\pi/2}) + \sqrt{2}/2$

77. $v = 20 \cos 4t$

79. 50π in.³/s

Exercise Set 3.6

1. -108

3. $1/8$

5. $-8\sqrt{3}/3$

7. $2\sqrt{3}$ cm/s

9. 144π cm³/s

11. $5/(8\pi)$ m/min

13. 0.8 cm/s

15. 6.25 m/min

17. 0.23 m/s

19. 4 persons per day

21. 68 km/h

23. $9\sqrt{3}/8$ cm/s

25. a. $(-\sqrt{2}/2, \sqrt{2}/2), (\sqrt{2}/2, -\sqrt{2}/2)$
b. $(\sqrt{2}/2, \sqrt{2}/2), (-\sqrt{2}/2, -\sqrt{2}/2)$

27. $10\pi \approx 31.416$

29. $\dfrac{25}{36\pi} \approx 0.22$ cm/s

31. $\dfrac{80\pi}{3} \approx 83.8$ m/s

33. $0.15\dfrac{\cos \alpha}{\cos \beta}$ rad/s

35. a. $\dfrac{4\alpha}{25}$ cm/min²

b. $-\dfrac{2\alpha}{25}$ cm/s²

Exercise Set 3.7

1. a. $2 + \frac{1}{4}\Delta x$ **b.** 2.025

3. a. $\frac{1}{2} - \frac{1}{4}\Delta x$ **b.** $\frac{11}{26}$

5. a. $\frac{1}{8} + \frac{3}{4}\Delta x$ **b.** $\frac{11}{100}$

7. a. $2\Delta x$ **b.** $\pi/45$

9. a. $\dfrac{\sqrt{3}}{2} - \dfrac{1}{2}\Delta x$ **b.** $\dfrac{\sqrt{3}}{2} - \dfrac{\pi}{360}$

11. $97/14$

13. $39/160$

15. $1/2 + \sqrt{3}\pi/360$

17. 26.19

19.

Exercise	11	13	15	17
Actual Value	6.928203	0.243902	0.515038	26.198073
Approximation	6.928571	0.24375	0.515115	26.19
Relative Error	0.000053	0.000625	0.000149	0.000308
Percentage Error	0.0053%	0.0625%	0.0149%	0.0308%

21. $0.0014, 0.14\%$

23. 5.05 **25.** 0.4π cm³

27. a. -32θ **b.** $-16\pi/3$ **c.** 0.047

29. a. $53/7$ **b.** $121/16$ **c.** $53/7 > 121/16$
d.

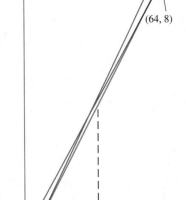

e. $53/7 > 121/16 > \sqrt{57}$

Exercise Set 3.8

1. $\frac{4}{3}(x + 2)^{1/3}$ **3.** $\frac{1}{2}x^{-1/2} - \frac{1}{2}x^{-3/2}$

5. $\frac{2}{3}x^{-1/3} - \frac{2}{3}x^{-5/3}$

7. $-8x(x^2 + 1)(x^2 - 1)^{-3}$

9. $1/(6x^{2/3}\sqrt{1 + x^{1/3}})$ **11.** $\dfrac{-2x^{1/3}}{3(x^{1/3} + x)^2}$

13. $\frac{1}{2}x^{-1/2} + \frac{1}{4}x^{-3/4} + \frac{1}{8}x^{-7/8}$

15. $\dfrac{4x(x^2 + 3)}{3(x^2 + 1)^{1/3}(1 - x^2)^{7/3}}$

17. $\frac{2}{3}x \cos x^2(\sin x^2)^{-2/3}$

19. $(\sin^3 x - \sqrt{x})^{-4/3}[(\sin^3 x - \sqrt{x})(\frac{19}{12}x^{7/12} - \frac{5}{4}x^{1/4})$
$\qquad\qquad -(x^{19/12} - x^{5/4})\frac{1}{3}(3\sin^2 x \cos x - \frac{1}{2}x^{-1/2})]$

21. $\frac{5}{3}x^{-1/3} \cos(x^{2/3}) \sin^{3/2}(x^{2/3})$

23. $\frac{1}{3}x^{-2/3} \cos \sqrt[3]{x} + \frac{1}{3}\cos x(\sin x)^{-2/3}$

25. $\frac{3}{4}(3x + 1)^{-3/4} \sec(\sqrt[4]{3x + 1}) \tan(\sqrt[4]{3x + 1})$

27. $-x/y$ **29.** $-\sqrt{y}/\sqrt{x}$

31. $\cos^2 y$, which equals $\dfrac{1}{1 + x^2}$ **33.** $\dfrac{\sin y + y \sin x}{\cos x - x \cos y}$

35. $\dfrac{2\sqrt{xy}(y - 1) - y}{x(1 - 2\sqrt{xy})}$, or equivalently $\dfrac{(y - 1)^2}{1 - 2xy + 2x}$

37. $\dfrac{-3x^2 - 2xy - y^2}{x^2 + 2xy + 3y^2}$

39. $\dfrac{2(y-1)\sqrt{x+y}-1}{1-2x\sqrt{x+y}}$

41. $\dfrac{6x-\csc^2(x+y)}{\csc^2(x+y)-\csc^2 y}$

43. $-y/x$

45. $5x^4/4y^3$

47. $\dfrac{\sec y}{2y-x\sec y\tan y}$

49. $\dfrac{1-2y\cos(2xy)}{2x\cos(2xy)}$

51. $x+y=6$

53. $x+y-4=0$

55. $3x-5y=0$

57. $y=3x-3$

61. $y=\sqrt{2}$
$y=-\sqrt{2}$
$x=\sqrt{2}$
$x=-\sqrt{2}$

63. $x=1$
$x=-1$

65. $y=-1$
$y=-3$
$x=0$
$x=-2$

67. $a=\pm 1$

71. $(\sqrt{2},-\sqrt{2}),\ (-\sqrt{2},\sqrt{2})$

Exercise Set 3.9

1. $12x$

3. $20x^3-18x^{-4}$

5. $2(1+x)^{-3}$

7. $2\cos(2x)$

9. $(2-\cos x)(1+\cos x)^{-2}$

11. $2\sec^4 x+4\sec^2 x\tan^2 x$

13. $-2\sec x\tan x-x\sec x\tan^2 x-x\sec^3 x$

15. $30x(x^3+1)^3(7x^3+1)$

17. $6x^{-4}-40x^{-6}$

19. $2\sec^3 x-\sec x$

21. $6x\cos x-6x^2\sin x-x^3\cos x$

23. $(2+6x^2)(1-x^2)^{-3}$

25. $-4(x+2)^{-3}$

27. $48(35x^4-30x^2+3)$

29. $\dfrac{dy}{dx}=\dfrac{2}{y},\ \dfrac{d^2y}{dx^2}=\dfrac{-4}{y^3}$

31. $\dfrac{dy}{dx}=\dfrac{y}{2y-x}$
$\dfrac{d^2y}{dx^2}=\dfrac{2y(y-x)}{(2y-x)^3}$

35. a. $v(t)=3t^2-12t-30$
b. $2-\sqrt{14}<t<2+\sqrt{14}$
c. $a(t)=6t-12$
d. $t>2$

37. $v(t)=-5\sin t,\ a(t)=-5\cos t$

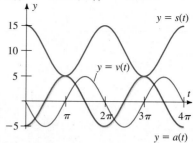

$s(t)$ gets larger when $v(t)>0$.
$v(t)$ gets larger when $a(t)>0$.

39. $-\cos x$

41. $-x\cos x-50\sin x$

43. $f'(x)=\begin{cases}2x & \text{if } |x|>1 \\ -2x & \text{if } -1<x<1\end{cases}$ $f''(x)=\begin{cases}2 & \text{if } |x|>1 \\ -2 & \text{if } -1<x<1\end{cases}$

45. $6-8x+3x^2$

Exercise Set 3.10

1. 0.229309

3. 0.618034

5. 2.30278

7. -1.49535

9. 2.30278

11. 6.0828

13. 1.7826

15. $-1.6180,\ 0.6180,\ 1.75$

17. $-2.6180,\ -1.6180,\ -0.38197,\ 0.6180$

19. 1.0144

21. $x_{10}\approx 0.00387$

Review Exercises—Chapter 3

1. $-7(3x-7)^{-2}$

3. $(-36x^4+9x^3-12x^2+x-2)x^{-3}(1+3x^2)^{-2}$

5. $\dfrac{1}{2\sqrt{t}}\sin t+\sqrt{t}\cos t$

7. $(x^3+x^2+x)^{-2}(-3x^2-2x-1)$

9. $(t^{-2}-t^{-3})^{-2}(2t^{-3}-3t^{-4})$

11. $2x\sin x^2+2x^3\cos x^2$

13. $-3(x^4+4x^2)^{-4}(4x^3+8x)$

15. $(2+7x)(x^2+1)^{-3/2}$

17. $30x(x^2+1)^2[(x^2+1)^3-7]^4$

19. $(-3x^2+18x+9)/[2\sqrt{3x-9}(x^2+3)^{3/2}]$

21. $-3\cot^2 s\csc^2 s$

23. $(\cos^2 x\sqrt{1 + \sin x})^{-1}(\frac{1}{2}\cos^2 x + \sin x + \sin^2 x)$

25. $\frac{9}{2}t^{7/2} - 15t^{3/2}$

27. $-2t\sin(2t^2 - 2)$

29. $\frac{1}{2}x^{-11/2}(12x^6 - 8x^5 - 4x^4 - 12x^3 + 20x^2 - 9)$

31. $f'(x) = \sin x + x\cos x,\quad f''(x) = 2\cos x - x\sin x$

33. $f'(x) = 2\sec^3 x - \sec x,\quad f''(x) = (6\sec^2 x - 1)\sec x\tan x$

35. $v = 2t + \frac{1}{2}t^{-1/2},\quad a = 2 - \frac{1}{4}t^{-3/2}$

37. $v = t/\sqrt{2 + t^2},\quad a = 2(2 + t^2)^{-3/2}$

39. 6

41. $1/(2\sqrt{x + 1})$

43. $2x + 1$

45. $(12x - y)/(x + 8y)$

47. $1/(1 + 2y + 3y^2)$

49. $-y/x$

51. $\sin y/(1 - x\cos y)$

53. $-(y^2 + x^2)/y^3$, which equals $-4/y^3$

55. $x + y - 1 = 0$

57. $2x + y - 1 - \pi/2 = 0$

59. $y = 3\sqrt{2}(x - \pi/4) + \sqrt{2}$

61. $(\sin x + x\cos x)\,dx$

63. $x(1 + x^2)^{-1/2}\,dx$

65. $\dfrac{-\sin\sqrt{x}}{2\sqrt{x}}\,dx$

67. 4.0667

69. 0.88

71. Not differentiable

73. Not differentiable

75. a. (v)   **b.** (viii)   **c.** (i)

77. $60\sqrt{3}$ cm²/5

79. $7/4$ m²/s

83. $y = -\dfrac{\sqrt{2\pi}}{\pi}x + \sqrt{2}$

85. $y = x/2$

87. $8/\pi$, 4

89. $2\sqrt{2} + \dfrac{\sqrt{2}}{8}\Delta x$

91. $1 + 2\Delta x$

93. 10

95. $0.00176 = 0.176\%$

97. $(1, 4), (-1, 4)$

99. $\frac{8}{5}x^{5/4}$

101. a. $v(t) = t^3 - 9t^2 + 24t - 20$ **b.** $a(t) = 3t^2 - 18t + 24$
   c. $t > 5$ **d.** $t < 2$ or $2 < t < 5$
   e. 2, 5   **f.** 5
   g. 2, 4

103. 1

Chapter 4

Exercise Set 4.1

1. a. x_1, x_2 **b.** a, b **c.** x_2 **d.** a, x_2

3. a. x_1, x_2, x_3 **b.** a, b **c.** x_3 **d.** x_3, b

5. Minimum

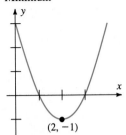
$(2, -1)$

7. Minimum

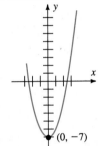
$(0, -7)$

9. Minimum

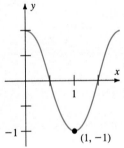
$(1, -1)$

	Critical numbers	Minimum	Maximum
11.	0, 8/3	$f(8/3) = -256/27$	$f(0) = 0$
13.	2	$f(1) = f(3) = -1/3$	$f(2) = -1/4$
15.	0, 4/3	$f(4/3) = -32/27$	$f(0) = 0$
17.	0	$f(-\pi/4) = f(\pi/4) = 0$	$f(0) = 1$
19.	$\pi/4$	$f(\pi) = -1$	$f(\pi/4) = \sqrt{2}$
21.	1	$f(1) = 2$	$f(1/2) = f(2) = 5/2$
23.	0	$f(0) = 1$	$f(-\pi/4) = f(\pi/4) = \sqrt{2}$
25.	0	$f(-1) = -1$	$f(1) = 1/3$
27.	0, 1	$f(8) = -16$	$f(1) = 6$

29. max: 4,   min: 0

31. max: 1/2,   min: 0

33. a. They are the same if $c \neq 0$.
   b. The critical numbers for h are the critical numbers for f moved left c units.
   c. The critical numbers for k are the critical numbers for f divided by c.

35. $b = -6$, $c = 2$

37. a. max: $\sqrt{2}$,   min: $-\sqrt{2}$
   b. max: 2,   min: -2
   c. max: 1,   min: -1

39. max: $(-2, 5)$,   min: $(3, 0)$

41. max: $(5, 14)$,   min: $(3, 0)$, $(-2, 0)$

43. max: $(-2, 1)$,   min: $(0, 0)$

Exercise Set 4.2

1. a. $f(x) = x(20 - x)$, $[0, 20]$, $(0, 20)$
   b. $f(x) = x^2 + (20 - x)^2$, $[0, 20]$, $(10, 10)$
   c. $f(x) = x^2 + (20 - x)^2$, $[0, 20]$, $(0, 20)$

3. $f(x) = x(20 - 2x)$, $[0, 10]$, $5\,\mathrm{m} \times 10\,\mathrm{m}$

5. $f(\ell) = 2\ell(20 - \ell)$, $[0, 20]$, $\ell = 10\,\mathrm{m}$, $w = 20\,\mathrm{m}$

7. $f(x) = 2x(12 - x)(8 - x)$, $[0, 8]$, $17.72\,\mathrm{cm} \times 4.86\,\mathrm{cm} \times 3.14\,\mathrm{cm}$

9. $V(h) = 2h(5 - h)(8 - h)$, $[0, 5]$, $2\,\mathrm{cm} \times 6\,\mathrm{cm} \times 6\,\mathrm{cm}$

11. $A(x) = 8x(1 - x/3)$, $[0, 3]$, $3\,\mathrm{cm} \times 2\,\mathrm{cm}$

13. $A(x) = 2x(4 - x^2)$, $[0, 2]$, $4\sqrt{3}/3 \times 8/3$

15. $A(r) = \dfrac{r^2}{2}\left(\dfrac{P}{r} - 2\right)$, $[0, P/2]$, $\theta = 2$ rad, $r = P/4$

17. $Y(t) = (25 + t)(495 - 15t)$, $[0, 33]$, 4 additional trees

19. $A(x) = x\sqrt{64 - x^2}$, $[0, 8]$, $4\sqrt{2} \times 4\sqrt{2}$

21. $V(r) = 5\pi r^2(1 - r/3)$, $[0, 3]$, $r = 2\,\mathrm{cm}$, $h = 5/3\,\mathrm{cm}$

23. $G(t) = 3 + \dfrac{2\pi}{3}\cos\left(\dfrac{\pi t}{3}\right)$, $[0, 9]$, max: $t = 0, 6$, min: $t = 3, 9$

25. $A(x) = 4x\sqrt{1 - x^2/4}$, $[0, 2]$, $2\sqrt{2} \times \sqrt{2}$

27. $A(x) = (50 - x)^2/16 + x^2/4\pi$, $[0, 50]$
   a. 50 cm
   b. $50\pi/(4 + \pi)$ cm

29. $T(x) = \sqrt{100^2 + x^2}/3 + (300 - x)/5$, $[0, 300]$, 75 m

31. $F(p) = 3000p - 100p^2 - 20{,}500$, $[0, 20]$, \$15

33. $R(p) = (200 - 4p)(40 + p)$, $[0, 50]$, \$45 per room

35. $S(x) = (0.1)x/(1 + (0.2x)^2)$, $[0, 10]$, 5 m

37. a. $V(\rho) = (\rho - 100)^2/100$, $[0, 100]$, 0
   b. $q(\rho) = \rho(\rho - 100)^2/100$, $[0, 100]$, 33.33 autos/km

Exercise Set 4.3

1. 1

3. $\pi/6$, $\pi/2$, $5\pi/6$, $7\pi/6$, $3\pi/2$, $11\pi/6$

5. All c with $0 < c < 3$ **7.** π

9. 2   **11.** $-3/2$

13. $(-1 + 2\sqrt{13})/3$

23. The distance travelled is between 450 miles and 540 miles.

29. 0.48166

Exercise Set 4.4

1. dec: $x \leq 2$
   abs min: $(2, 2)$
   inc: $x \geq 2$

3. inc: $x \leq -1$, $x \geq 2$
   dec: $-1 \leq x \leq 2$
   rel max: $(-1, 7)$
   rel min: $(2, -20)$

5. dec: $x \leq 0$
   abs min: $(0, 4)$
   inc: $x \geq 0$

7. inc: $-\infty < x < \infty$

9. dec: $x \leq -2$
   abs min: $(2, 0)$
   inc: $-2 \leq x \leq 0$
   rel max: $(0, 4)$
   dec: $0 \leq x \leq 2$
   abs min: $(2, 0)$
   inc: $x \geq 2$

11. inc: $0 \leq x \leq \pi/4$
   abs max: $(\pi/4, 1)$
   dec: $\pi/4 \leq x \leq 5\pi/4$
   abs min: $(5\pi/4, -1)$
   inc: $5\pi/4 \leq x \leq 2\pi$

13. inc: $0 \leq x < \infty$
   dec: $-\infty < x \leq 0$
   abs min: $(0, -1)$

15. dec: $-\infty < x < -1$, $-1 < x < \infty$

17. inc: $-\infty < x \leq 0$, $1 \leq x < \infty$
   dec: $0 \leq x \leq 1$
   rel min: $(1, 3)$,   rel max: $(0, 4)$

19. inc: $-\infty < x \leq -2$; $1/2 \leq x < \infty$
   dec: $2 \leq x \leq 1/2$
   rel max: $(-2, 35)$,   rel min: $(1/2, 15/4)$

21. inc: $-\infty < x < \infty$

23. inc: $-\infty < x < -1$, $-1 < x < \infty$

25. inc: $-\infty < x \leq 0$
   dec: $0 \leq x < \infty$
   abs max: $(0, 1)$

27. inc: $-3 \leq x < \infty$
   dec: $-\infty < x \leq -3$
   abs min: $(-3, 0)$

29. inc: $-\infty < x \leq -1$, $7 \leq x < \infty$
   dec: $-1 \leq x \leq 7$
   rel : $(-1, 26/3)$
   rel min: $(7, -230/3)$

31. inc: $-\infty < x \le -3$, $-1 \le x < \infty$
dec: $-3 \le x \le -1$
rel max: $(-3, 1)$
rel min: $(-1, -3)$

33. inc: $-\infty < x < \infty$

35. inc: $\dfrac{-3 - \sqrt{5}}{2} \le x \le \dfrac{-3 + \sqrt{5}}{2}$, $0 \le x < \infty$
dec: $\dfrac{-3 + \sqrt{5}}{2} \le x \le 0$, $-\infty < x \le \dfrac{-3 - \sqrt{5}}{2}$
rel max: $\left(\dfrac{-3 + \sqrt{5}}{2}, 1.09\right)$
rel min: $(0, 1)$
abs min: $\left(\dfrac{-3 - \sqrt{5}}{2}, -10.09\right)$

37. inc: $7 \le x < \infty$
dec: $0 \le x \le 7$
abs min: $(7, 15 - 196\sqrt{7})$
endpoint rel max: $(0, 15)$

39. inc: $0 \le x < \infty$
dec: $-\infty < x \le 0$
abs min: $(0, 0)$

41. dec: $-\infty < x < -8$, $-8 < x < 8$, $8 < x < \infty$

43. inc: $0 \le x < \pi/2$, $\pi \le x < 3\pi/2$
dec: $\pi/2 < x \le \pi$, $3\pi/2 < x \le 2\pi$
abs min: $(0, 1)$, $(\pi, 1)$, $(2\pi, 1)$

45. inc: $k\pi \le x \le k\pi + \pi/2$
dec: $k\pi - \pi/2 \le x \le k\pi$
abs max: $(k\pi + \pi/2, 1)$
abs min: $(k\pi, 0)$

47. inc: $-\infty < x \le 0$, $2 \le x < \infty$
dec: $0 \le x \le 2$
rel min: $(2, 3 - 3\sqrt[3]{4})$
rel max: $(0, 3)$

49. inc: $-1 \le x \le 2$, $5 \le x < \infty$
dec: $-\infty < x \le -1$, $2 \le x \le 5$
abs min: $(-1, -57/4)$, $(5, -57/4)$
rel max: $(2, 6)$

51. inc: $0 \le x \le 2\pi$
rel max: $(2\pi, 2\pi)$
rel min: $(0, 0)$

53. inc: $2 \le x < \infty$
dec: $-\infty < x \le 2$
abs min: $(2, -64)$

55. 3   **57.** -15 **61.** 4

63. a. 10   **b.** positive

65. b. $[0, 6]$ **c.** 36

67. a. $[0, 3]$;   to the right.   **b.** $t = 3$

69. $\sqrt[3]{V} \times \sqrt[3]{V} \times \sqrt[3]{V}$

71. $2\sqrt[3]{9}$ m $\times\ 2\sqrt[3]{9}$ m $\times\ 8\sqrt[3]{9}/3$ m

75. $8\sqrt{2}$ ft

77. $r = 3\sqrt{2}$ cm, $h = 6$ cm

93. -1.386

95. $-0.295, 0.5, 1.295$

Exercise Set 4.5

1. a. inc: $[1, 2]$, $[4, 6]$, $[8, 9]$
  b. concave up: $[3, 6]$, $[6, 9]$
  c. rel min: $x = 1, 4, 8$
  d. rel max: $x = 2, 6, 9$
  e. infl pt: 3

3. concave up: $-\infty < x < \infty$

5. concave up: $-\infty < x \le -1$, $1 \le x < \infty$
  concave down: $-1 \le x \le 1$
  infl pt: $(-1, 0)$, $(1, 0)$

7. concave up: $\pi/2 \le x \le 3\pi/2$
  concave down: $0 \le x \le \pi/2$, $3\pi/2 \le x \le 2\pi$
  infl pt: $(\pi/2, 0)$, $(3\pi/2, 0)$

9. concave up: $-3 \le x < \infty$
  concave down: $-\infty < x \le -3$
  infl pt: $(-3, 0)$

11. concave down: $-2 \le x < \infty$

13. concave up: $-\infty < x < -1$
  concave down: $-1 < x < \infty$

15. concave up: $-1/2 \le x < \infty$
  concave down: $-\infty < x \le -1/2$
  infl pt: $(-1/2, 0)$

17. concave up: $1/2 \le x < \infty$
  concave down: $-\infty < x \le 1/2$
  infl pt: $(1/2, -7/2)$

19. concave up: $-\infty < x \le -3$, $2 \le x < \infty$
  concave down: $-3 \le x \le 2$
  infl pt: $(-3, -375)$, $(2, -70)$

21. concave up: $-\infty < x \le (-3 - \sqrt{5})/2$,
        $(-3 + \sqrt{5})/2 \le x < \infty$
  concave down: $(-3 - \sqrt{5})/2 \le x \le (-3 + \sqrt{5})/2$
  infl pt: $[(-3 + \sqrt{5})/2, -3.02]$, $[(-3 - \sqrt{5})/2, -49.98]$

23. concave up: $2 - \sqrt{2} \le x \le 1$, $2 + \sqrt{2} \le x < \infty$
  concave down: $-\infty < x \le 2 - \sqrt{2}$, $1 \le x \le 2 + \sqrt{2}$
  infl pt: $(2 - \sqrt{2}, -159 + 106\sqrt{2})$, $(1, -19)$,
        $(2 + \sqrt{2}, -159 - 106\sqrt{2})$

25. concave up: $-\infty < x \le -1$
  concave down: $-1 \le x < 0$, $0 < x < \infty$
  infl pt: $(-1, 120)$

27. concave up: $-1 \le x < 0, \quad 0 < x < \infty$
concave down: $-\infty < x \le -1$
infl pt: $(-1, -3)$

29. rel max

31. no extremum

33. rel min

35. rel min

37. rel max

39. rel min: $(2, 2)$
rel max: $(0, 6)$
concave up: $1 \le x < \infty$
concave down: $-\infty < x \le 1$

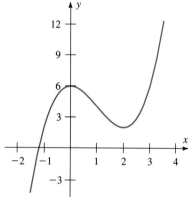

41. abs min: $(-3, 0), (3, 0)$
rel max: $(0, 9)$
concave up: $-\infty < x \le -3, \quad 3 \le x < \infty$
concave down: $-3 \le x \le 3$

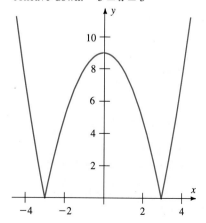

43. $\dfrac{20}{\sqrt[3]{2} + 1}$ from weaker source

45. inc: $[1, 3], [5, 6]$
concave down: $[2, 4], [7, 8]$
rel min: $x = 1, 5, 8$; rel max: $x = 0, 3, 6$

47. $a > 0$ and b, c are arbitrary

49. $f(x) = x^4$ with $b = 0$, for example

51. None; one; at most $k - 2$

53. $f(x) = -x^{1/3}$, for example

55. One possible graph:

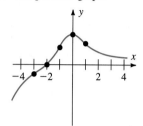

Exercise Set 4.6

1. $3/10$

3. $-1/3$

5. 0

7. 0

9. 0

11. 0

13. 1

15. 0

17. $+\infty$

19. Does not exist.

21. $+\infty$

23. $+\infty$

25. Does not exist.

27. -1

29. $26/10$

31. 10

33. $+\infty$

35. $+\infty$

37. $+\infty$

39. $y = 0$

41. $y = 2$

43. $y = 0$

45. $y = 3$

47. $y = -3\sqrt{2}/2$

49. $a = -3$

51. $x = -2$

53. $x = 1$

55. $x = 1$

57. $r = 2/3$, $a = 6$

59. $a = -8$, $b = 15$

63. One possible graph:

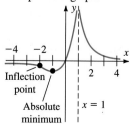

Exercise Set 4.7

1.

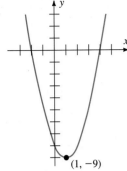

$(1, -9)$

3.

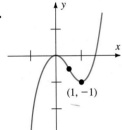

$(1, -1)$

5.

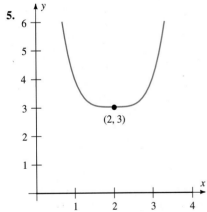

$(2, 3)$

7.

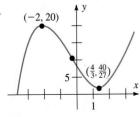

$(-2, 20)$

$\left(\frac{4}{3}, \frac{40}{27}\right)$

9.

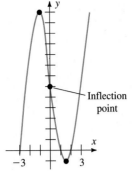

Inflection point

-3 3

11.

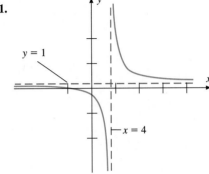

$y = 1$

$x = 4$

13.

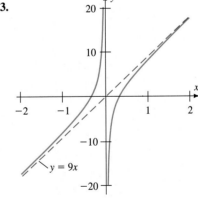

20

10

-2 -1 1 2

-10

$y = 9x$

-20

15.

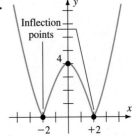

Inflection points

4

-2 $+2$

17.

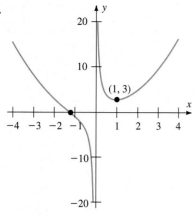

19.

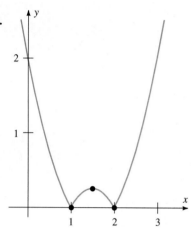

21.

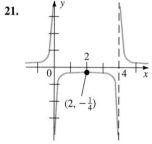

23.

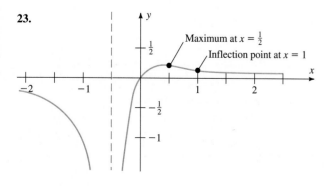

25.

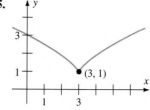

27.

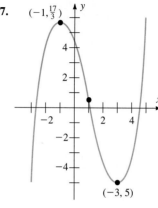

29.

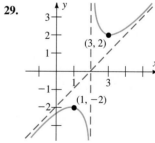

31.

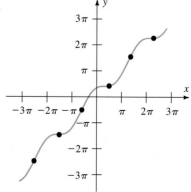

33.

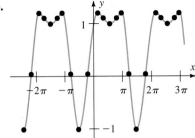

35.

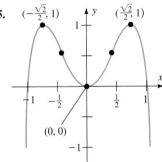

37.

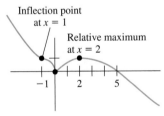

39.

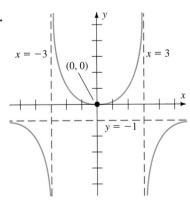

41.

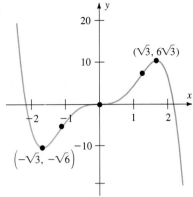

43.

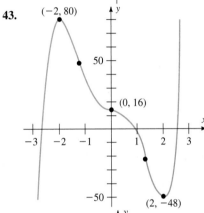

45.

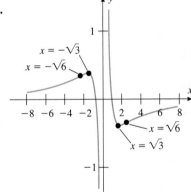

47.

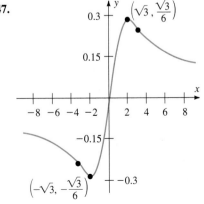

Exercise Set 4.8

1. approx: 4.08
M: 2
error: 0.0004

3. approx: 3.88
M: 2
error: 0.0009

5. approx: 16.96
M: 49.45
error: 0.0223

7. approx: $\sqrt{2}/2(1 + \pi/180)$
M: $23\pi/90$
error: 0.0001

9. approx: $\pi/36$
M: $\pi/36$
error: 0.0003

11. approx: 73/12
M: 1/864
error: 0.0005

15. $-0.215 < x < 0.215$

9.
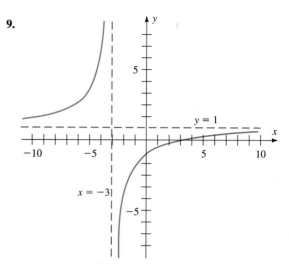

Review Exercises—Chapter 4

1.
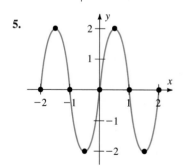

3.
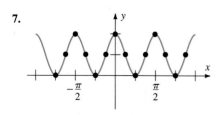

5.
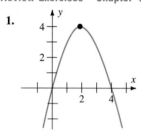

7.
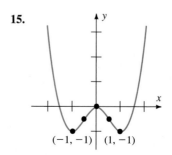

11.
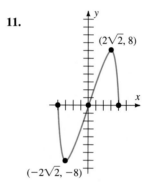

13.
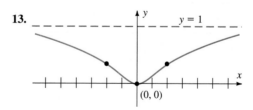

15.

17.

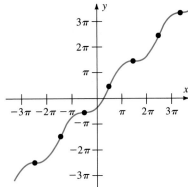

19.

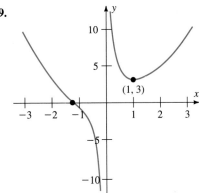

21.

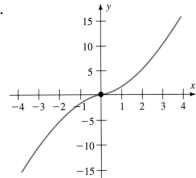

23.

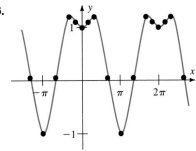

	Minimum	Maximum
25.	$f(-1) = f(1) = -1/3$	$f(0) = -1/4$
27.	$y(-1) = -3$	$y(0) = 1$
29.	$f(-1) = f(0) = 0$	$f(1) = 2$
31.	$y(-\sqrt{2}/2) = -\sqrt{2}$	$y(1) = 1$
33.	$f(-1) = -5$	$f(-2) = -3$
35.	$y(-1) = y(1) = -1$	$y(-2) = y(2) = 8$
37.	$f(-\pi) = f(\pi) = 1/3$	$f(0) = 3$
39.	$y(2) = 4\sqrt{5}/5$	$y(1/2) = \sqrt{5}$
41.	$f(0) = f\left(\dfrac{-1 + \sqrt{5}}{2}\right) = 0$	$f(1) = 1$

43. 3   **45.** $-\infty$ **47.** Does not exist

49. $-\infty$ **51.** 1   **53.** 1

55. One possible graph:

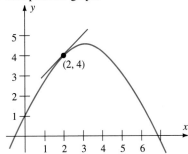

57. One possible graph:

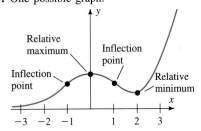

59. One possible graph:

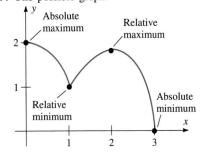

61. One possible graph:

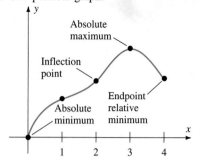

63. 5.4056 m

65. $40 per year

69. 1/9

73. $r = \sqrt{S}$, $\theta = 2$

75. $(1/4, \sqrt{17}/4)$, $(1/4, -\sqrt{17}/4)$

77. $\sqrt{2}$

79. $r = \sqrt[3]{\dfrac{3V}{28\pi}}$, $h = 8r$

83. $w = 4\sqrt[3]{4}$ cm, $l = h = 8\sqrt[3]{4}$ cm

85. -4

87. 6 times per year

89. 8 times per year

Chapter 5

Exercise Set 5.1

1. $\frac{2}{3}x^3 + x + C$

3. $\frac{3}{5}x^{5/3} + \frac{2}{7}x^{7/2} + C$

5. $2\sin t + \cos t + C$

7. $-\cos x + \tan x + C$

9. $\frac{2}{5}t^{5/2} + 2t^{1/2} + C$

11. $\frac{1}{4}t^4 + \frac{2}{3}t^3 + \frac{1}{2}t^2 + 2t + C$

13. $\frac{3}{2}x^{2/3} + C$

15. $\tan x + C$

17. $x^3 + \cos x + 2\tan x + C$

19. $\sin t + 2$

21. $\frac{2}{3}t^3 + 4$

23. $\frac{2}{5}t^{5/2} + \frac{8}{3}t^{3/2}$

25. $t^2/2 - 16t + 33$

27. $v(t) = 2t + 3$, $s(t) = t^2 + 3t$

29. $v(t) = \frac{3}{2}t^2$, $s(t) = \frac{1}{2}t^3 + 20$

31. $v(t) = 2t^2 + 4t + 8$, $s(t) = \frac{2}{3}t^3 + 2t^2 + 8t + 12$

33. $v(t) = -\cos t + 1$, $s(t) = -\sin t + t + 2$

35. $v(t) = 2\cos t - 1$
$s(t) = 2\sin t - t - \pi$

37. $v(t) = 1$
$s(t) = t + 1$

39. a. 1.53 s   **b.** 3.06 s

41. a. 28 m/s   **b.** 6 m/s

43. $\frac{1}{2}$ m/s^2

45. a. *ii*

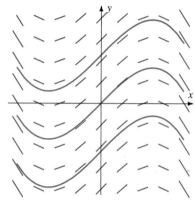

b. *i*

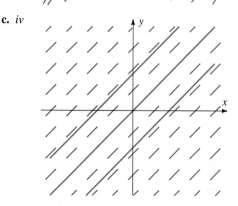

c. *iv*

d. *iii*

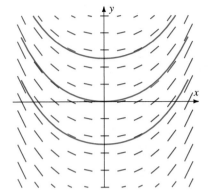

47. $x - x^2/2 + C$

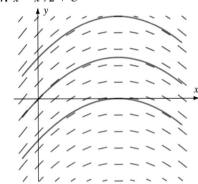

49. $-\cos x + C$

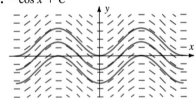

51.

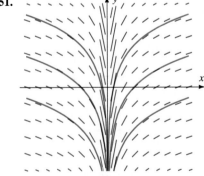

53. $f(x) = 3x - 2$

55. $f(x) = -2 \sin x + bx + c$

57.

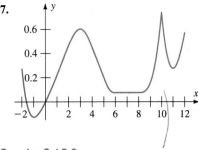

Exercise Set 5.2

1. $-\frac{1}{6} \cos(x^6 + 2) + C$

3. $2 \sin\sqrt{x} + C$

5. $\frac{2}{5}(x + 2)^{5/2} - \frac{4}{3}(x + 2)^{3/2} + C$

7. $\frac{1}{3}(x^2 + 1)^{3/2} + C$

9. $\frac{1}{2} \sec(2\theta) + C$

11. $-\frac{1}{2} \cot(x^2) + C$

13. $-\frac{2}{27}(3t^3 - 9t + 9)^{3/2} + C$

15. $\tan(x - \pi) + C$

17. $\frac{1}{3}(2\sqrt{x} + 3)^3 + C$

19. $\frac{1}{5}x^5 + \frac{2}{3}x^3 + x + C$

21. $\frac{1}{3}(x^2 - 1)^{3/2} + C$

23. $-\frac{1}{2} \sec(\pi - x^2) + C$

25. $\frac{1}{21}(x^6 - 3x^4)^{7/2} + C$

27. $\frac{1}{3}x^3 - \frac{2}{7}x^7 + \frac{1}{11}x^{11} + C$

29. $\frac{2}{3}\sqrt{t^3 + 6t} + C$

31. $-\frac{3}{4}(1 - 4x)^{1/3} + C$

33. $\frac{1}{5}(x^2 + 1)^{5/2} - \frac{1}{3}(x^2 + 1)^{3/2} + C$

35. For **a.–c.,** $u = 3x + 4$
   a. $\frac{1}{33}(3x + 4)^{11} + C$
   b. $\frac{2}{9}(3x + 4)^{3/2} + C$
   c. $-\frac{1}{3} \cos(3x + 4) + C$
   d. $u = \sin(3x + 4)$
   $\frac{1}{9} \sin^3(3x + 4) + C$

37. **a.** $u = x^2 + 6, \quad \frac{1}{44}(x^2 + 6)^{22} + C$
   b. $u = x^2 + 6, \quad \sqrt{x^2 + 6} + C$
   c. $u = x^2 + 6, \quad \frac{1}{2} \tan(x^2 + 6) + C$

Exercise Set 5.3

1. $y = \frac{1}{2}x^2 - x + C$

3. $y = \frac{1}{3}x^3 + x^{-1} + C$

5. $y = \frac{1}{3}(1 + \sin^2 x)^{3/2} + C$

7. $y = \frac{1}{2}x^2 - \frac{2}{3}x^{3/2} + C$

9. $y^2 = x^2 + C$

11. $y = \dfrac{1}{2x^2 + C}$

13. $y^2 = C - x^2$

15. $y = \sqrt{x^2 + 15}$

17. $y = \sqrt{4 - x^2}$

19. $y = \dfrac{-1}{3(1 + x^3)} + \dfrac{5}{6}$

21. $y = \dfrac{-3}{x^3 + 3x - 3}$

23. $y = \frac{1}{12}x^4 + 4x + 2$

25. $y = \sin 4x - 3x + 4\pi + 1$

27. $f(x) = \frac{1}{2}x^2 - \frac{2}{3}x^{3/2} + \frac{13}{6}$

29. $11/2$

31. $P(t) = 10\sqrt[3]{2.7t + 1}$

33. $P(t) = (2t + \sqrt{10})^2$

Review Exercises—Chapter 5

1. $2x^3 - x^2 + x + C$

3. $-2 \cos \sqrt{t} + C$

5. $2x^{3/2} + 6x^{1/2} + C$

7. $\frac{1}{2} \tan(x^2) + C$

9. $\dfrac{x^3}{3} - \dfrac{7}{2}x^2 + 6x + C$

11. $\dfrac{(t^2 + 1)^{7/2}}{7} - \dfrac{2(t^2 + 1)^{5/2}}{5} + \dfrac{(t^2 + 1)^{3/2}}{3} + C$

13. $\frac{1}{2} \sin(1 + x^2) + C$

15. $\frac{4}{3}x^3 - x + C$

17. $\begin{cases} x^2/2 + C, x \geq 0 \\ -x^2/2 + C, x < 0 \end{cases}$

19. $x + \sin x + 3$

21. $\frac{3}{2}x^2 + 3x + 4$

23. $\frac{1}{2} \sec x^2 + 1 - \sqrt{2}/2$

25. $y = x^2 + 5$

27. 122.5 m

29. $y = x + 1$

31. $y = x + 1$

33. a. -4 **b.** To the right   **c.** $t = 1$ and $t = 3$ **d.** Twice

Chapter 6

Exercise Set 6.1

1. 55 **3.** 6084 **5.** 1 **7.** 1

9. $\displaystyle\sum_{j=1}^{5} 2j$ **11.** $\displaystyle\sum_{j=2}^{6} j^2$ **13.** 6 **15.** 40

17. $\dfrac{\pi}{4}\sqrt{2}$ **19.** 17.96 **21.** 12 **23.** 85

25. $\dfrac{\pi}{4}(2 + \sqrt{2})$ **27.** 27.06 **31.** 420 **33.** 275

35. $2n^3 + n^2 - n$ **37.** 11 **39.** 5 **41.** $14{,}848$

43. 1 **45.** 2

47. a. 8.5 **b.** 9.94 **c.** $10 - 6/n$ **d.** 10
     e. 10.06 **f.** $10 + 6/n$ **g.** 10 **h.** Area is 10

49. a. $\dfrac{b^3}{6n^2}(n - 1)(2n - 1)$

  b. $\frac{1}{3}b^3$

  c. $\dfrac{b^3 - a^3}{3} - \dfrac{(b - a)^2(b + a)}{2n} + \dfrac{(b - a)^3}{6n^2}$

  d. $\dfrac{b^3 - a^3}{3}$

  f. $124/3$

51. $S_n = 26 - \dfrac{15}{2n} + \dfrac{1}{2n^2}$

  Area $= 26$

53. $S_n = 7 - \dfrac{3}{2n} + \dfrac{1}{2n^2}$

  Area $= 7$

55. $S_n = \begin{cases} 2 - 1/n, & \text{if } n \text{ is even} \\ 2 - 6/n, & \text{if } n \text{ is odd} \end{cases}$

  Area $= 2$

57. c. $S_3 \approx 3.38$
     d. $S_6 \approx 4.09$

Exercise Set 6.2

1. 17

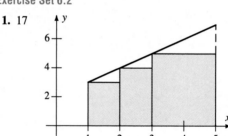

3. $6\frac{3}{4}$

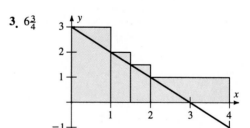

5. 26

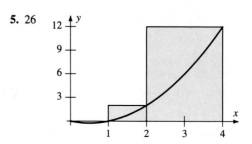

7. $\pi/2$

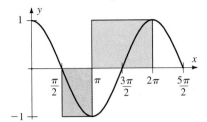

9. 28

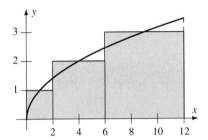

11. 0

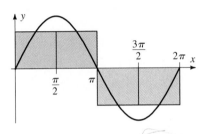

13. Largest = 14; smallest = 6

15. Largest = π; smallest = $-\pi$

17. 30   **19.** 6   **21.** -20

23. $-10/3$ **25.** 11/6   **27.** 112/3

29. -99

31. a. -3 **b.** 1   **c.** 3
    d. 6   **e.** -1 **f.** -3

33. a. 3   **b.** -22 **c.** -8
    d. 16   **e.** 12   **f.** 25

35. 9

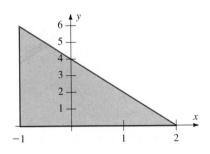

37. 12

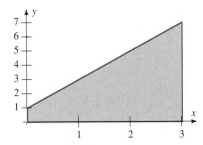

39. $9\pi/4$

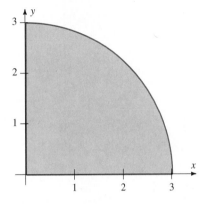

41. $6\frac{1}{2}$

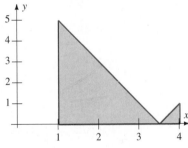

43. $\pi/2$

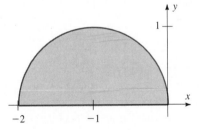

45. 3

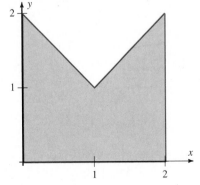

47. $6 - \pi$

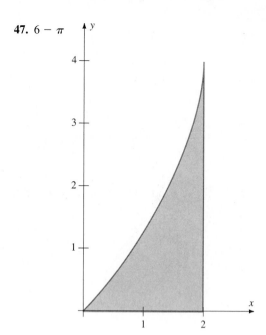

49. 11/6

51. −1/6

57. b. $\displaystyle\int_0^1 x\,dx = 1/2$

59. $\displaystyle\int_0^1 \sqrt{1 - x^2}\,dx = \pi/4$

71. $\frac{142}{105} \approx 1.35238$

73. $(3 + \sqrt{35} + \sqrt{133} + \sqrt{351})/\sqrt{8} \approx 13.8535$

75. $4(\frac{1}{5} + \frac{1}{13} + \frac{1}{29} + \frac{1}{53}) \approx 1.3211$

77. 1.08779

79. 2.55176

81.

b	0.0	0.5	1.0	1.5	2.0	2.5	3.0
Midpoint approx.	0.0	0.0411	0.3084	0.7853	0.8243	0.3677	0.8896

Exercise Set 6.3

1. −4

3. 33/4

5. $91\frac{1}{15}$

7. −1/8

9. $2 - \dfrac{8\sqrt{2}}{5}$

11. 5

13. 3

15. −209

17. $13\frac{1}{6}$

19. 2

21. $2\sqrt{3}$

23. 2

25. $\pi \sin 1$

27. 11/6

29. $4/\pi$

31. $F(x) = \dfrac{x^2 - 1}{2}$

$F'(x) = x$

33. $F(x) = \sin x$

$F'(x) = \cos x$

35. $x^2 \sin x^2$

37. $-x^3 \cos^2 x$

39. $3\sqrt{1 + \sin 3x}$

41. $\cos^3(x + 1) - 2x \cos^3(x^2 + 1)$

43. $\dfrac{2x^2}{(x^4 + 1)^3} + \displaystyle\int_3^{x^2} \dfrac{1}{(t^2 + 1)^3}\,dt$

45. a. $f(x)$ **b.** $-f(x)$ **c.** 0
d. $f(x)$ **e.** $f(b) - f(a)$ **f.** $f(x) + C$

47. a. $1\frac{1}{6}$ or $2\frac{5}{6}$
b. $\pm\sqrt{7/3}$

49. $b = 7$

51. a. $[0, 4]$ and $[7, 8]$
b. Yes; on $[6, 7]$
c. At $x = 4$ and $x = 8$
d. $A(0) = 0$

57. $\displaystyle\int_0^x \sin t\,dt = 1 - \cos x$

Exercise Set 6.4

1. 1/2

3. 1/2

5. $\dfrac{2^{51} - 3^{51}}{153}$

7. $\dfrac{\pi}{4} - \dfrac{1}{2}$

9. 7/3

11. $\sqrt{3}$

13. $\dfrac{1}{2} - \dfrac{\sqrt{2}}{4}$

15. 0

17. 2

19. 0

21. 2

23. 6/49

25. $\dfrac{1}{14}(2^{14} - 1) + \dfrac{2}{13}(2^{13} + 1) = 2430\dfrac{123}{182}$

27. 1/12

29. $\pi/2$

31. 2/3

33. 1/8

35. 1/3

37. $\dfrac{5\sqrt{5} - 1}{6}$

39. $\dfrac{\sqrt{2/\pi}}{2}$

41. $F(x) = \sqrt{1 + x^2} - \sqrt{2}$

$F'(x) = \dfrac{x}{\sqrt{1 + x^2}}$

43. $F(x) = \dfrac{(\cos^3 x) - 1}{3}$

$F'(x) = -\cos^2 x \sin x$

Exercise Set 6.5

1. 14   **3.** $2\sqrt{3} - 2$ **5.** 5/48

7. 52/3

9. 6

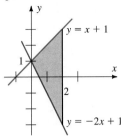

11. 80/3

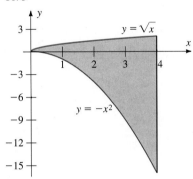

13. $4\sqrt{2}$

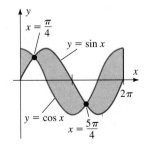

15. 52/3

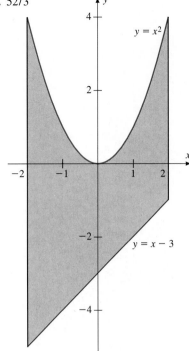

17. 44/3

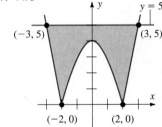

19. 1

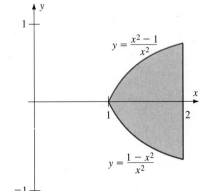

21. 125/6

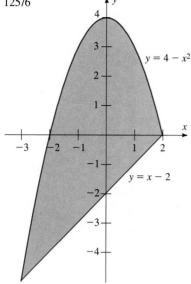

$y = 4 - x^2$

$y = x - 2$

23. 1/12

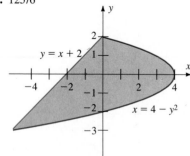

$y = x^2$

$y = x^3$

25. 125/6

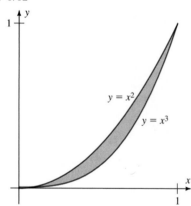

$y = x + 2$

$x = 4 - y^2$

27. 16/3

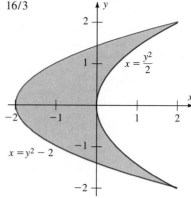

$x = \dfrac{y^2}{2}$

$x = y^2 - 2$

29. 8/15

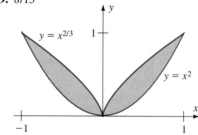

$y = x^{2/3}$

$y = x^2$

31. a. 4   **b.** 5

33. a. $-2/3$ **b.** $\frac{2}{3}(2\sqrt{2} - 1)$

35. a. 27   **b.** 27

37. 4/5   **39.** 3

41. $a = 4 - 4^{2/3}$

43. a. 8   **b.** 22

45. 8 subdivisions: -12.143
      50 subdivisions: -12.5273

Exercise Set 6.6

1. 1.59984   **3.** 72.9559

5. 15.0486   **7.** 1.62897

9. 75.2359   **11.** 1.61085

13. 22.4904   **15.** 0.795925

17. 0.784981   **19.** 1.40221

21. 0.69315

23. Midpoint: 3.14243
     Trapezoid: 3.13993

25. Lower bd: 72.9559
     Upper bd: 75.2359

27. 432.5 ft^2 **29.** 1280

Review Exercises—Chapter 6

1. a. 57/4   **b.** 75/4

3. a. 272/195   **b.** 337/195

5. a. $\sqrt{2}/4$ **b.** $(2 + \sqrt{2})/4$

7. a. 3.98   **b.** 4.02

9. Smallest: $\dfrac{\pi}{4}\left(\dfrac{\sqrt{2}}{2} - 1\right)$

   Largest: $\dfrac{\pi}{4}\left(\dfrac{\sqrt{2}}{2} + 2\right)$

11. Smallest: 39/8
   Largest: 59/8

13. 8

15. 18

17. $-1/15$

19. 9/20

21. 7/2

23. 200/3

25. 14/3

27. 8/3

29. 196/3

31. $1 - \sqrt{2}/2$

33. 1/2

35. $\dfrac{6^{11} - 5^{11}}{11} \approx 28{,}542{,}630.09$

37. 4/3

39. 3/4

41. 1

43. 1/30

45. $-6011/21 \approx -286.2381$

47. $-2/9$

49. 10/3

51. $2(\sqrt{2} - 1)$

53. $5\frac{2}{3}$

55. $\frac{4}{15}(1 + \sqrt{2})$

57. 86

59. 16/3

61. 83/15

63. 7/3

65. 23/3

67. 26/3

69. $4\sqrt{2}/3$

71. $\frac{243}{5}c^2 + 54c + 27$

73. 32/3

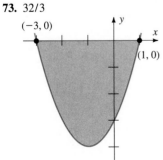

75. 8

77. 8

79. 32/3

81. 125/6

83. $2\sqrt{2} - 2$

85. 9/2

87. 8/15

89. 4/3

91. 31/2

93. $1/\pi$

95. $(\sqrt{5} - 1)/2$

97. 0

101. a. 0   **b.** $x^2\sqrt{1 + x}$
     c. 18   **d.** $4x^2\sqrt{1 + 2x}$

103. False   **105.** True   **107.** $-2\sec(2x)$

109. $\dfrac{[f(b)]^{n+1} - [f(a)]^{n+1}}{n + 1}$ if $n \neq -1$

111. $\pi/2$

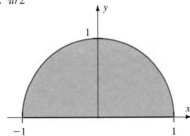

113. 18

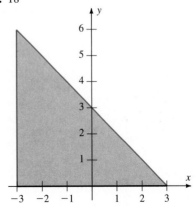

115. $6 + \pi/2$

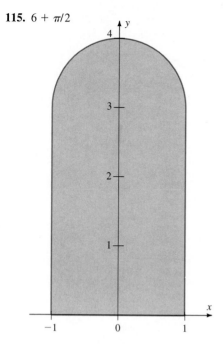

117. 9.37411   **119.** 87.1

121. $\frac{104}{105}\sqrt{2} \approx 1.40074$

Chapter 7

Exercise Set 7.1

1. a. $\displaystyle\int_1^4 \pi(2x + 1)^2 \, dx$

  b. 117π

3. a. $\displaystyle\pi\int_0^\pi \sin x \, dx$

  b. 2π

5. a. $\displaystyle\int_0^1 \pi x^2(x^3 - 1)^4 \, dx$

  b. $\dfrac{\pi}{15}$

7. a. $\displaystyle\int_1^2 \pi(4 - x^2) \, dx$

  b. $\dfrac{5\pi}{3}$

9. $\pi\left(1 - \dfrac{\pi}{4}\right)$ **11.** π **13.** 105π

15. a. $\displaystyle\int_{-2}^2 \pi[(6 - x^2)^2 - 4] \, dx$

  b. 76.8π

17. a. $\displaystyle\int_0^4 \pi\left(x^2 - \dfrac{x^4}{16}\right) dx$

  b. $\dfrac{128\pi}{15}$

19. π

21. a. $\displaystyle\int_0^4 \pi(4 - y) \, dy$

  b. 8π

23. a. $\displaystyle\int_0^8 \pi(4 - y^{2/3}) \, dy$

  b. $\dfrac{64\pi}{5}$

25. $\dfrac{64\pi}{3}$

27. $\dfrac{128\pi}{3}$ **29.** $\dfrac{4\pi r^3}{3}$

31. $\dfrac{512}{3}$ **33.** $\dfrac{2048}{15}$

35. $\dfrac{1}{3}\pi h(R^2 + rR + r^2)$ **37.** $\dfrac{5}{32\pi}$ m/min

39. $\dfrac{136\pi}{15}$ **41.** $\dfrac{225}{2}\pi$

43. $V = \dfrac{4}{3}\pi a^2 b$ **45.** $\dfrac{2}{3}r^3 \tan\theta$

47. ≈ 4.16 **49.** $12\pi^2 \approx 118.24$

Exercise Set 7.2

1. a.

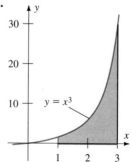
$y = 1 - x$

  b. $\displaystyle\int_0^1 2\pi x(1 - x) \, dx$ **c.** $\pi/3$

3. a.

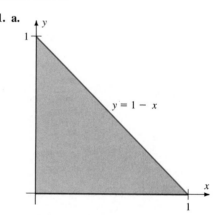
$y = x^3$

  b. $\displaystyle\int_1^3 2\pi x^4 \, dx$ **c.** $\dfrac{484\pi}{5}$

5. a.

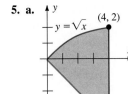

b. $\displaystyle\int_0^4 2\pi x(\sqrt{x} + x)\, dx$

c. $\dfrac{1024\pi}{15}$

7. a.

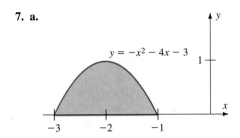

b. $\displaystyle\int_{-3}^{-1} 2\pi x(x^2 + 4x + 3)\, dx$

c. $\dfrac{16\pi}{3}$

9. a.

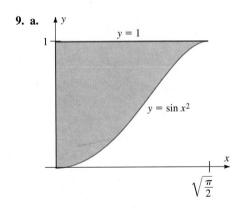

b. $\displaystyle\int_0^{\sqrt{\pi/2}} 2\pi x(1 - \sin x^2)\, dx$

c. $\dfrac{\pi^2}{2} - \pi$

11. a.

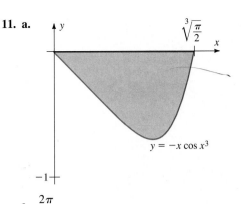

b. $\dfrac{2\pi}{3}$

13. a.

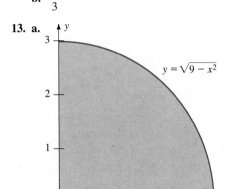

b. 18π

15. a.

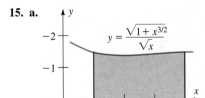

b. $24\pi - \dfrac{16\sqrt{2}\,\pi}{9}$

17. $\dfrac{1088}{15}\pi$ **19.** $\dfrac{49}{30}\pi$ **21.** $\dfrac{8}{15}\pi$

23. 56π **25.** $\frac{4}{3}\pi a^2 b$

27. Approx 19.7392   **29.** Approx 49.3449

Exercise Set 7.3

1. a. $\displaystyle\int_{-3}^2 \sqrt{1 + 4}\, dx$

 b. $5\sqrt{5}$

3. a. $\displaystyle\int_{-2}^2 \sqrt{1 + \frac{9}{4}(x + 2)}\, dx$

 b. $\frac{8}{27}(10\sqrt{10} - 1)$

5. a. $\int_0^3 \sqrt{1 + (x + 1)}\, dx$

b. $\frac{2}{3}(5\sqrt{5} - 2\sqrt{2})$

7. 12   **9.** 87/8   **11.** 227/24

13. a. $\int_1^3 2\pi(2x + 3)\sqrt{5}\, dx$

b. $28\pi\sqrt{5}$

15. a. $\int_0^2 2\pi y^3\sqrt{1 + 9y^4}\, dy$

b. $\frac{\pi}{27}(145\sqrt{145} - 1)$

17. $\frac{\pi}{6}(27 - 5\sqrt{5})$ **19.** $\frac{7819\pi}{162}$

21. 10/3   **25.** 40.3316

27. 14.4972   **29.** 74.0003

Exercise Set 7.4

1. 18   **3.** 2   **5.** 155/3

7. $2/\pi$ **9.** 1/3   **11.** $\frac{3}{2}\sqrt[3]{2}$

13. $\frac{8}{3}(5 + \sqrt{2})$ **15.** 1/2   **17.** 1/3

19. $\frac{7^6 + 5^6}{12}$

21. a. Displacement: -6, distance: 10
b. Displacement: $-40/3$, distance: $\frac{8}{3}(5 + 2\sqrt{2})$
c. Displacement: 1, distance: 1
d. Displacement: 1/2, distance: 1/2

23. a. $-\frac{1}{3}t^3 + \frac{5}{2}t^2 - 6t + 4$
b. $-2t + 5$
c. $t = 2, 3$
d. $2 < t < 3$
e. $t < 2$ or $t > 3$
f. 29/6
g. $-9/2$

25. a. $-9.8t + 12$
b. $-4.9t^2 + 12t + 10$
c. $\frac{60}{49}$ s
d. $t = \frac{1}{49}(60 + 10\sqrt{85})$
e. Displacement: -10 m, distance: $\frac{1210}{49}$ m
f. $-2\sqrt{85}$ m/s

27. 17.55 km

Exercise Set 7.5

1. 421.2 lb   **3.** 166.4 lb

5. 340,787.2 lb   **7.** 665.6 lb

9. 571.9 lb   **11.** 3 lb

13. 52,477.2 lb   **15.** 157.083 lb

17. 2995.2 lb

Exercise Set 7.6

1. 5 ft-lb   **3.** 15 ft-lb

5. 25 ft-lb   **7.** True

9. a. 5 lb/ft   **11.** $74,880\pi$ ft-lb
b. 5 lb
c. 22.5 ft-lb

13. $1,248,000\pi$ ft-lb   **15.** $74,880\pi$ ft-lb

17. 3833.9π ft-lb   **19.** 400 ft-lb

21. a. 11,600 ft-lb
b. 12,200 ft-lb

23. a. $\frac{160}{q^2}$ **25.** $\frac{91}{120}K$
b. 40 N-m

Exercise Set 7.7

1. $\frac{100}{39}$ **3.** 4.8 ft

9. $x_3 = 2,\ y_3 = \frac{-5}{3}$ **11.** $\left(\frac{3}{5}, \frac{12}{35}\right)$

13. $\left(\frac{4}{3}, \frac{4}{3}\right)$ **15.** $\left(0, \frac{8}{3\pi}\right)$

17. $\left(0, \frac{4}{5}\right)$ **21.** $\left(\frac{4}{5}, \frac{2}{7}\right)$

23. $\left(\frac{8}{3}, 0\right)$ **25.** $\left(\frac{49}{22}, \frac{933}{220}\right)$

27. $6\frac{4}{21}$ cm from the light end

29. $7\frac{44}{103}$ cm from the light end

31. On the axis, 2.5 cm from the base

33. $\left(\frac{5}{3}, 0\right)$ **37.** $\left(0, \frac{54}{55}\right)$

Exercise Set 7.8

1. $90\pi^2$ **3.** $3\pi/10$

5. $8\pi A$ **7.** 19,968 lb

9. 1291.65 lb   **11.** $4\pi^2$

Review Exercises—Chapter 7

1. $\dfrac{4}{3}\pi r^3$

3. $\dfrac{3\pi}{4}$

5. $\dfrac{3\pi^2}{4}$

7. $3\pi\sqrt[3]{4}$

9. $\pi(\sqrt{2}-1)$

11. $\pi\left(\dfrac{12}{7}\sqrt[3]{2}+\dfrac{48}{5}\sqrt[3]{4}+14\right)$

13. $\dfrac{8}{27}(10\sqrt{10}-1)$

15. $\dfrac{22\sqrt{22}-13\sqrt{13}}{27}$

17. $\dfrac{169}{24}$

19. $\dfrac{1}{3}Ah$

21. 31.25%

23. The exact answer is 14.424 to 3 decimal places.

25. 25/4 N-m; 11/4 N-m

27. $207{,}667\pi$ ft-lb

29. 4000 ft-lb

31. 32,700 ft-lb

33. 36.4 ft-lb

37. 12 cm from the light end

39. $\dfrac{21}{11}$ m ≈ 1.9091 m from the small end

41. $\left(-\dfrac{3}{4},1\right)$

43.

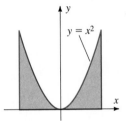

45. $(0, 6.4)$

47. $\left(3, \dfrac{18}{5}\right)$

49. $\left(3, \dfrac{67}{15}\right)$

51. $\left(\dfrac{5}{3}, \dfrac{5}{3}\right)$

53. a. $\dfrac{32\pi}{3}$ **b.** $8\pi\left(\dfrac{4}{3}+\pi\right)$ **c.** $8\pi\left(3\pi-\dfrac{4}{3}\right)$

Chapter 8

Exercise Set 8.1

1. 3

3. 1/3

5. −3

7. 3/2

9. 0

11. 2

13. 1/4

15. $3^{1/6}$

17. 9

19. 1

21. 1/8

23. $f^{-1}(x) = \dfrac{x+2}{3}$

25. $f^{-1}(x) = \dfrac{1}{x} - 2$

27. $f^{-1}(x) = \dfrac{3x}{1-x}$

29. $f^{-1}(x) = x^2 - 8x + 17$

31. $f^{-1}(x) = (x-3)^{3/5}$

33. 25/6

35. 1/2

39. a. $f^{-1}(x) = \sqrt{x} - 3$ **b.** $f^{-1}(x) = -1 - \sqrt{x}$
c. $f^{-1}(x) = \sqrt{x} - 2$ **d.** $f^{-1}(x) = 1 - \sqrt{x+4}$

41. a. 12   **b.** $f^{-1}(x) = x^3 - 5$ **c.** $(f^{-1})'(x) = 3x^2$

43. a. $\sqrt{2}$ **b.** 1   **c.** 2   **d.** $2\sqrt{3}/3$

45. a. 0   **b.** 1   **c.** 1   **d.** 1/4

47. a. 0   **b.** −1/35   **c.** 1   **d.** −1/40   **e.** −36

53. $I = I_0 10^{kx}$

55. False if $0 < a < 1$.

Exercise Set 8.2

1. a. 1/2   **b.** 2   **c.** $\sqrt{2}$ **d.** 0
e. e^2 **f.** 2   **g.** e^2 **h.** 2

3. True

5. $1 + \ln x$

7. $\dfrac{3x^2 - 1}{2(x^3 - x)}$

9. $\dfrac{\cos(\ln x)}{x}$

11. $\dfrac{\ln x}{(1 + \ln x)^2}$

13. $\dfrac{162(\ln \sqrt{x})^3}{x}$

15. $\dfrac{\dfrac{b}{a+bt}\ln(c+dt) - \dfrac{d}{c+dt}\ln(a+bt)}{\ln^2(c+dt)}$

17. $-\tan x$

19. $\dfrac{(\ln t - 1)^2}{(1 + \ln^2 t)^2}$

21. $\dfrac{x}{y}$

23. $\dfrac{xy - xy\ln y - y^2}{x^2 + xy\ln x}$

25. $-\dfrac{1}{2x}$

27.

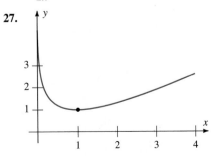

29.

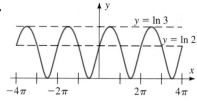

31. $\ln|x + 3| + C$

33. $-\ln|1 - x| + C$

35. $\frac{1}{2}\ln|x^2 - 2x| + C$

37. $\frac{1}{3}\ln|x^3 + 9x| + C$

39. $-\frac{1}{2}\ln|4 + 2\cos t| + C$

41. $\frac{1}{3}(\ln x)^3 + C$

43. $-2\ln|1 - \sqrt{x}| + C$

45. $\dfrac{x^4}{4} - \dfrac{x^3}{3} + 2x^2 - 3x + 4\ln|x + 1| + C$

47. $\ln 2$

49. $\left[\frac{1}{2}\ln|\ln x|\right]_e^{e^2} = \frac{1}{2}\ln 2$

51. $2\ln 3 - 3\ln 2$

53. $\ln 2$

55. $y = 0$

57. $6 - 8\ln 2$

59. True

61. $\dfrac{\pi(\ln 2)^3}{3}$

63. $\dfrac{1}{e - 1}$

65. $y = \ln Cx$

67. $y = \ln|\sec x| + C$

69. $y = \dfrac{1}{C - \frac{1}{2}\ln|2x + 1|}$

71. a. $\dfrac{dP}{dT} = \dfrac{2000P}{T^2}$ **b.** $\dfrac{dT}{dP} = \dfrac{T^2}{2000P}$

75. 1.59984

77. $1.59984 < \ln 5 < 1.62897.$

Exercise Set 8.3

1. a. 2   **b.** $\dfrac{1}{4}$ **c.** $\dfrac{x}{y}$ **d.** $-x^2$

   e. $\sqrt{x}$ **f.** 2^x **g.** $\dfrac{1}{x}$ **h.** x^4

   i. $x^2 + \ln x$ **j.** $\dfrac{e^x}{x}$

3. $y = x - \ln(x + 3)$

5. $3e^{3x}$

7. $\dfrac{e^{\sqrt{t}}}{2\sqrt{t}}$

9. $(2x - 1)e^{x^2 - x}$

11. $e^x(\sin x + \cos x)$

13. $\dfrac{e^x}{e^x + 1} - \dfrac{1}{x + 1}$

15. $-6xe^{x^2}(2 - e^{x^2})^2$

17. $\frac{1}{2}(e^x - e^{-x})$

19. $\left(\dfrac{e^{\sqrt{x}}}{2\sqrt{x}}\right)\left(\ln\sqrt{x} + \dfrac{1}{\sqrt{x}}\right)$

21. $\dfrac{1 - xy}{x^2}$, which equals $\dfrac{1 - \ln x}{x^2}$

23. $\dfrac{1}{e^y(x + 2y) - 2}$

25. True

27. True

29. $-e^{-x} + C$

31. $\frac{1}{2}e^{x^2 + 3} + C$

33. $\frac{1}{8}(1 + e^{2x})^4 + C$

35. $-e^{1/x} + C$

37. $\ln(1 + e^x) + C$

39. $\dfrac{-1}{3 + e^x} + C$

41. $2e^{\sqrt{x}} + e^{2\sqrt{x}} + C$

43. $\dfrac{1}{e} - \dfrac{1}{4}$

45. $e - 1$

47. 2

49. $e^2 - 1$

51. $\frac{1}{2}(e^{3/4} - 1)$

53. max: $(-1, 1)$, $(1, 1)$; min: $(0, 0)$

55. a. $[-\frac{1}{2}, \infty)$
   b. $(-\infty, -\frac{1}{2}]$
   c. min: $(-\frac{1}{2}, -\frac{1}{2}e^{-1}) \approx (-\frac{1}{2}, -0.184)$
   d. $[-1, \infty)$
   e. $(-\infty, -1]$
   f. $(-1, -e^{-2}) \approx (-1, -0.135)$

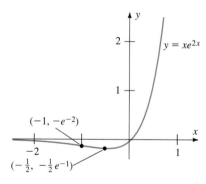

57. 3/4

59. $y = Ce^{x^2}$

61. $(-1, -e^{-1})$

63. $2\left(e + \dfrac{1}{e} - 2\right)$

65. a. $y = e^t$
   b. $y = -2e^{\pi t}$
   c. $y = 2e^{-3t}$

69. -4

73. 0.407263, 0.933326

75.

x	e^x	$P_1(x)$	$P_2(x)$	$P_3(x)$	$P_4(x)$
1	2.71828	2	2.5	2.66667	2.70833
0.5	1.64872	1.5	1.625	1.64583	1.64844
2	7.38906	3	5	6.3333	7
−2	0.135335	−1	1	−0.3333	0.3333

Exercise Set 8.4

1. a. $e^{\ln 2x}$ **b.** $e^{3 \ln \pi}$ **c.** $e^{(\ln 7)(\ln x)}$
d. $e^{\sqrt{2} \ln 4}$ **e.** $e^{(\ln 3)(\sin x)}$ **f.** $e^{(2-x)\ln 2}$

3. $10^a = e^{a \ln(10)}$, $e^b = 10^{b \log_{10}(e)}$

5. $y' = (\ln 3)3^x$

7. $\dfrac{2}{(2x-1)\ln 10}$ **9.** $x(\ln 10)(\ln x)$

11. $2t \log_2 t + \dfrac{t}{\ln 2}$ **13.** $(\ln \pi)\pi^x + \pi x^{\pi - 1}$

15. $x^x(\ln x + 1)$ **17.** $x^{\sqrt{x}-1/2}\left(1 + \dfrac{\ln x}{2}\right)$

19. $(\cos x)^{\sin x}\left[(\cos x)\ln(\cos x) - \dfrac{\sin^2 x}{\cos x}\right]$

21. $\dfrac{5^x}{\ln 5} + C$ **23.** $\dfrac{2}{\ln \pi}\pi^{\sqrt{x}} + C$

25. $\dfrac{x^{1+2\ln a}}{1 + 2\ln a} + C$ **27.** $\dfrac{40}{\ln 3}$

29. $\dfrac{39}{\ln 3}$ **31.** True

35. e^2 **37.** $\dfrac{4\pi}{\ln 3}$

39. $y = \dfrac{x-1}{\ln 10}$

41. $5000(\ln 2)$ dollars/year

43. $\dfrac{1}{3}\left(\dfrac{1}{x+2} - \dfrac{1}{x+3}\right)\sqrt[3]{\dfrac{x+2}{x+3}}$

45. $\left[\dfrac{6x}{x^2+2} + \dfrac{5}{x-1} - \dfrac{1}{x} - \dfrac{1}{2x+2} - \dfrac{1}{2x+4}\right] \cdot$
$\dfrac{(x^2+2)^3(x-1)^5}{x\sqrt{x+1}\sqrt{x+2}}$

47. $(x^2+1)^x\left(\ln(x^2+1) + \dfrac{2x^2}{x^2+1}\right)$

Exercise Set 8.5

1. e^{2t} **3.** $e^{5(t-1)}$ **5.** e^{-2t}

7. False

9. a. $v(t) = v_0 e^{(-c/m)t}$
b. $v(t) = 10e^{-4t}$
c. 0

11. a. 4 h   **b.** 800

13. Sell after 0.83 years. The original price does not affect this answer.

15. 23.3 h   **17.** 8.1 μg   **21.** 21.2 min

23. 8.68 min; 0.024°C   **25.** 2.63 yr

27. $\ln(1.08) \approx 0.070 = 7.70\%$

29. a. 10.25%   **b.** 10.38%
c. 10.516%   **d.** 10.517%

33. False. $P'(t) = \ln(1.05)P(t)$.

Review Exercises—Chapter 8

1. a. 4   **b.** $\dfrac{1}{6^5} = \dfrac{1}{7776}$ **c.** 32   **d.** $\dfrac{\sqrt{2}}{4}$

3. a. $f^{-1}(x) = x - 2$ **b.** $g^{-1}(x) = \sqrt{x}$
c. $y = \frac{1}{2}(x-1)$ **d.** $y = x^2$

5. $2x \ln(x-a) + \dfrac{x^2}{x-a}$ **7.** $\dfrac{2}{x \ln x}$

9. $2e^{x^2}(x \tan 2x + \sec^2 2x)$ **11.** $\dfrac{(1+t+e^t)e^t}{(1+e^t)^2}$

13. $\dfrac{(t^2+b)\ln(t^2+b) - 2t(t-a)\ln(t-a)}{(t-a)(t^2+b)[\ln(t^2+b)]^2}$

15. $\left(\dfrac{\sqrt{t}-2}{2t^2}\right)e^{\sqrt{t}}$ **17.** $\dfrac{-1}{3x^2} - \dfrac{1}{2(1-2x)} - \dfrac{1}{8x}$

19. $\dfrac{-(2xy \ln y + y^2)}{(2x^2 + xy \ln x)}$ **21.** $\dfrac{2x + ye^{xy}}{2y - xe^{xy}}$

23. $\dfrac{-y}{x}$ **25.** $\dfrac{4\sqrt{x}e^{2x} + \cos\sqrt{x}}{4\sqrt{x}(e^{2x} + \sin\sqrt{x})}$

27. $-\frac{1}{2}\ln|1 - x^2| + C$ **29.** $\frac{1}{4}\ln|4x + 2x^2| + C$

31. $2e^{x/2} + C$ **33.** $2(\sqrt{2} - 1)$

35. $\dfrac{5}{6} - 2\ln 2$ **37.** $\dfrac{e^4 + 3}{2}$

39. $e^x - e^{3x} + \frac{3}{5}e^{5x} - \frac{1}{7}e^{7x} + C$

41. $\ln|x^3 + x^2 - 7| + C$ **43.** $-\frac{1}{2}\ln 2$

45. $\frac{1}{2}(\ln x + 2)x^{\sqrt{x}-(1/2)}$

47. $\left(\dfrac{\sin^2 x}{x} + 2 \sin x \cos x \ln x\right) x^{\sin^2 x}$

49. min: $\left(\dfrac{1}{\sqrt{e}}, -\dfrac{1}{2e}\right)$ **51.** $(-1, 1)$

53. $2 \ln 5 + 2$ **55.** $\ln 5$

57. $2\pi \ln 2$

59.

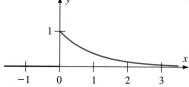

61.

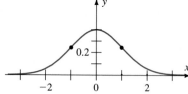

63. $y = -\dfrac{1}{3e}x + \dfrac{1 + 3e^2}{3}$

65. $y = 4x - 2e$; approx. by $10.8 - 2e \approx 5.36344$

69. 0.88623

71. a. 1.38284 **b.** 1.60321 **c.** -0.692835

73. $y = e^{2x}$ **75.** $y = \pi e^{-2x}$

77. $y \equiv 1$ **79.** $\dfrac{e}{2\sqrt{\ln 2}}$

81. $\left(\dfrac{54}{26 - 3 \ln 3}, \dfrac{358}{5(26 - 3 \ln 3)}\right)$

83. a. 273.1 million   **b.** 299.6 million

85. $100(5)^{4/10} \approx 190$ **87.** 4.54 hours

Chapter 9

Exercise Set 9.1

1. $\frac{1}{3} \sin 3x + C$ **3.** $\left[\frac{1}{2} \sec 2x\right]_0^{\pi/8} = \frac{1}{2}(\sqrt{2} - 1)$

5. $\tan x + C$

7. $\frac{1}{2} \ln |\sec x^2 + \tan x^2| + C$

9. $\frac{1}{6} \ln |\sec(3x^2 - 1)| + C$

11. $\left[\frac{1}{4} \tan^4 x\right]_{\pi/4}^{\pi/3} = 2$ **13.** $\left[e^{\tan x}\right]_{\pi/6}^{\pi/4} = e - e^{1/\sqrt{3}}$

15. $\ln |x + \tan x| + C$

17. $\frac{1}{2} \cot(1 - x^2) + \frac{1}{2}(1 - x^2) + C$

19. $2 \tan x + 2 \sec x - x + C$

21. $-\frac{1}{2} \ln |\sec(x^2 - 2x) + \tan(x^2 - 2x)| + C$

23.
$$\begin{cases} e^{\sin x} + C, & \text{if } \cos x \geq 0 \\ -e^{\sin x} + C, & \text{if } \cos x < 0 \end{cases}$$

25. $-2 \cot \sqrt{x} + 2 \csc \sqrt{x} + C$

27. $\frac{1}{2}\sqrt{1 + \sin 4x} + C$

29. $2\sqrt{\tan \theta + 1} + C$ **31.** $\frac{1}{2} \ln(2 + \sqrt{3})$

33. $-\dfrac{6}{\pi} \ln(2\sqrt{3} - 3)$ **35.** $\frac{1}{2} \tan^2 x + C_1, \quad \frac{1}{2} \sec^2 x + C_2$

37. $-\ln(2\sqrt{3} - 3)$ **39.** $\dfrac{2\pi}{3}$

41. $\dfrac{4\pi}{3} - 2 \ln(2 + \sqrt{3})$ **43.** $\dfrac{\pi \ln \sqrt{2}}{2}$

Exercise Set 9.2

1. $\frac{1}{2}x - \frac{1}{8} \sin 4x + C$

3. $\left[-\frac{1}{3} \cos^3 x\right]_{\pi/4}^{\pi/2} = \sqrt{2}/12$

5. $-\cos x + \frac{2}{3} \cos^3 x - \frac{1}{5} \cos^5 x + C$

7. $\dfrac{2}{15}$

9. $\frac{1}{5} \sec^5 x - \frac{1}{3} \sec^3 x + C$

11. $\dfrac{5\pi}{16}$

13. $-\frac{1}{2} \cot^2 x - \ln |\sin x| + C$

15. $\dfrac{1}{4}$

17. $\left[x - \frac{1}{2} \cos 2x\right]_{\pi/2}^{\pi} = \dfrac{\pi}{2} - 1$

19. $\frac{1}{5} \tan^5 x + \frac{1}{7} \tan^7 x + C$

21. $\frac{1}{7} \sec^7 x - \frac{2}{5} \sec^5 x + \frac{1}{3} \sec^3 x + C$

23. $\frac{1}{2} \sin x - \frac{1}{6} \sin 3x + C$

25. $2(\sin x)^{1/2} - \frac{2}{5}(\sin x)^{5/2} + C$

27. $\dfrac{-1}{3(1 + \tan x)^3} + C$

29. $\frac{1}{16} \sin 8x + \frac{1}{4} \sin 2x + C$

31. $\frac{1}{2} \cos^2 \theta - \ln |\cos \theta| + C$

33. $\frac{2}{3}(\tan x)^{3/2} + C$

35. $-\dfrac{1}{4}\dfrac{\cos(a+b+c)x}{a+b+c} - \dfrac{1}{4}\dfrac{\cos(a+b-c)x}{(a+b-c)}$

$\qquad -\dfrac{1}{4}\dfrac{\cos(a-b+c)x}{(a-b+c)} - \dfrac{1}{4}\dfrac{\cos(a-b-c)x}{(a-b-c)} + C$

37. 0   **39.** L **41.** 0   **49.** 1/2

51. $\pi/5$ **53.** $\ln(\sqrt{2}+1)$ **55.** $\dfrac{3-\sqrt{3}}{5}$

Exercise Set 9.3

1. $\dfrac{\pi}{6}$ **3.** 0   **5.** π

7. $\dfrac{\pi}{2}$ **9.** $\dfrac{1}{\sqrt{3}}$ **11.** 2

13. $\dfrac{-1}{2}$ **15.** 1   **17.** 0

19. $\sqrt{3}$ **21.** $\dfrac{x}{\sqrt{1-x^2}}$ **23.** $2x\sqrt{1-x^2}$

25. $\dfrac{\sqrt{x^2-1}}{x}$ **27.** $\dfrac{1}{x}$ **29.** $\dfrac{\sqrt{x^2-1}}{x}$

31. False   **33.** $\theta = \text{Tan}^{-1}\dfrac{x}{100}$ **35.** $\dfrac{\pi}{6}$

37. $\theta = \text{Tan}^{-1}\dfrac{x}{20}$

39. $\dfrac{\pi}{4} - \text{Sin}^{-1}\left(\dfrac{1}{2\sqrt{2}}\right) \approx 0.424$

Exercise Set 9.4

1. $\dfrac{3}{\sqrt{1-9x^2}}$ **3.** $\dfrac{1}{2\sqrt{t-t^2}}$ **5.** $\dfrac{-e^{-x}}{\sqrt{1-e^{-2x}}}$

7. $\dfrac{-1}{2\sqrt{(\text{Cos}^{-1}x)(1-x^2)}}$ **9.** $\dfrac{1}{(1+x^2)\,\text{Tan}^{-1}x}$

11. $\dfrac{-1}{\sqrt{1-x^2}}$ for $0 < x < 1$ **13.** $\dfrac{1-2x\,\text{Tan}^{-1}x}{(1+x^2)^2}$

15. $\dfrac{1}{2}\text{Tan}^{-1}\left(\dfrac{x}{2}\right) + C$ **17.** $\dfrac{1}{3}\text{Tan}^{-1}3x + C$

19. $\left[\dfrac{1}{8}\text{Sec}^{-1}\dfrac{x}{4}\right]^{8}_{8\sqrt{3}/3} = \dfrac{\pi}{48}$ **21.** $-\sqrt{1-x^2} + C$

23. $\text{Sec}^{-1}e^x + C$ **25.** $2\,\text{Tan}^{-1}e^{\sqrt{x}} + C$

27. $\dfrac{1}{3}\text{Tan}^{-1}x^3 + C$ **29.** $\dfrac{\pi^2}{32}$

31. $-\dfrac{1}{8}(\text{Cos}^{-1}2x)^4 + C$ **33.** $\pi/6$

35. $\dfrac{1}{5}\left[\text{Sec}^{-1}\left(\dfrac{12}{5}\right) - \text{Sec}^{-1}\left(\dfrac{8}{5}\right)\right]$

37. $\pi/6$ **39.** $y = \cos x$

41. 2 m   **43.** $\dfrac{1}{5}$ radians/s; $2\sqrt{2}$ ft/s

47. $\dfrac{\pi}{12}$ **49.** $\dfrac{\pi^2}{6}$

51. $2\pi\,\text{Tan}^{-1}(2\sqrt{3})$

Exercise Set 9.5

1. 0   **3.** 5/4

5. 3/5   **7.** 17/15

9. $\ln(1+\sqrt{2})$ **11.** $2\cosh 2x$

13. $(\sinh x)(1 + \text{sech}^2 x)$

15. $2(\cosh(4x))^{-1/2}\sinh(4x)$

17. $-\text{sech}\,x\tanh x$

19. $-2xe^x\,\text{csch}\,x^2\coth x^2 + e^x\,\text{csch}\,x^2$

21. $\dfrac{2}{\sqrt{4x^2+1}}$ **23.** $\dfrac{2s}{1-s^4}$

25. $\dfrac{\pi}{(\cosh^{-1}\pi x)\sqrt{\pi^2 x^2 - 1}}$

27. $\sinh x\cosh x^2 + 2x\cosh x\sinh x^2$

29. $\dfrac{2\cosh(\ln x^2)}{x} = x + x^{-3}$

31. $\dfrac{x\,\text{sech}^2(\sqrt{1+x^2})}{\sqrt{1+x^2}}$ **33.** $\ln\left(\dfrac{2+\sqrt{5}}{1+\sqrt{2}}\right)$

35. $x - \dfrac{1}{2}\tanh(2x) + C$

37. $\dfrac{\sinh^2 2}{2} = \dfrac{(e^2 - e^{-2})^2}{8}$

39. $\cosh^{-1}3 - \cosh^{-1}2 = \ln\left(\dfrac{3+2\sqrt{2}}{2+\sqrt{3}}\right)$

41. $\dfrac{1}{3}\sinh^{-1}\dfrac{3}{5} = \dfrac{1}{3}\ln(\dfrac{3}{5} + \dfrac{1}{5}\sqrt{34})$

43. $\ln(\cosh x) + C$ **45.** $2(\sinh x)^{1/2} + C$

47. $\dfrac{1}{3}\tanh^3 x + C$ **49.** False

59. $y_1 = C_1\sinh k\pi x + C_2\cosh k\pi x$
$\quad\; y_2 = C_1 e^{k\pi x} + C_2 e^{-k\pi x}$

61. $\ln\dfrac{25}{17}$

63. No relative extrema

65. min: $(\dfrac{1}{2}\ln\dfrac{7}{3}, \sqrt{21})$

67. $y'' - 9y = 0$

69. Even: $\cosh x$ Odd: $\sinh x$
$\quad\quad\quad$ sech x $\quad\quad\quad$ tanh x
$\quad\quad\quad\quad\quad\quad\quad\quad$ coth x
$\quad\quad\quad\quad\quad\quad\quad\quad$ csch x

73. $[0, +\infty)$

75. $\tanh x$: $\quad -\infty < x < \infty$
$\quad\,$ coth x: $\quad x \in (-\infty, 0) \cup (0, \infty)$
$\quad\,$ sech x: $\quad x \geq 0$
$\quad\,$ csch x: $\quad x \in (-\infty, 0) \cup (0, \infty)$

Review Exercises—Chapter 9

1. $\pi/3$ $\quad\quad\quad$ **3.** $\pi/3$ $\quad\quad\quad$ **5.** $\pi/4$ $\quad\quad\quad$ **7.** 1

9. $3/4$ $\quad\quad\quad\quad\quad\quad\quad\quad\quad$ **11.** $\dfrac{1}{2\sqrt{x(1-x)}}$

13. $\dfrac{2t}{1 + (1 - t^2)^2}$

15. $2x \sinh(1 - x) - x^2 \cosh(1 - x)$

17. $6[\csc(\cot 6x)][\cot(\cot 6x)](\csc^2 6x)$

19. $\dfrac{2}{(2x + 1)\sqrt{1 - [\ln(2x + 1)]^2}}$

21. $\dfrac{1 - 2x \operatorname{Tan}^{-1} x}{(1 + x^2)^2}$

23. $\dfrac{x}{(x^2 + 4)(\sqrt{x^2 + 3})}$

25. $-\dfrac{\operatorname{sech}^2 x}{(\pi + \tanh x)^2}$

27. $\dfrac{1}{5}(\sinh x + \operatorname{Cos}^{-1} 2x)^{-4/5}\left(\cosh x - \dfrac{2}{\sqrt{1 - 4x^2}}\right)$

29. $\dfrac{1}{(\operatorname{Sin}^{-1} x)\sqrt{1 - x^2}}$

31. $\tanh^{-1}(\ln x) + \dfrac{1}{1 - (\ln x)^2}$

33. $\dfrac{x}{(1 - x^4) \tanh^{-1}(x^2)}$

35. $\frac{1}{2}\cosh^{-1}(2x) + C = \frac{1}{2}\ln|2x + \sqrt{4x^2 - 1}| + C$

37. $\frac{1}{10}\sin^5 2x + C$

39. $\dfrac{\pi + 2}{4}$ $\quad\quad\quad$ **41.** $\pi/2$ $\quad\quad\quad$ **43.** $2/3$

45. $\tan 2x - x + C$ $\quad\quad\quad$ **47.** $1/6$

49. $\operatorname{Sin}^{-1}\left(\dfrac{x}{2}\right) + C$ $\quad\quad\quad$ **51.** $-\frac{1}{2}(\operatorname{Cos}^{-1} x)^2 + C$

53. $2\operatorname{Tan}^{-1}(e^{\sqrt{x}}) + C$

55. $\frac{2}{3}(\operatorname{Sin}^{-1} x)^{3/2} + C$

57. $\dfrac{(\operatorname{Sec}^{-1} 3x)^2}{4} + C$

59. $\dfrac{1}{6}\left(\sec^3 \dfrac{2\pi}{9} - 1\right)$

61. $2\sqrt{3}$

63. $\dfrac{-\sqrt{3}}{16}$

65. $\frac{2}{7}\cos^{7/2} x - \frac{2}{3}\cos^{3/2} x + C$

67. $\cos x + \sec x + C$

69. $\ln|1 + \sinh x| + C$

71. $\frac{1}{3}\cosh^3 x - \cosh x + C$

73. $\pi/2$ $\quad\quad\quad$ **75.** $\pi(1 - \pi/4)$

77. $\pi \operatorname{Tan}^{-1} 4$

79. $\dfrac{2xy}{\sqrt{x^4 y^2 - 1} - x^2}$

81. $\left[2\sinh x\right]_0^1 = e - \dfrac{1}{e}$ $\quad\quad\quad$ **83.** 2

87. -48 radians/h

89. $(-2, \operatorname{Cot}^{-1}(-2))$ $\quad$ and $\quad$ $(2, \operatorname{Cot}^{-1}(2))$

91. $3t - \dfrac{t^2}{2} + \operatorname{Tan}^{-1} t$

Chapter 10

Exercise Set 10.1

1. $xe^x - e^x + C$

3. $\left[x \sin x + \cos x\right]_{\pi/2}^{\pi} = -1 - \dfrac{\pi}{2}$

5. $x \operatorname{Tan}^{-1} x - \frac{1}{2}\ln(1 + x^2) + C$

7. $\dfrac{x}{\pi}\tan \pi x - \dfrac{1}{\pi^2}\ln|\sec \pi x| + C$

9. $\dfrac{x}{2}\sin(\ln x) - \dfrac{x}{2}\cos(\ln x) + C$

11. $27{,}761/18 \approx 1542.28$

13. $\left[x(\ln x)^2 - 2x \ln x + 2x\right]_e^{e^3} = 5e^3 - e$

15. $\left[\dfrac{x}{2}e^{2x} - \dfrac{1}{4}e^{2x}\right]_0^{\ln 2} = -\dfrac{3}{4} + 2\ln 2$

17. $\dfrac{e^2 + 3}{8}$

19. $\dfrac{3e^2 - 7}{4}$

21. $x \cosh x - \sinh x + C$

23. $\left[\frac{2}{5}e^{2x}\cos x + \frac{1}{5}e^{2x}\sin x\right]_0^\pi = -\frac{2}{5}(1 + e^{2\pi})$

25. $-2x - 2\sqrt{1 - x^2}\, \text{Cos}^{-1} x + x(\text{Cos}^{-1} x)^2 + C$

27. $\dfrac{(1 + x^2)^{3/2}}{15}(3x^2 - 2) + C$

29. $\dfrac{e^{ax^2}}{2a^2}(ax^2 - 1) + C$

31. $\frac{1}{4}\sec^3 x \tan x - \frac{1}{8}\sec x \tan x - \frac{1}{8}\ln|\sec x + \tan x| + C$

33. $\dfrac{2x^2 + 1}{4}\sinh^{-1} x - \dfrac{x}{4}\sqrt{1 + x^2} + C$

35. $-4(x^{3/2} + 3x + 6\sqrt{x} + 6)e^{-\sqrt{x}} + C$

37. $1/\pi$ $\qquad\qquad$ **39.** $\pi(e - 2)$

41. $2\pi^2$

43. $\dfrac{\pi}{1 + \pi^2}\left(1 + \dfrac{1}{e}\right)$ $\qquad$ **45.** $4/\pi$

47. $62.4\pi[12.5(\ln 5)^2 - 37.5(\ln 5) + 34]$

49. $\left(\dfrac{4e^3 - 3e^2 - 7}{6e^2 - 18}, \dfrac{e^3 - 3e + 5}{3e^2 - 9}\right)$

51. $-\frac{1}{2}\sin x \cos x + \frac{1}{2}x + C$

Exercise Set 10.2

1. $\dfrac{x\sqrt{4 - x^2}}{2} + 2\,\text{Sin}^{-1}\dfrac{x}{2} + C$

3. $\left[\text{Sin}^{-1}\left(\dfrac{x}{3}\right)\right]_0^{3/2} = \dfrac{\pi}{6}$

5. $\frac{1}{3}(x^2 + 8)\sqrt{x^2 - 4} + C$

7. $\left[\ln\left|\dfrac{1 - \sqrt{1 - x^2}}{x}\right| + \sqrt{1 - x^2}\right]_{1/2}^1$
$\qquad\qquad = -(\ln(2 - \sqrt{3}) + \sqrt{3}/2)$

9. $\dfrac{1}{16}\text{Sec}^{-1}\left(\dfrac{x}{2}\right) + \dfrac{\sqrt{x^2 - 4}}{8x^2} + C$

11. $\left[\dfrac{1}{128}\text{Tan}^{-1}\left(\dfrac{x}{4}\right) + \dfrac{x}{32(16 + x^2)}\right]_0^4 = \dfrac{2 + \pi}{512}$

13. $\dfrac{x}{\sqrt{1 - x^2}} - \text{Sin}^{-1} x + C$

15. $\left[-\frac{1}{15}(3x^2 + 2)(1 - x^2)^{3/2}\right]_0^1 = \frac{2}{15}$

17. $\ln\left|\dfrac{\sqrt{x^2 + 3} + x}{\sqrt{3}}\right| - \dfrac{x}{\sqrt{x^2 + 3}} + C$

19. $\dfrac{x\sqrt{x^2 + a^2}}{2} - \dfrac{a^2 \ln(x + \sqrt{x^2 + a^2})}{2} + C$

21. $\ln\left|\dfrac{x + \sqrt{x^2 - a^2}}{a}\right| - \dfrac{\sqrt{x^2 - a^2}}{x} + C$

23. $\dfrac{-1}{\sqrt{x^2 + a^2}} + C$

25. $\dfrac{1}{a}\text{Sec}^{-1}\left(\dfrac{x}{a}\right) + C$

27. $\dfrac{-\sqrt{x^2 + a^2}}{x} + \ln(x + \sqrt{x^2 + a^2}) + C$

29. $\dfrac{x\sqrt{x^2 + 9}}{2} - \dfrac{3}{2}\ln\left|\dfrac{x + \sqrt{x^2 + 9}}{3}\right| + C$

31. $-\sqrt{9 - x^2} - 4\,\text{Sin}^{-1}\left(\dfrac{x}{3}\right) + \dfrac{5}{3}\ln\left|\dfrac{3 - \sqrt{9 - x^2}}{x}\right| + C$

33. $20 - 9\ln 3$ $\qquad$ **35.** $\ln 2 - \frac{27}{40}$

37. $32\pi^2$ $\qquad\qquad$ **39.** $\dfrac{\pi}{3} - \dfrac{\pi^2}{12}$

41. 62.65% $\qquad\qquad$ **43.** $\dfrac{\sqrt{2} + \ln(\sqrt{2} + 1)}{4}$

Exercise Set 10.3

1. $\dfrac{-1}{x - 2} + C$

3. $\ln\left|\dfrac{x + 3 + \sqrt{x^2 + 6x + 13}}{2}\right| + C$

5. $\ln\left|\dfrac{(x - 3) + \sqrt{x^2 - 6x}}{3}\right| + C$

7. $-\sqrt{6x - x^2} + 3\,\text{Sin}^{-1}\left(\dfrac{x - 3}{3}\right) + C$

9. $\left[2\sqrt{x^2 + 6x + 18} - 5\ln\left|\dfrac{\sqrt{x^2 + 6x + 18} + (x + 3)}{3}\right|\right]_0^1$
$\qquad\qquad = 10 - 6\sqrt{2} + 5\ln\left(\dfrac{1 + \sqrt{2}}{3}\right)$

11. $\left[\frac{1}{2}\ln|x^2 - 8x - 7|\right]_{-3}^{-2} = -\ln\sqrt{2}$

13. $\left[-\dfrac{x + 2}{2(x^2 + 2x + 2)} - \dfrac{1}{2}\text{Tan}^{-1}(x + 1)\right]_0^1$
$\qquad\qquad = \dfrac{1}{5} + \dfrac{\pi}{8} - \dfrac{\text{Tan}^{-1} 2}{2}$

15. $\dfrac{-\sqrt{7}}{49}\ln\left|\dfrac{\sqrt{88 - 18x + x^2} - \sqrt{7}}{x - 9}\right|$
$\qquad\qquad - \dfrac{1}{7\sqrt{88 - 18x + x^2}} + C$

17. $\frac{1}{2}\left(x + \frac{5}{2}\right)\sqrt{4x^2 + 20x + 29}$

$$- \ln\left|\frac{\sqrt{4x^2 + 20x + 29}}{2} + \left(x + \frac{5}{2}\right)\right| + C$$

19. $\left[\frac{1}{2}\tan^{-1}\left(\frac{x - 2}{2}\right)\right]_2^4 = \frac{\pi}{8}$

21. $\left[\pi\sin^{-1}\left(\frac{x - 4}{4}\right)\right]_2^4 = \frac{\pi^2}{6}$

Exercise Set 10.4

1. $\ln\left|\frac{x}{x + 1}\right| + C$

3. $\left[3\ln|x - 3| - \ln|x - 1|\right]_4^6 = \ln\frac{81}{5}$

5. $-3\ln|1 - x| + \ln|1 + x| + C$

7. $\frac{1}{2}\ln\left|\frac{1 + x}{1 - x}\right| + C$

9. $\left[\ln|x| - \frac{1}{2}\ln(x^2 + 1)\right]_e^{e^2} = 1 + \ln\sqrt{\frac{1 + e^2}{1 + e^4}}$

11. $\frac{1}{2}\ln|x + 1| - \frac{1}{4}\ln(x^2 + 1) + \frac{1}{2}\tan^{-1}x + C$

13. $\left[2\ln|x - 2| + \frac{1}{2}\ln(x^2 + 1)\right]_3^4 = \ln(4\sqrt{\frac{17}{10}})$

15. $3\ln|x + 2| + \frac{1}{2}\ln(x^2 + 1) - \tan^{-1}x + C$

17. $\left[\ln|x| - \frac{3}{x + 1}\right]_e^{e^3} = 2 + \frac{3}{1 + e} - \frac{3}{1 + e^3}$

19. $2\ln|x| + \frac{1}{x + 1} + C$

21. $\left[\frac{1}{2}\ln|x| - \ln|x + 1| + \frac{1}{2}\ln|x + 2|\right]_1^2 = \ln\left(\frac{4\sqrt{2}}{3\sqrt{3}}\right)$

23. $\ln|x| + 2\ln|x - 3| + \ln(x^2 + 1) + C$

25. $4\ln|x| - 2\ln(x^2 + 1) + C$

27. $\frac{1}{4a^3}\ln\left|\frac{a + x}{a - x}\right| + \frac{1}{2a^2}\frac{x}{a^2 - x^2} + C$

29. $x^2 + 3\ln|x - 4| + 6\ln|x + 2| + C$

31. $3\ln|x - 1| + \ln|x^2 + x + 1| + C$

33. $2\ln|x| + 6\ln|x - 2| + \ln|x + 4| + C$

35. $4\ln 3 - \ln 2$

37. $3\pi\ln 5 - 4\pi\ln 3$

39. $\pi(\ln 5 + \frac{3}{2}\ln 13)$

43. $P = \frac{100}{1 + Me^{-2t}}$, $k = 100$

45. 1

Exercise Set 10.5

1. $x - 2\sqrt{x} + 2\ln(1 + \sqrt{x}) + C$

3. $2\sqrt{x} - 6\sqrt[3]{x} + 24\sqrt[6]{x} - 48\ln(\sqrt[6]{x} + 2) + C$

5. $\left[\frac{3}{5}(x + 1)^{5/3} - \frac{3}{2}(x + 1)^{2/3}\right]_0^7 = \frac{141}{10}$

7. $\frac{2\sqrt{3}}{3}\tan^{-1}\left(\frac{2\tan(x/2) + 1}{\sqrt{3}}\right) + C$

9. $-\cot\frac{x}{2} + C$

11. $\left[\frac{3}{8}(1 + x)^{8/3} - \frac{6}{5}(1 + x)^{5/3} + \frac{3}{2}(1 + x)^{2/3}\right]_0^2$

$$= \frac{51(3)^{2/3} - 27}{40}$$

13. $\frac{\sqrt{2}}{2}\ln\left|\frac{\tan(x/2) - 1 + \sqrt{2}}{\tan(x/2) - 1 - \sqrt{2}}\right| + C$

15. $\frac{1}{56}(1 - 2x)^{7/2} - \frac{3}{40}(1 - 2x)^{5/2}$

$$+ \frac{1}{8}(1 - 2x)^{3/2} - \frac{1}{8}(1 - 2x)^{1/2} + C$$

17. $\sin^{-1}x - \sqrt{1 - x^2} + C$

19. $\begin{cases} \dfrac{1}{\sqrt{a}}\ln\left|\dfrac{\sqrt{a + bx} - \sqrt{a}}{\sqrt{a + bx} + \sqrt{a}}\right| + C, & \text{if } a > 0 \\[3mm] \dfrac{2}{\sqrt{-a}}\tan^{-1}\sqrt{\dfrac{a + bx}{-a}} + C, & \text{if } a < 0 \end{cases}$

21. $\left[\frac{2}{3}(2 + x)^{3/2} - 4(2 + x)^{1/2}\right]_0^2 = \frac{8}{3}(\sqrt{2} - 1)$

Exercise Set 10.6

1. $\frac{1}{5}\ln\left(\frac{e^{5x}}{1 + e^{5x}}\right) + C$

3. $\left[\frac{4}{729}\ln\left(\frac{9 + 2x}{x}\right) - \frac{9 + 4x}{81x(9 + 2x)}\right]_1^2$

$$= \frac{151}{23,166} + \frac{4}{729}(\ln 13 - \ln 2 - \ln 11)$$

5. $\frac{1}{10}\ln(2 + 5x^2) + \frac{3}{\sqrt{10}}\tan^{-1}\left(\frac{\sqrt{5}x}{\sqrt{2}}\right) + C$

7. $\frac{1}{\pi}\left[\frac{1}{3}\tan^3\pi x - \tan\pi x + \pi x\right]_0^{1/4} = \frac{1}{4} - \frac{2}{3\pi}$

9. $\frac{\sin 3x}{6} - \frac{\sin 9x}{18} + C$

11. $\frac{-1}{2x} - \frac{3}{4}\ln\left|\frac{2 - 3x}{x}\right| + C$

13. $\frac{9}{20}\ln\left|\frac{2x + 5}{2x - 5}\right| + C$

15. $\frac{1}{10}\tan^2 5x + \frac{1}{5}\ln|\cos 5x| + C$

17. $\frac{4}{5}(3x^2 - 24x + 288)\sqrt{6+x} + C$

19. $\dfrac{3x}{7\sqrt{14-x^2}} + C$

21. $\left[-\dfrac{\pi}{8}\tan\left(\dfrac{\pi}{4} - x\right)\right]_0^{\pi/6} = \dfrac{\pi}{8}\left[1 - \tan\left(\dfrac{\pi}{12}\right)\right]$

23. $\dfrac{-3}{4}\ln\left|\dfrac{4 + \sqrt{x^2+16}}{x}\right| + C$

25. $\left[\dfrac{1}{3}\text{Tan}^{-1}(3x+2)\right]_{-1}^{-2/3} = \pi/12$

27. $\dfrac{-7}{2}\ln\left|\dfrac{2 + \sqrt{x^2+8x+20}}{x+4}\right| + C$

29. $9\,\text{Sin}^{-1}(x/4) - x\sqrt{16-x^2} + C$

Review Exercises—Chapter 10

1. $\frac{1}{3}(x^2+9)^{3/2} + C$

3. $\left[\dfrac{-x^2\cos 2x}{2} + \dfrac{x\sin 2x}{2} + \dfrac{\cos 2x}{4}\right]_0^{\pi} = -\dfrac{\pi^2}{2}$

5. $\dfrac{x}{2}\sqrt{x^2+a^2} + \dfrac{a^2}{2}\ln\left|\dfrac{x + \sqrt{x^2+a^2}}{a}\right| + C$

7. $\left[\ln|x(x+3)|\right]_1^3 = \ln(9/2)$

9. $\left[\dfrac{x}{\pi}\sin \pi x + \dfrac{1}{\pi^2}\cos \pi x\right]_{-1/2}^{1/2} = 0$

11. $\left[\dfrac{x\sin(2x+1)}{2} + \dfrac{\cos(2x+1)}{4}\right]_{-1/2}^{(\pi-1)/2} = -\dfrac{1}{2}$

13. $\left[-e^{-5x}\left(\dfrac{x^2}{5} + \dfrac{2x}{25} + \dfrac{2}{125}\right)\right]_0^1 = \dfrac{2}{125} - \dfrac{37}{125}e^{-5}$

15. $\dfrac{x^2}{2}\ln^2 x - \dfrac{x^2}{2}\ln x + \dfrac{x^2}{4} + C$

17. $\dfrac{\sqrt{5}}{5}\text{Tan}^{-1}\left(\dfrac{x-2}{\sqrt{5}}\right) + C$

19. $2\sqrt{x+4} + 2\ln\left|\dfrac{\sqrt{x+4}-2}{\sqrt{x+4}+2}\right| + C$

21. $\dfrac{32 + 16\sqrt{2}}{15}$ **23.** $\text{Sin}^{-1}\left(\dfrac{x}{a}\right) + C$

25. $\dfrac{-1}{x} + 3\ln|x+1| + C$ **27.** $1 - \dfrac{\pi}{4}$

29. $\dfrac{a^2}{2}\text{Sin}^{-1}\left(\dfrac{x}{a}\right) + \dfrac{x}{2}\sqrt{a^2-x^2} + C$

31. $\left[4\ln|x-2| + \text{Tan}^{-1}x\right]_0^1 = \dfrac{\pi}{4} - 4\ln 2$

33. $\frac{3}{5}(x+1)^{5/3} + C$

35. $\dfrac{x}{2}\sqrt{x^2-a^2} - \dfrac{a^2}{2}\ln\left|\dfrac{x + \sqrt{x^2-a^2}}{a}\right| + C$

37. $\ln|x+2| + \dfrac{2\sqrt{3}}{3}\text{Tan}^{-1}\left(\dfrac{2x+1}{\sqrt{3}}\right) + C$

39. $\frac{1}{4}(x^3 - 10x)\sqrt{x^2-4} + 6\ln|x + \sqrt{x^2-4}| + C$

41. $2\ln|x| + \dfrac{2}{3}\ln|x+1|$
$\qquad + \dfrac{1}{6}\ln(x^2 - x + 1) + \dfrac{1}{\sqrt{3}}\text{Tan}^{-1}\dfrac{2x-1}{\sqrt{3}} + C$

43. $-\frac{1}{2}\csc x \cot x + \frac{1}{2}\ln|\csc x - \cot x| + C$

45. $\dfrac{1}{3}\ln\left|\dfrac{\sqrt{9+x^2}-3}{x}\right| + C$

47. $\dfrac{1}{\sqrt{2}}\text{Tan}^{-1}\dfrac{x}{\sqrt{2}} - 3\ln|x-4| + C$

49. $\dfrac{4}{\sqrt{3}}\text{Tan}^{-1}\left[\sqrt{3}\tan\left(\dfrac{x}{2}\right)\right] - x + C$

51. $\left[\dfrac{1}{2}\ln\left|\dfrac{e^x - 1}{e^x + 1}\right|\right]_{\ln 2}^{\ln 3} = \ln\sqrt{3/2}$

53. $\frac{1}{2}\ln(a^2 + x^2) + C$

55. $\left[3\ln|x-1| + \ln(x^2+4)\right]_2^3 = \ln 13$

57. $-\ln(e^{-x} + 1) + C$

59. $3\sqrt{x^2+8x} - 11\ln|(x+4) + \sqrt{x^2+8x}| + C$

61. $\frac{1}{5}(x^2+1)^{5/2} - \frac{2}{3}(x^2+1)^{3/2} + \sqrt{x^2+1} + C$

63. $\dfrac{x^2}{2} - 2\ln(x^2+2) - \dfrac{2}{x^2+2} + C$

65. $\dfrac{1}{2}\ln\left|\dfrac{2x + \sqrt{4x^2-25}}{5}\right| + C$

67. $\left[x\ln^2 x - 2x\ln x + 2x\right]_1^e = e - 2$

69. $\dfrac{3}{2\sqrt{5}}\ln\left|\dfrac{x - \sqrt{5}}{x + \sqrt{5}}\right| + C$

71. $\dfrac{1}{3}\ln\left|\dfrac{\sqrt{9-\pi x}-3}{\sqrt{9-\pi x}+3}\right| + C$

73. $\dfrac{10}{49}\ln\left|\dfrac{x}{7-2x}\right| - \dfrac{5}{7x} + C$

75. $\frac{1}{2}(9+2x)^{3/2} - \frac{27}{2}(9+2x)^{1/2} + C$

77. $-\dfrac{\sqrt{8 - x^2}}{8x} + C$

79. $\dfrac{2\sqrt{65}}{65} \, \text{Tan}^{-1}\!\left[\sqrt{\dfrac{5}{13}}\,\tan\!\left(\dfrac{x}{2}\right)\right] + C$

81. $\dfrac{x^4}{4}\ln x - \dfrac{x^4}{16} + C$

83. $\left[-xe^{-x} - e^{-x} - \dfrac{x^2}{2e}\right]_0^1 = 1 - \dfrac{5}{2e}$

85. $\pi\!\left[\dfrac{1}{x} + \dfrac{2}{5}\ln\dfrac{x-5}{x}\right]_6^7 = \pi\!\left(-\dfrac{1}{42} + \dfrac{2}{5}\ln\dfrac{12}{7}\right)$

87. $\left[\text{Sin}^{-1} x\right]_0^{1/2} = \pi/6$

89. $\bar{x} = 2$

Chapter 11

Exercise Set 11.1

1. $-\dfrac{1}{e}$ **3.** 0   **5.** $\dfrac{1}{2}$

7. $-\infty$ **9.** 2   **11.** 1

13. $+\infty$ **15.** $+\infty$ **17.** 0

19. $\dfrac{e}{2}$ **21.** 0   **23.** 1

25. Does not exist

27. 0   **29.** 1   **31.** 1

33. 3   **35.** 1   **37.** $\dfrac{3}{4}$ **39.** 0

41.

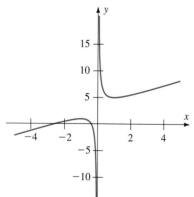

43.

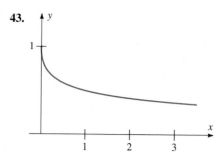

Exercise Set 11.2

1. $-\infty$ **3.** 1   **5.** 0   **7.** $+\infty$

9. 1   **11.** 1   **13.** 0   **15.** 0

17. e^6 **19.** 1   **21.** e^{-1} **23.** ln 2

25. $+\infty$ **27.** $\ln\frac{1}{2}$ **29.** e^{-1}

31.

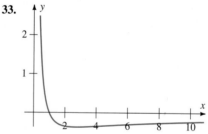

33.

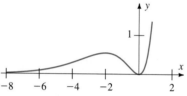

35. $r < 0$

Exercise Set 11.3

1. Diverges   **3.** Diverges   **5.** 1

7. 1/2   **9.** $e^4/2$ **11.** 1

13. Diverges   **15.** $\pi/2$ **17.** $-1/4$

19. $2e^3/9$ **21.** Diverges   **23.** 1/4

25. 1   **27.** 1/4   **31.** $p > -1$

35. Diverges   **37.** Converges   **39.** Diverges

43. a. $S = \displaystyle\int_1^{\infty} \dfrac{2\pi}{x}\sqrt{1 + \dfrac{1}{x^4}}\,dx > \int_1^{\infty}\dfrac{2\pi}{x}\,dx$, which diverges

45. 1/3   **47.** 1   **49.** $2\pi + 4\pi/e$

51. $\bar{x} = 2,\ \bar{y} = 1/5$

55. a. \$4,121,000   **b.** \$12,500,000

59. $2\displaystyle\lim_{t \to r}\int_0^t 2\pi r\,dx = \lim_{t \to r} 4\pi tr = 4\pi r^2$

Review Exercises—Chapter 11

1. 3/4   **3.** $+\infty$ **5.** 1   **7.** 0   **9.** 1

11. 0   **13.** 0   **15.** 1   **17.** 1   **19.** 2/e

21. Diverges   **23.** 0

25. Diverges   **27.** $\pi/2$

29. Horizontal asymptote $y = 3$ as $x \to \pm\infty$

31. Horizontal asymptote $y = 0$ as $x \to +\infty$

33. 7/2   **35.** $4\pi^2$

Chapter 12

Exercise Set 12.1

1. $\{\frac{1}{3}, \frac{2}{5}, \frac{3}{7}, \frac{4}{9}, \ldots\}$. Converges to $\frac{1}{2}$.

3. $\{-1, -\frac{1}{3}, -\frac{1}{11}, 0, \ldots\}$. Converges to 0.

5. $\{e, e^{1/2}, e^{1/3}, e^{1/4}, \ldots\}$. Converges to 1.

7. $\{\sqrt{5}, \sqrt{5}, \sqrt{5}, \sqrt{5}, \ldots\}$. Converges to $\sqrt{5}$.

9. $\{1, 0, -1, 0, \ldots\}$. Diverges.

11. $\left\{10, \dfrac{40}{1 + \sqrt{2}}, \dfrac{60}{1 + \sqrt{3}}, \dfrac{80}{3}, \ldots\right\}$. Diverges.

13. $\left\{\dfrac{2}{3}, \dfrac{3 + \sqrt{2}}{4}, \dfrac{3 - \sqrt{3}}{5}, \dfrac{5}{6}, \ldots\right\}$. Converges to 0.

15. $\{\sqrt{2}, \frac{1}{2}\sqrt{6}, \frac{2}{3}\sqrt{3}, \frac{1}{2}\sqrt{5}, \ldots\}$. Converges to 1.

17. $\{1, \cos\frac{1}{4}, \cos\frac{2}{9}, \cos\frac{3}{16}, \ldots\}$. Converges to $\cos 0 = 1$.

19. $\left\{\dfrac{3}{2}, \dfrac{1 + \sqrt{2}}{2\sqrt{2}}, \dfrac{2 + 3\sqrt{3}}{6\sqrt{3}}, \dfrac{5}{8}, \ldots\right\}$. Converges to $\dfrac{1}{2}$.

21. $\{\frac{1}{2}, \frac{1}{6}, \frac{1}{12}, \frac{1}{20}, \ldots\}$. Converges to

$$\lim\left(\frac{1}{n} - \frac{1}{n+1}\right) = \lim\left(\frac{1}{n(n+1)}\right) = 0.$$

23. $\{\sqrt{2} - 1, \sqrt{3} - \sqrt{2}, 2 - \sqrt{3}, \sqrt{5} - 2, \ldots\}$. Converges to 0.

25. $\left\{\sqrt{3}, \dfrac{3}{2}, \dfrac{\sqrt{19}}{3}, \dfrac{\sqrt{33}}{4}, \ldots\right\}$. Converges to $\sqrt{2}$.

27. $\{\text{Tan}^{-1} 2, \text{Tan}^{-1}\frac{3}{2}, \text{Tan}^{-1}\frac{4}{3}, \text{Tan}^{-1}\frac{5}{4}, \ldots\}$. Converges to $\pi/4$.

29. $\{-1, \frac{1}{2}, -\frac{1}{3}, \frac{1}{4}, \ldots\}$. Converges to 0.

31. $\{\ln\frac{3}{2}, \ln\frac{5}{2}, \ln\frac{7}{2}, \ln\frac{9}{2}, \ldots\}$. Diverges.

33. $\{0, \frac{1}{4}, \frac{8}{27}, \frac{81}{256}, \ldots\}$. Converges to $1/e$.

43. $\{(-1)^n\}$

Exercise Set 12.2

1. 2   **3.** 1   **5.** 0

7. Diverges   **9.** 1   **11.** e^{-3}

13. 0   **15.** 0   **17.** 0

19. e^3 **21.** Diverges

31. $a_n = 2n - 5$ **33.** $\{n\}$

Exercise Set 12.3

1. $-\frac{1}{2} + \frac{1}{4} - \frac{1}{8} + \frac{1}{16} - \cdots$

3. $\frac{3}{5} + \frac{5}{11} + \frac{9}{29} + \frac{17}{83} + \cdots$

5. $\ln\frac{1}{2} + \ln\frac{2}{3} + \ln\frac{3}{4} + \ln\frac{4}{5} + \cdots$

7. Converges to $\frac{7}{6}$ **9.** Diverges

11. Converges; $\frac{27}{23}$

13. Converges to $\dfrac{1}{1 - \left(\dfrac{1}{2 + x}\right)} = \dfrac{x + 2}{x + 1}$

15. Converges to $-\frac{1}{2}$ **17.** Diverges

19. Converges to $\frac{1}{3}$ **21.** Converges to $1/3$

23. Diverges   **25.** Converges to $\dfrac{3}{3 - \sqrt{2}}$

27. Diverges

29. **a.** $\displaystyle\sum_{k=1}^{\infty} \frac{3}{10^k}$ **b.** $\dfrac{1}{3}$

31. **a.** $\displaystyle\sum_{k=1}^{\infty} \frac{92}{100^k}$ **b.** $\dfrac{92}{99}$

33. **a.** $\displaystyle\sum_{k=1}^{\infty} \frac{412}{1000^k}$ **b.** $\dfrac{412}{999}$

35. $\displaystyle\sum_{k=0}^{\infty} (-1)^k x^k = \sum_{k=0}^{\infty} (-x)^k = \frac{1}{1 - (-x)} = \frac{1}{1 + x}$, if $|x| < 1$.

37. $\displaystyle\sum_{k=0}^{\infty} \frac{x^k}{y^k} = \sum_{k=0}^{\infty}\left(\frac{x}{y}\right)^k = \frac{1}{1 - x/y} = \frac{y}{y - x}$, if $|x| < |y|$.

39. $h + 2h\displaystyle\sum_{k=1}^{\infty}\left(\frac{2}{3}\right)^k = 5h$.

41. No, if $c = 0$.

45. $S_n = S_{n-1} + a_n$.

47.

S_5	S_{10}	S_{20}	S_{50}	S_{100}
2.736625514	2.96531694	2.999398543	2.999999997	3.

The actual sum is $\dfrac{1}{1-(2/3)} = 3$.

49.

S_5	S_{10}	S_{20}
$-.37186886$	-1.1286830	-1.2955650

S_{50}	S_{100}
-1.3015408	-1.3015410

The actual sum is $-\dfrac{3125}{2401}$.

51. $\dfrac{\sqrt{3}}{3}$

Exercise Set 12.4

1. Diverges

3. Diverges

5. Diverges

7. Converges

9. Diverges

11. Converges

13. Diverges

15. Diverges

17. Converges

19. Diverges

21. Diverges

23. Converges

25. Converges

27. $\displaystyle\int_{11}^{\infty} \frac{1}{x^2}\,dx \le R_n \le \int_{10}^{\infty} \frac{1}{x^2}\,dx$, so $\frac{1}{11} \le R_n \le \frac{1}{10}$

Exercise Set 12.5

1. Converges

3. Diverges

5. Converges

7. Diverges

9. Converges

11. Diverges

13. Converges

15. Converges. All terms are 0; hence, the sequence of partial sums converges.

17. Diverges. The Limit Comparison Test.

19. Converges. Comparison with geometric series.

21. Diverges. The Limit Comparison Test.

23. Diverges. The Limit Comparison Test.

25. Converges. Comparison with $\sum \dfrac{\pi/2}{k^2}$.

27. Diverges. kth Term Test.

29. Diverges. kth Term Test.

31. Converges. Comparison with $\sum \dfrac{\sqrt{k}}{k^2-1}$.

Exercise Set 12.6

1. Converges. The Ratio Test.

3. Diverges. The Ratio Test or kth Term Test.

5. Converges. The Root Test.

7. Converges. The Ratio Test.

9. Converges. The Limit Comparison Test.

11. Converges. The Root Test.

13. Diverges. The kth Term Test.

15. Diverges. The Ratio Test.

17. Converges. The Ratio Test.

19. Converges. The Ratio Test.

21. Converges. The Ratio Test.

23. Converges. The Root Test.

25. Diverges. The Integral Test.

27. All positive numbers x such that $x < 1$.

29. All positive numbers x.

33. $\displaystyle\lim_{k\to\infty} a_k = \infty$

Exercise Set 12.7

1. Absolutely

3. Absolutely

5. Absolutely

7. Conditionally

9. Conditionally

11. Absolutely

13. Diverges

15. Conditionally

17. Absolutely

19. Absolutely

21. Diverges

23. Absolutely

25. Conditionally

27. Absolutely

29. Absolutely

31. All x

33. All x

35. $n \ge 98$

Review Exercises—Chapter 12

1. Diverges

3. Diverges

5. Converges to 0

7. Diverges

9. Diverges

11. $\frac{6}{5}$

13. 2

15. $37 \displaystyle\sum_{k=1}^{\infty} \frac{1}{100^k} = \frac{37}{99}$

17. $\frac{1}{48}$

19. $\frac{10}{9}$

21. Converges. A telescoping series.

23. Diverges. The Ratio Test.

25. Converges. The Ratio Test.

27. Converges. The Alternating Series Test.

29. Converges. The Limit Comparison Test.

31. Converges. The Comparison Test.

33. Diverges. The Comparison and Integral Tests.

35. Converges. The Ratio Test.

37. Diverges. The Limit Comparison Test.

39. Converges. The Alternating Series Test.

41. Converges. The Root Test.

43. Converges. The Ratio Test.

45. Converges. All terms are 0.

47. Diverges. The Ratio Test.

49. Converges. The Alternating Series Test.

51. Converges. The Ratio Test.

53. Converges. The Alternating Series Test.

55. Converges. The Root Test.

57. Converges. The Root Test.

59. Converges conditionally

61. Diverges   **63.** Converges absolutely

65. Diverges   **67.** Converges absolutely

69. $\frac{1}{13}$ **71.** $\frac{1}{11}$

73. $-1 < c \le 1$ **75.** Σb_k must diverge

77. a. 90.6193 meters   **79. a.** Converges
    b. 110 meters   **b.** Converges
                    c. Converges

Chapter 13

Exercise Set 13.1

1. $1 - x + \dfrac{x^2}{2} - \dfrac{x^3}{6} + \dfrac{x^4}{24}$

3. $\dfrac{\sqrt{2}}{2} - \dfrac{\sqrt{2}}{2}\left(x - \dfrac{\pi}{4}\right) - \dfrac{\sqrt{2}}{4}\left(x - \dfrac{\pi}{4}\right)^2$
$+ \dfrac{\sqrt{2}}{12}\left(x - \dfrac{\pi}{4}\right)^3 + \dfrac{\sqrt{2}}{48}\left(x - \dfrac{\pi}{4}\right)^4$
$- \dfrac{\sqrt{2}}{240}\left(x - \dfrac{\pi}{4}\right)^5 - \dfrac{\sqrt{2}}{1440}\left(x - \dfrac{\pi}{4}\right)^6$

5. $x - \dfrac{x^2}{2} + \dfrac{x^3}{3} - \dfrac{x^4}{4}$

7. $x + \dfrac{x^3}{3}$

9. $-\pi x + \dfrac{\pi^3}{6}x^3 - \dfrac{\pi^5}{120}x^5$

11. $1 + x^2$

13. $1 - \dfrac{x}{2} - \dfrac{x^2}{8} - \dfrac{x^3}{16}$

15. $\sqrt{2} + \sqrt{2}\left(x - \dfrac{\pi}{4}\right) + \dfrac{3\sqrt{2}}{2}\left(x - \dfrac{\pi}{4}\right)^2$
$\qquad\qquad + \dfrac{11\sqrt{2}}{6}\left(x - \dfrac{\pi}{4}\right)^3$

17. $x - \dfrac{x^3}{3}$

19. $\frac{2}{3} + \frac{2}{9}(x - 1) - \frac{10}{27}(x - 1)^2$

21. $3(x + 1)^3 - 9(x + 1)^2 + 11(x + 1) - 4$

25. True

Exercise Set 13.2

1. $e^{-x} = 1 - x + \dfrac{x^2}{2} - \dfrac{x^3}{6} + \dfrac{e^{-c}}{24}x^4$; c is between 0 and x

3. $\cos x = 1 - \dfrac{x^2}{2} + \dfrac{x^4}{24} - \dfrac{\sin c}{120}x^5$; c is between 0 and x

5. $\text{Tan}^{-1} x = x - \dfrac{x^3}{3} + \dfrac{c - c^3}{(1 + c^2)^4}x^4$; c is between 0 and x

7. $\dfrac{1}{1 + x^2} = \dfrac{1}{2} - \dfrac{1}{2}(x - 1) + \dfrac{1}{4}(x - 1)^2$
$\qquad\qquad + 4\dfrac{c - c^3}{(1 + c^2)^4}(x - 1)^3$; c is between 1 and x

9. $\sec x = \sqrt{2} + \sqrt{2}\left(x - \dfrac{\pi}{4}\right) + \dfrac{3\sqrt{2}}{2}\left(x - \dfrac{\pi}{4}\right)^2$
$\qquad + \dfrac{(\sin c)(5 + \sin^2 c)}{6\cos^4 c}\left(x - \dfrac{\pi}{4}\right)^3$; c is between $\dfrac{\pi}{4}$ and x

11. $\sinh x = x + \dfrac{x^3}{6} + \dfrac{\cosh c}{5!}x^5$; c is between 0 and x

13. $\cosh x = \dfrac{5}{4} + \dfrac{3}{4}(x - \ln 2) + \dfrac{5}{8}(x - \ln 2)^2 + \dfrac{1}{8}(x - \ln 2)^3$
$\qquad + \dfrac{1}{24}(\cosh c)(x - \ln 2)^4$; c is between $\ln 2$ and x

21. $P_n(x) = 1 - x^2 + x^4 - x^6 + \ldots$, using all terms of degree $\le n$.

Exercise Set 13.3

Note: In Problems 1–23, there are many ways to handle the R_n and make accuracy estimates.

1. $\ln(1.5) \approx \frac{5}{12}$,
   error $< \frac{1}{64}$

3. $\sin 80° \approx 0.984769$,
   error < 0.0000387

5. $e^{0.2} \approx 1.22133$,
   error < 0.0001

7. $\sqrt{9.2} \approx 3.03315$,
   error < 0.0000021

9. $\sqrt[3]{10} \approx \frac{155}{72}$,
   error < 0.001929

11. $|R_2(x)| \le 0.0000208$

13. $|R_1(x)| \le 0.00125$

15. $|R_1(x)| < 0.0000724$

17. $|R_1(x)| < 0.00005$

19. $n = 6$

21. $n = 2$

23. $n = 3$

25. 2.7183, $n = 8$

27. $|R_1(x)| < 0.005483$

Exercise Set 13.4

1. $[-1, 1)$

3. $(-\infty, \infty)$

5. $(-\infty, \infty)$

7. $[-1, 1)$

9. $(-1, 1]$

11. $(-1, 1)$

13. $[-1, 1]$

15. $(-\infty, \infty)$

17. Converges only for $x = 1$.

19. $(\frac{1}{3}, 1]$

21. $(-1, 1)$

23. $[2, 4]$

25. $[-\frac{1}{2}, \frac{1}{2}]$

27. $(2 - e, 2 + e)$

29. $(-\frac{3}{7}, \frac{1}{7})$

31. $(-1, 1)$

33. $[-2, 4)$

35. $[-e, e)$

37. $(-\infty, \infty)$

39. $[-1, 1]$

Exercise Set 13.5

1. $\sum_{k=0}^{\infty} 2^k x^k$; $r = \frac{1}{2}$

3. $\sum_{k=0}^{\infty} x^{k+2}$; $r = 1$

5. $\sum_{k=0}^{\infty}(-1)^k 4^k x^{2k}$; $r = \frac{1}{2}$

7. $\sum_{k=0}^{\infty}(-1)^k x^{2k+1}$; $r = 1$

9. $\sum_{k=0}^{\infty} x^{4k}$; $r = 1$

11. $\sum_{k=0}^{\infty} \frac{x^{2k}}{9^{k+1}}$; $r = 3$

13. $-1 + \sum_{k=1}^{\infty}(-1)^{k+1} 2 x^k$; $r = 1$

15. $\sum_{k=1}^{\infty}(-1)^{k+1}(2k) x^{k-1}$; $r = 1$

17. $\sum_{k=1}^{\infty} k(-1)^{k+1} x^{2k-1}$; $r = 1$

19. $\sum_{k=0}^{\infty}(2k + 1) x^{2k}$; $r = 1$

21. $\sum_{k=1}^{\infty} 2k(-1)^{k+1} 4^k x^{2k-1}$; $r = \frac{1}{2}$

23. $\sum_{k=1}^{\infty} 2k(-1)^{k-1} x^{k-1}$; $r = 1$

25. $-\sum_{k=0}^{\infty} \frac{x^{k+1}}{k + 1}$; $r = 1$

27. $2\sum_{k=0}^{\infty} \frac{(-1)^k 4^k x^{2k+1}}{2k + 1}$; $r = \frac{1}{2}$

29. $\ln 4 + \sum_{k=0}^{\infty} \frac{(-1)^k x^{k+1}}{4^{k+1}(k + 1)}$; $r = 4$

31. $\sum_{k=0}^{\infty}(-1)^k \frac{x^{2k+2}}{k + 1}$; $r = 1$

Exercise Set 13.6

1. $\sum_{k=0}^{\infty} \frac{2^k x^k}{k!}$; all x

3. $\sum_{k=0}^{\infty} \frac{(-1)^k x^{2k+1}}{(2k + 1)!}$; all x

5. $\frac{\sqrt{2}}{2} - \frac{\sqrt{2}}{2}\left(x - \frac{\pi}{4}\right) - \frac{\sqrt{2}}{2(2!)}\left(x - \frac{\pi}{4}\right)^2$
   $+ \frac{\sqrt{2}}{2(3!)}\left(x - \frac{\pi}{4}\right)^3 + \frac{\sqrt{2}}{2(4!)}\left(x - \frac{\pi}{4}\right)^4 - - + \cdots$
   $\sum_{k=0}^{\infty}\left(\frac{\sqrt{2}}{2}\right)\frac{(-1)^{k(k+1)/2}}{k!}\left(x - \frac{\pi}{4}\right)^k$; all x

7. $5 + 4(x - 2) + (x - 2)^2$; all x

9. $\sum_{k=0}^{\infty}(-1)^k x^k$; $(-1, 1)$

11. $\sum_{k=0}^{\infty} \frac{(-1)^k 2^{2k+1} x^{2k+2}}{(2k + 1)!}$; all x

13. $\sum_{k=0}^{\infty} \frac{x^k (\ln 2)^k}{k!}$; all x

15. $1 + \sum_{k=1}^{\infty} \frac{\frac{3}{2}(\frac{3}{2} - 1)\cdots(\frac{3}{2} - k + 1) x^k}{k!}$; $|x| < 1$

17. $2 + \frac{1}{4}(x - 4) - \frac{1}{(2^5)2!}(x - 4)^2 + \ldots$; $(0, 8)$

19. $\sum_{k=0}^{\infty} \frac{(-1)^k (x - 2)^k}{2^{k+1}}$; $(0, 4)$

21. $\sum_{k=0}^{\infty} \frac{(-1)^k x^{k+2}}{k!}$; all x

23. $\displaystyle\sum_{k=0}^{\infty}\frac{(-1)^k x^{k+3}}{k+1}$; $(-1,1]$

25. $\displaystyle\frac{1}{2}\left[1-\sum_{k=0}^{\infty}\frac{(-1)^k 2^{2k}x^{2k}}{(2k)!}\right]$;   all x

27. $1+x^2$. Use $\dfrac{d}{dx}(\tan x)=\sec^2 x$.

29. $\displaystyle\sum_{k=0}^{\infty}\frac{(-1)^k x^{4k}}{(2k)!}$

31. $\displaystyle\sum_{k=0}^{\infty}\frac{x^{2k+1}}{(2k)!}$

33. 7.389   **35.** 0.157   **37.** 0.946   **39.** 0.747

41. $\displaystyle 1+x-\sum_{k=2}^{\infty}(-1)^k\frac{1\cdot 3\cdots(2k-3)}{k!}x^k$; $|x|<\frac{1}{2}$

43. $\displaystyle 27+\sum_{k=1}^{\infty}\frac{\frac{3}{2}(\frac{3}{2}-1)\cdots(\frac{3}{2}-k+1)x^k}{3^{k-3}\,k!}$; $|x|<3$

Review Exercises—Chapter 13

1. $2x-\frac{4}{3}x^3+\frac{4}{15}x^5$

3. $\dfrac{\pi\sqrt{2}}{8}+\dfrac{\sqrt{2}}{2}\left(1-\dfrac{\pi}{4}\right)\!\left(x-\dfrac{\pi}{4}\right)$
   $-\dfrac{\sqrt{2}}{4}\left(2+\dfrac{\pi}{4}\right)\!\left(x-\dfrac{\pi}{4}\right)^2+\dfrac{\sqrt{2}}{12}\left(\dfrac{\pi}{4}-3\right)\!\left(x-\dfrac{\pi}{4}\right)^3$

5. $1+x^2$

7. $\dfrac{\pi}{4}+\dfrac{1}{2}(x-1)-\dfrac{1}{4}(x-1)^2+\dfrac{1}{12}(x-1)^3$

9. $1+(\ln 2)x+\dfrac{(\ln 2)^2}{2}x^2+\dfrac{(\ln 2)^3}{6}x^3$

11. $\left|R_3\!\left(\dfrac{17\pi}{90}\right)\right|\le 6.999\cdot 10^{-7}$

13. $|R_2(35)|\le 0.000079$

15. $|R_3(\tfrac{1}{4})|<0.00021$

21. $(2,4)$ **23.** $[1,3)$ **25.** $(-1,1]$

27. $\displaystyle\sum_{k=0}^{\infty}x^{2k+1}$; $r=1$

29. $\displaystyle\ln 4+\sum_{k=0}^{\infty}\frac{(-1)^k x^{k+1}}{4^{k+1}(k+1)}$; $r=4$

31. $\displaystyle\sum_{k=0}^{\infty}\frac{(-1)^k(4k+3)x^{4k+2}}{(2k+1)!}$; $r=\infty$

33. Replace $k-1$ by k on the right.

35. $a_n=\dfrac{c^n a_0}{n!}(-1)^n$; $a_0 e^{-cx}$ **37.** $\Sigma_{k=0}^{\infty}(-1)^k x^{4k}$; $|x|<1$

39. $\displaystyle 1+\sum_{k=1}^{\infty}\frac{\frac{1}{3}(\frac{1}{3}-1)\cdots(\frac{1}{3}-k+1)x^{2k}}{k!}$

41. $1-\dfrac{1}{2\pi^2}\left(x-\dfrac{\pi^2}{4}\right)^2+\dfrac{1}{\pi^4}\left(x-\dfrac{\pi^2}{4}\right)^3$

43. $\displaystyle\sum_{k=0}^{\infty}\frac{x^{2k+2}}{k!}$ **45.** $e^{1/2}\approx 1.649$

Chapter 14

Exercise Set 14.1

	Vertex	Focal Length	Axis	Focus
1.	$(0,0)$	$\frac{1}{8}$	$x=0$	$(0,\frac{1}{8})$
3.	$(0,2)$	$-\frac{1}{8}$	$x=0$	$(0,\frac{15}{8})$
5.	$(1,0)$	$\frac{1}{12}$	$y=0$	$(\frac{13}{12},0)$
7.	$(0,\frac{1}{4})$	-2	$x=0$	$(0,-\frac{7}{4})$
9.	$(2,1)$	2	$x=2$	$(2,3)$
11.	$(-3,2)$	$\frac{3}{2}$	$y=2$	$(-\frac{3}{2},2)$
13.	$(-2,1)$	1	$y=1$	$(-1,1)$
15.	$(-1,-2)$	$\frac{3}{2}$	$x=-1$	$(-1,-\frac{1}{2})$
17.	$(-3,1)$	$\frac{5}{2}$	$x=-3$	$(-3,\frac{7}{2})$
19.	$(1,-2)$	$\frac{1}{2}$	$y=-2$	$(\frac{3}{2},-2)$

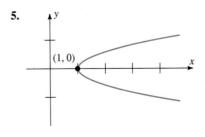

7.

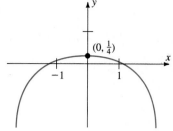

$(0, \frac{1}{4})$

9.

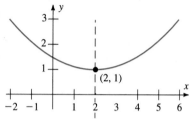

$(2, 1)$

11.

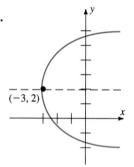

$(-3, 2)$

13.

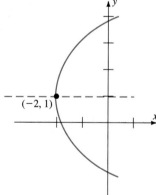

$(-2, 1)$

15.

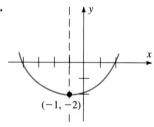

$(-1, -2)$

17.

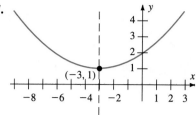

$(-3, 1)$

19.

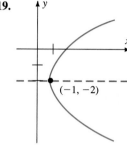

$(-1, -2)$

21. $x^2 = 12y$ **23.** $(x - 1)^2 = 32(y - 3)$

25. $(y + 6)^2 = 16(x + 1)$ **27.** $x^2 = 16(y - 1)$

33. $\frac{8}{3}$ **35.** $\frac{32}{2}$

37. $-\frac{41}{48}$

Exercise Set 14.2

	Center	Major Axis	Minor Axis	Foci
1.	$(0, 0)$	6	2	$(\pm 2\sqrt{2}, 0)$
3.	$(0, 0)$	8	4	$(\pm 2\sqrt{3}, 0)$
5.	$(0, 0)$	10	6	$(0, \pm 4)$
7.	$(2, 1)$	4	$2\sqrt{3}$	$(2, 0), (2, 2)$
9.	$(-3, 1)$	8	4	$(-3 \pm 2\sqrt{3}, 1)$
11.	$(1, -1)$	6	4	$(1, -1 \pm \sqrt{5})$
13.	$(-3, 1)$	10	6	$(1, 1), (-7, 1)$
15.	$(1, -1)$	$2\sqrt{3}$	$2\sqrt{2}$	$(0, -1), (2, -1)$

1.

$(0, 1)$ $(3, 0)$

3.

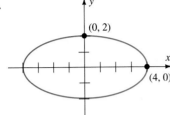

5.

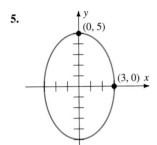

7.

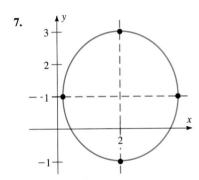

9.

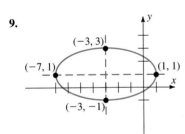

11.

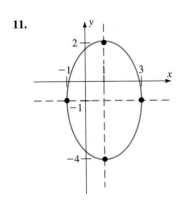

13.

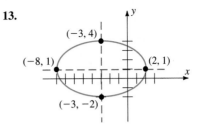

15.

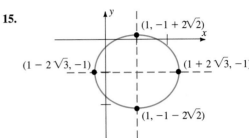

17. $\dfrac{x^2}{16} + \dfrac{y^2}{9} = 1$ **19.** $\dfrac{x^2}{25} + \dfrac{y^2}{9} = 1$

21. $\dfrac{(x-4)^2}{16} + \dfrac{(y-2)^2}{25} = 1$

23. $\dfrac{(x+2)^2}{16} + \dfrac{(y-6)^2}{25} = 1$

25. 50π **27.** $\dfrac{2b^2}{a}$ **29.** True. If $a = b$, we have a circle.

33. $\dfrac{(x+1)^2}{36} + \dfrac{(y-3)^2}{27} = 1$

35. 48π **37.** $1{,}809{,}450$ ft-lb

Exercise Set 14.3

	Center	Asymptotes
1.	$(0, 0)$	$y = \pm\frac{3}{2}x$
3.	$(0, 0)$	$y = \pm\dfrac{3}{\sqrt{5}}x$
5.	$(0, 0)$	$y = \pm 2x$
7.	$(0, 0)$	$y = \pm\dfrac{\sqrt{7}}{2}x$
9.	$(-3, 4)$	$y - 4 = \pm\frac{3}{2}(x + 3)$
11.	$(2, -3)$	$y + 3 = \pm\frac{3}{2}(x - 2)$
13.	$(0, 0)$	$y = \pm\frac{1}{2}x$
15.	$(\sqrt{2}, \sqrt{2})$	$y - \sqrt{2} = \pm\sqrt{2}(x - \sqrt{2})$
17.	$(-2, 3)$	$y - 3 = \pm\frac{4}{5}(x + 2)$
19.	$(-5, 7)$	$y - 7 = \pm\frac{4}{3}(x + 5)$

1.

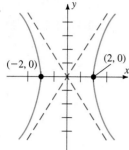

3.

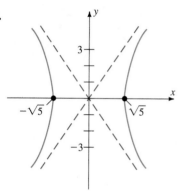

5.

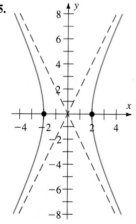

7.

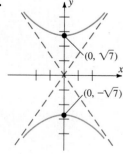

9.

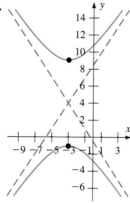

11.

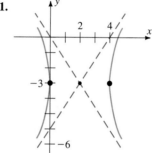

13.

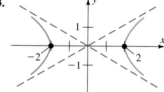

15.

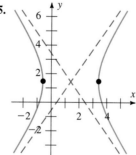

17.

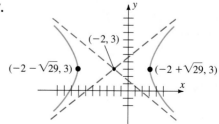

19.

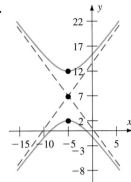

21. $x^2 - \dfrac{y^2}{3} = 1$

23. $\dfrac{(y-2)^2}{4} - x^2 = 1$

25. $\dfrac{(x+1)^2}{16} - \dfrac{(y-2)^2}{64} = \dfrac{1}{5}$

27. $\dfrac{y^2}{16} - \dfrac{x^2}{9} = 1$

29. They have the same asymptotes: $y = \pm\dfrac{b}{a}x$.

31. $b^2 x_0 x - a^2 y_0 y = a^2 b^2$

Exercise Set 14.4

1. $\theta = \dfrac{\pi}{8}$

3. $\theta = \dfrac{\pi}{4}$

5. $\theta = \dfrac{1}{2}\text{Tan}^{-1}\dfrac{4}{3}$

7. $\theta = \dfrac{\pi}{4}, \quad \dfrac{x'^2}{4} + \dfrac{y'^2}{16} = 1$

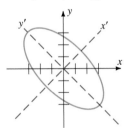

9. $\theta = \dfrac{\pi}{6}, \quad x'^2 = -(y' + 4)$

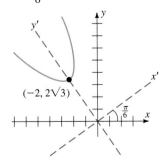

$(-2, 2\sqrt{3})$

11. $\theta = \text{Sin}^{-1}\dfrac{3}{5}, \quad y'^2 = 2(x' - 2)$

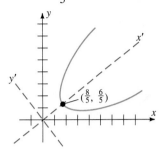

$\left(\dfrac{8}{5}, \dfrac{6}{5}\right)$

13. $\theta = -\dfrac{\pi}{3}, \quad \dfrac{x'^2}{4} + \dfrac{y'^2}{2} = 1$

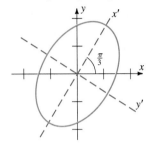

$\dfrac{\pi}{3}$

15. $\theta = \dfrac{\pi}{4}, \quad x'^2 = y'$

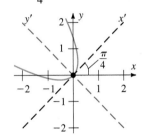

$\dfrac{\pi}{4}$

17. $\left(\mp\dfrac{3}{2}, \pm\dfrac{3\sqrt{3}}{2}\right)$

19. $\left(\pm\dfrac{3}{2}, \pm\dfrac{3\sqrt{3}}{2}\right)$

Review Exercises—Chapter 14

1. $(x-3)^2 + (y+4)^2 = 36$

3. $(x-2)^2 + (y-3)^2 = 16$

5. $(y-4)^2 = 8x$

7. $y = 3x^2 - 6x + 5$

9. $\dfrac{(x+1)^2}{9} + \dfrac{(y-1)^2}{5} = 1$

11. Outside

13. $(-3, -5)$ and $(2, 0)$

15. $(2, 2)$ and $(-2, 2)$

17. $(x - 1)^2 + (y + 3)^2 = 8$,   circle

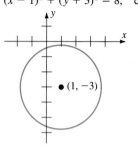

19. $\dfrac{y^2}{16} - \dfrac{x^2}{7} = 1$,   hyperbola

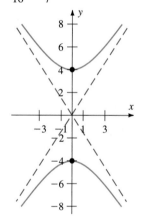

21. $\dfrac{x^2}{12} + \dfrac{y^2}{16} = 1$,   ellipse

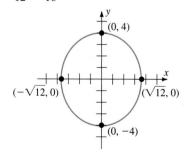

23. $\dfrac{y^2}{4} - \dfrac{x^2}{2} = 1$,   hyperbola

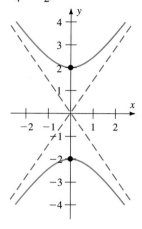

25. $\dfrac{x^2}{9} - \dfrac{y^2}{4} = 1$,   hyperbola

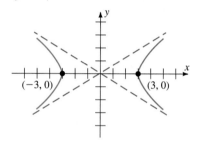

27. $y^2 = -(x - 2)$,   parabola

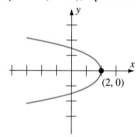

29. $(x + 7)^2 + (y - 5)^2 = 4$,   circle

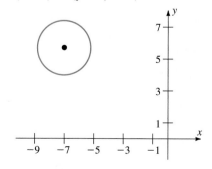

31. $\dfrac{x^2}{8} - \dfrac{y^2}{9} = 1$,   hyperbola

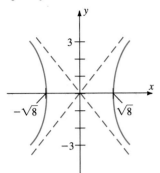

33. $\dfrac{x^2}{25} + \dfrac{y^2}{4} = 1$,   ellipse

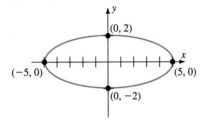

35. $\dfrac{y^2}{4} - \dfrac{x^2}{4} = 1$,   hyperbola

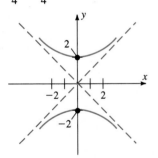

37. $(x - a)^2 + (y + 2a)^2 = 1$,   circle (assume $a > 0$)

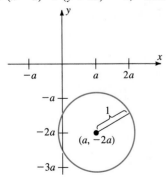

39. $y^2 - x^2 = 2$,   hyperbola

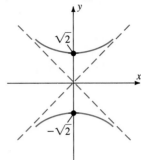

41. $(x - 1)^2 + (y - 3)^2 = 7$,   circle

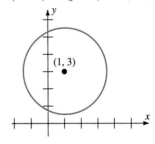

43. $\dfrac{y^2}{25} - \dfrac{x^2}{9} = 1$,   hyperbola

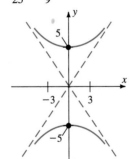

Chapter 15

Exercise Set 15.1

1. $P = (0, 1)$ **3.** $P = (0, 0)$

5. $P = (1, -1)$ **7.** $P = (3, 0)$

9. $P = (\sqrt{2}, \pi/4)$ **11.** $P = (-3, 0)$

13. $P = (-2, 2\pi/3)$ **15.** $P = (-2\sqrt{2}, 15\pi/4)$

17. Symmetric about the x-axis

19. Symmetric about the y-axis

21. x-axis, origin, y-axis

23. x-axis, y-axis, origin

25. $r = 2$

27. $r^2 = \dfrac{4}{1 + 3\cos^2\theta}$

29. $r^2 = (1 + \sin^2\theta)^{-1}$

31. $r^2 = \sec 2\theta$

33. $r = \dfrac{6}{\cos\theta} = 6\sec\theta$

35. $r = -2\sin\theta$

37. $\tan\theta = 1/2$

39. $x^2 + y^2 - 4y = 0$

41. $x^4 + x^2y^2 - y^2 = 0$

43. $x = 4$

45. $y^2 = 1 + 2x$

47. $x^4 + x^2y^2 - 16 = 0$

55.

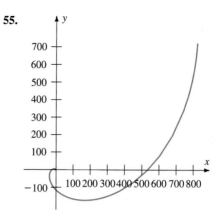

49.

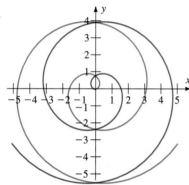

(blue: r > 0; red: r < 0)

57.

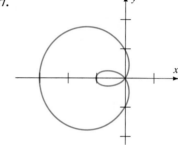

59. Center $(-1, 1)$ and radius $\sqrt{2}$

61. Center $(a/2, 0)$ and radius $a/2$

65. $9\pi/4$

67. $(0, 0), (4, 0), (4, 4)$

Exercise Set 15.2

51.

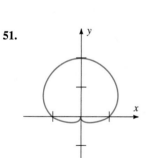

1.

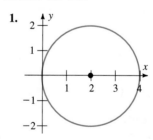

53.

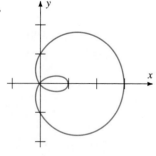

3.

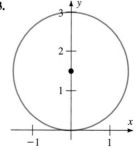

5.

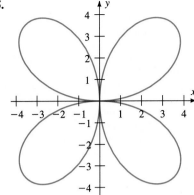

7.

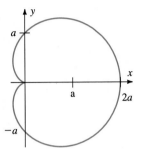

9.

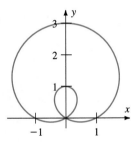

11.

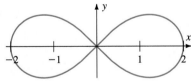

13.

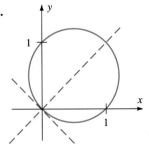

15. $(0, \theta)$, $(\sqrt{2}, \pi/4)$

17. $(3/2, \pi/6)$, $(3/2, 5\pi/6)$, $(0, \theta)$

19. $(3a/2, \pi/3)$, $(3a/2, 5\pi/3)$, $(0, \theta)$

21. $(0, \theta)$, $(1, 0)$, $(1, \pi)$

23. $(\sqrt{2}/2, \pi/8)$, $(\sqrt{2}/2, -\pi/8)$

25. $(0, \theta)$, $(1, \pi/2)$

29. $r = 1/(4 + 4 \sin \theta)$

31. $r = 2/(2 - \cos \theta)$

33. $y^2 = 1 - 2x$

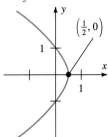

35. $\dfrac{(x + 4/3)^2}{(8/3)^2} + \dfrac{y^2}{(4/\sqrt{3})^2} = 1$ (ellipse)

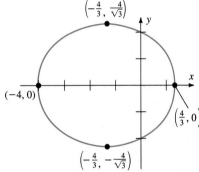

37. $y^2 = 4x + 4$ (parabola)

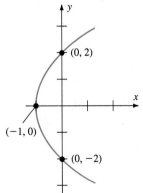

Exercise Set 15.3

1. $3\pi/4$

3. π

5. $\frac{9}{4}\pi + 4$

7. $\frac{33}{8}\pi + 4\sqrt{2} + \frac{1}{4}$

9. π

13. 8π

17. $\pi a^2/2$

21. $\pi/2$

25. $\dfrac{\pi - 2}{2}$

27. $\pi/6 + 1 - \sqrt{3}/2$

29. $11\pi/2$

31. $8/3$

11. $3\pi/2$

15. $\pi/2$

19. 2

23. $\frac{3}{2}a^2\pi - 4a^2$

Exercise Set 15.4

1. $y = x + 6$

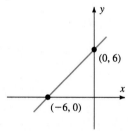

3. $y = \frac{2}{3}x + \frac{4}{3}$

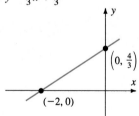

5. $y = x - 2; \quad x \geq 1$

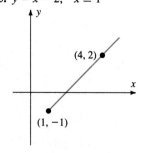

7. $y = x^3; \quad x > 0$

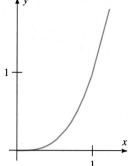

9. $x^2 - y^2 = 1$

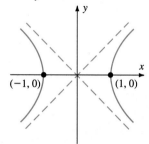

11. $y^2 = 4x^2(1 - x^2)$

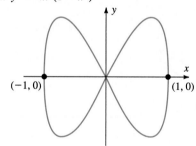

13. Slope $= 4, \quad y = 4x - 3$

15. Slope $= -\sqrt{3}, \quad y = -\sqrt{3}x + 2$

17. Slope $= -48, \quad y = -48x + 29$

19. Slope $= -1, \quad y = -x + 2$

21. Slope $= 1/3, \quad y = x/3 + 4/3$

23. Slope $= 2/e, \quad y = \dfrac{2}{e}x - 2$

25. 0 **27.** $5/\sqrt{3}$ **29.** $-2/\pi$

31. Vertical tangent: $(1, 0), (-1, 0)$
Horizontal tangent: $(0, 1), (0, -1)$

33. Vertical tangent at $(4, 2)$
Horizontal tangent at $(5, -1)$

35. Vertical tangent at $(6, 0)$
Horizontal tangent at $(27/4, 1/4)$

37. $(1, 5)$

39. $x(\theta) = f(\theta) \cos \theta = a \sin 3\theta \cos \theta$
$y(\theta) = f(\theta) \sin \theta = a \sin 3\theta \sin \theta$

41. $(3/2, \pi/6), (3/2, 5\pi/6)$

43. Vertical tangent at $(2, 0), (1/2, 2\pi/3), (1/2, 4\pi/3)$

45. a. $y = -(x^2 - 8x + 8)$
$y = x + 2$
b. Yes. $(2, 4)$ and $(5, 7)$.
c. The particles collide at $t = -2$ and at $t = 1$ at the points $(2, 4)$ and $(5, 7)$.

47. a. $x^2 + y^2 = 1$
$x = 0, -1 \le y \le 1$
b. Yes. $(0, -1)$ and $(0, 1)$.
c. They do not collide.

Exercise Set 15.5

1. $\frac{8}{27}(2^{3/2} - 1)$ **3.** $2(3^{3/2} - 1)$ **5.** 36

7. 3   **9.** $\sqrt{2}(e^\pi - 1)$

11. $(5^{3/2} - 8^{3/2}) + (12\sqrt{2} - 3\sqrt{5} + 12 \ln(\sqrt{5} - 1)$
$- 12 \ln(\sqrt{8} - 2))$

13. $\frac{1}{3}(17^{3/2} - 5^{3/2})$ **15.** $7\sqrt{2}/3$

17. $\frac{16\pi}{3}(8^{3/2} - 8)$ **19.** $\frac{\pi}{6}(27 - 5^{3/2})$

21. $\pi[133\sqrt{5} - \frac{7}{2}\sqrt{2} + \frac{3}{2}\ln(\sqrt{5} + 2) - \frac{3}{2}\ln(\sqrt{2} + 1)]$

23. 4   **25.** $\frac{\sqrt{5}}{2}(e^{2\pi} - 1)$ **27.** 4π

29. 3π **31.** $\sqrt{2}(e^{2\pi} - 1)$ **33.** $\frac{2}{5}\sqrt{2}\pi(e^{2\pi} + 1)$

35. $S = \int_a^b 2\pi x(t)\sqrt{(x'(t))^2 + (y'(t))^2}\, dt, \quad a \le t \le b$

37. $\frac{\pi}{64}[204\sqrt{2} - 36\sqrt{5} - \ln(17 + 12\sqrt{2}) + \ln(9 + 4\sqrt{5})]$

Review Exercises—Chapter 15

1. $x^2 + y^2 = 25$

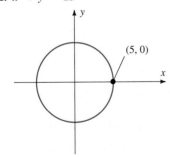

3. $r = \theta/\pi$

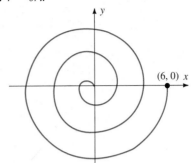

5. $y = 2x\sqrt{1 - x^2}$

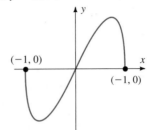

7. $y = x \tan \pi\sqrt{x^2 + y^2}$ **9.** $y^2 = 4x^2(1 - x^2)$

11. $\begin{cases} x(t) = 3 \cos t \\ y(t) = 2 \sin t \end{cases}$ **13.** $\begin{cases} x(t) = 2t^2 + 4t + 5 \\ y(t) = t \end{cases}$

15.

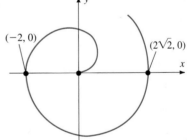

17. $x = \frac{y^2}{9} - 1$ (parabola)

19.

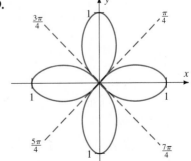

21.

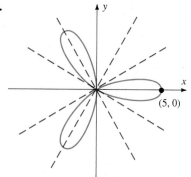

23.

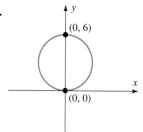

25.

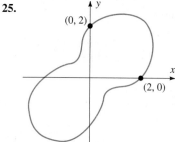

27. The y-axis

29. $r = 2a \cos \theta$

31. $r = -2a \sin \theta$

33. 6π

35. $\frac{3}{2}a^2\pi$

37. 9π

39. $\frac{1}{27}(22^{3/2} - 13^{3/2})$

41. $2[\sqrt{2} + \ln(1 + \sqrt{2})]$

43. π

45. $\pi/2$

47. $2\sqrt{2}$

49. $\dfrac{\pi}{8} - \dfrac{1}{4}$

51. 4π

53. $\dfrac{8\pi}{3} - 2\sqrt{3}$

55. $y = -2x - 9$

57. They cross at $(0, 0)$ and $(-1/4, -1/2)$.

They collide at $(0, 0)$.

As $t \to \infty$, $p \to (-\infty, -\infty)$ and $q \to (-\infty, -\infty)$.

As $t \to -\infty$, $p \to (+\infty, +\infty)$ and $q \to (-\infty, +\infty)$.

Chapter 16

Exercise Set 16.1

1.

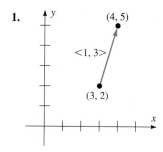

3.

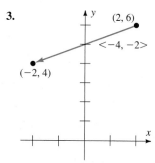

5.

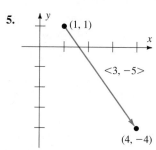

7. $\langle -1, 2 \rangle$

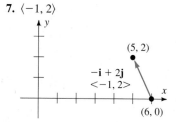

9. $\langle -3, 2 \rangle$

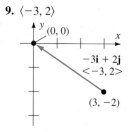

11. $\langle -3, -3 \rangle$

13. $\langle -3, -14 \rangle$

15. $\langle -6, -6 \rangle$

17. $\langle -1, 9 \rangle$

19. $\langle 10, 14 \rangle$

21. $\langle -1, 5 \rangle$

23. $\langle 7, 11 \rangle$

25. $\langle 4, 6 \rangle$

27. $\langle 19, 19 \rangle$

29. $\langle 12, 18 \rangle$

31.

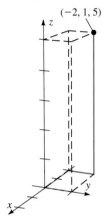

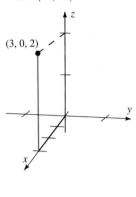

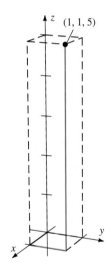

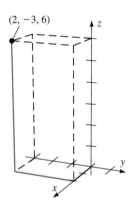

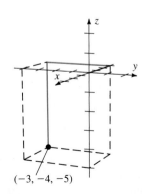

33. $\langle -16, 29, 5 \rangle$

35. $\langle 12, -13, 15 \rangle$

37. $9\sqrt{6}$

39. 5

41. 1

43. 5

45. $4\sqrt{10}$

47. $D = (1, -2, 1)$

49. a. $\sqrt{14}$ **b.** $\sqrt{14}$ **c.** $\sqrt{26}$ **d.** $\sqrt{6}$
   e. $\sqrt{26}$ **f.** $2\sqrt{26}$ **g.** $\sqrt{50}$

51. $-\mathbf{i}$

53. $-(1/2)\mathbf{i} + (\sqrt{3}/2)\mathbf{j}$

55. $(1/\sqrt{26})\mathbf{i} + (4/\sqrt{26})\mathbf{j} + (3/\sqrt{26})\mathbf{k}$

57. $-12\mathbf{i} + 1\mathbf{j} - 7\mathbf{k}$

59. $\left(\dfrac{3}{\sqrt{17}} + \dfrac{5}{\sqrt{35}} \right)\mathbf{i} - \left(\dfrac{2}{\sqrt{17}} + \dfrac{1}{\sqrt{35}} \right)\mathbf{j} + \left(\dfrac{2}{\sqrt{17}} + \dfrac{3}{\sqrt{35}} \right)\mathbf{k}$

61. $(1/\sqrt{10})\mathbf{i} + (3/\sqrt{10})\mathbf{j}, \quad -(1/\sqrt{10})\mathbf{i} - (3/\sqrt{10})\mathbf{j}$

63. a. $D = (2, 3)$
   b. $D = (0, -1)$ or $(10, 3)$

69. $(-3, 3, -3)$

73. a. $(2/3)\mathbf{i} - (4/3)\mathbf{j} + (4/3)\mathbf{k}$ and
       $-(2/3)\mathbf{i} + (4/3)\mathbf{j} - (4/3)\mathbf{k}$
   b. $a = -3$
   c. $-\dfrac{5\sqrt{14}}{14}\mathbf{i} + \dfrac{5\sqrt{14}}{7}\mathbf{j} - \dfrac{15\sqrt{14}}{14}\mathbf{k}$
   d. $a = -3/2, \quad b = 1/2$

77. 5 min

79. a. $400\mathbf{N} + 60\mathbf{E}$
   b. $20\sqrt{409}$ km
   c. $\mathrm{Sin}^{-1}\left(\dfrac{3}{20} \right)$ to the west of north

81. $50\sqrt{2}$ lb

Exercise Set 16.2

1. $x(t) = 1 + t, \quad y(t) = 2 + t, \quad z(t) = 3 - t$

3. $x(t) = 3t, \quad y(t) = -t, \quad z(t) = 5t$

5. $x(t) = -4 + t, \quad y(t) = 2 + 3t, \quad z(t) = 1 + 2t$

7. $x(t) = \pi t, \quad y(t) = 2t, \quad z(t) = \sqrt{2}t$

9. $\dfrac{x - 7}{1} = \dfrac{y + 6}{4} = \dfrac{z - 3}{-2}$

11. $\dfrac{x - 0}{1} = \dfrac{y + 6}{8} = \dfrac{z - 4}{4}$

13. $\dfrac{x - 3}{3} = \dfrac{y + 1}{-1} = \dfrac{z - 5}{5}$

15. $x(t) = 3t, \quad y(t) = 2t, \quad z(t) = 5t$

17. $x(t) = 4t - 4, \quad y(t) = -2t + 2, \quad z(t) = 3t - 3$

19. $\mathbf{i} + 4\mathbf{j} - 2\mathbf{k}$

21. $4\mathbf{i} - 2\mathbf{j} + 3\mathbf{k}$

23. $(2, 1, 2)$

25. $(4, -4, 8)$

27. They do not intersect.

29. $x(t) = t + 1$, $y(t) = 8t + 6$, $z(t) = 4t - 5$

31. $\dfrac{x - 1}{3} = \dfrac{y - 2}{2} = \dfrac{z - 1}{5}$

33. $\mathbf{r}(t) = \langle 1, 2, 4 \rangle + t\langle 1, 4, -2 \rangle$

35. ℓ_1 and ℓ_4 are the same line.
ℓ_3 and ℓ_5 are parallel.

Exercise Set 16.3

1. 1   **3.** -28 **5.** -3

7. 2   **9.** $-\dfrac{7}{\sqrt{170}}$ **11.** $5\sqrt{3}/9$

13. a. 14   **b.** -28 **c.** -14
d. 42   **e.** 78   **f.** -56
g. $\dfrac{14}{\sqrt{34}}$ **h.** $-\dfrac{1}{\sqrt{13}}$ **i.** $\langle -\tfrac{7}{5}, \tfrac{14}{5} \rangle$
j. $\langle \tfrac{16}{13}, \tfrac{24}{13} \rangle$

15. a. $-4/3$ **b.** 2
c. $3/4$ **d.** $-4/3$

17. 1   **19.** $\tfrac{18}{19}\mathbf{i} + \tfrac{18}{19}\mathbf{j} + \tfrac{6}{19}\mathbf{k}$

21. $6\sqrt{19}/19$ **23.** $2\mathbf{i} - \mathbf{j} + 5\mathbf{k}$

25. 23   **27.** $\tfrac{23}{7}\mathbf{i} - \tfrac{23}{35}\mathbf{j} + \tfrac{69}{35}\mathbf{k}$

29. $-\sqrt{6}/2$

31. **b** and **c** are orthogonal pairs.

33. $(\sqrt{5}/5)\mathbf{i} - (2\sqrt{5}/5)\mathbf{j}$, $-(\sqrt{5}/5)\mathbf{i} + (2\sqrt{5}/5)\mathbf{j}$

35. **a** and **c**

37. $x + 2y - z = 8$

39. $2x - 2y + 3z = 19$

41. $x + 2y + z = 1$

43. $\pm\sqrt{6}(\tfrac{1}{6}\mathbf{i} + \tfrac{1}{3}\mathbf{j} + \tfrac{1}{6}\mathbf{k})$

45. $(\tfrac{4}{5}\mathbf{i} - \tfrac{2}{5}\mathbf{j}) + (\tfrac{1}{5}\mathbf{i} + \tfrac{2}{5}\mathbf{j})$

47. $4\sqrt{2}$

49. $3x - 6y + 2z = 31$

51. $\theta = \mathrm{Cos}^{-1}(4/\sqrt{42})$

53. $27\sqrt{38}/38$

55. $x(t) = 2 + 2t$, $y(t) = 4 + 3t$, $z(t) = -3 - 7t$

57. $1000\sqrt{2}$

59. a. $\sqrt{14}/7$ **b.** $\sqrt{41}/41$ **c.** $-4\sqrt{21}/21$

61. $(\sqrt{2}/2)\mathbf{i} + (1/2)\mathbf{j} \pm (1/2)\mathbf{k}$

69. $\left(x - \dfrac{a_1 + a_2}{2}\right)^2 + \left(y - \dfrac{b_1 + b_2}{2}\right)^2 =$
$$\dfrac{(a_1 + a_2)^2 + (b_1 + b_2)^2 - 4(a_1a_2 + b_1b_2)}{4}$$

71. σ_1 and σ_5 are parallel.
σ_2 and σ_4 are equal.

Exercise Set 16.4

1. $-7\mathbf{i} + 8\mathbf{j} + 11\mathbf{k}$ **3.** $5\mathbf{i} - 4\mathbf{j} - \mathbf{k}$

5. 12   **7.** $3\mathbf{i} + 10\mathbf{j} + \mathbf{k}$

9. $-16\mathbf{i} + 8\mathbf{k}$ **11.** $37\mathbf{i} + 30\mathbf{j} - 11\mathbf{k}$

13. 40   **15.** $5\sqrt{3}$

17. 8   **19.** $\tfrac{5}{2}\sqrt{2}$

21. $\dfrac{5\sqrt{10}}{2}$ **23.** $\dfrac{23}{2}$

25. **k** or $-\mathbf{k}$

27. $\dfrac{\mathbf{i}}{\sqrt{3}} - \dfrac{\mathbf{j}}{\sqrt{3}} - \dfrac{\mathbf{k}}{\sqrt{3}}$ and $-\dfrac{\mathbf{i}}{\sqrt{3}} + \dfrac{\mathbf{j}}{\sqrt{3}} + \dfrac{\mathbf{k}}{\sqrt{3}}$

29. $\dfrac{x - 1}{2} = \dfrac{2 - y}{2}$, $z = 2$

31. $\dfrac{\sqrt{210}}{3}$ **33.** $\dfrac{\sqrt{1270}}{3}$

35. $\dfrac{5\sqrt{146}}{146}$ **37.** $29x - 6y - 15z = -16$

39. $x - y = -2$ **41.** $2\mathbf{i} + 11\mathbf{j} - 7\mathbf{k}$

43. $x = 7t + 1$, $y = 5t - 1$, $z = 6t - 3$

45. **a** and **c** are coplanar.

47. $y = z$ **51.** 53

Review Exercises—Chapter 16

1. $\sqrt{86}$

3. a. 2   **b.** $2\mathbf{i} - 3\mathbf{j} + \mathbf{k}$ **c.** $-2\mathbf{i} + 3\mathbf{j} - \mathbf{k}$
d. $\sqrt{2}$ **e.** $\mathbf{w} = \mathbf{j} - \mathbf{k}$ **f.** -4

5. $\dfrac{-1}{\sqrt{77}}$

7. $x(t) = 23t + 2$, $y(t) = -t + 1$, $z(t) = -9t - 3$

9. $7x - 9y + 3z = -22$

11. $8\mathbf{i} + \mathbf{j} - 2\mathbf{k}$

13. $\mathbf{r}(t) = (2 - 12t)\mathbf{i} + 11t\mathbf{j} + 9t\mathbf{k}$

17. $\dfrac{x - 1}{1} = \dfrac{y - 2}{1} = \dfrac{z + 1}{2}$

19. $-3x + 4y + z = 2$

21. $3x - 4y + 2z = 2$

23. $\sqrt{1581/30} \approx 7.259$

25. $(\sqrt{26}/26)\mathbf{i} - (3\sqrt{26}/26)\mathbf{j} + (2\sqrt{26}/13)\mathbf{k}$

27. $\dfrac{x - 1}{3} = \dfrac{y - 2}{7} = \dfrac{z - 3}{-1}$

29. a. $10\sqrt{11}/11$
  b. $\frac{10}{7}\mathbf{i} + \frac{15}{7}\mathbf{j} - \frac{5}{7}\mathbf{k}$

31. $5\sqrt{3}/3$

33. $\dfrac{x + 3}{-2} = \dfrac{y - 7}{3} = -z$

35. $3x + 2y - 2z = -1$

37. $\pm\dfrac{2\sqrt{3}}{15}(-5\mathbf{i} + 7\mathbf{j} + \mathbf{k})$

Chapter 17

Exercise Set 17.1

1.

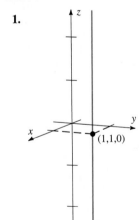

(1,1,0)

3.

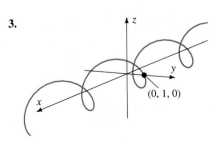

(0, 1, 0)

5.

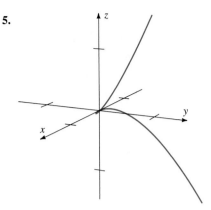

7.

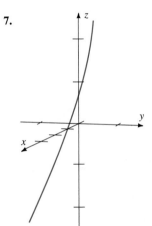

9. $t \geq 0$ **11.** $t < 1$

13. $t \neq \pi/2 + n\pi,\ n = 0,\ \pm 1,\ \ldots$ **15.** $0 < t \leq 3$

17. $(t + \sin t)\mathbf{i} + (3t + \cos t)\mathbf{j} + (t^2 + 1)\mathbf{k}$

19. $(3t + 2 \sin t)\mathbf{i} + (9t + 2 \cos t)\mathbf{j} + (3t^2 + 2)\mathbf{k}$

21. $t \sin t + 3t \cos t + t^2$

23. $(1 + t + 3\sqrt{t})\mathbf{i} + (4 - 3t - t^2)\mathbf{j} + (t + 3e^t)\mathbf{k}$

25. $(1 + t - \sqrt{t}) \sin t\,\mathbf{i} + (t - t^2) \sin t\,\mathbf{j} + (t - e^t) \sin t\,\mathbf{k}$

27. $\sqrt{t}(1 + t) + (1 - t)(1 - t^2) + te^t$

29. $\dfrac{\sqrt{2}}{2}\mathbf{i} + \mathbf{j} + \dfrac{\sqrt{2}}{2}\mathbf{k}$ **31.** $\mathbf{i} + \mathbf{k}$ **33.** Does not exist.

35. $6\mathbf{i} + 7\mathbf{j} + 14\mathbf{k}$

37. Does not exist.

39. $(-\pi/2 + n\pi, \pi/2 + n\pi),\ n = 0,\ \pm 1,\ \ldots$

41. $[0, 1)$ and $(1, \infty)$ **43.** $[0, 1)$

45. $(-\infty, 1)$ and $(1, \infty)$

47. $2k\pi \leq t \leq \pi + 2k\pi$

49. $600{,}000\pi$. One complete revolution takes $t = \pi/500$ s.

Exercise Set 17.2

1. $(0, \infty)$

3. $(-3, -2)$ and $(-2, \infty)$

5. $\mathbf{i} + \dfrac{1}{2\sqrt{t}}\mathbf{j}$

7. $\dfrac{1}{2\sqrt{t}}\mathbf{i} - \dfrac{3}{2}t^{-5/2}\mathbf{j} + \dfrac{2}{2t-1}\mathbf{k}$

9. $[e^t(t^2 + 2t - 1)]\mathbf{i} + (2t\sin t + t^2\cos t)\mathbf{j} + \frac{1}{2}(t^3 - 2t - 1)^{-1/2}(3t^2 - 2)\mathbf{k}$

11. $\dfrac{1}{\sqrt{1-t^2}}\mathbf{i} + \dfrac{t}{\sqrt{1+t^2}}\mathbf{j} - 3t^2 e^{-t^3}\mathbf{k}$

13. $-\dfrac{1}{4}t^{-3/2}\mathbf{i} + \dfrac{15}{4}t^{-7/2}\mathbf{j} - \dfrac{4}{(2t-1)^2}\mathbf{k}$

17. $(3\cos t - 2)\mathbf{i} + (6t + 2\sin t)\mathbf{k}$

19. $-\sin t\mathbf{i} + (3t^2 + \sin^2 t - \cos^2 t)\mathbf{j} - \mathbf{k}$

21. $3e^{3t}\cos e^{3t}\mathbf{i} + 6e^{6t}\mathbf{k}$

23. $5\mathbf{i} + 14\mathbf{j} + 4\mathbf{k}$

25. 138

27. 1

31. $\frac{2}{3}t^{3/2}\mathbf{i} + \frac{1}{2}e^{2t}\mathbf{j} + \ln|t|\mathbf{k} + \mathbf{C}$

33. $(t\ln t - t)\mathbf{i} + \ln|\ln t|\mathbf{j} + \mathbf{C}$

35. $(3 + t)\mathbf{i} + \left(5 + \dfrac{t^3}{3}\right)\mathbf{j}$

37. $\left(3 + \dfrac{t}{2} + \dfrac{\sin 2t}{4}\right)\mathbf{i} + \left(\dfrac{t}{2} - \dfrac{\sin 2t}{4} - 6\right)\mathbf{j} + (3 - e^{-t})\mathbf{k}$

39. $e^{-4t}\mathbf{i} + 4e^{-4t}\mathbf{k}$

41. $\left(\dfrac{b^2 - a^2}{2}\right)\mathbf{i} + \left(\dfrac{b^3 - a^3}{3}\right)\mathbf{j} - \left(\dfrac{b^4 - a^4}{4}\right)\mathbf{k}$

43. $\dfrac{\sqrt{2}}{2}\mathbf{i} + \dfrac{1}{2}\mathbf{j} + \left(\dfrac{\pi}{8} + \dfrac{1}{4}\right)\mathbf{k}$

47. False

Exercise Set 17.3

1. $48\mathbf{i} + \frac{1}{4}\mathbf{j}$

3. $\sqrt{2}\mathbf{i} + 2\mathbf{j}$

5. $-2\mathbf{i} + \mathbf{j} + 6\mathbf{k}$

7. $\frac{4}{5}\mathbf{i} + \frac{3}{5}\mathbf{j}$

9. $(1/\sqrt{10})\mathbf{i} + (3/\sqrt{10})\mathbf{k}$

11. 0

13. $-\sqrt{2}$

15. $x(t) = t, \quad y(t) = 1, \quad z(t) = 1 + 3t$

19. $(2, -8)$

21. $\frac{2}{5}x + 2y + \frac{3}{2}z = \ln(5^{2/5}2^{13/2})$

23. $-2\pi x + z = 1$

25. $\dfrac{\sqrt{362}}{11\sqrt{3}}$

27. Point: $(0, 0, -1)$
Angle: $\text{Cos}^{-1}(-\sqrt{3}/3)$

29. Point: $(1, \pi/2, 1)$
Angle: 0

31. $\pi/2$

33. $\frac{1}{2}\sqrt{6} + \frac{1}{2}\ln(2 + \sqrt{6}) - \frac{1}{4}\ln 2$

35. $61/3$

37. $\dfrac{1}{\sqrt{17}}(\mathbf{i} + 4\mathbf{j})$

39. $\dfrac{x-2}{-2} = y + 5 = \dfrac{z+6}{6}$

41. $\alpha = \sqrt{2}/2$

45. $-\cos t\mathbf{i} - \sin t\mathbf{j}$

47. $\sqrt{2}\left(1 - \dfrac{\omega}{12}\right)\mathbf{i} + \sqrt{2}\left(1 - \dfrac{\omega}{12}\right)\mathbf{j} + \dfrac{\pi}{2}\mathbf{k}$

Exercise Set 17.4

1. $\mathbf{v} = 2\mathbf{j}, \quad \mathbf{a} = 0$

3. $\mathbf{v} = -3\sin 3t\mathbf{i} + 3\cos 3t\mathbf{j}, \quad \mathbf{a} = -9\cos 3t\mathbf{i} - 9\sin 3t\mathbf{j}$

5. $\mathbf{v} = \dfrac{2}{t}\mathbf{i} - 2\sin 2t\mathbf{j} - \dfrac{1}{t^2}\mathbf{k}, \quad \mathbf{a} = \dfrac{-2}{t^2}\mathbf{i} - 4\cos 2t\mathbf{j} + \dfrac{2}{t^3}\mathbf{k}$

7. $\mathbf{v} = e^t(\cos t - \sin t)\mathbf{i} + e^t(\sin t + \cos t)\mathbf{j} - \sin t\mathbf{k}$,
$\mathbf{a} = -2e^t\sin t\mathbf{i} + 2e^t\cos t\mathbf{j} - \cos t\mathbf{k}$

9. $|\mathbf{v}| = 2\sqrt{41}$

11. $(3 + \sin t)\mathbf{i} + (3 - \cos t)\mathbf{j}$

13. $(\text{Tan}^{-1} t - 2)\mathbf{i} + (\frac{1}{2}e^{t^2} + \frac{1}{2})\mathbf{j} + 4\mathbf{k}$

15. $\mathbf{v} = 2t\mathbf{i} + t\mathbf{k}, \quad \mathbf{r} = (t^2 + 3)\mathbf{i} - \mathbf{j} + \left(\dfrac{t^2}{2} + 4\right)\mathbf{k}$

17. $\mathbf{v} = \sin t\mathbf{i} + (1 - \cos t)\mathbf{j} + \mathbf{k}$,
$\mathbf{r} = (4 - \cos t)\mathbf{i} + (-1 + t - \sin t)\mathbf{j} + (t + 2)\mathbf{k}$

21. Speed $= |\alpha| \cdot$ radius

23. a. $50\sqrt{3}\,t\mathbf{i} + (50t - \frac{1}{2}gt^2)\mathbf{j}$
b. $\dfrac{1250}{9.8}$ m
c. $\dfrac{100}{9.8}$ s
d. $\dfrac{5000\sqrt{3}}{9.8}$ m
e. 100 m/s

27. a. $\displaystyle\int_0^{50\sqrt{3}/g} \sqrt{2500 - 50\sqrt{3}gt + g^2t^2}\, dt$
b. 304.6

Exercise Set 17.5

1. $T(s) = \dfrac{1}{2}i + \dfrac{\sqrt{3}}{2}j,$ $r''(s) = 0,$ $\kappa(s) = 0$

3. $T(s) = \dfrac{-\sqrt{2}}{2}\sin(s)i + \dfrac{\sqrt{2}}{2}\cos(s)j + \dfrac{\sqrt{2}}{2}k,$

$r''(s) = \dfrac{-\sqrt{2}}{2}\cos(s)i - \dfrac{\sqrt{2}}{2}\sin(s)j,$ $\kappa(s) = \dfrac{\sqrt{2}}{2}$

5. 0

7. $\dfrac{e^t}{(1 + e^{2t})^{3/2}}$

9. $\dfrac{1}{2}$

11. $\dfrac{2}{(1 + 4x^2)^{3/2}}$

13. $\dfrac{|6x|}{(1 + 9x^4)^{3/2}}$

15. $\dfrac{|\cos x|}{(1 + \sin^2 x)^{3/2}}$

17. $\dfrac{2}{(1 + 4t^2)^{3/2}}$

19. $\dfrac{2e^t(1 + e^{2t})^{1/2}}{(2e^{2t} + 1)^{3/2}}$

21. $(\pi/2, 0)$

23. a. $\dfrac{2}{5^{3/2}}$ **b.** $(7/2, -4)$ **c.** 2

27. $\dfrac{3}{2^{3/2}(1 + \sin\theta)^{1/2}}$

29. $\dfrac{3}{2^{3/2}(1 - \cos\theta)^{1/2}}$

35. $a_T = \dfrac{2t}{\sqrt{t^2 + 1}},$ $a_N = \sqrt{\dfrac{16t^6 + 16t^4 + 4}{t^2 + 1}}$

Exercise Set 17.6

1. $C = (0, 0, 2),$ $r = 3$

3. $C = (2, -1, 3),$ $r = 4$

5. $C = (3, -1, -2),$ $r = 5$

	Equation	Name	Center	yz-sections	xz-sections	xy-sections
7.	$x^2 + y^2 + z^2 = 10$	sphere	$(0, 0, 0)$	circle*	circle*	circle*
9.	$\dfrac{x^2}{4} + \dfrac{y^2}{9} = z^2$	elliptic cone	$(0, 0, 0)$	hyperbola**	hyperbola**	ellipse*
11.	$-\dfrac{x^2}{9} - \dfrac{y^2}{4} + z^2 = 1$	hyperboloid of two sheets	$(0, 0, 0)$	hyperbola	hyperbola	ellipse*
13.	$\dfrac{(x + 2)^2}{3} - \dfrac{(y - 1)^2}{2} = z$	hyperbolic paraboloid	none	parabola	parabola	hyperbola**
15.	$\dfrac{(x + 3)^2}{4} + \dfrac{(y - 1)^2}{2} - \dfrac{z^2}{2} = 1$	hyperboloid of one sheet	$(-3, 1, 0)$	hyperbola	hyperbola	ellipse
17.	$\dfrac{x^2}{9} + \dfrac{(y + 2)^2}{6} = z$	elliptic paraboloid	none	parabola	parabola	ellipse*
19.	$\dfrac{z^2}{4} + \dfrac{y^2}{9} = x$	elliptic paraboloid	none	ellipse*	parabola	parabola
21.	$-x^2 + y^2 + z^2 = 1$	hyperboloid of one sheet	$(0, 0, 0)$	circle	hyperbola	hyperbola
23.	$(x - 3)^2 + (y + 1)^2 + (z + 3)^2 = 1$	sphere	$(3, -1, -3)$	circle*	circle*	circle*
25.	$(x + 2)^2 + (y - 1)^2 = 36(z - \frac{85}{18})$	paraboloid of revolution	none	parabola	parabola	circle*

*Section may degenerate to a point or the empty set.
**Section may degenerate to two lines.

27. Plane parallel to yz-plane

Direction of Ruling	Cross-section
29. z-axis	line
31. z-axis	hyperbola
33. z-axis	hyperbola
35. x-axis	parabola
37. y-axis	hyperbola
39. x-axis	parabola
41. y-axis	parabola

43. $y^2 = 4(x^2 + z^2)$

45. $y = x^2 + z^2$

47. a. S_1 in Q_2
  S_2 in Q_2
  S_3 in Q_1

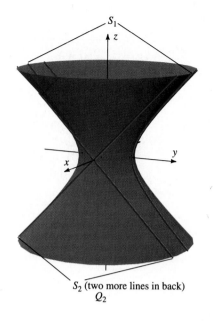

S_1

S_2 (two more lines in back)
Q_2

b.

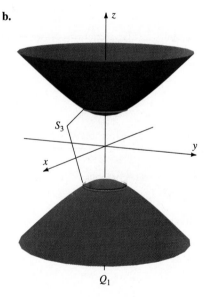

S_3

Q_1

Exercise Set 17.7

1. a. $(\sqrt{2}, \pi/4, 0)$ **b.** $(2, \pi/6, 3)$
  c. $(\sqrt{2}, 3\pi/4, -2)$ **d.** $(2, 2\pi/3, 4)$

3. a. $(\sqrt{2}, \sqrt{2}, -3)$ **b.** $(\sqrt{3}/2, 1/2, 4)$
  c. $(2, 2\sqrt{3}, -5)$ **d.** $(-1, -\sqrt{3}, 2)$

5. a. $(1, 0, \pi/2)$ **b.** $(2, \pi/4, \pi/4)$
  c. $(2, 7\pi/4, \pi/4)$ **d.** $(2\sqrt{2}, \pi/3, \pi/4)$

7. a. $(\sqrt{6}/2, \sqrt{6}/2, 1)$ **b.** $(0, 0, -1)$
  c. $(-1, 1, \sqrt{2})$ **d.** $(0, -3\sqrt{3}/2, -3/2)$

9. a. $(1, 0, \pi/2)$ **b.** $(2, -\pi/4, \pi/4)$
  c. $(2\sqrt{2}, \pi/3, \pi/4)$ **d.** $(2\sqrt{2}, \pi/4, \pi/4)$

11. a. $(2, \pi/4, 0)$ **b.** $(\sqrt{2}, \pi/2, \sqrt{2})$
  c. $(3, 2\pi/3, 0)$ **d.** $(\sqrt{3}/2, \pi/2, -1/2)$

13. $z = 3$

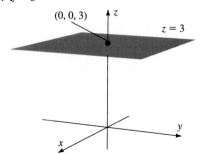

$(0, 0, 3)$ $z = 3$

15. $z^2 = 4(x^2 + y^2)$

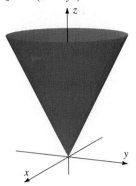

17. $x^2 + y^2 + z^2 = 4$

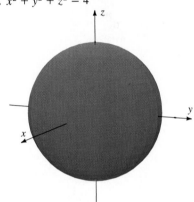

19. $x^2 - y^2 = a^2$

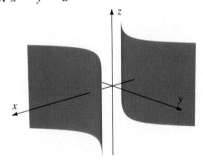

21. $r^2 = 9$ **23.** $r^2 = 9z$

25. $r^2 \cos^2 \theta + z^2 = 4$

27. $r^2 \cos \theta \sin \theta = 4 + ar \cos \theta$

29. $\rho^2 \sin^2 \phi = 9$

31. $\rho \sin^2 \phi = 9 \cos \phi$

33. $\rho^2 = \dfrac{4}{\sin^2 \phi \cos^2 \theta + \cos^2 \phi}$

35. $\rho^2 \sin^2 \phi \cos \theta \sin \theta - a\rho \sin \phi \cos \theta = 4$

Review Exercises—Chapter 17

1. $\dfrac{x^2}{a^2} + \dfrac{x^2}{b^2} = 1$ **3.** $[-1, 0) \cup (0, 1]$

5. $t = 2n\pi, \quad n = 0, 1, 2, \ldots$

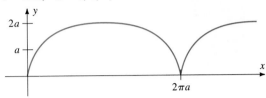

7. $3\mathbf{i} + \sqrt{2}\mathbf{j} + 2\mathbf{k}$ **9.** $t = \pi/2, 3\pi/2$

11. $t = \pm 3, \pm 2$

13. $\mathbf{r}'(t) = e^t(t + 1)\mathbf{i} + e^{-t}(1 - t)\mathbf{j}$
$\mathbf{r}''(t) = e^t(t + 2)\mathbf{i} + e^{-t}(t - 2)\mathbf{j}$

15. $\mathbf{r}'(t) = t^{-1}\mathbf{i} + \frac{1}{2}t^{-1/2}e^{\sqrt{t}}\mathbf{j}$
$\mathbf{r}''(t) = -t^{-2}\mathbf{i} + \frac{1}{4}(t^{-1} - t^{-3/2})e^{\sqrt{t}}\mathbf{j}$

17. $\mathbf{i} + (1 - 2e^{-1})\mathbf{j}$

19. $(4\mathbf{i} + 3\mathbf{j}) \cosh 3t + \mathbf{C} \sinh 3t$ for any constant vector $\mathbf{C}$

21. $\mathbf{r}(t) = 5 \cos (\frac{1}{5}t)\mathbf{i} + 5 \sin(\frac{1}{5}t)\mathbf{j}$

23. Sphere with center $(3, -2, 1)$ and radius $r = 2$

25.

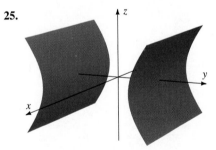

27. $(\sqrt{6}/2, \sqrt{6}/2, 1)$

29.

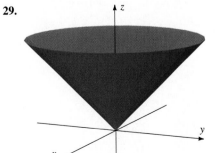

31.

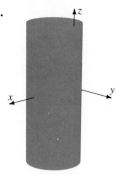

33. a. $\kappa(t) = 1/2$

b. $\mathbf{T}(t) = \dfrac{\sqrt{2}}{2}\mathbf{i} - \dfrac{\sqrt{2}}{2}\sin t\,\mathbf{j} + \dfrac{\sqrt{2}}{2}\cos t\,\mathbf{k}$

35. $3\sqrt{13}$

37. $\dfrac{12}{(145)^{3/2}}$

Equation	Name	Center	yz-section	xz-section	xy-section
41. $x^2 + \dfrac{y^2}{6} - \dfrac{z^2}{3} = 1$	hyperboloid of one sheet	$(0, 0, 0)$	hyperbola	hyperbola	ellipse
43. $\dfrac{y^2}{9} - \dfrac{x^2}{9} = z$	hyperbolic paraboloid	none	parabola	parabola	hyperbola*
45. $\dfrac{(x+1)^2}{4} + \dfrac{y^2}{4} - \dfrac{(z+2)^2}{9} = 1$	hyperboloid of one sheet	$(-1, 0, 2)$	hyperbola	hyperbola	circle

*Section may degenerate to two lines.

49. $\rho(0) = 1,\quad x^2 + y^2 = 1$

51. $s = \sqrt{1 + t^2}$

53. $\dfrac{20\sqrt{3} - g}{6}$

Chapter 18

Exercise Set 18.1

1. $\{(x, y) \mid (x, y) \neq (0, 0)\}$

3. $\{(x, y) \mid y \geq x\}$

5. All $(x, y) \in \mathbb{R}^2$

7. $\{(x, y) \mid x^2 + y^2 \leq 1 \quad \text{and} \quad (x, y) \neq (0, 0)\}$

9. All $(x, y, z) \in \mathbb{R}^3$

11. $\{(x, y, z) \mid xyz \neq 0\}$

13. $\{(x, y, z) \mid x \neq 0,\ y > 0,\ z \neq 0 \quad \text{and} \quad x^2yz^2 \neq 1\}$

15. a. $h(x, y) = \sqrt{x + y^2}$
   b. $\{(x, y) \mid x + y^2 \geq 0\}$

17. $S(r, h) = 2\pi r^2 + 2\pi rh$

19. $V(r, h) = \dfrac{\pi}{3}r^2 h + \dfrac{2\pi}{3}r^3$

21. -5 **23.** $1/\sqrt{2}$ **25.** 0

27. 0 **29.** 1 **31.** (iii), (b)

33. (ii), (d)   **35.** (v), (a)

37.

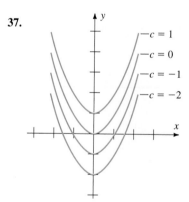

39.

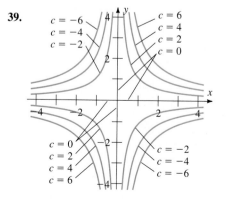

41.

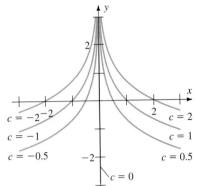

$c = -2$ $c = 2$

$c = -1$ $c = 1$

$c = -0.5$ $c = 0.5$

$c = 0$

43.

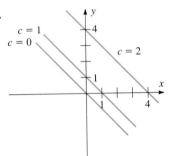

$c = 1$
$c = 0$ $c = 2$

45.

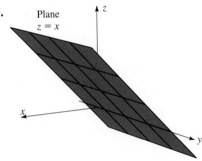

Plane
$z = x$

47.

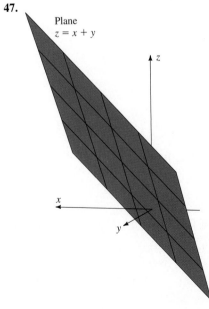

Plane
$z = x + y$

49.

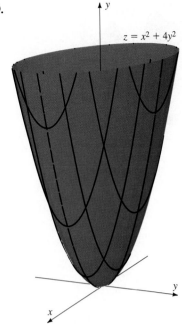

$z = x^2 + 4y^2$

51.

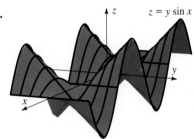

$z = y \sin x$

53.

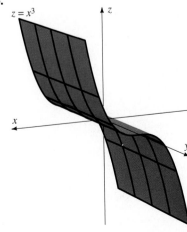

$z = x^3$

55.

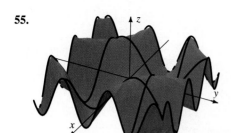

$z = \sin xy$

57. Does not exist

59. No

61. Yes

63. 0

65. False

Exercise Set 18.2

1. $\dfrac{\partial f}{\partial x} = y, \dfrac{\partial f}{\partial y} = x$

3. $\dfrac{\partial f}{\partial x} = \tan y^2, \dfrac{\partial f}{\partial y} = 2xy \sec^2 y^2$

5. $\dfrac{\partial f}{\partial x} = 2xe^{x^2+y^2}, \dfrac{\partial f}{\partial y} = 2ye^{x^2+y^2}$

7. $\dfrac{\partial f}{\partial r} = \cos \theta, \dfrac{\partial f}{\partial \theta} = -r \sin \theta$

9. $\dfrac{\partial f}{\partial x} = \dfrac{y^2 + 1}{xy^2 + x - y}, \dfrac{\partial f}{\partial y} = \dfrac{2xy - 1}{xy^2 + x - y}$

11. $\dfrac{\partial f}{\partial x} = yx^{y-1}, \dfrac{\partial f}{\partial y} = x^y \ln x$

13. $\dfrac{\partial f}{\partial r} = 2r \cos \theta, \dfrac{\partial f}{\partial \theta} = -r^2 \sin \theta$

15. $\dfrac{\partial f}{\partial x} = y^3, \dfrac{\partial f}{\partial y} = 3xy^2 - z^2, \dfrac{\partial f}{\partial z} = -2yz$

17. $\dfrac{\partial f}{\partial x} = \dfrac{2yz}{(x + y)^2}\left(\dfrac{x - y}{x + y}\right)^{z-1}$

$\dfrac{\partial f}{\partial y} = \dfrac{-2xz}{(x + y)^2}\left(\dfrac{x - y}{x + y}\right)^{z-1}$

$\dfrac{\partial f}{\partial z} = \left(\dfrac{x - y}{x + y}\right)^z \ln\left(\dfrac{x - y}{x + y}\right)$

19. $\dfrac{\partial f}{\partial r} = \dfrac{s \ln t \,(s^2 + t)}{2r^{1/2}\,(s^2 - 2r + t)^{3/2}}$

$\dfrac{\partial f}{\partial s} = \dfrac{r^{1/2} \ln t \,(-2r + t)}{(s^2 - 2r + t)^{3/2}}$

$\dfrac{\partial f}{\partial t} = \dfrac{r^{1/2} s \,(s^2 - 2r + t - \frac{1}{2}t \ln t)}{t \,(s^2 - 2r + t)^{3/2}}$

21. 125

23. $\dfrac{1}{2\sqrt{3}}$

25 $\dfrac{\partial^2 f}{\partial r^2} = 2 \cos \theta, \dfrac{\partial^2 f}{\partial r\, \partial \theta} = -2r \sin \theta, \dfrac{\partial^2 f}{\partial \theta^2} = -r^2 \cos \theta$

27. $\dfrac{2}{x^2 + y^2 + z^2}$

29. a. $v_2 = \dfrac{A_1 v_1}{A_2}$ **b.** $\dfrac{\partial v_2}{\partial A_2} = \dfrac{-A_1 v_1}{(A_2)^2}$

c. $\dfrac{-100 \text{ cm/s}}{9 \text{ cm}^2}$ **d.** $\dfrac{3 \text{ cm}^2}{20 \text{ cm/s}}$

31. $\rho \sin \phi \cos \theta$ **33.** $\sin \phi \cos \theta$ **35.** $-\rho \sin \phi$

37. a. $\cos(x + y)^2 - \cos x^2$ **b.** $\cos(x + y)^2$

41. a. $\dfrac{\partial V}{\partial a} = \dfrac{a\sigma}{2\epsilon_0 \sqrt{a^2 + r^2}}$

b. $\dfrac{\partial V}{\partial r} = \dfrac{\sigma}{2\epsilon_0}\left(\dfrac{r}{\sqrt{a^2 + r^2}} - 1\right)$

47. $f(x, y) = \sin(xy)$, for example

51. Slope $= 3$,   intersection $= (-2/3, 2, 0)$

Exercise Set 18.3

1. $2x + 6y - z = 10$ **3.** $x + 5y + z = -3$

5. $x - 2y - z = 4$ **7.** $x + y - z = 2$

9. $y + 2z = 2$ **11.** $y - z = 1$

13. $\dfrac{\sqrt{2}}{2}y - z = \dfrac{\sqrt{2}}{2}\left(\dfrac{\pi}{4} - 1\right)$ **15.** $2x + ez = 0$

17. $\mathbf{r}(t) = 3\mathbf{i} - \mathbf{j} + 4\mathbf{k} + t(4\mathbf{i} + 4\mathbf{j} - \mathbf{k})$

19. $\mathbf{r}(t) = \mathbf{i} - 2\mathbf{j} + 2\mathbf{k} + t(-\mathbf{i} + 2\mathbf{j} - 2\mathbf{k})$

21. $\mathbf{r}(t) = \mathbf{i} + \mathbf{j} + t(\mathbf{j} - \mathbf{k})$

23. $(2, 3, 20)$

25. $\mathbf{r}(t) = \mathbf{i} + t\mathbf{j} + t(4\mathbf{j})$

29. $18x + 16y - z = 25$

31. $x_0 x + y_0 y + z_0 z = r^2$

33. $-x - 2y + z + \pi/2 = 0$

35. $\pi/2$ **37.** 0

Exercise Set 18.4

1. Relative minimum: $(0, -2, 0)$

3. Saddle point: $(-3, 2, 0)$

5. Saddle point: $(0, 0, 9)$

7. Relative minimum: $(1, 3, 1)$

9. Saddle point: $(0, 0, 0)$

11. Saddle point: $(0, 0, 0)$

13. Saddle point: $(0, 0, 0)$
Relative maximum: $\left(-\frac{4}{3}, -\frac{4}{3}, \frac{64}{27}\right)$

15. Saddle points: $(0, \pi/2 + n\pi, 0)$

17. Absolute maximum: $x - y = \pi/2 + 2k\pi$
Absolute minimum: $x - y = -\pi/2 + 2$

19. Absolute minimum: $(0, 0, 0)$

21. Absolute maximum: $(0, 0, e)$

25. The graph is unbounded.

29. $l = 2$ m, $w = 2$ m, $h = 4$ m

31. $\left(\dfrac{ad}{a^2 + b^2 + c^2}, \dfrac{bd}{a^2 + b^2 + c^2}, \dfrac{cd}{a^2 + b^2 + c^2} \right)$

33. $l = 28''$, $w = h = 14''$

35. Absolute minimum: $-2\sqrt{2} + 4$
Absolute maximum: $2\sqrt{2} + 4$

37. Absolute minimum: 3
Absolute maximum: 9

39. $x = 75$, $y = 100$

43. 86

45. a. $y = \frac{23}{20}x - \frac{19}{4}$
b.

c. $y(4) = -3/20$

Exercise Set 18.5

1. $2xy^4 \, dx + 4x^2y^3 \, dy$

3. $\dfrac{x}{\sqrt{x^2 + y^4}} \, dx + \dfrac{2y^3}{\sqrt{x^2 + y^4}} \, dy$

5. $\dfrac{e^{\sqrt{x}}}{2\sqrt{x}} \cos y \, dx - e^{\sqrt{x}} \sin y \, dy$

7. $2xyz^3 \, dx + x^2z^3 \, dy + 3x^2yz^2 \, dz$

9. $\dfrac{1}{y + 3^z} \, dx - \dfrac{x}{(y + 3^z)^2} \, dy - \dfrac{x3^z \ln 3}{(y + 3^z)^2} \, dz$

11. $\dfrac{y^2 + z^2 - x^2 + 2xy}{(x^2 + y^2 + z^2)^2} \, dx + \dfrac{y^2 - x^2 - z^2 - 2xy}{(x^2 + y^2 + z^2)^2} \, dy$
$\qquad\qquad + \dfrac{2zy - 2zx}{(x^2 + y^2 + z^2)^2} \, dz$

13. a. $f(2, 3) = 48$
$\quad f(2.1, 3.2) = 57.12$
b. $\Delta f = 9.12$
c. 8.6
d. 0.0091, 0.91%

15. 5.076

19. 26.8

23. 25.28π

27. 0.079 s

17. 0.7441

21. 55.44

25. -3%

29. 10%

Exercise Set 18.6

1. $4x - 4yt = -16t + 4t^3$

3. $-y^2 \sin t + 2xy \cos t = -\sin^3 t + 2 \sin t \cos^2 t$

5. $\dfrac{1}{4\sqrt{tx + ty}} + \dfrac{te^{t^2}}{\sqrt{x + y}}$

7. $2t(y - 1) - 2x + 2z \cos t = -6t^2 - 2t + 2 \sin t \cos t$

9. $-2axz \sinh t + 2byz \cosh t - 2e^{-2t}(y^2 - x^2)$

11. $-2xy^3 \sin t + 3x^2y^2 (\sin t + t \cos t)$

13. a. $2xte^{st} + 2yt - 2z$
b. $2xse^{st} + 2ys + 2z$

15. a. $[y^2 \cos xy^2 - 2xy](2s - t) + [2xy \cos xy^2 - x^2](2t^2s)$
b. $[y^2 \cos xy^2 - 2xy](-s) + [2xy \cos xy^2 - x^2](2s^2t)$

17. a. $y^3z^2 \sin t - 3xy^2z^2t \sin s - 4sxy^3z$
b. $sy^3z^2 \cos t + 3xy^2z^2 \cos s + 4txy^3z$

19. a. $2ue^{u^2}y^2 - 4xye^{2v}$
b. $4xyue^{u^2} - 2x^2e^{2v}$

21. $6rt^2(1 - \cos \theta) + \dfrac{r^2t \sin \theta}{\sqrt{1 + t^2}}$

23. 116π cm³/s

25. 8.5 in.²/h

27. $\dfrac{208\pi\sqrt{109}}{109}$

29. a. $D(t) = \sqrt{1 + t^2}$
b. $D'(t) = \dfrac{t}{\sqrt{1 + t^2}}$

33. a. $12s^2t^2 - 8y^3 + 48s^2y^2$
b. $2s^4 + 8y^3 + 48t^2y^2$

35. $2s^2t^2e^{rst} - 4s^2t^2e^{2rst} + rs^3t^3e^{rst}$

37. $e^{y^2}[-2yr \sin^2 \theta - \sin \theta + 2yr \cos^2 \theta$
$\qquad\qquad + x(2 + 4y^2)r \sin \theta \cos \theta + 2xy \cos \theta]$

Exercise Set 18.7

1. $6\mathbf{i} + 9\mathbf{j}$

3. $\dfrac{\sqrt{2}}{2}\left(1 - \dfrac{\pi}{2}\right)\mathbf{i} + \dfrac{\sqrt{2}}{4} \pi\mathbf{j}$

5. $6\mathbf{i} + \mathbf{j} + 4\mathbf{k}$

7. $5\mathbf{i} - 3\mathbf{j} - \mathbf{k}$

9. $\left(e + \dfrac{1}{e}\right)\mathbf{i} + \dfrac{1}{4}\left(e + \dfrac{1}{e}\right)\mathbf{j} + 2\left(e - \dfrac{1}{e}\right)\mathbf{k}$

11. $-30\sqrt{2}$ **13.** $(2\sqrt{3} - 1)/18$

15. $2/\sqrt{3}$ **17.** $3\sqrt{3}/\sqrt{11}$

19. $\dfrac{13\sqrt{2}}{6} + \dfrac{1}{3}\left(\dfrac{1}{e} - e\right) + \dfrac{\pi}{12}\left(\dfrac{1}{e} + e\right)$

21. N: $2\mathbf{i} + 3\mathbf{j} + t(16\mathbf{i} - 6\mathbf{j})$
  T: $2\mathbf{i} + 3\mathbf{j} + t(6\mathbf{i} + 16\mathbf{j})$

23. N: $4\mathbf{i} + 4\mathbf{j} + t(\mathbf{i} + \mathbf{j})$
  T: $4\mathbf{i} + 4\mathbf{j} + t(\mathbf{i} - \mathbf{j})$

25. N: $\dfrac{\pi}{2}\mathbf{i} + \dfrac{\pi}{2}\mathbf{j} + t(\mathbf{j})$

  T: $\dfrac{\pi}{2}\mathbf{i} + \dfrac{\pi}{2}\mathbf{j} + t(\mathbf{i})$

27. N: $\mathbf{i} + e^{1/2}\mathbf{j} + t(\mathbf{i} + 2e^{-1/2}\mathbf{j})$
  T: $\mathbf{i} + e^{1/2}\mathbf{j} + t(2e^{-1/2}\mathbf{i} - \mathbf{j})$

29. $\sqrt{5}$ **31.** -1 **33.** $\dfrac{(4 + \pi)\sqrt{2}}{2\sqrt{16 + \pi^2}}$ **35.** $\dfrac{\sqrt{2}}{4}$

37. a. $(3, 1 + \sqrt{2}), (3, 1 - \sqrt{2})$
  b. $(1, 1), (5, 1)$

39. Along the curve $x = y \ln y$

41. N: $3\mathbf{i} + 6\mathbf{j} + 2\mathbf{k}$
  T: $3x + 6y + 2z = 18$

43. N: $5\mathbf{i} - \mathbf{j} - \mathbf{k}$
  T: $5x - y - z = 4$

45. N: $\mathbf{j}$
  T: $y = 1$

47. N: $\dfrac{1}{\sqrt{51}}(5\mathbf{i} + 5\mathbf{j} - \mathbf{k})$

57. $2\mathbf{i} + \mathbf{j}$

59. $-\mathbf{i} + \mathbf{j}$

Exercise Set 18.8

Maximum	Minimum
1. $f(0, \pm 1) = 4$	$f(\pm 1, 0) = 2$
3. None	$f(1, -1) = 2$

5. $f\left(\dfrac{1}{\sqrt{2}}, \dfrac{1}{\sqrt{2}}\right)$ $f\left(\dfrac{1}{\sqrt{2}}, \dfrac{-1}{\sqrt{2}}\right) = f\left(\dfrac{-1}{\sqrt{2}}, \dfrac{1}{\sqrt{2}}\right)$
  $= f\left(\dfrac{-1}{\sqrt{2}}, \dfrac{-1}{\sqrt{2}}\right) = \dfrac{1}{2}$ $= -\dfrac{1}{2}$

7. $f(2, 0) = 12$ $f(-2, 0) = -4$

9. $f\left(-\dfrac{2}{3}, \pm \dfrac{\sqrt{5}}{3}\right) = \dfrac{48}{9}$ $f(1, 0) = -3$

11. $f\left(\dfrac{1}{\sqrt{6}}, \dfrac{2}{\sqrt{6}}, -\dfrac{1}{\sqrt{6}}\right)$ $f\left(-\dfrac{1}{\sqrt{6}}, -\dfrac{2}{\sqrt{6}}, \dfrac{1}{\sqrt{6}}\right)$
  $= \sqrt{6}$ $= -\sqrt{6}$

13. $f(2, 2, 2) = 6$ $f(-2, -2, -2) = -6$

15. None   None

17. None $f(1, 0, 0) = 1$

19. $f\left(2, -1 + \dfrac{\sqrt{2}}{2}, \dfrac{3}{2}\right) = f\left(2, -1 - \dfrac{\sqrt{2}}{2}, \dfrac{3}{2}\right) = \dfrac{3}{4}$

21. $\left(3 - \dfrac{9\sqrt{13}}{13}, -2 + \dfrac{6\sqrt{13}}{13}\right)$

23. $(1, 0, 1)$

25. $(2, -1, 1)$

27. $\dfrac{2r}{\sqrt{3}} \times \dfrac{2r}{\sqrt{3}} \times \dfrac{2r}{\sqrt{3}}$

29. $r = \sqrt[3]{1/\pi},\ h = 2\sqrt[3]{1/\pi}$

31. $x = 2,\ y = 3,\ z = 1$

33. $x = y = z = 20$

Exercise Set 18.9

1. $x - y + C$ **3.** None

5. None   **7.** $x^3 \cos y - x + y^2 + C$

9. $x^2 e^{xy} - \sqrt{y} + C$ **11.** $x^2 y^2 + y \cos x + C$

Review Exercises—Chapter 18

1. No

3.

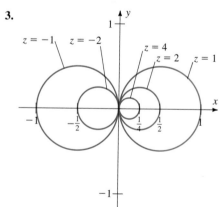

5. $-5\mathbf{i} - 7\mathbf{j}$ **7.** $-\mathbf{i} - \mathbf{k}$

9. $\dfrac{\partial f}{\partial x}(x, y) = \sin \sqrt{x^2 + y^2} + \dfrac{x^2 \cos \sqrt{x^2 + y^2}}{\sqrt{x^2 + y^2}}$,
  $\dfrac{\partial f}{\partial y}(x, y) = \dfrac{xy \cos \sqrt{x^2 + y^2}}{\sqrt{x^2 + y^2}}$

11. $\dfrac{784}{15}$

13. $\dfrac{\partial f}{\partial r} = 18x^2 + 4y^2 + 4xy - 2y,$

$\dfrac{\partial f}{\partial s} = 45x^2 + 10y^2 - 16sxy + 8sy$

1. $-\dfrac{1}{6}$

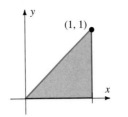

15. $5x + 4y = 10$

19. $-\dfrac{36}{5}\mathbf{i} - \dfrac{12}{5}\mathbf{j} - (2 - \text{Tan}^{-1}2)\mathbf{k}$

23. $\dfrac{2\sqrt{3}}{3} \times 2\sqrt{3} \times \dfrac{4\sqrt{3}}{3}$ **25.** Saddle point: $(0, 0, 0)$ **3.** $\dfrac{13}{24}$

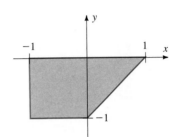

27. Saddle point: $(0, 0, 1)$ **29.** Saddle point: $(0, 0, e)$

31. a. 1
  b. 3
  c. $\mathbf{i} + 3\mathbf{j}$

33. $f(x, y) = x^2e^y + \cos y + C$

35. R increases by 10%

5. 0

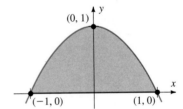

39. Maximum: $5 + 6\sqrt{2}$ **41.** No
  Minimum: $5 - 6\sqrt{2}$

43. $\left(\dfrac{4}{\sqrt{14}}, \dfrac{8}{\sqrt{14}}, \dfrac{-12}{\sqrt{14}}\right)$ **49.** $\left(\dfrac{3}{7}, \dfrac{-9}{7}, \dfrac{6}{7}\right)$

51. $x + 2y + z = 6$ **53.** $3 - \frac{1}{2}\sqrt{3}$

Chapter 19

7. $\dfrac{1}{20}$

1. 1   **3.** 7

5. 1/2   **7.** 8/3

9. $\dfrac{\pi}{2} + \dfrac{1}{3}$ **11.** $\dfrac{e^8 - 9}{8}$

13. 1/2   **15.** $4\pi/3$

17. 5/6   **19.** 8

21. $\dfrac{\ln 2}{4}$ **23.** $\dfrac{e^2 - 3}{2}$ **9.** $\dfrac{1}{3}$

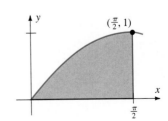

25. $\frac{32}{3}(3\sqrt{3} - 4\sqrt{2} + 1)$ **27.** 6

29. 1/2   **31.** $8e(e - 1)$

33. 8/3   **35.** $16\sqrt{2}/3$

11. $\dfrac{1}{2}$

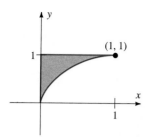

13. $\dfrac{1}{2}$

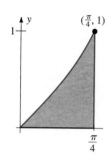

15. 2/15

17. $\frac{1}{3}(1 + e)^{3/2} - \frac{4}{3}2^{3/2} + 2\sqrt{e}\sqrt{1 + e} + 2 \ln \dfrac{\sqrt{e} + \sqrt{1 + e}}{1 + \sqrt{2}}$

19. 1/24

21. $\dfrac{3e^2 + e^{-2}}{4}$

23. 1/3

25. $\ln(2 + \sqrt{3})$

27. $e^4 - 4e$

29. 2

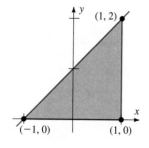

31. $\dfrac{1}{10}$

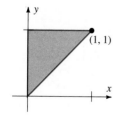

33. $\displaystyle\int_0^1 \int_{e^y}^e f(x, y) \, dx \, dy$

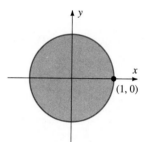

35. 9/2

37. $\sqrt{2} - 1$

39. 1/3

41. $\dfrac{\pi}{4} - \ln 2$

43. 1

45. 2/3

Exercise Set 19.3

1. $\dfrac{2\pi}{3}$

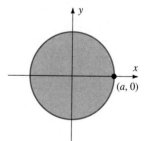

3. $\pi(e^{a^2} - 1)$

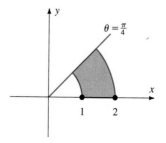

5. $\dfrac{45\pi}{16}$

7. $\dfrac{4}{27}$

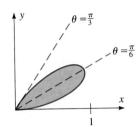

9. $\pi/4$

11. $2 + \pi/4$

13. $25\left(\dfrac{\pi}{3} + \dfrac{\sqrt{3}}{2}\right)$

15. $9\pi/2$

17. $\pi/2$

19. $\dfrac{\pi}{4}(e - 1)$

21. $\dfrac{\pi}{3} + \dfrac{\sqrt{3}}{2}$

23. 2

25. 0

27. 16π

29. $4\pi\sqrt{3}$

31. 3π

33. $\dfrac{\pi}{3}r^2h$

35. 16π

37. $\pi/4$

39. $16\pi/3$

Exercise Set 19.4

1. 3

3. 8

5. $\sqrt{2} - 1$

7. $\pi/4$

9. 2π

11. $(\bar{x}, \bar{y}) = (\frac{11}{9}, \frac{5}{9})$

13. $(\bar{x}, \bar{y}) = (3/\pi, 3/\pi)$

15. $(\bar{x}, \bar{y}) = (\frac{17}{54}, \frac{13}{9})$

17. $(\bar{x}, \bar{y}) = (0, \frac{8}{5})$

19. $(\bar{x}, \bar{y}) = (\frac{13}{5}, \frac{17}{10})$

21. $(\bar{x}, \bar{y}) = \left(\dfrac{4a}{3\pi}, \dfrac{4a}{3\pi}\right)$

23. $(\bar{x}, \bar{y}) = (\frac{8}{5}, 2)$

25. 4λ

Exercise Set 19.5

1. $\sqrt{3}$

3. $\dfrac{\pi}{6}(13^{3/2} - 1)$

5. $2\sqrt{3}$

7. $3\sqrt{2}\pi$

9. $\sqrt{2} + \ln(1 + \sqrt{2})$

11. $\dfrac{\sqrt{2}}{2} + \dfrac{1}{2}\ln(1 + \sqrt{2})$

13. $\dfrac{5\sqrt{2}}{12} + \dfrac{1}{4}\ln(3 + 2\sqrt{2})$

15. $\dfrac{\pi}{6}(7^3 - 29^{3/2})$

17. $4\pi(2 - \sqrt{3})$

19. $\dfrac{\pi}{12}(101^{3/2} - 1)$

21. 28π

23. $A \approx 7.9$ 4

25. $4\pi r^2$ **27.** $\dfrac{1 + \sqrt{5}}{12} + \dfrac{\ln(2 + \sqrt{5})}{4}$ **29.** $\dfrac{\pi}{24}(17^{3/2} - 1)$

Exercise Set 19.6

1. $1/8$

3. $1/2$

5. $14/3$

7. 0

9. $\displaystyle\int_0^2 \int_{y/2}^1 \int_0^{x+y} f(x, y, z)\, dz\, dx\, dy$

11. $\displaystyle\int_{-2}^2 \int_{-\sqrt{4-x^2}}^{\sqrt{4-x^2}} \int_{x^2+y^2}^4 f(x, y, z)\, dz\, dy\, d$

13.

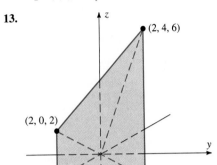

15. $2\pi/3$

17. 27π

19. 0

21. 128π

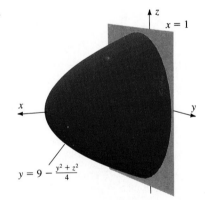

23. π

25. $\dfrac{256}{15}$

27. $405\pi/8$

29. 12λ

31. $(\bar{x}, \bar{y}, \bar{z}) = (0, \frac{3}{4}, \frac{15}{8})$

33. $(\bar{x}, \bar{y}, \bar{z}) = (\frac{3}{4}, \frac{3}{4}, \frac{3}{4})$

35. $\displaystyle\int_0^2 \int_0^{1-y/2} \int_0^{4-4x-2y} (x - y + z)\, dz\, dx\, dy,$

$\displaystyle\int_0^1 \int_0^{4-4x} \int_0^{2-2x-z/2} (x - y + z)\, dy\, dz\, dx,$

$\displaystyle\int_0^4 \int_0^{2-z/2} \int_0^{1-y/2-z/4} (x - y + z)\, dx\, dy\, dz,$

$$\int_0^2 \int_0^{4-2y} \int_0^{1-y/2-z/4} (x - y + z) \, dx \, dz \, dy,$$

$$\int_0^4 \int_0^{1-z/4} \int_0^{2-2x-z/2} (x - y + z) \, dy \, dx \, dz$$

Exercise Set 19.7

1. $81\pi/2$

3. $(\bar{x}, \bar{y}, \bar{z}) = (0, 0, 3)$

5. $64\alpha\pi/3$

7. $18\pi + 24$

9. $2\sqrt{2}\, 4^4/3^2 \approx 80$

11. $\pi/20$

13. $\pi/6$

15. $9\pi - 12$

17. $\frac{4}{3}\pi(R^3 - r^3)$

19. π

21. $6a\pi$

23. $(0, 3/8, 3/8)$

25. 2π

27. $\frac{\pi}{2}\left(1 - \frac{\sqrt{2}}{2}\right)$

29. $V = \frac{16}{3}R^3$

31. $9\pi(1 - \sqrt{3}/3)$

33. $\frac{32\sqrt{2}\,\pi}{15}$

35. $\frac{40\pi}{3}$

37. $\frac{104\pi}{3}\left(1 - \frac{\sqrt{2}}{2}\right)$

Exercise Set 19.8

1.

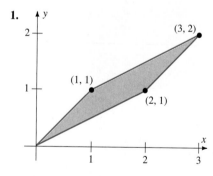

3.

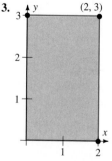

5.

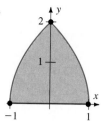

7. $21/4$

9. 48π

11. 12π

13. $3\pi/4$

Review Exercises—Chapter 19

1. $\dfrac{1}{15}$

3. $2\left(1 - \dfrac{1}{\sqrt{e}}\right)$

5. 64

7. $16(\pi/2 - 1)$

9. 3π

11. $abc/6$

13. 36π

15. $(\bar{x}, \bar{y}, \bar{z}) = (0, 0, 3/4)$

17. $\dfrac{2a^3}{9}(3\pi - 4)$

19. $4\sqrt{17}\pi$

21. $\dfrac{32\pi}{3}(8 - 3\sqrt{3})$

23. $\dfrac{1017}{20}$

25. $(\bar{x}, \bar{y}) = \left(0, \dfrac{16}{3\pi}\right)$

27. $\dfrac{2\pi}{3}(e - 1)$

29. $(\bar{x}, \bar{y}) = (5/6, 0)$

31. $81\pi/4$

33. $32\pi\rho$

35. $\dfrac{\pi}{2}(\sqrt{2} - 1)$

37. $256/9$

39. The integral is 1.9904 to 4 places

41. 36π

43. $8\sqrt{3}\pi$

Chapter 20

Exercise Set 20.1

1.

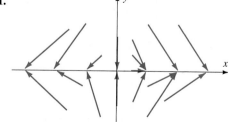

3.

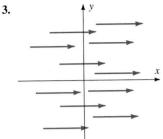

5.

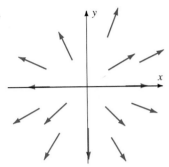

7.

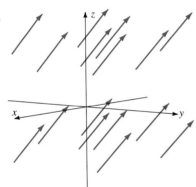

9.

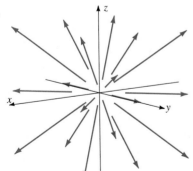

11. 4, 5, 9

13. $\mathbf{F} = 2x\mathbf{i} - 2y\mathbf{j}$

15. $\dfrac{1}{\sqrt{x^2 + y^2 + z^2}}(x\mathbf{i} + y\mathbf{j} + z\mathbf{k})$

17. $e^{x-y}(z\mathbf{i} - z\mathbf{j} + \mathbf{k})$

19. $\phi = x\tan(xy) + C$

21. Not a gradient

23. $\phi = e^{xyz} + C$

25. $\phi = x\sqrt{y}\cosh z + C$

27. $\phi = e^{x^2 y}\ln z - \mathrm{Tan}^{-1}(1 + z) + C$

31. a. $\mathbf{F}(0, 0, 0) = 4\mathbf{i}$
 b. $\mathbf{F}(-2, 0, 0) = -\frac{26}{9}\mathbf{i}$
 c. and **d.**

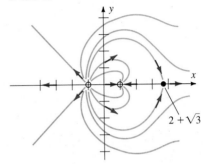

Exercise Set 20.2

1. 16

3. 1708/3

5. 328/15

7. $\pi + \pi^2/4$

9. -18

11. 12

13. $-\pi$

15. -4

17. $\dfrac{\pi}{2}\left(\dfrac{\pi}{2} - 1\right)$

19. 1672/3

21. -1

23. 3

25. $\dfrac{1}{6} + \dfrac{\pi}{16}$

27. $-442/3$

29. 4/3

33. $70\sqrt{5}/3$

35. $\frac{2}{3}(5^{3/2} - 1)$

37. 0

39. 5π

Exercise Set 20.3

For Exercises 1–10, $\mathbf{F} = \nabla f$ where

	f	$\int_c \mathbf{F} \cdot d\mathbf{r}$
1.	xy	3
3.	e^{xy}	$e^2 - 1$
5.	$x^2 + y^2$	0
7.	xe^{y^2}	$2e$
9.	$e^x \sin y$	$e\sqrt{2}/2$

11. 649

13. $\pi/6$

15. $-a$

23. No, the field is not conservative.

Exercise Set 20.4

1. $1/2$ **3.** $2/3$ **5.** $\pi/8$

7. $-11/60$ **9.** 0 **11.** -5π

13. 0 **15.** $-9\pi/2$ **17.** $1/6$

19. $3a^2\pi/8$

Exercise Set 20.5

1. $23\sqrt{6}/4$ **3.** $27\sqrt{14}/5$

5. 0 **7.** $48\pi(9 - 3\sqrt{5})$

9. $32\pi/3$ **11.** 2π

13. 16π **15.** $4\pi a^3/3$

17. 3

19. No, not the same: $4\pi/R$.

21. $(\bar{x}, \bar{y}, \bar{z}) = (0, 0, a/2)$

23. $(\bar{x}, \bar{y}, \bar{z}) = (1/2, 1/2, 1/2)$

Exercise Set 20.6

1. 0

3. $z\cos(yz)\mathbf{i} + xy\mathbf{j} + [y\sin(xy) - xz]\mathbf{k}$

5. Conservative   **7.** Not conservative

9. Both integrals are π.   **11.** Both integrals are $-2/3$.

13. Both integrals are 4.   **15.** -2π

17. Both integrals are 0.

Exercise Set 20.7

1. $2x + 2y + 2z$ **3.** $1 + x + xy$

5. $-y\sin(xy) + xze^{xyz} + xy\cos(xz)$

7. 0 **9.** -150

11. $11/6$ **13.** 0

15. $512\pi/15$ **17.** $80\pi/3$

23. a. Yes **b.** No **c.** No

29. False

Review Exercises—Chapter 20

1. a. $-y\sin x \ln z\mathbf{i} + \cos x \ln z\mathbf{j} + \dfrac{y\cos x}{z}\mathbf{k}$

  b. $[3x^2yz^2 + y\cos(xy)]\mathbf{i} + [x^3z^2 + x\cos(xy)]\mathbf{j} + 2x^3yz\mathbf{k}$

3. $\mathbf{F} = \nabla(-\text{Tan}^{-1}(y/x))$

7. Yes, since $\mathbf{F} = \nabla(xy^2z)$

13. True

17. $-4/3$

21. $-3/2$

25. $\sqrt{6}$

29. $-15\pi/2$

33. $8\pi(a + b + c)$

37. 0

5. -2π

9. $1/15$

15. $(a + b + c)V$

19. π

23. $2/5$

27. $1023\sqrt{2}\pi/5$

31. 0

35. $-164/3$

39. 0

Chapter 21

Exercise Set 21.1

1. $y = 2e^t - 1$

3. $y = \dfrac{-\pi}{2}e^{-t} + \pi$

5. $y = Ce^{-2t} + 2$

7. $y = \frac{1}{3}t^5 + Ct^2$

9. $y = e^{2t}(-\cos t + C)$

11. $y = \frac{1}{2}e^t + Ce^{-t}$

13. $y = \frac{1}{2}t + Ct^{-1}$

15. $y = ke^{ax^2/2}$

17. a. $dp/dt = -\frac{3}{200}p$
  b. 74.08 kg
  c. $p = 0$

19. a. $dp/dt = 0.10p + 100$
  b. \$1215.13

21. $I = I_0 e^{-t/RC}$

23. $52.8°$ C

25. $p = 6 - 4e^{-.02t}$

27. 6.93 years

29. $y = Cs^k$

Exercise Set 21.2

1. $x^2y = C$ **3.** $\sqrt{x^2 + y^2} = C$

5. Not exact   **7.** Not exact

9. Not exact   **11.** $x^2y = 3$

13. $xy^2 - x^2y = 2$

Exercise Set 21.3

1. $y = Ae^{-3t} + Be^{-2t}$

3. $y = Ae^t + Be^{-t}$

5. $y = Ae^{2t} + Be^{-5t}$

7. $y = Ae^{-t} + Bte^{-t}$

9. $y = Ae^{-3t} + Bte^{-3t}$

11. $y = A \sin 2t + B \cos 2t$

13. $y = A \sin \sqrt{5}t + B \cos \sqrt{5}t$

15. $y = Ae^{5t} + Be^{-2t}$

17. $y = Ae^{(2+\sqrt{3})t} + Be^{(2-\sqrt{3})t}$

19. $y = e^t(A \sin\sqrt{5}t + B \cos \sqrt{5}t)$

21. $y = Ae^{4t} + Be^{-3t}$

23. $y = e^{-t}(A \sin\sqrt{3}t + B \cos\sqrt{3}t)$

25. $y = e^{3t} - e^{-2t}$

27. $y = 3e^{-t} + 4te^{-t}$

29. $y = -\sin 3t + 3 \cos 3t$

31. $x = A \sin 2t + B \cos 2t$
$y = -2A \cos 2t + 2B \sin 2t$

33. $x(t) = \frac{1}{10} \cos 2t$

35. a. Oscillating with increasing frequency
b. Exponential function with increasingly large exponent
c. and d. Approach constant solution $x(t) = 1$.

Exercise Set 21.4

1. $y = C_1 e^{(-1+\sqrt{2})t} + C_2 e^{(-1-\sqrt{2})t} - \frac{5}{41} \sin 2t - \frac{4}{41} \cos 2t$

3. $y = C_1 \sin t + C_2 \cos t + \frac{1}{2}e^{-t}$

5. $y = C_1 e^{2t} + C_2 e^{-2t} - t$

7. $y = C_1 e^{5t} + C_2 e^{-2t} - \frac{5}{12} e^{2t}$

9. $y = (C_1 + \frac{1}{6}t) \sin 3t + C_2 \cos 3t$

11. $y = C_1 e^{4t} + (C_2 - \frac{2}{5}t)e^{-t}$

13. $y = (C_1 + C_2 t)e^{-t} + \frac{3}{4}e^t + t^2 - 4t + 6$

15. $y = (C_1 + C_2 t)e^t - \frac{11}{25} \sin 2t - \frac{2}{25} \cos 2t + \frac{1}{2}e^{-t}$

17. $y = e^{-t/2}(C_1 \sin \frac{1}{2}\sqrt{3}t + C_2 \cos \frac{1}{2}\sqrt{3}t) + t^3 - t^2 - 4t + 6$

23. $y = Ce^{3t} - 2t + 4$

25. $y = Ce^{4t} + t^2 + \frac{1}{2}t - \frac{1}{8}$

27. $y = Ce^{4t} - 2e^{3t}$

29. $y = Ce^{3t} - 2e^t - 2t + 4$

31. $y = 3e^{-4t}$

33. $y = 6e^{-3t} + e^{2t}$

Exercise Set 21.5

1. $y = a_0 \sum_{k=0}^{\infty} \frac{(-1)^k x^k}{k!} = a_0 e^{-x}$

3. $y = a_0 \sum_{k=0}^{\infty} \frac{6^k x^k}{k!} = a_0 e^{6x}$

5. $y = a_0 \sum_{k=0}^{\infty} \frac{x^{2k}}{k!} = a_0 e^{x^2}$

7. $y = \sum_{k=0}^{\infty} \frac{(-1)^k 4^k x^k}{k!} = e^{-4x}$

9. $y = a_1 \sum_{k=1}^{\infty} \frac{(-1)^k x^k}{k[(k-1)!]^2}$

11. $y = \sum_{k=0}^{\infty} \frac{(-1)^k x^{2k}}{(2k)!} = \cos x$

Exercise Set 21.6

1.

t_j	0	.25	.5	.75	1.0
y_j	1	1.5	2.25	3.375	5.0625
$y(t_j)$	1	1.6487	2.7183	4.4817	7.3891

$$y = e^{2t}$$

3.

t_j	0	.25	.50	.75	1.0
y_j	1	1.5	1.75	1.875	1.9375
$y(t_j)$	1	1.3934	1.6321	1.7769	1.8647

$$y = 2 - e^{-2t}$$

5.

t_j	0	.25	.5	.75	1.0
y_j	1	1	1.125	1.3906	1.8389
$y(t_j)$	1	1.0635	1.2663	1.6496	2.2974

$$y = 2e^{x^2/2} - 1$$

7.

t_j	0	.25	.50	.75	1.0
y_j	1.5	1.5	1.375	1.1563	0.9102
$y(t_j)$	1.5	1.4394	1.2788	1.0698	0.8679

$$y = e^{-t^2} + \frac{1}{2}$$

9.

t_j	0	.25	.5	.75	1.0
y_j	1.5	2.25	3.25	4.625	6.5625
$y(t_j)$	1.5	2.3987	3.7183	5.7317	8.8890

$$y = t + \tfrac{1}{2} + e^{2t}$$

11. See answer to Exercise 3.

Review Exercises—Chapter 21

1. $y = \tfrac{1}{2}x^2 - 3x + C$

3. $y = 2 \sec \sqrt{t} + C$

5. $y = \dfrac{-2}{3t^2 + C}$

7. $y = Ce^{(1+x)^2/2} - 2$

9. $y = Ce^{-2t} + 2$

11. $y = \dfrac{4}{Ce^{-4t} - 1}$

13. $y = Ce^{-\sin t}$

15. $y = Ce^{4t} - 2$

17. $y = e^{-2t}(C_1 + C_2 t)$

19. $y = C_1 e^{5t} + C_2 e^{-3t} - \frac{4}{377} \sin 2t - \frac{19}{377} \cos 2t$

21. $y = C_1 e^{3t} + C_2 e^{-3t} - \frac{5}{9} - \frac{1}{5} e^{2t}$

23. $y = Ce^{1/t}$

25. $y = C_1 e^{7t} + C_2 e^{-2t} - \frac{1}{14} t - \frac{23}{196}$

27. $y = Cx^{-2/3}$

29. $y = \dfrac{2x}{C + x}$ or $y = 0$

31. $y = 2 + \sqrt{x^2 + 1}$

33. $y = 3e^{t^2/2}$

35. $y = 3 \sin 3t + 3 \cos 3t$

37. $y = 6te^{3t}$

39. a. $y' = 2y$
b. $y'' - y = 0$

41. $y = e^{-t/5}$

43. 58° F

45. $y = C_1\left(x + \dfrac{x^2}{2} + \dfrac{x^3}{3!2!} + \dfrac{x^4}{4!3!} + \dfrac{x^5}{5!4!} + \cdots\right)$

47. $y = e^{2t}\left(A \sin \sqrt{2}t + B \cos \sqrt{2}t\right)$

Index

Quick Reference to Core Concepts

Quick Reference to Core Concepts

(This reference list begins on the previous page.)

Transcendental Functions

Techniques of Integration

Infinite Sequences, Series, and Power Series

Geometry